P9-CFB-356

# Get Ready for a Whole
## New Mastering Experience

**NEW!** **Ready-to-Go Teaching Modules** help instructors find the best assets to use before, during, and after class to teach the toughest topics in A&P. Created by teachers for teachers, these curated sets of teaching tools save you time by highlighting the most effective and engaging animations, videos, quizzing, coaching and active learning activities from MasteringA&P.

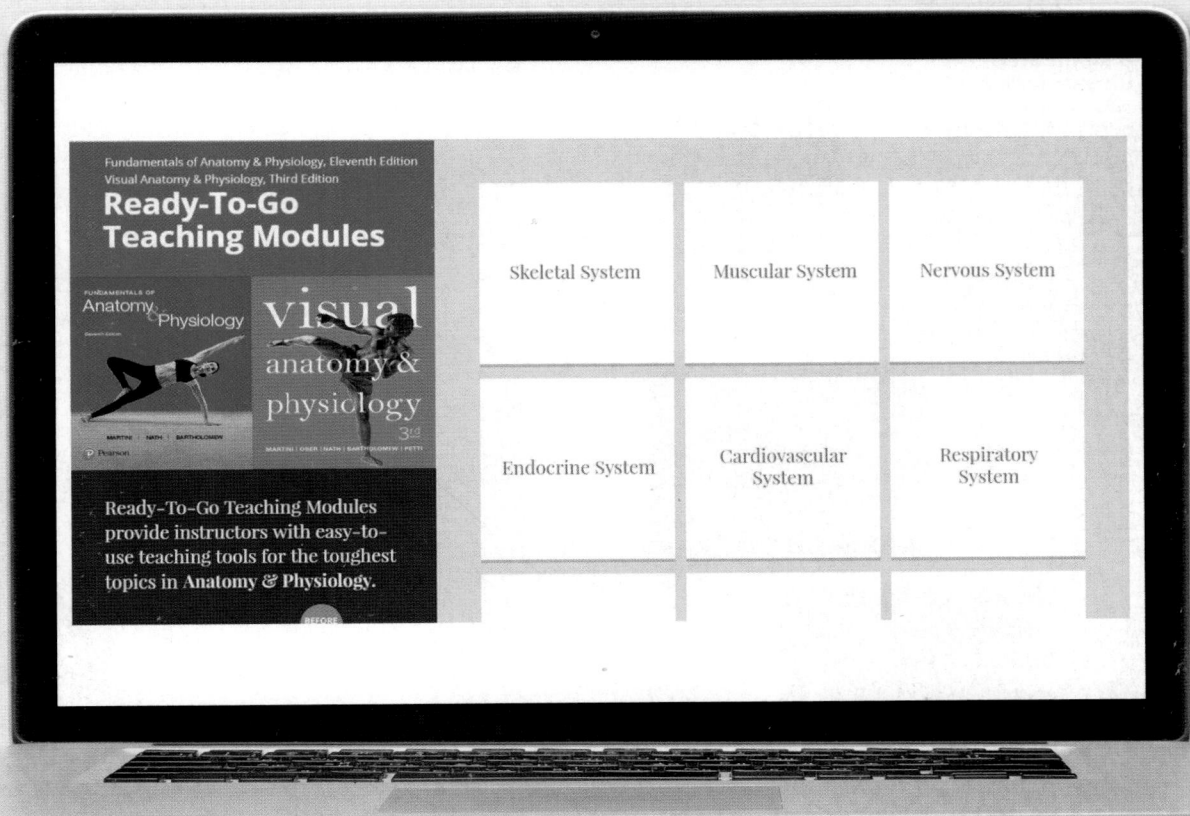

# Help Students Use Art
## More Effectively

## Module 1.2

### Comprehending the art is essential to understanding A&P

Think back to your first childhood book. You most likely began with a "picture book." Then, as you learned the alphabet and developed speech, you progressed to "word books." The next step was "chapter books." Somewhere along the way, you quit looking at pictures (art) and focused solely on the words (text). Maybe the shift to text-based reading without looking at the pictures happened in high school. You began reading words—paragraph after paragraph, page after page of words. Many of your books may have been colorful and filled with pictures, but you quickly decided that most were decorative and that the real information was to be found in the words. To succeed in a college science course, you need to break this pattern, shift your focus, and integrate information presented in the art as well as the words.

**1** In college, you are faced with many new terms, abstract concepts, and unfamiliar images. That's great, because college is intended to increase your knowledge and expand your horizons. But research has shown that undergraduate students have a tendency to simply read the text (also called the narrative) without paying attention to the art (referred to as figures or diagrams in your book). While you can certainly learn from this approach, there is abundant research showing that paying attention to the art while reading the text improves learning.

For example, researchers have set up studies that use eye-tracking equipment to measure eye movement and interaction with images on pages. They found that students spent very little time looking at the art while they were reading the text. When students were trained to read the text and view the art together, their time spent looking at the art increased. This attention to the art was critical: *Learning and comprehension levels were greater for students who studied both the art and the text together than for students who only read the text.*

Each student was given 30 minutes to study the material on a computer screen and then was tested to see how well they understood the content.

Eye-tracking equipment measured how long students focused their attention on either the written material or the visual material.

**2** Although reading text and viewing art side by side may sound like common sense, most students do not do that. We wrote this book to make it easier for you to study the text and the associated art at the same time and on the same page, without the page-flipping required by a traditional textbook format. In this book, the text and the art go hand in hand. Please continue reading as we walk you through the process of using this textbook to enhance your learning.

**? REVIEW**
A. What do eye-tracking studies tell us about the most effective way to learn?

**3** Here is an example of text in a narrative form and an illustration of the structure of a pencil. In most textbooks a figure will be as close to its descriptive lines of text as possible, but in some cases that may mean turning forward or backward a page or two to find it. You're expected to find and inspect a figure when you see a callout for it. The figure callout usually looks like this: (see Figure 1 or Figure 1). They are often intentionally color-coded so you can stop reading, look at the figure, and then find your place again when you go back to reading the text.

A pencil is a writing implement or art tool used to make marks on paper, canvas, or other surfaces. It is a long, round, slender instrument consisting of a thin stick of graphite enclosed in a long thin piece of wood. A typical pencil is approximately 8″ long, 3/8″ in diameter, and hexagonal in cross-section. On one end of the pencil is an eraser that is used to remove marks on paper. The opposite end of the pencil is pointed. The point consists of a semi-soft material called graphite that forms the core of the pencil shaft. Graphite is a form of carbon that easily rubs off from the pencil to the paper forming a mark. The pencil is held between the thumb and forefinger. **See Figure 1**

Eraser

Graphite tip

**Figure 1. Structure of a typical pencil**

**4** This example shows a different approach. Here the text and art are combined so key written information is integrated into the art itself. This **text-art integration** means you never have to flip pages or hunt for an illustration associated with specific text. Educational research shows such integration works best for comprehension. This book was written using this approach. After you answer the review question below, turn the page to see how text-art integration works with a topic about anatomy and physiology.

**? REVIEW**
B. Try this experiment. Pick one of the two examples, cover up the text, and focus solely on the art. What did you discover?

A pencil is a writing implement or art tool used to make marks on paper, canvas, or other surfaces. It is a long, round, slender instrument consisting of a thin stick of graphite enclosed in a long thin piece of wood.

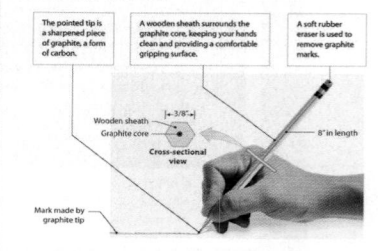

The pointed tip is a sharpened piece of graphite, a form of carbon.

A wooden sheath surrounds the graphite core, keeping your hands clean and providing a comfortable gripping surface.

A soft rubber eraser is used to remove graphite marks.

Wooden sheath
Graphite core
3/8″
Cross-sectional view
8″ in length

Mark made by graphite tip

**⤷ LEARNING OUTCOME**
Describe how to use the text and art together to master learning.

## Motor Units and Recruitment

**5** A typical skeletal muscle contains thousands of muscle fibers. Although some motor neurons control just a few muscle fibers, most control hundreds of them. The amount of tension produced is controlled at the subconscious level through variations in the number of muscle fibers stimulated.

A **motor unit** is a motor neuron and all the muscle fibers that it controls. The size of a motor unit indicates how fine, or precise, a movement can be. In the muscles of the eye, where precise control is extremely important, a motor neuron may control 4–6 muscle fibers. We have much less precise control over our leg muscles, where a single motor neuron may control 1000–2000 muscle fibers.

The muscle fibers of each motor unit are intermingled with those of other motor units. As a result, the direction of pull exerted on the tendon does not change when the number of activated motor units changes. When you decide to perform a specific movement, the contraction begins with the activation of the smallest motor units in the stimulated muscle. As the movement continues, larger motor units containing faster and more powerful muscle fibers are activated, and tension rises steeply. The smooth but steady increase in muscular tension produced by increasing the number of active motor units is called **recruitment**.

**? REVIEW**
**C.** Describe the relationship between the number of fibers in a motor unit and the precision of body movements.

KEY
Motor unit 1
Motor unit 2
Motor unit 3

Spinal cord
Cell bodies of motor neurons
Axons of motor neurons
Motor nerve

**? REVIEW**
**B.** Define *motor unit*.

**6** During a sustained contraction, motor units are activated on a rotating basis, allowing some to rest and recover while others are actively contracting. In this "relay team" approach, called **asynchronous motor unit summation**, each motor unit can recover somewhat before it is stimulated again. As a result, when your muscles contract for sustained periods, they produce slightly less than maximal tension.

Tension in tendon
Motor unit 1  Motor unit 2  Motor unit 3
Tension
Time

**Lo LEARNING OUTCOME**
Discuss the factors that affect peak tension production during the contraction of an entire skeletal muscle, and explain the significance of the motor unit in this process.

🏃 Everyday Physiology

### What Is Muscle Tone?

A variable number of motor units is always active, even when the entire muscle is not contracting. Their contractions do not produce enough tension to cause movement, but they do tense and firm the muscle. This resting tension in a skeletal muscle is called **muscle tone**, and it is regulated at the subconscious level. The activity level of each motor neuron changes constantly, and individual muscle fibers can relax while a constant tension is maintained in the attached tendon. Activated muscle fibers use energy, so the greater the muscle tone, the higher the "resting" rate of metabolism. Elevated muscle tone increases resting energy consumption by a small amount, but the effects are cumulative, and they continue 24 hours per day.

## Module 19.5

# Pressure, resistance, and venous return affect cardiac output

Go to
MasteringA&P™
**iP2**
Cardiac Output

You are now familiar with the factors that regulate cardiac output (see Module 18.16). Both cardiac output and the distribution of blood within the pulmonary and systemic circuits must constantly be adjusted to meet the demands of active tissues. The sites and mechanisms for such adjustments are previewed here.

### Cardiac Output

To maintain **cardiac output**, the heart must generate enough pressure to force blood through thousands of miles of peripheral capillaries, most no larger in diameter than a single red blood cell.

**? REVIEW**

**A.** Neural and endocrine regulatory mechanisms influence which factors?

### Venous Return

Venous pressure, although low, determines **venous return**—the amount of blood arriving at the right atrium each minute. Venous return has a direct impact on cardiac output.

### Arterial Pressure

**Arterial pressure** is the force exerted by blood against the walls of systemic arteries. Arterial pressure is much higher than venous pressure because it must push blood a greater distance and through smaller and smaller arteries and then through innumerable capillary networks.

### Blood Flow Regulation

**Autoregulation** acts locally. **Central regulation** involves neural and endocrine mechanisms to make coordinated adjustments in heart rate, stroke volume, peripheral resistance, and venous pressure so that cardiac output is sufficient to meet the needs of peripheral tissues.

### Venous Pressure

**Venous pressure** is the force exerted by blood within the venous system. As blood moves toward the heart, the veins become larger, resistance drops, and blood flow velocity increases. Muscular compression and the respiratory pump are two factors that assist in propelling blood toward the heart at low venous pressure.

### Peripheral Resistance

**Resistance** is a force that opposes movement. The **peripheral resistance** is the resistance of the arterial system as a whole. Resistance increases as the arterial branches become smaller and thinner.

**? REVIEW**

**B.** Which is greater: arterial pressure or venous pressure?

### Capillary Pressure

**Capillary pressure** is the force exerted by blood against capillary walls. It is normally low, allowing time for diffusion between the blood and surrounding interstitial fluid. This process, called **capillary exchange**, plays a key role in homeostasis.

Capillary exchange

Interstitial fluid

**? REVIEW**

**C.** Why is it beneficial for capillary pressure to be very low?

In this section we discuss the functioning of the cardiovascular system as a whole and learn how neural and hormonal mechanisms regulate cardiovascular function.

**Lo LEARNING OUTCOME**

Explain the effects of pressure, resistance, and venous return on cardiac output.

# and Practical Application

**Clinical Modules** throughout the book help students connect what they learn in class to the real-world. They may encounter similar examples in their future career.

**NEW! Everyday Physiology** features appear throughout the text to help students see connections to real-life applications.

## Everyday Physiology

### What Is Muscle Tone?

A variable number of motor units is always active, even when the entire muscle is not contracting. Their contractions do not produce enough tension to cause movement, but they do tense and firm the muscle. This resting tension in a skeletal muscle is called **muscle tone**, and it is regulated at the subconscious level. The activity level of each motor neuron changes constantly, and individual muscle fibers can relax while a constant tension is maintained in the attached tendon. Activated muscle fibers use energy, so the greater the muscle tone, the higher the "resting" rate of metabolism. Elevated muscle tone increases resting energy consumption by a small amount, but the effects are cumulative, and they continue 24 hours per day.

## Chapter Integration • Applying what you have learned

### A helmet-to-helmet collision causes a "stinger"

Dominic is a defenseman on his high school's varsity lacrosse team. He plays an aggressive style of defense and initiates extensive physical contact. As the final seconds of a recent game counted down, an opposing player was about to take a shot on goal that would have tied the game. Dominic ran full speed into the shooter. Their helmets collided dead-on, and Dominic's head snapped hard to the left. As his opponent fell backward onto the ground, Dominic felt an intense pain from the right side of his neck that radiated down his entire right upper limb to the tips of his fingers. The referee blew the whistle, indicating the end of the game. As the winning players and coaches congregated, the team's athletic trainer asked Dominic how he was feeling after delivering such a significant hit. Dominic replied that upon impact he felt a severe pain on the right side of his neck and entire right upper limb that was both numbing and burning, but that it lasted only a few moments. The trainer informed him that he likely experienced a *stinger* and that although the pain was quite intense, he would not likely have a lasting injury.

From what you have just learned about spinal nerves, answer the following questions.

1. Why did Dominic's pain radiate down only the right upper limb and not both upper limbs? Why was the pain also in his neck?

2. Which spinal nerves would you suspect to be involved?

3. Why do you think he experienced the numbing, burning pain, and why do you think the pain extended all the way to his fingertips?

**Chapter Integration** sections ask students to aggregate what they have learned so far in the context of real world situations. These scenarios are presented in a friendly, story format followed by critical thinking questions.

# Continuous Learning
## Before, During, and After Class

**Dynamic Study Modules** enable students to study more effectively on their own. With the Dynamic Study Modules mobile app, students can quickly access and learn the concepts they need to be more successful on quizzes and exams.

**NEW!** Instructors can now select which questions to assign to students.

**NEW! SmartArt Videos** help students navigate some of the complex figures in the text. They are accessible via QR code in the book and are assignable in MasteringA&P.

# with MasteringA&P™

**Learning Catalytics** is a "bring your own device" (laptop, smartphone, or tablet) engagement, assessment, and classroom intelligence system. Students use their device to respond to open-ended questions and then discuss answers in groups based on their responses.

*"My students are so busy and engaged answering Learning Catalytics questions during lecture that they don't have time for Facebook."*

*Declan De Paor, Old Dominion University*

# MasteringA&P™

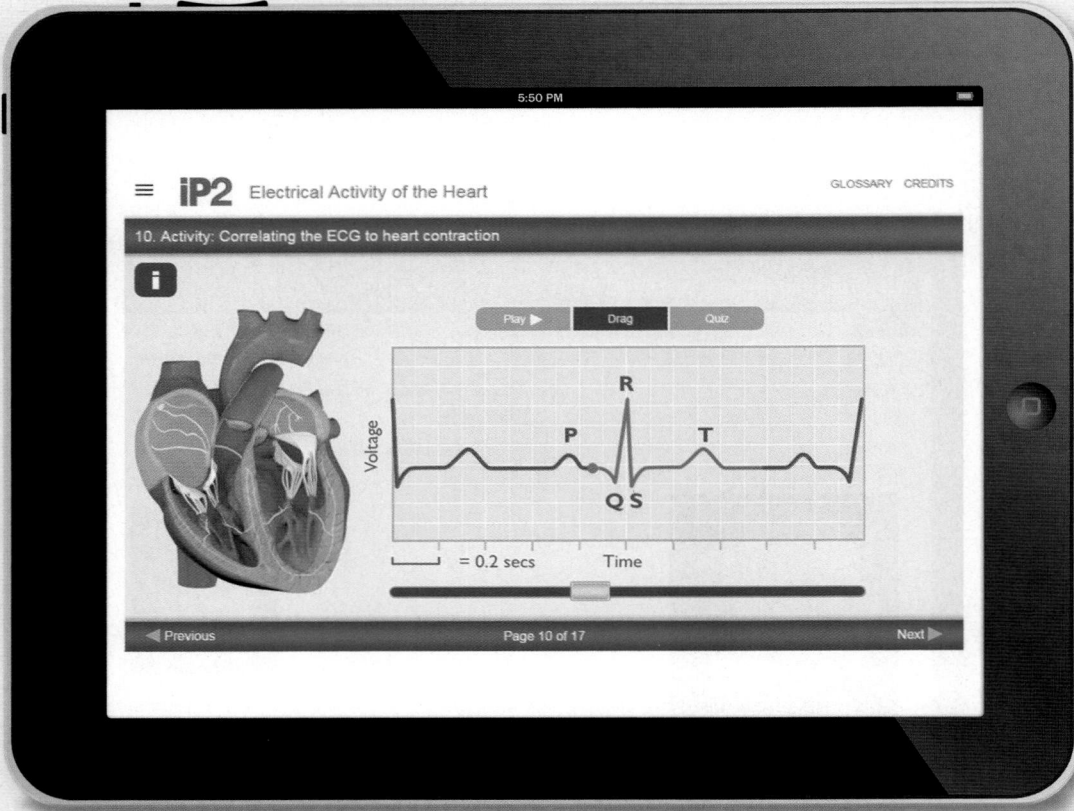

## NEW IP 2.0 modules include:

- Resting Membrane Potential
- Electrical Activity of the Heart
- Cardiac Output
- Factors Affecting Blood Pressure
- Generation of an Action Potential
- Cardiac Cycle

## Coming soon:

- Glomerular Filtration
- Neuromuscular Junction
- Tubular Reabsorption and Secretion
- Excitation Contraction Coupling

# More Practice, More Learning

Transverse tubules

**A&P Flix Coaching Activities** bring interactivity to these popular 3D movie-quality animations by asking students to answer questions related to the video.

## Additional assignable MasteringA&P activities include:

- Tough Topic Coaching Activities
- Bone & Dissection Video Coaching Activities
- Clinical Case Studies
- And More!

**NEW!** Beginning Fall 2017, all of the assignments from Sarikas's *Visual A&P Lab Manual,* 2e can be accessed in your *Visual A&P* Mastering course! Only one MasteringA&P code is needed to access these assignments.

Place the events that occur during excitation-contraction coupling in the correct order from left to right.

Reset   Help

| AP travels down T tubules to triads | AP propagates along sarcolemma | Ca²⁺ levels in sarcoplasm increase | Sarcoplasmic reticulum releases Ca²⁺ | Voltage-sensitive proteins open Ca²⁺ channels |

AP generated by motor neuron                    Contraction of skeletal muscle fiber

Submit   Hints   My Answers   Give Up   Review Part

**Incorrect; Try Again**
The AP causes the voltage-sensitive proteins located in the T tubules to change shape.

# MasteringA&P Lab Practice

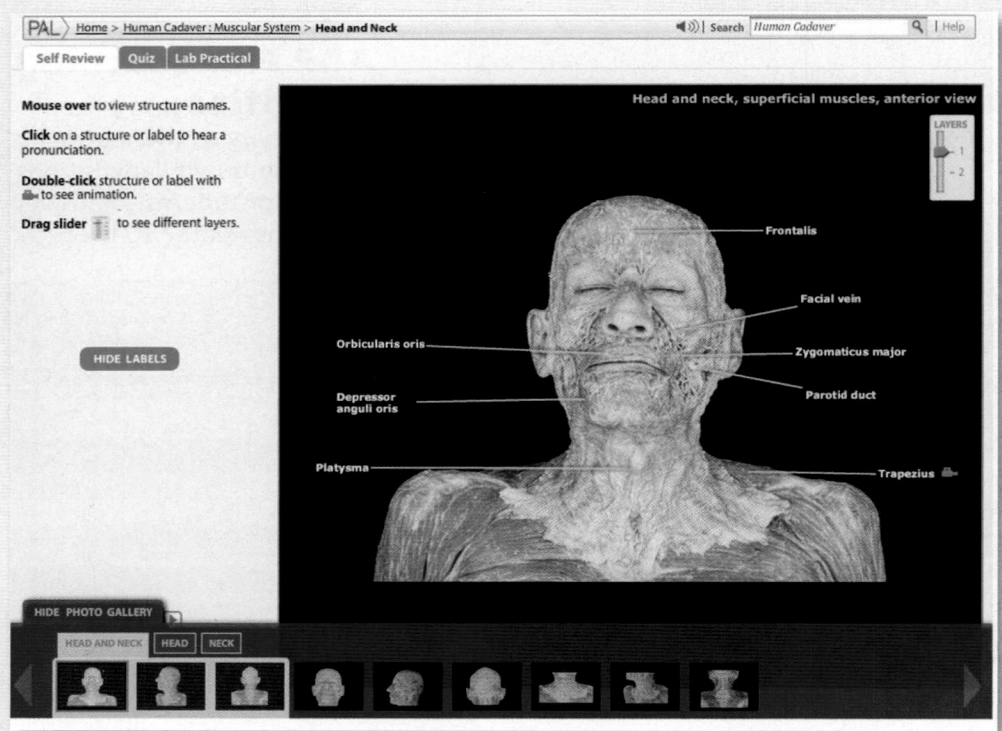

**Practice Anatomy Lab (PAL™ 3.0)** is a virtual anatomy study and practice tool that gives students 24/7 access to the most widely used lab specimens, including the human cadaver, anatomical models, histology, cat, and fetal pig. PAL 3.0 is easy to use and includes built-in audio pronunciations, rotatable bones, and simulated fill-in-the-blank lab practical exams.

## PhysioEx 9.1

is an easy-to-use lab simulation program that allows students to conduct experiments that are difficult in a wet lab environment because of time, cost, or safety concerns. Students are able to repeat labs as often as they like, perform experiments without animals, and are asked to frequently stop and predict within the labs.

# Access the Complete Textbook
## On or Offline with eText 2.0

**NEW! The Third Edition** is available in Pearson's fully-accessible eText 2.0 platform.*

**NEW! The eText 2.0 mobile app** offers offline access and can be downloaded for most iOS and Android phones and tablets from the iTunes or Google Play stores.

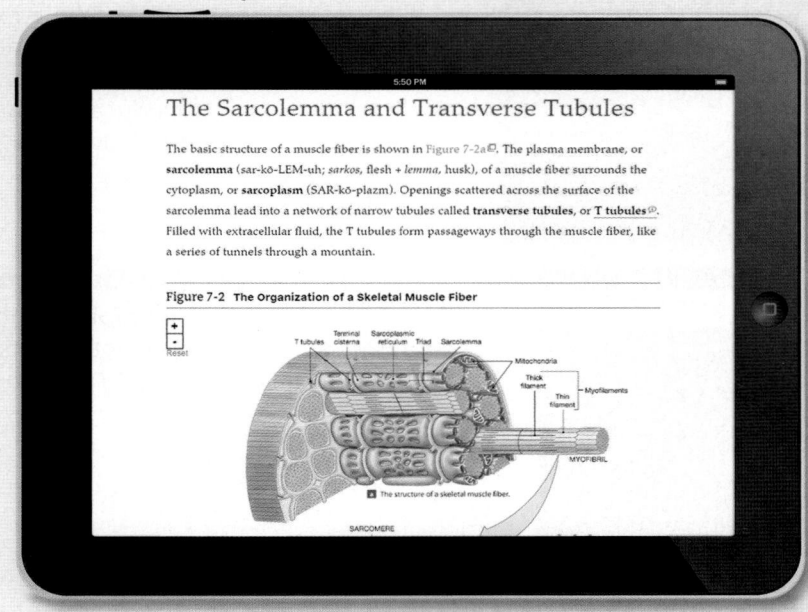

**Powerful interactive and customization functions** include instructor and student note-taking, highlighting, bookmarking, search, and links to glossary terms.

*The eText 2.0 edition will be live for Fall 2017 classes.

# Instructor and Student Support

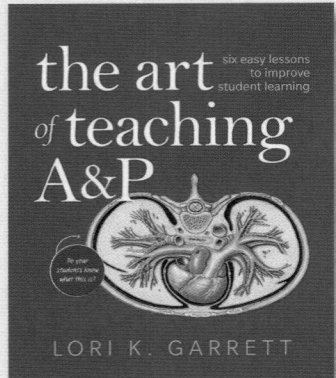

## NEW! The Art of Teaching A&P: Six Easy Lessons to Improve Student Learning by Lori K. Garrett

978-0-134-46951-5 / 0-134-46951-8

Author Lori Garrett (*Get Ready for A&P*) explores some of the most common challenges she's encountered in her classroom when using art to teach anatomy and physiology. From describing the challenge to researching why it occurs and proposing solutions to address it, Lori provides insight into how students look at images. She presents ideas for how educators can best use figures and illustrations to teach complex concepts without overwhelming or discouraging their students.

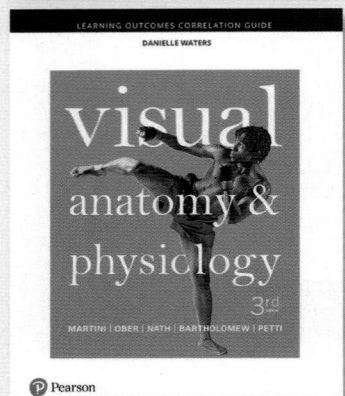

## NEW! The Learning Outcomes Correlation Guide

This guide ties each A&P topic and learning outcome established by HAPS (Human Anatomy & Physiology Society) to its corresponding module in the third edition of Martini's *Visual Anatomy & Physiology*.

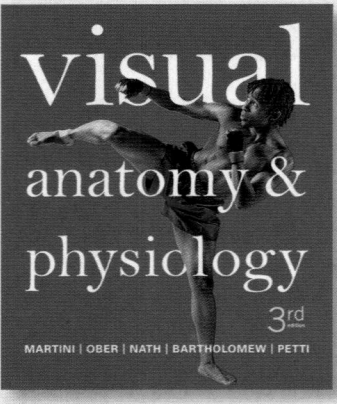

## A complete package of instructor resources includes:

- Customizable PowerPoint slides (with **NEW! Annotations on how to present complex art during lecture**)
- All figures from the book in JPEG format
- A&P Flix 3D movie-quality animations on tough topics
- Test Bank
- And more!

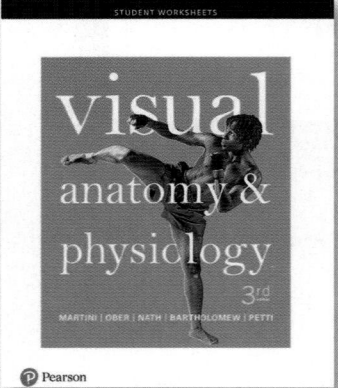

## Student Worksheets For Visual Anatomy & Physiology by Frederic H. Martini, William C. Ober, Judi L. Nath, Edwin F. Bartholomew, and Kevin Petti

978-0-134-48649-9 / 0-134-48649-8

This booklet contains all of the Section Review pages from the book for students who would prefer to mark their answers on separate pages rather than in the book itself.

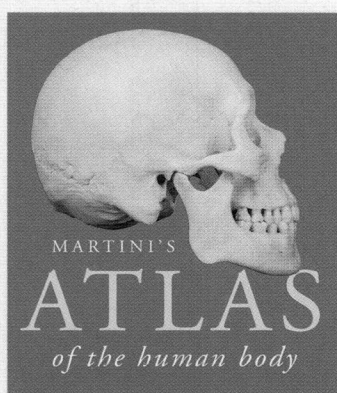

## Martini's Atlas of the Human Body by Frederic H. Martini
978-0-321-94072-8 / 0-321-94072-5

The Atlas offers an abundant collection of anatomy photographs, radiology scans, and embryology summaries, helping students visualize structures and become familiar with the types of images seen in a clinical setting. Free when packaged with the textbook.

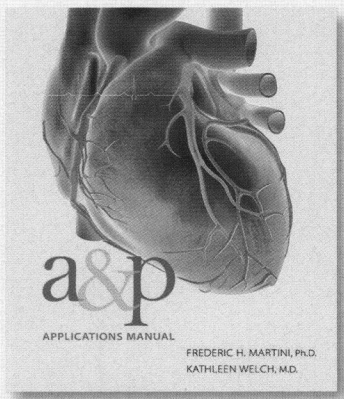

## A&P Applications Manual by Frederic H. Martini and Kathleen Welch
978-0-321-94973-8 / 0-321-94973-0

This manual contains extensive discussions on clinical topics and disorders to help students apply the concepts of anatomy and physiology to daily life and their future health professions. Free when packaged with the textbook.

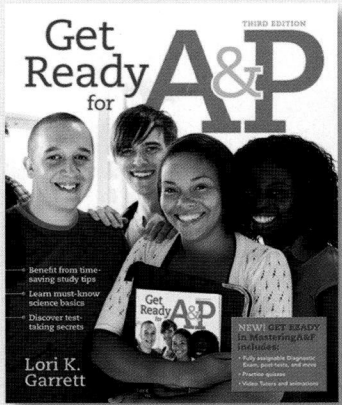

## Get Ready for A&P by Lori K. Garrett
978-032181336-7 / 0-321-81336-7

This book and online component were created to help students be better prepared for their A&P course. Features include pre-tests, guided explanations followed by interactive quizzes and exercises, and end-of-chapter cumulative tests. Also available in the Study Area of MasteringA&P. Free when packaged with the textbook.

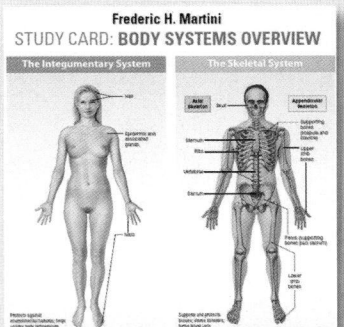

## Study Card for Martini: Body Systems Overview
0-321-92930-6 / 978-0-321-92930-3

A six-panel laminated card showing all body systems and their organs and functions. Free when packaged with the textbook.

# visual
## anatomy &
## physiology

**3rd** edition

**Frederic H. Martini, Ph.D.**
University of Hawaii at Manoa

**William C. Ober, M.D.**
Washington and Lee University

**Judi L. Nath, Ph.D.**
Lourdes University

**Edwin F. Bartholomew, M.S.**

**Kevin Petti, Ph.D.**
San Diego Miramar College

**Claire E. Ober, R.N.**
*Illustrator*

**Kathleen Welch, M.D.**
*Clinical Consultant*

**Ralph T. Hutchings**
*Biomedical Photographer*

 Pearson

330 Hudson Street, NY NY 10013

Editor-in-Chief: *Serina Beauparlant*
Courseware Portfolio Manager: *Cheryl Cechvala*
Content Producer: *Arielle Grant*
Managing Producer: *Nancy Tabor*
Courseware Director, Content Development: *Barbara Yien*
Courseware Sr. Analysts: *Suzanne Olivier, Alice Fugate*
Development Editors: *Molly Ward, Tanya Martin*
Courseware Editorial Assistant: *Kimberly Twardochleb*
Director of Digital Product Development: *Lauren Fogel*
Executive Content Producer: *Laura Tommasi*
Rich Media Content Producer: *Lauren Chen*
Content Developer, A&P and Microbiology: *Cheryl Chi*
Senior Mastering Media Producer: *Katie Foley*
Associate Mastering Media Producer: *Kristen Sanchez*
Full-Service Vendor: *Cenveo Publisher Services*

Full-Service Vendor Project Manager: *Mary Tindle*
Copyeditor: *Sally Peyrefitte*
Compositor: *Cenveo Publisher Services*
Art House: *Imagineering*
Contributing Illustrators: *imagineeringart.com, Anita Impagliazzo*
Art Coordinators: *Wynne Auyeung, Alicia Elliot*
Interior Designer: *Gibson Design Associates*
Design Manager: *Mark Ong*
Cover Designer: *Tamara Newnam*
Rights & Permissions Project Manager: *Kathleen Zander, Jason Perkins*
Rights & Permissions Management: *Cenveo*
Photo Researcher: *Kristin Piljay*
Manufacturing Buyer: *Stacey Weinberger*
Executive Marketing Manager: *Allison Rona*

Cover Photo Credit: *Nanettegrebe/Fotolia*

**Library of Congress Cataloging-in-Publication Data**
Names: Martini, Frederic | Ober, Claire E. | Welch, Kathleen (Kathleen Martini) | Hutchings, R. T., illustrator.
Title: Visual anatomy & physiology.
Other titles: Visual anatomy and physiology
Description: 3rd edition / Frederic H. Martini, Ph.D., University of Hawaii at Manoa [and four others]; Claire E. Ober, R.N., illustrator, Kathleen Welch, M.D., clinical consultant, Ralph T. Hutchings, biomedical photographer. | New York, NY : Pearson, [2018] | Includes bibliographical references and index.
Identifiers: LCCN 2016045395 | ISBN 0134394690
Subjects: LCSH: Human anatomy. | Human physiology. | Visual learning.
Classification: LCC QP31.2 .M373 2018 | DDC 612–dc23 LC record available at https://lccn.loc.gov/2016045395

2   17

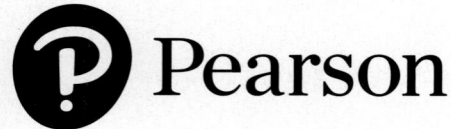

www.pearsonhighered.com

ISBN 10: 0-13-439469-0 (Student edition)
ISBN 13: 978-0-13-439469-5 (Student edition)
ISBN 10: 0-13-449973-5 (Instructor's Review Copy)
ISBN 13: 978-0-13-449973-4 (Instructor's Review Copy)

To my son, PK, for convincing me it was time to look at teaching and learning in new ways, and to the A&P students and instructors who helped shape the resulting text.

— RIC MARTINI

To my sons, Todd and Carl, whose warmth and humor have enriched my life in countless ways.

— BILL OBER

To my students and students everywhere, who make writing textbooks worthwhile. And, as always and in all ways, to my husband, Mike.

— JUDI NATH

To my daughters Ivy and Kate, grandchildren Awley, Rhyan, Finna, and Raya, and former students, who have given me the opportunity to touch the future.

— ED BARTHOLOMEW

To Coreen, my bride of over 25 years, and to Olivia and Dominic, the light of my life.

— KEVIN PETTI

**Frederic (Ric) H. Martini, Ph.D.**
Author

Dr. Martini received his Ph.D. from Cornell University in comparative and functional anatomy for work on the pathophysiology of stress. In addition to professional publications that include journal articles and contributed chapters, technical reports, and magazine articles, he is the lead author of 10 undergraduate texts on both anatomy and anatomy and physiology. Dr. Martini is currently affiliated with the University of Hawaii at Manoa and has a long-standing association with the Shoals Marine Laboratory, a joint venture between Cornell University and the University of New Hampshire. He has been active in the Human Anatomy and Physiology Society (HAPS) for over 24 years and was a member of the committee that established the course curriculum guidelines for A&P. He is now a President Emeritus of HAPS after serving as President-Elect, President, and Past-President over 2005–2007. Dr. Martini is also a member of the American Physiological Society, the American Association of Anatomists, the Society for Integrative and Comparative Biology, the Australia/New Zealand Association of Clinical Anatomists, the Hawaii Academy of Science, the American Association for the Advancement of Science, and the International Society of Vertebrate Morphologists.

**Judi L. Nath, Ph.D.**
Author

Dr. Judi Nath is a biology professor and the writer-in-residence at Lourdes University, where she teaches at both the undergraduate and graduate levels. Primary courses include anatomy, physiology, pathophysiology, medical terminology, and science writing. She received her bachelor's and master's degrees from Bowling Green State University, which included study abroad at the University of Salzburg in Austria. Her doctoral work focused on autoimmunity, and she completed her Ph.D. at the University of Toledo. Dr. Nath is devoted to her students and strives to convey the intricacies of science in captivating ways that are meaningful, interactive, and exciting. She has won the Faculty Excellence Award—an accolade recognizing effective teaching, scholarship, and community service—multiple times and in 2013 was named as an Ohio Memorable Educator. She is active in many professional organizations, notably the Human Anatomy and Physiology Society (HAPS), where she has served several terms on the board of directors. Dr. Nath is a coauthor of *Fundamentals of Anatomy & Physiology, Visual Essentials of Anatomy & Physiology, Anatomy & Physiology,* and *Human Anatomy* (published by Pearson), and she is the sole author of *Using Medical Terminology* and *Stedman's Medical Terminology* (published by Wolters Kluwer). Her favorite charities are those that have significantly affected her life, including the local Humane Society, the Cystic Fibrosis Foundation, and the ALS Association. In 2015, she and her husband established the Nath Science Scholarship at Lourdes University to assist students pursuing science-based careers. When not working, Dr. Nath enjoys spending days filled with family life, bicycling, and hanging with the dogs.

**Edwin F. Bartholomew, M.S.**
Author

Edwin F. Bartholomew received his undergraduate degree from Bowling Green State University in Ohio and his M.S. from the University of Hawaii. Mr. Bartholomew has taught human anatomy and physiology at both the secondary and undergraduate levels. In addition, he has taught a range of other science courses (from botany to zoology) at Maui Community College (now the University of Hawaii Maui College). For many years, he taught at historic Lahainaluna High School (LHS), the oldest high school west of the Rockies, where he assisted in establishing an LHS Health Occupations Students of America (HOSA) chapter. He is a coauthor of *Fundamentals of Anatomy & Physiology, Essentials of Anatomy & Physiology, Visual Essentials of Anatomy & Physiology, Structure and Function of the Human Body,* and *The Human Body in Health and Disease* (all published by Pearson). Mr. Bartholomew is a member of the Human Anatomy and Physiology Society (HAPS), the National Association of Biology Teachers, the National Science Teachers Association, and the American Association for the Advancement of Science.

**Kevin Petti, Ph.D.**
Author

Dr. Petti is a professor at San Diego Miramar College, where he teaches courses in human anatomy and physiology, human dissection, and health education. He is President Emeritus of the Human Anatomy and Physiology Society (HAPS) and holds a doctorate from the University of San Diego. As a dual U.S./Italian citizen, he also teaches courses in Italy that focus on the genesis of anatomy as a science and its influence on the Renaissance masters, a story unique to the Italian peninsula. His students range from anatomy professors pursuing continuing education to undergraduates in study-abroad programs. Dr. Petti is often invited to speak about the connection between art and anatomy in medieval and Renaissance Italy at museums, conferences, and universities. The Italian government has invited him to speak at their Cultural Institutes in Los Angeles and New York City, and the University of Palermo in Sicily included him in a seminar series celebrating its 210th anniversary.

### William C. Ober, M.D.
**Art Coordinator and Illustrator**

Dr. Ober received his undergraduate degree from Washington and Lee University and his M.D. from the University of Virginia. He also studied in the Department of Art as Applied to Medicine at Johns Hopkins University. After graduation, Dr. Ober completed a residency in family practice and later was on the faculty at the University of Virginia in the Department of Family Medicine and in the Department of Sports Medicine. He also served as Chief of Medicine of Martha Jefferson Hospital in Charlottesville, Virginia. He is currently a Visiting Professor of Biology at Washington and Lee University, where he has taught several courses and led student trips to the Galápagos Islands. He was on the Core Faculty at Shoals Marine Laboratory for 24 years, where he taught Biological Illustration every summer. Dr. Ober has collaborated with Dr. Martini on all of his textbooks in every edition.

### Claire E. Ober, R.N.
**Illustrator**

Claire E. Ober, R.N., B.A., practiced family, pediatric, and obstetric nursing before turning to medical illustration as a full-time career. She returned to school at Mary Baldwin College, where she received her degree with distinction in studio art. Following a five-year apprenticeship, she has worked as Dr. Ober's partner in Medical & Scientific Illustration since 1986. She was on the Core Faculty at Shoals Marine Laboratory and co-taught the Biological Illustration course with Dr. Ober for 24 years. The textbooks illustrated by Medical & Scientific Illustration have won numerous design and illustration awards.

### Kathleen Welch, M.D.
**Clinical Consultant**

Dr. Welch received her B.A. from the University of Wisconsin–Madison and her M.D. from the University of Washington in Seattle, and she completed her residency in family practice at the University of North Carolina in Chapel Hill. Participating in the Seattle WWAMI rural medical education program, she studied in Fairbanks, Anchorage, and Juneau, Alaska, with time in Boise, Idaho, and Anacortes, Washington, as well. For two years, she served as Director of Maternal and Child Health at the LBJ Tropical Medical Center in American Samoa and subsequently was a member of the Department of Family Practice at the Kaiser Permanente Clinic in Lahaina, Hawaii, and on the staff at Maui Memorial Hospital. She was in private practice from 1987 until her retirement in 2012. Dr. Welch is a Fellow of the American Academy of Family Practice and a member of the Hawaii Medical Association, the Maui County Medical Association, and the Human Anatomy and Physiology Society (HAPS). With Dr. Martini, she has coauthored both a textbook on anatomy and physiology and the *A&P Applications Manual.* She and Dr. Martini were married in 1979, and they have one son.

### Ralph T. Hutchings
**Biomedical Photographer**

Mr. Hutchings was associated with the Royal College of Surgeons for 20 years. An engineer by training, he has focused for years on photographing the structure of the human body. The result has been a series of color atlases, including *Color Atlas of Human Anatomy, Color Atlas of Surface Anatomy,* and *The Human Skeleton* (all published by Mosby-Yearbook Publishing). For his anatomical portrayal of the human body, the International Photographers Association has chosen Mr. Hutchings as the best photographer of humans in the 20th century. He lives in North London, where he tries to balance the demands of his photographic assignments with his hobbies of early motor cars and airplanes.

**V**isual Anatomy & Physiology is a comprehensive textbook for the two-semester A&P course. It combines a visual approach with a modular organization to deliver subject matter in an easy-to-use and time-efficient manner that uniquely meets the needs of today's students—without sacrificing the coverage of A&P topics required for careers in nursing and other allied health professions.

For the Third Edition, prior to revising or creating a module, we asked ourselves three questions: (1) How can we best make this information meaningful, manageable, and comprehensible? (2) Does the module spark interest and encourage students to read it? (3) Will students be able to answer "Why is this important?" after the module?

In essence, we want students to be excited about learning human anatomy and physiology. During the revision process, our team of content experts, medical illustrators, award-winning teaching professionals, academic authors, and publishing specialists worked together to write and design this academic text. We scrutinized every sentence, visual, and layout, ensuring that the narrative made sense, the content was accurate, and the combinations of text and visuals flowed together seamlessly over the one- and two-page module presentations. We read countless reviews and listened to our own students in the classroom. This end product is the culmination of the very best all involved had to offer.

To help improve future editions, we encourage you to send any pertinent information and remarks about the organization or content of this textbook to us directly, using the e-mail addresses below. We warmly welcome comments and suggestions and will carefully consider them in the preparation of the Fourth Edition.

## New to the Third Edition of *Visual Anatomy & Physiology*

### Global

- **A NEW emphasis on using art more effectively** informs multiple changes to layout and figure organization, as well as a new system of integrated figure prompts and questions. These help students view and navigate the art more efficiently and effectively to enhance learning.
- **NEW Smart Art with QR codes.** This new feature, which appears adjacent to select figures, gives students access to videos that help them navigate tough topics and reinforces the pedagogy of our art.
- **NEW Modules 1.1 through 1.5** introduce students to the importance of studying the art in the book and then guide them in how to study the figures in the text.
- **NEW Module Review and Module Integration questions. Module Review questions** appear adjacent to their relevant figures to encourage and prompt students to read the text and view the art together. **Module Integration questions at the end of a module encourage the student to engage in higher order learning skills.**
- **NEW Everyday Physiology features** are included throughout the text to add interest and help students see connections to real-life applications.
- The color palette has been enhanced to make the art more vibrant.
- Chapter 15 has been revised to place the section on vision before the section on equilibrium and hearing.
- The topics in Chapters 18 and 19 have been reversed: the heart and cardiovascular function are addressed before blood vessels and circulation. This arrangement provides a stronger foundation for understanding the structural and physiological factors that affect cardiac output and blood flow throughout the body.
- Terms have been standardized to match *Terminologia Anatomica*, *Terminologia Histologica*, and *Terminologia Embryologica*. *Stedman's Medical Dictionary* was used for terms not found in the preceding books.

# Chapter-by-Chapter Changes in the Third Edition

## Chapter 1: An Introduction to Anatomy & Physiology

- New Module 1.1: Using your textbook effectively is key to your success.
- New Module 1.2: Comprehending the art is essential to understanding A&P.
- New Module 1.3: Break down the art in step-wise fashion to learn the topic.
- New Module 1.4: Orient yourself to all art in the same way.
- New Module 1.5: The learning outcomes correspond by number to the chapter's modules and indicate what you should be able to do after completing the chapter.
- Revised Module 1.7 (formerly 1.2) contains a new chart on the characteristics of living organisms and a new illustrated chart on the processes of life.
- Revised Module 1.9 (formerly 1.4) includes a new Everyday Physiology box that relates principles of physics and chemistry to biology.
- Revised Modules 1.10 (formerly 1.5) and 1.13 (formerly 1.8) include updated art detailing the integration of organ systems at the organism level.
- Revised Module 1.17 includes a new flowchart demonstrating the regulation of temperature to maintain homeostasis.
- Revised Module 1.18 (formerly 1.13) includes a new flowchart of the regulation of body temperature by negative feedback.
- Revised Module 1.22 (formerly 1.17) includes updated axial skeleton art that provides points of reference to the body cavities of the trunk.

## Chapter 2: Chemical Level of Organization

- Revised Module 2.2 contains a new Everyday Physiology box discussing radioisotopes.
- Revised Module 2.4 contains a new Clinical Note discussing free radical damage.
- Revised Module 2.9 contains a new illustration and text describing the relationship between monomers and polymers.
- Revised Module 2.17 contains a new Clinical Note discussing protein denaturation.
- Revised Module 2.19 includes a revised illustration and additional text to include ATPase and water in the hydrolytic breakdown of ATP.

## Chapter 3: Cellular Level of Organization

- Revised Module 3.6 relocates the text boxes describing the functions of the Golgi apparatus and lysosomes to relate more closely to the art depicting them.
- Revised Module 3.12 includes updated art with additional details of the small ribosomal subunit and of the EPA sites on the large ribosomal subunit.
- Revised Module 3.15 includes a new Clinical Note describing osmolarity and tonicity in medicine.
- Revised Module 3.17 includes updated art to include the role of clathrin in receptor-mediated endocytosis.
- Revised Chapter Review contains new images in the Chapter Integration section.

## Chapter 4: Tissue Level of Organization

- Chapter art contains labels for micrographs of different tissue types.
- Revised Module 4.3 contains new art illustrating epithelia and glands.
- Revised Module 4.4 uses the term *basal lamina* instead of *clear layer* and the term *reticular lamina* instead of *dense layer*. The module also contains a new Everyday Physiology box describing the avascularity of epithelia.
- Revised Module 4.5 includes updated art depicting the endothelium lining the inside of the heart and provides a description of keratin.
- Revised Module 4.6 provides the magnification of the light micrograph depicting simple cuboidal epithelium (650×) and the LM of the stratified cuboidal epithelium (500×).
- Revised Module 4.7 provides the magnification of the LM of the pseudostratified columnar epithelium (350×).
- Revised Module 4.9 (formerly 4.8) differentiates between the terms *mucous cell* and *goblet cell*.
- Revised Module 4.14 (formerly 4.13) includes updated art that incorporates nerves.
- Revised Module 4.15 (formerly 4.14) uses the term *tissue membrane* and states that deep fascia consists of dense regular connective tissue.
- Revised Module 4.16 (formerly 4.15) contains new art of muscle tissue types.
- Revised Module 4.17 (formerly 4.16) contains a new Everyday Physiology box describing link between neural activity and thought processes.

## Chapter 5: The Integumentary System

- The text now uses *subcutaneous layer* as the primary term and *hypodermis* as the secondary term.
- Revised Module 5.4 uses the term *bulbous corpuscle* instead of *Ruffini corpuscle* and the term *tension lines* instead of *cleavage lines*. The module also contains a new Everyday Physiology text box describing subcutaneous fat accumulation.
- Revised Module 5.8 contains a new micrograph showing a sebaceous gland.

## Chapter 6: Bones and Bone Structure

- The chapter has a new title (formerly titled Osseous Tissue and Bone Structure).
- Revised Module 6.2 uses *bone markings* as the primary term and *surface features* as the secondary term.
- Revised Module 6.3 contains an expanded discussion of the periosteum.
- Revised Module 6.5 includes updated art that depicts the location of nerves within bone and includes the term *trabecular bone*.

- Revised Module 6.6 includes updated art depicting the location of blood vessels and nerves in relation to bone.
- Revised Module 6.7 defines the term *interstitial growth* and contains a new Clinical Note on the epiphyseal line in x-rays.
- Revised Module 6.8 contains a new illustration and description of diploë.
- Revised Module 6.9 contains a new image depicting acromegaly.
- Revised Module 6.11 contains a new description of the role of calcitonin.
- Revised Module 6.12 contains new art of a broken and healing tibia (formerly humerus).

### Chapter 7: The Skeleton
- Revised Module 7.4 uses the term *forehead* instead of *frons* and clarifies the locations of the zygomatic process and temporal process.
- Revised Module 7.5 describes foramina of the skull by the bone in which they are located.
- Revised Module 7.7 describes landmarks of the skull by which bone they are a part of and hyphenates the terms *supra-orbital* and *infra-orbital*.
- Revised Module 7.8 describes the function of the mental foramen.
- Revised Module 7.9 uses the term *posterior fontanelle* instead of *occipital fontanelle* and contains a new illustration and description comparing the skulls of a fetus, newborn, and adult.
- Revised Module 7.11 contains new art of the 12 thoracic vertebrae.
- Revised Module 7.12 describes the functions of vertebral processes.
- Revised Module 7.13 (formerly 7.12) hyphenates the term *sacro-iliac*.
- Revised Module 7.17 (formerly 7.16) hyphenates the terms *humero-ulnar* and *radio-ulnar*.
- Revised Module 7.18 contains a description on the arrangement of the pelvis.
- New Module 7.21 summarizes the differences between the male and female skeletons.
- Revised Module 7.23 (formerly 7.21) elaborates on the difference between the medial and lateral parts of the longitudinal arch and contains a description of flatfeet.

### Chapter 8: Joints
- Revised Module 8.2 contains a description of the joint cavity and a new Clinical Note on dislocations.
- Revised Module 8.3 contains descriptions of joints based on the number of axes they move around and new art of the axes; uses the term *plane joint* instead of *gliding joint*; and contains an updated chart that describes each type of synovial joint and the movement of each type.
- Revised Module 8.7 has a new title and describes the three types of joints within the vertebral column.

- Revised Module 8.8 (formerly 8.7) describes intervertebral disc disease.
- Revised Module 8.9 (formerly 8.8) uses *ligament of the femoral head* as the primary term and *ligamentum teres* as the secondary term.
- Revised Chapter Review contains new questions, 21 and 26.

### Chapter 9: Skeletal Muscle Tissue
- Revised Module 9.1 contains new art of the types of muscle tissue.
- Revised Module 9.6 defines *synaptic cleft*.
- Revised Module 9.10 contains a new Everyday Physiology box explaining muscle tone.

### Chapter 10: The Muscular System
- Revised Module 10.2 contains new art illustrating the different types of levers.
- Revised Module 10.11 clarifies the perineal region.
- Revised Module 10.16 contains new art illustrating supination and pronation.
- Revised Module 10.18 contains a new Clinical Note on trigger finger.

### Chapter 11: Nervous Tissue
- Chapter title has been changed (formerly titled Neural Tissue).
- Revised Module 11.1 includes the enteric nervous system (ENS) as a third division of the nervous system; simplifies the description of sensory receptors; includes *afferent, efferent, voluntary nervous system*, and *involuntary nervous system* as secondary terms; and includes the parasympathetic and sympathetic divisions.
- Revised Module 11.2 contains a new Clinical Note on the loss of neurons.
- Revised Module 11.3 includes an updated flowchart to include the parasympathetic and sympathetic divisions.
- Revised 11.5 contains new art showing the myelination of an axon in the PNS and a new Clinical Note on nerve regeneration.
- Revised Module 11.10 contains new art showing the axon hillock and initial segment.

### Chapter 12: The Spinal Cord, Spinal Nerves, and Spinal Reflexes
- Uses the terms *posterior* and *anterior* in reference to spinal roots, ganglion, and rami instead of *dorsal* and *ventral*.
- Revised Module 12.2 uses the term *lumbosacral enlargement* instead of *lumbar enlargement*.
- Revised Module 12.3 clarifies the term *rootlets*.
- Revised Module 12.4 contains a new Clinical Note on the clinical importance of gray matter organization.
- Revised Module 12.5 contains a new Clinical Note on shingles.
- Revised Module 12.7 includes the term *lumbosacral plexus* and an updated chart elaborating on the nerves and distribution of the cervical plexus.

- Revised Module 12.9 (formerly 12.8) includes an updated chart elaborating on the brachial plexus and a new Clinical Note on locating nerve injuries in the hand.
- Revised Module 12.10 (formerly 12.9) includes updated charts elaborating on the lumbar and sacral plexuses and a new Clinical Note on locating nerve injuries in the foot.

## Chapter 13: The Brain, Cranial Nerves, and Sensory and Motor Pathways
- Revised Module 13.3 uses the term *dural venous sinus* instead of *dural sinus*.
- Revised Module 13.4 includes updated art color-coded to clarify points of interest and updated charts clarifying the parts of the medulla oblongata and the pons.
- Revised Module 13.6 (formerly 13.5) includes updated art color-coded to clarify points of interest and includes a new Clinical Note on ataxia.
- Revised Module 13.8 (formerly 13.7) uses the term *nuclei* instead of *group*; includes updated art that specifies the regions the thalamus projects to; and includes an updated chart on the hypothalamus.
- Revised Module 13.9 (formerly 13.8) contains new charts that elaborate on the parts of the limbic system.
- Revised Module 13.10 (formerly 13.9) includes an updated chart that elaborates on the functions of the parts of the basal nuclei.
- Revised Module 13.12 (formerly 13.11) uses the terms *somatosensory* instead of *somatic sensory* and *Wernicke's area* instead of *general interpretive area*.
- Revised Module 13.13 (formerly 13.12) elaborates on projection fibers.
- Revised Module 13.15 (formerly 13.14) updates terminology of the branches of the trigeminal and vestibulo-cochlear cranial nerves.
- Revised Module 13.16 (formerly 13.15) includes an updated flowchart of the sensory pathway.
- Revised Module 13.18 (formerly 13.17) uses the term *lamellar corpuscle* instead of *lamellated corpuscle* and the term *bulbous corpuscle* instead of *Ruffini corpuscle*.
- Revised Module 13.19 (formerly 13.18) uses the term *somatotropy* instead of *sensory homunculus*.
- Revised Module 13.21 (formerly 13.20) uses the term *premotor cortex* instead of *motor association areas*.

## Chapter 14: The Autonomic Nervous System
- Revised Module 14.9 includes updated art.
- Revised Module 14.11 includes updated art with a key.

## Chapter 15: The Special Senses
- Revised Module 15.6 (formerly 15.13) uses the term *canthus* instead of *angle of the eye* and the term *bulbar* instead of *ocular*. The module also contains a new Clinical Note on conjunctivitis.

- Revised Module 15.8 (formerly 15.15) contains new art to orient a close-up illustration, and it uses the term *dilator pupillae* instead of *pupillary dilator* and the term *sphincter pupillae* instead of *pupillary constrictor*.
- Revised Module 15.9 (formerly 15.16) elaborates on the effect of distance on light refraction.
- Revised Module 15.11 (formerly 15.18) contains a new Clinical Note on color blindness.
- Revised Module 15.16 (formerly 15.6) introduces the term *pinna*; elaborates on otitis media; and describes hair within the external acoustic meatus.
- Revised Module 15.18 (formerly 15.8) uses the term *ampullary crest* instead of *crista ampullaris* and the term *ampullary cupula* instead of *cupula*, and it differentiates between the maculae of the utricle and saccule.
- Revised Module 15.19 (formerly 15.9) states the magnification of the light micrograph depicting the cochlear section ($60\times$).

## Chapter 16: The Endocrine System
- In revised Module 16.1, the chart describing mechanisms of intercellular communications includes a new row featuring autocrine communication. The module also includes a new text box illuminating the similarities between the nervous and endocrine systems.
- Revised Module 16.7 (formerly 16.6) includes updated art and flowchart clarifying the negative feedback mechanism that controls secretions of the hypothalamus, pituitary gland, and endocrine target organs.
- Revised Module 16.9 (formerly 16.8) uses the term *principal cells* instead of *chief cells* and includes a new flowchart elucidating the regulation of blood calcium.
- Revised Module 16.11 (formerly 16.10) uses the term *pancreatic polypeptide cells* instead of *F cells* and includes a new flowchart elucidating the regulation of blood glucose.
- Revised Module 16.15 (formerly 16.14) includes a new flowchart elucidating the regulation of blood pressure and volume.
- Section 2 Review includes updated art and corresponding terms for the Labeling section.

## Chapter 17: Blood
- Revised module 17.2 includes updated art of the composition of blood.
- Revised Module 17.3 includes updated art highlighting the differentiation of the lymphocyte lineage as well as the types of blast cells.
- Revised Module 17.5 contains a new Everyday Physiology box that discusses a red blood cell's ability to carry oxygen.
- Revised Module 17.6 includes updated art clarifying the sequence red blood cell production and recycling.

- Revised Module 17.7 includes updated art of shapes of anti-A and anti-B antibodies; anti-Rh replaces anti-D; added "clumping" or "no clumping" under test results for clarification).
- Revised Module 17.10 discusses the role of thrombin and a positive feedback loop in blood clotting.
- Revised Section 2 Review contains a new Concept Map and a new Matching section.

## Chapter 18: The Heart and Cardiovascular Function
- The chapter uses *mitral valve* as the primary term and *left atrioventricular valve* as the secondary term.
- Revised Module 18.1 (formerly 19.1) introduces the four-chambered structure of the heart and contains a new illustration of the systemic and pulmonary circuits.
- Revised Module 18.2 (formerly 19.3) contains a new Clinical Note describing cardiac tamponade.
- Revised Module 18.3 (formerly 19.2) includes an updated chart clarifying the layers of the pericardium (uses *parietal layer of serous pericardium* as primary term replacing *parietal pericardium* and *visceral layer of serous pericardium* as primary term and *epicardium* as the secondary term)
- Revised Module 18.7 (formerly 19.7) contains a new Clinical Note discussing surgical replacement of damaged heart valves.
- Revised Module 18.11 (formerly 19.12) contains new illustrations of a skeletal muscle fiber and a cardiac muscle cell.
- Revised Module 18.12 (formerly 19.11) contains new ECG tracings paired with events of the cardiac cycle and conducting system.
- Revised Module 18.16 (formerly 19.15) includes an updated flowchart of factors affecting stroke volume.
- Revised Chapter Review contains new questions 10, 13, 14, and 15.

## Chapter 19: Blood Vessels and Circulation
- Revised Module 19.1 (formerly 18.1) includes new art to present the circulatory system more realistically and incorporates the terminology *afferent vessels* and *efferent vessels*.
- Revised Module 19.2 (formerly 18.2) contains new art of an artery portraying a thicker tunica media.
- Revised Module 19.3 (formerly 18.3) contains a new micrograph of a capillary bed.
- Revised Module 19.4 (formerly 18.4) discusses that because veins are distensible they can act as blood reservoirs.
- Revised Module 19.5 (formerly 19.17) elaborates on the relationship between venous return, venous pressure, and cardiac output, and it distinguishes between autoregulation and central regulation of blood flow.
- Revised Module 19.7 (formerly 19.19) includes updated art that shows the relationship between vessel luminal diameter and cross-sectional area.

- Revised Module 19.8 (formerly 19.20) includes updated art clarifying fluid movements across a capillary.
- Revised Module 19.9 (formerly 19.21) contains new art depicting the autoregulation of blood volume and pressure and new art depicting the baroreceptor reflex.
- Revised Module 19.10 (formerly 19.22) contains new art depicting the response to decreasing blood pressure and volume and the response to increasing blood pressure and volume.
- Revised Module 19.11 (formerly 19.23) contains new art depicting chemoreceptor reflexes.
- Revised Module 19.13 (formerly 19.25) contains new art depicting the short-term and long-term mechanisms that compensate for a reduction in blood volume.
- Revised Section 3 Review contains a new Matching section linked to new art.
- Revised Module 19.14 (formerly 18.5) defines *blood island*, distinguishes the terms *hemangioblast* and *angioblast*, and contains new art detailing the yolk sac and vasculogenesis.
- Revised Module 19.16 (formerly 18.7) contains a new Everyday Physiology box discussing the functionality of dual venous drainage in the neck and limbs.
- Revised Module 19.19 (formerly 18.10) includes *confluence of sinuses*.
- New Module 19.23 provides flowcharts summarizing the systemic arterial and venous circuits.

## Chapter 20: The Lymphatic System and Immunity
- Revised Module 20.1 describes the immune system as a functional system.
- Revised Module 20.2 notes that small to medium-sized lymphatics contain valves.
- Revised Module 20.3 contains a new Clinical Note describing lymphedema.
- Revised Module 20.4 includes an updated flowchart that describes regulatory and memory T cells.
- Revised Module 20.5 uses the term *paracortex* instead of *deep cortex* and includes updated art that shows the medulla of a lymph node.
- Revised Module 20.6 contains a new Clinical Note describing myasthenia gravis.
- Revised Module 20.7 contains a new Clinical Note describing the implications of a ruptured spleen.
- Revised Module 20.11 (formerly 20.10) includes updated charts on the function of NK cells and immunological escape.
- Revised Module 20.12 (formerly 20.11) contains new art and descriptions of the three pathways of complement action.
- Revised Module 20.13 (formerly 20.12) contains new descriptions of aspects of innate immunity.
- Section 2 Review contains new questions 13 and 14.
- Revised Module 20.14 (formerly 20.13) uses the term *acquired* instead of *induced*.
- Revised Module 20.16 (formerly 20.15) uses the term *regulatory T cells* instead of *suppressor T cells*.

- Revised Module 20.18 (formerly 20.17) uses *haptens* as the primary term and *partial antigens* as the secondary term.
- Revised Module 20.22 (formerly 20.21) uses the term *transplant rejection* instead of *graft rejection* and clarifies the functioning of HIV.

## Chapter 21: The Respiratory System
- Revised Module 21.2 uses the term *mucociliary escalator* instead of *mucus escalator,* and contains a new description of mucous glands and a new Clinical Note describing cystic fibrosis.
- Revised Module 21.3 contains a new Everyday Physiology box describing how the nasal mucosa warms and humidifies the air entering the nasal cavity. The module uses the term *dorsum of nose* instead of *bridge of the nose* and the term *nostrils* instead of *external nares.*
- Revised Module 21.5 contains new art of the trachea and esophagus.
- Revised Module 21.7 uses the term *blood air barrier* instead of *respiratory membrane.*
- Revised Module 21.11 contains the equation for anatomic dead space.
- Revised Module 21.12 contains new art to present the circulatory system more realistically.
- Revised Module 21.13 contains a new Clinical Note on the time limitations of storing blood in a blood bank.
- Revised Module 21.17 contains a new flowchart of the regulation of arterial $P_{CO_2}$.
- Revised Module 21.18 contains new art.

## Chapter 22: The Digestive System
- Revised Module 22.2 uses the term *muscular layer* instead of *muscularis externa* and the term *submucosal neural plexus* instead of *submucosal plexus*).
- Revised Module 22.4 contains a new Clinical Note describing congenital megacolon.
- Revised Module 22.6 clarifies the locations of the palatine tonsils and the palatoglossal and palatopharyngeal arches, and it describes ankyloglossia.
- Revised Module 22.7 uses the term *cement* instead of *cementum*; defines *dentition*; and contains a new Clinical Note describing an impacted tooth.
- Revised Module 22.10 describes the pyloric orifice.
- Revised Module 22.12 contains new art and descriptions of Paneth, stem, and epithelial cells.
- Revised Module 22.14 contains a new description of enterocrinin.
- Revised Module 22.15 contains new descriptions of the local and neural responses of the gastric phase and of the hormonal responses of the intestinal phase.
- Revised Module 22.17 includes updated art of the defecation reflex.
- Revised Module 22.21 uses *portal triad* as the primary term and *portal area* as the secondary term, the term *stellate macrophage* instead of *Kupffer cell*, and contains a new Clinical Note on portal hypertension.
- Revised Module 22.22 uses the term *bile duct* instead of *common bile duct.*

## Chapter 23: Metabolism and Energetics
- Revised Module 23.3 (formerly part of 23.7) on glycolysis now precedes discussion of the citric acid cycle (formerly 23.3).
- Revised Module 23.5 (formerly 23.4) defines *oxidation*, *reduction*, and *chemiosmosis* and labels protein complexes of the electron transport chain by roman numerals.
- Revised Module 23.6 (formerly part of 23.7) describes total ATP yield from metabolsm of a glucose molecule based on recent values of ATP yield per NADH (2.5 ATP vs. previous 3 ATP) and $FADH_2$ (1.5 ATP vs. previous 2 ATP).
- Revised Module 23.14 (formerly 23.12) replaces the term *vitamin $D_3$* with *vitamin D.*

## Chapter 24: The Urinary System
- Revised Module 24.4 contains a new micrograph of nephron loops.
- Revised Module 24.5 contains a new Everyday Physiology box describing the innervation of the kidneys.
- Revised Module 24.7 contains new descriptions of the parts of a nephron and new illustrations of renal structures.
- Revised Module 24.8 uses the term *capsular layer* instead of *parietal layer.* The parts of the juxtaglomerular complex are now labeled.
- Revised Module 24.9 contains a new flowchart of the regulation of the glomerular filtration rate and a new Everyday Physiology box on the reabsorption of glomerular filtrate.
- Revised Module 24.10 includes updated art of the reabsorption of the proximal convoluted tubule.
- Revised Module 24.11 includes updated art of the nephron loop.
- Revised Module 24.13 includes a new step 8 discussing papillary duct permeability to urea and new art showing urea transporter.
- Revised Module 24.16 describes the detrusor of the urinary bladder and includes updated art showing the blood supply to the kidneys.
- Revised Module 24.17 contains new art describing urinary storage and voiding.

## Chapter 25: Fluid, Electrolyte, and Acid-Base Balance
- Revised Module 25.1 defines *intracellular fluid* and *extracellular fluid.*
- Revised Module 25.2 uses the term *dietary intake* instead of *dietary input* or *ingestion.*
- Revised Module 25.3 discusses sports drinks.
- Revised Module 25.4 contains new flowcharts of the regulation of sodium concentration and ECF volume.
- Revised Module 25.6 uses the term *metabolic acid* instead of *organic acid.*
- Revised Module 25.10 contains new flowcharts of the regulation of normal acid-base balance.
- Section 2 Review contains a new Labeling section.

## Chapter 26: The Reproductive System

- The chapter uses the term *sperm* instead of *spermatozoa*.
- Revised Module 26.1 includes a new description of the male reproductive system in terms of *internal genitalia* and *external genitalia*.
- Revised Module 26.4 uses the term *interstitial endocrine cells* instead of *interstitial cells* and contains an expanded description of the histology of a testis.
- Revised Module 26.6 contains a new Clinical Note on impotence.
- Revised Module 26.8 clarifies the description of the female reproductive system and defines the mons pubis.
- Revised Module 26.9 hyphenates the terms *retro-uterine* and *vesico-uterine*.
- Revised Module 26.11 describes peg cells.
- Revised Module 26.12 uses the term *basal layer* instead of *basilar zone* the term and *functional layer* instead of *functional zone*.
- Revised Module 26.13 contains a new Everyday Physiology box discussing breast size.
- Revised Module 26.15 includes an updated chart that depicts the GnRH pulse frequency. Text in Follicular Phase of the Ovarian Cycle box changed to reflect that one tertiary follicle from a group becomes dominant; *Tertiary ovarian follicle development* label replaces *Follicle development* label; temperature ranges changed for both Celsius and Fahrenheit scales; and Menses label changed to Menstrual Phase.

## Chapter 27: Development and Inheritance

- Revised Module 27.1 defines the term *pregnancy*.
- Revised Module 27.2 fertilization step titles and text in step art and clarified when DNA synthesis occurs; added a new Clinical Note on male sterility.
- Revised Module 27.3 includes updated art that shows implantation occurring over 6-9 days after fertilization, and uses *cytotrophoblast* instead of *cellular trophoblast* and *syncytiotrophoblast* instead of *syncytial trophoblast*.
- Revised Module 27.4 contains a new Clinical Note describing gestational trophoblastic neoplasia.
- Revised Module 27.5 uses the term *extra-embryonic* instead of *extraembryonic*.
- Revised Module 27.8 contains new art depicting the embryo after 3 weeks of development.
- Revised Module 27.9 contains a new Clinical Note describing the correlation between maternal age and medical risks during pregnancy.
- Revised Module 27.10 contains a new Clinical Note on the implications of premature labor.
- Revised Module 27.14 uses the term *autosomes* for autosomal chromosomes.
- Revised Module 27.16 (formerly 27.15) discusses incomplete dominance.
- Revised Module 27.17 (formerly 27.16) uses the term *sickle cell disease* instead of *sickle cell anemia* and defines *epigenetics*.

# Acknowledgments

This textbook represents a group effort, and we would like to acknowledge the people who worked together with us to create this Third Edition.

Foremost on our thank-you list are the instructors who offered invaluable suggestions throughout the revision process. We thank them for their participation and list their names and affiliations below.

We also express our appreciation to the editors and support staff at Pearson Science.

We owe special thanks to Senior Acquisitions Editor Cheryl Cechvala for shepherding this project from start to finish. Her ability to manage every detail with such fervor and interest is commendable. She also has an incredible command of the English language coupled with exceptional oratory ability—having candid conversations with Cheryl is always fun. Although this was her first year as our editor, she possesses the skills of a seasoned veteran. She is our biggest advocate and is always willing to champion our cause, despite the challenges of working with authors. We appreciate all her efforts on our behalf.

Content Producer Arielle Grant was incredibly skilled at keeping this project moving forward. Throughout every iteration, she kept track of the files, ensured we were on task, and maintained her high standards. Working with authors can be challenging, and Arielle was up for the task! Editorial Coordinator Kimberly Twardochleb was always available and answered every question we had with speed and accuracy. Having two highly skilled professionals working with us eased our burden. Thanks for not only preparing our material for publication, but also making sure it was the best it could possibly be. This past year could not have happened without them.

We wish to thank several Development Editors. Molly Ward started this project and provided suggestions for many of the early chapters. Alice Fugate and Tanya Martin picked up "midstream," using their collective skills to truly improve readability, and Suzanne Olivier came on board to review selected chapters to ensure student accessibility. Thanks to this editorial entourage!

We are grateful to Sally Peyrefitte for her very careful attention to detail and consistency in copyediting the text and art.

A special note of thanks is warranted for "Lt Dan" Lykins at the University of Colorado. Whenever he sent a suggestion, we knew it would be good. Thanks for appreciating our approach and wanting to make the book better with us.

The authors would also like to thank a few talented people who work in our virtual world. We especially thank Anita Impagliazzo for her continued creativity on this edition. Your artwork is superb! We are also appreciative of the work done by Jim Gibson and Christopher Gibson. Their attention to detail and running dialogue with us were instrumental in getting the pages right. Kudos Anita, Jim, and Christopher.

This book would not exist without the extraordinary dedication of the Production team. We are thankful for Mary Tindle at Cenveo, who skillfully led this book through the various phases of publishing.

We also wish to thank Allison Rona, Executive Marketing Manager, and the dedicated Pearson Science sales representatives for their continuing support of this project. Special thanks to Editor-in-Chief Serina Beauparlant, who took over the reins and worked closely with our new editor, Cheryl Cechvala, to ensure we had the resources necessary to publish what students need to succeed and what professors want in a textbook. And a round of applause and a backflip go out to Derek Perrigo, Senior A&P Specialist, our biggest cheerleader.

To help improve future editions, we encourage you to send any pertinent information, suggestions, or comments about the organization or content of this textbook to us directly, using the e-mail addresses to the right. We warmly welcome comments and suggestions and will carefully consider them in the preparation of the Fourth Edition.

*Frederic (Ric) H. Martini*
*Haiku, Hawaii*
martini@pearson.com

*Judi L. Nath*
*Sandusky, Ohio*
nath@pearson.com

*Edwin F. Bartholomew*
*Lahaina, Hawaii*
bartholomew@pearson.com

# Contents

# 6 Bones and Bone Structure   210

# 7 The Skeleton   240

## 8   Joints   288

## 9   Skeletal Muscle Tissue   314

# 17 Blood 630

# 18 The Heart and Cardiovascular Function 658

# 19 Blood Vessels and Circulation 696

# 20 The Lymphatic System and Immunity 750

## 22 The Digestive System 838

## 21 The Respiratory System 796

# 23 Metabolism, Nutrition, and Energetics 892

# 24 The Urinary System 932

# 27 Development and Inheritance 1036

# visual

## anatomy &

## physiology

### 3rd edition

**MARTINI | OBER | NATH | BARTHOLOMEW | PETTI**

## LEARNING OUTCOMES

These Learning Outcomes correspond by number to this chapter's modules and indicate what you should be able to do after completing the chapter. They also appear at the end of each module.

# 1

# An Introduction to Anatomy & Physiology

## Module 1.1
# Using your textbook effectively is key to your success

Welcome to the field of human anatomy and physiology—known simply as A&P! In this textbook we will introduce you to the inner workings of the human body, giving information about both its structure (anatomy) and its function (physiology). Many students who use this book are preparing for jobs in health-related fields, but regardless of your career choice, you will find the information within these pages relevant to your future.

**1** We will focus on the human body, but the principles you will learn apply to other living things as well. Our world contains an enormous diversity of living organisms, which vary widely in appearance and lifestyle. One aim of *biology*—the study of life—is to discover the unity and the patterns that underlie this diversity. As we study human A&P, we'll discover the interrelationship between structure and function—structure determines function. In much the same way, your textbook was intentionally structured so that it functions to enhance your learning.

**?  REVIEW**
**A.** What is human A&P?
**B.** Define the term *biology*.

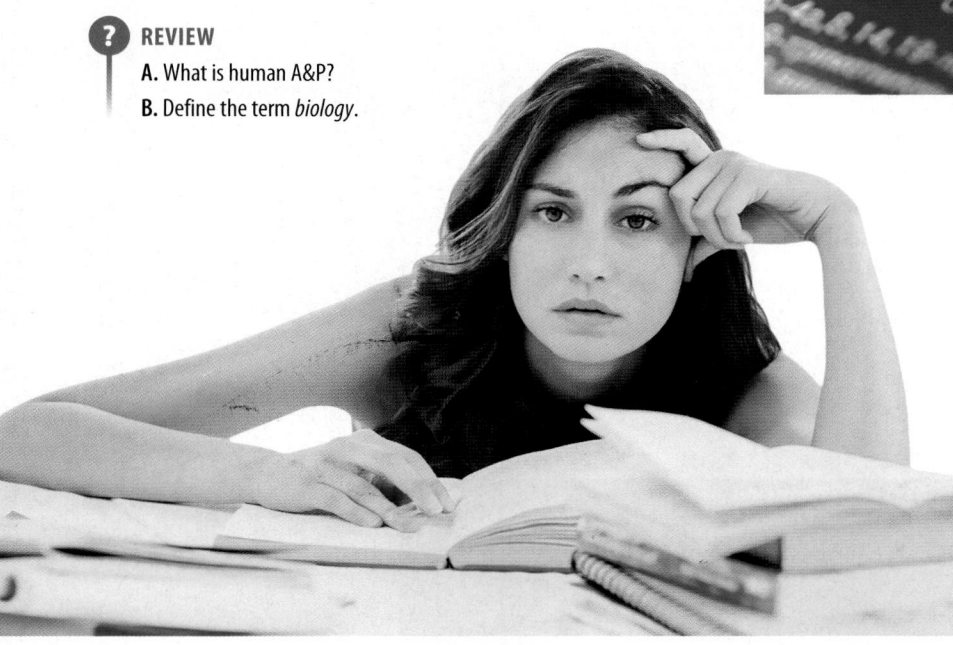

**2** Before we begin with the science of human A&P, let's take a few minutes to examine how you can best use this textbook. Research shows that many students underuse or even fail to use their textbooks. This is unfortunate, because when used effectively, your book can play a key role by helping you understand the concepts needed to pass your course. You will find that the illustrations are especially important, much more so than in other textbooks you have used in the past.

This introductory section sets the stage for your success in this course and introduces you to the basic principles of learning based on current research studies. Just as there are underlying concepts in A&P, there are basic strategies to using your A&P textbook effectively. If you master these techniques now, they will assist you not just in this course, but throughout your college career. Let's get started.

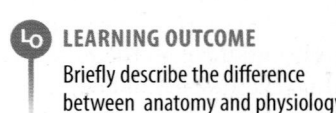 **LEARNING OUTCOME**
Briefly describe the difference between anatomy and physiology.

# Module 1.2
# Comprehending the art is essential to understanding A&P

Think back to your first childhood book. You most likely began with a "picture book." Then, as you learned the alphabet and developed speech, you progressed to "word books." The next step was "chapter books." Somewhere along the way, you quit looking at pictures (art) and focused solely on the words (text). Maybe the shift to text-based reading without looking at the pictures happened in high school. You began reading words—paragraph after paragraph, page after page of words. Many of your books may have been colorful and filled with pictures, but you quickly decided that most were decorative and that the real information was to be found in the words. To succeed in a college science course, you need to break this pattern, shift your focus, and integrate information presented in the art as well as the words.

**1** In college, you are faced with many new terms, abstract concepts, and unfamiliar images. That's great, because college is intended to increase your knowledge and expand your horizons. But research has shown that undergraduate students have a tendency to simply read the text (also called the narrative) without paying attention to the art (referred to as figures or diagrams in your book). While you can certainly learn from this approach, there is abundant research showing that paying attention to the art while reading the text improves learning.

For example, researchers have set up studies that use eye-tracking equipment to measure eye movement and interaction with images on pages. They found that students spent very little time looking at the art while they were reading the text. When students were trained to read the text and view the art together, their time spent looking at the art increased. This attention to the art was critical: *Learning and comprehension levels were greater for students who studied both the art and the text together than for students who only read the text.*

Each student was given 30 minutes to study the material on a computer screen and then was tested to see how well they understood the content.

Eye-tracking equipment measured how long students focused their attention on either the written material or the visual material.

**2** Although reading text and viewing art side by side may sound like common sense, most students do not do that. We wrote this book to make it easier for you to study the text and the associated art at the same time and on the same page, without the page-flipping required by a traditional textbook format. In this book, the text and the art go hand in hand. Please continue reading as we walk you through the process of using this textbook to enhance your learning.

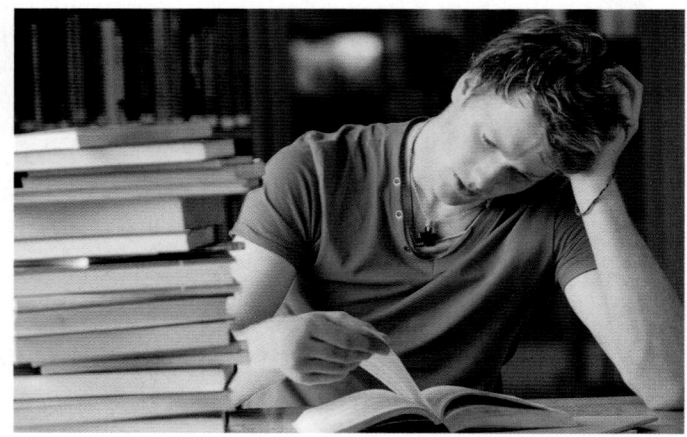

**? REVIEW**

**A.** What do eye-tracking studies tell us about the most effective way to learn?

**3** Here is an example of text in a narrative form and an illustration of the structure of a pencil. In most textbooks a figure will be as close to its descriptive lines of text as possible, but in some cases that may mean turning forward or backward a page or two to find it. You're expected to find and inspect a figure when you see a callout for it. The figure callout usually looks like this: (see Figure 1 or Figure 1). They are often intentionally color-coded so you can stop reading, look at the figure, and then find your place again when you go back to reading the text.

A pencil is a writing implement or art tool used to make marks on paper, canvas, or other surfaces. It is a long, round, slender instrument consisting of a thin stick of graphite enclosed in a long thin piece of wood. A typical pencil is approximately 8" long, 3/8" in diameter, and hexagonal in cross-section. On one end of the pencil is an eraser that is used to remove marks on paper. The opposite end of the pencil is pointed. The point consists of a semi-soft material called graphite that forms the core of the pencil shaft. Graphite is a form of carbon that easily rubs off from the pencil to the paper forming a mark. The pencil is held between the thumb and forefinger. **See Figure 1**

Eraser

Graphite tip

**Figure 1.** Structure of a typical pencil

**4** This example shows a different approach. Here the text and art are combined so key written information is integrated into the art itself. This **text-art integration** means you never have to flip pages or hunt for an illustration associated with specific text. Educational research shows such integration works best for comprehension. This book was written using this approach. After you answer the review question below, turn the page to see how text-art integration works with a topic about anatomy and physiology.

A pencil is a writing implement or art tool used to make marks on paper, canvas, or other surfaces. It is a long, round, slender instrument consisting of a thin stick of graphite enclosed in a long thin piece of wood.

The pointed tip is a sharpened piece of graphite, a form of carbon.

A wooden sheath surrounds the graphite core, keeping your hands clean and providing a comfortable gripping surface.

A soft rubber eraser is used to remove graphite marks.

Wooden sheath
Graphite core
|←3/8"→|
**Cross-sectional view**
8" in length

Mark made by graphite tip

**?** **REVIEW**

**B.** Try this experiment. Pick one of the two examples, cover up the text, and focus solely on the art. What did you discover?

**Lo** **LEARNING OUTCOME**

Describe how to use the text and art together to master learning.

# Module 1.3
# Break down the art in step-wise fashion to learn the topic

**1** If you were asked to calculate the area of a complex geometric shape like the one pictured here, with no further instruction, you might become overwhelmed and give up. But if you break down the figure into several smaller, easier-to-calculate shapes, your task becomes much easier.

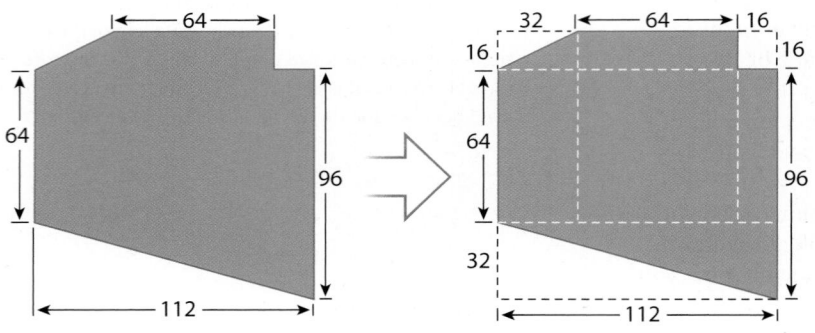

**? REVIEW**

**A.** Calculate the area in blue. All numbers are in centimeters (cm). If you need help, scan the QR code in the upper right corner on the opposite page.

**2** Nephrons (NEF-ronz) are the functional units of kidneys, and understanding what's happening in them is quite complicated. This figure shows a nephron, and it is part of a larger figure found in Chapter 24. Using what you just learned, examine this figure. What did you do first?

Let's break it down. Look at the figure, four blocks of grayed-out text, and the variety of arrows. Because you rely on both the art and the text, look at the key below and read the explanations for the various types of arrows; then come back and read the next paragraph.

After studying the key, begin at circled number 1 and follow in numerical order. Read through the associated text, and look at the adjacent figure part. Follow the numbered boxes in sequence. Although the nephron as a whole can be overwhelming, it becomes easier to understand when you break it down into separate parts. Boxed labels provide more than a specific structural term. This is the highlight of text-art integration.

Make sure you understand the meanings, similarities, and differences among these arrows.

**2** Proximal Convoluted Tubule

**4** Distal Convoluted Tubule

Renal tubule

Descending limb of loop begins

Ascending limb of loop ends

Ascending limb

Descending limb

Glomerular capsule

Capsular space

Glomerulus

**1** Renal Corpuscle

**3** Nephron Loop

Thin limb

Thick limb

**KEY**
- Filtrate
- Water reabsorption
- Variable water reabsorption
- Solute reabsorption or secretion
- Variable solute reabsorption or secretion
- Path of fluid within nephron

**3** A similar step-by-step approach will help you navigate other tough topics in A&P. Just like any textbook written in English, there is a purposeful flow to a page. While you are already used to reading left-to-right, column-by-column, the layouts in this book were created by removing the traditional white spaces and opening up the pages. This arrangement allows the narrative to be broken apart into blocks of text that are positioned next to the relevant art. There is no stumbling around to find the art that goes along with the text.

**QR codes.** When you see QR codes, scan them with either a smart phone or tablet to access additional material about a subject.

SmartArt

**Every new one- or two-page module starts with a title.** This is the main topic of the module. From there, you can proceed logically—just like when using your new cell phone or computer.

**Look for these numbered red blocks.** Each one indicates a text-art unit (indicated here by wide blue outlines).

- Read the text describing the image.
- Study the image and its related text describing key structures or functions.
- Answer any related "Review" questions.
- Move on to the next numbered red block.

## Module 13.1
# The brain develops from a hollow neural tube

**1**

Spinal cord

Neurocoel

**?** REVIEW

**2**

**Prosencephalon**

**Rhombencephalon**

Spinal cord

**?** REVIEW

Cerebrum   Diencephalon   Midbrain

Pons

Medulla oblongata

Cerebellum

Spinal cord

**3**

**Integration question.** After completing all of the text-art units in a module, answer any integration questions.

**?** INTEGRATION

 **LEARNING OUTCOME**
Explain how to approach complex concepts with multiple parts.

# Module 1.4
## Orient yourself to all art in the same way

**1** Anatomists and clinicians use the anatomical position as a standard point of reference. In the anatomical position, a person is standing erect, with the face gazing forward (anteriorly), arms at the side, and palms facing anteriorly. A person lying face upward is in the supine position. It is used for orienting and describing body structures and cross sections (views of internal organs).

Anatomical position

Supine position

**? REVIEW**
**A.** Look at the woman standing. On which wrist is her bracelet found?

**2** Always put yourself in the position of the image. Find your point of reference, and use a common language so everybody is talking about the same thing. The picture on the left shows an entire apple with a worm poking out of the side. If we section the apple, we can better see the interior structure of the apple and the locations of the sliced worm inside the apple.

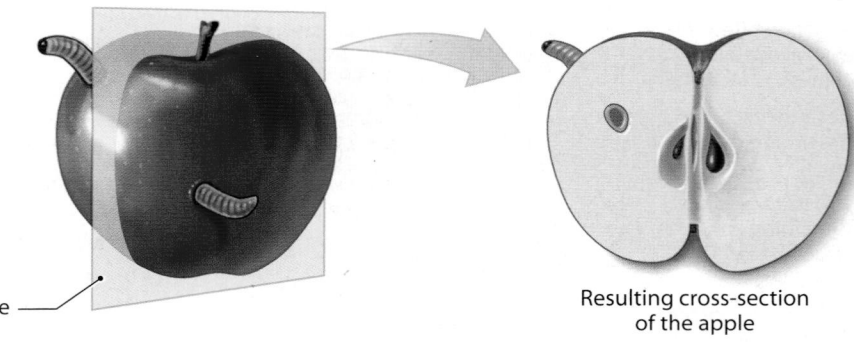

Cutting plane

Resulting cross-section of the apple

**3** All body cross sections in this text are oriented from the same perspective. Cross-sectional views are presented as they would appear in a CT or MRI scan in real life. To read them, orient yourself as though you are standing at the feet of a person who is in the supine position and you are looking toward their head.

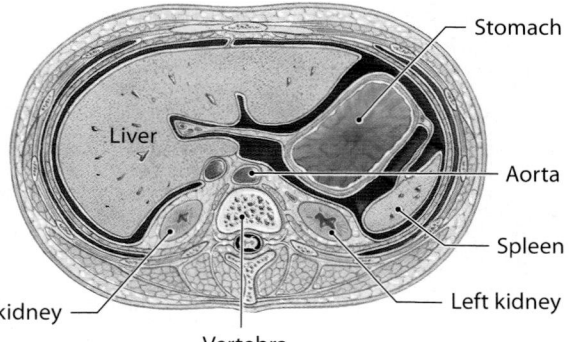

Stomach
Liver
Aorta
Spleen
Right kidney
Left kidney
Vertebra

Orientation of viewer

**? REVIEW**
**B.** What is the structure marked with an X in the MRI scan?

MRI scan
Liver
X

**Lo LEARNING OUTCOME**
Describe the anatomical position and how you should view sectional images.

# Module 1.5
## The learning outcomes correspond by number to the chapter's modules and indicate what you should be able to do after completing the chapter

Each chapter in this textbook is broken down into *sections* containing a series of one- or two-page *modules* with *text-art integration* and specific *learning outcomes*. The learning outcomes for each module are based on a *learning classification scheme*. Let's first explain these terms.

**1** A **module** is an independent, self-contained unit about a specific topic. The modules build on previously learned topics. When these modules are placed in a series, they become interdependent learning units called **sections**. The sectional layout promotes logical, efficient navigation through the material, and callouts to figures integrate the text with the art. Remember, text-art integration is crucial to understanding: Stop reading, look at the figure, and study the image along with the text.

**?** **REVIEW**
**A.** Define *module*, and state where the learning outcomes appear.

**2** **Learning outcomes** are educational objectives that use key verbs to target knowledge, comprehension, and specific skills. They indicate what you should be able to do after completing a chapter. The learning outcomes appear at the beginning of each chapter and in the lower right-hand corner of each module. Your instructor will test you on these learning outcomes, which are also tied to the questions in MasteringA&P™. Before moving on to the next module, be sure you can accomplish the task given in the learning outcome.

**3** Pay attention to these learning outcomes because they are tied directly to testing and summarize what you should be able to do after reading the text and studying the images of each module. These learning outcomes are based on a **learning classification scheme**, which identifies the fundamental levels of learning, from lower-order thinking skills to higher-order skills. You'll see these terms used in your listed learning outcomes, and you'll see some overlap within the scheme. (Here is where you can practice using what you just learned: Look at this figure, think about it, and then return to this text.) If you practice these basic strategies—that is, read the text, study the image, and pay attention to the learning outcomes—you are well on your way to success!

Module 19.4
**The boundaries between the chambers of the heart can be identified on its external surface**

Aortic arch
Ascending aorta
Superior vena cava
Right atrium
Right ventricle
Pulmonary trunk
Auricle of left atrium
Fat and vessels in the anterior interventricular sulcus
Left ventricle
Anterior surface

**Lo LEARNING OUTCOME**
Describe the cardiac chambers and the heart's external anatomy.

**Lo LEARNING OUTCOME**
Describe the cardiac chambers and the heart's external anatomy.

Knowing and remembering — Lower-order skills
Comprehending and understanding
Applying and describing
Synthesizing and creating
Analyzing and explaining
Evaluating and measuring
Higher-order skills

**?** **REVIEW**
**B.** Describe *learning classification scheme*.

**Lo LEARNING OUTCOME**
Explain the significance of learning outcomes in acquiring knowledge and skills.

# Section 1 Review
## An Introduction to Studying the Human Body

**Matching:** Match each lettered term with the most closely related description.

a. chapter module

b. chapter section

c. text-art integration

d. learning outcomes

e. learning classification scheme

**1** Educational objectives that use key verbs to target knowledge, comprehension, and specific skills

**2** An independent learning unit consisting of modules placed in a series

**3** A classification of the fundamental levels of learning, which includes lower-order and higher-order skills

**4** An independent, self-contained unit about a specific topic

**5** A textbook design where key information is found in both the text and art, and which are meant to be studied together

**1** _____

**2** _____

**3** _____

**4** _____

**5** _____

**Fill-in:** Fill in the missing word or phrase in the following statements.

**6** Learning and comprehension are greater for students who look at _____ instead of just reading the text.

**7** If symbols appear in an image that you don't understand (such as colored arrows pointing to different areas), a good place to look for an explanation besides reading the text would be the _____ .

**8** All images are oriented in the _____ , which means you are looking at a person who is standing erect, with the face gazing forward (anteriorly), arms at the side, and palms facing anteriorly.

**9** The learning outcomes can be found at the _____ and at the lower right-hand corner at the end of each module.

**Vocabulary:** Write each of the following terms under the proper heading.

• creating
• comprehending
• knowing
• analyzing
• applying
• remembering
• evaluating

**10** Lower-order skill

_____
_____
_____
_____
_____

**11** Higher-order Skill

_____
_____
_____
_____
_____

**Short answer:** In most A&P textbooks, figures are referenced somewhere within the narrative (such as See Figure 1-1) and too often appear on a different page from the related text. How does this description contrast with this textbook?

**12** _____
_____
_____
_____

# Module 1.6
# Focused study is important for learning anatomy and physiology

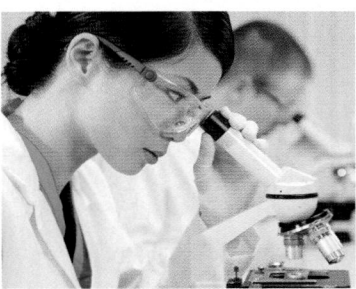

Human anatomy and physiology, or A&P, is about how the human body performs the functions that keep you alive and alert. You will learn many interesting and important facts about the human body as we proceed. However, the approach to learning those facts and the attitude you develop will be at least as important as the facts you memorize.

The basic approach in A&P can be summed up as "What is that structure, and how does it work?" The complexity of the answer depends on the level of detail you need. In science, if we know what something does but not how it works, it's usually called a "Black Box." The more you learn, the smaller (and more numerous) those Black Boxes become. That is, the more you learn, the more you realize how much you don't know.

We will devote considerable time to explaining how the body responds to normal and abnormal conditions and maintains **homeostasis**, a relatively stable internal environment. As we proceed, you will see how your body's anatomy and physiology work together to cope with injury, disease, or anything else that threatens homeostasis.

## Tips on How to Succeed in Your A&P Course

- **Approach the information in different ways.** For example, you might visualize the information with your own diagrams, talk it over with or "teach" a fellow student, or spend additional time in lab asking questions of your lab instructor.
- **Set up a study schedule** and stick to it.
- **Devote a block of time each day** to your A&P course.
- **Practice memorization.** Memorization is an important skill and an integral part of the course. You are going to have to memorize all sorts of things—among them muscle names, directional terms, and the names of bones and brain parts. Realize that this is an important study skill and that the more you practice, the better you will be at remembering terms and definitions. We will try to give you handles and tricks along the way to help you keep the information in mind.
- **Avoid shortcuts.** Actually there are no shortcuts. (Sorry.) You won't get the grade you want if you don't put in the time and do the work. This requires preparation throughout the term.
- **Attend all lectures, labs, and study sessions.** Ask questions and participate in discussions.
- **Read your lecture and lab assignments** before coming to class.
- **Do not procrastinate!** Do not do all your studying the night before the exam! Actually STUDY the material several times throughout the week. Marathon study sessions are often counterproductive. There is no easy learning button; you must push yourself.
- **Seek assistance as soon as possible if you have a problem understanding the material.** Do not wait until the end of the term, when it is too late to salvage your grade.

 **REVIEW**

**A.** What do scientists mean when they use the term "Black Box"?

**B.** Identify several strategies for success in this course.

 **LEARNING OUTCOME**
Describe homeostasis, and identify basic study skill strategies to use in this course.

# Module 1.7
# Organisms share common characteristics and processes

The world around us contains a variety of living organisms that look and behave differently. Despite this diversity, all living organisms share common characteristics.

## Common Characteristics of Life

**1** How do we know something is alive? Life is difficult to define; however, there are common characteristics that are regarded as essential to living things. Knowledge of these characteristics provides a framework from which to make predictions about the structure and function of organisms yet to be discovered. In our world, these common characteristics include the following:

| Characteristics of Life | Importance |
|---|---|
| **Cells** | Cells are the smallest structural and functional units of organisms. All living things are composed of one or more cells. Unicellular organisms are made up of one cell; multicellular organisms consist of more than one cell. |
| **Organization** Large molecules / Heart cell | All living things have a structural arrangement more complex than non-living things. This organization ranges from the chemical level to the more complex organism level. At the chemical level, combinations of large molecules such as carbohydrates, fats, proteins, and nucleic acids form cellular structures that make up tissues, organs, organ systems, and organisms. |
| **Responsiveness** Painful stimulus | Living things have the ability to detect and respond to stimuli (things that cause a reaction) in their immediate environment. For example, plants orient to the sun, or you react when touching a hot object. |
| **Regulation** Body temperature homeostasis / Normal range | Living things have mechanisms to maintain a relatively stable internal environment as external conditions fluctuate. This stabilizing process is called *homeostasis*. For example, your body keeps your temperature within a narrow range despite how cold or hot it is outside. |
| **Growth and Development** | Over time, organisms grow and develop. Growth is an increase in size, and it occurs as part of development. Development is the natural progression in physical maturation, such as a seed becoming a plant. Your brain began as a hollow tube that developed into a complex organ that can think about itself. |
| **Reproduction** Sperm / Egg | The process by which organisms produce offspring is called reproduction. Reproduction creates subsequent generations of the same kind of organisms. |
| **Metabolism** Catabolic chemical reaction | Metabolism refers to all the essential chemical processes that take place in living cells and organisms. Metabolism includes both anabolic (building up) processes, such as the synthesis of large molecules, and catabolic (breaking down) processes, such as the breakdown of large molecules to release energy. |

**? REVIEW**

**A.** List the common characteristics shared by all living things.

**? REVIEW**

**B.** Distinguish between growth and development.

## Basic Processes in Humans and Other Animals

**2** Just as there are common characteristics of life, there are basic processes that support the demands of living organisms. These processes require energy, which must be continually replaced as it is used. The basic processes in humans and other animals include the following:

**? REVIEW**
**C.** Describe the basic processes in humans and other animals.

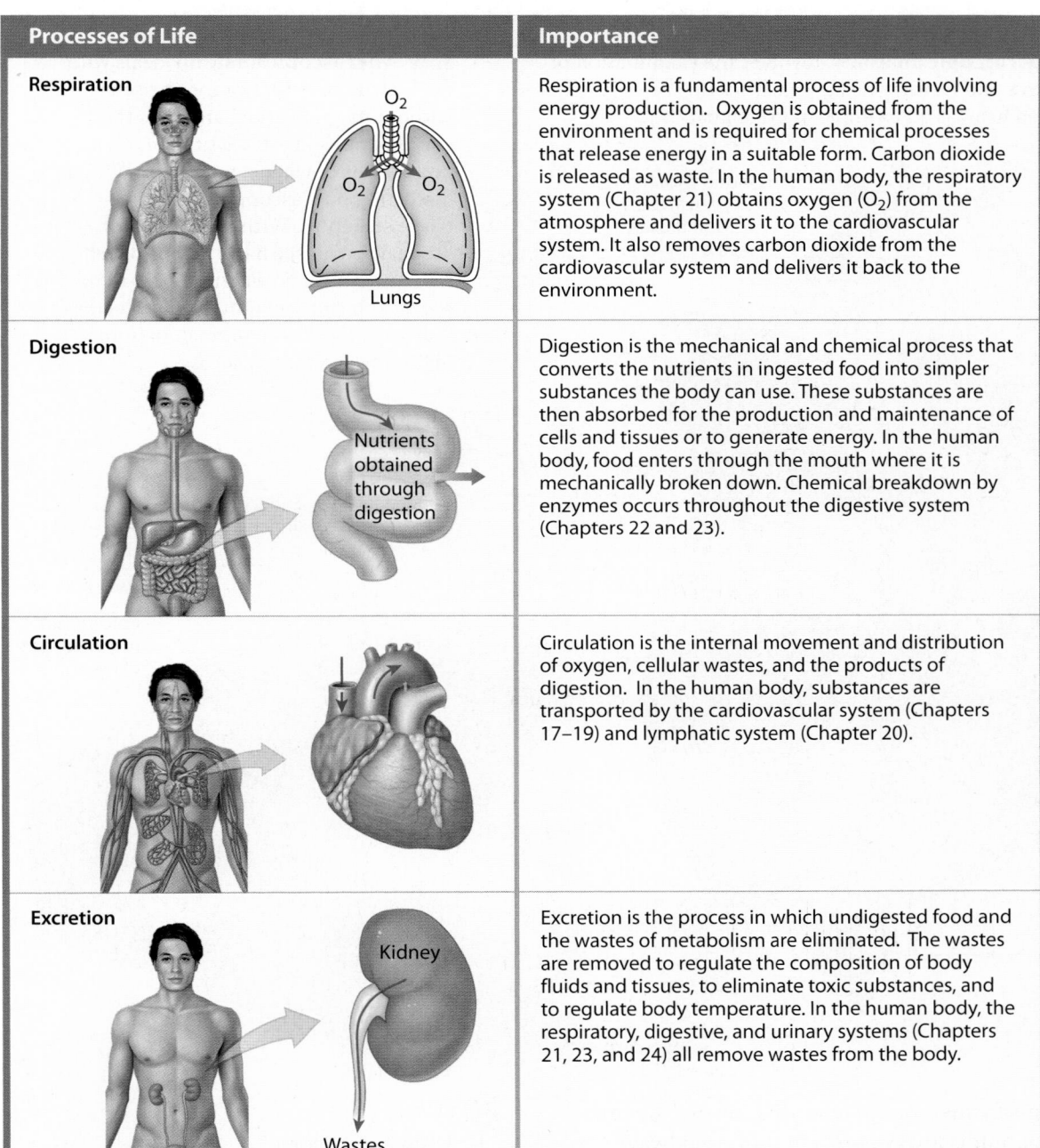

| Processes of Life | Importance |
|---|---|
| **Respiration** | Respiration is a fundamental process of life involving energy production. Oxygen is obtained from the environment and is required for chemical processes that release energy in a suitable form. Carbon dioxide is released as waste. In the human body, the respiratory system (Chapter 21) obtains oxygen ($O_2$) from the atmosphere and delivers it to the cardiovascular system. It also removes carbon dioxide from the cardiovascular system and delivers it back to the environment. |
| **Digestion** | Digestion is the mechanical and chemical process that converts the nutrients in ingested food into simpler substances the body can use. These substances are then absorbed for the production and maintenance of cells and tissues or to generate energy. In the human body, food enters through the mouth where it is mechanically broken down. Chemical breakdown by enzymes occurs throughout the digestive system (Chapters 22 and 23). |
| **Circulation** | Circulation is the internal movement and distribution of oxygen, cellular wastes, and the products of digestion. In the human body, substances are transported by the cardiovascular system (Chapters 17–19) and lymphatic system (Chapter 20). |
| **Excretion** | Excretion is the process in which undigested food and the wastes of metabolism are eliminated. The wastes are removed to regulate the composition of body fluids and tissues, to eliminate toxic substances, and to regulate body temperature. In the human body, the respiratory, digestive, and urinary systems (Chapters 21, 23, and 24) all remove wastes from the body. |

You'll notice that these basic processes are the central themes throughout this textbook. As you work your way through each chapter, pay close attention to the interrelationships among these topics. In the next 26 chapters we will consider the mechanics of each of these vital processes. Although our focus is on the processes of the human body specifically, the basic concepts have broad application in biology.

**Lo LEARNING OUTCOME**
Describe the common characteristics of life and the basic processes in humans and other animals.

# Module 1.8

# Anatomy is the study of structure …

**Anatomy**, which means "a cutting open," is the study of internal and external structures of the body and the physical relationships among body parts. Here is an overview of the anatomy of the heart, with the walls opened so that you can see the complexity of its internal structure.

**1** **Gross anatomy**, or **macroscopic anatomy**, involves the examination of relatively large structures and features usually visible with the unaided eye. This illustration of a dissected heart is an example of gross anatomy.

**2** **Microscopic anatomy** deals with structures that cannot be seen without magnification, and thus the equipment used establishes the boundaries of what can be seen. With a dissecting microscope, you can see tissue structure. With a light microscope, you can see basic details of cell structure. With an electron microscope, you can see individual molecules that are only a few nanometers (nm; billionths of a meter) across.

Blood with low oxygen and high carbon dioxide levels

Blood with high oxygen and low carbon dioxide levels

Pulmonary trunk

Superior vena cava

Ascending aorta

Left atrium

Right atrium

Left ventricle

Right ventricle

Inferior vena cava

Descending aorta

Endocardium (inner lining of heart)

Myocardium (heart muscle)

Epicardium (outer surface of heart)

All specific functions are performed by specific structures. The link between structure and function is always present, but not always understood. For example, although the anatomy of the heart was clearly described in the 15th century, almost 200 years passed before the heart's pumping action was demonstrated.

**?** **REVIEW**

**A.** What are the differences between gross anatomy and microscopic anatomy?

# ... and physiology is the study of function

**Physiology** is the study of function and how living organisms perform their vital functions. These functions are complex and much more difficult to examine than most anatomical structures. A physiologist looking at the heart focuses on its functional properties, such as the timing and sequence of the heartbeat, and its effects on blood pressure in the major arteries.

**3** The heartbeat is coordinated by electrical events within the heart muscle. Those electrical events can be detected by monitoring electrodes placed on the body surface. A record of these electrical events is called an electrocardiogram, or ECG.

**4** As the heart beats, pressure rises and falls within the major arteries and the chambers of the heart. Blood pressure in the major arteries must be maintained within normal limits to prevent vessel damage (from high pressures) or vessel collapse (from low pressures).

**? INTEGRATION**

**B.** Explain the link between anatomy and physiology.

**Lo LEARNING OUTCOME**

Define anatomy and physiology, and describe macroscopic and microscopic anatomy.

# Module 1.9
# Structure and function are interrelated

Physiology and anatomy are closely interrelated both theoretically and practically. Anatomical details are significant only because each has an effect on function, and physiological mechanisms can be fully understood only in terms of the underlying structural relationships.

**1** This relationship is easily understood at the gross anatomical level. You are well aware that your elbow joint functions like a hinge. It lets your forearm move toward or away from your shoulder, but it does not allow twisting at the joint. These functional limits are imposed by the internal structure of the joint.

Radius

Ulna

Humerus

The elbow joint functions as a hinge that permits movement only in one plane.

The ulna is the larger of the two bones of the forearm. It has a broad, deep depression into which the end of the humerus fits. Note the corresponding projections that help hold the humerus in position and prevent twisting.

The end of the humerus, the upper arm bone, has a cylindrical surface that is part of the elbow joint.

Ligaments

Humerus

With this interlocking arrangement, which is stabilized by ligaments and surrounding muscles, only hinge-like movement is permitted.

Ulna

 **REVIEW**

**A.** Compare the functioning of the elbow joint with a door on a hinge.

**B.** Predict what would happen to the function of a structure if its anatomy were altered.

**2** The relationship between structure and function also applies at the chemical level. Cells throughout your body communicate with one another through the use of chemical messengers, which you will learn more about in later chapters. Cellular communication usually involves the attachment of the chemical messenger released by one cell to a receptor molecule at another cell. That attachment depends on the three-dimensional shapes of the messenger and the receptor and on how well they fit together.

Each chemical compound has a specific size and three-dimensional shape. Chemical messengers come in a variety of sizes and shapes. In general, they are released by one cell to affect other cells.

Receptor molecules can be on the outer surface or inside the cell. This figure shows one receptor that extends across the plasma membrane (boundary of a cell). The plasma membrane is also known as the cell membrane.

Chemical messengers are detected when they attach, or bind, to a receptor that has the proper shape. Binding creates a new structure—messenger and receptor—and the entire complex often changes shape as a result. This can change the function of the receptor. In this case, a passageway opens up through the receptor molecule, permitting substances to cross the plasma membrane.

🏃 Everyday Physiology

### Physics and chemistry applied to biology

It's important to realize that no mysterious forces are involved in the workings of the body. Although our knowledge is incomplete, it is quite clear that living systems are subject to the same laws of physics and chemistry as buildings, oceans, and mountain ranges. In fact, many advances in our understanding of the human body came only after advances in one of the physical or applied sciences. For example, the action and purpose of the heart valves remained a mystery until the 1600s, when pumps containing valves were developed to remove the water from flooded coal mines. An English physician, William Harvey, was then astute enough to demonstrate that those design principles explained the function of the heart and the circulation of the blood.

**? INTEGRATION**

C. What features are common to forearm movement at an elbow joint and the opening of a cell membrane passageway?

**Lo LEARNING OUTCOME**

Explain the relationship between structure and function.

# Section 2 Review
## A&P in Perspective

**Vocabulary:** For each of the following descriptions, write the appropriate characteristic or process of living organisms on the corresponding blank.

**1** Usually refers to the absorption and utilization of oxygen and the generation and release of carbon dioxide

**2** Produces generations of the same type of organism

**3** Homeostatic mechanisms maintain relatively stable internal conditions

**4** Movement of fluid within the body; may involve a pump and a network of special vessels

**5** Elimination of chemical wastes generated by the body

**6** Mechanical and chemical process that converts food into simpler substances that the body can use.

**7** Includes anabolic and catabolic processes.

**8** Indicates that the organism detects and responds to stimuli in the immediate environment

**1** _respiration_

**2** _reproduction_

**3** _"regulation" - homeostasis_

**4** _circulation_

**5** _excretion_

**6** _digestion_

**7** _metabolism_

**8** _responsiveness_

Write each of the following terms under the proper heading.

- Right atrium
- Myocardium
- Valve to aorta opens
- Left ventricle
- Valve between left atrium and left ventricle closes
- Pressure in left atrium
- Electrocardiogram
- Endocardium
- Superior vena cava
- Heartbeat

**9 Anatomy**

_right atrium_
_myocardium_
_left ventricle_
_endocardium_
_superior vena cava_

**10 Physiology**

_valve closes_
_pressure_
_ECG_
_HR_
_valve opens_

**Short answer:** Briefly describe how the relationship of structure and function of a house key and its front door lock are both similar to and different from a chemical messenger and its receptor molecule.

**11** _Only certain chemical messengers, or proteins can 'enter' the cell. Same concept w/ the key. Only a certain type of key can open the door._

**Section integration:** How might a large organism's survival be affected by an inadequate internal circulation network?

**12** _if the organism was unable to circulate waste/nutrients it would not be able to maintain homeostasis. w/o homeostasis the mammal or organism could not live._

## Module 1.10

# The human body has multiple interdependent levels of organization

The apparent complexity of the human body represents multiple **levels of organization**. Each level is more complex than the underlying one, but all can be broken down into similar chemical and cellular components. The descriptions below highlight the organization of the cardiovascular system.

**Organism Level.** An **organism**—in this case, a human—is the highest level of organization. All organ systems of an organism's body work together to maintain life and health.

**Organ System Level.** (Chapters 5–27) Organs interact in organ systems. Each time it contracts, the heart pumps and pushes blood into a network of blood vessels. Together, the heart, blood, and blood vessels form the cardiovascular system, one of 11 **organ systems** in the body.

**Organ Level.** An **organ** consists of two or more tissues working to perform several functions. Layers of heart muscle tissue, in combination with connective tissue (another type of tissue), form the bulk of the wall of the heart, a hollow internal organ.

**Tissue Level.** (Chapter 4) A **tissue** is composed of similar cells (and their products) working together to perform one or more specific functions. Heart muscle cells, or cardiac muscle cells (*cardium*, heart), form cardiac muscle tissue.

**Cellular Level.** (Chapter 3) **Cells** are the smallest living units in the body. Their functions depend on organelles, intracellular structures composed of various complex molecules. Each organelle has a specific function; for example, one type provides the energy that powers the contractions of muscle cells in the heart.

**Chemical Level.** (Chapter 2) **Atoms**, the smallest stable units of matter, can combine to form **molecules** with complex shapes. The functional properties of a particular molecule are determined by its unique shape and atomic components.

Atoms in combination form molecules

Small molecules join to form complex protein molecules

Protein subunits form contractile filaments

 **REVIEW**

**A.** Name the simplest level of organization that includes the smallest living units in the body.

**B.** Define *organ*.

 **INTEGRATION**

**C.** Our understanding of how the human body works is based on a knowledge of which level(s) of organization?

 **LEARNING OUTCOME**

Describe the various levels of organization in the human body.

# Module 1.11
# Cells are the smallest units of life

Free-living **cells** are the smallest living structures, with all of the basic characteristics of life described earlier in the chapter. Most of the plants and animals you are familiar with are multicellular, consisting of thousands to billions of cells. These cells do not exist independently. They work together, each with its own characteristic structures and functions. Cells are the living building blocks of our bodies. There are literally trillions of cells in your body, but there are only an estimated 200 different types of cells. Nevertheless, those 200 types show remarkable diversity in form and function.

**1** The cells of the human body vary greatly in size. The dimensions of cells are usually given in terms of micrometers (μm). One micrometer is one-millionth of a meter, or approximately 1/25,000th of an inch. All the cells illustrated here are shown with the dimensions they would have if each were magnified about 1500 times. The skeletal muscle cells that give you the ability to move around are too large to illustrate here. At this magnification a large skeletal muscle cell would have the diameter of a small dinner plate and be over 300 m long.

| Smooth muscle cells | Blood cells | Bone cells | Fat cells |
|---|---|---|---|
|  | Red blood cell    White blood cell  |  |  |
| Smooth muscle cells, found in many organs, are long and slender. Their contractions in the walls of the stomach and intestines support the digestion and movement of ingested solids and liquids. | Blood cells are either flattened discs (red blood cells) or roughly spherical (white blood cells). Red blood cells transport oxygen and carbon dioxide in the bloodstream. White blood cells fight off infection and combat disease. | Bone cells reside within small cavities inside the mass of a bone. These cells maintain bone and recycle the calcium and phosphate stored there. | Fat cells are spherical storage containers. Whenever we take in more energy than we expend, the excess energy obtained from the food is stored as fat. As a result, these cells become larger and more numerous. |

**2** The importance of cells is apparent in the **cell theory**, one of the foundations of modern biology. The basic principles of the cell theory are as follows:

### Basic Principles of the Cell Theory

- Cells are the structural building blocks of all plants and animals.
- Cells are produced by the divisions of pre-existing cells.
- Cells are the smallest structural units that perform all vital functions.

**? REVIEW**

**A.** Name and define the unit used to measure cell size.

**B.** List the three basic principles of the cell theory.

## Cells lining the digestive tract

Cells lining the digestive tract are relatively delicate. These cells absorb the nutrients, vitamins, minerals, and water we need.

## Reproductive cells (sex cells)

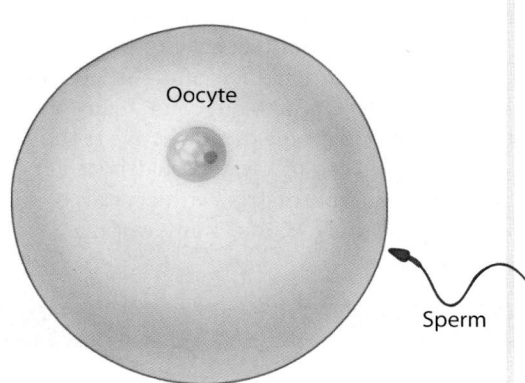

Oocyte

Sperm

Cells involved in sexual reproduction are called sex cells. Women produce relatively large oocytes (immature eggs) in very small numbers, usually at monthly intervals. Males continuously produce relatively tiny sperm in enormous numbers.

## Nerve cells (neurons)

Nerve cells, or neurons, are the equivalent of computer chips—they process information. Thought, memory, consciousness, and muscle control are all based on the actions of, and interactions among, neurons. There are many different types and shapes of neurons. This is a neuron from a part of the brain involved with the control of balance and movement. The extensive branching provides a huge surface area for receiving communications from other neurons.

The cells of the body work together, and our lives ultimately depend on their actions. If they don't do the right thing at the right time, we're in trouble. If our cells can't survive, we're doomed. Yet each individual cell remains unaware of its role in the "big picture"—it simply responds and adapts to changes in its local environment. How the responses of cells in different parts of the body are coordinated and controlled is obviously a key question, and we will spend considerable time in later chapters considering the answers.

**? INTEGRATION**

**C.** A red blood cell has a diameter of about 8 µm. Use that information to estimate the diameter of the oocyte in micrometers (µm) and millimeters (mm). (1 mm = 1000 µm)

**Lo LEARNING OUTCOME**

Describe various types of cells in the human body, and explain the basic principles of the cell theory.

# Module 1.12

# Tissues are specialized groups of cells and cell products

The roughly 200 different cell types in the body combine to form **tissues**, collections of similar cells and cell products that perform specific functions. **Histology** (*histos*, tissue) is the study of tissues. This module introduces the four **primary tissue types**—epithelial tissue, connective tissue, muscle tissue, and nervous tissue—that, in various combinations, form the organs of the body.

**1**    The most common type of **epithelial** (ep-i-THĒ-lē-ul) **tissue** is a layer of cells that forms a barrier with specific properties. Epithelia cover every exposed body surface; line the digestive, respiratory, reproductive, and urinary tracts; surround internal cavities such as the chest cavity or the fluid-filled chambers in the brain, eye, and inner ear; and line the inner surfaces of the blood vessels and heart.

**2**    **Connective tissue** is quite diverse in appearance. All forms of connective tissue contain cells and an extracellular **matrix** that consists of protein fibers and a liquid known as the ground substance. The amount and consistency of the matrix depend on the particular type of connective tissue. In blood, the cells are suspended in a watery matrix called plasma. Bone has a more durable matrix, with crystals of calcium salts organized around a fibrous framework, and very little ground substance.

### EPITHELIAL TISSUE

- Covers and protects exposed surfaces
- Lines internal passageways and chambers
- Produces glandular secretions

### CONNECTIVE TISSUE

- Fills internal spaces
- Provides structural support
- Stores energy

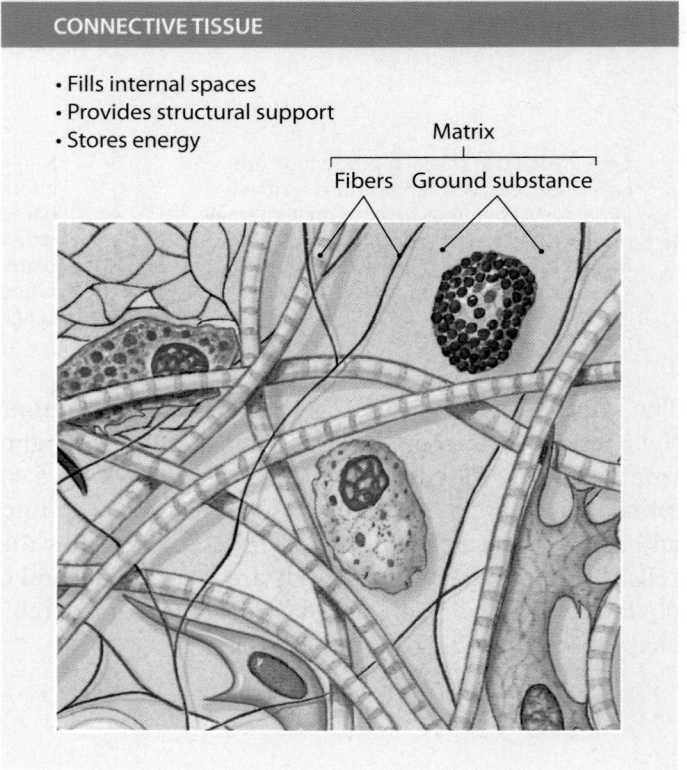

Matrix — Fibers — Ground substance

**3** **Muscle tissue** is unique because individual muscle cells can contract forcefully. Major functions of muscle tissue include skeletal movement, soft tissue support, maintenance of blood flow, movement of materials along internal passageways, and the stabilization of normal body temperature. There are three different types of muscle tissue.

## MUSCLE TISSUE

• Contracts to produce active movement

Nuclei

**Skeletal muscle tissue** is usually directly or indirectly attached to the skeleton. When it contracts, it moves or stabilizes the position of bones or internal organs.

Muscle cell    Nucleus

**Cardiac muscle tissue** is found in the heart, where its coordinated contractions propel blood through the blood vessels.

**Smooth muscle tissue** is found in the walls of blood vessels, within glands, and along the respiratory, circulatory, digestive, and reproductive tracts.

**4** **Nervous tissue** is specialized to carry information or instructions from one place in the body to another. Two basic types of cells are present: nerve cells, or **neurons** (NOO-rons; *neuro*, nerve), and supporting cells, or **neuroglia** (noo-RŌG-lē-uh; *glia*, glue). Neurons transmit information in the form of electrical impulses. Neuroglia isolate and protect neurons while forming a supporting framework. The nervous tissue in the body can be divided on anatomical grounds into the **central nervous system**, which consists of the brain and spinal cord, and the **peripheral nervous system**, which includes the nerves connecting the central nervous system with other tissues and organs.

## NERVOUS TISSUE

• Conducts electrical impulses
• Carries information

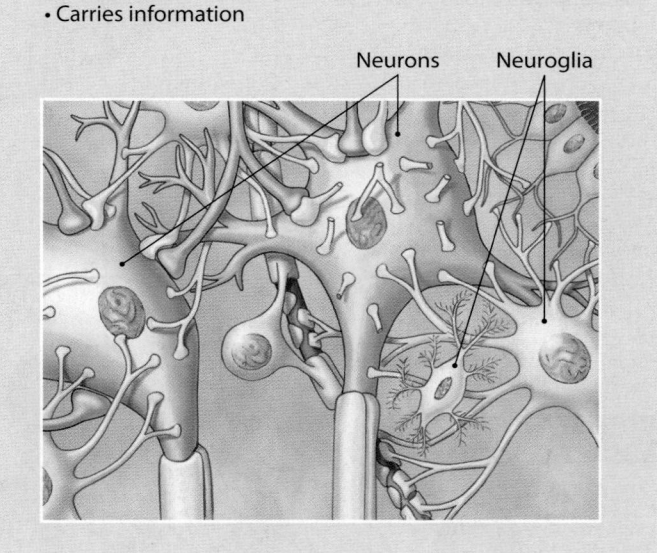

Neurons    Neuroglia

**? REVIEW**
**A.** Describe the general roles of the different types of muscle tissue.

**? INTEGRATION**
**B.** Identify the four primary tissue types, and explain the functions of each.

**Lo LEARNING OUTCOME**
Define histology, and explain the interrelationships among the various types of tissues.

# Module 1.13
# Organs and organ systems perform vital functions

**1** An **organ** is a functional unit composed of more than one tissue type. The particular combination and organization of tissues within an organ both determines and limits the organ's functions. For example, an organ with a flattened shape can provide protection (like a sheet of cardboard on a table surface), and an organ with a three-dimensional shape can house additional structures (like a cardboard box and its contents). An **organ system** consists of organs that interact to perform a specific range of functions, often in a coordinated fashion.

The heart, for example, is an organ that contains cardiac muscle tissue, epithelial tissue, connective tissues, and nervous tissue. The interconnections between cardiac muscle cells ensure that the contractions are coordinated, producing a heartbeat; the nervous tissue adjusts the heart rate. When the heart beats, the internal anatomy of the heart, largely composed of cardiac muscle tissue and connective tissue, allows it to function as a pump.

**Organism level**

**Organ level**

**Cardiovascular System**

**Endocrine**

**Nervous**

**Muscular**

**Skeletal**

**Organ system level**

**Integumentary**

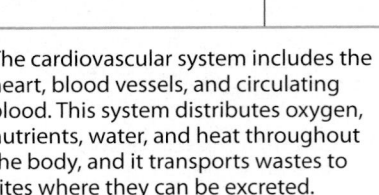

The cardiovascular system includes the heart, blood vessels, and circulating blood. This system distributes oxygen, nutrients, water, and heat throughout the body, and it transports wastes to sites where they can be excreted.

**2** The table at right lists the 11 organ systems in the human body. Although this categorization is a convenient way to organize information, the concept of separate "organ systems" is artificial and somewhat misleading. Nothing in the body functions in isolation—not cells, not tissues, not organs, and certainly not organ systems. Organs and organ systems are interdependent, and something that affects one organ will affect the functioning of the body as a whole. For example, the heart cannot pump blood effectively after massive blood loss. If the heart cannot pump and blood cannot flow, oxygen and nutrients cannot be distributed. Very soon, cardiac muscle tissue begins to break down as individual muscle cells die from oxygen and nutrient starvation. These changes will not be restricted to the cardiovascular system. All cells, tissues, and organs in the body will be damaged, with potentially fatal results.

**? REVIEW**

**A.** Which two organ systems are involved with circulation within the body?

| Organ Systems | Major Functions |
|---|---|
| Integumentary system | Protects against environmental hazards; helps control body temperature |
| Skeletal system | Provides support; protects tissues; stores minerals; forms blood cells |
| Muscular system | Produces movement; provides support; generates heat |
| Nervous system | Directs immediate responses to stimuli, usually by coordinating the activities of other organ systems |
| Endocrine system | Directs long-term changes in other organ systems |
| Cardiovascular system | Transports cells and dissolved materials in the bloodstream, including nutrients, wastes, and gases |
| Lymphatic system | Defends against infection and disease; returns tissue fluid to the bloodstream |
| Respiratory system | Delivers air to sites where gas exchange occurs between the air and circulating blood; produces sound |
| Digestive system | Processes food and absorbs nutrients (carbohydrates, fats, proteins, vitamins, minerals, and water) |
| Urinary system | Eliminates excess water, salts, and wastes; controls pH |
| Reproductive system | Produces sex cells and hormones; supports embryonic development from fertilization to birth (female) |

1

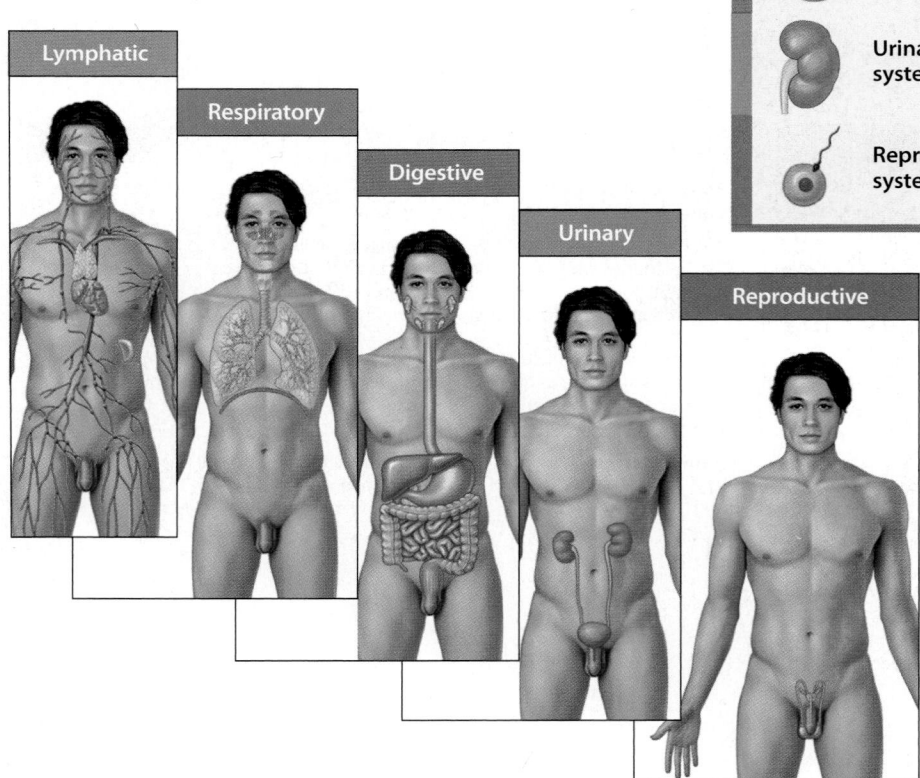

Lymphatic

Respiratory

Digestive

Urinary

Reproductive

**? INTEGRATION**

**B.** Using the table as a reference, describe how falling down a flight of stairs could affect at least six of the organ systems.

**Lo LEARNING OUTCOME**

Identify the 11 organ systems of the human body, and describe the major functions of each.

# Module 1.14

# Organs of the integumentary, skeletal, and muscular systems support and move the body ...

## Integumentary System

Protects against environmental hazards; helps control body temperature

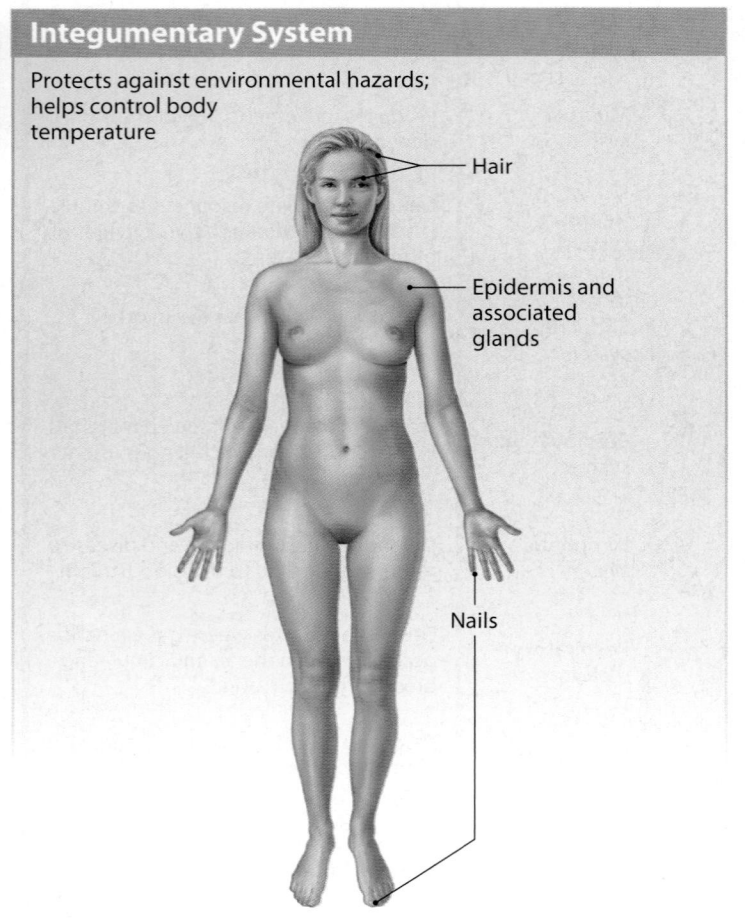

Hair

Epidermis and associated glands

Nails

## Skeletal System

Provides support; protects tissues; stores minerals; forms blood cells

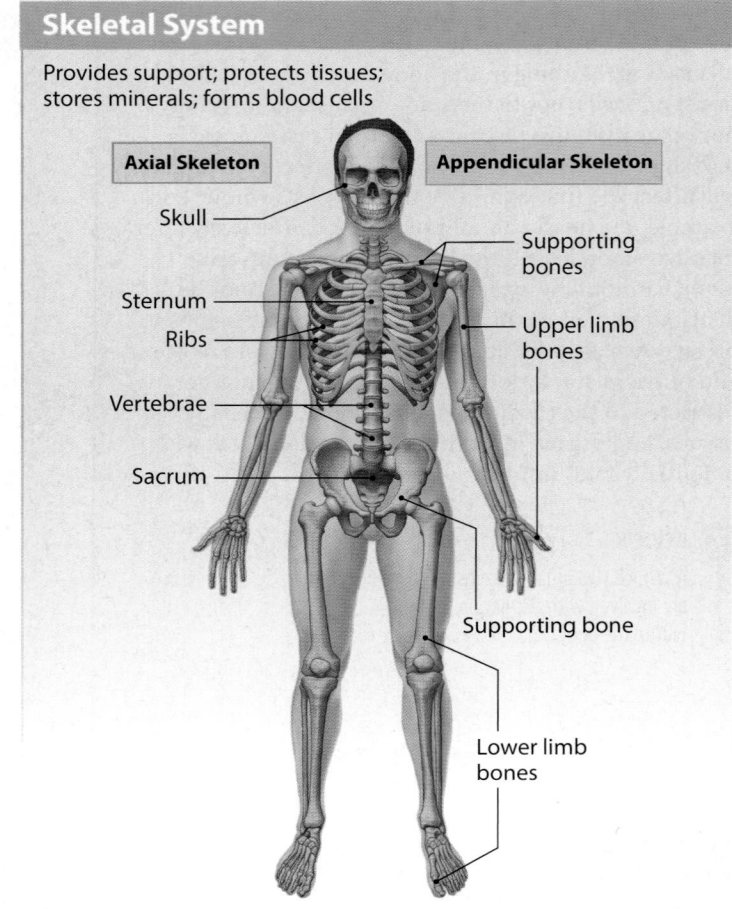

Axial Skeleton

Appendicular Skeleton

Skull

Supporting bones

Sternum

Ribs

Upper limb bones

Vertebrae

Sacrum

Supporting bone

Lower limb bones

| Organ/Structure | Primary Function |
| --- | --- |
| **Cutaneous Membrane** | |
| Epidermis | Covers surface; protects deeper tissues |
| Dermis | Nourishes epidermis; provides strength; contains glands |
| **Hair Follicles** | Produce hair; innervation provides sensation |
| Hairs | Provide protection for head |
| Sebaceous glands | Secrete lipid coating that lubricates hair shaft and epidermis |
| **Sweat Glands** | Produce perspiration for evaporative cooling |
| **Nails** | Protect and stiffen distal tips of digits |
| **Sensory Receptors** | Provide sensations of touch, pressure, temperature, pain |
| **Hypodermis** | Stores lipids; attaches skin to deeper structures |

| Organ/Structure | Primary Function |
| --- | --- |
| **Bones, Cartilages, and Joints** | Support, protect soft tissues; bones store minerals |
| Axial skeleton (skull, vertebrae, sacrum, coccyx, sternum, supporting cartilages and ligaments) | Protects brain, spinal cord, sense organs, and soft tissues of thoracic cavity; supports the body weight over lower limbs |
| Appendicular skeleton: limbs and supporting bones and ligaments | Provides internal support and positioning of the limbs; supports and moves axial skeleton |
| **Bone Marrow** | Primary site of blood cell production (red marrow); stores of energy in fat cells (yellow marrow) |

# ... and organs of the nervous system provide rapid control and regulation

## Muscular System

Produces movement; provides support; generates heat

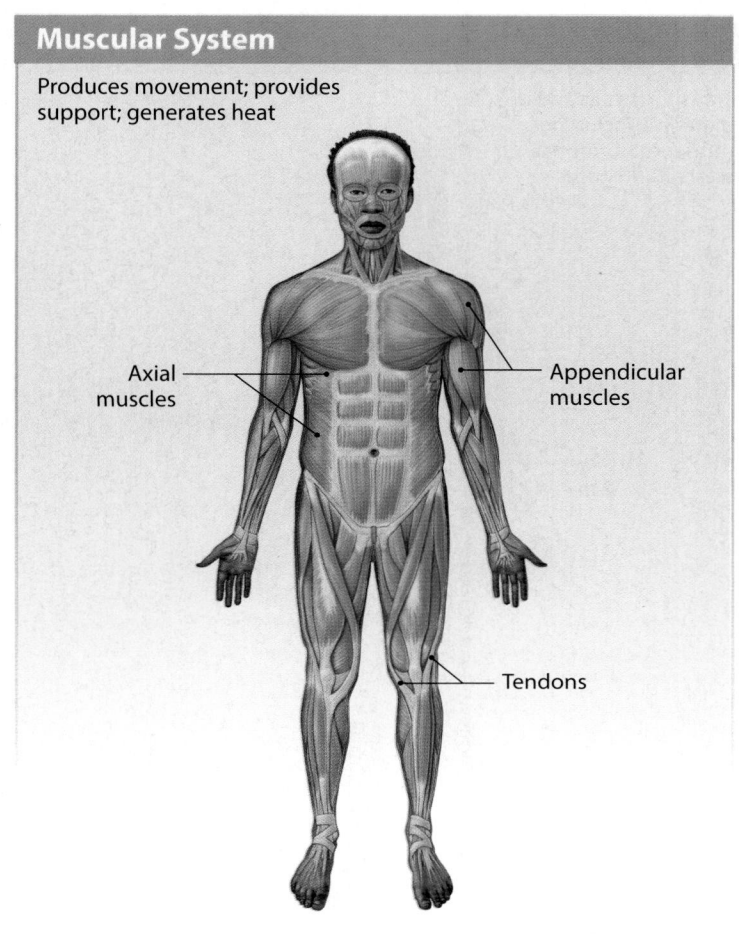

Axial muscles

Appendicular muscles

Tendons

## Nervous System

Directs immediate responses to stimuli, usually by coordinating the activities of other organ systems

**Central Nervous System**

Brain

Spinal cord

**Peripheral Nervous System**

Peripheral nerves

| Organ/Structure | Primary Functions |
|---|---|
| **Skeletal Muscles** | Provide skeletal movement; control entrances to digestive and respiratory tracts and exits from digestive and urinary tracts; produce heat; support skeleton; protect soft tissues |
| Axial muscles | Support and position axial skeleton |
| Appendicular muscles | Support, move, and brace limbs |
| **Tendons, Aponeuroses** | Use forces of contraction to perform specific tasks |

| Organ/Structure | Primary Function |
|---|---|
| **Central Nervous System (CNS)** | Acts as control center for nervous system; processes information; provides short-term control over activities of other systems |
| Brain | Performs complex integrative functions; controls both voluntary and involuntary activities |
| Spinal cord | Relays information to and from brain; performs less complex integrative activities |
| Special senses | Provide sensory input to the brain relating to sight, hearing, smell, taste, and equilibrium |
| **Peripheral Nervous System (PNS)** | Links CNS with other systems and with sense organs |

 **INTEGRATION**

**A.** How would a nervous system disorder affect the muscular system?

**LO** **LEARNING OUTCOME**

Describe the major organs of the integumentary, skeletal, muscular, and nervous systems, and briefly describe their functions.

# Organs of the endocrine system secrete chemicals that are carried by organs of the cardiovascular system …

## Endocrine System

Directs long-term changes in other organ systems with hormones (chemical messengers)

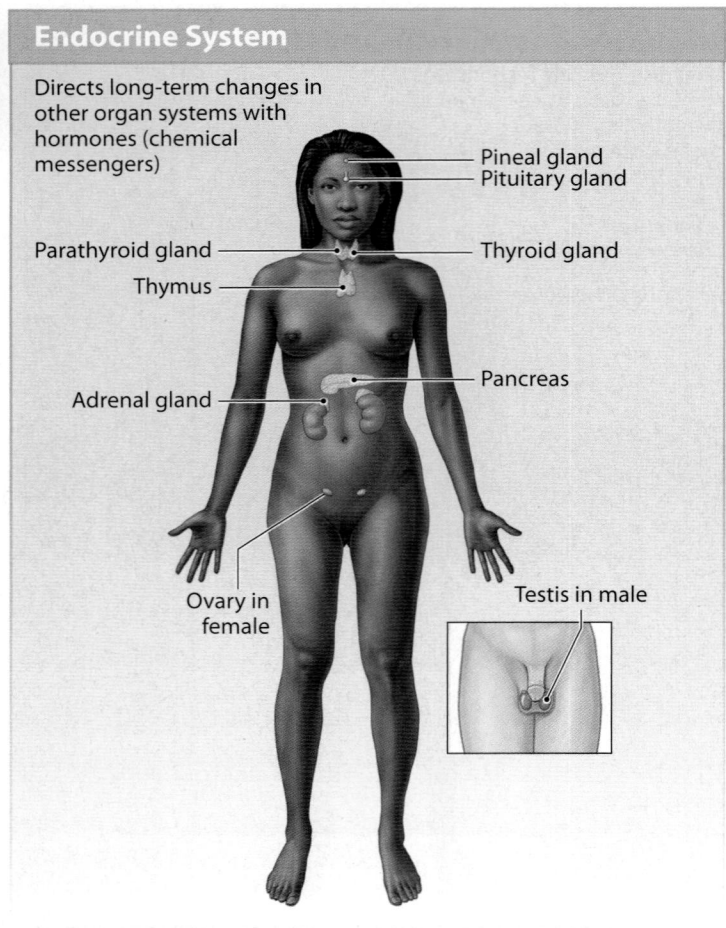

- Pineal gland
- Pituitary gland
- Parathyroid gland
- Thyroid gland
- Thymus
- Pancreas
- Adrenal gland
- Ovary in female
- Testis in male

## Cardiovascular System

Transports cells and dissolved materials, including nutrients, hormones, wastes, and gases

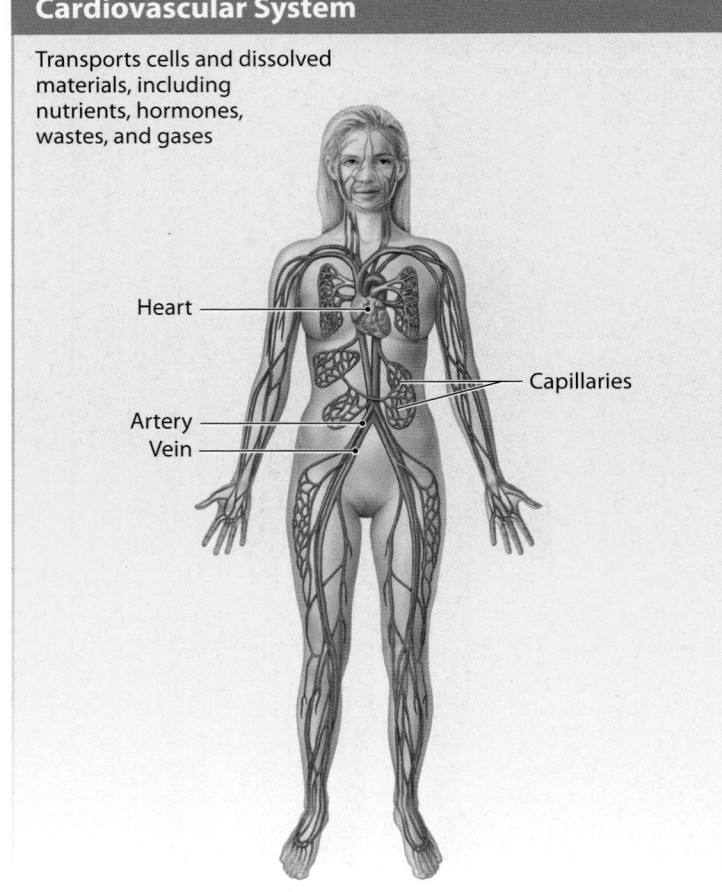

- Heart
- Capillaries
- Artery
- Vein

| Organ/Structure | Primary Function |
|---|---|
| **Pineal Gland** | May control timing of reproduction and set day-night rhythms |
| **Pituitary Gland** | Controls other endocrine glands; regulates growth and fluid balance |
| **Thyroid Gland** | Controls tissue metabolic rate; regulates calcium levels |
| **Parathyroid Glands** | Regulate calcium levels (with thyroid gland) |
| **Thymus** | Controls maturation of lymphocytes |
| **Adrenal Glands** | Adjust water balance, tissue metabolism, cardiovascular and respiratory activity |
| **Kidneys** | Control red blood cell production, elevate blood pressure, and assist in calcium homeostasis |
| **Pancreas** | Regulates blood glucose levels |
| **Gonads** | |
| Testes | Support male sexual characteristics and reproductive functions (see Module 1.16) |
| Ovaries | Support female sexual characteristics and reproductive functions (see Module 1.16) |

| Organ/Structure | Primary Function |
|---|---|
| **Heart** | Propels blood; maintains blood pressure |
| **Blood Vessels** | Distribute blood around the body |
| Arteries | Carry blood from the heart to capillaries |
| Capillaries | Permit diffusion between blood and interstitial fluids |
| Veins | Return blood from capillaries to the heart |
| **Blood** | Transports oxygen, carbon dioxide, and blood cells; delivers nutrients and hormones; removes wastes; assists in temperature regulation and defense against disease |

# ... organs of the lymphatic system defend the body, and organs of the respiratory system exchange vital gases

## Lymphatic System

Defends against infection and disease; returns tissue fluid to the bloodstream

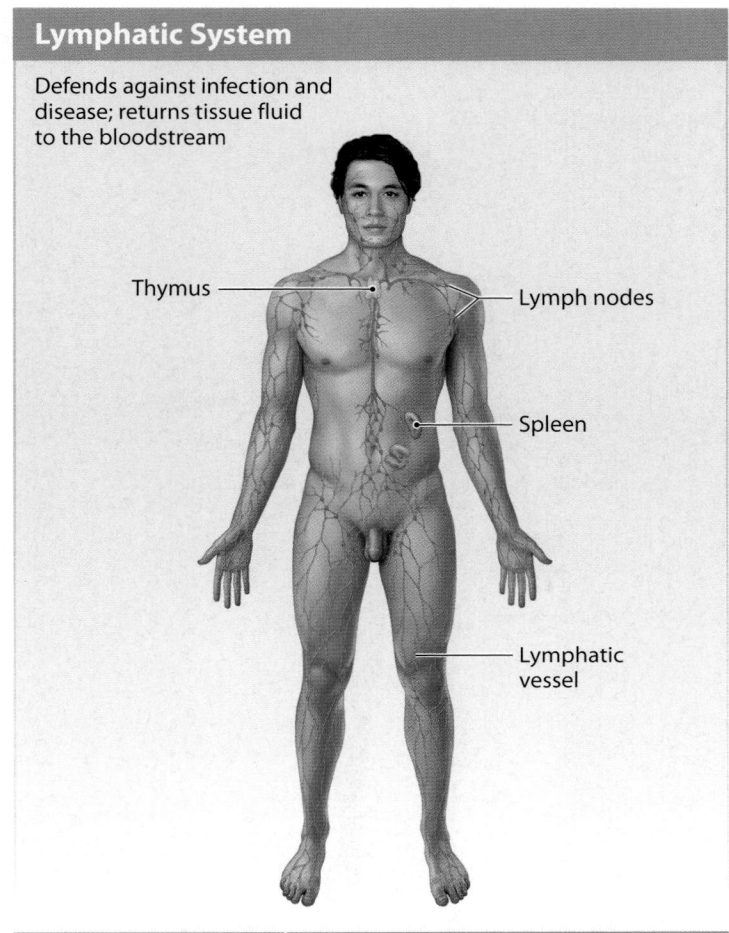

Thymus

Lymph nodes

Spleen

Lymphatic vessel

## Respiratory System

Delivers air to sites where gas exchange occurs between the air and circulating blood; produces sound

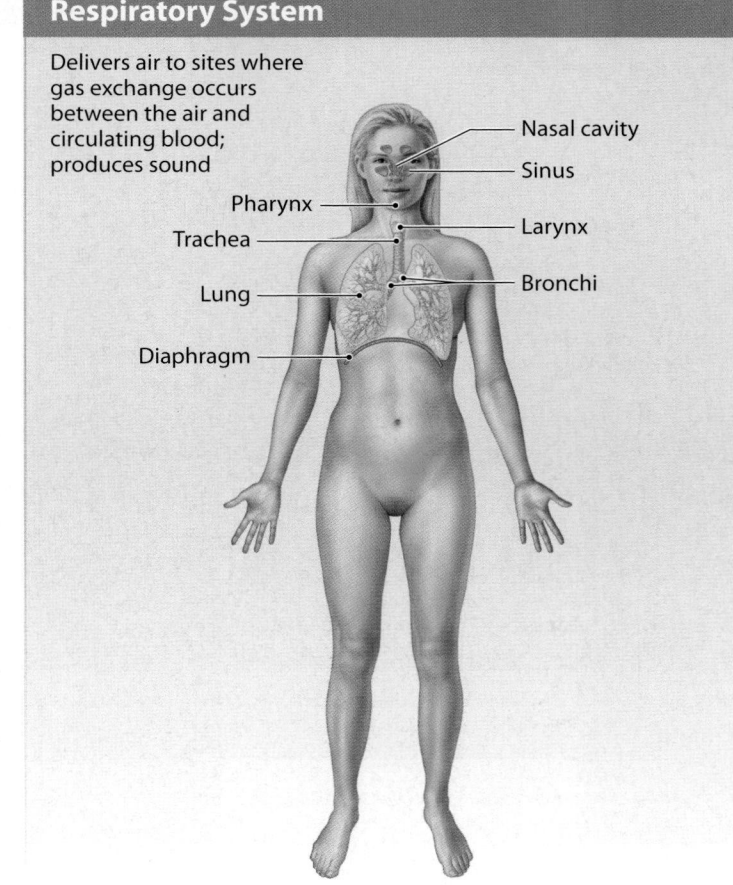

Nasal cavity

Sinus

Pharynx

Larynx

Trachea

Lung

Bronchi

Diaphragm

| Organ/Structure | Primary Function |
|---|---|
| Lymphatic Vessels | Carry lymph (water and proteins) and lymphocytes from peripheral tissues to veins of the cardiovascular system |
| Lymph Nodes | Monitor the composition of lymph; engulf pathogens; stimulate immune response |
| Spleen | Monitors circulating blood; engulfs pathogens and recycles red blood cells; stimulates immune response |
| Thymus | Controls development and maintenance of one class of lymphocytes (T cells) |

| Organ/Structure | Primary Function |
|---|---|
| Nasal Cavities and Paranasal Sinuses | Filter, warm, humidify air; detect smells |
| Pharynx | Conducts air to larynx; a chamber shared with the digestive tract |
| Larynx | Protects opening to trachea and contains vocal cords |
| Trachea | Filters air; cartilages keep airway open |
| Bronchi | Conducts air between trachea and lungs |
| Lungs | Responsible for air movement; alveoli within the lungs are sites of gas exchange between air and blood |

 **INTEGRATION**

**A.** How would a respiratory system disease affect the cardiovascular system?

 **LEARNING OUTCOME**

Describe the major organs of the endocrine, cardiovascular, lymphatic, and respiratory systems, and briefly describe their functions.

# Organs of the digestive system make nutrients available and, with the urinary system, excrete wastes ...

## Digestive System

Processes food and absorbs nutrients

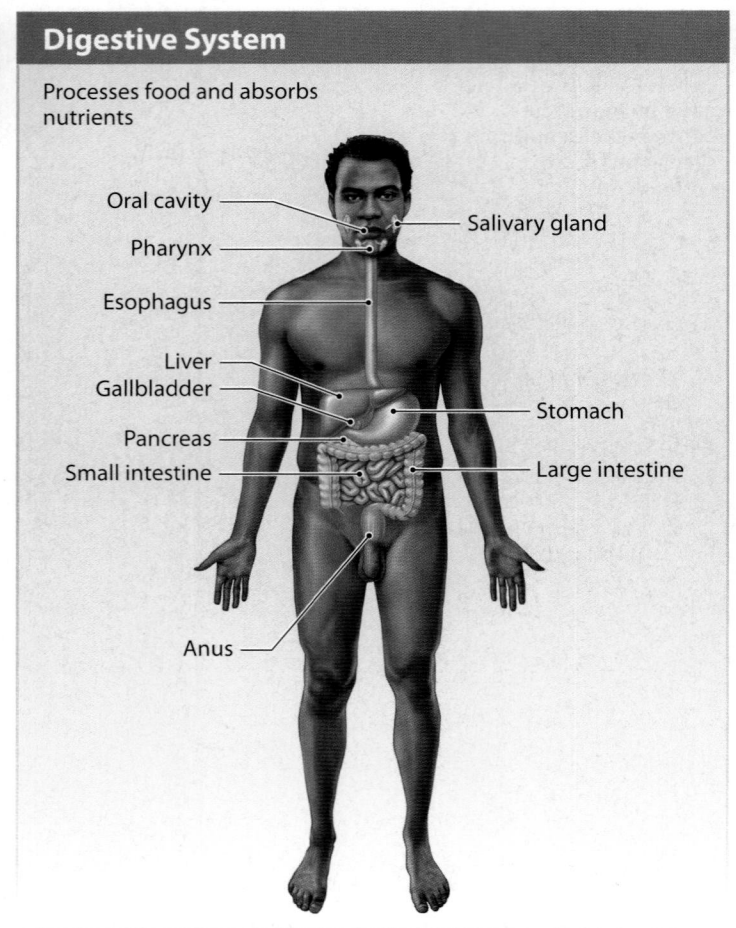

- Oral cavity
- Salivary gland
- Pharynx
- Esophagus
- Liver
- Gallbladder
- Stomach
- Pancreas
- Small intestine
- Large intestine
- Anus

## Urinary System

Eliminates excess water, salts, and wastes

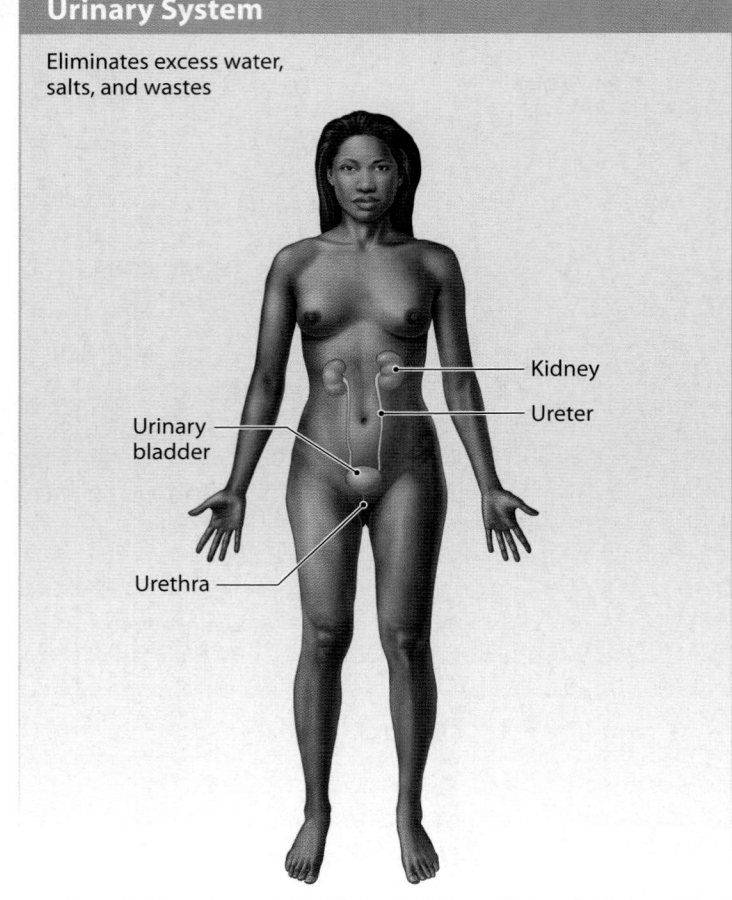

- Kidney
- Ureter
- Urinary bladder
- Urethra

| Organ/Structure | Primary Function |
| --- | --- |
| Oral Cavity (Mouth) | Cavity for food; works with associated structures (teeth, tongue) to break up food and pass food and liquids to pharynx |
| Salivary Glands | Provide buffers and lubrication; produce enzymes that begin digestion |
| Pharynx | Conducts solid food and liquids to esophagus; chamber shared with respiratory tract |
| Esophagus | Delivers food to stomach |
| Stomach | Secretes acids, enzymes, and hormones |
| Small Intestine | Secretes digestive enzymes, buffers, and hormones; absorbs nutrients |
| Liver | Secretes bile; regulates nutrient composition of blood |
| Gallbladder | Stores and concentrates bile for release into small intestine |
| Pancreas | Secretes digestive enzymes and buffers; contains endocrine cells |
| Large Intestine | Removes water from feces; stores wastes |

| Organ/Structure | Primary Function |
| --- | --- |
| Kidneys | Form and concentrate urine; regulate blood pH and ion concentrations; perform endocrine functions |
| Ureters | Conduct urine from kidneys to urinary bladder |
| Urinary Bladder | Stores urine for eventual elimination |
| Urethra | Conducts urine to exterior |

# … and organs of the male and female reproductive systems provide for the continuity of life

## Female Reproductive System

Produces sex cells and hormones; supports embryonic development from fertilization to birth

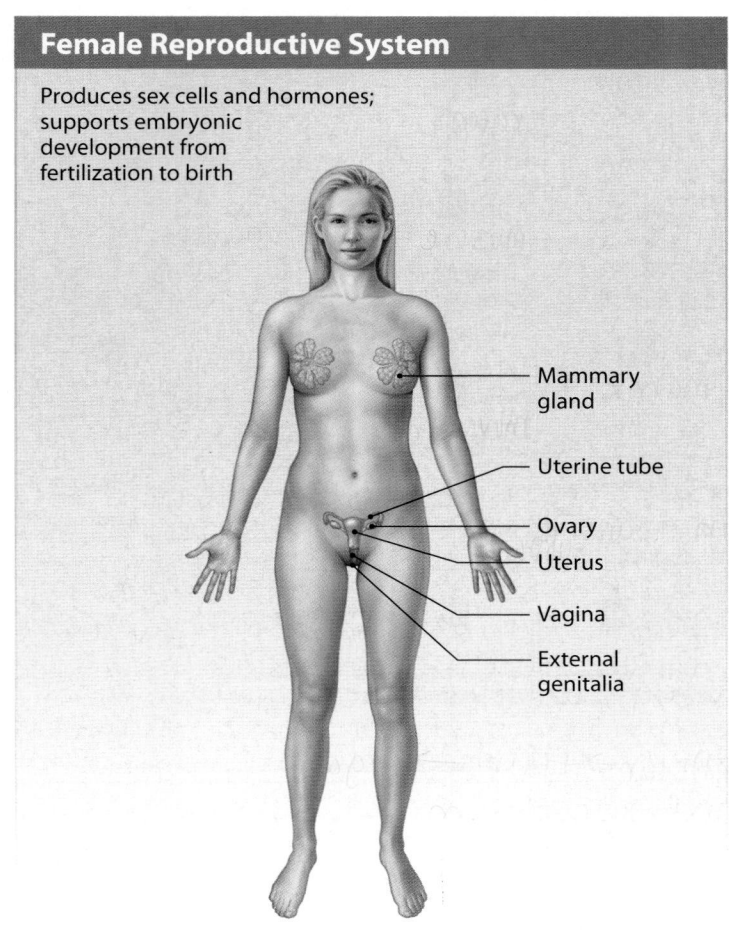

- Mammary gland
- Uterine tube
- Ovary
- Uterus
- Vagina
- External genitalia

## Male Reproductive System

Produces sex cells and hormones

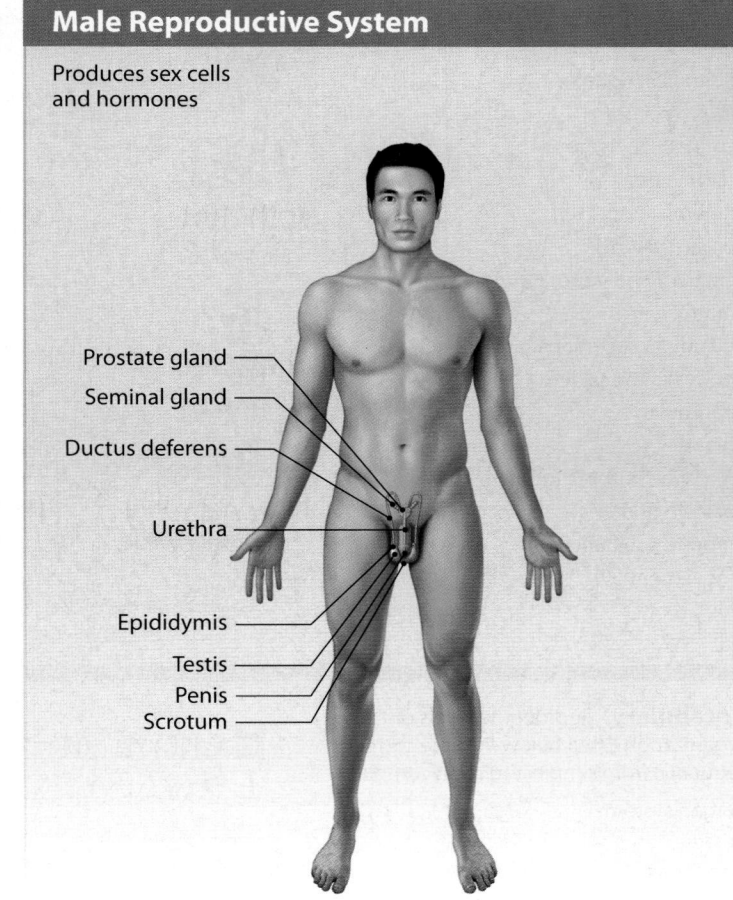

- Prostate gland
- Seminal gland
- Ductus deferens
- Urethra
- Epididymis
- Testis
- Penis
- Scrotum

| Organ/Structure | Primary Function |
|---|---|
| **Ovaries** | Produce oocytes and hormones |
| **Uterine Tubes** | Deliver oocyte or embryo to uterus; normal site of fertilization |
| **Uterus** | Site of embryonic development and exchange between maternal and fetal bloodstreams; sheds lining during menstruation |
| **Vagina** | Site of sperm deposition; acts as a birth canal during delivery; provides passageway for fluids during menstruation |
| **External Genitalia** | |
| Clitoris | Contains erectile tissue; provides pleasurable sensations during sexual activities |
| Labia | Contain glands that lubricate entrance to vagina |
| **Mammary Glands** | Produce milk that nourishes newborn infant |

| Organ/Structure | Primary Function |
|---|---|
| **Testes** | Produce sperm and hormones |
| **Accessory Organs** | |
| Epididymis | Acts as site of sperm maturation in each testis |
| Ductus deferens | Conducts sperm from the epididymis |
| Seminal glands | Secrete fluid that contributes to semen |
| Prostate gland | Secretes fluid and enzymes |
| Urethra | Conducts semen to exterior |
| **External Genitalia** | |
| Penis | Deposits sperm in vagina of female |
| Scrotum | Surrounds the testes and controls their temperature |

**? INTEGRATION**

**A.** How would a reproductive system disorder affect the urinary system?

**Lo LEARNING OUTCOME**

Describe the major organs of the digestive, urinary, and reproductive systems, and briefly describe their functions.

# Section 3 Review
## Levels of Organization

**Concept map:** Use each of the following terms once to fill in the blank boxes to correctly complete the map.

- organs
- epithelial tissue
- cells
- connective tissue
- muscle tissue
- nervous tissue
- organ systems
- external and internal surfaces
- matrix
- glandular secretions
- bones of the skeleton
- neuroglia
- blood
- materials within digestive tract
- protein fibers
- ground substance
- movement

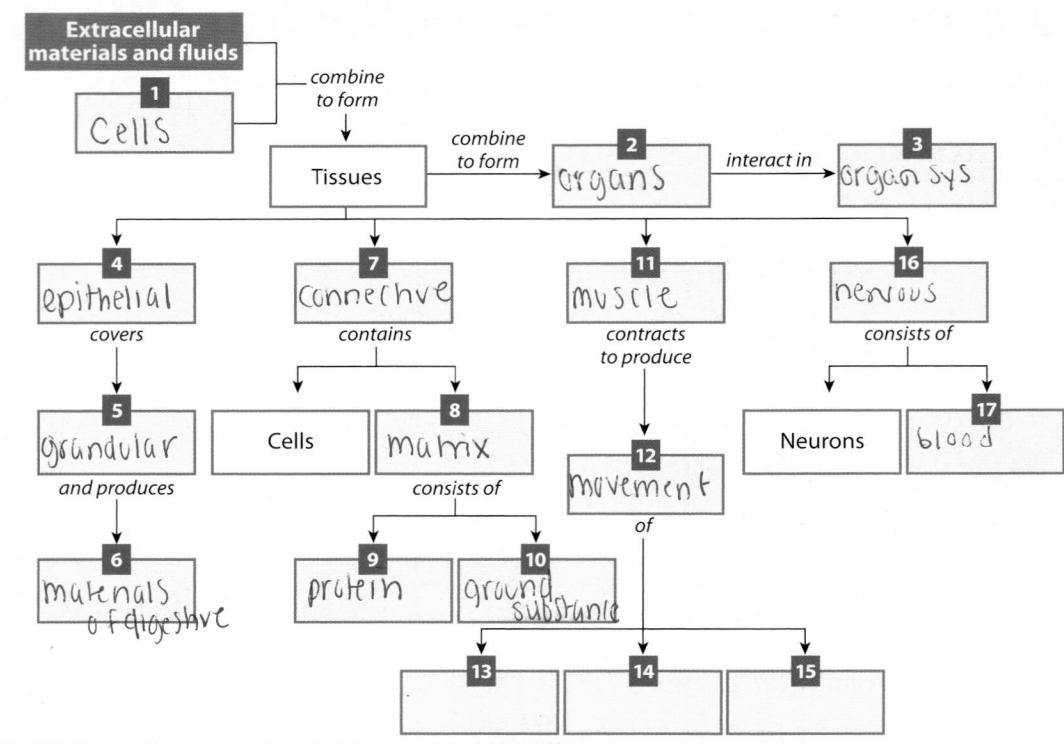

**Extracellular materials and fluids**

1. Cells — combine to form → Tissues — combine to form → 2. organs — interact in → 3. organ sys

Tissues →
- 4. epithelial — *covers*
- 7. connective — *contains*
- 11. muscle — *contracts to produce*
- 16. nervous — *consists of*

4. epithelial →
- 5. grandular — *and produces* → 6. materials of digestive

7. connective — *contains* →
- Cells
- 8. matrix — *consists of* → 9. protein, 10. ground substance

11. muscle → 12. movement — *of*

16. nervous — *consists of* →
- Neurons
- 17. blood

12. movement → 13. , 14. , 15.

**Vocabulary:** Reorder the levels of organization listed below into the correct sequence from simplest to most complex.

- organ system
- tissue
- organism
- organ
- chemical
- cellular

18. Chemical → cellular → tissue → organ → organ system → organism

**Short answer:** Summarize the major functions of each of the following organ systems.

| | | | |
|---|---|---|---|
| 19 | integumentary | 19 | protect |
| 20 | skeletal | 20 | generate heat / framework |
| 21 | muscular | 21 | provide movement |
| 22 | nervous | 22 | chemical signals |
| 23 | endocrine | 23 | hormone production |
| 24 | cardiovascular | 24 | movement of O₂, nutrients + waste |

| | | | |
|---|---|---|---|
| 25 | lymphatic | 25 | build tolerance against germs |
| 26 | respiratory | 26 | → O₂, eliminate CO₂ |
| 27 | digestive | 27 | absorb nutrients |
| 28 | urinary | 28 | eliminate waste |
| 29 | reproductive | 29 | produce offspring |

**Section integration:** For five different organ systems in the human body, identify a specialized cell type found in that system.

30. neurons → nervous        connective → lymphatic
Cardiac muscle cell → cardiovascular
Skeletal muscle cell → skeletal
epithelial → respiratory

# Module 1.17

# Homeostatic regulation relies on a receptor, a control center, and an effector

**Homeostasis** (hō-mē-ō-STĀ-sis; *homeo*, similar, unchanging + *stasis*, standing) is the presence of a stable internal environment. Maintaining homeostasis is absolutely vital to an organism's survival. Failure to maintain homeostasis soon leads to illness or even death. The principle of homeostasis is the central theme of this text and the foundation of modern physiology. **Homeostatic regulation** is the adjustment of physiological systems to preserve homeostasis in environments that are often inconsistent, unpredictable, and potentially dangerous. An understanding of homeostatic regulation is crucial to making accurate predictions about the body's responses to both normal and abnormal conditions.

**REVIEW**
**A.** Why is homeostatic regulation important to an organism?

**1** Maintaining a relatively constant temperature in your living space is a familiar example of homeostasis. Like all homeostatic regulatory mechanisms, it consists of (1) a **receptor** or sensor—in this case, a thermometer—that is sensitive to a particular environmental change, or stimulus; (2) a **control center** or integration center—in this case, a thermostat—that receives and processes the information supplied by the receptor and that sends out commands; and (3) an **effector**—in this case, an air conditioner—that responds to these commands by opposing the stimulus. The net effect is that any variation outside normal limits triggers a response that restores normal conditions.

The setting on a thermostat establishes the **set point**, or desired value, which in this case is the temperature you select. (In our example, the set point is 22°C, or about 72°F.) The function of the thermostat is to keep room temperature within acceptable limits, usually within a degree or so of the set point.

**REVIEW**
**B.** Describe the three parts necessary for homeostatic regulation.

**Control Center**

**2** Thermostat

**22°C**

Information affects

Sends commands to

**1** **Receptor**

Thermometer

**3** **Effector**

Air conditioner

Homeostasis **DISTURBED BY**

INCREASING

room temperature

STIMULUS

RESTORED

Homeostasis **RESTORED BY**

DECREASING

room temperature

**HOMEOSTASIS**

NORMAL ROOM TEMPERATURE

**2** Homeostatic control is not precise—it maintains a normal range rather than an absolute value. The same is true for controlling room temperature—a house may have the thermostat on one wall of one room and have the air conditioning outlets at multiple locations. Over time, the temperature in the house will oscillate around the set point.

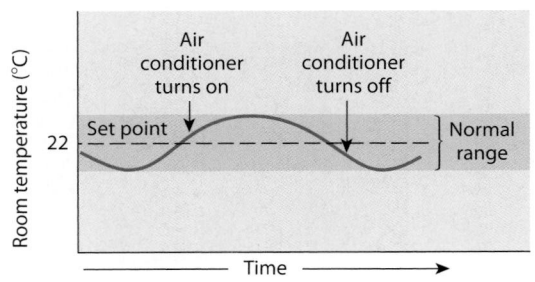

Air conditioner turns on

Air conditioner turns off

Set point

Room temperature (°C)

22

Normal range

Time

**INTEGRATION**
**C.** The Appendix contains tables of the normal physiological values regarding dissolved materials in various body fluids. Most of these values are listed as ranges, rather than averages. Why?

**LEARNING OUTCOME**
Describe the mechanisms of homeostatic regulation.

# Module 1.18
# Negative feedback provides stability ...

**Feedback** occurs when receptor stimulation triggers a response that changes the environment at the receptor. In the case of temperature control by a thermostat, temperature variation outside the desired range triggers an automatic response that corrects the situation. This method of homeostatic regulation is called **negative feedback** because an effector activated by the control center opposes, or negates, the original stimulus. Negative feedback thus tends to minimize change, keeping variation in key body systems within limits compatible with our long-term survival.

**1** Negative feedback is the primary mechanism of homeostatic regulation, and it provides long-term control over the body's internal conditions and systems. Homeostatic mechanisms using negative feedback normally ignore minor variations, and they maintain a normal range rather than a fixed value. The regulatory process itself is dynamic because the set point may vary with changing environments or differing activity levels. For example, when you are asleep, your thermoregulatory set point is lower, whereas when you work outside on a hot day (or when you have a fever), it is higher. Thus, body temperature can vary from moment to moment or from day to day for any person because of either small oscillations around the set point or changes in the set point. Comparable variations occur in all other aspects of physiology.

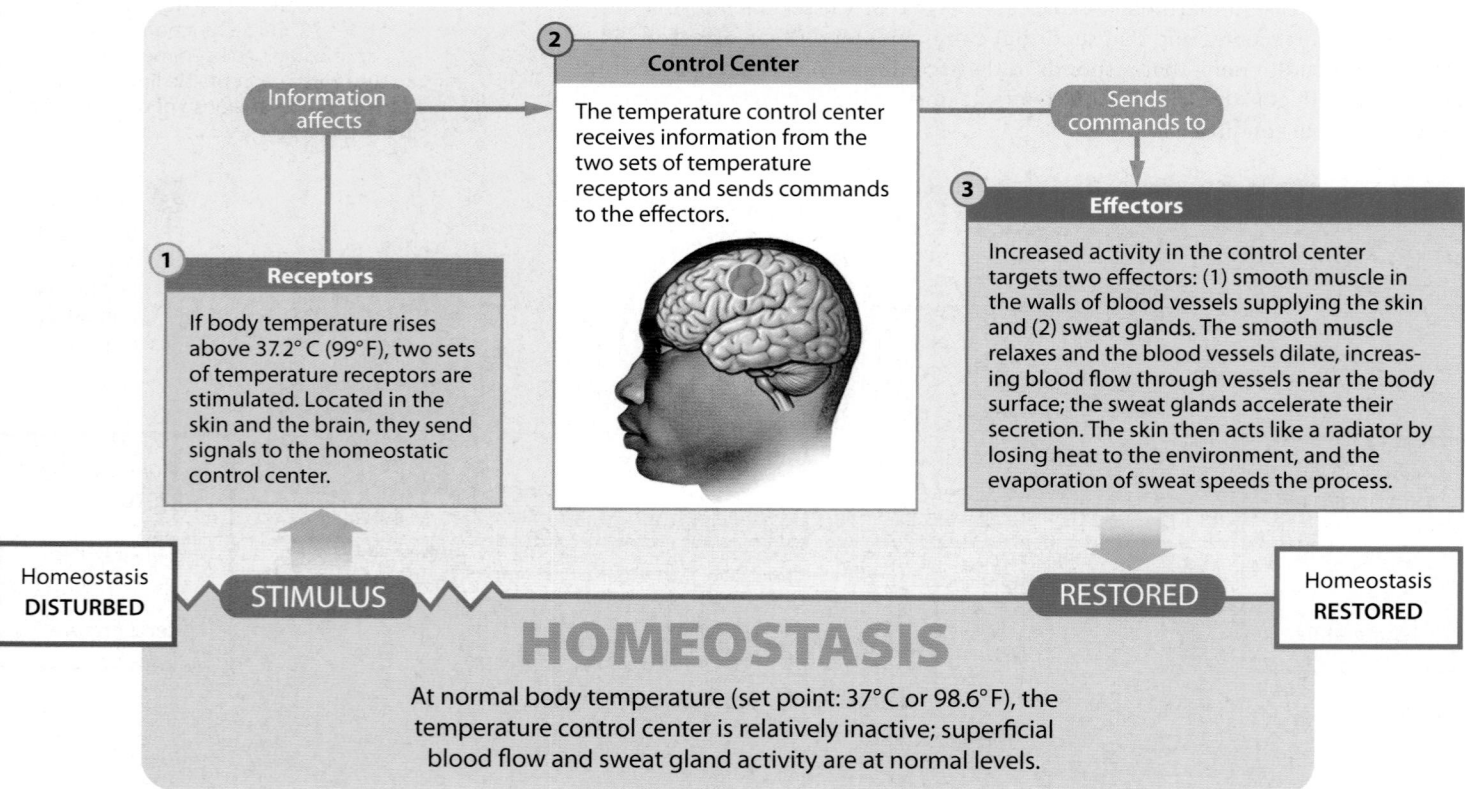

**2** Control Center

The temperature control center receives information from the two sets of temperature receptors and sends commands to the effectors.

Information affects

Sends commands to

**1** Receptors

If body temperature rises above 37.2° C (99°F), two sets of temperature receptors are stimulated. Located in the skin and the brain, they send signals to the homeostatic control center.

**3** Effectors

Increased activity in the control center targets two effectors: (1) smooth muscle in the walls of blood vessels supplying the skin and (2) sweat glands. The smooth muscle relaxes and the blood vessels dilate, increasing blood flow through vessels near the body surface; the sweat glands accelerate their secretion. The skin then acts like a radiator by losing heat to the environment, and the evaporation of sweat speeds the process.

Homeostasis **DISTURBED**    STIMULUS    RESTORED    Homeostasis **RESTORED**

## HOMEOSTASIS

At normal body temperature (set point: 37°C or 98.6°F), the temperature control center is relatively inactive; superficial blood flow and sweat gland activity are at normal levels.

**2** In this graph of body temperature over time in a warm environment, note that body temperature declines past the set point as the sweat already secreted continues to evaporate.

Body temperature (°C)

Vessels dilate, sweating increases

Vessels constrict, sweating decreases

37.2
37
36.7

Set point

Normal range

Time

**?** REVIEW

**A.** Explain the function of negative feedback systems, and give an example of homeostatic regulation by negative feedback in the body.

# … and positive feedback accelerates a process to completion

In **positive feedback**, an initial stimulus produces a response that exaggerates or enhances the change in the original conditions, rather than opposing it. You seldom encounter positive feedback in your daily life, simply because it tends to produce extreme responses. For example, suppose that the thermostat in your house were accidentally connected to a heater rather than to an air conditioner as shown in Module 1.17. Now, when room temperature exceeds the set point, the thermostat turns on the heater, causing a further rise in room temperature. Room temperature will continue to increase until someone switches off the thermostat, turns off the heater, or intervenes in some other way. This kind of escalating cycle is often called a **positive feedback loop**.

A break in a blood vessel wall causes bleeding

Clotting accelerates

*Positive feedback loop*

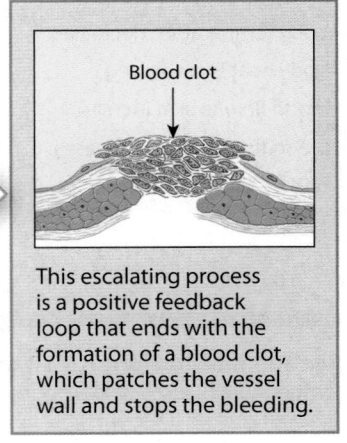

| Chemicals | | Chemicals | Blood clot |

Damage to cells in the blood vessel wall releases chemicals that begin the process of blood clotting.

The chemicals start chain reactions in which cells, cell fragments, and soluble proteins in the blood begin to form a clot.

As clotting continues, each step releases chemicals that further accelerate the process.

This escalating process is a positive feedback loop that ends with the formation of a blood clot, which patches the vessel wall and stops the bleeding.

**3** In the body, positive feedback loops are typically found when a potentially dangerous or stressful process must be completed quickly before homeostasis can be restored. For example, the immediate danger from a severe cut is blood loss, which can lower blood pressure and reduce the heart's efficiency.

**? REVIEW**

**B.** Why is positive feedback helpful in blood clotting but unsuitable for regulating body temperature?

**LO LEARNING OUTCOME**

Discuss the roles of negative feedback and positive feedback in maintaining homeostasis.

# Section 4 Review
## Homeostasis

**Vocabulary:** Write the term for each of the following descriptions in the space provided.

**1** Mechanism that increases a deviation from normal limits after an initial stimulus

**2** Adjustment of physiological systems to preserve homeostasis

**3** The maintenance of a relatively constant internal environment

**4** Homeostatic regulatory part that detects changes

**5** Corrective mechanism that opposes or cancels a variation from normal limits

1 _____
2 _____
3 _____
4 _____
5 _____

Indicate whether each of the following processes represents negative feedback or positive feedback.

**6** A rise in the level of calcium dissolved in the blood stimulates the release of a hormone that causes bone cells to deposit more of the calcium in bone.

**7** Labor contractions become increasingly forceful during childbirth.

**8** An increase in blood pressure triggers a nervous system response that lowers the blood pressure.

**9** Blood vessel cells damaged by a break in the vessel release chemicals that accelerate the blood clotting process.

6 _____
7 _____
8 _____
9 _____

**Short answer:** Assuming a normal body temperature range of 36.7–37.2°C (98–99°F), identify from the graph below what would happen if there were an increase or decrease in body temperature beyond the normal limits. Use the following descriptive terms to explain what would happen at (10) and (11) on the graph.

- body surface cools
- shivering occurs
- sweating increases
- body temperature decreases
- body heat is conserved
- blood flow to skin increases
- blood flow to skin decreases
- body temperature increases

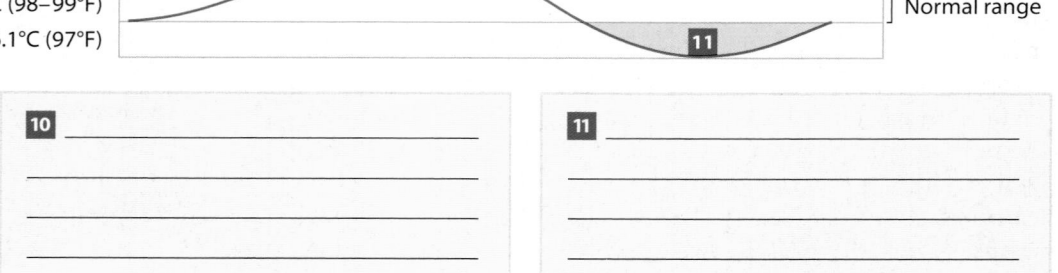

37.8°C (100°F)
36.7–37.2°C (98–99°F)
36.1°C (97°F)
] Normal range

10 _____
_____
_____
_____

11 _____
_____
_____
_____

**Section integration:** It is a warm day, and you feel a little chilled. On checking your temperature, you find that your body temperature is 1.5 °C below normal. Suggest some possible reasons for this situation.

12 _____
_____
_____
_____
_____
_____

36

## Module 1.19
# Anatomical terms have a long and varied history

Early anatomists created maps of the human body, and we still rely on maps for orientation. The landmarks are prominent anatomical structures; distances are measured in centimeters or inches; and we use specialized directional terms. In effect, anatomy uses a special language that must be learned almost at the start. Many terms are based on Latin or Greek words used by ancient anatomists. However, Latin and Greek terms are not the only ones that have been imported into the anatomical vocabulary over the centuries, and the vocabulary continues to expand. Many anatomical structures and clinical conditions were initially named after either the discoverer or, in the case of diseases, the most famous victim. Most of these commemorative names, or **eponyms**, have been replaced by more precise terms, but a few persist.

**1** The table below lists some eponyms you may already know, along with their equivalent anatomical or medical terms.

| Eponym | Equivalent Term |
|---|---|
| Achilles tendon | Calcaneal tendon |
| Broca's area | Speech center |
| Eustachian tube | Auditory tube |
| Krebs cycle | Citric acid cycle |

The study of cadaver-based human anatomy by medical professionals in a university can be traced back to medieval Europe. Founded in 1088, the University of Bologna in Italy is the first university in Europe and one of the oldest in the world. The most famous anatomist at Bologna was Mondino dei Liuzzi. He wrote the *Anatomia,* which is perhaps the first anatomy text for university students. Mondino was so revered that images of him presiding over dissections were included in other anatomy texts for many years.

Anatomy as a science was dramatically improved during the Renaissance at the University of Padua. Established in 1222 in Italy's Venetian Republic, Padua's most famous anatomist was Andreas Vesalius, who conducted frequent, detailed dissections. The result was a visually beautiful and amazingly accurate anatomical text titled *De Humani Corporis Fabrica.* Published in 1543, this text corrected the mistakes of previous anatomists and served as an early model for modern anatomy education. As you begin your anatomy studies, consider yourself as the latest generation of college and university anatomy students. You are studying an area of science that is centuries old and rich with history.

**2** Frontispiece to John of Ketham's *Fasciculus Medicinae,* 1493, showing Mondino dei Liuzzi, the lofty figure adorned with the ornate collar, presiding over a dissection.

**? REVIEW**

**A.** Which languages are the source of many modern anatomical terms?

**B.** Define the word *eponym*.

**C.** In what country was cadaver-based anatomy established as a discipline studied by medical professionals?

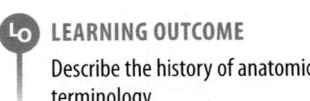

**LEARNING OUTCOME**
Describe the history of anatomical terminology.

# Module 1.20

# Superficial anatomy and regional anatomy indicate locations on or in the body

**1** This illustration shows the body in the **anatomical position**. In this position, the hands are at the sides with the palms facing forward, and the feet are together. Unless otherwise noted, all descriptions in this text refer to the body in the anatomical position. A person lying down is said to be **supine** (sū-PĪN) when face up, and **prone** when face down.

**? REVIEW**

**A.** Describe a person in the anatomical position.

**B.** A massage therapist often begins a massage by asking clients to lie face down with their arms at their sides. Which anatomical term describes that position?

**2** Why is it important to learn these terms? We need common anatomical terms to communicate effectively in a medical setting. For example, stating that a patient has a "bump on the back" does not give very precise information about its location. So, anatomists created maps of the body, naming superficial anatomical structures and identifying regional landmarks to help locate the exact point of that "bump on the back."

**Frontal** (forehead)

**Nasal** (nose)

**Cranial** (skull)

**Ocular, orbital** (eye)

**Cephalic** (head)

**Otic** (ear)

**Buccal** (cheek)

**Facial** (face)

**Cervical** (neck)

**Oral** (mouth)

**Mental** (chin)

**Thoracic** (thorax, chest)

**Mammary** (breast)

**Axillary** (armpit)

**Brachial** (arm)

**Abdominal** (abdomen) — **Trunk**

**Antecubital** (front of elbow)

**Umbilical** (navel)

**Antebrachial** (forearm)

**Pelvic** (pelvis)

**Carpal** (wrist)

**Palmar** (palm)

**Manual** (hand)

**Pollex** (thumb)

**Inguinal** (groin)

**Digits (phalanges)** (fingers)

**Pubic** (pubis)

**Patellar** (kneecap)

**Femoral** (thigh)

**Crural** (leg)

**Tarsal** (ankle)

**Digits (phalanges)** (toes)

**Pedal** (foot)

**Hallux** (great toe)

**Body regions: Anterior view**

TUMMY ACHE?

MAYBE AN EENSY KIDNEY STONE?

...OR, WAY INSIDE, DOWN AROUND THE BELLY BUTTON. SOME KIND OF BOO-BOO...

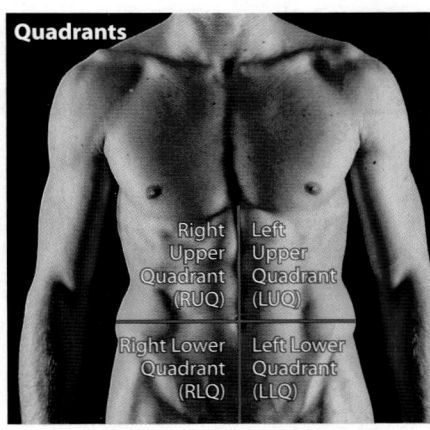

**Quadrants**

Right Upper Quadrant (RUQ)
Left Upper Quadrant (LUQ)
Right Lower Quadrant (RLQ)
Left Lower Quadrant (LLQ)

**3** Clinicians refer to four **abdominopelvic quadrants** formed by a pair of imaginary perpendicular lines that intersect at the umbilicus (navel). This simple method provides useful references for the description of aches, pains, and injuries. The location can help physicians determine the possible cause.

**4** Anatomists prefer more precise terms to describe the location and orientation of internal organs. Anatomists use nine **abdominopelvic regions**.

**Regions**

Right hypochondriac region
Epigastric region
Left hypochondriac region
Right lumbar region
Umbilical region
Left lumbar region
Right inguinal region
Hypogastric (pubic) region
Left inguinal region

**5** The image at the lower right shows the relationships among quadrants, regions, and internal organs.

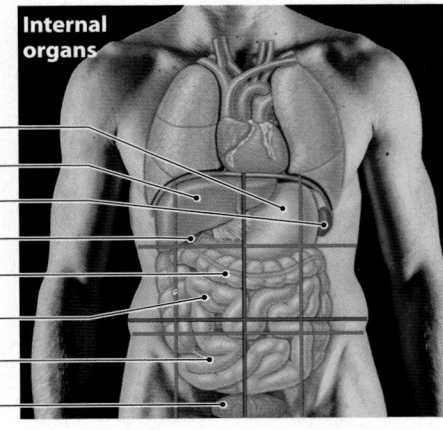

**Internal organs**

Stomach
Liver
Spleen
Gallbladder
Large intestine
Small intestine
Appendix
Urinary bladder

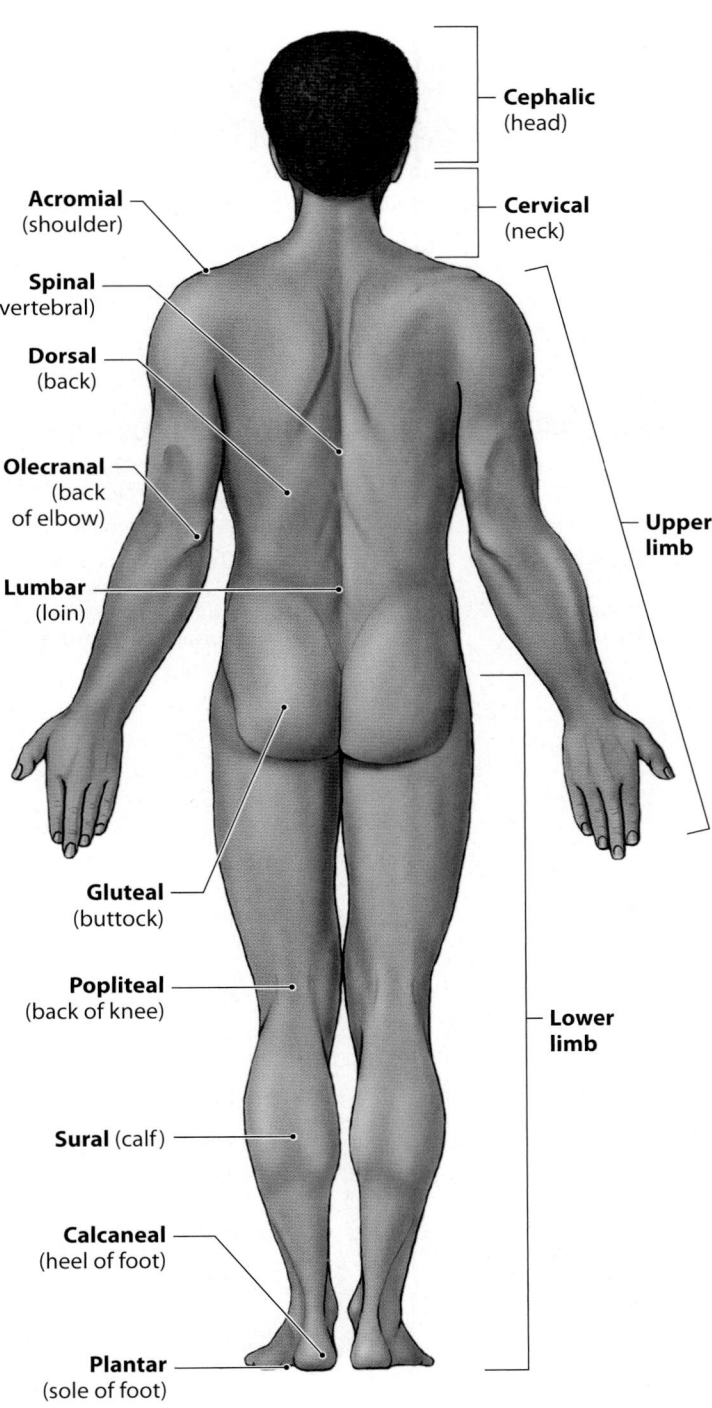

**Cephalic** (head)

**Cervical** (neck)

**Acromial** (shoulder)

**Spinal** (vertebral)

**Dorsal** (back)

**Olecranal** (back of elbow)

**Lumbar** (loin)

**Upper limb**

**Gluteal** (buttock)

**Popliteal** (back of knee)

**Lower limb**

**Sural** (calf)

**Calcaneal** (heel of foot)

**Plantar** (sole of foot)

**Body regions: Posterior view**

**? REVIEW**

**C.** Contrast the descriptions used by clinicians and anatomists when referring to the positions of injuries or internal organs of the abdomen and pelvis.

**? INTEGRATION**

**D.** What is the purpose of anatomical terms?

**Lo LEARNING OUTCOME**

Use correct anatomical terms to describe superficial and regional anatomy.

# Module 1.21

# Directional terms and sectional planes describe specific points of reference

**1** The figures on this page introduce the principal directional terms and examples of their use. There are many different terms, and some can be used interchangeably. As you learn these directional terms, it is important to remember that all anatomical directions utilize the anatomical position as the standard point of reference and that *left* and *right* always refer to the left and right sides of the subject, not the observer.

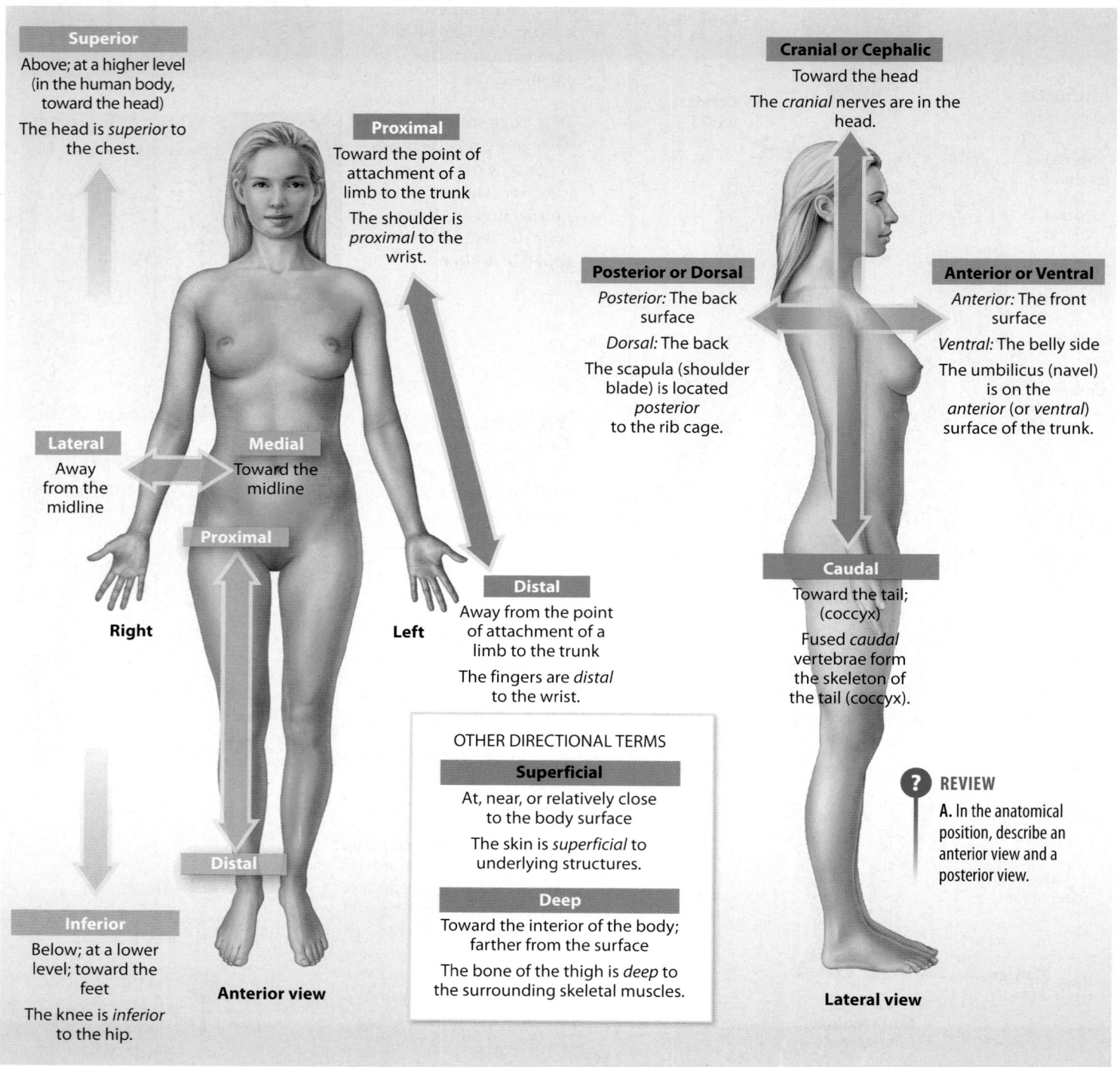

**Superior**
Above; at a higher level (in the human body, toward the head)

The head is *superior* to the chest.

**Proximal**
Toward the point of attachment of a limb to the trunk

The shoulder is *proximal* to the wrist.

**Lateral**
Away from the midline

**Medial**
Toward the midline

**Proximal**

**Right**

**Left**

**Distal**
Away from the point of attachment of a limb to the trunk

The fingers are *distal* to the wrist.

**Distal**

**Inferior**
Below; at a lower level; toward the feet

The knee is *inferior* to the hip.

**Anterior view**

OTHER DIRECTIONAL TERMS

**Superficial**
At, near, or relatively close to the body surface

The skin is *superficial* to underlying structures.

**Deep**
Toward the interior of the body; farther from the surface

The bone of the thigh is *deep* to the surrounding skeletal muscles.

**Cranial or Cephalic**
Toward the head

The *cranial* nerves are in the head.

**Posterior or Dorsal**
*Posterior:* The back surface

*Dorsal:* The back

The scapula (shoulder blade) is located *posterior* to the rib cage.

**Anterior or Ventral**
*Anterior:* The front surface

*Ventral:* The belly side

The umbilicus (navel) is on the *anterior* (or *ventral*) surface of the trunk.

**Caudal**
Toward the tail; (coccyx)

Fused *caudal* vertebrae form the skeleton of the tail (coccyx).

**Lateral view**

**?** REVIEW

**A.** In the anatomical position, describe an anterior view and a posterior view.

**2** A presentation in sectional view is sometimes the only way to illustrate the relationships between the parts of a three-dimensional object. The development of medical imaging techniques has made it even more important to understand sectional planes and terms.

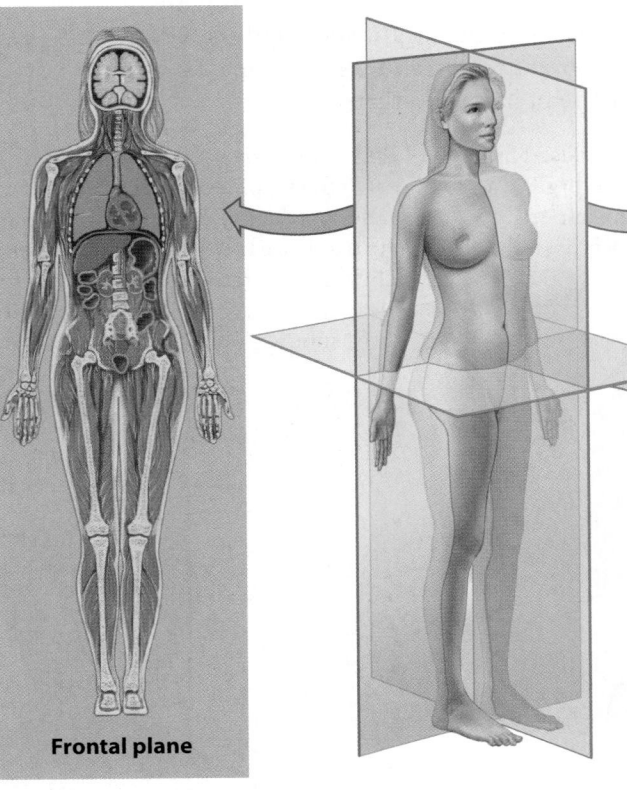

### Frontal, or coronal, plane

Plane is oriented parallel to long axis

A *frontal*, or *coronal*, *section* separates anterior and posterior portions of the body. *Coronal* usually refers to sections passing through the skull.

Directional term: frontally or coronally

**Frontal plane**

### Sagittal plane

Plane is oriented parallel to long axis

A *sagittal section* separates right and left portions. You examine a sagittal section, but you section sagittally.

A *midsagittal section* passes through the midline. It separates the body into equal right and left sides.

A *parasagittal section* misses the midline. It separates the body into unequal right and left sides.

**Midsagittal plane**

Directional term: sagittally

### Transverse, or horizontal, plane

Plane is oriented perpendicular to long axis

A *transverse*, or *horizontal*, section separates superior and inferior portions of the body. A cut in this plane is also called a **cross section**.

**Transverse plane** (inferior view)

Directional term: transversely or horizontally

**3** Sectional planes are used for visualization purposes. Here, we are serially sectioning a bent tube, which looks like a piece of elbow macaroni. Notice how the sectional views change as you approach the curve. Keep in mind these effects of sectioning when you are looking at slides under the microscope. These sectional views also affect the appearance of internal organs when seen in a CT or MRI scan. For example, the small intestine is a simple tube, but it can look like a pair of tubes, a dumbbell, an oval, or a solid, depending on where the section was taken.

**REVIEW**

**B.** What type of section would separate
(1) the two eyes?
(2) the nose and two ears?
(3) the neck and navel?

**INTEGRATION**

**C.** What is the purpose of directional and sectional terms?

**LEARNING OUTCOME**

Use correct directional terms and sectional planes to describe relative positions and relationships among body parts.

# Module 1.22

# Body cavities protect internal organs and allow them to change shape

The interior of the trunk of the body is often subdivided into regions established by the body wall. For example, everything deep to the chest wall is considered to be within the **thoracic cavity**, and all of the structures deep to the abdominal and pelvic walls are said to lie within the **abdominopelvic cavity**. Many vital internal organs within these regions are partially or fully enclosed by fluid-filled chambers called **body cavities**. True body cavities are lined by a serous membrane and share a common embryonic origin. Such body cavities have two essential functions: (1) They protect delicate organs from shocks and impacts; and (2) they permit significant changes in the size and shape of internal organs.

**? REVIEW**

**A.** Describe two distinctive features and two essential functions of true body cavities.

**1** The internal organs that are partially or completely enclosed by body cavities are called **viscera** (VIS-e-ruh) or visceral organs. Viscera do not float within the body cavities—they remain connected to the rest of the body. To understand the physical relationships, we examine the **heart**, **pericardium**, and **pericardial cavity**.

**2** The body cavities of the trunk contain the organs of the respiratory, cardiovascular, digestive, urinary, and reproductive systems. The two major body cavities of the trunk are the thoracic and abdominopelvic cavities. They are separated by the diaphragm. The boundaries of these cavities are indicated in red. Note that there are three subdivisions of the thoracic cavity.

The **pericardial cavity** is the fluid-filled space between the outer and inner layers of the serous pericardium. Imagine pushing your fist into a balloon. The wrist corresponds to the base of the heart, and the balloon represents the pericardium.

The **pericardium** (pehr-ih-KAR-dē-um) surrounds the heart and consists of an outer fibrous pericardium and an inner serous pericardium.

During each beat, the heart changes size and shape. The fluid within the pericardial cavity acts as a lubricant, reducing friction between the pericardial membranes as the heart beats.

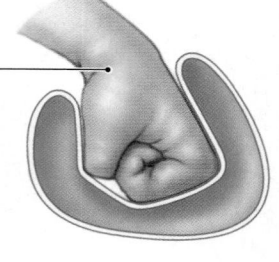

Cardiac muscle of the heart wall

A **serous membrane** covers the viscera and lines the true body cavities of the trunk.

A watery fluid secreted by the serous membranes coats the walls of these internal cavities and covers the surfaces of the enclosed viscera. It keeps the surfaces moist and reduces friction.

**BODY CAVITIES OF THE TRUNK**

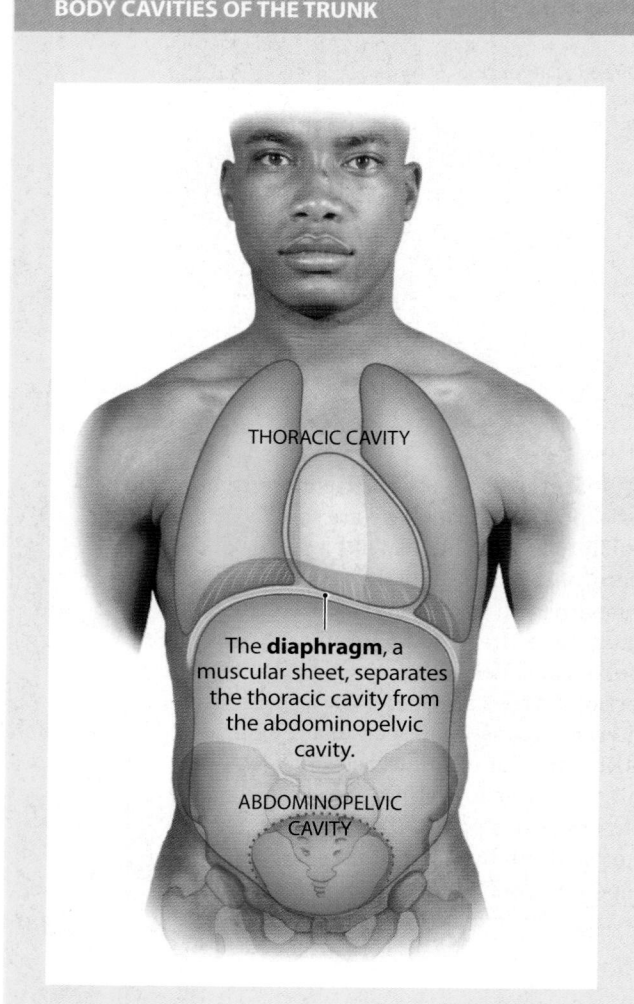

THORACIC CAVITY

The **diaphragm**, a muscular sheet, separates the thoracic cavity from the abdominopelvic cavity.

ABDOMINOPELVIC CAVITY

**3** The thoracic cavity contains the lungs, heart, and other structures. Its boundaries are established by the chest wall and diaphragm.

## THORACIC CAVITY

Each lung is enclosed within a **pleural cavity**, lined by a shiny, slippery serous membrane called the **pleura** (PLUR-uh).

Right lung

Heart

Left lung

Note the orientation of the section. Unless otherwise noted, all cross sections are shown as if the viewer were standing at the feet of a supine person and looking toward the head.

A horizontal section through the thoracic cavity shows the relationship between its three subdivisions.

The heart is embedded within the **mediastinum**, a mass of connective tissue that separates the two pleural cavities and stabilizes the positions of embedded organs and blood vessels.

## ABDOMINOPELVIC CAVITY

The abdominopelvic cavity encloses the **peritoneal** (per-i-tō-NĒ-al) **cavity**, a chamber lined by a serous membrane known as the **peritoneum** (per-i-tō-NĒ-um). A few organs, such as the kidneys and pancreas, lie between the peritoneal lining and the muscular wall of the abdominal cavity. Those organs are said to be **retroperitoneal** (re-trō-per-i-tō-NĒ-al; *retro*, behind). The peritoneum covers the ovaries and the uterus in females, as well as the superior portion of the urinary bladder in both sexes. Organs such as the urinary bladder and the distal portions of the ureters and large intestine, which extend inferior to the peritoneal cavity, are said to be **infraperitoneal**.

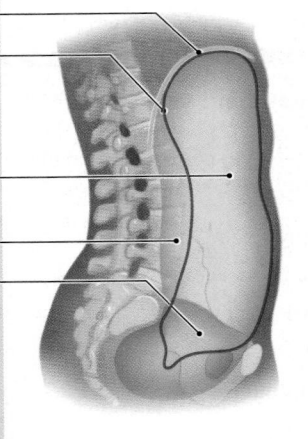

Diaphragm

Peritoneum (red) showing the boundaries of the peritoneal cavity

The abdominal cavity contains many digestive glands and organs

Retroperitoneal area

The pelvic cavity contains the urinary bladder, reproductive organs, and the last portion of the digestive tract. Many of these structures lie posterior to, or inferior to, the peritoneal cavity.

**4** The boundaries of the abdominopelvic cavity are established by the diaphragm, the muscles of the abdominal wall, the trunk muscles and inferior portions of the vertebral column, and the bones and muscles of the pelvis. It may be subdivided into the **abdominal cavity** and the **pelvic cavity**.

 **REVIEW**

**B.** Identify the body cavities of the trunk.

 **INTEGRATION**

**C.** If a surgeon makes an incision just inferior to the diaphragm, what body cavity will be opened?

**LEARNING OUTCOME**

Identify the major body cavities of the trunk and the subdivisions of each.

# Section 5 Review
## Anatomical Terms

**Labeling:** Label the directional terms in the figures at right.

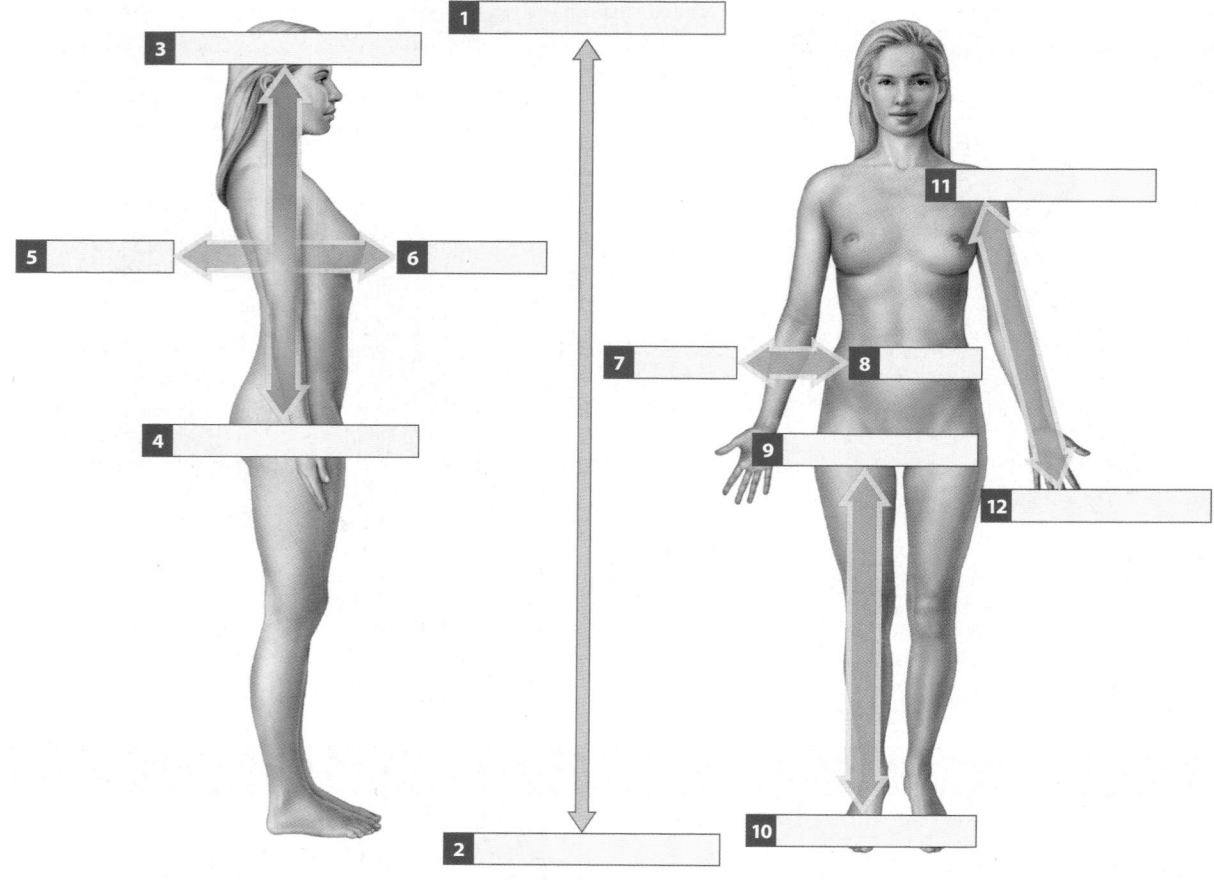

**Concept map:** Use each of the following terms once to fill in the blank boxes to correctly complete the body cavities concept map.

- digestive glands and organs
- abdominopelvic cavity
- thoracic cavity
- heart
- mediastinum
- diaphragm
- pelvic cavity
- trachea, esophagus
- reproductive organs
- left lung
- peritoneal cavity

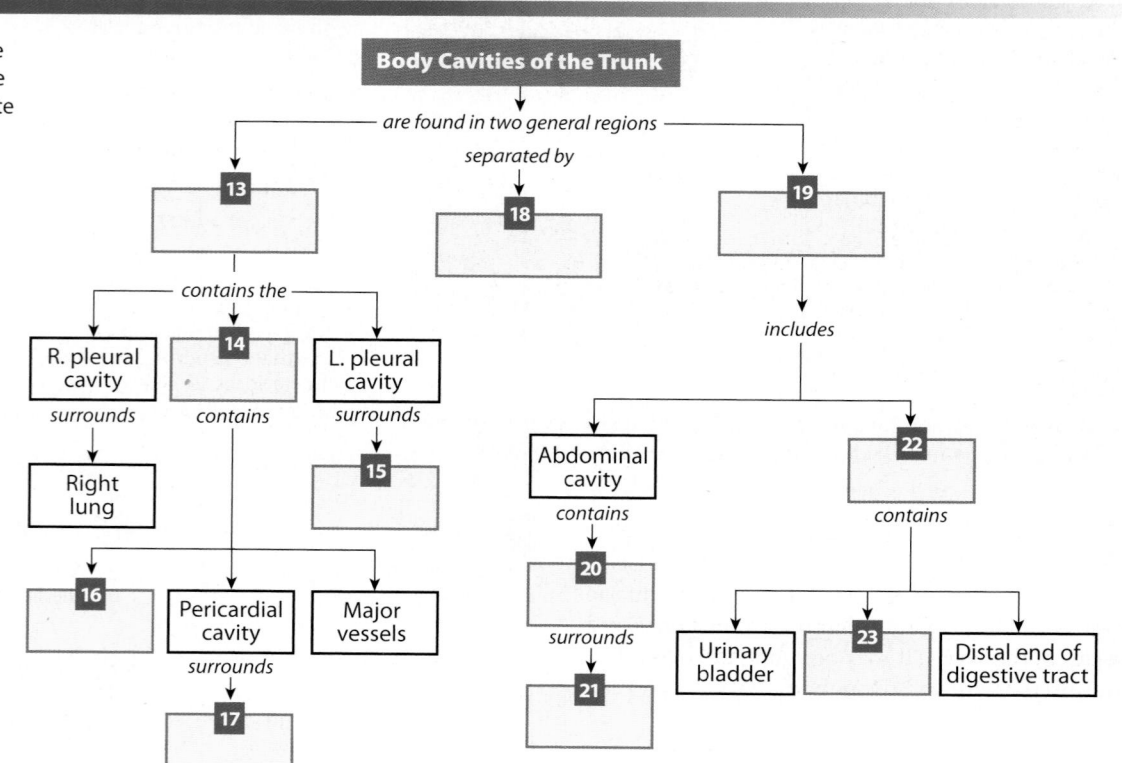

# Study Outline

## Section 1 • An Introduction to Studying the Human Body

### Module 1.1
**Using your textbook effectively is key to your success** p. 3

1. Many students don't use their textbook enough or at all.
2. The strategies in this section give you the tools to get the most out of your textbook.

### Module 1.2
**Comprehending the art is essential to understanding A&P** p. 4

3. Studying the art and text together, known as **text-art integration**, provides better learning and comprehension than studying just the text.
4. Key information is integrated into the art, not just found in the text.

### Module 1.3
**Break down the art in step-wise fashion to learn the topic** p. 6

5. Illustrations of complicated processes can be broken down to make the topic easier to understand.
6. Numbered red boxes divide each module, integrate text and art, and include review and integration questions.

### Module 1.4
**Orient yourself to all art in the same way** p. 8

7. The standard anatomical reference for the human body is the anatomical position.

### Module 1.5
**The learning outcomes correspond by number to the chapter's modules and indicate what you should be able to do after completing the chapter** p. 9

8. Each chapter is broken down into **sections**, which contain multiple **modules** about specific topics.
9. To effectively use text-art integration, stop reading, look at the figure, and study the image along with the text.
10. **Learning outcomes**, found at the beginning of each chapter and at the end of each module, state specific educational objectives and can address lower-order thinking skills to higher-order skills based on a **learning classification scheme**.

## Section 2 • A&P in Perspective

### Module 1.6
**Focused study is important for learning anatomy and physiology** p. 11

11. The basic approach in A&P can be summed up as "What is that structure, and how does it work?"
12. **Homeostasis** is a relatively constant internal environment.

### Module 1.7
**Organisms share common characteristics and processes** p. 12

13. All living things share these common characteristics and functions: cells, organization, responsiveness, regulation, growth and development, reproduction, and metabolism.

*The many different kinds of cells in the human body each have their own characteristics and functions. This nerve cell (neuron) is from a part of the brain involved with the control of balance and movement. The extensive branching provides a huge surface area for communication with other neurons.*

14. Basic processes in humans and other animals include respiration, digestion, circulation, and excretion.

### Module 1.8
**Anatomy is the study of structure, and physiology is the study of function** p. 14

15. **Gross**, or **macroscopic**, **anatomy** examines relatively large structures that are visible with the unaided eye.
16. **Microscopic anatomy** deals with structures that cannot be seen without magnification.
17. **Physiology** is the study of the complex functions of the human body.

### Module 1.9
**Structure and function are interrelated** p. 16

18. An anatomical example of how structure and function are interrelated is how the shapes of the humerus and ulna bones interlock and permit the elbow to function as a hinge joint.
19. An example of a physiological mechanism is how chemical messengers depend on their three-dimensional shapes to bind with cellular receptors.

## Section 3 • Levels of Organization

### Module 1.10
**The human body has multiple interdependent levels of organization** p. 19

20. Progressing from smallest to most complex, the **levels of organization** are **chemical**, **cellular**, **tissue**, **organ**, **organ system**, and **organism**.

### Module 1.11
**Cells are the smallest units of life** p. 20

21. **Cells** are the smallest independent organisms, and they are remarkably diverse.
22. **Cell theory** states that cells are the structural building blocks of plants and animals. They arise from divisions of preexisting cells, and they are the smallest structural units that perform all vital functions.

### Module 1.12
**Tissues are specialized groups of cells and cell products** p. 22

23. **Histology** is the study of **tissues**, which are groups of similar cells that perform specific functions.
24. There are four **primary tissue types**: **epithelial**, **connective**, **muscle**, and **nervous**.

### Module 1.13
**Organs and organ systems perform vital functions** p. 24

25. An **organ** is a functional unit composed of more than one tissue type. An **organ system** consists of organs that interact to perform a specific range of functions.

26. The 11 organ systems of the body are: **integumentary**, **skeletal**, **muscular**, **nervous**, **endocrine**, **cardiovascular**, **lymphatic**, **respiratory**, **digestive**, **urinary**, and **reproductive**.

#### Module 1.14

**Organs of the integumentary, skeletal, and muscular systems support and move the body, and organs of the nervous system provide rapid control and regulation** p. 26

27. The **integumentary system** protects against environmental hazards and helps control body temperature.

28. The **skeletal system** provides support, protects tissues, stores minerals, and forms blood cells.

29. The **muscular system** produces movement, provides support, and generates heat.

30. The **nervous system** directs immediate responses to stimuli by coordinating the activities of the other organ systems.

#### Module 1.15

**Organs of the endocrine system secrete chemicals that are carried by organs of the cardiovascular system, organs of the lymphatic system defend the body, and organs of the respiratory system exchange vital gases** p. 28

31. The **endocrine system** directs long-term changes in other organ systems.

32. The **cardiovascular system** transports cells and dissolved materials, including nutrients, wastes, and gases.

33. The **lymphatic system** defends against infection and disease and returns tissue fluid to the bloodstream.

34. The **respiratory system** delivers air to sites where gas exchange occurs between the air and circulating blood, and produces sound.

*The lymphatic system*

#### Module 1.16

**Organs of the digestive system make nutrients available and, with the urinary system, excrete wastes, and organs of the male and female reproductive systems provide for the continuity of life** p. 30

35. The **digestive system** processes food and absorbs nutrients.

36. The **urinary system** eliminates excess water, salts, and wastes.

37. The **female reproductive system** produces sex cells and hormones and supports embryonic development from fertilization to birth.

38. The **male reproductive system** produces sex cells and hormones.

### Section 4 • Homeostasis

#### Module 1.17

**Homeostatic regulation relies on a receptor, a control center, and an effector** p. 33

39. **Homeostasis** is the presence of a stable internal environment.

40. **Homeostatic regulation** is the adjustment of physiological systems to preserve homeostasis and consists of three necessary

parts: a **receptor** or sensor, a **control center**, and an **effector**. Homeostasis functions around a desired value, or **set point**.

#### Module 1.18

**Negative feedback provides stability, and positive feedback accelerates a process to completion** p. 34

41. In **negative feedback**, a stimulus produces a response that opposes, or negates, change.

42. In **positive feedback**, a stimulus produces a response that exaggerates change, creating a **positive feedback loop**. Positive feedback is seldom encountered in our daily lives because it produces extreme responses.

### SECTION 5 • Anatomical Terms

#### Module 1.19

**Anatomical terms have a long and varied history** p. 37

43. Many anatomical structures are named after the discoverer, whereas many diseases are named after the most famous victim of that disease. Such names, called **eponyms**, have been replaced with more precise terms.

44. Anatomy as a discipline studied by medical professionals began at the University of Bologna in Italy, which is considered the first modern university.

#### Module 1.20

**Superficial anatomy and regional anatomy indicate locations on or in the body** p. 38

45. In the **anatomical position**, the hands are at the sides with the palms facing forward and the feet together. A person lying down in the anatomical position is said to be **supine** when face up and **prone** when face down.

46. To describe the location of internal organs, clinicians refer to four **abdominopelvic quadrants**. Anatomists use nine, more precise, **abdominopelvic regions**.

#### Module 1.21

**Directional terms and sectional planes describe specific points of reference** p. 40

47. Directional terms utilize the anatomical position as a point of reference. The principal directional terms are **superior**, **inferior**, **proximal**, **distal**, **medial**, **lateral**, **cranial (or cephalic)**, **caudal**, **anterior (or ventral)**, **posterior (or dorsal)**, **superficial**, and **deep**.

48. The sectional planes used to illustrate the relationship between body structures are **frontal (coronal)**, **sagittal**, and **transverse (horizontal)**.

#### Module 1.22

**Body cavities protect internal organs and allow them to change shape** p. 42

49. Body structures deep to the chest wall are within the **thoracic cavity**; body structures deep to the abdominal and pelvic walls lie within the **abdominopelvic cavity**.

50. True **body cavities** are lined by a serous membrane and share a common embryonic origin. They function to protect organs from shocks and impacts and to permit significant changes in the size and shape of internal organs.

51. **Viscera** are internal organs that are partially or completely enclosed by body cavities. The double-layered **pericardium** surrounds the heart. The **pericardial cavity** is the fluid-filled space between the outer and inner layers.

52. The thoracic cavity contains the lungs, heart, and other structures. Its boundaries are the chest wall and a muscular sheet called the **diaphragm**.

53. The thoracic cavity consists of two **pleural cavities**, each surrounding a lung and each lined with a thin membrane known as the **pleura**. A tissue mass known as the **mediastinum** separates the pleural cavities. Within the mediastinum is the pericardial cavity that surrounds the heart.

54. The boundaries of the abdominopelvic cavity are the diaphragm, the abdominal wall muscles, the trunk muscles and inferior portions of the vertebral column, and the bones and muscles of the pelvis. It can be subdivided into the **abdominal cavity** and **pelvic cavity**.

55. The abdominal cavity and pelvic cavity contain the **peritoneal cavity**, a chamber lined by the **peritoneum**, a **serous membrane**.

56. Some organs lie between the peritoneal lining and the muscular wall of the abdominal cavity, and are said to be **retroperitoneal**. Other organs extend inferior to the peritoneal cavity and are said to be **infraperitoneal**.

1

## Chapter Review Questions

**Labeling:** Identify the body cavities and the organs they enclose, as well as the structure separating the superior and inferior body cavities.

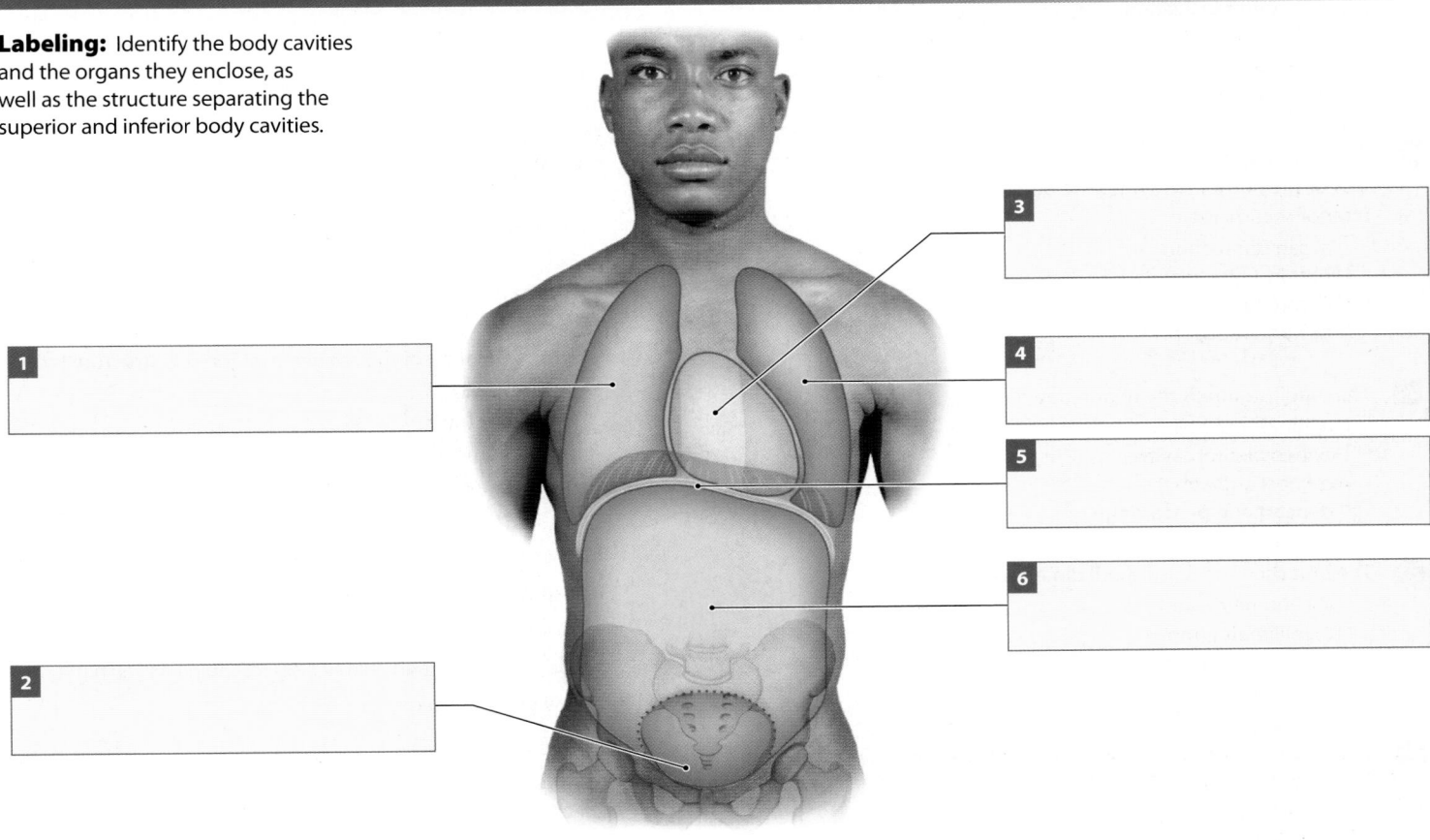

1

2

3

4

5

6

**Matching:** Match each lettered term with the most closely related description.

a. cytology
b. physiology
c. histology
d. anatomy
e. homeostasis
f. muscle
g. heart
h. endocrine
i. temperature regulation
j. labor and delivery
k. supine
l. prone
m. abdominopelvic cavity
n. pericardium

7 _____ Study of tissues
8 _____ Constant internal environment
9 _____ Face-up position
10 _____ Study of functions
11 _____ Positive feedback
12 _____ Organ system
13 _____ Study of cells

14 _____ Negative feedback
15 _____ Serous membrane
16 _____ Study of internal and external body structures
17 _____ Diaphragm tissue
18 _____ Peritoneal cavity
19 _____ Organ
20 _____ Face-down position

**Multiple choice:** Select the correct answer from the list provided.

**21** What is the correct order, from simplest to most complex, of the six levels of organization that make up the human body?
- ☐ a) cell, chemical, tissue, organ, organ system, organism
- ☐ b) chemical, cell, tissue, organ, organ system, organism
- ☐ c) chemical, cell, tissue, organ system, organ, organism
- ☐ d) chemical, cell, organ, tissue, organ system, organism
- ☐ e) cell, tissue, chemical, organ, organism, organ system

**22** The increasingly forceful labor contractions during childbirth are an example of
- ☐ a) receptor activation.
- ☐ b) effector shutdown.
- ☐ c) negative feedback.
- ☐ d) positive feedback.

**23** A plane through the body that passes perpendicular to the long axis of the body and divides the body into a superior and an inferior section is a
- ☐ a) sagittal section.
- ☐ b) transverse section.
- ☐ c) coronal section.
- ☐ d) frontal section.

**24** The mediastinum is the region between the
- ☐ a) lungs and heart.
- ☐ b) two pleural cavities.
- ☐ c) chest and abdomen.
- ☐ d) heart and pericardium.

**25** The unit used to measure cell dimensions is the
- ☐ a) centimeter (cm).
- ☐ b) millimeter (mm).
- ☐ c) micrometer (μm).
- ☐ d) kilometer (km).

**26** The two major body cavities of the trunk are the
- ☐ a) pleural cavity and pericardial cavity.
- ☐ b) pericardial cavity and peritoneal cavity.
- ☐ c) pleural cavity and peritoneal cavity.
- ☐ d) thoracic cavity and abdominopelvic cavity.

**27** Which of the following is *not* a characteristic of life?
- ☐ a) responsiveness
- ☐ b) movement
- ☐ c) manipulation of external environment
- ☐ d) reproduction

**28** Which sectional plane would divide the body so that the face remains intact?
- ☐ a) sagittal plane
- ☐ b) frontal (coronal) plane
- ☐ c) midsagittal plane
- ☐ d) parasagittal plane

**29** The _____ provides the means for understanding art presenting physiological processes.
- ☐ a) key
- ☐ b) module
- ☐ c) higher-order skill
- ☐ d) strategy

**30** The determination of the relative position of a structure is called
- ☐ a) metabolism.
- ☐ b) homeostasis.
- ☐ c) orientation.
- ☐ d) characteristics of living orgaisms.

**31** Put these steps in the correct order for maximum learning.
1. Read the text describing the image
2. Move to the next numbered red block
3. Study the image and related text describing its key parts
4. Answer any related review questions

- ☐ a) 3, 4, 1, 2
- ☐ b) 1, 3, 4, 2
- ☐ c) 2, 1, 4, 3
- ☐ d) 4, 2, 3, 1

**Short answer**

**32** Define *anatomy*. Define *physiology*.

**33** Describe the three basic principles of the cell theory.

**34** In which body cavity would each of the following organs be enclosed? Heart, small intestine, large intestine, lung, kidneys.

**35** Identify each of the four primary tissue types, and give an example of where in the body that tissue would be found.

**36** The hormone calcitonin is released from the thyroid gland in response to increased levels of calcium ions in the blood. If this hormone is controlled by negative feedback, what effect would calcitonin have on blood calcium levels?

**37** A stroke occurs when there is a disruption in blood flow to the brain, causing brain cells to die. Predict the kind of symptoms a stroke patient would have. Apply your knowledge of organ system function.

## Chapter Integration • Applying what you have learned

### The amazing experience of studying the human body firsthand

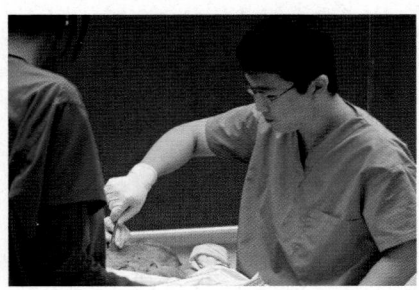

Your A&P studies are just beginning. The reading workload is massive, the terms seem like a foreign language, and the instructor seems to think that your entire life should revolve around her class. But you really do love anatomy and are lucky enough to be able to experience the study of the human body firsthand. At your last class you were introduced to the human anatomy lab, which you now approach with a little trepidation, as you have never had to study a deceased human body. Body donation for scientific purposes is a noble gesture on the part of the deceased, and you fully understand your instructor's expectations for showing complete respect to the cadaver.

Your instructor probably has you put on your gloves, apron, and eye protection before presenting the cadaver to you. Your fears prove groundless as this amazing experience begins. Your instructor explains that this cadaver was the victim of a gunshot wound to the lower right abdomen, with no exit wound. No autopsy was performed, so the anatomical/physiological cause of death has yet to be determined. The information on the body donor card provides the following information: male, 45 years old, no known medical history, and occupation unspecified. As a beginning anatomy student, use your knowledge to answer the following questions.

**1** Upon superficial examination, you discover that the bullet entered about 2 cm inferior to the umbilicus (navel). Using anatomical terminology, describe the location.

**2** Before opening the abdominal cavity, predict what organs may have been affected.

**3** Propose a plausible cause of death.

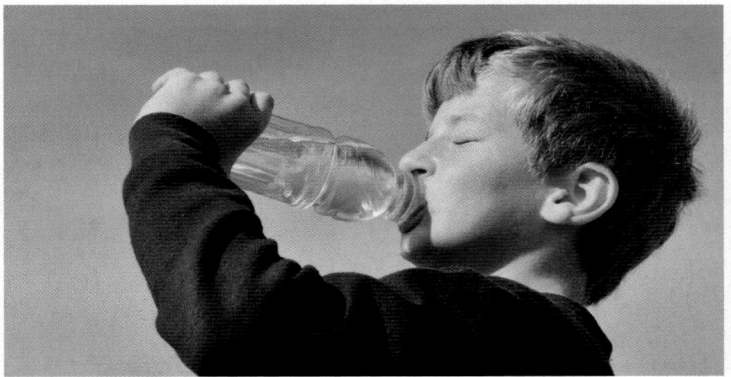

### Aaron has been extremely tired, thirsty, and urinating more than normal

Aaron is an active and apparently healthy 14-year-old boy. Lately, however, he has been extremely tired and thirsty and has been urinating more than normal. His parents take him to the doctor, and Aaron is diagnosed as having type 1 (insulin-dependent) diabetes. The physician explains to Eddie and Patty, his concerned parents, that Aaron's blood glucose (sugar) is much higher than it should be. She goes on to explain that when his blood glucose rises after meals, it should be absorbed into his body's tissues. Insulin, she continues, is the hormone responsible for this absorption, and it is normally produced by the pancreas. Aaron is no longer producing insulin, and even though his blood glucose levels are high, glucose is not going into his tissues. His body is being starved of glucose, explaining his lack of energy. Furthermore, the kidneys flush the excess glucose out of the blood and into the urine, resulting in more urine production. Considering what you have read in this chapter, can you answer any of these questions?

**4** Which one of Aaron's organ systems is failing? List other systems that could be affected.

**5** Applying what you learned about homeostasis (see Module 1.17) and negative feedback (see Module 1.18), predict the process that normally controls blood sugar, and then conclude which physiological mechanism is failing.

**6** Which dietary and medical interventions would you judge most effective in treating his type 1 diabetes?

## LEARNING OUTCOMES

These Learning Outcomes correspond by number to this chapter's modules and indicate what you should be able to do after completing the chapter. They also appear at the end of each module.

# 2

# Chemical Level of Organization

# Module 2.1
# Atoms are the basic particles of matter

Go to
MasteringA&P™

Get Ready
for A&P

Chemistry

Our study of the human body begins at the chemical level of organization. Chemistry is the science that studies the structure of **matter**, which is defined as anything that takes up space and has mass. **Mass** is the quantity of matter in an object. The more matter an object contains, the greater its mass. Within Earth's gravitational field, the mass of an object determines its weight. However, the two are not always equivalent: In orbit beyond Earth's gravity, you would be weightless, but your mass would remain unchanged.

**1** **Atoms** are the smallest stable units of matter. They are composed of **subatomic particles**, only three of which are important for understanding the basic chemical properties of matter.

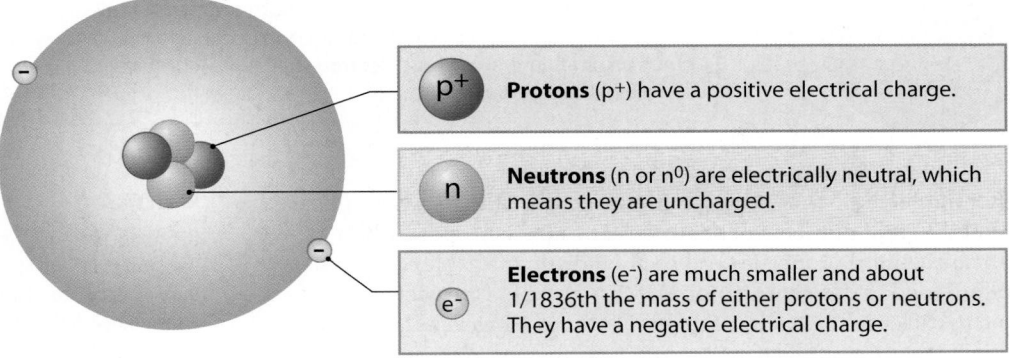

**Protons** (p⁺) have a positive electrical charge.

**Neutrons** (n or n⁰) are electrically neutral, which means they are uncharged.

**Electrons** (e⁻) are much smaller and about 1/1836th the mass of either protons or neutrons. They have a negative electrical charge.

**? REVIEW**

**A.** What is the relationship between an atom and matter?

**B.** Which subatomic particles have a positive charge? Which are uncharged?

**C.** Describe the subatomic particle not in the nucleus.

**2** Atoms can be subdivided into the nucleus and the electron cloud.

The **nucleus** of an atom lies at its center. The nucleus contains one or more protons and may contain neutrons as well. Protons and neutrons are similar in size and mass. The mass of the atom is mainly determined by the numbers of protons and neutrons in the nucleus.

The electrons in the atom whirl around the nucleus, tracing out a volume of space called an **electron cloud**.

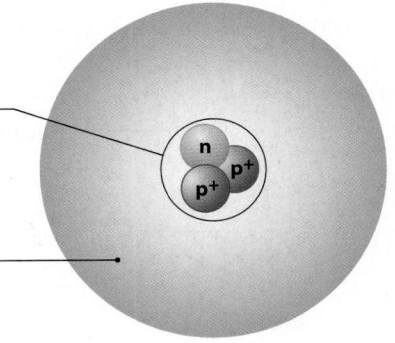

**3** Atoms interact by means of their electrons and produce larger, more complex structures.

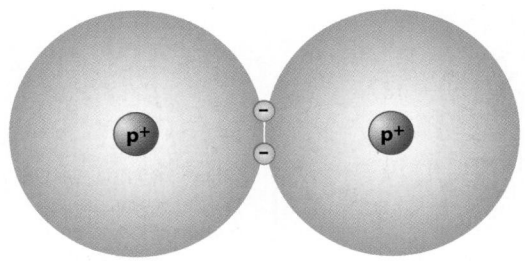

In summary, everything around us is composed of atoms in varying combinations. The unique characteristics of each object, living or nonliving, result from the types of atoms involved and the ways those atoms combine and interact.

**? INTEGRATION**

**D.** The gravitational field of the moon is 17% of Earth's. How would the weight and mass of a 100-pound astronaut change on the moon?

**LO LEARNING OUTCOME**

Define an atom, and describe the properties of its subatomic particles.

# Module 2.2
# Typical atoms contain protons, neutrons, and electrons

Atoms normally contain equal numbers of protons and electrons. The number of protons in an atom is known as the **atomic number**; the total number of both protons and neutrons is its **mass number**. An **element** is a pure substance consisting only of atoms with the same atomic number.

**?** **REVIEW**

**A.** Which is larger: an element's atomic number or mass number?

**1** All atoms are electrically neutral; they contain equal numbers of protons and electrons. **Hydrogen (H)** is the simplest atom, with an atomic number of 1. Thus, an atom of hydrogen contains one proton and one electron. A hydrogen atom's proton is located in the center of the atom and forms the nucleus.

The electrons whirl around the nucleus in a space known as the **electron cloud**. One reason an electron tends to remain in the electron cloud is that the negatively charged electron is attracted to the positively charged proton. The attraction between opposite electrical charges is an example of an electrical force.

Electrons are often shown in a fixed orbit around the nucleus, when in fact their movements are much more complex. In this simple two-dimensional model, the electron occupies a circular **electron shell**. The outermost electron shell represents the surface of the electron cloud.

**2** The atoms of a single element can differ in the number of neutrons in the nucleus. Atoms whose nuclei contain the same number of protons but different numbers of neutrons are called **isotopes**. Different isotopes of an element have essentially identical chemical properties and so are indistinguishable except on the basis of mass. The mass number is therefore used to designate isotopes. Mass numbers are useful because they tell us the number of subatomic particles in the nuclei of different atoms. However, they do not tell us the actual mass of the atoms. For example, they do not take into account the masses of the electrons or the slight difference between the mass of a proton and that of a neutron. The actual mass of an atom of a specific isotope is known as its **atomic mass**. The unit used to express atomic mass is the **atomic mass unit (amu)**, or dalton. By international agreement, 1 amu is equal to one-twelfth the mass of a carbon-12 atom.

**?** **REVIEW**

**B.** Carbon-12 ($^{12}$C) is the most common form of the element carbon. How is the isotope carbon-13 ($^{13}$C) similar to and different from $^{12}$C?

**Electron shell model**

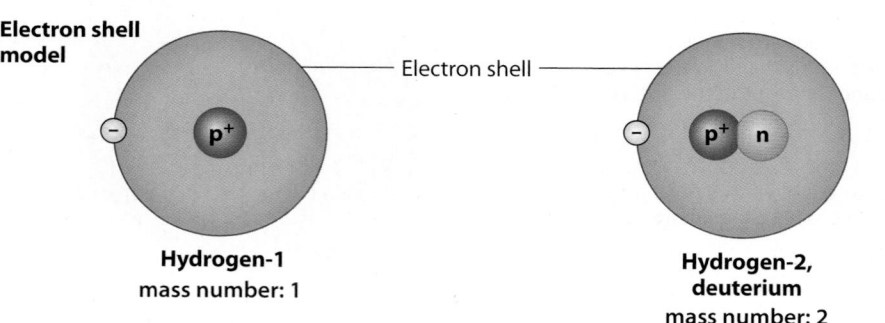

Electron shell

**Hydrogen-1**
mass number: 1

**Hydrogen-2, deuterium**
mass number: 2

**Hydrogen-3, tritium**
mass number: 3

**3** The **atomic weight** of an element is an average of the different atomic masses and proportions of its different isotopes. As a result, the atomic weight of an element is very close to the mass number of the most common isotope of that element. For example, the mass number of the most common isotope of hydrogen is 1, but the atomic weight of hydrogen is 1.0079, primarily because some hydrogen atoms (0.015 percent) have a mass number of 2, and even fewer have a mass number of 3. (In most cases, the periodic table of the elements in the Appendix shows the atomic weight of each element rounded to its nearest one-hundredth decimal place. For example, the atomic weight of hydrogen is shown as 1.01, rather than 1.0079).

**Atomic weight of hydrogen = 1.0079**

**?** **REVIEW**

**C.** How is it possible for two samples of hydrogen to contain the same number of atoms yet have different weights?

Go to
MasteringA&P™

Get Ready
for A&P

Atomic
Structure

2

## Principal Elements of the Human Body

| Element (% of total body weight) | Significance |
| --- | --- |
| **Oxygen, O** (65) | A component of water and other compounds (substances composed of two or more different elements); as a gas, essential for respiration |
| **Carbon, C** (18.6) | Found in all organic compounds |
| **Hydrogen, H** (9.7) | A component of water and most other compounds in the body |
| **Nitrogen, N** (3.2) | Found in proteins, nucleic acids, and other organic compounds |
| **Calcium, Ca** (1.8) | Found in bones and teeth; important for plasma membrane function, nerve impulses, muscle contraction, and blood clotting |
| **Phosphorus, P** (1.0) | Found in bones and teeth, nucleic acids, and high-energy compounds |
| **Potassium, K** (0.4) | Important for plasma membrane function, nerve impulses, and muscle contraction |
| **Sodium, Na** (0.2) | Important for blood volume, plasma membrane function, nerve impulses, and muscle contraction |
| **Chlorine, Cl** (0.2) | Important for blood volume, plasma membrane function, and water absorption |
| **Magnesium, Mg** (0.06) | A cofactor* for many enzymes |
| **Sulfur, S** (0.04) | Found in many proteins |
| **Iron, Fe** (0.007) | Essential for oxygen transport and energy capture |
| **Iodine, I** (0.0002) | A component of hormones of the thyroid gland |
| **Trace elements:** silicon (Si), fluorine (F), copper (Cu), manganese (Mn), zinc (Zn), selenium (Se), cobalt (Co), molybdenum (Mo), cadmium (Cd), chromium (Cr), tin (Sn), aluminum (Al), boron (B), and vanadium (V) | Some function as cofactors*; the functions of many trace elements are poorly understood |

* A cofactor is a mineral or nonprotein compound. It acts with proteins called enzymes to speed up chemical reactions in living things.

 **REVIEW**

**D.** Describe trace elements.

**4** Our bodies consist of many elements, and the 13 most abundant elements are listed in this table. The human body also contains atoms of another 14 elements—called **trace elements**—that are present in very small amounts. Only 92 elements exist in nature, although about two dozen additional elements have been created through nuclear reactions in research laboratories. Every element has a **chemical symbol**, an abbreviation recognized by scientists everywhere. Most of the symbols are easily connected with the English names of the elements (O for oxygen, N for nitrogen, C for carbon, and so on), but a few are abbreviations of their names in other languages. For example, the symbol for sodium, Na, comes from the Latin word *natrium*.

### Everyday Physiology

## Why are some isotopes unstable?

The nuclei of some isotopes are unstable, or radioactive. That is, they spontaneously break down and give off *radiation* (energy in the form of moving subatomic particles or waves). Such isotopes are called **radioisotopes**. The breakdown process is called *radioactive decay*. The decay rate of a radioisotope is commonly expressed as its **half-life**: the time required for half of a given amount of the isotope to decay. Radioisotopes differ in how rapidly they decay; their half-lives range from fractions of a second to billions of years.

**?** **INTEGRATION**

**E.** List the chemical symbols of the six most abundant elements in the human body and their total percentage contribution to total body weight.

**Lo** **LEARNING OUTCOME**

Describe an atom and how atomic structure affects the mass number and atomic weight of the various chemical elements.

# Module 2.3
# Electrons occupy various energy levels

Recall that atoms are electrically neutral; every positively charged proton is balanced by a negatively charged electron. Thus, with each increase in atomic number there is a comparable increase in the number of electrons traveling around the nucleus. Within the electron cloud, electrons occupy an orderly series of energy levels. Although the electrons in an energy level may travel in complex patterns around the nucleus, for our purposes the patterns are diagrammed as a series of concentric electron shells. The first electron shell (the one closest to the nucleus) corresponds to the lowest energy level. The number of electrons in the outermost energy level determines the atom's ability to participate in chemical reactions.

### Reactive elements

**1** The outermost energy level of an atom, called the **valence shell**, forms the atom's "surface." Electrons in the valence shell of an atom can be transferred to or shared with another atom. Atoms with unfilled energy levels, such as hydrogen and lithium, will react with other atoms, usually in ways that give them full valence shells. An atom with a filled valence shell is stable and does not readily react with other atoms.

### Noble gases

**2** Elements that do not readily participate in chemical processes are said to be **inert**. Inert elements, such as helium and neon, have filled outermost energy levels, and their atoms neither react with one another nor combine with atoms of other elements. Helium and neon are two of six such elements, called **noble gases**.

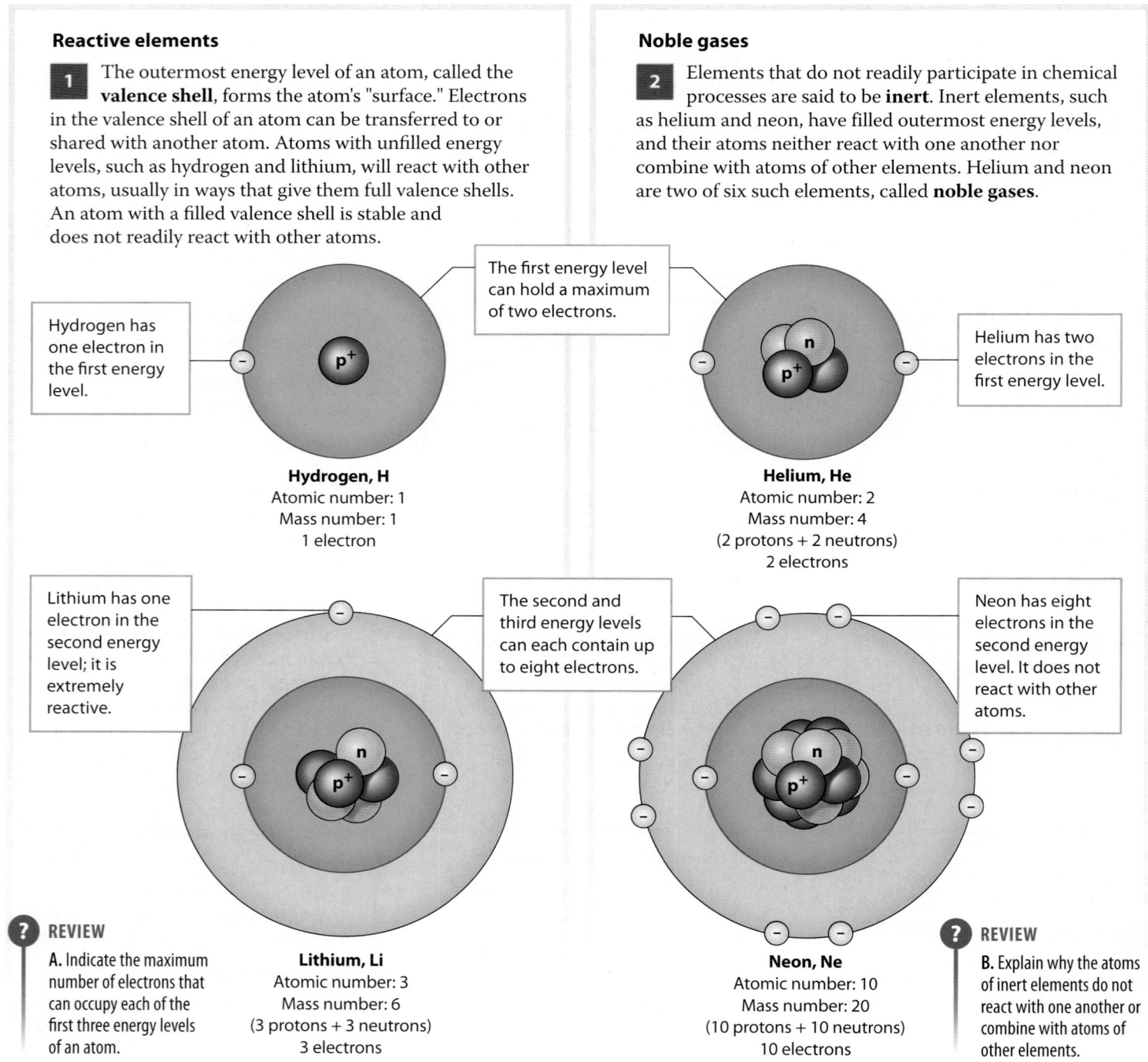

The first energy level can hold a maximum of two electrons.

Hydrogen has one electron in the first energy level.

Helium has two electrons in the first energy level.

**Hydrogen, H**
Atomic number: 1
Mass number: 1
1 electron

**Helium, He**
Atomic number: 2
Mass number: 4
(2 protons + 2 neutrons)
2 electrons

Lithium has one electron in the second energy level; it is extremely reactive.

The second and third energy levels can each contain up to eight electrons.

Neon has eight electrons in the second energy level. It does not react with other atoms.

**Lithium, Li**
Atomic number: 3
Mass number: 6
(3 protons + 3 neutrons)
3 electrons

**Neon, Ne**
Atomic number: 10
Mass number: 20
(10 protons + 10 neutrons)
10 electrons

**?** **REVIEW**
**A.** Indicate the maximum number of electrons that can occupy each of the first three energy levels of an atom.

**?** **REVIEW**
**B.** Explain why the atoms of inert elements do not react with one another or combine with atoms of other elements.

**3** Elements with unfilled valence shells, such as hydrogen, lithium, or sodium, are called **reactive** because they readily interact or combine with other atoms. Reactive atoms become stable by gaining, losing, or sharing electrons to fill their outermost energy level. When atoms gain or lose electrons, they are no longer electrically neutral, and they become **ions**. Atoms that lose electrons from the outer energy level have more protons than electrons. The atom has a net positive charge and is called a positive ion or **cation**. A single missing electron gives the ion a charge of +1. Some ions carry charges of +2, +3, or +4, depending on how many electrons are lost to become stable.

**Sodium atom, Na** (reactive)　　　　　　　　　　**Sodium ion, Na⁺** (stable)

**4** At other times, atoms become stable by filling their outer energy level with electrons obtained from other atoms. This also creates an atom that is no longer electrically neutral—it has more electrons than protons. The atom now has a net negative charge and is called a negative ion or **anion**. A single extra electron gives the ion a charge of −1. Some ions carry charges of −2, −3, or −4, depending on how many electrons are needed to become stable.

 **REVIEW**

**C.** Explain how cations and anions form.

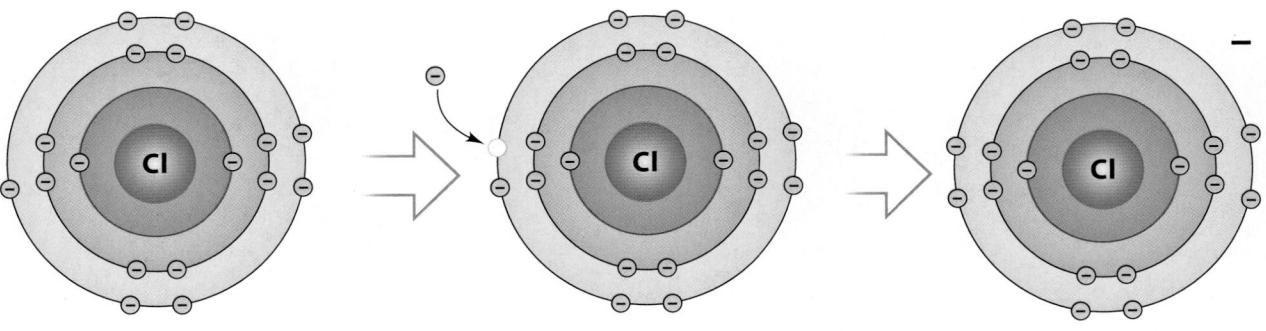

**Chlorine atom, Cl** (reactive)　　　　　　　　　**Chloride ion, Cl⁻** (stable)

The interactions that stabilize the outer energy levels of atoms often result in **chemical bonds**. These bonds hold the participating atoms together when the reaction has ended.

 **INTEGRATION**

**D.** Cations are smaller in diameter than their electrically neutral atom. Why?

**LEARNING OUTCOME**

Explain the relationship between electrons and energy levels.

# Module 2.4
# The most common chemical bonds are ionic bonds and covalent bonds

Go to
MasteringA&P™

Get Ready
for A&P

Chemical
Bonding

Chemical bonding forms new chemical entities called compounds and molecules. A **compound** is a chemical substance made up of atoms of two or more different elements in a fixed proportion, regardless of the type of bond joining them.

**1** **Ionic bonds** are chemical bonds created by the electrical attraction between cations (positive ions) and anions (negative ions). Ionic bond formation involves the transfer of one or more electrons from an atom that can lose them to achieve stability, to another atom that can gain them to achieve stability. Here we consider formation of the ionic compound sodium chloride (table salt).

**? REVIEW**
**A.** Describe why table salt is a compound.

| Step 1 | Step 2 |
|---|---|
| Sodium and chloride ion formation. The sodium atom loses an electron to the chlorine atom. This produces two stable ions with filled outer energy levels. | Ionic bond formation. Because these ions form close together and have opposite charges, they are attracted to one another. This creates NaCl, an ionic compound. |

Sodium atom

Na

Chlorine atom

Sodium ion ($Na^+$)

Na
+

Cl
−

Chloride ion ($Cl^-$)

Na
+

Cl
−

Sodium chloride (NaCl)

**2** A crystal of sodium chloride contains a large number of sodium and chloride ions packed closely together. The packed orientation of the sodium and chloride ions forms cube-shaped crystals.

Chloride ions ($Cl^-$)

Sodium ions ($Na^+$)

Sodium chloride crystals          × 0.5

**3** Some atoms can complete their outer electron shells not by gaining or losing electrons, but by sharing electrons with other atoms. Such sharing creates **covalent** (kō-VĀ-lent) **bonds** between the atoms involved. A **molecule** is a chemical structure consisting of atoms of one or more elements held together by covalent bonds.

**? REVIEW**

**B.** How many electrons are shared by the oxygen atoms in an oxygen molecule?

**2**

| Molecule | | Description |
|---|---|---|
| **Hydrogen** ($H_2$) |  | Hydrogen atoms are not found as single atoms. They exist as molecules, each containing a pair of hydrogen atoms. When the 2 hydrogen atoms share their electrons, each electron whirls around both nuclei. The sharing of one pair of electrons creates a **single covalent bond**. |
| **Oxygen** ($O_2$) | 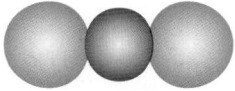 | An oxygen atom has 6 electrons in its outer energy level. By sharing two pairs of electrons, it forms a **double covalent bond** with another oxygen atom, and an oxygen molecule is created with a stable outer energy level. |
| **Carbon dioxide** ($CO_2$) | | A carbon atom has 4 electrons in its outer energy level, so it needs to gain another 4 from other atoms to achieve stability. In a molecule of carbon dioxide, a carbon atom shares a pair of electrons with each of 2 oxygen atoms and forms 2 double covalent bonds. |

**4** These are space-filling models of oxygen and carbon dioxide molecules. The spherical diameters of the atoms are shown in proportion to each other in these models. In nonpolar covalent bonds, the participating atoms share the electrons *equally*, and there is no electrical charge on the molecule. Because of this lack of electrical charge, such molecules are called **nonpolar molecules**.

Oxygen ($O_2$)

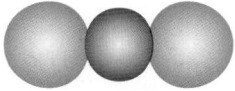

Carbon dioxide ($CO_2$)

**5** Some molecules, however, are formed by covalent bonds that involve an *unequal* sharing of electrons. In a water molecule, the electron clouds of the hydrogen atoms are distorted because the 8 protons in the oxygen atom have a much stronger attraction for the electrons than do the single protons of the hydrogen atoms. As a result, each hydrogen atom carries a slightly positive charge ($\delta^+$), and the oxygen atom carries a slightly negative charge ($2\delta^-$). (The Greek delta symbol [$\delta$] is used to denote partial charges.) This creates an asymmetrical **polar molecule**. Covalent bonds that produce polar molecules are called **polar covalent bonds**.

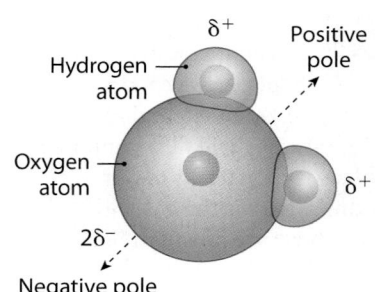

Hydrogen atom — $\delta^+$ — Positive pole

Oxygen atom — $2\delta^-$ — $\delta^+$

Negative pole

**? REVIEW**

**C.** Describe the kind of bonds that hold the atoms in a water molecule together.

### Clinical Note

## Free radicals damage or destroy vital compounds

Covalent bonds usually form molecules in which the valence shells of the atoms involved are filled. An atom, ion, or molecule that contains unpaired electrons in its valence shell is called a *free radical*. Free radicals are highly reactive. Almost as fast as it forms, a free radical enters additional reactions that are typically destructive. For example, free radicals can damage or destroy vital compounds, such as proteins. Evidence suggests that the cumulative damage from free radicals inside and outside our cells is a major factor in the aging process. Free radicals sometimes form in the course of normal metabolism, but cells have several methods of removing or inactivating them.

**? INTEGRATION**

**D.** Explain why we can use the term *molecule* for the smallest particle of water but not for that of table salt.

**Lo LEARNING OUTCOME**

Compare the ways in which atoms combine to form molecules and compounds.

# Module 2.5
# Matter may exist as a solid, a liquid, or a gas

Most matter in our environment exists in one of three states: solid, liquid, or gas. Whether a particular substance is a solid, a liquid, or a gas depends on the degree of motion or vibration due to its thermal energy (temperature) and the types of interactions among its atoms or molecules. The particles of a solid are held tightly together, while those of a liquid are less so; and the particles of a gas are independent of each other.

**1** Solids maintain their volume and their shape at ordinary temperatures and pressures. A lump of granite, a brick, and a textbook are solid objects.

**2** Liquids have a constant volume but no fixed shape. The shape of a liquid is determined by the shape of its container. Water, brewed coffee, and soda are liquids.

**3** A gas has neither a constant volume nor a fixed shape. Gases can be compressed or expanded; unlike liquids, they will fill a container of any size. The air of our atmosphere is the gas with which we are most familiar.

**?** **REVIEW**
**A.** Describe the different states of matter in terms of shape and volume.

**4** Water is the only substance that occurs as a solid (ice), a liquid (water), and a gas (water vapor) at temperatures compatible with life. Water exists as a liquid over a broad range of temperatures primarily because of hydrogen bonding among the water molecules.

**5** The small positive charges on the hydrogen atoms of one polar molecule can be attracted to the negative charges on another polar molecule, and this can change the shapes of the molecules or pull adjacent molecules together. This weak attractive force is called a **hydrogen bond**. In the liquid state, hydrogen bonds between adjacent water molecules are continually forming and breaking. When water freezes into ice, the hydrogen bonds lock the molecules in a lattice that occupies more space than did the liquid water. This accounts for the expansion of water during freezing, and because ice is less dense than liquid water, it floats. Water becomes a vapor or gas when all of the hydrogen bonds between adjacent water molecules are broken.

Water

**KEY**

● Hydrogen

● Oxygen

——— Hydrogen bond

- - - - - Hydrogen bond (breaking or reforming)

Ice

**6** At the water surface, the hydrogen bonds between water molecules slow the rate of evaporation and form what is known as **surface tension**. Surface tension acts as a barrier that keeps small objects from entering the water. For example, it allows insects, such as these water striders, to walk across the surface of a pond or puddle. Similarly, the surface tension in a layer of tears on the eye prevents dust particles from touching the surface of the eye.

**?** **REVIEW**

**B.** By what means are water molecules attracted to each other?

**?** **REVIEW**

**C.** Explain why small insects can walk on the surface of a pond and why tears protect the surface of the eye from dust particles.

**7** The polar charges on water molecules give water the ability to disrupt the ionic bonds of a variety of inorganic compounds and cause them to dissolve. Almost all naturally occurring elements are found in seawater, and at least 29 elements are dissolved in our body fluids.

**?** **INTEGRATION**

**D.** Describe the relationship between thermal energy (temperature) and stability of the hydrogen bonds between water molecules in ice, in liquid water, and as a gas.

**Lo** **LEARNING OUTCOME**

Describe the three states of matter and the importance of hydrogen bonds in liquid water.

# Section 1 Review
## Atoms , Molecules, and Compounds

**Matching:** Match each lettered term with the most closely related description.

a. atomic number
b. electrons
c. protons
d. neutrons
e. isotopes
f. ions
g. ionic bond
h. covalent bond
i. mass number
j. element
k. compound
l. hydrogen bond

**1** Atoms that have gained or lost electrons
**2** Subatomic particles in the nucleus; have no electrical charge
**3** Atoms of two or more different elements bonded together in a fixed proportion
**4** The number of protons in an atom
**5** Attractive force between water molecules
**6** Type of chemical bond within a water molecule
**7** The number of subatomic particles in the nucleus
**8** Substance composed only of atoms with same atomic number
**9** Subatomic particles in the nucleus; have an electrical charge
**10** Atoms of the same element with different masses
**11** Type of chemical bond in table salt
**12** Subatomic particles outside the nucleus; have an electrical charge

**1** F
**2** D
**3** K
**4** A
**5** l
**6** H
**7** i
**8** j
**9** C
**10** E
**11** G
**12** B

**Fill-in:** Fill in the missing information in the following table.

| Element | Number of protons | Number of electrons | Number of neutrons | Mass number |
|---|---|---|---|---|
| Helium | **13** 2 | 2 | 2 | **14** |
| Hydrogen | 1 | **15** | **16** | 1 |
| Carbon | 6 | **17** | 6 | **18** |
| Nitrogen | **19** | 7 | **20** | 14 |
| Calcium | **21** | **22** | 20 | 40 |

Indicate which of the following molecules are also compounds.

 $H_2$ (hydrogen)   $H_2O$ (water)    $O_2$ (oxygen)   CO (carbon monoxide)

**23** [ ]   **24** [ X ]   **25** [ ]   **26** [ X ]

**Section integration:** Describe how the following pairs of terms concerning atomic interactions are similar and how they are different.

**27** Inert element/reactive element

**28** Polar molecules/nonpolar molecules

**29** Covalent bond/ionic bond

**27** _____
_____
_____

**28** _____
_____
_____

**29** _____
_____
_____

# Module 2.6
# Chemical reactions and energy transfer are essential to cellular functions

Cells remain alive and functional by controlling chemical reactions. In a chemical reaction, new chemical bonds form between atoms, or existing bonds between atoms are broken. These changes take place as atoms in the reacting substances, called **reactants**, are rearranged to form different substances, or **products**. All of the reactions under way in the cells and tissues of the body at any given moment make up its **metabolism** (me-TAB-ō-lizm).

**1** In effect, each cell is a chemical factory. Growth, maintenance and repair, secretion, and contraction all involve complex chemical reactions. Cells also use chemical reactions to provide the energy they need to maintain homeostasis and to perform essential functions.

Typical cell

**Essential activities**
- Maintenance and repair
- Growth
- Division
- Special functions

**Energy transfer and use**

**Substances absorbed**

**Chemical reactions**

**?** REVIEW
**A.** Describe how cells are chemical factories.

**2** **Work** is the movement of an object or a change in the physical structure of matter. In your body, work includes movements such as walking or running, and also the synthesis of molecules and the conversion of liquid water to water vapor (evaporation).

    **Energy** is the capacity to perform work, and movement or physical change cannot take place without energy. **Kinetic energy** is the energy of motion, energy that can be transferred to another object and do work. **Potential energy** is stored energy, energy that has the potential (capability) to do work. It may derive from an object's position (you standing on a ladder) or from its physical or chemical structure (a stretched spring or a charged battery).

    Cells do work as they synthesize complex molecules and move materials into, out of, and within the cell. The cells of a skeletal muscle at rest, for example, contain potential energy in the form of the positions of protein filaments and the covalent bonds between molecules within the cells. When a muscle contracts, it performs work, and potential energy is converted into kinetic energy. Such a conversion is never 100 percent efficient. Each time an energy exchange or transfer occurs, some of the energy is released as heat. The amount of heat is proportional to the amount of work done. As a result, when you exercise, your body temperature rises.

**?** REVIEW

**B.** Compare and contrast the terms *work, energy, potential energy,* and *kinetic energy.*

**C.** Relate the terms *work, energy, potential energy,* and *kinetic energy* to a muscle contraction at the cellular level.

**Lo** LEARNING OUTCOME

Define metabolism, and distinguish among work, kinetic energy, and potential energy.

**2**

# Module 2.7

# Chemical notation is a concise method of describing chemical reactions

Before we can consider the specific compounds found in the human body, we must be able to describe chemical compounds and reactions effectively. Using sentences to describe chemical structures and events often leads to confusion. Chemists have developed a simple form of "chemical shorthand" known as **chemical notation** that makes communication much more efficient.

 **REVIEW**

**A.** Name the participants in a chemical reaction.

**B.** How are chemical reactions represented?

**1** Chemical notation allows us to describe complex events briefly and precisely. It is relatively easy to use chemical notation to calculate the weights of the reactants involved in a particular reaction.

## VISUAL REPRESENTATION

## CHEMICAL NOTATION

### Atoms

The symbol of an element indicates one atom of that element. A number preceding the symbol of an element indicates more than one atom of that element.

(H) one atom of hydrogen    (O) one atom of oxygen

(H) (H) two atoms of hydrogen    (O) (O) two atoms of oxygen

H one atom of hydrogen    O one atom of oxygen

2 H two atoms of hydrogen    2 O two atoms of oxygen

### Molecules

A subscript following the symbol of an element indicates a molecule with that number of atoms of that element.

(H)(H) hydrogen molecule composed of two hydrogen atoms

(O)(O) oxygen molecule composed of two oxygen atoms

 water molecule composed of two hydrogen atoms and one oxygen atom

$H_2$ hydrogen molecule    $O_2$ oxygen molecule

$H_2O$ water molecule

### Chemical Reactions

In a description of a chemical reaction, the participants at the start of the reaction are called reactants, and the reaction generates one or more products. Chemical reactions are represented by chemical equations. An arrow indicates the direction of the reaction, from reactants (usually on the left) to products (usually on the right). In the following reaction, two atoms of hydrogen combine with one atom of oxygen to produce a single molecule of water.

  →

reactants      product

Chemical reactions neither create nor destroy atoms; they merely rearrange atoms into new combinations. Therefore, the numbers of atoms of each element must always be the same on both sides of the equation for a chemical reaction. When this is the case, the equation is balanced.

$2 H + O \longrightarrow H_2O$
Balanced equation

$2 H + 2 O \longrightarrow H_2O$
Unbalanced equation

### Ions

A superscript plus or minus sign following the symbol of an element indicates an ion. A single plus sign indicates a cation with a charge of +1. (The original atom has lost one electron.) A single minus sign indicates an anion with a charge of –1. (The original atom has gained one electron.) If more than one electron has been lost or gained, the charge on the ion is indicated by a number preceding the plus or minus sign.

 sodium ion the sodium atom has lost one electron

 chloride ion the chlorine atom has gained one electron

 calcium ion the calcium atom has lost two electrons

$Na^+$ $Cl^-$ $Ca^{2+}$ sodium ion   chloride ion   calcium ion

**2** A **mole** (abbreviated mol) is a quantity with a weight in grams equal to an element's atomic weight. One mole of a given element always contains the same number of atoms as 1 mole of any other element. The atomic weight of oxygen is 16, and the atomic weight of hydrogen is 1. So a mole of oxygen will weigh 16 grams and contain the same number of atoms as a mole of hydrogen, which weighs 1 gram.

1 mol of oxygen          1 mol of hydrogen

**3** The **molecular weight** of a molecule or compound is the sum of the atomic weights of its component atoms. For ionic compounds (which do not form molecules), the term **formula weight** is used instead. Molecular and formula weights are important because you can neither handle individual molecules or ions nor easily count the billions of molecules or ions involved in chemical reactions in the body. To simplify our calculations of moles and molecular weights, we are rounding the atomic weights to the nearest whole number.

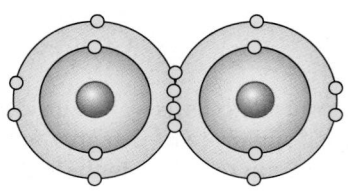

Molecular weight of $O_2 = 32$        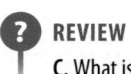 Molecular weight of $H_2 = 2$

**? REVIEW**
**C.** What is formula weight?

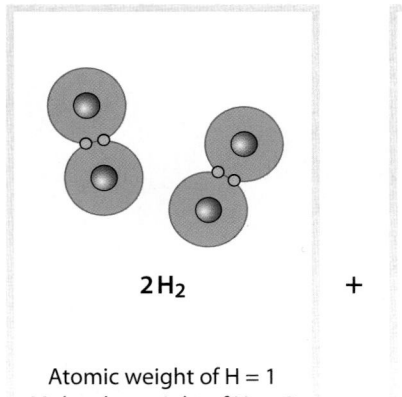

**2 H₂**

**Atomic weight of H = 1**
**Molecular weight of $H_2 = 2$**
**Two moles of $H_2$ weigh 4 g**

**+**

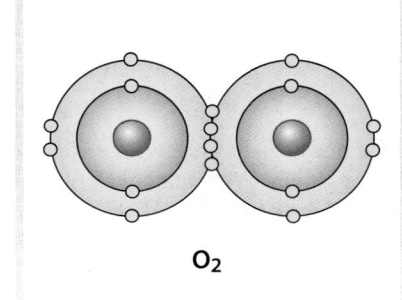

**O₂**

**Atomic weight of O = 16**
**Molecular weight of $O_2 = 32$**
**One mole of $O_2$ weighs 32 g**

**⟶**

**2 H₂O**

**Molecular weight of 2 $H_2O$:**
**2 x (2 + 16)**
**2 x 18 = 36**
**Two moles of $H_2O$ weigh 36 g**

**4** Using molecular weights, you can calculate the quantities of reactants needed for a specific reaction and determine the amount of product generated. For example, suppose you want to form water from hydrogen and oxygen according to this chemical equation: $2 H_2 + O_2 \longrightarrow 2 H_2O$. The first step is to calculate the molecular weights involved. As detailed above, combining 4 g of hydrogen with 32 g of oxygen yields 36 g of water. Although by convention we use grams, you could also work with ounces, pounds, or tons, as long as the proportions remained the same. Notice that when you do the calculation correctly, the molecular weights of reactants and products are balanced.

**? INTEGRATION**

**D.** Using chemical notation, write the molecular formula for glucose, a compound composed of 6 carbon (C) atoms, 12 hydrogen (H) atoms, and 6 oxygen (O) atoms.

**E.** Calculate the weight of 1 mol of glucose. (The atomic weight of carbon = 12.)

**Lo LEARNING OUTCOME**

Use chemical notation to symbolize chemical reactions.

# Module 2.8

# Three basic types of chemical reactions are important for understanding physiology

**1** A **decomposition** reaction breaks a molecule into smaller fragments.

A simple decomposition reaction is diagrammed below:

$$AB \longrightarrow A + B$$

Decomposition reactions occur outside cells as well as inside them. For example, a typical meal contains molecules of fats, sugars, and proteins that are too large and too complex to be absorbed and used by your body. Decomposition reactions in the digestive tract break these molecules down into smaller fragments so the body can absorb them.

Decomposition reactions involving water are important in the breakdown of complex molecules in the body. In **hydrolysis** (hī-DROL-i-sis; *hydro-*, water + *lysis*, a loosening), one of the bonds in a complex molecule is broken, and the components of a water molecule (H and OH) are added to the resulting fragments:

$$AB + H_2O \longrightarrow AH + BOH$$

The decomposition reactions of complex molecules within the body's cells and tissues are referred to collectively as **catabolism** (ka-TAB-ō-lizm; *katabole*, a throwing down). When a covalent bond—a form of potential energy—is broken, it releases kinetic energy that can perform work. By harnessing the energy released in this way, cells carry out vital functions such as growth, movement, and reproduction.

$$CD \longrightarrow C + D + ENERGY$$

**2** **Synthesis** (SIN-the-sis) is the opposite of decomposition. A synthesis reaction assembles smaller molecules into larger molecules.

A simple synthetic reaction is diagrammed here:

$$A + B \longrightarrow AB$$

Synthesis reactions may involve combining atoms or molecules to form even larger products. Water formation from hydrogen and oxygen molecules is a synthesis reaction. Synthesis always forms new chemical bonds, whether the reactants are atoms or molecules.

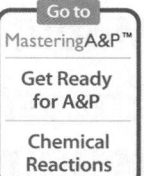

Go to
MasteringA&P™

Get Ready
for A&P

Chemical
Reactions

**Dehydration synthesis**, or condensation reaction, forms a complex molecule by removing a water molecule:

$$AH + BOH \longrightarrow AB + H_2O$$

Dehydration synthesis is the opposite of hydrolysis. We will see examples of both reactions in later sections.

**? REVIEW**

**A.** Compare the role of water in hydrolysis and dehydration synthesis reactions.

2

Synthesis of new molecules within the body's cells and tissues is known collectively as **anabolism** (a-NAB-ō-lizm; *anabole*, a throwing upward). Anabolism is usually considered an "uphill" process because it takes energy to create a chemical bond (just as it takes energy to push something uphill). Cells must balance their energy budgets, with catabolism providing the energy to support anabolism and other vital functions.

Chemical reactions are reversible (at least theoretically), so if

$$A + B \longrightarrow AB, \text{ then } AB \longrightarrow A + B$$

Many important biological reactions are freely reversible. Such reactions can be represented as an equation:

$$A + B \rightleftharpoons AB$$

> This equation indicates that, in a sense, two reactions are taking place at the same time. One is synthesis and the other decomposition. At **equilibrium**, the rates at which the two reactions proceed are in balance: As fast as one molecule of AB forms, another degrades into A + B.

**3** In an **exchange reaction**, parts of the reacting molecules are shuffled around to produce new products:

A simple exchange reaction is diagrammed here:

$$AB + CD \longrightarrow AD + CB$$

The reactants and products contain the same components (A, B, C, and D), but those components are present in different combinations. In an exchange reaction, the reactant molecules AB and CD must break apart (a decomposition) before they can interact with each other to form AD and CB (a synthesis).

**? REVIEW**

**B.** Identify and describe three types of chemical reactions important in human physiology.

**? INTEGRATION**

**C.** What is the source of energy that converts glucose, a six-carbon molecule, into two three-carbon molecules in cells?

**Lo LEARNING OUTCOME**

Distinguish among the major types of chemical reactions that are important for studying physiology.

# Module 2.9
# Enzymes lower the activation energy requirements of chemical reactions

Most chemical reactions do not take place spontaneously, or if they do, they occur so slowly that they would be of little value to cells. Before a reaction can proceed, enough energy must be provided to activate the reactants. The amount of energy required to start a reaction is called the **activation energy**. Many reactions can be activated by changes in temperature or acidity (low pH), but such changes are deadly to cells. Instead, your cells use special proteins called **enzymes** to perform most of the complex synthesis and decomposition reactions in your body.

**1** Enzymes promote chemical reactions by lowering the activation energy required. In doing so, they make it possible for chemical reactions, such as the breakdown of sugars, to proceed under conditions compatible with life. Enzymes belong to a class of substances called **catalysts** (KAT-uh-lists; *katalysis*, dissolution), compounds that speed up chemical reactions without themselves being permanently changed or consumed. Enzymatic reactions, which are generally reversible, proceed until an equilibrium is reached.

**? REVIEW**
**A.** What is an enzyme?
**B.** Why do our cells need enzymes?

In the external environment (outside the body), extreme conditions can provide the activation energy. For example, complex sugars can be broken down in a laboratory by boiling them in an acidic solution.

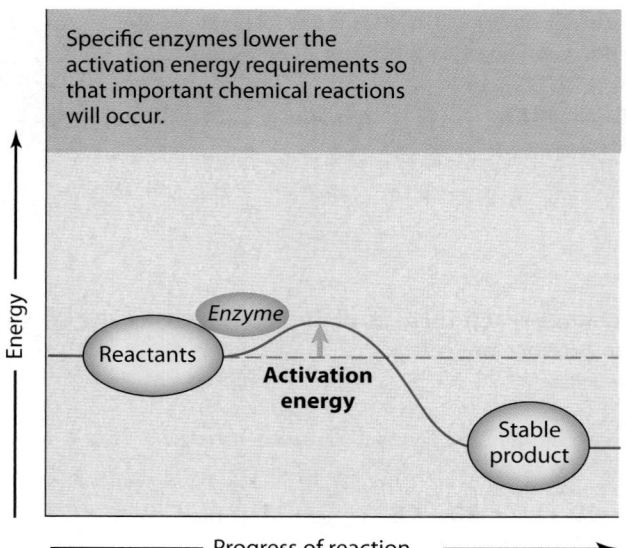

Specific enzymes lower the activation energy requirements so that important chemical reactions will occur.

**2** The complex reactions that support life proceed in a series of interlocking steps, each controlled by a specific enzyme. Such a reaction sequence is called a **metabolic pathway**. A synthetic pathway can be diagrammed as:

A — *enzyme 1* / Step 1 → B — *enzyme 2* / Step 2 → C — *enzyme 3* / Step 3 → and so on.

It takes activation energy to start a chemical reaction, but once it has begun, the reaction as a whole may absorb or release energy as it proceeds to completion. Reactions that release energy are said to be **exergonic** (*exo-*, outside). If more energy is required to begin the reaction than is released as it proceeds, the reaction is called **endergonic** (*endo-*, inside). Exergonic reactions are relatively common in the body. They generate the heat that maintains your body temperature.

**? REVIEW**
**C.** What is an important by-product of exergonic reactions?

**3** Enzymatic reactions are essential to the processing of **metabolites** (me-TAB-ō-līts; *metabole*, change), which are substances that can be synthesized or broken down by chemical reactions inside our bodies. **Nutrients** are essential metabolites that are normally obtained from the diet. Nutrients and metabolites can be broadly categorized as either *organic* or *inorganic*.

| Nutrients and metabolites |
| --- |

*include*

**? REVIEW**

**D.** Explain the differences between metabolites and nutrients.

2

| Organic compounds |
| --- |

**Organic compounds** always contain carbon and hydrogen atoms as their primary structural ingredients. Organic molecules can be much larger and more complex than inorganic compounds.

Examples: carbohydrates, lipids, proteins, and nucleic acids

| Inorganic compounds |
| --- |

**Inorganic compounds** generally do not contain carbon and hydrogen atoms as their primary structural ingredients.

Examples: carbon dioxide, oxygen, water, acids, bases, and salts

**4** A **macromolecule** is a large molecule, such as a carbohydrate, lipid, protein, or nucleic acid, made up of monomer subunits. A **monomer** (MON-ō-mer; *mono-* single + *-mer*, member of a group) is a molecule that can be bonded to other identical molecules to form a **polymer**. Repeating monomers join together through *dehydration synthesis* reactions to form polymers. The monomers of macromoleulces are separated, or released, through *hydrolysis reactions*.

**? INTEGRATION**

**E.** Explain why enzymes are often called organic catalysts.

**Lo LEARNING OUTCOME**

Describe the crucial role of enzymes in metabolism.

## Section 2 Review
## Chemical Reactions

**Matching:** Match each lettered term with the most closely related description.

a. exergonic

b. activation energy

c. organic compounds

d. exchange reaction

e. hydrolysis

f. endergonic

g. reactants

h. enzyme

**1** Catalyst

**2** Starting substances in a chemical reaction

**3** Chemical reaction involving water

**4** Reactions that absorb energy

**5** Shuffles parts of reactants

**6** Primary components are carbon and hydrogen

**7** Reactions that release energy

**8** Requirement for starting a chemical reaction

**1** H

**2** G

**3** E

**4** F

**5** D

**6** C

**7** A

**8** B

**Short answer:** Using chemical notation, write the formula of each of the following:

**9** One molecule of hydrogen

**10** Two atoms of hydrogen

**11** Six molecules of water

**12** One molecule of sucrose (in this order: 12 atoms of carbon, 22 atoms of hydrogen, and 11 atoms of oxygen)

**9** $H_2$

**10** $2H$

**11** $6H_2O$

**12** $C_{12}H_{22}O_{11}$

Write the chemical equation for the following chemical reaction: 1 molecule of glucose combined with 6 molecules of oxygen produce 6 molecules of carbon dioxide and 6 molecules of water.

**13** _____

Indicate which of the following is a hydrolysis reaction and which is a dehydration synthesis reaction.

**14** $A\text{-}B + H_2O \longrightarrow A\text{-}H + B\text{-}OH$

**15** $A\text{-}H + B\text{-}OH \longrightarrow A\text{-}B + H_2O$

**14** Hydrolysis

**15** dehydration synthesis

**Section integration:** In a metabolic pathway that consists of four steps, how would decreasing the amount of enzyme that catalyzes the second step affect the amount of product at the end of the pathway?

**16** _____

## Module 2.10
# Water has several important properties

Water ($H_2O$) is the most important substance in the body. It accounts for up to two-thirds of total body weight. A major change in the body's water content can be fatal because virtually all physiological systems will be affected. Although water is familiar to everyone, it has some highly unusual properties.

### Important Properties of Water

#### Lubrication

Water is an effective lubricant because there is little friction between water molecules. Thus even a thin layer of water between two opposing surfaces will greatly reduce friction between them; water reduces friction within joints and in body cavities.

#### Chemical reactant

In our bodies, chemical reactions occur in water, and water molecules are also participants in some reactions, including hydrolysis and dehydration synthesis.

#### High heat capacity

**Heat capacity** is the quantity of heat required to raise the temperature of a unit mass of a substance 1°C. Water has an unusually high heat capacity because water molecules in the solid and the liquid state are

attracted to one another through hydrogen bonding. Some features of a high heat capacity are listed at right:

- Water temperature must be high for individual molecules to develop enough kinetic energy to break all the hydrogen bonds to become water vapor, a gas.

- Water carries a great deal of heat away with it when it changes from a liquid to a gas. This feature accounts for the cooling effect of perspiration on the skin.

- A large mass of water changes temperature very slowly. This property is called **thermal inertia**.

#### Solvent

A remarkable number of inorganic and organic molecules will dissolve in water. The individual particles become dispersed within the water, and the result is a **solution**—a uniform mixture of two or more substances. The liquid in which other atoms, ions, or molecules are distributed is called the **solvent**. The dissolved substances are the **solutes**. In **aqueous solutions**, water is the solvent.

 **INTEGRATION**

**A.** Predict how water plays a role as a lubricant, reactant, coolant, and solvent during exercise.

**Lₒ LEARNING OUTCOME**

Describe four important properties of water and their significance in the body.

# Module 2.11
# Physiological systems depend on water

Many inorganic compounds are held together partially or completely by ionic bonds. In water, these compounds undergo **dissociation** (di-sō-sē-Ā-shun) or **ionization** (ī-on-ī-ZĀ-shun). In this process, ionic bonds are broken as the individual ions interact with the positive or negative poles of polar water molecules.

**1** A water molecule is said to be polar because it has positive and negative poles. This polarity is due to the asymmetrical positions of the hydrogen atoms that are attached by polar covalent bonds.

**2** In an aqueous solution, an ionic compound dissociates as water molecules break them apart. The anions are surrounded by the positive poles of water molecules, and the cations are surrounded by the negative poles of water molecules. The sheath of water molecules around an ion in solution is called a **hydration sphere**.

**3** Hydration spheres also form around an organic molecule containing polar covalent bonds. If the molecule binds water strongly, as does glucose, it will be carried into solution—in other words, it will dissolve. Molecules that interact readily with water molecules in this way are called **hydrophilic** (hī-drō-FIL-ik; *hydro-*, water + *philos*, loving). Glucose, an important soluble sugar, is one example.

**Sodium chloride crystal**

Negative pole

$2\delta^-$

O

H

$\delta^+$

$\delta^+$

Positive pole

Hydration spheres

Cl⁻

Na⁺

**NaCl in solution**

Glucose molecule

**Glucose molecule in solution**

**?** **REVIEW**
**A.** Explain how the ionic compound sodium chloride dissolves in water.

**4** An aqueous solution containing anions and cations will conduct an electrical current. Soluble inorganic substances whose ions will conduct an electrical current in solution are called **electrolytes** (e-LEK-trō-līts). Sodium chloride is an important electrolyte in body fluids. In an electrical field, cations in solution will move toward the negative side, or negative terminal, and anions will move toward the positive terminal. Electrical forces across plasma membranes affect the functioning of all cells. Small electrical currents carried by ions are essential to muscle contraction and nerve function, two topics that are discussed in later chapters.

+

−

Cl⁻

Na⁺

**?** **REVIEW**
**B.** Define *electrolytes*.

**5** The table at right lists the most important electrolytes and their dissociation products in the body. The dissociation of electrolytes in blood and other body fluids releases a variety of ions. Changes in the concentrations of electrolytes in body fluids will disturb almost every vital function. For example, declining potassium levels will lead to a general muscular paralysis, and rising concentrations will cause weak and irregular heartbeats. The concentrations of ions in body fluids are carefully regulated, primarily by the coordination of activities at the kidneys (ion excretion), the digestive tract (ion absorption), and the skeletal system (ion storage or release).

### Important Electrolytes That Dissociate in Body Fluids

| Electrolyte | | Ions Released |
|---|---|---|
| **NaCl** (sodium chloride) | $\longrightarrow$ | $Na^+ + Cl^-$ |
| **KCl** (potassium chloride) | $\longrightarrow$ | $K^+ + Cl^-$ |
| **CaPO$_4$** (calcium phosphate) | $\longrightarrow$ | $Ca^{2+} + PO_4^{2-}$ |
| **NaHCO$_3$** (sodium bicarbonate) | $\longrightarrow$ | $Na^+ + HCO_3^-$ |
| **MgCl$_2$** (magnesium chloride) | $\longrightarrow$ | $Mg^{2+} + 2\,Cl^-$ |
| **Na$_2$HPO$_4$** (sodium hydrogen phosphate) | $\longrightarrow$ | $2\,Na^+ + HPO_4^{2-}$ |
| **Na$_2$SO$_4$** (sodium sulfate) | $\longrightarrow$ | $2\,Na^+ + SO_4^{2-}$ |

**6** A solution containing dispersed proteins or other large molecules is called a **colloid**. The particles or molecules in a colloid will remain in solution indefinitely. Liquid Jell-O is a familiar, viscous (thick) colloid. In contrast, a **suspension** contains large particles in solution, but if undisturbed, its particles will settle out of solution because of the force of gravity. Whole blood is a temporary suspension because the blood cells are suspended in the blood plasma. If clotting is prevented, the cells in a blood sample will gradually settle to the bottom of the container.

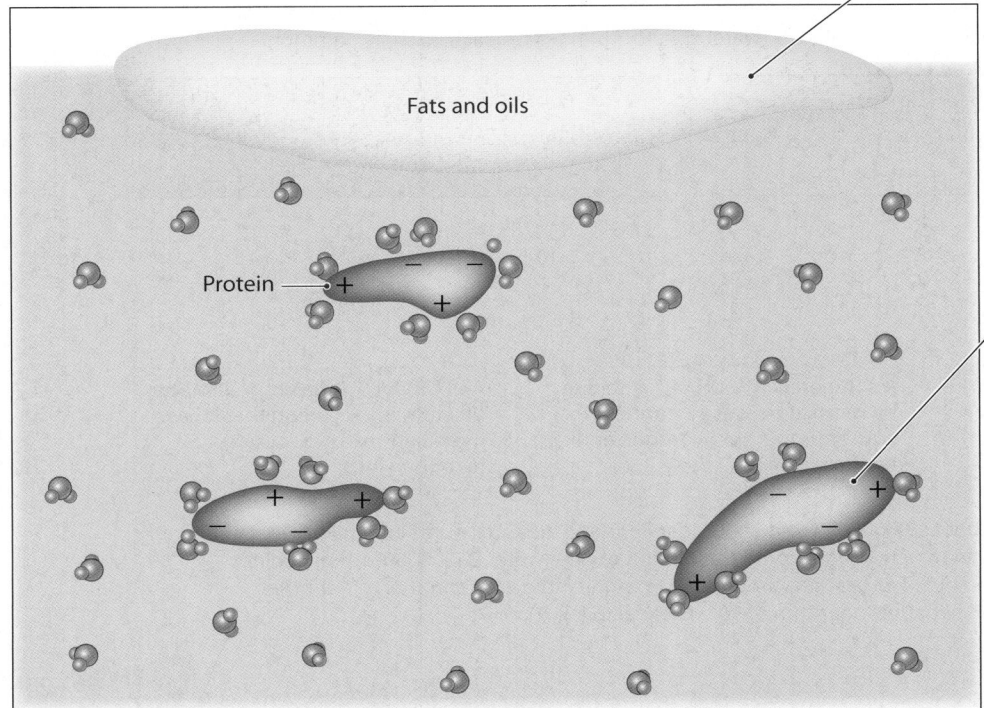

Fats and oils

Protein

Many organic molecules either lack polar covalent bonds or have very few. Such molecules do not have positive and negative poles and are thus nonpolar. When nonpolar molecules are exposed to water, hydration spheres do not form, and the molecules do not dissolve. Molecules that do not readily interact with water are called **hydrophobic** (hī-drō-FŌ-bik; *hydro-*, water + *phobos*, fear). Fats and oils of all kinds are some of the most familiar hydrophobic molecules.

Body fluids typically contain large and complex organic molecules, such as proteins, that are held in solution by their association with water molecules.

**?** **INTEGRATION**

**C.** Distinguish between hydrophilic and hydrophobic molecules.

**Lo** **LEARNING OUTCOME**

Explain how the chemical properties of water affect the solubility of inorganic and organic molecules.

# Module 2.12
# Regulation of body fluid pH is vital for homeostasis

A hydrogen atom involved in a chemical bond or participating in a chemical reaction can easily lose its electron to become a **hydrogen ion**, **H⁺**. Hydrogen ions are extremely reactive in solution. In excessive numbers, they will break chemical bonds, change the shapes of complex molecules, and generally disrupt cell and tissue functions. As a result, the concentration of hydrogen ions in body fluids must be regulated precisely.

**1** A few hydrogen ions are normally present even in a sample of pure water because some of the water molecules dissociate spontaneously, releasing a hydrogen ion, $H^+$, and a **hydroxide** (hī-DROK-sīd) **ion, OH⁻**.

**2** The hydrogen ion concentration in body fluids is so important to physiological processes that a special shorthand is used to express it. The **pH** of a solution is defined as the negative logarithm* of the hydrogen ion concentration in moles per liter (mol/L). For common liquids, the pH scale ranges from 0 to 14.

**Blood**

The pH of blood normally ranges from 7.35 to 7.45. Abnormal fluctuations in pH can damage cells and tissues by breaking chemical bonds, changing the shapes of proteins and altering cellular functions. **Acidosis** is an abnormal physiological state caused by low blood pH (below 7.35). A pH below 7 can produce coma. **Alkalosis** results from an abnormally high pH (above 7.45). A blood pH above 7.8 generally causes uncontrollable and sustained skeletal muscle contractions.

A solution with a pH below 7 is **acidic** (a-SI-dik), meaning that it contains more hydrogen ions than hydroxide ions.

A solution with a pH of 7 is said to be **neutral** because it contains equal numbers of hydrogen and hydroxide ions.

A pH above 7 is **basic**, or **alkaline** (AL-kuh-lin), meaning that it has more hydroxide ions than hydrogen ions.

\* One liter of pure water contains about 0.0000001 mol of hydrogen ions and an equal number of hydroxide ions. In other words, the concentration of hydrogen ions in a solution of pure water is 0.0000001 mol per liter. This can be written as $[H^+] = 10^{-7}$ mol/L. The brackets around the $H^+$ signify "the concentration of," another example of chemical notation. The negative logarithm of $10^{-7}$ is equal to $(-)(-7) = +7$.

**REVIEW**

**A.** Define *pH*.

## Acids and Bases

**3** An **acid** is any solute that dissociates in solution and releases hydrogen ions, thereby lowering the pH. Because a hydrogen atom that loses its electron consists solely of a proton, hydrogen ions are often referred to simply as protons, and acids as proton donors. A strong acid dissociates completely in solution, and the reaction occurs essentially in one direction only. **Hydrochloric acid (HCl)** is a representative strong acid; in water, it ionizes as follows:

$$HCl \longrightarrow H^+ + Cl^-$$

**? REVIEW**

**B.** A hydrogen ion is the same as what subatomic particle?

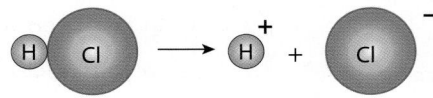

**4** A **base** is a solute that removes hydrogen ions from a solution and thereby acts as a proton acceptor. In solution, many bases release a hydroxide ion. Hydroxide ions have a strong affinity for hydrogen ions and react quickly with them to form water molecules. A strong base dissociates completely in solution. **Sodium hydroxide (NaOH)** is a strong base. In solution, it releases sodium ions and hydroxide ions:

$$NaOH \longrightarrow Na^+ + OH^-$$

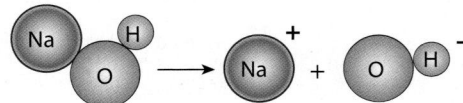

**5** Weak acids and weak bases fail to dissociate completely. At equilibrium, a significant number of molecules remain intact in the solution. For a given number of molecules in solution, weak acids and weak bases therefore have a lesser impact on pH than do strong acids and strong bases. **Carbonic acid (H$_2$CO$_3$)** is a weak acid found in body fluids. In solution, carbonic acid reversibly dissociates into a hydrogen ion and a **bicarbonate ion (HCO$_3^-$)**.

$$H_2CO_3 \rightleftharpoons H^+ + HCO_3^-$$

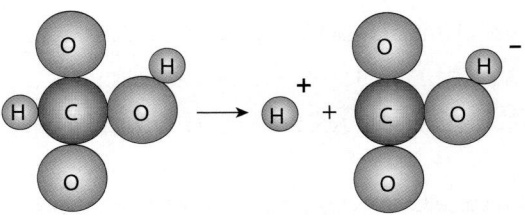

## Salts

**6** A **salt** is an ionic compound consisting of any cation except a hydrogen ion and any anion except a hydroxide ion. Because they are held together by ionic bonds, many salts dissociate completely in water, releasing cations and anions. For example, sodium chloride (table salt) dissociates immediately in water, releasing Na$^+$ and Cl$^-$, the most abundant ions in body fluids. The ionization of sodium chloride does not affect the local concentrations of hydrogen ions or hydroxide ions, so NaCl, like many salts, is a "neutral" solute. Other salts may indirectly affect the concentrations of H$^+$ and OH$^-$, making a solution slightly acidic or slightly basic.

$$NaCl \longrightarrow Na^+ + Cl^-$$

**? REVIEW**

**C.** Explain the differences among an acid, a base, and a salt.

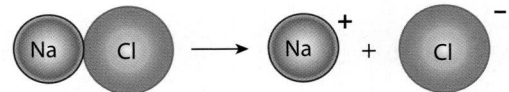

## Buffers

**7** **Buffers** are compounds that stabilize the pH of a solution by removing or replacing hydrogen ions. **Buffer systems** typically involve a weak acid and its related salt, which functions as a weak base. For example, the body's carbonic acid–bicarbonate buffer system consists of carbonic acid (H$_2$CO$_3$) and sodium bicarbonate, (NaHCO$_3$), otherwise known as baking soda. Buffers and buffer systems in body fluids help maintain pH within normal limits.

**? INTEGRATION**

**D.** What is the relationship between buffers and pH in physiological systems?

**Lo LEARNING OUTCOME**

Discuss the importance of pH and the role of buffers in body fluids.

# Water in the Body

**Matching:** Match each lettered term with the most closely related description.

a. solvent

b. water

c. buffers

d. hydrophilic

e. inorganic compounds

f. hydrophobic

g. acid

h. solute

i. alkaline

j. salt

| | | |
|---|---|---|
| **1** | HCl, NaOH, and NaCl | **1** E |
| **2** | A dissolved substance | **2** H |
| **3** | A solution with a pH greater than 7 | **3** I |
| **4** | Molecules that readily interact with water | **4** D |
| **5** | Fluid medium of a solution | **5** A |
| **6** | Ionic compound not containing hydrogen ions or hydroxide ions | **6** J |
| **7** | Compounds that stabilize pH in body fluids | **7** C |
| **8** | Solution with a pH of 6.5 | **8** G |
| **9** | Molecules that do not interact with water | **9** F |
| **10** | Makes up two-thirds of human body weight | **10** B |

**Short answer:** List four properties of water important to the functioning of the human body.

**11** chemical reactions (hydrolysis, dehydration synthesis)

**12** lubricant (low friction due to hydrogen bonds)

**13** universal solvent

**14** high heat capacity (takes alot of energy to change thermal temp)

**Short answer:** Identify the regions 15, 16, and 17 on the pH scale below:

**15** Acidic

**16** neutral

**17** basic

pH   0   1   2   3   4   5   6   7   8   9   10   11   12   13   14

[H$^+$] $10^0$   $10^{-1}$   $10^{-2}$   $10^{-3}$   $10^{-4}$   $10^{-5}$   $10^{-6}$   $10^{-7}$   $10^{-8}$   $10^{-9}$   $10^{-10}$   $10^{-11}$   $10^{-12}$   $10^{-13}$   $10^{-14}$

(mol/L)*

**18** How much more or less acidic is a solution of pH 3 compared to one with a pH of 6?

**18** _____

**19** Describe three negative effects of abnormal pH fluctuations in the human body.

**19** _____

**Section integration:** The addition of table salt to pure water does not result in a change in its pH. Why?

**20** _____

# Module 2.13

# All organic compounds contain carbon and hydrogen atoms

**Organic compounds** always contain the elements carbon and hydrogen, and generally oxygen as well. Many organic molecules are made up of long chains of carbon atoms linked by covalent bonds. The carbon atoms typically form additional covalent bonds with hydrogen or oxygen atoms and, less commonly, with nitrogen, phosphorus, sulfur, iron, or other elements. Many organic molecules are soluble in water. Although organic compounds are diverse, certain groupings of atoms occur again and again, even in very different types of molecules. These **functional groups** greatly influence the chemical (acid or base) and physical (solubility) properties of any molecule in which they occur.

 **REVIEW**

**A.** List the elements that make up organic compounds.

**B.** What is a functional group?

**C.** Identify the important functional groups of organic compounds.

## Important Functional Groups of Organic Compounds

| Functional Group | Structural Formula * | | Importance | Examples |
|---|---|---|---|---|
| **Amino group, $-NH_2$** | R—N with H, H | R—N with H, H | Acts as a base, accepting $H^+$, depending on pH; can form bonds with other molecules | • Amino acids |
| **Carboxyl group, $-COOH$** | R—C with O (double bond), OH | R—C with O, O—H | Acts as an acid, releasing $H^+$ to become $R-COO^-$ | • Fatty acids<br>• Amino acids |
| **Hydroxyl group, $-OH$** | R—O—H | R—O—H | May link molecules through dehydration synthesis; hydrogen bonding between hydroxyl groups and water molecules affects solubility | • Carbohydrates<br>• Fatty acids<br>• Amino acids<br>• Alcohols |
| **Phosphate group, $-PO_4^{2-}$** | R—O—P with $O^-$, $O^-$, O (double bond) | R—O—P with O, O, O, O | May link other molecules to form larger structures; may store energy | • Phospholipids<br>• Nucleic acids<br>• High-energy compounds |

\* A structural formula shows the covalent bonds within a molecule or functional group. For example, a single covalent bond is represented by a single line (—), a double bond by two parallel lines (=). Although not shown here, a triple bond is represented by three parallel lines (≡). **R** represents the organic molecule to which a functional group is attached.

 **INTEGRATION**

**D.** Describe the functional groups that are considered acidic or basic.

 **LEARNING OUTCOME**

Describe the common elements of organic compounds and how functional groups modify the properties of organic compounds

In this section we will introduce the major classes of organic compounds. We will also consider how enzymes facilitate essential reactions within living cells and how cells capture and transfer energy with special **high-energy compounds**.

## Module 2.14

# Carbohydrates contain carbon, hydrogen, and oxygen, usually in a 1:2:1 ratio

A **carbohydrate** is an organic molecule that contains carbon, hydrogen, and oxygen in a ratio near 1:2:1. Familiar carbohydrates include the sugars and starches that make up about half of the typical U.S. diet. Carbohydrates typically account for less than 1.5 percent of total body weight. Although they may have other functions, carbohydrates are most important as energy sources that are catabolized rather than stored.

 **REVIEW**

**A.** What is the most important function of carbohydrates?

**B.** Which of the structural representations of glucose shown below is more common in the body?

## Monosaccharides

**1** A simple sugar, or **monosaccharide** (mon-ō-SAK-uh-rīd; *mono-*, single + *sakcharon*, sugar), is a carbohydrate containing from 3 to 7 carbon atoms. A monosaccharide can be called a triose (three-carbon), tetrose (four-carbon), pentose (five-carbon), hexose (six-carbon), or heptose (seven-carbon). The hexose **glucose** (GLŪ-kōs) is the most important metabolic "fuel" in the body. Monosaccharides are the monomer building blocks of more complex carbohydrates.

The atoms in a glucose molecule may form either a straight chain or a ring. In the body, the ring form is more common.

**2** The three-dimensional structure of an organic molecule is an important characteristic because it usually determines the molecule's fate or function. Some molecules have the same molecular formula—in other words, the same types and numbers of atoms—but different structures. Such molecules are called **isomers**. The body usually treats different isomers as distinct molecules.

Glucose

Fructose

The monosaccharides glucose and **fructose** are isomers. Fructose is a hexose found in many fruits. Although its chemical formula, $C_6H_{12}O_6$, is the same as that of glucose, the arrangement of its atoms differs from that of glucose. (Not all of the carbon atoms are shown. C atoms also lie at the bond intersections.)

| Carbohydrates in the Body | | | |
| --- | --- | --- | --- |
| **Structural Class** | **Examples** | **Primary Function** | **Remarks** |
| **Monosaccharides** (simple sugars) | Glucose, fructose | Energy source | Manufactured in the body and obtained from food; distributed in body fluids |
| **Disaccharides** | Sucrose, lactose, maltose | Energy source | Sucrose is table sugar, lactose is in milk, and maltose is malt sugar; all must be broken down to monosaccharides before absorption |
| **Polysaccharides** | Glycogen | Glucose storage | Glycogen is in animal cells; other starches and cellulose are within or around plant cells |

## Disaccharides

Two monosaccharides joined together form a **disaccharide** (dī-SAK-uh-rīd; *di-*, two). Disaccharides such as **sucrose** (table sugar) have a sweet taste and, like monosaccharides, are quite soluble in water. The formation of sucrose involves dehydration synthesis.

**3** During dehydration synthesis, 2 molecules are joined by the removal of a water molecule.

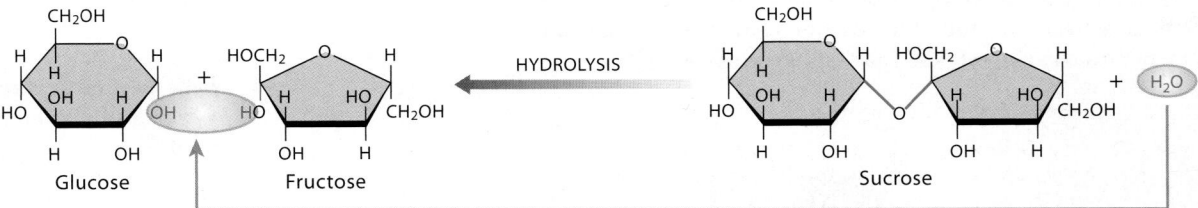

**4** Hydrolysis reverses the steps of dehydration synthesis. A complex molecule is broken down by the addition of a water molecule.

## Polysaccharides

**5** More complex carbohydrates result when repeated dehydration synthesis reactions add additional monosaccharides or disaccharides. These large molecules (polymers) are called **polysaccharides** (pol-ē-SAK-uh-rīdz; *poly-*, many). **Starches** are large polysaccharides formed from glucose molecules. Your digestive tract can break these molecules into monosaccharides. Starches such as those in potatoes and grains are a major dietary energy source.

The polysaccharide **glycogen** (GLĪ-kō-jen), or animal starch, has many side branches consisting of chains of glucose molecules. Muscle cells make and store glycogen. When these cells have a high demand for glucose, glycogen molecules are broken down. When the need is low, they absorb glucose from the bloodstream and rebuild glycogen reserves.

**? REVIEW**

**C.** List the three structural classes of carbohydrates, and give an example of each.

**? INTEGRATION**

**D.** Predict the reactants and the type of chemical reaction involved when muscle cells make and store glycogen.

**Lo LEARNING OUTCOME**

Discuss the structures and functions of carbohydrates.

# Module 2.15

# Lipids often have a carbon-to-hydrogen ratio of 1:2

Like carbohydrates, **lipids** (*lipos*, fat) contain carbon, hydrogen, and oxygen, and the carbon-to-hydrogen ratio is typically near 1:2. However, lipids contain much less oxygen than do carbohydrates with the same number of carbon atoms. The hydrogen-to-oxygen ratio is therefore very large. For example, a representative lipid, such as lauric acid, has a formula of $C_{12}H_{24}O_2$. Lipids may also contain small quantities of phosphorus, nitrogen, or sulfur. Familiar lipids include fats, oils, and waxes. Most lipids are insoluble in water, but special transport mechanisms carry them into the bloodstream.

 **REVIEW**

**A.** Describe lipids in terms of their elemental composition and solubility in water.

**B.** List examples of representative lipids in the body.

## Fatty Acids

**1** **Fatty acids** are long carbon chains with hydrogen atoms attached. One end of the carbon chain, called the head, is attached to a **carboxyl** (kar-BOK-sil) **group**: —COOH.

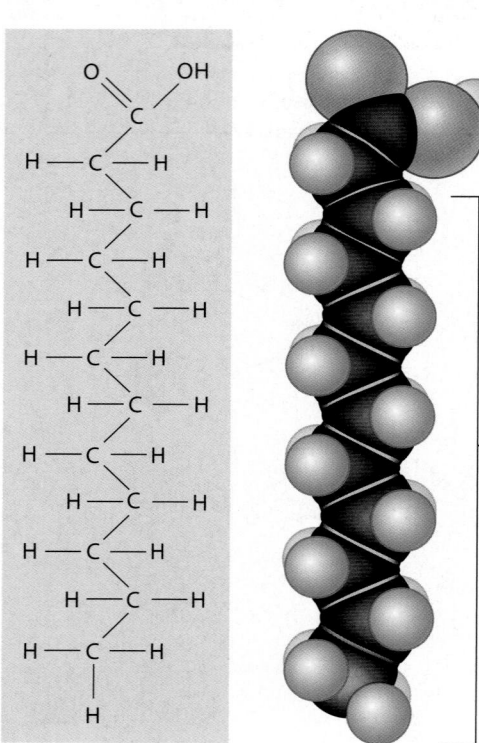

Lauric acid ($C_{12}H_{24}O_2$)

The carbon chain attached to the carboxyl group is known as the hydrocarbon tail of the fatty acid. The hydrocarbon tail is hydrophobic, so fatty acids have very limited solubility in water. In general, the longer the hydrocarbon tail, the lower the solubility of the molecule.

**2** In a **saturated fatty acid**, each carbon atom in the tail has four single covalent bonds.

**3** In an **unsaturated fatty acid**, one or more of the single covalent bonds between carbon atoms has been replaced by a double covalent bond. As a result, each carbon atom involved will bind only one hydrogen atom rather than two. This changes both the shape of the hydrocarbon tail and the way the fatty acid is metabolized. A monounsaturated fatty acid has a single double bond in the hydrocarbon tail. A polyunsaturated fatty acid contains multiple double bonds.

Double covalent bond →

**REVIEW**

**C.** Describe the structures of saturated and unsaturated fatty acids.

## Representative Lipids in the Body

| Lipid Type | Examples | Primary Functions | Remarks |
|---|---|---|---|
| **Fatty acids** | Lauric acid | Energy sources | Absorbed from food or synthesized in cells; transported in the blood |
| **Glycerides** | Monoglycerides, diglycerides, triglycerides | Energy sources, energy storage, insulation, and physical protection | Stored in fat deposits; must be broken down to fatty acids and glycerol before they can be used as an energy source |
| **Eicosanoids** (see Module 2.16) | Prostaglandins, leukotrienes | Chemical messengers coordinating local cellular activities | Prostaglandins are produced in most body tissues |
| **Steroids** (see Module 2.16) | Cholesterol | Structural components of cell membranes, hormones, digestive secretions in bile | All steroids have the same carbon ring framework |
| **Phospholipids, glycolipids** (see Module 2.16) | Lecithin (a phospholipid) | Structural components of plasma membranes | Derived from fatty acids and nonlipid components |

## Glycerides

**4** Individual fatty acids cannot be strung together in a chain by dehydration synthesis. But they can be attached to another compound, **glycerol** (GLIS-er-ol), through a similar reaction. The result is a lipid known as a **glyceride** (GLIS-er-īd).

**5** Lipids form essential components of all cells. In addition, lipid deposits are important as energy reserves. On average, lipids provide twice as much energy as carbohydrates do, gram for gram, when broken down in the body. Lipids normally account for 12–18 percent of total body weight of adult men, and 18–24 percent in adult women. The human body cannot synthesize all the lipids it needs, and several fatty acids must be obtained from the diet.

Dehydration synthesis can produce a **monoglyceride** (mon-ō-GLI-ser-īd), consisting of glycerol + one fatty acid. Subsequent reactions can yield a **diglyceride** (glycerol + two fatty acids) and then a **triglyceride** (glycerol + three fatty acids). Triglycerides are important lipid polymers. They are also known as triacylglycerols or neutral fats. Hydrolysis breaks the glycerides into fatty acids and glycerol.

**? REVIEW**
**D.** In the hydrolysis of a triglyceride, what are the reactants and the products?

**? INTEGRATION**
**E.** Summarize the functions of lipids in the body.

**Lo LEARNING OUTCOME**
Discuss the structures and functions of lipids.

# Module 2.16
## Eicosanoids, steroids, phospholipids, and glycolipids have diverse functions

Lipids are important as chemical messengers and as components of cellular structures. **Structural lipids** help form and maintain a cell's surrounding membrane and its intracellular membranes. At the cellular level, membranes are sheets or layers composed primarily of hydrophobic lipids. Functionally, a lipid membrane is an effective barrier that can separate two aqueous solutions of differing composition.

### Eicosanoids

1 **Eicosanoids** (ī-KŌ-sa-noydz) are lipids derived from arachidonic (ah-rak-i-DON-ik) acid, a fatty acid that must be absorbed from food because it cannot be synthesized by the body. **Leukotrienes** (lū-kō-TRĪ-ēnz) are produced primarily by cells involved with coordinating the responses to injury or disease, and they are considered in later chapters.

**? REVIEW**
**A.** Describe the structure and role of prostaglandins.

**Prostaglandins** (pros-tuh-GLAN-dinz) are short-chain fatty acids in which five of the carbon atoms are joined in a ring. These compounds are released by cells to coordinate or direct local cellular activities, and they are extremely powerful, even in minute quantities. Small amounts of prostaglandins released by damaged tissues, for example, stimulate nerve endings and produce the sensation of pain.

### Steroids

2 **Steroids** are large lipid molecules that share a distinctive carbon-ring framework. They differ in the functional groups that are attached to the basic ring structure.

**Cholesterol**

The outer boundary of all animal cells, the plasma membrane, contains **cholesterol** (kōh-LES-ter-ol; *chole-*, bile + *stereos*, solid). Cells need cholesterol to maintain their plasma membranes and for cell growth and division.

**? REVIEW**
**B.** Why is cholesterol necessary in the body?

**Estrogen**          **Testosterone**

Steroid hormones are involved in regulating sexual function. The sex hormones estrogen and testosterone are examples of steroid hormones.

## Phospholipids and Glycolipids

**3** **Phospholipids** (FOS-fō-lip-idz) and **glycolipids** (GLĪ-kō-lip-idz) are structurally related, and our cells can synthesize both types of lipids, primarily from fatty acids. Cholesterol, phospholipids, and glycolipids are considered structural lipids because of their roles in membrane structure.

In a phospholipid, a phosphate group links a diglyceride to a nonlipid group.

- Nonlipid group
- Phosphate group
- Glycerol
- Fatty acids

In a glycolipid, a carbohydrate is attached to a diglyceride.

- Carbohydrate
- Glycerol
- Fatty acids

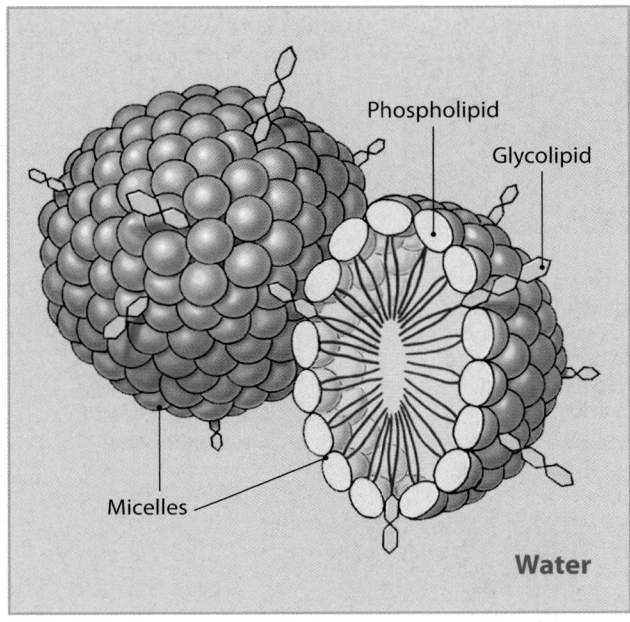

- Phospholipid
- Glycolipid
- Micelles
- Water

**4** The long hydrocarbon tails of phospholipids and glycolipids are hydrophobic, but the opposite ends, the nonlipid heads, are hydrophilic. In water, large numbers of these molecules tend to form droplets, or **micelles** (mī-SELZ), with the hydrophilic portions on the outside. Most meals contain a mixture of lipids and other organic molecules, and micelles form as the food breaks down in the digestive tract.

- Hydrophilic heads
- Hydrophobic tails

**Phospholipid**      **Glycolipid**

**? REVIEW**

**C.** Describe the orientations of phospholipids and glycolipids when they form a micelle.

**? INTEGRATION**

**D.** What do cholesterol, phospholipids, and glycolipids have in common?

**Lo LEARNING OUTCOME**

Discuss the structures and diverse functions of eicosanoids, steroids, phospholipids, and glycolipids.

# Module 2.17
# Proteins are formed from amino acids

**Proteins** are the most abundant organic components of the human body and, in many ways, the most important. The human body contains many different proteins, and they account for about 20 percent of total body weight. Proteins consist of long chains of organic molecules called **amino acids**. Twenty different amino acids are used as the building blocks of proteins in the human body. All amino acids contain carbon, hydrogen, oxygen, and nitrogen. Sulfur is also present in two amino acids, cysteine and methionine. A typical protein contains 1000 amino acids. The largest protein complexes have 100,000 or more.

**?** REVIEW

A. Describe proteins.

**1** Amino acids are simple organic compounds (monomers) that combine to form proteins (polymers). Every amino acid consists of a central carbon atom to which four different groups are attached: a hydrogen atom, an amino group, a carboxyl group, and a variable side group designated as R. The carboxyl group can act as an acid by releasing a hydrogen ion to become a carboxyl ion ($COO^-$). The amino group can act as a base by accepting a hydrogen ion, to become an amino ion ($-NH_3^+$). The result is a molecule, with both positive and negative charges, but a net charge of zero. Such molecules are called *zwitterions*, derived from the German word that means "hybrid." The different atoms of the R groups distinguish one amino acid from another, giving each its own chemical properties. For example, different R groups are polar, nonpolar, or electrically charged.

Amino group
Central carbon
Carboxyl group
R group (variable side chain of one or more atoms)

**2** Two amino acids can be linked together by dehydration synthesis.

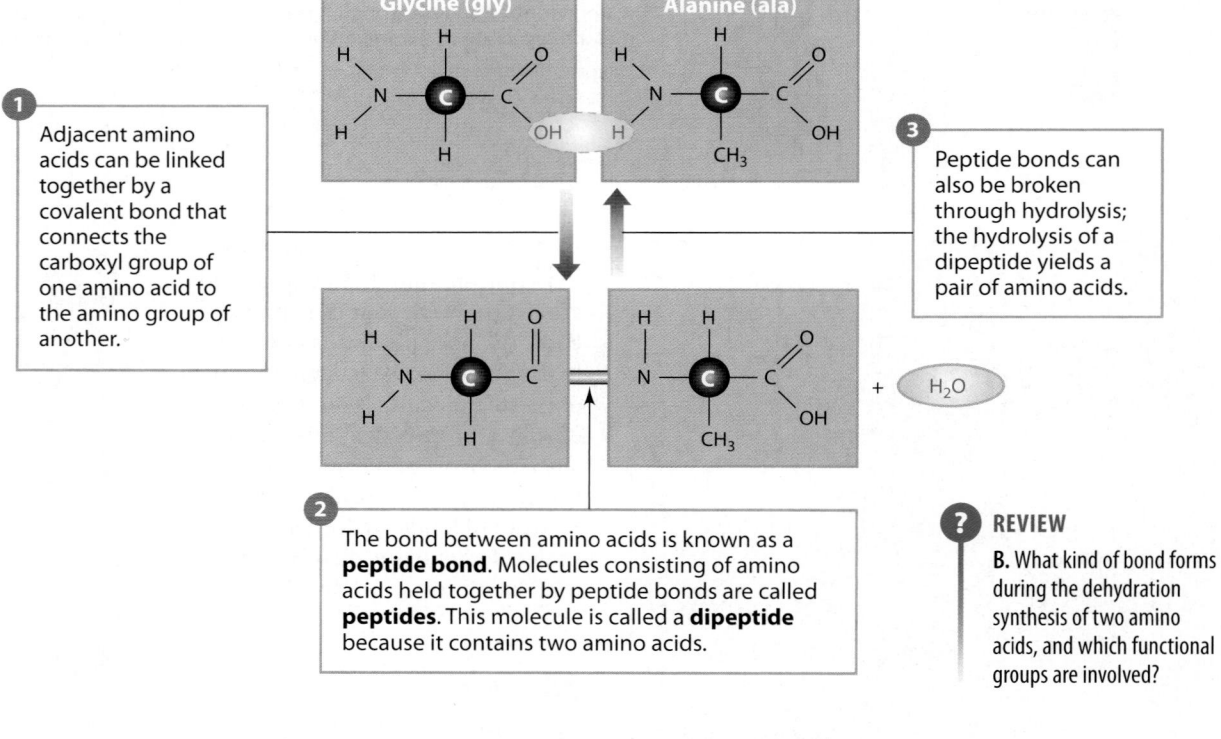

**1** Adjacent amino acids can be linked together by a covalent bond that connects the carboxyl group of one amino acid to the amino group of another.

**3** Peptide bonds can also be broken through hydrolysis; the hydrolysis of a dipeptide yields a pair of amino acids.

**2** The bond between amino acids is known as a **peptide bond**. Molecules consisting of amino acids held together by peptide bonds are called **peptides**. This molecule is called a **dipeptide** because it contains two amino acids.

**?** REVIEW

B. What kind of bond forms during the dehydration synthesis of two amino acids, and which functional groups are involved?

**3** The chain can be lengthened by the addition of more amino acids with peptide bonds. Attaching a third amino acid produces a tripeptide.

**4** Tripeptides and larger peptide chains are called **polypeptides**. Polypeptides containing more than 100 amino acids are called proteins, which can have up to four levels of structural complexity. The first level of complexity is called the primary structure.

### Primary Structure

**Primary structure** results from the sequence of amino acids bonded together in a linear chain.

### Secondary Structure

Hydrogen bond

**Alpha helix**

Hydrogen bond

or

**Beta sheet**

**Secondary structure** results from bonds between atoms at different parts of the polypeptide chain. Hydrogen bonding, for example, may create either a simple spiral, called an alpha helix (α helix) or a flat pleated sheet known as a beta sheet (β sheet). The alpha helix is the most common form, but a given polypeptide chain may have both helical and pleated sections.

### Tertiary Structure

Alpha helix

Heme unit

**Tertiary structure** results from the complex coiling and folding that gives a protein its final three-dimensional shape. Tertiary structure results primarily from interactions between the polypeptide chain and the surrounding water molecules and, to a lesser extent, from interactions between the R groups of amino acids in different parts of the molecule. This subunit of hemoglobin contains a heme unit that can bind to oxygen.

### Quaternary Structure

Heme unit

or

**Hemoglobin** (globular protein)

**Collagen** (fibrous protein)

**Quaternary structure** results from the interaction between individual polypeptide chains to form a protein complex. The protein **hemoglobin** contains four polypeptide subunits. Hemoglobin is found within red blood cells, where it binds and transports oxygen. It is an example of a globular protein. In **collagen**, three linear subunits intertwine, forming a fibrous protein. The three-dimensional shape of a protein plays an essential role in determining its functional properties.

---

⚕️ **Clinical Note**

## High temperature, proteins, and cell death

As temperatures rise, protein shape changes and enzyme function deteriorates. Eventually the protein undergoes **denaturation**, a change in tertiary or quaternary structure that makes it nonfunctional. Death occurs at very high body temperatures (above 43°C, or 110°F) because the denaturation of structural proteins and enzymes causes irreparable damage to organs and organ systems.

**?** **INTEGRATION**

**C.** Why does boiling a protein affect its functional properties?

**Lo** **LEARNING OUTCOME**

Discuss protein structure and the essential functions of proteins within the body.

# Module 2.18
# Enzymes are proteins with important regulatory functions

Almost everything that happens inside the human body does so because a specific enzyme makes it possible. Enzymes are organic catalysts that speed up cellular reactions (see Module 2.9). The reactants in enzymatic reactions are called **substrates**. As in other types of chemical reactions, the interactions among substrates yield specific products. Before an enzyme can function as a catalyst, substrates must bind to a specific region of the enzyme.

**? REVIEW**

**A.** What are the reactants in an enzymatic reaction called?

**1** Substrate binding occurs at the **active site**, typically a groove or pocket into which one or more substrates nestle, like a key fitting into a lock. Weak electrical attractive forces, such as hydrogen bonding, reinforce the physical fit. The tertiary or quaternary structure of the enzyme molecule determines the shape of the active site. Each enzyme catalyzes only one type of reaction, a characteristic called **specificity**. An enzyme's specificity is determined by the ability of its active sites to bind only to substrates with particular shapes and charges.

**? REVIEW**

**B.** Relate an enzyme's structure to its reaction specificity.

**2** Substrate binding produces an **enzyme-substrate complex**. Each cell contains an assortment of enzymes, and any particular enzyme may be active under one set of conditions and inactive under another. Virtually anything that changes the tertiary or quaternary shape of an enzyme can turn it "on" or "off" by changing the properties of the active site and preventing formation of an enzyme-substrate complex. Because the change is immediate, enzyme activation or inactivation is an important method of short-term control over reaction rates and pathways.

**3** Substrate binding typically results in a temporary, reversible change in the enzyme's shape. This change may further the reaction by placing physical stresses on the substrate molecules. The enzyme then promotes product formation. In some cases, the change in enzyme shape that accompanies substrate binding is sufficient to catalyze the reaction. In other cases, an external source must provide the activation energy required.

**4** The completed product then detaches from the active site, and the enzyme is free to repeat the process. When every enzyme molecule is cycling through its reaction sequence at top speed, further increases in substrate concentration will not affect the rate of reaction. The substrate concentration required for the maximum reaction rate is called the **saturation limit**.

**Lo LEARNING OUTCOME**

Explain how enzymes function within the body.

# Module 2.19

# High-energy compounds may store and transfer a portion of energy released during enzymatic reactions

**1** Enzymes may catalyze synthesis, decomposition, or exchange reactions. When product formation requires an energy donor, that donor is typically a **high-energy compound**. High-energy compounds contain **high-energy bonds**, covalent bonds whose breakdown releases energy under controlled conditions. The most common high-energy compound is **adenosine triphosphate**, or **ATP**.

ATP and related high-energy compounds provide the energy to power many vital functions, including muscle contraction and the enzymatic reactions responsible for synthesizing proteins, carbohydrates, and lipids.

**? REVIEW**

**A.** Describe ATP.

**B.** Compare AMP with ADP.

**C.** Where do cells obtain the energy needed for their vital functions?

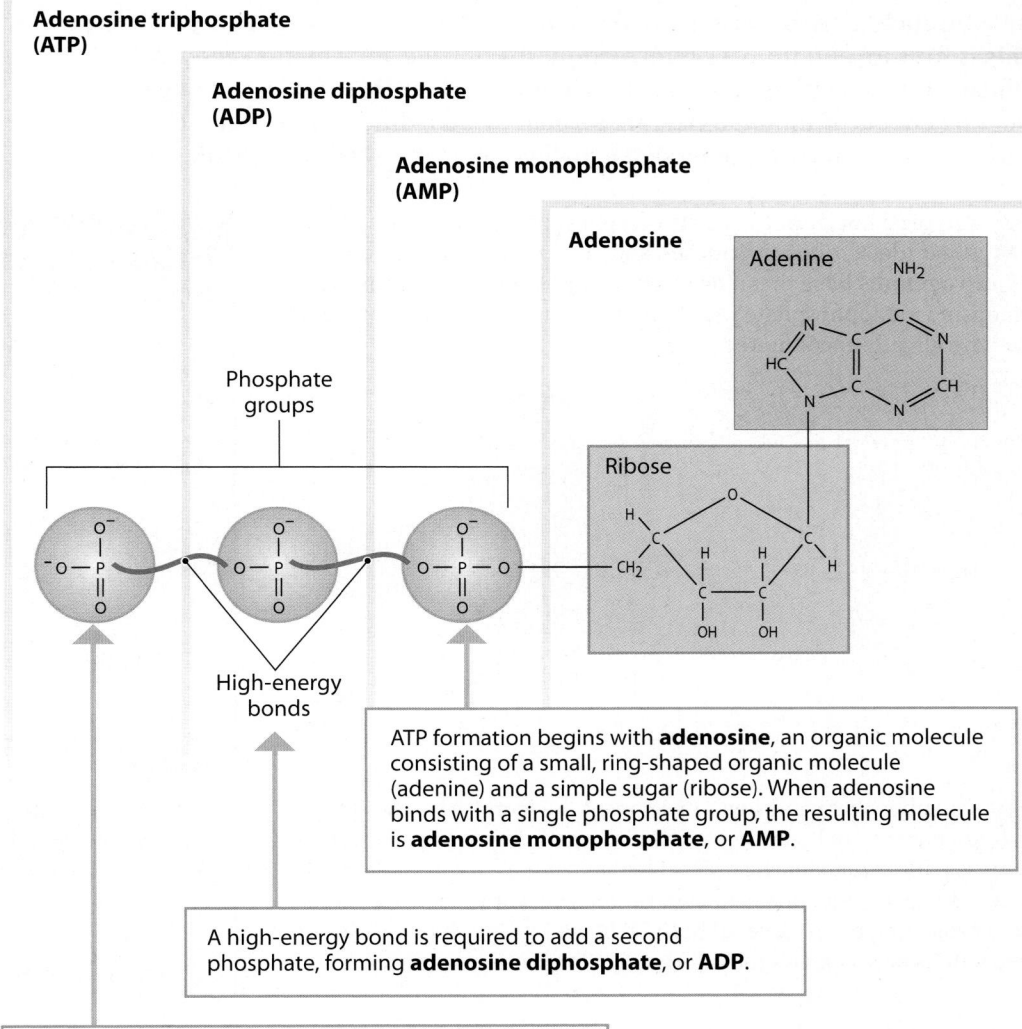

**Adenosine triphosphate (ATP)**

**Adenosine diphosphate (ADP)**

**Adenosine monophosphate (AMP)**

**Adenosine**

Adenine

Ribose

Phosphate groups

High-energy bonds

ATP formation begins with **adenosine**, an organic molecule consisting of a small, ring-shaped organic molecule (adenine) and a simple sugar (ribose). When adenosine binds with a single phosphate group, the resulting molecule is **adenosine monophosphate**, or **AMP**.

A high-energy bond is required to add a second phosphate, forming **adenosine diphosphate**, or **ADP**.

Additional energy must be provided to attach the third phosphate and create **adenosine triphosphate**, or **ATP**.

**2** The formation of ATP from ADP is a reversible reaction. The energy stored when ATP forms is released when it breaks down to ADP. Cells can synthesize ATP in one location and then break it down in another, harnessing the energy released to power essential activities. The hydrolytic breakdown of ATP requires an enzyme called **adenosine triphosphatase** (**ATPase**).

ADP + ENERGY + P $\rightleftharpoons$ ATP + $H_2O$
_ATPase_

**? INTEGRATION**

**D.** What are the products of ATP hydrolysis?

**LO LEARNING OUTCOME**

Discuss the structure and function of high-energy compounds.

# Module 2.20
# DNA and RNA are nucleic acids

**Nucleic** (noo-KLĀ-ik) **acids** are large organic molecules composed of carbon, hydrogen, oxygen, nitrogen, and phosphorus. The two classes of nucleic acid molecules are **deoxyribonucleic** (dē-oks-ē-rī-bō-noo-KLĀ-ik) **acid**, or **DNA**, and **ribonucleic** (rī-bō-noo-KLĀ-ik) **acid**, or **RNA**. The primary role of nucleic acids is to store and transfer information, specifically, information essential to cellular protein synthesis. A nucleic acid is a polymer that consists of one or two long chains of subunits that are formed by dehydration synthesis. The individual subunits (monomers) of a nucleic acid are called **nucleotides**.

**? REVIEW**

**A.** Describe and identify the two classes of nucleic acids.

**1** A typical nucleotide consists of a phosphate group, a 5-carbon sugar (deoxyribose or ribose), and an organic molecule known as a **nitrogenous base** that may be either a **purine** or a **pyrimidine**. Adenosine monophosphate (AMP) is an example of a nucleotide that you have already encountered.

**2** The phosphate and sugar of adjacent nucleotides can be strung together by dehydration synthesis, creating the long chains that comprise functional nucleic acids. The "backbone" of this molecule is a linear sugar-to-phosphate-to-sugar sequence, with the nitrogenous bases projecting to one side. In both DNA and RNA, this sequence of nitrogenous bases carries the information for protein synthesis.

## Nitrogenous bases

### Purines

The purines **adenine** and **guanine** are found in both DNA and RNA.

### Pyrimidines

There are three important pyrimidines. DNA and RNA both contain **cytosine**. **Thymine** is found only in DNA, and **uracil** is found only in RNA.

## A Comparison of DNA with RNA

| Characteristic | DNA | RNA |
| --- | --- | --- |
| **Sugar** | Deoxyribose | Ribose |
| **Nitrogenous bases** | Adenine (A), guanine (G), cytosine (C), thymine (T) | Adenine, guanine, cytosine, uracil (U) |
| **Number of nucleotides in typical molecule** | Always more than 45 million | Varies from fewer than 100 to about 50,000 |
| **Molecular shape** | Paired strands coiled in a double helix | Varies with hydrogen bonding along the length of the strand of each of the three main types (mRNA, tRNA, rRNA) |
| **Function** | Stores genetic information that controls protein synthesis | Performs protein synthesis as directed by DNA |

**3** A DNA molecule consists of a pair of nucleotide chains. Hydrogen bonding between opposing nitrogenous bases (a purine on one strand and a pyrimidine on the other) holds the two strands together. The shapes of the nitrogenous bases allow adenine to bond only to thymine, and cytosine to bond only to guanine. As a result, the combinations adenine–thymine (A–T) and cytosine–guanine (C–G) are known as **complementary base pairs**, and the two nucleotide chains of the DNA molecule are known as **complementary strands**.

**4** A molecule of RNA consists of a single chain of nucleotides. Its shape, and thus its function, depends on the order of the nucleotides and the interactions among them. Our cells have three types of RNA: (1) **messenger RNA** (**mRNA**), (2) **transfer RNA** (**tRNA**), and (3) **ribosomal RNA** (**rRNA**). Their specific functions are discussed in Chapter 3.

The two strands of DNA twist around one another in a double helix that resembles a spiral staircase.

Phosphate group
Deoxyribose
Hydrogen bond
Adenine
Thymine
DNA strand 1
DNA strand 2
Hydrogen bond
Cytosine
Guanine

**? REVIEW**
**B.** Explain how the complementary strands of DNA are held together.

Single-strand sugar-phosphate backbone
Nitrogenous bases

**? INTEGRATION**
**C.** Compare and contrast the nucleotides of DNA and RNA.

**Lo LEARNING OUTCOME**
Compare and contrast the structures and functions of DNA and RNA.

# Organic Compounds

**Concept map:** Use each of the following terms once to fill in the blank boxes to correctly complete the map.

- lipids
- carbohydrates
- nucleic acids
- disaccharides
- RNA
- fatty acids
- phosphate groups
- glycerol
- polysaccharides
- proteins
- monosaccharides
- ATP
- amino acids
- DNA
- nucleotides

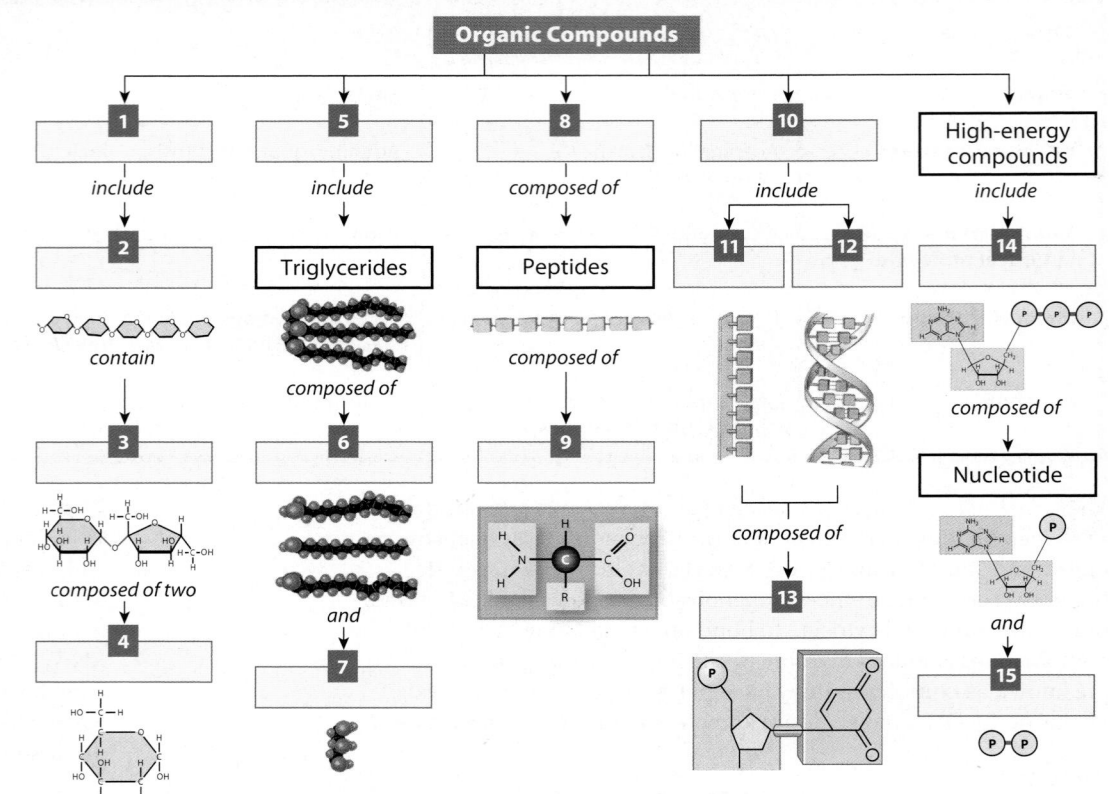

**Vocabulary:** In the space provided, write the boldfaced terms introduced in this section that contain the indicated word part.

| 16 | poly- *(many)* | 16 | _____ |
| 17 | tri- *(three)* | 17 | _____ |

| 18 | di- *(two)* | 18 | _____ |
| 19 | glyco- *(sugar)* | 19 | _____ |

**Matching:** Match each lettered term with the most closely related description.

a. monosaccharide

b. ATP

c. polyunsaturated

d. glycerol

e. cholesterol

f. isomers

g. glycogen

h. active site

i. nucleotide

j. RNA

k. peptide

| 20 | Polysaccharide with an energy-storage role in animal tissues | 20 | G |
| 21 | Molecules with same chemical formula but different structure | 21 | F |
| 22 | A fatty acid with more than one C-to-C double covalent bond | 22 | C |
| 23 | The region of an enzyme that binds the substrate | 23 | H |
| 24 | Three-carbon molecule that combines with fatty acids | 24 | ____ |
| 25 | A steroid essential to plasma membranes | 25 | ____ |
| 26 | A high-energy compound consisting of adenosine and three phosphate groups | 26 | ____ |
| 27 | A nucleic acid that contains the sugar ribose | 27 | ____ |
| 28 | The covalent bond between the carboxylic acid and amino groups of adjacent amino acids | 28 | ____ |
| 29 | Organic molecule consisting of a sugar, a phosphate group, and a nitrogenous base | 29 | ____ |
| 30 | A polysaccharide monomer | 30 | ____ |

# Study Outline

## SECTION 1 • Atoms, Molecules, and Compounds

### Module 2.1

### Atoms are the basic particles of matter    p. 51

1. **Matter** is anything that has mass and occupies space. **Mass** is the amount of material in matter.

2. **Atoms** are the smallest stable units of matter. They are composed of **subatomic particles**, called **protons**, **neutrons**, and **electrons**. Protons and neutrons reside in the **nucleus**. Electrons whirl around the nucleus, forming an **electron cloud**. A molecule forms when atoms interact to form larger, more complex structures.

### Module 2.2

### Typical atoms contain protons, neutrons, and electrons    p. 52

3. The number of protons in an atom is known as the **atomic number**. The total combined number of protons and neutrons in an atom is its **mass number**. **Isotopes** are atoms of the same **element** whose nuclei contain different numbers of neutrons.

4. Electrons occupy an orderly series of energy levels, also called **electron shells**.

### Module 2.3

### Electrons occupy various energy levels    p. 54

5. The electrons in the outermost energy level, or **valence shell**, determine the chemical property of an element.

6. **Reactive** elements have atoms with an unfilled outermost energy level. **Inert** elements have filled outermost energy levels and are also called **noble gases**.

7. An **ion** is an atom with an electrical charge. A **cation** has a positive charge. An **anion** has a negative charge.

### Module 2.4

### The most common chemical bonds are ionic bonds and covalent bonds    p. 56

8. A **compound** is a chemical substance made up of atoms of two or more different elements in a fixed proportion.

9. **Ionic bonds** are formed by the electrical attraction between cations and anions.

10. Atoms that share electrons to complete their outer electron shells create **covalent bonds**. A sharing of one pair of electrons is a **single covalent bond**.

11. A **molecule** is a chemical structure consisting of atoms of one or more elements held together by covalent bonds.

12. A bond with unequal sharing of electrons is a **polar covalent bond**.

### Module 2.5

### Matter may exist as a solid, a liquid, or a gas    p. 58

13. Solids maintain their volume, liquids have a constant volume but no fixed shape, and gases have neither constant volume nor a fixed shape.

14. Water is the only substance that occurs as a solid, liquid, or gas at temperatures consistent with life.

15. A **hydrogen bond** is a weak attractive force between polar molecules and gives water many of its properties, such as **surface tension**.

## SECTION 2 • Chemical Reactions

### Module 2.6

### Chemical reactions and energy transfer are essential to cellular functions    p. 61

16. A chemical reaction occurs when **reactants** are rearranged to form one or more **products**. **Metabolism** is all the chemical reactions in the body.

17. **Work** is the movement of an object or a change in the physical structure of matter. **Energy** is the capacity to perform work. **Kinetic energy** is energy in motion. **Potential energy** is stored energy.

### Module 2.7

### Chemical notation is a concise method of describing chemical reactions    p. 62

18. The rules of **chemical notation** are used to describe chemical reactions.

19. A **mole** (abbreviated as mol) is a quantity with a weight in grams equal to an element's atomic weight.

*One mole of a given element always contains the same number of atoms as one mole of any other element.*

20. The **molecular weight** of a molecule or compound is the sum of the atomic weights of its component atoms.

### Module 2.8

### Three basic types of chemical reactions are important for understanding physiology    p. 64

21. A chemical reaction is classified as a **decomposition**, a **synthesis**, or an **exchange reaction**.

22. In **hydrolysis**, one of the bonds in a complex molecule is broken, and the components of a water molecule are added to the resulting fragments. Decomposition reactions within the body's cells and tissues are called **catabolism**.

23. **Dehydration synthesis** is the formation of a complex molecule by the removal of a water molecule. The synthesis of new molecules within the body's cells and tissues is called **anabolism**.

24. At **equilibrium**, the rates at which the two reactions proceed are in balance.

### Module 2.9

### Enzymes lower the activation energy requirements of chemical reactions    p. 66

25. **Activation energy** is the amount of energy required to start a chemical reaction. **Enzymes** are **catalysts**, compounds that speed up chemical reactions without themselves being permanently changed or consumed. Enzymes promote chemical reactions, or **metabolic pathways**, by lowering the activation energy requirements.

26. **Exergonic** reactions release energy; **endergonic** reactions require more energy than is released.

27. **Metabolites** are molecules that are synthesized or broken down inside our bodies. **Nutrients** are the essential metabolites obtained from the diet. Nutrients and metabolites are broadly categorized as either **organic** or **inorganic compounds**. **Macromolecules** (large organic compounds) are **polymers** made up of identical **monomer** subunits.

## SECTION 3 • Water in the Body

### Module 2.10
**Water has several important properties**   p. 69

28. Water is the most important substance in the body. The important properties of water are **lubrication, chemical reactant, high heat capacity**, and being a **solvent**.

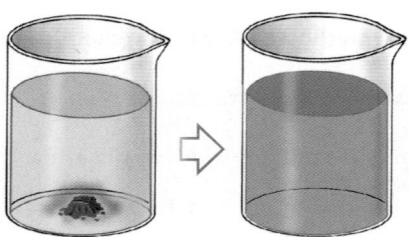

*A remarkable number of inorganic and organic molecules will dissolve in water. The individual particles become dispersed within the water, and the result is a solution—a uniform mixture of two or more substances.*

### Module 2.11
**Physiological systems depend on water**   p. 70

29. Many inorganic compounds, called **electrolytes**, undergo **dissociation**, or **ionization**, in water to form ions.

30. Molecules that interact readily with water molecules are called **hydrophilic**; those that do not are called **hydrophobic**.

31. A solution containing dispersed proteins or other large molecules is called a **colloid**, and these particles will remain in solution indefinitely. A **suspension** contains large particles in solution but if undisturbed will settle out due to gravity.

### Module 2.12
**Regulation of body fluid pH is vital for homeostasis**   p. 72

32. The **pH** of a solution indicates the concentration of **hydrogen ions ($H^+$)** it contains. Solutions are classified as **neutral, acidic**, or **basic (alkaline)** on the basis of pH. An **acid** releases hydrogen ions in solution; a **base** removes hydrogen ions from a solution.

33. A **salt** is an electrolyte whose cation is not a hydrogen ion ($H^+$) and whose anion is not a **hydroxide ion ($OH^-$)**. **Buffers** remove or replace hydrogen ions in solution. **Buffer systems** maintain pH within normal limits.

## SECTION 4 • Organic Compounds

### Module 2.13
**All organic compounds contain carbon and hydrogen atoms**   p. 75

34. Although **organic compounds** are diverse, certain groupings of atoms called **functional groups** repeatedly occur. The important functional groups are **amino groups, carboxyl groups, hydroxyl groups**, and **phosphate groups**.

### Module 2.14
**Carbohydrates contain carbon, hydrogen, and oxygen, usually in a 1:2:1 ratio**   p. 76

35. The three major types of carbohydrates are **monosaccharides** (simple sugars), **disaccharides**, and **polysaccharides**. Disaccharides and polysaccharides form from monosaccharides by dehydration synthesis.

### Module 2.15
**Lipids often have a carbon-to-hydrogen ratio of 1:2**   p. 78

36. **Fatty acids** are long carbon chains with hydrogen atoms attached. Fatty acids can be **saturated** or **unsaturated fatty acids**.

37. Long chains of fatty acids can be strung together by **glycerol**, resulting in a lipid known as a **glyceride**.

### Module 2.16
**Eicosanoids, steroids, phospholipids, and glycolipids have diverse functions**   p. 80

38. **Structural lipids** form and maintain intracellular and plasma membranes. Four classes of lipids are **eicosanoids, steroids, phospholipids**, and **glycolipids**.

### Module 2.17
**Proteins are formed from amino acids**   p. 82

39. **Proteins** are formed by 20 different **amino acids** arranged in long chains and complex shapes. The four levels of protein structure are **primary structure** (amino acid sequence), **secondary structure** (amino acid interactions by hydrogen bonds), **tertiary structure** (complex coiling and folding), and **quaternary structure** (formation of protein complexes from individual subunits).

*Proteins can have up to four levels of structural complexity. Hydrogen bonding in this secondary structure forms an alpha helix.*

### Module 2.18
**Enzymes are proteins with important regulatory functions** p. 84

40. The reactants in an enzymatic reaction, called **substrates**, interact to yield a product by binding to the enzyme's **active site**. Each enzyme catalyzes only one type of reaction, a characteristic called **specificity**.

### Module 2.19
**High-energy compounds may store and transfer a portion of energy released during enzymatic reactions**   p. 85

41. Cells store and transfer energy in the **high-energy bonds** of **high-energy compounds**. The most common high-energy compound is **adenosine triphosphate (ATP)**.

42. ATP formation begins by adding a phosphate group to **adenosine**, forming **adenosine monophosphate (AMP)**. Additional energy is required to attach a second phosphate group to form **adenosine diphosphate (ADP)**. Even more energy is required to attach a third phosphate group to form ATP.

### Module 2.20
**DNA and RNA are nucleic acids**   p. 86

43. **Nucleic acids** are chains of **nucleotides**. Each nucleotide contains a sugar, a phosphate group, and a **nitrogenous base**. The sugar is ribose in **RNA** and deoxyribose in **DNA**.

44. DNA is a two-stranded double helix containing the nitrogenous bases **adenine**, **guanine**, **cytosine**, and **thymine**. RNA consists of a single strand and contains **uracil** instead of thymine.

45. A DNA molecule consists of a pair of nucleotide chains. Hydrogen bonds hold together opposing nitrogenous bases. The **complementary strands** combine by means of **complementary base pairing**: adenine–thymine (A–T) and cytosine–guanine (C–G).

46. Human cells contain three types of RNA: **messenger RNA (mRNA)**, **transfer RNA (tRNA)**, and **ribosomal RNA (rRNA)**.

*A DNA molecule consists of a pair of nucleotide chains. Hydrogen bonding between opposing nitrogenous bases holds the two strands together.*

# Chapter Review Questions

**Multiple choice:** Select the correct answer from the list provided.

**1** If a polypeptide contains 10 peptide bonds, how many amino acids does it contain?

- ☐ a) 9
- ☒ b) 10
- ☒ c) 11
- ☐ d) 12

**2** A dehydration synthesis reaction between glycerol and a single fatty acid would yield a(n)

- ☐ a) micelle.
- ☐ b) omega-3 fatty acid.
- ☒ c) monoglyceride.
- ☒ d) diglyceride.

**3** An atom of calcium has 20 protons and 20 neutrons. What is its atomic number?

- ☐ a) 10
- ☒ b) 20
- ☐ c) 40
- ☐ d) 60

**4** In an exergonic reaction,

- ☐ a) large molecules are broken down into smaller ones.
- ☐ b) small molecules are assembled into larger ones.
- ☐ c) molecules are rearranged to form new molecules.
- ☐ d) molecules move from reactants to products and back.
- ☒ e) energy is released during the reaction.

**5** The hydrogen bonding that occurs in water is responsible for all of the following *except*

- ☐ a) the high boiling point of water.
- ☐ b) the low freezing point of water.
- ☒ c) the ability of water to dissolve nonpolar substances.
- ☐ d) the ability of water to dissolve inorganic salts.
- ☐ e) the high surface tension of water.

**6** The subatomic particle with the least mass

- ☒ a) carries a negative charge.
- ☐ b) carries a positive charge.
- ☐ c) plays no part in the atom's chemical reactions.
- ☐ d) is found only in the nucleus.

**7** A(n) _____ forms when atoms interact to produce larger, more complex structures.

- ☐ a) isotope
- ☐ b) enzyme
- ☒ c) molecule
- ☐ d) nucleus

**8** Isotopes of an element differ from each other in the number of

- ☐ a) protons in the nucleus.
- ☒ b) neutrons in the nucleus.
- ☐ c) electrons in the outer shells.
- ☐ d) a, b, and c are all correct.

**9** The number and arrangement of electrons in an atom's outer energy level determine the atom's

- ☐ a) atomic weight.
- ☐ b) atomic number.
- ☐ c) molecular weight.
- ☒ d) chemical properties.

**10** A _____ is a quantity with a weight in grams equal to an element's atomic weight.

- ☒ a) mole
- ☐ b) molecule
- ☐ c) compound
- ☐ d) synthesis

**2**

**11** Energy of motion is called
- [ ] a) end product.
- [×] b) kinetic.
- [ ] c) transfer work.
- [ ] d) potential.

**12** All organic compounds in the human body contain all of the following elements *except*
- [ ] a) hydrogen.
- [ ] b) oxygen.
- [ ] c) carbon.
- [×] d) calcium.

**13** All the chemical reactions that occur in the human body are collectively referred to as
- [ ] a) anabolism.
- [ ] b) catabolism.
- [×] c) metabolism.
- [ ] d) homeostasis.

**14** A pH of 7.8 in the human body typifies a condition referred to as
- [ ] a) acidosis.
- [×] b) alkalosis.
- [ ] c) dehydration.
- [ ] d) homeostasis.

**15** A(n) _____ is a solute that dissociates to release hydrogen ions, and a(n) _____ is a solute that removes hydrogen ions from solution.
- [ ] a) base, acid
- [ ] b) salt, base
- [ ] c) acid, salt
- [×] d) acid, base

**16** Special organic catalysts that control chemical reactions in the human body are called
- [×] a) enzymes.
- [ ] b) cytozymes.
- [ ] c) cofactors.
- [ ] d) activators.
- [ ] e) cytochromes.

**17** Complementary base pairing in DNA includes the pairs
- [ ] a) adenine–uracil and cytosine–guanine.
- [×] b) adenine–thymine and cytosine–guanine.
- [ ] c) adenine–guanine and cytosine–thymine.
- [ ] d) guanine–uracil and cytosine–thymine.

**18** When the energy stored in ATP is released, it is broken down into
- [ ] a) adenosine + energy.
- [ ] b) AMP + P + energy.
- [ ] c) P + P + P + energy.
- [×] d) ADP + P + energy.

## Short answer

**19** What are the three stable subatomic particles in atoms?

**20** What four major classes of organic compounds are found in the body?

**21** List three functions performed by lipids in the body.

**22** Identify the structural characteristics of a protein.

**23** a) What three components make up a nucleotide of DNA?
b) What three components make up a nucleotide of RNA?

**24** Explain how enzymes function in chemical reactions.

**25** What is a salt? How does a salt differ from an acid or a base?

**26** Explain the differences among nonpolar covalent bonds, polar covalent bonds, and ionic bonds.

**27** An organic molecule has the following constituents: carbon, hydrogen, oxygen, nitrogen, and phosphorus. Is the molecule more likely to be a carbohydrate, a lipid, a protein, or a nucleic acid?

**28** Explain how an insect can walk across the top of a pond without falling through the surface.

**29** A student eats a dinner of tomato salad with vinegar dressing and a glass of wine. Shortly thereafter he complains of an upset stomach. Considering what you know about pH and foods, can you predict why his stomach is upset? Could you give any suggestions as to the kinds of foods he could eat to alleviate his symptoms?

## Sketching exercise

**30** An oxygen atom has eight protons. a) Sketch in the arrangement of electrons around the nucleus of the oxygen atom. b) How many more electrons will it take to fill the outermost energy level (valence shell)?

Oxygen atom

# Chapter Integration • Applying what you have learned

## The chemistry of an after-dinner drink

Claire has just graduated from college, and her mother, Leslie, takes her to Italy as a graduation present. In Tuscany they enjoy a meal of vegetables, pasta, and lean meats at Massimo's family-owned osteria (a simple restaurant). Massimo provides his customers, locals and tourists alike, a culinary experience rich in cultural tradition. He takes pride in serving dinner slowly and offering dishes that are prepared simply and feature fresh ingredients.

When the women have finished their meals, Massimo offers them a *digestivo,* explaining that this after-dinner drink is part of the dining experience and necessary for proper digestion. Leslie orders an anise-flavored liqueur called sambuca, while Claire requests an amaro, a bitter, herbal drink. The two slowly sip their spirits and enjoy the flavorful end to their meal.

After they finish their drinks, they continue to sit and chat about their wonderful dining experience. Leslie comments that despite having eaten quite a bit of food, she doesn't feel overly full. Claire, while removing her jacket, agrees that her dinner, too, is indeed digesting comfortably.

Based on what you have learned about basic chemistry, answer the following questions about this scenario.

**1** If Massimo's *digestivo* was responsible for assisting the women's digestion, which type of chemical reaction is occurring?

**2** Explain why Claire might be removing her jacket.

**3** If the *digestivo* is assisting in the chemistry of digestion, which type of molecule would you predict to be present in the drink?

**4** The reactions facilitated by the molecules in the *digestivo* are breaking down the complex molecules of the meals into simpler molecules. Using your knowledge of organic compounds, identify the simple molecules to which these nutrients will be reduced.

## LEARNING OUTCOMES

These Learning Outcomes correspond by number to this chapter's modules and indicate what you should be able to do after completing the chapter. They also appear at the end of each module.

### Introduction to Cells

**3.1** Describe the cell theory and the process of cellular differentiation. p. 95

**3.2** Describe a body cell and its organelles, including the structure and function of each. p. 96

**3.3** Describe the structural and functional features of the plasma membrane. p. 98

**3.4** Differentiate among the structures and functions of the cytoskeleton. p. 100

**3.5** Describe the ribosome and smooth and rough endoplasmic reticula, and indicate their specific functions. p. 102

**3.6** Describe the Golgi apparatus, and indicate its specific functions. p. 104

**3.7** Describe the structure of a mitochondrion, and explain the significance of mitochondria to cellular function. p. 106

### Structure and Function of the Nucleus

**3.8** Describe the role of the nucleus in maintaining homeostasis at the cellular level. p. 109

**3.9** Describe the functions of the cell nucleus, and distinguish between chromatin and a chromosome. p. 110

**3.10** Discuss the nature of the genetic code, and summarize the process of protein synthesis. p. 112

**3.11** Summarize the process of transcription. p. 114

**3.12** Summarize the process of translation. p. 116

### How Substances Enter and Leave the Cell

**3.13** Contrast permeable, selectively permeable, and impermeable membranes. p. 119

**3.14** Explain the process of diffusion, and identify its significance in the body. p. 120

**3.15** Explain the process of osmosis, and identify its significance in the body. p. 122

**3.16** Describe carrier-mediated transport and its role in the absorption and removal of specific substances. p. 124

**3.17** Describe vesicular transport as a mechanism for facilitating the absorption or removal of specific substances from cells. p. 126

### Cell Life Cycle

**3.18** Distinguish between interphase and cell division in the cell cycle. p. 129

**3.19** Describe interphase, and explain its significance. p. 130

**3.20** Describe the process of mitosis and its role in the cell life cycle. p. 132

**3.21** **Clinical Module:** Discuss the relationship between cell division and cancer. p. 134

# 3

# Cellular Level of Organization

# Module 3.1
# Cellular differentiation produces specialized cells

**1** Objects smaller than 0.1 mm in diameter cannot be seen by the human eye. Because cells are smaller than this, no one could examine the structure of a cell until relatively effective microscopes were invented in the 17th century. Research over time has produced the **cell theory**, which is summarized at right:

**?** **REVIEW**
**A.** Describe the cell theory.

**Cells are the building blocks of all plants and animals.**

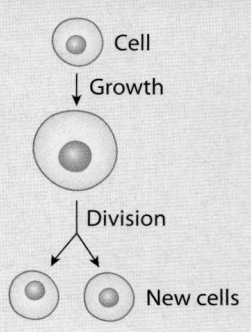

**All new cells come from the division of preexisting cells.**

Cell
↓ Growth

Division

New cells

**Cells are the smallest living units that carry out all vital physiological functions.**

Nutrients    $O_2$

Wastes    $CO_2$

3

**2** Each cell maintains homeostasis at the cellular level, but at higher levels of organization, achieving homeostasis requires the combined and coordinated actions of many cells. Although cells of the human body vary widely in size, shape, and function, all are the descendants of a single cell: the fertilized ovum.

At fertilization, the fertilized ovum, or zygote (ZĪ-gōt), contains the genetic potential to become any cell in the body.

**?** **REVIEW**
**B.** Identify the cell from which all the cells of your body are descendants.

As the first cell divisions occur, the new cells do not grow, but subdivide the ovum cytoplasm into smaller parcels. These parcels differ from one another because there were regional differences in the composition of the ovum cytoplasm at fertilization.

The cytoplasmic differences affect the DNA of the new cells, turning specific genes on or off. The descendant, or daughter, cells begin to develop specialized structural and functional characteristics. This process of gradual specialization is called **cellular differentiation**.

Cellular differentiation produces the specialized cells that form the tissues of the body.

Epithelial tissue      Connective tissue      Muscle tissue      Neural tissue

**?** **REVIEW**
**C.** Define *cellular differentiation*.

**Lo** **LEARNING OUTCOME**
Describe the cell theory and the process of cellular differentiation.

# Module 3.2
# Cells are the smallest living units of life

**1** Our body cells are surrounded by a watery medium known as the **extracellular fluid**. The extracellular fluid in most tissues is called **interstitial** (in-ter-STISH-ul) **fluid** (*interstitium*, space between cells in a tissue). This module introduces the major components of our cells, although not all of these components are found in every cell of the body.

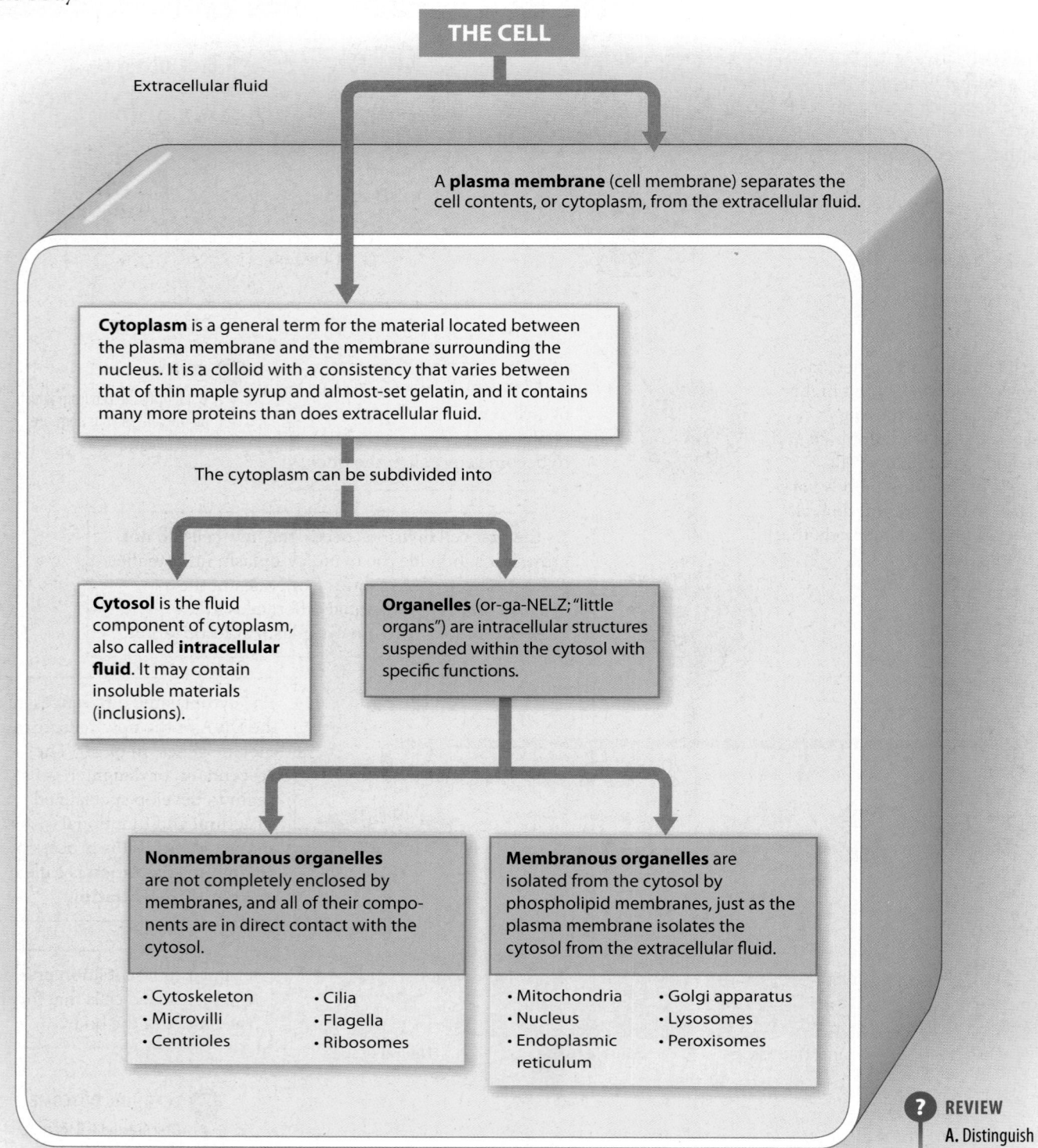

**THE CELL**

Extracellular fluid

A **plasma membrane** (cell membrane) separates the cell contents, or cytoplasm, from the extracellular fluid.

**Cytoplasm** is a general term for the material located between the plasma membrane and the membrane surrounding the nucleus. It is a colloid with a consistency that varies between that of thin maple syrup and almost-set gelatin, and it contains many more proteins than does extracellular fluid.

The cytoplasm can be subdivided into

**Cytosol** is the fluid component of cytoplasm, also called **intracellular fluid**. It may contain insoluble materials (inclusions).

**Organelles** (or-ga-NELZ; "little organs") are intracellular structures suspended within the cytosol with specific functions.

**Nonmembranous organelles** are not completely enclosed by membranes, and all of their components are in direct contact with the cytosol.

- Cytoskeleton
- Microvilli
- Centrioles
- Cilia
- Flagella
- Ribosomes

**Membranous organelles** are isolated from the cytosol by phospholipid membranes, just as the plasma membrane isolates the cytosol from the extracellular fluid.

- Mitochondria
- Nucleus
- Endoplasmic reticulum
- Golgi apparatus
- Lysosomes
- Peroxisomes

**? REVIEW**

**A.** Distinguish between the cytoplasm and cytosol.

**Peroxisome**
STRUCTURE: Vesicles (membranous sacs) containing degradative enzymes. FUNCTION: Breakdown of organic compounds; neutralization of toxic compounds generated in the process.

**Lysosome**
STRUCTURE: Vesicles containing digestive enzymes. FUNCTION: Breakdown of organic compounds and damaged organelles or pathogens.

**Microvilli**
STRUCTURE: Membrane extensions containing microfilaments. FUNCTION: Increase surface area to facilitate absorption of extracellular materials.

**Golgi apparatus**
STRUCTURE: Stacks of flattened membranes (cisternae) containing chambers. FUNCTION: Stores, alters, and packages synthesized products.

Centrosome

**Nucleus**
STRUCTURE: A fluid nucleoplasm containing enzymes, proteins, DNA, and nucleotides; surrounded by a double membrane, the **nuclear envelope.** FUNCTION: Controls metabolism; stores and processes genetic information; controls protein synthesis.

**Endoplasmic reticulum (ER)**
STRUCTURE: Network of membranous sheets and channels extending throughout the cytoplasm. FUNCTION: Synthesis of secretory products; intracellular storage and transport; detoxification of drugs or toxins.

- **Smooth ER**, which has no attached ribosomes, synthesizes lipids and carbohydrates.

- **Rough ER**, which has ribosomes bound to the membranes, modifies and packages newly synthesized proteins.

**Ribosomes**
STRUCTURE: RNA and proteins; fixed ribosomes bound to rough ER; free ribosomes scattered in cytoplasm. FUNCTION: Protein synthesis.

**Mitochondrion**
STRUCTURE: Double membrane, with inner membrane folds enclosing important metabolic enzymes. FUNCTION: Produces 95% of the ATP required by the cell.

Plasma membrane

**Cytoskeleton**
STRUCTURE: Proteins organized in fine filaments (microfilaments) or slender tubes (microtubules); organizing center at the **centrosome**, a cytoplasmic region that contains a pair of centrioles. FUNCTION: Strengthens and supports cell; aids in movement of cellular structures and materials.

 **REVIEW**

**B.** Identify the membranous organelles, and describe their functions.

**C.** Describe the functions of the cytoskeleton.

 **INTEGRATION**

**D.** Describe the external environment of most of the body's cells.

 **LEARNING OUTCOME**

Describe a body cell and its organelles, including the structure and function of each.

# Module 3.3

## The plasma membrane isolates the cell from its environment and performs varied functions

**1** The **plasma membrane** is a physical barrier that separates the inside of the cell from the surrounding extracellular fluid. It is a selectively permeable barrier that controls the entry of ions and nutrients, such as glucose; the elimination of wastes; and the release of secretions.

Superficial membrane carbohydrates form a layer known as the **glycocalyx** (glī -kō-KĀ-liks; *calyx*, cup). Carbohydrates account for about 3 percent of the weight of a plasma membrane. They are components of complex molecules such as proteoglycans (carbohydrate with some protein attached), glycoproteins (protein with some carbohydrate attached), and glycolipids (lipids with carbohydrates attached). The glycocalyx is important in cell recognition, binding to extracellular structures, and lubrication of the cell surface.

EXTRACELLULAR FLUID

Glycocalyx (extracellular carbohydrates)

Glycolipid

Integral protein with channel

CYTOPLASM

= 2 nm

Integral glycoproteins

Cytoskeleton (microfilaments)

**Integral proteins** are part of the membrane structure and cannot be removed without damaging or destroying the membrane. Most integral proteins span the width of the membrane one or more times and are therefore known as **transmembrane proteins**. Some contain pores or channels through which water and solutes may pass.

**Peripheral proteins** are bound to the inner or outer surface of the membrane and (like sticky notes) are easily separated from it. Integral proteins greatly outnumber peripheral proteins, which may have regulatory or enzymatic functions.

**2** The plasma membrane is extremely thin (6–10 nm) and very delicate. It is called a **phospholipid bilayer** because the different phospholipid molecules in it form two layers. In each half of the bilayer, the phospholipids lie with their hydrophilic heads at the membrane surface and their hydrophobic tails on the inside, just as in a micelle (see Module 2.16). The hydrophobic layer in the center of the membrane isolates the cytoplasm from the extracellular fluid. Such isolation is important because the composition of cytoplasm is very different from that of extracellular fluid.

Hydrophilic heads

Hydrophobic tails

**Cholesterol** is an important component of plasma membranes, with almost one cholesterol molecule for each phosholipid molecule. Like phospholipid molecules, cholesterol is an **amphipathic** molecule, which means that it has both hydrophobic and hydrophilic portions. Its polar hydroxyl group aligns with the hydrophilic heads of the phospholipid molecule, and its rigid steroid ring and nonpolar tail lie within the hydrophobic layer. Cholesterol "stiffens" the plasma membrane, making it less fluid and less permeable.

**? REVIEW**

**A.** Which structural component of the plasma membrane is mostly responsible for isolating a cell from its external environment?

**3** The general functions of the plasma membrane include physical isolation, regulation of exchange with the environment, sensitivity to the environment, and structural support. The lipid bilayer provides isolation, and the membrane proteins perform most of the other functions. The five major functional classes of plasma membrane proteins are described in the table below.

---

**Functional Classes of Membrane Proteins**

**Anchoring proteins** attach the plasma membrane to other structures and stabiliize its position. Inside the cell, membrane proteins are bound to the cytoskeleton, a network of supporting filaments in the cytoplasm.

**Recognition proteins** are detected by cells of the immune system.

**Enzymes** in plasma membranes may be integral or peripheral proteins.

**Receptor proteins** bind to specific extracellular molecules called **ligands** (LĪ-gandz). Such receptor-ligand binding can affect a cell's activities. A ligand can be anything from a small ion such as calcium, to a relatively large and complex hormone such as insulin.

**Carrier proteins** bind solutes and transport them across the plasma membrane.

**Channels** are integral proteins containing a central pore (channel) that forms a passageway completely through the plasma membrane. The channel permits the passage of water and small solutes that cannot otherwise cross the lipid bilayer of the plasma membrane.

---

**? REVIEW**

**B.** List the general functions of the plasma membrane.

**C.** Which type of integral protein allows water and small ions to pass through the plasma membrane?

**? INTEGRATION**

**D.** What characteristic of phospholipids accounts for their packing into a double layer?

**Lo LEARNING OUTCOME**

Describe the structural and functional features of the plasma membrane.

# Module 3.4
# The cytoskeleton plays both a structural and a functional role

The **cytoskeleton** functions as the cell's skeleton. It provides an internal protein framework that gives the cytoplasm strength and flexibility.

**1** The cytoskeleton of all cells includes microfilaments, intermediate filaments, and microtubules.

**Microvilli** (singular, *microvillus*) are finger-shaped extensions of the plasma membrane of some cells. A core of microfilaments stiffens each microvillus and anchors it to the cytoskeleton at the terminal web. Microvilli greatly increase the surface area of the cell and enhance its ability to absorb materials from the extracellular fluid.

The smallest of the cytoskeletal elements are the **microfilaments**. These protein strands are generally less than 6 nm in diameter. Typical microfilaments are composed of the protein **actin**. They are common in the periphery of the cell but relatively rare in the region immediately surrounding the nucleus.

Plasma membrane

The **terminal web** is a layer of microfilaments just inside the plasma membrane at the exposed surface of a cell that forms a layer or lining, such as that of the intestinal tract.

**Intermediate filaments**, which range from 7 to 11 nm in diameter, are the strongest and most durable cytoskeletal elements. Their protein composition varies among cell types.

**Microtubules** are the largest components of the cytoskeleton, with diameters of about 25 nm. They are hollow tubes built from the globular protein **tubulin**. Microtubules extend outward into the periphery of the cell from a region near the nucleus called the centrosome (see Module 3.2).

Secretory vesicle

Mitochondrion    Endoplasmic reticulum

**Microvilli**    SEM × 30,000

 **REVIEW**
**A.** List the three basic components of the cytoskeleton.

**2** This table summarizes the general components of the cytoskeleton.

| The Cytoskeleton | | | |
|---|---|---|---|
| **Structure** | **Remarks** | **Location** | **Functions** |
| **Microfilaments** | Present in most cells; best organized in skeletal and cardiac muscle cells | In bundles beneath the plasma membrane and throughout the cytoplasm | Provide strength, alter cell shape, bind the cytoskeleton to the plasma membrane, tie cells together, involved in muscle contraction |
| **Intermediate filaments** | Present in most cells; at least five types known | In cytoplasm | Provide strength, move materials through cytoplasm |
| **Thick filaments** | Found in skeletal and cardiac muscle cells | In cytoplasm | Interact with actin microfilaments to produce muscle contraction |
| **Microtubules** | Present in most cells | In cytoplasm radiating away from centrosome | Provide strength, move organelles |
| **Centrioles** | Nine groups of microtubule triplets form a short cylinder | In centrosome near nucleus | Organize microtubules in the spindle to move chromosomes during cell division |
| **Cilia** | Nine groups of microtubule doublets form a cylinder. Motile cilia have a central pair (9 + 2) | Extensions of plasma membrane | Multiple motile cilia propel fluids or solids across a cell surface. The single primary cilium of a cell detects environmental stimuli |
| **Flagella** | Nine groups of microtubule doublets form a cylinder around a central pair (9 + 2) | Extension of plasma membrane | A flagellum propels sperm |

**3** **Centrioles** are cylindrical structures composed of short microtubules. The microtubules form nine groups, three in each group. Each of these nine "triplets" is connected to its nearest neighbors on either side. Two centrioles are located in a region known as the **centrosome**. During cell division, the centrioles are associated with the movement of DNA strands. Cells that lack centrioles, such as red blood cells and skeletal muscle cells, cannot divide.

**4** **Cilia** (singular, *cilium*) are long, slender extensions of the plasma membrane. Multiple **motile cilia** are found on cells lining portions of the respiratory tract and the reproductive tract. Motile cilia resemble centrioles but have nine pairs of microtubules (rather than triplets) surrounding a central pair. The microtubules are anchored to a compact **basal body** just beneath the cell surface. A **primary cilium** lacks a central pair of microtubules and functions as a sensor of a cell's surroundings. **Flagella** (singular, *flagellum*) have the same microtubule structure as motile cilia but are much longer and beat in a wavelike fashion. The only human cell with a flagellum is a sperm.

**5** Motile cilia beat rhythmically to move fluids or secretions across the cell surface. The ciliated cells lining the trachea beat their cilia in synchronized waves to move sticky mucus and trapped dust particles toward the throat and away from delicate respiratory surfaces.

Power stroke

Return stroke

Microtubules in centriole

Plasma membrane

Microtubules

Basal body

**? REVIEW**

**B.** Which cytoskeletal component is common to both centrioles and cilia?

**C.** What is the function of motile cilia?

**? REVIEW**

**D.** Which cytoskeletal structure is found only in males?

**Lo LEARNING OUTCOME**

Differentiate among the structures and functions of the cytoskeleton.

## Module 3.5
# Ribosomes are responsible for protein synthesis and are often associated with the endoplasmic reticulum

**Ribosomes** are the organelles responsible for protein synthesis. The more proteins a cell synthesizes, the more ribosomes it has. Liver cells have many; fat cells have relatively few. Free ribosomes are scattered throughout the cytoplasm. The proteins they manufacture enter the cytosol. Ribosomes synthesizing proteins with destinations other than the cytosol become temporarily bound, or fixed, to the endoplasmic reticulum (ER), a membranous organelle.

**? REVIEW**

**A.** Describe the immediate cellular destinations of newly synthesized proteins from free ribosomes and fixed ribosomes.

**1** A functional ribosome consists of two subunits that are normally separate and distinct. These subunits contain special proteins and ribosomal RNA (rRNA), one of the RNA types introduced in Module 2.20. Before protein synthesis can begin, large and small ribosomal subunits must join together, as shown here.

Large ribosomal subunit

Small ribosomal subunit

**2** The **endoplasmic reticulum** (en-dō-PLAZ-mik re-TIK-ū-lum), or **ER**, is a network of intracellular membranes continuous with the nuclear envelope, which surrounds the nucleus.

The ER forms hollow tubes, flattened sheets, and chambers called **cisternae** (sis-TUR-nē; singular, *cisterna*, a reservoir for water).

Nuclear envelope

**3** The **smooth endoplasmic reticulum**, or **SER**, lacks ribosomes, and the cisternae are often tubular.

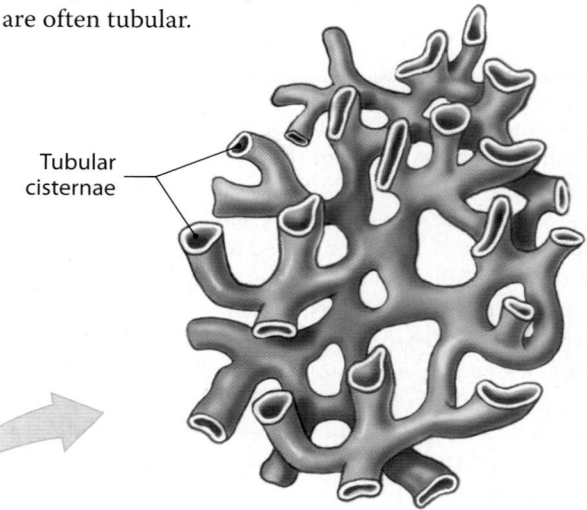

Tubular cisternae

### The Functions of the SER

- Synthesize the phospholipids and cholesterol needed for maintenance and growth of the plasma membrane, ER, nuclear envelope, and Golgi apparatus

- Synthesize steroid hormones, such as testosterone and estrogen (the dominant sex hormones in males and in females, respectively) in the reproductive organs

- Synthesize and store glycerides, especially triglycerides, in liver cells and fat cells

- Synthesize and store glycogen in skeletal muscle cells and liver cells

**4** The **rough endoplasmic reticulum (RER)** functions as a combination workshop and shipping warehouse. It is where many newly synthesized proteins are chemically modified and packaged for export to their next destination, the Golgi apparatus, described in Module 3.6.

? **REVIEW**

**B.** Compare and contrast the structure of SER and RER.

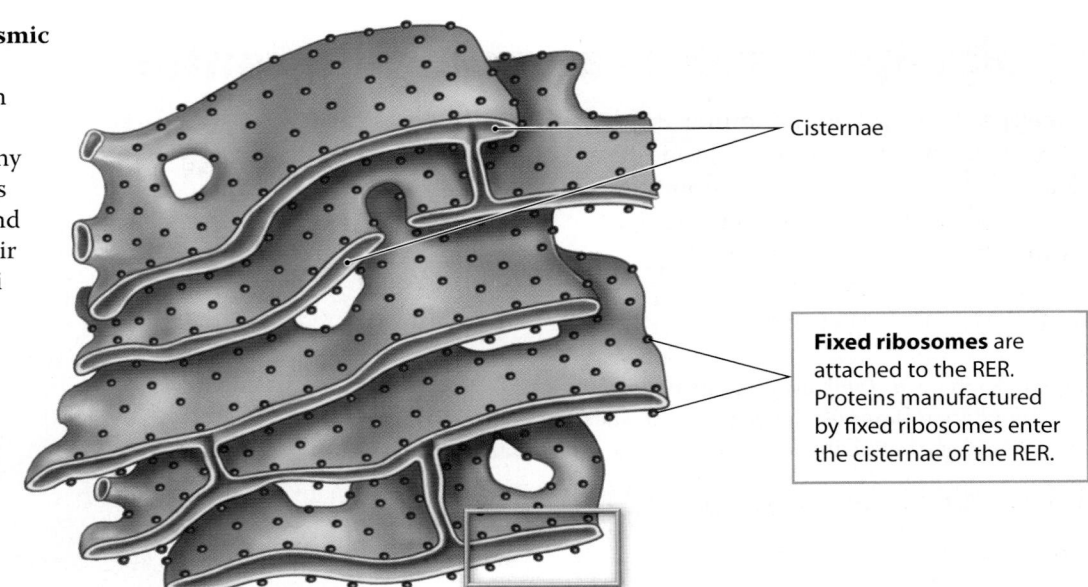

Cisternae

**Fixed ribosomes** are attached to the RER. Proteins manufactured by fixed ribosomes enter the cisternae of the RER.

mRNA strand

Ribosome

Transport vesicles

Growing polypeptide

Enzyme

Protein

Glycoprotein

| As a polypeptide is synthesized on a ribosome, the growing chain enters the cisterna of the RER. | The polypeptide assumes its secondary and tertiary structure. | The completed protein may become an enzyme or a glycoprotein. | Glycoproteins, proteins, and enzymes are packaged in **transport vesicles**. | Transport vesicles deliver proteins, enzymes, and glycoproteins to the Golgi apparatus. |

The amount of endoplasmic reticulum and the proportion of RER to SER vary with the type of cell and its ongoing activities. For example, pancreatic cells that manufacture digestive enzymes contain an extensive RER, but their SER is small. The situation is just the reverse in reproductive system cells that synthesize steroid hormones.

? **REVIEW**

**C.** Why do certain cells in the ovaries and testes contain large amounts of SER?

? **INTEGRATION**

**D.** The ER is connected to and continuous with what other organelle in the cell?

**Lo** **LEARNING OUTCOME**

Describe the ribosome and smooth and rough endoplasmic reticula, and indicate their specific functions.

# Module 3.6
# The Golgi apparatus is a packaging center

The **Golgi apparatus**, or **Golgi complex**, (1) renews or modifies the plasma membrane; (2) modifies and packages secretions, such as hormones or enzymes, for release outside the cell through a transport process called exocytosis; and (3) packages special enzymes within vesicles (lysosomes) for use in the cytosol.

**1** The Golgi apparatus typically consists of five or six flattened membranous discs called cisternae. A single cell may contain more than one Golgi apparatus, typically located near the nucleus.

Extracellular fluid

Cytosol

**Membrane renewal**

**Membrane renewal vesicles** add to the surface area of the plasma membrane. At the same time, other areas of the plasma membrane are being removed and recycled. So, the Golgi apparatus can change the properties of the plasma membrane, which can profoundly alter the sensitivity and functions of the cell.

**Secretion**

**Secretory vesicles** contain products that will be discharged from the cell by **exocytosis**. In this process, vesicles fuse with the plasma membrane and empty their contents into the extracellular environment.

**Enzymes for cytosol**

**Lysosomes** (LĪ-sō-sōmz; *lyso-*, a loosening + *soma*, body) are special vesicles that provide an isolated environment for potentially dangerous chemical reactions. These vesicles, produced by the Golgi apparatus, contain digestive enzymes whose varied functions are described on the facing page.

**Membrane renewal**

**Secretion**

Secretory vesicle

**Enzymes for cytosol**

Lysosome

Cisternae

*trans* face

**5** Ultimately, the product arrives at the *trans* face ("shipping" side), which is usually oriented toward the free surface of the cell.

**4** Small transport vesicles return resident Golgi proteins to cisternae of the *cis* face for reuse.

**3** Further modification and packaging occur as the cisternae migrate to the *trans* face.

**2** Multiple transport vesicles combine to form cisternae on the *cis* face. Inside the Golgi apparatus, enzymes modify the arriving proteins and glycoproteins. For example, the enzymes may change the carbohydrate structure of a glycoprotein, or they may attach a phosphate group, sugar, or fatty acid to a protein.

*cis* face

Transport vesicle

**Start** **1** Some proteins and glycoproteins synthesized in the rough endoplasmic reticulum (RER) are delivered to the Golgi apparatus by **transport vesicles**. The vesicles generally arrive at a cisterna known as the *cis* **face** ("receiving" side).

**2** Cells often need to break down and recycle large organic molecules and even complex structures such as organelles. The breakdown process requires powerful enzymes and often generates toxic chemicals that could damage or kill the cell. Lysosomes isolate those chemical reactions from the rest of the cytoplasm.

## Three basic functions of lysosomes

**Destruction of foreign material**

A lysosome may fuse with a vesicle containing fluids or solids from outside the cell.

Waste products and debris are ejected from the cell when the vesicle fuses with the plasma membrane through the process of exocytosis.

Extracellular solid or fluid

Vesicles containing fluids or solids may form at the surface of the cell.

Lysosomal enzymes are activated by fusion with another vesicle or organelle.

Lysosomes initially contain inactive enzymes.

As the materials or pathogens are broken down by lysosomal enzymes, released nutrients are absorbed.

As digestion occurs, nutrients are reabsorbed for recycling.

**Autolysis**

The lysosomal membrane may break down following injury to, or death of, the cell. The digestive enzymes become active and act upon the cytoplasm in a destructive process known as **autolysis**. For this reason, lysosomes are sometimes called "suicide packets."

**Removal of damaged organelles**

A lysosome may fuse with the membrane of another organelle, such as a mitochondrion. This activates the enzymes and begins the digestion of the lysosomal contents.

Golgi apparatus

**? REVIEW**

**A.** List the three major functions of the Golgi apparatus.

**B.** What do lysosomes contain?

**C.** Describe three functions of lysosomes.

With the exception of mitochondria, all membranous organelles in the cell are either interconnected or in communication through the movement of vesicles. This continuous movement and exchange is called **membrane flow**. In an actively secreting cell, an area equal to the entire membrane surface may be replaced *each hour*. Membrane flow is an example of the dynamic nature of cells. It gives cells a way to change the characteristics of their plasma membranes—the lipids, receptors, channels, anchoring proteins, and enzymes—as they grow, mature, or respond to a specific environmental stimulus.

 **LEARNING OUTCOME**

Describe the Golgi apparatus, and indicate its specific functions.

# Mitochondria are the powerhouses of the cell

The cells of all living things require energy to carry out the functions of life. The organelles that produce energy are the **mitochondria** (mī-tō-KON-drē-uh; singular, mitochondrion; *mitos*, thread + *chondrion*, granule).

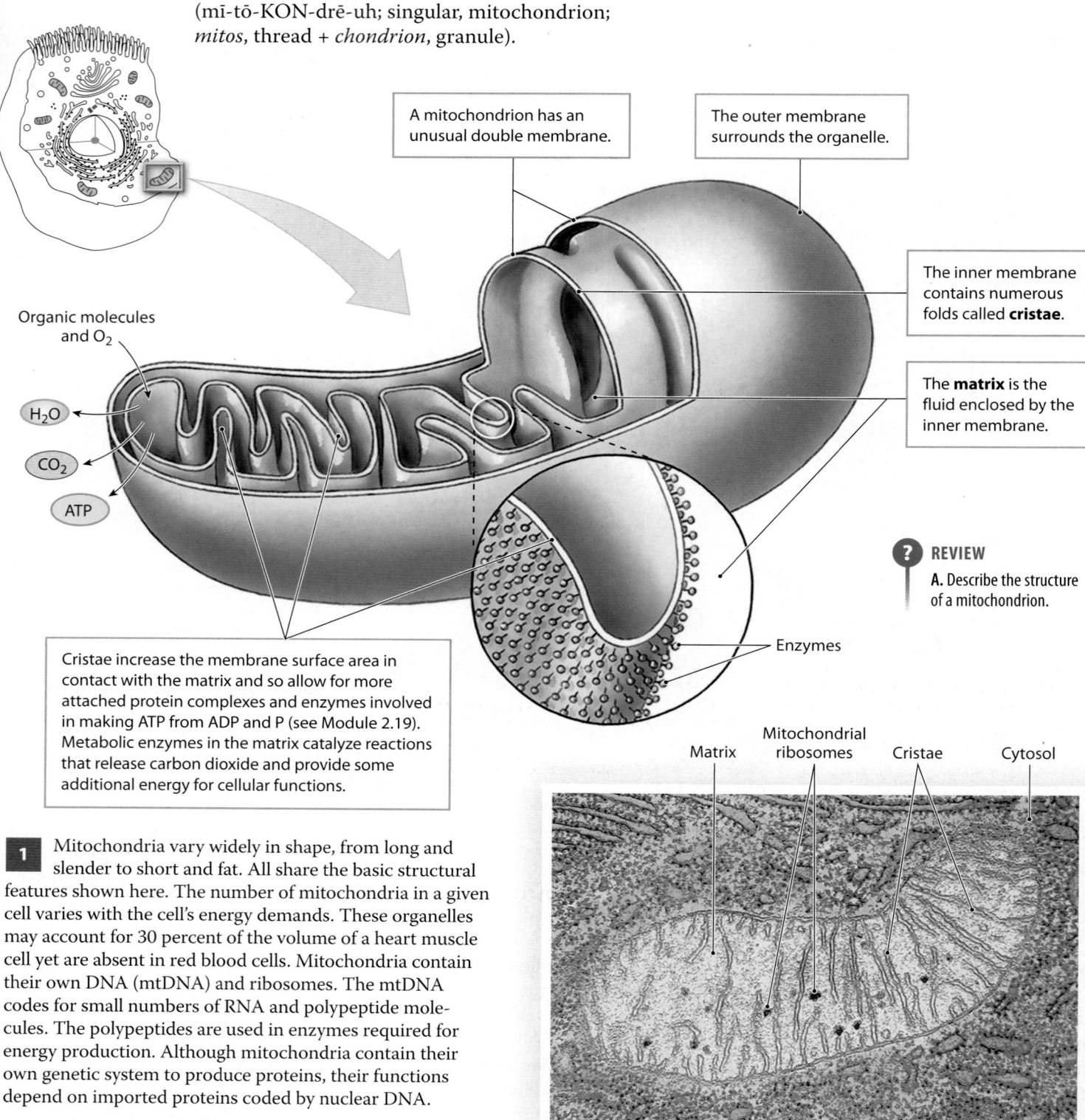

A mitochondrion has an unusual double membrane.

The outer membrane surrounds the organelle.

The inner membrane contains numerous folds called **cristae**.

Organic molecules and $O_2$

$H_2O$

$CO_2$

ATP

The **matrix** is the fluid enclosed by the inner membrane.

Cristae increase the membrane surface area in contact with the matrix and so allow for more attached protein complexes and enzymes involved in making ATP from ADP and P (see Module 2.19). Metabolic enzymes in the matrix catalyze reactions that release carbon dioxide and provide some additional energy for cellular functions.

Enzymes

**? REVIEW**
**A.** Describe the structure of a mitochondrion.

**1** Mitochondria vary widely in shape, from long and slender to short and fat. All share the basic structural features shown here. The number of mitochondria in a given cell varies with the cell's energy demands. These organelles may account for 30 percent of the volume of a heart muscle cell yet are absent in red blood cells. Mitochondria contain their own DNA (mtDNA) and ribosomes. The mtDNA codes for small numbers of RNA and polypeptide molecules. The polypeptides are used in enzymes required for energy production. Although mitochondria contain their own genetic system to produce proteins, their functions depend on imported proteins coded by nuclear DNA.

Matrix

Mitochondrial ribosomes

Cristae

Cytosol

**Mitochondrion**

TEM × 50,000

**2** Most cells generate ATP and other high-energy compounds as they break down carbohydrates, especially glucose. The major steps in the process are shown here.

**1** Although most ATP production occurs inside mitochondria, the first steps take place in the cytosol. In this reaction sequence, called **glycolysis** (*glycos*, sugar + *-lysis*, a loosening), each glucose molecule is broken down into 2 molecules of pyruvate. The mitochondria then absorb the pyruvate molecules.

**4** The energy released during these steps indirectly supports the enzymatic conversion of ADP to ATP. The ATP then leaves the mitochondrion.

CYTOSOL

Glucose

2 Pyruvate

$CO_2$

MITOCHONDRION

ADP + P

ATP

Citric acid cycle

H

Enzymes and coenzymes of cristae

$O_2$

MATRIX

$H_2O$

**2** In the mitochondrial matrix, a $CO_2$ molecule is removed from each absorbed pyruvate molecule. The remainder enters the **citric acid cycle**, or TCA (tricarboxylic acid) cycle, an enzymatic pathway that systematically breaks down the absorbed pyruvate remnant into carbon dioxide and hydrogen atoms. The carbon dioxide molecules diffuse into the cytoplasm.

**3** The hydrogen atoms are delivered to enzymes and coenzymes of the cristae, which catalyze ATP synthesis from ADP and P. At the end of this process, oxygen combines with the hydrogen atoms to form water molecules.

**?** **REVIEW**

**B.** Most of a cell's ATP is produced within its mitochondria. What gas do mitochondria require to produce ATP, and what gas results?

The amount of ATP generated during glycolysis is very small compared with the amount produced by the breakdown of pyruvate within mitochondria. Because mitochondrial activity requires oxygen, this method of ATP production is known as **aerobic metabolism** (*aer*, air + *bios*, life), or *cellular respiration*. Aerobic metabolism in mitochondria produces about 95 percent of the ATP needed to keep a cell alive. (Enzymatic reactions in the cytoplasm produce the rest.) Note that although the chemical reactions that release energy occur in the mitochondria, most of the cellular activities that require energy occur in the surrounding cytoplasm. So mitochondria are like little cellular batteries or fuel cells that provide the energy needed to power cellular functions.

**?** **INTEGRATION**

**C.** What does the presence of many mitochondria imply about a cell's energy requirements?

**Lo** **LEARNING OUTCOME**

Describe the structure of a mitochondrion, and explain the significance of mitochondria to cellular function.

## Section 1 Review
# Introduction to Cells

**3**

**Vocabulary:** In the space provided, write the boldfaced terms introduced in this section that contain the indicated word part.

1  glycos- *(sugar)*

2  aero- *(air)*

3  micro- *(small)*

4  lyso- *(a loosening)*

1  _____

2  _____

3  _____

4  _____

**Short answer:** Correctly label the indicated structures on the cell diagram below, and describe the functions of each.

5

6

7

8

9

10

11

12

13

**Section integration:** What is the advantage of having some of the cellular organelles enclosed by a membrane similar to the plasma membrane?

14  _____

# Module 3.8
# The nucleus is the control center for cellular homeostasis

The nucleus is usually the largest and most conspicuous structure in a cell. If you look at a human cheek cell under a light microscope, it is often the only organelle visible. The nucleus is the control center for cellular operations. A single nucleus stores all the information needed to direct the synthesis of more than 100,000 different proteins in the human body. This genetic information is coded in the sequence of nucleotides in DNA. The nucleus determines the structure of the cell and what functions it can perform by controlling which proteins are synthesized, under what circumstances, and in what amounts. Most cells in the body contain a single nucleus, but exceptions exist. For example, skeletal muscle cells have many nuclei, whereas mature red blood cells have none. Because a cell without a nucleus cannot repair itself, a red blood cell will disintegrate within 3 or 4 months.

**?** REVIEW
**A.** How is genetic information coded in the cell?
**B.** How many nuclei do most body cells contain?

**1** This is an overview of the role the nucleus plays in preserving homeostasis at the cellular level. When the extracellular environment changes, the cell has mechanisms that can provide rapid short-term adjustments. If the extracellular fluid (ECF) quickly returns to normal, those short-term adjustments may be sufficient to save the cell. But if the environmental stresses persist, the cell must adapt, and this involves long-term changes in its biochemical and physical characteristics. Dramatic, long-term changes in cell structure and function also occur as part of growth, development, and aging.

EXTRACELLULAR FLUID

Plasma membrane

Changes in ECF composition

Binding to membrane receptors

Movement of materials, such as ions, through membrane channels

Binding to nuclear receptors that alter genetic activity

DNA in nucleus

CYTOPLASM

**SHORT-TERM ADJUSTMENTS**

Enzyme activation or inactivation

**LONG-TERM ADJUSTMENTS**

Changes in the biochemical processes under way in the cell resulting from the synthesis of additional enzymes, fewer enzymes, or different enzymes

Changes in the physical structure of the cell due to alterations in the rates or types of structural proteins synthesized

In this section we consider topics of extreme importance to the rest of this text: the structure of the nucleus, the nature of the genetic material, how protein synthesis is controlled and directed, and how genetic information is divided and passed on during cell division.

**?** INTEGRATION
**C.** Describe why the nucleus is said to be the control center for a cell.

**Lo** LEARNING OUTCOME
Describe the role of the nucleus in maintaining homeostasis at the cellular level.

# Module 3.9

# The nucleus contains DNA, RNA, organizing proteins, and enzymes

The **nucleus** is often the most prominent and visible organelle. In standard histological sections and micrographs, the nucleus stains darkly and is easily identified.

**1** The nucleus is the control center for cellular processes. The surrounding nuclear envelope, with its nuclear pores, controls chemical communication between the cytoplasm and the nucleus.

The **nuclear envelope**, which surrounds the nucleus and separates it from the cytoplasm, is a double membrane.

The narrow **perinuclear space** (*peri-*, around) separates the two layers of the nuclear envelope.

**Nuclear pores**, which account for about 10 percent of the surface of the nucleus, are passageways that permit chemical communication between the nucleus and the cytosol. Proteins at the pores regulate the movement of ions and small molecules, and neither proteins nor DNA can freely cross the nuclear envelope.

**Nucleoplasm** is the fluid, gel-like substance of the nucleus. It contains the nuclear matrix, a network of fine filaments that provides structural support and may be involved in regulating genetic activity. The nucleoplasm also contains ions, enzymes, RNA and DNA nucleotides, small amounts of RNA, and DNA.

**Nucleoli** (noo-KLĒ-ō-lī; singular, *nucleolus*) are transient nuclear organelles that synthesize ribosomal RNA. They also assemble the ribosomal subunits, which enter the cytoplasm by carrier-mediated transport at the nuclear pores. Nucleoli are composed of RNA, enzymes, and proteins called **histones**. When the instructions for producing ribosomal proteins and RNA are being carried out, nucleoli form around the portions of DNA containing those instructions. Nucleoli are most prominent in cells that manufacture large amounts of proteins, such as liver, nerve, and muscle cells.

Nucleoplasm

Nuclear envelope

Nucleolus

Nuclear pore

**Nucleus**          TEM × 34,800

**?** **REVIEW**

**A.** Describe the contents and the structure of the nucleus.

**2** The DNA in the nucleus stores the instructions for protein synthesis. In the nucleus, the DNA strands are coiled, rather than straight, allowing a great deal of DNA to be packaged in a relatively small space. The coils wrap around histone molecules, forming complexes known as **nucleosomes**. The coiling can be tight or loose, and the entire chain of nucleosomes may coil around other proteins.

In cells that are not dividing, the nucleosomes are loosely coiled, forming a tangle of fine filaments known as **chromatin**.

2 nanometers (nm)

Nucleosome

Histones

DNA double helix

Nucleus of nondividing cell

**?** **REVIEW**

**B.** What molecule in the nucleus contains instructions for making proteins?

**3** At the beginning of cell division, DNA coiling becomes tighter and more complex, forming distinct structures called **chromosomes** (*chroma*, color).

At this stage, the chromosomes have already duplicated, and two copies of each chromosome are held together at a location known as the **centromere** (SEN-trō-mēr).

Supercoiled region

Dividing cell

Visible chromosome

In humans, the nuclei of somatic cells (general body cells, as opposed to sperm and oocytes, which are called sex cells) contain 23 pairs of chromosomes. One member of each pair is derived from the mother and one from the father. The DNA in these chromosomes carries the instructions for synthesizing both proteins and RNA. In addition, some DNA segments have a regulatory function, and others have no currently known function.

**?** **REVIEW**

**C.** How many chromosomes are contained within a typical somatic cell?

**?** **REVIEW**

**D.** The total length of the DNA within a human cell nucleus is approximately 2 meters. How does the DNA fit into the relatively small space of a human nucleus, which ranges some 6–10 μm in diameter?

**LO** **LEARNING OUTCOME**

Describe the functions of the cell nucleus, and distinguish between chromatin and a chromosome.

# Module 3.10

## Protein synthesis involves DNA, enzymes, and three types of RNA

In this module we introduce the key components and events of protein synthesis, a process we examine in detail in Modules 3.11 and 3.12.

**1** DNA consists of long, parallel chains of nucleotides. The two adjacent DNA chains are held together by hydrogen bonding between complementary base pairs. The nitrogenous bases involved are adenine (A), thymine (T), cytosine (C), and guanine (G). Information is stored in the sequence of base pairs. The chemical "language" the cell uses is known as the **genetic code**.

A triplet is a sequence of three nitrogenous bases along a DNA strand.

A **gene** is the functional unit of heredity. It contains all the DNA nucleotides needed to produce specific proteins. The number of nucleotides in a gene depends on the size of the polypeptide represented. A short polypeptide chain might need fewer than 300 nucleotides, but the instructions for building a large protein might involve 3000 or more nucleotides.

**? REVIEW**
**A.** What is a gene?

**KEY**

| | |
|---|---|
| C | Cytosine |
| G | Guanine |
| A | Adenine |
| T | Thymine |

DNA

**2** The genetic code is also called a **triplet code** because a sequence of three nitrogenous bases, or **triplet**, specifies the identity of a single amino acid. We discuss this in further detail in Module 3.11.

**? REVIEW**
**B.** Why is the genetic code described as a triplet code?

### Examples of the Genetic Code

| DNA Triplet Template Strand | DNA Triplet Coding Strand | mRNA Codon | tRNA Anticodon | Amino Acid |
|---|---|---|---|---|
| AAA | TTT | UUU | AAA | Phenylalanine |
| AAT | TTA | UUA | AAU | Leucine |
| ACA | TGT | UGU | ACA | Cysteine |
| CAA | GTT | GUU | CAA | Valine |
| TAC | ATG | AUG | UAC | Methionine |
| TCG | AGC | AGC | UCG | Serine |
| GGC | CCG | CCG | GGC | Proline |
| CGG | GCC | GCC | CGG | Alanine |

**3** Before a gene can affect a cell, the portion of the DNA molecule containing that gene must be uncoiled, and the histones temporarily removed. The factors controlling this process, called **gene activation**, are only partially understood. We know, however, that every gene contains segments responsible for regulating its own activity.

The two DNA strands separate in the region containing the activated gene, exposing the triplets to the nucleoplasm.

Paired DNA strands

**4** Enzymes then assemble a strand of **messenger RNA (mRNA)** by interconnecting complementary RNA nucleotides (A, G, C, U). This produces an RNA strand that contains information in triplets of RNA nucleotides. Those triplets are called **codons**. Each codon codes for a specific amino acid. The different codons and their respective amino acids are shown in the Appendix.

Enzyme

The mRNA strand containing the complementary codons passes through a nuclear pore and enters the cytoplasm.

Codon on mRNA

**5** At a ribosome in the cytoplasm, nucleotide triplets called **anticodons** on **transfer RNA (tRNA)** bind to the mRNA codons. There are many different types of tRNA, and each type carries a specific amino acid.

Amino acid — A1   A2   A3   A4

tRNA attaches to mRNA

Anticodon

U A C   G G C   U C G   A U U
A U G   C C G   A G C   U A A

mRNA strand

Codon

**6** The **ribosomal RNA (rRNA)** of the ribosome then strings amino acids together to form a polypeptide.

Polypeptide — A1 A2 A3 A4

methionine-proline-serine-leucine

**? REVIEW**
**C.** List the three types of RNA involved in protein synthesis.

So to summarize:

| The DNA triplets determine the sequence of mRNA codons. | → | The mRNA codons determine the sequence of tRNAs. | → | The sequence of tRNAs determines the sequence of amino acids in the polypeptide or protein. |

**? INTEGRATION**
**D.** Which type of RNA links the genetic information in the nucleus with the cytoplasmic sites of protein synthesis?

**Lo LEARNING OUTCOME**
Discuss the nature of the genetic code, and summarize the process of protein synthesis.

# Transcription encodes genetic instructions on a strand of RNA

A gene not only contains the instructions for the synthesis of proteins and nucleic acids, it also contains triplets that say "do read this message" or "do not read this message," "message starts here," or "message ends here." **Transcription** is the synthesis of RNA from a DNA template. The term *transcription* makes sense, as it means "to copy" or "rewrite." All three types of RNA are formed through DNA transcription, but we focus here on the transcription of mRNA, which carries the information needed to synthesize proteins.

**REVIEW**

**A.** What is transcription?

---

**1** Gene activation begins with the temporary disruption of the hydrogen bonds between the nitrogenous bases of the two DNA strands and the removal of the histone that "guards" the **control segment** (promoter), the first segment of a gene. Only the DNA template strand is used as the basis for synthesizing RNA.

Gene

DNA

The **template strand** is the DNA strand that will be used to synthesize RNA.

The segment at the start of the gene is known as the control segment, or promoter.

Triplet 1

Triplet 2

Triplet 3

Triplet 4

The enzyme **RNA polymerase** binds to the exposed control segment and, using the triplets as a guide, assembles a strand of mRNA.

Complementary triplets

**Coding strand** of DNA

**KEY**

A⊢ Adenine    U⊢ Uracil (RNA)

G⊢ Guanine    T⊢ Thymine (DNA)

C⊢ Cytosine

**REVIEW**

**B.** Define *DNA template strand.*

## Clinical Note

### Mutations

**Mutations** are permanent changes in a cell's DNA that affect the nucleotide sequence of one or more genes. The simplest is a *point mutation*, a change in a single nucleotide that affects one codon. A point mutation that produces a codon specifying a different amino acid changes the structure of the completed protein and can prove fatal. Certain cancers and two potentially lethal blood disorders, *thalassemia* and *sickle cell anemia*, result from changes in a single nucleotide.

**2** RNA polymerase promotes hydrogen bonding between the nitrogenous bases of the DNA template strand and complementary RNA nucleotides in the nucleoplasm. The complementary nucleotides are then strung together by covalent bonding.

**3** At the "stop" signal, the enzyme and the mRNA strand detach from the DNA strand, and transcription ends. The complementary DNA strands now reassociate as hydrogen bonding occurs between complementary base pairs.

**4** Each gene includes a number of triplets that are not needed to build a functional protein. As a result, the mRNA strand assembled during transcription, sometimes called **immature mRNA** or pre-mRNA, must be "edited" before it leaves the nucleus to direct protein synthesis. In this RNA processing, noncoding intervening sequences, called **introns**, are snipped out, and the remaining coding segments, or **exons**, are spliced together.

The process creates a shorter, functional strand of mRNA that then enters the cytoplasm through a nuclear pore. Intron removal is extremely important and tightly regulated. By changing the editing instructions and removing different introns, a single gene can produce mRNAs that code for several different proteins. Some introns, however, act as enzymes to catalyze their own removal. How this variable editing is regulated is unknown.

Codon 1

RNA nucleotide

RNA polymerase

mRNA strand

Codon 1

Codon 2

Codon 3

Codon 4 (stop codon)

RNA polymerase

RNA polymerase works only on RNA nucleotides—it can attach adenine, guanine, cytosine, or uracil, but never thymine. If the DNA triplet is TAC, the corresponding mRNA codon will be AUG.

**Immature mRNA**

| A | B | C | D | E | F | G | H | I |

D    H

B    F

Introns removed

| A | C | E | G | I |

Exons spliced together to form mature mRNA

**? REVIEW**

**C.** Name the substrates and product in the enzymatic reaction catalyzed by RNA polymerase.

**? INTEGRATION**

**D.** What process would be affected if a cell could not synthesize the enzyme RNA polymerase?

**Lᴏ LEARNING OUTCOME**

Summarize the process of transcription.

# Module 3.12

# Translation builds polypeptides as directed by an mRNA strand

Protein synthesis is the assembling of functional polypeptides in the cytoplasm. Protein synthesis takes place through **translation**, the formation of a linear chain of amino acids, using the information from an mRNA strand. Again, the name makes sense: To translate is to present the same information in a different language. In this case, a message written in the "language" of nucleic acids (the sequence of nitrogenous bases) is translated by ribosomes into the "language" of proteins (the sequence of amino acids in a polypeptide chain). Each mRNA codon designates a particular amino acid to be inserted into the polypeptide chain. There are three phases to translation: initiation, elongation, and termination.

**? REVIEW**
**A.** What is translation?

NUCLEUS — DNA

mRNA

After transcription and RNA processing take place, the mRNA leaves the nucleus through a nuclear pore. Translation then occurs in the cytoplasm.

**KEY**

 A  Adenine

 G  Guanine

 C  Cytosine

 U  Uracil

## Initiation

1 — Amino acid

tRNA

Small ribosomal subunit

U A C — Anticodon

Start codon

mRNA strand

tRNA binding sites (**E**, **P**, and **A**)

Large ribosomal subunit

E  P  A

**1** The **initiation** phase of translation begins when the mRNA strand binds to a small ribosomal subunit. An initiator tRNA carrying a specific amino acid, methionine, then binds to the start codon on the mRNA strand. Binding occurs between the nucleotides of the start codon and three complementary nucleotides in a segment of the tRNA strand known as the **anticodon**.

**? REVIEW**

**B.** The nucleotide sequence of three mRNA codons is AUU-GCA-CUA. What is the complementary anticodon sequence for the second codon?

**2** The small and large ribosomal subunits join together and enclose the mRNA and tRNA. The large ribosomal subunit has three sites for tRNA binding, called the **E** site (Exit), **P** site (Polypeptide building), and **A** site (Arrival). The **initiation complex** is now complete, and protein synthesis can proceed. The tRNA in the P site holds what will become the first amino acid of a peptide chain. The adjacent A site is where an additional tRNA can bind to the mRNA strand. More than 20 kinds of tRNA exist, each with a different nucleotide sequence in the anticodon. Each tRNA carries an amino acid, and there is at least one tRNA anticodon that corresponds to each of the amino acids used in protein synthesis.

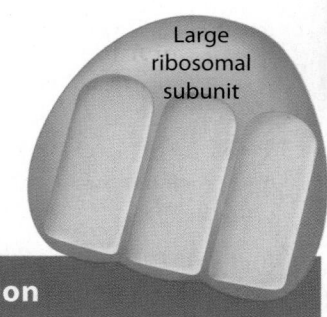

Large ribosomal subunit

## Elongation

Peptide bond

Stop codon

## Termination

Small ribosomal subunit

Completed polypeptide

mRNA strand

**3** In **elongation**, amino acids are added one by one to the growing polypeptide chain. When a complementary tRNA binds to the A site, ribosomal enzymes remove the amino acid from the tRNA at the P site and attach it to the amino acid delivered to the A site. The ribosome then moves one codon farther along the mRNA strand, and the tRNA that moves from the P site to the E site is released into the cytoplasm. The released tRNA can now bind another amino acid of the same type and repeat the cycle. Elongation ends when the ribosome reaches the stop codon at the end of the mRNA strand.

**4** **Termination** occurs as a protein-releasing factor recognizes the stop codon. A ribosomal enzyme then breaks the bond between the polypeptide and the tRNA in the P site, releasing the polypeptide. Other ribosomal enzymes separate the ribosomal subunits and free the intact strand of mRNA. The freed mRNA strand can interact with other ribosomes and create additional copies of the same polypeptide chain or protein.

Translation proceeds swiftly, producing a typical protein in about 20 seconds. The mRNA strand remains intact, and it can interact with other ribosomes to create additional copies of the same polypeptide chain. The process does not continue indefinitely, however, because after a few minutes to a few hours, mRNA strands are broken down, and the nucleotides recycled. However, large numbers of protein chains can be produced during that time. Although only two mRNA codons are "read" by a ribosome at any one time, many ribosomes can bind to a single mRNA strand, and a multitude of identical proteins may be quickly and efficiently produced.

**? INTEGRATION**

**C.** During the process of transcription, a nucleotide was deleted from an mRNA sequence that coded for a protein. What effect will this deletion have on the amino acid sequence of the protein?

**Lo LEARNING OUTCOME**

Summarize the process of translation.

# Section 2 Review
## Structure and Function of the Nucleus

**Matching:** Match each lettered term with the most closely related description.

a. introns

b. transcription

c. tRNA

d. chromosomes

e. exons

f. genetic information

g. nucleus

h. thymine

i. mRNA

j. gene

k. uracil

l. nuclear envelope

m. nuclear pore

n. nucleoli

**1** DNA strands and histones

**2** DNA nitrogenous base

**3** Double membrane

**4** mRNA noncoding regions

**5** RNA nitrogenous base

**6** Assemble ribosomal subunits

**7** Passageway for functional mRNA

**8** mRNA formation

**9** Functional unit of heredity

**10** mRNA coding regions

**11** Codon

**12** Anticodon

**13** Cell control center

**14** DNA nucleotide sequence

**1** _____

**2** _____

**3** _____

**4** _____

**5** _____

**6** _____

**7** _____

**8** _____

**9** _____

**10** _____

**11** _____

**12** _____

**13** _____

**14** _____

**Short answer:** The sequence of DNA bases at right are from a protein-coding gene. Use the sequence as the basis for answering the following questions.

T A C A A A A C A C G G C G G A A T

**15** Provide the corresponding mRNA base sequence, and insert a slash mark (/) between the codons.

**15** A U G / U U U / U G U / G C C / G C C / U U A

**16** Convert the mRNA codons you decoded above into tRNA anticodons.

**16** U A C / A A A / A C A / C G G / C G G / A A U

1  2  3  4  5  6

**17** Using the genetic code table in Module 3.10, translate the anticodon sequence in question 16 into the amino acid sequence of this polypeptide.

**17** Methionine / Phenylalanine / Cysteine
Alanine / Alanine / Leucine

**Section integration:** Explain the role of the nucleus in maintaining cellular homeostasis.

**18** _____

_____

_____

# Module 3.13
# The plasma membrane is a selectively permeable membrane

3

Conditions inside the cell can be much different from conditions outside the cell because the plasma membrane is an effective barrier. However, the barrier cannot be perfect, because cells are not self-sufficient, and their activities must be coordinated. In this section we consider how the plasma membrane selectively regulates the movement of materials into and out of the cell.

**1** **Permeability** is the property of the plasma membrane that determines precisely which substances can enter or leave the cytoplasm.

**? REVIEW**
**A.** Define *permeability*.
**B.** Identify three different types of membranes based on permeability.

**Freely permeable membranes**

**Freely permeable membranes** allow any substance to pass without difficulty. No living cell has a freely permeable plasma membrane.

**Selectively permeable membranes**

**Selectively permeable membranes** permit the passage of some materials and prevent the passage of others. Plasma membranes allow lipids and water free passage.

**Impermeable membranes**

Nothing can pass through **impermeable membranes**. Cells may be impermeable to specific substances, but no living cell has an impermeable plasma membrane.

**2** **Selectively permeable membranes** permit the free passage of some materials and restrict the passage of others. The distinction may be based on size, molecular shape, lipid solubility, electrical charge, or other factors. Cells differ in their permeabilities, depending on what lipids and proteins are present in the plasma membrane and how these components are arranged.

**? REVIEW**
**C.** Distinguish between passive and active processes of membrane passage.

**? INTEGRATION**
**D.** What kinds of molecules are involved in both active and passive processes of membrane passage?

**Lo LEARNING OUTCOME**
Contrast permeable, selectively permeable, and impermeable membranes.

## Module 3.14

# Diffusion is passive movement driven by concentration differences

Ions and molecules in liquids and gases are constantly in motion, colliding and bouncing off one another and off obstacles in their paths. Driven by thermal energy, the movement is random: A molecule can bounce in any direction. One result of this continuous random motion is that, over time, the molecules in any given space will tend to become evenly distributed. This distribution process, the net movement of a substance from an area of higher concentration to an area of lower concentration, is called **diffusion**. The difference between the high and low concentrations is a **concentration gradient**. After the gradient has been eliminated, the molecular motion continues, but net movement no longer occurs in any particular direction.

**1** Diffusion in air and water is slow, and it is most important over very short distances. A simple demonstration—observing a colored sugar cube placed in a beaker of water—can give you a good mental picture of what happens at the cellular level. However, compared to a cell, a beaker of water is enormous, and additional factors (which we will ignore) account for dye and sugar distribution over distances of centimeters as opposed to micrometers.

**?** **REVIEW**
**A.** Define *diffusion*.

**?** **REVIEW**
**B.** Describe the colliding molecules in the figure below.

## Diffusion of a Colored Sugar Cube through Water

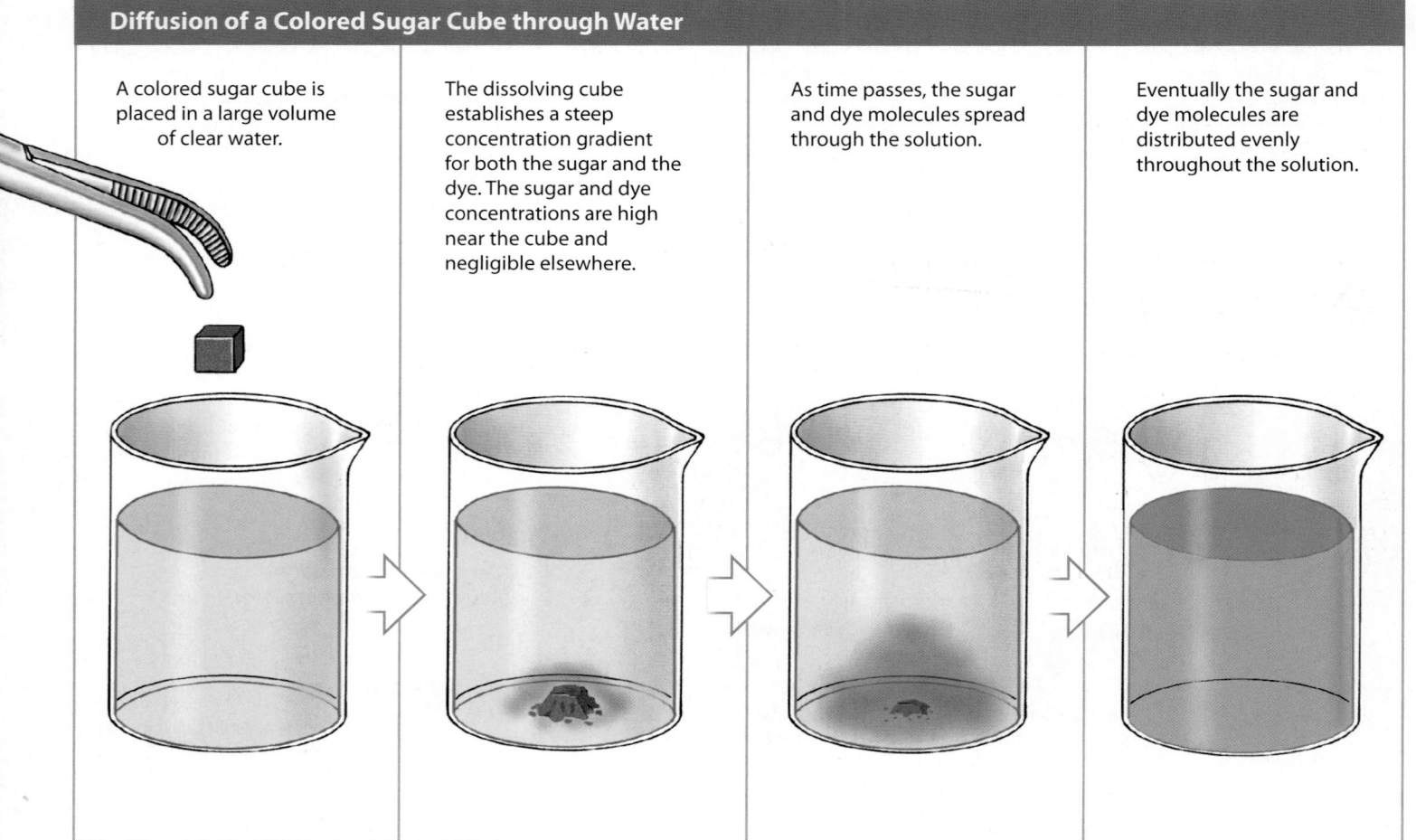

A colored sugar cube is placed in a large volume of clear water.

The dissolving cube establishes a steep concentration gradient for both the sugar and the dye. The sugar and dye concentrations are high near the cube and negligible elsewhere.

As time passes, the sugar and dye molecules spread through the solution.

Eventually the sugar and dye molecules are distributed evenly throughout the solution.

**2** In extracellular fluids, water and dissolved solutes diffuse freely. A plasma membrane, however, acts as a barrier that selectively restricts diffusion: Some substances pass through easily, but others cannot penetrate the membrane. An ion or a molecule can diffuse across a plasma membrane only by (1) crossing the lipid portion of the membrane or (2) passing through a membrane channel.

EXTRACELLULAR FLUID

Lipids, lipid-soluble molecules, water, and soluble gases ($O_2$ and $CO_2$) can diffuse across the lipid bilayer of the plasma membrane.

Water, small water-soluble molecules, and ions diffuse through membrane channel proteins that vary in shape, size, and specificity.

Plasma membrane

Channel protein

Large molecules that cannot fit through the membrane channels and cannot diffuse through the membrane lipids can cross the plasma membrane only when transported by a carrier mechanism.

CYTOPLASM

The diffusion of nutrients, waste products, and dissolved gases must keep pace with the demands of active cells. Important factors that influence diffusion rates are listed below.

- **Distance:** The shorter the distance, the more quickly concentration gradients are eliminated. In the human body, few cells are farther than 25 μm from a blood vessel.
- **Molecule or ion size:** The smaller the molecule or ion size, the faster the diffusion rate. Ions and small organic molecules, such as glucose, diffuse more rapidly than do large proteins.
- **Temperature:** The higher the temperature, the faster the diffusion rate.
- **Concentration gradient:** The steeper the concentration gradient, the faster diffusion proceeds.
- **Electrical forces:** Opposite electrical charges (+ and −) attract each other; like charges (+ and + or − and −) repel each other. Electrical attraction or repulsion can accelerate or reduce the ion diffusion rate.

**? REVIEW**

**C.** Identify factors that influence diffusion rates.

**? INTEGRATION**

**D.** How would a decrease in the oxygen concentration in the lungs affect oxygen diffusion into the blood?

**Lo LEARNING OUTCOME**

Explain the process of diffusion, and identify its significance in the body.

# Module 3.15
# Osmosis is the diffusion of water molecules across a selectively permeable membrane

Intracellular and extracellular fluids are solutions with a variety of dissolved materials. Each solute diffuses as though it were the only material in solution. If we ignore individual solute identities and simply count ions and molecules, we find that the total concentration of dissolved ions and molecules on either side of the plasma membrane stays the same. This state of equilibrium persists because a typical plasma membrane is freely permeable to water. The diffusion of water across a selectively permeable membrane is so important that it is given a special name: **osmosis** (oz-MŌ-sis; *osmos*, a push).

**REVIEW**
A. Describe osmosis.

**1** The movement of water driven by osmosis is called **osmotic flow**. The greater the initial difference in solute concentrations, the stronger the osmotic flow. The **osmotic pressure** of a solution is an indication of the force with which pure water moves into that solution as a result of its solute concentration. We can measure a solution's osmotic pressure in several ways. For example, an opposing pressure can prevent the osmotic flow of water into the solution.

**REVIEW**
B. Describe osmotic pressure, and state in which solution below it is greater.

## Water movement through a selectively permeable membrane

A selectively permeable membrane separates these two solutions, which have different solute concentrations. Water molecules (small blue dots) begin to cross the membrane toward solution B, the solution with the higher concentration of solutes (larger pink circles). The larger solute molecules cannot diffuse across the membrane.

At equilibrium, the solute concentrations on the two sides of the membrane are equal. Note that the volume of solution B has increased at the expense of the volume of solution A.

Pushing against a fluid generates **hydrostatic pressure**. The osmotic pressure of solution B is equal to the amount of hydrostatic pressure, indicated by the weight, required to stop the osmotic flow.

A
B
Water molecules
Solute molecules
Selectively permeable membrane

Volume decreased
Original level
Volume increased

Volumes equal
Applied force

**2** The total solute concentration in an aqueous solution is the solution's **osmolarity**, or **osmotic concentration**. The nature of the solutes, however, is often as important as the total osmolarity. When we describe the effects of various osmotic solutions on cells, we usually use the term **tonicity** instead of *osmolarity*. Although often used interchangeably, the terms *osmolarity* and *tonicity* do not always mean the same thing. *Osmolarity* refers to the solute concentration of the solution, whereas *tonicity* is a description of how the solution affects a cell. Tonicity may have one of three effects, as illustrated here.

**?** **REVIEW**

**C.** Contrast the effects of a hypotonic solution and a hypertonic solution on a red blood cell.

**3**

| Isotonic solution | Hypotonic solution | Hypertonic solution |
|---|---|---|
| A solution that does not cause an osmotic flow of water into or out of a cell is called **isotonic** (*iso-*, same + *tonos*, tension). The scanning electron micrograph (SEM) shows a red blood cell (RBC) that has been stored in an isotonic solution. | A **hypotonic** solution is one that causes osmotic water flow into the cell. If a red blood cell is placed in a hypotonic solution, it will swell up like a balloon. The cell may eventually burst, releasing its contents. This event is **hemolysis** (*hemo-*, blood + *lysis*, a loosening). | Exposure to a **hypertonic** solution results in the osmotic movement of water out of the cell. As water moves out, the cell shrivels and dehydrates. The shrinking of red blood cells is called **crenation**, and the cells are said to be crenated. |

Water molecules

Solute molecules

*SEM of a normal RBC in an isotonic solution*

*SEM of RBC in a hypotonic solution*

*SEM of crenated RBCs in a hypertonic solution*

**Clinical Note**

## Osmolarity and tonicity in medicine

It is often necessary to give patients large volumes of fluid to combat severe blood loss or dehydration. At such times the difference between osmolarity and tonicity is critically important. Consider a solution that has the same osmolarity as the intracellular fluid, but a higher concentration of one or more individual ions. If any of those ions can cross the plasma membrane and diffuse into the cell, the osmolarity of the intracellular fluid will gradually increase, and that of the extracellular solution will gradually decrease. Osmosis will then occur, moving water into the cell. If the process continues, the cell will gradually inflate like a water balloon. In this case, the extracellular solution and the intracellular fluid were initially equal in osmolarity, but the extracellular fluid was hypotonic rather than isotonic. In clinical emergencies, the fluid often administered is a 0.9 percent (0.9 g/dL) solution of sodium chloride (NaCl). This isotonic solution is called **normal saline**.

**?** **INTEGRATION**

**D.** Some pediatricians recommend using a 10 percent salt solution to relieve nasal congestion in infants. Explain the effects this treatment would have on the cells lining the nasal cavity. Would it be effective?

**Lo** **LEARNING OUTCOME**

Explain the process of osmosis, and identify its significance in the body.

# Module 3:16

# In carrier-mediated transport, integral proteins facilitate membrane passage

Watch

MasteringA&P™

A&PFlix

Membrane Transport

Nutrients that are insoluble in lipids and too large to fit through membrane channels may be transported across the plasma membrane by **carrier proteins**. Many carrier proteins move a specific substance in one direction only, either into or out of the cell. Some carrier proteins simultaneously move more than one substance in the same direction. This process is called **cotransport**. If a carrier protein simultaneously moves two substances in opposite directions, the process is called **countertransport,** and the carrier protein is called an **exchange pump**.

**? REVIEW**

**A.** Describe the process of carrier-mediated transport.

**B.** What two factors limit the rate of facilitated diffusion across a plasma membrane?

## Facilitated Diffusion

**1** Many substances can be passively transported across the plasma membrane in a process called **facilitated diffusion**. No ATP is expended in facilitated diffusion. The molecules simply move from an area of higher concentration to one of lower concentration. However, the molecules involved cannot cross the plasma membrane by diffusion through the membrane lipids or through open membrane channels. The number of suitable carrier proteins limits the rate of transport into the cell at any given moment. Once all the carrier proteins are saturated, the rate of transport cannot increase, regardless of further increases in the concentration of that substance in the ECF.

Facilitated diffusion begins when a specific molecule, such as glucose, binds to a receptor site on the integral carrier protein. In this case, the glucose is at a higher concentration in the extracellular fluid.

The shape of the protein then changes, moving the molecule across the plasma membrane. The carrier protein then releases the transported molecule into the cytoplasm. Note that this takes place without ever creating a continuous open channel between the cell's exterior and interior.

## Active Transport

**2** In **active transport**, ATP (or another high-energy compound) provides the energy needed to move ions or molecules across the plasma membrane. Active transport offers one great advantage: It is not dependent on a concentration gradient. As a result, the cell can import or export specific substances, regardless of their intracellular or extracellular concentrations. All cells contain carrier proteins called **ion pumps**, which actively transport the cations sodium ($Na^+$), potassium ($K^+$), calcium ($Ca^{2+}$), and magnesium ($Mg^{2+}$) across their plasma membranes. Specialized cells can transport additional ions, such as iodide ($I^-$), chloride ($Cl^-$), and iron ($Fe^{2+}$).

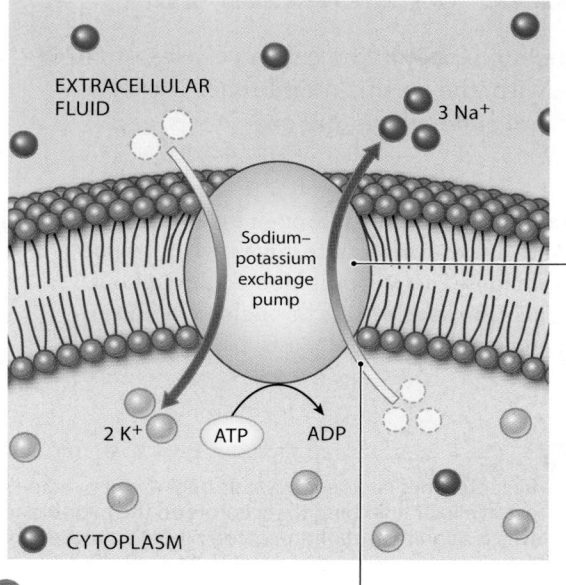

**1** Sodium ion concentrations are high in extracellular fluid but low in the cytoplasm. Potassium ion distribution is just the opposite: low in extracellular fluid and high in the cytoplasm. As a result, sodium ions slowly diffuse into the cell, and potassium ions diffuse out through leak channels. Homeostasis within the cell depends on the ejection of sodium ions and the recapture of lost potassium ions. The carrier protein for the sodium–potassium exchange pump is called *sodium–potassium ATPase.*

**2** On average, for each ATP molecule consumed, 3 sodium ions are ejected and the cell reclaims 2 potassium ions. The energy demands are impressive: Sodium–potassium ATPase may use up to 40 percent of the ATP produced by a resting cell!

**? REVIEW**

**C.** What do the transport processes of facilitated diffusion and active transport have in common?

## Secondary Active Transport

**3** In **secondary active transport**, the transport mechanism itself does not require energy from ATP, but the cell often needs to expend ATP at a later time to preserve homeostasis. Movement across the plasma membrane follows an existing concentration gradient for one of two substances transported, typically sodium ions, so transport does not require energy.

To preserve homeostasis, the cell must then expend ATP to pump the arriving sodium ions out of the cell by using the sodium–potassium exchange pump. It thus "costs" the cell one ATP for every three glucose molecules it transports into the cell.

A sodium ion and a glucose molecule bind to receptor sites on the carrier protein.

The carrier protein then changes shape, opening a path to the cytoplasm and releasing the transported materials. It then resumes its original shape and is ready to repeat the process.

**? INTEGRATION**

**D.** During digestion, the concentration of hydrogen ions ($H^+$) in the stomach contents increases to many times that in cells lining the stomach. Which transport process could be responsible?

**Lo LEARNING OUTCOME**

Describe carrier-mediated transport and its role in the absorption and removal of specific substances.

# Module 3.17

# In vesicular transport, vesicles selectively carry materials into or out of the cell

In **vesicular transport**, materials move into or out of the cell in vesicles, small membranous sacs that form at, or fuse with, the plasma membrane. The two major categories of vesicular transport are endocytosis and exocytosis, and both require energy in the form of ATP.

**1** **Endocytosis** is the importing of extracellular substances through the formation of vesicles at the cell surface. These vesicles are known as **endosomes**. Many important substances, including cholesterol and iron ions ($Fe^{2+}$), are distributed throughout the body attached to special transport proteins. Too large to pass through membrane pores, these proteins can enter cells by receptor-mediated endocytosis.

**? REVIEW**

**A.** Describe endocytosis.

**2** **Receptor-mediated endocytosis** produces vesicles containing a specific target molecule.

**Start**

**1** Receptor-mediated endocytosis begins when materials in the extracellular fluid bind to receptors on the membrane surface, which migrate to clathrin-coated pits. Most receptor molecules are glycoproteins, and each binds to a specific target molecule, or ligand, such as a transport protein or a hormone.

**2** Receptors bound to ligands cluster together. Once an area of the plasma membrane has become covered with ligands, it forms deep pockets that move to one area of the cell and then pinch off to form an endosome.

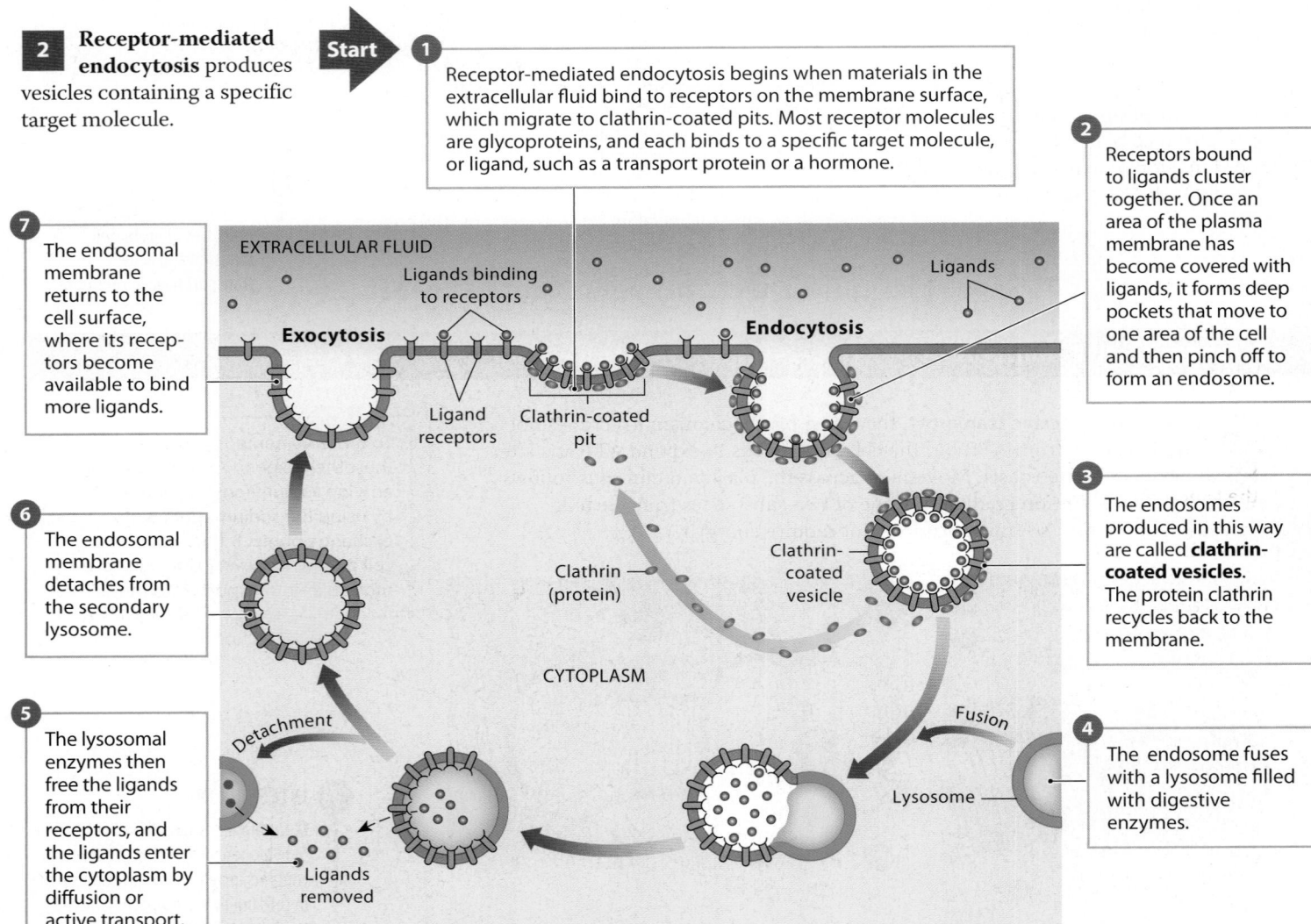

**7** The endosomal membrane returns to the cell surface, where its receptors become available to bind more ligands.

**6** The endosomal membrane detaches from the secondary lysosome.

**5** The lysosomal enzymes then free the ligands from their receptors, and the ligands enter the cytoplasm by diffusion or active transport.

**3** The endosomes produced in this way are called **clathrin-coated vesicles**. The protein clathrin recycles back to the membrane.

**4** The endosome fuses with a lysosome filled with digestive enzymes.

EXTRACELLULAR FLUID
Ligands binding to receptors
Ligands
**Exocytosis**
**Endocytosis**
Ligand receptors
Clathrin-coated pit
Clathrin (protein)
Clathrin-coated vesicle
CYTOPLASM
Fusion
Lysosome
Detachment
Ligands removed

**3** "Cell drinking," or **pinocytosis** (pi-nō-sī-TŌ-sis), is the formation of endosomes filled with extracellular fluid. This process is not as selective as receptor-mediated endocytosis, because no receptor proteins are involved. The target appears to be the fluid contents in general, rather than specific ligands. In a few specialized cells, pinocytosis produces vesicles on one side of the cell that are discharged on the other. This method of bulk transport is common in cells lining small blood vessels.

Bloodstream

Cytoplasm

Surrounding tissue

**Pinocytosis**                TEM × 20,000

Pinocytosis begins with the formation of deep grooves or pockets that then pinch off and enter the cytoplasm. The steps are similar to those of receptor-mediated endocytosis, except that ligand binding is not involved.

Endosome

Plasma membrane

**4** "Cell eating," or **phagocytosis** (fag-ō-sī-TŌ-sis), produces **phagosomes** containing solid objects that may be as large as the cell itself. Although most cells perform pinocytosis, only specialized cells perform phagocytosis. These cells are called **phagocytes** or **macrophages**.

Bacterium

Pseudopodium

**Phagocytosis**

Lysosome

Golgi apparatus

**Exocytosis**

Phagocytosis begins when cytoplasmic extensions called **pseudopodia** (soo-dō-PŌ-dē-ah; *pseudo-*, false + *podon*, foot; singular, *pseudopodium*) surround the object.

The pseudopodia then fuse at their tips to form a phagosome containing the targeted material.

This vesicle then fuses with many lysosomes. After that, lysosomal enzymes digest its contents.

Released nutrients diffuse into the surrounding cytoplasm.

The residue is then ejected from the cell through exocytosis.

**5** **Exocytosis** (ek-sō-sī-TŌ-sis) is the functional opposite of endocytosis. In exocytosis, a vesicle formed inside the cell fuses with, and becomes part of, the plasma membrane. When this takes place, the vesicle contents are released into the extracellular environment. The ejected material may be waste products (such as those accumulating in lysosomes) or secretory products (such as mucins or hormones).

**?** **REVIEW**

**B.** Describe the three types of endocytosis.

**C.** Describe exocytosis.

**?** **INTEGRATION**

**D.** Some white blood cells engulf bacteria and bring then into the cell. What is this process called?

**Lo** **LEARNING OUTCOME**

Describe vesicular transport as a mechanism for facilitating the absorption or removal of specific substances from cells.

# Section 3 Review
## How Substances Enter and Leave the Cell

**Concept map:** Use each of the following terms once to fill in the blank boxes to correctly complete the map.

- exocytosis — 8
- diffusion
- "cell eating" — 10
- molecule or ion size
- pinocytosis — 9
- facilitated diffusion
- vesicular transport — 7
- net diffusion of water — 4
- active transport
- specificity

**Membrane permeability** — *types*

Passive processes | No ATP energy | ATP energy | Active processes

*include* ... 1 ... Carrier-mediated transport ... 7 *include*

Carrier-mediated transport *includes*: Facilitated diffusion | 5 | Secondary active transport

7 *types*: Endocytosis | 8

1 *types*: Simple diffusion | 2 | Osmosis

*characteristics*: 6 | Saturation limits | Regulation

Simple diffusion *is affected by*:

**3**
Lipid solubility
Distance
Temperature
Concentration gradient
Electrical forces

Osmosis *is*: 4

8 *example*: Hormone secretion

Endocytosis *types*: Receptor-mediated | 9 | Phago-cytosis

9 *also called*: "Cell drinking"

Phago-cytosis *also called*: 10

---

**Short answer:** Classify each of the following situations as an example of diffusion, osmosis, or neither.

11. You walk into a room and smell a pine balsam aroma.

12. Water flows through a garden hose.

13. Grass in the yard wilts after being exposed to excess chemical fertilizer.

14. A sugar cube placed in a cup of hot tea dissolves.

15. After soaking several hours in water containing sodium chloride, a stalk of celery weighs less than before it was placed in the salty water.

11. _____
12. _____
13. _____
14. _____
15. _____

# Module 3.18

# Interphase and cell division make up the life cycle of a cell

Go to
MasteringA&P™
PAL 3.0
Histology→
Cytology

The period between fertilization and physical maturity involves tremendous changes in organization and complexity. At fertilization, a single cell is all there is. At maturity, your body has about 75 trillion cells. **Cell division**, a form of cellular reproduction, makes this transformation possible.

Even when development is complete, cell division continues to be essential to survival. Physical wear and tear, toxic chemicals, temperature changes, and other environmental stresses damage cells. And, like individuals, cells age. The life span of a cell varies from hours to decades, depending on the type of cell and the stresses involved. Many cells apparently self-destruct after a certain period of time as a result of the activation of specific "suicide genes" in the nucleus. This genetically controlled cell death is called **apoptosis** (ap-op-TŌ-sis or ap-ō-TŌ-sis; *apo-*, separated from + *ptosis*, a falling).

There are two different forms of cell division. **Mitosis** (mī-TŌ-sis), the focus of this section, produces two daughter cells, each containing a complete set of 46 chromosomes. **Meiosis** (mī-Ō-sis), which we examine in Chapter 26, produces sex cells (sperm or oocytes) containing only 23 chromosomes.

**? REVIEW**

**A.** Explain why cell division is important.

**B.** Define *apoptosis*.

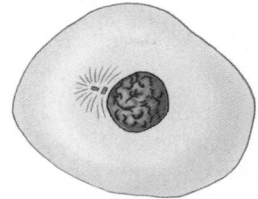

**Interphase** is the period in which the cell is performing normal functions and not actively engaged in cell division. Some cells are in interphase indefinitely. Others are always either dividing or preparing to divide.

During interphase in a cell preparing to divide, the chromosomes of the cell are duplicated, and associated proteins are synthesized.

**1** The division of a single cell produces a pair of **daughter cells**, each one half the size of the original. Before they themselves divide, each of the daughter cells will grow to the size of the original cell.

Original cell

Cell division

Daughter cells

**2** The diagram at right details the life cycle of a typical cell. The cycle ends only when the cell dies.

Cell division begins with mitosis. In mitosis, identical copies of the original chromosomes are distributed to each daughter cell.

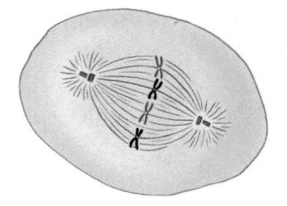

Cell division ends as **cytokinesis**, the division of the cytoplasm, physically separates the two daughter cells.

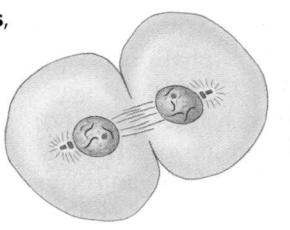

**INTERPHASE**

**THE CELL CYCLE**

**MITOSIS**

**CYTOKINESIS**

**? REVIEW**

**C.** When does cell division begin and end?

  **LEARNING OUTCOME**

Distinguish between interphase and cell division in the cell cycle.

# Module 3.19
# During interphase, the cell prepares for cell division

Most cells spend only a small part of their time actively engaged in cell division. **Somatic** (*soma*, body) **cells** spend most of their functional lives in a state known as **interphase**. During interphase, a cell performs all its normal functions and, if necessary, prepares for cell division.

**1** In a cell preparing to divide, interphase can be divided into the $G_1$, **S**, and $G_2$ **phases.**

When the G1 phase is complete, the cell enters the S phase. Over the next 6–8 hours, the cell duplicates its chromosomes. This involves DNA replication and the synthesis of histones and other proteins in the nucleus.

A cell that is ready to divide first enters the $G_1$ phase. In this phase, the cell makes enough mitochondria, cytoskeletal elements, endoplasmic reticula, ribosomes, Golgi membranes, and cytosol for two functional cells. Centriole replication begins in $G_1$ and commonly continues until $G_2$. In a cell dividing at top speed, $G_1$ may last just 8–12 hours. Such a cell pours all its energy into mitosis, and all other activities cease. If $G_1$ lasts for days, weeks, or months, preparation for mitosis occurs as the cell performs its normal functions.

When DNA replication ends, there is a brief (2- to 5-hour) $G_2$ phase devoted to last-minute protein synthesis and to completing centriole replication.

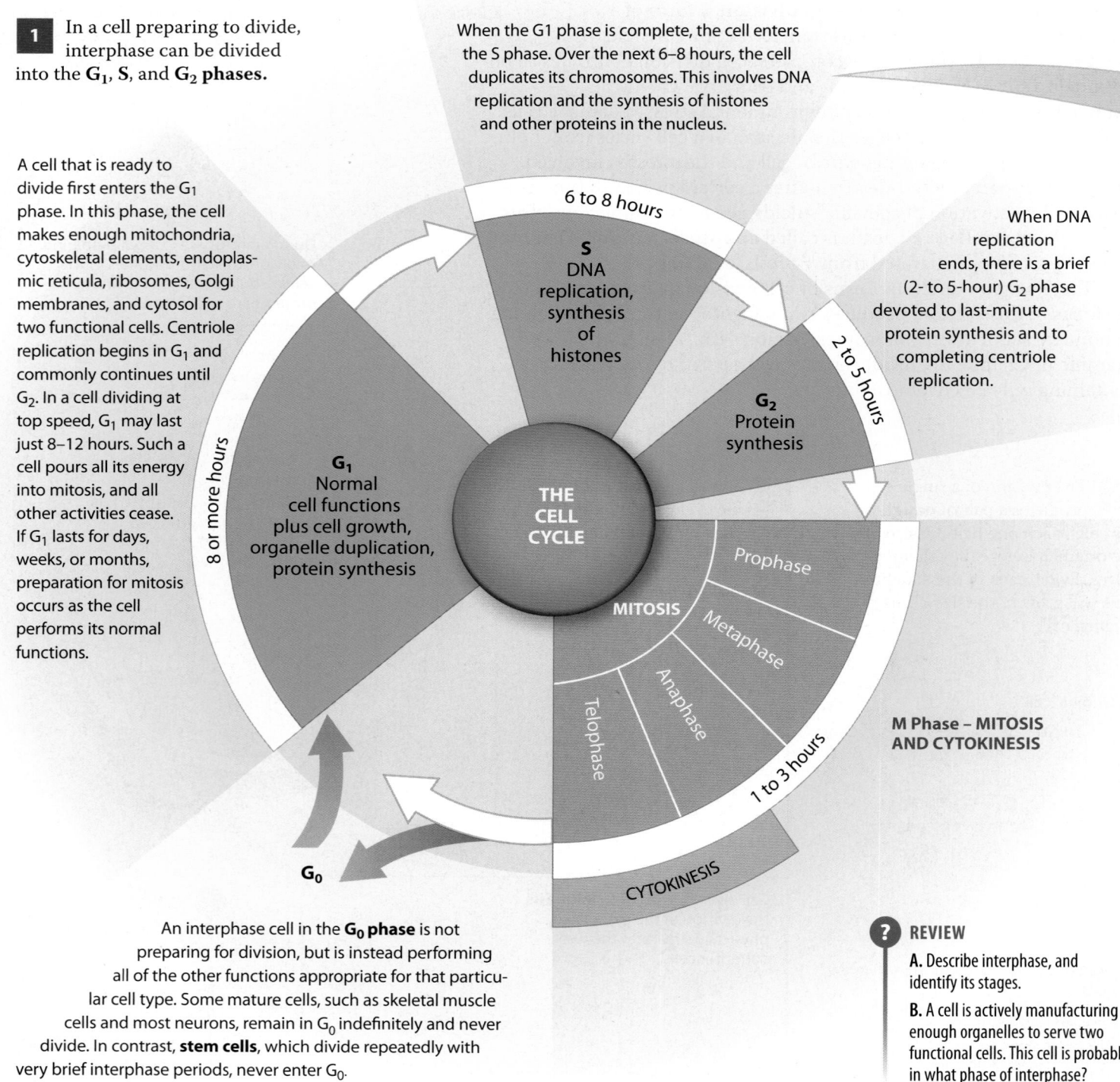

**THE CELL CYCLE**

6 to 8 hours

**S** DNA replication, synthesis of histones

2 to 5 hours

$G_2$ Protein synthesis

8 or more hours

$G_1$ Normal cell functions plus cell growth, organelle duplication, protein synthesis

**MITOSIS**

Prophase

Metaphase

Anaphase

Telophase

1 to 3 hours

CYTOKINESIS

**M Phase – MITOSIS AND CYTOKINESIS**

$G_0$

An interphase cell in the $G_0$ **phase** is not preparing for division, but is instead performing all of the other functions appropriate for that particular cell type. Some mature cells, such as skeletal muscle cells and most neurons, remain in $G_0$ indefinitely and never divide. In contrast, **stem cells**, which divide repeatedly with very brief interphase periods, never enter $G_0$.

**?** **REVIEW**

**A.** Describe interphase, and identify its stages.

**B.** A cell is actively manufacturing enough organelles to serve two functional cells. This cell is probably in what phase of interphase?

**2** During the S phase of the cell cycle, DNA is replicated. The goal of DNA replication is to copy the genetic information in the nucleus. The process occurs in cells preparing to undergo either mitosis or meiosis.

**1** **DNA replication** begins when **DNA helicase** enzymes unwind the strands and disrupt the hydrogen bonds between the bases. As the strands unwind, molecules of **DNA polymerase** bind to the exposed nitrogenous bases. This enzyme (1) promotes bonding between the nitrogenous bases of the DNA strand and complementary DNA nucleotides in the nucleoplasm and (2) links the nucleotides by covalent bonds.

Leading strand template

Leading strand

**2** As the two original strands separate, DNA polymerase binds to them. DNA polymerase can work in only one direction along a strand of DNA, but the two strands in a DNA molecule are oriented in opposite directions. The DNA polymerase bound to the upper leading strand template adds nucleotides to make a single, continuous complementary copy called the **leading strand** that grows toward the "zipper."

Segment 2

DNA nucleotide

Segment 1

**3** DNA polymerase on the lower lagging strand template can work only away from the zipper. So the first DNA polymerase to bind to this strand must add nucleotides and build a complementary DNA strand moving from left to right. As the two original strands continue to unzip, additional nucleotides are continuously being exposed to the nucleoplasm. The first DNA polymerase on this strand cannot go into reverse. It can only continue to elongate the strand it already started.

Lagging strand template

**KEY**

⬛▸ Adenine

⬛▸ Guanine

⬛▸ Cytosine

⬛▸ Thymine

**4** Thus, a second DNA polymerase must bind closer to the point of unzipping and assemble a complementary copy (segment 2) that grows until it "bumps into" segment 1 created by the first DNA polymerase. Enzymes called **DNA ligases** (LĪ-gās-ez; *liga*, to tie) then splice together the two DNA segments into a strand called the **lagging strand**.

**?** **REVIEW**
**C.** What enzymes must be present for DNA replication to proceed normally?

**3** Eventually, the unzipping completely separates the original strands. The copying ends, the last splicing is done, and two identical DNA molecules have formed. Once the DNA is replicated, the centrioles duplicated, and the necessary enzymes and proteins synthesized, the cell leaves interphase and is ready to proceed to mitosis.

Duplicated DNA double helices

**?** **INTEGRATION**
**D.** DNA replication occurs during what two cellular processes?

**Lo** **LEARNING OUTCOME**
Describe interphase, and explain its significance.

# Module 3.20

# Mitosis distributes chromosomes before cytokinesis separates the daughter cells

The M phase of the cell cycle includes mitosis and cytokinesis. **Mitosis** is a series of events during which the duplicated chromosomes of a cell separate and migrate into two identical nuclei. The four stages of mitosis are prophase, metaphase, anaphase, and telophase.

The term *mitosis* specifically refers to division and duplication of the cell's nucleus alone. The cytoplasm—the rest of the cell—divides into two distinct new cells in a separate, but related, process known as **cytokinesis** (sī-tō-ki-NĒ-sis; *cyto-*, cell + *kinesis*, motion).

## MITOSIS

The centrioles have replicated, and the pairs now move to opposite sides of the nucleus.

Microtubules extend outward from each pair of centrioles: **Astral rays** extend into the cytoplasm, whereas **spindle fibers** form between the centriole pairs.

Chromatids

Kinetochore

The nuclear envelope (nuclear membrane) disintegrates.

The kinetochore of each chromatid attaches to a spindle fiber.

Centrioles in centrosome

Nucleus

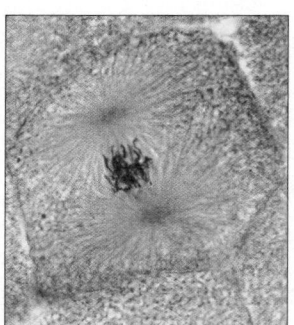

**1** During interphase, the DNA strands are loosely coiled, and chromosomes cannot be seen.

**2** **Prophase** (PRŌ-fāz; *pro*, before) begins when the chromosomes coil so tightly that they become visible as individual structures under a light microscope. As a result of DNA replication during the S phase, two copies of each chromosome now exist. Each copy is called a **chromatid**, and the paired chromatids are connected at a region known as the centromere (see Module 3.9). A raised region on the centromere, called the **kinetochore** (ki-NE-tō-kor), is where the spindle fibers attach to the chromatids.

 **REVIEW**

**A.** Define *mitosis*, and list its four stages.

**B.** What is a chromatid, and how many are present during normal mitosis in a human cell?

## CYTOKINESIS

3

The two chromatids are now pulled apart and drawn to opposite ends of the cell along the **spindle apparatus** (the complex of spindle fibers). Anaphase ends when the chromatids arrive near the centrioles at opposite ends of the cell.

As the chromatids approach the ends of the spindle apparatus, the cytoplasm constricts along the plane of the metaphase plate, forming a **cleavage furrow**.

Daughter cells

Metaphase plate

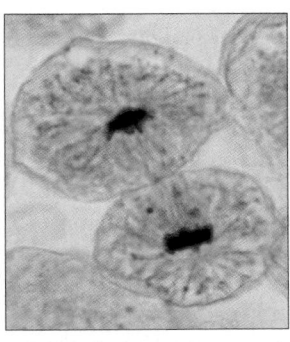

**3** **Metaphase** (MET-a-fāz; *meta*, after) begins as the chromatids move to a narrow central zone called the **metaphase plate**. Metaphase ends when all the chromatids are aligned in the plane of the metaphase plate.

**4** **Anaphase** (AN-a-fāz; *ana-*, apart) begins when the centromere of each chromatid pair splits and the chromatids separate. The chromatids are now pulled along the spindle fibers toward opposite sides of the dividing cell.

**5** During **telophase** (TĒL-ō-fāz; *telo-*, end), each new cell prepares to return to interphase. The nuclear envelopes re-form, the nuclei enlarge, and the chromosomes gradually uncoil to the chromatin state. This stage marks the end of mitosis.

**6** **Cytokinesis** usually begins with the formation of a cleavage furrow during anaphase and continues throughout telophase. The completion of cytokinesis marks the end of cell division.

**? INTEGRATION**

**C.** What would happen if spindle fibers failed to form in a cell during mitosis?

**Lo LEARNING OUTCOME**

Describe the process of mitosis and its role in the cell life cycle.

# Module 3.21

## Tumors and cancer are characterized by abnormal cell growth and division

When the rates of cell division and growth exceed the rate of cell death, a tissue begins to enlarge. The enlargement often results from the divisions of a single abnormal cell that no longer responds to the factors that regulate the rate of cell division. **Cancer** is an illness characterized by mutations—permanent changes in DNA nucleotide sequence and function—that disrupt normal control mechanisms that regulate the rates of cell division.

Abnormal cell

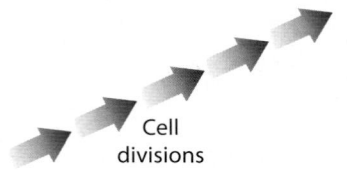

Cell divisions

**1** Cancer usually begins with a single abnormal cell. Every time a cell divides, there is a chance that something will go wrong with the control mechanism. As a result, cancers are most common in tissues where cells are dividing rapidly and continuously, such as the epithelium of the skin or the intestinal lining.

**?** **REVIEW**
**A.** Define *cancer*.
**B.** What is a benign tumor?

**2** A **tumor**, or **neoplasm**, is a mass or swelling produced by abnormal cell growth and division. In a **benign tumor**, the cells usually remain within the originating tissue. Such a tumor seldom threatens a person's life and usually can be removed surgically if its size or position disturbs tissue function.

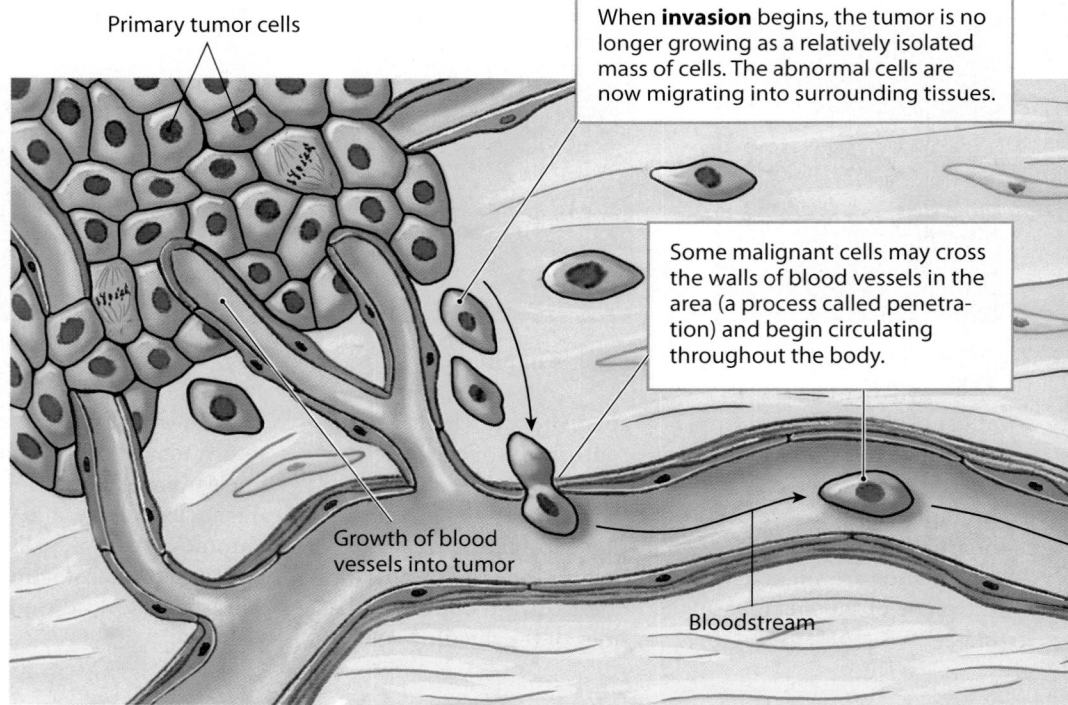

Primary tumor cells

When **invasion** begins, the tumor is no longer growing as a relatively isolated mass of cells. The abnormal cells are now migrating into surrounding tissues.

Some malignant cells may cross the walls of blood vessels in the area (a process called penetration) and begin circulating throughout the body.

Growth of blood vessels into tumor

Bloodstream

**3** Cells in a **malignant tumor** divide very rapidly, releasing chemicals that stimulate the growth of new blood vessels (**angiogenesis**) into the area. The availability of additional nutrients accelerates tumor growth, and malignant cells can then migrate into surrounding tissues and nearby blood vessels. This process—**metastasis** (me-TAS-tuh-sis)—can produce secondary tumors in tissues remote from the primary tumor site.

**4** Malignant cells may no longer perform their original functions, or they may perform normal functions in an abnormal way. For example, endocrine cancer cells may produce normal hormones, but in excessively large amounts. This photo shows a patient with a malignant tumor of the thyroid gland. Excessive amounts of thyroid hormone produced dramatic changes in metabolic activity throughout his body.

Responding to cues that are as yet unknown, cancer cells in the bloodstream ultimately escape out of blood vessels to establish secondary tumors at other sites. These tumors are extremely active metabolically, and their presence stimulates the growth of new blood vessels into the area. The increased blood supply provides additional nutrients to the cancer cells and further accelerates tumor growth and metastasis.

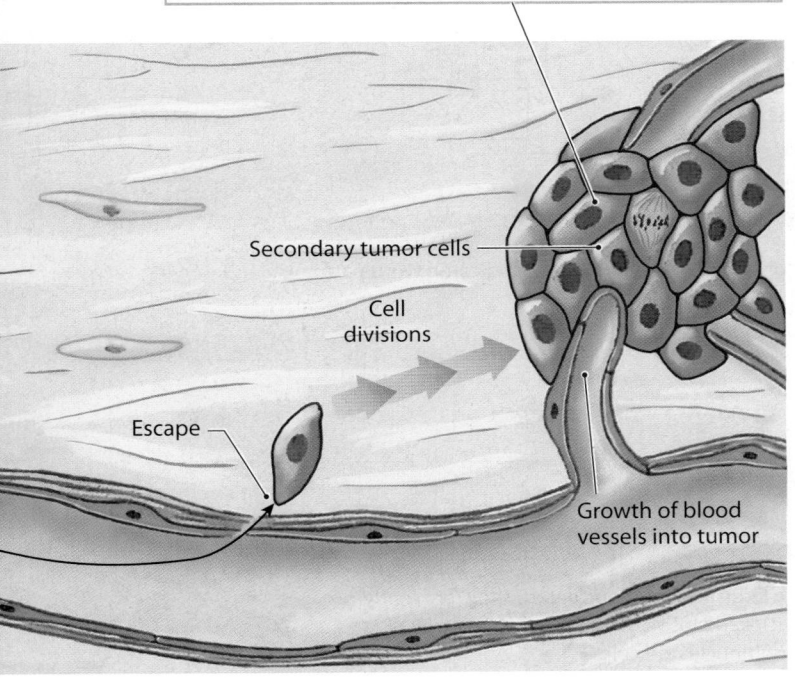

Secondary tumor cells

Cell divisions

Escape

Growth of blood vessels into tumor

Cancer cells do not use energy very efficiently. They grow and multiply at the expense of healthy tissues, competing with normal cells for space and nutrients. This competition contributes to the starved appearance of many patients in the late stages of cancer. Death may occur as a result of the compression of vital organs when nonfunctional cancer cells have killed or replaced the healthy cells in those organs, or when the cancer cells have starved normal tissues of essential nutrients. We discuss cancer further in later chapters that deal with specific body systems.

 **REVIEW**
**C.** Define *metastasis*.

 **INTEGRATION**
**D.** How does angiogenesis aid tumor growth?

 **LEARNING OUTCOME**
Discuss the relationship between cell division and cancer.

# Section 4 Review
## Cell Life Cycle

**Concept map:** Use each of the following terms once to fill in the blank boxes to correctly complete the map.

- metaphase
- DNA replication
- somatic cells
- G<sub>2</sub> phase
- telophase
- mitosis
- cytokinesis
- G<sub>1</sub> phase

| Cell cycle |

*occurs in*

**1**

*includes 2 major stages*

Interphase

**G<sub>0</sub> phase** — **2** — **S phase** — **3** — **5**

*time*
- Indefinite time
- 8 or more hours
- 6–8 hours
- 2–5 hours

*activity*
- Functioning cell not preparing for cell division
- Cell makes cytosol and organelles
- Cell duplicates chromosomes
- Protein synthesis

**4**

Preparation for cell division

2 new cells — *return to* / *resulting in*

*4 stages*

Prophase
- Chromatin to chromosomes
- Spindle fibers form
- Nuclear envelope disappears

**6**
- Chromatids move to metaphase plate

Anaphase
- Chromatid pairs separate
- Daughter chromosomes move toward opposite ends of cells

**7**
- Reverse of prophase
- End of mitosis

*result*

Daughter cells separate

**8**

---

**Vocabulary:** In the space provided, write the boldfaced terms introduced in this section that contain the indicated word part.

**9**  telo- *(end)*

**9** _____

**10**  pro- *(before)*

**10** _____

**11**  centro- *(in the middle)*

**11** _____

---

**Section integration:** The muscle cells that make up skeletal muscle tissue are large and multinucleated. They form early in development as groups of embryonic cells fuse together, contributing their nuclei and losing their individual plasma membranes. Describe an alternate mechanism that would also result in a large, multinucleated cell.

**12** _____

# Study Outline

## SECTION 1 • Introduction to Cells

### Module 3.1
### Cellular differentiation produces specialized cells    p. 95

1. **Cell theory** states that cells are the building blocks of all plants and animals, cells come from the division of pre-existing cells, and cells are the smallest units that perform all vital physiological functions.
2. All human body cells are descendants of a single fertilized ovum (zygote).
3. Cells undergo a process of gradual specialization called **cellular differentiation**.

### Module 3.2
### Cells are the smallest living units of life    p. 96

4. Cells are surrounded by **extracellular fluid** that in most cases is **interstitial fluid**.
5. The **cytoplasm** contains the fluid **cytosol** and the **organelles** suspended in the cytosol.
6. **Nonmembranous organelles** are not enclosed by membranes, and their components are in contact with the cytosol. They include the **cytoskeleton**, **microvilli**, **centrioles**, **cilia**, **flagella**, and **ribosomes**.
7. **Membranous organelles** are isolated from the cytosol by phospholipid membranes. They include the **mitochondria**, **nucleus**, **endoplasmic reticulum**, **Golgi apparatus**, **lysosomes**, and **peroxisomes**.

### Module 3.3
### The plasma membrane isolates the cell from its environment and performs varied functions    p. 98

8. The **plasma membrane** is a **phospholipid bilayer** that separates the inside of the cell from the surrounding extracellular fluid.
9. The general functions of the plasma membrane include physical isolation, regulation of exchange with the environment, sensitivity to the environment, and structural support.
10. Membrane proteins are functionally classified as **anchoring proteins**, **recognition proteins**, **receptor proteins**, **carrier proteins**, and **channels**.

*The plasma membrane is composed of two layers of phospholipid molecules with their hydrophilic heads at the membrane surface and their hydrophobic tails on the inside. The hydrophobic layer in the center of the membrane isolates the cytoplasm from the very different extracellular fluid.*

### Module 3.4
### The cytoskeleton plays both a structural and a functional role    p. 100

11. The **cytoskeleton** provides an internal protein framework that gives the cytoplasm strength and flexibility.
12. Components of the cytoskeleton are **microfilaments**, **intermediate filaments**, **thick filaments**, **microtubules**, **centrioles**, **cilia**. Flagella propel sperm.

### Module 3.5
### Ribosomes are responsible for protein synthesis and are often associated with the endoplasmic reticulum    p. 102

13. **Ribosomes** are organelles responsible for protein synthesis.
14. **Endoplasmic reticulum** (**ER**) is a network of intracellular membranes forming sheets or tubes called **cisternae**. **Smooth ER** is the site for the synthesis of phospholipids, cholesterol, steroid hormones, glycerides, and glycogen.

*The rough endoplasmic reticulum (RER) functions as a combination workshop for chemical modification of proteins and shipping warehouse to send them to their next destination.*

15. **Fixed ribosomes** are attached to **rough endoplasmic reticulum** (**RER**). Proteins are synthesized at the ribosomes and then modified and packaged in the RER.

### Module 3.6
### The Golgi apparatus is a packaging center    p. 104

16. The **Golgi apparatus**, or **Golgi complex**, consists of five or six flattened membranous discs called cisternae.
17. The Golgi apparatus functions to renew or modify the plasma membrane; modify and package secretions for release through exocytosis; and package special enzymes within vesicles for use in the cytosol.

### Module 3.7
### Mitochondria are the powerhouses of the cell    p. 106

18. **Mitochondria** have a double membrane and contain their own DNA. The inner membrane contains numerous folds called **cristae**.
19. Mitochondria are responsible for ATP production through **aerobic metabolism**.

# SECTION 2 • Structure and Function of the Nucleus

## Module 3.8

### The nucleus is the control center for cellular homeostasis p. 109

20. The nucleus contains DNA and determines the structure and function of the cell by controlling the synthesis of proteins.

## Module 3.9

### The nucleus contains DNA, RNA, organizing proteins, and enzymes   p. 110

21. The **nucleus** serves as the control center for cellular processes. It is surrounded by a **nuclear envelope** (a double membrane with a **perinuclear space**), through which it communicates with the cytosol by way of **nuclear pores**.

22. The nucleus contains a supportive **nucleoplasm** and one or more **nucleoli**, which synthesize ribosomal RNA.

23. The DNA in the nucleus stores the instructions for protein synthesis.

*The DNA in the nucleus is in coiled strands, allowing a great deal of DNA to be packaged in a relatively small space. The coils wrap around histone molecules to form nucleosomes.*

24. The DNA in a cell not dividing is loosely coiled forming a tangle of fine filaments known as **chromatin**. The DNA in dividing cells is coiled tightly forming a complex of distinct structures called **chromosomes**.

## Module 3.10

### Protein synthesis involves DNA, enzymes, and three types of RNA   p. 112

25. DNA consists of long, parallel chains of nucleotides that are held together by hydrogen bonding between complementary base pairs. The nitrogenous bases involved are adenine (A), thymine (T), cytosine (C), and guanine (G). The information stored in the sequence of base pairs is called the **genetic code**.

26. The genetic code is called a **triplet code** because a sequence of three nitrogenous bases, or **triplet**, specifies a single amino acid.

27. A **gene** is the functional unit of heredity. It contains all the DNA nucleotides needed to produce specific proteins.

28. As **gene activation** begins, a portion of the DNA molecule containing that gene uncoils.

29. A strand of **messenger RNA** (**mRNA**) is assembled that contains information in triplets, or **codons**.

30. **Transfer RNA** (**tRNA**) molecules bring amino acids to a ribosome. The amino acids form a polypeptide chain or protein.

## Module 3.11

### Transcription encodes genetic instructions on a strand of RNA   p. 114

31. **Transcription** is the synthesis of RNA from a DNA template. After transcription, a strand of mRNA carries instructions from the nucleus to the cytoplasm.

## Module 3.12

### Translation builds polypeptides as directed by an mRNA strand   p. 116

32. **Translation** is the formation of a linear chain of amino acids using the information provided by an mRNA strand. By complementary base pairing of tRNA **anticodons** to mRNA codons, amino acids are brought to the ribosome in the proper sequence. The three phases of translation are **initiation**, **elongation**, and **termination.**

# SECTION 3 • How Substances Enter and Leave the Cell

## Module 3.13

### The plasma membrane is a selectively permeable membrane   p. 119

33. The plasma membrane selectively regulates the movement of materials into and out of the cell. **Selectively permeable membranes** permit the free passage of some materials and restrict the passage of others.

34. Materials may cross the plasma membrane through active processes (require ATP) or passive processes (do not require ATP).

35. Vesicular transport is an active process. Diffusion is a passive process. Carrier-mediated processes may be either passive or active.

## Module 3.14

### Diffusion is passive movement driven by concentration differences   p. 120

36. **Diffusion** is the net movement of material from an area of higher concentration to an area of lower concentration. This occurs until the **concentration gradient** is eliminated.

37. Important factors that influence diffusion rates include distance, molecule and ion size, temperature, concentration gradient, and electrical forces.

*A plasma membrane acts as a barrier that selectively restricts diffusion. Some substances pass through easily, but others cannot penetrate the membrane.*

### Module 3.15

**Osmosis is the diffusion of water molecules across a selectively permeable membrane** p. 122

38. **Osmosis** is the net diffusion of water across a membrane. The movement of water driven by osmosis is called **osmotic flow**. The **osmotic pressure** of a solution is the force of water movement into that solution resulting from its solute concentration.

39. **Tonicity** is the effects of osmotic solutions on cells. A solution that does not cause an osmotic flow is **isotonic**. A solution that causes water to flow into a cell is **hypotonic** and can lead to **hemolysis**. A solution that causes water to flow out of a cell is **hypertonic** and can lead to **crenation**.

### Module 3.16

**In carrier-mediated transport, integral proteins facilitate membrane passage** p. 124

40. Substances that are insoluble in lipids and too large to fit through membrane channels may be transported across the plasma membrane by **carrier proteins**.

41. **Facilitated diffusion** is a passive process, **active transport** is an active process, and **secondary active transport** does not require ATP (although the cell may require ATP later to maintain homeostasis).

*Facilitated diffusion allows many substances to be passively transported across the plasma membrane.*

### Module 3.17

**In vesicular transport, vesicles selectively carry materials into or out of the cell** p. 126

42. In **vesicular transport**, materials move into or out of the cell in small membranous sacs called vesicles. The two major categories are **endocytosis** and **exocytosis**. Both require energy in the form of ATP.

43. In **receptor-mediated endocytosis**, specific target molecules are importend into a cell. **Pinocytosis** or "cell drinking" and **phagocytosis** or "cell eating" are other examples of endocytosis.

## SECTION 4 • Cell Life Cycle

### Module 3.18

**Interphase and cell division make up the life cycle of a cell** p. 129

44. **Cell division** of a single cell produces a pair of **daughter cells**.

45. **Interphase** is the period in which a cell is performing normal functions and not actively engaged in cell division.

46. **Apoptosis** is the genetically controlled death of cells. **Mitosis** is the nuclear division of somatic cells. Sex cells (sperm or oocytes) are produced by **meiosis**.

### Module 3.19

**During interphase, the cell prepares for cell division** p. 130

47. **Somatic cells** spend the majority of their functional lives in interphase, which includes the $G_1$, **S**, and $G_2$ **phases**.

### Module 3.20

**Mitosis distributes chromosomes before cytokinesis separates the daughter cells** p. 132

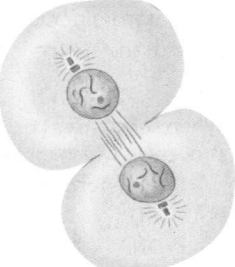

48. **Mitosis** specifically refers to the division and duplication of the cell's nucleus. It proceeds in four stages: **prophase**, **metaphase**, **anaphase**, and **telophase**.

49. Division of the cytoplasm into two distinct new cells involves a separate process known as **cytokinesis**.

### Module 3.21

**Tumors and cancer are characterized by abnormal cell growth and division** p. 134

50. A **tumor**, or **neoplasm**, can be **benign** or **malignant**.

51. Malignant cells may spread locally by **invasion**, or to distant tissues and organs through **metastasis**.

# Chapter Review Questions

**Multiple choice:** Select the correct answer from the list provided.

**1** Which of the following is *not* a characteristic of the cell theory?
- ☐ a) Cells are the building blocks of life.
- ☒ b) Cells combine to form tissues.
- ☐ c) All new cells come from the division of pre-existing cells.
- ☐ d) Cells are the smallest units that perform all vital physiological functions.

**2** The process of gradual cell specialization is called
- ☐ a) fertilization.
- ☐ b) meiosis.
- ☒ c) cellular differentiation.
- ☐ d) metastasis.

**3** The _____ is/are responsible for producing 95 percent of the ATP required by the cell.
- ☐ a) Golgi apparatus
- ☒ b) mitochondria
- ☐ c) rough endoplasmic reticulum
- ☐ d) smooth endoplasmic reticulum

**4** Somatic cell nuclei contain _____ pairs of chromosomes.
- ☐ a) 8
- ☒ b) 23
- ☐ c) 46
- ☐ d) 92

**3**

**5** The construction of a functional polypeptide by using the information in an mRNA strand is called
☒ a) translation.
☐ b) transcription.
☐ c) replication.
☐ d) differentiation.

**6** Genetically controlled cell death is called
☐ a) mitosis.
☒ b) apoptosis.
☐ c) metastasis.
☐ d) mutation.

**7** The _____ phase of protein synthesis encodes genetic instructions on a strand of mRNA.
☒ a) transcription
☐ b) translation
☐ c) transmigration
☐ d) transference

**8** When substances pass through the plasma membrane by active processes, which molecule is required?
☐ a) RNA polymerase
☒ b) ATP
☐ c) DNA
☐ d) tRNA

**9** The sodium–potassium exchange pump
☐ a) is an example of facilitated diffusion.
☐ b) does not require the input of cellular energy in the form of ATP.
☐ c) moves the sodium and potassium ions along their concentration gradients.
☒ d) is composed of a carrier protein located in the plasma membrane.

**10** If a cell lacked ribosomes, it would not be able to
☐ a) produce complex carbohydrates.
☒ b) synthesize proteins.
☐ c) divide.
☐ d) produce ATP.

**11** Suppose a DNA segment has the following nucleotide sequence: CTC/ATA/CGA/TTC/AAG/TTA. Which nucleotide sequence would a complementary mRNA strand have?
☐ a) GAG/UAU/GAU/AAC/UUG/AAU
☐ b) GAG/TAT/GCT/AAG/TTC/AAT
☒ c) GAG/UAU/GCU/AAG/UUC/AAU
☐ d) GUG/UAU/GGA/UUG/AAG/GGU

**12** How many amino acids are coded in the DNA segment in the previous question?
☐ a) 18
☐ b) 9
☒ c) 6
☐ d) 3

**Short answer**

**13** Identify the five functional classes of membrane proteins.

**14** Distinguish between mitosis and cytokinesis.

**15** Differentiate between diffusion and osmosis.

**16** Which organelle contains its own DNA, and what is its function?

**17** List the nonmembranous organelles, and then list the membranous organelles.

**18** List the stages of mitosis, and for each stage briefly describe the events that occur.

**19** If a cell has microvilli on its plasma membrane, in which activity is it likely to be actively engaged?

**20** If a red blood cell were immersed in a hypotonic solution, in which direction would water flow, and what effect would it have on the cell?

**21** Order the following steps of protein synthesis into the proper sequence:
a) mRNA exits nucleus through nuclear pore.
b) Introns are snipped from mRNA.
c) Amino acids form peptide bonds.
d) Ribosomal subunits detach from mRNA.
e) Transcription of DNA forms mRNA.
f) tRNA anticodons bind to mRNA codons.
g) Ribosomal subunits bind to mRNA.

**22** Describe the steps involved in phagocytosis.

**23** Why does mitosis produce cells containing 46 chromosomes, whereas meiosis produces cells containing only 23 chromosomes?

**24** Steroid hormones such as estrogen and testosterone are lipid molecules. Explain how these molecules cross the plasma membrane and enter the cell.

**25** Describe the process used by malignant tumors to accelerate their growth.

3

## Chapter Integration • Applying what you have learned

### Analyzing a stomach tumor

Vito has been experiencing persistent indigestion, heartburn, and loss of appetite. After many weeks of this discomfort, he finally schedules an appointment with his physician. The doctor performs a diagnostic procedure in which he passes a tube through Vito's mouth into his stomach. Through this tube the doctor can use a scope to view the tissue lining Vito's stomach. Upon examination, the doctor notices a tumor, or mass of abnormal cells, on the stomach wall. The doctor excises a small clump of tissue and sends it to the pathology laboratory for analysis. From what you have just learned about cells, answer the following questions about Vito's condition.

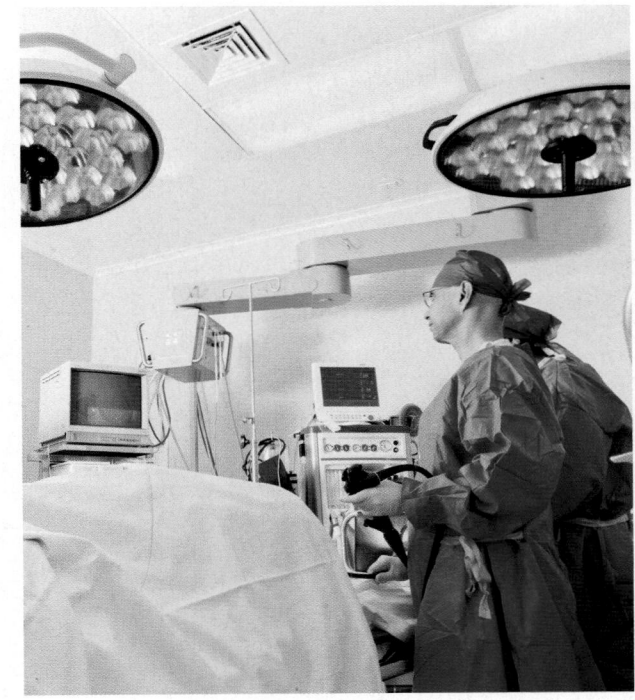

**1**   Which word describing the tumor do Vito and his doctor hope to read in the lab report? Why? Which word do they *not* want to read in the report? Why?

**2**   Explain the cellular processes that likely caused the tumor.

**3**   Besides the obvious threat to Vito's stomach, does this tumor pose a threat to other organs?

**4**   Which cellular process would Vito and his doctor want to switch on in these tumor cells if they were able to do so?

### Exercising muscle

Anna is a bodybuilder, certified personal trainer, and owner of her own fitness center. Five days per week she lifts weights for 3 hours per day. When working as a trainer, she often cites the benefits of electrolyte-laden sports drinks, claiming that they are good for fatigued cells.

**5**   Anna tells her clients that the muscle cells of people training for endurance sports, such as the marathon, have the potential for generating more energy molecules and improving muscle performance because those cells contain increased numbers of mitochondria. Is this true? Why or why not?

**6**   Using the principles of tonicity, explain the benefits of consuming energy drinks after strenuous exercise.

## LEARNING OUTCOMES

These Learning Outcomes correspond by number to this chapter's modules and indicate what you should be able to do after completing the chapter. They also appear at the end of each module.

# 4

# Tissue Level of Organization

## Module 4.1
# Four types of tissue make up the body

Our perspective has gradually changed over the preceding chapters. Atoms and molecules can only be examined through special imaging techniques and experimental procedures. Cellular details often escape detection unless an electron microscope is used. Tissue structure, however, can be examined with a light microscope, and based on your experiences in the laboratory, you may already be able to identify some tissues without using a microscope.

4

**1** The human body contains trillions of cells, but only about 200 different *types* of cells. To work efficiently, several different types of cells must coordinate their efforts. Cells working together form **tissues**—collections of similar cells and cell products that perform a limited number of specialized functions. The study of tissues is called **histology**. Histologists recognize four basic types of tissue: **epithelial tissue**, **connective tissue**, **muscle tissue**, and **nervous tissue**. This chapter examines each of these tissue types and sets the stage for our study of each organ system and their various organs.

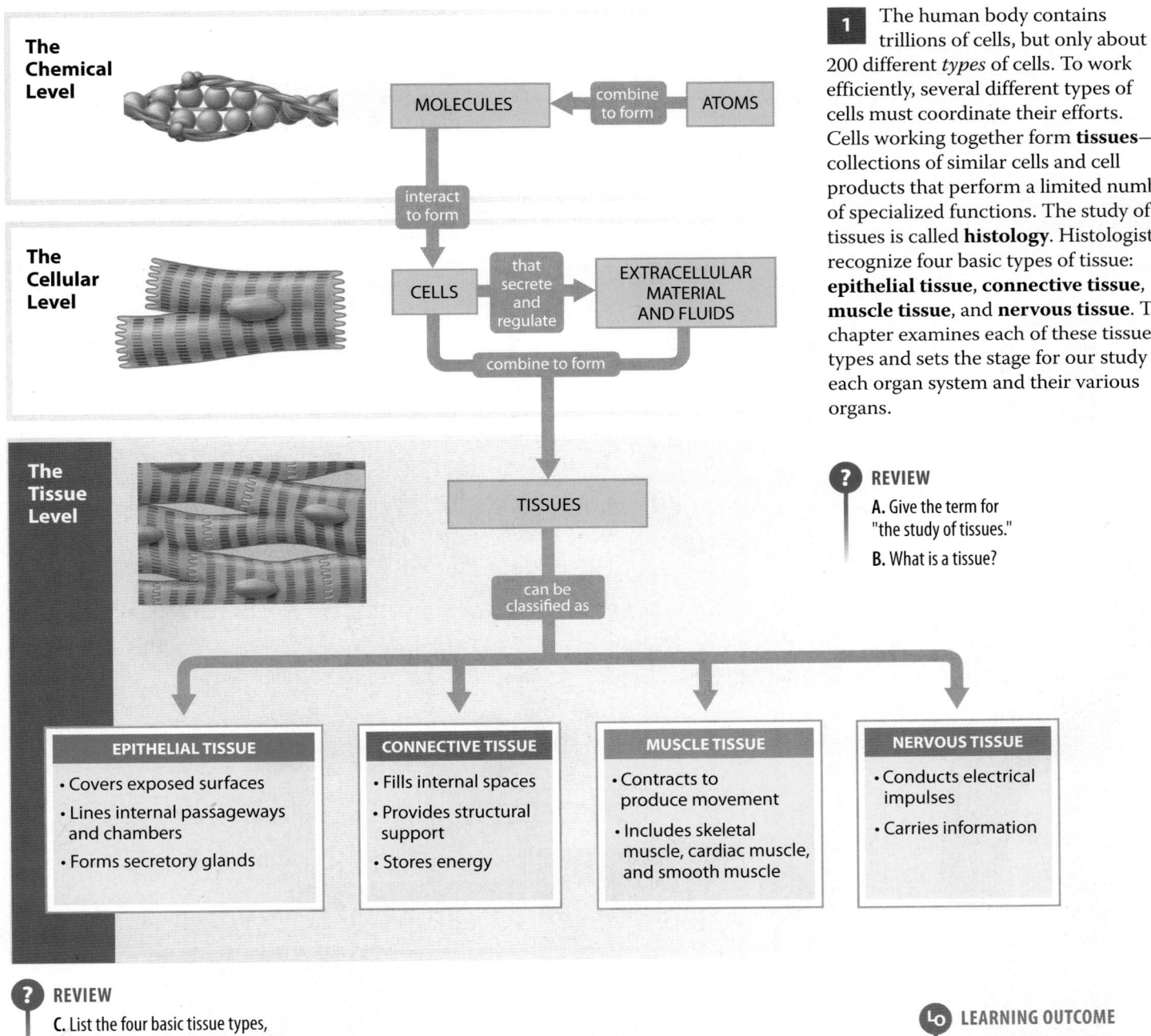

**The Chemical Level**

MOLECULES ← combine to form ← ATOMS

interact to form

**The Cellular Level**

CELLS → that secrete and regulate → EXTRACELLULAR MATERIAL AND FLUIDS

combine to form

**The Tissue Level**

TISSUES

can be classified as

| EPITHELIAL TISSUE | CONNECTIVE TISSUE | MUSCLE TISSUE | NERVOUS TISSUE |
|---|---|---|---|
| • Covers exposed surfaces<br>• Lines internal passageways and chambers<br>• Forms secretory glands | • Fills internal spaces<br>• Provides structural support<br>• Stores energy | • Contracts to produce movement<br>• Includes skeletal muscle, cardiac muscle, and smooth muscle | • Conducts electrical impulses<br>• Carries information |

**?** REVIEW
**A.** Give the term for "the study of tissues."
**B.** What is a tissue?

**?** REVIEW
**C.** List the four basic tissue types, and describe the functions of each.

**Lo** LEARNING OUTCOME
Identify the four types of tissue in the body, and describe their roles.

# Module 4.2
# Microscopes are used to study cells and tissues

Much of what is studied in anatomy and physiology cannot be seen with the naked eye or a simple hand lens. **Microscopy** (the use of microscopes) began about 400 years ago with the inventions of **simple microscopes** with one lens and **compound microscopes** with more than one lens. However, these early light microscopes magnified objects only 10 to 20 times actual size. The light microscope in your lab likely magnifies up to 1000 times, the maximum when using light. Modern electron microscopes using beams of electrons instead of light can magnify over 1 million times!

**?** **REVIEW**
**A.** How do early microscopes compare with modern microscopes?

**1** Anatomy can be studied at different scales. The amount of fine detail that can be seen in an image, called resolution, varies with magnification and the type of microscope being used. The figure below shows the magnification ranges and degree of resolution achieved by different types of microscopes.

| | meters (m) | | | millimeters (mm) | | | micrometers (µm) | | | nanometers (nm) | | |
|---|---|---|---|---|---|---|---|---|---|---|---|---|
| **Size** | 1 m | 100 mm | 10 mm | 1 mm | 100 µm | 10 µm | 1 µm | 100 nm | 10 nm | 1 nm |
| **Approximate Magnification** | | | | x 20 | x 100 | x 1000 | x 10,000 ($10^4$) | x 100,000 ($10^5$) | x $10^6$ | x $10^7$ |

Human heart • Fingertip (width) • Large protozoan • Human oocyte • Red blood cell • Bacteria • Mitochondrion • Viruses • Ribosomes • Proteins • DNA (diameter) • Amino acids

Unaided human eye

Compound light microscope

Scanning electron microscope

Transmission electron microscope

**2** A **compound light microscope** is an optical instrument that directs visible light through a thin section of tissue. The specimen is first magnified with an objective lens and focused in the tube of the microscope. A **revolving nosepiece** holds several objective lenses of progressive magnifying power. A second lens in the eyepiece, the ocular lens, magnifies the image further. The **total magnification** is calculated by multiplying the magnification of the objective lens times that of the ocular lens. The **resolution** of a microscope is the ability to distinguish between two separate points, or objects. The wavelengths of visible light limit the maximum resolution of a light microscope to about 200 nm (0.2 μm). Thus, objects closer together than 200 nm would appear as one object.

**3** A **transmission electron microscope** uses magnets to direct a beam of electrons through the surface of a finely sectioned object onto a photographic plate. The wavelength of the electron beam is about 0.00001 that of ordinary white light, which allows for far greater magnification and resolution. A transmission electron microscope has a maximum resolution of about 0.2 nm (0.0002 μm).

**4** The specimens prepared for **scanning electron microscopy** are not sectioned. The whole specimen is coated with an electron-dense material and then bombarded with electron beams. Because some of the electrons are reflected back, a three-dimensional image of the specimen is created. A scanning electron microscope has a maximum resolution of about 10 nm (0.01 μm).

**5** You will probably be using a compound light microscope during your laboratory experiences. Begin by using the objective lens with the lowest magnifying power to focus on your specimen. Then carefully rotate objective lenses with greater magnification into position. For each tissue you examine, open your text to an image of the same tissue. The images will look most similar if you focus your microscope close to the magnification of the image in the book.

Cilia

**Epithelium (respiratory tract)**    LM × 400

Cilia

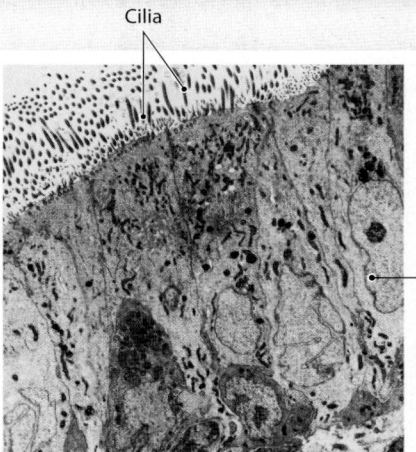

**Epithelium (respiratory tract)**    TEM × 3000

Cilia

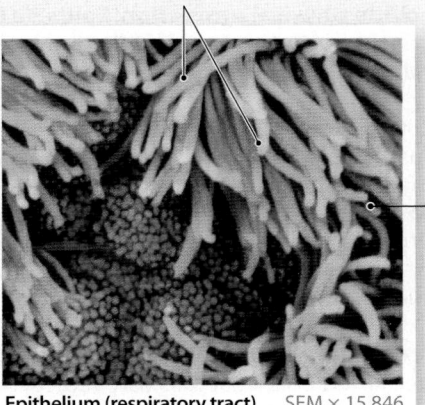

**Epithelium (respiratory tract)**    SEM × 15,846

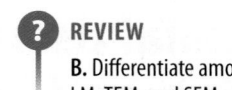

**? REVIEW**

**B.** Differentiate among LM, TEM, and SEM.

This photographic image of cells from the epithelial lining of the respiratory tract was taken with a light microscope and is called a light micrograph (LM).

Every micrograph in this book has a label indicating its type and magnification.

This transmission electron microscope image is of cells from the epithelial lining of the respiratory tract. The cell and organelle details are much sharper. This type of image is called a transmission electron micrograph (TEM).

This dramatic image of the surface of the epithelial lining of the respiratory tract was taken with a scanning electron microscope and then color enhanced. The type of image is called a scanning electron micrograph (SEM).

**? REVIEW**

**C.** The LM at the top of this page is magnified 400 times (400×). If the ocular lens used to make this image has a magnification of 10×, what is the magnification of the objective lens?

**Lo LEARNING OUTCOME**

Describe three microscopy techniques.

# Module 4.3

# Epithelial tissue covers surfaces, lines cavities, and forms secretory glands

**4**

**1** Let's begin our discussion of tissue types with **epithelial tissue** because it includes a very familiar feature: the surface of your skin. Epithelial tissue includes *epithelia* and *glands*.

## EPITHELIAL TISSUE

Includes

### Epithelia

**Epithelia** are cellular layers of different types that cover exposed surfaces and line internal cavities and passageways. They do not have blood vessels and often contain secretory cells, or gland cells, scattered among the other cell types.

### Glands

**Glands** are organized groups of cells or organs that contain epithelial-derived cells that synthesize substances for secretion. There are two types:

### Exocrine Glands

**Exocrine glands** secrete onto external surfaces or into internal passageways (ducts) that connect to the exterior.

### Endocrine Glands

**Endocrine glands** secrete hormones (chemical messengers) or their inactive precursors (prohormones) into the interstitial fluid that then enter the bloodstream for distribution.

**2** The table at right describes four essential functions of epithelial tissue.

## Functions of Epithelial Tissue

- **Provide physical protection**: Epithelia protect exposed and internal surfaces from abrasion, dehydration, and destruction by chemical or biological agents.

- **Control permeability**: Any substance that enters or leaves the body must cross an epithelium. Some epithelia are relatively impermeable, whereas others are permeable to compounds as large as proteins. Most are capable of selective absorption or secretion. The epithelial barrier can be regulated and modified in response to various stimuli. For example, think of the calluses that form on your hands when you do rough work for a period of time.

- **Provide sensation**: Sensory nerves extensively innervate most epithelia. Specialized epithelial cells can detect changes in the environment and convey information about such changes to the nervous system. For example, touch receptors respond to pressure by stimulating adjacent sensory nerves. A **neuroepithelium** is a sensory epithelium found in special sense organs that provide the sensations of smell, taste, sight, equilibrium, and hearing.

- **Produce specialized secretions**: Epithelial cells that produce secretions are called gland cells. Individual gland cells are often scattered among other cell types in an epithelium that may have many other functions. In a glandular epithelium, most or all of the epithelial cells produce secretions, and those secretions are the primary function of the tissue.

**?** **REVIEW**

**A.** List four essential functions of epithelial tissue.

**3** The cells of any epithelium share a number of basic features. An epithelium has an **apical** (Ā-pi-kal) **surface**, which faces the exterior of the body or some internal space, and a **base**, which is attached to adjacent tissues. The term **polarity** refers to the presence of structural and functional differences between the exposed and attached surfaces.

Microvilli are often found on the apical surfaces of epithelial cells that line internal passageways of the digestive, urinary, and reproductive tracts.

The apical surface is the region of the cell exposed to an internal or external environment. When the epithelium lines a tube, such as the intestinal tract, the apical surfaces of the epithelial cells are exposed to the space inside the tube, a passageway called the **lumen** (LOO-men).

The **basolateral surfaces** include both the base, where the cell attaches to underlying epithelial cells or deeper tissues, and the sides, where the cell contacts its neighbors.

Motile cilia cover the apical surfaces of cells in portions of the respiratory and reproductive tracts. A typical ciliated cell has about 250 cilia that beat in a coordinated manner (see Module 3.4).

### Membranous Organelles

Most epithelial cells have membranous organelles comparable to those of other cell types (see Module 3.2).

Golgi apparatus (facing apical surface)

Endoplasmic reticulum (often extensive around the nucleus)

Nucleus (in a tall cell, located closer to the base than the apical surface)

Mitochondria (may be apical or basal, depending on cell functions)

**4** Epithelial cells have three basic shapes: **squamous**, **cuboidal**, and **columnar**. For classification purposes, we look at the superficial cells in a section perpendicular to both the exposed surface and the basal surface. In sectional view, squamous cells appear thin and flat, cuboidal cells look like little boxes, and columnar cells look like tall, relatively slender rectangles. If only one layer of cells is present, that layer is a **simple epithelium**. In contrast, a **stratified epithelium** contains several layers of cells. Stratified epithelia are generally located in areas that need protection from physical or chemical stresses. The surface of the skin and the lining of the mouth are two examples.

**? REVIEW**
**B.** Summarize the classification of an epithelium based on cell shape and number of cell layers.

**Simple epithelia**

Squamous          Cuboidal          Columnar

**Stratified epithelia**

Squamous          Cuboidal          Columnar

**? INTEGRATION**
**C.** What function is served by motile cilia on epithelial cell surfaces?

**Lo LEARNING OUTCOME**
Describe epithelial tissues, including cell shape, layers, and functions.

# Module 4.4

## Epithelial cells are extensively interconnected, both structurally and functionally

To be effective as a barrier, an epithelium must form a complete cover or lining and be able to replace lost or damaged cells through stem cell division. The physical integrity of an epithelium depends on intercellular connections and attachment to adjacent tissues.

**1** The detailed structure of each form of intercellular connection, or attachment, demonstrates the linkage between structure and function at all levels.

Microvilli

APICAL SURFACE

### Intercellular Connections

**Tight** (occluding) **junctions** are characteristic of epithelial cells lining the intestinal tract, where they form a barrier that isolates the basolateral surfaces and deeper tissues from the contents of the lumen.

An **adhesion belt** locks together the terminal webs of neighboring cells, strengthening the apical region and preventing distortion and leakage at the tight junctions.

**Gap junctions** permit chemical communication (diffusion of ions and small molecules) that coordinates the activities of adjacent cells.

**Desmosomes** (DEZ-mō-sōms; *desmos*, ligament + *soma*, body) provide firm attachment between neighboring cells by interlocking their cytoskeletons.

BASE

**2** **Hemidesmosomes** attach the deepest epithelial cells to the basement membrane. At a hemidesmosome, the basal cytoskeleton is locked to peripheral proteins and to transmembrane proteins that are firmly attached to a layer of extracellular protein filaments and fibers.

Intermediate filaments of the cytoskeleton

Hemidesmosome

### Basement Membrane

The **basement membrane** is a noncellular structure produced by the basal surface of the epithelium and the underlying connective tissue.

The **basal lamina** is an amorphous layer that contains glycoproteins. It is produced by the adjacent layer of epithelial cells.

The **reticular lamina** contains reticular protein fibers and ground substance secreted by underlying connective tissue cells (discussed further in Module 4.11). It gives the basement membrane its strength and acts as a filter that restricts diffusion between adjacent tissues and the epithelium.

**3** At a tight junction, the attachment is so tight that it prevents water and solutes from passing between cells.

 **REVIEW**
**A.** Identify the various types of epithelial intercellular connections.

At a tight junction, the lipid portions of the two plasma membranes are tightly bound together by interlocking membrane proteins.

**4** A continuous adhesion belt forms a band that encircles each cell and binds it to its neighbors. The bands are dense proteins that are attached to the microfilaments of the terminal web.

**5** At a gap junction, two cells are held together by interlocking transmembrane proteins called **connexons**. Gap junctions between epithelial cells are common where the movement of ions helps coordinate functions such as secretion or cilia movement. Gap junctions are also common in other tissues. For example, gap junctions in cardiac muscle tissue and smooth muscle tissue are essential in coordinating muscle cell contractions.

Connexons are channel proteins that form a narrow passageway and let small molecules and ions pass from cell to cell. Each connexon is composed of six connexin proteins that form a cylinder with a central channel. A number of human diseases are caused by mutations in the genes that code for different connexin proteins.

**REVIEW**
**B.** What is the functional significance of gap junctions?

**6** At a desmosome, the opposing plasma membranes are locked together. Desmosomes are very strong and durable and help epithelial cells resist mechanical stresses such as stretching and twisting.

**Cell adhesion molecules (CAMs)** are transmembrane proteins that bind to each other and to extracellular materials. CAMs on the basolateral surface of an epithelium help bind the cell to the underlying basement membrane.

The membranes of adjacent cells may also be bonded by a thin layer of proteoglycans that contain polysaccharide derivatives, most notably **hyaluronic acid**.

🏃 **Everyday Physiology**

### Epithelia lack blood vessels

Because epithelia lack blood vessels, they are said to be **avascular** (*a*, without + *vas*, vessel). Epithelial cells must obtain nutrients by diffusion or absorption across either the exposed or the attached epithelial surface. The cells forming the deepest layer of an epithelium must remain firmly attached to underlying connective tissues because the blood vessels in those tissues nourish the entire epithelium.

 **REVIEW**
**C.** How do epithelial tissues obtain needed nutrients?

 **INTEGRATION**
**D.** What two types of tissues contribute to the formation and maintenance of the basement membrane?

🔵 **LEARNING OUTCOME**
Discuss the types and functions of intercellular connections between epithelial cells.

# Module 4.5

## The cells in a squamous epithelium are flat and irregularly shaped

The cells in a **squamous epithelium** (SKWĀ-mus; *squama*, plate or scale) are thin, flat, and somewhat irregularly shaped, like pieces of a jigsaw puzzle. From the surface, the cells resemble fried eggs laid side by side. In sectional view, the disc-shaped nucleus occupies the thickest portion of each cell.

**1** A **simple squamous epithelium** is the body's most delicate type of epithelium. This type of epithelium is located in protected regions where absorption or diffusion takes place or where a slick, slippery surface reduces friction. Simple squamous epithelia are found along passageways in the kidneys, inside the eye, and at the gas exchange surfaces (alveoli) of the lungs. Simple squamous epithelia in certain locations have special names. A **mesothelium** lines the pericardial, pleural, and peritoneal body cavities. The simple squamous epithelium lining the inner surface of the heart and all blood vessels is called an **endothelium**.

**? REVIEW**

**A.** What do a mesothelium and an endothelium have in common?

An endothelium lines the inside of the heart and the blood vessels.

A mesothelium lines the pericardial cavity and the peritoneal cavity.

Connective tissue

Sectioned epithelial cell

The surface of a simple squamous epithelium

**Surface and sectional view of the peritoneum**

Nucleus

Cytoplasm

Nuclei in supporting underlying connective tissue

**Simple squamous epithelia**        LM × 270

**2** A **stratified squamous epithelium** is generally located where physical or chemical stresses are severe. The cells form a series of layers, like the layers in a sheet of plywood. Stratified squamous epithelia form the surface of the skin and line the mouth, throat, esophagus, rectum, anus, and vagina.

**? REVIEW**

**B.** Why do the pharynx, esophagus, anus, and vagina have a similar epithelial organization?

Squamous superficial cells

Stem cells

Basement membrane

Connective tissue

**Surface of the tongue**

**Stratified squamous epithelium** LM × 400

**3** On exposed body surfaces, where physical stress and dehydration are potential problems, apical layers of epithelial cells are packed with filaments of the protein keratin. As a result, superficial layers are both tough and water resistant. Such an epithelium is said to be **keratinized**. A **nonkeratinized** stratified squamous epithelium resists abrasion but will dry out and deteriorate unless kept moist. Nonkeratinized stratified squamous epithelia are found in the mouth, pharynx, esophagus, anus, and vagina.

**? REVIEW**

**C.** What properties are common to keratinized epithelia?

Keratin fibers

Keratinized skin cells

Surface of human skin

Keratin is a fibrous protein that consists of two alpha-helical polypeptides wound together like the strands of a rope. It is a tough, water-resistant protein at the surface of the skin and in nails and hair.

**? INTEGRATION**

**D.** Under a light microscope, a tissue appears as a simple squamous epithelium. Can this be a sample of the skin surface? Why or why not?

**LO LEARNING OUTCOME**

Describe the structure and function of squamous epithelium.

# Module 4.6

# Cuboidal and transitional epithelia line several passageways and chambers connected to the exterior

## Cuboidal Epithelium

**1** The cells of a **cuboidal epithelium** resemble hexagonal boxes. (In sectional view, they appear square.) The spherical nuclei are near the center of each cell, and the distance between adjacent nuclei is roughly equal to the height of the epithelium.

Connective tissue

Basement membrane

Simple cuboidal cells

Lumen of duct

Nucleus

**Simple cuboidal epithelium** LM × 650

**2** A **simple cuboidal epithelium** provides limited protection and occurs where secretion or absorption takes place. Such an epithelium lines portions of the kidney tubules. They also line secretory chambers in the thyroid gland.

**? REVIEW**

**A.** Describe the appearance of simple cuboidal epithelial cells in sectional view.

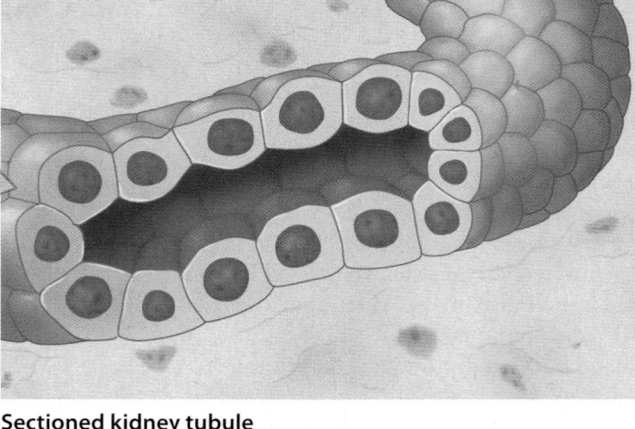

**Sectioned kidney tubule**

**3** **Stratified cuboidal epithelia** are relatively rare. They are located along the ducts of sweat glands and in the larger ducts of the mammary glands.

Lumen of duct

Stratified cuboidal cell

Basement membrane

Nucleus

Connective tissue

**Sweat gland duct**

**Stratified cuboidal epithelium** LM × 500

# Transitional Epithelium

**4** A **transitional epithelium** is an unusual stratified epithelium. Unlike most epithelia, it tolerates repeated cycles of stretching without damage. It is called *transitional* because the appearance of the epithelium changes as stretching occurs. Transitional epithelia line the urinary bladder, the ureters, and the urine-collecting chambers within the kidneys, where large changes in volume occur.

**REVIEW**

**B.** Identify the epithelium that lines the urinary bladder, and describe its unusual functional characteristic.

## Transitional Epithelium in an Empty Bladder

In an empty urinary bladder, the lining is not stretched to its limit. The superficial cells are plump and cuboidal with a dome-shaped surface.

Empty bladder

Epithelium (not stretched)

Basement membrane

Connective tissue and smooth muscle layers

LM × 400

## Transitional Epithelium in a Full Bladder

When the urinary bladder is full, the volume of urine has stretched the lining to its limits. The epithelium appears flattened and more like a stratified squamous epithelium.

Stretched bladder

Epithelium (stretched)

Basement membrane

Connective tissue and smooth muscle layers

LM × 400

**REVIEW**

**C.** Describe the changes in appearance of the transitional epithelium lining the urinary bladder as stretching occurs.

**INTEGRATION**

**D.** What functions are associated with a simple cuboidal epithelium and a transitional epithelium?

**LO** **LEARNING OUTCOME**

Describe the structure, function, and locations of cuboidal and transitional epithelia.

# Module 4.7

## Columnar epithelia absorb substances and protect the body from digestive chemicals

In a typical sectional view, the cells of a **columnar epithelium** appear rectangular. In reality, the densely packed cells are hexagonal, but they are taller and more slender than cells in a cuboidal epithelium. The elongated nuclei are crowded into a narrow band close to the basement membrane. The height of the epithelium is several times the distance between adjacent nuclei.

**1** A **simple columnar epithelium** is typically found where absorption or secretion takes place, as in the small intestine. These epithelia also line the stomach, gallbladder, uterine tubes, and ducts within the kidneys. These cells may have microvilli, which increase surface area for absorption, or motile cilia that move substances across the apical surface. In the stomach and intestinal tract, simple columnar epithelia secretions protect underlying tissues against chemicals involved with digestion.

**?** **REVIEW**

**A.** Describe the appearance of simple columnar epithelial cells in a sectional view.

Microvilli

Cytoplasm

Nucleus

Basement membrane

Connective tissue

**Intestinal lining**

**Simple columnar epithelium**  LM × 350

**2** A **pseudostratified columnar epithelium** includes several types of cells with varying shapes and functions. The distances between the cell nuclei and the exposed surface vary, so the epithelium appears to be layered, or stratified. It is not truly stratified, however, because every epithelial cell contacts the basement membrane. Pseudostratified columnar epithelial cells typically have motile cilia. They line the nasal cavities, the trachea (windpipe), and larger airways of the lungs and are also found along portions of the male reproductive tract.

 **REVIEW**

**B.** Explain why a pseudostratified columnar epithelium is not truly stratified.

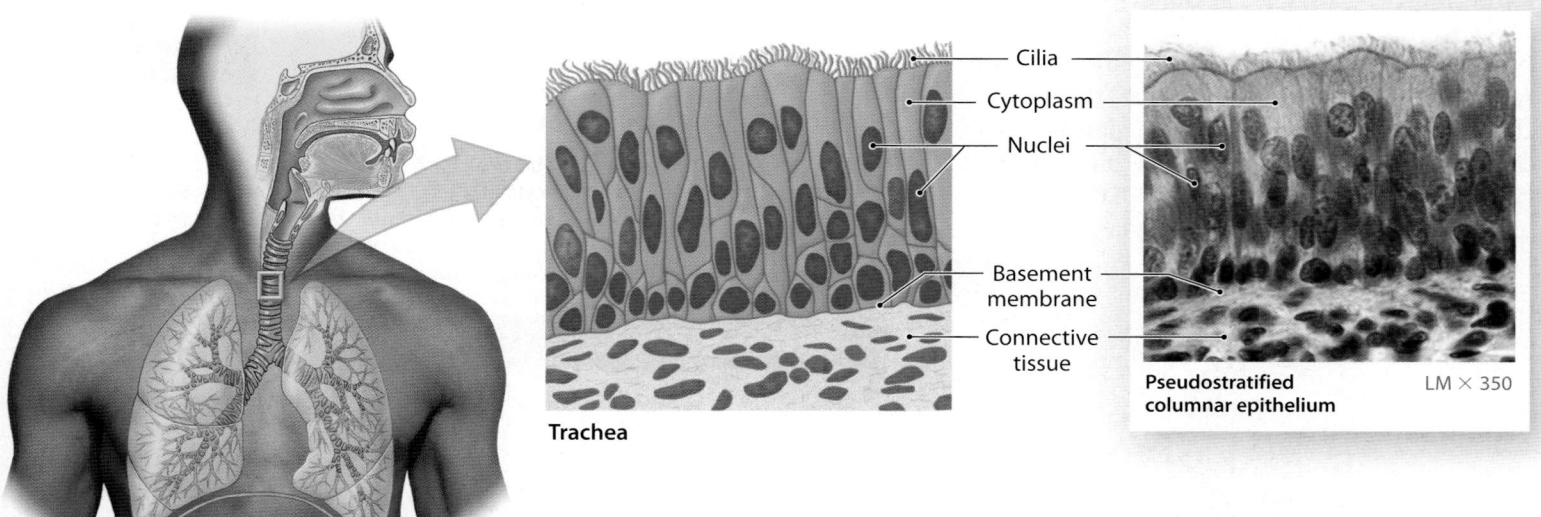

Trachea

Cilia
Cytoplasm
Nuclei
Basement membrane
Connective tissue

**Pseudostratified columnar epithelium** LM × 350

**3** **Stratified columnar epithelia** are not widely distributed in the body. These epithelia may have two or more layers. In the latter case, only the superficial cells are columnar in shape. Stratified columnar epithelia are most often found lining large ducts such as those of the salivary glands or pancreas.

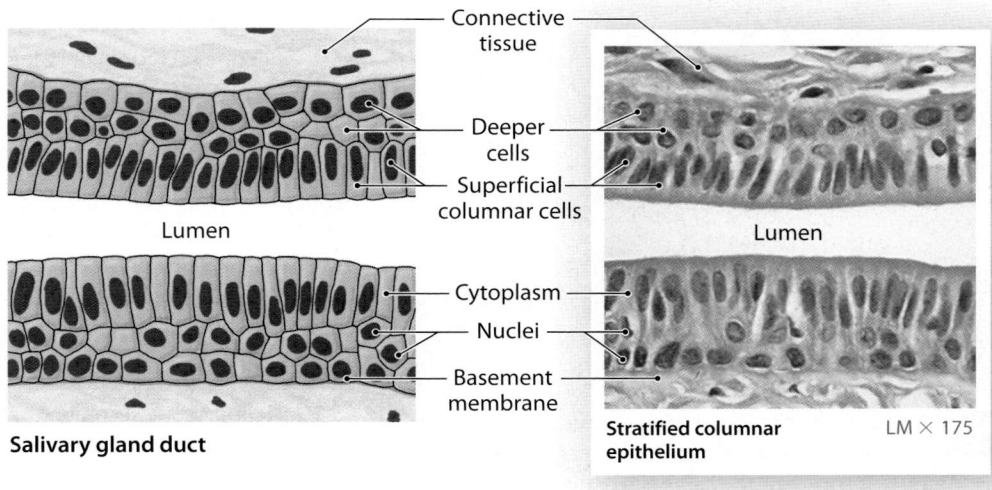

Connective tissue
Deeper cells
Superficial columnar cells
Lumen
Cytoplasm
Nuclei
Basement membrane

Lumen

**Salivary gland duct**

**Stratified columnar epithelium** LM × 175

**? INTEGRATION**

**C.** Describe the structures found on the surfaces of simple columnar and pseudostratified columnar epithelia.

**Lo LEARNING OUTCOME**

Describe the structure, function, and locations of columnar epithelia.

# Module 4.8
# Glandular epithelia are specialized for secretion

Many epithelia contain gland cells that are specialized for secretion. Collections of epithelial cells (or structures derived from epithelial cells) that produce secretions are called **glands**. They range from scattered cells to complex glandular organs. Some of these glands, called **endocrine glands**, release their secretions into the interstitial fluid. Others, known as **exocrine glands**, release their secretions into passageways called **ducts** that open onto an epithelial surface. We consider exocrine glands here. Endocrine glands are discussed in Chapter 16.

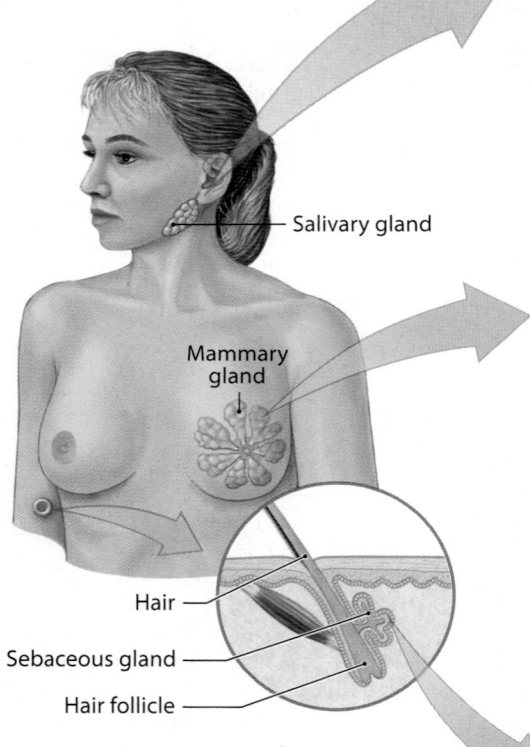

Salivary gland

Mammary gland

Hair

Sebaceous gland

Hair follicle

**1** Secretory cells of exocrine glands release their secretions into ducts using one of three methods: merocrine, apocrine, or holocrine secretion.

**? REVIEW**

**A.** Describe the two primary types of glands.

**B.** By what three methods do various secretory cells of exocrine glands release their secretions?

## Merocrine secretion

Mucin

Secretory vesicle (containing mucin)

Golgi apparatus

Nucleus

TEM × 3120

In **merocrine secretion** (MER-u-krin; *meros*, part), the product is released from secretory vesicles by exocytosis. This is the most common mode of exocrine secretion. **Mucin** is a merocrine secretion that mixes with water to form mucus. **Mucus** is an effective lubricant, a protective barrier, and a sticky trap for foreign particles and microorganisms.

## Apocrine secretion

Cytoplasm breaks down

Regrowth

Secretion

Golgi apparatus

**1** **2** **3** **4**

**Apocrine secretion** (AP-ō-krin; *apo-*, off) involves the loss of cytoplasm as well as the secretory product. The apical portion of the cytoplasm becomes packed with secretory vesicles and is then shed. Milk production in the mammary glands involves a combination of merocrine and apocrine secretions.

## Holocrine secretion

**3** Cells burst, releasing cytoplasmic contents.

**2** Cells produce secretion, increasing in size.

**1** Cell division replaces lost cells.

Stem cell

**Holocrine secretion** (HOL-ō-krin; *holos*, entire) destroys the gland cell. The entire cell becomes packed with secretory products and then bursts, releasing the secretion and killing the cell. Further secretion depends on replacement of destroyed cells through stem cell division. Sebaceous glands, associated with hair follicles, produce an oily hair coating by means of holocrine secretion.

**Lo LEARNING OUTCOME**

Compare the three different methods of exocrine secretion by glandular epithelia.

# Module 4.9
# Exocrine glands can be classified by structure

Three characteristics are used to describe the structure of **multi-cellular exocrine glands**: the structure of the duct, the shape of the secretory area of the gland (tubular or sac-like pockets), and the relationship between the duct (simple or compound) and the secretory areas (branched or not).

**1** A **simple gland** has a single duct that does not divide on its way to the gland cells.

In a **branched gland**, several secretory areas (tubular or acinar) share a duct. Note that "branched" refers to the glandular areas, not to the duct.

| SIMPLE TUBULAR | SIMPLE COILED TUBULAR | SIMPLE BRANCHED TUBULAR | SIMPLE ALVEOLAR (ACINAR) | SIMPLE BRANCHED ALVEOLAR |
|---|---|---|---|---|
| Examples:<br>• Intestinal glands | Examples:<br>• Merocrine sweat glands | Examples:<br>• Gastric glands<br>• Mucous glands of esophagus, tongue, duodenum | Examples:<br>• A stage in the embryonic development of simple branched glands; not found in adults | Examples:<br>• Sebaceous (oil) glands |

Glands whose glandular cells form tubes are **tubular**; the tubes may be straight or coiled.

Glands whose glandular cells form sac-like pockets are **alveolar** (al-VĒ-ō-lar; *alveolus*, sac) or **acinar** (AS-i-nar; *acinus*, chamber).

**2** In a **compound gland**, the duct divides one or more times on its way to the gland cells.

Glands whose secretory cells form both tubes and sacs are called **tubuloalveolar**.

| COMPOUND TUBULAR | COMPOUND ALVEOLAR (ACINAR) | COMPOUND TUBULOALVEOLAR |
|---|---|---|
| Examples:<br>• Mucous glands in mouth<br>• Bulbo-urethral glands in male reproductive system<br>• Seminiferous tubules of testes | Examples:<br>• Mammary glands | Examples:<br>• Salivary glands<br>• Glands of respiratory passages<br>• Pancreas |

**3** The only **unicellular exocrine glands** in the body are **goblet cells**. Goblet cells and mucous cells secrete mucin that hydrates to form mucus. Mucus protects the epithelial linings. Goblet cells are scattered among the absorptive cells in the epithelium of intestines. Mucous cells are scattered throughout other epithelial tissues.

Mucin

Golgi apparatus

Nucleus

Goblet cell

**? REVIEW**
**A.** What three characteristics are used to describe multicellular exocrine glands?

**? INTEGRATION**
**B.** Describe the simplest type of multicellular exocrine gland.

**LO LEARNING OUTCOME**
Explain how multicellular exocrine glands are classified by their structure.

# Section 1 Review
## Epithelial Tissue

**Labeling:** Label the types of epithelial tissues shown here.

| 1 | 2 | 3 | 4 | 5 | 6 |

**Concept map:** Use each of the following terms once to fill in the blank boxes to correctly complete the map.

- endocrine glands
- ducts
- goblet cells
- apocrine secretion
- interstitial fluid
- mucus
- merocrine secretion
- epithelial surfaces
- mucin
- exocrine glands
- holocrine secretion

**Vocabulary:** Write the term for each of the following descriptions in the space provided.

**18** A term meaning "no blood vessels"

**19** A gland whose glandular cells form sac-like pockets

**20** A type of epithelium that withstands stretching and that changes in appearance as stretching occurs

**21** The cell junction formed by the partial fusion of the lipid portions of two plasma membranes

**22** The complex structure attached to the basal surface of an epithelium

**23** A gland that has a single duct

**24** The type of epithelium lining the pericardial, pleural, and peritoneal body cavities

**18** _____
**19** _____
**20** _____
**21** _____
**22** _____
**23** _____
**24** _____

**Short answer:** Fill in the missing boxes with the epithelium type or example of an epithelial structure (or organ).

| Type of Epithelium | Structure (or Organ) |
| --- | --- |
| **25** | Lining of the trachea |
| Transitional epithelium | **26** |
| **27** | Surface of the skin |
| **28** | Lining of the small intestine |

| Type of Epithelium | Structure (or Organ) |
| --- | --- |
| Simple squamous epithelium | **29** |
| Cuboidal epithelium | **30** |
| **31** | Ducts of sweat glands |

## Module 4.10
# A matrix surrounds connective tissue cells

Go to
MasteringA&P™
PAL™3.0
Histology →
Connective
Tissue

4

Connective tissue varies widely in appearance and function, but all forms share three basic components: (1) specialized cells, (2) extracellular protein fibers, and (3) a fluid known as **ground substance**. Together, the extracellular fibers and ground substance make up the matrix that surrounds the cells. Whereas cells make up the bulk of epithelial tissue, the matrix typically accounts for most of the volume of connective tissue.

**Functions of Connective Tissue**

- Establish a structural framework for the body
- Transport fluids and dissolved materials
- Protect delicate organs
- Support, surround, and interconnect other types of tissue
- Store energy, especially in the form of triglycerides
- Defend the body from invading microorganisms

 **REVIEW**

**A.** Identify the three basic components of connective tissue.

**B.** Summarize the functions of connective tissue.

---

## Connective Tissue

The various types of connective tissue occur throughout the body but are never exposed to the outside environment. Many types of connective tissue are highly vascular (that is, they have many blood vessels) and contain sensory receptors that detect pain, pressure, temperature, and other stimuli.

### Connective Tissue Proper

**Connective tissue proper** includes those connective tissues with many types of cells and extracellular fibers in a syrupy ground substance.

### Fluid Connective Tissues

**Fluid connective tissues** have distinctive populations of cells suspended in a watery matrix that contains dissolved proteins.

### Supporting Connective Tissues

**Supporting connective tissues** differ from connective tissue proper in having a less diverse cell population and a matrix containing much more densely packed fibers. Supporting connective tissues protect soft tissues and support the weight of part or all of the body.

| **Loose** | **Dense** | **Blood** | **Lymph** | **Cartilage** | **Bone** |
|---|---|---|---|---|---|
| Fibers create loose, open framework | Fibers densely packed | Flows within cardiovascular system | Flows within lymphatic system | Solid, rubbery matrix | Solid, crystalline matrix |
| • areolar tissue | • dense regular | | | • hyaline cartilage | |
| • adipose tissue | • dense irregular | | | • elastic cartilage | |
| • reticular tissue | • elastic | | | • fibrocartilage | |

 **REVIEW**

**C.** Distinguish among connective tissue proper, fluid connective tissues, and supporting connective tissues.

 **LEARNING OUTCOME**

Describe the general structure of connective tissue.

# Module 4.11
# Loose connective tissues support other tissue types

Connective tissue proper contains extracellular protein fibers, a viscous (syrupy) ground substance, and two classes of cells. Fixed cells are permanent residents and are involved primarily with local maintenance, repair, and energy storage. Wandering cells primarily defend and repair damaged tissues. The number of cells at any given moment varies depending on local conditions. The body contains three types of **loose connective tissues:** areolar tissue, adipose tissue, and reticular tissue.

 **REVIEW**

**A.** Identify the types of cells found in connective tissue proper.

**B.** Describe the role of fibroblasts in connective tissue.

**1** **Areolar tissue** is the least specialized and most common form of connective tissue proper in adults. It is the general packing material in the body. All of the cell types found in other forms of connective tissue proper can be found in areolar tissue.

## Fibers

**Reticular fibers** are strong and form a branching network.

**Collagen fibers** are thick, straight, or wavy and often form bundles. They are very strong and resist stretching.

**Elastic fibers** are slender, branched, and very stretchy. They recoil to their original length after stretching or distortion.

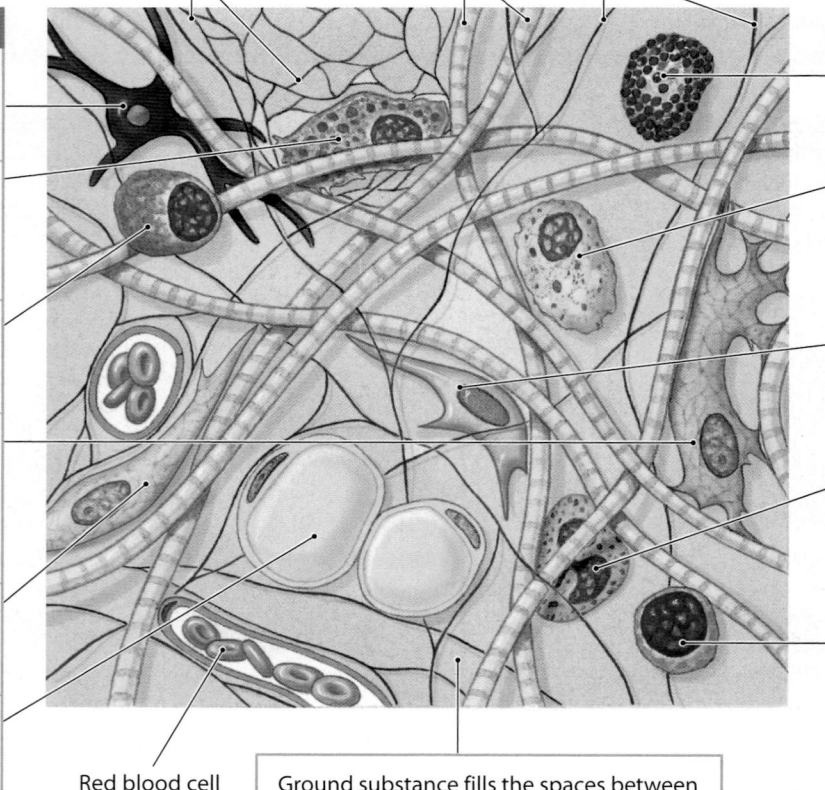

## Fixed Cells

**Melanocytes** are pigment cells that synthesize melanin, a brownish yellow pigment.

**Fixed macrophages** are stationary phagocytic cells that engulf cell debris and pathogens.

**Mast cells** are fixed cells that stimulate local inflammation and mobilize tissue defenses.

**Fibroblasts** secrete hyaluronic acid and proteins that form the ground substance and create the extracellular fibers.

**Fibrocytes** differentiate from fibroblasts and maintain the extracellular fibers.

**Adipocytes** (fat cells) are fixed cells that store lipids in large intracellular vesicles.

## Wandering Cells

**Plasma cells** are active, mobile immune cells that produce antibodies.

**Free macrophages** are wandering phagocytic cells that patrol the tissue, engulfing debris or pathogens.

**Mesenchymal cells** are mobile stem cells that repair damaged tissues.

**Neutrophils** and **eosinophils** are small, mobile, phagocytic blood cells that enter tissues during infection or injury.

**Lymphocytes** are mobile cells of the immune system.

Red blood cell in vessel

Ground substance fills the spaces between cells and surrounds connective tissue fibers. In all forms of connective tissue proper, ground substance is clear, colorless, and viscous because of the presence of proteoglycans and glycoproteins.

**2** **Adipose tissue** is found deep to the skin, especially at the flanks, buttocks, and breasts. It also forms a layer that provides padding within the orbit of the eyes, in the abdominopelvic cavity, and around the kidneys. The distinction between areolar tissue and adipose tissue is somewhat arbitrary. Adipocytes account for most of the volume of adipose tissue, but only a fraction of the volume of areolar tissue.

**? REVIEW**

**C.** Which type of loose connective tissue contains primarily lipids?

Adipocytes

Nuclei

**Adipose tissue**    LM × 340

**3** **Reticular tissue** is found in the liver, kidney, spleen, lymph nodes, and bone marrow, where it forms a tough, flexible network that provides support and resists distortion. In reticular tissue, reticular fibers create a complex three-dimensional supporting network known as a **stroma**. Fixed macrophages and fibroblasts are present in reticular tissues, but these cells are seldom visible because the stroma is dominated by cells that are unique to the organ under consideration and have specialized functions.

**? REVIEW**

**D.** What term means the fibrous supporting network formed of reticular fibers?

Reticular fibers

**Reticular tissue from liver**

**Reticular tissue**    LM × 375

**? INTEGRATION**

**E.** What types of phagocytic cells are present in connective tissue proper?

**Lo LEARNING OUTCOME**

Describe the structure, function, and locations of areolar tissue, adipose tissue, and reticular tissue.

# Module 4.12

# Dense connective tissues are dominated by extracellular fibers …

Extracellular fibers make up most of the volume of **dense connective tissues**. The body has three types of dense connective tissues: dense regular connective tissue, dense irregular connective tissue, and elastic tissue.

**1** **Dense regular connective tissue** is found in cords (tendons) or sheets connecting skeletal muscles to bone, and in cords (ligaments) that interconnect bones or stabilize the positions of internal organs. The forces applied to these cords and sheets arrive from a consistent direction: parallel to the long axis of the collagen fibers.

Collagen fiber

Fibroblast nuclei

**Tendon from triceps muscle**

**Dense regular connective tissue**    LM × 440

**2** In **dense irregular connective tissue**, the fibers form an interwoven meshwork in no consistent pattern. These tissues strengthen and support areas subjected to stresses from many directions. Dense irregular connective tissue forms (1) a covering, or capsule, that sheathes visceral organs; (2) a superficial layer covering bones, cartilages, and peripheral nerves; and (3) a thick supporting layer in the skin (the dermis).

Collagen fiber bundles

**Dermis**

**Dense irregular connective tissue**    LM × 111

**3** Dense regular and dense irregular connective tissues contain varying amounts of elastic fibers. When elastic fibers outnumber collagen fibers, the tissue is springy and resilient, allowing it to tolerate cycles of extension and recoil. This **elastic tissue** is found between vertebrae of the spinal column, in the walls of large blood vessels, in ligaments supporting transitional epithelia, and the erectile tissues of the penis.

Elastic fibers

Fibroblast nuclei

**Elastic ligament between vertebrae**

**Elastic tissue**    LM × 887

# ... whereas fluid connective tissues have an aqueous matrix

**4** **Fluid connective tissues** have an aqueous matrix that includes many types of suspended proteins that under normal conditions do not form insoluble fibers. In **blood**, this watery matrix is called **plasma**. Blood cells and platelets, collectively known as **formed elements**, are suspended in the plasma.

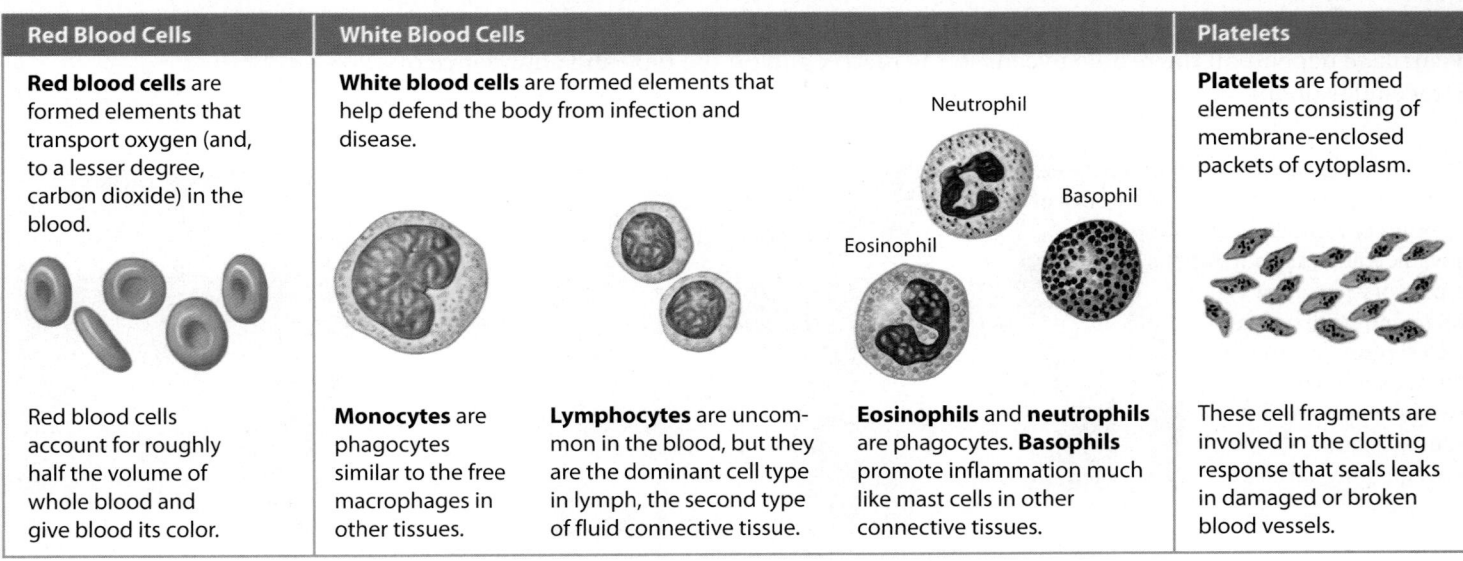

| Red Blood Cells | White Blood Cells | | | Platelets |
|---|---|---|---|---|
| **Red blood cells** are formed elements that transport oxygen (and, to a lesser degree, carbon dioxide) in the blood. | **White blood cells** are formed elements that help defend the body from infection and disease. Neutrophil, Eosinophil, Basophil | | | **Platelets** are formed elements consisting of membrane-enclosed packets of cytoplasm. |
| Red blood cells account for roughly half the volume of whole blood and give blood its color. | **Monocytes** are phagocytes similar to the free macrophages in other tissues. | **Lymphocytes** are uncommon in the blood, but they are the dominant cell type in lymph, the second type of fluid connective tissue. | **Eosinophils** and **neutrophils** are phagocytes. **Basophils** promote inflammation much like mast cells in other connective tissues. | These cell fragments are involved in the clotting response that seals leaks in damaged or broken blood vessels. |

## Extracellular Fluid Circulation

**1** **Arteries** carry blood away from the heart and into the tissues of the body.

**2** **Capillaries** are the smallest and most delicate blood vessels. All exchange between the blood and interstitial fluid occurs at capillaries.

**3** At capillary networks, blood pressure forces water and small solutes out of the bloodstream to form interstitial fluid in the surrounding tissues.

Heart

Water and solutes from bloodstream

**4** Lymph forms as interstitial fluid enters lymphatic vessels.

**5** Lymphatic vessels form a network that returns lymph to large veins near the heart.

**6** **Veins** carry blood from capillary beds to the heart.

**5** Extracellular fluid includes plasma, interstitial fluid, and lymph. Blood is normally confined to the vessels of the cardiovascular system, and contractions of the heart keep it in motion. As blood flows through body tissues, water and solutes move from the plasma into the surrounding interstitial fluid. **Lymph**, the second type of fluid connective tissue, is formed as interstitial fluid drains into lymphatic vessels that begin in peripheral tissues and empty into the venous system. The continuous recirculation of extracellular fluid is essential to homeostasis. It helps eliminate local differences in the levels of nutrients, wastes, or toxins; maintains blood volume; and alerts the immune system to infections that may be under way in peripheral tissues.

 **REVIEW**

**B.** Summarize the role of extracellular fluid in maintaining homeostasis.

**? INTEGRATION**

**C.** Lack of vitamin C in the diet interferes with the ability of fibroblasts to produce collagen. How might this affect connective tissue function?

**Lo LEARNING OUTCOME**

Describe the structure, function, and locations of dense connective tissues and fluid connective tissues.

# Module 4.13
# Cartilage provides a flexible support for body structures

In cartilage, the matrix is a firm gel that contains polysaccharide derivatives called **chondroitin sulfates** (kon-DROY-tin; *chondros*, cartilage). Chondroitin sulfates form complexes with proteins in the ground substance, producing proteoglycans. Cartilage cells, or **chondrocytes** (KON-drō-sīts), are the only cells in the avascular cartilage matrix. They occupy small chambers known as **lacunae** (la-KOO-nē; *lacus*, lake). The physical properties of cartilage depend on the proteoglycans in the matrix and on the type and abundance of extracellular fibers.

**?** **REVIEW**
**A.** Which connective tissue fiber is characteristic of the cartilage supporting the ear?

**1** **Hyaline cartilage** is found between the tips of the ribs and the bones of the sternum, covering bone surfaces at movable joints, supporting the respiratory passageways, and forming part of the nasal septum. It provides stiff but somewhat flexible support and reduces friction between bony surfaces.

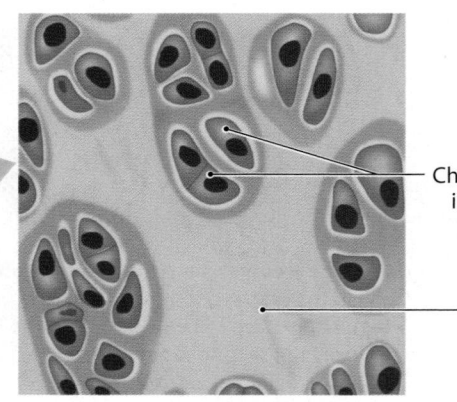

Chondrocytes in lacunae

Matrix

**Hyaline cartilage from shoulder joint**

Hyaline cartilage — LM × 500

**2** **Elastic cartilage** supports the external ear and a number of smaller internal structures. Its numerous elastic fibers allow it to distort without damage and return to its original shape.

Chondrocyte in lacuna

Elastic fibers in matrix

**Elastic cartilage from external ear**

Elastic cartilage — LM × 358

**3** **Fibrocartilage** is extremely durable and tough because it has little ground substance and its matrix is dominated by densely interwoven collagen fibers. Fibrocartilage pads are found within the knee joint, between the pubic bones of the pelvis, and in the intervertebral discs of the vertebral column. It resists compression, prevents bone-to-bone contact, and limits relative movement.

Collagen fibers in matrix

Chondrocytes

**Fibrocartilage from intervertebral disc**

Fibrocartilage — LM × 400

**4** Cartilage is generally set apart from surrounding tissues by a covering called a **perichondrium** (per-i-KON-drē-um; *peri-*, around). The perichondrium contains two distinct layers: an outer, fibrous layer of dense irregular connective tissue containing fibroblasts, and an inner, cellular layer. The fibrous layer gives mechanical support and protection and attaches the cartilage to other structures. The cellular layer is important to the growth and maintenance of the cartilage. Blood vessels in the perichondrium provide oxygen and nutrients to the underlying chondrocytes.

**Hyaline cartilage**   LM × 250

**? REVIEW**

**B.** Describe the two layers making up the perichondrium.

4

**5** Both appositional and interstitial growth occur during development before birth and through adolescence, but interstitial growth predominates. Normally, in adults cartilage growth no longer occurs. However, in unusual circumstances—such as after cartilage has been slightly damaged or subjected to excessive hormonal stimulation—some repair may occur by appositional growth. After severe damage, cartilage is not repaired, and a fibrous patch replaces the damaged area..

**? REVIEW**

**C.** Contrast appositional and interstitial growth of cartilage.

## Appositional growth

**Appositional growth** adds new layers of cartilage to the surface.

Cells in the cellular layer of the perichondrium differentiate into **chondroblasts** (immature chondrocytes).

The chondroblasts secrete new matrix and become separated from each other.

Dividing stem cells in the perichondrium continuously produce additional chondroblasts. Meanwhile, chondroblasts completely surrounded by matrix gradually mature into chondrocytes.

## Interstitial growth

**Interstitial growth** enlarges the cartilage from within.

Chondrocyte undergoes division within a lacuna surrounded by cartilage matrix.

As daughter cells secrete additional matrix, they move apart, expanding the cartilage from within.

**LO LEARNING OUTCOME**

Describe the structure, function, and locations of cartilage.

# Module 4.14
# Bone provides a strong framework for the body

**Bone tissue**, also called **osseous** (OS-ē-us; *os*, bone) **tissue**, contains a small volume of ground substance. About two-thirds of the bone matrix consists of a mixture of calcium salts—primarily calcium phosphate, with lesser amounts of calcium carbonate. The rest of the matrix is dominated by collagen fibers. This combination gives bone truly remarkable properties. By themselves, calcium salts are hard but rather brittle, whereas collagen fibers are strong and relatively flexible. In bone, the presence of the minerals surrounding the collagen fibers produces a strong, somewhat flexible combination that is highly resistant to shattering. In its overall properties, bone can compete with the best steel-reinforced concrete. In essence, the collagen fibers in bone act like the steel reinforcing rods, and the mineralized matrix acts like the concrete.

Compact bone also has a superficial layer of bone that was deposited during appositional growth of the bone. Because the matrix is solid and calcified, interstitial growth cannot occur in bone.

Unlike cartilage, bone is highly vascular. Large vessels outside the bone are connected to smaller vessels that supply areas of compact bone and the soft tissues that fill the interior cavity.

In compact bone, the matrix is organized in concentric layers around branches of blood vessels within the bone.

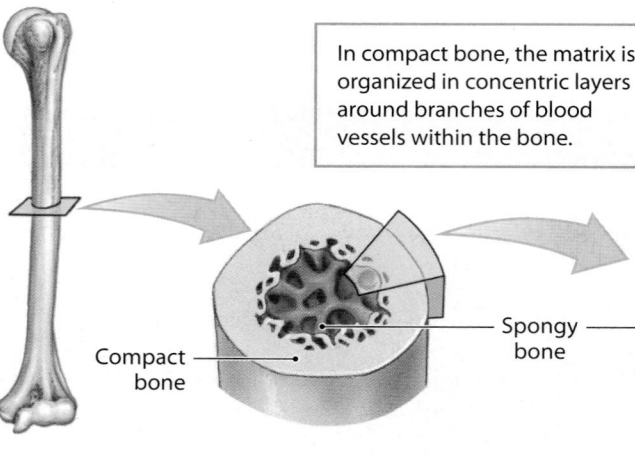

Compact bone

Spongy bone

Nerve

**1** A typical long bone is hollow, and its walls contain two different types of bone. The weight-bearing outer layer consists of well-organized **compact bone**, whereas a finer network of **spongy bone** lines the internal cavity. We consider the structural and functional differences between the two bone types in Chapter 6.

**2** Cartilage and bone share a number of functions, but these two supporting connective tissues are organized very differently.

### A Comparison of Cartilage and Bone

| Characteristic | Cartilage | Bone |
|---|---|---|
| Cells | Chondrocytes in lacunae | Osteocytes in lacunae |
| Ground substance | Chondroitin sulfate (in proteoglycan) and water | A small volume of liquid surrounding insoluble crystals of calcium salts |
| Fibers | Collagen, elastic, and reticular fibers in varying proportions | Collagen fibers predominate |
| Vascularity | No internal blood vessels | Extensive blood vessels |
| Covering | Perichondrium (two layers) | Periosteum (two layers) |
| Strength | Limited: bends easily, but hard to break | Strong: resists distortion until breaking point |

 **REVIEW**

**A.** Describe bone matrix.

**3** Although the arrangement of the layers of a bone may vary, the superficial and deeper layers share common structural features.

The fibrous layer attaches a bone to surrounding tissues and to associated tendons and ligaments.

The cellular layer functions in appositional bone growth and participates in repairs after an injury.

Except in joint cavities, where they are covered by a layer of hyaline cartilage, bone surfaces are sheathed by a **periosteum** (per-ē-OS-tē-um) composed of fibrous (outer) and cellular (inner) layers.

Lacunae in the matrix contain **osteocytes** (OS-tē-ō-sīts), or bone cells. The lacunae are typically organized around blood vessels that branch through the bony matrix.

Layers of matrix separate the lacunae. These layers are oriented along the main axis of the bone.

**Canaliculi** (kan-a-LIK-ū-lē; little canals) are fine passageways that form a branching network for the exchange of materials between blood vessels and osteocytes. This is important because diffusion cannot occur through the calcified matrix of bone.

A **central canal** at the center of an osteon contains sensory nerves and the blood vessels that provide oxygen and nutrients to the osteocytes.

**4** The functional unit of compact bone is called an **osteon**.

Osteon

Matrix

Lacunae

Canaliculi

**Compact bone**          LM × 320

**? REVIEW**

**B.** What are mature bone cells in lacunae called?

**C.** What is the functional unit of compact bone?

**? INTEGRATION**

**D.** Distinguish between the two types of supporting connective tissues with respect to their characteristic fibers.

**Lo LEARNING OUTCOME**

Describe the structure and function of bone.

Unlike cartilage, bone undergoes extensive remodeling throughout life, and complete repairs can be made even after severe damage has occurred. Bones also respond to the stresses placed on them, growing thicker and stronger with exercise and becoming thin and brittle with inactivity.

# Tissue membranes are physical barriers, and fasciae support and surround organs

A **membrane** is a physical barrier. There are many different types of anatomical membranes—you encountered plasma membranes in Chapter 3, and you will find many other kinds of membranes in later chapters. Here we consider **tissue membranes** that line or cover body surfaces. These membranes typically consist of an epithelium supported by connective tissue. Four types of tissue membranes occur in the body: mucous membranes, serous membranes, the cutaneous membrane, and synovial membranes.

**? REVIEW**

**A.** Which cavities in the body are lined by serous membranes?

## Tissue Membranes

**1** **Mucous membranes**, or *mucosae* (mū-KŌ-sē), line passageways and chambers that communicate with the exterior, including those in the digestive, respiratory, reproductive, and urinary tracts. These epithelial surfaces must be kept moist to reduce friction and, in many cases, to facilitate absorption or secretion. The epithelial surfaces are lubricated either by mucus (produced by goblet cells or mucous cells) or by fluids such as urine or semen.

Mucous secretion

Epithelium

The areolar tissue component of a mucous membrane is called the **lamina propria** (PRŌ-prē-uh).

**2** **Serous membranes**, or *serosae* (se-RŌ-sē), consist of a mesothelium supported by areolar tissue. They are extremely delicate and never directly connected to the exterior. Three serous membranes line the body cavities of the trunk: (1) the **pleura**, which lines the pleural cavities and covers the lungs; (2) the **pericardium**, which lines the pericardial cavity and covers the heart; and (3) the **peritoneum**, which lines the peritoneal cavity and covers the surfaces of visceral organs.

A watery **serous fluid** diffusing from underlying tissues coats the surfaces of a serous membrane and prevents friction.

Mesothelium

Areolar tissue

**3** The **cutaneous membrane** covers the surface of the body. It consists of a stratified squamous epithelium and a layer of areolar tissue reinforced by underlying dense irregular connective tissue. In contrast to serous and mucous membranes, the cutaneous membrane is thick, relatively waterproof, and usually dry. We take a closer look at the cutaneous membrane in Chapter 5.

Epithelium

Areolar tissue

Dense irregular connective tissue

**4** A **synovial membrane** lines freely movable joint cavities but does not cover the opposing joint surfaces. Although the covering of the synovial membrane is often called an epithelium, it differs from true epithelia in four respects: (1) It develops within a connective tissue, (2) no basement membrane is present, (3) gaps of up to 1 mm may separate adjacent cells, and (4) the synovial fluid and capillaries in the underlying connective tissue are continuously exchanging fluid and solutes.

Hyaline cartilage covers the bone surface within the joint cavity.

**Synovial** (si-NŌ-vē-ul) **fluid**, which fills the joint cavity, is produced and regulated by the synovial membrane.

| Synovial Membrane |
| --- |
| Epithelium |
| Areolar tissue |

Capillary

Adipocytes

Fibrous capsule

**? REVIEW**

**B.** Name the four types of tissue membranes found in the body.

## Fasciae

**5** **Fasciae** (FASH-ē-ē; singular, fascia) are connective tissue layers that support and surround organs. The fasciae consist of three types of layers: the **superficial fascia**, the **deep fascia**, and the **subserous fascia**.

Body wall

Body cavity

Skin

**Connective Tissue Framework of the Body**

| Superficial Fascia | Deep Fascia | Subserous Fascia |
| --- | --- | --- |
| • Lies between the skin and underlying organs<br>• Consists of areolar tissue and adipose tissue | • Forms a strong, fibrous internal framework<br>• Consists of dense regular connective tissue<br>• Is continuous with or bound to capsules, ligaments, and other connective tissue structures | • Lies between serous membranes and deep fascia<br>• Consists entirely of areolar tissue |

Serous membrane

Rib

Cutaneous membrane of the skin

**? INTEGRATION**

**D.** Which of the four tissue membranes is relatively waterproof and usually dry?

**Lo LEARNING OUTCOME**

Describe the arrangements of epithelial and connective tissues in the four types of tissue membranes, and describe the structures and locations of the three types of fasciae.

**? REVIEW**

**C.** Name the three layers of fascia and their types of connective tissue.

# Section 2 Review
# Connective Tissue

**Concept map:** Use each of the following terms once to fill in the blank boxes to correctly complete the map.

- loose connective tissue
- chondrocytes in lacunae
- fluid connective tissue
- tendons
- ligaments
- regular
- hyaline
- blood
- adipose
- bone

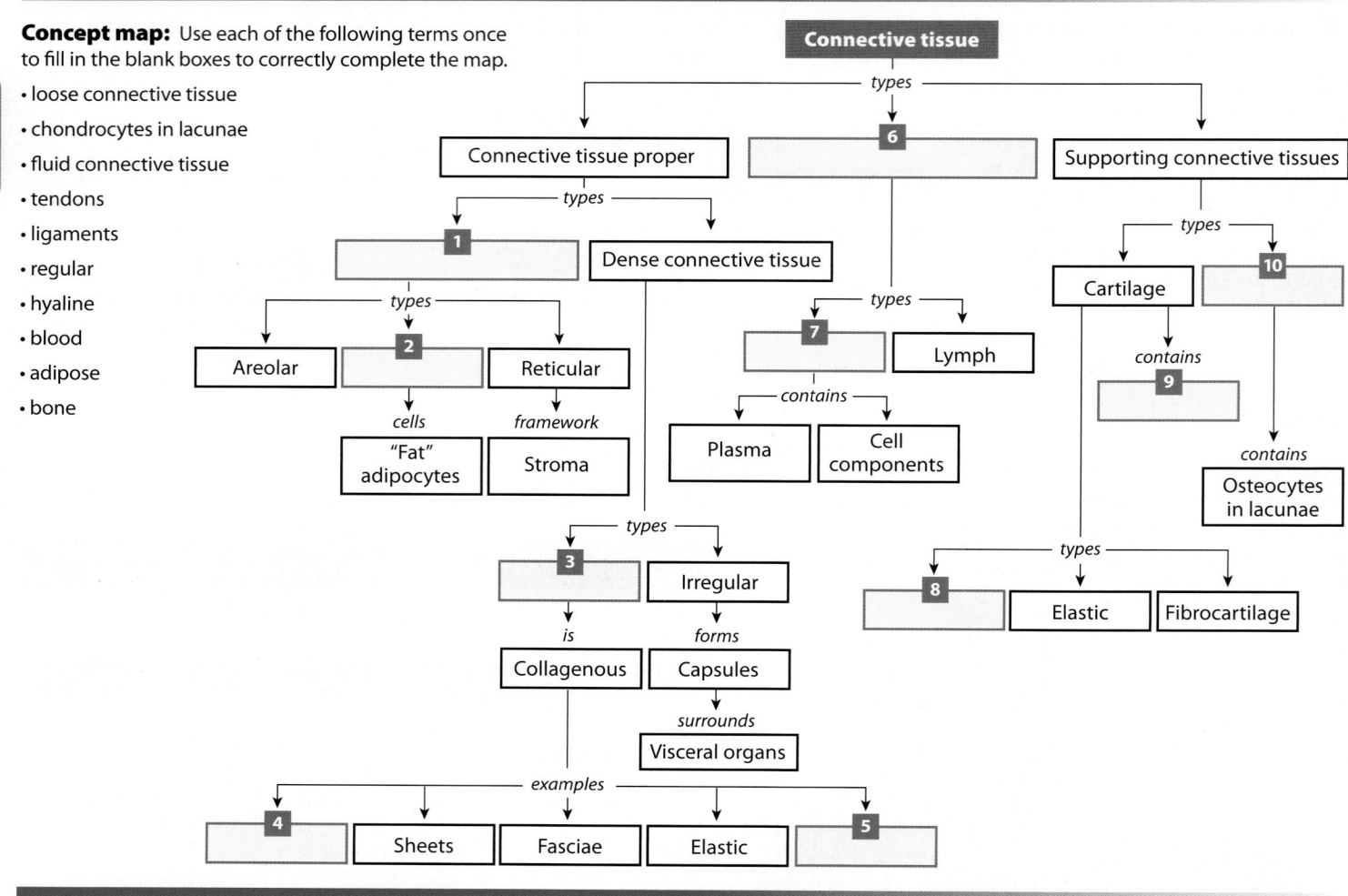

**Vocabulary:** In the space provided, write the boldfaced terms introduced in this section that contain the indicated word part.

11 peri- *(around)*
12 os- *(bone)*
13 chondro- *(cartilage)*
14 inter- *(between)*
15 lacus- *(lake)*

11 _____
12 _____
13 _____
14 _____
15 _____

Write the term for each of the following descriptions in the space provided.

16 A cartilage cell
17 Bone tissue
18 A type of cartilage that has a matrix with little ground substance and large amounts of collagen fibers
19 Cells that store lipid reserves
20 The tissue membrane that lines freely movable joint cavities
21 The tissue membrane that covers the surface of the body
22 Separates cartilage from surrounding tissues

16 _____
17 _____
18 _____
19 _____
20 _____
21 _____
22 _____

## Module 4.16
# Muscle tissue outweighs nervous tissue by 25:1

As we have seen, epithelia cover surfaces and line passageways, and connective tissues support and interconnect parts of the body. Together, these tissues provide a strong, interwoven framework within which the organs of the body can function. When we consider the collective weight of the four types of tissue in the body, connective tissues make up about 45 percent of the total body weight. Muscle tissue contributes the most, and nervous tissue the least, to body weight.

**? REVIEW**

**A.** What is the relative percentage of body weight from each of the four tissue types?

**4**

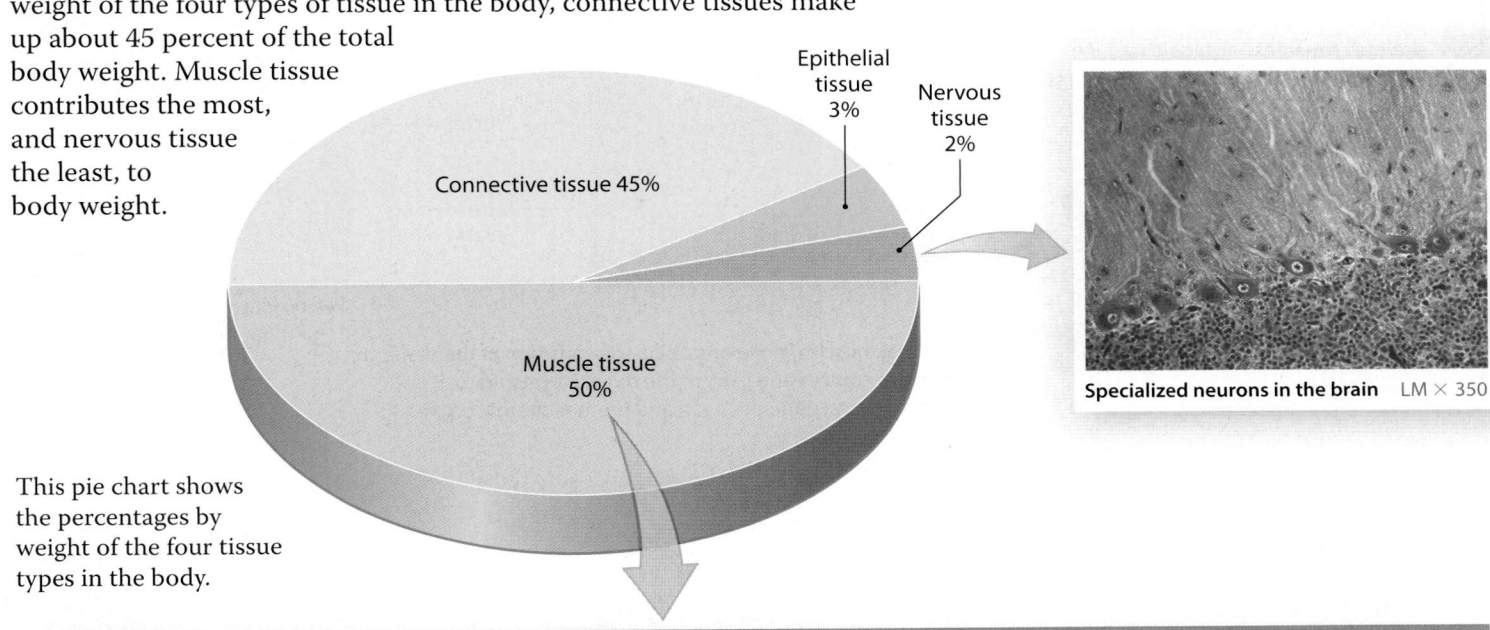

Epithelial tissue 3%

Nervous tissue 2%

Connective tissue 45%

Muscle tissue 50%

**Specialized neurons in the brain** LM × 350

This pie chart shows the percentages by weight of the four tissue types in the body.

**Muscle Tissue**

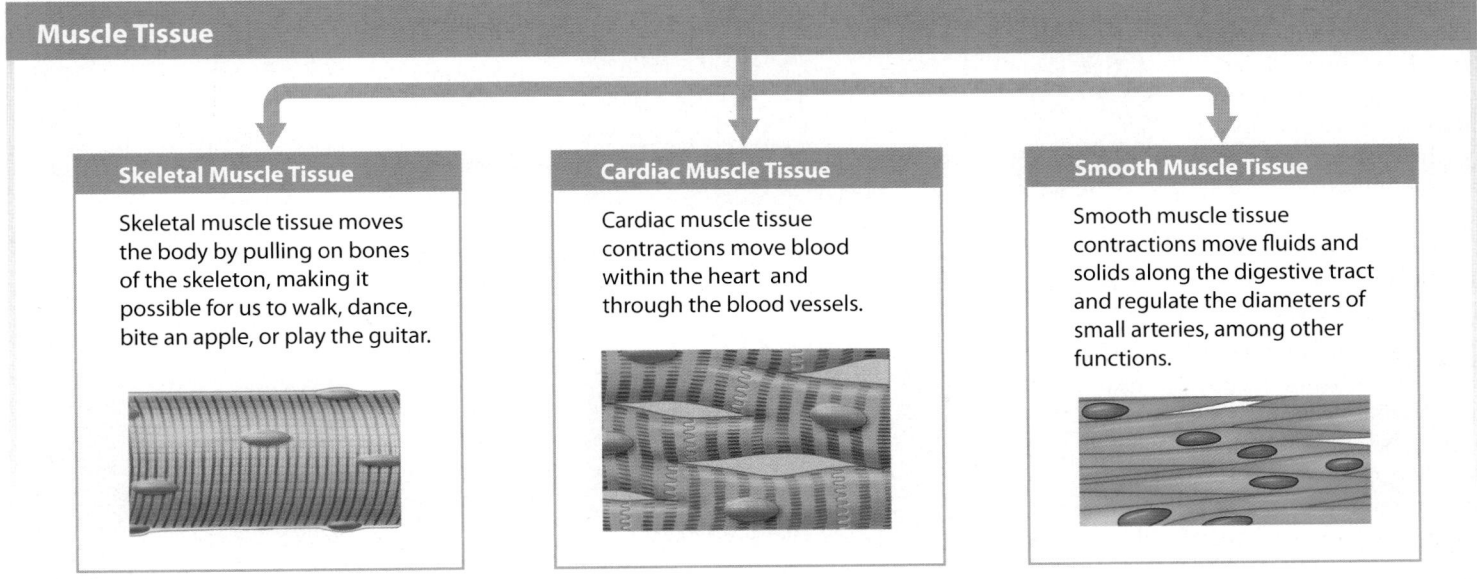

### Skeletal Muscle Tissue

Skeletal muscle tissue moves the body by pulling on bones of the skeleton, making it possible for us to walk, dance, bite an apple, or play the guitar.

### Cardiac Muscle Tissue

Cardiac muscle tissue contractions move blood within the heart and through the blood vessels.

### Smooth Muscle Tissue

Smooth muscle tissue contractions move fluids and solids along the digestive tract and regulate the diameters of small arteries, among other functions.

In this section we consider the histology of muscle tissue and nervous tissue. We only provide an overview of these tissue types because additional details will be given in later chapters dealing with the muscular system, nervous system, cardiovascular system, and digestive system.

**? REVIEW**

**B.** List the three classifications of muscle tissue, and describe a function for each type.

 **LEARNING OUTCOME**

Describe the relative proportions of muscle tissue and nervous tissue in the body.

# Module 4.17
# Muscle tissue is specialized for contraction ...

Many vital functions involve movement of one kind or another—movement of materials along the digestive tract, movement of blood around the cardiovascular system, or movement of the body from one place to another. Movement is produced by **muscle tissue**, which is specialized for contraction. There are three types of muscle tissue: skeletal muscle, cardiac muscle, and smooth muscle.

Go to
MasteringA&P™
PAL™3.0
Histology →
Muscle &
Nervous
Tissue

**? REVIEW**
**A.** Which type of muscle tissue regulates blood vessel diameter?

**1** **Skeletal muscle tissue** is found in skeletal muscles, organs that also contain connective tissues and nervous tissue. The cells are long, cylindrical, and banded (**striated**), and they have multiple nuclei (**multinucleate**).

Striations
Nuclei
Muscle fiber

**Skeletal muscle**   LM × 180

Skeletal muscles move or stabilize the position of the skeleton; guard entrances and exits to the digestive, respiratory, and urinary tracts; generate heat; and protect internal organs.

**2** **Cardiac muscle tissue** is found in the heart. Cardiac muscle cells are short, branched, and striated, usually with a single nucleus. These cells are interconnected at specialized intercellular junctions called **intercalated discs**. They help synchronize cardiocyte contractions.

Nuclei
Cardiac muscle cells
Intercalated discs
Striations

**Cardiac muscle**   LM × 450

Cardiac muscle moves blood and maintains blood pressure.

**3** **Smooth muscle tissue** is found throughout the body. For example, smooth muscle is found in the skin, in the walls of blood vessels, and in many digestive, respiratory, urinary, and reproductive organs. The cells are short, spindle shaped, and nonstriated, and they have a single, central nucleus.

Smooth muscle cells
Nuclei

**Smooth muscle**   LM × 235

Smooth muscle moves food, urine, and reproductive tract secretions; controls diameter of respiratory passageways; and regulates diameter of blood vessels.

# … and nervous tissue is specialized for communication

**Nervous tissue** is specialized to conduct electrical impulses from one region of the body to another. Ninety-eight percent of the nervous tissue in the body is in the brain and spinal cord, which are the control centers of the nervous system. Nervous tissue contains two types of cells: (1) **neurons** (NOOR-onz; *neuro*, nerve) and (2) several kinds of supporting cells, collectively called **neuroglia** (noo-ROG-lē-uh or noo-rō-GLĒ-uh), or **glial cells** (*glia*, glue).

**REVIEW**

**B.** Distinguish between neurons and neuroglia.

4

**4** Neurons transfer information from place to place and process information like living computer chips. Their sizes and shapes vary widely. The longest cells in your body are neurons. Many are as long as a meter (39 in.)!

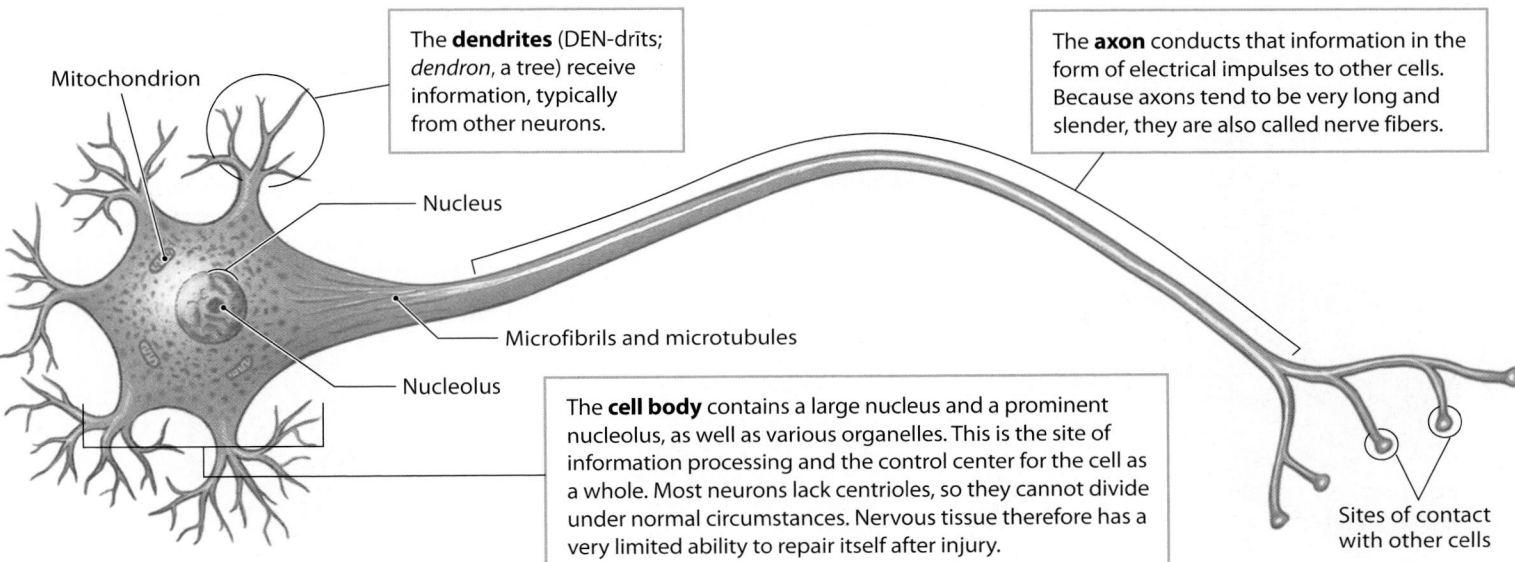

Mitochondrion

The **dendrites** (DEN-drīts; *dendron*, a tree) receive information, typically from other neurons.

The **axon** conducts that information in the form of electrical impulses to other cells. Because axons tend to be very long and slender, they are also called nerve fibers.

Nucleus

Microfibrils and microtubules

Nucleolus

The **cell body** contains a large nucleus and a prominent nucleolus, as well as various organelles. This is the site of information processing and the control center for the cell as a whole. Most neurons lack centrioles, so they cannot divide under normal circumstances. Nervous tissue therefore has a very limited ability to repair itself after injury.

Sites of contact with other cells

**5** There are several different types of neuroglia, each with specific functions. Each type has a distinctive appearance related to its primary function. In general, neuroglia protect, support, and repair nervous tissue and maintain the nutrient supply to neurons. Under a microscope, you can identify neuroglia by looking for the non-neuronal cells occupying space between neurons (see Module 11.4).

## Functions of Neuroglia

- Maintain physical structure of nervous tissue
- Repair nervous tissue framework after injury
- Perform phagocytosis
- Provide nutrients to neurons
- Regulate the composition of the interstitial fluid surrounding neurons

### 🏃 Everyday Physiology

### Our thought processes reflect neural activity in the brain

Our conscious and unconscious thought processes reflect the communication among neurons in the brain. Such communication involves the propagation of electrical impulses that are generated at the cell body and travel along the axon to reach other cells.

**INTEGRATION**

**C.** Organs are made up of different tissues. What tissues are found in skeletal muscles?

**LEARNING OUTCOME**

Specify the functions of muscle tissue and nervous tissue.

# Module 4.18

# The response to tissue injury involves inflammation and regeneration

Let's consider what happens after an injury, focusing on the interaction among different tissues. Tissues are not isolated from each other; they combine to form organs with diverse functions. Therefore, any injury affects several types of tissue simultaneously. These tissues must respond in a coordinated way to preserve homeostasis. The body utilizes two related processes, inflammation and regeneration, to restore homeostasis after an injury.

 **REVIEW**

**A.** Identify the two processes in the response to tissue injury.

## ① Injury

When a tissue is injured, the body activates a general defense mechanism.

### Exposure to Pathogens and Toxins

An injured tissue contains an abnormal concentration of pathogens, toxins, waste products, and the chemicals from injured cells.

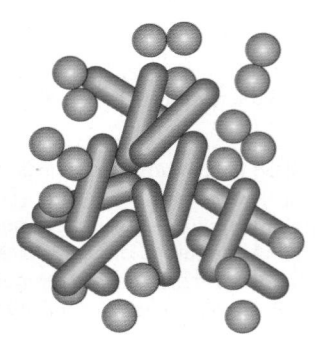

stimulates

### Mast Cell Activation

When an injury damages connective tissue, mast cells release a variety of chemicals. This process, called **mast cell activation**, stimulates inflammation.

Histamine

Mast cell

Heparin          Prostaglandins

Regeneration inhibits further mast cell activation

## ③ Regeneration

**Regeneration** is the repair that occurs after the damaged tissue has been stabilized and the inflammation has subsided.

As tissue conditions return to normal, fibroblasts move into the area, laying down a network of collagen fibers that stabilizes the injury site. This process produces a dense, collagenous framework known as **scar tissue**. Over time, scar tissue is usually "remodeled" and gradually assumes a more normal appearance.

Scar tissue

Normal tissue conditions restored

**2**

## Inflammation

**Inflammation** produces several familiar indications of injury: swelling, redness, heat (warmth), and pain. Inflammation may also result from the presence of pathogens, such as harmful bacteria, within the tissues. The presence of these pathogens constitutes an **infection**. Because all organs have connective tissues, inflammation can occur anywhere in the body.

**4**

### Increased Blood Flow

In response to the released chemicals, the smooth muscle tissue that surrounds local blood vessels relaxes, and the vessels **dilate**, or enlarge in diameter. This dilation increases blood flow through the damaged tissue.

### Increased Vessel Permeability

The dilation is accompanied by an increase in the permeability of the capillary walls. Plasma, including blood proteins, now diffuses into the injured tissue, so the area becomes swollen.

### Pain

The combination of abnormal conditions within the tissue and the chemicals released by mast cells stimulates nerve endings that produce the sensation of pain.

**PAIN**

### Increased local temperature
The increased blood flow and permeability cause the tissue to become warm and red.

### Increased oxygen and nutrients
Vessel dilation, increased blood flow, and increased vessel permeability result in enhanced delivery of oxygen and nutrients.

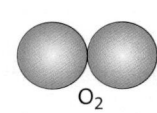

$O_2$

### Increased phagocytosis
Phagocytes in the tissue are activated, and they begin engulfing tissue debris and pathogens. Additional phagocytes migrate into the tissue from the bloodstream, drawn to the site by the abnormal local conditions.

### Removal of toxins and wastes
The enhanced circulation that increases the volume of interstitial fluid also carries away toxins and waste products, distributing them to the kidneys for excretion or to the liver for inactivation.

**Toxins and wastes**

Over a period of hours to days, the cleanup process generally succeeds in eliminating the inflammatory stimuli.

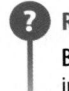 **REVIEW**

**B.** What are the four indications of inflammation that occur following an injury?

Each organ has a different ability to regenerate after injury. This ability is directly linked to the pattern of tissue organization in the injured organ. Epithelia, connective tissues (except cartilage), and smooth muscle tissue usually regenerate well, but skeletal and cardiac muscle tissues and nervous tissue regenerate relatively poorly, if at all. Damaged areas that do regenerate are often replaced by scar tissue. The permanent replacement of normal tissue by scar tissue is called **fibrosis** (fī-BRŌ-sis). Fibrosis in muscle and other tissues may occur in response to injury, disease, or aging.

 **INTEGRATION**

**C.** Why can inflammation occur in any organ in the body?

 **LEARNING OUTCOME**

Describe the roles of inflammation and regeneration in response to tissue injury.

## Section 3 Review
# Muscle Tissue and Nervous Tissue

**Concept map:** Use each of the following terms once to fill in the blank boxes to correctly complete the map.

- cardiac
- cell body
- dendrites
- maintain physical structure of nervous tissue
- multinucleate
- neuroglia
- nonstriated

- perform phagocytosis
- provide nutrients to neurons
- regulate composition of interstitial fluid surrounding neurons
- repair nervous tissue framework after injury
- smooth

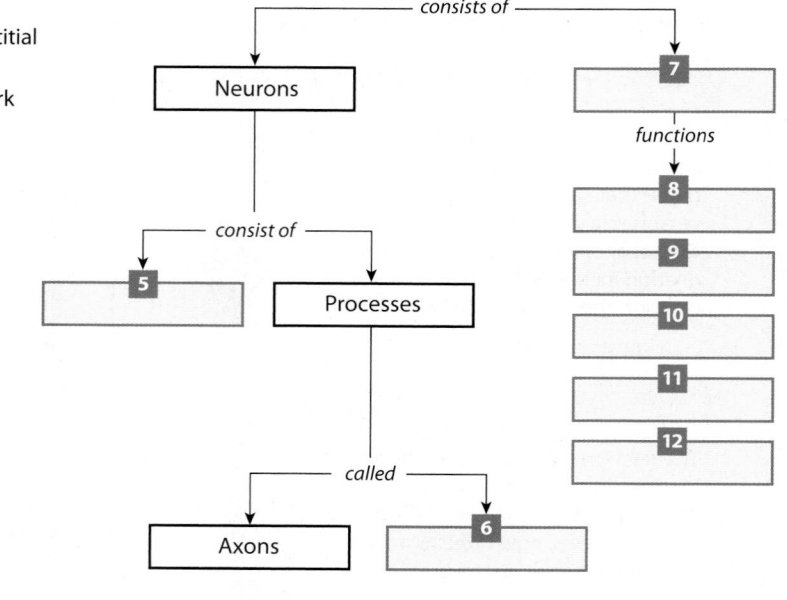

**Vocabulary:** Write the term for each of the following descriptions in the space provided.

**13** A single structure that extends from the cell body of a neuron and carries information to other cells

**14** A specialized intercellular junction between cardiac muscle cells

**15** The supporting cells found in nervous tissue

**16** Muscle tissue that contains large, multinucleate, striated cells

**17** Muscle tissue that regulates the diameter of blood vessels and respiratory passageways

**18** The repair process that occurs after inflammation has subsided

**19** The first process in a tissue's response to injury

**13** _____

**14** _____

**15** _____

**16** _____

**17** _____

**18** _____

**19** _____

**Section integration:** During inflammation, both blood flow and blood vessel permeability increase in the injured area. Describe how these responses aid the cleanup process and eliminate the inflammatory stimuli in the injured area.

**20** _____

_____

_____

_____

# Study Outline

## SECTION 1 • Epithelial Tissue

### Module 4.1

### Four types of tissue make up the body    p. 143

1. **Tissues** are collections of cells and cell products that perform a limited number of specialized functions.

2. The study of tissues is called **histology**. The four basic tissue types recognized by histologists are **epithelial tissue**, **connective tissue**, **muscle tissue**, and **nervous tissue**.

### Module 4.2

### Microscopes are used to study cells and tissues    p. 144

3. The earliest microscopes required a light source and could magnify only 10 or 20 times. Today's light microscopes can magnify up to 1000 times. **Simple microscopes** have one lens; **compound microscopes** feature more than one lens.

4. Modern microscopes use beams of electrons instead of light and can magnify over 1 million times. Common microscopes used in scientific study today are the **compound light microscope**, the **transmission electron microscope**, and the **scanning electron microscope.**

### Module 4.3

### Epithelial tissue covers surfaces, lines cavities, and forms secretory glands    p. 146

5. **Epithelial tissue** includes **epithelia** and **glands**. Epithelia cover exposed surfaces and line internal passageways. Glands are secretory structures and come in two types: **endocrine glands** and **exocrine glands**.

6. Endocrine glands secrete hormones or precursors into interstitial fluid, usually for distribution into the bloodstream. Exocrine glands secrete onto external surfaces or into ducts that connect to the exterior. The functions of epithelial tissue are to provide physical protection, control permeability, provide sensation, and produce specialized secretions.

7. An epithelium has a **base** that is attached to adjacent tissues, and an **apical surface** that faces the body exterior or internal space. Epithelial cells have three basic shapes: **squamous**, **cuboidal**, and **columnar**. A **simple epithelium** is a single layer of cells; several layers compose a **stratified epithelium**.

### Module 4.4

### Epithelial cells are extensively interconnected, both structurally and functionally    p. 148

8. The physical integrity of an epithelium depends on intercellular connections and attachment to underlying tissues.

*Interconnected epithelial cells*

9. **Intercellular attachments** include **tight** (occluding) **junctions**, **adhesion belts**, **gap junctions**, and **desmosomes**.

10. The **basement membrane** of the cell consists of two layers: the **basal lamina** and the **reticular lamina**. **Hemidesmosomes** attach the epithelium to the basement membrane.

11. Epithelial tissue is **avascular**; that is, it lacks blood vessels.

### Module 4.5

### The cells in a squamous epithelium are flat and irregularly shaped    p. 150

12. **Simple squamous epithelium** is a single layer of irregularly shaped cells that fit together like a jigsaw puzzle. This tissue is delicate, reduces friction, and is found along passageways and in areas where absorption takes place.

13. Simple squamous epithelium has special names in certain body locations. **Mesothelium** lines the pericardial, pleural, and peritoneal body cavities, and **endothelium** lines the inner surface of the heart and blood vessels.

14. **Stratified squamous epithelium** has several layers and is located where mechanical or chemical stresses are severe. This tissue may be **keratinized** or **nonkeratinized**.

### Module 4.6

### Cuboidal and transitional epithelia line passageways and chambers connected to the exterior    p. 152

15. **Cuboidal epithelium** resembles hexagonal boxes. **Simple cuboidal epithelium** is found where absorption occurs, as in the kidney tubules, or where secretion takes place, as in the thyroid gland.

*Cuboidal epithelia*

**Stratified cuboidal epithelia** are rare and located along the ducts of certain exocrine glands.

16. **Transitional epithelium** is an unusual stratified epithelium. It allows for stretching without damage. It lines the urinary bladder and other regions of the urinary system where volume changes.

### Module 4.7

### Columnar epithelia absorb substances and protect the body from digestive chemicals    p. 154

17. **Columnar epithelium** appears rectangular. **Simple columnar epithelium** is found where absorption or secretion takes place, such as the small intestine, stomach, and gallbladder.

18. **Pseudostratified columnar epithelia** are often ciliated, and they line the air passageways of the nasal cavities, trachea, and lungs. **Stratified columnar epithelia** are rare and line large ducts, such as those in the salivary gland or pancreas.

### Module 4.8

### Glandular epithelia are specialized for secretion    p. 156

19. Collections of epithelial cells that produce secretions are called **glands**. **Endocrine glands** release their secretions directly into interstitial fluid. **Exocrine glands** release their secretions into **ducts** or onto an epithelial surface.

4

20. Glandular epithelial cells release their secretions by merocrine, apocrine, or holocrine methods. In **merocrine secretion**, the most common method, the product is released by exocytosis. **Apocrine secretion** involves the loss of secretory product and cytoplasm. In **holocrine secretion**, gland cells become packed with secretions and then burst.

**Module 4.9**

**Exocrine glands can be classified by structure**    p. 157

21. Exocrine glands are classified based on the structure of the duct, the shape of the gland, and the relationship between the duct and the secretory areas. Forms of **simple** exocrine glands are **simple tubular**, **simple coiled tubular**, **simple branched tubular**, **simple alveolar (acinar)**, or **simple branched alveolar**. Forms of **compound** exocrine glands are **compound tubular**, **compound alveolar (acinar)**, or **compound tubuloalveolar**.

## SECTION 2 • Connective Tissue

**Module 4.10**

**A matrix surrounds connective tissue cells**    p. 159

22. Connective tissue varies widely, but all forms contain specialized cells, extracellular protein fibers, and **ground substance** fluid. Connective tissues provide a structural framework, transport fluids and materials, protect organs, support and interconnect other tissues, store energy, and defend the body.

23. The three broad categories of connective tissue are **connective tissue proper**, which includes **loose** and **dense connective tissues**; **fluid connective tissues** (**blood** and **lymph**); and **supporting connective tissues** (**cartilage** and **bone**).

**Module 4.11**

**Loose connective tissues support other tissue types**    p. 160

24. Connective tissue proper contains extracellular protein fibers, a viscous ground substance, and cells that are either fixed or wandering.

25. The three types of **loose connective tissues** are **areolar tissue** (the most common), **adipose tissue**, and **reticular tissue**.

**Module 4.12**

**Dense connective tissues are dominated by extracellular fibers, whereas fluid connective tissues have an aqueous matrix** p. 162

26. Extracellular fibers make up most of the volume of **dense connective tissues**.

27. The body has three types of dense connective tissues: **dense regular connective tissue**, **dense irregular connective tissue**, and **elastic tissue**.

28. **Fluid connective tissues** have a fluid matrix containing many types of suspended proteins. In **blood**, the watery matrix is called **plasma**. **Red blood cells**, **white blood cells**, and **platelets**, collectively known as **formed elements**, are suspended in the plasma. White blood cells in plasma include **monocytes**, **lymphocytes**, **eosinophils**, **neutrophils**, and **basophils**. **Lymph** is formed from interstitial fluid and has roles in maintaining blood volume and alerting the immune system to infections.

**Module 4.13**

**Cartilage provides a flexible support for body structures** p. 164

29. In **cartilage**, the matrix is a firm gel. Cartilage cells, or **chondrocytes**, occupy small chambers known as **lacunae** and are the only cells in the avascular matrix.

30. There are three types of cartilage: **hyaline cartilage**, **elastic cartilage**, and **fibrocartilage**. Cartilage is separated from surrounding tissues by a structure known as the **perichondrium**.

31. Cartilage grows by two mechanisms. **Appositional growth** adds new layers of cartilage to the surface, whereas **interstitial growth** enlarges the cartilage from within.

**Module 4.14**

**Bone provides a strong framework for the body**    p. 166

32. **Bone tissue** or **osseous tissue**, contains a small volume of ground substance. The matrix of bone is mostly calcium salts (calcium phosphate and calcium carbonate) and collagen fibers.

33. A typical long bone is hollow, and its walls contain either well-organized **compact bone** or a network of **spongy bone**.

**Module 4.15**

**Tissue membranes are physical barriers, and fasciae support and surround organs**    p. 168

34. **Tissue membranes** line or cover body surfaces. **Mucous membranes** line passageways to the exterior, **serous membranes** line the cavities of the body trunk, the **cutaneous membrane** covers the surface of the body, and **synovial membranes** line freely movable joint cavities.

35. **Fasciae** are connective tissue layers that support and surround organs. Fasciae consist of three types of layers: the **superficial fascia**, the **deep fascia**, or the **subserous fascia**.

## SECTION 3 • Muscle Tissue and Nervous Tissue

**Module 4.16**

**Muscle tissue outweighs nervous tissue by 25:1**    p. 171

36. As a percentage of body weight, the four tissue types contribute the following: muscle tissue 50 percent, connective tissue 45 percent, epithelial tissue 3 percent, and nervous tissue 2 percent.

**Module 4.17**

**Muscle tissue is specialized for contraction, and nervous tissue is specialized for communication**    p. 172

37. Movement is produced by **muscle tissue** that is specialized for contraction. There are three types: **skeletal muscle tissue**, **cardiac muscle tissue**, and **smooth muscle tissue**.

38. Skeletal muscle tissue composes skeletal muscles, has long cylindrical cells, is **striated**, and is **multinucleate**. Cardiac muscle tissue forms the heart, has cells that are short and branched, is striated, and contains **intercalated discs**. Smooth muscle tissue is found throughout the body and has cells that are spindle shaped, are nonstriated, and have a single, central nucleus.

39. **Nervous tissue** is specialized to conduct electrical impulses. Nervous tissue contains two basic cell types. **Neurons** are the cells that conduct electrical impulses and transfer and process information. **Neuroglia**, or **glial cells**, protect, support, and repair nervous tissue.

**Module 4.18**

**The response to tissue injury involves inflammation and regeneration**    p. 174

40. Restoration of homeostasis after an injury involves two processes: inflammation and regeneration.

41. **Inflammation** is indicated by swelling, redness, warmth, and pain. The inflammatory process removes damaged cells and tissue and fights pathogens. **Regeneration** is the repair that occurs after the damaged tissue has been stabilized and the inflammation has subsided.

# Chapter Review Questions

## Labeling

Identify the three categories of cartilage shown below.

**1** _____  **2** _____  **3** _____

Identify the three categories of muscle shown below.

**4** _____  **5** _____  **6** _____

Identify the six connective tissue types shown below.

**7** _____  **8** _____  **9** _____  **10** _____  **11** _____  **12** _____

**Multiple choice:** Select the correct answer from the list provided.

**13** Tissue that is specialized for contraction is
- a) epithelial tissue.
- b) connective tissue.
- c) muscle tissue.
- d) nervous tissue.

**14** The specialized intercellular junctions found in cardiac muscle are
- a) hemidesmosomes.
- b) intercalated discs.
- c) desmosomes.
- d) microvilli.

**15** Collections of specialized cells and cell products that perform a limited number of functions are called
- a) cellular aggregates.
- b) organs.
- c) tissues.
- d) organ systems.

**16** Axons, dendrites, and a cell body are characteristic of cells located in
- a) nervous tissue.
- b) connective tissue.
- c) muscle tissue.
- d) epithelial tissue.

**17** Specialized cells, extracellular protein fibers, and ground substance are characteristic of which type of tissue?
- a) muscle tissue
- b) epithelial tissue
- c) nervous tissue
- d) connective tissue

**18** Which of the following epithelia most easily permits diffusion?
- a) stratified squamous epithelium
- b) simple squamous epithelium
- c) transitional epithelium
- d) simple columnar epithelium

**19** The tissue lining the peritoneal cavity is an example of a
- ☐ a) mucous membrane.
- ☐ b) serous membrane.
- ☐ c) cutaneous membrane.
- ☐ d) synovial membrane.

**20** The addition of new layers of cartilage to the outer surface is typical of which type of growth?
- ☐ a) appositional
- ☐ b) interstitial

**21** Which of the following is the most common form of connective tissue proper?
- ☐ a) reticular tissue
- ☐ b) areolar tissue
- ☐ c) adipose tissue
- ☐ d) elastic tissue

**22** Which of the following membranes line movable joints?
- ☐ a) mucous membranes
- ☐ b) serous membranes
- ☐ c) cutaneous membranes
- ☐ d) synovial membranes

## Short answer

**23** Why is stratified squamous epithelium well suited to form the surface of the skin and the lining of the mouth and throat?

**24** Differentiate between endocrine glands and exocrine glands.

**25** Which two cell populations make up nervous tissue? What are the functions of each?

**26** Why are infections always a serious threat after a severe skin burn or abrasion?

**27** Describe the unique characteristics you would expect to see when examining cardiac muscle tissue under a light microscope.

**28** During a lab practical, Jason examines a tissue that is composed of densely packed protein fibers that are running parallel. There are no striations, but small nuclei are visible. Jason identifies the tissue as skeletal muscle. Why is Jason's choice wrong, and which tissue is he probably observing?

**29** While jogging through campus, Emily trips off a curb and falls onto the street. Her hands and exposed knees slide across the rough asphalt, resulting in deep abrasions. Outline the tissue injury response Emily's body will undergo.

---

# MasteringA&P™

Access more chapter study tools online in the MasteringA&P Study Area:
- Chapter Quizzes, Chapter Practice Test, and Clinical Case Studies

- Practice Anatomy Lab™ 3.0 — PAL™3.0
- Interactive Physiology™ — iP2™
- A&P Flix™ — A&PFlix™
- PhysioEx™ 9.1 — PhysioEx 9.1

# Chapter Integration • Applying what you have learned

## Some confusion in the histology lab

Alexis is a pre-nursing student enrolled in an anatomy and physiology class. Although she is familiar with basic microscope use, she is having difficulty identifying tissues in a histology lab. She has placed a slide labeled "simple columnar epithelium" onto the microscope stage and has focused her microscope with the lowest objective. As a source of reference she opens her text, *Visual Anatomy & Physiology,* to Module 4.7 and closely examines the light micrograph of simple columnar epithelium. To her dismay, the image in her microscope and the image in the text do not look at all alike.

**1**   Explain the discrepancy.
**2**   If you were her lab partner, how would you help her?

*A 350 × light micrograph of simple columnar epithelium as shown in Module 4.7.*

*What Alexis is seeing through her microscope.*

# 5

# The Integumentary System

## Module 5.1

# The integumentary system consists of the skin and various accessory structures

Go to
MasteringA&P™
PAL 3.0
Anatomical
Models→
Integumentary

The **integumentary system** is the most accessible and often the least appreciated organ system. Often referred to simply as the *skin* or **integument** (in-TEG-ū-ment), this system makes up about 16 percent of your total body weight. Its surface, 1.5–2 m² in area, is constantly worn away, attacked by microorganisms, irradiated by sunlight, and exposed to environmental chemicals. The integumentary system is the place where you and the outside world meet—and your body's first line of defense against an often hostile environment.

5

### Cutaneous Membrane

The **epidermis** (*epi*, above) consists of a stratified squamous epithelium.

Dermis ⎧ Papillary layer
⎩ Reticular layer

The **dermis** consists of a papillary layer of areolar tissue and a reticular layer of dense irregular connective tissue.

The **subcutaneous layer** (hypodermis) separates the integument from the fascia around deeper organs. Note that this tissue layer is not part of the integument.

Subpapillary plexus

Fat

### Accessory Structures

Hair shaft

Pore of sweat gland duct

Tactile (Meissner's) corpuscle

Sebaceous gland

Arrector pili muscle

Sweat gland duct

Hair follicle

Lamellar (Pacinian) corpuscle

Nerve fibers

Sweat gland

Artery

Vein

The network of arteries and veins connected to smaller vessels servicing the tissues of the integumentary system is the **cutaneous plexus**.

**1** Two main parts of the integumentary system are the **cutaneous membrane** and the **accessory structures**. The epidermis and dermis make up the *cutaneous membrane*. Components with supporting functions make up the *accessory structures*.

### Functions of the Integumentary System

- Protect underlying tissues and organs against impact, abrasion, fluid loss, and chemical attack
- Excrete salts, water, and organic wastes by integumentary glands
- Maintain normal body temperature through either insulation or evaporative cooling, as needed
- Produce melanin, which protects underlying tissue from ultraviolet radiation
- Produce keratin, which protects against abrasion and serves as a water repellent
- Synthesize vitamin $D_3$, a steroid that is subsequently converted to calcitriol, a hormone important to normal calcium metabolism
- Store lipids in adipocytes in the dermis and in adipose tissue in the subcutaneous layer
- Detect touch, pressure, pain, and temperature stimuli, and relay that information to the nervous system

**? REVIEW**

**A.** Identify the two major components of the cutaneous membrane.

**B.** Identify the major functions of the integumentary system.

**Lo LEARNING OUTCOME**

Describe the tissue structure of the integument and the functions of the integumentary system.

# Module 5.2
## The epidermis is composed of strata (layers) that have various functions

Go to
MasteringA&P™
PAL™ 3.0
Histology →
Integumentary
System

The epidermis is dominated by **keratinocytes** (ke-RAT-i-nō-sīts), the body's most abundant epithelial cells. These cells form several layers, or **strata**. Keratinocytes are continuously produced by stem cell divisions in the deepest layers and are shed at the exposed surface.

**1** The deeper layers of the epidermis form **epidermal ridges**, which extend into the dermis and are adjacent to dermal projections called **dermal papillae** (singular, *papilla*; small, nipple-like process) that project into the epidermis. These ridges and papillae are significant because they greatly increase the surface area for attachment, firmly binding the epidermis to the dermis.

Epidermis

Epidermal ridge

Dermal papilla

Dermis

**2** **Thin skin**, which covers most of the body surface, contains four strata. Thin skin is about as thick as the wall of a plastic sandwich bag (roughly 0.08 mm).

Epidermis

Dermis

**Thin skin**          LM × 225

**3** **Thick skin**, found on the palms of the hands and the soles of the feet, contains a fifth stratum. This skin is about as thick as a standard paper towel (roughly 0.5 mm). Note that the terms "thick" and "thin" refer to the relative thickness of just the epidermis, not to the cutaneous membrane as a whole.

Epidermis

Dermal papilla

Epidermal ridge

**Thick skin**          LM × 225

**Thick skin**

**4** This scanning electron micrograph of the tip of an index finger shows the ridge patterns in the thick skin that produce fingerprints. These ridge shapes are determined by genes and by the intrauterine environment during fetal development. Your epidermal ridge pattern is unique and does not change during your lifetime.

**Thick skin**          SEM × 25

**? REVIEW**
**A.** Compare thin skin with thick skin.

Papillary layer of dermis

LM × 470

**5** Like all other epithelia, the epidermis lacks local blood vessels. Epidermal cells rely on the diffusion of nutrients and oxygen from capillaries within the dermis. For this reason, the cells with the highest metabolic demand are closest to the underlying dermis.

## Layers of the Epidermis

The **stratum corneum** (STRA-tum KOR-nē-um; *cornu*, horn) is at the exposed surface of both thick skin and thin skin. It normally contains 15 to 30 layers of keratinized cells. Keratinization is the formation of protective, superficial layers of cells filled with the protein **keratin** (KER-a-tin; *keros*, horn). The dead cells in each layer of the stratum corneum remain tightly interconnected by desmosomes. It takes 7 to 10 days for a cell to move from the stratum basale to the stratum corneum. The dead cells generally remain in the exposed stratum corneum for an additional 2 weeks before they are shed or washed away. The stratum corneum is water resistant, but not waterproof. Water from interstitial fluid slowly diffuses to the surface and evaporates into the surrounding air. You lose about 500 mL (about 1 pt) of water in this way each day. The process is called **insensible perspiration** because you are unable to see or feel (sense) the water loss. In contrast, you are usually very aware of the **sensible perspiration** produced by active sweat glands.

In the thick skin of the palms and soles, a **stratum lucidum** ("clear layer") separates the stratum corneum from deeper layers. The cells in this layer are flattened, densely packed, largely without organelles, and filled with the proteins keratin and **keratohyalin** (ker-a-tō-HĪ-a-lin). These proteins were synthesized while the cells were in the underlying stratum granulosum. By the time they reach the stratum lucidum, they are dead and undergoing dehydration.

The **stratum granulosum** ("grainy layer") consists of three to five layers of keratinocytes. By the time cells reach this layer, most have stopped dividing and have started making large amounts of keratin and keratohyalin. As these protein fibers develop, the cells grow thinner, and their membranes thicken and become less permeable.

The **stratum spinosum** ("spiny layer") consists of 8 to 10 layers of keratinocytes bound together by desmosomes. Its name refers to the fact that the cells look like miniature pincushions in standard histological sections. The stratum spinosum also contains **dendritic (Langerhans) cells**, which participate in the immune response by stimulating a defense against (1) microorganisms that manage to penetrate the superficial layers of the epidermis and (2) superficial skin cancers.

The **stratum basale** (STRA-tum buh-SAHL-āy) is the basal layer of the epidermis. Hemidesmosomes attach the cells of this layer to the basement membrane that separates the epidermis from the areolar tissue in the adjacent papillary layer of the dermis. **Basal cells** dominate the stratum basale. Basal cells are stem cells whose divisions replace the more superficial keratinocytes that are lost or shed at the epithelial surface. Merkel cells involved with touch sensations are scattered among the basal cells. A Merkel cell along with a sensory nerve terminal is called a *tactile disc*. (An overview of tactile discs and other touch receptors of the skin is found in Module 5.4.)

**? REVIEW**

**B.** Identify the five layers of the epidermis (from superficial to deep).

**C.** Dandruff is caused by excessive shedding of cells from the outer layer of skin on the scalp. So, dandruff is composed of cells from which epidermal layer?

 **INTEGRATION**

**D.** A splinter that penetrates to the third layer of the epidermis of the palm is lodged in which layer?

**LO LEARNING OUTCOME**

Describe the main structural features of the epidermis, and explain the functional significance of each feature.

## Module 5.3
# Factors influencing skin color include epidermal pigmentation and dermal circulation

The color of one's skin is genetically determined. Skin color is influenced by the presence of pigments in the skin, the degree of dermal circulation, and the thickness and degree of keratinization in the epidermis. Increased pigmentation, or tanning, can result in response to ultraviolet (UV) radiation.

**1** The primary pigments involved in skin coloration are carotene and melanin. The micrograph at right shows a section through the thin skin of a light-skinned person. The ratio of melanin-producing melanocytes to basal cells ranges between 1:4 and 1:20, depending on the body region. The skin covering most areas of the body has about 1000 melanocytes per square millimeter. Skin color differences are due to the amount of melanin produced, not the number of melanocytes. Light-skinned people produce less melanin than dark-skinned people. Absence of melanin production leads to a disorder called **albinism**. People with this disorder have a normal distribution of melanocytes, but the cells lack the enzyme needed to produce melanin. In the figure below, begin with the center, text describing a melanocyte.

Melanocytes in stratum basale

Melanin pigment

Basement membrane

**Thin skin**          LM × 400

Keratinocyte

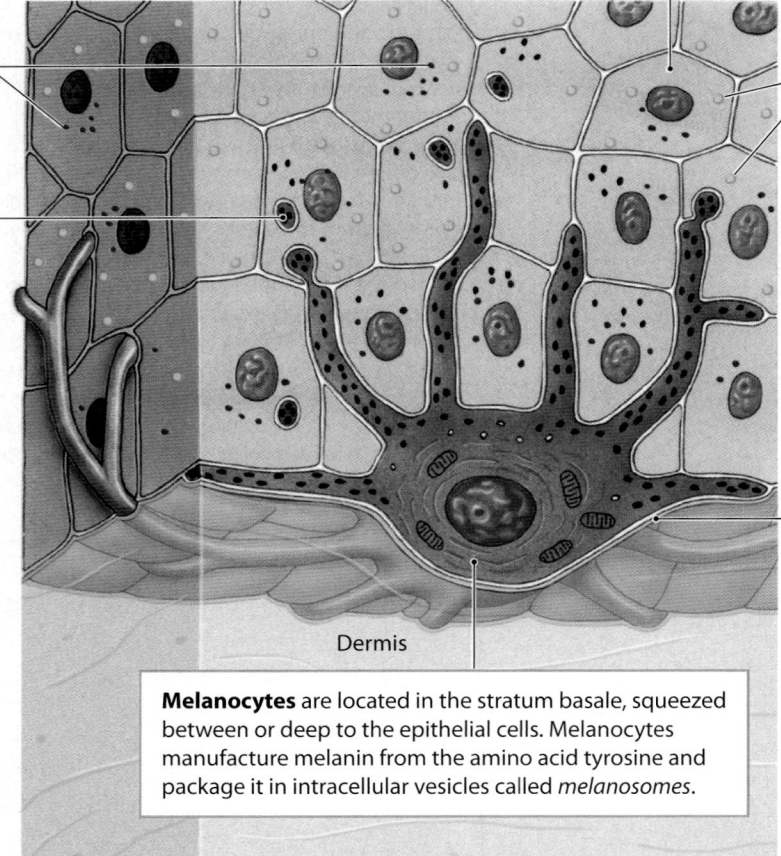

**Melanin** is a brown, yellow-brown, or black pigment produced by melanocytes.

**Melanosomes** travel within the processes of melanocytes and are transferred intact to keratinocytes. The transfer of pigmentation colors the keratinocyte temporarily, until the melanosomes are destroyed by fusion with lysosomes. In people with pale skin, this transfer occurs in the stratum basale and stratum spinosum, and the cells of more superficial layers lose their pigmentation. In dark-skinned people, the melanocytes are more active, melanosomes are larger and more numerous, and the transfer may occur in the stratum granulosum as well. As a result, skin pigmentation is darker and more persistent.

**Carotene** (KAR-uh-tēn) is an orange-yellow pigment that normally accumulates in epidermal cells. It is most apparent in the stratum corneum cells of light-skinned people, but it also accumulates in fatty tissues in the deep dermis and hypodermis. Carotene is found in orange vegetables, such as carrots and some squashes.

Basement membrane

Dermis

**Melanocytes** are located in the stratum basale, squeezed between or deep to the epithelial cells. Melanocytes manufacture melanin from the amino acid tyrosine and package it in intracellular vesicles called *melanosomes*.

 **REVIEW**

**A.** Name the two pigments contained in the epidermis.

**B.** Why does exposure to sunlight or sunlamps darken skin?

**2** The blood supply affects skin color because blood contains red blood cells filled with the red pigment hemoglobin. When bound to oxygen, hemoglobin is bright red, giving capillaries in the dermis a reddish tint that is most apparent in lightly pigmented people. If those vessels are dilated, the red tones become much more pronounced. For example, your skin may become flushed and red when your body temperature rises because the superficial blood vessels dilate so that the skin can act like a radiator and lose heat. When its blood supply is temporarily reduced, the skin may become pale. A light-skinned person who is frightened may "turn white" because of a sudden drop in blood supply to the skin. During a sustained reduction in blood flow, the oxygen levels in the tissues decline. Under these conditions, hemoglobin releases oxygen and turns a much darker red. Seen from the surface, the skin then takes on a bluish coloration called **cyanosis** (sī-uh-NŌ-sis; *kyanos*, blue). In people of any skin color, cyanosis is easiest to see in areas of very thin skin, such as the lips or beneath the nails.

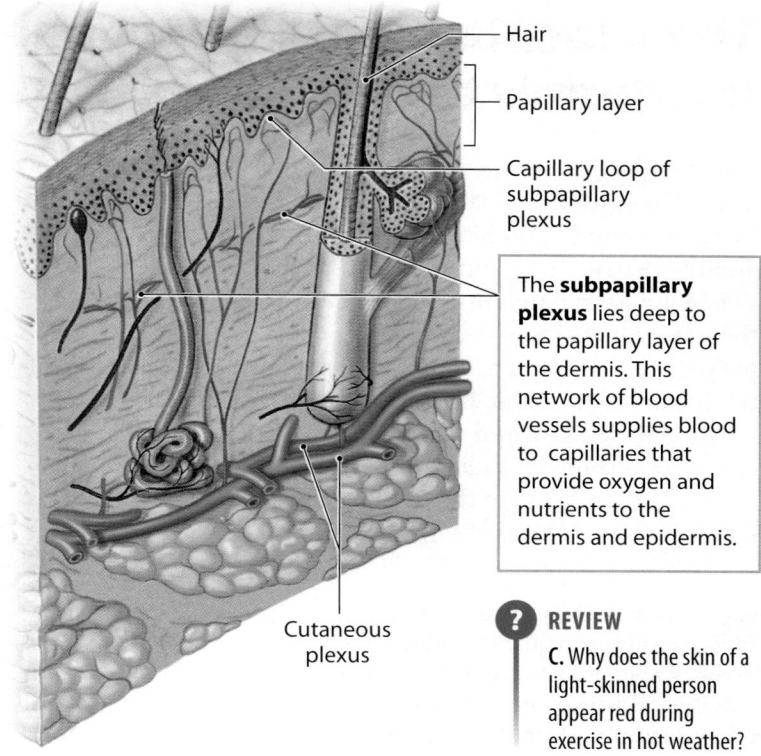

Hair

Papillary layer

Capillary loop of subpapillary plexus

The **subpapillary plexus** lies deep to the papillary layer of the dermis. This network of blood vessels supplies blood to capillaries that provide oxygen and nutrients to the dermis and epidermis.

Cutaneous plexus

5

**? REVIEW**
**C.** Why does the skin of a light-skinned person appear red during exercise in hot weather?

**3** Skin cancers are the most common types of cancers. The most common form of skin cancer is a **basal cell carcinoma**. This cancer originates in the stratum basale as a result of mutations caused by overexposure to the UV radiation in sunlight. Metastasis virtually never occurs in basal cell carcinomas, and most people survive these cancers. The melanin in keratinocytes provides some protection against the effects of UV radiation because the melanosomes are concentrated around the nucleus, where they act like sunshades for the enclosed DNA.

**4** In contrast, a **malignant melanoma** (mel-a-NŌ-muh) is extremely dangerous. In this condition, cancerous melanocytes grow rapidly and metastasize through the lymphatic system. The outlook for long-term survival is in many cases determined by how early the condition is diagnosed. If the cancer is detected early, while it is still localized, the affected area can be surgically removed, and the 5-year survival rate is 99 percent. If the condition is not detected until extensive metastasis has occurred, the 5-year survival rate drops to 14 percent.

**? REVIEW**
**D.** Why is basal cell carcinoma considered less dangerous than malignant melanoma?

**? INTEGRATION**
**E.** Explain why a healthy 6-month-old might have yellow-orange skin.

**Lo LEARNING OUTCOME**
Explain what accounts for individual differences in skin color, and compare basal cell carcinoma with malignant melanoma.

# Module 5.4

# The subcutaneous layer connects the dermis to underlying tissues

The **dermis** lies between the epidermis and the subcutaneous layer. The presence of two types of fibers enables the dermis to tolerate limited stretching. Collagen fibers are very strong and resist stretching, but they are easily bent or twisted. Elastic fibers permit stretching and then recoil to their original length. To prevent tissue damage, collagen fibers limit the flexibility of elastic fibers. Aging, hormonal changes, and the destructive effects of ultraviolet radiation permanently reduce the elasticity of the dermis. The results are wrinkles and sagging skin.

**1** The **papillary layer** of the dermis consists of highly vascularized areolar tissue. It contains the capillaries, lymphatic vessels, and sensory nerve fibers that supply the surface of the skin. The papillary layer gets its name from the dermal papillae that project between the epidermal ridges.

Capillary loop · Dermal papillae · Epidermal ridges

Papillary layer

Reticular layer

Subpapillary plexus

Nerve fiber · Cutaneous plexus · Lymphatic vessel

**Reticular layer of dermis**     SEM × 1500

Collagen fibers

Elastic fibers

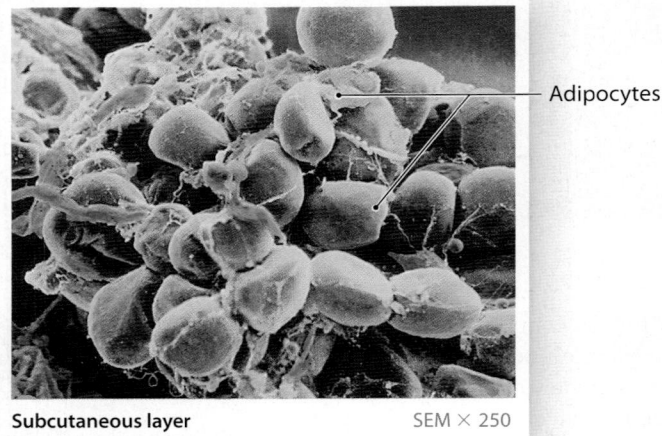

**Subcutaneous layer**     SEM × 250

Adipocytes

**2** The **reticular layer** consists of an interwoven meshwork of dense irregular connective tissue containing both collagen and elastic fibers. Bundles of collagen fibers extend superficially beyond the reticular layer to blend into those of the papillary layer, so the boundary between the two layers is indistinct. Collagen fibers of the reticular layer also extend into the deeper subcutaneous layer. In addition, the reticular and papillary layers of the dermis contain networks of blood vessels, lymphatic vessels, nerve fibers, and accessory organs such as hair follicles and sweat glands.

**3** The **subcutaneous layer** is not part of the skin. It separates the skin from deeper structures. This layer stabilizes the skin's position relative to underlying tissues while permitting independent movement. Often dominated by adipocytes, the subcutaneous fat in this layer stores energy. Because there are no vital organs and capillaries are sparse, this is a common site for subcutaneous injections with hypodermic needles.

**REVIEW**

**A.** Describe the layers of the dermis.

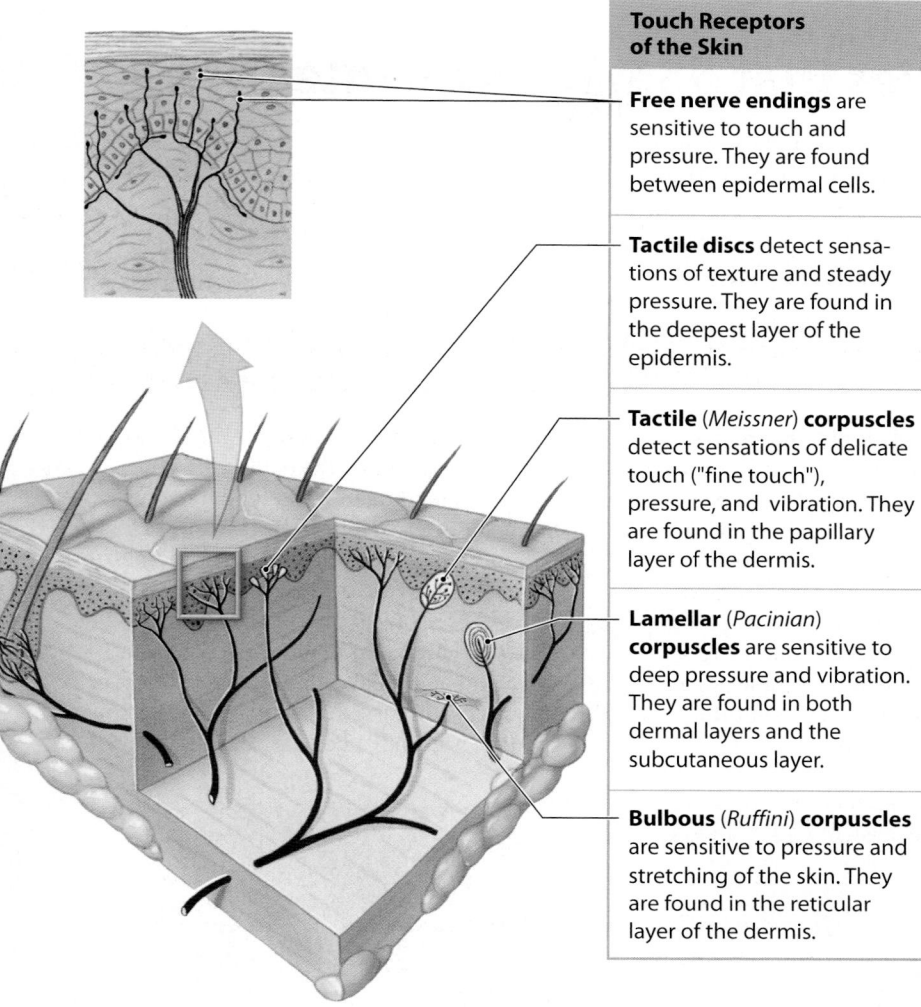

### Touch Receptors of the Skin

**Free nerve endings** are sensitive to touch and pressure. They are found between epidermal cells.

**Tactile discs** detect sensations of texture and steady pressure. They are found in the deepest layer of the epidermis.

**Tactile** (*Meissner*) **corpuscles** detect sensations of delicate touch ("fine touch"), pressure, and vibration. They are found in the papillary layer of the dermis.

**Lamellar** (*Pacinian*) **corpuscles** are sensitive to deep pressure and vibration. They are found in both dermal layers and the subcutaneous layer.

**Bulbous** (*Ruffini*) **corpuscles** are sensitive to pressure and stretching of the skin. They are found in the reticular layer of the dermis.

**4** The integument contains many sensory receptors, and anything that comes in contact with the skin—from the lightest touch of a mosquito to the weight of a loaded backpack—initiates a nerve impulse that can reach our conscious awareness. Each square centimeter of skin contains approximately 400 centimeters of nerve fibers! Here, we focus on the main *touch receptors* found in the epidermis and dermis. Those found in the epidermis are free nerve endings and tactile discs. The dermis contains tactile corpuscles, lamellar corpuscles, and bulbous corpuscles.

---

### 🏃 Everyday Physiology

### Subcutaneous fat accumulates beginning at puberty

Beginning at puberty, men accumulate subcutaneous fat at the neck, arms, lower back, and buttocks. In contrast, women accumulate it at the breasts, buttocks, hips, and thighs. In adults of either gender, the subcutaneous layer of the backs of the hands and the upper surfaces of the feet contains few fat cells, but fat can accumulate in the abdominal region, producing a "potbelly." An excessive amount of such abdominal fat, called central adiposity, is strongly correlated with cardiovascular disease.

| Anterior | Posterior |

**5** The banding shown here indicates the **tension** (*cleavage*) **lines** in the skin. Most of the collagen and elastic fibers at any location are arranged in parallel bundles oriented to resist the forces applied to the skin during normal movement. The resulting pattern of fiber bundles establishes tension lines. These lines are clinically significant: A cut parallel to a tension line will usually remain closed and heal with little scarring. A cut at right angles to a tension line will be pulled open as severed elastic fibers recoil, resulting in greater scarring. For these reasons, surgeons choose to make neat incisions parallel to the lines of cleavage.

**?** **INTEGRATION**

**B.** Predict the degree of scarring after the healing of a horizontal cut on the forehead just above the eyebrow.

**Lo** **LEARNING OUTCOME**

Describe the structures and functions of the dermis and subcutaneous layer.

# Module 5.5
# Burns are significant injuries that damage skin integrity

Burns are significant injuries because they can damage the integrity of large areas of the skin and compromise many essential functions. For example, dehydration and electrolyte imbalance can result, leading to kidney impairment and circulatory shock. Burns result from skin exposure to heat, friction, radiation, electrical shock, or strong chemical agents. The severity of the burn depends on the depth of penetration and the total area affected.

## Partial-Thickness Burns

**1** In a **first-degree burn**, only the surface of the epidermis is affected. In this type of burn, which includes most sunburns, the skin reddens and can be painful. The redness, a sign called **erythema** (er-i-THĒ-muh), results from inflammation of the sun-damaged tissues.

**2** In a **second-degree burn**, the entire epidermis and perhaps some of the dermis are damaged. Accessory structures such as hair follicles and glands are generally not affected, but blistering, pain, and swelling occur. If the blisters rupture, infection can easily develop. Healing typically takes 1 to 2 weeks, and some scar tissue may form. The gray shaded areas represent the depth of damaged skin.

## Full-Thickness Burns

**3** **Third-degree burns**, or *full-thickness burns*, destroy the epidermis and dermis, extending into the subcutaneous layer. Despite swelling, these burns are less painful than second-degree burns, because sensory nerves are destroyed. Extensive third-degree burns cannot repair themselves, because granulation tissue cannot form and epithelial cells are unable to cover the injury (discussed in Module 5.12). Skin grafting is usually necessary.

Erythema

**4** Burns that cover more than 20 percent of the skin surface threaten critical homeostatic functions of the skin.

## Skin Functions Affected by Burns

- **Fluid and Electrolyte Balance**. Even areas with partial-thickness burns lose their effectiveness as barriers to fluid and electrolyte losses. In full-thickness burns, the rate of fluid loss through the skin may reach five times the normal level.

- **Thermoregulation**. Increased fluid loss causes increased cooling by evaporation. As a result, more energy must be expended to keep body temperature within acceptable limits.

- **Protection from Infection**. The dampness of the epidermal surface, resulting from uncontrolled fluid loss, encourages bacterial growth. If the skin is broken, infection is likely. Widespread bacterial infection, or **sepsis** (*sepsis*, rotting), is the leading cause of death in burn victims.

 **REVIEW**

**A.** Distinguish among a first-degree, second-degree, and third-degree burn.

**B.** Which type of burn usually requires skin grafting? Why?

**5** Each year in the United States, roughly 4000 people die in fires or die of burn injuries. In evaluating burns in a clinical setting, two key areas must be determined: the percentage of skin that has been burned and the depth of the burns.

The illustration to the right shows a simple method of estimating the percentage of the surface area affected by burns. This method is called the **rule of nines** because the surface area in adults is divided into multiples of 9. The rule must be modified for children because their body proportions are different.

The depth of the burn can be quickly assessed with a pin. Because loss of sensation is characteristic of a full-thickness burn, the absence of a reaction to a pin prick indicates the presence of third-degree damage.

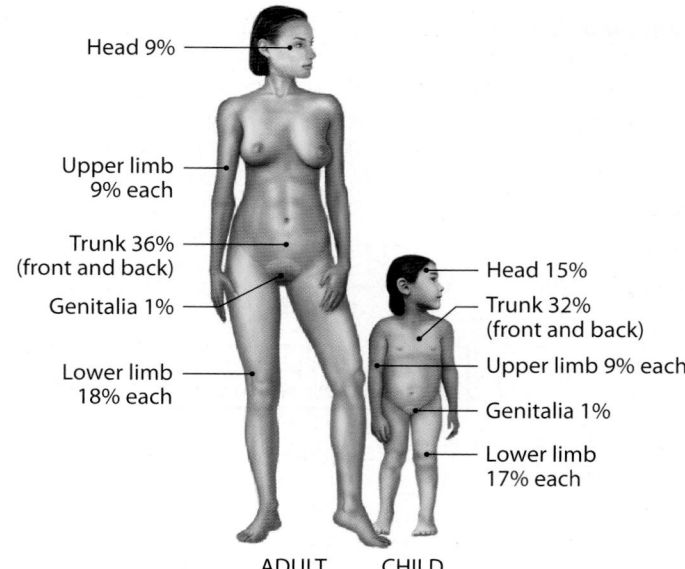

Head 9%
Upper limb 9% each
Trunk 36% (front and back)
Genitalia 1%
Lower limb 18% each

Head 15%
Trunk 32% (front and back)
Upper limb 9% each
Genitalia 1%
Lower limb 17% each

ADULT    CHILD

**6** Because of the loss of critical homeostatic functions, emergency treatment of burns focuses on the following procedures:

- Replacing lost fluids and electrolytes.
- Providing sufficient nutrients to meet increased metabolic demands for thermoregulation and healing.
- Preventing infection by cleaning and covering the burn while administering antibiotic drugs.
- Assisting tissue repair. Because large full-thickness burns cannot heal unaided, surgical procedures are necessary to encourage healing. In a **skin graft**, areas of intact skin are transplanted to cover the site of the burn. A split-thickness graft involves a transfer of the epidermis and superficial portions of the dermis. A full-thickness graft involves the epidermis and both layers of the dermis. A skin graft made with a patient's own undamaged skin is called an **autograft**. If the patient doesn't have enough undamaged skin, then an **allograft** (using frozen skin from a cadaver) or **xenograft** (using animal skin) may be necessary. Allografts and xenografts are temporary because the patient's immune system will reject them, usually within 7–21 days.

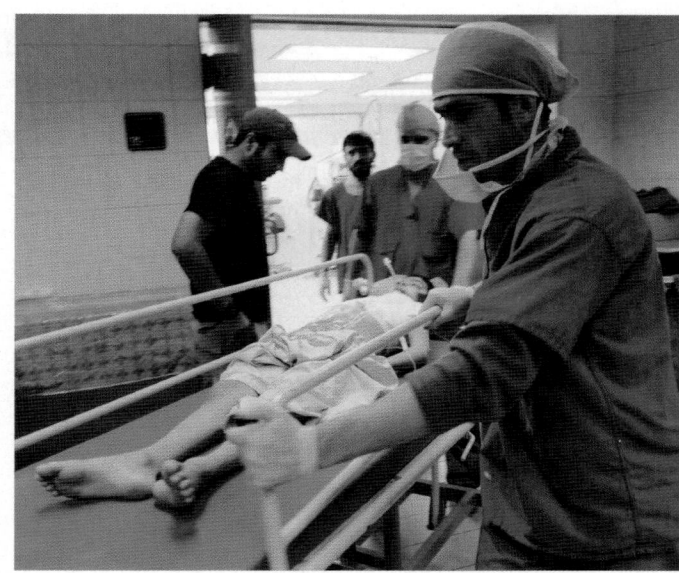

**? REVIEW**

**C.** Describe the three types of skin grafts. Which one is best? Why?

Fluid-replacement therapies, infection control methods, and grafting techniques provide young patients with burns over 80 percent of the body about a 50 percent chance of recovery. Recent advances in cell culturing may further improve survival rates. After a square centimeter or so of undamaged epidermis is removed and grown in the laboratory, basal cell divisions produce large sheets of epidermal cells—up to several square meters in area—that can be transplanted to cover the burn area. Although questions remain about the strength and flexibility of the repairs, skin cultivation is a breakthrough in the treatment of serious burns.

**? INTEGRATION**

**D.** A 32-year-old woman is admitted to the hospital with third-degree burns on her entire right leg, entire right arm, and the back of her trunk. Estimate the percentage of her body surface area affected by these burns.

**LO LEARNING OUTCOME**

Describe the classification of burns and the types of skin grafts.

# Functional Anatomy of the Skin

**Concept map:** Use each of the following terms once to fill in the blank boxes to correctly complete the map.

- accessory structures
- collagen
- connective
- epidermis
- fat
- granulosum
- hypodermis
- nerves
- papillary layer
- reticular layer

Integument

consists of

Cutaneous membrane | **1**

is superficial to

components

**2**

consists of

Stratified squamous epithelial cells of thin skin

consists of

- Stratum basale
- Stratum spinosum
- Stratum **3**
- Stratum corneum

Dermis

regions

Superficial

called

**4**

consists of

- Loose connective tissue
- Capillaries **5**
- 

Deeper

called

**6**

consists of

- Dense, irregular connective tissue
- **7** fibers
- 

Subcutaneous layer

also called

**8**

consists of

- **9** tissue
- **10** cells

**Short answer:** Identify and describe the parts of the cutaneous membrane and the underlying layer of loose connective tissue in the diagram at right.

**11**

**12**

**13**

**14**

**15**

**Section integration:** Describe why melanocyte malignancies are often fatal.

**16**

A firefighter is in the emergency room with serious burns. He tells the doctors that they really don't hurt much. Should they be concerned or relieved by this comment? Why?

**17**

## Module 5.6

# Hair follicles, exocrine glands, and nails are also components of the integumentary system

Hair and several other structures—hair follicles, exocrine glands (sebaceous and sweat glands), and nails—are considered **accessory structures** of the integument. These structures originate from the epidermis during embryonic development, so they are also known as *epidermal derivatives*. Embedded in the dermis, they project up to or above the surface of the epidermis. The accessory structures of the integumentary system and their functions are shown below.

**1** Hair follicles, sebaceous glands, and sweat glands are derived from **epithelial columns**, cords of epidermal cells that grow into the dermis.

**Hair Follicles**

Epithelial column

- Produce hairs that protect skull
- Produce hairs that provide delicate touch sensations on general body surface

**Exocrine Glands**

Epithelial column

**Sweat glands**
- Assist in thermoregulation
- Excrete wastes

**Sebaceous glands**
- Lubricate epidermis

**2** Nails begin as thickenings of the epidermis near the tips of the fingers and toes. These thickened areas settle into the dermis and form a distinct epidermis-dermis borderline and **nail field** containing basal cells.

**Nails**

Nail field

- Protect and support tips of fingers and toes

**? REVIEW**
**A.** What are epidermal derivatives?
**B.** Which exocrine glands are in the integument?

**Lo LEARNING OUTCOME**
Describe the main functions of the accessory structures of the integumentary system.

# Hair is composed of dead, keratinized cells produced in a specialized hair follicle

**Hairs** project above the surface of the skin almost everywhere, except over the sides and soles of the feet, the palms of the hands, the sides of the fingers and toes, the lips, and portions of the external genitalia. The human body has about 2.5 million hairs, and 75 percent of them are on the general body surface, not on the head. Hairs are nonliving structures produced in organs called hair follicles.

**1** A **hair follicle** is a complex structure composed of epithelial and connective tissues that forms a single hair. Adults have two types of hair in the integument that differ in texture. **Terminal hairs** are large, coarse, often darkly pigmented hairs such as those found on the scalp or the armpits. **Vellus hairs** are smaller, shorter, and more delicate hairs found on the general body surface.

**2** Hair formation begins at the expanded base of a hair follicle called the **hair bulb**. The mass of epithelial cells making up the hair bulb forms a cap that surrounds a small hair papilla, a peg of connective tissue containing capillaries and nerves.

A **sebaceous gland** produces secretions that coat the hair and the adjacent skin surface.

The **hair shaft**, which we see on the surface, begins deep within the hair follicle.

The **hair root**—the portion that anchors the hair into the skin—extends from the base of the hair follicle, where hair production begins, to the point where the hair shaft loses its connection with the walls of the follicle.

A connective tissue sheath surrounds the epithelial cells of the hair follicle.

A **root hair plexus** of sensory nerves surrounds the base of each hair follicle. As a result, you can feel the movement of the shaft of even a single hair.

The **cuticle** of the hair consists of daughter cells produced at the edges of the hair matrix. The cuticle forms the surface of the hair.

The **cortex** of the hair is an intermediate layer of daughter cells deep to the cuticle.

The **medulla** of the hair consists of daughter cells formed at the center of the hair matrix.

The **hair matrix** consists of a layer of basal cells at the base of the hair bulb in contact with the hair papilla. These actively dividing cells produce the hair. As these cells divide, daughter cells are gradually pushed toward the surface. Melanocytes are also scattered among the cells of the hair matrix.

The **hair papilla** is a small connective tissue peg filled with blood vessels and nerves.

Hair bulb

The **arrector pili** (a-REK-tor PĪ-lī) **muscles** are bundles of smooth muscles attached to the hair follicle. Contraction causes the hair to stand erect, producing "goose bumps."

**? REVIEW**

**A.** Describe a typical strand of hair.

**B.** What happens when an arrector pili muscle contracts?

**3** A cross section through a hair follicle reveals the internal structure of both the hair and the follicle. We can change our hair color by using chemicals that disrupt the cuticle, permitting dyes to enter and stain the cortex and medulla. These color treatments damage the hair's protective cuticle layer, dehydrating and weakening the hair shaft. As a result, the hair becomes more thin and brittle.

## Hair Structure

The medulla, or core, of the hair contains a flexible **soft keratin**.

The cortex contains thick layers of **hard keratin**, which give the hair its stiffness.

The cuticle, although thin, is very tough and contains hard keratin.

## Hair Follicle Structure

The **internal root sheath** surrounds the hair root and the deeper portion of the shaft. It is also produced from the hair matrix. The cells of this sheath disintegrate quickly, and this layer does not extend the entire length of the hair follicle.

The **external root sheath** consists of epithelial cells and extends from the skin surface to the hair matrix.

The **glassy membrane** is a thickened, clear basement membrane.

Connective tissue sheath

**4** Hairs grow and are shed according to a **hair growth cycle**. A hair in the scalp grows for 2 to 5 years, at a rate of about 0.33 mm per day. Variations in the growth rate and in the duration of the hair growth cycle account for individual differences in the length of uncut hair. Healthy adults with a full head of hair typically lose about 100 head hairs each day.

**2** The follicle then begins to undergo *regression* and transitions to the resting phase.

**?** **REVIEW**
C. Describe the four phases of the hair growth cycle.

**1** During the *active phase,* the hair grows continuously.

**3** During the *resting phase,* the hair loses its attachment to the follicle and becomes a **club hair**.

**4** When follicle *reactivation* occurs, the club hair is shed, and the hair matrix begins producing a replacement hair.

Variations in hair color reflect differences in structure and variations in the pigment produced by melanocytes at the hair matrix. Different forms of melanin give a dark brown, yellow-brown, or red color to the hair. As pigment production decreases with age, hair color lightens. Genes determine these structural and biochemical characteristics, but hormonal and environmental factors also influence your hair's condition. White hair results from the combination of a lack of pigment and the presence of air bubbles in the medulla of the hair shaft.

**?** **INTEGRATION**
D. Why is pulling a hair painful, yet cutting a hair is not?

**Lo** **LEARNING OUTCOME**
Describe the mechanisms of hair production, and explain the structural basis for hair texture and color.

# Module 5.8

## Sebaceous glands and sweat glands are exocrine glands in the skin

In this module, we examine the structure and function of the accessory structures that produce exocrine secretions, with particular attention to sebaceous glands and sweat glands. These exocrine glands assist in thermoregulation, excrete wastes, and lubricate the epidermis.

**1** **Sebaceous** (se-BĀ-shus) **glands**, or oil glands, discharge an oily lipid secretion into hair follicles by holocrine secretion (see Module 4.8). Sebaceous glands that communicate with a single follicle share a duct and thus are classified as simple branched alveolar glands (see Module 4.9). The lipids released from gland cells enter the lumen (open passageway) of the gland. Arrector pili muscle contractions may force sebaceous gland secretions along the hair follicle and onto the skin surface.

 **REVIEW**
**A.** Identify and describe the general functions of the exocrine glands found in the skin.

---

### Sebaceous Glands

Sebaceous glands secrete **sebum** (SĒ-bum), which coats the hair shaft and surrounding epidermal surfaces. Sebum is a mixture of triglycerides, cholesterol, proteins, and electrolytes. Sebum lubricates and keeps the hair shaft from becoming dry and brittle. It also moisturizes the surrounding skin and inhibits the growth of bacteria.

#### Typical Sebaceous Gland

Secretes sebum into hair follicles

- Lumen (hair removed)
- Wall of hair follicle
- Basement membrane
- Discharge of sebum
- Lumen
- Breakdown of cell membranes
- Mitosis and growth
- Basal cells

**Sebaceous gland**  LM × 150

- Sebaceous gland

#### Sebaceous Follicle

Secretes sebum onto the skin surface; located on the face, back, chest, nipples, and external genitalia

**REVIEW**
**B.** Describe the secretory method and functions of sebum.

## Sweat Glands

Sweat glands produce a watery solution by merocrine secretion, flush the epidermal surface, and perform other special functions.

### Apocrine Sweat Glands

**2** **Apocrine sweat glands** are found in the armpits (axillae), around the nipples, and in the pubic region, where they secrete into hair follicles. These glands produce a sticky, cloudy, and potentially odorous secretion. Despite their name, these glands rely on merocrine secretion.

Connective tissue of dermis

Lumen

Apocrine gland cells

**Section of apocrine sweat gland**      LM × 375

- Limited distribution (axillae, groin, nipples)
- Produce a viscous secretion of complex composition
- Possible function in olfactory communication
- Strongly influenced by hormones
- Include ceruminous glands of the external ear and mammary glands that produce milk

### Eccrine (Merocrine) Sweat Glands

**3** The adult integument also contains 2–5 million **eccrine sweat glands**, also called **merocrine sweat glands**, that discharge their secretions directly onto the surface of the skin. The palms and soles have the highest numbers; the palms have an estimated 500 eccrine sweat glands per square centimeter (3000 per square inch).

**Myoepithelial cells** (*myo-*, muscle) are contractile cells that squeeze the gland and discharge the accumulated secretion. They are also found in other exocrine glands: mammary glands, lacrimal (tear) glands, and salivary glands.

Eccrine gland cells

Lumina

**Section of eccrine sweat gland**      LM × 210

- Found in most areas of the skin
- Produce watery secretions containing electrolytes
- Merocrine secretion method
- Controlled primarily by nervous system
- Important in thermoregulation and excretion
- Some antibacterial action

Sweat pore

 **REVIEW**

**C.** Deodorants are used to mask the effects of secretions from which type of skin gland?

 **INTEGRATION**

**D.** Name the type of cell that assists the discharge of merocrine secretions, and list the glands in which such cells are present.

**LEARNING OUTCOME**

Describe the various kinds of exocrine glands in the skin and discuss the secretions of each.

# Nails are thick sheets of keratinized epidermal cells that protect the tips of fingers and toes

Nails protect the exposed dorsal surfaces of the tips of the fingers and toes. They also help limit distortion of the digits when they are subjected to mechanical stress—for example, when you run or grasp objects.

**1** These superficial and sectional views illustrate common landmarks of nail structure. The **nail body**, the visible portion of the nail, is made up of dead, tightly compressed cells packed with keratin. The nail body is recessed deep to the level of the surrounding epithelium and is bounded on either side by **lateral nail grooves** (depressions) and **lateral nail folds** (ridges).

The nail body covers an area of epidermis called the **nail bed**.

Nail body

Lateral nail fold

Lateral nail groove

Phalanx (bone of fingertip)

Direction of growth

Free edge of nail body

Underlying blood vessels give the nail its characteristic pink color.

Near the root, the dermal blood vessels are obscured by a thickened stratum basale in the nail bed, leaving a pale crescent known as the **lunula** (LŪ-nū-la; *luna*, moon).

Proximal nail fold

Eponychium

A portion of the stratum corneum of the nail root extends over the exposed nail, forming the **eponychium** (ep-ō-NIK-ē-um; *epi*, over + *onyx*, nail), or **cuticle**.

Proximal nail fold

Lunula    Nail body

The free edge of the nail—the distal portion that continues past the nail bed—extends over the **hyponychium** (hī-pō-NIK-ē-um), an area of thickened stratum corneum.

Nail production occurs at the **nail root**, an epidermal fold not visible from the surface. The deepest portion of the nail root lies very close to the bone of the fingertip.

Epidermis   Dermis   Phalanx

**? REVIEW**
**A.** Where does nail production occur?
**B.** Define *hyponychium*.

**2** The cells producing the nails can be affected by conditions that alter body metabolism, so changes in the shape, structure, or appearance of the nails can provide useful diagnostic information. For example, the nails may become pitted and distorted as a result of **psoriasis** (a condition marked by rapid stem cell division in the stratum basale), and concave as a result of some blood disorders.

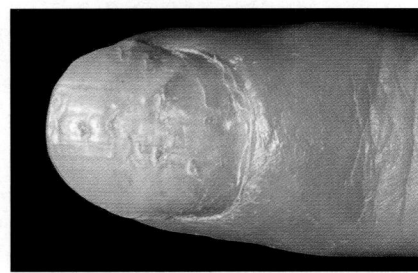

**? INTEGRATION**
**C.** Why does the free edge of a nail appear white?

**Lo LEARNING OUTCOME**
Describe the structure of a typical nail.

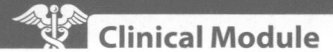 

# Module 5.10
# Age-related changes affect the integument

Aging affects all of the structures in the integument. The major structural and functional changes are summarized in this illustration. The gradual changes in the superficial appearance of the skin provide visual cues that we use unconsciously to estimate a person's age. This accounts for the popularity of cosmetic facial surgery.

**Fewer Melanocytes**

Melanocyte activity declines, and in light-skinned people the skin becomes very pale. With less melanin in the skin, people become more sensitive to sun exposure and more likely to experience sunburn.

**Drier Epidermis**

Sebaceous gland secretion decreases, and the skin becomes dry and often scaly.

**Thinning Epidermis**

The epidermis thins as basal cell activity declines. Connections between the epidermis and dermis weaken, making older people more prone to injury, skin tears, and skin infections. The skin's metabolic activity also declines. Reduced vitamin $D_3$ production (see Module 5.11) leads to muscle weakness and brittle bones.

**Diminished Immune Response**

The number of dendritic (Langerhans) cells decreases to about half the levels seen at maturity (roughly age 21). This reduction may decrease the sensitivity of the immune response and further encourage skin damage and infection.

**Thinning Dermis**

The dermis becomes thinner and has fewer elastic fibers, making the integument weaker and less resilient. The results—sagging and wrinkling—are most pronounced in body regions with the most exposure to the sun.

**Decreased Perspiration**

Eccrine sweat glands become less active, and with impaired perspiration, older people cannot lose heat as fast as younger people. So the elderly are at greater risk of overheating in warm environments.

**Altered Hair and Fat Distribution**

As sex hormone levels decrease with age, gender differences with respect to hair distribution and body fat distribution begin to fade.

**Fewer Active Follicles**

Hair follicles stop functioning or produce thinner, finer hairs. Decreased melanocyte activity makes these hairs gray or white.

**Slower Skin Repair**

Skin repairs proceed more slowly. Repairs to an uninfected blister might take 3 to 4 weeks in a young adult, but the same repairs could take 6 to 8 weeks at ages 65–75.

**Reduced Blood Supply**

A reduction in dermal blood supply cools the skin, which can stimulate thermoreceptors and make a person feel cold even in a warm room. Reduced circulation and sweat gland activity in the elderly lessens their ability to lose body heat, which can allow body temperatures to soar dangerously high with overexertion.

 **REVIEW**
**A.** Why does hair turn white or gray with age?
**B.** Why do people tolerate summer heat less well and become more susceptible to heat-related illness when they become older?

 **LEARNING OUTCOME**
Summarize the effects of aging on the skin.

# Module 5.11

# The integument responds to circulating hormones and has endocrine functions that are stimulated by ultraviolet radiation

## Circulating Horomones (Chemical Messengers)

**1** The integumentary system is physically and functionally tied to all other body systems. As a result, the state of the skin can be an indication of a person's health. Specialized cells of the integument respond to circulating levels of hormones in the body.

### Glucocorticoids

Steroid hormones called **glucocorticoids** are released during times of stress. These hormones loosen the connections between keratinocytes and reduce the effectiveness of the epidermis as a barrier to infection.

### Thyroid Hormones

**Thyroid hormones** maintain normal blood flow to the subpapillary plexus.

### Sex Hormones

**Sex hormones** stimulate epidermal cell divisions, increasing epidermal thickness and accelerating wound repair. These hormones also increase the number of dendritic cells that defend against cancer cells and pathogens.

### Growth Factors

**Growth factors** are compounds produced in the body that stimulate cell growth and cell division. Dozens of growth factors have been identified, but in many cases their origins and mechanisms of regulation are unknown.

**Epidermal growth factor (EGF)** is a peptide growth factor that has widespread effects on epithelia throughout the body. This peptide is produced by the salivary glands and glands of the duodenum, the initial portion of the small intestine.

EGF plays a role in:
- Promoting basal cell division in the stratum basale and stratum spinosum
- Accelerating keratin production in maturing keratinocytes
- Stimulating epidermal development and epidermal repair after injury
- Stimulating synthetic activity and secretion by epithelial glands

### Growth Hormone

**Growth hormone (GH)** stimulates fibroblast activity and collagen synthesis. Acting through intermediary compounds, GH also stimulates basal cell divisions, thickens the epidermis, and promotes wound repair.

**? REVIEW**

**A.** List some hormones that are necessary for maintaining a healthy integument.

# Vitamin D₃ Production

**2** Although too much sunlight can damage epithelial cells and deeper tissues, limited exposure to sunlight is beneficial because UV radiation plays a vital role in the synthesis of an important vitamin: **cholecalciferol** (kō-le-kal-SIF-er-ol), or **vitamin D₃**.

**Sources of Vitamin D₃**

**Sunlight:** When exposed to ultraviolet radiation, epidermal cells in the stratum spinosum and stratum basale convert a cholesterol-related steroid into cholecalciferol. This vitamin then diffuses across the basement membrane and enters capillaries of the subpapillary plexus.

**Diet:** Cholecalciferol can be obtained from the diet, but few foods contain it other than fish, fish oils, and shellfish, and even there its presence and quantities vary. Today, many food products are "fortified with vitamin D"—most notably milk, soy milk, and orange juice.

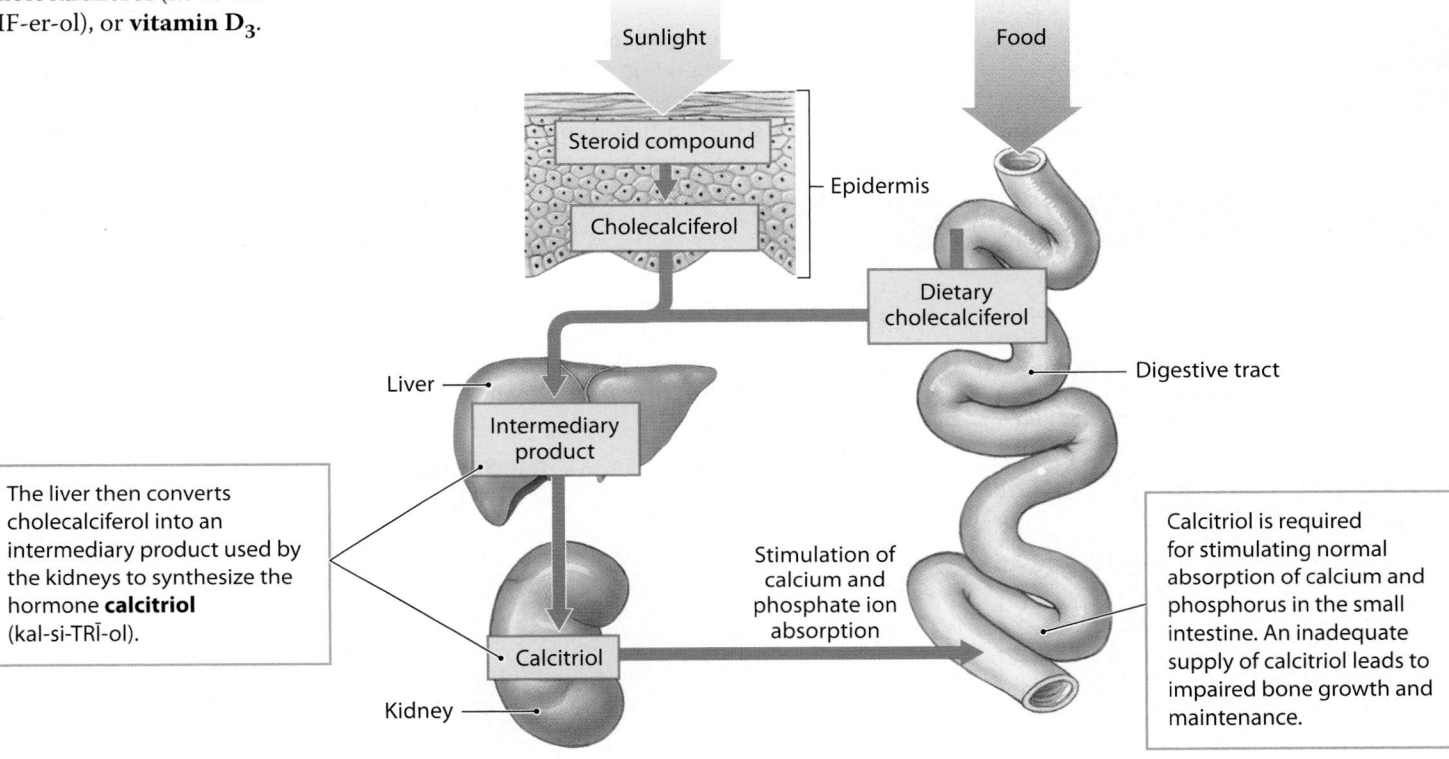

Sunlight

Food

Steroid compound

Cholecalciferol

Epidermis

Dietary cholecalciferol

Digestive tract

Liver

Intermediary product

The liver then converts cholecalciferol into an intermediary product used by the kidneys to synthesize the hormone **calcitriol** (kal-si-TRĪ-ol).

Stimulation of calcium and phosphate ion absorption

Calcitriol is required for stimulating normal absorption of calcium and phosphorus in the small intestine. An inadequate supply of calcitriol leads to impaired bone growth and maintenance.

Calcitriol

Kidney

**3** The legs shown here are characteristic of **rickets**, a condition that results in flexible, poorly mineralized bones. Rickets develops in a growing child whose skin is not exposed to sunlight and whose diet does not include a source of cholecalciferol (vitamin D₃). The bones have the proper shape, but they lack rigidity because the bone matrix contains inadequate amounts of calcium and phosphate. Rickets has largely been eliminated in the United States because dairy companies are required to add cholecalciferol, usually identified as "vitamin D," to the milk sold in grocery stores. However, cholecalciferol is required throughout life. Adults are less likely to develop rickets from cholecalciferol deficiency, but bone density will decrease, increasing the risk for fractures and slowing the healing process. The risks are especially acute for the elderly because skin production of cholecalciferol decreases by about 75 percent, even with exposure to sunlight.

**? REVIEW**

**B.** Explain the relationship between sunlight exposure and vitamin D₃.

**? INTEGRATION**

**C.** In some cultures, females must be covered from head to toe when they go outdoors. Explain why these women are at increased risk of developing bone problems later in life.

**Lo LEARNING OUTCOME**

Describe the interaction between sunlight and endocrine functioning as they relate to the skin.

# Module 5.12

# The integument can often repair itself, even after extensive damage

The integumentary system often functions independently, responding directly and automatically to local influences without involving the nervous or endocrine systems. A dramatic display of local regulation can be seen after an injury to the skin. There are four phases in skin regeneration after injury: **inflammation**, **migration**, **proliferation**, and **scarring**.

**1** An initial injury to the skin results in mast cell activation, which stimulates inflammation. Inflammation produces swelling, redness, heat (warmth), and pain (see Module 4.18).

**2** After several hours, a scab has formed, and cells of the stratum basale are migrating along the edges of the wound. Macrophages are removing debris, and more of these phagocytes are arriving via the enhanced circulation in the area. Clotting around the edges of the affected area partially isolates the region from adjacent undamaged tissues.

## Inflammation Phase

Immediately after the injury, mast cells in the region trigger an inflammatory response.

Bleeding occurs at the site of injury.

Epidermis

Dermis

## Migration Phase

The blood clot, or **scab**, that forms at the surface temporarily restores the integrity of the epidermis and restricts the entry of additional microorganisms into the area. The scab is red because of the presence of trapped red blood cells.

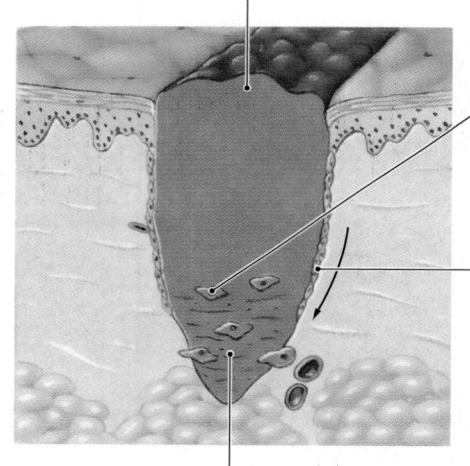

Macrophages patrol the damaged area of the dermis, phagocytizing debris and pathogens.

Cells of the stratum basale divide rapidly and migrate along the edges of the wound to replace the missing epidermal cells.

If the wound occupies an extensive area or involves a region covered by thin skin, dermal repairs must be under way before epithelial cells can cover the surface. Divisions by fibroblasts and mesenchymal cells produce mobile cells that invade the deeper areas of injury. Endothelial cells of damaged blood vessels also begin to divide, and new capillaries grow in behind the fibroblasts, enhancing circulation. The combination of blood clot, fibroblasts, and an extensive capillary network is called **granulation tissue**.

**? REVIEW**

**A.** Identify the first step in skin repair.

**B.** Describe granulation tissue.

**3** One week after the injury, the scab has been undermined by epidermal cells migrating over a meshwork produced by fibroblast activity. Phagocytic activity around the site has almost ended, and the blood clot is disintegrating.

**4** After several weeks, the scab has been shed, and the epidermis is complete. A shallow depression marks the injury site, but fibroblasts in the dermis continue to create scar tissue that will gradually elevate the overlying epidermis.

### Proliferation Phase

Over time, deeper portions of the clot dissolve, and the number of capillaries declines. Continued proliferation and activity by the fibroblasts produces collagen fibers and ground substance. The repairs do not restore the integument to its original condition, however, because the dermis will contain an abnormally large number of collagen fibers and relatively few blood vessels.

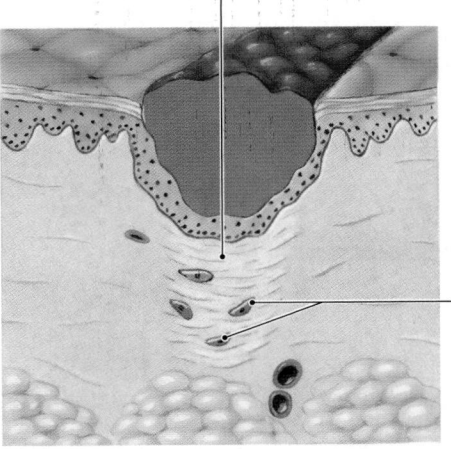

Fibroblasts

### Scarring Phase

Severely damaged hair follicles, sebaceous or sweat glands, muscle cells, and nerves are seldom repaired, and they too are replaced by fibrous tissue. The formation of this rather inflexible, fibrous, noncellular **scar tissue** completes the repair process but fails to restore the tissue to its original condition.

Scar tissue

**5** In some adults, most often those with dark skin, scar tissue formation may continue beyond the requirements of tissue repair. The result is a **keloid** (KĒ-loyd), a raised, thickened mass of scar tissue that begins at the site of injury and grows into the surrounding dermis. Keloids are covered by a shiny, smooth epidermal surface. Keloids can develop anywhere on the body where the tissue has been injured, including surgical sites. They are harmless and considered by some to be a cosmetic problem. In some cultures, however, keloids are produced intentionally as a form of body decoration.

**? REVIEW**

**C.** Why can skin regenerate effectively even after considerable damage?

 **LEARNING OUTCOME**

Explain how the skin responds to injury and is able to repair itself.

# Accessory Structures of the Skin

**Labeling:** Label the structures of a typical nail in the accompanying figures.

| 1 | | 7 | | 10 | |
| 2 | | 8 | | 11 | |
| 3 | | 9 | | 12 | |
| 4 | | | | 13 | |
| 5 | | | | 14 | |
| 6 | | | | 15 | |

Label the structures of a hair follicle in the accompanying figure.

| 16 | |
| 17 | |
| 18 | |
| 19 | |
| 20 | |

**Matching:** Match each lettered term with the most closely related description.

| | | | | | |
|---|---|---|---|---|---|
| a. malignant melanoma | | **21** | Produced by epidermal cells stimulated by UV radiation | **21** | _____ |
| b. keloid | | **22** | Epithelial fold not visible from the surface | **22** | _____ |
| c. nail root | | **23** | Found in the armpit | **23** | _____ |
| d. sebum | | **24** | Peptide produced by salivary glands | **24** | _____ |
| e. apocrine sweat glands | | **25** | Site of hair production | **25** | _____ |
| f. eponychium | | **26** | Excessive scar tissue | **26** | _____ |
| g. EGF | | **27** | Oily lipid secretion | **27** | _____ |
| h. vitamin $D_3$ | | **28** | Melanocytes metastasize through the lymphatic system | **28** | _____ |
| i. reticular layer of dermis | | **29** | Abundant in the palms and soles | **29** | _____ |
| j. eccrine sweat glands | | **30** | Cuticle | **30** | _____ |

**Section integration:** Many people change the natural appearance of their hair, either by coloring it or by altering the degree of curl in it. Which layers of the hair do you suppose are affected by the chemicals added during these procedures? Why are the effects of the procedures not permanent?

**31** _____

# Study Outline

## SECTION 1 • Functional Anatomy of the Skin

### Module 5.1

**The integumentary system consists of the skin and various accessory structures**   p. 183

1. The **integument**, or **integumentary system**, is composed of the **cutaneous membrane**, or *skin*, and the **accessory structures**.

2. The cutaneous membrane includes the **epidermis** (made of stratified squamous epithelium) and **dermis** (containing papillary and reticular layers).

3. Beneath the dermis lies the **subcutaneous layer**, or hypodermis. This tissue layer is not part of the integument.

4. Functions of the integument include protection, excretion, temperature regulation, vitamin $D_3$ synthesis, nutrient storage, and sensory detection.

### Module 5.2

**The epidermis is composed of strata (layers) that have various functions**   p. 184

5. The epidermis contains **keratinocytes**, arranged in **strata**, or layers. These cells are continuously produced in the deepest layers and shed at the exposed surface.

6. **Epidermal ridges** extend down into **dermal papillae** of the dermis. This arrangement firmly binds the epidermis to the dermis.

*Epidermal ridges and dermal papillae greatly increase the surface area for attachment, firmly binding the epidermis to the dermis.*

7. **Thin skin** contains four strata, whereas **thick skin**, which covers the palms of the hands and soles of the feet, contains five strata.

8. From deep to superficial, the layers of the epidermis are **stratum basale, stratum spinosum, stratum granulosum, stratum lucidum** (thick skin only), and **stratum corneum**.

### Module 5.3

**Factors influencing skin color include epidermal pigmentation and dermal circulation**   p. 186

9. **Melanin** and **carotene** are the primary pigments involved in skin coloration.

10. **Melanocytes**, located in the stratum basale, produce melanin, a brown, yellow-brown, or black pigment. Carotene, an orange-yellow pigment found in vegetables such as carrots and some

squashes, accumulates in the fatty tissues in the deep dermis and subcutaneous layer.

11. Blood supply also affects skin color because it contains red blood cells filled with the red pigment hemoglobin. During exercise, superficial blood vessels dilate to cool the body, and the skin becomes flushed (red). During a sustained reduction in circulation, oxygen levels decline, giving the skin a bluish coloration called **cyanosis**.

12. The most common types of cancers are skin cancers. **Basal cell carcinoma** is the most common form of skin cancer, yet it is often survived. **Malignant melanoma** is extremely dangerous.

### Module 5.4

**The subcutaneous layer connects the dermis to underlying tissues**   p. 188

13. Lying between the epidermis and subcutaneous layer, the **dermis** contains collagen and elastic fibers. The collagen fibers provide strength; the elastic fibers allow for stretching.

14. The dermis consists of the superficial **papillary layer** and the deeper **reticular layer**. The papillary layer contains highly vascularized areolar tissue, capillaries, lymphatic vessels, and sensory neurons. The reticular layer is a meshwork of dense irregular connective tissue containing collagen and elastic fibers.

15. The **subcutaneous layer** (hypodermis) is not part of the integument. It separates the skin from deeper structures.

16. The *touch receptors* of the integument are **free nerve endings**, **tactile corpuscles**, **tactile discs**, **lamellar corpuscles**, and **bulbous corpuscles**.

17. Most of the collagen and elastic fibers in any one area are arranged in parallel bundles to resist forces applied to the skin. Cuts parallel to the **tension lines** will heal with little scarring, but not a cut at right angles.

*These tension lines are clinically significant. A cut parallel to these lines will usually remain closed and heal with little scarring.*

### Module 5.5

**Burns are significant injuries that damage skin integrity**   p. 190

18. **First-degree** and **second-degree burns** are called *partial-thickness burns*. First-degree burns damage only the surface of the epidermis, whereas second-degree burns damage the entire epidermis and perhaps some of the dermis.

19. **Third-degree burns** are called *full-thickness burns* and extend into the subcutaneous layer. These burns cannot heal unaided and often require a **skin graft**.

20. The **rule of nines** is used to estimate the percentage of the surface area of the body affected by burns.

## SECTION 2 • Accessory Structures of the Skin

### Module 5.6
**Hair follicles, exocrine glands, and nails are also components of the integumentary system** p. 193

21. Hair follicles, exocrine glands, and nails are *epidermal derivatives*.

22. **Hair follicles** produce hairs that protect the skin and provide touch sensations to the body surface.

23. **Exocrine glands** assist in thermoregulation, excrete wastes, and lubricate the epidermis.

24. **Nails** protect and support the tips of the toes and fingers.

### Module 5.7
**Hair is composed of dead, keratinized cells produced in a specialized hair follicle** p. 194

25. A **hair follicle** is a complex structure made of connective and epithelial tissues that is responsible for the production of a single hair.

26. Hair formation begins at the base of the follicle at the **hair bulb**.

27. Each hair has a **hair root** and a **hair shaft**. At the base of the root is a **hair papilla**, surrounded by a hair bulb, and a **root hair plexus** of sensory nerves. Hairs have a central **medulla** of **soft keratin** surrounded by a **cortex** of **hard keratin**. The **cuticle** is a superficial layer of cells that protect the hair.

28. Composed of smooth muscle, each **arrector pili** muscle when contracted can make a single hair stand erect.

29. Hairs grow and shed according to the **hair growth cycle**. A typical scalp hair grows for 2–5 years and is then shed.

*A hair follicle is a complex structure. The human body has about 2.5 million of them.*

### Module 5.8
**Sebaceous glands and sweat glands are exocrine glands in the skin** p. 196

30. **Sebaceous glands** secrete **sebum**, which coats and protects the hair shaft.

31. The two types of sweat glands are **apocrine** and **eccrine** (**merocrine**) **sweat glands**. Apocrine sweat glands are found in the axillae, nipples, and pubic regions, where they secrete into hair follicles. Eccrine sweat glands discharge their secretion directly onto the skin surface.

### Module 5.9
**Nails are thick sheets of keratinized epidermal cells that protect the tips of fingers and toes** p. 198

32. The **nail body** covers the **nail bed**. Nail production occurs at the **nail root**. A portion of the stratum corneum of the nail root extends over the nail, forming the **eponychium**, or **cuticle**. The free edge of the nail extends over the **hyponychium**.

*The integumentary system is physically and functionally tied to all other body systems. Much of the communication between the skin and the rest of the body is by chemical communication with hormones.*

### Module 5.10
**Age-related changes affect the integument** p. 199

33. With aging, sebaceous and sweat gland activity decreases, the integument thins, blood flow decreases, cellular activity decreases, and repairs occur more slowly.

### Module 5.11
**The integument responds to circulating hormones and has endocrine functions that are stimulated by ultraviolet radiation** p. 200

34. **Glucocorticoids**, **thyroid hormones**, **sex hormones**, and **growth hormone** influence integument structure and function.

35. Limited exposure to sunlight is beneficial because UV radiation plays a role in the synthesis of **cholecalciferol**, or **vitamin $D_3$**. A lack of vitamin $D_3$ in children can result in **rickets**.

### Module 5.12
**The integument can often repair itself, even after extensive damage** p. 202

36. The skin can regenerate itself after damage. The process begins with bleeding and includes four phases: **inflammation, migration, proliferation**, and **scarring**.

37. In some people, **scar tissue** formation may continue beyond the requirements of tissue repair, resulting in **keloid** formation.

# Chapter Review Questions

**Labeling:** Identify the layers of the epidermis in this image.

1 _____

2 _____

3 _____

4 _____

5 _____

**True/False:** Indicate whether each statement is true or false.

6  Wrinkles and sagging skin are caused by damage to collagen fibers due to aging and exposure to ultraviolet radiation.

7  Another name for the subcutaneous layer is the hypodermis.

8  The perspiration you experience during intense exercise is called insensible perspiration.

9  The dermis is dominated by adipose tissue and therefore is an important site for energy storage.

10  Lamellar (Pacinian) corpuscles are sensitive to deep pressure.

11  Bluish skin as a result of decreased oxygen in blood is called cyanosis.

12  The root hair plexus is a smooth muscle whose contraction pulls on the follicle, forcing the hair to stand erect.

13  The primary pigments involved in skin coloration are carotene and melanin.

6  _____

7  _____

8  _____

9  _____

10  _____

11  _____

12  _____

13  _____

**Multiple choice:** Select the correct answer from the list provided.

14  Which portion of the hair follicle produces the hair?
- a) hair shaft
- b) hair matrix
- c) hair root plexus
- d) arrector pili

15  Which of the following glands discharges an oily secretion that coats the hair and the adjacent surface of the skin?
- a) apocrine gland
- b) sebaceous gland
- c) eccrine gland
- d) sweat gland

**16** Fine touch, pressure, and vibration are perceived by which nerve endings?

☐ a) free nerve endings
☐ b) tactile corpuscles
☐ c) tactile discs
☐ d) bulbous corpuscles

**17** What is the primary function of sensible perspiration?

☐ a) to get rid of wastes
☐ b) to protect the skin from dryness
☐ c) to maintain electrolyte balance
☐ d) to reduce body temperature

**18** In the elderly, blood supply to the dermis is reduced, and sweat glands are less active. What is most affected by this combination of factors?

☐ a) ability to thermoregulate
☐ b) ability to heal injured skin
☐ c) ease with which skin is injured
☐ d) ability to grow hair

**19** Which term is applied to the stratum corneum of the nail root that extends over the exposed nail?

☐ a) hyponychium
☐ b) eponychium
☐ c) lunula
☐ d) cerumen

**20** Which factor is associated with darker skin color?

☐ a) more melanocytes present in skin
☐ b) more layers of epidermis
☐ c) more melanin produced by melanocytes
☐ d) more superficial blood vessels

**21** Which anatomical feature is responsible for fingerprints?

☐ a) ridge patterns in the thick skin covering the fingertips
☐ b) dermal papillae
☐ c) the thickness of stratum lucidum
☐ d) the release of melanosomes by melanocytes

---

**Fill-in:** Fill in the following blanks in the spaces provided to the right.

**22** From superficial to deep, the two layers of the cutaneous membrane are the _____ and the _____ .

**23** The _____ layer separates the integument from the fascia around deeper organs.

**24** _____ sweat glands are found in the axillae, nipples, and pubic region.

**25** The most common form of skin cancer is _____ .

**26** A deficiency or absence of _____ production leads to a disorder known as albinism.

**22** _____ , _____

**23** _____

**24** _____

**25** _____

**26** _____

---

## Short answer

**27** In which layer(s) of the epidermis does cell division occur?

**28** What is the function of the arrector pili muscles?

**29** Explain how hair color can vary from person to person, and over a person's lifetime.

**30** What are the risks from having inadequate levels of vitamin $D_3$ in the body?

**31** Explain why when you warm your hands by a campfire, your face feels the heat more than your hands.

**32** Why is it important for a surgeon to choose an incision pattern according to the tension lines of the skin?

**33** Describe the functions of the skin that are compromised by extensive burns.

# Chapter Integration • Applying what you have learned

## Thinking about the science of tattoos

Tattoos are becoming increasingly visible in mainstream society. As tattoos become more fashionable, people of all ages and social classes are engaging in this age-old form of body art. Once considered a practice of only sailors, circus sideshow performers, or criminals, an increasingly diverse population is "sporting ink." Everyone from movie stars and athletes to doctors and soccer moms has tattoos these days. It's safe to say that most of you reading this text are either tattooed or know someone who is tattooed.

The tattooing process has remained essentially the same for thousands of years. Originally, a sharpened bone tool was dipped in dye and then tapped into the skin with a small hammer to create the tattoo. Today's tattooing uses an electrical tool to deposit pigment through the epidermis and into the dermis. Similar to a sewing machine, a motor-driven cluster of needles punctures the skin many times each second. The artist progressively shades and colors the tattoo in a process that can take many hours.

In light of the popularity and prevalence of tattooing, use your knowledge of the integument to answer the following questions.

**1**   Why does the tattoo ink stay in place?

**2**   Why do tattoos tend to fade over time?

**3**   Explain why it is somewhat painful to be tattooed.

**4**   Does tattooing carry a significant risk of infection?

**LEARNING OUTCOMES**

These Learning Outcomes correspond by number to this chapter's modules and indicate what you should be able to do after completing the chapter. They also appear at the end of each module.

# 6

# Bones and Bone Structure

## Module 6.1
# The skeletal system is made up of the axial and appendicular divisions

Go to
MasteringA&P™
PAL™ 3.0
Anatomical Models→
Axial Skeleton

The skeletal system includes the bones of the skeleton and the cartilages, ligaments, and other connective tissues that stabilize or interconnect them. This chapter expands on the introduction to bone tissue presented in Chapter 4. It considers the gross anatomy of bones and the processes involved in bone growth, remodeling, and repair.

**1** The adult skeletal system includes about 206 separate bones and a number of associated cartilages. This body system is divided into the **axial skeleton** and the **appendicular skeleton**.

### Axial Skeleton
(80 Bones)

The axial skeleton consists of the bones of the skull, thorax, and vertebral column. These elements form the longitudinal axis of the body.

**? REVIEW**

**A.** Describe the axial and appendicular divisions of the skeleton.

### Appendicular Skeleton (126 Bones)

The appendicular skeleton includes the bones of the limbs and the pectoral and pelvic girdles that attach the limbs to the axial skeleton.

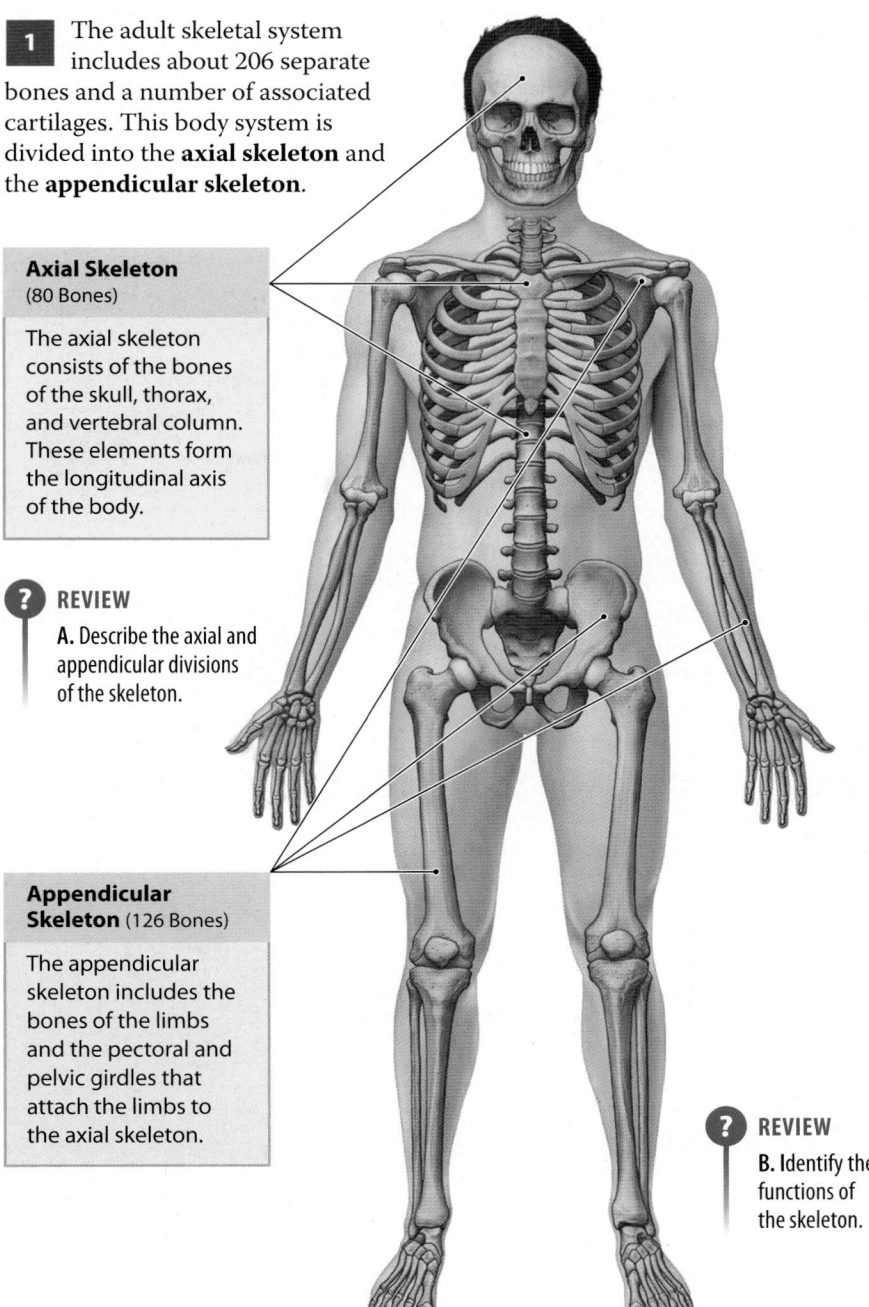

**? REVIEW**

**B.** Identify the functions of the skeleton.

**2** The vital functions of our skeleton are summarized in the table below. All of these functions ultimately depend on the unique and dynamic properties of bone tissue. The bone specimens that you study in lab or that you are familiar with from skeletons of dead animals are only the dry remains of this living tissue. They have the same relationship to the bone in a living organism as kiln-dried lumber does to a living tree.

### Functions of the Skeletal System

- **Support:** The skeletal system provides structural support for the entire body. Individual bones or groups of bones provide a framework for the attachment of soft tissues and organs.

- **Store minerals and lipids:** The calcium salts in bone are a valuable mineral reserve that maintains normal concentrations of calcium and phosphate ions in blood. Calcium is the most abundant mineral in the human body. Bones contain adipose tissue, which stores lipids as energy reserves.

- **Produce blood cells:** Red blood cells, white blood cells, and platelets are produced in the red bone marrow, which fills the internal cavities of many bones.

- **Protection:** The skeleton surrounds delicate tissues and organs. The ribs protect the heart and lungs, the skull encloses the brain, the vertebrae shield the spinal cord, and the pelvis cradles digestive and reproductive organs.

- **Leverage:** Many bones of the skeleton function as levers that can change the magnitude and direction of the forces skeletal muscles generate. The movements produced range from the fine motions of a fingertip to powerful changes in the position of the entire body.

**Lo**  **LEARNING OUTCOME**
Describe the two main divisions of the skeleton, and list the major functions of the skeletal system.

**6**

# Module 6.2

# Bones are classified according to shape and structure and have varied bone markings

**1** The 206 major bones in a typical adult human skeleton can be divided into six broad categories based on their shape. The skeleton also contains a variable number of minor bones (usually sutural bones and sesamoid bones) whose numbers vary from individual to individual.

**6**

### Flat Bones

**Flat bones** have thin, roughly parallel surfaces. Flat bones form the roof of the skull, the sternum, the ribs, and the scapulae. They protect underlying soft tissues and have an extensive surface area for the attachment of skeletal muscles.

Parietal bone

### Sutural Bones

**Sutural bones**, or Wormian bones, are small, flat, irregularly shaped bones between the flat bones of the skull. There are individual variations in the number, shape, and position of sutural bones. Their borders are like pieces of a jigsaw puzzle, and they range in size from a grain of sand to a quarter.

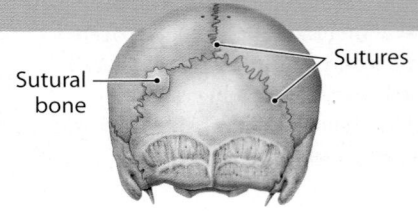
Sutures
Sutural bone

### Long Bones

**Long bones** are elongated and slender. They are located in the arm and forearm, thigh and leg, palms, soles, fingers, and toes. The femur, the long bone of the thigh, is the largest and heaviest bone in the body.

Humerus

### Irregular Bones

**Irregular bones** have complex shapes with short, flat, notched, or ridged surfaces. The spinal vertebrae, the bones of the pelvis, and several skull bones are irregular bones.

Vertebra

### Sesamoid Bones

**Sesamoid bones** are generally small, flat, and shaped somewhat like a sesame seed. They develop inside tendons and are most commonly located near joints at the knees, the hands, and the feet. Except for the patellae (pa-TEL-ē; singular, *patella*, a small shallow dish), or kneecaps, there are individual variations in the location and number of sesamoid bones. These variations, along with varying numbers of sutural bones, account for individual differences in the total number of bones in the skeleton.

Patella

**?** **REVIEW**
**A.** Identify the six broad categories for classifying a bone according to shape.

### Short Bones

**Short bones** are small and boxy. Examples of short bones include bones in the wrists (carpal bones) and in the ankles (tarsal bones).

Carpal bones

**2** The surfaces of bones have characteristic **bone markings** (*surface features*), which are projections, depressions, and openings related to particular functions. Projections are bumps that form where muscles, tendons, and ligaments attach and where adjacent bones form joints. Depressions and openings are sites where blood vessels or nerves lie alongside or penetrate the bone.

## Bone Markings of the Skull

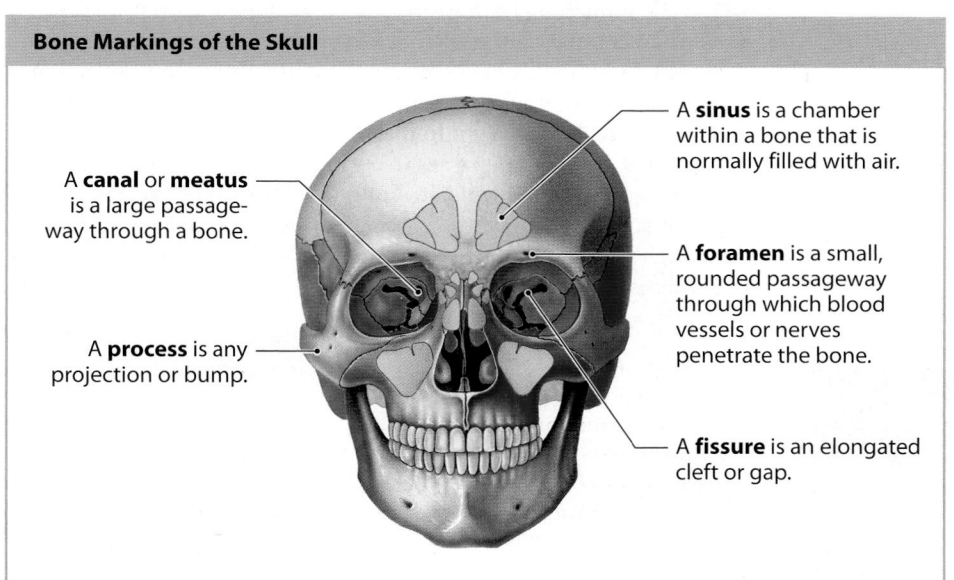

A **canal** or **meatus** is a large passage-way through a bone.

A **process** is any projection or bump.

A **sinus** is a chamber within a bone that is normally filled with air.

A **foramen** is a small, rounded passageway through which blood vessels or nerves penetrate the bone.

A **fissure** is an elongated cleft or gap.

## Bone Markings of the Humerus

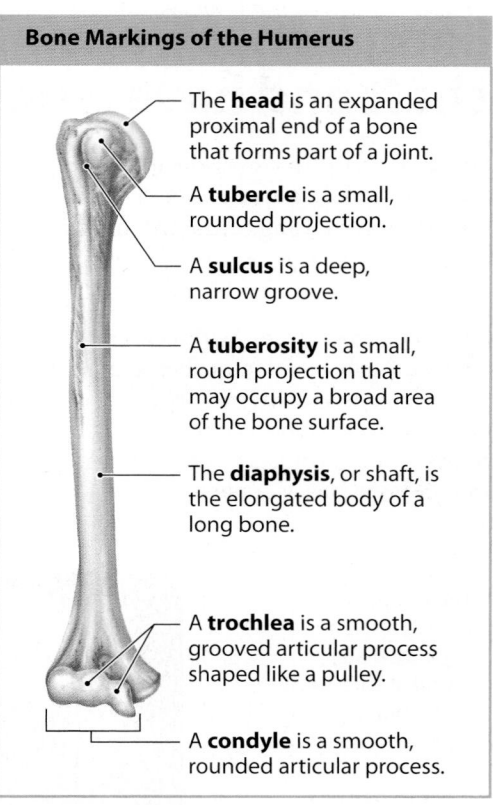

The **head** is an expanded proximal end of a bone that forms part of a joint.

A **tubercle** is a small, rounded projection.

A **sulcus** is a deep, narrow groove.

A **tuberosity** is a small, rough projection that may occupy a broad area of the bone surface.

The **diaphysis**, or shaft, is the elongated body of a long bone.

A **trochlea** is a smooth, grooved articular process shaped like a pulley.

A **condyle** is a smooth, rounded articular process.

## Bone Markings of the Femur

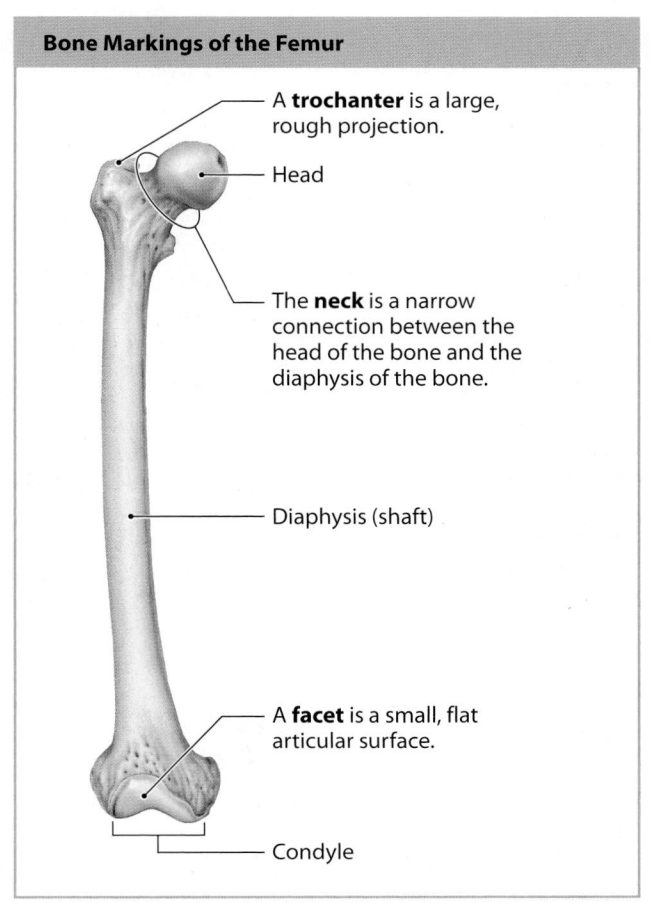

A **trochanter** is a large, rough projection.

Head

The **neck** is a narrow connection between the head of the bone and the diaphysis of the bone.

Diaphysis (shaft)

A **facet** is a small, flat articular surface.

Condyle

## Bone Markings of the Pelvis

A **crest** is a prominent ridge.

A **fossa** is a shallow depression or recess in the surface of the bone.

A **line** is a low ridge, more delicate than a crest.

A **spine** is a pointed or narrow process.

A **ramus** is a thick projection that makes an angle with the main body of the bone.

**? REVIEW**
**B.** Define *bone markings*.

**? INTEGRATION**
**C.** List the different terms used to describe projections.

**Lo LEARNING OUTCOME**
Classify bones according to their shapes, identify the major types of bone markings, and explain the functional significance of bone markings.

6

# Module 6.3

# Long bones transmit forces along the shaft and have a rich blood supply

Here we examine several aspects of the functional anatomy of a representative long bone, the femur.

**1** This frontal (coronal) section through a left femur shows its major regions and internal organization. Recall that typical long bones consist of both compact bone and spongy bone (see Module 4.14)

**2** The branching framework of spongy bone makes it weaker than compact bone when it is compressed from a single direction, but it is able to resist forces applied from many directions. At the femur, the spongy bone of the proximal epiphysis channels and directs the body weight to the compact bone of the diaphysis. The compact bone then passes these forces to the spongy bone of the distal epiphysis, which distributes the body weight across the articular surface at the knee. The medullary cavity does not reduce the bone's strength because the center of the bone is not subjected to compression or tension.

The **epiphysis** (e-PIF-i-sis) is an expanded area found at each end of the bone.

The **metaphysis** (me-TAF-i-sis; *meta*, between) is a narrow zone that connects the epiphysis to the shaft of the bone.

The **diaphysis** (shaft) is long and tubular.

Metaphysis

Epiphysis

The epiphysis consists largely of **spongy bone**, also called **trabecular** (tra-bek-YŪ-lar) **bone**. Spongy bone consists of an open network of struts and plates that resembles latticework with a thin covering of compact bone.

Relatively dense **compact bone** makes up the wall of the diaphysis. It forms a sturdy protective layer around a central space.

The **medullary cavity** (*medulla*, innermost part), or marrow cavity, is a space within the hollow shaft. In life, it is filled with two types of bone marrow. **Red bone marrow** is highly vascular and involved in the production of blood cells. **Yellow bone marrow** is adipose tissue important as an energy reserve.

Body weight (applied force)

Tension on lateral side of shaft

Compression on medial side of shaft

**? REVIEW**

**A.** List the major parts of a long bone.

**B.** Describe the function of the medullary cavity.

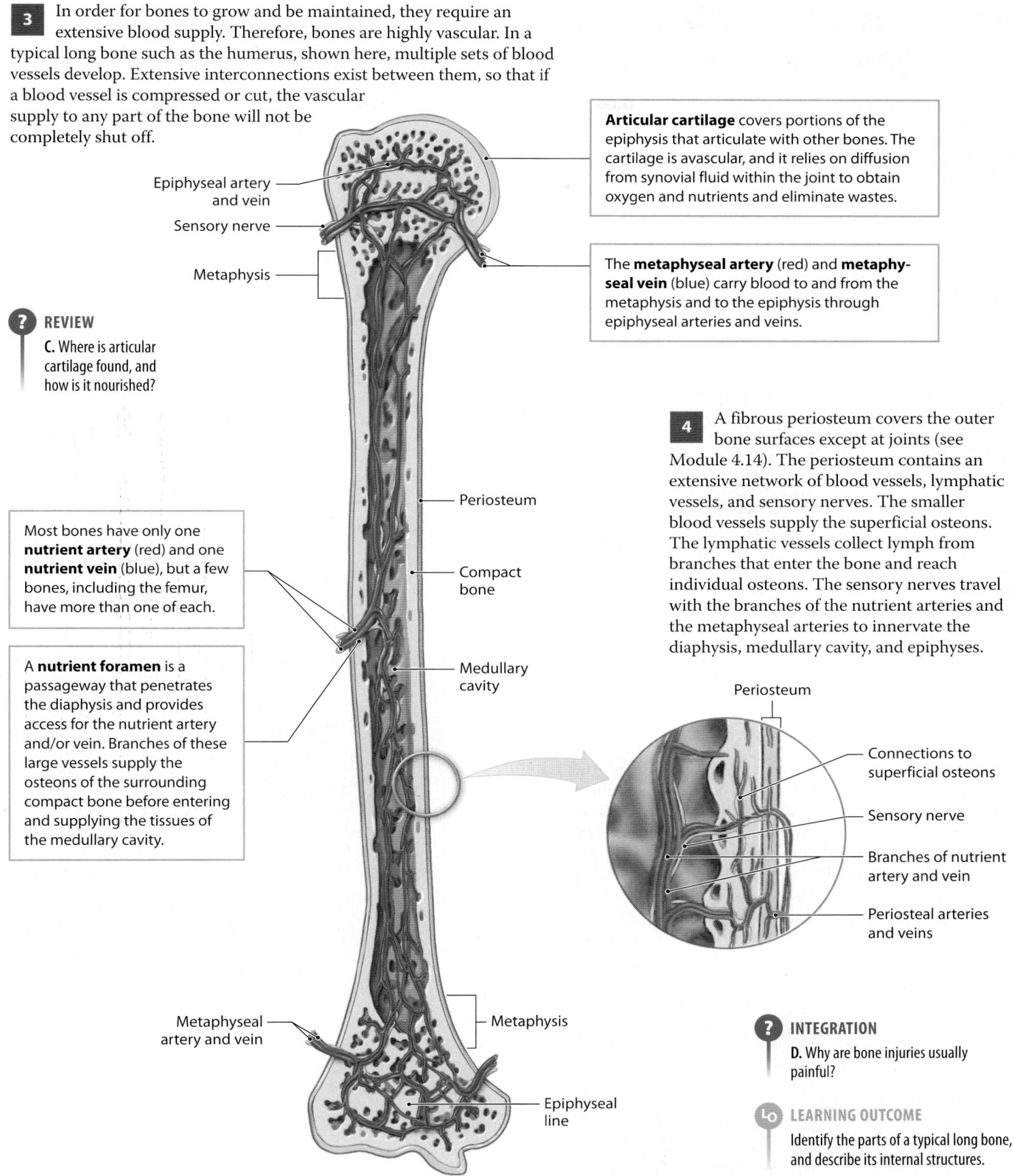

**3** In order for bones to grow and be maintained, they require an extensive blood supply. Therefore, bones are highly vascular. In a typical long bone such as the humerus, shown here, multiple sets of blood vessels develop. Extensive interconnections exist between them, so that if a blood vessel is compressed or cut, the vascular supply to any part of the bone will not be completely shut off.

Epiphyseal artery and vein

Sensory nerve

Metaphysis

**? REVIEW**
**C.** Where is articular cartilage found, and how is it nourished?

Most bones have only one **nutrient artery** (red) and one **nutrient vein** (blue), but a few bones, including the femur, have more than one of each.

A **nutrient foramen** is a passageway that penetrates the diaphysis and provides access for the nutrient artery and/or vein. Branches of these large vessels supply the osteons of the surrounding compact bone before entering and supplying the tissues of the medullary cavity.

**Articular cartilage** covers portions of the epiphysis that articulate with other bones. The cartilage is avascular, and it relies on diffusion from synovial fluid within the joint to obtain oxygen and nutrients and eliminate wastes.

The **metaphyseal artery** (red) and **metaphyseal vein** (blue) carry blood to and from the metaphysis and to the epiphysis through epiphyseal arteries and veins.

**4** A fibrous periosteum covers the outer bone surfaces except at joints (see Module 4.14). The periosteum contains an extensive network of blood vessels, lymphatic vessels, and sensory nerves. The smaller blood vessels supply the superficial osteons. The lymphatic vessels collect lymph from branches that enter the bone and reach individual osteons. The sensory nerves travel with the branches of the nutrient arteries and the metaphyseal arteries to innervate the diaphysis, medullary cavity, and epiphyses.

Periosteum

Compact bone

Medullary cavity

Periosteum

Connections to superficial osteons

Sensory nerve

Branches of nutrient artery and vein

Periosteal arteries and veins

Metaphyseal artery and vein

Metaphysis

Epiphyseal line

**? INTEGRATION**
**D.** Why are bone injuries usually painful?

**Lo LEARNING OUTCOME**
Identify the parts of a typical long bone, and describe its internal structures.

# Module 6.4

# Bone has a calcified matrix maintained and altered by osteogenic cells, osteoblasts, osteocytes, and osteoclasts

Both compact bone and spongy bone contain the same four types of cells.

**1** Bone contains small numbers of mesenchymal cells called **osteogenic** (os-tē-ō-JEN-ik) **cells**, or *osteoprogenitor* (os-tē-ō-prō-JEN-i-tor) *cells* (see Module 4.11). These stem cells divide to produce daughter cells that differentiate into osteoblasts. Osteogenic cells maintain populations of osteoblasts and so are important in the repair of a *fracture* (a break or a crack in a bone). Osteogenic cells are found in the inner, cellular layer of the periosteum. They are also found in a cellular layer, or *endosteum*, that lines medullary cavities and passageways for blood vessels that penetrate the matrix of compact bone.

Endosteum

Osteogenic cell

Medullary cavity

**2** **Osteoblasts** (OS-tē-ō-blasts; *-blast,* immature precursor cell) produce new bone matrix in a process called **ossification**, or **osteogenesis** (os-tē-ō-JEN-e-sis; *genesis,* production). Osteoblasts make and release the proteins and other organic components of the matrix. Before calcium salts are deposited, this "pre-bone" organic matrix is called **osteoid** (OS-tē-oyd). Osteoblasts also help increase local concentrations of calcium phosphate to the point where this calcium salt is deposited in the organic matrix. This process converts osteoid to bone. Osteocytes develop from osteoblasts that have become completely surrounded by bone matrix.

Osteoblast

Osteoid

Medullary cavity

**? REVIEW**

**A.** Describe the functions of osteogenic cells and osteoblasts.

**3**  **Osteocytes** (*osteo-*, bone + *-cyte*, cell) are mature bone cells that maintain the bone matrix. These cells continually recycle the protein and minerals of the surrounding matrix. Osteocytes secrete chemicals that dissolve the adjacent matrix, and the released minerals enter the bloodstream. They then rebuild the matrix, stimulating the deposition of mineral crystals. The turnover rate varies from bone to bone. Osteocytes also take part in the repair of damaged bone.

**?** **REVIEW**

**B.** Describe the functions of osteocytes.

The thin layers of matrix are called **lamellae** (lah-MEL-lē; singular, *lamella*, a thin plate).

Osteocytes make up most of the cell population in bone. Each osteocyte occupies a **lacuna**, a pocket sandwiched between layers of matrix. Osteocytes cannot divide, and a lacuna never contains more than one osteocyte.

Processes of the osteocytes extend into narrow passageways called **canaliculi** that penetrate the matrix. The canaliculi interconnect the lacunae and reach vascular passageways, providing a route for nutrient diffusion.

**4**  **Osteoclasts** (OS-tē-ō-clasts; *klastos,* broken) are cells that remove bone matrix. They are large cells with 50 or more nuclei. Osteoclasts secrete acids and protein-digesting enzymes that dissolve the matrix and release the stored minerals. This process, called **osteolysis** (os-tē-OL-i-sis; *lysis,* a loosening) or *resorption*, is important in regulating calcium and phosphate ion concentrations in blood and other body fluids. Osteoclasts are not related to osteogenic cells or their descendants. Instead, they are derived from the same stem cells that produce monocytes and macrophages, cells involved in the body's defense mechanisms.

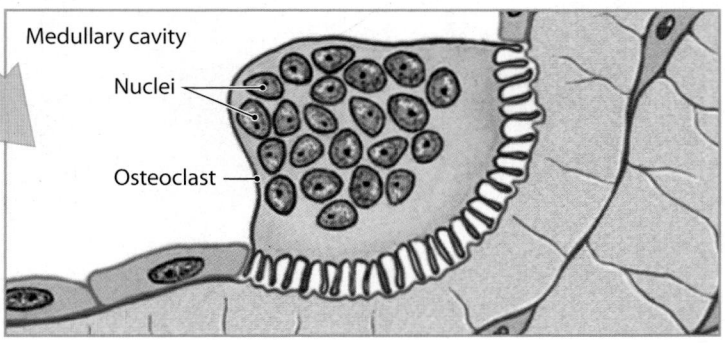

Medullary cavity

Nuclei

Osteoclast

**5**  A bone without a calcified matrix looks normal, but is very flexible. Roughly one-third of the weight of bone is due to collagen fibers. Bone cells account for only 2 percent of the weight of a typical bone. A calcium salt, calcium phosphate, $Ca_3(PO_4)_2$, makes up almost two-thirds of the weight of bone. Calcium phosphate interacts with calcium hydroxide, $Ca(OH)_2$, to form crystals of **hydroxyapatite**, $Ca_{10}(PO_4)_6(OH)_2$. As they form, these crystals incorporate other calcium salts, such as calcium carbonate ($CaCO_3$), and ions such as sodium, magnesium, and fluoride. Hydroxyapatite is very hard but inflexible and brittle. It can withstand compression but is likely to shatter with twisting, bending, or sudden impacts. Collagen fibers are strong and flexible. When subjected to tension (pull), they are stronger than steel, but when compressed, they bend. The protein–crystal combination in bone is strong, somewhat flexible, and highly resistant to breaking or shattering under compression. In its overall properties, bone is on a par with the best steel-reinforced concrete. In fact, bone is far superior to concrete because it can undergo remodeling (cycles of bone formation and resorption) as needed and can repair itself after injury.

**?** **REVIEW**

**C.** How would the compressive strength of a bone be affected if the ratio of collagen to hydroxyapatite increased?

**?** **INTEGRATION**

**D.** If osteoclast activity exceeds osteoblast activity in a bone, how will bone mass be affected?

**Lo** **LEARNING OUTCOME**

Identify the types of cells in bone, and list their major functions.

• Derived From stem cells That produce macrophages

# Compact bone consists of parallel osteons, and spongy bone consists of a network of trabeculae

The basic functional unit of mature compact bone is the **osteon** (OS-tē-on), or Haversian system. Many of the important features of osteons were introduced in Module 4.14.

**1** The lamellae of each osteon form a series of nested cylinders around the central canal. In transverse section, these **concentric lamellae** resemble a target, with the **central canal** as the bull's-eye. You might think of a single osteon as a drinking straw with very thick walls: When you attempt to push the ends of the straw together or to pull them apart, the straw is quite strong. But if you hold the ends and push from the side, the straw will bend sharply with relative ease. The osteons in the diaphysis of a long bone are parallel to the long axis of the shaft. Thus, the shaft does not bend, even when extreme forces are applied to either end. The femur can withstand 10–15 times the body's weight without breaking. Yet a much smaller force applied to the side of the shaft can break any long bone, even the femur.

Nerve

Capillary and venule (small vein)

Central canal

Concentric lamellae

Canaliculi radiating through the lamellae interconnect the lacunae of the osteons with one another and with the central canal.

Endosteum

**2** A section cut from the shaft of a long bone shows the organization of osteons and blood vessels and allows a comparison between compact and spongy bone.

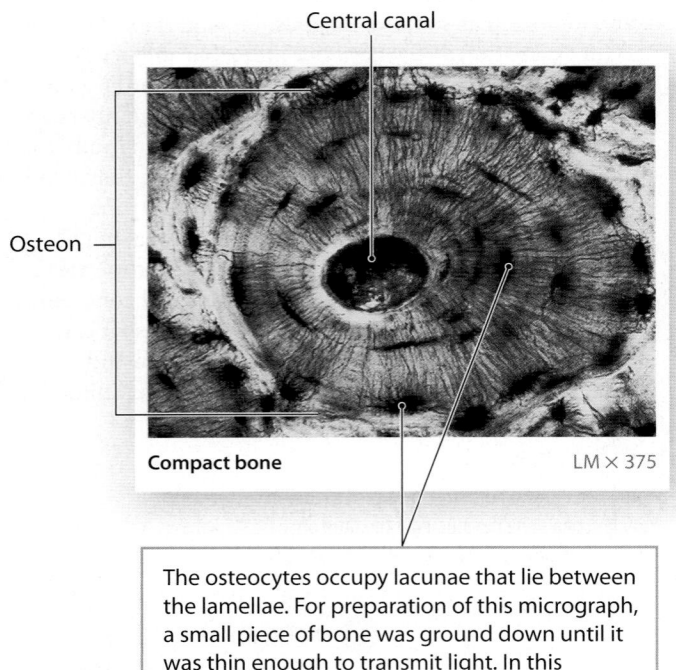

Central canal

Osteon

**Compact bone**          LM × 375

The osteocytes occupy lacunae that lie between the lamellae. For preparation of this micrograph, a small piece of bone was ground down until it was thin enough to transmit light. In this process, the lacunae and canaliculi are filled with bone dust and thus appear black.

Periosteum

Nerve

Vein

Artery

**?** **REVIEW**

**A.** Define *osteon*.

**3** Spongy bone, or *trabecular bone*, is found where bones are not heavily stressed or where stresses arrive from many directions (see Module 6.3). Below is the head of the femur, which transmits body weight to the shaft of that bone. The trabeculae are oriented along stress lines and are cross-braced extensively. In addition, spongy bone is much lighter than compact bone. Spongy bone thus reduces the weight of the skeleton, making it easier for muscles to move the bones. Spongy bone makes up most of the volume of short, flat, and irregular bones and the epiphyses of long bones.

**4** In spongy bone, lamellae are not arranged in osteons. The matrix in spongy bone forms struts and plates called **trabeculae**. The thin trabeculae branch, creating an open network. There are no capillaries or venules in the matrix of spongy bone. Nutrients reach the osteocytes by diffusion along canaliculi that open onto the surfaces of trabeculae. Spongy bone within the epiphyses of long bones, such as the femur, and the interior of other large bones such as the sternum and ilium (a large bone of the pelvis), contains red bone marrow. At other sites, spongy bone may contain yellow bone marrow. Blood vessels within this tissue deliver nutrients to the trabeculae and remove wastes generated by the osteocytes.

Trabeculae of spongy bone | Canaliculi opening on surface | Endosteum | Lamellae

**Circumferential lamellae** (*circum-*, around + *ferre*, to bear) are found at the outer and inner surfaces of the bone, where they are covered by the periosteum and endosteum, respectively. These lamellae are produced during the growth and maintenance of the bone.

**Interstitial lamellae** fill in the spaces between the osteons in compact bone. These lamellae are remnants of osteons whose matrix components have been almost completely recycled by osteoclasts.

Central canal

Perforating canal

Central canals generally run parallel to the surface of the bone. Other passageways, known as **perforating canals**, extend perpendicular to the surface. Blood vessels in these canals supply blood to osteons deeper in the bone and to tissues of the medullary cavity.

**? REVIEW**

**B.** Compare compact bone and spongy bone.

**? INTEGRATION**

**C.** A sample of bone has lamellae that are not arranged in osteons. Is the sample more likely from the epiphysis or from the diaphysis?

**LO LEARNING OUTCOME**

Compare the structures and functions of compact bone and spongy bone.

# Module 6.6

# Appositional bone growth involves the periosteum and the endosteum

The appositional growth of cartilage was detailed in Module 4.13. Appositional bone growth involves two bone coverings: the periosteum on the outside and the endosteum on the inside.

**1** The diameter of a bone enlarges through **appositional growth** at the outer surface. In this process, osteogenic cells in the inner layer of the periosteum produce daughter cells that differentiate into osteoblasts and add bone matrix to the surface. This adds successive layers of circumferential lamellae to the outer surface of the bone. Osteoblasts trapped between these lamellae differentiate into osteocytes. Over time, the deeper lamellae are recycled and replaced with the osteons typical of compact bone.

Additional circumferential lamellae are deposited, and the bone continues to increase in diameter.

**? REVIEW**
**A.** Define *appositional growth.*

Periosteum

**2** As bone is being added to the outer surface through appositional growth, osteoclasts are removing and recycling lamellae at the inner surface. As a result, the medullary cavity gradually enlarges as the bone increases in diameter.

Bone matrix is removed by osteoclasts

Infant    Child    Young adult    Adult

Bone deposited by superficial osteoblasts

**? REVIEW**
**B.** As a bone increases in diameter, what happens to the medullary cavity?

**3** Except within joint cavities, the superficial layer of compact bone that covers all bones is wrapped by a **periosteum**, which has a fibrous outer layer and a cellular inner layer. The periosteum (1) isolates the bone from surrounding tissues, (2) provides a route for the blood and nervous supply, and (3) actively participates in bone growth and repair. As the bone grows, the collagen fibers from tendons, ligaments, and joint capsules are cemented into the circumferential lamellae by osteoblasts from the cellular layer of the periosteum. These are called **perforating fibers**, and this method of attachment is extremely strong. An excessive pull on a tendon or ligament will usually break a bone rather than snap the collagen fibers at the bone surface.

**4** The **endosteum** is an incomplete cellular layer that lines the medullary cavity. This layer is active during bone growth, repair, and remodeling. It covers the trabeculae of spongy bone and lines the inner surfaces of the central canals. The endosteum consists of a simple flattened layer of osteogenic cells that covers the bone matrix, generally without any intervening connective tissue fibers. Where the cellular layer is incomplete, the matrix is exposed. At these exposed sites, osteoclasts and osteoblasts can remove or deposit matrix components. The osteoclasts generally occur in shallow depressions, called **osteoclastic crypts** or Howship's lacunae, that they have eroded into the matrix. As a result of their activities, the inner circumferential lamellae are often incomplete or interrupted.

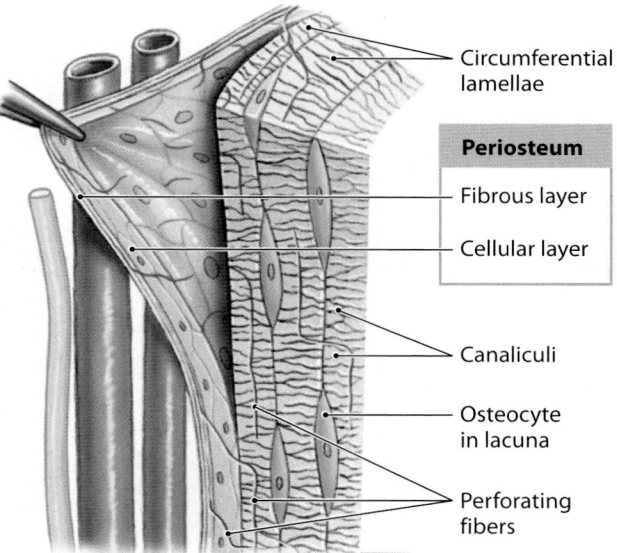

Circumferential lamellae

**Periosteum**

Fibrous layer

Cellular layer

Canaliculi

Osteocyte in lacuna

Perforating fibers

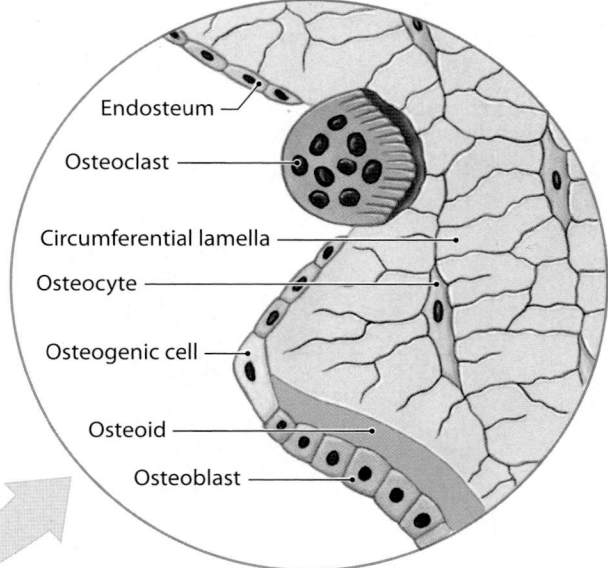

Endosteum

Osteoclast

Circumferential lamella

Osteocyte

Osteogenic cell

Osteoid

Osteoblast

**?** **REVIEW**

**C.** Distinguish between the periosteum and the endosteum.

Appositional bone growth is important in increasing the diameters of existing bones, but it does not form the original bones. Bone formation begins roughly 6 weeks after fertilization, when the embryo is 12 mm (0.5 in.) long. Two major processes are involved: endochondral ossification and intramembranous ossification.

**LEARNING OUTCOME**

Describe the process of appositional bone growth.

# Endochondral ossification replaces a cartilage model with bone

When bone formation begins in the embryo some 6 weeks after fertilization, all existing skeletal elements are made of hyaline cartilage. These cartilages are gradually replaced by bone through the process of **endochondral** (en-dō-KON-drul; *endo-*, inside + *chondros*, cartilage) **ossification**. The key steps in endochondral ossification are illustrated below. This process begins with a small cartilage that is basically a miniature model of the corresponding bone of the adult skeleton. As it forms, the bone grows in length (*interstitial growth*) and in diameter. Recall that the increase in bone diameter involves appositional bone growth (see Module 6.6).

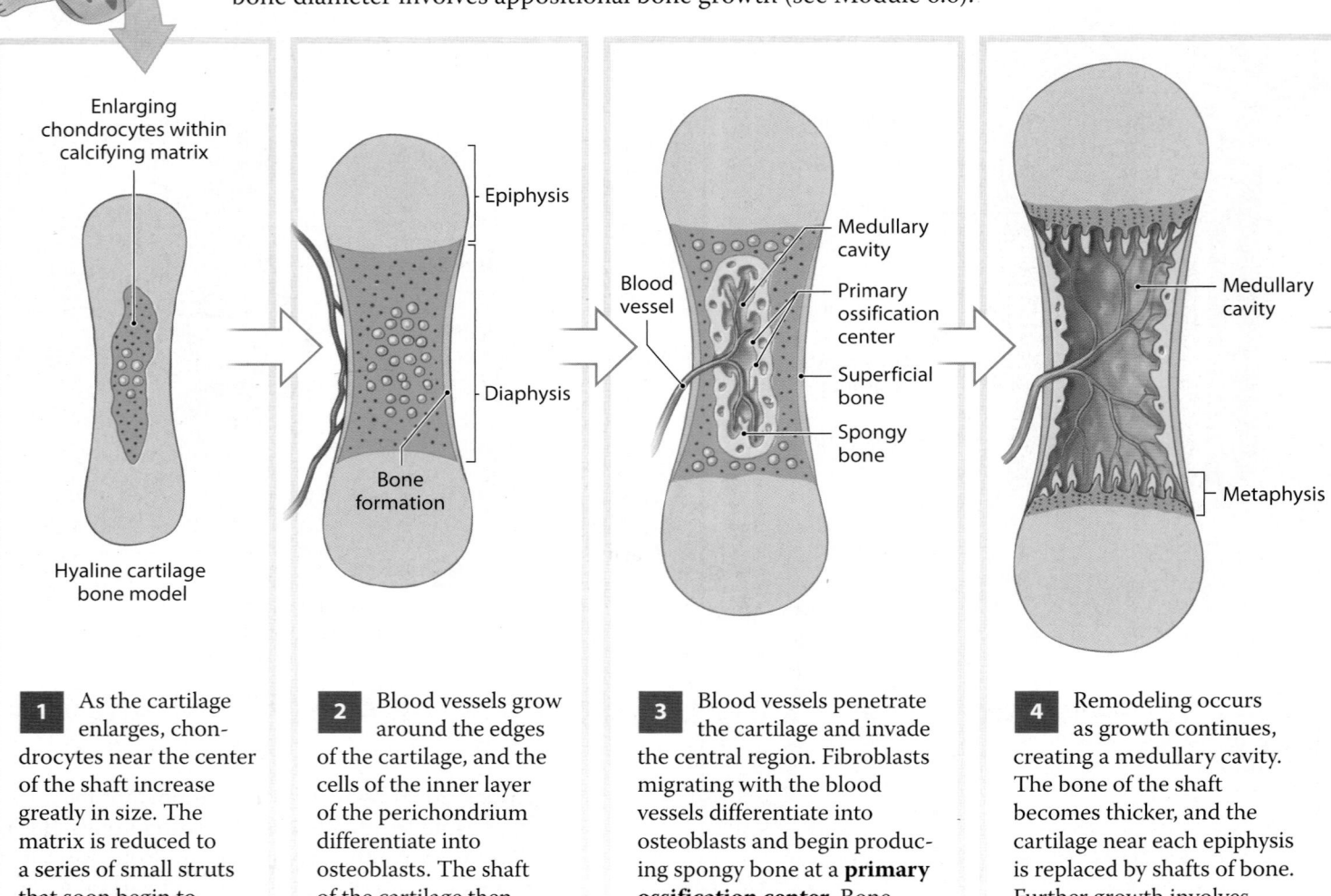

Enlarging chondrocytes within calcifying matrix

Hyaline cartilage bone model

Epiphysis

Diaphysis

Bone formation

Blood vessel

Medullary cavity

Primary ossification center

Superficial bone

Spongy bone

Medullary cavity

Metaphysis

**1** As the cartilage enlarges, chondrocytes near the center of the shaft increase greatly in size. The matrix is reduced to a series of small struts that soon begin to calcify. The enlarged chondrocytes then die and disintegrate, leaving cavities within the cartilage.

**2** Blood vessels grow around the edges of the cartilage, and the cells of the inner layer of the perichondrium differentiate into osteoblasts. The shaft of the cartilage then becomes ensheathed in a superficial layer of bone.

**3** Blood vessels penetrate the cartilage and invade the central region. Fibroblasts migrating with the blood vessels differentiate into osteoblasts and begin producing spongy bone at a **primary ossification center**. Bone formation then spreads along the shaft toward both ends.

**4** Remodeling occurs as growth continues, creating a medullary cavity. The bone of the shaft becomes thicker, and the cartilage near each epiphysis is replaced by shafts of bone. Further growth involves increases in length and diameter.

*Perichondrium - surrounding cartilage has fibrous & cellular layers, cellular layer has fibroblasts that can differentiate into chondrocytes*

**? REVIEW**

**A.** Define *endochondral ossification*.

**B.** In endochondral ossification, what is the original source of osteoblasts?

## The ephipyseal line in x-rays

In adults, the former location of the epiphyseal cartilage is often detectable in x-rays as a distinct **epiphyseal line**, which remains after epiphyseal growth has ended. In this x-ray of the left shoulder, the epiphyseal line of the head of the humerus is clearly visible.

Epiphyseal line

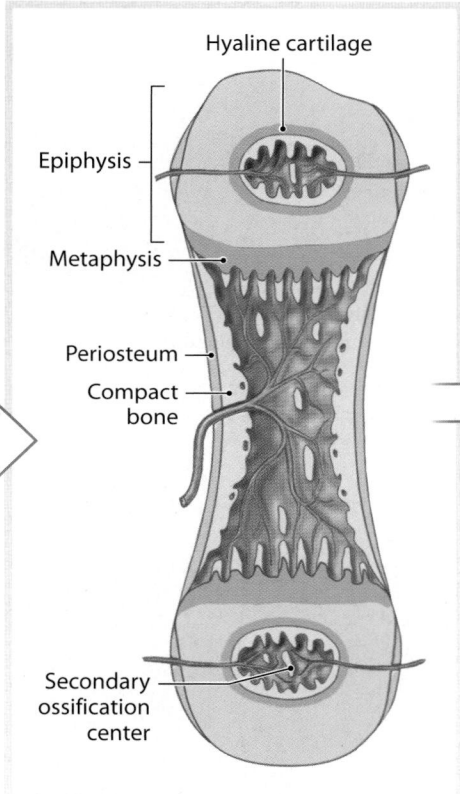

Hyaline cartilage

Epiphysis

Metaphysis

Periosteum

Compact bone

Secondary ossification center

**5** Capillaries and osteoblasts migrate into the epiphyses, creating **secondary ossification centers**. Interstitial growth takes place at each metaphysis.

Articular cartilage

Spongy bone

Epiphyseal cartilage

Diaphysis

**6** Soon the epiphyses are filled with spongy bone. Articular cartilage remains exposed to the joint cavity; over time it will be reduced to a thin superficial layer. At each metaphysis, an **epiphyseal cartilage**, or **epiphyseal plate**, separates the epiphysis from the diaphysis.

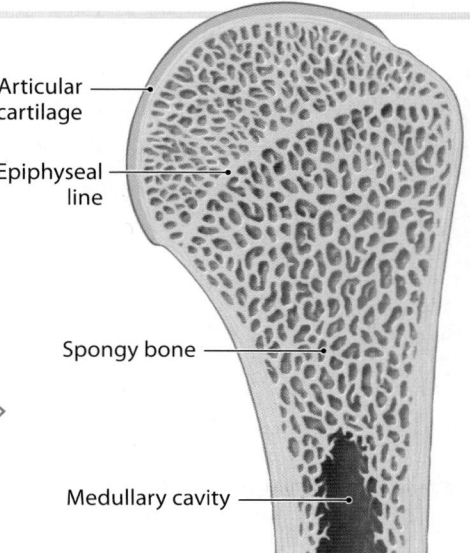

Articular cartilage

Epiphyseal line

Spongy bone

Medullary cavity

**8** At puberty, the combination of rising levels of sex hormones, growth hormone, and thyroid hormones stimulates bone growth dramatically. Osteoblasts now begin producing bone faster than chondrocytes are producing new cartilage. As a result, the osteoblasts "catch up," and the epiphyseal cartilage gets narrower and narrower until it ultimately disappears. The completion of epiphyseal growth is called **epiphyseal closure**.

Chondrocytes at the epiphyseal side of the cartilage continue to divide and enlarge.

Chondrocytes degenerate at the diaphyseal side.

Osteoblasts migrate upward from the diaphysis, and cartilage is gradually replaced by bone.

**7** The chondrocytes within the epiphyseal cartilage are arranged into zones. On the epiphyseal side, chondrocytes continually add new cartilage. On the diaphyseal side, chondrocytes are degenerating. Osteoblasts are continuously invading this zone and replacing it with bone. The osteoblasts are therefore moving toward the epiphysis, which is being pushed away by the continued production of new cartilage.

**? INTEGRATION**

**C.** How could x-rays of the femur be used to determine whether a person has reached full height?

 **LEARNING OUTCOME**

Describe the process of endochondral ossification.

# Intramembranous ossification forms bone without a prior cartilage model

**Intramembranous** (in-tra-MEM-bra-nus) **ossification** begins when mesenchymal cells within embryonic or fibrous connective tissue differentiate into osteoblasts. This process normally occurs in the deeper layers of the dermis, and the bones that result are often called **dermal bones** or *membrane bones*. Examples of dermal bones include the roofing bones of the skull, the lower jaw (mandible), and the collarbone (clavicle). Sesamoid bones are membrane bones that form within tendons. The patella (kneecap) is a sesamoid bone (see Module 6.2).

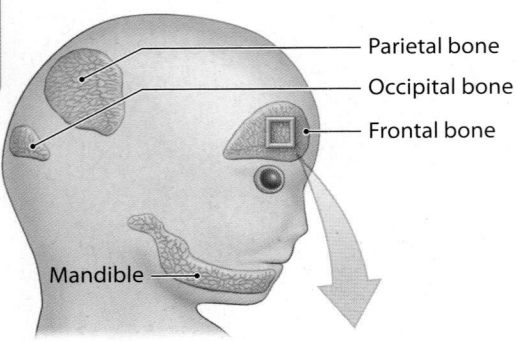

- Parietal bone
- Occipital bone
- Frontal bone
- Mandible

- Osteoblast layer
- Osteoid
- Osteocytes in lacunae
- Blood vessels
- Bone matrix
- Mesenchymal cells in embryonic connective tissue

**Intramembranous bone**    LM × 32

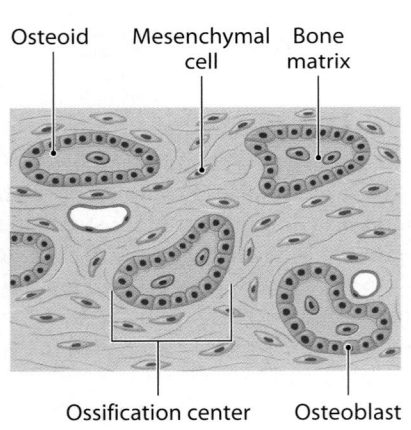

Osteoid    Mesenchymal cell    Bone matrix

Ossification center    Osteoblast

**1** Mesenchymal cells first cluster together, differentiate into osteoblasts, and start to secrete the organic components of the matrix. The resulting osteoid then becomes mineralized with calcium salts forming bone matrix. The location in a tissue where ossification begins is called an **ossification center**.

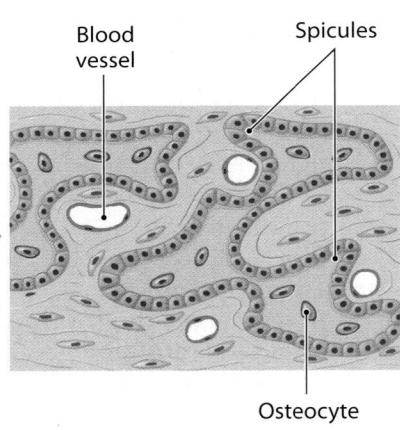

Blood vessel    Spicules

Osteocyte

**2** The developing bone grows outward from the ossification centers in small struts called **spicules**. As ossification proceeds, it traps some osteoblasts inside bony pockets; these cells differentiate into osteocytes. Meanwhile, mesenchymal cell divisions continue to produce additional osteoblasts.

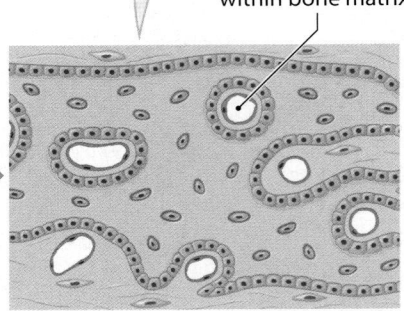

Blood vessel trapped within bone matrix

**3** Bone growth is an active process, and osteoblasts require oxygen and a reliable supply of nutrients. Blood vessels begin to grow into the area. As spicules meet and fuse together, some of these blood vessels become trapped within the developing bone.

 **REVIEW**

**A.** Define *intramembranous ossification*.

**B.** During intramembranous ossification, bone replaces which type of tissue?

**7** Intramembranous ossification starts approximately during the eighth week of embryonic development. These photos show the extent of intramembranous and endochondral ossification that occurs between 10 and 16 weeks of development. At 10 weeks, bone formation is under way, but the skeleton is incomplete. At 16 weeks, most of the bones of the adult skeleton can be identified.

**10 weeks of development**

**16 weeks of development**

Flat bones of the skull

Intramembranous ossification centers that produce the roofing bones of the skull

Future hip bone

Primary ossification centers of the long bones of the lower limb

Long bones of the limbs

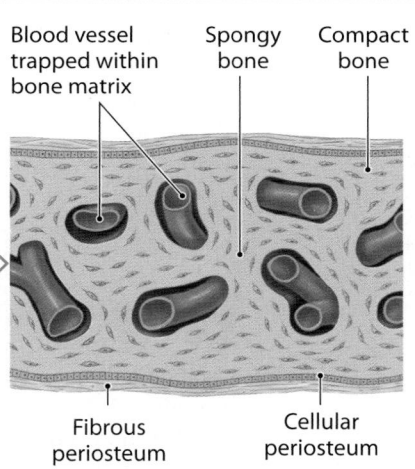

Osteocytes in lacunae

Osteoblast layer

Blood vessels

**4** Continued deposition of bone by osteoblasts located close to blood vessels results in a plate of spongy bone with blood vessels weaving throughout.

Blood vessel trapped within bone matrix

Spongy bone

Compact bone

Fibrous periosteum

Cellular periosteum

**5** Initially, the intramembranous bone consists of spongy bone only. Subsequent remodeling around trapped blood vessels produces osteons typical of compact bone. As the rate of growth slows, the connective tissue around the bone organizes into the fibrous layer of the periosteum. The osteoblasts closest to the bone surface become less active but remain as the inner, cellular layer of the periosteum.

Compact bone

Diploë

**6** Areas of spongy bone are remodeled, forming a layer called **diploë** (DIP-lō-ē; diplous, twofold) "sandwiched" between two layers of compact bone.

**?** **INTEGRATION**

**C.** Explain the primary difference between endochondral ossification and intramembranous ossification.

**Lo** **LEARNING OUTCOME**

Describe the process of intramembranous ossification.

# Module 6.9

## Abnormalities of bone growth and development produce recognizable physical signs

A variety of endocrine (hormonal) or metabolic problems can result in atypical skeletal growth. Here we consider several conditions that affect the skeleton as a whole.

**1** In **pituitary growth failure**, inadequate production of growth hormone (GH) leads to reduced epiphyseal cartilage activity and abnormally short bones. This condition is becoming increasingly rare in the United States because children can be treated with synthetic human GH.

**2** **Achondroplasia** (ā-kon-drō-PLĀ-zē-uh) results from abnormal epiphyseal activity. In this case the epiphyseal cartilages of the long bones grow unusually slowly and are replaced by bone early in life. As a result, the individual develops short, stocky limbs. Although other skeletal abnormalities occur, the trunk is normal in size, and sexual and mental development remain unaffected.

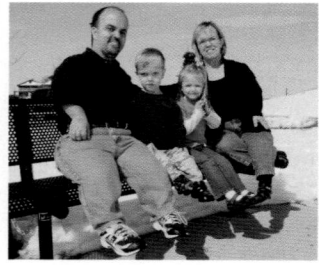

**3** Several inherited metabolic conditions that affect many systems influence the growth and development of the skeletal system. These conditions produce characteristic variations in body proportions. For example, many people with **Marfan syndrome** are very tall and have long, slender limbs because of excessive cartilage formation at the epiphyseal cartilages. Although this is an obvious physical distinction, the characteristic body proportions are not in themselves dangerous. However, the underlying mutation, which affects the structure of connective tissue throughout the body, commonly causes life-threatening cardiovascular problems.

**? REVIEW**

**A.** Why is pituitary growth failure less common today in the United States?

**B.** Describe Marfan syndrome.

**4** **Congenital talipes equinovarus** (TAL-i-pēz e-kwī-nō-VA-rus) (clubfoot) results from an inherited developmental abnormality that affects 2 in 1000 births. Boys are affected roughly twice as often as girls. One or both feet may be involved, and the condition may be mild, moderate, or severe. The underlying problem is abnormal muscle development that distorts growing bones and joints. In most cases, the tibia, ankle, and foot are affected, and the feet are turned medially and inverted. If both feet are involved, the soles face one another. Prompt treatment with casts or other supports in infancy helps alleviate the problem, and fewer than half of the cases require surgery.

**5** **Gigantism** (also called *giantism*) results from an overproduction of GH before puberty. Individuals can reach heights of over 2.7 m (8 ft 11 in.) and weights of over 200 kg (440 lb). Puberty is often delayed, and the facial features in adults resemble those of acromegaly (see **7** ). The most common cause is a pituitary tumor, which may be treated by surgery, radiation, or drugs that suppress GH release.

**6** Under abnormal conditions, mesenchymal stem cells in any connective tissue can develop into osteoblasts that begin producing bone. The person below (left) has **fibrodysplasia ossificans progressiva (FOP)**, a rare single gene mutation disorder that involves bone deposits around skeletal muscles. The skeleton below (right) shows the extent of abnormal ossification that can occur. Bones that develop in unusual places are called **heterotopic** (*hetero*, place), or **ectopic** (*ektos*, outside), **bones**. There is no effective treatment for this painful and debilitating condition, and patients seldom survive into their 40s.

**7** If GH levels rise abnormally after epiphyseal cartilages close, the skeleton does not grow longer, but bones get thicker, especially those in the face, jaw, and hands. Cartilage growth and alterations in soft-tissue structure lead to changes in physical features, such as the contours of the face. These physical changes occur in the disorder called **acromegaly**. Early diagnosis and treatment with pituitary surgery and/or drugs that reduce GH levels may prevent irreversible physical changes and extend life expectancy in both acromegaly and gigantism.

**?** **REVIEW**
**C.** Compare gigantism with acromegaly.

**Lo** **LEARNING OUTCOME**
Discuss various abnormalities of bone formation and growth.

# Introduction to the Structure and Growth of Bones

**Concept map:** Use each of the following terms once to fill in the blank boxes to correctly complete the map.

- lacunae ✓
- osteocytes ✓
- osteoid ✓
- intramembranous ossification ✓
- compact bone ✓
- periosteum
- hyaline cartilage ✓

6

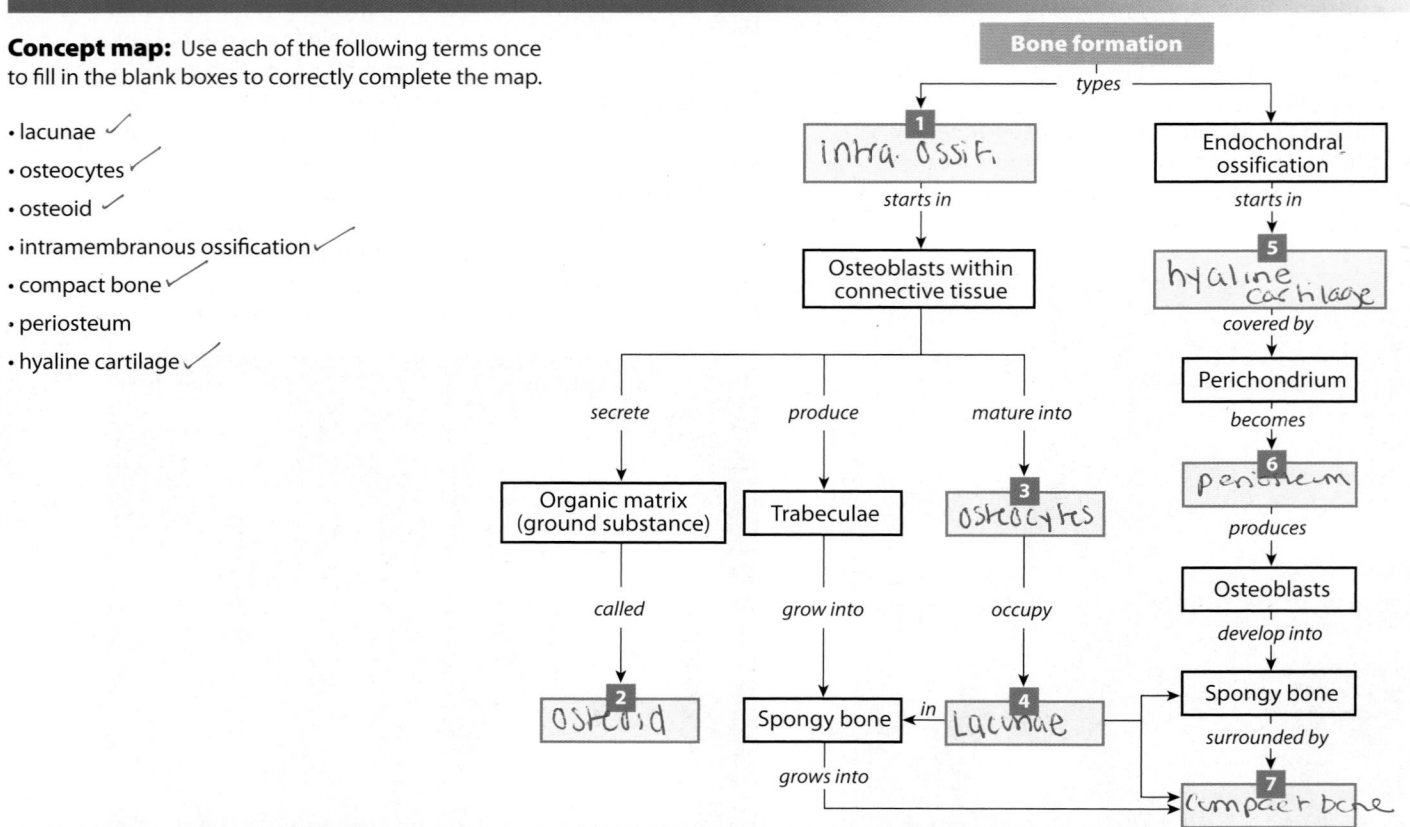

**Bone formation**

types

**1** intra. ossif.

Endochondral ossification

starts in

Osteoblasts within connective tissue

starts in

**5** hyaline cartilage

covered by

Perichondrium

becomes

**6** periosteum

produces

secrete

Organic matrix (ground substance)

produce

Trabeculae

mature into

**3** osteocytes

Osteoblasts

develop into

called

**2** osteoid

grow into

Spongy bone

occupy

**4** Lacunae

in

Spongy bone

surrounded by

grows into

**7** Compact bone

**Vocabulary:** Write the term for each of the following descriptions in the space provided.

**8**  Bones with complex shapes

**9**  The expanded ends of a long bone

**10**  A shallow depression in the surface of a bone

**11**  The marrow-filled space within a bone

**12**  The strut- and plate-shaped matrix of spongy bone

**13**  Cells that remove bone matrix

**14**  Bones that develop in tendons

**15**  The process that forms new bone matrix

**16**  The basic functional unit of compact bone

**17**  Type of bone growth that increases bone diameter

**18**  Process by which cartilage is replaced by bone

**8** Irregular bones
**9**
**10**
**11**
**12**
**13**
**14**
**15**
**16**
**17**
**18**

**Section integration:** While playing on her swing set, 10-year-old Rebecca falls and breaks her right leg. At the emergency room, the doctor tells her parents that the proximal end of the tibia where the epiphysis meets the diaphysis is fractured. The fracture is properly set and eventually heals. During a routine physical when she is 18, Rebecca learns that her right leg is 2.54 cm (1 in.) shorter than her left. What might account for this difference?

**19**

## Module 6.10
# Bones play an important role as mineral reservoirs

In this section, we consider the dynamic relationship between calcium concentrations in the blood and calcium reserves in the skeletal system. We also examine the role of hormones in regulating calcium balance in the body.

**1** This chemical analysis shows the importance of bones as mineral reservoirs. Minerals are inorganic ions that contribute to the osmotic balance of body fluids and are also vital in many physiological processes. In this section, we focus on the homeostatic regulation of calcium ion concentrations in blood. We consider other minerals in later chapters.

| Bone Composition | | Bone Contains |
|---|---|---|
| | Calcium 39% | 99% of the body's calcium |
| | Potassium 0.2% | 4% of the body's potassium |
| | Sodium 0.7% | 35% of the body's sodium |
| Organic compounds (mostly collagen) 33% | Magnesium 0.4% | 50% of the body's magnesium |
| | Carbonate 9.7% | 80% of the body's carbonate |
| | Phosphate 17% | 99% of the body's phosphate |
| | Total inorganic 67% components | |

6

In the intestines, calcium and phosphate ions are absorbed from the diet. The absorption rate is hormonally regulated.

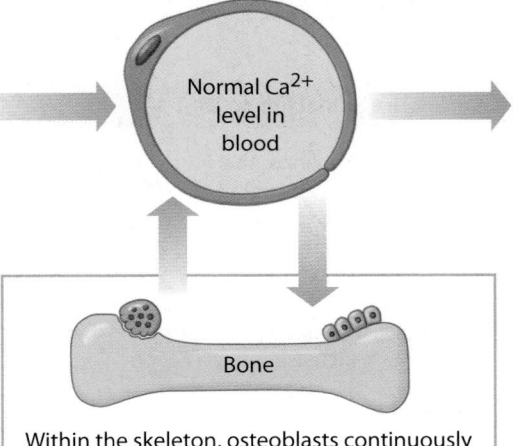

Normal Ca²⁺ level in blood

Bone

In the kidneys, the levels of calcium and phosphate ions lost in the urine are hormonally regulated.

**2** **Calcium** is the most abundant mineral in the human body. The typical human body contains 1–2 kg (2.2–4.4 lb) of calcium, with roughly 99 percent of it deposited in the skeleton. Calcium ions play a role in a variety of physiological processes, such as muscle contraction, generating nerve impulses, and blood coagulation. The homeostatic regulation of calcium ion level is a juggling act that balances activities under way in the intestines, bones, and kidneys.

Within the skeleton, osteoblasts continuously deposit new bone matrix. At the same time, osteoclasts erode existing matrix, releasing calcium and phosphate ions into the circulation. The balance between osteoblast and osteoclast activity is hormonally regulated.

 **REVIEW**

**A.** What is the ratio of organic compounds to inorganic components in the composition of bone?

**B.** Which three organ systems coordinate to maintain normal blood calcium level?

Even small variations from the normal calcium concentration affect cellular operations. If the calcium concentration of blood increases or decreases by more than 30–35 percent, neuron and muscle cell function is disrupted, with potentially lethal results. Calcium ion concentration is so closely regulated, however, that daily fluctuations of more than 10 percent are highly unusual.

 **INTEGRATION**

**C.** What physical signs would be expected in a person whose blood calcium was abnormally low?

**LEARNING OUTCOME**

List the minerals stored in the bones, and identify the organs involved in calcium homeostasis.

# The primary hormones regulating calcium ion metabolism are parathyroid hormone, calcitriol, and calcitonin

Calcium ion homeostasis is maintained by hormones that target the skeletal system, the digestive tract, and the kidneys.

**1** When blood calcium ion concentration falls below normal, cells of the **parathyroid glands**, embedded in the thyroid gland in the neck, release **parathyroid hormone (PTH)** into the bloodstream. Parathyroid hormone has multiple effects that increase blood calcium level.

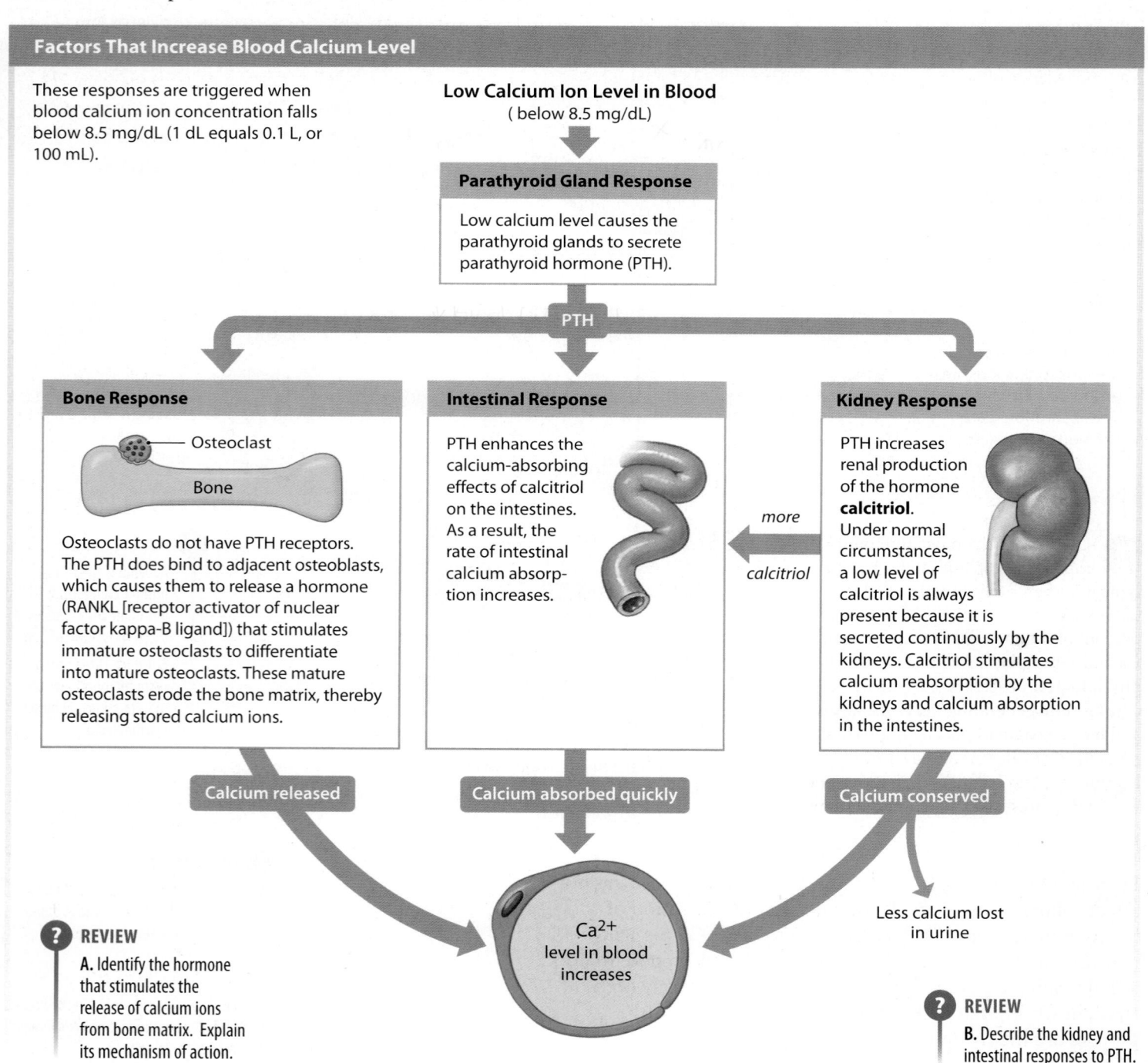

### Factors That Increase Blood Calcium Level

These responses are triggered when blood calcium ion concentration falls below 8.5 mg/dL (1 dL equals 0.1 L, or 100 mL).

**Low Calcium Ion Level in Blood**
( below 8.5 mg/dL)

**Parathyroid Gland Response**

Low calcium level causes the parathyroid glands to secrete parathyroid hormone (PTH).

PTH

**Bone Response**

Osteoclast

Bone

Osteoclasts do not have PTH receptors. The PTH does bind to adjacent osteoblasts, which causes them to release a hormone (RANKL [receptor activator of nuclear factor kappa-B ligand]) that stimulates immature osteoclasts to differentiate into mature osteoclasts. These mature osteoclasts erode the bone matrix, thereby releasing stored calcium ions.

**Intestinal Response**

PTH enhances the calcium-absorbing effects of calcitriol on the intestines. As a result, the rate of intestinal calcium absorption increases.

*more*

*calcitriol*

**Kidney Response**

PTH increases renal production of the hormone **calcitriol**. Under normal circumstances, a low level of calcitriol is always present because it is secreted continuously by the kidneys. Calcitriol stimulates calcium reabsorption by the kidneys and calcium absorption in the intestines.

Calcium released

Calcium absorbed quickly

Calcium conserved

$Ca^{2+}$ level in blood increases

Less calcium lost in urine

**? REVIEW**

**A.** Identify the hormone that stimulates the release of calcium ions from bone matrix. Explain its mechanism of action.

**? REVIEW**

**B.** Describe the kidney and intestinal responses to PTH.

**2** When blood calcium ion concentration rises above normal, **C cells** of the **thyroid gland** secrete **calcitonin**. Calcitonin was discovered more than 50 years ago, but its role in human physiology remains unclear. Although calcitonin secretion is stimulated by increased blood calcium, its effects are restricted to osteoclasts, whose activity is decreased. In pregnant and nursing women, both calcitonin and vitamin $D_3$ are increased, suggesting that these increases may be important in transferring maternal calcium to the fetus and for preserving and recovering bone mineral density in the mother. (Interestingly, pharmacological calcitonin used as a treatment for osteoporosis in postmenopausal women can inhibit osteoclast activity.)

## Factors That Decrease Blood Calcium Level

Intestinal and kidney responses are triggered when blood calcium ion concentration rises above 11 mg/dL.

**High Calcium Ion Level in Blood**
(above 11 mg/dL)

**Thyroid Gland Response**

C cells in the thyroid gland secrete calcitonin.

Calcitonin

**Bone Response**

Bone

Calcitonin decreases osteoclast activity but does not affect osteoblasts, which continue to deposit calcium ions within the bone matrix.

**Intestinal Response**

Decreasing PTH or calcitriol level results in a decrease in the rate of calcium ion absorption by the intestines.

**Kidney Response**

*less calcitriol*

Increased calcitonin level has an inhibitory effect on the kidneys and suppresses calcium ion reabsorption.

Calcium release slowed

Calcium absorbed slowly

Calcium excreted

More calcium lost in urine

$Ca^{2+}$ level in blood decreases

**? REVIEW**

**C.** How does calcitonin act to lower blood calcium?

As a calcium reserve, the skeleton plays the primary role in blood calcium homeostasis. This function has a direct effect on the shape and strength of the bones throughout the skeleton. When large numbers of calcium ions are released into the blood, the bones become weaker; when calcium salts are deposited, the bones become denser and stronger.

**LEARNING OUTCOME**

Discuss the effects of hormones on bone development, and explain the homeostatic mechanisms involved.

# Module 6.12

# A fracture is a crack or a break in a bone

Despite its mineral strength, bone can crack or even break if subjected to extreme loads, sudden impacts, or stresses from unusual directions. The damage produced constitutes a **fracture**. Most fractures heal even after severe damage, provided that the blood supply and the cellular components of the endosteum and periosteum survive.

## Steps in Fracture Repair

| Fracture Hematoma Formation | Callus Formation | Spongy Bone Formation | Compact Bone Formation |
| --- | --- | --- | --- |

Fracture hematoma

Dead bone    Bone fragments

Spongy bone of internal callus    Cartilage of external callus

Spongy bone of external callus

Periosteum

Internal callus    External callus

External callus

**1** Immediately after the fracture, extensive bleeding occurs. Over a period of several hours, a large blood clot, or **fracture hematoma**, soon closes off the injured vessels and leaves a fibrous meshwork in the damaged area.

**2** Next, cells of the intact endosteum and periosteum undergo rapid cell division, and the daughter cells migrate into the fracture zone. An **internal callus** (*callum*, hard skin) forms as a network of spongy bone unites the inner edges of the fracture. An **external callus** of cartilage and bone encircles and stabilizes the outer edges of the fracture.

**3** As the repair continues, osteoblasts replace the central cartilage of the external callus with spongy bone, which then unites the broken ends. Bone fragments and areas of dead bone closest to the break are removed and replaced. The ends of the fracture are now held firmly in place and can withstand normal stresses from muscle contractions.

**4** A swelling initially marks the location of the fracture. Over time, this region will be remodeled by osteoblasts and osteoclasts, and little evidence of the fracture will remain.

**?** **REVIEW**

**A.** List the steps involved in fracture repair, beginning just after the fracture occurs.

## Types of Fractures

**Transverse fractures**, such as this fracture of the ulna, break a bone shaft across its long axis.

**Spiral fractures**, such as this fracture of the femur, are produced by twisting stresses that spread along the length of the bone.

**Displaced fractures** produce new and abnormal bone alignments. **Nondisplaced fractures** retain the normal alignment of the bones or fragments.

**Compression fractures** occur in vertebrae subjected to extreme stresses, such as the forces produced when you land on your buttocks in a fall. Compression fractures are often associated with osteoporosis.

In a **greenstick fracture**, such as this fracture of the radius, only one side of the shaft is broken, and the other is bent. This type of fracture generally occurs in children, whose long bones have yet to ossify fully.

**Comminuted fractures**, such as this fracture of the femur, shatter the affected area into a multitude of bony fragments.

**Epiphyseal fractures**, such as this fracture of the femur, tend to occur where the bone matrix is undergoing calcification and chondrocytes are dying. A clean transverse fracture along this line generally heals well. Unless carefully treated, fractures between the epiphysis and the epiphyseal cartilage can permanently stop growth at this site.

A **Pott's fracture**, also called a bimalleolar fracture, occurs at the ankle and affects both the medial malleolus of the distal tibia and the lateral malleolus of the distal fibula.

A **Colles fracture**, a break in the distal portion of the radius, is typically the result of reaching out to cushion a fall.

**5** Fractures are named using various criteria, including their external appearance, their location, and the nature of the crack or break in the bone. Important types of fractures are illustrated here by representative x-rays. The broadest general categories are closed fractures and open fractures. Fractures are named by four basic either/or categories, each with two groupings: (1) position of bone ends (nondisplaced or displaced); (2) completeness of break (complete or incomplete); (3) orientation of break to the bone's long axis (linear or transverse); and (4) whether or not the broken bone penetrates the skin (open or closed). In an **open** (*compound*) **fracture**, the bone pierces the skin. **Closed** (*simple*) **fractures** are completely internal. Open fractures tend to be more dangerous than closed because of an increased risk of infection or uncontrolled bleeding. In closed fractures, the bone ends can only be seen on x-ray because the skin remains intact.

**REVIEW**

**B.** Define *open fracture* and *closed fracture*.

 **LEARNING OUTCOME**

Describe the types of fractures, and explain how fractures heal.

**Concept map:** Use each of the following terms once to fill in the blank boxes to correctly complete the map.

- ↓ Ca²⁺ intestinal absorption ✓
- homeostasis ✓
- release of stored Ca²⁺ from bone
- ↓ blood Ca²⁺ level ✓
- parathyroid glands ✓
- calcitonin ✓
- ↑ Ca²⁺ intestinal absorption ✓

6

### Regulation of Blood Calcium Level

by

**Thyroid gland**

releases

**1** Calcitonin

causes

**2** ↓ Ca²⁺ absorption

↑ Excretion of Ca²⁺ by kidneys

results in

**3** ↓ Blood Ca²⁺ level

**4** Parathyroid

release

**Parathyroid hormone**

causes

**5** ↑ Ca²⁺ intestinal absorption

Enhanced reabsorption of Ca²⁺ by kidneys

**6** release of stored Ca²⁺ from bone

results in

↑ Blood Ca²⁺ level

results in

stimulates ← ↑ Blood Ca²⁺ level — **7** homeostasis — ↓ Blood Ca²⁺ level → stimulates

---

**Short answer:** Identify the type of fracture and the bones involved in each of the following x-ray images.

**8** Spiral

**9** transverse

**10** greenstick

**11** comminuted

**12** Compressed

**13** Colles fracture

# Study Outline

## SECTION 1 • Introduction to the Structure and Growth of Bones

### Module 6.1

**The skeletal system is made up of the axial and appendicular divisions** p. 211

1. The **axial skeleton** has 80 bones and consists of the bones of the skull, thorax, and vertebral column.

2. The **appendicular skeleton** has 126 bones and consists of the limb bones, and the pectoral and pelvic girdles that attach the limbs to the axial skeleton.

3. The functions of the skeletal system are support, storage of minerals and lipids, blood cell production, protection, and leverage.

### Module 6.2

**Bones are classified according to shape and structure and have varied bone markings** p. 212

4. Bones can be divided into six broad categories based on their shapes: **flat bones**, **sutural bones**, **long bones**, **irregular bones**, **sesamoid bones**, and **short bones**.

5. Each bone has characteristic **bone markings** (*surface features*), including projections, depressions, and openings.

### Module 6.3

**Long bones transmit forces along the shaft and have a rich blood supply** p. 214

6. A representative long bone has an **epiphysis**, **metaphysis**, **diaphysis**, and a **medullary cavity**. The epiphyses consist mostly of **spongy bone** (also called **trabecular bone**), whereas the diaphysis is made of **compact bone**.

7. The medullary cavity is filled with **red bone marrow** (for blood cell production) and **yellow bone marrow** (for lipid storage).

8. **Articular cartilage** covers the portion of the epiphysis that articulates with other bones. It is avascular and relies on the diffusion of synovial fluid for its metabolism.

*The growth and maintenance of bones require an extensive blood supply. Therefore, bones are highly vascular.*

### Module 6.4

**Bone has a calcified matrix maintained and altered by osteogenic cells, osteoblasts, osteocytes, and osteoclasts** p. 216

9. **Osteogenic**, or *osteoprogenitor,* **cells** are mesenchymal cells present in bone. They reside in the inner layer of the periosteum, the endosteum, and in vascular passageways.

10. **Osteoblasts** produce new bone matrix in a process called **ossification** or **osteogenesis**. They develop into osteocytes once they have been surrounded by bone matrix.

11. **Osteocytes** are mature bone cells that account for most of the cell population in bone. These cells maintain the protein and mineral content of the bone matrix. They occupy a **lacuna** and have processes that extend into **canaliculi**.

12. **Osteoclasts** are cells that remove bone matrix.

### Module 6.5

**Compact bone consists of parallel osteons, and spongy bone consists of a network of trabeculae** p. 218

13. The basic functional unit of compact bone is the **osteon,** or Haversian system. Each osteon forms a series of **concentric lamellae** surrounding a **central canal**.

*Compact bone consists of parallel osteons. You might think of a single osteon as a drinking straw with very thick walls. Osteons are parallel to the long axis of a bone and give it great strength even when extreme force is applied to either end.*

14. Spongy bone is found in areas of less stress, as in the epiphyses of long bone. Spongy bone is formed by an open network of struts and plates called **trabeculae**.

15. **Circumferential lamellae** surround the outer and inner surfaces of bone. **Interstitial lamellae** fill in the spaces between the osteons of compact bone. **Perforating canals** run perpendicular to the long axis of bone.

### Module 6.6

**Appositional bone growth, involves the periosteum and endosteum,** p. 220

16. The diameter of a long bone enlarges through **appositional growth** at the outer surface as layers of bone are added. During this process, osteoclasts remove and recycle lamellae at the inner surface, increasing the diameter of the medullary cavity.

17. **Periosteum** covers the superficial layer of compact bone except within the joint cavities. **Endosteum** lines the medullary cavity.

## Module 6.7

### Endochondral ossification replaces a cartilage model with bone  p. 222

18. **Endochondral ossification** begins with a hyaline cartilage model that is gradually replaced with bone. A **primary ossification center** forms at the diaphysis; **secondary ossification centers** form at the epiphyses.

*Intramembranous ossification starts approximately during the eighth week of embryonic development.*
*At 10 weeks (left), bone formation is under way but the skeleton is incomplete. At 16 weeks, most of the bones of the adult skeleton can be identified.*

19. All the cartilage is replaced except at the **epiphyseal cartilage** of the metaphysis and the articular cartilage. Growth in length (*interstitial growth*) occurs at the metaphysis and involves the epiphyseal cartilage. Eventually, **epiphyseal closure** will signal the end of epiphyseal growth.

## Module 6.8

### Intramembranous ossification forms bone without a prior cartilage model  p. 224

20. **Intramembranous ossification** begins when mesenchymal cells differentiate into osteoblasts within embryonic or fibrous connective tissue. This process produces **dermal bones**, such as the roofing bones of the skull, the mandible, clavicle, and sesamoid bones such as the patella.

## Module 6.9

### Abnormalities of bone growth and development produce recognizable physical signs  p. 226

21. A variety of endocrine or metabolic problems can result in atypical skeletal growth. **Pituitary growth failure** and **achondroplasia** result in abnormally short bones. **Marfan syndrome** and **gigantism** cause abnormal tallness. Other conditions include **congenital talipes equinovarus**, **acromegaly**, and **fibrodysplasia ossificans progressiva**.

## SECTION 2 • Physiology of Bones

## Module 6.10

### Bones play an important role as mineral reservoirs  p. 229

22. **Calcium** is the most abundant mineral in the body; 99 percent of it is located in the skeleton.

23. The intestines, the skeleton, and the kidneys function to coordinate calcium homeostasis in blood.

## Module 6.11

### The primary hormones regulating calcium ion metabolism are parathyroid hormone, calcitonin, and calcitriol  p. 230

24. Low calcium ion level in blood triggers **parathyroid hormone** (**PTH**) release from the **parathyroid glands**.
PTH raises calcium ion levels in blood by indirectly stimulating osteoclasts and enhancing calcium absorption in the intestines and calcium reabsorption by the kidneys. Calcitriol release from the kidneys is also stimulated by PTH and enhances intestinal calcium absorption.

25. High calcium ion level in blood stimulates the **C cells** of the **thyroid gland** to release **calcitonin**. Calcitonin lowers blood calcium by decreasing osteoclast activity and decreasing calcitriol release, intestinal absorption of calcium ions, and calcium reabsorption by the kidneys.

## Module 6.12

### A fracture is a crack or a break in a bone  p. 232

26. A break or crack in a bone is a **fracture**. The repair of a fracture involves the formation of a **fracture hematoma**, an **internal callus**, and an **external callus**.

27. There are many types of bone fractures depending on the forces causing the damage.

28. In an **open** (*compound*) **fracture**, the bone pierces the skin. **Closed** (*simple*) **fractures** are completely internal, do not pierce the skin, and can only be seen on x-ray. **Open** (*compound*) or **compound fractures** project through the skin. These are especially dangerous because of the possibility of infection or uncontrolled bleeding.

*Bone can crack or break if subjected to extreme loads, sudden impacts, or stresses from unusual directions.*

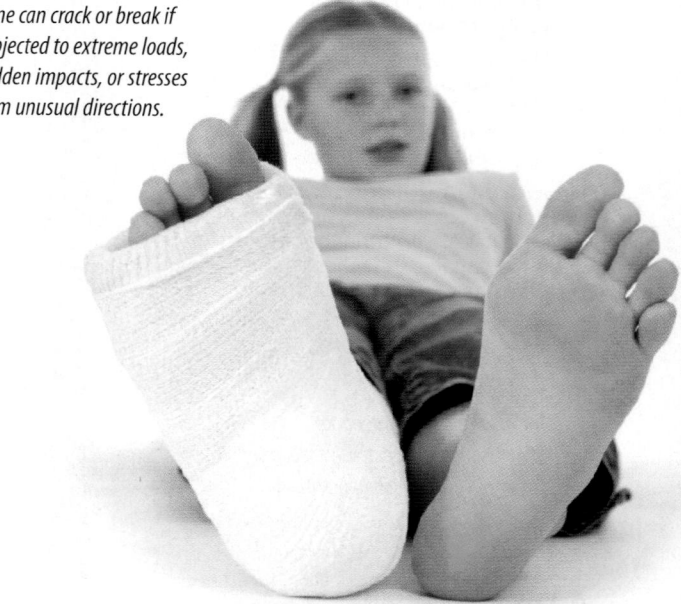

# Chapter Review Questions

**Labeling:** Identify the anatomical regions and structures in this image of a long bone.

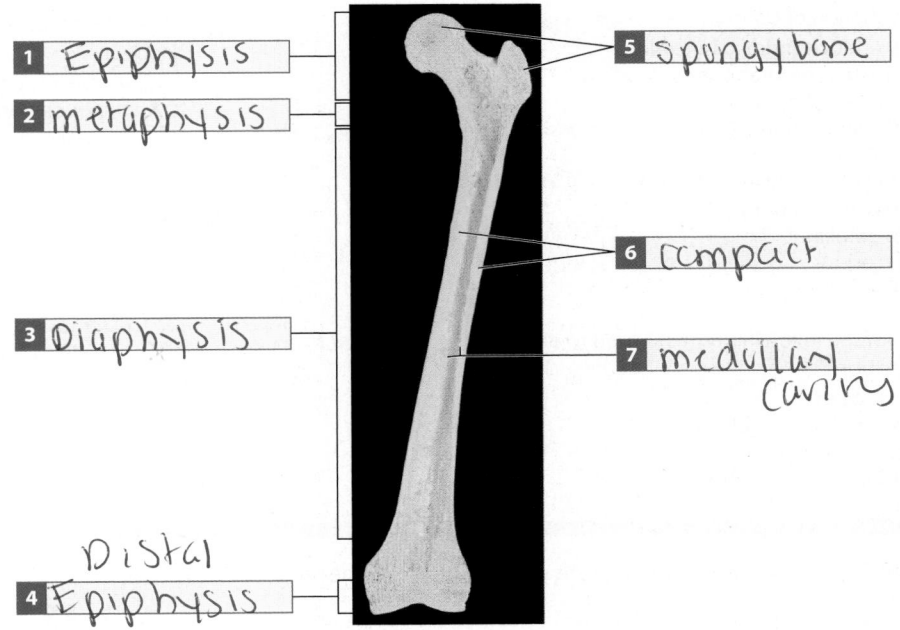

1. Epiphysis
2. metaphysis
3. Diaphysis
4. Distal Epiphysis
5. spongy bone
6. compact
7. medullary cavity

---

**True/False:** Indicate whether each statement is true or false.

8. A broken leg that results in a bone piercing the skin is called a compound fracture.
9. The primary ossification centers are localized in the epiphyses.
10. Osteoblastic activity tends to deposit calcium into the bone.
11. Parathyroid hormone (PTH) is released in response to low blood calcium level.
12. Both trabecular bone and compact bone have osteons.
13. A foramen is a small rounded passageway through which blood vessels or nerves penetrate.
14. Bone has a rich blood supply but lacks sensory innervation.
15. Appositional growth increases bone diameter.

8. T
9. F
10. T
11. T
12. F
13. T
14. F
15. T

---

**Multiple choice:** Select the correct answer from the list provided.

16. Blood cell formation occurs in
    - a) yellow bone marrow.
    - b) red bone marrow.
    - ☒ c) the matrix of bone tissue.
    - d) the ground substance of bones.

17. Which of the following hormones is involved with blood calcium ion homeostasis and is produced by the kidneys?
    - ☒ a) calcitriol
    - b) parathyroid hormone
    - c) prolactin
    - d) calcitonin

**18** The presence of an epiphyseal line indicates
- ☒ a) epiphyseal growth has ended.
- ☐ b) the bone is fractured at that location.
- ☐ c) the bone will continue increasing in length for many years.
- ☐ d) vitamin D3 insufficiency.

**19** The formation of bone without a prior cartilage model is
- ☐ a) endochondral ossification.
- ☒ b) intramembranous ossification.
- ☐ c) achondroplasia.
- ☐ d) fibrodysplasia ossificans progressiva.

**20** Which of the following are examples of irregular bones?
- ☐ a) Wormian bones
- ☐ b) patellae
- ☐ c) carpals
- ☒ d) vertebrae

**21** Which of the following terms is used to describe the structure of spongy bone?
- ☐ a) osteon
- ☐ b) circumferential lamellae
- ☒ c) trabeculae
- ☐ d) Haversian system

**22** Bone marrow is located in the
- ☒ a) medullary cavity.
- ☐ b) sinus cavity.
- ☐ c) facets.
- ☐ d) fossa.

**23** The external covering of bones, except at the joint cavity, is the
- ☐ a) endosteum.
- ☒ b) periosteum.
- ☐ c) osteon.
- ☐ d) interstitial lamellae.

**24** Which of the following is *not* formed by endochondral ossification?
- ☐ a) mandible
- ☐ b) patella
- ☐ c) femur
- ☒ d) roofing bones of the skull

**True/False:** Fill in the following blanks in the spaces provided to the right.

**25** The number of bones in the axial skeleton is _____ bones, whereas the appendicular skeleton contains _____ bones.

**26** _____ bones develop inside tendons and are commonly located near the joints of the knees, hands, and feet.

**27** An air-filled chamber within a bone is called a(an) _____.

**28** The functional unit of mature compact bone is called a(an) _____.

**29** _____ are cells that remove and recycle bone matrix.

**25** _____ , _____

**26** _____

**27** _____

**28** _____

**29** _____

## Short answer

**30** What are the five primary functions of the skeletal system?

**31** Which kind of fracture would you expect to occur in a patient whose shinbone broke when her foot stuck to the floor while she was dancing a pirouette?

**32** Compare and contrast pituitary growth failure and achondroplasia.

**33** What is the primary difference between endochondral ossification and intramembranous ossification?

**34** Describe why epiphyseal fractures are of particular concern.

**35** Why are impacts perpendicular to the shaft of a long bone more dangerous than stress applied parallel to the long axis of the bone?

**36** Distinguish between acromegaly and gigantism.

## MasteringA&P™

Access more chapter study tools online in the MasteringA&P Study Area:

- Chapter Quizzes, Chapter Practice Test, and Clinical Case Studies

- Practice Anatomy Lab™ 3.0 — PAL™3.0
- Interactive Physiology™ — iP2™
- A&P Flix™ — A&PFlix™
- PhysioEx™ 9.1 — PhysioEx™ 9.1

# Chapter Integration • Applying what you have learned

## Assess the damage to a construction worker's spine from a one-story fall

A construction worker falls from the second floor of a building. He forcefully lands directly on his buttocks in a position as if he were sitting. He immediately feels intense pain in his lower back and is unable to get onto his feet. While waiting for the ambulance, his coworkers perform a quick assessment and notice there are no bones protruding through his skin. Upon arrival at the emergency room, a series of diagnostic tests are performed. He is informed that the trauma he has suffered is two severely damaged vertebrae. From what you have just learned about osseous tissue and bone structure, answer the following questions.

**1** Taking into account the manner in which he fell and the forces his vertebral column experienced, which kind of fracture would you anticipate he has suffered? Is it likely to be a closed or open fracture?

**2** An orthopedic surgeon was called to evaluate his case and she has recommended surgery. She informed the patient that the damage to the vertebrae was so severe that she would like to fuse them to the neighboring healthy vertebrae in order to stabilize his vertebral column. She mentions a technique called bone grafting, where she will shave bone from his iliac crest (hip), and lay these shavings between the damaged and healthy bones. Describe the cellular activity you would predict to occur that would result in neighboring bones fusing into a single bone.

**3** What kind of processes would you expect to occur in the region of his hip where the bone for the graft was removed?

7

# The Skeleton

## Module 7.1

# The axial skeleton includes bones of the head, vertebral column, and trunk

The **axial skeleton** forms the longitudinal axis of the body. This division of the skeletal system includes the *skull* and associated bones, the *thoracic cage*, the *vertebral column*, and various supplemental cartilages. There are typically 80 bones in the axial skeleton—roughly 40 percent of the bones in the human body.

Go to
MasteringA&P™
PAL™3.0
Anatomical
Models→
Axial Skeleton

**? REVIEW**
**A.** How many bones comprise the skull and its associated bones?

7

| SKELETAL SYSTEM | 206 |

| Skull and associated bones | 29 |
| --- | --- |

Skull
| Cranium | 8 |
| Face | 14 |

Associated bones
| Auditory ossicles | 6 |
| Hyoid | 1 |

**AXIAL SKELETON** | 80

| Thoracic cage | 25 |
| Sternum | 1 |
| Ribs | 24 |

| Vertebral column | 26 |
| Vertebrae | 24 |
| Sacrum | 1 |
| Coccyx | 1 |

APPENDICULAR SKELETON
(*see Section 2*) | 126

Articular cartilage (covers bone surfaces in freely movable joints)

Costal cartilages (cartilages of ribs)

Intervertebral discs (cartilage)

**1** The axial skeleton provides a framework that supports and protects the brain, the spinal cord, and the organs in the body cavities of the trunk. It also provides an extensive surface area for the attachment of muscles that (1) adjust the positions of the head, neck, and trunk; (2) perform breathing movements; and (3) stabilize or position parts of the appendicular skeleton that support the limbs. The joints of the axial skeleton have limited movement, but they are very strong and heavily reinforced by ligaments.

**? REVIEW**
**B.** What are the primary functions of the axial skeleton?
**C.** Describe the general roles of the muscles that attach to the axial skeleton.

**LO LEARNING OUTCOME**
List the four major components of the axial skeleton, and describe its major functions.

# Module 7.2

# The skull has cranial and facial components that are usually bound together by sutures

The **skull** contains 22 bones: 8 form the *cranium*, or braincase, and 14 form the face. Seven additional bones are associated with the skull: six auditory ossicles are located within the temporal bones of the cranium, and the hyoid bone is connected to the inferior surfaces of the temporal bones by a pair of ligaments.

| SKULL | | |
|---|---|---|

| FACE | 14 |
|---|---|
| Maxillae | 2 |
| Palatine bones | 2 |
| Nasal bones | 2 |
| Inferior nasal conchae | 2 |
| Zygomatic bones | 2 |
| Lacrimal bones | 2 |
| Vomer | 1 |
| Mandible | 1 |

| CRANIUM | 8 |
|---|---|
| Occipital bone | 1 |
| Parietal bones | 2 |
| Frontal bone | 1 |
| Temporal bones | 2 |
| Sphenoid | 1 |
| Ethmoid | 1 |

| ASSOCIATED BONES (see Module 7.8) | 7 |
|---|---|
| Hyoid bone | 1 |
| Auditory ossicles enclosed in temporal bones | 6 |

**? REVIEW**

**A.** Identify the bones of the cranium.

**1** **Facial bones** protect and support the entrances to the digestive and respiratory tracts. They also provide areas for the attachment of muscles that control facial expressions and assist in manipulating food.

**Facial Bones**

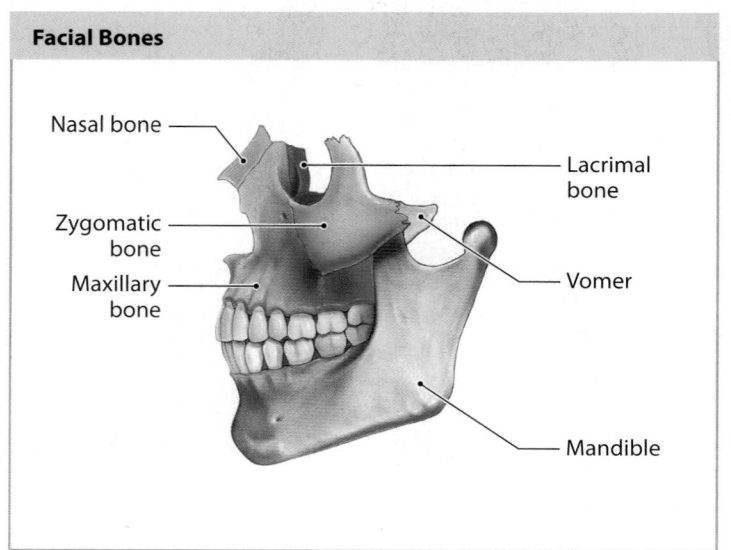

Nasal bone
Zygomatic bone
Maxillary bone
Lacrimal bone
Vomer
Mandible

**2** The **cranial bones** form the **cranium**, which protects the brain. It encloses the **cranial cavity**, a fluid-filled chamber that cushions and supports the brain. Membranes that stabilize the position of the brain are attached to the inner surface of the cranium. Its outer surface provides an extensive area for the attachment of muscles that move the eyes, jaws, and head.

**Cranial Bones**

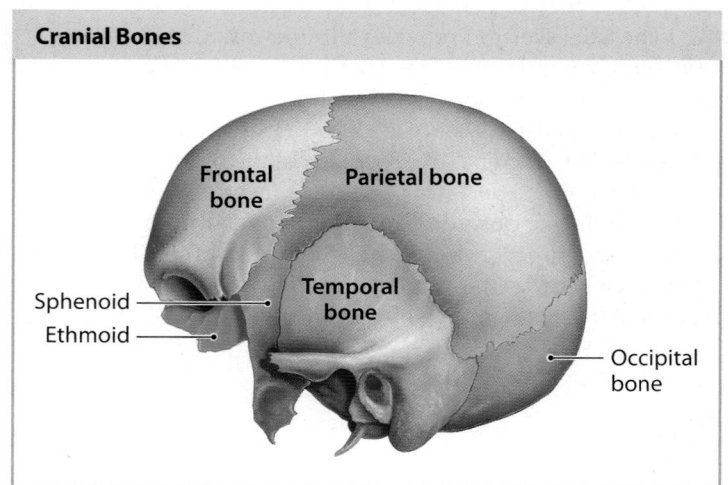

Frontal bone
Parietal bone
Sphenoid
Ethmoid
Temporal bone
Occipital bone

**? REVIEW**

**B.** Describe the functions of the facial bones.

**3** Joints, or *articulations*, form where two bones interconnect. Except where the mandible contacts the cranium, the connections between the skull bones of adults are immovable joints called **sutures**. At a suture, bones are tied firmly together with dense fibrous connective tissue. Each suture of the skull has a name, but at this point you need to know only four major sutures.

Watch

MasteringA&P™

Bone and Dissection Videos

Cranium

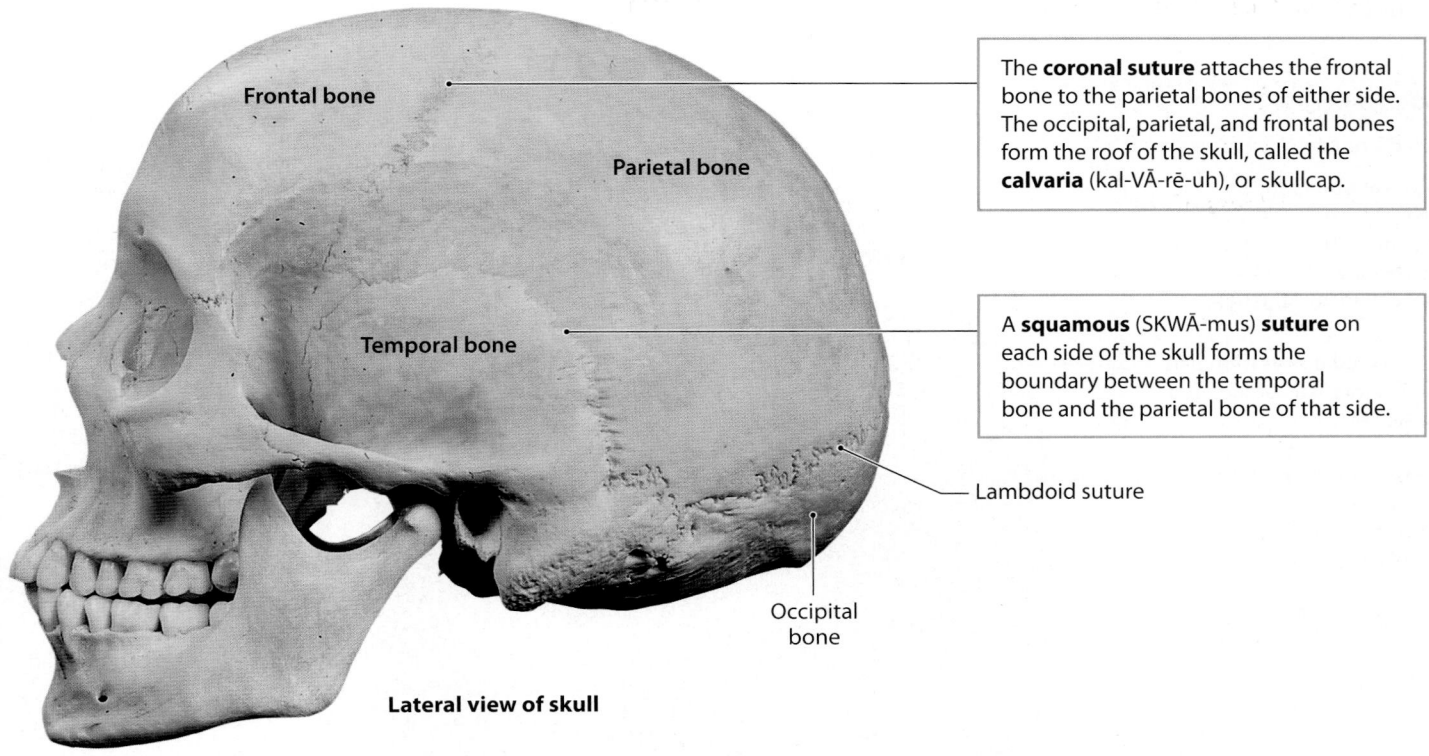

**Frontal bone**

**Parietal bone**

**Temporal bone**

Lambdoid suture

Occipital bone

**Lateral view of skull**

The **coronal suture** attaches the frontal bone to the parietal bones of either side. The occipital, parietal, and frontal bones form the roof of the skull, called the **calvaria** (kal-VĀ-rē-uh), or skullcap.

A **squamous** (SKWĀ-mus) **suture** on each side of the skull forms the boundary between the temporal bone and the parietal bone of that side.

7

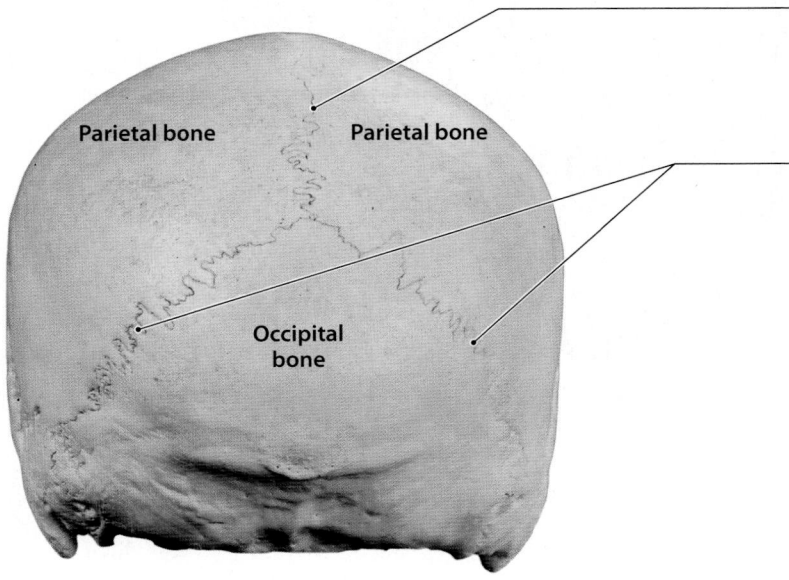

**Parietal bone**       **Parietal bone**

**Occipital bone**

**Posterior view of skull**

The **sagittal suture** extends from the lambdoid suture to the coronal suture, between the parietal bones.

The **lambdoid** (LAM-doyd) **suture** arches across the posterior surface of the skull. This suture separates the occipital bone from the two parietal bones. Occasionally, one or more sutural bones (Wormian bones) may be present along the lambdoid suture (see Module 6.2).

**?** INTEGRATION

**D.** Name the sutures that mark the boundaries of each parietal bone.

**?** REVIEW

**C.** Define *suture*.

**Lo** LEARNING OUTCOME

Identify the bones of the cranium and face, and locate and identify the cranial sutures.

# Facial bones dominate the anterior aspect of the skull, and cranial bones dominate the posterior surface

**1** We begin by examining the skull in anterior view. If you consider the cranium as the home of the brain, the facial bones form the front porch.

## Facial Bones

### Paired bones

The **nasal bones** support the superior portion of the bridge of the nose. They are connected to cartilages that support the distal portions of the nose.

The **lacrimal bones** form part of the medial wall of the orbits (eye sockets).

The **palatine bones** form the posterior portion of the hard palate and contribute to the floor of each orbit.

The **zygomatic bones** contribute to the rim and lateral wall of the orbit and form the prominent cheekbones.

The **maxillae** support the upper teeth and form the inferior orbital rim, the lateral margins of the external nares, the upper jaw, and most of the hard palate.

The **inferior nasal conchae** create turbulence in air passing through the nasal cavity and increase the epithelial surface area to warm and humidify inhaled air.

### Single bones

The **vomer** forms the inferior portion of the bony nasal septum.

The **mandible** forms the lower jaw.

## Cranial Bones

Parietal bone

The **frontal bone** forms the anterior portion of the cranium and the roof of the orbits. Mucus from the frontal sinuses within this bone helps to flush the nasal cavity surfaces.

The **sphenoid** forms part of the floor of the cranium, unites the cranial and facial bones, and acts as a cross-brace that strengthens the sides of the skull.

The **ethmoid** forms the anteromedial floor of the cranium, the roof of the nasal cavity, and part of the nasal septum and medial orbital wall.

**Anterior view**

**? REVIEW**

**A.** Identify the facial bones.

**B.** Identify the following bones as either a facial bone or a cranial bone: vomer, ethmoid, sphenoid, temporal, and inferior nasal conchae.

Watch
MasteringA&P™
Bone and
Dissection
Videos
Facial Bone

**2** Whereas the anterior view is dominated by facial bones, the posterior view is dominated by cranial bones. Several prominent bone markings on the occipital and temporal bones are identified here.

### Cranial Bones

The **parietal bone** on each side forms part of the superior and lateral surfaces of the cranium.

The **occipital bone** contributes to the posterior, lateral, and inferior surfaces of the cranium.

The **temporal bone** on either side (1) forms part of the lateral wall of the cranium and articulates with facial bones, (2) forms an articulation with the mandible, (3) surrounds and protects the sense organs of the internal ear, and (4) is an attachment site for muscles that close the jaws and move the head.

- The **mastoid process** is an attachment site for muscles that rotate or extend the head.

- The **styloid** (STĪ-loyd; *stylos*, pillar) **process**, near the base of the mastoid process, is attached to ligaments that support the hyoid bone and to the tendons of several muscles.

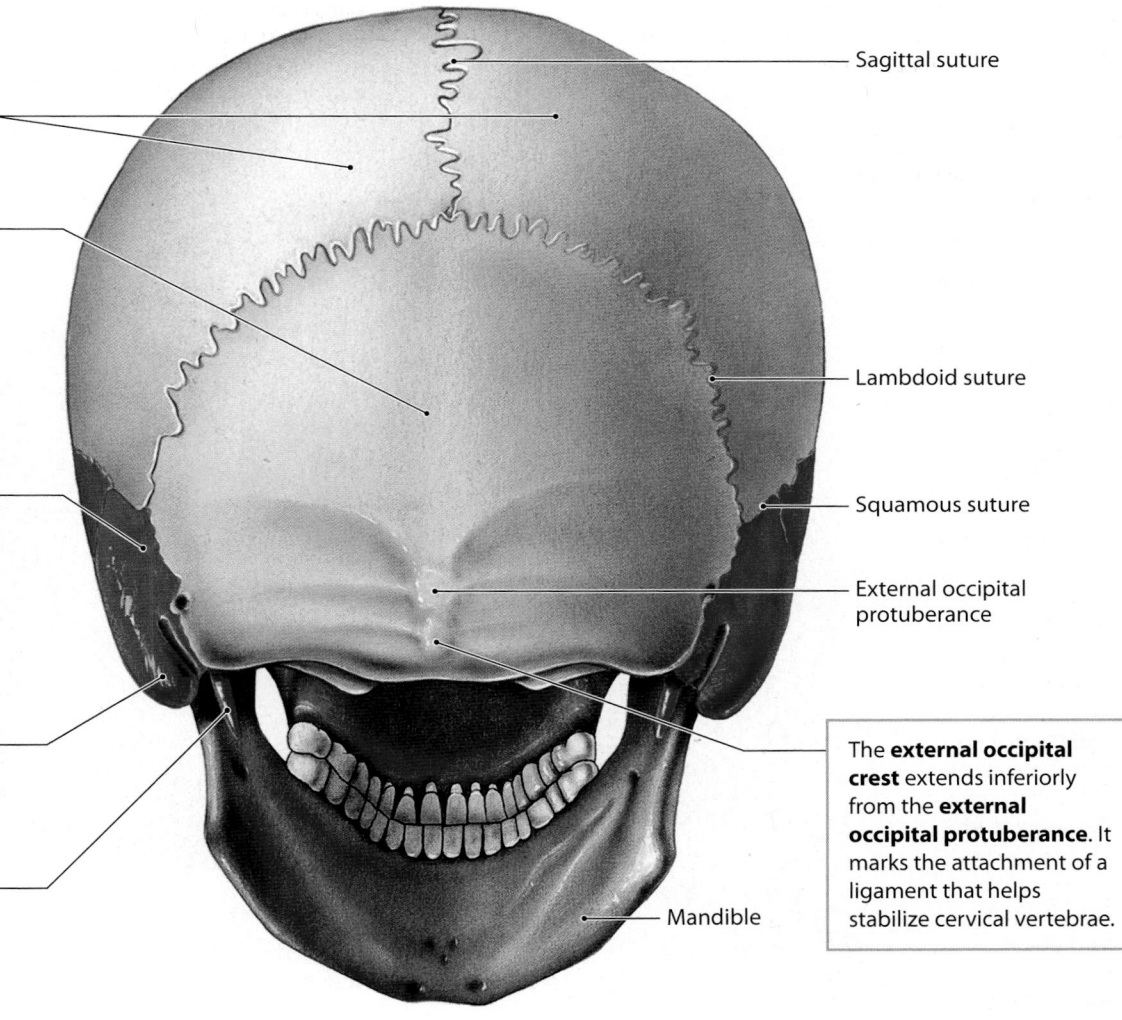

Sagittal suture

Lambdoid suture

Squamous suture

External occipital protuberance

The **external occipital crest** extends inferiorly from the **external occipital protuberance**. It marks the attachment of a ligament that helps stabilize cervical vertebrae.

Mandible

**Posterior view**

 **REVIEW**

**C.** The mandible articulates with which other cranial bones?

? **INTEGRATION**

**D.** Quincy suffers a hit to the skull that fractures the right superior lateral surface of his cranium. Which bone is fractured?

**Lo** **LEARNING OUTCOME**

Explain the significance of the markings and locations of the anterior and posterior aspects of the facial and cranial bones.

# The lateral and medial aspects of the skull share many bone markings

**1** This lateral view of the skull shows how the large bones interconnect and reveals the bone markings, or surface features, of the individual cranial and facial bones (see Module 6.2).

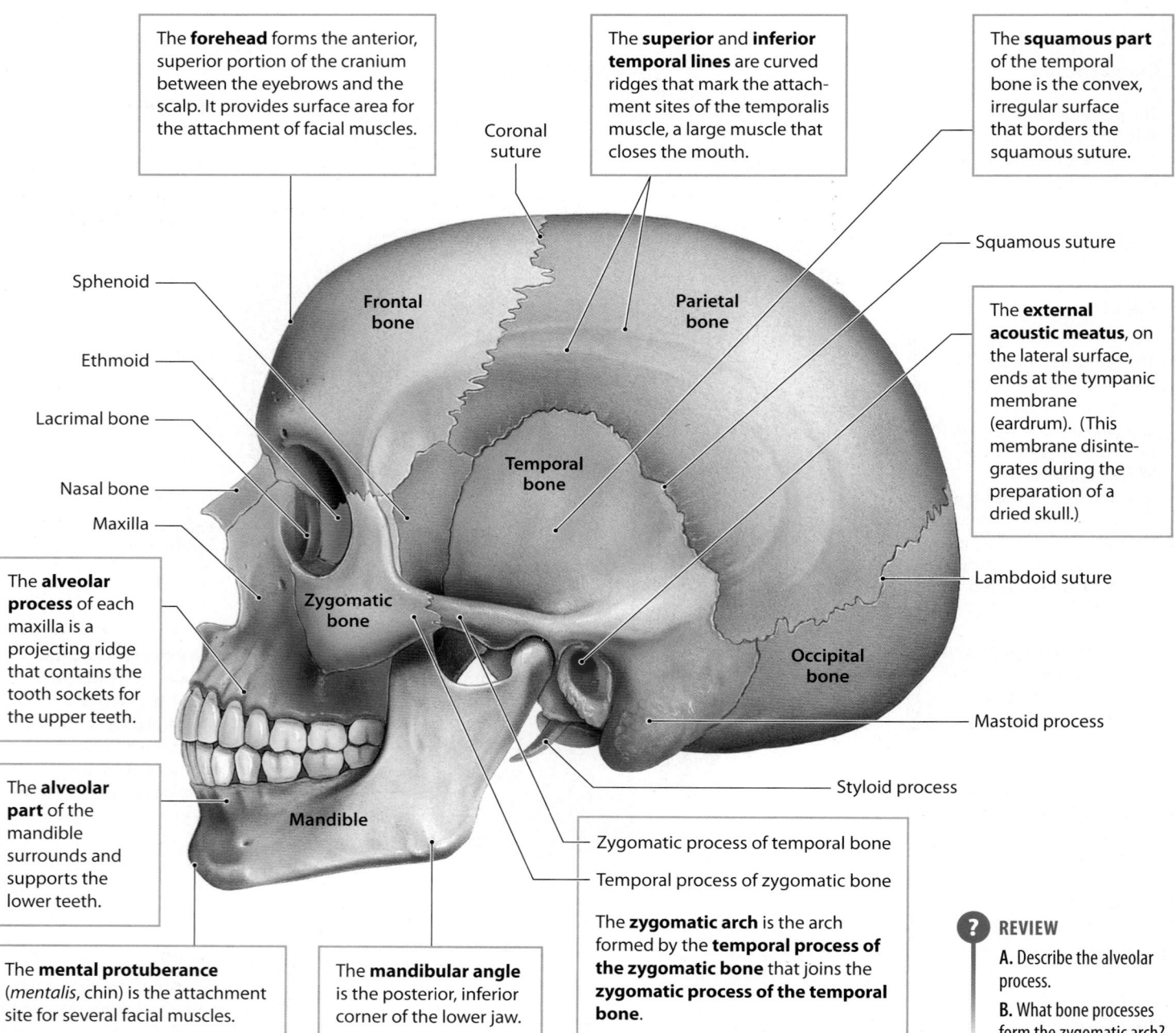

The **forehead** forms the anterior, superior portion of the cranium between the eyebrows and the scalp. It provides surface area for the attachment of facial muscles.

The **superior** and **inferior temporal lines** are curved ridges that mark the attachment sites of the temporalis muscle, a large muscle that closes the mouth.

The **squamous part** of the temporal bone is the convex, irregular surface that borders the squamous suture.

Coronal suture

Squamous suture

Sphenoid

Frontal bone

Parietal bone

The **external acoustic meatus**, on the lateral surface, ends at the tympanic membrane (eardrum). (This membrane disintegrates during the preparation of a dried skull.)

Ethmoid

Lacrimal bone

Nasal bone

Temporal bone

Maxilla

Lambdoid suture

The **alveolar process** of each maxilla is a projecting ridge that contains the tooth sockets for the upper teeth.

Zygomatic bone

Occipital bone

The **alveolar part** of the mandible surrounds and supports the lower teeth.

Mastoid process

Styloid process

Mandible

Zygomatic process of temporal bone

Temporal process of zygomatic bone

The **mental protuberance** (*mentalis*, chin) is the attachment site for several facial muscles.

The **mandibular angle** is the posterior, inferior corner of the lower jaw.

The **zygomatic arch** is the arch formed by the **temporal process of the zygomatic bone** that joins the **zygomatic process of the temporal bone**.

**? REVIEW**

**A.** Describe the alveolar process.

**B.** What bone processes form the zygomatic arch?

**2** This sagittal section passes slightly to the left of the midline, leaving the vomer and the perpendicular plate of the ethmoid intact. Note that the sinuses are named for the bone in which they occur.

Sphenoid

Frontal bone

Coronal suture

Sphenoidal sinus (right)

Hypophyseal fossa

The **petrous part** of the temporal bone encloses the structures of the internal ear—sense organs associated with hearing and balance. The auditory ossicles are located in the middle ear, a cavity within the petrous part.

The **frontal sinuses** are hollow spaces in the bone and vary greatly in size and time of appearance.

Parietal bone

Squamous suture

Lambdoid suture

Nasal bone

Ethmoid

Temporal bone

Inferior nasal concha (right)

The **internal acoustic meatus** begins on the medial surface of the petrous part of the temporal bone. It serves as a passageway for blood vessels and the facial and vestibulocochlear nerves.

Vomer

Palatine bone

Occipital bone

Maxilla

Mandible

Styloid process

The **hypoglossal canal** begins at the lateral base of each occipital condyle and ends on the inner surface of the occipital bone near the foramen magnum. The hypoglossal nerves—cranial nerves that control the tongue muscles—pass through these canals.

**? REVIEW**

**C.** Name each meatus found in the temporal bone.

**D.** What is the function of the internal acoustic meatus?

 **LEARNING OUTCOME**

Explain the significance of the markings and locations of the lateral and medial aspects of the facial and cranial bones.

# The foramina on the inferior surface of the skull mark the passageways for nerves and blood vessels

**1** In this inferior view you can see many of the important *foramina* (singular, *foramen*), or openings, that serve as passageways for blood vessels and nerves, as well as the large foramen used by the spinal cord.

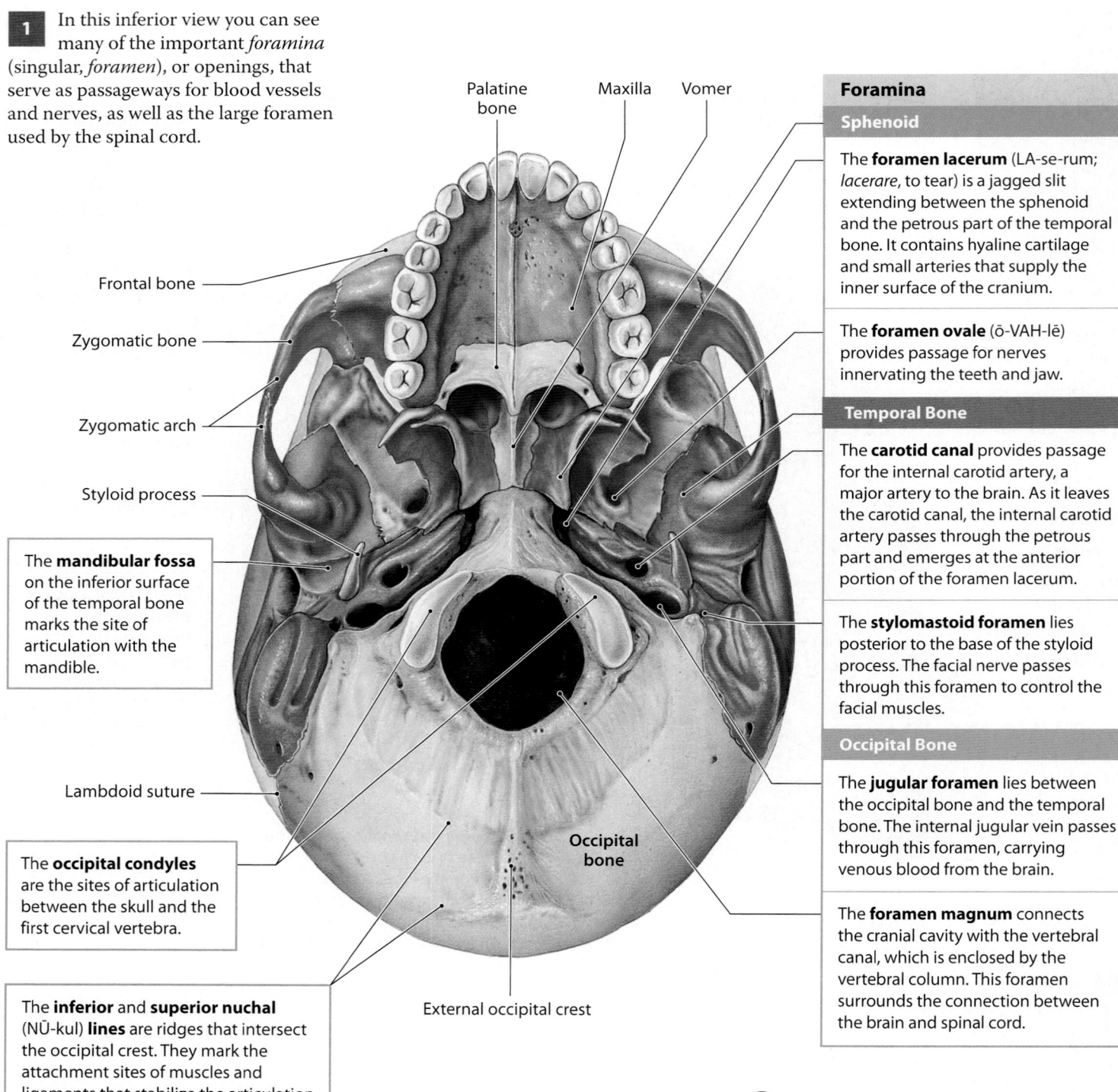

Palatine bone

Maxilla

Vomer

Frontal bone

Zygomatic bone

Zygomatic arch

Styloid process

The **mandibular fossa** on the inferior surface of the temporal bone marks the site of articulation with the mandible.

Lambdoid suture

The **occipital condyles** are the sites of articulation between the skull and the first cervical vertebra.

The **inferior** and **superior nuchal** (NŪ-kul) **lines** are ridges that intersect the occipital crest. They mark the attachment sites of muscles and ligaments that stabilize the articulation at the occipital condyles and balance the weight of the head over the cervical vertebrae.

External occipital crest

Occipital bone

### Foramina

#### Sphenoid

The **foramen lacerum** (LA-se-rum; *lacerare*, to tear) is a jagged slit extending between the sphenoid and the petrous part of the temporal bone. It contains hyaline cartilage and small arteries that supply the inner surface of the cranium.

The **foramen ovale** (ō-VAH-lē) provides passage for nerves innervating the teeth and jaw.

#### Temporal Bone

The **carotid canal** provides passage for the internal carotid artery, a major artery to the brain. As it leaves the carotid canal, the internal carotid artery passes through the petrous part and emerges at the anterior portion of the foramen lacerum.

The **stylomastoid foramen** lies posterior to the base of the styloid process. The facial nerve passes through this foramen to control the facial muscles.

#### Occipital Bone

The **jugular foramen** lies between the occipital bone and the temporal bone. The internal jugular vein passes through this foramen, carrying venous blood from the brain.

The **foramen magnum** connects the cranial cavity with the vertebral canal, which is enclosed by the vertebral column. This foramen surrounds the connection between the brain and spinal cord.

**? REVIEW**

**A.** Identify the bone containing the carotid canal, and name the structure that runs through this passageway.

**B.** Which foramen provides a passageway for nerves innervating the jaw?

**2** Compare this superior view of the floor of the cranial cavity with the inferior view of the skull at left.

Nasal bones

Frontal bone

Crista galli

Cribriform plate

Sella turcica

Parietal bone

**Occipital bone**

The **internal occipital crest** is a ridge that is important in anchoring blood vessels and membranes that stabilize the position of the brain.

## Foramina

### Ethmoid

The **olfactory foramina** are a series of small openings in the cribriform plate that permit passage of the olfactory nerves.

### Sphenoid

The **optic canals** permit the optic nerves to pass from the eyes to the brain.

The **foramen rotundum** permits the passage of a branch of the trigeminal nerve.

Foramen ovale

Foramen lacerum

The **foramen spinosum** permits the passage of blood vessels to membranes that surround the central nervous system.

### Temporal Bone

Carotid canal

Internal acoustic meatus

### Occipital Bone

Jugular foramen

The **hypoglossal canal** permits the passage of the hypoglossal nerve.

Foramen magnum

7

**REVIEW**

**C.** In which bone is the foramen magnum located, and what is significant about this opening?

**LEARNING OUTCOME**

Explain the significance of the markings and locations of the inferior and interior aspects of the facial and cranial bones.

# The shapes and markings of the sphenoid, ethmoid, and palatine bones are best seen in the isolated bones

**1** The **sphenoid** is an irregularly shaped bone that forms part of the floor of the cranium, unites the cranial and facial bones, and acts as a cross-brace that strengthens the sides of the skull.

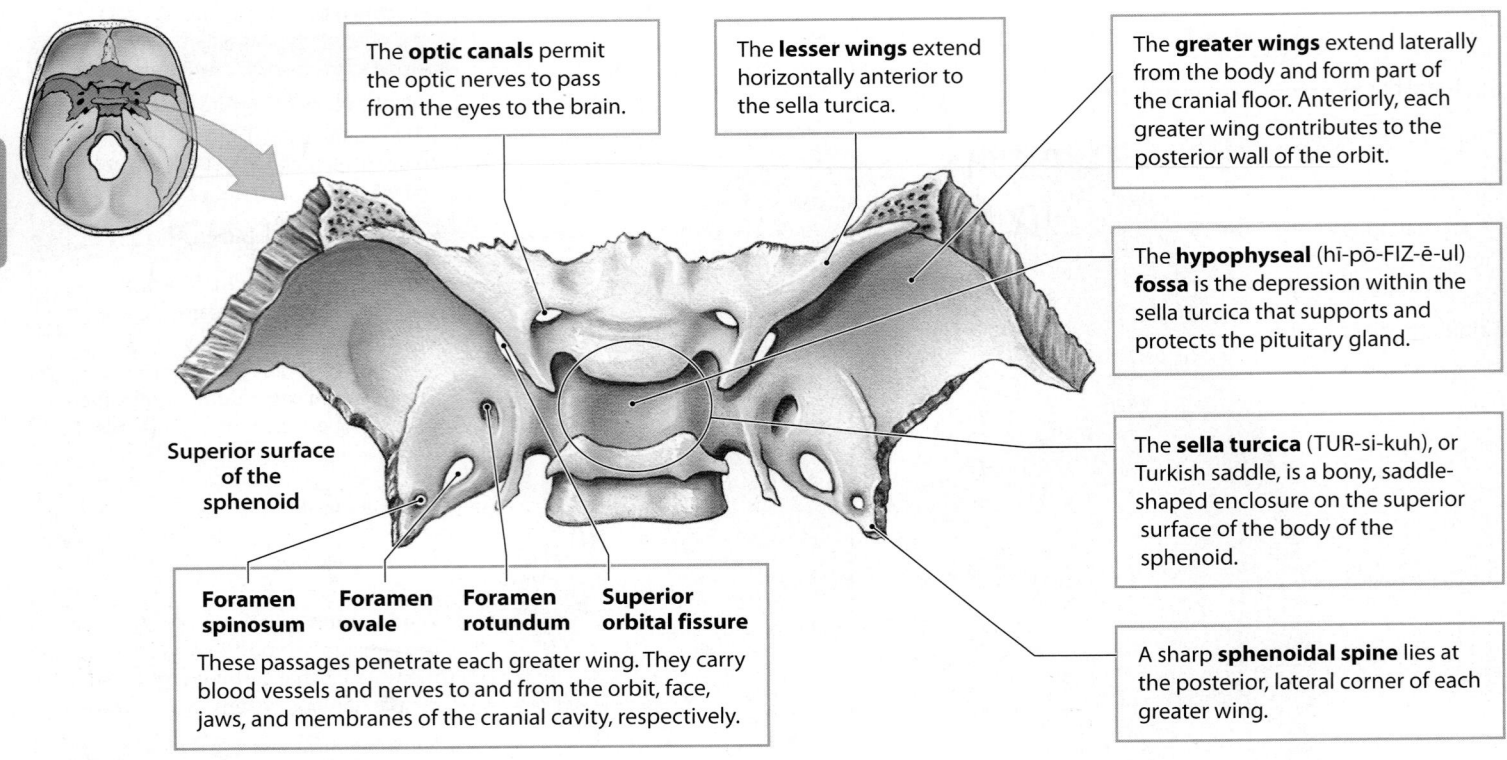

The **optic canals** permit the optic nerves to pass from the eyes to the brain.

The **lesser wings** extend horizontally anterior to the sella turcica.

The **greater wings** extend laterally from the body and form part of the cranial floor. Anteriorly, each greater wing contributes to the posterior wall of the orbit.

The **hypophyseal** (hī-pō-FIZ-ē-ul) **fossa** is the depression within the sella turcica that supports and protects the pituitary gland.

Superior surface of the sphenoid

The **sella turcica** (TUR-si-kuh), or Turkish saddle, is a bony, saddle-shaped enclosure on the superior surface of the body of the sphenoid.

**Foramen spinosum**  **Foramen ovale**  **Foramen rotundum**  **Superior orbital fissure**

These passages penetrate each greater wing. They carry blood vessels and nerves to and from the orbit, face, jaws, and membranes of the cranial cavity, respectively.

A sharp **sphenoidal spine** lies at the posterior, lateral corner of each greater wing.

Superior orbital fissure

The **sphenoidal sinuses** are on either side of the body, inferior to the sella turcica.

Greater wing

Lesser wing

Orbital surface of greater wing

Anterior surface of the sphenoid

Foramen rotundum

The **pterygoid** (TER-i-goyd; *pterygos*, wing) **processes** are vertical projections that originate on either side of the body.

Each pterygoid process forms a pair of **pterygoid plates**. They are attachment sites for muscles that move the mandible and soft palate.

Lateral plate

Medial plate

The **body** forms the central axis of the sphenoid.

**? REVIEW**

**A.** Identify the bone containing the optic canals, and cite the structures using this passageway.

**B.** Which bone contains the sella turcica? What structure is found within the sella turcica depression?

**2**    The **ethmoid** forms part of the floor of the anterior cranium, the roof of the nasal cavity, and part of the nasal septum and medial orbital wall. The ethmoid has three parts: (1) the cribriform plate, (2) the ethmoidal labyrinth on each side, and (3) the perpendicular plate.

 **REVIEW**

**C.** Identify the bone containing the cribriform plate. What is significant about this structure?

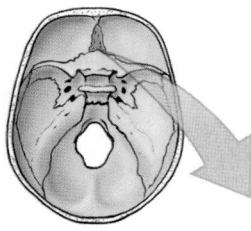

The **cribriform plate** (*cribrum*, sieve) forms the anteromedial floor of the cranium and the roof of the nasal cavity. The olfactory foramina in the cribriform plate permit passage of the olfactory nerves, which provide the sense of smell.

The **crista galli** (*crista*, crest + *gallus*, rooster, cock; cock's comb) is a bony ridge that projects superior to the cribriform plate. The falx cerebri, a membrane that stabilizes the position of the brain, attaches to this ridge.

**7**

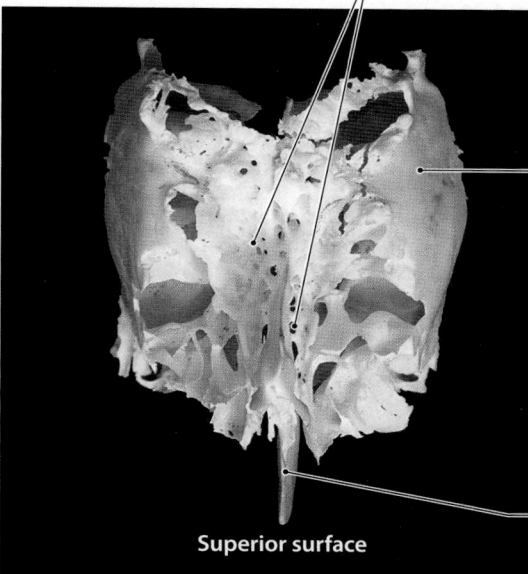

The **ethmoidal labyrinth** consists of **ethmoidal cells**, interconnected air-filled cavities, or sinuses. An ethmoidal labyrinth opens into the nasal cavity on each side.

The **superior nasal conchae** (KONG-kē; singular, *concha*, a snail shell) and the **middle nasal conchae** are delicate projections of the lateral masses.

The **perpendicular plate** forms part of the nasal septum, along with the vomer and a piece of hyaline cartilage.

**Superior surface**

**Posterior surface**

**3**    The **palatine bones** form the posterior portion of the hard palate and contribute to the floor of each orbit.

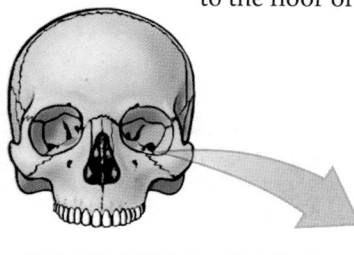

The **orbital process** forms part of the floor of the orbit. This process contains a small sinus that usually opens into the sphenoidal sinus.

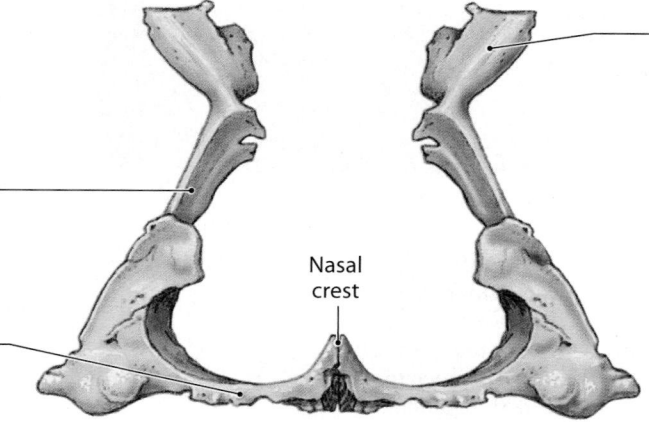

The **perpendicular plate** of the palatine bone is a flat portion that extends from the horizontal plate to the orbital process.

The **horizontal plate** forms the posterior part of the hard palate.

Nasal crest

**Anterior view of the two palatine bones**

 **INTEGRATION**

**D.** Which bone acts as a uniting bridge between the cranial and facial portions of the skull?

 **LEARNING OUTCOME**

Describe and locate the bone markings of the sphenoid, ethmoid, and palatine bones.

# Each orbital complex contains one eye, and the nasal complex encloses the nasal cavities

The facial bones not only protect and support the openings of the digestive and respiratory systems, but also protect the delicate sense organs for vision and smell. Together, certain cranial bones and facial bones form an *orbital complex* containing one eye and the *nasal complex* that surrounds the nasal cavities.

**1** The **orbits** are the bony recesses that contain the eyes. Each orbit is formed by the seven bones of the **orbital complex**. The frontal bone forms the roof, the zygomatic bone forms the lateral wall, and the maxilla provides most of the orbital floor. The orbital rim and the first portion of the medial wall are formed by the maxilla, the lacrimal bone, and the ethmoidal labyrinth of the ethmoid. The ethmoidal labyrinth articulates with the sphenoid and a small process of the palatine bone.

The **lacrimal fossa** on the superior and lateral surface of the orbit is a shallow depression in the frontal bone that marks the location of the lacrimal (tear) gland, which lubricates the eye surface.

The **supra-orbital margin** is a thickening of the frontal bone that helps protect the eye.

Either a **supra-orbital notch** or a fully enclosed supra-orbital foramen provides passage for blood vessels that supply the eyebrow, eyelids, and frontal sinuses.

Frontal bone

Palatine bone

Ethmoid

Sphenoid

Temporal bone

Zygomatic bone

The **lacrimal sulcus**, a groove along the anterior, lateral surface of the lacrimal bone, marks the location of the lacrimal sac. The lacrimal sulcus leads to the nasolacrimal canal.

The **nasolacrimal canal**, formed by a maxilla and lacrimal bone, protects the lacrimal sac and the nasolacrimal duct, which carries tears from the orbit to the nasal cavity.

Middle nasal concha

Inferior nasal concha

Maxilla

The **zygomaticofacial foramen** on the anterior surface of each zygomatic bone carries a sensory nerve that innervates the cheek.

The **infra-orbital foramen** marks the path of a major sensory nerve that reaches the brain by the foramen rotundum of the sphenoid.

**? REVIEW**

**A.** Identify the bones of the orbital complex.

**2** The **nasal complex** includes the bones that enclose the nasal cavities and the **paranasal sinuses**, air-filled chambers connected to the nasal cavities. The sphenoid, ethmoid, frontal bone, palatine bone, and maxillae contain the paranasal sinuses. (The tiny palatine sinuses, not shown, generally open into the sphenoidal sinuses.) The paranasal sinuses lighten the weight of the skull bones, allow the voice to resonate, and provide an extensive area of mucous epithelium. Inflammation of the sinuses due to viral, bacterial, or fungal infection is called **sinusitis**.

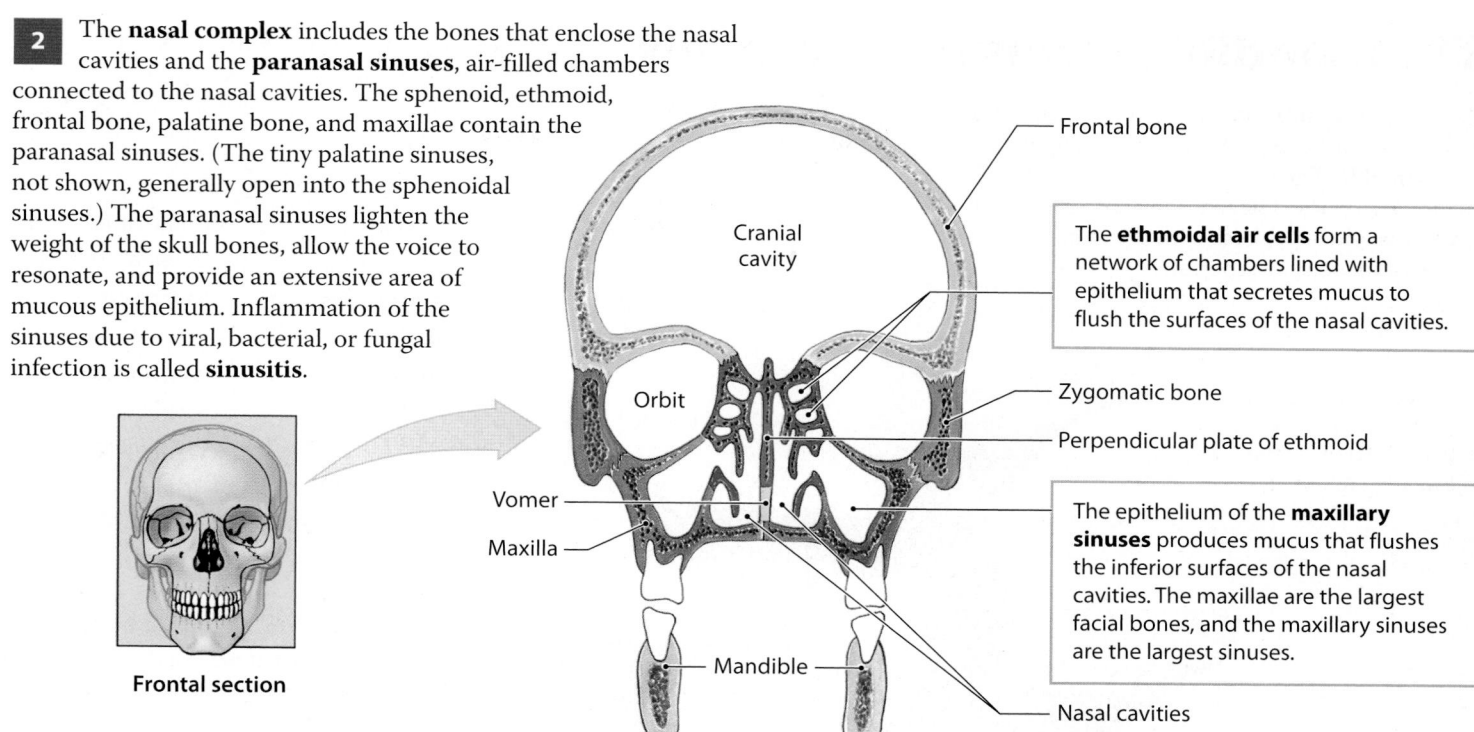

**Frontal section**

Frontal bone

Cranial cavity

The **ethmoidal air cells** form a network of chambers lined with epithelium that secretes mucus to flush the surfaces of the nasal cavities.

Orbit

Zygomatic bone

Perpendicular plate of ethmoid

Vomer

Maxilla

The epithelium of the **maxillary sinuses** produces mucus that flushes the inferior surfaces of the nasal cavities. The maxillae are the largest facial bones, and the maxillary sinuses are the largest sinuses.

Mandible

Nasal cavities

**3** This is a sagittal section of the skull with the nasal septum removed. The frontal bone, sphenoid, and ethmoid form the superior wall of the nasal cavities. The lateral walls are formed by the maxillae and the lacrimal bones, the ethmoid (the superior and middle nasal conchae), and the inferior nasal conchae. Much of the anterior margin of the nasal cavity is formed by the soft tissues of the nose, but the bridge of the nose is supported by the maxillae and nasal bones.

**Frontal Bone**

The **frontal sinuses** generally appear after age 6, but some people never develop them.

**Ethmoid**

Superior nasal concha

Middle nasal concha

**Sphenoid**

The paired **sphenoidal sinuses** are found on either side of the body of the sphenoid, inferior to the sella turcica. They vary in size.

**Palatine Bone**

Nasal bone

Lacrimal bone

Inferior nasal concha

Perpendicular plate

Horizontal plate

**Maxilla**

Hard palate

Pterygoid plates

**?** **REVIEW**
**B.** What are the functions of the paranasal sinuses?

**?** **INTEGRATION**
**C.** Name the complex—nasal, orbital, or both—where you find each of the following bones: frontal, maxilla, palatine, and nasal bones.

**Lo** **LEARNING OUTCOME**
Describe the structure of the orbital complex and nasal complex and the functions of their individual bones.

# The mandible forms the lower jaw ...

**1** The **mandible** can be subdivided into the horizontal body and the ascending rami (singular, *ramus*) of the mandible. The lower teeth are supported by the mandibular body. The mandible forms the entire lower jaw and articulates with the mandibular fossae of the temporal bones.

**Lateral view**

| FACE | 14 | CRANIUM | 8 |
|---|---|---|---|
| Maxillae | 2 | Occipital bone | 1 |
| Palatine bones | 2 | Parietal bones | 2 |
| Nasal bones | 2 | Frontal bone | 1 |
| Inferior nasal conchae | 2 | Temporal bones | 2 |
| Zygomatic bones | 2 | Sphenoid | 1 |
| Lacrimal bones | 2 | Ethmoid | 1 |
| Vomer | 1 | | |
| Mandible | 1 | | |

**SKULL**

Teeth (molars)

The **condylar process** articulates with the temporal bone at the *temporomandibular joint*.

The **alveolar part** of the mandible supports the lower teeth. The *dental alveoli* are the sockets for the roots of the teeth.

The **mandibular notch** is the depression that separates the condylar and coronoid processes.

The **mental foramen** is an opening for sensory nerves that carry sensory information from the lips and chin to the brain.

Mandibular angle

The **coronoid** (KOR-ō-noyd) **process** is the insertion point for the temporalis muscle, a powerful muscle that closes the jaws.

The **ramus** of the mandible is the ascending part that begins at the mandibular angle on either side. The masseter muscle, which is involved with chewing, is attached on its lateral surface.

The **body** of the mandible is the horizontal portion of that bone.

**2** The medial surface of the mandible has prominent surface markings.

**Sagittal section**

The **mylohyoid line** marks the insertion of the mylohyoid muscle, which supports the floor of the mouth.

Coronoid process

Condylar process

Head

Alveolar part

The **mandibular foramen** is the entrance to the mandibular canal, a passageway for blood vessels and nerves that service the lower teeth. Dentists typically anesthetize the sensory nerve within this canal before they work on the lower teeth.

A prominent depression on the medial surface marks the position of the submandibular salivary gland.

**? REVIEW**
**A.** Name the foramina of the mandible.

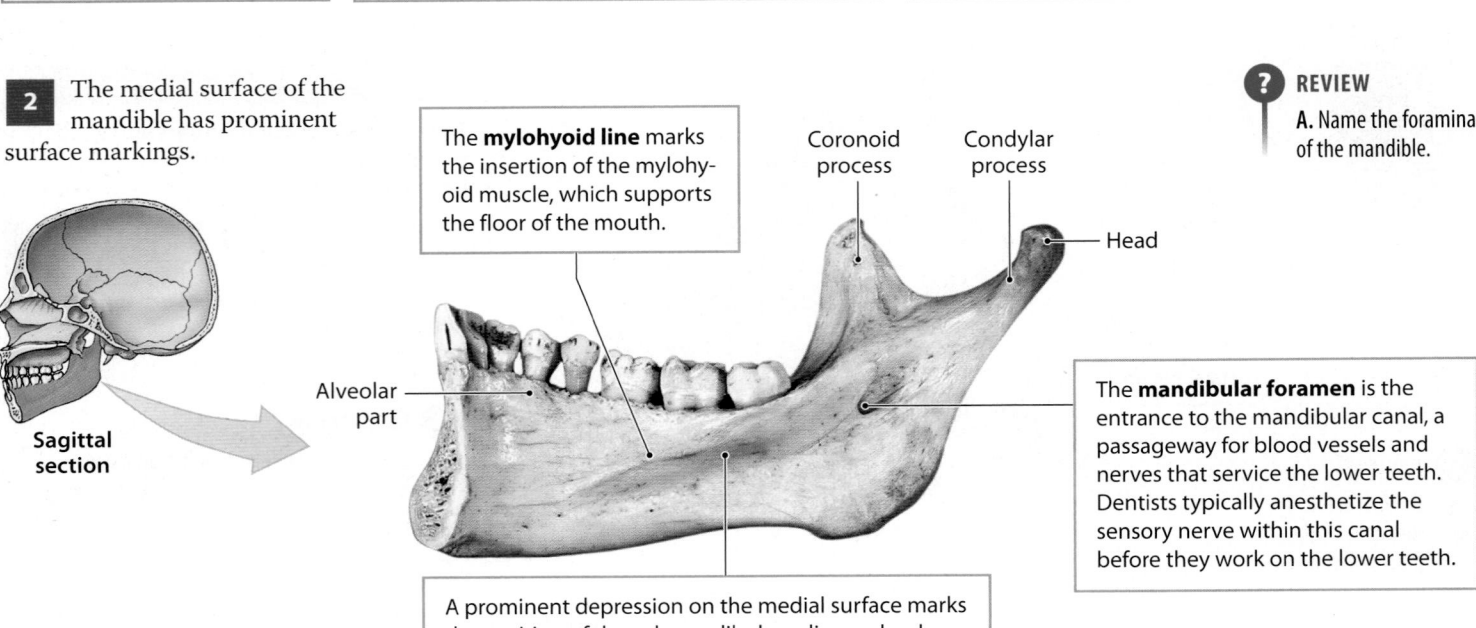

# … and the associated bones of the skull perform specialized functions

**ASSOCIATED BONES** 7

**Hyoid bone** 1

**Auditory ossicles enclosed in temporal bones (see Module 15.6)** 6

**?** **REVIEW**
**C.** Describe the location and function of the auditory ossicles.

7

**3** The **hyoid bone**—one of the seven associated bones of the skull—supports the larynx and is the attachment site for muscles of the larynx, pharynx, and tongue.

**?** **REVIEW**
**B.** Explain why your lab partner is correct when she claims that the hyoid bone does not join directly with any other bone.

**4** Also associated with the skull are six tiny bones called **auditory ossicles**. Three bones are located within each middle ear cavity, enclosed in the petrous part of the temporal bone. They play a key role in hearing by conducting the vibrations produced by sound waves from the tympanic membrane to the receptor complex of the internal ear.

The body of the hyoid is an attachment site for muscles of the larynx, tongue, and pharynx.

**Malleus** **Incus** **Stapes**

The **greater horn**, or greater cornu (plural, *cornua*), helps support the larynx and is attached to muscles that move the tongue.

**Superior view**

The **lesser horn**, or lesser cornu, is attached to ligaments that suspend the hyoid and larynx beneath the skull, like a child's swing from the limb of a tree.

**LO** **LEARNING OUTCOME**
Describe the mandible and the associated bones of the skull.

# Module 7.9
# Fontanelles permit cranial growth in infants and small children

The skull organizes around the developing brain. As birth approaches, the brain enlarges rapidly. The skull bones are also growing, but they fail to keep pace with the brain's growth. At birth, the cranial bones are connected by areas of flexible fibrous connective tissue that allow the skull's shape to change without damage and ease the infant's passage through the birth canal. The largest fibrous areas between the cranial bones are known as **fontanelles** (fon-tuh-NELZ; sometimes spelled *fontanels*).

**1** The **anterior fontanelle** is often referred to as the "soft spot" on newborns and is often the only fontanelle easily recognized by new parents. It generally persists until the child is nearly 2 years old. The anterior fontanelle is also of clinical importance. Normally it is taut, but it is shallow in a dehydrated infant.

The **anterior fontanelle** is the largest fontanelle. It lies at the intersection of the frontal, sagittal, and coronal sutures in the anterior portion of the skull.

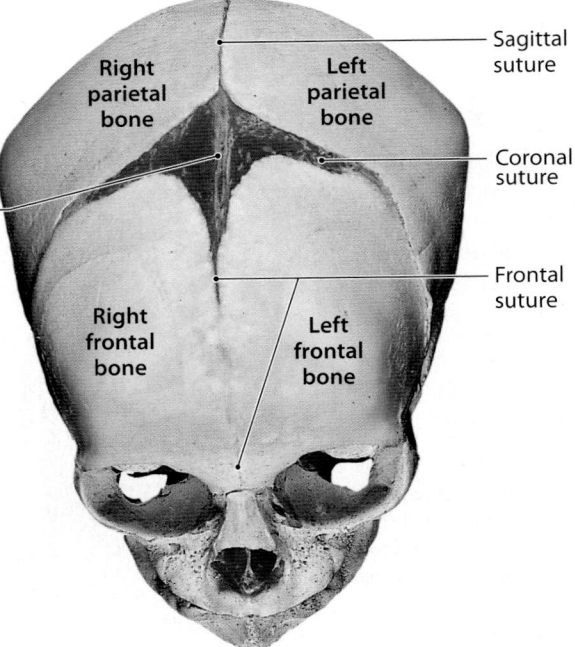

Sagittal suture

Right parietal bone

Left parietal bone

Coronal suture

Frontal suture

Right frontal bone

Left frontal bone

**Anterior/superior view**

**2** Over time, the sutures gradually narrow, and the fontanelles become smaller and smaller. In this posterior view of an infant skull, the **posterior fontanelle** has almost disappeared.

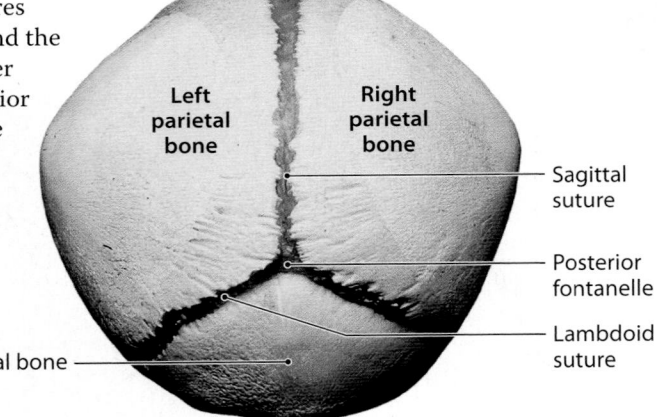

Left parietal bone

Right parietal bone

Sagittal suture

Posterior fontanelle

Lambdoid suture

Occipital bone

**Posterior view**

**3** The skulls of infants and adults differ in terms of the shape and structure of cranial elements, and this difference accounts for variations in both proportion and size. The most significant growth in the brain, about 70 percent, occurs during the fetal period. A newborn's brain will increase another 25 percent during infancy and the preschool years. As a result, when compared with the skull as a whole, the cranium of a young child is larger than that of an adult.

Developing brain

3 month fetus

Newborn

Adult

**4** This lateral view shows how small an infant's facial bones are compared with the cranium.

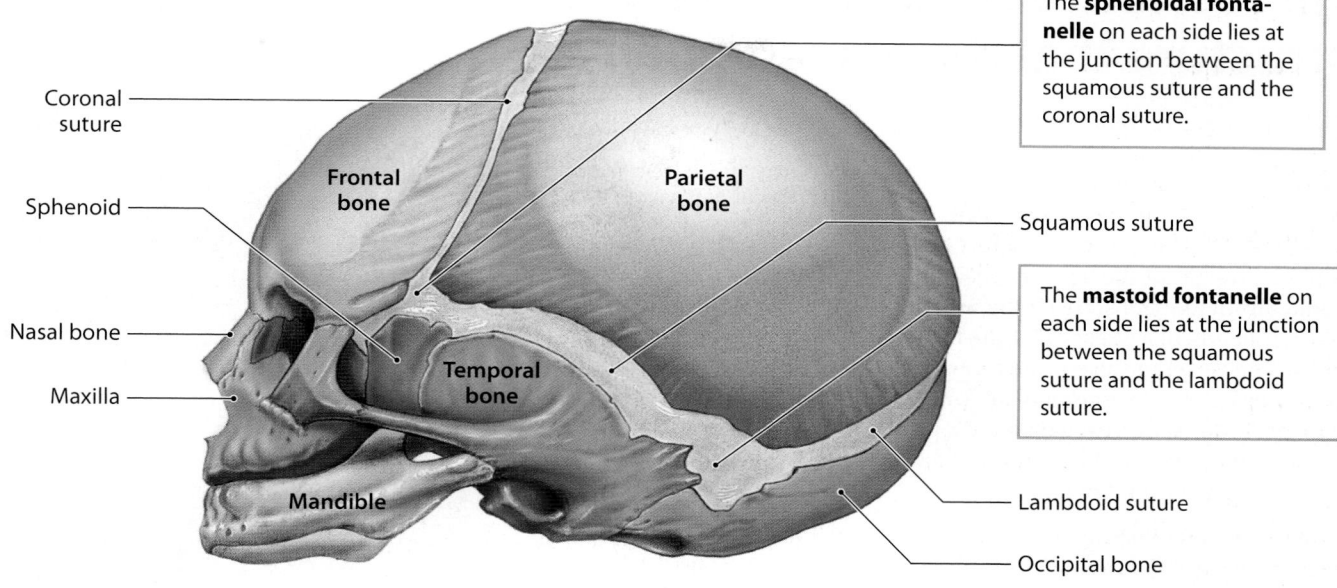

Coronal suture

Sphenoid

Nasal bone

Maxilla

**Frontal bone**

**Parietal bone**

**Temporal bone**

**Mandible**

The **sphenoidal fontanelle** on each side lies at the junction between the squamous suture and the coronal suture.

Squamous suture

The **mastoid fontanelle** on each side lies at the junction between the squamous suture and the lambdoid suture.

Lambdoid suture

Occipital bone

**5** This superior view shows the size of the anterior fontanelle relative to the width of the fibrous areas destined to become the sutures of the adult skull. Because it is composed of fibrous connective tissue and covers a major blood vessel, the anterior fontanelle pulses as the infant heart beats.

**? REVIEW**
**A.** Identify the major fontanelles.
**B.** What purposes do fontanelles serve?

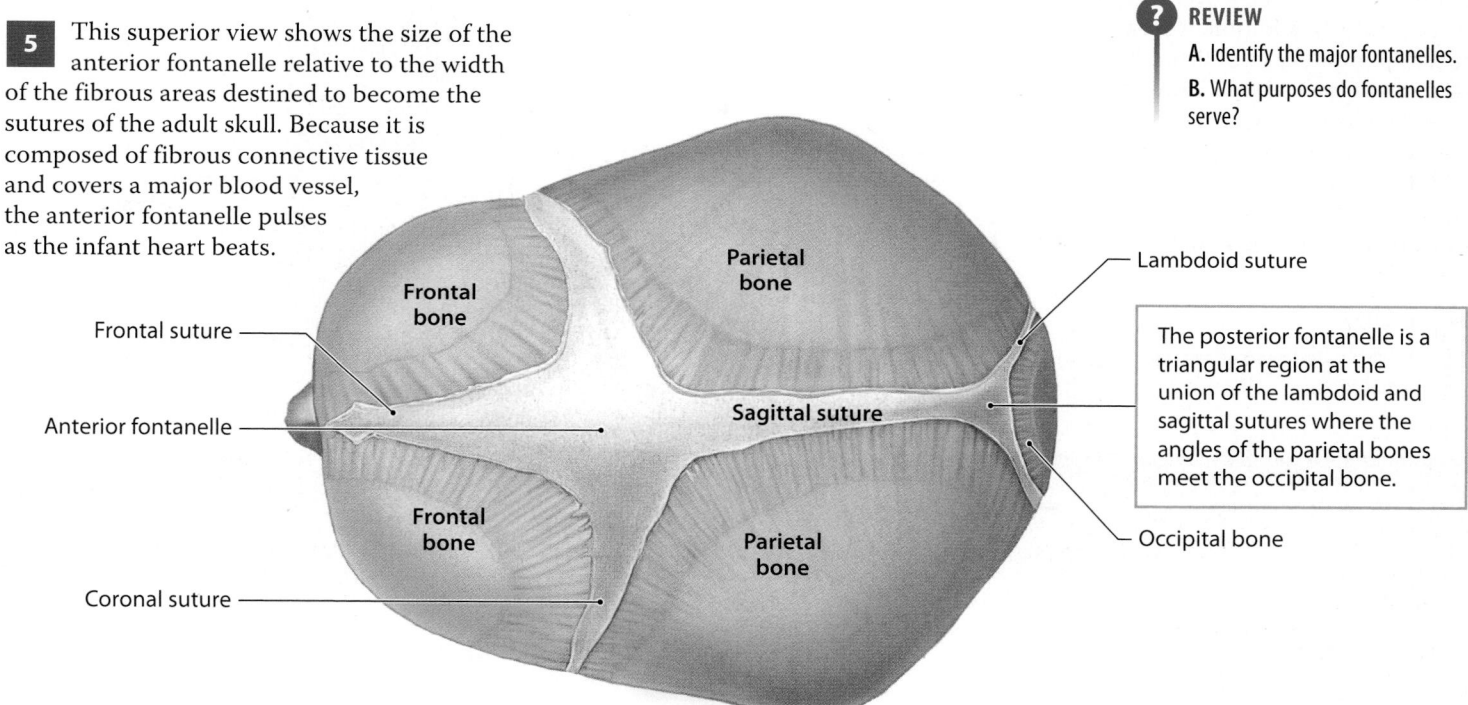

Frontal suture

Anterior fontanelle

Coronal suture

**Frontal bone**

**Frontal bone**

**Parietal bone**

**Parietal bone**

Sagittal suture

Lambdoid suture

The posterior fontanelle is a triangular region at the union of the lambdoid and sagittal sutures where the angles of the parietal bones meet the occipital bone.

Occipital bone

The posterior, sphenoidal, and mastoid fontanelles disappear within a month or two after birth. Even after the fontanelles disappear, the bones of the skull remain separated by fibrous connections. The most significant growth in the skull occurs before age 5, because at that time the brain stops enlarging and the cranial sutures ossify.

**? INTEGRATION**
**C.** Why are an infant's facial bones so small compared with its cranium?

 **LEARNING OUTCOME**
Describe key structural differences among the skulls of infants, children, and adults.

# The vertebral column has four spinal curves, and vertebrae share a basic structure that differs regionally

**Watch**
MasteringA&P™

**Bone and Dissection Videos**

Typical Vertebra

**1** The adult **vertebral column** (or spine) consists of 26 bones: the 24 *vertebrae*, the *sacrum*, and the *coccyx* (KOK-siks), or tailbone. The stacked vertebrae form a column of support that bears the weight of the head, neck, and trunk, ultimately transferring the weight to the appendicular skeleton of the lower limbs. The vertebrae also protect the spinal cord and help maintain an upright body position, as in sitting or standing. The length of the vertebral column of an adult averages 71 cm (28 in.).

Note that when referring to a specific vertebra, we use the capital letters C, T, L, S, and Co to indicate the cervical, thoracic, lumbar, sacral, and coccygeal regions, respectively. In addition, we use a subscript number to indicate the relative position of the vertebra within that region, with 1 indicating the vertebra closest to the skull. For example, $C_1$ is in contact with the skull and $C_3$ is the third cervical vertebra. Similarly, $T_{12}$ is in contact with $L_1$, and $L_4$ is the fourth lumbar vertebra.

**? REVIEW**

**A.** What is the importance of the secondary curves of the spine?

### Spinal Curves

Primary curves develop before birth, and secondary curves after birth.

The **cervical curve**, a secondary curve, develops as the infant learns to balance the weight of the head on the cervical vertebrae.

The **thoracic curve**, a primary curve, accommodates the thoracic organs.

The **lumbar curve**, a secondary curve, balances the weight of the trunk over the lower limbs. It develops with the ability to stand.

The **sacral curve**, a primary curve, accommodates the abdominopelvic organs.

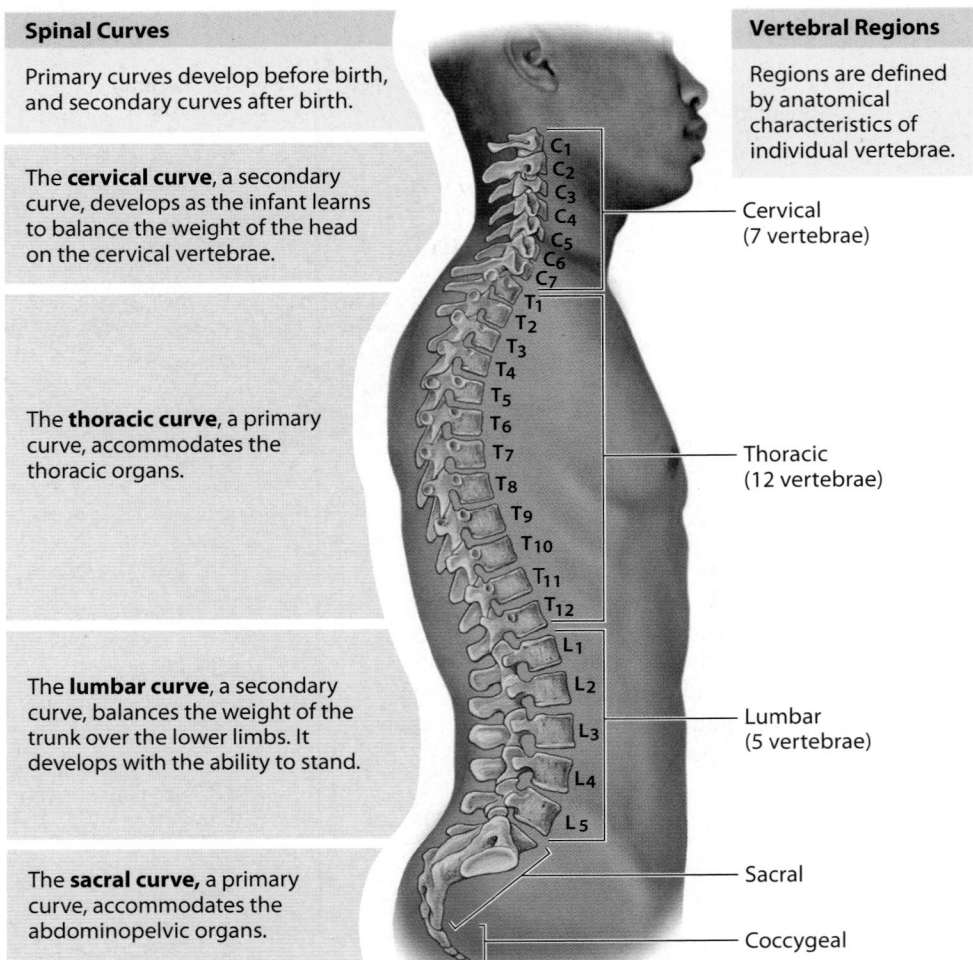

### Vertebral Regions

Regions are defined by anatomical characteristics of individual vertebrae.

Cervical (7 vertebrae)

Thoracic (12 vertebrae)

Lumbar (5 vertebrae)

Sacral

Coccygeal

## Regional Comparison of Vertebral Structure and Function

| | **Cervical Vertebrae (7)** | **Thoracic Vertebrae (12)** | **Lumbar Vertebrae (5)** |
|---|---|---|---|
| **Location** | Neck | Chest | Inferior portion of back |
| **Vertebral Body** | Small, oval, curved faces | Medium, heart-shaped, flat faces; facets for rib articulations | Massive, oval, flat faces |
| **Vertebral Foramen** | Large | Smaller | Smallest |
| **Spinous Process** | Long; split tip; points inferiorly | Long, slender; not split; points inferiorly | Blunt, broad; points posteriorly |
| **Transverse Processes** | Have transverse foramina | All but two have facets for rib articulations | Short; no articular facets or transverse foramina |
| **Functions** | Support skull, stabilize relative positions of brain and spinal cord, and allow controlled head movements | Support weight of head, neck, upper limbs, and chest; articulate with ribs to allow changes in volume of thoracic cage | Support weight of head, neck, upper limbs, and trunk |

**2** Each vertebra consists of three basic parts: (1) articular processes, (2) a vertebral arch, and (3) a vertebral body.

**Parts of a Vertebra**

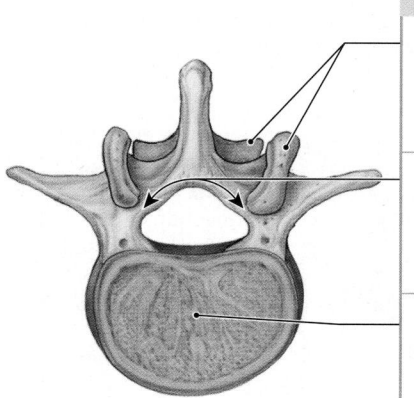

**Superior view**

The **articular processes** extend superiorly and inferiorly to articulate with adjacent vertebrae.

The **vertebral arch** forms the posterior and lateral margins of the vertebral foramen.

The **vertebral body** is the part of a vertebra that transfers weight along the axis of the vertebral column.

**?** **REVIEW**
**B.** Name the major components of a typical vertebra.

**3** The vertebral arch forms the posterior margin of each vertebral foramen. It consists of a pair each of laminae and pedicles and supports various processes: articular, spinous, and transverse.

The **vertebral foramen** is framed by the vertebral body and the vertebral arch.

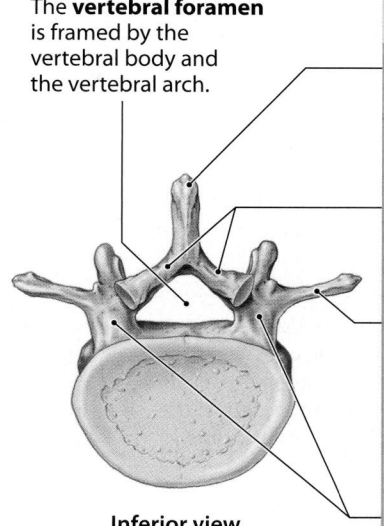

**Inferior view**

**The Vertebral Arch**

A **spinous process** projects posteriorly from the point where the vertebral laminae fuse to complete the vertebral arch.

The **laminae** (LAM-i-nē; singular, *lamina*, a thin plate) form the "roof" of the vertebral foramen.

**Transverse processes** project laterally on both sides from the point where the laminae join the pedicles. These processes are sites of muscle attachment and may also articulate with the ribs.

The **pedicles** (PED-i-kulz) form the sides, or "walls," of the vertebral arch and connect it to the vertebral body.

**7**

**4** This lateral and sectional view of three articulated vertebrae shows how they combine to form the vertebral canal.

Pedicle

Vertebral body

The bodies of adjacent vertebrae are interconnected by ligaments but are separated by pads of fibrocartilage, the **intervertebral discs**.

The spaces that are formed between successive pedicles are called **intervertebral foramina**. Nerves and blood vessels running to or from the spinal cord pass through these foramina.

Together, the vertebral foramina of successive vertebrae form the **vertebral canal**, which encloses the spinal cord.

**?** **REVIEW**
**C.** To which part of the vertebra do the intervertebral discs attach?

**5** Each articular process has a smooth, concave surface called an **articular facet**. This facet is the portion of the articular process that forms the joint with the adjacent vertebra. This posterior view shows that the vertebral arches form the roof of the vertebral canal.

Articular facet

The **superior articular processes** articulate with the inferior articular processes of a superior vertebra.

The **inferior articular processes** articulate with the superior articular processes of an inferior vertebra (or the sacrum, in the case of the last lumbar vertebra).

**L◦** **LEARNING OUTCOME**
Identify and describe the curves of the spinal column and their functions, and identify the vertebral regions.

# There are seven cervical vertebrae …

The seven **cervical vertebrae**, the smallest in the vertebral column, extend from the occipital bone of the skull to the thorax.

**1** The vertebral foramen of a cervical vertebra is very large. At this level, the spinal cord still contains most of the axons that connect the brain to the rest of the body. However, cervical vertebrae support only the weight of the head, so the vertebral body can be relatively small and light. As you continue toward the sacrum, the load increases, and the vertebral bodies gradually enlarge.

In a typical cervical vertebra ($C_2$–$C_6$), the tip of each spinous process bears a prominent notch. Such a process is a **bifid** (BĪ-fid) spinous process.

Vertebral arch

The transverse process is short and stumpy.

Vertebral foramen

Vertebral body

The **transverse foramen**, formed by a connection of the costal and transverse processes, protects the vertebral arteries and vertebral veins that service the brain.

A slender **costal process** extends anterolaterally from either side of the vertebral body.

**Cervical vertebra (superior view)**

**2** The first two cervical vertebrae are specialized to support and stabilize the cranium while permitting head movement. The **atlas**, $C_1$, has no vertebral body and no spinous process, but it has a large, round vertebral foramen bounded by anterior and posterior arches. The **axis**, $C_2$, resembles inferior cervical vertebrae except for the presence of the prominent dens on the superior surface of the body.

During development, the body of the atlas fuses to the body of the axis, where it forms the prominent **dens** (DENZ; *dens*, tooth), or **odontoid** (ō-DON-toyd; *odontos*, tooth) **process**.

Anterior arch of atlas

The atlas holds up the skull, articulating with the occipital condyles of the skull. This vertebra is named after Atlas, who, according to Greek myth, holds the world on his shoulders.

The articulation between the skull's occipital condyles and the atlas is a joint that permits nodding (such as when you indicate "yes").

Axis

Posterior arch of atlas

A transverse ligament binds the dens to the inner surface of the anterior arch of the atlas. This articulation permits rotation (as when you shake your head to indicate "no").

**3** The last cervical vertebra has a robust spinous process that ends in a solid tubercle that can easily be felt through the skin. For this reason, $C_7$ is known as the **vertebra prominens**, or prominent vertebra. The **ligamentum nuchae** (lig-uh-MEN-tum NŪ-kē; *nucha*, nape), a stout elastic ligament, begins at the vertebra prominens and extends to an insertion along the external occipital crest of the skull. When your head is upright, this ligament acts like the string on a bow to maintain the cervical curvature without muscular effort.

$C_1$
$C_2$
$C_3$
$C_4$
$C_5$
$C_6$
$C_7$

Vertebra prominens

**?** **REVIEW**

**A.** Joe suffered a hairline fracture at the base of the dens. Identify the bone and fracture site.

**B.** In which region of the vertebral column would you find a vertebra with a large foramen and two smaller foramina within the transverse processes?

# ... and twelve thoracic vertebrae

C<sub>7</sub>

**4** There are 12 **thoracic vertebrae**. From $T_1$ to $T_{12}$, each vertebral body is slightly larger and more massive than the one superior to it, because the weight being transmitted along the vertebral column is steadily increasing.

**5** A typical thoracic vertebra has a distinctive heart-shaped body that is much larger and more massive than that of a cervical vertebra, and the vertebral foramen is considerably smaller.

**6** Each thoracic vertebra articulates with ribs at **costal facets** on the dorsolateral surfaces of the vertebral body. The transverse processes of vertebrae $T_1$–$T_{10}$ also contain transverse costal facets for rib articulation. Thus, rib pairs 1 through 10 contact their vertebrae at two points: a costal facet and a transverse costal facet. Ribs 11 and 12 originate at costal facets on vertebrae $T_{11}$ and $T_{12}$. Because these ribs do not contact transverse processes, they contact their vertebrae at only one point.

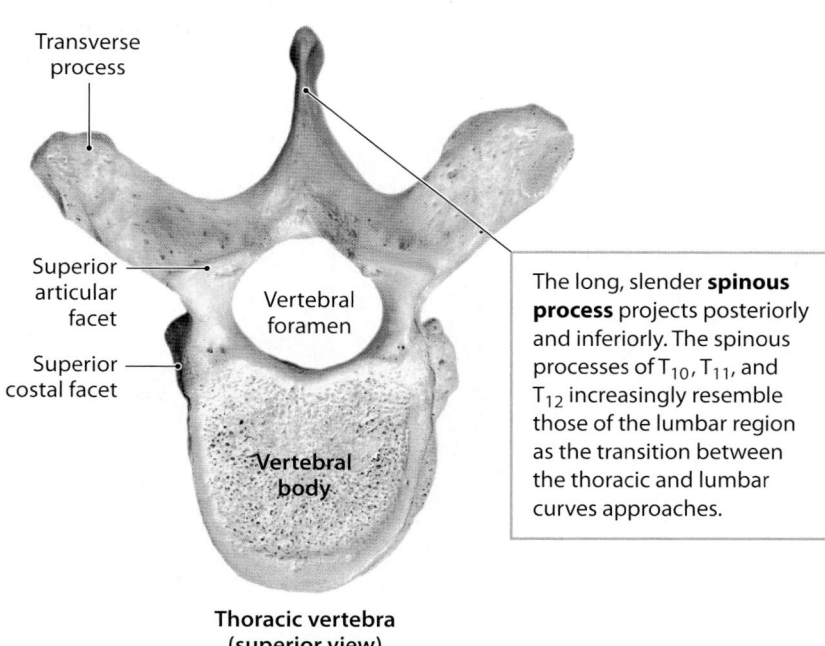

Transverse process

Superior articular facet

Superior costal facet

Vertebral foramen

Vertebral body

The long, slender **spinous process** projects posteriorly and inferiorly. The spinous processes of $T_{10}$, $T_{11}$, and $T_{12}$ increasingly resemble those of the lumbar region as the transition between the thoracic and lumbar curves approaches.

**Thoracic vertebra (superior view)**

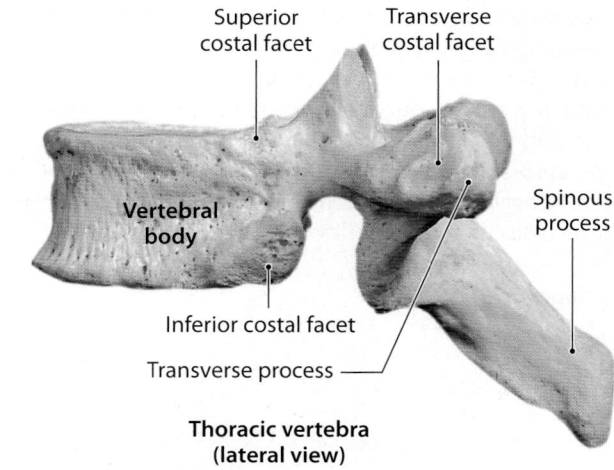

Superior costal facet

Transverse costal facet

Vertebral body

Spinous process

Inferior costal facet

Transverse process

**Thoracic vertebra (lateral view)**

**? REVIEW**

**C.** When you run your finger down the middle of a person's spine, what part of each vertebra are you feeling just beneath the skin?

 **LEARNING OUTCOME**

Describe the distinctive structural and functional characteristics of the cervical and thoracic vertebrae.

# Module 7.12
# There are five lumbar vertebrae

**1** The five **lumbar vertebrae** ($L_1$–$L_5$) are the largest vertebrae, and they withstand the most weight. Compression fractures of these vertebrae may occur at any age, but they occur most often after aging and osteoporosis have reduced the strength of bones throughout the body.

**? REVIEW**
**A.** How many vertebrae are in the lumbar region of the vertebral column?

L₁
L₂
L₃
L₄
L₅
Sacrum
Coccyx

**? REVIEW**
**B.** Why are the bodies of the lumbar vertebrae so large?

**2** The body of a typical lumbar vertebra is thicker than that of a thoracic vertebra, and the superior and inferior surfaces are oval rather than heart shaped. Other noteworthy features are that (1) lumbar vertebrae do not have costal facets; (2) the slender transverse processes, which lack transverse costal facets, project dorsolaterally; (3) the vertebral foramen is triangular; (4) the blunt spinous processes project posteriorly; (5) the superior articular processes face medially ("up and in"); and (6) the inferior articular processes face laterally ("down and out"). The massive spinous processes provide surface area for the attachment of lower back muscles and the orientation of the articular processes increase stability of the lumbar curve.

Spinous process
Lamina
Transverse process
Superior articular facet
Superior articular process
Vertebral foramen
Pedicle
Vertebral body

**Superior view**

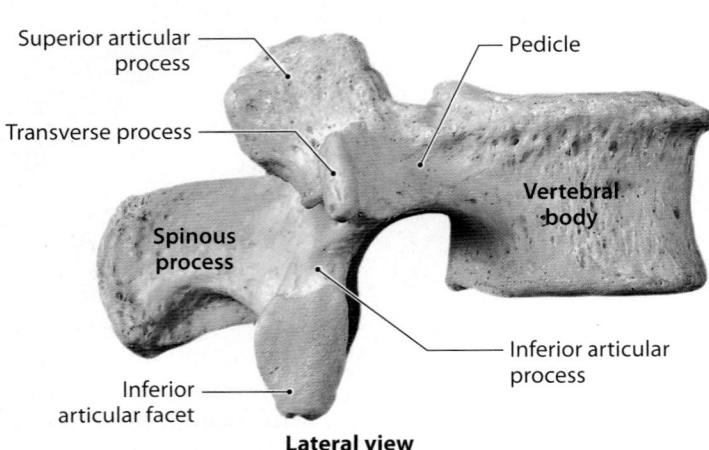

Superior articular process
Transverse process
Spinous process
Inferior articular facet
Pedicle
Vertebral body
Inferior articular process

**Lateral view**

**ʟᴏ LEARNING OUTCOME**
Describe the distinctive structural and functional characteristics of the lumbar vertebrae.

# Module 7.13

## The sacrum and coccyx consist of fused vertebrae

The **sacrum** is a single bone that consists of the fused components of five sacral vertebrae. These vertebrae begin fusing shortly after puberty and, in general, are completely fused at age 25–30. The sacrum protects the reproductive, digestive, and urinary organs. Through paired joints, the sacrum attaches the axial skeleton to the paired hip bones of the appendicular skeleton. The coccyx provides an attachment site for ligaments and a muscle that constricts the anal opening.

 The anterior surface of the sacrum is concave, and the posterior surface is convex. The degree of curvature is more pronounced in males than in females.

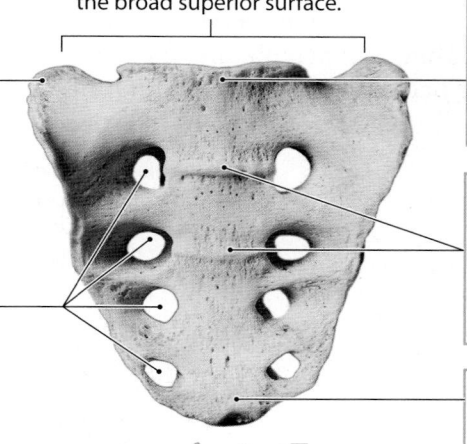

The **base** of the sacrum is the broad superior surface.

At the base of the sacrum, a broad sacral **ala**, or wing, extends on either side. The anterior and superior surfaces of each ala provide an extensive area for muscle attachment.

The **sacral promontory** is an important surface marking in females during pelvic examinations and during labor and delivery.

Prominent transverse lines mark the former boundaries of individual vertebrae that fused during the formation of the sacrum.

Four pairs of **sacral foramina** extend between the posterior and anterior surfaces. The intervertebral foramina of the fused sacral vertebrae open into these passageways.

The **apex** is the narrow, inferior portion of the sacrum.

Coccyx

**? REVIEW**
**A.** Which bone of the axial skeleton joins with the hip bones of the appendicular skeleton?

---

**2** A posterior view of the sacrum shows the broad surface that provides an extensive area for the attachment of muscles.

The **sacral canal** is a passageway that extends the length of the sacrum. Nerves and membranes that line the vertebral canal in the spinal cord continue into the sacral canal.

The **sacral tuberosity** is a roughened area posterior to the auricular surface. It marks the attachment site of ligaments that stabilize the sacro-iliac joint.

The **superior articular process** of the sacrum articulates with the last lumbar vertebra.

The **auricular surface** is a thickened, flattened area lateral and anterior to the superior portion of the lateral sacral crest. The auricular surface forms the *sacro-iliac joint* with the hip bone of the pelvic girdle.

The **median sacral crest** is a ridge formed by the fused spinous processes of the sacral vertebrae.

The **lateral sacral crest** is a ridge that represents the fused transverse processes of the sacral vertebrae.

The **sacral hiatus** (hī-Ā-tus) is the opening at the inferior end of the sacral canal.

Sacral cornu

**? REVIEW**
**B.** Which regions of the vertebral column do not consist of individual vertebrae?

The small **coccyx** begins as three to five separate coccygeal vertebrae. They typically begin fusing into a single bone by age 26. Each **coccygeal cornu** (plural, *cornua*) curves to meet the sacral cornu superior to it.

Coccygeal cornu

**Lo LEARNING OUTCOME**
Describe the distinctive structural and functional characteristics of the sacrum and coccyx.

# The thoracic cage protects organs in the chest and provides sites for muscle attachment

The skeleton of the chest, or **thoracic cage**, provides bony support to the walls of the thoracic cavity. It consists of the thoracic vertebrae, the *ribs*, and the *sternum* (breastbone). The ribs and the sternum form the **rib cage**, whose movements are important in breathing. The thoracic cage protects the heart, lungs, thymus, and other structures in the thoracic cavity, and serves as an attachment point for muscles involved in (1) breathing, (2) maintaining the position of the vertebral column, and (3) moving the pectoral girdles and upper limbs.

**1** An anterior view of the thoracic cage shows the sternum, the costal cartilages, and the major classes of ribs.

The **jugular notch**, located between the clavicular articulations, is a shallow indentation on the superior surface of the manubrium.

### Ribs

The **ribs** reinforce the posterior and lateral walls of the thoracic cavity. Ribs 1–7 gradually increase in length and curvature radius. Although the ribs are quite mobile and are among the most flexible of bones, they can be broken by a sharp blow or crushing impact. Fortunately, they are so stabilized by connective tissues and surrounding muscles that open fractures are rare, and splinting is unnecessary.

**Vertebrosternal ribs** (ribs 1–7), or **true ribs**, are connected to the sternum by individual *costal cartilages*.

The **vertebrochondral ribs** (ribs 8–10) are connected to the sternum by shared costal cartilages.

The last two pairs of ribs (11 and 12) are called **floating ribs** because they have no connection with the sternum, or **vertebral ribs** because they are attached only to the vertebrae and muscles of the body wall. Collectively, ribs 8–12 are also called **false ribs**.

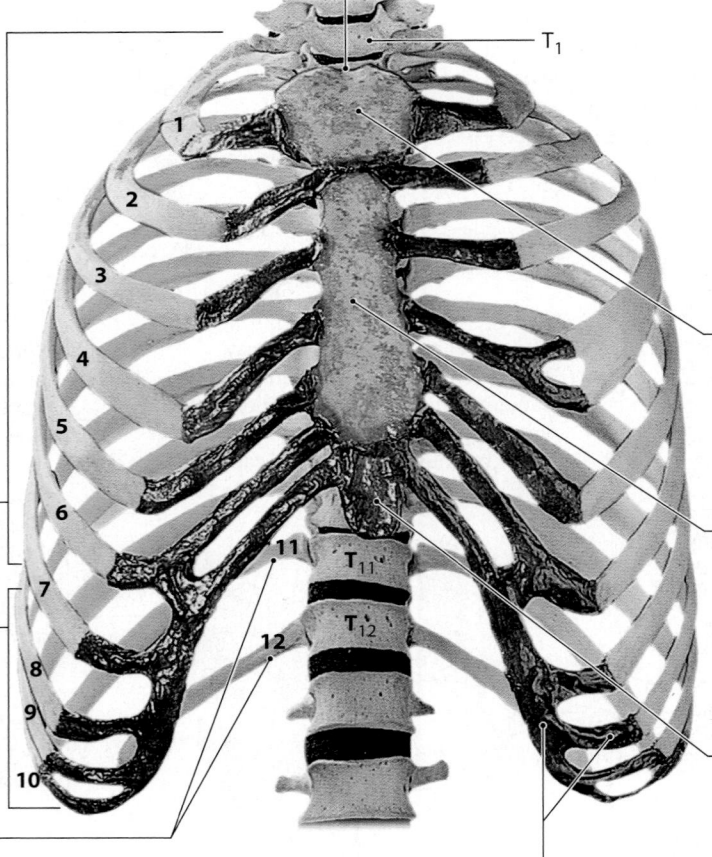

T₁ — $T_1$

### Sternum

The adult **sternum**, or breastbone, is a flat bone that forms in the anterior midline of the thoracic wall. It has three distinct regions that usually fuse together during adulthood.

**Manubrium:** The broad, trapezoid-shaped manubrium (ma-NŪ-brē-um) articulates with the clavicles (collarbones) and the cartilages of the first pair of ribs.

**Body:** The body attaches to the inferior surface of the manubrium and extends inferiorly along the midline. Individual costal cartilages from rib pairs 2–7 are attached to this portion of the sternum.

**Xiphoid process:** The xiphoid (ZĪ-foyd) process, the smallest part of the sternum, is attached to the inferior surface of the body of the sternum.

**Costal cartilages** connect ribs to the sternum, either individually or in groups.

**? REVIEW**
**A.** How are vertebrosternal ribs distinguished from vertebrochondral ribs?

**? REVIEW**
**B.** Why are ribs 11 and 12 called *floating ribs*?

**2** Ribs are long, curved, flattened bones that originate on or between the thoracic vertebrae and end in the wall of the thoracic cavity. Each of us, regardless of sex, has 12 pairs of ribs. Major bone markings on a representative rib (ribs 2–9) are shown in this posterior view.

The **tubercle** contains an articular facet, the *tubercular facet*, that contacts the transverse process of a thoracic vertebra.

Articular facets on head

The vertebral end of the rib articulates with the vertebral column at the **head**, or **capitulum** (ka-PIT-ū-lum), where there are two **articular facets**.

The bend, or **angle**, of the rib is the site where the tubular body, or **shaft**, begins curving toward the sternum.

The superficial surface is convex and provides an attachment site for muscles of the pectoral girdle and trunk. The intercostal muscles, which move the ribs, are attached to the superior and inferior surfaces.

Neck

Shaft

**Posterior view**

The **costal groove** along the inferior border marks the path of nerves and blood vessels.

**3** This superior view of a representative rib shows sites of articulations with a thoracic vertebra. The heads of ribs 2–9 articulate with costal facets on two adjacent vertebrae, and their tubercular facets articulate with the transverse costal facets of the inferior vertebra. The heads of ribs 1, 10, 11, and 12 articulate with individual vertebrae at single costal facets. The tubercular facets of ribs 1 and 10 articulate with the costal facets of vertebrae $T_1$ and $T_{10}$, respectively. Ribs 11 and 12 articulate only at their heads, and there are no tubercular facets.

Transverse process

Transverse costal facet

The tubercular facet of the rib articulates with the transverse costal facet of the inferior vertebra.

The superior articular facet of the rib articulates with the superior vertebra.

The inferior articular facet of the rib articulates with the superior costal facet of the vertebra.

**4** A typical rib acts like a bucket handle at a position just below horizontal. Lowering the handle swings it inward; pulling it up swings it outward. Because of the curvature of the ribs, the same movements change the position of the sternum. Depression of the ribs pulls the sternum inward, whereas elevation moves it outward. As a result, movements of the ribs affect both the width and the depth of the thoracic cage, increasing or decreasing its volume accordingly.

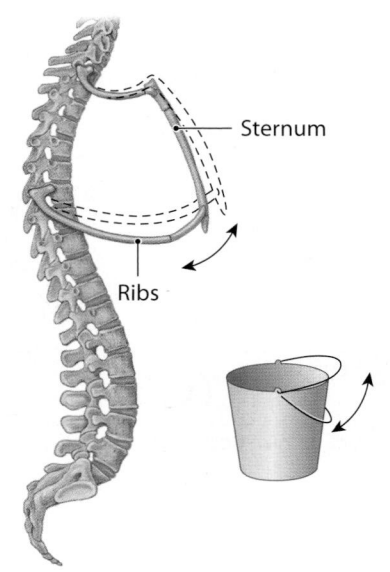

Sternum

Ribs

**? INTEGRATION**

**C.** Along with the ribs and sternum, what other bones make up the thoracic cage?

**Lo LEARNING OUTCOME**

Explain the significance of the articulations between the thoracic vertebrae and the ribs, and between the ribs and the sternum.

# Axial Skeleton

**Labeling:** In the image below, label the bones of the orbital complex.

**1** Frontal

**2** Sphenoid

**3** zygomatic Bone

**4** Ethmoid

**5** Lacrimal

**6** palatine

**7** Maxilla

In the images below, label the sutures and fontanelles of the infant skull.

**8** Sphenoidal fontanelle

**9** Anterior?

**10** Posterior

**11** mastoid

**12** Anterior

**13**

**14** Posterior

**Short answer:** Identify the vertebral region and regional characteristics for each vertebra.

**15** Region: Thoracic

**16** Characteristics:

**17** Region: Neck

**18** Characteristics:

**19** Region: Lumbar

**20** Characteristics:

## Module 7.15

# The appendicular skeleton includes the limb bones and the pectoral and pelvic girdles

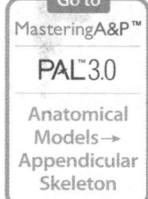

Go to
MasteringA&P™
PAL™3.0
Anatomical
Models→
Appendicular
Skeleton

The **appendicular skeleton** includes the bones of the limbs and the supporting bone girdles that connect them to the trunk. The descriptions in this section emphasize bone markings that either have functional importance (such as the attachment sites for skeletal muscles and the paths of major nerves and blood vessels) or provide landmarks that define areas and locate structures of the body.

 **REVIEW**

**A.** How many bones are in the appendicular skeleton?

| SKELETAL SYSTEM | 206 |
|---|---|

| AXIAL SKELETON | 80 |
|---|---|

7

| | |
|---|---|
| Clavicle | 2 |
| Scapula | 2 |

| Pectoral girdles | 4 |
|---|---|

| | |
|---|---|
| Humerus | 2 |
| Radius | 2 |
| Ulna | 2 |
| Carpal bones | 16 |
| Metacarpal bones | 10 |
| Phalanges (proximal, middle, distal) | 28 |

| Upper limbs | 60 |
|---|---|

| APPENDICULAR SKELETON | 126 |
|---|---|

| | |
|---|---|
| Hip bone (coxal bone) | 2 |

| Pelvic girdle | 2 |
|---|---|

 **REVIEW**

**B.** What is the function of the pectoral and pelvic girdles?

| | |
|---|---|
| Femur | 2 |
| Patella | 2 |
| Tibia | 2 |
| Fibula | 2 |
| Tarsal bones | 14 |
| Metatarsal bones | 10 |
| Phalanges | 28 |

| Lower limbs | 60 |
|---|---|

**LEARNING OUTCOME**

List the four major components of the appendicular skeleton.

# Module 7.16

# The pectoral girdles—the clavicles and scapulae—connect the upper limbs to the axial skeleton

Each arm articulates (forms a joint) with the trunk at the **pectoral girdle**, or shoulder girdle. The pectoral girdles consist of two S-shaped **clavicles** (KLAV-i-kulz; collarbones) and two broad, flat **scapulae** (SKAP-ū-lē; singular, *scapula*, SKAP-ū-luh; shoulder blades).

**1** Each clavicle originates at the superior, lateral border of the manubrium of the sternum, lateral to the jugular notch. It articulates with the sternum at the clavicular notch, forming the **sternoclavicular joint**. These sternoclavicular joints are the only articulations between the pectoral girdles and the axial skeleton.

**?** REVIEW

**A.** Name the bones of the pectoral girdles.

Acromioclavicular joint — Clavicle — Sternoclavicular joint — Jugular notch — Acromion (of scapula) — Clavicular notch — Scapula — Humerus

**Anterior view**

**2** From the sternoclavicular joint at the pyramid-shaped **sternal end**, each clavicle curves laterally and posteriorly for roughly half its length. It then forms a smooth posterior curve to articulate with a process of the scapula, the **acromion** (a-KRŌ-mē-on), at the **acromioclavicular joint**. Stabilizing ligaments attach to the conoid tubercle and costal tuberosity.

**?** REVIEW

**B.** How would a broken clavicle affect the mobility and stability of the scapula?

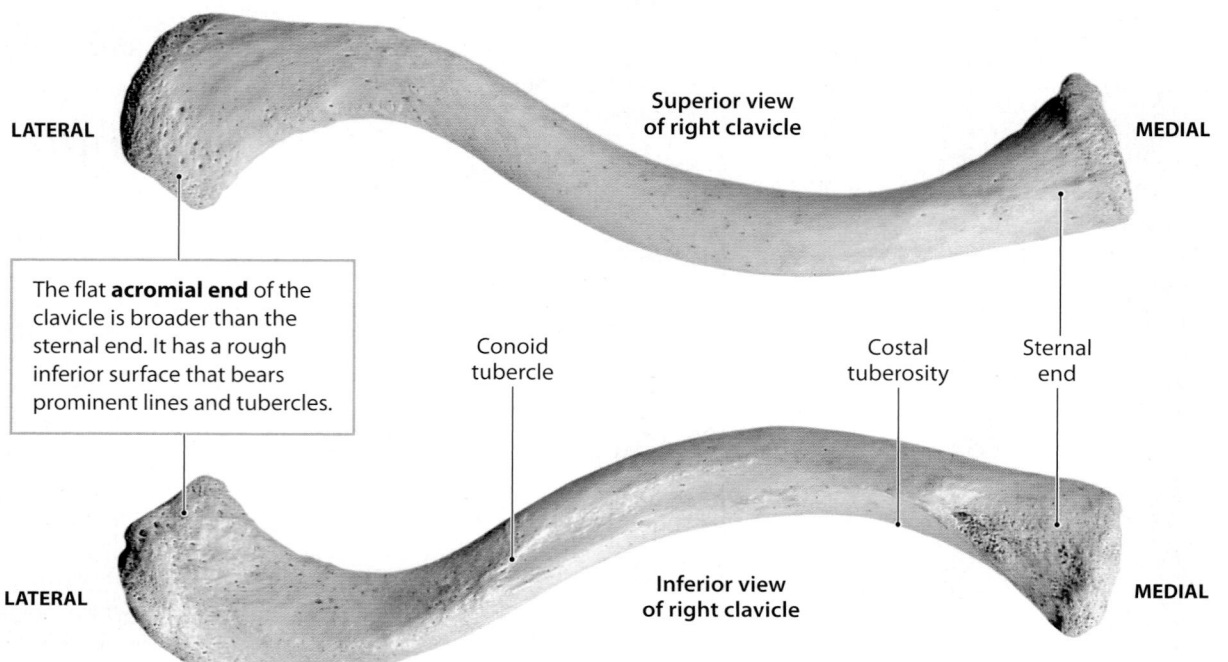

**LATERAL**

**Superior view of right clavicle**

**MEDIAL**

The flat **acromial end** of the clavicle is broader than the sternal end. It has a rough inferior surface that bears prominent lines and tubercles.

Conoid tubercle

Costal tuberosity

Sternal end

**LATERAL**

**Inferior view of right clavicle**

**MEDIAL**

**3** The anterior surface of the **body** of each scapula forms a broad, smooth triangle. The three sides of the triangle are the **superior border**; the **medial border**, or vertebral border; and the **lateral border**, or axillary (*axilla*, armpit) border. Muscles that position the scapula attach along these edges. The corners of the triangle are called the **superior angle**, the **inferior angle**, and the **lateral angle**. The depression in the anterior surface is called the **subscapular fossa**.

**4** The posterior surface of the scapula is convex and has prominent ridges and processes for muscle attachment.

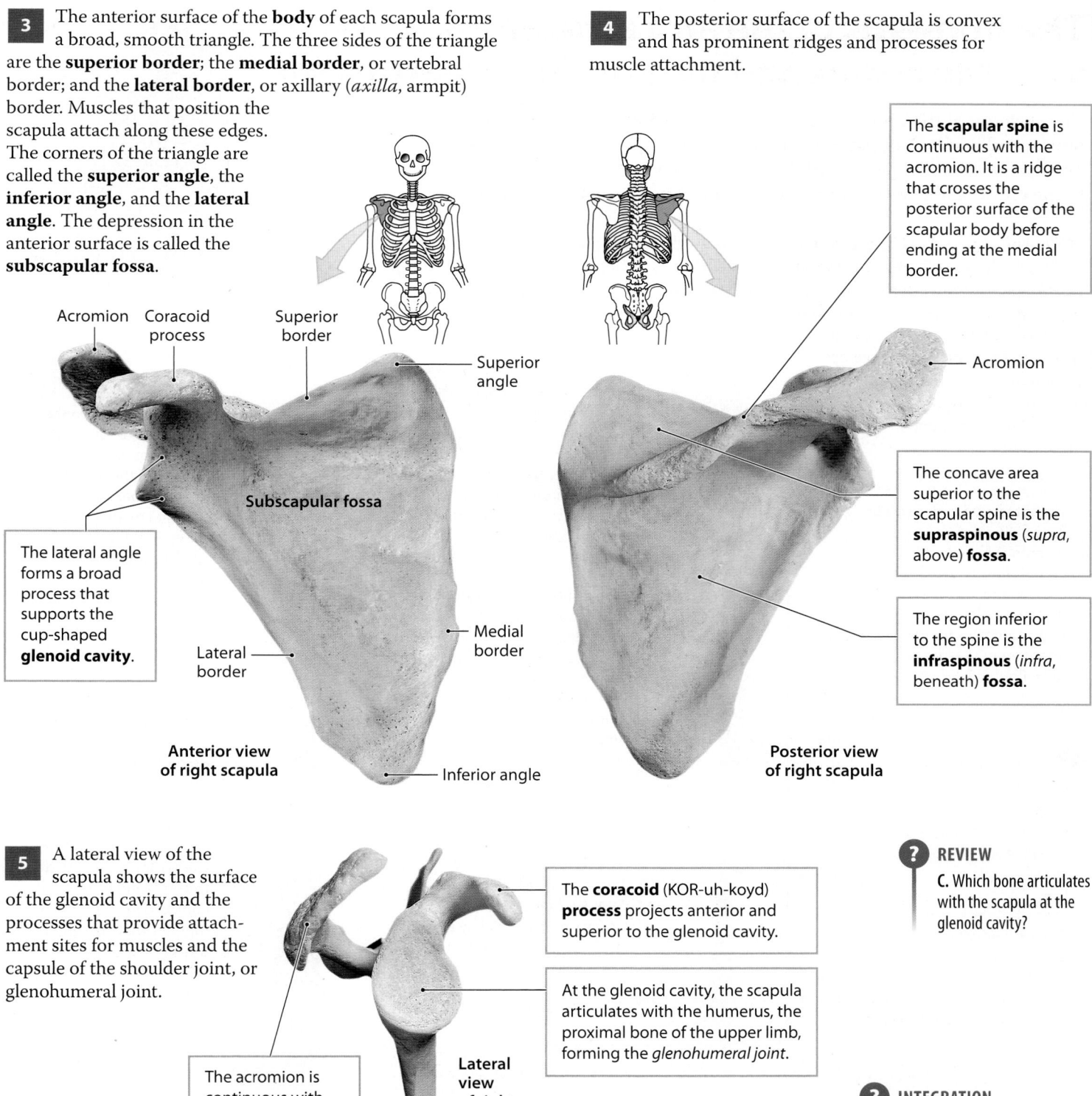

The **scapular spine** is continuous with the acromion. It is a ridge that crosses the posterior surface of the scapular body before ending at the medial border.

Acromion   Coracoid process   Superior border

Superior angle

Subscapular fossa

The lateral angle forms a broad process that supports the cup-shaped **glenoid cavity**.

Lateral border

Medial border

**Anterior view of right scapula**

Inferior angle

Acromion

The concave area superior to the scapular spine is the **supraspinous** (*supra*, above) **fossa**.

The region inferior to the spine is the **infraspinous** (*infra*, beneath) **fossa**.

**Posterior view of right scapula**

**5** A lateral view of the scapula shows the surface of the glenoid cavity and the processes that provide attachment sites for muscles and the capsule of the shoulder joint, or glenohumeral joint.

The **coracoid** (KOR-uh-koyd) **process** projects anterior and superior to the glenoid cavity.

At the glenoid cavity, the scapula articulates with the humerus, the proximal bone of the upper limb, forming the *glenohumeral joint*.

The acromion is continuous with the scapular spine. It is a large process that extends laterally, posterior and superior to the glenoid cavity.

**Lateral view of right scapula**

**? REVIEW**

**C.** Which bone articulates with the scapula at the glenoid cavity?

**? INTEGRATION**

**D.** How are the pectoral girdles of the appendicular skeleton attached to the axial skeleton?

**Lo LEARNING OUTCOME**

Identify the bones that form the pectoral girdles, their functions, and their superficial features.

# The humerus of the arm articulates with the radius and ulna of the forearm

The skeleton of the upper limbs consists of the bones of the arms, forearms, wrists, and hands. Note that in anatomical descriptions, the term *arm* refers only to the proximal portion of the upper limb (from shoulder to elbow), not to the entire limb. We will examine the bones of the right upper limb.

**1** The arm, or **brachium**, contains one bone, the **humerus**, which extends from the scapula to the elbow. Near the distal articulation with the bones of the forearm, the shaft expands to either side at the **medial** and **lateral epicondyles**. Epicondyles are processes that develop proximal to an articulation. They provide additional surface area for muscle attachment.

**Anterior view**

**Posterior view**

Greater tubercle

The **intertubercular sulcus** (also called the *intertubercular groove* or *bicipital groove*) lies between the greater and lesser tubercles. Both tubercles are important sites for muscle attachment. A large tendon runs along the groove.

The round **head** at the proximal end of the humerus articulates with the glenoid cavity of the scapula.

The **lesser tubercle** is a smaller projection that lies on the anterior medial surface of the epiphysis.

The **anatomical neck** marks the extent of the joint capsule.

The prominent **greater tubercle** is a rounded projection on the lateral surface of the epiphysis, near the margin of the humeral head. The greater tubercle establishes the lateral contour of the shoulder.

Shaft

The **surgical neck** corresponds to the metaphysis of the growing bone. The name reflects the fact that fractures typically occur at this site.

**? REVIEW**

**A.** Identify the two rounded projections on either side of the elbow, and state to which bone they belong.

The **deltoid tuberosity** is a large, rough elevation on the lateral surface of the shaft, approximately halfway along its length. It is named after the deltoid muscle, which attaches to it.

The **radial groove** crosses the inferior end of the deltoid tuberosity. This depression marks the path of the radial nerve, a large nerve that provides both sensory information from the posterior surface of the limb and motor control over the large muscles that straighten the elbow.

The **radial fossa** accommodates a portion of the head of radius (radial head) when the forearm approaches the humerus as the elbow bends.

Coronoid fossa

Olecranon fossa

Lateral epicondyle

Medial epicondyle

Trochlea

The rounded **capitulum** forms the lateral surface of the **condyle**. At the condyle, the humerus articulates with the radius and the ulna of the forearm, forming the *radiohumeral joint*. The condyle is divided into two articular regions.

The **trochlea** (*trochlea*, a pulley) is the spool-shaped medial portion of the condyle. The trochlea extends from the **olecranon** (ō-LEK-ruh-non) **fossa** on the posterior surface to the **coronoid** (*corona*, crown) **fossa** on the anterior surface. These depressions accept projections from the ulna as the elbow, or *humero-ulnar joint*, approaches the limits of its range of motion.

**2** The **ulna** and **radius** are parallel bones that support the forearm, or **antebrachium**. In the anatomical position, the ulna lies medial to the radius.

**REVIEW**

**B.** Which bone of the forearm is positioned laterally when in the anatomical position?

**Posterior view**

The disc-shaped **head of radius** articulates with the capitulum of the humerus. During flexion, the radial head swings into the radial fossa of the humerus.

The **neck** of the radius extends from the radial head to the radial tuberosity.

The **radial tuberosity** marks the attachment site of the biceps brachii muscle, a large muscle on the anterior surface of the arm.

The **olecranon**, the proximal end of the ulna, is the point of the elbow.

Proximal radio-ulnar joint

**Anterior view**

The **trochlear notch** of the ulna articulates with the trochlea of the humerus at the elbow joint.

The **coronoid process** forms the inferior lip of the trochlear notch. During flexion, the coronoid process fits into the coronoid fossa of the humerus.

The smooth **radial notch** accommodates the head of the radius at the *proximal radio-ulnar joint*.

Ulna —————— Radius

Radius ————

The **interosseous membrane** of the forearm is a fibrous sheet that connects the shafts of the ulna and radius.

—— Ulna

The **ulnar notch** of the radius marks the site of articulation with the head of the ulna.

The lateral surface of the ulnar head articulates with the distal end of the radius to form the *distal radio-ulnar joint*.

The **head of ulna**, or ulnar head, is slender and rounded.

The **radial styloid process** helps stabilize the wrist joint. The distal end of the radius articulates with the bones of the wrist.

—— Head of ulna

A short **ulnar styloid process** is attached to the posterior, lateral surface of the head. A triangular articular disc attaches to the styloid process; this cartilage separates the ulnar head from the bones of the wrist.

**REVIEW**

**C.** Name four different bone markings associated with the proximal portion of the ulna.

**LEARNING OUTCOME**

Identify the bones of the arm and forearm, their functions, and their bone markings.

# The wrist consists of carpal bones …

**1** The *carpus*, or wrist, contains eight **carpal bones** arranged in two rows: a row of four proximal carpal bones and a row of four distal carpal bones.

## Proximal Carpal Bones

| | | | |
|---|---|---|---|
| The **scaphoid** (*skaphe*, boat) is the proximal carpal bone on the lateral border of the wrist. It is the carpal bone closest to the styloid process of the radius. | The comma-shaped **lunate** (*luna*, moon) lies medial to the scaphoid and, like the scaphoid, articulates with the radius. | The **triquetrum** (*triquetrus*, three-cornered) is a small pyramid-shaped bone medial to the lunate. The triquetrum articulates with the articular disc that separates the ulnar head from the wrist. | The small, pea-shaped **pisiform** (PIS-i-form; *pisum*, pea) sits anterior to the triquetrum. |

**Right wrist and hand, anterior (palmar) view**

Radius   Ulna

I   II   III   IV   V

Metacarpal bones

Proximal phalanx

Middle phalanx

Distal phalanx

## Distal Carpal Bones

| | | | |
|---|---|---|---|
| The **trapezium** (*trapezion*, table) is the lateral bone of the distal row. Its proximal surface articulates with the scaphoid. | The wedge-shaped **trapezoid** lies medial to the trapezium. Like the trapezium, it has a proximal articulation with the scaphoid. | The **capitate** (*caput*, head) is the largest carpal bone. It sits between the trapezoid and the hamate. | The **hamate** (*hamatum*, hooked) is the medial distal carpal bone and has a prominent hook projecting anteriorly. |

It may help you to identify the eight carpal bones if you remember the sentence "Sam Likes To Push The Toy Car Hard." In lateral-to-medial order, the first four words stand for the proximal carpal bones (scaphoid, lunate, triquetrum, pisiform), and the last four stand for the distal carpal bones (trapezium, trapezoid, capitate, hamate).

**REVIEW**

**A.** Name the carpal bones.

# … and the hand consists of metacarpal bones and phalanges

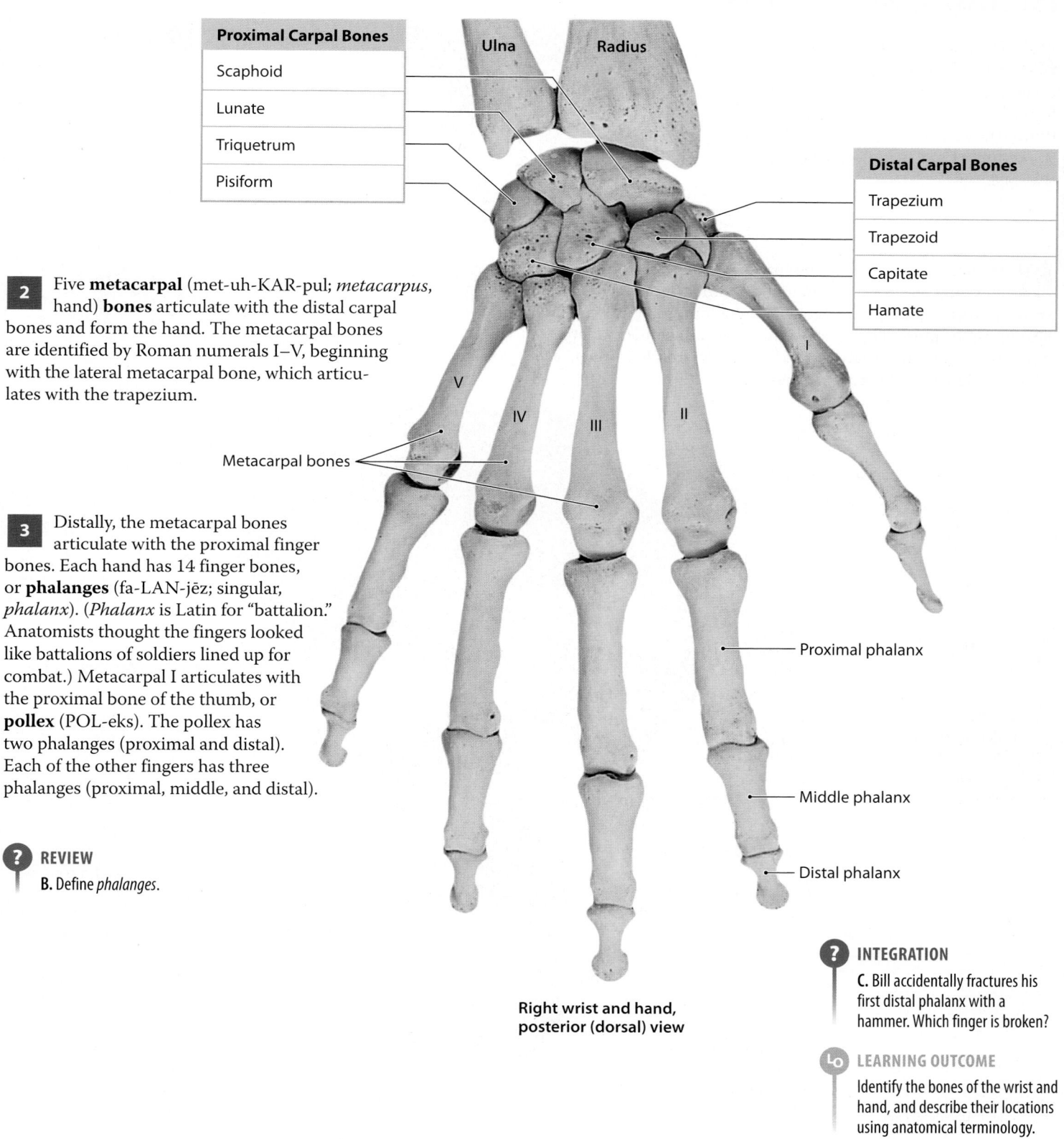

**Proximal Carpal Bones**

| |
|---|
| Scaphoid |
| Lunate |
| Triquetrum |
| Pisiform |

Ulna

Radius

**Distal Carpal Bones**

| |
|---|
| Trapezium |
| Trapezoid |
| Capitate |
| Hamate |

**2** Five **metacarpal** (met-uh-KAR-pul; *metacarpus*, hand) **bones** articulate with the distal carpal bones and form the hand. The metacarpal bones are identified by Roman numerals I–V, beginning with the lateral metacarpal bone, which articulates with the trapezium.

Metacarpal bones

V   IV   III   II   I

**3** Distally, the metacarpal bones articulate with the proximal finger bones. Each hand has 14 finger bones, or **phalanges** (fa-LAN-jēz; singular, *phalanx*). (*Phalanx* is Latin for "battalion." Anatomists thought the fingers looked like battalions of soldiers lined up for combat.) Metacarpal I articulates with the proximal bone of the thumb, or **pollex** (POL-eks). The pollex has two phalanges (proximal and distal). Each of the other fingers has three phalanges (proximal, middle, and distal).

Proximal phalanx

Middle phalanx

Distal phalanx

**? REVIEW**
**B.** Define *phalanges*.

**Right wrist and hand, posterior (dorsal) view**

**? INTEGRATION**
**C.** Bill accidentally fractures his first distal phalanx with a hammer. Which finger is broken?

**Lo LEARNING OUTCOME**
Identify the bones of the wrist and hand, and describe their locations using anatomical terminology.

# The hip bone forms by the fusion of the ilium, ischium, and pubis

**1** The **pelvic girdle** consists of the paired **hip bones**, which are also called the **coxal bones**. Each hip bone forms by the fusion of three bones: an **ilium** (IL-ē-um; plural, *ilia*), an **ischium** (IS-kē-um; plural, *ischia*), and a **pubis** (PŪ-bis).

Ilium

POSTERIOR ANTERIOR

Pubis

Ischium

**2** In lateral view, the ilium dominates the hip bone. Bone markings along the margin of the ilium include the **iliac spines**, which mark the attachment sites of important muscles and ligaments; the **gluteal lines**, which mark the attachment of large hip muscles; and the **greater sciatic** (sī-AT-ik) **notch**, through which the large sciatic nerve reaches the lower limb.

The **iliac crest** is an important ridge for muscle attachment.

| Gluteal Lines |
|---|
| Anterior |
| Inferior |
| Posterior |

Anterior superior iliac spine

Posterior superior iliac spine

Anterior inferior iliac spine

Posterior inferior iliac spine

Greater sciatic notch

The **lunate surface** is the smooth, cup-shaped articular surface.

The **acetabulum** (as-e-TAB-ū-lum; *acetabulum*, vinegar cup), a concave socket, articulates with the head of the femur. The ilium, ischium, and pubis meet inside the acetabulum, in an arrangement resembling a pie sliced into three pieces.

The prominent **ischial spine** projects superior to the *lesser sciatic notch*, which marks where blood vessels, nerves, and a small muscle pass across the surface.

Lesser sciatic notch

Ischial ramus

The **ischial tuberosity**, a roughened projection, bears the body's weight when you are seated.

The **acetabular notch** is a gap in the bony rim of the acetabulum.

**Lateral view of right hip bone**

**3** A medial view shows the locations of the ilium, ischium, and pubis in the right hip bone. The photo below shows the roughened articular surfaces that stabilize the pelvis and many additional features of the ischium and pubis.

Ilium

**ANTERIOR**

**POSTERIOR**

Pubis

Ischium

Iliac crest

The **iliac fossa** is a shallow depression that helps support the abdominal organs and provides additional area for muscle attachment.

Anterior inferior iliac spine

The **arcuate line** of the ilium is continuous with the pectineal line of the pubis.

The **pectineal line** is a ridge that ends at the **pubic tubercle**, a small, elevated area anterior and lateral to the pubic symphysis.

Superior pubic ramus

Pubis

The **iliac tuberosity** is a roughened area superior to the auricular surface. Ligaments originating here stabilize the sacro-iliac joint.

The **auricular surface** of the ilium articulates with the auricular surface of the sacrum at the sacro-iliac joint.

Greater sciatic notch

The **obturator** (OB-tū-rā-tor) **foramen** is a space that is closed by a sheet of collagen fibers whose inner and outer surfaces provide a firm base for the attachment of muscles of the hip. This foramen is bounded by the **ischial ramus**, the **inferior pubic ramus**, and the **superior pubic ramus**.

Ischial ramus

Inferior pubic ramus

**Medial view of right hip bone**

**4** At the **pubic symphysis**, the right and left pubic bones are attached to a median pad of fibrocartilage. Posteriorly, the two ilia articulate with the sacrum. Together, the hip bones and sacrum form the pelvis (the topic of the next module).

Sacrum

Pubic symphysis

**Anterior view of pelvis**

**? INTEGRATION**

**D.** Which bones articulate at the sacro-iliac joint?

 **LEARNING OUTCOME**

Describe the hip bones that form the pelvic girdle, their functions, and their bone markings.

*Section 2: Appendicular Skeleton •* **275**

# Module 7.20

# The pelvis consists of the two hip bones, the sacrum, and the coccyx

**1** The **pelvis** contains bones of the axial skeleton (the sacrum and coccyx) and the appendicular skeleton (the hip bones). An extensive network of ligaments connects the lateral borders of the sacrum with the iliac crest, the ischial tuberosity, the ischial spine, and the arcuate line. Other ligaments tie the ilia to the posterior lumbar vertebrae. These interconnections increase the stability of the pelvis.

Sacrum

Coccyx

**Hip Bone**

Ilium

Pubis

Ischium

Iliac crest

L5

Iliac fossa

Ilium

Each **sacro-iliac joint**, an articulation between the sacrum and the adjacent ilium, is supported by an extensive network of ligaments. These joints form the union between the axial skeleton and the appendicular skeleton.

Sacrum

Acetabulum

Pubic tubercle

The **pubic symphysis** is a fibrocartilage pad that forms the articulation between the two pubic bones.

Obturator foramen

Ischium

**Anterior view**

**? REVIEW**

**A.** Name the bones of the pelvis.

**B.** The pubic bones are joined anteriorly by what structure?

**2** The pelvis may be divided into the **true** (lesser) **pelvis** (shown in purple) and the **false** (greater) **pelvis.** The true pelvis encloses the pelvic cavity, a subdivision of the abdominopelvic cavity (see Module 1.22). The superior limit of the true pelvis is a line that extends from either side of the base of the sacrum, along the arcuate line and pectineal line to the pubic symphysis.

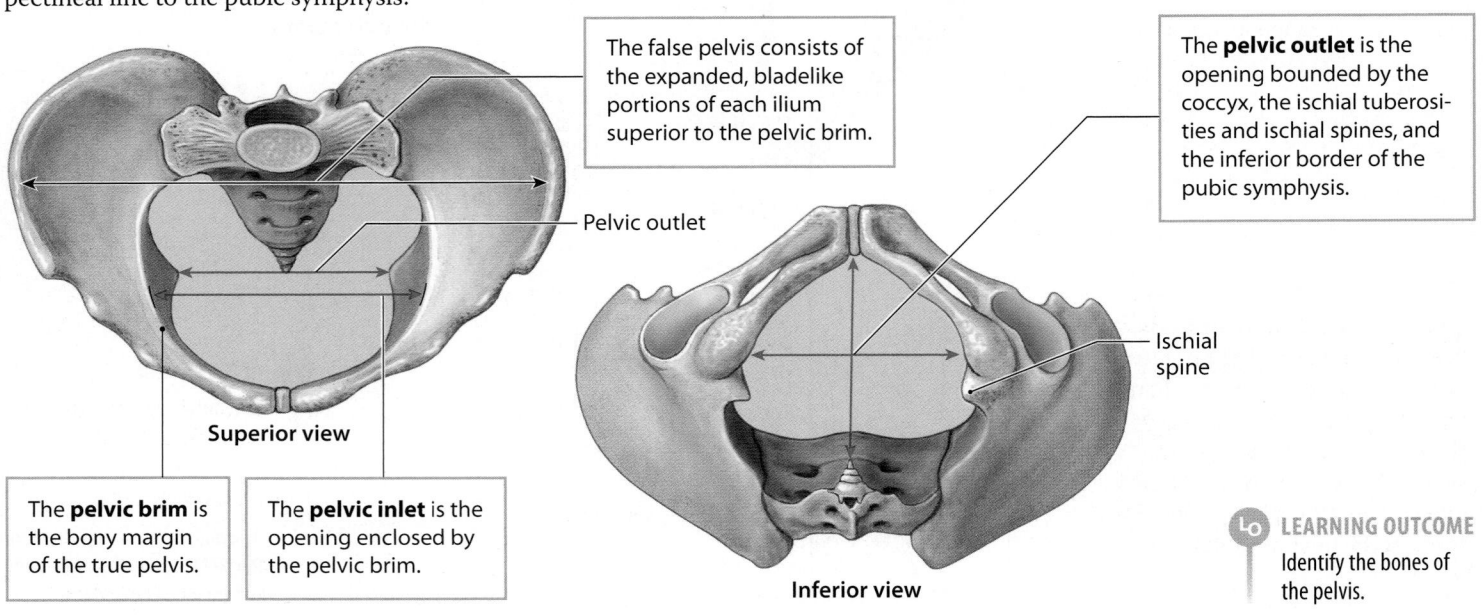

The false pelvis consists of the expanded, bladelike portions of each ilium superior to the pelvic brim.

The **pelvic outlet** is the opening bounded by the coccyx, the ischial tuberosities and ischial spines, and the inferior border of the pubic symphysis.

Pelvic outlet

Ischial spine

**Superior view**

The **pelvic brim** is the bony margin of the true pelvis.

The **pelvic inlet** is the opening enclosed by the pelvic brim.

**Inferior view**

**LEARNING OUTCOME**

Identify the bones of the pelvis.

# Module 7.21

# The adult male and female skeletons have significant differences

**1** The sex of an adult human skeleton can be determined most easily by examining the skull and pelvis. Many female pelvic differences appear to be adaptations to help support the weight of the fetus and its passage through the pelvic outlet during delivery. Not every skeleton shows every feature clearly, but this chart contrasts relative differences between the sexes.

**? REVIEW**

**A.** How does the female skull differ from the male skull?

| MALE | SKULL | FEMALE |
|---|---|---|
| Heavier, rougher | **General appearance** | Lighter, smoother |
| About 10% larger | **Cranium** | About 10% smaller |
| More sloping | **Forehead** | More vertical |
| Larger | **Sinuses** | Smaller |
| Larger | **Teeth** | Smaller |
| Larger, squarer, more robust | **Mandible** | Smaller, narrower, less robust |

| MALE | PELVIS | FEMALE |
|---|---|---|
| Narrower, rougher, more robust | **General appearance** | Broader, smoother, less robust |
| More vertical; extends further superior to sacro-iliac joint | **Ilium** | Less vertical; less extension superior to sacral articulation |
| Long, narrow triangle with pronounced sacral curvature | **Sacrum** | Broad, short triangle with less sacral curvature |
| Deeper | **Iliac fossa** | Shallower |
| Narrower, heart shaped | **Pelvic inlet** | Open, circular shaped |
| Narrow | **Pelvic outlet** | Wide |
| Points anteriorly | **Coccyx** | Points inferiorly |
| Directed laterally | **Acetabulum** | Faces slightly anteriorly |
| Oval | **Obturator foramen** | Triangular |
| < 90° | **Pubic angle** | ≥100° |

| MALE | OTHER | FEMALE |
|---|---|---|
| Greater | **Bone mass and density** | Lesser |
| More prominent | **Bone markings** | Less prominent |

**Lo LEARNING OUTCOME**

Discuss the differences between the male and female skeletons.

# The femur, tibia, and patella meet at the knee

The skeleton of each lower limb consists of a *femur* (thigh), a *patella* (kneecap), a *tibia* and a *fibula* (leg), and the *tarsal bones*, *metatarsal bones*, and *phalanges* of the foot. Once again, anatomical terminology differs from common usage. In anatomical terms, *leg* refers only to the distal portion of the limb, not to the entire lower limb. Thus, we will use *thigh* and *leg*, rather than *upper leg* and *lower leg*. The functional anatomy of the lower limbs differs from that of the upper limbs, primarily because the lower limbs transfer the body weight to the ground.

**1** The **femur** is the longest and heaviest bone in the body. It articulates with the hip bone at the hip joint and with the tibia of the leg at the knee joint. Major bone markings are shown here on the anterior and posterior surfaces of the right femur.

**Anterior view**

**Posterior view**

The **greater trochanter** is a large, rough projection that extends laterally from the junction of the head and shaft. The trochanters develop where large tendons attach to the femur.

At the **fovea capitis**, a small pit in the center of the head of the femur, a ligament attaches the acetabulum to the femur.

The rounded **head** of the femur articulates with the pelvis at the acetabulum.

The raised **intertrochanteric** (in-ter-trō-kan-TER-ik) **line** marks the edge of the articular capsule on the anterior surface of the femur.

The **lesser trochanter** projects posteriorly and medially from the junction of the head and shaft.

The **neck** of the femur joins the **shaft** at an angle of about 125° (formed by lines drawn through the axes of the shaft and neck of the femur).

Intertrochanteric crest

The **gluteal tuberosity** marks the attachment site of the gluteus maximus muscle.

The **linea aspera** is a rough ridge that marks the attachment site of powerful hip muscles.

Shaft

Medial supracondylar ridge

Lateral supracondylar ridge

The **adductor tubercle** marks the attachment site of the adductor magnus muscle.

The **popliteal surface** (*poples*, hollow of the knee) is a flattened triangular area.

Patellar surface

Lateral epicondyle

Lateral condyle

Medial epicondyle

Medial condyle

Intercondylar fossa

Lateral epicondyle

Lateral condyle

**2** At the distal end of the femur, the **medial** and **lateral condyles** are part of the knee joint. On the anterior and inferior surfaces, the two condyles are separated by the **patellar surface**, a smooth articular surface over which the patella glides. On the posterior surface, the medial and lateral condyles are separated by a deep **intercondylar fossa** that does not extend onto the anterior surface.

 **REVIEW**

**A.** Identify the bones of the lower limb.

**B.** Which structure articulates with the acetabulum?

**3** The **patella** is a large sesamoid bone that forms within the tendon of the quadriceps femoris, a group of muscles that extend (straighten) the knee. The right patella is shown here.

Base of patella

Attachment area for quadriceps tendon

Attachment area for the patellar ligament, which attaches the patella to the tibia

Apex of patella

**Anterior view**

**Lateral facet,** for lateral condyle of femur

**Medial facet,** for medial condyle of femur

Articular surface of patella

**Posterior view**

**4** The **tibia** (TIB-ē-uh), or shinbone, is the large medial bone of the leg. At the proximal end of the tibia, the **medial** and **lateral tibial condyles** articulate with the medial and lateral condyles of the femur. The slender **fibula** (FIB-ū-luh) parallels the lateral border of the tibia but does not participate in the knee joint and bears no weight. However, the fibula is important as a site for the attachment of muscles that move the foot and toes. In addition, the distal tip of the fibula extends lateral to the ankle, providing important stability to that joint.

Lateral tibial condyle

Superior tibiofibular joint

The **head** of the fibula articulates with the tibia. The articular facet is located on the anterior, inferior surface of the lateral tibial condyle.

Medial tibial condyle

The **tibial tuberosity** marks the attachment of the patellar ligament.

The **interosseous membrane** helps stabilize the positions of the tibia and fibula and provides additional surface area for muscle attachment.

The **anterior margin** is a ridge that begins at the tibial tuberosity and extends distally along the anterior tibial surface.

**Anterior view**

Fibula

The **lateral malleolus** of the fibula provides lateral stability to the ankle joint.

Tibia

Inferior articular surface

The **medial malleolus** (ma-LĒ-o-lus; *malleolus*, hammer) is a large tibial process familiar to you as the medial projection at the ankle. The medial malleolus provides medial support for the ankle joint.

Inferior tibiofibular joint

Lateral malleolus (fibula)

Articular surface of medial tibial condyle

The **intercondylar eminence** is a ridge that separates the condyles.

Articular surface of lateral tibial condyle

Lateral tibial condyle

Head of fibula

**Posterior view**

**? REVIEW**

**C.** Identify the sesamoid bone of the lower limb.

Fibula

**? INTEGRATION**

**D.** The fibula neither participates in the knee joint nor bears weight. Yet when it is fractured, walking becomes difficult. Why?

**⌐ LEARNING OUTCOME**

Identify the bones of the thigh and leg, their functions, and their bone markings.

# The ankle and foot consist of tarsal bones, metatarsal bones, and phalanges

**1** The bones of the *tarsus*, or ankle, accept the body weight from the leg and transfer it to the ground by distributing it through the bones of the foot. The combination of ankle and foot must be strong, yet flexible enough to deal with the changes in loading that occur during walking, running, and jumping.

### The Ankle (Tarsus)

The ankle consists of seven **tarsal bones**.

The **calcaneus** (kal-KĀ-nē-us), or heel bone, is the largest of the tarsal bones. When you stand normally, most of your weight is transmitted from the tibia, to the talus, to the calcaneus, and then to the ground.

The large **talus** transmits the weight of the body from the tibia toward the toes.

The **navicular** is anterior to the talus, on the medial side of the ankle. It articulates with the talus and with the three cuneiform (kū-NĒ-i-form) bones.

The **cuboid** articulates with the anterior surface of the calcaneus.

The three **cuneiform bones** are arranged in a row, with articulations between them. They are named according to their relative positions: medial, intermediate, and lateral.

### Metatarsals

The distal surfaces of the cuboid and the cuneiform bones articulate with the **metatarsal bones** of the foot. The five long metatarsal bones form the distal portion of the foot.

The metatarsal bones are identified by Roman numerals I–V, proceeding from medial to lateral across the sole. Proximally, metatarsal bones I–III articulate with the three cuneiform bones, and metatarsal bones IV and V articulate with the cuboid. Distally, each metatarsal bone articulates with a different proximal phalanx.

### Phalanges

The **phalanges**, or toe bones, have the same anatomical organization as the fingers. The toes contain 14 phalanges. The **hallux**, or great toe, has two phalanges (proximal and distal), and the other four toes each have three phalanges (proximal, middle, and distal).

The articulation between the talus and the tibia occurs across the superior and medial surfaces of the **trochlea**, a spool- or pulley-shaped articular process.

Proximal phalanx

Distal phalanx

Hallux

**Superior view**

**? REVIEW**

**A.** Identify the tarsal bones.

**B.** Which foot bone transmits the weight of the body from the tibia toward the toes?

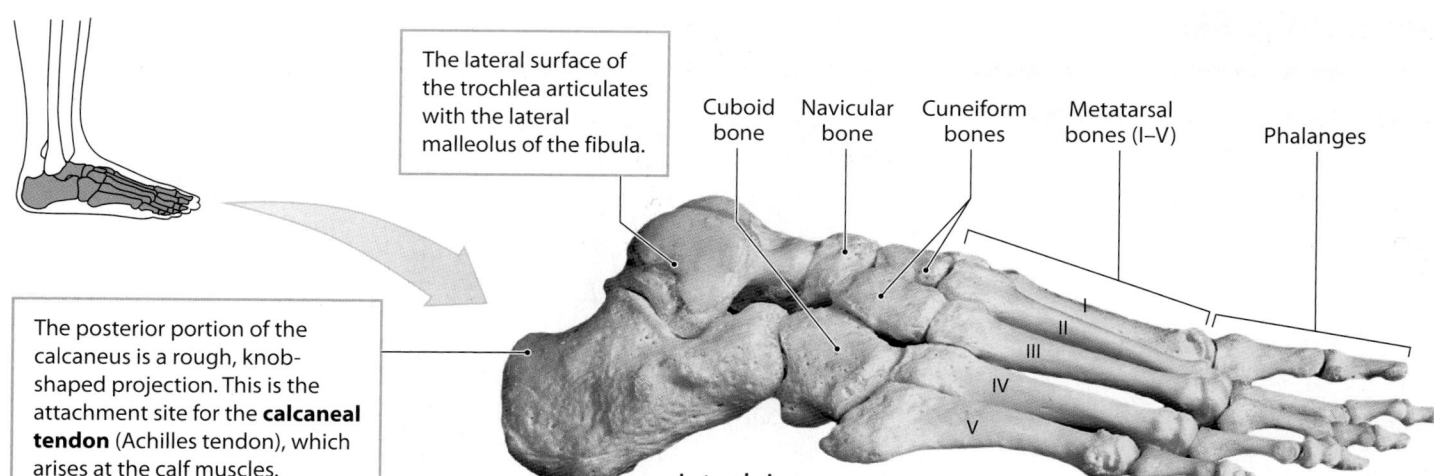

The lateral surface of the trochlea articulates with the lateral malleolus of the fibula.

Cuboid bone

Navicular bone

Cuneiform bones

Metatarsal bones (I–V)

Phalanges

**Lateral view**

The posterior portion of the calcaneus is a rough, knob-shaped projection. This is the attachment site for the **calcaneal tendon** (Achilles tendon), which arises at the calf muscles.

Phalanges

Metatarsal bones

Medial cuneiform bone

Navicular bone

Talus

Calcaneus

**Medial view**

**2** Weight transfer occurs along the **longitudinal arch** of the foot. The amount of weight transferred forward depends on the position of the foot and the placement of one's body weight. When you "dig in your heels" all your body weight rests on the calcaneus, but when you are on tiptoes all of your weight is transferred to the metatarsal bones and phalanges. When you stand normally, your body weight is distributed evenly between the calcaneus and the distal ends of the metatarsal bones.

The **main longitudinal arch** ties the calcaneus to the distal portions of the metatarsal bones. As a result, the medial plantar surface of the foot remains elevated, so that the muscles, nerves, and blood vessels that supply the inferior surface are not squeezed between the metatarsal bones and the ground. Note that the *lateral part* of the longitudinal arch has much less curvature than the *medial part*. Part of the reason is that the medial part has more elasticity. This elasticity absorbs the shocks from sudden changes in weight loading, such as the stresses that running or dancing place on the toes.

The **transverse arch** describes the degree of curvature from the medial border to the lateral border of the foot. However, in the condition known as *flatfeet*, normal arches are lost ("fall") or never form.

**3** In a **dancer's fracture**, the diaphysis of the fifth metatarsal is broken. Most such cases occur while the body weight is being supported by the longitudinal arch of the foot (as in ballet dancing). A sudden shift in weight from the medial portion of the arch to the lateral, less elastic border breaks the fifth metatarsal close to the small toe.

**? INTEGRATION**

**C.** Ten-year-old Joey jumps off the back porch, lands on his right heel, and breaks his foot. Which foot bone is most likely broken?

**Lo LEARNING OUTCOME**

Identify the bones of the ankle and foot, and describe their locations using anatomical terminology.

# Section 2 Review
## Appendicular Skeleton

**Labeling:** Label the bones of the appendicular skeleton in the diagram at right.

| 1 | |
| 2 | |
| 3 | |
| 4 | |
| 5 | |
| 6 | |
| 7 | |
| 8 | |

| 9 | |
| 10 | |
| 11 | |
| 12 | |
| 13 | |
| 14 | |
| 15 | |
| 16 | |

**Short answer:** In the pelvis diagrams below, identify the sex and the differences between the following: a. general appearance, b. ilium, c. iliac fossa, d. pelvic outlet, e. obturator foramen, f. pubic angle.

| 17 | Sex: |
|----|------|

| 18 | Differences: |
|----|--------------|
| a. | d. |
| b. | e. |
| c. | f. |

| 19 | Sex: |
|----|------|

| 20 | Differences: |
|----|--------------|
| a. | d. |
| b. | e. |
| c. | f. |

# Study Outline

## SECTION 1 • Axial Skeleton

### Module 7.1

**The axial skeleton includes bones of the head, vertebral column, and trunk    p. 241**

1. The **axial skeleton** forms the longitudinal axis of the body and includes the *skull* and associated bones, the *thoracic cage*, the *vertebral column*, and various supplemental cartilages.

2. The axial skeleton supports and protects the brain, the spinal cord, and the organs in the body cavities of the trunk. It also provides an extensive surface area for the attachment of muscles.

### Module 7.2

**The skull has cranial and facial components that are usually bound together by sutures    p. 242**

3. The **skull** contains 22 bones: 8 form the cranium, and 14 form the face. Seven other bones are associated with the skull: six auditory ossicles and the hyoid bone.

4. **Facial bones** protect and support the entrances to the digestive and respiratory tracts. They also provide areas for the attachment of facial muscles.

5. **Cranial bones** form the **cranium** and enclose the **cranial cavity**.

6. Except where the mandible contacts the cranium, the connections between the skull and bones of adults are immovable joints called **sutures**. The major sutures of the skull are the **coronal suture**, the **squamous suture**, the **sagittal suture**, and the **lambdoid suture**.

### Module 7.3

**Facial bones dominate the anterior aspect of the skull, and cranial bones dominate the posterior surface    p. 244**

7. The facial bones visible on the anterior aspect of the skull are the paired **nasal**, **lacrimal**, **palatine**, **zygomatic**, **maxillae**, **inferior nasal conchae**, and single **vomer** and **mandible bones**.

8. The cranial bones visible on the anterior aspect of the skull are the **parietal**, the **frontal**, **sphenoid**, and **ethmoid**.

9. The posterior view of the skull displays cranial bones only: **parietal**, **occipital**, and **temporal**.

### Module 7.4

**The lateral and medial aspects of the skull share many bone markings    p. 246**

10. The cranial bones visible from a lateral view of the skull are the frontal bone, parietal bone, occipital bone, temporal bone, sphenoid, and ethmoid. Prominent features are the **superior** and **inferior temporal lines**, the **squamous part** of the temporal bone, and the **external acoustic meatus**.

11. The facial bones visible from a lateral view are the zygomatic bone, mandible, maxilla, nasal bone, and lacrimal bone. Important features are the **zygomatic arch**, the **mandibular angle**, the **mental protuberance**, and the **alveolar process** of each maxilla and **alveolar part** of the mandible.

12. A sagittal section of a skull demonstrates the **frontal sinuses**, **petrous part** of the temporal bone, the **internal acoustic meatus**, and the **hypoglossal canal**.

### Module 7.5

**The foramina on the inferior surface of the skull mark the passageways for nerves and blood vessels    p. 248**

13. Visible in an inferior view of a skull are passageways for blood vessels, nerves, and the spinal cord. Important foramina include the **foramen lacerum**, **foramen ovale**, **carotid canal**, **jugular foramen**, **stylomastoid foramen**, and the **foramen magnum**. Other landmarks include the **mandibular fossa**, **occipital condyles**, and the **inferior** and **superior nuchal lines**.

14. In horizontal section, many of the foramina are visible as they open into the cranial cavity. This section also demonstrates the crista galli, cribriform plate, sella turcica, and the **internal occipital crest**.

### Module 7.6

**The shapes and markings of the sphenoid, ethmoid, and palatine bones are best seen in the isolated bones    p. 250**

15. The **sphenoid** forms part of the floor of the cranium, unites the cranial and facial bones, and acts as a cross-brace that ladder strengthens the sides of the skull.

16. The **ethmoid** forms part of the floor of the anterior cranium, the roof of the nasal cavity, a portion of the nasal septum and the medial orbital wall. It has three parts: the **cribriform plate**, the paired **ethmoidal labyrinths**, and the **perpendicular plate**. The **superior** and **middle nasal conchae** are projections of the lateral masses.

17. The **palatine bones** form the posterior portion of the hard palate and contribute to the floor of each orbit.

### Module 7.7

**Each orbital complex contains one eye, and the nasal complex encloses the nasal cavities    p. 252**

18. The orbits are bony recesses that contain the eyes and are formed by the seven bones of the **orbital complex**: frontal bone, zygomatic bone, maxilla, lacrimal, ethmoid, sphenoid, and palatine bone.

19. The **nasal complex** includes the bones of the nasal cavities and the **paranasal sinuses**. Sphenoid, ethmoid, frontal bone, palatine bone, and maxillae contain the paranasal sinuses.

### Module 7.8

**The mandible forms the lower jaw, and the associated bones of the skull perform specialized functions    p. 254**

20. The **mandible** can be subdivided into the **body** and the **rami**. Its most important features are the **coronoid process**, **alveolar part** of the mandible, **condylar process**, **mandibular notch**, **mandibular foramen**, and **mylohyoid line**.

21. The **hyoid bone** supports the larynx and is the attachment site for muscles. The **auditory ossicles** are enclosed in the petrous part of the temporal bone and are important in hearing.

### Module 7.9

**Fontanelles permit cranial growth in infants and small children    p. 256**

22. At birth, cranial bones are connected by flexible fibrous connective tissue called **fontanelles**.

23. The important fontanelles are the **anterior fontanelle**, **sphenoidal fontanelle**, **mastoid fontanelle**, and **posterior fontanelle**.

## Module 7.10

**The vertebral column has four spinal curves, and vertebrae share a basic structure that differs regionally    p. 258**

24. There are 26 bones in the adult **vertebral column**: the 24 *vertebrae*, the *sacrum*, and the *coccyx*.

25. The primary spinal curves develop before birth and remain in the adult as the **thoracic** and **sacral curves**. The secondary curves develop after birth. The **cervical curve** develops as the infant balances the weight of the head, and the **lumbar curve** develops with the ability to stand.

26. Each vertebra consists of **articular processes**, a **vertebral arch**, and a **vertebral body**. Vertebrae have structural differences in each region.

## Module 7.11

**There are seven cervical vertebrae and twelve thoracic vertebrae    p. 260**

27. Regional characteristics of the seven **cervical vertebrae** are **transverse foramina** and **bifid** spinous processes. Specialized cervical vertebrae are $C_1$ **(atlas)** and $C_2$ **(axis)**, which allow for head movement. The **dens**, or **odontoid process** of $C_2$, allows for head rotation.

28. Regional characteristics of the 12 **thoracic vertebrae** are **costal facets** for rib articulation and inferiorly projecting **spinous processes**.

## Module 7.12

**There are five lumbar vertebrae    p. 262**

29. The five lumbar vertebrae are the largest vertebrae. They do not have costal facets and have stumpier processes.

## Module 7.13

**The sacrum and coccyx consist of fused vertebrae    p. 263**

30. The **sacrum** is a single bone consisting of five fused sacral vertebrae. It begins fusing to the **coccyx** by age 26. Bone markings of the sacrum are the **base**, **sacral promontory**, **sacral foramina**, **apex**, **ala**, **sacral canal**, **sacral tuberosity**, **auricular surface**, **superior articular process**, **lateral** and **median sacral crests**, and **sacral hiatus**.

## Module 7.14

**The thoracic cage protects organs in the chest and provides sites for muscle attachment    p. 264**

31. The **thoracic cage** consists of the thoracic vertebrae, the **ribs**, and the **sternum**.

32. Ribs 1–7 are called **vertebrosternal ribs** because they are connected to the sternum by **costal cartilages**. Ribs 8–10, or **vertebrochondral ribs**, connect to the sternum by shared costal cartilages. Ribs 11 and 12 are called **floating ribs** because they have no connection with the sternum.

33. The sternum has three distinct regions: the **manubrium**, **body**, and **xiphoid process**.

34. Ribs contain a **tubercle**, **capitulum**, **articular facets**, and a **costal groove**. Ribs assist with breathing by changing the volume of the thoracic cavity.

## SECTION 2 • Appendicular Skeleton

## Module 7.15

**The appendicular skeleton includes the limb bones and the pectoral and pelvic girdles    p. 267**

35. The **appendicular skeleton** contains 126 bones. It contains the pectoral girdle and upper limbs, and the pelvic girdle and lower limbs.

## Module 7.16

**The pectoral girdles—the clavicles and the scapulae—connect the upper limbs to the axial skeleton    p. 268**

36. The **clavicles** originate on the superior lateral border of the **manubrium** and articulate laterally with the acromion of the **scapula**.

37. The important landmarks for muscle attachment to the scapula are the **superior**, **medial**, and **lateral borders**, **superior**, **inferior**, and **lateral angles**, and **subscapular**, **supraspinous**, and **infraspinous fossae**. A prominent spine on the posterior scapula ends laterally as the **acromion**. The **glenoid cavity** articulates with the head of the humerus to form the shoulder joint.

## Module 7.17

**The humerus of the arm articulates with the radius and ulna of the forearm    p. 270**

38. The upper limbs consist of the bones of the arms, forearms, wrists, and hands.

39. The proximal **humerus** contains the **head**, **greater** and **lesser tubercles**, **intertubercular sulcus**, and the **anatomical** and **surgical necks**. The shaft has an elevation called the **deltoid tuberosity**, where the deltoid muscle attaches. The distal humerus has the **trochlea**, **capitulum**, **radial fossa**, and **coronoid fossa**.

40. The **ulna** and **radius** are the parallel bones of the forearm. The elbow is formed by the **olecranon** of the ulna. The **head of ulna** and **ulnar styloid process** are on the distal ulna. The **head of radius** is on the proximal aspect of the bone and articulates with the capitulum of the humerus. The **radial styloid process** helps stabilize the wrist.

## Module 7.18

**The wrist consists of carpal bones, and the hand consists of metacarpal bones and phalanges    p. 272**

41. Eight **carpal bones** in two rows of four form the wrist. The carpals of the proximal row are the **scaphoid**, **lunate**, **pisiform**, and **triquetrum**. The distal row of carpals is composed of the **trapezium**, **trapezoid**, **capitate**, and **hamate**.

42. Five **metacarpal bones** form the hand. Laterally to medially, they are identified by Roman numerals I–V.

43. Each hand has 14 finger bones, or **phalanges**. Each finger contains a proximal, middle, and distal phalanx. The thumb, or **pollex**, however, has only a proximal and distal phalanx.

## Module 7.19

**The hip bone forms by the fusion of the ilium, ischium, and pubis    p. 274**

44. The **pelvic girdle** consists of the paired **hip**, or **coxal**, **bones**. Each hip bone is formed by the fusion of the **ilium**, **ischium**, and **pubis**.

45. In lateral view, the dominant features of a hip bone are the **iliac crest**, the **iliac** and **ischial spines**, **gluteal lines**, **ischial**

tuberosity, and the socket, or **acetabulum**. Visible from a medial view are the **iliac fossa**, **arcuate line**, **pectineal line**, **pubic tubercle**, **obturator foramen**, **ischial ramus**, and the **superior** and **inferior pubic rami**.

### Module 7.20

**The pelvis consists of the two hip bones, the sacrum, and the coccyx    p. 276**

46. An extensive network of ligaments connects the sacrum with the posterior aspects of the two hip bones. The **pubic symphysis** is a pad of fibrocartilage that forms the anterior articulation between the two hip bones.

47. The **pelvis** can be divided into a **true** (lesser) and **false** (greater) **pelvis**.

### Module 7.21

**The adult male and female skeletons have significant differences    p. 277**

48. The sex of adult human skeletons can be determined most easily by examining the skull and pelvis.

49. Adaptations for childbearing cause the female pelvis to be shaped differently from the male pelvis.

### Module 7.22

**The femur, tibia, and patella meet at the knee    p. 278**

50. Each lower limb consists of a *femur* (thigh), a *patella* (kneecap), a *tibia* and *fibula* (leg), and the *tarsal bones*, *metatarsal bones*, and *phalanges* of the foot.

51. The proximal femur has a **head, fovea capitis, neck, greater** and **lesser trochanters**, intertrochanteric crest, and **intertrochanteric line**. The distal femur contains **medial** and **lateral condyles**, medial and lateral epicondyles, a **patellar surface**, and an **intercondylar fossa**. The **linea aspera** runs the length of the posterior shaft.

52. The **patella** is a large sesamoid bone that forms within the quadriceps femoris tendon.

53. The proximal tibia has **medial** and **lateral tibial condyles** and an **intercondylar eminence**. The **head** of the fibula articulates with the proximal tibia. The projections of the ankle are formed by the **medial malleolus** of the tibia and the **lateral malleolus** of the fibula.

### Module 7.23

**The ankle and foot consist of tarsal bones, metatarsal bones, and phalanges    p. 280**

54. The ankle consists of seven **tarsal bones:** the **calcaneus, talus, navicular, cuboid**, and the three **cuneiform bones**.

55. Five **metatarsal bones** form the foot. Medially to laterally, they are identified by Roman numerals I–V.

56. Each foot has 14 toe bones, or **phalanges**. Each toe contains a proximal, middle, and distal phalanx. The great toe, or **hallux**, however, has only a proximal and distal phalanx.

57. Arches assist with transferring the weight of the body to the feet. The **longitudinal arch** runs from the calcaneus to the distal metatarsals. The **transverse arch** runs perpendicular to the longitudinal arch.

## Chapter Review Questions

**Labeling:** Identify the cranial and facial bones in the diagrams below.

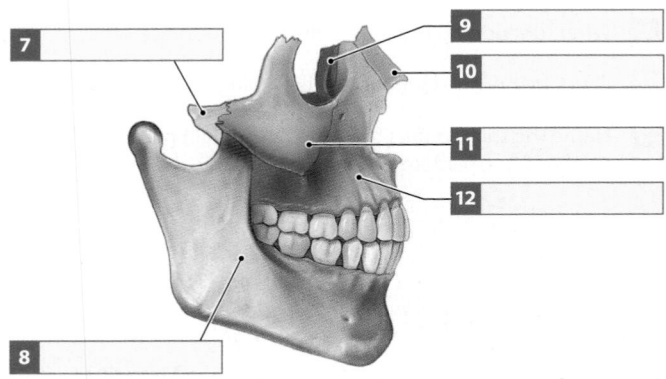

**Ture/False:** Indicate whether each statement is true or false.

13  The patella is a sesamoid bone.

14  The adductor tubercle is a small pit in the center of the head of the femur.

15  The lateral border of the scapula is also called the vertebral border.

16  The last two pairs of ribs are also known as floating ribs.

13  T

14  ___

15  ___

16  T

**Multiple choice:** Select the correct answer from the list provided.

**17** When you move your head as if to indicate no,
- ☐ a) the atlas rotates on the occipital condyles.
- ☐ b) $C_1$ and $C_2$ rotate on the adjoining cervical vertebrae.
- ☒ c) the atlas rotates on the dens of the axis.
- ☐ d) the skull rotates with both $C_1$ and $C_2$.

**18** Which of the following pairs of bones make up the bony nasal septum?
- ☐ a) inferior nasal conchae and vomer
- ☐ b) perpendicular plate of ethmoid and inferior nasal conchae
- ☒ c) vomer and perpendicular plate of ethmoid
- ☐ d) inferior nasal conchae and middle nasal conchae of ethmoid

**19** The unpaired facial bones include the
- ☐ a) lacrimal and nasal.
- ☒ b) vomer and mandible.
- ☐ c) maxilla and mandible.
- ☐ d) zygomatic and palatine.

**20** The joint between the frontal and parietal bones is correctly called the
- ☐ a) parietal suture.
- ☐ b) lambdoid suture.
- ☐ c) squamous suture.
- ☒ d) coronal suture.

**21** Which part of the ulna forms the point of the elbow?
- ☐ a) styloid process
- ☒ b) olecranon
- ☐ c) coronoid process
- ☐ d) trochlear notch

**22** Which of the following bones is *not* part of the orbital complex?
- ☐ a) lacrimal
- ☒ b) nasal
- ☐ c) ethmoid
- ☐ d) sphenoid

**23** The head of the femur articulates with the pelvis at the
- ☐ a) true pelvis.
- ☒ b) acetabulum.
- ☐ c) obturator foramen.
- ☐ d) pubic symphysis.

**24** What is the name of the flexible sheet that interconnects the radius and ulna (and the tibia and fibula)?
- ☒ a) interosseous membrane
- ☐ b) obturator foramen
- ☐ c) linea aspera
- ☐ d) intercondylar eminence

## Short answer

**25** What function do the fontanelles serve during birth?

**26** Describe how ribs function in breathing.

**27** Name the three bones that fuse to form a hip bone. Where do these bones meet?

**28** Distinguish between the primary and secondary curves of the vertebral column.

**29** Describe the arches of the feet.

# MasteringA&P ™

Access more chapter study tools online in the MasteringA&P Study Area:

- Chapter Quizzes, Chapter Practice Test, and Clinical Case Studies

- Practice Anatomy Lab™ 3.0    PAL˙3.0
- Interactive Physiology™    iP2˙
- A&P Flix™    A&PFlix™
- PhysioEx™ 9.1    PhysioEx˙ 9.1

# Chapter Integration • Applying what you have learned

## An illegal check from behind leaves a college hockey player with multiple injuries

Donny plays right wing on his college hockey team. He grew up in a hockey family in Peterborough, Canada. His dad played Juniors and had hopes of an NHL career, until marriage and children caused him to give up his dream. But he put skates on Donny as soon as Donny could walk and taught him the game. After high school, Donny followed in his father's footsteps by choosing to play college hockey. He, too, hopes to have a career in professional hockey.

During a collegiate hockey game against his archrivals, Donny was handling the puck well and outskating the defense, when he was checked from behind. (Checking is a maneuver in which an opponent is neutralized through the proper use of one's body or stick; checking from behind is illegal.) While lying on the ice being examined by the team's athletic trainer, Donny was experiencing chest and shoulder pain. His teammates helped him to the locker room, where a thorough physical exam revealed a badly bruised chest; dyspnea (difficulty breathing); and immobility of his left arm, especially in the anterior direction. X-rays subsequently taken at a nearby hospital revealed two fractured ribs.

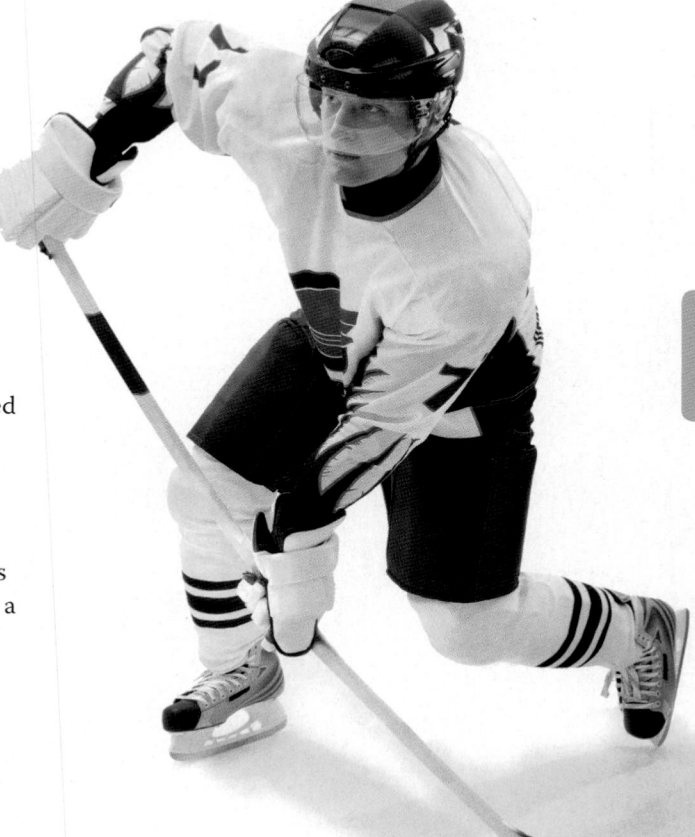

**1** Why might Donny be experiencing difficulty breathing?

**2** What is the probable reason for Donny's inability to move his left arm?

**3** State whether Donny's injuries are related to the axial skeleton or to the appendicular skeleton.

## Examining a baby's cleft palate

There are occasions when certain sutures of the skull do not form properly in fetal development. An example of this is cleft palate, in which a permanent gap remains between the bones of the hard palate. From what you have just learned about the bones of the skull, answer the following questions.

**4** In cleft palate, which bones do not properly form into a suture? How many bones are involved?

**5** If you were to look into the mouth of a baby with cleft palate, what would you expect to see?

**6** What kinds of difficulties do you think a baby with cleft palate might have?

# 8

# Joints

# Module 8.1

# Joints are classified according to structure and movement

The bones of the skeleton are relatively inflexible, so movements can occur only at **joints**, or **articulations**, where two bones interconnect.

**1** The anatomical structure of a joint determines the type and amount of movement the joint can execute. Each joint reflects a compromise between the need for strength and the need for mobility. The amount of movement at a joint is known as **range of motion (ROM)**. Joints are often categorized by their ROM and are further divided into subgroups based on anatomical structure.

**? REVIEW**
**A.** Define *range of motion (ROM)*.

## Functional and Structural Classifications of Joints

| Functional Category | Structural Category and Type | | Description |
|---|---|---|---|
| **Synarthrosis** (no movement)<br><br>At a synarthrosis (sin-ar-THRŌ-sis; *syn-*, together + *arthrosis*, joint), the bony edges are close together and may interlock. These extremely strong joints are located where movement between the bones must be prevented. | **Fibrous**<br>- Suture | | A **suture** (*sutura*, a sewing together) is a synarthrotic joint located only between the bones of the skull. The edges of the bones are interlocked and bound together at the suture by dense fibrous connective tissue. |
| | - Gomphosis | | A **gomphosis** (gom-FŌ-sis; *gomphos*, bolt) is a synarthrosis that binds the teeth to bony sockets in the maxillae and mandible. The fibrous connection between a tooth and its socket is a periodontal (per-ē-ō-DON-tal) ligament (*peri*, around + *odontos*, tooth). |
| | **Cartilaginous**<br>- Synchondrosis | | A **synchondrosis** (sin-kon-DRŌ-sis; *syn*, together + *chondros*, cartilage) is a rigid, cartilaginous bridge between two articulating bones. The cartilaginous connection between the ends of the first pair of ribs and the sternum is a synchondrosis. |
| | **Bony**<br>- Synostosis | | A **synostosis** (sin-os-TŌ-sis) is a totally rigid, immovable joint formed when two bones fuse and the boundary between them disappears. The frontal suture of the frontal bone and the epiphyseal lines of mature long bones are synostoses. |
| **Amphiarthrosis** (little movement)<br><br>An amphiarthrosis (am-fē-ar-THRŌ-sis; *amphi-*, on both sides) permits more movement than a synarthrosis but is much stronger than a freely movable joint. The articulating bones are connected by collagen fibers or cartilage. | **Fibrous**<br>- Syndesmosis | | At a **syndesmosis** (sin-dez-MŌ-sis; *desmos*, a band or ligament), bones are connected by a ligament. One example is the distal joint between the tibia and fibula. |
| | **Cartilaginous**<br>- Symphysis | | At a **symphysis**, the articulating bones are separated by a wedge or pad of fibrocartilage. The joint between the two pubic bones (the pubic symphysis) is an example of a symphysis. |
| **Diarthrosis** (free movement)<br><br>A diarthrosis (dī-ar-THRŌ-sis; *dia-*, through) permits the widest range of movement. | **Synovial** | | **Synovial** (si-NŌ-vē-ul) **joints** have the greatest range of motion compared to other joints. They are typically located at the ends of long bones, such as those of the upper and lower limbs. |

8

 **REVIEW**
**B.** Which structural category of joints allows for the greatest range of motion?

 **LEARNING OUTCOME**
Contrast the major categories of joints, and explain the relationship between structure and function for each category.

# Module 8.2

# Synovial joints are freely movable and lined with a synovial membrane

In this module we review the general structure of a synovial joint, which was introduced in our earlier discussion of synovial membranes (see Module 4.15). We will focus on the key structures of synovial joints.

**1** Under normal conditions, the opposing bony surfaces within a synovial joint cannot contact one another because these surfaces are covered by special *articular* cartilages, which are slick and smooth. The articular cartilages alone can reduce friction during movement at the joint. However, even when pressure is applied across a joint, the smooth articular cartilages do not touch one another, because they are separated by a thin film of synovial fluid within the joint cavity.

**? REVIEW**

**A.** Describe the key components of a synovial joint, and identify their functions.

## Synovial Joint Components

**Articular cartilage** covers bones at a joint. It is like hyaline cartilage, but it has no perichondrium and its matrix contains more water than other cartilages.

The **joint capsule**, or **articular capsule**, is a sac that encloses the articulating ends of the bones in the joint. It is made up of an outer fibrous layer and an inner synovial membrane. It may be reinforced with accessory structures such as tendons or ligaments.

The **synovial membrane** lines the interior of the joint capsule and secretes synovial fluid into the joint cavity.

The **joint cavity** contains synovial fluid. This fluid lubricates, cushions shocks, prevents abrasion, and supports the chondrocytes of the articular cartilages. Even in a large joint such as the knee, the total volume of synovial fluid in a joint is normally less than 3 mL.

Medullary cavity

The periosteum of each bone is continuous with the capsule of the joint. This integration adds strength and mobility to the joint.

Synovial fluid

Ligament

Spongy bone of epiphysis

Compact bone

**2** This table details the major functions of synovial fluid. It is a clear, straw-colored, viscous fluid with the consistency of raw egg white. Its normal viscosity is due to a high concentration of hyaluronic acid (see Module 4.4). Synovial fluid originates in the areolar tissue of the synovial membrane. During normal movement, synovial fluid circulates from the areolar tissue into the joint cavity and percolates through the articular cartilage, providing oxygen and nutrients to the chondrocytes and carrying away their metabolic wastes.

## Synovial Fluid Functions

- **Lubrication**. When part of an articular cartilage is compressed during movement, some of the synovial fluid squeezes out of the cartilage and into the space between the opposing surfaces. This thin layer of fluid markedly reduces friction between moving surfaces, just as a thin film of water reduces friction between a car's tires and a highway.

- **Nutrient Distribution**. Synovial fluid in a joint must circulate continuously to provide nutrients and dispose of waste for the chondrocytes in the articular cartilage. It circulates whenever the joint moves, and compression and reexpansion of the articular cartilage pump synovial fluid into and out of the cartilage matrix.

- **Shock Absorption**. Synovial fluid cushions shocks in joints that are subjected to sudden compression or impact. This cushioning effect occurs because the viscosity of synovial fluid increases with increasing pressure. For example, when you jog, your knees are severely compressed, and the viscosity of the synovial fluid increases. As the pressure lessens, viscosity decreases, and synovial fluid is again a lubricant.

**3** In complex synovial joints, such as the knee, a variety of accessory structures provide support and additional stability. Several of these structures are shown in this diagram of a sagittal section of the knee.

The tendon of the quadriceps muscles attaches to the base of the patella. Although not part of the joint itself, tendons passing across or around a joint can limit the joint's range of motion and provide mechanical support for it.

Patella

Synovial membrane

Joint capsule

Joint cavity

Articular cartilage

Femur

Tibia

### Accessory Structures Supporting the Knee

A **bursa** (BUR-sa; a pouch; plural, *bursae*) is a small, thin, fluid-filled pocket that forms in connective tissue outside a joint capsule. It contains synovial fluid and is lined by a synovial membrane. Bursae often form where a tendon or ligament rubs against other tissues. Located around most synovial joints, including the knee joint, bursae reduce friction and act as shock absorbers.

**Fat pads** are localized masses of adipose tissue covered by a layer of synovial membrane. They are commonly superficial to the joint capsule. Fat pads protect the articular cartilage and act as packing material for the joint. When the bones move, the fat pads fill in the spaces created as the joint cavity changes shape.

A **meniscus** (me-NIS-kus; a crescent; plural, *menisci*) is a pad of fibrocartilage between opposing bones within a synovial joint. Menisci are like articular discs that subdivide a synovial cavity, channel the flow of synovial fluid, or allow for variations in the shapes of the articular surfaces.

**Accessory ligaments** support, strengthen, and reinforce synovial joints. **Capsular ligaments**, or *intrinsic ligaments*, are localized thickenings of the joint capsule. **Extrinsic ligaments** are separate from the joint capsule. Extrinsic ligaments may pass outside or inside the joint capsule, and are called *extracapsular* or *intracapsular ligaments*, respectively.

The **patellar ligament** extends from the apex of the patella to the tibial tuberosity. This is an example of an **extracapsular ligament**.

The **cruciate** (KRŪ-shē-āt) **ligaments** that run through the interior of the knee joint are examples of **intracapsular ligaments**.

---

### Clinical Note

### Excessive force applied to a joint can cause dislocation

A joint cannot be both highly mobile and very strong. The greater the range of motion at a joint, the weaker it becomes. A synarthrosis, the strongest type of joint, has no movement. In contrast, a diarthrosis, such as the shoulder joint, is far weaker but has a broad range of motion. Any diarthrosis will be damaged by movement beyond its normal range of motion. When reinforcing structures cannot protect a joint from extreme stresses, a **dislocation**, or **luxation** (luk-SĀ-shun), results. In a dislocation, the articulating surfaces are forced out of position. The displacement can damage the articular cartilage, tear ligaments, or distort the joint capsule. Although the inside of a joint has no pain receptors, sensitive nerves monitor the joint capsule, ligaments, and tendons, so dislocations are very painful.

**? REVIEW**

**B.** Describe the accessory structures of complex synovial joints and identify their functions.

**? INTEGRATION**

**C.** Why would improper circulation of synovial fluid cause degeneration of articular cartilages in the affected joint?

**LO LEARNING OUTCOME**

Describe the basic structure of a synovial joint, and describe common accessory structures and their functions.

# Module 8.3

# Anatomical organization determines the motion at synovial joints

An accurate description of the functions of a joint includes terms that indicate the types of motion permitted. To show the general types of movement, we will use a simple model that you can try for yourself.

**1** Take a pencil and stand it upright on the surface of a desk. The pencil represents a bone, and the desktop represents an articular surface. A little imagination and a lot of twisting, pushing, and pulling will demonstrate that there are only three ways to move the pencil. Considering them one at a time gives a frame of reference for us to analyze complex movements.

**2** If you hold the pencil upright, without securing the point, you can push the pencil point across the surface. This kind of motion, **gliding**, is an example of linear motion. You could slide the point forward or backward, from side to side, or diagonally. However you move the pencil, the motion can be described by using two lines of reference (*axes*). One line represents forward–backward motion, the other left–right movement.

**3** If you hold the point in position, you can move the free (eraser) end of the pencil forward and backward, from side to side, or at some intermediate angle. These movements, which change the angle between the pencil shaft and the desktop, are examples of **angular motion**.

**4** A special term is used to describe a complex angular movement. Grasp the pencil eraser, and move the pencil in any direction until it is no longer vertical. Now swing the eraser through a complete circle. This movement, which corresponds to the path of your arm when you draw a large circle, is very difficult to describe. Anatomists avoid the problem by using the term **circumduction** (sir-kum-DUK-shun; *circum,* around) for this type of angular motion.

**5** If you keep the shaft vertical and the point at one location, you can still spin the pencil around its longitudinal axis. This movement is called **rotation**. Several joints permit partial rotation, but none can rotate freely without limit because such a movement would hopelessly tangle the blood vessels, nerves, and muscles that cross the joint.

Gliding

Superior–inferior axis

Lateral–medial axis

Anterior–posterior axis

**6** Movement of joints can also be described by the number of axes that they can rotate around. A joint that permits movement around one axis is called **monaxial**, a joint that permits movement around two axes is called **biaxial**, and one that permits movement around three axes is called **triaxial**.

Angular motion

Circumduction

Rotation

**? REVIEW**

**A.** Describe the types of motion possible at a synovial joint.

**7** This visual summary gives representative examples of the various anatomical classes of synovial joints based on the shapes of the articulating surfaces. It relates articular structure to simplified joint models and lists examples of each group of joints.

| Types of Synovial Joints | Models of Joint Motion | Examples |
|---|---|---|
| **Plane joint**<br><br>Plane joints, or gliding joints, have flattened or slightly curved surfaces that slide across one another, but the movement is very slight.<br><br>*Clavicle*, *Manubrium* | **Movement:** Gliding, slight monaxia | • Acromioclavicular and sternoclavicular joints<br>• Intercarpal and intertarsal joints<br>• Vertebrocostal joints<br>• Sacro-iliac joints |
| **Hinge joint**<br><br>Hinge joints permit angular motion in a single plane, like the opening and closing of a door.<br><br>*Humerus*, *Ulna* | **Movement:** Angular, monaxial | • Elbow joints<br>• Knee joints<br>• Ankle joints<br>• Interphalangeal joints |
| **Pivot joint**<br><br>Pivot joints permit only rotation.<br><br>*Atlas*, *Axis* | **Movement:** Rotation, monaxial | • Atlanto-axial joint<br>• Proximal radio-ulnar joints |
| **Condylar joint**<br><br>Condylar joints have an oval articular face nestled within a depression on the opposing surface.<br><br>*Scaphoid bone*, *Radius*, *Ulna* | **Movement:** Angular, biaxial | • Radiocarpal joints<br>• Metacarpophalangeal joints II–V<br>• Metatarsophalangeal joints |
| **Saddle joint**<br><br>Saddle joints have articular faces that fit together like a rider in a saddle. Each face is concave along one one axis and convex along the other.<br><br>*II*, *Metacarpal bone of thumb*, *Trapezium* | **Movement:** Angular, biaxial | • First carpometacarpal joints |
| **Ball-and-socket joint**<br><br>In a ball-and-socket joint, the round head of one bone rests within a cup-shaped depression in another.<br><br>*Scapula*, *Humerus* | **Movement:** Angular, circumduction, and rotation; triaxial | • Shoulder joints<br>• Hip joints |

**?** **REVIEW**

**B.** Identify the types of synovial joints.

**C.** Which type of synovial joint permits the greatest range of motion?

**8**

**?** **INTEGRATION**

**D.** Name the type of synovial joint for each of the following: shoulder, elbow, ankle, and thumb.

 **LEARNING OUTCOME**

Describe how the anatomical and functional properties of synovial joints permit movements of the skeleton.

# Module 8.4

# Specific terms are used to describe movements with reference to the anatomical position

It is easiest to understand the terms used to describe body movements when you see the actions under way. You should become very familiar with the descriptive terms presented here and in Module 8.5 because we will use them when we consider both the functions of specific joints and the actions of skeletal muscles. (The red dots you see in this module and the next indicate the locations of the joints involved in the illustrated movements.)

## Flexion and Extension

**1** *Flexion* and *extension* usually describe the movements at the hinge joints of the long bones of the limbs. However, they are also used to describe movements of the axial skeleton. For example, when you bring your head toward your chest, you flex the intervertebral joints of the neck.

**Lateral flexion** occurs when your vertebral column bends to the side. This movement is most pronounced in the cervical and thoracic regions.

**Flexion** (FLEK-shun) is movement in the anterior–posterior (sagittal) plane that decreases the angle between the articulating bones. Here you see flexion at the neck, the elbow, and the hip.

**Extension** occurs in the same plane as flexion, but it increases the angle between the articulating bones. When a person is in the anatomical position, all of the major joints of the axial and appendicular skeletons (except the ankle) are at full extension.

Extension past the anatomical position is called **hyperextension**.

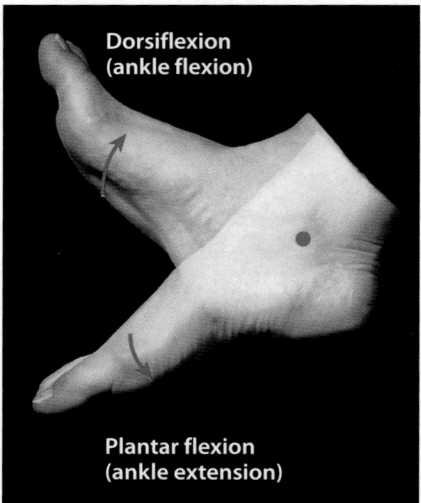

**Dorsiflexion** is upward movement of the foot or toes, as when you dig in your heel. **Plantar flexion** (*planta*, sole), the opposite movement, extends the ankle joint and bends the foot or toes, as when you stand on tiptoe. It is also acceptable (and simpler) to use "flexion and extension at the ankle," rather than "dorsiflexion and plantar flexion."

**? REVIEW**

**B.** Compare dorsiflexion to plantar flexion.

**? REVIEW**

**A.** Which movements are possible at hinge joints?

## Abduction and Adduction

**2** *Abduction* and *adduction* always refer to the movements of the appendicular skeleton, not to those of the axial skeleton.

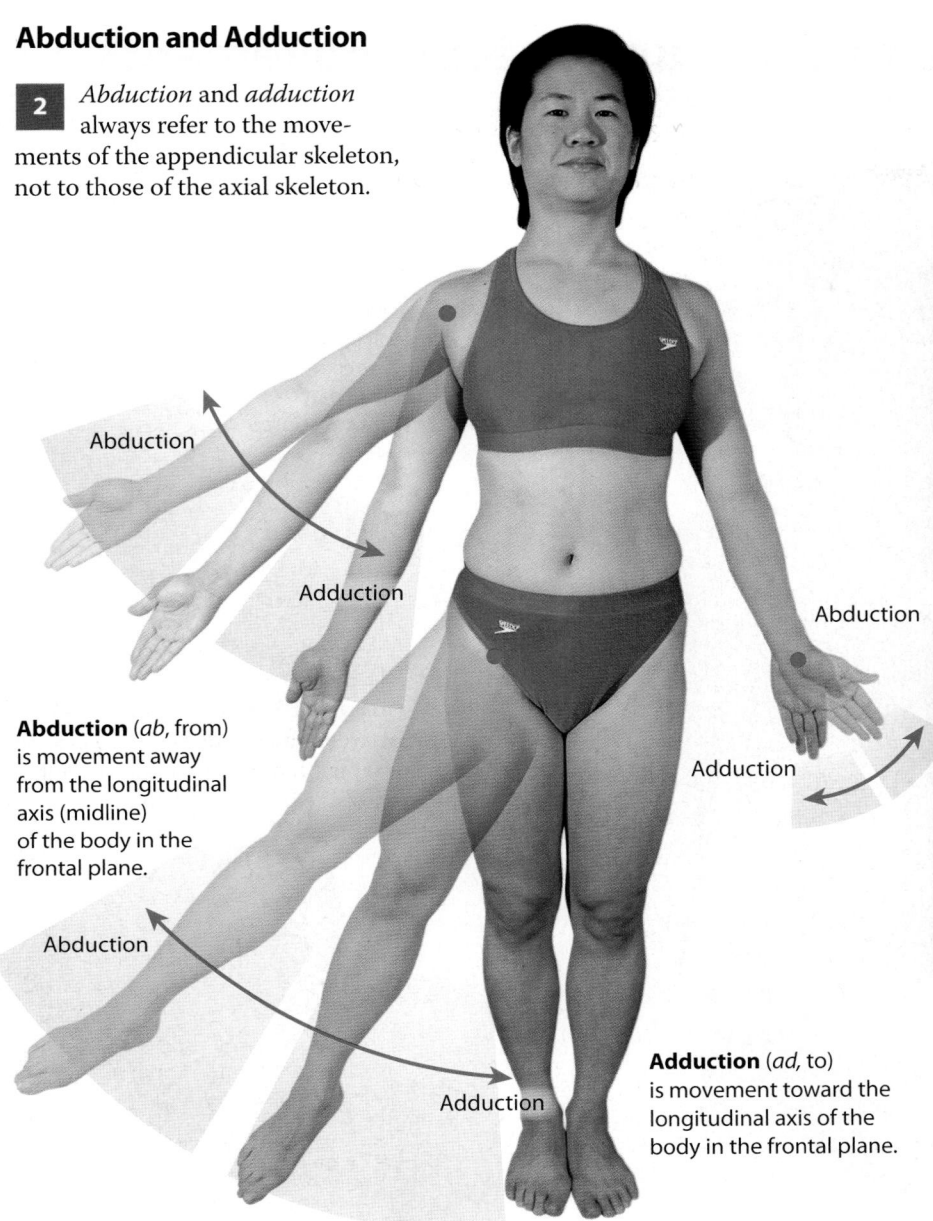

**Abduction** (*ab*, from) is movement away from the longitudinal axis (midline) of the body in the frontal plane.

Abduction

Adduction

Abduction

Abduction

Adduction

Abduction

Adduction

Adduction

**Adduction** (*ad*, to) is movement toward the longitudinal axis of the body in the frontal plane.

Spreading the fingers or toes apart abducts them, because they move away from a central digit. Bringing them together constitutes adduction. (Fingers move toward or away from the middle finger; toes move toward or away from the second toe.)

Central digit

Adduction

Abduction

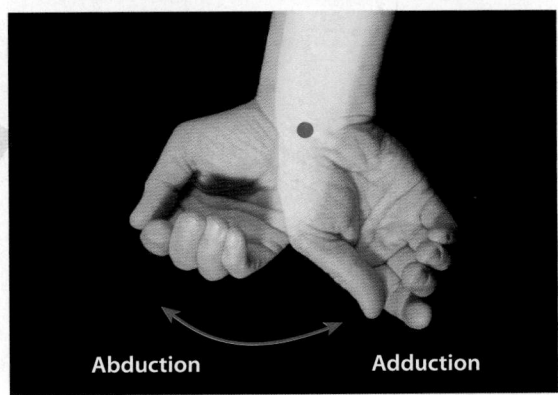

Abduction

Adduction

**? REVIEW**

**C.** When a person does jumping jacks, which limb movements are necessary?

## Circumduction

**3** Moving your arm as if to draw a big circle on the wall is **circumduction**. In this movement your hand moves in a circle, but your arm does not rotate.

**Lo LEARNING OUTCOME**

Describe flexion/extension, abduction/adduction, and circumduction movements of the skeleton.

# Specific terms describe rotation and special movements

## Rotation

**1** Head and trunk rotation is described as *right* or *left rotation*. Limb rotation can be described as *medial* or *lateral rotation*.

Right rotation

Left rotation

Atlanto-axial joint

Rotation is also described with reference to a figure in the anatomical position. Rotation of the head, which occurs around the atlanto-axial joint, may involve **left rotation** or **right rotation**.

Lateral (external) rotation

Medial (internal) rotation

**2** Special terms are used to describe forearm rotation. The proximal joint between the radius and the ulna (the proximal radio-ulnar joint) permits rotation of the head of the radius.

Limb rotation is the turning of a limb about its axis of rotation. During **medial rotation**, also known as *internal rotation* or *inward rotation*, the anterior surface of a limb turns toward the midline of the body. The reverse movement is called **lateral rotation**, *external rotation*, or *outward rotation*.

Proximal radio-ulnar joint

Distal radio-ulnar joint

As the shaft of the radius rotates, the distal epiphysis of the radius rolls across the anterior surface of the ulna. This movement, called **pronation** (prō-NĀ-shun), turns the wrist and hand from palm facing front to palm facing back (posteriorly). The opposing movement, in which the palm is turned anteriorly, is **supination** (sū-pi-NĀ-shun).

Supination

Pronation

**? REVIEW**

**A.** What movements are made possible by the rotation of the head of the radius?

# Special Movements

**3** Several specific terms apply to specific joints or unusual types of movement.

**Opposition**

**Opposition** is movement of the thumb toward the surface of the palm or the pads of other fingers. Opposition enables you to grasp and hold objects between your thumb and palm. It involves movement at the first carpo-metacarpal and metacarpophalangeal joints. Flexion at the fifth metacarpophalangeal joint can assist this movement.

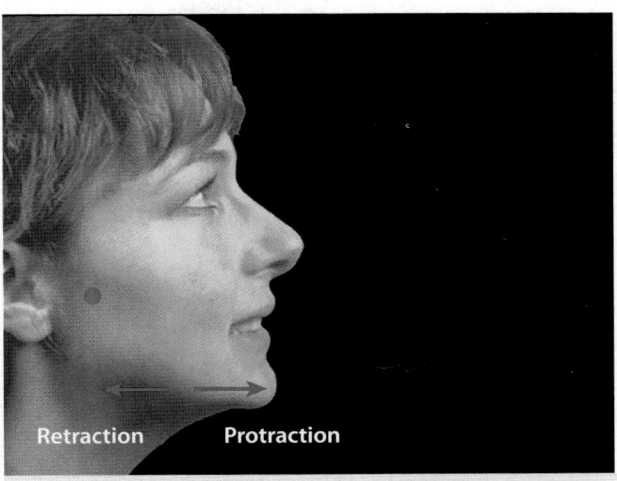

**Retraction** **Protraction**

**Protraction** is moving a part of the body anteriorly in the horizontal plane. **Retraction** is the reverse movement. You protract your jaw when you jut it forward, and you retract your jaw when you return it to its normal position. Although not shown here, you protract your clavicles when you cross your arms.

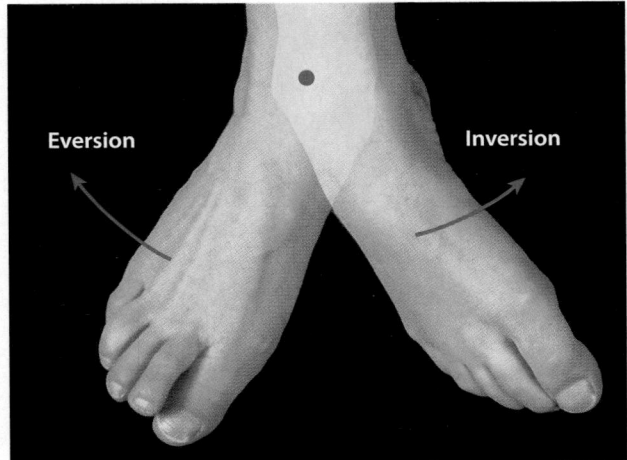

**Eversion** **Inversion**

**Inversion** (*in*, into + *vertere*, to turn) is a twisting motion of the foot that turns the sole inward, elevating the medial edge of the sole. The opposite movement is called **eversion** (ē-VER-zhun).

**Depression** **Elevation**

**Depression** takes place when a body part moves inferiorly. **Elevation** is lifting a body structure superiorly. You depress your mandible when you open your mouth, and you elevate your mandible as you close your mouth.

**? REVIEW**

**B.** Snapping your fingers involves what movement of the thumb?

**? INTEGRATION**

**C.** What hand movements occur when a person wriggles into tight-fitting gloves?

 **LEARNING OUTCOME**

Describe rotational and special movements of the skeleton.

# Joint Structure and Movement

**Labeling:** Label the structures in this diagram of a synovial joint.

| 1 | medullary cavity |
| 2 | spongy of epiphysis |
| 3 | Periosteum |
| 4 | synovial membrane |
| 5 | Articular cartilage |
| 6 | synovial fluid |
| 7 | Joint cavity |
| 8 | compact |

Identify each of the following movements.

| 9 | | 10 | | 11 | | 12 | | 13 | | 14 | | 15 | | 16 | | 17 | | 18 | | 19 | | 20 |

**Matching:** Match each lettered term with the most closely related description.

a. amphiarthrosis ✓
b. synarthrosis ✓
c. dislocation ✓
d. pronation/supination
e. diarthrosis ✓
f. shoulder ✓
g. articular discs ✓
h. fluid-filled pocket ✓

| 21 | Freely movable joint |
| 22 | Movements of forearm bones |
| 23 | Ball-and-socket joint |
| 24 | Menisci |
| 25 | Immovable joint |
| 26 | Luxation |
| 27 | Bursa |
| 28 | Slightly movable joint |

| 21 | E |
| 22 | D |
| 23 | F |
| 24 | G |
| 25 | B |
| 26 | C |
| 27 | H |
| 28 | A |

# Module 8.6

# Axial joints have less range of motion than appendicular joints

**1** Most joints in the axial skeleton are strong joints that permit very little movement. This section begins by examining the structure of the intervertebral joints, which permit limited but important movements of the vertebral column. Joints are typically named from the bones involved.

**2** A typical joint in the appendicular skeleton has an extensive range of motion. Because a joint cannot be both strong and highly mobile, these joints are often weaker than those of the axial skeleton. We will examine two joints in each limb: the shoulder and elbow in the upper limb, and the hip and knee in the lower limb.

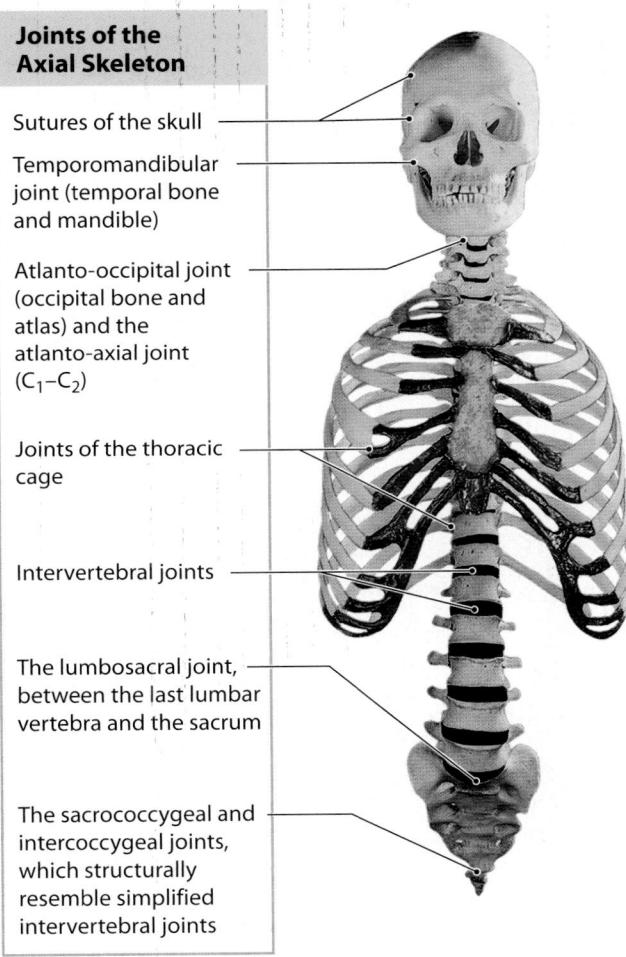

### Joints of the Axial Skeleton

Sutures of the skull

Temporomandibular joint (temporal bone and mandible)

Atlanto-occipital joint (occipital bone and atlas) and the atlanto-axial joint ($C_1$–$C_2$)

Joints of the thoracic cage

Intervertebral joints

The lumbosacral joint, between the last lumbar vertebra and the sacrum

The sacrococcygeal and intercoccygeal joints, which structurally resemble simplified intervertebral joints

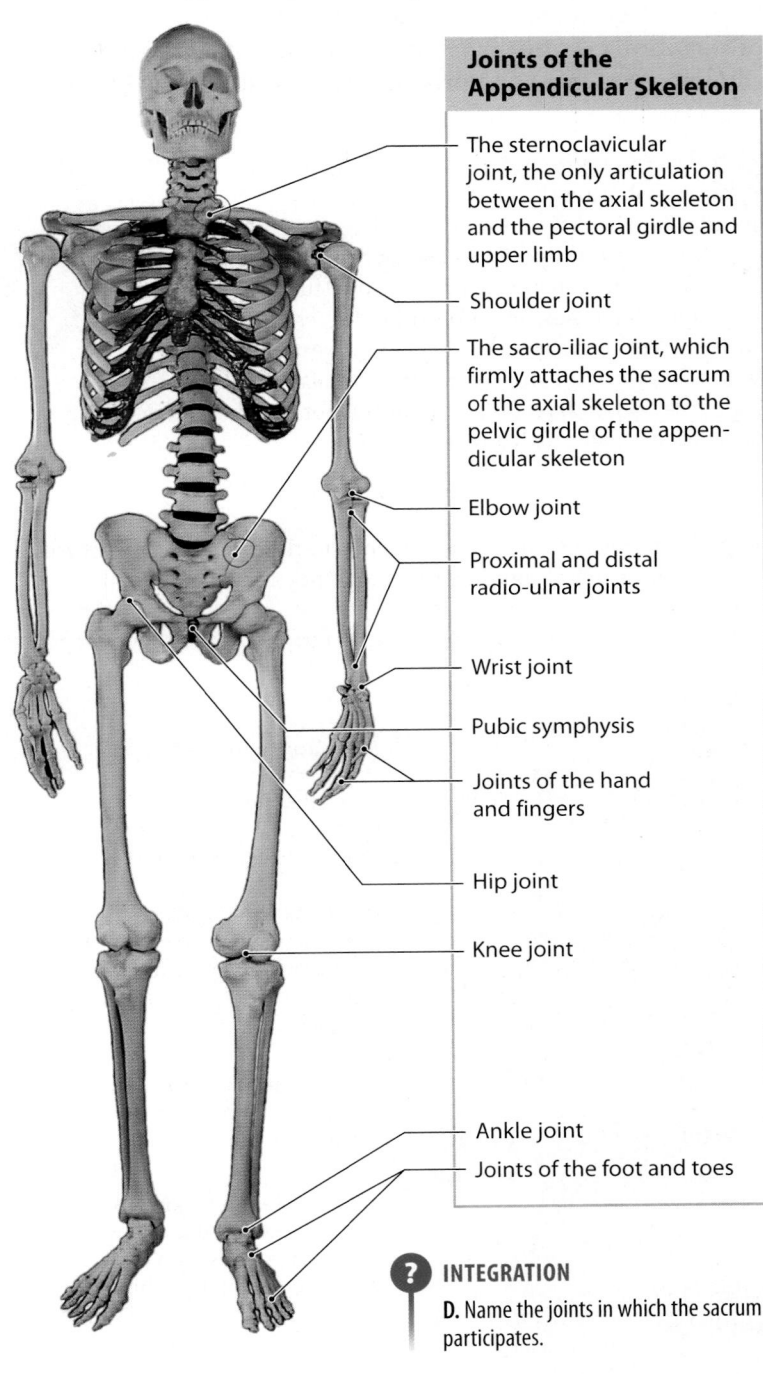

### Joints of the Appendicular Skeleton

The sternoclavicular joint, the only articulation between the axial skeleton and the pectoral girdle and upper limb

Shoulder joint

The sacro-iliac joint, which firmly attaches the sacrum of the axial skeleton to the pelvic girdle of the appendicular skeleton

Elbow joint

Proximal and distal radio-ulnar joints

Wrist joint

Pubic symphysis

Joints of the hand and fingers

Hip joint

Knee joint

Ankle joint

Joints of the foot and toes

**? REVIEW**

**A.** Describe the relationship between joint strength and mobility.

**B.** Which division of the skeleton has the greater range of motion?

**C.** Which joint attaches the pectoral girdle and upper limb to the axial skeleton?

**? INTEGRATION**

**D.** Name the joints in which the sacrum participates.

**LO LEARNING OUTCOME**

Compare the general relationship between joint stability and range of motion for axial and appendicular joints.

# The vertebral column includes three types of joints

There are three types of joints within the vertebral column: syndesmoses, synchondroses, and synovial. The fibrous joints uniting the vertebrae, which includes the vertebral ligaments, make up the **syndesmoses of vertebral column**. The intervertebral joints uniting the vertebral bodies with fibrocartilage, forming the **intervertebral discs**, make up the **synchondroses of vertebral column**. The numerous joints between the bony processes of adjoining vertebrae comprise the **vertebral synovial joints**.

**1** The intervertebral discs make a significant contribution to a person's height: they account for roughly one-fourth the length of the vertebral column superior to the sacrum. As we grow older, the water content of the nucleus pulposus (see right) decreases, and it becomes less effective as a cushion. This increases the chances for vertebral injury. Water loss from the intervertebral discs also causes shortening of the vertebral column, accounting for the characteristic decrease in height with advancing age.

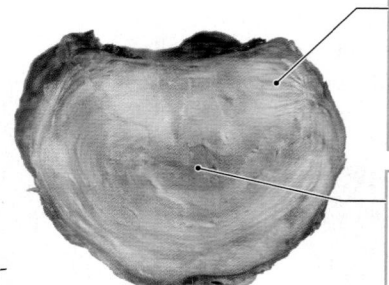

**Superior view**

Each intervertebral disc has a tough outer ring of fibrocartilage, the **anulus fibrosus** (AN-ū-lus fi-BRŌ-sus). The collagen fibers of this layer attach the disc to the bodies of adjacent vertebrae.

The anulus fibrosus surrounds the **nucleus pulposus** (pul-PŌ-sus), a soft, elastic, gelatinous core. The nucleus pulposus gives the disc resilience and enables it to absorb shocks.

**2** Numerous ligaments are attached to the bodies and processes of all vertebrae, binding them together and stabilizing the vertebral column. Ligaments making up the syndesmoses of the vertebral column are identified in the anterior view (left) and lateral section view (right).

**? REVIEW**

**A.** Describe the nucleus pulposus and anulus fibrosus of an intervertebral disc.

**B.** Name the primary vertebral ligaments.

### Vertebral Ligaments

Paired **ligamenta flava** (singular, *ligamentum flavum*) connect the laminae of adjacent vertebrae.

The **posterior longitudinal ligament** parallels the anterior longitudinal ligament and connects the posterior surfaces of adjacent vertebral bodies.

The **interspinous ligament** connects the spinous processes of adjacent vertebrae.

The **supraspinous ligament** interconnects the tips of the spinous processes from the sacrum to vertebra C$_7$. The ligamentum nuchae extends from vertebra C$_7$ to the base of the skull.

The **anterior longitudinal ligament** connects the anterior surfaces of adjacent vertebral bodies.

### Intervertebral Disc

- Anulus fibrosus
- Nucleus pulposus

Spinal cord

Spinal nerve

Posterior longitudinal ligament

**Lo LEARNING OUTCOME**

Describe the joints between the vertebrae of the vertebral column.

# Module 8.8
# Intervertebral disc disease and osteoporosis are common age-related health problems

**1** **Intervertebral disc disease** (**IVDD**) is a painful condition associated with disc degeneration that affects the spinal nerves. Causes include a bulging disc or herniated disc. If the posterior longitudinal ligaments weaken, as often occurs with age, the compressed nucleus pulposus may distort the anulus fibrosus, forcing it partway into the vertebral canal. This condition, seen here in lateral view, is called a **bulging disc**.

Normal intervertebral disc

Bulging disc

T₁₂

L₁

L₂

**?** **REVIEW**
**A.** Compare a bulging disc with a herniated disc.

**2** If the nucleus pulposus breaks through the anulus fibrosus, it too may protrude into the vertebral canal. This condition, shown here in superior view, is called a **herniated disc**. When a disc herniates, it compresses spinal nerves. which may cause pain and nerve damage that may even lead to paralysis.

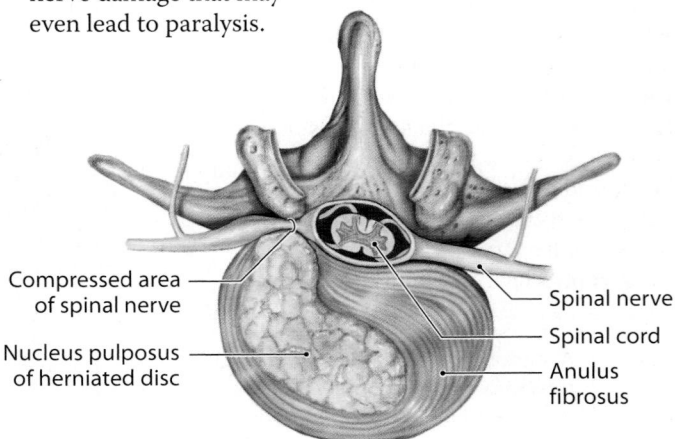

Compressed area of spinal nerve

Nucleus pulposus of herniated disc

Spinal nerve

Spinal cord

Anulus fibrosus

**3** The bones of the skeleton become thinner and weaker as a normal part of the aging process. Inadequate ossification is called **osteopenia** (os-tē-ō-PĒ-nē-uh; *penia*, lacking). This reduction in bone mass begins between the ages of 30 and 40 as osteoblast activity begins to decline, while osteoclast activity continues at previous levels. Thereafter, women lose roughly 8 percent of their skeletal mass every decade; men lose roughly 3 percent per decade. When the reduction in bone mass is sufficient to compromise normal function, the condition is known as **osteoporosis** (os-tē-ō-po-RŌ-sis; *porosus*, porous). The combination of osteopenia or osteoporosis and the reduced cushioning properties of the intervertebral discs makes vertebral fractures a common problem among the elderly.

**Clinical scan of a compression fracture in a lumbar vertebra**

**Normal spongy bone** SEM × 25

**Spongy bone with osteoporosis** SEM × 21

**?** **INTEGRATION**
**B.** What common age-related factors contribute to vertebral fractures in the elderly?

**Lo** **LEARNING OUTCOME**
Describe intervertebral disc disease and osteoporosis.

# The shoulder and hip are ball-and-socket joints

## Shoulder Joint

The **shoulder joint**, or **glenohumeral joint**, has the greatest range of motion of any joint. Because it is also the most frequently dislocated joint, it is an excellent example to show how stability is sacrificed for mobility. This joint is a ball-and-socket diarthrosis formed by the articulation of the head of the humerus with the glenoid cavity of the scapula. Another important joint in this area is the **acromioclavicular joint**, which is where the clavicle articulates with the scapula (see Module 7.16).

**1** Five major ligaments help stabilize the shoulder joint. As at other joints, bursae at the shoulder reduce friction where large muscles and tendons pass across the joint capsule. All or some of these ligaments may be injured in a shoulder dislocation (affecting the glenohumeral joint) or a shoulder separation (involving the acromioclavicular joint).

A tendon of the biceps brachii muscle runs through the shoulder joint. As it passes through the articular capsule, it is surrounded by a tubular bursa that is continuous with the joint cavity.

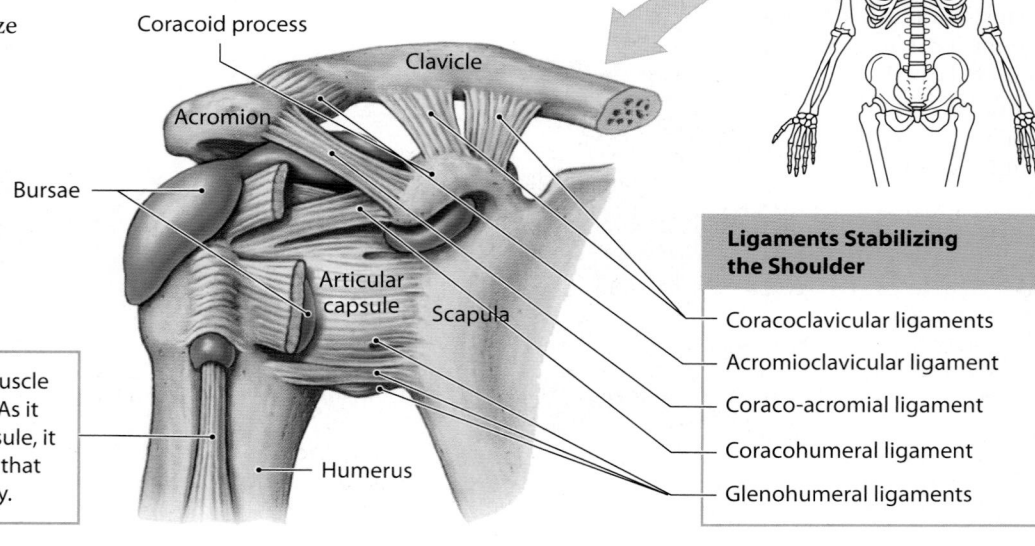

Coracoid process
Clavicle
Acromion
Bursae
Articular capsule
Scapula
Humerus

**Ligaments Stabilizing the Shoulder**

Coracoclavicular ligaments
Acromioclavicular ligament
Coraco-acromial ligament
Coracohumeral ligament
Glenohumeral ligaments

**2** This frontal section through the glenohumeral joint shows that the relatively loose articular capsule extends from the scapula to the anatomical neck of the humerus. The oversized articular capsule permits an extensive range of motion.

**3** In lateral view, you can see how small the articular cartilage of the glenoid cavity is compared with the articular capsule. The small articular cartilage is compensated by the glenoid labrum, surrounding rotator cuff muscles, associated tendons, and various ligaments that stabilize the joint.

Subdeltoid bursa
Articular capsule
Coraco-acromial ligament
Coracoclavicular ligaments
Clavicle
Humerus
Articular cartilages
Synovial membrane
Scapula

The area of the glenoid cavity is increased by a fibrocartilage **glenoid labrum** (*labrum*, lip or edge), which continues beyond the bony rim and deepens the socket.

Acromioclavicular ligament
Clavicle
Acromion
Articular capsule
Glenoid cavity
Articular cartilage
Scapula
Tendon of supraspinatus muscle
Tendon of infraspinatus muscle
Tendon of biceps brachii muscle
Coracohumeral ligament (cut)
Glenohumeral ligaments
Subscapularis muscle
Teres minor muscle

**? REVIEW**

**A.** Which structures provide most of the stability for the shoulder joint?

## Hip Joint

The **hip joint** is a sturdy ball-and-socket diarthrosis that allows flexion and extension, adduction and abduction, circumduction, and rotation.

**4** This lateral view shows the hip joint with the femur removed. Within the acetabulum, an articular cartilage pad extends like a horseshoe to either side of the acetabular notch. Two of the five ligaments that reinforce the articular capsule are shown here: the ligament of the femoral head and the transverse acetabular ligament.

Iliofemoral ligament

Articular cartilage

The **acetabular labrum**, a projecting rim of rubbery fibrocartilage, increases the depth of the joint cavity and helps to seal in synovial fluid.

The acetabulum, a deep fossa, accommodates the head of the femur.

Fat pad

The **ligament of the femoral head**, or **ligamentum teres** (*teres*, long and round), originates along the transverse acetabular ligament and attaches to the fovea capitis, a small pit at the center of the femoral head.

The **transverse acetabular ligament** crosses the acetabular notch, filling in the gap in the inferior border of the acetabulum.

**5** The articular capsule of the hip joint extends from the lateral and inferior surfaces of the pelvic girdle to the intertrochanteric line and intertrochanteric crest of the femur, enclosing both the head and neck of the femur. The three remaining broad ligaments that reinforce the articular capsule are shown in these views.

**? REVIEW**

**B.** At what site are the iliofemoral ligament, pubofemoral ligament, and ischiofemoral ligament located?

**Reinforcing Ligaments**

Pubofemoral ligament
Iliofemoral ligament
Ischiofemoral ligament

Greater trochanter

Intertrochanteric line

Lesser trochanter

**Anterior view**

Ischial tuberosity

**Posterior view**

**? INTEGRATION**

**C.** A football player is pushed out of bounds from behind. He falls onto his outstretched hand, pushing the humeral head forcefully upward. Which joints and ligaments are affected?

**LO LEARNING OUTCOME**

Describe the structure and function of the shoulder joint and hip joint.

Although the head of the femur is well supported, the ball-and-socket joint is not directly aligned with the weight distribution along the shaft. As a result, hip fractures (fractures of the femoral neck or between the greater and lesser trochanters of the femur) are much more common than hip dislocations.

# The elbow and knee are hinge joints

## Elbow Joint

**1** The **elbow joint** is a complex hinge joint that involves the humerus, radius, and ulna. It is extremely stable because (1) the bony surfaces of the humerus and ulna interlock, (2) a single, thick articular capsule surrounds both the humero-ulnar and proximal radio-ulnar joints, and (3) the articular capsule is reinforced by strong ligaments.

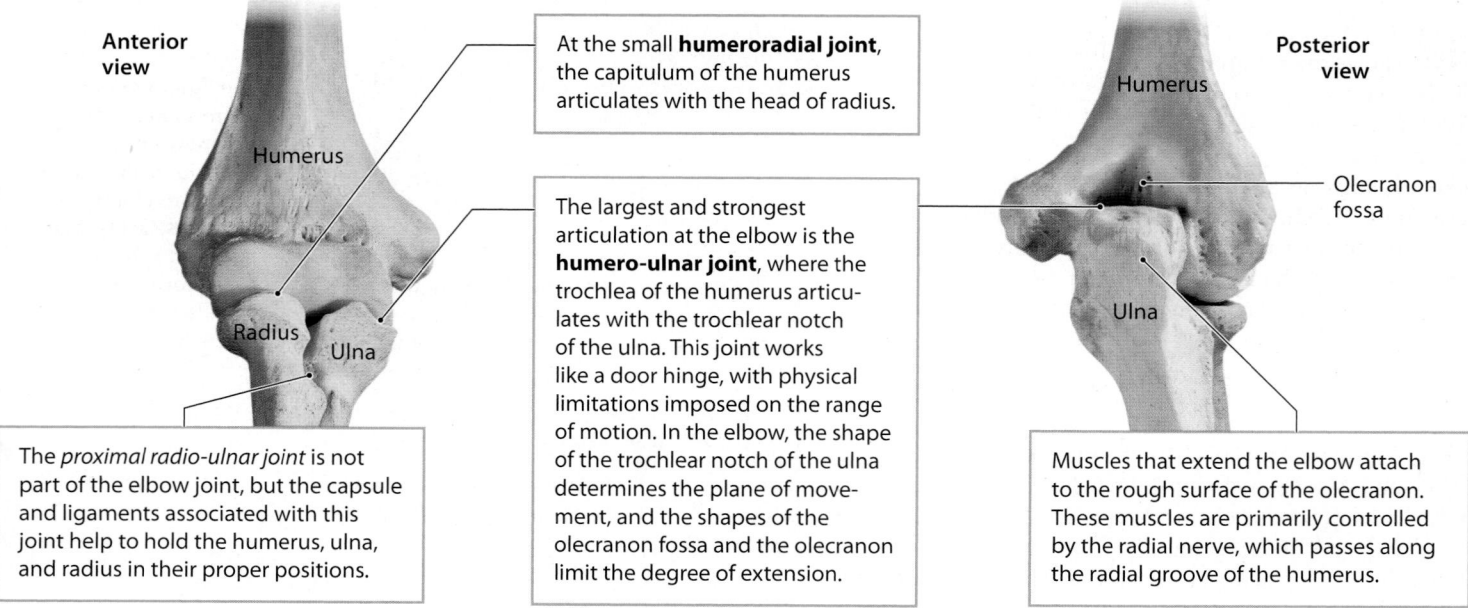

**Anterior view**

At the small **humeroradial joint**, the capitulum of the humerus articulates with the head of radius.

Humerus

Radius

Ulna

The *proximal radio-ulnar joint* is not part of the elbow joint, but the capsule and ligaments associated with this joint help to hold the humerus, ulna, and radius in their proper positions.

The largest and strongest articulation at the elbow is the **humero-ulnar joint**, where the trochlea of the humerus articulates with the trochlear notch of the ulna. This joint works like a door hinge, with physical limitations imposed on the range of motion. In the elbow, the shape of the trochlear notch of the ulna determines the plane of movement, and the shapes of the olecranon fossa and the olecranon limit the degree of extension.

**Posterior view**

Humerus

Olecranon fossa

Ulna

Muscles that extend the elbow attach to the rough surface of the olecranon. These muscles are primarily controlled by the radial nerve, which passes along the radial groove of the humerus.

**2** Although the elbow is extremely strong and stable, severe stresses can produce dislocations or other injuries, especially if epiphyseal growth has not been completed. For example, parents in a rush may drag a toddler along behind them. This exerts an upward, twisting pull on the elbow joint that can result in a partial dislocation of the head of radius from the annular ligament, known as **nursemaid's elbow**.

**? REVIEW**

**A.** Which ligaments stabilize the medial and lateral surfaces of the elbow joint?

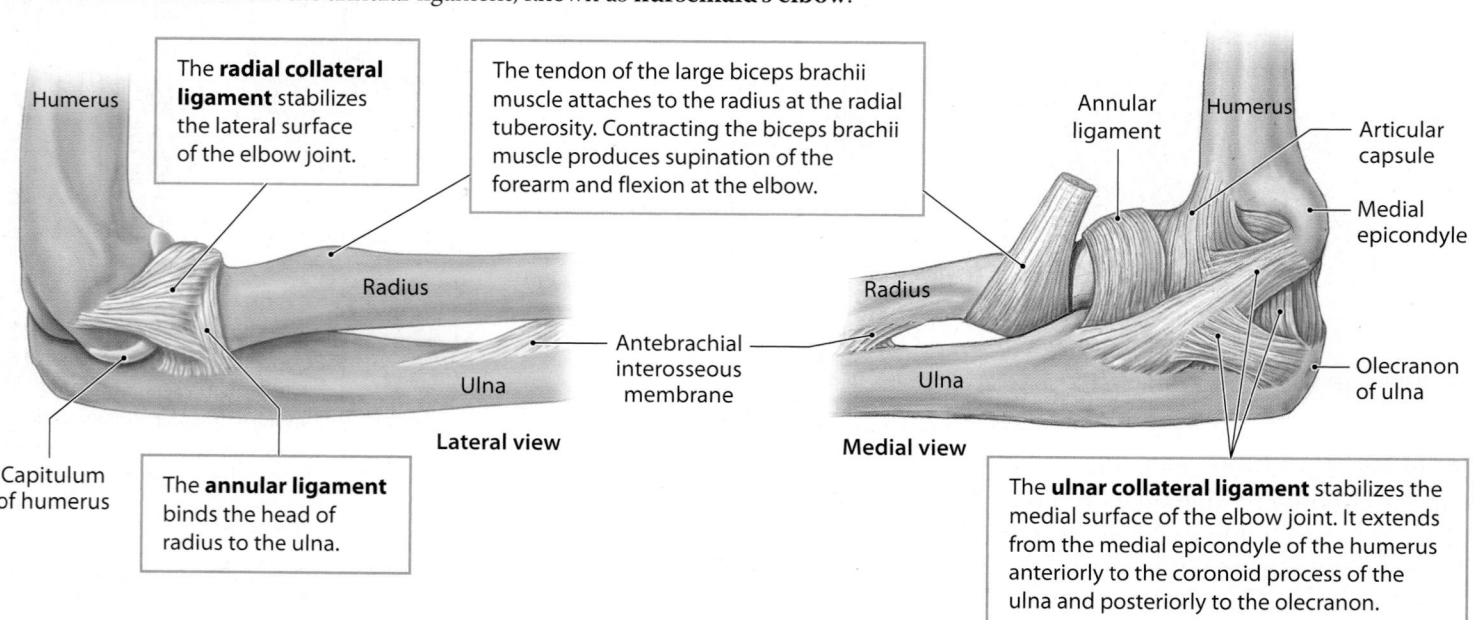

Humerus

The **radial collateral ligament** stabilizes the lateral surface of the elbow joint.

The tendon of the large biceps brachii muscle attaches to the radius at the radial tuberosity. Contracting the biceps brachii muscle produces supination of the forearm and flexion at the elbow.

Radius

Ulna

Antebrachial interosseous membrane

**Lateral view**

Capitulum of humerus

The **annular ligament** binds the head of radius to the ulna.

Annular ligament

Humerus

Radius

Ulna

**Medial view**

Articular capsule

Medial epicondyle

Olecranon of ulna

The **ulnar collateral ligament** stabilizes the medial surface of the elbow joint. It extends from the medial epicondyle of the humerus anteriorly to the coronoid process of the ulna and posteriorly to the olecranon.

# Knee Joint

The **knee joint** actually consists of three separate joints: two between the femur and tibia (medial condyle to medial condyle, and lateral condyle to lateral condyle) and one between the patella and the patellar surface of the femur. Together, these joints permit flexion, extension, and very limited rotation. Note that the fibula does not form part of the knee joint.

**3** The tendon from the quadriceps muscle passes over the anterior surface of the joint, enclosing the patella, and the patellar ligament then continues to its attachment on the anterior surface of the tibia. The patellar ligament and adjacent ligamentous bands support the anterior surface of the knee joint.

**4** The popliteal ligaments and several muscles that originate or insert on the femoral or tibial epiphyses reinforce the posterior surface of the joint.

Quadriceps tendon

Joint capsule

Bursa

Femur

Fibular collateral ligament

Tendon of biceps femoris muscle (cut)

Fibula

Tibia

The **fibular collateral ligament** provides lateral support. Clinically, it is referred to as the **lateral collateral ligament (LCL)**.

Patella

Patellar ligament

The **tibial collateral ligament** provides medial support. Clinically, it is referred to as the **medial collateral ligament (MCL)**.

Two **popliteal ligaments** extend between the femur and the heads of the tibia and fibula.

Fibula

Tibia

**Deep anterior view, flexed**

Patellar surface of femur

PCL

ACL

Fibular collateral ligament

Lateral condyle

Medial condyle

Tibia

Fibula

Crossing inside the capsule, the **anterior cruciate ligament (ACL)** and **posterior cruciate ligament (PCL)** attach the intercondylar area of the tibia to the condyles of the femur. These ligaments are named according to their position on the tibia.

Tibial collateral ligament

A pair of fibrocartilage pads, the **medial** and **lateral menisci**, lies between the femoral and tibial surfaces. The menisci act as cushions and provide lateral stability to the joint.

**Deep posterior view, extended**

ACL

PCL

Femur

Fibular collateral ligament

Medial condyle

Lateral condyle

Tibia

Fibula

Within the joint, the anterior cruciate (*cruciatus*, cross) ligament (ACL) and posterior cruciate ligament (PCL) limit the anterior and posterior movement of the tibia and maintain alignment of the femoral and tibial condyles. At full extension, a slight lateral rotation of the tibia tightens the ACL and forces the lateral meniscus between the tibia and femur. The knee joint is then locked in the extended position; unlocking requires medial rotation of the tibia or lateral rotation of the femur.

**? INTEGRATION**

**B.** What signs and symptoms would you expect in a person who has damaged the menisci of the knee joint?

**C.** Which ligament is a severely hyperextended knee more likely to damage: the ACL or the PCL?

**Lo LEARNING OUTCOME**

Describe the structure and function of the elbow joint and knee joint.

# Module 8.11
# Arthritis can disrupt normal joint structure and function

Joints are subjected to heavy use throughout our lifetime, and problems with joint function are relatively common, especially in older individuals. **Rheumatism** (RŪ-muh-tiz-um) is a general term that indicates pain and stiffness affecting the musculoskeletal system. **Arthritis** (ar-THRĪ-tis; *arthro,* joint + *itis,* inflammation) encompasses all the rheumatic diseases that affect synovial joints. Arthritis always involves damage to the articular cartilage, but the specific cause can vary. **Osteoarthritis** (os-tē-ō-ar-THRĪ-tis), also known as *degenerative arthritis* or *degenerative joint disease* (*DJD*), is the most common form of arthritis. It generally affects individuals age 60 or older. Osteoarthritis can result from the cumulative effects of wear and tear at joint surfaces or from genetic factors affecting collagen formation. In the U.S. population, 25 percent of women and 15 percent of men over age 60 show signs of this disease. Two other less common types of arthritis are **gouty arthritis** (see Module 23.14) and **rheumatoid arthritis** (see Module 20.21).

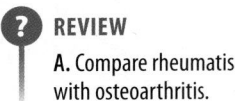 **REVIEW**
**A.** Compare rheumatism with osteoarthritis.

**1** This is a normal articular cartilage. A view through an arthroscope (red box 3) shows a smooth and slick surface. A sectional view reveals thick cartilage with a homogeneous matrix.

**2** This is an articular cartilage damaged by osteoarthritis. The exposed surfaces change from a slick, smooth-gliding surface to a surface composed of a rough feltwork of bristly collagen fibers. Such a change drastically increases friction at the joint, which then promotes further degeneration of the articular cartilage.

**Normal Joint**

Articular cartilage

LM × 180

Arthroscopic view of normal cartilage

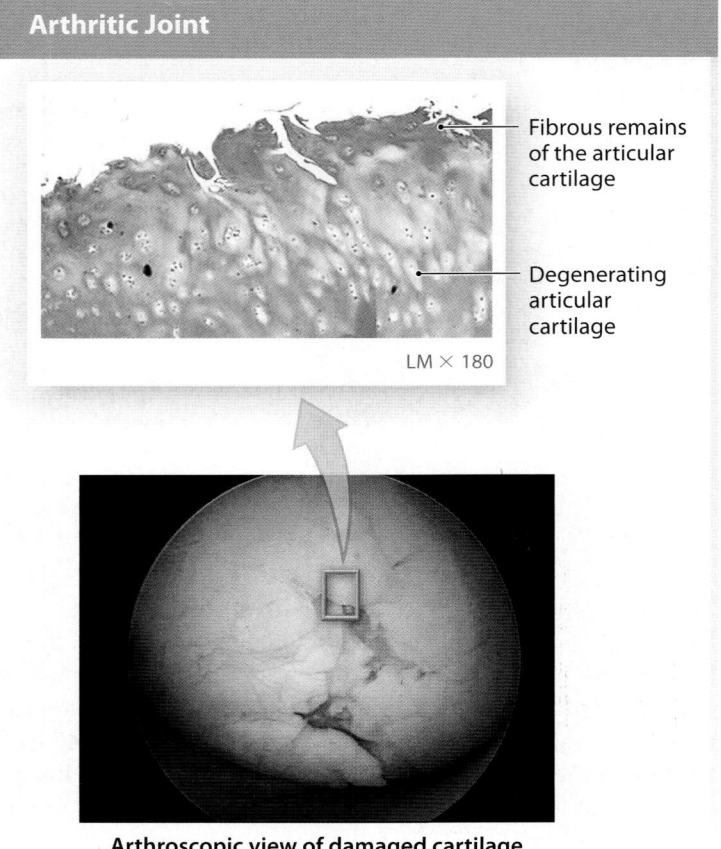

**Arthritic Joint**

Fibrous remains of the articular cartilage

Degenerating articular cartilage

LM × 180

Arthroscopic view of damaged cartilage

**3** There are several options available for examining the structure of problematic joints. This is an arthroscopic view of the interior of the left knee, showing injuries to the anterior and posterior cruciate ligaments. An **arthroscope** is a narrow, flexible fiberoptic scope containing a tiny camera that permits exploration of a joint without major surgery. Optical fibers are thin threads of glass or plastic that conduct light. The fibers can be bent around corners, so they can be introduced into a knee or other joint and moved around, enabling the physician to see inside the joint. If necessary, additional small incisions can be made to insert flexible instruments that permit surgery inside the joint, within view of the arthroscope. This procedure, called **arthroscopic surgery**, has greatly improved the treatment of knee and other joint injuries.

**4** An arthroscope cannot show soft tissue details outside the joint cavity, and repeated arthroscopy eventually leads to the formation of scar tissue and other joint problems. Magnetic resonance imaging (MRI) is a cost-effective and noninvasive method of viewing, without injury, and examining soft tissues around the joint. These are superior and posterior MRI views of the left knee joint. Note the clarity of the soft tissue detail.

**5** Implantation of artificial joints such as these may be the method of last resort when regular exercise, physical therapy, and anti-inflammatory drugs (such as aspirin) fail to slow the progress of arthritis. Artificial joints can restore mobility and relieve pain. People who undergo joint replacement surgery will have lifelong restrictions. For example, high-impact activities such as jogging, running, and playing football are no longer allowed after a hip or knee replacement. New knees and hips can last for more than 15 years, especially if the patient follows restrictions and avoids high-impact joint stress and strain.

**Artificial hip**

**Artificial shoulder**

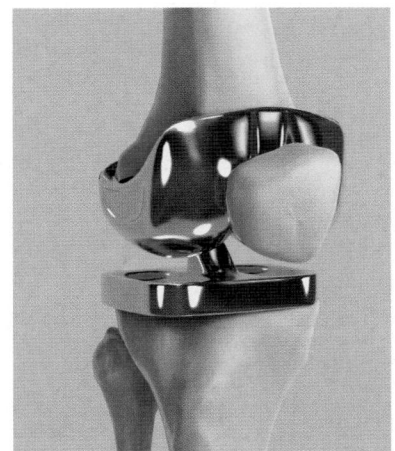

**Artificial knee**

**?** **INTEGRATION**

**B.** What can a person do to slow the progression of arthritis?

**Lo** **LEARNING OUTCOME**

Explain arthritis, and describe its effects on joint structure and function.

# Axial and Appendicular Joints

**Labeling:** Label each of the structures in the accompanying diagram of the shoulder joint.

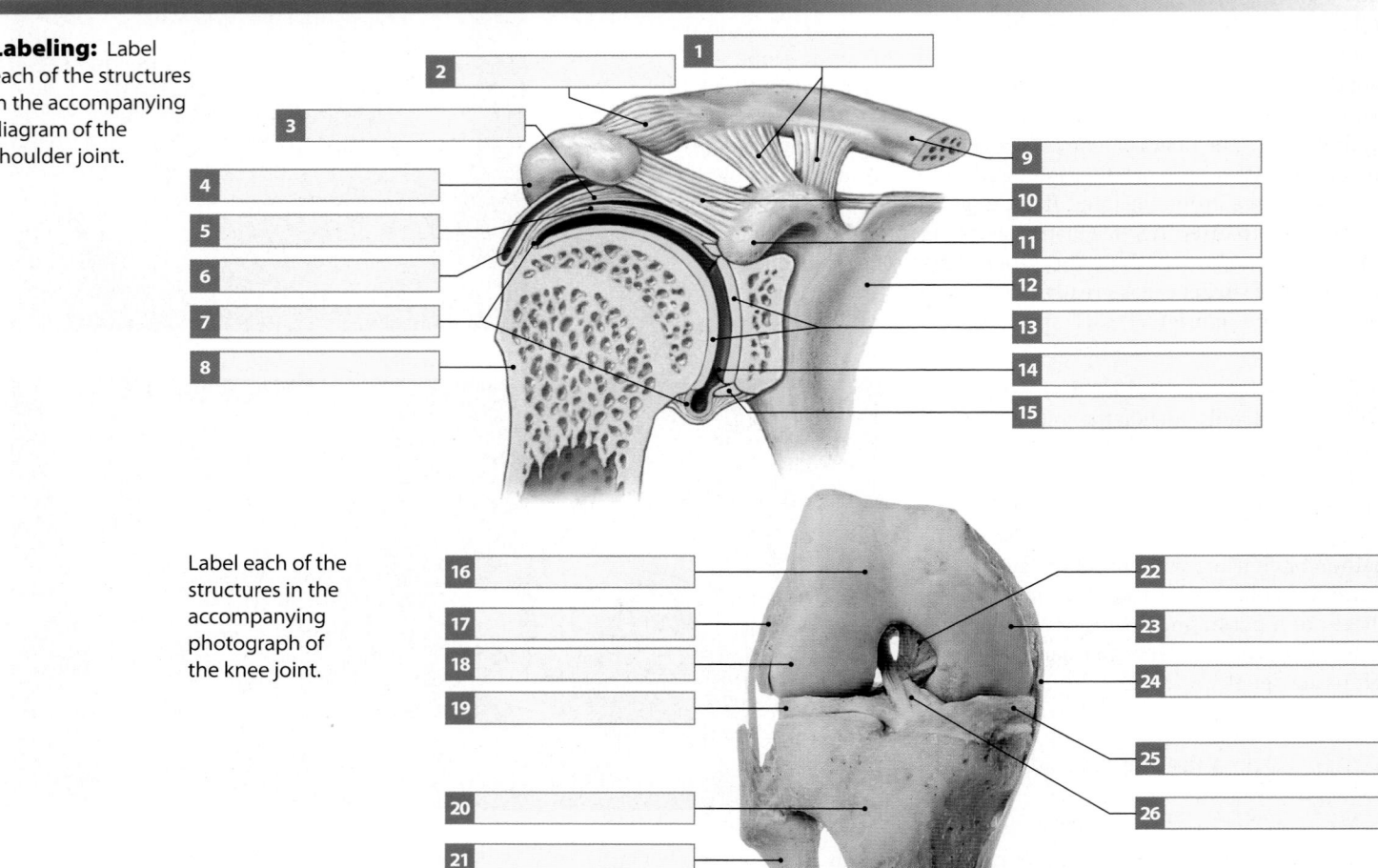

8

Label each of the structures in the accompanying photograph of the knee joint.

1 _____
2 _____
3 _____
4 _____
5 _____
6 _____
7 _____
8 _____
9 _____
10 _____
11 _____
12 _____
13 _____
14 _____
15 _____

16 _____
17 _____
18 _____
19 _____
20 _____
21 _____
22 _____
23 _____
24 _____
25 _____
26 _____

**Matching:** Match each lettered term with the most closely related description.

a. acetabulum
b. popliteal ligament
c. disc outer layer
d. dislocation
e. arthritis
f. disc inner layer
g. reinforce knee joint
h. osteoporosis

27 Knee joint posterior
28 Articular cartilage damage
29 Reduced bone mass
30 Cruciate ligaments
31 Anulus fibrosus
32 Deep fossa
33 Nucleus pulposus
34 Nursemaid's elbow

27 _____
28 _____
29 _____
30 _____
31 _____
32 _____
33 _____
34 _____

**Short answer:** Identify the joints that attach the pectoral and pelvic girdles to the axial skeleton.

35 _____

_____

# Study Outline

## SECTION 1 • Joint Structure and Movement

### Module 8.1

#### Joints are classified according to structure and movement p. 289

1. Movement of the skeleton occurs only at **joints** (**articulations**), which is where two bones meet.

2. A joint with no movement is called a **synarthrosis**, a joint with little movement is called an **amphiarthrosis**, and a freely movable joint is a **diarthrosis**.

3. The amount of movement at a joint is known as **range of motion (ROM)**.

### Module 8.2

#### Synovial joints are freely movable and lined with a synovial membrane p. 290

4. The components of a synovial joint are **articular cartilage**, a **joint capsule**, and a **synovial membrane**.

5. Synovial fluid lubricates the joint, provides nutrients for the chondrocytes of the articular cartilage, and helps to absorb shocks in the joints.

6. The accessory structures of a complex joint such as the knee include **bursae**, **fat pads**, **menisci**, and **accessory ligaments**.

7. Extreme stresses can cause joint **dislocation**, or **luxation**.

### Module 8.3

#### Anatomical organization determines the motion at synovial joints p. 292

8. Joint movement is determined by the structural arrangement of the involved bones.

9. The types of synovial joints include **plane**, **hinge**, **pivot**, **condylar**, **saddle**, and **ball-and-socket joints**.

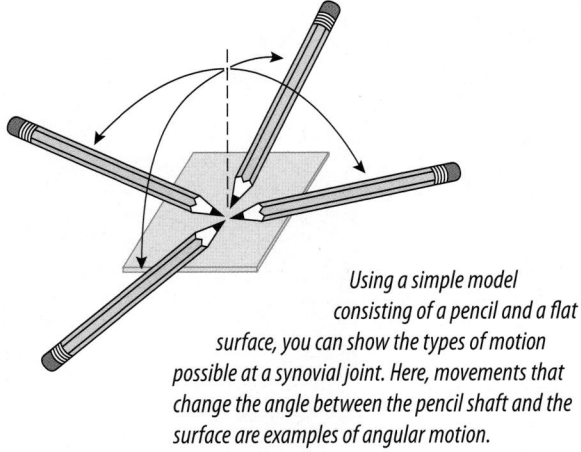

*Using a simple model consisting of a pencil and a flat surface, you can show the types of motion possible at a synovial joint. Here, movements that change the angle between the pencil shaft and the surface are examples of angular motion.*

### Module 8.4

#### Specific terms are used to describe movements with reference to the anatomical position p. 294

10. Terms used to describe joint movements include **flexion**, **extension**, **hyperextension**, **lateral flexion**, **dorsiflexion** (flexion at the ankle), **plantar flexion** (extension at the ankle), **abduction**, **adduction**, and **circumduction**.

*Abduction and adduction always refer to the movements of the appendicular skeleton, not to those of the axial skeleton.*

### Module 8.5

#### Specific terms describe rotation and special movements p. 296

11. Rotations include **right rotation**, **left rotation**, **lateral** (external) **rotation**, **medial** (internal) **rotation**, **pronation**, and **supination**.

12. Special movements include **opposition**, **protraction**, **retraction**, **inversion**, **eversion**, **depression**, and **elevation**.

## SECTION 2 • Axial and Appendicular Joints

### Module 8.6

#### Axial joints have less range of motion than appendicular joints p. 299

13. Most of the joints of the axial skeleton are strong and permit little movement.

14. Joints of the appendicular skeleton allow for extensive range of motion but are often weaker or less stable.

## Module 8.7
### The vertebral column includes three types of joints p. 300

15. There are three types of joints within the vertebral column: syndesmoses, synchondroses, and synovial.

16. The fibrous joints make up the **syndesmoses of vertebral column**. Fibrocartilage forms the **intervertebral discs** and makes up the **synchondroses of vertebral column**. Numerous joints between adjoining vertebrae comprise the **vertebral synovial joints**.

17. Vertebral stabilizing ligaments include the **ligamenta flava**, the **posterior longitudinal ligament**, the **interspinous ligament**, the **supraspinous ligament**, and the **anterior longitudinal ligament**.

## Module 8.8
### Intervertebral disc disease and osteoporosis are common age-related health problems p. 301

18. Abnormal intervertebral disc conditions include a **bulging disc** and a **herniated disc**.

19. Age-related loss of bone mass is **osteopenia**. **Osteoporosis** is a loss of bone mass that interferes with normal function.

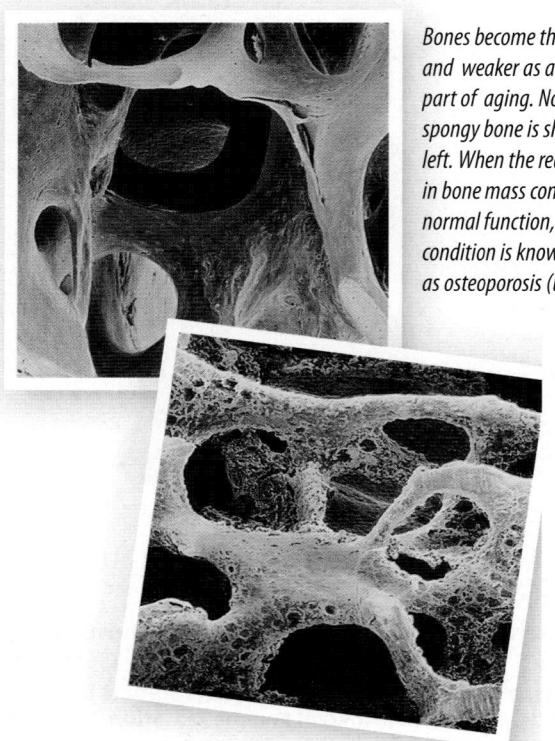

Bones become thinner and weaker as a normal part of aging. Normal spongy bone is shown left. When the reduction in bone mass compromises normal function, the condition is known as osteoporosis (below).

## Module 8.9
### The shoulder and hip are ball-and-socket joints p. 302

20. The **shoulder joint** (**glenohumeral joint**) is a ball-and-socket joint with the greatest range of motion.

21. The ligaments stabilizing the shoulder joint are the coracoclavicular, acromioclavicular, coraco-acromial, coracohumeral, and glenohumeral ligaments.

22. The area and depth of the glenoid cavity socket is increased by the **glenoid labrum**.

23. An **acetabular labrum**, the **ligament of the femoral head**, and the pubofemoral, the iliofemoral, and ischiofemoral ligaments stabilize the **hip joint**.

## Module 8.10
### The elbow and knee are hinge joints p. 304

24. The **elbow joint** includes the **humeroradial joint** and the **humero-ulnar joint**. The proximal radio-ulnar joint is not part of the elbow joint.

25. The **knee joint** includes three separate joints: two between the femur and tibia (medial condyle to medial condyle, and lateral condyle to lateral condyle) and one between the patella and the patellar surface of the femur.

26. The ligaments stabilizing the knee joint are the **fibular** and **tibial collateral ligaments**, the **anterior** and **posterior cruciate ligaments**, and the **medial** and **lateral menisci**.

## Module 8.11
### Arthritis can disrupt normal joint structure and function p. 306

27. **Rheumatism** is a general term indicating musculoskeletal pain and stiffness.

28. **Arthritis** is joint inflammation. **Osteoarthritis** is the most common form of arthritis and is the result of age-related wear and tear. Other familiar types are **gouty arthritis** and **rheumatoid arthritis**.

29. Treatments for problematic joints include **arthroscopic surgery** and joint replacement.

Joints are subjected to heavy use throughout our lifetimes, and problems with joint function are relatively common, especially in older individuals. This arthroscopic view of damaged cartilage shows how exposed surfaces change from a slick, smooth-gliding surface to a rough feltwork of bristly collagen fibers. This drastically increases friction at the joint and then promotes further degeneration.

# Chapter Review Questions

**True/False:** Indicate whether each statement is true or false.

**1** A bursa is a synovial fluid pocket outside a joint capsule.

**2** The acromioclavicular joint is an example of a condylar joint.

**3** Standing up from a chair involves hip joint extension and knee joint flexion.

**4** The ligamenta flava connect the laminae of adjacent vertebrae.

**5** The nucleus pulposus provides resilience and shock absorption to the intervertebral disc.

**6** The ligamentum teres attaches the head of the humerus into the glenoid cavity.

| | |
|---|---|
| 1 | _____ |
| 2 | _____ |
| 3 | _____ |
| 4 | _____ |
| 5 | _____ |
| 6 | _____ |

**Multiple choice:** Select the correct answer from the list provided.

**7** The synarthrosis formed between the bones of the skull is a
- a) gomphosis.
- b) synostosis.
- c) symphysis.
- d) suture.

**8** The joint between adjacent vertebral bodies is a
- a) synostosis.
- b) syndesmosis.
- c) symphysis.
- d) gomphosis.

**9** The joints typically located at the end of adjacent long bones are
- a) synarthroses.
- b) amphiarthroses.
- c) diarthroses.
- d) symphyses.

**10** The function of articular cartilage is to
- a) reduce friction.
- b) prevent bony surfaces from contacting one another.
- c) provide lubrication.
- d) do both a and b.

**11** Which of the following is *not* a function of synovial fluid?
- a) shock absorption
- b) nutrient distribution
- c) lubrication of articular surfaces
- d) maintenance of ionic balance

**12** A dislocation of an articulating surface is
- a) circumduction.
- b) hyperextension.
- c) luxation.
- d) supination.

**13** *Abduction* and *adduction* always refer to movements of the
- a) axial skeleton.
- b) appendicular skeleton.
- c) skull.
- d) vertebral column.

**14** Rotation of the radius that makes the palm face posteriorly is
- a) supination.
- b) pronation.
- c) protraction.
- d) inversion.

**15** A ballerina standing on her toes is performing which action of the ankle joint?
- a) dorsiflexion
- b) plantar flexion
- c) inversion
- d) eversion

**16** Increasing the angle between bones is
- a) flexion.
- b) extension.
- c) hyperextension.
- d) abduction.

**17** The hip is an extremely stable joint because it has

☐ a) a complete bony socket.

☐ b) a strong articular capsule.

☐ c) supporting ligaments.

☐ d) all of the above

**18** Movement of a limb away from the longitudinal axis of the body is

☐ a) abduction.

☐ b) adduction.

☐ c) hyperextension.

☐ d) protraction.

**19** Although the knee is considered to be only one joint, how many separate joints does it actually contain?

☐ a) two

☐ b) three

☐ c) four

☐ d) five

**20** The head of radius is attached to the ulna by the

☐ a) radial collateral ligament.

☐ b) ulnar collateral ligament.

☐ c) annular ligament.

☐ d) medial cruciate ligament.

**8**

## Short answer

**21** Explain why people tend to get shorter with advancing age.

**22** Explain how articular cartilage differs from other cartilage in the body.

**23** List the six different types of synovial joints, and give an example of each.

**24** Describe how a meniscus functions in a joint.

**25** Distinguish between osteopenia and osteoporosis.

**26** Which injury is more likely to result from playing a contact sport: a dislocated shoulder or a dislocated hip? Why?

# Chapter Integration • Applying what you have learned

## An avid 65-year-old golfer considers total knee replacement surgery to address chronic pain

Jon loved his golf. He loved the feel of just being out on a golf course. Even into his early 60s he still played; his only concession to age—and to a chronically aching knee—was pushing his bag on a wheeled cart instead of carrying his bag over his shoulder. He chalked the bum knee up to aging and managed the pain with wraps, ibuprofen, ice, and hot baths. For years his family physician had been urging him to consider a knee replacement because advances in the procedure made it almost miraculous in relieving pain and allowing near-normal activity. Finally realizing that the pain would only get worse and that he might be prevented from enjoying his golf, Jon did some research on knee replacement surgery, or knee arthroplasty.

Arthroplasty is used as a last resort when other treatments, such as physical therapy and anti-inflammatory drugs, do not relieve the signs and symptoms, and when the patient's discomfort becomes intense. In knee arthroplasty, a total knee dissection is performed, and damaged cartilage and bone, including the patella, are removed. The removed natural structures are replaced by synthetic structures made of metal and plastic—typically, a plastic patella rubs against the metal surfaces over the distal femur and proximal tibia. In the United States, the average age for a patient undergoing joint replacement is about 65–70 years. After reading about the surgery and speaking with several friends who had had the procedure, Jon decided to undergo knee arthroplasty. Here is an x-ray image of Jon's knees taken 6 months after surgery. He can now walk without pain and has returned to golf—although his game remains as bad as ever.

**1** Which bones compose the functioning knee joint?

**2** What purpose would physical therapy serve for a person suffering from joint pain?

**3** Classify the knee joint in terms of function and structure, and indicate the types of movements possible at this joint.

**4** Predict which activities may be restricted following a total knee arthroplasty.

## Assess a high-school quarterback's shoulder injury

Tommy is the quarterback on his high school football team. Trailing by 4 points late in the game, he calls for a long pass play. After taking the snap, he drops back and looks downfield for an open receiver. His linemen block as best they can, but they are not able to cover the blitzing linebacker. To avoid a sack, Tommy cocks back his arm and tries to rifle a pass to his tight end downfield. At the exact moment Tommy elevates the ball to throw, the linebacker is on him. The linebacker knocks the ball loose by chopping down onto Tommy's arm midway between his shoulder and elbow joints. When Tommy's arm is hit, he feels intense pain in his shoulder. As he runs to the sidelines in extreme discomfort, he is completely unable to move his shoulder joint. Using what you have just learned about joints and joint structure, answer the following questions.

**5** Do you think Tommy suffered a shoulder dislocation or a shoulder separation?

**6** Which bones, ligaments, and joint structures would you expect to be affected in Tommy's injury?

**7** Would you expect Tommy's injury to heal completely, or has he incurred a long-term joint instability?

## LEARNING OUTCOMES

These Learning Outcomes correspond by number to this chapter's modules and indicate what you should be able to do after completing the chapter. They also appear at the end of each module.

### Functional Anatomy of Skeletal Muscle Tissue

**9.1**   Describe the functions of skeletal muscle tissue. p. 315

**9.2**   Describe the organization of skeletal muscle at the tissue level. p. 316

**9.3**   Describe the structures of a sarcomere. p. 318

**9.4**   Describe the structures of a thin filament and a thick filament. p. 320

**9.5**   Describe a major characteristic of excitable membranes and its importance in generating an action potential. p. 322

**9.6**   Identify the components of the neuromuscular junction, and summarize the events involved in the control of skeletal muscles by motor neurons. p. 324

**9.7**   Describe the role of ATP in a muscle contraction, and explain the steps involved in the contraction of a skeletal muscle fiber. p. 326

### Functional Properties of Skeletal Muscle

**9.8**   Describe how muscle tension develops with respect to neural control and excitation-contraction coupling. p. 329

**9.9**   Describe the mechanism responsible for tension production in a muscle fiber, and discuss the factors that determine the peak tension developed during a contraction. p. 330

**9.10**   Discuss the factors that affect peak tension production during the contraction of an entire skeletal muscle, and explain the significance of the motor unit in this process. p. 332

**9.11**   Compare the different types of muscle contractions. p. 334

**9.12**   Describe the processes by which muscle fibers obtain the energy to power contractions. p. 336

**9.13**   Describe the factors that contribute to muscle fatigue, and discuss the processes involved in the muscle's subsequent recovery. p. 338

**9.14**   Relate the types of muscle fibers to muscle performance. p. 340

**9.15**   **Clinical Module:** Explain the physiological factors responsible for muscle hypertrophy, atrophy, and paralysis. p. 342

# 9

# Skeletal Muscle Tissue

# Module 9.1

# Skeletal muscle tissue enables body movement and other vital functions

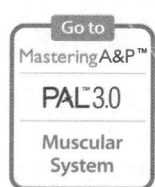

Go to
MasteringA&P™
PAL™3.0
Muscular
System

Muscle tissue, one of the four primary tissue types, consists chiefly of muscle cells that are highly specialized for contraction. Without the three types of muscle tissue—skeletal, cardiac, and smooth—nothing in the body would move, and the body itself could not move. In this chapter, we consider the structure and function of skeletal muscle tissue.

Each cell in skeletal muscle tissue is a single muscle fiber. **Skeletal muscles** are organs composed primarily of skeletal muscle tissue plus connective tissues, nerves, and blood vessels. Skeletal muscles are directly or indirectly attached to bones and have several functions, as detailed in the list of functions at right.

**Muscle Tissue**

### Skeletal Muscle Tissue

Skeletal muscle tissue contractions are under voluntary control. They move the body by pulling on bones, thus making it possible for us to be able to walk, dance, bite an apple, or play the piano.

### Cardiac Muscle Tissue

Cardiac muscle tissue is involuntary and is found in the heart. When it contracts, blood is pumped through the blood vessels.

*See Chapter 18.*

### Smooth Muscle Tissue

Smooth muscle tissue is involuntary and is found in the walls of hollow organs. In the digestive tract, it moves fluids and solids. As a lining in small arteries, it regulates vessel diameter.

*See Chapter 22.*

## Functions of Skeletal Muscle Tissue

- **Produce body movement**: Skeletal muscle contractions pull on tendons to move our bones. The effects range from simple motions such as extending the arm or breathing, to the highly coordinated movements of swimming, skiing, or piano playing.

- **Maintain posture and body position**: Tension in our skeletal muscles stabilizes joints to help maintain body posture— for example, holding your head still when you read a book, or balancing your body weight above your feet when you walk. Without constant muscular activity, you could neither sit upright nor stand.

- **Support soft tissues**: Layers of skeletal muscles make up the abdominal wall and the floor of the pelvic cavity. These muscles support the weight of our visceral organs and shield internal tissues from injury.

- **Guard body entrances and exits**: Skeletal muscles called sphincters encircle the openings of the digestive and urinary tracts. They give us voluntary control over swallowing, defecating, and urinating.

- **Maintain body temperature**: Muscle contractions use energy, and whenever energy is used in the body, some of it is converted to heat. The heat released by working muscles keeps body temperature in the range needed for normal functioning.

- **Store nutrients**: When our diet contains too few proteins or calories, the contractile proteins in skeletal muscles are broken down, and their amino acids are released into the circulation. The liver can use some of these amino acids to synthesize glucose, and others can be broken down to provide energy.

9

In this section, we examine the functional anatomy of a typical skeletal muscle. We place particular emphasis on the microscopic structural features that make contractions possible.

 **REVIEW**

**A.** Name the three types of muscle tissue, identify where they are found, and list their functions.

**B.** Which muscle types are voluntary, and which are involuntary?

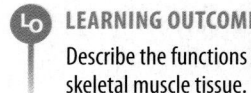 **LEARNING OUTCOME**

Describe the functions of skeletal muscle tissue.

# Skeletal muscle contains muscle tissue, connective tissues, blood vessels, and nerves

## Structure of a Skeletal Muscle

**1** A **skeletal muscle** is a complex organ containing *skeletal muscle fibers* (muscle cells), the connective tissues that harness the forces of contraction, the blood vessels that nourish the muscle fibers, and the nerves that control the contractions.

Perimysium    Muscle fascicle    Muscle fibers    Endomysium

Blood vessels    Nerve

The **epimysium** (ep-i-MIZ-ē-um; *epi-*, on + *mys*, muscle), a dense layer of collagen fibers, surrounds the entire muscle. The epimysium separates the muscle from surrounding tissues and organs. It is connected to the deep fascia, a dense connective tissue layer (see Module 4.15).

Epimysium
Blood vessels and nerves

**2** A **muscle fascicle** (FAS-i-kl; *fasciculus*, a bundle) is a bundle of muscle fibers. The fibrous perimysium layer separates the fascicles from each other.

Muscle fiber

The **perimysium** (per-i-MIZ-ē-um; *peri-*, around) is a fibrous layer that divides the skeletal muscle into a series of compartments. The perimysium contains collagen and elastic fibers as well as the blood vessels and nerves that supply the muscle fibers within the fascicles.

Endomysium

At the ends of a skeletal muscle, collagen fibers of the connective tissue layers merge to form either a bundle known as a **tendon** or a broad sheet called an **aponeurosis** (ap-ō-nū-RŌ-sis) (see Module 10.6). A tendon attaches the muscle to a specific point on a bone. An aponeurosis provides attachment over a broad area that may involve more than one bone. Where they contact the bone, collagen fibers extend into the bone matrix, providing a firm attachment. As a result, any contraction of the muscle pulls on the attached bone or bones.

**3** Individual **skeletal muscle fibers** are sheathed in a delicate endomysium that contains capillaries, myosatellite cells, and the axons of the neurons that control the muscle fibers.

Capillary    Myofibril

Axon of neuron

The **endomysium** (en-dō-MIZ-ē-um ) is a thin layer of areolar connective tissue that surrounds each muscle fiber. It loosely interconnects nearby muscle fibers. Each muscle fiber contains bundles of protein filaments called *myofibrils*.

**Myosatellite cells** are stem cells that function in the repair of damaged muscle tissue.

**? REVIEW**
**A.** Define *tendon* and *aponeurosis*.

**? REVIEW**
**B.** Describe the connective tissue layers associated with skeletal muscle tissue.

## Skeletal Muscle Development

**4** During embryonic development, groups of embryonic cells called **myoblasts** (*myo-*, muscle + *blastos*, formative cell) fuse, forming multinucleate cells. These large cells then develop into distinctive skeletal muscle fibers. Each nucleus in a skeletal muscle fiber represents the contribution of a single myoblast. Some myoblasts, however, do not fuse with developing muscle fibers. These unfused cells remain in the endomysium of adult skeletal muscle tissue as **myosatellite cells**. After an injury, myosatellite cells may enlarge, divide, and fuse with damaged muscle fibers, helping to repair the tissue.

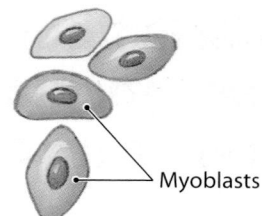

Myoblasts

Muscle fibers develop through the fusion of embryonic mesodermal cells called myoblasts.

Over time, most of the myoblasts fuse together to form larger multinucleate cells. However, a few myoblasts remain within the tissue as myosatellite cells, even in adults.

Myosatellite cell —

Nuclei —

Immature muscle fiber —

The multinucleate cells begin differentiating into skeletal muscle fibers as they enlarge and begin producing the proteins involved in muscle contraction.

**5** Compared to other cells, mature skeletal muscle fibers are enormous. A muscle fiber from a thigh muscle could have a diameter of 100 μm and be as long as 30 cm (12 inches). Each skeletal muscle fiber has hundreds of nuclei close to the plasma membrane. The genes in these nuclei control enzyme and structural protein production, which are needed for muscle contraction. The more gene copies, the faster the protein production.

Myosatellite cell —

Up to 30 cm in length

**6** Skeletal muscle fibers are so unusual that special terms are used to describe them. The plasma membrane is called the **sarcolemma** (sar-kō-LEM-uh; *sarkos*, flesh + *lemma*, husk), and the cytoplasm is called **sarcoplasm** (SAR-kō-plazm) (see Module 3.2). Most of the sarcoplasm consists of myofibrils.

**REVIEW**

**C.** What special terms are used to describe the plasma membrane and cytoplasm of a skeletal muscle fiber?

Myofibrils —

Sarcoplasm —

Mitochondria —

Sarcolemma —

— Striations

— Nuclei

**INTEGRATION**

**D.** How would severing a tendon attached to a muscle affect movement of that limb?

**LEARNING OUTCOME**

Describe the organization of skeletal muscle at the tissue level.

9

# Skeletal muscle fibers contain T tubules and sarcoplasmic reticula that surround contractile myofibrils made up of sarcomeres

**1** A **myofibril** is a cylindrical structure 1–2 μm in diameter and as long as the entire muscle fiber. The sarcoplasm of a single skeletal muscle fiber may contain hundreds to thousands of myofibrils. Because each myofibril has a banded appearance and the fiber is packed with myofibrils lying side by side, the entire muscle fiber appears to have light and dark bands, or **striations**.

Myofibril

Sarcolemma

Sarcoplasm

Skeletal muscle fiber

Striations

Nuclei

**2** Myofibrils consist of bundles of protein filaments called **myofilaments**. The most abundant myofilaments are **thin filaments** composed primarily of actin and **thick filaments** composed primarily of myosin.

Myofibril

Thin filament

Thick filament

Sarcolemma

Numerous mitochondria are scattered among the myofibrils.

**3** The myofilaments within each myofibril are arranged into repeating contractile units called **sarcomeres** (SAR-kō-mērz; *sarkos*, flesh + *meros*, part). Each myofibril consists of approximately 10,000 sarcomeres aligned end to end. Each sarcomere has a resting length of about 2 μm.

In a resting sarcomere, the **H band** is a lighter region on either side of the M line. The H band contains thick filaments, but no thin filaments.

The dense (dark) **A band** is the region that contains thick filaments.

The (light) **I band** contains thin filaments not overlapped by thick filaments.

Myofibril

Sarcomere

**REVIEW**

**A.** Describe the structures of a sarcomere.

The **M line** connects the central portion of each thick filament.

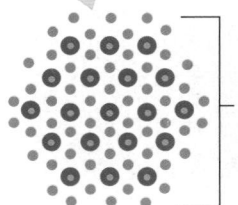

Within the A band, the **zone of overlap** has thin filaments arranged in a 6:1 ratio around each thick filament.

**Z lines** mark the boundary between adjacent sarcomeres. Z lines consist of proteins called *actinins*, which interconnect thin filaments of adjacent sarcomeres.

**4** The sarcolemma separates the sarcoplasm from the surrounding interstitial fluid. Negative electrical charges line the internal surface, and positive electrical charges line the external surface. This uneven distribution of electrical charges depends on the selective permeability of the plasma membrane. In skeletal muscle fibers, the distribution of electrical charges can change markedly with changes in membrane permeability. In fact, a sudden reversal in charge distribution is the first step that leads to a muscle contraction. These changes, triggered by the activity of a nerve cell (neuron) at one location on the sarcolemma, are swiftly propagated (spread) across the entire sarcolemma of the muscle fiber.

**Transverse tubules**, or **T tubules**, are narrow tubes that are continuous with the sarcolemma and extend into the sarcoplasm at right angles to the cell surface. T tubules form passageways through the muscle fiber, like a network of tunnels through a mountain.

Inside the sarcoplasm, T tubules encircle each sarcomere at the zones of overlap. Wherever a T tubule encircles a myofibril, the tubule is tightly bound to the membranes of the sarcoplasmic reticulum.

The **sarcoplasmic reticulum (SR)**, similar to the smooth endoplasmic reticulum of other cells, forms a tubular network around each individual myofibril. On either side of a T tubule, the tubules of the SR enlarge, fuse, and form expanded chambers called **terminal cisternae** (sis-TUR-nē).

Sarcolemma

Myofibril

Position of M line

Transverse tubule

Terminal cisternae

**?** **REVIEW**

**B.** Define *transverse tubules*.

9

The combination of a pair of terminal cisternae plus a T tubule is known as a **triad**. Although the membranes of the triad are tightly bound together, their fluid contents are separate and distinct.

Sarcoplasmic reticulum (SR)

Ca²⁺

Gated calcium channel (closed)

Cytosol

Calcium ion pump

**5** The sarcoplasmic reticulum is specialized for the storage and release of calcium ions. Its membrane contains ion pumps that actively transport calcium ions from the cytosol into the SR. Once in the SR, some of the calcium ions remain free in solution; the rest become reversibly bound to the protein **calsequestrin**. Including both the free calcium and the bound calcium, the total concentration of $Ca^{2+}$ within terminal cisternae in a resting (inactive) skeletal muscle fiber can be 40,000 times that of the surrounding cytosol. A muscle contraction begins when stored calcium ions are released into the cytosol through gated calcium channels.

**?** **INTEGRATION**

**C.** Within a resting skeletal muscle fiber, where is the greatest concentration of $Ca^{2+}$?

**Lo** **LEARNING OUTCOME**

Describe the structures of a sarcomere.

# Module 9.4

# The sliding of thin filaments past thick filaments produces muscle contraction

A longitudinal section of a sarcomere enables us to examine the structure and organization of the thin and thick filaments in greater detail.

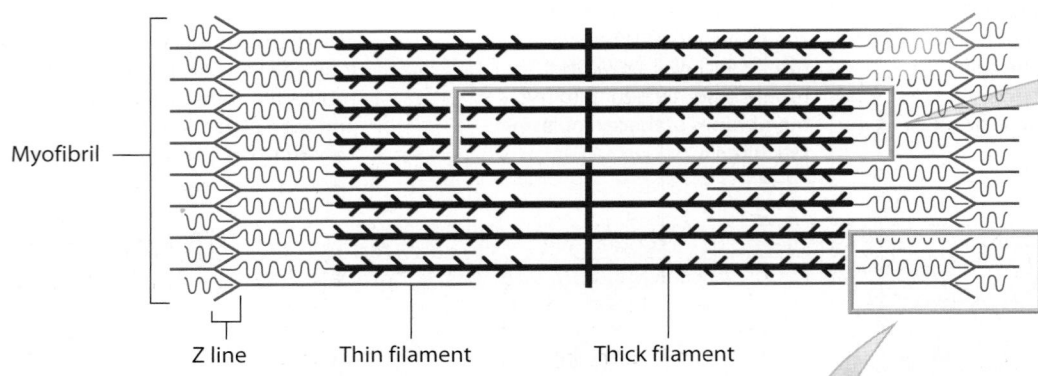

Myofibril

Z line    Thin filament    Thick filament

## Structure of Thin Filaments

**1** At either end of the sarcomere, the thin filaments are attached to the Z line. The protein **actinin** interconnects the thin filaments at the Z line. A typical thin filament is 5–6 nm in diameter and 1 μm long. As indicated in the magnified view below, each filament is composed of four main proteins: F-actin, nebulin, tropomyosin, and troponin.

Actinin    Z line

Active site

**F-actin** (filamentous actin) is a twisted strand composed of two rows of 300–400 individual molecules of G-actin (globular actin).

**Nebulin** is a large protein that extends along the F-actin strand in the cleft between the rows of G-actin molecules. It holds the F-actin strand together.

Each **G-actin** molecule contains an **active site** to which myosin can bind. This is much like a substrate molecule binding to an enzyme's active site.

Strands of **tropomyosin** (trō-pō-MĪ-ō-sin; *tropos*, turning) cover the active sites on G-actin and prevent actin–myosin interaction. A tropomyosin molecule is a double-stranded protein that is bound to one molecule of troponin midway along its length.

A **troponin** (TRŌ-pō-nin) molecule consists of three globular subunits. One subunit binds to tropomyosin, locking them together as a troponin–tropomyosin complex. A second subunit binds to one G-actin, holding the troponin–tropomyosin complex in position. And the third subunit has a receptor that binds two calcium ions.

**?** **REVIEW**

**A.** Compare F-actin with G-actin.

## Structure of Thick Filaments

**2** Thick filaments are 10–12 nm in diameter and 1.6 μm long. A thick filament contains about 300 **myosin** molecules, each made up of a pair of myosin subunits twisted around one another. All the myosin molecules are arranged with their tails pointing toward the M line. The myosin heads are arranged in a spiral, each facing one of the surrounding thin filaments. There are no myosin heads in a small area on either side of the M line.

Each thick filament has a core of an elastic protein called **titin**. Titin also extends from each end of the thick filament, across the I band to the Z line on that side. Titin has been described as a "molecular spring," because it recoils after stretching. In the normal resting sarcomere, the titin strands are completely relaxed; they become tense only when some external force stretches the sarcomere.

Thick filament     M line

**? REVIEW**
**B.** Name the proteins that make up a thick filament.

Myosin molecule

The **free head** has two globular protein subunits. During contraction, the myosin heads interact with the thin filaments.

Each long myosin **tail** is bound to other myosin molecules within the thick filament.

The connection between the head and the tail acts as a hinge that lets the head pivot at its base.

9

## Sliding Filament Theory

**3** When a skeletal muscle fiber contracts, thin filaments slide past the thick filaments. In this process, (1) the H bands and I bands get smaller, (2) the zones of overlap get larger, (3) the Z lines move closer together, and (4) the width of the A band remains constant. This explanation is known as the **sliding filament theory**.

**? REVIEW**
**C.** Summarize the sliding filament theory.

I band     A band

Z line     H band     Zone of overlap     Z line

**Sarcomere at rest**

I band     A band

Z line     H band     Zone of overlap     Z line

**Sarcomere contraction and filament sliding**

**4** During a contraction, sliding occurs in every sarcomere along a myofibril, so the myofibril gets shorter. Because myofibrils are attached to the sarcolemma at each Z line and at either end of the muscle fiber, when myofibrils shorten, so does the muscle fiber.

**? INTEGRATION**
**D.** Why is the zone of overlap an important region of the sarcomere?

**Lo LEARNING OUTCOME**
Describe the structures of a thin filament and a thick filament.

## Module 9.5

# Skeletal muscle fibers and neurons have excitable plasma membranes that produce and carry electrical impulses called action potentials

Here we discuss the electrical nature of cells, what an action potential is, and how it is propagated along the axon of a neuron or sarcolemma of a skeletal muscle fiber.

**1** Like miniature batteries with positive and negative electrical poles, all the cells of the body are polarized. A cell's polarization is due to an unequal distribution of positive and negative charges across its plasma membrane.

The charge separation represents potential energy, or a potential difference, and the cell is said to be **polarized**. We refer to the potential difference across the plasma membrane as the **membrane potential**. A cell's membrane potential is measured in millivolts (mV). It compares the inner membrane charge with the outer membrane charge. In undisturbed neurons and skeletal muscle fibers, typical **resting membrane potentials** are about –70 mV and –85 mV, respectively.

For a resting cell, the inner surface of its plasma membrane has a slight negative charge with respect to its outer surface. Although the positive and negative charges are attracted to each other, the plasma membrane keeps them apart.

**2** Plasma membranes are selectively permeable, and the cytosol and extracellular fluid (ECF) differ in composition.

The main contributors of positive electrical charges are sodium (Na⁺) and potassium (K⁺) ions. There is more Na⁺ outside a cell (in the ECF) than inside (in the cytosol), and more K⁺ inside a cell than outside.

There is a constant low rate of diffusion of Na⁺ and K⁺ down their concentration gradients across the membrane. Both require ion-specific "leak" membrane channels to cross the plasma membrane. The membrane is crossed more easily by K⁺ than Na⁺. However, sodium–potassium ion pumps constantly export 3 Na⁺ from the cell and import 2 K⁺ to maintain the cell's resting membrane potential (see Module 3.16).

Although there is an excess of negatively charged chloride ions (Cl⁻) in the ECF, little diffusion of these ions occurs across the plasma membrane.

The main contributors of negative charges within a cell are proteins that cannot cross the plasma membrane.

**?** **REVIEW**

A. Explain the function of sodium–potassium ion pumps.

**3** Neurons and skeletal muscle fibers have electrically excitable membranes. Such membranes contain **voltage-gated ion channels** that are activated (or inactivated) by changes in the membrane potential. Changes in the ion membrane permeability of a cell can temporarily reverse the distribution of electrical charges across the plasma membrane.

**? REVIEW**

**B.** Define *depolarization*, and describe the events that follow it.

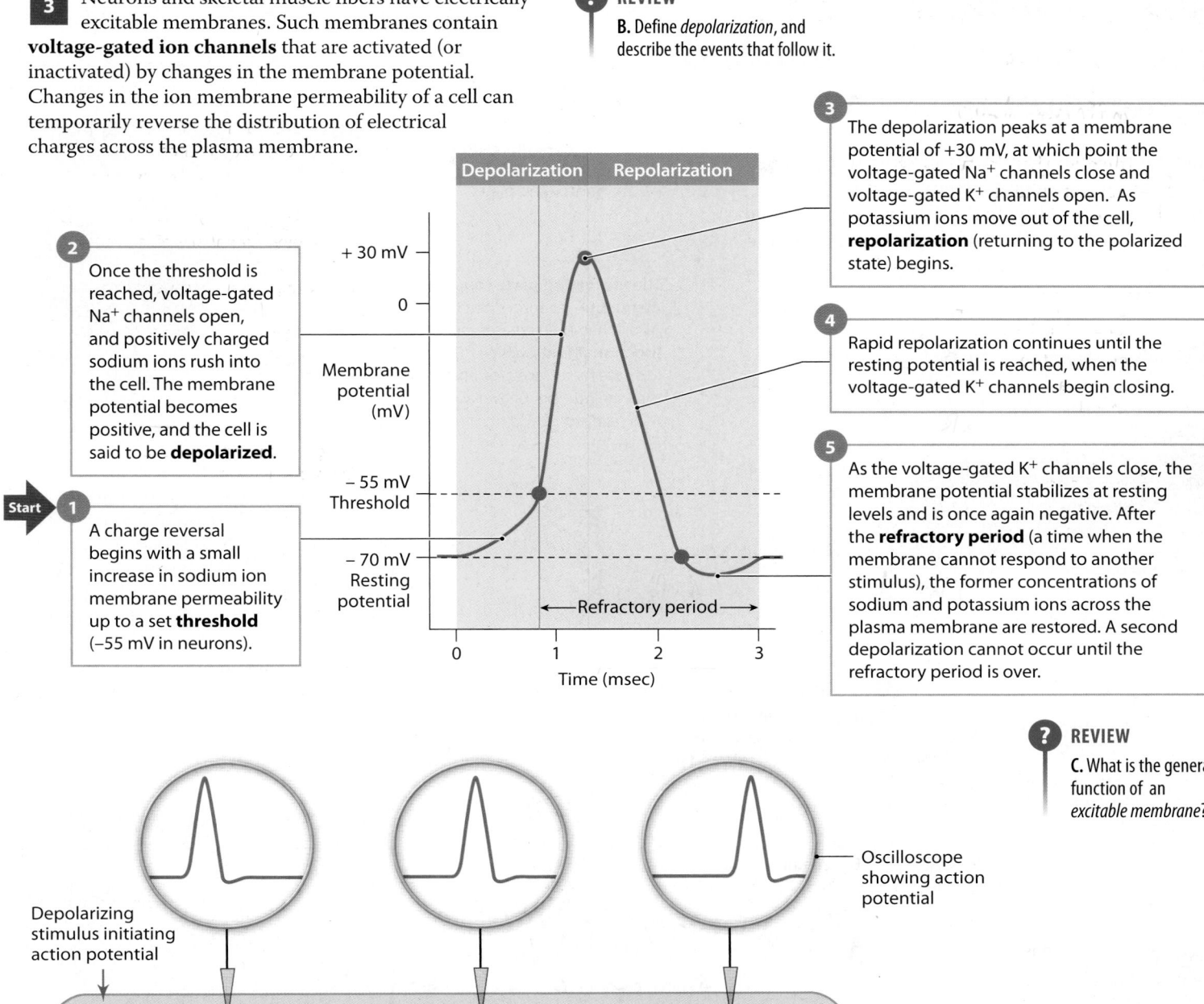

**2** Once the threshold is reached, voltage-gated Na⁺ channels open, and positively charged sodium ions rush into the cell. The membrane potential becomes positive, and the cell is said to be **depolarized**.

**1** A charge reversal begins with a small increase in sodium ion membrane permeability up to a set **threshold** (–55 mV in neurons).

**3** The depolarization peaks at a membrane potential of +30 mV, at which point the voltage-gated Na⁺ channels close and voltage-gated K⁺ channels open. As potassium ions move out of the cell, **repolarization** (returning to the polarized state) begins.

**4** Rapid repolarization continues until the resting potential is reached, when the voltage-gated K⁺ channels begin closing.

**5** As the voltage-gated K⁺ channels close, the membrane potential stabilizes at resting levels and is once again negative. After the **refractory period** (a time when the membrane cannot respond to another stimulus), the former concentrations of sodium and potassium ions across the plasma membrane are restored. A second depolarization cannot occur until the refractory period is over.

Depolarizing stimulus initiating action potential

Direction of action potential

Oscilloscope showing action potential

Plasma membrane of neuronal axon or skeletal muscle cell

**? REVIEW**

**C.** What is the general function of an *excitable membrane*?

**4** In most cells, the depolarization and repolarization of a plasma membrane is a localized change that does not spread across the entire plasma membrane. However, neurons and skeletal muscle fibers have electrically excitable membranes, and the initial depolarization and repolarization events produce an electrical impulse, or **action potential**, that is spread, or **propagated**, along their plasma membranes. The opening and closing of voltage-gated Na⁺ channels and the subsequent opening and closing of voltage-gated K⁺ channels generate an action potential in less than 2 msec. An action potential travels in one direction because the refractory period prevents it from propagating back in the direction from which it was initiated. Excitable membranes permit rapid communication between different parts of a cell.

**? INTEGRATION**

**D.** Explain why the propagation of action potentials along electrically excitable membranes occurs in only one direction.

**Lo LEARNING OUTCOME**

Describe a major characteristic of excitable membranes and its importance in generating an action potential.

# Module 9.6

# A skeletal muscle fiber contracts when stimulated by a motor neuron

Skeletal muscle fibers contract only under the control of the nervous system. Each skeletal muscle fiber is controlled by the nervous system at a specialized site known as a **neuromuscular junction** (**NMJ**), located midway along the muscle fiber's length.

Motor neuron

Path of electrical impulse (action potential)

Axon

Muscle fiber

Neuromuscular junction

Myofibril

Motor end plate

Sarcoplasmic reticulum

Myofibril

Axon terminal

Synaptic cleft

Motor end plate

**1** The NMJ is made up of an **axon terminal** (synaptic terminal) of a motor neuron, a *motor end plate* (a specialized region of the sarcolemma), and a *synaptic cleft* (an intervening space). A motor neuron is a nerve cell that propagates an electrical impulse from the nervous system to skeletal muscle fibers. A single axon may branch to control more than one skeletal muscle fiber, but each muscle fiber has only one NMJ. At the NMJ, the axon terminal of the neuron lies near the motor end plate of the muscle fiber.

The **synaptic cleft**, a narrow space, separates the axon terminal of the neuron from the opposing motor end plate.

The **motor end plate** contains membrane receptors that bind ACh. It has deep creases called **junctional folds**, which increase its surface area and thus the number of available ACh receptors.

Vesicles containing ACh (red)

Junctional fold of motor end plate

AChE

**2** The cytoplasm of the axon terminal contains vesicles filled with molecules of **acetylcholine** (as-e-til-KŌ-lēn), or **ACh**. Acetylcholine is a neurotransmitter, a chemical released by a neuron to change the permeability or other properties of another cell's plasma membrane. The synaptic cleft and the sarcolemma contain molecules of the enzyme **acetylcholinesterase** (**AChE**), which breaks down ACh.

T tubule

Sarcoplasm

Sarcoplasmic reticulum (SR)

$Ca^{2+}$   $Ca^{2+}$

**8** As the action potential sweeps down each T tubule and passes between the terminal cisternae, the SR's permeability changes, and calcium ions flood into the sarcomeres at the zones of overlap. This event, called **excitation-contraction coupling**, triggers the contraction of the muscle fiber (see Module 9.7).

**? REVIEW**

**A.** Describe the neuromuscular junction.

**324** • *Chapter 9: Skeletal Muscle Tissue*

9

Watch
MasteringA&P™
A&PFlix
Events at the Neuromuscular Junction

Arriving action potential

Junctional fold

**3** The stimulus for ACh release is the arrival of an electrical impulse, or action potential, at the axon terminal. An action potential is a sudden change in the membrane potential that travels along the length of the axon.

Sarcolemma of motor end plate

Exocytosis of ACh

**4** When the action potential reaches the neuron's axon terminal, changes in its membrane permeability trigger the exocytosis of ACh into the synaptic cleft. Exocytosis occurs as the synaptic vesicles fuse with the neuron's plasma membrane.

$Na^+$
$Na^+$
$Na^+$
$Na^+$

ACh-receptor membrane channel

**5** ACh molecules diffuse across the synaptic cleft and bind to ACh-receptor membrane channels. ACh binding alters the membrane's permeability to $Na^+$. Because the extracellular fluid contains a high concentration of $Na^+$, and $Na^+$ concentration inside the cell is very low, sodium ions rush into the sarcoplasm.

Action potential

Breakdown of ACh

AChE

**7** The action potential generated at the motor end plate now sweeps across the entire membrane surface. The effects are almost immediate because an action potential is an electrical event that flashes like a spark across the sarcolemmal surface. The effects are brief because the ACh has been removed, and no further stimulus acts upon the motor end plate until another action potential arrives at the axon terminal.

**6** The sudden inrush of sodium ions generates an action potential in the sarcolemma. ACh either diffuses away from the site or AChE quickly breaks it down in the synaptic cleft, thus closing the ACh-receptor membrane channels.

**? REVIEW**

**B.** How would a drug that blocks acetylcholine release affect muscle contraction?

**C.** What would happen if there were no AChE in the synaptic cleft?

**Lo LEARNING OUTCOME**

Identify the components of the neuromuscular junction, and summarize the events involved in the control of skeletal muscles by motor neurons.

# Module 9.7

# A muscle fiber contraction uses ATP in a cycle that repeats during the contraction

## Resting Sarcomere

**1** In the resting sarcomere, each myosin head is already "energized"—charged with the energy that will be used to power a contraction. Each myosin head points away from the M line. In this position, the myosin head is "cocked" like the spring in a mousetrap. Cocking the myosin head requires energy, which is obtained by breaking down ATP. In this process, the myosin head functions as ATPase, an enzyme that breaks down ATP. At the start of a contraction, the breakdown products, ADP and a phosphate (represented as P), remain bound to the myosin head.

## Contraction Cycle Begins

**2** The **contraction cycle** is a series of interrelated steps that begins with the arrival of calcium ions within the zone of overlap in a sarcomere.

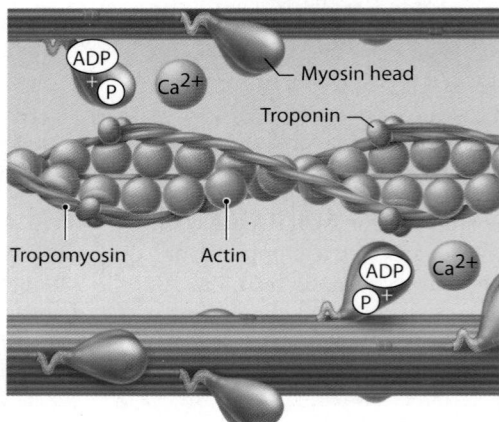

**? REVIEW**

**A.** What molecule supplies the energy for a muscle fiber contraction?

## Contracted Sarcomere

**8** The entire cycle repeats several times each second, as long as $Ca^{2+}$ concentrations remain elevated and ATP reserves are sufficient. Calcium ion levels will remain elevated only as long as action potentials continue to pass along the T tubules and stimulate the terminal cisternae. Once that stimulus is removed, the calcium channels in the SR close, and calcium ion pumps pull $Ca^{2+}$ from the cytosol and store it within the terminal cisternae. Troponin molecules then shift position, swinging the tropomyosin strands over the active sites and preventing further cross-bridge formation.

## Myosin Reactivates

**7** Myosin reactivates when the free myosin head splits ATP into ADP and P. The energy released is used to "recock" the myosin head.

## Active Sites Exposed

**3** Calcium ions bind to troponin, weakening the bond between actin and the troponin–tropomyosin complex. The troponin molecule then changes position, rolling the tropomyosin molecule away from the active sites on actin and allowing interaction with the energized myosin heads.

## Cross-Bridges Form

**4** Once the active sites are exposed, the energized myosin heads bind to them, forming **cross-bridges**.

## Cross-Bridges Detach

**6** When another ATP binds to the myosin head, the link between the myosin head and the active site on the actin molecule is broken. The active site is now exposed and able to form another cross-bridge.

## Myosin Heads Pivot

**5** After the cross-bridges form, the energy that was stored in the resting state is released as the myosin heads pivot toward the M line. This action is called the *power stroke*; when it occurs, the bound ADP and phosphate group are released.

**?** **REVIEW**

**B.** List the interrelated steps that occur once the contraction cycle begins.

**Lo** **LEARNING OUTCOME**

Describe the role of ATP in a muscle contraction, and explain the steps involved in the contraction of a skeletal muscle fiber.

# Section 1 Review
## Functional Anatomy of Skeletal Muscle Tissue

**Labeling:** Label the structures in this figure of a skeletal muscle fiber.

1. mitochondrea
2. Sarcolemma
3. MyoFibril
4. actin Thin filament
5. thick filament
6. Triad
7. S.R.
8. T-Tubule
9. sarcoplasmic reticulum / Terminal cisternae
10. Sarcoplasm
11. MyoFibril

M. HAIZ

Label the structures in this diagram of adjacent sarcomeres.

12.
13.
14. A band
15.
16.
17. Zone of overlap
18. M line
19.
20. thick filament
21.

**Fill-in:** In the spaces provided, reorder the following terms in their proper sequence from largest (22) to smallest (27).

- muscle fiber
- muscle fascicle
- sarcomere
- myofibril
- skeletal muscle
- myofilament

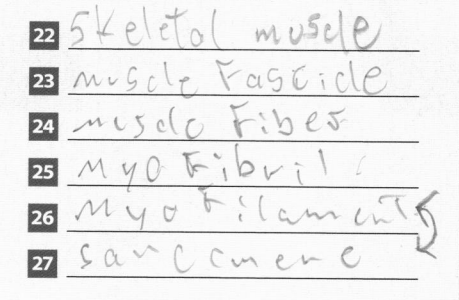

22. Skeletal muscle
23. muscle Fascicle
24. muscle Fiber
25. Myo Fibril
26. Myo Filaments
27. sarcomere

# Module 9.8

# Muscle tension develops from the events that occur during excitation-contraction coupling

**1** The link between the generation of an action potential in the sarcolemma and the start of a muscle contraction is called **excitation-contraction coupling.**

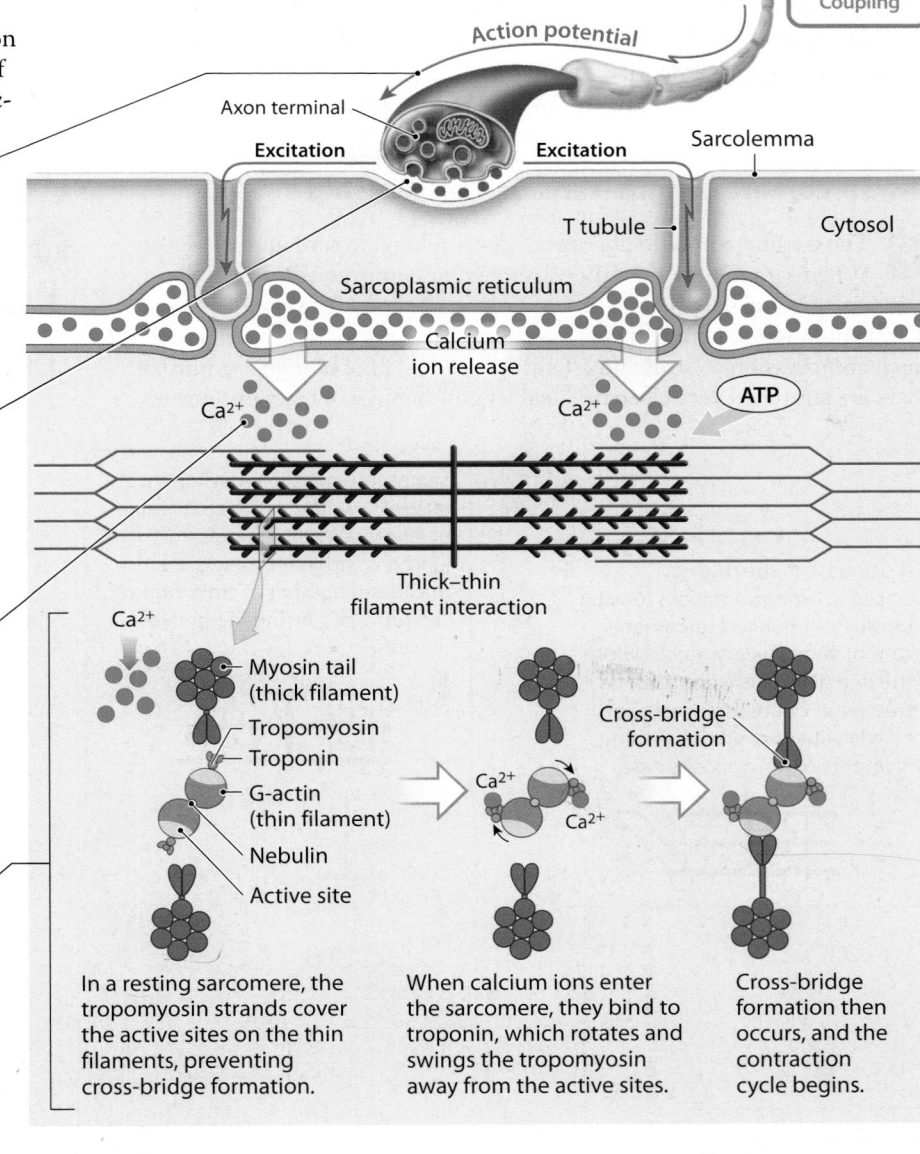

Action potential

Axon terminal

Excitation     Excitation     Sarcolemma

T tubule     Cytosol

Sarcoplasmic reticulum

Calcium ion release

$Ca^{2+}$     $Ca^{2+}$     ATP

Thick–thin filament interaction

$Ca^{2+}$

Myosin tail (thick filament)
Tropomyosin
Troponin
G-actin (thin filament)
Nebulin
Active site

$Ca^{2+}$     $Ca^{2+}$     Cross-bridge formation

In a resting sarcomere, the tropomyosin strands cover the active sites on the thin filaments, preventing cross-bridge formation.

When calcium ions enter the sarcomere, they bind to troponin, which rotates and swings the tropomyosin away from the active sites.

Cross-bridge formation then occurs, and the contraction cycle begins.

**1 Neural control**

A skeletal muscle fiber contracts when stimulated by a motor neuron at a neuromuscular junction. The stimulus arrives in the form of an action potential at the axon terminal.

**2 Excitation**

The action potential causes the release of ACh into the synaptic cleft, which leads to excitation—the generation of an action potential in the sarcolemma.

**3 Calcium ion release**

The muscle fiber action potential travels along the sarcolemma and into T tubules down to the triads. This triggers the release of $Ca^{2+}$ from the terminal cisternae of the sarcoplasmic reticulum.

**4 Contraction cycle begins**

The contraction cycle begins when $Ca^{2+}$ bind to troponin, exposing the active sites on the thin filaments. Cross-bridge formation begins—and continues—as long as ATP is available.

**5 Sarcomeres shorten**

As the thick and thin filaments interact, the sarcomeres shorten, pulling the ends of the muscle fiber closer together.

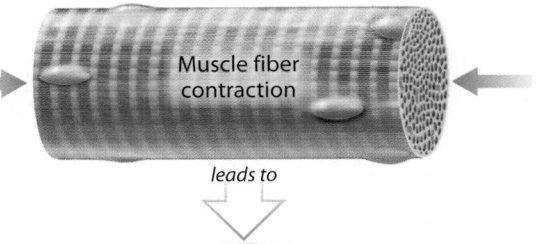

Muscle fiber contraction

leads to

**6 Muscle tension produced**

During the contraction, the entire skeletal muscle shortens and produces a pull, or **tension**, on the tendons at either end.

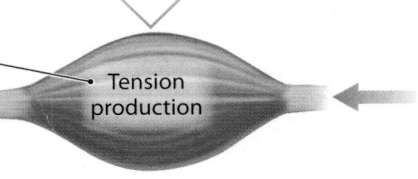

Tension production

**? REVIEW**

**A.** What causes calcium to be released from the sarcoplasmic reticulum?

**Lₒ LEARNING OUTCOME**

Describe how muscle tension develops with respect to neural control and excitation-contraction coupling.

# Module 9.9

# Tension is greatest when muscle fibers are stimulated at optimal length

There is no mechanism to regulate the amount of tension produced in a contraction by changing the number of contracting sarcomeres. When calcium ions are released, they are released from all triads in the muscle fiber. Thus, a muscle fiber is either "on" (producing tension) or "off" (relaxed). Tension produced at the individual muscle fiber does vary, however, depending on the fiber's resting length at the time of stimulation.

## The Sarcomere Length–Tension Relationship

**1** The tension a muscle fiber produces is related to sarcomere length. When sarcomeres are either stretched or compressed compared to optimal resting length, tension production declines. The arrangement of skeletal muscles, connective tissues, and bones normally prevents too much compression or stretching. During walking, for example, leg muscle fibers are stretched very close to "ideal length" before contractions occur.

**? REVIEW**

**A.** What sarcomere characteristic affects the amount of tension produced when a skeletal muscle fiber contracts?

**B.** Explain two key concepts of the sarcomere length–tension relationship.

**Start**

Sarcomeres produce tension most efficiently within an optimal range of lengths. When resting sarcomere length is within this range, the maximum number of cross-bridges can form, producing the greatest tension.

A decrease in the resting sarcomere length reduces tension because stimulated sarcomeres cannot shorten very much before the thin filaments extend across the center of the sarcomere and collide with or overlap the thin filaments of the opposite side.

An increase in sarcomere length reduces the tension produced by decreasing the size of the zone of overlap and the number of potential cross-bridge interactions.

Tension production falls to zero when the thick filaments are pressed against the Z lines and the sarcomere cannot shorten further.

When the zone of overlap is reduced to zero, thin and thick filaments cannot interact at all. The muscle fiber cannot produce any active tension, and no contraction can occur. Such extreme stretching of a muscle fiber is normally prevented by titin filaments (which tie the thick filaments to the Z lines) and by the surrounding connective tissues.

Tension (percent of maximum)
100
80
60
40
20
0

Normal range

1.2 μm    1.6 μm         2.6 μm              3.6 μm

Sarcomere length decreases

Sarcomere length increases

**Optimal resting length:** The normal range of sarcomere lengths in the body is 75 to 130 percent of the optimal length.

## Measuring Muscle Fiber Tension

**2** A **myogram** is a graphical representation of tension development in muscle fibers. This myogram shows three examples of a **twitch**, a single stimulus-contraction-relaxation sequence in a muscle fiber. Twitches vary in duration, depending on muscle type and location, internal and external environmental conditions, and other factors. We have all probably seen or felt an involuntary "muscle twitch" under the skin. Such a muscle twitch is called a **fasciculation** (fa-sik-yū-LĀ-shun). A fasciculation involves more than one muscle fiber. It is the synchronous contraction of a *motor unit*, that is, a group of skeletal muscle fibers all controlled by a single motor neuron (see Module 9.10).

**3** This myogram shows the phases of a 40-msec twitch in a muscle fiber from the gastrocnemius muscle, a prominent superficial muscle of the calf. Remember that a millisecond is one thousandth of a second! So these events happen rapidly—in less time than a finger snap.

Go to
MasteringA&P™
PhysioEx 9.1
Skeletal
Muscle
Length-
Tension
(Ex 2,6)

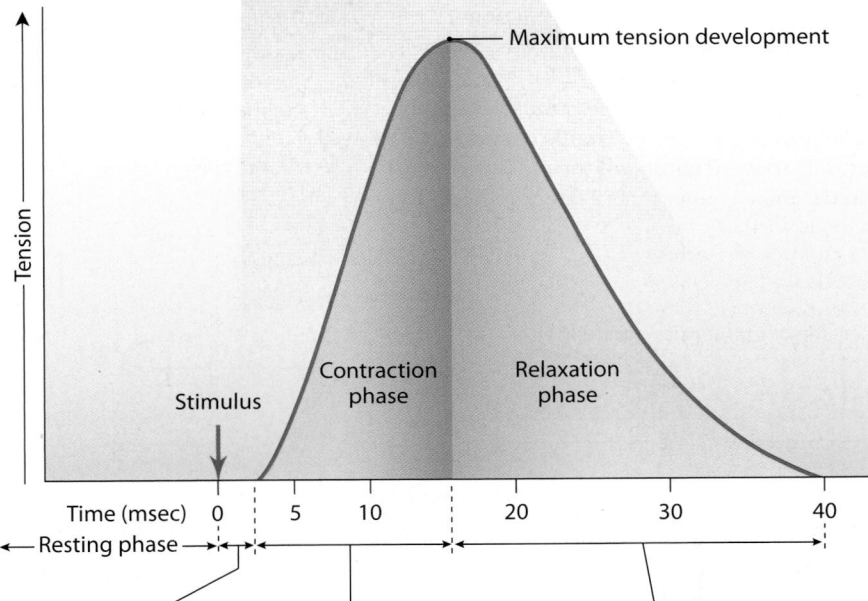

The **latent period** begins at stimulation and typically lasts about 2 msec. During this period, an action potential sweeps across the sarcolemma, and the SR releases calcium ions. The muscle fiber does not produce tension during the latent period, because the contraction cycle has yet to begin.

In the **contraction phase**, tension rises to a peak. For about 15 msec, calcium ions are binding to troponin, active sites on thin filaments are being exposed, and cross-bridge interactions are occurring.

The **relaxation phase** lasts about 25 msec. During this period, calcium levels are falling, active sites are being covered by tropomyosin, and the number of active cross-bridges is declining as they detach. As a result, tension returns to resting levels.

 **REVIEW**

**C.** Describe the events that occur during each phase of a twitch in a stimulated muscle fiber.

 **LEARNING OUTCOME**

Describe the mechanism responsible for tension production in a muscle fiber, and discuss the factors that determine the peak tension developed during a contraction.

# The peak tension developed by a skeletal muscle depends on the frequency of stimulation and the number of muscle fibers stimulated

Two factors determine the amount of tension produced by a skeletal muscle: (1) the amount of tension produced by each stimulated muscle fiber and (2) the total number of muscle fibers stimulated at a given moment.

 **REVIEW**

**A.** Compare incomplete tetanus with wave summation.

## Frequency of Stimulation and Muscle Fiber Tension

### Treppe

**1** Each time a skeletal muscle fiber is stimulated immediately after the relaxation phase has ended, the subsequent contraction will develop a slightly higher maximum tension than did the previous contraction. The increase in peak tension will continue over the first 30–50 stimulations. Because the tension rises like the steps in a staircase, this phenomenon is called **treppe** (TREP-eh, German for *staircase*). Most skeletal muscles do not demonstrate treppe. (However, treppe occurs in cardiac muscle tissue if stimuli of the same intensity are sent to the muscle fiber after a latent period.)

KEY

↓ = Stimulus

Maximum tension (in treppe)

Tension

Time

### Wave summation

**2** If a second stimulus arrives before the relaxation phase has ended, a second, more powerful contraction occurs. The addition of one twitch to another in this way is called **wave summation**. The duration of a single twitch determines the maximum time available to produce wave summation. For example, if a twitch lasts 20 msec (1/50 sec), a stimulus frequency of greater than 50 per second produces wave summation, whereas a stimulus frequency of less than 50 per second will produce individual twitches and treppe.

Tension

Time

### Incomplete tetanus

**3** A muscle producing almost peak tension during rapid cycles of contraction and relaxation is said to be in **incomplete tetanus** (*tetanos*, convulsive tension).

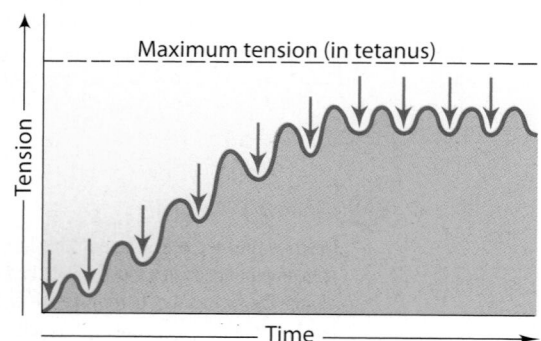

Maximum tension (in tetanus)

Tension

Time

### Complete tetanus

**4** **Complete tetanus** occurs when a higher stimulation frequency eliminates the relaxation phase. During complete tetanus, action potentials arrive so rapidly that the SR cannot reclaim calcium ions. The high cytosolic levels of $Ca^{2+}$ prolong the contraction, making it continuous. Tension plateaus at a maximum level. Although muscles can be forced into tetanus in the laboratory, they seldom if ever develop peak tension in the course of normal activities. In part, this is because normal movements require precise control over, and continuous variation in, the amount of tension produced.

Tension

Time

## Motor Units and Recruitment

**5** A typical skeletal muscle contains thousands of muscle fibers. Although some motor neurons control just a few muscle fibers, most control hundreds of them. The amount of tension produced is controlled at the subconscious level through variations in the number of muscle fibers stimulated.

A **motor unit** is a motor neuron and all the muscle fibers that it controls. The size of a motor unit indicates how fine, or precise, a movement can be. In the muscles of the eye, where precise control is extremely important, a motor neuron may control 4–6 muscle fibers. We have much less precise control over our leg muscles, where a single motor neuron may control 1000–2000 muscle fibers.

The muscle fibers of each motor unit are intermingled with those of other motor units. As a result, the direction of pull exerted on the tendon does not change when the number of activated motor units changes. When you decide to perform a specific movement, the contraction begins with the activation of the smallest motor units in the stimulated muscle. As the movement continues, larger motor units containing faster and more powerful muscle fibers are activated, and tension rises steeply. The smooth but steady increase in muscular tension produced by increasing the number of active motor units is called **recruitment**.

**KEY**

- Motor unit 1
- Motor unit 2
- Motor unit 3

Spinal cord

Cell bodies of motor neurons

Axons of motor neurons

Motor nerve

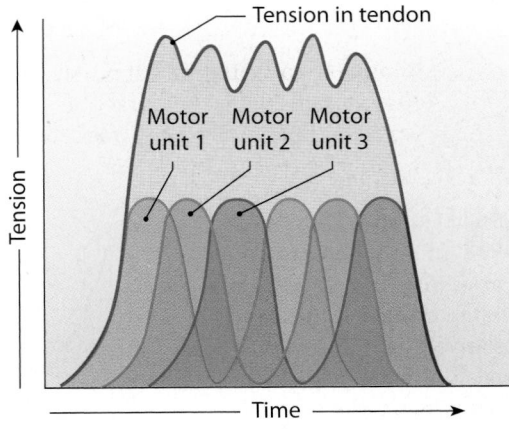

**?** **REVIEW**

**B.** Define *motor unit.*

 **?** **REVIEW**

**C.** Describe the relationship between the number of fibers in a motor unit and the precision of body movements.

**6** During a sustained contraction, motor units are activated on a rotating basis, allowing some to rest and recover while others are actively contracting. In this "relay team" approach, called **asynchronous motor unit summation**, each motor unit can recover somewhat before it is stimulated again. As a result, when your muscles contract for sustained periods, they produce slightly less than maximal tension.

 **Everyday Physiology**

### What Is Muscle Tone?

A variable number of motor units is always active, even when the entire muscle is not contracting. Their contractions do not produce enough tension to cause movement, but they do tense and firm the muscle. This resting tension in a skeletal muscle is called **muscle tone**, and it is regulated at the subconscious level. The activity level of each motor neuron changes constantly, and individual muscle fibers can relax while a constant tension is maintained in the attached tendon. Activated muscle fibers use energy, so the greater the muscle tone, the higher the "resting" rate of metabolism. Elevated muscle tone increases resting energy consumption by a small amount, but the effects are cumulative, and they continue 24 hours per day.

Tension in tendon

Motor unit 1   Motor unit 2   Motor unit 3

Tension

Time

 **LEARNING OUTCOME**

Discuss the factors that affect peak tension production during the contraction of an entire skeletal muscle, and explain the significance of the motor unit in this process.

# Muscle contractions may be isotonic or isometric; isotonic contractions may be concentric or eccentric

We can classify muscle contractions as *isotonic* or *isometric* on the basis of their pattern of tension production.

## Isotonic Contractions

In an **isotonic contraction** (*iso-*, equal + *tonos*, tension), tension rises to a constant level, and the skeletal muscle's length changes. Lifting an object off a desk, walking, and running involve isotonic contractions. There are two types of isotonic contractions: *concentric* and *eccentric*.

## Concentric Contractions

**1** In a **concentric contraction**, the muscle tension *exceeds* the load, and the muscle *shortens*. Consider an experiment with a skeletal muscle that is 1 cm$^2$ in cross-sectional area and can produce 4 kg (8.8 lb) of tension in complete tetanus. If we hang a load of 2 kg (4.4 lb) from that muscle and stimulate it, the muscle will shorten. An example of a concentric contraction is flexing the elbow joint from an extended position while holding a dumbbell. To lift the weight, the muscle must recruit enough motor units to overcome gravity and actively shorten.

**2** Before the muscle can shorten, the cross-bridges must produce enough tension to overcome the load—in this case, the 2-kg weight. At first, tension in the muscle rises until the tension in the tendon exceeds the load. Then, as the muscle shortens, its tension remains constant at a value that just exceeds the load. The term *isotonic* arose from this type of experiment.

**3** The speed, or rate, of muscle contraction varies inversely with the load on the muscle. If the load is relatively light, the muscle will contract quickly; if the load is heavy, the muscle will contract slowly or not at all. The speed of muscle contraction is fastest when the load equals zero.

**REVIEW**

**A.** Explain the relationship between load and speed of muscle contraction.

## Eccentric Contractions

**Go to**
MasteringA&P™
PhysioEx 9.1
Isotonic Contractions (Ex 2,7)

**4** In an **eccentric contraction**, the peak tension developed is *less than* the load, and the muscle *elongates* because of the contraction of another muscle or the pull of gravity. This figure shows what happens if we attach a 6-kg weight to the muscle and then stimulate the muscle (6 kg is 2 kg greater than the 4-kg maximum tension potential). To continue our example, think of lowering the dumbbell by extending your flexed elbow joint. To lower the weight, the muscle will decrease motor unit recruitment, the dumbbell will overcome the tension generated by the muscle, and the muscle will lengthen. In the example shown below, the muscle is still actively contracting, but not enough to lift the weight.

Support removed, eccentric contraction begins

6 kg

6 kg

**5** The rate of elongation during an eccentric contraction depends on the difference between the tension developed by the active muscle fibers and the size of the load. By varying the tension in an eccentric contraction, you can control the rate of elongation.

When the eccentric contraction ends, the unopposed load stretches the muscle until either the muscle tears, a tendon breaks, or the elastic recoil of the skeletal muscle is sufficient to oppose the load.

Muscle tension (kg)

Maximum tension produced

Support removed, contraction begins

Resting length

Muscle length (percent of resting length)

Time

**?  REVIEW**

**B.** Compare concentric and eccentric contractions.

## Isometric Contractions

In an **isometric contraction** (*metric*, measure), the muscle as a whole does not change length, and the tension produced never exceeds the load. Many of the reflexive muscle contractions that keep your body upright when you stand or sit involve isometric contractions.

**6** If the load equals the peak tension, the load won't move when the muscle contracts. The contracting muscle bulges, but not as much as it does during an isotonic contraction. In an isometric contraction, although the muscle as a whole does not shorten, the individual muscle fibers shorten as connective tissues stretch. The muscle fibers cannot shorten further, because the tension does not exceed the load.

Muscle contracts (isometric contraction)

6 kg

6 kg

Amount of load

Muscle tension (kg)

Muscle relaxes

Maximum potential tension produced

Contraction begins

Length unchanged

Muscle length (percent of resting length)

Time

**?  INTEGRATION**

**C.** Can a skeletal muscle contract without shortening? Why or why not?

**LO  LEARNING OUTCOME**

Compare the different types of muscle contractions.

# Module 9.12

# Muscle contraction requires large amounts of ATP that may be produced anaerobically or aerobically

## Energy Production

**1** Mitochondrial activity is the ultimate source of the energy required by active skeletal muscles.

**Glycolysis** is the anaerobic breakdown of glucose to pyruvate in the cytosol of a cell. It is an anaerobic process because it does not require oxygen. Glycolysis provides a net gain of 2 ATP molecules and generates 2 pyruvate molecules from each glucose molecule.

**Aerobic metabolism** normally provides 95 percent of the ATP demands of a resting cell. In this process, introduced in Module 3.7, mitochondria absorb oxygen, ADP, phosphate ions, and organic substrates (such as pyruvate) from the surrounding cytosol. Primarily through the activity of the electron transport chain, large amounts of energy are released and used to make ATP. The entire process is very efficient: For each molecule of pyruvate "fed" into the citric acid cycle, the cell gains 15 ATP molecules.

**? REVIEW**

**A.** What basic reactants do mitochondria absorb from the cytosol to synthesize ATP?

**2** This table characterizes the energy reserves of a typical skeletal muscle fiber. When energy demands are low and oxygen is abundant, a muscle fiber absorbs nutrients from the interstitial fluid and builds up its energy reserves. In addition to a small amount of free ATP in a cell, muscle fibers store a high-energy compound called **creatine phosphate (CP)**. (**Creatine [C]** is a compound assembled within skeletal muscles from catabolized amino acids.) However, muscle fibers store only a relatively small amount of energy in high-energy compounds. Most energy is stored as *glycogen*, which may account for 1.5 percent of total muscle weight and enables muscle contractions to continue for extended periods.

**? REVIEW**

**B.** Identify three sources of stored energy utilized by muscle fibers.

### Sources of Energy Stored in a Typical Muscle Fiber

| Energy Source | Initial Quantity | Utilization Process | Number of Twitches Supported by Each Energy Source Alone | Duration of Isometric Tetanic Contraction Supported by Each Energy Source Alone |
|---|---|---|---|---|
| Free ATP | 3 mmol | ATP ⟶ ADP + P | 10 | 2 sec |
| CP | 20 mmol | ADP + CP ⟶ ATP + C | 70 | 15 sec |
| Glycogen | 100 mmol | Glycolysis (anaerobic metabolism) | 670 | 130 sec |
| Glycogen | 100 mmol | Aerobic metabolism | 12,000 | 2400 sec (40 min) |

# Muscle Metabolism at Different Levels of Activity

## Muscle at rest

**3** In a resting skeletal muscle, the demand for ATP is low. More than enough oxygen is available for the mitochondria to meet that demand, and they produce a surplus of ATP. The extra ATP is used to build up reserves of CP and glycogen. Resting muscle fibers absorb fatty acids and glucose delivered by the bloodstream. The fatty acids are broken down in the mitochondria, generating ATP that is used to convert creatine to creatine phosphate and glucose to glycogen.

## Muscle at moderate activity levels

**4** At moderate levels of activity, the demand for ATP increases. This demand is met by the mitochondria, and the rate of oxygen consumption increases as a result. The skeletal muscle now relies primarily on the aerobic metabolism of pyruvate to generate ATP, and the pyruvate is provided by glycolysis, using glucose obtained from stored glycogen reserves. As long as mitochondrial activity can meet the demand for ATP, the muscle will not become fatigued until glycogen, lipid, and amino acid reserves are exhausted. This type of fatigue affects the muscles of endurance athletes, such as marathon runners, after hours of exertion.

## Muscle at peak activity levels

**5** At peak levels of activity, ATP demands are enormous and mitochondrial ATP production plateaus at a maximum rate determined by the availability of oxygen. At peak exertion, the mitochondria can provide only about one-third of the ATP needed. Glycolysis produces the rest. When glycolysis produces pyruvate faster than it can be utilized by the mitochondria, pyruvate levels rise in the cytosol. Under these oxygen-limited conditions, pyruvate is converted to a three-carbon **lactate** molecule. At the same time, the hydrolytic breakdown of ATP during muscle contraction continues to release hydrogen ions ($H^+$). During aerobic metabolism, these $H^+$ are absorbed by the mitochondria. However, when ATP production is relying on anaerobic metabolism (glycolysis), the $H^+$ are not being absorbed rapidly enough. At peak activity, then, both $H^+$ and lactate build up, resulting in an increased lactate blood level and metabolic acidosis, often referred to as **lactic acidosis**. After only a few seconds of peak activity, changes in pH will alter the functional characteristics of key enzymes so that the muscle fiber can no longer contract. Sprinters usually experience this type of muscle fatigue.

 **REVIEW**
**C.** When do muscle fibers produce lactate?

 **LEARNING OUTCOME**
Describe the processes by which muscle fibers obtain the energy to power contractions.

# Muscles fatigue and may need an extended recovery period

If a skeletal muscle can no longer perform at the required level of activity, it is **fatigued**. Many factors are involved in promoting muscle fatigue, but after peak activity a major factor is the decline in pH (from about 7.1 to 6.4) within the muscle fibers and the muscle as a whole, decreasing calcium ion binding to troponin and altering enzyme activities. In the **recovery period**, the conditions in muscle fibers are returned to normal, preexertion levels. After moderate activity, it may take several hours for muscle fibers to recover. After sustained activity at higher levels, complete recovery can take a week.

 **REVIEW**

**A.** How is skeletal muscle recovery different after moderate activity compared to sustained activity at higher levels?

**1** Glycolysis enables a skeletal muscle to continue contracting even when insufficient oxygen limits mitochondrial activity. Recall that glycolysis is not the most efficient way to generate ATP. It squanders the glucose reserves of the muscle fibers, and it is potentially dangerous because the dissociation of lactic acid can lower the pH of the blood and tissues. Glycolysis can produce ATP faster than aerobic metabolism, but only until glycogen reserves are depleted (1–2 minutes). In terms of energy efficiency, glycolysis captures some 4–6 percent of the energy as ATP in the conversion of glucose to pyruvate. The reactions of glycolysis also elevate body temperature, triggering increased sweat gland activity.

**2** During the recovery period, when oxygen is available, lactate can be recycled by conversion back to pyruvate. The pyruvate can then be used either by mitochondria to generate ATP or as a substrate for enzyme pathways that synthesize glucose and rebuild glycogen reserves. Aerobic metabolism is much more efficient than glycolysis—the muscle fiber can capture about 42 percent of the energy released. However, whenever they are using ATP, muscles are continuously releasing heat. About 70–80 percent of body heat is produced by resting skeletal muscles.

**3** The formation of lactate by skeletal muscle allows the continuation of ATP production during peak exertion. Much of the lactate released by muscle fibers during strenuous activity diffuses from the muscle tissue into the bloodstream. The liver absorbs and recycles it and releases glucose into the circulation. This shuffling of lactate to the liver and of glucose back to muscle cells is called the **Cori cycle**. Throughout the recovery period, the body's oxygen demand remains elevated above normal resting levels. The more ATP required, the more oxygen will be needed. The amount of oxygen required to restore normal, preexertion conditions is called the **oxygen debt**, or **excess postexercise oxygen consumption (EPOC)**. Most of the additional oxygen consumption occurs in skeletal muscle fibers, which must restore ATP, creatine phosphate, and glycogen concentrations to their former levels, and in liver cells, which generate the ATP needed to convert excess lactate to glucose.

**? REVIEW**

**B.** Define *oxygen debt* (excess postexercise oxygen consumption).

9

## Peak Activity

Much of the lactate produced during peak exertion diffuses out of the muscle fibers and into the bloodstream. The liver absorbs this lactate and begins converting it into pyruvate.

## Recovery

This process continues after exertion has ended because lactate levels within muscle fibers remain relatively high and lactate continues to diffuse into the bloodstream. After the absorbed lactate is converted to pyruvate in the liver, about 30 percent of the new pyruvate molecules are broken down in the mitochondria, providing the ATP needed to convert the remaining 70 percent of pyruvate molecules into glucose. The glucose molecules are then released into the circulation. Skeletal muscle fibers take up the glucose molecules and use them to rebuild their glycogen reserves.

20–30%

$O_2$

$H_2O$

LIVER

ATP        ADP

Pyruvate      70–80%      Glucose

Lactate

Lactate

MUSCLE
FIBER

Pyruvate

Glucose        Glucose

Glycogen reserves in muscle

**? REVIEW**

**C.** What happens to the lactate produced by skeletal muscle during peak activity?

**Lo LEARNING OUTCOME**

Describe the factors that contribute to muscle fatigue, and discuss the processes involved in the muscle's subsequent recovery.

# Fast, slow, and intermediate skeletal muscle fibers differ in size, internal structure, metabolism, and resistance to fatigue

The human body has three major types of skeletal muscle fibers: *fast fibers*, *slow fibers*, and *intermediate fibers*.

**1** Most of the skeletal muscle fibers in the body are called **fast fibers** because they can reach peak twitch tension in 0.01 second or less after stimulation. Fast fibers have large diameters and contain densely packed myofibrils, large glycogen reserves, and relatively few mitochondria. The tension produced by a muscle fiber is directly proportional to the number of myofibrils, so muscles dominated by fast fibers produce powerful contractions. However, fast fibers fatigue rapidly because their contractions use ATP in massive amounts, and they have relatively few mitochondria to generate ATP. As a result, prolonged activity is supported primarily by anaerobic metabolism. Skeletal muscles dominated by fast fibers appear pale and are often called *white muscles*.

**Fast fibers in cross section** LM × 171

**2** **Slow fibers** have only about half the diameter of fast fibers and take three times as long to reach peak tension after stimulation. Slow fibers are specialized to continue contracting for extended periods, long after a fast fiber would become fatigued. Slow muscle fibers are surrounded by a more extensive network of capillaries than fast muscle tissue, so they have a dramatically higher oxygen supply to support mitochondrial activity. Slow fibers also contain the red pigment **myoglobin** (MĪ-ō-glō-bin), a globular protein that is structurally related to hemoglobin, the red oxygen-carrying pigment in blood. Both myoglobin and hemoglobin reversibly bind oxygen molecules. Myoglobin is most abundant in slow fibers, so resting slow fibers contain substantial oxygen reserves that can be mobilized during a contraction. Because slow fibers have both an extensive capillary supply and a high concentration of myoglobin, skeletal muscles dominated by slow fibers appear dark red.

**Slow fibers in cross section** LM × 171

 **REVIEW**

**A.** Contrast fast fibers with slow fibers in terms of diameter, glycogen reserves, myoglobin content, and relative abundance of mitochondria.

**3** This transmission electron micrograph (a longitudinal section) shows that a fast muscle fiber (W, for white) and a slow muscle fiber (R, for red) have different sizes and densities and that slow fibers have more mitochondria (M) and a more extensive capillary supply. Our bodies also contain a third group of muscle fibers—**intermediate fibers**—that more closely resemble fast fibers, for they contain little myoglobin and are relatively pale. Intermediate fibers have a more extensive capillary network around them and are more resistant to fatigue than are fast fibers.

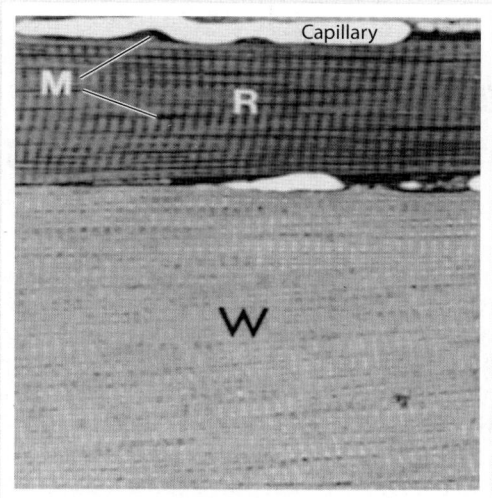

**Fast (W) and slow (R) fibers in longitudinal section**      TEM × 783

## Properties of Skeletal Muscle Fiber Types

| Property | Fast Fibers | Slow Fibers | Intermediate Fibers |
|---|---|---|---|
| **Cross-sectional diameter** | Large | Small | Intermediate |
| **Color** | White | Red | Pink |
| **Myoglobin content** | Low | High | Low |
| **Capillary supply** | Scarce | Dense | Intermediate |
| **Mitochondria** | Few | Many | Intermediate |
| **Time to peak tension** | Rapid | Prolonged | Medium |
| **Contraction speed** | Fast | Slow | Fast |
| **Fatigue resistance** | Low | High | Intermediate |
| **Glycolytic enzyme concentration in cytosol** | High | Low | High |
| **Substrates used for ATP generation during contraction (type of metabolism)** | Carbohydrates (anaerobic) | Lipids, carbohydrates, amino acids (aerobic) | Primarily carbohydrates (anaerobic) |
| **Alternative names** | Type II-B, FF (fast fatigue), white, fast-twitch glycolytic | Type I, S (slow), red, SO (slow oxidative), slow-twitch oxidative | Type II-A, FR (fast resistant), fast-twitch oxidative |

**4** This table compares the properties of fast, slow, and intermediate muscle fibers. The percentage of each fiber type in a skeletal muscle can vary greatly. Most human muscles contain a mixture of fiber types and so appear pink. However, there are no slow fibers in muscles of the eye or hand, where swift but brief contractions are required. Many back and calf muscles are dominated by slow fibers; these muscles contract almost continuously to maintain an upright posture. The percentage of fast versus slow fibers in each muscle is genetically determined, but the ratio of intermediate fibers to fast fibers can increase as a result of athletic training. For example, if a muscle is used repeatedly for endurance events, some of the fast fibers will develop the appearance and functional capabilities of intermediate fibers.

 **INTEGRATION**

**B.** Why would a sprinter experience muscle fatigue before a marathon runner would?

**C.** Which type of muscle fiber predominates in leg muscles of endurance athletes, such as long-distance runners?

**LEARNING OUTCOME**

Relate the types of muscle fibers to muscle performance.

# Module 9.15
# Many factors can result in muscle hypertrophy, atrophy, or paralysis

## Hypertrophy

**1** As a result of repeated, exhaustive stimulation, muscle fibers develop more mitochondria, a higher concentration of glycolytic enzymes, and larger glycogen reserves. Such muscle fibers have more myofibrils than do fibers that are less used, and each myofibril contains more thick and thin filaments. The net effect is **hypertrophy**, or an enlargement of the stimulated muscle. The number of muscle fibers does not change significantly, but the muscle as a whole enlarges because each muscle fiber increases in diameter. The muscle also becomes strong because tension production is proportional to the cross-sectional area of the muscle. The muscles of a bodybuilder are excellent examples of muscular hypertrophy. Although hypertrophy can be promoted by administering steroid hormones such as androgens (male sex hormones), exhaustive training is still required, and the side effects of steroid use can be very dangerous.

## Atrophy

Atrophy between 1st and 2nd metacarpals

**2** A skeletal muscle that is not regularly stimulated by a motor neuron loses muscle tone and mass. The muscle becomes flaccid, and the muscle fibers become smaller and weaker. This reduction in muscle size, tone, and power is called **atrophy**. A variable degree of muscle atrophy is a normal consequence of aging. Individuals of any age who are paralyzed by spinal injuries or other damage to the nervous system will gradually lose muscle tone and size in the areas affected. Even a temporary reduction in muscle use can lead to muscular atrophy; you can easily observe this effect by comparing "before and after" limb muscles in someone who has worn a cast. Muscle atrophy is initially reversible, but dying muscle fibers are not replaced. In extreme atrophy, the functional losses are permanent. That is why physical therapy is crucial for people who are temporarily unable to move normally.

**?** **REVIEW**
**A.** Define *muscle hypertrophy* and *muscle atrophy*.

## Muscular Dystrophy

**3** There are several different **muscular dystrophies** (DIS-trō-fēz), and all are inherited diseases that produce muscle weakness and deterioration. The most common and best understood are **Duchenne muscular dystrophy (DMD)** and **Becker muscular dystrophy** (**BMD**). These forms appear in childhood and generally affect only males. It is estimated that 1 in 3500 to 5000 newborn males worldwide is born with the condition. A progressive muscular weakness develops, and most individuals die as a result of respiratory paralysis. Skeletal muscles are primarily affected, although for some reason the facial muscles continue to function normally. In later stages of the disease, the facial muscles and cardiac muscle tissue may also become involved. The inheritance of DMD and BMD is sex-linked: Women carrying the defective gene are unaffected, but each of their male children will have a 50 percent chance of developing DMD or BMD. Genetic tests are available to determine whether or not a woman is carrying the defective gene, and prenatal tests can determine whether a fetus has the condition.

## Clinical Conditions That Affect Skeletal Muscles

**4** Because skeletal muscles depend on motor neurons for stimulation, disorders that impair the nervous system can indirectly affect the muscular system.

In **polio**, a virus attacks motor neurons in the spinal cord and brain, causing muscular atrophy and **paralysis** (loss of voluntary movement).

A very common bacterium, *Clostridium tetani*, causes the disease **tetanus**. *Clostridium* bacteria can thrive only in tissues with low oxygen levels. For this reason, a deep puncture wound, such as that from a nail, is much more likely to result in tetanus than a shallow, open cut that bleeds freely and flushes the site. When active in body tissues, these bacteria release a powerful toxin that suppresses the mechanism that inhibits motor neuron activity of skeletal muscles throughout the body. Severe tetanus has a 40–60 percent mortality rate. Thanks to an effective immunization program, only about 100 cases occur in the United States each year.

Neuromuscular communication can be disrupted by exposure to toxins that block ACh release or prevent ACh binding at the motor end plate. **Botulism**, caused by consuming food contaminated with toxins produced by *Clostridium botulinum*, paralyzes skeletal muscles by preventing ACh release at neuromuscular junctions. **Myasthenia gravis**, an autoimmune disorder characterized by the loss of ACh receptors at the neuromuscular junctions, results in progressive muscular weakness.

When skeletal muscles are deprived of nutrients and oxygen for an extended period, they begin to deteriorate. Within a few hours, the skeletal muscle fibers have run out of ATP, and the sarcoplasmic reticulum becomes unable to pump $Ca^{2+}$ out of the cytosol. Calcium ions diffusing into the cytosol from the extracellular fluid or leaking out of the sarcoplasmic reticulum then trigger a sustained contraction. Without ATP, the cross-bridges cannot detach from the active sites, so the affected muscle fibers lock up in the contracted state. This can occur locally after acute injury or after death.

**5** Shortly after death, a generalized skeletal muscle contraction, called **rigor mortis**, occurs throughout the body, beginning with the smaller muscles of the face, neck, and arms. As the SR deteriorates, calcium ions are released, and a sustained contraction begins. As ATP reserves are exhausted, the muscles become locked in the contracted state. Because all skeletal muscles are involved during rigor mortis, the individual becomes "stiff as a board." It typically begins 2–7 hours after death and disappears after 1–6 days or when decomposition begins, but the timing depends on environmental factors such as temperature. Forensic pathologists base time of death estimates on the degree of rigor mortis and environmental conditions.

**? REVIEW**

**B.** Explain how the flexibility or rigidity of a dead body can provide a clue about a murder victim's time of death.

**? INTEGRATION**

**C.** Six weeks after Fred broke his leg, the cast is removed. As he steps down from the exam table, his leg gives way, and he falls. Propose a logical explanation.

**LO LEARNING OUTCOME**

Explain the physiological factors responsible for muscle hypertrophy, atrophy, and paralysis.

**Labeling:** Use the following terms to correctly label the structures in the diagram representing a blood vessel and a skeletal muscle fiber at rest.

- creatine
- glycogen
- CP
- glucose
- $O_2$
- fatty acids

**Matching:** Match each lettered term with the most closely related description.

a. eccentric contraction

b. isometric contraction

c. isotonic contraction

d. concentric contraction

**7** Muscle does not change length during contraction

**8** Muscle changes length during contraction

**9** Peak tension less than load, and muscle elongates

**10** Peak tension greater than load, and muscle shortens

7 _____

8 _____

9 _____

10 _____

**Short answer:** Complete the following table by writing in the anatomical and physiological properties of the three types of skeletal muscle fibers.

| Property | Fast Fibers | Slow Fibers | Intermediate Fibers |
|---|---|---|---|
| Cross-sectional diameter | **11** | **12** | **13** |
| Color | **14** | Red | **15** |
| Myoglobin content | **16** | **17** | Low |
| Capillary supply | Scarce | **18** | **19** |
| Mitochondria | **20** | **21** | **22** |
| Time to peak tension | **23** | Prolonged | **24** |
| Contraction speed | **25** | **26** | **27** |
| Fatigue resistance | **28** | **29** | Intermediate |
| Glycolytic enzyme concentration in cytosol | High | **30** | **31** |

# Study Outline

## SECTION 1 • Functional Anatomy of Skeletal Muscle Tissue

### Module 9.1

**Skeletal muscle tissue enables body movement and other vital functions    p. 315**

1. The three types of muscle tissue (skeletal, cardiac, and smooth) consist chiefly of muscle cells that are highly specialized for contraction.

2. **Skeletal muscles** are complex organs that are attached directly or indirectly to bones.

3. Skeletal muscle functions to produce skeletal movement, maintain posture and body position, support soft tissue, guard entrances and exits of the body, maintain body temperature, and provide nutrient reserves.

### Module 9.2

**Skeletal muscle contains muscle tissue, connective tissues, blood vessels, and nerves    p. 316**

4. **Skeletal muscle** is surrounded by **epimysium**, a dense layer of collagen fibers.

5. A **muscle fascicle** is a bundle of muscle fibers surrounded by **perimysium**.

6. Individual muscle fibers are sheathed in **endomysium**.

7. Embryonic **myoblasts** fuse to form multinucleate cells that develop into distinct muscle fibers.

8. Skeletal muscle fibers have a plasma membrane called a **sarcolemma** and a cytoplasm called **sarcoplasm**.

### Module 9.3

**Skeletal muscle fibers contain T tubules and sarcoplasmic reticula that surround contractile myofibrils made up of sarcomeres    p. 318**

9. A single skeletal muscle fiber contains hundreds to thousands of banded, cylindrical structures called **myofibrils**.

10. Myofibrils consist of bundles of protein **myofilaments**. **Thin filaments** are composed mostly of actin, and **thick filaments** are composed primarily of myosin.

11. Myofilaments have repeating, light- and dark-banded functional units called **sarcomeres**.

12. The sarcolemma separates negative charges lining its internal surface from positive charges lining its external surface.

13. **Transverse tubules**, or **T tubules**, are extensions of the sarcolemma into the sarcoplasm. The combination of a T tubule plus a pair of **terminal cisternae** of the **sarcoplasmic reticulum (SR)** forms a **triad**.

14. Calcium ions ($Ca^{2+}$) are stored in the SR. A muscle contraction begins when the SR releases $Ca^{2+}$ into the cytosol.

### Module 9.4

**The sliding of thin filaments past thick filaments produces muscle contraction    p. 320**

15. The thin filaments are attached to the Z line by **actinin**.

16. **F-actin** (filamentous actin) is a twisted strand of individual **G-actin** (globular actin) molecules.

17. Strands of **tropomyosin** cover the active sites of G-actin. **Troponin** is bound to tropomyosin, and it also has one site bound to G-actin and two sites for calcium ions ($Ca^{2+}$).

18. Thick filaments contain **myosin molecules** arranged with their tails pointing toward the M line. The thick filaments are attached to the Z line by **titin**.

19. The myosin molecule has a **tail** bound to other myosin molecules, a **free head** that has two globular subunits, and a connection between the two that acts as a hinge.

20. During contraction, myosin heads form cross-bridges with thin filaments.

21. When a skeletal muscle fiber contracts, thin filaments slide past thick filaments. This is known as the **sliding filament theory**. This occurs in every sarcomere along a myofibril, so the myofibril gets shorter. When myofibrils get shorter, so does the muscle fiber.

### Module 9.5

**Skeletal muscle fibers and neurons have excitable plasma membranes that produce and carry electrical impulses called action potentials    p. 322**

22. At rest, a cell is **polarized**. The inner surface of its plasma membrane has a slight negative charge with respect to its outer surface.

23. In undisturbed neurons and skeletal muscle fibers, typical **resting membrane potentials** are –70 mV and –85 mV, respectively.

24. Open $Na^+$ and $K^+$ leak channels in the plasma membrane of cells allow sodium and potassium ions to move down their concentration gradients. Sodium–potassium ion pumps maintain the cell's resting potential by exporting 3 $Na^+$ from the cell, in exchange for 2 $K^+$.

25. Excitable membranes contain **voltage-gated ion channels** activated or inactivated by changes in membrane potential.

26. A cell is **depolarized** when a threshold potential is reached (–55 mV for skeletal muscle) and $Na^+$ rush into the cell. The depolarization peaks at +30 mV. $Na^+$ inflow then abruptly stops and **repolarization** begins as $K^+$ rush out of the cell. This loss of positive ions causes the membrane potential to become negative again.

27. In neurons and skeletal muscle fibers, the depolarization produces an electrical impulse, or **action potential**, that is spread, or **propagated**, along their excitable membranes.

### Module 9.6

**A skeletal muscle fiber contracts when stimulated by a motor neuron    p. 324**

28. Each skeletal muscle fiber is controlled by the nervous system at a specialized site called a **neuromuscular junction (NMJ)**.

29. The NMJ is made up of an **axon terminal**, a specialized region of the sarcolemma called the **motor end plate**, and a narrow space between the two called the **synaptic cleft**.

30. When the action potential reaches the synaptic terminal, it releases the neurotransmitter **acetylcholine (ACh)** into the synaptic cleft. The enzyme **acetylcholinesterase (AChE)** breaks down ACh.

31. ACh binds to ACh-membrane receptor channels in the motor end plate, changing the membrane permeability and causing $Na^+$ to rush into the sarcoplasm. This generates an action potential in the skeletal muscle fiber.

32. The action potential sweeps down each T tubule to the terminal cisternae, causing the SR to release $Ca^{2+}$. This event, called **excitation-contraction coupling**, triggers the contraction of the muscle fiber.

### Module 9.7

**A muscle fiber contraction uses ATP in a cycle that repeats during the contraction    p. 326**

33. Calcium ions ($Ca^{2+}$) bind to troponin, which causes the tropomyosin to move away from the actin active site, allowing interaction with the energized myosin heads. Binding of active sites to myosin heads forms **cross-bridges**.

34. The myosin head pivots as it releases the bound ADP and P.

35. The cross-bridge is broken when another ATP binds to the myosin head.

36. Myosin reactivation occurs when the free myosin head splits ATP into ADP and P.

37. The entire cycle is repeated several times each second, as long as $Ca^{2+}$ concentrations remain elevated and ATP reserves are sufficient.

## SECTION 2 • Functional Properties of Skeletal Muscle

### Module 9.8

**Muscle tension develops from the events that occur during excitation-contraction coupling    p. 329**

38. A skeletal muscle contracts when stimulated by a motor neuron at a neuromuscular junction.

39. **Excitation-contraction coupling** begins with the release of ACh at the axon terminal. This causes an action potential in the sarcolemma.

40. The action potential travels along the T tubules to the triads, where it triggers the release of $Ca^{2+}$ from the terminal cisternae of the SR.

41. Calcium ions ($Ca^{2+}$) permit thick and thin filaments to interact, causing the sarcomere to shorten, pulling the ends of the muscle fiber closer together.

42. The entire skeletal muscle shortens and produces **tension** on the tendons at either end.

### Module 9.9

**Tension is greatest when muscle fibers are stimulated at optimal length    p. 330**

43. The tension a muscle fiber produces is related to sarcomere length (**sarcomere length-tension relationship**).

44. Tension is produced most efficiently when the maximum number of cross-bridges can form. This sarcomere length is called **optimal resting length**.

45. A **twitch** is a single stimulus-contraction-relaxation sequence in a muscle fiber.

46. The phases of a twitch include the **latent period, contraction phase**, and **relaxation phase**.

### Module 9.10

**The peak tension developed by a skeletal muscle depends on the frequency of stimulation and the number of muscle fibers stimulated    p. 332**

47. The amount of tension produced by a skeletal muscle is determined by the amount of tension produced by each stimulated muscle fiber and by the total number of muscle fibers stimulated at a given moment.

48. **Treppe** is a progressive increase in twitch tension that is not demonstrated by most skeletal muscles.

49. Repeated stimulation before the relaxation phase ends may produce **wave summation**, in which one twitch is added to another.

50. A muscle producing almost peak tension during rapid cycles of contraction and relaxation is in **incomplete tetanus**. **Complete tetanus** occurs when the relaxation phase is eliminated.

51. A **motor unit** is a motor neuron and all the muscle fibers that it controls. Increasing the number of active motor units is called **recruitment**.

### Module 9.11

**Muscle contractions may be isotonic or isometric; isotonic contractions may be concentric or eccentric    p. 334**

52. In **isotonic contractions**, the tension in the muscle rises, and the length of the muscle changes. There are two types of isotonic contractions: concentric and eccentric contractions.

53. In **concentric contractions**, the muscle shortens. In **eccentric contractions**, the load is greater than peak tension, so the muscle lengthens.

54. In **isometric contractions**, the tension rises, but the length of the muscle does not change.

### Module 9.12

**Muscle contraction requires large amounts of ATP that may be produced anaerobically or aerobically    p. 336**

55. Mitochondrial activity is the ultimate source of the energy required by active skeletal muscles.

56. **Glycolysis** is the anaerobic breakdown of glucose to pyruvate in the cytosol of a cell. It is an anaerobic process because it does not require oxygen.

57. **Aerobic metabolism** normally provides 95 percent of the ATP demands of a resting cell. In this process, mitochondria absorb oxygen, ADP, phosphate ions, and pyruvate. For each molecule of pyruvate, the cell gains 15 ATP molecules.

58. In addition to a small amount of free ATP, muscle fibers store a high-energy compound called **creatine phosphate (CP)**.

59. In resting skeletal muscle fibers, the demand for ATP is low, and mitochondria produce a surplus of ATP.

60. At moderate levels of activity, ATP demand increases. Skeletal muscle now relies primarily on aerobic metabolism of pyruvate.

61. At peak activity levels, skeletal muscle relies heavily on glycolysis to generate ATP because the mitochondria cannot obtain enough oxygen to meet demand. The result is an accumulation of **lactate** in the muscle accompanied by a lowering of intracellular pH that contributes to fatigue.

## Module 9.13

### Muscles fatigue and may need an extended recovery period    p. 338

62. An active skeletal muscle is **fatigued** when it can no longer continue to perform.

63. In the **recovery period**, the conditions in muscle fibers are returned to normal, preexertion levels.

64. The **oxygen debt**, or **excess postexercise oxygen consumption (EPOC)**, resulting from muscle activity is the amount of oxygen required during the recovery period to restore the muscle to its normal condition.

## Module 9.14

### Fast, slow, and intermediate skeletal muscle fibers differ in size, internal structure, metabolism, and resistance to fatigue    p. 340

65. There are three major types of muscle fibers: fast, slow, and intermediate.

66. **Fast fibers** are large in diameter and contain densely packed myofibrils, large glycogen reserves, and relatively few mitochondria. They contract rapidly and powerfully, are supported by anaerobic metabolism, and fatigue easily.

67. **Slow fibers** are smaller in diameter, slower in contraction speed, contain more capillaries and mitochondria, and also contain the red oxygen-carrying pigment **myoglobin**.

68. **Intermediate fibers** are very similar to fast fibers but have greater resistance to fatigue.

## Module 9.15

### Many factors can result in muscle hypertrophy, atrophy, or paralysis    p. 342

69. **Hypertrophy** is the enlargement of stimulated muscle fibers as a result of development of more mitochondria, greater glycogen and enzyme reserves, and an increase in myofibrils.

70. **Atrophy** is a reduction in muscle size, tone, and power; it is often due to disuse or disease.

71. A variety of clinical conditions can affect skeletal muscles. These include **Duchenne muscular dystrophy (DMD)** and **Becker muscular dystrophy (BMD)**, **polio**, **tetanus**, **botulism**, and **myasthenia gravis**.

# Chapter Review Questions

**True/False:** Indicate whether each statement is true or false.

**1**  An increase in sarcomere length reduces the tension produced in a muscle fiber by reducing the size of the zone of overlap and the number of potential cross-bridge interactions.

**2**  Muscle hypertrophy is mostly explained by an increase in muscle fiber (cell) number as a result of exercise.

**3**  The Cori cycle describes the transfer of lactate to the liver and glucose back to the muscle cells during the recovery period.

**4**  In glycolysis, one molecule of glucose produces 2 molecules of pyruvate and 15 ATP.

**1** _____

**2** _____

**3** _____

**4** _____

**Matching:** Match each lettered term with the most closely related description.

a. myoblasts

b. sarcoplasmic reticulum

c. T tubules

d. tropomyosin

e. troponin

f. I band

g. A band

h. synaptic vesicles

i. motor end plate

j. triad

**5** Contains $Ca^{2+}$     **5** _____

**6** Contains thick filaments and thin filaments     **6** _____

**7** Contain ACh-receptor membrane channels     **7** _____

**8** Extensions of the sarcolemma into the sarcoplasm     **8** _____

**9** Embryonic cells that fuse to form muscle fibers     **9** _____

**10** Contains the neurotransmitter ACh     **10** _____

**11** Formed by T tubules and a pair of terminal cisternae     **11** _____

**12** Contains only thin filaments     **12** _____

**13** Has receptor sites for $Ca^{2+}$     **13** _____

**14** Cover the active sites of the G-actin     **14** _____

**9**

**Multiple choice:** Select the correct answer from the list provided.

**15** The connective tissue coverings of a skeletal muscle, listed from superficial to deep, are
- a) endomysium, perimysium, and epimysium.
- b) endomysium, epimysium, and perimysium.
- c) epimysium endomysium, and perimysium.
- d) epimysium, perimysium, and endomysium.

**16** The detachment of the myosin cross-bridges is directly triggered by
- a) the repolarization of T tubules.
- b) the attachment of ATP to the myosin heads.
- c) the hydrolysis of ATP.
- d) calcium ions.

**17** A muscle producing almost peak tension during rapid cycles of contraction and relaxation is said to be in
- a) incomplete tetanus.
- b) treppe.
- c) complete tetanus.
- d) a twitch.

**18** The type of contraction in which the muscle tension rises but the muscle does not change length is
- a) an isotonic contraction.
- b) an isometric contraction.
- c) a concentric contraction.
- d) an eccentric contraction.

**19** Which of the following statements about myofibrils is *not* correct?
- a) Each skeletal muscle fiber contains hundreds to thousands of myofibrils.
- b) Myofibrils contain repeating units called sarcomeres.
- c) Myofibrils extend the length of a skeletal muscle fiber.
- d) Filaments consist of bundles of myofibrils.

**20** Vesicles filled with acetylcholine (ACh) are found in the
- a) axon terminal.
- b) motor end plate.
- c) synaptic cleft.
- d) transverse tubule.

**21** The properties of slow fibers include all of the following *except*
- a) red in color.
- b) low myoglobin content.
- c) many mitochondria.
- d) small cross-sectional diameter.

**22** Which of the following activities involves eccentric muscle contractions?
- a) maintaining upright posture
- b) lifting a barbell over your head
- c) lowering yourself into a chair
- d) blinking your eyes

**23** A motor neuron and all the muscle fibers that it controls is called a
- a) motor unit.
- b) motor end plate.
- c) neuromuscular junction.
- d) cross-bridge.

**24** The region of the A band in a sarcomere that contains only thick filaments is the
- a) Z line.
- b) I band.
- c) zone of overlap.
- d) H band.

## Short answer

**25** What structural feature of a skeletal muscle fiber is responsible for carrying action potentials into the interior of the cell?

**26** Describe how it is possible that the muscles flexing the elbow can adjust their tension to lift a cup of coffee to your mouth or curl a dumbbell in the gym.

**27** Research in exercise physiology suggests that for several days after a bout of intense aerobic exercise, people consume oxygen at a rate greater than when they are normally at rest. Explain these results, and cite the process involved.

**28** What two processes are used to generate ATP from glucose in muscle cells?

**29** Explain the processes that cause a muscle to hypertrophy.

## MasteringA&P™

Access more chapter study tools online in the MasteringA&P Study Area:

- Chapter Quizzes, Chapter Practice Test, and Clinical Case Studies

- Practice Anatomy Lab™ 3.0 — PAL 3.0
- Interactive Physiology™ — iP2™
- A&P Flix™ — A&PFlix™
- PhysioEx™ 9.1 — PhysioEx™ 9.1

9

# Chapter Integration • Applying what you have learned

## Two childhood friends have become very different types of athletes

Sean and Stuart are neighborhood friends who have played sports together and competed against each other since early childhood. Although they are both outstanding athletes, they have noticed they each have their own individual skills in which they excel. Sean is a star soccer player. He can run the length of a soccer field for an entire game and does not experience the fatigue that his friend Stuart does when they compete. Stuart, a top prospect baseball player, is much more explosive in shorter-duration events. He always has much better times than his friend Sean when running from home plate to first base. From what you have just learned about skeletal muscle tissue, answer the following questions.

**1** Which muscle fiber type do you think Sean the soccer player has in greater abundance to explain his success in long-duration activities? How would this help him excel?

**2** Stuart the baseball player is likely to have a disproportionate amount of which fiber type? How do they assist him in explosive, short-duration activities such as running from home plate to first base?

**3** Do you think their athletic success is a result of training, genetics, or both?

**LEARNING OUTCOMES**

These Learning Outcomes correspond by number to this chapter's modules and indicate what you should be able to do after completing the chapter. They also appear at the end of each module.

# 10

# The Muscular System

## Module 10.1

# The axial and appendicular muscles have different functions

The skeletal muscles of the muscular system make up almost one-half the weight of your body. As we saw in Chapter 9, skeletal muscle tissue has multiple functions (see Module 9.1).

**1** The muscular system contributes more to body weight than does any other organ system.

Urinary system 0.7%
Respiratory system 1.7%
Nervous system 2%
Digestive system 6%
Lymphatic system 0.3%
Reproductive system 0.15%
Endocrine system 0.15%
Cardiovascular system 9%

Skeletal system 20%

Integumentary system 16%

Muscular system 44%

This section considers the functional organization of the muscular system. The human body contains approximately 700 skeletal muscles that differ widely in size, shape, and function. Although the individual skeletal muscle fibers contract the same way and to the same degree, skeletal muscle performance varies according to muscle fiber organization and how the muscle attaches to the skeleton.

**2** The **muscular system** is divided into axial and appendicular divisions. **Axial muscles** support and position the axial skeleton. **Appendicular muscles** support, move, and brace the limbs.

Axial muscles

Appendicular muscles

Tendons, which connect sketetal muscles to bones, conduct the forces of contraction to perform specific tasks.

**? REVIEW**

**A.** What are the divisions and functions of the muscular system?

**? REVIEW**

**B.** How many skeletal muscles are there in the body?

**C.** What factors influence the performance of a skeletal muscle?

**Lo LEARNING OUTCOME**

Describe the general function of the body's axial and appendicular muscles.

**351**

# Muscular power and range of motion are influenced by fascicle organization and leverage

## Fascicle Organization

**1** Recall that fascicles are bundles of muscle fibers within a skeletal muscle. In a **parallel muscle**, such as the biceps brachii, the fascicles are parallel to the long axis of the muscle. Most skeletal muscles in the body are parallel muscles. Some are flat bands with broad attachments (aponeuroses) at each end (see Module 9.2). Others are plump and cylindrical, with tendons at one or both ends. The muscle has a central **body**, also known as the *belly*. A skeletal muscle fiber can contract until it has shortened by about 30 percent. Because the muscle fibers in a parallel muscle are parallel to the long axis of the muscle, when those fibers contract together, the entire muscle shortens by about 30 percent. The tension developed during this contraction depends on the total number of myofibrils the muscle contains.

Fascicle

Body (belly)

**2** In a **convergent muscle**, such as the pectoralis major, muscle fascicles extending over a broad area converge on a common attachment site. A convergent muscle is versatile because the stimulation of different portions of the muscle can change the direction of pull. However, when the entire muscle contracts, the muscle fibers do not pull as hard on the attachment site as would a parallel muscle of the same size.

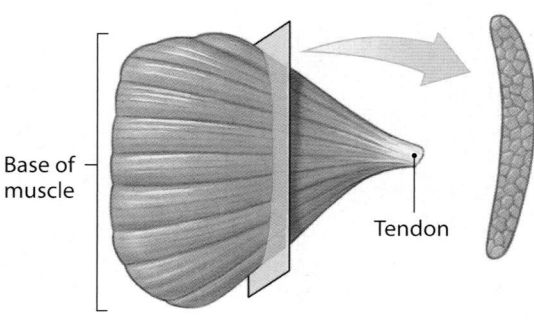

Base of muscle

Tendon

**3** In a **pennate muscle** (*penna*, feather), the fascicles form a common angle with the tendon. Because the muscle fibers pull at an angle, contracting pennate muscles do not move their tendons as far as parallel muscles do. But a pennate muscle contains more muscle fibers—and thus more myofibrils—than does a parallel muscle of the same size, so it produces more tension.

**?** **REVIEW**
**A.** Why does a pennate muscle generate more tension than does a parallel muscle of the same size?

**a** Extensor digitorum muscle

Extended tendon

In a **unipennate** muscle, all the muscle fibers are on the same side of the tendon.

**b** Rectus femoris muscle

In a **bipennate** muscle, fibers are on both sides of the tendon.

**c** Deltoid muscle

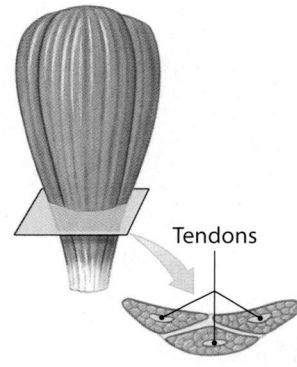

Tendons

In a **multipennate** muscle, the tendon branches within the muscle.

**4** In a **circular muscle**, or **sphincter** (SFINK-ter), the fascicles are concentrically arranged to encircle a duct, tube, or opening. When the muscle contracts, the diameter of the opening decreases (constricts).

Contracted          Relaxed

## Levers and Leverage

The force, speed, or direction of movement produced by muscle contraction can be modified by attaching the muscle to a lever. A **lever** is a rigid structure, such as a board or pry bar, used to lift or pry something that pivots on a fixed point called the **fulcrum**. A lever moves when an applied force is able to overcome any load that would otherwise oppose or prevent such movement. In the body, each bone is a lever and each joint is a fulcrum, and muscles crossing the joint provide the applied force.

**5** In a **first-class lever**, the fulcrum (F) lies between the applied force (AF) and the load (L). First-class levers are like a pry bar: the distance moved depends on the relative sizes of the force and the load and on how far each is from the fulcrum. Our bodies have few first-class levers, but neck extension is one example.

Example: Pry bar

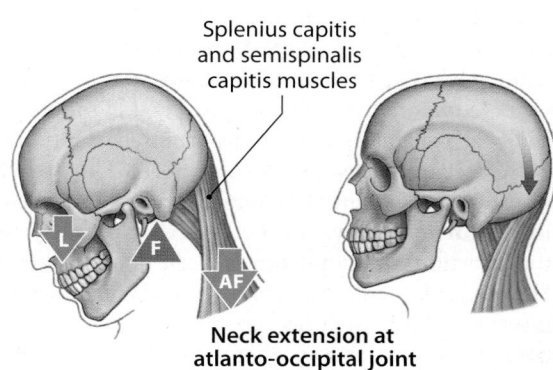

Splenius capitis and semispinalis capitis muscles

**Neck extension at atlanto-occipital joint**

**6** In a **second-class lever**, the load is located between the applied force and the fulcrum. A wheelbarrow is a familiar example of a second-class lever. Because the force is always farther from the fulcrum than the load is, a small force can move a larger weight. However, the load moves slowly and covers a short distance. Thus, the effective force is increased at the expense of speed and distance moved.

Example: Wheelbarrow

Gastrocnemius muscle

**Plantar flexion at the ankle joint**

**7** In **third-class levers** (the most common levers in the body), the force is applied between the load and the fulcrum. Barbecue tongs are a familiar example of a third-class lever. In contrast to second-class levers, speed and distance traveled are increased at the expense of effective force. In the biceps brachii muscle, the load is six times farther from the fulcrum than is the applied force. The effective force is reduced to the same degree.

Example: Tongs

Biceps brachii muscle

**Flexion at the elbow joint**

**? REVIEW**

**B.** Define *lever*, and describe the three classes of levers.

**? INTEGRATION**

**C.** The joint between the occipital bone of the skull and the first cervical vertebra (atlas) is which part of which class of lever system?

**Lo LEARNING OUTCOME**

Describe fascicle organization, and explain how levers affect muscle efficiency.

# Module 10.3

# The origins and insertions of muscles determine their actions, while ...

**1** In most cases, one end of a muscle is fixed in position, and the other end moves during a contraction. The place where the fixed end attaches is called the **origin** of the muscle. Most muscles originate at a bone, but some originate at a connective tissue sheath or band such as the intermuscular septa (components of the deep fascia that may separate adjacent skeletal muscles) or at the interosseous membranes of the forearm or leg (see Module 7.17). The site where the movable end attaches to another structure is called the **insertion** of the muscle. The origin is typically proximal to the insertion when the body is in the anatomical position. However, knowing which end is the origin and which is the insertion is ultimately less important than knowing where the two ends attach and what the muscle accomplishes when it contracts. When a muscle contracts, it produces a specific movement, or **action**. In general, we will describe actions in terms of movement at specific joints (as in the examples shown in Module 10.2).

Origins of biceps brachii muscle

Action

Insertion of biceps brachii muscle

**2** When complex movements occur, muscles commonly work in groups rather than individually. Their cooperation improves the efficiency of a particular movement. For example, large muscles of the limbs produce flexion or extension over an extended range of motion. Based on their functions, muscles may be described as agonists, antagonists, or synergists.

An **agonist**, or **prime mover**, is a muscle whose contraction is chiefly responsible for producing a particular movement. Determining which muscle in a group of muscles is the prime mover depends on the action under way and the relative positions of the articulating bones. In this simple example, the biceps brachii is an agonist that bends the elbow as when doing curls.

When a **synergist** (*syn-*, together + *ergon*, work) contracts, it helps a larger agonist work efficiently. Synergists may provide additional pull near the insertion or may stabilize the point of origin. The brachioradialis muscle assists in flexion and helps stabilize the elbow joint. Synergists that assist an agonist by preventing movement at another joint are called **fixators**.

An **antagonist** is a muscle whose action opposes that of a particular agonist. The triceps brachii muscle is an agonist that extends the elbow. It is therefore an antagonist of the biceps brachii muscle, and the biceps brachii is an antagonist of the triceps brachii.

Insertion of brachioradialis muscle

Origin of brachioradialis muscle

**?** **REVIEW**

**A.** Define the term *synergist* as it relates to muscle action.

**?** **REVIEW**

**B.** Muscle A abducts the humerus, and muscle B adducts the humerus. What is the relationship between these two muscles?

# ... their names can provide clues to appearance and/or function

**3** Familiarity with the terms in this table will help you identify and remember specific muscles. Except for the platysma and the diaphragm, the complete names of all skeletal muscles include the term "muscle." Although the full name, such as the biceps brachii muscle, will usually appear in the text, for simplicity only the descriptive name (biceps brachii) will be used in illustrations and tables.

## Muscle Terminology

| Terms Indicating Specific Regions of the Body* | Terms Indicating Position, Direction, or Fascicle Organization | Terms Indicating Structural Characteristics of the Muscle | Terms Indicating Actions |
|---|---|---|---|
| Abdominis (abdomen) | Anterior (front) | **Nature of Origin** | **General** |
| Ancon (elbow) | External (on the outside) | Biceps (two heads) | Abductor (movement away) |
| Auricular (auricle of ear) | Extrinsic (outside the structure) | Triceps (three heads) | Adductor (movement toward) |
| Brachial (brachium) | Inferior (below) | Quadriceps (four heads) | Depressor (lowering movement) |
| Capitis (head) | Internal (away from the surface) | | Extensor (straightening movement) |
| Carpi (wrist) | Intrinsic (within the structure) | **Shape** | Flexor (bending movement) |
| Cervicis (neck) | Lateral (on the side) | Deltoid (triangle) | Levator (raising movement) |
| Coccygeal (coccyx) | Medial (middle) | Orbicularis (circle) | Pronator (turning into prone position) |
| Costal (rib) | Oblique (slanting) | Pectinate (comblike) | Supinator (turning into supine position) |
| Cutaneous (skin) | Posterior (back) | Piriformis (pear-shaped) | Tensor (tensing movement) |
| Femoris (femur) | Profundus (deep) | Platy- (flat) | |
| Glossal (tongue) | Rectus (straight) | Pyramidal (pyramid) | |
| Hallucis (great toe) | Superficial (toward the surface) | Rhomboid (parallelogram) | **Specific** |
| Ilium (groin) | Superior (toward the head) | Serratus (serrated) | Buccinator (trumpeter) |
| Inguinal (groin) | Transverse (crosswise) | Splenius (bandage) | Risorius (laugher) |
| Lumbar (lumbar region) | | Teres (long and round) | Sartorius (like a tailor) |
| Nasalis (nose) | | Trapezius (trapezoid) | |
| Nuchal (back of neck) | | | |
| Ocular (eye) | | **Other Striking Features** | |
| Oris (mouth) | | Alba (white) | |
| Palpebra (eyelid) | | Brevis (short) | |
| Pollicis (thumb) | | Gracilis (slender) | |
| Popliteal (posterior to knee) | | Latae (wide) | |
| Psoas (loin) | | Latissimus (widest) | |
| Radial (forearm) | | Longissimus (longest) | |
| Scapular (scapula) | | Longus (long) | |
| Temporal (temple) | | Magnus (large) | |
| Thoracic (thorax) | | Major (larger) | |
| Tibial (tibia; shin) | | Maximus (largest) | |
| Ulnar (ulna) | | Minimus (smallest) | |
| | | Minor (smaller) | |
| | | Vastus (great) | |

*For other regional terms, refer to Module 1.20.

 **REVIEW**

**C.** What does the name *flexor carpi radialis longus* tell you about this muscle?

 **LEARNING OUTCOME**

Explain how the name of a muscle can help identify its location, appearance, or function.

# The skeletal muscles can be assigned to the axial division or the appendicular division based on origins and functions

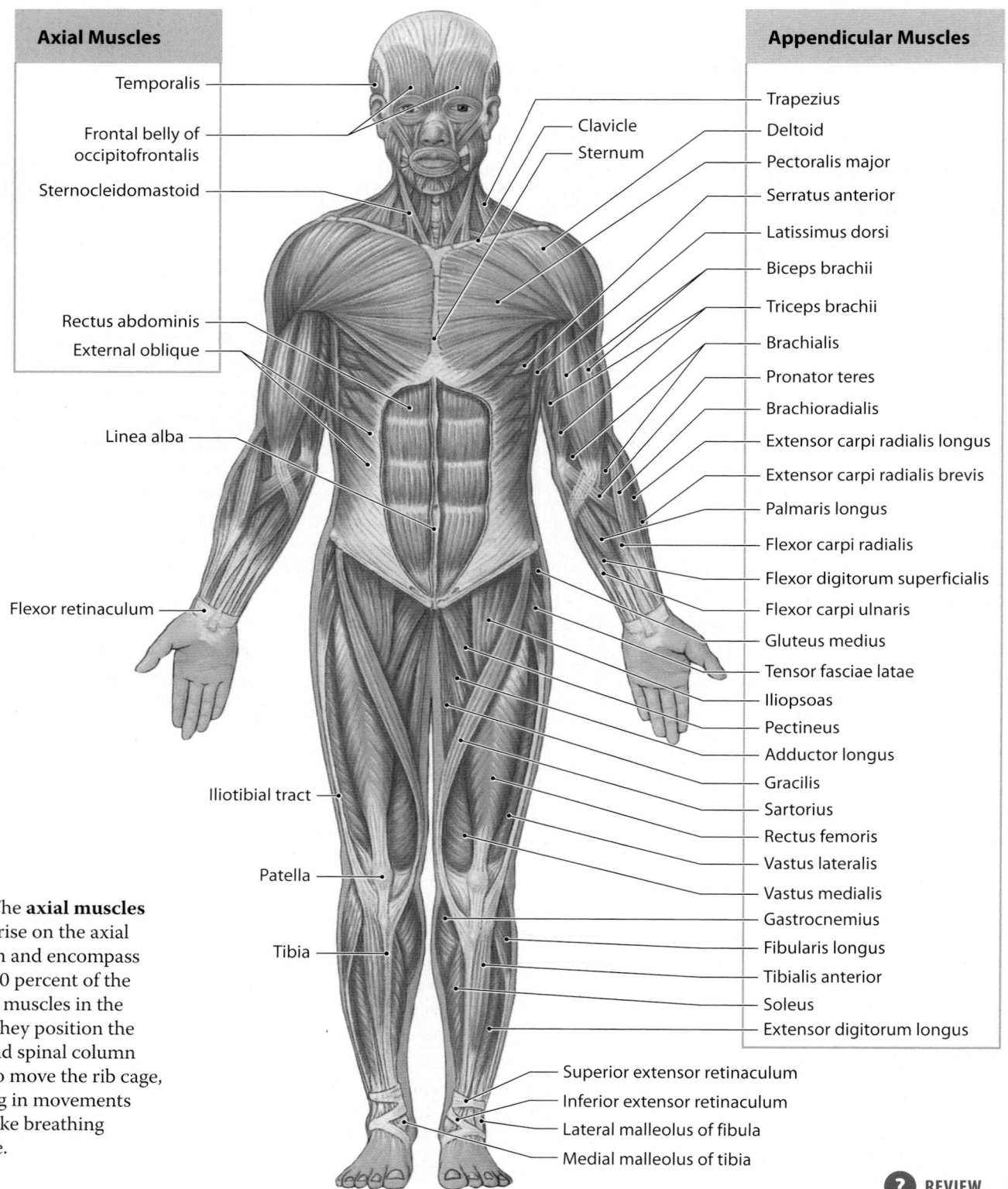

**Axial Muscles**

- Temporalis
- Frontal belly of occipitofrontalis
- Sternocleidomastoid
- Rectus abdominis
- External oblique
- Linea alba
- Flexor retinaculum
- Iliotibial tract
- Patella
- Tibia

Clavicle
Sternum

**Appendicular Muscles**

- Trapezius
- Deltoid
- Pectoralis major
- Serratus anterior
- Latissimus dorsi
- Biceps brachii
- Triceps brachii
- Brachialis
- Pronator teres
- Brachioradialis
- Extensor carpi radialis longus
- Extensor carpi radialis brevis
- Palmaris longus
- Flexor carpi radialis
- Flexor digitorum superficialis
- Flexor carpi ulnaris
- Gluteus medius
- Tensor fasciae latae
- Iliopsoas
- Pectineus
- Adductor longus
- Gracilis
- Sartorius
- Rectus femoris
- Vastus lateralis
- Vastus medialis
- Gastrocnemius
- Fibularis longus
- Tibialis anterior
- Soleus
- Extensor digitorum longus

Superior extensor retinaculum
Inferior extensor retinaculum
Lateral malleolus of fibula
Medial malleolus of tibia

**1** The **axial muscles** arise on the axial skeleton and encompass about 60 percent of the skeletal muscles in the body. They position the head and spinal column and also move the rib cage, assisting in movements that make breathing possible.

**? REVIEW**
**A.** What is the function of the axial muscles?

**Axial Muscles**

- Occipital belly of occipitofrontalis
- Sternocleidomastoid
- External oblique

**Appendicular Muscles**

- Trapezius
- Deltoid
- Infraspinatus
- Teres minor
- Teres major
- Rhomboid major
- Triceps brachii (long head)
- Triceps brachii (lateral head)
- Latissimus dorsi
- Brachioradialis
- Extensor carpi radialis longus
- Anconeus
- Flexor carpi ulnaris
- Extensor digitorum
- Extensor carpi ulnaris
- Gluteus medius
- Tensor fasciae latae
- Gluteus maximus
- Adductor magnus
- Semitendinosus
- Semimembranosus
- Gracilis
- Biceps femoris
- Sartorius
- Plantaris
- Gastrocnemius
- Soleus

- Iliotibial tract

- Calcaneal tendon
- Calcaneus

**10**

**? REVIEW**

**B.** Which structures labeled in the figures in this module are not muscles?

**2** The **appendicular muscles** stabilize or move the appendicular skeleton and include the remaining 40 percent of all skeletal muscles.

The modules that follow organize the muscles into functional groups. However, the muscular system, like the body itself, functions as an integrated whole, and more than one muscle group will usually be seen in one view.

**? INTEGRATION**

**C.** Identify the division (axial or appendicular) to which each of the following muscles belongs: biceps brachii, external oblique, temporalis, and vastus medialis.

**Lo LEARNING OUTCOME**

Describe the separation of muscles into axial and appendicular divisions.

# Functional Organization of the Muscular System

**Labeling:** Label each of the muscle types below according to its fascicle organization.

| 1 | | 2 | | 3 | | 4 | | 5 | |

Label each of the indicated superficial axial and appendicular muscles in the diagram to the right.

| 6 | |
| 7 | |
| 8 | |
| 9 | |
| 10 | |
| 11 | |
| 12 | |

| 18 | |
| 19 | |
| 20 | |

| 13 | |
| 14 | |
| 15 | |
| 16 | |
| 17 | |

| 21 | |
| 22 | |
| 23 | |
| 24 | |
| 25 | |

## Module 10.5
# There are four groups of axial muscles

The axial musculature stabilizes and positions the head, neck, and trunk. Based on location and/or function, we can divide the axial muscles into the four groups shown here. The groups do not always have distinct anatomical boundaries. For example, a function such as the extension of the vertebral column involves muscles along its entire length.

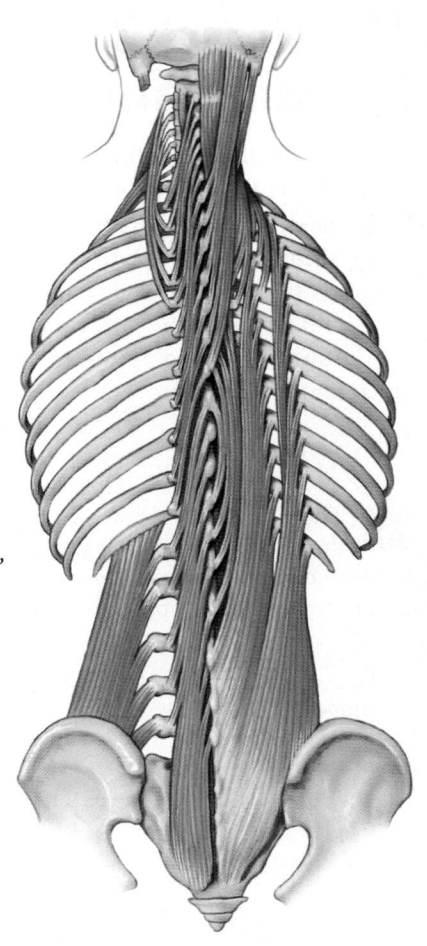

**1** The first group contains muscles of the head and neck that are not associated with the vertebral column. These muscles include the muscles of facial expression (see Module 10.6), the extrinsic eye muscles (see Module 10.7), and the muscles of the tongue, pharynx, and neck (see Module 10.8).

**2** The second group—the muscles of the vertebral column—includes numerous muscles of varied size that stabilize, flex, extend, or rotate the vertebral column (see Module 10.9).

**3** The third group consists of the oblique and rectus muscles of the trunk (see Module 10.10). These muscles are broad sheets or bands that form the muscular walls of the thoracic and abdominopelvic cavities.

**4** The fourth group—the muscles of the pelvic floor (see Module 10.11)—spans the pelvic outlet and supports the organs of the pelvis.

**? REVIEW**
**A.** The axial muscles stabilize and position which regions of the body?

**Lo LEARNING OUTCOME**
Describe the four groups of axial muscles and their general functions.

# The muscles of facial expression are important in eating and useful for communication

All the **muscles of facial expression** originate on the surface of the skull, except for the platysma of the neck. At their insertions, the fibers of the epimysium are woven into those of the superficial fascia (subcutaneous layer) and the dermis of the skin. Thus, when they contract, the skin moves, allowing you to have facial expressions.

**1** The lateral view below shows the major facial muscles. Note the abundance of muscles involved in movements of the lips.

The occipitofrontalis muscle, which forms the scalp, has two bellies that are connected by a collagenous sheet, the **epicranial aponeurosis**.

Occipital belly          Frontal belly

**2** This anterior view shows superficial muscles on the right side of the face and deeper muscles on the left side of the face.

Epicranial aponeurosis

Frontal belly of occipitofrontalis
Orbicularis oculi
Nasalis
Zygomaticus minor
Zygomaticus major
Orbicularis oris
Risorius
Platysma
Mentalis (cut)

Temporalis
Corrugator supercilii
Procerus
Levator labii superioris
Levator anguli oris
Masseter
Buccinator
Depressor anguli oris
Depressor labii inferioris
Thyroid cartilage of the larynx
Clavicle

Sternocleidomastoid

Epicranial aponeurosis

Orbicularis oculi
Nasalis

**Muscles of the Mouth and Cheek**

Levator labii superioris
Zygomaticus minor
Zygomaticus major
Buccinator
Orbicularis oris
Risorius
Mentalis (cut)
Depressor labii inferioris
Depressor anguli oris

Temporalis
Masseter

**? REVIEW**

**A.** Name the muscles associated with the mouth, and identify the one involved in whistling.

**3** This table groups the muscles of facial expression by region and summarizes their origins, insertions, and actions.

## Muscles of Facial Expression

| Region and Muscle | Origin | Insertion | Action |
|---|---|---|---|
| **Mouth** | | | |
| **Buccinator** | Alveolar process of maxilla and alveolar part of mandible | Blends into fibers of orbicularis oris | Compresses cheeks |
| **Depressor labii inferioris** | Mandible between the anterior midline and the mental foramen | Skin of lower lip | Depresses lower lip |
| **Levator labii superioris** | Inferior margin of orbit, superior to the infra-orbital foramen | Orbicularis oris | Elevates upper lip |
| **Levator anguli oris** | Maxilla below the infra-orbital foramen | Corner of mouth | Elevates the corner of the mouth |
| **Mentalis** | Incisive fossa of mandible | Skin of chin | Elevates and protrudes lower lip |
| **Orbicularis oris** | Maxilla and mandible | Lips | Compresses, purses lips |
| **Risorius** | Fascia surrounding parotid salivary gland | Angle of mouth | Draws corner of mouth to the side |
| **Depressor anguli oris** | Anterolateral surface of mandibular body | Skin at angle of mouth | Depresses corner of mouth |
| **Zygomaticus major** | Zygomatic bone near zygomaticomaxillary suture | Angle of mouth | Retracts and elevates corner of mouth |
| **Zygomaticus minor** | Zygomatic bone posterior to zygomaticotemporal suture | Upper lip | Retracts and elevates upper lip |
| **Eye** | | | |
| **Corrugator supercilii** | Orbital rim of frontal bone near nasal suture | Eyebrow | Pulls skin inferiorly and anteriorly; wrinkles brow |
| **Levator palpebrae superioris** | Tendinous band around optic foramen | Upper eyelid | Elevates upper eyelid |
| **Orbicularis oculi** | Medial margin of orbit | Skin around eyelids | Closes eye |
| **Nose** | | | |
| **Procerus** | Nasal bones and lateral nasal cartilages | Aponeurosis at bridge of nose and skin of forehead | Moves nose, changes position and shape of nostrils |
| **Nasalis** | Maxilla and alar cartilage of nose | Bridge of nose | Compresses bridge, depresses tip of nose; elevates corners of nostrils |
| **Scalp** | | | |
| **Occipitofrontalis** | | | |
| Frontal belly | Epicranial aponeurosis | Skin of eyebrow and bridge of nose | Raises eyebrows, wrinkles forehead |
| Occipital belly | Occipital and temporal bones | Epicranial aponeurosis | Tenses and retracts scalp |
| **Neck** | | | |
| **Platysma** | Superior thorax between cartilage of 2nd rib and acromion of scapula | Mandible and skin of cheek | Tenses skin of neck; depresses mandible and pulls lower lip inferiorly |

10

**? REVIEW**

**B.** State whether the following muscles involve the mouth, eye, nose, ear, scalp, or neck: buccinator, corrugator supercilii, mentalis, nasalis, platysma, procerus, and risorius.

**? INTEGRATION**

**C.** Explain how a person is able to consciously move the skin on the scalp but is not able to consciously move the skin of the thigh.

**Lo LEARNING OUTCOME**

Identify the facial expression muscles, and cite their origins, insertions, and actions.

# The extrinsic eye muscles position the eye ...

**1** Five of the six extrinsic eye muscles are visible in this lateral view of the right eye.

**2** One additional muscle, the medial rectus, can be seen in this medial view of the right eye.

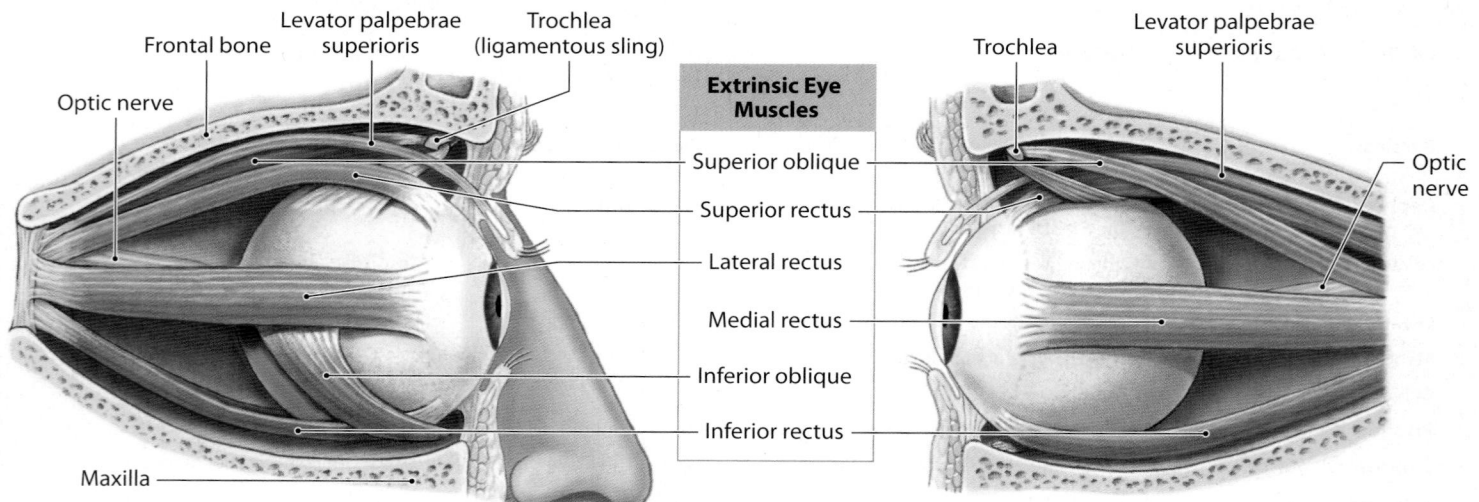

Frontal bone · Levator palpebrae superioris · Trochlea (ligamentous sling) · Optic nerve · Trochlea · Levator palpebrae superioris · Optic nerve · Maxilla

**Extrinsic Eye Muscles**

- Superior oblique
- Superior rectus
- Lateral rectus
- Medial rectus
- Inferior oblique
- Inferior rectus

**3** This anterior view of the right eye shows the direction of eye movements produced by the contraction of each extrinsic eye muscle operating independently.

Superior rectus · Trochlea · Superior oblique · Lateral rectus · Medial rectus · Inferior oblique · Inferior rectus

**4** This anterior view of the right orbit shows the origins of the extrinsic eye muscles.

Levator palpebrae superioris · Superior rectus · Oculomotor nerve (III) · Lateral rectus · Abducens nerve (VI) · Inferior oblique · Trochlear nerve (IV) · Trochlea · Superior oblique · Medial rectus · Optic nerve (II) · Inferior rectus

## Extrinsic Eye Muscles

| Muscle | Origin | Insertion | Action |
|---|---|---|---|
| **Inferior rectus** | Sphenoid around optic canal | Inferior, medial surface of eyeball | Eye looks inferiorly |
| **Medial rectus** | Sphenoid around optic canal | Medial surface of eyeball | Eye looks medially |
| **Superior rectus** | Sphenoid around optic canal | Superior surface of eyeball | Eye looks superiorly |
| **Lateral rectus** | Sphenoid around optic canal | Lateral surface of eyeball | Eye looks laterally |
| **Inferior oblique** | Maxilla at anterior portion of orbit | Inferior, lateral surface of eyeball | Eye rolls, looks superiorly and laterally |
| **Superior oblique** | Sphenoid around optic canal | Superior, lateral surface of eyeball | Eye rolls, looks inferiorly and laterally |

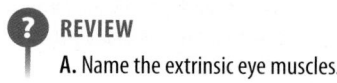

**REVIEW**

**A.** Name the extrinsic eye muscles.

# … and the muscles of mastication move the lower jaw

**5** This lateral view shows the largest superficial muscles of mastication (chewing) of the right side of the head.

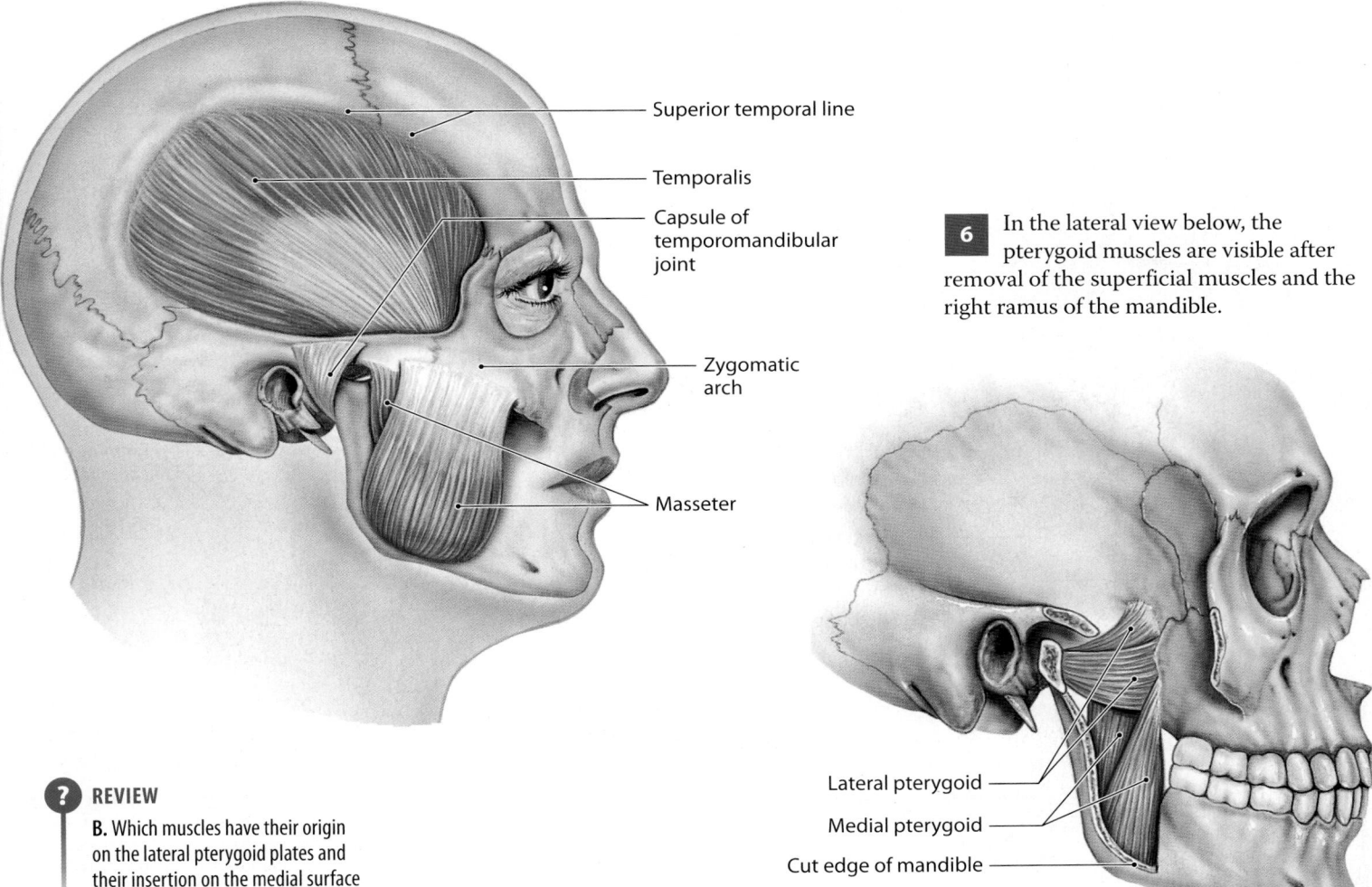

Superior temporal line

Temporalis

Capsule of temporomandibular joint

Zygomatic arch

Masseter

**6** In the lateral view below, the pterygoid muscles are visible after removal of the superficial muscles and the right ramus of the mandible.

10

Lateral pterygoid

Medial pterygoid

Cut edge of mandible

**? REVIEW**

**B.** Which muscles have their origin on the lateral pterygoid plates and their insertion on the medial surface of the ramus of the mandible?

| Muscles of Mastication | | | |
|---|---|---|---|
| **Muscle** | **Origin** | **Insertion** | **Action** |
| **Masseter** | Zygomatic arch | Lateral surface of ramus of mandible | Elevates mandible and closes the jaws |
| **Temporalis** | Along temporal lines of skull | Coronoid process of mandible | Elevates mandible |
| **Pterygoids** (**medial** and **lateral**) | Lateral pterygoid plate | Medial surface of ramus of the mandible | Medial: Elevates the mandible and closes the jaws, or slides the mandible from side to side<br><br>Lateral: Opens jaws, protrudes the mandible, or slides the mandible from side to side |

**? INTEGRATION**

**C.** If you were contracting and relaxing your masseter, what would you probably be doing?

**Lo LEARNING OUTCOME**

Identify the eye and jaw muscles, and cite their origins, insertions, and actions.

# The muscles of the tongue are closely associated with the muscles of the pharynx and neck

**1** This lateral view shows the tongue muscles after removal of the left half of the mandible.

**2** This lateral view shows the major groups of the muscles of the pharynx. Their roles in swallowing are discussed in Module 22.8.

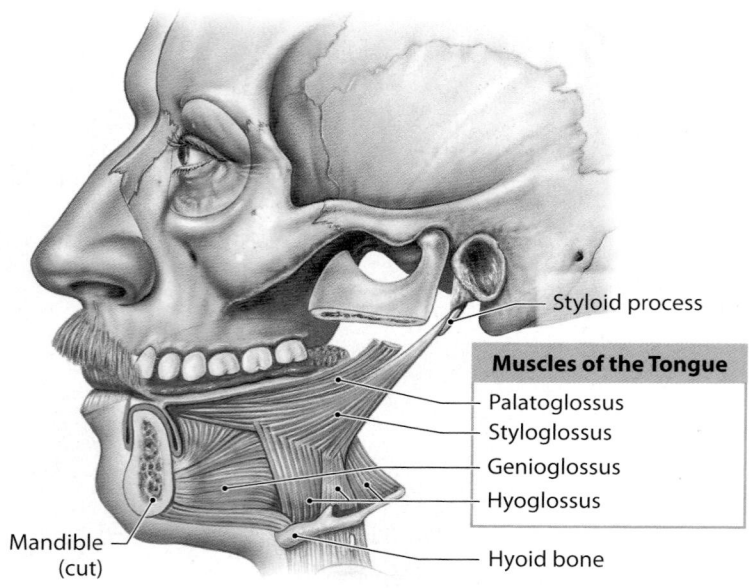

Styloid process

**Muscles of the Tongue**
Palatoglossus
Styloglossus
Genioglossus
Hyoglossus

Mandible (cut)

Hyoid bone

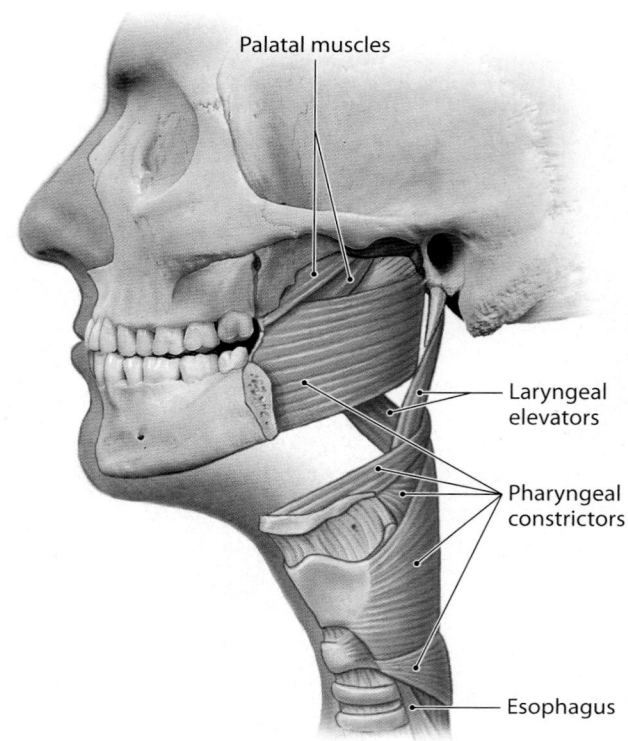

Palatal muscles

Laryngeal elevators

Pharyngeal constrictors

Esophagus

## Muscles of the Tongue

| Muscle | Origin | Insertion | Action |
|---|---|---|---|
| **Genioglossus** | Medial surface of mandible around chin | Body of tongue, hyoid bone | Depresses and protracts tongue |
| **Hyoglossus** | Body and greater horn of hyoid bone | Side of tongue | Depresses and retracts tongue |
| **Palatoglossus** | Anterior surface of soft palate | Side of tongue | Elevates tongue, depresses soft palate |
| **Styloglossus** | Styloid process of temporal bone | Along the side to tip and base of tongue | Retracts tongue, elevates side of tongue |

## Muscles of the Pharynx

| Muscle | Origin | Insertion | Action |
|---|---|---|---|
| **Pharyngeal constrictors** | Pterygoid process of sphenoid, medial surfaces of mandible, horns of hyoid bone, cricoid and thyroid cartilages of larynx | Median raphe attached to occipital bone | Constrict pharynx to propel an ingested food mass into the esophagus |
| **Laryngeal elevators** | Soft palate, cartilage around inferior portion of auditory tube, styloid process | Thyroid cartilage | Elevate larynx |
| **Palatal muscles** | Petrous part of temporal bone and adjacent soft tissues, sphenoidal spine and adjacent soft tissues | Soft palate | Elevate soft palate |

**? REVIEW**

**A.** List the muscles of the tongue.

**? REVIEW**

**B.** Which muscles elevate the soft palate?

**3** The anterior muscles of the neck are primarily involved in positioning the mandible, hyoid bone, and larynx.

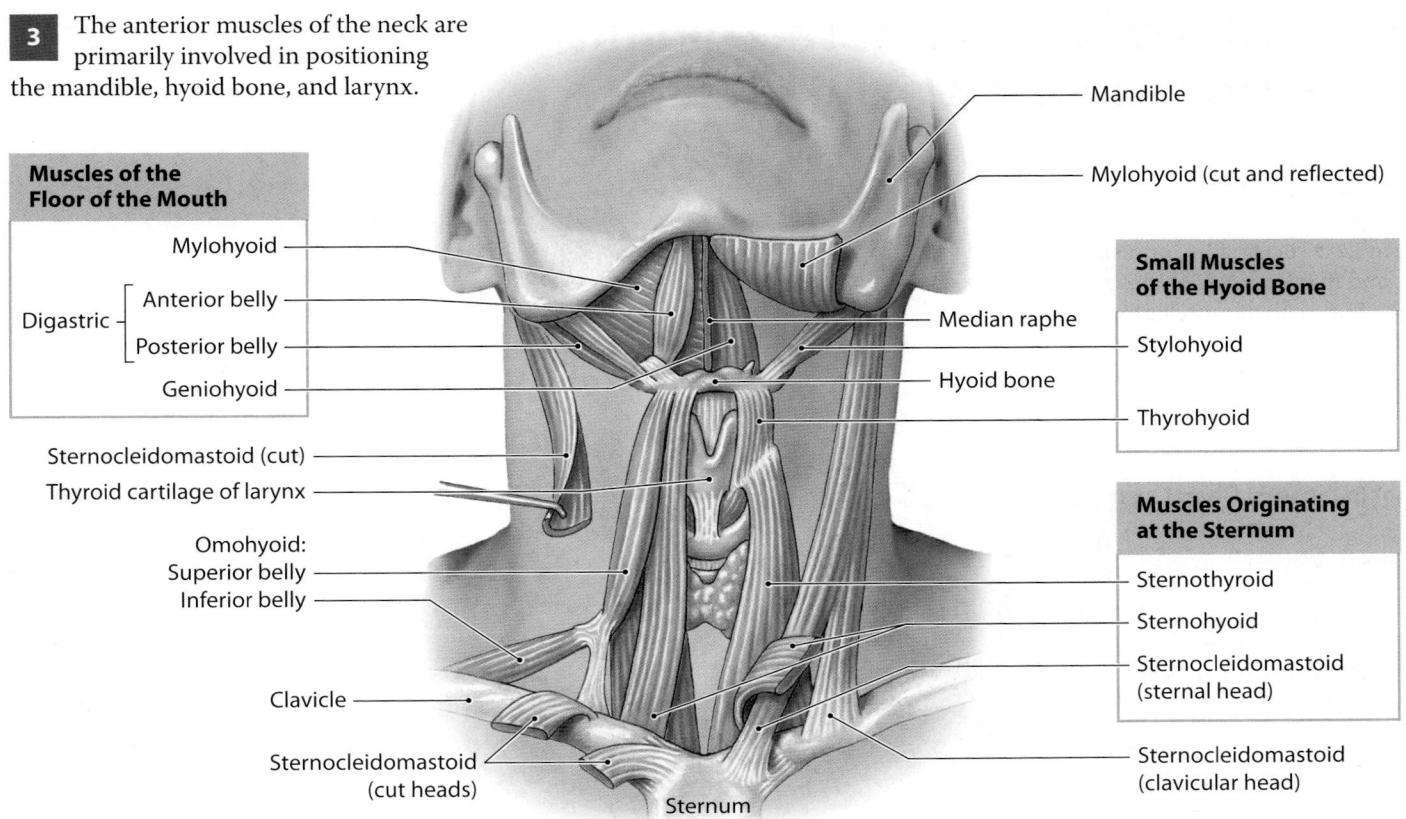

**Muscles of the Floor of the Mouth**

- Mylohyoid
- Digastric
  - Anterior belly
  - Posterior belly
- Geniohyoid

Sternocleidomastoid (cut)

Thyroid cartilage of larynx

Omohyoid:
- Superior belly
- Inferior belly

Clavicle

Sternocleidomastoid (cut heads)

Sternum

Mandible

Mylohyoid (cut and reflected)

Median raphe

Hyoid bone

**Small Muscles of the Hyoid Bone**

- Stylohyoid
- Thyrohyoid

**Muscles Originating at the Sternum**

- Sternothyroid
- Sternohyoid
- Sternocleidomastoid (sternal head)
- Sternocleidomastoid (clavicular head)

10

| Anterior Muscles of the Neck | | | |
|---|---|---|---|
| **Muscle** | **Origin** | **Insertion** | **Action** |
| **Digastric** | Inferior surface of mandible at chin and mastoid region | Hyoid bone | Depresses mandible or elevates larynx |
| **Geniohyoid** | Medial surface of mandible at chin | Hyoid bone | As above and pulls hyoid bone anteriorly |
| **Mylohyoid** | Mylohyoid line of mandible | Median raphe that runs to hyoid bone | Elevates hyoid bone or depresses mandible |
| **Omohyoid** | Superior border of scapula near scapular notch | Hyoid bone | Depresses hyoid bone and larynx |
| **Sternohyoid** | Clavicle and manubrium | Hyoid bone | Depresses hyoid bone and larynx |
| **Sternothyroid** | Manubrium and first costal cartilage | Thyroid cartilage of larynx | Depresses hyoid bone and larynx |
| **Stylohyoid** | Styloid process | Hyoid bone | Elevates larynx |
| **Thyrohyoid** | Thyroid cartilage of larynx | Hyoid bone | Elevates thyroid, depresses hyoid |
| **Sternocleidomastoid** | One head attaches to sternal end of clavicle; the other head attaches to manubrium | Mastoid region of skull and lateral portion of superior nuchal line | Flexes the neck; one alone bends head toward shoulder and rotates neck |

**4** This is a superior view of an isolated mandible. Several muscles extending from the hyoid bone to the mandible form the muscular floor of the mouth and support the tongue.

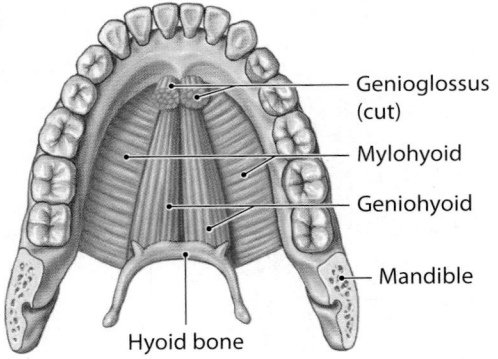

Genioglossus (cut)

Mylohyoid

Geniohyoid

Mandible

Hyoid bone

**? REVIEW**

**C.** Which muscles associated with the hyoid form the floor of the mouth?

**Lo LEARNING OUTCOME**

Identify the tongue, pharynx, and neck muscles, and cite their origins, insertions, and actions.

# Module 10.9

# The muscles of the vertebral column support and align the axial skeleton

**1** The **muscles of the vertebral column** are arranged in several layers. They include muscles originating or inserting on the ribs and the processes of the vertebrae. Although this mass of muscles extends from the sacrum to the skull, each muscle group is composed of numerous separate muscles of various lengths.

### Spinal Extensors, Superficial Layer

Splenius capitis

### Erector Spinae Muscles

Longissimus capitis

Spinalis cervicis

Longissimus cervicis

Iliocostalis cervicis

Iliocostalis thoracis

Longissimus thoracis

Spinalis thoracis

Iliocostalis lumborum

### Spinal Extensors, Deep Layer

#### Semispinalis Group

Semispinalis capitis

Semispinalis cervicis

Semispinalis thoracis

Multifidus

### Spinal Flexors

Quadratus lumborum

Longus capitis

Longus colli

**Muscles of the anterior cervical spine**

The **erector spinae** muscles are subdivided into **spinalis**, **longissimus**, and **iliocostalis muscle groups**. (The individual muscles are listed above.) These subdivisions are based on proximity to the vertebral column.

Thoracodorsal fascia

**Posterior view**

**? REVIEW**
A. List the spinal flexor muscles.

**? REVIEW**
B. Which muscles enable you to extend your neck?

**2** Note that this table lists many extensors of the vertebral column, but few flexors. Why? The vertebral column does not have a massive series of flexor muscles because (1) many of the large trunk muscles flex the vertebral column when they contract, and (2) most of the body weight lies anterior to the vertebral column, and (3) gravity tends to flex the spine when unopposed by the extensor muscles.

## Muscles of the Vertebral Column

| Group and Muscle | Origin | Insertion | Action |
|---|---|---|---|
| **Spinal Extensors — Superficial Layer** | | | |
| **Splenius** (splenius capitis, splenius cervicis) | Spinous processes and ligaments connecting inferior cervical and superior thoracic vertebrae | Mastoid process, occipital bone of skull, and superior cervical vertebrae | Together, the two sides extend neck; alone, each rotates and laterally flexes neck to that side |
| **Erector spinae** | | | |
| **SPINALIS GROUP** — **Spinalis cervicis** | Inferior portion of ligamentum nuchae and spinous process of $C_7$ | Spinous process of axis | Extends neck |
| **Spinalis thoracis** | Spinous processes of inferior thoracic and superior lumbar vertebrae | Spinous processes of superior thoracic vertebrae | Extends vertebral column |
| **LONGISSIMUS GROUP** — **Longissimus capitis** | Transverse processes of inferior cervical and superior thoracic vertebrae | Mastoid process of temporal bone | Together, the two sides extend head; alone, each rotates and laterally flexes neck to that side |
| **Longissimus cervicis** | Transverse processes of superior thoracic vertebrae | Transverse processes of middle and superior cervical vertebrae | Together, the two sides extend head; alone, each rotates and laterally flexes neck to that side |
| **Longissimus thoracis** | Broad aponeurosis and transverse processes of inferior thoracic and superior lumbar vertebrae; joins iliocostalis | Transverse processes of superior vertebrae and inferior surfaces of ribs | Extends vertebral column; alone, each produces lateral flexion to that side |
| **ILIOCOSTALIS GROUP** — **Iliocostalis cervicis** | Superior borders of vertebrosternal ribs near the angles | Transverse processes of middle and inferior cervical vertebrae | Extends or laterally flexes neck, elevates ribs |
| **Iliocostalis thoracis** | Superior borders of inferior seven ribs medial to the angles | Upper ribs and transverse process of last cervical vertebra | Stabilizes thoracic vertebrae in extension |
| **Iliocostalis lumborum** | Iliac crest, sacral crests, and spinous processes | Inferior surfaces of inferior seven ribs near their angles | Extends vertebral column, depresses ribs |
| **Spinal Extensors — Deep Layer** | | | |
| **SEMISPINALIS GROUP** — **Semispinalis capitis** | Articular processes of inferior cervical and transverse processes of superior thoracic vertebrae | Occipital bone, between nuchal lines | Together, the two sides extend head; alone, each extends and laterally flexes neck |
| **Semispinalis cervicis** | Transverse processes of $T_1$–$T_5$ or $T_6$ | Spinous processes of $C_2$–$C_5$ | Extends vertebral column and rotates toward opposite side |
| **Semispinalis thoracis** | Transverse processes of $T_6$–$T_{10}$ | Spinous processes of $C_5$–$T_4$ | Extends vertebral column and rotates toward opposite side |
| **Multifidus** | Sacrum and transverse processes of each vertebra | Spinous processes of the third or fourth more superior vertebrae | Extends vertebral column and rotates toward opposite side |
| **Spinal Flexors** | | | |
| **Longus capitis** | Transverse processes of cervical vertebrae | Base of the occipital bone | Together, the two sides flex the neck; alone, each rotates head to that side |
| **Longus colli** | Anterior surfaces of cervical and superior thoracic vertebrae | Transverse processes of superior cervical vertebrae | Flexes or rotates neck; limits hyperextension |
| **Quadratus lumborum** | Iliac crest and iliolumbar ligament | Last rib and transverse processes of lumbar vertebrae | Together, they depress ribs; alone, each side laterally flexes vertebral column |

10

**REVIEW**

**C.** What might account for a lack of massive flexor muscles?

**LEARNING OUTCOME**

Identify the vertebral column muscles, and cite their origins, insertions, and actions.

# Module 10.10

# The oblique and rectus muscles form the muscular walls of the trunk

**1** The oblique and rectus muscle groups share embryonic origins. As shown below, the scalene muscles extend from the cervical vertebrae in the neck to the first two ribs.

Scalenes

Rib 1

Rib 2

**Anterior view, frontal section**

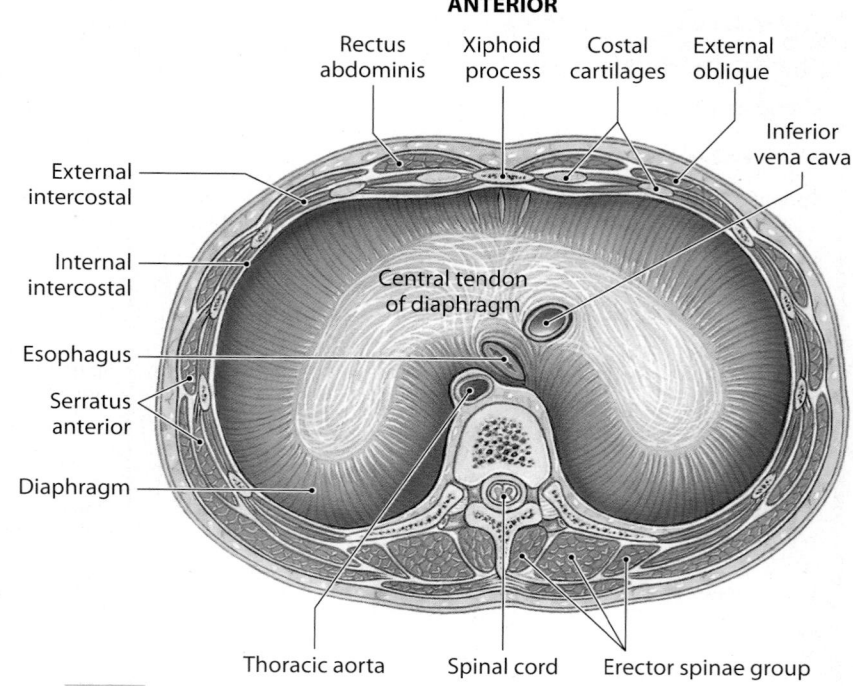

**ANTERIOR**

Rectus abdominis
Xiphoid process
Costal cartilages
External oblique

Inferior vena cava

External intercostal

Internal intercostal

Central tendon of diaphragm

Esophagus

Serratus anterior

Diaphragm

Thoracic aorta

Spinal cord

Erector spinae group

**Superior view of the diaphragm**

**2** The illustration below shows superficial muscles on the right side of the body and deeper muscles of the oblique and rectus groups on the left side.

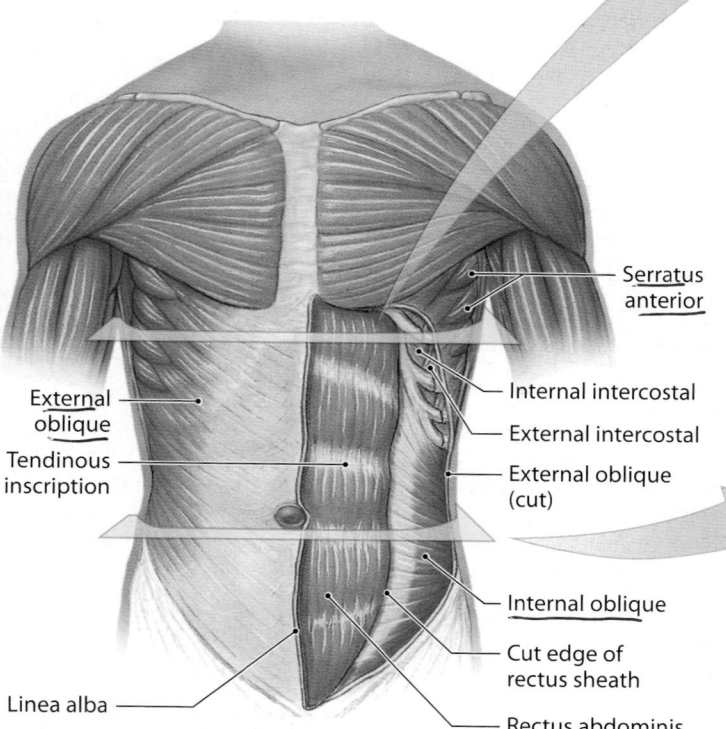

Serratus anterior

External oblique

Tendinous inscription

Internal intercostal

External intercostal

External oblique (cut)

Internal oblique

Cut edge of rectus sheath

Linea alba

Rectus abdominis

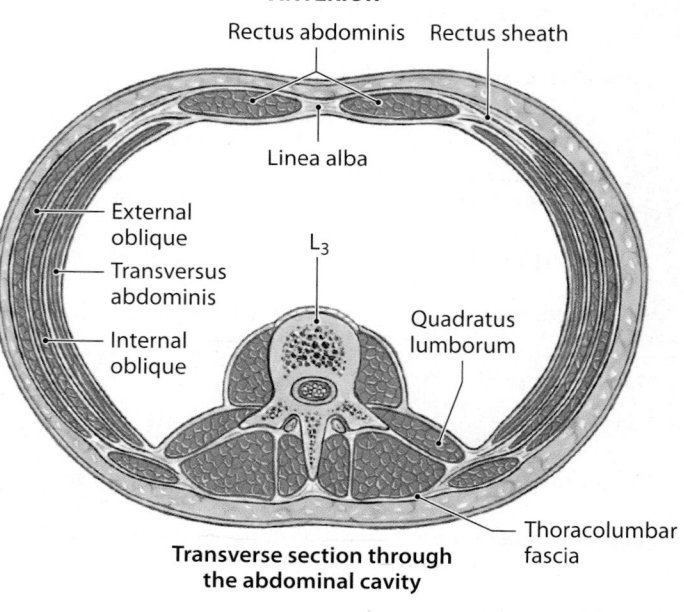

**ANTERIOR**

Rectus abdominis
Rectus sheath

Linea alba

External oblique

Transversus abdominis

Internal oblique

$L_3$

Quadratus lumborum

Thoracolumbar fascia

**Transverse section through the abdominal cavity**

**? REVIEW**

**A.** Which muscle forms the deepest layer of the abdominal wall muscles?

10

## Oblique and Rectus Muscles

| Group and Muscle | | Origin | Insertion | Action |
|---|---|---|---|---|
| **Oblique Group** | | | | |
| CERVICAL REGION | **Scalenes** | Transverse and costal processes of cervical vertebrae | Superior surfaces of first two ribs | Elevate ribs or flex neck |
| THORACIC REGION | **External intercostals** | Inferior border of each rib | Superior border of more inferior rib | Elevate ribs |
| | **Internal intercostals** | Superior border of each rib | Inferior border of the preceding rib | Depress ribs |
| | **Transversus thoracis** | Posterior surface of sternum | Cartilages of ribs | Depress ribs |
| ABDOMINAL REGION | **External oblique** | External and inferior borders of ribs 5–12 | Linea alba and iliac crest | Compresses abdomen, depresses ribs, flexes or bends spine |
| | **Internal oblique** | Lumbodorsal fascia and iliac crest | Inferior ribs, xiphoid process, and linea alba | Compresses abdomen, depresses ribs, flexes or bends spine |
| | **Transversus abdominis** | Cartilages of ribs 6–12, iliac crest, and lumbodorsal fascia | Linea alba and pubis | Compresses abdomen |
| **Rectus Group** | | | | |
| THORACIC REGION | **Diaphragm** | Xiphoid process, cartilages of ribs 4–10, and anterior surfaces of lumbar vertebrae | Central tendinous sheet | Contraction expands thoracic cavity, compresses abdominopelvic cavity |
| ABDOMINAL REGION | **Rectus abdominis** | Superior surface of pubis around symphysis | Inferior surfaces of costal cartilages (ribs 5–7) and xiphoid process | Depresses ribs, flexes vertebral column, compresses abdomen |

10

**3** Major landmarks of the inferior surface of the diaphragm.

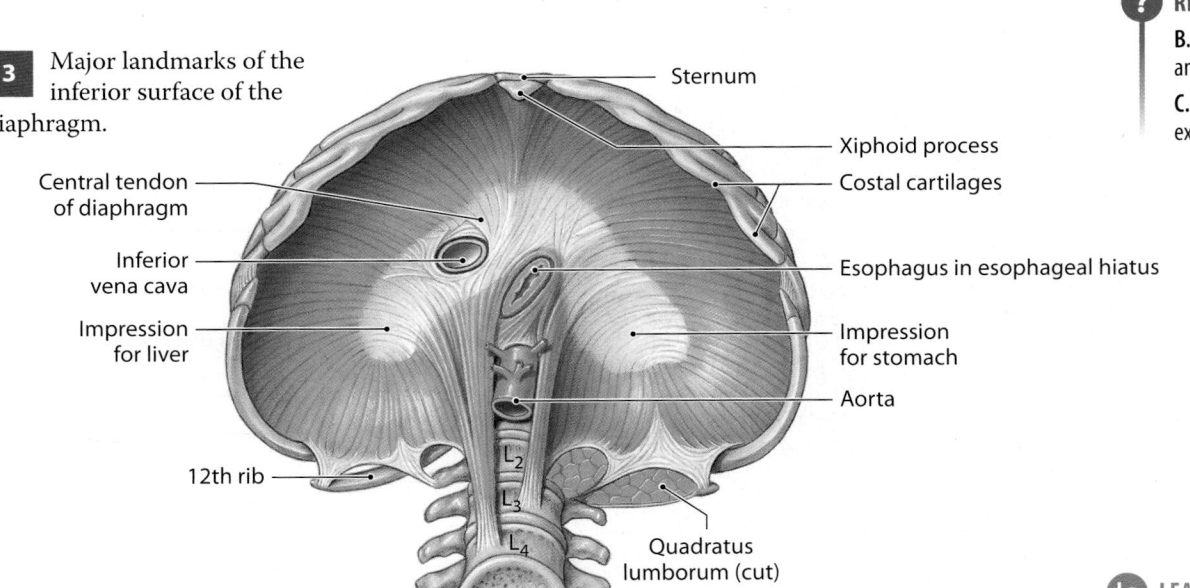

Sternum
Xiphoid process
Central tendon of diaphragm
Costal cartilages
Inferior vena cava
Impression for liver
Esophagus in esophageal hiatus
Impression for stomach
Aorta
12th rib
L₂
L₃
L₄
Quadratus lumborum (cut)

**Inferior view of the diaphragm**

**? REVIEW**

**B.** Which muscle connects the ribs and sternum to the pubic bones?

**C.** Describe the action of the external oblique.

**Lo LEARNING OUTCOME**

Identify the trunk muscles, and cite their origins, insertions, and actions.

# The muscles of the pelvic floor support the organs of the abdominopelvic cavity

**Superficial Dissections**

UROGENITAL TRIANGLE

**Deep Dissections**

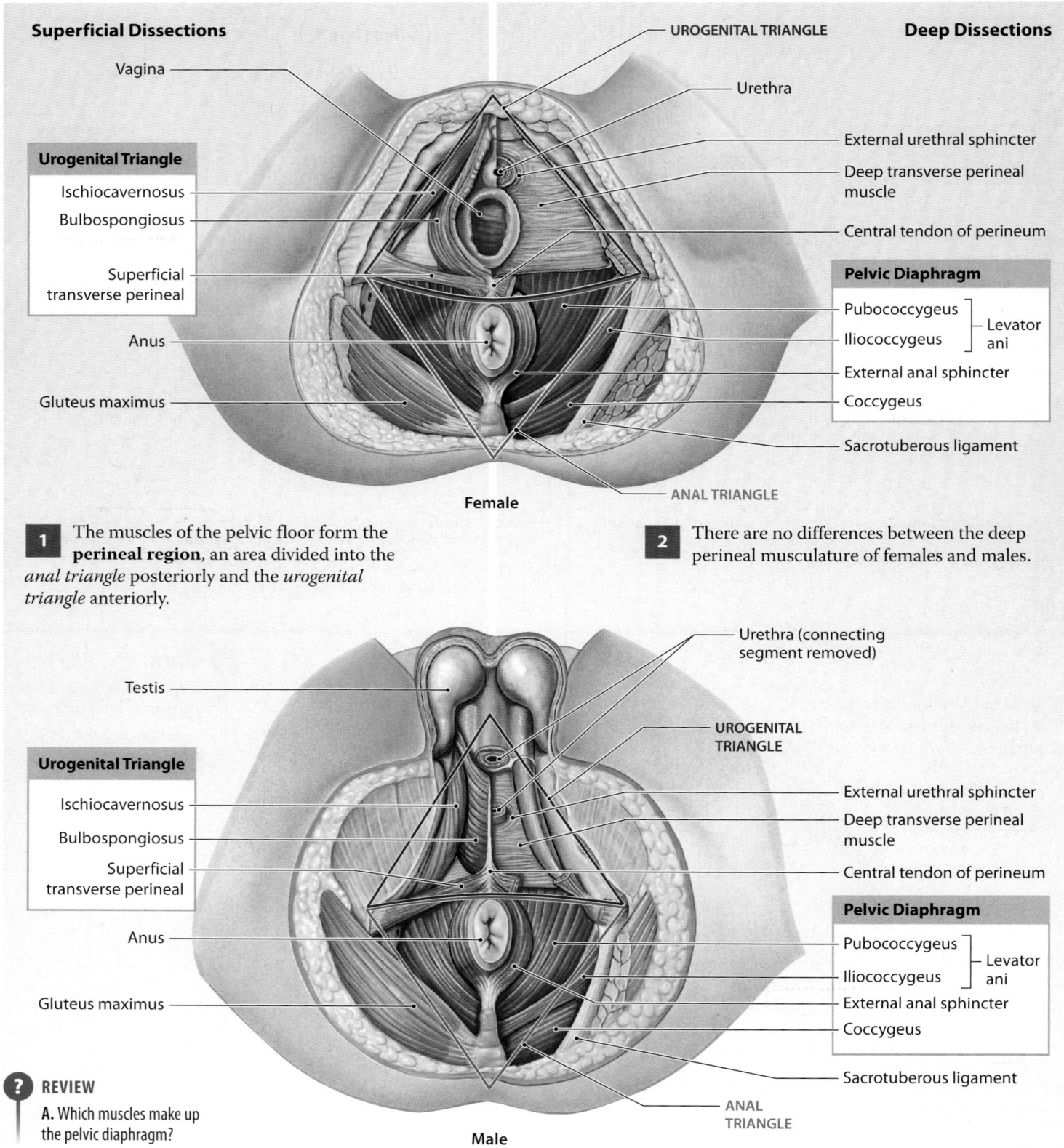

Vagina

Urethra

**Urogenital Triangle**

Ischiocavernosus

Bulbospongiosus

Superficial transverse perineal

Anus

Gluteus maximus

External urethral sphincter

Deep transverse perineal muscle

Central tendon of perineum

**Pelvic Diaphragm**

Pubococcygeus ⎤
           ⎟ Levator
Iliococcygeus ⎦ ani

External anal sphincter

Coccygeus

Sacrotuberous ligament

ANAL TRIANGLE

**Female**

**1** The muscles of the pelvic floor form the **perineal region**, an area divided into the *anal triangle* posteriorly and the *urogenital triangle* anteriorly.

**2** There are no differences between the deep perineal musculature of females and males.

Urethra (connecting segment removed)

Testis

**Urogenital Triangle**

Ischiocavernosus

Bulbospongiosus

Superficial transverse perineal

Anus

Gluteus maximus

UROGENITAL TRIANGLE

External urethral sphincter

Deep transverse perineal muscle

Central tendon of perineum

**Pelvic Diaphragm**

Pubococcygeus ⎤
           ⎟ Levator
Iliococcygeus ⎦ ani

External anal sphincter

Coccygeus

Sacrotuberous ligament

ANAL TRIANGLE

**Male**

**? REVIEW**

**A.** Which muscles make up the pelvic diaphragm?

## Muscles of the Pelvic Floor

| Group and Muscle | | Origin | Insertion | Action |
|---|---|---|---|---|
| **Urogenital Triangle** | | | | |
| **SUPERFICIAL MUSCLES** | **Bulbospongiosus** | | | |
| | Females | Collagen sheath at base of clitoris; fibers run on either side of urethral and vaginal opening | Central tendon of perineum | Compresses and stiffens clitoris; narrows vaginal opening |
| | Males | Collagen sheath at base of penis; fibers cross over urethra | Median raphe and central tendon of perineum | Compresses base and stiffens penis; ejects urine or semen |
| | **Ischiocavernosus** | Ischial ramus and tuberosity | Pubic symphysis anterior to base of penis or clitoris | Compresses and stiffens penis or clitoris |
| | **Superficial transverse perineal** | Ischial ramus | Central tendon of perineum | Stabilizes central tendon of perineum |
| **DEEP MUSCLES** | **Deep transverse perineal muscle** | Ischial ramus | Perineal body | Stabilizes central tendon of perineum |
| | **External urethral sphincter** | | | |
| | Females | Ischial and pubic rami | To median raphe; inner fibers encircle urethra | Closes urethra; compresses vagina and greater vestibular glands |
| | Males | Ischial and pubic rami | To median raphe at base of penis; inner fibers encircle urethra | Closes urethra; compresses prostate and bulbourethral glands |
| **Anal Triangle** | | | | |
| **Pelvic diaphragm** | | | | |
| **Coccygeus** | | Ischial spine | Lateral, inferior borders of sacrum and coccyx | Flexes coccygeal joints; tenses and supports pelvic floor |
| **Levator ani** | | | | |
| **Iliococcygeus** | | Ischial spine, pubis | Coccyx and median raphe | Tenses floor of pelvis; flexes coccygeal joints; elevates and retracts anus |
| **Pubococcygeus** | | Inner margins of pubis | Coccyx and median raphe | Tenses floor of pelvis; flexes coccygeal joints; elevates and retracts anus |
| **External anal sphincter** | | Coccyx | Encircles anal opening | Closes anal opening |

The pelvic diaphragm does not completely close the pelvic outlet, because the urethra, vagina, and anus pass through them to open at the exterior. Muscular sphincters surround their openings and permit voluntary control of urination and defecation. Muscles, nerves, and blood vessels also pass through the pelvic outlet as they travel to or from the lower limbs.

 **REVIEW**

**B.** In females, what is the action of the bulbospongiosus?

**C.** The coccygeus extends from the sacrum and coccyx to which structure?

 **LEARNING OUTCOME**

Identify the pelvic floor muscles, and cite their origins, insertions, and actions.

# Axial Muscles

**Labeling:** Label each of the indicated muscles of the face in the following diagram.

(frontal belly)

1 Occipitofrontalis

2 temporalis

C temporal fossa
I mandible
A elevate jaw

3 orbicularis oculi

4

5 zygomaticus minor

6 zygomaticus major

7

8 orbicularis oris

9

10

11

12 masseter

trapezius

platysma

Sternocleidomastoid

C clavicle/ Sternum
I mastoid process
A flexion of neck

Label each of the indicated muscles of the neck in the following diagram.

13

14

15

16

17

18

19

20

21

10

## Module 10.12
# The appendicular muscles stabilize, position, and support the limbs

In the following modules, we will group the appendicular muscles by their actions and origins. We can describe actions in two ways, one focused on the bone and one on the joint. The first way describes actions in terms of the region affected. For example, we say a muscle such as the biceps brachii performs "flexion of the forearm." The second way, used by specialists, such as kinesiologists and physical therapists, identifies the joint involved. In this approach, we say the action of the biceps brachii muscle is "flexion at (or of) the elbow." We will use this second way of describing muscle actions.

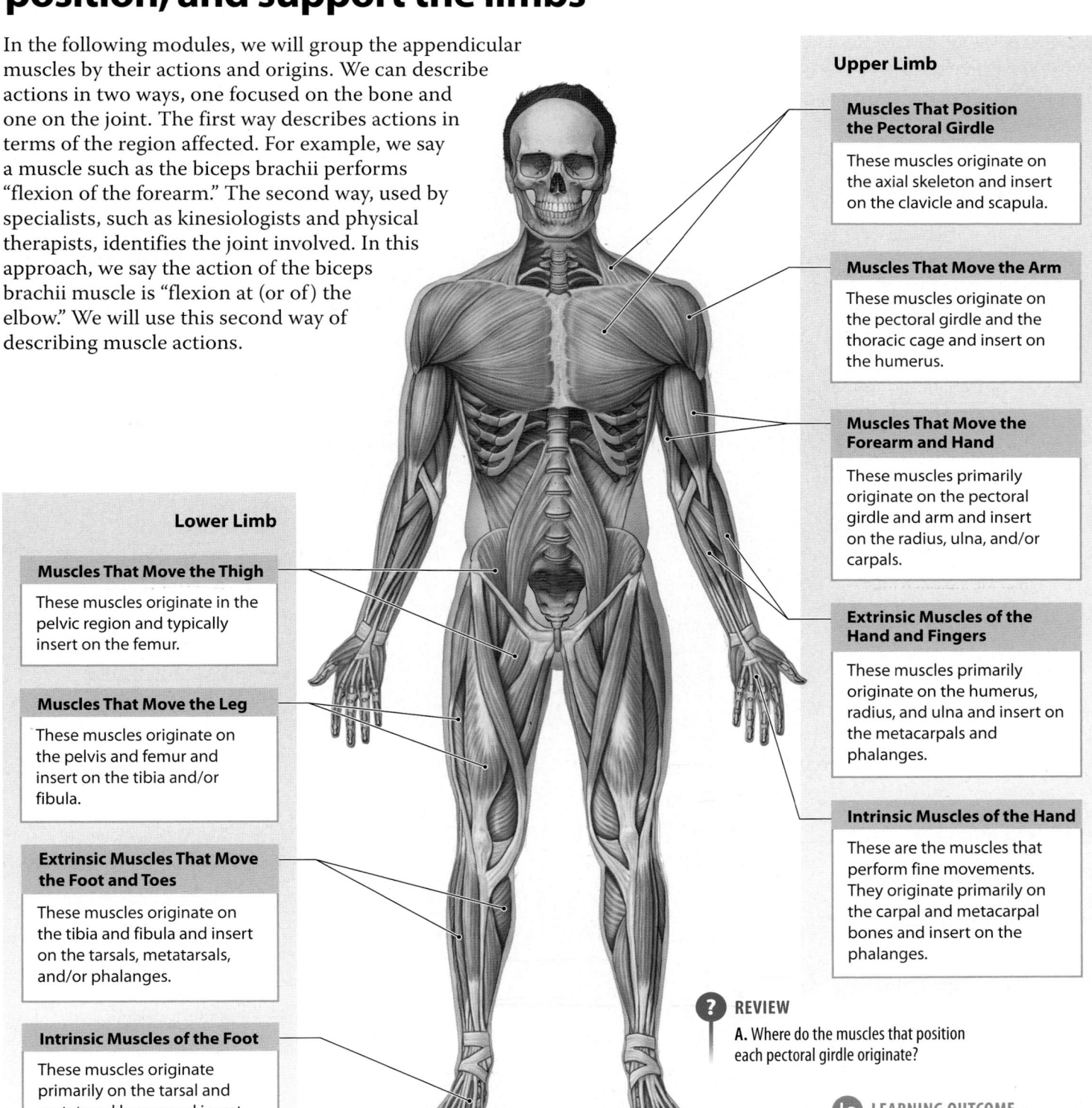

**Upper Limb**

**Muscles That Position the Pectoral Girdle**

These muscles originate on the axial skeleton and insert on the clavicle and scapula.

**Muscles That Move the Arm**

These muscles originate on the pectoral girdle and the thoracic cage and insert on the humerus.

**Muscles That Move the Forearm and Hand**

These muscles primarily originate on the pectoral girdle and arm and insert on the radius, ulna, and/or carpals.

**Extrinsic Muscles of the Hand and Fingers**

These muscles primarily originate on the humerus, radius, and ulna and insert on the metacarpals and phalanges.

**Intrinsic Muscles of the Hand**

These are the muscles that perform fine movements. They originate primarily on the carpal and metacarpal bones and insert on the phalanges.

**Lower Limb**

**Muscles That Move the Thigh**

These muscles originate in the pelvic region and typically insert on the femur.

**Muscles That Move the Leg**

These muscles originate on the pelvis and femur and insert on the tibia and/or fibula.

**Extrinsic Muscles That Move the Foot and Toes**

These muscles originate on the tibia and fibula and insert on the tarsals, metatarsals, and/or phalanges.

**Intrinsic Muscles of the Foot**

These muscles originate primarily on the tarsal and metatarsal bones and insert on the phalanges.

**? REVIEW**
**A.** Where do the muscles that position each pectoral girdle originate?

**LO LEARNING OUTCOME**
Describe the general functions of the muscles of the upper and lower limbs.

10

# Module 10.13

# The largest appendicular muscles originate on the trunk

**1** In general, muscles originating on the trunk control gross (large-scale) movements of the limbs. These muscles are often large and powerful. Distally, the limb muscles get smaller and more numerous, and their movements become more precise.

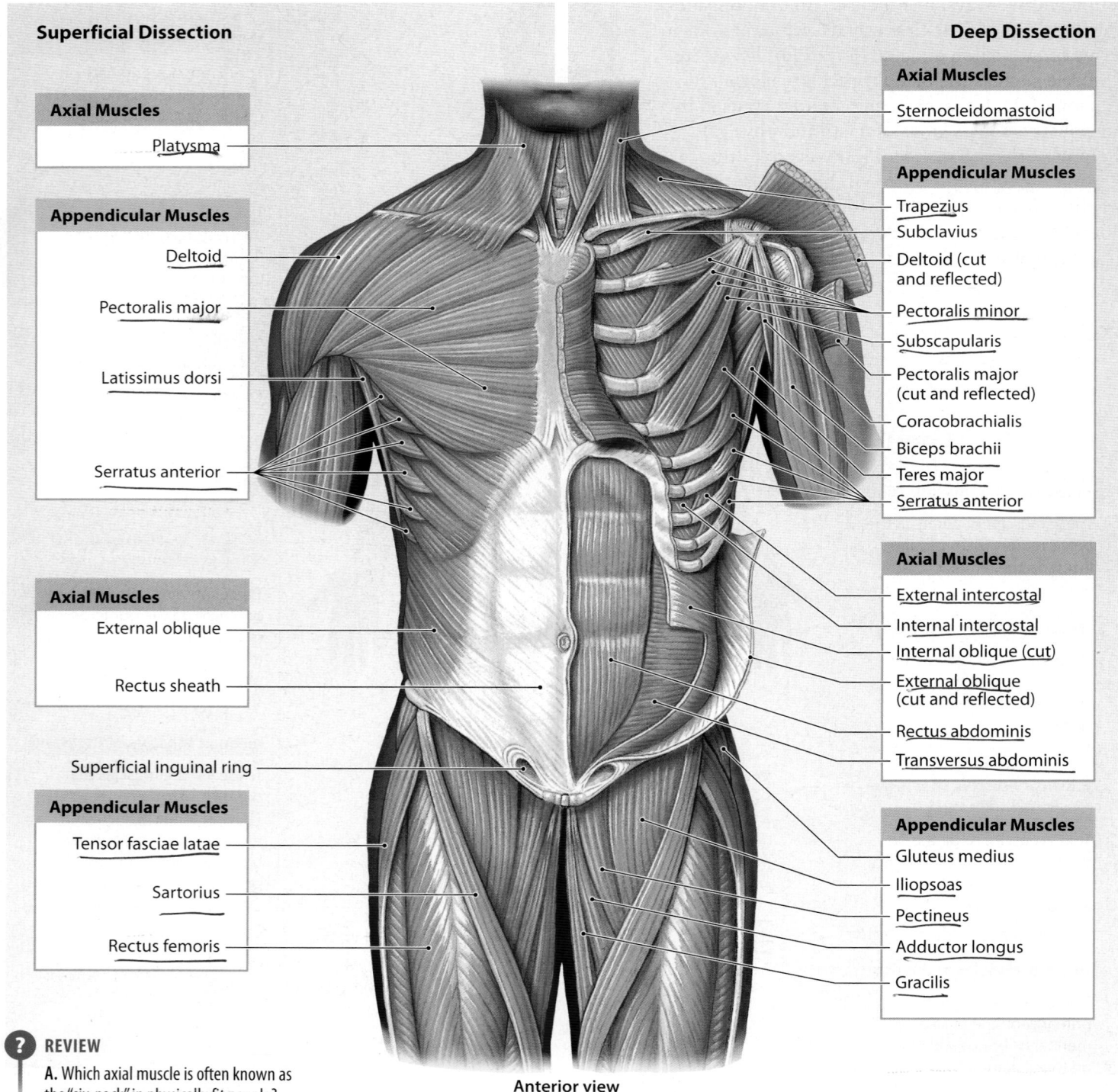

**Superficial Dissection**

**Axial Muscles**
- Platysma

**Appendicular Muscles**
- Deltoid
- Pectoralis major
- Latissimus dorsi
- Serratus anterior

**Axial Muscles**
- External oblique
- Rectus sheath
- Superficial inguinal ring

**Appendicular Muscles**
- Tensor fasciae latae
- Sartorius
- Rectus femoris

**Deep Dissection**

**Axial Muscles**
- Sternocleidomastoid

**Appendicular Muscles**
- Trapezius
- Subclavius
- Deltoid (cut and reflected)
- Pectoralis minor
- Subscapularis
- Pectoralis major (cut and reflected)
- Coracobrachialis
- Biceps brachii
- Teres major
- Serratus anterior

**Axial Muscles**
- External intercostal
- Internal intercostal
- Internal oblique (cut)
- External oblique (cut and reflected)
- Rectus abdominis
- Transversus abdominis

**Appendicular Muscles**
- Gluteus medius
- Iliopsoas
- Pectineus
- Adductor longus
- Gracilis

**Anterior view**

**? REVIEW**

**A.** Which axial muscle is often known as the "six-pack" in physically fit people?

**2** Appendicular muscles that originate on the large bones of the limb girdles and the proximal bones of the limbs dominate the posterior trunk.

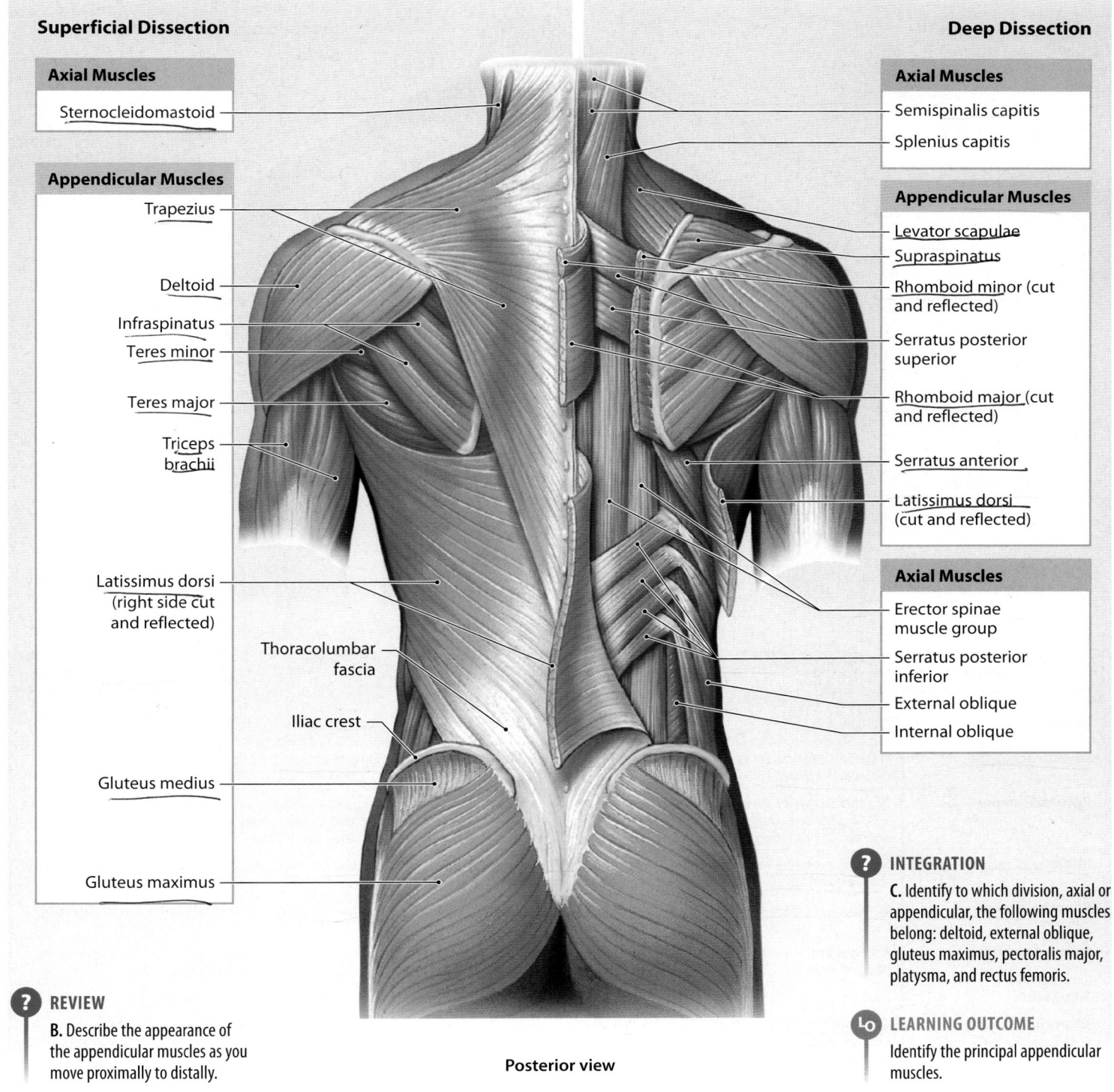

**Superficial Dissection**

**Deep Dissection**

**Axial Muscles**

Sternocleidomastoid

**Appendicular Muscles**

Trapezius

Deltoid

Infraspinatus

Teres minor

Teres major

Triceps brachii

Latissimus dorsi (right side cut and reflected)

Thoracolumbar fascia

Iliac crest

Gluteus medius

Gluteus maximus

**Axial Muscles**

Semispinalis capitis

Splenius capitis

**Appendicular Muscles**

Levator scapulae

Supraspinatus

Rhomboid minor (cut and reflected)

Serratus posterior superior

Rhomboid major (cut and reflected)

Serratus anterior

Latissimus dorsi (cut and reflected)

**Axial Muscles**

Erector spinae muscle group

Serratus posterior inferior

External oblique

Internal oblique

**Posterior view**

**? INTEGRATION**

**C.** Identify to which division, axial or appendicular, the following muscles belong: deltoid, external oblique, gluteus maximus, pectoralis major, platysma, and rectus femoris.

**Lo LEARNING OUTCOME**

Identify the principal appendicular muscles.

**? REVIEW**

**B.** Describe the appearance of the appendicular muscles as you move proximally to distally.

10

# Muscles that position each pectoral girdle originate on the occipital bone, superior vertebrae, and ribs

The muscles that position the pectoral girdles also anchor the pectoral girdles to the axial skeleton. Although these muscles have a smaller range of motion compared with other appendicular muscles, they help to increase upper limb mobility.

**1** Several deep muscles of the chest, notably the pectoralis minor and serratus anterior, work with the trapezius and levator scapulae of the back to position and stabilize each pectoral girdle.

**Muscles That Position the Pectoral Girdle**

- Trapezius
- Levator scapulae
- Subclavius
- Pectoralis minor

**Muscles That Position the Pectoral Girdle**

- Pectoralis minor (cut)
- Serratus anterior

- Internal intercostals
- External intercostals

T₁₂

**Anterior view**

## Muscles That Position the Pectoral Girdle

| Muscle | Origin | Insertion | Action |
|---|---|---|---|
| Levator scapulae | Transverse processes of first four cervical vertebrae | Vertebral border of scapula near superior angle | Elevates scapula |
| Pectoralis minor | Anterior-superior surfaces of ribs 2–4, 2–5, or 3–5, depending on anatomical variation | Coracoid process of scapula | Depresses and protracts shoulder; rotates scapula so glenoid cavity moves inferiorly (downward rotation); elevates ribs if scapula is stationary |
| Rhomboid major | Spinous processes of superior thoracic vertebrae | Vertebral border of scapula from spine to inferior angle | Adducts scapula and performs downward rotation |
| Rhomboid minor | Spinous processes of vertebrae C₇–T₁ | Vertebral border of scapula near spine | Adducts scapula and performs downward rotation |
| Serratus anterior | Anterior and superior margins of ribs 1–8 or 1–9 | Anterior surface of vertebral border of scapula | Protracts shoulder; rotates scapula so glenoid cavity moves superiorly (upward rotation) |
| Subclavius | First rib | Clavicle (inferior border) | Depresses and protracts shoulder |
| Trapezius | Occipital bone, ligamentum nuchae, and spinous processes of thoracic vertebrae | Clavicle and scapula (acromion and scapular spine) | Depends on active region and state of other muscles; may (1) elevate, retract, depress, or rotate scapula upward, (2) elevate clavicle, or (3) extend neck |

**2** The broad trapezius is the largest muscle in this group. The rhomboid major and rhomboid minor muscles and levator scapulae lie deep to the trapezius.

**?** **REVIEW**
**A.** Identify the largest of the superficial muscles that position the pectoral girdle.

**Superficial Dissection**

**Deep Dissection**

**Muscles That Position the Pectoral Girdle**

Trapezius

**Muscles That Position the Pectoral Girdle**

Levator scapulae

Rhomboid minor

Rhomboid major

Scapula

Serratus anterior

T$_{12}$ vertebra

**Posterior view**

The large, superficial trapezius muscles, commonly called the "traps," cover the back and portions of the neck, reaching to the base of the skull. These muscles are innervated by more than one nerve. For this reason, specific regions can be made to contract independently. As a result, their actions are quite varied.

**?** **REVIEW**
**B.** Which muscles enable you to shrug your shoulders?
**C.** Which muscle originates on the first rib and inserts on the inferior border of the clavicle?

**Lo** **LEARNING OUTCOME**
Identify the pectoral girdle muscles, and cite their origins, insertions, and actions.

# Muscles that move the arm originate on the clavicle, scapula, thoracic cage, and vertebral column

**Superficial Dissection**

**Deep Dissection**

Sternum

Clavicle

Ribs (cut)

**Muscles That Move the Arm**

Deltoid

Pectoralis major

**Muscles That Move the Arm**

Subscapularis*

Coracobrachialis

Teres major

(*Rotator cuff muscle)

**1** This anterior view shows muscles that move the arm and originate on the anterior surface of the chest and the anterior surface of the scapula.

**? REVIEW**

**A.** Which muscle originates on the anterior surface of the scapula and inserts on the lesser tubercle of the humerus?

Vertebra T$_{12}$

**Superficial Dissection**

**Deep Dissection**

Vertebra T$_1$

**Muscles That Move the Arm**

Supraspinatus*

Deltoid

Latissimus dorsi

(*Rotator cuff muscle)

Thoracolumbar fascia

**Muscles That Move the Arm**

Supraspinatus*

Infraspinatus*

Teres minor*

Teres major

(*Rotator cuff muscles)

**2** In a posterior view, the latissimus dorsi and deltoid muscles are the largest superficial muscles associated with arm movements. Many smaller muscles of this group originate on the scapula.

## Muscles That Move the Arm

| Muscle | Origin | Insertion | Action |
|--------|--------|-----------|--------|
| Deltoid | Clavicle and scapula (acromion and adjacent scapular spine) | Deltoid tuberosity of humerus | Whole muscle: abduction at shoulder; anterior part: flexion and medial rotation; posterior part: extension and lateral rotation |
| Supraspinatus* | Supraspinous fossa of scapula | Greater tubercle of humerus | Abduction at shoulder |
| Subscapularis* | Subscapular fossa of scapula | Lesser tubercle of humerus | Medial rotation at shoulder |
| Teres major | Inferior angle of scapula | Passes medially to reach the medial lip of intertubercular groove of humerus | Extension, adduction, and medial rotation at shoulder |
| Infraspinatus* | Infraspinous fossa of scapula | Greater tubercle of humerus | Lateral rotation at shoulder |
| Teres minor* | Lateral border of scapula | Passes laterally to reach the greater tubercle of humerus | Lateral rotation at shoulder |
| Coracobrachialis | Coracoid process | Medial margin of shaft of humerus | Adduction and flexion at shoulder |
| Pectoralis major | Cartilages of ribs 2–6, body of sternum, and inferior, medial portion of clavicle | Crest of greater tubercle and lateral lip of intertubercular groove of humerus | Flexion, adduction, and medial rotation at shoulder |
| Latissimus dorsi | Spinous processes of inferior thoracic and all lumbar vertebrae, ribs 8–12, and lumbodorsal fascia | Floor of intertubercular groove of the humerus | Extension, adduction, and medial rotation at shoulder |

\* Rotator cuff muscles

**3** The action of muscles positioning the arm can be understood by considering their direction of pull relative to the center of the glenoid cavity. The arrows indicate the line of force, or **line of action**, produced when a muscle contracts.

**? REVIEW**
**B.** Define *line of action*.

**? REVIEW**
**C.** Name the muscle that abducts the arm.

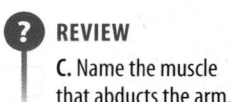

Extension

Abduction

Flexion

Deltoid

Tendons of biceps brachii

POSTERIOR

GLENOID CAVITY

ANTERIOR

Teres minor

Subscapularis

Triceps brachii

Lateral rotation

Teres major

Medial rotation

Adduction

Collectively, the supraspinatus, infraspinatus, teres minor, and subscapularis muscles and their associated tendons form the **rotator cuff**. The acronym SITS (representing the first letter of each muscle) can help you remember these four muscles. Sports that involve throwing a ball, such as baseball or football, place considerable strain on the rotator cuff, and rotator cuff injuries are relatively common.

**LEARNING OUTCOME**
Identify the muscles that move the arm, and cite their origins, insertions, and actions.

10

# Muscles that move the forearm and hand originate on the scapula, humerus, radius, or ulna

**1** This posterior view shows the superficial muscles involved in extension at the elbow and wrist.

**2** This anterior view shows the superficial muscles involved in flexion at the elbow and wrist.

Tricep brachii

brachioradialis

Extensor carpi radialis longus

**Elbow Extensors**

Triceps brachii, long head

Triceps brachii, lateral head

Anconeus

**Wrist Extensors**

Extensor carpi ulnaris

Extensor carpi radialis longus

Extensor carpi radialis brevis

Olecranon of ulna

Flexor carpi ulnaris

Radius

Ulna

abductor pollicis longus

Extensor digitorum (see Module 10.17)

extensor pollicis brevis

extensor pollicis longus

The **extensor retinaculum** (ret-i-NAK-yū-lum), a wide band of connective tissue, holds the tendons of the extensor muscles in place.

**Posterior view**

Coracoid process of scapula

Humerus

**Elbow Flexors**

Biceps brachii, long head

Biceps brachii, short head

Brachialis

Brachioradialis

**Elbow Extensors**

Triceps brachii, long head ✓

Triceps brachii, medial head

Medial epicondyle of humerus

**Wrist Flexors**

Flexor carpi radialis

Palmaris longus

Flexor carpi ulnaris

**Pronators and Supinators**

Pronator teres ✓

Supinator ✓

Pronator quadratus

The **flexor retinaculum**, a wide band of connective tissue, stabilizes the tendons of the flexor muscles.

**Anterior view**

**? REVIEW**

**A.** Define *retinaculum*.

**B.** Are the wrist extensors located on the anterior surface or the posterior surface of the forearm?

## Muscles That Move the Forearm and Hand

| Group and Muscle | Origin | Insertion | Action |
|---|---|---|---|
| **Action at the Elbow** | | | |
| FLEXORS — **Biceps brachii** | Short head from the coracoid process; long head from the supraglenoid tubercle (both on the scapula) | Tuberosity of radius *Radial tuberosity* | Flexion at elbow and shoulder; supination |
| FLEXORS — **Brachialis** | Anterior, distal surface of humerus | Tuberosity of ulna | Flexion at elbow |
| FLEXORS — **Brachioradialis** | Ridge superior to the lateral epicondyle of humerus | Lateral aspect of styloid process of radius | Flexion at elbow |
| EXTENSORS — **Anconeus** | Posterior, inferior surface of lateral epicondyle of humerus | Lateral margin of olecranon on ulna | Extension at elbow |
| EXTENSORS — **Triceps brachii** | Lateral head from the superior, lateral margin of humerus; long head from infraglenoid tubercle of scapula; and medial head from posterior surface of humerus inferior to radial groove | Olecranon of ulna | Extension at elbow, plus extension and adduction at the shoulder |
| PRONATORS / SUPINATORS — **Pronator quadratus** | Anterior and medial surfaces of distal portion of ulna | Anterolateral surface of distal portion of radius | Pronation |
| PRONATORS / SUPINATORS — **Pronator teres** | Medial epicondyle of humerus and coronoid process of ulna | Midlateral surface of radius | Pronation |
| PRONATORS / SUPINATORS — **Supinator** | Lateral epicondyle of humerus, annular ligament, and ridge near radial notch of ulna | Anterolateral surface of radius distal to the radial tuberosity | Supination |
| **Action at the Hand** | | | |
| FLEXORS — **Flexor carpi radialis** | Medial epicondyle of humerus | Bases of second and third metacarpal bones | Flexion and abduction at wrist |
| FLEXORS — **Flexor carpi ulnaris** | Medial epicondyle of humerus; adjacent medial surface of olecranon and anteromedial portion of ulna | Pisiform bone, hamate bone, and base of fifth metacarpal bone | Flexion and adduction at wrist |
| FLEXORS — **Palmaris longus** | Medial epicondyle of humerus | Palmar aponeurosis and flexor retinaculum | Flexion at wrist |
| EXTENSORS — **Extensor carpi radialis longus** | Lateral supracondylar ridge of humerus | Base of second metacarpal bone | Extension and abduction at wrist |
| EXTENSORS — **Extensor carpi radialis brevis** | Lateral epicondyle of humerus | Base of third metacarpal bone | Extension and abduction at wrist |
| EXTENSORS — **Extensor carpi ulnaris** | Lateral epicondyle of humerus; adjacent dorsal surface of ulna | Base of fifth metacarpal bone | Extension and adduction at wrist |

10

**3** The muscles involved in pronation and supination are the pronator teres, supinator, and pronator quadratus. Both the pronator teres and supinator originate on both the humerus and ulna and rotate the radius without either flexing or extending the elbow. The pronator quadratus muscle originates on the ulna and assists in pronation (see Module 8.5).

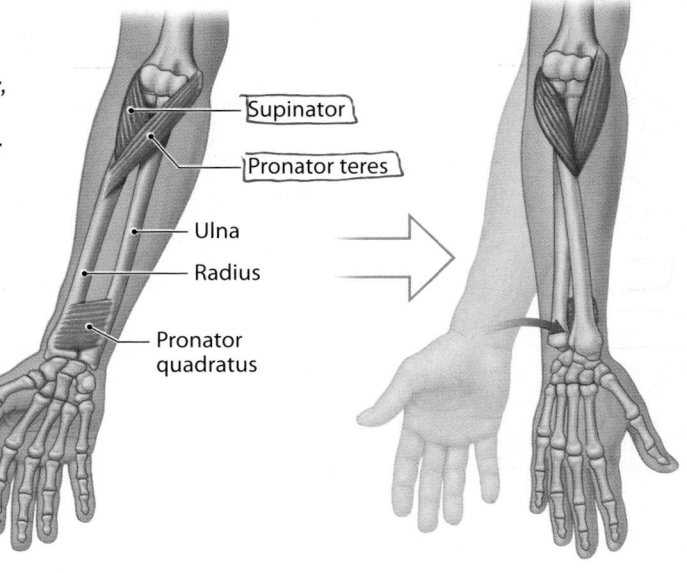

Supinator
Pronator teres
Ulna
Radius
Pronator quadratus

**Supination**

**Pronation**

**? INTEGRATION**

**C.** Which muscles are involved in turning a doorknob?

**Lo LEARNING OUTCOME**

Identify the forearm muscles, and cite their origins, insertions, and actions.

# Muscles that move the hand and fingers originate on the humerus, radius, ulna, and interosseous membrane

**1** In anterior view, the large flexor digitorum superficialis covers the smaller digital flexors.

**Muscles That Flex the Fingers and Thumb**
- Flexor digitorum superficialis
- Flexor digitorum profundus
- Flexor pollicis longus

Anterior view, middle layer

Anterior view, deepest layer

## Muscles That Move the Hand and Fingers

| Muscle | | Origin | Insertion | Action |
|---|---|---|---|---|
| **Abductor pollicis longus** | | Proximal dorsal surfaces of ulna and radius | Lateral margin of first metacarpal bone | Abduction at carpometacarpal joint of thumb and wrist |
| EXTENSORS | **Extensor digitorum** | Lateral epicondyle of humerus | Posterior surfaces of the phalanges, fingers 2–5 | Extension at finger joints and wrist |
| | **Extensor pollicis brevis** | Shaft of radius distal to origin of abductor pollicis longus | Base of proximal phalanx of thumb | Extension at carpometacarpal joint of thumb; abduction at wrist |
| | **Extensor pollicis longus** | Posterior and lateral surfaces of ulna and interosseous membrane | Base of distal phalanx of thumb | Extension at carpometacarpal joint of thumb; abduction at wrist |
| | **Extensor indicis** | Posterior surface of ulna and interosseous membrane | Posterior surface of phalanges of index finger (2), with tendon of extensor digitorum | Extension and adduction at joints of index finger |
| | **Extensor digiti minimi** | By extensor tendon to lateral epicondyle of humerus and from intermuscular septa | Posterior surface of proximal phalanx of little finger (5) | Extension at joints of little finger |
| FLEXORS | **Flexor digitorum superficialis** | Medial epicondyle of humerus; adjacent anterior surfaces of ulna and radius | Midlateral surfaces of middle phalanges of fingers 2–5 | Flexion at proximal interphalangeal, metacarpophalangeal, and wrist joints |
| | **Flexor digitorum profundus** | Medial and posterior surfaces of ulna, medial surface of coronoid process, and interosseus membrane | Bases of distal phalanges of fingers 2–5 | Flexion at distal interphalangeal joints and, to a lesser degree, proximal interphalangeal joints and wrist |
| | **Flexor pollicis longus** | Anterior shaft of radius, interosseous membrane | Base of distal phalanx of thumb | Flexion at carpometacarpal joint of thumb |

**2** In posterior view, the muscles that extend the fingers can only be seen after removal of the muscles involved in wrist movements. The deepest digital extensor muscles are those associated with movements of the thumb.

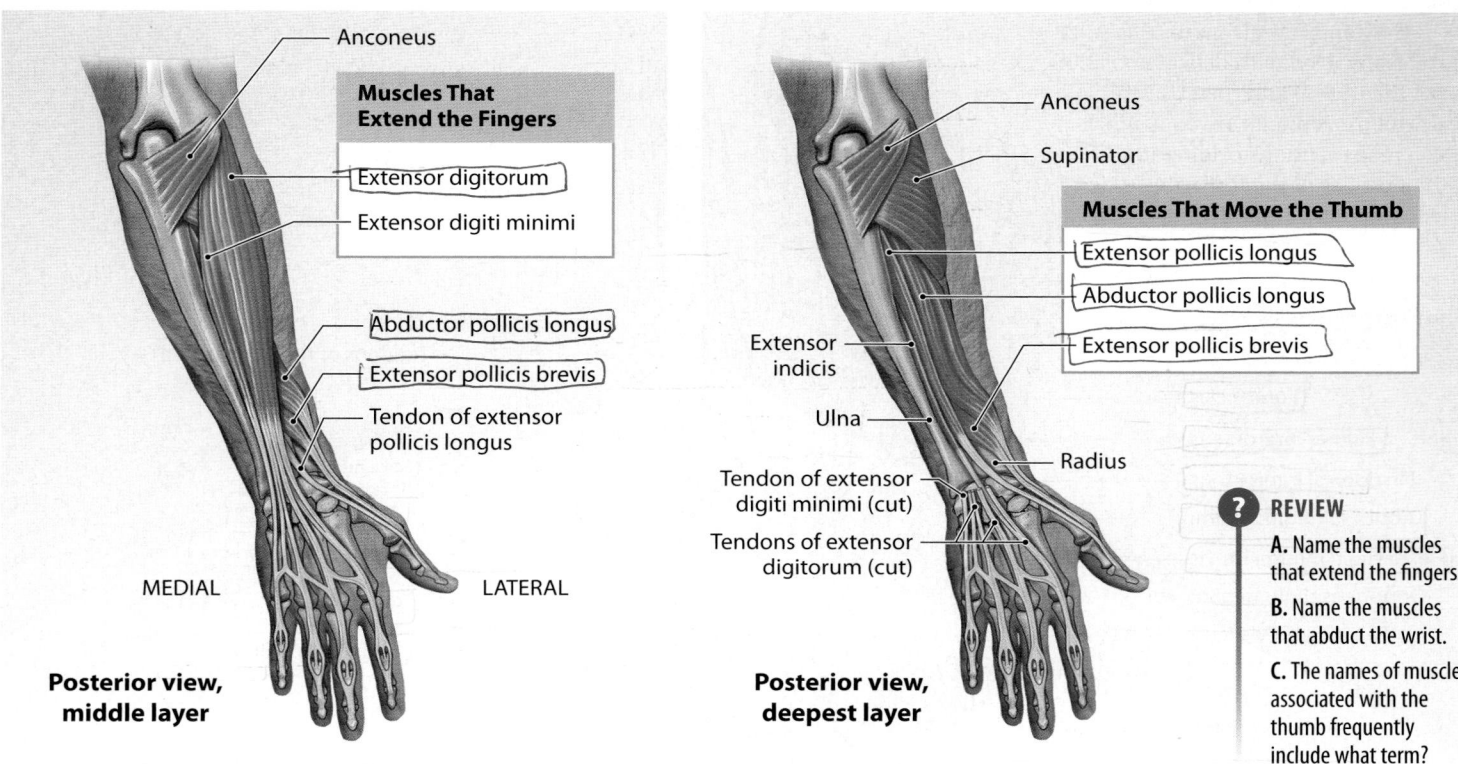

Anconeus

**Muscles That Extend the Fingers**

Extensor digitorum

Extensor digiti minimi

Abductor pollicis longus

Extensor pollicis brevis

Tendon of extensor pollicis longus

MEDIAL          LATERAL

**Posterior view, middle layer**

Anconeus

Supinator

**Muscles That Move the Thumb**

Extensor pollicis longus

Abductor pollicis longus

Extensor pollicis brevis

Extensor indicis

Ulna

Tendon of extensor digiti minimi (cut)

Tendons of extensor digitorum (cut)

Radius

**Posterior view, deepest layer**

**? REVIEW**

**A.** Name the muscles that extend the fingers.

**B.** Name the muscles that abduct the wrist.

**C.** The names of muscles associated with the thumb frequently include what term?

**3** **Synovial tendon sheaths** are tubular bursae that surround tendons where they cross bony surfaces. The tendons of the flexor muscles pass through such sheaths as they pass deep to the flexor retinaculum. Inflammation of the flexor retinaculum and synovial tendon sheaths can restrict movement and put pressure on the distal portions of the median nerve, a mixed (sensory and motor) nerve that innervates the hand. This condition, known as **carpal tunnel syndrome**, causes tingling, numbness, weakness, and chronic pain in the wrist and hand.

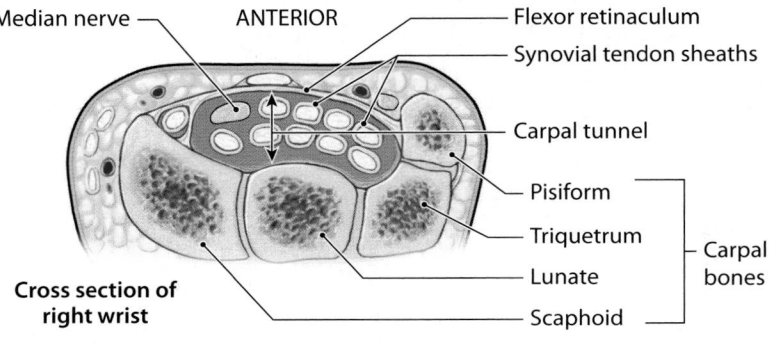

Median nerve          ANTERIOR          Flexor retinaculum

Synovial tendon sheaths

Carpal tunnel

Pisiform

Triquetrum          Carpal bones

Lunate

Scaphoid

**Cross section of right wrist**

As you study these muscles, you will notice that extensor muscles usually lie along the posterior and lateral surfaces of the forearm, whereas flexors are typically found on the anterior and medial surfaces. Remember, the limb must be in the anatomical position for this to be true. This information can be quite useful when you are trying to identify a particular muscle on a quiz or a lab practical.

**Lo LEARNING OUTCOME**

Identify the muscles of the hand and fingers, and cite their origins, insertions, and actions.

# The intrinsic muscles of the hand originate on the carpal and metacarpal bones and associated tendons and ligaments

**1** Dissection of the palmar surface of the hand reveals numerous small muscles involved with delicate positioning of the thumb and fingers. More powerful movements are controlled by the many tendons originating at the muscles of the forearm considered in Module 10.16.

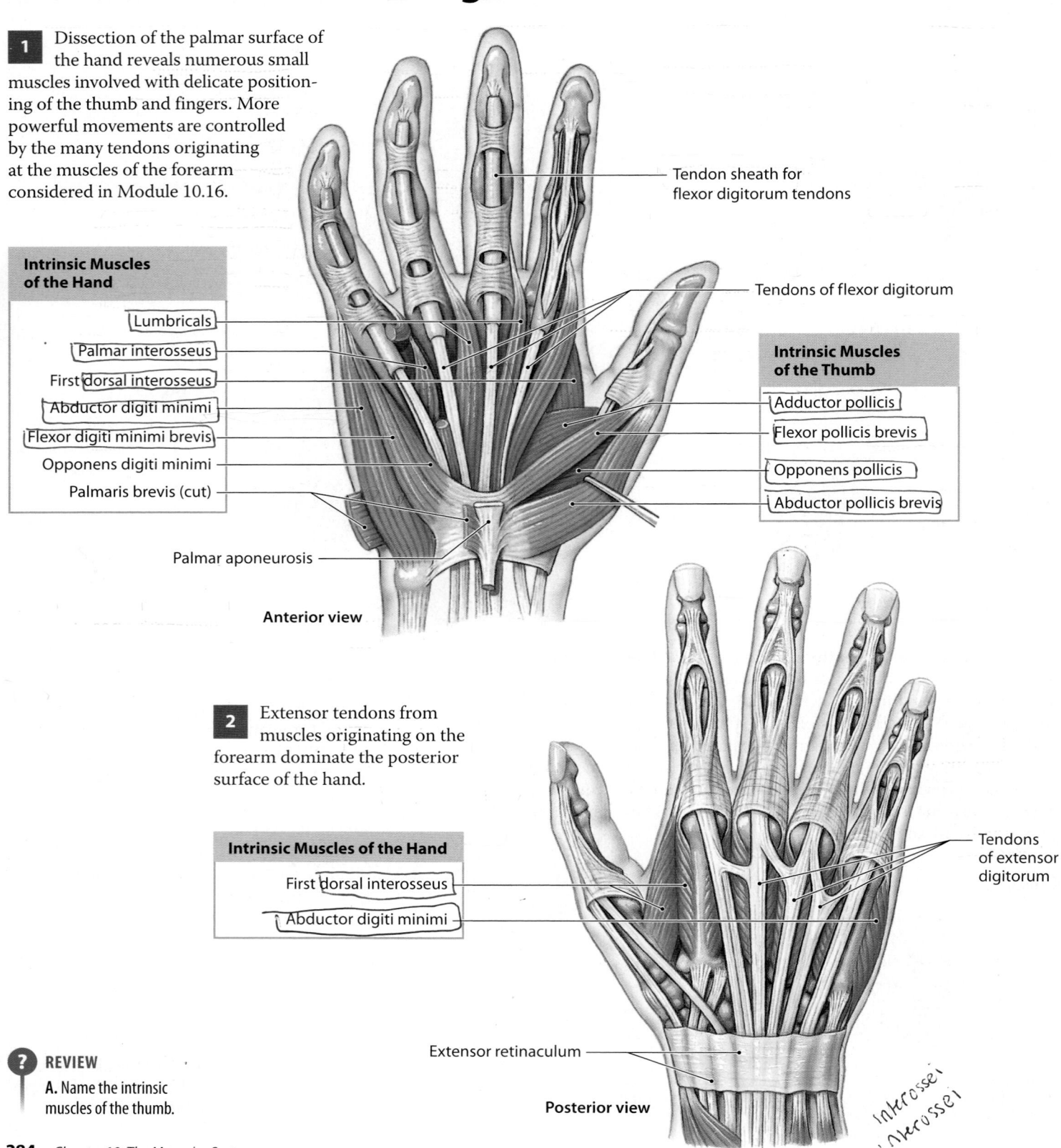

Tendon sheath for flexor digitorum tendons

Tendons of flexor digitorum

**Intrinsic Muscles of the Hand**

- Lumbricals
- Palmar interosseus
- First dorsal interosseus
- Abductor digiti minimi
- Flexor digiti minimi brevis
- Opponens digiti minimi
- Palmaris brevis (cut)

**Intrinsic Muscles of the Thumb**

- Adductor pollicis
- Flexor pollicis brevis
- Opponens pollicis
- Abductor pollicis brevis

Palmar aponeurosis

**Anterior view**

**2** Extensor tendons from muscles originating on the forearm dominate the posterior surface of the hand.

**Intrinsic Muscles of the Hand**

- First dorsal interosseus
- Abductor digiti minimi

Tendons of extensor digitorum

Extensor retinaculum

**Posterior view**

Interossei
Interossei

**? REVIEW**

**A.** Name the intrinsic muscles of the thumb.

## Intrinsic Muscles of the Hand

| Muscle | Origin | Insertion | Action |
|---|---|---|---|
| **Palmaris brevis** | Palmar aponeurosis | Skin of medial border of hand | Moves skin on medial border toward midline of palm |
| **Adductor pollicis** | Metacarpal and carpal bones | Proximal phalanx of thumb | Adduction of thumb |
| **Palmar interosseus (3–4)** | Sides of metacarpal bones II, IV, and V | Bases of proximal phalanges of fingers 2, 4, and 5 | Adduction at metacarpophalangeal joints of fingers 2, 4, and 5; flexion at metacarpophalangeal joints; extension at interphalangeal joints |
| **Abductor pollicis brevis** | Transverse carpal ligament, scaphoid bone, and trapezium | Radial side of base of proximal phalanx of thumb | Abduction of thumb |
| **Dorsal interosseus (4)** | Each originates from opposing faces of two metacarpal bones (I and II, II and III, III and IV, IV and V) | Bases of proximal phalanges of fingers 2–4 | Abduction at metacarpophalangeal joints of fingers 2 and 4; flexion at metacarpophalangeal joints; extension at interphalangeal joints |
| **Abductor digiti minimi** | Pisiform bone | Proximal phalanx of little finger | Abduction of little finger and flexion at its metacarpophalangeal joint |
| **Flexor pollicis brevis** | Flexor retinaculum, trapezium, capitate bone, and ulnar side of first metacarpal bone | Radial and ulnar sides of proximal phalanx of thumb | Flexion and adduction of thumb |
| **Lumbrical (4)** | Tendons of flexor digitorum profundus | Tendons of extensor digitorum to digits 2–5 | Flexion at metacarpophalangeal joints 2–5; extension at proximal and distal interphalangeal joints, digits 2–5 |
| **Flexor digiti minimi brevis** | Hamate bone | Proximal phalanx of little finger | Flexion at joints of little finger |
| **Opponens pollicis** | Trapezium and flexor retinaculum | First metacarpal bone | Opposition of thumb |
| **Opponens digiti minimi** | Trapezium and flexor retinaculum | Fifth metacarpal bone | Opposition of fifth metacarpal bone |

*Left side labels: ADDUCTION / ABDUCTION (spanning Adductor pollicis through Abductor digiti minimi); FLEXION (spanning Flexor pollicis brevis through Flexor digiti minimi brevis)*

**10**

---

**Clinical Note**

### Trigger finger

Trigger finger is a condition in which a finger gets stuck in a bent position and opens with a painful snap (like a trigger being released). It is caused by inflammation and thickening of the tendon sheath that covers the flexor digitorum tendon. The inflammation narrows the opening the tendon normally glides through. Treatment usually begins with rest and a hydrocortisone injection but may require surgery.

---

Fine control of the hand involves small intrinsic muscles that originate on the carpal and metacarpal bones. These intrinsic muscles are responsible for (1) flexion and extension of the fingers at the metacarpophalangeal joints, (2) abduction and adduction of the fingers at the metacarpophalangeal joints, and (3) opposition and reposition (relaxed position) of the thumb. No muscles originate on the phalanges, and only tendons extend across the distal joints of the fingers.

**? REVIEW**
**B.** Which muscles originate on the phalanges?

**? INTEGRATION**
**C.** If there are no muscles in the fingers, how are we able to move them?

**Lo LEARNING OUTCOME**
Identify the intrinsic hand muscles, and cite their origins, insertions, and actions.

# The muscles that move the thigh originate on the pelvis and associated ligaments and fasciae

**1** The **gluteal group** covers the posteri-or and lateral surfaces of the pelvis.

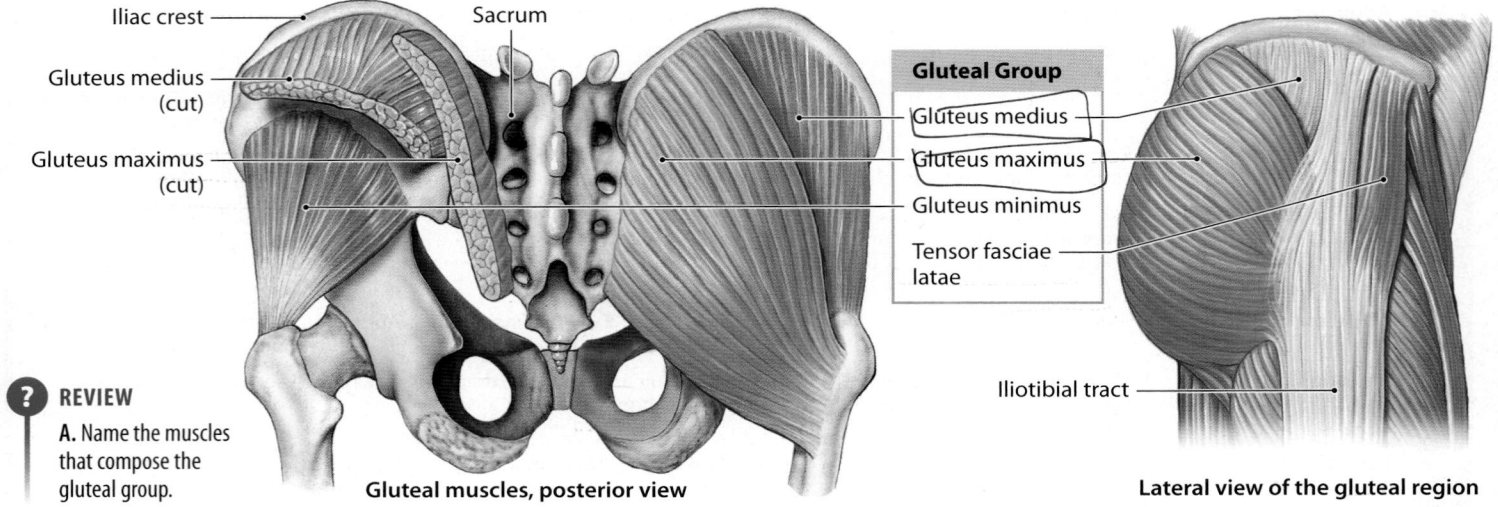

Iliac crest

Sacrum

Gluteus medius (cut)

Gluteus maximus (cut)

**Gluteal Group**

Gluteus medius

Gluteus maximus

Gluteus minimus

Tensor fasciae latae

Iliotibial tract

**? REVIEW**

**A.** Name the muscles that compose the gluteal group.

**Gluteal muscles, posterior view**

**Lateral view of the gluteal region**

**2** This dissection of the gluteal region in posterior view shows five of the six lateral rotators.

**Gluteal Group**

| Gluteus maximus (cut) | Gluteus medius (cut) | Gluteus minimus | Tensor fasciae latae |

**Lateral Rotator Group**

Piriformis

Superior gemellus

Obturator internus

Obturator externus

Inferior gemellus

Quadratus femoris

Ischial tuberosity

Iliotibial tract

**3** This anterior view shows the isolated iliopsoas muscle group and the adductor group.

**Iliopsoas Group**

Psoas major

Iliacus

L₅

Inguinal ligament

**Adductor Group**

Pectineus

Adductor brevis

Adductor longus

Adductor magnus

Gracilis

## Muscles That Move the Thigh

| Group and Muscle | Origin | Insertion | Action |
|---|---|---|---|
| **Gluteal Group** | | | |
| **Gluteus maximus** | Iliac crest, posterior gluteal line, and lateral surface of ilium; sacrum, coccyx, and thoracolumbar fascia | Iliotibial tract and gluteal tuberosity of femur | Extension and lateral rotation at hip |
| **Gluteus medius** | Anterior iliac crest of ilium, lateral surface between posterior and anterior gluteal lines | Greater trochanter of femur | Abduction and medial rotation at hip |
| **Gluteus minimus** | Lateral surface of ilium between inferior and anterior gluteal lines | Greater trochanter of femur | Abduction and medial rotation at hip |
| **Tensor fasciae latae** | Iliac crest and lateral surface of anterior superior iliac spine | Iliotibial tract | Extension of the knee and lateral rotation of the leg acting through the iliotibial tract; abduction** and medial rotation of the thigh |
| **Lateral Rotator Group** | | | |
| **Obturator** (**externus** and **internus**) | Lateral and medial margins of obturator foramen | Externus: trochanteric fossa of femur; internus: medial surface of greater trochanter | Lateral rotation at hip |
| **Piriformis** | Anterolateral surface of sacrum | Greater trochanter of femur | Lateral rotation and abduction at hip |
| **Gemellus** (**superior** and **inferior**) | Ischial spine and tuberosity | Medial surface of greater trochanter with tendon of obturator internus | Lateral rotation at hip |
| **Quadratus femoris** | Lateral border of ischial tuberosity | Intertrochanteric crest of femur | Lateral rotation at hip |
| **Adductor Group** | | | |
| **Adductor brevis** | Inferior ramus of pubis | Linea aspera of femur | Adduction, flexion, and medial rotation at hip |
| **Adductor longus** | Inferior ramus of pubis anterior to adductor brevis | Linea aspera of femur | Adduction, flexion, and medial rotation at hip |
| **Adductor magnus** | Inferior ramus of pubis posterior to adductor brevis and ischial tuberosity | Linea aspera and adductor tubercle of femur | Adduction at hip; superior part produces flexion and medial rotation; inferior part produces extension and lateral rotation |
| **Pectineus** | Superior ramus of pubis | Pectineal line inferior to lesser trochanter of femur | Flexion, medial rotation, and adduction at hip |
| **Gracilis** | Inferior ramus of pubis | Medial surface of tibia inferior to medial condyle | Flexion at knee; adduction and medial rotation at hip |
| **Iliopsoas Group*** | | | |
| **Iliacus** | Iliac fossa of ilium | Femur distal to lesser trochanter; tendon fused with that of psoas major | Flexion at hip |
| **Psoas major** | Anterior surfaces and transverse processes of vertebrae ($T_{12}$–$L_5$) | Lesser trochanter in company with iliacus | Flexion at hip or lumbar intervertebral joints |

\* The psoas major and iliacus are often considered collectively as the iliopsoas
\*\* Role in abduction is debatable.

One method for understanding the actions of these diverse muscles is to consider their orientation around the hip joint. Muscles originating on the surface of the pelvis and inserting on the femur will produce characteristic movements determined by their position relative to the acetabulum. Many of the muscles that act on the hip are very large, and they have insertions that extend over a broad area. As a result, these muscles often have more than one action line and therefore produce more than one action at the hip. For example, the action of the adductor magnus varies depending on what portion of the muscle is activated; when the entire muscle contracts, it produces a combination of flexion, extension, and adduction at the hip.

 **REVIEW**

**B.** Which leg movement would be impaired by injury to the obturator muscles?

 **LEARNING OUTCOME**

Identify the muscles that move the thigh, and cite their origins, insertions, and actions.

# The muscles that move the leg originate on the pelvis and femur

**1** **Flexors of the knee** originate on the pelvic girdle and extend along the posterior and medial surfaces of the thigh.

**2** Most of the **extensors of the knee** originate on the femoral surface and extend along the anterior and lateral surfaces of the thigh. Collectively, the four knee extensor muscles are called the **quadriceps femoris**.

Iliac crest
Gluteus medius
Tensor fasciae latae
Gluteus maximus
Adductor magnus
Gracilis
Iliotibial tract

**Flexors of the Knee**
Biceps femoris
Semitendinosus
Semimembranosus
Sartorius
Popliteus

**Posterior view**

Anterior superior iliac spine
Inguinal ligament
Iliacus ⎤
Psoas major ⎦ Iliopsoas
Pubic tubercle
Tensor fasciae latae
Pectineus
Adductor longus
Gracilis
Sartorius

**Extensors of the Knee (Quadriceps femoris)**
Rectus femoris
Vastus intermedius (lies deep to the rectus femoris and vastus lateralis)
Vastus lateralis
Vastus medialis
Quadriceps tendon
Patella
Patellar ligament

**Anterior view**

**? REVIEW**
A. Which muscles flex the knee?

**? REVIEW**
B. Name the muscles of the quadriceps femoris.

## Muscles That Move the Leg

| Group and Muscle | Origin | Insertion | Action |
|---|---|---|---|
| **Flexors of the Knee** | | | |
| **Biceps femoris** | Ischial tuberosity and linea aspera of femur | Head of fibula, lateral condyle of tibia | Flexion at knee; extension and lateral rotation at hip |
| **Semimembranosus** | Ischial tuberosity | Posterior surface of medial condyle of tibia | Flexion at knee; extension and medial rotation at hip |
| **Semitendinosus** | Ischial tuberosity | Proximal, medial surface of tibia near insertion of gracilis | Flexion at knee; extension and medial rotation at hip |
| **Sartorius** | Anterior superior iliac spine | Medial surface of tibia near tibial tuberosity | Flexion at knee; flexion and lateral rotation at hip |
| **Popliteus** | Lateral condyle of femur | Posterior surface of proximal tibial shaft | Medial rotation of tibia (or lateral rotation of femur); flexion at knee |
| **Extensors of the Knee** | | | |
| **Rectus femoris** | Anterior inferior iliac spine and superior acetabular rim of ilium | Tibial tuberosity by patellar ligament | Extension at knee; flexion at hip |
| **Vastus intermedius** | Anterolateral surface of femur and linea aspera (distal half) | Tibial tuberosity by patellar ligament | Extension at knee |
| **Vastus lateralis** | Anterior and inferior to greater trochanter of femur and along linea aspera (proximal half) | Tibial tuberosity by patellar ligament | Extension at knee |
| **Vastus medialis** | Entire length of linea aspera of femur | Tibial tuberosity by patellar ligament | Extension at knee |

**? REVIEW**

**C.** Identify the muscle whose origin is on the lateral condyle of the femur.

**3** This cross-sectional view shows the positions of the major thigh muscles relative to the femur. Together, the vastus muscles cradle the rectus femoris muscle the way a bun surrounds a hot dog. All four muscles insert on the patella via the quadriceps tendon.

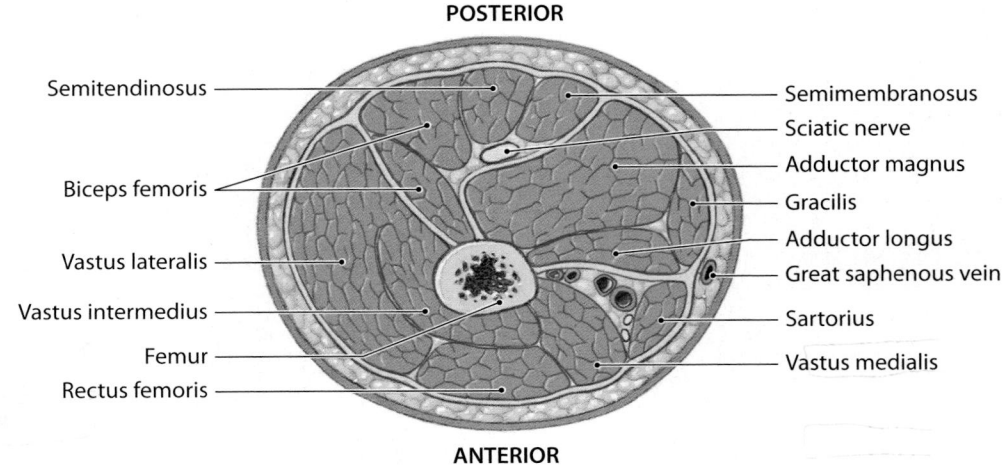

POSTERIOR

Semitendinosus — Semimembranosus
— Sciatic nerve
Biceps femoris — Adductor magnus
— Gracilis
Vastus lateralis — Adductor longus
— Great saphenous vein
Vastus intermedius — Sartorius
Femur — Vastus medialis
Rectus femoris —

ANTERIOR

**? INTEGRATION**

**D.** Which muscle of the leg crosses and produces actions at two joints?

**Lo LEARNING OUTCOME**

Identify the muscles that move the leg, and cite their origins, insertions, and actions.

10

# The extrinsic muscles that move the foot and toes originate on the tibia and fibula

**1** These views show the multiple muscle layers in the posterior aspect of the leg.

**Superficial Dissection** → **Deep Dissection**

**Ankle Extensors**
- Plantaris
- Gastrocnemius
- Soleus

Popliteus

Gastrocnemius (cut and removed)

Calcaneal tendon

Calcaneus

Head of fibula

**Ankle Extensors (Deep)**
- Tibialis posterior
- Fibularis longus
- Fibularis brevis

**Digital Flexors**
- Flexor digitorum longus
- Flexor hallucis longus

Tendon of flexor digitorum longus

Tendon of fibularis brevis

Tendon of fibularis longus

**2** These lateral, medial, and anterior views show the arrangement of the major superficial muscles.

**Lateral view**

Iliotibial tract

Head of fibula

**Ankle Extensors**
- Gastrocnemius
- Fibularis longus
- Soleus
- Fibularis brevis

**Ankle Flexors**
- Tibialis anterior

**Digital Extensors**
- Extensor digitorum longus
- Tendon of extensor hallucis longus

Superior extensor retinaculum

Calcaneal tendon

Inferior extensor retinaculum

Tendon of fibularis tertius

**Medial view**

Patella

Patellar ligament

Medial surface of tibial shaft

**Ankle Extensors**
- Gastrocnemius
- Soleus

Tibialis posterior

Superior extensor retinaculum

Calcaneal tendon

Inferior extensor retinaculum

Tendon of tibialis anterior

**Anterior view**

Patellar ligament

Fibularis longus

Tibialis anterior

Tibia

Extensor digitorum longus

Extensor hallucis longus

## Extrinsic Muscles That Move the Foot and Toes

| Group and Muscle | Origin | Insertion | Action |
|---|---|---|---|
| **Action at the Ankle** | | | |
| **FLEXORS (Dorsiflexors)** | | | |
| Tibialis anterior | Lateral condyle and proximal shaft of tibia | Base of first metatarsal bone and medial cuneiform bone | Flexion (dorsiflexion) at ankle; inversion of foot |
| Fibularis tertius | Distal anterior surface of fibula and interosseous membrane | Dorsal surface of fifth metatarsal bone | Flexion (dorsiflexion); eversion of foot |
| **EXTENSORS (Plantar flexors)** | | | |
| Gastrocnemius | Femoral condyles | Calcaneus by calcaneal tendon | Extension (plantar flexion) at ankle; inversion of foot; flexion at knee |
| Fibularis brevis | Midlateral margin of fibula | Base of fifth metatarsal bone | Eversion of foot and extension (plantar flexion) at ankle |
| Fibularis longus | Lateral condyle of tibia, head and proximal shaft of fibula | Base of first metatarsal bone and medial cuneiform bone | Eversion of foot and extension (plantar flexion) at ankle; supports longitudinal arch |
| Plantaris | Lateral supracondylar ridge | Posterior portion of calcaneus | Extension (plantar flexion) at ankle; flexion at knee |
| Soleus | Head and proximal shaft of fibula and adjacent posteromedial shaft of tibia | Calcaneus by calcaneal tendon (with gastrocnemius) | Extension (plantar flexion) at ankle |
| Tibialis posterior | Interosseous membrane and adjacent shafts of tibia and fibula | Tarsal and metatarsal bones | Adduction and inversion of foot; extension (plantar flexion) at ankle |
| **Action at the Toes** | | | |
| **DIGITAL FLEXORS** | | | |
| Flexor digitorum longus | Posteromedial surface of tibia | Inferior surfaces of distal phalanges, toes 2–5 | Flexion at joints of toes 2–5 |
| Flexor hallucis longus | Posterior surface of fibula | Inferior surface, distal phalanx of great toe | Flexion at joints of great toe |
| **DIGITAL EXTENSORS** | | | |
| Extensor digitorum longus | Lateral condyle of tibia, anterior surface of fibula | Superior surfaces of phalanges, toes 2–5 | Extension at joints of toes 2–5 |
| Extensor hallucis longus | Anterior surface of fibula | Superior surface, distal phalanx of great toe | Extension at joints of great toe |

 **REVIEW**

**A.** Name the muscles involved in extending the ankle.

**B.** Name the muscles involved in flexing the toes.

The largest muscles associated with ankle movement are the **gastrocnemius** and **soleus**. These muscles produce ankle extension (plantar flexion), a movement essential to walking and running (see Module 8.4). The muscles that move the toes are much smaller, and they originate on the surface of the tibia, fibula, or both. Large tendon sheaths surround the tendons of the tibialis anterior, extensor digitorum longus, and extensor hallucis longus muscles where they cross the ankle joint. The positions of these sheaths are stabilized by the **superior extensor retinaculum** and **inferior extensor retinaculum**, tough supporting bands of collagen fibers.

 **INTEGRATION**

**C.** How would a torn calcaneal tendon affect movement at the ankle?

 **LEARNING OUTCOME**

Identify the muscles that move the foot and toes, and cite their origins, insertions, and actions.

# The intrinsic muscles of the foot originate on the tarsal and metatarsal bones and associated tendons and ligaments

**1** This superior view introduces some of the **intrinsic muscles of the foot**. It also reveals the importance of the retinacula in stabilizing the positions of the tendons descending from the leg.

Superior extensor retinaculum

Medial malleolus of tibia

Tendon of tibialis anterior

Lateral malleolus of fibula

Inferior extensor retinaculum

Tendons of extensor digitorum longus

**Intrinsic Muscles of the Foot, Great Toe**

Extensor hallucis brevis

Abductor hallucis

Tendon of extensor hallucis longus

**Intrinsic Muscles of the Foot, Toes 2–5**

Dorsal interossei

Tendons of extensor digitorum brevis

**? REVIEW**
**A.** What are the functions of the superior and inferior retinacula of the foot?

**2** Intrinsic muscles are more numerous on the inferior surface of the foot and occur in several layers.

**Superficial Muscles of the Sole of the Foot**

Fibrous tendon sheaths

**Intrinsic Muscles of the Foot**

Lumbricals

Flexor hallucis brevis

Flexor digiti minimi brevis

Abductor hallucis

Quadratus plantae

Flexor digitorum brevis

Abductor digiti minimi

Plantar aponeurosis (cut)

Calcaneus

**Deep Muscles of the Sole of the Foot**

Tendons of flexor digitorum brevis

Tendon of flexor hallucis longus

Tendon of flexor digitorum longus

Tendon of tibialis posterior

Tendon of fibularis longus

**? REVIEW**
**B.** Identify the intrinsic muscle that flexes the great toe.

## Intrinsic Muscles of the Foot

| | Muscle | Origin | Insertion | Action |
|---|---|---|---|---|
| **FLEXION / EXTENSION** | **Flexor hallucis brevis** | Cuboid and lateral cuneiform bones | Proximal phalanx of great toe | Flexion at metatarsophalangeal joint of great toe |
| | **Flexor digitorum brevis** | Calcaneus (tuberosity on inferior surfaces) | Sides of middle phalanges, toes 2–5 | Flexion at proximal interphalangeal joints of toes 2–5 |
| | **Quadratus plantae** | Calcaneus (medial, inferior surfaces) | Tendon of flexor digitorum longus | Flexion at joints of toes 2–5 |
| | **Lumbrical (4)** | Tendons of flexor digitorum longus | Tendons of extensor digitorum longus, toes 2 to 5 | Flexion at metatarsophalangeal joints; extension at proximal interphalangeal joints of toes 2–5 |
| | **Flexor digiti minimi brevis** | Base of metatarsal bone V | Lateral side of proximal phalanx of toe 5 | Flexion at metatarsophalangeal joint of toe 5 |
| | **Extensor digitorum brevis** | Calcaneus (superior and lateral surfaces) | Dorsal surfaces of toes 1–4 | Extension at metatarsophalangeal joints of toes 1–4 |
| | **Extensor hallucis brevis** | Superior surface of anterior calcaneus | Dorsal surface of the base of proximal phalanx of great toe | Extension of great toe |
| **ADDUCTION / ABDUCTION** | **Adductor hallucis** | Bases of metatarsal bones II–IV and plantar ligaments | Proximal phalanx of great toe | Adduction at metatarsophalangeal joint of great toe |
| | **Abductor hallucis** | Calcaneus (tuberosity on inferior surface) | Medial side of proximal phalanx of great toe | Abduction at metatarsophalangeal joint of great toe |
| | **Plantar interosseus (3)** | Bases and medial sides of metatarsal bones | Medial sides of toes 3–5 | Adduction at metatarsophalangeal joints of toes 3–5 |
| | **Dorsal interosseus (4)** | Sides of metatarsal bones | Medial and lateral sides of toe 2; lateral sides of toes 3 and 4 | Abduction at metatarsophalangeal joints of toes 3 and 4 |
| | **Abductor digiti minimi** | Inferior surface of calcaneus | Lateral side of proximal phalanx, toe 5 | Abduction at metatarsophalangeal joint of toe 5 |

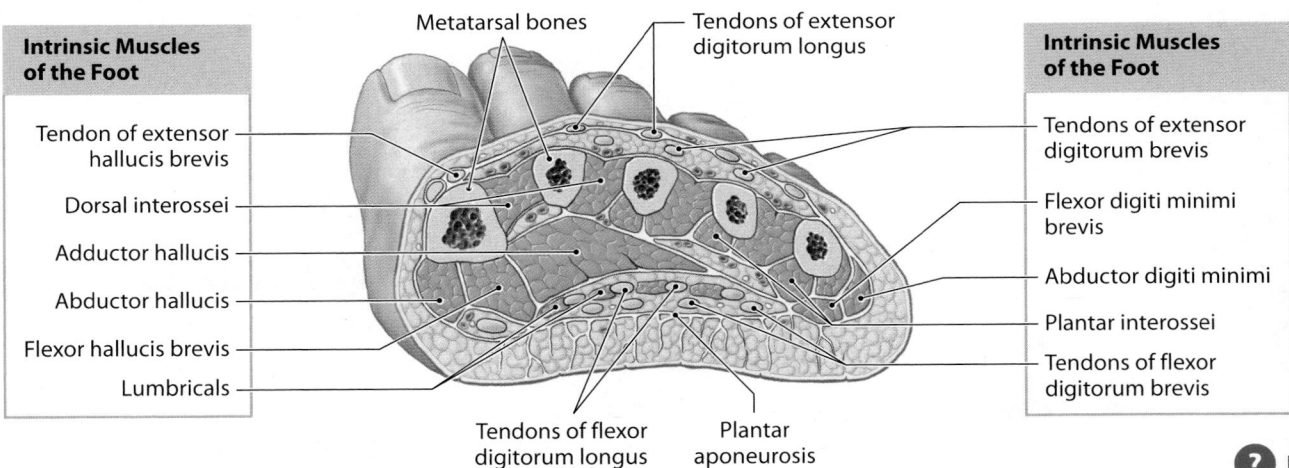

**Intrinsic Muscles of the Foot**

Tendon of extensor hallucis brevis
Dorsal interossei
Adductor hallucis
Abductor hallucis
Flexor hallucis brevis
Lumbricals

Metatarsal bones
Tendons of extensor digitorum longus

Tendons of flexor digitorum longus
Plantar aponeurosis

**Intrinsic Muscles of the Foot**

Tendons of extensor digitorum brevis
Flexor digiti minimi brevis
Abductor digiti minimi
Plantar interossei
Tendons of flexor digitorum brevis

**3** As you see in this cross section, most of the muscle mass in the foot lies inferior to the metatarsal bones. Many of these muscles are flexors that tense during ankle extension and help you "push off" when walking. This anatomical arrangement provides padding and assists in maintaining the arches of the foot.

**? REVIEW**

**C.** Which intrinsic foot muscles originate on and insert on tendons?

**Lo LEARNING OUTCOME**

Identify the intrinsic foot muscles, and cite their origins, insertions, and actions.

# The deep fascia divides the limb muscles into separate compartments

Fibrous partitions of the deep fascia in the limbs form **intermuscular septa** that separate muscles into sections called **compartments**. The muscles within each compartment have compatible functions (such as flexion or extension), and each has a characteristic blood supply and innervation. Because of this structural separation, infection or excess pressure is usually restricted to the affected compartment.

**1** Here are sectional views at intervals along the length of an upper and lower limb. In each relevant section the compartments are labeled, and representative muscles and important reference structures are indicated.

Deltoid

Triceps brachii

Medial intermuscular septum

Coracobrachialis

Biceps brachii

**Posterior Compartment**

Triceps brachii

Medial intermuscular septum

**Anterior Compartment**

Brachial artery and median nerve

Brachialis

Biceps brachii

Lateral intermuscular septum

**Posterior Compartment**

Extensor digitorum

Extensor carpi ulnaris

**Deep Anterior Compartment**

Flexor digitorum profundus

**Superficial Anterior Compartment**

Flexor digitorum superficialis

**Lateral Compartment**

Extensor carpi radialis brevis

Brachioradialis

ANTERIOR

**Anterior view**

10

POSTERIOR

Gluteus maximus

**Posterior Compartment**

Biceps femoris and
semitendinosus

Sciatic nerve

**Medial Compartment**

Adductor magnus

Adductor longus

**Anterior
Compartment**

Femoral artery, vein,
and nerve

Vastus medius

Rectus femoris

Vastus lateralis

**Posterior Compartment**

Biceps femoris

Sciatic nerve

**Anterior Compartment**

Vastus lateralis

Femoral artery,
vein, and nerve

Rectus femoris

**Medial Compartment**

Adductor magnus

Adductor longus

**? REVIEW**

**A.** Name the septa that divide
the arm into anterior and
posterior compartments.

**Lateral Compartment**

Fibularis longus

**Anterior Compartment**

Anterior tibial artery
and vein

Tibialis anterior

**Superficial Posterior
Compartment**

Gastrocnemius

Soleus

**Deep Posterior Compartment**

Posterior tibial artery
and vein

Tibialis posterior

**Anterior view**

ANTERIOR

Compartments are clinically important. For example, trauma to a
limb can cause bleeding, which elevates pressures and compresses
blood vessels and nerves within the compartment. A lack of blood
flow leads to "blood starvation," or *ischemia*. This condition, called
**compartment syndrome**, can lead to the paralysis or death of the
affected muscles if the pressure is not relieved within 2–4 hours.

 **INTEGRATION**

**B.** Propose a reason why compartment
syndrome can be life threatening.

 **LEARNING OUTCOME**

Describe the deep fascia and its relationship
to the various limb muscle compartments.

10

# Appendicular Muscles

**Labeling:** Label each of the indicated muscles that move the forearm and hand in the diagram at right.

1 tricep brachii

2 aponeurosis

3 extensor carpi ulnaris

4 extensor carpi ulna radialis brevi longus

5 extensor digitorum

6 flexor carpi ulnaris

Label each of the indicated structures or muscles that move the thigh and leg in the diagram below.

7 gluteus medius

8 tensor fasciae latae

9 gluteus maximus

Label each of the indicated muscles that move the foot and toes in the diagram below.

10 ?

11

12 Biceps femoris

13 Semitendinosus

14 Semimembranosus

15

25

16

17

18

19 Fibularis longus

20

21

22 Fibularis brevis

23

24

# Study Outline

## SECTION 1 • Functional Organization of the Muscular System

### Module 10.1
### The axial and appendicular muscles have different functions    p. 351

1. Skeletal muscle accounts for almost half the weight of your body.
2. The **muscular system** is divided into the **axial** and **appendicular muscles**.
3. The axial muscles support and position the axial skeleton, and the appendicular muscles support and move the limbs.

### Module 10.2
### Muscular power and range of motion are influenced by fascicle organization and leverage    p. 352

4. Muscle fascicles can be organized as **parallel**, **convergent**, **pennate** (**unipennate**, **bipennate**, or **multipennate**), or **circular** (**sphincter**).
5. A **lever** is a rigid structure that pivots on a fixed point called a **fulcrum**. In the body, bones act as levers, and joints act as fulcrums.
6. Levers are classified as **first-class**, **second-class**, and **third-class levers**. Third-class levers are the most common levers in the body.

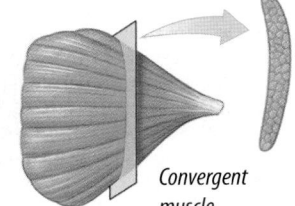

*Convergent muscle*

### Module 10.3
### The origins and insertions of muscles determine their actions, while their names can provide clues to appearance and/or function    p. 354

7. Each muscle can be identified by its **origin**, **insertion**, and **action**.
8. The site of attachment at the fixed end of the muscle is the **origin**; the site where the movable end of the muscle attaches to another structure is called the **insertion**. The movement produced when a muscle contracts is the **action**.
9. A muscle can be classified as an **agonist**, or **prime mover**; an **antagonist**; or a **synergist**.
10. Muscle terminology is associated with the location of the muscle, as well as its position, fascicle organization, structural characteristics, actions, and other features.

### Module 10.4
### The skeletal muscles can be assigned to the axial division or the appendicular division based on origins and functions    p. 356

11. About 60 percent of the skeletal muscles in the body are **axial muscles**. The remaining are **appendicular muscles**.

## SECTION 2 • Axial Muscles

### Module 10.5
### There are four groups of axial muscles    p. 359

12. The first group of axial muscles is the muscles of the head and neck that are not associated with the vertebral column. These are the muscles of the face, extrinsic eye, tongue, pharynx, and neck.
13. The second group of axial muscles is the muscles of the vertebral column. The third group of muscles is the muscles of the trunk, and the muscles of the pelvic floor form the fourth group.

### Module 10.6
### The muscles of facial expression are important in eating and useful for communication    p. 360

14. The **muscles of facial expression** originate on the surface of the skull. They insert on the superficial fascia and dermis of the skin.
15. The muscles of the mouth and cheek are **levator labii superioris**, **zygomaticus minor**, **zygomaticus major**, **buccinator**, **levator anguli oris**, **orbicularis oris**, **risorius**, **mentalis**, **depressor labii inferioris**, and **depressor anguli oris**.

### Module 10.7
### The extrinsic eye muscles position the eye, and the muscles of mastication move the lower jaw    p. 362

16. The extrinsic eye muscles are **inferior rectus**, **medial rectus**, **superior rectus**, **lateral rectus**, **inferior oblique**, and **superior oblique**.
17. The muscles of mastication are **masseter**, **temporalis**, **medial pterygoid**, and **lateral pterygoid**.

### Module 10.8
### The muscles of the tongue are closely associated with the muscles of the pharynx and neck    p. 364

18. The muscles of the tongue are **genioglossus**, **hyoglossus**, **palatoglossus**, and **styloglossus**.
19. The muscles of the pharynx are the **pharyngeal constrictors**, **laryngeal elevators**, and **palatal muscles**.
20. The anterior muscles of the neck are **digastric**, **geniohyoid**, **mylohyoid**, **omohyoid**, **sternohyoid**, **sternothyroid**, **stylohyoid**, **thyrohyoid**, and **sternocleidomastoid**.

### Module 10.9
### The muscles of the vertebral column support and align the axial skeleton    p. 366

21. The **erector spinae** muscles are subdivided into the **spinalis**, **longissimus**, and **iliocostalis muscle groups**.

### Module 10.10
### The oblique and rectus muscles form the muscular walls of the trunk    p. 368

22. The oblique muscles include the **scalenes**, **external** and **internal intercostals**, **transversus thoracis**, **external** and **internal obliques**, and **transversus abdominis**.
23. The rectus group includes the **diaphragm** and **rectus abdominis**.

**Module 10.11**

**The muscles of the pelvic floor support the organs of the abdominopelvic cavity    p. 370**

24. The muscles of the pelvic floor form the **perineum**, a muscular sheet that spans the pelvic outlet.

25. The muscles of the pelvic floor are divided into the muscles of the **urogenital triangle** and **anal triangle**.

## SECTION 3 • Appendicular Muscles

**Module 10.12**

**The appendicular muscles stabilize, position, and support the limbs    p. 373**

26. The upper limb muscles include muscles of the pectoral girdle, arm, forearm, hand, and fingers.

27. The lower limb muscles include muscles of the thigh, leg, foot, and toes.

**Module 10.13**

**The largest appendicular muscles originate on the trunk    p. 374**

28. Gross movements of the limbs are controlled by muscles that originate on the trunk. Limb muscles get smaller, more numerous, and more precise as they are located more distally on the limb.

**Module 10.14**

**Muscles that position each pectoral girdle originate on the occipital bone, superior vertebrae, and ribs    p. 376**

29. The muscles of the pectoral girdle are **trapezius**, **levator scapulae**, **subclavius**, **pectoralis minor**, **serratus anterior**, **rhomboid major**, and **rhomboid minor**.

**Module 10.15**

**Muscles that move the arm originate on the clavicle, scapula, thoracic cage, and vertebral column    p. 378**

30. The muscles that move the arm are **deltoid**, **pectoralis major**, **coracobrachialis**, **teres major**, **latissimus dorsi**, and the four muscles of the **rotator cuff: supraspinatus**, **infraspinatus**, **teres minor**, and **subscapularis**.

**Module 10.16**

**Muscles that move the forearm and hand originate on the scapula, humerus, radius, or ulna    p. 380**

31. The elbow extensors are **triceps brachii** and **anconeus**. The elbow flexors are **biceps brachii**, **brachialis**, and **brachioradialis**.

32. The wrist extensors are **extensor carpi ulnaris**, **extensor carpi radialis longus**, and **extensor carpi radialis brevis**. The wrist flexors are **flexor carpi radialis**, **palmaris longus**, and **flexor carpi ulnaris**.

33. The muscles of pronation are **pronator teres** and **pronator quadratus**. **Supinator** causes supination.

**Module 10.17**

**Muscles that move the hand and fingers originate on the humerus, radius, ulna, and interosseous membrane    p. 382**

34. The muscles that flex the fingers and the thumb are **flexor digitorum superficialis**, **flexor digitorum profundus**, and **flexor pollicis longus**. Finger extension is caused by **extensor, digitorum extensor pollicis brevis**, **extensor pollicis longus**, **extensor indicis**, and **extensor digiti minimi**.

35. The muscles that move the thumb are **extensor pollicis longus**, **abductor pollicis longus**, and **extensor pollicis brevis**.

36. The tendons of the flexor muscles pass through **synovial tendon sheaths**. Inflammation of the flexor retinaculum and the synovial tendon sheaths can put pressure on the median nerve, causing the pain known as **carpal tunnel syndrome**.

**Module 10.18**

**The intrinsic muscles of the hand originate on the carpal and metacarpal bones and associated tendons and ligaments    p. 384**

37. The intrinsic muscles of the hand perform flexion and extension of the fingers at the metacarpophalangeal joints; abduction and adduction of the fingers at the metacarpophalangeal joints; and opposition and reposition (relaxed position) of the thumb.

**Module 10.19**

**The muscles that move the thigh originate on the pelvis and associated ligaments and fasciae    p. 386**

38. The muscles of the **gluteal group** are **gluteus maximus**, **gluteus medius**, **gluteus minimus**, and **tensor fasciae latae**.

39. The iliopsoas group includes **psoas major** and **iliacus**.

40. The lateral rotator group muscles are **piriformis**, **superior** and **inferior gemellus**, **obturator internus** and **externus**, and **quadratus femoris**.

41. The adductor group muscles are **pectineus**, **adductor brevis**, **adductor longus**, **adductor magnus**, and **gracilis**.

**Module 10.20**

**The muscles that move the leg originate on the pelvis and femur    p. 388**

42. The **flexors of the knee** are **biceps femoris**, **semimembranosus**, **semitendinosus**, **sartorius**, and **popliteus**.

43. The **extensors of the knee** are **rectus femoris**, **vastus intermedius**, **vastus lateralis**, and **vastus medialis**.

**Module 10.21**

**The extrinsic muscles that move the foot and toes originate on the tibia and fibula    p. 390**

44. Plantar flexion is performed by **gastrocnemius**, **soleus**, **fibularis brevis**, **fibularis longus**, **plantaris**, and **tibialis posterior**. The **tibialis anterior** and **fibularis tertius** cause dorsiflexion.

45. Flexion of the toes is caused by **flexor digitorum longus** and **flexor hallucis longus**. Toe extension is caused by **extensor digitorum longus** and **extensor hallucis longus**.

## Module 10.22

**The intrinsic muscles of the foot originate on the tarsal and metatarsal bones and associated tendons and ligaments p. 392**

46. The **intrinsic muscles of the foot** stabilize the positions of the tendons descending from the leg. These muscles are more numerous on the inferior surface of the foot and occur in several layers.

47. These intrinsic muscles are responsible for flexion and extension of the interphalangeal joints, as well as abduction and adduction of the metatarsophalangeal joints.

## Module 10.23

**The deep fascia divides the limb muscles into separate compartments p. 394**

48. Fibrous partitions called **intermuscular septa** create **compartments** containing muscles with compatible functions as well as characteristic blood supply and innervation.

49. Trauma to a limb may cause bleeding into a compartment. This may elevate pressures and compress blood vessels and nerves within a compartment, a condition called **compartment syndrome**.

# Chapter Review Questions

**Labeling:** Identify the lever system in each of the images below.

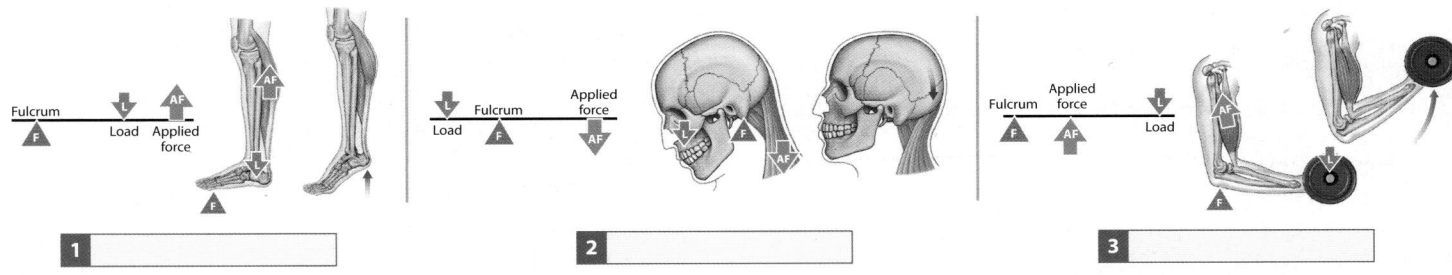

| 1 | | 2 | | 3 | |

**True/False:** Indicate whether each statement is true or false.

4 The pectoralis major is an example of a convergent muscle.

5 A wheelbarrow is an example of a third-class lever.

6 A muscle agonist can also be referred to as a prime mover.

7 The inferior rectus and superior oblique are examples of extrinsic eye muscles.

8 The esophageal hiatus is an opening in the transversus abdominis.

| 4 | _____ |
| 5 | _____ |
| 6 | _____ |
| 7 | _____ |
| 8 | _____ |

**Matching:** Match each lettered action with the most closely related muscle.

a. moves eye laterally
b. flexion at knee
c. extension at knee
d. abduction at shoulder
e. downward rotation of scapula
f. flexion at elbow
g. plantar flexion
h. mastication
i. medial rotation at shoulder
j. dorsiflexion

9 Tibialis anterior
10 Temporalis
11 Brachialis
12 Lateral rectus
13 Soleus
14 Biceps femoris
15 Rhomboid major
16 Vastus medialis
17 Deltoid
18 Subscapularis

| 9 | _____ |
| 10 | _____ |
| 11 | _____ |
| 12 | _____ |
| 13 | _____ |
| 14 | _____ |
| 15 | _____ |
| 16 | _____ |
| 17 | _____ |
| 18 | _____ |

**Multiple choice:** Select the correct answer from the list provided.

**19** The site where the more movable end of a muscle attaches is the
- ☐ a) origin.
- ☐ b) insertion.
- ☐ c) belly.
- ☐ d) fascicle.

**20** A muscle with a feather-shaped fascicle organization is called a
- ☐ a) parallel muscle.
- ☐ b) pennate muscle.
- ☐ c) convergent muscle.
- ☐ d) circular muscle.

**21** The most common lever system in the body is
- ☐ a) first-class.
- ☐ b) second-class.
- ☐ c) third-class.
- ☐ d) fourth-class.

**22** Which of the following muscles is an axial muscle?
- ☐ a) erector spinae
- ☐ b) trapezius
- ☐ c) deltoid
- ☐ d) flexor carpi radialis

**23** The major extensor of the elbow is
- ☐ a) triceps brachii.
- ☐ b) biceps brachii.
- ☐ c) deltoid.
- ☐ d) subscapularis.

**24** Inflammation of the retinaculum and synovial tendon sheaths resulting in pressure on the median nerve is called
- ☐ a) anterior compartment syndrome.
- ☐ b) rotator cuff syndrome.
- ☐ c) carpal tunnel syndrome.
- ☐ d) plantar fasciitis.

**25** When a person is doing a pull-up exercise, which of the following muscles is responsible for adduction at the shoulder joint?
- ☐ a) levator scapulae
- ☐ b) deltoid
- ☐ c) supraspinatus
- ☐ d) latissimus dorsi

**26** Which of the following muscles performs the hip and knee action required to kick a ball?
- ☐ a) rectus femoris
- ☐ b) pectineus
- ☐ c) gracilis
- ☐ d) biceps femoris

**27** Which of the following locations is an attachment site for the short head of the biceps brachii?
- ☐ a) tuberosity of ulna
- ☐ b) coracoid process of scapula
- ☐ c) acromion process of scapula
- ☐ d) supraglenoid tubercle of scapula

**28** Which of the following is *not* a muscle of mastication?
- ☐ a) masseter
- ☐ b) buccinator
- ☐ c) temporalis
- ☐ d) lateral pterygoid

**Short answer**

**29** What are the functions of the muscles of the pelvic floor?

**30** Identify the muscles of the rotator cuff.

**31** List the muscles of the quadriceps femoris, and describe their actions.

## MasteringA&P™

Access more chapter study tools online in the MasteringA&P Study Area:

- **Chapter Quizzes, Chapter Practice Test, and Clinical Case Studies**

- Practice Anatomy Lab™ 3.0 — PAL™3.0
- Interactive Physiology™ — iP2™
- A&P Flix™ — A&PFlix™
- PhysioEx™ 9.1 — PhysioEx™ 9.1

# Chapter Integration • Applying what you have learned

## Bodybuilding and lookin' good

Bodybuilders spend many hours in the gym lifting free weights to develop their muscles. Larger muscles and greater muscle definition are the goals, and looking "ripped" requires a lot of dedication. To sculpt their arms, for example, bodybuilders do a lot of biceps curls (flexion at the elbow holding weights in the anatomical position) and triceps curls (extension at the elbow).

As a 10-year-old, Jerry and his friends would go to the beach at Lion's Park, known locally as "Muscle Beach" because all the local bodybuilders would go

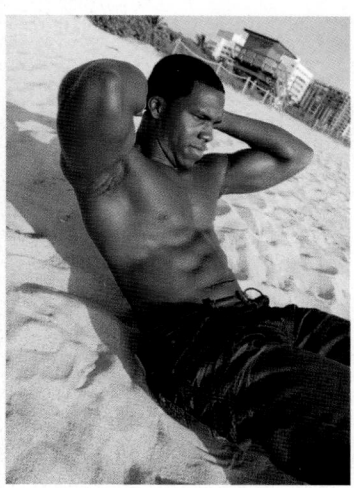

there in the summer to work out and show off for the girls. Some female bodybuilders even started going there. Jerry was always amazed by the size, shape, and strength of these musclemen and vowed that someday he would become one of them. As he reached puberty, Jerry became a fitness fanatic who worked out many hours a day. As he learned more about bodybuilding, he avoided the massive, heavily muscled look for one of a more athletic, lean, and well-defined musculature. Everyone came to admire his "six-pack abs," especially his girlfriend, DJ.

**1** Explain why doing both biceps curls and triceps curls helps achieve larger, well-toned arms.

**2** Which exercises would be best for shaping your abdominal muscles into "six-pack abs"?

## Sports, muscles, and joints

Jennifer is a high school freshman and an up-and-coming volleyball player on the junior varsity team. It is her intent to join the varsity team as a sophomore. One element of her game that she knows she needs to improve upon is her vertical jump. She has committed her offseason workouts to spending more time in the weight room strengthening the muscles involved in jumping. From what you have just learned about the muscular system, answer the following questions and design a weightlifting program to help Jennifer achieve her goals.

**3** Identify the actions involved at the hip, knee, and ankle joints when a person is jumping.

**4** Which muscles are involved in each of these actions?

**5** What kind of exercises would you suggest Jennifer perform in the weight room to increase the strength in these muscles?

11

# Nervous Tissue

# Module 11.1

# The nervous system has three divisions: the CNS, PNS, and ENS

The three divisions of the nervous system are the central nervous system (CNS), peripheral nervous system (PNS), and enteric nervous system (ENS). This chapter considers the CNS and PNS. Chapters 14 and 22 discuss the ENS, which is in the wall of the gastrointestinal tract and helps control digestive function.

**1** The major components and general functions of the CNS and PNS are described in this flowchart. It begins with the receptors of the PNS **1**.

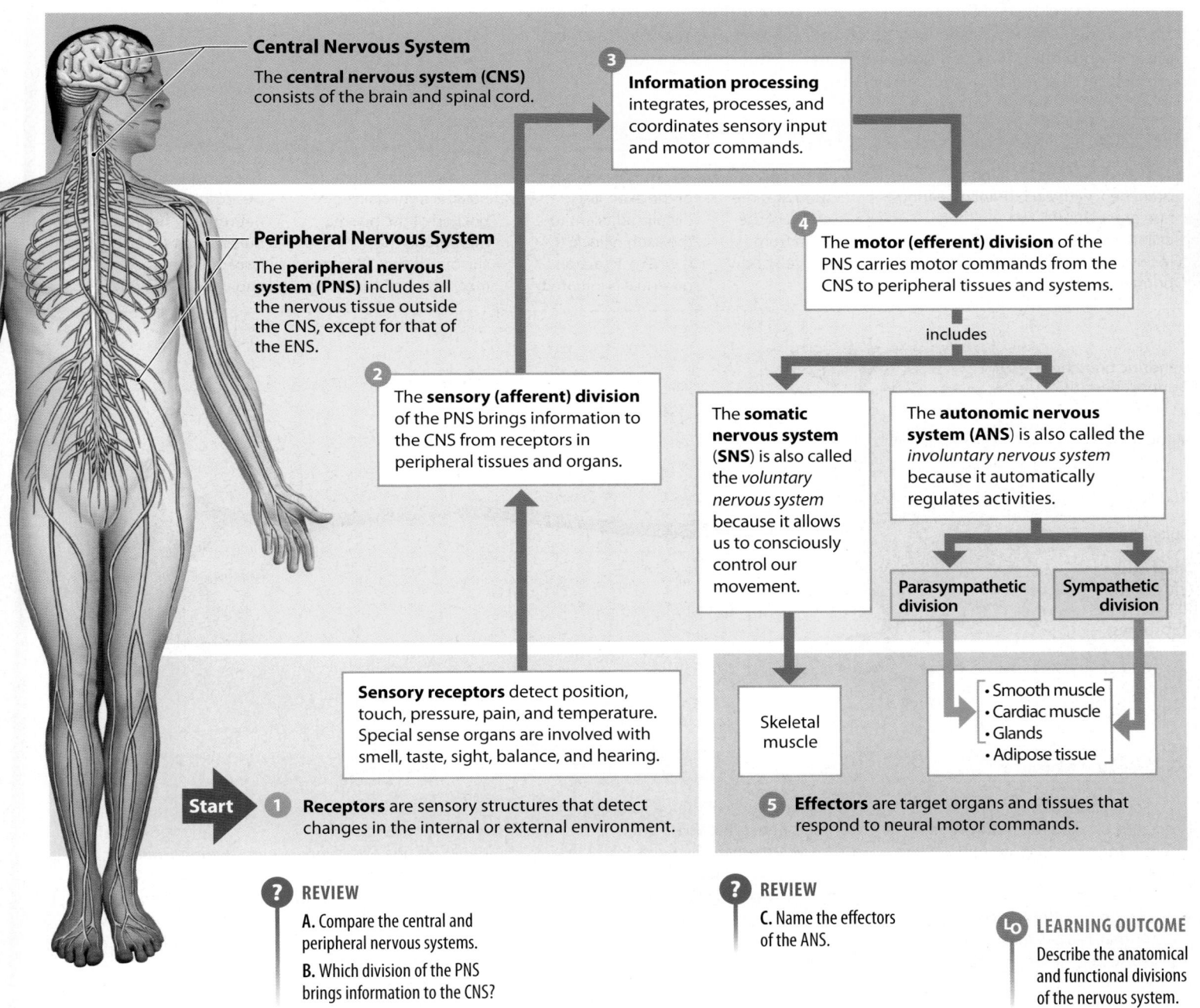

**Central Nervous System**

The **central nervous system (CNS)** consists of the brain and spinal cord.

**3** **Information processing** integrates, processes, and coordinates sensory input and motor commands.

**Peripheral Nervous System**

The **peripheral nervous system (PNS)** includes all the nervous tissue outside the CNS, except for that of the ENS.

**4** The **motor (efferent) division** of the PNS carries motor commands from the CNS to peripheral tissues and systems.

includes

**2** The **sensory (afferent) division** of the PNS brings information to the CNS from receptors in peripheral tissues and organs.

The **somatic nervous system (SNS)** is also called the *voluntary nervous system* because it allows us to consciously control our movement.

The **autonomic nervous system (ANS)** is also called the *involuntary nervous system* because it automatically regulates activities.

**Parasympathetic division**  **Sympathetic division**

**Sensory receptors** detect position, touch, pressure, pain, and temperature. Special sense organs are involved with smell, taste, sight, balance, and hearing.

Skeletal muscle

- Smooth muscle
- Cardiac muscle
- Glands
- Adipose tissue

**Start** **1** **Receptors** are sensory structures that detect changes in the internal or external environment.

**5** **Effectors** are target organs and tissues that respond to neural motor commands.

**? REVIEW**

**A.** Compare the central and peripheral nervous systems.

**B.** Which division of the PNS brings information to the CNS?

**? REVIEW**

**C.** Name the effectors of the ANS.

**Lo LEARNING OUTCOME**

Describe the anatomical and functional divisions of the nervous system.

11

# Module 11.2
# Neurons are nerve cells specialized for intercellular communication

Go to
MasteringA&P™
PAL 3.0
Histology →
Nervous
Tissue

In this module, we examine the structure of **neurons**. Recall from Chapter 4 that neurons have three general regions: *dendrites*, which receive stimuli from the environment or from other neurons; a *cell body*, which contains the nucleus and other organelles; and one *axon*, which carries information toward other cells. Specialized structures within each region contribute to intercellular communication.

**1** This is an illustration of a representative neuron. The **cell body** contains most of the organelles of a neuron. The **axon** is a long cytoplasmic process that extends away from the cell body. Many materials, including enzymes and lysosomes, travel the length of the axon along **neurotubules** (neuron microtubules) through a process called **axoplasmic transport**. Axoplasmic transport occurs in both directions. If debris or unusual chemicals appear in the axon terminal, **retrograde flow** soon delivers them to the cell body.

 **REVIEW**

**A.** Name the structural components of a typical neuron.

**11**

## Dendrites

Typical **dendrites** are highly branched, with each branch bearing fine processes 0.5 to 1 μm long called *dendritic spines*. CNS neurons receive most of their information primarily at the dendritic spines.

## Axon

The **axon hillock** is the origin of the axon from the cell body.

The **initial segment** of the axon lies distally adjacent to the axon hillock. It is where an action potential is initiated.

The **axolemma** (*lemma*, husk) is a specialized portion of the plasma membrane that surrounds the cytoplasm (*axoplasm*) of the axon.

The **axoplasm** (AK-sō-plazm) contains neurofibrils, neurotubules, small vesicles, lysosomes, mitochondria, and various enzymes.

Dendritic branches
(support dendritic
spines)

Mitochondrion

Nucleus

Nucleolus

**Nissl bodies**
(clusters of
RER and free
ribosomes)

## Cell Body

The cytoplasm surrounding the nucleus is called the **perikaryon** (per-i-KAR-ē-on; *peri*, around + *karyon*, nucleus). The perikaryon contains organelles that provide energy and synthesize the chemical neurotransmitters that are important in cell-to-cell communication.

The cytoskeleton of the perikaryon contains **neurofilaments** similar to the intermediate filaments in other cells. **Neurofibrils** are bundles of neurofilaments that extend into the dendrites and axon, providing internal support for these slender processes.

## Telodendria

The main axon trunk ends in a series of fine extensions, or **telodendria** (tel-ō-DEN-drē-uh; *telo-*, end + *dendron*, tree; singular: **telodendrion**).

The telodendria of an axon end at **axon terminals**, or **synaptic terminals**, where the neuron communicates with other cells.

**2** Each axon terminal is part of a **synapse**, a specialized site where the neuron communicates with another cell. Every synapse involves a **presynaptic cell** and a **postsynaptic cell.** Communication between those cells most commonly involves the release of chemicals called **neurotransmitters** into the **synaptic cleft**, a narrow space separating the two cells. Axon terminals receive (through axoplasmic transport) a continuous supply of neurotransmitters synthesized in the cell body. They also reabsorb and reassemble fragments of neurotransmitters broken down in the synaptic cleft.

**? REVIEW**

**B.** Describe a synapse.

Telodendrion of presynaptic cell

REPRESENTATIVE SYNAPSE

Mitochondrion

A simple, round axon terminal occurs where the postsynaptic cell is another neuron.

Neurotransmitters within the axon terminal are typically packaged in **synaptic vesicles**.

Presynaptic membrane

The narrow synaptic cleft separates the **presynaptic membrane**, where neurotransmitters are released, from the **postsynaptic membrane**, which has receptors for neurotransmitters.

Cytoplasm of postsynaptic cell

Postsynaptic membrane

**3** Except in a few special cases involving sensory receptors, a presynaptic cell is always a neuron. The postsynaptic cell can be either a neuron or another type of cell. Notice that axons may branch along their length, producing side branches known as **collateral branches**. Collateral branches enable a single neuron to communicate with several other cells.

**? REVIEW**

**C.** Compare presynaptic and postsynaptic cells.

**Synapses with Another Neuron**

Neuron 1

Dendrites

Collateral branch

Axolemma

Neuron 2

Synapses with another neuron

**Neuromuscular Junctions**

Neuron

Collateral branch

Neuromuscular junctions

Telodendria

Axon terminals

Skeletal muscle fibers

**Neuroglandular Synapses**

Neuron

Neuroglandular synapses

Gland cells

**Clinical Note**

## Neurons lost to injury are seldom replaced

Most CNS neurons lack centrioles and cannot divide. As a result, neurons lost to injury or disease are seldom replaced. Although neural stem cells persist in the adult nervous system, these cells are typically inactive (except in the epithelium responsible for our sense of smell, in the retina of the eye, and in the hippocampus, a portion of the brain involved with storing memories).

**? REVIEW**

**D.** Why is a CNS neuron not usually replaced after it is injured?

 **LEARNING OUTCOME**

Sketch and label the structure of a typical neuron, and describe the functions of each component.

# Neurons are classified on the basis of structure and function

There are four structural classifications of neurons. Functionally, neurons are classified as sensory neurons, interneurons, or motor neurons.

## Neuron Structure

**1** **Anaxonic** (an-aks-ON-ic) **neurons** are small and lack anatomical features that distinguish dendrites from axons; all the cell processes look alike. Anaxonic neurons are located in the brain and in special sense organs. Their functions are poorly understood.

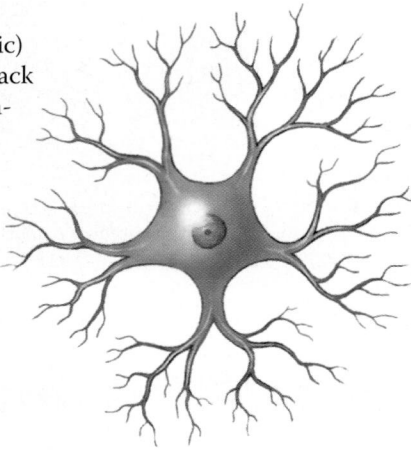

**2** **Bipolar neurons** have two distinct processes—one dendritic process that branches extensively at its distal tip, and one axon—with the cell body between the two. Bipolar neurons are rare but occur in special sense organs, where they relay information about sight, smell, or hearing from receptor cells to other neurons. Bipolar neurons are small; the largest measure less than 30 μm from end to end.

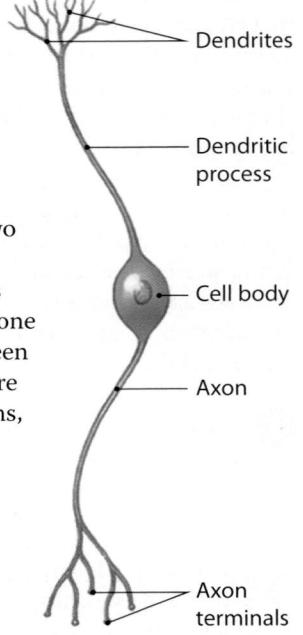

Dendrites

Dendritic process

Cell body

Axon

Axon terminals

**3** In a **unipolar neuron**, the dendrites and axon are continuous—basically, fused—and the cell body lies off to one side. In such a neuron, the **initial segment**, the first portion of the axon, lies where the dendrites converge. The rest of the process, which carries action potentials, is usually considered to be an axon. Most sensory neurons of the peripheral nervous system are unipolar. Their axons may extend a meter or more, ending at synapses in the central nervous system. The longest are those carrying sensations from the tips of the toes to the spinal cord.

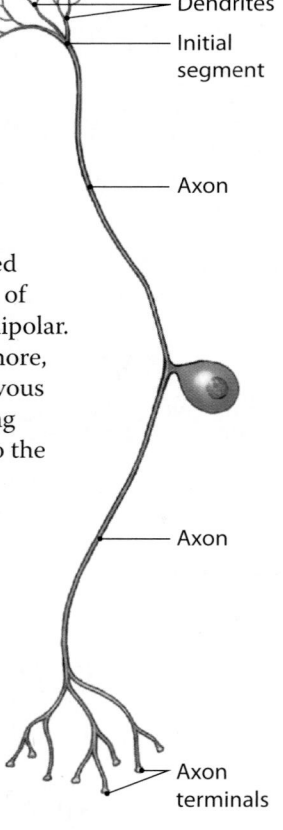

Dendrites

Initial segment

Axon

Axon

Axon terminals

**4** **Multipolar neurons** have two or more dendrites and a single axon. These are the most common neurons in the CNS. All motor neurons that control skeletal muscles, for example, are multipolar neurons. The axons of multipolar neurons can be as long as those of unipolar neurons. The longest carry motor commands from the spinal cord to small muscles that move the toes.

Dendrites

Axon

Axon terminals

**REVIEW**
**A.** Classify neurons based on their structure.

## Neuron Function within the CNS and PNS

**5** We have about 10 million sensory neurons, 20 billion interneurons, and half a million motor neurons. This flowchart describes the relationships among these neuron types. Begin reading at the red "Start" arrow at the bottom.

**?** **REVIEW**

**B.** Classify neurons based on their function.

### Central Nervous System (CNS)

One or more **interneurons** usually lie between sensory neurons and motor neurons. Interneurons receive sensory information from the PNS as well as input from other interneurons in the CNS. Interneurons are also responsible for higher functions, such as memory, planning, and learning.

**Somatic motor neurons** innervate skeletal muscles. You have conscious control over the activity of these neurons. The cell body of a somatic motor neuron lies in the CNS, and its axon extends within a **peripheral nerve** to innervate skeletal muscle fibers at neuromuscular junctions. A **nerve** is a bundle of axons in the PNS.

**Visceral motor neurons** innervate all peripheral effectors other than skeletal muscles—that is, smooth muscle, glands, cardiac muscle, and adipose tissue—throughout the body. Visceral motor neurons in the CNS control a second set of visceral motor neurons located in the PNS.

### Peripheral Nervous System (PNS)

**Sensory ganglia**

A **ganglion** is a collection of neuron cell bodies in the PNS. **Sensory neurons** are unipolar neurons whose cell bodies are located in **sensory ganglia** in the PNS. Sensory neurons form the afferent division of the PNS and deliver information from sensory receptors to the CNS. **Somatic sensory neurons** monitor the outside world and our position within it. **Visceral sensory neurons** monitor internal conditions and the organ systems.

**Efferent fibers** are axons that carry instructions from the CNS to peripheral tissues.

**Autonomic ganglia**

The second set of visceral motor neurons innervates peripheral effectors. These neuron cell bodies are found in peripheral **autonomic ganglia**.

| Parasympathetic division | Sympathetic division |
|---|---|

**Afferent fibers** are axons that carry sensory information to the CNS.

**Interoceptors** (*intero-*, inside) monitor the digestive, respiratory, cardiovascular, urinary, and reproductive systems and provide sensations of distension (stretching), deep pressure, and pain.

**Proprioceptors** (prō-prē-ō-SEP-torz) monitor the position and movement of skeletal muscles and joints.

**Exteroceptors** (*extero-*, outside) provide information about the external environment in the form of touch, temperature, or pressure sensations and the more complex senses of taste, smell, sight, equilibrium, and hearing.

**Somatic effectors**

Skeletal muscle fibers

**Visceral effectors**

Smooth muscles

Glands

Cardiac muscle

Adipose tissue

**Start** **Sensory receptors** are either cells monitored by sensory neurons or processes of specialized sensory neurons.

KEY
→ = Somatic (sensory & motor)
→ = Visceral (sensory & motor)

**?** **INTEGRATION**

**C.** Are unipolar neurons in a tissue sample of the PNS more likely to have a sensory or a motor function?

**Lo** **LEARNING OUTCOME**

Classify and describe neurons on the basis of their structure and function.

11

# Module 11.4

# Oligodendrocytes, astrocytes, ependymal cells, and microglia are neuroglia of the CNS

**Neuroglia** (nū-RŌG-lē-uh; *glia*, glue), or **glial cells**, support and protect neurons in the PNS and CNS. Neuroglia are abundant, diverse, and make up about half the volume of the nervous system.

**1** Nervous tissue in the CNS and PNS are organized differently, primarily because the CNS has a greater variety of glial cell types. This illustration summarizes information about neuroglia in the CNS.

### Neuroglia of the CNS

**Ependymal cells** form a simple cuboidal to columnar epithelium known as the **ependyma** (e-PEN-di-muh). The ependyma lines a fluid-filled passageway within the spinal cord and brain. The passageway narrows in the spinal cord and is called the *central canal*. In the brain, the passageway forms cavities called *ventricles*. **Cerebrospinal fluid** (**CSF**) fills these internal spaces and also surrounds the brain and spinal cord. Ependymal cells assist in producing, monitoring, and circulating CSF.

**Microglia** (mī-KRŌG-lē-uh) are developmentally related to monocytes and macrophages. They migrate into the CNS as the nervous system forms, and they persist as mobile cells, continuously moving through the nervous tissue, removing cellular debris, wastes, and pathogens by phagocytosis.

**Astrocytes** maintain the **blood–brain barrier** that isolates the CNS from the chemicals and hormones circulating in the blood. They also provide structural support within nervous tissue; regulate ion, nutrient, and dissolved gas concentrations in the interstitial fluid surrounding the neurons; absorb and recycle neurotransmitters that are not broken down or reabsorbed at synapses; and form scar tissue after CNS injury.

**Oligodendrocytes** (ol-i-gō-DEN-drō-sīts; *oligo-*, few) provide a structural framework within the CNS by stabilizing the positions of axons. They also produce **myelin** (MĪ-e-lin), a membranous wrapping around axons that increases the transmission speed of nerve impulses. When myelinating an axon, the tip of an oligodendrocyte process expands to form a wide pad that gets wound around the axon, forming concentric layers of plasma membrane. These layers create a **myelin sheath**.

Section of spinal cord including a portion of the central canal

Central canal

Neuron cell body

Neuron cell body

Neuron cell body

Gray matter

Myelinated axons

White matter

Myelin (cut)

Axon

Nodes

Capillary

**2** There are three types of ependymal cells: ependymocytes, tanycytes, and specialized CSF-producing ependymal cells (discussed in Chapter 13). **Ependymocytes** have motile cilia that help to circulate CSF. Ependymocytes have long slender basal processes that branch and make contact with neuroglia. **Tanycytes** are specialized nonciliated ependymal cells with microvilli on their apical surfaces. They are found in only one brain ventricle. It is thought that they transport substances between the CSF and the brain.

Cilia

Ependymocyte

Basal processes

Astrocyte processes

**3** Many oligodendrocytes cooperate in the formation of a myelin sheath along the length of an axon. Such an axon is said to be **myelinated**. Each oligodendrocyte myelinates segments of several axons. The relatively large areas of the axon that are thus wrapped in myelin are called **internodes** (*inter*, between). Internodes are typically 1–2 mm in length.

Internode

Myelin sheath

> The small unmyelinated gaps of a few micrometers that separate adjacent internodes are called **nodes** (or *nodes of Ranvier*; RAHN-vē-ā). In dissection, myelinated axons appear glossy white, primarily because of the lipids within the myelin. Regions dominated by myelinated axons make up the **white matter** of the CNS.

**4** Not all axons in the CNS are myelinated. **Unmyelinated axons** may not be completely covered by the processes of neuroglia. Such axons are common where relatively short axons and collaterals form synapses with densely packed neuron cell bodies. Areas containing neuron cell bodies, dendrites, and unmyelinated axons are a dusky gray color, and they make up the **gray matter** of the CNS.

Neuron cell bodies

Unmyelinated axon

**? REVIEW**

**A.** Name the neuroglia of the CNS.

**B.** Which neuroglia appear in increased numbers in a person with a brain infection?

**C.** Which glial cell protects the CNS from chemicals and hormones circulating in the blood?

 **INTEGRATION**

**D.** Contrast the white matter and gray matter in the CNS.

 **LEARNING OUTCOME**

Describe the locations and functions of neuroglia in the CNS.

11

# Module 11.5
# Schwann cells and satellite cells are the neuroglia of the PNS

**Schwann cells**, or *neurolemmocytes*, cover peripheral axons in two different ways and participate in axon repair after injury. *Satellite cells* surround peripheral cell bodies.

## Myelinated Axons

**1** Myelinating Schwann cells form a sheath around peripheral axons. Wherever a Schwann cell covers an axon, the outer surface of the Schwann cell is called the **neurilemma** (nū-ri-LEM-uh). Most axons in the PNS, whether myelinated or unmyelinated, are shielded from contact with interstitial fluid by Schwann cells.

## Unmyelinated Axons

**3** A single nonmyelinating Schwann cell encloses segments of many unmyelinated axons. It first surrounds the segments of a group of axons. The axons then come to lie in separate membrane folds at the periphery of the cell. This stabilizes the positions of these axons and isolates them from chemicals in the surrounding interstitial fluid. Note that nodes do not form between the adjacent nonmyelinating Schwann cells.

**Satellite cells** surround neuron cell bodies in PNS ganglia. They regulate the environment around the neurons, much as astrocytes do in the CNS.

**2** This illustration shows the steps in the myelination of an axon in the PNS. A single myelinating Schwann cell myelinates one internode of only one axon.

| **Step 1** | **Step 2** | **Step 3** |
|---|---|---|
| A Schwann cell first surrounds a portion of the axon within a groove of its cytoplasm. | The Schwann cell then begins to rotate around the axon. | As the Schwann cell rotates, myelin is wound around the axon in multiple layers, forming a tightly packed membrane. |

**?** **REVIEW**
**A.** Identify the neuroglia of the PNS.

# PNS Axon Injury and Repair

Site of injury

**4** Schwann cells assist in repairing damaged nerves (recall that nerves are bundles of axons in the PNS; see Module 11.3). The repair process, which often fails to restore full function, is known as **Wallerian** (waw-LĒ-rē-an) **degeneration**.

**? REVIEW**

**B.** In which part of a neuron does Wallerian degeneration occur?

**Step 1**
Distal to the injury site, the axon and myelin degenerate and fragment.

Axon  Myelin  Proximal stump  Distal stump

**Step 2**
The Schwann cells do not degenerate; instead, they proliferate along the path of the original axon. Over this period, macrophages move into the area and remove the degenerating debris distal to the injury site.

Macrophage

Cord of proliferating Schwann cells

**Step 3**
As the neuron recovers, its axon grows into the site of injury and then distally, along the path created by the newly divided Schwann cells.

**Step 4**
As the axon elongates, the Schwann cells wrap around it. If the axon reestablishes its normal synaptic contacts, normal function may be regained. However, if it stops growing or grows in some new direction, normal function will not return.

---

**Clinical Note**

## Nerve regeneration in the CNS is more limited than in the PNS

Limited regeneration may occur in the CNS, but there the situation is more complicated because (1) many more axons are likely to be involved, (2) astrocytes produce scar tissue that can prevent axon growth across the damaged area, and (3) astrocytes release chemicals that block the regrowth of axons.

**? INTEGRATION**

**C.** Describe the neurilemma of a myelinating and nonmyelinating Schwann cell.

**Lo LEARNING OUTCOME**

Describe the locations and functions of Schwann cells and satellite cells.

# Cellular Organization of the Nervous System

**Labeling:** Label each of the structures in the following diagram of a neuron.

1 dendritic spine
2 Nissl bodies
3 mitochondrion
4 nucleus
5 nucleolus
6 perikaryon
7 axon hillock
8 first segment
9 axolemma
10 axoplasm
11 telodendria
12 synaptic terminal

Label the anatomical classes of neurons shown below.

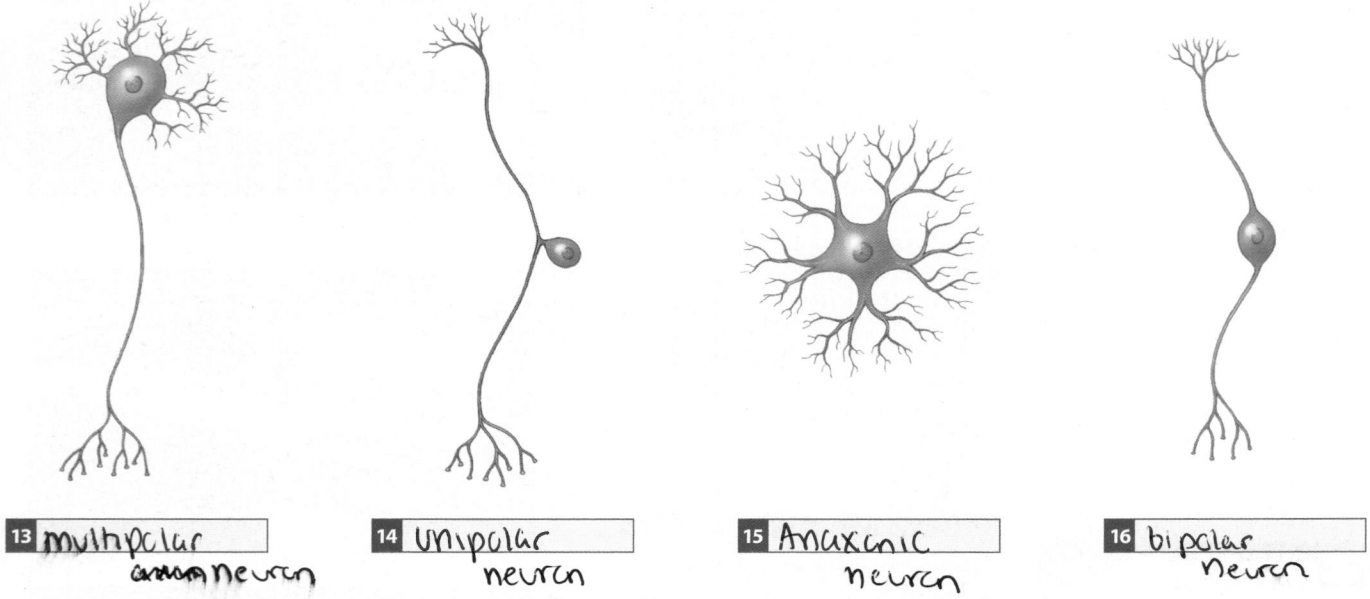

13 multipolar neuron
14 unipolar neuron
15 Anaxonic neuron
16 bipolar neuron

**Vocabulary:** In the space provided, write the boldfaced terms introduced in this section that contain the indicated word part.

| | | |
|---|---|---|
| 17 | neur- *(nerve)* | 17 neuron |
| 18 | dendr- *(tree)* | 18 dendrites |
| 19 | ef- *(away from)* | 19 Efferent |
| 20 | af- *(toward)* | 20 Afferent |

# Module 11.6
# Neuronal activity depends on changes in membrane potential

**1** Plasma membranes are selectively permeable, and the cytosol and extracellular fluid differ in composition. Under normal circumstances, the inner surface of the plasma membrane has a slight negative charge with respect to the outer surface. Recall that the unequal distribution of positive and negative electrical charges across the membrane surface represents potential energy, or a potential difference called the **membrane potential** (see Module 9.5). A membrane potential is characteristic of all living cells. It is the result of differences in the membrane's permeability to various ions and of active transport mechanisms.

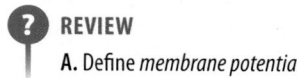

**2** The figure below provides an overview of the role of the membrane potential in neural activity. Changes in the membrane potential have many important functions; for example, they can trigger muscle contraction and gland secretion and transfer information in the nervous system.

**? REVIEW**
**A.** Define *membrane potential*.

| **1** The membrane potential of an undisturbed cell is called the **resting membrane potential**. All neural activities begin with a change in the resting membrane potential of a neuron. | **2** A typical stimulus produces a temporary, localized change in the resting membrane potential. This change, which decreases with distance away from the stimulus, is called a **graded potential**. | **3** If the graded potential is sufficiently large, it generates an action potential in the excitable membrane of the axon. An **action potential** is an electrical event that involves one location on the excitable membrane. Once an action potential develops in one location, it spreads along the surface of an axon toward the axon terminals. | **4** **Synaptic activity** then produces graded potentials in the plasma membrane of the postsynaptic cell. The process typically involves the release of neurotransmitters, such as ACh, by the presynaptic cell. These compounds bind to receptors on the postsynaptic plasma membrane, changing its permeability. | **5** The response of the postsynaptic cell depends on the degree of activity by the stimulated receptors and what other stimuli are also influencing the cell. The integration of stimuli at the level of the individual cell is the simplest form of **information processing** in the nervous system. |
|---|---|---|---|---|
| See Modules 11.7 and 11.8 | See Module 11.9 | See Modules 11.10 and 11.11 | See Module 11.12 | See Modules 11.13 and 11.14 |

11

**? REVIEW**
**B.** Compare a graded potential with an action potential.
**C.** Define *information processing*.

**LO LEARNING OUTCOME**
Describe the general role of membrane potential changes in neuronal activity.

# Differences in electrochemical gradients determine the resting membrane potential

The membrane potential of a cell is caused by the separation of positive and negative charges by the plasma membrane. The extracellular fluid (ECF) contains high concentrations of sodium ions (Na⁺) and chloride ions (Cl⁻), whereas the cytosol contains high concentrations of potassium ions (K⁺) and negatively charged proteins (Pr⁻). The ions cannot freely cross the lipid portions of the plasma membrane. They can enter or leave the cell only through membrane channels or by active transport mechanisms. The two ions we are most concerned with are potassium and sodium because they are the main factors that influence the membrane potential. This module discusses the roles of the differing membrane permeabilities for K⁺ and Na⁺, the *sodium–potassium exchange pump*, and the chemical and electrical gradients for K⁺ and Na⁺.

**1** The membrane potential exists primarily because plasma membranes contain passive **leak channels**, which are always open. Their size, shape, and structure determine which ions can pass through the membrane.

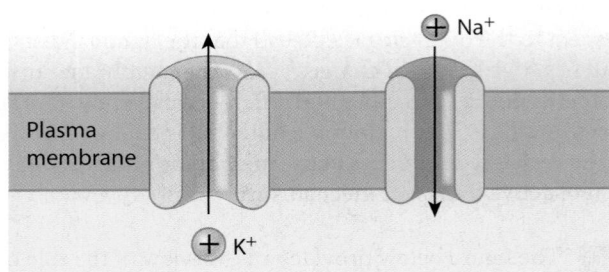

**2** An unstimulated, or "undisturbed," cell has a characteristic **resting membrane potential**. This is an overview of the events responsible for the resting membrane potential of a neuron. Both passive forces and active processes act across the plasma membrane to produce and maintain the resting membrane potential. The passive forces result from both chemical and electrical gradients. The Na⁺–K⁺ exchange pump is the active ATP-requiring process.

**?** **REVIEW**
**A.** Define *resting membrane potential*.

The resting membrane potential, a form of potential energy, is measured in millivolts (1 mV = one-thousandth of a volt). Different body cells have different membrane potentials, ranging from –5 mV to –100 mV. The minus sign indicates that the cell's interior is negatively charged. The resting membrane potential for a neuron is near –70 mV.

The cytosol contains an abundance of negatively charged proteins, whereas the extracellular fluid contains relatively few. Because they cannot cross the plasma membrane, they are an important electrical force involved in Na⁺ and K⁺ movement.

Potassium ions can diffuse out of the cell through potassium leak channels.

The **sodium–potassium exchange pump** ejects 3 Na⁺ for every 2 K⁺ recovered from the extracellular fluid. At a resting membrane potential of –70 mV, the rate of Na⁺ entry versus K⁺ loss is 3:2, and the exchange pump maintains a stable resting membrane potential.

Sodium ions can diffuse into the cell through sodium leak channels.

**?** **REVIEW**
**B.** What happens at the sodium–potassium exchange pump?

**3** The **chemical gradient** for any ion is its concentration gradient across the plasma membrane. The **electrical gradient** is created by the attraction between opposite charges or by the repulsion between like charges (+/+ or −/−). Together, these gradients determine an ion's **electrochemical gradient**, a form of potential energy. At a membrane potential known as the **equilibrium potential**, an ion's electrical and chemical gradients are in balance, and there is no net movement of ions across the membrane. The equilibrium potential indicates an ion's contribution to the resting membrane potential.

## Potassium Ion Gradients

At a neuron's resting membrane potential, the chemical and electrical gradients are opposed for potassium ions (K⁺). The net electrochemical gradient tends to drive potassium ions out of the cell.

If the plasma membrane were freely permeable to potassium ions, the outflow of K⁺ would continue until the equilibrium potential (−90 mV) was reached. Note how similar it is to the resting membrane potential.

## Sodium Ion Gradients

At a neuron's resting membrane potential, the chemical and electrical gradients for sodium ions (Na⁺) are combined. The net electrochemical gradient drives sodium ions into the cell.

If the plasma membrane were freely permeable to sodium ions, the influx of Na⁺ would continue until the equilibrium potential (+66 mV) was reached. Note how different it is from the resting membrane potential.

**? REVIEW**

**C.** What effect would decreasing the concentration of extracellular potassium ions have on the resting membrane potential of a neuron?

**LO LEARNING OUTCOME**

Explain how the resting membrane potential is created and maintained.

## Module 11.8

# Three types of gated ion channels change the permeability of the plasma membrane

Permeability changes are due to **gated ion channels** within the plasma membrane that open or close in response to specific stimuli. The three different types of gated ion channels are described in this module. Two types are especially important in neuron function.

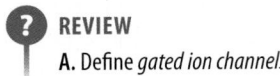

**? REVIEW**
**A.** Define *gated ion channels.*

## Chemically Gated Ion Channels

**1** **Chemically gated ion channels**, or **ligand-gated ion channels**, open when they bind specific chemicals. The receptors that bind *acetylcholine (ACh)* at the neuromuscular junction are chemically gated channels. Chemically gated channels are most abundant on the dendrites and cell body of a neuron, the areas where most synaptic communication occurs.

## Voltage-Gated Ion Channels

**2** **Voltage-gated ion channels** are characteristic of areas of excitable membrane. Such membranes are capable of generating and propagating, or spreading, an action potential (see Module 9.5). Voltage-gated channels open or close in response to changes in the membrane potential. The most important of such channels, for our purposes, are voltage-gated $Na^+$ channels, $K^+$ channels, and $Ca^{2+}$ channels. Sodium channels (shown here) have two gates that function independently: an activation gate that opens on stimulation, letting sodium ions into the cell, and an inactivation gate that closes to stop the entry of sodium ions.

**3** **Mechanically gated ion channels** open in response to mechanical stimuli that physically distort the neuron's membrane surface. Such channels are important in sensory receptors that respond to stretch, pressure, or vibration and cause the sense of touch and hearing.

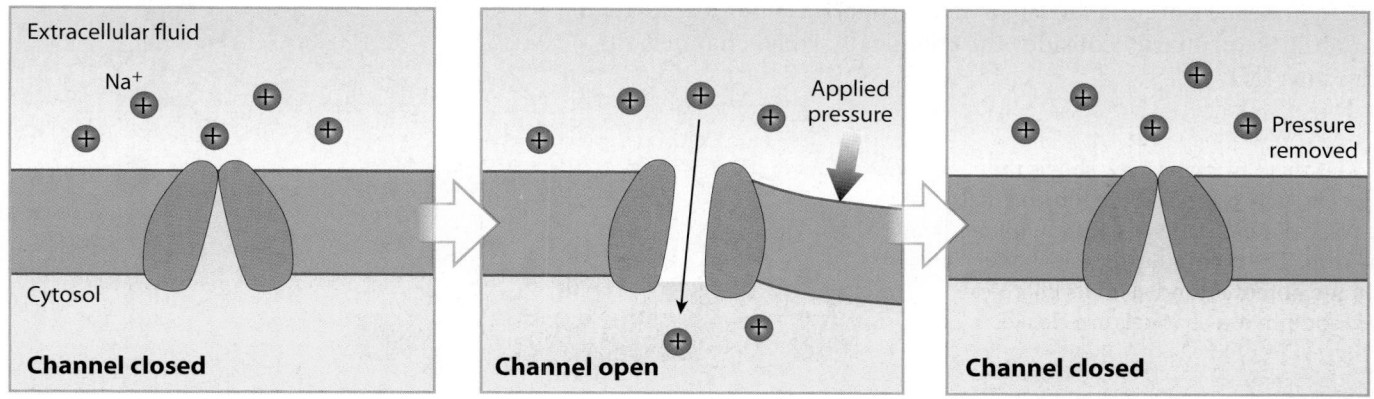

Extracellular fluid

Na+

**Channel closed**

Cytosol

Applied pressure

**Channel open**

Pressure removed

**Channel closed**

## Distribution of Gated Ion Channels on a Neuron

**4** Chemically gated ion channels are found on the neuron cell body and its dendrites. Voltage-gated Na+ channels and voltage-gated K+ channels are found along the axon of a neuron. As we will see, their distribution along an axon depends on whether it is myelinated or unmyelinated. Voltage-gated Ca²⁺ channels occur at axon terminals.

**? REVIEW**

**B.** Identify the three types of gated ion channels, and state the conditions under which each operates.

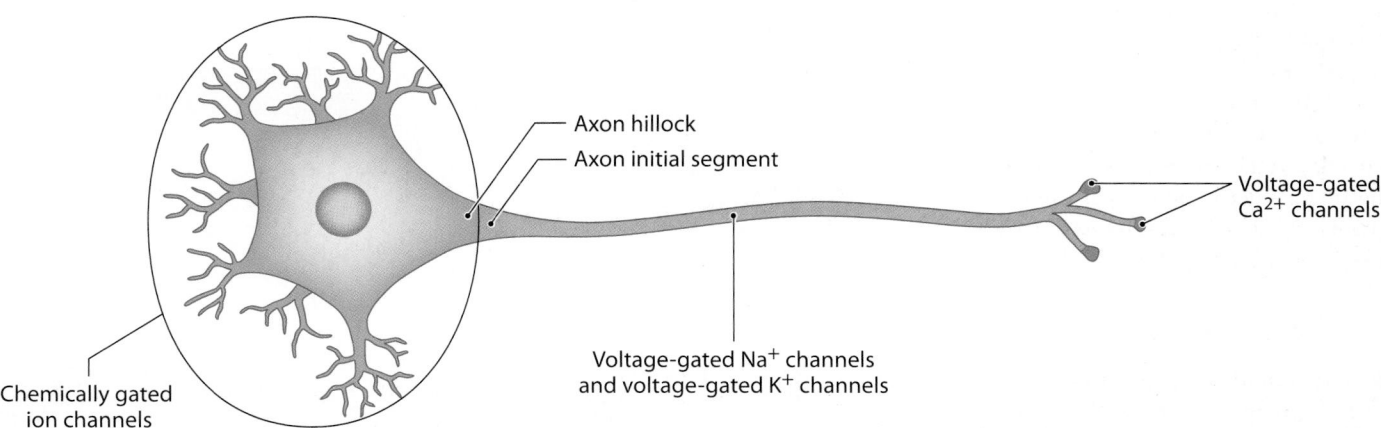

Axon hillock

Axon initial segment

Voltage-gated Ca²⁺ channels

Chemically gated ion channels

Voltage-gated Na+ channels and voltage-gated K+ channels

Most gated ion channels are closed at the resting membrane potential. When they open the rate of ion movement increases across the plasma membrane, and this changes the membrane potential.

**? INTEGRATION**

**C.** What effect would a chemical that blocks voltage-gated sodium ion channels in a neuron's plasma membrane have on its membrane potential?

**Lᴏ LEARNING OUTCOME**

Describe the functions of gated ion channels with respect to the permeability of the plasma membrane.

11

# Module 11.9

# Graded potentials are localized changes in the membrane potential

**Graded potentials**, or *local potentials*, are changes in the membrane potential that cannot spread far from the site of stimulation. Any stimulus that opens a gated ion channel will produce a graded potential. Note that when illustrating graded potentials, we can ignore the leak channels responsible for the resting membrane potential because their properties do not change. In this module we primarily consider the chemically gated channels for sodium ions ($Na^+$).

**1** This diagrammatic view shows the chemically gated sodium ion channels in the plasma membrane of the axon hillock of a neuron. The membrane is at its resting membrane potential, and all of its chemically gated sodium ion channels are closed.

**?** REVIEW
A. Define *graded potential*.

**2** When the membrane is exposed to a chemical that opens the chemically gated sodium channels, sodium ions enter the cell. This inrush of positive charges due to the large $Na^+$ electrochemical gradient reduces the membrane potential in this area. Any shift from the resting membrane potential toward a more positive value is called a **depolarization**. Depolarization applies to changes in potential from −70 mV to smaller negative values (toward 0 mV) as well as to membrane potentials above 0 mV.

**3** The sodium ions now inside the cell spread out, attracted by the negative charges along the inner surface of the membrane. The extent of depolarization spreads as well. As the plasma membrane depolarizes, extracellular sodium ions move toward the open channels, replacing ions that have entered the cell. This movement of positive charges parallel to the inner and outer surfaces of the membrane is called a **local current**.

11

**4** The degree of depolarization decreases with distance away from the stimulation site, primarily because the ions are entering only at one location but spreading in all directions. The impact on the membrane potential is proportional to the size of the stimulus, because the intensity of the stimulus determines the number of open sodium ion channels. The more open channels there are, the more sodium ions enter the cell, the greater the membrane area affected, and the greater the degree of depolarization.

Membrane potential (mV)

−70 mV
−65 mV
−60 mV
−65 mV
−70 mV

**5** This graph shows the changes in the membrane potential that occur over time when different chemical stimuli are applied to the axon hillock.

**?** **REVIEW**

**B.** Describe depolarization, repolarization, and hyperpolarization.

The presence of a chemical such as ACh opens chemically gated sodium ion channels. This leads to membrane depolarization.

When the chemical stimulus is removed, the membrane returns to its normal resting membrane potential as the excess sodium ions are transported out of the cytosol. This process is called **repolarization**.

It is also possible to expose the neuron to a chemical that opens chemically gated potassium channels. This results in an increasingly negative membrane potential as additional potassium ions leave the cytosol. A shift in the membrane potential past resting levels is called **hyperpolarization**.

Membrane potential (mV)

−60
−70
−80

Repolarization

Resting membrane potential

Depolarization

Chemical stimulus removed

Hyperpolarization

Return to resting membrane potential

Time

**6** This table summarizes the most important characteristics of graded potentials.

**?** **REVIEW**

**C.** What factors account for the local currents associated with graded potentials?

### Graded Potentials

Graded potentials, whether depolarizing or hyperpolarizing, share four basic characteristics:

**1.** The membrane potential is most affected at the site of stimulation, and the effect decreases with distance.

**2.** The effect spreads passively through local currents.

**3.** The graded change in membrane potential may involve either depolarization or hyperpolarization. The nature of the change is determined by the properties of the gated channels involved. For example, in a resting membrane, the opening of sodium ion channels will cause depolarization, whereas the opening of potassium ion channels will cause hyperpolarization. Thus, the change in membrane potential reflects whether positive charges enter or leave the cell.

**4.** The stronger the stimulus, the greater the change in the membrane potential, and the larger the area affected.

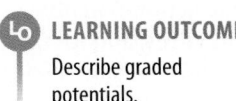 **LEARNING OUTCOME**
Describe graded potentials.

# Module 11.10

# Action potentials are all-or-none events for communication that begin with membrane potential reversal

Each neuron receives information in the form of graded potentials on its dendrites and cell body, and action potentials at the axon terminals trigger the release of neurotransmitters. However, the two ends of the neuron may be a meter apart, and even the largest graded potentials affect only a tiny area. Such relatively long-range communication requires a different mechanism—the action potential. **Action potentials** are propagated changes in the membrane potential that, once initiated, affect an entire excitable membrane. Whereas the resting membrane potential depends on leak channels and the graded potential depends on chemically gated ion channels, action potentials depend on voltage-gated ion channels. Recall that both neurons and skeletal muscle cells have areas of excitable membrane containing voltage-gated ion channels.

**Initial segment**

**Axon hillock**

**1** This figure shows the steps in the generation of an action potential at the initial segment of an axon. The first step is a graded depolarization caused by the opening of chemically gated sodium ion channels, usually at the axon hillock. Note that when illustrating action potentials, we can ignore both the leak channels and the chemically gated ion channels because their properties do not change.

**? REVIEW**

**A.** Define *action potential*.

**Resting Membrane Potential**

The axolemma contains both voltage-gated sodium ion channels and voltage-gated potassium ion channels. Both types are closed at the resting membrane potential.

**KEY**

■ Resting potential
■ Graded depolarization
■ Rapid depolarization
■ Repolarization

**Step 1**
**Depolarization to Threshold**

The stimulus that initiates an action potential is a graded depolarization large enough to open voltage-gated sodium channels. The opening of the channels occurs at a membrane potential known as the **threshold**.

**Step 2**
**Activation of Sodium Ion Channels and Rapid Depolarization**

When the sodium channel activation gates open, the plasma membrane becomes much more permeable to $Na^+$. Driven by the large electrochemical gradient, sodium ions rush into the cytosol, and rapid depolarization occurs. The inner membrane surface now contains more positive ions than negative ones, and the membrane potential has changed from −60 mV to a positive value.

In an excitable membrane, a graded depolarization is like the pressure on the trigger of a gun, and the action potential is like the firing of the gun. As long as the trigger is pulled hard enough, the gun fires, and it fires the same way every time. Similarly, all stimuli that bring the membrane to threshold generate identical action potentials. This concept is called the **all-or-none principle** because a given stimulus either triggers a typical action potential or triggers none at all.

**? REVIEW**

**B.** Describe the all-or-none principle.

**2** This graph shows the changes in the membrane potential over time at one location during the generation of an action potential.

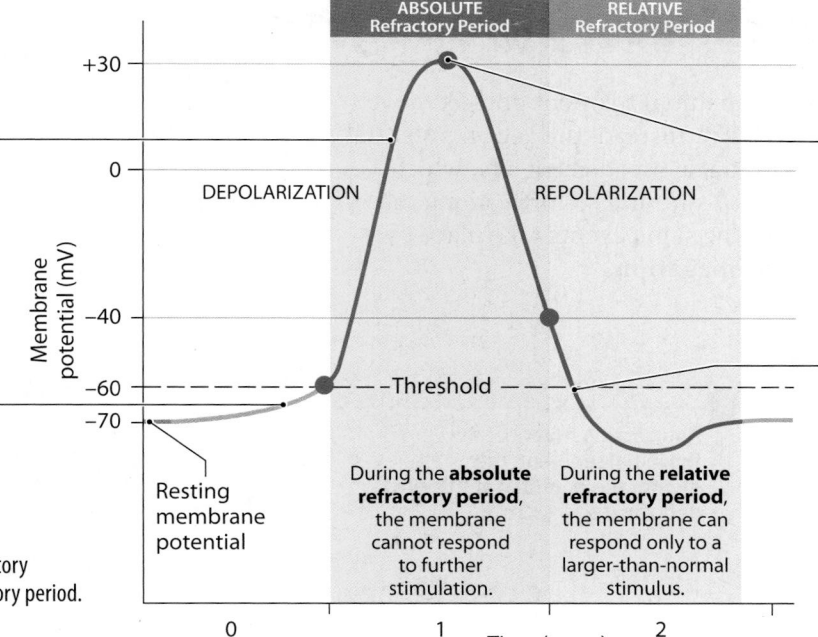

ABSOLUTE Refractory Period

RELATIVE Refractory Period

**Activation of Sodium Channels and Rapid Depolarization:** Voltage-gated sodium channels open, and sodium ions move into the cell. The membrane potential rises to +30 mV.

**Depolarization to Threshold:** A graded depolarization brings an area of excitable membrane to threshold (–60 mV).

DEPOLARIZATION

REPOLARIZATION

+30

0

–40

–60
–70

Membrane potential (mV)

Threshold

Resting membrane potential

During the **absolute refractory period**, the membrane cannot respond to further stimulation.

During the **relative refractory period**, the membrane can respond only to a larger-than-normal stimulus.

0          1          2
Time (msec)

**Inactivation of Sodium Channels and Activation of Potassium Channels:** Sodium channels close, voltage-gated potassium channels open, and potassium ions move out of the cell. Repolarization begins.

**Potassium Channels Close:** The voltage-gated potassium channels begin closing. Near threshold, the voltage-gated sodium channels begin reactivating, and the membrane soon returns to its normal resting state.

**? REVIEW**
**C.** Compare the absolute refractory period with the relative refractory period.

11

---

**Step 3**
Inactivation of Sodium Ion Channels and Activation of Potassium Ion Channels

+30 mV

As the membrane potential approaches +30 mV, the inactivation gates of the voltage-gated sodium channels close. This step is known as **sodium channel inactivation**, and it coincides with the opening of voltage-gated potassium channels. Positively charged potassium ions move out of the cytosol, shifting the membrane potential back toward resting levels. Repolarization now begins.

**Step 4**
Potassium Ion Channels Close

–90 mV

The voltage-gated sodium channels remain inactivated until the membrane has repolarized to near–threshold levels. At this time, they regain their normal status: closed but capable of opening. The voltage-gated potassium channels begin closing as the membrane reaches the normal resting membrane potential (about –70 mV). Until all of these potassium channels have closed, potassium ions continue to leave the cell. This produces a brief hyperpolarization.

**Resting Membrane Potential**

–70 mV

As the voltage-gated potassium channels close, the membrane potential returns to normal resting levels. The action potential is now over, and the membrane is once again at the resting membrane potential.

 **REVIEW**
**D.** List the events involved in the generation of an action potential.

 **LEARNING OUTCOME**
Describe the events involved in the generation of an action potential.

# Action potentials may affect adjacent portions of the plasma membrane through continuous propagation or saltatory propagation

An action potential generated at the initial segment doesn't move along the axon like a car on a highway. Instead, the action potential at one site triggers an action potential at an adjacent site, which triggers an action potential at a third site, and so forth, along the entire length of the axon. Because the same events take place over and over, this activity is termed **propagation**.

Initial segment
Segment 2
Segment 3

## Continuous Propagation

**1** In **continuous propagation**, the action potential appears to "move" along the axolemma of an axon in a series of tiny steps. Continuous propagation occurs along unmyelinated axons. This diagram shows how an action potential generated in one segment affects more distant portions of the axon. Each step takes only a millisecond, but the steps must be repeated along the entire axon, so propagation along an unmyelinated axon occurs only at a speed of about 1 meter per second.

**KEY**

| | |
|---|---|
| ▨ | Resting potential |
| ▨ | Graded depolarization |
| ▨ | Rapid depolarization |
| ▨ | Repolarization |

| Initial segment | Segment 2 | Segment 3 |
|---|---|---|

**1** As an action potential develops at the initial segment, the membrane potential at this site depolarizes to +30 mV.

Extracellular fluid

+30 mV    −70 mV    −70 mV

Na⁺

Plasma membrane    Cytosol

**2** As the sodium ions entering at the initial segment spread away from the open voltage-gated channels, a graded depolarization quickly brings the membrane in segment 2 to threshold.

−60 mV    −70 mV

Local current

**3** An action potential develops in segment 2.

The initial segment begins repolarization (and is now refractory).

+30 mV    −70 mV

Na⁺

**4** As the sodium ions entering at segment 2 spread laterally, a graded depolarization quickly brings the membrane in segment 3 to threshold.

The action potential can only move forward, not backward, because the membrane at the initial segment is in the absolute refractory period of repolarization.

−60 mV

Local current

**?** **REVIEW**
**A.** Define *continuous propagation* and *saltatory propagation*.

## Saltatory Propagation

**2** Continuous propagation cannot occur along a myelinated axon, because myelin blocks the flow of ions across the membrane. Ions can cross the plasma membrane only at the unmyelinated nodes (nodes of Ranvier). As a result, only the nodes can respond to a depolarizing stimulus. When an action potential appears at the initial segment of a myelinated axon, the local current skips the internode and depolarizes the closest node in a process called **saltatory propagation** (*saltare*, leaping).

Initial segment
Node 1
Node 2

**1** An action potential develops at the initial segment.

Initial segment — Node 1 — Node 2 — Extracellular fluid
+30 mV — −70 mV — −70 mV
Na⁺
Myelinated internode
Plasma membrane — Cytosol

**2** A local current produces a graded depolarization that brings the axolemma at node 1 to threshold.

−60 mV — −70 mV
Myelinated internode
Local current

**3** An action potential develops at node 1.

The initial segment begins repolarization (and is now refractory).

+30 mV — −70 mV
Na⁺
Myelinated internode

**4** A local current produces a graded depolarization that brings the axolemma at node 2 to threshold.

−60 mV
Myelinated internode
Local current

Saltatory propagation is much faster than continuous propagation, but there are differences among myelinated axons. That is because the speed varies with axon diameter. The larger the diameter, the lower the resistance to ion movement, and the faster the action potential travels.

**? REVIEW**

**B.** What is the relationship between myelin and the propagation speed of action potentials?

**Lo LEARNING OUTCOME**

Describe continuous propagation and saltatory propagation, and discuss the factors that affect the speed with which action potentials are propagated.

11

# Module 11.12

## At a synapse, information travels from the presynaptic cell to the postsynaptic cell

Watch
MasteringA&P™
Video Tutor
Cholinergic Synapse

In the nervous system, messages are transmitted from one location to another along axons in the form of action potentials, also known as "nerve impulses." To be effective, messages must be not only propagated along an axon but also transferred in some way to another neuron or an effector cell. That transfer occurs at a **synapse**. At a synapse involving two neurons, information is relayed from a presynaptic neuron to a postsynaptic neuron. The two types of synapses are chemical and electrical.

### Chemical Synapses

**1** **Chemical synapses**, which rely on neurotransmitter release, are by far the most abundant type of synapse. Most synapses between neurons, and all synapses between neurons and other types of cells, involve chemical synapses. Synapses that release acetylcholine (ACh) as a neurotransmitter are known as **cholinergic synapses**. This figure illustrates the events that occur at a cholinergic synapse when an action potential arrives at an axon terminal. Similar steps are involved in communication across a neuromuscular junction (see Module 9.6).

**? REVIEW**

**A.** Describe the components of a chemical synapse.

**B.** What event must occur in the postsynaptic cell before it generates an action potential?

**Step 1**
The normal stimulus for neurotransmitter release is the depolarization of the axon terminal by the arrival of an action potential.

**Step 2**
The depolarization of the axon terminal opens voltage-gated calcium ion channels. Calcium ions rush into the axon terminal and trigger the exocytosis of synaptic vesicles and the release of ACh into the synaptic cleft. The calcium ions that triggered exocytosis are rapidly removed, ending the release of ACh.

**Step 3**
The released ACh diffuses across the synaptic cleft and binds to the chemically gated $Na^+$ receptors on the postsynaptic membrane. The greater the amount of ACh released, the more receptors respond, and the larger the depolarization. If the depolarization is great enough to reach threshold, then an action potential will be generated in the postsynaptic neuron's membrane.

**Step 4**
The effects on the postsynaptic membrane are temporary because of the presence of the enzyme acetylcholinesterase (AChE) in the synaptic cleft. ACh molecules that bind to receptor sites are generally broken down within 20 msec of their arrival by AChE. Some other ACh molecules diffuse away from the synaptic cleft.

**2** At a cholinergic synapse, an arriving action potential depolarizes the axon terminal. Neurotransmitters are reabsorbed and recycled. However, if the neurotransmitters are released into the synaptic cleft at a faster rate than they can be reabsorbed or recycled, **synaptic fatigue** results, and the synapse will not function normally. Function returns when the ACh receptor is able to bind ACh again.

**REVIEW**
**C.** Describe synaptic fatigue.

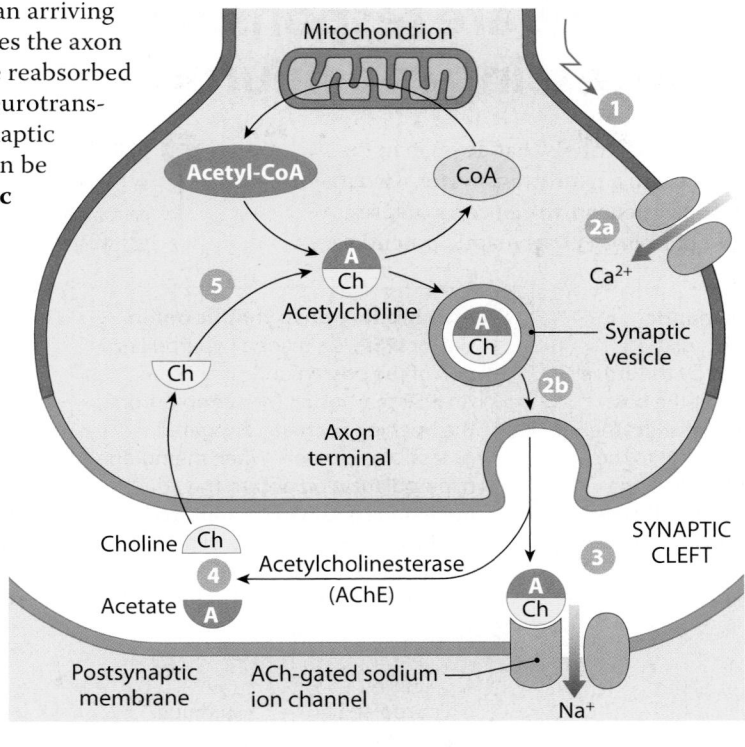

Mitochondrion

Acetyl-CoA

CoA

A Ch
Acetylcholine

A Ch
Synaptic vesicle

2a

Ca²⁺

2b

Axon terminal

Ch

Choline Ch

5

Acetylcholinesterase (AChE)

4

Acetate A

Postsynaptic membrane

ACh-gated sodium ion channel

A Ch

3

SYNAPTIC CLEFT

Na⁺

**Events Occurring at a Cholinergic Synapse**

**1** Action potential triggers the opening of voltage-gated calcium channels.

**2** (a) Calcium ions diffuse into the axon terminal, and (b) trigger synaptic vesicles to release ACh by exocytosis.

**3** ACh diffuses across synaptic cleft, binds to ACh-gated sodium ion channels, and produces a graded depolarizarion.

**4** Depolarization ends as ACh is broken down into acetate and choline by AChE.

**5** The axon terminal reabsorbs choline from the synaptic cleft and uses it to synthesize new molecules of ACh.

**11**

**3** A **synaptic delay** lasting 0.2–0.5 msec occurs between the arrival of the action potential at the axon terminal and the effect on the postsynaptic membrane. Although a delay of 0.5 msec is not very long, when information is being passed along a chain of interneurons the cumulative synaptic delay may be considerable. This is why reflexes are important for survival—they involve only a few synapses and thus provide rapid and automatic responses to stimuli. There is only one way to eliminate synaptic delay entirely—directly couple the presynaptic neuron to the postsynaptic cell at an electrical synapse.

## Electrical Synapses

**4** At an **electrical synapse**, the presynaptic and postsynaptic membranes are locked together by gap junctions. As a result, changes in the membrane potential of one cell will produce local currents that affect the other cell as if the two shared a common membrane. Electrical synapses are located in both the CNS and PNS, but they are extremely rare. They are present in some areas of the brain, in the eye, and in at least one pair of PNS ganglia (the ciliary ganglia). An action potential reaching an electrical synapse will always be propagated to the next cell, an arrangement that is efficient but not versatile. Much of the complexity and adaptability of the nervous system results from the fact that the responses of a postsynaptic cell can vary depending on the local chemical environment or the activities of synapses that release multiple neurotransmitters with varied effects.

Presynaptic neuron

Gap junctions connecting presynaptic and postsynaptic neurons

Postsynaptic neuron

**REVIEW**
**D.** Contrast an electrical synapse with a chemical synapse.

**Lo** **LEARNING OUTCOME**
Describe the general structure of synapses in the CNS and PNS, and discuss the events that occur at a chemical synapse.

# Module 11.13
# Postsynaptic potentials are responsible for information processing in a neuron

**1** **Postsynaptic potentials** are graded potentials that develop in the postsynaptic membrane in response to a neurotransmitter. Two major types of postsynaptic potentials develop at neuron-to-neuron synapses: excitatory postsynaptic potentials and inhibitory postsynaptic potentials.

**?** **REVIEW**

**A.** Define *excitatory postsynaptic potential (EPSP)* and *inhibitory postsynaptic potential (IPSP)*.

An **excitatory postsynaptic potential**, or **EPSP**, is a graded depolarization caused by the arrival of a neurotransmitter at the post-synaptic membrane that shifts the membrane potential closer to the threshold. When that occurs, the membrane is said to be **facilitated**. The larger the degree of facilitation, the smaller the additional stimulus needed to trigger an action potential.

An **inhibitory postsynaptic potential**, or **IPSP**, is a graded hyperpolarization of the postsynaptic membrane. An IPSP may result, for example, from the opening of chemically gated potassium channels. When membrane hyperpolarization occurs, the membrane is said to be **inhibited**, because a larger-than-usual depolarizing stimulus must be provided to bring the membrane potential to threshold.

**Summation** is the integration of the effects of graded potentials on a segment of the plasma membrane. EPSPs and IPSPs result from the activation of different types of chemically gated ion channels. If two different neurotransmitters arrive simultaneously, and both sets of channels open, the net effect may be no change in the membrane potential.

**2** A single neuron may receive information across thousands of synapses. Some of the neurotransmitters arriving at the postsynaptic cell at any moment may be excitatory, and others may be inhibitory. The net effect on the membrane potential at the axon hillock determines how the neuron responds from moment to moment. This is because (1) the axon hillock is closest to the initial segment, and graded potentials are what trigger the action potential, and (2) the threshold at the boundary between the axon hillock and the initial segment is lower than it is elsewhere on the cell body. Thus it is the axon hillock that integrates the excitatory and inhibitory stimuli affecting the cell body and dendrites at any given moment and determines the rate of action potential generation at the initial segment. This integration is the simplest level of information processing in the nervous system.

Axon hillock

Initial segment

Glial cell processes    Dendrite    Axon terminals

**3** An individual EPSP or IPSP has a small effect on the membrane potential. A typical EPSP produces a depolarization of about 0.5 mV at the postsynaptic membrane. Before an action potential will arise in the initial segment, local currents must depolarize that region by at least 10 mV. Therefore, a single EPSP will not result in an action potential, even if the synapse is on the axon hillock. However, individual EPSPs combine through the process of summation. There are two forms of summation that can generate an action potential: temporal summation and spatial summation.

## Temporal Summation

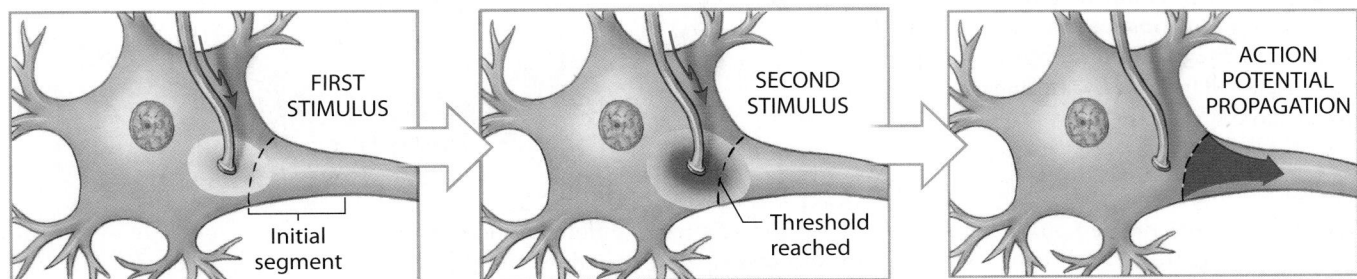

**Temporal summation** (*tempus*, time) occurs when a single synapse is active repeatedly. For example, a typical EPSP lasts about 20 msec, but under maximum stimulation an action potential can reach the axon terminal each millisecond. When a second EPSP arrives before the effects of the first EPSP have disappeared, another group of vesicles discharges ACh into the synaptic cleft, more ACh molecules arrive at the postsynaptic membrane, and the degree of depolarization increases. In this way, a series of small steps can eventually bring the initial segment to threshold.

## Spatial Summation

**Spatial summation** involves multiple synapses that are active at the same time. The effects on the membrane potential are cumulative because each active synapse opens gated channels that allow ion movement into or out of the cell. In this example, the active synapses are generating EPSPs and opening gated channels that let sodium ions enter the cell. The effects are cumulative, and the degree of depolarization at the initial segment depends on (1) how many excitatory synapses are active at any given moment and (2) how far they are from the initial segment. As in temporal summation, an action potential results when the membrane potential at the initial segment reaches threshold.

Each neuron integrates the information arriving across synapses and also monitors and responds to a variety of factors in the local environment. Neurons are dynamic, active cells, and their sensitivity to stimulation changes in response to variations in body temperature, oxygen or nutrient availability, or the presence of abnormal chemicals.

 **REVIEW**

**B.** Compare temporal summation with spatial summation.

 **INTEGRATION**

**C.** If a single EPSP depolarizes the initial segment from a resting membrane potential of −70 mV to −65 mV, and threshold is at −60 mV, will an action potential be generated? Explain your answer.

 **LEARNING OUTCOME**

Discuss the significance of postsynaptic potentials, including the roles of excitatory postsynaptic potentials and inhibitory postsynaptic potentials.

# Module 11.14

## Information processing involves interacting groups of neurons and various neurotransmitters that encode information as action potential frequency

**1** Cellular information processing occurs at the postsynaptic membrane as the membrane potential at the axon hillock rises and falls. Higher levels of information processing involve **regulatory neurons** that facilitate or inhibit the activities of presynaptic neurons by affecting the membrane of the cell body or by altering the sensitivity of axon terminals.

Regulatory neurons

Presynaptic neuron

Postsynaptic neuron

**? REVIEW**

A. Describe the role of regulatory neurons.

**2** One of the reasons the nervous system is so complex and adapted to different functions is that it uses more than 100 different neurotransmitters that each work in different ways. This table introduces a few of the important neurotransmitters. As you see, they may have indirect effects as well as direct effects on ion channels. The indirect effects usually involve binding to membrane receptors coupled with **G proteins**, a diverse family of enzyme complexes. Neurotransmitters that bind to G-coupled membrane receptors represent **first messengers**. Activated G proteins trigger the formation or release of ions or molecules called **second messengers** that alter conditions or activities of the postsynaptic cell.

### Selected Neurotransmitters

| Neurotransmitter | Chemical Structure | Mechanism of Action | Location | Comments |
|---|---|---|---|---|
| Acetylcholine (ACh) | $CH_3-N-CH_2-CH_2-O-C-CH_3$ (with $CH_3$ groups and $O$) | Primarily direct, through binding to chemically gated ion channels | CNS: Synapses throughout brain and spinal cord<br>PNS: Neuromuscular junctions, neuroglandular junctions, and synapses in autonomic ganglia | Widespread in CNS and PNS; best known and most studied of the neurotransmitters |
| Norepinephrine (NE) | $NH_2-CH_2-CH$ (with OH, OH, OH) | Indirect, through G proteins and second messengers | CNS: Cerebral cortex, hypothalamus, brainstem, cerebellum, and spinal cord<br>PNS: Most neuromuscular and neuroglandular junctions of sympathetic division of ANS | Involved in attention and consciousness, control of body temperature, and regulation of pituitary gland secretion |
| Epinephrine (E) | $CH_2-NH-CH_2-CH$ (with OH, OH, OH) | Indirect: G proteins and second messengers | CNS: Thalamus, hypothalamus, midbrain, and spinal cord | Generally excitatory effect along autonomic pathways |
| Serotonin | $NH_2-CH_2-CH_2$ (indole ring with OH, N) | Primarily indirect: G proteins and second messengers | CNS: Hypothalamus, limbic system, cerebellum, spinal cord, and retina | Important in emotional states, moods, and body temperature; several illicit hallucinogenic drugs, such as Ecstasy, target serotonin receptors |
| Glutamate | $HO-C-CH-CH_2-CH_2-C-OH$ (with $O$, $NH_2$, $O$) | Indirect: G proteins and second messengers<br>Direct: opens calcium/sodium ion channels | CNS: Cerebral cortex and brainstem | Important in memory and learning; most important excitatory neurotransmitter in the brain |
| Gamma-aminobutyric acid (GABA) | $NH_2-CH_2-CH_2-CH_2-C-OH$ (with $O$) | Direct or indirect (G proteins), depending on type of receptor | CNS: Cerebral cortex, cerebellum, interneurons throughout brain and spinal cord | Direct inhibitory effects: opens $Cl^-$ channels; indirect effects: opens $K^+$ channels and blocks entry of $Ca^{2+}$ |

11

**Twitch contractions**

Maximum tension (in tetanus)

**Incomplete tetanus**

**Tetanus**

Muscle fiber tension →

**KEY**

↓ = Arrival of action potential

Time →

**3** In the nervous system, complex information is translated into action potentials that are propagated along axons. On arrival, the message is often interpreted solely on the basis of the frequency of action potentials. Together, these figures show how the rate of action potentials arriving at a neuromuscular junction determines the tension and nature of the resulting contraction of a muscle fiber.

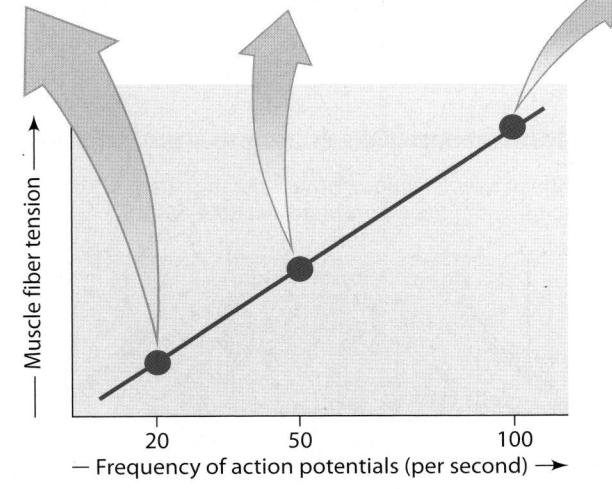

Muscle fiber tension →

20        50        100

— Frequency of action potentials (per second) →

**4** This table summarizes the key concepts for understanding information processing in the nervous system.

### Information Processing

- The neurotransmitters released at a synapse may have either excitatory or inhibitory effects. The effect on the axon's initial segment reflects a summation of the stimuli arriving at any moment. The frequency of action potentials generated is an indication of the degree of sustained depolarization at the axon hillock.

- Neurons may be facilitated or inhibited by extracellular chemicals other than neurotransmitters.

- The response of a postsynaptic neuron to the activation of a pre-synaptic neuron can be altered by (1) the presence of chemicals that cause facilitation or inhibition at the synapse, (2) activity under way at other synapses affecting the postsynaptic cell, and (3) modification of the rate of neurotransmitter release through facilitation or inhibition by regulatory neurons.

- Information is relayed in the form of action potentials. In general, the degree of sensory stimulation or the strength of the motor response is proportional to the frequency of action potentials.

**? REVIEW**

**B.** What determines the frequency of action potential generation?

**C.** The greater the degree of sustained depolarization at the axon hillock, the _____ (higher or lower) the frequency of action potentials generated.

**? INTEGRATION**

**D.** Compare the effects of the neurotransmitters acetylcholine and serotonin on ion channels.

**Lo LEARNING OUTCOME**

Discuss the interactions that make information processing in nervous tissue possible.

# Section 2 Review
## Neurophysiology

**Vocabulary:** Write the term for each of the following descriptions in the space provided.

**1** A propagated change in the membrane potential

**2** A synapse in which the presynaptic and postsynaptic neuronal membranes are locked together by gap junctions

**3** The membrane potential of an unstimulated cell

**4** Ion channels that open or close in response to specific stimuli

**5** Chemical synapses that release acetylcholine

**6** A shift in the membrane potential from $-70$ mV to $-85$ mV

**7** Movement of positive charges parallel to the inner and outer plasma membrane surfaces

**8** A shift in the membrane potential from $-70$ mV to $+30$ mV

**1** _____

**2** _____

**3** _____

**4** _____

**5** _____

**6** _____

**7** _____

**8** _____

---

**Short answer:** For the following diagram of a cholinergic synapse, write the names of components 9–14 in the boxes at left, and then fill in the table at right with descriptions of the events represented by 15–20.

**Components**

**9** _____

**10** _____

**11** _____

**12** _____

**13** _____

**14** _____

| Events Occurring at Synapse |
| --- |
| **15** |
| **16** |
| **17** |
| **18** |
| **19** |
| **20** |

---

**Section integration:** Guillain-Barré *(ghē-YAN bah-RĀ)* syndrome is a degeneration of myelin sheaths that ultimately may result in paralysis. Propose a mechanism by which myelin sheath degeneration can cause muscular paralysis.

**21** _____

_____

_____

# Study Outline

### SECTION 1 • Cellular Organization of the Nervous System

**Module 11.1**

**The nervous system has three divisions: the CNS, PNS, and ENS p. 403**

1. The **central nervous system (CNS)** consists of the brain and spinal cord. It is responsible for integrating, processing, and coordinating sensory data and motor commands.

2. The **peripheral nervous system (PNS)** includes all the nervous tissue outside the CNS, except for that of the ENS.

3. **Receptors** detect changes in the internal and external environment. The **sensory (afferent) division** of the PNS brings information from sensory receptors to the CNS.

4. The **motor (efferent) division** of the PNS carries motor commands from the CNS to the **effectors**, or target organs.

**Module 11.2**

**Neurons are nerve cells specialized for intercellular communication p. 404**

5. Neurons have three general regions: **dendrites** that receive stimuli; at *dendritic spines*; a **cell body** that contains the nucleus and other organelles; and an **axon** that carries information to other cells.

6. The **telodendria** of an axon end at **axon terminals**. Axon terminals are part of the **synapse**, where the neuron communicates with another cell.

7. Within the axon terminals are **synaptic vesicles** containing neurotransmitters.

*The cell body of a neuron contains most of its organelles.*

**Module 11.3**

**Neurons are classified on the basis of structure and function p. 406**

8. The four major anatomical classes of neurons are **anaxonic**, **bipolar**, **unipolar**, and **multipolar neurons**.

9. Functionally, neurons are classified as **sensory neurons**, **interneurons**, or **motor neurons** (**somatic** or **visceral**).

**Module 11.4**

**Oligodendrocytes, astrocytes, ependymal cells, and microglia are neuroglia of the CNS p. 408**

10. **Neuroglia**, or **glial cells**, support and protect neurons.

11. **Ependymal cells** are associated with **cerebrospinal fluid** production and circulation. **Microglia** remove cellular debris and pathogens. **Astrocytes** maintain the **blood–brain barrier**.

12. **Oligodendrocytes** form the **myelin sheath** that surrounds axons making up **white matter**. **Gray matter** is unmyelinated neuron cell bodies, dendrites, and **unmyelinated axons**.

**Module 11.5**

**Schwann cells and satellite cells are the neuroglia of the PNS p. 410**

13. **Schwann cells** form a myelin sheath around or enclose segments of peripheral axons.

14. **Satellite cells** surround neuron cell bodies in ganglia.

15. In the PNS, repair of axons in nerves may follow **Wallerian degeneration**, a process that often fails to restore full function.

*Schwann cells form a sheath around peripheral axons.*

### SECTION 2 • Neurophysiology

**Module 11.6**

**Neuronal activity depends on changes in membrane potential p. 413**

16. **Membrane potential** is the unequal charge distribution between the inner and outer surfaces of the plasma membrane, where there is a slight negative charge inside the plasma membrane with respect to the outside.

17. All neural activities begin with a change in the **resting membrane potential** of the neuron. If localized changes in resting membrane potential, called **graded potentials**, are sufficient, they can generate an **action potential**.

18. An action potential at the axon terminal causes release of neurotransmitter by the presynaptic cell and graded potentials in the postsynaptic cell. This entire process is called **synaptic activity**.

**Module 11.7**

**Differences in electrochemical gradients determine the resting membrane potential p. 414**

19. Passive **leak channels** allow the movement of $Na^+$ into the cell and $K^+$ out of the cell.

20. The **sodium-potassium exchange pump**, an active ATP-requiring process, ejects 3 $Na^+$ for every 2 $K^+$.

21. Potassium ion gradients force $K^+$ out of the cell, and sodium ion gradients drive $Na^+$ into the cell.

22. The resulting resting membrane potential for a neuron is approximately −70 mV.

**Module 11.8**

**Three types of gated ion channels change the permeability of the plasma membrane p. 416**

23. Resting membrane potential is stable until the cell is disturbed. When disturbed, the cell undergoes ion permeability changes due to **gated ion channels** within the plasma membrane.

24. **Chemically gated ion channels**, or **ligand-gated ion channels**, open when specific chemicals such as ACh bind to them.

11

25. **Voltage-gated ion channels** open or close in response to changes in membrane potential and are characteristic of excitable membranes. These membranes are capable of generating and propagating action potentials.

26. **Mechanically gated ion channels** open in response to physical distortion of the membrane surface. They are found in sensory receptors that respond to stretch, pressure, or vibration and cause sense of touch and hearing.

## Module 11.9

### Graded potentials are localized changes in the membrane potential   p. 418

27. **Graded potentials** cannot spread far from the site of stimulation.

28. Any shift from resting membrane potential toward a more positive value is called **depolarization**.

29. A return to normal resting membrane potential is called **repolarization** and results when excess Na⁺ leaves the cell.

30. **Hyperpolarization** is a shift in membrane potential more negative than resting membrane potential.

## Module 11.10

### Action potentials are all-or-none events for communication that begin with membrane potential reversal   p. 420

31. A graded potential that reaches **threshold** ($-60$ mV) initiates an **action potential**.

32. Once threshold is reached, voltage-gated Na⁺ channels open, Na⁺ enters the cell, and membrane potential rises to +30 mV.

33. Repolarization begins when Na⁺ channels close and then voltage-gated K⁺ channels open, allowing K⁺ to move out of the cell.

34. Repolarization continues as K⁺ channels close and both Na⁺ and K⁺ channels return to their normal state.

35. During the **absolute refractory period**, the membrane cannot respond to further stimulation. During the **relative refractory period**, the membrane can respond only to a larger-than-normal stimulus.

## Module 11.11

### Action potentials may affect adjacent portions of the plasma membrane through continuous propagation or saltatory propagation   p. 422

36. In **continuous propagation**, the action potential moves in a series of tiny steps along an unmyelinated axon.

37. In **saltatory propagation**, depolarization occurs only at the nodes (nodes of Ranvier) of a myelinated axon. Action potentials travel much faster in saltatory propagation than in continuous propagation.

38. Propagation speed along myelinated axons also depends on axon diameter. The larger the diameter, the lower the resistance to ion movement, and the faster the action potential travels.

## Module 11.12

### At a synapse, information travels from the presynaptic cell to the postsynaptic cell   p. 424

39. **Chemical synapses**, the most abundant type of synapse, release neurotransmitter into the synaptic cleft. **Cholinergic synapses** release ACh.

40. When an action potential depolarizes an axon terminal, calcium ions enter the cell, and ACh is released through exocytosis of synaptic vesicles from the presynaptic cell.

41. ACh binds to chemically gated Na⁺ channels on the postsynaptic membrane, producing a graded depolarization.

42. Depolarization ends as ACh is broken down into acetate and choline by AChE.

43. The axon terminal reabsorbs choline from the synaptic cleft and uses it to synthesize new ACh molecules.

44. In **electrical synapses**, the presynaptic and postsynaptic membranes of neurons are locked together by gap junctions. These structures allow for changes in membrane potential to be transferred from one cell to the other. These rare synapses are located in both the PNS and CNS.

## Module 11.13

### Postsynaptic potentials are responsible for information processing in a neuron   p. 426

45. An **excitatory postsynaptic potential**, or **EPSP**, is a graded depolarization that shifts membrane potential toward threshold.

46. When the membrane potential shifts toward threshold, the membrane is **facilitated**. The larger the facilitation, the smaller the additional stimulus needed to trigger an action potential.

47. An **inhibitory postsynaptic potential**, or **IPSP**, is a graded depolarization that opens chemically gated K⁺ channels resulting in hyperpolarization and an **inhibited** membrane.

48. **Summation** is the integration of EPSP and IPSP. **Temporal summation** occurs when a single synapse is active repeatedly. **Spatial summation** involves multiple synapses that are active simultaneously.

## Module 11.14

### Information processing involves interacting groups of neurons and various neurotransmitters that encode information as action potential frequency   p. 428

*A single neuron may receive information across thousands of synapses.*

49. Neurotransmitter release at a synapse may be either excitatory or inhibitory, depending on the summation of the stimuli, extracellular chemicals other than neurotransmitters, and other complex factors.

50. There are over 100 different neurotransmitters that each work in different ways.

# Chapter Review Questions

**Labeling:** Label the structures in the following diagram.

| 1 | |
| 2 | |
| 3 | |
| 4 | |
| 5 | |
| 6 | |
| 7 | |
| 8 | |

**True/False:** Indicate whether each statement is true or false.

9   Somatic sensory receptors monitor internal organs.

10  Synaptic vesicles contain neurotransmitters.

11  Microglia maintain the blood–brain barrier.

12  Schwann cells form the neurilemma.

13  The resting membrane potential for a neuron is near −70 mV.

| 9 | _____ |
| 10 | _____ |
| 11 | _____ |
| 12 | _____ |
| 13 | _____ |

**Matching:** Match each lettered term with the most closely related description.

a. relative refractory period

b. voltage-gated ion channel

c. oligodendrocyte

d. chemically gated ion channel

e. mechanically gated ion channel

f. Schwann cell

g. absolute refractory period

h. astrocyte

14  Produces myelin in the CNS

15  Opens in response to physical distortion

16  A time when an excitable membrane can respond only to a larger-than-normal stimulus

17  Opens or closes in response to changes in membrane potential

18  Produces myelin in the PNS

19  A time when an excitable membrane cannot respond to further stimulation

20  Maintains the blood–brain barrier

21  Opens in response to neurotransmitters

| 14 | _____ |
| 15 | _____ |
| 16 | _____ |
| 17 | _____ |
| 18 | _____ |
| 19 | _____ |
| 20 | _____ |
| 21 | _____ |

**Multiple choice:** Select the correct answer from the list provided.

**22** In the CNS, a neuron typically receives information from other neurons at its
- a) axon.
- b) Nissl bodies.
- c) dendrites.
- d) nucleus.

**23** The neural cells between sensory neurons and motor neurons are
- a) neuroglia.
- b) interneurons.
- c) sensory ganglia.
- d) autonomic ganglia.

**24** Any shift from resting membrane potential toward a more positive value is called
- a) repolarization.
- b) depolarization.
- c) membrane potential.
- d) hyperpolarization.

**25** Which of the following statements is true regarding plasma membrane $Na^+$ and $K^+$ leak channels?
- a) $Na^+$ passively moves through leak channels to exit cell.
- b) $Na^+$ actively moves through leak channels to exit cell.
- c) $K^+$ passively moves through leak channels to exit cell.
- d) $K^+$ actively moves through leak channels to exit cell.

**26** Receptors that bind to ACh at the postsynaptic membrane are
- a) voltage-gated channels.
- b) mechanically gated channels.
- c) passive channels.
- d) chemically gated channels.

**27** If the resting membrane potential of a neuron is $-70$ mV, and threshold is $-60$ mV, a membrane potential of $-55$ mV
- a) will produce an action potential.
- b) will hyperpolarize the membrane.
- c) is referred to as the absolute refractory period.
- d) will cause an IPSP.

**28** Repolarization does *not* include
- a) closing of voltage-gated $Na^+$ channels.
- b) opening of voltage-gated $Na^+$ channels.
- c) closing of voltage-gated $K^+$ channels.
- d) return of $Na^+$ and $K^+$ channels to their normal states.

**Short answer**

**29** What are the major components of the central nervous system (CNS) and the peripheral nervous system (PNS)?

**30** What three functional classes of neurons are found in the nervous system? What is the function of each type of neuron?

**31** Distinguish between continuous propagation and saltatory propagation of action potentials.

**32** How does a graded potential differ from an action potential?

**33** What is the difference between temporal summation and spatial summation of EPSPs?

MasteringA&P™

Access more chapter study tools online in the MasteringA&P Study Area:

- Chapter Quizzes, Chapter Practice Test, and Clinical Case Studies

- Practice Anatomy Lab™ 3.0 — PAL™3.0
- Interactive Physiology™ — iP2™
- A&P Flix™ — A&PFlix™
- PhysioEx™ 9.1 — PhysioEx™ 9.1

# Chapter Integration • Applying what you have learned

## Multiple sclerosis is a progressive, debilitating, demyelinating disease

Multiple sclerosis (MS) is a progressive, debilitating autoimmune disease in which the body's immune system attacks myelinated portions of the central nervous system, leading to demyelination of affected axons. The disease is so named because scleroses— also known as scars, plaques, or lesions—form in many places within myelinated regions (white matter). The age at onset is most commonly between 20 and 40 years. The cause of the disease is unknown but may involve some combination of environmental agents, genetic factors, and viral infections. Common signs and symptoms include partial loss of vision and problems with speech, balance, and general motor coordination, including loss of bowel and urinary bladder control. The incidence among women is about twice that of men. Individuals with MS experience unpredictable, recurrent cycles of deterioration, remission, and relapse. There is no cure for MS, although drugs that alter the sensitivity or responses of the immune system can slow the progression of the disease.

**1** Define *demyelination*.

**2** Why would individuals with MS experience generalized motor coordination dysfunction?

**3** Which glial cells are affected in MS?

## LEARNING OUTCOMES

These Learning Outcomes correspond by number to this chapter's modules and indicate what you should be able to do after completing the chapter. They also appear at the end of each module.

# 12

# The Spinal Cord, Spinal Nerves, and Spinal Reflexes

## Module 12.1

# The spinal cord can function independently from the brain

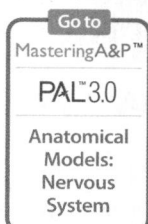

Organization is usually the key to success in any complex environment. In a large corporation, for example, only the most important problems reach the desk of the president. The nervous system works in much the same way. We are consciously aware of only a small fraction of the day-to-day activities monitored by sensory receptors and controlled by motor neurons. Throughout our lives, input pathways are routing sensations, processing centers are prioritizing and distributing information, and motor centers are directing responses to stimuli, most often outside of our awareness. This is possible only because the nervous system is so highly organized. Because our primary interest here is how the nervous system functions, we will consider the system from a functional perspective. Our approach is represented in the diagram below.

**? REVIEW**

**A.** Describe the direction of sensory input and motor commands relative to the spinal cord.

**Chapter 13: The Brain, Cranial Nerves, and Sensory and Motor Pathways**

Sensory receptors → Sensory input over cranial nerves → Reflex centers in brain → Motor output over cranial nerves → **Effectors**

Muscles

**Chapter 12: The Spinal Cord, Spinal Nerves, and Spinal Reflexes**

Sensory receptors → Sensory input over spinal nerves → Reflex centers in spinal cord → Motor output over spinal nerves

Glands

Adipose tissue

This chapter considers the spinal cord, spinal nerves, and spinal reflexes. We will begin with the **spinal cord**, the simplest part of the CNS, because the functional and structural relationships are relatively easy to understand. A **reflex** is a rapid, automatic response triggered by specific stimuli. **Spinal reflexes** are controlled in the spinal cord. They can function without any input from the brain. Chapter 13 examines the brain, cranial nerves, cranial reflexes, and the major pathways that interconnect the brain and spinal cord.

**? REVIEW**

**B.** What is a reflex?

**C.** Define *spinal reflex*.

**LEARNING OUTCOME**

Describe how the spinal cord can function without input from the brain.

# The spinal cord has 31 segments with 31 pairs of nerves

The adult spinal cord is approximately 45 cm (18 in.) long, with a maximum width of roughly 14 mm (0.55 in.). Note that the cord itself is not as long as the vertebral column—instead, the adult spinal cord ends between vertebrae $L_1$ and $L_2$. In sectional view, it has an outer layer of *white matter* and an inner layer of *gray matter* surrounding a small central canal. The amount of gray matter is greatest in segments of the spinal cord dedicated to the sensory and motor control of the limbs. These segments are expanded, forming the cervical and thoracic enlargements of the spinal cord.

**1** *Spinal nerves* are bundles of axons on the PNS that are connected to the spinal cord (see Module 11.3). There are 31 pairs of spinal nerves, each identified by its association with adjacent vertebrae. Spinal nerves in the cervical region are named by the adjacent inferior vertebrae until $C_8$. The first pair of cervical spinal nerves, $C_1$, passes between the skull and the first cervical vertebra. Cervical nerve $C_2$ precedes vertebra $C_2$, and the same naming system is used through $C_7$. A transition in naming occurs between the last (seventh) cervical vertebra and the first thoracic vertebra. The spinal nerves at this location are designated $C_8$. Therefore, although there are only 7 cervical vertebrae, there are 8 pairs of cervical spinal nerves. The remaining thoracic, lumbar, and sacral spinal nerves are named by the adjacent superior vertebrae. For example, spinal nerve $T_1$ emerges immediately inferior to vertebra $T_1$, spinal nerve $T_2$ follows vertebra $T_2$, and the pattern continues.

**KEY**

Spinal cord regions

☐ = Cervical
☐ = Thoracic
☐ = Lumbar
☐ = Sacral

**? REVIEW**

**A.** A typical spinal cord has how many pairs of spinal nerves, and where does the spinal cord end?

**B.** Name the pair of spinal nerves that emerges between vertebrae $C_7$ and $T_1$.

Cervical spinal nerves — $C_1$, $C_2$, $C_3$, $C_4$, $C_5$, $C_6$, $C_7$, $C_8$

$T_1$, $T_2$, $T_3$, $T_4$, $T_5$, $T_6$, $T_7$, $T_8$

Thoracic spinal nerves — $T_9$, $T_{10}$, $T_{11}$, $T_{12}$

$L_1$, $L_2$

Lumbar spinal nerves — $L_3$, $L_4$, $L_5$

Sacral spinal nerves

Coccygeal nerve ($Co_1$)

The **cervical enlargement** innervates (supplies nerves to) the shoulder and upper limbs.

Posterior median sulcus

The **lumbosacral enlargement** innervates structures of the pelvis and lower limbs.

The **conus medullaris** is the tapered, conical portion of the spinal cord inferior to the lumbar enlargement.

Inferior tip of spinal cord

Because the adult spinal cord ends at the level of the first or second lumbar vertebra, the posterior and anterior roots of spinal segments $L_2$ to $S_5$ extend inferiorly. When seen in gross dissection, the filum terminale and the long anterior and posterior roots resemble a horse's tail. Hence, this complex is called the **cauda equina** (KAW-duh eh-KWĪ-nuh; *cauda*, tail + *equinus*, horse).

The **filum terminale** ("terminal thread") is a slender strand of fibrous tissue that extends from the inferior tip of the conus medullaris to the second sacral vertebra. It provides longitudinal support to the spinal cord as a component of the coccygeal ligament.

**2** The spinal cord is divided into 31 segments, each of which gives rise to a pair of spinal nerves. There are 8 cervical, 12 thoracic, 5 lumbar, 5 sacral, and 1 coccygeal spinal segments.

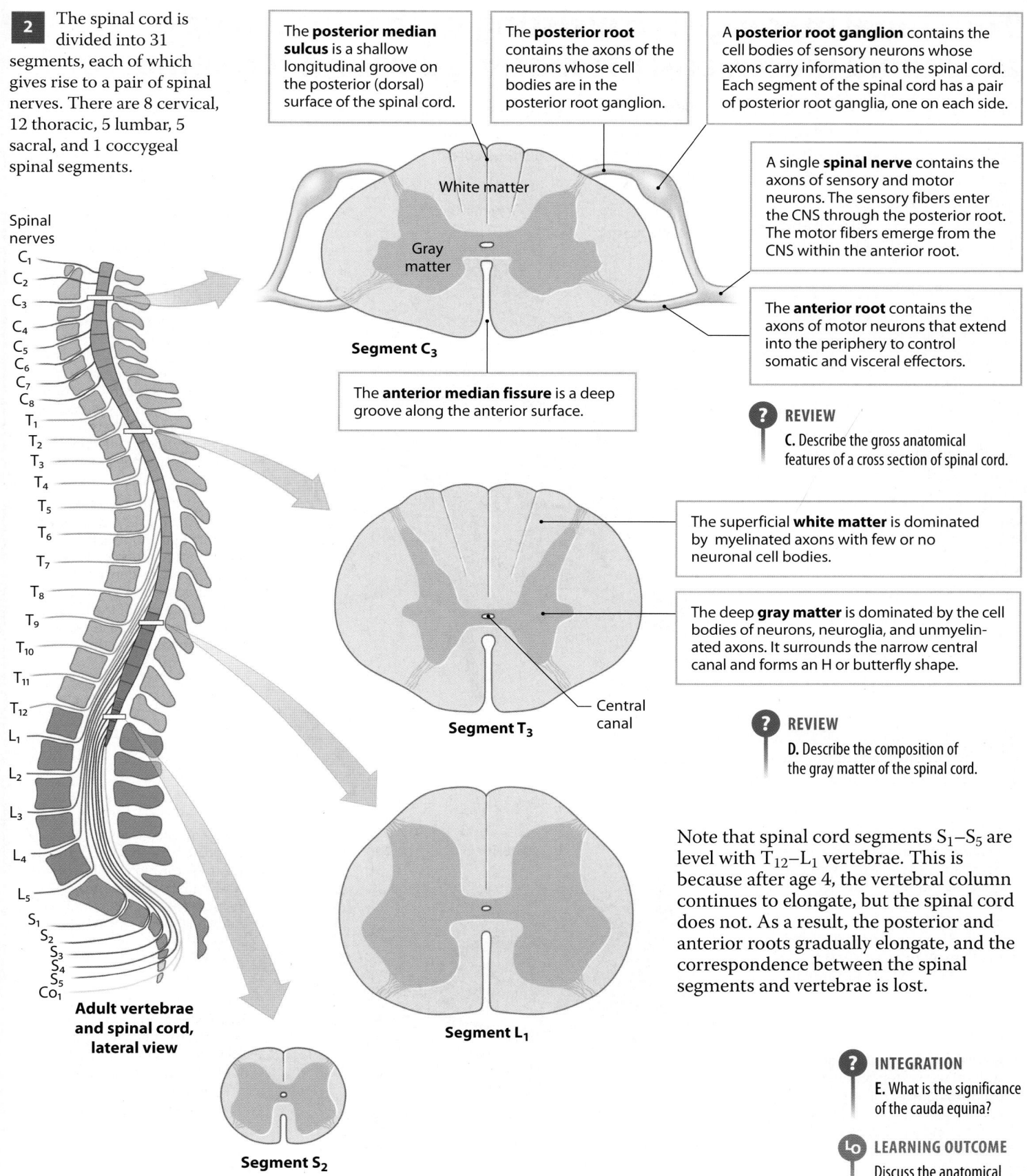

The **posterior median sulcus** is a shallow longitudinal groove on the posterior (dorsal) surface of the spinal cord.

The **posterior root** contains the axons of the neurons whose cell bodies are in the posterior root ganglion.

A **posterior root ganglion** contains the cell bodies of sensory neurons whose axons carry information to the spinal cord. Each segment of the spinal cord has a pair of posterior root ganglia, one on each side.

A single **spinal nerve** contains the axons of sensory and motor neurons. The sensory fibers enter the CNS through the posterior root. The motor fibers emerge from the CNS within the anterior root.

The **anterior root** contains the axons of motor neurons that extend into the periphery to control somatic and visceral effectors.

White matter

Gray matter

**Segment C₃**

The **anterior median fissure** is a deep groove along the anterior surface.

**Spinal nerves**

C₁, C₂, C₃, C₄, C₅, C₆, C₇, C₈, T₁, T₂, T₃, T₄, T₅, T₆, T₇, T₈, T₉, T₁₀, T₁₁, T₁₂, L₁, L₂, L₃, L₄, L₅, S₁, S₂, S₃, S₄, S₅, Co₁

**Adult vertebrae and spinal cord, lateral view**

**Segment T₃**

Central canal

The superficial **white matter** is dominated by myelinated axons with few or no neuronal cell bodies.

The deep **gray matter** is dominated by the cell bodies of neurons, neuroglia, and unmyelinated axons. It surrounds the narrow central canal and forms an H or butterfly shape.

**? REVIEW**
**C.** Describe the gross anatomical features of a cross section of spinal cord.

**12**

**? REVIEW**
**D.** Describe the composition of the gray matter of the spinal cord.

Note that spinal cord segments S₁–S₅ are level with T₁₂–L₁ vertebrae. This is because after age 4, the vertebral column continues to elongate, but the spinal cord does not. As a result, the posterior and anterior roots gradually elongate, and the correspondence between the spinal segments and vertebrae is lost.

**Segment L₁**

**? INTEGRATION**
**E.** What is the significance of the cauda equina?

**LO LEARNING OUTCOME**
Discuss the anatomical features of the spinal cord.

**Segment S₂**

# The spinal meninges, consisting of the dura mater, arachnoid mater, and pia mater, surround the spinal cord

Nervous tissues must be protected from shocks, including damaging contact with the surrounding bony walls of the vertebral canal. The **spinal meninges** (me-NIN-jēz; singular, *meninx*, membrane), a series of specialized membranes surrounding the spinal cord, provide the necessary physical stability and shock absorption. Blood vessels branching within these layers deliver oxygen and nutrients to the spinal cord. The spinal meninges consist of three layers: (1) the dura mater, (2) the arachnoid mater, and (3) the pia mater. At the foramen magnum of the skull, the spinal meninges are continuous with the cranial meninges, which surround the brain.

**1** The basic relationships among the spinal meninges can be seen in this posterior view of the dissected spinal cord.

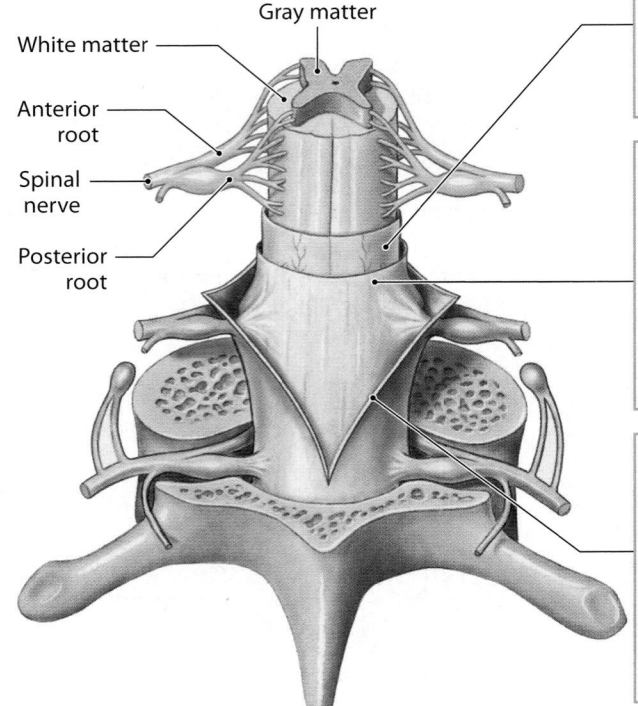

White matter

Gray matter

Anterior root

Spinal nerve

Posterior root

The **pia mater** (PĒ-uh MĀ-ter; *pia*, delicate + *mater*, mother) consists of a meshwork of elastic and collagen fibers that is firmly bound to the underlying neural tissue.

The **arachnoid** (a-RAK-noyd; *arachne*, spider) **mater** is the middle meningeal layer. It includes a simple squamous epithelium, called the *arachnoid membrane*, and the *subarachnoid space* that extends between the arachnoid membrane and the outer surface of the pia mater.

The tough, fibrous **dura mater** (DŪ-ruh; *dura*, hard) is the outermost covering of the spinal cord. It contains dense collagen fibers that are oriented along the longitudinal axis of the cord. A narrow subdural space separates the dura mater from the arachnoid mater.

**2** A cross-sectional view provides additional information about the surrounding structures and the spaces between the meningeal layers.

The **subarachnoid space** contains the *arachnoid trabeculae*, a network of collagen and elastic fibers that attaches the arachnoid mater to the pia mater. It is filled with **cerebrospinal fluid (CSF)**, which acts as a shock absorber and a diffusion medium for dissolved gases, nutrients, chemical messengers, and wastes.

Between the dura mater and the walls of the vertebral canal lies the **epidural space**. This region contains areolar tissue, blood vessels, and a protective padding of adipose tissue.

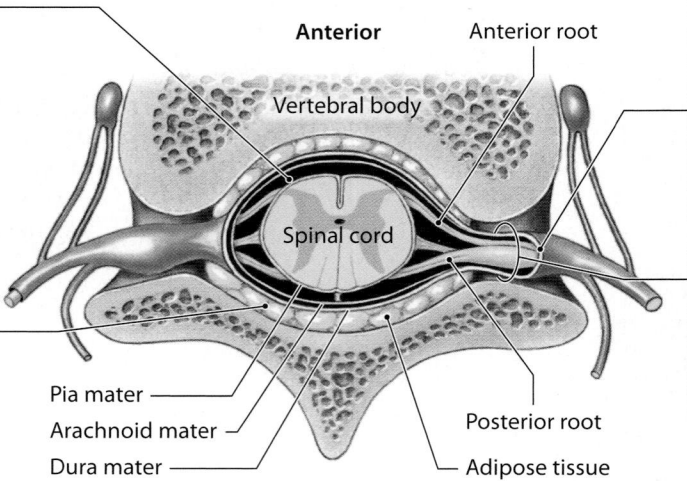

**Anterior**

Anterior root

Vertebral body

Spinal cord

Pia mater

Arachnoid mater

Dura mater

Posterior root

Adipose tissue

The spinal meninges accompany the posterior and anterior roots as they pass through the intervertebral foramina. The meningeal membranes are continuous with the connective tissues that surround the spinal nerves and their peripheral branches.

The posterior root ganglion lies between the pedicles of the adjacent vertebrae. The spinal nerve passes through the intervertebral foramen.

**?** **REVIEW**

**A.** Where is spinal CSF located?

**3** This anterior view of the cervical spinal cord shows the actual appearance of the meninges, supporting ligaments, and the roots of the spinal nerves.

Along the length of the spinal cord, paired **denticulate ligaments** extend from the pia mater through the arachnoid mater to the dura mater. Denticulate ligaments prevent lateral movement of the spinal cord. Together, the dural connections at the foramen magnum and the coccygeal ligament at the sacrum prevent superior–inferior movement of the spinal cord.

The blood vessels servicing the spinal cord run along the surface of the spinal pia mater, within the subarachnoid space.

Spinal cord

Anterior median fissure

Pia mater

Spinal nerve roots divide like a fan, forming **rootlets**

Posterior root

Anterior root, formed by several rootlets from one cervical segment

Arachnoid mater (cut and reflected back)

Dura mater (cut and reflected back)

**4** This is an x-ray of the inferior lumbar vertebrae. The vertebral canal inferior to vertebra $L_1$ contains the cauda equina bathed in the CSF of the subarachnoid space.

$L_2$
$L_3$
$L_4$
$L_5$

Cauda equina

**5** In adults, cerebrospinal fluid can be safely withdrawn in a procedure known as a **lumbar puncture**, or *spinal tap*. A needle is inserted into the subarachnoid space in the lumbar region inferior to the tip of the conus medullaris.

Dura mater
Epidural space

Body of third lumbar vertebra

Interspinous ligament

Lumbar puncture needle with tip in subarachnoid space

Cauda equina in subarachnoid space

**?** REVIEW

**B.** Name the structures and spinal coverings that are penetrated during a lumbar puncture procedure.

 **LEARNING OUTCOME**

Describe the three meningeal layers that surround the spinal cord.

12

# Module 12.4

# Gray matter integrates sensory and motor functions, and white matter carries information

**1** This micrograph shows the major landmarks in and around the spinal cord. Compare this cross section with the diagrammatic sectional views in this module.

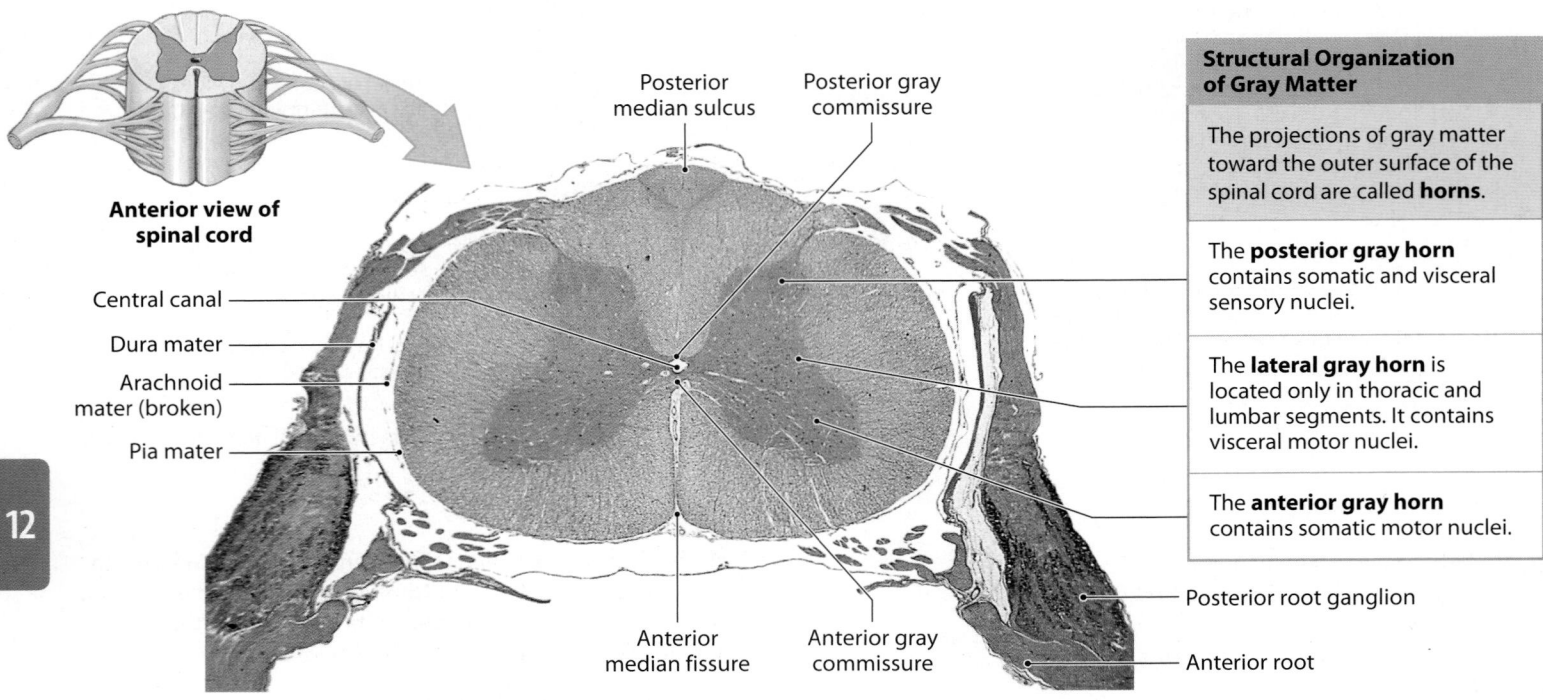

**Anterior view of spinal cord**

Central canal
Dura mater
Arachnoid mater (broken)
Pia mater

Posterior median sulcus
Posterior gray commissure

Anterior median fissure
Anterior gray commissure

### Structural Organization of Gray Matter

The projections of gray matter toward the outer surface of the spinal cord are called **horns**.

The **posterior gray horn** contains somatic and visceral sensory nuclei.

The **lateral gray horn** is located only in thoracic and lumbar segments. It contains visceral motor nuclei.

The **anterior gray horn** contains somatic motor nuclei.

Posterior root ganglion

Anterior root

**2** This diagrammatic view focuses on the organization of the gray matter of the spinal cord.

A frontal section along the length of the central canal (dashed line) of the spinal cord separates the sensory (posterior, or dorsal) nuclei from the motor (anterior) nuclei.

The **gray commissures** (*commissura*, a joining together) are posterior and anterior to the central canal. They contain axons that cross from one side of the cord to the other before they reach a destination in the gray matter.

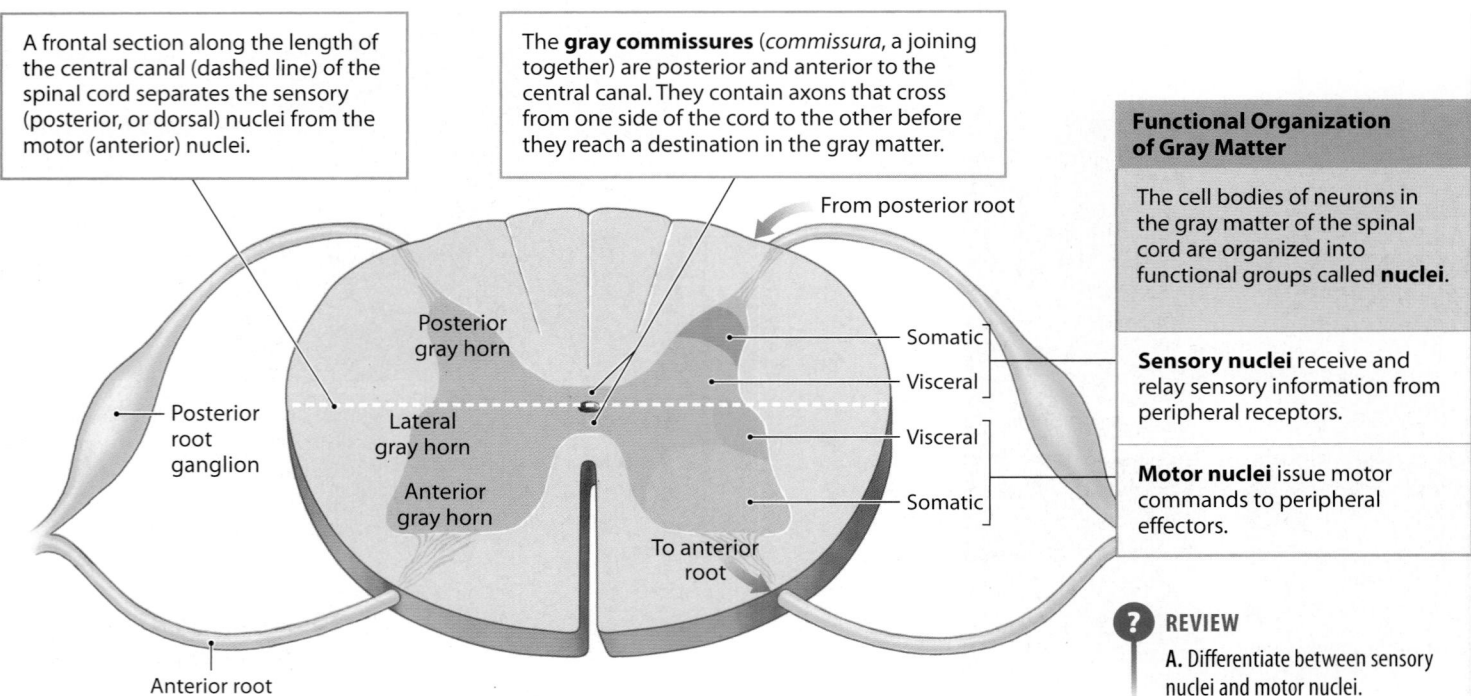

From posterior root

Posterior gray horn

Posterior root ganglion

Lateral gray horn

Anterior gray horn

Somatic
Visceral

Visceral

Somatic

To anterior root

Anterior root

### Functional Organization of Gray Matter

The cell bodies of neurons in the gray matter of the spinal cord are organized into functional groups called **nuclei**.

**Sensory nuclei** receive and relay sensory information from peripheral receptors.

**Motor nuclei** issue motor commands to peripheral effectors.

**? REVIEW**

**A.** Differentiate between sensory nuclei and motor nuclei.

**3** Like the gray horns, the white matter is organized according to the region of the body innervated. The white matter on each side of the spinal cord is divided into three regions called **columns**. These columns contain tracts. A **tract** is a bundle of axons in the CNS that is relatively uniform with respect to diameter, myelination, and propagation speed. All the axons within a tract relay the same type of information (sensory or motor) in the same direction. **Ascending tracts** carry sensory information toward the brain, and **descending tracts** convey motor commands to the spinal cord (discussed in Chapter 13).

### Organization of Tracts in the Posterior White Column

The posterior white column contains ascending tracts providing sensations from the trunk and limbs.

Leg
Hip
Trunk
Arm

### Structural and Functional Organization of White Matter

The **posterior white column** lies between the posterior gray horns and the posterior median sulcus.

The **lateral white column** includes the white matter on either side of the spinal cord, between the anterior and posterior columns.

The **anterior white column** lies between the anterior gray horns and the anterior median fissure.

Flexors/Extensors

The **anterior white commissure** interconnects the anterior white columns. This is where axons cross from one side of the spinal cord to the other.

Trunk  Shoulder  Arm  Forearm  Hand

In the cervical enlargement, which contains neurons involved with sensations and motor control of the upper limbs, the motor nuclei of the anterior gray horn are grouped by region. The motor neurons controlling flexor muscles are medial to those controlling extensor muscles.

 **Clinical Note**

## Gray matter organization can help localize spinal cord damage or disease

The high degree of organization of the spinal cord is clinically important. Damage to a specific area of gray matter allows clinicians to predict which muscles will be affected and what the cause might be. For example, *acute poliomyelitis* is caused by a virus that selectively damages neurons in the anterior horn, particularly in the cervical and lumbosacral enlargements of the spinal cord. Such damage leads to a characteristic muscle paralysis.

**?** **INTEGRATION**

**B.** A person with polio has lost the use of his leg muscles. In which area of his spinal cord would you expect the virus-infected motor neurons to be?

**C.** A disease that damages myelin sheaths would affect which portion of the spinal cord?

**Lo** **LEARNING OUTCOME**

Explain the roles of gray matter and white matter in processing and relaying sensory information and motor commands.

# Spinal nerves have a similar anatomical structure and distribution pattern

**1** Every segment of the spinal cord is connected to a pair of spinal nerves. Surrounding each spinal nerve is a series of connective tissue layers continuous with those of their associated peripheral nerves. These layers, best seen in sectional view, are comparable with those associated with skeletal muscles.

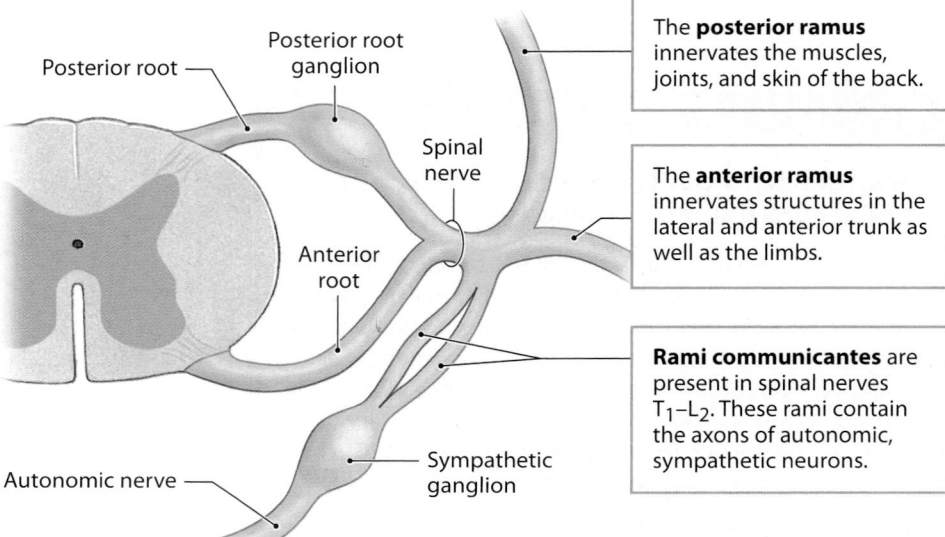

Perimysium

### Connective Tissue Layers of a Spinal Nerve

The **epineurium** (ep-i-NŪ-rē-um), or outermost covering of the nerve, consists of a dense network of collagen fibers.

The fibers of the **perineurium**, the middle layer, extend inward from the epineurium. These connective tissue partitions divide the nerve into a series of compartments that contain bundles of axons called **fascicles**.

The **endoneurium**, the innermost layer, consists of delicate connective tissues that extend from the perineurium and surround individual axons.

Arteries and veins penetrate the epineurium and branch within the perineurium. Capillaries leaving the perineurium branch in the endoneurium and supply the axons and Schwann cells of the nerve and the fibroblasts of the connective tissues.

Fascicle

Schwann cell

Myelinated axon

**2** Each spinal nerve divides to form **rami** (RĀ-mī; singular *ramus*, a branch). Some of these rami carry visceral motor fibers of the autonomic nervous system (ANS). Spinal nerves in the thoracic and upper lumbar segments of the spinal cord carry the motor output of the **sympathetic division** of the ANS (see Module 11.3). This division is responsible for the "fight or flight" response.

Posterior root

Posterior root ganglion

Spinal nerve

Anterior root

Autonomic nerve

Sympathetic ganglion

The **posterior ramus** innervates the muscles, joints, and skin of the back.

The **anterior ramus** innervates structures in the lateral and anterior trunk as well as the limbs.

**Rami communicantes** are present in spinal nerves $T_1$–$L_2$. These rami contain the axons of autonomic, sympathetic neurons.

**? REVIEW**

**A.** Identify the three layers of connective tissue of a spinal nerve, and identify the major peripheral branches of a spinal nerve.

**3** The specific bilateral sensory region of the skin surface monitored by a single pair of spinal nerves is known as a **dermatome**. Each pair of spinal nerves supplies its own dermatome, but the boundaries of adjacent dermatomes overlap to some degree. Spinal nerve $C_1$ typically lacks a sensory branch to the skin; when present, it innervates the scalp with $C_2$ and $C_3$. The face is monitored by the fifth pair of cranial nerves (CN V).

KEY
Spinal cord regions

☐ = Cervical
☐ = Thoracic
☐ = Lumbar
☐ = Sacral

**?** **REVIEW**
**B.** Describe a dermatome.

 **Clinical Note**

## Shingles

The same virus that causes chickenpox, the varicella-zoster virus (VZV), also causes **shingles** (derived from the Latin *cingulum,* girdle). This herpesvirus attacks neurons within the posterior roots of spinal nerves and sensory ganglia of cranial nerves. Shingles produces a painful rash and blisters whose distribution corresponds to that of the affected sensory nerves and follows its dermatome, as shown in the photo. Treatment for shingles typically involves large doses of antiviral drugs. Fortunately for people who are affected, attacks of shingles usually heal, often without treatment, and leave behind only unpleasant memories.

Anyone who has had chickenpox is at risk of developing shingles. After the initial encounter, the virus remains dormant within neurons of the anterior gray horns of the spinal cord. It is not known what triggers the reactivation of this pathogen. Many people who contract shingles suffer just a single episode in their adult lives. Some experience a painful condition known as *postherpetic neuralgia* after a bout of shingles. Moreover, the problem can recur in people with weakened immune systems, including the elderly, pregnant women, and people with AIDS or some forms of cancer. A vaccine to prevent shingles is available for people 50 years of age or older.

**?** **REVIEW**
**C.** Explain the cause of shingles.

**Lo** **LEARNING OUTCOME**
Describe the major components of a spinal nerve.

# Module 12.6

# Each ramus of a spinal nerve provides sensory and motor innervation to a specific region

**1** This diagram illustrates how a spinal nerve collects sensory information from peripheral structures and delivers it to sensory nuclei in the thoracic or superior lumbar segments of the spinal cord. The posterior ramus, anterior ramus, and white ramus communicans of each spinal nerve also contain sensory fibers. Somatic sensory information arrives over the posterior and anterior rami. Visceral sensory information reaches the posterior root through the posterior and anterior rami and white ramus communicans.

## Sensory Information

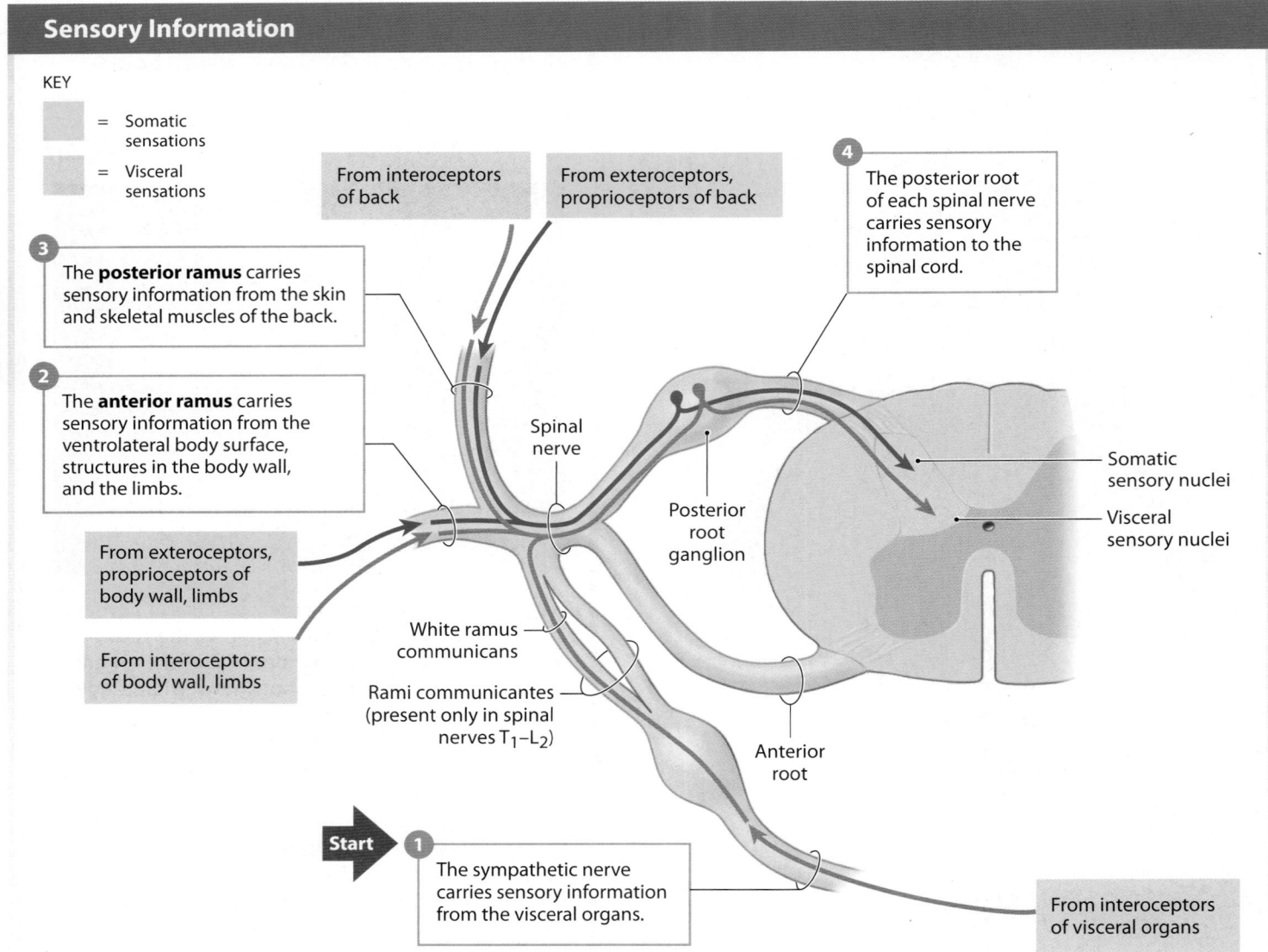

**KEY**

= Somatic sensations

= Visceral sensations

**From interoceptors of back**

**From exteroceptors, proprioceptors of back**

**4** The posterior root of each spinal nerve carries sensory information to the spinal cord.

**3** The **posterior ramus** carries sensory information from the skin and skeletal muscles of the back.

**2** The **anterior ramus** carries sensory information from the ventrolateral body surface, structures in the body wall, and the limbs.

From exteroceptors, proprioceptors of body wall, limbs

From interoceptors of body wall, limbs

Spinal nerve

Posterior root ganglion

White ramus communicans

Rami communicantes (present only in spinal nerves T$_1$–L$_2$)

Anterior root

Somatic sensory nuclei

Visceral sensory nuclei

**Start** **1** The sympathetic nerve carries sensory information from the visceral organs.

From interoceptors of visceral organs

**? REVIEW**

**A.** Which ramus carries sensory information from the limbs?

**B.** Through which root of a spinal nerve does visceral sensory information arrive?

**2** This diagram illustrates how a spinal nerve distributes motor commands that originate in motor nuclei of the thoracic or superior lumbar segments of the spinal cord. (Somatic and visceral motor neurons were introduced in Module 11.3.)

## Motor Commands

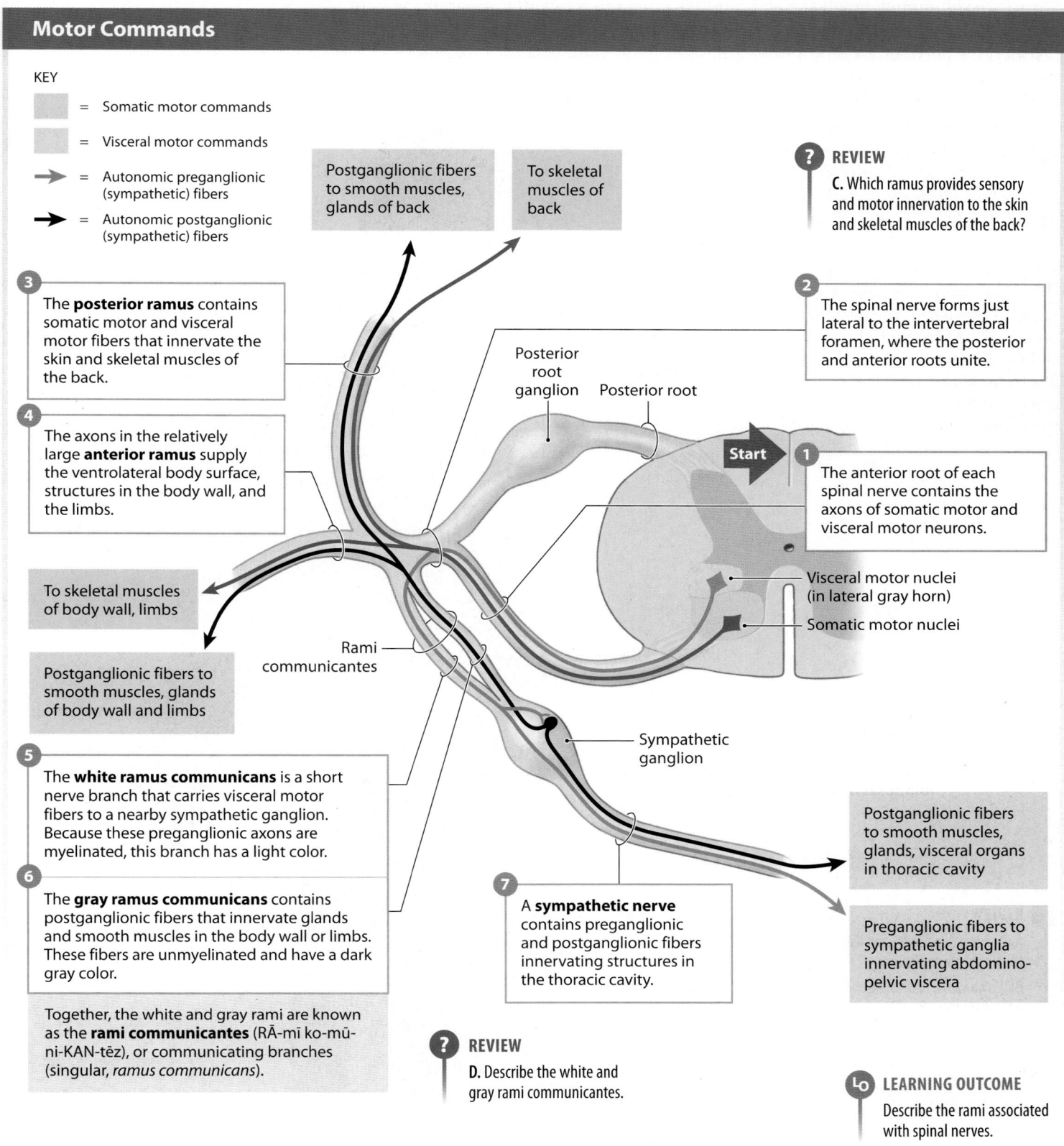

**KEY**

| | |
|---|---|
| = | Somatic motor commands |
| = | Visceral motor commands |
| → = | Autonomic preganglionic (sympathetic) fibers |
| ➤ = | Autonomic postganglionic (sympathetic) fibers |

Postganglionic fibers to smooth muscles, glands of back

To skeletal muscles of back

**3** The **posterior ramus** contains somatic motor and visceral motor fibers that innervate the skin and skeletal muscles of the back.

**4** The axons in the relatively large **anterior ramus** supply the ventrolateral body surface, structures in the body wall, and the limbs.

To skeletal muscles of body wall, limbs

Postganglionic fibers to smooth muscles, glands of body wall and limbs

**5** The **white ramus communicans** is a short nerve branch that carries visceral motor fibers to a nearby sympathetic ganglion. Because these preganglionic axons are myelinated, this branch has a light color.

**6** The **gray ramus communicans** contains postganglionic fibers that innervate glands and smooth muscles in the body wall or limbs. These fibers are unmyelinated and have a dark gray color.

Together, the white and gray rami are known as the **rami communicantes** (RĀ-mī ko-mū-ni-KAN-tēz), or communicating branches (singular, *ramus communicans*).

Rami communicantes

Posterior root ganglion

Posterior root

**Start**

**1**

Sympathetic ganglion

**7** A **sympathetic nerve** contains preganglionic and postganglionic fibers innervating structures in the thoracic cavity.

**REVIEW**
**C.** Which ramus provides sensory and motor innervation to the skin and skeletal muscles of the back?

**2** The spinal nerve forms just lateral to the intervertebral foramen, where the posterior and anterior roots unite.

**1** The anterior root of each spinal nerve contains the axons of somatic motor and visceral motor neurons.

Visceral motor nuclei (in lateral gray horn)

Somatic motor nuclei

Postganglionic fibers to smooth muscles, glands, visceral organs in thoracic cavity

Preganglionic fibers to sympathetic ganglia innervating abdomino-pelvic viscera

**REVIEW**
**D.** Describe the white and gray rami communicantes.

**LO LEARNING OUTCOME**
Describe the rami associated with spinal nerves.

12

# Module 12.7

## Spinal nerves form nerve plexuses that innervate the skin and skeletal muscles

During development, small skeletal muscles innervated by different anterior rami typically fuse to form larger muscles with compound origins. The anatomical distinctions between the component muscles may disappear, but separate anerior rami continue to provide sensory innervation and motor control to each part of the compound muscle. As they converge, the anterior rami of adjacent spinal nerves blend their fibers, producing a series of compound nerve trunks. This complex interwoven network of nerves is called a **nerve plexus** (PLEK-sus; *plexus*, braid).

**1** The anterior rami form four major plexuses: (1) the **cervical plexus**, (2) the **brachial plexus**, (3) the **lumbar plexus**, and (4) the **sacral plexus**. In this illustration, the spinal nerves are labeled on the left, and the major peripheral nerves are on the right.

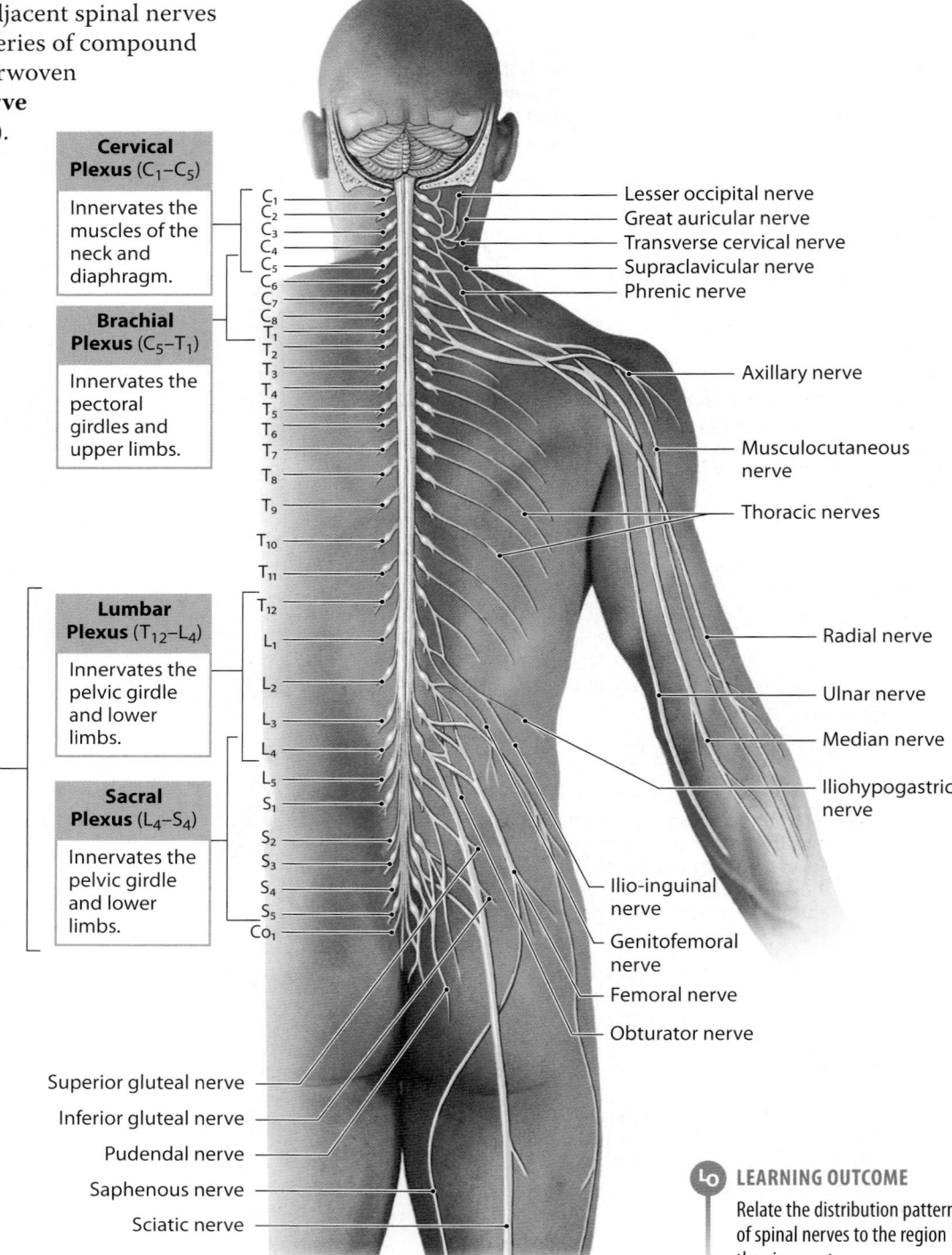

**Cervical Plexus ($C_1$–$C_5$)**

Innervates the muscles of the neck and diaphragm.

**Brachial Plexus ($C_5$–$T_1$)**

Innervates the pectoral girdles and upper limbs.

**Lumbar Plexus ($T_{12}$–$L_4$)**

Innervates the pelvic girdle and lower limbs.

**Sacral Plexus ($L_4$–$S_4$)**

Innervates the pelvic girdle and lower limbs.

The lumbar and sacral plexuses are often referred to as the **lumbosacral plexus**.

Lesser occipital nerve
Great auricular nerve
Transverse cervical nerve
Supraclavicular nerve
Phrenic nerve

Axillary nerve

Musculocutaneous nerve

Thoracic nerves

Radial nerve

Ulnar nerve

Median nerve

Iliohypogastric nerve

Ilio-inguinal nerve

Genitofemoral nerve

Femoral nerve

Obturator nerve

Superior gluteal nerve

Inferior gluteal nerve

Pudendal nerve

Saphenous nerve

Sciatic nerve

**? REVIEW**

**A.** Define *nerve plexus*, and list the major nerve plexuses.

**B.** Which spinal nerves are *not* part of a nerve plexus?

**Lo LEARNING OUTCOME**

Relate the distribution pattern of spinal nerves to the region they innervate.

# Module 12.8

## The cervical plexus innervates the muscles of the neck and diaphragm

**1** The **cervical plexus** consists of the anterior rami of spinal nerves $C_1$–$C_5$. The branches of the cervical plexus innervate the skin and muscles of the neck and extend into the thoracic cavity, where they control the diaphragm, a key muscle of breathing. The **phrenic nerve**, the major nerve of the cervical plexus, provides the entire nerve supply to the diaphragm. Other branches of this nerve plexus are distributed to the skin of the neck and the superior part of the chest.

 **REVIEW**

**A.** Damage to which nerve of the cervical plexus would interfere with the ability to breathe?

**Cranial Nerves***

- Accessory nerve (XI)
- Hypoglossal nerve (XII)

**Nerve Roots of Cervical Plexus**

- $C_1$
- $C_2$
- $C_3$
- $C_4$
- $C_5$

Clavicle

### Nerves of the Cervical Plexus

| Nerve and Distribution | Spinal Segments |
|---|---|
| **Great auricular** | |
| Skin over the posterior aspect of the ear and the neck | $C_2$–$C_3$ |
| **Lesser occipital** | |
| Skin of the neck and the scalp posterior and superior to the ear | $C_2$ |
| **Transverse cervical** | |
| Skin of the anterior triangle of the neck | $C_3$–$C_4$ |
| **Ansa cervicalis** | |
| Five of the extrinsic laryngeal muscles by way of the hypoglossal nerve (XII)* | $C_1$–$C_4$ |
| Geniohyoid | |
| Thyrohyoid | |
| Omohyoid | |
| Sternothyroid | |
| Sternohyoid | |
| **Phrenic** | |
| Diaphragm | $C_3$–$C_5$ |
| **Supraclaviculars** | |
| Skin of the neck and shoulder | $C_3$–$C_4$ |

* The twelve cranial nerves are indicated by Roman numerals (I–XII). We discuss cranial nerves in Module 13.15.

**REVIEW**

**B.** When an anesthetic blocks the function of the anterior rami of the cervical spinal nerves, which areas of the body will be affected?

**LO LEARNING OUTCOME**

Describe the cervical plexus.

# Module 12.9

## The brachial plexus innervates the pectoral girdles and upper limbs

**1** The **brachial plexus** innervates the pectoral girdles and upper limbs, with contributions from the anterior rami of spinal nerves $C_5$–$T_1$.

| **Brachial Plexus: Major Nerves of the Arm and Forearm** | |
|---|---|
| **Spinal Segments*** | **Nerve and Distribution** |
| | **Musculocutaneous** |
| $C_5$–$T_1$ | Flexor muscles on the arm (biceps brachii, brachialis, and coracobrachialis); sensory from skin over the lateral surface of the forearm through the lateral antebrachial cutaneous nerve |
| | **Radial** |
| $C_5$–$T_1$ | Many extensor muscles on the arm and forearm (triceps brachii, anconeus, extensor carpi radialis, extensor carpi ulnaris, and brachioradialis); and abductor pollicis muscle by the deep branch; sensory from skin over the postero-lateral surface of the limb through the posterior brachial cutaneus nerve, posterior antebrachial cutaneous nerve, and the superficial branch (radial half of the hand) |
| | **Median** |
| $C_6$–$T_1$ | Flexor muscles on the forearm (flexor carpi radialis and palmaris longus muscles); pronator quadratus and pronator teres; digital flexors (through the anterior interosseous nerve); sensory from skin over the anterolateral surface of the hand |
| | **Ulnar** |
| $C_8$–$T_1$ | Flexor carpi ulnaris, flexor digitorum profundus, adductor pollicis, and small digital muscles by the deep branch; sensory from skin over medial surface of the hand through the superficial branch |

\* The brachial plexus can also have fibers from $C_4$, $T_2$, or both.

Brachial plexus

Ulnar nerve

Lateral antebrachial cutaneous nerve

Superficial branch of radial nerve

Deep branch of radial nerve

Radial nerve

### Clinical Note

#### Nerve injury can be precisely located in the hand

The distribution of the cutaneous nerves of the wrist and hand is very important in clinical medicine. Nerve damage or injury in this region can be precisely localized by carefully testing the sensory function of the hand.

Radial nerve

Ulnar nerve

Median nerve

Anterior          Posterior

**? REVIEW**

**A.** Describe the brachial plexus.

**2** The nerves that form the brachial plexus originate from trunks and cords. **Trunks** are large bundles of axons from several spinal nerves. The trunks split to form **divisions**, which combine to form **cords**.

**? REVIEW**

**B.** Define a nerve plexus trunk and cord.

| Nerves of the Brachial Plexus | | Cords | Divisions | Trunks | Anterior Rami Roots |
|---|---|---|---|---|---|
| The nerves of the brachial plexus arise from one or more trunks or cords whose names indicate their positions relative to the axillary artery, a large artery supplying the upper limb. | | | | The anterior rami converge to form the superior, middle, and inferior trunks. | The brachial plexus is formed by the anterior rami of spinal nerves $C_5$–$T_1$. |
| **Spinal Segments** | **Nerve and Distribution** | | | | |
| | **Dorsal Scapular** | | | | |
| $C_5$ | Rhomboid and levator scapulae | | | | |
| | **Subclavian** | | | | |
| $C_5$–$C_6$ | Subclavius | | | | |
| | **Suprascapular** | | | | |
| $C_5$–$C_6$ | Supraspinatus and infraspinatus; sensory from shoulder joint and scapula | | | | |
| | **Pectoral (medial and lateral)** | | | | |
| $C_5$–$T_1$ | Pectoralis muscles | | | | |
| | **Subscapular nerves** | | | | |
| $C_5$–$C_6$ | Subscapularis and teres major | | | | |
| | **Thoracodorsal** | | | | |
| $C_6$–$C_8$ | Latissimus dorsi | | | | |
| | **Axillary** | | | | |
| $C_5$–$C_6$ | Deltoid and teres minor | | | | |
| | **Medial Antebrachial Cutaneous** | | | | |
| $C_8$–$T_1$ | Sensory from skin over anterior and medial surface of the arm and forearm | | | | |
| $C_5$–$T_1$ | **Musculocutaneous** | | | | |
| $C_6$–$T_1$ | **Median** | | | | |
| $C_8$–$T_1$ | **Ulnar** | | | | |
| $C_5$–$T_1$ | **Radial** | | | | |
| | **Long Thoracic** | | | | |
| $C_5$–$C_7$ | Serratus anterior | | | | |

Labels on figure: $C_5$, $C_6$, $C_7$, $C_8$, $T_1$, Superior trunk, Middle trunk, Inferior trunk, Lateral cord, Posterior cord, Medial cord, First rib

**12**

KEY
- Anterior rami roots
- Trunks
- Divisions
- Cords
- Peripheral nerves

 **REVIEW**

**C.** Name the major nerves associated with the brachial plexus.

**Lo LEARNING OUTCOME**

Relate the distribution pattern of the brachial plexus to its function.

# Module 12.10

# The lumbar and sacral plexuses innervate the skin and skeletal muscles of the trunk and lower limbs

The **lumbar plexus** and the **sacral plexus** arise from the lumbar and sacral segments of the spinal cord, respectively. The nerves arising at these plexuses innervate the pelvic girdle and lower limbs.

**1** These illustrations show the origins of the spinal nerves of the lumbar and sacral plexuses. Notice that the lumbosacral trunk provides some nerve input from the $L_4$ spinal nerve to the sacral plexus.

| Lumbar Plexus | |
|---|---|
| **Spinal Segments** | **Nerve and Distribution** |
| | **Iliohypogastric** |
| $T_{12}$–$L_1$ | External and internal oblique and transverse abdominis; skin over the inferior abdomen and buttocks |
| | **Ilio-inguinal** |
| $L_1$ | Abdominal muscles (with iliohypogastric nerve); skin over superior, medial thigh and portions of external genitalia |
| | **Genitofemoral** |
| $L_1$–$L_2$ | Skin over anteromedial thigh and portions of external genitalia |
| | **Lateral femoral cutaneous** |
| $L_2$–$L_3$ | Skin over anterior, lateral, and posterior thigh |
| | **Femoral** |
| $L_2$–$L_4$ | Quadriceps, sartorius, pectineus, and iliopsoas; skin of the anteromedial thigh and medial surface of the leg and foot |
| | **Obturator** |
| $L_2$–$L_4$ | Gracilis and adductor magnus, brevis, and longus; skin from the medial surface of the thigh |

**Lumbar plexus, anterior view**

| Sacral Plexus | |
|---|---|
| **Spinal Segments** | **Nerve and Distribution** |
| | **Superior gluteal** |
| $L_4$–$S_2$ | Gluteus minimus, gluteus medius, and tensor fasciae latae |
| | **Inferior gluteal** |
| $L_5$–$S_2$ | Gluteus maximus |
| | **Posterior femoral cutaneous** |
| $S_1$–$S_3$ | Skin over perineum and posterior thigh and leg |
| | **Sciatic** |
| $L_4$–$S_3$ | Semimembranosus, semitendinosus, and adductor magnus; branches into tibial and common fibular nerves |
| | **Pudendal** |
| $S_2$–$S_4$ | Muscles of the perineum; skin of external genitalia; bulbospongiosus and ischiocavernosus |

**Sacral plexus, anterior view**

**?** REVIEW

**A.** Describe the lumbar plexus and sacral plexus.

**2** In this anterior view of the lower trunk and lower limb, you can see the distribution of the nerves of both the lumbar and sacral plexuses.

**3** This posterior view of the lower limb shows the distribution of the nerves of the sacral plexus. The **sciatic (sī-AT-ik) nerve** is the largest and longest nerve in the body.

Iliohypogastric nerve
Ilio-inguinal nerve
Genitofemoral nerve
Lateral femoral cutaneous nerve
Femoral nerve
Obturator nerve
Superior gluteal nerve
Inferior gluteal nerve
Pudendal nerve
Posterior femoral cutaneous nerve (cut)
Sciatic nerve
Saphenous nerve
Common fibular nerve
Superficial fibular nerve
Deep fibular nerve

Superior gluteal nerve
Inferior gluteal nerve
Pudendal nerve
Posterior femoral cutaneous nerve
Sciatic nerve
Tibial nerve
Common fibular nerve
Sural nerve

**Clinical Note**

## Nerve injury can be precisely located in the foot

Just as in the hand, the distribution of the cutaneous nerves in the foot allows for precise location of nerve damage.

Saphenous nerve
Sural nerve
Common fibular nerve

Tibial nerve
Sural nerve
Saphenous nerve

Saphenous nerve
Sural nerve
Tibial nerve
Common fibular nerve

12

**?** **REVIEW**
**B.** List the major nerves of the sacral plexus.

**?** **REVIEW**
**C.** Which nerve divides into the tibial nerve and common fibular nerve?

 **LEARNING OUTCOME**

Relate the distribution patterns of the lumbar plexus and sacral plexus to their functions.

# Section 1 Review
## Functional Organization of the Spinal Cord

**Labeling:** Label each of the structures in the following cross-sectional diagram of the spinal cord.

Label the views (15, 19) and the nerves that innervate the indicated regions of the hands (16–18).

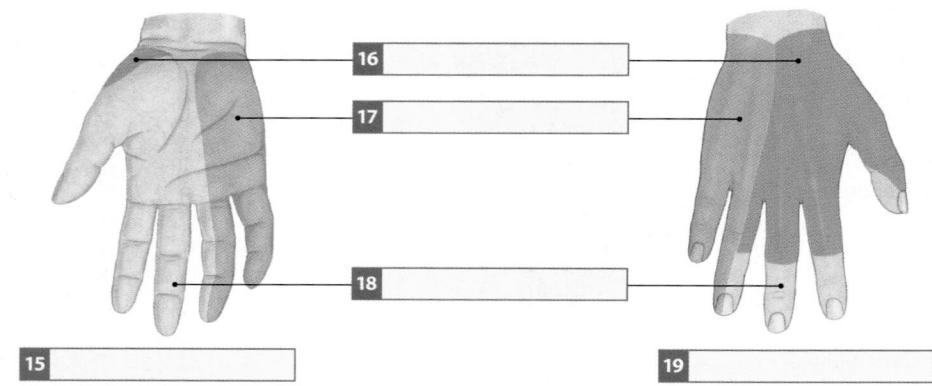

**Vocabulary:** Write the term for each of the following descriptions in the space provided.

20  The regions of white matter in the spinal cord

21  The tapered, conical portion of the spinal cord inferior to the lumbosacral enlargement

22  Bundles of axons in the PNS plus their associated blood vessels and connective tissues

23  Specialized membranes that provide stability and support for the spinal cord and brain

24  The complex made up of the filum terminale and the long posterior and anterior spinal nerve roots inferior to the spinal cord

25  The plexus from which the radial, median, and ulnar nerves originate

26  The outermost covering of the spinal cord

27  The connective tissue partition that separates adjacent bundles of nerve fibers in a spinal nerve or a peripheral nerve

28  A bundle of unmyelinated, postganglionic fibers that innervates glands and smooth muscles in the body wall or limbs

20  _____

21  _____

22  _____

23  _____

24  _____

25  _____

26  _____

27  _____

28  _____

12

# Module 12.11

# CNS neurons are grouped into neuronal pools, which form neural circuits

Your body has about 10 million sensory neurons, one-half million motor neurons, and 20 billion interneurons. The interneurons of the CNS are organized into a much smaller number of **neuronal pools**—functional groups of interconnected neurons. A neuronal pool may involve neurons in one specific location in the spinal cord or brain, or it may involve neurons in several regions of the brain. Estimates of the number of neuronal pools range between a few hundred and a few thousand.

**1** The pattern of interaction among neurons provides clues to the function of a neuronal pool. We refer to the "wiring diagram" as a **neural circuit**, like electrical circuits in the wiring of a house. Here are a few common circuit patterns.

**? REVIEW**

**A.** Differentiate between divergent and convergent neural circuits.

### Divergence

**Divergence** is the spread of information from one neuron to several neurons, or from one pool to multiple pools. Divergence permits the broad distribution of a specific input. Considerable divergence occurs when sensory neurons bring information into the CNS: The information is distributed to neuronal pools throughout the spinal cord and brain.

### Parallel Processing

**Parallel processing** occurs when several neurons or neuronal pools process the same information simultaneously. Divergence must take place before parallel processing can occur. As a result, many responses can occur simultaneously.

### Serial Processing

In **serial processing**, information is relayed in a stepwise fashion, from one neuron to another or from one neuronal pool to the next. This pattern occurs as sensory information is relayed from one part of the brain to another.

### Convergence

In **convergence**, several neurons synapse on a single postsynaptic neuron. Several patterns of activity in the presynaptic neurons can therefore have the same effect on the postsynaptic neuron. Through convergence, the same motor neurons can be subject to both conscious and subconscious control. For example, the movements of your diaphragm and ribs are now being controlled by your brain at the subconscious level. But you can also consciously control the same motor neurons, as when you take a deep breath and hold it.

### Reverberation

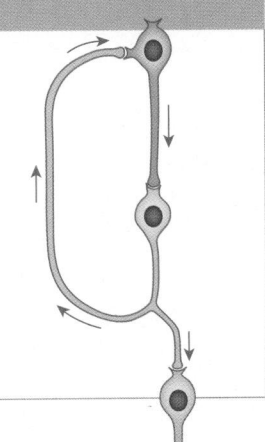

In **reverberation**, collateral branches of axons somewhere along the circuit extend back toward the source of an impulse and further stimulate the presynaptic neurons. Reverberation is like a positive feedback loop involving neurons: Once a reverberating circuit has been activated, it will continue to function until synaptic fatigue or inhibitory stimuli break the cycle.

The most complex neural processing occurs in the brain. The simplest circuits, which occur within the PNS and the spinal cord, control the automatic responses of reflexes. Reflexes are a bit like LEGO® bricks: Individually, they are quite simple, but they can be combined in a great variety of ways to create very complex motor responses (see Module 12.14).

**? REVIEW**

**B.** Which kind of neural circuit processes information in a stepwise fashion, one neuron to another?

**C.** Where does the simplest processing occur?

**Lo LEARNING OUTCOME**

Discuss the significance of neuronal pools, and describe their major patterns of neuron interaction.

# Module 12.12
# Reflexes are vital to homeostasis

**Reflexes** are rapid, automatic responses to specific stimuli. Reflexes preserve homeostasis by making rapid adjustments in the function of organs or organ systems. The response shows little variability: Each time a particular reflex is activated, it usually produces the same motor response. In a reflex, sensory fibers deliver information from peripheral receptors to an integration center in the CNS, and motor fibers carry motor commands to peripheral effectors.

**? REVIEW**
**A.** What are common characteristics of reflexes?

**1** The "wiring" or route of a single reflex is called a **reflex arc**. A reflex arc begins at a receptor and ends at a peripheral effector, such as a muscle fiber or a gland cell. There may or may not be intervening interneurons. This figure shows the five steps in a simple neural reflex.

**Step 2**
**Activation of a Sensory Neuron**
The stimulation of dendrites produces a graded polarization that leads to the formation and propagation of action potentials along the axons of the sensory neurons. This information reaches the spinal cord by way of a posterior root.

**Step 1**
**Stimulation of a Receptor**
A receptor is either a specialized cell or the dendrites of a sensory neuron. Receptors are sensitive to physical or chemical changes in the body or to changes in the external environment. In this example, leaning on a tack stimulates pain receptors in the hand. These receptors respond to stimuli that cause or accompany tissue damage.

**Step 3**
**Information Processing in the CNS**
Information processing begins when the sensory neuron releases excitatory neurotransmitters at the postsynaptic membrane of the interneuron.

Posterior root ganglion

To higher centers in the brain

REFLEX ARC

The neurotransmitter produces an excitatory postsynaptic potential (EPSP), which is integrated with other stimuli arriving at the postsynaptic cell at that moment (see Module 11.13).

Receptor

Stimulus

Effector

**Step 4**
**Activation of a Motor Neuron**
If the information processing leads to activation of the interneuron, motor neurons are stimulated and carry action potentials into the periphery. At the same time, collaterals from the interneuron may relay the pain sensations to other centers in the spinal cord and brain.

**Step 5**
**Response of a Peripheral Effector**
The release of neurotransmitters at axon terminals then leads to a response by a peripheral effector—in this case, a skeletal muscle whose contraction pulls your hand away from the tack. A reflex response generally removes or opposes the original stimulus. In this case, the contracting muscle pulls your hand away from a painful stimulus. This reflex arc is therefore an example of negative feedback.

**KEY**

— Sensory neuron (stimulated)

= Excitatory interneuron

— Motor neuron (stimulated)

**? REVIEW**
**B.** List the components of a reflex arc.

SmartArt

# Reflex Classifications

**2** Neural reflexes can be classified according to (1) their development, (2) the nature of the resulting motor response, (3) the complexity of the neural circuit involved, and (4) the site of information processing. These categories are not mutually exclusive. They represent different ways of describing a single reflex.

## Development

**Innate reflexes** result from the connections that form between neurons during development and are genetically or developmentally programmed. Such reflexes generally appear in a predictable sequence, from the simplest reflex responses (withdrawal from pain) to more complex motor patterns (chewing, suckling, or tracking objects with the eyes).

**Acquired reflexes,** or *conditioned reflexes*, are rapid and automatic, but they were learned rather than preestablished. Such reflexes are enhanced by repetition.

## Nature of the Response

**Somatic reflexes** provide for the involuntary control of skeletal muscles. (The withdrawal reflex at left is one example.) Somatic reflexes are essential because they are immediate. Somatic reflexes provide a rapid response that can later be supplemented by voluntary motor commands.

**Visceral reflexes**, or *autonomic flexes*, control or adjust the activities of smooth muscle, cardiac muscle, glands, and adipose tissues.

## Complexity of the Circuit

A **monosynaptic reflex** is the simplest reflex arc, involving only one synapse in the CNS. The sensory neuron synapses directly on a motor neuron, which serves as the processing center. Transmission across a chemical synapse always involves a synaptic delay, but with only one synapse, the delay between the stimulus and the response is minimized.

**Polysynaptic reflexes** involve at least one interneuron in addition to a sensory neuron and a motor neuron. These reflexes have a longer delay between stimulus and response, with the length of the delay proportional to the number of synapses. Polysynaptic reflexes can produce far more complicated responses than monosynaptic reflexes, because the interneurons can control motor neurons that activate several muscle groups simultaneously.

## Processing Site

In **spinal reflexes**, the important interconnections and processing events occur in nuclei of the spinal cord. In a simple segmental reflex, all of the processing occurs in a single spinal segment. **Intersegmental reflexes** involve multiple segments of the spinal cord, and they can produce very complex and coordinated responses.

In **cranial reflexes**, the important interconnections and processing events occur in nuclei (collections of neuron cell bodies) of the brain.

**? REVIEW**

**C.** Describe the various classifications of neural reflexes.

**12**

**LO LEARNING OUTCOME**

Describe the steps in a reflex.

# The stretch reflex is a monosynaptic reflex involving muscle spindles

**1** The best-known monosynaptic reflex is the **stretch reflex**. It automatically regulates skeletal muscle length. The **patellar**, or knee-jerk, **reflex** shown here is a well-known example.

**? REVIEW**

**A.** Define *stretch reflex*.

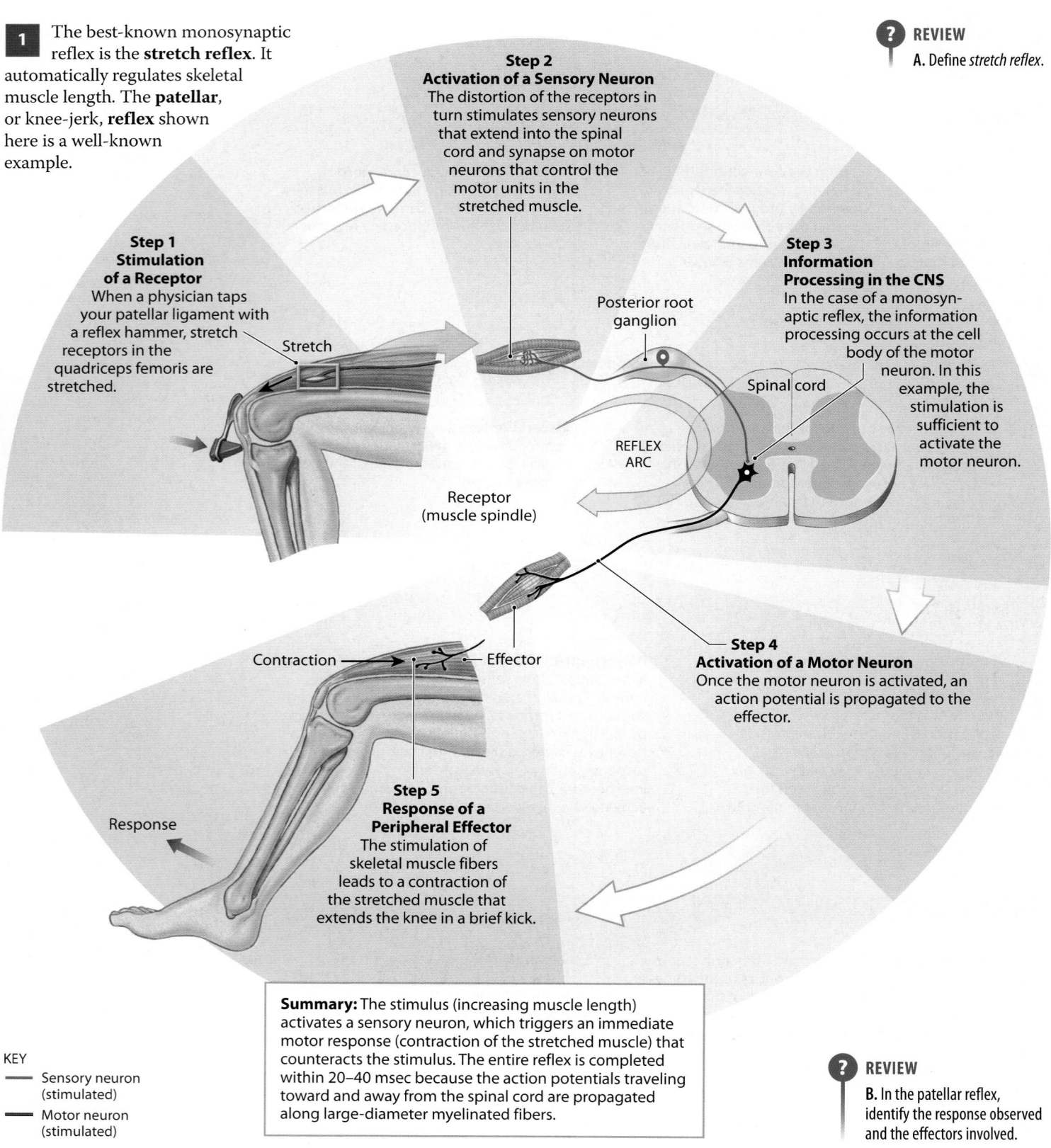

**Step 1**
**Stimulation of a Receptor**
When a physician taps your patellar ligament with a reflex hammer, stretch receptors in the quadriceps femoris are stretched.

Stretch

**Step 2**
**Activation of a Sensory Neuron**
The distortion of the receptors in turn stimulates sensory neurons that extend into the spinal cord and synapse on motor neurons that control the motor units in the stretched muscle.

Posterior root ganglion

**Step 3**
**Information Processing in the CNS**
In the case of a monosynaptic reflex, the information processing occurs at the cell body of the motor neuron. In this example, the stimulation is sufficient to activate the motor neuron.

Spinal cord

REFLEX ARC

Receptor (muscle spindle)

**Step 4**
**Activation of a Motor Neuron**
Once the motor neuron is activated, an action potential is propagated to the effector.

Contraction — Effector

**Step 5**
**Response of a Peripheral Effector**
The stimulation of skeletal muscle fibers leads to a contraction of the stretched muscle that extends the knee in a brief kick.

Response

**Summary:** The stimulus (increasing muscle length) activates a sensory neuron, which triggers an immediate motor response (contraction of the stretched muscle) that counteracts the stimulus. The entire reflex is completed within 20–40 msec because the action potentials traveling toward and away from the spinal cord are propagated along large-diameter myelinated fibers.

KEY
— Sensory neuron (stimulated)
— Motor neuron (stimulated)

**? REVIEW**

**B.** In the patellar reflex, identify the response observed and the effectors involved.

**2** The sensory receptors involved in the stretch reflex are **muscle spindles**. Each muscle spindle consists of a bundle of small, specialized skeletal muscle fibers called **intrafusal muscle fibers**. The muscle spindle is surrounded by the larger muscle fibers of the skeletal muscle responsible for resting muscle tone and, at greater levels of stimulation, for the contraction of the entire muscle. Both sensory and motor neurons innervate each intrafusal fiber.

**?** **REVIEW**

**C.** In the patellar reflex, how does stimulation of the muscle spindle by gamma motor neurons affect sensitivity and reaction time?

Skeletal muscle fibers

Intrafusal fibers

Muscle spindle

Axon of gamma motor neuron in CNS

The sensory neuron is always active, conducting impulses to the CNS. Its axon enters the CNS in a posterior root and synapses on motor neurons in the anterior gray horn of the spinal cord.

The dendrites of the sensory neuron spiral around the intrafusal fibers in a central sensory region.

Each muscle spindle is innervated by **gamma motor neurons** that control its sensitivity by altering tension in the intrafusal fibers. In this way the CNS can increase or decrease muscle tone when movements are under way.

**3** Stretching the central portion of the intrafusal fiber distorts the dendrites and stimulates the sensory neuron. This in turn stimulates the motor neurons, increasing muscle tone. Compressing the central portion inhibits the sensory neuron, which reduces stimulation of the motor neurons, and muscle tone decreases.

| Sensory Region | Action Potential Frequency in Sensory Neuron | Effect on Skeletal Muscle |
|---|---|---|
| Resting length | | Normal muscle tone persists |
| Stretched | | Muscle tone increases |
| Compressed | | Muscle tone decreases |

Many stretch reflexes are **postural reflexes**—reflexes that help us maintain a normal upright posture. Standing, for example, requires cooperation among many muscle groups. Some of these muscles work in opposition to one another, exerting forces that keep the body's weight balanced over the feet. If your body leans forward, stretch receptors in your calf muscles are stimulated. Those muscles then respond by increasing muscle tone, returning your body to an upright position. Postural muscles generally have a firm muscle tone and extremely sensitive stretch receptors. As a result, very fine adjustments are continually being made, and you are not aware of the cycles of contraction and relaxation that occur.

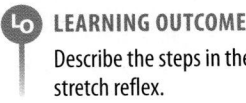

**LEARNING OUTCOME**
Describe the steps in the stretch reflex.

# Module 12.14

# Withdrawal reflexes and crossed extensor reflexes are polysynaptic reflexes

## Withdrawal Reflexes

**Withdrawal reflexes** move affected parts of the body away from a stimulus. The strongest withdrawal reflexes are triggered by painful stimuli, but these reflexes are sometimes initiated by the stimulation of touch receptors or pressure receptors. The **flexor reflex**, a representative withdrawal reflex, affects the muscles of a limb.

 **REVIEW**

**A.** Describe the flexor reflex.

**1** If you accidentally grab a hot pan on the stove, a dramatic flexor reflex will occur. When the pain receptors in your hand are stimulated, the sensory neurons activate interneurons in the spinal cord that stimulate motor neurons in the anterior gray horns. The result is a contraction of flexor muscles that yanks your hand away from the stove.

Distribution within gray horns to other segments of the spinal cord

Posterior root ganglion

Painful stimulus

Flexors stimulated

Extensors inhibited

When a specific muscle contracts, opposing muscles must relax to permit the movement. This is called **reciprocal inhibition**. The term *reciprocal* refers to the fact that the system works both ways. When the flexors contract, the extensors relax; when the extensors contract, the flexors relax.

Withdrawal reflexes show tremendous versatility because the sensory neurons activate many pools of interneurons. The distribution of the effects and the strength and character of the motor responses depend on the intensity and location of the stimulus. Mild discomfort might provoke a brief contraction in muscles of your hand and wrist. More powerful stimuli would produce coordinated muscular contractions affecting the positions of your hand, wrist, forearm, and arm. Severe pain would also stimulate contractions of your shoulder, trunk, and arm muscles. These contractions could persist for several seconds, due to the activation of reverberating circuits.

KEY

——— Sensory neuron (stimulated)

===== Excitatory interneuron

- - - - Inhibitory interneuron

——— Motor neuron (stimulated)

– – – Motor neuron (inhibited)

# Crossed Extensor Reflexes

The stretch reflexes and withdrawal reflexes involve **ipsilateral reflex arcs** (*ipsi*, same + *lateral*, side): The sensory stimulus and the motor response occur on the same side of the body. The **crossed extensor reflex** involves a **contralateral reflex arc** (*contra*, opposite), because an additional motor response occurs on the side opposite the stimulus.

**2** When you step on a tack, the flexor reflex pulls the affected foot away from the ground as excitatory interneurons stimulate the flexor muscles of that limb and inhibitory interneurons relax the extensor muscles that would otherwise oppose that movement. The crossed extensor reflex occurs simultaneously because collaterals of the excitatory and inhibitory interneurons cross to the other side of the spinal cord to motor neurons controlling muscles in the uninjured leg. The excitatory interneurons stimulate motor neurons controlling extensor muscles while the inhibitory interneurons relax the flexor muscles. As a result, your opposite leg straightens to support the shifting weight.

**? REVIEW**
**B.** During a withdrawal reflex of the foot, what happens to the limb on the side opposite the stimulus? What is this response called?

To motor neurons in other segments of the spinal cord

Extensors inhibited

Flexors stimulated

Extensors stimulated

Flexors inhibited

Painful stimulus

KEY

— Sensory neuron (stimulated)

══ Excitatory interneuron

---- Inhibitory interneuron

— Motor neuron (stimulated)

---- Motor neuron (inhibited)

**3** Polysynaptic reflexes are responsible for the automatic actions involved in complex movements such as walking and running. All polysynaptic reflexes share the basic characteristics summarized in this table.

**? REVIEW**
**C.** Identify the basic characteristics of polysynaptic reflexes.

## Properties of Polysynaptic Reflexes

- **They involve pools of interneurons**: Processing takes place in pools of interneurons before motor neurons are activated. The result may be excitation or inhibition.

- **They involve more than one spinal segment**: The interneuron pools extend across spinal segments and may activate muscle groups in many parts of the body.

- **They involve reciprocal inhibition**: Reciprocal inhibition coordinates muscular contractions and reduces resistance to movement.

- **They have reverberating circuits, which prolong the reflexive motor response**: Positive feedback between interneurons that innervate motor neurons and the processing pool maintains the stimulation even after the initial stimulus has faded.

- **Several reflexes may cooperate to produce a coordinated, controlled response**: As a reflex movement gets under way, antagonistic reflexes are inhibited. In complex polysynaptic reflexes, commands may be distributed along the length of the spinal cord, producing a well-coordinated response.

**Lo LEARNING OUTCOME**
Explain withdrawal reflexes and crossed extensor reflexes and the responses produced by each.

# Module 12.15

## The brain can inhibit or facilitate spinal reflexes, and reflexes can be used to determine the location and severity of damage to the CNS

Activities in the brain can have a profound effect on the performance of a reflex by facilitating or inhibiting the motor neurons or interneurons involved. The facilitation of motor neurons involved in reflexes is called **reinforcement**. For example, a method used to overemphasize the patellar reflex is a distractive technique called the Jendrassik maneuver. To do this, the person hooks the hands together by interlocking the fingers and then tries to pull the hands apart while a light tap is applied to the patellar ligament. This reinforcement produces a big kick rather than a twitch. A larger reflex response occurs even if the person realizes it is just a distraction.

**REVIEW**
**A.** Define *reinforcement* as it pertains to spinal reflexes.
**B.** What purpose does reflex testing serve?

**1** The **biceps reflex**, **triceps reflex**, and **ankle-jerk reflex** are stretch reflexes often tested during a physical exam. Each reflex is controlled by specific segments of the spinal cord. As a result, testing these reflexes provides information about the status of the corresponding spinal segments. The table on the right presents information on several reflexes commonly used in physical exams.

**Biceps reflex**

**Triceps reflex**

**Ankle-jerk reflex**

**2** Descending fibers may also have an inhibitory effect on spinal reflexes. Stroking an infant's foot on the lateral side of the sole produces an extension of the hallux and a fanning out of the other toes known as the **Babinski reflex**. This response disappears as descending motor pathways develop because those pathways inhibit this reflex response.

**3** In normal adults, stroking the lateral side of the sole produces a curling of the toes (plantar flexion), called a **plantar reflex** (or negative Babinski reflex). If either the higher centers of the brain or the descending tracts are damaged, the Babinski reflex will reappear in an adult. As a result, this reflex is often tested if CNS injury is suspected.

**Babinski reflex**

**Plantar reflex**

**4** The **abdominal reflex** depends on facilitation, rather than inhibition, of the descending tracts. In this reflex, seen in normal adults, a light stroking of the skin produces a reflexive twitch in the abdominal muscles that moves the navel toward the stimulus. The absence of an abdominal reflex may indicate damage to the descending tracts.

**Abdominal reflex**

## Reflexes Used in Diagnostic Testing

| Reflex | Stimulus | Afferent Nerve(s) | Spinal Segment | Efferent Nerve(s) | Normal Response |
|---|---|---|---|---|---|
| **Superficial Reflexes** | | | | | |
| **Abdominal reflex** | Light stroking of skin of abdomen | $T_7$–$T_{12}$ depending on region stroked | $T_7$–$T_{12}$ at level of arrival | Same as afferent | Contractions of abdominal muscles that pull navel toward the stimulus |
| **Cremasteric reflex** | Stroking of skin of upper thigh | Femoral nerve | $L_1$ | Genitofemoral nerve | Contraction of cremaster, elevation of scrotum |
| **Plantar reflex** | Longitudinal stroking of lateral side of sole of foot | Tibial nerve | $S_1$, $S_2$ | Tibial nerve | Flexion at toe joints |
| **Anal reflex** | Stroking of region around the anus | Pudendal nerve | $S_4$, $S_5$ | Pudendal nerve | Constriction of external anal sphincter |
| **Stretch Reflexes** | | | | | |
| **Biceps reflex** | Tap to tendon of biceps brachii near its insertion | Musculocutaneous nerve | $C_5$, $C_6$ | Musculocutaneous nerve | Flexion at elbow |
| **Triceps reflex** | Tap to tendon of triceps brachii near its insertion | Radial nerve | $C_6$, $C_7$ | Radial nerve | Extension at elbow |
| **Brachioradialis reflex** | Tap to forearm near styloid process of the radius | Radial nerve | $C_5$, $C_6$ | Radial nerve | Flexion at elbow, supination, and flexion at finger joints |
| **Patellar reflex** | Tap to patellar ligament | Femoral nerve | $L_2$–$L_4$ | Femoral nerve | Extension at knee |
| **Ankle-jerk reflex** | Tap to calcaneal tendon | Tibial nerve | $S_1$, $S_2$ | Tibial nerve | Extension (plantar flexion) at ankle |

 **INTEGRATION**

**C.** After injuring her back, 22-year-old Tina exhibits a Babinski reflex. What does this imply about her injury?

**LO** **LEARNING OUTCOME**

Explain the value of reflex testing and how the brain may control and modify reflex responses.

# Introduction to Reflexes

**Labeling:** Label the neural circuit patterns in the following diagrams.

| 1 | 2 | 3 | 4 | 5 |

Label the indicated structures in the accompanying diagram of a reflex arc.

REFLEX ARC

Stimulation of a

| 6 |

Activation of a

| 7 |

| 8 |

Information processing in

| 9 |

Activation of a

| 10 |

Response by an

| 11 |

---

**Vocabulary:** Write the term for each of the following descriptions in the space provided.

| 12 | A reflex in which the sensory stimulus and motor response occur on the same side of the body | 12 | _____ |
| 13 | Reflexes that move affected parts of the body away from a stimulus | 13 | _____ |
| 14 | Controls the sensitivity of a muscle spindle | 14 | _____ |
| 15 | Withdrawal reflex that affects the muscles of a limb | 15 | _____ |
| 16 | Autonomic reflexes that adjust nonskeletal muscles, glands, and adipose tissue | 16 | _____ |
| 17 | Type of reflexes enhanced by repetition | 17 | _____ |
| 18 | A reflex in which the motor response occurs on the side opposite to the stimulus | 18 | _____ |
| 19 | Process involved in preventing opposing muscles from contracting during a reflex | 19 | _____ |
| 20 | The enhancement of reflexes due to facilitation of motor neurons | 20 | _____ |

---

**Short answer:** Classify the withdrawal reflex illustrated in the reflex arc labeling exercise above according to development, response, complexity of circuit, and processing site.

| 21 | _____ |

# Study Outline

## SECTION 1 • Functional Organization of the Spinal Cord

### Module 12.1
### The spinal cord can function independently from the brain p. 437

1. The **spinal cord** receives sensory input and sends motor commands on spinal nerves.

2. A **reflex** is a rapid, automatic response triggered by specific stimuli.

3. **Spinal reflexes** are controlled in the spinal cord.

### Module 12.2
### The spinal cord has 31 segments with 31 pairs of nerves p. 438

4. There are 31 spinal segments, each giving rise to a pair of spinal nerves. There are 8 cervical, 12 thoracic, 5 lumbar, 5 sacral, and 1 coccygeal spinal segments.

5. The **cervical enlargement** supplies nerves to the shoulder and upper limb. The **lumbosacral enlargement** innervates the structures of the pelvis and lower limb.

6. The **conus medullaris** is the inferior tip of the spinal cord. Spinal segments $L_2$–$S_5$ extend as the **cauda equina**. The **filum terminale** extends from the conus medullaris to vertebra $S_2$.

7. Each spinal nerve connects to the spinal cord by a **posterior root** and an **anterior root**.

8. The posterior root contains axons of the **posterior root ganglion**. The posterior root ganglion contains the bodies of sensory neurons whose axons bring information to the spinal cord.

9. The anterior root contains somatic and visceral motor neuron axons.

10. The **white matter** of the spinal cord contains myelinated and unmyelinated axons. The **gray matter** contains cell bodies, neuroglia, and unmyelinated axons and surrounds a central canal.

### Module 12.3
### The spinal meninges, consisting of the dura mater, arachnoid mater, and pia mater, surround the spinal cord p. 440

11. The **spinal meninges** are a series of specialized membranes that surround the spinal cord, providing stability and shock absorption.

12. From superficial to deep, the meninges are (1) **dura mater**, (2) **arachnoid mater**, and (3) **pia mater**.

13. Between the dura mater and the walls of the vertebral canal is the **epidural space** containing protective adipose tissue padding. The **subarachnoid space** is filled with **cerebrospinal fluid (CSF)**, which acts as a shock absorber and a medium for the diffusion of gases, nutrients, and wastes.

14. CSF is withdrawn in a procedure known as a **lumbar puncture**, or *spinal tap*.

*The dissected spinal cord*

### Module 12.4
### Gray matter integrates sensory and motor functions, and white matter carries information p. 442

15. Projections of gray matter are called **horns**. The **posterior gray horn** contains somatic and visceral sensory nuclei. The **lateral gray horn** (present only in thoracic and lumbar spinal segments) contains visceral motor nuclei. The **anterior gray horn** contains somatic motor nuclei.

16. The **gray commissures** surround the central canal and contain axons that cross the spinal cord.

17. White matter is divided into **columns**: the **posterior white column**, the **lateral white column**, and the **anterior white column**.

18. Columns contain **tracts**, which are bundles of axons uniform in structure and function.

19. **Ascending tracts** carry sensory information to the brain. **Descending tracts** convey motor commands to the spinal cord.

### Module 12.5
### Spinal nerves have a similar anatomical structure and distribution pattern p. 444

20. **Epineurium**, a dense network of collagen fibers, surrounds spinal nerves. **Perineurium** divides the axons into bundles called **fascicles**. **Endoneurium** surrounds individual axons.

21. Each spinal nerve divides to form **rami** that carry visceral and motor fibers of the autonomic nervous system (ANS). Thoracic and lumbar spinal segments carry motor output for the **sympathetic division** of the ANS.

*Sectional view of a spinal nerve*

22. The specific region of the skin surface monitored by a single pair of spinal nerves is a **dermatome**. The varicella-zoster virus, the virus that causes chickenpox, can cause skin eruptions called **shingles** along a dermatome.

### Module 12.6
### Each ramus of a spinal nerve provides motor and sensory innervation to a specific region p. 446

23. The anterior root of each spinal nerve contains axons of somatic motor and visceral motor neurons. The posterior root of each spinal nerve carries sensory information to the spinal cord.

24. The **posterior ramus** contains somatic motor and visceral motor fibers that innervate the skin and skeletal muscles of the back. The posterior ramus also carries sensory information from the skin and skeletal muscles of the back.

25. The **anterior ramus** axons supply the ventrolateral body surface, structures in the body wall, and the limbs. The anterior ramus also carries sensory information from the ventrolateral body surface, structures in the body wall, and the limbs.

26. Each **white ramus communicans** and **gray ramus communicans** are together known as the **rami communicantes** and contain fibers of the ANS. They occur only with spinal nerves $T_1$–$L_2$.

**12**

## Module 12.7

### Spinal nerves form nerve plexuses that innervate the skin and skeletal muscles    p. 448

27. The anterior rami form four major plexuses: (1) the **cervical plexus**, (2) the **brachial plexus**, (3) the **lumbar plexus**, and (4) the **sacral plexus**.

## Module 12.8

### The cervical plexus innervates the muscles of the neck and diaphragm    p. 449

28. The **cervical plexus** consists of the anterior rami of spinal nerves $C_1$–$C_5$.

29. The cervical plexus innervates the muscles of the neck, the diaphragm, and the skin of the neck and superior chest.

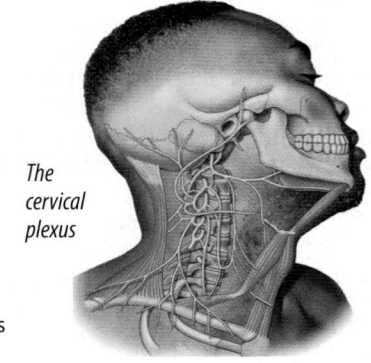

*The cervical plexus*

30. The **phrenic nerve**, a major nerve of the cervical plexus, supplies the diaphragm.

## Module 12.9

### The brachial plexus innervates the pectoral girdles and upper limbs    p. 450

31. The **brachial plexus** originates from the anterior rami of spinal nerves $C_5$–$T_1$.

32. **Trunks** (large bundles of axons from several spinal nerves) split into **divisions**, which regroup and combine into **cords** to form the brachial plexus.

## Module 12.10

### The lumbar and sacral plexuses innervate the skin and skeletal muscles of the trunk and lower limbs    p. 452

33. The **lumbar plexus** arises from spinal cord segments $T_{12}$–$L_4$. The **sacral plexus** arises from spinal cord segments $L_4$–$S_4$.

34. The largest and longest nerve in the body is the **sciatic nerve**, which arises from the sacral plexus.

## SECTION 2 • Introduction to Reflexes

## Module 12.11

### CNS neurons are grouped into neuronal pools, which form neural circuits    p. 455

35. The interneurons of the CNS are organized into functional groups of interconnected neurons called **neuronal pools**.

36. **Neural circuits** have varying patterns, termed **divergence**, **parallel processing**, **serial processing**, **convergence**, and **reverberation**.

## Module 12.12

### Reflexes are vital to homeostasis    p. 456

37. **Reflexes** are rapid, automatic responses to specific stimuli. The "wiring" of a single reflex is called a **reflex arc**.

38. The five steps in a simple neural reflex are (1) **stimulation of a receptor**, (2) **activation of a sensory neuron**, (3) **information processing in the CNS**, (4) **activation of a motor neuron**, and (5) **response of a peripheral effector**.

39. Neural reflexes are classified according to their development, the nature of the resulting motor response, the complexity of the neural circuit, and the information processing site.

## Module 12.13

### The stretch reflex is a monosynaptic reflex involving muscle spindles    p. 458

40. The **stretch reflex** is a monosynaptic reflex regulating skeletal muscle length. The **patellar reflex** is an example.

41. **Muscle spindles** are the sensory receptors involved in a stretch reflex. Each muscle spindle consists of specialized skeletal muscle fibers called **intrafusal muscle fibers**.

42. Stretching the intrafusal fiber stimulates a sensory neuron, which in turn stimulates the motor neurons.

43. Many stretch reflexes are **postural reflexes** that help us to maintain normal upright posture.

## Module 12.14

### Withdrawal reflexes and crossed extensor reflexes are polysynaptic reflexes    p. 460

44. **Withdrawal reflexes** move affected parts of the body away from a stimulus. Painful stimuli trigger the strongest withdrawal reflexes.

45. The **flexor reflex** affects the muscles of the limb. When a muscle contracts in a flexor reflex, opposing muscles relax to permit the movement. This is called **reciprocal inhibition**.

*A dramatic example of a flexor reflex in action*

46. Stretch reflexes and withdrawal reflexes involve **ipsilateral reflex arcs**, meaning the sensory stimulus and motor response occur on the same side of the body.

47. The **crossed extensor reflex** involves a **contralateral reflex arc** because an additional motor response occurs on the side opposite the stimulus.

48. **Polysynaptic reflexes** are responsible for the automatic actions involved in complex movements such as walking and running.

## Module 12.15

### The brain can inhibit or facilitate spinal reflexes, and reflexes can be used to determine the location and severity of damage to the CNS    p. 462

49. **Reinforcement** is the facilitation of motor neurons involved in reflexes.

50. Physical examinations test the status of spinal segments because each reflex is controlled by specific segments of the spinal cord.

51. The **biceps reflex**, **triceps reflex**, and **ankle-jerk reflex** are reflexes commonly tested in physical exams.

52. The **Babinski reflex** is present in infants but disappears in adults. Its presence in adults is a sign of damage to either the higher centers of the brain or descending tracts.

53. The **plantar reflex** and the **abdominal reflex** are normally present in adults.

# Chapter Review Questions

**Labeling:** Label the structures in the following diagram.

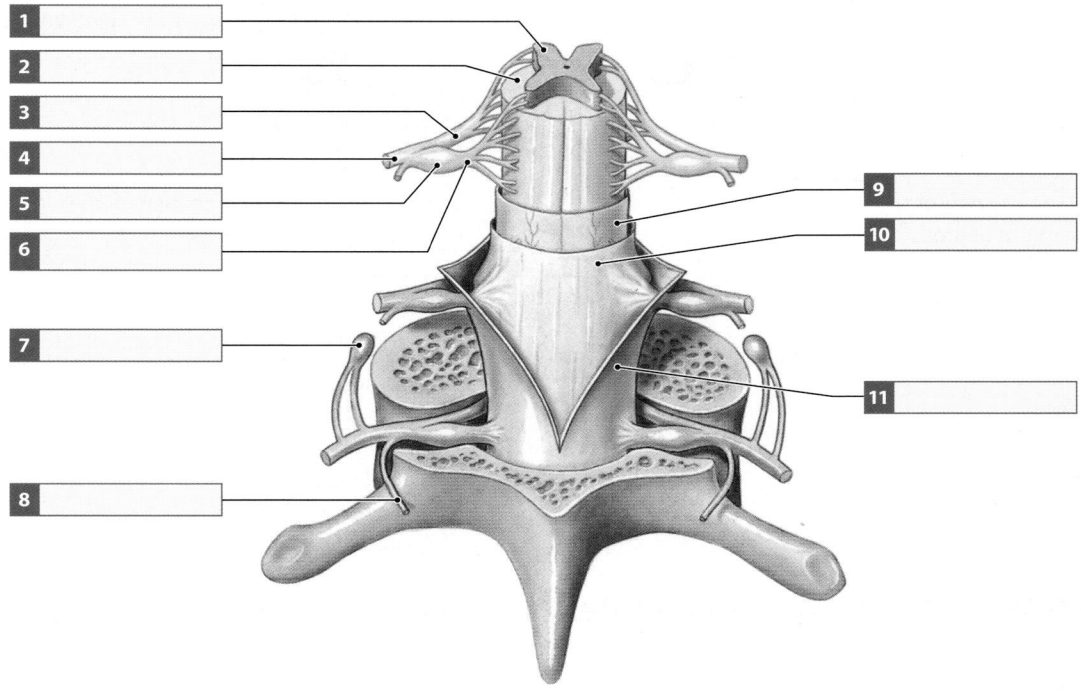

| | |
|---|---|
| 1 | |
| 2 | |
| 3 | |
| 4 | |
| 5 | |
| 6 | |
| 7 | |
| 8 | |
| 9 | |
| 10 | |
| 11 | |

**True/False:** Indicate whether each statement is true or false.

**12** Sensory input travels toward the spinal cord, and motor commands travel away from the spinal cord.

**13** Endoneurium is the outermost covering of a spinal nerve.

**14** The posterior root of each spinal nerve carries motor information away from the spinal cord.

**15** There are eight cervical spinal nerves.

**16** Motor nuclei are located in the posterior gray horn of the spinal cord.

**12** _____

**13** _____

**14** _____

**15** _____

**16** _____

**Multiple choice:** Select the correct answer from the list provided.

**17** The following steps are involved in a simple neural reflex: (1) activation of a sensory neuron; (2) activation of a motor neuron; (3) response of a peripheral effector; (4) stimulation of a receptor; (5) information processing. The proper sequence of these steps is

☐ a) 1, 3, 4, 5, 2.
☐ b) 4, 5, 3, 1, 2.
☐ c) 4, 1, 5, 2, 3.
☐ d) 4, 3, 1, 5, 2.

**18** The layers of the spinal meninges from superficial to deep are

☐ a) pia mater, arachnoid mater, dura mater.
☐ b) arachnoid mater, pia mater, dura mater.
☐ c) dura mater, arachnoid mater, pia mater.
☐ d) dura mater, pia mater, arachnoid mater.

**19** Fascicles containing bundles of axons are surrounded by
- [ ] a) epineurium.
- [ ] b) perineurium.
- [ ] c) endoneurium.
- [ ] d) dermatome.

**20** The adult spinal cord extends only to the
- [ ] a) coccyx.
- [ ] b) sacrum.
- [ ] c) third or fourth lumbar vertebra.
- [ ] d) first or second lumbar vertebra.

**21** The anterior root of each spinal nerve contains axons of
- [ ] a) visceral sensory neurons.
- [ ] b) somatic motor neurons only.
- [ ] c) visceral motor neurons only.
- [ ] d) both somatic motor and visceral motor neurons.

**22** A neuronal circuit that spreads information from one neuron to several neurons is
- [ ] a) divergence.
- [ ] b) convergence.
- [ ] c) serial processing.
- [ ] d) parallel processing.

**23** Reflex arcs in which the sensory stimulus and the motor response occur on the same side of the body are
- [ ] a) contralateral.
- [ ] b) ipsilateral.
- [ ] c) monosynaptic.
- [ ] d) crossed extensor.

**24** The tapered conical portion of the spinal cord is the
- [ ] a) cauda equina.
- [ ] b) conus medullaris.
- [ ] c) filum terminale.
- [ ] d) coccygeal nerve.

**25** The median nerve is a component of the
- [ ] a) cervical plexus.
- [ ] b) brachial plexus.
- [ ] c) lumbar plexus.
- [ ] d) sacral plexus.

**26** Stroking an infant's foot on the lateral side of the sole produces extension of the hallux and a fanning of the toes known as the
- [ ] a) Babinski reflex.
- [ ] b) plantar reflex.
- [ ] c) cremasteric reflex.
- [ ] d) ankle jerk reflex.

**12**

**Short answer**

**27** Describe the cause and symptoms of shingles.

**28** A student forcefully bumps the posterior medial aspect of her elbow on her desktop and feels pain in her hand. Which nerve is involved, and where in her hand is she experiencing pain?

**29** Why do cervical spinal nerves outnumber cervical vertebrae?

**30** Where is cerebrospinal fluid located, and what are its functions?

**MasteringA&P**™

Access more chapter study tools online in the MasteringA&P Study Area:

- **Chapter Quizzes, Chapter Practice Test, and Clinical Case Studies**

- **Practice Anatomy Lab™ 3.0**     PAL™3.0
- **Interactive Physiology™**     iP2™
- **A&P Flix™**     A&PFlix™
- **PhysioEx™ 9.1**     PhysioEx™ 9.1

# Chapter Integration • Applying what you have learned

## A helmet-to-helmet collision causes a "stinger"

Dominic is a defenseman on his high school's varsity lacrosse team. He plays an aggressive style of defense and initiates extensive physical contact. As the final seconds of a recent game counted down, an opposing player was about to take a shot on goal that would have tied the game. Dominic ran full speed into the shooter. Their helmets collided dead-on, and Dominic's head snapped hard to the left. As his opponent fell backward onto the ground, Dominic felt an intense pain from the right side of his neck that radiated down his entire right upper limb to the tips of his fingers. The referee blew the whistle, indicating the end of the game. As the winning players and coaches congregated, the team's athletic trainer asked Dominic how he was feeling after delivering such a significant hit. Dominic replied that upon impact he felt a severe pain on the right side of his neck and entire right upper limb that was both numbing and burning, but that it lasted only a few moments. The trainer informed him that he likely experienced a *stinger* and that although the pain was quite intense, he would not likely have a lasting injury.

From what you have just learned about spinal nerves, answer the following questions.

**1** Why did Dominic's pain radiate down only the right upper limb and not both upper limbs? Why was the pain also in his neck?

**2** Which spinal nerves would you suspect to be involved?

**3** Why do you think he experienced the numbing, burning pain, and why do you think the pain extended all the way to his fingertips?

## A really bad day

Karen had just finished her A&P midterm exam and was hurrying out of the classroom to get home and finish packing for her first-ever spring break trip to Cancun, Mexico. She was excited about her upcoming trip and equally excited about how well she was doing in her first year of nursing school. She thought back to how she had struggled with every high school science class she'd had, but her experience in her A&P class was completely different. Her professor was wonderful at presenting the information in an understandable manner and in guiding students to numerous ancillary tools to help them comprehend a seemingly overwhelming subject.

Seeing the crowd of students waiting for the elevator, Karen chose to take the stairs. In her haste to get home, she slipped on the edge of a step and fell down a flight of stairs. During the fall, she suffered lumbar and sacral spinal cord damage due to a hyperextension of her back. The injury resulted in edema around the central canal that compressed the anterior horn of the lumbar region.

An additional consequence of her injury was the loss of the ability to control her bowels and urinary bladder because control of these functions involves spinal reflex arcs located

in the sacral region of the spinal cord. In both instances, two sphincter muscles—an inner sphincter of smooth muscle and an outer sphincter of skeletal muscle—control the passage of wastes (whether feces or urine) out of the body.

**4** What signs and symptoms would you expect Karen to exhibit during her neurological evaluation?

**5** How would a transection of the spinal cord at the $L_1$ level affect a person's bowel and urinary bladder control?

**6** Compare the effects of a transection of the spinal cord at $L_1$ in an adult with a newborn's inability to control urination.

## LEARNING OUTCOMES

These Learning Outcomes correspond by number to this chapter's modules and indicate what you should be able to do after completing the chapter. They also appear at the end of each module.

# 13

# The Brain, Cranial Nerves, and Sensory and Motor Pathways

# Module 13.1
# The brain develops from a hollow neural tube

The first section of this chapter introduces the functional anatomy of the brain and cranial nerves. A "typical" brain has a volume of 1200 mL (71 in.$^3$). However, brain size varies among individuals. The brains of males are, on average, about 10 percent larger than those of females because of differences in average body size. No correlation exists between brain size and intelligence. Individuals with the smallest brains (750 mL) and the largest brains (2100 mL) are functionally normal.

**1** This lateral view of the brain of an embryo after 4 weeks of development shows the **neural tube**, the hollow cylinder that is the beginning of the central nervous system (CNS). At the cephalic (rostral) portion of the neural tube, three swellings called the **primary brain vesicles** develop. They are named for their relative positions: prosencephalon, mesencephalon, and rhombencephalon.

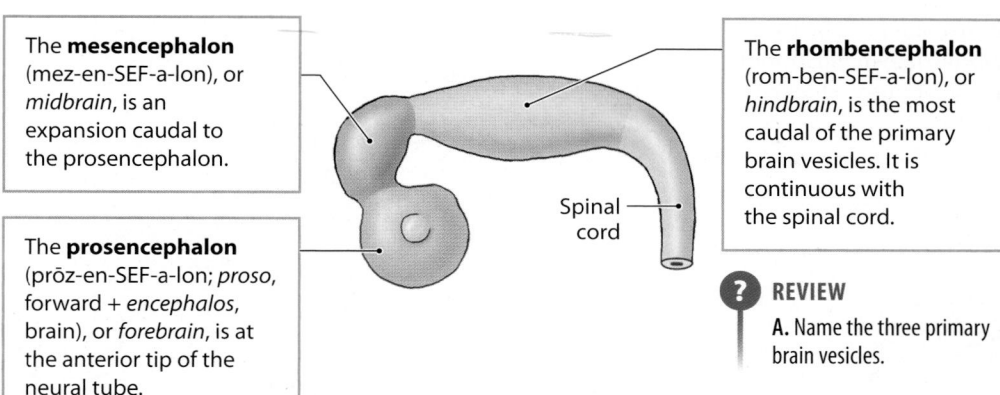

The **mesencephalon** (mez-en-SEF-a-lon), or *midbrain*, is an expansion caudal to the prosencephalon.

The **prosencephalon** (prōz-en-SEF-a-lon; *proso*, forward + *encephalos*, brain), or *forebrain*, is at the anterior tip of the neural tube.

Spinal cord

The **rhombencephalon** (rom-ben-SEF-a-lon), or *hindbrain*, is the most caudal of the primary brain vesicles. It is continuous with the spinal cord.

**? REVIEW**
**A.** Name the three primary brain vesicles.

13

**2** By week 5 of development, the primary brain vesicles have changed position, and the prosencephalon and rhombencephalon have subdivided, forming **secondary brain vesicles**.

| Prosencephalon | Rhombencephalon | |
|---|---|---|
| The **diencephalon** (dī-en-SEF-a-lon) becomes the major relay and processing center for information headed to and from the cerebrum. | The **metencephalon** (met-en-SEF-a-lon; *meta*, after) is caudal to the **midbrain**. This region will form the **cerebellum** and the **pons** of the adult brain. | The **myelencephalon** (mī-el-en-SEF-a-lon; *myelon*, spinal cord) will become the **medulla oblongata**. |
| The **telencephalon** (tel-en-SEF-a-lon; *telos*, end) begins to expand rapidly, eventually becoming the **cerebrum**, the largest part of the adult brain. | | |

Spinal cord

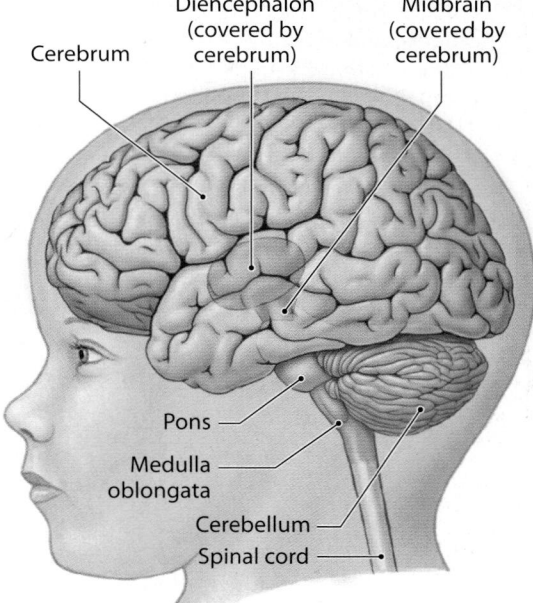

Cerebrum

Diencephalon (covered by cerebrum)

Midbrain (covered by cerebrum)

Pons

Medulla oblongata

Cerebellum

Spinal cord

**3** As development continues, the cerebrum enlarges to the point where it covers other regions of the brain.

**? REVIEW**
**B.** Which structures form the secondary brain vesicles?

**? REVIEW**
**C.** Which embryonic brain vesicle develops into the largest region of the adult brain?

**Lo LEARNING OUTCOME**
Describe the origins of the different regions of the brain from the embryonic neural tube.

# Module 13.2
# Each region of the brain has distinct structural and functional characteristics

**1** This diagrammatic view of the brain introduces the four major regions of the brain, their distinct structures, and their general functions. The regions and their structures are color coded, and these colors will be used as appropriate in illustrations throughout the chapter.

### Cerebrum

The **cerebrum** (se-RĒ-brum) of the adult brain is divided into left and right **cerebral hemispheres**. Their surfaces are highly folded and covered by a superficial layer of gray matter, 1.5 to 4.5 mm thick, called the **cerebral cortex** (*cortex*, rind or bark). Cerebral functions include conscious thought, memory storage and processing, sensory processing, and regulating skeletal muscle contractions.

**Fissures** are deep grooves that subdivide each cerebral hemisphere.

**Gyri** (JĪ-rī; singular, *gyrus*) are folds in the cerebral hemispheres that increase its surface area.

**Sulci** (SUL-sī; singular, *sulcus*) are shallow grooves in the cerebral hemispheres that separate adjacent gyri.

### Diencephalon

The **diencephalon** is the structural and functional link between the cerebral hemispheres and the rest of the CNS.

The **thalamus** (THAL-a-mus) contains relay and processing centers for sensory information.

The **hypothalamus** (*hypo-*, below), or floor of the diencephalon, contains centers involved with emotions, autonomic function, and hormone production.

Spinal cord

### Cerebellum

The **cerebellum** (ser-e-BEL-um) is the second largest region of the brain. It is partially hidden by the cerebral hemispheres. The functions of the cerebellum include coordinating and modulating motor commands from the cerebral cortex. The cerebellum has only one-tenth of the brain's volume, but over one-half of the brain's neurons.

 **REVIEW**
**A.** Describe the role of the medulla oblongata.

### Brainstem

The **brainstem** includes the midbrain, pons, and medulla oblongata.

The **midbrain** contains nuclei that process visual and auditory information and control reflexes triggered by these stimuli. It also contains centers that help maintain consciousness.

The **pons** (*pons*, bridge) connects the cerebellum to the brainstem. In addition to tracts and relay centers, the pons also contains nuclei that function in somatic and visceral motor control.

The **medulla oblongata** relays sensory information to other portions of the brainstem and to the thalamus. The medulla oblongata also contains major centers that regulate autonomic functions, such as heart rate and blood pressure.

**2** During development, the neural tube within the cerebral hemispheres, diencephalon, metencephalon, and medulla oblongata expands to form chambers called **ventricles** (VEN-trih-kls). The ventricles are filled with cerebrospinal fluid and lined by ependymal cells (see Module 11.4).

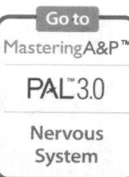

**?** **REVIEW**
**B.** Which region of the brain contains two ventricles?

Cerebral hemispheres

Pons

Medulla oblongata

Spinal cord

Central canal

**Ventricular system of the brain, lateral view**

| Ventricular System of the Brain |
| --- |
| Each cerebral hemisphere contains a large **lateral ventricle**. |
| Each lateral ventricle communicates with the third ventricle through an **interventricular foramen**. |
| The **third ventricle** is located in the diencephalon. |
| The **cerebral aqueduct** is a slender canal within the midbrain that connects the third ventricle to the fourth ventricle. |
| The **fourth ventricle** begins in the metencephalon and extends into the superior portion of the medulla oblongata. It then narrows and becomes the central canal of the spinal cord. |

Cerebral hemispheres

Central canal

Cerebellum

**Ventricular system of the brain, anterior view**

13

**3** The interconnections between the ventricles are seen in this frontal section of the brain.

Lateral ventricles

Interventricular foramen

Third ventricle

Inferior tip of lateral ventricle

Cerebral aqueduct

Fourth ventricle

Central canal

**Frontal section**

Cerebellum

The **corpus callosum** is a thick tract of white matter that interconnects the two cerebral hemispheres.

The **septum pellucidum** is a thin partition that separates the two lateral ventricles.

**?** **REVIEW**
**C.** Compare the corpus callosum with the septum pellucidum.

**?** **INTEGRATION**
**D.** Which region of the brain is enclosed or hidden by the cerebrum?

**Lo** **LEARNING OUTCOME**
Name the four major regions of the brain, and describe their functions.

# The cranial meninges and cerebrospinal fluid protect and support the brain

The cranial bones, **cranial meninges** (membranous brain coverings), and the *cerebrospinal fluid* protect the delicate brain tissues. In addition, the nervous tissue of the brain is biochemically isolated from the general circulation by the *blood brain barrier* (see Module 11.4).

## Meninges

**1** The three layers that make up the cranial meninges—the cranial **dura mater**, **arachnoid mater**, and **pia mater**—are continuous with those of the spinal meninges (see Module 12.3).

**? REVIEW**
**A.** From superficial to deep, name the layers that make up the cranial meninges.

**1 Dura Mater**

The cranial dura mater consists of outer and inner fibrous layers. The outer layer is fused to the periosteum of the cranial bones. As a result, there is no epidural space. The outer (periosteal) and inner (meningeal) layers of the cranial dura mater are typically fused together. Where separated, they contain tissue fluids and blood vessels, including several large dural venous sinuses. The dural venous sinuses collect blood from the veins of the brain.

Subdural space

Cranium (skull)

**2 Arachnoid Mater**

The arachnoid mater is a weblike membrane that covers the brain, providing a smooth surface that does not follow the underlying folds. The subarachnoid space extends between the arachnoid mater and the pia mater. Fibrous strands of arachnoid trabeculae traverse the subarachnoid space.

Periosteal cranial dura
Dural venous sinus
Meningeal cranial dura

Arachnoid membrane
Subarachnoid space
Arachnoid trabeculae

Cerebral cortex

**3 Pia Mater**

Astrocyte processes bind the pia mater to the surface of the brain. The pia mater sticks to the surface of the brain. It extends into every fold and accompanies the branches of cerebral blood vessels as they penetrate the surface of the brain to reach internal structures.

**2** At several sites, the inner layer of the dura mater extends into the cranial cavity, forming **dural folds**—sheets that dip inward and then return. These stabilize and support the brain. **Dural venous sinuses** are large collecting veins located within the dural folds. There are three large dural folds.

Inferior sagittal sinus

The **superior sagittal sinus** is the largest dural sinus.

The **tentorium cerebelli** (ten-TŌ-rē-um ser-e-BEL-ī; *tentorium*, a tent) separates the cerebral hemispheres from the cerebellum.

The **falx cerebri** (FALKS SER-e-brī; *falx*, sickle shaped) is a fold of dura mater between the cerebral hemispheres. Its inferior portions attach anteriorly to the crista galli of the ethmoid and posteriorly to the internal occipital crest of the occipital bone. The superior and inferior sagittal sinuses lie within this dural fold.

The **falx cerebelli** separates the two cerebellar hemispheres along the midsagittal line inferior to the tentorium cerebelli.

## Cerebrospinal Fluid

**3** **Cerebrospinal fluid (CSF)** surrounds and bathes the exposed surfaces of the CNS. CSF has three important functions: (1) supporting the weight of the brain, (2) cushioning the brain and spinal cord against physical trauma, and (3) transporting nutrients, chemical messengers, and wastes.

Each ventricle contains an area of **choroid plexus** (*choroid*, vascular coat; *plexus*, network) involved in producing and maintaining CSF. Each choroid plexus consists of specialized ependymal cells with tight junctions and capillaries.

The CSF circulates from the choroid plexuses through the ventricles and fills the central canal of the spinal cord. As it circulates, materials diffuse between the CSF and the interstitial fluid of the CNS across the ependymal cells.

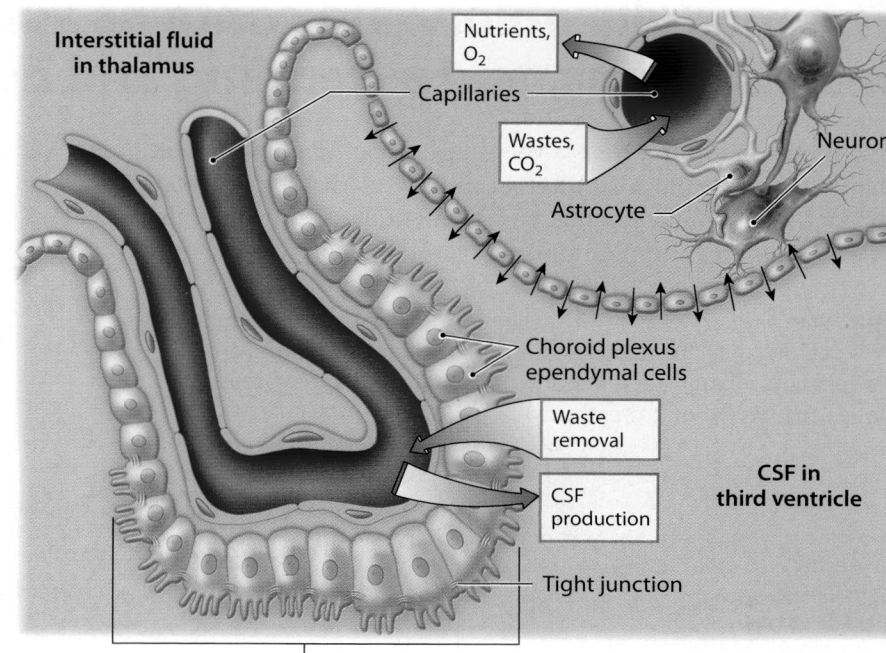

Interstitial fluid in thalamus

Nutrients, O₂

Capillaries

Wastes, CO₂

Neuron

Astrocyte

Choroid plexus ependymal cells

Waste removal

CSF production

**CSF in third ventricle**

Tight junction

Choroid plexus

---

Superior sagittal sinus

Third ventricle

Cerebral aqueduct

The CSF reaches the sub-arachnoid space through two **lateral apertures** and a single **median aperture** in the roof of the fourth ventricle.

The CSF then flows through the subarachnoid space surrounding the brain, spinal cord, and cauda equina.

The choroid plexuses produce CSF at a rate of about 500 mL/day. The total volume of CSF at any moment is approximately 150 mL, and the entire volume of CSF is replaced about every 8 hours.

Central canal of spinal cord

Dura mater

Arachnoid

Subarachnoid space

Cauda equina

Filum terminale

---

**4** Fingerlike extensions of the arachnoid membrane penetrate the meningeal layer of the dura mater and extend into the superior sagittal sinus. In adults, these extensions form large **arachnoid granulations**, sites where CSF is absorbed into the venous circulation.

Dura mater

Superior sagittal sinus

Cranium

Arachnoid granulation

CSF movement

Subdural space

Arachnoid membrane

Cerebral cortex

Pia mater

**? INTEGRATION**

**B.** What brain tissues have tight junctions?

**C.** How would decreased diffusion across the arachnoid granulations affect the volume of cerebrospinal fluid in the ventricles?

**Lo LEARNING OUTCOME**

Explain how the brain is protected and supported and how cerebrospinal fluid forms and circulates.

13

# The medulla oblongata contains autonomic reflex centers, relay stations, and ascending and descending tracts

The **medulla oblongata** is a very busy place: All communication between the brain and spinal cord travels along tracts that ascend or descend through the medulla oblongata. The medulla oblongata is also a center that coordinates complex autonomic reflexes and visceral functions.

**1** Descending tracts cover the anterior surface of the medulla oblongata. The medulla oblongata contains **autonomic centers** controlling vital functions and contains **relay stations** along sensory and motor pathways.

Pons

The **pyramids** contain tracts of motor fibers that originate at the cerebral cortex.

Some of the pyramidal fibers cross to the opposite side of the medulla oblongata as they descend into the spinal cord. This crossing over is called a **decussation** (de-kuh-SĀ-shun; *decussation*, crossing over).

The **olive** is a prominent olive-shaped bulge along the anterolateral surface of the medulla oblongata. It follows the contours of the olivary nucleus.

**Autonomic Centers**
- Reticular formation
- Cardiovascular centers
- Solitary nuclei

**Relay Stations**
- Olivary nucleus
- Cuneate nucleus
- Gracile nucleus

**Anterior view**

**? REVIEW**
**A.** Describe the pyramids of the medulla oblongata and a decussation.

**2** Major components and functions of the medulla oblongata are summarized in this table. The medulla oblongata also contains the nuclei associated with five cranial nerves.

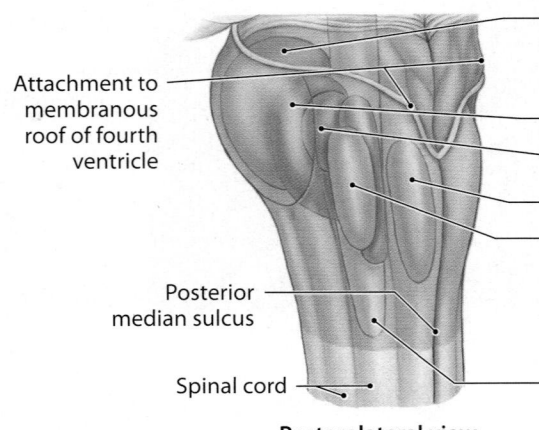

Attachment to membranous roof of fourth ventricle

Posterior median sulcus

Spinal cord

**Posterolateral view**

### The Medulla Oblongata

| Region/Nucleus | Function |
|---|---|
| **GRAY MATTER** | |
| **Inferior olivary complex** | Relays information from the red nucleus, other midbrain centers, and the cerebral cortex to the cerebellum |
| **Reflex centers** | |
| Cardiovascular centers | Regulate heart rate and force of contraction |
| Respiratory rhythmicity centers | Set the basic pace of respiratory movements |
| **Gracile nucleus** **Cuneate nucleus** | Relay somatic information to the thalamus |
| **Other nuclei/centers:** Cranial nerves VIII (in part), IX, X, XI (in part), and XII* | Sensory and motor nuclei of five cranial nerves; nuclei relay ascending information from the spinal cord to higher centers |
| **Reticular formation** | Contains nuclei and centers that regulate vital autonomic functions; extends into the pons and midbrain |
| **WHITE MATTER** | |
| **Ascending and descending tracts within funiculi** | Link the brain with the spinal cord |

**? REVIEW**
**B.** Which medulla oblongata components relay somatic sensory information to the thalamus?
**C.** What is the function of the ascending and descending tracts in the medulla oblongata?

*The full names and functions of the cranial nerves are discussed in Module 13.15.

 **LEARNING OUTCOME**
List the main components of the medulla oblongata, and specify their functions.

13

# Module 13.5

# The pons links the cerebellum to the brain and spinal cord and has vital autonomic reflex centers

**1** The **pons** links the cerebellum with the midbrain, diencephalon, cerebrum, medulla oblongata, and spinal cord. The pons contains four groups of structures: (1) ascending, descending, and transverse tracts that interconnect other portions of the CNS; (2) nuclei involved with the control of respiration; (3) sensory and motor nuclei of cranial nerves; and (4) nuclei and tracts that process and relay information sent to and from the cerebellum.

**? REVIEW**

**A.** List the four groups of pontine structures.

- Midbrain
- Pons
- Medulla oblongata
- Spinal cord

**Lateral view of the brainstem**

Cerebellum

Fourth ventricle

Medulla oblongata

Inferior olivary complex

**Lateral view**

The **reticular formation** is a loosely organized mass of gray matter that contains embedded nuclei and centers that regulate vital autonomic functions. It extends from the medulla oblongata to the midbrain.

| The Pons | |
|---|---|
| **Region/Nucleus** | **Function** |
| **WHITE MATTER** | |
| Descending tracts | Carry motor commands from higher centers of the brain to nuclei of cranial or spinal nerves |
| Ascending tracts | Carry sensory information from brainstem nuclei to the thalamus |
| Transverse pontine fibers | Link pontine nuclei with the cerebellum of the opposite side |
| **GRAY MATTER** | |
| Apneustic and pneumotaxic centers | Adjust activities of the respiratory rhythmicity centers in the medulla oblongata and pons |
| Reticular formation | Automatic processing of incoming sensations and outgoing motor commands |
| Nuclei of cranial nerves V, VI, VII, and VIII (in part) | Relay sensory information and issue somatic motor commands to jaw, face, and extrinsic eye muscles |
| Other nuclei/relay centers | Relay sensory and motor information to the cerebellum |

**Lo LEARNING OUTCOME**

List the main components of the pons, and specify their functions.

# Module 13.6

# The cerebellum coordinates learned and reflexive patterns of muscular activity at the subconscious level

The cerebellum is an automatic processing center that monitors proprioceptive, visual, tactile, balance, and auditory sensations. It has two primary functions:

- **Adjusting the postural muscles of the body.** The cerebellum coordinates rapid, automatic adjustments that maintain balance and equilibrium. The cerebellum makes these adjustments by modifying the activities of motor centers in the brainstem.

- **Programming and fine-tuning movements controlled at the conscious and subconscious levels.** The cerebellum refines learned movement patterns such as riding a bicycle or playing the piano. It performs this function indirectly by regulating activity along motor pathways at the cerebral cortex, basal nuclei, and motor centers in the brainstem. The cerebellum compares the motor commands with proprioceptive information (perception of muscle and joint position) and stimulates any adjustments needed to make the movement smooth.

Cerebellum

**Posterior view**

**1** As seen in this view of the posterior superior surface, the cerebellum has large **anterior** and **posterior lobes**. Like the cerebrum, the cerebellum has two hemispheres, and the surface is covered with a convoluted surface of gray matter called the **cerebellar cortex**.

Anterior lobe

Posterior lobe

**Left hemisphere of cerebellum**

**Right hemisphere of cerebellum**

The **vermis** (VER-mis; worm) is a narrow band of cortex that separates the cerebellar hemispheres.

The anterior and posterior lobes are separated by the **primary fissure**.

The **folia** (FŌ-lē-uh; leaves) are folds of the cerebellar surface. They are less prominent than the folds in the surfaces of the cerebral hemispheres.

**2** The cerebellar cortex has an outer molecular layer, an intermediate Purkinje layer with stacked Purkinje cells, and an inner granular layer. The **Purkinje** (pur-KIN-jē) **cells** are highly branched and create a layer of neuron cell bodies each with a single axon. Most axons carrying sensory information to the cerebellum do not synapse in the cerebellar nuclei but pass through the deeper layers of the cerebellum on their way to the dendrites of Purkinje cells. Information about the motor commands issued at the conscious and subconscious levels reaches the Purkinje cells after being relayed by nuclei in the pons or by the cerebellar nuclei. The only axons that leave the cerebellar cortex are those of the Purkinje cells.

**Purkinje cells**      LM × 400

The extensive dendrites of each Purkinje cell receive input from up to 200,000 synapses, more than any other type of brain cell.

Cell body of Purkinje cell

Purkinje cell axons project into the white matter of the cerebellum.

**3** This sagittal section through the vermis shows the internal organization of the cerebellum. The white matter of the cerebellum forms a branching array that in sectional view resembles a tree. It connects the cerebellar cortex and nuclei with tracts of white matter that form the cerebellar peduncles. Anatomists call it the **arbor vitae** (AR-bor VĪ-tē), or "tree of life." The three **cerebellar peduncles** (*pedunculus*, foot) on each side contain tracts that link the cerebellum with the brainstem, cerebrum, and spinal cord. The middle and inferior peduncles carry most of the input to the cerebellum, and the superior peduncle carries most of the cerebellar output.

**? REVIEW**

**A.** Identify the components of the cerebellar gray matter.

Anterior lobe

Midbrain

Pons

Choroid plexus of the fourth ventricle

Medulla oblongata

Spinal cord

Posterior lobe

**Lateral view**

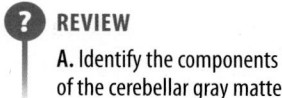

| The Cerebellum | |
|---|---|
| **Region/Nuclei** | **Function** |
| **GRAY MATTER** | |
| **Cerebellar cortex** | Involuntary coordination and control of ongoing body movements |
| **Cerebellar nuclei** | Involuntary coordination and control of ongoing body movements |
| **WHITE MATTER** | |
| **Arbor vitae** | Connects cerebellar cortex and nuclei with cerebellar peduncles |
| **Cerebellar peduncles** | |
| **Superior** | Link cerebellum with midbrain, diencephalon, and cerebrum |
| **Middle** | Carry communications between the cerebellum and pons |
| **Inferior** | Link the cerebellum with the medulla oblongata and spinal cord |
| **Transverse pontine fibers** | Interconnect pontine nuclei with the opposite cerebellar hemisphere |

**13**

**? REVIEW**

**B.** Describe the arbor vitae, including its makeup, location, and function.

**? REVIEW**

**C.** Describe ataxia.

**Clinical Note**

## Damage or disruption of the cerebellum results in loss of coordination

The cerebellum can be permanently damaged by trauma or stroke or be temporarily affected by drugs such as alcohol. The result is **ataxia** (a-TAK-sē-uh; *ataxia*, lack of order), an inability to coordinate muscle movement. In severe ataxia, the person cannot sit or stand without assistance.

**LO LEARNING OUTCOME**

List the main components of the cerebellum, and specify the functions of each.

# Module 13.7

# The midbrain regulates auditory and visual reflexes and controls alertness

The **midbrain** is the most complex and integrative portion of the brainstem. Working independently or with the cerebellum, the midbrain can direct complex motor patterns at the subconscious level. It also influences the level of activity in the entire nervous system.

Posterior view of brainstem and diencephalon

**1** This posterior view of the midbrain shows the major superficial landmarks as well as underlying nuclei.

### Corpora Quadrigemina

The **corpora quadrigemina** (KOR-pŏr-uh kwa-dri-JEM-ih-nuh) are two pairs of sensory nuclei located in the roof of the midbrain.

Each **superior colliculus** (ko-LIK-ū-lus; *colliculus*, hill) receives visual inputs from the thalamus and controls the reflex movements of the eyes, head, and neck in response to visual stimuli, such as a bright light.

Each **inferior colliculus** receives auditory input from nuclei in the medulla oblongata and pons and controls reflex movements of the head, neck, and trunk in response to auditory stimuli, such as a loud noise.

Pineal gland

Thalamus

The **reticular activating system (RAS)** is a specialized part of the reticular formation. Stimulation of the RAS makes you more alert and attentive; damage to the RAS produces unconsciousness.

The **red nucleus** contains numerous blood vessels, which give it a red color. It receives information from the cerebrum and cerebellum and issues subconscious motor commands that affect upper limb position and background muscle tone.

The **substantia nigra** (NĪ-gruh; *nigra*, black) contains darkly pigmented cells that inhibit activity in the basal nuclei of the cerebrum.

The **cerebral peduncles** are nerve fiber bundles on the ventrolateral surfaces of the midbrain. They contain (1) descending fibers that reach the cerebellum by way of the pons and (2) descending fibers that carry voluntary motor commands issued by the cerebral hemispheres.

**? REVIEW**

**A.** Identify the sensory nuclei contained within the corpora quadrigemina.

**B.** Give the functions of the superior colliculi and inferior colliculi.

**2** This superior view of a horizontal section through the midbrain shows the internal subdivisions relative to the cerebral aqueduct.

Posterior view of brainstem and diencephalon

**Posterior**

Inferior colliculus

Superior colliculus

Red nucleus

Substantia nigra

Cerebral peduncle

**Anterior**

The **tectum**, or roof of the midbrain, is the region posterior to the cerebral aqueduct.

Cerebral aqueduct

The **tegmentum** is the area anterior to the cerebral aqueduct.

**3** These views show the relationships between brainstem structures and the cranial nerves that arise from them.

**? REVIEW**
**C.** Which cranial nerves arise from the brainstem?

**Cranial Nerves of Brainstem**

III
IV

V

Pons

VI
VII
VIII
IX
X
XI
XII

Spinal nerve C₁

Spinal nerve C₂

Anterior root

**Lateral view**

Posterior root

**Midbrain**

Cerebral peduncle
Superior colliculus
Inferior colliculus

**Cerebellar Peduncles**

Superior cerebellar peduncle
Middle cerebellar peduncle
Inferior cerebellar peduncle

Medulla oblongata

Spinal cord

IV

Medulla oblongata

Spinal cord

Choroid plexus in roof of fourth ventricle

Posterior roots of spinal nerves C₁ and C₂

**Posterior view**

13

| Components of the Midbrain and Their Functions | | |
|---|---|---|
| **Component** | **Region/Nuclei** | **Function** |
| **GRAY MATTER** | | |
| **Tectum (roof)** | **Superior colliculi** | Integrate visual information with other sensory inputs; initiate reflex responses to visual stimuli |
| | **Inferior colliculi** | Relay auditory information to medial geniculate body; initiate reflex responses to auditory stimuli |
| **Walls and floor** | **Red nuclei** | Provide subconscious control of upper limb position and background muscle tone |
| | **Substantia nigra** | Secretes the neurotransmitter *dopamine* and inhibits activity in the basal nuclei |
| | **Reticular formation (RAS headquarters)** | Processes incoming sensations and outgoing motor commands automatically; can initiate involuntary motor responses to stimuli; helps maintain consciousness (RAS) |
| | **Other nuclei/centers** | Are associated with cranial nerves III and IV |
| **WHITE MATTER** | | |
| | **Cerebral peduncles** | Connect primary motor cortex with motor neurons in brain and spinal cord; carry ascending sensory information to thalamus |

**? REVIEW**
**D.** What part of the midbrain influences the activity of the entire nervous system?

**Lo LEARNING OUTCOME**
List the main components of the midbrain, and specify the functions of each.

# The diencephalon consists of the epithalamus, thalamus, and hypothalamus

## Epithalamus

**1** The **epithalamus** is the roof of the diencephalon superior and posterior to the third ventricle. As seen in this sagittal section of the brain, the anterior portion of the epithalamus contains an extensive choroid plexus that extends through the interventricular foramina.

The anterior limit of the diencephalon is marked by the **anterior commissure**, a tract that interconnects the cerebral hemispheres, and the **optic chiasm**, where the optic nerves (II) connect to the brain.

Choroid plexus

Epithalamus (anterior portion)

A projection of gray matter called an **interthalamic adhesion** extends into the third ventricle from the left and right sides of the thalamus, although no fibers cross the midline. It is absent in about one-fifth of human brains.

The **pineal gland** lies in the posterior portion of the epithalamus. It is an endocrine structure that secretes the hormone **melatonin**. Melatonin is important in regulating day–night cycles and reproductive functions.

Hypothalamus

Thalamus

## Thalamus

**2** The left and right sides of the **thalamus** sit superior to the midbrain. The important landmarks in this lateral view can be seen only after the cerebral hemispheres and cerebral peduncles have been removed.

**? REVIEW**

**A.** Damage to the lateral geniculate bodies of the thalamus would interfere with what particular function?

Thalamus

Optic chiasm

Optic tract

Cerebral peduncle (midbrain)

**Lateral view of the left side of thalamus and midbrain**

The **lateral geniculate** (je-NIK-ū-lāt; *genicula*, little knee) **body** of each thalamus receives visual information over the **optic tract** and sends signals to both the midbrain and the occipital lobe of the cerebral hemisphere on that side.

The **medial geniculate body** of each thalamus relays auditory information from specialized receptors of the internal ear to the appropriate area of the cerebral cortex.

**3** The thalamus is the final relay point for ascending sensory information that will be relayed, or **projected**, to the cerebral cortex. The thalamus acts as a filter, passing on only a small portion of the arriving sensory information to the cerebral hemispheres. Each region of the thalamus contains nuclei, groups of nuclei, or bodies that connect to specific regions of the cerebral cortex (see table at the upper right).

*Note: Colors indicate the associated areas of the cerebral cortex*

Limbic system

Frontal lobes

Parietal lobe and cingulate gyrus

Association areas of cerebral cortex

Medial nuclei

Anterior nuclei

Lateral nuclei

Pulvinar

Ventral nuclei

Auditory input

Medial geniculate body

Basal nuclei

Cerebellum

General sensory input

Visual input

Lateral geniculate body

## Components of the Thalamus and Their Functions

| Component | Function |
|---|---|
| **Anterior nuclei** | Part of the limbic system (see Module 13.9) |
| **Medial nuclei** | Integrate sensory information for projection to the frontal lobes of the cerebral hemispheres |
| **Ventral nuclei** | Project sensory information to the primary sensory cortex; relay information from cerebellum and basal nuclei to the motor area of the cerebral cortex |
| **Pulvinar nuclei** | Integrate sensory information for projection to association areas of the cerebral cortex |
| **Lateral geniculate body** | Projects visual information to the visual cortex: integrates sensory information and influences emotional states |
| **Medial geniculate body** | Projects auditory information to the auditory cortex: integrates sensory information and influences emotional states |

## Hypothalamus

**4** The **hypothalamus** contains important control and integrative centers that are shown in this sagittal section. Hypothalamic centers may be stimulated by (1) sensory information from the cerebrum, brainstem, and spinal cord; (2) changes in the composition of the CSF and interstitial fluid; or (3) chemicals in the circulating blood that rapidly enter the hypothalamus because this region lacks a blood brain barrier.

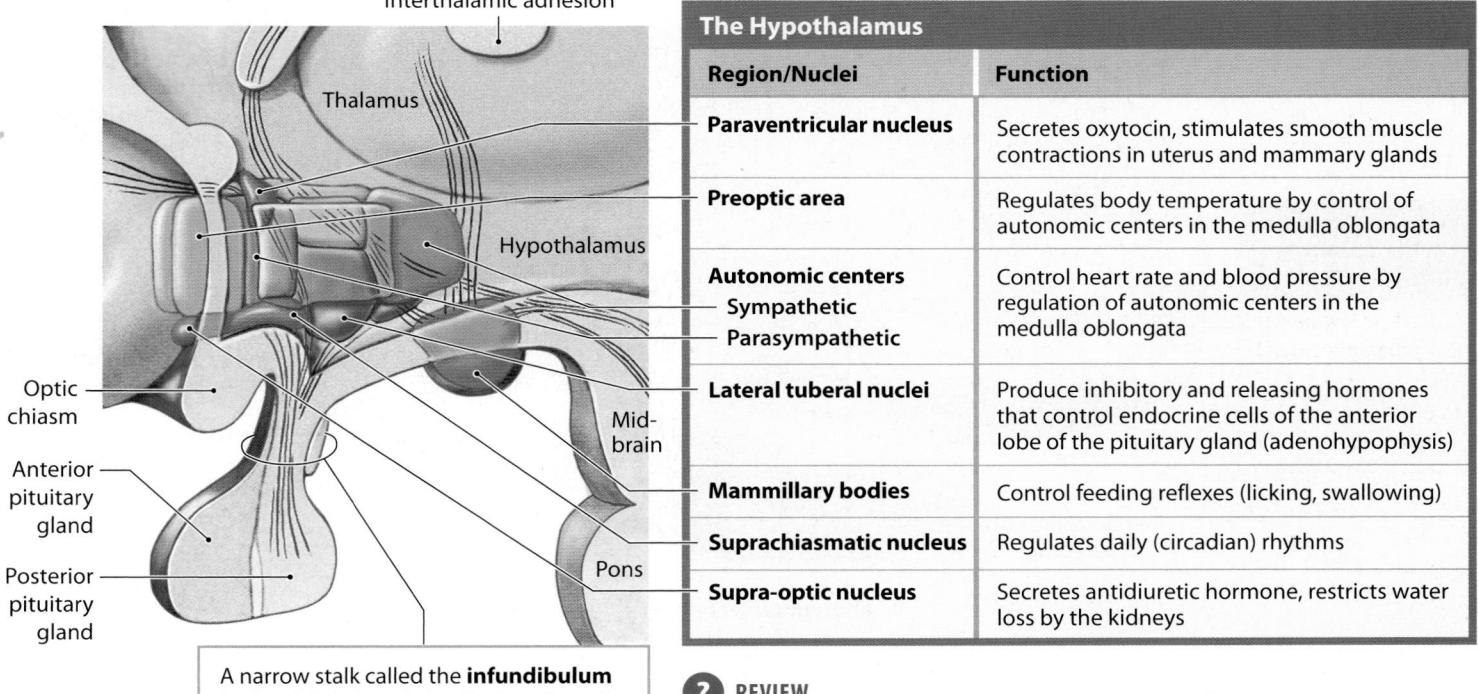

Interthalamic adhesion

Thalamus

Hypothalamus

Optic chiasm

Anterior pituitary gland

Posterior pituitary gland

Mid-brain

Pons

A narrow stalk called the **infundibulum** (in-fun-DIB-ū-lum; *infundibulum*, funnel) extends inferiorly, connecting the floor of the hypothalamus to the pituitary gland.

### The Hypothalamus

| Region/Nuclei | Function |
|---|---|
| **Paraventricular nucleus** | Secretes oxytocin, stimulates smooth muscle contractions in uterus and mammary glands |
| **Preoptic area** | Regulates body temperature by control of autonomic centers in the medulla oblongata |
| **Autonomic centers** Sympathetic Parasympathetic | Control heart rate and blood pressure by regulation of autonomic centers in the medulla oblongata |
| **Lateral tuberal nuclei** | Produce inhibitory and releasing hormones that control endocrine cells of the anterior lobe of the pituitary gland (adenohypophysis) |
| **Mammillary bodies** | Control feeding reflexes (licking, swallowing) |
| **Suprachiasmatic nucleus** | Regulates daily (circadian) rhythms |
| **Supra-optic nucleus** | Secretes antidiuretic hormone, restricts water loss by the kidneys |

**? REVIEW**

**B.** Which component of the diencephalon is stimulated by changes in body temperature?

**Lo LEARNING OUTCOME**

List the main components of the diencephalon, and specify the functions of each.

# Module 13.9

# The limbic system is a functional group of tracts and nuclei located in the cerebrum and diencephalon

The **limbic system** (*limbus*, border) includes nuclei and tracts along the border between the cerebrum and diencephalon. This system is a functional grouping rather than an anatomical one. Functions of the limbic system include (1) establishing emotional states; (2) linking the conscious, intellectual functions of the cerebral cortex with the unconscious and autonomic functions of the brainstem; (3) facilitating memory storage and retrieval; and (4) affecting motivation.

**?** **REVIEW**

**A.** List the primary functions of the limbic system.

**1** The limbic system has components that include nuclei and tracts along the border between the cerebrum and diencephalon. This diagrammatic section shows the position and orientation of its major components.

Corpus callosum · Fornix · Central sulcus · Pineal gland

**Components of the Limbic System in the Diencephalon**

- Anterior thalamic nuclei
- Hypothalamus
- Mammillary body

Temporal lobe of cerebrum

**Midsagittal section**

**Components of the Limbic System in the Cerebrum**

The region of the cerebral hemisphere shown in green is known as the **limbic lobe**. It contains three gyri.

**Cingulate gyrus** (superior portion of limbic lobe)

**Parahippocampal gyrus** (inferior portion of limbic lobe)

**Dentate gyrus** (posterior portion of limbic lobe)

## The Limbic System

### Diencephalic Components

**Thalamus**

The anterior thalamic nuclei relay information from the mammillary body (of the hypothalamus) to the cingulate gyrus on that side.

**Hypothalamus**

Hypothalamic nuclei

Mammillary body

### Other Components

**Reticular formation** (not shown)

The boundaries between the hypothalamic nuclei of the limbic system are often poorly defined, but experimental stimulation has identified a number of important hypothalamic centers responsible for the emotions of rage, fear, pain, sexual arousal, and pleasure. The stimulation of specific regions of the hypothalamus can also produce heightened alertness and general-ized excitement or generalized lethargy and sleep. These responses are caused by the stimulation or inhibition of the reticular formation.

Corpus callosum

Olfactory tract

Dentate gyrus

Parahippocampal gyrus

**?** **REVIEW**

**B.** What are some functions of the amygdaloid body?

**?** **REVIEW**

**C.** Which region of the limbic system is particularly important for the storage and retrieval of long-term memories?

## The Limbic System

### Cerebral Components

**Limbic lobe cortical areas**

**Cingulate gyrus** (superior portion of limbic lobe)

**Dentate gyrus** (posterior portion of limbic lobe)

**Parahippocampal gyrus** (inferior portion of limbic lobe)

### Tracts

The **fornix** (FOR-niks, arch) is a tract of white matter that connects the hippocampus with the hypothalamus.

### Nuclei

The **amygdaloid** (ah-MIG-da-loyd; *amygdale*, almond) **body**, or *amygdala*, of the limbic lobe appears to act as an interface among the limbic system, the cerebrum, and various sensory systems. It plays a role in the regulation of heart rate, in the control of the "fight or flight" response, and in linking emotions with specific memories.

The **hippocampus** is an elongated nucleus whose shape reminded early anatomists of a sea horse (*hippocampus*). It is impor-tant in learning, especially in the storage and retrieval of long-term memories.

The cerebral cortex (gray matter of the cerebrum) enables you to perform complex tasks, but it is largely the limbic system that makes you *want* to do them. For this reason, the limbic system is also known as the *motivational system*.

**Lo** **LEARNING OUTCOME**

Identify the main components of the limbic system, and specify the locations and functions of each.

13

# Module 13.10

# The basal nuclei of the cerebrum adjust and refine ongoing voluntary movements

The **basal nuclei** are masses of gray matter that lie within each cerebral hemisphere deep to the floor of the lateral ventricle. The basal nuclei provide subconscious control of skeletal muscle tone and help coordinate learned movement patterns. Under normal conditions, these nuclei do not initiate particular movements. But once a movement is under way, the basal nuclei provide the general pattern and rhythm, especially for movements of the trunk and proximal limb muscles.

**1** The basal nuclei consist of the **caudate nucleus** and the **lentiform** (lens-shaped) **nucleus**. The lentiform nucleus is subdivided into a medial **globus pallidus** (GLŌ-bus PAL-i-dus; pale globe) and a lateral **putamen** (pū-TĀ-men). The axon bundles that link the cerebral cortex to the diencephalon and brainstem pass between and around the basal nuclei. Together these fibers form the *internal capsule* of the cerebrum.

**?** **REVIEW**

**A.** Define *basal nuclei.*

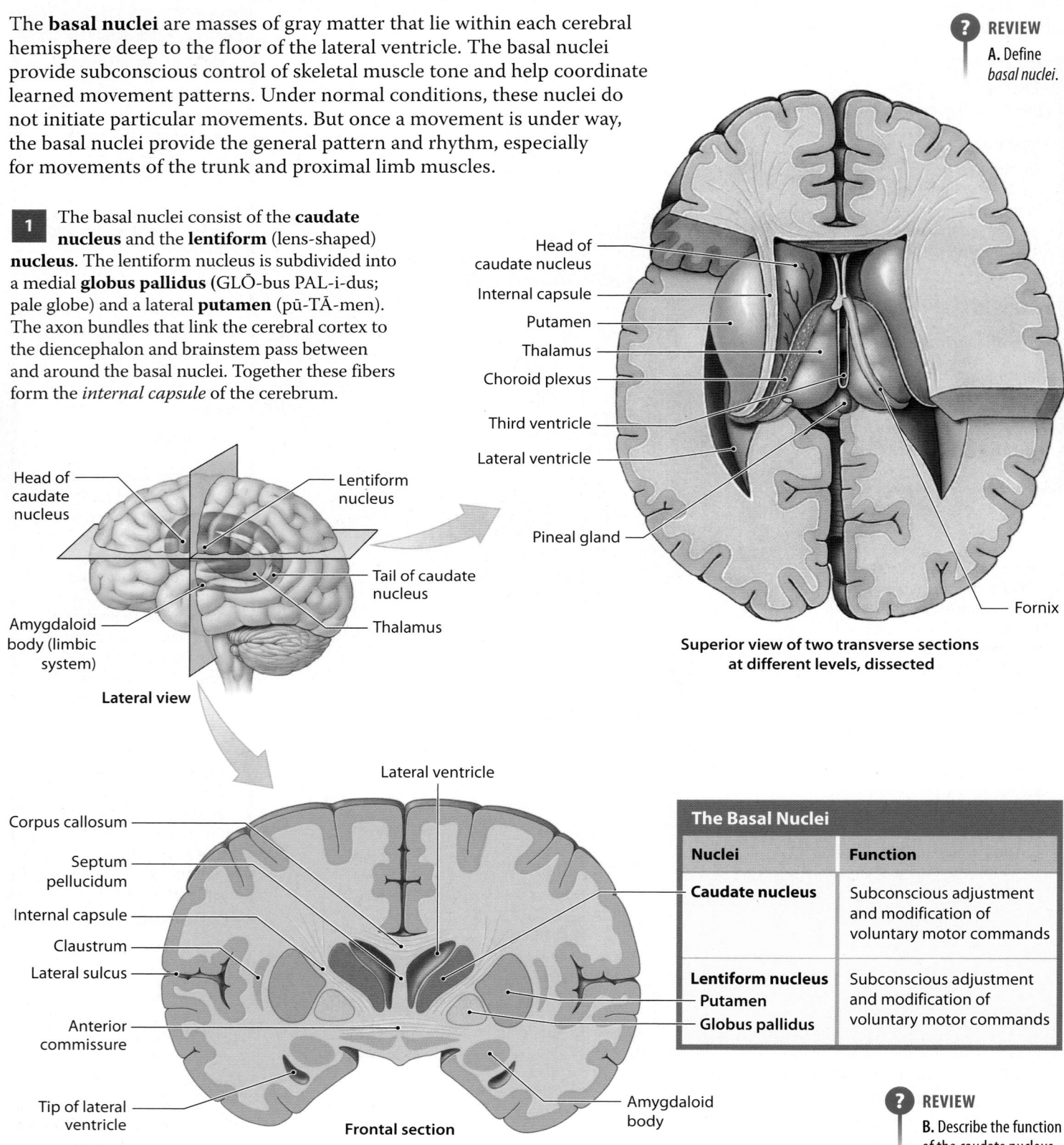

**Superior view of two transverse sections at different levels, dissected**

Head of caudate nucleus
Internal capsule
Putamen
Thalamus
Choroid plexus
Third ventricle
Lateral ventricle
Pineal gland
Fornix

**Lateral view**

Head of caudate nucleus
Lentiform nucleus
Tail of caudate nucleus
Thalamus
Amygdaloid body (limbic system)

**Frontal section**

Lateral ventricle
Corpus callosum
Septum pellucidum
Internal capsule
Claustrum
Lateral sulcus
Anterior commissure
Tip of lateral ventricle
Amygdaloid body

| The Basal Nuclei | |
|---|---|
| **Nuclei** | **Function** |
| **Caudate nucleus** | Subconscious adjustment and modification of voluntary motor commands |
| **Lentiform nucleus** Putamen Globus pallidus | Subconscious adjustment and modification of voluntary motor commands |

**?** **REVIEW**

**B.** Describe the function of the caudate nucleus.

**2** This diagram indicates the roles of the basal nuclei in modifying ongoing movements.

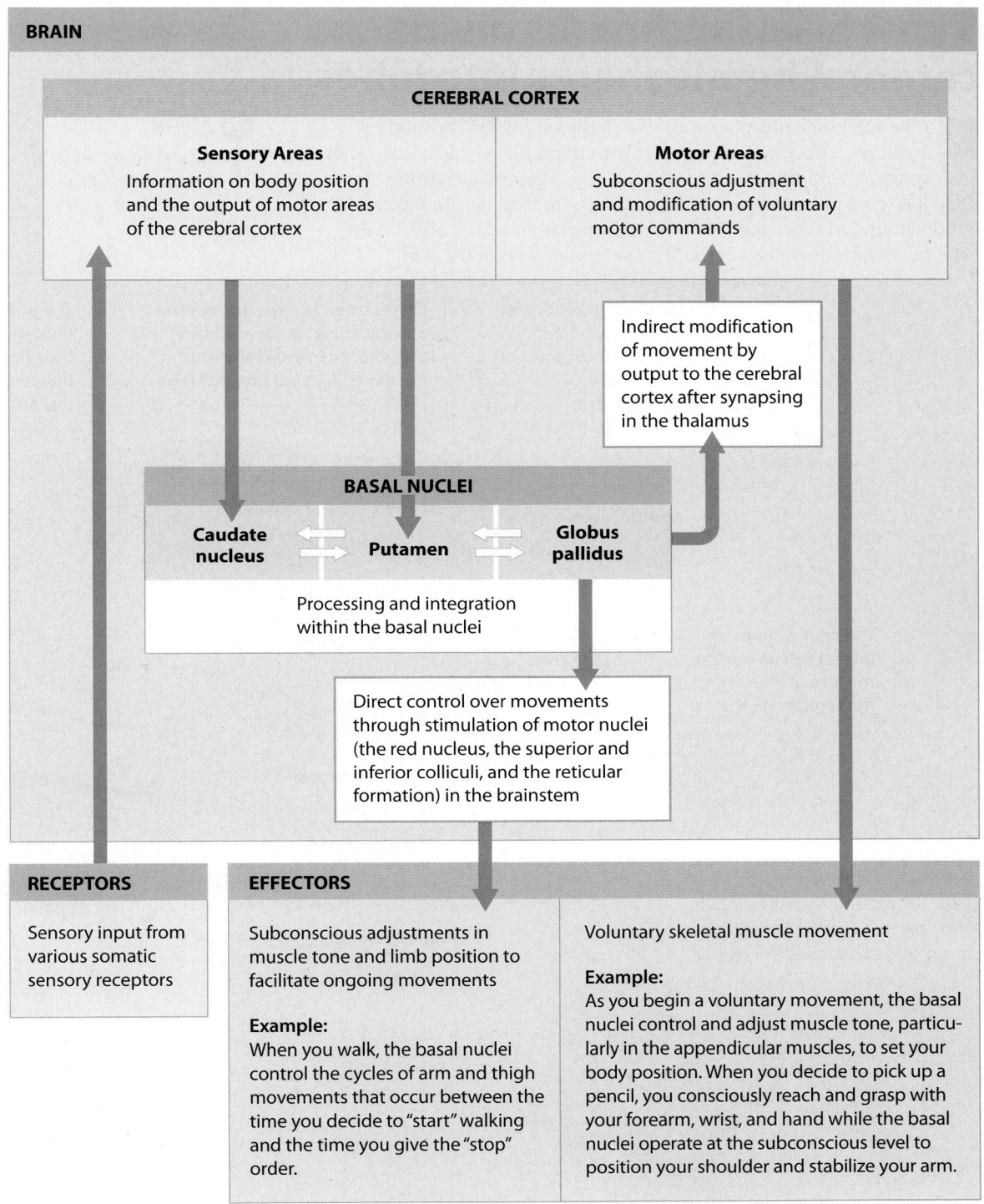

BRAIN

**CEREBRAL CORTEX**

**Sensory Areas**

Information on body position and the output of motor areas of the cerebral cortex

**Motor Areas**

Subconscious adjustment and modification of voluntary motor commands

Indirect modification of movement by output to the cerebral cortex after synapsing in the thalamus

**BASAL NUCLEI**

**Caudate nucleus** ⇄ **Putamen** ⇄ **Globus pallidus**

Processing and integration within the basal nuclei

Direct control over movements through stimulation of motor nuclei (the red nucleus, the superior and inferior colliculi, and the reticular formation) in the brainstem

**RECEPTORS**

Sensory input from various somatic sensory receptors

**EFFECTORS**

Subconscious adjustments in muscle tone and limb position to facilitate ongoing movements

**Example:**
When you walk, the basal nuclei control the cycles of arm and thigh movements that occur between the time you decide to "start" walking and the time you give the "stop" order.

Voluntary skeletal muscle movement

**Example:**
As you begin a voluntary movement, the basal nuclei control and adjust muscle tone, particularly in the appendicular muscles, to set your body position. When you decide to pick up a pencil, you consciously reach and grasp with your forearm, wrist, and hand while the basal nuclei operate at the subconscious level to position your shoulder and stabilize your arm.

**?** **INTEGRATION**

**C.** What signs and symptoms would be present in a person who has basal nuclei damage?

**Lo** **LEARNING OUTCOME**

Describe the structure and function of the basal nuclei of the cerebrum.

13

# Module 13.11

# Superficial landmarks divide the cerebral hemispheres into lobes

**1** The cerebral hemispheres consist of the cerebral cortex and its associated fiber system together with the deeper-lying basal nuclei. Each cerebral hemisphere can be divided into regions called **lobes**. Your brain has a unique pattern of sulci and gyri, as individual as a fingerprint, but the boundaries between lobes are reliable landmarks. Lobes on the external surfaces are named after the overlying bones of the skull.

 **REVIEW**
**A.** Identify the lobes of the cerebrum, and indicate the basis for their names.

The **precentral gyrus**, anterior to the central sulcus, contains the primary motor cortex. Motor neurons control voluntary movements.

On each hemisphere, the **central sulcus**, a deep groove, divides the anterior **frontal lobe** from the more posterior **parietal lobe**.

The **postcentral gyrus**, posterior to the central sulcus, contains the primary somatosensory cortex that receives sensory information that reaches our conscious awareness.

Frontal lobe

Parietal lobe

The nearly horizontal **lateral sulcus** separates the frontal lobe from the **temporal lobe**.

Occipital lobe

Temporal lobe

Cerebellum

Pons

Medulla oblongata

**2** Retracting the superficial cerebral cortex along the lateral sulcus exposes the *insula*.

The **insula** (IN-sū-luh; *insula*, island), an "island" of cortex, lies medial to the lateral sulcus.

? **REVIEW**
**B.** Describe the insula.

**Lateral view, cadaver brain**

**3** This midsagittal view indicates the inner boundaries of the lobes and highlights the way the cerebral hemispheres cover the rest of the brain. For clarity, structures and regions outside the cerebrum are labeled with gray lettering.

Precentral gyrus — Central sulcus — Postcentral gyrus

Limbic lobe

Frontal lobe

Parietal lobe

> The **parieto-occipital sulcus** visible on the medial surface separates the parietal lobe from the **occipital lobe**.

Corpus callosum

Fornix

Thalamus

Occipital lobe

Pineal gland

Hypothalamus

Corpora quadrigemina

Cerebral aqueduct

Optic chiasm

Fourth ventricle

Pituitary gland

Pons

Cerebellum

Temporal lobe

Mammillary body

Medulla oblongata

It's important to remember the following general facts about the cerebral hemispheres:

- Each cerebral hemisphere receives sensory information from, and sends motor commands to, the opposite side of the body. Thus the motor areas of the left cerebral hemisphere control muscles on the right side, and motor areas of the right cerebral hemisphere control muscles on the left side. This crossing over, which occurs in the brainstem and spinal cord, has no known functional significance.

- Even though the two hemispheres may look identical and have many similar functions, important differences exist.

- The correspondence between a specific function and a specific region of the cerebral hemisphere is imprecise. The boundaries are not distinct and have considerable overlap, and some cortical functions, such as consciousness, cannot easily be assigned to any single region. However, we know that normal individuals use all portions of the brain.

**Midsagittal section, cadaver brain**

**? INTEGRATION**

**C.** What effect would damage to the left postcentral gyrus produce?

**Lo LEARNING OUTCOME**

Identify the major superficial landmarks of the cerebrum, and cite the location of each.

13

# Module 13.12

# The lobes of the cerebral cortex have regions with specific functions

**1** The cerebral cortex is divided into six functional categories. The *primary motor cortex* issues voluntary commands to skeletal muscles, and the *primary somatosensory cortex* receives general somatic sensory information. The special senses of sight, sound, smell, and taste reach other portions of the cerebral cortex. Each sensory and motor region of the cortex is connected to a nearby association area. **Association areas** are regions of the cortex that interpret incoming data or coordinate a motor response.

**? REVIEW**
**A.** Where is the primary motor cortex located?

## Motor Cortex

Neurons of the **primary motor cortex** direct voluntary movement by controlling somatic motor neurons in the brainstem and spinal cord.

The **premotor cortex**, or somatic motor association area, coordinates learned movements.

## Gustatory Cortex

The **gustatory cortex** of the insula receives information from taste receptors.

## Olfactory Cortex

The **olfactory cortex** receives sensory information from the olfactory (smell) receptors.

## Auditory Cortex

The **primary auditory cortex** is responsible for monitoring auditory (sound) information.

The **auditory association area** monitors sensory activity in the auditory cortex and recognizes sounds, such as spoken words.

Precentral gyrus   Central sulcus   Postcentral gyrus

**Parietal lobe**

**Frontal lobe**

**Occipital lobe**

Lateral sulcus

**Temporal lobe**

## Sensory Cortex

Neurons in the **primary somatosensory cortex** receive somatic sensory information from receptors for touch, pressure, pain, vibration, or temperature.

The **somatosensory association area** monitors activity in the primary somatosensory cortex. It allows you to recognize a touch as light as a mosquito landing on your arm.

## Visual Cortex

The **primary visual cortex** receives information from the lateral geniculate bodies.

The **visual association area** monitors the patterns of activity in the visual cortex and interprets the results. When you see the symbols c, a, and r, your visual association area recognizes that they form the word "car."

**? REVIEW**
**B.** Which senses are affected by damage to the temporal lobes?

**2** **Integrative centers** receive information from association areas and direct motor activities as well as performing analytical functions. They are located in the lobes and cortical areas of both cerebral hemispheres. **Language areas** are primarily associated with the left cerebral hemisphere.

**Broca's area (motor speech area)** is associated with speech production. It regulates breathing and vocalization patterns needed for normal speech. A person with damage to Broca's area can make sounds but not words.

The **prefrontal cortex** integrates information from sensory association areas and performs intellectual functions, such as predicting the consequences of possible responses.

The **frontal eye field** controls learned eye movements, such as when you scan these lines of text.

**Wernicke's area** is associated with language comprehension. This analytical center receives information from the somatosensory association areas and plays an important role in personality by integrating sensory information and coordinating access to visual and auditory memories.

13

**3** Each of the two cerebral hemispheres is responsible for specific functions that are not ordinarily performed by the opposite hemisphere. This regional specialization is called **hemispheric lateralization**.

### Left Cerebral Hemisphere

In most people, the left hemisphere contains the specialized language areas and is responsible for language-based skills. Reading, writing, and speaking, for example, depend on processing done in the left cerebral hemisphere. In addition, the premotor cortex that controls hand movements is larger on the left side for right-handed people than for left-handed people. The left hemisphere is also important for analytical tasks, such as mathematics and logic.

### Right Cerebral Hemisphere

The right cerebral hemisphere analyzes sensory information and relates the body to the sensory environment. Interpretive centers in this hemisphere enable you to identify familiar objects by touch, smell, sight, taste, or feel. For example, the right hemisphere plays a dominant role in recognizing faces and in understanding three-dimensional relationships. It is also important in analyzing the emotional context of a conversation—for instance, distinguishing between the threat "Get lost!" and the question "Get lost?"

LEFT HAND

RIGHT HAND

Prefrontal cortex

Prefrontal cortex

Speech center

Anterior commissure

CORPUS CALLOSUM

Writing

Analysis by touch

Auditory cortex (right ear)

Auditory cortex (left ear)

Language and mathematical calculation

Spatial visualization and analysis

Visual cortex (right visual field)

Visual cortex (left visual field)

Left-handed people represent about 9 percent of the population. In most cases, the primary motor cortex of the right hemisphere controls motor function for the dominant left hand. However, the centers involved with speech and analytical function are in the left hemisphere, just as they are in right-handed people. Interestingly, an unusually high percentage of musicians and artists are left-handed. Additionally, the more a person favored one hand over the other, the stronger the connection was with the other side of the brain.

 **INTEGRATION**

**C.** Which brain region has been affected in a stroke victim who is unable to speak?

 **LEARNING OUTCOME**

Identify the locations of the motor, sensory, and association areas of the cerebral cortex, and discuss the functions of each.

# White matter connects the cerebral hemispheres and the lobes of each hemisphere and links the cerebrum to the rest of the brain

The interior of the cerebral hemispheres consists primarily of white matter (myelinated axons or fibers) organized into groups that share common functions.

**1** **Association fibers** interconnect areas of cerebral cortex within a single cerebral hemisphere.

**Lateral view**

### Association Fibers

The shortest association fibers are called **arcuate** (AR-kyū-āt) **fibers**, because they curve in an arc to pass from one gyrus to another.

Longer association fibers are organized into discrete bundles, or *fasciculi*. The **longitudinal fasciculi** connect the frontal lobe to the other lobes of the same cerebral hemisphere.

**? REVIEW**
**A.** What is the function of the longitudinal fasciculi?

**2** **Commissural** (kom-ih-SŪR-al; *commissura*, crossing over) **fibers** connect the cerebral hemispheres. **Projection fibers** link the cerebral cortex to the diencephalon, brainstem, cerebellum, and spinal cord. All projection fibers must pass through the diencephalon, where axons heading to sensory areas of the cerebral cortex pass among the axons descending from motor areas of the cortex. In gross dissection, the ascending fibers and descending fibers look alike, and the entire mass is known as the **internal capsule** (see Module 13.10).

Longitudinal fissure

**Anterior view**

### Commissural Fibers

The **corpus callosum** is the most important band of commissural fibers because it allows communication and coordination between the left and right cerebral hemispheres. It contains more than 200 million axons carrying some 4 billion impulses per second!

The **anterior commissure** is a smaller tract of commissural fibers that provides another route for communication between the cerebral hemispheres. Its importance increases if the corpus callosum is damaged.

### Projection Fibers

Projection fibers of the internal capsule connect the cerebral cortex to the diencephalon, brainstem, cerebellum, and spinal cord.

**? REVIEW**
**B.** What special names are given to axons in the white matter of the cerebral hemispheres?

 **LEARNING OUTCOME**
Discuss the significance of the white matter of the cerebral hemispheres.

# Module 13.14

# Brain activity can be monitored using external electrodes; the record is called an electroencephalogram, or EEG

**1** Neural function depends on electrical events of the plasma membranes of neurons. The brain contains billions of neurons, and their activity generates an electrical field that can be measured by placing electrodes on the scalp. The electrical activity changes constantly as nuclei and cortical areas are stimulated or quieted down. A printed recording of the electrical activity of the brain is called an **electroencephalogram (EEG)**. The electrical patterns observed are called **brain waves**.

**Alpha waves** occur in the brains of healthy, awake adults who are resting with their eyes closed. Alpha waves disappear during sleep, but they also vanish when the individual begins to concentrate on some specific task.

**Beta waves** are higher-frequency waves that appear in people who are either concentrating on a task, under stress, or in a state of psychological tension.

**Theta waves** may appear transiently during sleep in normal adults but are most often observed in children and in intensely frustrated adults. The presence of theta waves under other circumstances may indicate the presence of a brain disorder, such as a tumor.

**Delta waves** are large-amplitude, low-frequency waves. They are normally seen during deep sleep in people of all ages. Delta waves are also seen in the brains of infants (whose brains are still developing) and in awake adults when a tumor, vascular blockage, or inflammation has damaged portions of the brain.

0  Seconds   1   2   3   4

Electrical activity in the two hemispheres is generally synchronized by a "pacemaker" mechanism that seems to involve the thalamus. Asynchrony between the hemispheres can indicate localized damage or other cerebral abnormalities. For example, a tumor or injury affecting one hemisphere typically changes the pattern in that hemisphere, and the patterns of the two hemispheres are no longer aligned. A **seizure** is a temporary cerebral disorder accompanied by abnormal movements, unusual sensations, inappropriate behavior, or some combination of these signs and symptoms. Clinical conditions characterized by seizures are known as seizure disorders, or **epilepsies**. Seizures of all kinds are accompanied by a marked change in the pattern of electrical activity recorded in an EEG. The change begins in one portion of the cerebral cortex but may spread across the entire cortical surface, like a wave on the surface of a pond.

 **REVIEW**

**A.** Name and describe the four wave types associated with an EEG.

**B.** Differentiate between a *seizure* and *epilepsy*.

 **LEARNING OUTCOME**

Discuss the origin and significance of the major categories of brain waves seen in an electroencephalogram (EEG).

# The twelve pairs of cranial nerves are classified as sensory, special sensory, motor, or mixed nerves

**1** Cranial nerves are PNS components that connect directly to the brain.

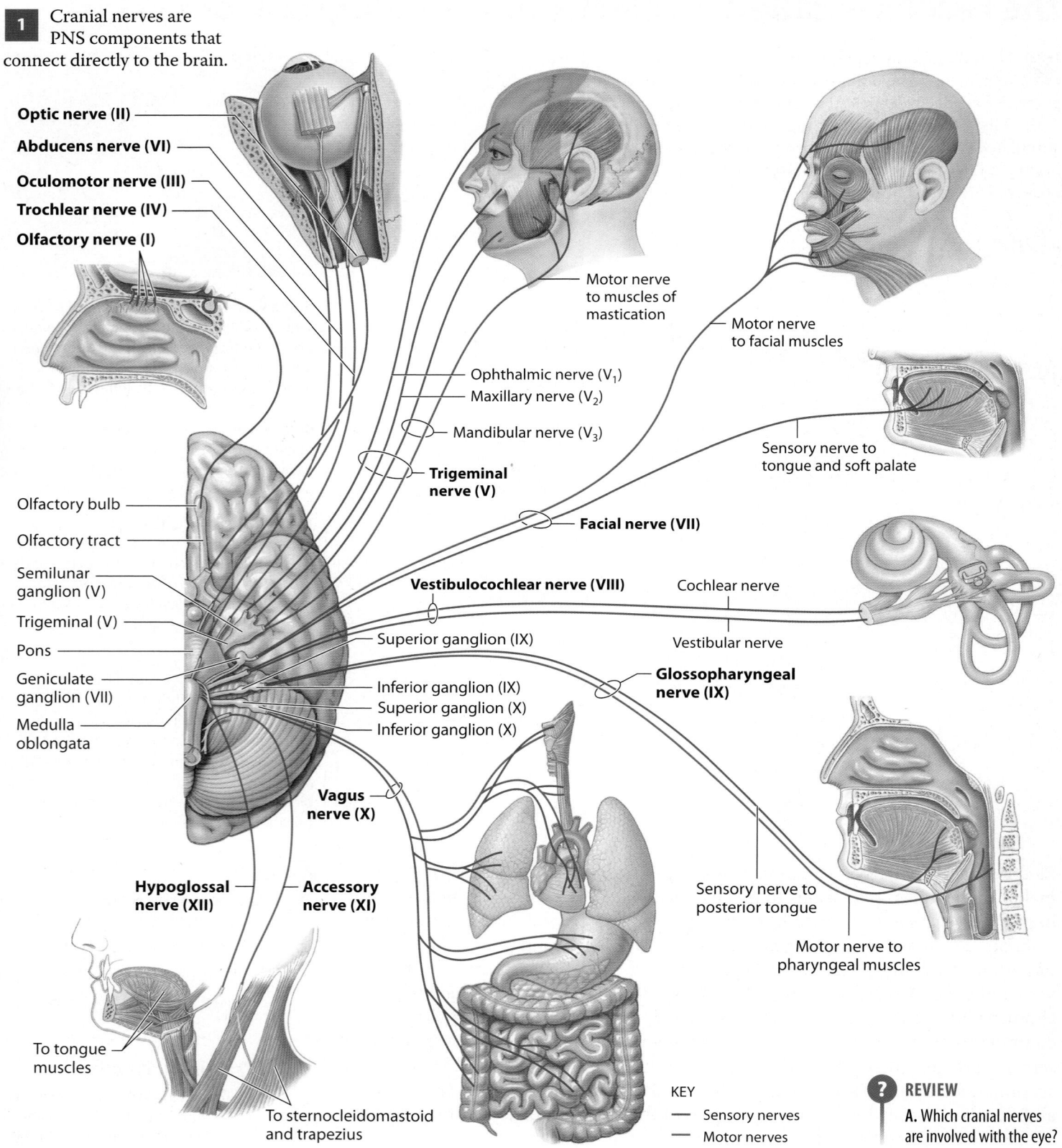

**Optic nerve (II)**

**Abducens nerve (VI)**

**Oculomotor nerve (III)**

**Trochlear nerve (IV)**

**Olfactory nerve (I)**

Motor nerve to muscles of mastication

Motor nerve to facial muscles

Ophthalmic nerve (V₁)

Maxillary nerve (V₂)

Mandibular nerve (V₃)

**Trigeminal nerve (V)**

Sensory nerve to tongue and soft palate

Olfactory bulb

Olfactory tract

**Facial nerve (VII)**

Semilunar ganglion (V)

Trigeminal (V)

Pons

Geniculate ganglion (VII)

Medulla oblongata

**Vestibulocochlear nerve (VIII)**

Superior ganglion (IX)

Inferior ganglion (IX)

Superior ganglion (X)

Inferior ganglion (X)

Cochlear nerve

Vestibular nerve

**Glossopharyngeal nerve (IX)**

**Vagus nerve (X)**

**Hypoglossal nerve (XII)**

**Accessory nerve (XI)**

Sensory nerve to posterior tongue

Motor nerve to pharyngeal muscles

To tongue muscles

To sternocleidomastoid and trapezius

KEY
— Sensory nerves
— Motor nerves

**?** **REVIEW**

**A.** Which cranial nerves are involved with the eye?

**2** The number assigned to a cranial nerve corresponds to its position along the longitudinal axis of the brain. Roman numerals preceded by CN are usually used.

## Branches and Functions of Cranial Nerves

| Cranial Nerve (Number) | Sensory Ganglion | Branch | Primary Function | Foramen | Innervation |
|---|---|---|---|---|---|
| **Olfactory (I)** | | | Special sensory | Olfactory foramina of ethmoid | Olfactory epithelium |
| **Optic (II)** | | | Special sensory | Optic canal | Retina of eye |
| **Oculomotor (III)** | | | Motor | Superior orbital fissure | Inferior, medial, superior rectus, inferior oblique, and levator palpebrae superioris muscles; intrinsic eye muscles |
| **Trochlear (IV)** | | | Motor | Superior orbital fissure | Superior oblique muscle |
| **Trigeminal (V)** | Semilunar | | Mixed | Superior orbital fissure | Areas associated with the jaws |
| | | Ophthalmic nerve ($V_1$) | Sensory | Superior orbital fissure | Orbital structures, nasal cavity, skin of forehead, upper eyelid, eyebrows, and part of nose |
| | | Maxillary nerve ($V_2$) | Sensory | Foramen rotundum | Lower eyelid; superior lip, gums, and teeth; cheek, part of nose, palate, and part of pharynx |
| | | Mandibular nerve ($V_3$) | Mixed | Foramen ovale | *Sensory:* inferior gums, teeth, lips, part of palate, and part of tongue<br>*Motor:* muscles of mastication |
| **Abducens (VI)** | | | Motor | Superior orbital fissure | Lateral rectus muscle |
| **Facial (VII)** | Geniculate | | Mixed | Internal acoustic meatus to facial canal; exits at stylomastoid foramen | *Sensory:* taste receptors on anterior two-thirds of tongue<br>*Motor:* muscles of facial expression, lacrimal gland, submandibular gland, and sublingual salivary glands |
| **Vestibulocochlear (acoustic) (VIII)** | | Vestibular nerve<br><br>Cochlear nerve | Special sensory | Internal acoustic meatus | Vestibule (receptors for motion and balance)<br>Cochlea (receptors for hearing) |
| **Glossopharyngeal (IX)** | Superior and inferior | | Mixed | Jugular foramen | *Sensory:* posterior third of tongue; pharynx and part of palate; receptors for blood pressure, pH, oxygen, and carbon dioxide concentrations<br>*Motor:* pharyngeal muscles and parotid salivary gland |
| **Vagus (X)** | Superior and inferior | | Mixed | Jugular foramen | *Sensory:* pharynx, auricle and external acoustic canal, diaphragm, visceral organs in thoracic and abdominopelvic cavities<br>*Motor:* palatal and pharyngeal muscles and visceral organs in thoracic and abdominopelvic cavities |
| **Accessory (XI)** | | Internal branch | Motor | Jugular foramen | Skeletal muscles of palate, pharynx, and larynx (with vagus nerve) |
| | | External branch | Motor | Jugular foramen | Sternocleidomastoid and trapezius muscles |
| **Hypoglossal (XII)** | | | Motor | Hypoglossal canal | Tongue musculature |

 **REVIEW**
B. Which cranial nerves have motor functions only?

 **LEARNING OUTCOME**
Identify the cranial nerves by name and number, and cite the functions of each.

13

# Functional Anatomy of the Brain and Cranial Nerves

**Labeling:** Label the structures in the accompanying figure of a lateral view of the human brain.

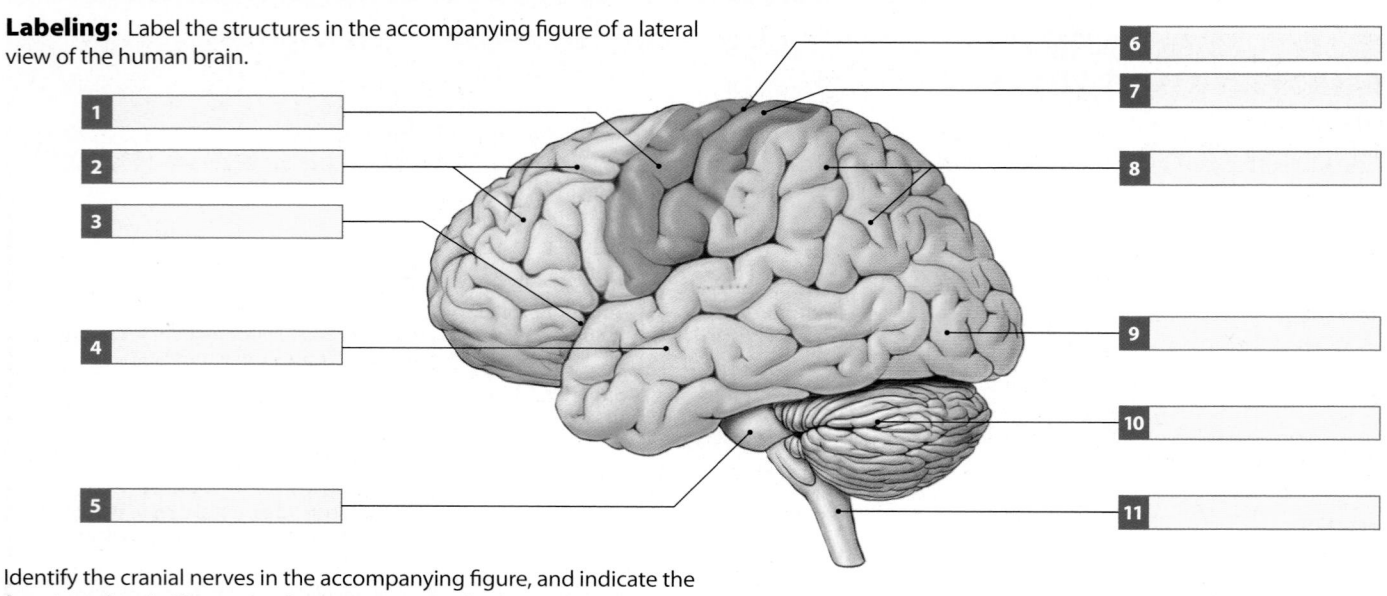

1 [_____]
2 [_____]
3 [_____]
4 [_____]
5 [_____]

6 [_____]
7 [_____]
8 [_____]
9 [_____]
10 [_____]
11 [_____]

Identify the cranial nerves in the accompanying figure, and indicate the function of each: M = motor, S = sensory, or B = both motor and sensory.

12 [_____] ○M ⊗S ○B

13 [_____] ⊗M ○S ○B
14 [_____] ○M ○S ⊗B
15 [_____] ○M ○S ⊗B
16 [_____] ○M ○S ⊗B
17 [_____] ○M ○S ⊗B

18 [_____] ○M ⊗S ○B
19 [_____] ⊗M ○S ○B
20 [_____] ⊗M ○S ○B
21 [_____] ○M ⊗S ○B
22 [_____] ○M ○S ⊗B
23 [_____] ○M ○S ⊗B

**Vocabulary:** Write the term for each of the following descriptions in the space provided.

24 Forms the walls of the diencephalon

25 The shortest association fibers in the CNS white matter

26 The tract of white matter that connects the hippocampus with the hypothalamus

27 The fibers that permit communication between the two cerebral hemispheres

28 The nuclei made up of the caudate nuclei and the lentiform nuclei

24 _____
25 _____
26 _____
27 _____
28 _____

**Section integration:** Smelling salts may restore consciousness after a person has fainted. The active ingredient of smelling salts is ammonia, and it acts by irritating the lining of the nasal cavity. Propose a mechanism by which smelling salts would raise a person from the unconscious state to the conscious state.

29 _____
_____
_____

# Module 13.16

# Sensations carried by sensory pathways to the CNS begin with transduction at a sensory receptor

Go to
MasteringA&P™

**iP**™

Nervous System: Anatomy Review

**Sensory receptors** are specialized cells or cell processes that inform your CNS about conditions inside or outside the body. The term **general senses** is used to describe our sensitivity to temperature, pain, touch, pressure, vibration, and *proprioception* (position). General sensory receptors are distributed throughout the body, and they are relatively simple in structure. **Sensory pathways** begin at peripheral receptors and end within the CNS. Much of the information carried by a sensory pathway never reaches the primary sensory cortex and our awareness. The information carried by a sensory pathway is called a **sensation**, and the conscious awareness of a sensation is called a **perception**.

**?** **REVIEW**
**A.** Define the term *general senses*.

**1** The area monitored by a single receptor cell is its **receptive field**. The larger the receptive field, the poorer your ability to localize a stimulus. A touch receptor on the general body surface, for example, may have a receptive field 7 cm (2.5 in.) in diameter. As a result, you can describe a light touch there as affecting only a general area, not an exact spot. On the tongue or fingertips, where the receptive fields are less than a millimeter in diameter, you can be very precise about the location of a stimulus.

Receptive field 1    Receptive field 2

**2** This diagram presents the basic events that occur along sensory and motor pathways.

**?** **REVIEW**
**B.** Relate receptive field size to stimulus localization.

13

Arriving stimulus

**Sensory Pathway**

**Depolarization of Receptor**

The process begins when a stimulus results in a graded change in the membrane potential of a receptor cell. The conversion of a stimulus to a change in membrane potential is called **transduction**.*

**Action Potential Generation**

If the stimulus depolarizes the receptor cell to threshold, action potentials develop in the initial segment. The longer the depolarization is sustained the higher the frequency of action potentials.

**Propagation over Labeled Line**

A **labeled line** consists of axons carrying information about one type of stimulus (touch, pressure, light). The CNS interprets the stimulus according to which labeled line is activated. That's why you see lights when you rub your eyes—the stimulated receptor sends information over axons that normally carry visual information.

**CNS Processing**

Information processing occurs at every synapse along the labeled line. The line may branch repeatedly, distributing the sensory information to multiple nuclei and centers in the spinal cord and brain.

**Immediate Involuntary Response**

Processing centers in the spinal cord or brainstem may direct an immediate reflex response even before sensations reach the cerebral cortex.

**Motor Pathway (involuntary)**

**Voluntary Response**

The voluntary response, which is not immediate, can moderate, enhance, or supplement the relatively simple involuntary reflexive response.

**Motor Pathway (voluntary)**

**Perception**

Only about 1 percent of arriving sensations are relayed to the primary somatosensory cortex.

* Note: A *transducer* is a device that converts a physical signal into an electrical signal, or the reverse. For example, a digital scale has a transducer that converts a force into an electrical signal proportional to the applied force.

 **LEARNING OUTCOME**
Describe the basic events that occur along a sensory pathway.

# Module 13.17

# Receptors are classified by function or response to the stimulus

**1** The simplest receptors are the dendrites of sensory neurons. As shown at the right, the branching tips of these dendrites, called *free nerve endings*, are not protected by accessory structures. Free nerve endings extend through a tissue the way grass roots extend into the soil. They can be stimulated by many different stimuli and therefore exhibit little receptor specificity. For example, free nerve endings that respond to tissue damage by providing pain sensations may be stimulated by chemical stimulation, pressure, temperature changes, or trauma. The sensitivity and specificity of a free nerve ending may be altered by its location, the presence of accessory structures, or both.

Free nerve endings

**2** As shown in this concept map, general sensory receptors can be classified according to the nature of the primary stimulus.

## A Functional Classification of General Sensory Receptors

### Nociceptors

**Nociceptors** are pain receptors that do not adapt quickly.  These free nerve endings have large receptive fields. Two types of axons—type A and type C fibers—carry pain sensations. Type A fibers propagate action potentials faster and respond to less intense stimuli than do C fibers.

### Thermoreceptors

**Thermoreceptors** are temperature receptors that adapt quickly. They are free nerve endings located in the dermis, skeletal muscles, liver, and hypothalamus. Cold receptors are three or four times more numerous than warm receptors. There are no structural differences between cold and warm thermoreceptors.

### Mechanoreceptors

**Mechanoreceptors** are sensitive to stimuli that distort their plasma membranes. These membranes contain mechanically gated ion channels whose gates open or close in response to stretching, compression, twisting, or other distortions of the membrane.

### Chemoreceptors

**Chemoreceptors** respond to water-soluble and lipid-soluble substances that are dissolved in body fluids (interstitial fluid, blood, and CSF).

**? REVIEW**

**A.** List the four types of general sensory receptors based on function, and identify the type of stimulus that excites each type.

---

Myelinated **Type A fibers** carry sensations of **fast pain**, or prickling pain. An injection or a deep cut produces this type of pain. These sensations quickly reach the CNS, where they often trigger somatic reflexes. They are also relayed to the primary somatosensory cortex and so receive conscious attention. In most cases, the arriving information permits the stimulus to be localized to an area several centimeters in diameter.

Slower, unmyelinated **Type C fibers** carry sensations of **slow pain**, or burning and aching pain. These sensations cause a generalized activation of the reticular formation and thalamus. The person becomes aware of the pain but has only a general idea of the area affected.

**Proprioceptors** monitor the positions of joints and muscles. They are the most structurally and functionally complex of the general sensory receptors. One example is the muscle spindle, discussed in Module 12.13.

**Baroreceptors** (bar-ō-rē-SEP-torz; *baro-*, pressure) detect pressure changes in the walls of blood vessels and in portions of the digestive, respiratory, and urinary tracts.

**Tactile receptors** provide the sensations of touch, pressure, and vibration. Touch sensations provide information about shape or texture, whereas pressure sensations indicate the degree and frequency of mechanical distortion. Extremely sensitive **fine touch and pressure receptors** provide detailed information about a stimulation, whereas **crude touch and pressure receptors** provide poor localization and give little information.

**? REVIEW**

**B.** Describe the three classes of mechanoreceptors.

**3** Receptors can also be categorized based on the nature of their response to stimulation.

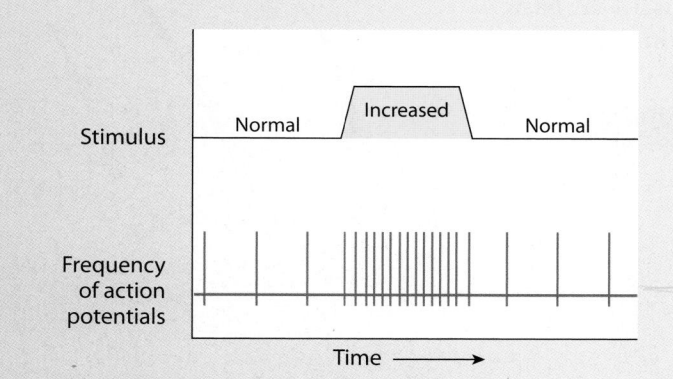

**Tonic receptors**, slow-adapting receptors, are always active and generate action potentials at a frequency that reflects the background level of stimulation. When the stimulus increases or decreases, the rate of action potential generation changes accordingly.

**Phasic receptors**, fast-adapting receptors, are normally inactive but become active for a short time in response to a change in the conditions they are monitoring.

**4** A reduction in sensitivity in the presence of a constant stimulus is called **adaptation**. You seldom notice the rumble of the tires when you ride in a car, or the background noise of the air conditioner, because your nervous system quickly adapts to stimuli that are painless and constant. Adaptation may be peripheral or central.

**? REVIEW**

**C.** Explain adaptation, and differentiate between peripheral adaptation and central adaptation.

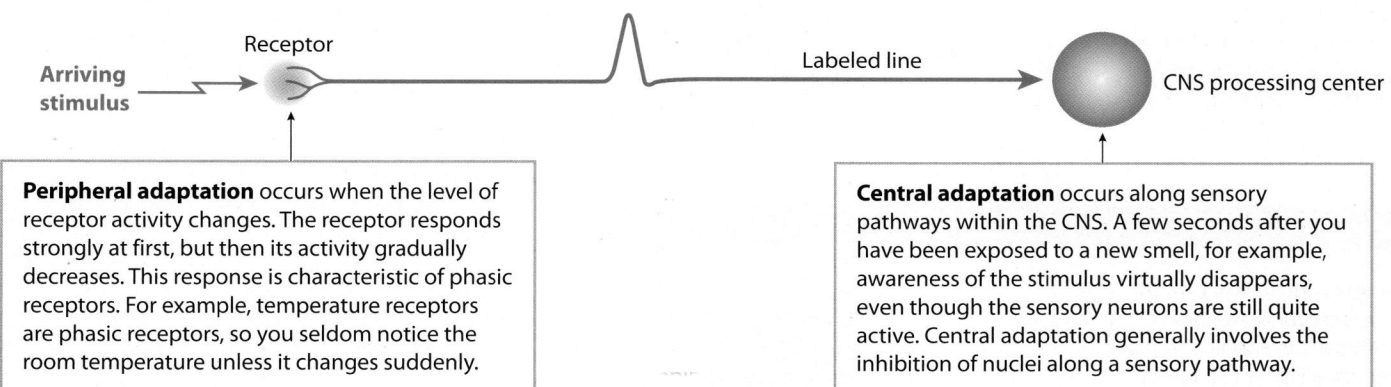

**Peripheral adaptation** occurs when the level of receptor activity changes. The receptor responds strongly at first, but then its activity gradually decreases. This response is characteristic of phasic receptors. For example, temperature receptors are phasic receptors, so you seldom notice the room temperature unless it changes suddenly.

**Central adaptation** occurs along sensory pathways within the CNS. A few seconds after you have been exposed to a new smell, for example, awareness of the stimulus virtually disappears, even though the sensory neurons are still quite active. Central adaptation generally involves the inhibition of nuclei along a sensory pathway.

Peripheral adaptation reduces the amount of information that reaches the CNS. Central adaptation at the subconscious level further restricts the amount of detail that arrives at the cerebral cortex. Most of the incoming sensory information is processed in centers along the spinal cord or brainstem at the subconscious level. This processing can produce reflexive motor responses, but you are seldom consciously aware of either the stimuli or the responses.

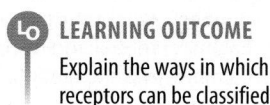

**LEARNING OUTCOME**

Explain the ways in which receptors can be classified.

# Module 13.18

# Tactile receptors have a simple structure and are abundant in the skin

Tactile sensory receptors are mechanoreceptors that provide the sensations of touch, pressure, and vibration. Not surprisingly, the greatest diversity is found in the skin, which is in constant contact with the external environment and its associated hazards and threats to homeostasis. This module discusses the basic structure and function of the important tactile receptors in the skin.

Hair

## Free Nerve Endings

**Free nerve endings** are the branching tips of sensory neurons. They are not protected by any accessory structures, and they are nonspecific: They can respond to touch and pressure, pain, and temperature stimuli. Free nerve endings are the most common receptors in the skin.

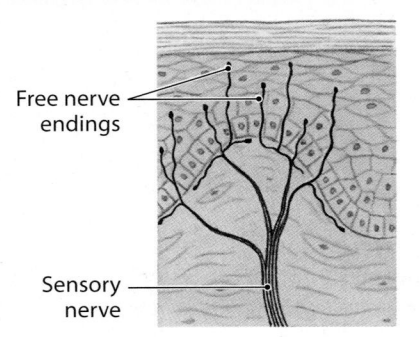

Free nerve endings

Sensory nerve

## Root Hair Plexuses

Wherever hairs are located, the nerve endings of the **root hair plexus** monitor distortions and movements across the body surface (see Module 5.7). When a hair is displaced, the movement of the follicle distorts the sensory dendrites and produces action potentials. These receptors adapt rapidly, so they are best at detecting initial contact and subsequent movements.

Hair shaft

Root hair plexus

Sensory nerves

## Tactile Discs

**Tactile discs** are fine touch and pressure receptors. They are extremely sensitive tonic receptors with very small receptive fields. The dendritic processes of many nerve terminals branching from a single myelinated afferent fiber make close contact with **Merkel cells**, unusually large epithelial cells in the stratum basale, the deepest layer of the epidermis. Each Merkel cell and its nerve terminal make up a tactile disc.

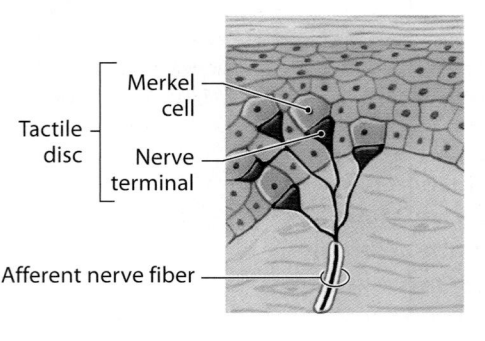

Tactile disc

Merkel cell

Nerve terminal

Afferent nerve fiber

## Tactile Corpuscles

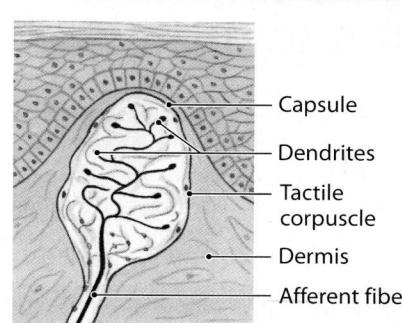

Capsule

Dendrites

Tactile corpuscle

Dermis

Afferent fiber

**Tactile corpuscles**, or *Meissner's* (MĪS-nerz) *corpuscles*, provide sensations of fine touch and pressure and low-frequency vibration. They adapt to stimulation within a second after contact. Tactile corpuscles are fairly large structures, measuring roughly 100 µm long by 50 µm wide. These receptors are most abundant in the eyelids, lips, fingertips, nipples, and external genitalia. The dendrites are highly coiled and interwoven, and they are surrounded by modified Schwann cells. A fibrous capsule surrounds the entire complex and anchors it within the papillary dermis.

## Lamellar Corpuscles

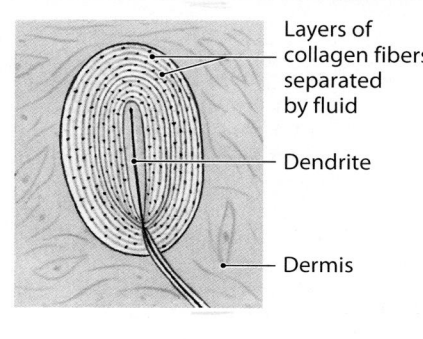

Layers of collagen fibers separated by fluid

Dendrite

Dermis

**Lamellar** (LAM-eh-lar; *lamella*, thin plate) **corpuscles**, or *Pacinian* (pa-SIN-ē-an) *corpuscles*, are sensitive to deep pressure. Because they are fast-adapting receptors, they are most sensitive to pulsing or high-frequency vibrating stimuli. A single dendrite lies within concentric layers of collagen fibers and specialized fibroblasts. The entire corpuscle may reach 4 mm long and 1 mm in diameter. The concentric layers, separated by interstitial fluid, shield the dendrite from virtually every source of stimulation other than direct pressure. Somatic sensory information is provided by lamellar corpuscles located throughout the dermis, notably in the fingers, mammary glands, and external genitalia; in the superficial and deep fasciae; and in joint capsules. Visceral sensory information is provided by lamellar corpuscles in mesenteries, in the pancreas, and in the walls of the urethra and urinary bladder.

**13**

## Bulbous Corpuscles

Capsule

Dendrites

Afferent fiber

**Bulbous corpuscles**, or *Ruffini* (roo-FĒ-nē) *corpuscles*, are sensitive to pressure and distortion of the reticular (deep) dermis, as when the skin is stretched. These receptors are tonic and show little if any adaptation. A capsule surrounds a core of collagen fibers that are continuous with those of the surrounding dermis. Within the capsule, a network of dendrites is intertwined with the collagen fibers. Any tension or distortion of the dermis tugs or twists the capsular fibers, stretching or compressing the attached dendrites and altering the activity in the myelinated afferent fiber.

 **INTEGRATION**

**B.** Which is likely to be more sensitive to continuous deep pressure: a lamellar corpuscle or a bulbous corpuscle?

 **LEARNING OUTCOME**

List the types of tactile receptors, and specify the functions of each.

 **REVIEW**

**A.** Which types of tactile receptors are located only in the dermis?

# Three major somatic sensory pathways carry information from the skin and muscles to the CNS

**1** Two tracks within the **spinothalamic pathway** provide conscious sensations of poorly localized ("crude") touch, pressure, pain, and temperature. In this pathway, axons of **first-order neurons** enter the spinal cord and synapse on **second-order neurons** within the posterior gray horns. The axons of these interneurons cross to the opposite side of the spinal cord before ascending to the thalamus. **Third-order neurons** synapse in the primary somatosensory cortex.

## Spinothalamic Pathway

The **anterior spinothalamic tracts** of the spinothalamic pathway carry crude touch and pressure sensations.

The **lateral spinothalamic tracts** of the spinothalamic pathway carry pain and temperature sensations.

A **somatotopy** is a functional map of the somatosensory cortex. This **sensory homunculus**, or "little man in the brain," looks like a caricature, showing that the relative size of cortex devoted to a particular body region is proportional to the density of sensory input received from that body region.

The second-order neurons synapse in the ventral nuclei of the thalamus.

Midbrain

Medulla oblongata

Anterior spinothalamic tract

Spinal cord

**Anterior views**

Crude touch and pressure sensations from right side of body

Midbrain

Medulla oblongata

Lateral spinothalamic tract

Spinal cord

Pain and temperature sensations from right side of body

**? REVIEW**
**A.** Define *somatotopy*.

KEY

→ Axon of first-order neuron

⇨ Second-order neuron

➡ Third-order neuron

**2** The **posterior column pathway** carries sensations of highly localized ("fine") touch, pressure, vibration, and proprioception. This pathway begins at a peripheral receptor and ends at the primary somatosensory cortex of the cerebral hemispheres.

## Posterior Column Pathway

Primary Somatosensory Cortex

Ventral nuclei in thalamus

Midbrain

The **medial lemniscus** is a tract leading from the gracile nucleus and cuneate nucleus to the thalamus.

Gracile nucleus and cuneate nucleus

Medulla oblongata

The sensory axons ascend in the medial gracile fasciculus and lateral cuneate fasciculus.

Spinal cord

Posterior root ganglion

Fine touch, vibration, pressure, and proprioception sensations from right side of body

**3** The cerebellum receives proprioceptive information about the position of skeletal muscles, tendons, and joints along the **spinocerebellar pathway**. The posterior spinocerebellar tracts contain axons that do not cross over to the opposite side of the spinal cord. These axons reach the cerebellar cortex by the inferior cerebellar peduncle of that side. The anterior spinocerebellar tracts are dominated by axons that have crossed over to the opposite side of the spinal cord.

## Spinocerebellar Pathway

Pons

Cerebellum

Medulla oblongata

Spinocerebellar pathway

Posterior spinocerebellar tract

Spinal cord

Anterior spinocerebellar tract

Proprioceptive input from muscle spindles, tendons, and joint capsules

**4** This cross section through the spinal cord shows the locations of the somatic sensory pathways.

**Posterior**

Posterior column pathway

Spinocerebellar pathway

Spinothalamic pathway

**Anterior**

**?** **INTEGRATION**

**B.** Which spinal tracts carry action potentials generated by nociceptors?

**C.** Which cerebral hemisphere receives impulses conducted by the right gracile fasciculus?

**LO** **LEARNING OUTCOME**

Identify and describe the major sensory pathways.

13

# Module 13.20

# The somatic nervous system controls skeletal muscles through upper and lower motor neurons

Somatic motor pathways always involve at least two motor neurons: an **upper motor neuron**, whose cell body lies in a CNS processing center, and a **lower motor neuron**, whose cell body lies in a nucleus of the brainstem or spinal cord. The upper motor neuron synapses on the lower motor neuron, which in turn innervates a single motor unit in a skeletal muscle.

**1** The **corticospinal pathway** provides voluntary control over skeletal muscles. This pathway is sometimes called the *pyramidal system* because it begins at the *pyramidal cells* of the primary motor cortex. The axons of these upper motor neurons descend into the brainstem and spinal cord to synapse on lower motor neurons that control skeletal muscles.

**? REVIEW**

**A.** Describe the role of the corticobulbar tracts.

**B.** Define *corticospinal tracts*.

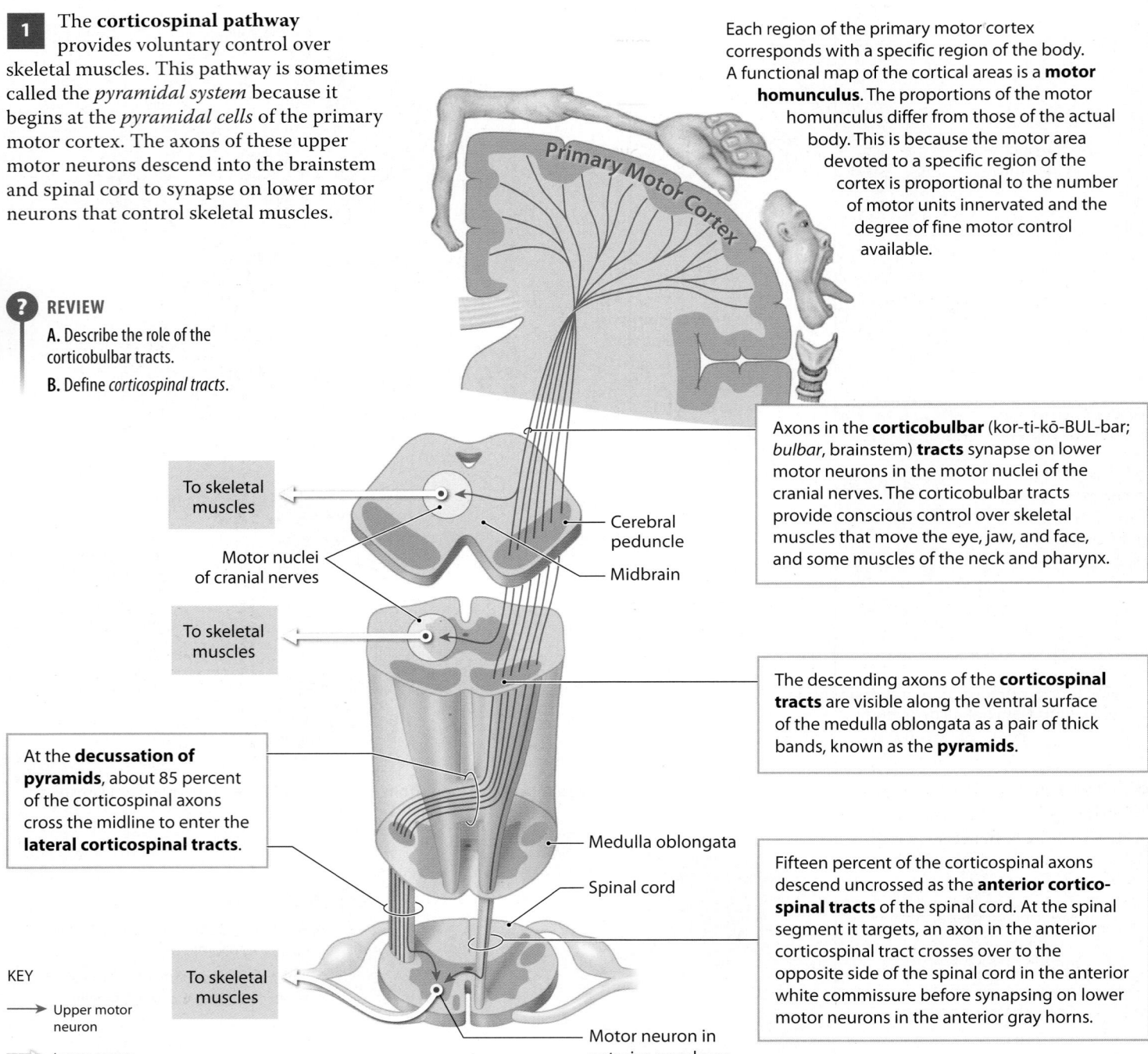

Each region of the primary motor cortex corresponds with a specific region of the body. A functional map of the cortical areas is a **motor homunculus**. The proportions of the motor homunculus differ from those of the actual body. This is because the motor area devoted to a specific region of the cortex is proportional to the number of motor units innervated and the degree of fine motor control available.

Primary Motor Cortex

To skeletal muscles

Motor nuclei of cranial nerves

To skeletal muscles

Cerebral peduncle

Midbrain

Axons in the **corticobulbar** (kor-ti-kō-BUL-bar; *bulbar*, brainstem) **tracts** synapse on lower motor neurons in the motor nuclei of the cranial nerves. The corticobulbar tracts provide conscious control over skeletal muscles that move the eye, jaw, and face, and some muscles of the neck and pharynx.

The descending axons of the **corticospinal tracts** are visible along the ventral surface of the medulla oblongata as a pair of thick bands, known as the **pyramids**.

At the **decussation of pyramids**, about 85 percent of the corticospinal axons cross the midline to enter the **lateral corticospinal tracts**.

Medulla oblongata

Spinal cord

Fifteen percent of the corticospinal axons descend uncrossed as the **anterior corticospinal tracts** of the spinal cord. At the spinal segment it targets, an axon in the anterior corticospinal tract crosses over to the opposite side of the spinal cord in the anterior white commissure before synapsing on lower motor neurons in the anterior gray horns.

**KEY**

To skeletal muscles

→ Upper motor neuron

⇨ Lower motor neuron

Motor neuron in anterior gray horn

**Anterior view**

**2** Several centers in the cerebrum, diencephalon, and brainstem may issue somatic motor commands as a result of processing performed at a subconscious level. The locations of those centers are indicated in the diagram to the right. The components of the **medial pathway** help control gross movements of the trunk and proximal limb muscles, whereas those of the **lateral pathway** help control the distal limb muscles that perform more precise movements.

**?** **REVIEW**

**C.** What effect would increased stimulation of the motor neurons of the red nucleus have on muscle tone?

**3** The locations of the medial and lateral pathways are indicated in this cross section of the spinal cord.

The **red nucleus**, the primary nucleus of the lateral pathway, receives information from the cerebrum and cerebellum and adjusts upper limb position and background muscle tone.

| Nuclei of the Medial Pathway |
| --- |
| Superior and inferior colliculi |
| Reticular formation |
| Vestibular nucleus |

Motor cortex — Thalamus — Basal nuclei — Cerebellar nuclei — Medulla oblongata

**Corticospinal Pathway**

Anterior corticospinal tract    Lateral corticospinal tract

**Medial Pathway**

The medial pathway is primarily concerned with the control of muscle tone and gross movements of the neck, trunk, and proximal limb muscles. The upper motor neurons of the medial pathway are located in the vestibular nuclei, the superior and inferior colliculi, and the reticular formation.

The **reticulospinal tracts** contain the axons of upper motor neurons in the reticular formation. The reticular formation receives input from almost every ascending and descending pathway. It also has extensive connections with the cerebrum, the cerebellum, and brainstem nuclei.

The **vestibulospinal tracts** begin at the vestibular nuclei of cranial nerve VIII. These nuclei receive sensory information from the internal ear about the position and movement of the head. They respond to changes in the orientation of the head by issuing motor commands that alter the muscle tone and position of the neck, eyes, head, and limbs.

The **tectospinal tracts** contain the axons of upper motor neurons in the superior and inferior colliculi of the midbrain. Axons in the tectospinal tracts direct reflexive changes in the position of the head, neck, and upper limbs in response to bright lights, sudden movements, or loud noises.

**Lateral Pathway**

The lateral pathway is primarily concerned with the control of muscle tone and the more precise movements of the distal parts of the limbs.

The upper motor neurons of the lateral pathway lie within the red nuclei of the midbrain. Axons of these motor neurons cross to the opposite side of the brain and descend into the spinal cord in the **rubrospinal tracts** (*ruber*, red). In humans, the rubrospinal tracts are small and extend only to the cervical spinal cord, where they provide motor control over distal muscles of the upper limbs.

**Lo** **LEARNING OUTCOME**

Describe the components, processes, and functions of the somatic motor pathways.

13

# Module 13.21
# There are multiple levels of somatic motor control

**1** In the last two chapters you encountered many nuclei in the spinal cord and brain that play roles in skeletal muscle contractions. In general, the closer a motor center is to the cerebral cortex, the more complex and variable the motor activities are. The cerebellum is involved in coordinating the motor activities at multiple levels.

**Cerebral Cortex**

Plans and initiates voluntary motor activity

**Basal Nuclei**

Modify voluntary and reflexive motor patterns at the subconscious level

**Hypothalamus**

Controls reflex motor patterns related to eating, drinking, and sexual activity; modifies respiratory reflexes

**Thalamus and Midbrain**

Control reflexes in response to visual and auditory stimuli

**Pons and Medulla Oblongata**

Control balance reflexes and more complex respiratory reflexes

**Cerebellum**

Coordinates complex motor patterns through feedback loops involving the cerebral cortex and basal nuclei as well as nuclei of the medial and lateral pathways

**Brainstem and Spinal Cord**

Control simple cranial and spinal reflexes

| Level of Motor Complexity | |
|---|---|
| **Region** | **Complexity** |
| Cerebral cortex | More complex |
| Cerebellum | |
| Basal nuclei | |
| Thalamus and midbrain | |
| Hypothalamus | |
| Pons and medulla oblongata | |
| Brainstem and spinal cord | Less complex |

**? REVIEW**

**A.** Which brain regions control reflexes in response to visual and auditory stimuli that are experienced while viewing a movie?

**B.** The basic reflex motor patterns related to eating and drinking are controlled by which region of the brain?

## Preparing for Movement

**2**  When you make a conscious decision to perform a specific movement, information is relayed from the frontal lobes to the premotor cortex. These areas in turn relay the information to the cerebellum and basal nuclei.

## Performing a Movement

**3**  As the movement begins, the premotor cortex sends instructions to the primary motor cortex. Feedback from the basal nuclei and cerebellum modifies those commands, and output along the medial and lateral pathways directs involuntary adjustments in position and muscle tone.

The basal nuclei adjust patterns of movement in two ways:
1. They alter the sensitivity of the pyramidal cells to adjust the output along the corticospinal tract.
2. They change the excitatory or inhibitory output of the medial and lateral pathways.

If the primary motor cortex is damaged, the person loses fine motor control over skeletal muscles. However, some voluntary movements can still be controlled by the basal nuclei based on information from the prefrontal cortex concerning planned movements. But because the corticospinal pathway is inoperative, the cerebellar feedback cannot fine-tune the ongoing movements. A person in this condition can stand, maintain balance, and even walk, but all movements are hesitant, awkward, and poorly controlled.

Premotor cortex

Decision in frontal lobes

Basal nuclei

Cerebellum

Cerebral cortex

Premotor cortex

Primary motor cortex

Basal nuclei

Other nuclei of the medial and lateral pathways

Cerebellum

Cerebral cortex

Corticospinal pathway

Lower motor neurons

Motor activity

As the movement proceeds, the cerebellum monitors proprioceptive and vestibular information (balance and equilibrium) and compares the arriving sensations with those experienced during previous movements. It then adjusts the activities of the upper motor neurons involved.

**? INTEGRATION**

**C.** During a tennis match, you decide how and where to hit the ball. Explain how the premotor cortex is involved in your decisions.

**Lo LEARNING OUTCOME**

Describe the levels of information processing involved in motor control.

# Module 13.22
# Nervous system disorders may result from problems with neurons, pathways, or a combination of the two

Heart

## Referred Pain

**1** **Referred pain** is the sensation of pain in a part of the body other than its actual source. A familiar example is the pain of a heart attack, which is frequently felt in the left arm. Strong visceral pain sensations arriving at a segment of the spinal cord can stimulate interneurons that are part of the spinothalamic pathway. Activity in these interneurons leads to the stimulation of the primary sensory cortex, so the person feels pain in a specific part of the body surface.

**? REVIEW**
**A.** Define *referred pain*.

## Parkinson's Disease

Normal substantia nigra    Diminished substantia nigra in Parkinson's patient

**2** **Parkinson's disease** results when neurons of the substantia nigra are damaged or secrete less dopamine. The basal nuclei become more active, which raises skeletal muscle tone and produces rigidity and stiffness. People who have Parkinson's disease have difficulty starting voluntary movements because opposing muscle groups do not relax; they must be overpowered. Once a movement is under way, every aspect must be voluntarily controlled through intense effort and concentration.

## Rabies

**3** **Rabies** is a dramatic example of a clinical condition directly related to retrograde flow in peripheral axons. A bite from a rabid animal injects the rabies virus into peripheral tissues, where virus particles quickly enter axon terminals. Retrograde flow then carries the virus into the CNS, with potentially fatal results. Many toxins (including heavy metals), some pathogenic bacteria, and other viruses also bypass CNS defenses by exploiting axoplasmic transport (see Module 11.2).

**? REVIEW**
**B.** Describe how rabies is contracted.

Rabies viruses

Retrograde flow

Axon terminal

## Cerebral Palsy

**4** **Cerebral palsy (CP)** refers to a number of disorders that affect voluntary motor movement. The motor dysfunction is nonprogressive, appears during infancy or childhood, and persists throughout the person's lifetime. The cause may be trauma associated with premature or unusually stressful birth, maternal exposure to drugs (including alcohol), or a genetic defect that causes improper development of motor pathways.

## Amyotrophic Lateral Sclerosis

**5** **Amyotrophic lateral sclerosis (ALS)** is a progressive, degenerative disorder that affects motor neurons in the spinal cord, brainstem, and cerebral hemispheres. The degeneration affects both upper and lower motor neurons. A defect in axonal transport is thought to underlie the disease. Because a motor neuron and its dependent muscle fibers are so intimately related, the destruction of CNS neurons causes atrophy of the associated skeletal muscles. ALS is commonly known as *Lou Gehrig's disease*, named after the famous New York Yankees player who died of the disorder. Noted physicist Stephen Hawking is also afflicted with this condition.

**?** **REVIEW**
**C.** Describe amyotrophic lateral sclerosis (ALS).

## Alzheimer's Disease

**6** **Alzheimer's disease (AD)** is a progressive disorder characterized by the loss of higher-order cerebral functions. It is the most common cause of **senile dementia**, or senility. Signs and symptoms may appear at 50–60 years of age or later, although the disease occasionally affects younger people. An estimated 2 million people in the United States—including roughly 15 percent of those over age 65, and nearly half of those over age 85—have some form of the condition, and it causes approximately 100,000 deaths each year. Microscopic examination of the brains of AD patients reveals intracellular and extracellular abnormalities in brain regions that are specifically associated with memory processing, such as the hippocampus.

Abnormal dendrites, axons, and extracellular proteins form complexes known as Alzheimer's plaques.

## Multiple Sclerosis

**7** **Multiple sclerosis** (skler-Ō-sis; *sklerosis*; hardness), or **MS**, is a disease of the CNS characterized by recurrent incidents of demyelination that affects axons in the optic nerve, brain, and spinal cord. Common signs and symptoms include partial loss of vision and problems with speech, balance, and general motor coordination, including bowel and urinary bladder control. The time between incidents and the degree of recovery vary from case to case. In about one-third of all cases, the disorder is progressive, and functional impairment increases following each new incident. The first attack typically occurs at 30–40 years of age; the incidence among women is 1.5 times that among men.

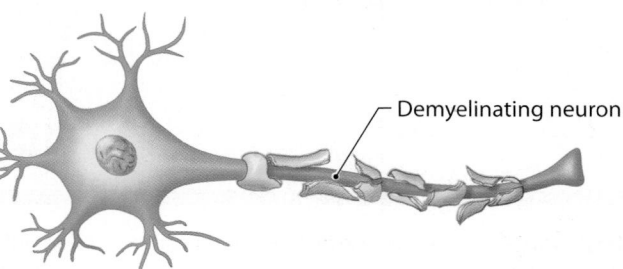

Demyelinating neuron

**Lo** **LEARNING OUTCOME**
Describe the roles of the nervous system in referred pain, Parkinson's disease, rabies, cerebral palsy, amyotrophic lateral sclerosis, Alzheimer's disease, and multiple sclerosis.

# Sensory and Motor Pathways

**Labeling:** Label each type of tactile receptor found in the skin.

1 _____

2 _____

3 _____

4 _____

5 _____

6 _____

## Fill-in

The general organization of the spinal cord is such that motor tracts are   7 _____   (anterior or posterior), and sensory tracts are   8 _____   (anterior or posterior).

**13**

**Short answer:** Identify the descending and ascending tracts and pathways in the accompanying sectional diagram of the spinal cord, and then describe the general functions of the tracts of each pathway.

9 _____

10 _____

11 _____

12 _____

13 _____

14 _____

15 _____

16 _____

17 _____

18 _____

19 _____

**Section integration:** A person whose primary motor cortex has been injured retains the ability to walk, maintain balance, and perform other voluntary and involuntary movements. Even though the movements lack precision and are awkward and poorly controlled, why is the ability to walk and maintain balance possible?

20 _____

# Study Outline

## SECTION 1 • Functional Anatomy of the Brain and Cranial Nerves

### Module 13.1
### The brain develops from a hollow neural tube    p. 471

1. A hollow cylinder called the **neural tube** develops into the brain and spinal cord.

2. The cephalic portion of the neural tube enlarges into three divisions called the **primary brain vesicles**: the **prosencephalon**, **mesencephalon**, and **rhombencephalon**.

3. **Secondary brain vesicles** develop from the prosencephalon (**diencephalon** and **telencephalon**) and rhombencephalon (**metencephalon** and **myelencephalon**).

### Module 13.2
### Each region of the brain has distinct structural and functional characteristics    p. 472

4. The **cerebrum** is divided into **cerebral hemispheres**, and is covered by a superficial layer of gray matter called **cerebral cortex**.

5. The functions of the cerebrum include thought, memory, sensory processing, and regulating skeletal muscle contractions.

6. Visible on the cerebrum are **fissures** (deep grooves), **gyri** (folds), and **sulci** (shallow depressions).

7. The **diencephalon** links cerebral hemispheres with the rest of the CNS. It is composed of the **thalamus** and **hypothalamus**.

8. The **brainstem** includes the **midbrain**, **pons**, and **medulla oblongata**.

9. The **cerebellum** controls motor commands from the cerebral cortex.

10. Chambers within the brain called **ventricles** are filled with cerebrospinal fluid and lined by ependymal cells. The ventricles of the brain contain two **lateral ventricles**, a **third ventricle**, and a **fourth ventricle**. Each lateral ventricle is connected to the third ventricle through an **interventricular foramen**, and the **cerebral aqueduct** connects the third ventricle to the fourth ventricle.

11. The **corpus callosum** interconnects the two cerebral hemispheres. The **septum pellucidum** is a thin partition separating the two lateral ventricles.

### Module 13.3
### The cranial meninges and cerebrospinal fluid protect and support the brain    p. 474

12. The cranial **dura mater**, **arachnoid mater**, and **pia mater** are continuous with the layers of the spinal meninges.

13. The cranial dura mater is made of an outer periosteal layer that is fused to the periosteum of the cranial bones, and typically separated from the inner meningeal layer by the **dural venous sinuses**.

14. The cranial arachnoid mater consists of the arachnoid membrane and the arachnoid trabeculae that connect it to the pia mater.

15. The pia mater adheres to the surface of the brain by astrocyte processes.

16. **Dural folds** extend into the cranial cavity. These are the **falx cerebri**, **tentorium cerebelli**, and **falx cerebelli**. The **superior sagittal sinus** is the largest dural sinus.

17. **Cerebrospinal fluid (CSF)** is produced by the **choroid plexuses** and bathes the surfaces of the CNS. CSF reaches the subarachnoid space through two **lateral apertures** and a single **median aperture** in the roof of the fourth ventricle.

18. CSF is absorbed into the venous circulation at **arachnoid granulations**.

### Module 13.4
### The medulla oblongata contains autonomic reflex centers, relay stations, and ascending and descending tracts    p. 476

19. All communication between the brain and spinal cord involves tracts that ascend or descend through the **medulla oblongata**. It is also a center for autonomic reflexes and visceral functions.

20. In the medulla oblongata, **pyramids** contain tracts of motor fibers. Some of these fibers cross over to the opposite side of the medulla. This crossing over is called a **decussation**.

21. The medulla also contains **autonomic centers** and **relay stations**.

### Module 13.5
### The pons links the cerebellum to the brain and spinal cord and has vital autonomic reflex centers    p. 477

22. The **pons** links the cerebellum with the midbrain, diencephalon, cerebrum, medulla oblongata, and spinal cord.

### Module 13.6
### The cerebellum coordinates learned and reflexive patterns of muscular activity at the subconscious level    p. 478

23. The **cerebellum** functions to (1) adjust the postural muscles of the body and (2) fine-tune movements controlled at the conscious and subconscious levels.

24. The cerebellar hemispheres are each made of an **anterior** and **posterior lobe**, and separated by the **vermis**. The **folia** are folds on the cerebellar surface.

25. The three **cerebellar peduncles** on each side are tracts that link the cerebellum with the brainstem, cerebrum, and spinal cord.

### Module 13.7
### The midbrain regulates auditory and visual reflexes and controls alertness    p. 480

26. The **midbrain** is the most complex and integrative portion of the brainstem.

27. Major landmarks of the midbrain are the **corpora quadrigemina** (formed by the paired **superior** and **inferior colliculi**), the **red nucleus**, the **substantia nigra**, and the **cerebral peduncles**.

28. The **tectum** and **tegmentum** surround the cerebral aqueduct.

### Module 13.8
### The diencephalon consists of the epithalamus, thalamus, and hypothalamus    p. 482

29. The **epithalamus** is the roof of the diencephalon. Its anterior portion contains an extensive choroid plexus that extends through the interventricular foramina.

30. The anterior aspect of the diencephalon is marked by the **anterior commissure** and the **optic chiasm**.

31. The **pineal gland** is an endocrine structure that secretes **melatonin**, an important regulator of day–night cycles and reproductive functions.

32. Important structures of the **thalamus** are the **lateral geniculate body**, **medial geniculate body**, and the **optic tract**.

33. The **hypothalamus** contains important control and integrative centers. Hypothalamic nuclei include the **preoptic area** and **suprachiasmatic nucleus**.

34. The **infundibulum** connects the floor of the hypothalamus to the pituitary gland.

### Module 13.9

**The limbic system is a functional group of tracts and nuclei located in the cerebrum and diencephalon    p. 484**

35. The **limbic system** is a functional grouping rather than an anatomical one. It functions to establish emotional states, link conscious and unconscious functions, and facilitate memory.

36. The components of the limbic system in the diencephalon are the anterior group of thalamic nuclei, hypothalamus, and mammillary body.

37. The **cingulate gyrus**, **parahippocampal gyrus**, and **dentate gyrus** form the **limbic lobe**, the region of the limbic system in the cerebrum.

38. The **amygdaloid body** of the limbic lobe plays a role in the regulation of heart rate, the "fight or flight" response, emotions, and memory.

39. The **hippocampus** is important in learning and long-term memory, and is connected to the hypothalamus by the **fornix**.

### Module 13.10

**The basal nuclei of the cerebrum adjust and refine ongoing voluntary movements    p. 486**

40. The **basal nuclei** are masses of gray matter in the floor of the lateral ventricles. They provide subconscious control of skeletal muscle tone and help coordinate learned movement patterns.

41. The basal nuclei consist of the **caudate nucleus** and **lentiform nucleus**. The lentiform nucleus is subdivided into the **globus pallidus** and **putamen**.

### Module 13.11

**Superficial landmarks divide the cerebral hemispheres into lobes    p. 488**

42. Each cerebral hemisphere can be divided into **lobes** that are named after the overlying bones of the skull.

43. Important landmarks of the cerebrum are the **precentral gyrus**, **central sulcus**, **postcentral gyrus**, and the **lateral sulcus**.

44. Retraction of the cerebral cortex along the lateral sulcus exposes the **insula**.

45. Each cerebral hemisphere receives information from and sends commands to the opposite side of the body. Although each hemisphere has similar appearances, they have different functions. The correspondence between function and region in the cerebral cortex is imprecise.

### Module 13.12

**The lobes of the cerebral cortex have regions with specific functions    p. 490**

46. The **primary motor cortex** of the precentral gyrus sends voluntary commands to skeletal muscles. The **premotor cortex**, or somatic association area, coordinates learned movements.

47. The **primary somatosensory cortex** receives general somatic sensory information.

48. **Association areas** interpret incoming data or coordinate a motor response.

49. The **gustatory cortex** of the insula receives information from taste receptors, and the **olfactory cortex** of the insula receives sensory information from the olfactory receptors.

50. The **auditory cortex**, located in the temporal lobe, contains the **primary auditory cortex** and the **auditory association area**.

51. The **visual cortex** in the occipital lobe contains the **primary visual cortex** and the **visual association area**.

52. **Integrative centers** receive information from association areas, direct motor activities, and perform analyses. They include **Broca's area**, the **prefrontal cortex**, the **frontal eye field**, and **Wernicke's area**.

53. Each cerebral hemisphere is responsible for specific functions not performed by the opposite hemisphere. This regional specialization is called **hemispheric lateralization**.

### Module 13.13

**White matter connects the cerebral hemispheres and the lobes of each hemisphere and links the cerebrum to the rest of the brain    p. 492**

54. **Association fibers** such as the **arcuate fibers** and the **longitudinal fasciculi** interconnect areas of the neural cortex within a single cerebral hemisphere.

55. **Commissural fibers** such as the **corpus callosum** and **anterior commissure** interconnect the cerebral hemispheres.

### Module 13.14

**Brain activity can be monitored using external electrodes; the record is called an electroencephalogram, or EEG    p. 493**

56. **EEG** electrical patterns are called **brain waves**.

57. **Alpha waves** appear in awake adults, **beta waves** are visible when a person is concentrating or under stress, **theta waves** appear mostly in children, and **delta waves** are normally seen during sleep.

### Module 13.15

**The twelve pairs of cranial nerves are classified as sensory, special sensory, motor, or mixed nerves    p. 494**

58. The sensory cranial nerves are the ophthalmic and maxillary nerves of the trigeminal nerve (V).

59. The special sensory cranial nerves are the **olfactory (I)**, **optic (II)**, and **vestibulocochlear (VIII) nerves**.

60. The motor cranial nerves are the **oculomotor (III)**, **trochlear (IV)**, **abducens (VI)**, **accessory (XI)**, and **hypoglossal (XII) nerves**.

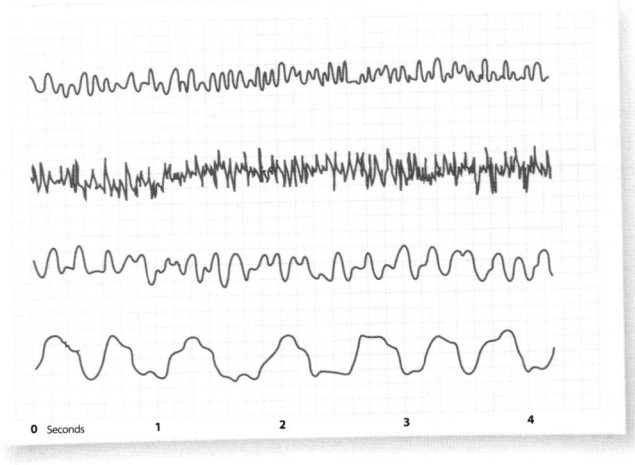

0  Seconds    1    2    3    4

61. The mixed cranial nerves are the **trigeminal (V)**, **facial (VII)**, **glossopharyngeal (IX)**, and **vagus (X) nerves**.

## SECTION 2 • Sensory and Motor Pathways

### Module 13.16

**Sensations carried by sensory pathways to the CNS begin with transduction at a sensory receptor    p. 497**

62. The **general senses** detect temperature, pain, touch, pressure, vibration, and proprioception. The information carried by a sensory pathway is a **sensation**, and the conscious awareness of a sensation is a **perception**.

63. The area monitored by a single receptor cell is its **receptive field**. The larger a receptive field is, the poorer your ability to localize a stimulus.

### Module 13.17

**Receptors are classified by function or the response to the stimulus    p. 498**

64. The simplest receptors are the dendrites of sensory neurons. These free nerve endings are not protected by accessory structures.

65. The general sensory receptors are **nociceptors**, **thermoreceptors**, **chemoreceptors**, and **mechanoreceptors**.

### Module 13.18

**Tactile receptors have a simple structure and are abundant in the skin    p. 500**

66. The greatest diversity of tactile receptors is found in the skin.

67. The important tactile receptors of the skin are **free nerve endings**, **root hair plexuses**, **tactile discs**, **tactile corpuscles**, **lamellar corpuscles**, and **bulbous corpuscles**.

### Module 13.19

**Three major somatic sensory pathways carry information from the skin and muscles to the CNS    p. 502**

68. The **spinothalamic pathway** has two tracts. The **anterior spinothalamic tracts** carry crude touch and pressure sensations. The **lateral spinothalamic tracts** carry pain and temperature sensations.

69. In the spinothalamic pathway, axons of **first-order neurons** enter the spinal cord and synapse with **second-order neurons** within the posterior gray horns. The axons of these interneurons cross to the opposite side of the spinal cord before ascending to the thalamus. **Third-order neurons** synapse in the primary sensory cortex.

70. A **somatotopy** is a functional map illustrating the area of primary somatosensory cortex devoted to a particular body region.

71. The **posterior column pathway** carries sensations of fine touch, pressure, vibration, and proprioception.

72. The **spinocerebellar pathway** sends proprioceptive information to the cerebellum.

### Module 13.20

**The somatic nervous system controls skeletal muscles through upper and lower motor neurons    p. 504**

73. Somatic motor pathways involve at least two motor neurons: an **upper motor neuron**, whose cell body lies in a CNS processing center, and a **lower motor neuron**, whose cell body lies in a nucleus of the brainstem or spinal cord.

74. The **corticospinal pathway** provides voluntary control over skeletal muscles.

75. A **motor homunculus** is functional map illustrating the primary motor cortex area devoted to a particular body region.

76. The **medial pathway** controls gross movements and proximal limb muscles. The **lateral pathway** controls the distal limb muscles that perform precise movements.

### Module 13.21

**There are multiple levels of somatic motor control    p. 506**

77. The closer a motor center is to the cerebral cortex, the more complex the motor activities will be.

78. Increasing levels of complexity begin at the brainstem and spinal cord, and continue to the pons and medulla oblongata, hypothalamus, thalamus and midbrain, basal nuclei, and cerebral cortex. The cerebellum is involved at multiple levels.

79. Voluntary movement begins with a conscious decision in the frontal lobes, and then information is relayed to the premotor cortex.

80. As the movement begins, the premotor cortex send instructions to the primary motor cortex.

### Module 13.22

**Nervous system disorders may result from problems with neurons, pathways, or a combination of the two    p. 508**

81. **Referred pain** is pain in a part of the body other than its actual source. **Parkinson's disease** is caused by a decrease in dopamine secretion by the substantia nigra. **Rabies** damages the CNS through retrograde flow beginning at the axon terminals.

82. **Cerebral palsy (CP)** refers to a number of disorders that affect voluntary motor movement. **Amyotrophic lateral sclerosis (ALS)** is a progressive degeneration of CNS motor neurons.

83. **Alzheimer's disease (AD)** is a progressive disorder characterized by the loss of higher-order cerebral functions. **Multiple sclerosis (MS)** results from demyelination of the axons of the optic nerve, brain, and spinal cord.

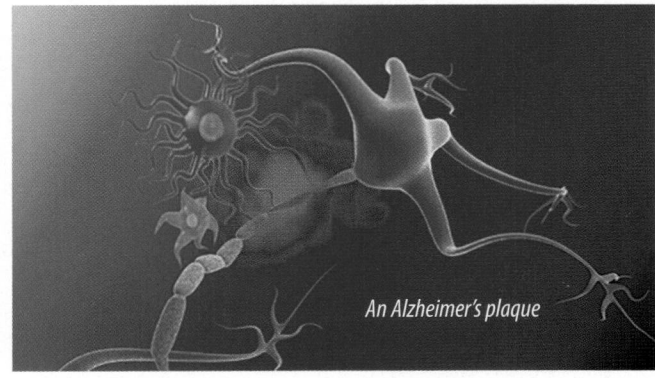

*An Alzheimer's plaque*

13

# Chapter Review Questions

**Labeling:** Label the structures of the brain indicated in the following diagram.

| 9 | |
| 10 | |
| 11 | |
| 1 | |
| 12 | |
| 2 | |
| 13 | |
| 3 | |
| 14 | |
| 4 | |
| 15 | |
| 16 | |
| 5 | |
| 6 | |
| 7 | |
| 17 | |
| 8 | |
| 18 | |

**13**

**True/False:** Indicate whether each statement is true or false.

| 19 | The special sensory cranial nerves are I, II, VIII, and X. | 19 _____ |
| 20 | Proprioceptors are the general sensory receptors that monitor the positions of joints and muscles. | 20 _____ |
| 21 | Parkinson's disease is characterized by demyelination that affects axons of the optic nerve, brain, and spinal cord. | 21 _____ |
| 22 | Lamellar corpuscles are often referred to as free nerve endings. | 22 _____ |
| 23 | The posterior column pathway carries sensations of touch, pressure, vibration, and proprioception. | 23 _____ |

**Multiple choice:** Select the correct answer from the list provided.

**24** The adult cerebrum develops from the
- ☐ a) diencephalon.
- ☐ b) telencephalon.
- ☐ c) mesencephalon.
- ☐ d) metencephalon.

**25** The cerebral aqueduct connects the
- ☐ a) lateral ventricles to the third ventricle.
- ☐ b) third ventricle to the fourth ventricle.
- ☐ c) fourth ventricle to the central canal.
- ☐ d) lateral ventricles to the interventricular foramen.

**26** The fold of dura mater that projects between the cerebral hemispheres is the
- ☐ a) falx cerebri.
- ☐ b) falx cerebelli.
- ☐ c) tentorium cerebelli.
- ☐ d) superior sagittal sinus.

**27** The site where CSF is absorbed into venous circulation is the
- ☐ a) lateral apertures.
- ☐ b) median aperture.
- ☐ c) choroid plexuses.
- ☐ d) arachnoid granulations.

**28** The superior and inferior colliculi collectively form the
- [ ] a) thalamus.
- [ ] b) corpora quadrigemina.
- [ ] c) limbic system.
- [ ] d) reticular activating system.

**29** The centers in the pons that adjust the respiratory rhythmicity centers in the medulla oblongata are the
- [ ] a) apneustic and pneumotaxic centers.
- [ ] b) inferior and superior peduncles.
- [ ] c) cardiac and vasomotor centers.
- [ ] d) gracile nucleus and cuneate nucleus.

**30** The final relay point for ascending sensory information that will be projected to the primary sensory cortex is the
- [ ] a) hypothalamus.
- [ ] b) thalamus.
- [ ] c) spinal cord.
- [ ] d) pons.

**31** The brain waves normally seen during deep sleep in people of all ages are called
- [ ] a) alpha waves.
- [ ] b) beta waves.
- [ ] c) theta waves.
- [ ] d) delta waves.

### Short answer

**32** Cerebral meningitis is a condition in which the meninges of the brain become inflamed as the result of viral or bacterial infection. This condition can be life threatening. Why?

**33** Julia is about to have a cavity in her tooth filled. The tooth in need of attention is on the posterior right side of her bottom jaw. Which nerve do you think the dentist will block when he gives her an injection to control the pain? Is this a sensory nerve, motor nerve, or mixed nerve? Besides stopping the pain from the tooth, what other sensory or motor activities would you expect to be affected by the injection?

---

**13**

## MasteringA&P™

Access more chapter study tools online in the MasteringA&P Study Area:

- **Chapter Quizzes, Chapter Practice Test, and Clinical Case Studies**

- **Practice Anatomy Lab™ 3.0** — PAL™3.0
- **Interactive Physiology™** — iP2™
- **A&P Flix™** — A&PPFlix™
- **PhysioEx™ 9.1** — PhysioEx™ 9.I

---

# Chapter Integration • Applying what you have learned

## A frightening awakening

Upon awakening one morning, John feels an unusual sensation on the left side of his face. He looks in the mirror and sees that half of his face is drooping. He tries to make a variety of facial expressions such as smiling and frowning, but can do so only on his right side. He is completely unable to control the facial muscles on his left side. Extremely alarmed at this condition, he drives himself to his physician's office right away.

John's doctor asks him a few questions about how his symptoms developed, and then gives him a physical and neurological examination. After asking a few more questions, the doctor tells John that his condition is called *Bell's palsy*. He assures John that although his symptoms are alarming, he has not had a stroke or a heart attack, and the Bell's palsy is likely to last only a few weeks.

Using what you have learned about cranial nerves, answer the following questions.

**1** Which cranial nerve do you think is most likely to be involved in John's Bell's palsy?

**2** Besides the inability to move his facial muscles, which other symptoms would you expect John to be experiencing?

**3** What advice would you give John to help him manage his condition?

**LEARNING OUTCOMES**

These Learning Outcomes correspond by number to this chapter's modules and indicate what you should be able to do after completing the chapter. They also appear at the end of each module.

# 14

# The Autonomic Nervous System

# Module 14.1
# Ganglionic neurons of the ANS control visceral effectors

Go to
MasteringA&P™
PAL™3.0
Autonomic
Nervous
System

In Chapters 12 and 13, we considered the organization of the **somatic nervous system (SNS)**. The SNS provides conscious and subconscious control over the skeletal muscles of the body. In this chapter, we consider the **autonomic nervous system (ANS)**, which controls visceral function mostly outside our awareness, such as adjustment of heart rate.

**?** **REVIEW**

**A.** Compare the SNS with the ANS.

**1** In the SNS, motor neurons of the central nervous system exert direct control over skeletal muscles. The lower motor neurons may be controlled by reflexes based in the spinal cord or brain, or by upper motor neurons whose cell bodies lie within nuclei of the brain or the primary motor cortex (see Module 13.20).

**2** In the ANS, motor neurons of the central nervous system synapse on visceral motor neurons in autonomic ganglia, and these ganglionic neurons control **visceral effectors**. The integrative centers for autonomic activity are located in the hypothalamus. The neurons in these autonomic centers are comparable to the upper motor neurons in the SNS. However, instead of a single lower motor neuron, there are two motor neurons in series. Visceral motor neurons whose cell bodies lie in the brainstem and spinal cord are known as **preganglionic neurons**. These neurons are part of visceral reflex arcs, and most of their activities represent direct reflex responses, rather than responses to commands from the hypothalamus. The axons of preganglionic neurons leave the CNS and synapse on **ganglionic neurons**—visceral motor neurons in peripheral ganglia. These ganglia, which contain hundreds to thousands of ganglionic neurons, are called **autonomic ganglia**. Ganglionic neurons innervate visceral effectors such as cardiac muscle, smooth muscle, glands, and adipose tissue.

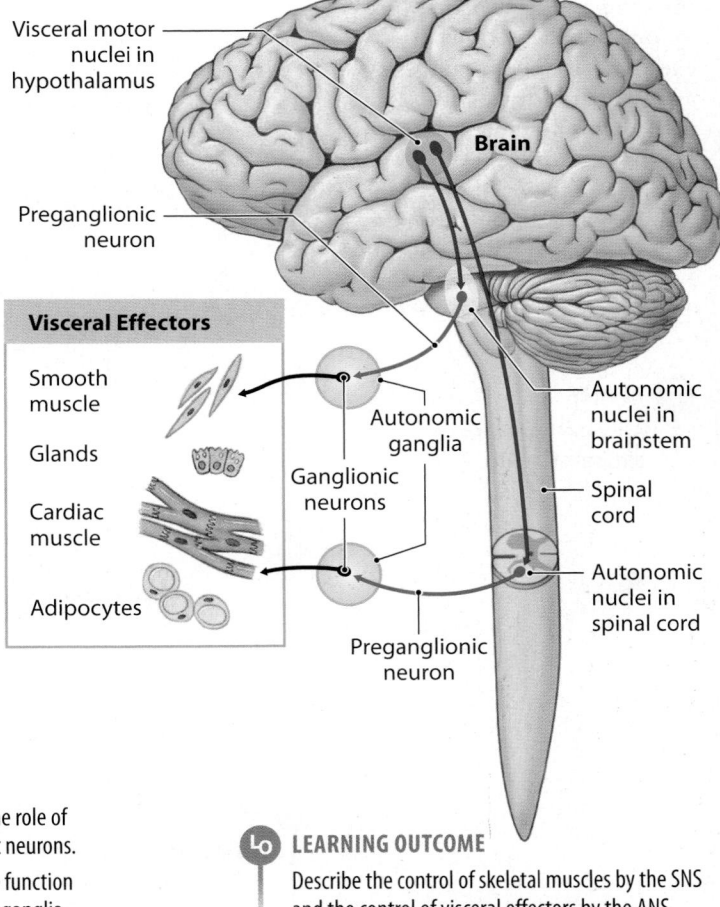

**?** **REVIEW**

**B.** Describe the role of preganglionic neurons.

**C.** Explain the function of autonomic ganglia.

**Lo** **LEARNING OUTCOME**

Describe the control of skeletal muscles by the SNS and the control of visceral effectors by the ANS.

# Module 14.2
# The ANS consists of sympathetic and parasympathetic divisions

**1** The ANS contains two major divisions whose names are probably already familiar to you: the **sympathetic division** and the **parasympathetic division**. Most often, these two divisions have opposing effects: If the sympathetic division causes excitation, the parasympathetic causes inhibition. However, this is not always the case. The two divisions may work independently, because some structures are innervated by only one division; or the two divisions may work together, each controlling a stage of a complex process. In general, the sympathetic division "kicks in" only during periods of exertion, stress, or emergency, and the parasympathetic division predominates under resting conditions. Both primary divisions of the ANS influence the third nervous system division, known as the *enteric nervous system*.

**? REVIEW**
**A.** Identify two major divisions of the ANS.

## Autonomic Nervous System

### Sympathetic Division

In the sympathetic division, or **thoracolumbar** (thor-a-kō-LUM-bar) **division**, axons emerge from the cell bodies within the lateral gray horns of thoracic and superior lumbar segments of the spinal cord ($T_1$–$L_2$). These axons innervate ganglia relatively close to the spinal cord.

### Parasympathetic Division

In the parasympathetic division, or **craniosacral** (krā-nē-ō-SĀ-krul) **division**, axons emerge from the brainstem nuclei and the lateral gray horns of the sacral segments of the spinal cord. These axons innervate ganglia very close to (or within) target organs.

III

VII

IX

X

Cranial nerves (III, VII, IX, and X)

Sympathetic chain ganglia

$T_1$
$T_2$
$T_3$
$T_4$
$T_5$
$T_6$
$T_7$
$T_8$
$T_9$
$T_{10}$
$T_{11}$
$T_{12}$

Thoracic nerves

Lumbar nerves ($L_1$, $L_2$ only)

$L_1$
$L_2$

$S_2$
$S_3$
$S_4$

Sacral nerves ($S_2$, $S_3$, $S_4$ only)

**? REVIEW**
**B.** Compare the anatomy of the sympathetic division with that of the parasympathetic division.

**2** The **enteric nervous system (ENS)** is a third division of the nervous system (see Module 11.1). Previously considered a third division of the ANS (along with the sympathetic and parasympathetic divisions), it is an extensive network of neurons and nerve networks located in the walls of the digestive tract. Although the sympathetic and parasympathetic divisions influence ENS activities, the ENS initiates and coordinates many complex visceral reflexes locally, without instructions from the CNS. The ENS has about 100 million neurons—at least as many as the spinal cord. It also uses the same neurotransmitters found in the brain. The ENS is discussed with the digestive system in Chapter 22.

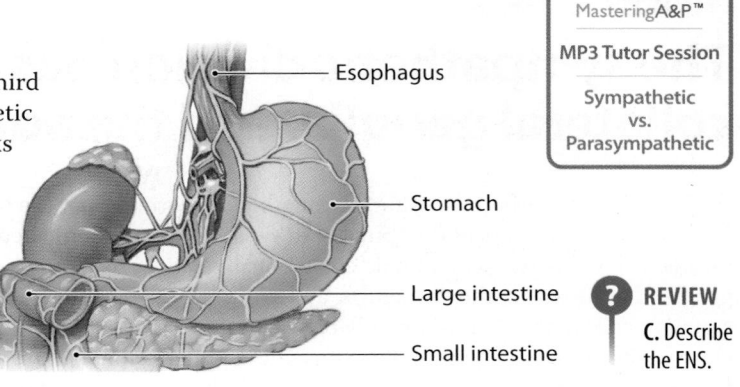

Esophagus

Stomach

Large intestine

Small intestine

**?** **REVIEW**
**C.** Describe the ENS.

**3** In this chapter, we discuss the general organization of the sympathetic and parasympathetic divisions. In dissection, the sympathetic and parasympathetic nerves are not clearly segregated. Instead, they form a complex of intertwining nerves and nerve plexuses (see Module 12.7).

Aortic arch

Right vagus nerve

Trachea

Left vagus nerve

**Autonomic Plexuses and Ganglia**

Cardiac plexus

Thoracic spinal nerves

Pulmonary plexus

Esophagus

Thoracic sympathetic chain ganglia

Esophageal plexus

Splanchnic nerves

Celiac plexus and ganglion

Superior mesenteric ganglion

Diaphragm

Superior mesenteric artery

Inferior mesenteric plexus and ganglion

Inferior mesenteric artery

Hypogastric plexus

Pelvic sympathetic chain ganglia

**?** **INTEGRATION**
**D.** Which division of the ANS is responsible for the physiological changes you experience when startled by a loud noise?

**Lo** **LEARNING OUTCOME**
List the divisions of the ANS and the general functions of each.

# Module 14.3

# The sympathetic division has chain ganglia, collateral ganglia, and the adrenal medullae …

**1** In the sympathetic division, preganglionic neurons from the thoracic and superior lumbar segments of the spinal cord synapse on ganglionic neurons that may be located (1) within the sympathetic chain of ganglia near the spinal cord, (2) in collateral ganglia that lie within the thoracic or abdominopelvic cavities, or (3) within modified ganglion cells in the adrenal medullae. Because the ganglionic neurons are relatively close to the vertebral column, the axons of the preganglionic neurons (**preganglionic fibers**) are short compared to axons of the ganglionic neurons (**postganglionic fibers**).

**KEY**

➡ Preganglionic fibers

➡ Postganglionic fibers

➡ Hormones released into circulation

Ganglionic neurons in the sympathetic chain and collateral ganglia exert their effects through innervation of peripheral target organs.

**Preganglionic Neurons**

Lateral gray horns of spinal segments $T_1$–$L_2$

**Ganglionic Neurons**

**Sympathetic chain ganglia** lie on both sides of the vertebral column.

The **collateral ganglia** innervate abdominopelvic tissues and include the **celiac**, **superior mesenteric**, and **inferior mesenteric ganglia**.

The **adrenal medullae** are modified sympathetic ganglia that secrete neurotransmitters directly into the bloodstream.

Ganglionic neurons release neurotransmitters that function as hormones into the general circulation.

**Target Organs**

Visceral effectors in thoracic cavity, head, body wall, and limbs

Visceral effectors in abdomino-pelvic cavity

Organs and systems throughout the body

The sympathetic division prepares the body to deal with emergencies. When fully activated, this division produces what is known as the "fight or flight" response, which readies the body for a crisis that may require sudden, intense physical activity. The general pattern of responses to increased levels of sympathetic activity includes the following changes: (1) heightened mental alertness, (2) increased metabolic rate, (3) decreased digestive and urinary functions, (4) activation of energy reserves, (5) dilation of respiratory passageways and increased respiratory rate, (6) increased heart rate and blood pressure, and (7) activation of sweat glands.

**? REVIEW**

**A.** Starting in the spinal cord, trace the path of a nerve impulse through the sympathetic ANS to its target organ in the abdominopelvic cavity.

# ... whereas the parasympathetic division has terminal or intramural ganglia

**2** In the parasympathetic division, a typical preganglionic fiber synapses on six to eight ganglionic neurons. These neurons may be located in **terminal ganglia**, located near the target organ, or in **intramural ganglia** (*murus*, wall), which are embedded in the tissues of the target organ. Terminal ganglia are usually paired. Examples of terminal ganglia include the parasympathetic ganglia associated with the cranial nerves. Intramural ganglia typically consist of interconnected masses and clusters of ganglion cells.

**? REVIEW**

**B.** Describe an intramural ganglion.

KEY

➤ Preganglionic fibers

➤ Postganglionic fibers

14

| Preganglionic Neurons | | Ganglionic Neurons | Target Organs |
|---|---|---|---|
| | CN III | **Ciliary ganglia** | Intrinsic eye muscles (pupil and lens shape) |
| The midbrain, pons, and medulla oblongata contain parasympathetic nuclei associated with cranial nerves III, VII, IX, and X. | CN VII | **Pterygopalatine** and **submandibular ganglia** | Nasal glands, tear glands, and salivary glands |
| | CN IX | | |
| | CN X | **Otic ganglia** | Parotid salivary glands |
| | | Intramural ganglia | Visceral organs of neck, thoracic cavity, and most of abdominal cavity |
| Parasympathetic nuclei lie in the lateral gray horns of sacral spinal cord segments $S_2$–$S_4$. | Pelvic nerves | Intramural ganglia | Visceral organs in inferior portion of abdominopelvic cavity |

**? REVIEW**

**C.** List general responses to increased sympathetic activity and to parasympathetic activity.

The parasympathetic division is concerned with conserving energy and regulating visceral function. It is known as the "rest and digest" system. The overall pattern of responses to increased levels of parasympathetic activity includes the following changes: (1) decreased metabolic rate, (2) decreased heart rate and blood pressure, (3) increased secretion by salivary and digestive glands, (4) increased motility and blood flow in the digestive tract, and (5) stimulation of urination and defecation.

**Lo LEARNING OUTCOME**

Describe the structures and functions of the sympathetic and parasympathetic divisions of the ANS.

# Module 14.4

# The two ANS divisions innervate many of the same structures, but the innervation patterns are different

**1** The distribution of sympathetic fibers is the same on both sides of the body. For simplicity, the left side of this image shows the distribution to the skin (and to skeletal muscles and other tissues of the body wall), whereas the right side depicts the innervation of visceral organs.

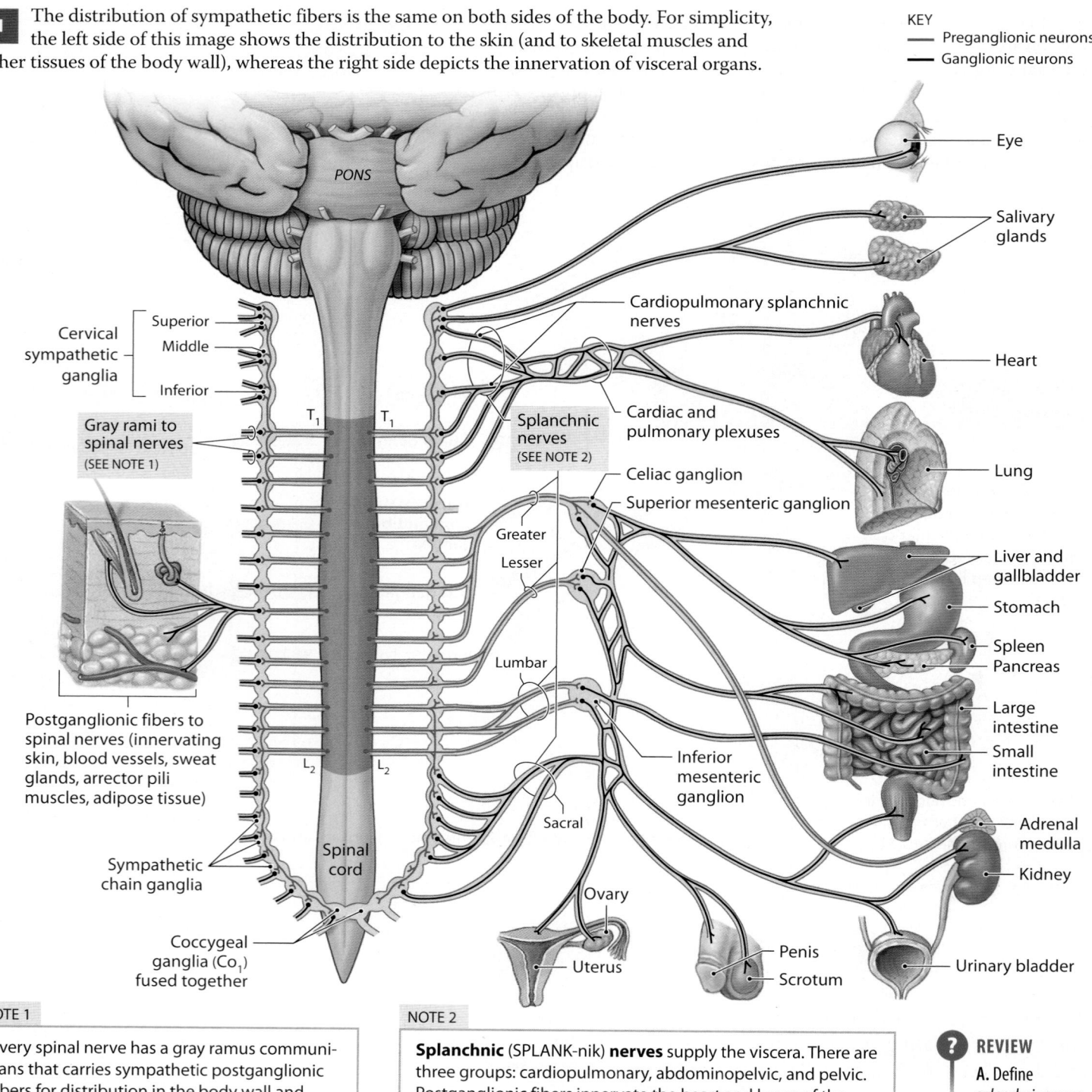

**KEY**
— Preganglionic neurons
— Ganglionic neurons

Eye

Salivary glands

PONS

Cardiopulmonary splanchnic nerves

Heart

Cervical sympathetic ganglia
Superior
Middle
Inferior

$T_1$    $T_1$

Gray rami to spinal nerves (SEE NOTE 1)

Splanchnic nerves (SEE NOTE 2)

Cardiac and pulmonary plexuses

Celiac ganglion
Superior mesenteric ganglion

Greater
Lesser

Lung

Liver and gallbladder

Stomach

Spleen
Pancreas

Postganglionic fibers to spinal nerves (innervating skin, blood vessels, sweat glands, arrector pili muscles, adipose tissue)

$L_2$    $L_2$

Lumbar

Inferior mesenteric ganglion

Large intestine
Small intestine

Sympathetic chain ganglia

Spinal cord

Sacral

Adrenal medulla

Kidney

Coccygeal ganglia ($Co_1$) fused together

Ovary
Uterus

Penis
Scrotum

Urinary bladder

---

**NOTE 1**

Every spinal nerve has a gray ramus communicans that carries sympathetic postganglionic fibers for distribution in the body wall and limbs. In the head and neck, postganglionic sympathetic fibers leaving the superior cervical sympathetic ganglia supply the regions innervated by cranial nerves III, VII, IX, and X.

**NOTE 2**

**Splanchnic** (SPLANK-nik) **nerves** supply the viscera. There are three groups: cardiopulmonary, abdominopelvic, and pelvic. Postganglionic fibers innervate the heart and lungs of the thoracic cavity; preganglionic fibers synapse in collateral ganglia (celiac, superior mesenteric, and inferior mesenteric) and innervate the viscera of the abdominopelvic cavity; and postganglionic fibers inferior to $L_2$ innervate the pelvic cavity.

**? REVIEW**
**A.** Define *splanchnic nerves.*

**2** This image shows the parasympathetic innervation on one side of the body; the innervation on the opposite side (not shown) follows the same pattern.

**KEY**
— Preganglionic neurons
— Ganglionic neurons

Pterygopalatine ganglion

Occulomotor (III)

Lacrimal gland

Eye

PONS

Ciliary ganglion

Facial (VII)

Submandibular ganglion

Salivary glands

Glossopharyngeal (IX)

Otic ganglion

Heart

The vagus nerve (X) alone provides roughly 75 percent of all parasympathetic outflow. The numerous branches of the vagus nerve intermingle with sympathetic preganglionic and postganglionic fibers, forming nerve plexuses comparable to those formed by spinal nerves innervating the limbs.

Cardiac plexus

Lungs

Celiac plexus

Liver and gallbladder

Stomach

Spleen

Inferior mesenteric plexus

Pancreas

Hypogastric plexus

Large intestine

Small intestine

Rectum

Spinal cord

**? REVIEW**
**B.** Name the plexuses innervated by the vagus nerve.

Kidney

The preganglionic fibers in the sacral segments of the spinal cord carry the sacral parasympathetic output. These fibers do not join the anterior roots of the spinal nerves. Instead, the preganglionic fibers form distinct **pelvic nerves**, which innervate intramural ganglia in the walls of the kidneys, urinary bladder, terminal portions of the large intestine, and sex organs.

$S_2$
$S_3$
$S_4$

Penis

Scrotum

Urinary bladder

Uterus    Ovary

**? REVIEW**
**C.** Which nerve carries most of the parasympathetic outflow?

**Lo LEARNING OUTCOME**
Describe the innervation patterns of the sympathetic and parasympathetic divisions of the ANS.

14

# Module 14.5

# The functional differences between the two ANS divisions reflect their divergent anatomical and physiological characteristics

**1** Unlike the parasympathetic division, the entire sympathetic division can be quickly activated. This table summarizes the effects of sympathetic activation.

**2** This illustration and the table below it summarize the anatomical characteristics of the sympathetic division of the ANS.

### Effects of Sympathetic Activation

The sympathetic division can change the activities of tissues and organs by releasing norepinephrine (NE) at peripheral synapses and by distributing epinephrine (E) and NE throughout the body in the bloodstream. The visceral motor fibers that target specific effectors, such as smooth muscle fibers in blood vessels of the skin, can be activated in reflexes that do not involve other visceral effectors.

In a crisis, however, the entire division responds. This event, called **sympathetic activation**, is controlled by sympathetic centers in the hypothalamus. The effects are not limited to peripheral tissues; sympathetic activation also alters CNS activity. During sympathetic activation, the following changes occur:

- Increased alertness by stimulation of the reticular activating system, causing the person to feel "on edge."

- A feeling of energy and euphoria, often associated with a disregard for danger and a temporary insensitivity to painful stimuli.

- Increased activity in the cardiovascular and respiratory centers of the pons and medulla oblongata, which increases blood pressure, heart rate, breathing rate, and depth of respiration.

- A general increase in muscle tone through stimulation of the medial and lateral pathways, so the person *looks* tense and may begin to shiver.

- Mobilization of energy reserves, through the accelerated breakdown of glycogen in muscle and liver cells and the release of lipids by adipose tissues.

These changes, plus the peripheral changes already noted, complete the preparations necessary for the person to cope with a stressful situation.

 **REVIEW**

**A.** What physiological changes are typical in a tense or anxious person?

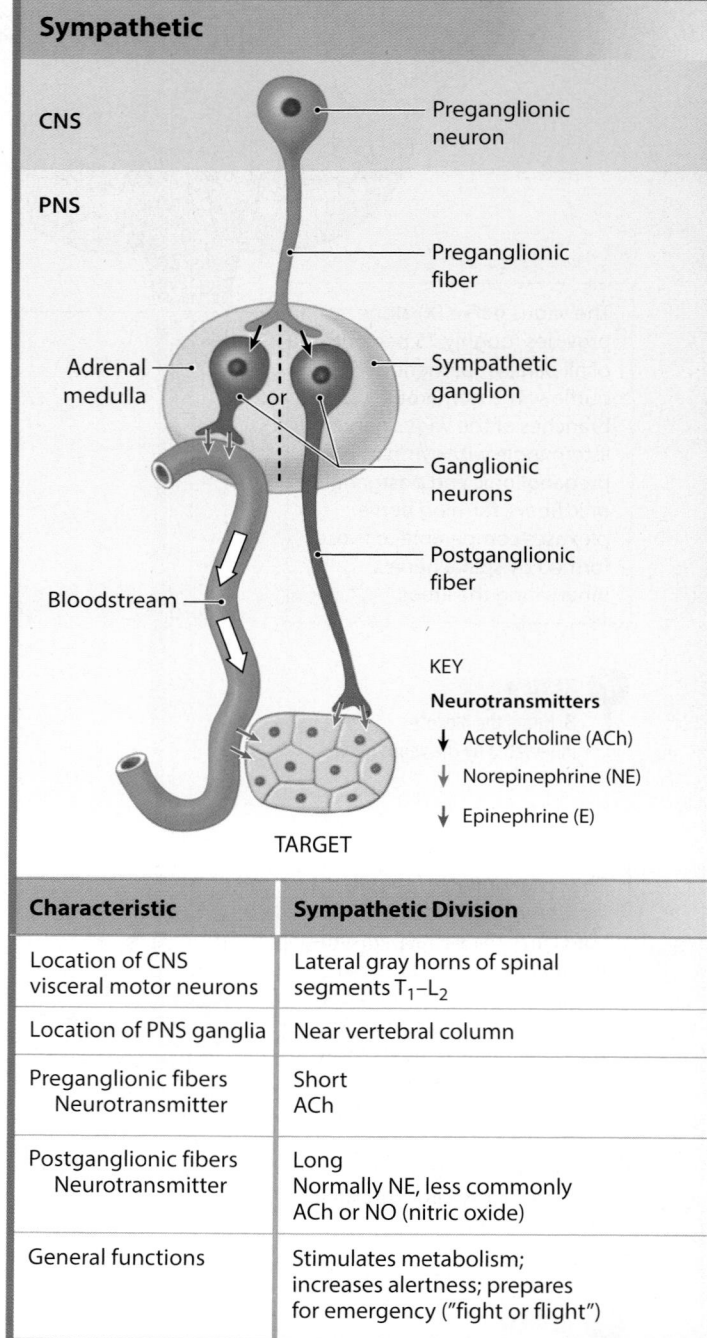

**Sympathetic**

CNS — Preganglionic neuron

PNS

Preganglionic fiber

Adrenal medulla

or

Sympathetic ganglion

Ganglionic neurons

Postganglionic fiber

Bloodstream

**KEY**

**Neurotransmitters**

↓ Acetylcholine (ACh)

↓ Norepinephrine (NE)

↓ Epinephrine (E)

TARGET

| Characteristic | Sympathetic Division |
|---|---|
| Location of CNS visceral motor neurons | Lateral gray horns of spinal segments $T_1$–$L_2$ |
| Location of PNS ganglia | Near vertebral column |
| Preganglionic fibers Neurotransmitter | Short ACh |
| Postganglionic fibers Neurotransmitter | Long Normally NE, less commonly ACh or NO (nitric oxide) |
| General functions | Stimulates metabolism; increases alertness; prepares for emergency ("fight or flight") |

**3** This illustration and the table below it summarize the anatomical characteristics of the parasympathetic division of the ANS.

### Parasympathetic

Preganglionic neuron

CNS

PNS

Preganglionic fiber

Ganglionic neuron

Postganglionic fiber

Parasympathetic ganglion

**KEY**

**Neurotransmitters**

↓ Acetylcholine

TARGET

| Characteristic | Parasympathetic Division |
|---|---|
| Location of CNS visceral motor neurons | Brainstem and spinal segments $S_2$–$S_4$ |
| Location of PNS ganglia | Typically intramural |
| Preganglionic fibers Neurotransmitter | Relatively long ACh |
| Postganglionic fibers Neurotransmitter | Relatively short ACh |
| General functions | Promotes relaxation, nutrient uptake, energy storage ("rest and digest") |

**4** The parasympathetic division does not release neurotransmitters directly into the bloodstream, so it does not undergo a division-wide activation. This table summarizes the effects of parasympathetic stimulation.

### Effects of Parasympathetic Activation

The parasympathetic division is responsible for conserving the body's energy and maintaining the resting metabolic rate. Although it is active continuously, the activities are reflex responses to conditions within specific structures or regions. Examples of the major effects produced by the parasympathetic division include the following:

- Constriction of the pupils (to restrict the amount of light that enters the eyes) and focusing of the lenses of the eyes on nearby objects.

- Secretion by digestive glands, including salivary glands, gastric glands, duodenal glands, intestinal glands, the pancreas (exocrine and endocrine), and the liver.

- Secretion of hormones that promote the absorption and utilization of nutrients by peripheral cells.

- Changes in blood flow and glandular activity associated with sexual arousal.

- Increased smooth muscle activity along the digestive tract.

- Stimulation and coordination of defecation.

- Contraction of the urinary bladder during urination.

- Constriction of the respiratory passageways.

- Reduction in heart rate and force of contraction.

These functions center on relaxation, food processing, and energy absorption. The parasympathetic division has been called the **anabolic system** (*anabole*, a raising up) because its stimulation leads to a general increase in the nutrient content of the blood. Cells throughout the body respond to this increase by absorbing nutrients and using them to support growth, cell division, and the creation of energy reserves in the form of lipids or glycogen.

 **REVIEW**

**B.** What neurotransmitter is released by all parasympathetic neurons?

**C.** Why is the parasympathetic division called the *anabolic system*?

 **LEARNING OUTCOME**

Describe the mechanisms of neurotransmitter release in the ANS, and explain the effects of neurotransmitters on target organs and tissues.

# Module 14.6

# Membrane receptors at target organs mediate the effects of sympathetic and parasympathetic stimulation

## Neurotransmitter Release in the Sympathetic Division

The effects of sympathetic stimulation result primarily from the interactions of norepinephrine (NE) and epinephrine (E) with **adrenergic receptors** in the plasma membrane. There are two classes of adrenergic receptors: **alpha receptors** and **beta receptors**. In general, NE stimulates alpha receptors to a greater degree than it does beta receptors, whereas E stimulates both classes of receptors. Localized sympathetic activity involving the release of NE at sympathetic terminals primarily affects alpha receptors near those terminals. The effects typically persist for a few seconds. By contrast, generalized sympathetic activation and release of E and NE by the adrenal medullae affect alpha and beta receptors throughout the body. Tissue concentrations of E and NE released by the adrenal medullae remain elevated for as long as 30 seconds, and the metabolic effects may persist for several minutes. Because the adrenal medullae release three times as much E as they do NE, during sympathetic activation the effects of beta receptors predominate.

**1** Stimulation of alpha (α) receptors activates their associated G proteins on the cytoplasmic side of the plasma membrane. Depending on receptor type, their activation triggers different activities in the cell. Stimulation of alpha-1 receptors generally excites the target cell, whereas stimulation of alpha-2 receptors generally inhibits the target cell.

**2** Beta (β) receptors are located on the plasma membranes of cells in many organs, including skeletal muscles, the lungs, the heart, and the liver. Stimulation of beta receptors and G protein activation triggers changes in the metabolic activity of the target cell. The three types of beta receptors are beta-1 ($\beta_1$), beta-2 ($\beta_2$), and beta-3 ($\beta_3$).

**? REVIEW**

**A.** Compare and contrast alpha and beta receptors.

## Neurotransmitter Release in the Parasympathetic Division

Although all the synapses and neuromuscular or neuroglandular junctions of the parasympathetic division use the same transmitter—acetylcholine (ACh)—two types of **cholinergic receptors** occur on the postsynaptic membranes: nicotinic receptors and muscarinic receptors.

**3** **Nicotinic** (nik-ō-TIN-ik) **receptors** are found on all postganglionic neurons, adrenal medullae cells, and at the neuromuscular junctions of skeletal muscle fibers. ACh always excites the ganglionic neuron or muscle fiber by opening chemically gated ion channels in the postsynaptic membrane. Nicotinic receptors are also stimulated by *nicotine*, a powerful toxin that can be obtained from a variety of sources, including tobacco.

**4** **Muscarinic** (mus-ka-RIN-ik) **receptors** are G protein–coupled receptors. They occur at cholinergic neuromuscular or neuroglandular junctions in the parasympathetic division and at the few cholinergic junctions in the sympathetic division. Muscarinic receptor stimulation and G protein activation produce longer-lasting effects than does the stimulation of nicotinic receptors. The response, which reflects the activation or inactivation of specific enzymes, can be either excitatory or inhibitory. Muscarinic receptors are stimulated by *muscarine*, a toxin produced by some poisonous mushrooms.

14

**? REVIEW**

**B.** Compare nicotinic receptors with muscarinic receptors.

**? INTEGRATION**

**C.** A person with high blood pressure (hypertension) is prescribed a drug that blocks beta receptors. How could this medication alleviate hypertension?

**Lo LEARNING OUTCOME**

Describe the types of sympathetic and parasympathetic receptors and their associated neurotransmitters.

# Functional Anatomy of the Autonomic Nervous System (ANS)

**Labeling:** Fill in the missing labels in this diagram of the sympathetic division of the ANS.

**Concept map:** Use each of the following terms once to fill in the blank boxes to correctly complete the map.

- sympathetic division
- craniosacral division
- cranial nerves III, VII, IX, X
- thoracolumbar division
- parasympathetic division
- sacral nerves
- thoracic nerves
- enteric nervous system
- lumbar nerves

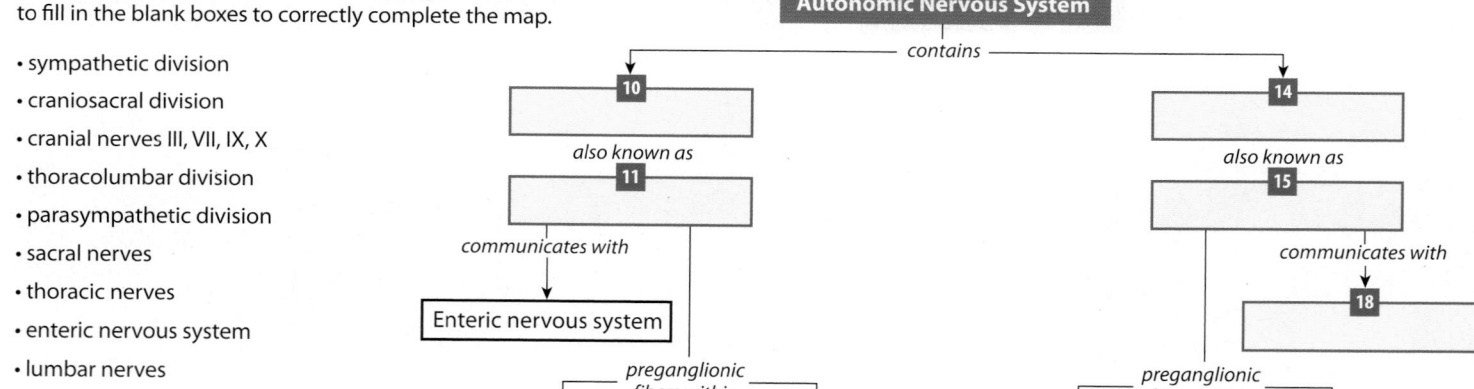

**Matching:** Match each lettered term with the most closely related description.

a. nicotinic, muscarinic
b. secrete norepinephrine
c. receptors
d. alpha, beta
e. cholinergic
f. splanchnic nerves
g. acetylcholine
h. parasympathetic activation

| 19 | Collateral ganglia |
| 20 | Parasympathetic neurotransmitter |
| 21 | Sexual arousal |
| 22 | Adrenal medullae |
| 23 | Cholinergic receptors |
| 24 | Determines neurotransmitter effects |
| 25 | All parasympathetic neurons |
| 26 | Adrenergic receptors |

| 19 | _____ |
| 20 | _____ |
| 21 | _____ |
| 22 | _____ |
| 23 | _____ |
| 24 | _____ |
| 25 | _____ |
| 26 | _____ |

# Module 14.7

# The ANS adjusts visceral motor responses to maintain homeostasis

If all consciousness were eliminated, vital physiological processes would continue virtually unchanged; after all, a night's sleep is not a life-threatening event. Longer, deeper states of unconsciousness are not necessarily more dangerous, as long as nourishment and other basic care are provided. People who have suffered severe brain injuries can survive in a coma for decades. Over that period, the ANS continues to adjust activities of the digestive, cardiovascular, respiratory, and reproductive systems to maintain homeostasis without instructions or interference from the conscious mind.

**? REVIEW**
**A.** Explain the significance of the ANS to homeostasis.

**1** This flowchart gives a general overview of the way sensory information is distributed and motor commands are issued by the nervous system.

**? REVIEW**
**B.** Name the two types of motor pathways.
**C.** Identify somatic effectors and visceral effectors.

The output of the ANS has an impact on virtually every body system. This section considers those effects and how autonomic output is coordinated and directed.

**Lo LEARNING OUTCOME**
Describe the role of the ANS in maintaining homeostasis during unconsciousness.

14

# Module 14.8
# The ANS provides precise control over visceral functions

**1** This table lists the effects of sympathetic and parasympathetic innervation on major organs and systems of the body. Even in the absence of stimuli, autonomic motor neurons maintain a continuous level of spontaneous activity called **autonomic tone**. Many vital organs receive **dual innervation**, receiving instructions from both the sympathetic and parasympathetic divisions. The effects may be opposing or complementary. In cases where only the sympathetic division provides innervation, the response may still vary widely depending on the type of receptor stimulated.

 **REVIEW**

**A.** Define *dual innervation*.

**B.** Explain autonomic tone and its significance in controlling visceral function.

## A Functional Comparison of the Sympathetic and Parasympathetic Divisions of the ANS

| Structure | Sympathetic Effects (receptor or synapse type) | Parasympathetic Effects (all muscarinic) |
|---|---|---|
| **Eye** | | |
| | Dilation of pupil ($\alpha_1$); accommodation for distance vision ($\beta_2$) | Constriction of pupil; accommodation for close vision |
| **Lacrimal glands** | None (not innervated) | Secretion |
| **Skin** | | |
| **Sweat glands** | Increased secretion, palms and soles ($\alpha_1$); generalized increase in secretion (cholinergic) | None (not innervated) |
| **Arrector pili muscles** | Contraction; erection of hairs ($\alpha_1$) | None (not innervated) |
| **Cardiovascular System** | | |
| **Blood vessels** | | |
| To skin | Dilation ($\beta_2$ and cholinergic); constriction ($\alpha_1$) | None (not innervated) |
| To skeletal muscles | Dilation ($\beta_2$ and cholinergic) | None (not innervated) |
| To heart | Dilation ($\beta_2$); constriction ($\alpha_1$, $\alpha_2$) | None (not innervated) |
| To lungs | Dilation ($\beta_2$); constriction ($\alpha_2$) | None (not innervated) |
| To digestive viscera | Constriction ($\beta_2$); dilation ($\alpha_2$) | None (not innervated) |
| To kidneys | Constriction, decreased urine production ($\alpha_1$, $\alpha_2$); dilation, increased urine production ($\beta_1$, $\beta_2$) | None (not innervated) |
| To brain | Dilation (cholinergic and nitroxidergic) | None (not innervated) |
| **Veins** | Constriction ($\alpha_1$, $\beta_2$) | None (not innervated) |
| **Heart** | Increased heart rate, force of contraction, and blood pressure ($\alpha_1$, $\beta_1$) | Decreased heart rate, force of contraction, and blood pressure |
| **Endocrine System** | | |
| **Adrenal gland** | Secretion of epinephrine, norepinephrine by adrenal medulla | None (not innervated) |
| **Neurohypophysis** | Secretion of ADH ($\beta_1$) | None (not innervated) |
| **Pancreas** | Decreased insulin secretion ($\alpha_2$) | Increased insulin secretion |
| **Pineal gland** | Increased melatonin secretion ($\beta_1$, $\beta_2$) | Inhibition of melatonin synthesis |
| **Respiratory System** | | |
| **Airways** | Increased airway diameter ($\beta_2$) | Decreased airway diameter |
| **Secretory glands** | Mucous secretion ($\alpha_1$) | None (not innervated) |
| **Digestive System** | | |
| **Salivary glands** | Production of viscous secretion ($\alpha_1$, $\beta_1$) containing mucins and enzymes | Production of copious, watery secretion |
| **Sphincters** | Constriction ($\alpha_1$) | Dilation |
| **General level of activity** | Decreased ($\alpha_2$, $\beta_2$) | Increased |
| **Secretory glands** | Inhibition ($\alpha_2$) | Stimulation |
| **Liver** | Glycogen breakdown, glucose synthesis and release ($\alpha_1$, $\beta_2$) | Glycogen synthesis |
| **Pancreas** | Decreased exocrine secretion ($\alpha_1$) | Increased exocrine secretion |

## A Functional Comparison of the Sympathetic and Parasympathetic Divisions of the ANS (Continued)

| Structure | Sympathetic Effects (receptor type) | Parasympathetic Effects (all muscarinic) |
|---|---|---|
| **Muscular System** | | |
| | Increased force of contraction, glycogen breakdown ($\beta_2$) | None (not innervated) |
| | Facilitation of ACh release at neuromuscular junction ($\alpha_2$) | None (not innervated) |
| **Adipose Tissue** | | |
| | Lipolysis, fatty acid release ($\alpha_1$, $\beta_1$, $\beta_3$) | None (not innervated) |
| **Urinary System** | | |
| **Kidneys** | Secretion of renin ($\beta_1$) | Uncertain effects on urine production |
| **Urinary bladder** | Constriction of internal sphincter; relaxation of urinary bladder ($\alpha_1$, $\beta_2$) | Tensing of urinary bladder, relaxation of internal sphincter to eliminate urine |
| **Male Reproductive System** | | |
| | Increased glandular secretion and ejaculation ($\alpha_1$) | Erection |
| **Female Reproductive System** | | |
| | Increased glandular secretion; contraction of pregnant uterus ($\alpha_1$) | Variable (depending on hormones present) |
| | Relaxation of nonpregnant uterus ($\beta_2$) | Variable (depending on hormones present) |

**2** This graph shows the effects of dual innervation on the heart. The heart consists of cardiac muscle tissue, and its contractions are triggered by specialized pacemaker cells. The two autonomic divisions have opposing effects on pacemaker function. ACh released by the parasympathetic division decreases the heart rate, whereas NE released by the sympathetic division increases the heart rate. Because autonomic tone exists, small amounts of both neurotransmitters are released continuously. However, under resting conditions, the effects of parasympathetic innervation predominate.

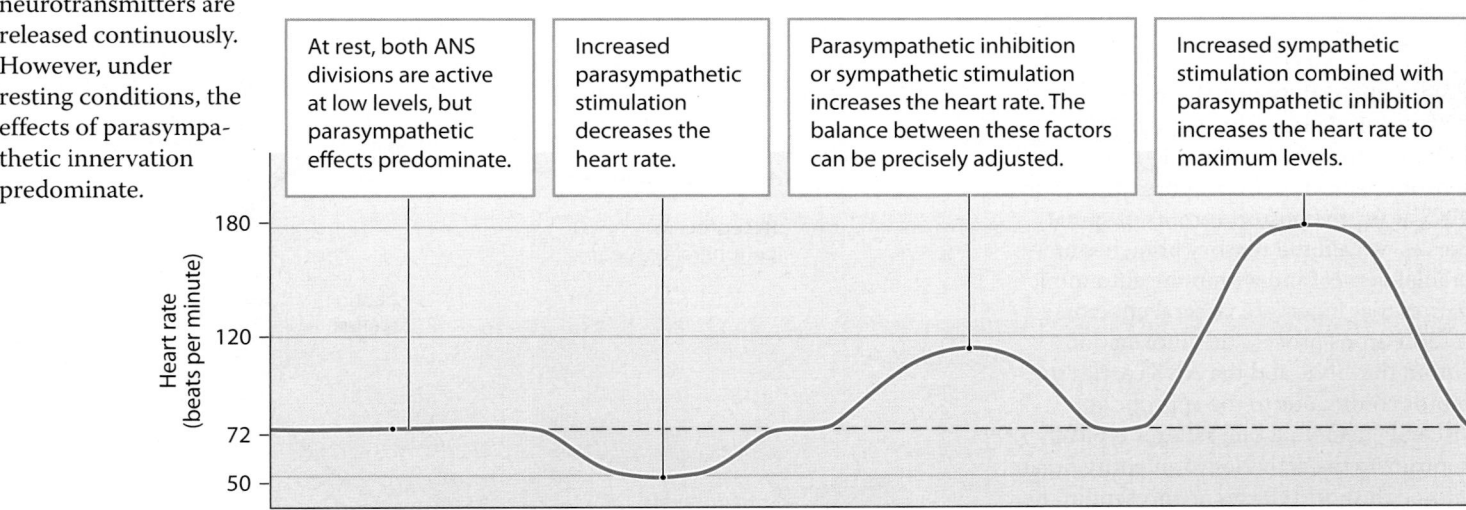

At rest, both ANS divisions are active at low levels, but parasympathetic effects predominate.

Increased parasympathetic stimulation decreases the heart rate.

Parasympathetic inhibition or sympathetic stimulation increases the heart rate. The balance between these factors can be precisely adjusted.

Increased sympathetic stimulation combined with parasympathetic inhibition increases the heart rate to maximum levels.

**? INTEGRATION**

**C.** You go outside on a cold day, and blood flow to your skin is reduced, conserving body heat. You become angry, and your face turns red. Explain these changes.

**Lo LEARNING OUTCOME**

Discuss the relationship between the two divisions of the ANS and the significance of dual innervation.

## Module 14.9

# Most visceral functions are controlled by visceral reflexes

**Visceral reflexes** are autonomic reflexes initiated in the viscera. They provide automatic motor responses that can be modified, facilitated, or inhibited by higher centers, especially those of the hypothalamus. Each visceral reflex arc consists of a receptor, a sensory neuron, a processing center (one or more interneurons), and one or two visceral motor neurons. All visceral reflexes are polysynaptic. They can be short reflexes or long reflexes.

**?** **REVIEW**

**A.** Define *visceral reflex*.

**1** **Short reflexes** bypass the CNS entirely. They involve sensory neurons and interneurons whose cell bodies are located within autonomic ganglia. The interneurons synapse on ganglionic neurons, and the motor commands are then distributed to effectors by postganglionic fibers. Short reflexes control very simple motor responses with localized effects. In general, short reflexes control activity in one small part of a target organ. Short reflexes predominate in the enteric nervous system, which operates largely outside of the awareness and control of the CNS.

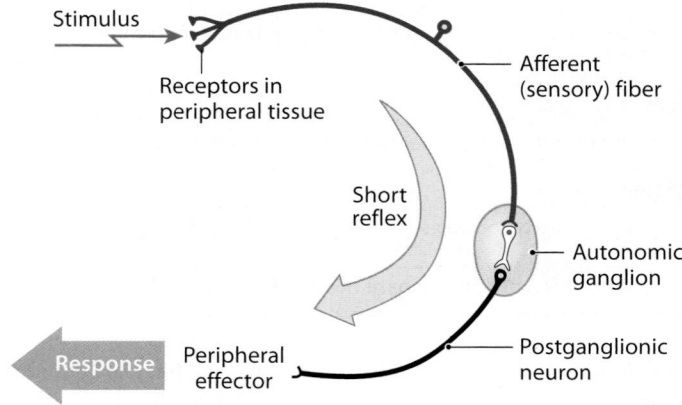

**2** **Long reflexes** are the autonomic equivalents of the polysynaptic reflexes introduced in Module 12.12. Visceral sensory neurons deliver information to the CNS along the posterior roots of spinal nerves, within the sensory branches of cranial nerves, and within the autonomic nerves that innervate visceral effectors. Interneurons process the information within the CNS, and the ANS carries the motor commands to the appropriate visceral effectors. Long reflexes typically coordinate the activities of an entire organ. Although short reflexes are present in the sympathetic and parasympathetic divisions, long reflexes predominate, and they are responsible for coordinating responses involving multiple organ systems.

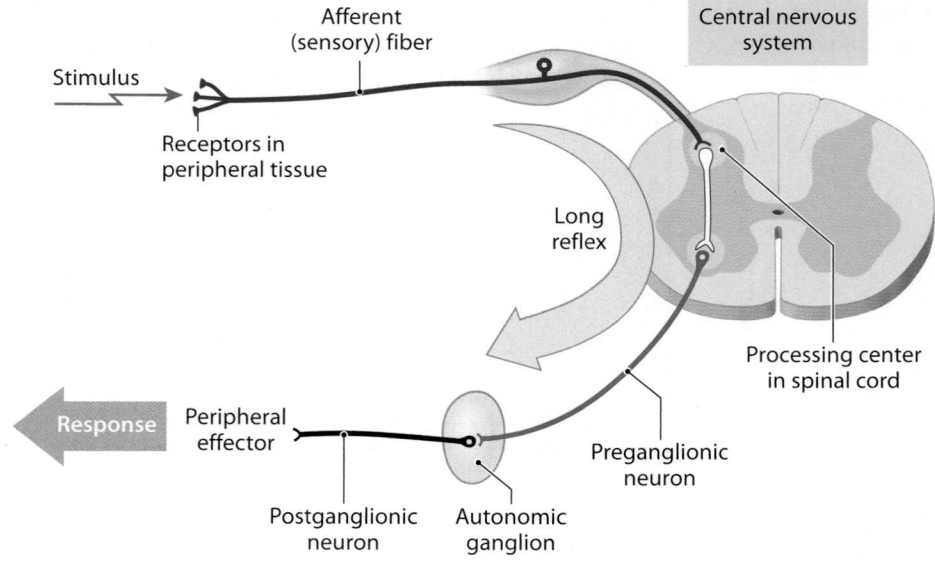

**?** **REVIEW**

**B.** Compare short reflexes with long reflexes.

**3** Visceral sensory information is collected by interoceptors monitoring visceral tissues and organs, primarily within the thoracic and abdominopelvic cavities. These interoceptors include nociceptors, thermoreceptors, tactile receptors, baroreceptors, and chemoreceptors, although none of them are as numerous as they are in somatic tissues. Almost all of the processing of this sensory information is done at the subconscious level, in nuclei in the spinal cord and the **solitary nuclei** in the brainstem.

**REVIEW**

**C.** Describe the solitary nuclei.

| Interoceptors provide visceral sensory information. | The axons of the visceral sensory neurons usually travel in company with autonomic motor fibers innervating the same visceral structures. | The visceral sensory information is delivered to the solitary nucleus. |
|---|---|---|

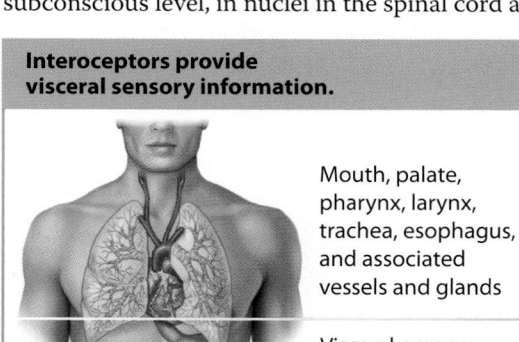

Mouth, palate, pharynx, larynx, trachea, esophagus, and associated vessels and glands → **Cranial nerves V, VII, IX, and X**

Visceral organs located between the diaphragm and the pelvic cavity → **Posterior roots of spinal nerves T₁–L₂**

Organs in the inferior portion of the pelvic cavity → **Posterior roots of spinal nerves S₂–S₄**

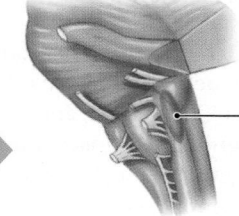

A solitary nucleus is located on each side of the medulla oblongata.

Each solitary nucleus is a major processing and sorting center for visceral sensory information. It has extensive connections with the various cardiovascular and respiratory centers as well as with the reticular formation. The information is seldom relayed to higher centers, so we remain unaware of the sensations or the resulting motor responses.

**14**

## Common Visceral Reflexes

| | Reflex | Stimulus | Response | Comments |
|---|---|---|---|---|
| **SYMPATHETIC** | **Cardioacceleratory reflex** | Sudden decrease in carotid blood pressure | Increase in heart rate and force of contraction | Coordinated in cardiac centers of medulla oblongata |
| | **Vasomotor reflexes** | Changes in blood pressure in major arteries | Changes in diameter of peripheral vessels | Coordinated in vasomotor center in medulla oblongata |
| | **Pupillary light reflex** | Light reaches visual receptors | Changes in pupil diameter | Allows eye to adjust to light; lack of light inhibits reflex |
| | **Ejaculation (in males)** | Erotic stimuli (primarily tactile) | Skeletal muscle contractions ejecting semen | Process results in propulsion of semen |
| **PARASYMPATHETIC** | **Gastric and intestinal reflexes** | Pressure and physical contact | Smooth muscle contractions that propel food materials and mix them with secretions | Initiated by vagus nerve and controlled locally by ENS |
| | **Defecation** | Distension of rectum | Relaxation of internal anal sphincter | Requires voluntary relaxation of external anal sphincter |
| | **Urination** | Distension of urinary bladder | Contraction of walls of urinary bladder; relaxation of internal urethral sphincter | Requires voluntary relaxation of external urethral sphincter |
| | **Consensual light response** | Bright light shining in eye(s) | Constriction of pupils of both eyes | Reflex observed on one side of the body when the other side is stimulated |
| | **Swallowing reflex** | Movement of food and liquids into pharynx | Smooth muscle and skeletal muscle contractions | Coordinated by medullary swallowing center |
| | **Coughing reflex** | Irritation of respiratory tract | Sudden explosive ejection of air | Coordinated by medullary coughing center |
| | **Baroreceptor reflex** | Sudden increase in carotid blood pressure | Reduction in heart rate and force of contraction | Coordinated in cardiac centers of medulla oblongata |
| | **Sexual arousal** | Erotic stimuli (visual or tactile) | Increased glandular secretions, sensitivity, erection | Heightened sensitivity and blood flow to sexual organs |

**LEARNING OUTCOME**

Define a visceral reflex, and explain the significance of such reflexes.

# Module 14.10

# Baroreceptors and chemoreceptors initiate important autonomic reflexes involving visceral sensory pathways

Baroreceptors and chemoreceptors play key roles in the control of visceral function by the autonomic nervous system.

**1** **Baroreceptors** are stretch receptors that monitor changes in pressure. The receptor consists of free nerve endings that branch within the elastic tissues in the walls of hollow organs, blood vessels, and tubes in the respiratory, digestive, and urinary tract. When the pressure changes, the elastic walls of these structures stretch or recoil. These changes in shape distort the receptor's dendritic branches and alter the rate of action potential generation. Baroreceptors monitor blood pressure in the walls of major vessels, including the carotid artery (at the **carotid sinus**) and the aorta (at the **aortic sinus**). The information provided by these baroreceptors plays a major role in regulating cardiac function and adjusting blood flow to vital tissues. Baroreceptors in the lungs monitor the degree of lung expansion. This information is relayed to the respiratory rhythmicity centers in the medulla oblongata, which sets the pace of respiration. Baroreceptors in the urinary and digestive tracts trigger a variety of visceral reflexes, such as those associated with urination and defecation.

**Baroreceptors of Carotid Sinus and Aortic Sinus**

Provide information on blood pressure to cardiovascular and respiratory control centers

**Baroreceptors of Lungs**

Provide information on lung expansion to respiratory rhythmicity centers for control of respiratory rate

**Baroreceptors of Digestive Tract**

Provide information on volume of contents within tract segments; trigger reflex movement of materials along tract

**Baroreceptors of Bladder Wall**

Provide information on volume of urinary bladder; trigger urination reflex

**Baroreceptors of Colon**

Provide information on volume of fecal material in colon; trigger defecation reflex

 **REVIEW**

**A.** Define *baroreceptors*. Where are baroreceptors located within the body?

**2** **Chemoreceptors** are specialized neurons that detect small changes in the concentrations of specific chemicals or compounds. Chemoreceptors are found (1) within the medulla oblongata and elsewhere in the brain; (2) in the **carotid bodies**, near the origin of the internal carotid arteries on each side of the neck; and (3) in the **aortic bodies** between the major branches of the aortic arch. (The carotid bodies and aortic bodies are small tissue masses of receptors and supporting cells.) The neurons in the respiratory centers in the medulla oblongata monitor the pH and $P_{CO_2}$ in cerebrospinal fluid. The carotid and aortic bodies monitor the pH and levels of carbon dioxide ($P_{CO_2}$) and oxygen ($P_{O_2}$) in arterial blood. (The symbol P refers to the partial pressure of a single gas in a mixture of gases, such as air. As we discuss in Chapter 21, differences in partial pressure drive the diffusion of gases in the lungs and body tissues.) These chemoreceptors play an important role in the reflexive control of respiration and cardiovascular function.

**?** **REVIEW**

**B.** Which type of receptor is sensitive to changes in blood pH?

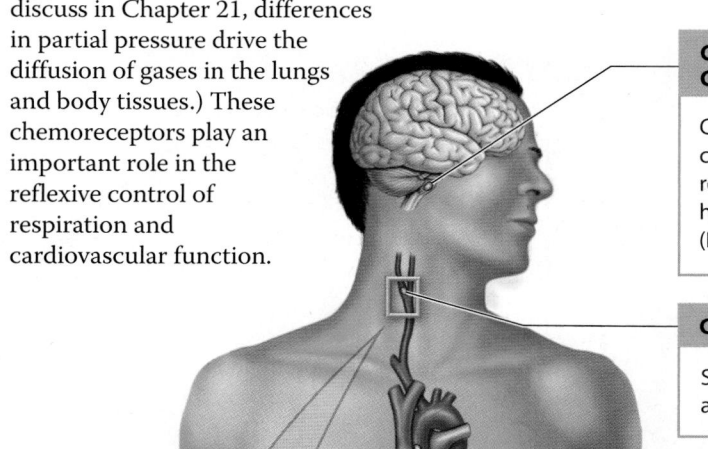

**Chemoreceptors in Respiratory Centers in the Medulla Oblongata**

Chemoreceptors within the respiratory centers of the medulla oblongata respond to the concentration of hydrogen ions (pH) and carbon dioxide ($P_{CO_2}$) in cerebrospinal fluid.

Trigger reflexive adjustments in depth and rate of respiration

**Chemoreceptors of Carotid Bodies**

Sensitive to changes in the pH, $P_{CO_2}$, and $P_{O_2}$ in arterial blood

By cranial nerve IX

Trigger reflexive adjustments in respiratory and cardiovascular activity

**Chemoreceptors of Aortic Bodies**

Sensitive to changes in the pH, $P_{CO_2}$, and $P_{O_2}$ in arterial blood

By cranial nerve X

**14**

**3** Enlarged view of the carotid sinus and the location of the carotid body.

Branch of cranial nerve IX
Internal carotid
External carotid

Carotid body
Carotid sinus
Common carotid

Chemoreceptors

Blood vessel

**Carotid body**          LM × 400

**Lo** **LEARNING OUTCOME**

Explain the roles of baroreceptors and chemoreceptors in homeostasis.

# Module 14.11

# The autonomic nervous system has multiple levels of motor control

The levels of activity in the sympathetic and parasympathetic divisions of the ANS are controlled by centers in the brainstem that regulate specific visceral functions. As in the SNS, simple ANS reflexes based in the autonomic ganglia and the spinal cord provide relatively rapid and automatic responses to stimuli. More complex sympathetic and parasympathetic reflexes are coordinated by processing centers in the medulla oblongata. In addition to the cardiovascular and respiratory centers, the medulla oblongata contains centers and nuclei involved with salivating, swallowing, digestive secretions, peristalsis, and urinary function. These centers are in turn subject to regulation by the hypothalamus. Because the hypothalamus interacts with all other portions of the brain, activity in the limbic system, thalamus, or cerebral cortex can have dramatic effects on autonomic function. For example, when you become angry, your heart rate accelerates, your blood pressure rises, and your respiratory rate increases; when you consider your next meal, your stomach "growls", and your mouth waters.

**1** This diagram shows the pathways and levels of ANS control. The hypothalamus acts as the "headquarters" of both the sympathetic and parasympathetic divisions, and hypothalamic output is indicated by solid red arrows. Nuclei in the spinal cord and brainstem control basic visceral reflex patterns. There is also continual feedback between higher brain centers and the hypothalamus and brainstem; this subconscious communication is indicated by the dashed red arrows.

**?** **REVIEW**
**A.** What brain structure is considered to be the headquarters for the ANS?

**Cerebral Cortex**
Communication at subconscious level

**Limbic System**
Emotions, memory, and behavior

**Thalamus**
Relay of sensory information

**Hypothalamus**
Sympathetic and para-sympathetic headquarters

**Pons**
Higher levels of respiratory control

**Includes**
- Cardiac centers
- Vasomotor center
- Swallowing center
- Coughing center
- Respiratory centers

**Medulla Oblongata**
Processing centers for complex visceral reflexes (sympathetic and parasympathetic)

**Spinal Cord $T_1$–$L_2$**
Neurons controlling sympathetic visceral reflexes

**Sacral Spinal Cord $S_2$–$S_4$**
Neurons controlling parasympathetic visceral reflexes (such as defecation and urination)

**2** The SNS and the ANS have parallel organization and are integrated at the level of the brainstem. This diagram shows the basic patterns of organization and integration. Blue arrows indicate ascending sensory information; red arrows, descending motor commands; dashed lines, pathways of communication and feedback among higher centers. The activities of the SNS and those of the ANS are integrated at many levels. Although we have considered somatic and visceral motor pathways separately, the two have many parallels in terms of both organization and function. Sensory pathways may carry information distributed to both the SNS and the ANS, triggering integrated and compatible reflexes. Higher levels of integration involve the brainstem, and both systems are influenced, if not controlled, by higher brain centers.

KEY

→ Sensory information

→ Motor commands

←- -→ Communication and feedback

**Central Nervous System**

Cerebral cortex

Limbic system

Thalamus

Hypothalamus

Somatosensory —→

—— Visceral sensory

Relay and processing centers in brainstem

**Somatic reflexes**

**Long reflexes**

**Peripheral Nervous System**

Lower motor neuron

Preganglionic neuron

Sensory pathways

**SNS**

**ANS**

**Short reflexes**

Ganglionic neuron

Skeletal muscles

Sensory receptors

Visceral effectors

Stimulus

**? REVIEW**

**B.** What brain structure relays somatosensory information?

**? INTEGRATION**

**C.** Harry has a brain tumor that is pressing against his hypothalamus. Explain the effect on autonomic function.

**Lo LEARNING OUTCOME**

Describe the hierarchy of interacting levels of control in the ANS, beginning with the hypothalamus.

14

## Section 2 Review
# Autonomic Regulation and Control Mechanisms

**Concept map:**

Use each of the following terms once to fill in the blank boxes to correctly complete the map.

- respiratory
- pons
- spinal cord $T_1$–$L_2$
- vasomotor
- coughing
- hypothalamus
- sympathetic visceral reflexes
- parasympathetic visceral reflexes
- complex visceral reflexes
- limbic system and thalamus

**Fill-in:** Fill in S (sympathetic) or P (parasympathetic) to indicate the ANS division responsible for each of the following effects.

| | | |
|---|---|---|
| 11 | Decreased metabolic rate | 11 ○S ○P |
| 12 | Increased salivary and digestive secretions | 12 ○S ○P |
| 13 | Increased metabolic rate | 13 ○S ○P |
| 14 | Stimulation of urination and defecation | 14 ○S ○P |
| 15 | Activation of sweat glands | 15 ○S ○P |
| 16 | Heightened mental alertness | 16 ○S ○P |
| 17 | Decreased heart rate and blood pressure | 17 ○S ○P |
| 18 | Mobilization of energy reserves | 18 ○S ○P |
| 19 | Increased heart rate and blood pressure | 19 ○S ○P |
| 20 | Reduced digestive and urinary functions | 20 ○S ○P |
| 21 | Increased motility and blood flow in the digestive tract | 21 ○S ○P |
| 22 | Increased respiratory rate and dilation of respiratory passages | 22 ○S ○P |
| 23 | Constriction of the pupils and focus of the eyes on nearby objects | 23 ○S ○P |

**Short answer:** Recent surveys show that about one-third of the American adult population is involved in some type of exercise program. What contributions does sympathetic activation make to help the body adjust to changes that occur during exercise and still maintain homeostasis?

24 _____

_____

# Study Outline

## SECTION 1 • Functional Anatomy of the Autonomic Nervous System

### Module 14.1
### Ganglionic neurons of the ANS control visceral effectors p. 517

1. The **autonomic nervous system (ANS)** controls visceral functions outside our conscious awareness.

2. In the **somatic nervous system (SNS)**, neurons control skeletal muscles. The ANS controls the **visceral effectors**: smooth muscle, glands, cardiac muscle, and adipocytes.

3. ANS motor neurons synapse with visceral motor neurons in autonomic ganglia. These **ganglionic neurons** control visceral effectors.

4. The visceral motor neurons whose cell bodies lie in the brainstem and spinal cord are known as **preganglionic neurons**. The axons of these neurons leave the CNS and synapse with ganglionic neurons at **autonomic ganglia**.

### Module 14.2
### The ANS consists of sympathetic and parasympathetic divisions p. 518

5. The most familiar divisions of the ANS are the **sympathetic division** and the **parasympathetic division**.

6. While it is not always the case, the sympathetic division causes excitation, and the parasympathetic division causes inhibition.

7. The two divisions may work independently, because some structures are innervated by only one division, or they may work together during complex processes.

8. Both divisions influence a third division of the nervous system, the **enteric nervous system**.

9. The enteric nervous system is an extensive network of neurons in the walls of the digestive tract.

### Module 14.3
### The sympathetic division has chain ganglia, collateral ganglia, and the adrenal medullae, whereas the parasympathetic division has terminal or intramural ganglia p. 520

10. The preganglionic neurons of the sympathetic division have their cell bodies in the lateral gray horns of spinal segments $T_1$–$L_2$.

11. They synapse with ganglionic neurons located at the **sympathetic chain ganglia**, **collateral ganglia**, or modified ganglion cells in the **adrenal medullae**.

12. The axons of preganglionic neurons (**preganglionic fibers**) are relatively short compared with the axons of ganglionic neurons (**postganglionic fibers**).

13. The sympathetic division prepares the body for heightened somatic activity, such as the "fight or flight" response.

14. The preganglionic neurons of the parasympathetic division have their cell bodies in cranial nerves III, VII, IX, and X and in the lateral gray horns of sacral spinal segments $S_2$–$S_4$.

15. Axons of preganglionic parasympathetic neurons are relatively long. They synapse with ganglionic neurons in **terminal ganglia** (located near the target organ) or in **intramural ganglia** (embedded in the wall of the target organ).

16. The parasympathetic division is known as the "rest and digest" system because it regulates visceral function and energy conservation.

### Module 14.4
### The two ANS divisions innervate many of the same structures, but the innervation patterns are different p. 522

17. Every spinal nerve has a gray ramus communicans that carries sympathetic postganglionic fibers for distribution in the body wall and limbs.

18. Postganglionic sympathetic fibers of the cervical sympathetic ganglia supply the regions innervated by cranial nerves III, VII, IX, and X.

19. **Splanchnic nerves** supply the viscera. There are three groups: cardiopulmonary, abdominopelvic (greater, lesser, and lumbar), and pelvic (sacral).

20. The vagus nerve (X) carries 75 percent of parasympathetic efferent commands.

21. The preganglionic fibers of the sacral segments carry sacral parasympathetic output and form distinct **pelvic nerves**. These nerves innervate intramural ganglia in the kidneys, urinary bladder, terminal portions of the large intestine, and sex organs.

### Module 14.5
### The functional differences between the two ANS divisions reflect their divergent anatomical and physiological characteristics p. 524

22. In crisis, the entire sympathetic division responds. This is called **sympathetic activation**.

23. During sympathetic activation, the following changes occur: increased alertness, feelings of energy and euphoria, temporary insensitivity to painful stimuli, increased heart rate, increased blood pressure, increased breathing rate and depth of respiration, elevated muscle tone, and a mobilization of energy reserves.

24. The major effects produced by the parasympathetic division include constriction of the pupils, focusing of the lens on near objects, secretion of digestive glands, secretion of hormones promoting the absorption of nutrients, blood flow and glandular activity in response to sexual arousal, control of defecation, control of urination, constriction of the respiratory passageways, and reduction of heart rate and force of contraction.

25. The parasympathetic division has been called the **anabolic system** because its stimulation increases nutrient content in blood.

### Module 14.6
### Membrane receptors at target organs mediate the effects of sympathetic and parasympathetic stimulation p. 526

26. Sympathetic stimulation is a result of the interactions of the neurotransmitters norepinephrine (NE) and epinephrine (E) with **adrenergic receptors** in the plasma membrane.

27. There are two classes of adrenergic receptors: **alpha receptors** and **beta receptors**.

28. There are two types of alpha receptors: alpha-1 and alpha-2.

29. Stimulation of alpha-1 receptors generally excites the target cell. Stimulation of alpha-2 receptors generally inhibits the target cell.

30. There are three types of beta receptors: beta-1, beta-2, and beta-3.

31. Beta receptor stimulation changes the metabolic activity of the target cell.

32. All parasympathetic stimulation uses the neurotransmitter ACh that binds to **cholinergic receptors**.

33. There are two types of cholinergic receptors: **nicotinic receptors** and **muscarinic receptors**.

34. ACh always excites nicotinic receptors. ACh can excite or inhibit muscarinic receptors.

*Muscarinic receptors are stimulated by muscarine, a toxin produced by some poisonous mushrooms.*

## SECTION 2 • Autonomic Regulation and Control Mechanisms

### Module 14.7

**The ANS adjusts visceral motor responses to maintain homeostasis** p. 529

35. The ANS adjusts activities of the digestive, cardiovascular, respiratory, and reproductive systems to maintain homeostasis without instructions or interference from the conscious mind.

36. General sensory receptors are stimulated, sending sensory information toward the CNS, where it is processed.

37. Motor commands are issued from the CNS toward either somatic or visceral effectors.

### Module 14.8

**The ANS provides precise control over visceral functions p. 530**

38. Even in the absence of stimuli, autonomic motor neurons show a continuous level of spontaneous activity called **autonomic tone**.

39. Many organs receive instructions from both sympathetic and parasympathetic divisions. This is called **dual innervation**, and the effects may be opposing or complementary.

### Module 14.9

**Most visceral functions are controlled by visceral reflexes p. 532**

40. **Visceral reflexes** provide automatic motor responses that can be influenced by higher centers in the brain.

41. All visceral reflexes are polysynaptic reflex arcs that consist of a receptor, a sensory neuron, a processing center, and one or two visceral motor neurons.

42. **Short reflexes** control simple motor responses with localized effects and operate outside the awareness and control of the CNS.

43. **Long reflexes** are the autonomic equivalents of the polysynaptic reflexes. Visceral sensory neurons send information to the CNS along the dorsal roots of spinal nerves, sensory branches of cranial nerves, and autonomic nerves that innervate visceral effectors.

44. Long reflexes are more common than short reflexes, and they are responsible for controlling multiple organ systems.

45. Visceral sensory information is collected by interoceptors that monitor visceral tissues and organs. These include nociceptors, thermoreceptors, tactile receptors, baroreceptors, and chemoreceptors.

46. The processing of sensory information is performed at the subconscious level in the nuclei in the spinal cord and the solitary nuclei in the brainstem.

### Module 14.10

**Baroreceptors and chemoreceptors initiate important autonomic reflexes involving visceral sensory pathways p. 534**

47. **Baroreceptors** are stretch receptors that monitor changes in pressure.

48. Baroreceptors monitor blood pressure in the walls of major vessels such as the **carotid sinus** and **aortic sinus**. This allows for the regulation of cardiac function and blood flow to vital tissues.

49. Baroreceptors in the lung monitor lung expansion to control respiratory rhythmicity, and those in the bladder and colon control urination and defecation.

50. **Chemoreceptors** detect changes in the concentration of specific chemicals or compounds.

51. Chemoreceptors are located in the medulla oblongata, the **carotid bodies**, and the **aortic bodies**.

52. The chemoreceptors of the medulla oblongata respond to changes in pH and $P_{CO_2}$ in cerebrospinal fluid to regulate depth and rate of respiration.

53. The chemoreceptors of the carotid bodies and aortic bodies respond to changes in pH, $P_{CO_2}$, and $P_{O_2}$ in arterial blood to adjust respiratory and cardiovascular activity.

### Module 14.11

**The autonomic nervous system has multiple levels of motor control** p. 536

54. The hypothalamus acts as the headquarters of both ANS divisions.

55. Nuclei in the spinal cord and brainstem control basic visceral reflex patterns.

56. There is also continued feedback between higher brain centers and the hypothalamus and brainstem.

57. The somatic nervous system and autonomic nervous system have parallel organization and are integrated at the level of the brainstem.

58. Sensory pathways may distribute information to both the somatic nervous system and autonomic nervous system, triggering integrated and compatible reflexes.

# Chapter Review Questions

**Labeling:** Label the anatomical characteristics of the ANS shown in the diagram below.

**True/False:** Indicate whether each statement is true or false.

10  Ganglionic neurons have their cell bodies in the CNS.

11  The enteric nervous system is influenced by both the sympathetic and parasympathetic divisions.

12  The celiac ganglion is associated with sympathetic regulation of the cardiovascular and respiratory systems.

13  All effector organs receive dual innervation.

14  Chemoreceptors of the carotid and aortic bodies are sensitive to changes in blood pressure.

15  Nicotinic receptors can be found at the neuromuscular junctions of skeletal muscle fibers.

10 _____
11 _____
12 _____
13 _____
14 _____
15 _____

**Multiple choice:** Select the correct answer from the list provided.

**16** Which of the following is *not* a visceral effector?
- a) smooth muscle
- b) glands
- c) skeletal muscle
- d) adipocytes

**17** Sympathetic division axons emerge from
- a) the brainstem and the sacral segments of the spinal cord.
- b) cranial nerves III, VIII, IX, and X.
- c) only sacral nerves.
- d) the thoracic nerves and lumbar nerves $L_1$ and $L_2$.

**18** Which of the following ganglionic neurons synapse with preganglionic fibers of the parasympathetic division?
- a) ciliary ganglia
- b) sympathetic chain
- c) adrenal medulla
- d) inferior mesenteric ganglia

**19** The autonomic nervous system directs
- a) voluntary motor activity.
- b) conscious control of skeletal muscles.
- c) unconscious processes that maintain homeostasis.
- d) sensory input from the skin.

**20** The division of the autonomic nervous system that prepares the body for activity and stress is the
- a) sympathetic division.
- b) parasympathetic division.
- c) enteric division.
- d) somatomotor division.

**21** The two types of cholinergic receptors are
- a) alpha-1 and alpha-2 receptors.
- b) beta-1 and beta-2 receptors.
- c) nicotinic and muscarinic receptors.
- d) alpha-1 and beta-1 receptors.

**22** Which receptors would you block to decrease heart rate?
- a) alpha-1 receptors
- b) beta-1 receptors
- c) nicotinic receptors
- d) muscarinic receptors

**23** Where are the chemoreceptors that are sensitive to changes in pH and $P_{CO_2}$ in arterial blood located?
- a) carotid and aortic bodies
- b) carotid and aortic sinuses
- c) baroreceptors of the lungs
- d) respiratory centers of the medulla oblongata

**24** A drug that stimulates which receptor would relax the smooth muscle passages and increase airway diameter of the respiratory tract?
- a) beta-1
- b) beta-2
- c) beta-3
- d) muscarinic

**25** Which of the following responses is *not* characteristic of sympathetic activation?
- a) increased alertness
- b) elevation of muscle tone
- c) erection
- d) increased activity in the cardiovascular and respiratory centers of the pons and medulla oblongata

**26** Which neurotransmitter is released by all preganglionic fibers of the autonomic nervous system?
- a) acetylcholine
- b) epinephrine
- c) norepinephrine
- d) serotonin

**27** Dual innervation refers to situations in which vital organs
- a) receive information from both sympathetic and parasympathetic fibers.
- b) receive information from both the somatic and autonomic nervous systems.
- c) are receptive to multiple neurotransmitters.
- d) are under both conscious and unconscious control.

**Short answer**

**28** Identify the components of a visceral reflex arc, and distinguish between short reflexes and long reflexes.

**29** Identify at least three differences between the somatic and autonomic nervous systems.

**30** Which cranial and sacral nerves are associated with the parasympathetic division?

## Chapter Integration • Applying what you have learned

### A patch for the ocean

Andrew, a recent high school graduate who has spent his entire life in his hometown of Tulsa, Oklahoma, is very excited to have been accepted to a university in San Diego. One of the first social events organized by his dormitory is an all-day whale-watching tour. Although he had fished from a boat in lakes as a young boy, Andrew has never been out on the ocean. Despite being apprehensive, he signed up.

Prior to the tour, Andrew visited the campus health center to ask about seasickness medicine. The doctor recommended that Andrew use a patch that he can place behind one of his ears. The doctor said that even though it's not fully understood how the patch's active ingredient, scopolamine, works to prevent seasickness, his patients have had good results with it. Andrew decided to give it a try.

As he applied the patch on the morning of the whale-watching tour, he read the package label. It described scopolamine as antimuscarinic and anticholinergic. Andrew had no idea what that meant. Using what you have just learned about the autonomic nervous system, answer the following questions.

**1** Which division of the autonomic nervous system would be affected if muscarinic receptors were blocked?

**2** Blocking muscarinic receptors would affect which neurotransmitter?

**3** Why does the label also describe scopolamine as anticholinergic?

**4** Predict some side effects Andrew might experience while wearing the patch.

### Bee careful in the garden!

Gregor is outside gardening when he is stung on his finger by a bee. Shortly afterward, he experiences pain, swelling, redness, and itching in the immediate area of the sting; his throat begins to swell, and breathing becomes more difficult.

An ambulance is called, and Gregor is taken to the emergency room. Drugs are administered to prevent anaphylaxis, a severe, life-threatening allergic response. Anaphylaxis is characterized by the contraction of smooth muscles along the respiratory passageways and a peripheral vasodilation that produces a fall in blood pressure, which would affect multiple organ systems.

**5** Would acetylcholine or epinephrine be more helpful in relieving his condition? Why?

**6** If the condition progressed to the point where there is a fall in blood pressure and inadequate blood flow to body tissues, describe what ANS responses would attempt to restore normal blood flow and blood pressure.

## LEARNING OUTCOMES

These Learning Outcomes correspond by number to this chapter's modules and indicate what you should be able to do after completing the chapter. They also appear at the end of each module.

# 15

# The Special Senses

# Module 15.1

# A generator potential is a depolarization of the membrane

The five special senses are *olfaction* (smell), *gustation* (taste), *vision, equilibrium* (balance), and *hearing.* These diverse and complex sensations all originate at sensory receptor cells. They may be neurons or specialized receptor cells that communicate with sensory neurons.

**1** Olfactory receptors are the dendrites of specialized neurons. When dissolved odorants bind to olfactory receptors, a depolarization called a **generator potential** occurs. If the stimulus is strong enough, the generator potential triggers action potentials that travel the axon to the CNS.

Specialized olfactory neuron

**? REVIEW**
**A.** Where do the special senses originate?

**? REVIEW**
**B.** What is a generator potential?

**2** The receptors for the senses of taste, vision, equilibrium, and hearing are specialized cells that have inexcitable membranes and form synapses with the processes of sensory neurons. When stimulated, the membrane of the receptor cell undergoes a graded depolarization that triggers the release of neurotransmitters at the synapse. These neurotransmitters then depolarize the sensory neuron, inducing a generator potential capable of producing action potentials that are propagated to the CNS. Because a synapse is involved, there is a slight synaptic delay. However, this arrangement permits modification of the sensitivity of the receptor cell by presynaptic facilitation or inhibition.

In the following modules we will consider the organization of receptors and the sensory pathways for each of the special senses. We begin with the special senses of olfaction and gustation.

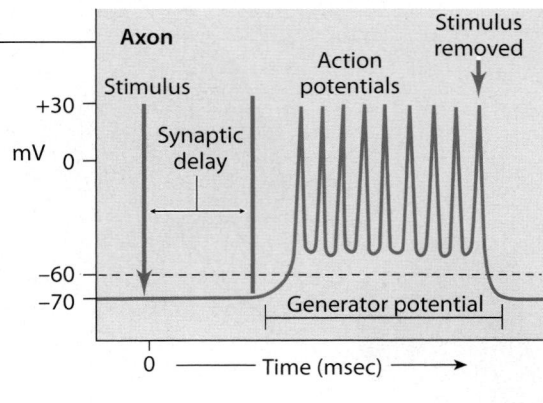

**? REVIEW**
**C.** Compare olfactory receptors with receptors for the other special senses.

**Lo LEARNING OUTCOME**
Explain the roles of generator potentials and depolarization in sensory neurons and receptor cells.

# Olfaction involves specialized chemoreceptive neurons and delivers sensations directly to the cerebrum

**1** The sense of smell, called **olfaction**, is made possible by paired **olfactory organs**. These olfactory organs, located in the nasal cavity on each side of the nasal septum, contain olfactory receptor cells. The epithelium of the olfactory organs covers the inferior surface of the cribriform plate, the superior portion of the perpendicular plate, and the superior nasal conchae of the ethmoid (see Module 7.6).

**? REVIEW**
**A.** Describe olfaction.

## Olfactory Pathway to the Cerebrum

| | | | | |
|---|---|---|---|---|
| Chemicals in the air stimulate the *olfactory receptor cells*, sensory neurons within the olfactory organ. | Axons leaving the olfactory epithelium collect into 20 or more bundles that penetrate the cribriform plate of the ethmoid bone. | The first synapse occurs in the **olfactory bulb**, which is located just superior to the cribriform plate. | Axons leaving the olfactory bulb travel along the **olfactory tract** to reach the olfactory cortex, the hypothalamus, and portions of the limbic system. | The distribution of olfactory information to the limbic system and hypothalamus explains how certain smells trigger profound emotional and behavioral responses, as well as memories. |

Cribriform plate of ethmoid bone

Olfactory epithelium

Superior nasal concha

**Olfactory organ on the right side of the nasal septum**

**? REVIEW**
**B.** Trace the olfactory pathway, beginning at the olfactory epithelium.

15

**2** The olfactory organs are made up of two layers: the **olfactory epithelium** and the **lamina propria**. The **olfactory receptor cells** are specialized neurons within the epithelium. The exposed tip of each receptor cell forms a knob that projects beyond the epithelial surface. The knob is a base for up to 20 cilia-shaped dendrites that extend into the surrounding mucus and lie parallel to the epithelial surface, where they are exposed to dissolved chemicals. Between 10 and 20 million olfactory receptors are packed into an area of roughly 5 cm². If we take into account the exposed dendritic surfaces, the actual sensory area probably approaches that of the entire body surface.

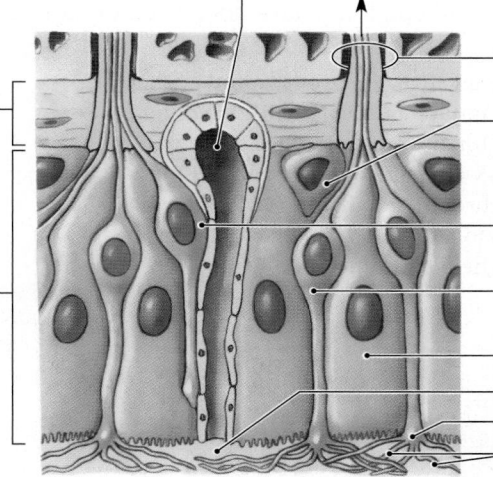

Olfactory gland

To olfactory bulb

The underlying lamina propria consists of areolar tissue, numerous blood vessels, and nerves. This layer also contains **olfactory glands** whose secretions absorb water and form a thick, pigmented mucus.

The olfactory epithelium contains the olfactory receptor cells, **supporting cells**, and regenerative **basal cells** (stem cells). The olfactory receptor population undergoes considerable turnover; new receptor cells are produced by the division and differentiation of basal cells. This turnover is one of the few examples of replacement of neurons in adult humans.

Olfactory nerve fibers

Basal cell: divides to replace worn-out olfactory receptor cells

Developing olfactory receptor cell

Olfactory receptor cell

Supporting cell

Mucous layer

Knob

Olfactory dendrites: surfaces contain receptor proteins

**3** Olfactory reception occurs on the dendrites of the olfactory receptor cells. **Odorants**—dissolved chemicals that stimulate olfactory neurons—interact with membrane receptors called *odorant-binding proteins* on the membrane surface. In general, odorants are small, airborne, water-soluble and lipid-soluble substances. As few as four odorant molecules can activate an olfactory receptor cell.

**? REVIEW**

**C.** Describe the events leading up to the depolarization and generation of an action potential by an olfactory receptor cell.

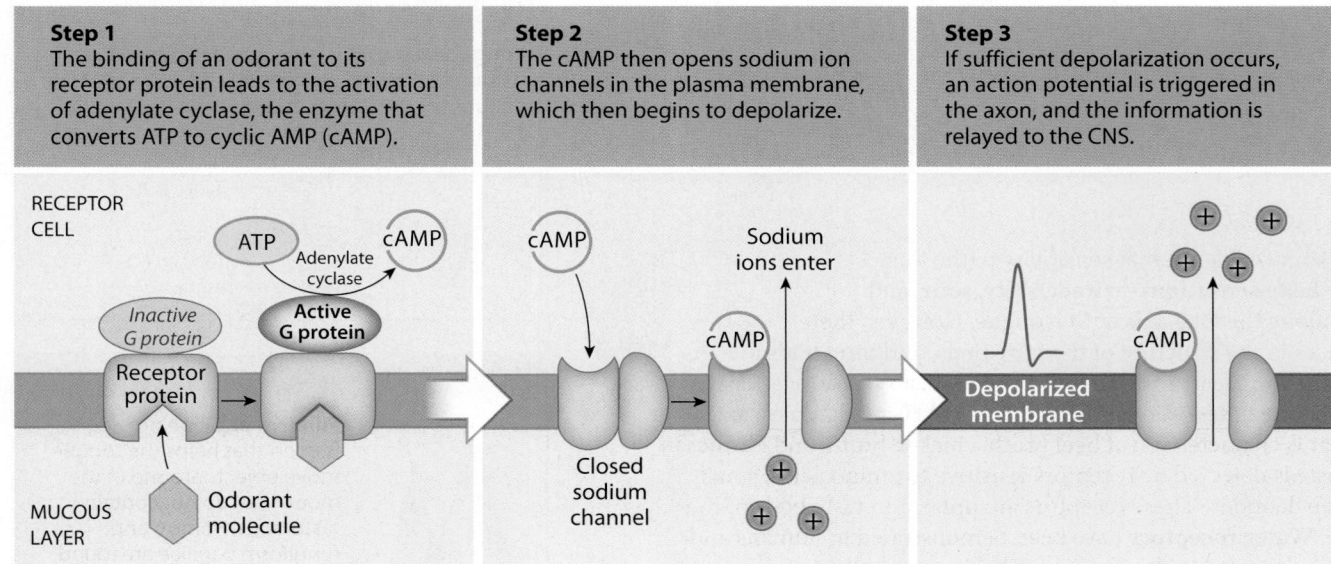

**Step 1**
The binding of an odorant to its receptor protein leads to the activation of adenylate cyclase, the enzyme that converts ATP to cyclic AMP (cAMP).

**Step 2**
The cAMP then opens sodium ion channels in the plasma membrane, which then begins to depolarize.

**Step 3**
If sufficient depolarization occurs, an action potential is triggered in the axon, and the information is relayed to the CNS.

RECEPTOR CELL

ATP
Adenylate cyclase
cAMP
*Inactive G protein*
**Active G protein**
Receptor protein

MUCOUS LAYER
Odorant molecule

cAMP
Sodium ions enter
cAMP
Closed sodium channel

Depolarized membrane
cAMP

 **LEARNING OUTCOME**
Describe the sensory organs of smell, trace the olfactory pathways to their destinations in the cerebrum, and explain how olfactory perception occurs.

15

# Gustation involves epithelial chemoreceptor cells located in taste buds

**Gustation**, the sense of taste, provides information about the foods and liquids we eat and drink. *Taste receptors*, or *gustatory* (GUS-ta-tor-ē) *receptor cells*, are distributed over the superior surface of the tongue and adjacent portions of the pharynx and larynx. The most important taste receptors are on the tongue. By the time we reach adulthood, the taste receptors on the pharynx, larynx, and epiglottis have decreased in number.

**1** The superior surface of the tongue has numerous variously shaped epithelial projections called **lingual papillae** (pa-PIL-ē; *papilla*, a nipple-shaped mound). The human tongue has four types of lingual papillae: (1) vallate (VAL-āt; *vallum*, wall) papillae, (2) foliate (FŌ-lē-āt) papillae, (3) fungiform (*fungus*, mushroom) papillae, and (4) filiform (*filum*, thread) papillae. The vallate, foliate, and fungiform papillae contain taste receptors sensitive to chemical stimuli and specialized epithelial cells in sensory structures called **taste buds**. An adult has about 5000 taste buds.

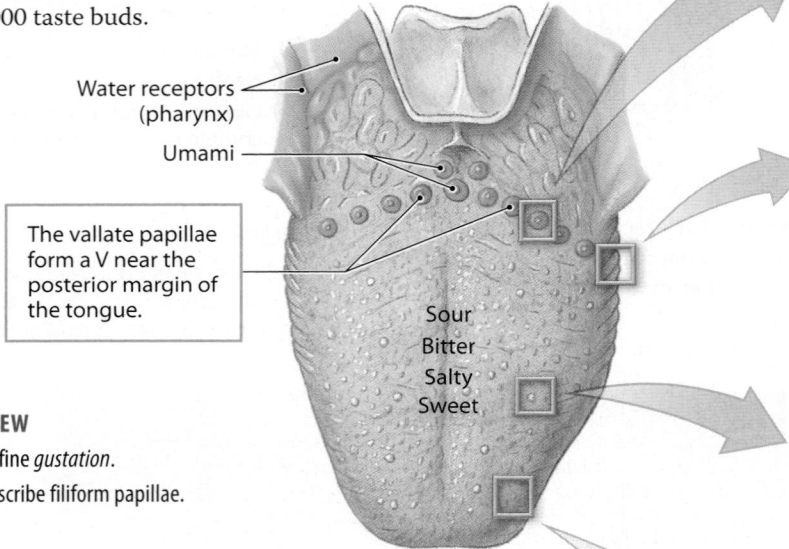

Water receptors (pharynx)

Umami

The vallate papillae form a V near the posterior margin of the tongue.

Sour
Bitter
Salty
Sweet

**? REVIEW**
**A.** Define *gustation*.
**B.** Describe filiform papillae.

**2** There is some evidence that sensitivity to the four **primary taste sensations—sweet, salty, sour,** and **bitter**—varies along the long axis of the tongue. However, there are no differences in the structure of the taste buds, and taste buds in all portions of the tongue provide all four primary taste sensations. There are also two other taste sensations. **Umami** (oo-MAH-mē) is a pleasant, savory taste that is characteristic of beef broth, chicken broth, and Parmesan cheese. This taste is detected by receptors sensitive to amino acids, small peptides, and nucleotides. These receptors are present in taste buds of the vallate papillae. **Water receptors** have been demonstrated in humans and appear to be concentrated in the pharynx. The sensory output of these receptors is processed in the hypothalamus and affects water balance and the regulation of blood volume. For example, drinking stimulates water receptors, which in turn stimulate the hypothalamus to secrete antidiuretic hormone, which ultimately prevents overingestion of water.

## Vallate Papillae

**Vallate papillae** are relatively large, shaped like the tip of a pencil eraser, and surrounded by deep epithelial folds. Each papilla contains as many as 100 taste buds.

Taste buds

## Foliate Papillae

**Foliate papillae** are found on the lateral margins of the posterior region of the tongue.

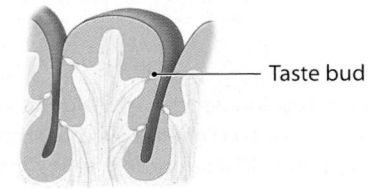

Taste bud

## Fungiform Papillae

**Fungiform papillae** are shaped like small buttons within shallow depressions. Each papilla contains about five taste buds.

Taste bud

## Filiform Papillae

**Filiform papillae** provide friction that helps the tongue move objects around in the mouth but do not contain taste buds. Filiform and fungiform papillae are found on the anterior two-thirds of the superior surface of the tongue.

15

**3** Taste buds are recessed into the surrounding epithelium, isolated from the unprocessed contents of the mouth. Each taste bud contains 40–100 gustatory receptor cells and many small stem cells.

Each **gustatory receptor cell** extends slender microvilli, sometimes called *taste hairs*, into the surrounding fluids through the **taste pore**, a narrow opening. A typical gustatory receptor cell survives about 10 days before it is replaced.

Axons of sensory neurons

**Basal cells** are stem cells that divide to produce *transitional cells* that mature into gustatory receptor cells.

Transitional cell

Taste hairs (microvilli)

Taste pore

**Diagrammatic view of a taste bud**

Taste buds

**Taste buds**  LM × 280

**Taste bud**  LM × 650

15

**4** The threshold for taste receptor stimulation varies for each of the primary taste sensations, and taste receptors respond more readily to unpleasant than to pleasant stimuli. For example, we are 100,000 times more sensitive to bitter substances and almost 1000 times more sensitive to acids (sour) than to either sweet or salty chemicals. This sensitivity has survival value: Acids can damage the mucous membranes of the mouth and pharynx, and many potent biological toxins have an extremely bitter taste.

Our tasting abilities change with age. We begin life with more than 10,000 taste buds, but the number begins declining dramatically by age 50. The sensory loss becomes especially significant because aging people also experience a decline in the number of olfactory receptors. As a result, many elderly people find that their food tastes bland and unappetizing, whereas children tend to find the same foods too spicy.

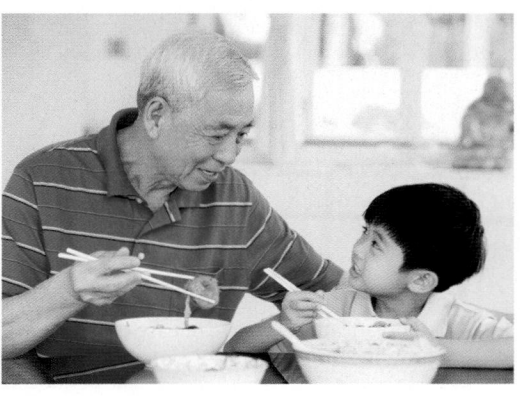

**? REVIEW**

**C.** Which taste receptors offer a survival advantage when tasting something for the first time?

**Lo LEARNING OUTCOME**
Describe the sensory organs of gustation.

# Module 15.4

# Gustatory reception relies on membrane receptors and ion channels, and sensations are carried by facial, glossopharyngeal, and vagus nerves

**1** Gustatory reception, like olfactory reception, is stimulated by dissolved chemicals. Dissolved chemicals contacting the taste hairs may either diffuse through plasma membrane leak channels or bind to receptor proteins of the gustatory receptor cell. Some 90 percent of the gustatory receptor cells respond to two or more different taste stimuli. The different tastes involve different receptor mechanisms.

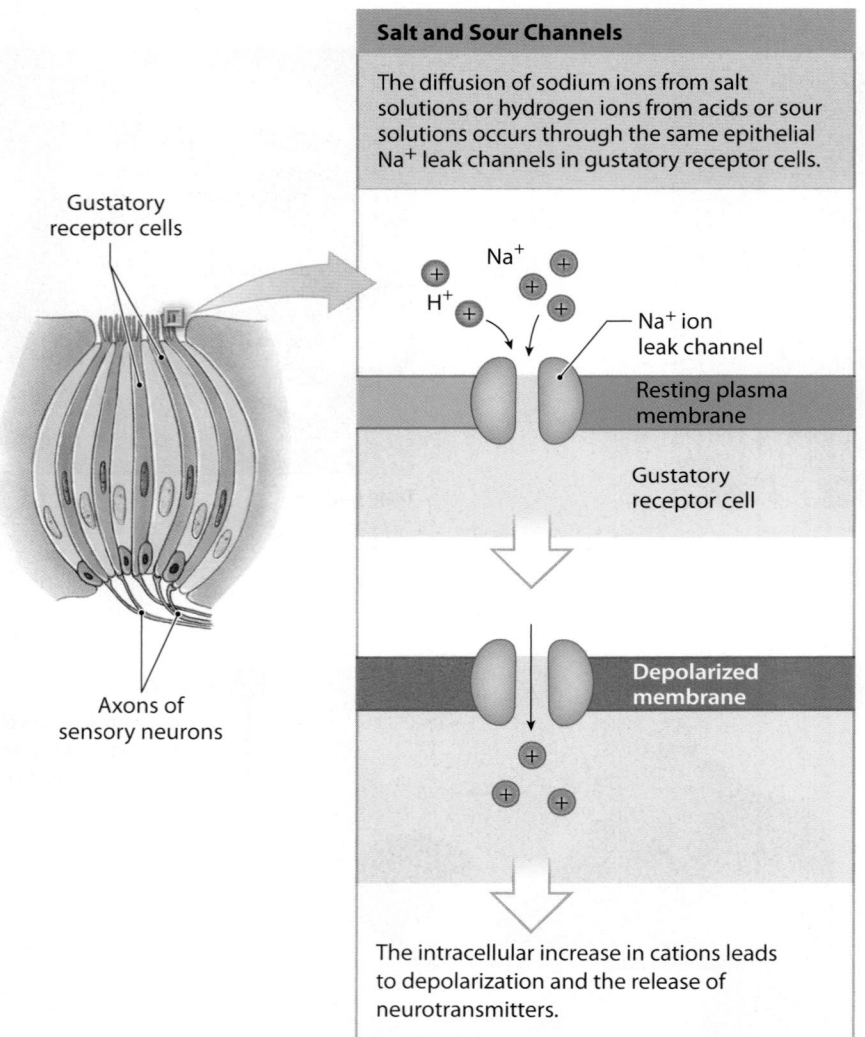

Gustatory receptor cells

Axons of sensory neurons

**Salt and Sour Channels**

The diffusion of sodium ions from salt solutions or hydrogen ions from acids or sour solutions occurs through the same epithelial $Na^+$ leak channels in gustatory receptor cells.

$Na^+$

$H^+$

$Na^+$ ion leak channel

Resting plasma membrane

Gustatory receptor cell

Depolarized membrane

The intracellular increase in cations leads to depolarization and the release of neurotransmitters.

**Sweet, Bitter, and Umami Receptors**

Membrane receptors responding to stimuli that produce sweet, bitter, and umami sensations are linked to G proteins called **gustducins** (GUST-dus-inz)—protein complexes that use second messengers to produce their effects.

Sweet, bitter, or umami

Membrane receptor

Inactive G protein

Active G protein

Depolarized membrane

Active G protein

Active 2nd messenger

Inactive 2nd messenger

Activation of second messengers stimulates release of neurotransmitters.

The released neurotransmitters enter synapses with the dendrites of adjacent sensory neurons. Depolarization of sensory neurons leads to a generator potential and the propagation of action potentials along the gustatory pathway to the CNS.

 **REVIEW**

**A.** What are *gustducins*?

**2** This illustration follows the **gustatory pathway** from the gustatory receptor cells to the cerebral cortex. A conscious perception of taste is produced by processing at the primary somatosensory cortex as the information received from the taste buds is integrated with other sensory data. Information about the texture of food, along with taste-related sensations such as "peppery" or "burning hot," comes from sensory afferents in the trigeminal nerve (V). Taste receptors adapt slowly, but central adaptation quickly reduces your sensitivity to a new taste (see Module 13.17). In addition, the level of stimulation from the olfactory receptors plays an overwhelming role in taste perception. Thus, you are several thousand times more sensitive to "tastes" when your olfactory organs are fully functional. By contrast, when you have a cold and your nose is stuffed up, airborne molecules cannot reach your olfactory receptors, so meals taste dull and unappealing. This reduction in taste perception occurs even though the taste buds are responding normally.

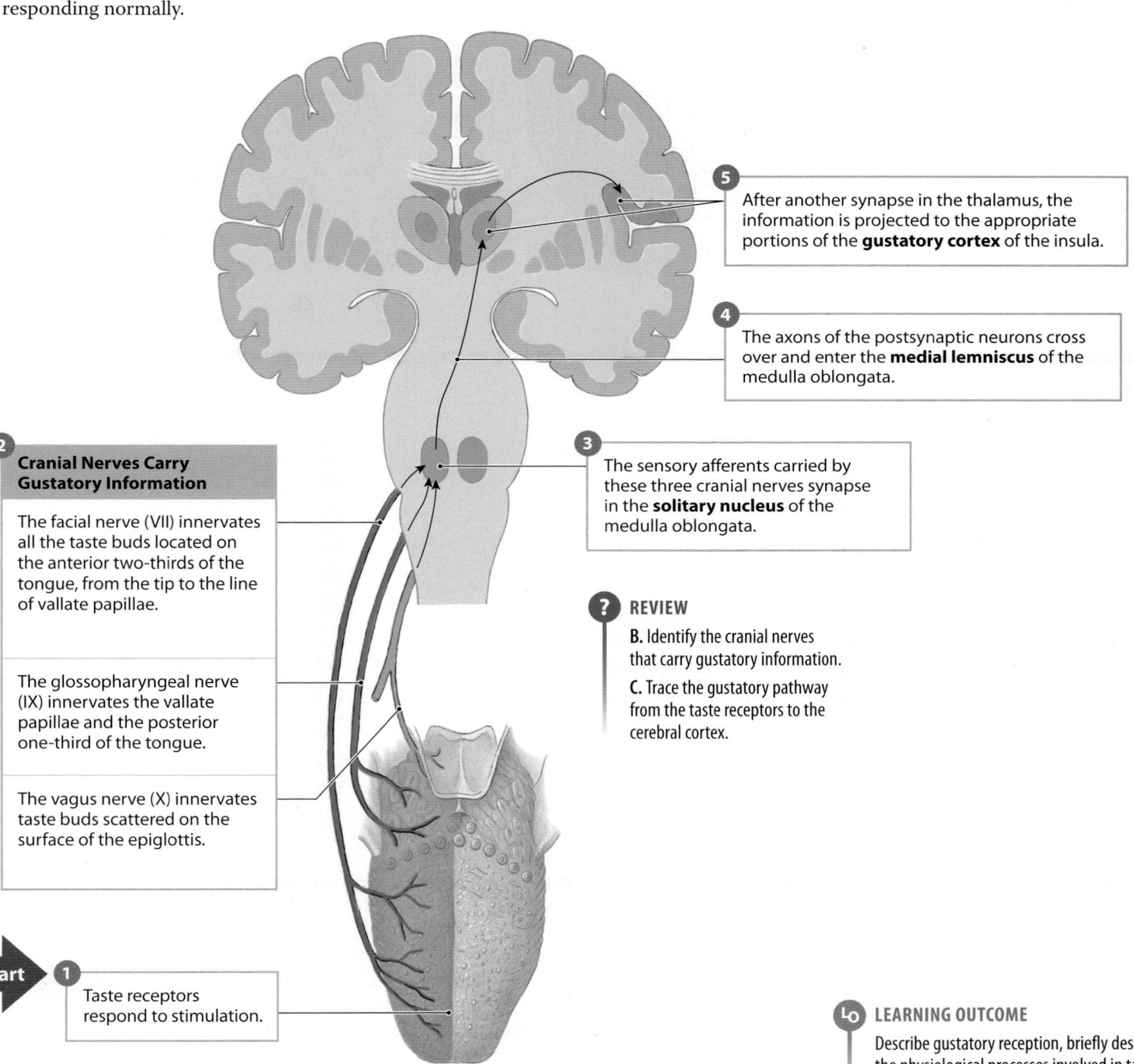

**5** After another synapse in the thalamus, the information is projected to the appropriate portions of the **gustatory cortex** of the insula.

**4** The axons of the postsynaptic neurons cross over and enter the **medial lemniscus** of the medulla oblongata.

**2 Cranial Nerves Carry Gustatory Information**

The facial nerve (VII) innervates all the taste buds located on the anterior two-thirds of the tongue, from the tip to the line of vallate papillae.

The glossopharyngeal nerve (IX) innervates the vallate papillae and the posterior one-third of the tongue.

The vagus nerve (X) innervates taste buds scattered on the surface of the epiglottis.

**3** The sensory afferents carried by these three cranial nerves synapse in the **solitary nucleus** of the medulla oblongata.

**? REVIEW**

**B.** Identify the cranial nerves that carry gustatory information.

**C.** Trace the gustatory pathway from the taste receptors to the cerebral cortex.

**Start 1** Taste receptors respond to stimulation.

**Lo LEARNING OUTCOME**

Describe gustatory reception, briefly describe the physiological processes involved in taste, and trace the gustatory pathway.

15

**Labeling:** Label taste sensations associated with areas along the tongue axis (1–5) and the four types of lingual papillae (6–9).

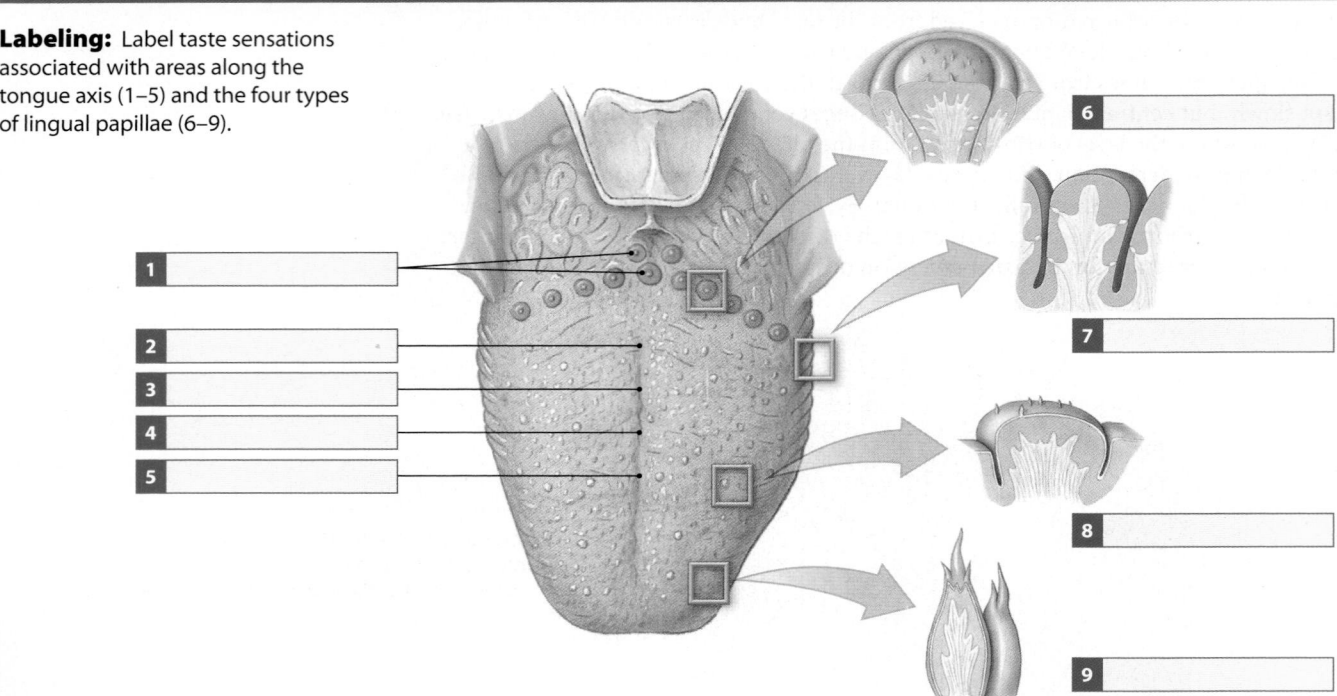

| | |
|---|---|
| 1 | |
| 2 | |
| 3 | |
| 4 | |
| 5 | |
| 6 | |
| 7 | |
| 8 | |
| 9 | |

**Matching:** Match each lettered term with the most closely related description.

a. gustation
b. depolarization
c. lamina propria
d. lingual papillae
e. G proteins
f. stem cells
g. bitter
h. olfactory dendrites
i. olfaction
j. taste bud
k. cerebral cortex
l. olfactory bulb
m. odorant

10  Sweet, bitter, and umami sensations
11  Sense of smell
12  Basal cells
13  Chemical stimulus
14  Sense of taste
15  Olfactory glands
16  Receives all special senses stimuli
17  Cluster of gustatory receptors
18  Site of first synapse by olfactory receptors
19  Contain olfactory receptor proteins
20  Most sensitive taste sensation
21  Produces generator potential
22  Epithelial projections of tongue

10 _____
11 _____
12 _____
13 _____
14 _____
15 _____
16 _____
17 _____
18 _____
19 _____
20 _____
21 _____
22 _____

**Section integration:** Contrast the sensory receptors for olfaction with the sensory receptors for taste, vision, equilibrium, and hearing.

23 _____
_____
_____

15

# Module 15.5
# The eyes form early in embryonic development

This section will examine the anatomy and physiology of the adult eyes. The eyes, which give us the sense of vision, are our most complex sense organs. A tremendous amount of sensory integration, feedback, and processing occurs within the eye before any sensations are relayed to the CNS. This functional complexity far exceeds that of the receptor complexes responsible for olfaction, gustation, equilibrium, and hearing. That complexity is directly related to the way the eyes form during embryonic development. In many ways, the eyes resemble detached CNS nuclei that are totally devoted to processing visual stimuli.

**Week 4 embryo**

**1** The first indication of eye development appears as a pair of bulges called **optic vesicles** form in the lateral walls of the prosencephalon. These bulges extend to either side like a pair of dumbbells, each containing a cavity continuous with the neural tube (see Module 13.1).

Neural tube

Optic vesicle

**? REVIEW**
**A.** What are the first structures that form during eye development?

**2** The lateral bulges become indented, forming a pair of **optic cups** that remain connected to the diencephalon by slender stalks. The epidermis overlying the optic cup responds by forming a pocket that later pinches off and develops into the lens of the eye.
The inner and outer layers of the optic cup develop into the retina. The retina will house the light-sensing *photoreceptor cells*.

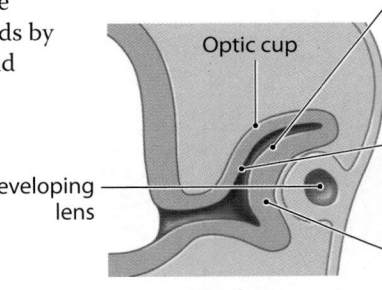

Optic cup

Developing lens

**Week 5**

**Layers of the Developing Retina**

The ependymal cells in this region of the outer layer of the optic cup develop into the photoreceptors.

The ependymal cells on the inner layer of the optic cup develop into pigment cells that absorb light that has passed through the photoreceptor layer. The separation gradually decreases until the photoreceptors and pigment cell layers are in contact.

This nervous tissue of the outer layer of the optic cup forms layers of neurons, ganglion cells, and specialized glial cells that process and integrate visual information.

**? REVIEW**
**B.** Which structures develop into the retina?
**C.** Which cells develop into the photoreceptors?

**3** Embryonic cells aggregating around the optic cup form supporting layers of connective tissue that isolate the nervous tissues from the rest of the body. The eye develops interior chambers filled with fluid that is continuously generated and reabsorbed. Modules within this section will examine each of these components of the eye in greater detail.

The optic nerve (II) follows the path of the original stalk connected to the optic cup.

Lens

Eyelids

Fluid-filled chambers

Connective tissue layers

**Week 6**

**Lo LEARNING OUTCOME**
Outline the embryonic development of the eye.

15

# Accessory structures of the eye provide protection while allowing light to reach the interior of the eye

**1** **Accessory structures** of the eye include the eyelids, eyelashes, the superficial epithelium of the eye, and the structures associated with the production, secretion, and removal of tears.

The **cornea** is a transparent area on the anterior surface of the eye.

Laterally, the two eyelids are connected at the **lateral angle of the eye**. *canthus*

Light enters the eye by passing through the cornea and then through the **pupil**, an opening at the center of the colored **iris**.

### Eyelids and Eyelashes

The **eyelashes**, along the margins of the eyelids, are very robust hairs that help prevent foreign matter from reaching the surface of the eye.

The **eyelid**, or *palpebra* (PAL-peh-bruh), is a continuation of the skin. The continual blinking of the eyelids keeps the surface of the eye lubricated and removes dust and debris. The eyelids can also close firmly to protect the delicate surface of the eye.

The **palpebral fissure** is the gap that separates the free margins of the upper and lower eyelids.

Medially, the two eyelids are connected at the **medial angle of the eye**.

The **lacrimal caruncle** (KAR-ung-kul) is a small, reddish body at the medial angle of the eye. It produces the thick secretions that cause the gritty deposits that sometimes appear after a good night's sleep.

**?** **REVIEW**

**A.** Identify the accessory structures of the eye.

**2** The epithelium covering the inner surfaces of the eyelids and the outer surface of the eye is called the **conjunctiva** (kon-junk-TĪ-vuh). It is a mucous membrane covered by a specialized stratified squamous epithelium.

*palpebrae*

Modified sebaceous glands called **tarsal glands**, or *Meibomian* (mī-BŌ-mē-an) *glands*, are along the inner margin of the lid. These glands secrete a lipid-rich product that helps keep the eyelids from sticking together.

### Conjunctiva

The **palpebral conjunctiva** covers the inner surface of the eyelids.

The **bulbar conjunctiva**, or *ocular conjunctiva*, covers the anterior surface of the eye.

The pocket created where the palpebral conjunctiva becomes continuous with the bulbar conjunctiva is known as the **fornix** of the eye.

The bulbar conjunctiva is continuous with the very delicate corneal epithelium that covers the surface of the transparent cornea.

15

**3** A constant flow of tears keeps conjunctival surfaces moist and clean. Tears reduce friction, remove debris, prevent bacterial infection, and provide nutrients and oxygen to portions of the conjunctival epithelium. The **lacrimal apparatus** produces, distributes, and removes tears. The lacrimal apparatus of each eye consists of (1) a *lacrimal gland* (or tear gland) with associated ducts, (2) paired *lacrimal canaliculi*, (3) a *lacrimal sac*, and (4) a *nasolacrimal duct*.

**? REVIEW**

**B.** Which layer of the eye would be the first affected by inadequate tear production?

### Components of the Lacrimal Apparatus

The almond-shaped **lacrimal gland** is about 12–20 mm (0.5–0.75 in.) long. Each day it produces about 1 mL of watery, slightly alkaline tears. The tears lubricate, nourish, and oxygenate the corneal cells. Lacrimal gland secretions contain the antibacterial enzyme **lysozyme** and antibodies that attack pathogens before they enter the body.

There are 10–12 **tear ducts** that deliver tears from the lacrimal gland to the space behind the upper eyelid.

The **lacrimal puncta** (singular, *punctum*) are two small pores that drain the *lacrimal lake*, or "lake of tears."

The **lacrimal canaliculi** are small canals that connect the lacrimal puncta to the lacrimal sac.

The **lacrimal sac** is a small chamber that nestles within the lacrimal sulcus of the orbit.

The **nasolacrimal duct** begins at the inferior tip of the lacrimal sac. It passes through the nasolacrimal canal to deliver tears to the nasal cavity.

The nasolacrimal duct empties into the *inferior meatus*, a narrow passageway inferior and lateral to the inferior nasal concha.

Superior rectus

Upper eyelid

Lower eyelid

A layer of orbital fat posterior to the eyeball provides padding while permitting eye movements.

Inferior rectus

Inferior oblique

**15**

---

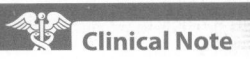 **Clinical Note**

## Conjunctivitis is an inflammation of the conjunctiva

**Conjunctivitis**, or pinkeye, is an inflammation of the conjunctiva. The most obvious sign, redness, is due to the dilation of blood vessels deep to the conjunctival epithelium. This condition may be caused by pathogenic infection or by physical, allergic, or chemical irritation of the conjunctival surface.

**? REVIEW**

**C.** Describe conjunctivitis.

 **LEARNING OUTCOME**

Identify the accessory structures of the eye, and explain their functions.

# Module 15.7

# The hollow eyeball has a layered wall and fluid-filled anterior and posterior cavities

**1** The wall of the eye has three layers, formerly called tunics: (1) an outer *fibrous layer*, (2) an intermediate *vascular layer*, and (3) a deep *inner layer (retina)*.

## Fibrous Layer

The outer **fibrous layer** of the eye consists of the cornea and the white **sclera** (SKLER-uh). These two components are continuous, and the border between the two is called the **corneoscleral junction**, or *corneal limbus*. The fibrous layer (1) supports and protects the eye, (2) is an attachment site for the extrinsic eye muscles, and (3) contains the transparent cornea. Light first enters the eye through the cornea; its curvature aids in the focusing process.

## Vascular Layer

The **vascular layer**, or *uvea* (YŪ-vē-uh), contains numerous blood vessels, lymphatic vessels, and the intrinsic (smooth) muscles of the eye. The functions of this middle layer include (1) providing a route for blood vessels and lymphatics that supply tissues of the eye; (2) regulating the amount of light that enters the eye; (3) secreting and reabsorbing the fluid called *aqueous humor* that circulates within the chambers of the eye; and (4) controlling the shape of the lens, an essential part of the focusing process.

The colored **iris** is visible through the transparent corneal surface. It contains blood vessels, pigment cells (melanocytes), and two layers of smooth muscle fibers. When these muscles contract, they change the diameter of the pupil.

The **ciliary body** is a thickened region that bulges into the interior of the eye. A ring of fibers extends from the ciliary body to the lens, holding it in position posterior to the pupil.

The **choroid** (KOR-oyd) is a vascular layer that is covered by the sclera. The choroid has an extensive capillary network that delivers oxygen and nutrients to the neural tissue within the neural layer.

Sclera  Corneoscleral junction  Cornea

Lens

Optic nerve

## Inner Layer

The **inner layer**, or **retina**, is the innermost layer of the eye. The retina consists of a thin, outer deep layer (the *pigmented layer*) that absorbs light, and a thick, inner superficial layer (the *neural layer*) that contains the **photoreceptors**—the cells that are sensitive to light.

**? REVIEW**

**A.** Name the three layers of the eye.

**2** This sectional view shows that the ciliary body and the lens divide the interior of the eye into a small **anterior cavity** and a large **posterior cavity**.

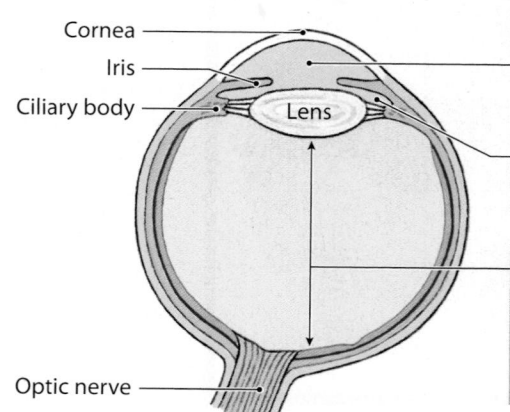

Cornea

Iris

Ciliary body

Lens

Optic nerve

## Anterior Cavity

The **anterior chamber** extends from the cornea to the iris.

The **posterior chamber** extends between the iris and the ciliary body and lens.

## Posterior Cavity

Most of the posterior cavity's volume is taken up by a gelatinous substance known as the **vitreous body**. *Vitreous humor* is the fluid part of the vitreous body.

15

## The Anterior Cavity

**3** A view of the anterior cavity at greater magnification reveals additional details about the structure of the iris and ciliary body. Eye color is determined by genes that influence the density and distribution of melanocytes on the anterior surface and interior of the iris, as well as the density of the pigmented epithelium. When the connective tissue of the iris contains few melanocytes, light passes through it and bounces off the pigmented epithelium; the eye then appears blue. Individuals with green, brown, or black eyes have increasing numbers of melanocytes in the body and surface of the iris. The eyes of people with albinism appear a very pale gray or blue-gray.

## The Aqueous Humor

**4** **Aqueous humor** is a fluid that circulates within the anterior cavity, passing from the posterior to the anterior chamber through the pupil. It also freely diffuses through the vitreous body and across the surface of the retina, where it is called **vitreous humor**. Circulation of the aqueous humor provides an important route for nutrient and waste transport, in addition to forming a fluid cushion. The pressure exerted by the aqueous humor helps retain the eye's shape; it also stabilizes the position of the retina, pressing the neural layer against the pigmented layer. A procedure called **tonometry** measures the eye's intra-ocular pressure—the fluid pressure within the anterior chamber. Normal intra-ocular pressure ranges from 12 to 21 mm Hg.

The body of the iris consists of a highly vascular, pigmented loose connective tissue. The anterior surface has no epithelial covering; instead, it has an incomplete layer of fibroblasts and melanocytes. The posterior surface is covered by a pigmented epithelium that is part of the neural layer.

Aqueous humor forms through active secretion by epithelial cells of the ciliary body's ciliary processes at a rate of 1–2 μL per minute. The epithelial cells regulate its composition, which is similar to cerebrospinal fluid.

Aqueous humor leaves the eye through the **scleral venous sinus** (canal of Schlemm), a passageway that extends completely around the eye at the level of the corneoscleral junction. Collecting channels deliver the aqueous humor from this canal to veins in the sclera. The rate of removal normally keeps pace with the rate of secretion, and aqueous humor is removed and recycled within a few hours of its formation.

The bulk of the ciliary body consists of the **ciliary muscle**, a smooth muscular ring that projects into the interior of the eye.

The **ciliary processes** are folds of epithelium covering the ciliary muscle.

The **ora serrata** (Ō-ra ser-RA-tuh; serrated mouth) is the jagged anterior edge of the neural layer of the retina. The outer pigmented layer of the retina continues anteriorly across the posterior surface of the iris.

The fibers making up the **ciliary zonule** (*suspensory ligament*) attach to the tips of the ciliary processes. These connective tissue fibers hold the lens posterior to the iris and centered on the pupil. As a result, any light passing through the pupil will also pass through the lens.

Cornea

Anterior chamber

Iris

Lens

Ciliary body

Conjunctiva

Posterior cavity (vitreous chamber)

**?** **REVIEW**
**C.** Where in the eye is aqueous humor located?

**?** **REVIEW**
**B.** What gives eyes their characteristic color?

 **LEARNING OUTCOME**
Describe the layers of the eye wall and the anterior and posterior cavities of the eye.

# Module 15.8

# The structures of the eye direct light along a visual axis to the fovea centralis of the retina

**1** The superior sectional view below presents key aspects of eye anatomy that are associated with positioning the eye and allowing light to reach the photoreceptors of the retina.

**?** **REVIEW**

**A.** Which eye structure does *not* contain blood vessels?

### Sectional Anatomy of the Eye

The cornea allows light to enter the eye, so its transparency and clarity are vital to eye function. The cornea consists primarily of a dense matrix containing multiple layers of collagen fibers, organized so as not to interfere with the passage of light. The cornea is avascular; the superficial epithelial cells must obtain oxygen and nutrients from the tears that flow across their free surfaces.

The lens lies posterior to the cornea, held in place by the ciliary zonule that originates on the ciliary body. The lens consists of concentric layers of cells surrounded by a dense fibrous capsule. The cells are slender, long, and filled with transparent proteins called **crystallins**, which are responsible for both the clarity and the focusing power of the lens. Around the edges of the lens, its capsular fibers intermingle with those of the ciliary zonule. The primary function of the lens is to focus the visual image on the photoreceptors. The lens does this by changing shape.

Tension in the series of fibers making up the ciliary zonule resists the tendency of the lens to assume a spherical shape.

The ciliary body supports the lens and controls its shape.

The retina contains the photoreceptors, pigment cells, supporting cells, and neurons.

The blood vessels of the choroid directly or indirectly provide nutrients to all structures within the eye.

The sclera, or "white of the eye," consists of dense fibrous connective tissue containing both collagen and elastic fibers. This layer is thickest over the posterior surface of the eye, near the exit of the optic nerve, and thinnest over the anterior surface. The sclera stabilizes the shape of the eye during eye movements. The six extrinsic eye muscles insert on the sclera, blending their collagen fibers with those of the sclera.

The **optic nerve** (II) carries visual information to the brain.

15

**2** The amount of light entering the eye and passing through the lens is controlled by the two pupillary muscles of the iris. When these smooth muscles contract, they change the diameter of the pupil. Both muscle layers are controlled by the autonomic nervous system. Parasympathetic activation in response to bright light causes the pupils to constrict, whereas sympathetic activation in response to dim light causes the pupils to dilate.

**?** **REVIEW**
**B.** What happens to the pupils when light intensity decreases?

Sphincter pupillae

Dilator pupillae

The **dilator pupillae** extend radially away from the edge of the pupil. Contraction of these muscles enlarges the pupil.

The **sphincter pupillae** form a series of concentric circles around the pupil. When these sphincter muscles contract, the diameter of the pupil decreases.

Decreased light intensity
Increased sympathetic stimulation

Increased light intensity
Increased parasympathetic stimulation

**3** Light passing through the center of the cornea and the center of the lens strikes a specific location that contains the highest density of photoreceptors anywhere in the eye.

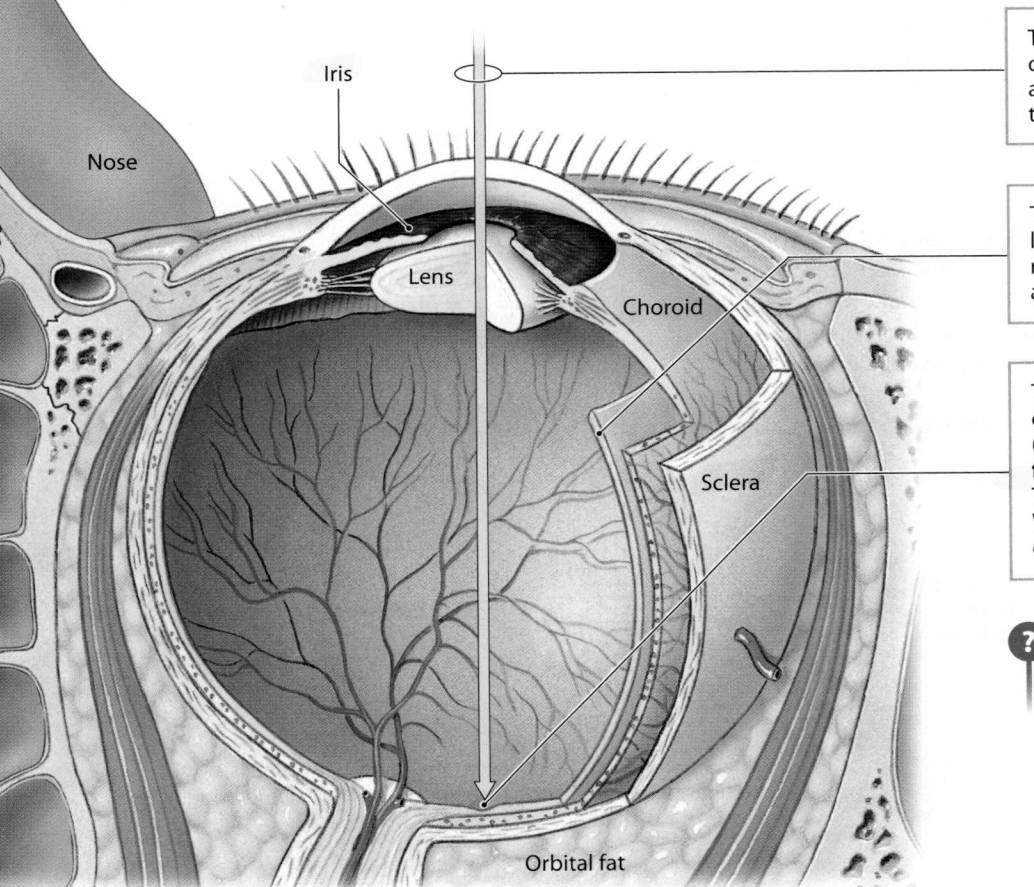

Iris

Nose

Lens

Choroid

Sclera

Orbital fat

The **visual axis** of the eye is an imaginary line drawn from the center of an object you are looking at, directly through the center of the cornea and the center of the lens to the retina.

The photoreceptors are located in the inner, neural layer of the retina. The type and density of receptors vary from one portion of the retina to another.

The very highest concentration of photoreceptors occurs at the center of an area called the **macula** (MAK-yū-luh; spot). This central area is called the **fovea centralis** (FŌ-vē-uh; shallow depression). The fovea centralis is the point of sharpest vision: When you look directly at an object, its image falls on this portion of the retina.

**?** **REVIEW**
**C.** Light passing through the eye along the visual axis strikes what part of the retina?

 **LEARNING OUTCOME**
Explain how light is directed to the fovea centralis of the retina.

# Focusing of light produces a sharp image on the retina

The eye is often compared to a camera. To provide useful information, the lens of the eye, like a camera lens, must focus the arriving image. For an image to be "in focus," the rays of light arriving from an object must strike the sensitive surface of the retina in precise order, forming a miniature image of the object. If the rays are not perfectly focused, the image is blurry. Focusing typically occurs in two steps: as the light passes through the cornea and as it passes through the lens.

**1** Light is **refracted**, or bent, when it passes from one medium to another medium with a different density. In the human eye, the greatest amount of refraction occurs when light passes from the air into the corneal tissues, which have a density close to that of water. You cannot vary the amount of refraction that occurs at the cornea. Additional refraction takes place when the light passes from the aqueous humor into the relatively dense lens. The lens provides the extra refraction needed to focus the light rays from an object toward a **focal point**—a specific point of intersection on the retina. The distance between the center of the lens and its focal point is the **focal distance** of the lens. Whether in the eye or in a camera, the focal distance is determined by the distance from the object to the lens and by the shape of the lens.

**?** REVIEW

**A.** Define *focal point*.

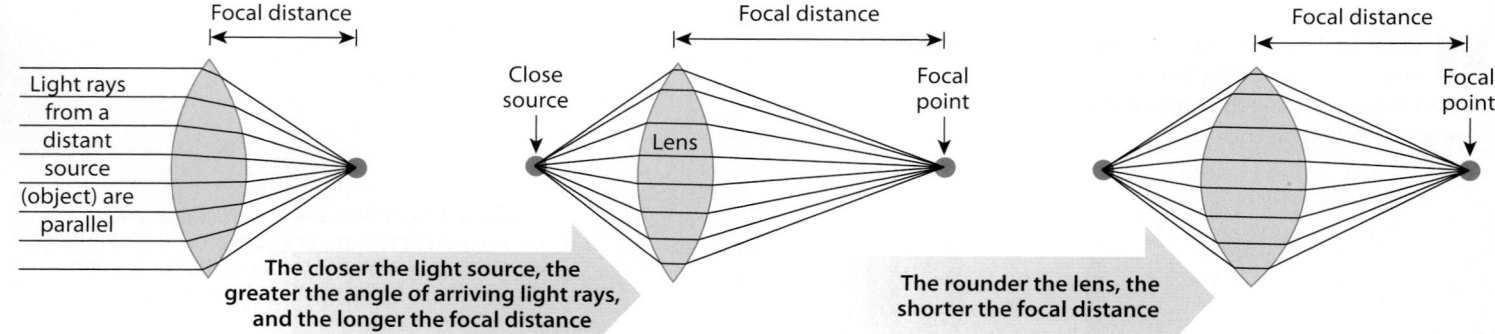

Focal distance

Light rays from a distant source (object) are parallel

**The closer the light source, the greater the angle of arriving light rays, and the longer the focal distance**

Focal distance

Close source

Lens

Focal point

**The rounder the lens, the shorter the focal distance**

Focal distance

Focal point

**2** A camera focuses an image by moving the lens toward or away from the digital image sensor. This method of focusing cannot work in our eyes, because the distance from the lens to the retina cannot change. We focus images on the retina by changing the shape of the lens to keep the focal distance constant and give us clear vision, a process called **accommodation**.

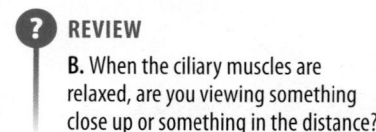

**?** REVIEW

**B.** When the ciliary muscles are relaxed, are you viewing something close up or something in the distance?

**For Close Vision: Ciliary Muscle Contracted, Lens Rounded**

Focal point on fovea centralis

When the ciliary muscle contracts, the ciliary body moves toward the lens, thereby reducing the tension in the ciliary zonule. The elastic capsule of the lens then pulls it into a more spherical shape that increases the refractive power of the lens, enabling it to bring light from nearby objects into focus on the retina.

**For Distant Vision: Ciliary Muscle Relaxed, Lens Flattened**

Focal point on fovea centralis

When the ciliary muscle relaxes, the ciliary zonule pulls at the circumference of the lens, making the lens flatter and bringing the image of a distant object into focus on the retina.

**3** An object in view is not a single point. The image consists of a large number of individual points, like the pixels on a computer screen. In the eye, light from each point is focused on the retina, creating a miniature image of the original. However, the image is inverted and reversed. The brain compensates for this image reversal, and we are not aware of any difference between the orientation of the image on the retina and that of the object. The compensation is learned by experience—a person wearing glasses that reverse and invert the visual image can adapt to the change relatively quickly.

Here we consider a sagittal section through an eye that is looking at a telephone pole.* The image of the top of the pole lands at the bottom of the retina, and the image of the bottom hits the top of the retina.

Now consider the same sagittal section through an eye that is looking at a picket fence.* The image of the left edge of the fence falls on the right side of the retina, and the image of the right edge falls on the left side of the retina.

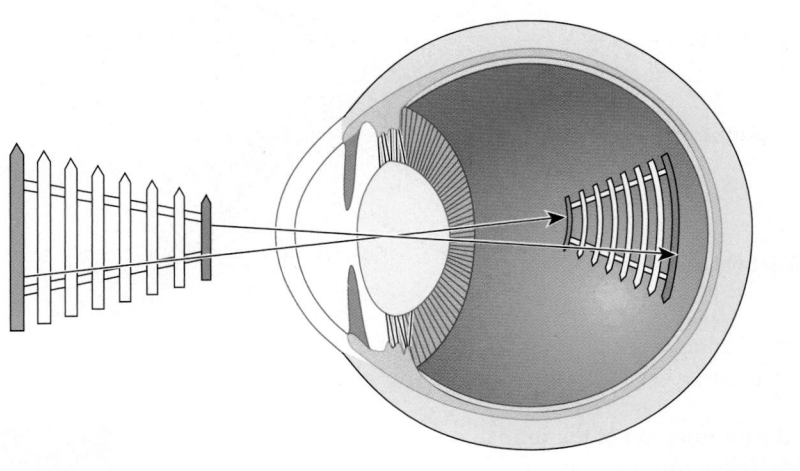

*Note: These illustrations are not drawn to scale because the fovea centralis occupies a small area of the retina, and the projected images are very tiny. As a result, the crossover of light rays is shown in the lens, but it actually occurs very close to the fovea centralis.

The greatest amount of refraction is required to view objects that are very close to the lens. The inner limit of clear vision for close objects, known as the **near point of vision**, is determined by the degree of elasticity in the lens. Children can usually focus on something 7–9 cm (3–4 in.) from the eye, but over time the lens tends to become stiffer and less responsive. A young adult can usually focus on objects 15–20 cm (6–8 in.) away. As we age, this distance gradually increases. The near point at age 60 is typically about 83 cm (33 in.).

 **REVIEW**

**C.** Why does the near point of vision typically increase with age?

 **LEARNING OUTCOME**

Describe the process by which images are focused on the retina.

15

# The neural layer of the retina contains multiple layers of specialized photoreceptors, neurons, and supporting cells

**1** This diagrammatic sectional view through the eye shows the retina near the origin of the optic nerve.

**Pigmented Layer of the Retina**

The pigmented layer of the retina absorbs light that passes through the neural layer, preventing light from bouncing back and producing visual "echoes." The pigment cells also have important biochemical interactions with the photoreceptors located in the neural layer of the retina.

**Neural Layer of the Retina**

The neural layer of the retina contains photoreceptors, supporting cells, and neurons that perform preliminary processing and integration of visual information.

The photoreceptors of the neural layer are the cells closest to the pigmented layer of the retina.

**Ganglion cells** form the innermost cell layer in the neural layer of the retina.

The axons of the ganglion cells converge at the **optic disc** to form the optic nerve, which carries visual information to the brain. The optic disc has no photoreceptors. Because an image falling on this portion of the retina cannot be detected, it is called the **blind spot**.

Central retinal artery

Optic nerve

Blood vessels enter and leave the interior of the eye within the optic nerve. They radiate across the inner surface of the eye, servicing the inner layers of cells in the neural layer of the retina.

Central retinal vein

Sclera

Choroid

**2** This is a photograph of the retinal surface, taken through the cornea, pupil, and lens of the right eye.

Optic disc (blind spot)

Fovea centralis (dense area at the center of the lighter macula)

Macula

Central retinal artery and vein emerging from center of optic disc

**? REVIEW**

**A.** What is the eye's blind spot?

**3** This sectional view shows that the retina contains multiple layers of specialized cells, including two types of photoreceptors: rods and cones.

**?** **REVIEW**

**B.** Compare rods with cones.

Pigmented layer of the retina

### Photoreceptors of the Retina

**Rods** do not discriminate among colors of light and provide vision in black and white. Highly sensitive, they enable us to see in dimly lit rooms, at twilight, and in pale moonlight.

**Cones** provide us with color vision. They give us sharper, clearer images than the rods, but cones require more intense light.

Rods and cones synapse with neurons called **bipolar cells**.

Bipolar cells synapse on ganglion cells.

### Horizontal and Amacrine Cells

These cells can facilitate or inhibit communication between photoreceptors and ganglion cells, thereby altering the sensitivity of the retina. These cells play an important role in the eye's adjustment to dim or brightly lit environments.

A network of **horizontal cells** extends across the deep portion of the retina at the level of the synapses between photoreceptors and bipolar cells.

A layer of **amacrine** (AM-a-krin) **cells** occurs where bipolar cells synapse with ganglion cells.

LIGHT

15

**4** The two color bands in this illustration of the retinal surface indicate the relative densities of cones and rods on either side of a horizontal line passing through the fovea centralis and optic disc of the right eye. When you look directly at an object, the image falls on the fovea centralis, the center of color vision and image sharpness. However, in very dim light, cones cannot function. That is why you cannot see a dim star if you stare directly at it, but you can see it if you shift your gaze to one side or the other so that the image falls on the more sensitive rods.

**?** **REVIEW**

**C.** If you enter a dimly lit room, will you be able to see clearly? Why or why not?

Fovea centralis · Optic disc

Low Density of Cones
High Density of Cones

The retina of each eye contains approximately 6 million cones. The density of cones reaches its maximum at the fovea centralis of the macula, where there are no rods.

Low Density of Rods
High Density of Rods

The retina contains approximately 125 million rods. The density of rods is highest at the periphery of the retina, where there are very few cones.

Visual acuity →

Lateral border · Fovea centralis · Blind spot · Nasal border

This graph plots the sharpness of vision, or **visual acuity**, along the axis of the horizontal line. There is a direct correlation between visual acuity and cone density.

 **LEARNING OUTCOME**

Describe the structure and function of the retina's layers of cells, and explain the distribution of rods and cones and their relation to visual acuity.

# Photoreception occurs in the outer segment of rod and cone cells

The rods and cones of the retina are called *photoreceptors* because they detect photons, the basic units of visible light. Light energy is a form of radiant energy that travels in waves with a characteristic wavelength (distance between wave peaks). Our eyes are sensitive to wavelengths of 400–700 nm, the spectrum of visible light. A nanometer (nm) is one billionth of a meter.

**1** This illustration shows the major structural features of rods and cones and the adjacent pigmented epithelium and bipolar cells. The pigmented epithelium forms the pigmented layer of the retina (see Module 15.10).

**? REVIEW**
**A.** Describe the structure of a photoreceptor.

**Structure of Cones** **Structure of Rods**

**Pigmented Epithelium**

The pigmented epithelium absorbs photons that are not absorbed by visual pigments. It also phagocytizes old discs shed from the tip of the outer segment.

In a cone, the discs are infoldings of the plasma membrane, and the outer segment tapers to a blunt point.

Melanin granules

In a rod, each disc is an independent entity, and the outer segment forms an elongated cylinder.

**Outer Segment**

The outer segment of a photoreceptor contains flattened membranous plates, or **discs**, that contain thousands of **visual pigment** molecules.

Discs

Connecting stalks

**Inner Segment**

The inner segment contains the photoreceptor's major organelles and is responsible for all cell functions other than photoreception.

Mitochondria

Golgi apparatus

Nuclei

Cone        Rods

Rhodopsin molecule

Retinal

Opsin

Each photoreceptor synapses with a bipolar cell.

Bipolar cell

LIGHT

**? REVIEW**
**B.** How could a diet deficient in vitamin A affect vision?

**2** Visual pigments are derivatives of the compound **rhodopsin** (rō-DOP-sin), or visual purple, the visual pigment found in rods. Rhodopsin consists of a protein, **opsin**, bound to the pigment **retinal** (RET-i-nal), which is synthesized from vitamin A. The type of opsin present determines the wavelength of light that can be absorbed by retinal.

**3** All rods contain the same type of opsin. It is most sensitive to blue-green wavelengths of light. There are three types of cones: **blue cones**, **green cones**, and **red cones**. Each type has a different form of opsin that is sensitive to a different range of wavelengths. Their stimulation in various combinations is the basis for color vision. In a person with normal color vision, the cone population consists of 16 percent blue cones, 10 percent green cones, and 74 percent red cones. Although their wavelength sensitivities overlap, each type is most sensitive to a specific portion of the visual spectrum. If all three cone populations are stimulated, we perceive the color as white. We also perceive white if rods (but not cones) are stimulated, which is why everything appears black and white when we enter dimly lit surroundings or walk by starlight.

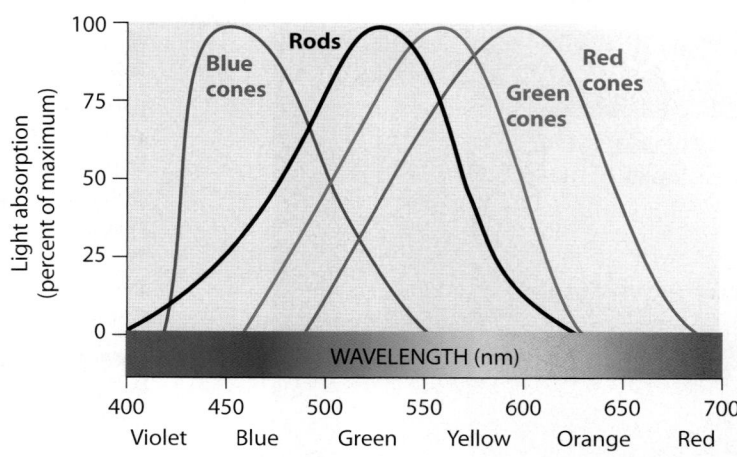

**? REVIEW**

**C.** Identify the three types of cones.

15

## Color blindness is caused by nonfunctioning cones

People who cannot distinguish certain colors have a form of **color blindness**. The standard tests for color vision involve picking numbers or letters out of a complex colored picture such as the 12 in the one to the right. Color blindness occurs when one or more types of cones are nonfunctional. The cones may be absent, or they may be present but unable to manufacture the necessary visual pigments. In the most common type of color blindness (red–green color blindness), the red cones are missing, so the person cannot distinguish red light from green light. Inherited color blindness involving one or two cone pigments is not unusual. Ten percent of all males show some color blindness, but only about 0.67 percent of all females are color blind. Total color blindness is extremely rare. Only 1 person in 300,000 does not manufacture any cone pigments.

🦯 **Clinical Note**

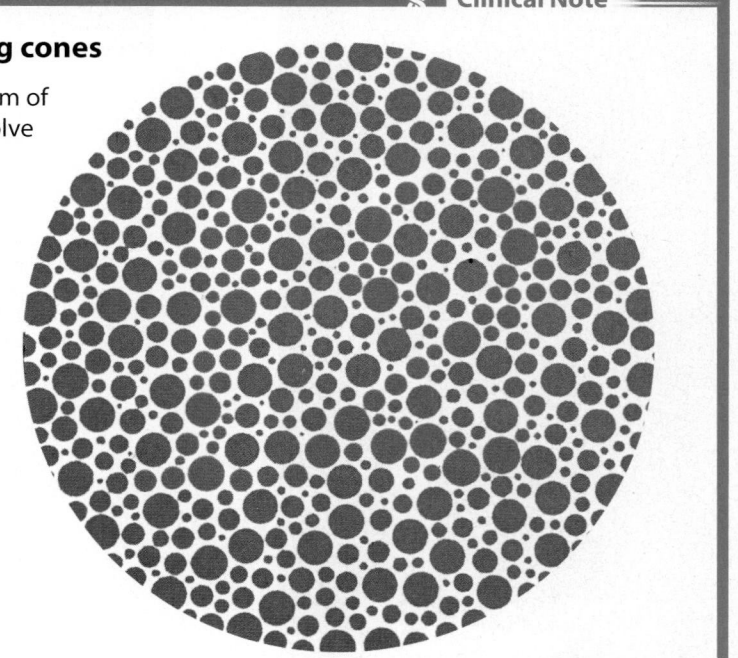

**Lo LEARNING OUTCOME**

Describe the structure of the photoreceptors and how we are able to distinguish colors.

# Module 15.12

# Photoreception involves activation, bleaching, and reassembly of visual pigments

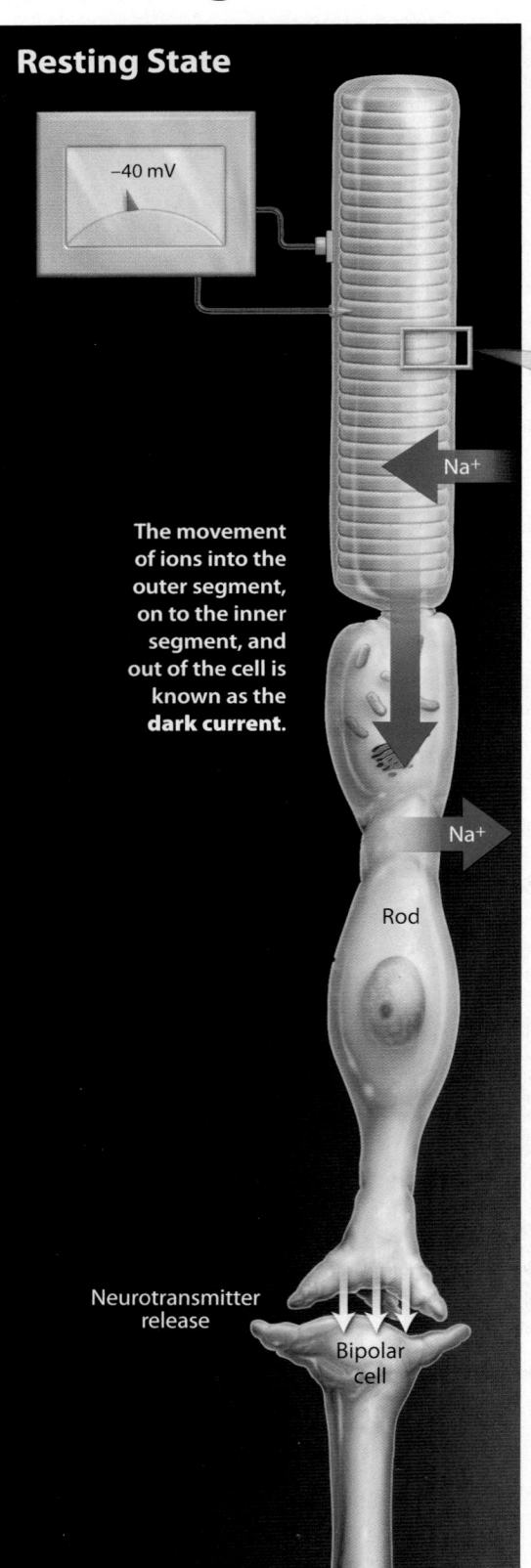

**Resting State**

−40 mV

The movement
of ions into the
outer segment,
on to the inner
segment, and
out of the cell is
known as the
**dark current.**

Na⁺

Na⁺

Rod

Neurotransmitter
release

Bipolar
cell

**1** The plasma membrane in the outer segment of the photoreceptor contains chemically gated sodium ion channels. In darkness, these gated channels are kept open in the presence of cGMP (cyclic guanosine monophosphate), a derivative of the high-energy compound guanosine triphosphate (GTP). Because the channels are open, the membrane potential is approximately −40 mV, rather than the −70 mV typical of resting neurons. At the −40 mV membrane potential, the photoreceptor is continuously releasing neurotransmitters across synapses to bipolar cells. The inner segment also continuously pumps sodium ions (Na⁺) out of the cytosol.

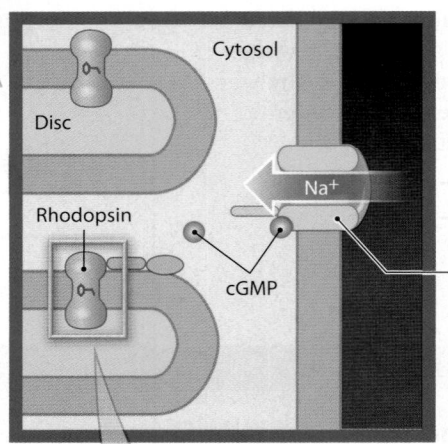

Cytosol

Disc

Rhodopsin

Na⁺

cGMP

Gated Na⁺
channel

**2** The bound retinal molecule in rhodopsin has two possible configurations: the bent or curved 11-*cis* form and the more linear 11-*trans* form. Normally, in the dark, the molecule is in the 11-*cis* form. On absorbing light, it changes to the 11-*trans* form. This change activates the opsin molecule.

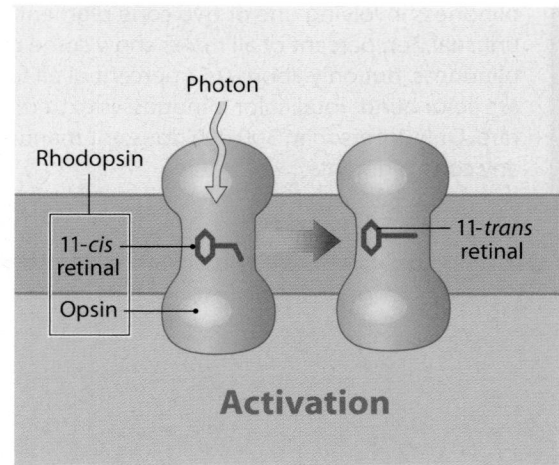

Photon

Rhodopsin

11-*cis*
retinal

Opsin

11-*trans*
retinal

**Activation**

**? REVIEW**

**A.** What are the
two configurations
of retinal?

**3** Opsin then activates **transducin**, a G protein bound to the disc membrane. The transducin in turn activates the enzyme **phosphodiesterase (PDE)**.

**4** Phosphodiesterase is an enzyme that breaks down cGMP. The removal of cGMP from the gated sodium channels results in their inactivation. The rate of $Na^+$ entry into the cytosol then decreases.

## Active State

**5** The decrease in the rate of $Na^+$ entry reduces the dark current. At the same time, active transport continues to export $Na^+$ from the cytosol. When the sodium channels close, the membrane potential drops toward −70 mV. As the plasma membrane hyperpolarizes, the rate of neurotransmitter release decreases. This decrease signals the adjacent bipolar cell that the photoreceptor has absorbed a photon.

**6** Rhodopsin cannot respond to additional photons until its retinal component regains its original shape. It does not spontaneously revert to the 11-*cis* form. Instead, the entire rhodopsin molecule must be broken down into retinal and opsin in a process called **bleaching**. The retinal is then converted to its original *cis* shape. This conversion requires energy in the form of ATP. Opsin and 11-*cis* retinal are reassembled, and the rhodopsin molecule is ready to repeat the cycle.

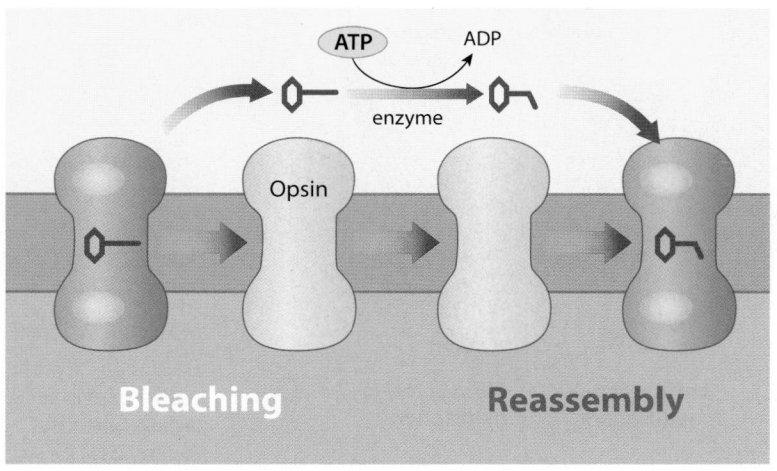

**?** **REVIEW**

**B.** Visual pigments undergo which three changes during photoreception?

**C.** When during photoreception is ATP required?

**Lo** **LEARNING OUTCOME**

Explain photoreception and how visual pigments are activated.

# The visual pathways distribute visual information from each eye to both cerebral hemispheres

**1** The visual images from the left and right eyes overlap, and the visual cortex of each cerebral hemisphere receives information from both eyes. The information is sorted, however, so that the left visual cortex gets information on the right half of the visual field, and the right visual cortex receives information on the left half of the visual field. **Depth perception** is the ability to judge depth or distance by interpreting the three-dimensional relationships among objects in view. Your brain perceives depth by comparing the relative positions of objects within the images received by both eyes. The map in the visual cortex is upside down and backward, duplicating the orientation of the visual image at the retina.

**Combined Visual Field**

Left side | Right side

Left eye only | Binocular vision | Right eye only

Retina
Optic disc

Collaterals from the fibers synapsing in the lateral geniculate bodies continue to subconscious processing centers in the diencephalon and brainstem. The pupillary reflexes and reflexes that control eye movement, for example, are triggered by collaterals carrying information to the superior colliculi.

Diencephalon and brainstem

Optic tract

Superior colliculus

Lateral geniculate body

Left cerebral hemisphere | Right cerebral hemisphere

## The Visual Pathways

### Photoreceptors in the retina

The visual pathways begin at the photoreceptors in the retina. Each photoreceptor monitors a specific receptive field of vision and, when stimulated, passes the information through a bipolar cell to a ganglion cell.

### Optic nerve (II)

Axons from the approximately 1 million ganglion cells converge on the optic disc, penetrate the wall of the eye, and proceed toward the diencephalon as the optic nerve (II).

### Optic chiasm

The two optic nerves, one from each eye, reach the diencephalon at the **optic chiasm.**

### Optic tracts to the lateral geniculate bodies

From the optic chiasm, about half the fibers proceed toward the lateral geniculate body of the same side of the brain, whereas the other half cross over to reach the lateral geniculate body of the opposite side.

### Projection fibers (optic radiation)

From each lateral geniculate body, visual information travels to the occipital cortex of the cerebral hemisphere on that side. The bundle of projection fibers linking each lateral geniculate body with the visual cortex is known as the **optic radiation.**

### Visual cortex of cerebral hemispheres

The perception of a visual image reflects the integration of information that arrives at the visual cortex of the occipital lobes. Each eye receives a slightly different visual image because (1) the foveae centrales are 5–7.5 cm (2–3.0 in.) apart, and (2) the nose and eye socket block a portion of the view of the opposite side.

15

# Module 15.14

# Refractive problems result from abnormalities in the cornea or lens or in the shape of the eye

**1** In the normal healthy eye, when the ciliary muscle is relaxed and the lens is flattened, the image of a distant object will be focused on the retina's surface. This condition is called **emmetropia** (*emmetro-*, proper + *opia*, vision), or *normal vision*.

**Emmetropia**
(normal vision)

**? REVIEW**

**A.** Define *emmetropia*.

**2** Small variations in the performance of the lens or the structure of the eye commonly occur. These can be corrected with external lenses (glasses or contact lenses).

**3** Variable success at correcting myopia and hyperopia has been achieved by surgery that reshapes the cornea.

## Myopia (Nearsightedness)

If the eyeball is too deep or the resting curvature of the lens is too great, the image of a distant object is projected in front of the retina. Such individuals are said to be *nearsighted* because vision at close range is clear but distant objects are blurry and out of focus. Their condition is more formally termed **myopia** (*myein*, to shut + *ops*, eye).

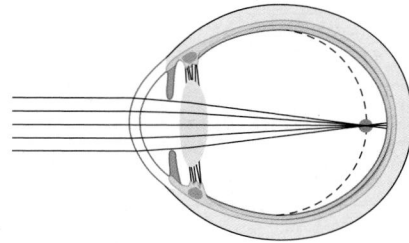

Myopia can be treated by placing a diverging lens in front of the eye. Diverging lenses have at least one concave surface and spread the light rays apart as if the object were closer to the viewer.

Diverging,
concave lens

## Hyperopia (Farsightedness)

If the eyeball is too shallow or the lens is too flat, **hyperopia** results. The ciliary muscle must contract to focus even a distant object on the retina, and at close range the lens cannot provide enough refraction to focus an image on the retina. Hyperopic people are said to be *farsighted* because they can see distant objects most clearly.

Hyperopia can be corrected by placing a converging lens in front of the eye. Converging lenses have at least one convex surface and provide the additional refraction needed to bring nearby objects into focus.

Converging
convex lens

## Surgical Correction

In **photorefractive keratectomy (PRK)**, a computer-guided laser shapes the cornea to exact specifications. Tissue is removed only to a depth of 10–20 μm—no more than about 10 percent of the cornea's thickness. The entire procedure can be done in less than a minute. A variation on PRK is called **laser-assisted in-situ keratomileusis (LASIK)**. In this procedure, the interior layers of the cornea are reshaped and covered by a flap of the normal corneal epithelium. Roughly 70 percent of LASIK patients achieve normal vision; it is now the most common form of refractive surgery. Each year, an estimated 100,000 people undergo PRK therapy in the United States. Corneal scarring is rare, and approximately 10 million Americans have had corneal refractive surgery. However, many still need reading glasses, and both immediate and long-term visual problems can occur.

**? REVIEW**

**B.** Which type of lens would correct hyperopia?

 **LEARNING OUTCOME**

Describe various refraction problems associated with the cornea, lens, or shape of the eye.

# Section 2 Review
## Vision

**Labeling:** Label the structures in this diagram of a sagittal section of the left eye.

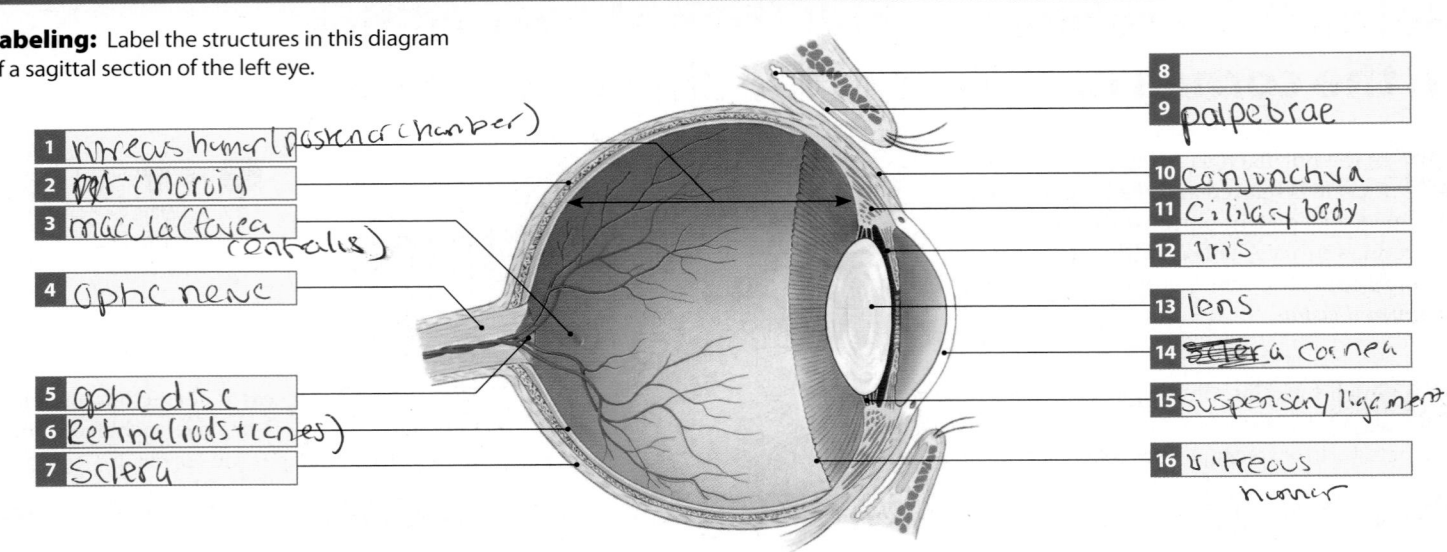

1. vitreous humer (posterior chamber)
2. choroid
3. macula (fovea contralis)
4. Optic nerve
5. Optic disc
6. Retinal (rods + cones)
7. Sclera

8. 
9. palpebrae
10. Conjunctiva
11. Ciliary body
12. Iris
13. lens
14. sclera cornea
15. suspensory ligament
16. vitreous humer

**Matching:** Match each lettered term with the most closely related description.

a. ganglion cells
b. cones
c. vascular layer
d. rods
e. optic disc
f. posterior chamber
g. crystallins
h. sclera
i. retina
j. occipital lobe
k. palpebrae
l. rhodopsin
m. fovea centralis
n. posterior cavity
o. pupil

17. Visual pigment
18. Eyelids
19. Transparent proteins in the cells of a lens
20. White of the eye
21. Opening surrounded by the iris
22. Inner layer
23. Site of vitreous body
24. Photoreceptors that enable vision in dim light
25. Extends between the iris and the ciliary body and lens
26. Sharpest vision
27. Visual cortex
28. Photoreceptors that provide the perception of color
29. Their axons form the optic nerves
30. Iris, ciliary body, and choroid
31. Region of retina called the "blind spot"

17. _____
18. _____
19. _____
20. _____
21. _____
22. _____
23. _____
24. _____
25. _____
26. _____
27. _____
28. _____
29. _____
30. _____
31. _____

**Section integration:** A bright flash of light from nearby exploding fireworks blinds Rachel's eyes. The result is a "ghost" image that temporarily remains on her retinas. What might account for the images and their subsequent disappearance?

32. _____

## Module 15.15
# Equilibrium and hearing involve the internal ear

**1** In olfaction, the sensory receptors are specialized sensory neurons. In gustation, sensory receptor cells communicate with sensory neurons. In vision, sensory receptors called photoreceptors respond to light. In each case, the sensory receptor cells are located within epithelia exposed to the external environment, and the information is routed directly to the CNS for processing.

Olfactory receptor cell (specialized sensory neuron)

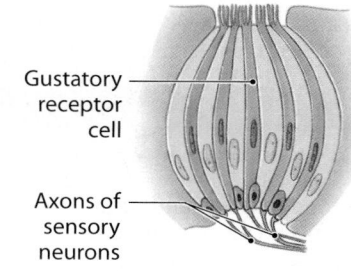

Gustatory receptor cell

Axons of sensory neurons

**2** The sensory receptors for the sensations of equilibrium and hearing are isolated and protected from the external environment. These sensory receptor cells are located within the **internal ear**, a complex sense organ, where the sensory information is integrated and organized before it is forwarded to the CNS.

Internal ear

**? REVIEW**

**A.** Contrast the olfactory and gustatory sensory receptors with those of equilibrium and hearing.

15

**3** The sensory receptor cells of the internal ear are called **hair cells** because their free surfaces are covered with specialized processes similar to the cilia and microvilli of other cells. They are basically mechanoreceptors sensitive to contact or movement, and they are always surrounded by supporting cells and monitored by the dendrites of sensory neurons. When an external force pushes against the processes of a hair cell, their movement distorts the plasma membrane and alters the rate at which the hair cell releases neurotransmitters. In this way, hair cells provide information about the direction and strength of mechanical stimuli.

**? REVIEW**

**B.** Describe the sensory receptors of the internal ear.

**C.** What kind of stimuli can the internal ear sense?

Displacement in this direction stimulates hair cell

Extracellular fluid

Displacement in this direction inhibits hair cell

Certain hair cells involved with balance contain a **kinocilium** (ki-nō-SIL-ē-um), a single large cilium. Hair cells do not actively move their kinocilia (or stereocilia).

The free surface of every hair cell supports 80–100 long **stereocilia**, which resemble very long microvilli.

Hair cell

Dendrite of sensory neuron

Supporting cell

The internal ear has a complex three-dimensional structure that determines what stimuli can reach the hair cells in each region. For example, hair cells in one region can respond only to gravity or acceleration, whereas those in other regions respond only to rotation or only to sound. In this section you will learn how the anatomy of the internal ear enables a single receptor cell type to provide such a diversity of information.

**LO LEARNING OUTCOME**
Describe the sensory receptors of the internal ear.

# The ear is divided into the external ear, the middle ear, and the internal ear

**1** The ear is divided into three anatomical regions: the external ear, the middle ear, and the internal ear.

| External Ear | Middle Ear | Internal Ear |
|---|---|---|
| The **external ear**—the visible portion of the ear—collects and directs sound waves toward the middle ear. | The **middle ear**, or tympanic cavity, is an air-filled chamber separated from the external acoustic meatus by the tympanic membrane. It is connected to the pharynx by the *auditory tube* and contains the three auditory ossicles. | The **internal ear** contains the sensory organs for hearing and equilibrium. It receives amplified sound waves from the middle ear. |

Elastic cartilage keeps the external acoustic meatus open and makes the **auricle**, or *pinna*, flexible.

The **auditory ossicles** are three tiny bones within the middle ear. These bones connect the tympanic membrane with one of the receptor complexes of the internal ear.

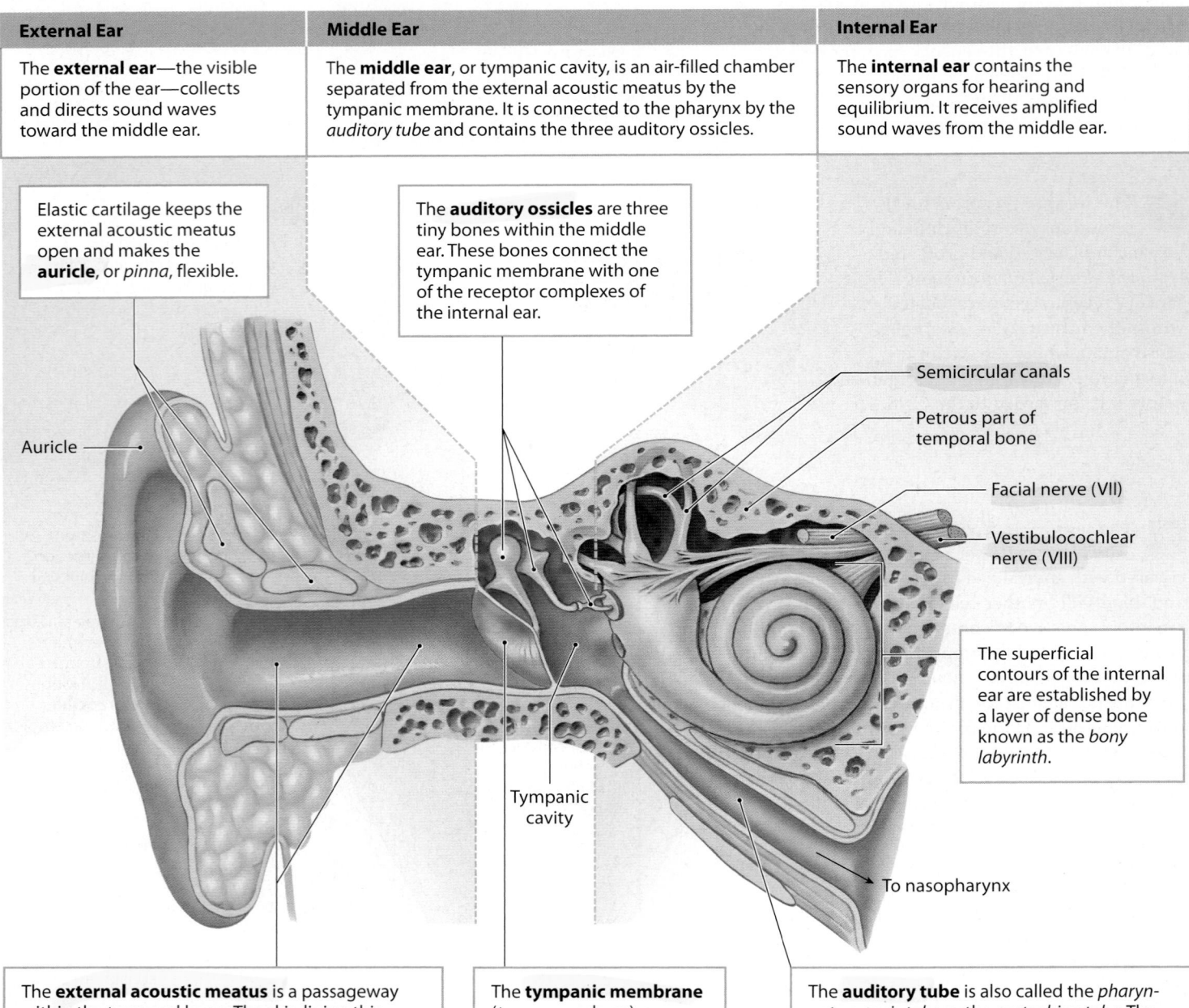

Auricle

Semicircular canals

Petrous part of temporal bone

Facial nerve (VII)

Vestibulocochlear nerve (VIII)

The superficial contours of the internal ear are established by a layer of dense bone known as the *bony labyrinth*.

Tympanic cavity

To nasopharynx

The **external acoustic meatus** is a passageway within the temporal bone. The skin lining this passage contains **ceruminous glands** that secrete a waxy material, **cerumen** (or earwax), that helps keep out foreign objects or small insects. Cerumen also slows the growth of microorganisms and so reduces the chances of infection. In addition, the canal is lined with many small, outwardly projecting hairs. These hairs trap debris and also provide increased tactile sensitivity through their root hair plexuses.

The **tympanic membrane** (*tympanon*, drum), or eardrum, lies at the end of the external acoustic meatus. The tympanic membrane is a thin, semitransparent sheet that separates the external ear from the middle ear.

The **auditory tube** is also called the *pharyngotympanic tube* or the *eustachian tube*. The auditory tube permits pressure equalization on either side of the tympanic membrane. Unfortunately, the auditory tube can also allow microorganisms to travel from the nasopharynx into the middle ear. Invasion by microorganisms can lead to an unpleasant middle ear infection known as **otitis media**. It can cause temporary hearing loss and, if untreated, may progress to the internal ear.

**2** The middle ear communicates both with the *nasopharynx* (the superior portion of the pharynx), through the auditory tube, and the mastoid air cells, through a number of small connections. The three auditory ossicles in the middle ear are the malleus, incus, and stapes. The articulations between the auditory ossicles are the smallest synovial joints in the body. Each ossicle has a tiny capsule and supporting extracapsular ligaments.

## Auditory Ossicles

| | | |
|---|---|---|
| The **malleus** (*malleus*, hammer) attaches at three points to the interior surface of the tympanic membrane. | The **incus** (*incus*, anvil), the middle ossicle, attaches the malleus to the stapes. | The edges of the base of the **stapes** (STĀ-pēz; *stapes*, stirrup) are bound to the edges of the *oval window*, an opening in the bone that surrounds the internal ear. |

**? REVIEW**

**A.** Name the three tiny bones located in the middle ear, from lateral to medial.

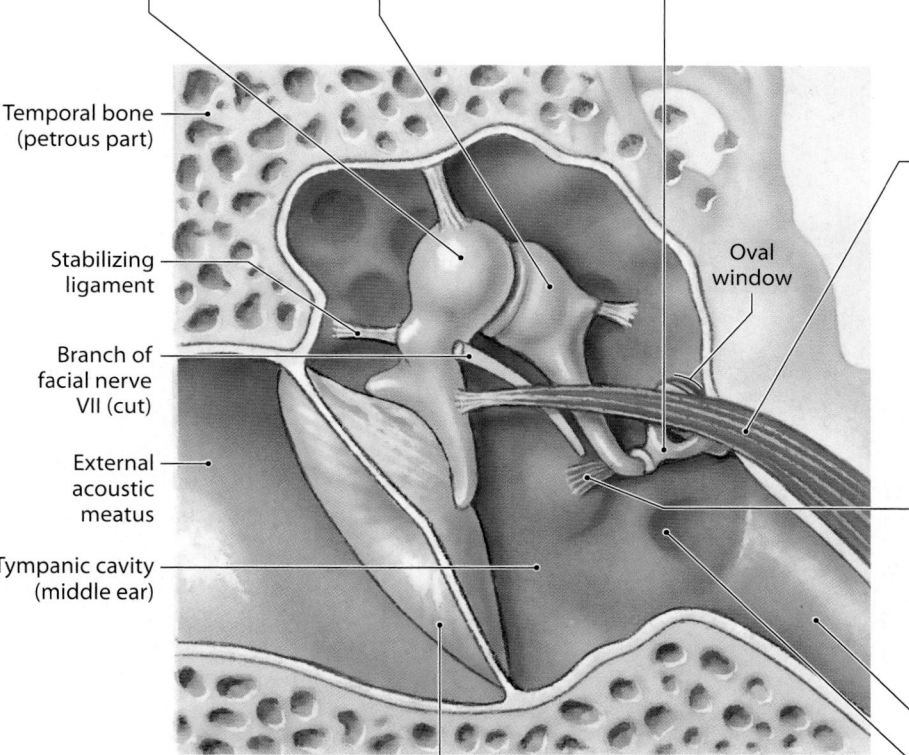

Temporal bone (petrous part)

Stabilizing ligament

Branch of facial nerve VII (cut)

External acoustic meatus

Tympanic cavity (middle ear)

Oval window

Auditory tube

Round window

## Muscles of the Middle Ear

The **tensor tympani** (TEN-sor tim-PAN-ē) is a short ribbon of muscle that originates on the petrous part of the temporal bone and the auditory tube and inserts on the "handle" of the malleus. When the tensor tympani contracts, it pulls the malleus medially, stiffening the tympanic membrane. This increased stiffness reduces the amount of movement, or vibration, in response to a sound. The tensor tympani is innervated by motor fibers of the mandibular nerve of the trigeminal nerve (V).

The **stapedius** (sta-PĒ-dē-us), innervated by the facial nerve (VII), originates from the posterior wall of the middle ear and inserts on the stapes. Contraction of the stapedius pulls the stapes, reducing movement of the stapes at the oval window.

Arriving sound waves vibrate the tympanic membrane, thereby converting the waves into mechanical movements. The auditory ossicles conduct those vibrations to the internal ear because they are connected in such a way that an in–out movement of the tympanic membrane produces a rocking motion of the stapes. The ossicles thus function as a lever system that collects the force applied to the tympanic membrane and focuses it on the oval window. Because the tympanic membrane is 22 times larger and heavier than the oval window, considerable amplification occurs, so we can hear very faint sounds. But that degree of amplification can be a problem when we are exposed to very loud noises. The tensor tympani and stapedius contract to protect the tympanic membrane and ossicles from violent movements under very noisy conditions.

**? REVIEW**

**B.** What is the function of the auditory tube?

**LEARNING OUTCOME**

Describe the structures of the external, middle, and internal ear, and explain how they function.

*malleus*
*incus*
*stapes*

**15**

# In the internal ear, the bony labyrinth protects the membranous labyrinth and its receptors

**1** The internal ear is a winding passageway called a **labyrinth** (*labyrinthos*, network of canals). It can be subdivided into an outer bony labyrinth and an inner membranous labyrinth.

## Bony Labyrinth

The **bony labyrinth** is a shell of dense bone that surrounds and protects the membranous labyrinth. Between the bony and membranous labyrinths flows the **perilymph** (PEHR-ih-limf), a liquid that closely resembles cerebrospinal fluid. The bony labyrinth consists of three parts: the **semicircular canals**, the **vestibule** (VES-tih-byūl), and the **cochlea** (KOK-lē-a; *cochlea*, snail shell).

Semicircular canals  Vestibule  Cochlea

Receptor areas

## Membranous Labyrinth

The inner contours of the bony labyrinth are closely followed by the contours of the **membranous labyrinth**, a delicate, interconnected network of fluid-filled tubes. The receptors for equilibrium and hearing are found within these tubes. The membranous labyrinth contains a fluid called **endolymph** (EN-dō-limf).

The **semicircular ducts** lie within the semicircular canals. Their receptors are stimulated by rotation of the head.

The vestibule contains a pair of membranous sacs, the **utricle** (YŪ-trih-kul) and the **saccule** (SAK-yūl), which provide sensations of gravity and linear acceleration.

The **cochlear duct** is a slender, elongated portion of the membranous labyrinth that is contained within the cochlea. The cochlear duct sits sandwiched between a pair of perilymph-filled chambers, and the entire complex makes turns around a central bony hub in a spiral arrangement that resembles a snail shell.

**2** This cross section of a semicircular canal shows the relationships among the bony labyrinth, perilymph, membranous labyrinth, and endolymph.

Bony labyrinth

Perilymph

Membranous labyrinth

Endolymph

KEY

Membranous labyrinth

Bony labyrinth

**? REVIEW**

**A.** Identify the structures of the bony labyrinth.

**3** This concept map summarizes the regions and functions of the membranous labyrinth.

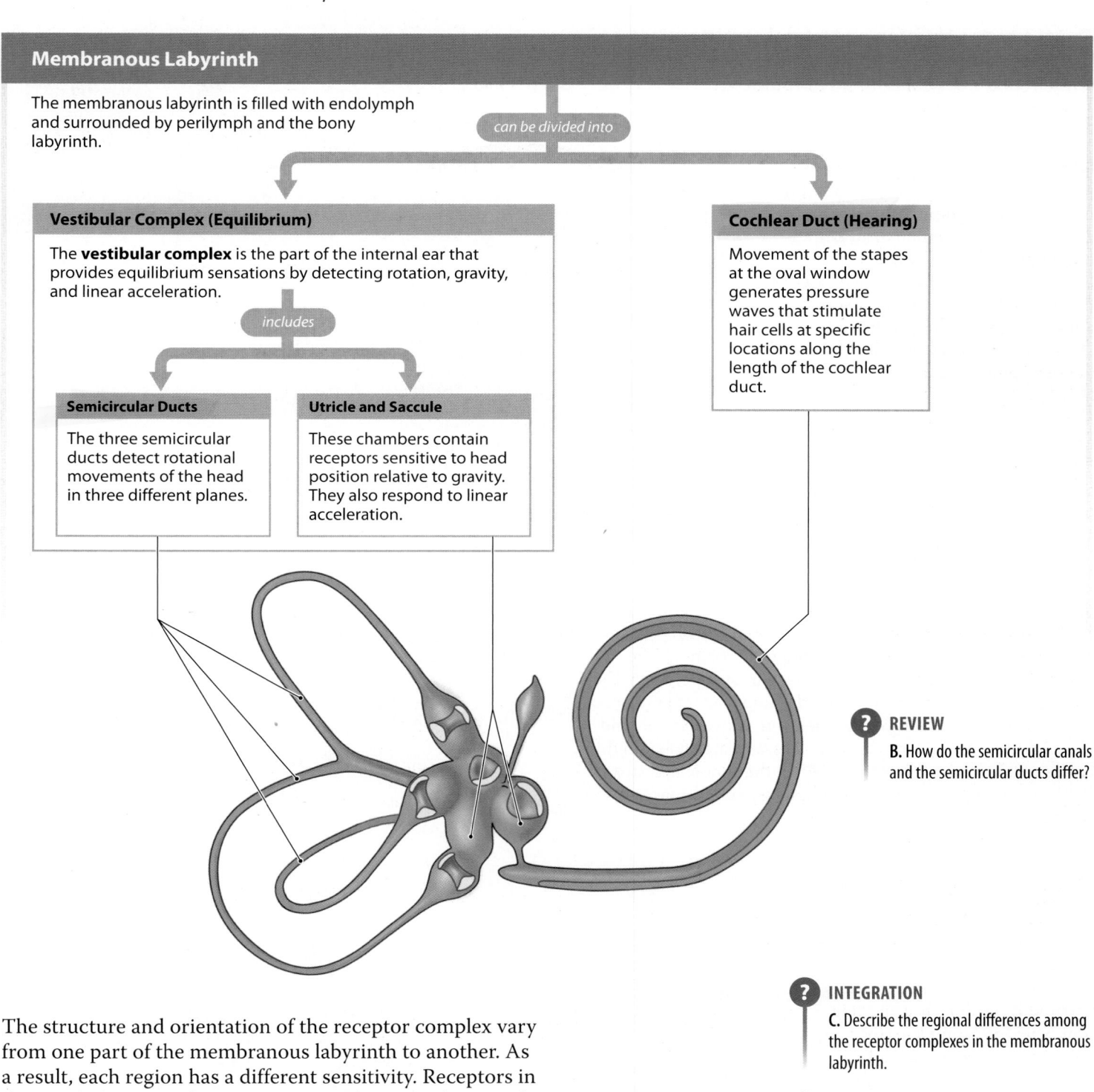

**Membranous Labyrinth**

The membranous labyrinth is filled with endolymph and surrounded by perilymph and the bony labyrinth.

*can be divided into*

**Vestibular Complex (Equilibrium)**

The **vestibular complex** is the part of the internal ear that provides equilibrium sensations by detecting rotation, gravity, and linear acceleration.

*includes*

**Semicircular Ducts**

The three semicircular ducts detect rotational movements of the head in three different planes.

**Utricle and Saccule**

These chambers contain receptors sensitive to head position relative to gravity. They also respond to linear acceleration.

**Cochlear Duct (Hearing)**

Movement of the stapes at the oval window generates pressure waves that stimulate hair cells at specific locations along the length of the cochlear duct.

The structure and orientation of the receptor complex vary from one part of the membranous labyrinth to another. As a result, each region has a different sensitivity. Receptors in the vestibule respond to gravity or acceleration, those in the semicircular ducts respond only to rotation, and hair cells in the cochlear duct are sensitive only to sound.

**? REVIEW**

**B.** How do the semicircular canals and the semicircular ducts differ?

**? INTEGRATION**

**C.** Describe the regional differences among the receptor complexes in the membranous labyrinth.

**Lo LEARNING OUTCOME**

Describe the structures and functions of the bony labyrinth and membranous labyrinth.

# Hair cells in the semicircular ducts respond to rotation; hair cells in the utricle and saccule respond to gravity and linear acceleration

The **anterior**, **posterior**, and **lateral semicircular ducts** are continuous with the utricle. Each semicircular duct contains an **ampulla**, an expanded region that contains the receptors.

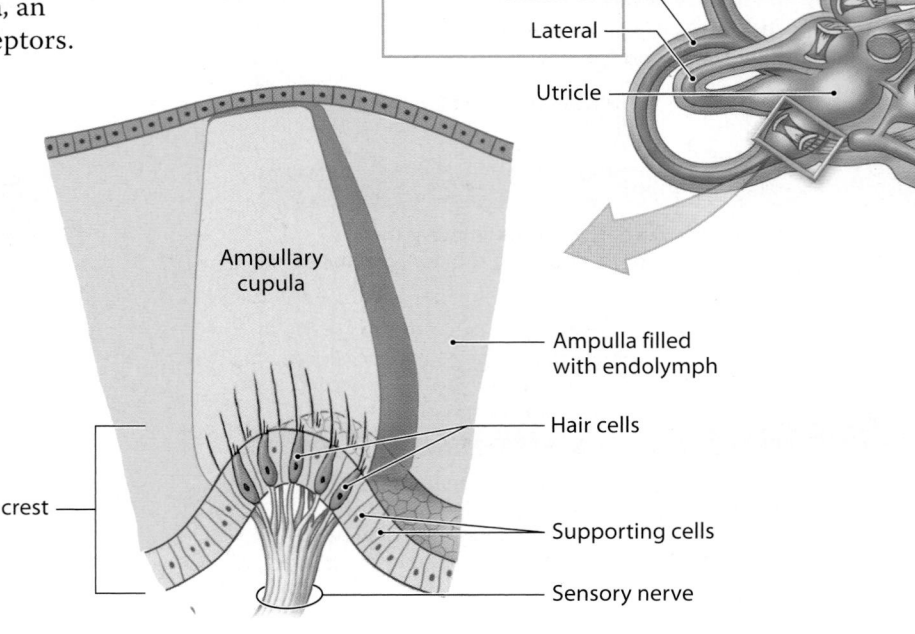

| Semicircular Ducts | |
|---|---|
| Anterior | |
| Posterior | |
| Lateral | |
| Utricle | |

Ampulla

**1** The region in the wall of the ampulla that contains the receptors is known as the **ampullary crest**. Each ampullary crest is bound to an **ampullary cupula** (KŪ-pū-luh), a flexible, elastic, gelatinous structure that extends the full width of the ampulla. At an ampullary crest, the kinocilia and stereocilia of the hair cells are embedded in the ampullary cupula.

Ampullary cupula

Ampulla filled with endolymph

Hair cells

Ampullary crest

Supporting cells

Sensory nerve

**2** The ampullary cupula has a density very close to that of the surrounding endolymph, so it essentially floats above the receptor surface. When your head rotates in the plane of a semicircular duct, the movement of endolymph along the length of the duct pushes the ampullary cupula to the side, distorting the receptor processes. Movement of fluid in one direction stimulates the hair cells, and movement in the opposite direction inhibits them. When the endolymph stops moving, the elastic nature of the ampullary cupula makes it return to its normal position.

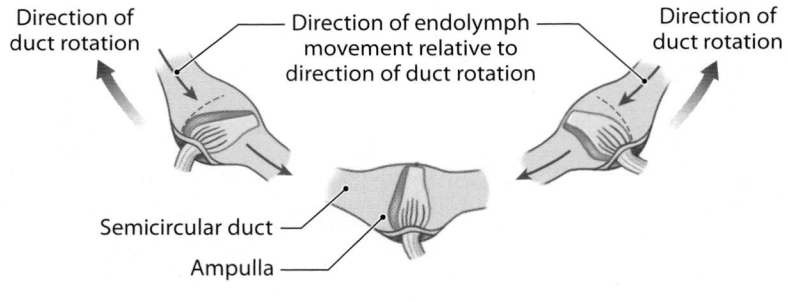

Direction of duct rotation

Direction of endolymph movement relative to direction of duct rotation

Direction of duct rotation

Semicircular duct

Ampulla

**3** Even the most complex angular movements can be analyzed in terms of motion in three rotational planes. Each semicircular duct responds to one of these rotational movements. A horizontal rotation, as in shaking your head "no," stimulates the hair cells of the lateral semicircular duct. Nodding "yes" excites the anterior duct, and tilting your head from side to side activates receptors in the posterior duct.

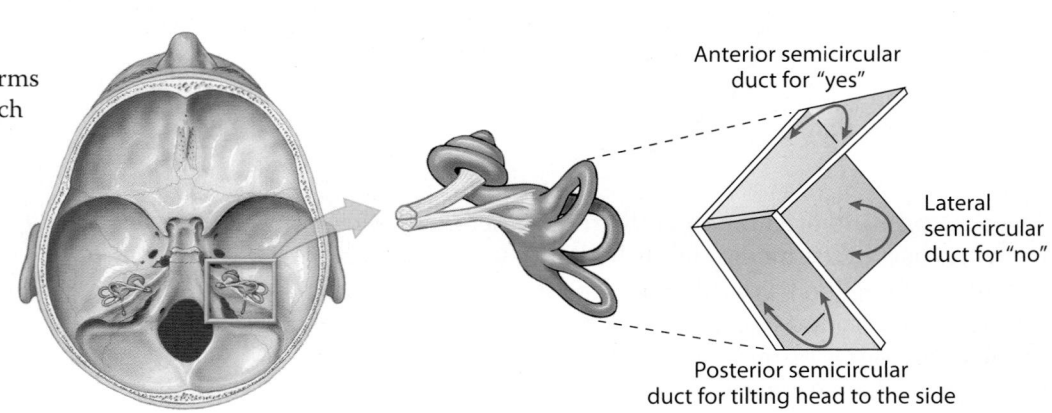

Anterior semicircular duct for "yes"

Lateral semicircular duct for "no"

Posterior semicircular duct for tilting head to the side

**4** The utricle and saccule provide equilibrium sensations when the body is moving in a straight line or is stationary. These two chambers are connected by a slender passageway that is continuous with the narrow **endolymphatic duct**, which ends in a blind pouch called the **endolymphatic sac**. This sac projects into the subarachnoid space, where it is surrounded by a capillary network. Portions of the endolymphatic duct secrete endolymph continuously, and endolymph returns to the general circulation at the endolymphatic sac.

**5** As in the ampullae, the hair cell processes are embedded in a gelatinous structure, here the **otolithic membrane**. This membrane's surface contains densely packed calcium carbonate crystals. These crystals are called **otoliths** ("ear stones").

The hair cells of the utricle and saccule are clustered in oval structures called **maculae** (MAK-yū-lē; *macula*, spot). The **macula of utricle** is sensitive to changes in horizontal movement. The **macula of saccule** is sensitive to changes in vertical movement.

**? REVIEW**
**A.** Define *otoliths*.

**6** Changes in the position of the head cause distortion of the hair cell processes in the maculae and send signals to the brain.

When your head is in the normal, upright position, the otoliths sit atop the otolithic membrane of macula of utricle. Their weight presses on the macular surface, pushing the hair cell processes down rather than to one side or another.

When your head is tilted, the pull of gravity on the otoliths shifts them to the side, thereby distorting the hair cell processes and stimulating the macular receptors. The change in receptor activity tells the CNS that your head is no longer level. This mechanism also accounts for your perception of linear acceleration, as when your car speeds up suddenly. The otoliths lag behind, and the effect on the hair cells is comparable to tilting your head back.

Gravity

Gravity

Receptor output increases

Otolith moves "downhill," distorting hair cell processes

**? REVIEW**
**B.** Cite the functions of sensory receptors in the saccule and utricle.

**Lo LEARNING OUTCOME**
Describe the functions of hair cells in the semicircular ducts, utricle, and saccule.

# The cochlear duct contains the hair cells of the spiral organ that function in hearing

**KEY**

➡ From oval window to tip of cochlea

➡ From tip of cochlea to round window

**1** The **cochlear duct**, or **scala media**, lies between a pair of perilymphatic chambers: the **scala vestibuli** (SKĀ-luh ves-TIB-yū-lē), or *vestibular duct*, and the **scala tympani,** or *tympanic duct*. The outer surfaces of these ducts are encased by the bony labyrinth everywhere except at the **oval window** (the base of the scala vestibuli) and the **round window** (the base of the scala tympani). Because the vestibuli and tympani scalae are interconnected at the tip of the spiral-shaped cochlea, they really form one long and continuous perilymphatic chamber. This chamber begins at the oval window, extends through the scala tympani, and proceeds around the top of the cochlea and along the scala tympani; it ends at the round window.

Round window

Stapes at oval window

Scala vestibuli

Cochlear duct

Scala tympani

| Vestibulocochlear nerve (VIII) | Cochlear nerve | Vestibular nerve |
|---|---|---|

Semicircular canals

From oval window

Temporal bone (petrous part)

Vestibular membrane

Basilar membrane

Scala vestibuli

Cochlear duct

Spiral organ

Scala tympani

Spiral ganglion

Cochlear nerve

To round window

Vestibulocochlear nerve (VIII)

**Cochlear section**   LM × 60

**2** The cochlear duct is a long, coiled tube suspended between the scala vestibuli and the scala tympani. The **vestibular membrane** separates the cochlear duct from the scala vestibuli, and the **basilar membrane** separates the cochlear duct from the scala tympani. The hair cells of the cochlear duct are located in a structure called the **spiral organ**, or *organ of Corti*. This sensory structure sits on the basilar membrane.

**? REVIEW**

**A.** Where is the spiral organ located?

**3** This sectional view shows a single turn of the cochlea. The scala vestibuli and scala tympani are filled with perilymph, whereas the cochlear duct, which contains the spiral organ, is filled with endolymph.

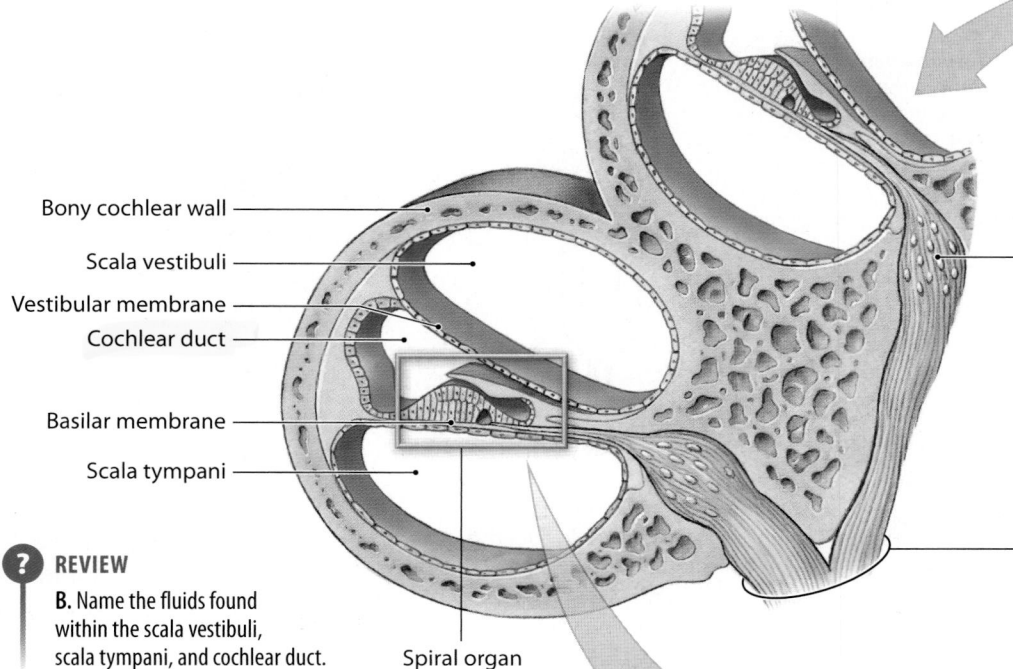

Bony cochlear wall

Scala vestibuli

Vestibular membrane

Cochlear duct

Basilar membrane

Scala tympani

The **spiral ganglion** contains the cell bodies of sensory neurons that monitor the adjacent hair cells of the spiral organ.

The cochlear nerve of the vestibulocochlear nerve (VIII) contains the axons of the neurons of the spiral ganglion.

**? REVIEW**
**B.** Name the fluids found within the scala vestibuli, scala tympani, and cochlear duct.

Spiral organ

**4** The hair cells of the spiral organ are arranged in a series of longitudinal rows. They lack kinocilia, and their stereocilia are in contact with the overlying **tectorial** (tek-TOR-ē-al; *tectum*, roof) **membrane**. This membrane is firmly attached to the inner wall of the cochlear duct. When a portion of the basilar membrane bounces up and down in response to pressure waves within the perilymph, the stereocilia of the hair cells are pressed against the tectorial membrane and distorted. If the amount of movement increases, more hair cells—and more rows of hair cells—are stimulated. These pressure waves within the perilymph are triggered by sound waves that originally arrived at the tympanic membrane.

Tectorial membrane

Outer hair cell

Basilar membrane    Inner hair cell    Nerve fibers

**? REVIEW**
**C.** When the basilar membrane moves, what happens to the hair cells of the spiral organ?

At rest

Shear

Pressure wave in perilymph

**Lo LEARNING OUTCOME**
Describe the structures and functions of the spiral organ.

15

# Module 15.20

# Sound waves lead to movement of the basilar membrane in the process of hearing

**1** Hearing is the perception of sound, but what is sound? It consists of waves of pressure conducted through a medium such as air or water. In air, each *pressure wave* consists of a region where the air molecules are crowded together and an adjacent zone where they are farther apart.

The **wavelength** of sound is the distance between two adjacent wave crests (peaks) or the distance between two adjacent wave troughs.

Tuning fork

Air molecules

Tympanic membrane

**2** Sound waves can be graphed as S-shaped curves that repeat in a regular pattern. Sound waves travel through the air at about 1235 km/h (768 mph). The **frequency** is the number of waves that pass a fixed reference point—such as the tympanic membrane— in a given time. Physicists use the term *cycles* rather than *waves*. Hence, the frequency of a sound is measured in terms of the number of cycles per second (cps), a unit called **hertz (Hz)**. Wavelength and frequency are inversely related. What we perceive as the *pitch* of a sound is our sensory response to its frequency. A high-pitched sound (high frequency, short wavelength) might have a frequency of 15,000 Hz or more; a very low-pitched sound (low frequency, long wavelength) could have a frequency of 100 Hz or less.

Because all sound waves travel at the same speed, as the frequency increases the wavelength must become shorter.

The **amplitude** is determined by the amount of energy carried by the wave. The greater the amplitude, the louder the sound.

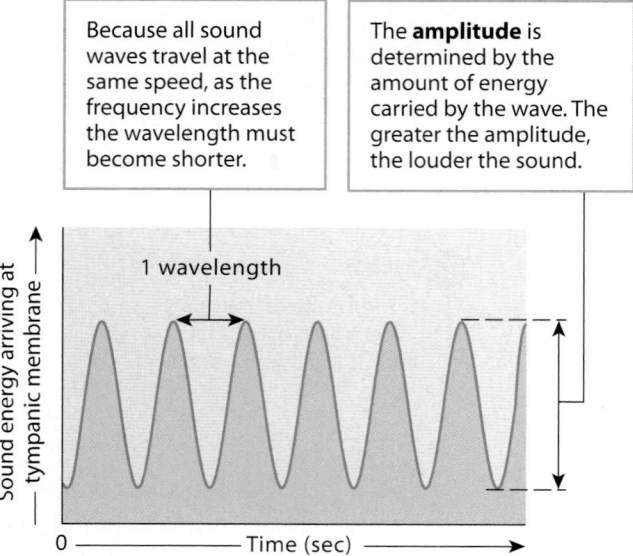

Sound energy arriving at tympanic membrane

1 wavelength

0 ————— Time (sec) —————→

**3** This table lists the **intensity**—or energy in sound waves—of familiar sounds. Intensity determines how loud it seems. The greater the energy content, the larger the amplitude, and the louder the sound. Sound energy is reported in **decibels** (DES-i-belz) (**dB**).

 **REVIEW**

**A.** Define *decibel*.

| Intensity of Representative Sounds | | |
|---|---|---|
| **Typical Decibel Level** | **Example** | **Dangerous Time Exposure** |
| 0 | Lowest audible sound | |
| 30 | Quiet library; soft whisper | |
| 40 | Quiet office; living room; bedroom away from traffic | |
| 50 | Light traffic at a distance; refrigerator; gentle breeze | |
| 60 | Air conditioner from 20 feet; conversation; sewing machine in operation | |
| 70 | Busy traffic; noisy restaurant | Some damage if continuous |
| 80 | Subway; heavy city traffic; alarm clock at 2 feet; factory noise | More than 8 hours |
| 90 | Truck traffic; noisy home appliances; shop tools; gas lawn mower | Less than 8 hours |
| 100 | Chain saw; boiler shop; pneumatic drill | 2 hours |
| 120 | "Heavy metal" rock concert; sandblasting; thunderclap nearby | Immediate danger |
| 140 | Gunshot; jet plane | Immediate danger |
| 160 | Rocket launching pad | Hearing loss inevitable |

**4** The energy of sound waves is a physical pressure. When sound waves strike a flexible object, the object responds to that pressure. Given the right combination of frequencies and amplitudes, the object will begin to vibrate at the same frequency as the sound, a phenomenon called **resonance**. The higher the amplitude, the greater the amount of vibration. For you to be able to hear a sound, your tympanic membrane must vibrate in resonance with the sound waves. Pressure waves at the tympanic membrane generate movement of the stapes at the oval window. The flexibility of the basilar membrane varies along its length, so pressure waves of different frequencies affect different parts of the membrane. This diagram depicts the movement produced by sound waves with a frequency of 6000 Hz. When the stapes moves inward, the basilar membrane distorts toward the round window, which bulges into the middle ear cavity. When the stapes moves outward, the basilar membrane rebounds and distorts toward the oval window. The location of the vibration is interpreted as **pitch** ( **2** ). The number of stimulated hair cells is interpreted as **volume**.

**5** This illustration summarizes the events involved in hearing.

**Events Involved in Hearing**

| Sound waves arrive at the tympanic membrane. | Movement of the tympanic membrane displaces the auditory ossicles. | Movement of the stapes at the oval window produces pressure waves in the perilymph of the scala vestibuli. | The pressure waves distort the basilar membrane on their way to the round window of the scala tympani. | Vibration of the basilar membrane causes hair cells to vibrate against the tectorial membrane. | Information about the region and the intensity of stimulation is relayed to the CNS over the cochlear nerve of cranial nerve VIII. |

Round window

Scala tympani (contains perilymph)
Basilar membrane
Cochlear duct (contains endolymph)
Vestibular membrane
Scala vestibuli (contains perilymph)

Movement of sound waves

Tympanic membrane

**? INTEGRATION**

**C.** How would sound perception be affected if the round window could not bulge out as a result of increased perilymph pressure?

**? REVIEW**

**B.** Beginning at the external acoustic meatus, list, in order, the structures involved in hearing.

**LO LEARNING OUTCOME**

Explain the anatomical and physiological basis for pitch and volume sensations for hearing.

# The vestibulocochlear nerve carries equilibrium and hearing sensations to the brainstem

## Equilibrium

**1** The receptors for the sense of equilibrium (balance) are the hair cells of the vestibule and the semicircular ducts. The information the receptors collect is passed along the vestibular nerve of cranial nerve VIII to the vestibular nuclei, which in turn distribute the information throughout the CNS.

**1** Hair cells of the vestibule and semicircular ducts monitor body position and motion.

**2** Sensory neurons located in adjacent vestibular ganglia carry information from the hair cells. These sensory fibers form the **vestibular nerve** of the vestibulocochlear nerve (VIII).

**3** The vestibular nuclei in the medulla oblongata integrate sensory information from both ears and relay that information to the cerebral cortex, cerebellum, and motor nuclei in the brainstem and spinal cord.

**4** The automatic movements of the eyes that occur in response to sensations of motion are directed by the superior colliculi. These movements attempt to keep your gaze focused on a specific point in space, despite changes in body position and orientation.

The reflexive motor commands issued by the vestibular nuclei are distributed to the motor nuclei for the cranial nerves involved with eye, head, and neck movements (III, IV, VI, and XI).

The vestibular nuclei relay information about position and balance to the cerebellum.

Instructions descending in the vestibulospinal tracts of the spinal cord adjust peripheral muscle tone and complement the reflexive movements of the head or neck.

Vestibular ganglion

Semicircular canals

Vestibule

CN III

CN IV

CN VI

CN XI

Cochlear nerve

Vestibulocochlear nerve (VIII)

**?** **REVIEW**

**A.** Where are the hair cell receptors for equilibrium located?

**B.** Which cranial nerves are involved with eye, head, and neck movements?

**2** The table at right summarizes the primary functions of the vestibular nuclei as they process information arriving from the equilibrium receptors in both ears.

### Functions of the Vestibular Nuclei

1. Integrating sensory information about equilibrium that arrives from both ears
2. Relaying information from the vestibular complex to the cerebellum
3. Relaying information from the vestibular complex to the cerebral cortex, providing a conscious sense of head position and movement
4. Sending commands to motor nuclei in the brainstem and spinal cord

# Hearing

**3** Hair cells along the basilar membrane are the receptors for the sense of hearing. Information they collect is passed along the cochlear nerve of the vestibulocochlear nerve (VIII) to the cochlear nuclei, which in turn distribute that information to the superior olivary nuclei, inferior colliculi, thalamus, and auditory cortex.

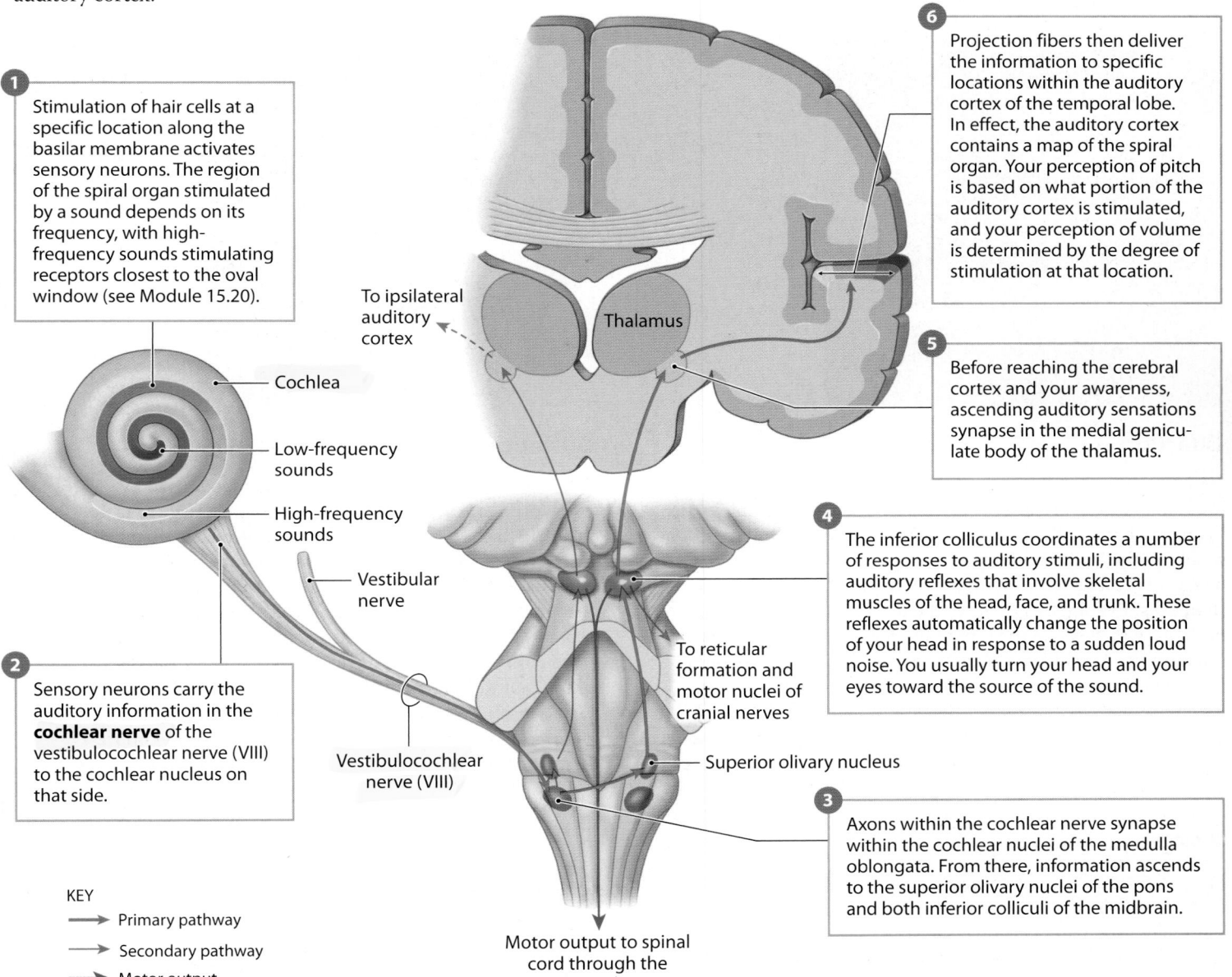

**1** Stimulation of hair cells at a specific location along the basilar membrane activates sensory neurons. The region of the spiral organ stimulated by a sound depends on its frequency, with high-frequency sounds stimulating receptors closest to the oval window (see Module 15.20).

**2** Sensory neurons carry the auditory information in the **cochlear nerve** of the vestibulocochlear nerve (VIII) to the cochlear nucleus on that side.

**6** Projection fibers then deliver the information to specific locations within the auditory cortex of the temporal lobe. In effect, the auditory cortex contains a map of the spiral organ. Your perception of pitch is based on what portion of the auditory cortex is stimulated, and your perception of volume is determined by the degree of stimulation at that location.

**5** Before reaching the cerebral cortex and your awareness, ascending auditory sensations synapse in the medial geniculate body of the thalamus.

**4** The inferior colliculus coordinates a number of responses to auditory stimuli, including auditory reflexes that involve skeletal muscles of the head, face, and trunk. These reflexes automatically change the position of your head in response to a sudden loud noise. You usually turn your head and your eyes toward the source of the sound.

**3** Axons within the cochlear nerve synapse within the cochlear nuclei of the medulla oblongata. From there, information ascends to the superior olivary nuclei of the pons and both inferior colliculi of the midbrain.

To ipsilateral auditory cortex

Thalamus

Cochlea

Low-frequency sounds

High-frequency sounds

Vestibular nerve

To reticular formation and motor nuclei of cranial nerves

Vestibulocochlear nerve (VIII)

Superior olivary nucleus

**KEY**
→ Primary pathway
→ Secondary pathway
→ Motor output

Motor output to spinal cord through the tectospinal tracts

Most of the auditory information from one cochlea is projected to the auditory cortex of the cerebral hemisphere on the opposite side of the brain. However, each auditory cortex also receives information from the cochlea on that side. These interconnections play a role in localizing sounds (left/right). They can also reduce the functional impact of damage to a cochlea or ascending pathway.

**? INTEGRATION**

**C.** What is your reflexive response to hearing a loud noise, such as a firecracker?

**Lo LEARNING OUTCOME**

Trace the pathways for the sensations of equilibrium and hearing to their respective destinations in the brain.

15

# Module 15.22

# Aging is associated with many disorders of the special senses; trauma, infection, and abnormal stimuli may cause problems at any age

**? REVIEW**

**A.** Which cranial nerves provide taste sensations from the tongue?

## Olfaction

**1** Disorders of the sense of smell may result from a head injury or from normal, age-related changes. If an injury to the head damages the olfactory nerves (I), then the sense of smell may be impaired. Unlike other populations of neurons, olfactory receptor cells are regularly replaced by the division of stem cells. Despite this process, the total number of receptors declines with age, and the remaining receptors become less sensitive. As a result, elderly people have difficulty detecting odors in low concentrations. This explains why your grandmother may overdo her perfume and why your grandfather's aftershave may seem so strong—they must use more of the odorous solution to be able to smell it themselves.

## Gustation

**2** Disorders of the sense of taste can be caused by problems with olfactory receptors, damage to taste buds, damage to cranial nerves, and age-related changes. The sense of smell also makes a large contribution to our sense of taste, so conditions that affect the olfactory receptors—such as the common cold—can also dull your sense of taste. A reduced sense of taste may also result from damage to taste buds by inflammation or infections of the mouth. Alternatively, the cranial nerves carrying taste sensations—facial (VII), glossopharyngeal (IX), and vagus (X)—may be damaged through trauma or compression by a tumor.

## Vision

**3** Age-related eye disorders include senile cataracts and presbyopia. **Senile cataracts**, which are a natural consequence of aging, result in the loss of lens transparency. As the condition advances, the person needs brighter and brighter light for reading, and visual acuity may eventually decrease to the point of blindness. Cataracts may be corrected by surgery. **Presbyopia** is age-related farsightedness and is caused by loss of lens elasticity.

**Normal eye**

**Eye with cataract**

## Equilibrium

**4** **Vertigo** is a feeling that you are dizzily spinning or that things are dizzily turning about you. Vertigo is usually caused by conditions that alter the function of the internal ear receptor complex, the vestibular nerve of the vestibulocochlear nerve, or sensory nuclei and pathways in the central nervous system. Any event that sets endolymph into motion can stimulate the equilibrium receptors and produce vertigo; flushing the external auditory canal with cold water may chill the endolymph in the outermost portions of the labyrinth and establish a temperature-related circulation of fluid that produces mild and temporary vertigo. Although vertigo is usually caused by a problem with the internal ear, it can also be due to vision problems or drug intake. Excessive consumption of alcohol or exposure to certain drugs can also produce vertigo by changing endolymph composition or disturbing hair cells. Perhaps the most common cause of vertigo is **motion sickness**. Its unpleasant signs and symptoms include headache, sweating, flushing of the face, nausea, and vomiting. The drugs commonly administered to prevent motion sickness appear to depress activity in the brainstem.

 **REVIEW**
**B.** What causes vertigo?

## Hearing

**5** In the United States, approximately 37.5 million adults report having some hearing trouble, and about 30 million people aged 12 years or older report hearing loss in both ears. Hearing loss is typically of two kinds: conductive and sensorineural.
**Conductive hearing loss** is due to a problem conducting sound waves through the external ear to the auditory ossicles. People have a reduced ability to hear faint sounds. Causes include impacted earwax, infection, and perforated tympanic membrane.
**Sensorineural hearing loss** results from damage to the cochlea or to the nerve pathways from the internal ear to the brain. Causes include exposure to loud noise, head trauma, and aging. With age, the tympanic membrane becomes less flexible, the articulations between the auditory ossicles stiffen, and the round window may begin to ossify.

 **REVIEW**
**C.** Identify two common classes of hearing-related disorders.

**Lo** **LEARNING OUTCOME**
Describe age-related disorders of olfaction, gustation, vision, equilibrium, and hearing.

## Section 3 Review
# Equilibrium and Hearing

**Labeling:** Label the structures in the following diagram of the right ear.

Label the structures in the following micrograph of the internal ear.

**Section integration:** For a few seconds after you ride an express elevator from the 25th floor to the ground floor, you still feel as if you are descending, even though you have come to a stop. Why?

21 _____

# Study Outline

## SECTION 1 • Olfaction and Gustation

### Module 15.1

### A generator potential is a depolarization of the membrane    p. 545

1. Receptors for olfaction, or the sense of smell, are the dendrites of specialized neurons. Dissolved chemicals contact these dendrites, causing a **generator potential**.

2. Receptors for taste, vision, equilibrium, and hearing are specialized cells with inexcitable membranes. When stimulated, a graded depolarization triggers chemical transmitter release at the synapse with a sensory neuron.

3. Depolarization of the sensory neuron creates a generator potential and action potentials that are propagated to the CNS.

### Module 15.2

### Olfaction involves specialized chemoreceptive neurons and delivers sensations directly to the cerebrum    p. 546

4. The sense of smell, or **olfaction**, is provided by paired **olfactory organs**.

5. The olfactory pathway to the cerebrum is as follows: Sensory neurons within the olfactory organ are stimulated by chemicals in the air; axons leave the olfactory epithelium and penetrate the cribriform plate of the ethmoid bone; synapses occur within the **olfactory bulb**; axons travel along the **olfactory tract** to the olfactory cortex, hypothalamus, and limbic system; distribution of olfactory information triggers not only the sensation of smell but also emotions, behaviors, and memories.

6. The olfactory organs are made up of the **olfactory epithelium** and the **lamina propria**.

7. The olfactory epithelium contains **olfactory receptor cells**, **supporting cells**, and **basal cells**. The exposed tip of each receptor cell forms a knob that serves as a base for dendrites that are exposed to dissolved compounds.

8. The lamina propria consists of areolar tissue, blood vessels, nerves, and **olfactory glands**.

9. Dissolved chemicals called **odorants** stimulate olfactory receptors at their dendrites by binding to specific receptors.

### Module 15.3

### Gustation involves epithelial chemoreceptor cells located in taste buds    p. 548

10. **Gustation**, or the sense of taste, provides information about the foods and liquids we eat and drink.

11. *Taste receptors* are distributed along the superficial tongue and adjacent portions of the pharynx and larynx.

12. The superior surface of the tongue has four types of **lingual papillae** that contain taste receptors and specialized epithelial cells in sensory structures called **taste buds**.

13. There are four **primary taste sensations: sweet**, **salty**, **sour**, and **bitter**. There are also receptors for **umami**, a savory taste characteristic of beef and chicken broth and Parmesan cheese. **Water receptors** are concentrated in the pharynx.

14. Taste buds are recessed deep into the surrounding epithelium and contain 40–100 receptor cells called **gustatory receptor cells**.

15. Each gustatory receptor cell extends microvilli (taste hairs) through **taste pores**.

*Vallate papillae are projections on the tongue that contain taste buds.*

16. **Basal cells** are stem cells within a taste bud that divide to produce daughter cells that mature into gustatory receptor cells.

17. Taste reception stimulation threshold varies by primary taste sensation. Receptors also respond more readily to unpleasant stimuli. The number of taste buds declines as we age.

### Module 15.4

### Gustatory reception relies on membrane receptors and ion channels, and sensations are carried by facial, glossopharyngeal, and vagus nerves    p. 550

18. Gustatory reception is stimulated by dissolved chemicals that contact the taste hairs and either diffuse through membrane channels or bind to receptor proteins of the gustatory cell.

19. Taste hairs stimulated by the entry of $Na^+$ or $H^+$ through leak channels are responsible for salt and sour sensations. Membrane receptors provide the sensations of sweet, bitter, and umami.

20. The **gustatory pathway** to the cerebral cortex begins with receptor stimulation. The cranial nerve involved is dependent on which region of the tongue the receptors are stimulated. The facial nerve (VII) innervates taste buds on the anterior two-thirds of the tongue. The glossopharyngeal nerve (IX) innervates taste buds on the posterior one-third of the tongue, and the vagus nerve (X) innervates the taste buds on the epiglottis.

21. Sensory afferents from the involved cranial nerve synapse in the **solitary nucleus** of the medulla oblongata. Axons then enter the **medial lemniscus** of the medulla oblongata and synapse in the thalamus. Information is then projected into the **gustatory cortex** of the insula.

## SECTION 2 • Vision

### Module 15.5

### The eyes form early in embryonic development    p. 553

22. The first indication of eye development is a pair of bulges called **optic vesicles** that appear in the 4th week of embryonic development.

23. **Optic cups** form from the optic vesicles. Ependymal cells on the outer layer of the optic cups develop into photoreceptors. Ependymal cells on the inner layer develop into pigment cells.

24. The lens develops from cells superficial to the optic cups. The eye develops interior chambers filled with fluid.

## Module 15.6

### Accessory structures of the eye provide protection while allowing light to reach the interior of the eye p. 554

25. The **accessory structures** of the eye include the **eyelids**, **eyelashes**, the superficial epithelium of the eye, and the structures associated with tears.

26. The two eyelids (*palpebrae*) are connected laterally by the **lateral angle of the eye** and medially by the **medial angle of the eye**. The **palpebral fissure** is the gap between the two eyelids.

27. The **palpebral conjunctiva** covers the inner surface of the eyelids, and the **bulbar conjunctiva** covers the anterior surface of the eye. **Conjunctivitis** is an inflammation of the **conjunctiva**.

28. The **lacrimal apparatus** of each eye produces, distributes, and removes tears. It consists of a **lacrimal gland** with associated **tear ducts**, paired **lacrimal canaliculi**, a **lacrimal sac**, and a **nasolacrimal duct**.

## Module 15.7

### The hollow eyeball has a layered wall and fluid-filled anterior and posterior cavities p. 556

29. The **layers** (tunics) of the wall of the eye are the **fibrous layer**, **vascular layer**, and **inner layer**.

30. The fibrous layer, the outermost layer of the eye, consists of the cornea and **sclera**. The fibrous layer supports and protects the eye, serves as an attachment site for extrinsic eye muscles, and contains the cornea, through which light passes to begin the focusing process.

31. The vascular layer, the middle layer of the eye, provides a route for blood and lymphatic vessels (at the **choroid**), regulates the light entering the eye through the pupil by contracting the smooth muscles of the **iris**, secretes and reabsorbs aqueous humor, and controls the shape of the lens in the focusing process by contracting and relaxing the **ciliary muscle** of the **ciliary body**.

32. The inner layer, or **retina**, is the innermost layer of the eye. It contains a thin deep layer (pigmented layer) that absorbs light, and a thick superficial layer (neural layer) that contains light-sensitive **photoreceptors**.

33. The eye is divided by the ciliary body and lens into an **anterior cavity** and a larger **posterior cavity**. The anterior cavity contains an **anterior chamber** and a **posterior chamber**.

34. **Aqueous humor** circulates within the anterior cavity. It is secreted by the ciliary body and reabsorbed at the **scleral venous sinus**. The gelatinous **vitreous body** occupies most of the posterior cavity. The **vitreous humor** is the fluid part of the vitreous body.

## Module 15.8

### The structures of the eye direct light along a visual axis to the fovea centralis of the retina p. 558

35. The transparaent cornea permits light to enter the eye and is avascular.

36. The lens focuses the visual image on the photoreceptors. The ciliary body supports the lens and controls its shape.

37. Blood vessels in the choroid provide nutrients to the eye.

38. The thickest layer of the eye is the sclera. It stabilizes the shape of the eye and is the site of insertion for the six extrinsic eye muscles.

39. The two layers of pupillary muscles control the amount of light entering the eye by changing the diameter of the pupil. These muscles are the **dilator pupillae** and **sphincter pupillae**.

40. The **visual axis** of the eye is an imaginary line from the object of direct sight through the center of the cornea and the center of the lens to the retina.

41. The highest concentration of photoreceptors is at the **fovea centralis** of the **macula**. This is the area of sharpest vision.

## Module 15.9

### Focusing of light produces a sharp image on the retina p. 560

42. Light rays must be focused on the retina, or the image will be blurry. Focusing occurs first as light passes through the cornea and then through the lens.

43. Light is bent, or **refracted**, by the cornea, aqueous humor, and then the lens toward a **focal point**.

44. **Accommodation** is the process by which the lens changes shape to focus an image on the retina.

45. Light is refracted in a way that results in an inverted image that the brain reverses.

46. Refraction is greatest for closer objects. The ability to focus on near objects, the **near point of vision**, decreases as we age.

## Module 15.10

**The neural layer of the retina contains multiple layers of specialized photoreceptors, neurons, and supporting cells p. 562**

47. The pigmented layer of the retina absorbs light that passes through the neural layer, preventing the light from bouncing back and causing "echoes."

48. The neural layer of the retina contains photoreceptors, supporting cells, and neurons that process and integrate visual information. **Ganglion cells** form the innermost portion of the neural layer of the retina.

49. Axons of the ganglion cells converge to form the optic nerve. This region—the **optic disc**—cannot detect an image because it has no photoreceptors and is therefore called the **blind spot**.

50. The photoreceptors of the retina are called rods and cones.

51. **Rods** are not color sensitive, but their high sensitivity allows for vision in dim light. **Cones** allow for color vision as well as sharp, clear vision. Rods and cones synapse with **bipolar cells**.

52. **Horizontal cells** and **amacrine cells** facilitate or inhibit communication between photoreceptor and ganglion cells.

53. **Visual acuity** is best at the fovea centralis, where cone density is greatest.

## Module 15.11

**Photoreception occurs in the outer segment of rod and cone cells p. 564**

54. The outer segments of rods and cones have membranous **discs** contiaining **visual pigments**. These pigments are derivatives of the compound **rhodopsin**.

55. Rhodopsin consists of the protein **opsin** bound to the pigment **retinal**. Retinal is synthesized from vitamin A.

56. There are three types of cones: **blue cones**, **green cones**, and **red cones**. Each is sensitive to a different range of light wavelength. Their stimulation in various combinations is the basis for color vision. A person with **color blindness** lacks one or more cone pigments.

## Module 15.12

**Photoreception involves activation, bleaching, and reassembly of visual pigments p. 566**

57. Lacking photon stimulation, a photoreceptor produces a **dark current** and constantly releases neurotransmitters.

58. On absorbing light, retinal changes shape and activates the opsin molecule. This activation leads to changes in the permeability of the outer segment to $Na^+$ that decrease the rate of neurotransmitter release by the inner segment at its synapse with a bipolar cell. Change in bipolar cell activity is detected by one or more ganglion cells. This determines the region of retina stimulation.

59. After absorbing a photon, rhodopsin breaks down into retinal and opsin in a process called **bleaching**. Retinal is converted back to its original shape in a process that requires ATP. Once retinal is converted, it can recombine with opsin and repeat the cycle.

## Module 15.13

**The visual pathways distribute visual information from each eye to both cerebral hemispheres p. 568**

60. The visual pathway begins at the photoreceptor in the retina. Each photoreceptor monitors a specific receptive field. Information is passed to bipolar and ganglion cells.

61. Ganglion cell axons converge at the optic disc and proceed to the diencephalon as the optic nerve (II). The two optic nerves meet at the diencephalon at the **optic chiasm**. At that point, about half the fibers continue to the lateral geniculate body of the same side of the brain, and the other half cross to the opposite side.

62. The **optic radiation** links each lateral geniculate body with the visual cortex of the occipital lobe.

63. The visual images from the left and rght eyes overlap, and the visual cortex of each cerebral hemisphere receives information from both eyes. The sorting of this information allows for **depth perception**.

## Module 15.14

**Refractive problems result from abnormalities in the cornea or lens or in the shape of the eye p. 569**

64. In a normal relaxed eye, the image will be focused on the retina. Normal vision is called **emmetropia**.

65. In **myopia**, or nearsightedness, the image is projected in front of the retina. This is corrected with a diverging lens.

66. In **hyperopia**, or farsightedness, the image cannot focus on the retina. This condition is corrected with a converging lens.

## SECTION 3 • Equilibrium and Hearing

### Module 15.15

**Equilibrium and hearing involve the internal ear p. 571**

67. The receptor cells for olfaction and gustation are exposed to the external environment and communicate with the CNS. By contrast, the receptors for equilibrium and hearing are isolated and protected from the external environment within the **internal ear**.

68. The receptors for hearing and equilibrium are called **hair cells** because their free surfaces are covered with processes similar to cilia and microvilli. Certain hair cells contain a **kinocilium**, a single large cilium, and many **stereocilia**.

69. These processes are sensitive to contact or movement. Forces pushing against the hair cell processes alter the rate of neurotransmitter release, thereby providing information about the direction and strength of the stimuli.

70. The complex three-dimensional structure of the internal ear can determine acceleration, rotation, or sound, depending on which region is stimulated.

### Module 15.16

**The ear is divided into the external ear, the middle ear, and the internal ear p. 572**

71. The ear is divided into three anatomical regions: the **external ear**, the **middle ear**, and the **internal ear**.

**15**

72. The external ear consists of the **auricle**, the **external acoustic meatus**, and the **ceruminous glands**. **Cerumen** is a waxy material secreted by the ceruminous glands that, along with hairs projecting from the skin, helps keep foreign objects from reaching deeper delicate structures.

73. The middle ear, or tympanic cavity, is an air-filled cavity that contains the three **auditory ossicles**: **malleus**, **incus**, and **stapes**.

74. The internal ear contains the sensory organs for hearing and equilibrium. It contains a layer of dense bone called the *bony labyrinth*.

75. The **auditory tube** connects the middle ear with the nasopharynx and allows pressure equalization on either side of the tympanic membrane.

76. An infection of the middle ear is known as **otitis media**.

**Module 15.17**

**In the internal ear, the bony labyrinth protects the membranous labyrinth and its receptors    p. 574**

77. The internal ear can be subdivided into a **bony labyrinth** and **membranous labyrinth**. It contains receptors for the special senses of equilibrium and hearing.

78. The membranous labyrinth is a collection of tubes and chambers that contain **endolymph**.

79. The **semicircular ducts** of the membranous labyrinth lie within the semicircular canals of the bony labyrinth and are stimulated by the rotation of the head.

80. The **cochlear duct** is an extension of the membranous labyrinth within the cochlea and is surrounded by a pair of perilymph-filled chambers. It is responsible for hearing.

*The bony labyrinth and membranous labyrinth of the internal ear*

81. The bony labyrinth surrounds and protects the membranous labyrinth. Between the bony and membranous labyrinths flows **perilymph**.

82. The bony labyrinth consists of the **semicircular canals**, the **vestibule**, and the **cochlea**.

83. The membranous labyrinth, filled with endolymph and surrounded by perilymph, contains the **vestibular complex**, which is responsible for equilibrium. The three semicircular ducts detect rotational movements of the head, and the utricle and saccule monitor head position relative to gravity and linear acceleration.

**Module 15.18**

**Hair cells in the semicircular ducts respond to rotation; hair cells in the utricle and saccule respond to gravity and linear acceleration    p. 576**

84. The **anterior**, **posterior**, and **lateral semicircular ducts** are continuous with the utricle. Each duct contains an **ampulla**, an extended region that contains receptors.

85. The **ampullary crest** is a region in the ampulla wall that contains receptors. Here the kinocilia and stereocilia are embedded in the **ampullary cupula**, a gelatinous structure that extends the full width of the ampulla.

86. Head rotation in the plane of a semicircular duct causes the endolymph in that duct to push the cupula to the side and distort the receptor processes.

87. Equilibrium sensations are provided by **maculae** in the utricle and saccule. Hair cell processes of the utricle and saccule are embedded in a gelatinous **otolithic membrane** containing **otoliths**.

88. Gravity holds the otoliths in place when the head is in the normal upright position. Changes in head position cause the otoliths to move and distort the hair cells. A similar mechanism can also detect linear acceleration.

**Module 15.19**

**The cochlear duct contains the hair cells of the spiral organ that function in hearing    p. 578**

89. The **cochlear duct**, or **scala media**, lies between two perilymph-filled chambers: the **scala vestibuli** and **scala tympani**. The scalae vestibuli and tympani are one long continuous duct interconnected at the tip of the cochlear spiral.

90. The outer surfaces of these three ducts are encased by the bony labyrinth except at the **oval window** (the base of the scala vestibuli) and the **round window** (base of the scala tympani).

91. The **vestibular membrane** separates the cochlear duct from the scala vestibuli, and the **basilar membrane** separates the cochlear duct from the scala tympani.

92. The hair cells of the cochlear duct are within the **spiral organ**, or *organ of Corti*, a structure atop the basilar membrane.

93. Movement of the basilar membrane in response to perilymph pressure changes causes the stereocilia of the hair cells to press against the **tectorial membrane**.

94. The **spiral ganglion** contains the cell bodies of sensory neurons that monitor the hair cells of the spiral organ.

**Module 15.20**

**Sound waves lead to movement of the basilar membrane in the process of hearing    p. 580**

95. Air pressure at various **frequencies** and **amplitudes** vibrates the tympanic membrane.

96. The location of the vibration along the basilar membrane is interpreted as **pitch**; the number of hair cells stimulated is interpreted as **volume**.

97. The events involved in hearing are as follows: Sound waves arrive at the tympanic membrane; auditory ossicles vibrate; the stapes at the oval window causes pressure waves in perilymph of scala vestibuli; pressure waves on their way to the round window distort the basilar membrane; hair cells vibrate against tectorial membrane; region and intensity of stimulation are relayed to CNS over the cochlear nerve of cranial nerve VIII.

## Module 15.21

### The vestibulocochlear nerve carries equilibrium and hearing sensations to the brainstem    p. 582

98. The receptors for equilibrium are the hair cells of the vestibule and the semicircular ducts. Sensory information is passed along the **vestibular nerve** of cranial nerve VIII to the vestibular nuclei and then throughout the CNS.

99. Hair cells of the basilar membrane are the receptors for hearing. Information is sent along the **cochlear nerve** of cranial nerve VIII to the cochlear nuclei, superior olivary nuclei, inferior colliculi, thalamus, and auditory cortex.

## Module 15.22

### Aging is associated with many disorders of the special senses; trauma, infection, and abnormal stimuli may cause problems at any age    p. 584

100. Aging and a variety of conditions can decrease the number of receptor cells, resulting in diminished senses. Both the number of olfactory receptors and the number of taste buds decline with age. Lens transparency lessens with age, leading to **senile cataracts**. The tympanic membrane loses some flexibility with age, leading to hearing loss. **Vertigo** and **sensorineural hearing loss** are examples of disorders that may result from damage to the cranial nerves and brain.

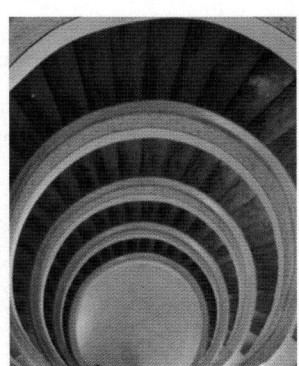

# Chapter Review Questions

**Labeling:** Label the structures in this sectional view through the back of the eye.

**True/False:** Indicate whether each statement is true or false.

10  Humans are most sensitive to the sweet taste.

11  The cochlear duct and scala tympani both contain perilymph, and the scala vestibuli contains endolymph.

12  Vibration of the basilar membrane causes vibration of hair cells against the tectorial membrane.

13  The accessory structures of the eye include the fibrous layer, vascular layer, and inner layer.

14  Rods are the photoreceptors that enable vision in dim light.

15  The retinal area of greatest visual acuity is the fovea centralis.

10 _____

11 _____

12 _____

13 _____

14 _____

15 _____

**Multiple Choice:** Select the correct answer from the list provided.

16  A blind spot occurs in the retina where
☐ a) the fovea centralis is located.
☐ b) ganglion cells synapse with bipolar cells.
☐ c) axons of ganglion cells converge at the optic nerve.
☐ d) rod cells are clustered to form the macula.

17  The choroid is found in the
☐ a) fibrous layer.
☐ b) vascular layer.
☐ c) inner layer.
☐ d) vitreous body.

18  Receptors in the saccule and utricle provide the sensation of
☐ a) gravity and linear acceleration.
☐ b) hearing.
☐ c) rotational movements of the head.
☐ d) vibration.

19  The equalization of pressure on either side of the tympanic membrane is achieved by the
☐ a) round window.
☐ b) oval window.
☐ c) auditory tube.
☐ d) scala tympani.

20  What accounts for the feeling you experience when you accelerate down rollercoaster tracks?
☐ a) Pressure waves in the perilymph distort the basilar membrane.
☐ b) The otoliths move horizontally and distort the hair cell processes, stimulating the macular receptors.
☐ c) The hair cells in the spiral organ press against the tectorial membrane.
☐ d) Movement of the tympanic membrane causes displacement of the auditory ossicles.

**Short answer**

21  Identify the four primary taste sensations. Name the two additional taste sensations that have been identified.

22  Describe senile cataracts.

23  Describe otitis media.

---

## MasteringA&P™

Access more chapter study tools online in the MasteringA&P Study Area:

■ Chapter Quizzes, Chapter Practice Test, and Clinical Case Studies

■ Practice Anatomy Lab™ 3.0 — PAL™3.0

■ Interactive Physiology™ — iP2™

■ A&P Flix™ — A&PFlix™

■ PhysioEx™ 9.1 — PhysioEx™ 9.1

# Chapter Integration • Applying what you have learned

## Curious complications from the common cold

Earl Drummond, 75 years old, visits his physician for a routine examination. During the history and physical, Mr. Drummond states that he has been experiencing difficulty seeing at a distance while driving and  that at times he feels dizzy, especially when he closes his eyes. Additionally, Mr. Drummond tells his physician that he has a slight "cold." The physician asks Mr. Drummond to stand with his feet together and arms extended forward. The doctor discovers that as long as Mr. Drummond keeps his eyes open, he exhibits very little swaying movement. However, when he closes his eyes, his body begins to sway a great deal, and his arms tend to drift together toward the left side of his body. After a complete workup, which includes a neurological evaluation and a basic vision-screening test, the preliminary diagnosis is a slight fever, myopia, and vertigo.

It is estimated that 20–25 percent of all adults in the United States have myopia, while 40 percent of people in the United States experience vertigo at least once during their lifetime. The prevalence for both conditions increases with age.

**1** Explain the diagnosis of myopia.

**2** What treatment is available for people with myopia?

**3** Based on the information given, provide an explanation for Mr. Drummond's vertigo.

**4** Why is Mr. Drummond's vertigo worse when his eyes are closed?

**5** Why might Mr. Drummond's arms move toward his left side when he has his eyes closed?

## Flight deck hearing loss

Daniel is a U.S. Navy veteran who spent several years regularly working 12-hour shifts on the flight deck of an aircraft carrier. He consistently wore earplugs as well as a helmet designed to provide ear protection. There were, however, occasions when he was exposed to the sound of jet engines that approached 140 dB with no ear protection at all. Such events were inevitable despite his best efforts to minimize his exposure. At the time of his discharge from the Navy, it was noted that he had suffered appreciable hearing loss. His physician informed Daniel that he is now partially deaf. Using what you have just learned about hearing, answer the following questions.

**6** What kind of deafness do you think Daniel has experienced from his prolonged exposure to the sound of jet engines?

**7** What has occurred at the cellular level that has resulted in his deafness?

**8** Do you think it is possible for Daniel to regain his hearing? Why or why not?

15

## LEARNING OUTCOMES

These Learning Outcomes correspond by number to this chapter's modules and indicate what you should be able to do after completing the chapter. They also appear at the end of each module.

# 16

# The Endocrine System

## Module 16.1

# The nervous and endocrine systems release chemical messengers that bind to target cells

**1** To preserve homeostasis, cellular activities must be coordinated throughout the body. The table below introduces the various ways cells communicate and coordinate their activities using chemical messengers or signals.

### Mechanisms of Intercellular Communication

| Mechanism | | Transmission | Chemical Signals | Distribution of Effects |
|---|---|---|---|---|
| **Direct communication** | | Through gap junctions | Ions, small solutes, lipid-soluble materials | Usually limited to adjacent cells of the same type that are interconnected by connexons |
| **Paracrine communication** | | Through extracellular fluid | Paracrines | Primarily limited to the local area, where paracrine concentrations are relatively high; target cells must have appropriate receptors (see Modules 3.3, 11.8) |
| **Autocrine communication** | | Through extracellular fluid | Autocrines | Limited to the cell that secretes the chemical signal |
| **Endocrine communication** | | Through the bloodstream | Hormones | Target cells are mainly in other distant tissues and organs and must have appropriate receptors |
| **Synaptic communication** | | Across synapses | Neurotransmitters | Limited to very specific area; target cells must have appropriate receptors |

**2** Viewed from a general perspective, the differences between the nervous and endocrine systems seem relatively clear. In fact, these broad organizational and functional differences are the basis for treating them as two separate systems. Yet when we consider them in detail, we see similarities in the ways they are organized.

### Similarities between the Nervous and Endocrine Systems

- Both systems rely on the release of chemicals that bind to specific receptors on their target cells.

- The two systems share many chemical messengers; for example, norepinephrine and epinephrine are called hormones when released into the bloodstream, but they are called neurotransmitters when released across synapses.

- Both systems are regulated primarily by negative feedback control mechanisms.

- The two systems share a common goal: to preserve homeostasis by coordinating and regulating the activities of other cells, tissues, organs, and systems.

**? REVIEW**

**A.** What is the common goal of the nervous and endocrine systems?

In this section we consider the structure and function of the endocrine system, hormones, and the integration of neural and endocrine activities.

 **LEARNING OUTCOME**

Describe the similarities between the endocrine and nervous systems and their specific modes of intercellular communication.

# Hormones may be amino acid derivatives, peptides, or lipid derivatives

**1** The hormones and paracrines and autocrines of the body can be divided into three groups on the basis of their chemical structure: (1) **amino acid derivatives**, (2) **peptide hormones**, and (3) **lipid derivatives**.

## Amino Acid Derivatives

This group of hormones includes (1) thyroid hormones, produced by the thyroid gland; (2) the compounds epinephrine (E), norepinephrine (NE), and dopamine, which are sometimes called **catecholamines** (kat-e-KŌ-la-mēnz); and (3) tryptophan derivatives, one of which is melatonin (mel-a-TŌ-nin), secreted by the pineal gland.

### Thyroid Hormones

Example: Thyroxine (T$_4$)

### Catecholamines

Example: Epinephrine

### Tryptophan Derivatives

Example: Melatonin

## Peptide Hormones

In general, peptide hormones are synthesized as **prohormones**—inactive precursor molecules that are converted to active hormones. Peptide hormones range from short polypeptide chains of amino acids, such as antidiuretic hormone (ADH) and oxytocin (9 amino acids apiece), to small proteins, such as insulin (51 amino acids), growth hormone (GH; 191 amino acids) and prolactin (PRL; 198 amino acids). This group includes all the hormones secreted by the hypothalamus, heart, thymus, digestive tract, and pancreas, and most of the hormones of the pituitary gland. Glycoproteins—polypeptides that have carbohydrate side chains—may also function as hormones. Examples include thyroid-stimulating hormone (TSH), luteinizing hormone (LH), and follicle-stimulating hormone (FSH) from the pituitary gland, and several hormones produced in other organs.

## Lipid Derivatives

These hormones consist of carbon rings and side chains built either from fatty acids (eicosanoids) or cholesterol (steroid hormones).

### Eicosanoids

Example: Prostaglandin E

**Eicosanoids** (Ī-kō-sa-noydz) are important paracrines that coordinate cellular activities and affect enzymatic processes (such as blood clotting) in extracellular fluids. Some eicosanoids, such as **leukotrienes** (loo-kō-TRĪ-ēns), have secondary roles as hormones. A second group of eicosanoids—**prostaglandins**—is involved primarily in coordinating local cellular activities.

### Steroid Hormones

Example: Estrogen

**Steroid hormones** are released by the reproductive organs (androgens by the testes in males, estrogens and progesterone by the ovaries in females), by the cortex of the adrenal glands (corticosteroids), and by the kidneys (calcitriol). Because circulating steroid hormones are bound to specific transport proteins in the blood, they remain in circulation longer than do secreted peptide hormones.

 **REVIEW**

**A.** Describe the structural classification of hormones.

 **LEARNING OUTCOME**

Explain the chemical classification of hormones.

# Module 16.3

# The endocrine system includes organs and tissues with primary and secondary hormone-secreting roles

Go to
MasteringA&P™
PAL™ 3.0
Endocrine
System

**1** The **endocrine system** includes those organs (indicated in orange in this figure) whose primary function is the production of hormones or paracrines. Many other organs contain tissues that secrete hormones, but their endocrine functions are secondary. Examples include the heart, kidneys, intestines, and reproductive organs. The endocrine functions of these organs will be considered primarily in later chapters.

**? REVIEW**

**A.** Define *endocrine system*.

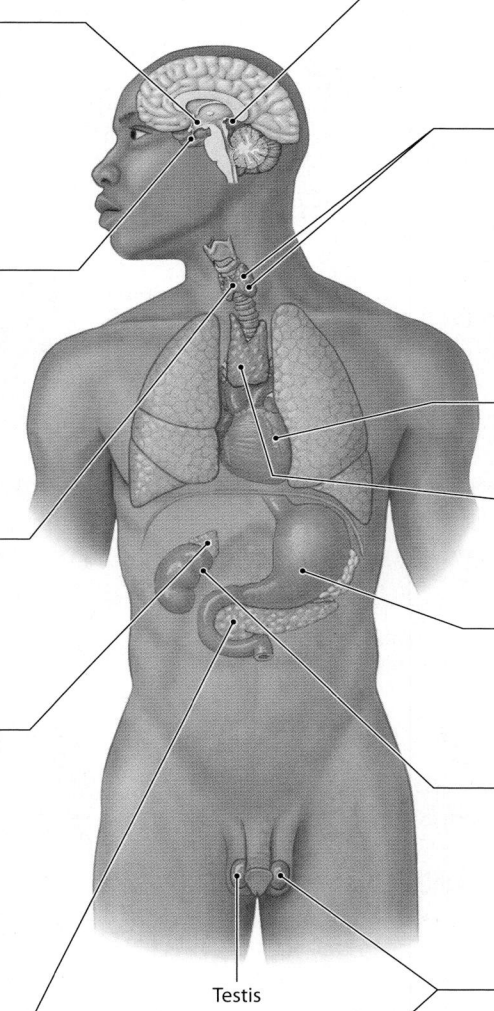

Testis

Ovary

### Hypothalamus

The **hypothalamus** secretes hormones involved with fluid balance, smooth muscle contraction, and the control of hormone secretion by the anterior lobe of the pituitary gland.

### Pituitary Gland

The **pituitary gland** secretes multiple hormones that regulate the endocrine activities of the adrenal cortex, thyroid gland, and reproductive organs, and a hormone that stimulates melanin production.

### Thyroid Gland

The **thyroid gland** secretes hormones that affect metabolic rate and calcium ion levels in body fluids.

### Adrenal Glands

The two **adrenal glands** secrete hormones involved with mineral balance, metabolic control, and resistance to stress. They also release epinephrine and norepinephrine during sympathetic activation.

### Pancreas (Pancreatic Islets)

The **pancreatic islets** secrete hormones regulating the rate of glucose uptake and utilization by body tissues.

### Pineal Gland

The **pineal gland** secretes melatonin, which affects reproductive function and helps establish circadian (day/night) rhythms.

### Parathyroid Glands

The **parathyroid glands** secrete a hormone important to the regulation of calcium ion levels in body fluids.

### Organs with Secondary Endocrine Functions

| | |
|---|---|
| Heart: Secretes hormones involved in regulating blood volume | See Chapter 18 |
| Thymus: Secretes hormones involved in stimulating and coordinating the immune response | See Chapter 20 |
| Digestive tract: Secretes numerous hormones involved in coordinating system functions, glucose metabolism, and appetite | See Chapter 22 |
| Kidneys: Secrete hormones that regulate red blood cell production, the rates of calcium and phosphate absorption by the intestinal tract, and an enzyme involved in regulating blood volume | See Chapter 24 |
| Gonads: Secrete hormones affecting growth, metabolism, and sexual characteristics, and other hormones that coordinate the activities of organs in the reproductive system | See Chapter 26 |

16

**? REVIEW**

**B.** Name the organs of the endocrine system.

 **LEARNING OUTCOME**

Identify the organs and tissues of the endocrine system and the key functions of the hormones they secrete.

# Module 16.4

# Hormones affect target cells after binding to receptors in the plasma membrane, cytoplasm, or nucleus

To affect a target cell, a hormone must first interact with an appropriate *receptor*—a protein molecule to which it binds strongly. Each cell has receptors for responding to several different hormones, but cells in different tissues have different combinations of receptors. This is one reason why hormones have differential effects on specific tissues. For every cell, the presence or absence of a specific receptor determines the cell's hormonal sensitivities. If a cell has a receptor that can bind a particular hormone, that cell will respond to the hormone's presence. If not, the hormone will have no effect on that cell. Hormone receptors are located either on plasma membranes (*extracellular receptors*) or within target cells (*intracellular receptors*).

**REVIEW**

**A.** Define *hormone receptor.*

**1** The receptors for catecholamines (E, NE, and dopamine) and peptide hormones, and most eicosanoids are on the plasma membranes of their target cells. However, binding of the hormone with its receptor on a target cell does not produce a direct effect on its intracellular activities. Instead, the hormone, or **first messenger**, uses an intracellular intermediary, or **second messenger**, to exert the hormone's effects in the cell. The second messenger may act as an enzyme activator, inhibitor, or cofactor. Regardless, the net result is a change in the rates of various metabolic reactions. The two most important second messengers are **cyclic AMP (cAMP)**, a derivative of ATP, and calcium ions ($Ca^{2+}$).

**REVIEW**

**B.** Differentiate between a first messenger and a second messenger.

The link between the first messenger and the second messenger generally involves a **G protein**, an enzyme complex coupled to a membrane receptor.

## Effects on cAMP Levels

Many G proteins, once activated, exert their effects by changing the concentration of cAMP, which acts as the second messenger within the cell.

If levels of cAMP increase, enzymes may be activated or ion channels may be opened, accelerating the metabolic activity of the cell.

In some instances, G protein activation results in decreased levels of cAMP in the cytoplasm. This decrease has an inhibitory effect on the cell.

## Effects on Ca²⁺ Levels

Some G proteins use $Ca^{2+}$ as a second messenger.

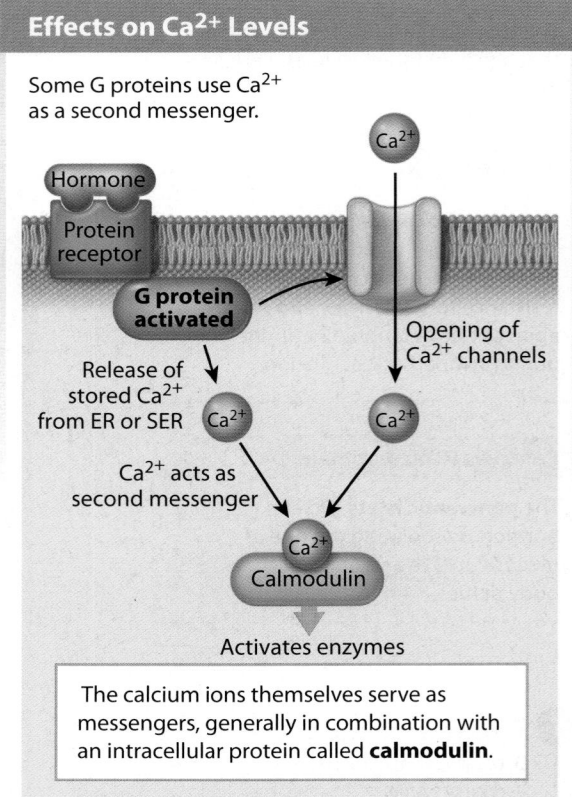

The calcium ions themselves serve as messengers, generally in combination with an intracellular protein called **calmodulin**.

**2** Steroid hormones diffuse across the lipid part of the plasma membrane and bind to intracellular receptors in the cytoplasm or nucleus. The hormone–receptor complexes then alter the activity of specific genes. By this mechanism, steroid hormones can alter the rate of DNA transcription in the nucleus, changing the pattern of protein synthesis. The resulting changes in the synthesis of enzymes or structural proteins directly affect the target cell's metabolic activity and structure. For example, in response to the sex hormone testosterone, skeletal muscle fibers increase production of enzymes and structural proteins, causing increases in muscle size and strength.

**3** Thyroid hormones are primarily transported across the plasma membrane by carrier-mediated processes. Once in the cells, these hormones bind to receptors on mitochondria and within the nucleus. Thyroid hormones bound to mitochondria increase the rate of ATP synthesis in the mitochondria. Hormone–receptor complexes in the nucleus activate specific genes or change the rate of transcription, which affects the metabolic activities of the cell by increasing or decreasing the concentrations of specific enzymes.

Watch

MasteringA&P™

**A&PFlix**

Mechanism of Hormone Action: Second Messenger cAMP

**Steroid hormone**

**Thyroid hormone**

16

**?** **REVIEW**

**C.** Which type of hormone diffuses across the plasma membrane and binds to receptors in the cytoplasm?

**Lo** **LEARNING OUTCOME**

Explain the general mechanisms of hormonal action.

# The hypothalamus exerts direct or indirect control over the activities of many endocrine organs

**1** The **hypothalamus** contains both brain centers and endocrine tissue. It provides the highest level of endocrine control by integrating the activities of the nervous and endocrine systems. The hypothalamus does this through three mechanisms.

Hypothalamic neurons synthesize two hormones—**antidiuretic hormone (ADH)** and **oxytocin (OXT)**—and transport them along axons within the infundibulum to the posterior lobe of the pituitary gland (see Module 13.8).

The hypothalamus secretes **regulatory hormones**. These regulatory hormones control the secretory activities of endocrine cells in the anterior lobe of the pituitary gland.

The hypothalamus contains autonomic centers that exert direct neural control over the endocrine cells of the adrenal medulla. When the sympathetic division is activated, the adrenal medulla is stimulated directly and immediately.

**? REVIEW**

**A.** Define *regulatory hormone*.

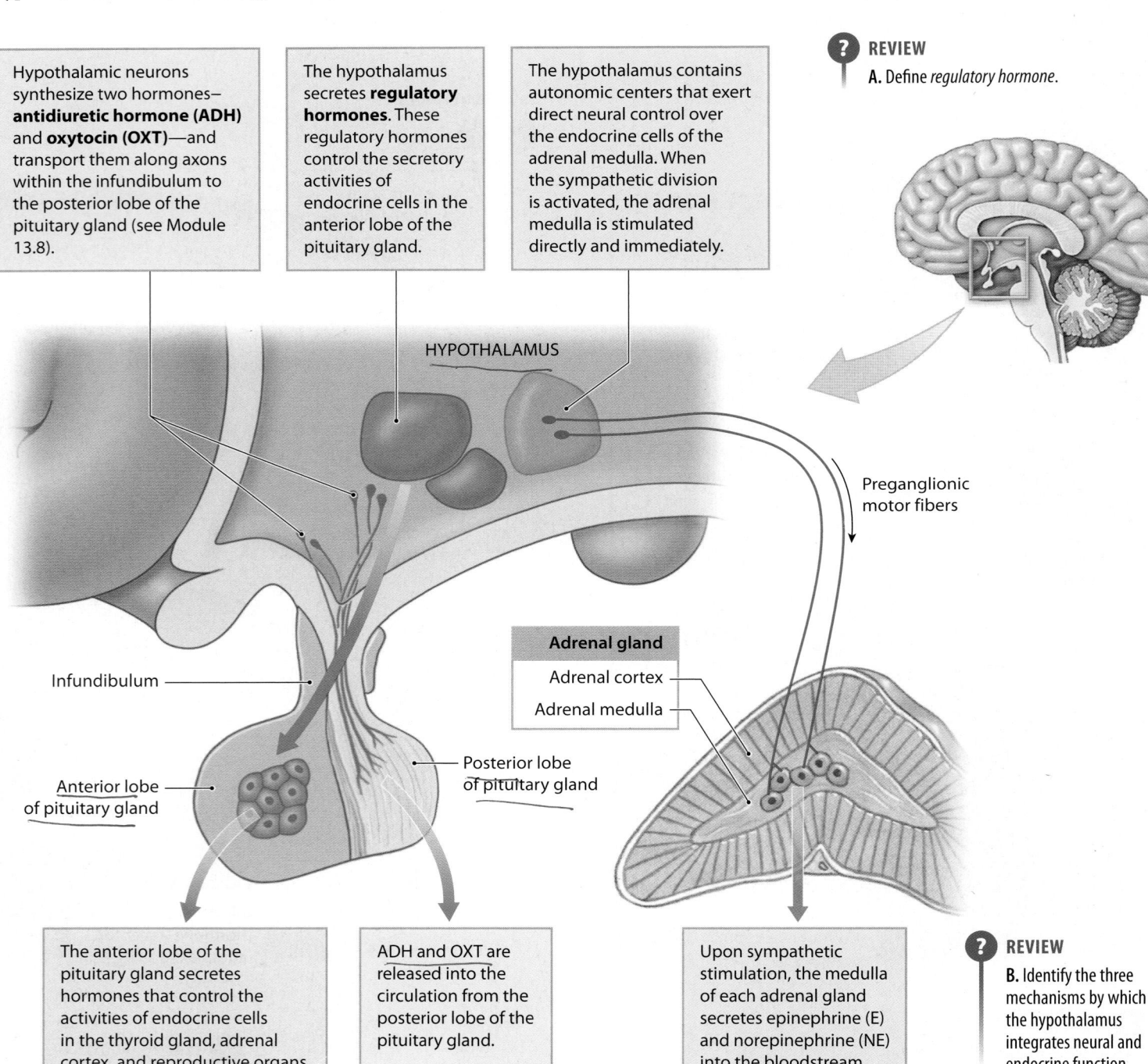

HYPOTHALAMUS

Preganglionic motor fibers

Infundibulum

**Adrenal gland**

Adrenal cortex

Adrenal medulla

Posterior lobe of pituitary gland

Anterior lobe of pituitary gland

The anterior lobe of the pituitary gland secretes hormones that control the activities of endocrine cells in the thyroid gland, adrenal cortex, and reproductive organs.

ADH and OXT are released into the circulation from the posterior lobe of the pituitary gland.

Upon sympathetic stimulation, the medulla of each adrenal gland secretes epinephrine (E) and norepinephrine (NE) into the bloodstream.

**? REVIEW**

**B.** Identify the three mechanisms by which the hypothalamus integrates neural and endocrine function.

16

**2** The hypothalamus controls the production of hormones in the anterior lobe of the pituitary gland by secreting specific regulatory hormones. At the **median eminence**, a swelling near the attachment of the infundibulum, hypothalamic neurons release regulatory factors into the surrounding interstitial fluids. These secretions enter the bloodstream quite easily because the endothelial cells lining the capillaries in this region are unusually permeable. Called **fenestrated** (FEN-es-trā-ted; *fenestra*, window) **capillaries**, these vessels allow relatively large molecules to enter or leave the bloodstream.

Neurons of the supra-optic and paraventricular nuclei synthesize ADH and OXT, respectively. These hormones are released by axon terminals at fenestrated capillaries in the posterior lobe of the pituitary gland.

Supra-optic nuclei | Paraventricular nuclei | Neurosecretory neurons

HYPOTHALAMUS

MEDIAN EMINENCE

Superior hypophyseal artery

Infundibulum

Inferior hypophyseal artery

Posterior lobe of pituitary gland

Endocrine cells

Anterior lobe of pituitary gland

Hypophyseal veins

**Hypophyseal Portal System**

The capillary networks and the interconnecting vessels make up a portal system. Portal systems are named after their destinations. This particular system is known as the **hypophyseal** (hī-pō-FIZ-ē-al) **portal system** (*hypophysis* is the Latin name for the pituitary gland). Portal systems are an efficient means of chemical communication. This one ensures that all the hypothalamic hormones entering the portal vessels will reach their target cells in the anterior lobe before being diluted through mixing with the general circulation. The communication is strictly one way, however. Any chemicals released by the cells "downstream" must do a complete circuit of the cardiovascular system before they reach the capillaries of the portal system.

The capillary networks in the median eminence are supplied by the superior hypophyseal artery. Before leaving the hypothalamus, the capillary networks unite to form a series of larger vessels that spiral around the infundibulum to reach the anterior lobe.

The vessels between the median eminence and the anterior lobe carry blood from one capillary network to another. Blood vessels that link two capillary networks are called **portal vessels**. In this case, they have the histological structure of veins, so they are called portal veins.

Once within the anterior lobe, these vessels form a second capillary network that branches among the endocrine cells.

The regulatory hormones secreted by the hypothalamus are transported directly to the anterior lobe by the hypophyseal portal system. Two classes of hypothalamic regulatory hormones exist: releasing hormones and inhibiting hormones. A **releasing hormone (RH)** stimulates the synthesis and secretion of one or more hormones at the anterior lobe, whereas an **inhibiting hormone (IH)** prevents the synthesis and secretion of hormones from the anterior lobe. Releasing hormones, inhibiting hormones, or some combination of the two may control an endocrine cell in the anterior lobe.

**? REVIEW**

**C.** Name and describe the characteristics and functions of the blood vessels that link the hypothalamus with the anterior lobe of the pituitary gland.

**LEARNING OUTCOME**

Describe how the hypothalamus controls endocrine organs.

# Module 16.6

# The anterior lobe of the pituitary gland produces and releases 7 tropic hormones, while the posterior lobe releases 2 hormones

**1** The **pituitary gland**, or **hypophysis** (hī-POF-i-sis), is a small, oval gland that lies nestled within the sella turcica, a depression in the sphenoid bone. Nine important peptide hormones are released by the pituitary gland—seven by the anterior lobe and two by the posterior lobe. All nine hormones bind to plasma membrane receptors, and all nine use cAMP as a second messenger. The hormones of the anterior lobe are also called **tropic hormones** (*trope*, a turning), because they "turn on" endocrine glands or support the functions of other organs. (Some sources call them *trophic hormones* [*trophe*, nourishment].)

The **infundibulum** (in-fun-DIB-ū-lum; funnel) is a stalk that connects the pituitary gland to the hypothalamus.

HYPOTHALAMUS

Median eminence

Optic chiasm

A fold of the dura mater encircles the base of the infundibulum, locking the pituitary gland in position and isolating it from the cranial cavity.

The **anterior lobe of the pituitary gland**, or **adenohypophysis** (ad-eh-nō-hī-POF-ih-sis), contains a variety of endocrine cells.

The **posterior lobe of the pituitary gland**, or **neurohypophysis** (noo-rō-hī-POF-ih-sis), contains the axons of hypothalamic neurons.

**? REVIEW**
**A.** Name the two lobes of the pituitary gland.

---

## Hormones of the Anterior Lobe

### TSH

Thyroid gland

**Thyroid-stimulating hormone (TSH)** targets the thyroid gland, where it triggers the release of thyroid hormones. TSH is released in response to **thyrotropin-releasing hormone (TRH)** from the hypothalamus. As circulating concentrations of thyroid hormones rise, the rate of TRH and TSH production falls— another example of negative feedback.

### ACTH

Adrenal gland

**Adrenocorticotropic hormone (ACTH)**, also known as *corticotropin*, stimulates the release of steroid hormones by the adrenal cortex, the outer portion of the adrenal gland. ACTH specifically targets cells that produce hormones that affect glucose metabolism. ACTH release occurs under the stimulation of **corticotropin-releasing hormone (CRH)** from the hypothalamus.

### Gonadotropins (FSH and LH)

The hormones called **gonadotropins** (gō-nad-ō-TRŌ-pinz) regulate the activities of the gonads. (These organs—the testes in males and ovaries in females—produce reproductive cells as well as hormones.) The production of gonadotropins occurs under stimulation by **gonadotropin-releasing hormone (GnRH)** from the hypothalamus.

**Follicle-stimulating hormone (FSH)** promotes ovarian follicle development in females and, in combination with luteinizing hormone, stimulates the secretion of **estrogens** (ES-trō-jenz) by ovarian cells. In males, FSH promotes the physical maturation of developing sperm. FSH production is inhibited by **inhibin**, a peptide hormone released by cells in the testes and ovaries.

Ovary

**Luteinizing** (LŪ-tē-in-ī-zing) **hormone (LH)** induces *ovulation*, the release of reproductive cells in females. It also promotes the secretion, by the ovaries, of estrogens and progesterone, which prepare the body for possible pregnancy. In males, this gonadotropin stimulates the production of sex hormones by the interstitial cells of the testes. These sex hormones are called **androgens** (AN-drō-jenz; *andros*, man); the most important one is *testosterone*.

Testis

**? REVIEW**
**B.** Identify the nine pituitary hormones and their target tissues.

## Hormones of the Posterior Lobe

### ADH

**Antidiuretic hormone (ADH)**, also known as *vasopressin*, is released in response to a variety of stimuli, most notably an increase in the solute concentration in the blood or a decrease in blood volume or blood pressure. An increase in the solute concentration stimulates specialized hypothalamic neurons. Because they respond to a change in the osmotic concentration of body fluids, these neurons are called **osmoreceptors**. The osmoreceptors then stimulate the neurosecretory neurons that release ADH.

Kidney

The primary function of ADH is to act on the kidneys to retain water and decrease urination. With losses minimized, any water absorbed from the digestive tract will be retained, reducing the concentrations of electrolytes in the extracellular fluid. In high concentrations, ADH also causes vasoconstriction, a narrowing of peripheral blood vessels that helps increase blood pressure. Alcohol inhibits ADH release, which explains why people find themselves making frequent trips to the bathroom after consuming alcoholic beverages.

### OXT

In women, **oxytocin** (*okytokos,* swift birth), or **OXT**, stimulates smooth muscle contraction in the wall of the uterus, promoting labor and delivery. After delivery, OXT promotes the ejection of milk by stimulating the contraction of myoepithelial cells around the secretory alveoli and the ducts of the mammary glands. Although the functions of OXT in sexual activity remain unclear, it is known that circulating concentrations of OXT rise during sexual arousal and peak at orgasm in both sexes. The secretion of OXT involves both neural and endocrine mechanisms. Such a pathway is called a *neuroendocrine response*. Another example is the secretion of E and NE (see Module 16.5).

Uterus

### GH

**Growth hormone (GH)** stimulates cell growth and reproduction by accelerating the rate of protein synthesis. Skeletal muscle cells and chondrocytes are particularly sensitive to GH. The production of GH is regulated by **growth hormone–releasing hormone (GH–RH)** and **growth hormone–inhibiting hormone (GH–IH)** from the hypothalamus. The stimulation of growth by GH involves two different mechanisms of control:

Musculo-skeletal system

The primary control mechanism is indirect. Liver cells respond to GH by synthesizing and releasing **somatomedins**, known also as **insulin-like growth factors (IGFs)**. These peptide hormones stimulate tissue growth by binding to receptors on a variety of plasma membranes. In skeletal muscle fibers, cartilage cells, and other target cells, somatomedins increase the uptake of amino acids and their incorporation into new proteins.

The direct actions of GH are more selective:
- In epithelia and connective tissues, GH stimulates stem cell divisions and the differentiation of daughter cells.
- In adipose tissue, GH stimulates the breakdown of stored triglycerides by adipocytes (fat cells), which then release fatty acids into the blood. Many tissues then stop breaking down glucose and use fatty acids instead to generate ATP. This is termed a **glucose-sparing effect**.
- In the liver, GH stimulates the breakdown of glycogen reserves, leading to the release of glucose into the bloodstream.

### PRL

**Prolactin** (*pro-*, before + *lac*, milk) **(PRL)** works with other hormones to stimulate mammary gland development. In pregnancy and during the nursing period that follows delivery, PRL also stimulates milk production by the mammary glands. PRL production is inhibited by **prolactin-inhibiting hormone** (PIH) and stimulated by **prolactin-releasing hormones** (PRH).

Mammary gland

### MSH

The *pars intermedia*, a narrow portion of the anterior lobe closest to the posterior lobe, may secrete **melanocyte-stimulating hormone (MSH)**. MSH stimulates the melanocytes of the skin to increase their production of melanin. In adults, this portion of the anterior lobe is virtually nonfunctional, and the circulating blood usually does not contain MSH.

 **INTEGRATION**

**C.** In a dehydrated person, how would the amount of ADH released by the posterior lobe of the pituitary change?

 **LEARNING OUTCOME**

Describe the location and structure of the pituitary gland, and identify pituitary hormones and their functions.

16

# Negative feedback mechanisms control the secretion rates of the hypothalamus and the pituitary gland

**1** Hormone secretion is typically controlled through negative feedback; the relationships among the hypothalamus, the pituitary, and the endocrine target organ provide many useful examples. In the common arrangement shown here, the hypothalamus produces a releasing hormone (or factor) that triggers the release of a hormone by the anterior lobe of the pituitary gland. The pituitary hormone stimulates release of a second hormone by the target organ. This second hormone suppresses secretion of both the hypothalamic releasing hormone and the pituitary hormone. The table below lists major hormones whose secretion is controlled in this way.

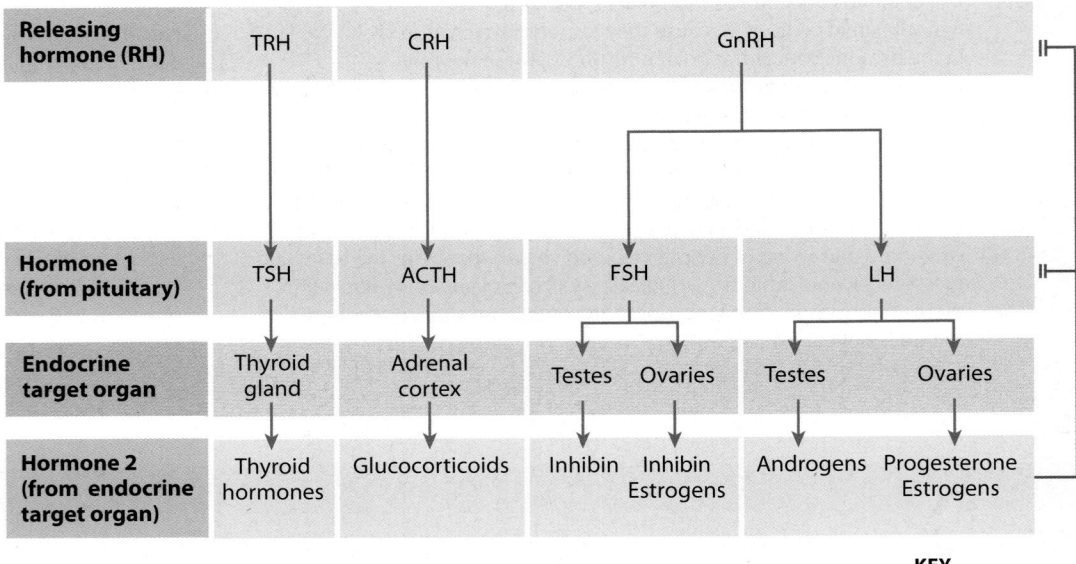

| | | | | | | |
|---|---|---|---|---|---|---|
| **Releasing hormone (RH)** | TRH | CRH | | GnRH | | |
| **Hormone 1 (from pituitary)** | TSH | ACTH | FSH | | LH | |
| **Endocrine target organ** | Thyroid gland | Adrenal cortex | Testes | Ovaries | Testes | Ovaries |
| **Hormone 2 (from endocrine target organ)** | Thyroid hormones | Glucocorticoids | Inhibin | Inhibin Estrogens | Androgens | Progesterone Estrogens |

**KEY**

⟶ Stimulation

⊢ Inhibition

**2** Growth hormone (GH) production involves both releasing and inhibiting hormones. The hypothalamic neurons releasing these hormones are sensitive to levels of **somatomedins**, which function like Hormone 2 in the general pattern shown above. However, the somatomedins not only suppress secretion of GH–RH but also stimulate secretion of GH–IH. This combination provides more rapid and precise regulation of GH levels.

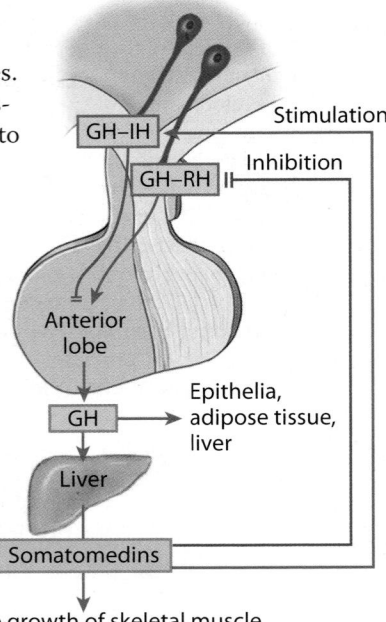

**3** Prolactin (PRL) production is also controlled by a pair of regulatory hormones. Secretion of prolactin-releasing hormone (PRH) is inhibited by PRL, whereas secretion of prolactin-inhibiting hormone (PIH) is stimulated by PRL. When one is stimulated, the other is inhibited.

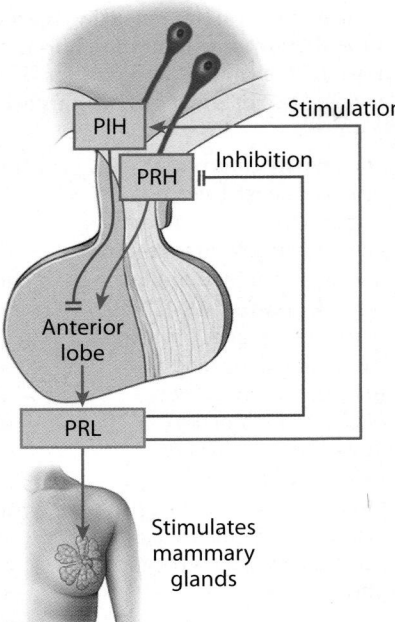

**4** This figure gives an overview of the hormonal relationships discussed in Modules 16.4 through 16.6.

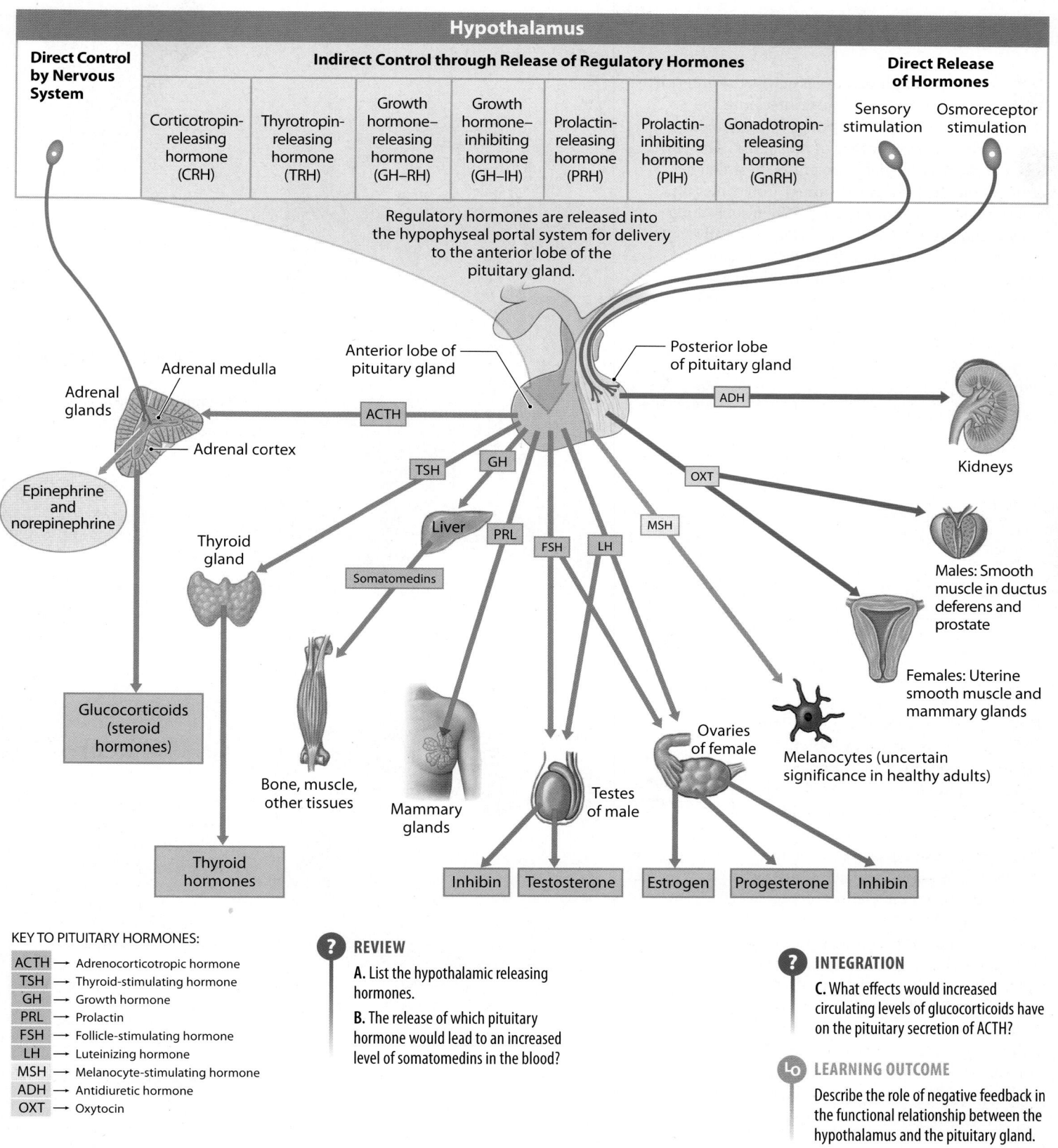

| Hypothalamus | | | | | | | | | |
|---|---|---|---|---|---|---|---|---|---|
| **Direct Control by Nervous System** | **Indirect Control through Release of Regulatory Hormones** | | | | | | | **Direct Release of Hormones** | |
| | Corticotropin-releasing hormone (CRH) | Thyrotropin-releasing hormone (TRH) | Growth hormone–releasing hormone (GH–RH) | Growth hormone–inhibiting hormone (GH–IH) | Prolactin-releasing hormone (PRH) | Prolactin-inhibiting hormone (PIH) | Gonadotropin-releasing hormone (GnRH) | Sensory stimulation | Osmoreceptor stimulation |

Regulatory hormones are released into the hypophyseal portal system for delivery to the anterior lobe of the pituitary gland.

Anterior lobe of pituitary gland

Posterior lobe of pituitary gland

Adrenal glands
Adrenal medulla
Adrenal cortex

ACTH

ADH → Kidneys

Epinephrine and norepinephrine

TSH    GH    OXT

Thyroid gland

Liver    PRL    FSH    LH    MSH

Somatomedins

Glucocorticoids (steroid hormones)

Bone, muscle, other tissues

Mammary glands

Testes of male

Ovaries of female

Melanocytes (uncertain significance in healthy adults)

Males: Smooth muscle in ductus deferens and prostate

Females: Uterine smooth muscle and mammary glands

Thyroid hormones

| Inhibin | Testosterone | Estrogen | Progesterone | Inhibin |

**KEY TO PITUITARY HORMONES:**

ACTH → Adrenocorticotropic hormone
TSH → Thyroid-stimulating hormone
GH → Growth hormone
PRL → Prolactin
FSH → Follicle-stimulating hormone
LH → Luteinizing hormone
MSH → Melanocyte-stimulating hormone
ADH → Antidiuretic hormone
OXT → Oxytocin

**? REVIEW**

**A.** List the hypothalamic releasing hormones.

**B.** The release of which pituitary hormone would lead to an increased level of somatomedins in the blood?

**? INTEGRATION**

**C.** What effects would increased circulating levels of glucocorticoids have on the pituitary secretion of ACTH?

**Lo LEARNING OUTCOME**

Describe the role of negative feedback in the functional relationship between the hypothalamus and the pituitary gland.

16

# The thyroid gland contains follicles and requires iodine to produce hormones that stimulate tissue metabolism

**1** The **thyroid gland** curves across the anterior surface of the trachea just inferior to the thyroid ("shield-shaped") cartilage, which forms most of the anterior surface of the larynx. The two lobes of the thyroid gland are united by a slender connection, the **isthmus** (IS-mus). You can easily feel the gland with your fingers. When something goes wrong with it, the thyroid gland typically becomes visible as it enlarges and distorts the surface of the neck. The size of the gland varies, depending on heredity and environmental and nutritional factors, but its average weight is about 34 g (1.2 oz). An extensive blood supply gives the thyroid gland a deep red color.

Thyroid cartilage

Internal jugular vein

Left lobe of thyroid gland

Isthmus of thyroid gland

Right lobe of thyroid gland

Common carotid artery

Trachea

Outline of clavicle

Outline of sternum

**2** The thyroid gland contains large numbers of **thyroid follicles**, hollow spheres lined by a simple cuboidal epithelium. The follicle cells surround a follicle cavity that holds a viscous *colloid*, a fluid containing large quantities of dissolved proteins. A network of capillaries surrounds each follicle, delivering nutrients and regulatory hormones to the glandular cells and accepting their secretory products and metabolic wastes. The follicle cells synthesize a globular protein called **thyroglobulin** (thī-rō-GLOB-ū-lin) and secrete it into the colloid of the thyroid follicles. Thyroglobulin molecules contain the amino acid tyrosine, the building block of thyroid hormones.

Simple cuboidal epithelium of follicle

Thyroid follicle

Thyroglobulin in colloid

**Section of thyroid gland**  LM × 260

A second population of endocrine cells lies sandwiched between the follicle cells and their basement membranes. These large, pale cells—called **C (clear) cells**—produce the hormone **calcitonin (CT)**, which helps to regulate $Ca^{2+}$ concentrations in body fluids.

 **REVIEW**

**A.** What thyroid hormone aids in calcium regulation?

**3** This figure shows the continuous process by which thyroid hormones are produced and then stored within thyroglobulin in thyroid follicles. Thyroid hormones are released into and removed from the circulation as needed.

Follicle cavity

**? REVIEW**

**B.** Name the hormones of the thyroid gland.

**3** Tyrosine molecules with attached iodine atoms become linked, forming molecules of thyroid hormones that remain in thyroglobulin. The hormone **thyroxine** (thī-ROK-sēn), also called *tetraiodothyronine* or **$T_4$**, contains four iodine atoms. A related molecule called **triiodothyronine ($T_3$)** contains three iodine atoms.

**2** The iodide ions diffuse to the apical surface of each follicle cell, where they lose an electron and are converted to an iodine atom ($I^0$) by the enzyme *thyroid oxidase*. This reaction also attaches one or two iodine atoms to the tyrosine portions of a thyroglobulin molecule within the follicle cavity.

FOLLICLE CAVITY

Thyroglobulin (contains $T_3$ and $T_4$)

Thyroglobulin

Iodine atoms ($I^0$)

Other amino acids

Tyrosine

Diffusion

FOLLICLE CELL

TSH-sensitive carrier

**Start** ▶ **1**

CAPILLARY

Iodide ions ($I^-$)

$T_4$ & $T_3$

**4** Follicle cells remove thyroglobulin from the follicle cavity by endocytosis.

Endocytosis

**5** Lysosomal enzymes break the thyroglobulin down, and the released amino acids and thyroid hormones enter the cytoplasm. The amino acids are then recycled and used to synthesize more thyroglobulin.

$T_4$ $T_3$

Diffusion

**6** The released molecules of $T_3$ and $T_4$ diffuse across the basement membrane and enter the bloodstream. About 90 percent of all thyroid secretions is $T_4$. $T_3$ is secreted in comparatively small amounts. However, its metabolic effects are much stronger than those of $T_4$.

**1** Iodide ions are absorbed from the diet and delivered by the bloodstream to the thyroid gland. TSH-sensitive carrier proteins in the basement membrane of the follicle cells actively transport iodide ions ($I^-$) into the cytoplasm.

**7** Roughly 75 percent of the $T_4$ molecules and 70 percent of the $T_3$ molecules entering the bloodstream become attached to transport proteins called **thyroid-binding globulins (TBGs)**. Most of the rest of the $T_4$ and $T_3$ in circulation is attached to **transthyretin** or to *albumin*, one of the plasma proteins. The transport proteins release thyroid hormones only gradually, and the bound thyroid hormones represent a substantial reserve: The bloodstream normally contains more than a week's supply of thyroid hormones.

## Effects of Thyroid Hormones on Peripheral Tissues

- Increased rates of oxygen consumption and energy consumption; in children, may cause a rise in body temperature
- Increased heart rate and force of contraction; generally results in a rise in blood pressure
- Increased sensitivity to sympathetic stimulation
- Maintenance of normal sensitivity of respiratory centers to changes in oxygen and carbon dioxide concentrations
- Stimulation of red blood cell formation and thus enhanced oxygen delivery
- Stimulation of activity in other endocrine tissues
- Accelerated turnover of minerals in bone

 **INTEGRATION**

**C.** Why do signs and symptoms of decreased thyroxine concentration not appear until about a week after a thyroidectomy (surgical removal of the thyroid gland)?

 **LEARNING OUTCOME**

Describe the location and structure of the thyroid gland, identify the hormones it produces, and specify the functions of those hormones.

# Parathyroid hormone, produced by the parathyroid glands, is the primary regulator of blood calcium ion levels

**1** Two pairs of **parathyroid glands** are embedded in the posterior surfaces of the thyroid gland. All together, the four parathyroid glands weigh a mere 1.6 g (0.06 oz).

**? REVIEW**

**A.** Describe the locations of the parathyroid glands.

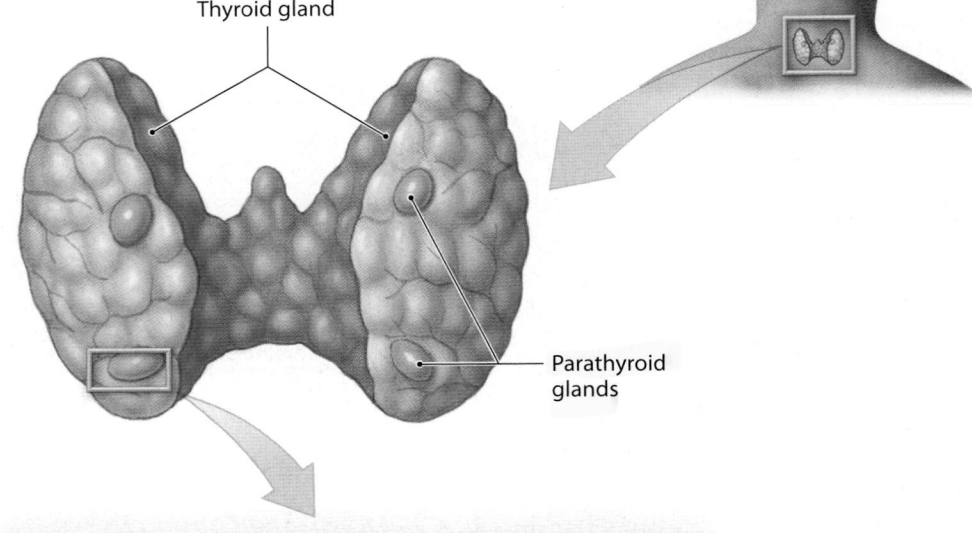

Thyroid gland

Parathyroid glands

**2** The parathyroid glands have at least two cell populations. The **parathyroid (principal) cells** produce parathyroid hormone. The functions of the other cells, called *parathyroid oxyphil cells*, are unknown. They do not appear until after puberty, and then their numbers increase with age. Like the C cells of the thyroid gland, the parathyroid cells monitor the circulating concentration of calcium ions. When the $Ca^{2+}$ concentration of the blood falls below normal, the parathyroid cells secrete **parathyroid hormone (PTH)**. The net result of PTH secretion is an increase in $Ca^{2+}$ concentration in body fluids.

Blood vessel

A dense fibrous capsule separates the cells of the parathyroid gland from those of the thyroid gland.

Thyroid follicles

**Parathyroid gland**     LM × 94

Parathyroid (principal) cells

Oxyphil cells

**Parathyroid cells and oxyphil cells**     LM × 600

16

**3** Parathyroid hormone and calcitonin (from the thyroid) have opposing effects on levels of calcium ions in body fluids. However, in healthy adults, PTH, aided by calcitriol secreted by the kidneys, is the primary regulator of blood calcium level (see Module 6.11). Removal of the thyroid gland seldom affects calcium ion homeostasis because dietary intake and metabolic demand are so closely balanced that increased blood calcium levels are very rare. However, calcitonin can be administered clinically to treat several metabolic disorders that cause increased calcium levels and excessive bone formation.

Secrete calcitonin → **Effectors** Kidneys, digestive tract → *respond with* → Increased excretion of calcium by kidneys / Calcium absorption from digestive tract prevented

**Receptors** C cells in thyroid gland

Homeostasis **DISTURBED BY** INCREASING blood calcium level

**STIMULUS**

Homeostasis **RESTORED BY** DECREASING blood calcium level

**RESTORED**

# HOMEOSTASIS
### NORMAL BLOOD CALCIUM RANGE
### (8.5–11 mg/dL)

Homeostasis **DISTURBED BY** DECREASING blood calcium level

**STIMULUS**

Homeostasis **RESTORED BY** INCREASING blood calcium level

**RESTORED**

**Receptors** Principal cells in parathyroid glands

Secrete parathyroid hormone (PTH) → **Effectors** Kidneys, bone, digestive tract → *respond with* → Increased reabsorption of calcium by kidneys / Calcium release from bone / Increased calcitriol production by kidneys causes $Ca^{2+}$ absorption by digestive tract

**16**

## Effects of Parathyroid Hormone on Peripheral Tissues

- PTH mobilizes calcium from bone by affecting osteoblast and osteoclast activity. PTH stimulates osteoblasts to secrete a growth factor known as RANKL. Osteoclasts have no PTH receptors, but both precursor and mature osteoclasts have RANKL receptors. This growth factor results in an increase in osteoclasts and osteoclast activity. With more osteoclasts, the rates of mineral turnover and $Ca^{2+}$ release accelerate. As bone matrix erodes, blood $Ca^{2+}$ rises.

- PTH enhances the reabsorption of $Ca^{2+}$ by the kidneys, reducing urinary losses.

- PTH stimulates the formation and secretion of calcitriol by the kidneys. In general, the effects of calcitriol complement or enhance those of PTH, but calcitriol also enhances $Ca^{2+}$ and $PO_4^{3-}$ absorption by the digestive tract.

**? REVIEW**

**B.** Explain how parathyroid hormone increases blood calcium level.

**C.** Decreased blood calcium level would result in increased secretion of which hormone?

 **LEARNING OUTCOME**

Describe the location of the parathyroid glands, and identify the functions of the hormone they produce.

# The adrenal hormones are involved in metabolic regulation, electrolyte balance, and stress responses

**1** A yellow, pyramid-shaped **adrenal gland**, or *suprarenal* (sū-pra-RĒ-nal; *supra-*, above + *ren*, kidney) *gland*, sits on the superior border of each kidney. The adrenal glands are retroperitoneal, as are the kidneys, and only their anterior surfaces are covered by a layer of parietal peritoneum. Like other endocrine glands, the adrenal glands are richly supplied with blood vessels. Each adrenal gland has three parts: a superficial *capsule*, an *adrenal cortex*, and an inner *adrenal medulla*.

**2** The **adrenal cortex** is yellowish as a result of stored lipids, especially cholesterol and various fatty acids. The adrenal cortex produces more than two dozen steroid hormones, collectively called **corticosteroids**. Like other steroid hormones, corticosteroids exert their effects by determining which genes in the nuclei of their target cells are transcribed, and at what rate. The resulting changes in the nature and concentration of enzymes in the cytoplasm affect cellular metabolism. Corticosteroids are vital: If the adrenal glands are destroyed or removed, the person will die unless corticosteroids are administered. The **adrenal medulla** synthesizes epinephrine and norepinephrine.

**The Adrenal Hormones**

**Adrenal gland** LM × 250

Capsule

Cortex

Medulla

Right adrenal gland

Adrenal arteries

Left adrenal gland

Right kidney

Left kidney

Abdominal aorta

Inferior vena cava

16

**3** Deep to the adrenal capsule are three distinct regions, or zones, in the adrenal cortex. Each zone synthesizes specific steroid hormones. Deep to the cortex lies the adrenal medulla, which contains cells similar to those in sympathetic ganglia.

| Region/Zone | Hormones | Primary Targets | Hormonal Effects | Regulatory Control |
|---|---|---|---|---|
| **ADRENAL CAPSULE** | | | | |
| **ADRENAL CORTEX** | | | | |
| The **zona glomerulosa** (glō-mer-ū-LŌ-suh) is the outer region of the adrenal cortex. | **Mineralocorticoids**, primarily **aldosterone** | Kidneys | Aldosterone increases renal reabsorption of $Na^+$ and water, (especially in the presence of ADH). It also accelerates urinary loss of $K^+$. | Mineralocorticoid secretion is stimulated by decreased blood pressure and volume through a kidney enzyme that activates RAAS (see Module 16.15). |
| The **zona fasciculata** (fa-sik-ū-LA-tuh; *fasciculus*, little bundle) is the large, central portion of the adrenal cortex. | **Glucocorticoids** (glū-kō-KOR-ti-koydz) are steroid hormones that affect glucose metabolism. The primary hormones are **cortisol** (KOR-ti-sol), also called *hydrocortisone*, and smaller amounts of the related steroid **corticosterone** (kor-ti-KOS-te-rōn). The liver converts some of the circulating cortisol to **cortisone**, another metabolically active glucocorticoid. | Most cells | Glucocorticoids increase rates of glucose and glycogen formation by the liver. They also stimulate the release of amino acids from skeletal muscles and release of lipids from adipose tissues. They also promote lipid catabolism within peripheral cells. These actions supplement the glucose-sparing effect of growth hormone (see Module 16.5). Cortisol also reduces inflammation (an **anti-inflammatory** effect). | Glucocorticoid secretion is stimulated by ACTH from the anterior lobe of the pituitary gland. |
| The **zona reticularis** (re-tik-ū-LAR-is; *reticulum*, network) forms a narrow band bordering the adrenal medulla. | Small quantities of androgens (male sex hormones) that may be converted to estrogens in the bloodstream | Skin, bones, and other tissues, but minimal effects in normal adults | Adrenal androgens stimulate the development of pubic hair in boys and girls before puberty. | Androgen secretion is stimulated by ACTH. |
| **ADRENAL MEDULLA** | | | | |
| | Epinephrine (E), norepinephrine (NE) | Most cells | Epinephrine and norepinephrine increase cardiac activity, blood pressure, glycogen breakdown, and blood glucose levels. | Epinephrine and norepinephrine secretion is stimulated by sympathetic preganglionic fibers during sympathetic activation. |

16

**? REVIEW**

**A.** Name the regions and zones of an adrenal gland from superficial to deep.

**B.** Cite the hormones secreted by each region of an adrenal gland.

**? INTEGRATION**

**C.** What effect would increased cortisol levels have on blood glucose level?

**Lo LEARNING OUTCOME**

Describe the location, structure, and functions of the adrenal glands, identify the hormones produced, and specify the functions of each hormone.

# The pancreatic islets secrete insulin and glucagon, which regulate glucose use by most cells

**1** The **pancreas** is mostly retroperitoneal and lies in the loop between the inferior border of the stomach and the proximal portion of the small intestine. It is a slender, pale organ with a nodular (lumpy) texture. In adults, the pancreas is 20–25 cm (8–10 in.) long and weighs about 80 g (2.8 oz). The **exocrine pancreas**, roughly 99 percent of the organ's volume, consists of clusters of gland cells and their attached ducts. Together, the gland and duct cells secrete large quantities of an alkaline, enzyme-rich fluid that reaches the lumen of the intestinal tract through one or two **pancreatic ducts**.

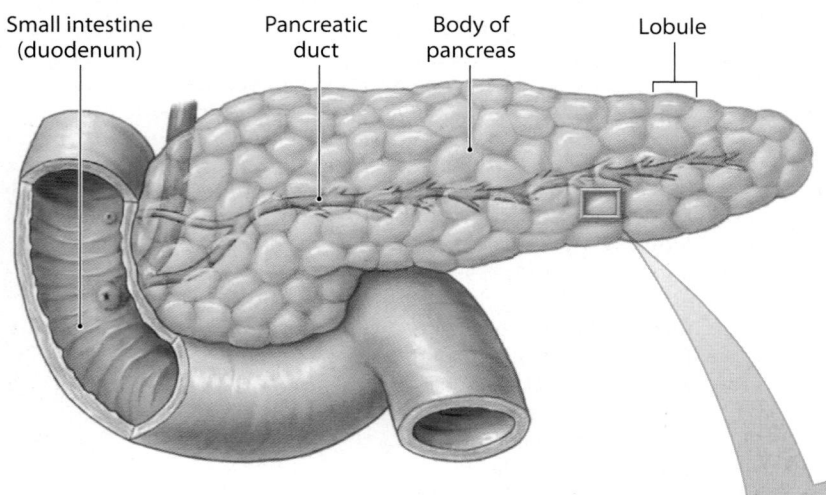

Small intestine (duodenum)  Pancreatic duct  Body of pancreas  Lobule

**2** The **endocrine pancreas** consists of small groups of cells scattered among the exocrine cells. The endocrine clusters are known as **pancreatic islets**, or the *islets of Langerhans* (LANGH-er-hanz). Pancreatic islets account for only about 1 percent of all cells in the pancreas. Nevertheless, a typical pancreas contains about 2 million pancreatic islets, and their secretions are vital to our survival.

The exocrine cells form small clusters that secrete into a lumen continuous with a pancreatic duct. The glandular clusters are called **pancreatic acini** (singular, *acinus*).

**Alpha cells** produce the hormone **glucagon** (GLŪ-ka-gon). Glucagon raises blood glucose levels by increasing the rates of glycogen breakdown and glucose release by the liver.

**Beta cells** produce the hormone **insulin** (IN-suh-lin). Insulin lowers blood glucose levels by increasing the rate of glucose uptake and use by most body cells, and by increasing glycogen synthesis in skeletal muscles and the liver.

Pancreatic islet   Capillaries

**Delta cells** produce a peptide hormone identical to growth hormone–inhibiting hormone (GH–IH). GH–IH suppresses the release of glucagon and insulin by other islet cells and slows the rates of food absorption and enzyme secretion along the digestive tract.

**Pancreatic polypeptide cells** (**PP cells**) produce the hormone *pancreatic polypeptide (PP)*. PP inhibits gallbladder contractions and regulates the production of some pancreatic enzymes. It may also help control the rate of nutrient absorption by the digestive tract.

 **REVIEW**

**A.** Identify the types of cells in the pancreatic islets and the hormones produced by each.

**3** Insulin and glucagon are the primary hormones responsible for the regulation of the blood glucose level. When the blood glucose level increases, beta cells secrete insulin, which then stimulates the transport of glucose across plasma membranes and into target cells. When the blood glucose level decreases, alpha cells secrete glucagon, which stimulates glycogen breakdown and glucose release by the liver.

Secrete insulin → **Effectors** All body cells → *respond with* →

**Receptors** Beta cells in pancreatic islets

Pancreas

- Increased rate of glucose transport into target cells
- Increased rate of glucose use and ATP generation
- Increased conversion of glucose to glycogen
- Increased amino acid absorption and protein synthesis
- Increased triglyceride (fat) synthesis in adipose tissue

Homeostasis **DISTURBED BY** INCREASING blood glucose level

STIMULUS

Homeostasis **RESTORED BY** DECREASING blood glucose level

RESTORED

# HOMEOSTASIS
## NORMAL BLOOD GLUCOSE RANGE
### (70–110 mg/dL)

Homeostasis **DISTURBED BY** DECREASING blood glucose level

STIMULUS

Homeostasis **RESTORED BY** INCREASING blood glucose level

RESTORED

**Receptors** Alpha cells in pancreatic islets

Pancreas

Secrete glucagon → **Effectors** Liver, skeletal muscle, adipose cells → *respond with* →

- Increased breakdown of glycogen to glucose (in liver, skeletal muscle)
- Increased breakdown of fat to fatty acids (in adipose tissue)
- Increased synthesis and release of glucose (in liver)

**16**

**? REVIEW**

**B.** The secretion of which hormone lowers blood glucose concentration?

**? INTEGRATION**

**C.** What is the effect of increased glucagon levels on the amount of glycogen stored in the liver?

**Lo LEARNING OUTCOME**

Describe the location and structure of the pancreas, identify the hormones it produces, and specify the functions of those hormones.

# Module 16.12

# The pineal gland of the epithalamus secretes melatonin, which affects the circadian rhythm

**1** The **pineal gland**, the main part of the epithalamus, lies in the posterior portion of the roof of the third ventricle in the brain. The pineal gland contains neurons, neuroglia, and special secretory cells called **pinealocytes** (pin-Ē-al-ō-sīts). These cells synthesize the hormone **melatonin** from molecules of the neurotransmitter *serotonin*. Collaterals from the visual pathways enter the pineal gland and affect the rate of melatonin production. This rate is lowest during daylight hours and highest at night.

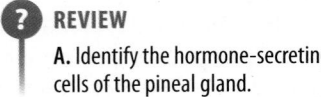
**? REVIEW**
**A.** Identify the hormone-secreting cells of the pineal gland.

Astrocyte

Pinealocytes

**Pineal gland**     LM × 450

**? REVIEW**
**B.** Increased amounts of light would inhibit the production of which hormone?

## Functions of Melatonin in Humans

- **Setting circadian rhythms.** Because pineal activity is cyclical, the pineal gland may also be involved with the maintenance of basic **circadian rhythms**—daily changes in physiological processes that follow a regular day/night pattern. Increased melatonin secretion in darkness has been suggested as a primary cause of *seasonal affective disorder (SAD)*. This condition can develop during the winter in people who live at high latitudes, where sunlight is scarce or lacking. It is characterized by changes in mood, eating habits, and sleeping patterns.

- **Inhibiting reproductive functions.** In some mammals, melatonin slows the maturation of sperm, oocytes, and reproductive organs by reducing the rate of GnRH secretion. The significance of this effect in humans remains unclear, but circumstantial evidence suggests that melatonin may play a role in the timing of human sexual maturation. Melatonin levels in the blood decline at puberty, and pineal tumors that eliminate melatonin production cause premature puberty in young children.

- **Protecting against damage by free radicals.** Melatonin is a very effective antioxidant. It may protect CNS neurons from **free radicals**, such as nitric oxide (NO) or hydrogen peroxide ($H_2O_2$), that may be formed in active neural tissue.

**? REVIEW**
**C.** List three functions suggested for melatonin in humans.

**LEARNING OUTCOME**
Describe the location of the pineal gland, and identify the functions of the hormone that it produces.

## Module 16.13

# Diabetes mellitus is an endocrine disorder characterized by an excessively high blood glucose level

### Diabetes Mellitus

**Diabetes mellitus** (mel-Ī-tus; *mellitum*, honey) is characterized by a glucose concentration high enough to overwhelm the reabsorption capabilities of the kidneys. The presence of abnormally high blood glucose levels is called **hyperglycemia** (hī-per-glī-SĒ-mē-ah). In diabetes mellitus, glucose appears in the urine (**glycosuria**; glī-kō-SOO-rē-a), and urine volume generally becomes excessive (**polyuria**). Diabetes mellitus can be caused by genetic abnormalities or mutations that result in inadequate insulin production, the synthesis of abnormal insulin molecules, or the production of defective insulin-receptor proteins. An estimated 25.8 million people in the United States have some form of diabetes.

 **REVIEW**
**A.** Define *diabetes mellitus*.

### Type 1 Diabetes

**Type 1 diabetes** is characterized by inadequate insulin production by the pancreatic beta cells. People with Type 1 diabetes must receive insulin to live—typically multiple injections daily or continuous infusion through an insulin pump or other device. Type 1 diabetes accounts for only about 5–10 percent of diabetes cases and often develops in children and young adults.

### Type 2 Diabetes

**Type 2 diabetes** is the most common form of diabetes mellitus. Most people with this form of diabetes produce normal amounts of insulin, at least initially, but their tissues do not respond properly—a condition known as insulin resistance. Type 2 diabetes is associated with obesity, and weight loss through diet and exercise can be an effective treatment, especially when coupled with drugs that alter rates of glucose synthesis and release by the liver.

**1** Untreated diabetes mellitus disrupts metabolic activities throughout the body. Clinical problems arise because the tissues involved are experiencing an energy crisis—in essence, most of the tissues are responding as they would during chronic starvation, breaking down lipids and even proteins because they are unable to absorb glucose from their surroundings. Problems involving abnormal changes in blood vessel structure are particularly dangerous.

 **REVIEW**
**B.** Describe the two types of diabetes mellitus.
**C.** Identify some clinical problems associated with diabetes mellitus.

### Clinical Problems Caused by Diabetes Mellitus

The proliferation of capillaries and hemorrhaging at the retina may cause partial or complete blindness. This condition is called **diabetic retinopathy**.

Degenerative blockages in cardiac circulation can lead to early heart attacks. For a given age group, heart attacks are three to five times more likely in people with diabetes than in people who do not have the condition.

Degenerative changes in the kidneys, a condition called **diabetic nephropathy**, can lead to kidney failure.

Abnormal blood flow to nervous tissues is probably responsible for a variety of problems with peripheral nerves, including abnormal autonomic function. As a group, these disorders are termed **diabetic neuropathy**.

Blood flow to the distal portions of the limbs is reduced, and peripheral tissues may be damaged as a result. A reduction in blood flow to the feet, for example, can lead to tissue death, ulceration, infection, and the loss of toes or a major portion of one or both feet.

**Lo LEARNING OUTCOME**
Explain diabetes mellitus: its types, clinical manifestations, and treatments.

**16**

# Hormones and Intercellular Communication

**Concept map:** Use each of the following terms once to fill in the blank boxes to correctly complete the map.

- steroid hormones
- tryptophan derivatives
- glycoproteins
- short polypeptides
- catecholamines
- peptide hormones
- thyroid hormones
- transport proteins
- lipid derivatives
- small proteins
- eicosanoids

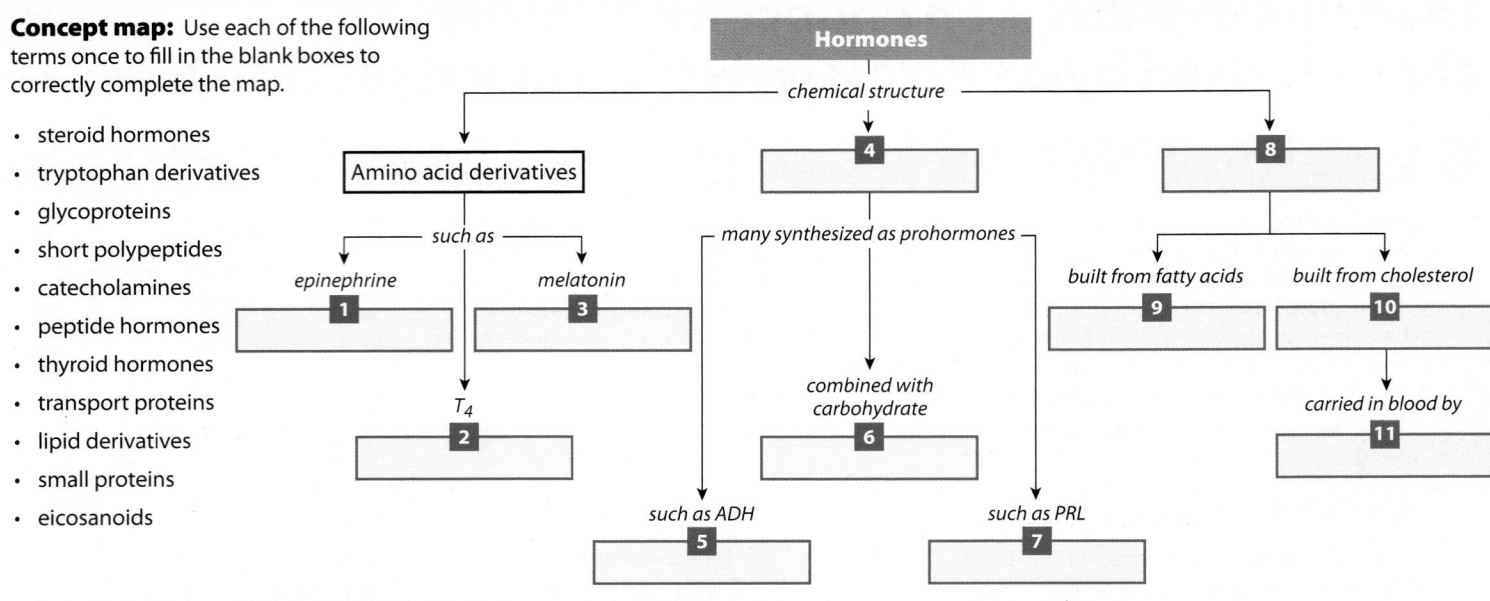

**Matching:** Match each lettered term with the most closely related description.

| | | | |
|---|---|---|---|
| a. | induces ovulation | **12** Calcitonin | **12** _____ |
| b. | secretes androgens | **13** Adrenal medulla | **13** _____ |
| c. | C cells | **14** Prostaglandin | **14** _____ |
| d. | secrete parathyroid hormone | **15** Pituitary gland | **15** _____ |
| e. | sympathetic activation | **16** Second messenger | **16** _____ |
| f. | coordinates local cell activities | **17** Hypothalamus | **17** _____ |
| g. | secretes tropic hormones | **18** Zona reticularis | **18** _____ |
| h. | secretes regulatory hormones | **19** Thyroxine | **19** _____ |
| i. | contains iodine | **20** Luteinizing hormone | **20** _____ |
| j. | cyclic AMP | **21** Principal cells | **21** _____ |

**Short answer:** Identify the endocrine gland—or endocrine cells—based on the major effects produced by its/their secreted hormone(s).

**22** Stimulates and coordinates the immune response

**22** _____

**23** Establishes circadian rhythms

**23** _____

**24** Secretes insulin and regulates glucose uptake and utilization

**24** _____

**25** Controls hormone secretion of the pituitary gland

**25** _____

**26** Regulates red blood cell production and calcium and phosphate absorption by the digestive tract

**26** _____

**27** Regulates mineral balance, metabolic control, and responses to stress

**27** _____

**28** Regulates secretions of adrenal cortex, thyroid gland, and reproductive organs

**28** _____

**29** Affects growth, metabolism, and sexual characteristics

**29** _____

16

## Module 16.14

# Hormones interact to produce coordinated physiological responses

Go to
MasteringA&P™
PhysioEx™ 9.I

Ex 4:
Endocrine
System
Physiology

**1** Extracellular fluids contain a mixture of hormones whose concentrations change daily or even hourly. Because cells have more than one type of receptor, they respond to multiple hormones simultaneously. When a cell receives instructions from two hormones at the same time, there are four possible outcomes.

**? REVIEW**

**A.** Define *synergistic effect*.

### Antagonistic Effects

Two hormones may have **antagonistic** (opposing) **effects**, as in the negative feedback case of insulin and glucagon (see Module 16.11). The net result depends on the balance between the two hormones. In general, when antagonistic hormones are present, the observed effects are weaker than those produced by either hormone acting unopposed.

### Additive Effects

Two hormones may have **additive effects**, so that the net result is greater than the effect that each would produce acting alone. In some cases, the net result is greater than the sum of the hormones' individual effects. This interaction is a **synergistic effect** (sin-er-JIS-tik; *synairesis*, a drawing together). An example is the enhanced glucose-sparing action of GH in the presence of glucocorticoids (see Module 16.10).

### Permissive Effects

One hormone may have a **permissive effect** on another. In such cases, the first hormone is needed for the second to produce its effect. For example, epinephrine does not change the rate of energy consumption in a tissue unless thyroid hormones are also present in normal concentrations.

### Integrative Effects

Two hormones may produce different but complementary effects in specific tissues and organs. These **integrative effects** are important in coordinating the activities of diverse physiological systems. The differing effects of calcitriol and parathyroid hormone on tissues involved in calcium metabolism are an example (see Module 16.9).

In this section, we will consider the ways hormones interact to preserve homeostasis and control short-term or long-term processes.

**? REVIEW**

**B.** Which of the four hormonal effects are involved in a negative feedback response?

**? INTEGRATION**

**C.** What kind of effect do hormones have if they produce different but complementary effects?

**LO LEARNING OUTCOME**

Explain how hormones interact to produce coordinated physiological responses.

16

# Regulation of blood pressure and blood volume involves hormones from primary endocrine organs and from endocrine tissues in the heart and kidneys

The long-term regulation of blood pressure and blood volume involves not only the pituitary and adrenal glands, but also endocrine cells in the heart and kidneys.

**1** The endocrine cells of the heart are located in the heart walls. If blood volume becomes too great, these cells are stretched excessively and begin to secrete **natriuretic peptides** (nā-trē-ū-RET-ik; *natrium*, sodium + *ouresis*, urination). Natriuretic peptides promote the loss of $Na^+$ and water by the kidneys, suppress thirst, and inhibit *renin* release by the kidneys and the secretion of ADH and aldosterone. The net result is a decrease in both blood volume and blood pressure, thereby reducing the stretching of the heart walls.

**? REVIEW**
**A.** Name the hormones secreted by the heart.

Releases natriuretic peptides → **Effectors** Kidneys and blood vessels → *response to natriuretic peptides*

**Receptor** Cardiac muscle cells

Increased $Na^+$ and water loss in urine

Reduced thirst

Inhibition of aldosterone and ADH secretion

Dilation of blood vessels

Decreased blood pressure and blood volume

Homeostasis **DISTURBED BY** INCREASING blood pressure and volume

Homeostasis **RESTORED BY** DECREASING blood pressure and volume

STIMULUS — RESTORED

# HOMEOSTASIS
## NORMAL BLOOD PRESSURE AND VOLUME

Homeostasis **DISTURBED BY** DECREASING blood pressure and volume

STIMULUS — RESTORED

Homeostasis **RESTORED BY** INCREASING blood pressure and volume

**Receptor** Specialized kidney endocrine cells

Erythropoietin (EPO) is released → Increased red blood cell production

Renin is released

Increase in blood pressure and volume

Increased fluid intake and retention

Aldosterone secreted
ADH secreted
Stimulation of thirst

Angiotensinogen (plasma protein) → Angiotensin I — **ACE** → Angiotensin II

**? REVIEW**
**B.** Name the hormone and enzyme secreted by the kidneys.

**2** When blood flow to the kidneys decreases, endocrine cells in the kidneys release a hormone, **erythropoietin (EPO)**, and an enzyme, **renin**, which activates the **renin-angiotensin-aldosterone system (RAAS).** Angiotensin-converting enzyme (ACE) in the lungs modifies angiotensin I to the hormone angiotensin II. The responses to angiotensin II increase fluid intake and fluid retention.

 **LEARNING OUTCOME**
Describe the functions of the hormones produced by the kidneys and the heart.

# Module 16.16

# Normal growth requires the cooperation of many endocrine organs

**1** Several hormones—GH, thyroid hormones, insulin, PTH, calcitriol, and reproductive hormones—are especially important in normal growth, although many others have secondary effects on growth. The circulating levels of these hormones are regulated independently. Every time the hormonal mixture changes, metabolic operations are modified to some degree. The modifications vary in duration and intensity, producing unique growth patterns in different people.

### Insulin

Growing cells need adequate supplies of energy and nutrients. Without insulin, the passage of glucose and amino acids across plasma membranes is drastically reduced or eliminated.

### Parathyroid Hormone and Calcitriol

Parathyroid hormone (PTH) and calcitriol promote the absorption of calcium salts for later deposition in bone. Without adequate levels of both hormones, bones can still enlarge, but they will be poorly mineralized, weak, and flexible.

### Thyroid Hormones

Normal growth requires appropriate levels of thyroid hormones. If these hormones are absent during fetal development or the first year of life, the nervous system will not develop normally, and mental retardation will result. If thyroid hormone levels decline before puberty, normal skeletal development will not continue.

### Reproductive Hormones

The activity of osteoblasts in key locations and the growth of specific cell populations are affected by the presence or absence of reproductive hormones (androgens in males, estrogens in females). These sex hormones stimulate cell growth and differentiation in their target tissues. The targets differ for androgens and estrogens, and the differential growth induced by each accounts for sex-related differences in skeletal proportions and secondary sex characteristics.

| | | | |
| 5 ft | | | |
| 4 ft | | | |
| 3 ft | | | |
| 2 ft | | | |
| 1 ft | | | |

Newborn    Child (6 years)    Adult

### Growth Hormone

| In Children | In Adults |
| --- | --- |
| The effects of GH on protein synthesis and cellular growth are most apparent in children. GH supports their muscular and skeletal development. | In adults, GH helps to maintain normal blood glucose concentrations and to mobilize lipid reserves stored in adipose tissue. It is not the primary hormone involved, however, and an adult with a GH deficiency but normal levels of thyroid hormones, insulin, and glucocorticoids will have no physiological problems. |

**? REVIEW**

**A.** Identify several hormones necessary for normal growth and development.

 **LEARNING OUTCOME**

Explain the roles of other endocrine organs and hormones in normal growth and development.

16

# Module 16.17

# The stress response is a predictable response to any significant threat to homeostasis

Any condition—whether it is physical or emotional—that threatens homeostasis is a form of **stress**. Many stresses are opposed by specific homeostatic adjustments. For example, a drop in body temperature leads to shivering or changes in the pattern of blood flow, which can restore body temperature to normal. In addition, the body has a general response to stress that can occur while other, more specific responses are under way. Exposure to a wide variety of stress-causing factors will produce the same general pattern of hormonal and physiological adjustments. These responses are part of the **stress response**, also known as the **general adaptation syndrome** (**GAS**). The stress response is divided into three phases: (1) alarm, (2) resistance, and (3) exhaustion.

## Alarm Phase ("Fight or Flight")

**1** During the **alarm phase**, an immediate response to the stress occurs. This response is directed by the sympathetic division of the autonomic nervous system. In the alarm phase, (1) energy reserves are mobilized, mainly in the form of glucose, and (2) the body prepares to deal with the stress-causing factor by "fight or flight" responses. Epinephrine is the dominant hormone of the alarm phase. Its secretion is part of a generalized sympathetic activation.

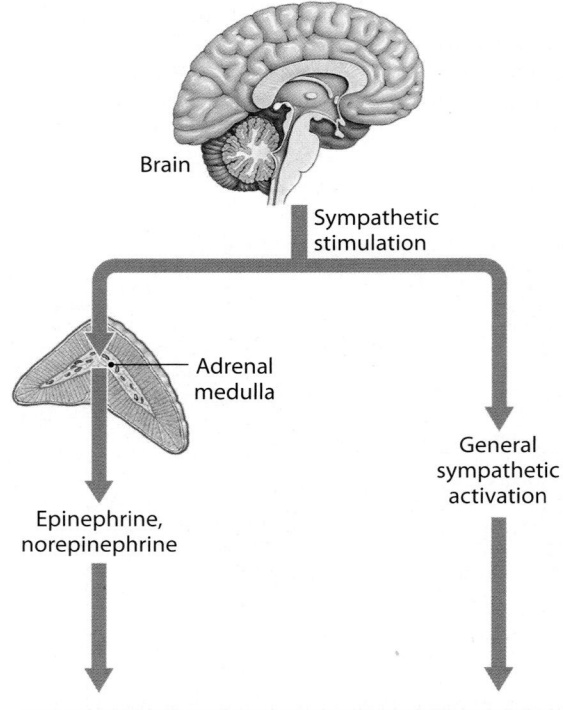

Brain

Sympathetic stimulation

Adrenal medulla

General sympathetic activation

Epinephrine, norepinephrine

**Immediate Short-Term Responses to Crises**

- Increased mental alertness
- Increased energy use by all cells
- Mobilization of glycogen and lipid reserves
- Changes in circulation
- Reduction in digestive activity and urine production
- Increased sweat gland secretion
- Increased heart rate and respiratory rate

**REVIEW**

**A.** List the three phases of the stress response.

16

## Resistance Phase

**2** If a stress lasts longer than a few hours, the person enters the **resistance phase** of the stress response. Glucocorticoids are the dominant hormones of the resistance phase. Epinephrine, GH, and thyroid hormones are also involved. Energy demands in the resistance phase remain higher than normal because of the combined effects of these hormones. Neural tissue has a high demand for energy, and neurons must have a reliable supply of glucose. Glycogen reserves are adequate to maintain normal glucose concentrations during the alarm phase but are nearly exhausted after several hours. The hormones of the resistance phase mobilize lipids and amino acids, thus shifting tissue metabolism away from glucose, so that whatever glucose becomes available it can be used by neural tissues.

Brain

Kidney

Sympathetic stimulation

ACTH

Pancreas

Renin-angiotensin-aldosterone system

Adrenal cortex

Mineralocorticoids (with ADH)

Glucocorticoids

Glucagon

Growth hormone

### Long-Term Metabolic Adjustments

- Mobilization of remaining energy reserves: Lipids are released by adipose tissue; amino acids are released by skeletal muscle
- Conservation of glucose: Peripheral tissues (except nervous) break down lipids to obtain energy
- Increased blood glucose concentrations: Liver synthesizes glucose from other carbohydrates, amino acids, and lipids
- Conservation of salts and water, loss of $K^+$ and $H^+$

**? REVIEW**

**B.** Describe the resistance phase.

## Exhaustion Phase

**3** The body's lipid reserves are sufficient to maintain the resistance phase for a period of weeks or even months. But the resistance phase cannot be sustained indefinitely. When the resistance phase ends, homeostatic regulation breaks down, and the **exhaustion phase** begins. Unless corrective actions are taken almost immediately, the failure of one or more organ systems will prove fatal. Mineral imbalances contribute to the existing problems with major systems. The production of aldosterone throughout the resistance phase results in a conservation of $Na^+$ at the expense of $K^+$. As the body's $K^+$ content declines, a variety of cells—notably neurons and muscle fibers—begin to malfunction. Although a single cause (such as heart failure) may be listed as the cause of death, the underlying problem is the body's inability to sustain the endocrine and metabolic adjustments of the resistance phase.

### Factors That Can Trigger the Exhaustion Phase

- Exhaustion of lipid reserves and the breakdown of structural proteins as the body's primary energy source, damaging vital organs
- Infections that develop as a result of the suppression of inflammation and of the immune response, a secondary effect of the glucocorticoids that are essential to the metabolic activities of the resistance phase
- Cardiovascular damage and complications resulting from the ADH- and aldosterone-related elevations in blood pressure and blood volume
- Inability of the adrenal cortex to continue producing glucocorticoids, which results in a failure to maintain acceptable blood glucose concentration
- Failure to maintain adequate fluid and electrolyte balance

**? INTEGRATION**

**C.** During which phase of the general adaptation syndrome is there a collapse of vital systems?

**Lo LEARNING OUTCOME**

Define the general adaptation syndrome, and compare homeostatic responses with stress responses.

16

# Module 16.18

# Overproduction or underproduction of hormones can cause endocrine disorders

Endocrine disorders may develop for a variety of reasons, including abnormalities in the endocrine gland, the endocrine or neural regulatory mechanisms, or the target tissues. For example, a hormone level may rise because its target organs are becoming less responsive, because a tumor has formed among the gland cells, or because something has interfered with the normal feedback control mechanism. When naming endocrine disorders, clinicians use the prefix *hyper-* when referring to excessive hormone production and *hypo-* when referring to inadequate hormone production. Endocrine tumors can result in **hypersecretion**, but its incidence is relatively rare. Most endocrine disorders are the result of problems within the endocrine gland that result in **hyposecretion**, the production of inadequate levels of a particular hormone.

 **REVIEW**

**A.** Define the prefixes *hyper-* and *hypo-* in the context of endocrine disorders.

**1**    **Acromegaly** results from the overproduction of growth hormone after the epiphyseal plates have fused. Bone shapes change, and cartilaginous areas of the skeleton enlarge. Signs of acromegaly include broad facial features and enlarged lower jaw.

## Common Causes of Hormone Hyposecretion

| Metabolic Factors | Physical Damage | Congenital Disorders |
|---|---|---|
| Hyposecretion may result from a deficiency in some key substance needed to synthesize the hormone. For example, hypothyroidism can be caused by inadequate levels of iodine in the diet. | Any condition that interrupts the normal circulatory supply to endocrine cells or that physically damages those cells may cause them to become inactive immediately or after an initial surge of hormone release. | A person may be unable to produce normal amounts of a particular hormone because (1) the gland is too small, (2) the required enzymes are abnormal, (3) the receptors that trigger secretion are relatively insensitive, or (4) the gland cells lack the receptors normally involved in stimulating secretory activity. |

**2**    An enlarged thyroid gland, or **goiter**, is usually associated with thyroid hyposecretion due to nutritional iodine deficiency.

 **REVIEW**

**B.** Identify three common causes of hormone hyposecretion.

Endocrine abnormalities can also be caused by the presence of abnormal hormonal receptors in target tissues. In such cases, the gland and the regulatory mechanisms involved are normal, but the peripheral cells are unable to respond to the circulating hormone. The best example of this type of abnormality is type 2 diabetes, in which peripheral cells do not respond normally to insulin.

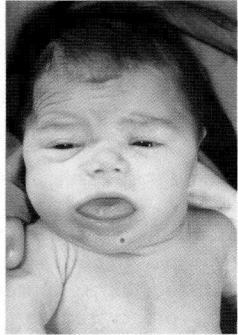

**3**    **Infantile hypothyroidism** is a congenital disorder due to thyroid hormone deficiency. It is characterized by mental disability, puffy face, and a thick tongue.

16

## An Overview of Endocrine Disorders

| Hormone | Results of Under-production or Tissue Insensitivity | Principal Signs and Symptoms | Results of Over-production or Tissue Hypersensitivity | Principal Signs and Symptoms |
|---|---|---|---|---|
| **Growth hormone (GH)** | Pituitary growth failure | Delayed growth, abnormal fat distribution, low blood glucose hours after a meal | Gigantism, acromegaly | Excessive growth |
| **Antidiuretic hormone (ADH)** | Diabetes insipidus | Polyuria, dehydration, thirst | Syndrome of inappropriate ADH secretion (SIADH) | Increased body weight and water content |
| **Thyroxine ($T_4$) and triiodothyronine ($T_3$)** | Hypothyroidism, infantile hypothyroidism, myxedema (in adults) | Low metabolic rate, low body temperature, impaired physical and mental development | Hyperthyroidism, Graves' disease | High metabolic rate and body temperature |
| **Parathyroid hormone (PTH)** | Hypoparathyroidism | Muscular weakness, neurological problems, formation of dense bones, tetany (muscular spasms due to low blood $Ca^{2+}$ level) | Hyperparathyroidism | Neurological, mental, and muscular problems due to high blood $Ca^{2+}$ concentrations; weak and brittle bones |
| **Insulin** | Diabetes mellitus (type 1) | High blood glucose, impaired glucose utilization, dependence on lipids for energy, glycosuria | Excess insulin production (or administering too much insulin) | Low blood glucose levels, possibly causing coma |
| **Mineralocorticoids (MCs)** Example: aldosterone | Hypoaldosteronism | Polyuria, low blood volume, high blood $K^+$ and low blood $Na^+$ concentrations | Aldosteronism | Increased body weight due to $Na^+$ and water retention, low blood $K^+$ concentration |
| **Glucocorticoids (GCs)** Example: cortisol | Addison's disease | Inability to tolerate stress, mobilize energy reserves, or maintain normal blood glucose concentrations | Cushing's disease | Excessive breakdown of tissue proteins and lipid reserves, impaired glucose metabolism |
| **Epinephrine (E) and norepinephrine (NE)** | None identified | None identified | Pheochromocytoma (benign tumor of the adrenal medulla) | High metabolic rate, blood pressure, and heart rate; increased blood glucose levels |
| **Estrogens (females)** | Hypogonadism | Sterility, lack of secondary sex characteristics | Adrenogenital syndrome | Overproduction of androgens by zona reticularis of adrenal cortex; leads to masculinization |
|  |  |  | Precocious puberty | Premature sexual maturation and related behavioral changes |
| **Androgens (males)** | Hypogonadism | Sterility, lack of secondary sex characteristics | Adrenogenital syndrome (gynecomastia) | Abnormal production of estrogens, sometimes due to adrenal or interstitial cell tumors; leads to breast enlargement |
|  |  |  | Precocious puberty | Premature sexual maturation and related behavioral changes |

**16**

**4** **Addison's disease** is caused by hyposecretion of corticosteroids, especially glucocorticoids. Pigment changes result from stimulation of melanocytes by ACTH, which is similar in structure to MSH.

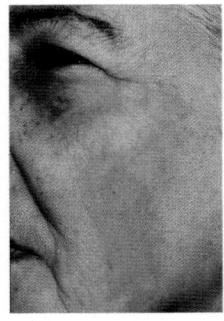

**5** **Cushing's disease** is caused by hypersecretion of glucocorticoids. Lipid reserves are mobilized, and adipose tissue accumulates in the cheeks and at the base of the neck.

 **REVIEW**

**C.** What condition is characterized by increased body weight due to $Na^+$ and water retention and a low blood $K^+$ concentration?

**Lo** **LEARNING OUTCOME**

Describe key endocrine disorders, citing their characteristic signs and symptoms.

# Hormones and System Integration

**Labeling:** Write the phrases below in the boxes to complete the diagram of the homeostatic regulation of blood pressure and volume.

- specialized kidney endocrine cells
- erythropoietin (EPO) released
- cardiac muscle cells
- aldosterone secreted
- reduced thirst
- decreasing blood pressure and volume
- increasing blood pressure and volume
- renin released
- ADH secreted
- release of natriuretic peptides
- increased Na⁺ and water loss in urine
- increased red blood cell production

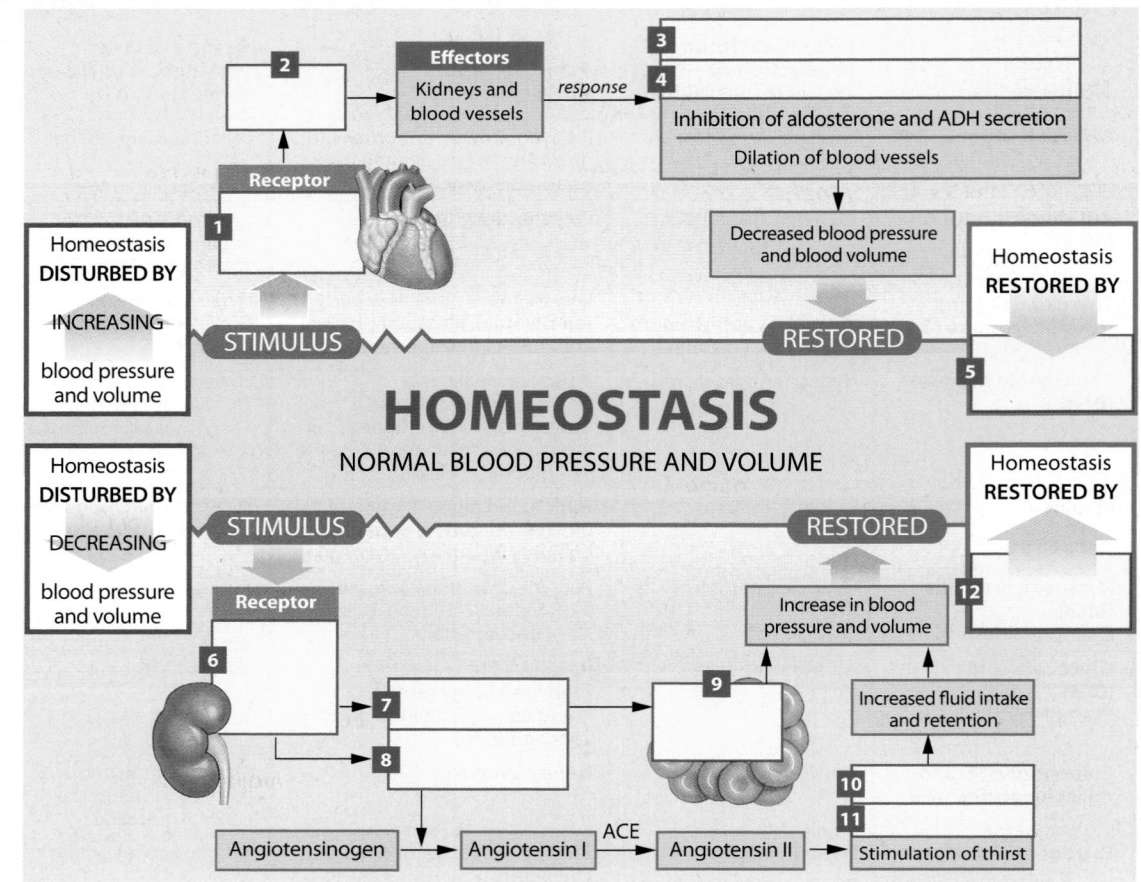

**16**

**Matching:** Match each lettered term with the most closely related description.

a. sympathetic activation
b. decreased blood pressure and volume
c. insulin and glucagon
d. GH and glucocorticoids
e. increased blood pressure and volume
f. homeostasis threat
g. glucocorticoids
h. PTH and calcitriol
i. gigantism
j. protein synthesis

| 13 | Antagonistic effect | 13 _____ |
| 14 | Resistance phase | 14 _____ |
| 15 | Alarm phase | 15 _____ |
| 16 | Renin and EPO effect | 16 _____ |
| 17 | Growth hormone (GH) | 17 _____ |
| 18 | Additive effect | 18 _____ |
| 19 | Excessive GH in children | 19 _____ |
| 20 | Natriuretic peptides effect | 20 _____ |
| 21 | Stress | 21 _____ |
| 22 | Integrative effect | 22 _____ |

**Section integration:** Describe, and give an example of, the four possible effects that may occur when a cell receives instructions from two different hormones.

23 _____

_____

# Study Outline

## SECTION 1 • Hormones and Intercellular Communication

### Module 16.1

### The nervous and endocrine systems release chemical messengers that bind to target cells    p. 595

1. Cells communicate in a variety of ways including through gap junctions, extracellular fluid (by paracrines and autocrines), the bloodstream (by hormones), or across synapses (by neurotransmitters).

2. The nervous and endocrine systems both rely on the release of chemicals that bind to receptors on target cells, share many chemical messengers, are regulated by negative feedback mechanisms, and preserve homeostasis.

### Module 16.2

### Hormones may be amino acid derivatives, peptides, or lipid derivatives    p. 596

3. Hormones and paracrines can be grouped as **amino acid derivatives**, **peptide hormones**, and **lipid derivatives**.

4. Amino acid derivatives include thyroid hormones and **catecholamines**.

5. Peptide hormones are synthesized as **prohormones** that are converted to active hormones.

6. Lipid derivatives are built from either fatty acids (**eicosanoids**) or cholesterol (**steroid hormones**).

### Module 16.3

### The endocrine system includes organs and tissues with primary and secondary hormone-secreting roles    p. 597

7. The **endocrine system** organs include the **hypothalamus**, **pituitary gland**, **thyroid gland**, **adrenal glands**, **pancreas (pancreatic islets)**, **pineal gland**, and **parathyroid glands**.

8. Organs with secondary endocrine functions are the heart, thymus, digestive tract, kidneys, and gonads.

### Module 16.4

### Hormones affect target cells after binding to receptors in the plasma membrane, cytoplasm, or nucleus    p. 598

9. Each cell has receptors for different hormones. Cells in different tissues have different combinations of receptors.

10. A hormone binding to plasma membrane receptors does not have a direct effect on cellular activities. This hormone, or **first messenger**, requires a **second messenger**.

11. The two most important second messengers are **cyclic AMP (cAMP)** and calcium ions ($Ca^{2+}$).

12. Steroid hormones diffuse across the plasma membrane and bind to receptors in the cytoplasm or nucleus.

13. Thyroid hormones are transported across the plasma membrane by carrier-mediated processes.

### Module 16.5

### The hypothalamus exerts direct or indirect control over the activities of many endocrine organs    p. 600

14. The **hypothalamus** regulates the nervous and endocrine systems through three mechanisms: synthesizing **antidiuretic hormone (ADH)** and **oxytocin (OXT)** and transporting them to the posterior lobe of the pituitary; secreting **regulatory hormones** that control endocrine cells in the anterior lobe of the pituitary gland; and controlling the endocrine cells of the adrenal medulla.

15. The **hypophyseal portal system** ensures that hypothalamic hormones reach their target cells in the anterior lobe of the pituitary gland before being diluted by the general circulation.

16. Hypothalamic regulatory hormones can be either a **releasing hormone** or an **inhibiting hormone**. Individually or in combination, these hormones control endocrine cells in the anterior lobe of the pituitary gland.

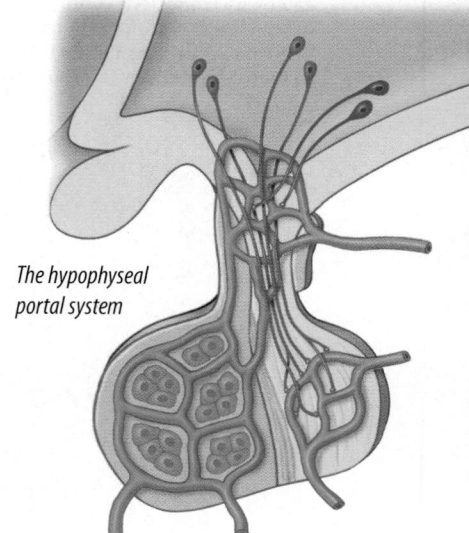

*The hypophyseal portal system*

### Module 16.6

### The anterior lobe of the pituitary gland produces and releases 7 tropic hormones, while the posterior lobe releases 2 hormones    p. 602

17. The **pituitary gland** is connected to the hypothalamus by the **infundibulum** and lies within the sella turcica of the sphenoid bone.

18. The pituitary gland releases nine hormones: seven from the **anterior lobe of the pituitary (adenohypophysis)**, two from the **posterior lobe of the pituitary (neurohypophysis)**.

19. The hormones of the anterior lobe of the pituitary are **thyroid-stimulating hormone (TSH)**, **adrenocorticotropic hormone (ACTH)**, **follicle-stimulating hormone (FSH)**, **luteinizing hormone (LH)**, **growth hormone (GH)**, **prolactin (PRL)**, and **melanocyte-stimulating hormone (MSH)**.

20. The hormones of the posterior lobe of the pituitary are **antidiuretic hormone (ADH)**, or vasopressin, and **oxytocin (OXT)**.

### Module 16.7

### Negative feedback mechanisms control the secretion rates of the hypothalamus and the pituitary gland    p. 604

21. Hormone secretion is controlled through negative feedback.

22. The hypothalamus produces a releasing hormone that triggers the release of a hormone from the anterior lobe of the pituitary gland.

23. The pituitary hormone stimulates the release of a second hormone by the target organ. This second hormone suppresses secretion of the hypothalamic and pituitary hormones.

### Module 16.8

### The thyroid gland contains follicles and requires iodine to produce hormones that stimulate tissue metabolism    p. 606

24. The **thyroid gland** is located inferior to the thyroid cartilage. The **isthmus** connects its two lobes.

25. The thyroid gland contains **thyroid follicles** that produce **thyroglobulin**, which is used to make thyroid hormones.

**16**

26. This process produces thyroid hormones: Iodine ions from the diet are absorbed and delivered to the thyroid by the bloodstream; these ions diffuse to the apical surface of each follicle cell and are converted to iodine atoms; thyroglobulin produces the thyroid hormones **thyroxine ($T_4$) and triiodothyronine ($T_3$)**; thyroglobulin is removed from a follicle cavity and recycled; $T_3$ and $T_4$ diffuse into the bloodstream, where the majority of thyroid hormones are bound to **thyroid-binding globulins (TBGs)**. Any remaining $T_4$ and $T_3$ is attached to **transthyretin** or to *albumin* (a plasma protein).

*The thyroid gland*

27. The effects of thyroid hormones on peripheral tissues include increased oxygen and energy consumption, increased heart rate and force of contraction, increased blood pressure, increased sensitivity to sympathetic stimulation, stimulation of red blood cell formation, increased activity of other endocrine tissues, and accelerated turnover of bone minerals.

### Module 16.9

### Parathyroid hormone, produced by the parathyroid glands, is the primary regulator of blood calcium ion levels    p. 608

28. Two pairs of **parathyroid glands** are located on the posterior surface of the thyroid gland.

29. **Parathyroid (principal) cells** produce parathyroid hormone and monitor the circulating concentration of calcium ions.

30. When $Ca^{2+}$ decreases below normal, the parathyroid (principal) cells secrete **parathyroid hormone (PTH)**, resulting in increased $Ca^{2+}$ concentration in body fluids.

31. PTH functions by increasing osteoclast number and activity, enhancing $Ca^{2+}$ reabsorption in the kidneys, and stimulating formation and release of calcitriol from the kidney, which increases calcium reabsorption by the digestive tract.

### Module 16.10

### The adrenal hormones are involved in metabolic regulation, electrolyte balance, and stress responses    p. 610

32. An **adrenal gland**, or *suprarenal gland*, sits on each kidney.

33. The adrenal cortex releases more than two dozen steroid hormones called **corticosteroids**. These hormones determine which genes are transcribed in the nuclei of their target cells, and they affect cellular metabolism.

34. The adrenal medulla synthesizes epinephrine and norepinephrine.

### Module 16.11

### The pancreatic islets secrete insulin and glucagon, which regulate glucose use by most cells    p. 612

35. The **pancreas** is retroperitoneal and lies in the loop between the inferior border of the stomach and the proximal small intestine.

36. The **exocrine pancreas** comprises approximately 99 percent of the organ's volume. It secretes an alkaline, enzyme-rich fluid into the digestive tract through **pancreatic ducts**.

37. The **endocrine pancreas** consists of small groups of endocrine cell clusters called **pancreatic islets**, or *islets of Langerhans*.

38. Pancreatic islets contain **alpha cells** that produce the hormone **glucagon** (which raises blood glucose), and **beta cells** that produce the hormone **insulin** (which lowers blood glucose).

39. **Delta cells** produce growth hormone–inhibiting hormone, and **pancreatic polypeptide cells (PP cells)** produce the hormone *pancreatic polypeptide (PP)*.

### Module 16.12

### The pineal gland of the epithalamus secretes melatonin, which affects the circadian rhythm    p. 614

40. The **pineal gland** lies in the posterior portion of the roof of the third ventricle in the brain.

41. Its **pinealocytes** secrete **melatonin**, which sets basic **circadian rhythms** and may function to inhibit reproductive functions and protect against damage by **free radicals**.

### Module 16.13

### Diabetes mellitus is an endocrine disorder characterized by an excessively high blood glucose level    p. 615

42. **Diabetes mellitus** is characterized by high blood glucose concentration (**hyperglycemia**), glucose in the urine (**glycosuria**), and excessive urine volume (**polyuria**).

43. **Type 1 diabetes** is characterized by inadequate insulin production by the pancreatic beta cells. People who have the disorder must receive insulin by injection or pump.

44. **Type 2 diabetes**, the most common form of diabetes mellitus, is the result of insulin resistance. It is associated with obesity.

45. Clinical problems caused by diabetes include blindness from **diabetic retinopathy**, heart attack from cardiac circulation blockage, kidney damage from **diabetic nephropathy**, problems with peripheral nerves from **diabetic neuropathy**, and a reduction in peripheral blood flow that can lead to foot amputation.

## SECTION 2 • Hormones and System Integration

### Module 16.14

**Hormones interact to produce coordinated physiological responses**   p. 617

46. Cells never respond to only one hormone; they respond to multiple hormones simultaneously.

47. When a cell receives instructions from multiple hormones, there are four possible outcomes: **antagonistic effects**, **additive (synergistic) effects**, **permissive effects**, or **integrative effects**.

### Module 16.15

**Regulation of blood pressure and blood volume involves hormones from primary endocrine organs and from endocrine tissues in the heart and kidneys**   p. 618

48. Blood pressure and volume are regulated not only by the pituitary gland and adrenal glands, but also by endocrine cells in the heart and kidney.

49. If blood volume is too high, the endocrine cells located in the heart walls secrete **natriuretic peptides**. The result is a decrease in blood volume and pressure by the loss of water and Na$^+$ at the kidneys, an inhibition of renin release, and the secretion of ADH and aldosterone.

50. If blood pressure or volume is too low, blood flow to the kidney decreases, endocrine cells in the kidney release the hormone **erythropoietin**, which increases red blood cell production, and the enzyme **renin**, which activates the **renin-angiotensin-aldosterone system (RAAS)**. Responses to angiotensin II from RAAS increase fluid intake and retention.

### Module 16.16

**Normal growth requires the cooperation of many endocrine organs**   p. 619

51. Normal growth is regulated by hormones including insulin, parathyroid hormone, calcitriol, thyroid hormones, reproductive hormones, and growth hormone.

### Module 16.17

**The stress response is a predictable response to any significant threat to homeostasis**   p. 620

52. **Stress** is a physical or emotional condition that threatens homeostasis.

*Stress is a physical or emotional condition that threatens homeostasis*

53. Stress-causing factors produce a pattern of responses called the **stress response** or the **general adaptation syndrome (GAS)**.

54. The three phases of the stress response are the **alarm phase** ("fight or flight"), the **resistance phase**, and the **exhaustion phase**.

### Module 16.18

**Overproduction or underproduction of hormones can cause endocrine disorders**   p. 622

55. Endocrine disorders may develop for a variety of reasons, including abnormalities of the endocrine gland, regulatory mechanisms, or the target tissues.

56. Endocrine disorders named with the prefix *hyper-* refer to excessive hormone production, or **hypersecretion**.

57. Endocrine disorders named with the prefix *hypo-* refer to inadequate hormone production, or **hyposecretion**.

# Chapter Review Questions

**True/False:**  Indicate whether each statement is true or false.

1   Intercellular communication through extracellular fluid is a characteristic of paracrines.

2   Catecholamines are lipid-derivative hormones.

3   Hormones that bind to receptors in the plasma membrane require a second messenger.

4   Rising levels of blood calcium cause the parathyroid glands to secrete PTH.

5   People with type 2 diabetes must receive insulin daily by injection or through an insulin pump.

| | |
|---|---|
| **1** T | |
| **2** F | |
| **3** T | |
| **4** F | |
| **5** F | |

## Fill-in

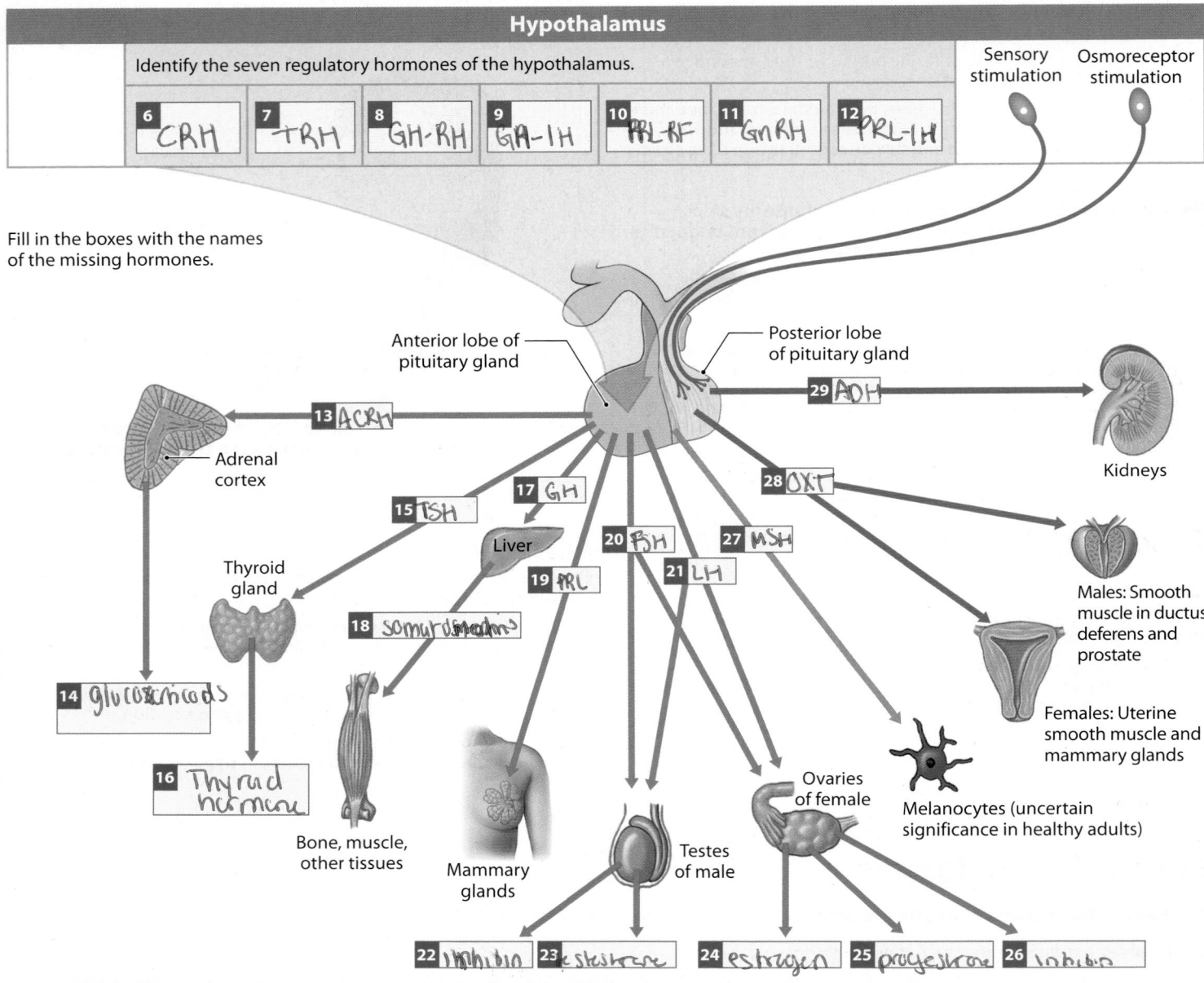

**Hypothalamus**

Identify the seven regulatory hormones of the hypothalamus.

| 6 CRH | 7 TRH | 8 GH-RH | 9 GH-IH | 10 PRL-RF | 11 GnRH | 12 PRL-IH |

Sensory stimulation  Osmoreceptor stimulation

Fill in the boxes with the names of the missing hormones.

Anterior lobe of pituitary gland

Posterior lobe of pituitary gland

29 ADH

Kidneys

13 ACRH

Adrenal cortex

28 OXT

17 GH

15 TSH

Liver

20 FSH

27 MSH

Males: Smooth muscle in ductus deferens and prostate

Thyroid gland

19 PRL

21 LH

18 somatomedins

14 glucocorticoids

Females: Uterine smooth muscle and mammary glands

16 Thyroid hormone

Ovaries of female

Melanocytes (uncertain significance in healthy adults)

Bone, muscle, other tissues

Mammary glands

Testes of male

22 inhibin   23 testosterone   24 estrogen   25 progesterone   26 inhibin

## Multiple choice: Select the correct answer from the list provided.

**30** Secretions from the pancreatic islets are
- ☐ a) glucagon from delta cells and insulin from beta cells.
- ☒ b) glucagon from alpha cells and insulin from beta cells. ✓
- ☐ c) glucagon from beta cells and insulin from alpha cells.
- ☐ d) glucagon from beta cells and insulin from delta cells.

**31** The pineal gland secretes
- ☐ a) corticosteroids.
- ☐ b) thyroxine.
- ☐ c) melanocyte-stimulating hormone.
- ☒ d) melatonin.

**32** The two most important second messengers are
- ☐ a) cAMP and G proteins.
- ☐ b) G proteins and calcium ions.
- ☐ c) catecholamines and eicosanoids.
- ☒ d) cAMP and calcium ions.

**33** FSH production in males supports
- ☒ a) physical maturation of developing sperm.
- ☐ b) development of muscle size.
- ☐ c) the release of androgens.
- ☐ d) an increased desire for sexual activity.

**34** Which of the following is *not* an effect of thyroid hormones on peripheral tissues?
- [ ] a) Increased rates of oxygen consumption and energy
- [x] b) Decreased heart rate
- [ ] c) Accelerated turnover of minerals in bone
- [ ] d) Increased sensitivity to sympathetic stimulation

**35** Steroid hormones bind to receptors in the
- [ ] a) plasma membrane only.
- [ ] b) plasma membrane or cytoplasm.
- [ ] c) plasma membrane or nucleus.
- [x] d) cytoplasm or nucleus.

## Short answer

**36** What are the functions of the hormones and enzyme secreted by the kidneys?

**37** What effects do calcitonin and parathyroid hormone have on blood calcium level?

---

# MasteringA&P™

Access more chapter study tools online in the MasteringA&P Study Area:

- Chapter Quizzes, Chapter Practice Test, and Clinical Case Studies

| | |
|---|---|
| ■ Practice Anatomy Lab™ 3.0 | PAL 3.0 |
| ■ Interactive Physiology™ | iP2™ |
| ■ A&P Flix™ | A&PFlix™ |
| ■ PhysioEx™ 9.1 | PhysioEx 9.1 |

---

# Chapter Integration • Applying what you have learned

## What's wrong with me?

Sherry, a 43-year-old mother of two, has been experiencing some slightly troublesome physical and mental problems that began about 8 months ago. She had been a physically fit 120-lb person with a bright, happy, and cheerful disposition all her adult life. Within the past few months, however, her husband has repeatedly commented that she overreacts to otherwise normal family situations and seems irritated by the slightest problems. She has also lost a considerable amount of weight, even though her diet and exercise levels haven't changed. Because Sherry just isn't feeling "right," she decides to make an appointment with her family physician.

Sherry tells her physician that she has been restless, anxious, and quite irritable lately. She is having a hard time sleeping, complains of diarrhea and weight loss, and has an increased appetite without weight gain. Visual examination of her neck reveals an enlargement of the right anterior neck. During the examination, her physician notices a higher-than-normal heart rate and a fine tremor in her outstretched fingers. Although further clinical tests are necessary, the doctor suspects hyperthyroidism.

**1** Why does the physician suspect hyperthyroidism?

**2** Explain hyperthyroidism.

**3** What tests could the physician order to make a positive diagnosis of Sherry's condition?

**4** Which of Sherry's signs and symptoms indicate nervous system involvement?

## 17

# Blood

## Module 17.1

# Blood is the fluid portion of the cardiovascular system

 **1** The **cardiovascular system** includes a fluid (**blood**), a series of conducting tubes (the **blood vessels**) that distribute the fluid throughout the body, and a pump (the **heart**) that keeps the fluid in motion.

### Components of the Cardiovascular System

The **HEART** propels blood and maintains blood pressure

**BLOOD VESSELS** distribute blood around the body

| | |
|---|---|
| **Arteries** | carry blood away from the heart to the capillaries |
| **Capillaries** | permit diffusion between blood and interstitial fluids |
| **Veins** | return blood from capillaries to the heart |

**BLOOD** distributes oxygen, carbon dioxide, and blood cells; delivers nutrients and hormones; transports waste products; and assists in temperature regulation and defense against disease.

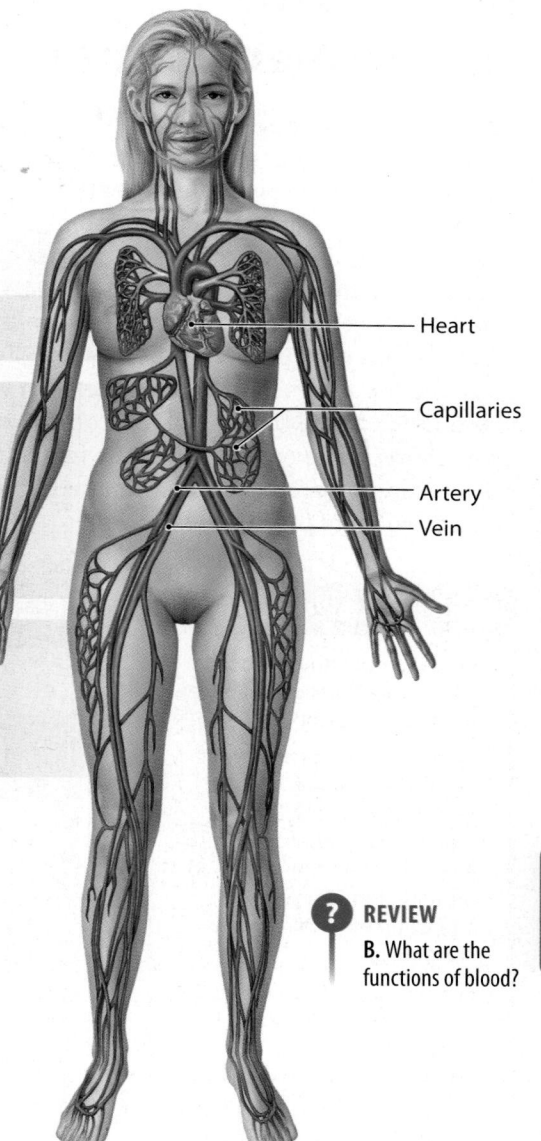

Heart

Capillaries

Artery
Vein

**?** **REVIEW**
**A.** Identify the components of the cardiovascular system.

**?** **REVIEW**
**B.** What are the functions of blood?

**17**

### Functions of Blood

- **Transport dissolved gases, nutrients, hormones, and metabolic wastes:** Circulating blood contains red blood cells that carry oxygen from the lungs to peripheral tissues and carry carbon dioxide from those tissues to the lungs. Blood distributes nutrients absorbed by the digestive tract or released from storage in adipose tissue or in the liver. It carries hormones from endocrine glands toward their target cells. It also absorbs cellular wastes and carries them to the kidneys for excretion.

- **Regulate the pH and ion composition of interstitial fluids:** Diffusion between interstitial fluids and blood eliminates local deficiencies or excesses of ions such as calcium or potassium. Blood also absorbs and neutralizes metabolic acids generated by active tissues.

- **Restrict fluid losses at injury sites:** Blood contains enzymes and other substances that respond to breaks in vessel walls by initiating the clotting process. A blood clot acts as a temporary patch that prevents further blood loss.

- **Defend against toxins and pathogens:** Blood transports *white blood cells*, specialized cells that migrate into peripheral tissues to fight infections or remove debris. Blood also transports *antibodies*, proteins that specifically attack invading organisms or foreign substances.

- **Stabilize body temperature:** Blood absorbs the heat generated by active skeletal muscles and redistributes it to other tissues. If body temperature is already high, that heat will be lost across the surface of the skin. If body temperature is too low, the warm blood is directed to the brain and to other temperature-sensitive organs.

This section describes the vital roles of blood and introduces its two main components: *plasma* and *formed elements*. In Section 2, we examine the structure and function of the formed elements.

 **LEARNING OUTCOME**
List the components of the cardiovascular system, and describe several important functions of blood.

# Module 17.2

# Blood is a fluid connective tissue containing plasma and formed elements

Go to
MasteringA&P™
PhysioEx™ 9.1
Ex: 11 Blood Analysis 1 Hematocrit

Blood is a fluid connective tissue with a unique composition. It consists of **plasma** (PLAZ-muh), a liquid matrix, and **formed elements** (cells and cell fragments). The cardiovascular system of an adult male contains 5–6 liters (5.3–6.4 quarts) of blood; that of an adult female contains 4–5 liters (4.2–5.3 quarts). The difference in blood volume between the sexes is due to differences in average body size. After blood is removed for analysis or storage, the term **whole blood** is used to indicate that the blood composition has not been altered. The components of whole blood can, however, be separated, or fractionated, if only one component is of interest.

### Properties of Whole Blood

- Blood temperature is about 38°C (100.4°F), slightly above normal body temperature.

- Blood is five times as viscous as water—that is, five times as resistant to flow. Blood's high viscosity results from interactions among dissolved proteins, formed elements, and water molecules in the plasma.

- Blood is slightly alkaline, with a pH between 7.35 and 7.45 (average: 7.4).

**Whole blood** *consists of*

## PLASMA

1  Plasma forms 55 percent of the volume of whole blood. In many respects, the composition of plasma resembles that of interstitial fluid. This similarity exists because water, ions, and small solutes are continuously exchanged between plasma and interstitial fluids across the walls of capillaries. The primary differences between plasma and interstitial fluid involve (1) the levels of respiratory gases (oxygen and carbon dioxide) due to the respiratory activities of tissue cells, and (2) the concentrations and types of dissolved proteins (because plasma proteins cannot cross capillary walls).

**Plasma**
**55%**
(Range 46–63%)

Plasma Proteins

Other Solutes

Water

+

**Formed Elements**
**45%**
(Range 37–54%)

Platelets

White Blood Cells

Red Blood Cells

## FORMED ELEMENTS

2  Formed elements are blood cells and cell fragments suspended in plasma. They make up 45 percent of the volume of whole blood. The **hematocrit** (hē-MAH-tō-krit) is the percentage of formed elements in a sample of whole blood. Red blood cells make up 99.9% of these formed elements. In adult males, the normal hematocrit, or **packed cell volume (PCV)**, averages 47 (range: 40–54); the average for adult females is 42 (range: 37–47). The difference in hematocrit between the sexes primarily reflects the fact that androgens (male hormones) stimulate red blood cell production, whereas estrogens (female hormones) do not.

? **REVIEW**

A. Identify the two components making up whole blood, and list the composition of each.

B. Define *hematocrit*.

17

## Plasma Proteins 7%

**Plasma proteins** are in solution rather than forming insoluble fibers like those in other connective tissues, such as loose connective tissue or cartilage. On average, each 100 mL of plasma contains 7.6 g of protein, almost five times the concentration in interstitial fluid. The large size and globular shapes of most blood proteins usually prevent them from leaving the bloodstream. The liver synthesizes and releases more than 90 percent of all plasma proteins.

**Albumins** (al-BŪ-minz) make up about 60 percent of the plasma proteins. As the most abundant plasma proteins, they are major contributors to the osmotic pressure of plasma.

**Globulins** (GLOB-ū-linz) make up approximately 35 percent of the proteins in plasma. Important plasma globulins include antibodies and transport globulins. **Antibodies**, also called **immunoglobulins** (i-mū-nō-GLOB-ū-linz), attack foreign proteins and pathogens. **Transport globulins** bind small ions, hormones, lipids, and other compounds.

**Fibrinogen** (fī-BRIN-ō-jen) functions in clotting and normally makes up about 4 percent of plasma proteins. Under certain conditions, fibrinogen molecules interact to form large, insoluble strands of **fibrin** (FĪ-brin) that form the basic framework for a blood clot.

The plasma also contains active and inactive enzymes and hormones whose concentrations vary widely.

## Other Solutes 1%

Other solutes are generally present in plasma in concentrations similar to interstitial fluid. However, arterial blood and venous blood can contain different concentrations of nutrients and wastes.

**Electrolytes:** Normal extracellular ion composition is essential for vital cellular activities. The major plasma electrolytes are $Na^+$, $K^+$, $Ca^{2+}$, $Mg^{2+}$, $Cl^-$, $HCO_3^-$, $HPO_4^-$, and $SO_4^{2-}$.

**Organic nutrients:** Organic nutrients are used for ATP production, growth, and cell maintenance. This category includes lipids (fatty acids, cholesterol, and glycerides), carbohydrates (primarily glucose), and amino acids.

**Organic wastes:** Wastes are carried to sites for breakdown or excretion. Examples of organic wastes include urea, uric acid, creatinine, bilirubin, and ammonium ions.

## Water 92%

## Platelets <0.1%

**Platelets** (PLĀT-lets) are small, membrane-bound cell fragments that contain enzymes and other substances important for clotting.

 "thrombocytes"

17

## White Blood Cells <0.1%

**White blood cells (WBCs)**, or **leukocytes** (LŪ-kō-sīts; *leukos*, white + *-cyte*, cell), play a role in the body's defense mechanisms. There are five classes of leukocytes, each with slightly different functions that will be explored later in the chapter.

 Neutrophils

Eosinophils

 Basophils

Lymphocytes

Monocytes

## Red Blood Cells 99.9%

**Red blood cells (RBCs)**, or **erythrocytes** (e-RITH-rō-sīts; *erythros*, red + *-cyte*, cell), are the most abundant blood cells. These specialized cells are essential for the transport of oxygen in the blood.

 **INTEGRATION**

**C.** Which specific plasma proteins would you expect to be elevated during an infection?

 **LEARNING OUTCOME**

Describe the important components and major properties of blood.

# Module 17.3

# Formed elements are produced by stem cells in red bone marrow

**1** The formed elements develop in red bone marrow in a process called **hemopoiesis** (hē-mō-poy-Ē-sis (*hemo-*, blood + *poiesis*, a making), or **hematopoiesis**.

Nutrient artery

Venous sinuses

Red bone marrow

### Hemocytoblasts

**Hemocytoblasts** form from *hematopoietic stem cells* (HSCs). For RBCs to be produced, hemocytoblasts in the red bone marrow must divide, producing two types of cells: lymphoid stem cells and myeloid stem cells. *Lymphoid stem cells* divide to produce white blood cells called lymphocytes. *Myeloid stem cells* divide to produce RBCs and several classes of white blood cells.

**?** REVIEW
**A.** Define *hemocytoblasts*.

### Lymphoid Stem Cells

**Lymphoid stem cells** are responsible for the production of *lymphocytes*, which are cells that function in the immune response. These stem cells originate in the red bone marrow. Some remain there, whereas others migrate to **lymphoid tissues**, including the thymus, spleen, and lymph nodes. Lymphocytes are then produced in these organs as well as in the red bone marrow.

**Colony-stimulating factors** are hormones released by activated lymphocytes and other cells during an immune response to stimulate blood cell formation.

Stimulation

### Myeloid Stem Cells

**Myeloid stem cells** are stem cells in red bone marrow that divide to give rise to all types of formed elements other than lymphocytes.

**2** The term *formed elements* is appropriate because platelets are cell fragments rather than specialized cells. The flowchart at right describes the origin of platelets from megakaryocytes. This table summarizes important information about platelets.

**?** REVIEW
**B.** Describe platelets and their origin.

| Structure and Function of Platelets | | | |
| --- | --- | --- | --- |
| Appearance in a Stained Blood Smear | Abundance (Average Number per μL) | Function | Remarks |
| Platelets are flattened discs that appear round when viewed from above and appear spindle shaped in section or in a blood smear.  | 350,000 (range: 150,000–500,000) | Platelets clump together and stick to damaged vessel walls, and they release chemicals that stimulate blood clotting. | Platelets are continuously replaced. Each platelet circulates for 9–12 days before being removed by phagocytes, mainly in the spleen. |

17

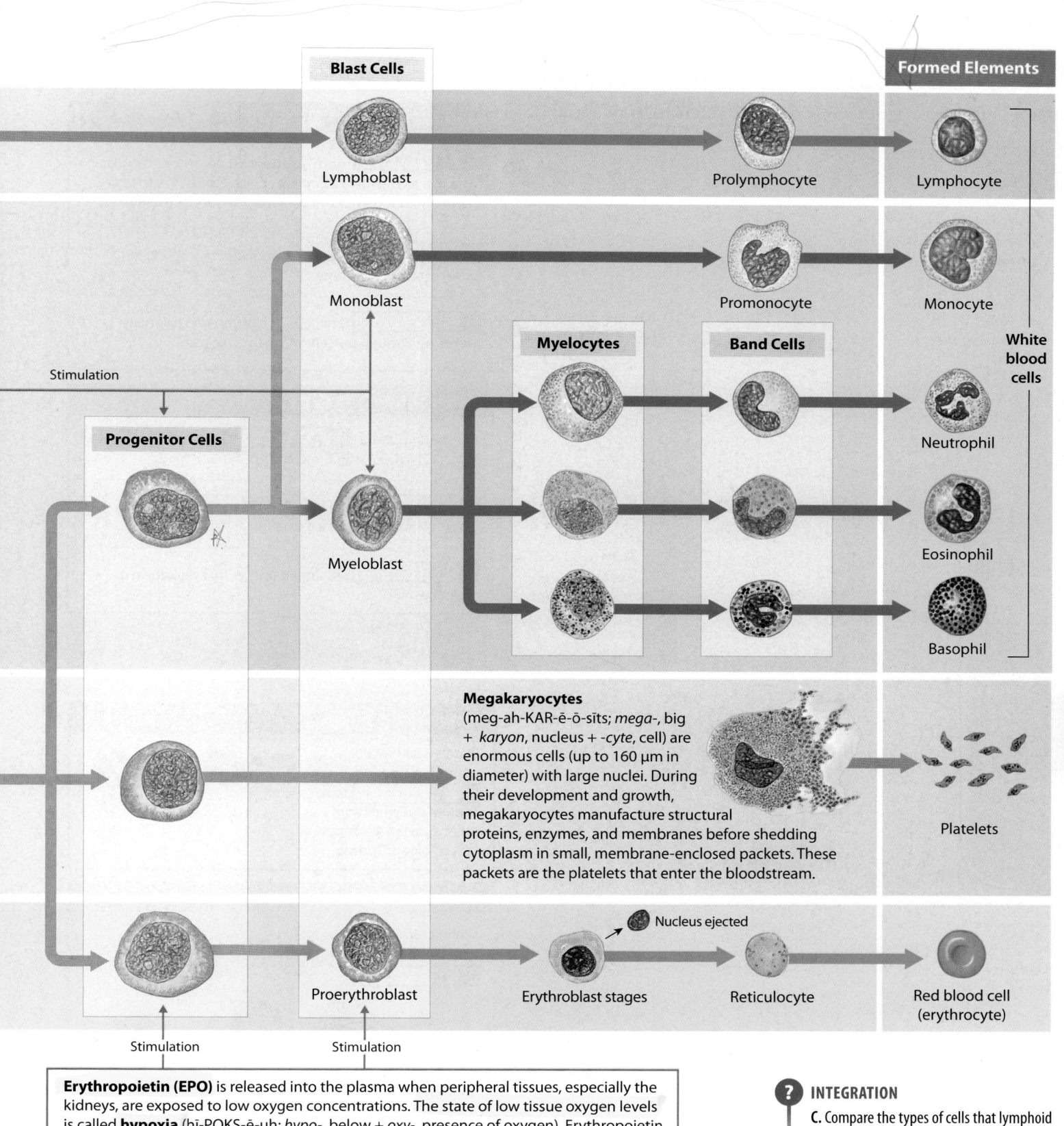

**Blast Cells**

**Formed Elements**

Lymphoblast

Prolymphocyte

Lymphocyte

Monoblast

Promonocyte

Monocyte

White blood cells

**Progenitor Cells**

Stimulation

**Myelocytes**

**Band Cells**

Neutrophil

Myeloblast

Eosinophil

Basophil

**Megakaryocytes** (meg-ah-KAR-ē-ō-sīts; *mega-*, big + *karyon*, nucleus + *-cyte*, cell) are enormous cells (up to 160 μm in diameter) with large nuclei. During their development and growth, megakaryocytes manufacture structural proteins, enzymes, and membranes before shedding cytoplasm in small, membrane-enclosed packets. These packets are the platelets that enter the bloodstream.

Platelets

Nucleus ejected

Stimulation

Proerythroblast

Stimulation

Erythroblast stages

Reticulocyte

Red blood cell (erythrocyte)

**Erythropoietin (EPO)** is released into the plasma when peripheral tissues, especially the kidneys, are exposed to low oxygen concentrations. The state of low tissue oxygen levels is called **hypoxia** (hī-POKS-ē-uh; *hypo-*, below + *oxy-*, presence of oxygen). Erythropoietin is released (1) during anemia; (2) when blood flow to the kidneys decreases; (3) when the oxygen content of air in the lungs decreases due to disease or high altitude; and (4) when the respiratory surfaces of the lungs are damaged. Once in the bloodstream, EPO is carried to areas of red bone marrow, where it stimulates stem cells and developing RBCs.

17

**?** **INTEGRATION**

**C.** Compare the types of cells that lymphoid stem cells and myeloid stem cells produce.

**LO** **LEARNING OUTCOME**

Explain the origins and differentiation of the formed elements.

# Section 1 Review
## Plasma and Formed Elements

**Matching:** Use the lettered terms below to fill in the blanks in the diagram.

a. electrolytes
b. globulins
c. 99.9%
d. oxygen transport
e. cell fragments
f. 7%
g. fibrinogen
h. defense mechanisms
i. organic nutrients
j. 92%
k. organic wastes
l. 1%
m. <0.1%
n. <0.1%
o. albumins

### Plasma Proteins [1] **F**

[2] **Albumins** make up about 60 percent of the plasma proteins. As the most abundant plasma proteins, they are major contributors to the osmotic pressure of plasma.

[3] **globulins** make up approximately 35 percent of the proteins in plasma.

[4] **G** functions in clotting and normally makes up about 4 percent of plasma proteins.

### Other Solutes [5] **i**

The major plasma [6] **A** are $Na^+$, $K^+$, $Ca^{2+}$, $Mg^{2+}$, $Cl^-$, $HCO_3^-$, $HPO_4^-$, and $SO_4^{2-}$.

[7] **i** are used for ATP production, growth, and maintenance of cells.

[8] **K** are carried to sites for breakdown or excretion.

### Water [9] **J**

O

B

fibnogen

**Plasma**

**55%**
(Range 46-63%)

**Whole blood** *consists of*

+

**Formed Elements**

**45%**
(Range 37–54%)

### Platelets [10] **e**

Platelets are small, membrane-bound [11] **E** that contain enzymes and other substances important to blood clotting.

### White Blood Cells [12] **n**

White blood cells play a role in the body's [13] **H**

### Red Blood Cells [14] **C**

Red blood cells are essential for [15] **D** in the blood.

636

## Module 17.4

# Hematology is the study of blood and blood-forming tissues

**Hematology** (hē-muh-TOL-ō-jē) is the study of blood, blood-forming tissues, and blood disorders. It is important as a branch of science because much can be learned about a person's health by evaluating blood. Blood provides a readily accessible medium for diagnosing and treating disease. Blood tests can detect disorders such as anemia, infection, and clotting disorders.

**?** **REVIEW**
**A.** What is hematology?

**1** Since blood courses through nearly every tissue of the body, **dyscrasias** (dis-KRĀ-zē-uz), or blood disorders, can have systemic effects. Blood tests are used for many reasons. Some are simple, such as determining blood type; others are done to evaluate the types and numbers of RBCs, WBCs, and platelets. One test commonly performed is a **complete blood count (CBC)**. A CBC determines the RBC count, WBC count, erythrocyte indices (such as hemoglobin content), hematocrit, and the platelet count in 1 cubic millimeter of blood (1 microliter, μL). A **WBC differential count** identifies the types and numbers for each white blood cell. Blood clotting tests and bleeding time give information on blood coagulation (see Module 17.10). Abnormal values may indicate underlying medical conditions. At right is a vial of whole blood and a representative medical report showing a complete blood count.

**2** Several common tests focus on RBCs and assess the number, size, shape, and maturity of circulating RBCs, indicating the erythropoietic activities under way (see Module 17.6). The tests can also be useful in detecting problems, such as internal bleeding, that may not produce obvious signs or symptoms. The table below lists examples of important RBC tests and terms associated with abnormal values.

**Complete Blood Count (CBC)**
**Hematocrit:** 42%
**Hemoglobin:** 14.9 g/dL
**MCH:** 30 μg/RBC
**MCV:** 90 μm³/cell
**Platelet count:** 350,000/mm³
**RBC:** 5.2 million/μL
**WBC:** 7000/μL

**WBC differential count:**
Neutrophils:  67%
Lymphocytes: 25%
Monocytes:  3%
Eosinophils:  4%
Basophils:  1%

**?** **REVIEW**
**B.** Describe a complete blood count (CBC)

### RBC Tests and Related Terminology

| Test | Determines | Terms Associated with Abnormal Values | |
| --- | --- | --- | --- |
| | | Elevated | Depressed |
| **Hematocrit (Hct)** | Percentage of formed elements in whole blood Normal = 37–54% | Polycythemia | Anemia |
| **Hemoglobin Concentration (Hb, Hgb)** | Concentration of hemoglobin in blood Normal = 12–18 g/dL | Polycythemia | Anemia |
| **Mean Corpuscular Hemoglobin Concentration (MCH)** | Average weight of Hb in one RBC Normal = 27–34 μg/RBC (normochromic) | Hyperchromic | Hypochromic |
| **Mean Corpuscular Volume (MCV)** | Average volume of one RBC Normal = 82–101 μm³/cell (normocytic) | Macrocytic | Microcytic |
| **RBC Count** | Number of RBCs per μL of whole blood Normal = 4.2–6.3 million cells/μL | Erythrocytosis/polycythemia | Anemia |
| **Reticulocyte Count (Retic.)** | Percentage of circulating reticulocytes Normal = 0.8% | Reticulocytosis | Diminished erythropoiesis |

This section examines the structure and function of the formed elements of blood, *hemostasis* (how bleeding is stopped), and various blood disorders.

**?** **REVIEW**
**C.** Which condition would a patient have if she had a depressed hematocrit level?

 **LEARNING OUTCOME**
Define hematology, describe the elements of a complete blood count (CBC), and give examples of red blood cell lab tests.

**17**

# Module 17.5

# Red blood cells, the most common formed elements, contain hemoglobin that transports respiratory gases

**1** The human body contains an enormous number of red blood cells (RBCs). A standard blood test—the **red blood cell count**—reports the number of RBCs per microliter (μL) of whole blood. In adult males, 1 μL, or 1 cubic millimeter (mm³), of whole blood contains 4.5–6.3 million RBCs; in adult females, 1 μL contains 4.2–5.5 million. A single drop of whole blood contains approximately 260 million RBCs, and the blood of an average adult has 25 trillion RBCs. RBCs thus account for roughly one-third of all cells in the human body.

**2** Each RBC is a biconcave disc with a thin central region and a thicker outer margin. The given range dimensions shown here are for a normal RBC.

**Stained blood smear** LM × 450

**RBCs** Colorized SEM × 2100

7.2–8.4 μm

0.45–1.16 μm    2.31–2.85 μm

**3** The biconcave shape and flexible plasma membrane of an RBC are important for the three reasons detailed in this table.

## Functional Aspects of Red Blood Cells

- **Large surface area–to–volume ratio**. Each RBC carries oxygen bound to *hemoglobin*, an intracellular protein, and that oxygen must be absorbed or released quickly as the RBC passes through the capillaries. The greater the surface area per unit volume, the faster the exchange of materials between the RBC's interior and the surrounding plasma. The total surface area of all the RBCs in the blood of a typical adult is about 3800 square meters, roughly 2000 times the total surface area of the body.

- **RBCs can form stacks**. Like dinner plates, RBCs can form stacks that ease the flow through narrow blood vessels. An entire stack can pass along a blood vessel only slightly larger than the diameter of a single RBC, whereas individual cells would bump the walls, bang together, and form logjams that could restrict or prevent blood flow.

- **Flexibility**. The cytoplasmic surface of an RBC plasma membrane is a meshwork of flexible proteins. As a result, RBCs can bend and flex when entering small capillaries as narrow as 4 μm.

Stacks of RBCs known as **rouleaux** (rū-LŌ) form and dissociate repeatedly without affecting the cells involved.

Blood vessels (viewed in longitudinal section)

Red blood cell

Nucleus of endothelial cell

**Sectional view of capillaries** LM × 1430

**? REVIEW**

**A.** Describe the functional aspects of RBCs.

17

**4** A red blood cell is very different from a "typical cell." As our RBCs develop, they lose most of their organelles, including nuclei; they retain only the cytoskeleton. Because mature RBCs lack nuclei, a condition called **anucleate**, and also lack ribosomes, they cannot divide or synthesize structural proteins or enzymes. As a result, RBCs cannot repair themselves, and their life span is normally less than 120 days. In effect, a developing RBC loses any organelle not directly associated with this primary function: the transport of respiratory gases. That function is performed by molecules of **hemoglobin (Hb, Hgb)**, which account for more than 95 percent of an RBC's intracellular proteins. The hemoglobin content of whole blood is reported in grams of Hb per deciliter (100 mL) of whole blood (g/dL). Normal ranges are 14–18 g/dL in males and 12–16 g/dL in females.

**5** Hemoglobin has a complex quaternary protein structure. Each Hb molecule has two **alpha (α) chains** and two **beta (β) chains** of polypeptides. Each chain is a globular protein subunit that resembles the myoglobin in skeletal and cardiac muscle cells. Like myoglobin, each Hb chain contains a single molecule of **heme**, a nonprotein pigment complex.

β chain 2

Heme

α chain 2

α chain 1

β chain 1

**? REVIEW**

**B.** Describe hemoglobin.

**6** Each heme unit holds an iron ion in such a way that the iron can interact with an oxygen molecule, forming **oxyhemoglobin**, **HbO$_2$**. Blood containing RBCs filled with oxyhemoglobin is bright red. The iron–oxygen interaction is very weak, and the two can easily dissociate without damaging the heme unit or the oxygen molecule. The binding of an oxygen molecule to the iron in a heme unit is therefore completely reversible. A hemoglobin molecule whose iron is not bound to oxygen is called **deoxyhemoglobin**, or reduced hemoglobin. Blood containing RBCs filled with deoxyhemoglobin is dark red—almost burgundy.

**? REVIEW**

**C.** Compare oxyhemoglobin with deoxyhemoglobin.

**Heme**

**🏃 Everyday Physiology**

## How much oxygen can a red blood cell carry?

Each red blood cell contains about 280 million Hb molecules. Because one Hb molecule contains four heme units, each RBC can potentially carry more than a billion molecules of oxygen at a time. Roughly 98.5% of the oxygen carried by the blood travels through the bloodstream bound to Hb molecules inside RBCs. The rest is dissolved in the plasma.

**LEARNING OUTCOME**

List the characteristics and functions of red blood cells, and describe the structure and functions of hemoglobin.

17

# Module 17.6
# Red blood cells are continually produced …

About 1 percent of the circulating RBCs are replaced each day, and in the process approximately 3 million new RBCs enter the bloodstream each second! Such a rapid rate of replacement is necessary because a typical RBC has a relatively short life span. After it travels about 700 miles in 120 days, either its plasma membrane ruptures, or it is engulfed by macrophages in the spleen, liver, or bone marrow. The continual elimination of RBCs usually goes unnoticed, as long as new ones enter the bloodstream at a comparable rate. RBC maturation is shown in (1), while (2)–(5) detail RBC recycling.

**1**

**Events Occurring in the Red Bone Marrow**

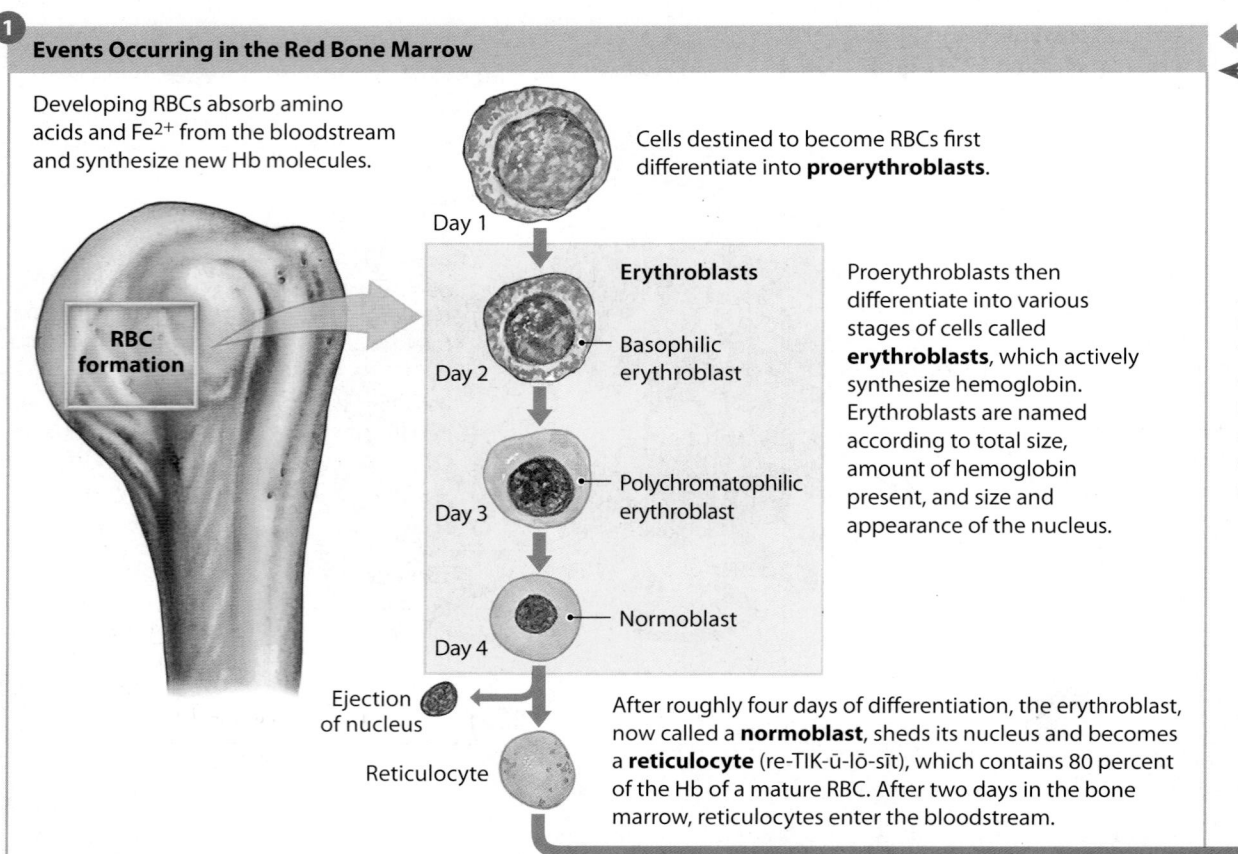

Developing RBCs absorb amino acids and $Fe^{2+}$ from the bloodstream and synthesize new Hb molecules.

**RBC formation**

Day 1

**Erythroblasts**

Basophilic erythroblast
Day 2

Polychromatophilic erythroblast
Day 3

Normoblast
Day 4

Ejection of nucleus

Reticulocyte

Cells destined to become RBCs first differentiate into **proerythroblasts**.

Proerythroblasts then differentiate into various stages of cells called **erythroblasts**, which actively synthesize hemoglobin. Erythroblasts are named according to total size, amount of hemoglobin present, and size and appearance of the nucleus.

After roughly four days of differentiation, the erythroblast, now called a **normoblast**, sheds its nucleus and becomes a **reticulocyte** (re-TIK-ū-lō-sīt), which contains 80 percent of the Hb of a mature RBC. After two days in the bone marrow, reticulocytes enter the bloodstream.

After 24 hours in circulation, the reticulocytes complete their maturation and resemble other mature RBCs.

Reticulocytes account for about 0.8% of the circulating RBCs.

In adults, red blood cell formation, or **erythropoiesis** (e-rith-rō-poy-Ē-sis), occurs only in red bone marrow, or **myeloid** (MĬ-e-loyd) **tissue** (*myelos*, marrow). This tissue is located in portions of the vertebrae, sternum, ribs, skull, scapulae, pelvis, and proximal limb bones. Other marrow areas contain a fatty tissue known as yellow bone marrow. Under extreme stimulation, such as severe and sustained blood loss, areas of yellow marrow can convert to red marrow, increasing the rate of RBC formation.

**5**

**Events Occurring in the Kidney**

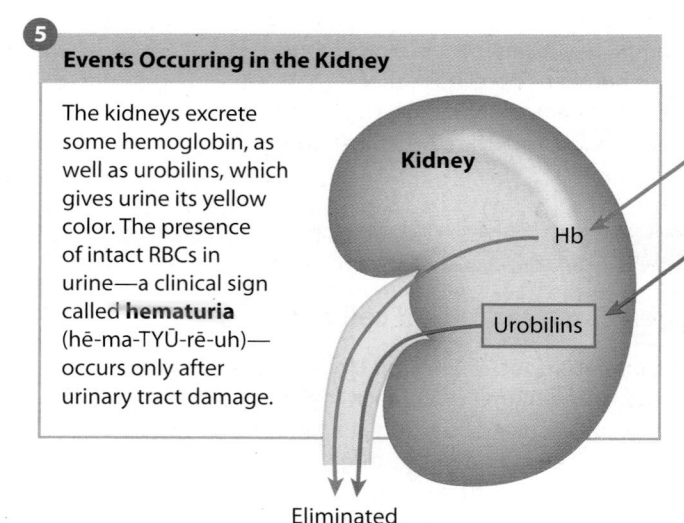

The kidneys excrete some hemoglobin, as well as urobilins, which gives urine its yellow color. The presence of intact RBCs in urine—a clinical sign called **hematuria** (hē-ma-TYŪ-rē-uh)— occurs only after urinary tract damage.

Kidney

Hb

Urobilins

Eliminated in urine

Wed Feb 7

# ... and their components recycled or eliminated

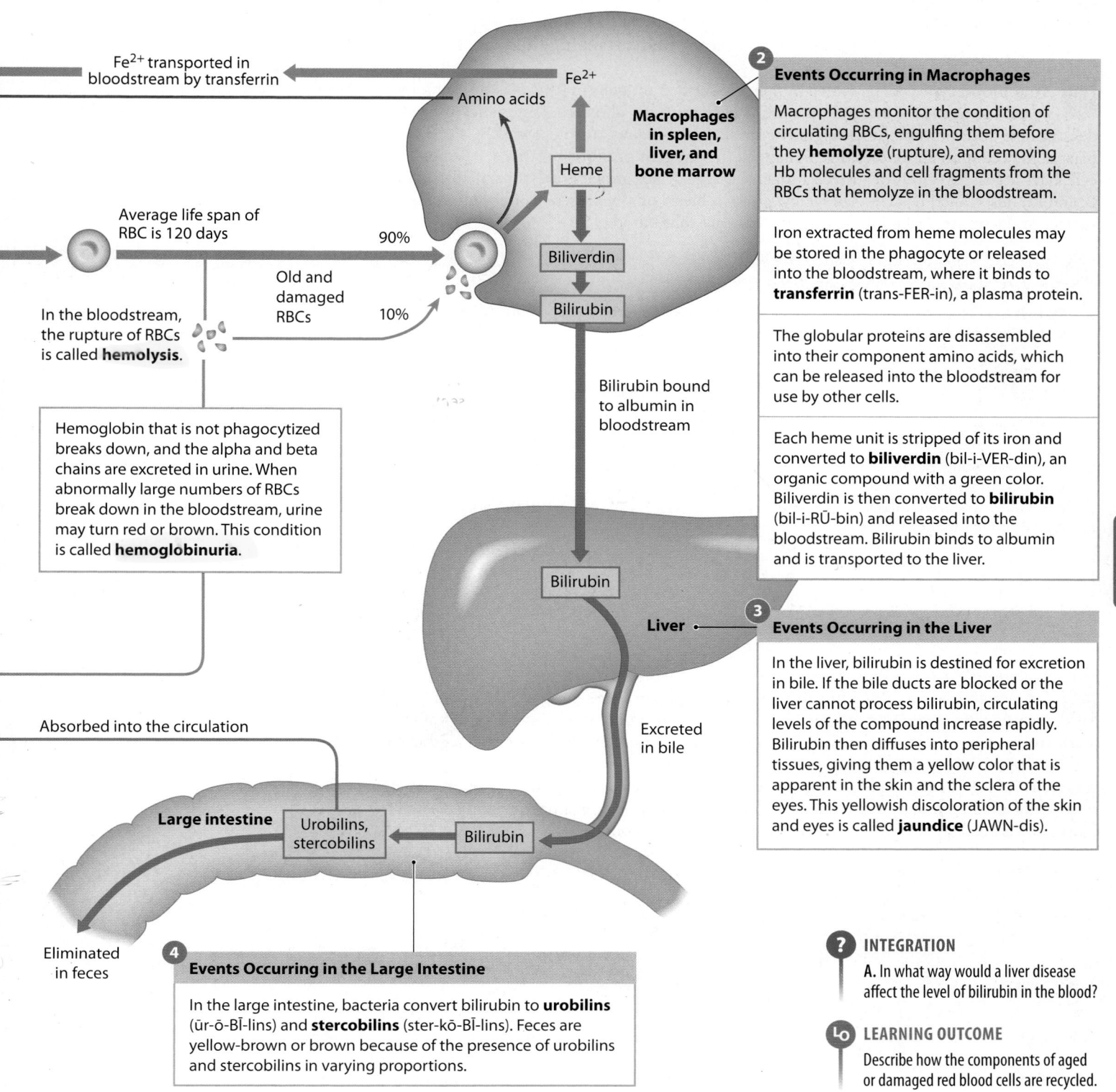

Fe²⁺ transported in bloodstream by transferrin

Fe²⁺

Amino acids

**Macrophages in spleen, liver, and bone marrow**

Heme

**2** **Events Occurring in Macrophages**

Macrophages monitor the condition of circulating RBCs, engulfing them before they **hemolyze** (rupture), and removing Hb molecules and cell fragments from the RBCs that hemolyze in the bloodstream.

Iron extracted from heme molecules may be stored in the phagocyte or released into the bloodstream, where it binds to **transferrin** (trans-FER-in), a plasma protein.

The globular proteins are disassembled into their component amino acids, which can be released into the bloodstream for use by other cells.

Each heme unit is stripped of its iron and converted to **biliverdin** (bil-i-VER-din), an organic compound with a green color. Biliverdin is then converted to **bilirubin** (bil-i-RŪ-bin) and released into the bloodstream. Bilirubin binds to albumin and is transported to the liver.

Average life span of RBC is 120 days

90%

Old and damaged RBCs

10%

In the bloodstream, the rupture of RBCs is called **hemolysis**.

Biliverdin

Bilirubin

Bilirubin bound to albumin in bloodstream

Hemoglobin that is not phagocytized breaks down, and the alpha and beta chains are excreted in urine. When abnormally large numbers of RBCs break down in the bloodstream, urine may turn red or brown. This condition is called **hemoglobinuria**.

Bilirubin

**3** **Events Occurring in the Liver**

Liver

In the liver, bilirubin is destined for excretion in bile. If the bile ducts are blocked or the liver cannot process bilirubin, circulating levels of the compound increase rapidly. Bilirubin then diffuses into peripheral tissues, giving them a yellow color that is apparent in the skin and the sclera of the eyes. This yellowish discoloration of the skin and eyes is called **jaundice** (JAWN-dis).

Absorbed into the circulation

Excreted in bile

**Large intestine**

Urobilins, stercobilins

Bilirubin

Eliminated in feces

**4** **Events Occurring in the Large Intestine**

In the large intestine, bacteria convert bilirubin to **urobilins** (ūr-ō-BĪ-lins) and **stercobilins** (ster-kō-BĪ-lins). Feces are yellow-brown or brown because of the presence of urobilins and stercobilins in varying proportions.

**?** **INTEGRATION**

**A.** In what way would a liver disease affect the level of bilirubin in the blood?

**Lᴏ** **LEARNING OUTCOME**

Describe how the components of aged or damaged red blood cells are recycled.

17

# Module 17.7

# Blood type is determined by the presence or absence of specific surface antigens on RBCs

Go to
MasteringA&P™
PhysioEx 9.1
Ex 11: Blood Analysis 4 Blood Typing

**Antigens** are substances that can trigger a protective defense mechanism called an **immune response**. Most antigens are proteins, but some other types of organic molecules are antigens as well. The plasma membranes of your cells contain **surface antigens**, substances your immune system recognizes as "normal," or "self." In other words, your immune system ignores these substances rather than attacking them as "foreign."

**1** Your **blood type** is determined by specific surface antigens in RBC plasma membranes. These surface antigens are integral membrane glycoproteins whose characteristics are genetically determined. Your immune system ignores the surface antigens on your own RBCs. However, your plasma may contain antibodies that will attack the antigens on "foreign" RBCs. Although red blood cells have at least 50 kinds of surface antigens, three surface antigens are of particular importance: A, B, and Rh (also called D). The *ABO blood group* is based on the presence or absence of the A and B surface antigens. According to this group, there are four blood types.

**? REVIEW**

**A.** What is determined by the surface antigens on RBCs?

---

## Antigens and Antibodies by Blood Type

**Type A** blood has RBCs with surface antigen A only.

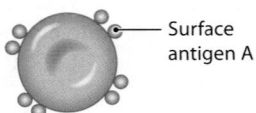
Surface antigen A

If you have type A blood, your plasma contains anti-B antibodies, which will attack type B surface antigens.

**Type B** blood has RBCs with surface antigen B only.

Surface antigen B

If you have type B blood, your plasma contains anti-A antibodies.

**Type AB** blood has RBCs with both A and B surface antigens.

If you have type AB blood, your plasma has neither anti-A nor anti-B antibodies.

**Type O** blood has RBCs without both A and B surface antigens.

If you have type O blood, your plasma contains both anti-A and anti-B antibodies.

---

**2** Surface antigens on RBCs are also known as **agglutinogens** (ah-glu-TIN-ō-jenz), because if antigens of one blood type are exposed to the corresponding antibodies (also known as **agglutinins**) from another type, the RBCs will clump together in a process called **agglutination**. The cells may also undergo hemolysis. Such a **cross-reaction** is very dangerous because clumps and fragments of RBCs can plug small blood vessels in the kidneys, lungs, heart, or brain, damaging or destroying affected tissues. Accidental cross-reactions may occur if a person being treated for severe blood loss is accidentally given a transfusion of the wrong blood type.

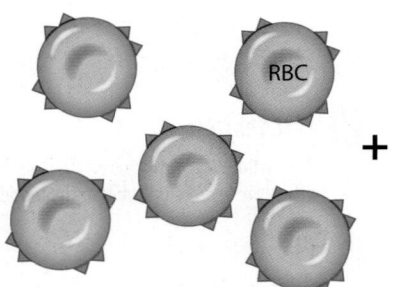
RBCs with surface antigen B

+

Anti-B antibodies

Agglutination (clumping)

Hemolysis

**3** The *Rh blood group* is based on the presence or absence of the Rh surface antigen. The term **Rh positive (Rh+)** indicates the presence of the **Rh** surface antigen. It is referred to as "Rh" because this antigen was first discovered in *Rh*esus monkeys before it was found in humans. The absence of this antigen is indicated as **Rh negative (Rh−)**. When the complete blood type is recorded, "Rh" is usually omitted and blood type is reported as O negative (O−), A positive (A+), and so on.

> **? REVIEW**
> **B.** What is the most common blood type in the United States?

### Differences in Blood Type Distribution

| Population | Percentage with Each Blood Type | | | | |
|---|---|---|---|---|---|
| | **O** | **A** | **B** | **AB** | **Rh+** |
| **United States** | | | | | |
| Black American | 49 | 27 | 20 | 4 | 95 |
| Caucasian | 45 | 40 | 11 | 4 | 85 |
| Chinese American | 42 | 27 | 25 | 6 | 100 |
| Filipino American | 44 | 22 | 29 | 6 | 100 |
| Hispanic | 57 | 31 | 10 | 2 | 92 |
| Japanese American | 31 | 39 | 21 | 10 | 100 |
| Korean American | 32 | 28 | 30 | 10 | 100 |
| **Native North American** | 79 | 16 | 4 | <1 | 100 |
| **Native South American** | 100 | 0 | 0 | 0 | 100 |
| **Australian Aborigine** | 44 | 56 | 0 | 0 | 100 |

**4** Shown here are the results of blood typing tests on blood samples from four people. Drops of each person's blood are mixed with solutions containing antibodies to the surface antigens A, B, and Rh. Clumping (agglutination) occurs when the sample contains the corresponding surface antigen(s). The blood type for each person is shown at right. Because cross-reactions, or **transfusion reactions**, are so dangerous, care must be taken to ensure that the blood types of a blood donor and the recipient are **compatible**—that is, the donor's blood cells and the recipient's plasma will not cross-react.

> **? REVIEW**
> **C.** Which blood type(s) can be safely transfused into a person with type O− blood?

| Anti-A Antibodies | Anti-B Antibodies | Anti-Rh Antibodies | Blood type | Can receive blood from |
|---|---|---|---|---|
| Clumping | No clumping | Clumping | A+ | A+, O+, O− |
| No clumping | Clumping | Clumping | B+ | B+, O+, O− |
| Clumping | Clumping | Clumping | AB+ | A+, B+, AB+, O+, O− (universal recipient) |
| No clumping | No clumping | No clumping | O− | O− (universal donor) |

The presence of anti-A and/or anti-B antibodies is genetically determined and remains constant throughout life, regardless of whether the person has ever been exposed to foreign RBCs. In contrast, the plasma of an Rh-negative person does not necessarily contain anti-Rh antibodies. These antibodies are present only if the person has been sensitized by previous exposure to Rh-positive RBCs.

> **? INTEGRATION**
> **D.** Why can't a person with type A blood safely receive blood from a person with type B blood?

> **LO LEARNING OUTCOME**
> Explain the importance of blood typing and the basis for ABO and Rh incompatibilities.

## Module 17.8

# Hemolytic disease of the newborn is an RBC-related disorder caused by a cross-reaction between fetal and maternal blood types

Genes controlling the presence or absence of any surface antigen in the plasma membrane of a red blood cell are provided by both parents, so a child can have a blood type different from that of either parent. During pregnancy, when fetal and maternal vascular systems are closely intertwined, the mother's antibodies against RBC surface antigens may cross the placenta, attacking and destroying fetal RBCs. The resulting condition, called **hemolytic disease of the newborn** (**HDN**), has many forms, some quite dangerous and others so mild as to remain undetected.

Rh⁻
mother

Rh⁺
fetus

### First Pregnancy of an Rh– Mother with an Rh+ Infant

In its most common form, hemolytic disease of the newborn develops after an Rh⁻ woman has carried an Rh⁺ fetus. Unlike anti-A and anti-B antibodies, anti-Rh antibodies are able to cross the placenta and enter the fetal bloodstream.

**1** During a first pregnancy, problems seldom develop because very few fetal cells enter the maternal bloodstream, and thus the mother's immune system is not stimulated to produce anti-Rh antibodies.

Placenta

**During First Pregnancy**

Maternal blood supply and tissue

Fetal blood supply and tissue

### Second Pregnancy of an Rh– Mother with an Rh+ Infant

**4** If a future pregnancy involves an Rh⁺ fetus, maternal anti-Rh antibodies produced after the first delivery cross the placenta and enter the fetal bloodstream. These antibodies destroy fetal RBCs, producing a dangerous anemia. The fetal demand for red blood cells increases, and they leave the bone marrow and enter the bloodstream before completing their development. Because these immature RBCs are erythroblasts, HDN is also known as **erythroblastosis fetalis** (e-rith-rō-blas-TŌ-sis fē-TAL-is).

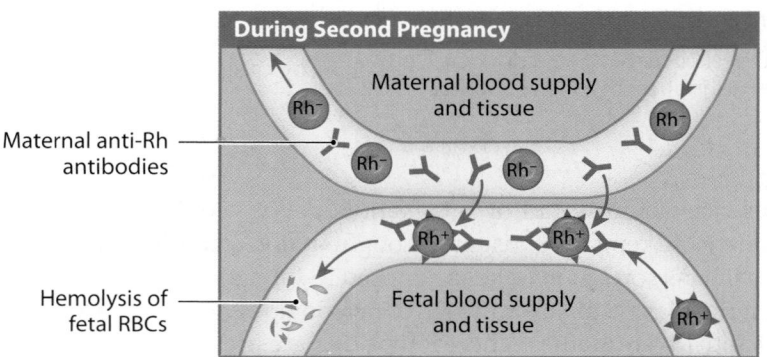

**During Second Pregnancy**

Maternal anti-Rh antibodies

Maternal blood supply and tissue

Hemolysis of fetal RBCs

Fetal blood supply and tissue

**❓ REVIEW**

**A.** Define *hemolytic disease of the newborn* (HDN).

**2** Exposure to fetal red blood cell antigens generally occurs during delivery, when bleeding takes place at the placenta and uterus. Such mixing of fetal and maternal blood can stimulate the mother's immune system to produce anti-Rh antibodies, leading to **sensitization**.

**3** About 20 percent of Rh⁻ mothers who carried Rh⁺ children become sensitized within 6 months of delivery. Because the anti-Rh antibodies are not produced in significant amounts until after delivery, a woman's first infant is not affected.

**Hemorrhaging at Delivery**

Maternal blood supply and tissue

Maternal RBC

Fetal blood supply and tissue

Rh antigen on fetal red blood cells

**Maternal Antibody Production**

Maternal blood supply and tissue

Maternal antibodies to Rh antigen

Without treatment, a fetus with erythroblastosis fetalis may die before delivery or shortly thereafter. A newborn with severe HDN is anemic, and the high concentration of circulating bilirubin produces jaundice (see Module 17.6). Because the maternal antibodies remain active in the newborn for 1–2 months after delivery, the infant's entire blood volume may require replacement to remove the maternal anti-Rh antibodies, as well as the damaged RBCs. Fortunately, the mother's anti-Rh antibody production can be prevented if anti-Rh antibodies (available under the name RhoGAM) are administered to the mother in weeks 26–28 of pregnancy and during and after delivery. These antibodies destroy any fetal RBCs that cross the placenta before they can stimulate a maternal immune response. Because maternal sensitization does not occur, no anti-Rh antibodies are produced. In the United States, this relatively simple procedure has almost entirely eliminated HDN mortality caused by Rh incompatibilities.

**? REVIEW**

**B.** Why is RhoGAM administered to pregnant Rh⁻ women?

**? INTEGRATION**

**C.** Does an Rh⁺ mother carrying an Rh⁻ fetus require a RhoGAM injection? Explain your answer.

**Lo LEARNING OUTCOME**

Describe hemolytic disease of the newborn, explain the clinical significance of the cross-reaction between fetal and maternal blood types, and cite preventive measures.

## Module 17.9

# The various types of white blood cells contribute to the body's defenses

Go to
MasteringA&P™
PAL 3.0
Cardiovascular System

White blood cells (WBCs), or **leukocytes**, have nuclei and other organelles and lack hemoglobin. Their common properties are described below. A phrase to help you remember the five types of WBCs is "Never Let Monkeys Eat Bananas" where "N" stands for neutrophils, "L" for lymphocytes, "M" for monocytes, "E" for eosinophils, and "B" for basophils.

RBC

Stained blood smear    LM × 1500

**1** **Granular leukocytes**, or *granulocytes*, have abundant cytoplasmic granules (secretory vesicles and lysosomes) that absorb histological stains, such as Wright stain or Giemsa stain.

**GRANULAR LEUKOCYTES**

Neutrophils

Eosinophils

Basophils

*WBCs can be divided into two classes*

**AGRANULAR LEUKOCYTES**

Monocytes

Lymphocytes

**2** **Agranular leukocytes**, or *agranulocytes*, contain secretory vesicles and lysosomes smaller than those in granulocytes. However, few, if any, of these cytoplasmic granules absorb histological stain.

### Shared Properties of WBCs

- WBCs circulate for only a short portion of their life span, using the bloodstream primarily to travel between organs and to rapidly reach areas of infection or injury. WBCs spend most of their time migrating through loose and dense connective tissues throughout the body.

- All WBCs can migrate out of the bloodstream. When circulating WBCs in the bloodstream become activated, they contact and adhere to the vessel walls and squeeze between adjacent endothelial cells to enter the surrounding tissue. This process is called **emigration**, or **diapedesis** (dī-ah-peh-DĒ-sis; *dia*, through + *pedesis*, a leaping).

- All WBCs are attracted to specific chemical stimuli. This characteristic, called **positive chemotaxis** (kē-mō-TAK-sis), guides WBCs to invading pathogens, damaged tissues, and other active WBCs.

- Neutrophils, eosinophils, and monocytes are capable of phagocytosis. They can engulf pathogens, cell debris, or other materials. Macrophages are monocytes that move from the bloodstream into peripheral tissues (see Module 4.11).

**?** **REVIEW**

**A.** Identify the five types of white blood cells.

**B.** How do basophils respond to tissue damage?

17

## White Blood Cells

| Cell Type | Quantity (Average Number per µL) | Appearance in a Stained Blood Smear | Functions | Remarks |
|---|---|---|---|---|
| **Neutrophils** | 4150 (range: 1800–7300) Differential count: 50–70% | Round cell; nucleus lobed and may resemble a string of beads; cytoplasm contains large, pale inclusions | Phagocytic: engulf pathogens or debris in injured or infected tissues; release cytotoxic enzymes and chemicals | Move into tissues after several hours; may survive for minutes to days, depending on tissue activity; produced in red bone marrow |
| **Eosinophils** | 165 (range: 0–700) Differential count: 2–4% | Round cell; nucleus generally has two lobes; cytoplasm contains large granules that generally stain bright red | Phagocytic: engulf antibody-labeled materials; release cytotoxic enzymes; reduce inflammation; increase in abundance in allergies and parasitic infections | Move into tissues after several hours; may survive for minutes to days, depending on activity in tissues; produced in red bone marrow |
| **Basophils** | 44 (range: 0–150) Differential count: <1% | Round cell; nucleus generally cannot be seen through dense, blue-stained granules in cytoplasm | Enter damaged tissues and release histamine and other chemicals that promote inflammation | Survival time unknown; assist mast cells of tissues in producing inflammation; produced in red bone marrow |
| **Monocytes** | 456 (range: 200–950) Differential count: 2–8% | Very large cell; nucleus kidney bean–shaped; abundant cytoplasm | Enter tissues and become macrophages; engulf pathogens or debris | Move into tissues after 1–2 days; survive for months or longer; produced primarily in red bone marrow |
| **Lymphocytes** | 2185 (range: 1500–4000) Differential count: 20–40% | Generally round cell, slightly larger than RBC; round nucleus; very little cytoplasm | Cells of lymphatic system; provide defense against specific pathogens or toxins | Survive for months to decades; circulate from blood to tissues and back; produced in red bone marrow and lymphoid tissues |

17

A variety of conditions, including infections, inflammation, and allergic reactions, cause characteristic changes in the populations of circulating WBCs. Examining a stained blood smear can provide a window on such changes. A **WBC differential count** identifies the types and numbers of white blood cells in a sample of blood. The values reported indicate the number, as a percentage, of each type of cell in a sample of 100 WBCs (see Module 17.4).

 **INTEGRATION**

**C.** Which type of white blood cell would you find in the greatest numbers in an infected cut?

 **LEARNING OUTCOME**

Categorize the various types of white blood cells on the basis of their structures and functions.

# Module 17.10
# The clotting response is a complex cascade of events that reduces blood loss

The process of **hemostasis** (*haima*, blood + *stasis*, halt), the stopping of bleeding, halts blood loss through the walls of damaged vessels. At the same time, it establishes a framework for tissue repairs. Hemostasis has three phases—*vascular*, *platelet*, and *coagulation*—in which many events happen at once, and all of them interact to some degree.

**? REVIEW**
**A.** Define *hemostasis*.

## Vascular Phase

The **vascular phase** of hemostasis lasts for roughly 30 minutes after the injury occurs. It is dominated by the response of the endothelial cells and the contraction, or **vascular spasm**, of smooth muscle of the vessel walls.

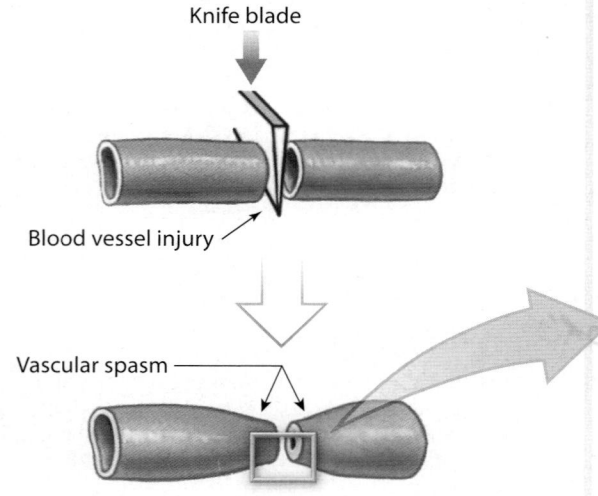

### Events of the Vascular Phase

- The endothelial cells contract and expose the underlying basement membrane to the bloodstream.

- The endothelial cells begin releasing chemical factors and local hormones. Endothelial cells also release **endothelins**, peptide hormones that (1) stimulate smooth muscle contraction and promote vascular spasms and (2) stimulate the division of endothelial cells, smooth muscle cells, and fibroblasts to accelerate the repair process.

- The endothelial plasma membranes become "sticky." A tear in the wall of a small artery or vein may be partially sealed off by the attachment of endothelial cells on either side of the break. In small capillaries, endothelial cells on opposite sides of the vessel may stick together and prevent blood flow along the damaged vessel. The stickiness also aids the attachment of platelets as the platelet phase gets under way.

## Platelet Phase

The **platelet phase** of hemostasis begins with the attachment of platelets to sticky endothelial surfaces, the basement membrane, exposed collagen fibers, and each other.

### Chemicals Released by Activated Platelets

- Adenosine diphosphate (ADP), which stimulates platelet aggregation and secretion

- Several chemicals that stimulate vascular spasms

- **Platelet factors**, proteins that play a role in blood clotting

- **Platelet-derived growth factor (PDGF)**, a peptide that promotes vessel repair

- Calcium ions ($Ca^{2+}$), which are required for platelet aggregation and in several steps in the clotting process

## Coagulation Phase

The **coagulation** (cō-ag-ū-LĀ-shun) **phase** of hemostasis does not start until 30 seconds or more after vessel damage. **Coagulation**, or blood clotting, involves intertwined cascades called the extrinsic, intrinsic, and common pathways. These pathways lead to the conversion of circulating fibrinogen (a soluble protein) into **fibrin** (an insoluble protein). As the fibrin network grows, blood cells and additional platelets become trapped, forming a **blood clot** that seals off the damaged portion of the vessel. **Procoagulants** (clotting factors) in the plasma play a key role in this phase. Essential clotting factors include $Ca^{2+}$ and 11 different proteins (identified by Roman numerals). Many clotting factors are **proenzymes** (inactive enzymes), which, when converted to active enzymes, direct essential reactions in the clotting response. The activation of one proenzyme commonly creates a **cascade**, or chain reaction of successive events.

### Common Pathway

The **common pathway** begins when enzymes from either the extrinsic or intrinsic pathway activate factor X. Activated factor X activates a complex called prothrombin activator. Prothrombin activator converts the proenzyme **prothrombin** into the enzyme **thrombin** (THROM-bin). Thrombin then completes the clotting process by converting fibrinogen to fibrin.

### Extrinsic Pathway

The **extrinsic pathway** begins with the release of **tissue factor** (factor III) by damaged endothelial cells or peripheral tissues. The greater the damage, the more tissue factor is released, and the faster clotting occurs. Tissue factor then combines with $Ca^{2+}$ and another clotting factor to form an enzyme complex capable of activating factor X, the first step in the common pathway.

Thrombin from the common pathway also speeds up clotting by stimulating additional tissue factor formation and platelet factor release. This is a positive feedback loop (see Module 1.18).

### Intrinsic Pathway

The **intrinsic pathway** begins with the activation of proenzymes exposed to collagen fibers at the injury site. This pathway proceeds with the assistance of **PF-3**, a platelet factor released by aggregating platelets. After a series of linked reactions, activated clotting factors combine to form an enzyme complex capable of activating factor X.

Factor X

Tissue factor complex

Prothrombin activator

Factor VII

Prothrombin → Thrombin

Factor X activator complex

Factors VIII and IX

$Ca^{2+}$

$Ca^{2+}$

Fibrin ← Fibrinogen

Factor III

Platelet factor (PF-3)

Activated proenzymes (usually factor XII)

Tissue damage

Contracted smooth muscle cells

Blood clot containing trapped RBCs — SEM × 1200

**17**

### Clot Retraction

Once the fibrin meshwork has formed, platelets and red blood cells stick to the fibrin strands. The platelets then contract, and the entire clot begins to undergo **clot retraction**, a process that continues over a period of 30–60 minutes and pulls the cut edges together.

**REVIEW**

**C.** Describe the events that follow the coagulation phase.

As repairs proceed, the clot gradually dissolves. This process, called **fibrinolysis** (fī-bri-NOL-i-sis), begins with the activation of the proenzyme **plasminogen** by thrombin, produced by the common pathway, and **tissue plasminogen activator** (**t-PA**), released by damaged tissues. The activation of plasminogen produces the enzyme **plasmin** (PLAZ-min), which erodes the foundation of the clot.

**REVIEW**

**B.** Briefly describe the vascular, platelet, and coagulation phases of hemostasis.

**LEARNING OUTCOME**

Discuss the mechanisms that control blood loss after an injury, and describe the reaction sequences responsible for blood clotting.

Mon.

# Module 17.11

# Blood disorders can be classified by their origins and the changes in blood characteristics

**1** The procedure called **venipuncture** (VĒN-i-punk-chur; *vena*, vein + *punctura*, a piercing) is crucial in diagnosing blood disorders. Fresh whole blood is generally collected from a superficial vein, such as the median cubital vein on the anterior surface of the elbow. Venipuncture is commonly used because (1) superficial veins are easy to locate, (2) the walls of veins are thinner than those of comparably sized arteries, and (3) blood pressure in the venous system is relatively low, so the puncture wound seals quickly. The most common clinical procedures examine venous blood.

**2** Blood disorders can have a variety of causes. The major categories of blood disorders are introduced here.

## Nutritional Blood Disorders

In **iron deficiency anemia**, normal hemoglobin synthesis cannot occur because either iron reserves or the dietary intake of iron is inadequate. Because developing RBCs cannot synthesize functional hemoglobin, they are unusually small (**microcytic**). Women are especially dependent on a normal dietary supply of iron because their iron reserves are about one-half that of a typical man.

A deficiency in **vitamin B$_{12}$** prevents normal stem cell divisions in the bone marrow, which can result in **pernicious** (per-NISH-us) **anemia**. Fewer red blood cells are produced, and those that are produced are abnormally large (**macrocytic***) and may develop bizarre shapes. Pernicious anemia can also result from a lack of intrinsic factor, a mucoprotein secreted by the stomach that is necessary for adequate vitamin B$_{12}$ absorption.

Calcium ions and **vitamin K** affect almost every aspect of the clotting process. All three pathways (intrinsic, extrinsic, and common) require Ca$^{2+}$, so any disorder that lowers blood Ca$^{2+}$ levels can impair blood clotting. Additionally, the liver requires adequate amounts of vitamin K for synthesizing four of the clotting factors, including prothrombin.

**? REVIEW**
**A.** Compare pernicious anemia with iron deficiency anemia.

## Congenital Blood Disorders

**Sickle cell disease (SCD)** is a group of inherited RBC disorders. All are due to mutations in the chains of the Hb molecule. Affected RBCs take on a sickled shape when they release bound oxygen, making the RBCs fragile and easily damaged. Moreover, a sickled RBC can become stuck in a capillary, blocking the flow of blood and oxygen to nearby tissues.

To develop the most severe SCD, **sickle cell anemia**, a person must have two copies of the sickling gene—one from each parent. If only one sickling gene is present, the person has the **sickling trait**. In such cases, most of the hemoglobin is of the normal form, and the RBCs

Normal and sickled RBCs

function normally. However, having the sickling trait gives the person some resistance to malaria, a deadly mosquito-borne parasitic disease. Infection of RBCs by the parasites induces sickling, and sickled cells containing the parasites are engulfed and destroyed by macrophages.

**Thalassemias** (thal-ah-SĒ-mē-uhs) are forms of SCD. They are also caused by an inability to produce adequate amounts of normal protein subunits of hemoglobin. The severity of different types of thalassemias depends on which and how many protein subunits are abnormal.

**Hemophilia** (hē-mō-FĒ-lē-a) is an inherited bleeding disorder. About 1 person in 10,000 is a hemophiliac, and of those, 80–90 percent are males. In most cases, hemophilia is caused by the reduced production of a single clotting factor. The severity of hemophilia varies, depending on how little clotting factor is produced. In severe cases, extensive bleeding occurs with relatively minor contact, and bleeding occurs at joints and around muscles.

17

## Blood Infections

Blood is normally free of microorganisms, but they can enter the blood through a wound or infection. **Bacteremia** is a condition in which bacteria circulate in blood but do not multiply there. **Viremia** is a similar condition associated with viruses. **Sepsis** (SEP-sis) is a widespread pathogenic infection of body tissues. Sepsis of the blood, or **septicemia** (formerly known as "blood poisoning"), results if pathogens are present and multiplying in the blood and spreading throughout the body.

Infant with septicemia from meningococcal bacteria

Mosquito biting a human

Red blood cell infected with *Plasmodium*

**Malaria** is a parasitic disease caused by several species of the protozoan *Plasmodium*. It is one of the most severe diseases in tropical countries, killing 1.5–3 million people per year, of which up to half are children under the age of 5. Malaria is transmitted from person to person by a mosquito. The parasite initially infects liver cells, then enlarges and fragments into smaller forms that infect red blood cells. Periodically, at intervals of 2–3 days, all of the infected RBCs rupture simultaneously and release more parasites that infect additional RBCs. The release and reinfection of RBCs corresponds to the cycles of fever and chills that characterize malaria. Dead RBCs can block blood vessels leading to vital organs, such as the kidney and brain, resulting in tissue death.

## Blood Cell Cancers

**Leukemias** are cancers of blood-forming tissues. The cancerous cells of leukemia do not form a compact tumor, but instead are spread throughout the body from their origin in red bone marrow. Both of the two types of leukemia—myeloid and lymphoid— are characterized by elevated levels of circulating WBCs. **Myeloid leukemia** is characterized by the presence of abnormal granulocytes (neutrophils, eosinophils, and basophils) or other cells of the bone marrow. **Lymphoid leukemia** involves lymphocytes and their stem cells. The first symptoms appear when immature and abnormal white blood cells appear in the bloodstream. Untreated leukemia is invariably fatal.

Abnormal WBCs (∗) seen in a blood smear of a patient with myeloid leukemia

**? REVIEW**

**B.** Identify the two types of leukemia.

## Degenerative Blood Disorders

In **disseminated intravascular coagulation (DIC)**, bacterial toxins activate several steps in the coagulation process that converts fibrinogen to fibrin within the circulating blood. Although much of the fibrin is removed by phagocytes or is dissolved by plasmin, small clots may block small vessels and damage nearby tissues. If the liver cannot produce enough circulating fibrinogen to keep pace with the rate at which fibrinogen is being removed, clotting abilities decrease, and uncontrolled bleeding may occur.

 **INTEGRATION**

**C.** Explain why venipuncture is a common clinical procedure for obtaining blood for examination.

 **LEARNING OUTCOME**

Explain how blood disorders are detected, and describe examples of the various categories of blood disorders.

# Structure and Function of Formed Elements

**Concept map:** Use each of the following terms once to fill in the blank boxes to correctly complete the map.

- iron-deficiency anemia
- reticulocyte count
- lymphoid leukemia
- congenital
- sepsis
- pernicious anemia
- MCV
- hematocrit
- nutritional
- RBC count
- MCH
- malaria
- Hb concentration
- infections
- myeloid leukemia
- hemophilia
- leukemias
- sickle cell disease

**Hematology** *studies*

**Blood tests** — *measure*
- **1** 37–54%
  - **2** 82–101 μm$^3$
- **3** 12–18 g/dL
  - **4** 27–34 μg
- **5** 4.2–6.3 million cells/μL
  - **6** 0.8%

**Blood-forming tissue** — *cancer*
- **7** — *such as*
  - **8** granulocytes
  - **9** agranulocytes

**Blood disorders** — *types*
- **10** — *such as*
  - **13** lack vitamin B$_{12}$
  - **14** lack iron
- **11** — *such as*
  - **15** abnormal Hb
  - **16** abnormal clotting
- **12** — *such as*
  - **17** "blood poisoning"
  - **18** *Plasmodium*

**Matching:** Match each lettered description with the most closely related term.

a. blood disorders
b. basophil
c. jaundice
d. venipuncture
e. red bone marrow
f. mature RBCs
g. parasitic disease
h. blood clotting elements
i. erythropoietin
j. cross-reaction
k. macrophages
l. lymphocyte

**19** Myeloid tissue          19 _____
**20** Anucleated              20 _____
**21** Dyscrasias              21 _____
**22** Monocytes               22 _____
**23** Granulocyte             23 _____
**24** Agglutination           24 _____
**25** Agranulocyte            25 _____
**26** Bilirubin               26 _____
**27** Median cubital vein     27 _____
**28** Malaria                 28 _____
**29** Hormone                 29 _____
**30** Platelets               30 _____

**Section integration:** Why are mature red blood cells in humans incapable of protein synthesis and mitosis?

**31** _____

17

# Study Outline

## SECTION 1 • Plasma and Formed Elements

### Module 17.1

**Blood is the fluid portion of the cardiovascular system** p. 631

1. The **cardiovascular system** includes the heart, blood vessels, and blood.

2. The **heart** propels blood and maintains blood pressure, and the **blood vessels** distribute blood around the body.

3. **Blood** functions to transport gases, nutrients, hormones, and waste products; regulate pH and ion composition of interstitial fluid; restrict fluid loss at injury sites; provide defense against disease; and assist in temperature regulation.

### Module 17.2

**Blood is a fluid connective tissue containing plasma and formed elements** p. 632

4. Blood is composed of a liquid matrix called **plasma**. Suspended in the plasma are blood cells and cell fragments called **formed elements**.

5. Plasma forms 55 percent of blood volume; formed elements are 45 percent of blood volume.

6. Plasma is primarily composed of water (92 percent), which functions in the transportation of organic and inorganic molecules, formed elements, and heat.

7. The remaining composition of plasma is 7 percent **plasma proteins** (**albumins**, **globulins**, **fibrinogen**, enzymes, and hormones), and 1 percent other solutes (**electrolytes**, **organic nutrients**, and **organic wastes**).

8. Formed elements are 99.9 percent **red blood cells (RBCs)**, or **erythrocytes**, which transport oxygen in blood.

9. The remaining formed elements are **platelets** (cell fragments important in blood clotting) and **white blood cells (WBCs)**, or **leukocytes**. There are five classes of leukocytes that function in body defense mechanisms.

### Module 17.3

**Formed elements are produced by stem cells in red bone marrow** p. 634

10. **Hemopoiesis** is the process by which the formed elements develop.

11. All formed elements are produced by **hemocytoblasts**.

12. Hemocytoblasts give rise to **lymphoid stem cells** and **myeloid stem cells**. Lymphoid stem cells give rise to lymphocytes. Myeloid stem cells give rise to all other types of formed elements.

13. **Erythropoietin (EPO)** stimulates RBC development. EPO is released into the plasma by the kidneys during **hypoxia** (low oxygen levels in the tissues).

## SECTION 2 • Structure and Function of Formed Elements

### Module 17.4

**Hematology is the study of blood and blood-forming tissues** p. 637

14. **Hematology** is the study of blood, blood-forming tissues, and blood disorders.

15. Blood circulates through nearly every tissue of the body, so **dyscrasias**, or blood disorders, can have systemic effects.

16. Various diagnostic blood tests measure changes in the proportions and quantities of formed elements.

17. A **complete blood count (CBC)** is a common blood test that determines the RBC count, WBC count, erythrocyte indices (hemoglobin content), hematocrit, platelet count, and WBC differential count.

18. Several common RBC tests assess the number, size, shape, and maturity of red blood cells. Various terms are used to describe abnormal measured values.

### Module 17.5

**Red blood cells, the most common formed elements, contain hemoglobin that transports respiratory gases** p. 638

19. The **red blood cell count** is a blood test that reports the number of RBCs per microliter of whole blood.

20. A red blood cell has the shape of a biconcave disc and a flexible plasma membrane. As a result, RBCs have a large surface-to-area ratio, stack into **rouleaux**, and bend to fit narrow capillaries.

21. Mature RBC's do not have a nucleus and are **anucleate**. They have also lost most of their other organelles (including ribosomes) and have a life span of less than 120 days.

22. Ninety-five percent of RBC intracellular protein is **hemoglobin (Hb)**, which is responsible for transporting respiratory gases. Each Hb molecule contains two **alpha (α) chains** and two **beta (β) chains** of polypeptides. Each chain contains a single **heme** molecule.

*Hemoglobin has a complex quaternary structure.*

**17**

23. Each heme holds an iron ion and interacts with an oxygen molecule to form **oxyhemoglobin**, or **HbO₂**. This is a weak interaction that is easily reversed.

24. If a hemoglobin molecule is not bound to oxygen, it is called **deoxyhemoglobin**, or reduced hemoglobin.

## Module 17.6

### Red blood cells are continually produced and their components recycled or eliminated    p. 640

25. About 1 percent of the circulating RBCs are replaced each day.

26. Red blood cell formation, or **erythropoiesis**, occurs only in red bone marrow, or **myeloid tissue**, within the vertebrae, sternum, ribs, skull, scapulae, pelvis, and proximal limb bones.

*Red blood cells are produced in the red bone marrow.*

27. Developing RBCs absorb amino acids and Fe²⁺ and synthesize hemoglobin (Hb) molecules.

28. **Proerythroblasts** differentiate into the various stages of **erythroblasts**. After four days of differentiation and synthesizing hemoglobin, the erythroblast becomes a **normoblast** and sheds its nucleus. It is now a **reticulocyte** and contains 80 percent of the Hb of a mature RBC. After two days in the bloodstream, reticulocytes complete their maturation.

29. After an average life span of 120 days, RBCs are engulfed by macrophages. Ten percent of RBCs will **hemolyze** (rupture) before this, and macrophages will engulf their cell fragments and Hb molecules.

30. The iron extracted from the heme molecules is either stored in the macrophage or transported back to bone marrow by **transferrin**. The globular proteins are broken down into their amino acids and used by other cells. The heme unit is stripped of iron, converted to **biliverdin**, which is converted to **bilirubin**, and released into the bloodstream.

31. Bilirubin binds to albumin and is transported to the liver for excretion in bile. In the large intestine, bacteria convert bilirubin to **urobilins** and **stercobilins**, which are eliminated in feces. The kidneys excrete some hemoglobin and urobilins.

32. Hemoglobin that is not phagocytized breaks down, and the alpha and beta chains are eliminated in urine. The resulting red or brown urine is called **hemoglobinuria**. The presence of intact RBCs in urine is called **hematuria**.

## Module 17.7

### Blood type is determined by the presence or absence of specific surface antigens on RBCs    p. 642

33. **Blood type** is a classification determined by specific, genetically determined **surface antigens** in RBC plasma membranes. These surface antigens are recognized as "self" and ignored by your immune system, but your plasma may contain antibodies that will attack the antigens on "foreign" RBCs.

34. There are four blood types based on the presence or absence of the A and B surface antigens (**agglutinogens**).

35. **Type A** blood has surface antigen A and contains the anti-B antibodies, which will attack type B surface antigens.

36. **Type B** blood has surface antigen B and contains the anti-A antibodies, which will attack type A surface antigens.

37. **Type AB** blood has both A and B surface antigens and neither anti-A nor anti-B antibodies.

38. **Type O** blood has neither A nor B surface antigens and has both anti-A and anti-B antibodies.

39. RBCs will **agglutinate**, or clump, if the surface antigens on RBCs of one blood type are exposed to the corresponding antibodies (**agglutinins**) from another blood type. Such a **cross-reaction** is dangerous because the RBC clumps and fragments can plug small blood vessels and damage many organs and tissues. The RBCs may undergo hemolysis.

40. **Rh positive (Rh⁺)** indicates the presence of the Rh surface antigen. The absence of this surface antigen is indicated as **Rh negative (Rh⁻)**.

*RBCs will agglutinate if the surface antigens on RBCs of one blood type are exposed to the corresponding antibodies of another blood type.*

17

41. A **transfusion reaction** is so dangerous that blood from the donor and blood from the recipient must be **compatible**. A blood type test can determine whether the donor's RBCs are compatible with the recipient's blood plasma.

### Module 17.8

**Hemolytic disease of the newborn is an RBC-related disorder caused by a cross-reaction between fetal and maternal blood types p. 644**

42. Genes that determine blood type are contributed by both parents, so a child can have a blood type different from that of either parent.

43. **Hemolytic disease of the newborn (HDN)**, or **erythroblastosis fetalis**, can result if the mother's antibodies cross the placenta and attack and destroy fetal RBCs.

44. The most common form of HDN develops after an Rh⁻ woman has carried an Rh⁺ fetus. During delivery of the first pregnancy, fetal RBC antigens will stimulate the mother's immune system to produce anti-Rh antibodies, leading to **sensitization**. The process of sensitization can take months so the woman's first infant is not affected.

Rh⁺ fetus

45. If a future pregnancy involves an Rh⁺ fetus, maternal anti-Rh antibodies can cross the placenta and cause HDN. Without treatment the fetus will die before delivery or shortly thereafter.

46. Injections of RhoGAM during and after a first pregnancy can prevent the mother from producing anti-Rh antibodies.

### Module 17.9

**The various types of white blood cells contribute to the body's defenses p. 646**

47. The shared properties of **white blood cells (WBCs)**, or **leukocytes**, are circulation in bloodstream for only a short period of time; **emigration**, or **diapedesis**, out of the bloodstream and into areas of tissue or organ infection or injury; attraction to pathogens or tissue damage through **positive**

chemotaxis; and phagocytosis (neutrophils, eosinophils, and monocytes).

48. The granular leukocytes are neutrophils, eosinophils, and basophils.

49. The agranular leukocytes are monocytes and lymphocytes.

50. A **WBC differential count** can indicate a variety of pathogenic conditions, such as infections, inflammation, and allergic reactions.

### Module 17.10

**The clotting response is a complex cascade of events that reduces blood loss p. 648**

51. **Hemostasis** is the process of stopping blood loss through damaged blood vessels.

52. The **vascular phase** of hemostasis is dominated by endothelial cells and contraction, or **vascular spasm**, of smooth muscle of the vessel walls. The endothelial cells release **endothelins** that stimulate cell division, and their plasma membranes become sticky, which helps close the vessel and aid platelet attachment.

Blood clot containing trapped RBCs    SEM × 1200

53. The **platelet phase** of hemostasis begins when platelets attach to the sticky endothelial surfaces, to exposed collagen fibers, and to each other.

54. The **coagulation phase** of hemostasis includes **coagulation** (blood clotting), a complex process leading to the conversion of fibrinogen to **fibrin**, forming a fibrin meshwork of platelets and RBCs stuck to the fibrin strands. The platelets contract in **clot retraction** to pull the cut edges together.

55. The clot gradually dissolves in a process called **fibrinolysis**.

### Module 17.11

**Blood disorders can be classified by their origins and the changes in blood characteristics p. 650**

56. Diagnosis of blood disorders requires **venipuncture**, usually from the median cubital vein, to collect fresh whole blood.

57. Blood disorders can be classified as nutritional, congenital, infectious, cancerous, or degenerative.

**17**

# Chapter Review Questions

**Labeling:** Label the blood cells shown here.

1 _____  2 _____  3 _____  4 _____  5 _____  6 _____

**True/False:** Indicate whether each statement is true or false.

7   Globulins make up 60 percent of the plasma proteins.

8   Monocytes enter damaged tissues and release histamines that promote inflammation.

9   Red blood cells have an average life span of 120 days.

10  The first phase of hemostasis is the coagulation phase.

11  Neutrophils are the most abundant white blood cells.

| | |
|---|---|
| 7 | F |
| 8 | F |
| 9 | T |
| 10 | F |
| 11 | T |

**Multiple choice:** Select the correct answer from the list provided.

12  Plasma percentage of the whole blood volume; formed elements percentage of whole blood volume.
- [X] a) 55, 45
- [ ] b) 45, 55
- [ ] c) 92, 8
- [ ] d) 63, 37

13  A hemoglobin molecule is composed of
- [ ] a) two protein chains and two heme units.
- [ ] b) four protein chains and two heme units.
- [X] c) four protein chains and four heme units.
- [ ] d) one protein chain and one heme unit.

14  The correct order of erythrocyte formation is
- [X] a) hemocytoblast, myeloid stem cell, progenitor cell, proerythroblast, erythroblast stages, reticulocyte, erythrocyte.
- [ ] b) hemocytoblast, progenitor cell, myeloid stem cell, proerythroblast, reticulocyte, erythroblast stages, erythrocyte.
- [ ] c) hemocytoblast, myeloid stem cell, progenitor cell, reticulocyte, proerythroblast, erythroblast stages, erythrocyte.
- [ ] d) myeloid stem cell, hemocytoblast, progenitor cell, proerythroblast, reticulocyte, erythroblast stages, erythrocyte.

15  Megakaryocytes shed cytoplasm in small, membrane-enclosed packets that enter the bloodstream as
- [ ] a) neutrophils.
- [ ] b) lymphocytes.
- [ ] c) monocytes.
- [X] d) platelets.

16  Which of the following blood types is the "universal donor"?
- [ ] a) O⁺
- [X] b) O⁻
- [ ] c) AB⁺
- [ ] d) AB⁻

17  When an RBC is engulfed by a macrophage, the hemoglobin heme units are stripped of their iron and converted to
- [ ] a) bilirubin.
- [X] b) biliverdin.
- [ ] c) urobilins.
- [ ] d) stercobilins.

18  The percentage of formed elements in a sample of whole blood is called
- [ ] a) hemoglobinuria.
- [X] b) hematocrit.
- [ ] c) hematuria.
- [ ] d) oxyhemoglobin.

**Short answer**

19  Describe the various types of leukemias.

20  What is the role of blood in stabilizing and maintaining body temperature?

21  Identify the RBC surface antigens and plasma antibodies for blood types A, B, AB, and O.

## MasteringA&P™

Access more chapter study tools online in the MasteringA&P Study Area:

■ **Chapter Quizzes, Chapter Practice Tests, and Clinical Case Studies**

■ Practice Anatomy Lab™ 3.0 — PAL™3.0

■ Interactive Physiology™ — iP2™

■ A&P Flix™ — A&PFlix™

■ PhysioEx™ 9.1 — PhysioEx™ 9.1

# Chapter Integration • Applying what you have learned

## The dangerous search for better cycling

Piero is an elite competitive cyclist who aspires to compete on the world stage. Despite his intense training in the past year along the coast in his hometown of Miami, he has been unable to achieve "the next level" of performance. He has heard about a drug that endurance athletes can take that could improve his athletic performance, but he has concerns because he also heard there could be serious side effects from using this drug.

Using what you have just learned about blood, answer the following questions.

**1** Which hormone do you think this drug is performing like?

**2** How is it that this drug could help him in his cycling?

**3** Which serious side effects do you think Piero would be at risk for if he were to use this drug?

**4** Are there training techniques other than drugs that Piero could try instead?

## An unwelcome change in blood cells

Ursula is a 27-year-old single female with no known medical history. A normally active person with a well-balanced diet, she is rarely ill. One weekend, Ursula felt as though she was "coming down with something." On Saturday, she was fatigued but had no trouble eating. By Sunday, she was extremely tired and simply could not eat. Monday afternoon, her lethargy was so profound that she was unable to get out of bed, so she called a family member to take her to the physician's office.

While her physician was performing a physical exam, her blood pressure plummeted to 60/20, her heart rate increased to 190 beats per minute, and her body temperature was 3 degrees below normal (95.6°F). She was rushed to the hospital, where blood was drawn and tested. Obtaining blood using a venipuncture was nearly impossible because Ursula was in the beginning stages of vascular collapse. Her gums and skin were pale.

When viewed through a microscope, Ursula's red blood cells were smaller than normal and spherically shaped and thus were characterized as "spherocytes." Antibodies in her plasma were attacking her red blood cells, causing them to hemolyze and transform into "spherocytes." Additionally, her blood's hemoglobin levels were well below normal. Ultimately, Ursula was diagnosed with immune-mediated hemolytic anemia, and crisis treatment—a blood transfusion and the administration of corticosteroids and chemotherapeutic agents—was begun.

**5** Describe the shape of normal red blood cells, and define *hemolysis*.

**6** Explain why Ursula's body temperature had dropped significantly.

**7** Provide a plausible explanation for Ursula's fatigue.

17

# 18

**LEARNING OUTCOMES**

These Learning Outcomes correspond by number to this chapter's modules and indicate what you should be able to do after completing the chapter. They also appear at the end of each module.

# The Heart and Cardiovascular Function

## Module 18.1

# The heart has four chambers that pump and circulate blood through the pulmonary and systemic circuits

Go to
MasteringA&P™
PAL™3.0
Cardiovascular

Recall that the cardiovascular system consists of the *heart* and a network of *blood vessels* transporting blood between the heart and peripheral tissues (see Module 17.1). Blood vessels of the *pulmonary circuit* carry blood to and from the gas exchange surfaces of the lungs, while blood vessels of the *systemic circuit* transport blood to and from the rest of the body.

**1** The heart is located near the anterior chest wall, directly behind the sternum.

**?** **REVIEW**

**A.** Describe the location and position of the heart.

**B.** Compare the base of the heart with the apex.

The **base** of the heart is at its superior border, where the largest veins and arteries are attached. The base sits posterior to the sternum at the level of the third costal cartilage, centered about 1.2 cm (0.5 in.) to the left side.

Ribs

The inferior, pointed tip of the heart is the free **apex** (Ā-peks). A typical adult heart measures approximately 12.5 cm (5 in.) from the base to the apex. The apex reaches the fifth intercostal space, approximately 7.5 cm (3 in.) to the left of the midline.

**2** The **heart** has four muscular chambers, two associated with each circuit—in essence, the heart is two side-by-side pumps. The *right atrium* (Ā-trē-um; entry chamber; plural, *atria*) receives blood from the systemic circuit and passes it to the *right ventricle* (VEN-tri-kel; little belly), which then pumps blood into the pulmonary circuit. The *left atrium* collects blood from the pulmonary circuit and empties it into the *left ventricle*, which pumps blood into the systemic circuit. When the heart beats, first the atria contract, and then the ventricles contract. The two ventricles contract at the same time and eject equal volumes of blood into the pulmonary and systemic circuits.

**?** **REVIEW**

**C.** Name the four chambers of the heart.

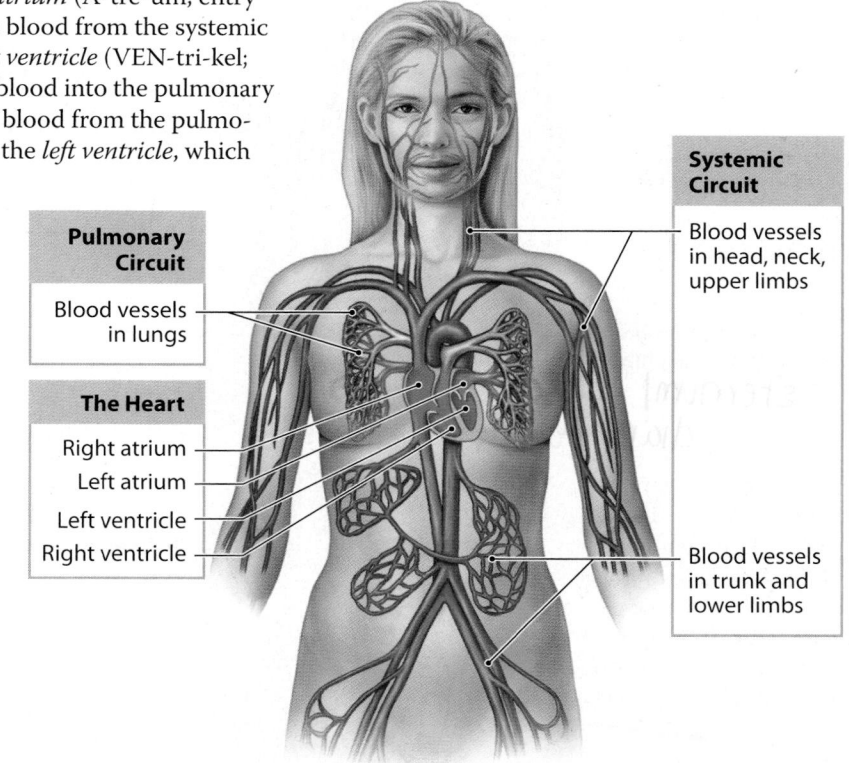

**Pulmonary Circuit**

Blood vessels in lungs

**The Heart**

Right atrium
Left atrium
Left ventricle
Right ventricle

**Systemic Circuit**

Blood vessels in head, neck, upper limbs

Blood vessels in trunk and lower limbs

**?** **INTEGRATION**

**D.** Compare the volume of blood each circuit receives from contraction of the ventricles.

**Lo** **LEARNING OUTCOME**

Describe the heart's location, shape, its four chambers, and the pulmonary and systemic circuits.

This section examines the anatomy of the heart. We then consider the regulation of cardiac function in Section 2.

**659**

# Module 18.2

# The heart is located in the mediastinum and is enclosed by the pericardial cavity

**1** This anterior view of the chest shows the position of the heart and major blood vessels relative to the ribs, lungs, and diaphragm. The heart sits in the anterior portion of the **mediastinum** (mē-dē-as-TĪ-num), the region between the two pleural cavities. This mass of connective tissue stabilizes the positions of the great vessels, thymus, esophagus, and trachea.

**? REVIEW**

**A.** Define *mediastinum*.

Trachea

Thyroid gland

First rib (cut)

Base of heart

Right lung

Left lung

Apex of heart

Diaphragm

**Anterior view of chest cavity**

**2** The heart is surrounded by the **pericardium**. The pericardium consists of a *fibrous pericardium* and a *serous pericardium*. A dense network of stabilizing fibrous bands extend from the fibrous pericardium to the sternum and diaphragm. The two-layered serous pericardium is made up of a parietal layer and a visceral layer. These layers are separated by a potential fluid-filled space called the *pericardial cavity*.

The pericardial cavity contains 15–50 mL of **pericardial fluid**, secreted by the pericardial serous membranes. This fluid acts as a lubricant that reduces friction between the opposing surfaces as the heart beats. Pathogens can infect the pericardium, producing the condition **pericarditis**. The inflamed pericardial surfaces rub together, producing a distinctive scratching sound that can be heard through a stethoscope.

Base of heart

Fibrous pericardium

Cut edge of visceral layer of serous pericardium

Cut edge of parietal layer of serous pericardium

| **Parietal Layer of Serous Pericardium** |
| --- |
| Areolar tissue |
| Mesothelium |

Fibrous attachment to sternum and diaphragm

*sternum/ diaphragm*

Apex of heart

**3** To visualize the relationship between the heart, pericardium, and the pericardial cavity, imagine pushing your fist toward the center of a large, partially inflated balloon. The balloon represents the pericardium, and your fist is the heart. Your wrist, where the balloon folds back on itself, corresponds to the base of the heart, to which the **great vessels**, the largest veins and arteries in the body, are attached. The air space inside the balloon corresponds to the pericardial cavity.

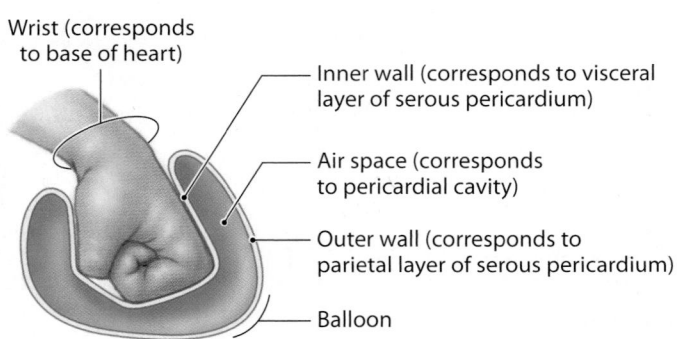

Wrist (corresponds to base of heart)

Inner wall (corresponds to visceral layer of serous pericardium)

Air space (corresponds to pericardial cavity)

Outer wall (corresponds to parietal layer of serous pericardium)

Balloon

**4** This is a diagrammatic superior view of a partial dissection of the thoracic cavity. The image illustrates the position of the pericardial cavity and the physical relationships among the components in the mediastinum. Also shown are three of the great vessels attached to the heart—the aorta, superior vena cava, and pulmonary trunk.

**REVIEW**

**B.** Describe the heart's location in the body.

Esophagus

Posterior mediastinum

Aorta (arch segment removed)

Left pulmonary artery

Right pleural cavity

*Right lung*

*Left lung*

Left pleural cavity

Bronchus of lung

Left pulmonary vein

Right pulmonary artery

Aortic arch

Pulmonary trunk

Right pulmonary vein

Left atrium

Superior vena cava

Left ventricle

Right atrium

Pericardial cavity

Visceral layer of the serous pericardium

Parietal Pericardium

Right ventricle

Anterior mediastinum

---

### Clinical Note

## Fluid accumulation in the pericardial cavity can cause cardiac tamponade

As you can see in this image, the heart does not have a lot of empty space around it—the thoracic cavity is very crowded. Traumatic injuries that damage the pericardium or chest wall can result in fluid accumulation within the pericardial cavity, which can restrict the movement of the heart. This condition, called **cardiac tamponade** (tam-po-NĀD; *tampon*, plug), can also be caused by acute pericarditis.

**INTEGRATION**

**C.** Why can cardiac tamponade be a life-threatening condition?

**LO** **LEARNING OUTCOME**

Describe the location and general features of the heart.

# The heart wall contains concentric layers of cardiac muscle tissue

**1** This is a view of a section taken from the wall of the heart and the surrounding pericardium. The heart wall contains three layers: pericardium, myocardium, and endocardium.

**? REVIEW**

**A.** From superficial to deep, name the layers of the heart wall.

## The Pericardium

The **pericardium** (pehr-ih-KAR-dē-um) surrounds the heart. It consists of an outer fibrous pericardium and a two-layered **serous pericardium**.

## Parietal Layer of Serous Pericardium

The **parietal layer** lines the outer wall of the pericardial cavity. The parietal layer is in contact with dense fibrous tissue of the fibrous pericardium.

Connective tissues

Pericardial cavity (contains pericardial fluid)

Dense fibrous tissue

Areolar tissue

Mesothelium ⎤ Parietal layer

## Myocardium

The **myocardium**, or muscular wall of the heart, forms both atria and ventricles. This middle layer contains cardiac muscle tissue, blood vessels, and nerves. The myocardium consists of concentric layers of cardiac muscle tissue.

## Visceral Layer of Serous Pericardium

The **visceral layer**, or **epicardium**, covers the outer surface of the heart. This portion of the serous pericardium consists of an exposed mesothelium and an underlying layer of areolar tissue that is attached to the myocardium (see Module 4.5).

## Endocardium

The inner surfaces of the heart, including those of the heart valves, are covered by the **endocardium**, a simple squamous epithelium and underlying areolar tissue. The squamous epithelial lining of the cardiovascular system is called an *endothelium*. The endothelium of the heart is continuous with the endothelium of the attached great vessels.

Endothelium

Areolar tissue

Heart wall

Mesothelium ⎤ Visceral layer

Areolar tissue

**? REVIEW**

**B.** Describe the tissue layers of the pericardium.

**2** The atrial myocardium contains muscle bundles that wrap around the atria and encircle the great vessels. Superficial ventricular muscles wrap around both ventricles; deeper muscle layers spiral around and between the ventricles toward the apex.

Atrial musculature forms bands that wrap around the atria in a figure-eight pattern.

Ventricular musculature forms bands that spiral around the ventricles.

**3** This light micrograph shows cardiac muscle tissue. Histological characteristics that distinguish cardiac muscle tissue from skeletal muscle tissue: (1) small cell size, (2) a single, centrally located nucleus, (3) branching interconnections between cells called *intercalated discs*, and (4) specialized intercellular connections.

Each cardiac muscle cell is connected to several others at specialized sites known as **intercalated** (in-TER-ka-lā-ted) **discs**.

**Cardiac muscle tissue**     LM × 575

**4** Cardiac muscle cells are found only in the heart. Like skeletal muscle fibers, cardiac muscle cells contain organized myofibrils, and the presence of many aligned sarcomeres gives the cells a striated appearance. They are almost totally dependent on aerobic metabolism to obtain the energy they need to continue contracting. The sarcoplasm of a cardiac muscle cell contains many mitochondria and abundant reserves of myoglobin that store oxygen. Because these cells are metabolically very active and have a high demand for oxygen and nutrients, cardiac tissues are richly supplied with capillaries.

Cardiac muscle cells are relatively small, averaging 10–20 μm in diameter and 50–100 μm in length.

Intercalated disc (sectioned)

Nucleus

Mitochondria

Bundles of myofibrils

Intercalated disc

**5** At an intercalated disc, the plasma membranes of two adjacent cardiac muscle cells are extensively intertwined and bound together by gap junctions and desmosomes (see Module 4.4). These connections help stabilize the positions of adjacent cells. The gap junctions allow ions and small molecules to move from one cell to another. This creates a direct electrical connection between the two muscle cells. An action potential can travel across an intercalated disc, moving quickly from one cardiac muscle cell to another. Myofibrils in the two interlocking muscle cells are firmly anchored by Z lines to the membrane at the intercalated disc and can "pull together" with maximum efficiency. Because the cardiac muscle cells are mechanically, chemically, and electrically connected to one another, the entire tissue resembles a single, enormous muscle cell. For this reason, cardiac muscle has been called a **functional syncytium** (sin-SISH-ē-um; a fused mass of cells).

Gap junction

**Intercalated Disc**

Z lines bound to opposing plasma membranes

Desmosomes

**? INTEGRATION**

**C.** Why is it important that cardiac tissue contain many mitochondria and capillaries?

**Lo LEARNING OUTCOME**

Describe the structure of the pericardium and explain its functions, identify the layers of the heart wall, and describe the structures and functions of cardiac muscle.

# The boundaries between the four chambers of the heart can be identified on its external surface

**1** The four cardiac chambers can easily be identified in a superficial view of the anterior surface of the heart. The left and right atria have relatively thin muscular walls and are highly expandable. When not filled with blood, the outer portion of each atrium deflates and becomes a lumpy, wrinkled flap. Shallow grooves, or **sulci** (singular, *sulcus*), mark the boundaries between the atria and ventricles and between the left and right ventricles. The connective tissue of the epicardium generally contains substantial amounts of fat, especially along the sulci. In fresh or preserved hearts, this fat must be stripped away to expose the underlying grooves. These sulci also contain the coronary arteries and veins that carry blood to and from the cardiac muscle.

Base

Aortic arch

Ascending aorta

Superior vena cava

The **ligamentum arteriosum** attaches the pulmonary trunk of the pulmonary circuit to the aortic arch of the systemic circuit. It is a fibrous remnant of the fetal connection between the aorta and the pulmonary trunk (see Module 19.24).

Pulmonary trunk

Auricle of left atrium

The **auricle** (AW-ri-kl; *auris*, ear) of the atrium is an expandable pouch. It has this name because it reminded early anatomists of the external ear.

Right atrium

Fat and vessels in the anterior interventricular sulcus

Right ventricle

Left ventricle

The **coronary sulcus**, a deep groove, marks the border between the atria and the ventricles. It contains fat and vessels.

The **anterior interventricular sulcus** is a shallow depression on the anterior surface that marks the boundary between the left and right ventricles.

**Anterior surface**

apex

**? REVIEW**

**A.** The anterior view of the heart is dominated by which chambers?

**2** This view of the posterior surface of the heart shows the left atrium and its connection to the pulmonary veins of the pulmonary circuit. It also shows the right atrium and its connection to the superior and inferior venae cavae of the systemic circuit and coronary veins.

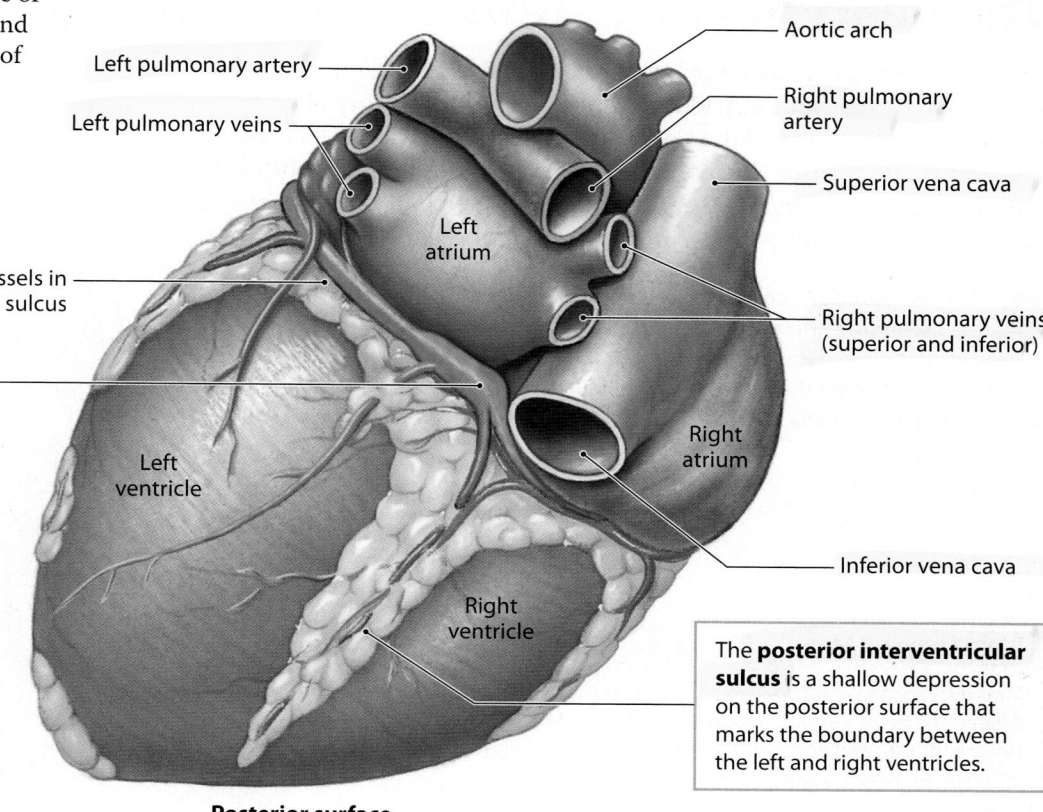

The **coronary sinus** carries blood collected from the myocardium by numerous coronary veins and conveys the blood to the right atrium.

Left pulmonary artery

Left pulmonary veins

Fat and vessels in the coronary sulcus

Left atrium

Left ventricle

Right ventricle

Aortic arch

Right pulmonary artery

Superior vena cava

Right pulmonary veins (superior and inferior)

Right atrium

Inferior vena cava

**Posterior surface**

The **posterior interventricular sulcus** is a shallow depression on the posterior surface that marks the boundary between the left and right ventricles.

**?** **REVIEW**

**B.** Which structures collect blood from the myocardium, and into which heart chamber does this blood flow?

**?** **REVIEW**

**C.** Name and describe the shallow depressions and grooves found on the heart's external surface.

**3** As you see in this anterior view, a dissected heart from a preserved cadaver is not conveniently color-coded for identification of its structures. However, the removal of fat does provide an unobstructed view of coronary blood vessels, the topic of the next module.

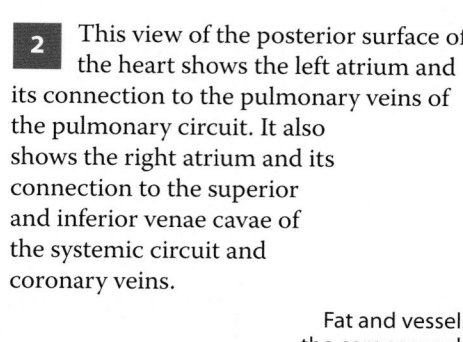

Aortic arch

Ascending aorta

Superior vena cava

Auricle of right atrium

Right atrium

Right coronary artery

Coronary sulcus

Marginal branch of right coronary artery

Left subclavian artery

Left common carotid artery

Brachiocephalic trunk

Ligamentum arteriosum

Left pulmonary artery

Pulmonary trunk

Auricle of left atrium

Left coronary artery

Anterior interventricular sulcus

Left ventricle

Right ventricle

Anterior interventricular artery

**Cadaver dissection, anterior view**

**LO** **LEARNING OUTCOME**
Describe the cardiac chambers and the heart's external anatomy.

# Module 18.5
# The heart has an extensive blood supply

The heart works continuously, so cardiac muscle cells require reliable supplies of oxygen and nutrients. Although a great volume of blood flows through the chambers of the heart, the myocardium needs its own, separate blood supply. The coronary circulation supplies that blood to the muscle tissue of the heart. During maximum exertion, blood flow to the myocardium may increase to nine times that of resting levels.

**1** The left and right **coronary arteries** originate at the base of the ascending aorta, where blood pressure is the highest in the systemic circuit. However, myocardial blood flow is not steady: It peaks while the heart muscle is relaxed and almost ceases while it contracts.

### Left Coronary Artery

The **left coronary artery** supplies blood to the left ventricle, left atrium, and the *interventricular septum* separating the left and right ventricles.

Circumflex artery

The large **anterior interventricular** (left anterior descending) **artery** runs along the surface within the anterior interventricular sulcus.

### Right Coronary Artery

The **right coronary artery**, which follows the coronary sulcus around the heart, supplies blood to the right atrium, portions of both ventricles, and portions of the electrical conducting system of the heart, which controls and coordinates the heartbeat.

**Marginal arteries** from the right coronary artery supply the surface of the right ventricle.

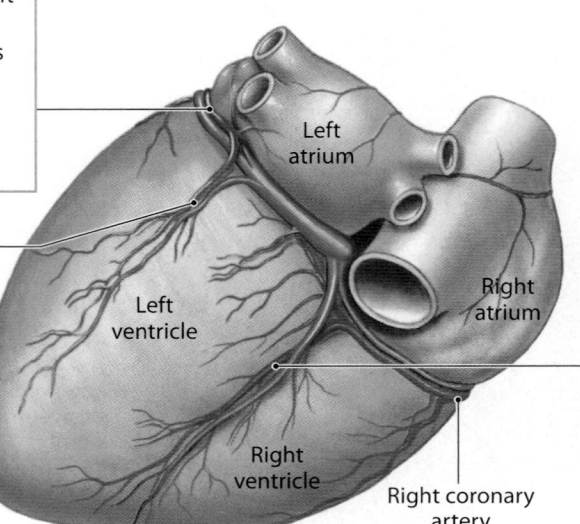

**Anterior view**

Pulmonary trunk
Aortic arch
Ascending aorta
Left atrium
Right atrium
Right ventricle
Left ventricle

Arterial anastomoses between the anterior and posterior interventricular arteries maintain a fairly continuous blood flow despite pressure fluctuations in the left and right coronary arteries.

**2** Branches of the left and right coronary arteries continue onto the posterior surface of the heart.

The **circumflex artery** is a branch of the left coronary artery that curves to the left around the coronary sulcus, eventually meeting and fusing with small branches of the right coronary artery. A marginal artery branches from the circumflex artery to supply the posterior surface of the left ventricle.

Marginal artery

Left atrium

Left ventricle

Right atrium

Right ventricle

Right coronary artery

**Posterior view**

The right coronary artery continues across the posterior surface of the heart, supplying the **posterior interventricular** (posterior descending) **artery**, which runs toward the apex within the posterior interventricular sulcus. This vessel supplies blood to the interventricular septum and adjacent portions of the ventricles.

**? REVIEW**

**A.** Describe the areas of the heart supplied by the right and left coronary arteries.

**3** This view identifies the major collecting vessels on the anterior surface of the heart.

The **great cardiac vein** begins on the anterior surface of the ventricles, along the interventricular sulcus. This vein drains blood from the region supplied by the anterior interventricular artery. The great cardiac vein reaches the level of the atria and then curves around the left side of the heart within the coronary sulcus to empty into the coronary sinus.

The **anterior cardiac veins**, which drain the anterior surface of the right ventricle, empty directly into the right atrium.

**Anterior view**

**4** This view identifies the major collecting vessels on the posterior surface of the heart.

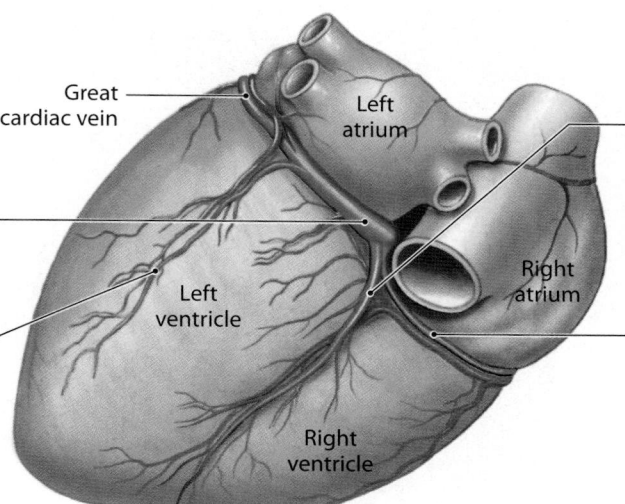

The **coronary sinus** is an expanded vein that opens into the right atrium near the base of the inferior vena cava.

The **posterior vein of left ventricle** drains the area supplied by the circumflex artery.

The **middle cardiac vein**, draining the area supplied by the posterior interventricular artery, empties into the coronary sinus.

The **small cardiac vein** receives blood from the posterior surfaces of the right atrium and ventricle. It empties into the coronary sinus along with the middle cardiac vein.

**Posterior view**

**? REVIEW**
**B.** Compare the anterior cardiac veins to the posterior vein of left ventricle.

**? REVIEW**
**C.** List the arteries and veins of the heart.

Each time the left ventricle contracts, it forces blood into the aorta. The arrival of additional blood at elevated pressures stretches the elastic walls of the aorta. When the left ventricle relaxes, pressure decreases, and the walls of the aorta recoil. This recoil, called **elastic rebound**, pushes blood both forward, into the systemic circuit, and backward, into the coronary arteries. Thus, the combination of blood pressure and elastic rebound ensures a continuous flow of blood to meet the demands of active cardiac muscle tissue.

**? REVIEW**
**D.** Describe what happens to blood flow during elastic rebound.

 **LEARNING OUTCOME**
Describe the major vessels supplying the heart, and cite their locations.

# Module 18.6
# Internal valves control the direction of blood flow between the heart chambers and great vessels

**1** In a sectional view, you can see that the right atrium communicates with the right ventricle, and the left atrium with the left ventricle. The atria are separated by the **interatrial septum**. The ventricles are separated by the much thicker **interventricular septum**. **Atrioventricular (AV) valves**, folds of fibrous tissue, extend into the openings between the atria and ventricles. These valves permit blood flow in one direction only: from the atria to the ventricles. There are also two **semilunar valves** between the ventricles and their great vessels, which ensure blood flows into these vessels.

### Right Atrium

The right atrium receives blood from the superior and inferior venae cavae. It also receives blood from the cardiac veins through the coronary sinus.

The **fossa ovalis** is an oval depression that is a remnant of the foramen ovale, an opening between the atria found in the fetal heart. The foramen ovale closes at birth and is permanently sealed off during the next 3 months.

The anterior atrial wall and the inner surface of the auricle contain prominent muscular ridges called the **pectinate muscles** (*pectin*, comb).

The opening of the coronary sinus carries blood from the cardiac veins.

### Right Ventricle

Blood travels from the right atrium into the right ventricle through a broad opening bordered by the **tricuspid** (trī-KUS-pid; *tri*, three) **valve**, also known as the **right atrioventricular valve**.

The free edge of each valve consists of three flaps, or **cusps**, attached to tendinous connective tissue fibers called the **chordae tendineae** (KOR-dē TEN-di-nē-ē; tendinous cords).

The fibers of the chordae tendineae originate at conical muscular projections called the **papillary** (PAP-i-ler-ē) **muscles**.

The superior portion of the right ventricle tapers toward the **pulmonary valve**, or pulmonary semilunar valve. Blood leaving the right ventricle passes through this valve to enter the pulmonary trunk.

Superior vena cava

Ascending aorta

Pulmonary trunk

Aortic arch

Inferior vena cava

Interventricular septum

### Left Atrium

The left atrium receives blood from the pulmonary veins.

Left pulmonary veins

### Left Ventricle

The left ventricle is much thicker than the right ventricle.

The **mitral** (MĪ-tral; *mitre*, a bishop's hat) **valve**, or *left atrioventricular valve*, is also known as the *bicuspid valve*. It permits the flow of blood from the left atrium into the left ventricle but prevents backflow during ventricular contraction. As the name *bicuspid* implies, the mitral valve contains two, not three, cusps.

Blood leaves the left ventricle by passing through the **aortic valve** (aortic semilunar valve) and into the ascending aorta.

The **trabeculae carneae** (tra-BEK-ū-lē KAR-nē-ē; *carneus*, fleshy) are a series of muscular ridges on the inner surfaces of the right and left ventricles.

The **moderator band** provides a rapid-conduction path that tenses the papillary muscles before the ventricular myocardium contracts. This prevents "slamming" of the cusps of the mitral valve.

**2** The function of an atrium is to collect blood that is returning to the heart and convey it to the attached ventricle. The functional demands on the right and left atria are similar, and the two chambers look almost identical. The demands on the right and left ventricles, however, are very different, and the two have significant structural differences. These differences are best seen in sectional view.

Posterior interventricular sulcus

In sectional view, the relatively thin wall of the right ventricle resembles a pouch attached to the massive wall of the left ventricle. Such a thin wall is adequate because the right ventricle normally does not need to work very hard to push blood through the pulmonary circuit— the lungs are close to the heart, and the pulmonary vessels are relatively short and wide.

The left ventricle has an extremely thick muscular wall and is round in cross section. It must develop four to six times as much pressure to push blood around the systemic circuit as the right ventricle develops to push blood around the pulmonary circuit.

Fat in anterior interventricular sulcus

**? REVIEW**

**A.** Why is the left ventricle more muscular than the right ventricle?

**3** These cross sections indicate the changes in ventricular shape that occur when the ventricles contract.

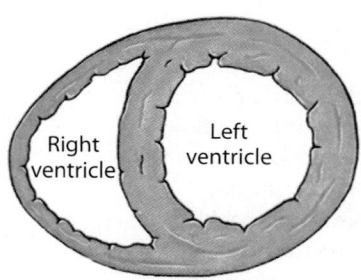

Right ventricle

Left ventricle

**Dilated (relaxed)**

When the left ventricle contracts, (1) the diameter of the ventricular chamber decreases, and (2) the distance between the base and apex decreases. The effect is similar to simultaneously squeezing and rolling up the end of a toothpaste tube. This motion also reduces the volume of the right ventricle, and a person whose right ventricular musculature has been severely damaged may still survive because the contraction of the left ventricle helps push blood into the pulmonary circuit.

When the right ventricle contracts, it acts like a bellows, squeezing the blood against the thick wall of the left ventricle. This mechanism moves blood very efficiently with minimal effort, but it develops relatively low pressures.

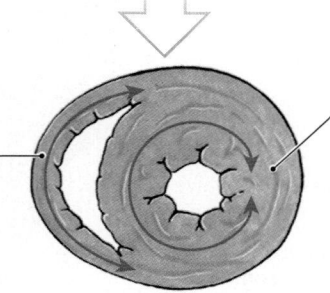

**Contracted**

**? INTEGRATION**

**B.** Damage to the semilunar valve on the right side of the heart would affect blood flow to which vessel?

**Lo LEARNING OUTCOME**

Trace blood flow through the heart, identifying the major blood vessels, chambers, and heart valves.

# When the heart beats, the AV valves close before the semilunar valves open, and the semilunar valves close before the AV valves open

**1** When the ventricles are relaxed, the chordae tendineae are loose, and the AV valves (tricuspid and mitral) offer no resistance to the flow of blood from the atria into the ventricles. Blood pressure in the pulmonary and systemic circuits keeps the semilunar (pulmonary and aortic) valves closed until the ventricles contract.

Pulmonary veins

Left atrium

Aortic valve (closed)

Mitral (left AV) valve (open)

Left ventricle (dilated)

Right ventricle

Tricuspid (right AV) valve (open)

Aortic valve (closed)

Pulmonary valve (closed)

**Superior view of cardiac valves**

Chordae tendineae (loose)

Papillary muscles (relaxed)

**KEY**

Oxygenated blood

Deoxygenated blood

**2** When the ventricles contract, blood moving back toward the atria pushes the cusps of the AV valves together, closing them and preventing backflow. At the same time, the contraction of the papillary muscles tenses the chordae tendineae, stopping the cusps before they swing into the atria. If the chordae tendineae are cut or the papillary muscles are damaged, **regurgitation** (backflow) of blood into the atria occurs each time the ventricles contract. As the ventricular pressures rise above those in the pulmonary and systemic circuits, the aortic and pulmonary valves open, and blood flows out of the ventricles.

The **aortic sinuses** are saclike dilations adjacent to each cusp of the aortic valve. The right and left coronary arteries originate at the aortic sinuses.

**?** **REVIEW**

**A.** Define cardiac *regurgitation*.

Aorta

Left atrium

Aortic valve (open)

Mitral valve (closed)

Left ventricle (contracted)

Tricuspid valve (closed)

Aortic valve (open)

Pulmonary valve (open)

**Superior view of cardiac valves**

Chordae tendineae (tense)

Papillary muscles (contracted)

Ventricular contraction

**Frontal section through left atrium and ventricle**

**3** The heart valves are encircled and supported by a flexible connective tissue frame known as the **cardiac skeleton** (fibrous skeleton). The cardiac skeleton consists of interconnected bands of dense connective tissue that encircle the heart valves and the bases of the pulmonary trunk and aorta. The cardiac skeleton stabilizes the positions of the heart valves and ventricular muscle cells and electrically isolates the ventricular myocardium from the atrial myocardium.

Cardiac skeleton

**Superior view**

? **REVIEW**
**B.** Describe the structural and functional roles of the cardiac skeleton.

**4** The pulmonary and aortic valves each consist of three semilunar (half-moon shaped) cusps of thick connective tissue. Unlike the AV valves, the semilunar valves do not require muscular braces, because the cusps are stable. When the semilunar valves close, the three symmetrical cusps support one another like the legs of a tripod.

Open                Closed
**Superior views**

? **REVIEW**
**C.** What do semilunar valves prevent?

---

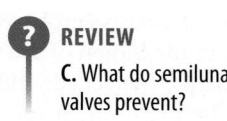 **Clinical Note**

## Damaged heart valves may require surgical replacement

Serious valve problems can interfere with cardiac function. If valve function deteriorates to the point at which the heart cannot maintain adequate blood flow, symptoms of **valvular heart disease** (**VHD**) appear. Congenital malformations may be responsible, but in many cases the condition develops as a consequence of **carditis**, an inflammation of the heart. In severe cases, the only option may be to replace the damaged valve with a prosthetic valve.

This is a superior view of a damaged aortic valve. The cusps are irregular in shape; they are also stiff and relatively inflexible. Such a valve would not open properly and cannot close completely.

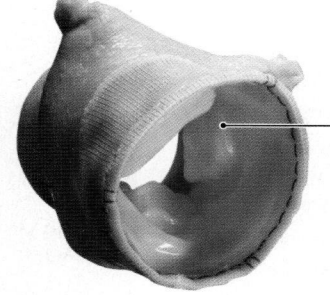

This bioprosthetic valve is an example of an artificial valve that uses the cusps from a pig's heart. Pig or cow valves do not stimulate the clotting system, but they may wear out after about 10 years.

 **LEARNING OUTCOME**
Describe the relationship between the AV and semilunar valves during a heartbeat.

# Module 18.8
# Arteriosclerosis can lead to coronary artery disease

**Arteriosclerosis** (ar-tē-rē-ō-skle-RŌ-sis; *arterio-*, artery + *sklerosis*, hardness) is a thickening and toughening of arterial walls. This condition may not sound life threatening, but complications related to arteriosclerosis account for roughly half of all deaths in the United States. The effects of arteriosclerosis are varied; for example, arteriosclerosis of coronary vessels is responsible for *coronary artery disease (CAD)*, and arteriosclerosis of arteries supplying the brain can lead to strokes.

**1** **Atherosclerosis** (ath-er-ō-skler-Ō-sis; *athero-*, fatty degeneration) is the formation of lipid deposits in the arterial tunica media; these deposits are associated with damage to the endothelial lining. Here is a histological view and a photograph of a dissected atherosclerotic artery. Atherosclerosis, the most common form of arteriosclerosis, tends to develop in people whose blood contains elevated levels of plasma lipids—specifically, cholesterol. The circulating lipids are removed and deposited within arterial walls, which become increasingly abnormal in structure. The result is an atherosclerotic **plaque**, a fatty mass of tissue that projects into the lumen of the vessel and restricts blood flow. Elderly people—especially elderly men—are most likely to develop atherosclerotic plaques. In addition to being older and male, other important risk factors for atherosclerosis include high blood cholesterol levels, high blood pressure, and cigarette smoking. About 20 percent of middle-aged men have all three of these risk factors; these men are four times as likely to experience a heart attack as other men in their age group. Although fewer women develop atherosclerotic plaques, elderly female smokers with high blood cholesterol and high blood pressure are at much greater risk than other women.

**?** **REVIEW**

**A.** Compare arteriosclerosis with atherosclerosis.

Tunica externa (outer layer of blood vessel wall)

Lipid deposits (plaque)

Tunica media (middle layer of blood vessel wall)

**Coronary artery**        LM × 6

Plaque in vessel wall

**2** Plaques can be treated by removing the damaged segment of the vessel and replacing it (often with a superficial vein removed from the leg), but such surgery can be difficult and dangerous. An alternative is **balloon angioplasty** (AN-jē-ō-plas-tē; *angeion*, vessel), in which the tip of a catheter contains an inflatable balloon. Once in position, the balloon is inflated, pressing the plaque against the vessel walls. This photo shows the catheter within an artery of a cadaver. The artery has been opened to show the orientation and relative sizes of the catheter and artery. Balloon angioplasty is most effective in treating small, soft plaques. Several factors make this a highly attractive treatment: (1) The mortality rate during surgery is only about 1 percent; (2) the success rate is over 90 percent; and (3) the procedure can be performed on an outpatient basis.

Catheter        Balloon        Arterial wall

**3** The term **coronary artery disease (CAD)** refers to areas of partial or complete blockage of coronary circulation. Cardiac muscle cells need a constant supply of oxygen and nutrients, so any reduction in blood flow to the heart muscle produces a corresponding reduction in cardiac performance. Such reduced circulatory supply, known as **coronary ischemia** (is-KĒ-mē-uh), generally results from partial or complete blockage of the coronary arteries. The usual cause is the formation of an atherosclerotic plaque in a coronary artery. The plaque, or an associated **thrombus** (blood clot), then narrows the passageway and reduces blood flow. Spasms in the smooth muscles of the vessel wall can further decrease or even stop blood flow. Plaques may be visible in clinical scans or high-resolution ultrasound images.

**? REVIEW**
**B.** What is coronary ischemia?

This is a color-enhanced **digital subtraction angiography (DSA)** scan of a normal heart. The major branches of the left and right coronary arteries are clearly visible.

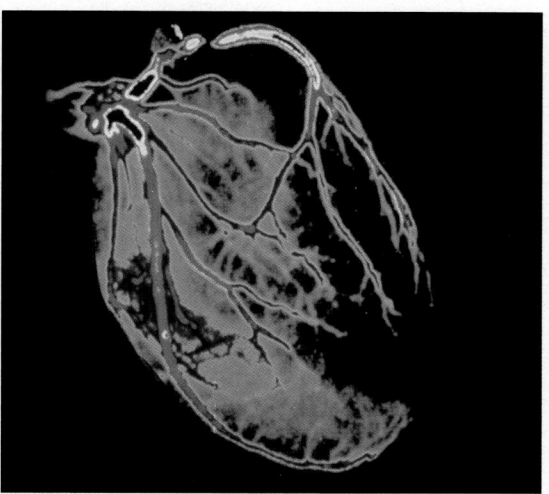

This is a color-enhanced DSA scan of the heart of a person with advanced CAD. Blood flow to the ventricular myocardium is severely restricted.

**4** Because plaques commonly redevelop after angioplasty, a fine wire-mesh tube called a **stent** may be inserted into the vessel. The stent pushes against the vessel wall, holding it open. This is a scan of a stent that has been placed in the anterior interventricular artery. (This imaging technique reveals the lumen of the blood vessels, rather than the vessel, so the stent appears to surround the vessel.) Stents are routinely used by many cardiac specialists because their long-term success rate and incidence of complications are significantly lower than those for balloon angioplasty alone. If the circulatory blockage is extensive, multiple stents can be inserted along the length of the vessel.

Stent

**? REVIEW**
**C.** Describe the purpose of a stent.

**Lo LEARNING OUTCOME**
Define arteriosclerosis, and explain its significance to health.

# Structure of the Heart

**Labeling:** Label each of the structures in this figure.

1 _____
2 _____
3 _____
4 _____
5 _____
6 _____
7 _____
8 _____
9 _____
10 _____
11 _____
12 _____
13 _____
14 _____
15 _____
16 _____
17 _____
18 _____
19 _____
20 _____
21 _____
22 _____
23 _____
24 _____

**Matching:** Match each lettered term with the most closely related description.

a. fossa ovalis

b. intercalated discs

c. serous membrane

d. tricuspid valve

e. aortic valve

f. endocardium

g. aorta

h. myocardium

i. bicuspid valve

j. coronary sinus

25 Blood to systemic arteries

26 Muscular wall of heart

27 Carries blood from coronary veins to right atrium

28 Depression in interatrial septum

29 Right atrioventricular valve

30 Alternate name for mitral valve

31 Cardiac muscle fiber connections

32 Pericardium

33 Inner surface of heart

34 Semilunar valve

25 _____
26 _____
27 _____
28 _____
29 _____
30 _____
31 _____
32 _____
33 _____
34 _____

**Short answer:** Beginning with the right atrium, list in order the heart chambers and valves through which (a) deoxygenated blood and (b) oxygenated blood flows.

35 _____
_____

# Module 18.9
# The cardiac cycle is a complete round of systole and diastole

**1** Each single cardiac contraction, or **heartbeat**, is followed by a brief resting phase, which allows time for the chambers to relax and prepare for the next heartbeat. The period between the start of one heartbeat and the beginning of the next is a single **cardiac cycle**. The heart rate is the number of beats per minute (bpm).

**2** Although we think of the heart as a pump, it is really four pumps that work in pairs, and thus a heartbeat is a complicated event. If all four chambers contracted at once, normal blood flow couldn't occur. Instead, the two atria contract first, pushing blood into the ventricles, and then the two ventricles contract, pushing blood through the pulmonary and systemic circuits and into the atria. The elaborate pacemaking and conducting systems within the heart normally provide the required spacing between atrial and ventricular contractions.

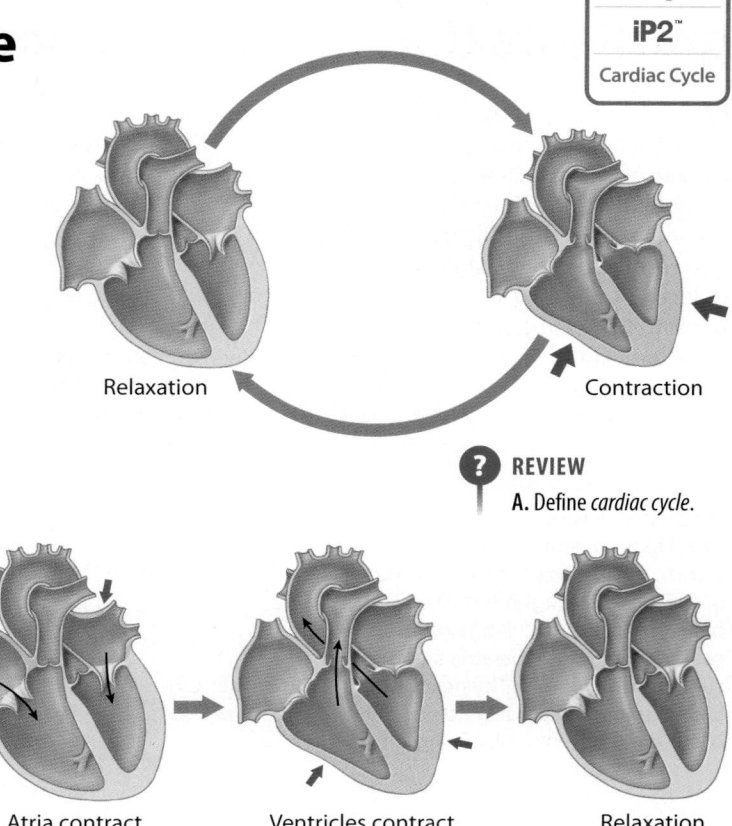

Relaxation          Contraction

**? REVIEW**
**A.** Define *cardiac cycle*.

Relaxation     Atria contract     Ventricles contract     Relaxation

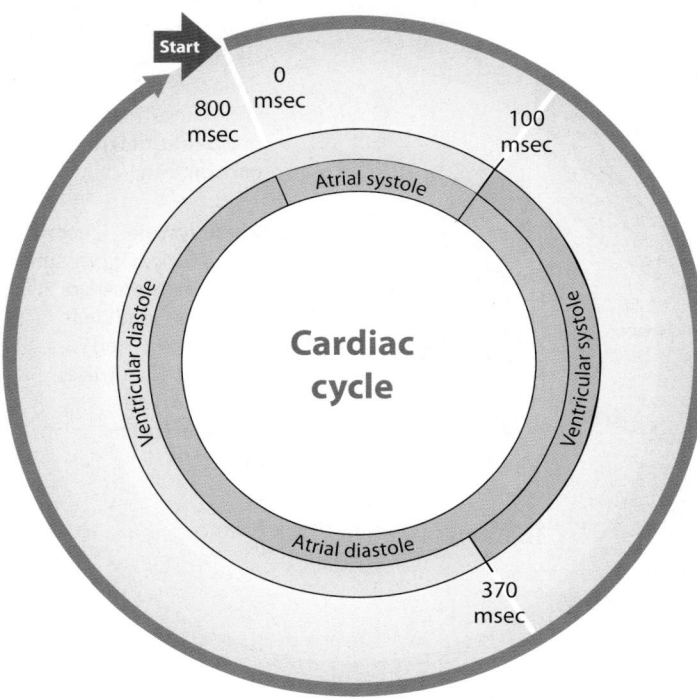

**3** For any one chamber in the heart, the cardiac cycle can be divided into two phases: (1) systole and (2) diastole. During **systole** (SIS-tō-lē), or contraction, the chamber contracts and pushes blood into an adjacent chamber or into an arterial trunk. Systole is followed by **diastole** (dī-AS-tō-lē), or relaxation. During diastole, the chamber fills with blood and prepares for the next contraction. At a representative heart rate of 75 beats per minute (bpm), a sequence of systole and diastole in either the atria or the ventricles lasts 800 msec. For convenience, we will assume that the cardiac cycle is determined by the atria and that it includes one cycle of atrial systole and atrial diastole. This section will examine the details of a representative cardiac cycle—how it is initiated and coordinated and how the pressures generated by the contracting chambers result in directional blood flow.

**? REVIEW**
**B.** Give the alternate terms for heart contraction and heart relaxation.

**? INTEGRATION**
**C.** Compare the duration of atrial and ventricular systole at a representative heart rate of 75 bpm.

**ʟᴏ LEARNING OUTCOME**
Explain the complete round of cardiac systole and diastole.

# Module 18.10

# The cardiac cycle creates pressure gradients that maintain blood flow

**1** The phases of the cardiac cycle are diagrammed here for a heart rate of 75 beats per minute (bpm). When the heart rate increases, all the phases of the cardiac cycle are shortened. The greatest reduction occurs in the length of time spent in diastole. When the heart rate climbs from 75 bpm to 200 bpm, the time spent in systole drops by less than 40 percent, but the duration of diastole is reduced by almost 75 percent. (The red-hued areas indicate which heart chambers are in systole.)

**Start**

**1** When the cardiac cycle begins, all four chambers are relaxed, and the ventricles are partially filled with blood.

**2** During **atrial systole**, the atria contract, completely filling the relaxed ventricles with blood. Atrial systole lasts 100 msec.

**3** Atrial systole ends, and **atrial diastole** begins and continues until the start of the next cardiac cycle.

**?** REVIEW
**A.** List the phases of the cardiac cycle.

As atrial systole ends, ventricular systole begins. This period, which lasts 270 msec, can be divided into two phases.

**4** **Ventricular systole—first phase:** Ventricular contraction pushes the AV valves closed but does not create enough pressure to open the semilunar valves. This is known as the period of **isovolumetric contraction**.

**5** **Ventricular systole—second phase:** As ventricular pressure rises and exceeds pressure in the arteries, the semilunar valves open, and blood is forced out of the ventricles. This is known as the period of **ventricular ejection**.

Ventricular diastole lasts 530 msec (the 430 msec remaining in this cardiac cycle, plus the first 100 msec of the next). Throughout the rest of this cardiac cycle, filling occurs passively, and both the atria and the ventricles are relaxed. The next cardiac cycle begins with atrial systole and the completion of ventricular filling.

**8** **Ventricular diastole—late:** All chambers are relaxed. The ventricles fill passively to about 70% of their final volume.

**7** Blood flows into the relaxed atria, but the AV valves remain closed. This is known as the period of **isovolumetric relaxation**.

**6** **Ventricular diastole—early:** As the ventricles relax, the pressure in them drops. Blood flows back against the cusps of the semilunar valves and forces them closed.

800 msec

0 msec

100 msec

370 msec

Atrial systole

Ventricular diastole

Ventricular systole

Atrial diastole

**Cardiac cycle**

**?** REVIEW
**B.** What are the two phases of ventricular systole?

KEY

← Blood flow

← Contraction

**2** This graph plots the pressure changes within the aorta, the left atrium, and the left ventricle during the cardiac cycle. Although pressures are lower in the right atrium and right ventricle, the same principles apply; both sides of the heart contract at the same time, and they eject equal volumes of blood.

| ATRIAL DIASTOLE | ATRIAL SYSTOLE | ATRIAL DIASTOLE | ATRIAL SYSTOLE |
|---|---|---|---|

| VENTRICULAR DIASTOLE | VENTRICULAR SYSTOLE | VENTRICULAR DIASTOLE |
|---|---|---|

Pressure (mm Hg)

**Aorta**

120

60

30

**Left atrium**
**Left ventricle**

0

Aortic valve opens

Ventricular ejection occurs **5**

**6** Aortic valve closes

When the aortic valve closes, pressure rises again as the elastic arterial walls recoil. This small rise produces a valley in the pressure tracing, called a **dicrotic** (dī-KROT-ik) **notch** (*dikrotos*, double beating).

**4** Isovolumetric contraction

**7** Isovolumetric relaxation occurs

Left atrium ejects blood into ventricles **2**

Mitral valve closes

Mital valve opens

0    100    200    300    400    500    600    700    800

**1** Atrial contraction begins

**3** Atrial systole ends; mitral valve closes

**8** Mitral valve opens; passive ventricular filling occurs

—— Time (msec) ——

Heart sounds

$S_4$   $S_1$   $S_2$   $S_3$   $S_4$

"Lubb"   "Dupp"

**3** There are four heart sounds, designated as $S_1$ through $S_4$. If you listen to your own heart with a stethoscope, you will clearly hear the first and second heart sounds. The first heart sound, known as "lubb" ($S_1$), lasts a little longer than the second, called "dupp" ($S_2$). $S_1$, which marks the start of ventricular contraction, is produced as the AV valves close. $S_2$ occurs when the semilunar valves close. Third and fourth heart sounds are usually very faint and seldom are audible in healthy adults. These sounds are associated with blood flowing into the ventricles ($S_3$) and atrial contraction ($S_4$), rather than with valve action.

**? INTEGRATION**

**C.** Is the heart always pumping blood when pressure in the left ventricle is rising? Explain.

**Lo LEARNING OUTCOME**

Explain the events of the cardiac cycle, and relate the heart sounds to specific events.

# Module 18.11

# Cardiac muscle cell contractions last longer than skeletal muscle fiber contractions primarily because of differences in calcium ion membrane permeability

## Skeletal Muscle Fiber and Cardiac Muscle Cell Contractions

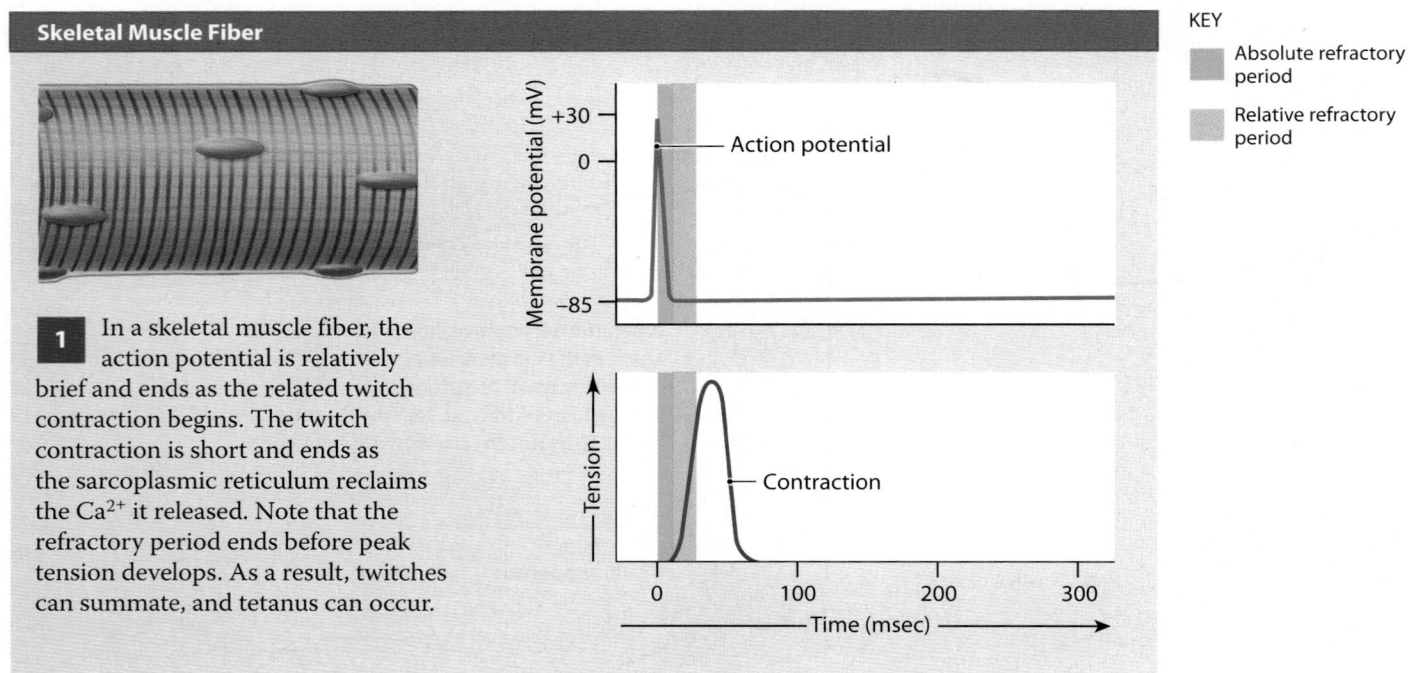

**Skeletal Muscle Fiber**

**KEY**

Absolute refractory period

Relative refractory period

**1** In a skeletal muscle fiber, the action potential is relatively brief and ends as the related twitch contraction begins. The twitch contraction is short and ends as the sarcoplasmic reticulum reclaims the $Ca^{2+}$ it released. Note that the refractory period ends before peak tension develops. As a result, twitches can summate, and tetanus can occur.

**Cardiac Muscle Cell**

**2** In a cardiac muscle cell, the action potential is prolonged because extracellular calcium ions enter the cell for an extended period. As a result, the period of active muscle cell contraction is extended. Note that the refractory period continues until relaxation is well under way. Thus, summation cannot occur, and tetanic contractions do not occur in normal cardiac muscle tissue. This feature is vital: A heart in tetany could not pump blood.

**? REVIEW**

**A.** Why does tetany not occur in cardiac muscle?

With a single twitch lasting 250 msec or longer, a normal cardiac muscle cell could reach 300–400 contractions per minute under maximum stimulation. The normal heart rate never gets that high because the stimulus for contraction cannot spread that quickly through the heart muscle.

Go to
MasteringA&P™
PhysioEx 9.1
Ex 6.1: Cardiac
Muscle
Refractory
Period

# Cardiac Action Potentials

**3** The action potential in a cardiac muscle cell can be divided into three stages.

### Rapid Depolarization

The stage of **rapid depolarization** in a cardiac muscle cell resembles that in a skeletal muscle fiber. At threshold, voltage-gated sodium channels open, and the membrane suddenly becomes permeable to Na$^+$. The result is a massive influx of sodium ions and the rapid depolarization of the sarcolemma. The channels involved are called **fast sodium channels** because they open quickly and remain open for only a few milliseconds.

### Plateau

During the **plateau**, the membrane potential remains near 0 mV. Two opposing factors are involved. As the membrane potential approaches +30 mV, the voltage-gated sodium channels close, and the cell begins actively pumping Na$^+$ out of the cell. However, as the sodium channels are closing, voltage-gated calcium channels are opening. These channels are called **slow calcium channels**. That is, they open slowly and remain open for a relatively long period—roughly 175 msec. While the slow calcium channels are open, the entry of Ca$^{2+}$ nearly balances the loss of Na$^+$, and the membrane potential hovers near 0 mV.

### Repolarization

After approximately 175 msec, slow calcium channels begin closing, and slow potassium channels begin opening. As the channels open, potassium ions (K$^+$) rush out of the cell, and the net result is a period of rapid **repolarization** that restores the resting potential.

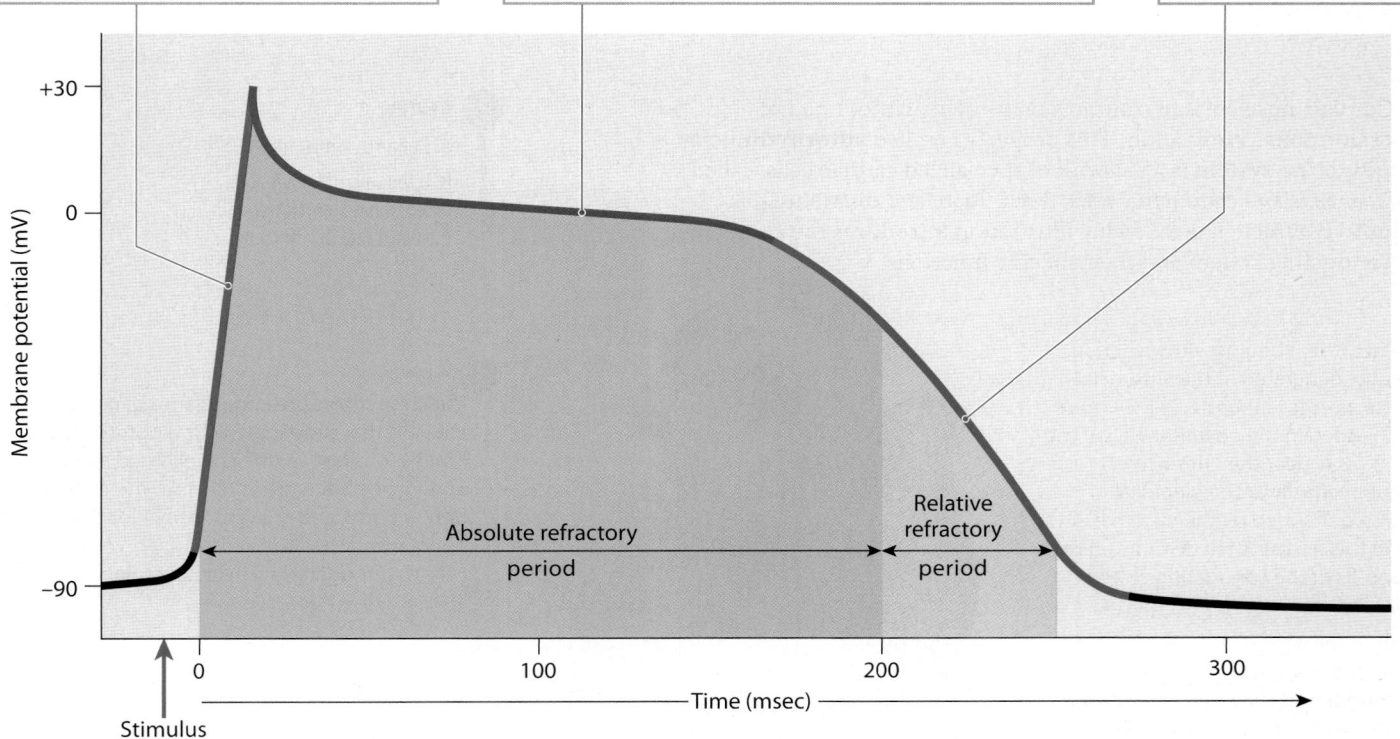

**REVIEW**

**B.** List the three stages of an action potential in a cardiac muscle cell.

**C.** Describe slow calcium channels and the significance of their activity.

**LEARNING OUTCOME**

Describe an action potential in cardiac muscle, and explain the role of calcium ions.

# Module 18.12
## Electrical events of pacemaker cells and conducting cells establish the heart rate

**1** The goal of cardiovascular regulation is to maintain adequate blood flow to vital tissues. The best overall indicator of peripheral blood flow is **cardiac output (CO)**, the amount of blood pumped by the left ventricle into the aorta each minute. Cardiac output depends on two factors: the **heart rate** and the **stroke volume**, the amount of blood pumped out of the ventricle during a single heartbeat.

**2** The calculation of cardiac output (CO) is very straightforward: you multiply the heart rate (HR) by the average stroke volume (SV). For example, if the heart rate is 75 bpm and stroke volume is 80 mL/beat, the CO is:

HR: 75 bpm × SV: 80 mL/beat = CO: 6000 mL/min

The body precisely adjusts cardiac output so that peripheral tissues receive an adequate circulatory supply under a variety of conditions. When necessary, the heart rate can increase by 250 percent, and stroke volume in a normal heart can almost double.

**3** Cardiac muscle tissue contracts on its own, without neural or hormonal stimulation. This property is called **autorhythmicity**. The **conducting system** is a network of specialized cardiac cells, called *pacemaker cells* and *conducting cells*, that initiate and distribute this stimulus to contract. The following illustration introduces the components of the conducting system and their specific functions.

**? REVIEW**
**A.** Define *autorhythmicity*.
**B.** If the cells of the SA node failed to function, how would the heart rate be affected?

**1** Each heartbeat begins with an action potential generated at the **sinoatrial (SA)** (sī-nō-Ā-trē-al) **node**. The SA node is embedded in the posterior wall of the right atrium, near the entrance of the superior vena cava. The electrical impulse generated by the pacemaker cells in this node is then distributed to other cells of the conducting system.

**2** In the atria, conducting cells form **internodal pathways**, which distribute the contractile stimulus to atrial muscle cells as the impulse travels toward the ventricles.

**3** The **atrioventricular (AV) node** is located at the junction between the atria and ventricles. The AV node also contains pacemaker cells, but they do not ordinarily affect the heart rate. However, if the SA node or internodal pathways are damaged, the heart will continue to beat because the AV node will generate impulses at a rate of 40–60 bpm. This is slower than the 60–100 impulses/min generated at the SA node.

Moderator band

**5** **Purkinje fibers** are large-diameter conducting cells that propagate action potentials very rapidly—as fast as small myelinated axons. Purkinje cells are the final link in the distribution network and are responsible for the depolarization of the ventricular myocardial cells that triggers ventricular systole.

**4** The AV node delivers the stimulus to the conducting cells of the **AV bundle**, located within the interventricular septum. The AV bundle is normally the only electrical connection between the atria and the ventricles.

The AV bundle leads to the right and left **bundle branches**, which are also made up of conducting cells. The left bundle branch, which supplies the massive left ventricle, is much larger than the right bundle branch. Both branches extend toward the apex of the heart, turn, and fan out deep to the endocardial surface.

**4** This sequence illustrates the distribution of the contractile stimulus and shows how the conducting system coordinates the contractions of the cardiac cycle. This conducting system allows the atria to contract before the ventricles. Although the cardiac cycle is continuous, we conventionally begin with atrial contraction, about 50 msec after an action potential is generated at the SA node.

**1** An action potential is generated at the SA node, and atrial activation begins.

SA node

Time = 0

**ECG Tracing**

The sequence of electrical events occurring in the conducting system during a contraction cycle can be measured by an electro-cardiogram (ECG). This will be covered in greater depth in Module 18.13.

**2** The stimulus spreads across the atrial surfaces by cell-to-cell contact within the internodal pathways and soon reaches the AV node.

AV node

Elapsed time = 50 msec

**P wave:** atrial depolarization

P

**3** A 100-msec delay occurs at the AV node. During this delay, atrial contraction occurs.

AV bundle

Bundle branches

Elapsed time = 150 msec

**P–R interval:** conduction through AV node and AV bundle

P

The cells of the AV node can conduct impulses at a maximum rate of 230 per minute. Because each impulse results in a ventricular contraction, this value is the maximum normal heart rate. Even if the SA node generates impulses at a faster rate, the ventricles will still contract at 230 bpm. Higher heart rates occur only when the heart or the conducting system has been damaged or stimulated by drugs.

**4** As atrial contraction is completed, the impulse travels along the interventricular septum within the AV bundle and the bundle branches to the Purkinje fibers and, by the moderator band, to the papillary muscles of the right ventricle. Ventricular contraction begins.

Moderator band

Elapsed time = 175 msec

**Q wave:** beginning of ventricular depolarization

P

Q

**5** The impulse is distributed by Purkinje fibers and relayed throughout the ventricular myocardium. Ventricular contraction reaches full force and proceeds to completion.

Elapsed time = 225 msec

Purkinje fibers

**QRS complex:** completion of ventricular depolarization

R

P

Q S

**? INTEGRATION**

**C.** Why is it important for impulses from the atria to be delayed at the AV node before they pass into the ventricles?

**Lo LEARNING OUTCOME**

Describe the components and functions of the conducting system of the heart.

# Module 18.13

## Normal and abnormal cardiac activity can be detected in an electrocardiogram

The electrical events occurring in the heart are powerful enough to be detected by electrodes on the surface of the body. A recording of these events over time is an **electrocardiogram** (e-lek-trō-KAR-dē-ō-gram), also called an **ECG** or **EKG**. Clinicians can use an ECG to assess the performance of specific nodal, conducting, and contractile components. When a portion of the heart has been damaged by a heart attack, for example, the ECG will reveal an abnormal pattern of impulse conduction.

**1** The appearance of the ECG varies with the placement of the monitoring electrodes, or leads. The photo shows the leads in one of the standard configurations, and the graph indicates the important features of an ECG obtained using that configuration.

**? REVIEW**

**A.** Define *electrocardiogram*.

---

The **P wave** represents depolarization of the atria, which causes atrial contraction.

The **QRS complex** corresponds to depolarization of the ventricles. This is a relatively strong electrical signal because the ventricular muscle is much more massive than that of the atria. It is also a complex signal, largely because of the complex pathway that the spread of depolarization takes through the ventricles. The ventricles begin contracting shortly after the peak of the **R wave**, the first positive (upward) deflection of the QRS complex.

The smaller **T wave** represents repolarization of the ventricles. A deflection corresponding to atrial repolarization is not apparent, because it occurs while the ventricles are depolarizing, and the electrical events are masked by the QRS complex.

**? REVIEW**

**B.** List the important features of the ECG, and indicate what each represents.

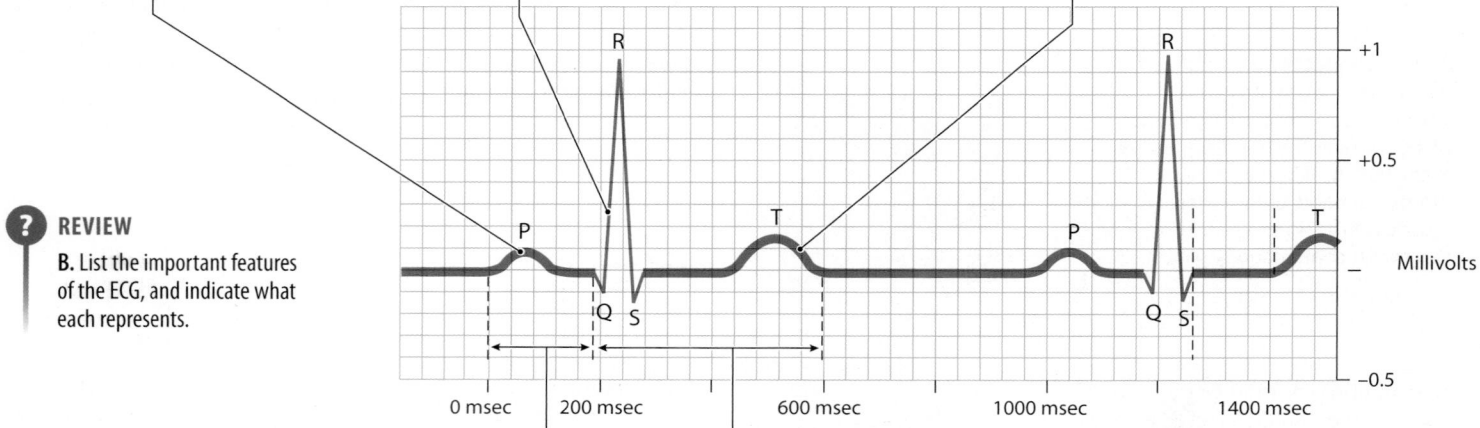

The **P–R interval** extends from the start of atrial depolarization to the start of the QRS complex (ventricular depolarization) rather than to R, because in abnormal ECGs the peak at R can be difficult to determine. Extension of the P–R interval to more than 200 msec can indicate damage to the conducting pathways or AV node.

The **Q–T interval** indicates the time from the beginning of the Q wave to the end of the T wave. It indicates the time required for the ventricles to undergo a single cycle of depolarization and repolarization. The Q–T interval can be lengthened by electrolyte disturbances, some medications, conduction problems, coronary ischemia, or myocardial damage.

**2** Despite the variety of sophisticated equipment available to assess or visualize cardiac function, in the majority of cases the ECG provides the most important diagnostic information. ECG analysis is especially useful in detecting and diagnosing **cardiac arrhythmias** (ă-RITH-mē-uz)—abnormal patterns of cardiac electrical activity. Momentary arrhythmias are not inherently dangerous; about 5 percent of healthy people experience a few abnormal heartbeats each day. Clinical problems appear when arrhythmias reduce the pumping efficiency of the heart. Several important examples of arrhythmias are described here.

SmartArt

### Premature Atrial Contractions (PACs)

**Premature atrial contractions (PACs)** often occur in healthy people. In a PAC, the normal atrial rhythm is momentarily interrupted by a "surprise" atrial contraction. Stress, caffeine, and various drugs may increase the incidence of PACs, presumably by increasing the permeabilities of the pacemaker cells in the SA node. The impulse spreads along the conduction pathway, and a normal ventricular contraction follows the atrial beat.

### Paroxysmal Atrial Tachycardia (PAT)

In **paroxysmal** (par-ok-SIZ-mal) **atrial tachycardia**, or **PAT**, a premature atrial contraction triggers a flurry of atrial activity. The ventricles are still able to keep pace, and the heart rate jumps to about 180 bpm.

### Atrial Fibrillation (AF)

During **atrial fibrillation** (fib-ri-LĀ-shun), the impulses move over the atrial surface at rates of perhaps 500 bpm. The atrial wall quivers instead of producing an organized contraction. The ventricular rate cannot follow the atrial rate and may remain within normal limits. Even though the atria are now nonfunctional, their contribution to the maximum amount of blood the ventricles can hold at the end of atrial contraction is so small that the condition may go unnoticed in older individuals.

### Premature Ventricular Contractions (PVCs)

**Premature ventricular contractions (PVCs)** occur when a Purkinje cell or ventricular myocardial cell depolarizes to threshold and triggers a premature contraction. Single PVCs are common and not dangerous. Because the pacemaker cell is no longer an SA nodal cell, the cell responsible is called an **ectopic** (ek-TOP-ik; out of place) **pacemaker**. The frequency of PVCs can be increased by exposure to epinephrine, to other stimulatory drugs, or to ionic changes that depolarize cardiac muscle plasma membranes.

### Ventricular Tachycardia (VT)

**Ventricular tachycardia** is defined as four or more PVCs without intervening normal beats. It is also known as **VT** or **V-tach**. Multiple PVCs and VT may indicate that serious cardiac problems exist.

### Ventricular Fibrillation (VF)

**Ventricular fibrillation (VF)**, or **V-fib**, is responsible for the condition known as **cardiac arrest**. VF is rapidly fatal because the ventricles quiver and stop pumping blood. The heart in VF is said to resemble a bag of squirming worms.

**REVIEW**
**C.** Why is ventricular fibrillation fatal?

**LEARNING OUTCOME**
Identify the electrical events shown on an electrocardiogram.

# Module 18.14

# The intrinsic heart rate can be altered by autonomic activity

**1** The pacemaker cells of the SA and AV nodes cannot maintain a stable resting membrane potential. After each repolarization, the membrane gradually drifts toward threshold. This gradual spontaneous depolarization is called a **pacemaker potential**. The rate of spontaneous depolarization is fastest at the SA node, which without neural or hormonal stimulation generates action potentials at a rate of 60–100 per minute. Because the SA node reaches threshold first, it establishes the heart rate— the impulse generated by the SA node brings the AV nodal cells to threshold before the prepotential of the AV nodal cells can do so.

**2** Any factor that changes the rate of spontaneous depolarization or the duration of repolarization will alter the heart rate by changing the time required to reach threshold. Acetylcholine released by parasympathetic neurons opens chemically gated K⁺ channels in the plasma membrane, thereby slowing the rate of spontaneous depolarization and also slightly extending the duration of repolarization. The result is a decline in heart rate.

**3** Norepinephrine (NE) released by sympathetic neurons binds to beta-1 receptors, leading to the opening of ion channels that increase the rate of depolarization and shorten the period of repolarization. Because the nodal cells reach threshold more quickly, the heart rate increases.

**4** Every person has a characteristic resting heart rate that varies with age, general health, and physical conditioning. However, there is a normal range of heart rates. The American Heart Association considers 60–100 bpm to be the normal range of resting heart rates.

**? REVIEW**

**A.** Compare bradycardia with tachycardia.

**Bradycardia** (brăd-ē-KAR-dē-uh; *bradys*, slow) is a condition in which the heart rate is slower than normal.

**60 bpm** — Normal range of resting heart rates — **100 bpm**

**Tachycardia** (tak-ē-KAR-dē-uh; *tachys*, swift) indicates a faster-than-normal heart rate.

**5** The cardiac centers of the medulla oblongata contain the autonomic headquarters for cardiac control. These centers innervate the heart by means of the cardiac plexus.

**? REVIEW**

**B.** Describe the sites and actions of the cardioinhibitory and cardioacceleratory centers.

The **cardioinhibitory center** controls the parasympathetic neurons that slow the heart rate.

The **cardioacceleratory center** controls sympathetic neurons that increase the heart rate.

Vagal nucleus

Medulla oblongata

Vagus nerve (X)

Spinal cord

**Sympathetic**

Sympathetic innervation arrives in postganglionic fibers within the cardiac nerves. These fibers innervate the nodes, the conducting system, and the atrial and ventricular myocardium.

Sympathetic preganglionic fiber

Sympathetic ganglia (cervical ganglia and superior thoracic ganglia [$T_1$–$T_4$])

Sympathetic postganglionic fiber

Cardiac nerve

**Parasympathetic**

Parasympathetic innervation arrives in the vagus nerve and synapses with ganglion cells in the cardiac plexus. Postganglionic fibers innervate the SA node, AV node, and atrial musculature. Innervation of the ventricular musculature is very limited.

Parasympathetic preganglionic fiber

Synapses in cardiac plexus

Parasympathetic postganglionic fibers

Reflex pathways and input from higher centers, especially the sympathetic and parasympathetic headquarters in the hypothalamus, regulate the cardiac centers. As both divisions are chronically active, the heart has a resting autonomic tone. The normal resting heart rate is somewhat slower than the intrinsic SA nodal stimulation rate of 80–100 bpm, because at rest the effects of parasympathetic innervation predominate.

**? INTEGRATION**

**C.** Caffeine has effects on conducting cells and contractile cells that are similar to those of NE. What effect would drinking large amounts of caffeinated beverages have on the heart rate?

 **LEARNING OUTCOME**

Describe the factors affecting the heart rate.

# Module 18.15
## Stroke volume depends on the relationship between end-diastolic volume and end-systolic volume

**1** The **stroke volume** of the heart can be compared to pumping water with a manual pump. The amount pumped varies with the amount of movement of the pump handle. Although the heart contains two large pumps—the right and left ventricles—we can use a single pump as a model because the two work in the same way and pump equal amounts of blood.

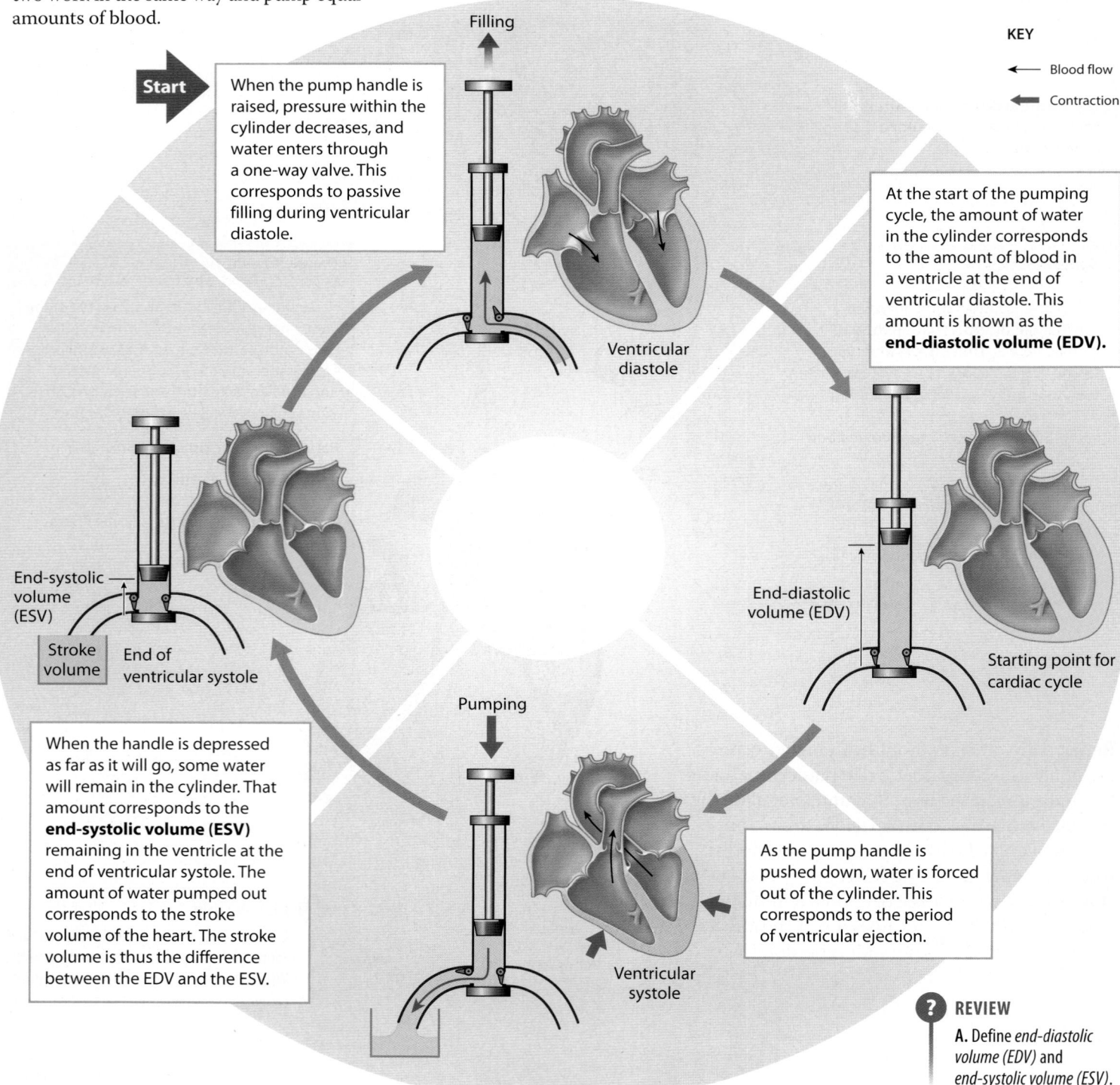

**Filling**

**Start**

When the pump handle is raised, pressure within the cylinder decreases, and water enters through a one-way valve. This corresponds to passive filling during ventricular diastole.

Ventricular diastole

KEY

← Blood flow
← Contraction

At the start of the pumping cycle, the amount of water in the cylinder corresponds to the amount of blood in a ventricle at the end of ventricular diastole. This amount is known as the **end-diastolic volume (EDV).**

End-diastolic volume (EDV)

Starting point for cardiac cycle

End-systolic volume (ESV)

Stroke volume | End of ventricular systole

When the handle is depressed as far as it will go, some water will remain in the cylinder. That amount corresponds to the **end-systolic volume (ESV)** remaining in the ventricle at the end of ventricular systole. The amount of water pumped out corresponds to the stroke volume of the heart. The stroke volume is thus the difference between the EDV and the ESV.

**Pumping**

As the pump handle is pushed down, water is forced out of the cylinder. This corresponds to the period of ventricular ejection.

Ventricular systole

**?** REVIEW

**A.** Define *end-diastolic volume (EDV)* and *end-systolic volume (ESV).*

**2** Note that in our water pump example, the amount of water pumped per stroke could be changed by altering the EDV, ESV, or both. In the resting heart, stroke volume is relatively stable, but it can increase substantially if necessary.

**? REVIEW**

**B.** What effect would an increase in venous return have on the stroke volume?

**1** The EDV is influenced by the venous return and the filling time.

**Venous return** is the amount of blood delivered to the right atrium by the venae cavae and coronary sinus. Blood volume, muscular activity, and the rate of blood flow through peripheral capillaries are the major factors that determine venous return. When blood volume goes down significantly, venous return is reduced. When skeletal muscles contract and compress adjacent veins, or when peripheral tissue activity increases (which accelerates regional blood flow), venous return increases.

The **filling time** is the duration of ventricular diastole. Assuming that venous return remains constant, slowing the heart rate increases the EDV, and increasing the heart rate decreases the EDV.

The amount of myocardial stretching is called the **preload**. The greater the EDV, the larger the preload. When the EDV is very small, the sarcomeres in the cardiac muscle cells are too short to develop much power. As the EDV increases and the sarcomeres approach optimal length, the ventricular muscle cells can contract more efficiently and produce more powerful contractions and eject more blood. This relationship—greater EDV = greater stroke volume— is called the *Frank-Starling law of the heart.*

**2** The ESV is influenced by the contractility and the afterload.

The **contractility** is the amount of force produced during a contraction at a given amount of preload. Sympathetic stimulation increases contractility. Many hormones—including epinephrine, norepinephrine, thyroid hormones, and glucagon—increase contractility to some degree. Many drugs used in clinical practice are intended to reduce contractility; examples include the "beta-blockers" such as propranolol or timolol and calcium channel blockers such as nifedipine or verapamil.

The **afterload** is the amount of tension the contracting ventricle must produce to force open the semilunar valve and eject blood. The greater the afterload, the longer the period of isovolumetric contraction, the shorter the duration of ventricular ejection, and the larger the ESV. In other words, as the afterload increases, the stroke volume decreases. Afterload is increased by any factor that restricts blood flow through the arterial system. For example, vasoconstriction (the narrowing of peripheral blood vessels) increases the afterload, whereas vasodilation (the relaxation and widening of peripheral blood vessels) decreases the afterload.

**? REVIEW**

**C.** What effect would an increase in sympathetic stimulation of the heart have on the end-systolic volume (ESV)?

 **LEARNING OUTCOME**

Describe the variables that influence stroke volume.

# Module 18.16

# Cardiac output is regulated by adjustments in heart rate and stroke volume

**1** As this figure indicates, cardiac output varies widely to meet metabolic demands. The body must closely monitor and tightly control those variations.

**2** This diagram summarizes the important information presented thus far concerning the factors that affect cardiac output.

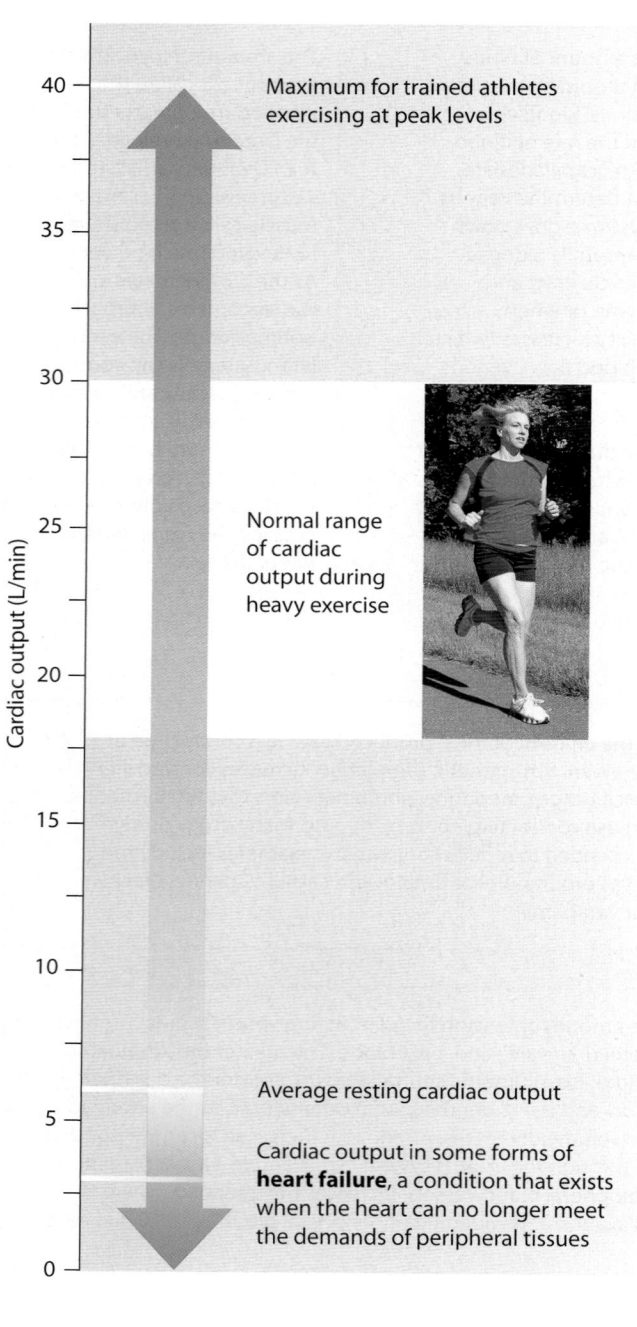

40 — Maximum for trained athletes exercising at peak levels

Normal range of cardiac output during heavy exercise

Average resting cardiac output

Cardiac output in some forms of **heart failure**, a condition that exists when the heart can no longer meet the demands of peripheral tissues

Cardiac output (L/min)

**?** **REVIEW**
**A.** Define *heart failure*.

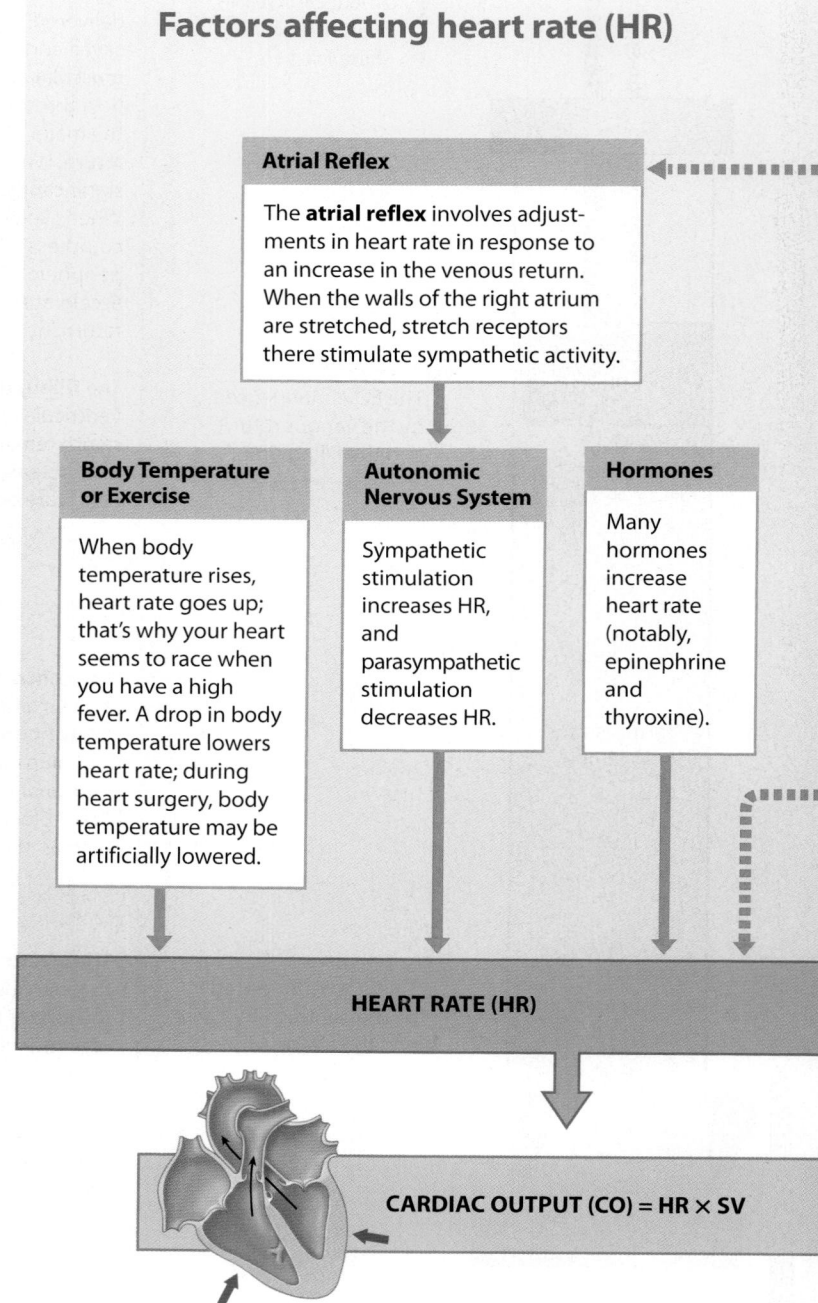

## Factors affecting heart rate (HR)

**Atrial Reflex**

The **atrial reflex** involves adjustments in heart rate in response to an increase in the venous return. When the walls of the right atrium are stretched, stretch receptors there stimulate sympathetic activity.

**Body Temperature or Exercise**

When body temperature rises, heart rate goes up; that's why your heart seems to race when you have a high fever. A drop in body temperature lowers heart rate; during heart surgery, body temperature may be artificially lowered.

**Autonomic Nervous System**

Sympathetic stimulation increases HR, and parasympathetic stimulation decreases HR.

**Hormones**

Many hormones increase heart rate (notably, epinephrine and thyroxine).

**HEART RATE (HR)**

**CARDIAC OUTPUT (CO) = HR × SV**

## Factors affecting stroke volume (SV)

**Exercise**

Muscular contractions compress veins and assist valves in directing venous blood toward the right atrium.

**Blood Volume**

Large reductions in blood volume due to bleeding or dehydration reduce venous return.

**Blood Flow**

Changes in peripheral blood flow patterns can increase or decrease venous return.

**Venous Return**

In general, when venous return increases, SV increases. When venous return decreases, SV decreases.

**Filling Time**

Increase in filling time increases the EDV.

**Autonomic Nervous System**

Sympathetic stimulation increases both heart rate and contractility. Parasympathetic action slows the heart rate but has little influence on contractility.

**Hormones**

Many hormones increase contractility in addition to their effects at other target organs.

**Preload**

The amount of preload affects the EDV by stretching the myocardium.

**Contractility**

The greater the contractility, the smaller the ESV.

Vasodilation decreases afterload.

Vasoconstriction increases afterload.

**End-Diastolic Volume (EDV)**

The EDV is influenced by the venous return, the filling time, and the preload.

**End-Systolic Volume (ESV)**

The ESV is influenced by the contractility and the afterload.

**Afterload**

The afterload is the resistance to blood flow out of the heart. The greater the afterload, the lower the pumping efficiency of the heart, and the larger the ESV.

↑ EDV = ↑ SV
↓ EDV = ↓ SV

**STROKE VOLUME (SV) = EDV − ESV**

↑ ESV = ↓ SV
↓ ESV = ↑ SV

**? INTEGRATION**

**B.** Compute Joe's stroke volume if his end-systolic volume (ESV) is 40 mL and his end-diastolic volume (EDV) is 125 mL.

**Lo LEARNING OUTCOME**

Explain how stroke volume and cardiac output are coordinated.

**Matching:** Match each lettered term with the most closely related description.

a. P wave

b. cardiac output

c. autorhythmicity

d. "lubb" sound

e. "dupp" sound

f. sympathetic neurons

g. stroke volume

h. tachycardia

i. bradycardia

j. parasympathetic neurons

**1** AV valves close

**2** Self-stimulated cardiac muscle contractions

**3** Atrial depolarization

**4** Amount of blood ejected by ventricle during a single beat

**5** HR × SV

**6** Decrease the heart rate

**7** Term for slower-than-normal HR

**8** Increase the heart rate

**9** Semilunar valve closes

**10** Term for faster-than-normal HR

**1** _____
**2** _____
**3** _____
**4** _____
**5** _____
**6** _____
**7** _____
**8** _____
**9** _____
**10** _____

**Short answer:** Refer to the accompanying graph of the cardiac cycle to answer the questions below.

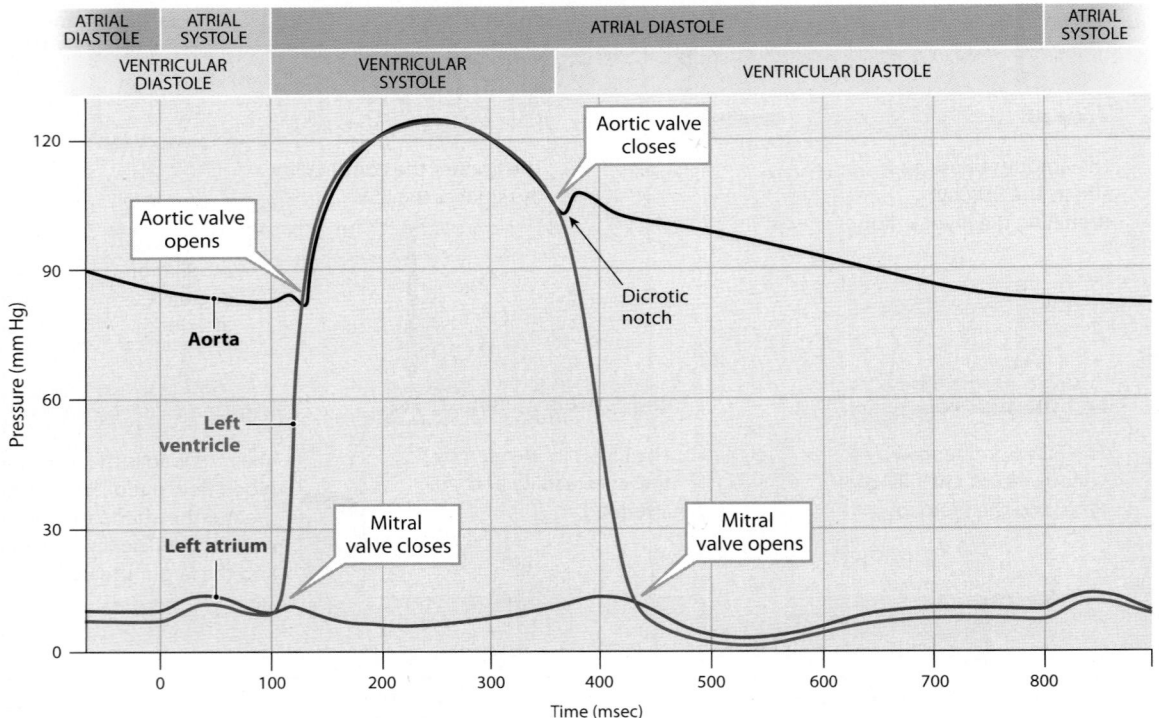

**11** What event occurs when the pressure in the left ventricle rises above that in the left atrium?

**12** During ventricular systole, is the blood volume in the **atria** increasing or decreasing?

**13** During ventricular systole, is the volume in the **ventricles** increasing or decreasing?

**14** During most of ventricular diastole, is the pressure in the left ventricle greater than, the same as, or less than the pressure in the left atrium?

**15** What event occurs when the pressure within the left ventricle becomes greater than the pressure within the aorta?

**16** During isovolumetric contraction, where is pressure the highest?

**17** During what part of the cardiac cycle is blood pressure highest in the large systemic arteries?

**11** _____
**12** _____
**13** _____
**14** _____
**15** _____
**16** _____
**17** _____

# Study Outline

## SECTION 1 • Structure of the Heart

### Module 18.1

**The heart has four chambers that pump and circulate blood through the pulmonary and systemic circuits    p. 659**

1. The **base** of the heart is at the superior end where the largest veins and arteries are attached.The **apex** is a pointed tip at the inferior aspect of the heart.

2. The four chambers of the heart are the *left atrium* and *left ventricle* and the *right atrium* and *right ventricle.*

3. The left atrium receives blood from the *pulmonary circuit,* and the left ventricle pumps blood into the *systemic circuit.* The right atrium receives blood from the systemic circuit, and the right ventricle pumps blood into the pulmonary circuit.

### Module 18.2

**The heart is located in the mediastinum and enclosed by the pericardial cavity    p. 660**

4. The heart sits in the anterior portion of the **mediastinum**. The mediastinum also contains the **great vessels,** thymus, esophagus, and trachea.

5. The pericardial cavity is the potential space between the visceral layer of the serous pericardium and parietal layer of the serous pericardium, and it contains 15–50 mL of **pericardial fluid.**

6. **Pericarditis** is an infection of the pericardium. **Cardiac tamponade** is fluid accumulation within the pericardial cavity.

### Module 18.3

**The heart wall contains concentric layers of cardiac muscle tissue    p. 662**

7. The heart wall contains three layers: epicardium, myocardium, and endocardium.

8. The **epicardium**, or **visceral layer of the serous pericardium**, is a serous membrane that covers the surface of the heart. The **parietal layer of the serous pericardium** is a serous membrane that forms the outer wall of the pericardial cavity. It forms part of the pericardium that surrounds the heart.

*The layers of the heart wall*

9. The **myocardium** contains cardiac muscle tissue, blood vessels, and nerves.

10. The **endocardium** is a simple squamous epithelium and underlying areolar tissue that covers the inner surface of the heart and heart valves.

11. Cardiac muscle cells are small; contain a single centrally located nucleus, branching interconnections between cells, and specialized intercellular connections known as **intercalated discs**; and are striated in appearance.

12. Cardiac muscle cells are almost totally dependent on aerobic metabolism and are richly supplied with capillaries. Gap junctions between these cells allow for ions and small molecules to move from one cell to another and create a direct electrical connection between them. Cardiac muscle is mechanically, chemically, and electrically connected such that it resembles a single, enormous muscle cell,  or **functional syncytium**.

### Module 18.4

**The boundaries between the four chambers of the heart can be identified on its external surface    p. 664**

13. The **auricle** of the atrium is an expandable pouch. The **coronary sulcus** is a deep groove between the atria and the ventricles. The **anterior interventricular sulcus** is a depression between the left and right ventricles. The **ligamentum arteriosum** is a remnant of the fetal connection between the aorta and pulmonary trunk.

14. The **coronary sinus** carries blood collected from the coronary veins to the right atrium. The **posterior interventricular sulcus** is a shallow depression between the left and right ventricles.

### Module 18.5

**The heart has an extensive blood supply    p. 666**

15. The left and right **coronary arteries** originate at the base of the ascending aorta.

16. The **right coronary artery** follows the coronary sulcus and supplies blood to the right atrium, portions of both ventricles, and portions of the conducting system of the heart. **Marginal arteries** from the right coronary artery supply the surface of the right ventricle.

17. The **left coronary artery** supplies blood to the left atrium, left ventricle, and interventricular septum. The **anterior interventricular artery** is a branch of the left coronary artery that lies within the anterior interventricular sulcus.

18. The **circumflex artery** is a branch of the left coronary artery that curves to the left around the coronary sulcus. The **posterior interventricular artery** is a branch of the right coronary artery.

19. The **anterior cardiac veins** drain the anterior surface of the right

*The major collecting vessels of the heart*

ventricle and empty directly into the right atrium. The **great cardiac vein** drains blood from regions supplied by the anterior interventricular artery.

20. The **posterior vein of left ventricle**, **small cardiac vein**, and **middle cardiac vein** all drain blood from the posterior aspect of the heart and empty into the **coronary sinus.**

### Module 18.6

**Internal valves control the direction of blood flow between the heart chambers and great vessels    p. 668**

21. The right atrium communicates with the right ventricle, and the left atrium communicates with the left ventricle.

22. The atria are separated by an **interatrial septum**, and a much thicker **interventricular septum** separates the two ventricles.

23. **Atrioventricular (AV) valves** permit the flow of blood in one direction only: from atria to ventricles. The **right atrioventricular valve** has three **cusps** and is also known as the **tricuspid valve**. The **left atrioventricular valve** has two cusps and is also called the **bicuspid valve** or **mitral valve**.

24. The free edge of each AV valve is attached to **chordae tendineae** that originate at **papillary muscles**.

25. Blood leaving the right ventricle passes through the **pulmonary valve** (semilunar) to enter the pulmonary trunk. Blood leaving the left ventricle passes through the **aortic valve** (semilunar) and into the ascending aorta.

26. The atria collect blood returning to the heart and convey it to their attached ventricles.

27. When the right ventricle contracts, it squeezes blood against the left ventricular wall.

28. When the left ventricle contracts, the diameter of the ventricular chamber decreases, and the distance between the apex and base decreases.

## Module 18.7

### When the heart beats, the AV valves close before the semilunar valves open, and the semilunar valves close before the AV valves open    p. 670

29. When the ventricles are relaxed, the chordae tendineae are loose, and the AV valves allow blood flow from the atria into the ventricles.

30. When the ventricles contract, the AV valves close to prevent **regurgitation** into the atria.

31. As the ventricular pressure rises above those in the pulmonary and systemic circuits, the aortic and pulmonary valves open, and blood flows out of the ventricles.

32. Interconnected bands of dense connective tissue called the **cardiac skeleton** encircle the heart valves and bases of the pulmonary trunk and aorta.

33. The semilunar valves do not require muscular braces.

34. A valve with **valvular heart disease (VHD)** cannot maintain adequate blood flow. This can be caused by congenital malformations or **carditis** (inflammation of the heart) and require valve replacement.

## Module 18.8

### Arteriosclerosis can lead to coronary artery disease    p. 672

35. **Arteriosclerosis** is a thickening and toughening of the arterial walls and can cause coronary artery disease and stroke.

36. **Atherosclerosis**, deposits of **plaque** in the vessel wall, is the most common form of arteriosclerosis.

37. **Coronary artery disease (CAD)** is partial or complete blockage of coronary circulation, known as **coronary ischemia**. Blood flow can be blocked by plaque, a **thrombus** (blood clot), or spasms in the vessel.

*A color-enhanced digital subtraction angiography (DSA) of a normal heart*

38. Plaques can be treated by **balloon angioplasty** or by surgical insertion of a wire-mesh tube called a **stent**.

## SECTION 2 • Cardiac Cycle

## Module 18.9

### The cardiac cycle is a complete round of systole and diastole    p. 675

39. A **heartbeat** is a single cardiac contraction. The period between the start of one heartbeat and the beginning of the next is a single **cardiac cycle**.

40. All four chambers do not contract at once. Instead, the two atria contract and push the blood into the ventricles, and then the two ventricles contract and push the blood into the pulmonary and systemic circuits.

41. The cardiac cycle can be divided into two phases: **systole** (contraction) and **diastole** (relaxation).

42. A representative heart rate is 75 beats per minute (bpm). A representative cardiac cycle lasts 800 msec.

## Module 18.10

### The cardiac cycle creates pressure gradients that maintain blood flow    p. 676

43. When the cardiac cycle begins, all four chambers are relaxed, and the ventricles are partially filled with blood.

44. During **atrial systole**, the atria contract and fill the ventricles. **Atrial diastole** now begins and continues until the next cardiac cycle.

45. **Ventricular systole** has two phases: **isovolumetric contraction**, when the ventricles contract, the AV valves close, but pressure is not great enough to open the semilunar valves; and **ventricular ejection**, when the pressure is great enough to open the semilunar valves and blood is ejected.

46. **Ventricular diastole—early** is when the pressure drops and blood flows back against the semilunar valve cusps to close them. Blood flows into the relaxed atria, but AV valves are still closed.

47. During **ventricular diastole—late**, all chambers are relaxed and filling. The ventricles fill passively to 70 percent of their final volume.

48. Four heart sounds can be heard with a stethoscope, and they are designated as **S₁** to **S₄**. $S_1$ is also known as "lubb" and is the sound of the AV valves closing. $S_2$ is "dupp" and is the sound of the semilunar valves closing. The two other sounds are associated with blood flow, are faint, and are seldom audible in healthy adults.

## Module 18.11

### Cardiac muscle cell contractions last longer than skeletal muscle fiber contractions primarily because of differences in calcium ion membrane permeability    p. 678

49. Skeletal muscle cell action potentials are brief and end as the muscle twitch begins.

50. Cardiac muscle fiber action potentials are prolonged because calcium ions enter the cell for an extended period. The result is a longer period of active contraction. There is also a long refractory period, which prevents tetanic contractions.

51. There are three stages of cardiac muscle cell action potential: **rapid depolarization**, **plateau**, and **repolarization**.

## Module 18.12

### Electrical events of pacemaker cells and conducting cells establish the heart rate    p. 680

52. **Cardiac output (CO)** is the amount of blood ejected by the left ventricle in 1 minute.

*Conducting system of the heart*

53. CO depends on **heart rate (HR)** and **stroke volume (SV),** which is the amount of blood pumped from the ventricle in a single heartbeat.

54. CO can be calculated as HR × SV. HR: 75 bpm × SV: 80 mL/beat = CO: 6000 mL/min.

55. The heart beats independently of neural or hormonal stimulation. This **autorhythmicity** is initiated and distributed by the **conducting system**.

56. Each heartbeat begins with an action potential at the **sinoatrial (SA) node**. The impulse continues along **internodal pathways**, the **atrioventricular (AV) node**, the **AV bundle**, **bundle branches**, and **Purkinje fibers**.

## Module 18.13

### Normal and abnormal cardiac activity can be detected in an electrocardiogram    p. 682

57. The electrical events in the heart can be recorded in an **electrocardiogram** (**ECG** or **EKG**).

58. The **P wave** represents atrial depolarization, the **QRS complex** represents ventricular depolarization, and the **T wave** represents ventricular repolarization. Repolarization of the atria is not visible because it is masked by the QRS complex.

59. An ECG can be useful in detecting and diagnosing many **cardiac arrhythmias**.

## Module 18.14

### The intrinsic heart rate can be altered by autonomic activity p. 684

60. The cells of the SA and AV node cannot maintain a stable resting potential. After repolarization, a gradual, spontaneous depolarization called a **pacemaker potential** begins. This is fastest at the SA node.

61. Acetylcholine released by the parasympathetic neurons will result in a slower heart rate. Norepinephrine released by sympathetic neurons will increase heart rate.

62. The normal resting heart rate range is 60–100 bpm. Age, health, and physical conditioning influence it.

63. A slower-than-normal heart rate is called **bradycardia**; a faster-than-normal heart rate is called **tachycardia**.

64. The **cardioinhibitory center** of the medulla oblongata controls the parasympathetic neurons that slow heart rate. The **cardioacceleratory center** controls the sympathetic neurons that increase heart rate.

## Module 18.15

### Stroke volume depends on the relationship between end-diastolic volume and end-systolic volume    p. 686

65. The amount of blood in a ventricle at the end of ventricular diastole is **end-diastolic volume** (**EDV**). After ventricular systole, some blood will remain in the ventricle. This is called **end-systolic volume** (**ESV**).

66. EDV is influenced by **venous return** and **filling time**.  The Frank-Starling law of the heart describes the relationship between EDV and stroke volume:  The greater the EDV, the greater the stroke volume. ESV is influenced by **contractility** and **afterload**.

## Module 18.16

### Cardiac output is regulated by adjustments in heart rate and stroke volume    p. 688

67. Cardiac output (CO) varies widely to meet metabolic demands. Important factors that affect heart rate (HR) are sympathetic and parasympathetic stimulation, body temperature, and hormones. Important factors that influence stroke volume (SV) are exercise, blood volume, blood flow, the autonomic nervous system, and hormones.

# Chapter Review Questions

**True/False:** Indicate whether each statement is true or false.

**1** The $S_1$, or "lubb," sound of the heart is produced by the closing of the AV valves.

**2** The T wave of an electrocardiogram represents repolarization of the atria.

**3** The circumflex artery supplies blood to the surface of the right ventricle.

**4** Isovolumetric contraction is the first phase of ventricular systole, when pressures close the AV valves but are not high enough to open the semilunar valves.

**5** Tachycardia is a faster-than-normal heart rate, usually above 100 bpm.

**1** T
**2** F
**3** F
**4** T
**5** T

**Multiple choice:** Select the correct answer from the list provided.

**6** Cardiac output (CO) is calculated by which of the following formulas?
- a) CO = HR × ESV
- b) CO = HR × EDV
- ☒ c) CO = HR × SV
- d) CO = HR × MAP

**7** During diastole, a chamber of the heart
- ☒ a) relaxes and fills with blood.
- b) contracts and pushes blood into the adjacent chamber.
- c) experiences a sharp increase in pressure.
- d) reaches a pressure of approximately 120 mm Hg.

**8** Which of the following is longest?
- a) the refractory period of a skeletal muscle fiber
- ☒ b) the refractory period of a cardiac muscle cell
- c) the action potential of a skeletal muscle fiber
- d) the contraction time of a skeletal muscle fiber

**9** If the papillary muscles fail to contract,
- a) the AV valves will not open.
- b) the semilunar valves will not open.
- ☒ c) the AV valves will not close properly.
- d) the semilunar valves will not close properly.

**10** The simple squamous epithelium and areolar tissue covering the heart valves is the
- a) epicardium.
- ☒ b) endocardium.
- c) myocardium.
- d) cardiac skeleton.

**11** Which of the following is a fibrous remnant of a fetal connection between the aorta and pulmonary trunk?
- a) fossa ovalis
- ☒ b) ligamentum arteriosum
- c) chordae tendineae
- d) trabeculae carneae

**12** The Frank-Starling law of the heart states that
- ☒ a) the greater the EDV, the greater the stroke volume.
- b) the greater the ESV, the greater the stroke volume.
- c) the greater the EDV, the lesser the stroke volume.
- d) the lesser the ESV, the lesser the stroke volume.

**13** The great cardiac vein drains blood from the heart muscle to the
- a) left ventricle.
- b) right ventricle.
- ☒ c) right atrium.
- d) left atrium.

**14** Which of the following is a function of the cardiac skeleton of the heart?
- a) It physically isolates the muscle fibers of the atria from those of the ventricles.
- ☒ b) It helps distribute the forces of cardiac contraction.
- c) It allows more rapid contraction of the ventricles.
- d) It strengthens and helps prevent overexpansion of the heart.

**15** During the cardiac cycle, the amount of blood ejected from the left ventricle when the semilunar valve opens is the
- ☒ a) stroke volume (SV).
- b) end-diastolic volume (EDV).
- c) end-systolic volume (ESV).
- d) cardiac output (CO).

## Short answer

**16** What role do the chordae tendineae and papillary muscles play in the normal function of the AV valves?

**17** Trace the normal pathway of an electrical impulse through the conducting system of the heart.

**18** Describe the three distinct layers that make up the heart wall.

**19** What effects do sympathetic and parasympathetic stimulation have on the heart?

**20** What are the sources and significance of the four heart sounds?

## MasteringA&P™

Access more chapter study tools online in the MasteringA&P Study Area:

■ Chapter Quizzes, Chapter Practice Test, and Clinical Case Studies

■ Practice Anatomy Lab™ 3.0    PAL 3.0

■ Interactive Physiology™    iP2™

■ A&P Flix™    A&PFlix™

■ PhysioEx™ 9.1    PhysioEx 9.1

# Chapter Integration • Applying what you have learned

## An abnormal click in a 12-year-old's heart

Brian is a 12-year-old boy about to enter the sixth grade. While his pediatrician is listening to his heart during a routine physical examination, she hears an abnormal click during ventricular systole. The doctor informs Brian's mother that she can hear a murmur, or irregular sound, during the beating of his heart and that the boy most likely has a mitral valve prolapse. She explains that his mitral valve is not functioning perfectly and that when it closes, it bulges, or prolapses, into the chamber it is attempting to prevent blood from entering. The good news is that she cannot hear any regurgitation.

Brian's mother became noticeably upset at the thought of her son having a heart condition. The doctor calmly explained that his condition is relatively common, is not life threatening, and often causes no symptoms. She will order an echocardiogram (an ultrasound of the heart) so she can examine his mitral valve and determine the extent of his condition, but she expects his condition to be relatively minor.

**1** The doctor heard the murmur during ventricular systole. Is the mitral valve closed or open at that time?

**2** Into which chamber is the valve prolapsing, and why is there no regurgitation?

**3** Predict what the echocardiogram could show to explain Brian's heart murmur.

## LEARNING OUTCOMES

These Learning Outcomes correspond by number to this chapter's modules and indicate what you should be able to do after completing the chapter. They also appear at the end of each module.

# 19

# Blood Vessels and Circulation

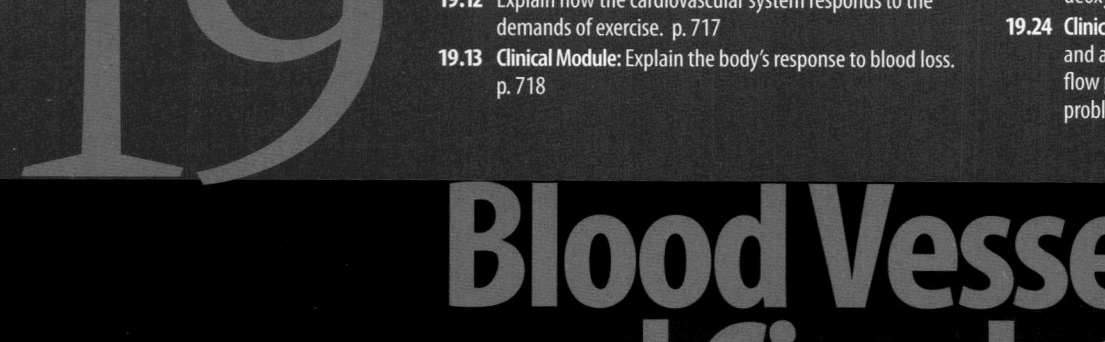

## Module 19.1

# The heart pumps blood, in sequence, through the arteries, capillaries, and veins of the pulmonary and systemic circuits

Blood flows through a network of blood vessels that extend between the heart and peripheral tissues. Those blood vessels are organized into a **pulmonary circuit**, which carries blood to and from the gas exchange surfaces of the lungs, and a **systemic circuit**, which transports blood to and from the rest of the body.

**2** **Pulmonary Circuit**

Pulmonary arteries

Capillaries in lungs

Pulmonary veins

**Start**

**1** The **right atrium** receives blood from the systemic circuit and passes it to the **right ventricle**, which pumps blood into the pulmonary circuit.

**4** **Systemic Circuit**

Capillaries in head, neck, upper limbs

Systemic arteries

**3** The **left atrium** collects blood from the pulmonary circuit and empties it into the **left ventricle**, which pumps blood into the systemic circuit.

Systemic veins

Capillaries in trunk and lower limbs

**1** Each circuit begins and ends at the heart, and blood travels through these circuits in sequence. Thus, blood returning to the heart from the systemic circuit must complete the pulmonary circuit before reentering the systemic circuit. Blood is carried away from the heart by **arteries** and returns to the heart by way of **veins**. Vessels that carry blood away from the heart (arteries) are also called *efferent vessels*. Vessels that carry blood toward the heart (veins) are also known as *afferent vessels*. Microscopic, thin-walled vessels called **capillaries** interconnect the smallest arteries and the smallest veins. Capillaries are called *exchange vessels* because their thin walls allow the exchange of nutrients, dissolved gases, and wastes between blood and the surrounding interstitial fluid.

**?** REVIEW

**C.** Which chamber of the heart receives blood from the systemic circuit?

**?** REVIEW

**A.** Describe the pulmonary circuit.

**B.** Describe the systemic circuit.

In this section we take a close look at the structure of the major blood vessels in the body.

**?** INTEGRATION

**D.** Distinguish among efferent vessels, afferent vessels, and exchange vessels.

**Lo** LEARNING OUTCOME

Distinguish between the pulmonary and systemic circuits, and identify afferent and efferent blood vessels.

# Module 19.2
## Arteries and veins differ in the structure and thickness of their walls

**1** The walls of arteries and veins contain three distinct layers: the *tunica intima*, *tunica media*, and *tunica externa*. These vessel walls are too thick to allow diffusion between the blood and tissues or even between blood and the vessel itself. Thus, large vessel walls contain small arteries and veins that supply the smooth muscle cells and fibroblasts of the tunica media and tunica externa. These blood vessels are called the **vasa vasorum** ("vessels of vessels").

Lumen of artery

Lumen of vein

**Artery and vein**  LM × 60

The **tunica intima** (IN-ti-muh), or *tunica interna*, is the innermost layer of a blood vessel. This layer includes the endothelial lining and an underlying layer of connective tissue containing elastic fibers. In arteries, the outer margin of the tunica intima contains a thick layer of elastic fibers called the **internal elastic membrane**.

The **tunica media**, the middle layer, contains concentric sheets of smooth muscle tissue in a framework of loose connective tissue. When these smooth muscles contract, the vessel decreases in diameter; this is called **vasoconstriction**. When the smooth muscles relax, the diameter increases; this is called **vasodilation**. The tunica media in arteries is separated from the surrounding tunica externa by a thin band of elastic fibers called the **external elastic membrane**.

The **tunica externa** (eks-TER-nuh), or *tunica adventitia*, the outermost layer of a blood vessel, is a connective tissue sheath. In arteries, this layer contains collagen fibers with scattered bands of elastic fibers. In veins, it is generally thicker than the tunica media and contains networks of elastic fibers and bundles of smooth muscle cells. The connective tissue fibers of the tunica externa typically blend into those of adjacent tissues, stabilizing and anchoring the blood vessel.

**Artery**

Endothelium

Internal elastic membrane

Smooth muscle

Elastic fiber

External elastic membrane

**Vein**

Endothelium

Smooth muscle

Tunica intima

Tunica media

Tunica externa

## Features of Typical (Medium-Sized) Arteries and Veins

| Feature | Typical Artery | Typical Vein |
|---|---|---|
| **General appearance in sectional view** | Usually round, with relatively thick wall Small lumen | Usually flattened or collapsed, with relatively thin wall Large lumen |
| **Tunica intima** | | |
| Endothelium | Usually rippled because of vessel constriction | Often smooth |
| Internal elastic membrane | Present | Absent |
| **Tunica media** | Thick, dominated by smooth muscle cells and elastic fibers | Thin, dominated by smooth muscle cells and collagen fibers |
| External elastic membrane | Present | Absent |
| **Tunica externa** | Collagen and elastic fibers | Collagen, elastic fibers, and smooth muscle cells |

**2** There are five general classes of blood vessels in the cardiovascular system. **Arteries** carry blood away from the heart. As arteries enter peripheral tissues, they branch repeatedly, and the branches decrease in diameter. The smallest arterial branches are called **arterioles** (ar-TĒR-ē-ōls). From the arterioles, blood moves into **capillaries**, where diffusion occurs between blood and interstitial fluid. From the capillaries, blood enters small **venules** (VEN-yūls), which unite to form larger **veins** that return blood to the heart.

**? REVIEW**

**A.** List the five general classes of blood vessels.

## Large Vein

**Large veins** include the superior and inferior venae cavae and their branches within the abdominopelvic and thoracic cavities. They have a thin tunica media and large lumen. Average luminal diameter is about 2 cm, with an average wall thickness of about 2 mm.

Tunica externa
Tunica media
Tunica intima

## Elastic Artery

**Elastic arteries** are large vessels that transport blood away from the heart and include the pulmonary trunk and aorta. These are resilient, elastic vessels capable of stretching and recoiling as the heart beats and arterial pressures change. They have an average luminal diameter of about 1.5 cm and wall thickness of about 1 mm.

Internal elastic membrane
Tunica intima
Endothelium
Tunica media
Tunica externa

## Medium-Sized Vein

**Medium-sized veins** correspond in general size to medium-sized arteries. In these veins, the tunica media is thin, and it contains relatively few smooth muscle fibers. Their luminal diameters range from 2 to 9 mm, and average wall thickness is about 4 mm.

Tunica externa
Tunica media
Endothelium
Tunica intima

## Muscular Artery

**Muscular arteries**, or medium-sized arteries, distribute blood to the body's skeletal muscles and internal organs. They have a thicker tunica media with a greater percentage of smooth muscle fibers than elastic arteries. They have an average luminal diameter of 4 mm and wall thickness of about 1 mm.

Tunica externa
Tunica media
Endothelium
Tunica intima

## Venule

**Venules** collect blood from capillary beds and are the smallest venous vessels. Venules smaller than 50 μm lack a tunica media and resemble expanded capillaries. They have an average luminal diameter of about 20 μm and wall thickness of about 1 μm.

Tunica externa
Endothelium

## Arteriole

**Arterioles** have a poorly defined tunica externa, and the tunica media consists of scattered smooth muscle fibers that may not form a complete layer. They have an average luminal diameter of about 30 μm and wall thickness of about 6 μm.

Smooth muscle cells
Endothelium

## Fenestrated Capillary

Pores
Endothelial cells
Basement membrane

## Capillaries

**Capillaries** lack a tunica externa and tunica media. They are the only blood vessels whose walls permit exchange between blood and the surrounding interstitial fluids. Because capillary walls are thin, diffusion distances are short, so exchange can occur quickly.

## Continuous Capillary

Endothelial cells
Basement membrane

**? REVIEW**

**B.** Describe a capillary.

**Lo LEARNING OUTCOME**

Distinguish among the types of blood vessels on the basis of their structure and function.

# Module 19.3
# Capillary structure and capillary blood flow affect the rates of exchange between the blood and interstitial fluid

A typical capillary consists of a tube of endothelial cells within a basement membrane. The tunica media and tunica externa are absent. The average diameter of a capillary is a mere 8 μm, very close to that of a single red blood cell. Two major types of capillaries exist: continuous capillaries and fenestrated capillaries.

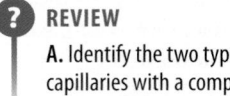

**? REVIEW**
**A.** Identify the two types of capillaries with a complete endothelium.

## Continuous Capillary

**1** In a **continuous capillary**, the endothelium is a complete lining. A cross section through a large continuous capillary cuts across several endothelial cells. In a small continuous capillary, a single endothelial cell may completely encircle the lumen. Continuous capillaries are located throughout the body, in all tissues except epithelia and cartilage. Continuous capillaries permit the diffusion of water, small solutes, and lipid-soluble materials into the surrounding interstitial fluid but prevent the loss of blood cells and plasma proteins. In addition, some exchange may occur between blood and interstitial fluid through selective vesicular transport (see Module 3.17). In specialized continuous capillaries throughout most of the central nervous system and in the thymus, the endothelial cells are bound together by tight junctions. Permeability in these capillaries is both restricted and precisely regulated.

## Fenestrated Capillary

**2** **Fenestrated** (FEN-es-trā-ted) **capillaries** (*fenestra*, window) also have a complete, or uninterrupted, endothelium. These capillaries contain "windows," or pores, that penetrate the endothelial lining. The pores (60–80 nm in diameter) permit the rapid exchange of water and solutes as large as small peptides between blood and interstitial fluid. Examples of fenestrated capillaries include the choroid plexuses of the brain and the capillaries of the hypothalamus, pituitary gland, pineal gland, and thyroid gland. Fenestrated capillaries are also located along absorptive areas of the intestinal tract and at filtration sites in the kidneys.

Basement membrane —
Endothelial cell —
Nucleus —

Fenestrations, or pores

Vesicles containing materials transported across the endothelial cell

Basement membrane

Boundary between endothelial cells

Boundary between endothelial cells

Basement membrane

**? REVIEW**
**B.** At what sites in the body are fenestrated capillaries located?

**3** Sinusoids (SĪ-nuh-soydz), or **sinusoidal capillaries**, resemble fenestrated capillaries that are flattened and irregularly shaped. In contrast to continuous and fenestrated capillaries, sinusoids have a discontinuous endothelium. Gaps 30–40 µm in diameter are present between adjacent endothelial cells, and the basement membrane is either thinner or absent. As a result, sinusoids permit the free exchange of water and solutes as large as plasma proteins between the slow-moving blood and interstitial fluid. Sinusoids occur in the liver, bone marrow, spleen, and many endocrine organs, including the pituitary and adrenal glands.

Gap between adjacent cells

**Sinusoids**

**4** Capillaries function as part of an interconnected network called a **capillary bed**. A capillary bed contains several relatively direct connections between arterioles and venules. Such a connection, which resembles a typical capillary in structure, is called a **thoroughfare channel**. Although blood normally flows at a constant rate from the arteriole to the venule across the capillary bed, the flow within each capillary is quite variable. Bands of smooth muscle at the entrance to each capillary—called **precapillary sphincters**—alternately contract and relax, perhaps a dozen times per minute. These rhythmic changes in vessel diameter are called **vasomotion**. As a result, the blood flow within the associated capillary occurs in pulses rather than as a steady and constant stream. The net effect is that blood may reach the venules by one route now and by a different route later.

Arteriole

Small artery

Capillary bed          LM × 125

More than one artery may supply blood to a capillary bed. The multiple arteries are called **collaterals**. They fuse before giving rise to arterioles. The fusion of two collateral arteries that supply a capillary bed is an example of an **arterial anastomosis**. (An *anastomosis* is the joining of blood vessels.) An arterial anastomosis acts like an insurance policy: If one artery is compressed or blocked, capillary circulation will continue.

A single arteriole generally gives rise to dozens of capillaries that empty into several venules.

A **precapillary sphincter** guards the entrance to each capillary. Contraction or relaxation of the smooth muscle cells changes the diameter of the capillary entrance, thereby controlling the flow of blood through the sphincter.

An **arteriovenous** (ar-tēr-ē-ō-VĒ-nus) **anastomosis** is a direct connection between an arteriole and a venule. When this anastomosis is dilated, blood will bypass the capillary bed and flow directly into the venous circulation. The pattern of blood flow through these anastomoses is regulated primarily by sympathetic innervation under the control of the cardiovascular centers of the medulla oblongata.

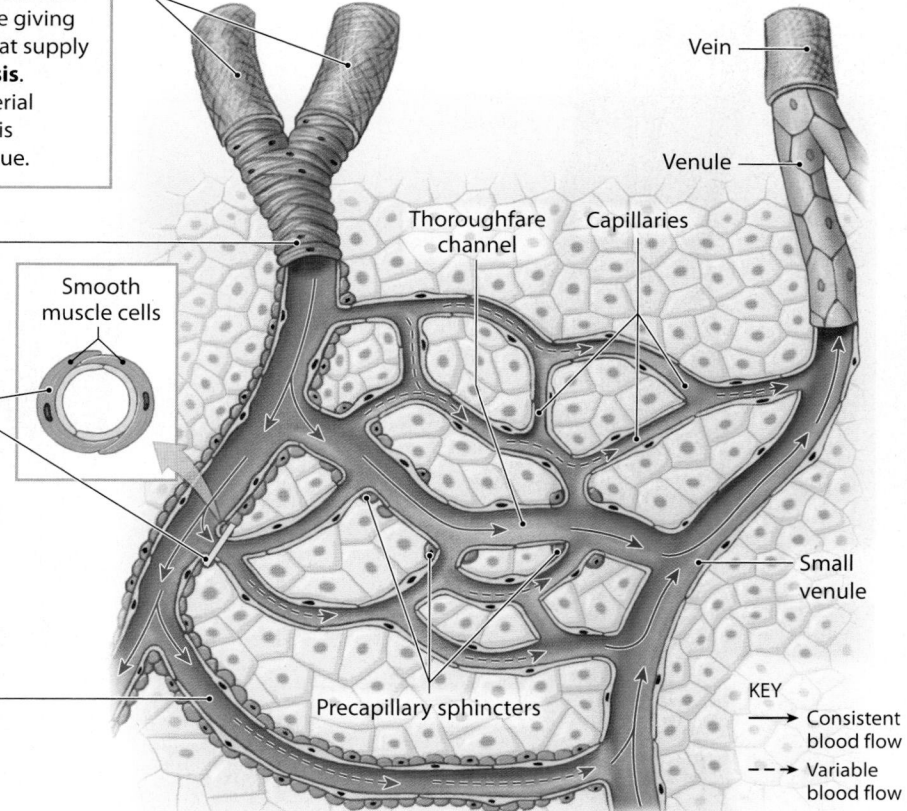

Vein

Venule

Thoroughfare channel

Capillaries

Smooth muscle cells

Small venule

Precapillary sphincters

KEY
→ Consistent blood flow
--→ Variable blood flow

**LEARNING OUTCOME**

Describe the structures of capillaries and their functions in the exchange of dissolved materials between blood and interstitial fluid.

# Module 19.4

# The venous system has low pressures and contains almost two-thirds of the body's blood volume

The arterial system is a high-pressure system: Almost all the force developed by the heart is required to push blood along the network of arteries and through miles of capillaries. Blood pressure in a peripheral venule is only about 10 percent of that in the ascending aorta, and pressures continue to fall along the venous system.

**1** The blood pressure in venules and medium-sized veins is so low that it cannot overcome the force of gravity. In the limbs, veins of this size contain **valves**, folds of the tunica intima that project from the vessel wall and point in the direction of blood flow. Venous valves (like valves in the heart) permit blood flow in one direction only, thereby preventing the backflow of blood toward the capillaries. If the walls of the veins near the valves weaken or become stretched and distorted, the valves may no longer work properly. Blood then pools in the veins, which become grossly distended. The effects range from mild discomfort and a cosmetic problem, as in superficial **varicose veins** in the thighs and legs, to painful distortion of adjacent tissues, as in the **hemorrhoids** that form in venous networks of the anal canal.

**REVIEW**

**A.** Why are valves located in veins but not in arteries?

**B.** How is blood pressure maintained in veins to counter the force of gravity?

When you are standing, venous blood from your feet must overcome the pull of gravity to ascend to the heart. Valves compartmentalize the blood within the veins, thereby dividing the weight of the blood between the compartments.

Valve closed

Valve closed

Any contraction of the surrounding skeletal muscles squeezes the blood toward the heart. Although you are probably not aware of it, when you stand, rapid cycles of contraction and relaxation are occurring within your leg muscles, helping to push blood toward the trunk.

Valves superior to the contracting muscle open, allowing blood to move toward the heart.

Valves inferior to the contracting muscle are forced closed, preventing backflow of blood to the capillaries.

**64%**

**Systemic
venous
system**

**9%**

**Pulmonary
circuit**

**Heart**

**7%**

**Systemic
arterial
system**

**13%**

**Systemic
capillaries**

**7%**

WHOLE
BLOOD

Affix
Donation
Number
Here

Expiry Date

For storage of whole blood.
Do not reuse if there is any visible sign of
deterioration. The fluid pathways of this
pack are sterile and non-pyrogenic.
Contents of unit pack must be used within 10 days
of opening
Do not use after.

ABO and RhD
Group

Do not vent.
Do not reuse this container

Centre Identity

VOLUNTEER DONOR

Date Bled

**2** This chart shows the distribution of the body's total blood volume at rest. The total blood volume is unevenly distributed among arteries, capillaries, and veins. The systemic venous system contains nearly two-thirds of total blood volume. In a man with a 6 L total blood volume, systemic veins contain roughly 3.5 L of whole blood. Of that amount, approximately 1 L is found in venous networks carrying blood from the digestive organs to the liver. Because veins are much more *distensible*, or expandable, than arteries, they can act as **blood reservoirs** and accommodate large changes in blood volume. The pulmonary circuit, the heart, and the systemic arteries and capillaries contain the rest of the blood volume (about 1.5 L).

**? REVIEW**
**C.** Describe the distribution of total blood volume in the body.

**3** If serious hemorrhaging (profuse blood loss) occurs, the body maintains blood volume within the arterial system at near-normal levels by reducing the volume of blood in the venous system. In the neural mechanism involved, the vasomotor center in the medulla oblongata stimulates sympathetic nerves innervating smooth muscle cells in the tunica media of medium-sized veins. The resulting contraction produces **venoconstriction** (vē-nō-kon-STRIK-shun), reducing the diameter of the veins and the amount of blood contained in the venous system. In addition, blood enters the general circulation from venous networks of slowly moving blood from the digestive organs that drain into the liver.

Sympathetic
nerves stimulated

Smooth
muscle
contracts

Vein
constricts

**? INTEGRATION**
**D.** What factors are involved in the formation of varicose veins?

**Lo LEARNING OUTCOME**
Describe the venous system, and indicate the distribution of blood within the cardiovascular system.

# Functional Anatomy of Blood Vessels

**Labeling:** Label the blood vessels and their components in the images below.

LM × 60

1 _____

2 _____

3 _____

4 _____

5 _____

6 _____

7 _____

8 _____

**Matching:** Match each lettered term with the most closely related description.

a. pulmonary circuit

b. vein

c. exchange vessels

d. continuous capillary

e. systemic circuit

f. right atrium

g. venous valves

h. fenestrated capillaries

i. artery

j. anastomosis

k. sinusoids

l. tunica media

| 9 | Left ventricle pumps blood into this |
| 10 | Relatively thick vessel wall |
| 11 | Joining of blood vessels |
| 12 | Present only in veins |
| 13 | Blood to and from the lungs |
| 14 | Present in liver, bone marrow, and spleen |
| 15 | Relatively thin vessel wall |
| 16 | Complete endothelial lining with pores |
| 17 | Smooth muscle tissue |
| 18 | Capillaries |
| 19 | Complete endothelial lining without pores |
| 20 | Receives blood from systemic circuit |

9 _____

10 _____

11 _____

12 _____

13 _____

14 _____

15 _____

16 _____

17 _____

18 _____

19 _____

20 _____

**Section integration:** What is the function of precapillary sphincters, and what role would you expect them to play during exercise and in cold temperatures?

21 _____

## Module 19.5

# Pressure, resistance, and venous return affect cardiac output

Go to
MasteringA&P™
**iP2™**
Cardiac
Output

You are now familiar with the factors that regulate cardiac output (see Module 18.16). Both cardiac output and the distribution of blood within the pulmonary and systemic circuits must constantly be adjusted to meet the demands of active tissues. The sites and mechanisms for such adjustments are previewed here.

**Cardiac Output**

To maintain **cardiac output**, the heart must generate enough pressure to force blood through thousands of miles of peripheral capillaries, most no larger in diameter than a single red blood cell.

**? REVIEW**

**A.** Neural and endocrine regulatory mechanisms influence which factors?

**Venous Return**

Venous pressure, although low, determines **venous return**—the amount of blood arriving at the right atrium each minute. Venous return has a direct impact on cardiac output.

**Arterial Pressure**

**Arterial pressure** is the force exerted by blood against the walls of systemic arteries. Arterial pressure is much higher than venous pressure because it must push blood a greater distance and through smaller and smaller arteries and then through innumerable capillary networks.

**Blood Flow Regulation**

**Autoregulation** acts locally. **Central regulation** involves neural and endocrine mechanisms to make coordinated adjustments in heart rate, stroke volume, peripheral resistance, and venous pressure so that cardiac output is sufficient to meet the needs of peripheral tissues.

**Venous Pressure**

**Venous pressure** is the force exerted by blood within the venous system. As blood moves toward the heart, the veins become larger, resistance drops, and blood flow velocity increases. Muscular compression and the respiratory pump are two factors that assist in propelling blood toward the heart at low venous pressure.

**Peripheral Resistance**

**Resistance** is a force that opposes movement. The **peripheral resistance** is the resistance of the arterial system as a whole. Resistance increases as the arterial branches become smaller and thinner.

**Capillary Pressure**

**Capillary pressure** is the force exerted by blood against capillary walls. It is normally low, allowing time for diffusion between the blood and surrounding interstitial fluid. This process, called **capillary exchange**, plays a key role in homeostasis.

Capillary
exchange

Interstitial fluid

**? REVIEW**

**B.** Which is greater: arterial pressure or venous pressure?

**? REVIEW**

**C.** Why is it beneficial for capillary pressure to be very low?

In this section we discuss the functioning of the cardiovascular system as a whole and learn how neural and hormonal mechanisms regulate cardiovascular function.

**Lo LEARNING OUTCOME**

Explain the effects of pressure, resistance, and venous return on cardiac output.

# Module 19.6

# Vessel luminal diameter is the main source of resistance within the cardiovascular system

For circulation to occur, the heart must develop sufficient pressure to overcome the **total peripheral resistance**—the resistance of the entire cardiovascular system. The total peripheral resistance of the cardiovascular system depends on three factors: *vascular resistance*, *viscosity*, and *turbulence*.

**?** REVIEW

**A.** List the factors that contribute to total peripheral resistance.

**1** **Vascular resistance**, the opposition to blood flow in vessels, is the largest component of total peripheral resistance. Vascular resistance results primarily from friction between blood and the vessel's luminal wall. The amount of friction depends on two factors: vessel length and vessel diameter.

### Friction and Vessel Length

Friction occurs between the moving blood and the walls of the vessel. The longer the vessel, the greater the surface area in contact with the blood, and the greater the resistance. You can easily blow the water out of a snorkel that is 25 cm (10 in.) long, but you cannot blow the water out of a 15 m garden hose of the same diameter because the total frictional resistance is too great. The most dramatic changes in blood vessel length occur between birth and maturity, as individuals grow to adult size. In adults, vessel length can be considered constant.

### Friction and Vessel Luminal Diameter

Friction also occurs between layers of fluid moving at different speeds. The layer of blood closest to the luminal wall of the vessel is slowed by friction with the endothelial surface. The adjacent layer of blood is slowed by friction with the more superficial layer. This effect diminishes as the distance from the wall increases. In a small-diameter vessel, all the blood is slowed to some degree, and resistance is relatively high. Blood near the center of a large-diameter vessel does not encounter friction with the luminal walls, so the resistance is fairly low.

### Vessel Length versus Vessel Luminal Diameter

Differences in luminal vessel diameter have a greater effect on resistance than do vessel length differences. If two vessels have an equal luminal diameter (or radius) but one is twice as long as the other, the longer vessel has twice as much resistance to blood flow. If two vessels have equal length, but one has twice the luminal diameter of the other, the narrower one has 16 times more resistance to blood flow. This relationship, expressed in terms of the vessel radius (r) and resistance (R), is summarized as $R \propto 1/r^4$. Thus, a small change in luminal vessel diameter produces a large change in resistance.

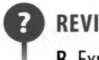 REVIEW

**B.** Explain the equation $R \propto 1/r^4$.

**2** **Viscosity** is a measure of the resistance to flow caused by interactions among molecules and suspended materials in a liquid. Liquids of low viscosity, such as water (viscosity 1.0), flow at low pressures. Thick, syrupy fluids, such as molasses (viscosity 300), flow only under higher pressures. Whole blood has a viscosity about five times that of water, because of the presence of plasma proteins and blood cells. Under normal conditions, the viscosity of blood remains stable, but disorders that affect the hematocrit or plasma composition also change blood viscosity and thus peripheral resistance.

**3** High flow rates, irregular surfaces, and sudden changes in vessel diameter upset the smooth flow of blood, creating eddies and swirls. This phenomenon, called **turbulence**, increases resistance and slows blood flow. Here you see the turbulence created by the presence of a plaque within a small artery. Turbulence also occurs in normal individuals when blood flows between the atria and the ventricles, and between the ventricles and the aortic and pulmonary trunks. Turbulent flow is responsible for the production of the third and fourth heart sounds (see Module 18.10).

Go to
MasteringA&P™
PhysioEx™ 9.1
Ex 5.1: Blood Vessel Radius and Blood Flow Rate

Relative viscosity

1000
100
10
0

300
100
40
10
5
1

**Molasses**

**Motor Oil**

**Maple Syrup**

**Blood**

**Water**

Turbulence

Plaque deposit

**? INTEGRATION**

**C.** Which would reduce peripheral resistance: an increase in vessel length or an increase in vessel diameter?

**Lo LEARNING OUTCOME**

Describe the factors that influence total peripheral resistance.

# Module 19.7

# Blood flow is determined by the interplay between arterial pressure and peripheral resistance

**Blood flow** is the volume of blood flowing per unit of time through a vessel or group of vessels. In general terms, blood flow is directly proportional to the **arterial pressure**, and inversely proportional to **peripheral resistance**. However, the absolute pressure is less important than the **pressure gradient**—the difference in pressure from one end of the vessel to the other. The largest pressure gradient is found between the base of the aorta and the proximal ends of peripheral capillary beds. Cardiovascular pressures are measured in millimeters of mercury (mm Hg).

**? REVIEW**

**A.** Define *blood flow*, and describe its relationship to blood pressure and peripheral resistance.

**1** The most important component of total peripheral resistance is vascular resistance (see Module 19.6). The two curves in the figure at right illustrate the inverse relationship between vessel luminal diameter and vessel cross-sectional area. The U-shaped curve shows the decrease in arterial diameters toward the capillaries and the increase in venous diameters toward the heart. The hump-shaped curve indicates the total cross-sectional areas of individual vessels. This second curve (black line) shows that the capillaries contribute most to vascular resistance. In effect, your blood moves from one big pipe (the aorta, with a cross-sectional area of 4.5 cm²) into countless tiny ones (the peripheral capillaries, with a total cross-sectional area of 5000 cm²), and then blood travels back to the heart through two large pipes (the venae cavae).

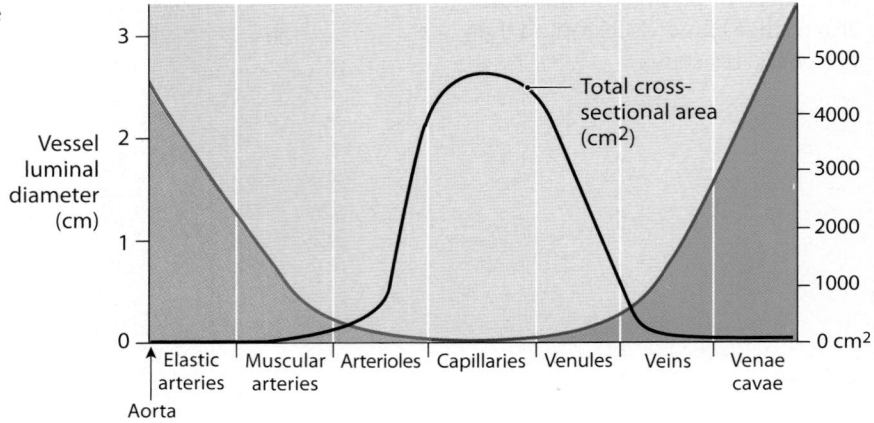

**2** The heart generates a pressure of around 120 mm Hg as it pushes blood into the aorta. At each branching of the arterial system, the arterial pressure drops as blood is pushed into ever-increasing numbers of smaller and smaller branches. At the start of peripheral capillaries, arterial pressure has fallen to approximately 35 mm Hg, and by the time blood reaches the venules, it has fallen to approximately 18 mm Hg.

**3** As blood pressure decreases, the blood moves more slowly. Blood flow is highest in the aorta, where blood pressure is high and the large luminal diameter provides for low resistance. Blood flow is slowest in the capillaries, whose luminal diameter is nearly the same as that of a single red blood cell. This is important because exchange with interstitial fluid occurs only in the capillaries, and diffusion is a relatively slow process. Blood flow speeds up in the venous system because although pressures are falling, the vessels are merging into large-diameter passages with very low resistance.

**4** Arterial pressure is not constant. It rises during ventricular systole and falls during ventricular diastole as the elastic arterial walls stretch and recoil. The peak blood pressure measured during ventricular systole is called **systolic pressure**, and the minimum blood pressure at the end of ventricular diastole is called **diastolic pressure**. In recording blood pressure, we separate systolic and diastolic pressures with a slash. Thus the blood pressure illustrated here is written as 120/90 (read as "one-twenty over ninety").

**?** **REVIEW**

**B.** In a healthy person, where is blood pressure greater: in the aorta or in the inferior vena cava? Explain.

Go to MasteringA&P™ **iP2**™ Factors Affecting Blood Pressure

The difference between the systolic and diastolic pressures is the **pulse pressure**.

The **mean arterial pressure (MAP)** is calculated by adding one-third of the pulse pressure to the diastolic pressure. For a systolic pressure of 120 mm Hg and a diastolic pressure of 90 mm Hg, MAP can be calculated as follows:

$$90 + (120 - 90)/3$$
or
$$90 + 10 = 100 \text{ mm Hg}$$

Systolic

Diastolic

Blood pressure (mm Hg): 120, 100, 80, 60, 40, 20, 0

Aorta | Elastic arteries | Muscular arteries | Arterioles | Capillaries | Venules | Medium-sized veins | Large veins | Venae cavae

Increasing distance from left ventricle →

**5** Capillary exchange refers to the bidirectional movement of substances between the blood plasma and interstitial fluid across capillary walls. It involves three processes: *filtration*, diffusion, and osmosis. Blood pressure within the capillary beds, called *capillary hydrostatic pressure* (*CHP*), provides the driving force for filtration. Along the length of a typical capillary, CHP pushes water and soluble molecules out of the bloodstream and into the interstitial fluid. Only small solutes can cross the endothelium; larger molecules, such as plasma proteins, remain in the bloodstream. This size-selective process is called **filtration**. We consider further details about the process of capillary exchange in the next module.

Capillary hydrostatic pressure (CHP)

Amino acid
Blood protein
Glucose
Ions
Interstitial fluid

Small solutes

Hydrogen bond

Water molecule

Endothelial cell 1

Endothelial cell 2

Interstitial fluid

Relatively small solutes are carried along as CHP pushes water into the interstitial fluid during filtration.

**?** **INTEGRATION**

**C.** Calculate the mean arterial pressure for a person whose blood pressure is 125/70.

**Lo** **LEARNING OUTCOME**

Describe the factors that determine blood flow.

# Module 19.8

# Capillary exchange is a dynamic process that includes diffusion, filtration, and reabsorption

**1** Diffusion is the net movement of ions or molecules from an area where their concentration is higher to an area where their concentration is lower. Diffusion occurs most rapidly when (1) the distances involved are short, (2) the concentration gradient is large, and (3) the ions or molecules involved are small (see Module 3.14). Diffusion occurs continuously across capillary walls, but different substances use different routes, as noted in the table below.

## Routes of Diffusion Across Capillary Walls

- Water, ions, and small organic molecules, such as glucose, amino acids, and urea, can usually enter or leave the bloodstream by diffusion either between adjacent endothelial cells or through the pores of fenestrated capillaries.

- Many ions, including sodium, potassium, calcium, and chloride, can diffuse across endothelial cells by passing through channels in plasma membranes.

- Large water-soluble compounds are unable to enter or leave the bloodstream except at fenestrated capillaries, such as those in the hypothalamus, the kidneys, many endocrine organs, and the intestinal tract.

- Lipids, such as fatty acids and steroids, and lipid-soluble materials, including soluble gases such as oxygen and carbon dioxide, can cross capillary walls by diffusion through the endothelial plasma membranes.

- Plasma proteins are normally unable to cross the endothelial lining anywhere except in sinusoids, such as those of the liver, where plasma proteins enter the bloodstream.

**2** In a capillary, blood pressure decreases as blood flows from the arterial end to the venous end. As a result, the rates of filtration and fluid reabsorption (by osmosis) gradually change as blood passes along the length of a capillary.

### Near Arteriole, Filtration Predominates

Filtration predominates at the arterial end of a capillary, where **capillary hydrostatic pressure (CHP)** is highest. As filtration occurs, the plasma osmolarity, or **blood colloid osmotic pressure (BCOP)**, increases. BCOP rises because water is leaving but the larger solutes, especially soluble plasma proteins, are remaining behind. The difference between CHP (which pushes water out) and BCOP (which draws water in) is known as the **net filtration pressure (NFP)**. At the start of the capillary, NFP is high, but as blood flows along the capillary, CHP decreases as water flows out, and BCOP rises as plasma protein concentrations increase.

Return to circulation

3.6 L/day flow into lymphatic vessels

Tissue cells

Arteriole

24 L/day

35 mm Hg   25 mm Hg

**NFP = +10 mm Hg**

Capillary

CHP > BCOP
Fluid forced out of capillary

KEY

**CHP** (Capillary hydrostatic pressure)

**BCOP** (Blood colloid osmotic pressure)

**NFP** (Net filtration pressure)

**3** Any conditions that affect either blood pressure or the osmotic pressures of the blood or interstitial fluid will shift the balance between hydrostatic and osmotic forces. We can then predict the effects on the basis of an understanding of capillary dynamics. Three examples are presented in the table to the right.

### Representative Variations in Capillary Exchange

- If hemorrhaging occurs, both blood volume and blood pressure decline. This decrease in CHP lowers the NFP and increases the amount of capillary reabsorption. The result is a reduction in the volume of interstitial fluid and an increase in the circulating plasma volume. This process is known as a **recall of fluids**.

- If dehydration occurs, the plasma volume decreases because of water loss, and the concentration of plasma proteins increases. The increase in BCOP accelerates reabsorption and a recall of fluids that delays the onset and severity of clinical problems caused by low blood volume and blood pressure.

- If the CHP rises or the BCOP decreases, fluid moves out of the blood in capillaries and builds up in peripheral tissues, an abnormal condition called **edema**.

**? REVIEW**

**B.** Define *edema*.

**No Net Movement**

At a point roughly two-thirds of the way along the capillary, NFP is zero, and there is no net movement of fluid into or out of the capillary.

**Near Venule, Reabsorption Predominates**

In the final segment of the capillary, CHP falls below BCOP, and water flows back into the capillary. Note that more water leaves the bloodstream during filtration than gets retrieved through reabsorption. The difference, approximately 3.6 L/day, flows through peripheral tissues and enters lymphatic vessels that eventually return it to the venous system.

**Venule**

20.4 L/day

No net fluid movement

| 25 mm Hg | 25 mm Hg |

**NFP = 0**

**Capillary**

| 18 mm Hg | 25 mm Hg |

**NFP = –7 mm Hg**

CHP = BCOP
No net fluid movement

BCOP > CHP
Fluid moves into capillary

**? REVIEW**

**A.** Under what general conditions would fluid move into a capillary?

**? INTEGRATION**

**C.** Identify the conditions that would shift the balance between hydrostatic and osmotic forces.

**Lo LEARNING OUTCOME**

Describe the movement of fluids between capillaries and interstitial spaces.

## Module 19.9

# Cardiovascular regulatory mechanisms respond to changes in blood pressure or blood chemistry

**1** Homeostatic mechanisms regulate cardiovascular activity to ensure that **tissue perfusion**—the blood flow through the tissues—meets the tissue demands for oxygen and nutrients. There are two regulatory pathways: one that involves local *autoregulation* and, if that is ineffective, a second that involves *central regulation* by neural and endocrine mechanisms.

**? REVIEW**
**A.** Define *tissue perfusion*.

---

### Autoregulation

**Autoregulation** involves local changes in the pattern of blood flow within capillary beds. These changes are due to precapillary sphincters that open and close in response to chemical changes in the interstitial fluid. Factors that promote the dilation of blood vessels are called **vasodilators**.

Inadequate local blood flow and blood pressure

**Effectors**
Precapillary sphincters

*Respond with* →

Local decrease in resistance and increase in blood flow

If autoregulation does not restore homeostasis

Homeostasis **DISTURBED BY**

- Chemical changes (decreased $O_2$ or pH, increased $CO_2$ or prostaglandins)
- Increased tissue activity
- Physical stress (trauma, high local temperature)

Homeostasis **RESTORED BY**

INCREASING

blood flow

**? REVIEW**
**B.** Describe autoregulation as it relates to cardiovascular function.

**STIMULUS**

**RESTORED**

# HOMEOSTASIS
### NORMAL BLOOD VOLUME AND PRESSURE

**RESTORED**

**Central regulation** involves both neural and endocrine mechanisms.

**Endocrine mechanisms** involve long-term increases in blood volume and blood pressure.

Homeostasis **RESTORED BY**

INCREASING

blood volume and pressure

Long-term increase in blood volume and blood pressure

Endocrine response (see Module 19.10)

**Effectors**
Kidneys

**Neural mechanisms** increase cardiac output and decrease blood flow to nonessential or inactive tissues.

Short-term increase in blood pressure by sympathetic stimulation of the heart and peripheral vasoconstriction

Activation of cardiovascular center in the medulla oblongata

Stimulation of receptors sensitive to changes in systemic blood pressure or chemistry

**Effector**
Nervous system

**Central Regulation: Neural and Hormonal Mechanisms**

**2** The **baroreceptor reflexes** (*baro-*, pressure) respond to changes in blood pressure. Arterial baroreceptors are located in the (1) carotid sinuses, expanded chambers near the base of each internal carotid artery of the neck, and (2) the aortic arch (see Module 14.10).

**Responses to Increased Baroreceptor Stimulation**

- Cardiac centers decrease cardiac output through parasympathetic stimulation and sympathetic inhibition.
- Widespread peripheral vasodilation due to the inhibition of the *vasomotor center* of the medulla oblongata.

**?** **REVIEW**

**C.** Explain the function of baroreceptor reflexes.

Baroreceptors stimulated

**Effector**

**Medulla oblongata**

Cardiovascular centers:
- Cardioinhibitory center stimulated
- Cardioacceleratory center inhibited

→ *Result in* → Decreased cardiac output

- Vasomotor center inhibited

→ *Results in* → Vasodilation

**Receptor**

Baroreceptors in carotid sinuses and aortic arch

Reduced blood pressure

Homeostasis **DISTURBED BY** INCREASING blood pressure

STIMULUS

RESTORED

Homeostasis **RESTORED BY** DECREASING blood pressure

# HOMEOSTASIS
## NORMAL RANGE OF BLOOD PRESSURE

Homeostasis **DISTURBED BY** DECREASING blood pressure

STIMULUS

RESTORED

Homeostasis **RESTORED BY** INCREASING blood pressure

**Receptor**

Baroreceptors in carotid sinuses and aortic arch

Increased blood pressure

Baroreceptors inhibited

**Effector**

**Medulla oblongata**

Cardiovascular centers:
- Vasomotor center stimulated

→ *Results in* → Vasoconstriction

- Cardioacceleratory center stimulated
- Cardioinhibitory center inhibited

→ *Result in* → Increased cardiac output

**Responses to Decreased Baroreceptor Stimulation**

- Cardiac centers increase cardiac output through the stimulation of sympathetic innervation to the heart. This results from the stimulation of the cardioacceleratory center and is accompanied by an inhibition of the cardioinhibitory center (see Module 18.14).
- Widespread peripheral vasoconstriction caused by the stimulation of the vasomotor center.
- In a crisis, such as a major blood loss sympathetic activation occurs; both norepinephrine and epinephrine enter the bloodstream. The net effect is an immediate increase in cardiac output and generalized vascular constriction, a combination that rapidly elevates blood pressure.

 **LEARNING OUTCOME**

Explain central regulation, autoregulation, and baroreceptor reflexes in response to changes in blood pressure and blood composition.

# Module 19.10
# Endocrine responses to low blood pressure and low blood volume ...

The endocrine system provides both short-term and long-term regulation of cardiovascular performance through the endocrine functions of the heart and the kidneys, in addition to the actions of antidiuretic hormone from the pituitary gland.

**1** When blood pressure and blood volume decrease below normal, the immediate response is the release of epinephrine (E) and norepinephrine (NE) from the adrenal medullae, stimulating cardiac output and peripheral vasoconstriction. Other hormones important in the long-term response include (1) antidiuretic hormone (ADH), (2) angiotensin II, (3) erythropoietin (EPO), and (4) aldosterone. Although ADH and angiotensin II also affect blood pressure, these two hormones are concerned primarily with the long-term regulation of blood volume.

**? REVIEW**
**A.** Identify the hormones responsible for short-term regulation of decreasing blood pressure and blood volume.

# HOMEOSTASIS
## NORMAL BLOOD PRESSURE AND VOLUME

Homeostasis **DISTURBED BY** **DECREASING** blood pressure and volume

STIMULUS

RESTORED

Homeostasis **RESTORED BY** **INCREASING** blood pressure and volume

**Short-term effects**

**Long-term effects**

**Combined Short-Term and Long-Term Effects**
Increased blood pressure
Increased blood volume

**Receptors**
Baroreceptors

**Receptors**
Kidneys

**Endocrine Response of Kidneys**
Erythropoietin (EPO) is released
Renin release leads to angiotensin II activation

Increased red blood cell formation

Sympathetic activation and release of adrenal hormones E and NE

**Angiotensin II Effects**
Antidiuretic hormone (ADH) released
Aldosterone secreted
Thirst stimulated

**Effectors**
Heart and blood vessels

*Respond with*

Increased cardiac output and peripheral vasoconstriction

**? REVIEW**
**B.** How does the kidney respond to vasoconstriction of the renal artery?

# … are very different from those to high blood pressure and high blood volume

**2** Excessive blood volume triggers a response through its effects on the walls of the heart. When the heart walls are abnormally stretched during diastole, cardiac muscle cells release **natriuretic peptides**. **Atrial natriuretic peptide** (nā-trē-ū-RET-ik; *natrium*, sodium + *ouresis*, urination), or **ANP**, is produced by cardiac muscle cells in the wall of the right atrium. A related hormone called **brain natriuretic peptide** (**BNP**) is produced by ventricular muscle cells. The homeostatic responses to ANP and BNP are summarized below. As blood volume and blood pressure decrease, the stresses on the walls of the heart are reduced, and natriuretic peptide production ceases.

Release of natriuretic peptides (ANP and BNP)

**Receptor**
Cardiac muscle cells

**Effectors**
Kidneys and blood vessels

*Response to ANP and BNP*

Increased Na⁺ loss in urine

Increased water loss in urine

Reduced thirst

Inhibition of ADH, aldosterone, epinephrine, and norepinephrine release

Peripheral vasodilation

Decreased blood pressure and blood volume

Homeostasis **DISTURBED BY** **INCREASING** blood pressure and volume

**STIMULUS**

**RESTORED**

Homeostasis **RESTORED BY** **DECREASING** blood pressure and volume

## HOMEOSTASIS
NORMAL BLOOD VOLUME AND PRESSURE

**REVIEW**
**C.** Describe the roles of the natriuretic peptides.

**LEARNING OUTCOME**
Explain the hormonal regulation of blood pressure and blood volume.

# Module 19.11

# Chemoreceptors monitor the chemical composition of the blood and cerebrospinal fluid

The **chemoreceptor reflexes** respond to changes in carbon dioxide, oxygen, or pH levels in blood and cerebrospinal fluid (CSF). The chemoreceptors involved are sensory neurons located in the **carotid bodies**, situated in the neck near the carotid sinus; in the **aortic bodies**, near the aortic arch; and on the ventrolateral surfaces of the medulla oblongata (see Module 14.10). Stimulation of these chemoreceptors triggers coordinated adjustments in cardiovascular and respiratory activity.

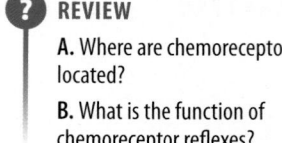

**REVIEW**

**A.** Where are chemoreceptors located?

**B.** What is the function of chemoreceptor reflexes?

| Effector |
| --- |
| **Medulla oblongata** |

Respiratory centers stimulated — *Result in* → Increased respiratory rate

Cardiovascular centers:

- Cardioacceleratory center stimulated
- Cardioinhibitory center inhibited

*Result in* → Increased cardiac output and blood pressure

- Vasomotor center stimulated

*Results in* → Vasoconstriction

Chemoreceptors stimulated

| Receptors |
| --- |
| Chemoreceptors in carotid and aortic bodies and medulla oblongata |

Increased pH and $O_2$ levels and decreased $CO_2$ levels

Homeostasis **DISTURBED BY**

INCREASING $CO_2$ levels and DECREASING pH and $O_2$ levels

STIMULUS

RESTORED

Homeostasis **RESTORED BY**

DECREASING $CO_2$ levels and INCREASING pH and $O_2$ levels

# HOMEOSTASIS

NORMAL pH, $O_2$, AND $CO_2$ LEVELS IN BLOOD AND CSF

**INTEGRATION**

**C.** What effect does an increase in the respiratory rate have on $CO_2$ levels?

**LEARNING OUTCOME**

Describe the role of chemoreceptor reflexes in adjusting cardiovascular activity.

# Module 19.12

# The cardiovascular center makes extensive adjustments to cardiac output and blood distribution during exercise

## At Rest

At rest, cardiac output averages around 5800 mL/min. Note the pattern of blood distribution to major organs, especially the skeletal muscles, brain, and abdominal viscera.

## Light Exercise

As you begin light exercise, three interrelated changes take place:

1. Vasodilation occurs, peripheral resistance drops, and blood flow through the capillaries increases.
2. The venous return increases as skeletal muscle contractions squeeze blood along the peripheral veins. At the same time, each inhalation creates a negative pressure in the thoracic cavity that pulls blood into the venae cavae from their branches. This mechanism is called the **respiratory pump**.
3. Cardiac output increases, primarily because of the increased venous return.

## Heavy Exercise

At higher levels of exertion, cardiac output increases toward maximal levels. Major changes in the peripheral distribution of blood allow a massive increase in blood flow to skeletal muscles while preventing a potentially disastrous decrease in systemic blood pressure. Under massive sympathetic stimulation, the vasomotor center severely restricts blood flow to "nonessential" organs, such as the digestive viscera. Although blood flow to most tissues is diminished, body temperature rises and skin perfusion increases to promote heat loss. Only the blood supply to the brain remains unchanged.

| | At Rest | Light Exercise | Heavy Exercise |
|---|---|---|---|
| Brain | 750 mL/min | 750 mL/min | 750 mL/min |
| Cardiac output | 5800 mL/min | 9500 mL/min | 17,500 mL/min |
| Kidney | 250 mL/min / 1100 mL/min | 350 mL/min / 900 mL/min | 750 mL/min / 600 mL/min |
| Abdominal viscera | 1400 mL/min | 1100 mL/min | 600 mL/min |
| Skin | 500 mL/min | 1500 mL/min | 1900 mL/min |
| Other tissues | 600 mL/min | 400 mL/min | 400 mL/min |
| Skeletal muscles | 1200 mL/min | 4500 mL/min | 12,500 mL/min |

Cardiovascular performance improves significantly with training. Trained athletes have bigger hearts and larger stroke volumes than do nonathletes. Because cardiac output is equal to the stroke volume times the heart rate, at a given cardiac output, the larger the stroke volume, the slower the heart rate. An athlete at rest can maintain normal blood flow to peripheral tissues at a heart rate as low as 32 beats per minute, and, when necessary, the cardiac output of an athlete in peak condition can increase to levels 50 percent higher than those of nonathletes.

**? REVIEW**

A. Describe the respiratory pump.

B. Describe the changes in cardiac output and blood flow during exercise.

 **LEARNING OUTCOME**

Explain how the cardiovascular system responds to the demands of exercise.

# Module 19.13

# Short-term and long-term mechanisms compensate for a reduction in blood volume

**1** When hemostasis fails to prevent significant blood loss, the entire cardiovascular system makes adjustments to maintain blood pressure and restore blood volume. The immediate problem is the maintenance of adequate blood pressure and peripheral blood flow. The long-term problem is the restoration of normal blood volume. These mechanisms can cope with blood losses equivalent to approximately 30 percent of total blood volume.

**? REVIEW**

**A.** Identify the compensatory mechanisms that respond to blood loss.

## HOMEOSTASIS
### NORMAL BLOOD PRESSURE AND VOLUME

Homeostasis **DISTURBED BY**

DECREASING

blood pressure and volume caused by extensive bleeding

**STIMULUS**

**RESTORED**

Homeostasis **RESTORED BY**

INCREASING

blood pressure and volume

**Responses directed by the nervous system**

**Responses coordinated by the endocrine system**

**Receptors**

Baroreceptors and chemoreceptors

**Receptors**

Kidneys

**Long-Term Hormonal Response**

ADH, angiotensin II, aldosterone, and EPO released

Elevation of blood volume

**Combined Short-Term and Long-Term Effects**

Increased blood pressure

Increased blood volume

**Cardiovascular Responses**

Peripheral vasoconstriction

Mobilization of venous blood reservoirs

Increased cardiac output

Pain, stress, anxiety, fear

**Higher Centers in Brain**

Stimulation of cardiovascular center

General sympathetic activation

## Short-Term Responses to Blood Loss

- Carotid and aortic reflexes increase cardiac output and cause peripheral vasoconstriction. Cardiac output is maintained by increasing the heart rate, typically to 180–200 bpm.

- The combination of stress and anxiety stimulates the sympathetic nervous system, which causes a further increase in vasomotor tone, constricting the arterioles and increasing blood pressure. At the same time, venoconstriction, the constriction of large and medium-sized veins, improves venous return and shifts blood into the arterial system.

- Sympathetic activation causes the secretion of E and NE by the adrenal medullae, increasing cardiac output and extending peripheral vasoconstriction. The release of ADH by the neurohypophysis and the production of angiotensin II enhance vasoconstriction; ADH also participates in the long-term response.

## Long-Term Responses to Blood Loss

- The decrease in capillary blood pressure triggers a recall of fluids from the interstitial spaces.

- Aldosterone and ADH promote fluid retention and reabsorption by the kidneys.

- Thirst increases, and additional water is obtained by absorption across the digestive tract.

- Erythropoietin (EPO) stimulates the maturation of red blood cells in the red bone marrow, resulting in increased blood volume and improved oxygen delivery to peripheral tissues.

**2** **Shock** is an acute cardiovascular crisis marked by low blood pressure (**hypotension**) and inadequate peripheral blood flow. Severe and potentially fatal signs and symptoms develop as vital tissues become starved for oxygen and nutrients. The most common cardiovascular causes of shock are a drop in cardiac output after a hemorrhage or damage to the heart, as in a heart attack. Shock develops when normal compensation mechanisms fail to maintain adequate cardiac output and peripheral blood flow. **Circulatory shock** involves a series of interlocking positive feedback loops that begin when blood loss exceeds about 35 percent of total blood volume. **Progressive shock** is an initial stage characterized by positive feedback loops that accelerate tissue damage. Unless the positive feedback loops are broken, the person will enter the fatal stage of **irreversible shock**.

### Progressive Shock

1 When blood volume declines by more than 35 percent, blood pressure remains abnormally low, venous return is reduced, and cardiac output is inadequate despite sustained vasoconstriction and the mobilization of the venous blood reservoir.

2 The first of several dangerous positive feedback loops begins when low cardiac output reduces blood flow to the heart. This damages the myocardium, which leads to a further reduction in cardiac output.

3 Reduced cardiac output accelerates oxygen starvation in peripheral tissues, and the resulting chemical changes promote clotting within vessels that further restricts peripheral blood flow.

4 Local pH changes increase capillary permeability, and this further reduces blood volume.

### Irreversible Shock

5 Carotid sinus baroreceptors trigger a massive activation of the vasomotor center. To preserve blood flow to the brain at any cost, the sympathetic output causes a sustained and maximal vasoconstriction. This decreases peripheral circulation to an absolute minimum, but it increases blood pressure to about 70 mm Hg and temporarily improves blood flow to cerebral vessels.

6 Unless prompt treatment is provided, blood pressure will again decrease. By now the heart muscle will be seriously damaged as blood flow to the brain decreases.

7 When arteriolar smooth muscles and precapillary sphincters become unable to contract despite the commands of the vasomotor center, **circulatory collapse** occurs. This condition causes widespread peripheral vasodilation, an immediate and fatal decrease in blood pressure, and the cessation of blood flow.

If feedback loops at 2–4 can be broken, the person may survive. If the mean arterial blood pressure (MAP) falls below 50 mm Hg, damage soon becomes irreversible

**? REVIEW**

**B.** Name the immediate and long-term problems related to hemorrhage.

**? REVIEW**

**C.** Describe circulatory shock, progressive shock, and irreversible shock.

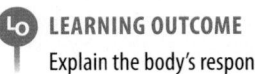 **LEARNING OUTCOME**

Explain the body's response to blood loss.

**Matching:** Match each lettered term with the most closely related description.

a. autoregulation

b. venous return

c. local vasodilators

d. natriuretic peptides

e. chemoreceptors

f. baroreceptors

g. turbulence

h. CHP>BCOP

i. medulla oblongata

j. edema

k. viscosity

l. CHP<BCOP

| 1 | Detect changes in pressure |
| 2 | Vasomotor center |
| 3 | Aided by thoracic pressure changes due to breathing |
| 4 | Causes immediate, local homeostatic responses |
| 5 | Decreased tissue $O_2$ and increased $CO_2$ |
| 6 | Carotid bodies |
| 7 | Reabsorption of water into a capillary |
| 8 | Opposite of smooth blood flow |
| 9 | Resistance to flow |
| 10 | Forces water out of a capillary |
| 11 | Peripheral vasodilation |
| 12 | Excess interstitial fluid accumulation |

1 _____

2 _____

3 _____

4 _____

5 _____

6 _____

7 _____

8 _____

9 _____

10 _____

11 _____

12 _____

Match each lettered item with the appropriate numbered blank box in the diagram below.

a. increased cardiac output and peripheral vasoconstriction

b. aldosterone secreted

c. increased blood pressure

d. blood pressure and volume

e. thirst stimulated

f. increased blood volume

g. renin released

h. antidiuretic hormone released

i. increased red blood cell formation

j. erythropoietin released

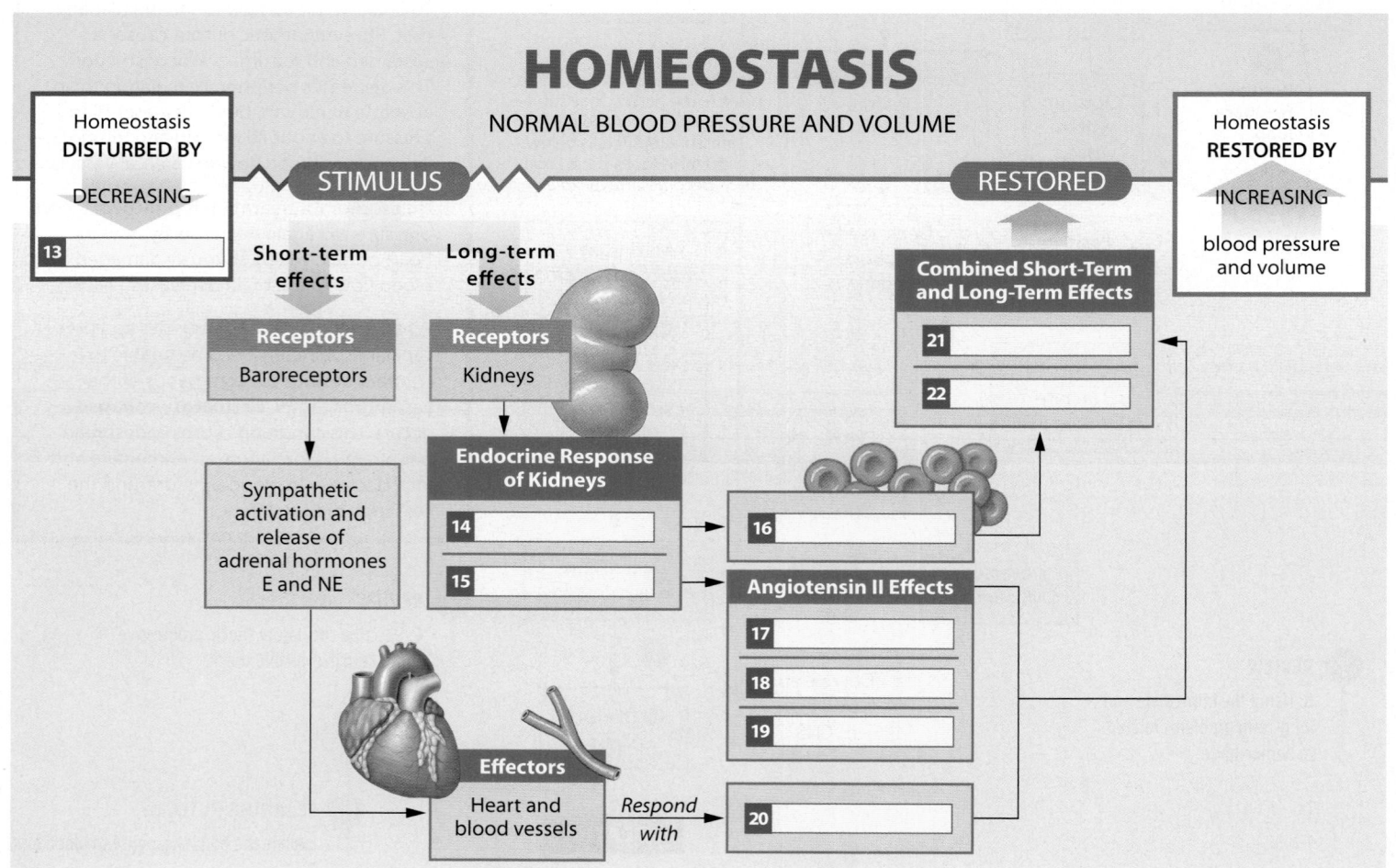

# HOMEOSTASIS
## NORMAL BLOOD PRESSURE AND VOLUME

Homeostasis **DISTURBED BY**

DECREASING

STIMULUS

13 _____

Homeostasis **RESTORED BY**

INCREASING

blood pressure and volume

RESTORED

**Short-term effects**

**Long-term effects**

**Receptors**

Baroreceptors

**Receptors**

Kidneys

**Combined Short-Term and Long-Term Effects**

21 _____

22 _____

Sympathetic activation and release of adrenal hormones E and NE

**Endocrine Response of Kidneys**

14 _____

15 _____

16 _____

**Angiotensin II Effects**

17 _____

18 _____

19 _____

**Effectors**

Heart and blood vessels

*Respond with*

20 _____

# Module 19.14
# New blood vessels form through vasculogenesis and angiogenesis

**1** At about day 13–15 of embryonic development, aggregates of cells scatter within the yolk sac forming **blood islands**, the first type of hemangiogenic (blood vessel-forming) tissue. As the embryo develops, these islands give rise to *hematopoietic stem cells* and *hemangioblasts*. **Vasculogenesis** is the formation of new vessels by hemangioblasts and precursor endothelial cells. **Angiogenesis** is the growth of new blood vessels from preexisting vessels. Angiogenic remodeling occurs with the migration of endothelial cells to form vascular networks.

**? REVIEW**
**A.** What are blood islands, and from which cells do they form?
**B.** What is the function of hemangioblasts?

Precursor endothelial cells

Yolk sac

**14-DAY EMBRYO**

**Vasculogenesis**

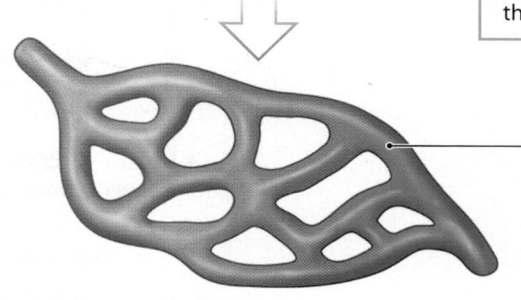

Blood island

An inner layer of embryonic **hematopoietic stem cells** gives rise to all other blood cells.

An outer layer of **hemangioblasts** forms embryonic blood cells and the vascular endothelium.

Hemangioblasts remodel the blood islands first into primitive capillary networks and then into larger arterial and venous networks.

**2** This diagram shows the main arteries and veins in a 4-week-old embryo when it is about 4 mm long. The *dorsal aorta* and *cardinal veins* are products of vasculogenesis. Further growth of blood vessels occurs through angiogenesis.

**Angiogenesis**

**Formation of the Aortic Arch and Its Branches**

The **aortic arches** are a series of arterial channels that ultimately form the carotid arteries, the aortic arch, and part of the pulmonary arteries. The **dorsal aorta** becomes the descending aorta.

Dorsal aorta

Aortic arches

Yolk sac

**Formation of the Venae Cavae**

The **cardinal veins** are a series of venous channels that ultimately form the superior and inferior venae cavae and vessels returning to the heart.

Anterior cardinal vein (from cephalic region)

Posterior cardinal vein (from caudal region)

**4-WEEK EMBRYO**

In this section we follow the distribution of the major blood vessels in the body.

 **LEARNING OUTCOME**
Distinguish between vasculogenesis and angiogenesis.

# Module 19.15

# The pulmonary circuit carries deoxygenated blood from the right ventricle to the lungs and returns oxygenated blood to the left atrium

 **1** This flowchart is an overview of the organization of the cardiovascular system. The **pulmonary circuit** is composed of arteries and veins that transport blood between the heart and the lungs. This circuit begins at the right ventricle and ends at the left atrium. From the left ventricle, the arteries of the **systemic circuit** transport oxygenated blood and nutrients to all organs and tissues, ultimately returning deoxygenated blood to the right atrium.

**? REVIEW**

**A.** Compare the oxygen content in the two circulatory circuits.

**2** The table below summarizes the general organization and naming of the body's blood vessels.

## General Patterns of Blood Vessel Organization

- The peripheral distributions of arteries and veins on the body's left and right sides are generally identical, except near the heart, where the largest vessels connect to the atria or ventricles.

- A single vessel may have several names as it crosses specific anatomical boundaries, making accurate anatomical descriptions possible when the vessel extends far into the periphery. For example, the external iliac artery becomes the femoral artery as it leaves the trunk and enters the lower limb.

- Tissues and organs are usually serviced by several arteries and veins. Often, anastomoses between adjacent arteries or veins reduce the impact of a temporary or even permanent **occlusion** (blockage) of a single blood vessel.

**? REVIEW**

**B.** Briefly describe general patterns of blood vessel organization.

Brain

Upper limbs

Pulmonary circuit (arteries)

Pulmonary circuit (veins)

Systemic circuit (arteries)

Systemic circuit (veins)

Lungs

Right atrium

Left atrium

Left ventricle

Right ventricle

Kidneys

Liver

Spleen

Digestive organs

Gonads

Lower limbs

**3** Arteries of the pulmonary circuit differ from those of the systemic circuit in that they carry deoxygenated blood. (This is why most color-coded diagrams show the pulmonary arteries in blue, the same color as systemic veins.) By convention, several large arteries are called **trunks**; the **pulmonary trunk** is one important example. As the pulmonary trunk curves over the superior border of the heart, it gives rise to the left and right **pulmonary arteries**. These large arteries enter the lungs before branching repeatedly, giving rise to smaller and smaller arteries. The smallest branches, the **pulmonary arterioles**, provide blood to the alveolar capillary network that surrounds small air pockets called **alveoli** (al-VĒ-ō-lī; singular, *alveolus*). Alveolar walls are thin enough for gas exchange between the capillary blood and inspired air; the blood absorbs oxygen and eliminates carbon dioxide. Oxygenated blood leaving the alveolar capillaries enters venules that in turn unite to form larger vessels carrying blood toward the **pulmonary veins**. These four veins, two from each lung, deliver oxygenated blood into the left atrium, completing the pulmonary circuit.

**?  REVIEW**

**C.** Trace a drop of blood through the lungs, beginning at the right ventricle and ending at the left atrium.

**Lo  LEARNING OUTCOME**

Identify the major arteries and veins of the pulmonary circuit, and name the areas each serves.

# Module 19.16

## The arteries and veins of the systemic circuit operate in parallel, and the major vessels often have similar names

**1** This figure is an overview of the major **systemic arteries**. Note that all the arteries of the systemic circuit originate from the **aorta**, the large elastic artery extending from the left ventricle of the heart. In each of the seven modules that follow, the left-hand page will illustrate and discuss the systemic arterial vessels and their branches. To reduce clutter, "artery" will not be repeated in every label. Because most of the major arteries are paired, with one artery of each pair on either side of the body, the terms "right" and "left" will appear in figure labels only when the arteries on both sides are labeled.

**?** **REVIEW**

**A.** Identify the largest artery in the body.

Vertebral

Common carotid

Subclavian

Brachiocephalic trunk

Axillary

Ascending aorta

Brachial

Radial

Ulnar

Digital arteries

Palmar arches

Popliteal

Fibular

Dorsalis pedis

Plantar arch

Aortic arch

Descending aorta

Diaphragm (muscle)

Celiac trunk

Renal

Gonadal

Lumbar

Common iliac

Internal iliac

External iliac

Deep femoral

Femoral

Descending genicular

Posterior tibial

Anterior tibial

**2** This figure is an overview of the major **systemic veins**. Note that all of the veins of the systemic circuit merge into two large veins: the **superior vena cava**, which collects systemic blood from the head, chest, and upper limbs, and the **inferior vena cava**, which collects systemic blood from all structures inferior to the diaphragm. In each of the seven modules that follow, the right-hand page will illustrate and discuss the systemic venous vessels and their branches. To reduce clutter, "vein" will not be repeated in every label. The same name often applies to both the artery and the vein servicing a particular structure or region. Additionally, "right" and "left" will appear in figure labels only when the veins on both sides are shown.

### Everyday Physiology

## Why are there both superficial and deep veins in the neck and limbs?

One significant difference between the arterial and venous systems concerns the distribution of major veins in the neck and limbs. Arteries in these areas are located deep beneath the skin, protected by bones and surrounding soft tissues. In contrast, the neck and limbs generally have two sets of peripheral veins, one superficial and the other deep. This dual venous drainage is important for controlling body temperature. In hot weather, venous blood flows through superficial veins, where heat loss take place. In cold weather, blood is routed to the deep veins to minimize heat loss.

**? REVIEW**

**B.** Name the two large veins that collect blood from the systemic circuit.

Labels (left side, top to bottom):
Vertebral
External jugular
Internal jugular
Subclavian
Brachiocephalic
Axillary
Cephalic
Brachial
Basilic
Radial
Median antebrachial
Ulnar
Palmar venous arches
Digital veins

Labels (right side, top to bottom):
Superior vena cava
Intercostal veins
Diaphragm (muscle)
Inferior vena cava
Renal
Gonadal
Lumbar veins
Common iliac
Internal iliac
External iliac
Deep femoral
Femoral
Posterior tibial
Anterior tibial

Labels (lower left):
Great saphenous
Popliteal
Small saphenous
Fibular
Plantar venous arch
Dorsal venous arch

**KEY**
▨ Superficial veins
■ Deep veins

**LO LEARNING OUTCOME**

Identify the major arteries and veins of the systemic circuit, and name the areas each serves.

**? REVIEW**

**C.** Besides containing valves, cite another major difference between the arterial and venous systems.

# Module 19.17
# The branches of the aortic arch supply structures ...

**1** This left side of the module presents the arteries in the same sequence in which arterial blood flow occurs.

**Start**

## Branches of the Aortic Arch

Three elastic arteries originate along the aortic arch and deliver blood to the head, neck, shoulders, and upper limbs:

| | | |
|---|---|---|
| The **brachiocephalic** (brā-kē-ō-se-FAL-ik) **trunk** is the first branch off the aortic arch. It ascends for a short distance before branching to form the **right subclavian artery** and the **right common carotid artery**. | The **left common carotid artery**, the second vessel arising from the aortic arch, supplies the left side of the head and neck. | The **left subclavian artery**, the third artery arising from the aortic arch, has the same distribution pattern as the right subclavian artery. |

## The Right Subclavian Artery

Two major branches arise before a subclavian artery leaves the thoracic cavity: (1) the **internal thoracic artery** (internal mammary), supplying the pericardium and anterior wall of the chest; and (2) the **vertebral artery**, which provides blood to the brain and spinal cord.

## Arteries of the Arm

After leaving the thoracic cavity and passing across the superior border of the first rib, the subclavian is called the **axillary artery**. This artery crosses the axilla to enter the arm, where it becomes the **brachial artery**, which supplies blood to the upper limb. The brachial artery gives rise to the deep brachial artery, which supplies deep structures on the posterior aspect of the arm, and the ulnar collateral arteries, which supply the area around the elbow.

## Arteries of the Forearm

As it approaches the coronoid fossa of the humerus, the brachial artery divides into the **radial artery**, which follows the radius, and the **ulnar artery**, which follows the ulna to the wrist. At the wrist, the radial and ulnar arteries fuse to form the superficial and deep **palmar arches**, which supply blood to the hand and to the **digital arteries** of the thumb and fingers.

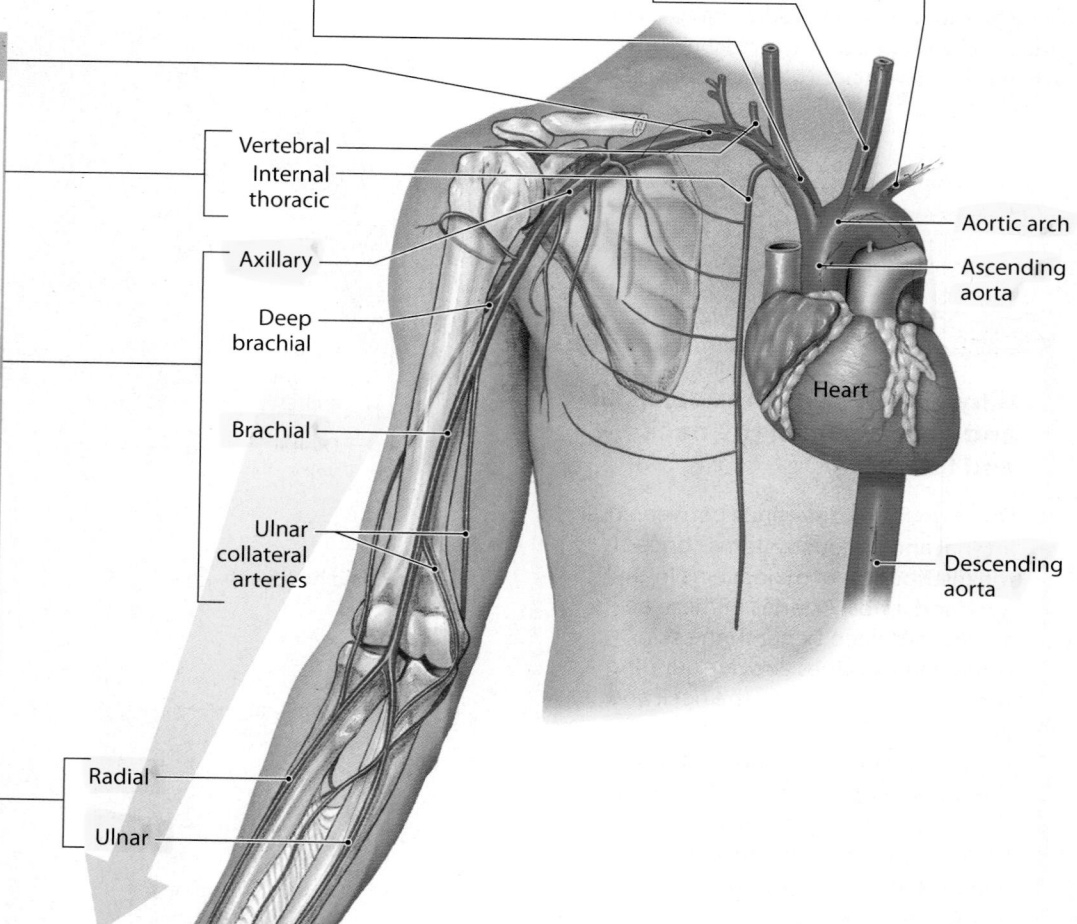

Vertebral
Internal thoracic
Axillary
Deep brachial
Brachial
Ulnar collateral arteries
Radial
Ulnar
Deep palmar arch
Superficial palmar arch
Digital arteries

Aortic arch
Ascending aorta
Heart
Descending aorta

## REVIEW

**A.** Name the two arteries formed by the division of the brachiocephalic trunk.

**B.** A blockage of which branch of the aortic arch would interfere with blood flow to the left arm?

# … that are drained by the superior vena cava

**2** This right side of the module presents the veins in the same sequence in which venous blood flow occurs.

**Veins of the Neck**

The **external jugular vein** drains superficial structures of the head and neck.

The **vertebral vein** drains the cervical spinal cord and the posterior surface of the skull.

The **internal jugular vein** drains deep structures of the head and neck.

**The Right Subclavian Vein**

The axillary vein is joined by the cephalic vein on the lateral surface of the first rib, forming the **subclavian vein**, which continues into the chest.

**Veins of the Arm**

As the **brachial vein** merges with the **basilic vein** it becomes the **axillary vein**, which enters the axilla.

The **cephalic vein** extends along the lateral side of the arm.

**Veins of the Forearm**

The **median cubital vein** interconnects the cephalic and basilic veins. (The median cubital is the vein from which venous blood samples are typically collected.)

The **ulnar vein** and the **radial vein** drain the **deep palmar arch**. Before crossing the elbow, these veins fuse to form the brachial vein.

The cephalic vein, the **median antebrachial vein**, and the basilic vein drain the **superficial palmar arch**.

**Start**

The **digital veins** empty into superficial and deep veins of the hand, which are interconnected to form the palmar venous arches.

The **brachiocephalic vein** forms as the jugular veins empty into the subclavian vein. It receives blood from the vertebral vein and the internal thoracic vein draining the anterior chest wall.

The **superior vena cava (SVC)** carries blood from the two brachiocephalic veins to the right atrium of the heart.

The **internal thoracic vein** collects blood from the intercostal veins and delivers it to the brachiocephalic vein.

*cephalic*

Brachial
Basilic

*median cubital vein*

Cephalic

Radial

Basilic

Ulnar

Deep palmar arch

Superficial palmar arch

**KEY**

Superficial veins

Deep veins

**? INTEGRATION**

**C.** Whenever Thor gets angry, a large vein bulges in the lateral region of his neck. Which vein is this?

**Lo LEARNING OUTCOME**

Identify the branches of the aortic arch and the branches of the superior vena cava, and name the areas each serves.

# Module 19.18

# The external carotid arteries supply the neck, lower jaw, and face, and the internal carotid and vertebral arteries supply the brain …

**1** The **common carotid arteries** ascend deep in the tissues of the neck and supply blood to the structures of the face, neck, and brain. To locate the carotid artery, gently press a finger along either side of the windpipe (trachea) until you feel a strong pulse. Details of the arterial supply of the brain are covered in Module 19.19.

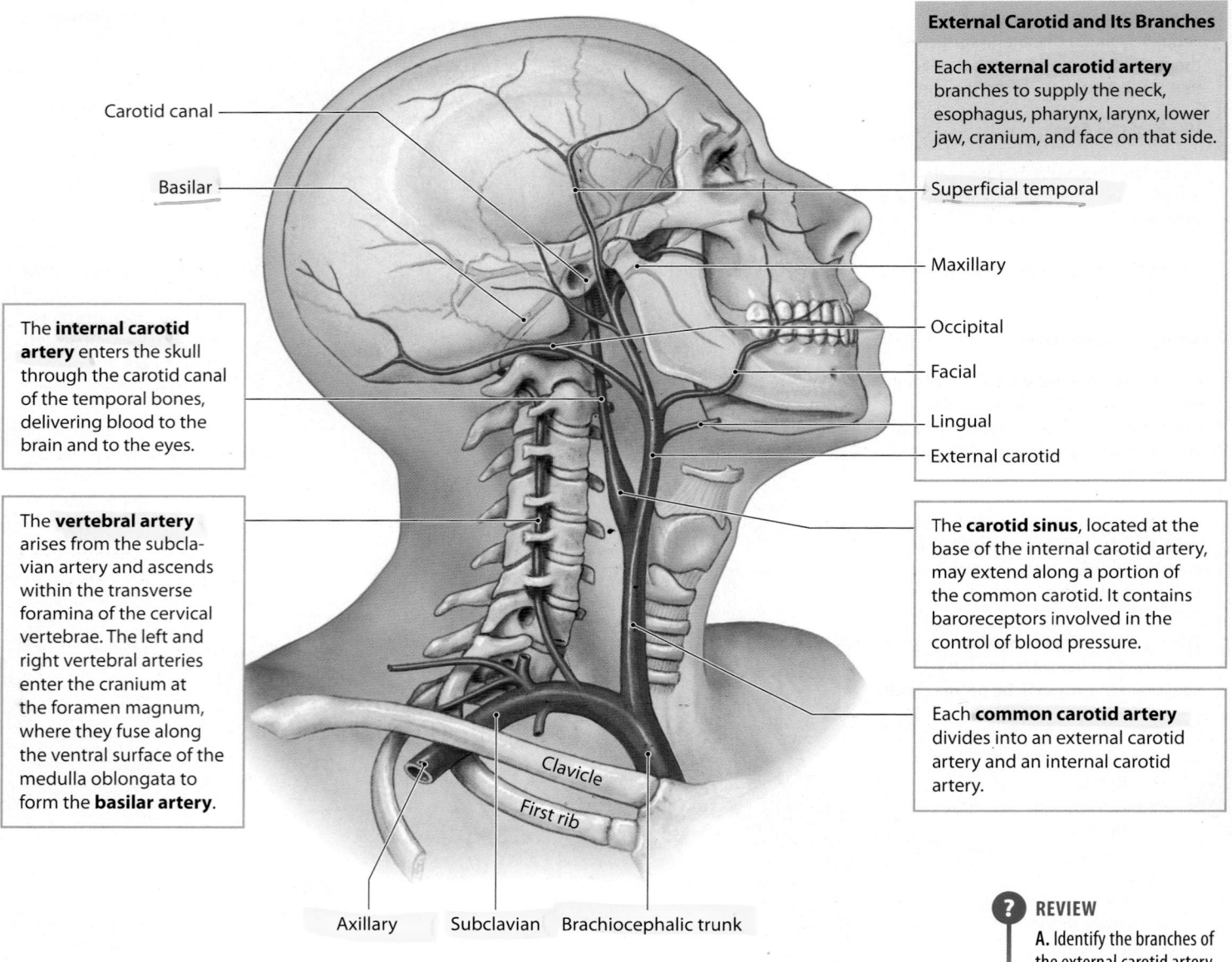

Carotid canal

Basilar

The **internal carotid artery** enters the skull through the carotid canal of the temporal bones, delivering blood to the brain and to the eyes.

The **vertebral artery** arises from the subclavian artery and ascends within the transverse foramina of the cervical vertebrae. The left and right vertebral arteries enter the cranium at the foramen magnum, where they fuse along the ventral surface of the medulla oblongata to form the **basilar artery**.

Clavicle

First rib

Axillary    Subclavian    Brachiocephalic trunk

**External Carotid and Its Branches**

Each **external carotid artery** branches to supply the neck, esophagus, pharynx, larynx, lower jaw, cranium, and face on that side.

Superficial temporal

Maxillary

Occipital

Facial

Lingual

External carotid

The **carotid sinus,** located at the base of the internal carotid artery, may extend along a portion of the common carotid. It contains baroreceptors involved in the control of blood pressure.

Each **common carotid artery** divides into an external carotid artery and an internal carotid artery.

**? REVIEW**

**A.** Identify the branches of the external carotid artery.

**B.** Name the arterial structure in the neck region that contains baroreceptors.

# ... while the external jugular veins drain the regions supplied by the external carotid arteries, and the internal jugular veins drain the brain

**2** The **external jugular veins** are formed by the maxillary and temporal veins, while the **internal jugular veins** drain the blood from the various venous sinuses within the cranium (see Module 19.19 for additional details). The external and internal jugular veins combine with the vertebral and subclavian to form the **brachiocephalic vein**.

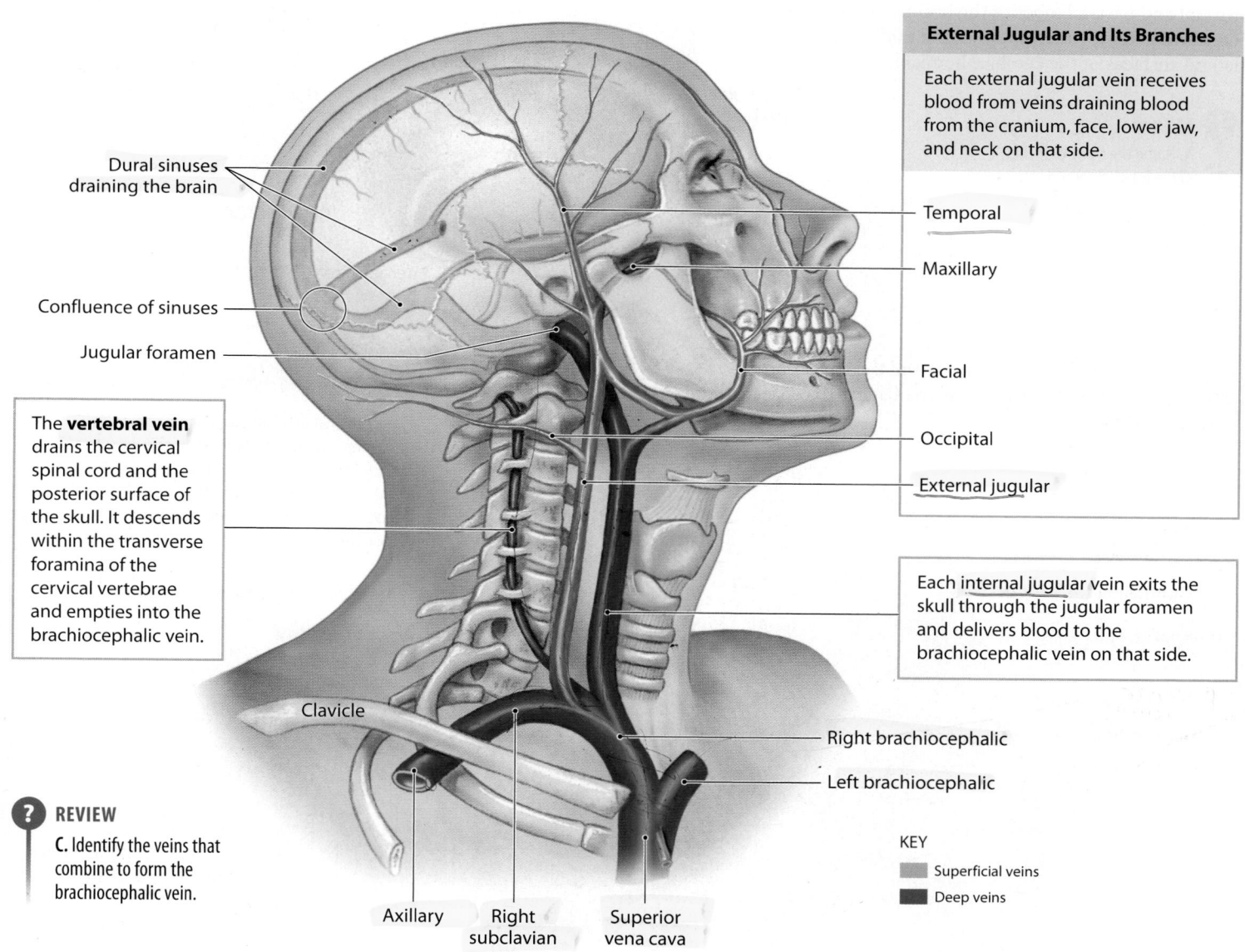

**Dural sinuses draining the brain**

**Confluence of sinuses**

**Jugular foramen**

The **vertebral vein** drains the cervical spinal cord and the posterior surface of the skull. It descends within the transverse foramina of the cervical vertebrae and empties into the brachiocephalic vein.

Clavicle

### External Jugular and Its Branches

Each external jugular vein receives blood from veins draining blood from the cranium, face, lower jaw, and neck on that side.

Temporal

Maxillary

Facial

Occipital

External jugular

Each internal jugular vein exits the skull through the jugular foramen and delivers blood to the brachiocephalic vein on that side.

Right brachiocephalic

Left brachiocephalic

**KEY**
- Superficial veins
- Deep veins

Axillary      Right subclavian      Superior vena cava

**? REVIEW**
**C.** Identify the veins that combine to form the brachiocephalic vein.

**LO LEARNING OUTCOME**
Identify the branches of the carotid arteries and the branches of the external jugular veins, and name the areas each serves.

# Module 19.19
# The internal carotid arteries and the vertebral arteries supply the brain …

**1** This lateral view shows the major arteries supplying the brain. The internal carotid arteries normally supply the arteries of the anterior half of the cerebrum, and the rest of the brain receives blood from the vertebral and basilar arteries. The internal carotid artery ascends to the level of the optic nerves, where each artery divides into three branches: (1) an **ophthalmic artery**, which supplies the eyes; (2) an **anterior cerebral artery**, which supplies the frontal and parietal lobes of the brain; and (3) a **middle cerebral artery**, which supplies the midbrain and the lateral surfaces of the cerebral hemispheres.

**? REVIEW**
**A.** Name the three branches of the internal carotid artery.

Middle cerebral
Anterior cerebral
Posterior cerebral
Basilar
Vertebral
Ophthalmic
Cerebral arterial circle
Internal carotid

**2** The internal carotid arteries and the basilar artery are interconnected in a ring-shaped anastomosis called the **cerebral arterial circle**.

Anterior cerebral
Ophthalmic
Internal carotid (cut)
Posterior communicating
Middle cerebral
Pituitary gland
Posterior cerebral
Cerebellar

*middle cerebral artery*

### Cerebral Arterial Circle

The cerebral arterial circle, or *circle of Willis*, encircles the infundibulum of the pituitary gland. This arrangement reduces the likelihood of a serious interruption of cerebral blood flow because the brain can receive blood from either the carotid or the vertebral arteries.

Anterior communicating
Anterior cerebral
Posterior communicating
Posterior cerebral

Basilar
Vertebral

Within the cranium, the vertebral arteries and the basilar artery supply blood to the spinal cord, medulla oblongata, pons, and cerebellum before dividing into the posterior cerebral arteries, which in turn branch off into the posterior communicating arteries.

**? REVIEW**
**B.** Describe the structure and function of the cerebral arterial circle.

# … which is drained by the dural sinuses and the internal jugular veins

**3** The superficial cerebral veins and small veins of the brainstem empty into a network of dural sinuses. Most of the deep cerebral veins converge within the brain to form the **great cerebral vein**, which delivers blood from the interior of the cerebral hemispheres and the choroid plexus to the **straight sinus**. The superior sagittal sinus, occipital sinus, straight sinus, and the left and right transverse sinuses connect at the *confluence of sinuses*. Numerous small veins from the orbit and other cerebral veins drain into the **cavernous sinus**.

The **superior sagittal sinus**, in the falx cerebri, is the largest dural sinus (see Module 13.3).

**Unpaired Median Sinuses**

Superior sagittal sinus
Inferior sagittal sinus
Straight sinus
Cavernous sinus

Occipital sinus

**Paired Lateral Sinuses**

Right transverse sinus
Right sigmoid sinus
Petrosal sinuses

Confluence of sinuses

Great cerebral vein

Internal jugular

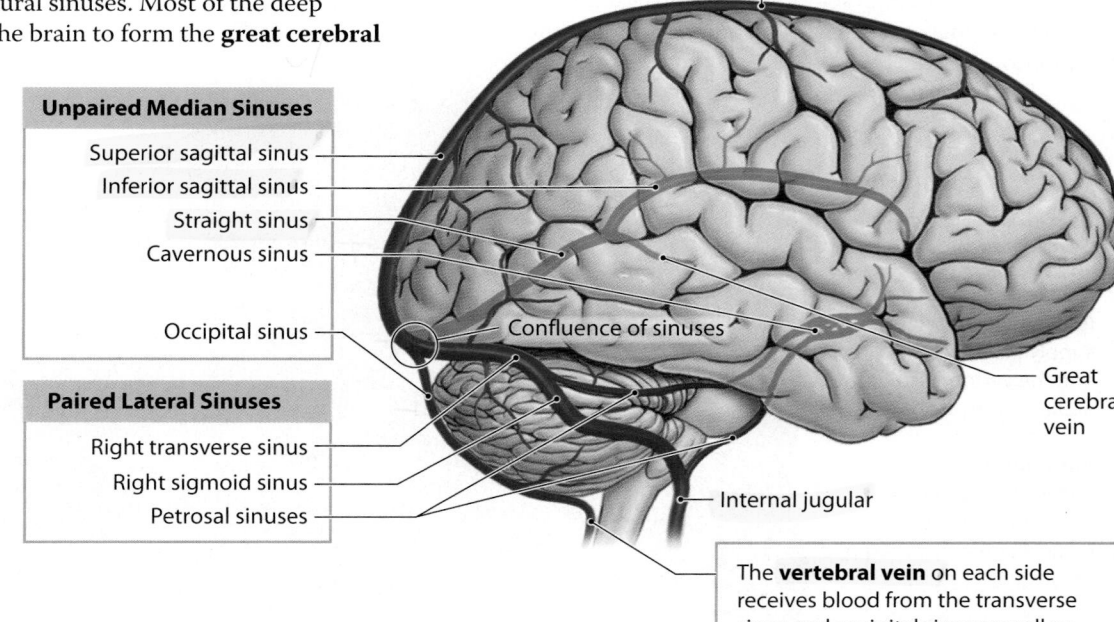

The **vertebral vein** on each side receives blood from the transverse sinus and occipital sinus as well as superficial veins of the skull and veins draining the cervical vertebrae.

**4** The cavernous sinus empties into the two **petrosal sinuses**, which in turn drain into the **transverse sinuses**. The transverse sinuses, the straight sinus, and the superior sagittal sinus converge to form the **sigmoid sinuses**, which penetrate the jugular foramina and leave the skull as the **internal jugular veins**.

Superior sagittal sinus (cut)

Cavernous sinus

Internal jugular

Cerebellar veins

Straight sinus

Confluence of sinuses

Cerebral veins

Petrosal sinus

Sigmoid sinus

Transverse sinus

**? REVIEW**

**C.** Name the veins that drain the dural sinuses of the brain.

**Lo LEARNING OUTCOME**

Identify the branches of the internal carotid and vertebral arteries and the branches of the internal jugular veins, and name the areas each serves.

# Module 19.20
# The regions supplied by the descending aorta …

**1** The descending aorta is continuous with the aortic arch. The diaphragm divides the descending aorta into a superior **thoracic aorta** and an inferior **abdominal aorta**. The figure details the branches of the thoracic aorta and introduces the branches of the abdominal aorta that are detailed further in the table below.

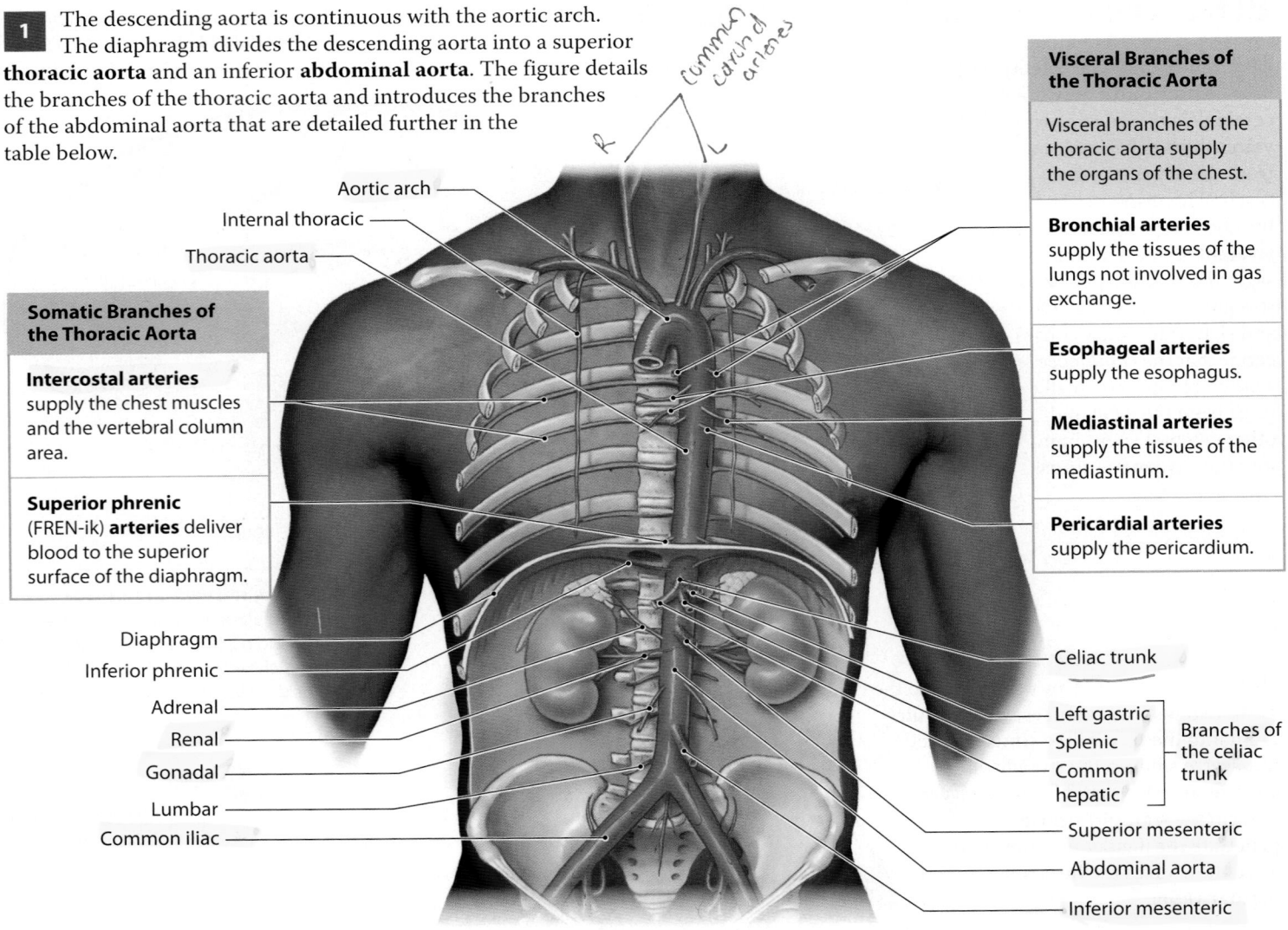

**Somatic Branches of the Thoracic Aorta**

**Intercostal arteries** supply the chest muscles and the vertebral column area.

**Superior phrenic** (FREN-ik) **arteries** deliver blood to the superior surface of the diaphragm.

**Visceral Branches of the Thoracic Aorta**

Visceral branches of the thoracic aorta supply the organs of the chest.

**Bronchial arteries** supply the tissues of the lungs not involved in gas exchange.

**Esophageal arteries** supply the esophagus.

**Mediastinal arteries** supply the tissues of the mediastinum.

**Pericardial arteries** supply the pericardium.

Labels: Aortic arch, Internal thoracic, Thoracic aorta, Diaphragm, Inferior phrenic, Adrenal, Renal, Gonadal, Lumbar, Common iliac, Celiac trunk, Left gastric, Splenic, Common hepatic — Branches of the celiac trunk, Superior mesenteric, Abdominal aorta, Inferior mesenteric

(handwritten) common arch of arteries, R, L

## Major Paired Branches of the Abdominal Aorta

- The **inferior phrenic arteries** supply the inferior surface of the diaphragm and the inferior portion of the esophagus.
- The **adrenal arteries** supply the adrenal glands, which cap the superior part of each kidney.
- The short **renal arteries** arise along the posterolateral surface of the abdominal aorta, just inferior to the superior mesenteric artery. We consider the branches of the renal arteries in Chapter 24.
- The **gonadal** (gō-NAD-al) **arteries** originate between the superior and inferior mesenteric arteries. In males, they are called *testicular arteries*; in females, they are termed *ovarian arteries*. The distribution of gonadal vessels (both arteries and veins) differs by sex; we examine the differences in Chapter 26.
- Small **lumbar arteries** arise on the posterior surface of the aorta and supply the vertebrae, spinal cord, and abdominal wall.

## Major Unpaired Branches of the Abdominal Aorta

- The **celiac** (SĒ-lē-ak) **trunk** divides into three branches: (1) the *left gastric artery*, which supplies the stomach and the inferior portion of the esophagus; (2) the *splenic artery*, which supplies the spleen and arteries to the stomach; and (3) the *common hepatic artery*, which supplies arteries to the liver, stomach, gallbladder, and the proximal portion of the small intestine.
- The **superior mesenteric** (mez-en-TER-ik) **artery** arises inferior to the celiac trunk to supply arteries to the pancreas and duodenum and to most of the large intestine.
- The **inferior mesenteric artery** arises just superior to the bifurcation of the aorta, and it delivers blood to the terminal portions of the colon and the rectum. We examine the branches of these vessels in Module 19.21.

# … are drained by the superior and inferior venae cavae

**2** The superior vena cava drains blood from the head, neck, shoulders, chest, and upper limbs. The chief collecting vessels of the thorax are the **azygos** (AZ-i-gos) **vein** and the **hemi-azygos vein**. These veins receive blood from (1) **intercostal veins**, which in turn receive blood from the chest muscles; (2) **esophageal veins**, which drain blood from the inferior portion of the esophagus; (3) **bronchial veins** draining the passageways of the lungs; and (4) **mediastinal veins** draining other mediastinal structures. The inferior vena cava collects most of the blood inferior to the diaphragm.

Brachiocephalic

Superior vena cava

Internal thoracic

Inferior vena cava

Hepatics

Phrenic

Adrenal

Renal

Gonadal

Lumbar

Common iliac

### The Azygos and Hemi-azygos Veins

The azygos vein is a major branch of the superior vena cava.

The smaller hemi-azygos vein drains into the azygos vein and may also drain into the left brachiocephalic vein.

Branches:
Esophageal, bronchial, and mediastinal veins
Intercostal veins

 **REVIEW**

**A.** Which vessel collects most of the venous blood inferior to the diaphragm?

## Major Branches of the Inferior Vena Cava

- **Lumbar veins** drain the lumbar portion of the abdomen, including the spinal cord and muscles of the body wall.

- **Gonadal** (ovarian or testicular) **veins** drain the ovaries or testes. The right gonadal vein empties into the inferior vena cava; the left gonadal vein generally drains into the left renal vein.

- **Hepatic veins** drain the sinusoids of the liver.

- **Renal veins**, the largest branches of the inferior vena cava, collect blood from the kidneys.

- **Adrenal veins** drain the adrenal glands. In most individuals, only the right adrenal vein drains into the inferior vena cava; the left adrenal vein drains into the left renal vein.

- **Phrenic veins** drain the diaphragm. Only the right phrenic vein drains into the inferior vena cava; the left drains into the left renal vein.

 **REVIEW**

**B.** Identify the major branches of the inferior vena cava.

**INTEGRATION**

**C.** Grace is in an automobile accident, and her celiac trunk is ruptured. Which organs will be affected most directly by this injury?

**LEARNING OUTCOME**

Identify the branches of the descending aorta and the branches of the venae cavae, and name the areas each serves.

# Module 19.21
# The viscera supplied by the celiac trunk and mesenteric arteries ...

**1** The abdominal aorta begins immediately inferior to the diaphragm. Three unpaired branches supply the abdominal viscera: the celiac trunk, the superior mesenteric artery, and the inferior mesenteric artery.

## The Celiac Trunk

The **celiac trunk** divides into the common hepatic artery, the left gastric artery, and the splenic artery.

The **common hepatic artery** branches to supply the liver, stomach, gallbladder, and the duodenum (the proximal segment of the small intestine).

The **left gastric artery** supplies the stomach. Its anastomosis with the right gastric artery ensures the stomach a continuous blood supply.

The **splenic artery** supplies the spleen and sends branches to the stomach and pancreas.

### Branches of the Common Hepatic Artery

Hepatic artery proper (liver)

Cystic (gallbladder)

Gastroduodenal (stomach and duodenum)

Right gastric (stomach)

Right gastro-epiploic (stomach and duodenum)

Superior pancreatico-duodenal (duodenum)

Ascending colon

### Superior Mesenteric Artery

The **superior mesenteric artery** branches to supply the pancreas and duodenum, small intestine, and most of the large intestine.

Inferior pancreatico-duodenal (pancreas and duodenum)

Right colic (large intestine)

Ileocolic (large intestine)

Middle colic (cut) (large intestine)

Intestinal arteries (small intestine)

### Branches of the Splenic Artery

Left gastro-epiploic (stomach)

Pancreatic (pancreas)

Spleen

### Inferior Mesenteric Artery

The **inferior mesenteric artery** delivers blood to the terminal portions of the colon and the rectum.

Left colic (colon)

Sigmoid (colon)

Rectal (rectum)

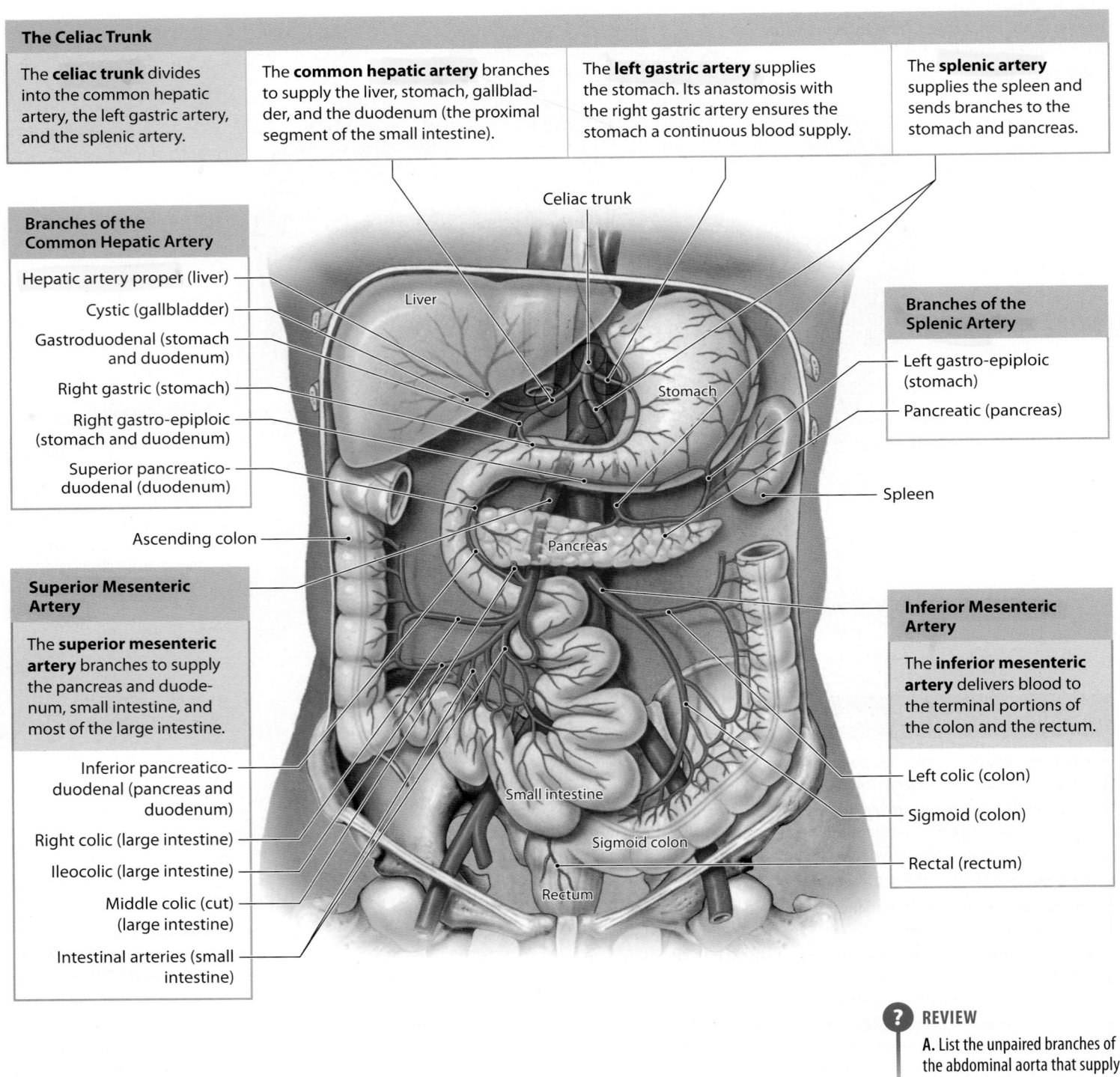

Celiac trunk

Liver

Stomach

Pancreas

Small intestine

Sigmoid colon

Rectum

**? REVIEW**

**A.** List the unpaired branches of the abdominal aorta that supply blood to the visceral organs.

# ... are drained by the branches of the hepatic portal vein

**2** The **hepatic portal vein** forms through the fusion of the **superior mesenteric, inferior mesenteric,** and **splenic veins**. The largest volume of blood (and most of the nutrients) flows through the superior mesenteric vein. The hepatic portal vein receives blood from the left and right **gastric veins**, which drain the medial border of the stomach, and from the **cystic vein**, coming from the gallbladder. This circulatory pattern is called the **hepatic portal system**. It directs blood with absorbed nutrients from the digestive system to the liver for processing.

**? REVIEW**

**B.** Identify the three veins that merge to form the hepatic portal vein.

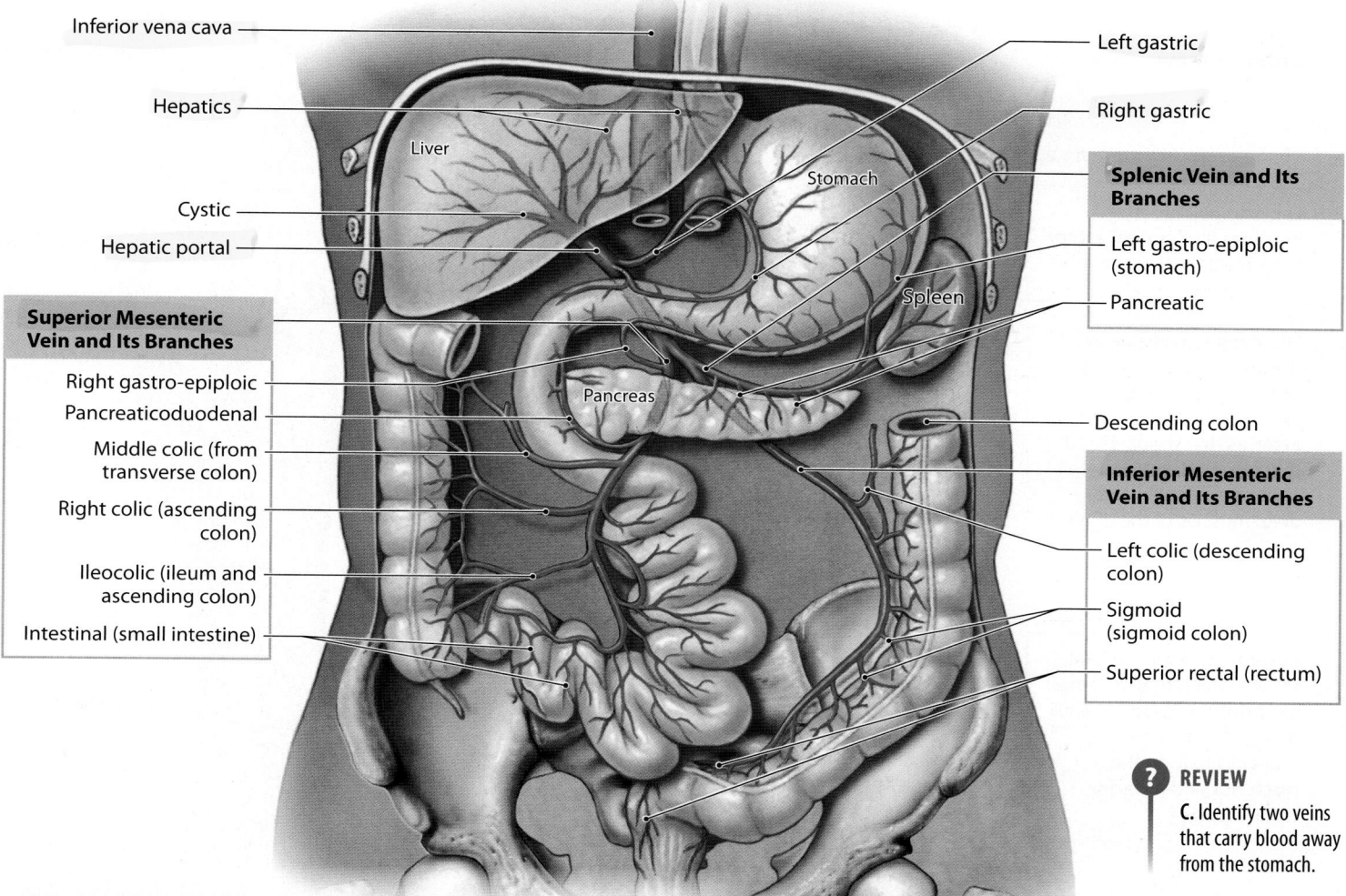

**? REVIEW**

**C.** Identify two veins that carry blood away from the stomach.

## Branches of the Hepatic Portal Vein

- The inferior mesenteric vein collects blood from capillaries along the inferior portion of the large intestine. It drains the left colic vein and the superior rectal veins, which collect venous blood from the descending colon, sigmoid colon, and rectum.

- The splenic vein is formed by the union of the inferior mesenteric vein and veins from the spleen, the lateral border of the stomach (left gastro-epiploic vein), and the pancreas (pancreatic veins).

- The superior mesenteric vein collects blood from veins draining the stomach (right gastro-epiploic vein), the small intestine (intestinal and pancreaticoduodenal veins), and two-thirds of the large intestine (ileocolic, right colic, and middle colic veins).

**? INTEGRATION**

**D.** Describe the function of the hepatic portal system, and name its primary vessel.

**Lo LEARNING OUTCOME**

Identify the branches of the visceral arterial vessels and the venous branches of the hepatic portal system, and name the areas each serves.

# Module 19.22

# The pelvis and lower limbs are supplied by branches of the common iliac arteries ...

**1** Near the level of vertebra $L_4$, the abdominal aorta divides to form a pair of elastic arteries: the **right** and **left common iliac** (IL-ē-ak) **arteries**. At the level of the lumbosacral joint, each common iliac divides to form an **internal iliac artery** and an **external iliac artery**.

**REVIEW**

**A.** Name the first two branches of the common iliac artery.

### Anterior View

Common iliac

External iliac

The external iliac artery becomes the **femoral artery** as it enters the lower limb.

The **deep femoral artery**, the first large branch of the femoral artery, forms the **femoral circumflex arteries** that supply blood to the ventral and lateral regions of the skin and deep muscles of the thigh.

Posterior to the knee joint, the femoral artery becomes the **popliteal** (pop-LIT-ē-al) **artery**, which then branches to form the **posterior** and **anterior tibial arteries**.

Anterior tibial

Posterior tibial

Fibular

### Internal Iliac and Its Branches

The internal iliac artery enters the pelvic cavity to supply the urinary bladder, the internal and external walls of the pelvis, the external genitalia, the medial side of the thigh, and, in females, the uterus and vagina. It has four major branches:

Lateral sacral

Internal pudendal

Obturator

Superior gluteal

Femoral

The descending genicular artery supplies the area around the knee.

### Arteries of the Foot

In the foot, the tibial and fibular arteries are interconnected by the anastomoses of the **dorsalis pedis**, **dorsal arch**, and **plantar arch**, which supply distal portions of the foot and the digital arteries of the toes.

Dorsalis pedis

Medial plantar

Lateral plantar

Dorsal arch

Plantar arch

### Posterior View

Right external iliac

Deep femoral

Femoral circumflex

Popliteal

Anterior tibial

Posterior tibial

The posterior tibial artery gives rise to the **fibular artery**, or **peroneal** (*perone*, fibula) **artery**, before continuing inferiorly along the posterior surface of the tibia.

# … and drained by branches of the common iliac veins

**2** The external iliac veins receive blood from the lower limbs, the pelvis, and the lower abdomen. As the left and right external iliac veins cross the inner surface of the ilium, they are joined by the internal iliac veins, which drain the pelvic organs. The internal iliac veins are formed by the fusion of the gluteal, internal pudendal, obturator, and lateral sacral veins. The union of external and internal iliac veins forms the **common iliac vein**.

**Anterior View**

**Posterior View**

- Common iliac
- External iliac
- Internal iliac
- Gluteal
- Internal pudendal
- Lateral sacral
- Obturator
- Femoral
- Femoral circumflex
- Deep femoral
- Femoral
- Great saphenous
- Popliteal
- Small saphenous
- Anterior tibial
- Posterior tibial
- Fibular
- Digital

**Femoral Vein and Its Branches**

Immediately before penetrating the abdominal wall, the femoral vein receives blood from (1) the great saphenous vein; (2) the **deep femoral vein**, which collects blood from deeper structures in the thigh; and (3) the **femoral circumflex vein**, which drains the region around the neck and head of the femur. The femoral vein penetrates the body wall and emerges in the pelvic cavity as the **external iliac vein**.

At the popliteal fossa, the small saphenous vein merges with the **popliteal vein** to form the **femoral vein**.

**? REVIEW**

**B.** The plantar venous arch carries blood to which three veins?

The **plantar venous arch** delivers blood to the deep veins of the leg: the **anterior tibial**, the **posterior tibial**, and the **fibular (peroneal) veins**.

The **dorsal venous arch** collects blood from capillaries on the superior surface of the foot and the digital veins of the toes. The dorsal venous arch is drained by two superficial veins: the **great saphenous** (sa-FĒ-nus) **vein** (*saphenes*, prominent) and the **small saphenous vein**.

**? INTEGRATION**

**C.** A blood clot that blocks the popliteal vein would interfere with blood flow in which other veins?

**Lo LEARNING OUTCOME**

Identify the branches of the common iliac arteries and the branches of the common iliac veins, and name the areas each serves.

# Module 19.23

# The systemic arterial circuit delivers oxygenated blood throughout the body. . .

**1** This flowchart summarizes the systemic arterial circulation as it delivers blood from the left ventricle to body tissues except for the lungs. (The *bronchial arteries* supply the tissues of the lungs not involved in gas exchange.)

**? REVIEW**

**A.** Trace the path of a drop of blood from the left ventricle to the right hip joint.

# . . . and the systemic venous circuit returns deoxygenated blood back to the heart

**2** This flowchart summarizes the systemic venous circulation as it returns blood back to the right atrium.

**Right external jugular**
Collects blood from neck, face, salivary glands, scalp

**Right vertebral**
Collects blood from cranium, spinal cord, vertebrae

**Right internal jugular**
Collects blood from cranium, face, and neck

**Left internal jugular**

**Left vertebral**

**Left external jugular**

**Right subclavian**

**Right brachiocephalic**

**Left brachiocephalic**

**Left subclavian**

**Right internal thoracic**
Collects blood from structures of anterior thoracic wall

**Mediastinal veins**
Collect blood from the mediastinum

**Superior vena cava**

Through most superior intercostal vein

**Left internal thoracic**

**RIGHT ATRIUM**

**Right axillary**

**Left axillary**
Collects blood from the left upper limb

**Right brachial**
Collects blood from forearm, wrist, and hand

**Right cephalic**
Collects blood from lateral surface of upper limb

**Right basilic**
Collects blood from medial surface of upper limb

**Azygos**

**Hemi-azygos**

**Right radial**

**Right ulnar**

Venous network of wrist and hand

Interconnected by median cubital vein and median antebrachial network

**Right intercostal veins**
Collect blood from vertebrae and body wall

**Esophageal veins**
Collect blood from the esophagus

**Left intercostal veins**
Collect blood from vertebrae and body wall

**Inferior vena cava**

**Hepatic veins**
Collect blood from the liver

**Phrenic veins**
Collect blood from the diaphragm

**Gonadal veins**
Collect blood from the gonads

**Adrenal veins**
Collect blood from the adrenal glands

**Lumbar veins** (paired, segmental)
Collect blood from spinal cord and body wall

**Renal veins**
Collect blood from the kidneys

**Right common iliac**

**Left common iliac**

**Right external iliac**

**Right internal iliac**

**Left internal iliac**
Collects blood from the left gluteal, internal pudendal, obturator, and lateral sacral veins

**Left external iliac**
Collects blood from the left lower limb

**Femoral**

**Gluteal veins**

**Internal pudendal vein**

**Obturator vein**

**Lateral sacral vein**

**Deep femoral**

**Great saphenous**

Collect blood from the pelvic muscles, skin, and urinary and reproductive organs of the right side of the pelvis

**Small saphenous**

**Popliteal**

**Posterior tibial**

**Fibular**

**Anterior tibial**
Collect blood from the foot and toes

**KEY**
☐ Superficial veins
☐ Deep veins

**? REVIEW**

**B.** Trace the path of a drop of blood from the right forearm to the right atrium.

**LO LEARNING OUTCOME**

Describe the pathways that oxygenated and deoxygenated blood take in the systemic arterial and venous systems.

# Module 19.24
# The pattern of blood flow through the fetal heart and the systemic circuit must change at birth

**1** Fetal blood flows to the *placenta* through a pair of **umbilical arteries**, which arise from the internal iliac arteries and enter the *umbilical cord*. Blood returns from the placenta in the single **umbilical vein**, bringing oxygen and nutrients to the developing fetus. The umbilical vein drains into the **ductus venosus**, a vascular connection to an intricate network of veins within the developing liver. The ductus venosus collects blood from the veins of the liver and from the umbilical vein and empties into the inferior vena cava. When the placental connection is broken at birth, blood stops flowing in the umbilical vessels, and they soon degenerate. However, remnants of these vessels persist throughout life as fibrous cords.

Prior to birth, the lungs are collapsed, and most blood bypasses the pulmonary circuit completely. The **foramen ovale**, or *interatrial opening*, allows blood to pass from the right atrium to the left atrium, but any backflow is prevented by a flap that acts as a one-way valve.

A second short circuit exists between the pulmonary trunk and the aorta. This connection, the **ductus arteriosus**, consists of a short, muscular vessel. Most of the blood that does reach the right ventricle flows through the ductus arteriosus and enters the systemic circuit rather than flowing through the pulmonary arteries into the lungs.

Placenta

Aorta

Umbilical vein

Liver

Umbilical cord

Pulmonary trunk

Inferior vena cava

Ductus venosus

Umbilical arteries

**Full-term fetus (before birth)**

**?** **REVIEW**
**A.** Describe the pattern of fetal blood flow to and from the placenta.

**2** At birth, the infant takes a first breath, inflating the lungs and expanding the pulmonary blood vessels. Blood rushes into the pulmonary vessels, and the resulting pressure changes at the heart close the foramen ovale. Within a few seconds, rising $O_2$ levels stimulate constriction of the ductus arteriosus, isolating the pulmonary and aortic trunks from one another. In adults, the heart has a shallow depression, the **fossa ovalis**, at the location of the fetal passageway. The remnants of the ductus arteriosus persist throughout life as a fibrous cord known as the **ligamentum arteriosum**.

Ductus arteriosus (closed)

Pulmonary trunk

Left atrium

Foramen ovale (closed)

Right atrium

Left ventricle

Right ventricle

Inferior vena cava

**After delivery**

**?** **REVIEW**
**B.** Identify the six structures that are vital to fetal circulation but cease to function at birth, and describe what becomes of these structures.

**3**    Although minor individual variations in the vascular network are quite common, congenital cardiovascular problems serious enough to threaten homeostasis are relatively rare. Most congenital heart problems result from abnormal formation of the heart or problems with the connections between the heart and the great vessels. If diagnosed early, most can be surgically corrected—sometimes prior to delivery.

## Ventricular Septal Defect

Ventricular septal defect

Ventricular septum

**Ventricular septal defects** are openings in the interventricular septum that separate the right and left ventricles. These defects are the most common congenital heart problems, affecting 0.12 percent of newborns. The opening between the two ventricles has an effect similar to a connection between the atria: When the more powerful left ventricle beats, it ejects blood into the right ventricle and pulmonary circuit.

## Patent Foramen Ovale and Patent Ductus Arteriosus

Patent ductus arteriosus

Patent foramen ovale

If the foramen ovale remains open, or **patent** (PĀ-tent), blood recirculates through the pulmonary circuit instead of entering the left ventricle. The movement, driven by the relatively high systemic pressure, is called a "left-to-right shunt." Arterial oxygen content is normal, but the left ventricle must work much harder than usual to provide adequate blood flow through the systemic circuit. Hence, pressures rise in the pulmonary circuit. If the pulmonary pressures rise enough, they may force blood into the systemic circuit through the ductus arteriosus. This condition—a **patent ductus arteriosus**—creates a "right-to-left shunt." Because the circulating blood is not adequately oxygenated, it develops a deep red color. The skin then develops the blue tones typical of cyanosis, and the infant is known as a "blue baby."

## Tetralogy of Fallot

Patent ductus arteriosus

Pulmonary stenosis

Ventricular septal defect

Enlarged right ventricle

The **tetralogy of Fallot** (fa-LŌ) is a complex group of heart and circulatory defects that affect 0.10 percent of newborn infants. In this condition, (1) the pulmonary trunk is abnormally narrow (pulmonary stenosis); (2) the interventricular septum is incomplete; (3) the aorta originates where the interventricular septum normally ends; and (4) the right ventricle is enlarged, and both ventricles thicken in response to the increased workload.

 **REVIEW**

**C.** Compare a ventricular septal defect with tetralogy of Fallot.

## Atrioventricular Septal Defect

Atrial defect

Ventricular defect

In an **atrioventricular septal defect**, both the atria and ventricles are incompletely separated. The results are quite variable, depending on the extent of the defect and the effects on the atrioventricular valves. This type of defect most commonly affects infants with Down syndrome, a disorder caused by the presence of an extra copy of chromosome 21.

## Transposition of the Great Vessels

Patent ductus arteriosus

Aorta

Pulmonary trunk

In the **transposition of the great vessels**, the aorta is connected to the right ventricle instead of to the left ventricle, and the pulmonary artery is connected to the left ventricle instead of to the right ventricle. This malformation affects 0.05 percent of newborn infants.

**LEARNING OUTCOME**

Describe the differences between fetal and adult circulation patterns, identify changes in blood flow patterns at birth, and list common congenital heart problems.

# Patterns of Blood Flow

**Labeling:** Label the major arteries in the diagram at right.

Label the major veins in the diagram below.

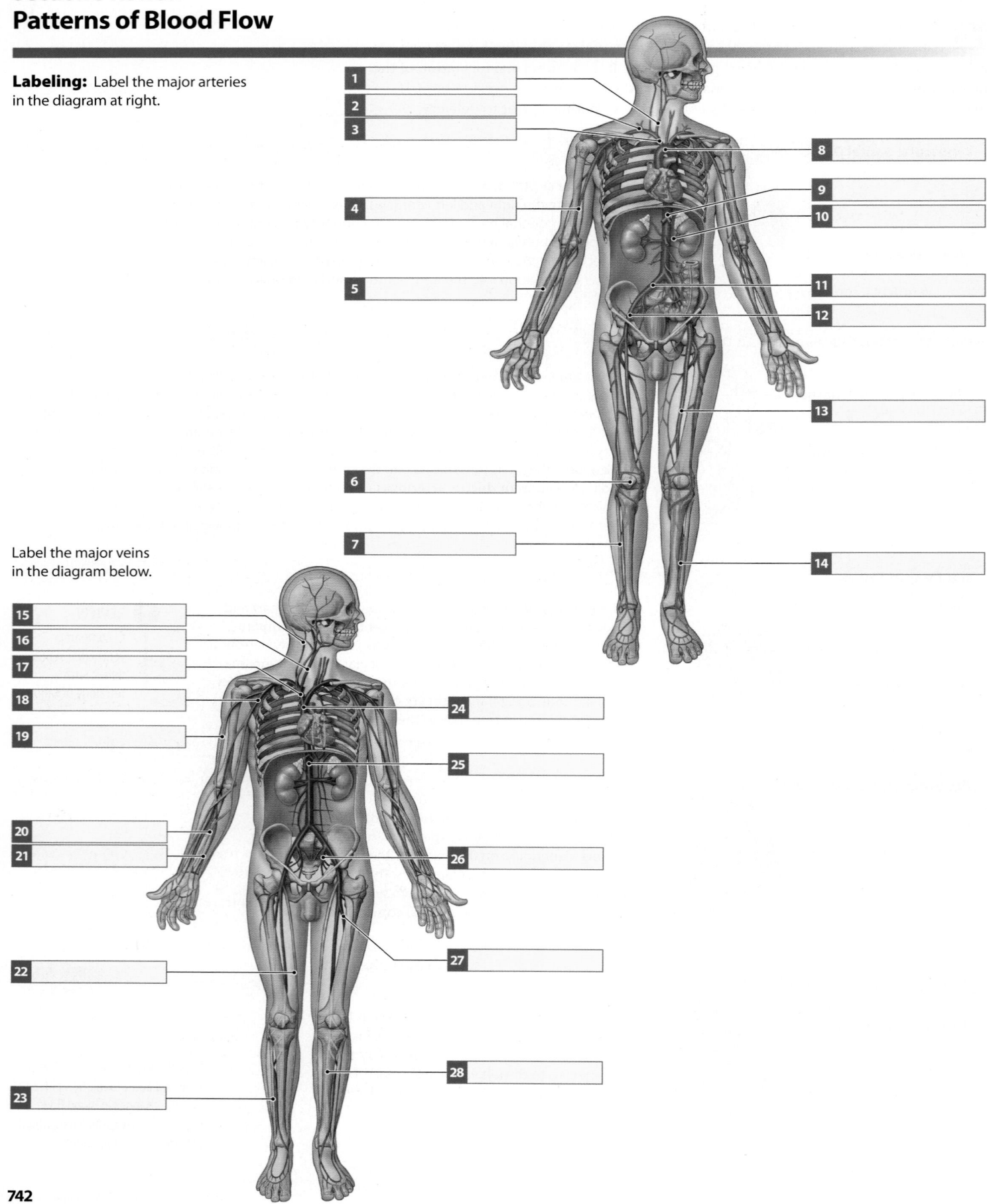

1

2

3

4

5

6

7

8

9

10

11

12

13

14

15

16

17

18

19

20

21

22

23

24

25

26

27

28

# Study Outline

## SECTION 1 • Functional Anatomy of Blood Vessels

### Module 19.1

### The heart pumps blood, in sequence, through the arteries, capillaries, and veins of the pulmonary and systemic circuits   p. 697

1. The **pulmonary circuit** carries blood to and from the lungs. The **systemic circuit** transports blood to the rest of the body.

2. Blood is carried away from the heart by **arteries** *(efferent vessels)*, and returns to the heart by **veins** *(afferent vessels)*.

3. Thin-walled **capillaries** interconnect the smallest arteries and the smallest veins. The thin walls of capillaries allow for the exchange of nutrients, dissolved gases, and wastes between the blood and tissues.

### Module 19.2

### Arteries and veins differ in the structure and thickness of their walls   p. 698

4. The walls of arteries and veins have three distinct layers: the tunica intima, tunica media, and tunica externa.

5. The **tunica intima** is the innermost layer, and it includes an endothelial lining and underlying layer of connective tissue containing elastic fibers. Arteries also have an **internal elastic membrane**.

6. The **tunica media** is the middle layer, and it contains concentric sheets of smooth muscle tissue in a framework of loose connective tissue. This layer allows for **vasoconstriction** and **vasodilation**.

7. The **tunica externa**, or *tunica adventitia*, is the outermost layer and is a connective tissue sheath.

8. The classes of blood vessels as they carry blood away from the heart are **arteries**, **arterioles**, and **capillaries**. Blood then returns to the heart by **venules** and **veins**.

### Module 19.3

### Capillary structure and capillary blood flow affect the rates of exchange between the blood and interstitial fluid   p. 700

9. A typical capillary is a tube of endothelial cells with a delicate basement membrane. The tunica media and tunica externa are absent.

10. In a **continuous capillary**, the endothelium is a complete lining. These capillaries are located throughout the body.

11. In **fenestrated capillaries**, there are pores that penetrate the complete endothelial lining that permit rapid exchange of water and solutes as large as small peptides. These are located in the choroid plexus of the brain and the hypothalamus, pituitary gland, pineal gland, and thyroid gland.

12. **Sinusoids** are discontinuous capillaries and have gaps between adjacent endothelial cells. They are present in the liver, bone marrow, spleen, and many endocrine organs.

13. Capillaries function as part of an interconnected **capillary bed**. Capillary beds are often supplied with blood as **collateral** arteries fuse in **arterial anastomoses**.

14. **Precapillary sphincters** at the entrance to capillaries continually contract and relax in a process called **vasomotion**.

### Module 19.4

### The venous system has low pressures and contains almost two-thirds of the body's blood volume   p. 702

64%

Systemic venous circuit

*The venous system contains almost two-thirds of the body's blood volume.*

15. Blood pressure in a peripheral vein is about 10 percent of that in the ascending aorta.

16. Folds of the tunica intima in some veins form **valves** that permit blood flow in only one direction.

17. Veins distended from weakness in their walls near valves can form **varicose veins** or **hemorrhoids**.

18. Blood volume is unevenly distributed among blood vessels in the following pattern: 9 percent, pulmonary circuit; 7 percent, heart; 13 percent, systemic arterial system; 7 percent, systemic capillaries; 64 percent, systemic venous system.

19. During hemorrhaging, **venoconstriction** (contraction of smooth muscle cells in medium-sized veins) maintains blood volume in the arterial system at near-normal levels. Veins are expandable and can act as **blood reservoirs**.

## SECTION 2 • Coordination of Cardiac Output and Blood Flow

### Module 19.5

### Pressure, resistance, and venous return affect cardiac output   p. 705

20. **Cardiac output** and the distribution of blood in the pulmonary and systemic circuits must constantly be adjusted to meet the demands of active tissues.

21. Factors regulating cardiac output are **arterial pressure, peripheral resistance, capillary pressure, venous pressure, venous return,** and blood flow regulation.

## Module 19.6

### Vessel luminal diameter is the main source of resistance within the cardiovascular system    p. 706

22. The heart must develop enough pressure to overcome **total peripheral resistance**.

23. **Vascular resistance** is the opposition to blood flow in vessels. The longer the vessel length, the greater the friction, and thus the greater the resistance. The larger the vessel luminal diameter, the lesser the friction, and thus the lesser the resistance.

24. A small change in vessel diameter produces a large change in resistance.

*Blood flow turbulence due to plaque*

25. **Viscosity** is the resistance to flow. Increased blood viscosity and **turbulence** due to plaque or other factors increase peripheral resistance.

## Module 19.7

### Blood flow is determined by the interplay between arterial pressure and peripheral resistance    p. 708

26. **Blood flow** is directly proportional to **blood pressure** and inversely proportional to **peripheral resistance**.

27. Absolute pressure is less important than the **pressure gradient**: the difference in pressure from one end of the vessel to the other.

28. As blood moves toward the capillaries, the diameter of the arteries decreases, blood pressure drops, and the blood moves more slowly.

29. As blood returns to the heart, the diameter of the veins increases, blood pressure continues to drop, but blood velocity accelerates.

30. Arterial pressure rises during ventricular systole and falls during ventricular diastole. The peak pressure measured is called **systolic pressure**, and the minimum pressure measured is called **diastolic pressure**.

31. **Pulse pressure** is the difference between systolic and diastolic pressures. **Mean arterial pressure** (**MAP**) is calculated by adding one-third of pulse pressure to diastolic pressure.

32. Capillary exchange involves a combination of **filtration**, diffusion, and osmosis. *Capillary hydrostatic pressure* is the blood pressure within capillary beds.

## Module 19.8

### Capillary exchange is a dynamic process that includes diffusion, filtration, and reabsorption    p. 710

33. Diffusion occurs most rapidly when the distances are short, the concentration gradient is large, and the ions or molecules are small.

*Capillary exchange involves filtration, diffusion, and osmosis.*

34. In a capillary, blood pressure declines as blood flows from the arterial end to the venous end. Filtration predominates at the arterial end, where **capillary hydrostatic pressure** (**CHP**) is highest, and reabsorption predominates at the venous end, where **blood colloid osmotic pressure** (**BCOP**) is highest.

35. More water leaves the bloodstream during filtration than is retrieved through reabsorption. The difference is approximately 3.6 L/day and is returned to the venous system by the lymphatic vessels.

## Module 19.9

### Cardiovascular regulatory mechanisms respond to changes in blood pressure or blood chemistry    p. 712

36. Homeostatic mechanisms regulate cardiovascular activity to ensure that **tissue perfusion** (blood flow through tissues) meets the demands for oxygen and nutrients.

37. The two regulatory pathways are autoregulation and central regulation.

38. **Autoregulation** involves changes in capillary blood flow in response to chemical changes in the interstitial fluid. Local **vasodilators** accelerate blood flow.

39. **Central regulation** involves neural (*cardiovascular center*) and endocrine mechanisms.

40. **Baroreceptor reflexes** respond to changes in blood pressure to maintain blood pressure within a normal range.

## Module 19.10

**Endocrine responses to low blood pressure and low blood volume are very different from those to high blood pressure and high blood volume** p. 714

41. When blood pressure and blood volume decrease, epinephrine and norepinephrine stimulate cardiac output and vasoconstriction. Other hormones in long-term response include ADH, angiotensin II, EPO, and aldosterone.

42. Increased blood volume abnormally stretches the heart walls, causing cardiac muscle cells to release the **natriuretic peptides**: **atrial natriuretic peptide** and **brain natriuretic peptide**. These hormones reduce blood volume and blood pressure.

## Module 19.11

**Chemoreceptors monitor the chemical composition of the blood and cerebrospinal fluid** p. 716

43. **Chemoreceptor reflexes** stimulated by chemoreceptors in the **carotid bodies**, **aortic bodies**, and medulla oblongata respond to changes in carbon dioxide, oxygen, or pH in blood and CSF.

44. Chemoreceptor reflexes trigger adjustments in cardiovascular and respiratory activity.

## Module 19.12

**The cardiovascular center makes extensive adjustments to cardiac output and blood distribution during exercise** p. 717

45. Cardiac output at rest averages around 5800 mL/min. At light exercise, it increases to 9500 L/min and can increase to 17,500 mL/min during heavy exercise.

46. Trained athletes have bigger hearts, larger stroke volumes, and lower resting heart rates than nonathletes. The cardiac output of an athlete in peak condition can be 50 percent higher than that of a nonathlete.

## Module 19.13

**Short-term and long-term mechanisms compensate for a reduction in blood volume** p. 718

47. When homeostasis fails to prevent significant blood loss, the entire cardiovascular system makes adjustments to maintain blood pressure and restore blood volume.

48. **Shock** is an acute cardiovascular crisis marked by low blood pressure (**hypotension**) and inadequate peripheral blood flow. It develops when normal compensation mechanisms fail to maintain adequate cardiac output and peripheral blood flow.

49. **Circulatory shock** begins with **progressive shock** (blood loss exceeding 35 percent) and can lead to the fatal stage of **irreversible shock** and **circulatory collapse**.

## SECTION 3 • Patterns of Blood Flow

## Module 19.14

**New blood vessels form through vasculogenesis and angiogenesis** p. 721

50. **Blood islands** are embryonic cells scattered within the yolk sac that form blood vessels and blood cells.

51. During embryonic development, blood islands give rise to **hematopoietic stem cells** and **hemangioblasts**.

52. Hemangioblasts remodel the blood islands into capillaries and then into arterial and venous networks in a process called **vasculogenesis**. **Angiogenesis** is the formation of new blood vessels from preexisting vessels.

## Module 19.15

**The pulmonary circuit carries deoxygenated blood from the right ventricle to the lungs and returns oxygenated blood to the left atrium** p. 722

53. The **pulmonary circuit** is composed of arteries and veins that transport blood between the heart and the lungs. This circuit begins at the right ventricle and ends at the left atrium.

*The pulmonary circuit*

54. Blood from the right ventricle enters the **pulmonary trunk**, then the left and right **pulmonary arteries**, **pulmonary arterioles**, and then the capillary network that surrounds the **alveoli**. This is where gas exchange occurs between the capillary blood and inspired air.

55. Oxygenated blood leaves the alveolar capillaries and enters venules that unite to form four **pulmonary veins** (two from each lung). These veins deliver blood to the left atrium, completing the pulmonary circuit.

## Module 19.16

**The arteries and veins of the systemic circuit operate in parallel, and the major vessels often have similar names** p. 724

56. All the major **systemic arteries** originate from the **aorta**.

57. All the veins of the systemic circuit merge into either the **superior vena cava** (which collects systemic blood from the head, chest, and upper limbs) or the **inferior vena cava** (which collects systemic blood from all structures inferior to the diaphragm).

## Module 19.17

**The branches of the aortic arch supply structures that are drained by the superior vena cava** p. 726

58. The branches of the aortic arch are the **brachiocephalic trunk**, **left common carotid artery**, and **left subclavian artery**. The brachiocephalic trunk branches to form the **right subclavian artery** and **right common carotid artery**.

59. Within the thoracic cavity, the subclavian artery branches to form the **internal thoracic artery** (which supplies the pericardium and anterior chest wall) and **vertebral artery** (which supplies the brain and spinal cord).

60. After leaving the thoracic cavity, the subclavian artery forms the **axillary artery** and then the **brachial artery**. The brachial artery divides into the **radial artery** and **ulnar artery**. These two arteries fuse at the wrist to form the superficial and deep **palmar arches**. These supply blood to the hand and to the **digital arteries** that supply the fingers.

*The branches of the aortic arch supply structures that are . . .*

*. . . drained by the superior vena cava.*

61. The **digital veins** collect blood from the palmar venous arches. The cephalic vein, **median antebrachial vein**, and the basilic vein drain the **superficial palmar arch**.

62. The **ulnar vein** and **radial vein** drain the **deep palmar arch**. These two veins join to form the brachial vein.

63. The **median cubital vein** interconnects the cephalic and basilic vein. This vein is often the site from which venous blood samples are collected.

64. The **cephalic vein** extends along the forearm. The **brachial vein** and **basilic vein** merge to become the **axillary vein**.

65. The **external jugular vein** drains the superficial head and neck. The **internal jugular vein** drains the deep head and neck. The **vertebral vein** drains the cervical spinal cord and the posterior surface of the skull.

66. The **brachiocephalic vein** forms as the jugular veins join the axillary vein. The **superior vena cava** carries blood from the two brachiocephalic veins to the right atrium.

## Module 19.18

**The external carotid arteries supply the neck, lower jaw, and face, and the internal carotid and vertebral arteries supply the brain, while the external jugular veins drain the regions supplied by the external carotid arteries, and the internal jugular veins drain the brain.    p. 728**

67. The **common carotid arteries** supply the face, neck, and brain. Each common carotid artery branches to form an external and internal carotid artery.

68. The **internal carotid artery** enters the skull through the carotid canal of the temporal bones to deliver blood to the brain and eyes. The **carotid sinus** is located at the base of the internal carotid artery.

69. The **vertebral artery** travels within the vertebral foramina of the cervical vertebrae. The left and right vertebral arteries fuse along the ventral surface of the medulla oblongata to form the **basilar artery**.

70. The **external carotid artery** branches to supply the neck, esophagus, pharynx, larynx, cranium, and face.

71. The maxillary and temporal veins form the **external jugular veins**. The **internal jugular veins** drain blood from the venous sinuses within the cranium. The external and internal jugular veins combine with the **vertebral vein** and subclavian vein to form the **brachiocephalic vein**.

## Module 19.19

**The internal carotid arteries and the vertebral arteries supply the brain, which is drained by the dural sinuses and the internal jugular veins    p. 730**

72. The internal carotid arteries normally supply the arteries of the anterior half of the cerebrum, and the rest of the brain receives blood from the vertebral and basilar arteries.

73. The internal carotid divides into an **ophthalmic artery** (which supplies the eyes), **anterior cerebral artery** (which supplies frontal and parietal lobes of the brain), and **middle cerebral artery** (which supplies the midbrain and lateral surfaces of cerebral hemispheres).

74. The internal carotid arteries and basilar artery are interconnected in a ring-shaped anastomosis called the **cerebral arterial circle** (*circle of Willis*). This arrangement reduces the likelihood of a serious interruption of cerebral blood flow.

75. The superficial cerebral veins and small veins of the brainstem empty into a network of dural sinuses. Most of the deep cerebral veins converge into the **great cerebral vein**, which delivers blood to the **straight sinus**. Veins from the orbit and other cerebral veins drain into the **cavernous sinus**.

76. The **superior sagittal sinus**, in the falx cerebri, is the largest dural sinus. Other sinuses are the **petrosal sinuses**, **transverse sinuses**, and **sigmoid sinuses**.

## Module 19.20

**The regions supplied by the descending aorta are drained by the superior and inferior venae cavae    p. 732**

77. The descending aorta is continuous with the aortic arch. The diaphragm divides the descending aorta into a **thoracic aorta** and an **abdominal aorta**.

78. The visceral branches of the thoracic aorta are the **bronchial arteries**, **esophageal arteries**, **mediastinal arteries**, and **pericardial arteries**.

79. The somatic branches of the thoracic aorta are the **intercostal** and **superior phrenic arteries**.

80. The major paired branches of the abdominal aorta are the **inferior phrenic**, **adrenal**, **renal**, **gonadal**, and **lumbar arteries**.

81. The major unpaired branches of the abdominal aorta are the **celiac trunk**, **superior mesenteric artery**, and **inferior mesenteric artery**.

82. The major collecting vessels of the thorax are the **azygos** and **hemi-azygos veins**. These veins receive blood from the **intercostal**, **esophageal**, **bronchial**, and **mediastinal veins**.

83. The major branches of the inferior vena cava are the **lumbar**, **gonadal** (ovarian or testicular), **hepatic**, **renal**, **adrenal**, and **phrenic veins**.

## Module 19.21

**The viscera supplied by the celiac trunk and mesenteric arteries are drained by the branches of the hepatic portal vein     p. 734**

84. The **celiac trunk** divides into the **common hepatic artery** (which supplies the liver, stomach, gallbladder, and duodenum), **left gastric artery** (which supplies the stomach), and **splenic artery** (which supplies the spleen, stomach, and pancreas).

85. The **superior mesenteric artery** branches to supply the pancreas and duodenum, small intestine, and most of the large intestine.

86. The **inferior mesenteric artery** delivers blood to the terminal portions of the colon and the rectum.

87. The **hepatic portal vein** forms through the fusion of the **superior mesenteric**, **inferior mesenteric**, and **splenic veins**.

88. The hepatic portal vein receives blood from the left and right **gastric veins** (medial border of stomach) and the **cystic vein** (from the gallbladder). This circulation directs blood with absorbed nutrients from the digestive system and is called the **hepatic portal system**.

## Module 19.22

**The pelvis and lower limbs are supplied by branches of the common iliac arteries and drained by branches of the common iliac veins     p. 736**

89. Near vertebra L$_4$, the abdominal aorta divides to form the **right** and **left common iliac arteries**. Near the lumbosacral joint, each common iliac divides to form an **internal iliac artery** and **external iliac artery**.

90. The external iliac artery becomes the **femoral artery** when it enters the lower limb. The first branch is the **deep femoral artery**, which branches into the **femoral circumflex arteries**.

91. Posterior to the knee joint, the femoral artery becomes the **popliteal artery**. This branches into the **posterior** and **anterior tibial arteries**.

92. The internal iliac artery supplies the organs of the pelvic cavity.

93. The tibial artery and **fibular (peroneal) artery** are interconnected by the anastomoses of the **dorsalis pedis**, **dorsal arch**, and **plantar arch**. These arteries supply the distal portions of the foot and digital arteries of the toes.

94. The external iliac veins receive blood from the lower limbs, pelvis, and lower abdomen. The internal iliac veins drain the pelvic organs. The union of the external and internal iliac veins forms the **common iliac vein**.

95. In the foot, the **dorsal venous arch** is drained by the **great saphenous vein** and **small saphenous vein**. The small saphenous vein merges with the **popliteal vein** to form the **femoral vein**, which also receives blood from the **deep femoral vein** and the **femoral circumflex vein**.

## Module 19.23

**The arteries of the systemic circuit deliver oxygenated blood throughout the body, and the veins of the systemic circuit return deoxygenated blood back to the heart     p. 738**

96. The systemic arterial circulation delivers blood from the left ventricle to the body.

97. The systemic venous circulation returns blood from the body back to the right atrium.

## Module 19.24

**The pattern of blood flow through the fetal heart and the systemic circuit must change at birth     p. 740**

98. Fetal blood flows into the placenta through a pair of **umbilical arteries** that arise from the internal iliac arteries. Blood returns from the placenta through a single **umbilical vein**.

99. The umbilical vein drains into the **ductus venosus**, which connects to a collection of veins in the developing liver, and empties into the inferior vena cava. At birth, blood stops flowing into the umbilical vessels, and they soon degenerate.

100. Prior to birth the lungs are collapsed, and most blood bypasses the pulmonary circuit. Blood passes from the right atrium to the left atrium through the **foramen ovale**. Blood also bypasses the lungs through the **ductus arteriosus**, a short vessel between the pulmonary trunk and aorta.

111. At birth the foramen ovale closes, forming the **fossa ovalis**. The ductus arteriosus forms the **ligamentum arteriosum**.

102. Most congenital heart problems result from abnormal formation of the heart or problems with the connections between the heart and the great vessels.

103. Common congenital heart problems are **ventricular septal defect**, **patent foramen ovale** and **patent ductus arteriosus**, **tetralogy of Fallot**, **atrioventricular septal defect**, and **transposition of the great vessels**.

# Chapter Review Questions

**Labeling:** Label the arteries of the brain shown in the image below.

**Multiple choice:** Select the correct answer from the list provided.

**9** The heart pumps blood into the systemic circuit from the
- a) right atrium.
- b) right ventricle.
- c) left atrium.
- ☒ d) left ventricle.

**10** The blood vessel layer containing smooth muscle fibers allowing for the processes of vasoconstriction and vasodilation is the
- a) tunica intima.
- ☒ b) tunica media.
- c) tunica adventitia.
- d) tunica externa.

**11** Capillaries containing pores that permit the passage of large molecules are
- a) continous capillaries.
- b) precapillary sphincters.
- c) collaterals.
- ☒ d) fenestrated capillaries.

**12** Most of the body's total blood volume is distributed within the
- a) pulmonary circuit.
- b) systemic arterial system.
- ☒ c) systemic venous system.
- d) systemic capillaries.

**13** The process of generating blood vessels from preexisting vessels is called
- a) angiogenesis.
- b) vasculogenesis.
- c) fenestration.
- d) arteriosus.

**14** Which of the following is the correct combination of celiac trunk branches?
- a) common hepatic artery, superior mesenteric artery, inferior mesenteric artery
- b) common hepatic artery, splenic artery, superior mesenteric artery
- c) common hepatic artery, left gastric artery, phrenic artery
- d) common hepatic artery, left gastric artery, splenic artery

**15** The vein that drains the venous sinuses of the brain is the
- a) internal jugular vein.
- b) external jugular vein.
- c) vertebral vein.
- d) great cerebral vein.

**16** The two branches from the brachiocephalic trunk are the
- a) left subclavian and left common carotid arteries.
- ☒ b) right subclavian and right common carotid arteries.
- c) left subclavian and left common carotid veins.
- d) right subclavian and right common carotid veins.

**17** The brachiocephalic trunk branches from the right side of the aortic arch. Why is there no such vessel on the left side?

**18** List the distribution of the body's total blood volume.

**19** Describe the function of the hepatic portal system.

---

## MasteringA&P™

Access more chapter study tools online in the MasteringA&P Study Area:

■ Chapter Quizzes, Chapter Practice Test, and Clinical Case Studies

■ Practice Anatomy Lab™ 3.0 — PAL™3.0

■ Interactive Physiology™ — iP2™

■ A&P Flix™ — A&PFlix™

■ PhysioEx™ 9.1 — PhysioEx™ 9.1

---

# Chapter Integration • Applying what you have learned

## Mapping cerebral circulation with angiography

Mr. Samuel was having bouts of light-headedness and slurred speech that alarmed his family, so they rushed him to the hospital. The attending physician was concerned that Mr. Samuel may be experiencing a stroke (damaged brain cells from lack of blood supply) due to a narrowed artery or blood clot in his brain. The physician ordered a cerebral angiogram, an x-ray image that would show the circulation in Mr. Samuel's head and neck. The physician explained to the family that the doctor performing the procedure would insert a tube called a catheter into Mr. Samuel's thigh and guide it through his arteries to his neck. There, a contrasting agent, or dye, would be released to generate the image.

The image obtained from the procedure showed that Mr. Samuel had suffered a minor stroke. The report said that he had an 80 percent blockage of the vertebral artery on his left side. The doctor told the family that Mr. Samuel was fortunate to have had the blockage where he did because he would have suffered a more serious stroke if the blockage had occurred elsewhere.

**1** The angiography required a catheter to be guided from arteries in Mr. Samuel's thigh all the way to the base of his neck. Name the vessels through which this catheter would travel.

**2** Where do you think the dye would be released? Why is that the best place for it to be released?

**3** Can you explain what the doctor meant when he said that Mr. Samuel was lucky to have had the blockage where he did? Where do you think a more serious blockage would be?

**LEARNING OUTCOMES**

These Learning Outcomes correspond by number to this chapter's modules and indicate what you should be able to do after completing the chapter. They also appear at the end of each module.

# 20

# The Lymphatic System and Immunity

## Module 20.1

# The lymphatic system consists of lymphatic vessels, nodes, and lymphoid tissue

The lymphatic system includes the cells, tissues, and organs responsible for providing **immunity** (the ability to resist infection and disease). It also maintains normal blood volume and the composition of interstitial fluid.

**1** **Lymphocytes** are the primary cells of the lymphatic system. These cells respond to the presence of invading pathogens (such as bacteria or viruses), abnormal body cells (such as virus-infected cells or cancer cells), and foreign proteins (such as the toxins released by some bacteria). Lymphocytes circulate within the blood and lymph. **Lymph** is interstitial fluid that has entered a lymphatic vessel.

— Lymphocyte

**2** The lymphatic system includes a network of **lymphatic vessels**, often called **lymphatics**, which begin in peripheral tissues and end at connections to veins near the heart. The lymphatic system also includes an array of lymphoid tissues and lymphoid organs scattered throughout the body. **Primary lymphoid tissues and organs** are sites where lymphocytes are formed and mature. They include red bone marrow and the thymus. **Secondary lymphoid tissues and organs** are where lymphocytes are activated and cloned (identical cellular copies are produced). These structures include the lymph nodes, tonsils, mucosa-associated lymphoid tissue (MALT), appendix, and spleen.

All the cells and tissues involved in producing immunity are part of the **immune system**. The immune system is a *functional* system, and it includes parts of the integumentary, skeletal, lymphatic, cardiovascular, respiratory, and digestive systems.

**? REVIEW**
**B.** List the components of the lymphatic system.

**Lymph Nodes and Lymphatic Vessels**

- Cervical lymph nodes
- Thoracic duct
- Right lymphatic duct
- Axillary lymph nodes
- Lymphatics of mammary gland
- Cisterna chyli
- Lymphatics of upper limb
- Lumbar lymph nodes
- Pelvic lymph nodes
- Inguinal lymph nodes
- Lymphatics of lower limb

**Other Lymphoid Tissues and Organs**

- Tonsil
- Thymus
- Spleen
- Mucosa-associated lymphoid tissue (MALT) in digestive, respiratory, urinary, and reproductive tracts
- Appendix
- Red bone marrow

**? REVIEW**
**A.** What is the difference between a lymphocyte and lymph?

This section considers the organization and functional anatomy of the lymphatic system.

**Lo LEARNING OUTCOME**
Identify the various components of the lymphatic system.

20

## Interstitial fluid flows continuously into lymphatic capillaries and exits tissues as lymph in lymphatic vessels

**20.2**

**Lymphatic vessels** carry lymph from peripheral tissues to the venous system (see Module 4.12). The lymphatic network begins with **lymphatic capillaries**, the smallest lymphatic vessels.

**?** **REVIEW**
**A.** What is the function of lymphatic vessels?

**1** Lymphatic capillaries are present in almost every tissue and organ in the body. As shown here, they are closely associated with blood capillary networks within tissues. Interstitial fluid flows continuously into lymphatic capillaries, at which point it is called lymph. *Lymph* is a fluid that resembles plasma but contains a much lower concentration of suspended proteins.

Arteriole

Smooth muscle

Endothelial cells

Lymphatic capillary

Blood capillaries

Areolar connective tissue

Venule

Interstitial fluid

Lymph flow

**2** Lymphatic capillaries differ from blood capillaries. They (1) are closed at one end rather than forming continuous tubes, (2) have larger diameters, (3) have thinner walls, and (4) typically have a flattened or irregular outline in sectional view.

**?** **REVIEW**
**B.** What is the function of overlapping endothelial cells in lymphatic capillaries?

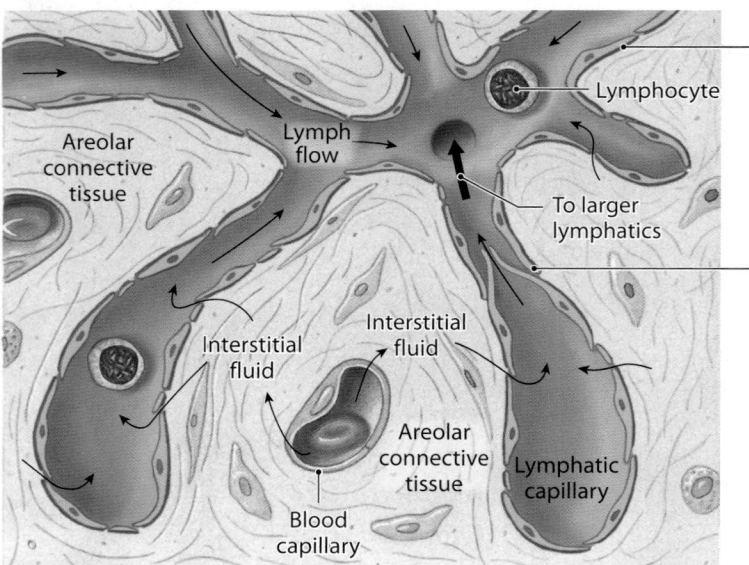

Areolar connective tissue

Lymph flow

Lymphocyte

To larger lymphatics

Interstitial fluid

Interstitial fluid

Areolar connective tissue

Interstitial fluid

Blood capillary

Lymphatic capillary

Although lymphatic capillaries are lined by endothelial cells, the basement membrane is incomplete or absent.

The endothelial cells of a lymphatic capillary overlap. The region of overlap acts as a one-way valve, permitting the entry of fluids and solutes (including proteins), as well as viruses, bacteria, and cell debris, but preventing their return to the intercellular spaces.

**Sectional view**

20

**3** From the lymphatic capillaries, lymph flows into larger lymphatic vessels that lead toward the body's trunk. These lymphatic vessels commonly occur in association with arteries and veins.

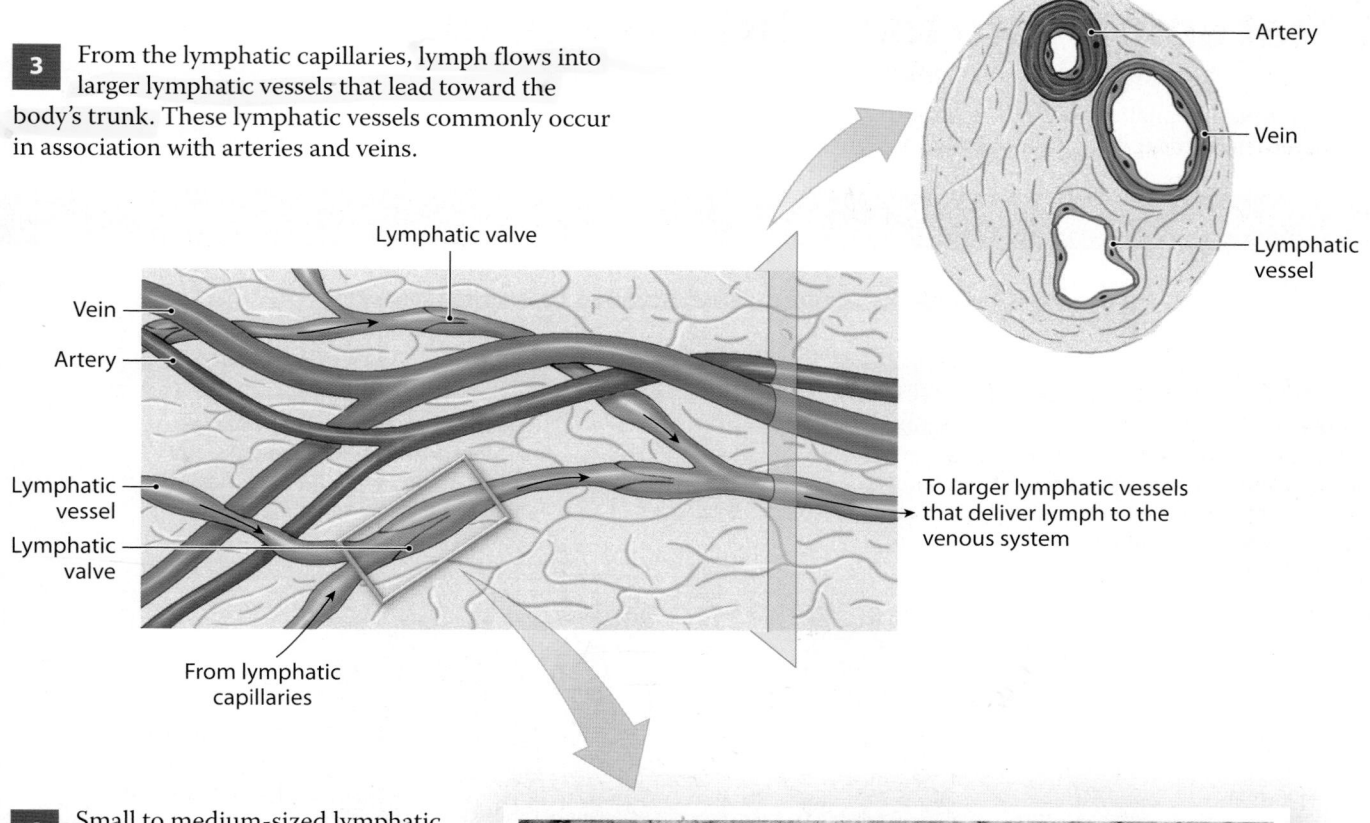

Lymphatic valve

Vein

Artery

Lymphatic vessel

Lymphatic valve

From lymphatic capillaries

To larger lymphatic vessels that deliver lymph to the venous system

Artery

Vein

Lymphatic vessel

**4** Small to medium-sized lymphatic vessels contain valves that prevent backflow. The valves are quite close together and produce noticeable bulges. As a result, some lymphatic vessels resemble a string of pearls. Pressures within the lymphatic system are minimal, and the valves are essential to maintaining normal lymph flow toward the thoracic cavity. Surrounding skeletal muscle contractions aid lymph flow.

Lymphatic valve

Lymphatic vessel

**Valve in lymphatic vessel**                    LM × 65

 **REVIEW**

**C.** What structure prevents the backflow of lymph in some lymphatic vessels?

Important lymphatic capillaries in the small intestine, called *lacteals*, transport lipids absorbed by the digestive tract. Lymphatic capillaries are absent in areas without a blood supply, such as the cornea of the eye.

**?** **INTEGRATION**

**D.** Compare lymphatic capillaries in the small intestine with those found elsewhere.

**Lo** **LEARNING OUTCOME**

Describe the structure and function of important lymphatic vessels.

20

# Module 20.3

# Small lymphatic vessels converge to form lymphatic ducts that empty into the subclavian veins

**1** The lymph from lymphatic capillaries flows into either superficial or deep lymphatic vessels.

## Lymphatic Vessels

### Superficial Lymphatics

**Superficial lymphatics** are located in the subcutaneous layer deep to the skin; in the areolar tissues of the mucous membranes lining the digestive, respiratory, urinary, and reproductive tracts; and in the areolar tissues of the serous membranes lining the pleural, pericardial, and peritoneal cavities.

### Deep Lymphatics

**Deep lymphatics** accompany deep arteries and veins supplying skeletal muscles and other organs of the neck, limbs, and trunk, and the walls of visceral organs.

Superficial inguinal lymph nodes and lymphatic vessels

Deep inguinal lymph nodes and lymphatic vessels

**2** Superficial and deep lymphatics converge to form larger vessels called **lymphatic trunks**, which in turn empty into two large collecting vessels: the *right lymphatic duct* and the *thoracic duct*. The smaller right lymphatic duct collects lymph from the right side of the body superior to the diaphragm. The thoracic duct collects lymph from the entire body inferior to the diaphragm but only from the left side of the body superior to the diaphragm.

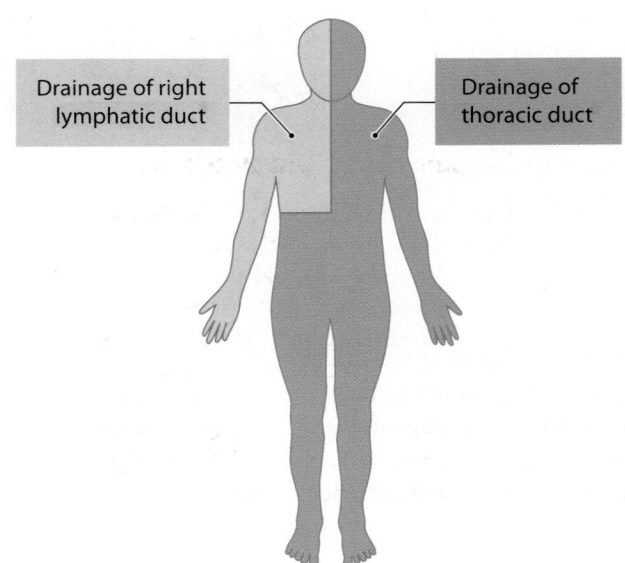

Drainage of right lymphatic duct

Drainage of thoracic duct

**?** **REVIEW**

**A.** Describe the drainage of the right lymphatic duct and the thoracic duct.

**3** This illustration shows the relationship between the lymphatic ducts and the venous system. The right lymphatic duct is about 2 cm long. In contrast, the thoracic duct is about 42 cm long and is the largest lymphatic vessel.

### Right Lymphatic Duct

The right lymphatic duct is formed by the merging of the **right jugular trunk**, **right subclavian trunk**, and **right bronchomediastinal trunk**. It then empties into the right subclavian vein.

Right jugular trunk

Right subclavian trunk

Right lymphatic duct entering right subclavian vein

Right bronchomediastinal trunk

Superior vena cava (cut)

Rib (cut)

Azygos vein

Diaphragm

Right internal jugular vein

Brachiocephalic veins

Left internal jugular vein

### Thoracic Duct

The thoracic duct ascends along the left side of the vertebral column, collecting lymph from the **left bronchomediastinal trunk**, the **left subclavian trunk**, and the **left jugular trunk**. It empties into the left subclavian vein.

Left jugular trunk

Left subclavian trunk

Thoracic duct entering left subclavian vein

Left bronchomediastinal trunk

Thoracic duct

Thoracic lymph nodes

Hemi-azygos vein

Parietal pleura (cut)

Intestinal trunk

**? REVIEW**

**B.** Name the two large lymphatic vessels into which the lymphatic trunks empty.

Inferior vena cava (cut)

Right lumbar trunk

Left lumbar trunk

The **cisterna chyli** (KĪ-lī) is an expanded, sac-like chamber at the base of the thoracic duct. The cisterna chyli receives lymph from the inferior part of the abdomen, the pelvis, and the lower limbs by way of the right and left **lumbar trunks** and the **intestinal trunk**.

**20**

---

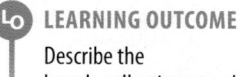 **Clinical Note**

## Blocked lymphatic vessels cause lymphedema

Blockage of the lymphatic drainage produces **lymphedema** (limf-e-DĒ-muh), a condition in which interstitial fluids accumulate and the affected area gradually becomes swollen and grossly distended. Lymphedema most often affects a limb, as in this photo, although it can occur in other locations. If the condition persists, the connective tissues lose their elasticity and the swelling becomes permanent. Because the interstitial fluids are basically stagnant, toxins and pathogens can accumulate and overwhelm local defenses without fully activating the immune system. The immune system is the body's defense against foreign substances.

**? REVIEW**

**C.** Explain lymphedema.

**Lo LEARNING OUTCOME**

Describe the lymph-collecting vessels, and explain lymphedema.

# Lymphocytes are responsible for the immune functions of the lymphatic system

Lymphocytes make up 20–40 percent of the circulating leukocyte population. However, circulating lymphocytes are only a small fraction of the total lymphocyte population. The body contains some $10^{12}$ lymphocytes, with a combined weight of more than one kilogram (2.2 lb).

**1** Three classes of lymphocytes circulate in blood. Although each class has distinctive biochemical and functional characteristics, each is sensitive to chemicals called **antigens**. Most antigens are pathogens, foreign proteins, or abnormal substances that can stimulate the immune system and induce a reaction to these agents known as the *immune response*.

**? REVIEW**

**A.** Identify the three main classes of lymphocytes.

## Classes of Lymphocytes

*(handwritten annotations: "Thymus", "Bone marrow")*

### T Cells

Approximately 80 percent of circulating lymphocytes are classified as **T cells** (T lymphocytes). There are many different types of T cells, but we will focus on four major categories.

#### Cytotoxic T Cells

**Cytotoxic T cells** attack foreign cells or body cells infected by viruses. Their attack commonly involves direct contact. These lymphocytes are the primary cells involved in the production of **cell-mediated immunity** (cellular immunity).

#### Helper T Cells

**Helper T cells** stimulate the responses of both T cells and B cells. Helper T cells are absolutely vital because they must activate B cells before the B cells can produce antibodies.

#### Regulatory T Cells

**Regulatory T cells** are a subset of T cells that moderate the immune response. The interplay between regulatory T and helper T cells helps establish and control the sensitivity of the immune response.

#### Memory T Cells

**Memory T cells** respond to antigens they have already encountered by cloning more lymphocytes to ward off the invader.

### B Cells

**B cells** account for 10–15 percent of circulating lymphocytes.

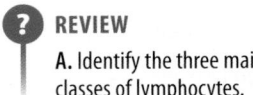

#### Plasma Cells

When stimulated, B cells can differentiate into **plasma cells**, which produce and secrete antibodies. B cells are said to be responsible for **antibody-mediated immunity** (humoral immunity) because antibodies circulate widely in body fluids and attack targets throughout the body.

### NK Cells

**Natural killer (NK) cells** make up 5–10 percent of circulating lymphocytes. These lymphocytes attack foreign cells, body cells infected with viruses, and cancer cells that appear in normal tissues. Their continuous monitoring of peripheral tissues is called **immune surveillance**.

**2** In adults, erythropoiesis (red blood cell formation) is normally confined to red bone marrow, but lymphocyte production, or **lymphocytopoiesis** (lim-fō-sī-tō-poy-Ē-sis), involves the red bone marrow, thymus, and peripheral lymphoid tissues. Red bone marrow plays the primary role in maintaining normal lymphocyte populations.

## Thymus

Lymphoid stem cells and their descendants are isolated from the general circulation by the *blood thymus barrier*, comparable to the blood brain barrier. Under the influence of thymic hormones, the lymphoid stem cells divide repeatedly, producing the various kinds of T cells. These T cells undergo a selection process to ensure they will not react to the body's own healthy cells and cellular products. Up to 98 percent of developing T cells are deselected and undergo apoptosis (programmed cell death).

Production and differentiation of surviving T cells

## Red Bone Marrow

The divisions of hemocytoblasts in the red bone marrow of adults generate the **lymphoid stem cells** that produce all types of lymphocytes.

One group of lymphoid stem cells migrates to the thymus.

A second group of lymphoid stem cells remains in the bone marrow and divides to produce immature B cells and NK cells.

B cells

NK cells

Cancer suppressers

As they mature, B cells and NK cells enter the bloodstream and migrate to peripheral tissues.

**?** REVIEW

**B.** What tissues are involved in lymphopoiesis?

When their maturation is nearing completion, T cells re-enter the bloodstream and travel to peripheral lymphoid tissues and organs.

## Peripheral Tissues

**Cell-mediated immunity**

Mature T cells are responsible for cell-mediated immunity.

**Antibody-mediated immunity**

Most of the B cells move into the lymph nodes, spleen, or other lymphoid tissues.

**Immune surveillance**

The NK cells migrate throughout the body, moving through peripheral tissues in search of abnormal cells.

20

The T cells and B cells that migrate from their sites of origin retain the ability to divide. Their divisions produce daughter cells of the same type. For example, a dividing B cell produces other B cells, but not T cells or NK cells. The ability of specific types of lymphocytes to increase in number is crucial to the success of the immune response.

**?** REVIEW

**C.** Which cells are responsible for antibody-mediated immunity?

 **LEARNING OUTCOME**

Identify the classes of lymphocytes, discuss their importance, and describe their distribution in the body.

# Module 20.5

# Lymphocytes aggregate within lymphoid tissues and lymphoid organs

**Lymphoid tissues** are connective tissues dominated by lymphocytes. In a **lymphoid nodule**, the lymphocytes are densely packed in an area of areolar tissue. The boundaries are not distinct, because no fibrous capsule surrounds the tissue. In many areas, lymphoid nodules form large clusters. In contrast, **lymphoid organs** (lymph nodes, the thymus, and the spleen) are separated from surrounding tissues by a fibrous connective tissue capsule.

**? REVIEW**

**A.** Name the lymphoid tissue that protects epithelia lining the digestive, respiratory, urinary, and reproductive tracts.

**1** Clusters of lymphoid nodules deep to the epithelial lining of the intestine are known as **aggregated lymphoid nodules**, or *Peyer's patches*. Each nodule often has a central zone called a **germinal center**, which contains dividing lymphocytes. The lymphoid tissues that protect epithelia of the digestive, respiratory, urinary, and reproductive tracts from pathogens and toxins form the **mucosa-associated lymphoid tissue (MALT)**.

**Aggregated lymphoid nodules** LM × 20

Intestinal lumen

Mucous membrane of intestinal wall

Germinal center

Aggregated lymphoid nodule in intestinal mucosa

Underlying connective tissue

**2** The **tonsils** are large lymphoid nodules in the walls of the pharynx. A single **pharyngeal tonsil**, often called the *adenoid*, lies in the posterior superior wall of the nasopharynx. Left and right **palatine tonsils** are located at the posterior, inferior margin of the oral cavity, along the boundary of the pharynx. A pair of **lingual tonsils** lie deep to the mucous epithelium covering the base (pharyngeal portion) of the tongue. Because of their location, the lingual tonsils are not usually visible unless they become infected and swollen. **Tonsillitis** is an inflammation of the tonsils (especially the palatines) although the other tonsils may also be affected. Tonsils reach their largest size by puberty and then begin to atrophy.

Germinal centers within nodules

Pharyngeal epithelium

Pharyngeal tonsil

Palate

Palatine tonsil

Lingual tonsil

**Pharyngeal tonsil** LM × 40

**? REVIEW**

**B.** Define *tonsil*, and name the five tonsils.

**3** **Lymph nodes** are small lymphoid organs ranging in diameter from 1 mm to 25 mm (about 1 in.). The shape of a typical lymph node resembles that of a kidney bean. The largest lymph nodes are located where the peripheral lymphatics from the neck and the limbs connect with the trunk, such as in the groin, the axillae, and the base of the neck. These nodes, often called **lymph glands**, detect pathogens before they reach the vital organs of the trunk. A lymph node functions like a kitchen water filter, purifying lymph before it reaches the venous circulation. As lymph flows through a lymph node, at least 99 percent of the antigens in the lymph are removed, and immune responses are stimulated as needed.

Lymph node

Lymph vessel

**Lymph nodes**

## Path of Lymph Flow through a Lymph Node

**6** **Efferent** (*efferens*, to bring out) **lymphatics** leave the lymph node at the hilum. These vessels collect lymph from the medullary sinus and carry it toward the venous circulation.

**5** Lymph continues into the **medullary sinus** at the core of the lymph node. This region contains B cells and macrophages.

**4** Lymph then flows through lymph sinuses in the **paracortex**, which is dominated by T cells.

**3** Lymph next flows into the **outer cortex**. The periphery of the cortex contains B cells within germinal centers similar to those of lymphoid nodules.

**2** The afferent vessels deliver lymph to the subcapsular space, a meshwork of reticular fibers, macrophages, and dendritic cells. **Dendritic cells** are involved in starting the immune response. Antigens are also carried by dendritic cells from peripheral tissues to lymph nodes.

**Start**

**1** **Afferent** (*afferens*, to bring to) **lymphatics** carry lymph to the lymph node from peripheral tissues. The afferent lymphatics penetrate the capsule of the lymph node on the side opposite the hilum.

Lymph node artery and vein

Blood vessels and nerves enter and leave the lymph node at the **hilum**, a shallow indentation.

Medulla

Capsule

Germinal center

Subcapsular space

**Trabeculae** (*trabecula*, a beam) are fibrous partitions extending inward from the capsule.

20

**?** REVIEW

**C.** Trace the path of lymph through a lymph node, beginning at the afferent lymphatics.

The digestive, respiratory, urinary, and reproductive tracts communicate with the exterior environment, which contains dangerous pathogens and toxins. MALT, which defends the exposed epithelia, contains much of the lymphoid tissue in the body. A variety of clinical disorders can result from infection and/or inflammation of MALT components. Tonsillitis is one familiar example; another is **appendicitis** (inflammation of the lymphoid tissues of the appendix).

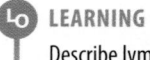 LEARNING OUTCOME

Describe lymphoid tissues, and trace lymph flow through a lymph node.

# The thymus is a lymphoid organ that produces functional T cells

The **thymus** produces several hormones that are important to the development of functional T cells and thus to the maintenance of normal immunological defenses. *Thymosin* (THĪ-mō-sin) is the name originally given to an extract from the thymus that promotes the development and maturation of lymphocytes. This extract actually contains several hormones collectively known as **thymosins**. Just before puberty, the thymus weighs about 40 g (1.4 oz). After puberty, it gradually shrinks and becomes increasingly fibrous and fatty, a process called **involution**. By the time a person reaches age 50, the thymus may weigh less than 12 g (0.3 oz). The gradual decrease in the size and secretory abilities of the thymus that occurs with age is correlated with an increased susceptibility to disease.

**1** In adults, the thymus is a pink, grainy organ located in the mediastinum, generally just posterior to the sternum.

**? REVIEW**
**A.** Where is the thymus located?

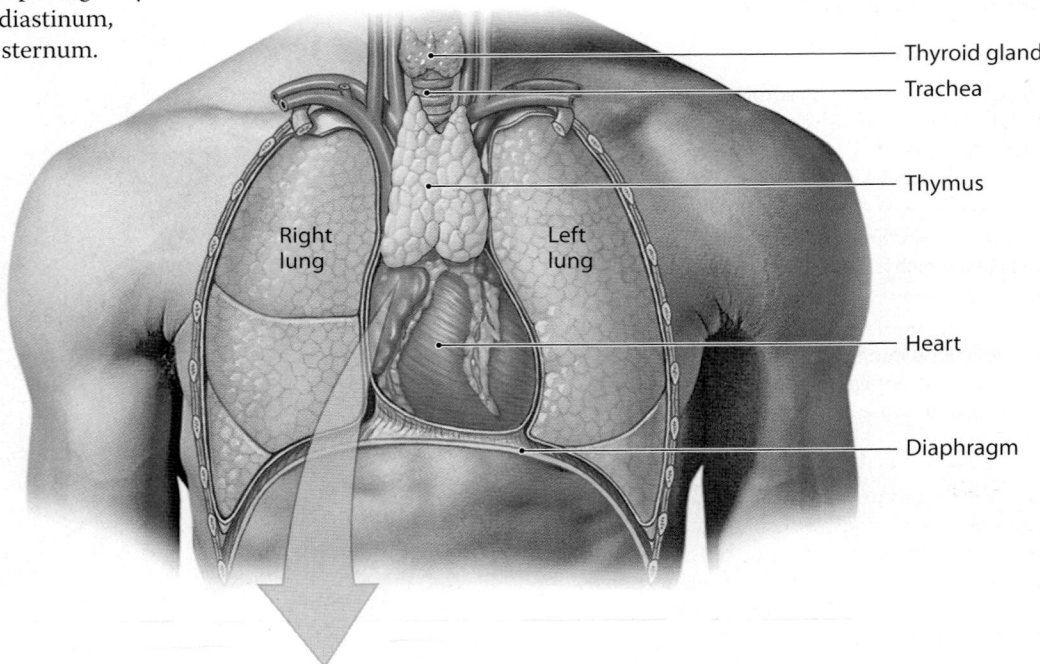

- Thyroid gland
- Trachea
- Thymus
- Right lung
- Left lung
- Heart
- Diaphragm

20

**2** The capsule that covers the thymus divides it into left and right **lobes**. Fibrous partitions called **septa** (singular, *septum*) originate at the capsule and divide the lobes into **lobules** 0.5–2 mm in diameter.

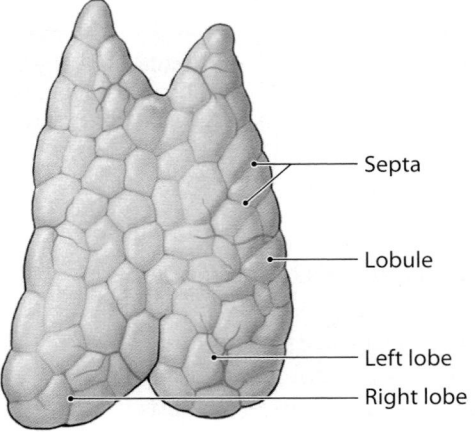

- Septa
- Lobule
- Left lobe
- Right lobe

**? REVIEW**
**B.** Describe the gross anatomy of the thymus.

**3** Each lobule consists of a dark outer **cortex** and a lighter central **medulla**. Lymphocytes divide and differentiate under controlled conditions in the cortex. Lymphocytes in the cortex are arranged in clusters that are completely surrounded by **epithelial reticular cells (ERCs)**. Epithelial reticular cells regulate T cell development and function. ERC processes also encircle the capillaries in the cortex and maintain the **blood thymus barrier**. Macrophages in the cortex engulf lymphocytes undergoing apoptosis (see Module 20.4).

Thymus — LM × 50

**4** After roughly 3 weeks, developing T cells leave the cortex and enter the medulla. Because there is no blood thymus barrier in the medulla, developing T cells may enter a medullary blood vessel. Instead of ensheathing blood vessels as they do in the cortex, the epithelial reticular cells in the medulla cluster together in concentric layers, forming distinctive structures known as **thymic corpuscles**, or *Hassal's corpuscles*. These structures are involved in the production of T regulatory cells. Within the medulla, developing T cells may also leave the thymus through an efferent lymphatic.

**?** **REVIEW**

**C.** Which cells maintain the blood thymus barrier?

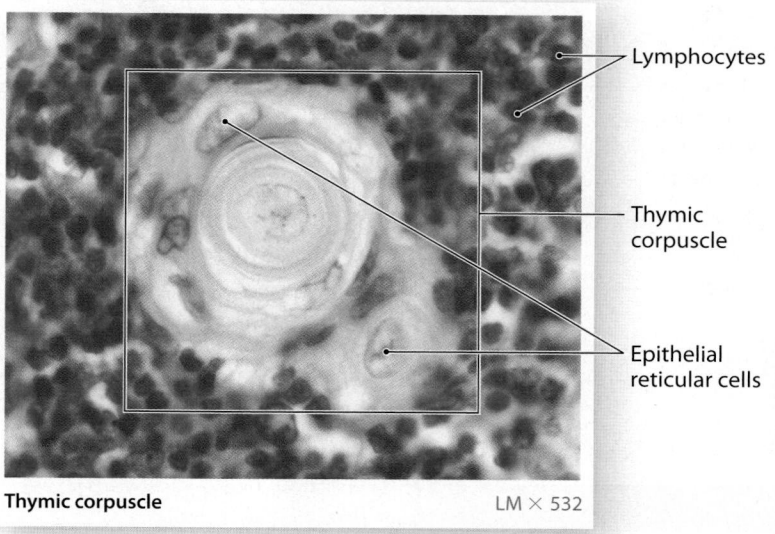

Thymic corpuscle — LM × 532

**20**

---

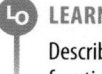 **Clinical Note**

## Myasthenia gravis is an autoimmune disease associated with an enlarged, hyperactive thymus

Myasthenia gravis is a disease characterized by skeletal muscle weakness. It is an autoimmune disease caused by a person's own antibodies blocking, altering, or destroying the receptors for acetylcholine (ACh) at the neuromuscular junction. Myasthenia gravis is associated with a hyperactive thymus and some patients have benign thymic tumors. It is thought that the thymus causes the production of abnormal antibodies that attack neuromuscular ACh receptors.

**Lo** **LEARNING OUTCOME**

Describe the structure and function of the thymus.

# The spleen, the largest lymphoid organ, responds to antigens in the bloodstream

The adult **spleen** contains the largest collection of lymphoid tissue in the body. In essence, the spleen performs the same functions for blood that lymph nodes perform for lymph. Functions of the spleen can be summarized as (1) removing abnormal blood cells and other blood components by phagocytosis, (2) storing iron recycled from red blood cells, and (3) initiating immune responses by B cells and T cells in response to antigens in circulating blood.

**?** **REVIEW**
**A.** What are the functions of the spleen?

Spleen

**1** The spleen lies along the curving lateral border of the stomach, extending between the 9th and 11th ribs on the left side. It is attached to the lateral border of the stomach by the **gastrosplenic ligament**, a broad band of mesentery.

Stomach
Rib
Liver
Pancreas
Inferior vena cava
Aorta
Kidneys

Diaphragm
Parietal peritoneum
Visceral peritoneum
Gastrosplenic ligament
Gastric area
Spleen

The **diaphragmatic surface** of the spleen is smooth and convex, conforming to the shape of the diaphragm and body wall.

Hilum
Renal area

Inferior view, transverse section

**Clinical Note**

### A ruptured spleen is a medical emergency

The spleen tears so easily that a seemingly minor impact to the left side of the abdomen can rupture the capsule. Because the spleen is relatively fragile, it is very difficult to repair surgically, so a severely ruptured spleen is removed, a process called a **splenectomy** (splē-NEK-tō-mē). A person can survive without a spleen but lives with an increased risk of bacterial infection (particularly involving pneumococcal bacteria).

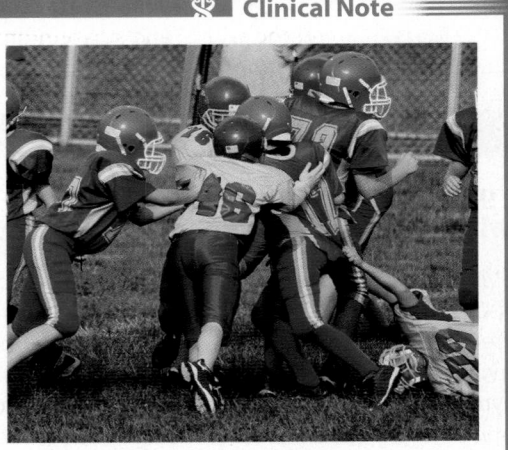

20

**2** The spleen is about 12 cm (5 in.) long and weighs, on average, nearly 160 g (5.6 oz). In gross dissection, it is deep red because of the blood it contains. The spleen has a soft texture, so its shape primarily reflects its association with the structures around it. The medial, **visceral surface** contains indentations that conform to the shape of the stomach (the **gastric area**) and that of the kidney (the **renal area**).

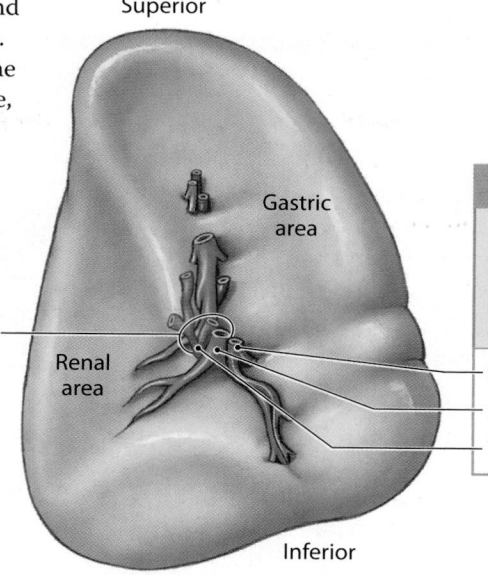

Superior

Gastric area

Hilum

Renal area

Inferior

**Landmarks on the visceral surface**

### Splenic Blood Vessels and Lymphatic Vessels

Splenic blood vessels and lymphatic vessels communicate with the spleen on the visceral surface at the **hilum**, a groove marking the border between the gastric and renal areas.

— Splenic artery

— Splenic vein

— Splenic lymphatic vessels

**3** The spleen is surrounded by a capsule containing collagen and elastic fibers. The cellular components within make up the **pulp** of the spleen. **Red pulp** contains large numbers of red blood cells, whereas **white pulp** resembles lymphoid nodules and contains lymphocytes.

White pulp    Red pulp

**The spleen**                      LM × 50

### Trabeculae

Fibrous **trabeculae** (partitions) radiate outward toward the capsule through the interior from the hilum. Blood vessels travel within the trabeculae.

— Capsule

— Trabecula

— Trabecular artery

— Central artery

**? REVIEW**

**B.** Describe red pulp and white pulp found in the spleen.

**20**

**4** An unusual circulatory arrangement within the spleen gives the phagocytes of the spleen an opportunity to identify and engulf any damaged or infected cells in circulating blood. Because macrophages are scattered throughout the red pulp, and lymphocytes and dendritic cells are found within and surrounding the white pulp, any pathogen or other antigen in the blood quickly triggers an immune response.

### Path of Blood Flow through the Spleen

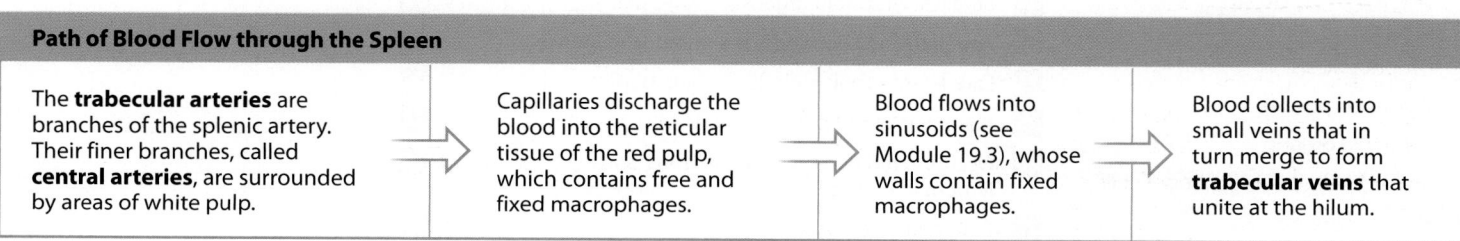

| | | | |
|---|---|---|---|
| The **trabecular arteries** are branches of the splenic artery. Their finer branches, called **central arteries**, are surrounded by areas of white pulp. | Capillaries discharge the blood into the reticular tissue of the red pulp, which contains free and fixed macrophages. | Blood flows into sinusoids (see Module 19.3), whose walls contain fixed macrophages. | Blood collects into small veins that in turn merge to form **trabecular veins** that unite at the hilum. |

 **REVIEW**

**C.** Beginning at the trabecular arteries, trace the path of blood through the spleen.

 **LEARNING OUTCOME**

Describe the structure and function of the spleen, and trace blood flow through it.

# Anatomy of the Lymphatic System

**Labeling:** Label the structures of the lymphatic system in the accompanying figure.

1
2
3
4
5
6
7
8
9
10
11
12
13
14
15
16
17

**Matching:** Match each lettered term with the most closely related description.

a.  B cells
b.  spleen
c.  lymphatic capillaries
d.  cytotoxic T cells
e.  epithelial reticular cells
f.  tonsils
g.  lymphopoiesis
h.  thymic corpuscles
i.  lymphoid organs
j.  helper T cells and regulatory T cells
k.  afferent lymphatics
l.  right subclavian vein
m.  lymph nodes

| 18 | Beginning of lymphatic system | 18 |
| 19 | Thymus medullary structures | 19 |
| 20 | Receives lymph from right lymphatic duct | 20 |
| 21 | Thymus, spleen, and lymph nodes | 21 |
| 22 | Smallest lymphoid organs | 22 |
| 23 | Regulate and coordinate the immune response | 23 |
| 24 | Largest mass of lymphoid tissue in body | 24 |
| 25 | Maintains blood thymus barrier | 25 |
| 26 | Occurs in red bone marrow, thymus, and lymphoid tissues | 26 |
| 27 | Cell-mediated immunity | 27 |
| 28 | Lymphoid nodules in walls of pharynx | 28 |
| 29 | Antibody-mediated immunity | 29 |
| 30 | Carry lymph to lymph nodes | 30 |

20

## Module 20.8

# Innate immunity is nonspecific and is not stimulated by specific antigens

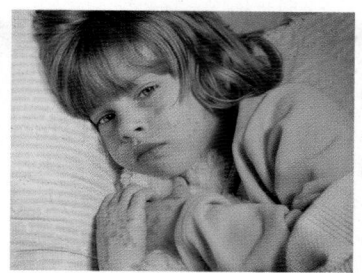

Immunity (i-MYŪ-ni-tē) is the body's ability to resist and defend against infectious organisms or other substances that could damage tissues and organs. The body's defensive reaction is known as the **immune** (*immune*, protected) **response**. Humans have two types of immunity. **Innate (nonspecific) immunity** uses defense mechanisms that either prevent, or slow the entry, of infectious organisms or attack them if they do enter. Section 3 discusses adaptive (specific) immunity.

## Immunity

### Innate (Nonspecific) Immunity

**Innate defenses** do not distinguish one type of threat from another. Their response is the same, regardless of the type of invading agent. These defenses, which are present at birth, provide a defensive capability known as **nonspecific resistance**.

**Physical barriers** such as the skin and mucous membranes keep hazardous organisms and materials outside the body. Such physical barriers are often called the first line of the body's defenses.

**Phagocytes** are cells that engulf pathogens and cell debris. Examples of phagocytes are the macrophages of peripheral tissues and the eosinophils and neutrophils of blood.

**Immune surveillance** is the destruction of abnormal cells by NK cells in peripheral tissues.

 Destruction of abnormal cells

**Interferons** are chemicals that coordinate the defenses against viral infections.

**Complement** is a system of circulating proteins that assists antibodies in the destruction of pathogens.

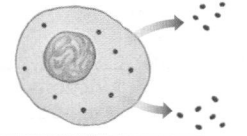 Lysed pathogen

**Inflammation** is a localized, tissue-level response that tends to limit the spread of an injury or infection.

 Inflammation

Mast cell

**Fever** is an elevation of body temperature that accelerates tissue metabolism and defenses.

### Adaptive (Specific) Immunity

**Adaptive defenses** protect against particular threats. For example, a specific defense may protect against infection by one type of bacterium but be ineffective against other bacteria and viruses. Adaptive immunity depends on the activities of specific lymphocytes. The body's specific defenses produce a state of protection known as **specific resistance**. We discuss adaptive immunity in Section 3.

**?** **REVIEW**

**A.** Distinguish between innate immunity and adaptive immunity.

**B.** How does innate immunity protect us from disease?

**?** **INTEGRATION**

**C.** A child falls off her bike and skins her knee. Which form of immunity will be activated immediately?

**Lo** **LEARNING OUTCOME**

Give an overview of the components of innate (nonspecific) immunity.

20

# Module 20.9

# Physical barriers prevent pathogens and toxins from entering body tissues

To cause trouble, a foreign substance or pathogen must enter body tissues. In other words, it must cross an epithelium—either at the skin or across a mucous membrane.

**1** The integumentary system provides the major physical barrier to the external environment.

### Secretions

Sebaceous gland

Most epithelia are protected by specialized accessory structures and secretions. The epidermal surface also receives the secretions of sebaceous and sweat glands. These secretions, which flush the surface to wash away microorganisms and chemical agents, may also contain bactericidal chemicals, destructive enzymes (lysozymes), and antibodies.

### Hair

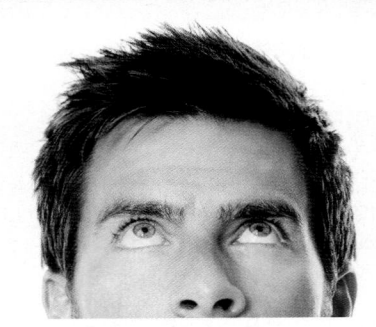

The hairs on most areas of your body protect against physical abrasion (especially on the scalp), and they often prevent hazardous materials or insects from touching your skin.

### Stratified squamous epithelium

Keratinized cells

Desmosomes

The epithelial covering of the skin has multiple layers, a coating of keratinized cells, and a network of desmosomes that lock adjacent cells together.

**2** The mucous membranes lining the digestive, respiratory, urinary, and reproductive tracts are an important barrier that protects against foreign substances and pathogens. MALT in these locations provides additional defenses (see Module 20.5).

Mucus bathes most surfaces of your digestive tract, and your stomach contains a powerful acid that can destroy many pathogens. Mucus moves across the lining of the respiratory tract, urine flushes the urinary passageways, and glandular secretions do the same for the reproductive tract. Special enzymes, antibodies, and an acidic pH add to the effectiveness of these secretions.

Along the more delicate internal passageways, epithelial cells are usually tied together by tight junctions and supported by a fibrous basement membrane.

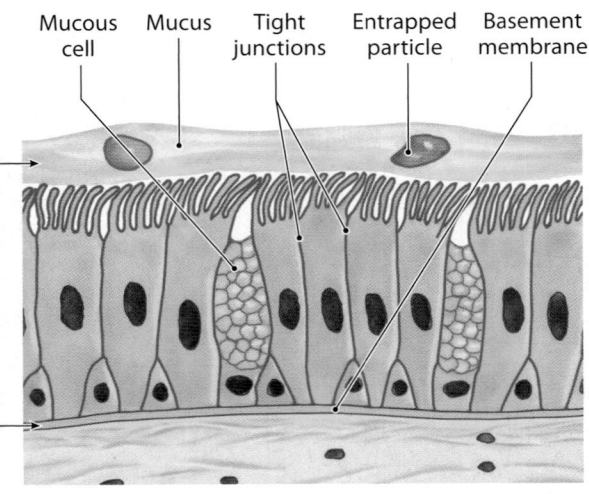

Mucous cell | Mucus | Tight junctions | Entrapped particle | Basement membrane

**? REVIEW**

**A.** How does the integumentary system protect the body?

**Lo LEARNING OUTCOME**

Explain how physical barriers play a role in innate immunity.

# Module 20.10
# Phagocytes respond to pathogen invasion

**1** **Phagocytes** serve as janitors and police in peripheral tissues. They remove cellular debris and respond to foreign substances or pathogens that have breached the body's physical barriers. Many phagocytes attack and remove microorganisms even before lymphocytes detect them. All phagocytic cells function in much the same way, although the target of phagocytosis may differ from one type of phagocyte to another.

**? REVIEW**

**A.** Identify the types of phagocytes in the body, and differentiate between fixed macrophages and free macrophages.

## Types of Phagocytes

| Neutrophils | Eosinophils | Macrophages |
|---|---|---|
| 12 µm | 8–10 µm | There are two major classes of **macrophages** derived from the monocytes of the circulating blood. This collection of phagocytic cells is called the **monocyte–macrophage system**, or the *reticuloendothelial system*.<br><br>Monocyte<br><br>**Fixed macrophages** are permanent residents of specific tissues and organs and are scattered among connective tissues. They are immobile and stay within these tissues. **Free macrophages** travel throughout the body, arriving at the site of an injury by migrating through adjacent tissues or by recruitment from the circulating blood. |
| **Neutrophils** are abundant, mobile, and quick to phagocytize cellular debris or invading bacteria. They circulate in the bloodstream and roam through peripheral tissues, especially at sites of injury or infection. | **Eosinophils**, which are less abundant than neutrophils, phagocytize foreign substances or pathogens that have been coated with antibodies. | |

**2** All phagocytes share several functional characteristics, which are summarized in the table below.

## Functional Characteristics of Phagocytes

- Phagocytes can leave capillaries by squeezing between adjacent endothelial cells, a process known as **emigration**.

- Phagocytes may be attracted to or repelled by chemicals in the surrounding fluids, a phenomenon called **chemotaxis**. They are particularly sensitive to chemicals released by body cells or pathogens.

- Phagocytosis always begins with **adhesion**, the attachment of the phagocyte to its target. In this process, receptors on the plasma membrane of the phagocyte bind to the surface of the target.

- After adhesion, the phagocyte may either destroy the target itself or promote its destruction by activating specific defenses.

**Phagocyte engulfing bacteria**   SEM × 2900

**? REVIEW**

**B.** Define *chemotaxis*.

**Lo LEARNING OUTCOME**

Explain the role phagocytes play in innate immunity.

# Module 20.11
# NK cells perform immune surveillance, detecting and destroying abnormal cells

The immune system generally ignores the body's own cells unless they become abnormal in some way. **Natural killer (NK)** cells are responsible for recognizing and destroying abnormal cells when they appear in peripheral tissues. The continuous "policing" of peripheral tissues by NK cells is called **immune surveillance**. The plasma membranes of cancer cells generally contain unusual proteins called **tumor-specific antigens**, which NK cells recognize as abnormal. The affected cells are then destroyed, preserving the tissue.

**? REVIEW**

**A.** Define *immune surveillance.*

**1** NK cells recognize bacteria, foreign cells, cells infected by viruses, and cancer cells. In each case, the steps leading to target cell destruction are similar.

| Step 1: Recognition and Adhesion | Step 2: Realignment of Golgi Apparatus | Step 3: Secretion of Perforins | Step 4: Lysis of Abnormal Cell |
|---|---|---|---|
| If a cell has unusual components in its plasma membrane, an NK cell recognizes that other cell as abnormal. Such recognition activates the NK cell, which then adheres to its target cell. | The Golgi apparatus moves around the nucleus until the maturing face points directly toward the abnormal cell. The Golgi apparatus then produces a flood of secretory vesicles that contain proteins called **perforins**. These vesicles move through the cytosol toward the cell surface. | The perforins are released at the cell surface by exocytosis and diffuse across the narrow gap separating the NK cell from its target. | As a result of the pores made by perforin molecules, the target cell can no longer maintain its internal environment, and it quickly disintegrates. |

NK cell · Golgi apparatus · Antigen · Abnormal cell

**? REVIEW**

**B.** How do NK cells detect cancer cells?

Perforin molecules

NK cell · Abnormal cell

When perforin molecules reach the opposing plasma membrane, they interact to form a network of pores. These pores are large enough to allow ions, proteins, and other intracellular materials to leave the abnormal cell.

20

**2** Cell division is a complicated process, and mistakes occasionally occur. The abnormal daughter cells usually cannot survive, and NK cells detect and destroy the few that can still function. That detection is important because a small number of those abnormal cells are potentially lethal cancer cells.

Stem cell

Abnormal cell

NK cell identifies and destroys abnormal cell

Daughter cells

**3** Cancer cells often mutate and can sometimes avoid detection by NK cells. This process of avoiding detection or neutralizing body defenses is called **immunological escape**.

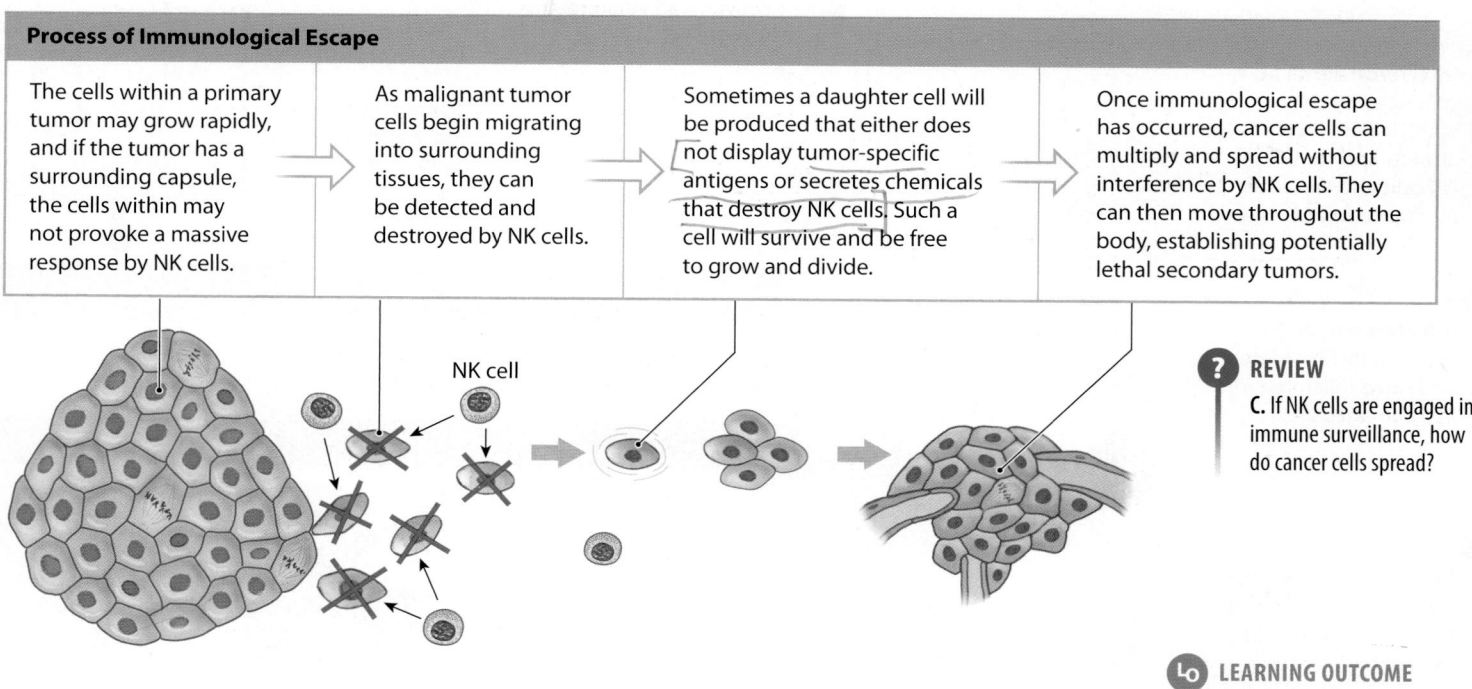

### Process of Immunological Escape

The cells within a primary tumor may grow rapidly, and if the tumor has a surrounding capsule, the cells within may not provoke a massive response by NK cells.

As malignant tumor cells begin migrating into surrounding tissues, they can be detected and destroyed by NK cells.

Sometimes a daughter cell will be produced that either does not display tumor-specific antigens or secretes chemicals that destroy NK cells. Such a cell will survive and be free to grow and divide.

Once immunological escape has occurred, cancer cells can multiply and spread without interference by NK cells. They can then move throughout the body, establishing potentially lethal secondary tumors.

NK cell

**? REVIEW**

**C.** If NK cells are engaged in immune surveillance, how do cancer cells spread?

**Lo LEARNING OUTCOME**

Describe immune surveillance, and explain the role of NK cells.

# Module 20.12

# Interferons and the complement system are distributed widely in body fluids

## Interferons

Interferons (in-ter-FĒR-onz) **(IFNs)** are small proteins released by activated lymphocytes and macrophages and by cells infected with viruses. An interferon binds to surface receptors on the membrane of a normal cell and, by second messengers, triggers the production of **antiviral proteins** in the cytoplasm. Antiviral proteins do not prevent viruses from entering the cell. Instead, they interfere with viral replication inside the cell. In addition to slowing the spread of viral infections, IFNs stimulate macrophages and NK cells.

**? REVIEW**
**A.** Define *interferons*.

**1** At least three types of interferons exist, each of which has additional specialized functions. Interferons are examples of **cytokines** (SĪ-tō-kīnz)— chemicals released by cells to coordinate local activities. Cytokines produced by most cells are used only for cell-to-cell communication within a given tissue. However, cytokines released by phagocytes and lymphocytes can also act as hormones, affecting cells and tissues throughout the body.

**1** **Interferon alpha (α)** is produced by cells infected with viruses. IFN alpha attracts and stimulates NK cells and enhances resistance to viral infection.

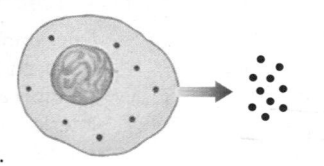

**2** **Interferon beta (β)** is secreted by fibroblasts and slows inflammation in a damaged area.

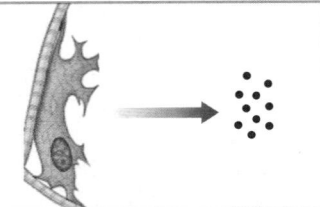

**3** **Interferon gamma (γ)** is secreted by T cells and NK cells and stimulates macrophage activity.

## a. Classical Pathway

The most rapid and effective activation of the complement system occurs through the classical pathway.

Antibodies

Bacterial cell wall

**Antibodies bind to bacterial cell wall**

C1

C1 must attach to two antibodies for its activation.

The attached C1 protein acts as an enzyme, catalyzing a series of reactions involving other complement proteins that split C3 into C3a and C3b. C3a diffuses away and activates an inflammatory response.

C2    C4
C3
C3a    C3b

**Activation and Cascade**

**C3b Attachment (classical pathway)**

C3b binds to bacterial surface and enhances phagocytosis.

C3b

## Results of Complement Activation

All three pathways split inactive C3 proteins to form activated C3b and C3a proteins.

### Killing of Pathogen (Cell Lysis)

Once an activated C3b protein has attached to the cell wall, additional complement proteins may form a **membrane attack complex (MAC)** in the membrane that destroys the integrity of the target cell.

C5-C9
MAC

Multiple pores in bacterium

**Cell lysis**

20

# Complement System

**2** Plasma contains over 30 special proteins that form the **complement system**. The term *complement* refers to the fact that this system complements the action of antibodies. The complement proteins interact with one another in chain reactions, or cascades, similar to the blood clotting system. There are three possible pathways for complement action: (a) **classical pathway**, (b) **lectin pathway**, and (c) **alternative pathway**.

## b. Lectin Pathway

The lectin pathway is activated by the protein mannose-binding lectin (MBL). MBL binds to carbohydrates on bacterial surfaces.

### Lectin binds to bacterial cell wall

MBL binds to carbohydrates on the bacterial surface.

### C3 Activation

Bound MBL forms a complex that splits C3 into C3a and C3b. C3a activates an inflammatory response.

### C3b Attachment (lectin pathway)

C3b binds to bacterial surface and enhances phagocytosis.

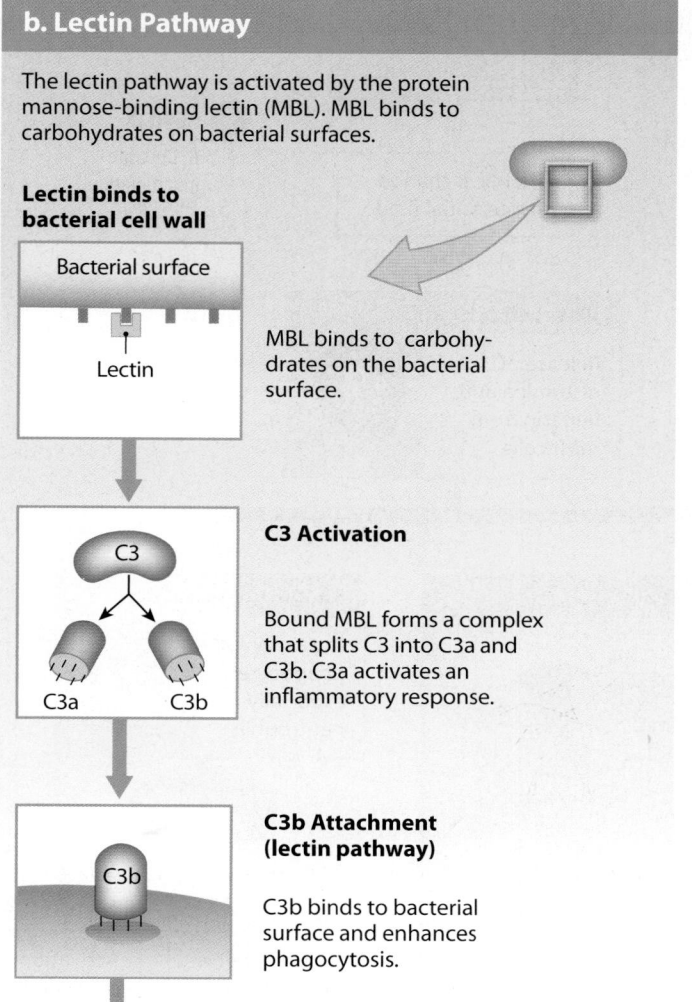

## c. Alternative Pathway

The alternative pathway is important in the defense against bacteria, some parasites, and virus-infected cells.

### Complement proteins interact in plasma

The alternative pathway begins when several complement proteins, notably properdin, interact in the plasma, causing C3 to split into C3a and C3b. This interaction can be triggered by exposure to foreign materials, such as the capsule of a bacterium. C3a activates an inflammatory response.

### C3b Attachment (alternative pathway)

C3b binds to bacterial surface and enhances phagocytosis.

## Enhanced Phagocytosis (Opsonization)

The attached C3b may also act to enhance phagocytosis. This enhancement of phagocytosis, a process called **opsonization**, occurs because macrophage membranes contain receptors that detect and bind to complement proteins and bound antibodies.

## Inflammation (Histamine Release)

Release of histamine by mast cells and basophils increases the degree of local inflammation, attracts and activates phagocytes, and accelerates blood flow to the region.

**? REVIEW**

**B.** Briefly explain the role of complement proteins.

**C.** What is the effect of histamine release?

**Lo LEARNING OUTCOME**

Describe the types of interferons, and explain the pathways of complement activation.

20

# Module 20.13

# Inflammation is a localized tissue response to injury; fever is a generalized response to tissue damage and infection

## Inflammation

**1** **Inflammation**, or the **inflammatory response**, is a localized tissue response to injury. Inflammation produces local redness, swelling, heat, and pain. These are called the *cardinal signs and symptoms of inflammation*. Sometimes a fifth is included: lost function. Many stimuli, including impact, abrasion, distortion, chemical irritation, infection by pathogens, and extreme temperatures (hot or cold), can produce inflammation. Each of these stimuli kills cells, damages connective tissue fibers, or injures the tissue in some other way. The changes alter the chemical composition of the interstitial fluid. Damaged cells release prostaglandins, proteins, and potassium ions, and the injury itself may have introduced foreign proteins or pathogens. The changes in the interstitial environment trigger the complex process of inflammation.

**Tissue Damage**

Chemical change in interstitial fluid

**Mast Cell Activation**

Release of histamine and heparin from mast cells

**? REVIEW**
**A.** Describe inflammation.

**Redness, Swelling, Heat, and Pain**

Dilation of blood vessels, increased blood flow, increased vessel permeability

Clot formation (temporary repair)

**Phagocyte Attraction**

Attraction of phagocytes, especially neutrophils

Release of cytokines

Removal of debris by neutrophils and macrophages; stimulation of fibroblasts

Activation of specific defenses

## Fever

**2** **Fever** is a body temperature greater than 37.2°C (99°F). Circulating fever-inducing proteins called **pyrogens** (PĪ-rō-jenz; *pyro-*, fever or heat + *-gen*, substance) can reset the temperature thermostat in the hypothalamus and raise body temperature. Within limits, a fever can be beneficial. High body temperatures may inhibit some viruses and bacteria, but the most likely beneficial effect is on body metabolism. For each 1°C rise in body temperature, metabolic rate increases by 10 percent. As a result, tissue defenses can be mobilized more rapidly, and the repair process speeds up.

**? REVIEW**
**B.** What effect do pyrogens have in the body?

**Tissue Repair**

Pathogen removal, clot erosion, scar tissue formation

# Summary of Innate Immunity (Nonspecific Defenses)

**3** This table summarizes the information on innate (nonspecific) immunity presented in this section.

## Physical Barriers

Keep hazardous organisms and materials outside the body

Secretions

Epithelium

Duct of eccrine sweat gland

Hair

## Phagocytes

Remove pathogens and debris

Neutrophil    Eosinophil    Monocyte    Free macrophage    Fixed macrophage

## Immune Surveillance

NK cells destroy abnormal cells in peripheral tissues

Natural killer cell

Lysed abnormal cell

## Interferons

Chemical messengers that coordinate the defenses against viral infections

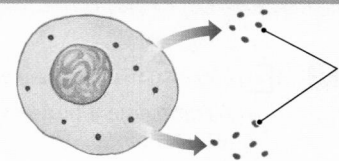

Interferons released by activated lymphocytes, macrophages, or virus-infected cells

## Complement System

Attacks and breaks down the surfaces of cells, bacteria, and viruses; attracts phagocytes; stimulates inflammation

Complement

Lysed pathogen

## Inflammation

A localized, tissue-level response that tends to limit the spread of an injury or infection

Mast cell

1. Increases blood flow
2. Activates macrophages
3. Increases capillary permeability
4. Activates complement
5. Stimulates regional clotting reaction
6. Increases regional temperature
7. Activates adaptive defenses

 **REVIEW**

**C.** A rise in the level of interferons in the body suggests what kind of infection?

## Fever

Mobilizes defenses; accelerates repairs; inhibits pathogens

Body temperature rises above 37.2°C in response to pyrogens

 **LEARNING OUTCOME**

Explain the significance of inflammation and fever as innate defense mechanisms.

20

# Innate Immunity

**Labeling:** Identify the type of innate immunity (nonspecific defense) described below.

**1** keep hazardous organisms and materials outside the body. For example, a mosquito that lands on your head may be unable to reach the surface of the scalp if you have a full head of hair.

**2** are cells that engulf pathogens and cell debris. Examples are the macrophages of peripheral tissues and the eosinophils and neutrophils of blood.

**3** is the destruction of abnormal cells by NK cells in peripheral tissues.

Destruction of abnormal cells

**4** are chemical messengers that coordinate the defenses against viral infections.

**5** is a system of circulating proteins that assists antibodies in the destruction of pathogens.

**6** is a localized, tissue-level response that tends to limit the spread of an injury or infection.

**7** is an elevation of body temperature that accelerates tissue metabolism and mobilizes defenses.

**1** _____

**2** _____

**3** _____

**4** _____

**5** _____

**6** _____

**7** _____

**Multiple choice:** Select the correct answer from the list provided.

**8** A physical barrier such as the skin provides a nonspecific body defense due to its makeup, which includes
- ☐ a) multiple layers.
- ☐ b) a coating of keratinized cells.
- ☐ c) a network of desmosomes locking adjacent cells together.
- ☐ d) all of these.

**9** NK cells sensitive to the presence of abnormal plasma membranes are primarily involved in
- ☐ a) defenses against specific threats.
- ☐ b) phagocytic activity for defense.
- ☐ c) complex, time-consuming defense mechanisms.
- ☐ d) immune surveillance.

**10** The nonspecific defense that breaks down cells, attracts phagocytes, and stimulates inflammation is
- ☐ a) the inflammatory response.
- ☐ b) the action of interferons.
- ☐ c) the complement system.
- ☐ d) immune surveillance.

**11** The protein(s) that interfere with the replication of viruses is (are)
- ☐ a) complement proteins.
- ☐ b) heparin.
- ☐ c) pyrogens.
- ☐ d) interferons.

**12** Circulating proteins that reset the thermostat in the hypothalamus, causing a rise in body temperature, are called
- ☐ a) pyrogens.
- ☐ b) interferons.
- ☐ c) lysosomes.
- ☐ d) complement proteins.

**13** Provide defense against pathogens that overcome the body's nonspecific physical barriers:
- ☐ a) phagocytes
- ☐ b) mucus
- ☐ c) hair
- ☐ d) interferon

**Short answer:** Can a fever be beneficial? Explain your answer.

**14** _____

**20**

# Module 20.14
# Adaptive immunity provides the body's specific defenses

**1** **Adaptive (specific) immunity** is [coordinated and produced by T cells and B cells.] As the flowchart below shows, specific immunity may be acquired in different ways. To prevent disease, the most common way is to induce immunity artificially by vaccination (immunization). A **vaccine** is a preparation designed to induce an immune response. It contains either a dead or an inactive pathogen, or antigens derived from that pathogen.

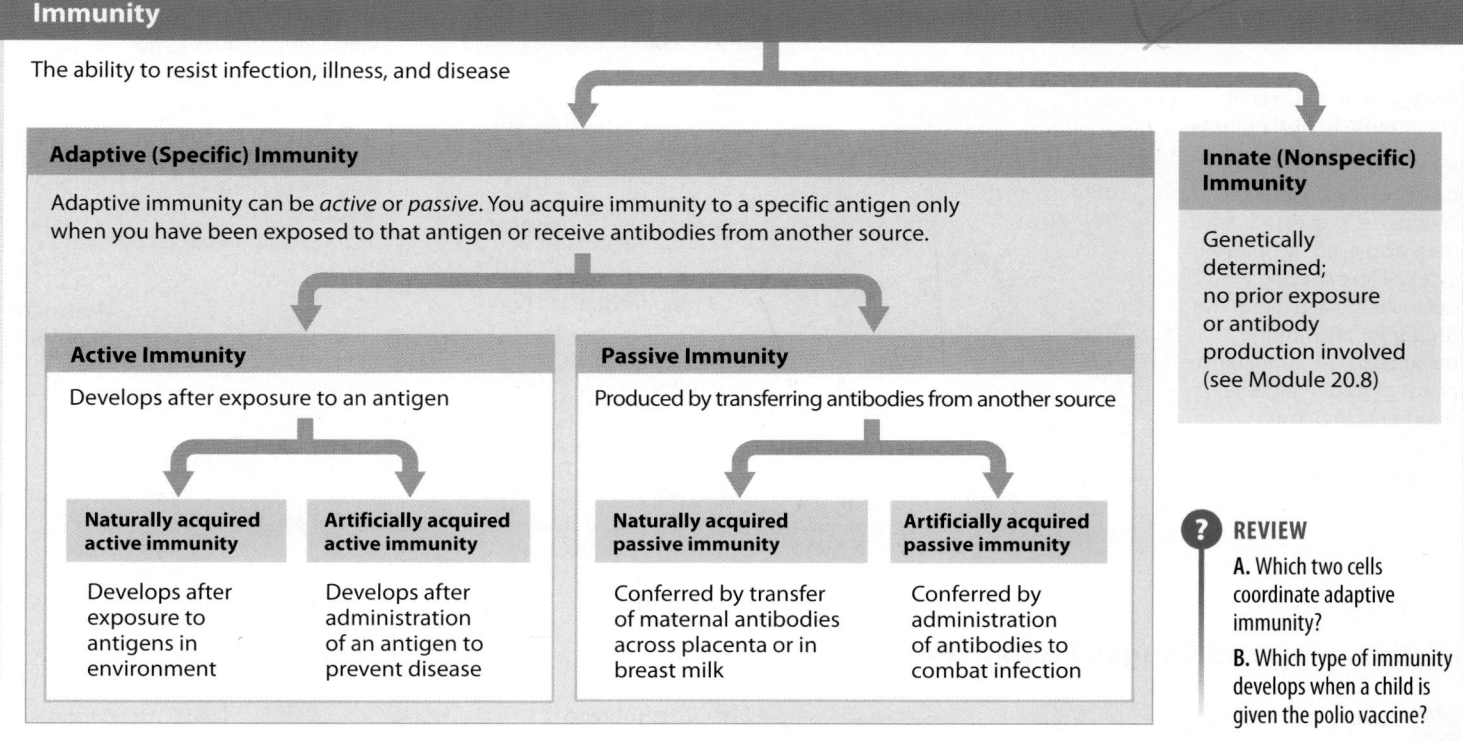

**Immunity**

The ability to resist infection, illness, and disease

**Adaptive (Specific) Immunity**

Adaptive immunity can be *active* or *passive*. You acquire immunity to a specific antigen only when you have been exposed to that antigen or receive antibodies from another source.

**Active Immunity**

Develops after exposure to an antigen

**Naturally acquired active immunity**

Develops after exposure to antigens in environment

**Artificially acquired active immunity**

Develops after administration of an antigen to prevent disease

**Passive Immunity**

Produced by transferring antibodies from another source

**Naturally acquired passive immunity**

Conferred by transfer of maternal antibodies across placenta or in breast milk

**Artificially acquired passive immunity**

Conferred by administration of antibodies to combat infection

**Innate (Nonspecific) Immunity**

Genetically determined; no prior exposure or antibody production involved (see Module 20.8)

**? REVIEW**

**A.** Which two cells coordinate adaptive immunity?

**B.** Which type of immunity develops when a child is given the polio vaccine?

**2** Regardless of the form, adaptive immunity exhibits four general properties, which are summarized in the following table.

**20**

**Properties of Adaptive Immunity**

- **Specificity** results from the activation of specific lymphocytes and the production of antibodies with targeted effects. Each T cell or B cell has receptors that bind to one specific antigen but ignore all others. The response of an activated T cell or B cell is equally specific and leaves other antigens unaffected.
- **Versatility** results from the large diversity of lymphocytes present in the body. There are millions of different lymphocyte populations, each sensitive to a different antigen. When activated by a specific antigen, a lymphocyte divides, producing more lymphocytes with the same specificity. All the cells produced by the division of an activated lymphocyte constitute a **clone**.

- **Memory** exists because cell divisions of activated lymphocytes produce two groups of cells: one group that attacks the invader immediately, and another that remains inactive unless it is exposed to the same antigen at a later date. These inactive **memory cells** enable your immune system to "remember" antigens it has previously encountered and to launch a faster, stronger, and longer-lasting counterattack if such an antigen reappears.
- **Tolerance** exists because the immune response ignores normal ("self") tissues but targets abnormal and foreign cells ("nonself") as well as toxins. Tolerance can also develop over time in response to chronic exposure to an antigen in the environment. Such tolerance generally lasts only as long as the exposure continues.

**Lo LEARNING OUTCOME**

Describe the types of adaptive immunity and four properties of adaptive immunity.

# Module 20.15

# Adaptive immunity is triggered by exposure of T cells and B cells to specific antigens

**1** This figure provides a "big picture" overview of the immune response. The presentation of specific antigens may stimulate (1) cell-mediated immunity involving attacks by T cells and (2) antibody-mediated immunity involving specific antibodies secreted by activated B cell descendants. We examine each of these processes more closely in later modules. This module focuses on how antigens trigger the body's specific defenses.

**Antigens or Antigenic Fragments in Body Fluids**

Most antigens must either be displayed on infected cells or be "processed" and then displayed by phagocytes before specific defenses are activated. The trigger for an immune response is the appearance of antigens or antigenic fragments in plasma membranes; this is called **antigen presentation**.

**Adaptive Immunity**

Antigen presentation triggers an immune response.

Antigens

**Cell-Mediated Immunity**

Phagocytes activated → T cells activated

Communication and feedback

**Antibody-Mediated Immunity**

Activated B cells give rise to cells that produce antibodies.

**Direct Physical and Chemical Attack**

Activated T cells find the pathogens and attack them through phagocytosis or the release of chemical toxins.

**Attack by Circulating Antibodies**

**Destruction of antigens**

## MHC Proteins and Antigen Presentation

**2** Antigen presentation occurs when an antigen-glycoprotein combination capable of activating T cells appears in a plasma membrane. The structure of these glycoproteins is genetically determined. The genes controlling their synthesis are located along one portion of chromosome 6, in a region called the **major histocompatibility complex (MHC)**. These membrane glycoproteins are called **MHC proteins. Class I MHC proteins** are in the plasma membranes of all nucleated cells.

**1** Viral or bacterial infection of a body cell triggers antigen presentation by class I MHC proteins.

Plasma membrane

Viral or bacterial pathogen

**2** The infection results in the appearance of abnormal peptides in the cytoplasm.

Infected cell

Transport vesicle

**3** The endoplasmic reticulum produces class I MHC proteins to which the abnormal peptides are attached.

Endoplasmic reticulum

Nucleus

**5** The abnormal peptides are displayed by class I MHC proteins on the plasma membrane.

**4** After export to the Golgi apparatus, the MHC proteins with their bound abnormal peptides are transported to the plasma membrane.

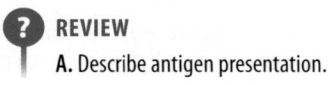

**REVIEW**

**A.** Describe antigen presentation.

**3** **Class II MHC proteins** are present only in the membranes of antigen-presenting cells and lymphocytes. **Antigen-presenting cells (APCs)** are specialized cells that include all the phagocytic cells of the monocyte–macrophage group and the dendritic cells of the skin and lymphoid organs. Class II MHC proteins appear in the plasma membrane only when the cell is processing antigens. The mechanism of antigen presentation is shown here for a phagocytic cell.

**1** Phagocytic APCs engulf the extracellular pathogens.

**2** Lysosomal action produces antigenic fragments.

Lysosome

Phagocytic cell

Plasma membrane

Nucleus

**5** Antigenic fragments are displayed by class II MHC proteins on the plasma membrane.

**4** Antigenic fragments are bound to class II MHC proteins.

**3** The endoplasmic reticulum produces class II MHC proteins.

Endoplasmic reticulum

**4** In this preparation, cells were exposed to fluorescent-tagged antibodies that bind to specific structures and glow when the cells are illuminated with fluorescent light. The green cell is an APC and the red cell is a lymphocyte; the bright yellow areas indicate where labeled antigens are being displayed on MHC proteins.

Lymphocyte

APC

**? REVIEW**

**B.** What is the major histocompatibility complex (MHC)?

**C.** Where are class I MHC proteins and class II MHC proteins found?

**20**

**? INTEGRATION**

**D.** What types of immunity are stimulated by antigen presentation?

**Lo LEARNING OUTCOME**

Explain how antigens trigger an immune response.

# Module 20.16

# Infected cells stimulate the formation and division of cytotoxic T cells, memory T$_C$ cells, and regulatory T cells

**1** Inactive T cells have receptors that recognize either class I or class II MHC proteins on other cells when those proteins are bound to specific antigens. If the MHC protein contains the antigen that the T cell is programmed to detect, binding occurs. This process is called **antigen recognition**.

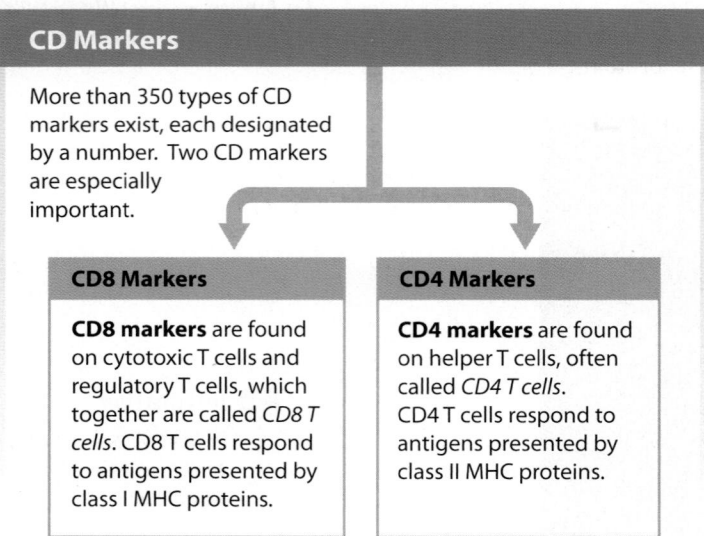

Inactive T cell

Antigen recognition protein

T cell receptor
Antigen
MHC protein

Infected body cell (including APCs)

**2** The membrane proteins involved in antigen recognition are members of a class of proteins called **CD** (cluster of differentiation) **markers**.

### CD Markers

More than 350 types of CD markers exist, each designated by a number. Two CD markers are especially important.

| CD8 Markers | CD4 Markers |
| --- | --- |
| **CD8 markers** are found on cytotoxic T cells and regulatory T cells, which together are called *CD8 T cells*. CD8 T cells respond to antigens presented by class I MHC proteins. | **CD4 markers** are found on helper T cells, often called *CD4 T cells*. CD4 T cells respond to antigens presented by class II MHC proteins. |

**? REVIEW**

**A.** Describe CD markers.

**3** Antigen recognition prepares a T cell for activation. A CD8 T cell can recognize antigens bound to class I MHC proteins. Full activation requires exposure of the cell to specific physical or chemical stimuli. This results in activation and cell division.

### Antigen Recognition

Antigen recognition occurs when a CD8 T cell encounters an appropriate antigen on the surface of another cell, bound to a class I MHC protein.

Infected cell

Viral or bacterial antigen

Inactive CD8 T cell

### Activation and Cell Division

Antigen recognition results in T cell activation and cell division, producing populations of three different types of CD8 T cells sensitive to the same antigen.

Active T$_C$ cells

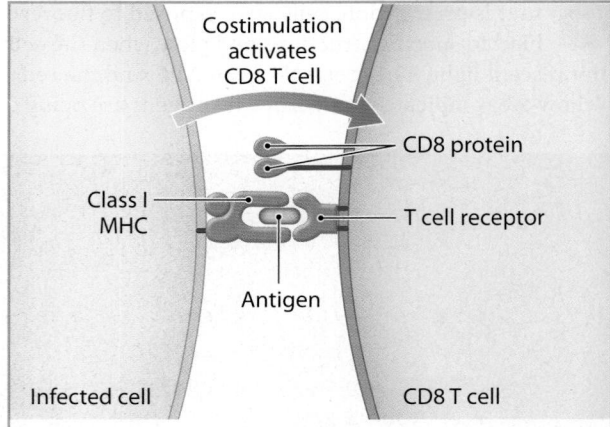

Costimulation activates CD8 T cell

CD8 protein

Class I MHC

T cell receptor

Antigen

Infected cell

CD8 T cell

### Costimulation

In order to proceed from recognition to activation, a T cell must also bind to the stimulating cell at a second site. This vital secondary binding process, called **costimulation**, essentially confirms the initial activation signal. Costimulation is like the safety switch on a gun: It helps prevent T cells from mistakenly attacking normal (self) tissues.

**4** Following activation, the cell divides repeatedly, and daughter cells differentiate into three types of CD8 T cells.

### Cytotoxic T Cells Seek Out Antigen-Bearing Cells

**Cytotoxic T (T$_c$) cells** seek out and destroy abnormal and infected cells. Cytotoxic T cells are highly mobile cells that roam throughout injured tissues. When a T$_c$ cell encounters its target antigens bound to class I MHC proteins, it attacks the target cell.

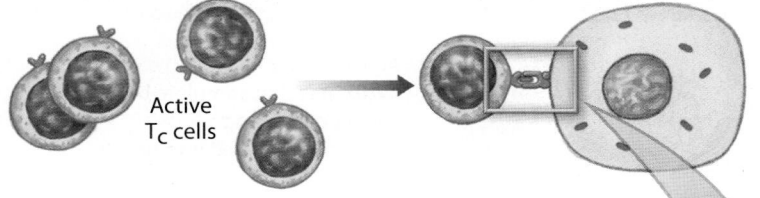

Active
T$_c$ cells

### Destruction of Target Cells

The T$_c$ cell destroys the antigen-bearing cell. As indicated below, it may use several different mechanisms to kill the target cell.

### Memory T$_c$ Cells Are Produced

**Memory T$_c$ cells** are produced by the same cell divisions that produce cytotoxic T cells. Thousands of these cells are produced the first time the antigen triggers an immune response. However, they remain inactive and do not differentiate further.

Memory T$_c$ cells
(inactive)

### Regulatory T Cells Provide a Delayed Suppression

**Regulatory T (T$_{reg}$) cells** moderate the responses of other T cells and of B cells by secreting inhibitory cytokines called *suppression factors*. Suppression does not occur immediately, because activation takes much longer for regulatory T cells than for other types of T cells. Regulatory T cells act *after* the initial immune response.

Regulatory
T cells

Perforin release — Destroys integrity of target cell membrane through the release of perforins

Cytokine release — Activates genes within the target cell nucleus that result in the self-destruction of the cell through a process called **apoptosis** (ap-op-TŌ-sis)

Lymphotoxin release — Disrupts cell metabolism through the release of **lymphotoxin** (lim-fō-TOK-sin)

The first time CD8 T cells encounter a specific antigen, the entire sequence of events—from the appearance of the antigen in a tissue to widespread cell destruction by cytotoxic T cells—takes 2 days or more. During this time, the damage or infection may spread, making it more difficult to control. However, if the same antigen appears a second time, memory T$_c$ cells will immediately differentiate into cytotoxic T cells, producing a prompt, effective cellular response that can overwhelm an invading organism before it becomes well established in the tissues.

**? REVIEW**

**B.** Identify the three major types of T cells activated by class I MHC proteins.

**C.** How do abnormal antigens attached to class I MHC proteins initiate an immune response?

**LEARNING OUTCOME**

Explain the events of antigen recognition and the roles of CD markers in T cell differentiation.

20

# Antigen-presenting cells can stimulate activation of CD4 T cells, producing helper T cells that promote B cell activation and antibody production

**1** Before they can initiate antibody-mediated immunity, inactive CD4 T cells must be exposed to antigens that are bound to class II MHC proteins. Costimulation then completes their activation. Upon activation, CD4 T cells undergo a series of divisions, and daughter cells differentiate into active **helper T cells (T$_H$ cells)** and **memory T$_H$ cells**.

**? REVIEW**
**A.** Explain the function of cytokines secreted by helper T cells.

**Antigen Recognition by CD4 T Cell**

Antigen-presenting cell (APC)

Foreign antigen

Inactive CD4 (T$_H$) cell

APC

Class II MHC

Costimulation

CD4 protein

Antigen

T cell receptor

T$_H$ cell

**CD4 T Cell Activation and Cell Division**

Cytokines

Active helper T cells secrete **cytokines** that stimulate both cell-mediated and antibody-mediated immunity.

Active helper T cells

The Golgi apparatus is packaging membrane receptors (red) to the APC-presented antigen that will be incorporated into the surface of CD4 T cells. These receptors are essential to the costimulation of B cells.

Memory T$_H$ cells (inactive)

**An activated helper T cell** — Fluorescent LM × 400

**2** When a B cell encounters its specific antigen, it prepares for activation. This preparatory process is called **sensitization**. During sensitization, antigens brought into the cell by endocytosis subsequently appear on the surface of the B cell, bound to class II MHC proteins.

**? REVIEW**
**B.** Define *sensitization*.

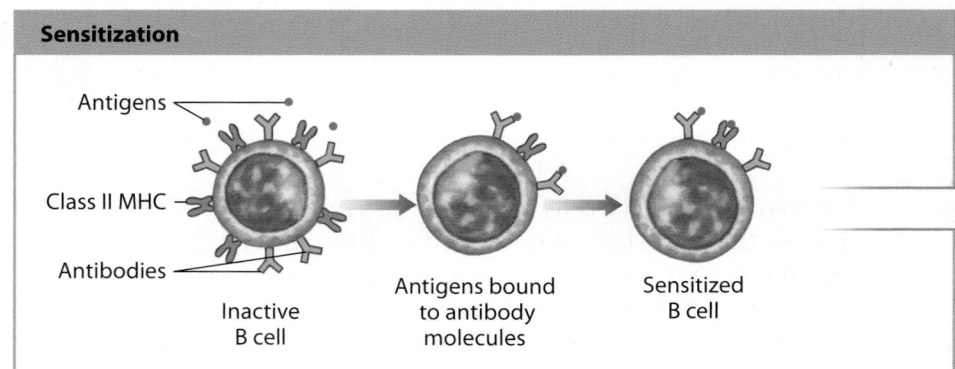

**Sensitization**

Antigens

Class II MHC

Antibodies

Inactive B cell

Antigens bound to antibody molecules

Sensitized B cell

**3**   To become fully activated, a sensitized B cell must encounter a helper T cell that was activated by exposure to the same antigen. When that happens, the helper T cell binds to the MHC complex of the sensitized B cell, recognizes the presence of the antigen, and begins secreting cytokines that promote B cell activation.

**4**   Under the continued stimulation of cytokines released by active helper T cells, an activated B cell undergoes a series of cell divisions. These divisions yield daughter cells with two different fates: memory B cells and plasma cells.

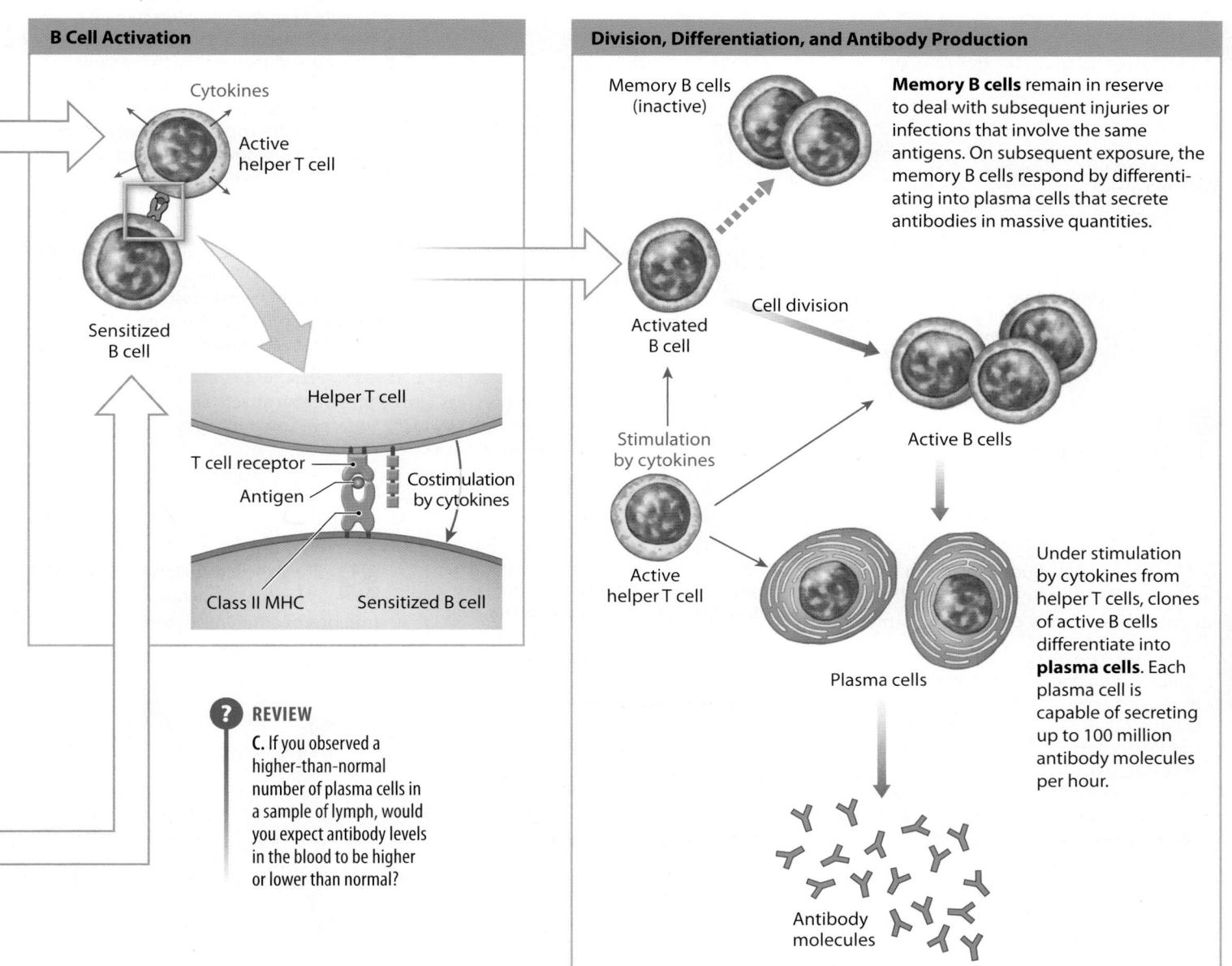

**B Cell Activation**

Cytokines

Active helper T cell

Sensitized B cell

Helper T cell

T cell receptor

Antigen

Costimulation by cytokines

Class II MHC    Sensitized B cell

**Division, Differentiation, and Antibody Production**

Memory B cells (inactive)

**Memory B cells** remain in reserve to deal with subsequent injuries or infections that involve the same antigens. On subsequent exposure, the memory B cells respond by differentiating into plasma cells that secrete antibodies in massive quantities.

Activated B cell

Cell division

Stimulation by cytokines

Active B cells

Active helper T cell

Plasma cells

Under stimulation by cytokines from helper T cells, clones of active B cells differentiate into **plasma cells**. Each plasma cell is capable of secreting up to 100 million antibody molecules per hour.

Antibody molecules

**?** **REVIEW**

**C.** If you observed a higher-than-normal number of plasma cells in a sample of lymph, would you expect antibody levels in the blood to be higher or lower than normal?

**Lo** **LEARNING OUTCOME**

Explain the sensitization and activation of B cells and the role of plasma cells.

20

# Module 20.18

# Antibodies are small soluble proteins that bind to specific antigens and whose abundance increases upon later antigen exposure

**1** An antibody molecule consists of two parallel pairs of polypeptide chains: one pair of **heavy chains** and one pair of **light chains**. Each chain contains both **constant segments** and **variable segments**. The constant segments of the heavy chains form the base of the antibody molecule.

The free tips of the two variable segments form the **antigen-binding sites** of the antibody molecule. These sites can interact with an antigen in the same way that the active site of an enzyme interacts with a substrate molecule. Small differences in the amino acid sequence of the variable segments affect the precise shape of the antigen-binding site. These differences account for differences in specificity among the antibodies produced by different B cells.

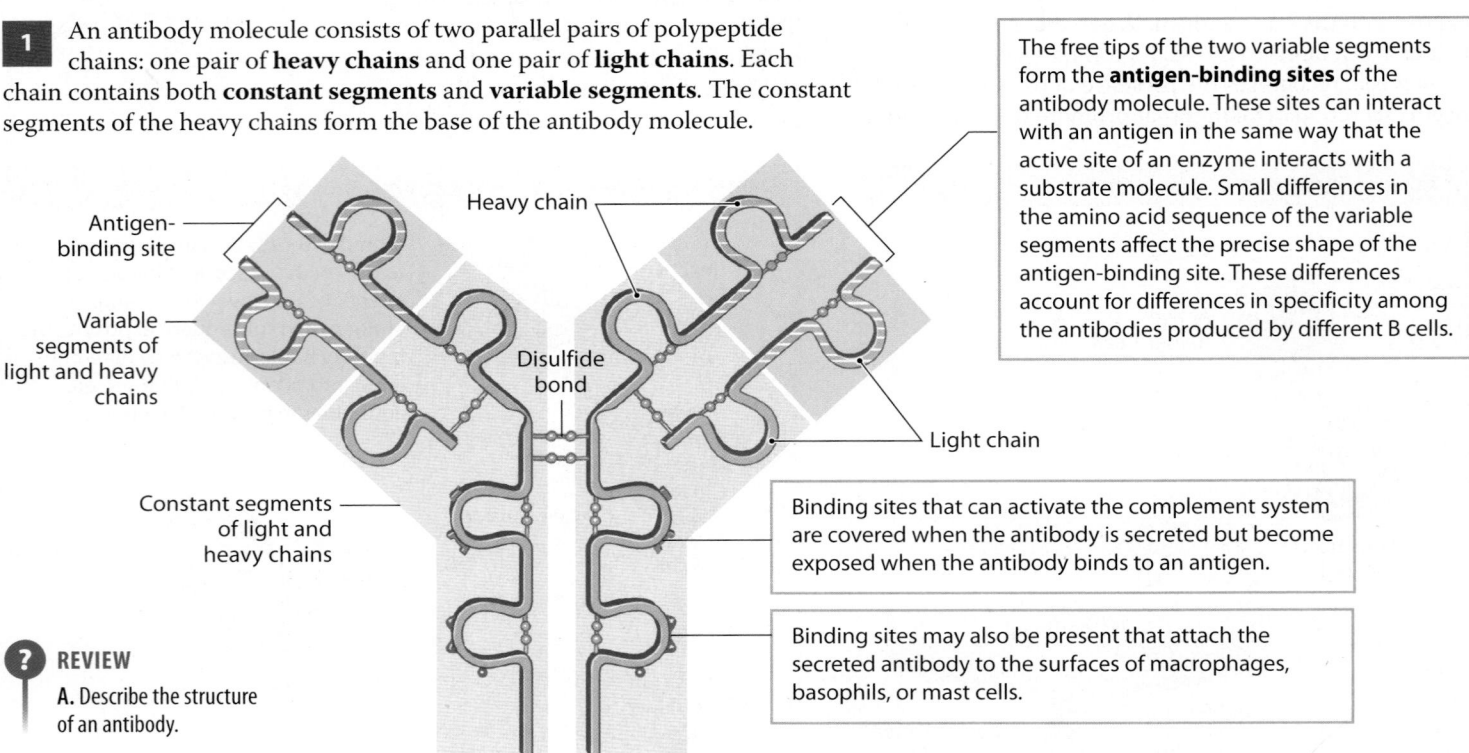

Binding sites that can activate the complement system are covered when the antibody is secreted but become exposed when the antibody binds to an antigen.

Binding sites may also be present that attach the secreted antibody to the surfaces of macrophages, basophils, or mast cells.

**? REVIEW**
**A.** Describe the structure of an antibody.

**2** When an antibody molecule binds to its corresponding antigen molecule, an **antigen-antibody complex** is formed.

Antibodies do not bind to the entire antigen, but to specific portions of its exposed surface—regions called **antigenic determinant sites**.

A **complete antigen** is an antigen with at least two antigenic determinant sites, one for each of the antigen-binding sites on an antibody molecule.

**Haptens**, or *partial antigens*, do not ordinarily cause B cell activation. However, they may become attached to carrier molecules, forming combinations that can function as complete antigens. The antibodies produced will attack both the hapten and the carrier molecule. If the carrier molecule is normally present in the tissues, the antibodies may begin attacking and destroying normal cells. This is the basis for several drug reactions, including allergies to penicillin.

**3** The exposed surface of something as large as a bacterium contains millions of antigenic determinant sites, and it may become carpeted with antibodies.

Antibody

Antigen-antibody complex

Carrier molecule

Hapten

Antibody

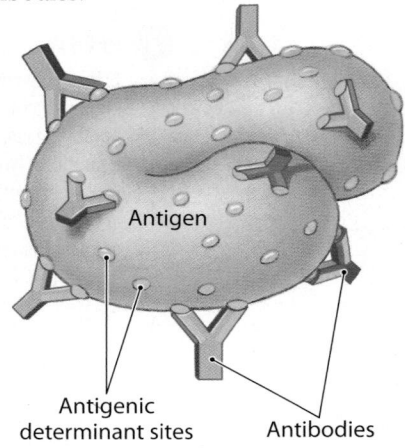

Antigen

Antigenic determinant sites

Antibodies

**? REVIEW**
**B.** Define *antigenic determinant site.*

**4** There are five classes of antibodies, or **immunoglobulins** (**Igs**). The classes are determined by differences in the structure of the heavy-chain constant segments and so have no effect on the antibody's specificity, which is determined by the antigen-binding sites.

### Classes of Antibodies

**IgG** antibodies account for 80 percent of all antibodies. IgG antibodies are responsible for resistance against many viruses, bacteria, and bacterial toxins.

**IgE** attaches as an individual molecule to the exposed surfaces of basophils and mast cells.

**IgD** is an individual molecule on the surfaces of B cells, where it can bind antigens in the extracellular fluid. This binding can play a role in the sensitization of the B cell involved.

**IgM** is the first class of antibody secreted after an antigen is encountered. IgM concentration declines as IgG production accelerates. The anti-A and anti-B antibodies responsible for the agglutination of incompatible blood types are IgM antibodies.

**IgA** is found primarily in glandular secretions such as mucus, tears, saliva, and semen. These antibodies attack pathogens before they gain access to internal tissues.

**5** The initial response to antigen exposure is called the **primary response**. Because the antigen must activate the appropriate B cells, which must then differentiate into antibody-secreting plasma cells, the primary response takes time to develop. During the primary response, the **antibody titer**, which is a laboratory test that measures the level of antibodies in a blood sample, does not peak until 1 or 2 weeks after the initial exposure. If the person is no longer exposed to the antigen, the antibody levels decrease.

**6** When an antigen is encountered a second time, it triggers a more extensive and prolonged **secondary response**. During the secondary response, antibody titers increase sooner and reach levels many times higher than they did in the primary response. This reflects the presence of large numbers of memory cells that are already primed for the arrival of the antigen. These memory B cells respond immediately—much faster than the B cells stimulated during the initial exposure. The secondary response appears even if the second exposure occurs years after the first, because memory cells may survive for 20 years or more.

**? INTEGRATION**

**C.** Which would be more affected by a lack of memory B cells and memory T cells: the primary response or the secondary response?

**Lo LEARNING OUTCOME**

Describe the structure of an antibody, discuss the types of antibodies, and explain the primary and secondary responses to antigen exposure.

# Antibodies use many different mechanisms to destroy target antigens

Go to
MasteringA&P™
PhysioEx 9.1
Ex 12.3: Indirect Enzyme-Linked Immunosorbent Assay (ELISA)

As a group, the antibodies of the various classes provide a versatile and effective defense against a variety of threats.

**1** The formation of an antigen-antibody complex may eliminate the antigen in seven different ways.

### Neutralization

Both viruses and bacterial toxins must bind to the plasma membranes of body cells before they can enter or injure those cells. Binding occurs at superficial sites on the bacteria or toxins. Antibodies may bind to those sites, making the virus or toxin incapable of attaching itself to a cell. This mechanism is known as **neutralization**.

### Prevention of Pathogen Adhesion

Antibodies dissolved in saliva, mucus, tears, and sweat coat epithelia, providing an additional layer of defense. A covering of antibodies makes it difficult for bacteria or viruses to adhere to and penetrate body surfaces.

### Activation of Complement

When bound to an antigen, portions of the antibody molecule change shape, exposing areas that bind complement proteins. The bound complement molecules then activate the complement system, which destroys the antigen.

### Opsonization

Phagocytes can bind more easily to antibodies and complement proteins on the surface of a pathogen than they can to an antibody-free and complement-free surface. As a result, a coating of antibodies and complement proteins increases the effectiveness of phagocytosis. This effect is called *opsonization* (see Module 20.12). Some bacteria have slick plasma membranes or capsules, and phagocytes must be able to hang onto their prey before they can engulf it.

### Precipitation and Agglutination

If antigens are close together, an antibody can bind to antigenic determinant sites on two different antigens. In this way, antibodies can tie large numbers of antigens together, creating an **immune complex**. When the antigen is a soluble molecule, this process may create complexes too large to remain in solution. The formation of these insoluble immune complexes is called **precipitation**. When the target antigen is on the surface of a cell or a virus, the formation of immune complexes is called **agglutination**. The clumping of erythrocytes that occurs when incompatible blood types are mixed is an agglutination reaction.

**Immune Complex**

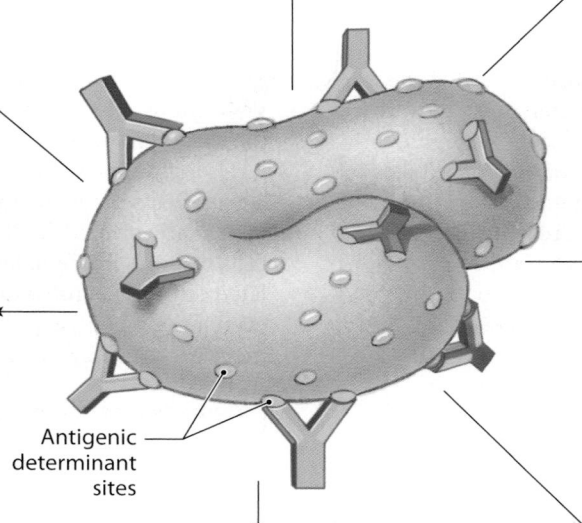

Antigenic determinant sites

### Stimulation of Inflammation

Antibodies may promote inflammation by stimulating the release of chemicals from basophils and mast cells.

### Attraction of Phagocytes

Antigens covered with antibodies attract eosinophils, neutrophils, and macrophages—cells that phagocytize pathogens and destroy foreign or abnormal plasma membranes.

 **REVIEW**

**A.** List the ways that antigen-antibody complexes can destroy target antigens.

**B.** Define *opsonization*.

**C.** Which cells are involved in inflammation?

 **LEARNING OUTCOME**

Explain the mechanisms by which antibodies destroy target antigens.

20

## Module 20.20

# Hypersensitivities are abnormal reactions to antigens

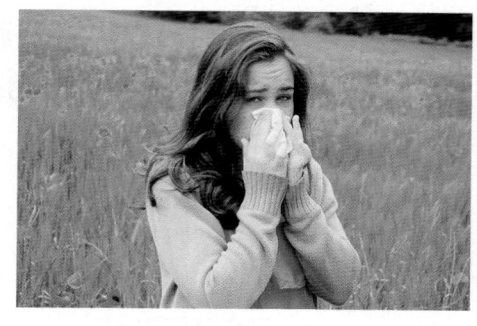

**Hypersensitivities**, commonly called *allergies*, are excessive immune responses to antigens. The sudden increase in cellular activity or antibody titers can have several unpleasant side effects. For example, neutrophils or cytotoxic T cells may destroy normal cells while attacking the antigen, or the antigen-antibody complex may trigger massive inflammation. Antigens that trigger allergic reactions are often called **allergens**.

**? REVIEW**

**A.** Describe hypersensitivities.

**1** Sensitization to an allergen during the initial exposure leads to the production of large quantities of IgE. The tendency to produce IgE antibodies in response to specific allergens may be genetically determined.

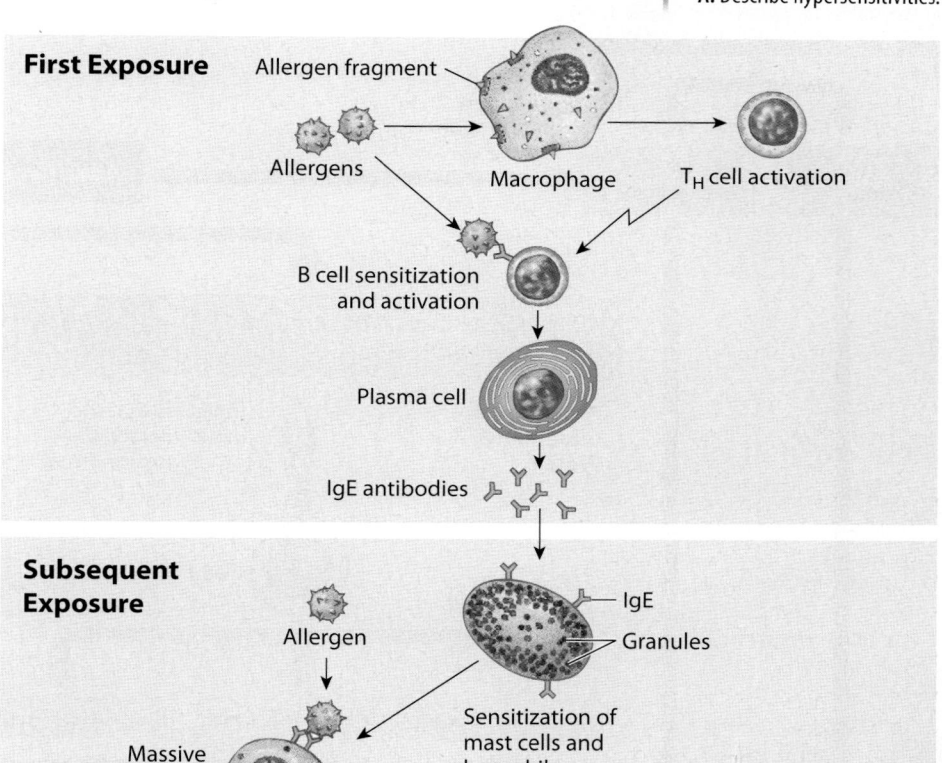

**First Exposure**

Allergen fragment

Allergens

Macrophage

T_H cell activation

B cell sensitization and activation

Plasma cell

IgE antibodies

**Subsequent Exposure**

Allergen

IgE

Granules

Sensitization of mast cells and basophils

Massive stimulation of mast cells and basophils

Release of histamines, leukotrienes, and other chemicals that cause pain and inflammation

**2** **Immediate hypersensitivity** (type I) is a rapid and especially severe response to the presence of an antigen. One form, **allergic rhinitis**, includes hay fever and other environmental allergies and may affect 15 percent of the U.S. population. Allergic rhinitis (inflammation of the nasal membranes) is one example of a hypersensitivity reaction that is restricted to the body surface. If the allergen enters the bloodstream, however, the response could be lethal. In **anaphylaxis** (an-a-fi-LAK-sis; *ana-*, again + *phylaxis*, protection), a circulating allergen affects mast cells throughout the body. In severe cases extensive peripheral vasodilation occurs, producing a fall in blood pressure that can lead to circulatory collapse. This response, called **anaphylactic shock**, is a life-threatening emergency.

**? REVIEW**

**C.** What is anaphylaxis?

**? REVIEW**

**B.** Which chemicals do mast cells and basophils release when stimulated in an allergic reaction?

| Localized Allergic Reactions | Systemic Allergic Reactions |
|---|---|
| If the allergen is at the body surface: localized inflammation, pain, and itching | If the allergen is in the bloodstream: itching, swelling, and difficulty breathing (due to airway constriction) |
| Example: urticaria (hives) | Example: anaphylaxis |

**Lo LEARNING OUTCOME**

Explain hypersensitivities, anaphylaxis, and the role of antibodies in each.

20

# Module 20.21

# Innate immunity and adaptive immunity work together to defeat pathogens

**1** This flowchart shows the relationships among the responses of innate and adaptive immunity to antigens.

**Antigens**

**? REVIEW**
**A.** Which T cells contain CD8 markers? CD4 markers?

Trigger

**Innate (Nonspecific) Immunity**

Complement system

NK cells macrophages

Antigen presentation by APCs

**Adaptive (Specific) Immunity**

**Activation by Class I MHC Proteins (Cell-Mediated)**

Antigen and Class I MHC Protein

Indicates that the cell is infected or otherwise abnormal

CD8 T cells

**Cytotoxic T Cells**
Attack and destroy infected and abnormal cells displaying antigen

**Memory T_C Cells**
Await reappearance of the antigen

**Regulatory T Cells**
Moderate immune response by T cells and B cells

**Activation by Class II MHC Proteins (Antibody-Mediated)**

Antigen and Class II MHC Protein

Indicates the presence of pathogens, toxins, or foreign proteins

CD4 T cells

**Helper T Cells**
Stimulate immune response by T cells and B cells

**Memory T_H Cells**
Await reappearance of the antigen

Inhibition

Inhibition

Direct physical and chemical attack

Direct physical and chemical attack

Activation of B cells

Production of memory B cells

Production of plasma cells

**? REVIEW**
**B.** Which cells produce antibodies?

**Destruction of Antigens**

Attack by circulating proteins

Secretion of antibodies

20

**2** This illustration gives an overview of the course of events responsible for overcoming a bacterial infection. The most effective defenses against bacteria involve phagocytosis and antigen presentation by APCs.

**3** Defenses against viruses involve direct contact with virus-infected cells by cytotoxic T cells and NK cells and antigen presentation by APCs.

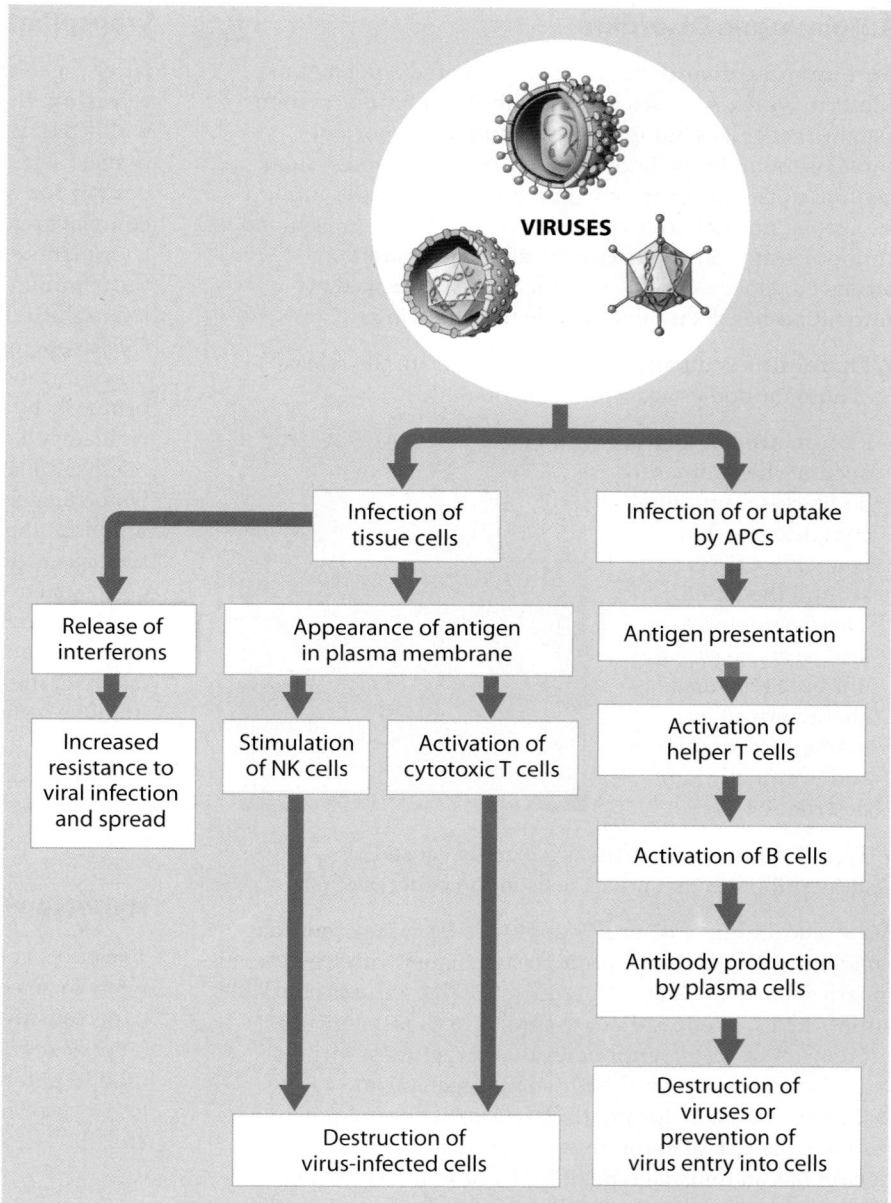

**? REVIEW**

**C.** Which cells can be activated by direct contact with virus-infected cells?

**LO LEARNING OUTCOME**

Summarize the integration of innate and adaptive immunity.

# Module 20.22
## Immune disorders involving either overactivity ...

### Autoimmune Disorders

**Autoimmune disorders** affect an estimated 5 percent of adults in North America and Europe. The immune system usually recognizes but ignores self-antigens—normal antigens found in the body. When the recognition system malfunctions, however, activated B cells may make antibodies against normal body cells and tissues. These "misguided" antibodies are called **autoantibodies**. The condition produced depends on the specific antigen attacked by autoantibodies. Examples include the following:

- **Thyroiditis** is inflammation resulting from the release of autoantibodies against thyroglobulin.

- **Rheumatoid arthritis** occurs when autoantibodies form immune complexes within connective tissues around the joints. These complexes lead to an excessive immune response characterized by marked inflammation and eventually joint destruction.

- **Type 1 diabetes mellitus** is generally caused by autoantibodies that attack cells in the pancreatic islets.

Many autoimmune disorders appear to be cases of mistaken identity. For example, proteins associated with the measles, Epstein–Barr, influenza, and other viruses contain amino acid sequences that are similar to those of myelin proteins. As a result, antibodies that target these viruses may also attack myelin sheaths. This mechanism is likely responsible for multiple sclerosis. For unknown reasons, the risk of autoimmune disorders increases if a person has an unusual type of MHC protein. At least 50 clinical conditions have been linked to specific variations in MHC structure.

**REVIEW**
**A.** Define *autoimmune disorders*.

### Transplant Rejection

After organ transplant surgery, the major problem is **transplant rejection**. In transplant rejection, T cells are activated by contact with MHC proteins on plasma membranes in the donated tissues. The cytotoxic T cells that develop then attack and destroy the foreign cells. **Immunosuppression**—partial or complete reduction of the immune response—can make transplants significantly more successful. An understanding of the communication among T cells, macrophages, and B cells has led to the development of drugs with more selective effects. **Cyclosporine** is an immunosuppressive medication used to prevent transplant rejection. It suppresses the immune response primarily by inhibiting helper T cell activity while leaving regulatory T cells relatively unaffected. Unfortunately, immunosuppressive agents can also destroy stem cells and lymphocytes and lead to a complete failure of the immune response.

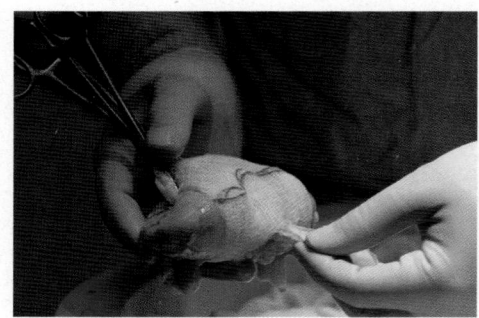

### Hypersensitivities

The effects of the many forms of hypersensitivities (allergies) range from mild to potentially lethal (see Module 20.20).

20

# … or underactivity can be harmful

## Inadequate Immune Response

### Immunodeficiency Diseases

**Immunodeficiency diseases** result from (1) problems with the embryonic development of lymphoid organs and tissues; (2) an infection with a virus that depresses immune function; or (3) treatment with, or exposure to, immunosuppressive agents, such as radiation or drugs. **Acquired immunodeficiency syndrome (AIDS)**, or *late-stage HIV disease*, is caused by the **human immunodeficiency virus** (**HIV**). This virus is a *retrovirus*: It carries its genetic information in RNA rather than in DNA. Specifically, the virus binds to CD4 T cells. HIV also infects several types of antigen-presenting cells, including those of the monocyte–macrophage line. It is the infection of helper T cells that leads to clinical problems.

The gradual destruction of helper T cells impairs the immune response because these cells play a central role in coordinating cell-mediated and antibody-mediated responses to antigens. To make matters worse, regulatory T cells are relatively unaffected by the virus. Over time, the excess of suppressing factors "turns off" the normal immune response. Circulating antibody levels decline, and cell-mediated immunity is reduced. The body is left with impaired defenses against a wide variety of microbial invaders. Microorganisms that would ordinarily be harmless can now initiate lethal *opportunistic infections*. The risk of cancer also increases because immune surveillance is depressed.

**HIV (green) budding from an infected T$_H$ cell**     SEM × 40,000

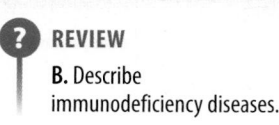 **REVIEW**

**B.** Describe immunodeficiency diseases.

### Age-Related Reductions in Immune Activity

As we age, the immune system becomes less effective at combating disease. T cells become less responsive to antigens, so fewer cytotoxic T cells respond to an infection. This effect may, at least in part, be associated with the gradual involution of the thymus and a reduction in circulating levels of thymic hormones. Because the number of helper T cells is also reduced, B cells are less responsive, so antibody levels do not rise as quickly after antigen exposure. The net result is an increased susceptibility to viral and bacterial infections. For this reason, vaccinations for infectious diseases such as the flu (influenza) and pneumococcal pneumonia, are strongly recommended for elderly people. The increased incidence of cancer in elderly people reflects the fact that immune surveillance declines, so tumor cells are not eliminated as effectively.

 **REVIEW**

**C.** Provide a plausible explanation for the increased incidence of cancer in the elderly.

**LEARNING OUTCOME**

Explain autoimmune disorders, transplant rejection, and immunodeficiency diseases, and describe age-related changes in the immune response.

**20**

**Matching:** Match each lettered term with the most closely related description.

a. opsonization

b. helper T cells

c. antibody

d. class II MHC

e. costimulation

f. IgM

g. class I MHC

h. IgG

i. passive immunity

j. anaphylaxis

k. CD4 markers

l. acquired immunity

m. B lymphocytes

| 1 | Two parallel pairs of polypeptide chains |
| 2 | Found on helper T cells |
| 3 | Active and passive |
| 4 | Transfer of antibodies |
| 5 | Attacked by HIV |
| 6 | Enhances phagocytosis |
| 7 | MHC proteins present in the plasma membranes of all nucleated cells |
| 8 | Differentiate into memory and plasma cells |
| 9 | MHC proteins present in the plasma membranes of all APCs and lymphocytes |
| 10 | Antibodies used to determine blood type |
| 11 | Secondary binding process required for T cell activation |
| 12 | Accounts for 80 percent of all immunoglobulins |
| 13 | Circulating allergen stimulates mast cells throughout body |

1 _____

2 _____

3 _____

4 _____

5 _____

6 _____

7 _____

8 _____

9 _____

10 _____

11 _____

12 _____

13 _____

Match each lettered term with the most closely related description.

a. cytotoxic T cells

b. viruses

c. B cells

d. antibodies

e. helper T cells

f. macrophages

g. natural killer (NK) cells

h. regulatory T cells

i. memory T cells and B cells

**20**

*The race is on:*

**14** _____

try to replicate before the immune system can mount a response. Two have already taken over cells in the body.

**15** _____

quickly recognize the viruses as a foreign threat. They begin destroying viruses by engulfing them.

*Stimulated by the release of cytokines from macrophages, helper T cells, and interferons,*

**16** _____

join the attack on virus-infected cells. They also fight cancer cells.

**19** _____

are Y-shaped proteins designed specifically to recognize a particular viral or bacterial invader; they bind to it and neutralize it.

**18** _____

mature into plasma cells, which in turn produce Y-shaped proteins.

**17** _____

(the battle managers of the immune system) emit signals to other lymphocytes to join the attack.

**20** _____

wage chemical warfare on virus-infected cells by releasing lethal proteins.

*As the body begins to conquer the viruses,*

**21** _____

help moderate the immune response. Otherwise, the immune response might attack the body.

*As the viruses are being defeated, the body creates*

**22** _____

to circulate permanently in the bloodstream, ensuring that next time, that particular virus will be swiftly conquered.

# Study Outline

## SECTION 1 • Anatomy of the Lymphatic System

### Module 20.1

**The lymphatic system consists of lymphatic vessels, nodes, and lymphoid tissue** p. 751

1. The lymphatic system includes cells, tissues, and organs that provide **immunity** (the ability to defend the body against infection, illness, and disease) and return interstitial fluid to the bloodstream.

2. **Lymphocytes** are the cells of the lymphatic system. They are surrounded by **lymph**, interstitial fluid that has entered a lymphatic vessel.

3. **Lymphatics (lymphatic vessels)** are vessels that begin in the peripheral tissues and end at veins.

4. **Primary lymphoid tissues and organs** are where lymphocytes form and mature and include the red bone marrow and the thymus.

5. **Secondary lymphoid tissues and organs** are where lymphocytes are activated and cloned and include the lymph nodes, tonsils, mucosa-associated lymphoid tissue (MALT), appendix, and spleen.

### Module 20.2

**Interstitial fluid flows continuously into lymphatic capillaries and exits tissues as lymph in lymphatic vessels** p. 752

6. **Lymphatic vessels** begin as **lymphatic capillaries** and carry lymph from the peripheral tissues to the venous system.

7. Lymphatic capillaries differ from blood capillaries. They originate at pockets rather than from continuous tubes, are larger in diameter, have thinner walls, and have a flattened or irregular outline in sectional view.

*Valve in a lymphatic vessel*

8. Small to medium-sized lymphatic vessels have valves.

9. Lacteals are prominent lymphatic capillaries in the small intestine.

### Module 20.3

**Small lymphatic vessels converge to form lymphatic ducts that empty into the subclavian veins** p. 754

10. The lymph from lymphatic capillaries flows into either **superficial lymphatics** or **deep lymphatics** that converge to form **lymphatic trunks**. The lymphatic trunks empty into either the **thoracic duct** or **right lymphatic duct**.

11. The **right jugular trunk**, **right subclavian trunk**, and **right bronchomediastinal trunk** merge to form the right lymphatic duct. It then empties into the right subclavian vein.

12. The **left bronchomediastinal trunk**, the **left subclavian trunk**, and the **left jugular trunk** merge to form the thoracic duct. It then empties into the left subclavian vein.

13. Blockage of lymphatic drainage can cause **lymphedema**.

### Module 20.4

**Lymphocytes are responsible for the immune functions of the lymphatic system** p. 756

14. Lymphocytes make up 20–40 percent of the circulating leukocyte population, yet circulating lymphocytes are only a fraction of the total lymphocyte population.

15. Three classes of lymphocytes circulate in the blood (**T cells**, **B cells**, **NK cells**), and they are all sensitive to **antigens** that can stimulate the immune system.

16. **Lymphopoiesis** (lymphocyte production) involves red bone marrow, thymus, and peripheral lymphoid tissues.

### Module 20.5

**Lymphocytes aggregate within lymphoid tissues and lymphoid organs** p. 758

17. **Lymphoid tissues** are connective tissues dominated by lymphocytes.

18. In a **lymphoid nodule**, the lymphocytes are densely packed in areolar tissues. In contrast, **lymphoid organs** (lymph nodes, thymus, and spleen) are surrounded by a fibrous connective tissue capsule.

19. **Aggregated lymphoid nodules**, or *Peyer's patches,* are found in the small intestine.

20. **Tonsils** are large lymphoid nodules. A single **pharyngeal tonsil** (*adenoid*) is located in the posterior wall of the nasopharynx. Paired **palatine** and **lingual tonsils** are located in the oral cavity.

21. **Lymph nodes**, or **lymph glands**, are small lymphoid organs shaped like a kidney bean. They detect pathogens before they reach vital organs.

22. **Afferent lymphatics** bring lymph to the lymph node on the side opposite the **hilum**. **Efferent lymphatics** exit at the hilum, along with blood vessels and nerves.

### Module 20.6

**The thymus is a lymphoid organ that produces functional T cells** p. 760

23. The **thymus** produces a collection of hormones called **thymosins** that are important to the development of T cells.

24. The thymus undergoes **involution** with aging. This process is associated with increased susceptibility to disease.

25. The thymus is located in the mediastinum, posterior to the sternum. It has right and left **lobes**, and **lobules** that are divided by **septa**.

26. The thymus has a dark outer **cortex** and lighter central **medulla**. Lymphocytes in the cortex are in clusters and surrounded by **epithelial reticular cells (ERCs)**. These cells maintain the **blood thymus barrier.**

## Module 20.7
**The spleen, the largest lymphoid organ, responds to antigens in the bloodstream   p. 762**

27. The **spleen** functions to (1) remove abnormal blood cells and other blood components by phagocytosis, (2) store iron recycled from red blood cells, and (3) initiate immune responses by B cells and T cells.

28. The spleen is located on the left side of the body at the level of the 9th and 11th ribs. It is attached to the stomach by the **gastrosplenic ligament**.

29. A **splenectomy** is the removal of the spleen due to injury.

30. The **visceral surface** (medial) contains indentations from the stomach (**gastric area**) and kidney (**renal area**).

31. **Red pulp** contains red blood cells, and **white pulp** resembles lymphoid nodules and contains lymphocytes.

## SECTION 2 • Innate Immunity

### Module 20.8
**Innate immunity is nonspecific and is not stimulated by specific antigens   p. 765**

32. Two complementary mechanisms protect the human body from disease: **innate (nonspecific) immunity** and adaptive (specific) immunity.

33. **Innate defenses** do not distinguish one type of threat from another and include **physical barriers**, **phagocytes**, **immune surveillance**, **interferons**, **complement**, **inflammation**, and **fever**.

34. **Adaptive defenses** protect against particular threats. This is known as adaptive immunity or **specific resistance**.

### Module 20.9
**Physical barriers prevent pathogens and toxins from entering body tissues   p. 766**

35. The integumentary system provides the major physical barrier to the external environment.

36. The epithelia lining the digestive, respiratory, urinary, and reproductive tracts also provide a barrier.

### Module 20.10
**Phagocytes respond to pathogen invasion   p. 767**

37. **Phagocytes** (**neutrophils**, **eosinophils**, and **fixed** and **free macrophages**) remove cellular debris and respond to foreign substances and pathogens.

## Module 20.11
**NK cells perform immune surveillance, detecting and destroying abnormal cells   p. 768**

38. **Natural killer (NK) cells** recognize and destroy abnormal cells. Constant monitoring by NK cells is called **immune surveillance**. Many cancer cells have **tumor-specific antigens** on their plasma membranes that NK cells recognize as abnormal.

39. Cancer cells often mutate and avoid NK cell detection, in a process called **immunological escape**.

## Module 20.12
**Interferons and the complement system are distributed widely in body fluids   p. 770**

40. **Interferons (IFNs)** are proteins released by activated lymphocytes and macrophages and by cells infected with viruses.

41. Interferons bind to surface receptors and, through second messengers, trigger the production of **antiviral proteins** in the cytoplasm.

42. Interferons are examples of **cytokines**, chemicals released by cells to coordinate local activities.

43. **Interferon alpha (α)** is produced by cells infected with viruses and attracts NK cells. **Interferon beta (β)** is secreted by fibroblasts and slows inflammation. **Interferon gamma (γ)** is secreted by T cells and NK cells and stimulates macrophage activity.

44. The **complement system** complements the actions of antibodies. There are three possible pathways: the **classical pathway**, the **lectin pathway**, and the **alternative pathway**.

## Module 20.13
**Inflammation is a localized tissue response to injury; fever is a generalized response to tissue damage and infection   p. 772**

45. **Inflammation**, or the **inflammatory response**, is a localized tissue response to injury that produces redness, swelling, heat, and pain.

46. **Fever**, a body temperature greater than 37.2°C (99°F), is caused by circulating proteins called **pyrogens** that reset the thermostat in the hypothalamus.

47. Fever can be beneficial in that it inhibits some viruses and bacteria; it also increases metabolic rate.

## SECTION 3 • Adaptive Immunity

### Module 20.14
**Adaptive immunity provides the body's specific defenses   p. 775**

48. **Adaptive (specific) immunity** is coordinated and produced by T cells and B cells. **Vaccines** stimulate an immune response by producing antibodies to a specific disease.

49. Two types of adaptive (specific) immunity are **active immunity** (which appears after exposure to an antigen) and **passive immunity** (produced by the transfer of antibodies from another source).

50. Adaptive immunity exhibits four general properties: **specificity**, **versatility**, **memory**, and **tolerance**.

51. **Memory cells** enable the immune system to "remember" previous target antigens.

## Module 20.15
### Adaptive immunity is triggered by exposure of T cells and B cells to specific antigens    p. 776

52. The first step in triggering an immune response is **antigen presentation**.

53. If a cell-mediated response is triggered, phagocytes and then T cells will be activated. The activated T cells will attack through phagocytosis or chemical toxins.

*The presence of antigens triggers specific defenses.*

54. If an antibody-mediated response is triggered, B cells will form antibody-producing cells whose products will destroy the antigen.

55. Antigen presentation occurs when an antigen is attached to a **class I MHC protein** or **class II MHC protein** displayed on the plasma membrane of an infected cell or **antigen-presenting cells (APCs)**, respectively.

## Module 20.16
### Infected cells stimulate the formation and division of cytotoxic T cells, memory T$_C$ cells, and regulatory T cells    p. 778

56. T cells have receptors for either class I or class II MHC proteins. **Antigen recognition** will occur only if the MHC protein presents the antigen to a T cell that can detect it.

57. There are many **CD** (cluster of differentiation) **markers**. **CD8** and **CD4 markers** are two important membrane proteins involved in antigen recognition.

58. CD8 markers are found on *CD8 cells*. These cells respond to antigens presented by class I MHC proteins. They provide cell-mediated immunity.

59. CD4 markers are found on *CD4 cells*. These cells respond to antigens presented by class II MHC proteins.

60. Before activation can occur, a T cell undergoes **costimulation**, a process that prevents T cells from mistakenly attacking normal body cells.

61. Following activation, CD8 T cells will divide and then differentiate into three types of CD8 T cells: **cytotoxic T (T$_C$) cells**, **memory T$_C$ cells**, and **regulatory T (T$_{reg}$) cells**.

## Module 20.17
### Antigen-presenting cells can stimulate activation of CD4 T cells, producing helper T cells that promote B cell activation and antibody production    p. 780

62. For CD4 T cells to initiate antibody-mediated immunity, they must be exposed to antigen-bearing class II MHC proteins. Costimulation then completes their activation.

63. Upon activation, CD4 T cells divide and differentiate into active **helper T cells (T$_H$ cells)** and **memory T$_H$ cells**.

64. B cells prepare for activation by presenting an antigen bound to class II MHC proteins in a process called **sensitization**.

65. A sensitized B cell must encounter a helper T cell that was activated by the same antigen. When these two cells bind, the active helper T cell secretes **cytokines**.

66. The cytokines continue stimulation of the activated B cell, causing it to divide into either **memory B cells** or **plasma cells**. The memory B cells remain in reserve to deal with subsequent infections. The plasma cells secrete antibodies.

## Module 20.18
### Antibodies are small soluble proteins that bind to specific antigens and whose abundance increases upon later antigen exposure    p. 782

67. An antibody molecule contains two parallel pairs of polypeptide chains: one pair of **heavy chains** and one pair of **light chains**. Each chain contains **constant segments** and **variable segments**.

68. An antibody molecule binds to an antigen at **antigenic determinant sites**, forming an **antigen-antibody complex**.

69. There are five different classes of antibodies, or **immunoglobulins (Igs)**, determined by structure. These classes are **IgG, IgE, IgD, IgM**, and **IgA**.

70. During the **primary response**, the concentration of antibodies in the blood (**antibody titer**) does not peak until 1 or 2 weeks after initial exposure.

71. When an antigen is encountered a second time, a more extensive and prolonged **secondary response** occurs. Antibody levels increase more rapidly and reach much higher levels than they did during the primary response.

## Module 20.19
### Antibodies use many different mechanisms to destroy target antigens    p. 784

72. An antigen-antibody complex may eliminate an antigen in seven ways: **neutralization**, **prevention of pathogen adhesion**, **activation of complement**, **opsonization**, **attraction of phagocytes**, **stimulation of inflammation**, and **precipitation** and **agglutination**.

## Module 20.20
### Hypersensitivities are abnormal reactions to antigens    p. 785

73. **Hypersensitivities** (*allergies*) are inappropriate or excessive immune responses to antigens. Antigens that trigger allergic reactions are called **allergens**.

74. An allergen entering the bloodstream could be lethal, as in **anaphylaxis**, when an allergen affects mast cells throughout the body. In severe cases, circulatory collapse is possible, in a response called **anaphylactic shock**.

## Module 20.21
### Innate immunity and adaptive immunity work together to defeat pathogens    p. 786

75. Elements of innate immunity and adaptive immunity work together in a coordinated response to antigens.

## Module 20.22
### Immune disorders involving either overactivity or underactivity can be harmful    p. 788

76. Immune disorders of excessive or misdirected response include **autoimmune disorders** (**thyroiditis, rheumatoid arthritis**, and **type 1 diabetes mellitus**), **transplant rejection**, and **hypersensitivities**.

77. Immune disorders of inadequate immune response include **immunodeficiency diseases** (such as **acquired immunodeficiency syndrome**, or **AIDS**, which is caused by the **human immunodeficiency virus**, or **HIV**), and age-related reductions in immune activity.

20

## Chapter Review Questions

**Labeling:** Label the antibody molecule shown here.

1 _____

2 _____

3 _____

4 _____

5 _____

6 _____

**True/False:** Indicate whether each statement is true or false.

**7** One difference between lymphatic vessels and veins is that lymphatic vessels do not have valves.

**8** Lymphopoiesis occurs exclusively in lymphoid tissue.

**9** Efferent lymphatics arise from the hilum of a lymph node.

**10** A decrease in circulating pyrogens results in an increase in body temperature.

**11** Antibody molecules consist of four pairs of heavy chains and four pairs of light chains.

**12** Under stimulation from cytokines, B cells can differentiate into plasma cells.

**7** _____

**8** _____

**9** _____

**10** _____

**11** _____

**12** _____

**Multiple choice:** Select the correct answer from the list provided.

**13** Lymph from the left arm, the left half of the head, the left side of the body superior to the diaphragm, and the entire body inferior to the diaphragm is received by the
- ☐ a) thoracic duct.
- ☐ b) cisterna chyli.
- ☐ c) azygos vein.
- ☐ d) left lymphatic duct.

**14** CD4 T cells respond to antigens presented by
- ☐ a) class I MHC.
- ☐ b) class II MHC.
- ☐ c) cytokines.
- ☐ d) plasma cells.

**15** Interferons increase resistance of cells to
- ☐ a) fungal infections.
- ☐ b) bacterial infections.
- ☐ c) viral infections.
- ☐ d) cancer.

**16** Which of the following immunoglobulins accounts for 80 percent of all antibodies?
- ☐ a) IgG
- ☐ b) IgE
- ☐ c) IgD
- ☐ d) IgM

**17** The human immunodeficiency virus (HIV) infects which of the following cells?

   ☐ a) regulatory T cells

   ☐ b) memory T cells

   ☐ c) helper T cells

   ☐ d) cytotoxic T cells

**18** Which of the following is *not* a phagocyte?

   ☐ a) neutrophil

   ☐ b) monocyte

   ☐ c) free macrophage

   ☐ d) natural killer cell

**19** The adenoid is also known as the

   ☐ a) pharyngeal tonsil.

   ☐ b) palatine tonsil.

   ☐ c) lingual tonsil.

   ☐ d) Peyer's patch.

**20** The gradual decrease in secretions and size of this organ with age is correlated with an increased susceptibility to disease.

   ☐ a) spleen

   ☐ b) thymus

   ☐ c) tonsils

   ☐ d) appendix

**Short answer**

**21** Describe transplant rejection.

**22** How does a cytotoxic T cell (T$_C$ cell) destroy its target cell?

**23** Why is it necessary for certain vaccinations, such as hepatitis B, to be administered as a series of injections over a period of months?

---

## MasteringA&P™

Access more chapter study tools online in the MasteringA&P Study Area:

■ Chapter Quizzes, Chapter Practice Test, and Clinical Case Studies

■ Practice Anatomy Lab™ 3.0    PAL 3.0

■ Interactive Physiology™    iP2™

■ A&P Flix™    A&PFlix™

■ PhysioEx™ 9.1    PhysioEx 9.1

---

# Chapter Integration • Applying what you have learned

## Catching childhood diseases in adulthood

The varicella-zoster virus (VZV) causes chickenpox (varicella) and shingles (herpes zoster). Chickenpox is a common childhood disease characterized by small, itchy blisters, fever, fatigue, and headache. Shingles is caused by a reactivation of VZV later in life. The Centers for Disease Control and Prevention (CDC) recommends that people aged 19–49 who had chickenpox earlier in life receive the chickenpox vaccine to prevent shingles. The CDC also recommends the shingles vaccine, Zostavax, for people 60 years old and older to prevent shingles.

A 20-year-old student discovers that she has been exposed to chickenpox by her twin nieces and is concerned that she might have contracted the disease. She goes to see her physician, who takes a blood sample and sends it to a lab for antibody titers. The results reveal an elevated level of IgM antibodies to the VZV but very few IgG antibodies to the virus.

**1** Has this student come down with chickenpox?

**2** Who is susceptible to contracting chickenpox?

**3** What is the causative agent for chickenpox and for shingles?

**4** Distinguish between IgM and IgG antibodies.

**5** Explain how antigenic specificity of antibodies is possible.

**6** Compare the primary and secondary immune responses as they pertain to IgM and IgG.

**7** Administration of the chickenpox vaccine confers what type of immunity?

# 21

# The Respiratory System

## Module 21.1
# The respiratory system has an upper and lower respiratory tract with different functions

The **respiratory system** is composed of structures involved in breathing, or **pulmonary ventilation** (airflow to and from the lungs), and gas exchange. In this section we consider this body system's structures and functions.

**1** The respiratory system is anatomically divided into an upper respiratory system and a lower respiratory system. The **respiratory tract** is a branching passageway that carries air to and from the gas exchange surfaces of the lungs. The tract is composed of a conducting portion and a respiratory portion. The **conducting portion** begins at the nasal cavity and extends through the pharynx, larynx, trachea, bronchi, and larger bronchioles. The **respiratory portion** includes the smallest, thinnest *respiratory bronchioles* and the air-filled sacs called **alveoli** (al-VĒ-ō-lī). Gas exchange between air and blood occurs at the alveoli.

**? REVIEW**
**A.** Where does gas exchange between air and the lungs occur?

Labels (handwritten): Pharynx, nasopharynx, uvula, laryngopharynx, oropharynx, superior nasal conchae, middle nasal conchae, anterior nares, external nares, inferior nasal conchae, Tongue

Labels: Esophagus, Clavicle, Ribs, Right lung, Left lung, Diaphragm

### Upper Respiratory System

The **upper respiratory tract** filters, warms, and humidifies incoming air—protecting the more delicate surfaces of the lower respiratory system—and reabsorbs heat and water from outgoing air.

- Nose
- Nasal cavity
- Paranasal sinuses
- Pharynx

### Lower Respiratory System

The **lower respiratory tract** conducts air to and from the gas exchange surfaces.

- Larynx
- Trachea
- Bronchus
- Bronchioles
- Respiratory bronchioles

- Alveoli

**2** This table summarizes the major functions of the respiratory system.

## Functions of the Respiratory System

- Providing an extensive surface area for gas exchange between air and circulating blood
- Moving air to and from the exchange surfaces of the lungs along the respiratory passageways
- Protecting respiratory surfaces from dehydration, temperature changes, and invasion by pathogens
- Producing sounds for speaking, singing, and other forms of communication
- Detecting odors by olfactory receptors in the superior portions of the nasal cavity

**? REVIEW**
**B.** Distinguish between the conducting portion and respiratory portion of the respiratory tract.

**Lo LEARNING OUTCOME**
Identify the structures of the respiratory system, and list its major functions.

21

# Module 21.2

## The respiratory defense system protects the respiratory mucosa

Debris or pathogens in inhaled air can severely damage the delicate exchange surfaces of the respiratory system. A series of filtration mechanisms that make up the **respiratory defense system** prevent such contamination.

**Go to**
MasteringA&P™
PAL™3.0
Respiratory System

**? REVIEW**
**A.** Define *respiratory defense system.*

**1** The **respiratory mucosa** (mū-KŌ-suh) lines the conducting portion of the respiratory tract. A pseudostratified ciliated columnar epithelium with numerous mucous cells lines the nasal cavity, the superior portion of the pharynx, the trachea, bronchi, and large bronchioles.

**? REVIEW**
**B.** What membrane lines the conducting portion of the respiratory tract?

Mucus

Debris

Mucous cell

Ciliated columnar epithelial cell

The beating of cilia sweeps mucus and any trapped debris or microorganisms toward the pharynx, where they will be coughed out or swallowed and exposed to the acids and enzymes of the stomach. This ciliary movement continuously cleans and protects the respiratory surfaces. The flow of mucus is often described as a **mucociliary escalator**. By the time air reaches the respiratory portion of the tract, particles larger than around 5 μm have been trapped and removed.

The **lamina propria** (LAM-in-nuh PRŌ-prē-uh) is the underlying layer of areolar tissue that supports the respiratory epithelium.

In the upper respiratory system, trachea, and bronchi, the lamina propria contains **mucous glands** that discharge secretions onto the epithelial surface.

**2** This sectional view of the respiratory mucosa shows the histological appearance of the respiratory epithelium.

Mucous cells in the epithelium and mucous glands in the lamina propria produce a sticky mucus that bathes exposed surfaces.

Nucleus of columnar epithelial cell

Cilia

Lamina propria

Basement membrane

Stem cells in the epithelium divide to replace damaged or aged cells.

**3** The structure of the respiratory epithelium changes markedly along the respiratory tract.

A respiratory mucosa, with mucous cells, lines the nasal cavity and the superior portion of the pharynx.

A stratified squamous epithelium lines the inferior portions of the pharynx, protecting the epithelium from abrasion and chemicals.

A typical respiratory mucosa lines the superior portion of the lower respiratory tract.

In the narrower bronchioles, the epithelium becomes cuboidal.

The gas exchange surfaces consist of a delicate simple squamous epithelium. Here the distance between the air and the blood in adjacent capillaries is generally less than 1 μm.

---

**Clinical Note**

## Cystic fibrosis is an inherited disease characterized by the production of thick and sticky mucus

**Cystic fibrosis (CF)** is the most common lethal inherited disease in individuals of Northern European descent. With CF, the respiratory mucosa produces thick mucus that the respiratory defense system cannot transport, making breathing difficult. The mucociliary escalator stops working, leading to frequent infections. Potentially lethal infections may develop in the respiratory passageways and lungs if bacteria such as *Pseudomonas aeruginosa* colonize the stagnant mucus. So, liquid medicines are made into a mist and then inhaled to thin the mucus and open the airways. The average lifespan for people with CF who live into adulthood is 37 years. Death generally results from heart failure associated with a massive chronic bacterial infection of the lungs.

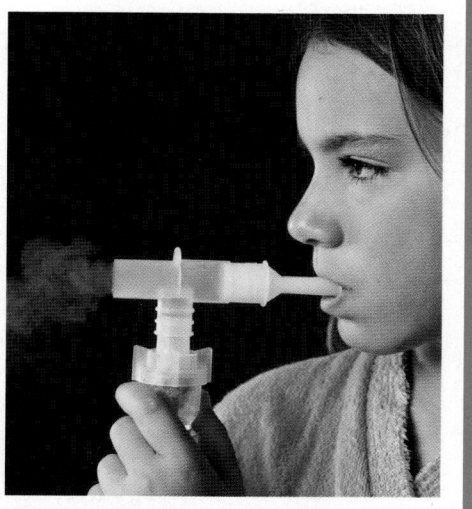

**? REVIEW**

**C.** Why can cystic fibrosis become lethal?

**L₀ LEARNING OUTCOME**

Explain how the delicate respiratory exchange surfaces are protected from pathogens, debris, and other hazards, and describe cystic fibrosis.

21

# Module 21.3
# The upper respiratory system includes the nose, nasal cavity, paranasal sinuses, and pharynx

**1** The nose is the primary passageway for air entering the respiratory system when you are resting and breathing quietly. This anterior view shows the nasal cartilages and external landmarks.

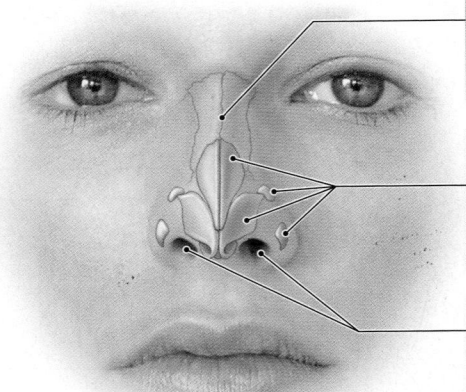

The two nasal bones form the external bridge, called the **dorsum of nose**. It is supported by hyaline cartilage.

Small, elastic **nasal cartilages** extend laterally from the bridge of the nose. These cartilages help to keep the nostrils open and prevent their collapse during a strong inhalation.

Air normally enters through the paired **nostrils**, or external **nares** (NĀ-res), that open into the nasal cavity.

**2** Several important features of the nasal cavity can be best seen in frontal section. The superior, middle, and inferior nasal conchae (singular, *concha*) project toward the nasal septum from the lateral walls of the nasal cavity. (Some references use the terms *superior*, *middle*, and *inferior turbinates*.) To pass from the nostrils to the *choanae* (KŌ-an-ē)—openings of the nasal cavity—air flows between adjacent conchae, through the **superior**, **middle**, and **inferior nasal meatuses** (mē-Ā-tus-ez;). These narrow grooves bounce incoming air off the conchal surfaces and churn it like a stream flowing over rocks. As the air swirls, small airborne particles become trapped in the mucus that coats the lining of the nasal cavity. In addition, the turbulence provides extra time for warming and humidifying incoming air. It also creates circular air currents that bring olfactory stimuli to the olfactory receptors.

 **REVIEW**

**A.** List the structures of the upper respiratory system.

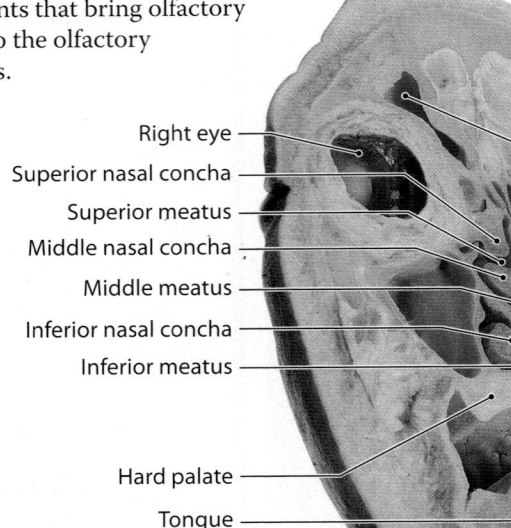

Right eye
Superior nasal concha
Superior meatus
Middle nasal concha
Middle meatus
Inferior nasal concha
Inferior meatus

Hard palate
Tongue

Frontal sinus
Ethmoidal air cell
Maxillary sinus

Perpendicular plate of ethmoid
Vomer

### Paranasal Sinuses

The maxillary, frontal, ethmoid, and sphenoid bones that form the lateral and superior walls of the nasal cavity contain **paranasal sinuses** (see Module 7.7). The mucus produced in these sinuses, aided by tears draining through the nasolacrimal ducts, keep the surfaces of the nasal cavity moist and clean.

### Nasal Septum

The **nasal septum**, formed by the fusion of the perpendicular plate of the ethmoid and the vomer, divides the nasal cavity into left and right sides.

21

**3** Other features of the upper respiratory system are best seen in sagittal section. The larynx marks the beginning of the lower respiratory system.

### Pharynx

The **pharynx** (FAR-inks) is a chamber shared by the digestive and respiratory systems. The curving superior and posterior walls of the pharynx are closely bound to the axial skeleton, but the lateral walls are flexible and muscular.

The **nasopharynx** (nā-zō-FAR-inks) is the superior portion of the pharynx. It is connected to the posterior portion of the nasal cavity through the choanae. Each auditory tube opens into the nasopharynx at the **pharyngeal opening of the auditory tube** on each side of the pharyngeal tonsil.

The **oropharynx** extends between the soft palate and the base of the tongue at the level of the hyoid bone. At the boundary between the nasopharynx and the oropharynx, the epithelium changes from pseudostratified columnar to stratified squamous epithelium.

The narrow **laryngopharynx** (la-rin-gō-FAR-inks) includes that portion of the pharynx between the hyoid bone and the entrance to the larynx and esophagus. Like the oropharynx, the laryngopharynx is lined with a stratified squamous epithelium.

The **trachea**, or windpipe, conducts air toward the lungs.

**?** REVIEW

**B.** Trace the pathway of air through the upper respiratory system.

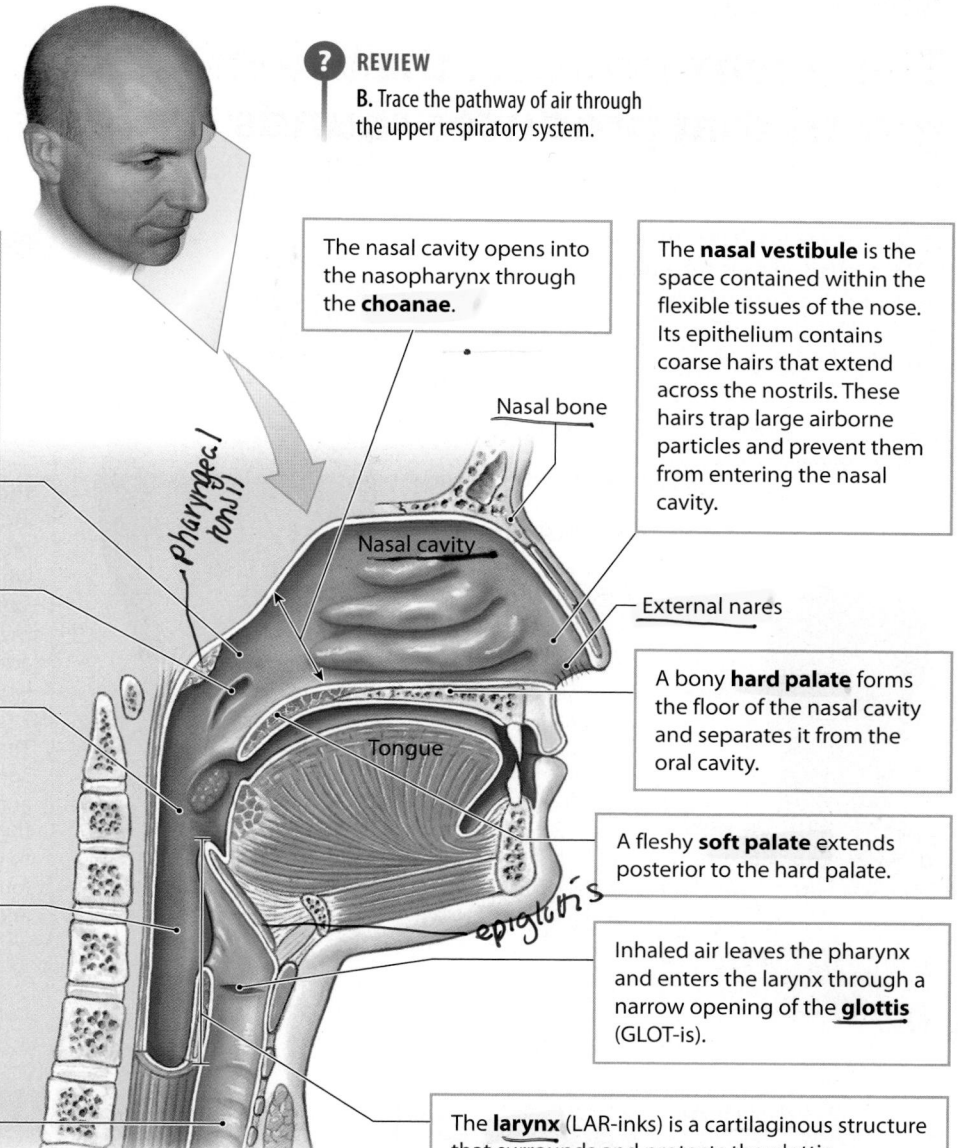

*pharyngeal tonsil*

*epiglottis*

The nasal cavity opens into the nasopharynx through the **choanae**.

Nasal bone

Nasal cavity

Tongue

The **nasal vestibule** is the space contained within the flexible tissues of the nose. Its epithelium contains coarse hairs that extend across the nostrils. These hairs trap large airborne particles and prevent them from entering the nasal cavity.

External nares

A bony **hard palate** forms the floor of the nasal cavity and separates it from the oral cavity.

A fleshy **soft palate** extends posterior to the hard palate.

Inhaled air leaves the pharynx and enters the larynx through a narrow opening of the **glottis** (GLOT-is).

The **larynx** (LAR-inks) is a cartilaginous structure that surrounds and protects the glottis.

**Everyday Physiology**

### How does the nasal mucosa adjust air temperature and moisture in the air we breathe?

Throughout much of the nasal cavity, the lamina propria contains an extensive network of highly expandable veins that can release heat like a radiator. As cool, dry air passes inward over the exposed surfaces of the nasal cavity, the air warms, and water in the mucus evaporates. Air moving from your nasal cavity to your lungs is thus heated almost to body temperature, and it is nearly saturated with water vapor. This protects more delicate respiratory surfaces from chilling or drying out. As air moves out of the respiratory tract, it again passes over the epithelium of the nasal cavity. The air now is warmer and more humid than it was when it entered. The nasal mucosa reabsorbs the heat and water from the outgoing air, thus reducing both heat loss and water loss to the environment. Breathing through your mouth eliminates much of the conditioning of inhaled air and increases heat and water loss at every exhalation.

**?** REVIEW

**C.** Why is the vascularization of the nasal cavity important?

*nasopharynx*

21

**Lo** LEARNING OUTCOME

Identify the organs and structures of the upper respiratory system, and describe their functions.

# Module 21.4
# The larynx protects the glottis that produces sounds

**1** The **larynx** (LAR-ingks) is a cartilaginous tube that surrounds and protects the glottis. The glottis is the "voicebox" of the larynx.

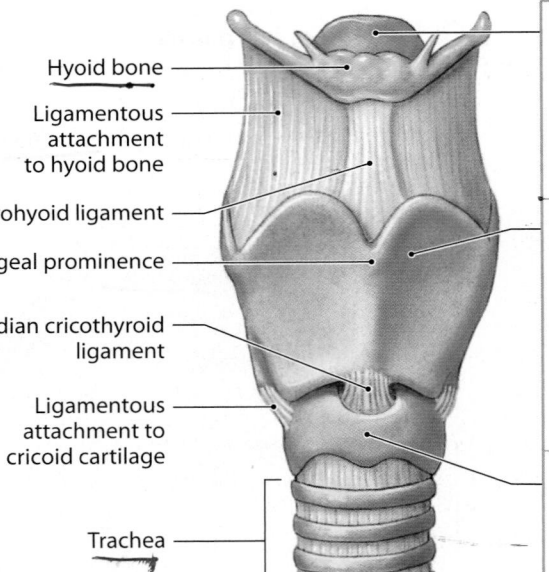

Hyoid bone

Ligamentous attachment to hyoid bone

Median thyrohyoid ligament

Laryngeal prominence

Median cricothyroid ligament

Ligamentous attachment to cricoid cartilage

Trachea

**Anterior view**

## Large Cartilages of the Larynx

The shoehorn-shaped **epiglottis** (ep-i-GLOT-is) projects superior to the glottis and forms a lid over it. During swallowing, the larynx is elevated, and the epiglottis folds back over the glottis, preventing both solids and liquids from entering the respiratory tract.

The **thyroid** (*thyroid*, shield-shaped) **cartilage** forms most of the anterior and lateral walls of the larynx. The prominent anterior surface of the thyroid cartilage, which you can easily see and feel, is called the *laryngeal prominence*, or "Adam's apple." The median thyrohyoid ligament and the surrounding ligaments attach to the hyoid bone, the epiglottis, and smaller laryngeal cartilages.

The **cricoid** (KRĪ-koyd; ring-shaped) **cartilage** is a complete ring of cartilage with a narrow anterior band and broad posterior portion. Together the cricoid and thyroid cartilages protect the glottis and the entrance to the trachea, and their broad surfaces provide sites for the attachment of important laryngeal muscles and ligaments.

Hyoid bone

Epiglottis

Thyroid cartilage

Corniculate cartilage

Arytenoid cartilage

Cricoid cartilage

Tracheal cartilages

Anterior    Posterior

**Sagittal section**

### Vestibular and Vocal Ligaments

The **vestibular ligaments** and the **vocal ligaments** extend between the thyroid cartilage and the arytenoid cartilages.

Vestibular ligament

Vocal ligament

Epiglottis

Thyroid cartilage

Arytenoid cartilage

Cricoid cartilage

Tracheal cartilages

**Posterior view**

**2** The cartilages of the larynx are isolated in this posterior view. Now the small laryngeal cartilages you can see in relationship to the larger, supporting cartilages.

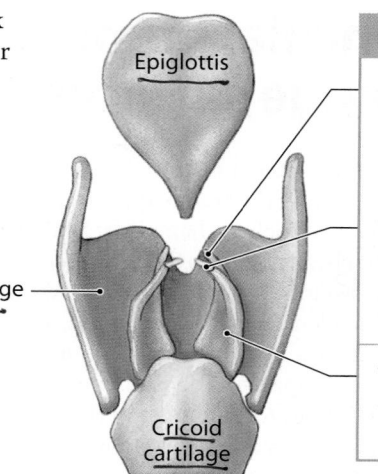

Epiglottis

Thyroid cartilage

Cricoid cartilage

**?** REVIEW

**A.** Identify the paired and unpaired cartilages that compose the larynx.

**Small Laryngeal Cartilages**

The **cuneiform** (kū-NĒ-i-form; wedge-shaped) **cartilages** are long and curved, and they lie within folds of tissue that extend between the lateral surface of each arytenoid cartilage and the epiglottis.

The **corniculate** (kor-NIK-ū-lat; horn-shaped) **cartilages** articulate with the arytenoid cartilages. The corniculate and arytenoid cartilages function in opening and closing the glottis and producing sound.

The small, paired **arytenoid** (ar-i-TĒ-noyd; ladle-shaped) **cartilages** articulate with the superior surface of the cricoid cartilage.

**3** Air enters or leaves the larynx through the glottis. The **glottis** (GLOT-is) is made up of the vocal folds and the rima glottidis. The opening or closing of the glottis involves rotational movements of the arytenoid cartilages. When the glottis is open, passing air vibrates its **vocal folds**, tissue folds that contain the elastic **vocal ligaments**. The vibration of the vocal folds produces sound waves, and the pitch of the sound produced depends on the diameter, length, and tension in the vocal folds. The tension is controlled by the contraction of voluntary muscles that reposition the arytenoid cartilages relative to the thyroid cartilage. These diagrammatic superior views show the glottis in the open and closed positions.

**?** REVIEW

**B.** Describe the structures of the glottis.

**Glottis**

The **rima glottidis** is the opening between the vocal folds and the arytenoid cartilages.

The **vocal folds**, which contain the vocal ligaments, lie inferior to the vestibular folds. The vocal folds are involved with sound production and are also known as the *vocal cords*.

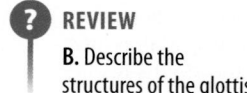

Corniculate cartilage
Cuneiform cartilage
Ary-epiglottic fold
Vestibular fold
Epiglottis
Root of tongue

Vocal folds of glottis

**Glottis (open)**

**Glottis (closed)**

The vestibular ligaments lie within a pair of relatively inelastic **vestibular folds**. These folds help prevent foreign objects from entering the glottis and contacting the more delicate vocal folds.

**4** Sound production at the larynx is called **phonation** (fō-NĀ-shun; *phone*, voice). Phonation is one component of speech. However, clear speech also requires **articulation**, the modification of those sounds by other structures, such as the tongue, teeth, and lips. In a stringed instrument, such as a guitar, the quality of the sound produced does not depend solely on the nature of the vibrating string. Rather, the entire instrument becomes involved as the walls vibrate and the composite sound echoes within the hollow body. Similar amplification and resonance occur within your pharynx, oral cavity, nasal cavity, and paranasal sinuses. The combination determines the particular and distinctive sound of your voice.

**?** REVIEW

**C.** Distinguish between phonation and articulation.

**Lo** LEARNING OUTCOME

Describe the structure of the larynx, and discuss its role in normal breathing and in the production of sound.

**21**

## Module 21.5

# The trachea, bronchi, and bronchial branches convey air to and from lung gas exchange surfaces

**1** The **trachea** (TRĀ-kē-uh), or windpipe, is a tough, flexible tube with a diameter of about 2.5 cm (1 in.) and a length of about 11 cm (4.33 in.). It begins anterior to vertebra C$_6$ and ends in the mediastinum, at the level of vertebra T$_5$, where it branches to form the right and left **main bronchi** (BRONG-ki; singular, *bronchus*).

**2** This is a sectional view of the trachea. An elastic ligament and the **trachealis**, a band of smooth muscle, connect the ends of each C-shaped tracheal cartilage. Contraction of the trachealis reduces the diameter of the trachea, increasing the resistance to airflow. This allows air to be more forcefully expelled, such as when you cough. The diameter of the trachea changes from moment to moment, primarily under the control of the sympathetic division of the ANS. Sympathetic stimulation increases the diameter of the trachea and makes it easier to move air along the respiratory passageways.

The trachea contains 15–20 **tracheal cartilages**, which stiffen the tracheal walls and protect the airway. They also prevent it from collapsing or overexpanding as pressures change in the respiratory system.

The right main bronchus is larger in diameter than the left, and it descends toward the lung at a steeper angle. Thus, most foreign objects that enter the trachea find their way into the right bronchus rather than the left.

Because the tracheal cartilages are incomplete posteriorly, the posterior tracheal wall can easily distort when large masses of food pass along the esophagus.

The mucosa of the trachea resembles that of the nasal cavity and the nasopharynx.

Hyoid bone

Larynx

Esophagus

Trachea

Trachealis

Lumen of trachea

Tracheal cartilage

Tracheal gland

Carina of trachea

Lobar bronchi

Left main bronchus

Lobar bronchi

**Right lung**

**Left lung**

**? REVIEW**

**A.** Compare the two main bronchi.

**B.** What function do the C-shaped tracheal cartilages allow?

**3** This flowchart indicates the general pattern of airflow in the conducting portion of the respiratory tract.

| Air-Conducting Passageways in the Lower Respiratory Tract | | | | |
|---|---|---|---|---|
| The trachea is a single conducting tube, extending from the larynx to the mediastinum. | The trachea branches to form two main bronchi, one for each lung. Unlike the C-shaped cartilaginous rings of the trachea, those of the main bronchi form complete rings. | After entering the lung, each main bronchus divides to form **lobar bronchi**. The right lung has three, and the left lung has two. The lobar bronchi are supported by small cartilage plates rather than rings. | Each lobar bronchus divides to form **segmental bronchi**. The cartilages in the walls of segmental bronchi resemble those of lobar bronchi. | Each segmental bronchus branches several times, giving rise to multiple **bronchioles**. |

21

**4** This diagrammatic view highlights important structural features of bronchi and bronchioles and the general pattern of airway distribution along the lower respiratory tract. At each new branch, the diameter decreases. Overall, there is a five- to eightfold decrease in diameter from the trachea to the terminal bronchioles, which have a diameter of 0.3–0.5 mm.

Cartilage

LEFT LUNG

Left main bronchus

Lobar bronchus

Lobar bronchus

Cartilage plates

Visceral pleura

Segmental bronchi

### Bronchiole

The walls of bronchioles lack cartilages but contain relatively thick layers of smooth muscle, and smooth muscle tone has a significant effect on bronchial diameter and the resistance to airflow. Resting tension in this smooth muscle commonly causes the bronchiolar mucosa to form a series of folds. Sympathetic activation leads to **bronchodilation**, the enlargement of airway diameter. Parasympathetic stimulation leads to **bronchoconstriction**, a reduction in the diameter of the airway. Extreme bronchoconstriction may occur during allergic reactions such as **asthma** (AZ-muh), in which swelling and bronchoconstriction can severely restrict or prevent airflow.

Respiratory epithelium

Smooth muscle

Bronchioles

Terminal bronchiole

Respiratory bronchiole

Pulmonary lobule

Each bronchiole branches further to form **terminal bronchioles**. Roughly 6500 terminal bronchioles arise from each segmental bronchus, and each terminal bronchiole supplies a single **pulmonary lobule**, where gas exchange occurs within alveoli.

 **REVIEW**

**C.** Trace the pathway of airflow along the passages of the lower respiratory tract.

**Lo LEARNING OUTCOME**

Discuss the structures and functions of the airways outside and inside the lungs.

21

# The lungs have lobes that are subdivided into bronchopulmonary segments

**1** The branching pattern of bronchi and bronchioles is often called the **bronchial tree**. Each segmental bronchus ultimately supplies air to a single **bronchopulmonary segment**, a specific region of one lung. The lungs have distinct lobes that are separated by deep fissures. Each lobe contains at least a pair of bronchopulmonary segments.

Right

Left

Bronchopulmonary segments of right superior lobe

Bronchopulmonary segments of right middle lobe

Bronchopulmonary segments of right inferior lobe

Bronchopulmonary segments of left superior lobe

Bronchopulmonary segments of left inferior lobe

**A diagrammatic view of the bronchial tree, lung lobes, and bronchopulmonary segments**

**?** REVIEW

**A.** Define *bronchopulmonary segment.*

**2** The left and right lungs are surrounded by the left and right pleural cavities, respectively. This photo shows the collapsed, preserved lungs in a cadaver. The lungs in a living, healthy nonsmoker have a lighter, pinkish color.

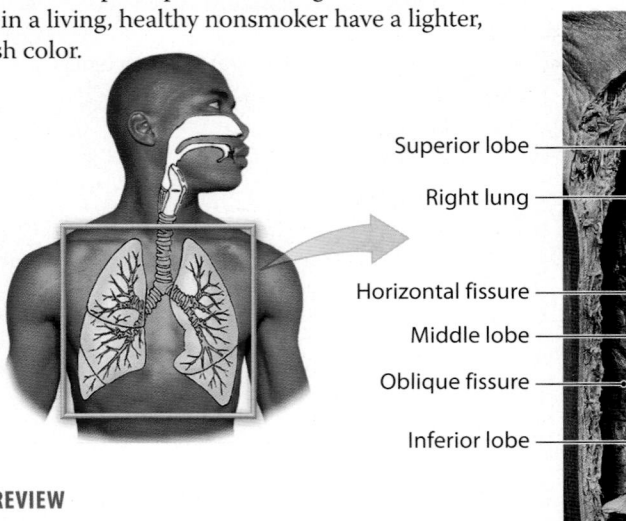

Superior lobe

Right lung

Horizontal fissure

Middle lobe

Oblique fissure

Inferior lobe

Boundary between right and left pleural cavities

Left lung

Superior lobe

Oblique fissure

Fibrous layer of pericardium

Inferior lobe

Falciform ligament

Cut edge of diaphragm

Liver, right lobe

Liver, left lobe

**?** REVIEW

**B.** Describe the location of the lungs within the thoracic cavity.

21

**3** Each lung is a blunt cone with a tip extending to the superior border of the first rib. The broad concave inferior portion of each lung rests on the superior surface of the diaphragm. You can see major superficial landmarks on the left and right lungs in these lateral and medial views.

**? REVIEW**

**C.** Describe the lung borders and landmarks.

## Lateral Views

The curving anterior and inferior borders of each lung follow the contours of the rib cage.

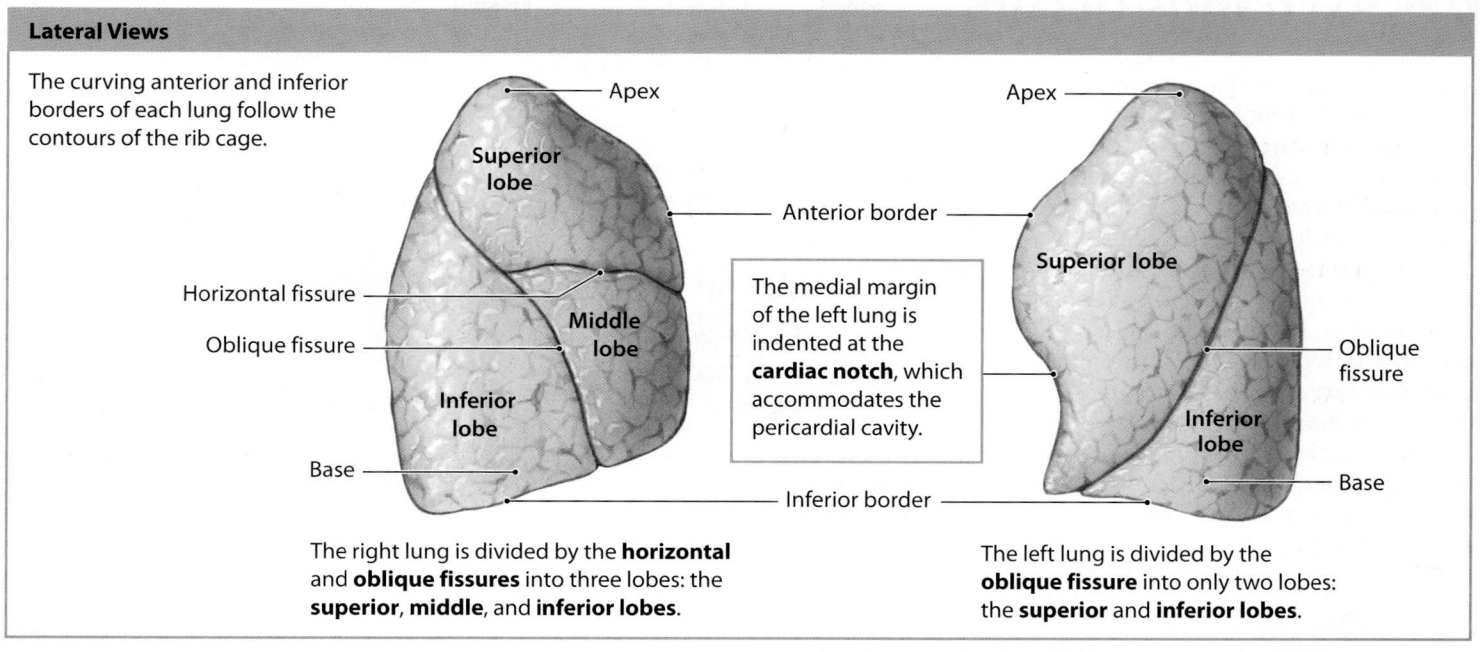

The medial margin of the left lung is indented at the **cardiac notch**, which accommodates the pericardial cavity.

The right lung is divided by the **horizontal** and **oblique fissures** into three lobes: the **superior**, **middle**, and **inferior lobes**.

The left lung is divided by the **oblique fissure** into only two lobes: the **superior** and **inferior lobes**.

## Medial Views

The mediastinal surface, which contains the hilum, has an irregular shape. Both lungs have grooves that mark the position of the great vessels and the heart.

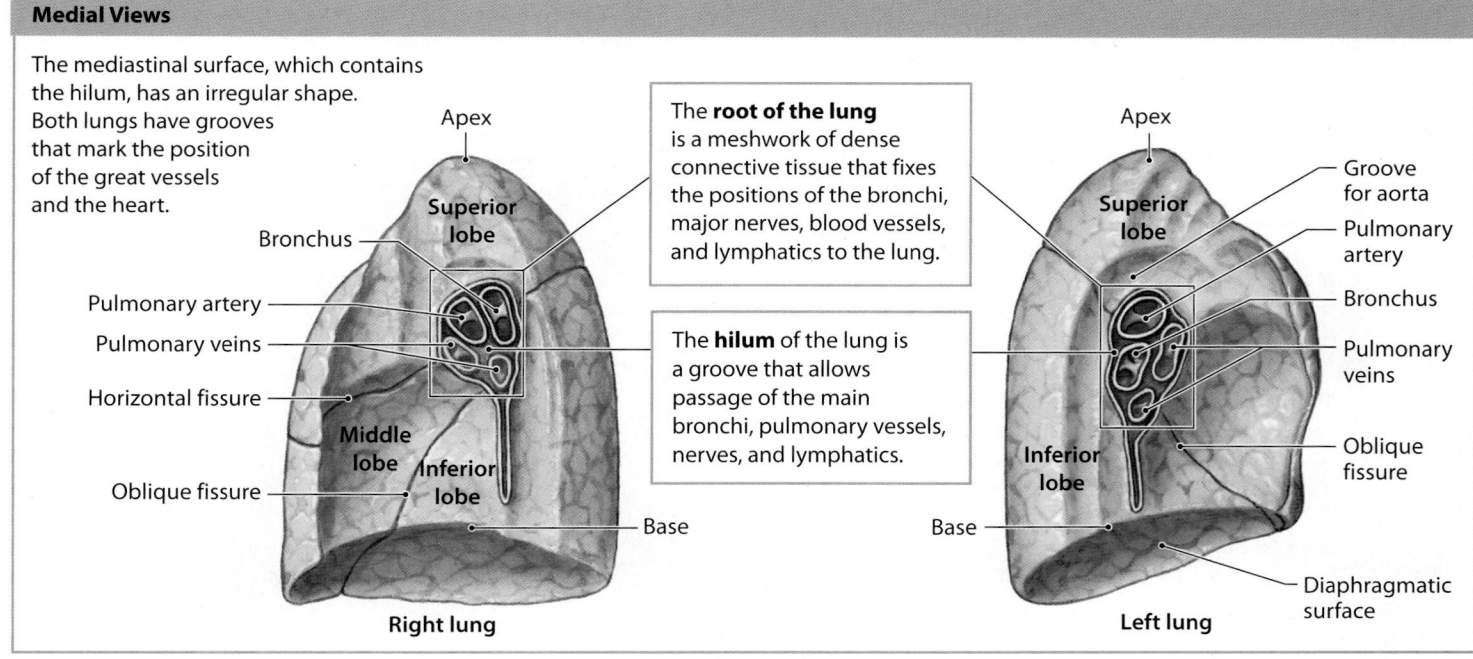

The **root of the lung** is a meshwork of dense connective tissue that fixes the positions of the bronchi, major nerves, blood vessels, and lymphatics to the lung.

The **hilum** of the lung is a groove that allows passage of the main bronchi, pulmonary vessels, nerves, and lymphatics.

**Right lung**

**Left lung**

**? REVIEW**

**D.** Name the lobes and fissures of each lung.

**Lo LEARNING OUTCOME**

Describe the superficial anatomy of the lungs.

21

# Module 21.7
# Pulmonary lobules contain alveoli, where gas exchange occurs

**1** The continuous passageways of the conducting and respiratory portions of the respiratory tract end in thin-walled sac-like structures called **pulmonary alveoli** (al-VĒ-ō-lī; singular, *alveolus*). Each lung contains about 150 million alveoli, and they give the lung an open, spongy appearance. Each alveolus is surrounded by an extensive capillary network that receives blood from a branch of a pulmonary artery and discharges blood into a branch of a pulmonary vein. A network of elastic fibers surrounds these capillaries. Recoil of these fibers during exhalation reduces the size of the alveoli and helps push air out of the lungs.

**? REVIEW**
**A.** Define *pulmonary lobule*.

**1** Each segmental bronchus delivers air to a single bronchopulmonary segment.

**2** Within a bronchopulmonary segment, bronchioles branch repeatedly, forming about 6500 terminal bronchioles.

**3** Each terminal bronchiole supplies a single **pulmonary lobule**.

Trachea
Left main bronchus
Lobar bronchus
Bronchioles
Respiratory bronchiole
Alveoli
Broncho-pulmonary segment

**4** A **respiratory bronchiole** forms by the branching of a terminal bronchiole inside a pulmonary lobule. This is the start of the respiratory portion of the respiratory tract.

**5** The walls of respiratory bronchioles open into alveoli. Gas exchange occurs across the walls of the alveoli.

**6** Respiratory bronchioles open into regions called **alveolar ducts**. An alveolar duct is a common passage that is connected to multiple individual alveoli.

**7** The alveolar ducts end in clusters of alveoli called **alveolar sacs** or **alveolar saccules**.

Branch of pulmonary artery
Bronchiole
Bronchial artery (red), nerve (yellow), and vein (blue)
Smooth muscle around terminal bronchiole
Terminal bronchiole
Bands of elastic fibers
Lymphatic vessel
Branch of pulmonary vein
Interlobular septum
Capillary beds
Alveoli

### The Pleurae

The **visceral pleura** covers the outer surfaces of the lungs.

The **pleural cavity**, a potential space between the pleural membranes, contains **pleural fluid** that coats the pleural surfaces to reduce friction.

The **parietal pleura** covers the inner surface of the thoracic wall and extends over the diaphragm and mediastinum.

**2** The alveolar epithelium is primarily a simple squamous epithelium. The surrounding capillaries differ functionally from other capillaries; they dilate when alveolar oxygen levels are high and constrict when alveolar oxygen levels are low. This response directs blood flow to the alveoli containing the most oxygen.

The squamous epithelial cells, called **pneumocytes type I**, are unusually thin and delicate and are the sites of gas diffusion.

Roaming **alveolar macrophages** patrol the epithelial surface, phagocytizing any particles that have eluded other respiratory defenses and reached the alveoli.

**Pneumocytes type II** are scattered among the squamous cells. These cells produce **surfactant** (sur-FAK-tant), an oily secretion containing a mixture of phospholipids and proteins that forms a superficial coating over a thin layer of water.

Capillary

Alveolar macrophage

Elastic fibers

Interconnection between adjacent alveoli

Endothelial cell of capillary

**3** Gas exchange occurs across the **blood air barrier** at each alveolus. Diffusion proceeds very rapidly because the distance is short and both oxygen and carbon dioxide are lipid soluble. In addition, the surface area of the blood air barrier is very large.

**The Blood Air Barrier**

The total distance separating alveolar air from blood can be as little as 0.1 μm; it averages about 0.5 μm.

Alveolar cell layer

Fused basement membranes of alveolar epithelium and capillary endothelium

Capillary endothelium

**Alveolar air space**

**Capillary lumen**

Red blood cell

Surfactant plays a key role in keeping the alveoli open. It reduces the surface tension in the water coating the alveolar surface. Alveolar walls, like soap bubbles, are very delicate; without surfactant, the normally high surface tension of water would collapse the alveoli.

Nucleus of endothelial cell

**?** **REVIEW**
**C.** What would happen to the alveoli if surfactant were not produced?

**?** **REVIEW**
**B.** Describe the structure and function of the blood air barrier.

21

**Lo** **LEARNING OUTCOME**
Describe the structure of a pulmonary lobule and the functional anatomy of the alveoli.

**Labeling:** Label each of the respiratory system structures in the figure below.

1. middle concha
2. hard palate
3. oropharynx
4. glottis
5. hyaline cartilage
6. Right lung
7. nares
8. Larynx
9. Primary bronch
10. 
11. 
12. 
13. 
14. 
15. 
16. 

**Matching:** Match each lettered term with the most closely related description.

a. respiratory bronchiole
b. respiratory mucosa
c. phonation
d. bronchodilation
e. terminal bronchiole
f. laryngeal prominence
g. pneumocytes type I
h. pneumocytes type II
i. cystic fibrosis
j. trachea
k. pharynx
l. blood air barrier
m. larynx
n. bronchoconstriction

17 Produce surfactant
18 Windpipe
19 Simple squamous epithelial cells
20 Sympathetic activation
21 Supplies a pulmonary lobule
22 Parasympathetic activation
23 Start of respiratory portion of respiratory tract
24 Gas exchange
25 Sound production at the larynx
26 Chamber shared by respiratory and digestive systems
27 Surrounds and protects the glottis
28 Lethal inherited respiratory disease
29 Lines the conducting portion of respiratory tract
30 Anterior surface of thyroid cartilage

17 _____
18 _____
19 _____
20 _____
21 _____
22 _____
23 _____
24 _____
25 _____
26 _____
27 _____
28 _____
29 _____
30 _____

21

# Module 21.8
# Respiratory physiology involves external and internal respiration

Go to
MasteringA&P™
PhysioEx 9.1
Ex 7: Repiratory System Mechanics

The general term **respiration** refers to two integrated processes: external respiration and internal respiration.

**1** This illustration provides an overview of respiration and shows relationships among external respiration, gas diffusion and transport, and internal respiration.

 **REVIEW**

**A.** Define *external respiration, gas diffusion,* and *internal respiration.*

## Respiration

### External Respiration

**External respiration** includes all the processes involved in the exchange of oxygen and carbon dioxide between blood, lungs, and the external environment. The purpose of external respiration, and the primary function of the respiratory system, is to meet the respiratory demands of the body's cells.

**Gas diffusion** occurs across the blood air barrier between alveolar air spaces and alveolar capillaries and across capillary walls between blood and other tissues.

### Internal Respiration

**Internal respiration** is the absorption of $O_2$ from blood and the release of $CO_2$ by tissue cells. We consider the biochemical pathways responsible for $O_2$ consumption and $CO_2$ generation (known collectively as *cellular respiration*) in Chapter 23.

**Pulmonary ventilation**, or breathing, involves the physical movement of air into and out of the lungs. The primary function of pulmonary ventilation is to maintain adequate **alveolar ventilation**—the movement of air into and out of the alveoli. Alveolar ventilation prevents the buildup of $CO_2$ in the alveoli and ensures a continuous supply of oxygen that keeps pace with its absorption by the bloodstream.

Lungs

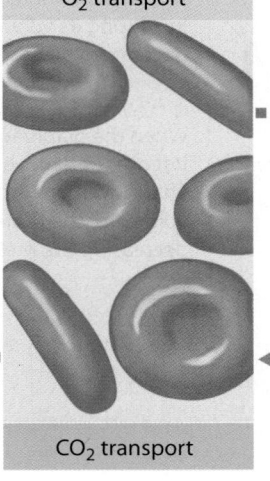

$O_2$ transport

Gas diffusion

Gas diffusion

$CO_2$ transport

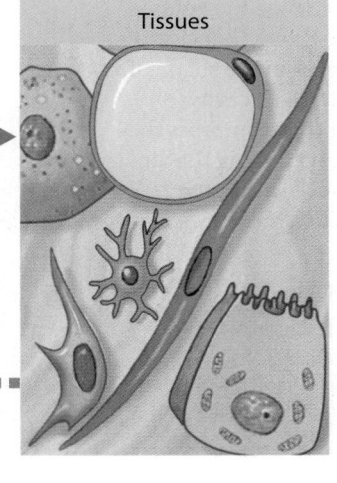

Tissues

Gas diffusion

Gas diffusion

**21**

In this section, we examine the integrated steps involved in external respiration. Abnormalities affecting any of the steps in external respiration ultimately affect the gas concentrations of interstitial fluids and, thus, cellular activities. If the oxygen content declines, the affected tissues will become starved for oxygen. **Hypoxia**, or low tissue oxygen levels, places severe limits on the metabolic activities of the affected area. If the supply of oxygen is cut off completely, **anoxia** (an-OK-sē-a; *a-*, without + *ox-*, oxygen) results. Much of the damage caused by strokes and heart attacks results from localized anoxia.

**?** **REVIEW**

**B.** How are hypoxia and anoxia different?

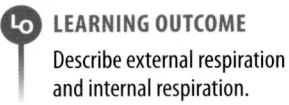 **LEARNING OUTCOME**

Describe external respiration and internal respiration.

# Module 21.9
## Pulmonary ventilation is driven by pressure changes within the pleural cavities

In a gas, such as air, the molecules bounce around as independent objects. At normal atmospheric pressures, gas molecules are much farther apart than the molecules in a liquid, so the density of air is relatively low. The pressure exerted by the enclosed gas results from the collision of gas molecules with the walls of the container. The greater the number of collisions, the higher the pressure.

**1** These figures show the relationships between pressure and volume in a gas. For a gas in a closed container and at a constant temperature, pressure ($P$) is inversely proportional to volume ($V$). That is, *if you decrease the volume of a gas, its pressure will increase. If you increase the volume of a gas, its pressure will decrease.* In particular, the relationship between pressure and volume is reciprocal: If you double the external pressure on a flexible container, its volume will drop by half, and if you reduce the external pressure by half, the volume of the container will double. This relationship, $P = 1/V$, is called **Boyle's law**.

If you decrease the volume of the container, collisions occur more frequently per unit time, increasing the pressure of the gas.

If you increase the volume, fewer collisions occur per unit time, because it takes longer for a gas molecule to travel from one wall to another. As a result, the gas pressure inside the container decreases.

**?** **REVIEW**
**A.** Define *Boyle's law*.

**2** Movements of the diaphragm and rib cage change the volume of the thoracic cavity. Changes in the shape of the thoracic cavity expand or compress the lungs, changing the air pressure within the respiratory tract.

Superior movement of the rib cage increases the depth and width of the thoracic cavity, increasing its volume and decreasing pressure within it.

When the diaphragm contracts, it tenses and moves inferiorly. This movement increases the volume of the thoracic cavity, decreasing the pressure within it.

**?** **REVIEW**
**B.** What physical changes affect the volume of the lungs?

**3** At the start of a breath, pressures inside and outside the thoracic cavity are identical, and no air moves into or out of the lungs.

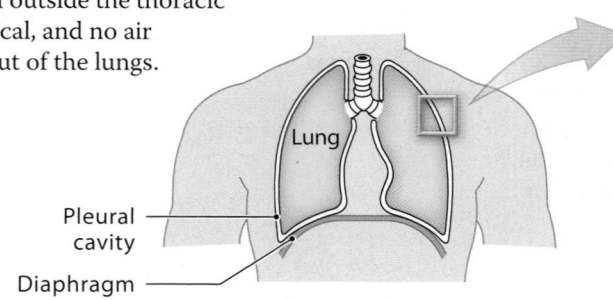

Lung

Pleural cavity

Diaphragm

$P_{outside} = P_{inside}$

Pressure outside and inside are equal, so no air movement occurs

Although separated by the pleural cavity, the layer of pleural fluid makes the lungs stick to the inner walls of the thorax. This kind of fluid bond is responsible for making a coaster stick to the bottom of a wet glass. The elastic tissues of the lungs are always trying to recoil and reduce lung volume to about 5 percent of its normal size. This collapse is prevented by the fluid bond between the parietal and visceral pleura. If an injury allows air into the pleural cavity, this bond is broken, and the lung collapses. This condition is called **atelectasis** (a-te-LEK-ta-sis).

Thoracic wall

Parietal pleura

Pleural fluid

Visceral pleura

Lung

**4** Air will flow from an area of higher pressure to an area of lower pressure. When the thoracic cavity enlarges and increases in volume during inhalation, pressure decreases inside the lungs, and air flows in. When the thoracic cavity decreases in volume during exhalation, pressure increases inside the lungs, forcing air out of the respiratory tract.

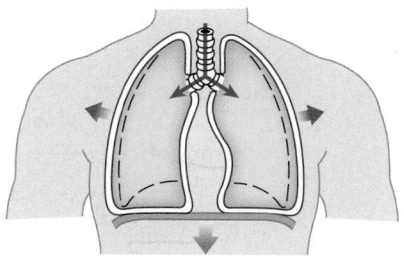

Inhalation: volume increases

$P_{outside} > P_{inside}$

Pressure inside decreases, so air flows in

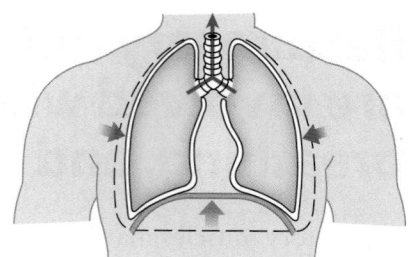

Exhalation: volume decreases

$P_{outside} < P_{inside}$

Pressure inside increases, so air flows out

**5** The direction of airflow is determined by the difference between atmospheric pressure and **intrapulmonary** (in-tra-PUL-mo-nār-ē) **pressure**, the pressure inside the respiratory tract (usually measured at the alveoli).

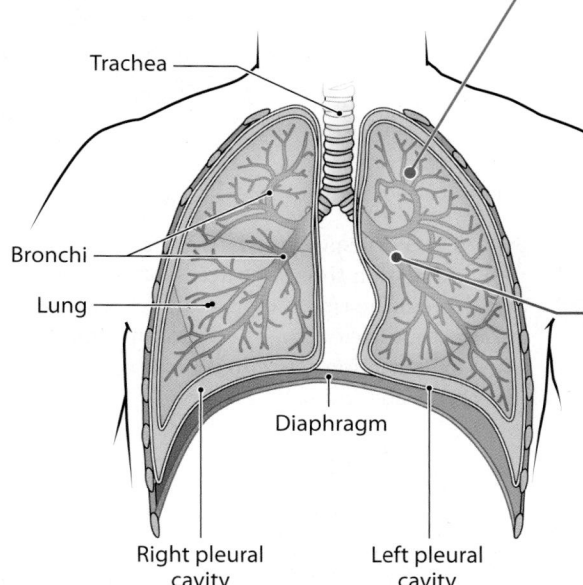

Trachea

Bronchi

Lung

Diaphragm

Right pleural cavity

Left pleural cavity

### Intrapulmonary Pressures Measured within Alveoli

A pressure differential of 0 mm Hg exists when atmospheric and intrapulmonary pressures are equal. Positive intrapulmonary pressures will push air out of the lungs; negative intrapulmonary pressures will pull air into the lungs.

### Airflow into and out of the Lungs

The **tidal volume ($V_T$)** is the amount of air moved into the lungs during inhalation and out of the lungs during exhalation. At rest, the tidal volume is approximately 500 mL.

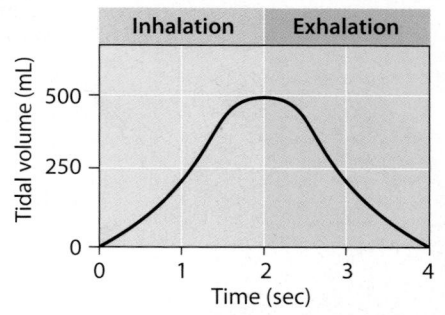

**6** Although we will use mm Hg to report pressures, other units are also used in clinical practice. The table below compares several units used to report gas pressures.

### Four Common Units for Reporting Gas Pressures

- **Millimeters of mercury (mm Hg):** This is the most common unit for reporting blood pressure and gas pressures. Normal atmospheric pressure is approximately 760 mm Hg.

- **Torr:** This unit of measurement is preferred by many respiratory therapists; it is also commonly used in Europe and in some technical journals. One torr is equivalent to 1 mm Hg; in other words, normal atmospheric pressure is equal to 760 torr.

- **Centimeters of water (cm $H_2O$):** In a hospital setting, anesthetic gas pressures and oxygen pressures are commonly measured in centimeters of water. One cm $H_2O$ is equivalent to 0.735 mm Hg; normal atmospheric pressure is 1033.6 cm $H_2O$.

- **Pounds per square inch (psi):** Pressures in compressed gas cylinders and other industrial applications are generally reported in psi. Normal atmospheric pressure at sea level is approximately 15 psi.

**?** **REVIEW**

**C.** What pressures determine the direction of airflow within the respiratory tract?

21

**Lₒ** **LEARNING OUTCOME**

Summarize the physical principles governing the movement of air into and out of the lungs.

# Module 21.10

# Respiratory muscles are involved with breathing, and ...

Respiratory muscles may be involved with either inhalation or exhalation. Those involved with inhalation are called **inspiratory muscles**; those involved in exhalation are called **expiratory muscles**.

**1** This anterior view introduces the **primary respiratory muscles** and **accessory respiratory muscles**. The primary muscles, the diaphragm and external intercostal muscles, are involved in inhalation. When you are breathing quietly, inhalation is active, but exhalation is passive—elastic forces and gravity are sufficient to decrease the volume of the lungs.

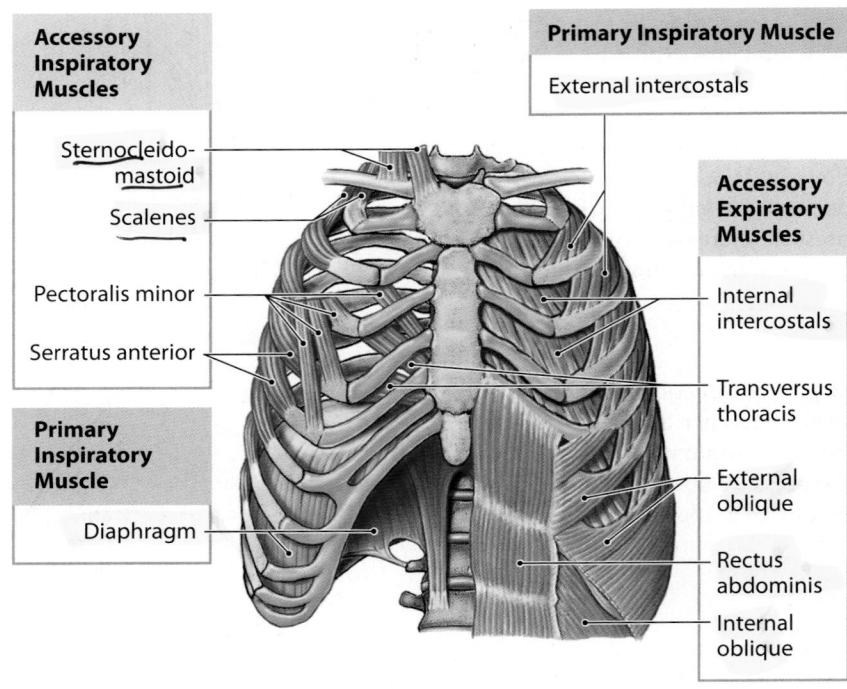

**Accessory Inspiratory Muscles**

Sternocleido-mastoid

Scalenes

Pectoralis minor

Serratus anterior

**Primary Inspiratory Muscle**

Diaphragm

**Primary Inspiratory Muscle**

External intercostals

**Accessory Expiratory Muscles**

Internal intercostals

Transversus thoracis

External oblique

Rectus abdominis

Internal oblique

**2** This lateral view during inhalation shows the inspiratory muscles that elevate the ribs and depress the diaphragm to enlarge the thoracic cavity.

**Accessory Inspiratory Muscles (active when needed)**

The contraction of accessory muscles assists the external intercostal muscles in elevating the ribs. The muscles increase the speed and amount of rib movement when the primary respiratory muscles are unable to move enough air to meet the oxygen demands of tissues.

**Primary Inspiratory Muscles**

Contraction of the external intercostal muscles elevates the ribs. This action contributes about 25 percent to the volume of air in the lungs at rest.

Contraction of the diaphragm flattens the floor of the thoracic cavity, increasing its volume and drawing air into the lungs. This is responsible for roughly 75 percent of the air movement in normal breathing at rest.

**?** **REVIEW**

**A.** Identify the primary inspiratory muscles.

**3** This corresponding lateral view during active exhalation shows the accessory expiratory muscles that depress the ribs and push the relaxed diaphragm into the thoracic cavity. The abdominal muscles that assist in exhalation are represented by a single muscle (the rectus abdominis).

**Accessory Expiratory Muscles (active when needed)**

The internal intercostals and transversus thoracis depress the ribs and reduce the width and depth of the thoracic cavity.

The abdominal muscles push the relaxed diaphragm into the thoracic cavity.

**?** **REVIEW**

**B.** When do the accessory respiratory muscles become active?

21

# ... pulmonary function tests determine lung performance

**4** Pulmonary function tests monitor respiratory function by measuring rates and volumes of air movement. Only a small volume of the air in our lungs is exchanged during a single quiet respiratory cycle (inhalation and exhalation). The total volume of the lungs is divided into a series of *volumes* and *capacities* (each the sum of various volumes), as shown in this spirogram, a tracing of respiratory movements. The red line indicates air volume within the lungs during breathing for an average male.

The **inspiratory reserve volume (IRV)** is the amount of air that you can breathe in over and above the tidal volume.

The **inspiratory capacity** is the amount of air that you can draw into your lungs after you have completed a quiet respiratory cycle. The inspiratory capacity is the sum of the tidal volume and the inspiratory reserve volume.

The **tidal volume (V$_T$)** is the amount of air you move into or out of your lungs during a single respiratory cycle under resting conditions.

The **vital capacity** is the maximum amount of air that you can move into or out of your lungs in a single respiratory cycle. The vital capacity is the sum of the expiratory reserve volume, the tidal volume, and the inspiratory reserve volume.

The **expiratory reserve volume (ERV)** is the amount of air that you can voluntarily expel after you have completed a normal, quiet respiratory cycle.

The **functional residual capacity (FRC)** is the amount of air remaining in your lungs after you have completed a quiet respiratory cycle. The FRC is the sum of the expiratory reserve volume and the residual volume.

The **minimal volume**, a component of the residual volume, is the amount of air that would remain in your lungs if they were allowed to collapse. The minimal volume ranges from 30 to 120 mL, but, unlike other volumes, it cannot be measured in a healthy person.

The **total lung capacity** is the total volume of your lungs, calculated by adding the vital capacity and the residual volume. The total lung capacity averages around 6000 mL in males and 4200 mL in females.

The **residual volume** is the amount of air that remains in your lungs even after a maximal exhalation.

Volume (mL): 6000, 2700, 2200, 1200, 0

Time →

21

| Sex Differences | | |
|---|---|---|
| | **Males** | **Females** |
| VT | 500 mL | 500 mL |
| IRV | 3300 mL | 1900 mL |
| ERV | 1000 mL | 700 mL |
| Residual volume | 1200 mL | 1100 mL |
| Total lung capacity | 6000 mL | 4200 mL |
| Vital capacity | 4800 mL | 3100 mL |
| Inspiratory capacity | 3800 mL | 2400 mL |
| Functional residual capacity | 2200 mL | 1800 mL |

 **REVIEW**
**C.** Name the various measurable pulmonary volumes.

 **LEARNING OUTCOME**
Name the respiratory muscles, describe their actions, and define the various pulmonary function tests.

# Module 21.11
# Pulmonary ventilation must be closely regulated to meet tissue oxygen demands

The respiratory system adjusts pulmonary ventilation over a broad range to meet the oxygen demands of the body. These adjustments involve varying both the number of breaths per minute and the amount of air moved in each breath. When you are exercising at peak levels, the amount of air moving into and out of your respiratory tract can be 50 times the amount moved at rest. The factors involved in pulmonary ventilation should remind you of the factors involved in regulating cardiovascular function.

## Respiratory Rate

**1** Your **respiratory rate** (*f*), or *respiratory frequency* (*f*), is the number of breaths you take each minute. As you read this, you are probably breathing quietly, with a low respiratory rate. The normal respiratory rate of a resting adult ranges from 12–18 breaths each minute, roughly one for every four heartbeats. Children breathe more rapidly, at rates of about 18–20 breaths per minute.

**?** **REVIEW**
**A.** Define *respiratory rate*.

## Respiratory Minute Volume

**2** We can calculate the volume of air moved each minute, symbolized $V_E$, by multiplying the respiratory rate, $f$, by the tidal volume, $V_T$. This value is called the **respiratory minute volume**. The respiratory rate at rest averages 12 breaths per minute, and the tidal volume at rest averages around 500 mL per breath. On that basis, we can calculate that respiratory minute volume at rest is approximately 6 L/min.

**Respiratory Minute Volume**

$$V_E = f \times V_T$$

$$\left(\begin{array}{c}\text{Volume of air moved}\\\text{each minute}\end{array}\right) = \left(\begin{array}{c}\text{Respiratory}\\\text{rate}\end{array}\right) \times \left(\begin{array}{c}\text{Tidal}\\\text{volume}\end{array}\right)$$

= 12 × 500 mL per minute

= 6000 mL per minute

= 6.0 liters per minute

In other words, the respiratory minute volume at rest is approximately 6 liters per minute.

**3** The factors involved in respiratory minute volume are easily diagrammed, but their functional relationships are complex. Increasing either respiratory rate or tidal volume will increase the respiratory minute volume. But what if the respiratory rate goes up, and the tidal volume goes down? The respiratory minute volume may remain the same, but the effects on respiratory performance are very different. We now consider the differences and their significance.

**Respiratory Rate**
Breaths per minute

**Respiratory Minute Volume**
Volume of air moved each minute

**Tidal Volume**
Volume of air moved with each breath

## Alveolar Ventilation

**4** **Alveolar ventilation**, symbolized $V_A$, is the amount of air reaching the alveoli each minute. The alveolar ventilation is less than the respiratory minute volume, because some of the air never reaches the alveoli but remains in the conducting portion of the respiratory system. This is known as the **anatomic dead space ($V_D$)**. To calculate the anatomic dead space, multiply the tidal volume by 0.3. The example below shows that with a tidal volume of 500 mL, anatomic dead space equals 150 mL.

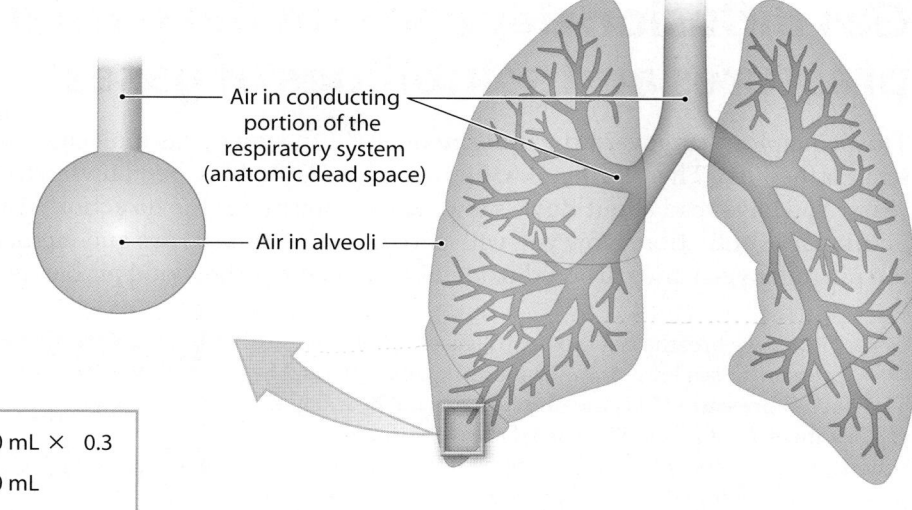

Air in conducting portion of the respiratory system (anatomic dead space)

Air in alveoli

$$V_D = V_T \times 0.3$$

Anatomic dead space $=$ Tidal volume $\times$ 0.3

$$V_D = 500 \text{ mL} \times 0.3$$
$$= 150 \text{ mL}$$

**5** Alveolar ventilation is calculated as the respiratory rate ($f$) multiplied by the difference between tidal volume and the anatomic dead space. The units are mL/min.

$$V_A = f \times V_T - V_D$$

Alveolar ventilation $=$ Breaths per minute $\times$ ( Tidal volume $-$ Anatomic dead space )

$$V_A = 12 \times (500 \text{ mL} - 150 \text{ mL})$$
$$= 12 \times 350 \text{ mL}$$
$$= 4200 \text{ mL/min}$$

However, the composition of the gas arriving in the alveoli is significantly different from that of the surrounding atmosphere because inhaled air always mixes with "used" air in the conducting passageways (the anatomic dead space) on its way to the exchange surfaces. Alveolar air thus contains less oxygen and more carbon dioxide than atmospheric air.

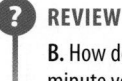 **REVIEW**

**B.** How does the respiratory minute volume differ from alveolar ventilation?

**6** Let's return to the effects of altering tidal volume and respiratory rate. If the respiratory rate increases to 20 breaths per minute but the tidal volume decreases to 300 mL, the respiratory minute volume will remain unchanged. However, the alveolar ventilation rate drops dramatically, falling from 4.2 L/min to 3 L/min.

? **REVIEW**

**C.** Which ventilates alveoli more effectively: slow, deep breaths or rapid, shallow breaths? Explain why.

**Respiratory Minute Volume**

$$V_E = f \times V_T$$
$$= 20 \times 300 \text{ mL per minute}$$
$$= 6.0 \text{ liters per minute}$$

**Alveolar Ventilation**

$$V_A = f \times (V_T - V_D)$$
$$= 20 \times (300 \text{ mL} - 150 \text{ mL}) \text{ per minute}$$
$$= 3.0 \text{ liters per minute}$$

This alveolar ventilation rate is almost 30 percent below its original value of 4.2 L/min, and if tissue demands are elevated as well, widespread tissue hypoxia could result. Thus, whenever the demand for oxygen increases, both the tidal volume *and* the respiratory rate must be increased.

 **LEARNING OUTCOME**

Explain how respiratory rate and tidal volume affect pulmonary and alveolar ventilation.

21

# Module 21.12
# Gas diffusion depends on the partial pressures and solubilities of gases

The principles that govern the movement and diffusion of gas molecules are relatively straightforward. These principles, known as **gas laws**, have been understood for about 250 years. You have read about Boyle's law, which determines the direction of air movement in pulmonary ventilation. Now you will learn about other gas laws and factors that determine the rate of oxygen and carbon dioxide diffusion across the blood air barrier.

**REVIEW**

**A.** Define *Dalton's law*.

**1** The air we breathe is a mixture of gases. Their total pressure at sea level, 1 atmosphere, equals 760 mm Hg. The **partial pressure (P)** is the pressure exerted by a single gas in a mixture of gases. The partial pressures for atmospheric gases are included in this table. All the partial pressures added together equal the total pressure exerted by the gas mixture; this is known as **Dalton's law**. As soon as air enters the respiratory tract, its characteristics begin to change. In passing through the nasal cavity, inhaled air becomes warmer, and the amount of water vapor increases. When it reaches the alveoli, the incoming air mixes with air remaining in the alveoli from the previous respiratory cycle. During the subsequent exhalation, the departing alveolar air mixes with air in the anatomic dead space, producing yet another mixture that differs from both atmospheric and alveolar air samples.

| Atmospheric Pressure = 760 mm Hg |
|---|
| $597(P_{N_2}) + 159(P_{O_2}) + 0.3(P_{CO_2}) + 3.7(P_{H_2O}) = 760$ mm Hg |

**Partial Pressures (mm Hg) and Normal Gas Concentrations (%) in Air**

| Source of Sample | Nitrogen ($N_2$) | Oxygen ($O_2$) | Carbon Dioxide ($CO_2$) | Water Vapor ($H_2O$) |
|---|---|---|---|---|
| Inhaled air (dry) | 597 (78.6%) | 159 (20.9%) | 0.3 (0.04%) | 3.7 (0.5%) |
| Alveolar air (saturated) | 573 (75.4%) | 100 (13.2%) | 40 (5.2%) | 47 (6.2%) |
| Exhaled air (saturated) | 569 74.8% | 116 (15.3%) | 28 (3.7%) | 47 (6.2%) |

**2** At a given temperature, the amount of a particular gas in solution is directly proportional to the partial pressure of that gas. This principle is known as **Henry's law**.

**21**

 **REVIEW**

**B.** What is the significance of Henry's law to the process of respiration?

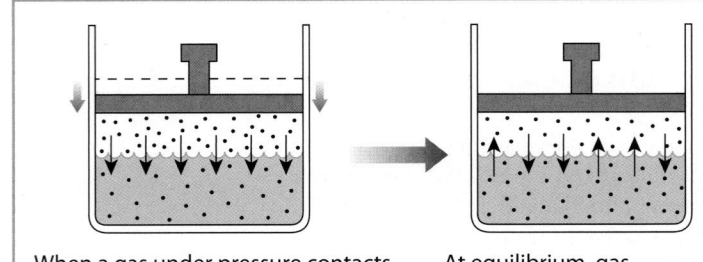

When a gas under pressure contacts a liquid, the pressure tends to force gas molecules into solution. At a given pressure, the number of dissolved gas molecules will increase until an equilibrium is established.

At equilibrium, gas molecules diffuse out of the liquid as quickly as they enter it, so the total number of gas molecules in solution remains constant.

**Example:**
Soda is put into the can under pressure, and the gas (carbon dioxide) is in solution at equilibrium.

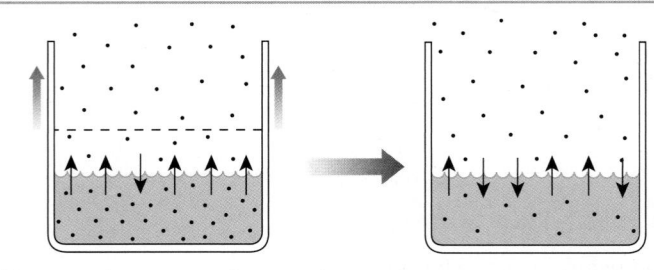

If the partial pressure is decreased, gas molecules will come out of solution.

Eventually a new equilibrium will be established.

When you open a soda can, the internal pressure decreases, and the gas molecules begin to come out of solution. The volume of the can is so small, and the volume of the atmosphere so great, that within a half hour or so virtually all the carbon dioxide comes out of solution, and you are left with "flat" soda.

**3** This illustration shows the partial pressures of $O_2$ and $CO_2$ during external respiration in the pulmonary circuit and during internal respiration in the systemic circuit.

Systemic circuit

Pulmonary circuit

## External Respiration

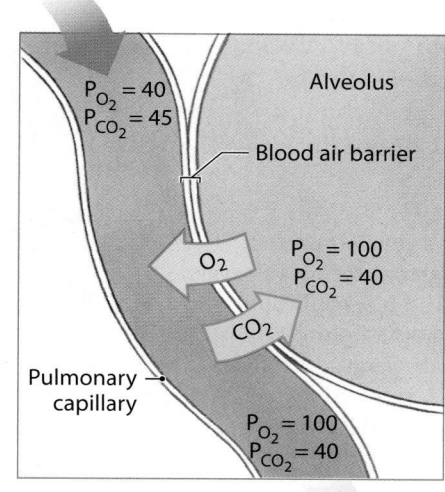

Alveolus

$P_{O_2} = 40$
$P_{CO_2} = 45$

Blood air barrier

$O_2$

$P_{O_2} = 100$
$P_{CO_2} = 40$

$CO_2$

Pulmonary capillary

$P_{O_2} = 100$
$P_{CO_2} = 40$

Blood arriving in the pulmonary arteries has a lower $P_{O_2}$ and a higher $P_{CO_2}$ than does alveolar air. Diffusion between the alveolar mixture and the pulmonary capillaries thus increases blood $P_{O_2}$ while decreasing its $P_{CO_2}$. By the time the blood enters the pulmonary venules, it has reached equilibrium with the alveolar air. Hence, blood departs the alveoli with a $P_{O_2}$ of about 100 mm Hg and a $P_{CO_2}$ of about 40 mm Hg. This is possible because each gas moves independently.

## Internal Respiration

Interstitial fluid

$P_{O_2} = 95$
$P_{CO_2} = 40$

$P_{O_2} = 40$
$P_{CO_2} = 45$

$O_2$

$CO_2$

$P_{O_2} = 40$
$P_{CO_2} = 45$

Systemic capillary

The partial pressure of oxygen in the pulmonary veins decreases to about 95 mm Hg as it mixes with venous blood from capillaries surrounding the conducting passageways rather than to the alveoli. This is the $P_{O_2}$ in the blood that arrives at peripheral capillaries. Normal interstitial fluid has a $P_{O_2}$ of 40 mm Hg. As a result, oxygen diffuses out of the capillaries until the capillary $P_{O_2}$ is the same as that in the adjacent tissues. Inactive peripheral tissues normally have a $P_{CO_2}$ of about 45 mm Hg, whereas blood entering peripheral capillaries normally has a $P_{CO_2}$ of 40 mm Hg. As a result, carbon dioxide diffuses into the blood as oxygen diffuses into the tissue.

The system just described is at equilibrium, and the $P_{O_2}$ and $P_{CO_2}$ are stable in the alveoli and in the tissues. Every oxygen molecule entering peripheral tissues is balanced by an oxygen molecule absorbed at the alveoli, and the absorbed oxygen molecule will be replaced in the next respiratory cycle. But if tissue oxygen demand accelerates, that equilibrium is disturbed, and the respiratory rate and tidal volume must then increase.

 **REVIEW**

**C.** Explain the decrease in $P_{O_2}$ from the pulmonary venules to the blood arriving in the peripheral capillaries of the systemic circuit.

 **LEARNING OUTCOME**

Summarize the physical principles governing the diffusion of gases into and out of the blood.

21

SmartArt

# Module 21.13
# Almost all the oxygen in blood is transported bound to hemoglobin within red blood cells

Each 100 mL of blood leaving the alveolar capillaries carries away roughly 20 mL of oxygen. Of this amount, only about 0.3 mL (1.5 percent) consists of oxygen molecules in solution. The rest of the oxygen molecules are bound to hemoglobin (Hb) molecules—specifically, to the iron ions in the center of **heme units**.

**1** A hemoglobin molecule consists of four globular protein subunits, each containing a heme unit that can bind an oxygen molecule. Thus, each hemoglobin molecule can reversibly bind up to four molecules of oxygen, forming **oxyhemoglobin (HbO$_2$)**. Carbon monoxide, CO, is a gas released by petroleum-burning engines and heaters. It is dangerous because it irreversibly binds to heme units, making them unavailable for oxygen transport.

Protein subunits

Iron ion

Heme unit

**?** **REVIEW**
**A.** Define *oxyhemoglobin*.

**2** The percentage of heme units containing bound oxygen at any given moment is called the **hemoglobin saturation**. If all the Hb molecules in the blood are fully loaded with oxygen, saturation is 100 percent. If, on average, each Hb molecule carries 2 O$_2$ molecules, saturation is 50 percent. This graph is an **oxygen-hemoglobin saturation curve**, which shows the saturation of hemoglobin at different partial pressures of oxygen. Notice that hemoglobin will be more than 90 percent saturated if exposed to a P$_{O_2}$ above 60 mm Hg.

Where the slope is steep, a very small change in blood P$_{O_2}$ will result in a large change in the amount of oxygen bound to Hb or released from HbO$_2$.

Blood entering the systemic circuit has a P$_{O_2}$ of 95 mm Hg, and the hemoglobin is about 97% saturated with oxygen.

Blood leaving peripheral tissues has an average P$_{O_2}$ of 40 mm Hg, so the hemoglobin drops from 97% to 75% saturation. It releases only 22% of its stored oxygen, so even venous blood contains substantial oxygen reserves.

The P$_{O_2}$ in active muscle tissue may drop to 15–20 mm Hg. Hemoglobin passing through these capillaries will go from 97% saturation to about 20% saturation. This means that an active tissue can receive 3.5 times more oxygen than an inactive tissue, even if the rate of blood flow remains the same.

Note: The curve has this shape because each arriving oxygen molecule increases the affinity of hemoglobin for the next oxygen molecule. Once the first oxygen molecule binds to the hemoglobin, the slope rises steeply until reaching a plateau near 100% saturation.

21

**3** Blood pH has a direct effect on the oxygen-hemoglobin saturation curve because the shape of hemoglobin molecules changes as the number of bound $O_2$ molecules increases and these changes affect its affinity for oxygen. The curves in this figure show how the slope of the oxygen-hemoglobin saturation curve changes when the pH of the blood shifts away from the normal average value of 7.4. This effect of pH on the hemoglobin saturation curve is called the **Bohr effect**. Carbon dioxide is the primary compound responsible for the Bohr effect.

If the pH increases, the saturation curve shifts to the left, and hemoglobin releases less oxygen.

The range of normal blood pH is 7.35–7.45.

If the pH decreases, the saturation curve shifts to the right, and hemoglobin releases more oxygen. At a pH of 7.4, hemoglobin saturation would be 75% at a $P_{O_2}$ of 40 mm Hg; if the pH shifts to 7.2, at the same $P_{O_2}$ hemoglobin saturation would be 60%. Thus, the pH shift triggered the release of an additional 15% of the bound oxygen.

**4** Temperature changes also affect the saturation curve. The higher the temperature, the more readily hemoglobin gives up its oxygen reserves. Blood temperature is about 38°C. Temperature effects are significant only in active tissues that are generating large amounts of heat. For example, active skeletal muscles generate heat, and the heat warms blood that flows through these organs. As the blood warms, the Hb molecules release more oxygen that can be used by the active muscle fibers.

 **REVIEW**

**B.** During exercise, hemoglobin releases more oxygen to active skeletal muscles than it does when those muscles are at rest. Why?

 **REVIEW**

**C.** Explain the relationship among BPG, oxygen, and hemoglobin.

---

Clinical Note

### Fresh whole blood has a limited storage time in a blood bank

Red blood cells (RBCs) do not contain mitochondria, so they can generate ATP only through glycolysis. The metabolic pathways involved in glycolysis also generate the compound **2,3-bisphosphoglycerate** (biz-fos-fō-GLIS-er-āt), or **BPG**. For any partial pressure of oxygen, the higher the concentration of BPG, the more oxygen will be released by the Hb molecules. BPG production decreases as RBCs age, and levels of BPG can determine how long a blood bank can store fresh whole blood. When BPG levels get too low, hemoglobin becomes firmly bound to the available oxygen. The blood is then useless for transfusions, because the RBCs will no longer release oxygen to peripheral tissues, even at a disastrously low $P_{O_2}$.

 **LEARNING OUTCOME**

Discuss the structure and function of hemoglobin, explain the oxygen-hemoglobin saturation curve, and describe the role of 2,3-bisphosphoglycerate.

21

# Carbon dioxide is transported three ways in the bloodstream

Carbon dioxide is generated by aerobic metabolism in peripheral tissues. After entering the bloodstream, a $CO_2$ molecule (1) dissolves in the plasma, (2) binds to hemoglobin within red blood cells, or (3) is converted to a bicarbonate ion ($HCO_3^-$) within red blood cells. All three reactions are completely reversible.

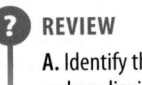 **REVIEW**

**A.** Identify three ways that carbon dioxide is transported in the bloodstream.

**1** We will consider the events that occur as blood enters peripheral tissues in which the $P_{CO_2}$ is 45 mm Hg.

$CO_2$ diffuses into the bloodstream.

$CO_2$ has limited solubility in plasma, and only ~7% remains in solution.

93% diffuses into RBCs.

Roughly 23% of the carbon dioxide carried by blood is bound to the globular protein portions of Hb molecules. These $CO_2$ molecules are reversibly attached to exposed amino groups ($NH_2$) of the Hb molecules. The resulting compound is called **carbaminohemoglobin** (kar-BAM-i-nō-hē-mō-glō-bin), **HbCO$_2$**.

Most of the carbon dioxide absorbed by blood (roughly 70% of the total) is converted to carbonic acid through the activity of the enzyme **carbonic anhydrase**.

The carbonic acid molecules immediately dissociate into a hydrogen ion and a bicarbonate ion. The net reaction is:

$$H_2CO_3 \rightleftharpoons H^+ + HCO_3^-$$

Most of the hydrogen ions bind to hemoglobin molecules, forming **HbH$^+$**. The Hb molecules thus function as pH buffers, tying up the released hydrogen ions before the ions can leave the RBCs and decrease the plasma pH.

$H^+$

$HCO_3^-$

**PLASMA**

**RBC**

$Cl^-$

The **bicarbonate ions** move into the surrounding plasma with the aid of a countertransport mechanism that exchanges intracellular bicarbonate ions ($HCO_3^-$) for extracellular chloride ions ($Cl^-$). This exchange, which trades one anion for another, does not require ATP. The result is a mass movement of chloride ions into the RBCs, an event known as the **chloride shift**.

21

**2** This illustration summarizes the transport of oxygen and carbon dioxide in the lungs (at left) and in peripheral tissues (at right). This system is at equilibrium, and the partial pressures of oxygen ($P_{O_2}$) and carbon dioxide ($P_{CO_2}$) are stable in the alveoli and in the tissues (see Module 21.12). Every oxygen molecule entering peripheral tissues is balanced by an oxygen molecule absorbed by the alveoli, and the absorbed oxygen molecule will be replaced in the next respiratory cycle.

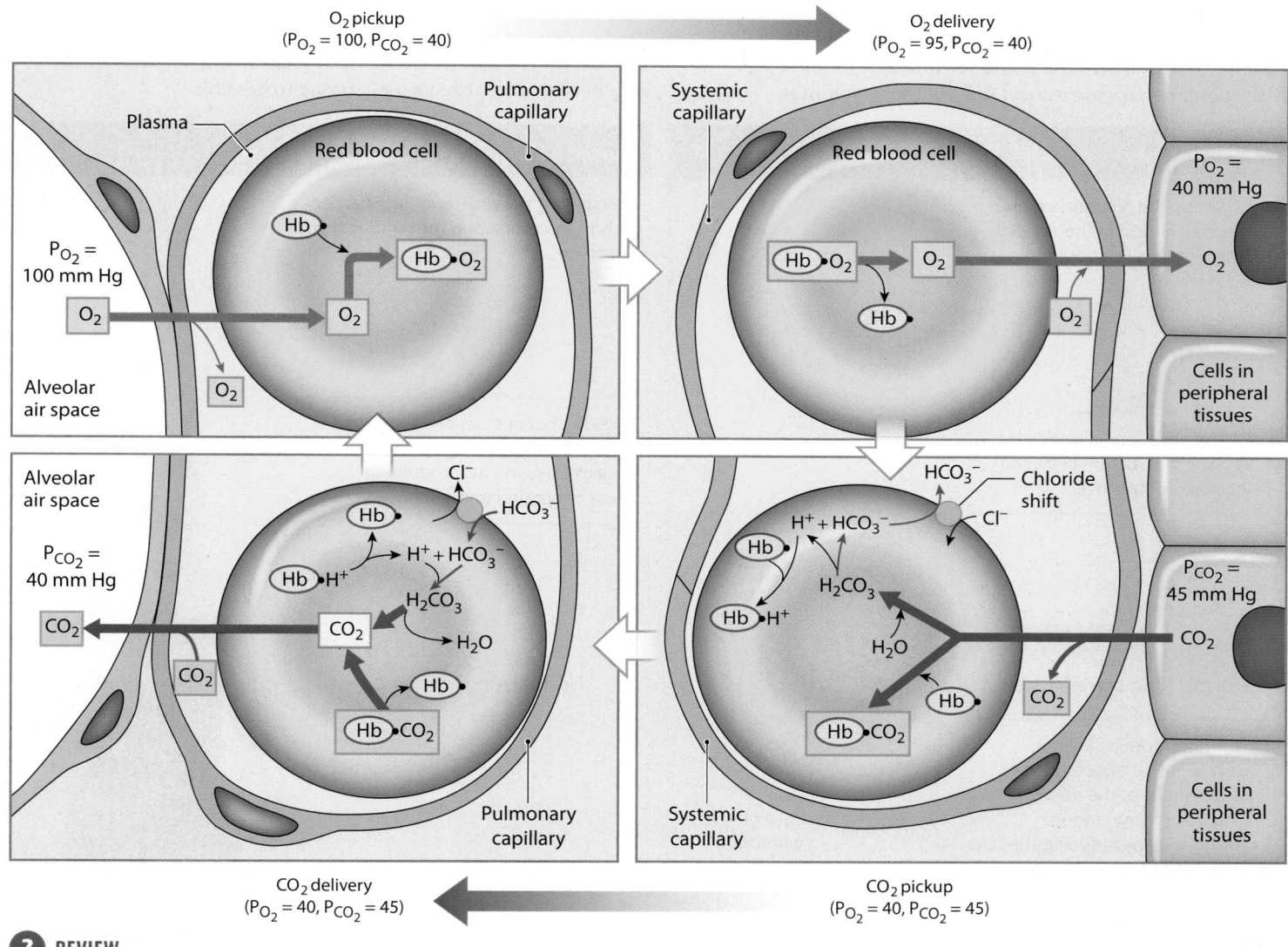

$O_2$ pickup
($P_{O_2} = 100$, $P_{CO_2} = 40$)

$O_2$ delivery
($P_{O_2} = 95$, $P_{CO_2} = 40$)

$CO_2$ delivery
($P_{O_2} = 40$, $P_{CO_2} = 45$)

$CO_2$ pickup
($P_{O_2} = 40$, $P_{CO_2} = 45$)

**? REVIEW**

**B.** Describe the forces that drive oxygen and carbon dioxide transport between the blood and peripheral tissues.

The equilibrium between oxygen absorption and oxygen use is disturbed when tissue oxygen demand increases. If the respiratory rate and tidal volume do not increase, the alveolar $P_{O_2}$ will steadily decrease, and the alveolar, blood, and tissue $P_{CO_2}$ will steadily increase. This is a particulary unpleasant combination that can lead to widespread hypoxia and a dangerous decrease in the pH of body fluids. In a clinical setting, pulmonary ventilation is therefore closely regulated by monitoring the $P_{O_2}$, $P_{CO_2}$, and pH of body fluids.

**? INTEGRATION**

**C.** How would blockage of the trachea affect blood pH?

**Lo LEARNING OUTCOME**

Describe how carbon dioxide is transported in the blood, and explain how oxygen is picked up, transported, and released into the bloodstream.

# Module 21.15
# Pulmonary disease can affect both lung elasticity and airflow

 **1** The **compliance** of the lungs is a measure of their expandability, or how easily the lungs expand in response to applied pressure. It is influenced by the internal structure of the lungs (elasticity and resilience) and the flexibility of the chest wall. It is determined by monitoring the intrapulmonary pressure at different lung volumes.

**2** The **resistance** of the lungs is an indication of how much force is required to inflate or deflate them. At rest, the muscular activity involved in pulmonary ventilation accounts for 3–5 percent of the resting energy demand. If resistance increases, that figure increases dramatically, and an individual may become exhausted simply trying to breathe.

## Compliance

The greater the compliance, the lower the tension in the walls of the lungs at a given volume.

The loss of supporting tissues and the destruction of alveoli increase compliance.

Normal lung compliance

Arthritis or other skeletal disorders that affect the articulations of the ribs or spinal column reduce compliance by reducing the elasticity of the chest wall.

Compliance = ease of expansion

**Respiratory distress syndrome** is caused by inadequate surfactant production. This leads to alveolar collapse on exhalation. As a result, compliance is reduced.

The lower the compliance, the greater the tension in the walls of the lungs at a given volume, and the harder it is for air to flow along the conducting passages.

## Resistance

The higher the resistance, the harder it is to force air along the conducting passages.

Bronchoconstriction increases resistance, as does excessive mucus production or airway inflammation.

Normal lung resistance

Sympathetic activation of bronchodilation reduces resistance.

Resistance = inflation force

The lower the resistance, the more easily air flows along the conducting passageways.

 **? REVIEW**
**A.** Define *compliance* and *resistance*.

**21**

# Chronic Obstructive Pulmonary Disease

**3** **Chronic obstructive pulmonary disease (COPD)** is a general term for a progressive disorder of the airways that restricts airflow and reduces alveolar ventilation. We highlight three different COPDs below. *Asthma*, or asthmatic bronchitis, is the term used when symptoms are acute and intermittent. The terms *chronic bronchitis* and *emphysema* are usually applied when signs and symptoms are long lasting and progressive with occasional crises due to infections.

## Asthma

**Asthma** (AZ-muh) is a condition characterized by conducting passageways that are extremely sensitive to irritation. The airways respond to irritation by constricting smooth muscles all along the bronchial tree. This is accompanied by edema and swelling of the mucosa of the respiratory passageways and the accelerated production of mucus. The combination makes breathing very difficult, and resistance is markedly increased. This can be caused by allergies, toxins, or exercise.

## Chronic Bronchitis

**Chronic bronchitis** (brong-KĪ-tis) is a long-term inflammation and swelling of the bronchial lining, leading to overproduction of mucus. The characteristic sign is frequent coughing with copious sputum production. This condition is most commonly related to cigarette smoking but also results from other environmental irritants, such as chemical vapors. Over time, the increased mucus production can block smaller airways, increasing resistance and reducing respiratory efficiency. Chronic bacterial infections can further damage lungs. People with chronic bronchitis may have signs of heart failure, including widespread edema. Their blood oxygenation is low, and their skin may have a bluish color (cyanosis). The combination of widespread edema and bluish coloration has led to the descriptive term **blue bloaters** for people with this condition.

## Emphysema

**Emphysema** (em-fi-ZĒ-muh) is a chronic, progressive condition characterized by shortness of breath and an inability to tolerate physical exertion. The underlying problem is the destruction of alveolar surfaces and inadequate surface area for oxygen and carbon dioxide exchange. The alveoli gradually expand, and adjacent alveoli merge to form larger air spaces supported by fibrous tissue without alveolar capillary networks. As elastic connective tissues are lost, compliance increases, but the loss of respiratory surface area restricts oxygen absorption, so the person becomes short of breath. The respiratory muscles work hard, and people with this condition, who use a lot of energy just breathing, tend to be thin. Chest x-rays, such as the one at right, show overexpanded lungs. Because their exaggerated respiratory movements typically maintain near normal blood oxygenation, the skin of pale-skinned emphysema patients is usually pink. The combination of heavy breathing and pink coloration has led to the descriptive term **pink puffers** for these people.

**21**

**?** **REVIEW**

**B.** Identify three chronic obstructive pulmonary diseases (COPDs).

**C.** Compare chronic bronchitis with emphysema.

**Lo** **LEARNING OUTCOME**

Explain how pulmonary disease affects compliance and resistance.

## Module 21.16
# Respiratory control mechanisms involve interacting centers in the brainstem

**1** Respiratory control involves multiple levels of regulation. Respiratory rate and rhythm are a function of a network of respiratory centers that includes centers in the pons and medulla. Most of the regulatory activities occur outside our awareness.

### Higher Centers

**Higher centers** in the hypothalamus, limbic system, and cerebral cortex can alter the activity of the pneumotaxic centers, but essentially normal respiratory cycles continue even if the brainstem superior to the pons has been severely damaged.

### Apneustic and Pneumotaxic Centers in the Pons

The **apneustic** (ap-NŪ-stik) **centers** and the **pneumotaxic** (nū-mō-TAK-sik) **centers** of the pons are paired nuclei that adjust the output of the respiratory rhythmicity centers.

**REVIEW**
**A.** Name the paired central nervous system nuclei that adjust the pace of respiration.

### Respiratory Rhythmicity Centers

The most basic level of respiratory control involves pacemaker cells in the medulla oblongata. These neurons generate cycles of contraction and relaxation in the diaphragm. The paired **respiratory rhythmicity centers** establish the pace of respiration by adjusting these pacemakers and coordinating additional respiratory muscle actions. Each rhythmicity center is subdivided into a **dorsal respiratory group (DRG)** and a **ventral respiratory group (VRG)**. The DRG modifies its activities in response to input from chemoreceptors that monitor $O_2$, $CO_2$, and pH in the blood and cerebrospinal fluid and from baroreceptors (stretch receptors) that monitor stretching in the walls of the lungs. Current research suggests that it is impossible to attribute respiratory rate and rhythm to a single area, aside from the basic pattern generator in the ventrolateral medulla known as the pre-Bötzinger complex. However, the DRG is mainly concerned with inspiration, and the VRG is primarily associated with expiration.

**Higher Centers**
- Cerebral cortex
- Limbic system
- Hypothalamus

The pneumotaxic centers inhibit the apneustic centers and promote passive or active exhalation. An increase in pneumotaxic output quickens the pace of respiration by shortening the duration of each inhalation. A decrease in pneumotaxic output slows the respiratory pace but increases the depth of respiration because the apneustic centers are more active.

Inhibition

The apneustic centers promote inhalation by stimulating the DRG. During forced breathing, the apneustic centers adjust the degree of stimulation in response to sensory information from the vagus nerve (X) concerning the amount of lung inflation.

Pons

Medulla oblongata

To diaphragm ←

To external intercostal muscles ←

The **inspiratory center** of the DRG contains neurons that control lower motor neurons innervating the external intercostal muscles and the diaphragm. This center functions in every respiratory cycle.

To accessory inspiratory muscles ←

To accessory expiratory muscles ←

The VRG has inspiratory and expiratory centers that function only when breathing demands increase and accessory respiratory muscles become involved.

In addition to the centers in the pons, DRG and VRG, the **pre-Bötzinger complex** in the medulla is essential to all forms of breathing. Its mechanisms are poorly understood.

**REVIEW**
**B.** Which brainstem centers generate the respiratory pace?

21

**2** This flowchart shows the events involved in **quiet breathing** in a person at rest.

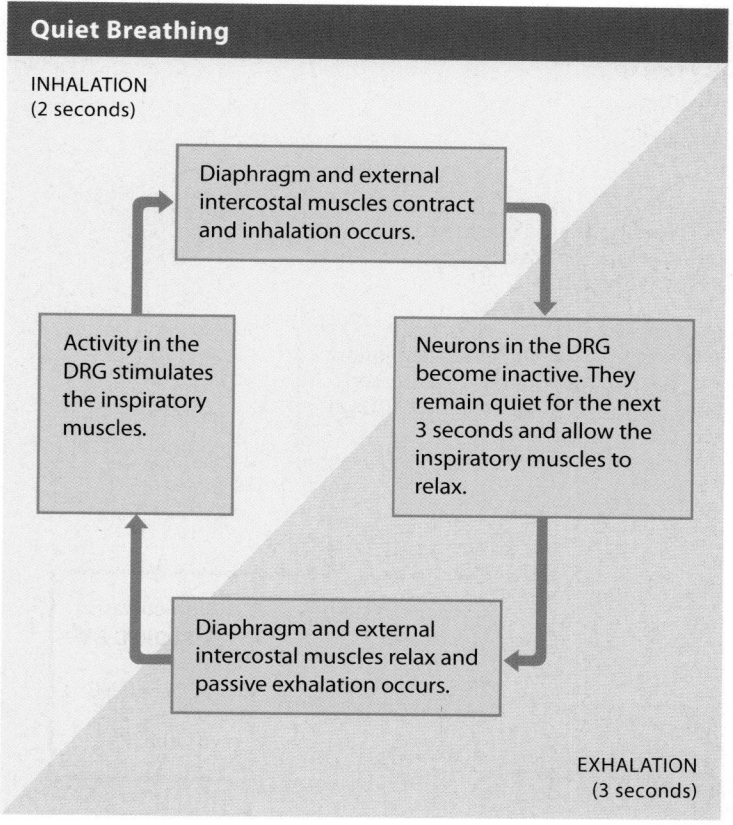

**Quiet Breathing**

INHALATION
(2 seconds)

Diaphragm and external intercostal muscles contract and inhalation occurs.

Activity in the DRG stimulates the inspiratory muscles.

Neurons in the DRG become inactive. They remain quiet for the next 3 seconds and allow the inspiratory muscles to relax.

Diaphragm and external intercostal muscles relax and passive exhalation occurs.

EXHALATION
(3 seconds)

**3** This flowchart shows what happens during **forced breathing**, when pulmonary ventilation increases toward maximal levels by involving accessory respiratory muscles.

**Forced Breathing**

INHALATION

Inspiratory muscles contract, and expiratory muscles relax. Inhalation occurs.

Increased activity in the DRG stimulates neurons of the VRG that in turn activate the accessory muscles involved in inhalation. The expiratory center of the VRG is inhibited.

DRG and inspiratory center of VRG are inhibited. Expiratory center of VRG is active.

After each inhalation, active exhalation occurs as the neurons of the expiratory center of the VRG stimulate the appropriate accessory muscles. Inspiratory muscles relax.

EXHALATION

**4** This table lists the various sensory stimuli that can modify the activities of the respiratory centers. These stimuli can trigger automatic responses known as **respiratory reflexes**, which are the focus of the next module.

**Stimuli That Trigger Respiratory Reflexes**

- Chemoreceptors sensitive to the pH, $P_{O_2}$, or $P_{CO_2}$ of the blood or cerebrospinal fluid alter the activities of the respiratory centers.
- Baroreceptors in the aortic or carotid sinuses sensitive to changes in blood pressure alter the activities of the respiratory centers.
- Stretch receptors that respond to changes in the volume of the lungs are responsible for inflation and deflation reflexes.
- Irritating physical or chemical stimuli in the nasal cavity, larynx, or bronchial tree initiate protective reflexes, such as coughing or sneezing.

**? REVIEW**

**C.** Which chemical factors in blood or cerebrospinal fluid stimulate the respiratory centers?

21

Respiratory control is an area of intense study, and our understanding of the interactions and mechanisms remains incomplete. Because of the brainstem's location, it is nearly impossible to study this mechanism in humans, and much of the work has been done in cats and newborn rodents.

 **LEARNING OUTCOME**

Describe the brainstem structures that influence the control of respiration.

# Module 21.17
# Respiratory reflexes provide rapid automatic adustments in pulmonary ventilation

## Chemoreceptor Reflexes

**1** This diagram shows the effect of changes in $P_{CO_2}$ on the respiratory rate. Under normal conditions, the $P_{CO_2}$ is the most important factor stimulating chemoreceptors and thereby influencing respiratory activity. An increase of just 10 percent in the arterial $P_{CO_2}$ causes the respiratory rate to double, even if the $P_{O_2}$ remains completely normal. In contrast, a decrease in arterial $P_{O_2}$ has little effect on the respiratory centers until the arterial $P_{O_2}$ decreases below 60 mm Hg.

**? REVIEW**
**A.** Are chemoreceptors more sensitive to blood $CO_2$ levels or blood $O_2$ levels?

An increase in the arterial blood $P_{CO_2}$ constitutes **hypercapnia**. The most common cause of hypercapnia is **hypoventilation**—when respiratory activity is insufficient to meet the demands for tissue oxygen delivery and carbon dioxide removal. Carbon dioxide then accumulates in the blood.

Chemoreceptors stimulated by increased $P_{CO_2}$ and decreased pH

**Effector**
Respiratory muscles stimulated

*Result in*

Increased respiratory rate with increased elimination of $CO_2$ at alveoli

**Receptor**
Chemoreceptors in arteries and medulla oblongata

Decreased arterial $P_{CO_2}$

Homeostasis **DISTURBED BY**
INCREASING
arterial $P_{CO_2}$ (hypercapnia)

**STIMULUS**

**RESTORED**

Homeostasis **RESTORED BY**
DECREASING
arterial $P_{CO_2}$

# HOMEOSTASIS
## NORMAL ARTERIAL $P_{CO_2}$

Homeostasis **DISTURBED BY**
DECREASING
arterial $P_{CO_2}$ (hypocapnia)

**STIMULUS**

**RESTORED**

Homeostasis **RESTORED BY**
INCREASING
arterial $P_{CO_2}$

**Receptor**
Chemoreceptors in arteries and medulla oblongata

Increased arterial $P_{CO_2}$

**? REVIEW**
**B.** Define *hypercapnia* and *hypocapnia*.

Chemoreceptors inhibited by decreased $P_{CO_2}$ and increased pH

**Effector**
Respiratory muscles inhibited

*Result in*

Decreased respiratory rate with decreased elimination of $CO_2$ at alveoli

**Hyperventilation**, when the rate and depth of respiration exceed the demands for oxygen delivery and carbon dioxide removal, gradually leads to **hypocapnia**, an abnormally low $P_{CO_2}$. Swimmers sometimes hyperventilate to extend their time underwater; this works because it is the $P_{CO_2}$ that stimulates respiratory activity. If the $P_{CO_2}$ is driven down too far, they may become unconscious from oxygen starvation in the brain without ever feeling the urge to breathe. This condition is **shallow water blackout**.

21

## Baroreceptor Reflexes

**2** Baroreceptors in the carotid and aortic sinuses are monitored by sensory nerves within the glossopharyngeal and vagus nerves, respectively. The cardiovascular effects of their stimulation were considered in Module 19.9. The sensory information is also distributed to the respiratory centers, triggering adjustments in the respiratory minute volume.

When arterial blood pressure decreases below its normal range, the respiratory centers are stimulated, causing the respiratory minute volume to increase.

— Normal range —

When arterial blood pressure increases above normal levels, the respiratory centers are inhibited, causing the respiratory minute volume to decrease.

## Inflation/Deflation Reflexes

**3** Inflation and deflation reflexes are activated by stretch receptors in the lungs during forced breathing when the tidal volume is large (1000 mL or more). This sensory information is distributed to the apneustic centers and the VRG.

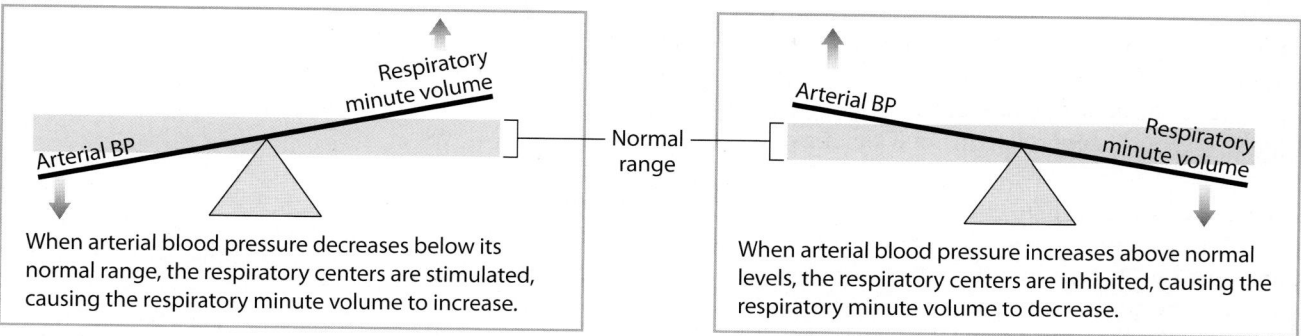

The **inflation reflex** prevents overexpansion of the lungs during forced breathing.

**2** Respiratory muscles are inhibited by the respiratory rhythmicity centers, and inhalation stops.

DRG
VRG

**1** Stretch receptors located in the smooth muscle around bronchioles are stimulated by lung expansion.

**3** The expiratory centers of the VRG are then stimulated, and forced exhalation begins.

The **deflation reflex** inhibits the expiratory centers and stimulates the inspiratory centers when the lungs are deflating.

**3** Forced inhalation now begins, with activation of the DRG and the inspiratory centers of the VRG.

DRG
VRG

**1** Stretch receptors located in the alveolar walls are stimulated as elastic fibers recoil and the alveolar volume is reduced.

**2** The expiratory centers of the VRG are inhibited. The inhibition continues until exhalation stops.

21

## Protective Reflexes

**4** **Protective reflexes** operate when you are exposed to irritants and include sneezing and coughing. Both reflexes involve **apnea** (AP-nē-uh), a period in which breathing has stopped, usually followed by a forceful expulsion of air to remove the offending stimulus. Air leaving the larynx can travel at 160 kph (99 mph), carrying mucus, particles, and irritating gases out of the respiratory tract through the nose or mouth.

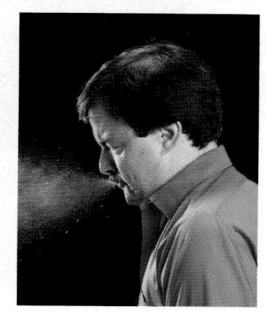

**? INTEGRATION**

**C.** Johnny is angry, so he tells his mom that he will hold his breath until he turns blue and dies. Explain whether this will likely happen.

**Lo LEARNING OUTCOME**

Identify and discuss reflex respiratory activity in pulmonary ventilation.

# Module 21.18

# Respiratory function decreases with age; smoking makes matters worse

All aspects of respiratory function are affected by aging. As elastic tissue deteriorates throughout the body, vital capacity decreases. As arthritic changes stiffen rib joints, compliance and maximum respiratory minute volume are decreased. These respiratory limitations make strenuous exercise difficult or impossible.

**1** In addition to skeletal and connective tissue changes associated with age, some degree of emphysema is normal in people over age 50. However, the extent varies widely with the lifetime exposure to cigarette smoke and other respiratory irritants. This graph compares the respiratory performance of people who have never smoked with those who have smoked for various periods of time. The message is quite clear: Although some decrease in respiratory performance is inevitable, you can prevent serious respiratory deterioration by stopping smoking or never starting.

**?** **REVIEW**

**A.** Name several age-related factors that affect the respiratory system.

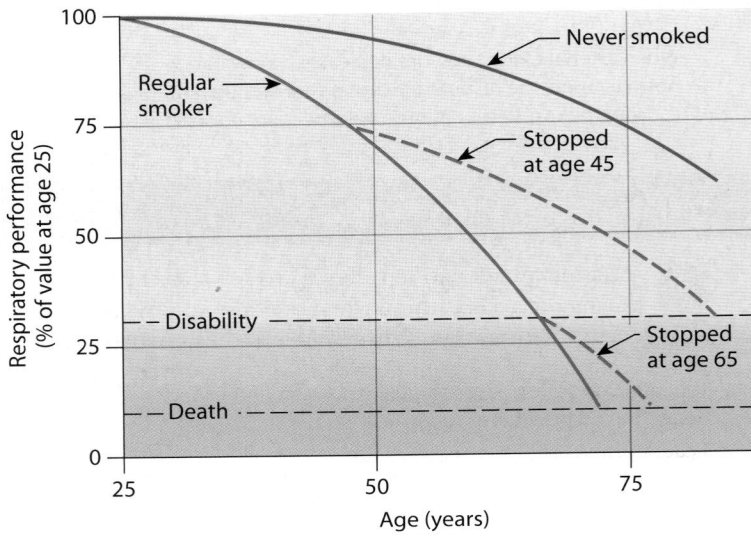

## Lung Cancer

**2** **Lung cancer** is an aggressive class of malignancies. These cancers affect the epithelial cells that line conducting passageways, mucous glands, or alveoli. Signs and symptoms generally do not appear until tumors restrict airflow or compress adjacent structures. Chest pain, shortness of breath, a cough or a wheeze, and weight loss are common. Evidence has shown that 85–90% of all lung cancers are the direct result of cigarette smoking. Treatment varies with the cellular organization of the tumor and whether *metastasis* (cancer cell migration) has occurred. Surgery, radiation, or chemotherapy may be involved. According to the CDC, more Americans die from lung cancer than any other type of cancer. The risk of getting lung cancer increases as we get older and is greater in men than women.

Normal lung       Lung cancer

**?** **REVIEW**

**B.** Describe lung cancer.

**3** Smoke is an irritant, and chemicals found in cigarette smoke include several carcinogens (cancer-causing substances). This figure shows the progression of events leading to lung cancer. The fact that cigarette smoking causes cancer is not surprising, given that the smoke contains carcinogens. What is surprising is that more smokers do not develop lung cancer. Evidence suggests that some smokers have a genetic predisposition to developing at least one form of lung cancer.

### Normal Respiratory Epithelium

The normal respiratory epithelium consists of ciliated columnar epithelium with an abundance of mucous cells, which produce mucus that helps clean inhaled air. The irritants and inhaled carcinogens contained in cigarette smoke begin a series of changes to this epithelium.

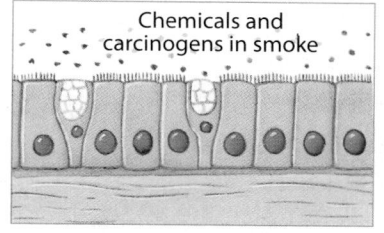

Chemicals and carcinogens in smoke

Reversible

### Dysplasia

In **dysplasia**, cells are damaged and their functional characteristics change. The cilia of respiratory epithelial cells are damaged and paralyzed by exposure to cigarette smoke. These changes cause the local buildup of mucus and reduce the effectiveness of the epithelium in protecting deeper, more delicate portions of the respiratory tract.

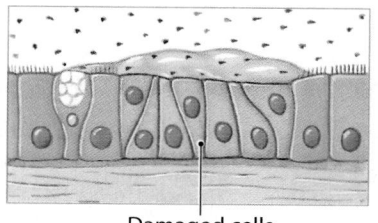

Damaged cells

Reversible

### Metaplasia

In **metaplasia**, a tissue changes its structure in response to injury or chemical stresses. In this case, the stressed respiratory surface converts to a stratified epithelium that protects underlying connective tissue but does nothing for other areas of the respiratory tract. Metaplasia may be reversed if the stressful stimulus is removed before further damage occurs.

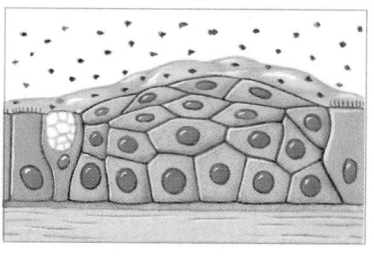

Irreversible

### Neoplasia and Anaplasia

In **neoplasia**, the growth of abnormal cells forms a cancerous tumor called a **neoplasm**. In **anaplasia**, the most dangerous stage, the cells become malignant and metastasize (spread) to other parts of the body. Neither neoplasia nor anaplasia is reversible, although the cancers may be treated by chemicals, radiation, and/or surgery.

**? REVIEW**

**C.** Compare dysplasia, metaplasia, neoplasia, and anaplasia.

21

 **LEARNING OUTCOME**

Describe age-related changes to, and the effects of cigarette smoking on, the respiratory system.

## Section 2 Review
# Respiratory Physiology

**Matching:** Match each lettered term with the most closely related description.

a. apnea

b. hemoglobin releases more $O_2$

c. decreases vital capacity

d. external intercostals

e. iron ion

f. Boyle's law

g. compliance

h. bicarbonate ion

i. anoxia

j. atelectasis

k. hypocapnia

l. pneumotaxic centers

m. apneustic centers

n. partial pressure

| 1 | Single gas in a mixture |
| 2 | A cause of tissue death |
| 3 | Inverse pressure/volume relationship |
| 4 | Expandability of lungs |
| 5 | $CO_2$ transport |
| 6 | Elastic tissue deterioration |
| 7 | Act to elevate ribs |
| 8 | Heme unit |
| 9 | Blood pH decreases |
| 10 | Promotes passive or active exhalation |
| 11 | Stimulate DRG and promote inhalation |
| 12 | Hyperventilation |
| 13 | Collapsed lung |
| 14 | Period of suspended respiration |

1 _____
2 _____
3 _____
4 _____
5 _____
6 _____
7 _____
8 _____
9 _____
10 _____
11 _____
12 _____
13 _____
14 _____

**Short answer:** Identify and describe the pulmonary volumes and capacities indicated in the spirogram below.

15 _____

16 _____

17 _____

18 _____

19 _____

20 _____

21 _____

22 _____

23 _____

**Section integration:** Compare and contrast external respiration, pulmonary ventilation, and internal respiration.

24 _____

_____

# Study Outline

## SECTION 1 • Anatomy of the Respiratory System

### Module 21.1

**The respiratory system has an upper and lower respiratory tract with different functions**   p. 797

1. The **respiratory system** is composed of structures for breathing, or **pulmonary ventilation** (airflow to and from lungs), and gas exchange.

2. The respiratory system is divided into an upper and lower respiratory system.

3. The **respiratory tract** is a passageway that carries air to and from the exchange surfaces of the lungs. It has a **conducting portion** and a **respiratory portion**.

4. The conducting portion extends from the nasal cavity down to the larger bronchioles of the lungs. The respiratory portion includes the smallest bronchioles and **alveoli** (air-filled sacs where gas exchange occurs).

5. The respiratory system provides an extensive surface area for gas exchange between the air and circulating blood, moves air to and from the gas exchange surfaces of the lungs, protects the respiratory surfaces, produces sounds for communication, and facilitates olfaction.

### Module 21.2

**The respiratory defense system protects the respiratory mucosa**   p. 798

6. The mucus and cilia of the **respiratory defense system** prevent contamination of the respiratory system.

7. The **respiratory mucosa** lines the conducting portion of the respiratory tract.

8. A pseudostratified ciliated columnar epithelium with numerous mucous cells lines the nasal cavity, the superior portion of the pharynx, and the trachea, bronchi, and large bronchioles.

9. The cilia move mucus, debris, and microorganisms toward the pharynx, where they will be swallowed. This is the **mucociliary escalator**.

10. **Cystic fibrosis (CF)** causes production of a thick and sticky mucus in the conducting portions of the respiratory tract. Potentially lethal infections may result as such dense mucus accumulates, because it cannot be transported.

### Module 21.3

**The upper respiratory system includes the nose, nasal cavity, paranasal sinuses, and pharynx**   p. 800

11. The two nasal bones form the external bridge called the **dorsum of nose**. Air enters through the **nostrils**.

12. Air then flows past the conchae, causing swirling that can trap particles and assist olfaction.

13. The maxillary, frontal, ethmoid, and sphenoid bones that form the lateral and superior walls of the nasal cavity contain **paranasal sinuses**. The mucus produced in these sinuses drains into the nasal cavity and helps keep it moist.

14. The **nasal septum** divides the nasal cavity and is formed by the perpendicular plate of the ethmoid bone and vomer.

15. The **pharynx** is divided into the **nasopharynx**, **oropharynx**, and **laryngopharynx**. Other important structures of the upper respiratory system are the **choanae**, **nasal vestibule**, **hard palate**, and **soft palate**.

16. The extensive network of veins in the nasal cavity delivers body heat to the nasal cavity, so inhaled air is warmed before it leaves the nasal cavity. The heat also evaporates moisture from the epithelium to humidify the incoming air.

### Module 21.4

**The larynx protects the glottis that produces sounds**   p. 802

17. The **larynx** is made of three unpaired cartilages: the **epiglottis** (prevents liquids and solid food from entering glottis when swallowing), **thyroid cartilage** (forms most of the anterior and lateral larynx), and **cricoid cartilage** (supports and protects the glottis).

18. The **vestibular ligaments** and the **vocal ligaments** extend between the thyroid cartilage and the **arytenoid cartilages**.

19. When the **glottis** is open, flowing air vibrates its **vocal folds**, or *vocal cords*, that contain the elastic vocal ligaments. Vibration of the vocal folds produces sound waves.

20. Sound production at the larynx is called **phonation**. Clear speech requires **articulation**.

### Module 21.5

**The trachea, bronchi, and bronchial branches convey air to and from lung gas exchange surfaces**   p. 804

21. The **trachea** is a flexible airway containing 15–20 C-shaped tracheal rings that support and protect its wall. The trachea branches to form left and right **main bronchi**.

22. The **trachealis** connects the ends of the C-shaped **tracheal cartilages.** Sympathetic stimulation of this muscle increases the diameter of the trachea to allow more airflow.

23. Air-conducting passageways in the lower respiratory tract occur in the following order: trachea, main bronchi, **lobar bronchi** (three to right lung, two to left lung), **segmental bronchi**, **bronchioles**, **terminal bronchioles**, and **pulmonary lobules**.

24. Bronchiole walls lack cartilages but contain smooth muscle. Sympathetic stimulation causes **bronchodilation**, and parasympathetic stimulation causes **bronchoconstriction**. Extreme bronchoconstriction may occur during allergic reactions such as **asthma**.

### Module 21.6

**The lungs have lobes that are subdivided into bronchopulmonary segments**   p. 806

25. The **bronchial tree** is the branching pattern of bronchi and bronchioles.

26. Each segmental bronchus supplies air to a single **bronchopulmonary segment**. Each lung lobe contains at least two bronchopulmonary segments.

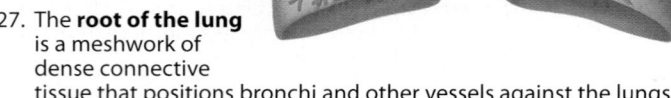

27. The **root of the lung** is a meshwork of dense connective tissue that positions bronchi and other vessels against the lungs.

28. The right lung has three **lobes** (**superior**, **middle**, and **inferior**), an **oblique fissure**, and a **horizontal fissure**. The left lung has two **lobes** (**superior** and **inferior**) and an **oblique fissure**.

29. A **hilum** is a groove on the medial surface of each lung and is where the bronchi and other vessels enter and exit the lung.

### Module 21.7

**Pulmonary lobules contain alveoli, where gas exchange occurs**   p. 808

30. Each lung contains 150 million **pulmonary alveoli**, or air sacs. They are at the end of the respiratory tract.

31. Each segmental bronchus delivers air to a single bronchopulmonary segment, where it repeatedly branches into terminal bronchioles.

32. Each terminal bronchiole forms a respiratory bronchiole as it supplies a single **pulmonary lobule**. This is the start of the respiratory portion of the respiratory tract.

33. A **respiratory bronchiole** opens into an **alveolar duct**. Each alveolar duct opens up into an **alveolar sac**. An alveolar sac is connected to many individual alveoli.

34. The alveolar epithelium is primarily a simple squamous epithelium of **pneumocytes type I**. **Pneumocytes type II** secrete **surfactant**, an oily secretion that prevents alveoli from collapsing. **Alveolar macrophages** patrol the epithelium.

35. Gas exchange occurs across the **blood air barrier** at each alveolus.

## SECTION 2 • Respiratory Physiology

### Module 21.8

**Respiratory physiology involves external and internal respiration**   p. 811

36. **Respiration** refers to external respiration and internal respiration.

37. **External respiration** involves the exchange of oxygen and carbon dioxide between the body and the external environment.

38. **Internal respiration** involves the absorption of oxygen and release of carbon dioxide by tissues.

39. **Pulmonary ventilation**, or breathing, involves the physical movement of air into and out of the lungs. It primarily functions to maintain **alveolar ventilation**, the movement of air into and out of the alveoli.

### Module 21.9

**Pulmonary ventilation is driven by pressure changes within the pleural cavities**   p. 812

40. The relationship between pressure ($P$) and volume ($V$) is inversely proportional. If the volume of a container is decreased by half, the pressure of the gas within it will double; if the volume is doubled, the pressure of the gas decreases by half. This is called **Boyle's law**.

41. Movements of the diaphragm and rib cage cause a change in volume of the thoracic cavity. At the start of a breath, no air moves, because the pressures inside and outside the thoracic cavity are identical.

42. Air will flow from higher pressure to lower pressure. As the thoracic cavity volume enlarges during inhalation, air flows into the lungs. When the thoracic cavity volume decreases in exhalation, increasing air pressure forces air out of the lungs.

43. The direction of airflow is determined by the difference between atmospheric pressure and **intrapulmonary pressure**.

44. The amount of air moved during inhalation and exhalation is called **tidal volume ($V_T$)**. At rest, this is approximately 500 mL.

### Module 21.10

**Respiratory muscles are involved with breathing, and pulmonary function tests determine lung performance**   p. 814

45. **Inspiratory muscles** are involved with inhalation, and **expiratory muscles** are involved with exhalation.

46. The **primary inspiratory muscles** are the diaphragm and external intercostals. The **accessory inspiratory muscles** are sternocleidomastoid, scalene, pectoralis minor, and serratus anterior.

47. The **accessory expiratory muscles** are internal intercostals, transversus thoracis, external oblique, rectus abdominis, and internal oblique.

48. Pulmonary volumes and capacities include **tidal volume ($V_T$)**, **inspiratory reserve volume (IRV)**, **inspiratory capacity**, **vital capacity**, **functional residual capacity (FRC)**, **residual volume**, **total lung capacity**, **minimal volume**, and **expiratory reserve volume (ERV)**.

### Module 21.11

**Pulmonary ventilation must be closely regulated to meet tissue oxygen demands**   p. 816

49. The respiratory system adjusts pulmonary ventilation to meet the oxygen demands of the body.

50. **Respiratory rate ($f$)** is the number of breaths per minute. The normal rate for a resting adult is 12–18 breaths per minute, and for children it is 18–20 breaths per minute.

51. The volume of air moved each minute is called **respiratory minute volume ($V_E$)**. It is calculated by multiplying respiratory rate ($f$) by tidal volume ($V_T$). Tidal volume at rest averages 500 mL per breath. Respiratory minute volume at rest is about 6 L/min.

52. **Alveolar ventilation ($V_A$)** is the amount of air reaching the alveoli each minute. Some of the air never reaches the alveoli because it remains in the conducting portion of the lungs. This is the **anatomic dead space ($V_D$)** and at rest is nearly 150 mL of the 500 mL of tidal volume.

53. Alveolar ventilation equals respiratory rate multiplied by tidal volume minus anatomic dead space.

54. Whenever the demand for oxygen increases, both tidal volume and respiratory rate must be increased.

**21**

## Module 21.12

### Gas diffusion depends on the partial pressures and solubilities of gases  p. 818

55. The **partial pressure (P)** of a gas is the pressure exerted by a single gas in a mixture of gases.

56. All the partial pressures added together equal the total pressure exerted by the gas mixture; this is known as **Dalton's law**.

57. A gas in solution is directly proportional to the partial pressure of that gas. This principle is known as **Henry's law**.

58. In external respiration, blood arriving at the pulmonary arteries has a lower $P_{O_2}$ and a higher $P_{CO_2}$ than does alveolar air. Hence, diffusion across the blood air barrier increases blood $P_{O_2}$ and decreases blood $P_{CO_2}$.

59. The values for pulmonary arterial blood are a $P_{O_2}$ of 40 mm Hg, and a $P_{CO_2}$ of 45 mm Hg. The values for alveolar air are a $P_{O_2}$ of 100 mm Hg, and a $P_{CO_2}$ of 40 mm Hg. Blood reaches equilibrium with alveolar air and departs the alveoli at a $P_{O_2}$ of 100 mm Hg, and a $P_{CO_2}$ of 40 mm Hg.

60. Partial pressure of oxygen drops to 95 mm Hg as it mixes with venous blood from the conducting passageways. Therefore, $P_{O_2}$ in the peripheral capillaries is 95 mm Hg. Normal interstitial fluid has a $P_{O_2}$ of 40 mm Hg.

61. In internal respiration, blood enters systemic capillaries at a $P_{O_2}$ of 95 mm Hg, and a $P_{CO_2}$ of 40 mm Hg. Interstitial fluid is a $P_{O_2}$ of 40 mm Hg, and a $P_{CO_2}$ of 45 mm Hg. Blood reaches equilibrium and departs systemic capillaries with a $P_{O_2}$ of 40 mm Hg, and a $P_{CO_2}$ of 45 mm Hg.

## Module 21.13

### Almost all the oxygen in the blood is transported bound to hemoglobin within red blood cells  p. 820

62. In the blood leaving the alveolar capillaries, only 1.5% of the oxygen molecules are in solution. The rest are bound to the **heme units** of hemoglobin (Hb) molecules.

63. A hemoglobin molecule has four globular protein subunits, each containing a heme unit. Thus, each hemoglobin molecule can reversibly bind up to four oxygen molecules, forming **oxyhemoglobin (HbO₂)**.

64. **Hemoglobin saturation** is the percentage of heme units containing bound oxygen. An **oxygen-hemoglobin saturation curve** shows the saturation of hemoglobin at different partial pressures of oxygen. Above 60 mm Hg, hemoglobin is 90 percent saturated.

65. The **Bohr effect** is the change of oxygen-hemoglobin saturation due to pH. Normal blood pH is 7.35–7.45. As pH increases, the saturation curve shifts to the left (hemoglobin releases less oxygen); as pH decreases, it shifts to the right (hemoglobin releases more oxygen).

66. The higher the temperature, the more readily hemoglobin releases oxygen.

## Module 21.14

### Carbon dioxide is transported three ways in the bloodstream  p. 822

67. After entering the bloodstream, a $CO_2$ molecule dissolves in the plasma (7%), binds to the protein portion of hemoglobin (23%) or (3) is converted to a bicarbonate ion (70%). All three reactions are reversible.

68. Only 7% of $CO_2$ is dissolved in plasma.

69. Some 23% of the $CO_2$ is bound to the globular protein portion of the Hb molecule in **carbaminohemoglobin (HbCO₂)**.

70. **Carbonic anhydrase** in RBCs converts 70% of the total $CO_2$ in blood to carbonic acid. The carbonic acid molecules immediately dissociate into a hydrogen ion and a bicarbonate ion. The **bicarbonate ions** move into the plasma in exchange for chloride ions in an event called the **chloride shift**.

## Module 21.15

### Pulmonary disease can affect both lung elasticity and airflow  p. 824

71. **Compliance** describes how easily lungs expand. **Resistance** describes how much force is required to inflate or deflate lungs.

72. **Chronic obstructive pulmonary disease (COPD)** is a general term for a disorder restricting airflow and decreasing alveolar ventilation.

73. **Asthma** is the term used when symptoms are acute and intermittent. The terms **chronic bronchitis** and **emphysema** are usually applied when signs and symptoms are chronic and progressive with occasional crises due to infections.

## Module 21.16

### Respiratory control mechanisms involve interacting centers in the brainstem  p. 826

74. Respiratory control involves the **respiratory rhythmicity centers** in the medulla oblongata, the **apneustic centers** and **pneumotaxic centers** in the pons, and the **higher centers**: the hypothalamus, limbic system, and cerebral cortex.

## Module 21.17

### Respiratory reflexes provide rapid automatic adjustments in pulmonary ventilation  p. 828

75. **Chemoreceptor reflexes** are triggered by stimulation of chemoreceptors. Normally, $P_{CO_2}$ is the most important factor stimulating chemoreceptors and thereby influencing respiratory factors.

76. An increase in arterial $P_{CO_2}$ (**hypercapnia**) is most commonly caused by hypoventilation. This occurs when respiratory activity is insufficient for $CO_2$ removal, causing $CO_2$ to increase in blood.

77. A decrease in arterial $P_{CO_2}$ (**hypocapnia**) occurs when the rate and depth of respiration exceed the demands for $O_2$ delivery and $CO_2$ removal, as in **hyperventilation**.

78. **Baroreceptor reflexes** are triggered by stimulation of baroreceptors. Baroreceptors in the carotid and aortic sinuses trigger adjustments in respiratory minute volume.

79. **Inflation reflexes** prevent overexpansion of the lungs. **Deflation reflexes** stimulate the inspiratory centers when the lungs are deflating.

80. **Protective reflexes** include sneezing and coughing. Both reflexes involve **apnea**, a period when breathing has stopped.

## Module 21.18

### Respiratory function decreases with age; smoking makes matters worse  p. 830

81. Respiratory function (vital capacity and compliance) decreases with age. Never having smoked, or stopping smoking, can prevent serious respiratory deterioration.

82. **Lung cancer** kills more people than any other type of cancer. Smoking causes 85–90% of all lung cancers.

# Chapter Review Questions

**Labeling:** Provide the values, in millimeters of mercury (mm Hg), on this illustration of the partial pressures of oxygen and carbon dioxide during external respiration in the pulmonary circuit and internal respiration in the systemic circuit.

**External Respiration**

$P_{O_2} =$ [1]
$P_{CO_2} =$ [2]

Alveolus

Blood air barrier

$P_{O_2} =$ [3]
$P_{CO_2} =$ [4]

$O_2$

$CO_2$

Pulmonary capillary

$P_{O_2} =$ [5]
$P_{CO_2} =$ [6]

**Internal Respiration**

Interstitial fluid

Systemic capillary

$P_{O_2} =$ [7]
$P_{CO_2} =$ [8]

$O_2$

$P_{O_2} =$ [11]
$P_{CO_2} =$ [12]

$CO_2$

$P_{O_2} =$ [9]
$P_{CO_2} =$ [10]

---

**Matching:** Match each lettered term with the most closely related description.

a. respiratory rate
b. heme unit of Hb molecule
c. left lung
d. increasing pH
e. pulmonary ventilation
f. vital capacity
g. total lung capacity
h. right lung
i. globular unit of Hb molecule
j. respiratory minute volume
k. resistance
l. alveolar ventilation

[13] Airflow to and from lungs
[14] Volume of air reaching alveoli each minute
[15] Has two lobes
[16] Binds $CO_2$
[17] Volume of air moved into and out of lungs each minute
[18] Binds $O_2$
[19] Number of breaths each minute
[20] Hb molecule releases less $O_2$
[21] Force required to inflate/deflate lungs
[22] Maximum amount of air moved into and out of lungs in one respiratory cycle
[23] Has three lobes
[24] Vital capacity plus residual volume

[13] _____
[14] _____
[15] _____
[16] _____
[17] _____
[18] _____
[19] _____
[20] _____
[21] _____
[22] _____
[23] _____
[24] _____

---

**Multiple choice:** Select the correct answer from the list provided.

[25] Surfactant
☐ a) protects the outer surface of the lungs.
☐ b) phagocytizes small particles.
☐ c) replaces mucus in the alveoli.
☐ d) helps prevent the alveoli from collapsing.

[26] One of the responses to an abnormally low $P_{CO_2}$ would be
☐ a) decreased respiratory rate.
☐ b) increased respiratory rate.
☐ c) hypercapnea.
☐ d) chronic obstructive pulmonary disease.

**27** The narrow opening in the larynx for inhaled air is called the

- ☐ a) epiglottis.
- ☐ b) rima glottidis.
- ☐ c) pharynx.
- ☐ d) internal nares.

**28** Which of the following muscle combinations are the primary muscles of inspiration?

- ☐ a) diaphragm and the internal intercostal muscles
- ☐ b) diaphragm and the external intercostal muscles
- ☐ c) abdominal muscles and the internal intercostal muscles
- ☐ d) abdominal muscles and the external intercostal muscles

**29** Boyle's law states that

- ☐ a) all the partial pressures in a gas mixture added together equals the total pressure exerted by the gas mixture.
- ☐ b) the amount of a particular gas in solution is directly proportional to the partial pressure of that gas.
- ☐ c) the shape of hemoblogin molecules changes as the number of bound $O_2$ molecules increases, and these changes affect its affinity for oxygen.
- ☐ d) if you reduce the volume of a flexible container by half, the pressure within it will double.

**30** $V_E = f \times V_T$ is the formula used to calculate

- ☐ a) Dalton's law.
- ☐ b) alveolar ventilation.
- ☐ c) respiratory minute volume.
- ☐ d) vital capacity.

## Short answer

**31** Name the three regions of the pharynx, and identify where each region is located.

**32** By what three mechanisms is carbon dioxide transported in the bloodstream?

**33** Why is breathing through the nasal cavity more desirable than breathing through the mouth?

## MasteringA&P™

Access more chapter study tools online in the MasteringA&P Study Area:

- Chapter Quizzes, Chapter Practice Test, and Clinical Case Studies

- Practice Anatomy Lab™ 3.0  — PAL 3.0
- Interactive Physiology™ — iP2™
- A&P Flix™ — A&PFlix™
- PhysioEx™ 9.1 — PhysioEx 9.1

# Chapter Integration • Applying what you have learned

## Spring break snorkeling danger

Jake is an active 22-year-old college student who enjoys water sports. In fact, he has an athletic scholarship to swim on the university swim team. While on spring break at an oceanfront hotel with both beach and pool access, Jake decides to go snorkeling. The sun is shining, the air temperature is 80°F, and the ocean is a refreshing 72°F. His normal alveolar ventilation rate (AVR) during mild exercise is 6.0 L/min, and his snorkel has a volume of 50 mL.

After snorkeling in the ocean all afternoon, Jake heads to the pool to meet up with some friends. Before diving in, he hyperventilates for several minutes. After he enters and begins swimming underwater, he blacks out and almost drowns. Fortunately, friends notice that something's wrong and pull him out of the pool to safety. Jake recovers within 15 seconds and does not require pulmonary resuscitation or medical treatment.

**1** Assuming a constant tidal volume of 500 mL and an anatomic dead space of 150 mL, what would Jake's respiratory rate have to be for him to maintain an AVR of 6.0 L/min while snorkeling?

**2** What caused Jake to black out and nearly drown?

21

## 22 The Digestive System

## Module 22.1
# The digestive system consists of the digestive tract and accessory organs

The **digestive system** consists of a muscular tube, the *digestive tract*—also called the *gastrointestinal (GI) tract* or *alimentary canal*—and various accessory organs. Food enters the mouth and passes along the length of the digestive tract. On the way, accessory organs mechanically process the food and produce secretions containing water, enzymes, buffers, and other components that assist in preparing organic and inorganic nutrients for absorption.

**1** The digestive system works with other systems to support tissues that have no direct connection with the outside environment and no other means of obtaining nutrients.

The digestive system provides the nutrients cells need for fuel, maintenance, and growth.

The respiratory system works with the cardio-vascular system to supply oxygen to cells and remove carbon dioxide.

$O_2$ and $CO_2$

Cardio-vascular system

Nutrients

Tissue cells

Wastes

The urinary system removes the organic wastes generated by cell activities.

**?  REVIEW**

**A.** Which other systems work with the digestive system to support the cells and tissues of the human body?

**2** The **digestive tract** begins at the oral cavity (mouth) and continues through the pharynx (throat), esophagus, stomach, small intestine and large intestine (referred to as the *small bowel* and *large bowel*), which opens to the exterior at the anus. Accessory digestive organs include the teeth, tongue, gallbladder, and *glandular organs*, such as the salivary glands, liver, and pancreas.

**?  REVIEW**

**B.** Starting at the mouth, identify the major organs of the digestive tract.

**C.** List the accessory organs of the digestive system.

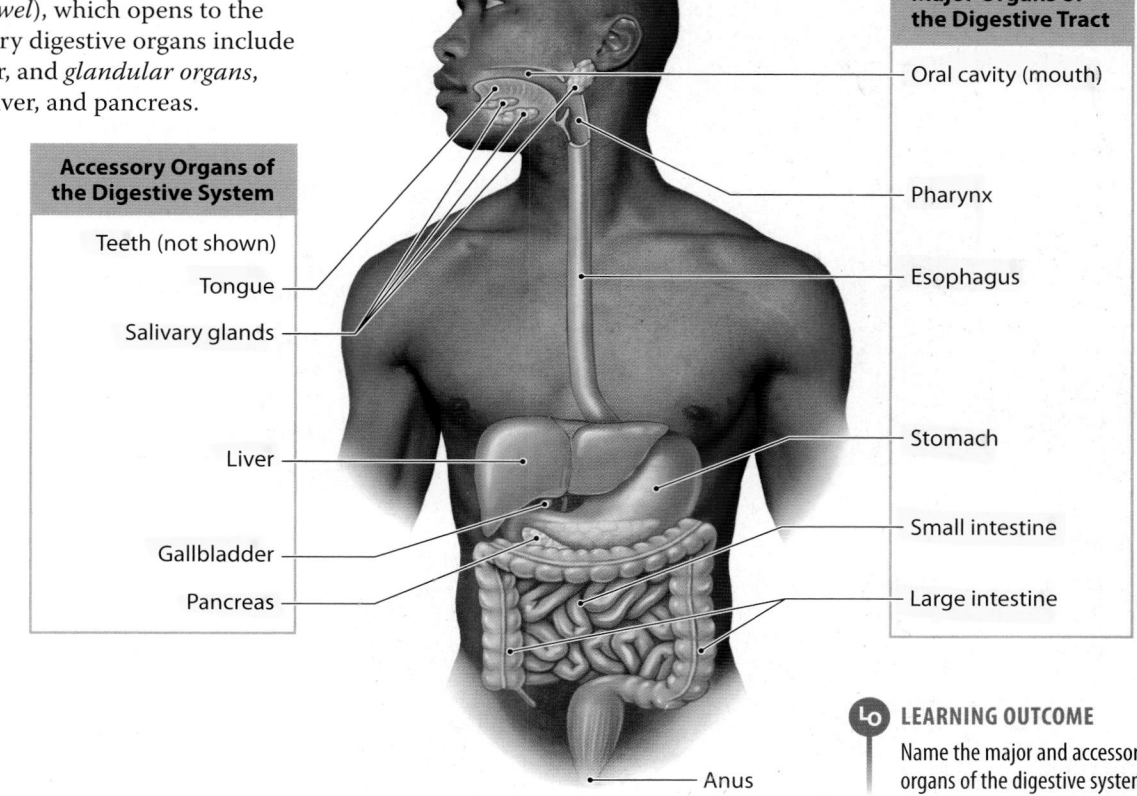

**Accessory Organs of the Digestive System**

Teeth (not shown)

Tongue

Salivary glands

Liver

Gallbladder

Pancreas

**Major Organs of the Digestive Tract**

Oral cavity (mouth)

Pharynx

Esophagus

Stomach

Small intestine

Large intestine

Anus

**Lo  LEARNING OUTCOME**

Name the major and accessory organs of the digestive system.

22

# Module 22.2

# The digestive tract is a muscular tube lined by a mucous epithelium

The structure of the digestive tract varies to some degree from one region of the digestive tract to another. The composite view shown in this module most closely resembles the small intestine, the longest segment of the digestive tract.

**1** The digestive tract is basically a long muscular tube lined by a mucous membrane. The lining often contains permanent ridges as well as temporary folds that disappear as the passageway fills. Together the ridges and folds dramatically increase the surface area available for absorption.

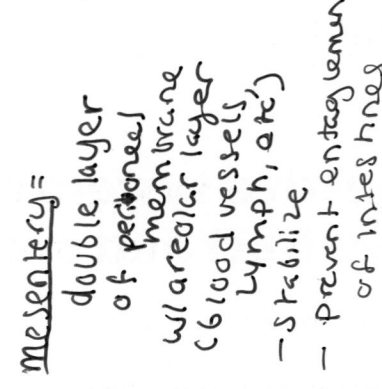

*[handwritten notes:]* Serous membrane = serous membrane " connecting/ panel • visceral peritoneum

*[handwritten notes:]* mesentery = double layer of peritoneal membrane w/ areolar layer (blood vessels, lymph, etc.) — stabilize — Prevent entanglement of intestines

### Mesentery

Portions of the digestive tract are suspended within the peritoneal cavity by sheets of serous membrane that connect the parietal peritoneum with the visceral peritoneum. These **mesenteries** are double sheets of peritoneal membrane. The areolar tissue between the mesothelial surfaces provides a route to and from the digestive tract for blood vessels, nerves, and lymphatic vessels. Mesenteries also stabilize the positions of the attached organs, thus preventing the intestines from becoming entangled during digestive movements or sudden changes in body position.

— Mesothelium
— Areolar tissue
— Mesothelium

Mesenteric artery and vein

Circular folds

Mesentery

Lumen of digestive tract

### Major Layers of the Digestive Tract

#### Mucosa

The inner lining, or **mucosa**, of the digestive tract is a mucous membrane consisting of an epithelium, which is moistened by glandular secretions, and a lamina propria of areolar tissue.

#### Submucosa

The **submucosa** is a layer of dense irregular connective tissue that surrounds the mucosa. The submucosa has large blood vessels and lymphatic vessels, and in some regions it also contains exocrine glands that secrete buffers and enzymes into the lumen of the digestive tract.

#### Muscular Layer

The **muscular layer** is dominated by smooth muscle cells in two layers, an inner circular layer and an outer longitudinal layer. These layers play an essential role in mechanical processing and in moving materials along the digestive tract.

#### Serosa

Along most portions of the digestive tract within the peritoneal cavity, the muscular layer is covered by a layer of visceral peritoneum known as the **serosa**. There is no serosa covering the muscular layer of the oral cavity, pharynx, esophagus, and rectum. Instead, a dense network of collagen fibers forms a sheath—called an **adventitia** (ad-ven-TISH-uh)—that firmly attaches the digestive tract to adjacent structures.

**22**

**? REVIEW**

**A.** What is the importance of the mesenteries?

**B.** Name the four layers of the digestive tract beginning from the lumen of the digestive tract.

**2** Each layer in the wall of the digestive tract has a different set of functions. The mucosa has several structural features that enhance its abilities to absorb nutrients from the lumen of the tract. Those features are shown in this diagrammatic view.

## Secretory Glands

The secretions of gland cells located in the mucosa and submucosa—or in accessory glandular organs—are carried to the epithelial surfaces by ducts.

Intestinal glands

Submucosal gland

Artery and vein

Lymphatic vessel

Mucosa

Submucosa

## Muscular Layer

Circular layer

Longitudinal layer

Serosa

**Circular folds** (plicae circulares) are permanent transverse folds in the intestinal lining.

## Components of the Mucosa

The mucosal epithelium of the oral cavity, pharynx, and esophagus are lined by a stratified squamous epithelium. The stomach, small intestine, and almost the entire length of the large intestine are lined by a simple columnar epithelium that contains mucin-secreting goblet cells (see Module 4.9).

**Villi** (singular, *villus*) are small mucosal projections, like tiny fingers, that stick into the lumen of the small intestine. These are permanent features that further increase the surface area available for absorption.

The **lamina propria** is a layer of areolar tissue that contains blood vessels, sensory nerve endings, lymphatic vessels, smooth muscle cells, scattered areas of lymphoid tissue, and, in some regions, mucous glands.

The **muscularis** (mus-kū-LAIR-is) **mucosae** (mū-KŌ-sē) consists of two concentric layers of smooth muscle. The inner *circular layer* encircles the lumen, and the outer *longitudinal layer* contains muscle cells oriented parallel to the long axis of the tract. Contractions in these muscular layers alter the shape of the lumen and move the circular folds and villi.

Superficial to the muscular layer, the submucosa contains the **submucosal neural plexus**, a nerve network that contains sensory neurons, parasympathetic ganglionic neurons, and sympathetic postganglionic fibers that innervate the mucosa and submucosa.

The muscular layer contains the **myenteric** (mī-en-TER-ik) **plexus** (*mys*, muscle + *enteron*, intestine). This network of parasympathetic ganglia, sensory neurons, interneurons, and sympathetic postganglionic fibers lies sandwiched between the circular and longitudinal muscle layers. The myenteric plexus and the submucosal plexus contain neurons of the enteric nervous system (ENS), and most digestive activities are locally controlled (see Module 14.2). In general, parasympathetic stimulation increases muscle tone and activity; sympathetic stimulation decreases muscle tone and activity.

 **REVIEW**

**C.** Compare the submucosal neural plexus with the myenteric plexus.

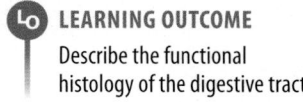 **LEARNING OUTCOME**

Describe the functional histology of the digestive tract.

22

# Module 22.3

# Smooth muscle tissue is found throughout the body, but it plays a particularly prominent role in the digestive tract

Smooth muscle tissue forms sheets, bundles, or sheaths around other tissues in almost every organ. Smooth muscles around blood vessels regulate blood flow through vital organs. In the digestive and urinary systems, smooth muscle sphincters regulate the movement of materials along internal passageways.

**1** Smooth muscle cells are relatively long and slender, ranging from 5 to 10 μm in diameter and 30 to 200 μm in length. Wherever smooth muscle tissue forms layers, the cells are aligned parallel to one another. In the digestive tract, there is usually an inner circular layer and an outer longitudinal layer. In a longitudinal section of the digestive tract, the muscle cells in the circular layer of the muscular layer look like little round balls, whereas those in the longitudinal layer look like long spindles.

 **REVIEW**

**A.** Describe the orientation of smooth muscle fibers in the muscular layer of the digestive tract.

Circular muscle layer

Longitudinal muscle layer

**Smooth muscle**　　　　　LM × 65

**2** As with skeletal and cardiac muscle cells, actin and myosin filaments are involved in smooth muscle contractions. However, the organization of those filaments differs from that of skeletal or cardiac muscle cells. A smooth muscle fiber has no T tubules, and the sarcoplasmic reticulum forms a loose network throughout the sarcoplasm. Smooth muscle cells also lack myofibrils and sarcomeres. As a result, this tissue has no striations and is called nonstriated muscle.

**Relaxed (sectional view)**

Dense body

The thin filaments in a smooth muscle cell are attached to **dense bodies**, structures distributed throughout the sarcoplasm. Dense bodies are similar to the Z lines of skeletal muscle tissue.

Thick filaments are scattered throughout the sarcoplasm, and there are more myosin heads per thick filament than in skeletal or cardiac muscle cells.

**Relaxed (superficial view)**

A network of intermediate filaments interconnects the dense bodies. Some of the dense bodies are firmly attached to the sarcolemma.

Adjacent smooth muscle cells are bound together at dense bodies, transmitting the contractile forces from cell to cell throughout the tissue.

**Contracted (superficial view)**

When sliding occurs between thin and thick filaments, the cell shortens. Because the dense bodies are not arranged in straight lines, the muscle cell twists like a corkscrew as it contracts.

 **REVIEW**

**B.** Identify the structural characteristics of smooth muscle fibers.

**3** Smooth muscle tissue can be divided into two subtypes based on distinctively different methods of innervation and control.

## Types of Smooth Muscle

### Multi-unit Smooth Muscle

**Multi-unit smooth muscle cells** are innervated in motor units comparable to those of skeletal muscles, but each smooth muscle cell may be connected to more than one motor neuron. Multi-unit smooth muscle tissue is located in the iris of the eye, where it regulates the diameter of the pupil; along portions of the male reproductive tract; within the walls of large arteries; and in the arrector pili muscles of the skin.

Eye

Smooth muscle cell

Neuron

KEY
⟶ = Stimulation by neurotransmitter release

### Visceral Smooth Muscle

Most **visceral smooth muscle cells** lack direct contact with any motor neuron. These muscle cells are arranged in sheets or layers, with adjacent muscle cells electrically connected by gap junctions and mechanically connected by dense bodies. As a result, whenever one muscle cell contracts, the stimulus for contraction can travel to adjacent smooth muscle cells, and the contraction spreads in a wave throughout the layer. A contraction can occur in response to neural, hormonal, or chemical stimuli. In addition, many visceral smooth muscle networks show rhythmic cycles of activity triggered by **pacesetter cells** that contract spontaneously at regular intervals. Visceral smooth muscle cells are located in the walls of the digestive tract, the gallbladder, the urinary bladder, and many other internal organs.

Small intestine

Gap junction

Neuro-transmitter

Pacesetter cell

Smooth muscle cell

KEY
⟶ = Stimulation across cell-to-cell connections

Because the thick and thin filaments of smooth muscle are scattered and are not organized into sarcomeres, tension development and resting length are not directly related. A stretched smooth muscle soon adapts to its new length and retains the ability to contract on demand. This ability to function over a wide range of lengths is called **plasticity**. Smooth muscle can contract over a range of lengths four times greater than that of skeletal muscle. This is extremely important in the stomach, intestines, urinary bladder, or uterus, which must undergo major changes in size and shape. These smooth muscle tissues have a normal background level of activity known as **smooth muscle tone**. Neural, hormonal, or local chemical factors can increase or decrease smooth muscle tone and alter the degree of tension in the wall of a muscular organ.

 **REVIEW**

**C.** Why can smooth muscle contract over a wider range of resting lengths than skeletal muscle?

22

 **LEARNING OUTCOME**

Describe the structural and functional features of smooth muscle tissue.

# Module 22.4

# Smooth muscle contractions produce motility of the digestive tract ...

Coordinated contractions of smooth muscle within the muscular layer play a vital role in digestive tract movement, or *motility*, by moving material along the digestive tract (peristalsis) and in mechanical digestion (segmentation).

## Peristalsis

**1** Food enters the digestive tract as a moist, compact mass known as a **bolus** (BŌ-lus). The muscular layer propels materials from one portion of the digestive tract to another by contractions known as **peristalsis** (per-i-STAL-sis). During peristalsis, a wave of contraction in the circular muscles forces the bolus forward.

**1**

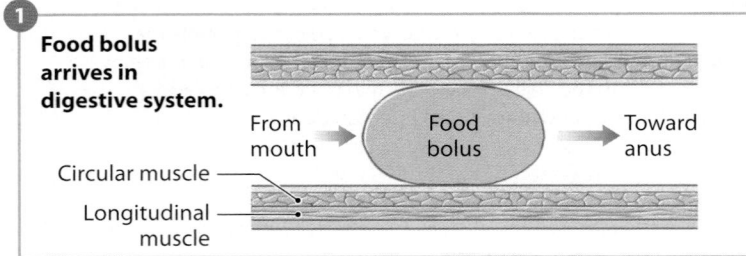

**Food bolus arrives in digestive system.**

From mouth → Food bolus → Toward anus

Circular muscle

Longitudinal muscle

**2**

**Circular muscles contract behind bolus.**

**3**

**Longitudinal muscles contract ahead of bolus.**

**4**

**A wave of contraction in circular muscle layer forces bolus forward.**

## Segmentation

**2** Most areas of the small intestine and some portions of the large intestine undergo cycles of contraction that churn and fragment the bolus, mixing the contents with intestinal secretions. This activity is called **segmentation**. These rhythmic cycles of contraction do not follow a set pattern and thus do not push materials along the tract in any one direction.

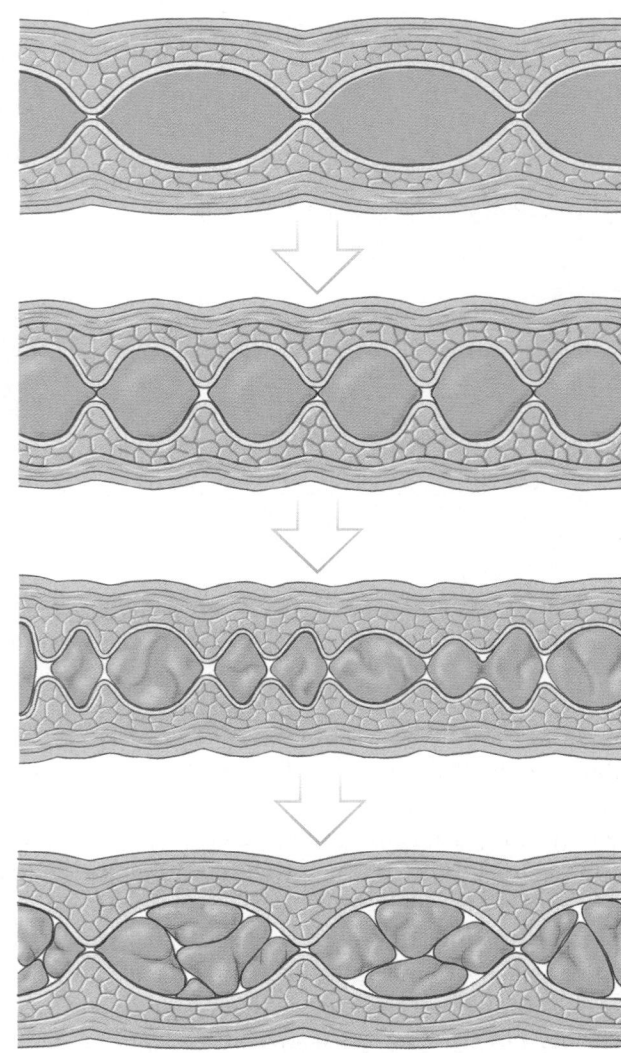

**? REVIEW**

**A.** Which is more efficient in propelling intestinal contents along the digestive tract: peristalsis or segmentation? Why?

22

# … and local factors interact with neural and hormonal mechanisms to regulate digestive activities

**3** Three major mechanisms regulate and control digestive activities. The primary stimuli involved are local factors that may in turn activate neural or hormonal control mechanisms.

**② Neural Control Mechanisms**

Neural control mechanisms include both short and long **myenteric reflexes** that result in peristalsis and segmentation. The movement of materials along the digestive tract, as well as many secretory functions, is primarily controlled by local factors. Short reflexes are triggered by chemoreceptors or stretch receptors in the walls of the digestive tract; the controlling neurons are located in the myenteric plexus. Long reflexes involving interneurons and motor neurons in the CNS provide a higher level of control over digestive and glandular activities, generally controlling large-scale peristalsis that moves materials from one region of the digestive tract to another. Long reflexes may involve parasympathetic motor fibers in the glossopharyngeal (IX), vagus (X), or pelvic nerves that synapse in the myenteric plexus (see Module 14.9).

CNS

**Long reflex**

Myenteric plexus

**Short reflex**

Stretch receptors, chemoreceptors

Peristalsis and segmentation movements

Buffers, acids, enzymes released

Secretory cells

**① Local Factors**

Local factors are the primary stimulus for digestive activities and include changes in the pH of the contents of the lumen, physical distortion of the wall of the digestive tract, or the presence of chemicals—either specific nutrients or chemical messengers released by cells of the mucosa.

Enteroendocrine cells

Bloodstream transport

Hormones released

**③ Hormonal Control Mechanisms**

The digestive tract produces at least 18 hormones that affect almost every aspect of digestive function, and some of them also affect the activities of other systems. These hormones are peptides produced by **enteroendocrine cells**, endocrine cells in the epithelium of the digestive tract. We consider these hormones as we proceed down the digestive tract.

 **REVIEW**

**B.** Cite the major mechanisms that regulate and control digestive activities.

**C.** Describe enteroendocrine cells.

22

---

Clinical Note

## Congenital megacolon results from missing nerve cells in the colon

**Congenital megacolon**, or *Hirschsprung disease*, is a congenital disorder characterized by absence (aganglionosis) or marked reduction (hypoganglionosis) in the number of ganglion cells in the myenteric plexus of the rectum. This causes paralysis of the smooth muscles and an absence of peristaltic movements. The result is abnormal dilation and hypertrophy of the colon. Patients complain of chronic constipation, bloating, and abdominal pain.

 **LEARNING OUTCOME**

Explain the processes by which materials move through the digestive tract.

# Organization of the Digestive System

**Labeling:** Label the structures of the digestive tract in the accompanying figure.

| | |
|---|---|
| 1 | |
| 2 | |
| 3 | |
| 4 | |
| 5 | |
| 6 | |
| 7 | |

**Matching:** Match each lettered term with the most closely related description.

a. lamina propria

b. peristalsis

c. pacesetter cells

d. esophagus

e. muscularis mucosa

f. segmentation

g. circular folds

h. sphincter

i. myenteric plexus

j. visceral smooth muscle cells

k. plasticity

l. liver

m. multi-unit smooth muscle cells

n. bolus

| 8 | Digestive tube between the pharynx and stomach |
|---|---|
| 9 | Moves circular folds and villi |
| 10 | Areolar tissue layer containing blood vessels, nerve endings, and lymphatic vessels |
| 11 | Permanent transverse folds in the digestive tract lining |
| 12 | Waves of muscular contractions that propel materials along digestive tract |
| 13 | Stimulate rhythmic cycles of activity along digestive tract |
| 14 | Nerve network within the muscular layer |
| 15 | Have direct contact with motor neurons |
| 16 | Digestive system accessory organ |
| 17 | Rhythmic muscular contractions that mix materials in digestive tract |
| 18 | Form of food entering the digestive tract |
| 19 | Lack direct contact with motor neurons |
| 20 | Ability of smooth muscle cells to function over varied lengths |
| 21 | Ring of muscle tissue |

| 8 | _____ |
|---|---|
| 9 | _____ |
| 10 | _____ |
| 11 | _____ |
| 12 | _____ |
| 13 | _____ |
| 14 | _____ |
| 15 | _____ |
| 16 | _____ |
| 17 | _____ |
| 18 | _____ |
| 19 | _____ |
| 20 | _____ |
| 21 | _____ |

**Section integration:** How would a decrease in smooth muscle tone affect the digestive processes and possibly cause constipation (infrequent bowel movements)?

| 22 | _____ |
|---|---|
| | _____ |
| | _____ |
| | _____ |

22

# Module 22.5
# The digestive tract begins with the mouth and ends with the anus

**1** The digestive tract is a muscular tube approximately 10 m (33 ft) long. It can be divided into regions that differ in histological structure and functional properties. In this section, we consider each of the major organs of the digestive tract and their functions.

| Major Organs of the Digestive Tract |
| --- |
| **Oral Cavity (Mouth)** |
| Mechanical digestion with accessory organs (teeth and tongue), moistening, mixing with salivary secretions |
| **Pharynx** |
| Muscular propulsion of materials into the esophagus |
| **Esophagus** |
| Transport of materials to the stomach |
| **Stomach** |
| Chemical breakdown of materials by acid and enzymes; mechanical digestion through muscular contractions |
| **Small Intestine** |
| Enzymatic digestion and absorption of water, organic substrates, vitamins, and ions |
| **Large Intestine** |
| Dehydration and compaction of indigestible materials in preparation for elimination |

| Accessory Organs of the Digestive System |
| --- |
| Teeth |
| Tongue |
| Salivary glands |
| Liver |
| Gallbladder |
| Pancreas |

Anus

**2** This table summarizes the general functions of the digestive tract. The lining of the digestive tract also protects surrounding tissues against digestive acids and enzymes, abrasion, and pathogens within the digestive tract.

| Functions of the Digestive Tract |
| --- |
| • **Ingestion**: Food and liquids enter the digestive tract through the mouth. |
| • **Mechanical digestion and propulsion**: Involves crushing and shearing ingested solids and then propelling them along the digestive tract. It includes tearing and mashing by the teeth and compressing by the tongue. Swirling, mixing, and churning motions of the stomach and intestines provide mechanical digestion after ingestion. |
| • **Chemical digestion**: The chemical and enzymatic breakdown of food into substances that can be absorbed by the digestive epithelium. |
| • **Secretion**: Secretions are produced along most of the digestive tract. In addition, accessory organs secrete acids, enzymes, and buffers. |
| • **Absorption**: The movement of nutrients across the digestive epithelium and into the bloodstream. |
| • **Defecation**: Indigestible food is compacted into material waste called **feces**. The discharge of feces from the body is called **defecation** (def-e-KĀ-shun). |

 **REVIEW**

**A.** Define *ingestion*.

**B.** Distinguish between chemical digestion and absorption.

**C.** Describe the function of the large intestine.

 **LEARNING OUTCOME**

Name the structures and primary functions of the digestive tract organs.

22

# The oral cavity is a space that contains the tongue, teeth, and gums

The **oral cavity** (mouth) is lined by the **oral mucosa**, which has a stratified squamous epithelium. A layer of keratinized cells covers regions exposed to severe abrasion, such as the superior surface of the tongue and the opposing surface of the hard palate. The epithelial lining of the cheeks, lips, and inferior surface of the tongue is relatively thin and nonkeratinized. Nutrients are not absorbed in the oral cavity, but digestion of carbohydrates and lipids begins here. The mucosa inferior to the tongue is thin enough and vascular enough to allow rapid absorption of lipid-soluble drugs, such as nitroglycerin, used to treat angina attacks.

**? REVIEW**
**A.** The oral cavity is lined by which type of epithelium?

**1** This sagittal section introduces the major structures forming the boundaries of the oral cavity (mouth).

**Superior Boundary of the Oral Cavity**

| | |
|---|---|
| The **hard palate** is formed by the palatine processes of the maxillary bones and the horizontal plates of the palatine bones. | The muscular **soft palate** lies posterior to the hard palate. |

**? REVIEW**
**B.** Name the structure that forms the roof of the mouth.

**Anterior and Lateral Boundary of the Oral Cavity**

Anteriorly, the mucosa of each cheek is continuous with that of the lips, or **labia** (LĀ-bē-uh; singular, *labium*).

The mucosae of the **cheeks**, or lateral walls of the oral cavity, are supported by pads of fat and the buccinator.

The **body** of the tongue is a muscular organ that forms part of the oral cavity floor.

**Inferior Boundary of the Oral Cavity**

The floor of the mouth inferior to the tongue receives extra support from the geniohyoid and mylohyoid (see Module 10.8).

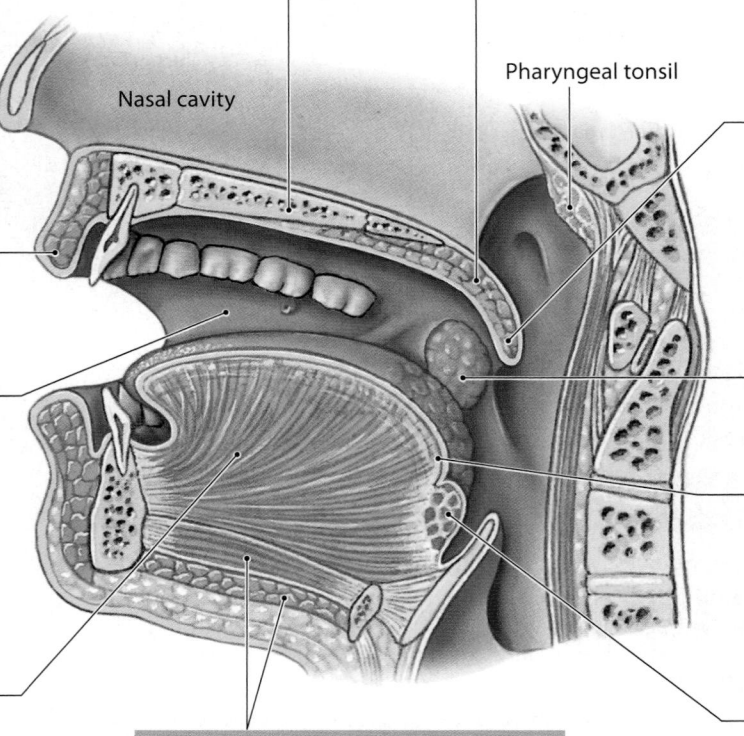

Nasal cavity

Pharyngeal tonsil

**Posterior Boundary of the Oral Cavity**

The posterior margin of the soft palate supports the **uvula** (Ū-vū-luh), a dangling process that helps prevent food from entering the pharynx prematurely and swings upwards during swallowing to prevent food from entering the nasopharynx.

A **palatine tonsil** lies on either side of the entrance to the *oropharynx*.

The **root** of the tongue is fixed and projects into the oropharynx. A V-shaped line of vallate papillae marks the boundary between the root and the body of the tongue (see Module 15.3).

A pair of **lingual tonsils** are embedded in the root of the tongue.

**22**

**2** An anterior view of the oral cavity shows additional details not visible in the sagittal section.

The retractors are placed in the **oral vestibule**, the space between the cheeks (or lips) and the teeth.

Anteriorly, a delicate **frenulum** (FREN-ū-lum; *frenulum*, a small bridle) **of upper lip** attaches the gums to the upper lip.

The mucosa covering the hard palate is thick and has complex ridges. When your tongue compresses food against the hard palate, these ridges provide traction.

Soft palate

Palatine tonsil

The **frenulum of tongue**, or *lingual frenulum*, connects the body of the tongue to the floor of the oral cavity.

The **frenulum of lower lip** attaches the gums to the lower lip.

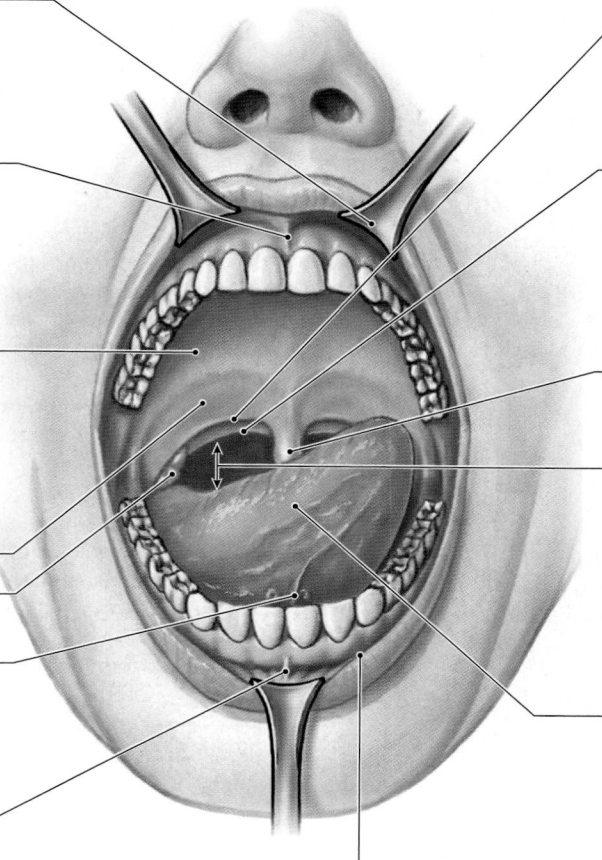

### Palatal Arches

On either side of the uvula are two pairs of muscular arches, the *palato-glossal* and the *palatopharyngeal*.

The more anterior **palatoglossal** (pal-a-tō-GLOS-al) **arch** extends between the soft palate and the base of the tongue.

The more posterior **palatopharyn-geal** (pal-ah-tō-fah-RIN-jē-al) **arch** extends from the soft palate to the pharyngeal wall. A palatine tonsil lies between the palatoglossal and palato-pharyngeal arches on each side.

Uvula

On each side, the ridge that connects the more anterior palato-glossal arch with the uvula forms the superior margin of the **fauces** (FAW-sēz), the space between the oral cavity and the oropharynx. (The root of the tongue forms the inferior margin.)

The **tongue** manipulates materials inside the mouth and may occasion-ally be used to bring foods into the oral cavity, as when licking an ice cream cone. The tongue's surface is flushed by the secretions of small glands that extend into the underly-ing lamina propria. These secretions contain water, mucin, and the enzyme **lingual lipase**, which starts digesting lipids even before they leave the oral cavity.

The **gingivae** (JIN-ji-vē), or gums, are ridges of oral mucosa that surround the base of each tooth on the alveolar processes of the maxillary bones and mandible. In most regions, the gingivae are firmly bound to the periostea of the underlying bones.

**? REVIEW**
**C.** Describe the location of the fauces.

22

✠ **Clinical Note**

## Tongue-tie is more than not finding the right words

**Ankyloglossia** (ang-ki-lō-GLOS-ē-uh), or *tongue-tie*, is a condition that is present at birth in which the frenulum of tongue is too short. People with ankyloglossia have difficulty sticking out their tongue, so it interferes with breastfeeding in newborns and learning to speak in toddlers.

**? INTEGRATION**
**D.** What effects might a shortened frenulum of tongue cause?

**Lo LEARNING OUTCOME**
Describe the anatomy of the oral cavity, and discuss the functions of its structures.

# Module 22.7
# Teeth in different regions of the jaws vary in size, shape, and function

**1** This is a sectional view through a representative adult tooth. The bulk of each tooth consists of a mineralized matrix similar to that of bone. This material, called **dentin**, differs from bone in that it does not contain cells. Instead, cytoplasmic processes extend into the dentin from cells making up the pulp in the central **pulp cavity**.

Dentin

Pulp cavity

The **crown** of the tooth projects into the oral cavity from the surface of the gingiva.

The **neck** of the tooth marks the boundary between the crown and the root.

The **root** of each tooth sits in a bony cavity called a *tooth socket*, or *tooth alveolus*.

**? REVIEW**
**A.** Name the three main parts of a typical tooth.

### Components of a Tooth

The portion of the crown that is used for crushing, slicing, or chewing is the **occlusal surface** of the tooth.

A layer of **enamel** covers the dentin of the crown. Enamel, which contains calcium phosphate in a crystalline form, is the hardest biologically manufactured substance. Adequate amounts of calcium, phosphates, and vitamin D₃ during childhood are essential if the enamel coating is to be complete and resistant to decay.

A shallow groove called the **gingival** (JIN-ji-val) **sulcus** surrounds the neck of each tooth. The epithelium of the gingiva is bound to the tooth at the base of the sulcus. This epithelial attachment prevents bacteria from entering the deeper tissues around the root.

A layer of **cement** (sē-MENT) covers the dentin of the root. Cement is less resistant to erosion than is dentin.

Collagen fibers of the **periodontal ligament** extend from the dentin of the root to the compact bone of the alveolus, creating a strong joint known as a **gomphosis** (see Module 8.1).

The pulp cavity receives blood vessels and nerves through the **root canal**, a narrow tunnel within the root of the tooth. The opening into the root canal is called the **apical foramen**.

**2** There are four different types of teeth, each with a distinctive shape and root pattern.

### Types of Teeth

| **Incisors** (in-SĪ-zerz) are blade-shaped teeth located at the front of the mouth. Incisors are useful for clipping or cutting. These teeth have a single root. | The **canines**, or *cuspids* (KUS-pidz), are conical, with a sharp ridgeline and a pointed tip. They are used for tearing or slashing. Canines have a single root. | **Premolars**, or *bicuspids* (bī-KUS-pidz), have flattened crowns with prominent ridges. They crush, mash, and grind. Premolars have one or two roots. | **Molars** have very large, flattened crowns with prominent ridges adapted for crushing and grinding. Molars in the upper jaw typically have three roots, whereas those in the lower jaw usually have two roots. |
|---|---|---|---|

Upper jaw —

Lower jaw —

22

**3** The teeth are collectively called the *dentition*. Two dentitions form during development: deciduous and permanent. The first to appear are the **deciduous teeth** (de-SID-yū-us; *deciduus*, falling off), also called *primary teeth*, *milk teeth*, or *baby teeth*. At 2 years of age, children have 20 deciduous teeth—five on each side of the upper and lower jaws: two incisors, one canine, and a pair of deciduous molars.

**4** As the jaw grows larger, the deciduous teeth are gradually replaced by the **permanent teeth**. Three additional molars appear on each side of the upper and lower jaws as the person ages, bringing the count of permanent teeth to 32. As replacement proceeds, the periodontal ligaments and roots of the deciduous teeth erode. The deciduous teeth either fall out or are pushed aside by the eruption (emergence) of the permanent teeth.

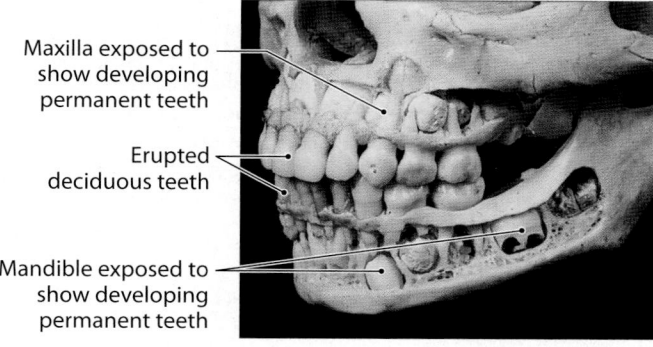

Central incisors (7.5 mo)
Lateral incisor (9 mo)
Canine (18 mo)
Deciduous 1st molar (14 mo)
Deciduous 2nd molar (24 mo)

Deciduous 2nd molar (20 mo)
Deciduous 1st molar (12 mo)
Lateral incisor (7 mo)
Canine (16 mo)
Central incisors (6 mo)

**Deciduous teeth, with the age at eruption given in months**

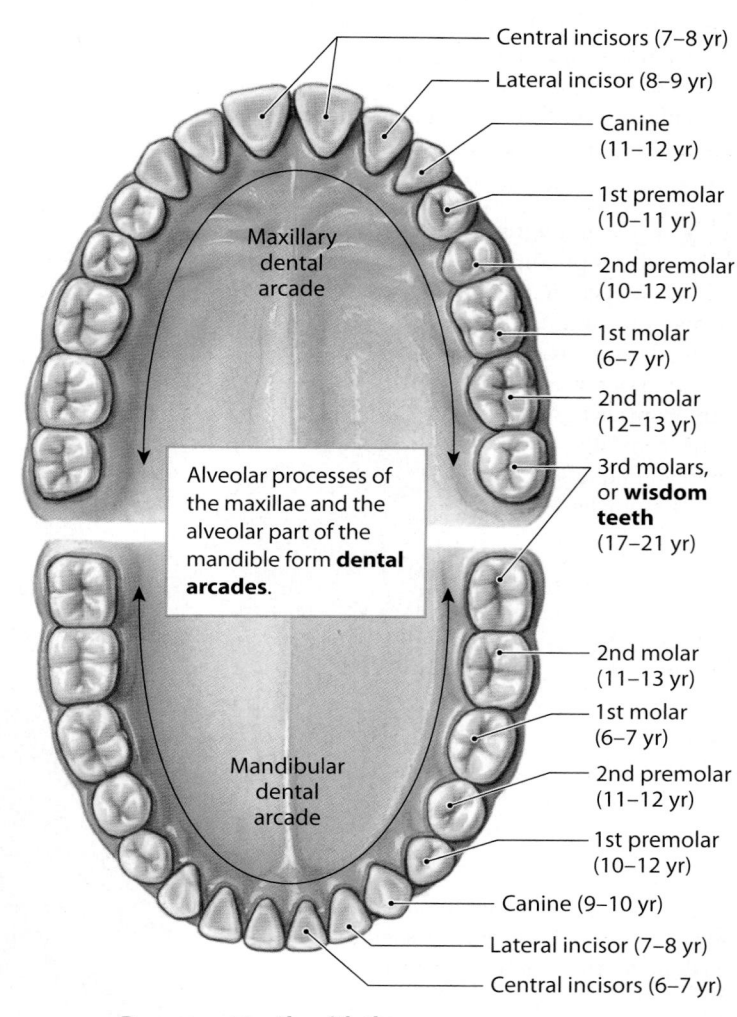

Central incisors (7–8 yr)
Lateral incisor (8–9 yr)
Canine (11–12 yr)
1st premolar (10–11 yr)
2nd premolar (10–12 yr)
1st molar (6–7 yr)
2nd molar (12–13 yr)
3rd molars, or **wisdom teeth** (17–21 yr)

Maxillary dental arcade

Alveolar processes of the maxillae and the alveolar part of the mandible form **dental arcades**.

2nd molar (11–13 yr)
1st molar (6–7 yr)
2nd premolar (11–12 yr)
1st premolar (10–12 yr)
Canine (9–10 yr)
Lateral incisor (7–8 yr)
Central incisors (6–7 yr)

Mandibular dental arcade

**Permanent teeth, with the age at eruption given in years**

Maxilla exposed to show developing permanent teeth
Erupted deciduous teeth
Mandible exposed to show developing permanent teeth

**Unerupted permanent teeth exposed in the skull of a child**

**? REVIEW**

**B.** What is the name sometimes given to the third set of molars?

**Lo LEARNING OUTCOME**

Describe the types of teeth, and differentiate between deciduous teeth and permanent teeth.

# Module 22.8

## The muscular walls of the pharynx and esophagus play a key role in swallowing

**1** The **pharynx** (FAR-ingks), or throat, is an anatomical space and a common passageway for food, liquids, and air. Food passes through the oropharynx and laryngopharynx on its way to the esophagus. This illustration shows the major landmarks and boundaries of the pharynx (see Module 21.3).

Internal nares
Oral cavity
Uvula
Epiglottis
Larynx
Esophagus

**Regions of the Pharynx**

Nasopharynx

Oropharynx

Laryngopharynx

**?** **REVIEW**
**A.** Name the regions and functions of the pharynx.

**2** The **esophagus** is a hollow muscular tube approximately 25 cm (10 in.) long and about 2 cm (0.80 in.) in diameter at its widest point. The primary function of the esophagus is to actively move food and liquids to the stomach.

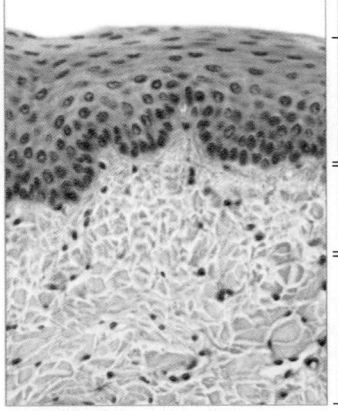

**3** The layers of the esophageal wall are comparable with those in other portions of the digestive tract. However, the shape of the lumen and the structure of the muscular layer are unique to the esophagus. There is no serosa, but an **adventitia** of connective tissue outside the muscular layer anchors the esophagus to the posterior body wall.

The mucosa and submucosa have large folds that extend the length of the esophagus. These folds allow for expansion during the passage of a large bolus. Muscle tone in the walls keeps the lumen closed, except when you swallow.

Mucosa

Submucosa

**4** This light micrograph illustrates the extreme thickness of the epithelium making up the esophageal mucosa layer.

Adventitia

In the superior third of the esophagus, the circular and longitudinal layers of the muscular layer contain skeletal muscle fibers; the middle third contains a mixture of skeletal and smooth muscle tissue; and along the inferior third, only smooth muscle occurs.

The mucosa of the esophagus contains a nonkeratinized, stratified squamous epithelium similar to that of the pharynx and oral cavity.

Lamina propria

The muscularis mucosae consists of an irregular layer of smooth muscle.

**?** **REVIEW**
**B.** Describe the muscular layer of the esophagus.

**Esophageal mucosa**    LM × 300

22

**5** Swallowing, or **deglutition** (dē-glū-TISH-un), is a complex process that can be started voluntarily but proceeds automatically once it begins. Although you take conscious control over swallowing when you eat or drink, swallowing is also controlled at the subconscious level. Each day you swallow approximately 2400 times. We divide swallowing into three phases.

## Buccal Phase

Hard palate
Soft palate
Bolus
Oropharynx
Tongue

The **buccal phase** begins with the compression of the bolus against the hard palate. Retraction of the tongue then forces the bolus into the oropharynx and assists in the elevation of the soft palate, thereby sealing off the nasopharynx. The buccal phase is strictly voluntary. Once the bolus enters the oropharynx, reflex responses are initiated and the bolus is moved toward the stomach.

## Pharyngeal Phase

Uvula
Tongue
Bolus
Epiglottis
Larynx

The **pharyngeal phase** begins when tactile receptors on the palatal arches and uvula are stimulated. In response, motor commands from the swallowing center in the medulla oblongata then direct a coordinated pattern of muscle contraction in the pharyngeal muscles. Elevation of the larynx and folding of the epiglottis results from contractions of the pharyngeal muscles, while the palatal muscles elevate the uvula and soft palate to block the entrance to the nasopharynx. Pharyngeal constrictors then force the bolus through the pharynx, past the closed glottis, and into the esophagus.

## Esophageal Phase

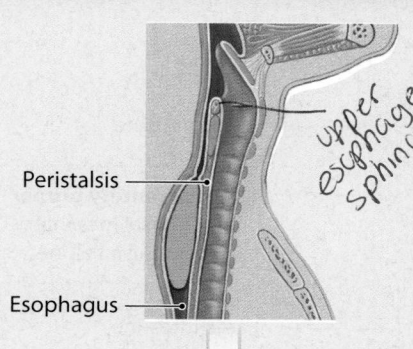

Peristalsis
Esophagus
*upper esophageal sphincter*
Thoracic cavity
Lower esophageal sphincter
Stomach

The **esophageal phase** of swallowing begins as the contraction of pharyngeal muscles forces the bolus through the entrance to the esophagus. Once in the esophagus, the bolus is pushed toward the stomach by peristalsis. The approach of the bolus triggers the opening of the lower esophageal sphincter, and the bolus then continues into the stomach. For a typical bolus, the entire trip takes about 9 seconds. Liquids may travel faster, flowing ahead of the peristaltic contractions with the assistance of gravity. A dry or poorly lubricated bolus travels much more slowly, and a series of local reflexes, called **secondary peristaltic waves**, may be required to push it all the way to the stomach.

The esophagus begins posterior to the cricoid cartilage. From this point, where it is at its narrowest, the esophagus descends toward the thoracic cavity posterior to the trachea. It passes inferiorly along the posterior wall of the mediastinum and enters the abdominopelvic cavity through the **esophageal hiatus** (hī-Ā-tus), an opening in the diaphragm. The esophagus is innervated by parasympathetic and sympathetic fibers from the esophageal plexus. Resting muscle tone in the circular muscle layer in the superior 3 cm of the esophagus normally prevents air from entering the esophagus. The band of smooth muscle involved functions as an **upper esophageal sphincter**. A comparable area of smooth muscle at the inferior end of the esophagus forms the **lower esophageal sphincter** (cardiac sphincter). It normally remains in a state of active contraction, preventing the backflow of materials from the stomach into the esophagus.

 **REVIEW**

**C.** Describe the major event in each of the three phases of swallowing.

22

 **LEARNING OUTCOME**

Describe the anatomy and functions of the pharynx and esophagus, and explain the swallowing process.

# Module 22.9

# The stomach and most of the intestinal tract are suspended by mesenteries and covered by the peritoneum

The **peritoneal cavity** is a serous membrane that lines the abdominopelvic cavity, covering most of its organs. The **peritoneal cavity** is a potential space between the parietal and visceral layers of the peritoneum. The **visceral peritoneum** covers the organs and the **parietal peritoneum** lines the abdominal walls.

**1** These diagrams are cross-sectional views from early embryos. During embryonic development, the digestive tract and accessory organs are enclosed by the peritoneal cavity and suspended by **dorsal** and **ventral mesenteries** (MEZ-en-ter-ēz).

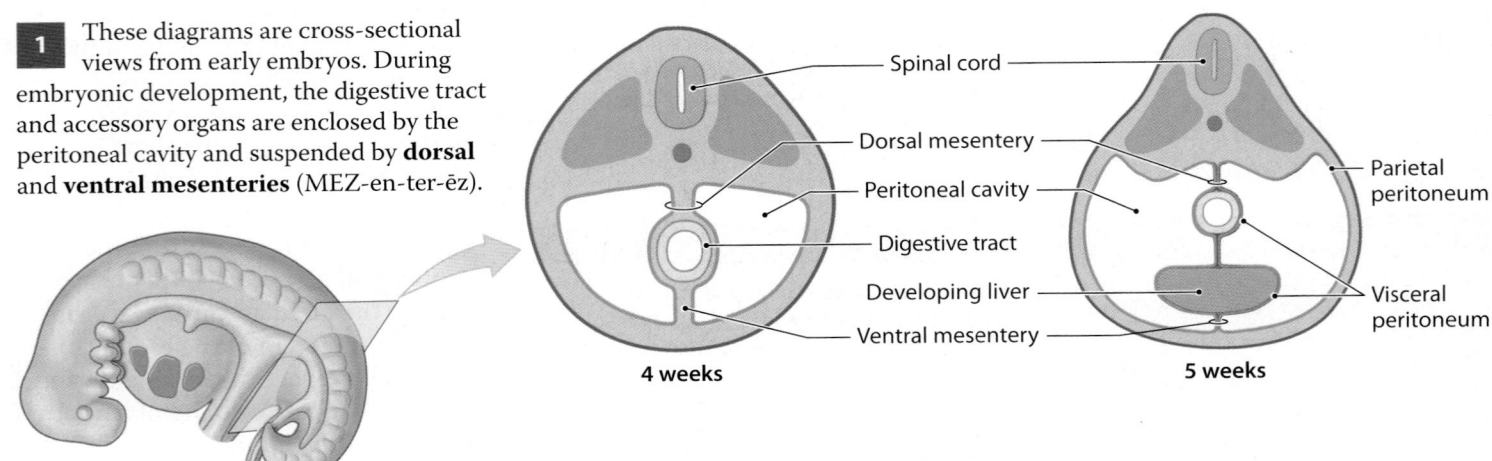

Spinal cord

Dorsal mesentery

Peritoneal cavity

Digestive tract

Developing liver

Ventral mesentery

Parietal peritoneum

Visceral peritoneum

4 weeks

5 weeks

**2** The diagrammatic cross-sectional view below and the lateral view to the right show the mesenteries supporting the digestive tract and accessory organs early in development. The ventral mesentery later disappears along most of the digestive tract, remaining in adults in only two places: between the stomach and the liver (the lesser omentum), and between the liver and the anterior abdominal wall (the falciform ligament).

Falciform ligament

Liver

Lesser omentum

Stomach

Greater omentum

Pancreas

Small intestine

Large intestine

Pancreas

Stomach

Liver

The **greater omentum** (ō-MEN-tum; *omentum*, fat skin) is the dorsal mesentery of the stomach.

The **lesser omentum** is the remnant of the ventral mesentery between the stomach and the liver.

The **mesentery proper** is the dorsal mesentery of the small intestine.

The **mesocolon** is the dorsal mesentery of the large intestine.

The **falciform** (FAL-si-form; *falx*, sickle + *forma*, form) **ligament** is the remnant of the ventral mesentery between the liver and the anterior wall of the peritoneal cavity.

**? REVIEW**

**A.** What is the falciform ligament?

**3** As the digestive tract elongates, the position of the mesenteries changes. Some segments of the tract contact the posterior wall and become fixed in position. As the small intestine coils and increases in length, segments of the mesentery proper come into contact and fuse together.

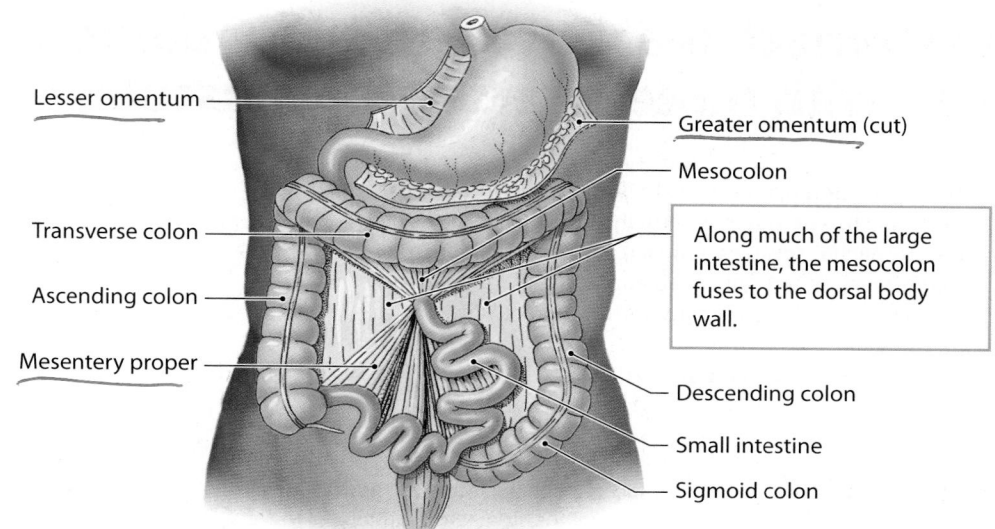

Lesser omentum

Transverse colon

Ascending colon

Mesentery proper

Greater omentum (cut)

Mesocolon

Along much of the large intestine, the mesocolon fuses to the dorsal body wall.

Descending colon

Small intestine

Sigmoid colon

**4** A sagittal section reveals the orientation of the mesenteries in an adult. The coils of the small intestine are suspended by the mesentery proper. The transverse colon and a small section near the end of the large intestine are suspended by remnants of the original mesocolon.

The mesentery associated with the initial portion of the small intestine and the pancreas fuses with the posterior abdominal wall, locking those structures in a retroperitoneal position.

Pancreas

Initial segment of small intestine (duodenum)

Mesentery proper

Mesocolon

Last segment of the large intestine (rectum)

Anus

Falciform ligament    Diaphragm

Liver

The lesser omentum stabilizes the position of the stomach and provides an access route for blood vessels and other structures entering or leaving the liver.

Stomach

Mesocolon

Transverse colon

The greater omentum forms an enormous pouch that extends inferiorly between the body wall and the anterior surface of the small intestine. Adipose tissue in the greater omentum conforms to the shapes of the surrounding organs, providing padding and protection across the anterior and lateral surfaces of the abdomen.

Parietal peritoneum

Small intestine

**? REVIEW**
**B.** What is the function of the lesser omentum?

22

The serous membrane lining the peritoneal cavity continuously produces peritoneal fluid, which provides essential lubrication. Because a thin layer of peritoneal fluid covers the parietal and visceral surfaces, sliding movements can occur without friction and resulting irritation. About 7 liters of fluid are secreted and reabsorbed each day, although the volume within the peritoneal cavity at any one time is about 50 mL. Liver disease, kidney disease, and heart failure can accelerate the rate at which fluids move into the peritoneal cavity. The accumulation of peritoneal fluid creates a characteristic abdominal swelling called **ascites** (a-SĪ-tēz).

**? REVIEW**
**C.** Explain the significance of peritoneal fluid.

**Lo LEARNING OUTCOME**
Explain the embryonic development of the mesenteries, and describe the mesenteries that remain in adulthood.

# The stomach is a muscular, expandable, J-shaped organ with three layers in the muscular layer

**1** This illustration presents the major surfaces and regions of the stomach. The shape is actually highly variable. When empty, the stomach resembles a muscular tube with a narrow, constricted lumen. When full, it can contain 1–1.5 liters of material. That material, which consists of food combined with saliva and the secretions of the gastric glands, is a viscous, highly acidic, soupy mixture called **chyme** (kīm).

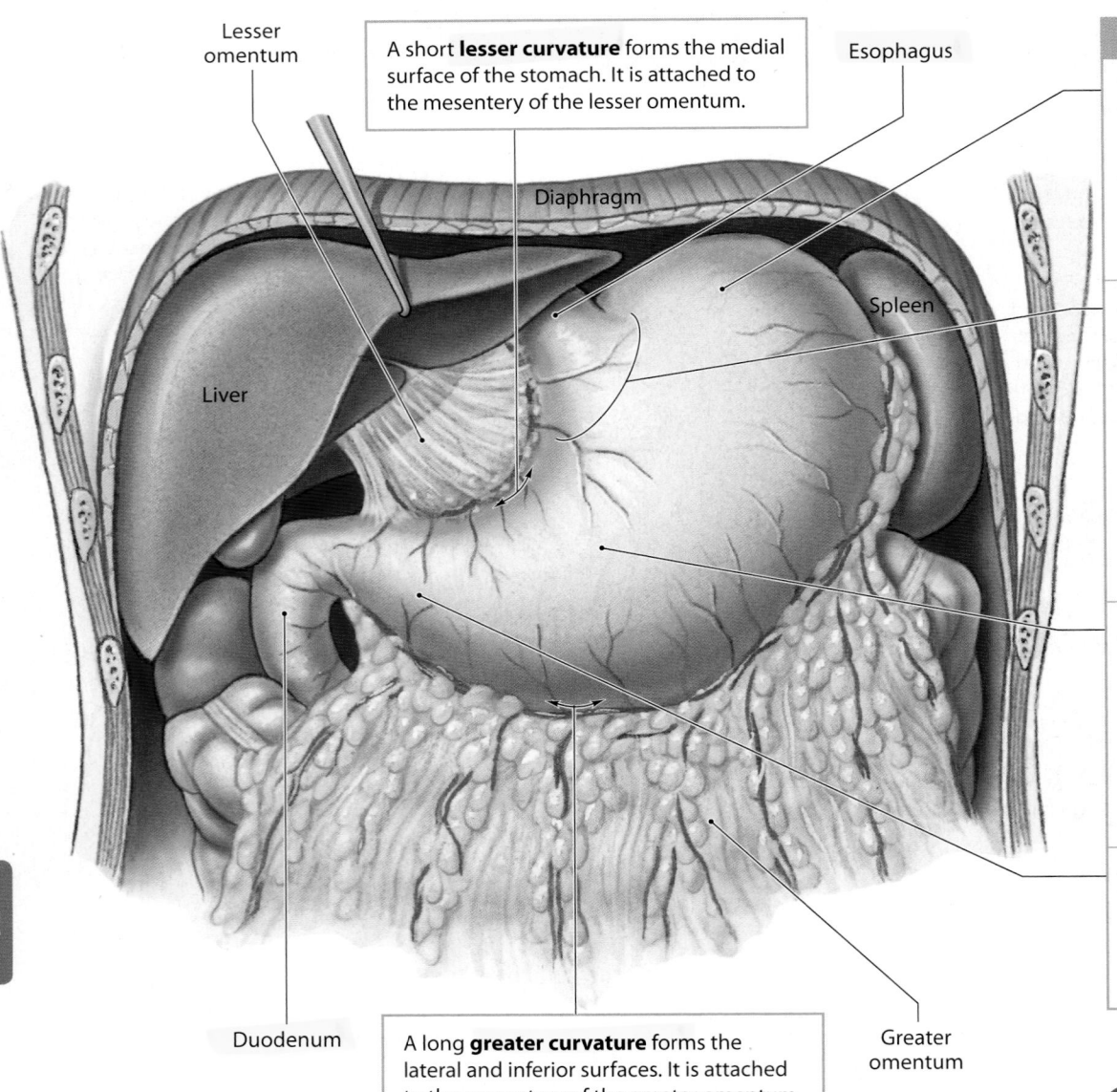

Lesser omentum

A short **lesser curvature** forms the medial surface of the stomach. It is attached to the mesentery of the lesser omentum.

Esophagus

Diaphragm

Liver

Spleen

Duodenum

A long **greater curvature** forms the lateral and inferior surfaces. It is attached to the mesentery of the greater omentum.

Greater omentum

### Regions of the Stomach

The **fundus** (FUN-dus) is the portion of the stomach that is superior to the junction between the stomach and the esophagus. The fundus contacts the inferior, posterior surface of the diaphragm.

The **cardia** (KAR-dē-uh) is the superior, medial portion of the stomach within 3 cm (1.2 in.) of the junction between the stomach and the esophagus. The cardia contains abundant mucous glands whose secretions coat the junction with the esophagus and help protect that tube from the acid and enzymes of the stomach.

The **body** is the area of the stomach between the fundus and the curve of the J. It is the largest region of the stomach. The body functions as a mixing bowl for ingested food and secretions produced in the stomach.

The **pyloric part** forms the sharp curve of the J. As mixing movements occur during digestion, the pyloric part frequently changes shape.

**? REVIEW**

**A.** Name the four major regions of the stomach in order from its junction with the esophagus to the small intestine.

22

**2** This cadaver dissection exposes the stomach and greater omentum.

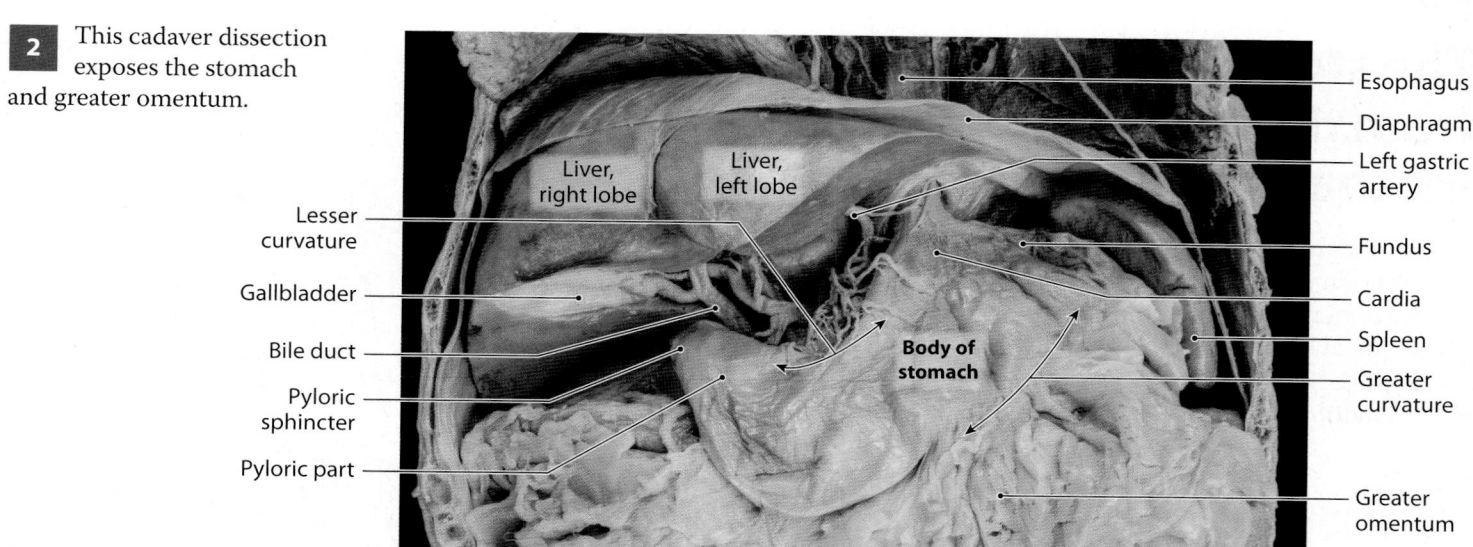

Esophagus

Diaphragm

Left gastric artery

Liver, right lobe

Liver, left lobe

Lesser curvature

Gallbladder

Bile duct

Pyloric sphincter

Pyloric part

Fundus

Cardia

Spleen

Greater curvature

**Body of stomach**

Greater omentum

**3** A partially sectioned and dissected stomach reveals additional details about the internal structure of this organ.

Fundus

Esophagus

Anterior surface

Cardia

Body

Lesser curvature

**The Pyloric Part**

The **pyloric antrum** (*antron*, cavity) is the region of the pyloric part connected to the body of the stomach.

The **pyloric canal** empties into the *duodenum*, the proximal segment of the small intestine.

Ther **pyloric orifice** is the stomach outlet. It opens into the duodenum.

A muscular **pyloric sphincter** regulates the release of chyme into the duodenum.

**Muscular Layer**

The muscular layer of the stomach includes an inner, oblique layer of smooth muscle. This additional layer of smooth muscle strengthens the stomach wall and assists in the mixing and churning to form chyme.

Longitudinal layer

Circular layer

Oblique layer

Greater curvature

**Rugae** (RŪ-gē; wrinkles) are prominent but temporary mucosal folds that allow the gastric lumen to expand. As the stomach fills, the rugae gradually flatten out until, at maximum distension, they almost disappear.

**?** **REVIEW**

**B.** What anatomical feature of the stomach allows the organ to form chyme?

22

**?** **REVIEW**

**C.** Describe the lining of the stomach.

**Lo** **LEARNING OUTCOME**

Describe the anatomy of the stomach and its histological features.

# Module 22.11

## The stomach receives food and liquids from the esophagus and aids in mechanical and chemical digestion

The **stomach** has four major functions:
(1) temporary storage of ingested food,
(2) mechanical digestion of ingested food,
(3) chemical digestion of food through the action of acid and enzymes, and
(4) production of intrinsic factor.

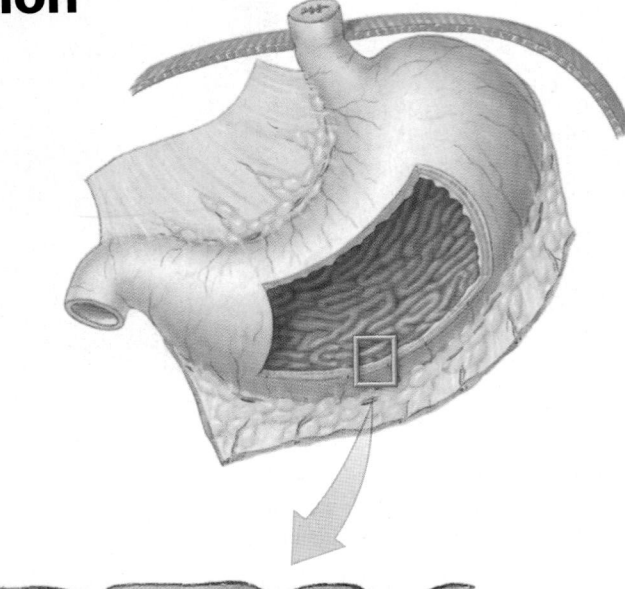

**1** The wall of the stomach is relatively thick and muscular, and its mucosa has deep folds that form gastric glands.

| Layers of the Stomach Wall |
| --- |
| **Mucosa** |
| A simple columnar epithelium lines all portions of the stomach. This epithelium produces a carpet of mucus that covers the interior surfaces of the stomach. The alkaline mucous layer protects epithelial cells against the acid and enzymes in the gastric lumen. Still, the environment is harsh, and a typical gastric epithelial cell has a life span of only 3 to 7 days. |
| **Submucosa** |
| **Muscular Layer** |
| **Serosa** |

Lamina propria
Muscularis mucosae

Oblique layer
Circular layer
Longitudinal layer

Lymphatic vessel

Artery and vein

Myenteric plexus

**? REVIEW**

**A.** Explain the significance of the alkaline mucous layer lining the interior surface of the stomach.

22

**2** **Gastric glands** in the fundus and body secrete most of the acid and enzymes involved in gastric digestion. The gastric glands in these areas are dominated by **parietal cells** and **chief cells**. Together, they secrete about 1500 mL (1.6 qt) of *gastric juice* each day. Gastric glands in the pyloric part secrete mucus and hormones involved in the coordination and control of digestive activity.

Shallow depressions called **gastric pits** open onto the gastric surface. Stem cells at the base, or neck, of each gastric pit actively divide, replacing superficial cells that are shed into the chyme.

Each gastric pit communicates with several gastric glands that extend deep into the lamina propria.

**?** **REVIEW**
**B.** What is the function of parietal cells?

Lamina propria
Mucous epithelial cells
Neck

### Cells of Gastric Glands

**Parietal cells** secrete **intrinsic factor**, a glycoprotein that helps absorb vitamin $B_{12}$ across the intestinal lining. This vitamin is necessary for erythropoiesis (see Module 17.11). Parietal cells also secrete hydrochloric acid (HCl).

**G cells** are enteroendocrine cells that produce a variety of hormones. These hormones and their functions will be considered in Module 22.14.

**Chief cells** secrete **pepsinogen** (pep-SIN-ō-jen), an inactive proenzyme. Acid (HCl) in the gastric lumen converts pepsinogen to **pepsin**, an active proteolytic (protein-digesting) enzyme. In addition, the stomachs of newborn infants (but not of adults) produce **rennin** and **gastric lipase**, enzymes important for the digestion of milk.

**3** Parietal cells indirectly secrete *hydrochloric acid* (HCl) to keep the stomach contents at pH 1.5–2.0. They do not produce this strong acid in the cytoplasm, because it would erode secretory vesicles and destroy the cell. Instead, $H^+$ and $Cl^-$ are transported and secreted independently, as diagrammed below. When gastric glands are actively secreting, enough bicarbonate ions enter the bloodstream to increase the pH of the blood significantly. This sudden influx of bicarbonate ions has been called the **alkaline tide**.

**?** **REVIEW**
**C.** Describe the alkaline tide.

**1** Hydrogen ions ($H^+$) are generated inside a parietal cell as the enzyme carbonic anhydrase converts $CO_2$ and $H_2O$ to carbonic acid ($H_2CO_3$), which then dissociates.

**2** An anion countertransport mechanism ejects the bicarbonate ions into the interstitial fluid and imports chloride ions into the cell.

**3** The chloride ions then diffuse across the cell and exit through open chloride channels into the lumen of the gastric gland.

Parietal cell

$CO_2 + H_2O$

*Carbonic anhydrase*

**4** The hydrogen ions are actively transported into the lumen of the gastric gland.

$H_2CO_3$

Dissociation

Interstitial fluid

$HCO_3^-$
$Cl^-$ → $Cl^-$

$HCO_3^- + H^+$

$Cl^-$

$H^+$

$Cl^-$

Alkaline tide

Enters bloodstream

Lumen of gastric gland

**KEY**

Diffusion

Carrier-mediated transport

Active transport

Countertransport

**ʟᴏ** **LEARNING OUTCOME**
Describe the anatomy of the stomach relating to its role in digestion and absorption.

22

# Module 22.12
# The intestinal tract is specialized to absorb nutrients

**1** The intestinal lining has a series of transverse folds called **circular folds** or *plicae* (PLĪ-sē; folds) *circulares*. Unlike the rugae in the stomach, these circular folds are permanent features that do not disappear when the small intestine fills. The small intestine contains roughly 800 circular folds, most of them within the jejunum. Their presence greatly increases the surface area available for absorption.

**2** The mucosa of the small intestine contains a series of fingerlike projections, the **intestinal villi**. If the small intestine were a simple tube with smooth walls, it would have a total absorptive area of nearly 3300 cm² (3.6 ft²). Instead, the mucosa contains circular folds, which support a forest of villi. Each villus is covered by epithelial cells whose exposed surfaces are covered with microvilli. This arrangement increases the total area for absorption by a factor of more than 600, to approximately 2 million cm² (more than 2200 ft²).

**3** This diagrammatic sectional view of the intestinal wall shows features common to all segments of the small intestine.

Circular fold

Villi

Villi | Submucosal artery and vein | Lacteal (lymphatic capillary) | Lymphoid nodule

**Layers of the Small Intestine**

Mucosa

Submucosa

Muscular layer

Serosa

Lymphatic vessel

At the bases of the villi are the entrances to the **intestinal glands** (*intestinal crypts*).

Contractions of the **muscularis mucosae** and smooth muscle cells within the intestinal villi move the villi back and forth, exposing the epithelial surfaces to the liquefied intestinal contents. This movement improves the efficiency of absorption by quickly eliminating local differences in nutrient concentration. Movements of the villi also squeeze the lacteals, which help move lymph out of the villi.

Submucosal plexus

Circular layer of smooth muscle

Myenteric plexus

Longitudinal layer of smooth muscle

**? REVIEW**

**A.** Name the layers of the small intestine from superficial to deep.

**4** Each villus has a complex internal structure. The lamina propria of each villus contains an extensive network of capillaries that originate in a vascular network within the submucosa. These capillaries carry absorbed nutrients to the hepatic portal system for delivery to the liver, which adjusts the nutrient concentrations in blood before the blood reaches the general systemic circulation.

**5** The surface of each villus consists of a simple columnar epithelium that is carpeted with microvilli. Because the microvilli project from the epithelium like the bristles on a brush, these cells are said to have a **brush border**. Brush border enzymes are integral membrane proteins located on the surfaces of intestinal microvilli. These enzymes break down materials that come in contact with the brush border. The epithelial cells then absorb the breakdown products.

Each villus contains a central lymphatic capillary called a **lacteal** (LAK-tē-ul; *lacteus*, milky). Lacteals transport materials that cannot enter blood capillaries. For example, absorbed fatty acids are assembled into protein–lipid packages that are too large to diffuse into the bloodstream. These packets, called *chylomicrons*, reach the venous circulation through the thoracic duct, which delivers lymph into the left subclavian vein (see Module 20.2).

Goblet cell

Nerve

Capillary network

Columnar epithelial cell

Lamina propria

Arteriole

Lymphatic vessel

Venule

Smooth muscle cell

Capillaries

Brush border

Goblet cells in the columnar epithelium secrete mucin onto the intestinal surfaces.

Lacteal

**Tip of villus**          LM × 250

**?** **REVIEW**

**B.** Describe the anatomy of the intestinal mucosa.

**C.** Explain the function of lacteals.

**6** Near the base of each intestinal gland, *stem cells* divide and produce new generations of *epithelial cells*, which are continuously displaced toward the intestinal surface. In a few days, the new cells reach the tip of a villus and are shed into the intestinal lumen. This ongoing process renews the epithelial surface. The disintegration of the shed cells adds enzymes to the lumen. **Paneth cells** at the base of the intestinal glands have a role in innate (nonspecific) immunity and release defensins and lysozyme. These secretions kill some bacteria and allow others to live, thereby establishing the microbiota of the intestinal lumen.

Epithelial cell

Stem cell

Paneth cell

**Lo** **LEARNING OUTCOME**

Describe the anatomy of the intestinal tract and its histological features.

# Module 22.13
# The small intestine is divided into the duodenum, jejunum, and ileum

The **small intestine**, or small bowel, plays the key role in nutrient digestion and absorption. Ninety percent of nutrient absorption occurs here; most of the rest occurs in the large intestine.

**1** This diagram illustrates the distribution of the three segments of the small intestine by abdominopelvic region. The small intestine fills much of the abdominopelvic cavity, and the mesentery proper stabilizes its position. The small intestine averages 6 m (19.7 ft) long and has a diameter ranging from 4 cm (1.6 in.) at the stomach to about 2.5 cm (1 in.) at its junction with the large intestine.

## Segments of the Small Intestine

The **duodenum** (dū-ō-DĒ-num), 25 cm (10 in.) long, is the segment closest to the stomach. This portion of the small intestine is a "mixing bowl." It receives chyme from the stomach and digestive secretions from the pancreas and liver. Except for the proximal 2.5 cm (1 in.), the duodenum is in a retroperitoneal position, firmly attached to the posterior body wall (see Module 22.9).

The **jejunum** (je-JŪ-num) is the segment between the duodenum and the ileum and is marked by a sharp bend at its beginning. At this junction, the small intestine reenters the peritoneal cavity, supported by a sheet of mesentery. The jejunum is about 2.5 meters (8.2 ft) long. The bulk of chemical digestion and nutrient absorption occurs there.

The **ileum** (IL-ē-um), the final segment of the small intestine, is also the longest, averaging 3.5 meters (11.5 ft) in length. The ileum ends at the **ileocecal** (il-ē-o-SĒ-kal) **valve**, a sphincter that controls the flow of material from the ileum into the *cecum* (SĒ-kum) of the large intestine. To help you remember the order of the small intestine segments, use this mnemonic: **d**on't **j**ump **i**n—**d**uodenum, **j**ejunum, and **i**leum.

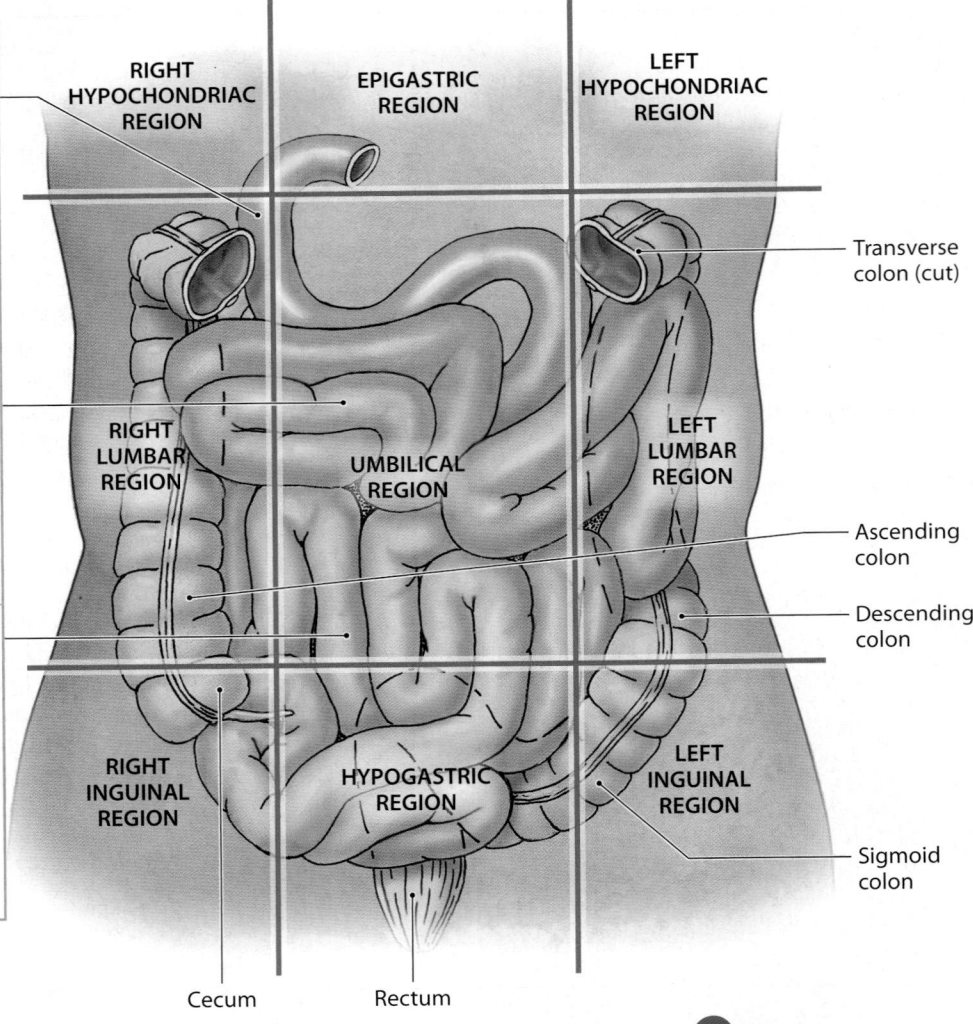

RIGHT HYPOCHONDRIAC REGION

EPIGASTRIC REGION

LEFT HYPOCHONDRIAC REGION

Transverse colon (cut)

RIGHT LUMBAR REGION

UMBILICAL REGION

LEFT LUMBAR REGION

Ascending colon

Descending colon

RIGHT INGUINAL REGION

HYPOGASTRIC REGION

LEFT INGUINAL REGION

Sigmoid colon

Cecum

Rectum

**22**

**? REVIEW**
A. Name the three segments of the small intestine from proximal to distal.

**? REVIEW**
B. Identify the segment of the small intestine found within the epigastric region.

**2** Each segment of the small intestine has characteristic features related to its primary functions. The transition from one region to another is gradual, and the boundaries are indistinct.

### Duodenum

Circular folds

Duodenal ampulla

The duodenum has few circular folds, and their villi are small. The primary function of the duodenum is to receive chyme from the stomach and neutralize its acids before they can damage the absorptive surfaces of the small intestine.

### Jejunum

Circular folds

The jejunum has numerous circular folds, and the villi are abundant and very long.

**Duodenum**

The submucosa of the duodenum is dominated by **duodenal glands** that produce mucous secretions.

Circular folds

**Jejunum**

Serosa

Muscular layer

Submucosa

Mucosa

Muscularis mucosae

Villi

### Ileum

Circular folds

The ileum has fewer circular folds than the jejunum, and those disappear distally. Its villi are relatively stumpy. The submucosa contains aggregated lymphoid nodules.

Aggregated lymphoid nodules of the submucosa protect the small intestine from bacteria that normally inhabit the large intestine.

**Ileum**

**? REVIEW**

**C.** What is the primary function of the duodenum?

**? INTEGRATION**

**D.** A traumatic injury to the umbilical region could affect which segments of the small intestine?

**Lo LEARNING OUTCOME**

Describe the anatomy and physiology of the small intestine.

22

# Several hormones regulate digestion

**1**   Several hormones that regulate digestion are produced by the duodenum, which receives partially digested materials from the stomach. The duodenum adjusts gastric activity and coordinates accessory digestive organ secretion depending on the composition of the arriving chyme.

## Major Hormones of the Duodenum

### Gastrin

**Gastrin** is secreted by G cells in the pyloric antrum and enteroendocrine cells in the duodenum. The pyloric antrum secretes gastrin when stimulated by the vagus nerves or when food arrives in the stomach. Duodenal cells release gastrin when they are exposed to large quantities of incompletely digested proteins. The functions of gastrin include increasing stomach motility and stimulating gastric acid and enzyme production.

### Secretin

**Secretin** is released when chyme arrives in the duodenum. Secretin's primary effect is an increase in the secretion of buffers by the pancreas, which increase the pH of the chyme. Among its secondary effects, secretin also stimulates secretion of bile by the liver and reduces gastric motility and gastric secretory rates.

### Gastric Inhibitory Peptide (GIP)

**Gastric inhibitory peptide (GIP)** is secreted when fats and carbohydrates—especially glucose—enter the small intestine. The inhibition of gastric activity is accompanied by the stimulation of insulin release at the pancreatic islets. GIP has several secondary effects, including stimulating submucosal duodenal gland activity, stimulating lipid synthesis in adipose tissue, and increasing glucose use by skeletal muscles.

### Cholecystokinin (CCK)

**Cholecystokinin (CCK)** is secreted when chyme arrives in the duodenum, especially when the chyme contains lipids and partially digested proteins. In the pancreas, CCK accelerates the production and secretion of all types of digestive enzymes. It also causes a relaxation of the hepatopancreatic sphincter and contraction of the gallbladder, resulting in the ejection of bile and pancreatic juice into the duodenum. Thus, the net effects of CCK are to increase the secretion of pancreatic enzymes and to push pancreatic secretions and bile into the duodenum. The presence of CCK in high concentrations has two additional effects: It inhibits gastric activity, and it appears to have CNS effects that reduce the sensation of hunger.

### Vasoactive Intestinal Peptide (VIP)

**Vasoactive intestinal peptide (VIP)** stimulates the secretion of intestinal glands, dilates regional capillaries, and inhibits acid production in the stomach. By dilating capillaries in active areas of the intestinal tract, VIP provides an efficient mechanism for removing absorbed nutrients.

### Enterocrinin

**Enterocrinin** is released by enteroendorine cells of the duodenum in response to stimulation by the vagus nerves before chyme reaches the pyloric sphincter and also when chyme enters the duodenum. It stimulates production of alkaline mucus by the duodenal submucosal glands, which protects the intestine from acidic chyme.

**? REVIEW**

**A.** Name the major hormones that regulate digestive activities.

**2** This flowchart summarizes the pattern of hormone release and the effects of those hormones within the digestive system.

Go to
MasteringA&P™
PhysioEx 9.1
Ex 8: Chemical and Physical Digestion

Ingested food

Food in stomach

Chyme in duodenum

Material arrives in jejunum

| Hormone | Action |
|---|---|
| Gastrin | Acid production by parietal cells |
| | Stimulation of gastric motility; mixing waves increase in intensity |
| GIP | Release of insulin from pancreas |
| Secretin and CCK | Release of pancreatic enzymes and buffers |
| | Bile secretion and ejection of bile from gallbladder |
| VIP | Dilation of intestinal capillaries |
| Enterocrinin | Alkaline mucus production neutralizes acidic chyme |

KEY
⊣ Inhibits
→ Stimulates

Pancreas

facilitates          facilitates

**Nutrient absorption** → **NUTRIENT UTILIZATION BY ALL TISSUES**

22

**? INTEGRATION**

**B.** How would the pH of the intestinal contents be affected if the small intestine did not produce secretin?

**C.** Does a high-fat meal raise or lower the level of cholecystokinin (CCK) in the blood?

**Lo LEARNING OUTCOME**

Discuss the major digestive hormones and their primary effects.

# Module 22.15
# Central and local mechanisms coordinate gastric ...

**1** The duodenum plays a key role in controlling digestive function because it monitors the contents of the chyme and adjusts the activities of the stomach and accessory glands to protect the delicate absorptive surfaces of the jejunum. This pivotal role of the duodenum is apparent when you consider the three phases of gastric secretion. The phases are named according to the location of the control center involved.

## Cephalic Phase

The **cephalic phase** of gastric secretion begins when you see, smell, taste, or think of food. This phase, which is directed by the CNS, prepares the stomach to receive food. The neural output proceeds by way of the parasympathetic division of the autonomic nervous system, and the vagus nerves (X) innervate the submucosal plexus of the stomach. Next, postganglionic parasympathetic fibers innervate mucous cells, chief cells, parietal cells, and G cells of the stomach. In response to stimulation, gastric juice production accelerates, reaching rates of about 500 mL/h. This phase generally lasts only minutes.

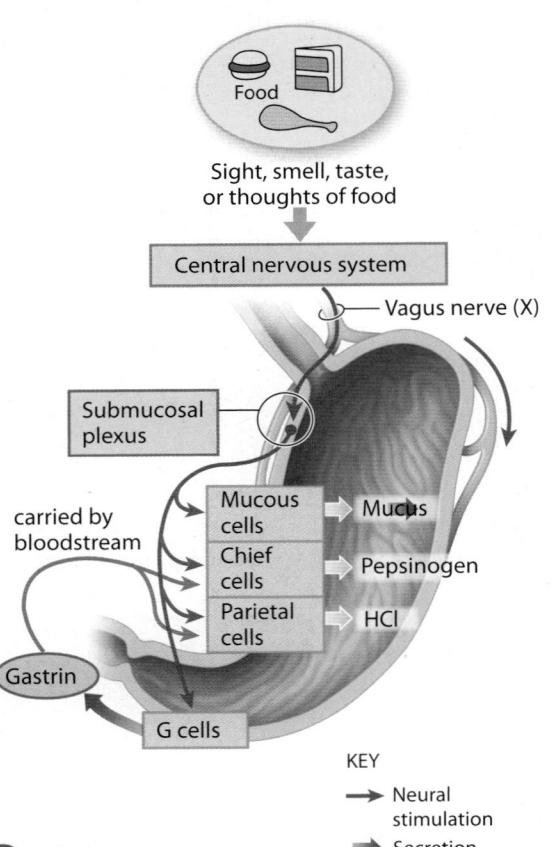

## Gastric Phase

The **gastric phase** begins when food arrives in the stomach and builds on the stimulation provided during the cephalic phase. The stimuli that initiate the gastric phase are (1) distension of the stomach, (2) an increase in the pH of the gastric contents (due to dilution of acid during mixing), and (3) the presence of undigested food in the stomach, especially proteins and peptides. The gastric phase may continue for 3 to 4 hours while the acid and enzymes process the ingested food. During this period, gastrin stimulates contractions in the muscular layer of the stomach and intestinal tract. After the first hour, the material in the stomach is churning like clothing in a washing machine. The gastric phase consists of the following mechanisms:

### Local Responses

Distension of the gastric wall stimulates the release of histamine in the lamina propria, which binds to receptors on the parietal cells and stimulates acid secretion.

### Neural Responses

The stimulation of stretch receptors and chemoreceptors triggers short reflexes coordinated in the submucosal and myenteric plexuses. This in turn activates the stomach's secretory cells. The stimulation of the myenteric plexus produces powerful contractions called **mixing waves** in the muscular layer.

### Hormonal Response

Neural stimulation and the presence of peptides and amino acids in chyme stimulate the secretion of the hormone gastrin. Gastrin travels by the bloodstream to parietal and chief cells, whose increased secretions reduce the pH of the gastric juice. In addition, gastrin also stimulates gastric motility.

**? REVIEW**

**A.** Name and briefly describe an important characteristic of each of the three phases of gastric secretion.

# … and intestinal activities

## Intestinal Phase

The **intestinal phase** of gastric secretion begins when chyme first enters the small intestine, usually after several hours of mixing contractions. The function of the intestinal phase is controlling the rate of gastric emptying to ensure that the secretory, digestive, and absorptive functions of the small intestine can proceed with reasonable efficiency. Although here we consider the intestinal phase as it affects stomach activity, the arrival of chyme in the small intestine also triggers other neural and hormonal events that coordinate the activities of the intestinal tract and the pancreas, liver, and gallbladder.

### Neural Responses

Chyme leaving the stomach decreases the distension of the stomach, thereby reducing the stimulation of stretch receptors. Distension of the duodenum by chyme stimulates stretch receptors and chemoreceptors that trigger the **enterogastric reflex**. This reflex inhibits both gastrin production and gastric contractions and stimulates the contraction of the pyloric sphincter, which prevents further discharge of chyme. At the same time, local reflexes at the duodenum stimulate mucus production, which helps protect the duodenal lining from the arriving acid and enzymes.

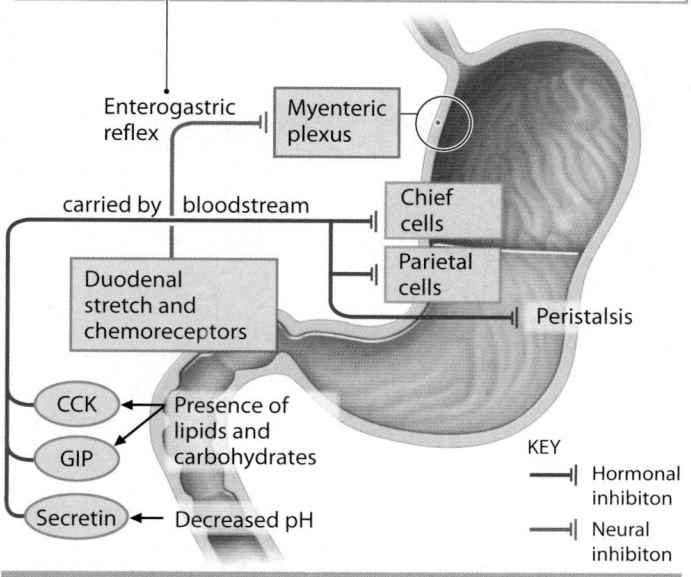

KEY
- ⊣ Hormonal inhibiton
- ⊣ Neural inhibiton

### Hormonal Responses

- Arrival of lipids and carbohydrates stimulates the secretion of cholecystokinin (CCK) and gastric inhibitory peptide (GIP).
- A decrease in pH below 4.5 stimulates the secretion of secretin.
- Partially digested proteins in the duodenum stimulate the secretion of gastrin, which speeds gastric processing.

**2** Two central reflexes are triggered by the stimulation of stretch receptors in the stomach wall as it fills. These reflexes accelerate movement along the small intestine while the enterogastric reflex controls the rate of chyme entry into the duodenum.

### Central Gastric Reflexes

The **gastroenteric reflex** stimulates motility and secretion along the entire small intestine.

The **gastroileal** (gas-trō-IL-ē-al) **reflex** triggers the opening of the ileocecal valve, allowing materials to pass from the small intestine into the large intestine.

The **ileocecal valve** regulates the passage of materials into the large intestine.

**? REVIEW**

**B.** Describe two central reflexes triggered by stimulation of the stretch receptors in the stomach wall.

In general, the rate of chyme movement into the small intestine is fastest when the stomach is greatly distended and the meal contains little protein. A big meal that contains small amounts of protein, large amounts of carbohydrates, alcohol, or caffeine will leave the stomach very quickly. One reason for this is that both alcohol and caffeine stimulate gastric secretion and motility.

**? INTEGRATION**

**C.** Why might severing the branches of the vagus nerves that supply the stomach provide relief for a person who suffers from chronic gastric ulcers (sores on the stomach lining)?

**Lo LEARNING OUTCOME**

Explain the regulation of gastric activity by central and local mechanisms.

22

# Module 22.16
# The large intestine stores and concentrates fecal material

The horseshoe-shaped **large intestine**, also known as the large bowel, has an average length of about 1.5 meters (4.9 ft) and a diameter of 7.5 cm (3 in.). The major functions of the large intestine include (1) reabsorbing water and compacting the intestinal contents into feces, (2) absorbing important vitamins generated by bacterial action, and (3) storing fecal material prior to defecation. The large intestine consists of three segments: the *cecum*, the *colon*, and the *rectum*.

Right colic (hepatic) flexure

> **? REVIEW**
> **A.** Name the major functions of the large intestine.

**1** Material arriving from the ileum first enters an expanded pouch called the **cecum** (SĒ-kum). The cecum collects and stores materials from the ileum and begins the process of **compaction** (the forming of feces by compression).

The **ascending colon** begins at the cecum and ascends along the right margin of the peritoneal cavity to the inferior surface of the liver. The colon then bends sharply to the left at the **right colic** (*hepatic*) **flexure**, and this marks the end of the ascending colon.

The ileum attaches to the medial surface of the cecum and opens into the cecum at the **ileocecal valve**.

Ileum

Cecum

Ileum

The slender, hollow **appendix** (also called the *vermiform appendix*) is attached to the cecum. The appendix is generally about 9 cm (3.6 in.) long, but its size and shape are quite variable. The mucosa and submucosa of the appendix are dominated by lymphoid nodules, and the appendix functions primarily as an organ of the lymphatic system. Inflammation of the appendix is known as **appendicitis**.

**2** The **colon** has a larger diameter and a thinner wall than the small intestine. We can subdivide the colon into four regions: the ascending colon, transverse colon, descending colon, and sigmoid colon. The ascending and descending colon are retroperitoneal and firmly attached to the abdominal wall. The transverse colon and sigmoid colon are suspended by remnants of the embryonic mesocolon.

Hepatic portal vein

Superior mesenteric vein

Superior mesenteric artery

Inferior mesenteric vein

Aorta

Superior mesenteric vein

**?** **REVIEW**
**B.** Identify the four regions of the colon.

The **transverse colon** crosses the abdomen from right to left. It is supported by the transverse mesocolon and is separated from the anterior abdominal wall by the layers of the greater omentum. As the transverse colon reaches the left side of the body, the colon makes a 90° turn at the **left colic** (splenic) **flexure**.

Left colic (splenic) flexure

Greater omentum (cut)

The serosa of the colon contains numerous teardrop-shaped sacs of fat called **omental** (fatty) **appendices**.

The **descending colon** proceeds inferiorly along the body's left side until reaching the iliac fossa. At the iliac fossa, the descending colon ends at the **sigmoid flexure**.

Three separate longitudinal bands of smooth muscle—called the **teniae coli** (TĒ-nē-ē KŌ-lē;)—run along the outer surfaces of the colon just deep to the serosa. These bands correspond to the outer longitudinal layer of the muscular layer in other portions of the digestive tract.

Muscle tone within the teniae coli creates **haustra** (HAWS-truh), a series of pouches in the wall of the colon. Cutting into the intestinal lumen reveals that the creases between the haustra affect the mucosal lining as well, producing a series of internal folds. Haustra permit the expansion and elongation of the colon, rather like the bellows that allow an accordion to lengthen.

The sigmoid flexure is the start of the **sigmoid** (SIG-moyd; *sigmeidos*, the Greek letter S) **colon**, an S-shaped segment that is about 15 cm (6 in.) long. The sigmoid colon empties into the rectum.

Powerful peristaltic contractions called **mass movements** occur a few times each day in response to distension of the stomach and duodenum. These contractions begin at the transverse colon and push materials along the distal portion of the large intestine.

The **rectum** (REK-tum), forms the last 15 cm (6 in.) of the digestive tract. It is an expandable organ for the temporary storage of feces. The movement of fecal material into the rectum triggers the urge to defecate.

**REVIEW**
**C.** Describe mass movements.

**LEARNING OUTCOME**
Describe the gross anatomy of the three segments of the large intestine.

22

# Module 22.17

# The large intestine compacts fecal material; the defecation reflex coordinates the elimination of feces

Teniae coli
Omental appendices
Haustrum

**1** The major characteristics of the wall of the large intestine are the lack of villi and the presence of distinctive intestinal glands dominated by mucin-secreting goblet cells. The mucosa of the large intestine does not produce enzymes; any digestion that occurs results from enzymes introduced into the small intestine or from bacterial action. The mucus lubricates the feces as it becomes drier and more compact.

**? REVIEW**
**A.** How does digestion occur in the large intestine?

Simple columnar epithelium

Intestinal gland

Goblet cells

Muscularis mucosae

Submucosa

Aggregated lymphoid nodule

**Muscular Layer**

Circular layer

Serosa

Longitudinal layer (teniae coli)

**2** This illustration shows the characteristic features of the rectum, the last segment of the digestive tract. The lamina propria and submucosa of the distal portion of the rectum contain a network of veins. If venous pressures there rise too high because of straining during defecation or pregnancy, the veins can become distended, producing **hemorrhoids**.

The distal portion of the rectum, the **anal canal**, contains small longitudinal folds called **anal columns**. The margins of these columns are joined by transverse folds that mark the boundary between the columnar epithelium of the proximal rectum and a stratified squamous epithelium like that in the oral cavity.

Rectum

Anal columns

The circular muscle layer of the muscular layer here forms the **internal anal sphincter**, which is composed of smooth muscle fibers and is not under voluntary control.

The **external anal sphincter** consists of a ring of skeletal muscle fibers that encircles the distal portion of the anal canal. This sphincter consists of skeletal muscle and is under voluntary control.

Rectum

**Rectum, sectioned**

**? REVIEW**
**B.** Define *hemorrhoids*.

The **anus** is the exit of the anal canal. Here, the epidermis becomes keratinized and identical to the surface of the skin.

**3** Less than 10 percent of the nutrient absorption within the digestive tract occurs in the large intestine. Nevertheless, absorption in this segment of the digestive tract is very important. In addition to preventing dehydration by reabsorbing water, the epithelium absorbs three vitamins (K, $B_5$, and biotin) produced by the normal bacteria living in the colon.

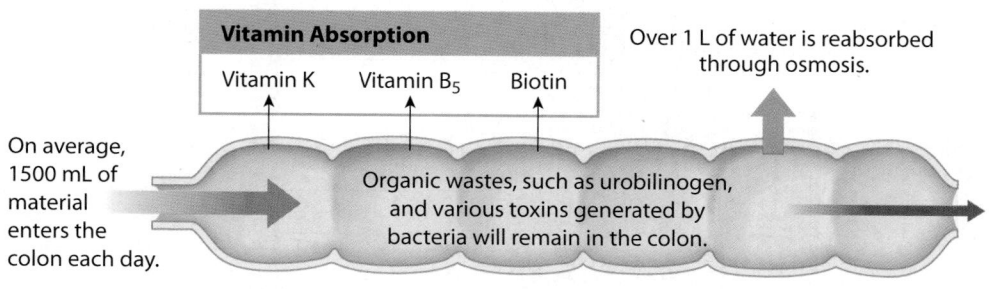

**Vitamin Absorption**

| Vitamin K | Vitamin $B_5$ | Biotin |

Over 1 L of water is reabsorbed through osmosis.

On average, 1500 mL of material enters the colon each day.

Organic wastes, such as urobilinogen, and various toxins generated by bacteria will remain in the colon.

Only 200 mL of feces is ejected. Fecal material is 75% water, 5% bacteria, and the rest a mixture of indigestible materials, inorganic matter, and the remains of epithelial cells. Bacteria produce several compounds that contribute to the odor of feces, including ammonia, *indole* and *skatole* (two nitrogen-containing compounds), and *hydrogen sulfide* ($H_2S$), a gas that produces a "rotten egg" odor.

**4** The rectal chamber is usually empty, except when a powerful peristaltic contraction forces feces out of the sigmoid colon. Distension of the rectal wall then starts the **defecation reflex**, an involuntary response to stimuli in the large bowel that results in a bowel movement. This reflex involves two positive feedback loops, a short *intrinsic myenteric defecation reflex* and a long *parasympathetic defecation reflex*, triggered by the stimulation of stretch receptors in the walls of the rectum. In addition, other consciously directed activities can increase intra-abdominal pressures and help force fecal material out of the rectum. These activities include the *Valsalva maneuver*: tensing the abdominal muscles and elevating intraabdominal pressures by exhaling forcibly with a closed glottis.

**4** Voluntary relaxation of the external anal sphincter allows defecation to occur at a convenient time.

**3** The **long reflex** is a spinal reflex coordinated by sacral parasympathetic motor neurons. This reflex stimulates mass movments that push feces toward the rectum from the descending colon and sigmoid colon. It further relaxes the internal anal sphincter.

**2** The first loop is a **short reflex** that triggers a series of peristaltic contractions in the rectum that move feces toward the anus. This reflex is mediated by the myenteric plexus in the sigmoid colon and rectum. The internal anal sphincter relaxes.

Spinal cord

**1** Feces move into rectum, causing distension and stimulation of stretch receptors

**Start**

DISTENSION OF RECTUM

Stretch receptors

Internal anal sphincter

External anal sphincter

**?** **REVIEW**

**C.** Describe the two positive feedback loops involved in the defecation reflex.

**Lo** **LEARNING OUTCOME**

Describe the large intestine's histology and role in fecal compaction, and explain the defecation reflex.

22

# Section 2 Review
## Digestive Tract

**Labeling:** Label the structures of a typical tooth in the accompanying figure.

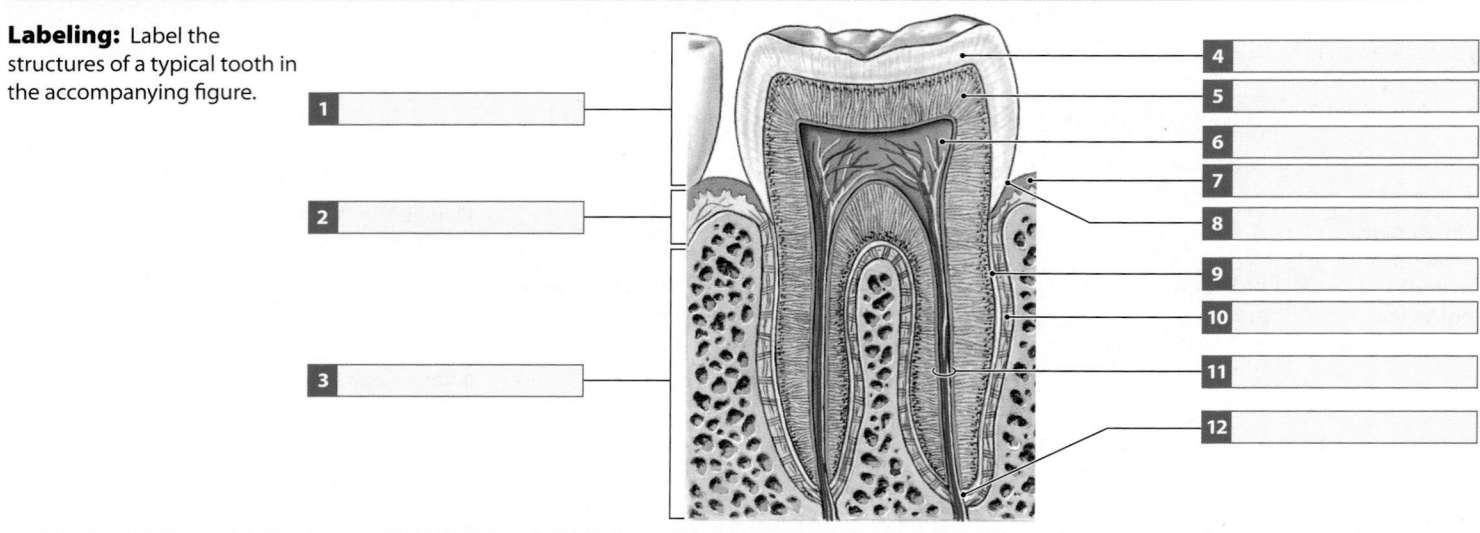

1 _____

2 _____

3 _____

4 _____

5 _____

6 _____

7 _____

8 _____

9 _____

10 _____

11 _____

12 _____

**Concept map:** Use each of the following terms once to fill in the blank boxes to correctly complete the map.

- acid production
- gastrin
- VIP
- insulin
- intestinal capillaries
- GIP
- material in jejunum
- gallbladder
- bile
- inhibits
- secretin and CCK
- nutrient utilization by tissues

**Hormones of Digestive Activity**

Food in stomach —*release of*→ [14] —*promotes*→ [19]

Food in stomach —*conveyed to*→ Chyme in duodenum

[14] → Gastric motility

[18] → Gastric motility

Chyme in duodenum —*secretion of*→ [15] —*release of*→ [20] from pancreas —*facilitates*→

[15] → [18]

[16] —*release of*→ Pancreatic enzymes and buffers

[16] —*secretion of*→ [21] —*from*→ [23]

[17] —*dilation of*→ [22]

Chyme in duodenum —*conveyed to*→ [13]

[22] —*facilitates*→ Nutrient absorption

[24]

**Short answer:** Briefly describe the similarities and differences between parietal cells and chief cells in the stomach wall.

25 _____

_____

22

## Module 22.18

# Some accessory digestive organs have secretory functions

The accessory digestive organs with secretory functions are the salivary glands, gallbladder, pancreas, and liver. The salivary glands and pancreas produce and store enzymes and buffers that are essential to normal digestive function. In addition to their roles in digestion, the salivary glands, liver, and pancreas have vital metabolic and endocrine functions.

| Accessory Digestive Organs |
| --- |
| **Salivary Glands** |
| Three major pairs of salivary glands produce secretions that contain mucin and enzymes. |
| **Gallbladder** |
| The gallbladder stores and concentrates bile secreted by the liver. |
| **Pancreas** |
| Exocrine cells secrete buffers and digestive enzymes; endocrine cells secrete insulin, glucagon, pancreatic polypeptide (PP), and GH–IH (see Module 16.11). |
| **Liver** |
| The liver has almost 200 known functions. Some of the most important are listed in the table below. |

### Digestive and Metabolic Functions of the Liver

- Synthesizing and secreting bile
- Storing glycogen and lipids
- Maintaining normal concentrations of glucose, amino acids, and fatty acids in the bloodstream
- Synthesizing and interconverting nutrient types (such as the conversion of carbohydrates to lipids)
- Synthesizing and releasing cholesterol bound to transport proteins
- Inactivating toxins
- Storing iron
- Storing fat-soluble vitamins

**Other Major Functions**

- Synthesizing plasma proteins
- Synthesizing clotting factors
- Phagocytizing damaged red blood cells (by stellate macrophages)
- Storing blood
- Absorbing and breaking down circulating hormones and immunoglobulins
- Absorbing and inactivating lipid-soluble drugs

 **REVIEW**

**A.** What is the function of the salivary glands?

**B.** Distinguish between the exocrine and endocrine secretions of the pancreas.

**C.** Which accessory organ of the digestive system is responsible for almost 200 known functions?

 **LEARNING OUTCOME**

Describe the functions of the accessory organs of the digestive system.

22

# Module 22.19

# Saliva lubricates, moistens, and protects the mouth and begins carbohydrate digestion

**1** Three major pairs of salivary glands secrete into the oral cavity. Each pair has a distinctive cellular organization and produces *saliva*, a mixture of glandular secretions with slightly different properties. Any object in your mouth can trigger a salivary reflex by stimulating receptors monitored by the trigeminal nerve (V) or taste buds innervated by cranial nerves VII, IX, or X. Parasympathetic stimulation accelerates secretion by all the salivary glands, resulting in the production of large amounts of saliva.

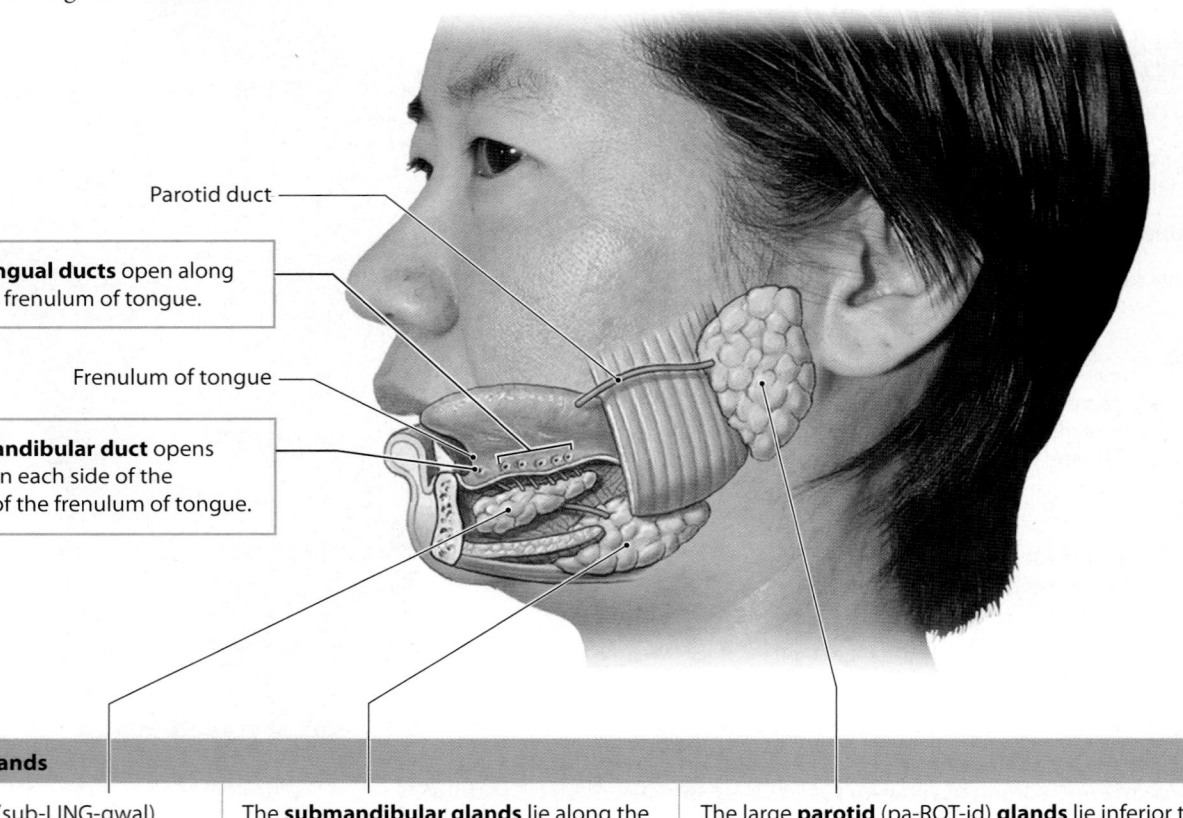

Parotid duct

Numerous **sublingual ducts** open along either side of the frenulum of tongue.

Frenulum of tongue

One large **submandibular duct** opens into the mouth on each side of the anterior margin of the frenulum of tongue.

### The Salivary Glands

The **sublingual** (sub-LING-gwal) **glands** lie under either side of the tongue, covered by the mucous membrane of the floor of the mouth. These glands produce a mucous secretion that acts as a buffer and lubricant.

The **submandibular glands** lie along the inner surface of the mandible within the mandibular groove. Cells of the submandibular glands secrete a mixture of buffers, mucin, and **salivary amylase**, an enzyme that breaks down starches (complex carbohydrates). The gland cells also transport antibodies (IgA) into the saliva to provide additional protection against pathogens in food.

The large **parotid** (pa-ROT-id) **glands** lie inferior to the zygomatic arch deep to the skin covering the lateral and posterior surface of the mandible. Each gland has an irregular shape, extending from the mastoid process of the temporal bone across the outer surface of the masseter. The parotid glands produce a serous secretion containing large amounts of salivary amylase. The secretions of each parotid gland are drained by a **parotid duct**, which empties into the oral vestibule at the level of the second upper molar.

 **REVIEW**

**A.** Name the three pairs of salivary glands.

 **REVIEW**

**B.** Damage to the parotid glands would affect the digestion of which nutrient?

22

**2**   Each submandibular gland contains a mixture of secretory cells, some specialized for mucous secretion and others specialized for enzyme production.

Ducts collect the secretions, and the duct cells assist in the secretion of buffers and antibodies (IgA).

**Mucous cells** secrete mucin, water, and buffers.

**Serous cells** secrete salivary amylase and **lysozyme**, an antibacterial enzyme. They also transport antibodies from the interstitial fluid into the saliva.

**Submandibular gland**

LM × 600

**3**   The **saliva** in the mouth is a clear, watery substance. About 70 percent originates in the submandibular glands, 25 percent from the parotid glands, and 5 percent from the sublingual glands. Collectively the salivary glands produce 1.0–1.5 L of saliva each day, and 99.4 percent of that volume is water. The remaining 0.6 percent includes electrolytes (mainly $Na^+$, $Cl^-$, and $HCO_3^-$), buffers, glycoproteins, antibodies, enzymes, and wastes. Salivary mucin lubricates the oral cavity.

A continuous background level of saliva secretion flushes the oral surfaces, helping keep them clean. Buffers in the saliva keep the pH of your mouth near 7.0 and prevent the buildup of acids produced by bacteria. In addition, saliva contains antibodies (IgA) and lysozyme, which help control populations of oral bacteria. Food usually remains in the mouth long enough for chewing (mastication) to mix it with saliva and break the combination into a relatively homogeneous, pulpy mass.

**? REVIEW**

**C.** Which pair of salivary glands contributes most to saliva production?

**? INTEGRATION**

**D.** Which pair of salivary glands secretes substances that reduce oral bacteria populations?

**Lo LEARNING OUTCOME**

Discuss the structure and functions of the salivary glands.

22

# Module 22.20

## The liver, the largest visceral organ, is divided into left, right, caudate, and quadrate lobes

**1** The **liver**, the largest visceral organ, is a firm, reddish brown organ weighing about 1.5 kg (3.3 lb). These two horizontal sections at vertebral levels $T_{11}$ and $T_{12}$ shown below give you an idea of the size and position of the liver relative to other visceral organs.

Liver

> The liver is wrapped in a tough fibrous capsule and is covered by a layer of visceral peritoneum.

Falciform ligament

> Afferent blood vessels and other structures reach the liver by traveling within the connective tissue of the lesser omentum. They converge at a region called the *porta hepatis* ("doorway to the liver").

Sternum

Left lobe of liver

Stomach

Lesser omentum

Right lobe of liver

Caudate lobe of liver

Inferior vena cava

Right pleural cavity

Cut edge of diaphragm

Aorta

Spleen

Peritoneal cavity

**Horizontal section at the level of vertebra $T_{11}$ (diagrammatic view)**

Falciform ligament

Cut edge of diaphragm

Right pleural cavity

Caudate lobe of liver

Inferior vena cava

Right lobe of liver

Left lobe of liver

Parietal peritoneum

Stomach

Aorta

Spleen

Left kidney

**? REVIEW**
**A.** Describe the outer covering of the liver.

**Horizontal section at the level of vertebra $T_{12}$ (cadaver)**

22

**2** The major anatomical landmarks and the four liver lobes are shown below.

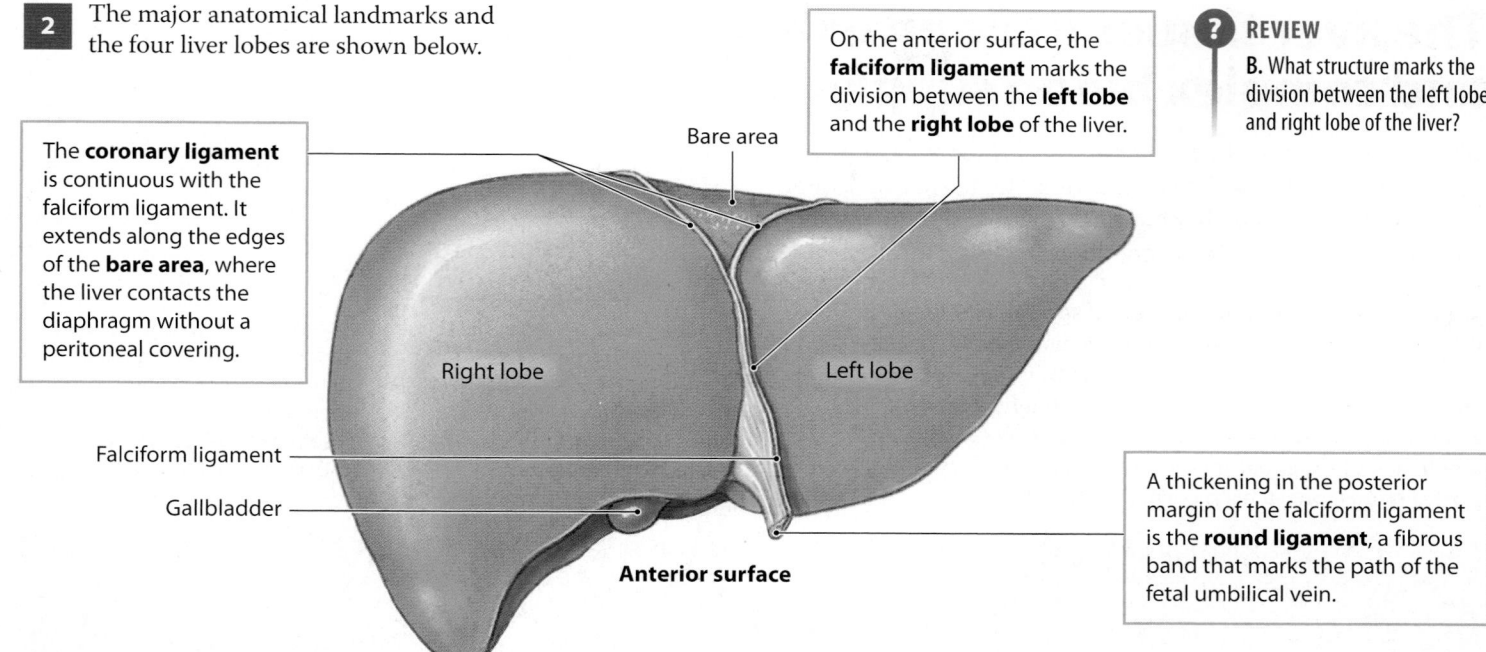

The **coronary ligament** is continuous with the falciform ligament. It extends along the edges of the **bare area**, where the liver contacts the diaphragm without a peritoneal covering.

On the anterior surface, the **falciform ligament** marks the division between the **left lobe** and the **right lobe** of the liver.

Bare area

Right lobe

Left lobe

Falciform ligament

Gallbladder

**Anterior surface**

A thickening in the posterior margin of the falciform ligament is the **round ligament**, a fibrous band that marks the path of the fetal umbilical vein.

**?** **REVIEW**
**B.** What structure marks the division between the left lobe and right lobe of the liver?

Left hepatic vein

Inferior vena cava

Coronary ligament

Bare area

Porta hepatis

Hepatic portal vein

Hepatic artery proper

The **bile duct** carries bile from the liver and gallbladder to the duodenum by way of the porta hepatis.

**Visceral surface**

The **gallbladder** temporarily stores bile produced by the liver.

**Lobes of the Liver**

Left lobe

The **caudate** (KAW-dāt) **lobe** lies on the posterior surface of the liver. The impression left by the inferior vena cava marks the division between the right lobe and the small caudate lobe.

Right lobe

The **quadrate lobe** is sandwiched between the left lobe and the gallbladder.

**?** **REVIEW**
**C.** Name the lobes of the liver.

**?** **REVIEW**
**D.** What is the function of the gallbladder?

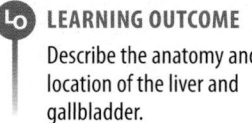

**LEARNING OUTCOME**
Describe the anatomy and location of the liver and gallbladder.

22

# The liver tissues have an extensive and complex blood supply

The lobes of the liver are divided by connective tissue into approximately 100,000 **liver lobules**, the basic functional units of the liver. The histological organization and structure of a typical liver lobule are the focus of this module.

**1** This is a diagrammatic view of several liver lobules. Each lobule is roughly 1 mm in diameter. Liver cells, called **hepatocytes** (HEP-a-tō-sīts), within these lobules adjust circulating levels of nutrients through selective absorption and secretion. In cross section, a typical liver lobule has a hexagonal shape and is surrounded by six **portal triads**, one at each corner of the lobule.

The hepatocytes in a liver lobule form a series of irregular plates arranged like the spokes of a wheel. The plates are only one cell thick, and exposed hepatocyte surfaces are covered with short microvilli.

The plates of hepatocytes are separated by **liver sinusoids**, blood vessels that lack a basement membrane but otherwise resemble large fenestrated capillaries (see Module 19.3).

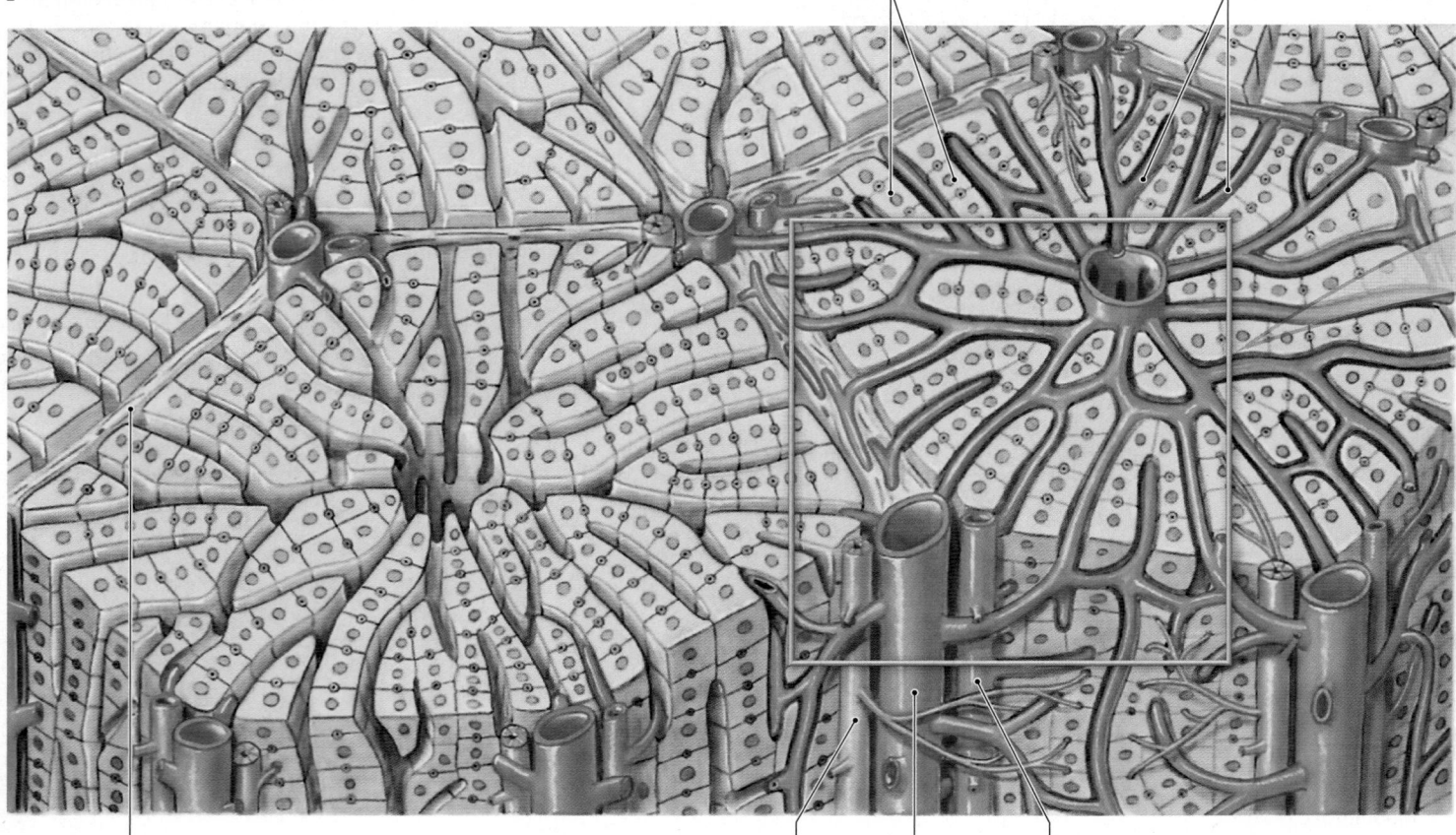

Adjacent lobules are separated from each other by an **interlobular septum**.

Interlobular bile duct    Interlobular vein    Interlobular artery

**REVIEW**

A. Define *hepatocyte*.

### Portal Triad

A **portal triad,** or portal area, contains three structures: (1) an interlobular vein, (2) an interlobular artery, and (3) an interlobular bile duct. Branches from the arteries and veins of each portal triad deliver blood to the liver sinusoids, or hepatic sinusoids, of adjacent liver lobules. The **hepatic portal system** is a venous portal system in which the hepatic portal vein receives blood from the capillaries of most of the abdominal viscera and delivers it to the liver sinusoids (see Module 19.21).

**REVIEW**

B. Describe a portal triad.

**2** This figure provides an enlarged view of one portion of a single liver lobule and shows additional details of its functional anatomy.

**6** Bile canaliculi merge to form **bile ductules** (DUK-tūlz), which carry bile to **interlobular bile ducts** in the nearest portal triad. Bile plays an important role in the digestion of fats in the small intestine.

**5** The hepatocytes secrete a fluid called **bile** into a network of narrow channels between the opposing membranes of adjacent liver cells. These passageways, called **bile canaliculi**, extend outward, away from the central vein. Bile contains acidic bile salts.

**4** The **central vein** collects blood from the sinusoids of the lobule. The central veins of all the lobules ultimately merge to form the hepatic veins, which then empty into the inferior vena cava.

**3** In addition to containing typical endothelial cells, the sinusoidal lining includes a large number of **stellate macrophages** (*Kupffer cells*). These phagocytic cells, part of the monocyte–macrophage system, engulf pathogens, cell debris, and damaged blood cells. They are also responsible for storing iron, some lipids, and heavy metals (such as tin or mercury) that are absorbed by the digestive tract.

**2** As blood flows through the liver sinusoids, hepatocytes adjacent to them regulate solute and nutrient levels and absorb or secrete molecules as large as plasma proteins.

**Start**

**1** Blood enters the liver sinusoids from small branches of the interlobular artery and interlobular vein. About one-third of the blood supply to the liver is arterial blood from the *hepatic artery proper*. The rest is venous blood from the *hepatic portal vein*, which begins in the capillaries of the esophagus, stomach, small intestine, and most of the large intestine.

Interlobular bile duct    Interlobular vein    Interlobular artery

**?** REVIEW

**C.** Define *stellate macrophages*, and indicate their functions.

---

🦅 **Clinical Note**

## Portal hypertension is caused by restricted blood flow to the liver

Liver diseases (such as the various forms of **viral hepatitis**) and conditions such as alcoholism can lead to degenerative changes in the liver tissue and constriction of blood flow. Pressures in the hepatic portal system are usually low, averaging 10 mm Hg or less. This pressure can increase markedly, however, if blood flow through the liver becomes restricted as a result of a blood clot or damage to the organ. Such a rise in portal pressure is called **portal hypertension**. As pressures rise, small peripheral veins and capillaries in the portal system become distended; if they rupture, potentially fatal bleeding can occur. Portal hypertension can also force fluid into the peritoneal cavity across the serosal surfaces of the liver and viscera, producing *ascites* (see Module 22.9).

**22**

 **LEARNING OUTCOME**
Describe the histological features of liver tissue.

# Module 22.22
# The gallbladder stores and concentrates bile

The **gallbladder** is a hollow, pear-shaped organ that stores and concentrates bile prior to its ejection into the small intestine. This muscular sac is located in a recess in the posterior surface of the liver's right lobe. The gallbladder is divided into three regions: the **fundus**, the **body**, and the **neck**.

**1** In this view, the liver has been pulled upward to show the gallbladder and its associated ducts.

The **right** and **left hepatic ducts** collect bile from all the bile ducts of the liver lobes.

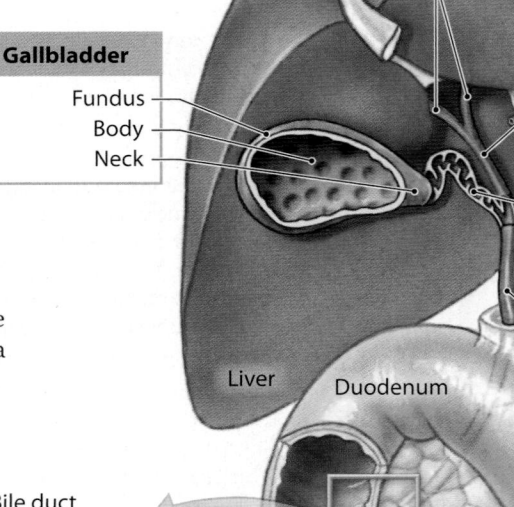

**Gallbladder**
- Fundus
- Body
- Neck

The hepatic ducts unite to form the **common hepatic duct**. Bile in the common hepatic duct either flows into the bile duct or enters the **cystic duct**, which leads to the gallbladder.

**2** The bile duct passes within the lesser omentum and penetrates the wall of the duodenum to meet the pancreatic duct at the entrance to the **duodenal ampulla** (am-PUL-a), a chamber within the **duodenal papilla** that projects into the intestinal lumen.

- Cystic duct
- Cut edge of lesser omentum
- Hepatic portal vein and hepatic artery proper

Liver

Duodenum

The cystic duct and the common hepatic duct unite to form the **bile duct**. It empties into the duodenum.

Stomach

Pancreas

Pancreas

Bile duct

Pancreatic duct

Intestinal lumen | Duodenal papilla | Duodenal ampulla

The muscular **hepatopancreatic sphincter** encircles the lumen of the bile duct, pancreatic duct, and duodenal ampulla. Resting tension in the sphincter prevents bile flow into the duodenum except at mealtimes.

**3** Bile can enter the duodenum only when the hepato-pancreatic sphincter is open. So for most of the day, bile flows into the gallbladder instead.

**1** The liver secretes bile continuously: roughly 1 liter per day.

**2** As it remains in the gallbladder, bile becomes more concentrated.

**3** The release of CCK by the duodenum triggers dilation of the hepatopancreatic sphincter and contraction of the gallbladder. This ejects bile into the duodenum through the duodenal ampulla.

Liver

**4** In the lumen of the digestive tract, bile salts break the lipid droplets apart in a process called **emulsification** (ē-mul-si-fi-KĀ-shun). This increases the surface area accessible to enzymes.

CCK

Lipid droplet

Bile duct

Duodenum

**REVIEW**

**A.** Define *emulsification*.

**B.** Trace a drop of bile from the hepatic ducts to the duodenal lumen.

**LEARNING OUTCOME**

Describe the structure, functions, and regulatory activities of the gallbladder.

22

# Module 22.23

## The pancreas has vital endocrine and exocrine functions

The **pancreas** lies posterior to the stomach, extending laterally from the duodenum toward the spleen. The pancreas is a slender, pinkish gray organ about 15 cm (6 in.) long and weighing about 80 g (3 oz).

**1** The broad **head** of the pancreas lies within the loop formed by the duodenum as it leaves the pyloric part. Like the duodenum, the pancreas is retroperitoneal and bound to the posterior wall of the abdominal cavity.

The large **pancreatic duct** delivers the exocrine secretions to the duodenum. The epithelial cells lining the duct and its smaller branches secrete water and ions that mix with the secretions of the exocrine gland cells to form a watery **pancreatic juice**.

Partitions of connective tissue divide the interior of the pancreas into distinct **pancreatic lobules**.

Bile duct

In 3–10 percent of the population, a small **accessory pancreatic duct** (Santorini duct) branches from the pancreatic duct and empties separately into the duodenum.

Body of pancreas

Tail of pancreas

Head of pancreas

Duodenal papilla

The pancreatic duct meets the bile duct at the entrance to the duodenal ampulla.

Duodenum

Pancreatic duct

Endocrine cells in pancreatic islet

**2** The pancreas is primarily an exocrine organ, and pancreatic tissue is dominated by the **pancreatic acini** (AS-i-nī; *berry*), which produce digestive enzymes and buffers.

**3** Each day, the pancreas secretes about 1000 mL (1 qt) of pancreatic juice containing a variety of enzymes and a watery buffer solution. This table introduces the primary pancreatic enzymes produced; their functions will be considered further in the next chapter.

**Pancreatic acinar cells** secrete pancreatic enzymes that do most of the digestive work in the small intestine, breaking down ingested materials into small molecules suitable for absorption.

### Major Pancreatic Enzymes

- **Pancreatic alpha-amylase** is a carbohydrase (kar-bō-HĪ-drās)—an enzyme that breaks down certain starches. Pancreatic alpha-amylase is almost identical to salivary amylase.
- **Pancreatic lipase** breaks down certain complex lipids, releasing products (such as fatty acids) that can be easily absorbed.
- **Nucleases** break down RNA or DNA.
- **Proteolytic enzymes** break proteins apart. They are secreted as inactive proenzymes that become active once they are in the duodenal lumen. The active enzymes include trypsin, chymotrypsin, carboxypeptidase, and elastase. Together, they break down proteins into a mixture of dipeptides, tripeptides, and amino acids.

**? REVIEW**

**A.** What is the primary digestive function of the pancreas?

**Lo LEARNING OUTCOME**

Describe the structure, functions, and regulatory activities of the pancreas.

22

# Module 22.24
# Disorders of the digestive system are diverse and relatively common

## Oral Cavity

**Periodontal disease**, the most common cause for the loss of teeth, occurs when dental plaque forms in the area between the gums and teeth. The bacterial activity may cause **gingivitis** (shown here), tooth decay, and, eventually, breakdown of the periodontal ligament and surrounding bone.

Gingivitis

## Salivary Glands

The **mumps virus** causes **mumps**, an infection of the salivary glands. The infection most often occurs in the parotid gland, as seen here, but it may also infect other salivary glands and other organs, including the gonads and the meninges. Infection typically occurs at 5 to 9 years of age. In postado-lescent males, the mumps virus infecting the testes may cause sterility. An effective mumps vaccine became available in 1967. The vaccine is usually combined with measles and rubella vaccines (to form the MMR vaccine) and is adminis-tered to infants after the age of 15 months.

Mumps in parotid gland

## Esophagus

**Esophagitis** (ē-sof-a-JĪ-tis) is an inflammation of the esophagus. This painful condition, which can be seen in this endoscopic view of the esophagus, usually results from the presence of stomach acids that leak through a weakened or permanently relaxed lower esophageal sphincter. Such backflow, or **gastro-esophageal reflux**, is responsible for the symptoms of heartburn.

Esophagitis seen on endoscopy

## Liver

Any condition that severely damages the liver is a threat to life. **Hepatitis**, an inflammation of the liver, can be caused by alcohol abuse, drugs, or infection. **Cirrhosis** (sir-RŌ-sis) is a form of hepatitis characterized by the degeneration of liver cells and their replacement with fibrous connective tissue (a process called scarring). The surviving liver cells divide, but the fibrous tissue prevents the reestablishment of normal tissues. As a result, liver function declines, and other complications develop. There are many different forms of viral hepatitis: The most common are **hepatitis A, B,** and **C**. The hepatitis viruses disrupt liver function by attacking and destroying liver cells. An infected person may develop a high fever, and the liver may become inflamed and tender. In a condition called **jaundice** (JAWN-dis), the skin and eyes develop a yellow color because the bilirubin normally excreted in the bile is accumulating in body fluids.

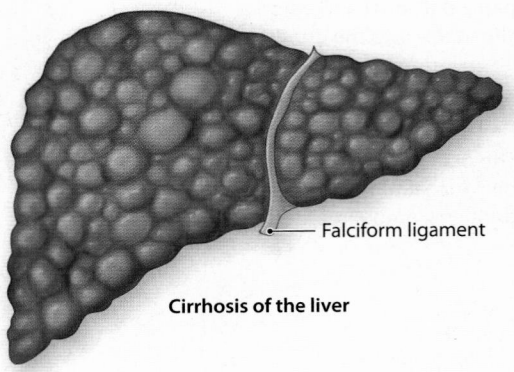

Falciform ligament

Cirrhosis of the liver

## Gallbladder

If bile becomes too concentrated, crystals of insoluble minerals and salts called **gallstones** form. Small gallstones are not a problem if they can be flushed through the bile duct and excreted. In **cholecystitis** (kō-lē-sis-TĪ-tis; *chole*, bile + *kystis*, bladder + *itis*, inflammation), the gallstones are so large that they damage the wall of the gallbladder or block the cystic duct or bile duct. In that case, the gallbladder may need to be surgically removed. This removal does not seriously impair digestion, because bile production continues at normal levels.

Gallstones

22

## Stomach

Inflammation of the mucous membrane lining the stomach is called **gastritis** (gas-TRĪ-tis). This condition may develop after ingesting drugs, including aspirin and alcohol. It may also appear after severe emotional or physical stress, bacterial infection of the gastric wall, or the ingestion of strong chemicals. Gastritis may lead to ulcer formation. A **peptic ulcer** develops when gastric enzymes and acids erode through the stomach or duodenal lining. Specifically, a peptic ulcer located in the stomach is a **gastric ulcer**, and one located in the duodenum is a **duodenal ulcer**. Infection by the bacterium *Helicobacter pylori* (HE-li-kō-bak-ter pī-LŌR-ī) is responsible for over 80 percent of peptic ulcers. Treatment for ulcers involves the administration of drugs, such as **cimetidine** (si-MET-i-dēn) (Tagamet), that inhibit acid production by gastric glands, combined with antibiotics if *Helicobacter pylori* is present.

**Gastric ulcer**

## Pancreas

**Pancreatitis** (pan-krē-a-TĪ-tis) is an inflammation of the pancreas. Factors that may cause this condition include blockage of the excretory ducts by viral infections, and toxic drugs, such as alcohol. Any of these stimuli may injure exocrine cells in a portion of the organ. Lysosomes then activate digestive enzymes within the cells, which begin to break down. In about one-eighth of cases, death results when the process does not stop, and the released lysosomal enzymes destroy the pancreas.

**Acute pancreatitis**

## Small Intestine

**Enteritis** is inflammation of the intestine (usually the small intestine). Enteritis typically causes watery bowel movements, or **diarrhea** (dī-a-RĒ-uh). One cause of diarrhea due to enteritis is the protozoan *Giardia lamblia* (shown here). **Dysentery** (dis-en-TER-ē) is inflammation of the small and large intestines that usually produces diarrhea containing blood and mucus. **Gastroenteritis** is inflammation of the stomach and the intestines due to bacterial, viral, protozoan, or parasitic worm infections. Most of these conditions are prevalent in areas that have poor sanitation and low water quality.

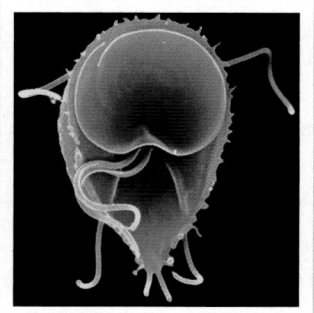
*Giardia lamblia*    SEM × 4000

## Large Intestine

**Colitis** (kō-LĪ-tis) is a general term referring to inflammation of the colon, often involving diarrhea or constipation. Diarrhea results when the lining of the colon is unable to reabsorb water normally, or when so much fluid enters the colon that its water reabsorption capacity is exceeded. **Constipation** is infrequent bowel movement (or defecation), generally involving dry, hard feces. It results when fecal material moves through the colon so slowly that excessive water reabsorption occurs.

**Colorectal cancer** is the third most common cancer in the United States, affecting both men and women. The death rate has declined over the past 20 years for both men and women. It is most common among persons over 50 years of age. Primary risk factors for colorectal cancer include a diet rich in animal fats and low in fiber. There are also a number of inherited disorders that promote epithelial tumor formation along the intestines. It is believed that most colorectal cancers begin as small, localized tumors, or **polyps** (POL-ips), that grow from the mucosa lining the intestinal wall. For individuals at increased risk because of family history, associated disease, or older age, visual inspection of the intestinal lumen by fiberoptic colonoscopy is prudent. Such examination can find polyps before they develop into cancers. The 5-year survival rate for people whose cancer is found at an early stage and treated immediately is greater than 90 percent.

**Colon polyp seen on colonoscopy**

 **REVIEW**

**A.** Describe hepatitis.

**B.** Describe cholecystitis.

**C.** What bacterium is responsible for most peptic ulcers?

22

 **LEARNING OUTCOME**
Briefly describe several digestive system disorders.

# Accessory Digestive Organs

**Labeling:** Label the structures of a liver lobule in the accompanying figure.

1. _____
2. _____
3. _____
4. _____
5. _____
6. _____
7. _____
8. _____
9. _____

**Matching:** Match each lettered term with the most closely related description.

a. lysozyme
b. emulsification
c. gallstones
d. stellate macrophages
e. pancreatic lipase
f. liver
g. starch
h. pancreas
i. submandibular glands
j. hepatocytes
k. gallbladder
l. mumps
m. bile duct
n. peptic ulcer

| 10 | Pancreatic alpha-amylase substrate | 10 | _____ |
| 11 | Retroperitoneal organ | 11 | _____ |
| 12 | Drains liver and gallbladder | 12 | _____ |
| 13 | Bile-secreting cells | 13 | _____ |
| 14 | Viral infection of salivary glands | 14 | _____ |
| 15 | Digestive epithelial damage by acids | 15 | _____ |
| 16 | Process of breaking lipid droplets apart | 16 | _____ |
| 17 | Pancreatic enzyme that breaks down complex lipids | 17 | _____ |
| 18 | Organ that secretes bile continuously | 18 | _____ |
| 19 | Antibacterial enzyme | 19 | _____ |
| 20 | Greatest producer of saliva | 20 | _____ |
| 21 | Phagocytize and store iron | 21 | _____ |
| 22 | Stores bile | 22 | _____ |
| 23 | Cholecystitis | 23 | _____ |

**Short answer:** Describe the beneficial roles of saliva.

24. _____

**Section integration:** Predict the consequences of a blockage of the duodenal ampulla by a tumor.

25. _____

# Study Outline

## SECTION 1 • Organization of the Digestive System

### Module 22.1

**The digestive system consists of the digestive tract and accessory organs    p. 839**

1.  The **digestive system** consists of the digestive tract plus accessory organs.

2.  The **digestive tract** begins at the oral cavity (mouth) and continues through the pharynx (throat), esophagus, stomach, small intestine, and large intestine.  It ends with the large intestine, which opens to the exterior at the anus.

3.  The accessory organs of the digestive system are the teeth, tongue, salivary glands, liver, gallbladder, and pancreas.

4.  The digestive system works with other systems to support tissues that have no direct connection with the outside environment.

### Module 22.2

**The digestive tract is a muscular tube lined by a mucous epithelium    p. 840**

5.  A double sheet of peritoneal membrane called **mesentery** stabilizes the positions of the digestive tract.

6.  The major layers of the digestive tract are the **mucosa** (inner lining), **submucosa** (surrounds the mucosa), **muscular layer** (circular and longitudinal layers of smooth muscle cells), and **serosa** (visceral peritoneum).

*Section of digestive tract*

7.  Secretions from gland cells located in the mucosa and submucosa, or in accessory glandular organs, are carried to the epithelial surfaces by ducts.

8.  **Circular folds** (plicae circulares) are permanent transverse folds in the intestinal lining. **Villi** are small mucosal projections into the lumen of the small intestine.

### Module 22.3

**Smooth muscle tissue is found throughout the body, but it plays a particularly prominent role in the digestive tract  p. 842**

9.  In the digestive tract, the muscular layer usually contains an inner circular layer and an outer longitudinal layer of smooth muscle.

10. **Multi-unit smooth muscle cells** are innervated by motor units and are located in the iris of the eye, the male reproductive tract, large arteries, and the arrector pili muscles of the skin.

11. **Visceral smooth muscle cells** lack a direct contact with motor neurons. They are arranged in sheets, and the layers contract in waves triggered by **pacesetter cells**.

12. **Plasticity** is the ability of a stretched smooth muscle to function over a wide range of lengths. **Smooth muscle tone** is the normal background level of activity.

### Module 22.4

**Smooth muscle contractions produce motility of the digestive tract, and local factors interact with neural and hormonal mechanisms to regulate digestive activities    p. 844**

13. Food enters the digestive tract as a moist, compact mass called a **bolus**.

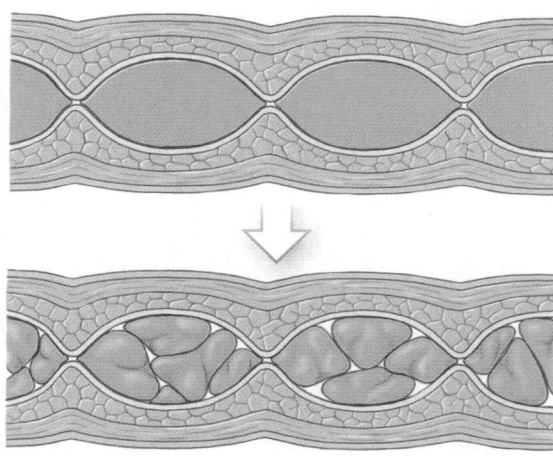

14. The muscular layer propels the bolus with wavelike contractions called **peristalsis**. **Segmentation** churns and fragments a bolus without any directional movement.

15. Three major mechanisms regulate and control digestive activities: **local factors**, **neural control mechanisms**, and **hormonal control mechanisms**.

## SECTION 2 • Digestive Tract

### Module 22.5

**The digestive tract begins with the mouth and ends with the anus   p. 847**

16. The major organs of the digestive tract are the **oral cavity, pharynx, esophagus, stomach, small intestine**, and **large intestine**. The accessory organs are the teeth, tongue, salivary glands, liver, gallbladder, and pancreas.

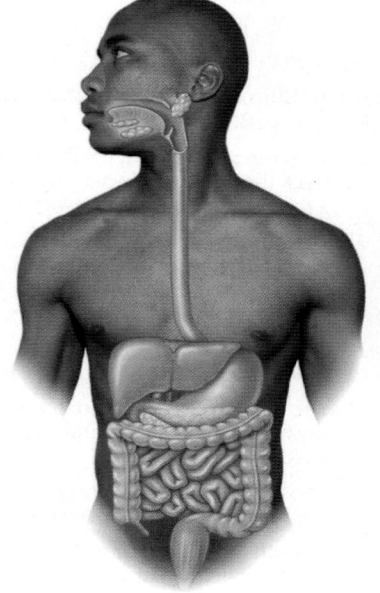

17. The general functions of the digestive tract are **ingestion, mechanical digestion and propulsion, chemical digestion, secretion, absorption**, and **defecation**.

22

## Module 22.6
### The oral cavity is a space that contains the tongue, teeth, and gums    p. 848

18. The superior boundary of the **oral cavity** is formed by the **hard palate** (palatine processes of the maxillary bones and the horizontal plates of the palatine bones) and the muscular **soft palate**.

*Sagittal secton of the oral cavity*

19. The posterior boundary of the oral cavity contains the **uvula**, the **palatine tonsils**, the lingual tonsils, and the **root** of the tongue.

20. The tongue forms the inferior boundary of the oral cavity. The **labia** (lips) and **cheeks** form the anterior and lateral boundaries.

21. The **oral vestibule** is the space between the cheeks and teeth. The **frenulum of upper lip** and **frenulum of lower lip** attach lips to the **gingivae**, or gums. The **frenulum of tongue** attaches the tongue to the floor of the mouth. The **palatal arches** are on either side of the uvula. The **fauces** is the space between the oral cavity and the oropharynx.

## Module 22.7
### Teeth in different regions of the jaws vary in size, shape, and function    p. 850

22. The components of a tooth are the **crown** (above the gums), **neck** (boundary between the crown and root), and **root** (within a *tooth socket*, or *tooth alveolus*).

23. The bulk of each tooth is made of **dentin**. **Enamel** covers the dentin of the crown. **Cement** covers the dentin of the root. The **periodontal ligament** extends from the dentin to the alveolus to form a joint called a **gomphosis**.

24. The pulp cavity receives blood vessels and nerves through the **apical foramen** into the **root canal**.

25. The four different types of teeth are **incisors**, **canines** (*cuspids*), **premolars** (*bicuspids*), and **molars**. Third molars are called **wisdom teeth**.

26. The teeth are collectively called the *dentition*. There are two dentitions: 20 **deciduous teeth** and 32 **permanent teeth**.

27. A tooth that does not erupt because of overcrowding of adjacent teeth is called an **impacted tooth**. Wisdom teeth are commonly impacted.

## Module 22.8
### The muscular walls of the pharynx and esophagus play a key role in swallowing    p. 852

28. The **pharynx** is a passageway for solid food, drinks, and air. Food passes through the oropharynx and laryngopharynx on its way to the **esophagus**, a hollow muscular organ leading to the stomach.

29. The shape of the lumen and structure of the muscular layer of the esophagus are unique to the rest of the digestive tract. There is no serosa, but an **adventitia** of connective tissue anchors it to the posterior body wall.

30. Swallowing, or **deglutition**, involves three phases: the **buccal phase**, the **pharyngeal phase**, and the **esophageal phase**.

31. The esophagus enters the abdominopelvic cavity through the **esophageal hiatus** of the diaphragm. The **upper esophageal sphincter** prevents air from entering the esophagus, and the **lower esophageal sphincter** prevents the backflow of food from the stomach into the esophagus.

## Module 22.9
### The stomach and most of the intestinal tract are suspended by mesenteries and covered by the peritoneum    p. 854

32. **Visceral peritoneum** covers the organs enclosed by the peritoneal cavity. **Parietal peritoneum** lines the inner surfaces of the body wall.

33. **Dorsal** and **ventral mesenteries** suspend the digestive tract and accessory organs during embryonic development.

34. The ventral mesentery remains in the adult as the **lesser omentum** and the **falciform ligament**.

35. The **greater omentum** is the dorsal mesentery of the stomach, the **mesentery proper** is the dorsal mesentery of the small intestine, and the **mesocolon** is the dorsal mesentery of the large intestine.

## Module 22.10
### The stomach is a muscular, expandable, J-shaped organ with three layers in the muscular layer    p. 856

36. The regions of the stomach are the **fundus, cardia, body**, and **pyloric part**.

37. The medial surface of the stomach forms the **lesser curvature**, and the lateral and inferior surfaces form the **greater curvature**.

38. The pyloric part is composed of the **pyloric antrum, pyloric canal, pyloric orifice,** and **pyloric sphincter**.

39. The muscular layer of the stomach has an inner oblique layer of smooth muscle and mucosal folds called **rugae**.

22

## Module 22.11

### The stomach receives food and liquids from the esophagus and aids in mechanical and chemical digestion    p. 858

40. The **stomach** stores, breaks down, and disrupts chemical bonds in food and produces intrinsic factor.

41. The lumen of the stomach is lined with shallow depressions called **gastric pits**, which communicate with several gastric glands. **Gastric glands** secrete most of the acid and enzymes involved in gastric digestion.

42. The cells of gastric glands are **parietal cells** that secrete **intrinsic factor**, which aids vitamin $B_{12}$ absorption in the small intestine, and HCl; **G cells**

*Layers of the stomach wall*

(enteroendocrine cells); and **chief cells** that secrete **pepsinogen**.

43. Pepsinogen is converted to **pepsin** by acid in the gastric lumen.

## Module 22.12

### The intestinal tract is specialized to absorb nutrients    p. 860

44. The intestinal lining has transverse folds called **circular folds** that greatly increase its surface area for absorption.

45. The mucosa of the small intestine contains fingerlike projections called **intestinal villi**. At the base of the villi are the entrances into the **intestinal glands**.

46. Each villus has a complex internal structure with a blood capillary network that carries absorbed nutrients to the liver via hepatic portal circulation, and a lymphatic capillary called a **lacteal** that absorbs materials that cannot enter the blood capillaries.

47. The microvilli of the simple columnar epithelium on the surface of each villus project to form a **brush border**.

*Layers of the small intestine*

## Module 22.13

### The small intestine is divided into the duodenum, jejunum, and ileum    p. 862

48. The **small intestine** accounts for 90 percent of nutrient absorption.

49. The regions of the small intestine are the **duodenum**, **jejunum**, and **ileum**. The ileum ends at the **ileocecal valve**.

50. Each segment of the small intestine has characteristic features. The transition from one region to another is gradual, and the boundaries are indistinct.

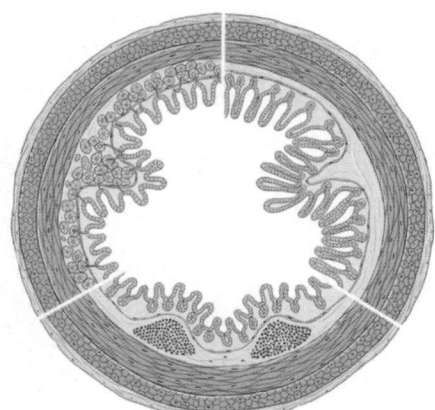

## Module 22.14

### Several hormones regulate digestion    p. 864

51. The duodenum produces several hormones that regulate digestion.

52. The hormones are **gastrin** (secreted by G cells in stomach; stimulates gastric motility), **secretin** (release of pancreatic enzymes and bile secretion), **gastric inhibitory peptide** (release of insulin from pancreas), **cholecystokinin** (**CCK**) (release of pancreatic enzymes and bile secretion), **vasoactive intestinal peptide (VIP)** (dilation of intestinal capillaries), and **enterocrinin** (stimulates alkaline mucus production by duodenal submucosal glands).

## Module 22.15

### Central and local mechanisms coordinate gastric and intestinal activities    p. 866

53. The duodenum plays a key role in controlling the three phases of gastric secretion: cephalic phase, gastric phase, and intestinal phase.

54. The **cephalic phase** of gastric secretion begins when you see, smell, taste, or think of food. The CNS is preparing the stomach to receive food.

55. The **gastric phase** is when food arrives in the stomach. It includes distension of the stomach, an increase in pH, and the presence of undigested materials in the stomach.

*The intestinal phase of gastric secretion*

56. The **intestinal phase** begins when chyme first enters the small intestine.

57. Two **central gastric reflexes** are stimulated by stretch receptors in the stomach. The **gastroenteric reflex** stimulates motility and secretion, and the **gastroileal reflex** opens the ileocecal valve.

22

## Module 22.16

### The large intestine stores and concentrates fecal material p. 868

58. The functions of the **large intestine** include reabsorbing water and compacting the intestinal contents into feces, absorbing important vitamins generated by bacteria, and storing fecal material prior to defecation.

59. Material from the ileum enters the **cecum** through the **ileocecal valve**. **Compaction** begins in the cecum. The slender **appendix** is attached to the cecum.

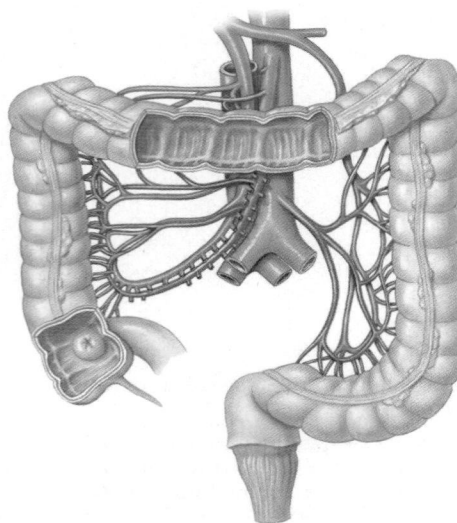

60. The three segments of the large intestine are the cecum, the **colon,** and the rectum. The anatomy of the colon begins at the cecum and continues to the **ascending colon**, **right colic** (*hepatic*) **flexure**, **transverse colon**, **left colic** (*splenic*) **flexure**, **descending colon**, **sigmoid flexure**, **sigmoid colon**, and **rectum**.

61. Along the length of the large intestine are pouches called **haustra** and longitudinal bands of smooth muscle called **teniae coli**.

62. Powerful peristaltic contractions called **mass movements** push materials toward the distal portion of the large intestine.

## Module 22.17

### The large intestine compacts fecal material; the defecation reflex coordinates the elimination of feces p. 870

63. The large intestine lacks villi but has distinct intestinal glands dominated by goblet cells. Digestion in the small intestine is carried out by enzymes secreted into the small intestine and by bacteria.

64. **Hemorrhoids** are distended veins in the rectum that result from high venous pressures.

65. The distal portion of the **anal canal** contains small longitudinal folds called **anal columns**.

66. The circular muscle layer of the muscular layer forms the **internal anal sphincter**, which is not under voluntary control.

67. The **external anal sphincter** is a ring of skeletal muscle fibers and is under voluntary control.

68. The colon absorbs three bacteria-produced vitamins: vitamin K, vitamin $B_5$, and biotin.

69. Distension of the rectal walls starts the **defecation reflex**, which involves two positive feedback loops: the **short reflex** (*intrinsic myenteric defecation reflex*) and the **long reflex** (*parasympathetic defecation reflex*).

## SECTION 3 • Accessory Digestive Organs

### Module 22.18

### Some accessory digestive organs have secretory functions p. 873

70. The accessory digestive organs with secretory functions are the **salivary glands**, **gallbladder**, **pancreas**, and **liver**. The liver has almost 200 known functions.

### Module 22.19

### Saliva lubricates, moistens, and protects the mouth and begins carbohydrate digestion p. 874

71. Three pairs of salivary glands secrete into the oral cavity: the **sublingual glands**, the **submandibular glands**, and the **parotid glands**.

72. **Saliva** is a mixture of glandular secretions: 70 percent originates from submandibular glands, 25 percent from parotid glands, 5 percent from sublingual glands.

73. One to 1.5 L of saliva is produced per day, of which 99.4 percent is water.

74. **Mucous cells** secrete mucin, water, and buffers. **Serous cells** secrete **salivary amylase** and **lysozyme**.

### Module 22.20

### The liver, the largest visceral organ, is divided into left, right, caudate, and quadrate lobes p. 876

75. The **liver** is wrapped in a tough fibrous capsule and covered by visceral peritoneum.

76. The **falciform ligament** marks the division between the **right lobe** and **left lobe**. A thickening of the posterior margin of the falciform ligament forms the **round ligament**.

77. The inferior vena cava lies between the **caudate lobe** and right lobe. The **quadrate lobe** lies between the left lobe and gallbladder.

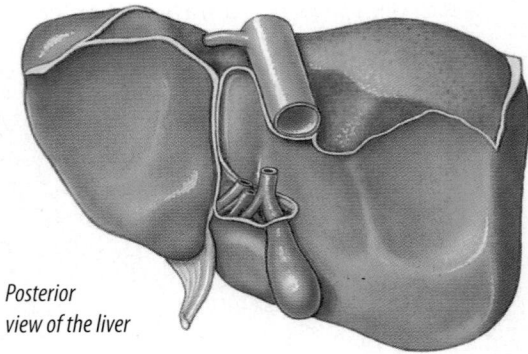

*Posterior view of the liver*

78. The **gallbladder** stores bile. The **bile duct** carries bile from the liver and gallbladder to the duodenum.

### Module 22.21

### The liver tissues have an extensive and complex blood supply p. 878

79. The lobes of the liver are divided by connective tissue into 100,000 **liver lobules**, the basic functional units of the liver. A typical lobule is hexagonal and surrounded by six **portal triads**, one at each corner.

80. A portal triad contains three structures: an interlobular vein (a branch of the hepatic portal vein), an interlobular artery (a branch of the hepatic artery proper), and an interlobular bile duct (a small branch of the bile duct).

81. Liver cells, or **hepatocytes**, within the lobules adjust circulating levels of nutrients. **Liver sinusoids** separate plates of hepatocytes. Hepatocytes secrete **bile** into **bile canaliculi**. Bile canaliculi merge to form **bile ductules**, which carry bile to the **interlobular bile ducts**.

82. The **central vein** collects blood from the sinusoids of the lobule. The central veins of all the lobules merge to form the hepatic veins, which empty into the inferior vena cava.

83. Sinusoids contain **stellate macrophages** (*Kupffer cells*), phagocytic cells that are part of the monocyte–macrophage system and engulf pathogens, cell debris, and damaged blood cells.

84. **Viral hepatitis**, alcoholism, and **portal hypertension** can damage the liver.

## Module 22.22

### The gallbladder stores and concentrates bile    p. 880

85. The **gallbladder** is located in a recess in the posterior surface of the liver's right lobe.  The three regions of the gallbladder are the **fundus**, **body**, and **neck**.

86. The **right** and **left hepatic ducts** collect bile from the liver, and transport the bile to the **common hepatic duct**. Bile enters either the bile duct or the **cystic duct**, which leads to the gallbladder. The **bile duct** is formed by the union of the cystic duct and common hepatic duct and enters the duodenum.

87. The bile duct penetrates the wall of the duodenum and meets with the pancreatic duct at the **duodenal ampulla**, a chamber within the **duodenal papilla** that projects into the intestinal lumen. The **hepatopancreatic sphincter** prevents bile flow into the duodenum except at mealtimes.

## Module 22.23

### The pancreas has vital endocrine and exocrine functions p. 881

88. The **head** of the **pancreas** lies within the loop formed by the duodenum. The **pancreatic duct** delivers **pancreatic juice** to the duodenum.

89. The pancreas is primarily an exocrine organ. **Pancreatic acini** produce digestive enzymes that break down starches, lipids, RNA, DNA, and proteins. Pancreatic acini also secrete buffers.

## Module 22.24

### Disorders of the digestive system are diverse and relatively common    p. 882

90. **Periodontal disease**, **mumps**, and **gastro-esophageal reflux** are common disorders of the digestive system. Other disorders include **hepatitis**, **gallstones**, **pancreatitis**, **gastritis**, and **enteritis**. **Colorectal cancer** is the third most common cancer in the United States, affecting both men and women.

# Chapter Review Questions

**Labeling:** Label the figure shown here.

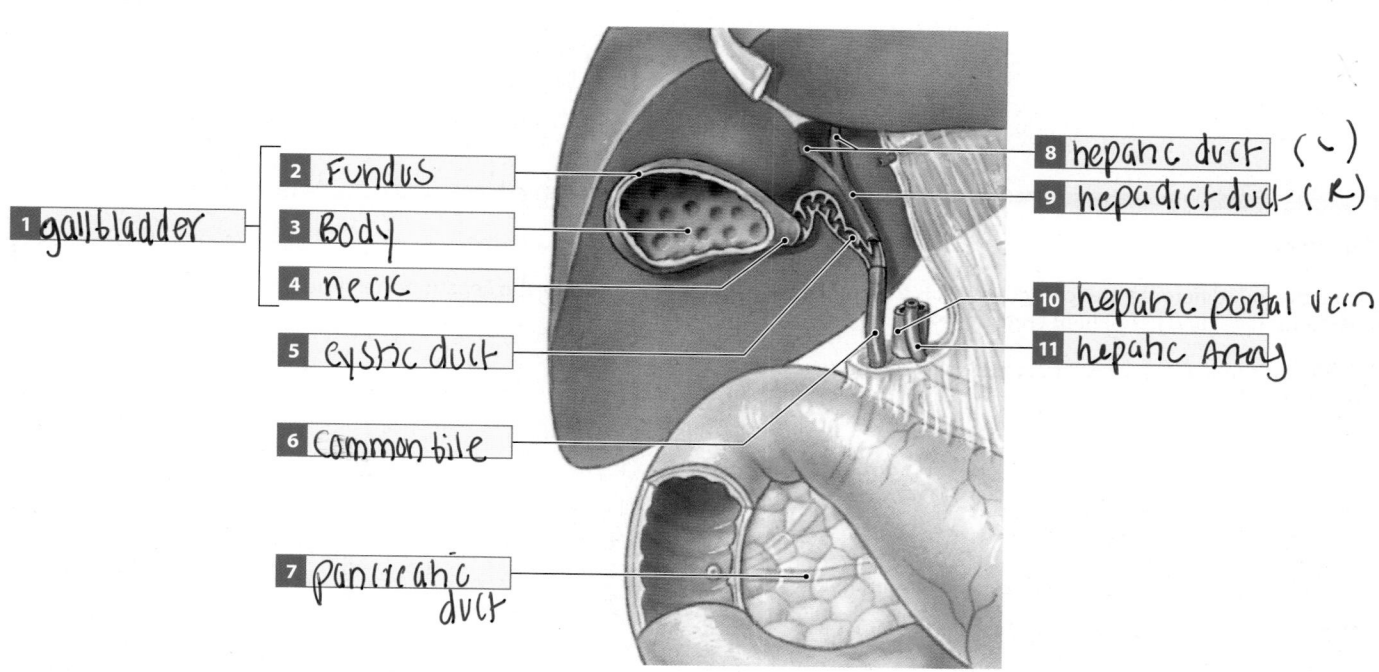

1 gallbladder
2 Fundus
3 Body
4 neck
5 cystic duct
6 Common bile
7 Pancreatic duct
8 hepatic duct (ᴸ)
9 hepatic duct (R)
10 hepatic portal vein
11 hepatic Artery

22

**True/False:** Indicate whether each statement is true or false.

12  Stellate macrophages are phagocytic cells within liver sinusoids.

13  The gallbladder produces almost as much bile on a daily basis as the liver.

14  Pancreatic islets produce the exocrine secretions for digestion.

15  Most saliva originates in the submandibular glands.

16  The correct order of small intestine segments is: duodenum, ileum, and jejunum.

17  Haustra are a series of pouches along the length of the colon.

| | |
|---|---|
| 12 | |
| 13 | False |
| 14 | False (endocrine) |
| 15 | True |
| 16 | False |
| 17 | true |

**Multiple choice:** Select the correct answer from the list provided.

18  The release of pancreatic secretions and bile into the duodenum is caused by
☒ a) cholecystokinin.
☐ b) gastric inhibitory peptide.
☐ c) gastrin.
☐ d) vasoactive intestinal peptide.

19  Which of the following is *not* a major function of the large intestine?
☐ a) reabsorbing water and compacting the intestinal contents into feces
☒ b) producing enzymes that assist pancreatic juice
☐ c) absorbing vitamins produced by bacterial activity
☐ d) storing fecal material prior to defecation

20  Fingerlike projections into the lumen of the small intestine are called
☐ a) microvilli.
☒ b) villi.
☐ c) circular folds.
☐ d) rugae.

21  A mesentery that forms a pouch between the body wall and the anterior surface of the small intestine is the
☒ a) greater omentum.
☐ b) lesser omentum.
☐ c) mesocolon.
☐ d) mesentery proper.

22  Pepsin is formed by a reaction between
☐ a) gastric lipase and acid.
☐ b) pepsinogen and intrinsic factors.
☐ c) bile and pancreatic juice.
☒ d) pepsinogen and acid.

23  Which of the following structures is *not* a component of a portal triad?
☐ a) interlobular vein
☐ b) central vein
☐ c) interlobular artery
☐ d) interlobular bile duct

**Short answer**

24  Name the four different types of teeth. What is the number of deciduous teeth and permanent teeth?

25  Identify the three phases of gastric secretion, and provide a brief description of each phase.

26  Describe the hepatic portal system.

# MasteringA&P™

Access more chapter study tools online in the MasteringA&P Study Area:

■ **Chapter Quizzes, Chapter Practice Test, and Clinical Case Studies**

| | |
|---|---|
| ■ Practice Anatomy Lab™ 3.0 | PAL™3.0 |
| ■ Interactive Physiology™ | iP2™ |
| ■ A&P Flix™ | A&PFlix™ |
| ■ PhysioEx™ 9.1 | PhysioEx™ 9.1 |

# Chapter Integration • Applying what you have learned

## Three forms of weight control surgery

Obesity is a medical condition in which excess body fat adversely affects the quality of life, leading to increased health problems and decreased life expectancy. *Obesity* refers to an excess amount of body fat. Both obesity and overweight (an excess amount of body weight that can come from muscles, bone, fat, and water) result from energy imbalance: When the number of calories eaten is greater than the number of calories used, there is an energy imbalance. So, if people eat and drink more calories than they use ("burn"), the balance tips toward weight gain. Over time, if this imbalance continues, the person becomes overweight. If the pattern continues further, obesity results. Weight management can be a complex issue because it involves complex interactions among behavior, diet, exercise, and genetics. Being overweight or obese can increase the risk for type 2 diabetes, hypertension, cardiovascular disease, and early death.

Because obesity is a leading preventable cause of death, more people have turned to surgery to help them lose weight. One form of weight control surgery involves gastric stapling.

In this procedure, a large portion of the gastric lumen is stapled shut, leaving only a small pouch in contact with the esophagus and duodenum.

Gastric bypass is another surgical procedure to induce weight loss. In this surgery, the proximal small intestine (duodenum) is connected to a small pouch formed by a superior portion of the stomach. Although this procedure seems more effective than gastric stapling, it involves more complicated surgery.

A third procedure, adjustable gastric band surgery (also known as lap-band surgery), is shown above. It involves placing an inflatable silicone device around the superior aspect of the stomach by laparoscopic surgery. This procedure is performed on obese patients with a body mass index (BMI) of 35–40 or greater. (The body mass index is a ratio calculated by dividing body weight in kilograms by the square of body height in meters; it is used to identify people who are overweight or underweight.)

After each type of surgery, the stomach will hold approximately 110–220 grams of food at each meal, compared to 1500 grams for a normal, distended stomach. Although surgical intervention is successful in many cases, the risks for potential complications are also high.

**1** How would gastric stapling result in weight loss?

**2** Explain the roles of the stretch receptors and the feeling of fullness after the gastric stapling procedure.

**3** Using your knowledge of smooth muscle tissue, what will happen to the gastric muscular layer over time after gastric stapling?

**4** How do gastric bypass surgery and lap-band surgery help achieve the goal of weight loss?

22

**LEARNING OUTCOMES**

These Learning Outcomes correspond by number to this chapter's modules and indicate what you should be able to do after completing the chapter. They also appear at the end of each module.

**Introduction to Cellular Metabolism**

**Digestion and Metabolism of Organic Nutrients**

**Energetics and Thermoregulation**

23

# Metabolism, Nutrition, and Energetics

# Module 23.1
# Metabolism is the sum of catabolic and anabolic reactions

**Metabolism** is the sum of all chemical and physical changes that occur in body tissues. It consists of catabolism (breakdown processes) and anabolism (synthesis processes). Catabolism converts large molecules into smaller ones and releases energy, and anabolism uses energy to convert small molecules into larger ones.

**REVIEW**

**A.** Compare catabolism and anabolism.

---

**An Overview of Cellular Metabolism**

**Catabolism** proceeds in a series of steps. In general, the first steps take place in the cytosol. Fats are broken down into fatty acids and glycerol; carbohydrates are broken down into glucose; and proteins are catabolized into amino acids. Different processes are involved in the catabolism of each nutrient.

*Nutrients*, the nourishing components in food, are necessary for normal physiology. The available nutrients are distributed in the blood, forming a **nutrient pool**, which can be used for anabolism or ATP production. **Anabolism** allows cells to continuously replace cellular components, which is part of *metabolic turnover*.

The cell "feeds" its mitochondria from its nutrient pool. In return, the cell gets the ATP it needs. About 40% of the energy released in catabolic reactions is captured and used to generate ATP from ADP and $P_i$ (inorganic phosphate). The other 60% escapes as heat that warms the interior of the cell and the surrounding tissues.

The ATP produced by mitochondria provides energy to support both anabolism and other cell functions.

---

**REVIEW**

**B.** Explain the process of metabolic turnover.

INTERSTITIAL FLUID

Plasma membrane

**Results of Anabolism**
- Maintenance and repair
- Growth
- Secretion
- Nutrient reserves

**CATABOLISM**

**ANABOLISM**

**Organic Molecules**
- Amino acids
- Lipids
- Simple sugars

**NUTRIENT POOL**

Anaerobic catabolism in the cytosol releases small amounts of ATP that are significant only under unusual conditions.

ATP

**HEAT**  60%

**Aerobic Metabolism** (in mitochondria)

40%  ATP

**Other ATP Expenses**
- Movement
- Contraction
- Intracellular transport
- Cytokinesis
- Endocytosis
- Exocytosis

CYTOSOL

23

**LEARNING OUTCOME**

Define metabolism, catabolism, and anabolism, and give an overview of cellular metabolism.

# Module 23.2
# Cells use nutrients from the nutrient pool for metabolism

Mitochondria accept only specific organic molecules. So, chemical reactions in the cytosol take whichever organic nutrients are available and break them down into smaller fragments that the mitochondria can process.

**1** Cells are continuously replacing membranes, organelles, enzymes, and structural proteins as well as performing their specialized functions. All of these functions require ATP. So, the cell's mitochondria must have a continuous supply of 2-carbon substrates to generate ATP. Substrates are the substances acted on by an enzyme (see Module 2.18). In this figure, arrow thickness indicates the relative importance of that pathway in a cell.

**?** **REVIEW**
**A.** List the reactants required and products generated by mitochondria.

Organic compounds that can be absorbed by cells are distributed to cells throughout the body by the bloodstream.

**Structural, functional, and storage components**

Triglycerides | Glycogen | Proteins

Nutrient pool — Fatty acids | Glucose | Amino acids

3-carbon chains

2-carbon chains

**MITOCHONDRIA**

Citric acid cycle → Coenzymes → Electron transport chain → ATP

$O_2$

$H_2O$

$CO_2$

**KEY**

↓ = Catabolic pathway

↑ = Anabolic pathway

**2** The nutrient pool of each cell contributes to the metabolic reserves of the body as a whole. When absorption across the digestive tract is unable to maintain normal nutrient levels in the blood, those reserves can be mobilized.

Nutrients obtained through digestion and absorption

Adipocytes convert excess fatty acids to triglycerides for storage. If absorption by the digestive tract and reserves in the liver cannot maintain normal nutrient levels, the triglycerides are broken down and the fatty acids released.

Liver cells store triglycerides and glycogen. If absorption by the digestive tract does not maintain normal nutrient levels, the triglycerides and glycogen are broken down, and the fatty acids and glucose are released.

**Nutrients distributed in the blood**

Skeletal muscles at rest metabolize fatty acids and use glucose to build glycogen reserves. Amino acids are used to increase the number of myofibrils. If the digestive tract, adipocytes, and liver cannot maintain normal nutrient levels, the contractile proteins can be broken down and amino acids released into the bloodstream for use by other tissues.

Nervous tissue requires a continuous supply of glucose. During starvation, other tissues shift to fatty acid or amino acid catabolism, conserving glucose for nervous tissue.

Cells in most tissues continuously absorb and catabolize glucose.

**L₀** **LEARNING OUTCOME**
Describe the role of the nutrient pool in cellular metabolism.

23

# Module 23.3

# Glycolysis is the first step in glucose catabolism

Glucose is the preferred molecule for catabolism and ATP production. **Glycolysis** is an anaerobic process that breaks down a 6-carbon glucose molecule into two 3-carbon molecules of pyruvate through a series of enzymatic steps. This diagram follows changes in the carbon chain. There is a net gain of 2 ATP molecules for each glucose molecule converted to 2 molecules of pyruvate. Also, cytosolic NAD (nicotinamide adenine dinucleotide) accepts hydrogen atoms to form NADH. Their fate depends on the degree to which glucose is catabolized. The complete catabolism of pyruvate begins with its entry into a mitochondrion.

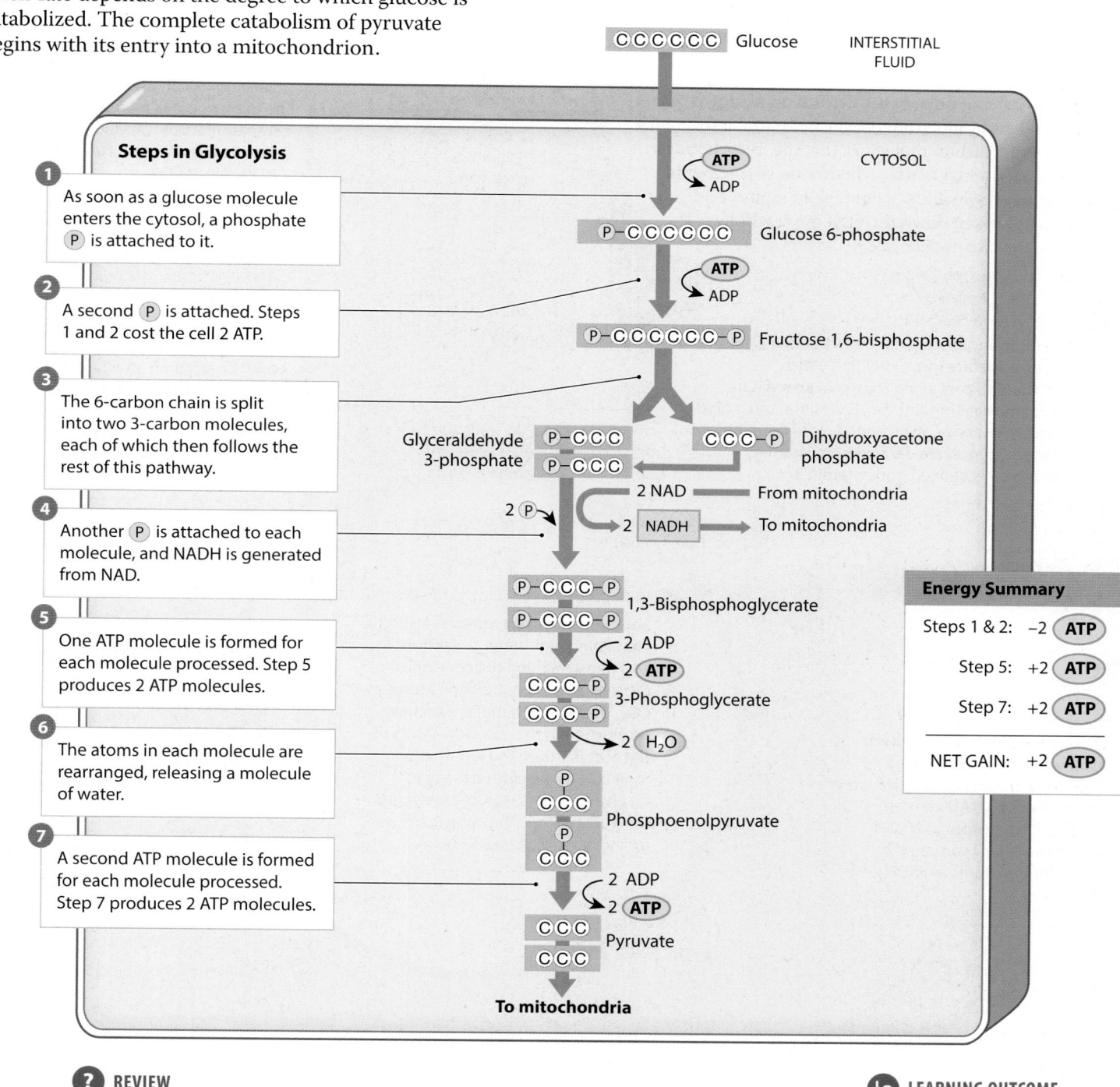

**Steps in Glycolysis**

1. As soon as a glucose molecule enters the cytosol, a phosphate Ⓟ is attached to it.

2. A second Ⓟ is attached. Steps 1 and 2 cost the cell 2 ATP.

3. The 6-carbon chain is split into two 3-carbon molecules, each of which then follows the rest of this pathway.

4. Another Ⓟ is attached to each molecule, and NADH is generated from NAD.

5. One ATP molecule is formed for each molecule processed. Step 5 produces 2 ATP molecules.

6. The atoms in each molecule are rearranged, releasing a molecule of water.

7. A second ATP molecule is formed for each molecule processed. Step 7 produces 2 ATP molecules.

**Energy Summary**

| | | |
|---|---|---|
| Steps 1 & 2: | −2 | **ATP** |
| Step 5: | +2 | **ATP** |
| Step 7: | +2 | **ATP** |
| NET GAIN: | +2 | **ATP** |

**? REVIEW**
A. List the products of glycolysis.

**Lo LEARNING OUTCOME**
Summarize the important events and products of glycolysis.

# Module 23.4
# The citric acid cycle transfers hydrogen atoms to coenzymes

Within the mitochondria, in the process of **aerobic metabolism**, pyruvate is broken down, and the energy released is used to produce a large amount of ATP. This oxygen-requiring process involves two pathways: the citric acid cycle and the electron transport chain. The **citric acid cycle** is also known as the *Krebs cycle* and the *tricarboxylic acid cycle*. Its function is to remove hydrogen atoms (H) from organic substrates entering from the cytosol and transfer them to the electron transport chain.

**?** **REVIEW**
**A.** Briefly describe the citric acid cycle, and explain its role.

**1** Chemical reactions in the cytosol break down organic nutrients from the nutrient pool into smaller 3-carbon and 2-carbon molecules that the mitochondria can process. It doesn't matter whether these fragments came from a carbohydrate, a lipid, or an amino acid. Everything is broken down through the catabolic pathway of the citric acid cycle.

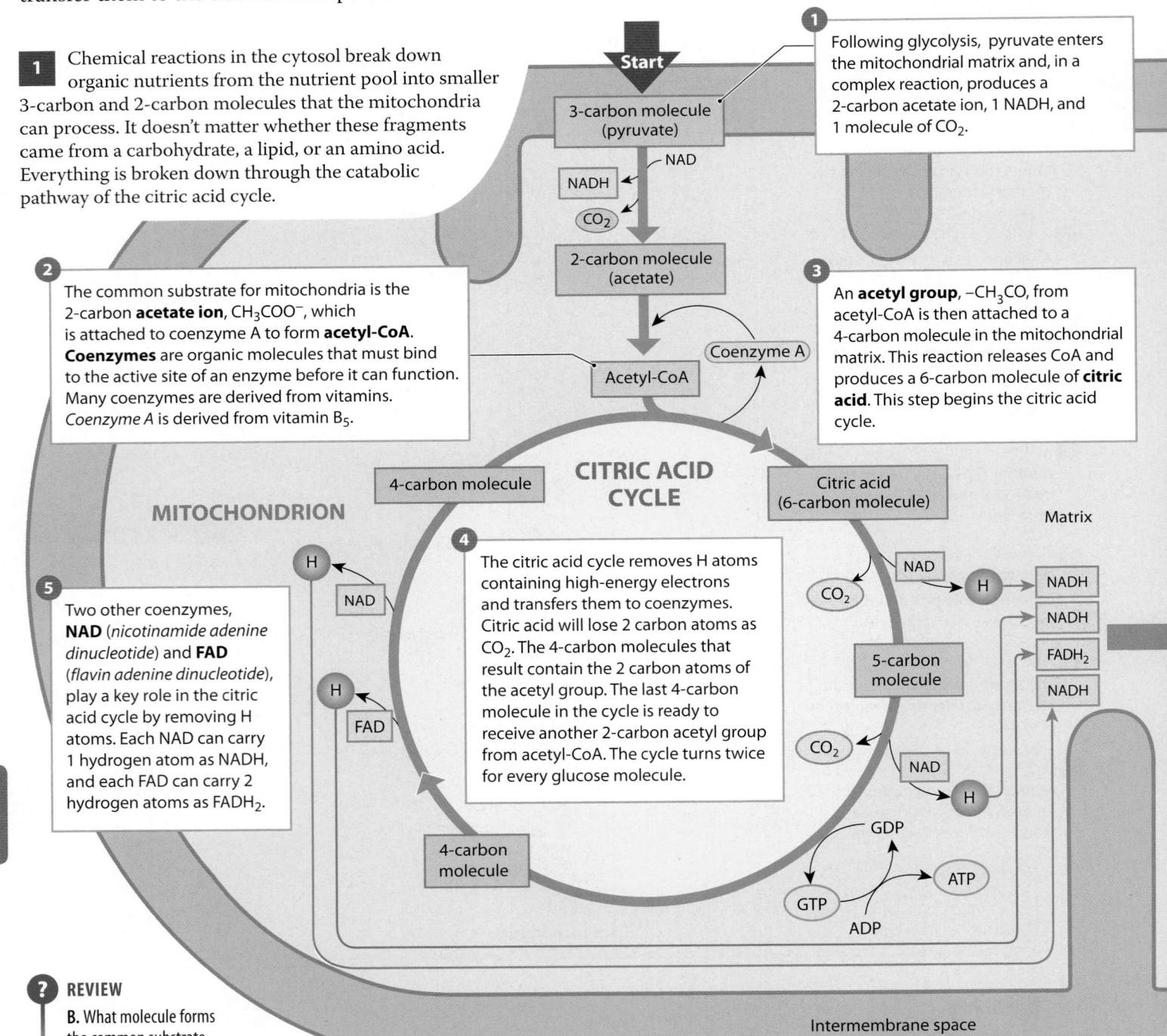

**2** The common substrate for mitochondria is the 2-carbon **acetate ion**, $CH_3COO^-$, which is attached to coenzyme A to form **acetyl-CoA**. **Coenzymes** are organic molecules that must bind to the active site of an enzyme before it can function. Many coenzymes are derived from vitamins. *Coenzyme A* is derived from vitamin $B_5$.

**1** Following glycolysis, pyruvate enters the mitochondrial matrix and, in a complex reaction, produces a 2-carbon acetate ion, 1 NADH, and 1 molecule of $CO_2$.

**3** An **acetyl group**, $-CH_3CO$, from acetyl-CoA is then attached to a 4-carbon molecule in the mitochondrial matrix. This reaction releases CoA and produces a 6-carbon molecule of **citric acid**. This step begins the citric acid cycle.

**5** Two other coenzymes, **NAD** (*nicotinamide adenine dinucleotide*) and **FAD** (*flavin adenine dinucleotide*), play a key role in the citric acid cycle by removing H atoms. Each NAD can carry 1 hydrogen atom as NADH, and each FAD can carry 2 hydrogen atoms as $FADH_2$.

**4** The citric acid cycle removes H atoms containing high-energy electrons and transfers them to coenzymes. Citric acid will lose 2 carbon atoms as $CO_2$. The 4-carbon molecules that result contain the 2 carbon atoms of the acetyl group. The last 4-carbon molecule in the cycle is ready to receive another 2-carbon acetyl group from acetyl-CoA. The cycle turns twice for every glucose molecule.

**CITRIC ACID CYCLE**

**MITOCHONDRION**

**?** **REVIEW**
**B.** What molecule forms the common substrate for the citric acid cycle?

23

**2** This figure is a more detailed view of the citric acid cycle showing the changes in the carbon chain as it works its way through the cycle. (At physiological pH, the acids of the citric acid cycle exist in their dissociated, ionized form. Thus, citric acid is called citrate.) For each acetyl-CoA molecule that enters the citric acid cycle, 5 hydrogen atoms are removed and transferred to coenzymes, 2 molecules of $CO_2$ are produced, and 2 molecules of water are consumed. Because the citric acid cycle turns twice for each glucose molecule, the net energy gain from two acetyl groups entering the citric acid cycle is 2 ATP molecules.

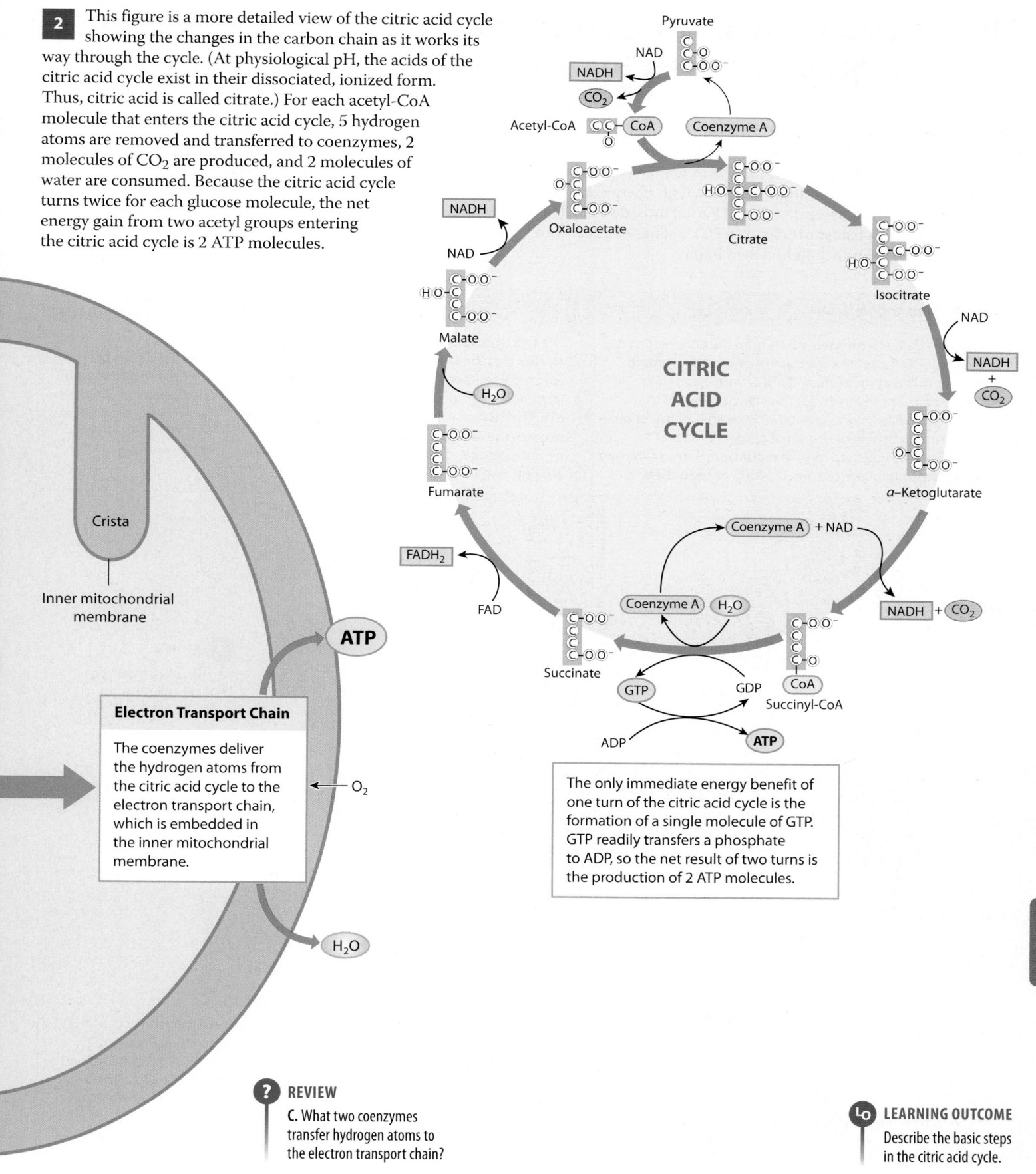

**Electron Transport Chain**

The coenzymes deliver the hydrogen atoms from the citric acid cycle to the electron transport chain, which is embedded in the inner mitochondrial membrane.

The only immediate energy benefit of one turn of the citric acid cycle is the formation of a single molecule of GTP. GTP readily transfers a phosphate to ADP, so the net result of two turns is the production of 2 ATP molecules.

**? REVIEW**

**C.** What two coenzymes transfer hydrogen atoms to the electron transport chain?

**Lo LEARNING OUTCOME**

Describe the basic steps in the citric acid cycle.

# Module 23.5
# The electron transport chain establishes a proton gradient used to make ATP

**1** **Oxidative phosphorylation** is the generation of ATP as the result of the transfer of electrons from the coenzymes NADH and $FADH_2$ to oxygen by a sequence of electron carriers within mitochondria. *Oxidation* refers to the transfer of electrons, and *phosphorylation* refers to the attachment of a high-energy phosphate. The process produces more than 90 percent of the ATP used by body cells. The key reactions take place in the **electron transport chain (ETC)**, a series of integral and peripheral proteins in the inner mitochondrial membrane.

**? REVIEW**

**A.** Define *oxidative phosphorylation*.

---

**Oxidative Phosphorylation**

**1** Coenzymes deliver hydrogen atoms from the citric acid cycle to the ETC, which contains a sequence of protein-pigment molecules called **cytochromes** (SĪ-tō-krōmz; *cyto-*, cell + *chroma*, color). These molecules transfer electrons of the hydrogen atoms from one to another in a series of *oxidation–reduction (redox) reactions*. A gain of oxygen, or loss of hydrogen or electrons, is a form of **oxidation**. A loss of oxygen, or gain of hydrogen or electrons, is a form of **reduction**.

**2** The first cytochrome passes electrons to the second, which passes them to the third, and so on, down the ETC. This stepwise passage releases the energy carried by the electrons in manageable amounts, not all at once.

**3** At several steps along the way, enough energy is released to pump hydrogen ions into the intermembrane space.

**4** The diffusion of hydrogen ions back into the matrix through hydrogen ion channels powers the production of ATP by the enzyme **ATP synthase**.

Crista

Matrix

Intermembrane space

$2e^-$

$2H^+$

NAD and FAD

Cytochromes of the electron transport chain

$2H^+$

ADP + P

ATP

$O^{-2}$ + $2H^+$

$H_2O$

**MITOCHONDRION**   Crista

Intermembrane space

**? REVIEW**

**B.** Compare oxidation and reduction.

**5** The final electron acceptor is oxygen. After accepting 2 electrons, an oxygen ion then reacts with 2 hydrogen ions in the matrix to form a water molecule.

23

**2** The basis of oxidative phosphorylation is the final acceptance of electrons by oxygen and the formation of water. The electron transport chain (ETC) consists of four respiratory complexes (I-IV), coenzyme Q, and cytochrome molecules (*b*, *c*, *a*, and $a_3$). Except for cytochrome *c* (Cyt *c*), these molecules are associated with respiratory complexes III and IV. The hydrogen atoms delivered by NADH and FADH$_2$ are first split into electrons (e$^-$) and protons (H$^+$). The high-energy electrons are passed along in a way that ensures that their energy is released in a series of small steps rather than all at once. The result is a proton (H$^+$) electrochemical gradient across the inner membrane. Despite this gradient, H$^+$ cannot diffuse into the matrix because they are not lipid soluble. However, H$^+$ channels in the inner membrane permit H$^+$ to diffuse into the matrix. The kinetic energy of passing hydrogen ions generates ATP in a process known as **chemiosmosis**. The red line with the arrow indicates the paths and destination of the electrons.

Mitochondrion

**CYTOSOL**

Outer membrane

Intermembrane space

Inner membrane

Electron transport chain establishes proton (H$^+$) gradient

Chemiosmosis uses proton (H$^+$) gradient

ATP synthase with proton (H$^+$) channel

H$^+$   H$^+$   H$^+$

Cyt *c*

Q   *b*   IV

I   III   *a*   $a_3$

II

H$^+$

NADH   H$^+$

NAD

FADH$_2$   FAD + H$^+$

H$^+$

H$^+$   e   O + 2 H$^+$   ADP + P

H$_2$O   ATP

**MITOCHONDRIAL MATRIX**

| The electrons from NADH pass from respiratory complex I to coenzyme Q. The energy released by the electrons is used to pump protons (H$^+$) from the matrix into the intermembrane space. | The electrons from FADH$_2$ are transferred to respiratory complex II and then directly to coenzyme Q. | The electrons are transferred from coenzyme Q to respiratory complex III. More electron energy is released and used to pump H$^+$ into the intermembrane space. | The electrons transferred to respiratory complex IV release their remaining energy to pump H$^+$ into the intermembrane space. The electrons are released into the matrix and combine with oxygen and H$^+$ to form water molecules. | H$^+$ pass through hydrogen ion channels into the matrix. Their movement provides the energy for the ATP synthase complex to combine ADP and P into ATP. |

Mitochondria are very efficient, but aerobic metabolism is dependent on the availability of oxygen that must be absorbed first by the cell and then by the mitochondria. If that supply of oxygen is restricted, the cytochromes cannot pass on their electrons, and the ETC stops working. When the ETC stops, NADH and FADH$_2$ can't drop off their hydrogen atoms, so the citric acid cycle halts as well. If the problem persists, the cell will die because it lacks the ATP to support vital functions. Poisons such as cyanide (used as a pesticide) bind to cytochromes and prevent the transfer of electrons to oxygen. As a result, cells die from energy starvation.

 **REVIEW**

**C.** Describe the role that hydrogen ion channels play in the generation of ATP.

 **LEARNING OUTCOME**

Describe the basic steps in the electron transport chain.

23

# Module 23.6
# Glucose catabolism yields 30–32 ATP

**1** This diagram summarizes the sources of ATP produced during the complete catabolism of a single glucose molecule. Two ATP are gained through glycolysis, whereas 30 are gained through aerobic metabolism in mitochondria. The energy yields of NADH and $FADH_2$ are different because they transfer their H atoms to different complexes in the ETC.

> The NADH generated in the cytosol yields fewer ATP than the NADH generated in the citric acid cycle because it "costs" the cell ATP to get the hydrogen atoms to the ETC within the mitochondria.

CYTOSOL

Glucose (6-carbon)

2 ATP

4 ADP

2 ADP

4 ATP

2 NAD

**GLYCOLYSIS**

2 NADH

2 NAD

2 Pyruvate (3-carbon)

2 $CO_2$

By intermediates

2 NAD

Coenzyme A

2 NADH

2

2 Acetyl-CoA

**ELECTRON TRANSPORT CHAIN**

5 ATP

8 NAD + 20 ATP

2 FAD + 3 ATP

4 $CO_2$

6 NAD

6 NADH

**Citric Acid Cycle (2 turns)**

2 FAD

2 $FADH_2$

6 $O_2$   12 $H_2O$

2 GDP

2 ATP

2 GTP

2 ADP

**MITOCHONDRIA**

## Energy Summary

**Glycolysis (Anaerobic):**

| 4 ATP | produced during enzymatic reactions in the cytosol |
| –2 ATP | used to initiate glycolysis |
| 2 ATP | net gain to cell |

**The Electron Transport Chain and Citric Acid Cycle (Aerobic):**

| 5 ATP | (3 ATP if 2 NADH transfer their electrons to FAD) |
| 20 ATP | from NADH generated in citric acid cycle |
| 3 ATP | from $FADH_2$ generated in citric acid cycle |
| 2 ATP | by GTP produced during enzymatic reactions |

| 30–32 ATP | net gain to cell from complete catabolism of one glucose molecule |

## Reasons Why Glucose Is the Primary Energy Source for Cells

- Glucose is a small, soluble molecule that is easily distributed through body fluids.
- Glucose can provide ATP anaerobically through glycolysis. Although only a small amount of ATP is produced, glycolysis is important during peak levels of physical activity, in red blood cells, or when a tissue is temporarily deprived of oxygen.
- Glucose can be stored as glycogen, which forms compact, insoluble granules.
- Glucose can be easily mobilized because the breakdown of glycogen (glycogenolysis) occurs very quickly and involves only a single enzymatic step. Mobilization of other intracellular reserves involves much more complex pathways and takes considerably more time.

**? REVIEW**
**A.** Identify when most of the $CO_2$ is released during the complete catabolism of glucose.

**? INTEGRATION**
**B.** Explain when glycolysis is important in cellular metabolism.

**Lo LEARNING OUTCOME**
Identify the sources of ATP production and energy yield at each source during glucose catabolism.

23

# Module 23.7
# Nutrient metabolism follows several pathways

**1** Neither the diet nor the nutrient pool provides everything needed to build every protein, carbohydrate, and lipid a cell might require. With few exceptions, this is not a problem because the cell also contains the enzymes necessary to synthesize what it needs from available substrates. This diagram is a general overview of the catabolic and anabolic pathways involved; these pathways will be the focus of the next section.

KEY

↓ = Catabolic pathway

↑ = Anabolic pathway

Triglycerides

Glycogen

Proteins

Fatty acids can be stored as triglycerides.

Stored triglycerides can be broken down into fatty acids and glycerol.

**Glycogenesis** is the formation of glycogen from glucose.

**Glycogenolysis** is the breakdown of glycogen to glucose.

The primary use of amino acids is the synthesis of proteins. Amino acids are seldom broken down if other energy sources are available. However, in starvation the proteins of muscle tissues are mobilized, releasing amino acids that can be catabolized by other tissues.

Nutrient pool

Fatty acids

Glucose

Amino acids

The breakdown of a fatty acid releases acetyl-CoA that can enter the citric acid cycle.

**Gluconeogenesis** is the formation of glucose from non-carbohydrates.

**Glycolysis** is the breakdown of glucose to 2 3-carbon molecules.

3-carbon chains

2-carbon chains

Fatty acid synthesis begins with acetyl-CoA. Because this is the common intermediary for all aerobic catabolic pathways, fatty acids can be synthesized from excess carbohydrates or amino acids.

MITOCHONDRIA

ATP

Citric acid cycle

Coenzymes

Electron transport chain

$O_2$

$H_2O$

$CO_2$

$O_2$ must be continuously provided by diffusion from the extracellular fluid. This requires normal respiratory function and adequate tissue perfusion.

23

**? REVIEW**

**A.** Compare glycogenesis and glycogenolysis.

$CO_2$ must leave the cytosol by diffusion into the extracellular fluid. The bloodstream must continuously absorb $CO_2$ in peripheral tissues and eliminate it at the lungs to prevent potentially dangerous changes in body fluid pH.

**? INTEGRATION**

**B.** Why do cells make new compounds?

**Lo LEARNING OUTCOME**

Define glycogenesis, glycogenolysis, and gluconeogenesis.

# Section 1 Review
## Introduction to Cellular Metabolism

**Matching:** Match each lettered term with the appropriate numbered blank in the cellular metabolism figure to the right.

a. glucose
b. electron transport chain
c. $O_2$
d. fatty acids
e. proteins
f. citric acid cycle
g. ATP
h. $CO_2$
i. $H_2O$
j. coenzymes
k. 2-carbon chains

Structural, functional, and storage components

Triglycerides   Glycogen

[ ] 3

Nutrient pool   Amino acids

[ ] 1
[ ] 2

Three-carbon chains

[ ] 4
[ ] 5
[ ] 6

MITOCHONDRIA

[ ] 7
[ ] 8
[ ] 9
[ ] 10
[ ] 11

Match each lettered term with the most closely related description.

a. coenzymes
b. cytochromes
c. citric acid
d. nutrients scarce
e. nutrient pool
f. anabolism
g. ATP
h. water
i. acetate
j. oxygen
k. citric acid cycle
l. catabolism
m. oxidative phosphorylation
n. nutrients abundant

| 12 | 6-carbon molecule | 12 |
| 13 | Collection of all the cell's organic substances | 13 |
| 14 | Synthesis of new organic molecules | 14 |
| 15 | Process that produces over 90 percent of ATP used by body cells | 15 |
| 16 | ETC proteins | 16 |
| 17 | Carry hydrogen atoms to the ETC | 17 |
| 18 | Final acceptor of electrons from the ETC | 18 |
| 19 | Breakdown of organic molecules | 19 |
| 20 | Condition when cells preferentially break down carbohydrates | 20 |
| 21 | Product of hydrogen ion diffusion within mitochondria | 21 |
| 22 | Source of mitochondrial $CO_2$ production | 22 |
| 23 | ETC by-product | 23 |
| 24 | Condition when cells preferentially break down lipids | 24 |
| 25 | Common substrate for mitochondrial ATP production | 25 |

**Section integration:** Nervous tissue requires a constant supply of glucose. What general shifts in cellular metabolism occur during fasting or starvation to meet that requirement?

26 _____

23

## Module 23.8

# Digestion involves a series of steps to make nutrients available to the body

**1** The food we eat has an organized physical structure, and the compounds it contains are large and often insoluble. During digestion the physical structure is broken down, and a combination of chemical and enzymatic action breaks the complex organic compounds into simpler components that can be absorbed by the digestive tract and then distributed by the bloodstream. Cells throughout the body rely on these organic molecules for energy production and to replenish the intracellular nutrient pool.

 **REVIEW**

**A.** Why is digestion important?

 **REVIEW**

**B.** Where does most nutrient absorption occur?

**C.** Most of the absorbed nutrients enter into which blood vessel?

This section provides an overview of the digestion, absorption, and fates of carbohydrates, lipids, and proteins. The modules emphasize general patterns that will make it easier for you to understand the specifics presented in biochemistry or nutrition courses. We focus attention on how nutrients are absorbed, stored or interconverted, or catabolized to yield the energy needed to support vital activities. We will not consider the major anabolic pathways here because Chapter 2 discussed the synthesis of carbohydrates, lipids, and proteins in sufficient detail.

### Steps in the Process of Digestion

In the oral cavity, saliva dissolves some organic nutrients, and mechanical processing with the teeth and tongue disrupts the physical structure of the material and provides access for digestive enzymes. Those enzymes begin the digestion of complex carbohydrates (polysaccharides) and lipids.

In the stomach, food is further broken down physically and chemically by stomach acid and by enzymes that can operate at an extremely low pH.

In the duodenum, buffers from the pancreas and liver moderate the pH of the arriving chyme. The pancreas also secretes various digestive enzymes that catalyze the catabolism of carbohydrates, lipids, proteins, and nucleic acids.

Nutrient absorption occurs in the small intestine, primarily in the jejunum, and the nutrients enter the bloodstream.

Indigestible food and wastes enter the large intestine, where water is reabsorbed and bacteria generate both organic nutrients and vitamins. Most of these organic products are absorbed before the residue is excreted as feces.

Most of the nutrients absorbed by the digestive tract are transported in a branch of the hepatic portal vein that ends at the liver. The liver absorbs nutrients as needed to maintain normal levels in the bloodstream.

Within peripheral tissues, cells absorb the nutrients needed to maintain their nutrient pool and ongoing operations.

**23**

 **LEARNING OUTCOME**

Outline the steps involved in digestion, and list the nutrients used by the body.

# Module 23.9

# Carbohydrates are usually the preferred substrates for catabolism and ATP production under resting conditions

**Carbohydrates ingested**

**1** In the mouth, chewing saturates the bolus with secretions from the salivary glands. **Salivary amylase** is an enzyme that breaks down complex carbohydrates into a mixture of disaccharides and trisaccharides.

**2** Salivary amylase continues to digest carbohydrates from the meal until the pH throughout the contents of the stomach falls below 4.5. This enzyme generally remains active for 1–2 hours after a meal.

**3** When chyme arrives in the duodenum, secretin stimulates the release of buffers that shift the duodenal pH from acidic to alkaline (all intestinal enzymes require an alkaline pH). At the same time, cholecystokinin (CCK) release triggers the secretion of pancreatic buffers and enzymes, including **pancreatic alpha-amylase**. This enzyme has the same functions as salivary amylase, which was deactivated by denaturation in the stomach.

**4** The arrival of chyme containing large amounts of carbohydrates triggers the release of gastric inhibitory peptide (GIP), which stimulates insulin release by the pancreas (see Module 22.14).

Hepatic portal vein

**5** The epithelial cells lining the jejunum finish the digestion of carbohydrates. The plasma membrane at the brush border contains the enzymes **maltase**, **sucrase**, and **lactase**, which break the disaccharides maltose (glucose + glucose), sucrose (glucose + fructose), and lactose (glucose + galactose) into simple sugars that are then absorbed. As these enzymes function, they transport the monosaccharides across the plasma membrane and release them into the cytosol.

The simple sugars that are transported into the cell at its apical surface diffuse through the cytosol and reach the interstitial fluid by facilitated diffusion across the basolateral surfaces. These monosaccharides then diffuse into the capillaries of the intestinal villi for eventual transport to the liver in the hepatic portal vein.

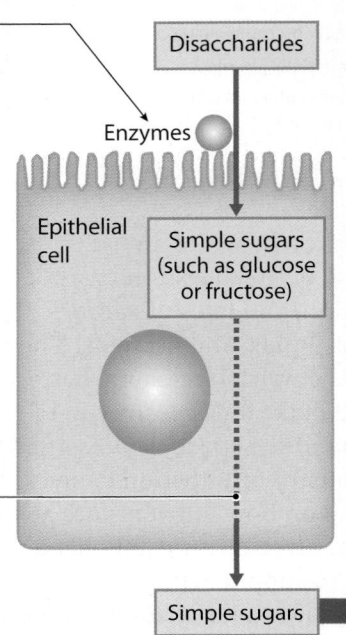

Disaccharides

Enzymes

Epithelial cell

Simple sugars (such as glucose or fructose)

Simple sugars

**6** Intestinal enzymes do not alter indigestible carbohydrates such as cellulose, so they arrive in the colon virtually intact. These carbohydrates provide a reliable nutrient source for bacteria in the colon, whose metabolic activities generate small quantities of **flatus**, or intestinal gas. Foods containing large amounts of indigestible carbohydrates (such as beans) stimulate bacterial gas production, leading to distension of the colon, cramps, and the frequent discharge of intestinal gases.

**?** REVIEW

A. Describe the source of intestinal gas.

**Liver**

Simple sugars are absorbed and converted to glucose as needed. Glucose is released into the blood sufficient to maintain normal glucose levels (~90 mg/dL).

Excess stored as glycogen or used in other ways

**Skeletal Muscle**

Resting skeletal muscles primarily break down fatty acids for energy and absorb glucose to build up their glycogen reserves.

Glycogen

**? REVIEW**

**B.** Explain the role of glycogen in cellular metabolism.

Carbohydrates (such as glucose) are usually preferred for catabolism because proteins and lipids are more important as structural components of cells and tissues.

**GLUCOSE**
(6-carbon)

In most tissues, the transport of glucose into the cell is dependent on the presence of a **carrier protein**.

Insulin signals the cell to recruit the carrier proteins to the cell surface.

Inside the cell, the glucose may be converted to another simple sugar, such as ribose, and used to build glycoproteins, other structural materials, or nucleic acids. Glucose may also be converted to glycerol for the synthesis of glycerides.

CYTOSOL

ATP

Other simple sugars

Pyruvate (3-carbon)

Pyruvate (3-carbon)

If needed to provide energy, the 6-carbon glucose molecule is broken down into two 3-carbon molecules of **pyruvate**. This anaerobic process, called glycolysis, yields a net gain of 2 ATP for every glucose molecule broken down. (See Module 23.3 for the details of glycolysis.)

Coenzyme A

$CO_2$

Each pyruvate molecule can then be used by mitochondria after conversion to acetyl-CoA.

Acetyl-CoA (2-carbon)

Citric acid cycle

Coenzymes

Electron transport chain

$O_2$

**ATP**

For each molecule of pyruvate processed by mitochondria, the cell gains 15 ATP, consumes 3 molecules of $O_2$, and generates 3 molecules of $CO_2$ and 6 molecules of water. Thus for each pair of pyruvate molecules catabolized, the cell gains 30–32 ATP (see Module 23. 6).

$H_2O$

**MITOCHONDRIA**

$CO_2$

**? REVIEW**

**C.** Explain why carbohydrates are preferred over proteins and lipids as an energy source.

**Lo LEARNING OUTCOME**
Describe carbohydrate metabolism.

23

# Lipids reach the bloodstream in chylomicrons; the cholesterol is then extracted and released as lipoproteins

**Lipids ingested** →

**Thoracic duct**

**1** Chewing in the mouth breaks material into smaller chunks and disrupts connective tissue organization. As this is under way, the bolus becomes saturated with saliva containing **lingual lipase**. This enzyme splits triglycerides, breaking them down into monoglycerides and fatty acids.

**2** Most dietary lipids are not water soluble. The mixing of chyme in the stomach creates large drops containing a variety of lipids. Lingual lipase continues to function in the acidic environment but can break down triglycerides only at the surfaces of these drops. As a result, only about 20% of the triglycerides have been broken down by the time the chyme leaves the stomach.

**3** When chyme arrives in the duodenum, CCK is released, triggering the secretion of pancreatic enzymes, including **pancreatic lipase**, and also stimulating gallbladder contraction and the ejection of bile into the duodenum. Bile contains **bile salts** that break the large lipid drops into tiny droplets, a process called **emulsification**, which provides better access for pancreatic lipase. Pancreatic lipase then breaks apart the triglycerides to form a mixture of fatty acids, monoglycerides, and glycerol. As these molecules are released, they interact with bile salts in the lumen to form small lipid–bile salt complexes called **micelles** (mī-SELZ).

**4** When a micelle contacts the intestinal epithelium, the lipids diffuse across the plasma membrane and enter the cytosol. The intestinal cells synthesize new triglycerides from the monoglycerides, fatty acids, and glycerol. These triglycerides, in company with absorbed cholesterol, phospholipids, and other lipid-soluble materials, are then coated with proteins, forming **chylomicrons** (kī-lō-MĪ-kronz; *chylos*, milky lymph + *mikros*, small). Chylomicrons are **lipoproteins**—lipid-protein complexes that contain insoluble lipids. The superficial coating of phospholipids and proteins makes the entire complex water soluble.

The intestinal cells then secrete the chylomicrons into interstitial fluid by exocytosis. The superficial protein coating of the chylomicrons keeps them suspended in the interstitial fluid, but their size generally prevents them from diffusing into capillaries.

Bile salts are released. Most are reabsorbed and returned to the liver before they leave the small intestine.

Micelle

Monoglycerides, fatty acids

Triglycerides + other lipids and proteins

Exocytosis

Most of the chylomicrons released diffuse into the intestinal lacteals of the lymphatic system, which lack basement membranes and have large gaps between adjacent endothelial cells.

**5** From the lacteals, the chylomicrons proceed along the lymphatic vessels and into the thoracic duct.

23

**6** The chylomicrons enter the bloodstream at the left subclavian vein, then pass through the pulmonary circuit before entering the systemic circuit.

**7** Capillary walls contain the enzyme **lipoprotein lipase**, which breaks down the chylomicrons and releases fatty acids and monoglycerides that can diffuse into the interstitial fluid.

Resting skeletal muscle fibers absorb fatty acids and break them down, using the ATP provided to power the contractions that maintain muscle tone and to convert glucose to glycogen.

Adipocytes absorb the monoglycerides and fatty acids and use them to synthesize triglycerides for storage.

**? REVIEW**
**A.** What is the difference between a micelle and a chylomicron?

## Lipoproteins and Lipid Transport and Distribution

**8** The liver absorbs chylomicrons and creates **low-density lipoproteins (LDLs)** and **very low density lipoproteins (VLDLs)**. To make LDLs, triglycerides are removed from the chylomicrons, cholesterol is added, and the surface proteins are altered. VLDLs contain triglycerides manufactured by the liver, plus small amounts of phospholipids and cholesterol. Some of the cholesterol is used by the liver to synthesize bile salts; excess cholesterol is excreted in the bile.

Chylomicrons

VLDL

**9** VLDLs transport triglycerides from the liver to muscle and adipose tissue.

**10** The LDLs enter the bloodstream and are delivered to peripheral tissues.

LDL

Cholesterol extracted

**14** The HDLs return the cholesterol to the liver, where it is extracted and packaged in new LDLs and VLDLs or excreted with bile salts in bile.

Excess cholesterol is excreted with the bile salts

**11** Once in peripheral tissues, the LDLs are absorbed.

LDL

HDL
High cholesterol

HDL
Low cholesterol

Peripheral cells

Lysosomal breakdown

Used in synthesis of membranes, hormones, other materials

**? REVIEW**
**B.** What does the liver do with the chylomicrons it receives?

HDL

Cholesterol release

**12** The cell extracts the cholesterol and uses it in various ways.

The total cholesterol level and the LDL:HDL ratio are often used to detect potentially serious cardiovascular problems. Elevated total cholesterol (above 200 mg/dL) plus a high LDL:HDL ratio means that (1) there is a lot of cholesterol in the circulation and (2) most of it is going out to the tissues (in LDLs and VLDLs) and staying there, rather than returning to the liver (in HDLs) for recycling or excretion. This is a potential problem because excess cholesterol in peripheral tissues tends to accumulate in atherosclerotic plaques (see Module 19.8) that can cause heart attacks and strokes.

**13** The cholesterol not used by the cell diffuses out of the cell across the plasma membrane and re-enters the bloodstream. **High-density lipoproteins (HDLs)**, proteins released by the liver, absorb excess cholesterol and transport it back to the liver for storage or excretion in the bile.

**? INTEGRATION**
**C.** Describe the roles of LDLs and HDLs.

**Lo LEARNING OUTCOME**
Describe the mechanisms of lipid transport and distribution.

23

# Module 23.11
# Fatty acids can be broken down to provide energy or converted to other lipids

**1** During lipid catabolism, or **lipolysis**, lipids are broken down into pieces that can be either converted to pyruvate or channeled directly into the citric acid cycle. We will consider triglyceride catabolism, as the principles apply to other lipids as well. A triglyceride is first split into its component parts by hydrolysis, yielding 1 molecule of glycerol and 3 fatty acid molecules. Glycerol is converted to pyruvate. The catabolism of fatty acids, a process called **beta-oxidation**, involves a completely different set of enzymes that that generate acetyl-CoA directly.

**?** **REVIEW**
**A.** Define *beta-oxidation*.

Triglyceride

Absorption through endocytosis

CYTOSOL

**Fatty acid (18-carbon)**

**1** Lysosomal enzymes break down triglyceride molecules into 1 glycerol molecule and 3 fatty acids.

Glycerol

**3** The complete catabolism of fatty acids does not occur until they have been absorbed by mitochondria.

Beta-oxidation

2 **ATP**

**2** In the cytosol, the glycerol is converted to pyruvate through the glycolysis pathway, yielding 2 ATP for each triglyceride broken down.

**Fatty acid (18-carbon)**

Coenzyme A
(attaches to the carboxyl group of the fatty acid)

FADH$_2$

Pyruvate

**Fatty acid (18-carbon) – CoA**

CoA

Coenzyme A
(this step requires a second coenzyme A)

NADH

Coenzyme A

CO$_2$

Acetyl-CoA

13 **ATP**

**Fatty acid (16-carbon) – CoA**

CoA + **Acetyl-CoA**

CoA

Citric acid cycle

Coenzymes

Electron transport chain

O$_2$

**4** An enzymatic reaction then breaks off the first two carbons as acetyl-CoA while leaving a shorter fatty acid bound to the second molecule of coenzyme A.

H$_2$O

For each step in beta-oxidation, the cell gains 13 ATP, and the process can be repeated until the entire fatty acid has been broken down. Note that this is much more efficient than glucose catabolism: The complete breakdown of a 6-carbon glucose molecule yields 30–32 ATP, whereas the complete breakdown of 6 carbons from a fatty acid yields 39–40 ATP.

CO$_2$

23

**2** The synthesis of lipids is known as **lipogenesis** (lip-ō-JEN-e-sis). Because most catabolic pathways are reversible, the cell can also synthesize most—but not all—lipids on demand. This diagram shows the major pathways for lipogenesis.

**4** The glycerol required for triglyceride production is synthesized from one of the intermediate products of glycolysis.

**3** All of the other structural and functional lipids can be synthesized from fatty acids.

**2** Fatty acid synthesis involves a reaction sequence quite distinct from that of beta-oxidation. As a result, body cells cannot build every fatty acid they can break down. For example, our cells lack the enzymes to insert double bonds in the proper locations to synthesize two 18-carbon fatty acids synthesized by plants: **linolenic acid** (an **omega-3** fatty acid) or **linoleic acid** (an **omega-6** fatty acid). However, these fatty acids are needed to synthesize prostaglandins and some of the phospholipids found in plasma membranes throughout the body. They are therefore called **essential fatty acids** because they must be included in your diet.

**Start ►** **1** The synthesis of most types of lipids, including nonessential fatty acids and steroids, begins with acetyl-CoA. Lipogenesis can use almost any organic substrate, because lipids, amino acids, and carbohydrates can be converted to acetyl-CoA.

CYTOSOL

Steroids

Triglycerides

Glucose

Glycerol

Cholesterol

Prostaglandins

Phospholipids

Glycolipids

Fatty acids

ADP

ATP

Pyruvate

Coenzyme A

$CO_2$

Acetyl-CoA

Citric acid cycle

MITOCHONDRIA

**? REVIEW**

**B.** Which molecule is a key reactant in both ATP production from fatty acids and lipogenesis?

**C.** Identify the fates of fatty acids.

KEY

▲ = Anabolic pathway    ▼ = Catabolic pathway

23

Lipids are useful as an energy reserve primarily because beta-oxidation is so efficient, and excess lipids can be stored as triglycerides. The problem is that triglyceride reserves form insoluble droplets that are difficult for water-soluble enzymes to access. As a result, stored lipids cannot provide large amounts of ATP very quickly—that's why carbohydrate catabolism is so important. Lipids, however, can be stored in much larger quantities, and they are well suited for meeting chronic energy demands during periods of stress or starvation.

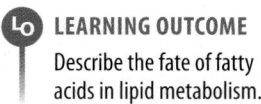

**LEARNING OUTCOME**
Describe the fate of fatty acids in lipid metabolism.

# Module 23.12

# An amino acid not needed for protein synthesis may be broken down or converted to a different amino acid

**Proteins ingested**

**1** Proteins have very complex structures, so protein digestion is both complex and time-consuming. The first step, breaking down solid food, begins in the mouth as the food is thoroughly chewed and mixed with saliva.

**2** Additional mechanical processing occurs in the stomach through churning and mixing. Exposure of the bolus to a strongly acidic environment kills pathogens and breaks down connective tissues and plant cell walls. Stomach acids also denature most proteins and disrupt tertiary and secondary protein structure, exposing peptide bonds to enzymatic action by the proteolytic enzyme **pepsin**. Chief cells secrete pepsinogen, an inactive proenzyme, but HCl in the stomach converts pepsinogen to pepsin. Protein digestion is not completed in the stomach, because time is limited and pepsin acts only on specific types of peptide bonds. However, pepsin generally has enough time to break down complex proteins into smaller peptide and polypeptide chains before the chyme enters the duodenum.

Hepatic portal vein

**3** When acidic chyme arrives in the duodenum, CCK stimulates production and release of pancreatic enzymes. These enzymes are secreted as inactive proenzymes that are then activated within the duodenum. **Enteropeptidase**, an enzyme released by the duodenal epithelium, begins the process, converting the proenzyme trypsinogen to the proteolytic enzyme **trypsin**. Trypsin then converts the other proenzymes to yield **chymotrypsin**, **carboxypeptidase**, and **elastase**. Each enzyme splits peptide bonds linking specific amino acids and ignores others. Together, they break down proteins into a mixture of dipeptides, tripeptides, and amino acids.

**? REVIEW**

**A.** Describe the role of CCK release and its effects on proteins.

**4** The epithelial surfaces of the small intestine contain several **peptidases**—enzymes that break peptide bonds—notably **dipeptidases** that break dipeptides apart and release individual amino acids. These amino acids, as well as those released by the action of pancreatic enzymes, are absorbed through both facilitated diffusion and cotransport mechanisms.

After diffusing to the basal surface of the cell, the amino acids are released into interstitial fluid by facilitated diffusion and cotransport. Once in the interstitial fluid, the amino acids enter intestinal capillaries for transport to the liver in the hepatic portal vein.

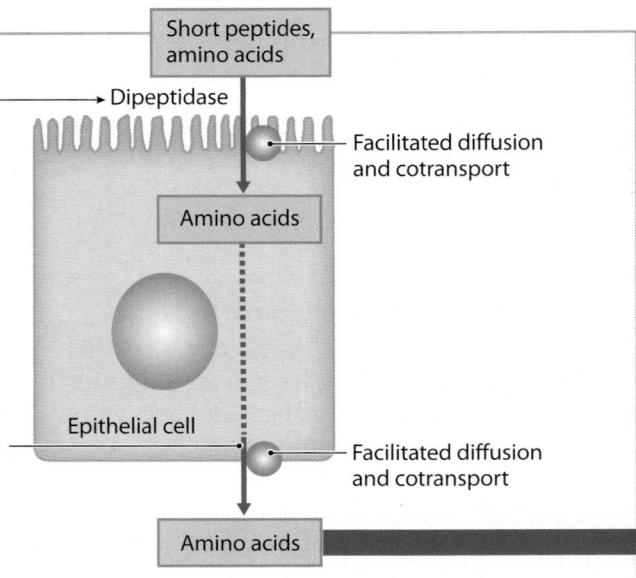

Short peptides, amino acids

Dipeptidase

Facilitated diffusion and cotransport

Amino acids

Epithelial cell

Facilitated diffusion and cotransport

Amino acids

23

**5**

The liver does not control circulating levels of amino acids as precisely as it does glucose concentrations. Blood amino acid levels normally range between 35 and 65 mg/dL, but they may become elevated after a protein-rich meal. The liver itself uses many amino acids for synthesizing plasma proteins, and it has all of the enzymes needed to synthesize, convert, or catabolize amino acids. In addition, amino acids that can be broken down to 3-carbon molecules can be used for gluconeogenesis when other sources of glucose are unavailable.

Note: Alanine aminotransferase (ALT) and aspartate aminotransferase (AST) are transaminases that leak from injured hepatocytes into the circulation. High levels in a blood test indicate liver disease or inflammation.

## Amino Acid Synthesis

Liver cells and other body cells can readily synthesize the carbon frameworks of about half of the amino acids needed to synthesize proteins. There are 10 **essential amino acids** that the body either cannot synthesize or cannot produce in amounts sufficient for growing children.

In **amination**, an ammonium ion ($NH_4^+$) is used to form an amino group that is attached to a molecule, yielding an amino acid.

$\alpha$–Ketoglutarate → Glutamic acid

In **transamination**, the amino group of one amino acid gets transferred to another molecule, yielding a different amino acid. The remaining carbon chain can then be broken down or used in other ways.

Glutamic acid + Organic acid 1 ⇌ (Transaminase) Organic acid 2 + Tyrosine

## Amino Acid Catabolism

The first step in amino acid catabolism is the removal of the amino group, leaving a carbon chain that can be converted to pyruvate, acetyl-CoA, or an intermediary in the citric acid cycle.

In **deamination**, the amino group is removed, and an ammonium ion is released.

Glutamic acid → (Deaminase) Organic acid + $NH_4^+$ Ammonium ion

Ammonium ions are highly toxic, even in low concentrations. Liver cells, the primary sites of deamination, have enzymes that use ammonium ions to synthesize **urea**, a relatively harmless water-soluble compound that is excreted in urine. The **urea cycle** is the reaction sequence responsible for the production of urea.

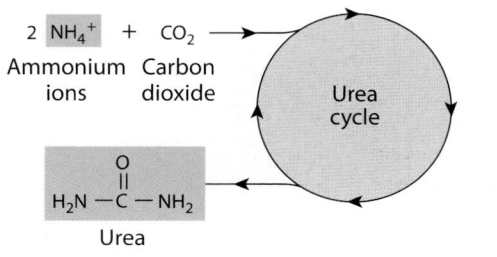

$2 \; NH_4^+ + CO_2 →$
Ammonium ions   Carbon dioxide → Urea cycle → Urea

**?** **REVIEW**

**B.** In amino acid metabolism, identify the processes by which the amino group is removed.

**C.** What happens to the ammonium ions that are removed from amino acids during deamination?

When broken down in the mitochondria, amino acids have an energy yield comparable with that of carbohydrates. However, a cell surviving by amino acid catabolism is like someone in a winter cabin burning the walls to stay warm. It can help temporarily, but it isn't a permanent solution, and in the long run it only makes matters worse.

**Lo** **LEARNING OUTCOME**

Summarize the main features of protein metabolism and the use of proteins as an energy source.

**23**

# Module 23.13

# There are two general patterns of metabolic activity: the absorptive and postabsorptive states

Metabolic activity changes from moment to moment, but there are broad patterns that follow a daily cycle. These cycles are maintained by many different hormones. The degree of response differs from tissue to tissue, but we will consider a representative cell that shares characteristics with all body cells.

**1** The **absorptive state** is the time following a meal, when nutrient absorption is under way. After a typical meal, the absorptive state continues for about 4 hours. If you are fortunate enough to eat three meals a day, you spend 12 out of every 24 hours in the absorptive state. Insulin is the primary hormone of the absorptive state, although various other hormones stimulate amino acid uptake (growth hormone) and protein synthesis (growth hormone, androgens, and estrogens).

**? REVIEW**

**A.** Define and describe the *absorptive state*.

**KEY**

↓ = Catabolic pathway   ⬆ = Anabolic pathway   ↑ = Stimulation

Insulin (binding to insulin membrane receptor recruits carrier proteins to the cell surface)

Blood glucose levels elevated

**LIPIDS**

Triglycerides

**CARBOHYDRATES**

Glycogen

Glucose ← Insulin

**PROTEINS**

Proteins

Insulin

← Insulin
← Androgens
← Estrogens
← Growth hormone

Glycolysis

ATP

Blood lipid levels elevated → Fatty acids | Glycerol ← | Amino acids ← Blood amino acids elevated

Insulin, Growth hormone

Pyruvate

**In the Absorptive State**

- Insulin stimulates (1) glucose uptake and glycogenesis, (2) amino acid uptake and protein synthesis, and (3) triglyceride synthesis.
- Androgens, estrogens, and growth hormone also stimulate protein synthesis.
- Glycolysis and aerobic metabolism provide the ATP needed to power cellular activities as well as the synthesis of lipids and proteins.

Insulin

$CO_2$

Acetyl-CoA

Citric acid cycle → Coenzymes → Electron transport chain

ATP

$O_2$ ← $O_2$

$H_2O$

**MITOCHONDRIA**

$CO_2$

23

**2** The **postabsorptive state** is the time when nutrient absorption is not under way and your body must rely on internal energy reserves to continue meeting its energy demands. You spend about 12 hours each day in the postabsorptive state, although a person who is skipping meals can extend that time considerably. Metabolic activity in the postabsorptive state is focused on the mobilization of energy reserves and the maintenance of normal blood glucose levels. These activities are coordinated by several hormones, including glucagon, epinephrine, glucocorticoids, and growth hormone.

Whe blood glucose levels are low, glucocorticoids in the liver stimulate the synthesis of glucose that is released into the circulation, for use primarily by neural tissues.

KEY

= Catabolic pathway

= Anabolic pathway

= Stimulation

Glucose released into the bloodstream by the liver

**LIPIDS** Triglycerides

**CARBOHYDRATES** Glycogen

**PROTEINS** Proteins

When blood lipid levels are low, fatty acids are released by adipocytes. These fatty acids can be absorbed and catabolized in other tissues.

Glucagon → Epinephrine →

Glucose

Glucagon, Epinephrine, Glucocorticoids, Growth hormone

When blood amino acid levels are low, skeletal muscle and other tissues release amino acids.

Glucocorticoids, Growth hormone

Fatty acids released into the bloodstream by adipocytes

Fatty acids

Glycerol

Amino acids

Amino acids released into the bloodstream

Gluconeogenesis

ATP

Glucagon

Glucocorticoids stimulate lipid catabolism in all tissues other than nervous tissue, which can metabolize only glucose. The shift in metabolic activity elsewhere preserves blood glucose for use by the nervous system.

Glucocorticoids, Growth hormone

Pyruvate

CO_2

Acetyl-CoA

ATP

Liver cells break down lipids and amino acids, and as the levels of acetyl-CoA increase, compounds called **ketone bodies** begin to form. Liver cells do not catabolize ketone bodies, which diffuse into the general circulation. Cells in peripheral tissues absorb them, convert them to acetyl-CoA, and use them as an energy source.

Citric acid cycle

Coenzymes

Electron transport chain

$O_2$ ← $O_2$

$H_2O$

**MITOCHONDRIA**

Ketone bodies

$CO_2$

**? REVIEW**
**B.** When and how do ketone bodies form?

23

**In the Postabsorptive State**

- Glucocorticoids stimulate the mobilization of lipid and protein reserves; these effects are enhanced by growth hormone.
- Glucagon stimulates glycogenolysis and gluconeogenesis, primarily in the liver. The release of glucose by the liver and the shift away from glucose metabolism by other tissues stabilizes blood glucose levels.
- Epinephrine is important in stimulating glycogenolysis in skeletal and cardiac muscle, and lipolysis in adipocytes.

**? INTEGRATION**
**C.** How do the absorptive and postabsorptive states maintain normal blood glucose levels?

**Lo LEARNING OUTCOME**
Differentiate between the absorptive and postabsorptive metabolic states, and summarize the characteristics of each.

# Module 23.14

# Vitamins are essential to the function of many metabolic pathways

The absorption of nutrients from food is called **nutrition**. **Vitamins** are organic compounds required in very small quantities but that play an essential role in specific metabolic pathways. **Minerals** are important inorganic nutrients that serve as electrolytes in body fluids. We discuss their roles in Module 25.3.

**? REVIEW**
**A.** Define *nutrition*.

## Fat-Soluble Vitamins

**1** The **fat-soluble vitamins** are vitamins A, D, E, and K. These vitamins are absorbed primarily from the digestive tract along with the lipid contents of micelles. Vegetables are potential sources of fat-soluble vitamins. However, when exposed to sunlight, your skin can synthesize small amounts of vitamin D, and intestinal bacteria produce some vitamin K.

**2** This table summarizes information concerning the fat-soluble vitamins. Because these vitamins are stored in lipid deposits throughout the body, your body contains a significant reserve of these vitamins, and normal metabolic operations can continue for several months after dietary sources have been cut off. For this reason, symptoms of **hypovitaminosis** (hī-pō-vī-ta-min-Ō-sis), or vitamin deficiency disease, rarely result from a dietary insufficiency of fat-soluble vitamins (except in the case of vitamin D, for reasons discussed in Module 5.11). Too much of a vitamin can also produce harmful effects. **Hypervitaminosis** (hī-per-vī-ta-min-Ō-sis) occurs when dietary intake exceeds the body's abilities to store, utilize, or excrete a particular vitamin. This condition most commonly involves one of the fat-soluble vitamins.

### The Fat-Soluble Vitamins

| Vitamin | Significance | Sources | Recommended Daily Allowance (RDA) in mg | Effects of Deficiency | Effects of Excess |
|---------|-------------|---------|------------------------------------------|----------------------|-------------------|
| A | Maintains epithelia; required for synthesis of visual pigments; supports immune system; promotes growth and bone remodeling | Leafy green and yellow vegetables, liver, dairy products, and fish | 0.7–0.9 | Retarded growth, night blindness, deterioration of epithelial membranes | Liver damage, skin paling, CNS effects (nausea, anorexia) |
| D | Required for normal bone growth, intestinal calcium and phosphorus absorption, and retention of these ions by the kidneys | Synthesized in skin exposed to sunlight; fortified dairy products and egg yolks | 0.005–0.015* | Rickets, skeletal deterioration | Calcium deposits in many tissues, disrupting functions |
| E | Prevents breakdown of vitamin A and fatty acids | Meat, milk, vegetables | 15 | Anemia, other problems suspected | Nausea, stomach cramps, blurred vision, fatigue |
| K | Essential for liver synthesis of prothrombin and other clotting factors | Vegetables; production by intestinal bacteria | 0.09–0.12 | Bleeding disorders | Liver dysfunction, jaundice |

## Water-Soluble Vitamins

**3** The **water-soluble vitamins** are the B vitamins and vitamin C. Most of them are components of coenzymes. The B vitamins are found in meat, eggs, and dairy products, while vitamin C is found in citrus fruits.

**4** This table summarizes information about water-soluble vitamins. These water-soluble vitamins are rapidly exchanged between the fluid compartments of the digestive tract and the circulating blood, and excessive amounts are readily excreted in urine. For this reason, hypervitaminosis involving water-soluble vitamins is unlikely unless you are taking megadoses of vitamin supplements.

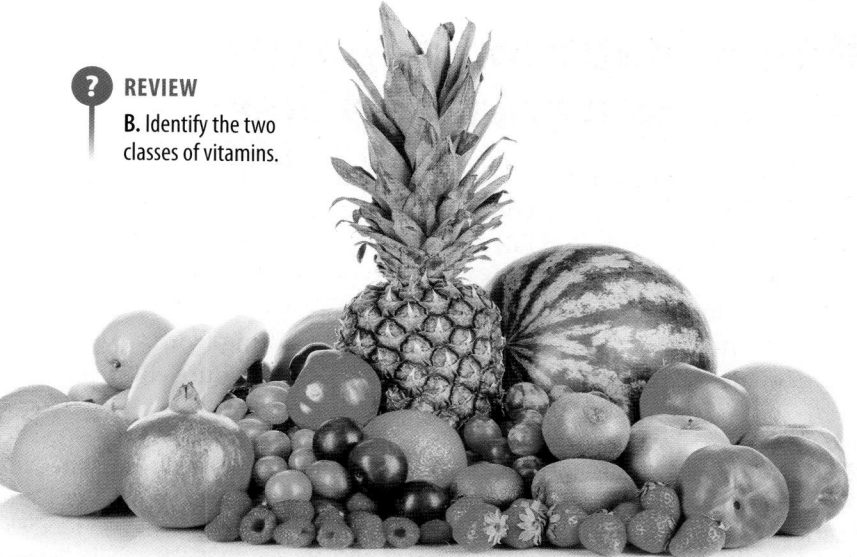

**?** REVIEW

**B.** Identify the two classes of vitamins.

### The Water-Soluble Vitamins

| Vitamin | Component or Precursor of | Sources | Recommended Daily Allowance (RDA) in mg | Effects of Deficiency | Effects of Excess |
|---|---|---|---|---|---|
| $B_1$ (thiamine) | Coenzyme in many pathways | Milk, meat, bread | 1.1–1.2 | Muscle weakness, CNS and cardiovascular problems, including heart disease; called *beriberi* | Hypotension |
| $B_2$ (riboflavin) | Part of FAD, involved in multiple pathways, including glycolysis and citric acid cycle | Milk, meat, eggs and cheese | 1.1–1.3 | Epithelial and mucosal deterioration | Itching, tingling |
| $B_3$ (niacin) | Part of NAD, involved in multiple pathways | Meat, bread, potatoes | 14–16 | CNS, GI, epithelial, and mucosal deterioration; called *pellagra* | Itching, burning; vasodilation; death after large dose |
| $B_5$ (pantothenic acid) | Coenzyme A, in multiple pathways | Milk, meat | 10 | Retarded growth, CNS disturbances | None reported |
| $B_6$ (pyridoxine) | Coenzyme in amino acid and lipid metabolism | Meat, whole grains, vegetables, orange juice, cheese and milk | 1.3–1.7 | Retarded growth, anemia, convulsions, epithelial changes | CNS alterations, perhaps fatal |
| $B_9$ (folic acid) | Coenzyme in amino acid and nucleic acid metabolism | Leafy vegetables, some fruits, liver, cereal and bread | 0.2–0.4 | Retarded growth, anemia, gastrointestinal disorders, developmental abnormalities | Few noted, except at massive doses |
| $B_{12}$ (cobalamin) | Coenzyme in nucleic acid metabolism | Milk, meat | 0.0024 | Impaired RBC production, causing pernicious anemia | Polycythemia |
| $B_7$ (biotin) | Coenzyme in many pathways | Eggs, meat, vegetables | 0.03 | Fatigue, muscular pain, nausea, dermatitis | None reported |
| C (ascorbic acid) | Coenzyme in many pathways | Citrus fruits | 75–90; smokers add 35 mg | Epithelial and mucosal deterioration; called *scurvy* | Kidney stones |

**23**

Only vitamins $B_{12}$ and C are stored in significant quantities, so insufficient intake of other water-soluble vitamins can lead to initial signs and symptoms of vitamin deficiency within a period of days to weeks. The bacteria living in your intestines help prevent deficiency diseases by producing small amounts of four of the nine water-soluble vitamins ($B_5$, $B_7$, $B_9$, and $B_{12}$). The intestinal epithelium can easily absorb all the water-soluble vitamins except $B_{12}$. The $B_{12}$ molecule is large, and it must be bound to **intrinsic factor** synthesized by the gastric mucosa before absorption can occur.

**?**  INTEGRATION

**C.** If vitamins do not provide a source of energy, what is their role in nutrition?

**LO**  LEARNING OUTCOME

Explain the role of fat-soluble vitamins and water-soluble vitamins in metabolic pathways.

# Module 23.15

# Proper nutrition depends on eating a balanced diet

A **balanced diet** contains all the ingredients needed to maintain homeostasis, including adequate substrates to produce ATP, essential amino acids and fatty acids, and vitamins. In addition, the diet must include minerals (electrolytes) and enough water to replace losses in urine, feces, and sweat. A balanced diet prevents **malnutrition**, an unhealthy state resulting from inadequate or excessive absorption of one or more nutrients.

**?** **REVIEW**
**A.** Define *balanced diet.*

**Know the Limits on Fats, Sugars, and Salt (Sodium)**

- Get most of your fat from fish, nuts, and vegetable oils, which are key sources of essential fatty acids.

- Limit solid fats like butter, shortening, and lard that are high in saturated fats and cholesterol.

- Check the Nutrition Facts label to keep saturated fats, trans fats, and sodium low.

- Choose food and beverages low in added sugars. Added sugars contribute to excessive caloric intake.

**1** How do you know if you're eating a balanced diet? Follow the updated food recommendations on the Website www.choosemyplate.gov. The United States Department of Agriculture created this diagram to offer personalized eating plans and food assessments based on the current Dietary Guidelines for Americans. The color-coded food group slices indicate the proportions of food we should consume from each of the five basic food groups: grains, vegetables, fruits, dairy, and proteins. In addition, use oils sparingly.

**Find Your Balance between Food and Physical Activity**

- Stay within your daily calorie needs.

- Adults should be physically active for at least 30 minutes most days of the week.

- About 60 minutes a day of physical activity may be needed to prevent weight gain.

- To sustain weight loss, at least 60 to 90 minutes a day of physical activity may be required.

- Children and teenagers should be physically active for 60 minutes most days.

ChooseMyPlate.gov

| GRAINS | VEGETABLES | FRUITS | OILS | DAIRY | PROTEINS |
|---|---|---|---|---|---|
| Make half your grains whole | Vary your veggies | Focus on fruits | | Get your calcium-rich foods | Go lean with proteins |

**2** In addition to considering the types of food you eat, think about how much energy that food contains relative to how much energy you use on a daily basis. We express the energy content of food in calories. One **calorie** is the energy needed to raise the temperature of 1 g of water by 1°C. We use the term **kilocalorie** (KIL-o-kal-o-re) (kcal) or Calorie (Cal) when talking about the metabolism of the entire body. One **Calorie** is the amount of energy needed to raise the temperature of 1 kilogram of water 1°C. The energy yield of the components of your diet differ; the values are 4.18 Cal/g for carbohydrates, 4.32 Cal/g for proteins, and 9.46 Cal/g for lipids. Depending on the level of activity, the average adult needs 2000 to 3000 Cal each day to maintain a stable weight. That's a lot of mitochondrial activity—1 Cal is roughly equivalent to 83 million trillion ATP.

### Energy Content of Some Food Types

| Food Type | Serving | Cal |
|---|---|---|
| Breakfast bar | 1 bar | 368 |
| Long-grain rice | 1.5 cup | 308 |
| Bread, whole wheat | 4 slices | 277 |
| Butter | 0.5 tbsp | 51 |
| Beer, regular | 12 fl oz | 160 |
| Cola, regular | 12 fl oz | 140 |

23

**3** Here is additional information about the various food groups. The food groups in themselves are less important than making intelligent choices about what (and how much) you eat. Poor choices can lead to malnutrition even if all five groups are represented.

## Five Basic Food Groups and Their Effects on Health

| Nutrient Group | Provides | Health Effects |
|---|---|---|
| **Grains** (recommended: at least half of total eaten should be whole grains) | Carbohydrates; vitamins E, thiamine, niacin, folate; calcium; phosphorus; iron; sodium; dietary fiber | Whole grains prevent rapid rise in blood glucose levels and consequent rapid rise in insulin levels |
| **Vegetables** (recommended: especially dark-green and orange vegetables) | Carbohydrates; vitamins A, C, E, folate; dietary fiber; potassium | Reduce risk of cardiovascular disease; protect against colon cancer (folate) and prostate cancer (lycopene in tomatoes) |
| **Fruits** (recommended: a variety of fruit each day) | Carbohydrates; vitamins A, C, E, folate; dietary fiber; potassium | Reduce risk of cardiovascular disease; protect against colon cancer (folate) |
| **Dairy** (recommended: low-fat or fat-free milk, yogurt, and cheese) | Complete proteins; fats; carbohydrates; calcium; potassium; magnesium; sodium; phosphorus; vitamins A, $B_{12}$, pantothenic acid, thiamine, riboflavin | Good source of calcium, which strengthens bones. Whole milk: High in calories, may cause weight gain; saturated fats associated with heart disease |
| **Meat, Fish, Eggs, and Legumes** (recommended: lean meats, fish, poultry, eggs, beans, peas, lentils) | Complete proteins; fats; calcium; potassium; phosphorus; iron; zinc; vitamins E, thiamine, $B_6$ | Fish and poultry lower risk of heart disease and colon cancer (compared to red meat). Consumption of up to one egg per day does not appear to increase incidence of heart disease; nuts and legumes improve blood cholesterol ratios, lower risk of heart disease and diabetes |

**4** Certain foods such as milk, yogurt, beef, fish, poultry, and eggs provide all the essential amino acids in sufficient quantities. They are said to contain **complete proteins**. (Recall that essential amino acids must come from the diet because they cannot be synthesized by the body. Nonessential amino acids are those the body can make on demand.)

**?** **REVIEW**

**B.** Distinguish between a complete protein and an incomplete protein.

**5** Many plants supply adequate *amounts* of protein, but they are **incomplete proteins**, which lack one or more of the essential amino acids. Vegetarians must become adept at varying their food choices to include combinations of ingredients that meet all their amino acid requirements. Even with a proper balance of amino acids, vegans, strict vegetarians who consume no animal products of any type, need to supplement with vitamin $B_{12}$ because it is obtained only from animal products or from fortified foods.

**?** **INTEGRATION**

**C.** Of these—carbohydrates, lipids, or proteins—which releases the greatest amount of energy per gram during catabolism?

**Lo** **LEARNING OUTCOME**

Explain what constitutes a balanced diet and why such a diet is important.

23

# Module 23.16
# Metabolic disorders may result from nutritional or biochemical problems

## Disorders Related to Diet and Digestion

### Eating Disorders

**Eating disorders** are psychological problems that result in either inadequate or excessive food consumption. There are many different forms; a common thread in these conditions is an obsessive concern about food and body weight.

#### Anorexia Nervosa
**Anorexia** (an-ō-REK-sē-uh) is the lack or loss of appetite. It may also accompany disorders that involve other systems. **Anorexia nervosa** is a form of self-induced starvation that appears to be the result of severe psychological problems. It is most common in adolescent Caucasian females whose weight is nearly 30 percent below normal levels. Although very obviously underweight, patients are convinced that they are too fat and refuse to eat normal amounts of food. Death rates from severe cases range from 10 to 15 percent.

#### Bulimia
In **bulimia** (bū-LĒM-ē-uh) a person goes on an "eating binge" that may involve a meal that lasts 1–2 hours and may include 20,000 or more calories. The meal is followed by induced vomiting, usually along with the use of laxatives (to promote the movement of material through the digestive tract) and diuretics (drugs that cause excessive urination). Bulimia is more common than anorexia nervosa and generally affects adolescent females. The health risks of bulimia result from (1) cumulative damage to the stomach, esophagus, oral cavity, and teeth from repeated exposure to stomach acids; (2) electrolyte imbalances resulting from the loss of sodium and potassium ions in gastric juices, diarrhea, and urine; (3) edema; and (4) cardiac arrhythmias.

 **REVIEW**
**A.** Identify and briefly define two eating disorders.

### Obesity

**Obesity** is defined as a condition of being 20 percent over ideal weight. It is at this point that serious health risks appear, such as diabetes, hypertension, and hypercholesterolemia. On that basis, the U.S. Centers for Disease Control and Prevention estimate that more than one-third of U.S. adults (35.7 percent) and about 17 percent of children aged 2–19 are obese. Basically, obese people take in more food energy than they are using. There are two major categories of obesity: regulatory obesity and metabolic obesity.

**Regulatory obesity**, the most common form, results from a failure to regulate food intake so that appetite, diet, and activity are in balance. In **metabolic obesity**, the condition is secondary to some underlying bodily malfunction that affects cell and tissue metabolism. Cases of metabolic obesity are relatively rare.

### Elevated Cholesterol Levels

Elevated cholesterol levels are associated with the development of atherosclerosis and coronary artery disease. Nutritionists currently recommend that you reduce cholesterol intake to under 300 mg per day. This amount represents a 40 percent reduction for the average American adult. In fasting people, triglycerides are usually present at levels of 40–150 mg/dL. A high total cholesterol value linked to a high LDL level spells trouble. In effect, an excessive amount of cholesterol is being exported to peripheral tissues. Problems can also exist in people with low HDL levels (below 35 mg/dL). In such cases, excess cholesterol delivered to the tissues cannot easily be returned to the liver for excretion. In either event, the amount of cholesterol in peripheral tissues—and especially in arterial walls—is likely to increase.

LDL cholesterol deposited in arterial walls, deforming the lumen of the blood vessel

# Nutritional/Metabolic Disorders

## Phenylketonuria

Several inherited metabolic disorders result from an inability to produce specific enzymes involved in amino acid metabolism. People with **phenylketonuria** (fen-il-kē-tō-NŪ-rē-uh), or **PKU**, for example, cannot convert phenylalanine to tyrosine. This reaction is an essential step in the synthesis of norepinephrine, epinephrine, dopamine, and melanin. If PKU is not detected in infancy, central nervous system development is inhibited, and severe brain damage results. The condition is common enough that a warning is printed on the packaging of products, such as diet drinks, that contain phenylalanine.

 **REVIEW**
**B.** Briefly describe phenylketonuria (PKU).

## Protein Deficiency Diseases

Regardless of the energy content of the diet, if it is deficient in essential amino acids, the person will be malnourished to some degree. In a **protein deficiency disease**, protein synthesis decreases throughout the body. As protein synthesis in the liver fails to keep pace with the breakdown of plasma proteins, plasma osmolarity falls. This reduced osmolarity results in a fluid shift as more water moves out of the capillaries and into interstitial spaces, the peritoneal cavity, or both. The longer the person remains in this state, the more severe the ascites and edema that result. It is estimated that more than 100 million children worldwide suffer from protein deficiency diseases.

**Kwashiorkor** (kwash-ē-OR-kor) occurs in children whose protein intake is inadequate, even if the caloric intake is acceptable. In each case, additional complications include damage to the developing brain. The term is from the Ghana language and its literal translation means "first-second", describing the development of the disease in an older child who had been weaned from his mother's breast when the younger sibling was born. Protein deficiency diseases are rare in the United States.

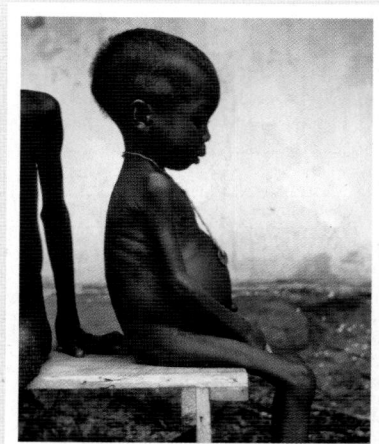
**Kwashiorkor**

## Ketoacidosis

When glucose supplies are limited, the breakdown of fatty acids and some amino acids in liver cells elevates acetyl-CoA levels and results in the production of small organic acids called **ketone bodies**. Most of these compounds diffuse out of the liver and accumulate in the bloodstream. One of the ketone bodies, acetone, has a "fruity" aroma that can be smelled in the breath of someone who has skipped one or more meals. This condition is called **ketosis** (ke-TŌ-sis), and over time ketone bodies can lower the pH of blood. This acidification of the blood is called **ketoacidosis** (kē-tō-as-i-DŌ-sis). In severe cases, the blood pH may drop below 7.05, and this may cause coma, cardiac arrhythmias, and death. A person with poorly controlled or undiagnosed diabetes mellitus is at serious risk of developing ketoacidosis because the liver responds as if the person is starving, catabolizing proteins and lipids and dumping ketone bodies into the circulation.

## Gout

When RNA is recycled as part of metabolic turnover, the purines (adenine and guanine) cannot be catabolized. Instead, they are deaminated and excreted as **uric acid**. Like urea, uric acid is a relatively nontoxic waste product, but it is far less soluble than urea. Urea and uric acid are called **nitrogenous wastes** because they are waste products that contain nitrogen atoms. Normal uric acid concentrations in plasma average 2.7–7.4 mg/dL, depending on sex and age. At concentrations over 7.4 mg/dL, body fluids become saturated with uric acid, and insoluble uric acid crystals may begin to form. The condition that then develops is called **gout**. Initially, the joints of the limbs, especially the metatarsal–phalangeal joint of the great toe, are likely to be affected. This intensely painful condition, called **gouty arthritis**, may persist for several days and then disappear for a period of days to years.

**Acute gouty arthritis of the great toe**

 **REVIEW**
**C.** Define *protein deficiency disease*, and cite an example.

 **LEARNING OUTCOME**
Describe several metabolic disorders resulting from nutritional or biochemical problems.

**23**

# Section 2 Review
## Digestion and Metabolism of Organic Nutrients

**Matching:** Match each lettered term with the most closely related description.

a. lipogenesis
b. anorexia
c. lipolysis
d. A, D, E, K
e. absorptive state
f. deamination
g. ketone bodies
h. calorie
i. uric acid
j. B complex and C
k. urea formation
l. lipoproteins
m. insulin
n. skeletal muscle

| 1 | Absorptive state hormone |
| 2 | Glycogen reserves |
| 3 | Water-soluble vitamins |
| 4 | Fat catabolism |
| 5 | Lipid synthesis |
| 6 | Amino acid catabolism |
| 7 | Fat-soluble vitamins |
| 8 | Lipid transport |
| 9 | Removal of amino group |
| 10 | Lipid breakdown |
| 11 | Gout |
| 12 | Unit of energy |
| 13 | Period following a meal |
| 14 | Lack or loss of appetite |

1 _____
2 _____
3 _____
4 _____
5 _____
6 _____
7 _____
8 _____
9 _____
10 _____
11 _____
12 _____
13 _____
14 _____

**Multiple choice:** Select the correct answer from the list provided.

**15** Intestinal absorption of nutrients occurs in the
- a) duodenum.
- b) ileocecum.
- c) ileum.
- d) jejunum.

**16** When blood glucose concentrations are elevated, the glucose molecules are
- a) catabolized for energy.
- b) used to build protein.
- c) used for tissue repair.
- d) all of these

**17** Most of the lipids absorbed by the digestive tract are immediately transferred to the
- a) liver.
- b) red blood cells.
- c) hepatocytes for storage.
- d) venous circulation by the thoracic duct.

**18** Hypervitaminosis involving water-soluble vitamins is relatively uncommon because
- a) the excess amount is stored in adipose tissue.
- b) the excess amount is readily excreted in the urine.
- c) the excess amount is stored in the bones.
- d) excess amounts are readily absorbed by skeletal muscle tissue.

## Short answer

**19** What is the difference between an essential amino acid and a non-essential amino acid?

**20** Describe four reasons why protein catabolism is an impractical source of quick energy.

**21** What is the primary difference between the absorptive and postabsorptive states?

**22** Why is the liver the focal point for metabolic regulation and control?

19 _____
_____

20 _____
_____

21 _____
_____

22 _____
_____

**Section integration:** Claudia suffers from anorexia nervosa. One afternoon she is rushed to the emergency room because of cardiac arrhythmias. Her breath smells fruity, and her blood and urine samples contain high levels of ketone bodies. Why does her breath have a fruity smell, and what would be the pH of her blood and urine?

23 _____

23

# Module 23.17

# Energetics is the study of energy changes, and thermoregulation involves heat balance

The amount of energy needed to support ongoing activities varies from moment to moment. The study of the flow of energy and its change(s) from one form to another is called **energetics**. A common benchmark in energetics studies is the **basal metabolic rate (BMR)**, the minimum resting energy expenditure of an awake, alert person.

**? REVIEW**
**A.** Define *energetics*.
**B.** What is basal metabolic rate?

**1** A direct method of determining the BMR involves monitoring respiratory activity. If we assume that average amounts and proportions of carbohydrates, lipids, and proteins are being catabolized, 4.825 Cal are expended per liter of oxygen consumed.

**2** An average person has a BMR of 70 Cal per hour, or about 1680 Cal per day. The actual energy consumption per day may be several times that amount, depending on a person's size, weight, and level of physical activity. This graph shows the approximate number of Calories expended per hour at various levels of physical exertion.

**Estimated Calories expended by a 154-lb (70-kg) person**

To maintain energy balance, food intake must be adequate to support the level of activity. This section begins with a look at the control of appetite. All of the reactions that generate ATP also generate heat; only about 40 percent of the energy released through catabolism can be used to form ATP, and the rest warms the surrounding cytoplasm. When activity levels increase, ATP production accelerates, and more heat is generated. Enzymes will function normally only over a relatively narrow range of temperatures, and metabolic pathways are at risk unless heat is lost as quickly as it is produced. So this section ends with the topic of **thermoregulation**—the homeostatic control of body temperature.

**? REVIEW**
**C.** Define *thermoregulation*.

**Lo LEARNING OUTCOME**
Explain energetics and the role of thermoregulation in maintaining homeostasis.

23

# Module 23.18
# The control of appetite is complex and involves both short-term and long-term mechanisms

The **feeding center** (related to hunger) and the **satiety center** (related to food satisfaction) are hypothalamic nuclei involved with controlling appetite. Multiple factors influence these centers, and the most important satiety factors are diagrammed here. In addition, social factors, psychological pressures, and dietary habits can play a role, although the mechanisms and pathways involved remain to be determined.

**Short-Term Regulation of Appetite**

**Stimulation of Satiety Center** ➡️

Elevated blood glucose levels depress appetite, and low blood glucose stimulates appetite. The likely mechanism is glucose entry stimulating the neurons of the satiety center.

Several hormones of the digestive tract, including CCK , suppress appetite during the absorptive state.

Stimulation of stretch receptors along the digestive tract, especially in the stomach, causes a sense of satiation and suppresses appetite.

**Stimulation of Feeding Center** ➡️

Several neurotransmitters have been linked to appetite regulation. **Neuropeptide Y (NPY)**, for example, is a hypothalamic neurotransmitter that (among other effects) stimulates the feeding center and increases appetite.

The hormone **ghrelin** (GREL-in), secreted by the gastric mucosa, stimulates appetite. Ghrelin levels are high when the stomach is empty and decline as the stomach fills. Some research suggests that leptin plays a role in regulating ghrelin, showing that these "hunger hormones" are complex and warrant further study.

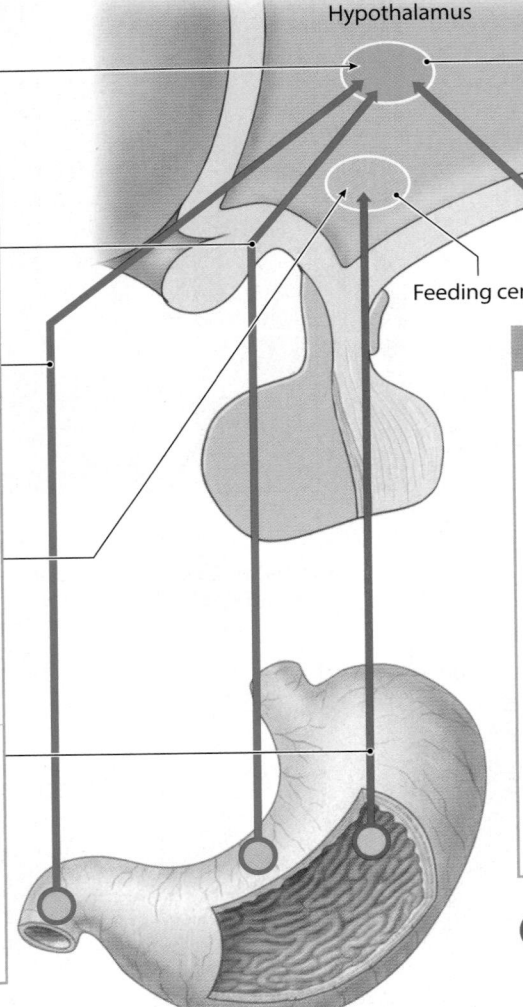

Hypothalamus

Satiety center

Feeding center

**Long-Term Regulation of Appetite**

When appetite outpaces energy usage, excess calories are stored as fat in adipose tissue. **Leptin** is a peptide hormone released by adipose tissues as they synthesize triglycerides. When leptin binds to CNS neurons that control emotions and appetite, this results in satiation and appetite suppression. Leptin levels are lower in thin people and higher in overweight people. Obese people have resistance to the appetite-suppressing effects of leptin.

**? REVIEW**
B. Describe leptin and its effect on appetite.

**? REVIEW**
A. Which hormone inhibits the satiety center and stimulates appetite in the short term?

**? INTEGRATION**
C. How might a lack of neuropeptide Y in the hypothalamus affect the control of appetite?

**Lo LEARNING OUTCOME**
Describe the roles of the satiety center and the feeding center in the regulation of food intake.

In general, activation of the satiety center causes inhibition of the feeding center, and vice versa. In addition, factors that primarily stimulate one center often have a secondary, inhibitory effect on the other.

23

# Module 23.19

# To maintain a constant body temperature, heat gain and heat loss must be in balance

Only about 40 percent of energy released by catabolism can be captured as ATP, and the rest is lost as heat that warms surrounding tissues. If body temperature is to remain constant, heat production and heat loss must be kept in balance despite wide variations in activity levels and environmental conditions.

**1** This illustration introduces the primary mechanisms of heat transfer between the body and the surrounding environment.

## Primary Mechanisms of Heat Transfer

**Radiation:** Objects warmer than the environment lose heat energy as infrared radiation. When you feel the heat from the sun, you are detecting that radiation. Your body loses heat the same way, but in proportionately smaller amounts. More than 50 percent of the heat you lose indoors is attributable to radiation; the exact amount varies with both body temperature and skin temperature.

**Evaporation:** When water evaporates, it changes from a liquid to a vapor. Evaporation absorbs energy—roughly 0.58 Cal per gram of water evaporated—and cools the surface where evaporation occurs. Each hour, 20–25 mL of water crosses epithelia and evaporates from the alveolar surfaces of the lungs and the surface of the skin. This **insensible perspiration** remains relatively constant; at rest, it accounts for about 20 percent of your body's average indoor heat loss. The sweat glands responsible for **sensible perspiration** have a tremendous scope of activity, ranging from virtual inactivity to secretory rates of 2–4 liters per hour.

**Convection:** Convection is heat loss to the cooler air that moves across the surface of your body. As your body loses heat to the air next to your skin, that air warms and rises, moving away from the surface of the skin. Cooler air replaces it, and as this air in turn becomes warmed, the pattern repeats. Convection accounts for nearly 15 percent of the body's heat loss indoors.

**Conduction:** Conduction is the direct transfer of energy through physical contact. Conduction is generally not an effective way for gaining or losing heat, and its impact depends on the temperature of the object and the amount of skin area it contacts. When you are standing, conductive losses are negligible.

**2** The physiological effects of a failure to control body temperature are indicated in the chart to the right.

**? REVIEW**

**A.** Define *insensible perspiration*.

**B.** What heat transfer process accounts for about one-half of a person's heat loss when indoors?

**C.** How is heat loss different between conduction and convection?

| Underlying physical or environmental condition | °F | °C | Thermoregulatory capabilities | Major physiological effects |
|---|---|---|---|---|
| CNS damage | 114 | 44 | Severely impaired | Death, Proteins denature |
| Heat stroke | 110 | 42 | Impaired | Convulsions, Cell damage |
| | 106 | 40 | | Disorientation |
| Disease-related fevers, Severe exercise, Active children | 102 | 38 | Effective | |
| **Normal range (oral)** | 98 | 36 | | Systems normal |
| Early mornings in cold weather | 94 | 34 | Impaired | Disorientation |
| Severe exposure | 90 | 32 | | Loss of muscle control |
| Hypothermia for open heart surgery | 86 | 30 | Severely impaired | Loss of consciousness |
| | 82 | 28 | | Cardiac arrest |
| | 78 | 26 | | |
| | 74 | 24 | Lost | Death |

**Lo LEARNING OUTCOME**

Discuss the mechanisms involved in heat gain and heat loss.

# Module 23.20

## Thermoregulatory centers in the hypothalamus adjust heat loss and heat gain

Heat loss and heat gain involve the activities of many systems. Those activities are coordinated by the **heat-loss center** and **heat-gain center** in the pre-optic area of the hypothalamus.

Pre-optic area

Heat-loss center

Heat-gain center

Radiation

Convection

**Responses to High Body Temperature Coordinated by the Heat-Loss Center**

**Behavioral changes:** A sense of discomfort leads to behavioral responses—getting into the shade, going into the water, or taking other steps that reduce body temperature.

**Vasodilation and shunting of blood to skin surface:** The inhibition of the vasomotor center causes peripheral vasodilation, and warm blood flows to the surface of the body. The skin takes on a reddish color, skin temperatures rise, and radiational and convective losses increase.

**Sweat production:** As blood flow to the skin increases, sweat glands are stimulated to increase sweating. The perspiration flows across the body surface, and evaporative heat losses accelerate. Maximal secretion, if completely evaporated, would remove 2320 Cal per hour.

**Respiratory heat loss:** The respiratory centers are stimulated, and the depth of respiration increases. Often, the person begins breathing through an open mouth rather than through the nasal passageways, increasing evaporative heat losses through the lungs.

**? REVIEW**

**A.** Predict the effect of peripheral vasodilation on a person's body temperature.

23

# When Body Temperature Falls

Adrenal gland

Radiation

Convection

Warm blood from trunk

Warm blood returns to trunk

37°C

36.5°– 37°C

Heat transfer

24°C

23°C

Cooled blood to distal capillaries

Cool blood returns to trunk

### Responses to Low Body Temperature Coordinated by the Heat-Gain Center

The heat-gain center responds to low body temperature in two ways:

#### Increased Generation of Body Heat

**Nonshivering thermogenesis** (ther-mō-JEN-e-sis) involves the release of hormones that increase the metabolic activity of all tissues. Sympathetic stimulation of the adrenal medullae releases epinephrine, which quickly increases the rate of glycogenolysis in the liver and skeletal muscle and the metabolic rate of most tissues. In children, but not usually in adults, the heat-gain center can increase the rate of thyrotropin-releasing hormone (TRH) release by the hypothalamus when body temperature is below normal. This stimulates the release of thyroid-stimulating hormone (TSH) by the anterior lobe of the pituitary gland, and as the rate of thyroid hormone release increases, so does the rate of catabolism throughout the body (see Module 16.8).

In **shivering thermogenesis**, a gradual increase in muscle tone increases the energy consumption of skeletal muscle tissue throughout your body. Both agonists and antagonists are involved, and muscle tone gradually increases to the point at which stretch receptor stimulation will produce brief, oscillatory contractions of antagonistic muscles. In other words, you begin to shiver. Shivering can elevate body temperature quite effectively, increasing the rate of heat generation by as much as 400 percent.

#### Conservation of Body Heat

The vasomotor center decreases blood flow to the dermis, thereby reducing losses by radiation and convection. The skin cools, and with blood flow restricted, it may take on a bluish or pale color. The epithelial cells are not damaged, because they can tolerate extended periods at temperatures as low as 25°C (77°F) or as high as 49°C (120°F).

The deep veins lie alongside the deep arteries, and heat is conducted from the warm blood flowing outward to the limbs to the cooler blood returning from the periphery. This arrangement traps the heat close to the body core and dramatically reduces heat loss. The transfer of heat, water, or solutes between fluids moving in opposite directions is called **countercurrent exchange**. The intermittent cycle of constriction and release when capillaries in our hands are exposed to cold temperature is called the *Lewis Wave* or the *hunter's response*. This is an autonomic response to reduce heat loss at the extremities.

23

**? REVIEW**

**B.** Describe the role of nonshivering thermogenesis in regulating body temperature.

**C.** Name the heat conservation mechanism that conducts heat from deep arteries to adjacent deep veins in the limbs.

 **LEARNING OUTCOME**

Discuss the homeostatic mechanisms that maintain a constant body temperature.

# Energetics and Thermoregulation

**Matching:** Match each lettered term with the most closely related description.

a. ghrelin
b. basal metabolic rate
c. 40 percent
d. leptin
e. insensible perspiration
f. inhibits feeding center
g. shivering thermogenesis
h. peripheral vasoconstriction
i. thermoregulation
j. sensible perspiration
k. neuropeptide Y
l. 60 percent
m. nonshivering thermogenesis
n. peripheral vasodilation

| 1 | Sweat gland activity; heat loss |
| 2 | Adipose tissue hormone |
| 3 | Homeostatic control of body temperature |
| 4 | General role of satiety center |
| 5 | Release of hormones; increased metabolism |
| 6 | Percent of catabolic energy released as heat |
| 7 | Stimulation of vasomotor center |
| 8 | Appetite-regulating neurotransmitter |
| 9 | Resting energy expenditure |
| 10 | Stomach hormone |
| 11 | Percent of catabolic energy captured as ATP |
| 12 | Inhibition of vasomotor center |
| 13 | Lung and epithelial water loss |
| 14 | Result of increased skeletal muscle tone |

1 _____
2 _____
3 _____
4 _____
5 _____
6 _____
7 _____
8 _____
9 _____
10 _____
11 _____
12 _____
13 _____
14 _____

**Multiple choice:** Select the correct answer from the list provided.

**15** A person's BMR is influenced by
- a) sex.
- b) body weight.
- c) age.
- d) all of these

**16** The processes involved in heat transfer between the body and the environment are
- a) sensible perspiration, insensible perspiration, heat loss, and heat gain.
- b) radiation, conduction, convection, and evaporation.
- c) physiological responses and behavioral modifications.
- d) sensible perspiration, insensible perspiration, hormones, and heat conservation.

**17** The primary mechanisms for increasing heat loss from the body include
- a) vasomotor and respiratory.
- b) sensible and insensible.
- c) physiological responses and behavioral modifications.
- d) acclimatization and vasomotor.

**18** All of the following are responses to an increase in body temperature *except*
- a) stimulation of the respiratory centers.
- b) stimulation of sweat glands.
- c) peripheral vasoconstriction.
- d) peripheral vasodilation.

**19** If daily intake exceeds total energy demands, the excess energy is stored primarily as
- a) triglycerides in adipose tissue.
- b) lipoproteins in the liver.
- c) glycogen in the liver.
- d) glucose in the bloodstream.

**20** All of the following factors suppress appetite *except*
- a) low blood glucose levels.
- b) high blood glucose levels.
- c) leptin.
- d) stimulation of stretch receptors along the digestive tract.

**23**

## Short answer

**21** How can a person's energy expenditure at rest be estimated by monitoring oxygen consumption?

**22** Describe the responses generated by the heat-gain center.

**23** Describe the heat-gain mechanisms involved in nonshivering thermogenesis.

21 _____
_____
22 _____
_____
23 _____
_____
_____

# Study Outline

## SECTION 1 • Introduction to Cellular Metabolism

### Module 23.1

**Metabolism is the sum of catabolic and anabolic reactions p. 893**

1. **Metabolism** is all the chemical reactions in the body. **Cellular metabolism** is the chemical reactions within cells.

2. **Catabolism** is the breakdown of organic substances; **anabolism** is the synthesis of new organic molecules.

3. *Metabolic turnover* is the process in which cells continuously replace their cellular components.

4. Organic molecules (amino acids, lipds, and simple sugars) in the blood form a **nutrient pool**.

5. In mitochondria, the catabolic reactions of aerobic metabolism release significant amounts of energy. Forty percent of this energy is captured and used to convert ADP to ATP. The remainder is lost as heat that warms the cell interior and surrounding tissues.

### Module 23.2

**Cells use nutrients from the nutrient pool for metabolism p. 894**

6. The **nutrient pool** is the source of substrates for both catabolism and anabolism.

7. Cellular activity requires ATP, so the cell's mitochondria must have a continuous supply of 2-carbon substrate molecules.

8. The nutrient pool of each cell contributes to the metabolic reserves of the body as a whole.

9. If absorption in the digestive tract does not provide adequate nutrient levels, liver cells, adipocytes, and skeletal muscles can mobilize nutrient reserves for use by other cells.

### Module 23.3

**Glycolysis is the first step in glucose metabolism p. 895**

10. **Glycolysis** is an anaerobic process that breaks down a 6-carbon glucose molecule into two 3-carbon molecules of pyruvate through a series of enzymatic steps.

11. There is a net gain of 2 ATP molecules for each glucose molecule converted to 2 molecules of pyruvate. In addition, NAD accepts hydrogen atoms that can be transferred to mitochondria.

### Module 23.4

**The citric acid cycle transfers hydrogen atoms to coenzymes p. 896**

12. Mitochondria provide almost all of the energy that supports cellular operations in the form of ATP by absorbing specific organic molecules, removing the hydrogen atoms from those molecules, and transferring them to **coenzymes**.

13. Organic nutrients from the nutrient pool are broken down into smaller 3-carbon and 2-carbon molecules that the mitochondria can process.

14. The process begins when a pyruvate enters the mitochondrial matrix and produces a 2-carbon **acetate ion**, 1 NADH, and 1 molecule of $CO_2$.

15. Acetate attaches to coenzyme A to form **acetyl-CoA**. The acetyl group from the acetyl-CoA attaches to a 4-carbon molecule in the mitochondrial matrix and produces a 6-carbon molecule of **citric acid**. This is the first step of the **citric acid cycle**.

16. The citric acid cycle removes H atoms containing high-energy electrons and transfers them to coenzymes for further processing.

17. **NAD** and **FAD** are coenzymes that play a key role in the citric acid cycle by removing H atoms and delivering them to the electron transport chain.

18. The energy gain from each acetyl group entering the citric acid cycle is 1 ATP molecule.

### Module 23.5

**The electron transport chain establishes a proton gradient used to make ATP p. 898**

19. **Oxidative phosphorylation** is the generation of ATP as the result of the transfer of electrons from the coenzymes NADH and $FADH_2$ to oxygen by a sequence of electron carriers within mitochondria. This process produces 90 percent of the ATP used by cells. The key reactions take place in the **electron transport chain (ETC)**.

20. Coenzymes deliver hydrogen atoms from the citric acid cycle to a sequence of molecules called **cytochromes**. The first cytochrome passes electrons to the second, which passes them down the ETC by a series of *oxidation-reduction (redox) reactions*. This passage gradually releases the energy carried by the electrons.

21. Along the way, hydrogen ions are pumped into the intermembrane space. The diffusion of hydrogen ions back into the matrix powers the production of ATP by the enzyme **ATP synthase**.

22. The final electron acceptor is oxygen. After accepting 2 electrons, an oxygen ion then reacts with a pair of hydrogen ions in the matrix to form a water molecule.

23. Aerobic metabolism is dependent on the availability of oxygen. If oxygen is severely restricted, the cytochromes cannot pass on their electrons, and the ETC stops working.

### Module 23.6

**Glucose catabolism yields 30–32 ATP p. 900**

24. Catabolism of a single glucose molecule yields 2 ATP through glycolysis and 28–30 ATP through aerobic metabolism.

### Module 23.7

**Nutrient metabolism follows several pathways p. 901**

25. Neither the diet nor the nutrient pool provides everything needed to build every protein, carbohydrate, and lipid a cell might require. Cells can synthesize what they need from available substrates.

## SECTION 2 • Digestion and Metabolism of Organic Nutrients

### Module 23.8

**Digestion involves a series of steps to make nutrients available to the body p. 903**

26. During digestion, a combination of chemical and enzymatic action breaks the complex organic compounds of food into simpler components that can be absorbed by the digestive tract and then distributed by the bloodstream.

27. Cells throughout the body rely on these organic molecules for energy production and to replenish the intracellular nutrient pool.

## Module 23.9

### Carbohydrates are usually the preferred substrates for catabolism and ATP production under resting conditions  p. 904

28. **Salivary amylase** in saliva is an enzyme that breaks down complex carbohydrates into disaccharides and trisaccharides. It remains active until pH falls below 4.5 in the stomach.

29. In the duodenum, buffers shift pH to alkaline. **Pancreatic alpha-amylase** now functions as salivary amylase did. Chyme with carbohydrates causes the release of gastric inhibitory peptide (GIP), which causes the release of insulin.

30. Epithelial cells of the jejunum finish the digestion of carbohydrates. The plasma membrane of the brush border contains the enzymes **maltase**, **sucrase**, and **lactase**, which break down the disaccharides into simple sugars that are absorbed.

31. Indigestible carbohydrates such as cellulose provide a nutrient source for bacteria in the colon. These bacteria produce intestinal gas, or **flatus**.

32. Simple sugars are absorbed in the liver and converted to glucose. Excess glucose is stored as glycogen in the liver and skeletal muscles.

33. Glucose is transported into cells by a **carrier protein** stimulated by insulin. Inside the cell, glucose is used to build structural materials or nucleic acids. If energy is needed, it is broken down into two molecules of **pyruvate** in an anaerobic process called glycolysis.

34. Glycolysis yields 2 ATP for every glucose molecule broken down. Each pyruvate molecule can be used by the mitochondria after conversion to acetyl-CoA.

35. For each molecule of pyruvate processed by mitochondria, the cell gains 15 ATP, consumes 3 molecules of $O_2$, and generates 3 molecules of $CO_2$ and 6 molecules of water. For each pair of pyruvate molecules catabolized, the cell gains 30 ATP.

## Module 23.10

### Lipids reach the bloodstream in chylomicrons; the cholesterol is then extracted and released as lipoproteins  p. 906

36. **Lingual lipase** in the saliva breaks down triglycerides into monoglycerides and fatty acids. Large drops containing a variety of lipids form in the stomach.

37. **Bile salts** in bile break the large drops of lipids into tiny droplets in a process called **emulsification**. This allows **pancreatic lipase** better access to break apart the triglycerides to a mixture of fatty acids, monoglycerides, and glycerol. These molecules form small lipid–bile salt complexes called **micelles**.

38. The lipids within the micelle diffuse across the intestinal epithelium. The intestinal cells synthesize triglycerides and other lipids that are coated with proteins in a complex known as **chylomicrons**, which are **lipoproteins**.

39. Most of the chylomicrons diffuse into the intestinal lacteals and proceed along the lymphatic vessels, the thoracic duct, the left subclavian vein, the pulmonary circuit, and then into the systemic circuit.

40. The liver absorbs chylomicrons and creates **low-density lipoproteins (LDLs)** and **very low density lipoproteins (VLDLs)**. VLDLs transport triglycerides from the liver to muscle and adipose tissue. The LDLs enter the bloodstream and are delivered to peripheral tissues.

41. The cholesterol not used by the cell re-enters the bloodstream. **High-density lipoproteins (HDLs)** absorb excess cholesterol and transport it back to the liver for storage or excretion in the bile.

## Module 23.11

### Fatty acids can be broken down to provide energy or converted to other lipids  p. 908

42. During lipid catabolism, or **lipolysis**, lipids are broken down into pieces that can be either converted to pyruvate or channeled directly into the citric acid cycle in a process called **beta-oxidation**.

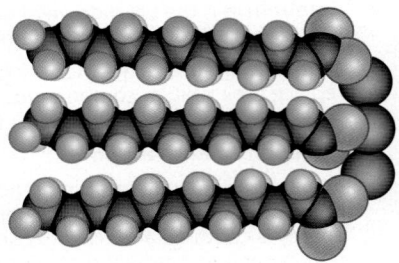

*The principles of triglyceride catabolism apply to other lipids as well.*

43. The synthesis of lipids, known as **lipogenesis**, begins with acetyl-CoA and results in fatty acids that can be used to synthesize other structural and functional lipids.

## Module 23.12

### An amino acid not needed for protein synthesis may be broken down or converted to a different amino acid  p. 910

44. Protein digestion begins when food is chewed and mixed with saliva and continues through the churning and mixing in the stomach. Stomach acids also denature most proteins, exposing them to enzymatic action by **pepsin**.

45. A variety of enzymes act on specific peptide bonds to break down proteins into a mixture of dipeptides, tripeptides, and amino acids.

46. The epithelial surfaces of the small intestine contain **peptidases** that break dipeptides apart and release individual amino acids, which are absorbed into the cell. Amino acids are then released into interstitial fluid, intestinal capillaries, hepatic portal vein, and then the liver.

47. The liver uses amino acids for synthesizing plasma proteins, and it has all of the enzymes needed to synthesize, convert, or catabolize amino acids. **Deamination** releases toxic ammonium ions, which liver cells use to synthesize relatively harmless **urea** that is excreted in urine.

## Module 23.13

### There are two general patterns of metabolic activity: the absorptive and postabsorptive states  p. 912

48. The **absorptive state** is the time following a meal, when nutrient absorption is under way. After a typical meal, the absorptive state continues for about 4 hours.

49. In the absorptive state, insulin stimulates glucose uptake and glycogenesis, protein synthesis is stimulated, and glycolysis and aerobic metabolism provide the ATP needed for cellular activities.

50. The **postabsorptive state** is the time when nutrient absorption is not under way and your body must rely on internal energy reserves to continue meeting its energy demands.

51. In the postabsorptive state, lipid and protein reserves are mobilized, blood glucose levels are stabilized by the release of glucose by the liver, skeletal and cardiac muscle stimulate glycogenolysis, and adipocytes stimulate lipolysis.

## Module 23.14

### Vitamins are essential to the function of many metabolic pathways    p. 914

52. **Nutrition** is the absorption of nutrients from food.

53. **Vitamins** are organic compounds required in very small quantities that play an essential role in specific metabolic pathways. The **fat-soluble vitamins** are vitamins A, D$_3$, E, and K. They are absorbed from the digestive tract along with the lipid contents of micelles. Vegetables are potential sources of fat-soluble vitamins.

54. The **water-soluble vitamins** are the B vitamins and vitamin C. The B vitamins are found in meat, eggs, and dairy products; vitamin C is found in citrus fruits.

## Module 23.15

### Proper nutrition depends on eating a balanced diet    p. 916

55. A **balanced diet** contains everything needed to maintain homeostasis, including adequate substrates to produce ATP, essential amino acids and fatty acids, vitamins, electrolytes, and water. A balanced diet helps prevent **malnutrition**.

56. In addition to considering the types of food we eat, we should consider the energy that food contains. The energy content of food is expressed in **calories**.

57. One calorie is the energy needed to raise the temperature of 1 g of water by 1°C. We use the term **kilocalorie** (kcal) or **Calorie** (Cal) when talking about the metabolism of the entire body. One Calorie is the amount of energy needed to raise the temperature of 1 kilogram of water 1°C.

58. The energy yield is different for components of the diet. The values are 4.18 Cal/g for carbohydrates, 4.32 Cal/g for proteins, and 9.46 Cal/g for lipids. Depending on activity level, the average adult needs between 2000 and 3000 Cal each day.

59. The five basic food groups are grains, vegetables, fruits, dairy, and meat and beans.

60. Some members of the dairy and meat and beans groups provide all the essential amino acids and are said to be **complete proteins**. Many plants supply adequate amounts of protein, but they are **incomplete proteins** because they lack one or more of the essential amino acids.

## Module 23.16

### Metabolic disorders may result from nutritional or biochemical problems    p. 918

61. **Eating disorders** are psychological problems that result in either inadequate or excessive food consumption. These include self-induced starvation called **anorexia nervosa** and binge eating and purging, called **bulimia**.

62. **Obesity** is defined as being 20 percent over ideal weight. Elevated cholesterol levels are associated with atherosclerosis and coronary artery disease.

63. **Phenylketonuria** is an inherited inability to produce the enzymes involved in amino acid metabolism and can result in brain damage if not detected in infancy.

64. **Kwashiorkor** is a **protein deficiency disease** in children whose protein intake is inadequate.

65. When glucose is limited, the breakdown of fatty acids and some amino acids results in the production of **ketone bodies**. **Ketosis** is an accumulation of ketone bodies in the bloodstream; it can often be detected on the breath because it has a fruity aroma. This condition can lower blood pH in a condition called **ketoacidosis**. This can occur in poorly controlled or undiagnosed diabetes mellitus and in cases of starvation.

66. An accumulation of a **nitrogenous waste**, **uric acid**, can cause the formation of uric acid crystals. This condition is called **gout** and can lead to **gouty arthritis** in the joints of the limbs.

## SECTION 3 • Energetics and Thermoregulation

## Module 23.17

### Energetics is the study of energy changes, and thermoregulation involves heat balance    p. 921

67. The study of the flow of energy and its change(s) from one form to another is called **energetics**. A common benchmark in energetics studies is the **basal metabolic rate (BMR)**, the minimum resting energy expenditure of an awake, alert person.

68. BMR may be determined by measuring respiratory activity and assuming that 4.825 Calories are expended per liter of oxygen consumed. An average person has a BMR of 70 Cal per hour, or about 1680 Cal per day.

69. All of the reactions that generate ATP also generate heat; only about 40 percent of the energy released through catabolism can be used to form ATP, and the rest warms the surrounding cytoplasm.

70. The homeostatic control of body temperature is **thermoregulation**.

## Module 23.18

### The control of appetite is complex and involves both short-term and long-term mechanisms    p. 922

71. The **feeding center** (involved with hunger) and the **satiety center** (involved with food satisfaction) are hypothalamic nuclei involved with the control of appetite.

72. Elevated blood glucose levels depress appetite, and low blood glucose stimulates appetite. Hormones of the digestive tract and stimulation of stretch receptors in the digestive tract also suppress appetite.

73. The feeding center is stimulated by **neuropeptide Y (NPY)** and **ghrelin**.

74. Excess calories are stored as fat in adipocytes. **Leptin**, a hormone released by adipose tissues as they synthesize triglycerides, suppresses appetite.

23

### Module 23.19

**To maintain a constant body temperature, heat gain and heat loss must be in balance    p. 923**

75. If body temperature is to remain constant, heat production and heat loss must be kept in balance despite wide variations in activity levels and environmental conditions.

76. The primary mechanisms for heat transfer are **radiation**, **evaporation**, **convection**, and **conduction**.

### Module 23.20

**Thermoregulatory centers in the hypothalamus adjust heat loss and heat gain    p. 924**

77. Heat loss and heat gain are regulated by the **heat-loss center** and the **heat-gain center**, respectively, in the pre-optic area of the hypothalamus.

78. Responses to high body temperature are coordinated by the heat-loss center. These responses include **behavioral change, vasodilation and shunting of blood to the skin surface, sweat production,** and **respiratory heat loss**.

79. Responses to low body temperature are coordinated by the heat-gain center. These responses include **nonshivering thermogenesis**, **shivering thermogenesis**, and the **conservation of body heat** using the **countercurrent exchange** mechanism as blood flow decreases to the surface.

## Chapter Review Questions

**True/False:** Indicate whether each statement is true or false.

**1** The chemical reactions within cells are collectively known as cellular anabolism.

**2** The breakdown of glucose into 2 3-carbon molecules is called glycolysis.

**3** The energy yield for carbohydrates is 4.18 Cal/g.

**4** Kwashiorkor occurs in adults when uric acid crystals form in body fluids, causing a specific type of arthritis.

**5** Leptin is a hormone secreted by the gastric mucosa that stimulates appetite.

**6** The average person has a BMR of 70 Cal/hour, or about 1680 Cal/day.

1 _____

2 _____

3 _____

4 _____

5 _____

6 _____

**Multiple choice:** Select the correct answer from the list provided.

**7** Catabolism refers to
- a) the creation of a nutrient pool.
- b) the sum total of all chemical reactions in the body.
- c) the production of organic compounds.
- d) the breakdown of organic substrates.

**8** The breakdown of glucose to pyruvate is
- a) anaerobic.
- b) aerobic.
- c) lipogenic.
- d) anabolic.

**9** The process that produces 90 percent of the ATP used by our cells is
- a) glycolysis.
- b) the citric acid cycle.
- c) oxidative phosphorylation.
- d) thermoregulation.

**10** Glucose synthesis from smaller carbon chains is known as
- a) glycolysis.
- b) gluconeogenesis.
- c) glycogenesis.
- d) glycogenolysis.

**11** The net result of one turn of the citric acid cycle is
- a) 1 ATP.
- b) 2 ATP.
- c) 15 ATP.
- d) 30 ATP.

**12** Which of the following is a protein deficiency disease?
- a) phenylketonuria
- b) kwashiorkor
- c) ketoacidosis
- d) gouty arthritis

**13** The enzyme that breaks down complex carbohydrates into a mixture of disaccharides and trisaccharides in the mouth is
- a) pepsin.
- b) lingual lipase.
- c) sucrase.
- d) salivary amylase.

**14** Which of the following is a water-soluble vitamin?
- a) vitamin A
- b) vitamin C
- c) vitamin D
- d) vitamin K

**15** The hormone released by adipose tissues as they synthesize triglycerides is

- ☐ a) leptin.
- ☐ b) ghrelin.
- ☐ c) neuropeptide Y.
- ☐ d) insulin.

**16** More than 50 percent of the heat you lose indoors is attributable to

- ☐ a) radiation.
- ☐ b) sensible perspiration.
- ☐ c) convection.
- ☐ d) conduction.

## Short answer

**17** Explain how catabolism and anabolism are "linked" by ATP.

**18** What level of total cholesterol is considered elevated, and what are the health risks of elevated cholesterol?

**19** What are the risks of too much vitamin intake, and which vitamins are most commonly involved?

**20** Distinguish between a calorie, a kilocalorie, and a Calorie.

---

## MasteringA&P™

Access more chapter study tools online in the MasteringA&P Study Area:

- Chapter Quizzes, Chapter Practice Test, and Clinical Case Studies

- Practice Anatomy Lab™ 3.0 — PAL 3.0
- Interactive Physiology™ — iP2™
- A&P Flix™ — A&P*Flix*™
- PhysioEx™ 9.1 — PhysioEx 9.1

---

# Chapter Integration • Applying what you have learned

## Finding balance in foods

Diet has a profound influence on a person's general health. Too many or too few nutrients, hypervitaminosis (too many vitamins) or hypovitaminosis (too few vitamins), and above-normal or below-normal concentrations of minerals (inorganic nutrients) can adversely affect health. Subtle, long-term problems can occur when the diet includes the wrong proportions or combinations of nutrients. The average diet in the United States contains too much sodium, too much sugar, and too many calories in general, and lipids (particularly saturated fats) provide too great a proportion of those calories. Poor diets increase the incidence of obesity, heart disease, atherosclerosis, hypertension, and diabetes in the U.S. population.

Mrs. Henderson, who has a family history of cardiovascular disease, is concerned about her health. A comprehensive blood test reveals elevated levels of LDL (so-called "bad cholesterol") and decreased levels of HDL (so-called "good cholesterol"), normal glucose, and adequate vitamin levels except vitamin $B_6$. The physician orders a consultation with the registered dietitian, who discovers that Mrs. Henderson's semi-vegetarian diet is high in carbohydrates and fats.

**1** Why are vitamins and minerals essential components of the diet?

**2** Explain the labeling of HDL as "good cholesterol" and LDL as "bad cholesterol."

**3** Why are HDLs considered beneficial?

**4** What process in the liver increases after a high-carbohydrate meal?

**5** What is the significance of a pyridoxine (vitamin $B_6$) deficiency?

23

**LEARNING OUTCOMES**

These Learning Outcomes correspond by number to this chapter's modules and indicate what you should be able to do after completing the chapter. They also appear at the end of each module.

# The Urinary System

## Module 24.1
# The urinary system organs are the kidneys, ureters, urinary bladder, and urethra

In this section, we consider the functional anatomy of the structures of the urinary system. Later sections consider how the kidneys remove metabolic wastes from the bloodstream to produce urine, how the process is regulated, and how urine is eliminated through the **urinary tract**, the passages from kidneys to the ureters, urinary bladder, and urethra.

**1** The **urinary system** eliminates excess water, salts, and metabolic wastes through the production of urine.

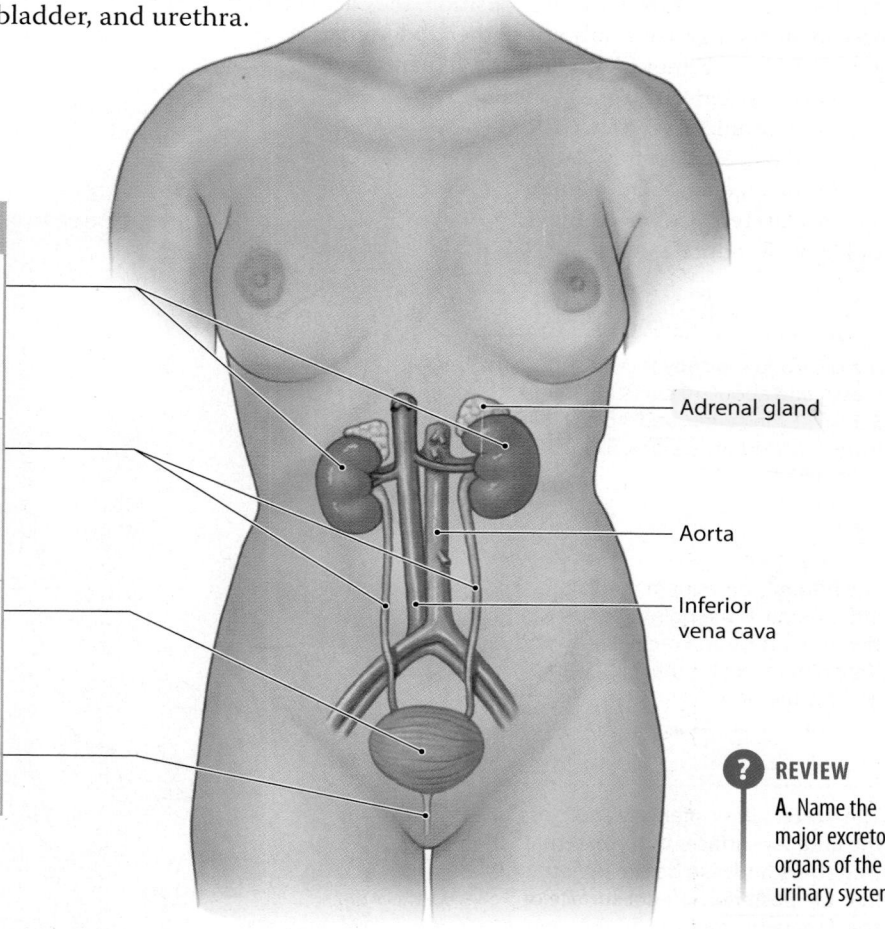

Adrenal gland

Aorta

Inferior vena cava

**Major Organs of the Urinary System**

The **kidneys** are metabolically very active. At rest they receive about 25 percent of the cardiac output. The two kidneys are the major excretory organs of the urinary system. They produce **urine**, a fluid containing water, ions, and small soluble substances.

The **ureters** receive the urine from the kidneys and conduct it to the urinary bladder. Urine movement involves a combination of gravity and the peristaltic contractions of smooth muscle in the walls of the ureters.

The **urinary bladder** receives and stores urine prior to its elimination from the body. Urine elimination, or **urination**, is driven by the contraction of smooth muscle layers in the walls of the urinary bladder.

The **urethra** is a passageway that conducts urine from the urinary bladder to the exterior.

**?** REVIEW

**A.** Name the major excretory organs of the urinary system.

**2** The urinary system removes wastes generated by cells throughout the body, but it has several other essential homeostatic functions. These functions are summarized in the table below.

**?** REVIEW

**B.** Describe the functions of the urinary system.

**Functions of the Urinary System**

- Adjusting blood volume and blood pressure
- Regulating blood plasma concentrations of sodium, potassium, chloride, and other ions
- Stabilizing blood pH
- Conserving valuable nutrients by preventing their loss in urine
- Removing drugs and toxins from the bloodstream

24

**L₀** LEARNING OUTCOME

Identify the organs of the urinary system, and cite a primary function of each.

# Module 24.2
# The kidneys are paired retroperitoneal organs

In this module, you will learn the location of the kidneys and their placement in relation to the axial skeleton and the organs of the abdominopelvic cavity. You will also see how the kidneys are supported, protected, and stabilized in position.

**1** The kidneys, ureters, urinary bladder, and the associated blood vessels are shown in this anterior view. Because the kidneys are in a retroperitoneal position, they are clearly visible in an anterior view only after other abdominal organs have been removed. To help you remember the organs positioned retroperitoneally, use the mnemonic **SAD PUCKER** for **S**uprarenal (adrenal) glands, **A**orta and inferior vena cava, **D**uodenum, **P**ancreas, **U**reters, **C**olon, **K**idneys, **E**sophagus, **R**ectum.

A typical adult **kidney** is reddish brown and about 10 cm (4 in.) long, 5.5 cm (2.2 in.) wide, and 3 cm (1.2 in.) thick. Each kidney weighs about 150 g (5.25 oz).

The **hilum**, a prominent medial indentation, is the point of entry for the renal artery and renal nerves and the point of exit for the renal vein and the ureter.

The **ureters** pass inferiorly and cross the anterior surfaces of the external iliac artery and vein before penetrating into the posterior, inferior surface of the urinary bladder.

This is the cut edge of the posterior peritoneum, which has been removed. The kidneys, adrenal glands, and ureters lie between the muscles of the posterior body wall and the parietal peritoneum, in a **retroperitoneal** position.

Right adrenal gland · Vena cava · Aorta · Esophagus (cut) · Left adrenal gland · Diaphragm · Right kidney · Left kidney · Rectum · Urinary bladder

 **REVIEW**

**A.** What structures enter and exit the kidney at the hilum?

24

**2** In this posterior view, you can see that the kidneys are located on each side of the vertebral column, between vertebrae $T_{12}$ and $L_3$. In this position the kidneys are protected by the visceral organs (anteriorly), the musculature of the body wall, and the 11th and 12th ribs (posteriorly and laterally). The left kidney lies slightly superior to the right kidney.

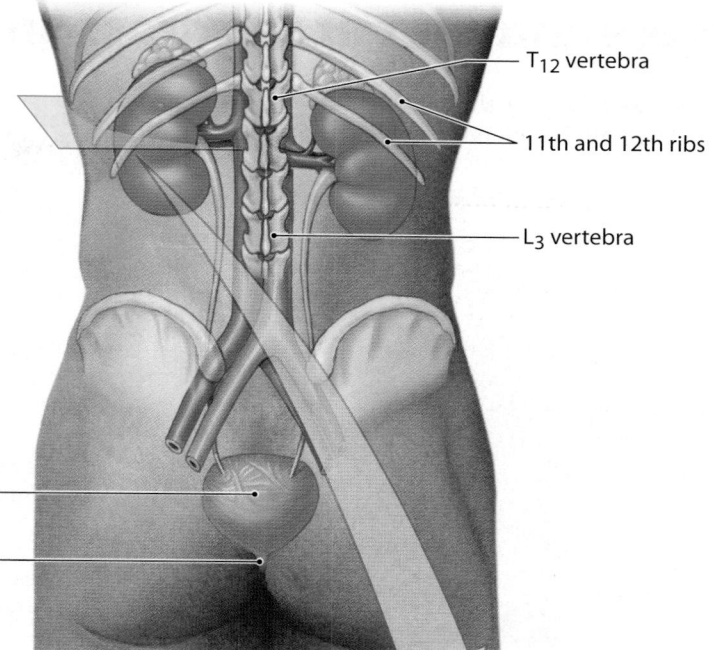

$T_{12}$ vertebra

11th and 12th ribs

$L_3$ vertebra

Urinary bladder

Urethra

Go to
MasteringA&P™
PAL™3.0
Urinary System

Anteriorly, the renal fascia forms a thick layer that fuses with the peritoneum.

**3** The position of the kidneys in the abdominal cavity is maintained by the overlying peritoneum, contact with adjacent visceral organs, and supporting connective tissues. As seen in this sectional view, three concentric layers of connective tissue protect and stabilize each kidney.

Stomach

Pancreas

Aorta

Vena cava

Spleen

Left kidney

## Connective Tissue Layers Supporting the Kidneys

### Fibrous Capsule

The **fibrous capsule** is a layer of collagen fibers that covers the outer surface of the entire organ.

### Perinephric Fat

The **perinephric fat** (perirenal fat capsule) is a cushioning thick layer of adipose tissue that surrounds the fibrous capsule.

### Renal Fascia

The **renal fascia** is a dense, fibrous outer layer that anchors the kidney to surrounding structures. Collagen fibers extend outward from the fibrous capsule through the perinephric fat to this layer. Posteriorly, it fuses with the deep fascia surrounding the muscles of the body wall. Anteriorly, it forms a thick layer that fuses with the peritoneum.

 **REVIEW**

**B.** Describe the concentric layers of connective tissue that protect and anchor the kidney.

**?** **INTEGRATION**

**C.** What would happen to a kidney's position if the perinephric fat layer were depleted and the collagen fibers of the fibrous capsule were to become detached?

**Lo** **LEARNING OUTCOME**

Describe the location and structural features of the kidneys.

24

# Module 24.3
# The kidneys are complex at the gross and microscopic levels

**1** This illustration of a sectioned kidney shows the major landmarks and features. The blood vessels servicing the kidney enter through the **hilum**.

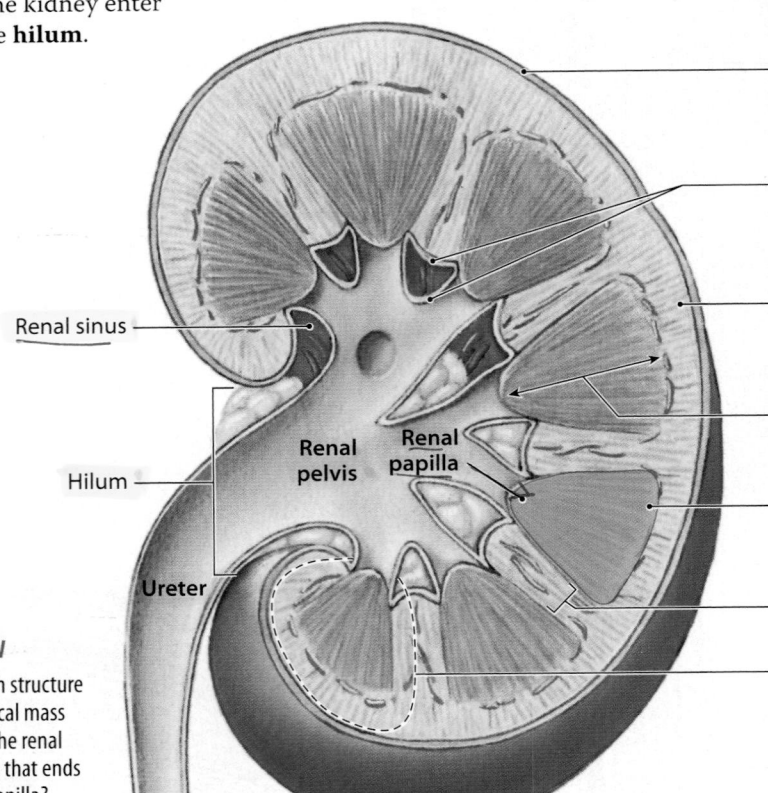

Renal sinus

Hilum

Renal pelvis

Renal papilla

Ureter

### Major Structural Landmarks of the Kidney

The ~~Renal~~ **fibrous capsule** covering the outer surface of the kidney also lines the **renal sinus**, an internal cavity within the kidney. The outer and inner linings are continuous at the hilum.

Within the renal sinus, the fibrous capsule stabilizes the positions of the ureter, the renal blood vessels, and renal nerves.

The **renal cortex** is the superficial region of the kidney, in contact with the fibrous capsule. The cortex is reddish brown and granular.

The **renal medulla** extends from the renal cortex to the renal sinus. It contains 6 to 18 renal pyramids.

A **renal pyramid** is a conical structure extending from the cortex to a tip called the **renal papilla**.

A **renal column** is a band of granular tissue that separates adjacent pyramids.

A **kidney lobe** consists of a renal pyramid, the overlying area of renal cortex, and adjacent tissues of the renal columns. Each kidney contains 6–18 kidney lobes. Urine is produced in the kidney lobes.

**? REVIEW**

**A.** Which structure is a conical mass within the renal medulla that ends at the papilla?

---

**2** Here is a frontal section of a human kidney. The labels follow the path of urine within the collecting system (discussed later) of the renal sinus, which communicates with the ureter at the hilum.

**? REVIEW**

**B.** Compare the minor and major calyces.

A **minor calyx** collects the urine produced by a single kidney lobe.

A **major calyx** forms through the fusion of 4–5 minor calyces (KÃ-li-sēz).

Hilum

The **renal pelvis** is a large, funnel-shaped chamber that collects urine from the major calyces. It fills most of the renal sinus and is connected to the ureter, which drains the kidney.

Ureter

**3** **Nephrons** are microscopic functional units of the kidney. Nephrons from different locations differ slightly in structure. Approximately 85 percent of all nephrons are **cortical nephrons**, located almost entirely within the superficial renal cortex. The remaining 15 percent of nephrons, termed **juxtamedullary** (juks-tuh-MED-yū-lar-ē; *juxta*, near) **nephrons**, have long **nephron loops** that extend deep into the renal medulla.

Renal cortex

Cortical nephrons are the most numerous nephrons. They are responsible for most of the regulatory functions of the kidneys.

Nephron loop of cortical nephron

Juxtamedullary nephrons form a small percentage of the total number of nephrons. They play a crucial role in establishing conditions in the renal medulla that are essential to water conservation and the production of concentrated urine.

Renal medulla

Nephron loop of juxtamedullary nephron

Renal papilla

Minor calyx

**? REVIEW**

**C.** Which type of nephron is essential for water conservation and concentrated urine production?

**Lo LEARNING OUTCOME**

Describe the gross structural features of the kidney, and distinguish between cortical and juxtamedullary nephrons.

24

# Module 24.4

# A nephron is divided into segments; each segment has specific functions

**1** The nephron consists of a renal corpuscle and a renal tubule. At the renal corpuscle, blood pressure forces water and dissolved solutes out of the glomerular capillaries and into a chamber—the **capsular space**—that is continuous with the lumen of the renal tubule. The **renal tubule** is a tubular passageway that may be 50 mm (1.97 in.) long. Filtration produces an essentially protein-free solution, known as a **filtrate**, that is similar to blood plasma. After modification by the renal tubule and collecting system, the filtrate leaves the kidneys as urine.

**? REVIEW**
**A.** Describe filtrate.

## Nephron

**2 Proximal Convoluted Tubule**

The **proximal convoluted tubule (PCT)** is the initial segment of the renal tubule. It reabsorbs nutrients from the filtrate, now known as **tubular fluid**.

**4 Distal Convoluted Tubule**

The **distal convoluted tubule (DCT)** is the last segment of the renal tubule. It makes further adjustments in the composition of the tubular fluid through secretion and reabsorption.

Renal tubule

**Start**

**1 Renal Corpuscle**

The **renal corpuscle** (KOR-pus-ul) is a spherical structure consisting of the **glomerular capsule** (*Bowman's capsule*), a cup-shaped chamber approximately 200 μm in diameter, and a capillary network known as the **glomerulus** (glo-MER-yū-lus).

Efferent arteriole

Afferent arteriole

Bowman's
Glomerular capsule

Capsular space

Glomerulus

Descending limb of loop begins

Ascending limb of loop ends

Thick ascending limb

**? REVIEW**
**B.** Identify the structures of the renal corpuscle.

**3 Nephron Loop**

The **nephron loop** is the middle segment of the renal tubule. It establishes an osmotic gradient in the renal medulla that promotes water reabsorption from tubular fluid in the nephron and collecting system. Each limb contains a thin segment and a thick segment. The terms *thin* and *thick* refer to the height of the epithelium. Thin limbs are lined with squamous epithelium. Thick limbs have cuboidal epithelium.

Descending thin limb

Ascending thin limb

**KEY**

⇨ Filtrate

⇨ Water reabsorption

⇢ Variable water reabsorption

⇨ Solute reabsorption or secretion

⇢ Variable solute reabsorption or secretion

Thin limb          Thick limb

**2** Each nephron empties into the **collecting system**, a series of tubes that carry tubular fluid away from the nephron. Collecting ducts receive this fluid from many nephrons. Each collecting duct begins in the renal cortex and descends into the renal medulla, carrying fluid to a papillary duct that drains into a minor calyx.

## Collecting System

The collecting system receives the urine from individual nephrons and makes final adjustments in urine volume and composition before delivering it to a minor calyx.

**5 Collecting Duct**

A **collecting duct** carries tubular fluid through the osmotic gradient in the renal medulla. There are two main types of cells in the collecting duct: intercalated cells and principal cells.

**Intercalated cells** are cuboidal cells with microvilli that play a role in secreting and reabsorbing both hydrogen and bicarbonate ions.

**Principal cells** are cuboidal cells that reabsorb water and secrete potassium ions.

**6 Papillary Duct**

A **papillary duct** collects tubular fluid from multiple collecting ducts and delivers it to a minor calyx.

**3** This light micrograph was taken from a section through the nephron loops and collecting ducts in the renal medulla.

Thick ascending limbs · Collecting duct · Descending thin limbs

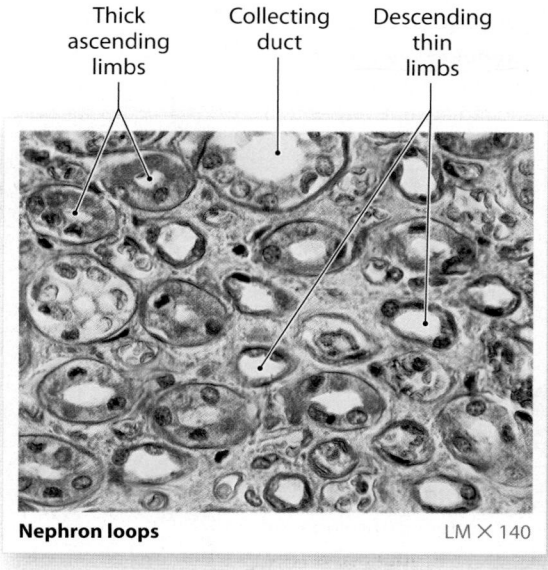

**Nephron loops** LM × 140

**? REVIEW**
**C.** Describe the structures of the collecting system.

As it travels along the renal tubule, tubular fluid gradually changes in composition. The composition of the urine that enters the minor calyx varies from moment to moment depending on the activities under way in each segment of the nephron and collecting system.

**24**

 **LEARNING OUTCOME**
Describe the segments of the nephron and collecting system, including their general functions and histological appearance.

# The kidneys are highly vascular, and the circulation patterns are complex

**1** These illustrations show the **arterial** system supplying the kidney (top) and the **venous** system draining the kidney (bottom).

**Arcuate arteries** originate at interlobar arteries and arch along the boundary between the renal cortex and renal medulla.

**Cortical radiate arteries** supply the cortical portions of adjacent kidney lobes.

**Afferent arterioles** that branch off the cortical radiate arteries supply blood to individual nephrons.

**Interlobar arteries** branch from the segmental arteries and radiate outward within the renal columns.

**Segmental arteries** form through the branching of the renal artery inside the renal sinus.

**Start**

Each kidney receives blood from a **renal artery**, which originates at the aorta near the origin of the superior mesenteric artery.

Each afferent arteriole delivers blood to a capillary knot called a **glomerulus**. Blood is then distributed to the capillaries of the nephron, as detailed in **2**.

**Cortical radiate veins** collect blood from the capillaries of the nephrons.

**Arcuate veins** collect blood from associated cortical radiate veins.

**Interlobar veins** collect blood from arcuate veins. They drain directly into the renal vein because there are no segmental veins.

The **renal vein** returns the blood to the inferior vena cava.

**?** **REVIEW**

**A.** Trace the pathway of blood from the renal artery to the renal vein.

**2** In a cortical nephron, the nephron loop is relatively short, and the *efferent arteriole* delivers blood to a network of *peritubular capillaries*, which surround the entire renal tubule. Both the nephrons and the peritubular capillaries are surrounded by interstitial fluid called **peritubular fluid**. These capillaries drain into small venules that carry blood to the cortical radiate veins.

**3** In a juxtamedullary nephron, the peritubular capillaries are connected to the **vasa recta** (*vasa*, vessels + *recta*, straight)—long, straight capillaries that parallel the limbs or the nephron loop.

**3** The **peritubular capillaries** collect water and solutes reabsorbed by the nephron and deliver other solutes to the nephron for secretion.

Glomerulus in renal corpuscle

**2** The **efferent arteriole** carries blood from the glomerulus to the peritubular capillaries.

**Start**

**1** The afferent arteriole delivers blood to the glomerulus, where filtration occurs.

To the cortical radiate vein

**? REVIEW**
**B.** Describe how blood enters and leaves the glomerulus.

**Cortical nephron**

Peritubular capillaries

Efferent arteriole

Afferent arteriole

To cortical radiate vein

Capillaries of the vasa recta collect and transport water and solutes within the renal medulla.

Collecting duct

Nephron loop

**? REVIEW**
**C.** Define the *vasa recta*.

**Juxtamedullary nephron**

## Everyday Physiology

### How is the kidney innervated?

Each kidney has approximately 1.25 million nephrons, with a combined length of about 145 km (85 miles). Both the cortical and the juxtamedullary nephrons are innervated by renal nerves that enter at the hilum and follow the branches of the renal arteries. Most of the nerve fibers involved are sympathetic postganglionic fibers from the celiac plexus and the inferior splanchnic nerves. Sympathetic innervation adjusts blood flow and blood pressure at the glomeruli and stimulates the release of renin.

**Lo LEARNING OUTCOME**
Trace the pathway of blood flow through a kidney, and compare the pattern of blood flow in cortical and juxtamedullary nephrons.

24

# Anatomy of the Urinary System

**Concept map:** Use each of the following terms once to fill in the blank boxes to correctly complete the map.

- ureter
- proximal convoluted tubule
- glomerulus
- urinary bladder
- renal tubules
- papillary ducts
- major calyces
- renal medulla
- renal sinus
- nephrons

**Short answer:** Label the kidney structures in the following diagram, and then provide a brief functional/anatomical description of each.

11 _____

12 _____

13 _____

14 _____

15 _____

16 _____

17 _____

18 _____

19 _____

20 _____

21 _____

22 _____

23 _____

24

# Module 24.6
# The kidneys maintain homeostasis by removing wastes and producing urine

Go to
MasteringA&P™
PhysioEx 9.1
Ex 9: Renal System Physiology

The primary function of urine production is to maintain homeostasis by regulating the volume and composition of blood. This process involves the excretion of solutes—specifically, metabolic wastes. Our bodies form three important metabolic wastes:

- **Urea**, the most abundant waste, is a by-product of the breakdown of amino acids in the liver.
- **Creatinine** is generated in skeletal muscle tissue through the breakdown of creatine phosphate, a high-energy compound that plays an important role in muscle contraction.
- **Uric acid** is a waste formed during the recycling of the nitrogenous bases of RNA molecules.

The kidneys are capable of producing concentrated urine with an osmotic concentration, or *osmolarity*, of 855–1335 mOsm/L, more than four times that of blood plasma. In a clinical setting, the osmolarity of urine is often reported as *osmolality*, which is milliosmoles per kilogram of water (mOsm/kg $H_2O$). In discussions of osmotic concentrations in body fluids, these terms are often used interchangeably.

**1** The kidneys produce fluid (urine) that is different from other body fluids. This table shows renal efficiency by comparing concentrations of typical substances found in blood plasma and urine.

| Normal Laboratory Values for Solutes in Plasma and Urine | | |
|---|---|---|
| **Solute** | **Blood Plasma** | **Urine** |
| **Ions (mEq/L)** | | |
| Sodium ($Na^+$) | 135–145 | 40–220 |
| Potassium ($K^+$) | 3.5–5.0 | 25–100 |
| Bicarbonate ($HCO_3^-$) | 20–28 | 1.9 |
| **Metabolites and Nutrients (mg/dL)** | | |
| Glucose | 70–110 | 0.009 |
| Lipids | 450–1000 | 0.002 |
| Proteins | 6.0–8.0 g/dL | 0.000 |
| **Nitrogenous Wastes (mg/dL)** | | |
| Urea | 8–25 | 1800 |
| Creatinine | 0.6–1.5 | 150 |
| Uric acid | 2–6 | 40 |
| Ammonia | <0.1 | 60 |

Note: The values indicated are typical ranges; specific numbers vary depending on the laboratory and testing methods used.

**2** Kidney function involves three distinct physiological processes.

In **filtration**, blood pressure forces water and solutes across the membranes of the glomerular capillaries and into the capsular space. Solute molecules small enough to pass through the filtration membrane are carried by the surrounding water molecules.

**Reabsorption** is the transport of water and solutes from the tubular fluid, across the tubular epithelium, and into the peritubular fluid.

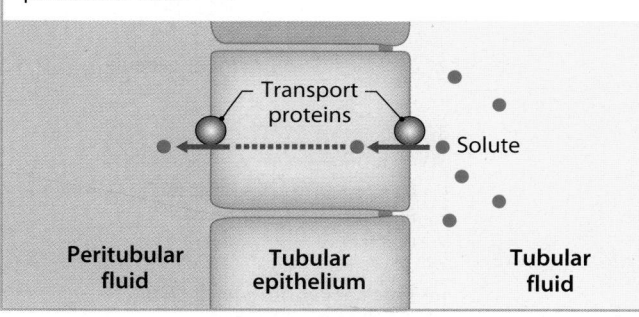

**Secretion** is the transport of solutes from the peritubular fluid, across the tubular epithelium, and into the tubular fluid.

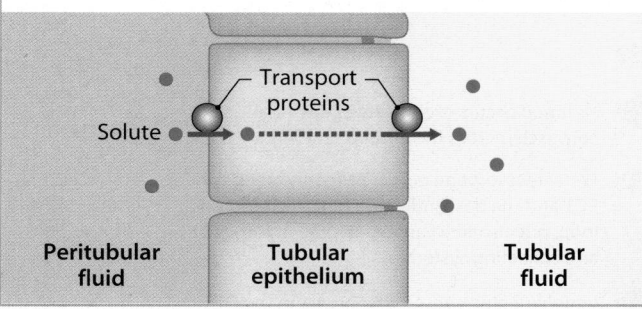

**? REVIEW**

**A.** In which direction do fluids and solutes move in each of the three kidney processes?

**Lo LEARNING OUTCOME**

Briefly describe how the kidneys maintain homeostasis and produce urine.

24

# Module 24.7
# Filtration, reabsorption, and secretion occur in specific segments of the nephron and collecting system

**1** This diagram shows where the processes of urine formation—filtration, reabsorption, and secretion—occur along the nephron and collecting system. Most segments perform both reabsorption and secretion, but the balance between the two processes varies from one segment to another. Regulation of the final volume and solute concentration of the urine results from the interaction between the collecting system and the nephron loops, especially the long loops of the juxtamedullary nephrons.

**? REVIEW**
**A.** Identify the three distinct processes of urine formation in the kidney.

**Proximal Convoluted Tubule (PCT)**
- Reabsorption of water, ions, and all organic nutrients

**Distal Convoluted Tubule (DCT)**
- Secretion of ions, acids, drugs, toxins
- Variable reabsorption of water, sodium ions, and calcium ions (under hormonal control)

**Start**

**Renal Corpuscle**
- Production of filtrate

Glomerulus

Glomerular capsule
Bowman's

Collecting duct

**Collecting Duct**
- Variable reabsorption of water
- Reabsorption or secretion of sodium, potassium, hydrogen, and bicarbonate ions

**Nephron Loop**
- **Descending thin limb (DTL)** Further reabsorption of water
- **Thick ascending limb (TAL)** Reabsorption of sodium and chloride ions

nephron loop

**Papillary Duct**
- Delivery of urine to minor calyx

Urine storage and elimination

KEY

Filtration occurs exclusively in the renal corpuscle, across the filtration membrane.

Water reabsorption occurs primarily along the PCT and the descending limb of the nephron loop, but also to a variable degree in the DCT and collecting system.

Variable water reabsorption occurs in the DCT and collecting system.

Solute reabsorption occurs along the PCT, the ascending limb of the nephron loop, the DCT, and the collecting system.

Variable solute reabsorption or secretion occurs at the PCT, the DCT, and the collecting system.

**? REVIEW**
**B.** Where does filtration exclusively occur in the kidney?

24

**2** The table below gives an overview of the functions of the various segments of the nephron and collecting system. We examine the roles of these nephron segments and surrounding blood vessels in greater detail in subsequent modules.

## Renal Structures and Their Functions

| Segment | General Functions | Specific Functions |
|---|---|---|
| **Renal corpuscle**  | *Filtration* of blood; generates approximately 180 L/day of filtrate similar in composition to blood plasma but without plasma proteins | *Filtration*<br>Water and inorganic and organic solutes from plasma<br><br>*Retention*<br>Plasma proteins and blood cells |
| **Proximal convoluted tubule (PCT)** | *Reabsorption* of 60–70% of the water (108–116 L/day), 99–100% of the organic substrates, and 60–70% of the sodium and chloride ions in the original filtrate | *Active reabsorption*<br>Glucose, other simple sugars, amino acids, vitamins, ions (including sodium, potassium, calcium, magnesium, phosphate, and bicarbonate)<br><br>*Passive reabsorption*<br>Urea, chloride ions, lipid-soluble materials, water<br><br>*Secretion*<br>Hydrogen ions, ammonium ions, creatinine, drugs, and toxins |
| **Nephron loop** | *Reabsorption* of 25% of the water (45 L/day) and 20–25% of the sodium and chloride ions in the original filtrate; creation of the concentration gradient in the renal medulla | *Reabsorption*<br>Sodium and chloride ions, water |
| **Distal convoluted tubule (DCT)** | *Reabsorption* of a variable amount of water (usually 5%, or 9 L/day) under antidiuretic hormone (ADH) stimulation, and a variable amount of sodium ions under aldosterone stimulation | *Reabsorption*<br>Sodium and chloride ions, sodium ions (variable), calcium ions (variable), water (variable)<br><br>*Secretion*<br>Hydrogen ions, ammonium ions, creatinine, drugs, and toxins |
| **Collecting system** | *Reabsorption* of a variable amount of water (usually 9.3%, or 16.8 L/day) under antidiuretic hormone stimulation, and a variable amount of sodium ions under aldosterone stimulation | *Reabsorption*<br>Sodium ions (variable), bicarbonate ions (variable), water (variable)<br><br>*Secretion*<br>Potassium and hydrogen ions (variable) |

## Blood Vessels

| Segment | General Functions | Specific Functions |
|---|---|---|
| **Peritubular capillaries** | *Redistribution* of water and solutes reabsorbed in the renal cortex | Return of water and solutes from the peritubular fluid to the general circulation |
| **Vasa recta** | *Redistribution* of water and solutes reabsorbed in the renal medulla, and stabilization of the concentration gradient of the renal medulla | Return of water and solutes from the peritubular fluid to the general circulation |

24

**? REVIEW**

**C.** Which segment of the nephron is solely involved in the reabsorption of water and sodium and chloride ions?

 **LEARNING OUTCOME**

Describe filtration, reabsorption, and secretion along each segment of the nephron and collecting system.

# Module 24.8
## Filtration occurs at the renal corpuscle

The glomerulus filters blood through the filtration membrane to produce filtrate.

Go to
MasteringA&P™
iP2™
Urinary System:
Glomerular
Filtration

**1** At the renal corpuscle, the capillary knot of the glomerulus projects into the capsular space like the heart projects into the pericardial sac. Like the pericardium, the glomerular capsule has an outer capsular layer and an inner visceral layer.

The **efferent arteriole** delivers blood to peritubular capillaries. It has a smaller diameter than the afferent arteriole, and this increases the blood pressure within the glomerulus.

### Juxtaglomerular Complex

The **juxtaglomerular complex** consists of specialized cells that secrete renin when glomerular blood pressure decreases.

Macula densa

Extraglomerular mesangial cells

Juxtaglomerular cells

The **afferent arteriole** delivers blood to the glomerulus from a cortical radiate artery.

The **capsular space** separates the capsular outer and visceral layers of the glomerular capsule.

### Glomerular Capsule

The **glomerular capsule** forms the outer wall of the renal corpuscle and covers the glomerular capillaries.

Capsular outer layer

Visceral layer

Glomerulus

DCT

Initial segment of renal tubule

**2** The visceral layer consists of large cells with complex processes, or "feet," that wrap around the specialized basement membrane of the glomerular capillaries. These cells are called **podocytes** (PŌ-dō-sīts; *podos*, foot + *-cyte*, cell), and their feet are known as **foot processes**, or *pedicels*. Substances passing out of the blood at the glomerulus must be small enough to pass between the narrow gaps, or **filtration slits**, between adjacent foot processes.

**A podocyte**    SEM × 2100

Filtration slits

Foot processes

Podocyte

### Filtration Membrane

Fenestrated capillary endothelium

Basement membrane

Foot processes of podocyte

**Intraglomerular mesangial cells** are supporting cells that lie between adjacent glomerular capillaries. These cells contract or relax to control capillary diameter and the rate of capillary blood flow.

RBC

Capsular space

24

**3** The glomerular capillaries are fenestrated capillaries containing large-diameter pores (see Module 19.3). The endothelium covers a *dense layer*, a specialized type of basement membrane. The **basement membrane** of the endothelium differs from that in other capillary networks in that it may encircle more than one capillary. Together, the fenestrated endothelium, the basement membrane, and the foot processes of the podocyte form the **filtration membrane**. Under normal circumstances, only a few plasma proteins—such as albumin molecules, with an average diameter of 7 nm—can cross the filtration membrane and enter the capsular space.

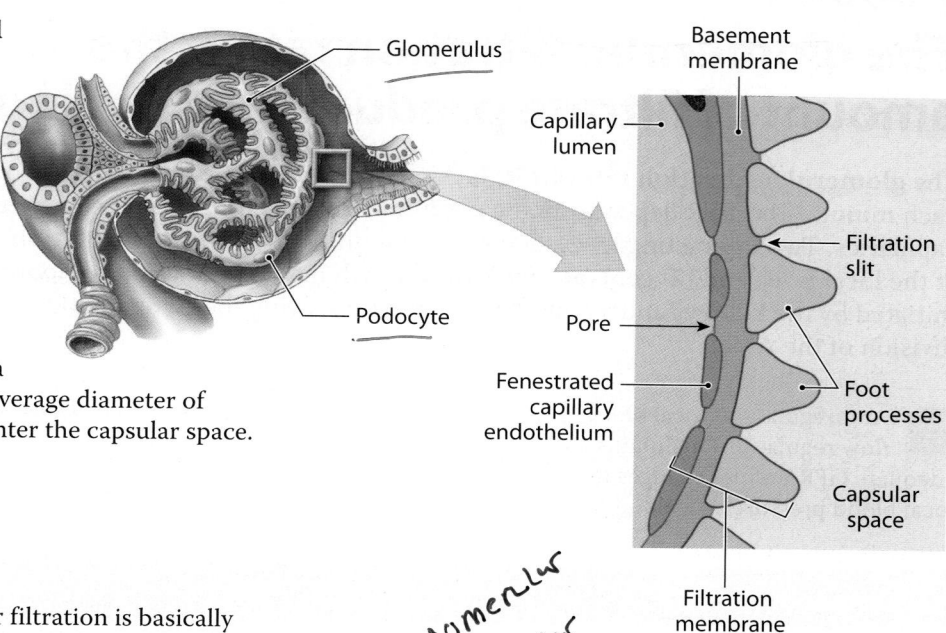

**? REVIEW**

**A.** What three elements form the filtration membrane?

**4** The primary factor involved in glomerular filtration is basically the same as that governing fluid and solute movement across capillaries throughout the body: the balance between **hydrostatic pressure** (fluid pressure) and **colloid osmotic pressure** (pressure due to materials in solution) on either side of the glomerular capillaries.

*Juxtaglomerular*
*Juxtaglomerular*

## Factors Controlling Glomerular Filtration

The **glomerular hydrostatic pressure (GHP)** is the blood pressure in the glomerular capillaries. This pressure tends to push water and solute molecules out of the plasma and into the filtrate. The GHP averages 50 mm Hg. It is significantly higher than capillary pressures elsewhere in the systemic circuit because of the different diameters of the afferent and efferent capillaries.

Filtrate in capsular space

Plasma proteins

Solutes

The **capsular colloid osmotic pressure** is usually 0 because few, if any, plasma proteins enter the capsular space.

The **blood colloid osmotic pressure (BCOP)** tends to draw water out of the filtrate and into the plasma; it thus opposes filtration. Over the entire length of the glomerular capillary bed, the BCOP averages about 25 mm Hg.

The **net filtration pressure (NFP)** is the pressure acting across the glomerular capillaries. It is the sum of the hydrostatic pressures and the colloid osmotic pressures. Under normal circumstances, the net filtration pressure is approximately 10 mm Hg. This is the average pressure forcing water and dissolved substances out of the glomerular capillaries and into the capsular space.

**Capsular hydrostatic pressure (CsHP)** opposes GHP. CsHP, which tends to push water and solutes out of the filtrate and into the plasma, results from the resistance of filtrate already present in the nephron that must be pushed toward the renal pelvis.

**? REVIEW**

**B.** Explain why blood pressure is higher in glomerular capillaries than in other systemic capillaries.

**? INTEGRATION**

**C.** Blood colloid osmotic pressure tends to draw water out of the filtrate and into the plasma. Why does this occur?

**Lo LEARNING OUTCOME**

Describe the structural features of a renal corpuscle, and explain the functions of the filtration membrane components.

24

# The glomerular filtration rate is the amount of filtrate produced each minute

The **glomerular filtration rate** (**GFR**) is the amount of filtrate the kidneys produce each minute. The GFR depends on the net filtration pressure across the glomerular capillaries. Two interacting levels of control stabilize GFR: (1) **autoregulation** at the local level, and (2) **central regulation**, which has an endocrine component initiated by the kidneys and a neural component involving the sympathetic division of the ANS.

**1** Autoregulation (local blood flow regulation) maintains an adequate GFR despite changes in local blood pressure and blood flow.

*if insufficient to restore normal GFR*

**Autoregulation**

Immediate local response in the kidney

Decreased filtration pressure; decreased filtrate and urine production

Contraction of intraglomerular mesangial cells

Constriction of efferent arterioles

Dilation of afferent arterioles

*if sufficient*

Increased glomerular blood pressure

Homeostasis **DISTURBED BY**

**DECREASING**

blood flow to kidneys

Homeostasis **RESTORED BY**

**INCREASING**

blood flow to kidneys

STIMULUS

RESTORED

# HOMEOSTASIS

NORMAL GLOMERULAR FILTRATION RATE (GFR)

**? REVIEW**

**A.** Describe autoregulation at the kidneys.

24

**2** If autoregulation is ineffective, the kidneys call for a wider response to increase the GFR. This response involves multiple systems and mechanisms.

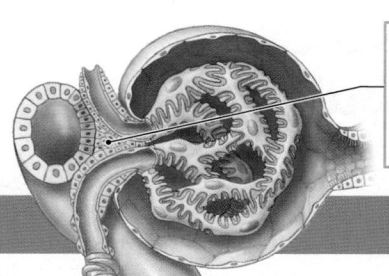

The juxtaglomerular complex plays a key role in coordinating responses to decreased GFR by initiating the renin-angiotensin-aldosterone system (RAAS).

## Central Regulation

### Endocrine response

| Effector |
|---|
| **Juxtaglomerular complex** |
| releases renin into the bloodstream |

Renin in the bloodstream triggers formation of angiotensin I, which is then activated to angiotensin II by **angiotensin-converting enzyme (ACE)** in the capillaries of the lungs

**Neural responses** triggered by angiotensin II

Angiotensin II triggers increased aldosterone secretion by the adrenal glands

| Effector |
|---|
| **Nervous system** |

**? REVIEW**
**B.** What does the juxtaglomerular complex do in response to decreased filtration pressure?

Angiotensin II constricts peripheral arterioles and further constricts the efferent arterioles

Aldosterone increases Na⁺ retention

Increased stimulation of thirst centers

Increased fluid consumption

Increased fluid retention

Increased antidiuretic hormone (ADH) production

Increased glomerular blood pressure

Increased systemic blood pressure

Increased blood volume

Constriction of systemic veins

Increased cardiac output

Increased sympathetic motor tone

Together, angiotensin II and sympathetic activation stimulate peripheral vasoconstriction

**? REVIEW**
**C.** Angiotensin II has what effect on the CNS?

 **Everyday Physiology**

## What percentage of glomerular filtrate is reabsorbed?

Each kidney contains about 6 m²—some 64 square feet—of filtration surface, and the GFR averages an astounding 125 mL per minute. This means that about 10 percent of the fluid delivered to the kidneys by the renal arteries leaves the bloodstream and enters the capsular spaces. In the course of a single day, the glomeruli generate about 180 liters (48 gal) of filtrate, about 70 times the total plasma volume. But as filtrate passes through the renal tubules, about 99 percent of it is reabsorbed.

**LO LEARNING OUTCOME**
Describe the factors that influence filtration pressure and the glomerular filtration rate.

24

# Module 24.10

# Reabsorption predominates along the proximal convoluted tubule …

The proximal convoluted tubule (PCT) is the first segment of the renal tubule. The entrance to the PCT lies almost directly opposite the point where the afferent and efferent arterioles connect to the glomerulus. The distal convoluted tubule (DCT), which forms the last segment of the nephron, makes final adjustments to the solute composition of the tubular fluid.

**1** Normally, before the tubular fluid enters the nephron loop, the PCT reabsorbs more than 99 percent of the glucose, amino acids, and other organic nutrients in the fluid. The PCT also reabsorbs sodium, potassium, bicarbonate, magnesium, phosphate, and sulfate ions. As reabsorption occurs, the solute concentration of tubular fluid decreases, and that of peritubular fluid and adjacent capillaries increases. Osmosis then pulls water out of the tubular fluid and into the peritubular fluid. Along the PCT, this mechanism results in the reabsorption of approximately 108 liters of water each day.

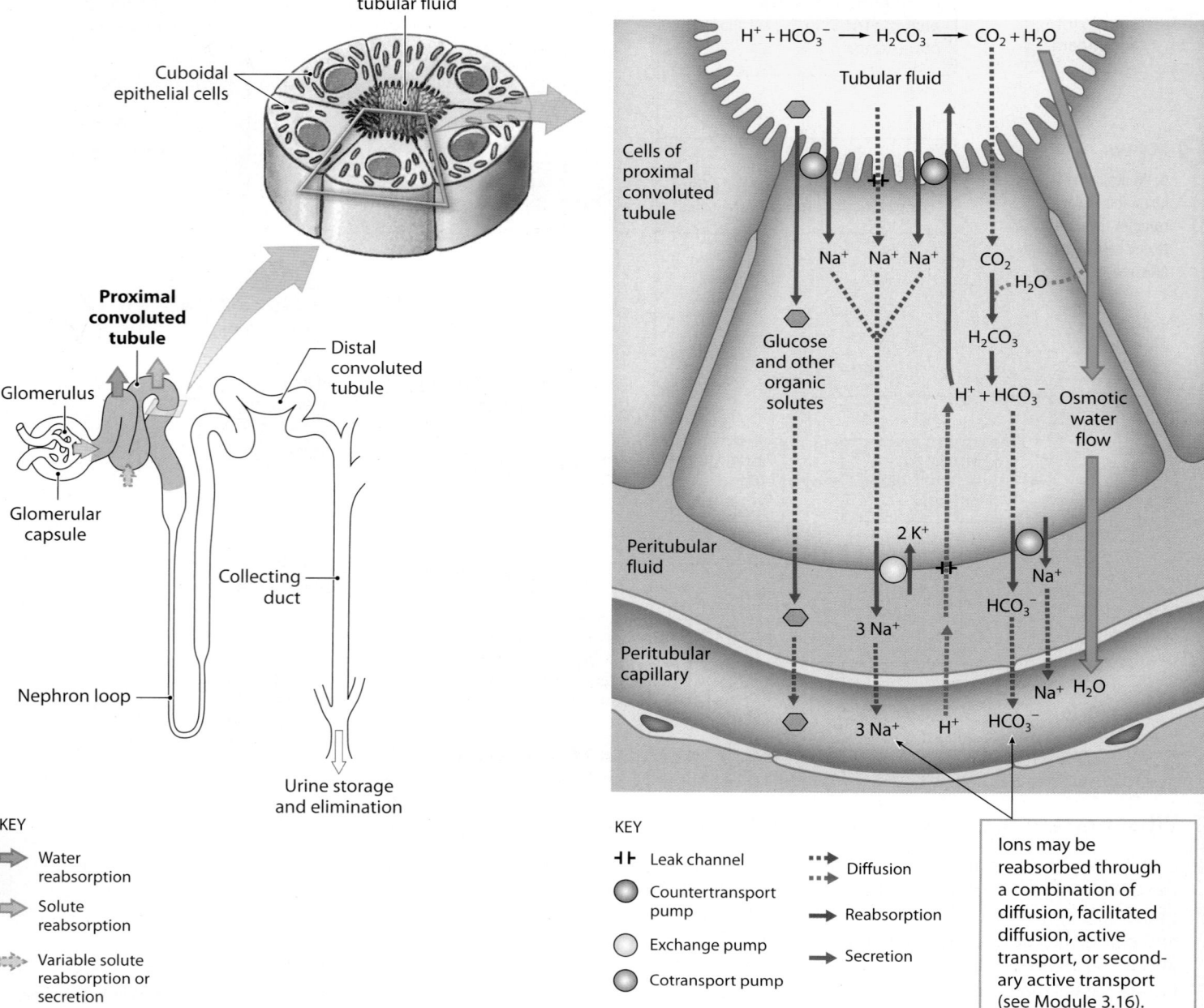

Lumen containing tubular fluid

Cuboidal epithelial cells

**Proximal convoluted tubule**

Glomerulus

Glomerular capsule

Distal convoluted tubule

Collecting duct

Nephron loop

Urine storage and elimination

$H^+ + HCO_3^- \longrightarrow H_2CO_3 \longrightarrow CO_2 + H_2O$

Tubular fluid

Cells of proximal convoluted tubule

$Na^+$  $Na^+$  $Na^+$    $CO_2$

$H_2O$

Glucose and other organic solutes

$H_2CO_3$

$H^+ + HCO_3^-$    Osmotic water flow

2 $K^+$

Peritubular fluid

$Na^+$

$HCO_3^-$

Peritubular capillary

3 $Na^+$

3 $Na^+$    $H^+$    $HCO_3^-$

$Na^+$  $H_2O$

**KEY**

⇨ Water reabsorption

⇨ Solute reabsorption

⇨ Variable solute reabsorption or secretion

**KEY**

⊣⊢ Leak channel

◯ Countertransport pump

◯ Exchange pump

◯ Cotransport pump

┈▶ Diffusion

┈▶ Diffusion

⟶ Reabsorption

⟶ Secretion

Ions may be reabsorbed through a combination of diffusion, facilitated diffusion, active transport, or secondary active transport (see Module 3.16).

24

# ... whereas reabsorption and secretion are often linked along the distal convoluted tubule

**? REVIEW**

**A.** Identify the segment of the nephron that makes final adjustments to the composition of tubular fluid.

**2** Only 15–20 percent of the initial filtrate volume reaches the DCT, and the concentrations of electrolytes and organic wastes in the arriving tubular fluid no longer resemble the concentrations in blood plasma. In the DCT, a combination of secretion and reabsorption further alters the solute composition of the tubular fluid.

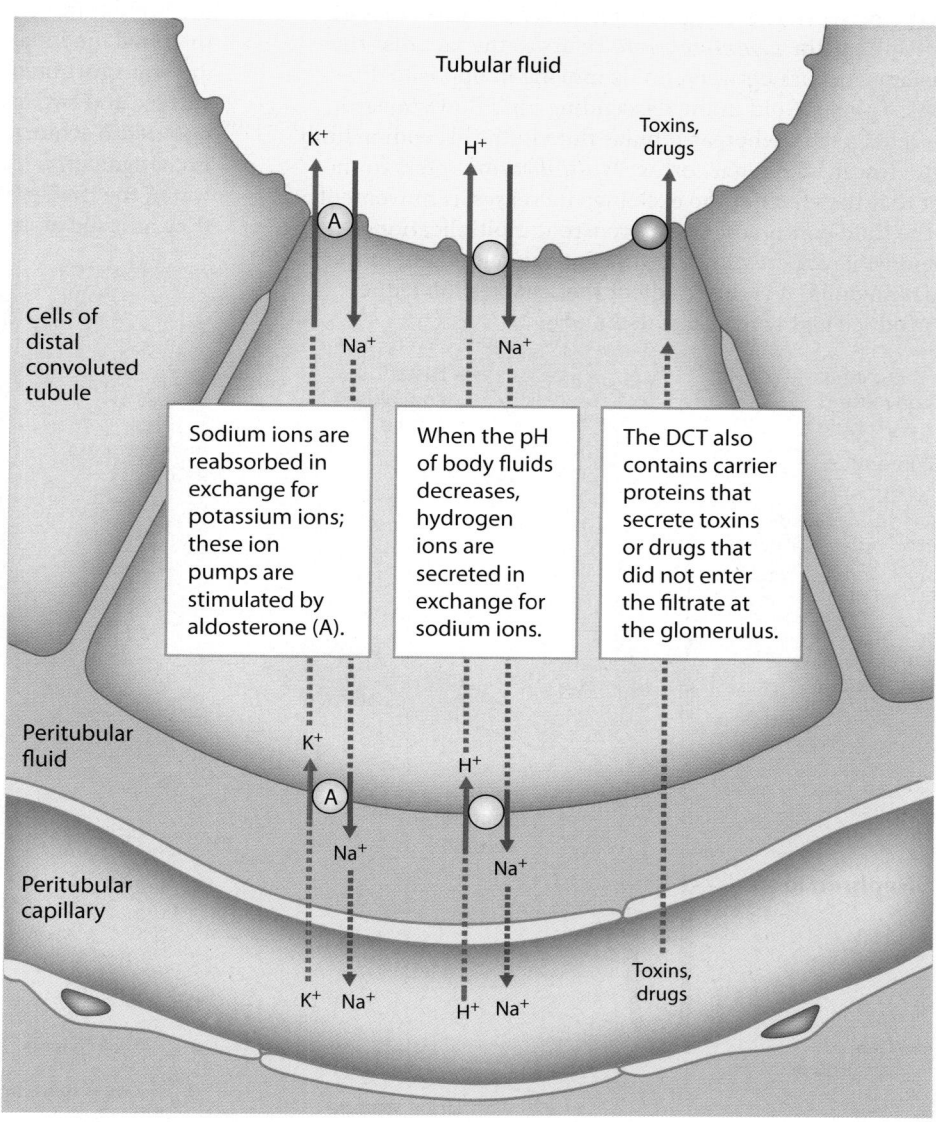

Sodium ions are reabsorbed in exchange for potassium ions; these ion pumps are stimulated by aldosterone (A).

When the pH of body fluids decreases, hydrogen ions are secreted in exchange for sodium ions.

The DCT also contains carrier proteins that secrete toxins or drugs that did not enter the filtrate at the glomerulus.

**KEY**

⬤ Carrier protein
◯ Exchange pump
Ⓐ Aldosterone-regulated pump
┈┈▶ Diffusion
──▶ Reabsorption
──▶ Secretion

**? REVIEW**

**B.** What effect would increased amounts of aldosterone have on the K+ concentration in urine?

**? INTEGRATION**

**C.** What effect would a decrease in the Na+ concentration of filtrate have on the pH of tubular fluid?

**Lo LEARNING OUTCOME**

Identify the types of transport mechanisms along the proximal and distal convoluted tubules of the nephron.

24

# Module 24.11
# Exchange between the limbs of the nephron loop creates an osmotic concentration gradient in the renal medulla

**1** The descending thin limb and the thick ascending limb of the nephron loop are very close together, separated only by peritubular fluid. The exchange that occurs between these segments is called **countercurrent multiplication**. *Countercurrent* refers to the fact that the exchange occurs between fluids moving in opposite directions: Tubular fluid in the descending limb flows toward the renal pelvis, whereas tubular fluid in the ascending limb flows toward the renal cortex. *Multiplication* refers to the fact that the effect of the exchange increases as movement of the fluid continues. Countercurrent multiplication is responsible for creating the concentration gradient in the renal medulla. It is this gradient that enables the kidney to produce highly concentrated urine.

**2** The key to this process is the transport activity performed by the cells of the thick ascending limb. Active transport at the apical surface moves sodium and chloride ions out of the tubular fluid, and they then enter the peritubular fluid of the renal medulla. The mechanism of sodium and chloride ion transport involves the $Na^+$–$K^+$/2 $Cl^-$ carrier at the apical surface and two carriers at the basal surface of the tubular cell: a potassium–chloride cotransport pump and a sodium–potassium exchange pump. Because the apical surface is impermeable to water, the transport of ions does not result in an osmotic water flow as it did along the PCT.

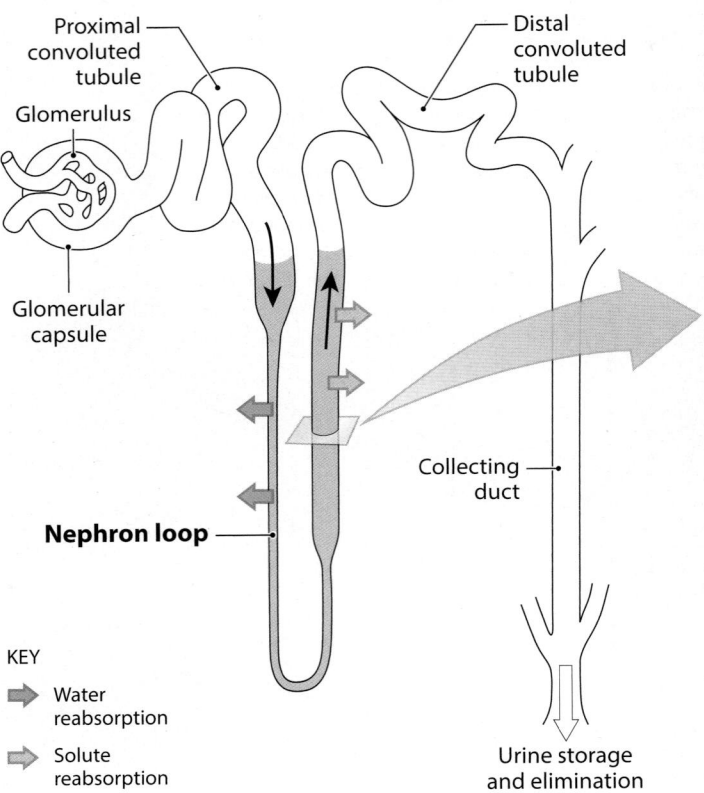

Proximal convoluted tubule

Glomerulus

Glomerular capsule

Distal convoluted tubule

Collecting duct

**Nephron loop**

KEY

Water reabsorption

Solute reabsorption

Urine storage and elimination

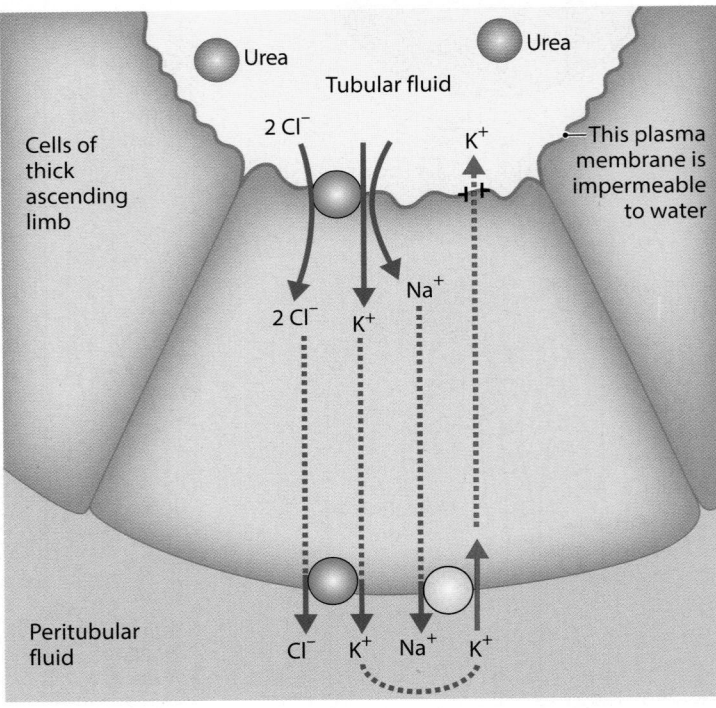

Urea        Urea

Tubular fluid

Cells of thick ascending limb

2 $Cl^-$        $K^+$

This plasma membrane is impermeable to water

2 $Cl^-$    $K^+$        $Na^+$

Peritubular fluid

$Cl^-$    $K^+$    $Na^+$    $K^+$

KEY

= Leak channel        = Reabsorption

= Cotransport pump        = Secretion

= Exchange pump        = Diffusion

? **REVIEW**

**A.** Define *countercurrent multiplication* as it occurs in the kidneys.

**3** The removal of sodium and chloride ions from the tubular fluid in the ascending limb increases the osmotic concentration of the peritubular fluid around the descending thin limb. The descending thin limb is permeable to water but impermeable to solutes; as tubular fluid travels deeper into the renal medulla along the descending thin limb, osmosis moves water into the peritubular fluid.

**4** Urea is an important solute that is eliminated in the urine. As water is reabsorbed along the DCT and collecting duct, the concentration of urea gradually increases in the tubular fluid. The tubular fluid reaching the papillary duct typically contains urea at a concentration of about 450 mOsm/L. Papillary duct permeability to urea makes up nearly one-third of the solutes in the deepest portions of the medulla.

The descending thin limb is permeable to water but impermeable to solutes. Solutes remain behind as water leaves, so the tubular fluid reaching the turn of the nephron loop has a higher osmotic concentration than it did at the start.

The pumping mechanism of the thick ascending limb is highly effective: Almost two-thirds of the sodium and chloride ions that enter it are pumped out of the tubular fluid before that fluid reaches the DCT.

The thick ascending limb is impermeable to water and other solutes and selectively permeable to Na⁺ and Cl⁻. In other tissues, differences in solute concentration are quickly resolved by osmosis. But since osmosis cannot occur across the impermeable wall of the thick ascending limb, the solute concentration in the tubular fluid decreases as Na⁺ and Cl⁻ are removed.

**1** The descending thin limb is permeable to water but impermeable to solutes, including urea.

**2** The thick ascending limb of the nephron loop is impermeable to water and solutes.

**3** The DCT and the collecting ducts are impermeable to urea but have variable permeability to water.

**4** The papillary duct is permeable to urea. As a result, urea tends to diffuse out of the tubular fluid and into the peritubular fluid in the deepest part of the medulla.

**KEY**

—— = Impermeable to solutes

—— = Impermeable to urea; variable permeability to water

- - - = Permeable to urea

—— = Impermeable to water

| 300 | = Fluid osmotic concentration (mOsm/L) |

○ = Na⁺

○ = Cl⁻

○ = Urea

**? REVIEW**

**B.** The thick ascending limb of the nephron loop actively pumps what substances into the peritubular fluid?

**? INTEGRATION**

**C.** An increase in sodium and chloride ions in the peritubular fluid affects the descending thin limb in what way?

**Lo LEARNING OUTCOME**

Explain the role of countercurrent multiplication in the formation of a concentration gradient in the renal medulla.

24

# Module 24.12
# Urine volume and concentration are hormonally regulated

Urine volume and osmotic concentration are regulated through the hormonal control of water reabsorption by antidiuretic hormone (ADH). The water permeabilities of the PCT and descending limb of the nephron loop cannot be adjusted, and water reabsorption occurs whenever the osmotic concentration of the peritubular fluid exceeds that of the tubular fluid. Because these water movements cannot be prevented, they represent **obligatory water reabsorption**. Obligatory reabsorption usually recovers 85 percent of the volume of filtrate produced. The volume of water lost in urine depends on how much of the water in the remaining tubular fluid (15 percent of the filtrate volume, or nearly 27 liters per day) is reabsorbed along the DCT and collecting system. The amount can be precisely controlled by a process called **facultative water reabsorption**.

**1** Without antidiuretic hormone (ADH), water is not reabsorbed in the distal collecting tubule or the collecting duct, so all the fluid reaching the DCT is lost in the urine. No facultative water reabsorption occurs, and the person then produces large amounts of very dilute urine.

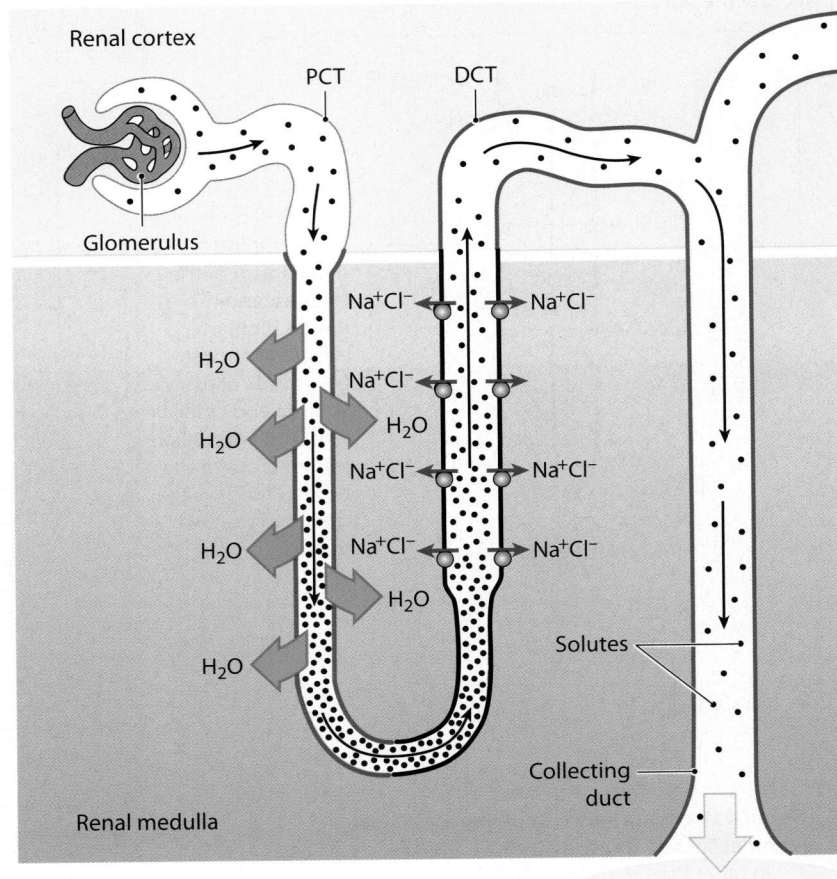

**Obligatory Water Reabsorption**

Glomerulus

Glomerular capsule

Proximal convoluted tubule

Nephron loop

**Facultative Water Reabsorption**

Distal convoluted tubule

Collecting duct

Urine storage and elimination

**KEY**

➡ = Water reabsorption

⇢ = Variable water reabsorption

Renal cortex

PCT        DCT

Glomerulus

$Na^+Cl^-$

$H_2O$

$Na^+Cl^-$

$H_2O$

$Na^+Cl^-$

$H_2O$

$Na^+Cl^-$

$H_2O$

$Na^+Cl^-$        $Na^+Cl^-$

$H_2O$        $H_2O$

$Na^+Cl^-$        $Na^+Cl^-$

$Na^+Cl^-$        $Na^+Cl^-$

Solutes

Collecting duct

Renal medulla

Large volume of dilute urine

**? REVIEW**

**A.** Can the water permeability of the PCT or DCT ever change? Explain.

**B.** Compare obligatory water reabsorption with facultative water reabsorption.

**2** ADH causes special water channels called **aquaporins** to appear in the apical surfaces of the plasma membranes lining the DCT and collecting duct. The aquaporins dramatically increase the rate of osmotic water movement. As ADH levels increase, the DCT and collecting system become more permeable to water, and the urine osmotic concentration increases. Under maximum ADH stimulation, the DCT and collecting system become so permeable to water that the osmotic concentration of the urine is equal to that of the peritubular fluid in the deepest portion of the renal medulla. The result is a small volume of concentrated urine.

**3** A healthy adult typically produces 1200 mL of urine per day (about 0.6 percent of the filtrate volume), with an osmotic concentration of about 1000 mOsm/L. However, normal values differ from person to person and from day to day, as the kidneys alter their function to maintain homeostatic conditions within body fluids.

KEY

- <image of arrow> = Water reabsorption
- <image of dashed arrow> = Variable water reabsorption
- <image of Na+/Cl- transport> = Na⁺/Cl⁻ transport
- **ADH** = Antidiuretic hormone

- ∙∙∙ = Solutes
- —— = Impermeable to solutes
- **——** = Impermeable to water
- —— = Variable permeability to water

Small volume of concentrated urine

| General Characteristics of Normal Urine | |
|---|---|
| **Characteristic** | **Normal Range** |
| **pH** | 4.5–8 (average: 6.0) |
| **Specific gravity** | 1.003–1.030 |
| **Osmotic concentration (osmolarity)** | 855–1335 mOsm/L |
| **Water content** | 93–97% |
| **Volume** | 700–2000 mL/day |
| **Color** | Pale yellow |
| **Odor** | Varies with composition |
| **Bacterial content** | None (sterile) |

**? REVIEW**

**C.** What effect does an increase in ADH levels have on the DCT?

**D.** When ADH levels in the DCT decrease, what happens to the urine osmotic concentration?

**? INTEGRATION**

**E.** What effect does ADH have on the apical plasma membranes lining the DCT and collecting ducts?

**Lo LEARNING OUTCOME**

Describe how antidiuretic hormone influences the volume and concentration of urine.

24

# Renal function is an integrative process involving filtration, reabsorption, and secretion

**Renal cortex**

**1** The filtrate produced by the renal corpuscle has the same osmotic concentration as plasma—about 300 mOsm/L. It has the same composition as plasma but does not contain plasma proteins.

**2** In the proximal convoluted tubule (PCT), the active removal of ions and organic nutrients results in the osmotic flow of water out of the tubular fluid. This decreases the volume of filtrate but keeps the solutions inside and outside the tubule isotonic. Between 60 and 70 percent of the filtrate volume is absorbed here.

**3** In the PCT and descending limb of the nephron loop, water moves into the surrounding peritubular fluid, leaving a small volume of highly concentrated tubular fluid. This reduction occurs by obligatory water reabsorption.

$H_2O$

Nutrients

$H_2O$

Electrolytes

Descending limb of nephron loop

**Renal medulla**

$H_2O$

Vasa recta

$H_2O$

Increasing osmolarity

$H_2O$

$H_2O$

**? REVIEW**

**A.** In the PCT, ions and organic substrates are actively removed, thus causing what to occur?

300
300
300
300
300
600
600
600
900
900
900
1200
1200
1200
1200

**KEY**

= Water reabsorption

= Variable water reabsorption

= $Na^+/Cl^-$ transport

Ⓐ = Aldosterone-regulated pump

Ⓤ = Urea transporter

= Impermeable to solutes

= Impermeable to water

= Variable permeability to water

300 = Fluid osmotic concentration (mOsm/L)

= Solutes

24

**4** The thick ascending limb is impermeable to water and solutes. The tubule cells actively transport Na$^+$ and Cl$^-$ out of the tubule, thereby decreasing the osmotic concentration of the tubular fluid. Because just Na$^+$ and Cl$^-$ are removed, urea makes up a higher proportion of the total osmotic concentration at the end of the nephron loop.

**5** Further adjustments in the composition of the tubular fluid occur in the DCT and the collecting system. The solute concentration of the tubular fluid can be adjusted through active transport (reabsorption or secretion).

**6** The final adjustments in the volume and osmotic concentration of the tubular fluid are made by controlling the water permeabilities of the distal portions of the DCT and the collecting system. ADH levels determine the final urine volume and concentration.

**7** The vasa recta absorb the solutes and water that were reabsorbed by the nephron loop and the collecting ducts. By transporting these solutes and water into the systemic circuit, the vasa recta maintain the concentration gradient of the renal medulla.

**8** The papillary duct is permeable to urea, where its concentration is about 450mOsm/L. This is the same value as in the deepest parts of the medulla.

**L$_o$** **LEARNING OUTCOME**

Summarize the major steps involved in water reabsorption and urine production.

24

# Module 24.14
# Renal failure is a life-threatening condition

**1**   **Renal failure** occurs when the kidneys cannot filter wastes from the blood and can no longer maintain homeostasis. When kidney filtration slows for any reason, urine production decreases. As the patient makes less and less urine, signs and symptoms of renal failure appear because water, ions, and metabolic wastes are retained. Virtually all systems in the body are affected. Blood pressure rises; anemia develops due to a decrease in erythropoietin production; and CNS problems can lead to sleeplessness, seizures, delirium, and even coma.

 **REVIEW**

**A.** Briefly explain the difference between chronic and acute renal failure.

## Renal Failure

### Chronic Renal Failure

In **chronic renal failure**, kidney function deteriorates gradually, and the associated problems accumulate over time. The management of chronic renal failure typically involves restricting water and salt intake and minimizing protein intake. This combination reduces strain on the urinary system by (1) minimizing the volume of urine produced and (2) preventing the generation of large quantities of nitrogenous wastes. Acidosis, a common problem in persons with renal failure, can be countered by ingesting bicarbonate ions.

Chronic renal failure generally cannot be reversed; its progression can only be slowed.

### Acute Renal Failure

**Acute renal failure** occurs when filtration slows suddenly or stops due to renal ischemia, urinary obstruction, trauma, or nephrotoxic drugs. The reduction in kidney function takes place over a few days and may persist for weeks. Sensitized individuals can also develop acute renal failure after an allergic response to antibiotics or anesthetics. People in acute renal failure may recover if they survive the incident. The kidneys may then regain partial or complete function. (With supportive treatment, the survival rate is approximately 50 percent.)

**2**   **Hemodialysis** (hē-mō-dī-AL-i-sis) is a treatment that "cleanses" the blood by substituting for normal kidney functioning. The basic principle involved in this process, called **dialysis**, is passive diffusion across a selectively permeable membrane. The patient's blood flows past an artificial dialysis membrane, which contains pores large enough to permit the diffusion of ions, nutrients, and organic wastes, but small enough to prevent the loss of plasma proteins. A special dialysis fluid flows on the other side of the membrane.

**?** **REVIEW**

**B.** Define *dialysis*.

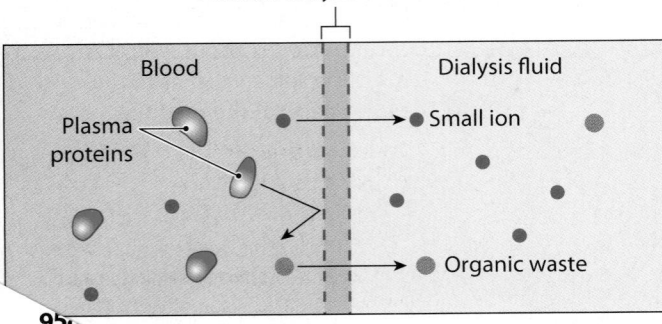

Artificial dialysis membrane

Blood     Dialysis fluid

Plasma proteins

Small ion

Organic waste

### Comparative Composition of Plasma and Dialysis Fluid

| Component | Plasma | Dialysis Fluid |
|---|---|---|
| **Electrolytes (mEq/L)** | | |
| Sodium ($Na^+$) | 135–145 | 136–140 |
| Potassium ($K^+$) | 3.5–5.0 | 0–3.0 |
| Calcium ($Ca^{2+}$) | 4.3–5.3 | 1.5 |
| Magnesium ($Mg^{2+}$) | 1.4–2.0 | 0.5–1.0 |
| Chloride ($Cl^-$) | 100–108 | 99–110 |
| Bicarbonate ($HCO_3^-$) | 21–28 | 27–39 |
| Phosphate ($PO_4^{2-}$) | 3 | 0 |
| Sulfate ($SO_4^{2-}$) | 1 | 0 |
| **Nutrients (mg/dL)** | | |
| Glucose | 70–110 | 100 |

Note: Although these values are representative, the precise dialysis fluid composition can be tailored to meet specific clinical needs. For example, if plasma potassium levels are too low, the dialysis fluid potassium ion concentration can be elevated to remedy the situation.

**3** The signs and symptoms of acute renal failure can be relieved by kidney dialysis, but this treatment is not a cure.

Thermometer

Dialysis fluid

Holding tank

Blood pump

Dialysis chamber

To drain

Flow meter

Air detector and clamp

Blood flowing in a tube of dialysis membrane

Artery

Vein

As diffusion takes place across the dialysis membrane, the composition of the blood changes. Potassium ions, phosphate ions, sulfate ions, urea, creatinine, and uric acid diffuse across the membrane into the dialysis fluid. Bicarbonate ions and glucose diffuse into the bloodstream. In effect, diffusion across the dialysis membrane takes the place of normal glomerular filtration, and the characteristics of the dialysis fluid ensure that important metabolites remain in the bloodstream rather than diffusing across the membrane.

Silicone rubber tubes called **shunts** are inserted into a medium-sized artery and vein. (The typical location is the forearm, although the lower leg is sometimes used.) The two shunts are then connected, forming a short circuit that does not impede the flow of blood. The shunts can then be used, like taps in a wine barrel, to draw a blood sample or to connect the patient to a dialysis machine. For long-term dialysis, a surgically created arteriovenous anastomosis provides access.

The only real cure for severe renal failure is a kidney transplant. This procedure involves implanting a new kidney obtained from a living or deceased donor. The recipient's nonfunctioning kidney(s) may be removed, especially if an infection is present. Patient survival is more than 90 percent at 2 years after the transplant. The use of kidneys from close relatives significantly improves the chances that the transplant will succeed. Immunosuppressive drugs are given to reduce tissue rejection (see Module 20.22). Unfortunately, this treatment also lowers the recipient's resistance to infection or cancer.

 **INTEGRATION**

**C.** Explain why patients on dialysis often receive Epogen or Procrit, synthetic forms of erythropoietin.

 **LEARNING OUTCOME**

Compare and contrast chronic and acute renal failure, and explain the process of hemodialysis.

# Overview of Renal Physiology

**Matching:** Match each lettered term with the most closely related description.

a. aquaporins
b. ADH
c. aldosterone
d. PCT
e. secretion
f. renal corpuscle
g. nephron loop
h. filtrate
i. BCOP
j. podocytes

| **1** | Site of plasma filtration |
| **2** | Glomerular epithelium |
| **3** | Protein-free solution |
| **4** | Opposes filtration |
| **5** | Countercurrent multiplication |
| **6** | Water channels |
| **7** | Primary method for eliminating drugs or toxins |
| **8** | Stimulates ion pump—Na$^+$ reabsorbed |
| **9** | Primary site of nutrient reabsorption in the nephron |
| **10** | Regulates passive reabsorption of water from urine in the collecting system |

1 _____
2 _____
3 _____
4 _____
5 _____
6 _____
7 _____
8 _____
9 _____
10 _____

**Short answer:** Identify the structures of the representative nephron and collecting system in the following diagram, and describe the functions of each.

11 _____

12 _____    13 _____

16 _____

Glomerulus

Efferent arteriole

Afferent arteriole

Glomerular capsule

14 _____

17 _____

Descending limb    Ascending limb

15 _____

18 _____

**Section integration:** Marissa has had a urinalysis that detected large amounts of plasma proteins and white blood cells in her urine. What condition might be responsible, and what effects would it have on her urine output?

19 _____
_____

## Module 24.15
# The urinary tract transports, stores, and eliminates urine

Filtrate modification and urine production end when the fluid enters the renal pelvis. The structures of the **urinary tract** transport, store, and eliminate urine.

**1** A **pyelogram** (PĪ-el-ō-gram) is an image of the urinary system. This image of the male urinary system was obtained by taking an x-ray of the kidneys after a radiopaque dye was administered intravenously. Such an image provides an orientation to the relative sizes and positions of the system's main structures.

 **REVIEW**
**A.** When does urine production end?
**B.** What is a pyelogram?

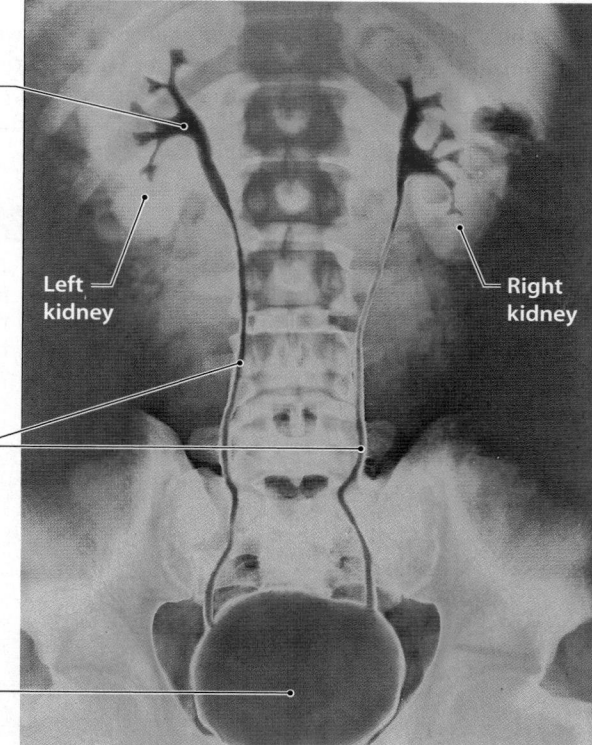

Renal pelvis

Left kidney

Right kidney

The **ureters** are a pair of muscular tubes that extend from the kidneys to the urinary bladder—a distance of about 30 cm (12 in.). The ureters are retroperitoneal and are firmly attached to the posterior abdominal wall. The paths taken by the ureters in men and women are different, because of variations in the nature, size, and position of the reproductive organs. In males, the base of the urinary bladder lies between the rectum and the pubic symphysis. In females, the base of the urinary bladder sits inferior to the uterus and anterior to the vagina.

The **urinary bladder** is a hollow, muscular organ that serves as a temporary reservoir for urine. A full urinary bladder can contain as much as a liter of urine.

**2** This is a sectional view of a male pelvis showing the locations of the lower sructures of the urinary tract.

**3** This is a sectional view of a female pelvis showing the locations of the lower sructures of the urinary tract.

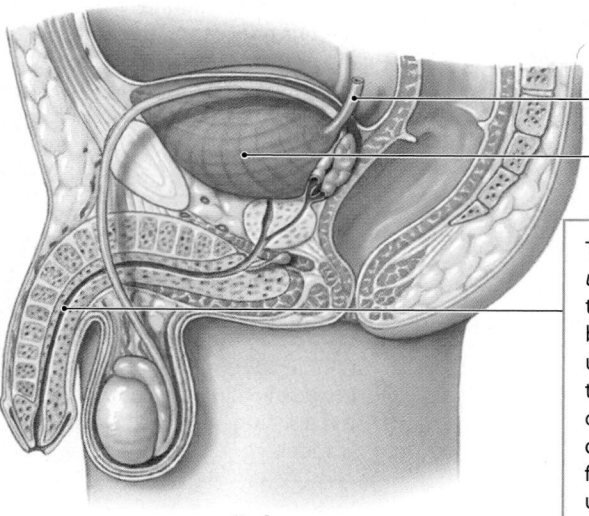

Ureter

Urinary bladder

The **urethra** (plural, *urethrae*) extends from the neck of the urinary bladder and transports urine to the exterior of the body. The urethrae of males and females differ in length and in function. The male urethra transports not only urine, but also semen.

**Male**

**Female**

24

**? REVIEW**
**C.** How does the urethra differ between males and females?

**Lo LEARNING OUTCOME**
Name the organs responsible for the transport, storage, and elimination of urine.

# Module 24.16

# The ureters, urinary bladder, and urethra are specialized to conduct urine

**1** The urinary bladder is filled by the ureters and drained by the urethra. Urinary bladder dimensions vary with its state of distension. A layer of peritoneum covers the superior surfaces of the urinary bladder. Several peritoneal folds assist in stabilizing its position, and tough ligamentous bands anchor it to the pelvic and pubic bones.

**? REVIEW**

**A.** Urine is transported by the _____, stored within the _____, and eliminated through the _____.

**Supporting Ligaments**

The **lateral umbilical ligaments** pass along the sides of the bladder to the umbilicus (navel). These fibrous cords are the vestiges of the two umbilical arteries, which supplied blood to the placenta during embryonic and fetal development.

The **middle umbilical ligament** extends from the anterior, superior border toward the umbilicus.

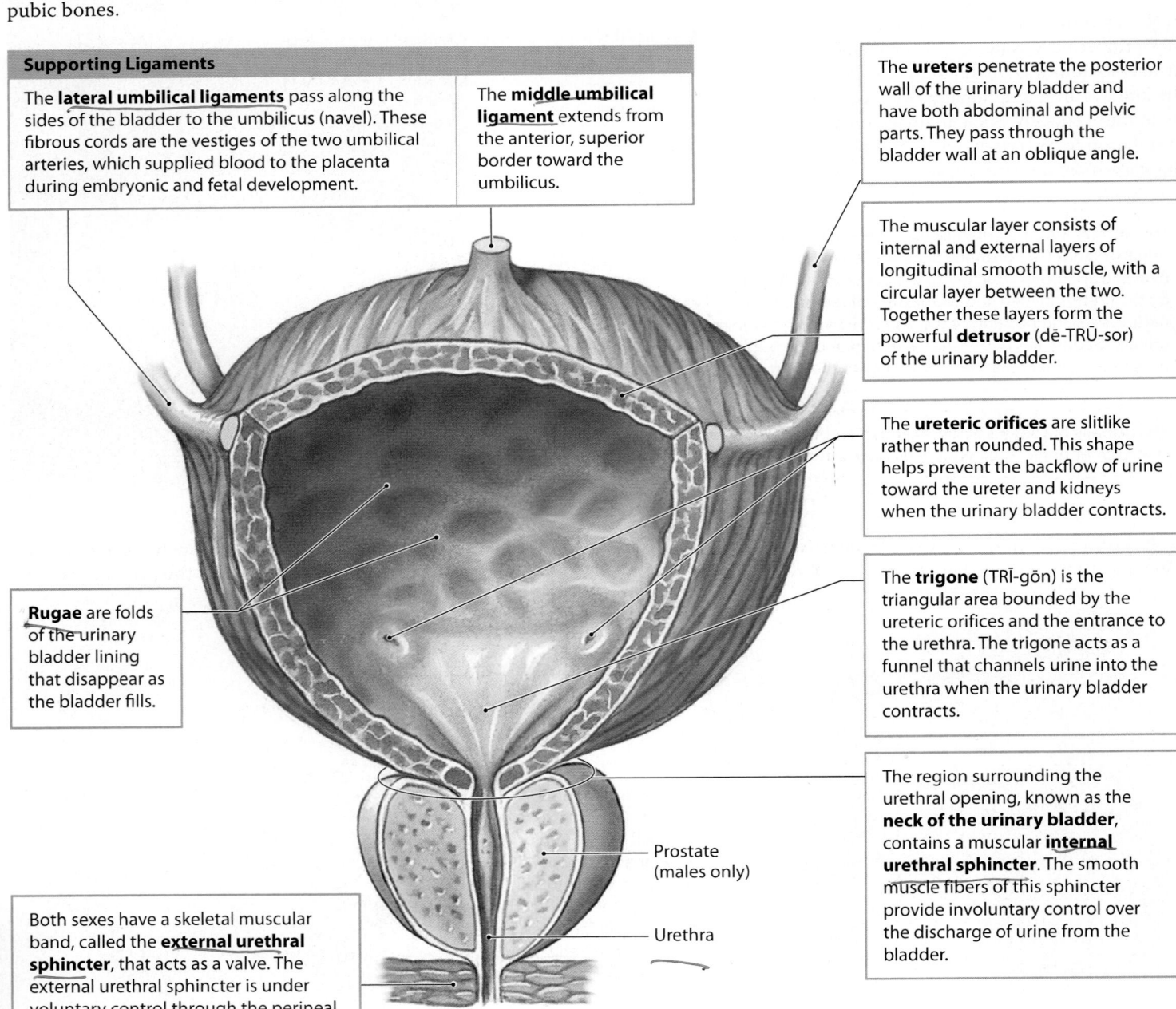

The **ureters** penetrate the posterior wall of the urinary bladder and have both abdominal and pelvic parts. They pass through the bladder wall at an oblique angle.

The muscular layer consists of internal and external layers of longitudinal smooth muscle, with a circular layer between the two. Together these layers form the powerful **detrusor** (dē-TRŪ-sor) of the urinary bladder.

The **ureteric orifices** are slitlike rather than rounded. This shape helps prevent the backflow of urine toward the ureter and kidneys when the urinary bladder contracts.

The **trigone** (TRĪ-gōn) is the triangular area bounded by the ureteric orifices and the entrance to the urethra. The trigone acts as a funnel that channels urine into the urethra when the urinary bladder contracts.

**Rugae** are folds of the urinary bladder lining that disappear as the bladder fills.

Prostate (males only)

Urethra

The region surrounding the urethral opening, known as the **neck of the urinary bladder**, contains a muscular **internal urethral sphincter**. The smooth muscle fibers of this sphincter provide involuntary control over the discharge of urine from the bladder.

Both sexes have a skeletal muscular band, called the **external urethral sphincter**, that acts as a valve. The external urethral sphincter is under voluntary control through the perineal branch of the pudendal nerve. This sphincter has a resting muscle tone and must be voluntarily relaxed to permit urination.

**? REVIEW**

**B.** What has to happen to the external urethral sphincter to allow urination?

24

**2** The wall of each ureter consists of three layers: (1) an inner mucosa made up of a transitional epithelium and the surrounding lamina propria; (2) a middle muscular layer made up of longitudinal and circular bands of smooth muscle; and (3) an outer connective tissue layer that is continuous with the fibrous capsule and peritoneum. About every 30 seconds, a peristaltic contraction begins at the renal pelvis and sweeps along the ureter, forcing urine toward the urinary bladder.

Smooth muscle

Outer connective tissue layer

**Mucosa**

Lamina propria

Transitional epithelium

**3** The urinary bladder wall has mucosa, submucosa, and muscular (detrusor) layers with a visceral peritoneum covering. When the detrusor contracts, the urinary bladder compresses and expels urine into the urethra.

**Mucosa**

Transitional epithelium

Lamina propria

Submucosa

Detrusor

Visceral peritoneum

**4** The urethral lining consists of a stratified epithelium that varies from transitional at the neck of the urinary bladder, to stratified columnar at the midpoint, to stratified squamous near the external urethral orifice. The lamina propria is thick and elastic, and the mucosa has longitudinal folds. Mucin-secreting cells are located in the epithelial pockets. Connective tissues of the lamina propria anchor the urethra to surrounding structures.

Lumen of urethra

Stratified squamous epithelium of mucosa

Lamina propria containing mucous epithelial glands

Smooth muscle

 **REVIEW**

**C.** Name the specialized smooth muscle of the urinary bladder, and describe its function.

**Lo LEARNING OUTCOME**

Describe the structures and functions of the ureters, urinary bladder, and urethra.

24

# Module 24.17
# Urinary reflexes coordinate urine storage and voiding

Urine reaches the urinary bladder by peristaltic contractions of the ureters. The urge to urinate generally appears when your bladder contains about 200 mL of urine. Whether or not we urinate depends on spinal reflexes, higher centers in the brain, and conscious control over urinating. Two spinal reflexes control urination, also called *micturition*: the **urine storage reflex** and the **urine voiding reflex**. These must be considered together because urine storage and release involve simultaneous, coordinated activities.

**? REVIEW**
**A.** What are the two reflexes that control urination?

**1** Urine storage occurs by spinal reflexes and the **pontine storage center**. When urine is stored, afferent impulses from stretch receptors in the bladder stimulate sympathetic outflow to the detrusor and internal urethral sphincter, inhibiting contraction of the detrusor and stimulating contraction of the internal urethral sphincter. The pontine storage center inhibits urination by decreasing parasympathetic activity and increasing somatic motor nerve activity at the external urethral sphincter. These spinal reflexes and pontine responses promote urine storage (continence).

**2** Urine voiding occurs by spinal reflexes and the **pontine micturition center**. The voiding reflex begins when afferent impulses from stretch receptors in the bladder initiate sacral spinal reflexes that (1) stimulate increased parasympathetic activity (detrusor contracts and internal urethral sphincter relaxes), (2) decrease sympathetic activity (detrusor contracts and internal urethral sphincter relaxes), and (3) decrease somatic motor nerve activity (external urethral sphincter relaxes). Ascending afferent input from the spinal cord also stimulates interneurons that relay sensations to the pontine micturition center. The pontine micturition center promotes urination by the same three spinal efferents in the sacral spinal reflexes.

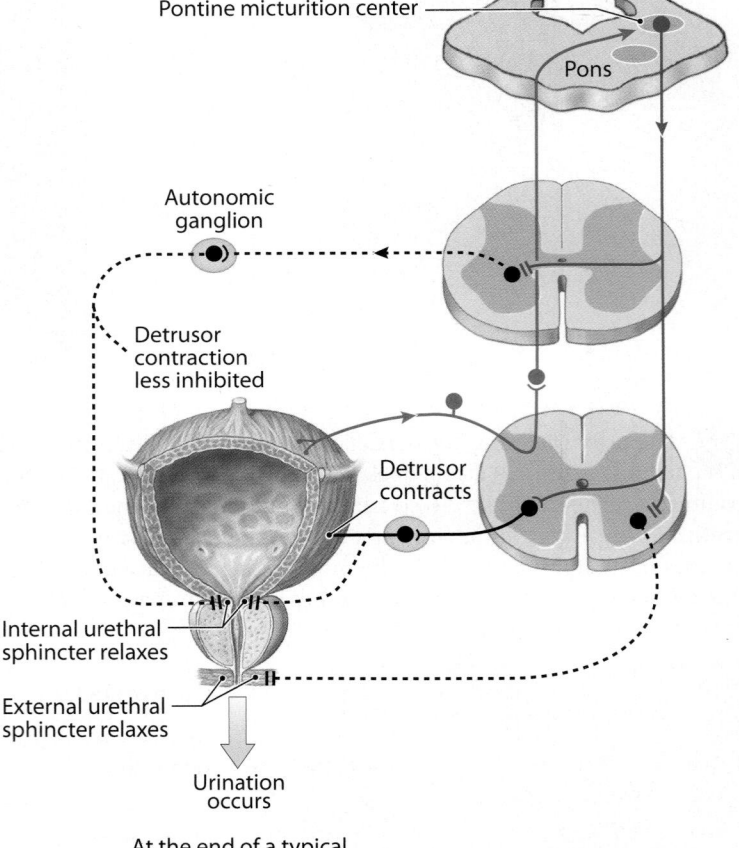

At the end of a typical micturition, less than 10 mL of urine remains in the bladder.

KEY
— Afferent
— Stimulated ⎫ Efferent
···‖ Inhibited ⎭
— Pontine efferent

**? REVIEW**
**B.** The ability to consciously control urination depends on your ability to control which muscle?

**Lo LEARNING OUTCOME**
Describe the urine storage and urine voiding reflexes.

# Module 24.18

# Urinary disorders can often be detected by physical examinations and laboratory tests

**1** The primary signs and symptoms of urinary system disorders include changes in volume and appearance of urine, frequency of urination, and pain. The nature and location of the pain can provide clues to the source.

Pain in the superior pubic region may be associated with urinary bladder disorders.

Pain in the superior lumbar region or in the flank that radiates to the right upper quadrant or left upper quadrant can be caused by kidney infections such as **pyelonephritis**, or by **renal calculi** (kidney stones).

**Dysuria** (painful or difficult urination) can occur with cystitis (bladder inflammation) or urethritis (ureter inflammation), or with urinary obstructions. In males, enlargement of the prostate can compress the urethra and lead to dysuria.

**?** REVIEW

**A.** What is the term for painful or difficult urination?

**2** Characteristic changes in urinary output or frequency give clues to underlying problems with this system or other systems.

## Abnormal Urine Output and Frequency

| Increased Urgency or Frequency | Changes in Urinary Output | Incontinence | Urinary Retention |
| --- | --- | --- | --- |
| An irritation of the lining of the ureters or urinary bladder can lead to **urgency** (strong desire to urinate) with **frequency** (increased need to urinate), although the total amount of urine produced each day remains normal. | Changes in the volume of urine produced by a person with average fluid intake indicate problems either with the kidneys or with the control of renal function. **Polyuria**, the production of excessive amounts of urine, results from hormonal or metabolic problems, such as those associated with diabetes or glomerulonephritis. **Oliguria** (a urine volume of 50–500 mL/day) and **anuria** (0–50 mL/day) are conditions that indicate serious kidney problems and potential renal failure. | **Incontinence**, an inability to control urination voluntarily, may involve **stress incontinence** (involuntary leakage), **urge incontinence** (inability to delay urination), or **overflow incontinence** (the continual, slow trickle of urine from an always-full bladder). | In **urinary retention**, renal function is normal, at least initially, but urination does not occur. Urinary retention in males commonly results from enlargement of the prostate and compression of the prostatic urethra. |

Important clinical signs of urinary system disorders include the following:

- **Edema**. Renal disorders often lead to **proteinuria** (protein in the urine) and if severe, result in generalized edema in peripheral tissues. Facial swelling, especially around the eyes, is common.

- **Fever**. A fever commonly develops with a *urinary tract infection* (*UTI*). **Cystitis** (urinary bladder inflammation) may result in a low-grade fever; kidney infections, such as pyelonephritis, can produce very high fevers.

**?** INTEGRATION

**B.** Why is urinary obstruction at the urethra more dangerous than at the ureter?

 **LEARNING OUTCOME**

Describe common urinary disorders related to output and frequency.

**24**

**Labeling:** Label the bladder structures in the diagram at right.

1 _____
2 _____
3 _____
4 _____
5 _____
6 _____
7 _____
8 _____

**Matching:** Match each lettered term with the most closely related description.

a. urethra
b. external urethral sphincter
c. detrusor
d. rugae
e. internal urethral sphincter
f. trigone
g. transitional epithelium
h. micturition
i. stratified squamous epithelium
j. external urethral orifice
k. middle umbilical ligament

9  The ring of smooth muscle in the neck of the urinary bladder
10 Triangular area within the urinary bladder
11 Relaxation of this muscle leads to urination
12 The external opening of the urethra
13 Folds lining the surface of the empty urinary bladder
14 Superior, supporting fibrous cord of the urinary bladder
15 Contraction of this smooth muscle compresses the urinary bladder
16 The type of epithelium that lines the ureters
17 Tube that transports urine to the exterior
18 Term for urination
19 Epithelium that lines the urethra

9  _____
10 _____
11 _____
12 _____
13 _____
14 _____
15 _____
16 _____
17 _____
18 _____
19 _____

**Short answer:** List four primary signs and symptoms of urinary disorders.

20 _____

Briefly describe the similarities and differences in the following pairs of terms.

21 cystitis/pyelonephritis
22 stress incontinence/overflow incontinence
23 polyuria/proteinuria

21 _____
22 _____
23 _____

24

# Study Outline

## SECTION 1 • Anatomy of the Urinary System

### Module 24.1

**The urinary system organs are the kidneys, ureters, urinary bladder, and urethra    p. 933**

1. The **urinary system** eliminates excess water, salts, and metabolic wastes through **urine**.

2. The **kidneys** perform the excretory functions of the urinary system. Urine is eliminated by the **urinary tract** (**ureters**, **urinary bladder**, and **urethra**).

3. The functions of the urinary system are adjusting blood volume and pressure, regulating blood plasma concentrations of ions, stabilizing blood pH, conserving nutrients by preventing their loss in urine, and removing drugs and toxins from the bloodstream.

### Module 24.2

**The kidneys are paired retroperitoneal organs    p. 934**

4. The **hilum** is a medial indentation on the kidney. It is the entry point for the renal artery and renal nerves, and an exit point for the renal vein and ureters.

5. The **ureters** pass inferiorly and cross the anterior surfaces of the external iliac artery and vein before emptying into the posterior, inferior surface of the urinary bladder.

6. The **kidneys**, adrenal glands, and ureters lie between the muscles of the posterior body wall and the parietal peritoneum, in a **retroperitoneal** position.

7. Visceral organs, the body wall musculature, and the 11th and 12th ribs protect the kidneys. The left kidney lies slightly superior to the right kidney.

8. A **fibrous capsule** of collagen fibers covers the outer surface of the kidney.

9. **Perinephric fat** surrounds the fibrous capsule. **Renal fascia** anchors the kidney to surrounding structures.

### Module 24.3

**The kidneys are complex at the gross and microscopic levels    p. 936**

10. The **fibrous capsule** also lines the **renal sinus**, an internal cavity within the kidney.

11. The **renal cortex** is the superficial region of the kidney. The **renal medulla** extends from the renal cortex to the renal sinus.

12. A **renal pyramid** is a conical structure extending from the cortex to a tip called the **renal papilla**. A **renal column** separates adjacent pyramids.

13. A **minor calyx** collects the urine produced by a single **kidney lobe**. A **major calyx** is a fusion of 4–5 **minor calyces**. The **renal pelvis** collects urine from the major calyces and is continuous with the ureters.

14. **Nephrons** are microscopic functional units of the kidney. Approximately 85 percent of nephrons are **cortical nephrons**. The remaining nephrons are **juxtamedullary nephrons** that have long **nephron loops**.

### Module 24.4

**A nephron is divided into segments; each segment has specific functions    p. 938**

15. A nephron consists of a **renal corpuscle** and a **renal tubule**.

16. At the renal corpuscle (which consists of the **glomerular capsule** and **glomerulus**), blood pressure forces water and dissolved solutes out of the glomerular capillaries and into the **capsular space**. Filtration produces **filtrate** that is similar to blood plasma.

17. After modification by the renal tubule and collecting system, the filtrate leaves the kidneys as urine.

18. The **proximal convoluted tubule** (**PCT**) reabsorbs nutrients from the **tubular fluid**. The **distal convoluted tubule** (**DCT**) further adjusts the tubular fluid.

19. The **nephron loop** establishes an osmotic gradient in the renal medulla that promotes water reabsorption from tubular fluid, permitting the production of concentrated urine.

*Renal corpuscle, PCT, nephron loop, DCT, and collecting system*

20. Each nephron empties into a **collecting system** of tubes (**collecting duct** and **papillary duct**) that performs final adjustments to urine volume and composition.

### Module 24.5

**The kidneys are highly vascular, and the circulation patterns are complex    p. 940**

21. The pathway of renal blood flow is: **renal artery** → **segmental arteries** → **interlobar arteries** → **arcuate arteries** → **cortical radiate arteries** → **afferent arterioles** → **glomerulus** → **efferent arterioles** → **peritubular capillaries** → **cortical radiate veins** → **arcuate veins** → **interlobar veins** → **renal vein**.

## SECTION 2 • Overview of Renal Physiology

### Module 24.6

**The kidneys maintain homeostasis by removing wastes and producing urine    p. 943**

22. Three solutes excreted in urine are **urea** (from liver breakdown of amino acids), **creatinine** (from the breakdown of creatine phosphate), and **uric acid** (from the recycling of the nitrogenous bases of RNA).

23. The three distinct processes of kidney function are **filtration**, **reabsorption**, and **secretion**.

24. The kidneys produce a fluid (urine) very different from other body fluids. Laboratory tests can measure the concentration of substances within the urine.

## Module 24.7

### Filtration, reabsorption, and secretion occur in specific segments of the nephron and collecting system p. 944

25. Filtration occurs at the renal corpuscle.

26. Most segments of the nephron perform a combination of reabsorption and secretion, but the balance between the two processes varies from one segment to another.

27. Final volume and solute concentration of the urine result from the interaction between the collecting system and the nephron loops.

## Module 24.8

### Filtration occurs at the renal corpuscle p. 946

28. The glomerulus filters blood through the filtration membrane.

29. The **afferent arteriole** delivers blood from a cortical radiate artery, while the **efferent arteriole** delivers blood to peritubular capillaries. The efferent arteriole has a smaller diameter than the afferent arteriole, and this elevates the blood pressure within the glomerulus.

30. The **juxtaglomerular complex** secretes renin when glomerular blood pressure decreases.

31. Substances passing out of the glomerulus must be small enough to pass through the **filtration slits** between the **foot processes** (*pedicels*) of the **podocytes**.

32. Glomerular capillaries are fenestrated capillaries. The endothelium covers a **basement membrane** that along with the foot processes of the podocyte forms the **filtration membrane**.

33. The primary factor involved in glomerular filtration is the balance between **hydrostatic pressure** and **colloid osmotic pressure**.

## Module 24.9

### The glomerular filtration rate is the amount of filtrate produced each minute p. 948

34. The **glomerular filtration rate (GFR)** is the amount of filtrate the kidneys produce each minute.

35. GFR is controlled by autoregulation at the local level and by central regulation, which has an endocrine component initiated by the kidneys and an autonomic component involving the sympathetic division of the ANS.

36. Through **autoregulation**, the kidneys adjust GFR when filtration rate decreases by dilation of the afferent arterioles, contraction of the intraglomerular mesangial cells and constriction of the efferent arterioles. The result is an increase in glomerular blood pressure and normal GFR.

37. If autoregulation is ineffective, **central regulation** responds by activating endocrine and neural mechanisms. If GFR does not return to normal, the juxtaglomerular complex increases renin production.

38. Renin triggers the formation of angiotensin I, which then is activated to angiotensin II by **angiotensin-converting enzyme (ACE)** in the capillaries of the lungs.

39. Angiotensin II triggers a series of **neural responses** to increase glomerular pressure, including stimulation of thirst centers, increased ADH and aldosterone production, increased cardiac output, and increased sympathetic motor tone.

## Module 24.10

### Reabsorption predominates along the proximal convoluted tubule, whereas reabsorption and secretion are often linked along the distal convoluted tubule p. 950

40. The proximal convoluted tubule (PCT) reabsorbs more than 99 percent of the glucose, amino acids, and other organic nutrients in the fluid. The PCT also reabsorbs sodium, potassium, bicarbonate, magnesium, phosphate, and sulfate ions.

41. Only 15 to 20 percent of the initial filtrate volume reaches the distal convoluted tubule (DCT). In the DCT, sodium ions are reabsorbed and exchanged for potassium ions by ion pumps that are stimulated by aldosterone. When pH decreases, hydrogen ions are exchanged for sodium ions, and the DCT also secretes toxins and drugs.

## Module 24.11

### Exchange between the limbs of the nephron loop creates an osmotic concentration gradient in the renal medulla p. 952

42. The exchange that occurs between the descending thin limb and the thick ascending limb of the nephron loop is called **countercurrent multiplication**.

43. *Countercurrent* refers to the fact that tubular fluid in the descending limb flows toward the renal pelvis, whereas tubular fluid in the ascending limb flows toward the renal cortex.

44. *Multiplication* refers to the fact that the effect of the exchange increases as fluid movement continues.

45. Active transport in the thick ascending limb moves sodium and chloride ions out of the tubular fluid and into the peritubular fluid of the renal medulla.

46. The descending thin limb is permeable to water but impermeable to solutes.

47. As water is reabsorbed along the DCT and collecting duct, the concentration of urea gradually increases in the tubular fluid.

## Module 24.12

### Urine volume and concentration are hormonally regulated p. 954

48. Water reabsorption may occur passively or by hormonal control. **Obligatory water reabsorption**, when water movements cannot be prevented, occurs in the PCT and descending limb of the nephron loop.

49. **Facultative water reabsorption**, which is controlled by antidiuretic hormone (ADH), occurs in the DCT and collecting system. **Aquaporins** in the apical surfaces of plasma membranes lining the DCT and collecting duct enhance the rate of osmotic water movement.

50. An adult typically produces 1200 mL of urine per day.

24

## Module 24.13

### Renal function is an integrative process involving filtration, reabsorption, and secretion    p. 956

51. The filtrate produced by the renal corpuscle has the same osmotic concentration as plasma, about 300 mOsm/L.

52. In the PCT, the active removal of ions and organic nutrients produces a continuous osmotic flow of water out of the tubular fluid. Obligatory water reabsorption in the PCT and descending limb moves water into the peritubular fluid leaving highly concentrated tubular fluid.

53. The thick ascending limb of the nephron loop is impermeable to water and solutes. The tubule cells actively transport $Na^+$ and $Cl^-$ out of the tubule.

54. Further adjustments in the composition of the tubular fluid occur in the DCT and the collecting system.

55. Final adjustments of the tubular fluid are made in the DCT and the collecting system where the level of ADH determines the final urine concentration.

56. The vasa recta absorb the solutes and water reabsorbed by the nephron loop, DCT, and the collecting ducts.

## Module 24.14

### Renal failure is a life-threatening condition    p. 958

57. **Renal failure** occurs when the kidneys cannot perform the excretory functions needed to maintain homeostasis.

58. In **chronic renal failure**, kidney function deteriorates gradually, and the associated problems accumulate over time.

59. **Acute renal failure** occurs when exposure to toxic drugs, renal ischemia, urinary obstruction, or trauma causes filtration to slow suddenly or stop.

60. In **hemodialysis** an artificial membrane regulates the composition of blood by means of a dialysis machine. The basic principle in this process, called **dialysis**, is passive diffusion across a selectively permeable membrane. Diffusion across the dialysis membrane takes the place of normal glomerular filtration.

*Hemodialysis machine*

## SECTION 3 • Urine Storage and Elimination

## Module 24.15

### The urinary tract transports, stores, and eliminates urine p. 961

61. The **urinary tract** is composed of the ureters, urinary bladder, and urethra. It transports, stores, and eliminates urine.

62. The **ureters** are a pair of muscular tubes that extend from the kidneys to the urinary bladder.

63. The **urinary bladder** is a hollow, muscular organ that serves as a temporary reservoir for urine.

64. The **urethra** extends from the neck of the urinary bladder and transports urine to the exterior of the body. The male urethra is longer, and transports semen as well as urine.

## Module 24.16

### The ureters, urinary bladder, and urethra are specialized to conduct urine    p. 962

65. The **lateral umbilical ligaments** and the **middle umbilical ligament** support the urinary bladder.

66. The **ureters** penetrate the posterior wall of the urinary bladder at an oblique angle without entering the peritoneal cavity. The **ureteric orifices** are slitlike to help prevent the backflow of urine as the bladder contracts.

67. The **trigone** is a triangular area bounded by the urethral openings and the entrance to the urethra. It acts like a funnel that channels urine toward the urethra. **Rugae** are folds of the urinary bladder lining that disappear as the bladder fills.

68. The **neck of the urinary bladder** is the region surrounding the urethral opening. It contains the smooth muscle **internal urethral sphincter**.

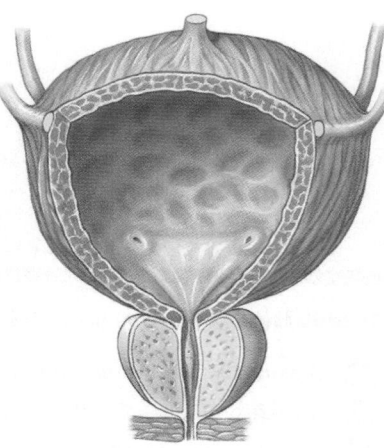

69. Each ureter consists of three layers: (1) an inner mucosa, made up of a transitional epithelium and the surrounding lamina propria; (2) a middle muscular layer made up of longitudinal and circular bands of smooth muscle; and (3) an outer connective tissue layer. Peristalsis moves urine toward the urinary bladder.

70. The urinary bladder wall contains mucosa, submucosa, and muscular layers. The muscular layer forms the **detrusor** that compresses the urinary bladder when it contracts to force urine into the urethra.

71. The urethral lining consists of a stratified epithelium that varies from transitional at the neck of the urinary bladder, to stratified columnar at the midpoint, to stratified squamous near the external urethral orifice.

## Module 24.17

### Urinary reflexes coordinate urine storage and voiding    p. 964

72. The **urine storage reflex** and the **urine voiding reflex** control *micturition* (urination). Urine storage occurs by spinal reflexes and the **pontine storage center**. Urine voiding occurs by spinal reflexes and the **pontine micturition center**.

73. Voluntary relaxation of the external urethral sphincter causes relaxation of the internal urethral sphincter. Because the local pathway already has increased pressures within the urinary bladder, relaxation of these sphincters leads to urination.

## Module 24.18

### Urinary disorders can often be detected by physical examinations and laboratory tests    p. 965

74. The primary signs and symptoms of urinary system disorders include changes in volume and appearance of urine, frequency of urination, and pain.

75. **Pyelonephritis** (kidney infection) or **renal calculi** (kidney stones) can cause pain in the superior lumbar region. **Dysuria** (painful or difficult urination) can be due to a variety of causes.

76. **Proteinuria** (protein in the urine), **edema**, **fever**, and **cystitis** (urinary bladder inflammation) are all important clinical signs of urinary system disorders.

24

# Chapter Review Questions

**Labeling:** Label the image of a renal corpuscle and associated structures shown here.

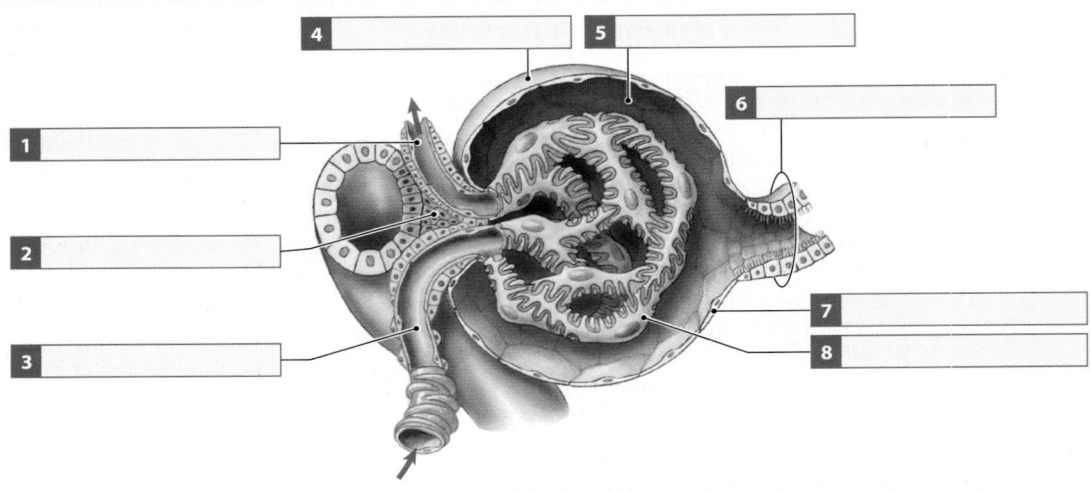

4 _____  5 _____

1 _____  6 _____

2 _____

3 _____  7 _____

8 _____

**True/False:** Indicate whether each statement is true or false.

9  The urinary tract includes the kidneys, ureters, urinary bladder, and urethra.

10  The left kidney lies slightly superior to the right kidney.

11  Most nephrons are cortical nephrons.

12  Tubular fluid is no longer modified beyond the DCT.

13  Contraction of the detrusor muscle moves urine along the ureters, toward the urinary bladder.

14  The process of urination is coordinated by the micturition reflex.

9  _____

10  _____

11  _____

12  _____

13  _____

14  _____

**Multiple choice:** Select the correct answer from the list provided.

15  The basic functional unit of the kidney is the
  - a) filtration unit.
  - b) nephron loop.
  - c) glomerulus.
  - d) nephron.

16  When ADH levels increase,
  - a) the amount of water reabsorbed increases.
  - b) the amount of water reabsorbed decreases.
  - c) the DCT becomes impermeable to water.
  - d) sodium ions are exchanged for potassium ions.

17  Which of the following conditions would cause an increase in the glomerular filtration rate?
  - a) constriction of the afferent arteriole
  - b) constriction of the efferent arteriole
  - c) decrease in systemic blood pressure
  - d) decrease in aldosterone and ADH production

18  Which of the following statements best describes the action of aldosterone at the DCT?
  - a) Aldosterone stimulates actions that cause the secretion of drugs and toxins from the peritubular fluid.
  - b) Sodium ions are reabsorbed in exchange for potassium ions by ion pumps stimulated by aldosterone.
  - c) Aldosterone increases water secretion.
  - d) Aldosterone causes glucose to be reabsorbed into the peritubular fluid.

19  Water reabsorption occurs primarily along the
  - a) entire length of the nephron loop.
  - b) ascending limb of the nephron loop and the DCT.
  - c) PCT and descending limb of the nephron loop.
  - d) DCT and collecting system.

20  In central regulation, decreased GFR is coordinated by the
  - a) renal corpuscle.
  - b) juxtamedullary complex.
  - c) renal pelvis.
  - d) cerebral cortex.

24

**21** Urea is a by-product of the breakdown of

⬜ a) amino acids by the liver.

⬜ b) creatine phosphate in skeletal muscle.

⬜ c) the recycling of the nitrogenous bases in RNA molecules.

⬜ d) glycogen to glucose.

**22** Which of the following is *not* a response to angiotensin II?

⬜ a) decreased ADH production

⬜ b) increased stimulation of the thirst centers

⬜ c) increased aldosterone secretion by the adrenal glands

⬜ d) constriction of peripheral arteries and further constriction of the efferent arterioles

---

**Short answer**

**23** Identify and describe the three distinct processes in urine production.

**24** Describe the primary process involved in glomerular filtration.

**25** What is countercurrent multiplication?

---

# MasteringA&P™

Access more chapter study tools online in the MasteringA&P Study Area:

■ Chapter Quizzes, Chapter Practice Test, and Clinical Case Studies

| ■ Practice Anatomy Lab™ 3.0 | PAL 3.0 |
| ■ Interactive Physiology™ | iP2™ |
| ■ A&P Flix™ | *A&PFlix*™ |
| ■ PhysioEx™ 9.1 | PhysioEx™ 9.1 |

---

# Chapter Integration • Applying what you have learned

## Melamine contamination and food product safety

For consumers, some troubling developments have turned the purity of food and food products, for humans and animals alike, from a matter of trust into a matter of concern.

In 2007, the unintentional adulteration of pet food made in China with the nitrogen-containing industrial chemical melamine caused pet illnesses and deaths in the United States and resulted in massive recalls of contaminated products. In 2008, milk intentionally contaminated with melamine—put there to skew laboratory tests that measure nitrogen as an index of the protein content in a food or drink powder—sickened at least 64,000 children in China, several of whom died. Clearly, vigilance concerning our food supply is crucial, even while standards and regulations are in place.

In the body, melamine contamination has two major effects: (1) the formation of crystalline masses in the filtrate and/or urine, and (2) acidification of the tubular fluid. Resulting clinical problems range from blood in the urine, to acid-base and electrolyte disorders, to urinary obstruction and (in severe cases) kidney failure.

**1** Propose a linkage between the stated effects of melamine poisoning and the clinical problems observed.

**2** Primary therapeutic options for melamine poisoning include infusion of fluids, dialysis, and medication. Explain how these options address specific problems caused by melamine poisoning, and suggest possible follow-up tests.

24

**LEARNING OUTCOMES**

These Learning Outcomes correspond by number to this chapter's modules and indicate what you should be able to do after completing the chapter. They also appear at the end of each module.

**Fluid and Electrolyte Balance**

**25.1** Name the body's fluid compartments, identify the solid components, and summarize their contents. p. 973

**25.2** Explain what is meant by fluid balance, and discuss its importance for homeostasis. p. 974

**25.3** Explain what is meant by mineral balance, and discuss its importance for homeostasis. p. 976

**25.4** Summarize the relationship between sodium and water in maintaining fluid and electrolyte balance. p. 978

**25.5** Clinical Module: Explain factors that control potassium balance, and discuss hypokalemia and hyperkalemia. p. 980

**Acid-Base Balance**

**25.6** Describe the three categories of acids in the body. p. 983

**25.7** Explain the role of buffer systems in maintaining acid-base balance and pH. p. 984

**25.8** Explain the role of buffer systems in regulating the pH of the intracellular fluid and the extracellular fluid. p. 986

**25.9** Describe the compensatory mechanisms involved in maintaining of acid-base balance. p. 988

**25.10** Clinical Module: Describe respiratory acidosis and respiratory alkalosis. p. 990

# 25

# Fluid, Electrolyte, and Acid-Base Balance

# Module 25.1

# Body composition may be viewed in terms of solids and two fluid compartments

Think about a small pond and the fish whose lives are totally dependent on its water quality. Your cells live in a pond, and they rely on water as a diffusion medium for gas, nutrient, and waste distribution. This chapter focuses on the homeostatic mechanisms that regulate body fluids and are critical to cells.

**1** The **extracellular fluid**, or **ECF**, is made up of the interstitial fluid of peripheral tissues and the plasma of circulating blood; the **intracellular fluid**, or **ICF**, consists of the cytosol inside cells. Minor components of the ECF include lymph, cerebrospinal fluid (CSF), synovial fluid, serous fluids (pleural, pericardial, and peritoneal fluids), aqueous humor, perilymph, and endolymph. The ECF and ICF are called **fluid compartments** because they commonly behave separately, maintaining different ionic compositions. These pie charts compare the total body composition of adult males and females. In both sexes, the ICF contains a greater proportion of total body water than the ECF.

**REVIEW**
**A.** Define *ECF* and *ICF*.
**B.** Describe the fluid compartments.

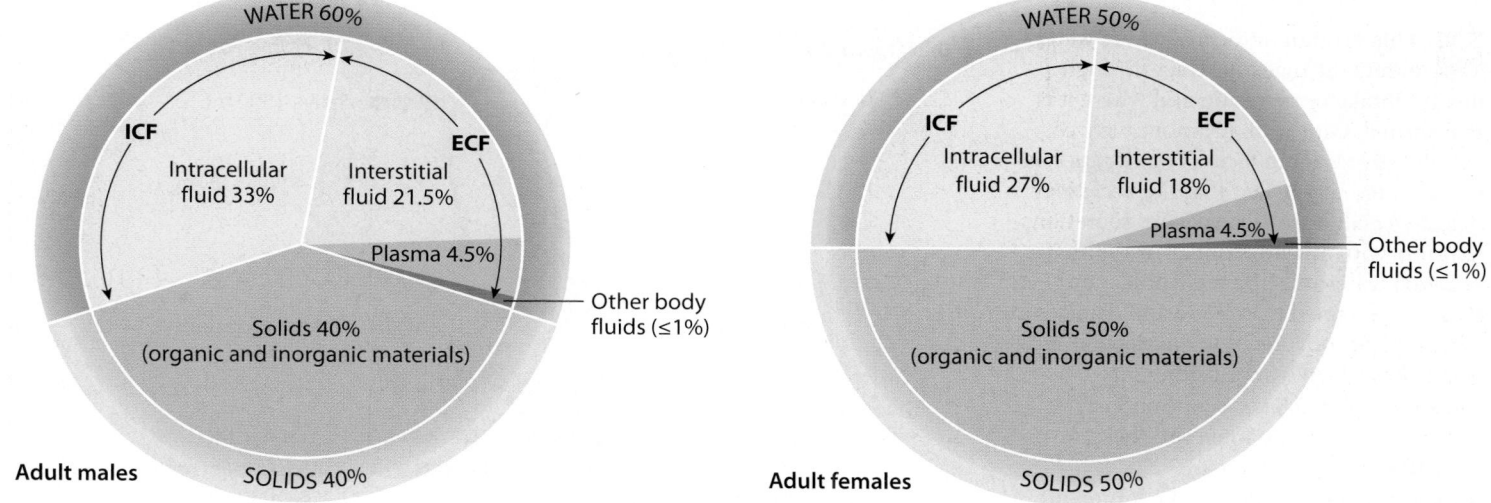

**Adult males**

WATER 60%
ICF — Intracellular fluid 33%
ECF — Interstitial fluid 21.5%
Plasma 4.5%
Other body fluids (≤1%)
Solids 40% (organic and inorganic materials)
*SOLIDS 40%*

**Adult females**

WATER 50%
ICF — Intracellular fluid 27%
ECF — Interstitial fluid 18%
Plasma 4.5%
Other body fluids (≤1%)
Solids 50% (organic and inorganic materials)
*SOLIDS 50%*

**2** Solid components make up only 40–50 percent of the body mass. This bar graph presents an overview of the solid components of a 70-kg (154-pound) person with a minimum of body fat. The distribution was obtained by averaging values for males and females ages 18–40 years.

**SOLID COMPONENTS** (31.5 kg; 69.3 lb)

Kg

15

10

5

0

Proteins  Lipids  Minerals  Carbohydrates  Miscellaneous

**REVIEW**
**C.** Which solid component makes up most of the body mass?

**LEARNING OUTCOME**
Name the body's fluid compartments, identify the solid components, and summarize their contents.

# Module 25.2

# Fluid balance exists when water gain equals water loss

**1** Your body is in **fluid balance** when its water content remains stable over time. Water gains occur primarily in the digestive tract. Water losses occur through many routes, but almost half of daily water loss occurs through urination.

**? REVIEW**

**A.** Identify routes of fluid loss from the body.

| Fluid Balance | |
|---|---|
| **Source** | **Daily Input (mL)** |
| Water content of food | 1000 |
| Water consumed as liquid | 1200 |
| Metabolic water produced during catabolism | 300 |
| **Total** | **2500** |
| **Method of Elimination** | **Daily Output (mL)** |
| Urination | 1200 |
| Evaporation at skin | 750 |
| Evaporation at lungs | 400 |
| Loss in feces | 150 |
| **Total** | **2500** |

**2** This diagram indicates where water enters the digestive tract through dietary intake or secretion and where it is reabsorbed. Only a small amount leaves the digestive tract in feces. The situation is complicated by the fact that the accessory digestive glands produce watery secretions that mix with arriving food. Most of that secreted water must be recovered along with water gained from food and drink. All of this water movement involves passive water flow down osmotic gradients. Intestinal epithelial cells continuously absorb nutrients and ions, and these activities gradually decrease the solute concentration in the lumen of the digestive tract and increase the solute concentration in the interstitial fluid of the lamina propria. As the solute concentration decreases in the lumen, water moves across the epithelium and into the interstitial fluid, maintaining osmotic equilibrium. Once within the interstitial fluid, the absorbed water is rapidly distributed throughout the ECF.

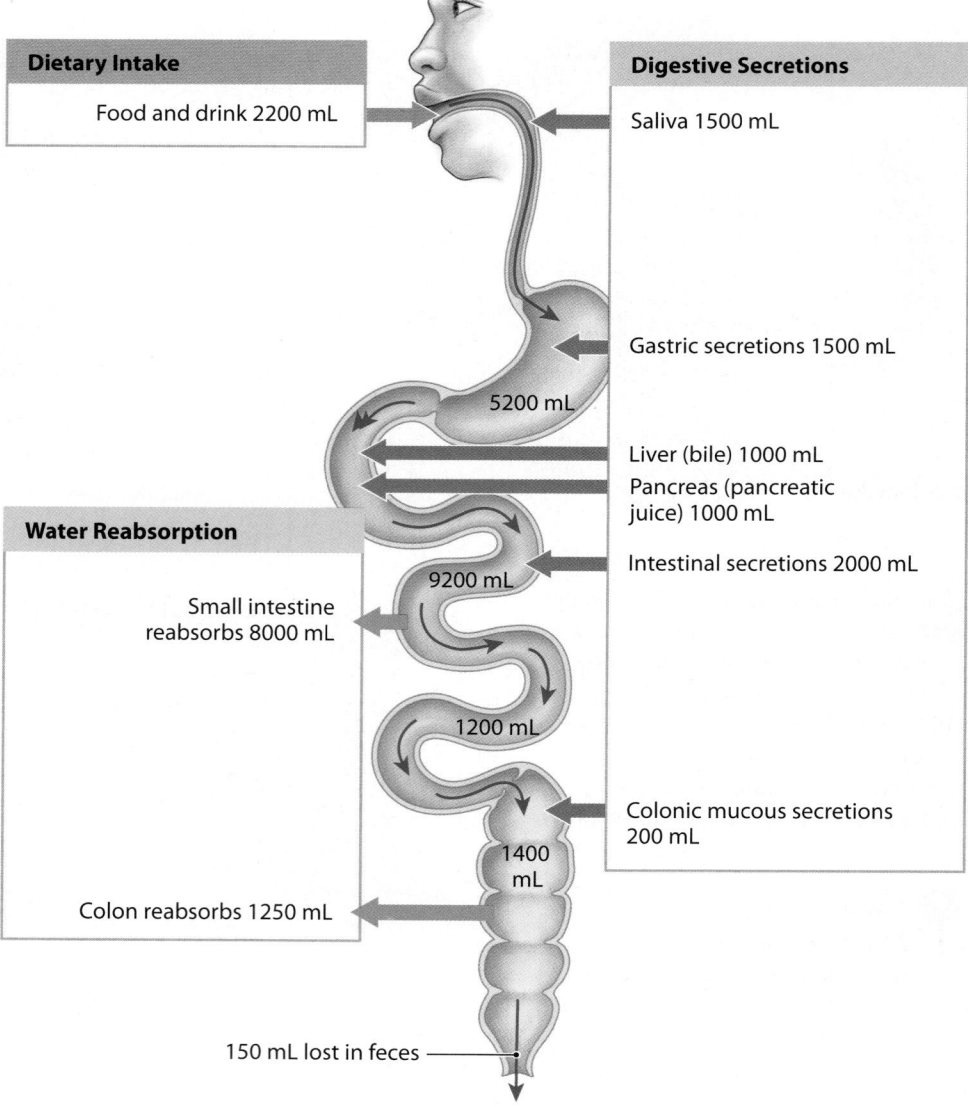

**Dietary Intake**

Food and drink 2200 mL

**Digestive Secretions**

Saliva 1500 mL

Gastric secretions 1500 mL

5200 mL

Liver (bile) 1000 mL

Pancreas (pancreatic juice) 1000 mL

**Water Reabsorption**

Intestinal secretions 2000 mL

9200 mL

Small intestine reabsorbs 8000 mL

1200 mL

Colonic mucous secretions 200 mL

1400 mL

Colon reabsorbs 1250 mL

150 mL lost in feces

**3** This diagram illustrates the major factors that affect ECF volume. Although the composition of the ECF and that of ICF are very different, the two are at osmotic equilibrium. Note that the volume of the ICF is larger than that of the ECF. The volume of water held within cells represents a significant reserve that can prevent sudden changes in the solute and water concentrations in the ECF.

A rapid water movement between the ECF and the ICF in response to an osmotic gradient is called a **fluid shift**. Fluid shifts occur rapidly in response to changes in the osmotic concentration of the ECF and reach equilibrium within minutes to hours. This can be an important factor when water intake is restricted but water losses are severe.

Water absorbed across digestive epithelium (2200 mL)

Water vapor lost at skin and lungs (1150 mL)

Water lost in feces (150 mL)

Water secreted by sweat glands (variable)

Metabolic water (300 mL)

ICF

ECF

Plasma membranes of tissue cells

Water lost in urine (1200 mL)

 **REVIEW**

**B.** Describe a fluid shift.

**4** **Dehydration** develops when water losses outpace water gains. When you lose water but retain electrolytes, the osmotic concentration of the ECF increases. Osmosis then moves water out of the ICF and into the ECF until the two solutions are again isotonic. At that point, both the ECF and ICF are somewhat more concentrated than normal, and both volumes are lower than they were before the fluid loss. Because the ICF is considerably larger than the ECF, the ICF is an effective water reserve. However, if the fluid imbalance continues unchecked, the loss of water from the ICF will produce severe thirst, dryness, and wrinkling of the skin. Eventually blood volume and blood pressure may decrease to the point at which circulatory shock develops.

The ECF and ICF are in balance, with the two solutions isotonic.

Intracellular fluid (ICF)

Extracellular fluid (ECF)

Water loss from ECF decreases volume and makes this solution hypertonic with respect to the ICF.

ECF water loss

An osmotic water shift from the ICF into the ECF restores osmotic equilibrium but decreases the ICF volume.

Decreased ICF volume

Increased ECF volume

 **REVIEW**

**C.** Explain dehydration and its effect on the osmotic concentration of blood.

 **LEARNING OUTCOME**

Explain what is meant by fluid balance, and discuss its importance for homeostasis.

# Module 25.3
## Mineral balance involves balancing electrolyte gain and loss

**1** Minerals and electrolytes are closely associated. **Minerals** are inorganic substances, and **electrolytes** are the ions released when mineral salts dissociate. **Mineral balance** exists when ion absorption and excretion are about the same.

 **REVIEW**

**A.** Define *mineral balance*.

**Ion Absorption**

Ion absorption occurs across the epithelial lining of the small intestine and colon.

Ion reserves (primarily in the skeleton)

Ion pool in body fluids

ICF    ECF

**Ion Excretion**

Sweat gland secretions (secondary site of ion loss)

Kidneys (primary site of ion loss)

**2** Within body fluids, the dissociated salts are *electrolyte solutions*. The movement of the electrolytes determines fluid flow.

 **REVIEW**

**B.** Identify the electrolytes absorbed by active transport.

| Movement of Electrolytes in Solution | |
|---|---|
| **Electrolyte** | **Mechanism** |
| Sodium ($Na^+$) | Channel-mediated diffusion, cotransport, or active transport |
| Calcium ($Ca^{2+}$) | Active transport |
| Potassium ($K^+$) | Channel-mediated diffusion |
| Magnesium ($Mg^{2+}$) | Active transport |
| Iron ($Fe^{2+}$) | Active transport |
| Chloride ($Cl^-$) | Channel-mediated diffusion or carrier-mediated transport |
| Iodide ($I^-$) | Channel-mediated diffusion or carrier-mediated transport |
| Bicarbonate ($HCO_3^-$) | Channel-mediated diffusion or carrier-mediated transport |
| Nitrate ($NO_3^-$) | Channel-mediated diffusion or carrier-mediated transport |
| Phosphate ($PO_4^{3-}$) | Active transport |
| Sulfate ($SO_4^{2-}$) | Active transport |

**3** The body contains substantial reserves of key minerals. This table summarizes the functions of various minerals and indicates the primary routes of ion excretion. The amount lost each day must equal the daily intake to maintain mineral balance.

## Minerals and Mineral Reserves

| Mineral | Functions | Total Body Content (Adults) | Primary Route of Excretion | Recommended Dietary Allowance (RDA) (in mg) |
|---|---|---|---|---|
| **Bulk Minerals** | | | | |
| **Sodium (Na)** | Major cation in body fluids; essential for normal membrane function | 110 g, primarily in body fluids | Urine, sweat, feces | 1500 |
| **Potassium (K)** | Major cation in cytosol; essential for normal membrane function | 140 g, primarily in cytosol | Urine | 4700 |
| **Chloride (Cl)** | Major anion in body fluids; functions in forming HCl | 89 g, primarily in body fluids | Urine, sweat | 2300 |
| **Calcium (Ca)** | Essential for normal muscle and neuron function and normal bone structure | 1.36 kg, primarily in skeleton | Urine, feces | 1000–1200 |
| **Phosphorus (P)** | In high-energy compounds, nucleic acids, and bone matrix (as phosphate) | 744 g, primarily in skeleton | Urine, feces | 700 |
| **Magnesium (Mg)** | Cofactor of enzymes, required for normal membrane functions | 29 g (skeleton, 17 g; cytosol and body fluids, 12 g) | Urine | 310–400 |
| **Trace Minerals** | | | | |
| **Iron (Fe)** | Component of hemoglobin, myoglobin, and cytochromes | 3.9 g (1.6 g stored as ferritin or hemosiderin) | Urine (traces) | 8–18 |
| **Zinc (Zn)** | Cofactor of enzyme systems, notably carbonic anhydrase | 2 g | Urine, hair (traces) | 8–11 |
| **Copper (Cu)** | Required as cofactor for hemoglobin synthesis | 127 mg | Urine, feces (traces) | 0.9 |
| **Manganese (Mn)** | Cofactor for some enzymes | 11 mg | Feces, urine (traces) | 1.8–2.3 |
| **Cobalt (Co)** | Cofactor for transaminations; mineral in vitamin $B_{12}$ (cobalamin) | 1.1 g | Feces, urine | 0.0001 |

## Everyday Physiology

### Which sports drink to choose?

Some sports drinks contain sugars and vitamins as well as electrolytes. During endurance events (marathons, ultramarathons, and distance cycling), solutions containing less than 10 g/dL of glucose may improve performance if consumed late in the event, when metabolic reserves are exhausted. However, high sugar concentrations (above 10 g/dL) can cause cramps, diarrhea, and other problems. The benefit of "glucose polymers" (often cornstarch) in sports drinks has yet to be proved. Drinking beverages "fortified" with vitamins is actively discouraged: Vitamins are not lost during exercise, and the consumption of these beverages in large volumes could, over time, cause hypervitaminosis.

 **INTEGRATION**

**C.** Explain the significance of two important body minerals: sodium and calcium.

 **LEARNING OUTCOME**

Explain what is meant by mineral balance, and discuss its importance for homeostasis.

# Module 25.4

# Water balance depends on sodium balance, and the two are regulated simultaneously

## Response to Changes in Sodium Levels

**1** **Sodium balance** exists when sodium gains equal sodium losses. The regulatory mechanism involved changes the ECF volume but keeps the Na$^+$ concentration relatively stable. When sodium gains exceed losses, the ECF volume increases; when sodium losses exceed gains, the volume of the ECF decreases. The changes in ECF volume occur without a significant change in the osmotic concentration of the ECF. These adjustments cause minor changes in ECF volume that do not cause adverse physiological effects.

ADH secretion increases

**Effector**
Kidney

Water reabsorption by kidneys

**Recall of Fluids**

Because the ECF osmolarity increased, water shifts out of the ICF, increasing ECF volume and decreasing Na$^+$ concentration

Thirst promotes water consumption

**Receptors**
Osmoreceptors in hypothalamus stimulated

Homeostasis **DISTURBED BY**

INCREASING

Na$^+$ levels in ECF

STIMULUS

RESTORED

Homeostasis **RESTORED BY**

DECREASING

Na$^+$ levels in ECF

# HOMEOSTASIS
## NORMAL Na$^+$ CONCENTRATION IN ECF

Homeostasis **DISTURBED BY**

DECREASING

Na$^+$ levels in ECF

STIMULUS

RESTORED

Homeostasis **RESTORED BY**

INCREASING

Na$^+$ levels in ECF

**Receptors**
Osmoreceptors in hypothalamus inhibited

Thirst is suppressed

Water loss decreases the ECF volume, the osmolarity increases and Na$^+$ ion concentration increases

ADH secretion decreases

**Effector**
Kidney

Water loss by kidney increases

**REVIEW**

**A.** What effect does inhibition of osmoreceptors have on ADH secretion and thirst?

## Response to Changes in ECF Volume

**2** If the changes in ECF volume caused by disturbances in sodium balance are extreme, the homeostatic mechanisms responsible for regulating blood volume and blood pressure will be activated. When ECF volume changes, so does plasma volume and, in turn, blood volume. If ECF volume increases, blood volume increases; if ECF volume decreases, blood volume decreases.

Sustained abnormalities in the Na$^+$ concentration in the ECF occur only when there are severe problems with fluid balance. When the Na$^+$ concentration of the ECF falls below 136 mEq/L, a state of **hyponatremia** (*natrium*, sodium) exists; this can be caused by excessive water intake (overhydration) or inadequate salt intake. When body water content decreases, the Na$^+$ concentration increases; when that concentration exceeds 145 mEq/L, **hypernatremia** exists; dehydration is the most common cause. Both conditions are serious and potentially life threatening.

**? REVIEW**

**B.** What effect does aldosterone have on sodium ion concentration in the ECF?

**Lo LEARNING OUTCOME**

Summarize the relationship between sodium and water in maintaining fluid and electrolyte balance.

# Module 25.5

# Disturbances of potassium balance are uncommon but extremely dangerous

**1** This diagram indicates the major factors involved in **potassium balance**. The key factors in maintaining potassium balance are the rate of K+ entry across the digestive epithelium and the rate of K+ loss into urine.

## Factors Controlling Potassium Balance

Approximately 100 mEq (1.9–5.8 g) of potassium ions are absorbed by the digestive tract each day.

About 98 percent of the potassium content of the human body is in the ICF, rather than the ECF.

The K+ concentration in the ECF is relatively low. The rate of K+ entry from the ICF through leak channels is balanced by the rate of K+ recovery by the Na+-K+ exchange pump.

When potassium balance exists, the rate of urinary K+ excretion matches the rate of digestive tract absorption.

K+                              K+

The potassium ion concentration in the ECF is approximately 5 mEq/L.

K+

Na+

**K+**

The potassium ion concentration in the ICF is approximately 135 mEq/L.

Renal K+ losses are approximately 100 mEq per day.

### KEY

→ = Absorption

→ = Secretion

- - → = Diffusion through leak channels

**2** Because dietary intake is relatively constant, the kidneys are the main factor determining the K+ concentration in the ECF. Ion pumps along the distal portions of the nephron and collecting systems control potassium loss by the kidneys. Aldosterone regulates these pumps. Aldosterone stimulates Na+ reabsorption while simultaneously accelerating K+ secretion (see Module 24.10).

The primary mechanism of potassium secretion involves an exchange pump that ejects potassium ions while reabsorbing sodium ions.

K+          K+

Na+          Na+

Tubular fluid          ECF

The sodium ions are then pumped out of the cell in exchange for potassium ions in the ECF. This is the same pump that ejects sodium ions entering the cytosol through leak channels.

### KEY

(A) = Aldosterone-sensitive exchange pump

○ = Sodium-potassium exchange pump

**? REVIEW**

**A.** Which organs are primarily responsible for regulating the potassium ion concentration in the ECF?

**3** Potassium balance does not occur in isolation. For example, aldosterone secretion increases in response to either high blood K⁺ concentrations or a decrease in ECF sodium levels. However, disturbances of acid-base balance can disrupt potassium balance by affecting the aldosterone-sensitive pumps (see Module 24.13).

**? REVIEW**

**B.** Identify factors that cause potassium excretion.

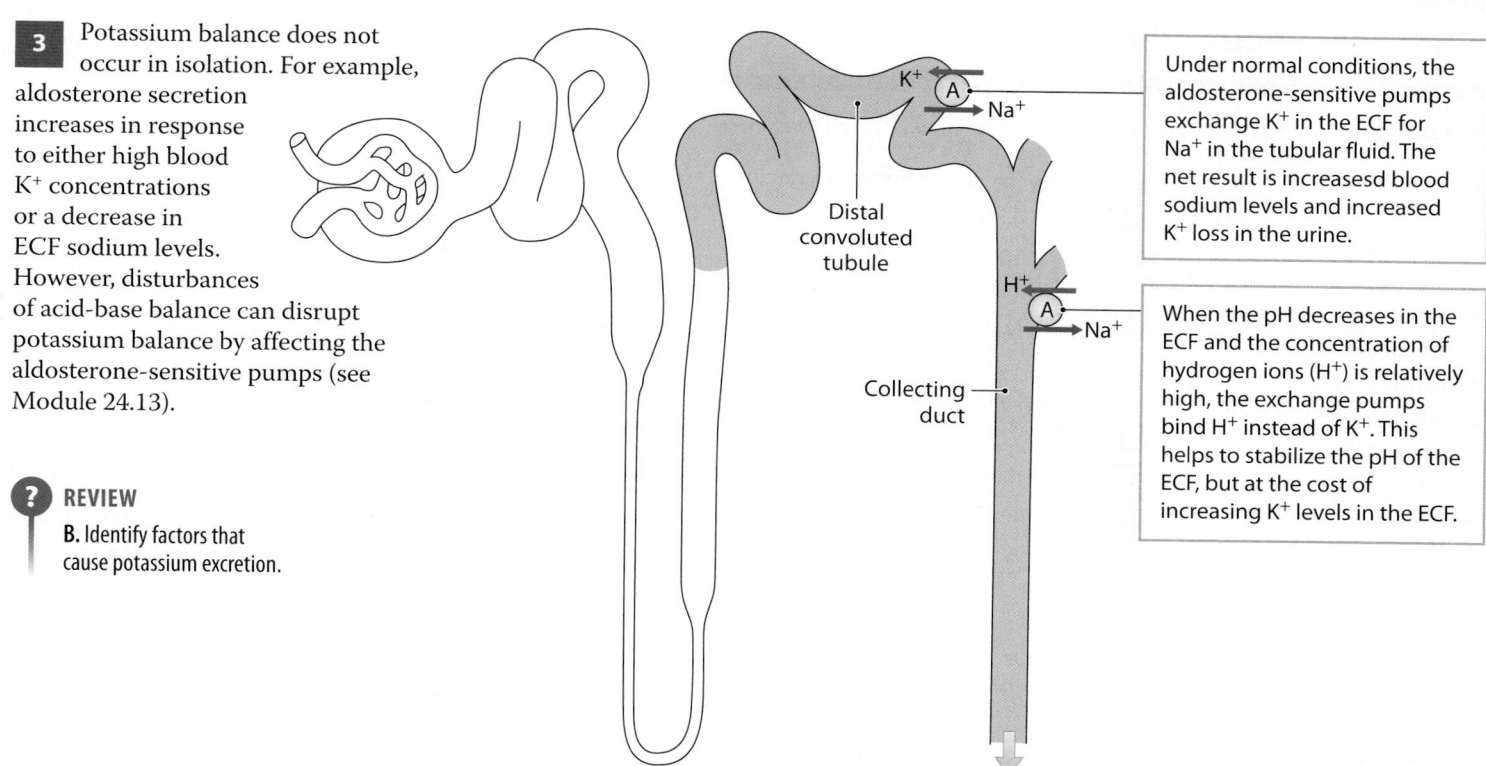

Distal convoluted tubule

Collecting duct

Under normal conditions, the aldosterone-sensitive pumps exchange K⁺ in the ECF for Na⁺ in the tubular fluid. The net result is increasesd blood sodium levels and increased K⁺ loss in the urine.

When the pH decreases in the ECF and the concentration of hydrogen ions (H⁺) is relatively high, the exchange pumps bind H⁺ instead of K⁺. This helps to stabilize the pH of the ECF, but at the cost of increasing K⁺ levels in the ECF.

**4** The diagram below indicates major factors involved in disturbances of potassium balance. The normal range of K⁺ concentration in the ECF is 3.5–5.0 mEq/L. Variations outside of that range are unusual and potentially dangerous.

When the blood concentration of potassium falls below 2 mEq/L, extensive muscular weakness develops, followed by eventual paralysis. This condition, called severe **hypokalemia** (*kalium*, potassium), is potentially lethal due to its effects on the heart.

Normal potassium levels in blood (3.5–5.0 mEq/L)

High K⁺ concentrations in the blood produce an equally dangerous condition known as **hyperkalemia**. Severe hyperkalemia occurs when K⁺ concentration exceeds 7 mEq/L, which results in cardiac arrhythmias.

| Factors Promoting Hypokalemia | |
|---|---|
| Several diuretics can produce hypokalemia by increasing the volume of urine produced. | The endocrine disorder called **aldosteronism**, characterized by excessive aldosterone secretion, results in hypokalemia by overstimulating sodium retention and potassium loss. |

| Factors Promoting Hyperkalemia | | |
|---|---|---|
| Chronically low blood pH promotes hyperkalemia by interfering with K⁺ excretion by the kidneys. | Kidney failure due to damage or disease will prevent normal K⁺ secretion and thereby produce hyperkalemia. | Several drugs promote urination by blocking Na⁺ reabsorption by the kidneys. When sodium reabsorption slows down, so does potassium secretion, and hyperkalemia can result. |

Treatment for hypokalemia generally includes increasing dietary intake of potassium by salting food with potassium salts (KCl) or by taking potassium tablets. Treatment for hyperkalemia typically includes diluting the ECF with a solution low in K⁺, stimulating K⁺ loss in urine by using diuretics, adjusting the pH of the ECF, and restricting dietary K⁺ intake. In cases resulting from renal failure, kidney dialysis may also be required.

 **REVIEW**

**C.** Define *hypokalemia* and *hyperkalemia*.

 **LEARNING OUTCOME**

Explain factors that control potassium balance, and discuss hypokalemia and hyperkalemia.

# Fluid and Electrolyte Balance

**Matching:** Match each lettered term with the most closely related description.

a. kidneys

b. potassium

c. fluid compartments

d. fluid balance

e. hypertonic blood plasma

f. dehydration

g. aldosterone

h. plasma, interstitial fluid

i. osmoreceptors

j. fluid shift

k. hypokalemia

l. ADH

m. hyponatremia

n. sodium

| **1** | Monitor blood osmotic concentration | **1** _____ |
| **2** | Water gain = water loss | **2** _____ |
| **3** | Major components of ECF | **3** _____ |
| **4** | Dominant cation in ECF | **4** _____ |
| **5** | Hormone that restricts water loss and stimulates thirst | **5** _____ |
| **6** | Caused by overhydration | **6** _____ |
| **7** | Dominant cation in ICF | **7** _____ |
| **8** | ICF and ECF | **8** _____ |
| **9** | Most important sites of sodium ion regulation | **9** _____ |
| **10** | Water movement between ECF and ICF | **10** _____ |
| **11** | Water moves from cells into ECF | **11** _____ |
| **12** | Result of aldosteronism | **12** _____ |
| **13** | Water losses greater than water gains | **13** _____ |
| **14** | Regulates sodium ion absorption along distal convoluted tubule and collecting system | **14** _____ |

**Multiple choice:** Select the correct answer from the list provided.

**15** Nearly two-thirds of the total body water content is

☐ a) extracellular fluid (ECF).

☐ b) intracellular fluid (ICF).

☐ c) tissue fluid.

☐ d) interstitial fluid.

**16** Electrolyte balance involves balancing the rates of absorption across the digestive tract with rates of loss at the

☐ a) heart and lungs.

☐ b) stomach and liver.

☐ c) kidneys and sweat glands.

☐ d) pancreas and gallbladder.

**17** If the ECF is hypertonic with respect to the ICF, water will move

☐ a) from the ECF into cells until osmotic equilibrium is restored.

☐ b) from cells into the ECF until osmotic equilibrium is restored.

☐ c) in both directions until osmotic equilibrium is restored.

☐ d) in response to the sodium–potassium exchange pump.

**18** When pure water is consumed, the ECF

☐ a) becomes hypotonic with respect to the ICF.

☐ b) becomes hypertonic with respect to the ICF.

☐ c) becomes isotonic with respect to the ICF.

☐ d) electrolytes become more concentrated.

**19** Physiological adjustments affecting fluid and electrolyte balance are mediated primarily by

☐ a) antidiuretic hormone.

☐ b) aldosterone.

☐ c) natriuretic peptides.

☐ d) all of the above.

**20** When water is lost but electrolytes are retained, the osmolarity of the ECF rises, and osmosis then moves water

☐ a) out of the ECF and into the ICF.

☐ b) back and forth between the ICF and ECF.

☐ c) out of the ICF and into the ECF.

☐ d) none of the above.

**Section integration:** Malia, a nursing student, has been caring for burn patients. She notices that they consistently show elevated levels of potassium in their urine and wonders why. What would you tell her?

**21** _____

_____

_____

# Module 25.6
# There are three categories of acids in the body

Your body is in **acid-base balance** when the production of hydrogen ions is precisely offset by their loss and when the pH of body fluids remains within normal limits.

**?** **REVIEW**
**A.** When is your body in acid-base balance?

**1** This diagram shows the major factors involved in maintaining acid-base balance. The primary challenge to homeostasis is that your body generates a variety of acids during normal metabolic operations, and a significant decrease in body fluid pH must be prevented. The respiratory and urinary systems, and buffer systems in the ICF and ECF, play important homeostatic roles in acid-base balance.

**?** **REVIEW**
**B.** What is the primary challenge to acid-base homeostasis?

### Respiratory System

The respiratory system plays a key role by eliminating carbon dioxide.

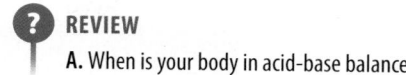

$CO_2$

### Active Tissue Cells

Active tissues continuously generate carbon dioxide ($CO_2$), which in solution forms carbonic acid ($H_2CO_3$) that immediately dissociates into a hydrogen ion and a bicarbonate ion ($H^+ + HCO_3^-$). Additional acids, such as pyruvic acid, are produced in the course of normal metabolic operations.

Tissue cells

$CO_2$

$H^+$

Normal blood pH (7.35–7.45)

### Kidneys

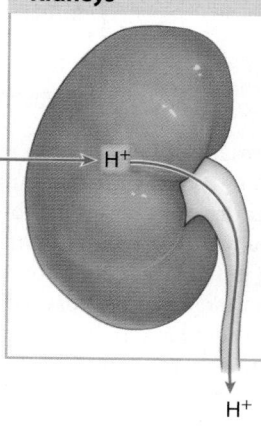

$H^+$

The kidneys play a major role by secreting hydrogen ions into the urine and generating buffers that enter the bloodstream. The rate of excretion rises and falls as needed to maintain normal blood pH. As a result, the normal pH of urine varies widely but averages 6.0—slightly acidic.

$H^+$

### Buffer Systems

Buffer systems can temporarily store hydrogen ions and thereby provide short-term pH stability.

**2** There are three categories of acids that can threaten pH balance.

## Categories of Acids

### Fixed Acids

**Fixed acids** are acids that do not leave solution; once produced, they remain in body fluids until they are eliminated by the kidneys. Sulfuric acid and phosphoric acid are the most important fixed acids in the body. They are generated in small amounts during the catabolism of amino acids and compounds that contain phosphate groups, including phospholipids and nucleic acids.

### Metabolic Acids

**Metabolic acids** take part in, or are produced by, cellular metabolism. Important metabolic acids include pyruvic acid, lactic acid, and ketone bodies (synthesized from acetyl-CoA). Under normal conditions, most metabolic acids are metabolized rapidly, so significant accumulations do not occur.

### Volatile Acids

**Volatile acids** can leave the body and enter the atmosphere via the lungs. Carbonic acid ($H_2CO_3$) is a volatile acid that forms through the interaction of water and carbon dioxide.

**?** **REVIEW**
**C.** Compare the three categories of acids.

**Lo** **LEARNING OUTCOME**
Describe the three categories of acids in the body.

# Module 25.7

# Potentially dangerous disturbances in acid-base balance are opposed by buffer systems

**1** Buffers in body fluids temporarily neutralize the acids produced by normal metabolic operations. In order to understand how buffers work, we need a bit of review of acids and bases. Key terms in this table were introduced in Chapter 2.

| A Review of Important Terms Relating to Acid-Base Balance | |
|---|---|
| **pH** | The negative exponent (negative logarithm) of the hydrogen ion concentration [H$^+$] in a solution ; pH is a measure of how acidic or basic a solution is |
| **Neutral** | A solution with a pH of 7; the solution contains equal numbers of hydrogen ions (H+) and hydroxide ions (OH$^-$) |
| **Acidic** | A solution with a pH below 7; in this solution, hydrogen ions predominate |
| **Basic (alkaline)** | A solution with a pH above 7; in this solution, hydroxide ions predominate |
| **Acid** | A substance that dissociates to release hydrogen ions, decreasing pH |
| **Base** | A substance that dissociates to release hydroxide ions or to remove hydrogen ions, increasing pH |
| **Salt** | An ionic compound consisting of a cation other than a hydrogen ion and an anion other than a hydroxide ion |
| **Buffer** | A substance that tends to oppose changes in the pH of a solution by removing or replacing hydrogen ions; in body fluids, buffers maintain blood pH within normal limits (7.35–7.45) |

**2** The pH of the ECF normally remains within relatively narrow limits, usually 7.35–7.45. Any deviation from the normal range is extremely dangerous because changes in H$^+$ concentrations disrupt the stability of plasma membranes, alter the structure of proteins, and change the activities of important enzymes. You could not survive for long with an ECF pH below 6.8 or above 7.7. In practice, decreases in pH are much more common than increases because several acids, including carbonic acid, are generated by normal cellular activities. Any shift in pH affects virtually all body systems, but the nervous and cardiovascular systems are particularly sensitive to pH fluctuations.

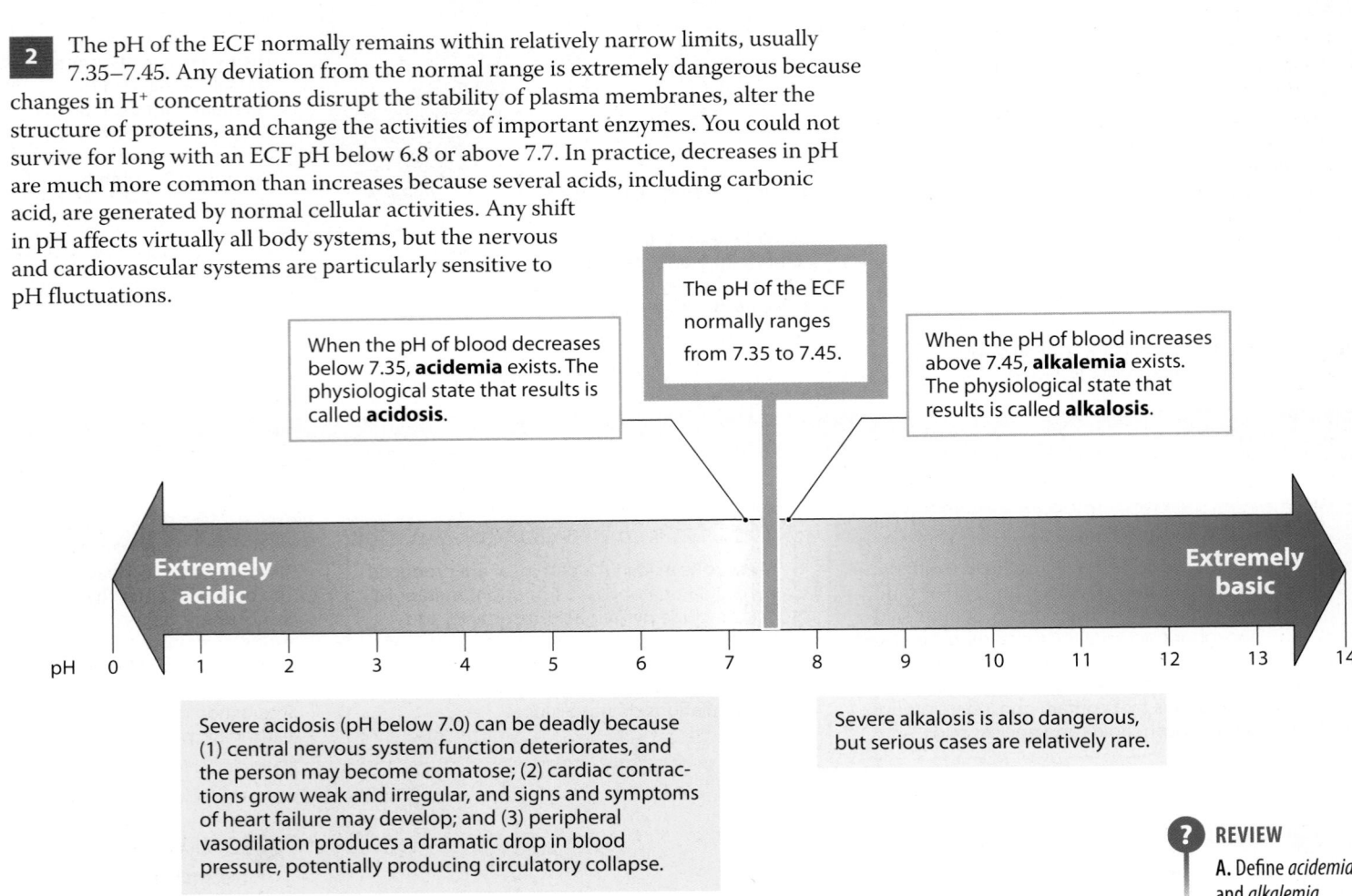

When the pH of blood decreases below 7.35, **acidemia** exists. The physiological state that results is called **acidosis**.

The pH of the ECF normally ranges from 7.35 to 7.45.

When the pH of blood increases above 7.45, **alkalemia** exists. The physiological state that results is called **alkalosis**.

Extremely acidic

Extremely basic

pH  0   1   2   3   4   5   6   7   8   9   10   11   12   13   14

Severe acidosis (pH below 7.0) can be deadly because (1) central nervous system function deteriorates, and the person may become comatose; (2) cardiac contractions grow weak and irregular, and signs and symptoms of heart failure may develop; and (3) peripheral vasodilation produces a dramatic drop in blood pressure, potentially producing circulatory collapse.

Severe alkalosis is also dangerous, but serious cases are relatively rare.

**? REVIEW**

**A.** Define *acidemia* and *alkalemia*.

**3** The partial pressure of carbon dioxide ($P_{CO_2}$) in blood is the most important factor affecting the pH of body tissues. However, its effect on pH is indirect. This is because carbon dioxide first combines with water to form **carbonic acid** ($H_2CO_3$). Most of the carbon dioxide in solution is converted to carbonic acid by carbonic anhydrase, and most of the carbonic acid dissociates into hydrogen ions and bicarbonate ions. For this reason, there is an inverse relationship between the $P_{CO_2}$ and pH.

 **REVIEW**

**B.** What intermediate compound formed from water and carbon dioxide directly affects the pH of the ECF?

| $P_{CO_2}$ 35–45 mm Hg | **HOMEOSTASIS** | pH 7.35–7.45 |

STIMULUS

Homeostasis **DISTURBED BY** INCREASING $P_{CO_2}$

Homeostasis **DISTURBED BY** DECREASING $P_{CO_2}$

$$H_2O + CO_2 \longrightarrow H_2CO_3 \longrightarrow H^+ + HCO_3^-$$

When blood carbon dioxide increases, more carbonic acid forms, additional hydrogen ions and bicarbonate ions are released, and the pH decreases.

$$H^+ + HCO_3^- \longrightarrow H_2CO_3 \longrightarrow H_2O + CO_2$$

When blood carbon dioxide decreases, carbonic acid dissociates into carbon dioxide and water. This removes $H^+$ ions from solution and increases the pH.

**REVIEW**

**C.** Summarize the relationship between $P_{CO_2}$ levels and pH.

**4** A **buffer system** in body fluids generally consists of a combination of a weak acid (HY) and the anion ($Y^-$) released by its dissociation. The anion functions as a weak base.

In solution, molecules of the weak acid exist in equilibrium with its dissociation products. In chemical notation, this relationship is represented as

$$HY \rightleftharpoons H^+ + Y^-$$

Adding $H^+$ to the solution upsets the equilibrium and results in the formation of additional molecules of the weak acid.

$$H^+ + Y^- \xrightarrow{H^+} H^+ + HY$$

Removing $H^+$ from the solution also upsets the equilibrium and results in the dissociation of additional molecules of HY. This releases $H^+$.

$$H^+ + HY \xrightarrow{\quad} H^+ + Y^- \atop H^+$$

**LEARNING OUTCOME**

Explain the role of buffer systems in maintaining acid-base balance and pH.

# Module 25.8

# Buffer systems can delay, but not prevent, pH shifts in the ICF and ECF

**1** The body has three major buffer systems, each with slightly different characteristics and distributions. Although buffer systems can bind excess H⁺, they provide only a temporary solution to an acid-base imbalance. The hydrogen ions are not eliminated, but merely rendered harmless. In the process, a buffer molecule is tied up, and the supply of buffers is limited.

## Buffer Systems

exist in

**Intracellular fluid (ICF)** ------- **Extracellular fluid (ECF)**

**Phosphate Buffer System**

The **phosphate buffer system** has an important role in buffering the pH of the ICF and of urine.

**Protein Buffer Systems**

**Protein buffer systems** contribute to the regulation of pH in the ECF and ICF. These buffer systems interact extensively with the other two buffer systems.

**Hemoglobin buffer system (RBCs only)** | **Amino acid buffers (all proteins)** | **Plasma protein buffers**

**Carbonic Acid-Bicarbonate Buffer System**

The **carbonic acid-bicarbonate buffer system** is most important in the ECF.

**2** The **hemoglobin buffer system** is the only intracellular buffer system that can have an immediate effect on the pH of body fluids. In the tissues, red blood cells absorb carbon dioxide from the plasma and convert it to carbonic acid. As the carbonic acid dissociates, the hydrogen ions are buffered by hemoglobin proteins. In the lungs, the entire reaction sequence proceeds in reverse, and the $CO_2$ diffuses into the alveoli for exhalation (see Module 21.14).

**? REVIEW**

**A.** Identify the body's three major buffer systems.

**B.** Which fluids are buffered by the phosphate buffer system?

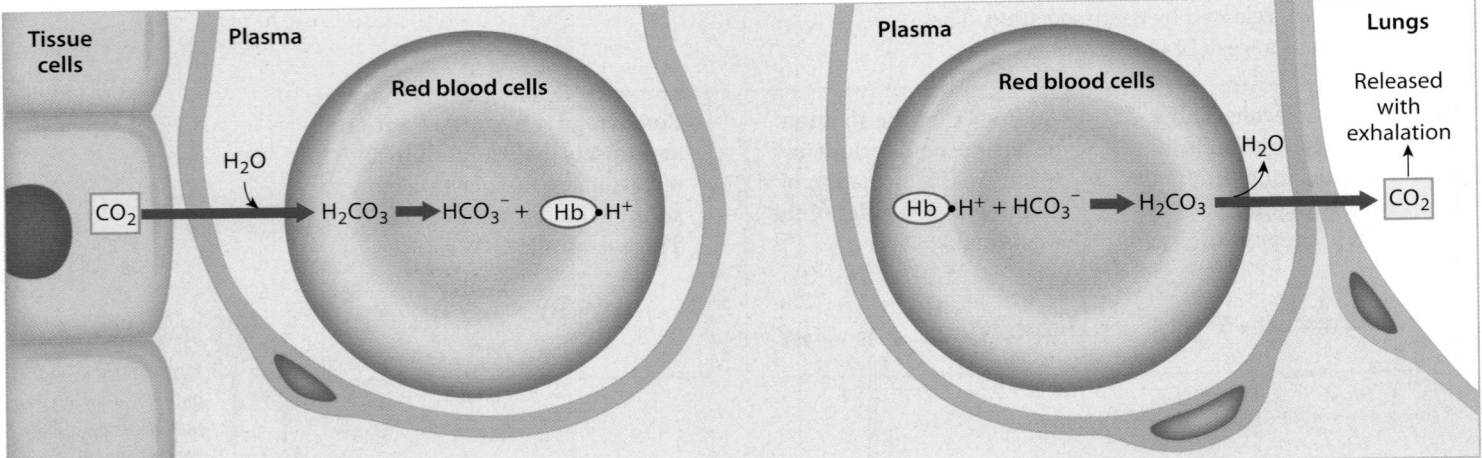

Tissue cells | Plasma | Red blood cells

$H_2O$

$CO_2 \rightarrow H_2CO_3 \rightarrow HCO_3^- + (Hb) \cdot H^+$

Plasma | Red blood cells | Lungs

$(Hb) \cdot H^+ + HCO_3^- \rightarrow H_2CO_3$

$H_2O$

Released with exhalation

$CO_2$

**3** Protein buffer systems prevent significant pH change, usually by binding excess hydrogen ions (H⁺). These buffer systems depend on the ability of amino acids to respond to pH changes by accepting or releasing H⁺. Each amino acid has a specific pH (usually less than 7) at which the carboxyl group has released a hydrogen ion and the amino group has bound one. This form, which has no net ionic charge, is called a *zwitterion*. In solutions above pH 7, the amino group (NH₂), carboxylate group (COO⁻), and side group (R) of most free amino acids can act as buffers. The underlying mechanism is shown here. However, in a protein, most of the carboxylate and amino groups in the main chain are tied up in peptide bonds. Only the exposed amino group and carboxylate group at either end of a protein are available as buffers. Thus, most of the buffering capacity of proteins is provided by the R-groups of the component amino acids.

Increasing acidity (decreasing pH)

**Normal pH**
(7.35–7.45)

Amino acid (zwitterion)

If pH decreases, the carboxylate group (COO⁻) and the amino group (–NH₂) of an amino acid can act as weak bases and accept additional hydrogen ions, forming a carboxyl group (–COOH) and an amino ion (–NH₃⁺), respectively. Many of the R-groups can also accept hydrogen ions, forming RH⁺.

At the normal pH of body fluids (7.35–7.45), neither the carboxylate groups (COO⁻) nor the amino groups (–NH₂) of most amino acids are bound to hydrogen ions.

**4** The carbonic acid-bicarbonate buffer system involves freely reversible reactions. A change in the concentration of any participant affects the concentrations of all other participants and shifts the direction of the reactions under way.

**BICARBONATE RESERVE**

Body fluids contain a large reserve of HCO₃⁻, primarily in the form of dissolved molecules of the weak base sodium bicarbonate (NaHCO₃). This readily available supply of HCO₃⁻ is known as the **bicarbonate reserve**.

**CARBONIC ACID-BICARBONATE BUFFER SYSTEM**

The primary function of the carbonic acid-bicarbonate buffer system is to protect against the effects of the metabolic and fixed acids generated through metabolic activity. In effect, it takes the H⁺ released by these acids and generates carbonic acid that dissociates into water and carbon dioxide, which can easily be eliminated by the lungs.

**Metabolic acid-base disorders** result from the production or loss of excessive amounts of fixed or metabolic acids. The primary role of the carbonic acid-bicarbonate buffer system is to protect against such disorders. **Respiratory acid-base disorders** result from an imbalance between the rate of CO₂ generation and the rate of CO₂ elimination by the lungs. The carbonic acid-bicarbonate buffer system cannot protect against respiratory disorders; the imbalances must be corrected by reflexive changes in the depth and rate of respiration.

**? REVIEW**

**C.** Describe the carbonic acid-bicarbonate buffer system.

**Lo LEARNING OUTCOME**

Explain the role of buffer systems in regulating the pH of the intracellular fluid and the extracellular fluid.

# Module 25.9

# The homeostatic responses to metabolic acidosis and alkalosis involve respiratory and renal mechanisms as well as buffer systems

## Responses to Metabolic Acidosis

**1** **Metabolic acidosis** develops when large numbers of hydrogen ions are released by metabolic or fixed acids and the pH decreases. For homeostasis to be preserved, the excess $H^+$ must either be permanently bound through the formation of water (linked to the formation of $CO_2$ that can be eliminated by the lungs), or removed from body fluids through secretion by the kidneys.

Addition of $H^+$

**CARBONIC ACID-BICARBONATE BUFFER SYSTEM**

**BICARBONATE RESERVE**

$CO_2 \leftarrow CO_2 + H_2O \leftarrow H_2CO_3$ (carbonic acid) $\leftarrow H^+ + HCO_3^-$ (bicarbonate ion) $\leftarrow HCO_3^- + Na^+ \leftarrow NaHCO_3$ (sodium bicarbonate)

Lungs

Generation of $HCO_3^-$

KIDNEYS

**Respiratory Response to Acidosis**

Increased respiratory rate decreases $P_{CO_2}$, effectively converting carbonic acid molecules to water.

Other buffer systems absorb $H^+$

**Renal Response to Acidosis**

Kidney tubules respond by secreting $H^+$ ions, removing $CO_2$, and reabsorbing $HCO_3^-$ to help replenish the bicarbonate reserve.

Secretion of $H^+$

**2** Renal tubule cells secrete $H^+$ into the tubular fluid along the proximal convoluted tubule (PCT), the distal convoluted tubule (DCT), and the collecting system. They also bolster the capabilities of the carbonic acid-bicarbonate buffer system by increasing the concentration of bicarbonate ions in the ECF, replacing those pulled from the bicarbonate reserve.

Tubular fluid

Renal tubule cells

ECF

**Steps in $H^+$ Removal and $HCO_3^-$ Production**

$CO_2$ generated by the tubule cell is added to the $CO_2$ diffusing into the cell from the urine and from the ECF.

Carbonic anhydrase converts $CO_2$ and water to carbonic acid, which then dissociates.

The chloride ions exchanged for bicarbonate ions are excreted in the tubular fluid.

Bicarbonate ions and sodium ions are transported into the ECF, adding to the bicarbonate reserve.

**?** **REVIEW**

**A.** Describe metabolic acidosis.

## Responses to Metabolic Alkalosis

**3** **Metabolic alkalosis** develops when large numbers of hydrogen ions are removed from body fluids, resulting in an increased pH. When this occurs, the rate of $H^+$ secretion by the kidneys decreases, tubule cells do not reclaim the bicarbonate ions in tubular fluid, and the collecting system transports $HCO_3^-$ into tubular fluid while releasing a strong acid (HCl) into the ECF.

**Respiratory Response to Alkalosis**

Decreased respiratory rate increases $P_{CO_2}$, effectively converting $CO_2$ molecules to carbonic acid.

Removal of $H^+$

**CARBONIC ACID–BICARBONATE BUFFER SYSTEM**

Lungs $CO_2 + H_2O \rightarrow$ $H_2CO_3$ (carbonic acid) $\rightarrow$ $H^+$ + $HCO_3^-$ (bicarbonate ion)

**BICARBONATE RESERVE**

$HCO_3^- + Na^+ \rightarrow NaHCO_3$ (sodium bicarbonate)

Other buffer systems release $H^+$

Generation of $H^+$

KIDNEYS

**Renal Response to Alkalosis**

Kidney tubules respond by conserving $H^+$ ions and secreting $HCO_3^-$.

Secretion of $HCO_3^-$

**4** Tubule cells secrete bicarbonate ions into the tubular fluid along the proximal convoluted tubule (PCT), the distal convoluted tubule (DCT), and the collecting system. They also move $H^+$ ions into the ECF.

**Steps in $HCO_3^-$ Removal and $H^+$ Production**

$CO_2$ generated by the tubule cell is added to the $CO_2$ diffusing into the cell from the tubular fluid and from the ECF.

Carbonic anhydrase converts $CO_2$ and water to carbonic acid, which then dissociates.

The hydrogen ions are actively transported into the ECF, accompanied by the diffusion of chloride ions.

$HCO_3^-$ is pumped into the tubular fluid in exchange for chloride ions that will diffuse into the ECF.

**?** **REVIEW**

**B.** Describe metabolic alkalosis.

**?** **INTEGRATION**

**C.** If the kidneys are conserving $HCO_3^-$ and eliminating $H^+$ in acidic urine, which is occurring: metabolic alkalosis or metabolic acidosis?

**Lo** **LEARNING OUTCOME**

Describe the compensatory mechanisms involved in maintaining acid-base balance.

# Module 25.10
# Respiratory acid-base disorders are the most common challenges to acid-base balance

Go to
MasteringA&P™
PhysioEx 9.1
Ex: 10 Acid-Base Balance

**Respiratory acid-base disorders** result from an imbalance between the rate of $CO_2$ generation in body tissues and the rate of $CO_2$ elimination by the lungs.

## Respiratory Acidosis

**1** Respiratory acid-base disorders cannot be corrected, even temporarily, by the carbonic acid-bicarbonate buffer system. If the rate of $CO_2$ generation exceeds the rate of $CO_2$ removal, **respiratory acidosis** develops. Respiratory acidosis is relatively common and may be life threatening.

**CARBONIC ACID-BICARBONATE BUFFER SYSTEM**

**BICARBONATE RESERVE**

$$CO_2 \quad CO_2 + H_2O \rightarrow \underset{\text{(carbonic acid)}}{H_2CO_3} \rightarrow \boxed{H^+} + \underset{\text{(bicarbonate ion)}}{HCO_3^-} \qquad HCO_3^- + Na^+ \underset{}{\rightleftharpoons} \underset{\text{(sodium bicarbonate)}}{NaHCO_3}$$

**Lungs**

**1** When respiratory activity does not keep pace with the rate of $CO_2$ generation, alveolar and blood $P_{CO_2}$ increases. This upsets the equilibrium and drives the reaction to the right, generating additional $H_2CO_3$, which releases $H^+$ and decreases blood pH.

**2** As bicarbonate ions and hydrogen ions are released through the dissociation of carbonic acid, the excess bicarbonate ions become part of the bicarbonate reserve.

To limit the pH effects of respiratory acidosis, the excess $H^+$ must either be "tied up" by other buffer systems or secreted by the kidneys. The underlying problem, however, cannot be eliminated without an increase in the respiratory rate.

**? REVIEW**
**A.** What would happen to the blood $P_{CO_2}$ of a patient who has an airway obstruction?

**2** This flowchart summarizes the integrated homeostatic responses to respiratory acidosis.

Homeostasis **DISTURBED BY** INCREASING $P_{CO_2}$ caused by hypoventilation

**STIMULUS**

**Receptors**
Arterial and CSF chemoreceptors stimulated

Increased $P_{CO_2}$ results in a decreased blood pH

**Responses to Acidosis**
Increased respiratory rate

Kidneys secrete $H^+$ and generate $HCO_3^-$

Buffer systems other than the carbonic acid-bicarbonate system accept $H^+$

Decreased $P_{CO_2}$

Decreased $H^+$ and increased $HCO_3^-$

**RESTORED**

Homeostasis **RESTORED BY** blood pH returning to normal

**HOMEOSTASIS**
NORMAL ACID-BASE BALANCE

## Respiratory Alkalosis

**3** If the rate of $CO_2$ elimination exceeds the rate of $CO_2$ generation, **respiratory alkalosis** develops. Respiratory alkalosis is relatively uncommon and rarely severe.

**CARBONIC ACID–BICARBONATE BUFFER SYSTEM**

**BICARBONATE RESERVE**

$$CO_2 \leftarrow CO_2 + H_2O \leftarrow H_2CO_3 \text{ (carbonic acid)} \leftarrow H^+ + HCO_3^- \text{ (bicarbonate ion)} \leftarrow HCO_3^- + Na^+ \rightleftharpoons NaHCO_3 \text{ (sodium bicarbonate)}$$

Lungs

**1** If respiratory activity exceeds the rate of $CO_2$ generation, alveolar and blood $P_{CO_2}$ decreases, and this disturbs the equilibrium and drives the reactions to the left, removing $H^+$ and increasing blood pH.

**2** As bicarbonate ions and hydrogen ions are removed in the formation of carbonic acid, the bicarbonate ions—but not the hydrogen ions—are replaced by the bicarbonate reserve.

**4** The homeostatic responses to respiratory alkalosis are summarized in this flowchart. Most cases are related to anxiety, and the resulting hyperventilation is self-limiting—the person often faints, and the respiratory rate then decreases to normal levels. One common treatment is to have the person breathe in and out of a paper bag; the rising $CO_2$ level in the recycled air increases blood $P_{CO_2}$ and eliminates the alkalosis.

# HOMEOSTASIS
### NORMAL ACID-BASE BALANCE

Homeostasis **DISTURBED BY**

DECREASING

$P_{CO_2}$

caused by hyperventilation

**STIMULUS**

Decreased $P_{CO_2}$ results in an increased blood pH

**Receptors**

Arterial and CSF chemoreceptors inhibited

**Responses to Alkalosis**

Kidneys generate $H^+$ and secrete $HCO_3^-$

Buffer systems other than the carbonic acid-bicarbonate system release $H^+$

Decreased respiratory rate

Increased $H^+$ and decreased $HCO_3^-$

Increased $P_{CO_2}$

**RESTORED**

Homeostasis **RESTORED BY**

blood pH returning to normal

**?** **INTEGRATION**

**B.** How would a decrease in the pH of body fluids affect the respiratory rate?

**Lo** **LEARNING OUTCOME**

Describe respiratory acidosis and respiratory alkalosis.

# Acid-Base Balance

**Labeling:** Use the following terms to label the boxes in the two flowcharts. Terms may be used more than once.

- arterial and CSF chemoreceptors
- blood pH decrease
- blood pH increase
- increasing $P_{CO_2}$
- decreasing $P_{CO_2}$
- increased
- decreased
- $H^+$
- $HCO_3^-$

Respiratory acidosis

**Receptors**

3 _____

stimulated

**Responses to Acidosis**

Respiratory rate

4 _____

Kidneys

secrete 5 _____

generate 6 _____

Buffer systems other than the carbonic acid-bicarbonate system accept $H^+$

7 _____ $P_{CO_2}$

8 _____ $H^+$

9 _____ $HCO_3^-$

Homeostasis **DISTURBED BY**

1 _____

caused by hypoventilation

Increased $P_{CO_2}$ results in

2 _____

STIMULUS

RESTORED

Homeostasis **RESTORED BY**

blood pH

returning to normal

# HOMEOSTASIS
## NORMAL ACID-BASE BALANCE

Homeostasis **DISTURBED BY**

10 _____

caused by hyperventilation

STIMULUS

RESTORED

Homeostasis **RESTORED BY**

blood pH

returning to normal

Decreased $P_{CO_2}$ results in an

11 _____

**Receptors**

12 _____

inhibited

**Responses to Alkalosis**

Kidneys

generate 13 _____

secrete 14 _____

Buffer systems other than the carbonic acid-bicarbonate system release $H^+$

Respiratory rate

15 _____

18 _____ $H^+$

17 _____ $HCO_3^-$

16 _____ $P_{CO_2}$

Respiratory alkalosis

**Chapter integration:** After falling into a deep lake and nearly drowning, a young boy is rescued. His rescuers assess his condition and find that his body fluids have high $P_{CO_2}$ and lactate levels and low $P_{O_2}$ levels. Identify the underlying problem, and recommend the necessary treatment to restore homeostatic conditions.

19 _____

# Study Outline

## SECTION 1 • Fluid and Electrolyte Balance

### Module 25.1

**Body composition may be viewed in terms of solids and two fluid compartments    p. 973**

1. Homeostatic mechanisms regulate body fluids that are critical to cells.

2. The **extracellular fluid**, or **ECF**, is made up of the interstitial fluid of peripheral tissues and the plasma of circulating blood. The **intracellular fluid**, or **ICF**, consists of the cytosol inside cells.

3. Intracellular fluid (ICF) and extracellular fluid (ECF) are called **fluid compartments**.

4. Solid components make up only 40–50 percent of body mass.

### Module 25.2

**Fluid balance exists when water gains equal water losses    p. 974**

5. **Fluid balance** in the body occurs when water content remains stable over time.

6. Water gains occur primarily in the digestive tract. Almost half of water loss occurs through urination.

7. Daily water input from water content in food, consumed liquids, and metabolic water produced during catabolism is approximately 2500 mL. A total of 2500 mL of water is eliminated through urination, evaporation at the skin and lungs, loss through feces, and some secretion by sweat glands.

The total body composition of an adult female

8. A rapid water movement between the ECF and the ICF in response to an osmotic gradient is called a **fluid shift**.

9. **Dehydration** develops when water losses outpace water gains.

### Module 25.3

**Mineral balance involves balancing electrolyte gain and loss    p. 976**

10. **Minerals** are inorganic substances, and **electrolytes** are the ions released when mineral salts dissociate. **Mineral balance** exists when ion absorption and excretion are about the same.

11. The body contains substantial reserves of key minerals. The amount lost each day must equal the daily intake if the person is to stay in mineral balance.

### Module 25.4

**Water balance depends on sodium balance, and the two are regulated simultaneously    p. 978**

12. **Sodium balance** exists when sodium gains equal sodium losses.

13. When sodium gains exceed losses, the ECF volume increases; when sodium losses exceed gains, the volume of the ECF decreases.

14. If ECF volume rises, blood volume goes up. If ECF volume drops, blood volume goes down

15. When the $Na^+$ concentration in the ECF falls below 136 mEq/L, a state of **hyponatremia** exists. This can be caused by excessive water intake (overhydration) or inadequate salt intake.

16. When body water content declines, the $Na^+$ concentration rises; when that concentration exceeds 145 mEq/L, **hypernatremia** exists; dehydration is the most common cause.

### Module 25.5

**Disturbances of potassium balance are uncommon but extremely dangerous    p. 980**

17. The key factors in maintaining **potassium balance** are the rate of $K^+$ entry across the digestive epithelium and the rate of $K^+$ loss into urine.

18. The kidneys are the main factor determining the $K^+$ concentration in the ECF.

19. Potassium loss is regulated by the kidneys through aldosterone stimulating $Na^+$ reabsorption while simultaneously accelerating $K^+$ excretion.

20. The endocrine disorder **aldosteronism** (excessive aldosterone secretion), results in **hypokalemia** (low potassium) from excess sodium retention and potassium loss. Diuretics can cause hypokalemia by increasing the volume of urine produced.

21. **Hyperkalemia** (excess potassium) can be caused by low blood pH, kidney failure, and diuretics that block $Na^+$ reabsorption (which also slows potassium ion secretion).

## SECTION 2 • Acid-Base Balance

### Module 25.6

**There are three categories of acids in the body**    p. 983

22. **Acid-base balance** occurs in the body when the production of $H^+$ ions is precisely offset by their loss and when the pH of body fluids remains within normal limits.

23. The primary challenge to acid-base homeostasis is that your body generates a variety of acids during normal metabolic operations, and a significant decrease in body fluid pH must be prevented.

24. Three classes of acids can threaten pH balance: **fixed acids** (acids that do not leave solution), **metabolic acids** (products of cellular metabolism), and **volatile acids** (acids leaving the body through the lungs).

### Module 25.7

**Potentially dangerous disturbances in acid-base balance are opposed by buffer systems**    p. 984

25. The pH of the ECF normally remains between 7.35 and 7.45.

26. Changes in $H^+$ concentrations from the normal range disrupt the stability of plasma membranes, alter the structure of proteins, and change the activities of important enzymes.

27. When the pH of blood decreases below 7.35, **acidemia** exists, resulting in **acidosis**. When the pH of blood increases above 7.45, **alkalemia** exists, resulting in **alkalosis**.

28. The most important factor affecting the pH of body tissues is $P_{CO_2}$ because carbon dioxide combines with water to form **carbonic acid** that dissociates into a hydrogen ion $H^+$ and a bicarbonate ion ($HCO_3^-$) in the reaction: $H_2O + CO_2 \rightarrow H_2CO_3 \rightarrow H^+ + HCO_3^-$. There is an inverse relationship between $P_{CO_2}$ and pH.

29. A **buffer system** in body fluids generally consists of a combination of a weak acid (HY) and the anion ($Y^-$) released by its dissociation.

### Module 25.8

**Buffer systems can delay, but not prevent, pH shifts in the ICF and ECF**    p. 986

30. There are three major buffer systems: the phosphate buffer system, the protein buffer systems, and the carbonic acid-bicarbonate buffer system.

31. The **phosphate buffer system** plays an important role in buffering the pH of the ICF and urine.

32. The **protein buffer systems** regulate the pH of the ECF and ICF and interact with the two other buffer systems. Protein buffer systems prevent significant pH change by binding excess $H^+$. This involves the amino, carboxylate, and R (side) groups of free amino acids or, in proteins, mostly exposed R-groups.

33. The **carbonic acid-bicarbonate buffer system** is most important in the ECF. It takes the $H^+$ released by metabolic activity and generates carbonic acid that dissociates into water and carbon dioxide, which can easily be eliminated by the lungs.

### Module 25.9

**The homeostatic responses to metabolic acidosis and alkalosis involve respiratory and renal mechanisms as well as buffer systems**    p. 988

34. **Metabolic acidosis** develops when metabolic or fixed acids release large numbers of $H^+$ and the pH decreases.

35. The excess $H^+$ must either be permanently bound up through the formation of water or removed from body fluids through secretion by the kidneys.

36. The respiratory response to acidosis is increased respiratory rate to decrease $P_{CO_2}$, resulting in carbonic acid and water. The renal response is for tubule cells to secrete $H^+$ into the tubular fluid along the PCT, the DCT, and the collecting system.

37. **Metabolic alkalosis** develops when large numbers of $H^+$ are removed from body fluids, resulting in a rise in pH.

38. The respiratory response to metabolic alkalosis is decreased respiratory rate. This elevates $P_{CO_2}$, resulting in $CO_2$ molecules converting into carbonic acid and adding $H^+$. The renal response is for the kidney tubules to conserve $H^+$ and secrete $HCO_3^-$.

### Module 25.10

**Respiratory acid-base disorders are the most common challenges to acid-base balance**    p. 990

39. **Respiratory acid-base disorders** result from an imbalance between the rate of $CO_2$ generation in body tissues and the rate of $CO_2$ elimination at the lungs. This cannot be corrected by the carbonic acid-bicarbonate buffer system.

40. If the rate of $CO_2$ generation exceeds the rate of $CO_2$ removal by hypoventilation, then **respiratory acidosis** develops. This cannot be eliminated without an increase in respiratory rate.

41. If the rate of $CO_2$ elimination exceeds the rate of $CO_2$ generation by hyperventilation, then **respiratory alkalosis** develops. This is relatively uncommon and rarely severe.

42. Respiratory alkalosis is often related to anxiety-induced hyperventilation and is corrected when the person faints, resulting in normal ventilation rates. One common treatment is to have the person breathe into a paper bag. The rising $CO_2$ level in the recycled air increases blood $P_{CO_2}$ and eliminates the alkalosis.

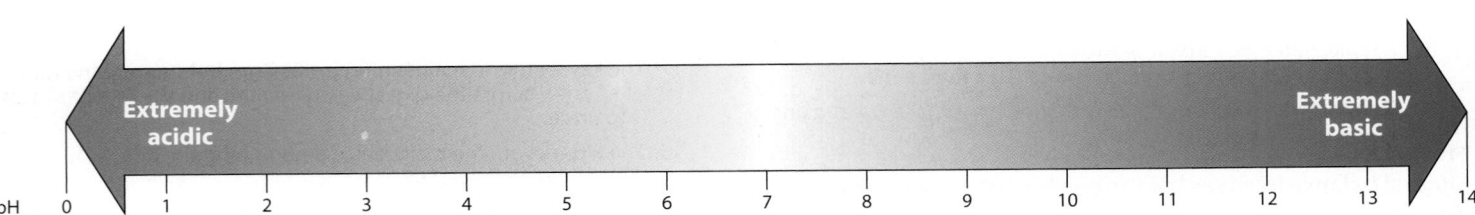

**Extremely acidic**

**Extremely basic**

| pH | 0 | 1 | 2 | 3 | 4 | 5 | 6 | 7 | 8 | 9 | 10 | 11 | 12 | 13 | 14 |

*Severe acidosis and severe alkalosis are both dangerous to the human body.*

# Chapter Review Questions

**True/False:** Indicate whether each statement is true or false.

**1** Metabolic acidosis develops when large numbers of hydrogen ions are removed from body fluids, resulting in an increase in pH.

**2** Proteins are the most abundant solid components of body mass.

**3** About half of daily water loss occurs through evaporation at the lungs.

**4** Respiratory alkalosis can be treated by simply having the person breathe in and out of a paper bag.

**5** Hypokalemia is caused by inadequate aldosterone secretion.

**6** Iron is classified as a bulk mineral.

**7** Carbonic acid is a volatile acid that forms through the interaction of water and carbon dioxide.

**8** The pH of ECF normally ranges from 7.0 to 7.30.

**9** The total body composition of adult males is 60 percent water and 40 percent solids, whereas for females, total body composition is 50 percent water and 50 percent solids.

**10** About 98 percent of the potassium content of the human body is in the ICF, rather than the ECF.

**1** _____

**2** _____

**3** _____

**4** _____

**5** _____

**6** _____

**7** _____

**8** _____

**9** _____

**10** _____

**Matching:** Match each lettered term with the most closely related description.

a. buffer

b. acid

c. basic, or alkaline

d. pH

e. acidic

f. base

g. neutral

h. salt

**11** The negative exponent of the hydrogen ion concentration in a solution

**12** A substance that dissociates to release hydroxide ions or to remove hydrogen ions, increasing pH

**13** A solution with a pH of 7; it contains equal numbers of hydrogen ions and hydroxide ions

**14** A substance that tends to oppose changes in the pH of a solution by removing or replacing hydrogen ions

**15** A solution with a pH above 7; hydroxide ions predominate

**16** A substance that dissociates to release hydrogen ions, decreasing pH

**17** An ionic compound consisting of a cation other than a hydrogen ion and an anion other than a hydroxide ion

**18** A solution with a pH below 7; hydrogen ions predominate

**11** _____

**12** _____

**13** _____

**14** _____

**15** _____

**16** _____

**17** _____

**18** _____

**Multiple choice:** Select the correct answer from the list provided.

**19** The most important factor affecting the pH of body tissues is the

- ☐ a) hydrochloric acid.
- ☐ b) ketone bodies.
- ☐ c) lactic acid.
- ☐ d) partial pressure of carbon dioxide.

**20** Respiratory acidosis develops when the pH of blood is

- ☐ a) increased due to decreased $P_{CO_2}$ level.
- ☐ b) decreased due to increased $P_{CO_2}$ level.
- ☐ c) increased due to increased $P_{CO_2}$ level.
- ☐ d) decreased due to decreased $P_{CO_2}$ level.

**21** Which of the following is the correct chemical equation of the reaction between water and carbon dioxide?

- ☐ a) $H_2CO_3 + H^+ \longrightarrow HCO_3^- + H_2O \longrightarrow CO_2$
- ☐ b) $H_2O + CO_2 \longrightarrow H_2CO_3 \longrightarrow H^+ + HCO_3^-$
- ☐ c) $H_2O + CO_2 \longrightarrow HCO_3^- \longrightarrow H_2CO_3 \longrightarrow H^+$
- ☐ d) $H_2O + CO_2 + H^+ \longrightarrow H_2CO_3 \longrightarrow HCO_3^-$

**22** Most of the buffering capacity of proteins is provided by the

- ☐ a) R-groups of the component amino acids.
- ☐ b) peptide bonds between adjacent amino acids.
- ☐ c) amino and carboxyl groups of the component amino acids.
- ☐ d) central carbon group of most amino acids.

**23** Which of the following is *not* an endocrine response to decreased blood volume and blood pressure?

- ☐ a) increased renin secretion and angiotensin II activation
- ☐ b) increased aldosterone release
- ☐ c) increased ADH release
- ☐ d) increased $Na^+$ in urine

**24** The condition of below-normal $Na^+$ concentration in the blood is called

- ☐ a) hyponatremia.
- ☐ b) hypernatremia.
- ☐ c) hypokalemia.
- ☐ d) hyperkalemia.

**25** Which of the following minerals functions as a component of myoglobin, hemoglobin, and cytochromes?

- ☐ a) zinc
- ☐ b) copper
- ☐ c) iron
- ☐ d) cobalt

**26** The primary site of ion loss is the

- ☐ a) sweat glands.
- ☐ b) kidneys.
- ☐ c) intestines.
- ☐ d) lungs.

**Short answer**

**27** Differentiate fluid balance, mineral balance, and acid-base balance.

**28** Why can prolonged vomiting produce metabolic alkalosis?

## MasteringA&P™

Access more chapter study tools online in the MasteringA&P Study Area:

- Chapter Quizzes, Chapter Practice Test, and Clinical Case Studies

- Practice Anatomy Lab™ 3.0 — PAL™3.0
- Interactive Physiology™ — iP2™
- A&P Flix™ — A&PFlix™
- PhysioEx™ 9.1 — PhysioEx™ 9.1

# Chapter Integration • Applying what you have learned

## Intestinal trouble in paradise

While visiting a foreign country with poor-quality water, Beth and Tom each had a drink with ice during dinner. Many travelers avoid drinking the water in foreign countries but forget that ice is often made with local tap water! As a result, both Beth and Tom contracted an intestinal disease that causes severe diarrhea. (Diarrhea is characterized by loose watery bowel movements that, if left untreated, can cause serious health complications.)

It is estimated that each year such "traveler's diarrhea" affects as many as 10 million Americans. High-risk destinations include the lesser-developed countries of Latin America, Africa, the Middle East, and regions of Asia. The Centers for Disease Control and Prevention (CDC) maintains a website to advise travelers.

After 2 days of experiencing uncontrolled bouts of diarrhea, the couple sought medical attention at a nearby medical clinic. Various tests revealed that the culprit was the bacterium *Escherichia coli*. Further tests revealed that both Beth and Tom were dehydrated, so intravenous fluids were prescribed. Unfortunately, the attending nurse became distracted and erroneously gave Tom a hypertonic glucose solution instead of the normal saline that Beth received.

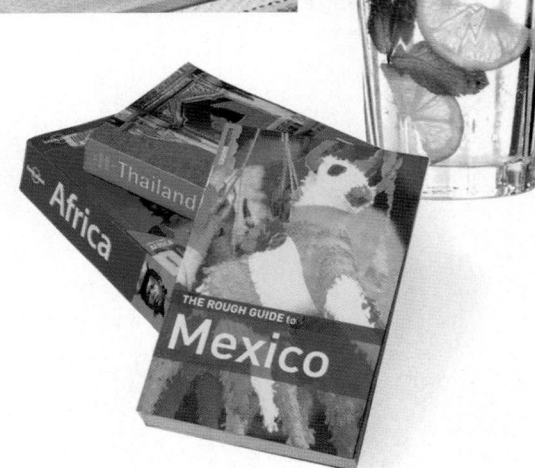

**1** How would Beth's and Tom's dehydrated states affect their blood pH, urine pH, and breathing pattern?

**2** What effect will the intravenous hypertonic glucose solution have on Tom's blood ADH levels and urine volume?

# 26

# The Reproductive System

## Module 26.1
# Male reproductive structures include the external genitalia and internal genitalia

Our **reproductive system** is the only system that is not essential to sustaining life, but its functions affect other systems and ensure species continuation. This chapter discusses how the male and female reproductive organs, called **gonads**, produce and store reproductive cells called **gametes**. Male gametes are **sperm**. We also look at how various reproductive organs secrete hormones that maintain normal sexual function, allowing those gametes to come together to form new individuals.

**?** **REVIEW**
**A.** What are the functions of the gonads?

**1** The male reproductive system is made up of the external genitalia and the internal genitalia. **Genitalia** are the organs of the reproductive system. The **male external genitalia** include the penis, urethra, and scrotum. **Male internal genitalia** include the ductus deferens, seminal gland, prostate, bulbo-urethral gland, epididymis, spermatic cord (not shown), and testis. Sperm travel along a duct system known as the *reproductive tract*, mixing with glandular secretions to form a fluid called *semen*.

**Male External Genitalia**

The **penis** contains erectile tissue, deposits sperm in the vagina of the female, and produces pleasurable sensations during sexual activities.

The **urethra** conducts semen to the exterior.

The **scrotum** surrounds the testes.

**Male Internal Genitalia**

Each **ductus deferens** (sperm duct) conducts sperm between the epididymis and prostate.

Each **seminal gland** secretes fluid that makes up much of the volume of semen.

The **prostate** secretes fluid and enzymes.

Each **bulbo-urethral gland** secretes fluids that lubricate the tip of the penis.

Each **epididymis** (plural, *epididymides*) is the site of sperm maturation.

Each **testis** (plural, *testes*) produces sperm and hormones.

 **REVIEW**
**B.** Name the structures of the male external genitalia.

 **REVIEW**
**C.** Name the structures of the male internal genitalia.

**Lo** **LEARNING OUTCOME**
Identify the structures and functions of the male external genitalia and internal genitalia.

# Module 26.2

# Sperm transport relies on ducts, glands, and related structures of the scrotum and testes

**1** The main structures of the male reproductive system are shown in this sagittal section. Starting from a testis, the sperm travel within the **epididymis** (ep-i-DID-i-mus), along the **ductus deferens** (DUK-tus DEF-e-renz), and then along the **ejaculatory duct** and the urethra before leaving the body. Accessory glands—the **seminal** (SEM-i-nal) **glands**, the **prostate** (PROS-tāt), and the **bulbo-urethral** (bul-bō-ū-RĒ-thral) **glands**—secrete various fluids into the ejaculatory duct and urethra. The distal part of the urethra passes through the penis, opening to the exterior.

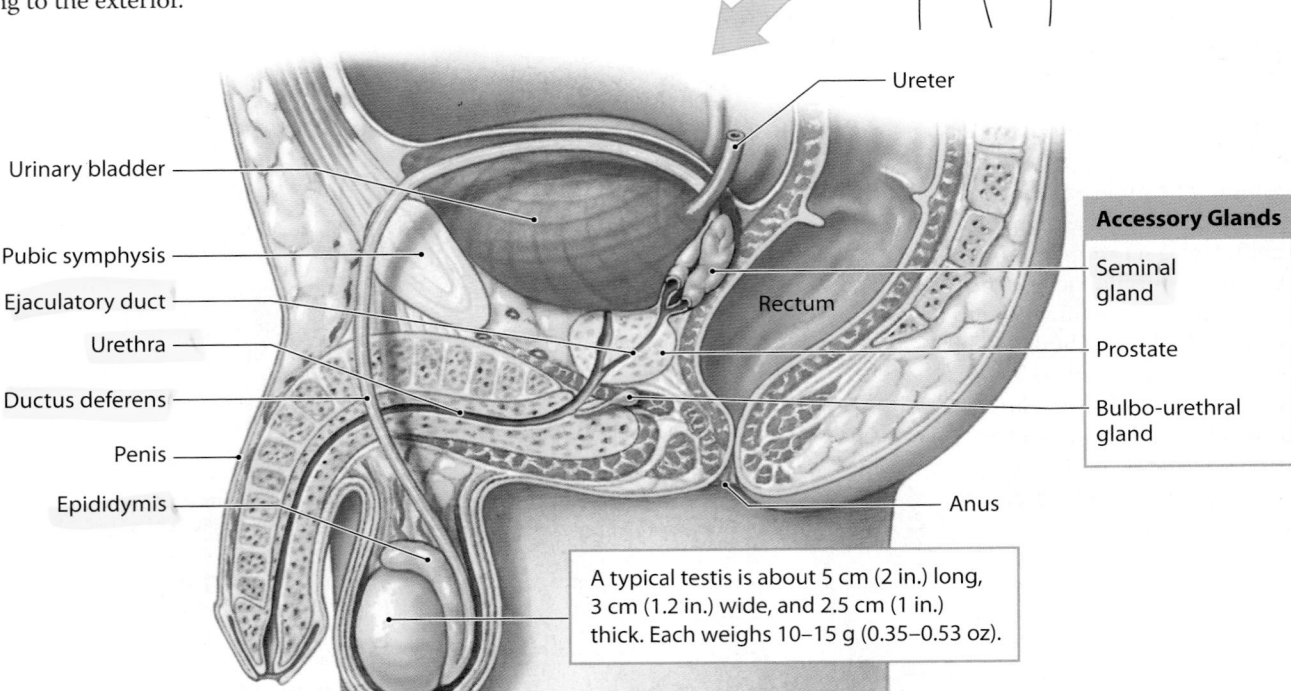

Ureter

Urinary bladder

Pubic symphysis

Ejaculatory duct

Urethra

Ductus deferens

Penis

Epididymis

Rectum

### Accessory Glands

Seminal gland

Prostate

Bulbo-urethral gland

Anus

A typical testis is about 5 cm (2 in.) long, 3 cm (1.2 in.) wide, and 2.5 cm (1 in.) thick. Each weighs 10–15 g (0.35–0.53 oz).

Scrotum

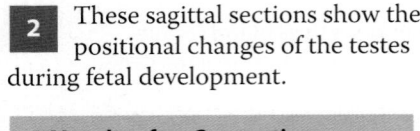

**2** These sagittal sections show the positional changes of the testes during fetal development.

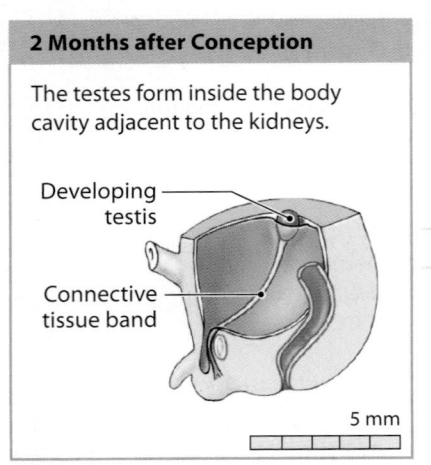

### 2 Months after Conception

The testes form inside the body cavity adjacent to the kidneys.

Developing testis

Connective tissue band

5 mm

### 3 Months after Conception

The connective tissue bands do not get longer as the fetus grows. As a result, the relative position of the testes changes as the body enlarges.

5 mm

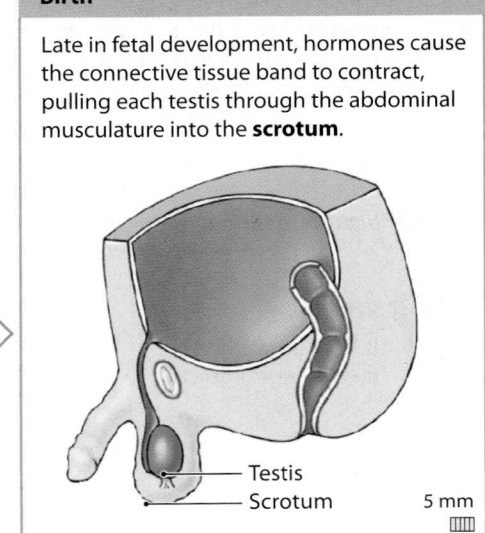

### Birth

Late in fetal development, hormones cause the connective tissue band to contract, pulling each testis through the abdominal musculature into the **scrotum**.

Testis

Scrotum

5 mm

**3** This view shows superficial and deeper features of the scrotum, testes, and related structures.

Inguinal ligament

The **superficial inguinal ring** penetrates the layers of abdominal muscles and forms the entrance to the inguinal canal.

The **cremaster** (krē-MAS-ter) lies deep to the dermis. When this muscle contracts during sexual arousal or in response to cool temperature, it tenses and pulls the testes closer to the body.

The scrotum consists of a thin layer of skin and the underlying superficial fascia. The dermis contains a layer of smooth muscle, the **dartos** (DAR-tōs) **muscle**. Resting muscle tone in the dartos muscle elevates the testes and causes the wrinkling of the scrotal surface.

The **inguinal canal** extends from the inguinal ring to the scrotal cavity. In normal adult males, the inguinal canals are closed, but the spermatic cords create weak points in the abdominal wall that remain throughout life. As a result, **inguinal hernias**—protrusions of visceral tissues or organs into the inguinal canal—are fairly common in males.

The **spermatic cords** extend through the inguinal canals between the abdominopelvic cavity and the testes. Each spermatic cord consists of layers of fascia and muscle enclosing the ductus deferens and the blood vessels, nerves, and lymphatic vessels that supply the testes.

Nerve
Artery
Venous plexus
Ductus deferens
*vas*

Scrotal septum
Scrotal cavity
Raphe of scrotum

The left and right **scrotal cavities** are separated by the **scrotal septum**. The partition between the two is marked by a raised thickening in the scrotal surface known as the **raphe** (RĀ-fē) **of scrotum**.

**4** This horizontal section through the scrotum shows the internal organization of the testes.

Epididymis   Efferent ductule   Straight tubule   Ductus deferens

**Scrotum**
Skin
Dartos muscle
Superficial fascia
Cremaster
Scrotal cavity

The **tunica vaginalis** (TU-nih-ka vaj-ih-NAL-is), a serous membrane, lines the scrotal cavity and reduces friction between the opposing parietal (outer) layer and visceral (inner) layer.

The **tunica albuginea** (al-bū-JIN-ē-uh), a tough fibrous capsule, covers the testis. This capsule is continuous with septa testis (fibrous partitions) that subdivide the testis into separate **lobules**.

Septa testis

Each testis is divided into about 250 lobules, and each lobule has 1 to 4 slender, coiled **seminiferous tubules**. Sperm production occurs in the seminiferous tubules. Smooth, muscle-like cells called *myoid cells,* contract to help transport sperm through the tubules.

The seminiferous tubules merge into straight tubules, which are connected to a maze of passageways known as the **rete** (RĒ-tē; *rete,* a net) **testis**. Fifteen to 20 large **efferent ductules** connect the rete testis to the epididymis.

**? INTEGRATION**

**B.** On a warm day, would the cremaster be contracted or relaxed? Why?

**Lo LEARNING OUTCOME**

Outline fetal descent of the testes, and describe the structures related to the scrotum and testes.

**? REVIEW**

**A.** Identify the complex network of channels that is connected to the seminiferous tubules.

# Module 26.3

## Spermatogenesis occurs in the testes and produces mature sperm

**Spermatogenesis**, or sperm production, involves (1) **mitosis** and cell division (cytokinesis), (2) **meiosis** (mī-Ō-sis), a special form of cell division involved in gamete production, and (3) **spermiogenesis**, the differentiation of immature male gametes into physically mature sperm.

**? REVIEW**

**A.** Define *spermatogenesis*.

**1** As you may recall from Chapter 3, somatic cells contain 23 pairs of chromosomes. Mitosis is part of the process of somatic cell division, producing two daughter cells each containing identical pairs of chromosomes. This diagram follows the fates of three representative chromosomes during mitosis and cell division. Because the daughter cells each contain 23 pairs of chromosomes, or two sets of chromosomes, they are called **diploid** (DIP-loyd; *diplo*, double) (2n) cells. Each mitotic division of a stem cell within the seminiferous tubules produces one daughter cell that remains a stem cell and one daughter cell that enters meiosis.

**2** Meiosis involves two cycles of cell division (**meiosis I** and **meiosis II**) and produces four cells, each containing 23 individual chromosomes. Because these reproductive cells (gametes) contain only one member of each pair of chromosomes, they are called **haploid** (HAP-loyd; *haplo*, single) (n) cells. The events in the nucleus are the same for the formation of sperm or oocytes (female gametes). In meiosis, the corresponding maternal (inherited from the mother) and paternal (inherited from the father) chromosomes now come together in an event known as **synapsis** (sih-NAP-sis). Synapsis involves 23 pairs of chromosomes, and each member of each pair consists of two chromatids (see Module 3.20). A matched set of four chromatids is called a **tetrad** (TET-rad; *tetras*, four).

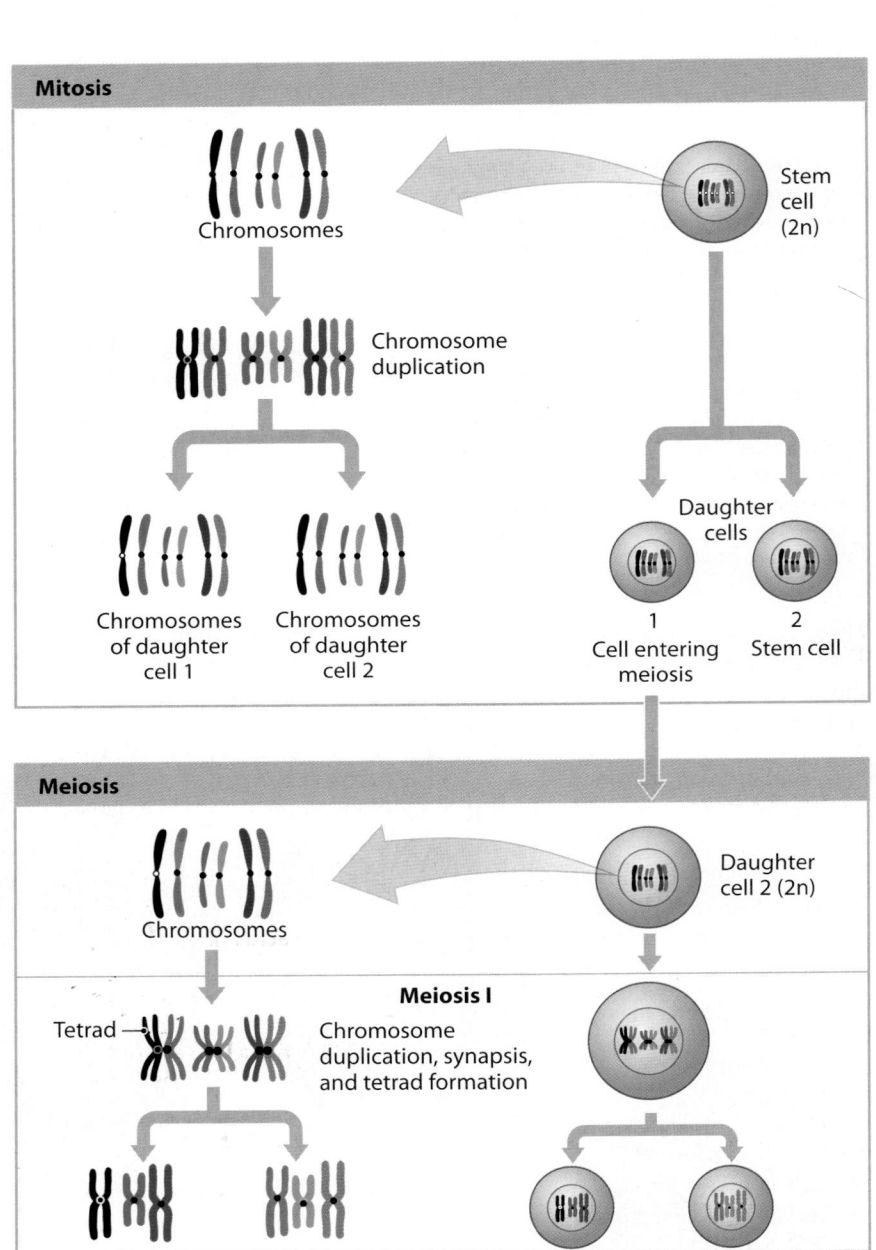

**Mitosis**

Chromosomes

Stem cell (2n)

Chromosome duplication

Chromosomes of daughter cell 1

Chromosomes of daughter cell 2

Daughter cells

1 Cell entering meiosis

2 Stem cell

**Meiosis**

Chromosomes

Daughter cell 2 (2n)

Tetrad

**Meiosis I**
Chromosome duplication, synapsis, and tetrad formation

**Meiosis II**

1 2 3 4
Chromosomes of gametes

1 2 3 4
Gametes (n)

**3** **Spermatogonia** (singular, *spermatogonium*) are the stem cells in the seminiferous tubules that divide to produce sperm in a process called **spermatogenesis**. The complete process of spermatogenesis occurs in four phases and takes about 64 days.

**Mature sperm**

**?** **REVIEW**

**B.** How many sperm will eventually be produced from each primary spermatocyte?

## Spermatogenesis

### Mitosis of a spermatogonium

Each division of a diploid spermatogonium produces two daughter cells. One is a spermatogonium that remains in contact with the basement membrane of the tubule, and the other is a **primary spermatocyte** that is displaced toward the lumen. The progression from spermatogonium to primary spermatocyte takes 16 days.

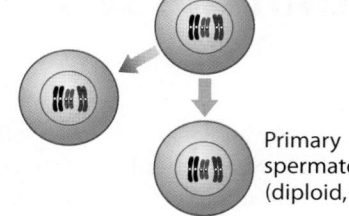

Primary spermatocyte (diploid, 2n)

### Meiosis I

As meiosis I begins, each primary spermatocyte contains 46 individual chromosomes. At the end of meiosis I, the daughter cells are called **secondary spermatocytes**. Every secondary spermatocyte contains 23 chromosomes, each with a pair of duplicate chromatids. This phase takes about 24 days.

DNA replication

Primary spermatocyte

Synapsis and tetrad formation

Secondary spermatocytes

### Meiosis II

The secondary spermatocytes soon enter meiosis II, which yields four haploid **spermatids**, each containing 23 chromosomes. For each primary spermatocyte that enters meiosis, four spermatids are produced. This phase lasts only a few hours.

Spermatids (haploid, n)

### Spermiogenesis (physical maturation)

In spermiogenesis, the last step of spermatogenesis, each spermatid matures into a single sperm. The process—from spermatids to sperm—takes 24 days.

Sperm (haploid, n)

---

**4** Here you can see the distinctive, specialized features of a **sperm**. Unlike other, less specialized cells, a mature sperm lacks an endoplasmic reticulum, a Golgi apparatus, lysosomes, peroxisomes, inclusions, and many other intracellular structures. The loss of these organelles reduces the cell's size and mass; it is essentially a mobile carrier for the enclosed chromosomes, and extra weight would slow it down.

## Structure of a Sperm

The **acrosome** (AK-rō-sōm) is a membranous compartment containing enzymes essential to fertilization. *nucleus cytoplasm*

The **head** is a flattened ellipse containing a nucleus with densely packed chromosomes.

The **neck** contains both centrioles of the original spermatid. The microtubules of the distal centriole are continuous with those of the middle piece and tail.

The **middle piece** contains mitochondria arranged in a spiral around the microtubules. Mitochondria provide the ATP required to move the tail.

The **tail** is a **flagellum**, a whiplike organelle that moves the sperm.

The tail of a sperm is the only flagellum in the human body. Whereas cilia beat in a predictable, wavelike fashion, the tail of a sperm has a whiplike, corkscrew motion.

**?** **REVIEW**

**C.** List the structures of a sperm.

*acrosome*
*spermatozoon*

**Lo** **LEARNING OUTCOME**

Summarize the events of spermatogenesis, and describe the functional anatomy of a mature sperm.

# Module 26.4
# Meiosis and early spermiogenesis occur within the seminiferous tubules

**1** This is a light micrograph of a transverse section (cross section) of a testis. Because the seminiferous tubules are tightly coiled, most tissue slides show them in transverse section. Note how the *rete testis* is a network of canals. The twisting, curved tubules within each lobule are the *seminiferous tubules*, where spermatogenesis occurs. The *septa testis* are the fibrous partitions that converge toward the region nearest the entrance to the epididymis.

Rete testis

Seminiferous tubules

Septa testis

**Testis**                                                    LM × 26

**2** Here is a higher-magnification view of several sections through one or more seminiferous tubules. Spermatogenesis and spermiogenesis together take approximately 9 weeks. If spermatogenesis were synchronized along the entire length of a tubule, sperm would be released once every nine weeks. Instead, each segment of a seminiferous tubule is at a different phase of the process of spermatogenesis. As a result, every seminiferous tubule is continuously producing sperm.

Seminiferous tubule containing late spermatids

Seminiferous tubule containing sperm

Seminiferous tubule containing early spermatids

**Seminiferous tubules**                                      LM × 75

**3** Each tubule is surrounded by a delicate connective tissue capsule, and areolar tissue fills the spaces between the tubules. Within those spaces are numerous blood vessels and large interstitial endocrine cells. **Interstitial endocrine cells** (Leydig cells) produce androgens, such as testosterone and androstenedione, the dominant sex hormones in males.

**? REVIEW**
**A.** What is the function of interstitial endocrine cells?

**4** Each seminiferous tubule contains spermatogonia, spermatocytes at various stages of meiosis, spermatids, sperm, and large **nurse cells** (Sertoli cells). Nurse cells are attached to the tubular capsule and extend to the lumen between the other types of cells. Developing spermatocytes undergoing meiosis, and spermatids undergoing spermiogenesis, are not free in the seminiferous tubules. Instead, they are surrounded by the cytoplasm of the nurse cells. As spermiogenesis proceeds, the spermatids gradually develop into mature sperm. At **spermiation**, a sperm loses its attachment to the nurse cell and enters the lumen of the seminiferous tubule.

**5** Nurse cells provide a microenvironment that supports spermatogenesis. One of their many important functions is to isolate the seminiferous tubules from the general circulation by creating a **blood testis barrier** comparable in function to the blood brain barrier. Nurse cells are joined by tight junctions, forming a layer that divides the seminiferous tubule into an outer **basal compartment**, which contains the spermatogonia, and an inner **luminal compartment**, where meiosis and spermiogenesis occur.

**? REVIEW**
**B.** Describe the process of spermiation.

**? REVIEW**
**C.** What is the role of nurse cells?

**Lo LEARNING OUTCOME**
Explain meiosis and early spermiogenesis within the seminiferous tubules.

# Module 26.5

# The male reproductive tract receives secretions from the seminal, prostate, and bulbo-urethral glands

The testes produce physically mature sperm. The other portions of the male reproductive system are responsible for the functional maturation, nourishment, storage, and transport of sperm. The sperm leaving the testes are physically mature but immobile and incapable of fertilizing an oocyte. To become motile (actively swimming) and fully functional, sperm must undergo a process called **capacitation**. Capacitation normally occurs in two steps: (1) Sperm become motile when they are mixed with secretions of the seminal glands, and (2) they become capable of successful fertilization when exposed to conditions in the female reproductive tract. The epididymis secretes a substance (as yet unidentified) that prevents premature capacitation.

**1** This posterior view shows the urinary bladder, prostate, and other structures of the male reproductive system.

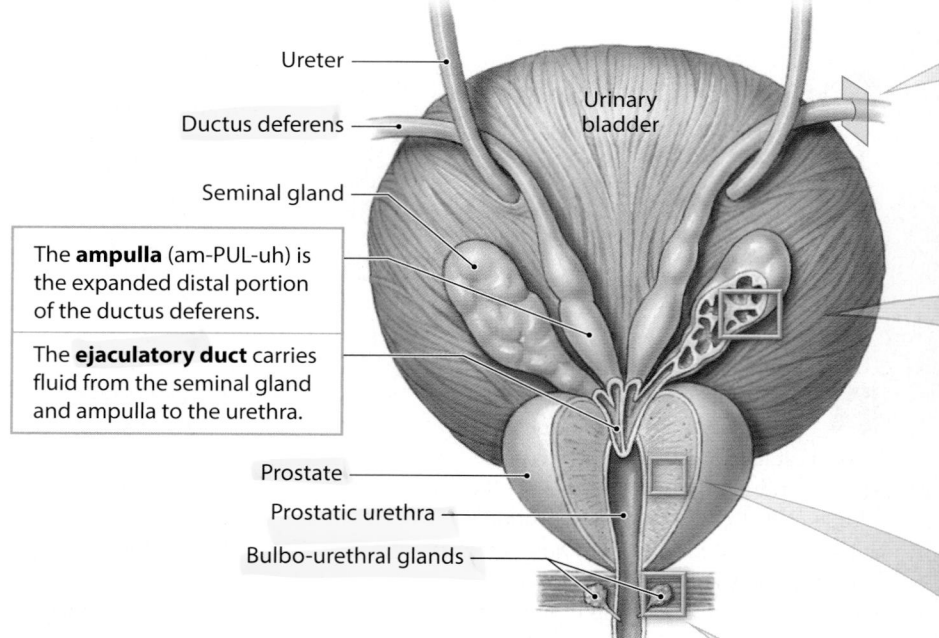

Ureter

Ductus deferens

Seminal gland

Urinary bladder

The **ampulla** (am-PUL-uh) is the expanded distal portion of the ductus deferens.

The **ejaculatory duct** carries fluid from the seminal gland and ampulla to the urethra.

Prostate

Prostatic urethra

Bulbo-urethral glands

**2** The **epididymis** (ep-i-DID-i-mis; *epi*, on + *didymos*, twin; plural, *epididymides*), the start of the male reproductive tract, is a coiled tube bound to the posterior border of each testis. A tubule almost 7 m (23 ft) long, the epididymis is coiled and twisted so as to take up very little space. It takes up to 2 weeks for a sperm to pass through the epididymis and complete its functional maturation. During this time, sperm exist in a sheltered environment that is precisely regulated by the surrounding epithelial cells.

**3** The pseudostratified columnar epithelial lining of the epididymis has extremely long **stereocilia**. These processes increase the surface area available for absorption from, and secretion into, the fluid in the tubule.

| Regions of the Epididymis |
| --- |
| The **head** receives sperm from the efferent ductules. |
| The **body** extends inferiorly along the posterior surface of the testis. |
| Near the inferior border of the testis, the number of coils decreases, marking the start of the **tail**. The tail recurves and ascends, connecting to the ductus deferens. Sperm are stored primarily within the tail. |

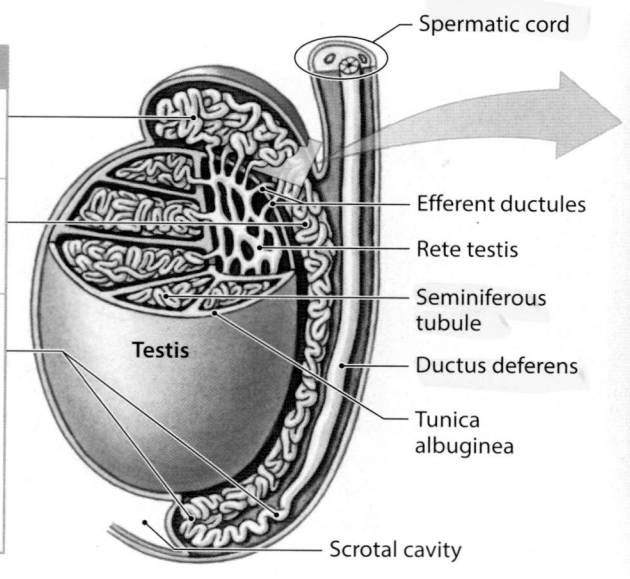

Spermatic cord

Efferent ductules

Rete testis

Seminiferous tubule

Ductus deferens

Tunica albuginea

Testis

Scrotal cavity

Epithelium

Sperm in lumen

Stereocilia

**Epididymis**

LM × 240

The wall of the ductus deferens contains a thick layer of smooth muscle. Peristaltic contractions in this layer propel sperm and fluid along the duct.

The lumen of the ductus deferens is lined by pseudo-stratified ciliated columnar epithelium.

**Ductus deferens** LM × 100

**4** Each **ductus deferens**, or *vas deferens*, is 40–45 cm (16–18 in.) long and begins at the tail of the epididymis. As part of the spermatic cord, it ascends through the inguinal canal, enters the abdominal cavity, and passes posterior to the urinary bladder to reach the prostate. In addition to transporting sperm from the epididymis to the ejaculatory duct, the ductus deferens can store sperm in the ampulla for several months. During this time, the sperm remain in a temporary state of inactivity with low metabolic rates.

**? REVIEW**

**A.** What is the function of the ductus deferens?

**Seminal gland** LM × 35

Lumen

Secretory pockets

Smooth muscle

**5** The **seminal glands**, also called the *seminal vesicles*, are embedded in connective tissue on either side of the midline, sandwiched between the posterior wall of the urinary bladder and the rectum. The seminal glands produce about 60 percent of the volume of **semen**. When mixed with the secretions of the seminal glands, previously inactive but functional sperm undergo the first step in capacitation and begin beating their flagella, becoming highly motile. The secretions are ejected by smooth muscle contractions controlled by the sympathetic division of the ANS. About 2–5 mL of semen is released in a typical ejaculation. Semen contains sperm and **seminal fluid**. In addition to seminal gland secretions, semen also contains secretions from nurse cells, the epididymis, the prostate, and the bulbo-urethral glands.

**Prostate** LM × 35

Connective tissue and smooth muscle

Tubulo-alveolar glands

**6** The **prostate** is a small, muscular, rounded organ about 4 cm (1.6 in.) in diameter. The prostate encircles the proximal portion of the urethra as it leaves the urinary bladder. The tubulo-alveolar glands of the prostate are surrounded by a thick blanket of smooth muscle fibers. The prostate produces 20–30 percent of the volume of semen. Prostatic secretions contain **seminalplasmin** (sem-i-nal-PLAZ-min), an antibiotic protein that may help prevent urinary tract infections in males. These secretions are ejected into the prostatic urethra by peristaltic contractions of the muscular prostate wall.

**Bulbo-urethral gland** LM × 140

Smooth muscle

Capsule

Mucous glands

Lumen

**7** The paired **bulbo-urethral glands** (Cowper's glands) are located at the base of the penis. The diameter of each is nearly 10 mm (less than 0.5 in.). The duct of each gland empties into the urethra. These glands secrete a thick, alkaline mucus that helps neutralize any urinary acids that may remain in the urethra, and it also lubricates the tip of the penis.

**? REVIEW**

**B.** What are the functions of the bulbo-urethral glands?

**? INTEGRATION**

**C.** Trace the ductal pathway from the epididymis to the urethra.

**Lo LEARNING OUTCOME**

Explain the roles played by the male reproductive tract and accessory glands in the functional maturation, nourishment, storage, and transport of sperm.

# Module 26.6
# The penis conducts urine and semen to the exterior

**1** The **penis** is a tubular organ for sexual intercourse and conducting urine to the exterior. Main structures include the root, body, and glans penis.

Pubic symphysis

Membranous urethra

Bulb of penis

Right crus of penis

Ischial ramus

Corpus spongiosum

Corpora cavernosa

Scrotum

Neck of glans

External urethral orifice

The **root** of the penis is the fixed portion that attaches the penis to the body wall. This connection occurs within the urogenital triangle immediately inferior to the pubic symphysis.

The **body of penis** (*shaft*), is the tubular, movable portion of the organ.

The **glans penis** (*head*) is the expanded distal end that surrounds the external urethral orifice. The **neck of glans** is the narrow portion of the penis between the shaft and the glans.

**2** This cross section shows that most of the body of the penis consists of three cylindrical columns of vascularized **erectile tissue**. Erectile tissue consists of a three-dimensional network with vascular spaces incompletely separated by partitions of elastic connective tissue and smooth muscle fibers. In the resting state, the arterial branches are constricted, and the muscular partitions are tense. This combination restricts blood flow into the erectile tissue. During **erection**, the penis stiffens and elevates to an upright position, but the flaccid (nonerect) penis hangs inferior to the pubic symphysis and anterior to the scrotum.

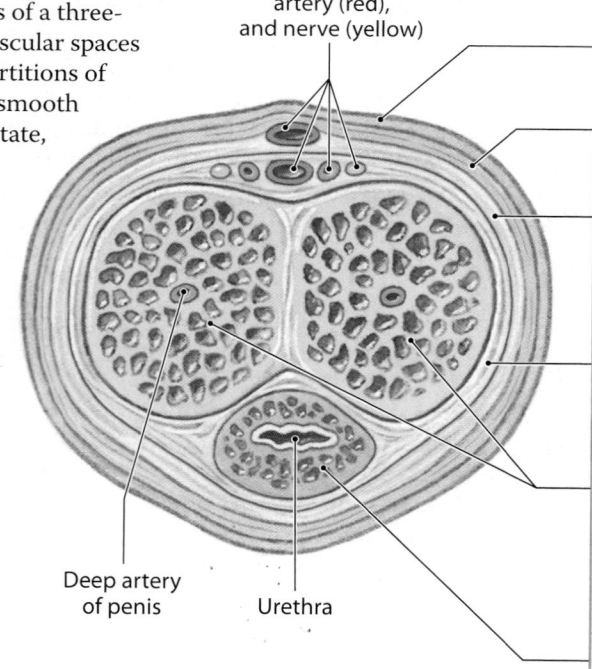

Dorsal veins (blue), artery (red), and nerve (yellow)

Deep artery of penis

Urethra

### Tissue Layers of the Penis

The skin overlying the penis resembles that of the the skin on the scrotum.

The dermis has a layer of smooth muscle that is a continuation of the dartos muscle of the scrotum.

The underlying areolar tissue allows the thin skin to move without distorting deeper structures. The areolar tissue also contains superficial arteries, veins, and lymphatic vessels.

Deep to the areolar tissue, a dense network of elastic fibers encircles the internal structures of the penis.

### Erectile Tissues

The anterior surface of the flaccid penis covers two cylindrical masses of erectile tissue: the **corpora cavernosa** (KOR-por-a ka-ver-NŌ-suh; singular, *corpus cavernosum*). The corpora cavernosa extend along the length of the penis as far as its neck.

The relatively slender **corpus spongiosum** (spon-jē-Ō-sum) surrounds the urethra. This erectile body extends to the tip of the penis, where it expands to form the glans penis.

**? REVIEW**
**A.** Name the three columns of erectile tissue in the penis.

**3** This view of the male reproductive tract provides a three-dimensional perspective on the relationships among the glands and passageways.

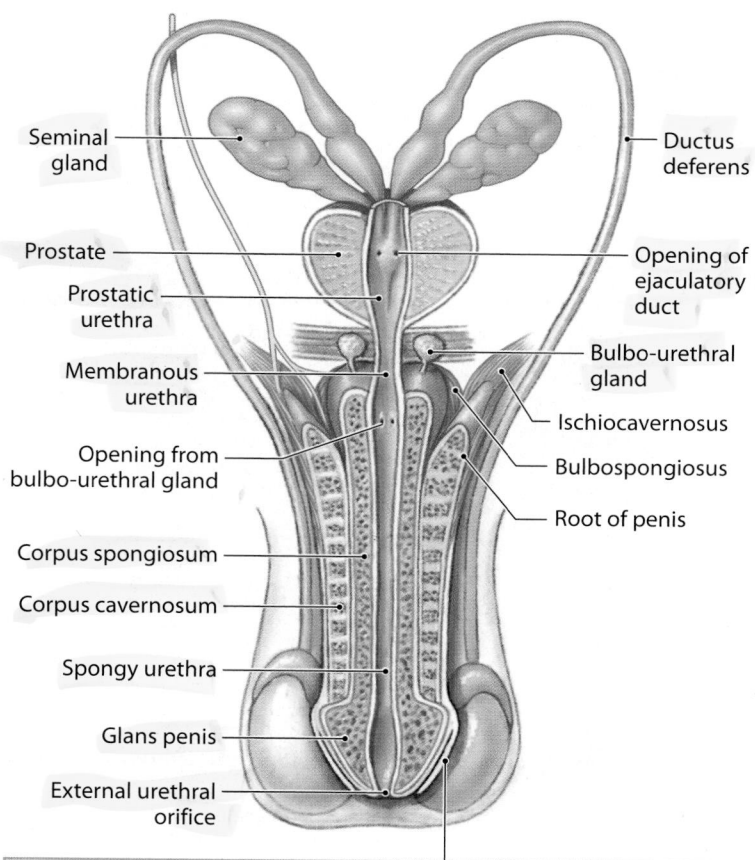

Seminal gland

Ductus deferens

Prostate

Prostatic urethra

Opening of ejaculatory duct

Membranous urethra

Bulbo-urethral gland

Opening from bulbo-urethral gland

Ischiocavernosus

Bulbospongiosus

Root of penis

Corpus spongiosum

Corpus cavernosum

Spongy urethra

Glans penis

External urethral orifice

A fold of skin called the **foreskin** or **prepuce** (PRĒ-pūs), surrounds the tip of the penis. The foreskin attaches to the relatively narrow neck of the penis and continues over the glans penis. *Preputial* (prē-PŪ-shē-al) *glands* in the skin of the neck and the inner surface of the prepuce secrete a waxy material known as **smegma** (SMEG-ma).

---

### ⚕ Clinical Note

## Impotence is an inability to achieve or maintain an erection

**Impotence**, or **erectile dysfunction** (**ED**), is an inability to achieve or maintain an erection. There may be a variety of physical causes, because erection involves vascular changes as well as neural commands. For example, low blood pressure in the arteries of the penis, due to a cardiovascular blockage such as a plaque, reduce the ability to attain an erection. Psychological factors such as depression or anxiety can also result in impotence. ED can be successfully treated with oral medications that enhance the effects of nitric oxide—the chemical that dilates blood vessels by relaxing vascular smooth muscles. These medications increase blood flow to the penis, causing an erection in response to sexual stimulation.

---

**4** Human sexual function refers to how the body reacts during the sexual response. Complex neural reflexes coordinate sexual functions.

### Arousal

During **arousal**, erotic thoughts or sensory nerve stimuli lead in the genital region leads to an increase in the parasympathetic outflow over the pelvic nerves.

The parasympathetic innervation of the penile arteries involves neurons that release **nitric oxide** (NO) at their axon terminals. The smooth muscles in the arterial walls relax when NO is released, at which time the vessels dilate, blood flow increases, the vascular channels become engorged with blood, and the penis becomes erect.

Arousal also stimulates the bulbo-urethral glands. Their secretion lubricates the tip of the glans penis.

### Emission

Further stimulation leads to sympathetic activation that causes **emission**.

Emission begins with peristaltic contractions in the ampullae of the ductus deferens. This pushes sperm into the prostatic urethra.

Contractions then begin in the walls of the seminal glands and the prostate, and their secretions now enter the urethra and mix with the sperm introduced by ampullary contractions to form **semen**.

### Ejaculation

**Ejaculation** occurs as powerful, rhythmic contractions take place in the ischiocavernosus and bulbospongiosus. The contractions of both muscles are controlled by somatic motor neurons in the inferior lumbar and superior sacral segments of the spinal cord. These contractions are associated with the pleasurable sensations known as **male orgasm**.

The bulbospongiosus wraps around the base of the penis. The contraction of this muscle pushes semen toward the external urethral orifice.

The ischiocavernosus inserts along the sides of the penis, and its contractions serve primarily to stiffen the erect penis.

**? INTEGRATION**

**B.** An inability to contract the ischiocavernosus and bulbospongiosus would interfere with which phase of the male sexual response?

**LO LEARNING OUTCOME**

Describe the structure and function of the penis.

# Module 26.7

# Testosterone plays a key role in establishing and maintaining male sexual function

Male sexual and reproductive function is regulated by a complex interaction of hormones from the hypothalamus, anterior lobe of the pituitary gland, and testes. **Testosterone** is produced primarily by the interstitial endocrine cells of the testes, and negative feedback systems keep testosterone levels within a relatively narrow range until late in life. This flowchart diagrams the hormonal interactions that regulate male sexual function.

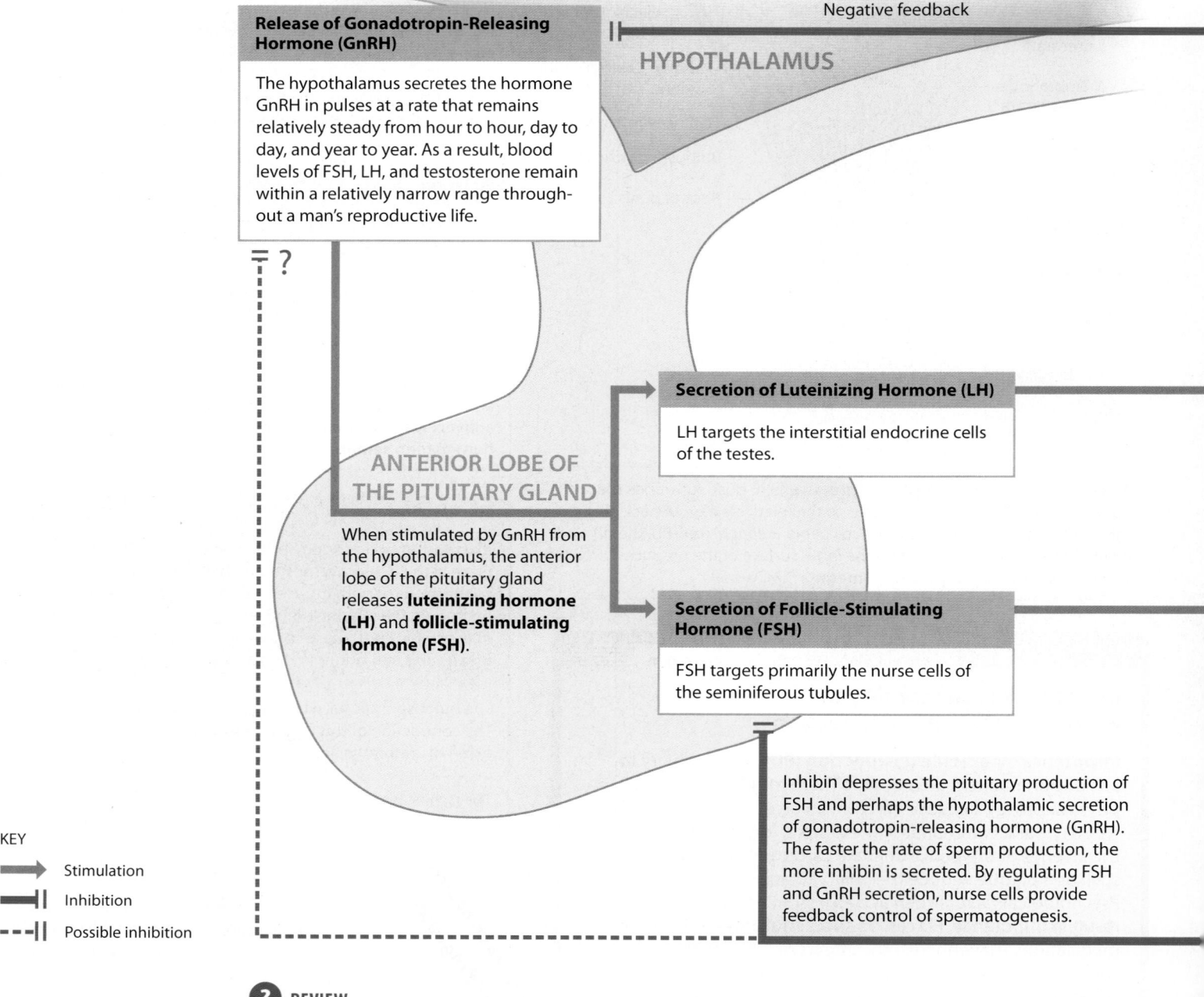

**Release of Gonadotropin-Releasing Hormone (GnRH)**

The hypothalamus secretes the hormone GnRH in pulses at a rate that remains relatively steady from hour to hour, day to day, and year to year. As a result, blood levels of FSH, LH, and testosterone remain within a relatively narrow range throughout a man's reproductive life.

Negative feedback

**HYPOTHALAMUS**

**ANTERIOR LOBE OF THE PITUITARY GLAND**

When stimulated by GnRH from the hypothalamus, the anterior lobe of the pituitary gland releases **luteinizing hormone (LH)** and **follicle-stimulating hormone (FSH)**.

**Secretion of Luteinizing Hormone (LH)**

LH targets the interstitial endocrine cells of the testes.

**Secretion of Follicle-Stimulating Hormone (FSH)**

FSH targets primarily the nurse cells of the seminiferous tubules.

Inhibin depresses the pituitary production of FSH and perhaps the hypothalamic secretion of gonadotropin-releasing hormone (GnRH). The faster the rate of sperm production, the more inhibin is secreted. By regulating FSH and GnRH secretion, nurse cells provide feedback control of spermatogenesis.

KEY

→ Stimulation

⊣| Inhibition

---⊣| Possible inhibition

**? REVIEW**

**A.** Identify important regulatory hormones that establish and maintain male sexual function.

In many target tissues, some of the arriving testosterone is converted to **dihydrotestosterone (DHT)**. A small amount of DHT diffuses back out of the cell and into the bloodstream, and DHT levels are usually about 10 percent of circulating testosterone levels. Dihydrotestosterone can also enter peripheral cells and bind to the same hormone receptors targeted by testosterone. In addition, some tissues (notably those of the external genitalia) respond to DHT rather than to testosterone, and other tissues (including the prostate) are more sensitive to DHT than to testosterone.

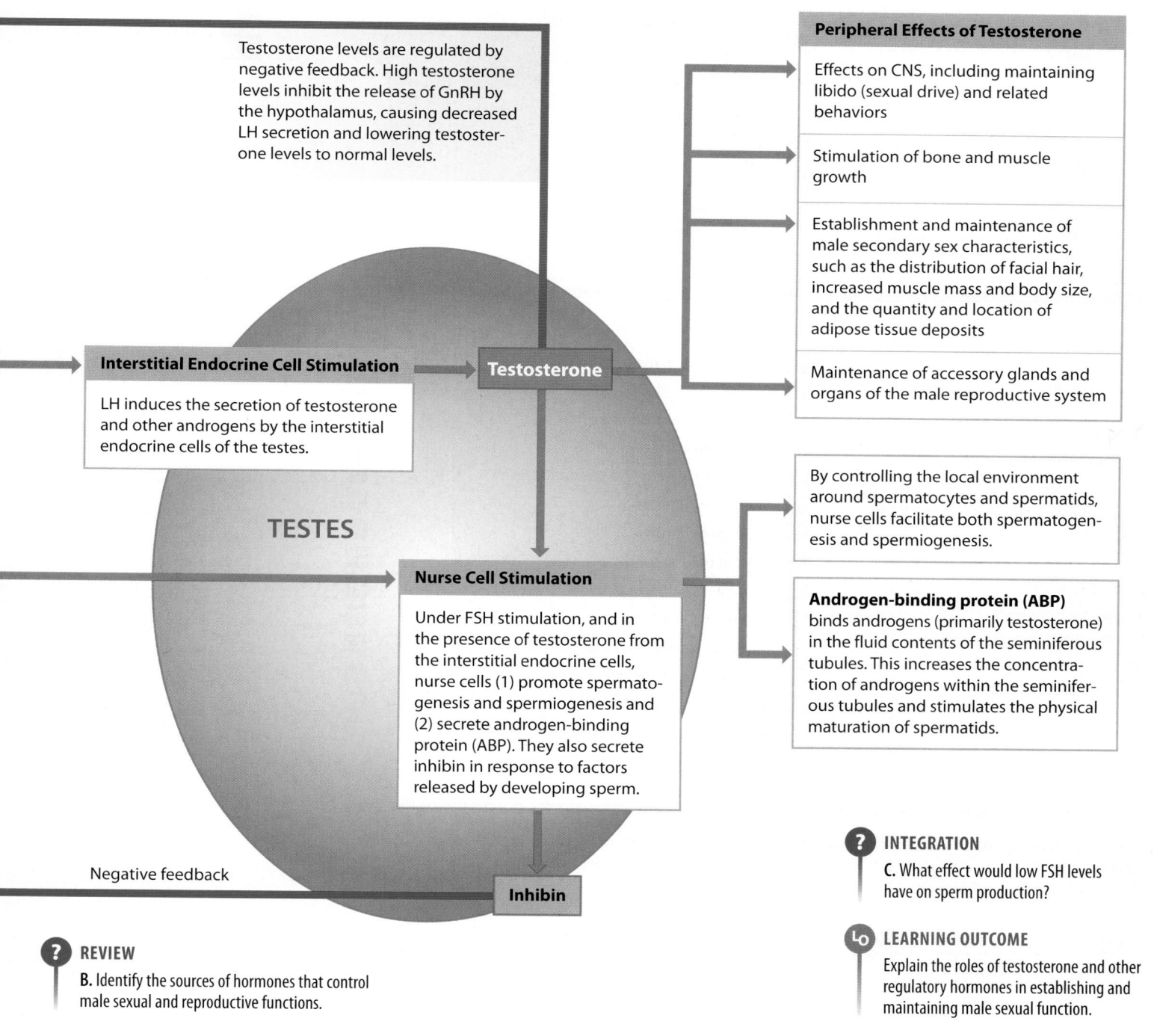

Testosterone levels are regulated by negative feedback. High testosterone levels inhibit the release of GnRH by the hypothalamus, causing decreased LH secretion and lowering testosterone levels to normal levels.

**Interstitial Endocrine Cell Stimulation**

LH induces the secretion of testosterone and other androgens by the interstitial endocrine cells of the testes.

**TESTES**

**Testosterone**

**Peripheral Effects of Testosterone**

Effects on CNS, including maintaining libido (sexual drive) and related behaviors

Stimulation of bone and muscle growth

Establishment and maintenance of male secondary sex characteristics, such as the distribution of facial hair, increased muscle mass and body size, and the quantity and location of adipose tissue deposits

Maintenance of accessory glands and organs of the male reproductive system

By controlling the local environment around spermatocytes and spermatids, nurse cells facilitate both spermatogenesis and spermiogenesis.

**Nurse Cell Stimulation**

Under FSH stimulation, and in the presence of testosterone from the interstitial endocrine cells, nurse cells (1) promote spermatogenesis and spermiogenesis and (2) secrete androgen-binding protein (ABP). They also secrete inhibin in response to factors released by developing sperm.

**Androgen-binding protein (ABP)** binds androgens (primarily testosterone) in the fluid contents of the seminiferous tubules. This increases the concentration of androgens within the seminiferous tubules and stimulates the physical maturation of spermatids.

Negative feedback

**Inhibin**

**? REVIEW**

**B.** Identify the sources of hormones that control male sexual and reproductive functions.

**? INTEGRATION**

**C.** What effect would low FSH levels have on sperm production?

**Lo LEARNING OUTCOME**

Explain the roles of testosterone and other regulatory hormones in establishing and maintaining male sexual function.

# Section 1 Review
# Male Reproductive System

**Labeling:** Label the structures of the male reproductive system in the accompanying diagram.

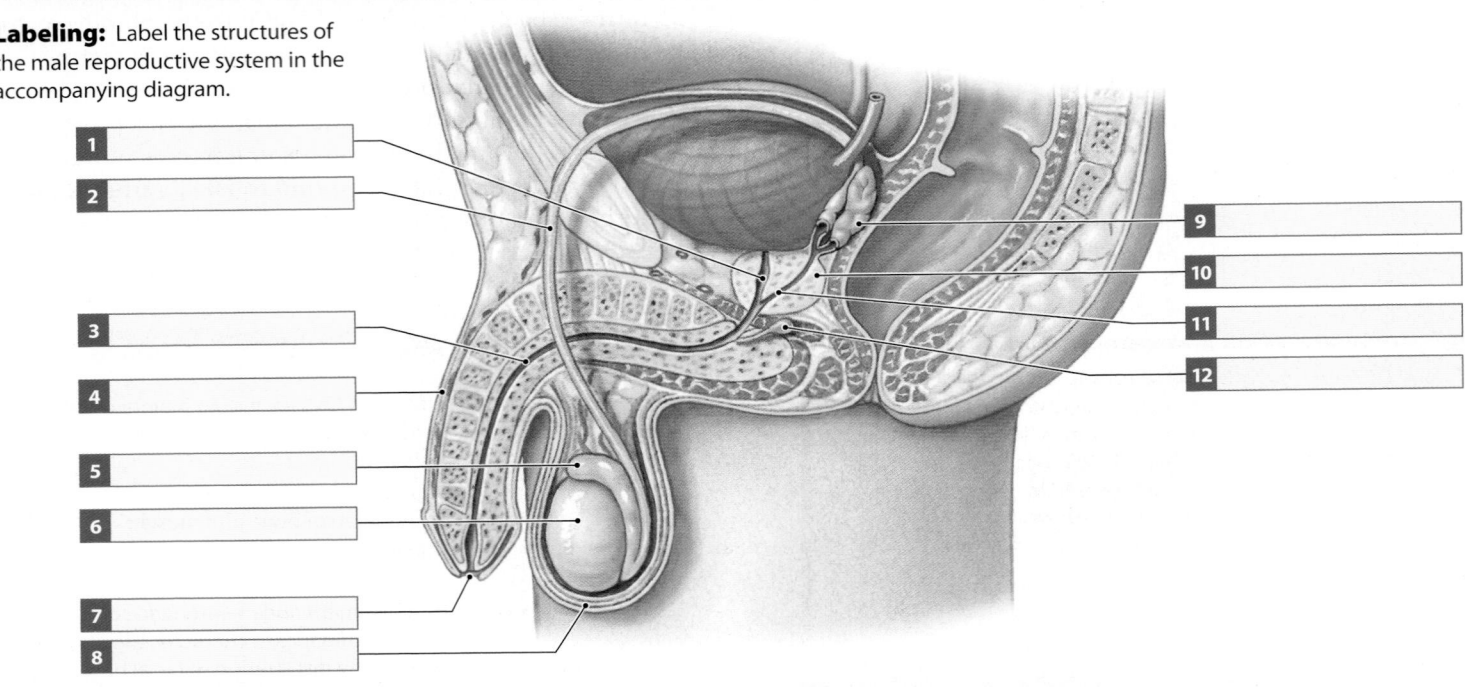

1 _____
2 _____
3 _____
4 _____
5 _____
6 _____
7 _____
8 _____
9 _____
10 _____
11 _____
12 _____

**Matching:** Match each lettered term with the most closely related description.

a. semen
b. epididymis
c. nurse cells
d. corpus spongiosum
e. luteinizing hormone (LH)
f. impotence
g. follicle-stimulating hormone (FSH)
h. spermatogonia
i. seminiferous tubules
j. dartos muscle
k. spermatogenesis
l. interstitial endocrine cells
m. spermiogenesis
n. penis and scrotum

13 Scrotal smooth muscle
14 Sperm stem cells
15 Sites of sperm production
16 Produce testosterone
17 Physical maturation of spermatids
18 Sperm production
19 External genitalia
20 Start of male reproductive tract
21 Maintain blood testis barrier
22 Sperm and seminal fluid
23 Inability to achieve or maintain an erection
24 Erectile tissue surrounding the urethra
25 Induces secretion of androgens
26 Hormone that targets nurse cells

13 _____
14 _____
15 _____
16 _____
17 _____
18 _____
19 _____
20 _____
21 _____
22 _____
23 _____
24 _____
25 _____
26 _____

**Section integration:** In males, the endocrine disorder hypogonadism is primarily due to the underproduction of testosterone or the lack of tissue sensitivity to testosterone, and it results in sterility. What are five primary functions of testosterone in males?

27 _____

_____

# Module 26.8
# Female reproductive structures include the external genitalia and internal genitalia

A woman's reproductive system produces sex hormones and functional gametes, and it must also be able to protect and support a developing embryo, maintain a growing fetus, and nourish a newborn infant. The female reproductive system is made up of the external genitalia and the internal genitalia. The **female external genitalia** include the vulva— a collective term describing the outer genitals—and its associated structures. The female urethra (not shown) is also part of the external genitalia. **Female internal genitalia** include the ovaries, uterine tubes, uterus, and vagina. Each **mammary gland** of the breast produces milk that nourishes the infant.

**? REVIEW**
**A.** Identify the main organs of the female reproductive system.

**1** This figure gives an overview of the structures of the female reproductive system. The female gonads, called *ovaries* (singular, *ovary*), produce immature female gametes called **oocytes** (Ō-ō-sīts), which later mature into **ova** (singular, *ovum*). Oocytes leave the ovary and then travel along the female duct system, or **female reproductive tract**. If fertilization occurs, it will normally occur in the uterine tubes, and further embryonic development will occur within the uterus.

**Mammary Glands**

The **mammary gland** is a milk-secreting structure within each breast.

**Major Structures of the Female Reproductive Tract**

Each **ovary** produces oocytes and hormones.

Each **uterine tube** delivers an oocyte or embryo to the uterus. These tubes are the normal sites of fertilization.

The **uterus** is the site of embryonic and fetal development and of exchange between the maternal and embryonic/fetal bloodstreams.

The **vagina** is a muscular tube leading to the cervix. It is an organ of sexual intercourse, serves as the birth canal, and is a passageway for menstrual fluids.

**External Genitalia**

The **mons pubis** is the pad of fatty tissue overlying the symphysis pubis.

The **clitoris** is located at the most anterior part of the vulva and contains erectile tissue.

The **labia** (*labium majus* and *labium minus*) border the vagina.

**? REVIEW**
**B.** Name the structures of the female external genitalia.

**? REVIEW**
**C.** Where does fertilization normally occur?

**ᴸᴼ LEARNING OUTCOME**
Identify the structures and functions of the female external genitalia, internal genitalia, and mammary glands.

# Module 26.9

# Major female reproductive organs are the ovaries, uterus, and their associated structures

**1** The main organs of the female reproductive system are the *ovaries*, *uterine tubes*, *uterus*, *vagina*, and components of the external genitalia—the *labia* and *clitoris*. This sagittal section through the pelvic cavity shows the location and orientation of the female reproductive organs.

The pocket formed between the posterior wall of the uterus and the anterior surface of the colon is the **recto-uterine** (rek-tō- Ū-ter-in) **pouch**.

The pocket formed between the uterus and the posterior wall of the urinary bladder is the **vesico-uterine** (ves-i-kō-Ū-ter-in) **pouch**.

Urinary bladder

Pubic symphysis

Urethra

Accessory glands lubricate the entrance to the vagina and the external genitalia.

Infundibulum

*Fimbriae*

Sigmoid colon

Rectum

Vaginal fornix

Anus

**? REVIEW**

**A.** Distinguish between the vesico-uterine and recto-uterine pouches.

## Major Female Reproductive Organs

The paired **ovaries** are small, lumpy, almond-shaped organs near the lateral walls of the pelvic cavity. The ovaries have three main functions: (1) producing immature female gametes, or **oocytes**; (2) secreting female sex hormones, including **estrogens** and **progesterone**; and (3) secreting **inhibin**, involved in the feedback control of pituitary FSH production. (Estrogens include estradiol, estriol, and estrone; estradiol is the most abundant of the three types.)

Each **uterine tube** begins with a funnel-shaped structure, called an **infundibulum**, that is open into the pelvic cavity along the medial surface of the ovary. The other end of the uterine tube opens into the uterine cavity.

The **uterus** sits inferior to the ovaries and covers the superior and posterior surfaces of the urinary bladder.

The **vagina** extends from the base of the uterus (cervix) to the exterior.

## External Genitalia

The **labia** are folds that surround the central space called the *vestibule*.

The **clitoris** contains erectile tissue that projects into the vestibule. Stimulating the clitoris produces pleasurable sensations associated with the female orgasm.

**? REVIEW**

**B.** What roles do the ovaries perform?

**2** Each ovary is stabilized by several thickened peritoneal folds called *ligaments*. This is a superior posterior view with the left uterine tube pulled away from the ovary to show the ligaments more clearly.

The **ovarian ligament** extends from the uterus, near the attachment of the uterine tube, to the medial surface of the ovary.

The **mesovarium** (mez-ō-VĀ-rē-um), a thickened fold of mesentery, supports and stabilizes the position of each ovary.

Ovarian artery and vein

Infundibulum

Fimbriae

Ovary

Uterine tube

Uterus

The **suspensory ligament** extends from the lateral surface of the ovary past the open end of the uterine tube to the pelvic wall. The major blood vessels of the ovary (the ovarian artery and ovarian vein) pass along this ligament to reach the mesovarium.

Broad ligament

Ureter

Uterosacral ligament

External os

Cervix

Vaginal rugae

Vaginal wall

**Posterior view**

The **broad ligament** is an extensive mesentery that encloses the ovaries, uterine tubes, and uterus. It attaches to the sides and floor of the pelvic cavity, where it becomes continuous with the parietal peritoneum. The broad ligament thus subdivides this part of the peritoneal cavity.

**3** This is a cross section taken through the mesovarium and the broad ligament. A typical ovary is about 5 cm long, 2.5 cm wide, and 8 mm thick (2 in. by 1 in. by 0.33 in.) and weighs 6–8 g (roughly 0.25 oz). Blood vessels enter and leave the ovary at the *ovarian hilum*, where the ovary attaches to the mesovarium. The interior of the ovary is divided into a superficial **cortex** and a deeper **medulla**. Gametes are produced in the cortex.

**? REVIEW**

**C.** Name the structures enclosed by the broad ligament, and describe the function of the mesovarium.

The visceral peritoneum, or **germinal epithelium**, covers the surface of each ovary and consists of a layer of columnar epithelial cells.

The germinal epithelium overlies a dense connective tissue layer called the **tunica albuginea**.

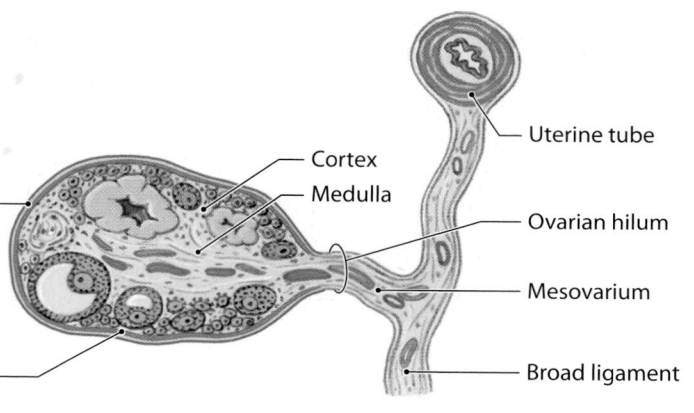

Cortex

Medulla

Uterine tube

Ovarian hilum

Mesovarium

Broad ligament

**LO LEARNING OUTCOME**

Describe the anatomy of the ovaries, uterus, and associated structures.

# Module 26.10

## Oogenesis occurs in the ovaries, and ovulation occurs during the 28-day ovarian cycle

**Oogenesis** (ō-ō-JEN-e-sis) is the formation and development of the oocyte. It begins before a woman's birth, accelerates at puberty, and ends at menopause. Between puberty and menopause, an oocyte matures on a monthly basis as part of the ovarian cycle.

**1** Although the nuclear events in the ovaries during meiosis are the same as those in the testes, the cytoplasm of the **primary oocyte** is unevenly distributed during the two meiotic divisions. Oogenesis produces one functional **secondary oocyte**, which contains most of the original cytoplasm, and two or three **polar bodies**, nonfunctional cells that later disintegrate. Another difference is that the ovary releases a secondary oocyte rather than a mature ovum. The secondary oocyte is suspended in metaphase of meiosis II. Meiosis will only be completed if fertilization occurs.

Why is the secondary oocyte considered haploid even though it has the same amount of DNA as a diploid primary oocyte? It is haploid because each set of duplicate chromatids originated from only *one* member of a chromosome pair in the primary oocyte.

### Everyday Physiology

#### What happens to most primary oocytes?

After 5 months of development, fetal ovaries contain an estimated 7 million primary oocytes. Not all survive. At birth the ovaries have approximately 2 million primary oocytes. However, most of these primary oocytes do not survive until puberty, degenerating in a process called *atresia* (ah-TRĒ-zē-uh). By puberty, their number has dropped to about 400,000.

### Oogenesis

#### Mitosis of oogonium

During fetal development, female reproductive stem cells called **oogonia** (ō-ō-GŌ-nē-uh singular, *oogonium*), undergo mitosis, producing diploid *primary oocytes* (2n).

Oogonium (stem cell)

Oogonium

Primary oocyte (diploid, 2n)

DNA replication

#### Meiosis I

Between the 3rd and 7th months of fetal development, the daughter cells, or primary oocytes (Ō-ō-sīts), prepare to undergo meiosis. They proceed as far as the prophase of meiosis I, but then the process comes to a halt.

The primary oocytes remain in a state of suspended development until puberty, when rising levels of FSH trigger the start of the **ovarian cycle**. Each month after the ovarian cycle begins, some of the primary oocytes are stimulated to undergo further development. Meiosis I is then completed, yielding a haploid secondary oocyte and a first polar body.

Tetrad

Primary oocyte

First polar body

First polar body may not complete meiosis II

Secondary oocyte (haploid, n)

#### Meiosis II

Each month after the ovarian cycle begins, usually one secondary oocyte leaves the ovary suspended in metaphase of meiosis II.

At the time of fertilization, a second polar body forms, and the fertilized oocyte is then called a mature **ovum**. (A cell in any of the preceding steps in oogenesis is sometimes called an immature ovum.)

Secondary oocyte released (ovulation) in metaphase of meiosis II

Second polar body

Sperm (n)

Nucleus of ovum (n)

**Fertilization**
(See Module 27.2)

**REVIEW**

**A.** Define *oocyte*.

**2** **Ovarian follicles** are structures where oocyte growth and meiosis I occur. Follicle development may take almost a year to complete. During the 28-day ovarian cycle, a *tertiary ovarian follicle* matures, releases its secondary oocyte, and degenerates. *Follicles do not physically move around the cortex; this figure shows the relative sizes and sequence of events.*

### Primordial Ovarian Follicles in Egg Nest

Primary oocytes are located in the outer portion of the ovarian cortex, near the tunica albuginea, in clusters called **egg nests**. An inactive primary oocyte is surrounded by a simple squamous layer of follicle cells, forming a **primordial follicle**.

Primary oocyte

Follicle cells

LM × 270

### Formation of Primary Ovarian Follicle

Follicular cells enlarge, divide, and form several layers of cells around an activated primary oocyte. These follicle cells are now called **granulosa cells**. A region around the oocyte develops, called the **zona pellucida** (ZŌ-na pe-LU-si-duh; *pellucidus*, translucent). As the granulosa cells enlarge and multiply, a layer of **thecal endocrine cells** (*theca*, a box) forms around the follicle. Thecal cells and granulosa cells work together to produce estrogens.

Granulosa cells

Primary oocyte

Thecal endocrine cells

LM × 270

### Formation of Secondary Ovarian Follicle

**Secondary follicles** develop as the wall of the follicle thickens and the deeper follicular cells begin secreting fluid that accumulates in small pockets. These pockets gradually expand and separate the inner and outer layers of the follicle.

Thecal cells

Zona pellucida

Nucleus of primary oocyte

Granulosa cells

LM × 245

### Formation of a Tertiary Ovarian Follicle

By days 10–14 of the ovarian cycle, usually only one follicle has enlarged to become a mature **tertiary follicle**, or graafian (GRAH-fē-an) follicle, roughly 15 mm in diameter. The oocyte projects into the **antrum** (AN-trum), or expanded central chamber of the follicle. The granulosa cells associated with the secondary oocyte form a protective layer known as the **corona radiata** (kō-RŌ-nuh rā-dē-AH-tuh).

Antrum containing follicular fluid

Secondary oocyte

Corona radiata

LM × 80

Released secondary oocyte

Corona radiata

Ruptured follicle

Ovary

### Formation of Corpus Albicans

If fertilization does not occur, after 12 days progesterone and estrogen levels fall markedly. Fibroblasts invade the nonfunctional corpus luteum, producing a knot of pale scar tissue called a **corpus albicans** (AL-bi-kanz). The disintegration, or involution, of the corpus luteum marks the end of the ovarian cycle. A new ovarian cycle then begins with the development of another dominant tertiary follicle.

### Formation of Corpus Luteum

The empty tertiary follicle initially collapses, and under LH stimulation the remaining granulosa cells proliferate to create the **corpus luteum** (LŪ-tē-um; *lutea*, yellow), which secretes progesterone (prō-JES-ter-ōn) and estrogens. Progesterone prepares the uterus for pregnancy by stimulating the maturation of the uterine lining and the secretions of uterine glands.

### Ovulation

At **ovulation**, the tertiary follicle releases the secondary oocyte and corona radiata into the pelvic cavity. Ovulation marks the end of the **follicular phase** of the ovarian cycle and the start of the **luteal phase**.

Secondary oocyte

LM × 170

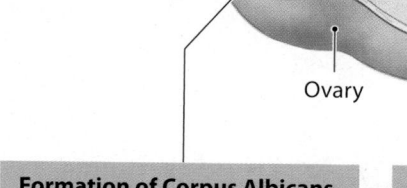

**? REVIEW**

**B.** List the important events in follicle development.

**? INTEGRATION**

**C.** What are the main differences in gamete production between males and females?

**LO LEARNING OUTCOME**

Outline the processes of meiosis and oogenesis in the ovaries.

# Module 26.11

# The uterine tubes are connected to the uterus, a hollow organ with thick muscular walls

 Each **uterine tube** (fallopian tube) is a hollow, muscular structure measuring approximately 13 cm (5.2 in.) in length. The distal portion of each uterine tube connects to the uterus. The **uterus** is a hollow, muscular organ that is about 7.5 cm (3 in.) long with a maximum diameter of 5 cm (2 in.). It weighs 30–40 g (1–1.4 oz). The sectional illustration below shows the internal structure of the uterine tube and the connection between the lumen of the uterine tube and the large uterine cavity within the uterus.

## Uterine Tube

| | | | |
|---|---|---|---|
| The **infundibulum** is a funnel-like expansion of the uterine tube. It has numerous fingerlike projections that extend into the pelvic cavity. The projections are called **fimbriae** (FIM-brē-ē). Fimbriae drape over the surface of the ovary, but there is no physical connection between the two structures. The cilia lining the infundibulum create a current that carries the secondary oocyte toward the uterine tube. | The **ampulla** is the middle region. The thickness of its smooth muscle layers gradually increases as the tube approaches the uterus. | The ampulla leads to the **isthmus** (IS-mus), a short region connected to the uterine wall. | The **uterine part** is embedded within the uterine wall and connects the uterine tube to the uterine cavity. |

**?** **REVIEW**
**A.** How do recently released secondary oocytes reach the uterine tube?

Fimbriae

Uterine cavity

Uterine artery and vein

Vagina

## Layers of the Uterine Wall

The outer surface of the uterus is an incomplete serosa called the **perimetrium** (per-i-mē-trē-um; *peri-*, around + *metra*, uterus). It is continuous with the peritoneal lining and covers most of the uterine surface.

The perimetrium covers a thick, muscular **myometrium** (mī-ō-MĒ-trē-um; *myo-*, muscle). The smooth muscle tissue of the myometrium provides much of the force needed to move a fetus out of the uterus and into the vagina during delivery.

The inner lining consists of a glandular **endometrium** (en-dō-MĒ-trē-um) whose tissue changes in the course of the monthly uterine cycle.

## The Uterine Cavity

The **uterine cavity** is the large, superior chamber that is continuous with the isthmus of the uterine tube on each side.

The **internal os** (*os*, an opening or mouth) is the opening that connects the uterine cavity to the cervical canal.

The **cervical canal** is a constricted passageway at the inferior end of the uterine cavity. It begins at the internal os and ends at the external os.

The **external os** is the curved vaginal opening into the uterus.

**?** **REVIEW**
**B.** Describe the three layers of the uterine wall.

**2** This cross-section shows the smooth muscle layers surrounding the uterine tube mucosa. A combination of ciliary movement and peristaltic contractions in the walls of the uterine tube transport oocytes toward the uterus.

Mucosa

Smooth muscle

**Isthmus of uterine tube** LM × 100

**3** This colorized SEM shows the ciliated epithelium of the uterine tube. The cilia beat toward the uterine cavity, establishing fluid currents that help transport the secondary oocyte after ovulation.

Cilia

**Epithelial surface** SEM × 4000

**4** The uterus is divided into three anatomical regions, as shown here. The uterus, which is capable of great changes in size and shape, provides mechanical protection, nutritional support, and waste removal for the developing **embryo** (weeks 1–8) and **fetus** (week 9 through delivery). In addition, contractions in the muscular wall of the uterus are important in delivering the fetus at birth.

Ovarian artery and vein

Suspensory ligament of ovary

Ovary

Broad ligament

**Regions of the Uterus**

The **fundus** is the upper rounded portion of the uterine body. It projects superior to the openings of the uterine tubes.

The uterine **body** is the largest portion of the uterus. The body ends at a constriction that encircles the internal os. It makes up about two-thirds of the organ.

The **cervix** (SER-viks) is the inferior portion of the uterus that surrounds the cervical canal and the external os and projects into the vagina.

Vaginal artery

Vagina

**? REVIEW**
**C.** Name the regions of the uterus.

It normally takes 3 to 4 days for a secondary oocyte to travel from the infundibulum to the uterine cavity. If fertilization is to occur, the secondary oocyte must encounter sperm during the first 12–24 hours of its passage along the uterine tube.

 **LEARNING OUTCOME**
Describe the structure, histology, and functions of the uterine tubes and uterus.

# Module 26.12

# The uterine (menstrual) cycle involves changes in the functional layer of the endometrium

**1** Within the myometrium, branches of the uterine arteries form **arcuate arteries**, which encircle the endometrium. From the arcuate arteries, **radial arteries** supply the endometrium.

Endometrium · Myometrium

Uterine cavity

Arcuate arteries

Perimetrium

Radial artery

Uterine artery

**2** The endometrium contains a **basal layer** adjacent to the myometrium, and a **functional layer**, the region closest to the uterine cavity. **Straight arteries** deliver blood to the basal layer, and **spiral arteries** supply the functional layer. The functional layer of the endometrium contains large tubular uterine glands.

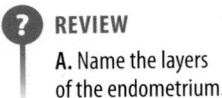 **REVIEW**
**A.** Name the layers of the endometrium.

Simple columnar epithelium · Uterine glands

Functional layer

Spiral artery

Basal layer

Straight artery

Myometrium

**3** The structure of the basal layer remains relatively constant over time, but that of the functional layer undergoes cyclical changes in response to sex hormone levels. These cyclical changes produce the characteristic histological features of the **uterine cycle**. The uterine cycle averages 28 days in length, but it can range from 21 to 35 days in healthy women of reproductive age.

| Menstrual Phase | Proliferative Phase | Secretory Phase |
|---|---|---|

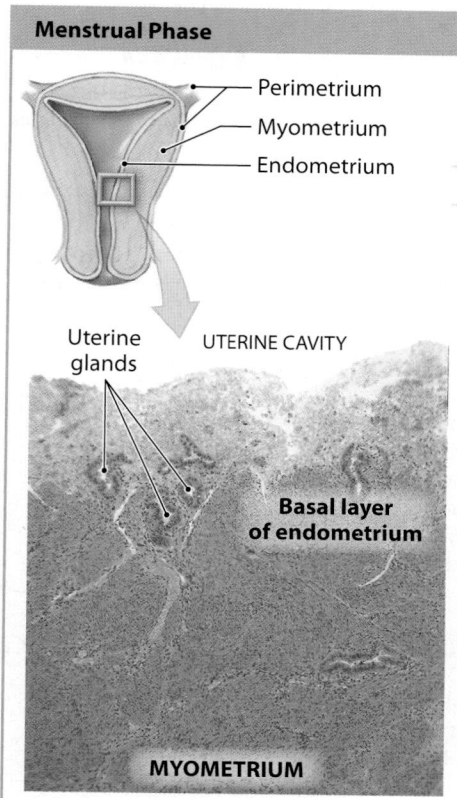

Perimetrium
Myometrium
Endometrium

Uterine glands
UTERINE CAVITY

**Basal layer of endometrium**

**MYOMETRIUM**

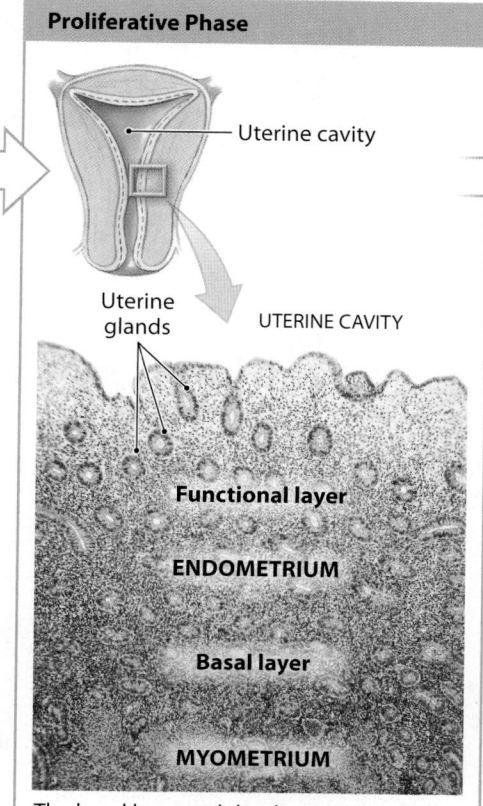

Uterine cavity

Uterine glands
UTERINE CAVITY

**Functional layer**

**ENDOMETRIUM**

**Basal layer**

**MYOMETRIUM**

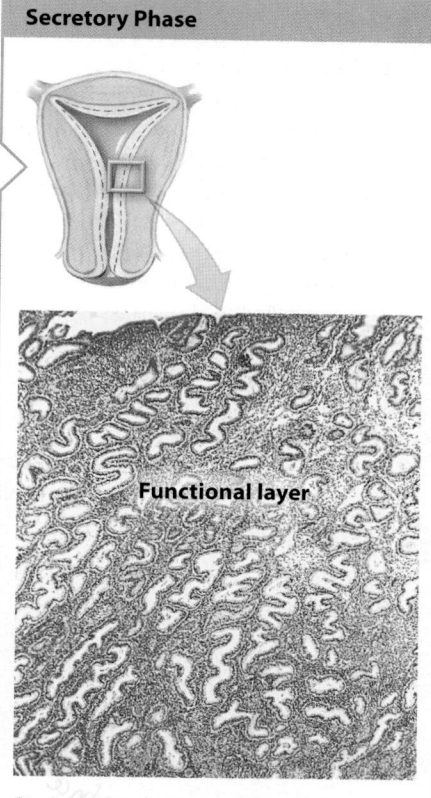

**Functional layer**

The uterine cycle begins with the **menstrual phase**. This phase is marked by the degeneration and sloughing (shedding) of the endometrial functional layer, leading to **menstruation** (men-strū-Ā-shun), also called **menses** (MEN-sēz). This degeneration is caused by constriction of the spiral arteries, which reduces endometrial blood flow. Eventually, the weakened arterial walls rupture, and blood pours into the connective tissues of the functional layer. Blood cells and degenerating tissues then break away and enter the uterine cavity, to be lost by passage through the external os and into the vagina. Menstruation generally lasts from 1 to 7 days. During this time, about 35 to 50 mL of blood are lost.

The basal layer and the deepest uterine glands survive menstruation intact. The epithelial cells of the uterine glands then multiply and spread across the endometrial surface, restoring the uterine epithelium. During this reorganization, the endometrium is in the **proliferative phase**. Estrogens secreted by the developing ovarian follicles stimulate and sustain the proliferative phase. By the time ovulation occurs, the functional layer is several millimeters thick, and prominent mucous glands extend to the border with the basal layer. At this time, the uterine glands are manufacturing a glycogen-rich mucus that can be metabolized by an early embryo.

During the **secretory phase**, the uterine glands enlarge, accelerating their rate of secretion, and the arteries that supply the uterine wall elongate and spiral through the tissues of the functional layer. This activity occurs under the combined stimulatory effects of progesterone and estrogens from the corpus luteum. The secretory phase begins at the time of ovulation and lasts as long as the corpus luteum remains intact. When the corpus luteum stops producing stimulatory hormones, a new uterine cycle begins with the onset of menstruation and the disintegration of the functional layer.

 **REVIEW**

**B.** How long does each menstruation typically last?

The uterine (menstrual) cycle begins at puberty. The first cycle, known as **menarche** (me-NAR-kē; *men*, month + *arche*, beginning), typically occurs at age 11–12. The cycles continue until **menopause** (MEN-ō-pawz), the end of menstruation, at age 45–55. Over the interim, regular uterine cycles are interrupted only by circumstances such as illness, stress, starvation, medications, contraceptives, or pregnancy.

 **REVIEW**

**C.** Describe the phases of the uterine cycle.

 **LEARNING OUTCOME**

Identify the phases and events of the uterine cycle.

# Module 26.13

## The vagina opens into the vestibule

**1** The **vagina** is an elastic, muscular tube extending between the cervix and the **vestibule**, a space bordered by the labia minora. The vagina is typically 7.5–9 cm (3–3.6 in.) long, but its diameter varies because it is highly distensible. The internal passageway is called the **vaginal canal**. The vagina serves as a passageway for the elimination of menstrual fluids; receives the penis during sexual intercourse and holds sperm prior to their passage into the uterus; and forms the inferior portion of the birth canal, through which the fetus passes during delivery.

At the proximal end of the vagina, the **cervix** projects into the vaginal canal. The shallow recess in the vagina surrounding the tip of the cervix is known as the **vaginal fornix** (FOR-niks).

External os
Vaginal fornix
Vaginal artery
Vaginal vein

Vaginal canal

**? REVIEW**
**A.** Describe the anatomy of the vagina.

In the relaxed state, the vaginal lining forms folds called **rugae**. The vaginal canal is lined by a nonkeratinized stratified squamous epithelium.

Throughout childhood, the vagina and vestibule are usually separated by the **hymen** (HĪ-men), an elastic epithelial fold that partially blocks the entrance to the vagina. An intact hymen is typically stretched or torn during first sexual intercourse, tampon use, pelvic exam, or physical activity. It is frequently absent, and its condition is not an indicator of sexual virginity.

Greater vestibular gland
Labium minus
Vestibule

**? REVIEW**
**B.** Identify the functions of the vagina.

*orifice*

The urethra opens into the vestibule just anterior to the vaginal entrance.

*orifice*

Vaginal entrance
Bulb of vestibule
Greater vestibular gland

Anus

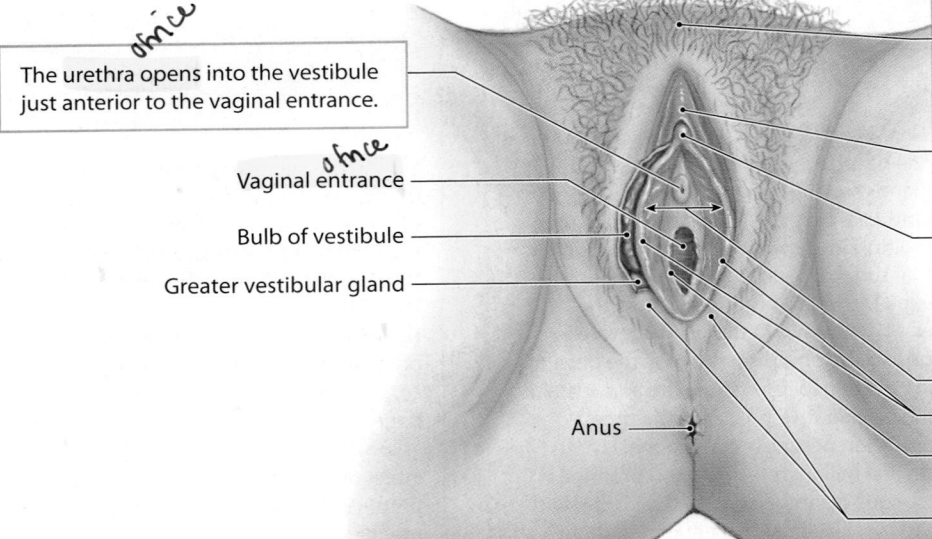

### Female External Genitalia

The bulge of the **mons pubis** is created by adipose tissue deep to the skin and superficial to the pubic symphysis.

Extensions of the **labia minora** encircle the body of the clitoris, forming its **prepuce**, or hood.

The **clitoris** (KLIT-ō-ris or kli-TŌR-is) projects into the vestibule. This small, rounded tissue projection contains erectile tissue comparable to the corpora cavernosa and corpus spongiosum of the penis.

Vestibule
Labia minora
Hymen (torn)

The **labia majora** (singular, *labium majus*) are prominent folds of skin that encircle and partially conceal the labia minora and adjacent structures.

**2** The area containing the female external genitalia is the **vulva** (VUL-vuh), or **pudendum** (pū-DEN-dum). The vagina opens into the vestibule, a central space bounded by small folds known as the **labia minora** (LĀ-be-uh mi-NOR-uh; singular, *labium minus*). A variable number of small **lesser vestibular glands** discharge their secretions onto the exposed surface of the vestibule, keeping it moist. During sexual arousal, a pair of ducts discharges the secretions of the **greater vestibular glands** (Bartholin's glands) into the vestibule. These mucous glands have the same embryonic origins as the bulbo-urethral glands of males. The **bulbs of vestibule** are masses of erectile tissue on either side of the vaginal entrance. These bulbs have the same embryonic origins as the corpus spongiosum of the penis in males.

**? INTEGRATION**
**C.** Cite the similarities that exist between certain structures in the reproductive systems of females and males.

**LO LEARNING OUTCOME**
Describe the structure, histology, and functions of the vagina.

# Module 26.14
# Each breast contains a mammary gland that secretes milk

A newborn infant cannot fend for itself, and several of its key systems are not fully developed. Over the initial period of adjustment to an independent life, the infant can gain nourishment from milk secreted by the maternal **mammary glands**. These organs are controlled mainly by hormones released by the reproductive system and the placenta, a temporary structure that provides the embryo and fetus with nutrients. The interaction of these hormones results in milk production, or **lactation** (lak-TĀ-shun).

**REVIEW**
**A.** Define *lactation*.

**1** Female **breasts** are two projections anterior to the pectoralis major. Each breast has a mammary gland, a variable amount of fat, a nipple, and an areola. Breasts are rudimentary in the male.

Pectoralis major

Lobules

Lobe

**Left breast**

### Structures of the Breast and Mammary Gland

On each side, a mammary gland lies in the subcutaneous **adipose tissue** (pectoral fat pad) deep to the skin of the chest.

Dense connective tissue surrounds the duct system and forms partitions that extend between the lobes and the lobules. These bands of connective tissue, the **suspensory ligaments of the breast**, originate in the dermis of the overlying skin.

The glandular tissue of the breast consists of separate **lobes**, each containing several secretory **lobules**. Each lobule is composed of many **secretory alveoli**.

Ducts leaving the lobules converge, giving rise to a single **lactiferous** (lak-TIF-er-us) **duct** in each lobe.

Near the nipple, each lactiferous duct enlarges, forming an expanded chamber called a **lactiferous sinus**.

Each breast has a **nipple**, a small conical projection where 15–20 lactiferous sinuses open onto the body surface.

The darker skin around each nipple is the **areola** (a-RĒ-ō-luh). Large sebaceous glands deep to the areolar surface give it a grainy texture.

*lactiferous duct → sinus*

**INTEGRATION**

**B.** Explain whether the blockage of a single lactiferous sinus would or would not interfere with the delivery of milk to the nipple.

**C.** Trace the route of milk from its site of production to the body surface.

**Lo** **LEARNING OUTCOME**

Discuss the structure and function of the breasts and mammary glands.

---

![Everyday Physiology]

## What determines breast size?

An active mammary gland is a tubulo-alveolar gland. An inactive, or *resting*, mammary gland is dominated by a duct system rather than by active glandular cells. Breast size in a nonpregnant woman reflects primarily the amount of adipose tissue present, not the amount of glandular tissue. The secretory apparatus normally does not complete its development unless pregnancy occurs.

# Module 26.15

# The ovarian and uterine cycles are regulated by hormones of the hypothalamus, pituitary gland, and ovaries

The ovarian and uterine cycles must operate in synchrony to ensure proper reproductive function. If the two cycles are not properly coordinated, infertility results. A female who doesn't ovulate cannot conceive, even if her uterus is perfectly normal. A female who ovulates normally but whose uterus is not ready to support an embryo will also be infertile.

**1** As in males, GnRH from the hypothalamus regulates reproductive function in females. However, in females, the pulse frequency of GnRH release and GnRH levels change throughout the course of the ovarian cycle.

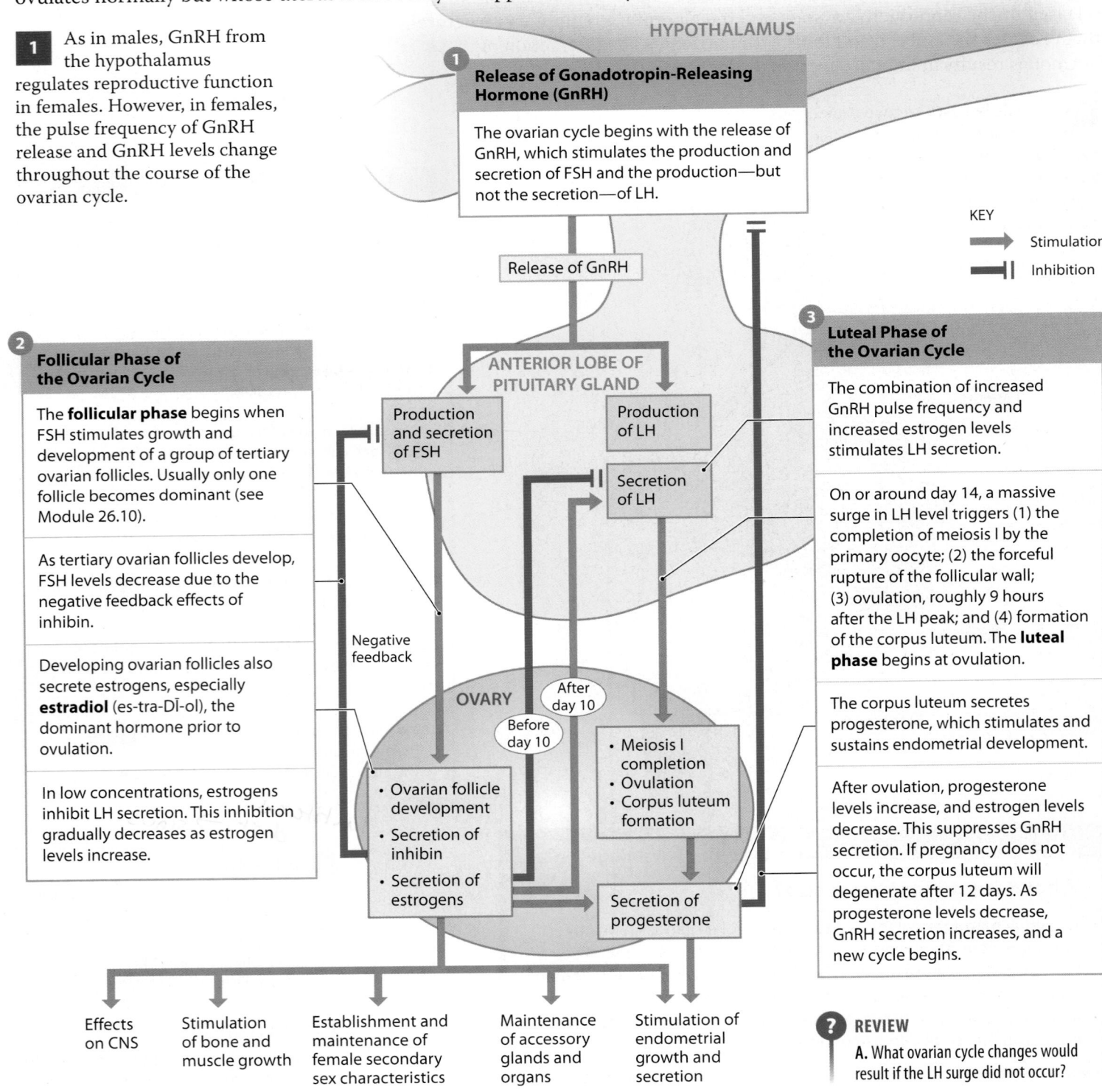

**HYPOTHALAMUS**

**1 Release of Gonadotropin-Releasing Hormone (GnRH)**

The ovarian cycle begins with the release of GnRH, which stimulates the production and secretion of FSH and the production—but not the secretion—of LH.

Release of GnRH

KEY
→ Stimulation
⊣ Inhibition

**2 Follicular Phase of the Ovarian Cycle**

The **follicular phase** begins when FSH stimulates growth and development of a group of tertiary ovarian follicles. Usually only one follicle becomes dominant (see Module 26.10).

As tertiary ovarian follicles develop, FSH levels decrease due to the negative feedback effects of inhibin.

Developing ovarian follicles also secrete estrogens, especially **estradiol** (es-tra-DĪ-ol), the dominant hormone prior to ovulation.

In low concentrations, estrogens inhibit LH secretion. This inhibition gradually decreases as estrogen levels increase.

**ANTERIOR LOBE OF PITUITARY GLAND**

Production and secretion of FSH

Production of LH

Secretion of LH

Negative feedback

**OVARY**

Before day 10

After day 10

• Ovarian follicle development
• Secretion of inhibin
• Secretion of estrogens

• Meiosis I completion
• Ovulation
• Corpus luteum formation

Secretion of progesterone

**3 Luteal Phase of the Ovarian Cycle**

The combination of increased GnRH pulse frequency and increased estrogen levels stimulates LH secretion.

On or around day 14, a massive surge in LH level triggers (1) the completion of meiosis I by the primary oocyte; (2) the forceful rupture of the follicular wall; (3) ovulation, roughly 9 hours after the LH peak; and (4) formation of the corpus luteum. The **luteal phase** begins at ovulation.

The corpus luteum secretes progesterone, which stimulates and sustains endometrial development.

After ovulation, progesterone levels increase, and estrogen levels decrease. This suppresses GnRH secretion. If pregnancy does not occur, the corpus luteum will degenerate after 12 days. As progesterone levels decrease, GnRH secretion increases, and a new cycle begins.

Effects on CNS

Stimulation of bone and muscle growth

Establishment and maintenance of female secondary sex characteristics

Maintenance of accessory glands and organs

Stimulation of endometrial growth and secretion

**? REVIEW**

**A.** What ovarian cycle changes would result if the LH surge did not occur?

**2** This illustration combines the key events in the ovarian and uterine cycles. The monthly hormonal fluctuations cause physiological changes that affect core body temperature. During the follicular phase—when estrogens are the dominant hormones—the **basal body temperature**, or the resting body temperature measured upon awakening in the morning, is about 0.3°C (0.5°F) lower than it is during the luteal phase, when progesterone dominates.

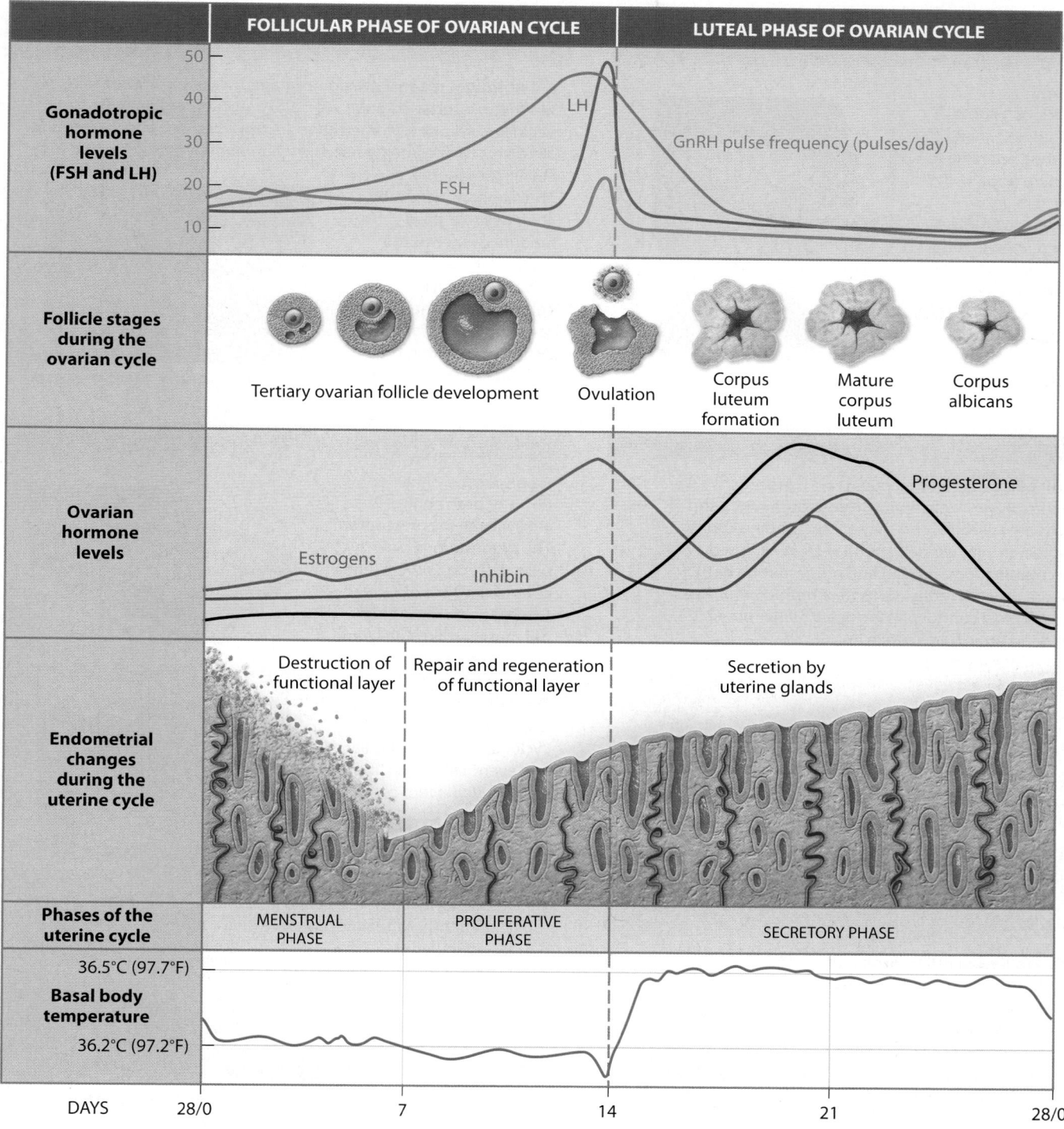

**FOLLICULAR PHASE OF OVARIAN CYCLE**

**LUTEAL PHASE OF OVARIAN CYCLE**

**Gonadotropic hormone levels (FSH and LH)**

LH

GnRH pulse frequency (pulses/day)

FSH

**Follicle stages during the ovarian cycle**

Tertiary ovarian follicle development

Ovulation

Corpus luteum formation

Mature corpus luteum

Corpus albicans

**Ovarian hormone levels**

Progesterone

Estrogens

Inhibin

**Endometrial changes during the uterine cycle**

Destruction of functional layer

Repair and regeneration of functional layer

Secretion by uterine glands

**Phases of the uterine cycle**

MENSTRUAL PHASE

PROLIFERATIVE PHASE

SECRETORY PHASE

36.5°C (97.7°F)

**Basal body temperature**

36.2°C (97.2°F)

DAYS 28/0     7     14     21     28/0

**? REVIEW**

**B.** What uterine cycle event occurs when estrogen and progesterone decrease?

**Lo LEARNING OUTCOME**

Summarize the hormonal regulation of the female reproductive cycles.

# Module 26.16
## Birth control strategies vary in effectiveness and associated risks

### Male Condom

**Male condoms** (prophylactics, or "rubbers") cover the glans and shaft of the penis during intercourse and keep sperm from reaching the female reproductive tract. Of all the strategies described in this module, only latex condoms protect against **sexually transmitted diseases (STDs)**, such as syphilis, gonorrhea, human papillomavirus (HPV), and HIV/AIDS.

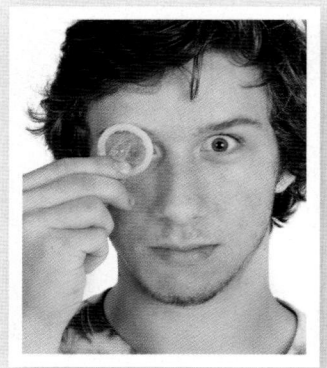

### Diaphragm with Spermicide

A **diaphragm** is a shallow, dome-shaped silicone cup with a flexible rim that is inserted into the vagina to prevent pregnancy. Because vaginas vary in size, women choosing this method must be individually fitted. Before intercourse, the diaphragm is inserted so that it covers the external os. The diaphragm must be coated with a small amount of spermicide (sperm-killing chemical) to be an effective contraceptive. The failure rate of a properly fitted and used diaphragm is estimated at 5–6 percent.

### Birth Control Pills—Combined (Estrogen and Progesterone)

**Birth control pills** (oral contraceptives) are medications taken in pill form that prevent conception. Numerous brands and combinations are available, and they are the leading contraceptive method among women aged 15–29. When used as prescribed, combination estrogen and progesterone birth control pills are the most effective form of birth control. Risks associated with birth control pills include a higher risk of blood clots, heart attack, and stroke in women who smoke, especially if they are over age 35. However, pregnancy itself has similar or higher risks.

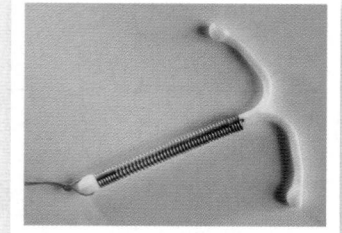

### Progesterone-Only Forms of Birth Control

Progesterone-only forms of birth control are available by prescripton: The progesterone-only pill and Depo-Provera (birth control shot) are examples. The **progesterone-only pill** must be taken daily, and skipping even one pill may result in pregnancy. **Depo-Provera** is injected every 3 months. Uterine cycles are initially irregular and eventually cease in roughly 50 percent of women using this product. The most common problems with this contraceptive method are a tendency to gain weight and a slow return to fertility (up to 18 months) after injections are discontinued.

### IUD (Intrauterine Device)

An **intrauterine device (IUD)** consists of a small plastic loop or a T that is inserted into the uterine cavity. Two types are available in the United States: copper and hormonal. The copper type affects sperm movement, effectively blocking them from reaching an oocyte. The hormonal type may prevent ovulation, and it also thickens cervical mucus, blocking sperm entry.

**? REVIEW**

**A.** Which birth control method provides some protection against sexually transmitted diseases?

**B.** The use of which birth control method often results in the cessation of the uterine cycle?

## The Rhythm Method

"Natural Family Planning," also called the **rhythm method** or *fertility awareness-based methods*, involves abstaining from sexual intercourse on the days ovulation might be occurring. The timing is estimated on the basis of previous patterns of menstruation; monitoring changes in indications of ovulation, including basal body temperature and cervical mucus texture; and, for some, urine tests for LH. Because of the irregularity of many women's uterine cycles, this method of contraception has a failure rate estimated to be 13–20 percent.

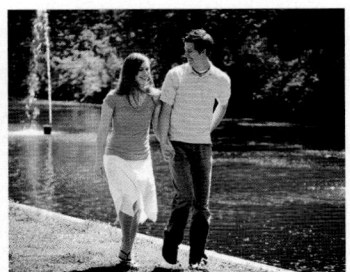

## Post-Coital Contraceptives (Plan B)

**Hormonal post-coital contraception**, or the emergency "morning after" pill, involves taking levonorgestrel contraceptive pills up to 72 hours (3 days) after unprotected intercourse. The pill is most effective if taken immediately after unprotected sex. It works by either preventing or delaying ovulation or interfering with oocyte fertilization. Some forms are effective for up to 120 hours (5 days) of unprotected sex. In the United States, it is available over-the-counter without age restrictions.

## Surgical Sterilization—Male

In a **vasectomy** (va-SEK-tō-mē), each ductus deferens is cut, and either a segment is removed (and the ends tied or cauterized), or silicone plugs are inserted (which makes it relatively easy to reverse the procedure). After a vasectomy, sperm cannot pass from the epididymides to the distal portions of the reproductive tract. The surgery can be performed in a physician's office in a matter of minutes; the failure rate (due to incomplete closure or blockage of either ductus deferens) is extremely low. After vasectomy, men experience normal sexual function because the secretions of the epididymides and testes normally account for only about 5 percent of the volume of semen. Sperm continue to develop, but they remain inactive and eventually degenerate.

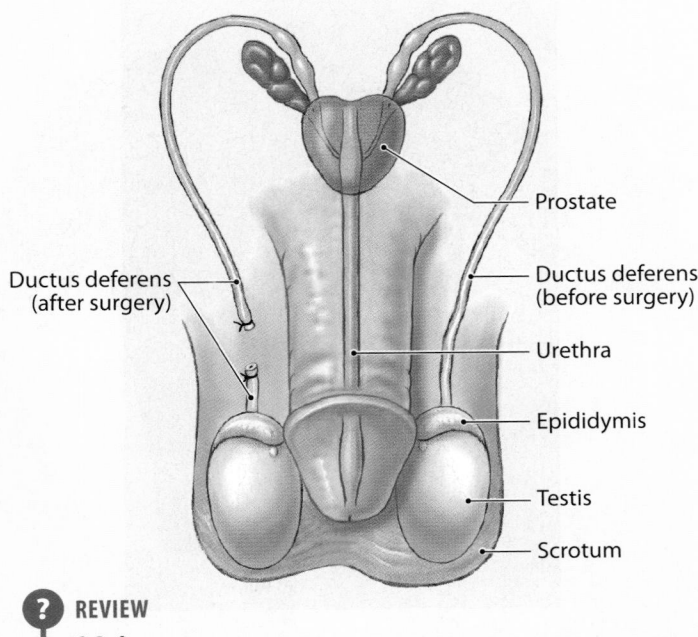

## Surgical Sterilization—Female

The uterine tubes can be blocked by a surgical procedure known as a **tubal ligation**. The failure rate for this procedure is also extremely low. Because the surgery requires that the abdomino-pelvic cavity be entered, most commonly by laparoscopy, a general anesthetic is required, and complications are more likely than with a vasectomy.

**REVIEW**

**C.** Define *vasectomy*.

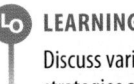 **LEARNING OUTCOME**

Discuss various birth control strategies and their associated risks.

# Module 26.17
# Reproductive system disorders are relatively common and often deadly

**1** Enlargement of the prostate, or **benign prostatic hypertrophy (BPH)**, typically occurs spontaneously in men over age 50. The increase in size occurs as testosterone production by the interstitial endocrine cells decreases. At the same time, the interstitial endocrine cells begin releasing small quantities of estrogens into the bloodstream. The combination of lower testosterone levels and the presence of estrogens probably stimulates prostatic growth. In severe cases, prostatic swelling constricts and blocks the urethra, producing urinary obstruction. **Prostate cancer**, a malignancy of the prostate, is the most common cancer among men and is the second most common cause of cancer deaths in males. Blood tests are often used for screening purposes. The most sensitive is a blood test for **prostate-specific antigen (PSA)**. Elevated levels of this antigen, normally present in low concentrations, may indicate the presence of prostate cancer. Screening with periodic PSA tests is now being recommended for some men over age 50. Treatment of localized prostate cancer often involves radiation therapy or surgical removal of the prostate—a **prostatectomy** (pros-ta-TEK-tō-mē).

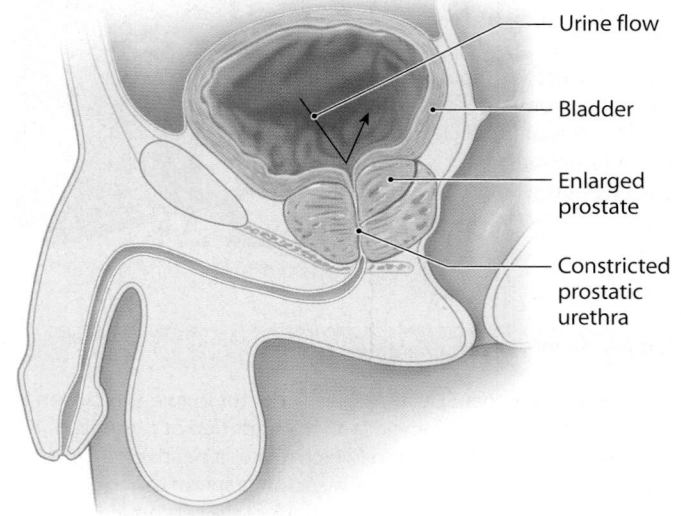

Urine flow

Bladder

Enlarged prostate

Constricted prostatic urethra

**2** **Testicular cancer** is cancer that starts in the testicles. It is not common and affects about 1 in every 263 males. The average age at the time of diagnosis is 33. More than 95 percent of testicular cancers result from abnormal spermatogonia or spermatocytes, rather than abnormal nurse cells, interstitial endocrine cells, or other testicular cells. Treatment generally consists of a combination of orchiectomy (removal of the testes) and chemotherapy. The 5-year survival rate for localized testicular cancer is 99%, primarily as a result of earlier diagnosis and improved treatment protocols.

**?** **REVIEW**

**A.** Compare benign prostatic hypertrophy with prostate cancer.

Cancerous tissue

Testis

**3** The mammary glands are stimulated by the changing levels of circulating reproductive hormones that accompany the uterine cycle, and, late in the ovarian cycle, occasional discomfort or even inflammation of mammary gland tissues can occur. If inflamed lobules become walled off by scar tissue, **cysts** are created. Clusters of cysts can be felt in the breast as discrete masses, a condition known as **fibrocystic disease**. Biopsy may be needed to distinguish between this benign condition and **breast cancer**. Breast cancer, a malignant, metastasizing tumor, is the most common cancer among American women, except for skin cancers. About one in eight women will develop invasive breast cancer during her lifetime. Notable risk factors include (1) a family history of breast cancer, (2) a first pregnancy after age 30, and (3) early menarche or late menopause. Treatment of breast cancer includes surgical removal of the tumor. Because in many cases cancer cells begin to spread before the condition is diagnosed, part or all of the affected mammary gland is surgically removed, and usually the axillary lymph nodes on that side are biopsied to detect signs of metastasis. A combination of chemotherapy, radiation treatments, and hormone treatments may be used to supplement the surgical procedures.

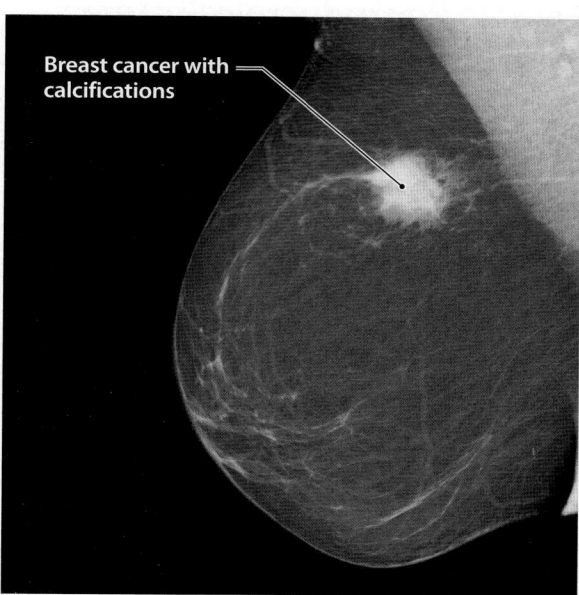

Breast cancer with calcifications

**4** **Ovarian cancer** begins in the ovaries and spreads. It accounts for more deaths than any other cancer of the female reproductive system. It is the most dangerous because it is seldom diagnosed in its early stages. The prognosis is relatively good for cancers that originate in the general ovarian tissues or from abnormal oocytes. These cancers respond well to some combination of chemotherapy, radiation, and surgery. However, 85 percent of ovarian cancers are **carcinomas** (cancers derived from epithelial cells), and sustained remission can be obtained in only about one-third of the cases of this type.

Ovarian carcinoma

Normal ovary

**5** **Cervical cancer** starts in the cells of the cervix. Cervical pre-cancers are diagnosed more often than invasive cervical cancers. It tends to occur in women younger than age 50. The cervical cancer death rate has decreased because of increased use of the Pap test, which detects the cancer in the early, most curable stage. Vaccines are available to protect against several types of **human papillomavirus (HPV)**, which cause about 75 percent of cervical cancers.

**? REVIEW**
**B.** Which pathogen is associated with most cases of cervical cancer?

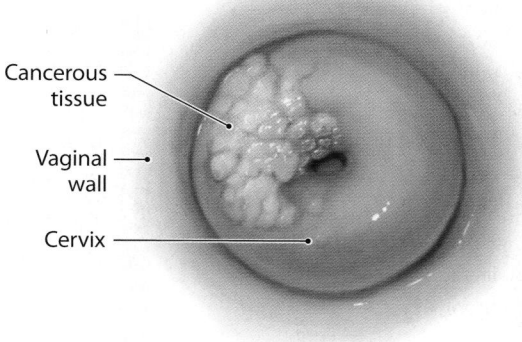

Cancerous tissue

Vaginal wall

Cervix

Sexual activity carries with it the risk of infection with a variety of microorganisms. The consequences of such an infection may range from merely inconvenient to potentially lethal. **Sexually transmitted diseases (STDs)/sexually transmitted infections (STIs)** are transferred from person to person, primarily by sexual contact. At least two dozen bacterial, viral, and fungal infections are currently recognized as STDs. The bacterium *Chlamydia* can cause pelvic inflammatory disease (PID) and infertility; HIV can be a deadly viral disease. Cases of chlamydia, gonorrhea, and syphilis have all increased in the United States during recent years. Poverty, intravenous drug use, prostitution, and the appearance of drug-resistant pathogens all contribute to the problem.

**? REVIEW**
**C.** Define *sexually transmitted disease*.

**Lo LEARNING OUTCOME**
Discuss several common reproductive system disorders.

**Labeling:** Label the structures of the female reproductive system in the accompanying diagram.

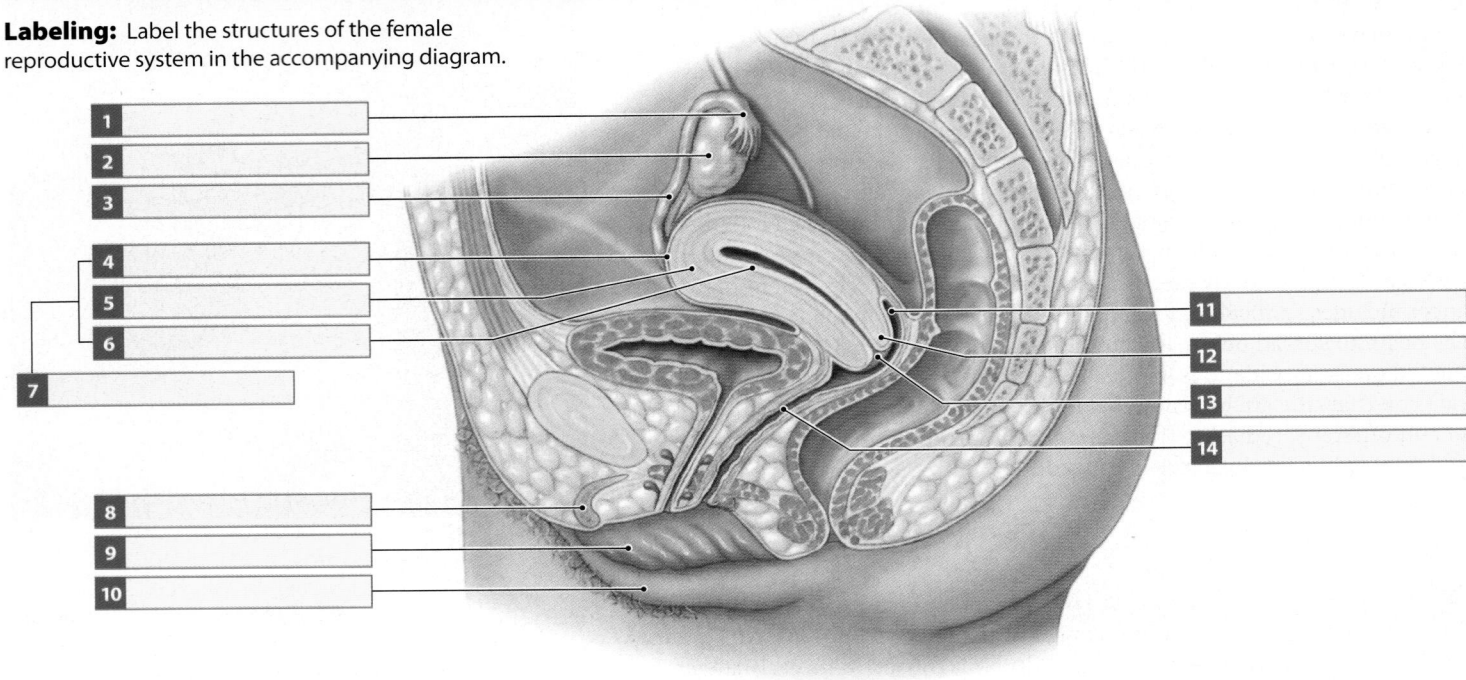

**Matching:** Match each lettered term with the most closely related description.

a. LH surge
b. recto-uterine pouch
c. tubal ligation
d. menarche
e. ovaries
f. corpus luteum
g. vulva
h. cervix
i. broad ligament
j. oocytes
k. GnRH
l. vesico-uterine pouch
m. lactation
n. uterine cycle

15 Immature female gametes
16 First menstrual cycle
17 Pocket anterior to the uterus
18 Encloses the ovaries, uterine tubes, and uterus
19 Pocket posterior to the uterus
20 Endocrine structure
21 Averages 28 days
22 Milk production
23 Oocyte and hormone production
24 Triggers ovulation
25 Inferior portion of the uterus
26 Female surgical sterilization
27 Stimulates FSH production and secretion
28 Contains female external genitalia

15 _____
16 _____
17 _____
18 _____
19 _____
20 _____
21 _____
22 _____
23 _____
24 _____
25 _____
26 _____
27 _____
28 _____

**Section integration:** In a condition known as endometriosis, endometrial cells migrate from the body of the uterus either into the uterine tubes or through the uterine tubes and into the peritoneal cavity, where they become established. Explain why periodic pain is a major symptom of endometriosis.

29 _____

_____

# Study Outline

## SECTION 1 • Male Reproductive System

### Module 26.1

### Male reproductive structures include the external genitalia and internal genitalia    p. 999

1. Male and female reproductive organs, called **gonads**, produce and store reproductive cells called **gametes**. Male gametes are **sperm**.

2. **Genitalia** are the organs of the reproductive system.

3. The **male external genitalia** include the penis, urethra, and scrotum.

4. **Male internal genitalia** include the ductus deferens (paired), seminal gland (paired), prostate, bulbo-urethral gland (paired), epididymis (paired), spermatic cord (paired), and testis (paired).

5. Sperm travel along a duct system known as the *reproductive tract*, mixing with glandular secretions to form a fluid called *semen*.

### Module 26.2

### Sperm transport relies on ducts, glands, and related structures of the scrotum and testes    p. 1000

6. Starting from the **epididymis**, sperm travel the **ductus deferens**, **ejaculatory duct**, and urethra before leaving the body.

7. The accessory organs (**seminal glands, prostate, and bulbo-urethral glands**) secrete fluids into the ejaculatory ducts and urethra.

8. The testes descend into the **scrotum**.

9. The **superficial inguinal ring** forms the entrance to the **inguinal canal**. The **spermatic cord** lies within the inguinal canal and encloses the ductus deferens, nerves, and blood and lymphatic vessels.

10. The **dartos muscle** is the smooth muscle of the scrotum. The two **scrotal cavities** are separated by the **scrotal septum**.

11. The **tunica albuginea** cover the testes, and are continuous with the septa that separate the **lobules**. Each lobule contains coiled **seminiferous tubules** where sperm production occurs. Contractions of smooth musclelike *myoid cells* transport sperm through the tubules.

12. The seminiferous tubules are connected to the **rete testis**, and **efferent ductules** connect to the epididymis.

### Module 26.3

### Spermatogenesis occurs in the testes and produces mature sperm    p. 1002

13. **Spermatogenesis** involves **mitosis**, **meiosis**, and **spermiogenesis**.

14. Each mitotic division of a stem cell produces one daughter cell that remains a stem cell and one daughter cell that enters meiosis. These **diploid** (2n) cells contain 23 pairs of chromosomes.

15. Meiosis involves two cycles of cell division (**meiosis I** and **meiosis II**) and produces four **haploid** (n) cells that each contain 23 individual chromosomes.

16. In **synapsis**, maternal and paternal chromosomes form 23 chromosome pairs. Each member of a pair consists of two duplicate chromatids, and the set of four chromatids is called a **tetrad**.

17. **Spermatogenesis** begins with the division of a **spermatogonium** (stem cell) into a daughter stem cell and **primary spermatocyte**.

18. In meiosis I, **secondary spermatocytes** form from a primary spermatocyte. Meiosis II yields four haploid **spermatids**.

19. In **spermiogenesis**, each spermatid matures into a single **sperm** that contains an **acrosome**, **head**, **neck**, **middle piece**, and **tail**, or **flagellum**.

### Module 26.4

### Meiosis and early spermiogenesis occur within the seminiferous tubules    p. 1004

20. Every seminiferous tubule is continuously producing sperm.

21. **Interstitial endocrine cells** (Leydig cells) produce testosterone and androstenedione, the dominant sex hormones in males.

22. **Nurse cells** (Sertoli cells) surround developing spermatocytes and spermatids. At **spermiation**, a sperm loses its attachment and enters the lumen of the seminiferous tubule.

23. Nurse cells are joined by tight junctions to isolate the seminiferous tubules from the general circulation by creating a **blood testis barrier**.

### Module 26.5

### The male reproductive tract receives secretions from the seminal, prostate, and bulbo-urethral glands    p. 1006

24. Sperm become motile and capable of fertilization only after undergoing **capacitation**.

25. Each **epididymis** consists of a **head**, **body**, and **tail**. It is here that sperm complete maturation.

26. Each **ductus deferens** is part of the spermatic cord. It transports and stores sperm.

27. The paired **seminal glands** produce about 60 percent of the volume of **semen**. Semen includes sperm and **seminal fluid**.

28. The **prostate** encircles the proximal portion of the urethra as it leaves the urinary bladder and produces 20–30 percent of the volume of semen.

29. The paired **bulbo-urethral glands** are located at the base of the penis. They secrete alkaline mucus that helps neutralize urinary acids remaining in the urethra and lubricates the tip of the penis.

### Module 26.6

### The penis conducts urine and semen to the exterior    p. 1008

30. The **penis** conducts urine to the exterior and introduces semen into the female's vagina during sexual intercourse.

31. The penis is attached to the body wall by the **root**. The **body of penis** (*shaft*) is the tubular portion of the organ. The **glans** is the expanded distal end that surrounds the external urethral orifice. The **neck of glans** is the narrow portion of the penis between the shaft and the glans. The **prepuce** surrounds the tip of the penis.

32. The penis consists of three cylindrical columns of **erectile tissue**. The two **corpora cavernosa** extend along the length of the penis as far as its neck. The **corpus spongiosum** surrounds the urethra.

33. The three phases of male sexual response are **arousal**, **emission**, and **ejaculation**.

34. **Impotence**, or **erectile dysfunction (ED)**, is the inability to achieve or maintain an erection due to physical or psychological causes.

### Module 26.7

**Testosterone plays a key role in establishing and maintaining male sexual function    p. 1010**

35. The hypothalamus releases gonadotropin-releasing hormone (GnRH), which causes the anterior pituitary to release **luteinizing hormone (LH)** and **follicle- stimulating hormone (FSH)**. LH stimulates the interstitial cells of the testes to secrete **testosterone**. FSH stimulates nurse cell activity.

36. Through negative feedback, testosterone decreases GnRH secretion, and inhibin secretion by nurse cells depresses FSH production.

37. The peripheral effects of testosterone are to maintain libido, stimulate bone and muscle growth, maintain male secondary sex characteristics, and maintain organs of the male reproductive system.

## SECTION 2 • Female Reproductive System

### Module 26.8

**Female reproductive structures include the external genitalia and internal genitalia    p. 1013**

38. The **female external genitalia** include the vulva—a collective term describing the outer genitals—and its associated structures.

39. **Female internal genitalia** include the ovary, uterine tube, uterus, and vagina.

40. The **mammary glands** are accessory structures that produce milk that nourishes the newborn and infant.

41. Female gametes are called **oocytes**. They travel the **female reproductive tract** through the uterine tubes to the uterus.

### Module 26.9

**Major female reproductive organs are the ovaries, uterus, and their associated structures    p. 1014**

42. The **ovaries** have three main functions: **oocyte** production, secretion of **estrogens** and **progesterone**, and secretion of **inhibin**.

43. Each **uterine tube** begins with an expanded funnel called an **infundibulum**, and ends at the uterine cavity.

44. The pocket formed between the uterus and the posterior wall of the urinary bladder is the **vesico-uterine pouch**. The pocket formed between the posterior wall of the uterus and the anterior surface of the colon is the **recto-uterine pouch**.

45. The **vagina** extends from the base of the uterus to the exterior.

46. The external genitalia include the **clitoris** and the **labia**.

47. Each ovary is stabilized by an **ovarian ligament** and **suspensory ligament**, as well as the **mesovarium**.

### Module 26.10

**Oogenesis occurs in the ovaries, and ovulation occurs during the 28-day ovarian cycle    p. 1016**

48. **Oogenesis** is the formation and development of the oocyte. Between puberty and menopause, it occurs on a monthly basis as part of the ovarian cycle.

49. Oogenesis produces one functional **secondary oocyte** and two or three **polar bodies** (nonfunctional cells that later disintegrate).

50. The **oogonia** complete their mitotic divisions before birth. Between the 3rd and 7th months of fetal development, primary oocytes undergo meiosis. The process stops at prophase of meiosis I. The meiotic process begins again at puberty when rising FSH triggers the **ovarian cycle**. Meiosis I is then completed, yielding a secondary oocyte and a polar body.

51. Each month after the ovarian cycle begins, one or more secondary oocytes leave the ovary suspended in metaphase of meiosis II.

52. At the time of fertilization, a second polar body forms, and the fertilized oocyte is then called a mature **ovum**.

53. **Ovarian follicles** are specialized structures in the cortex of the ovaries, where both oocyte growth and meiosis I occur.

54. An inactive primary oocyte is surrounded by a layer of follicle cells, forming a **primordial ovarian follicle**. Follicular cells are called **granulosa cells** when they enlarge, divide, and form several layers of cells around an activated primary oocyte.

55. A **zona pellucida** forms around the oocyte, and **thecal endocrine cells** form around the follicle.

56. Together, thecal cells and granulosa cells produce estrogens.

57. **Secondary ovarian follicles** develop into tertiary ovarian follicles. By days 10–14 of the ovarian cycle, one **tertiary ovarian follicle** has become dominant.

58. The oocyte projects into the **antrum**. The granulosa cells associated with the secondary oocyte form a protective layer known as the **corona radiata**.

59. At **ovulation**, the tertiary ovarian follicle releases the secondary oocyte with its corona radiata into the pelvic cavity. Ovulation marks the end of the **follicular phase** of the ovarian cycle and the start of the **luteal phase**.

60. The empty follicle initially collapses, and under LH stimulation forms the **corpus luteum**, which secretes progesterone. If fertilization does not occur, the nonfunctional corpus luteum forms a pale scar called a **corpus albicans**.

### Module 26.11

**The uterine tubes are connected to the uterus, a hollow organ with thick muscular walls    p. 1018**

61. The regions of the **uterine tube** are the funnel-like **infundibulum**, the fingerlike **fimbriae**, the **ampulla**, and **isthmus**.

62. The layers of the uterine wall are an outer **perimetrium**, a muscular **myometrium**, and an inner glandular **endometrium**.

63. The **uterine cavity** is continuous with the **internal os** that leads to the **cervical canal** and the **external os** that opens into the vagina.

64. The **fundus** is the rounded upper portion of the uterine body. The **body** is the largest portion of the uterus. The **cervix** is the inferior portion of the uterus.

### Module 26.12

**The uterine (menstrual) cycle involves changes in the functional layer of the endometrium    p. 1020**

65. Within the myometrium, branches of the uterine arteries form **arcuate arteries**. From the arcuate arteries, **radial arteries** supply the endometrium.

66. The endometrium contains a **basal layer** and a **functional layer**. **Straight arteries** deliver blood to the basal layer, and **spiral arteries** supply the functional layer.

67. The basal layer remains constant over time, but the functional layer changes in response to sex hormone levels.

68. The **uterine cycle** averages 28 days in length but can range from 21 to 35 days.

69. The uterine cycle begins when the functional layer of the endometrium degenerates, the weakened arterial walls rupture, and blood pours into the connective tissues of the functional layer. This is called **menses**. The process of endometrial shedding is called **menstruation** and lasts from 1 to 7 days.

70. Restoration of the endometrium occurs in the **proliferative phase**. During the **secretory phase**, the uterine glands enlarge, and the arteries that supply the uterine wall elongate and spiral through the functional layer. The secretory phase begins at the time of ovulation and lasts as long as the corpus luteum remains intact.

71. The uterine cycle, or **menstrual cycle**, begins at puberty. The first cycle is known as **menarche**. The cycles continue until **menopause**.

## Module 26.13

### The vagina opens into the vestibule  p. 1022

72. The **vagina** is an elastic, muscular tube extending between the cervix and the **vestibule**. The internal passageway is called the **vaginal canal**.

73. The vagina serves as a passageway for menstrual fluids, receives the penis during sexual intercourse, and forms the inferior portion of the birth canal.

74. The shallow recess in the vagina surrounding the tip of the **cervix** is known as the **vaginal fornix**. The vaginal lining forms folds called **rugae**.

75. Throughout childhood, the vagina and vestibule are usually separated by the **hymen**, but the hymen is typically stretched or torn during sexual intercourse, tampon use, or heavy physical exercise.

76. The **mons pubis** is a bulge of adipose tissue superficial to the pubic symphysis.

77. The **labia majora** are prominent folds of skin that encircle and partially conceal the **labia minora** and adjacent structures. Extensions of the labia minora encircle the body of the **clitoris**, forming its **prepuce**.

78. The area containing the female external genitalia is called the **vulva**, or **pudendum**.

79. **Lesser vestibular glands** discharge onto the exposed surface of the vestibule, keeping it moist. During sexual arousal, the **greater vestibular glands** secrete into the vestibule.

## Module 26.14

### Each breast contains a mammary gland that secretes milk p. 1023

80. **Breasts** lie directly over the pectoralis major within the subcutaneous **adipose tissue** (pectoral fat pad)

81. The **suspensory ligaments of the breast** surround the duct system and form partitions between the **lobes** and **lobules**.

82. The glandular tissue of the breast consists of lobes that contain lobules. Each lobule is composed of many **secretory alveoli**.

83. Ducts leaving the lobules converge to a single **lactiferous duct** that enlarges to a **lactiferous sinus**. At the **nipple**, 15 to 20 lactiferous sinuses open onto the body surface.

84. The reddish brown skin around each nipple is the **areola**.

## Module 26.15

### The ovarian and uterine cycles are regulated by hormones of the hypothalamus, pituitary gland, and ovaries  p. 1024

85. The ovarian and uterine cycles must operate in synchrony to ensure proper reproductive function.

86. GnRH levels from the hypothalamus change throughout the ovarian cycle. The ovarian cycle begins with the release of GnRH, which stimulates the release of FSH from the anterior lobe of the pituitary gland.

87. The **follicular phase** of the ovarian cycle begins when FSH stimulates the growth of a group of tertiary ovarian follicles. One becomes dominant. Follicles release inhibin and estrogens (especially **estradiol**, the dominant hormone prior to ovulation).

88. Inhibin decreases FSH. Estrogens affect the CNS, stimulate bone and muscle growth, maintain female secondary sex characteristics, maintain accessory glands, and stimulate endometrial growth and secretion.

89. GnRH and elevated estrogen levels stimulate LH secretion from the anterior pituitary. LH causes the completion of meiosis I and ovulation. The **luteal phase** of the ovarian cycle begins at ovulation.

90. The corpus luteum secretes progesterone, which stimulates and sustains endometrial development.

91. After ovulation, GnRH decreases because progesterone levels increase and estrogens decrease. If there is no pregnancy, the corpus luteum degenerates, progesterone decreases, the pulse frequency of GnRH secretion increases, and a new cycle begins.

## Module 26.16

### Birth control strategies vary in effectiveness and associated risks  p. 1026

92. Birth control methods include **male condoms**, the **diaphragm**, **birth control pills**, the **progesterone-only pill**, Depo-Provera injections, the **intrauterine device**, the **rhythm method**, **hormonal post-coital contraceptives**, and surgical sterilization (**vasectomy** and **tubal ligation**).

## Module 26.17

### Reproductive system disorders are relatively common and often deadly  p. 1028

93. **Benign prostatic hypertrophy** (BPH) occurs in men over age 50 as testosterone decreases. It can cause urinary obstruction. **Prostate cancer** is the second most common cause of cancer deaths in men and can be screened for with the **prostate-specific antigen** (PSA) blood test.

94. **Testicular cancer** is not common. The average age of diagnosis is 33. The survival rate is nearly 99%.

95. Benign **fibrocystic disease** and **breast cancer** are distinguished with a biopsy.

96. **Ovarian cancer** is dangerous because it is seldom diagnosed in its early stages.

97. **Cervical cancer** tends to occur in women under the age of 50. Vaccines are available to protect against several types of **human papillomavirus** (HPV), which cause about 75 percent of cervical cancers.

98. At least two dozen **sexually transmitted diseases** (STDs)/**sexually transmitted infections** (STIs) can be acquired primarily or exclusively by sexual intercourse.

# Chapter Review Questions

**Labeling:** Label the male and female reproductive structures shown here.

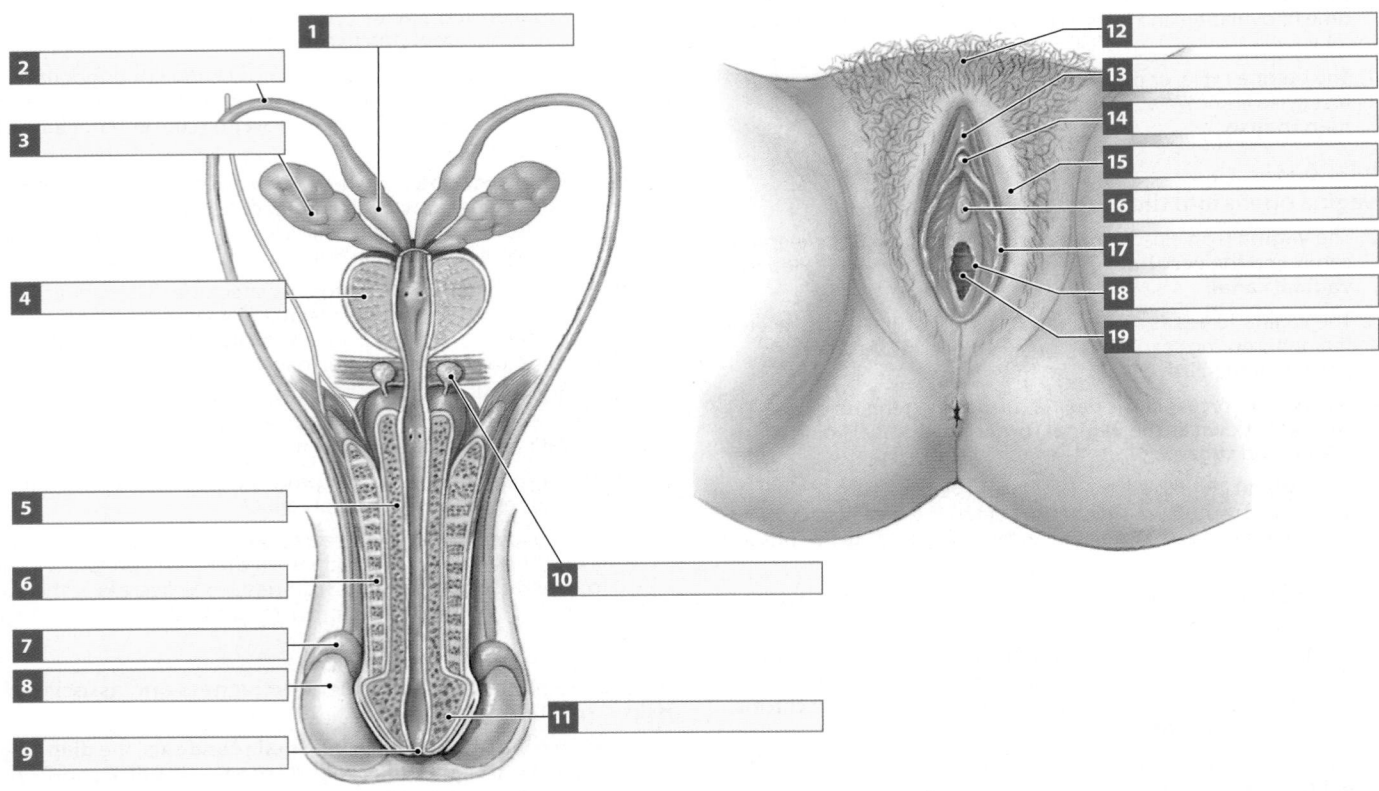

**True/False:** Indicate whether each statement is true or false.

20  The prostate is the site of sperm maturation.

21  Secondary spermatocytes contain 23 chromosomes.

22  In the male, LH induces the secretion of testosterone and other androgens by the interstitial endocrine cells of the testes.

23  At the end of meiosis II in females, four functional oocytes are produced.

24  A massive surge of LH on or around day 14 of the menstrual cycle triggers ovulation and formation of the corpus luteum.

25  Vaccines are available that protect against the virus that causes about 75 percent of cervical cancers.

20 _____

21 _____

22 _____

23 _____

24 _____

25 _____

**Multiple choice:** Select the correct answer from the list provided.

26  In females, meiosis II is not completed until
- a) birth.
- b) ovulation.
- c) fertilization.
- d) uterine implantation.

27  Progesterone secretion is performed by the
- a) primary ovarian follicle.
- b) secondary ovarian follicle.
- c) corpus luteum.
- d) corpus albicans.

**28** Spermatogenesis begins with the division of spermatogonia in the

- ☐ a) epididymis.
- ☐ b) seminiferous tubules.
- ☐ c) ductus deferens.
- ☐ d) bulbo-urethral glands.

**29** Erection of the penis is a result of

- ☐ a) parasympathetic neurons releasing nitric oxide at their axon terminals, causing the penile arteries to dilate, increasing blood flow to the penis.
- ☐ b) sympathetic neurons releasing nitric oxide at their axon terminals, causing the penile arteries to dilate, increasing blood flow to the penis.
- ☐ c) parasympathetic neurons releasing nitric oxide at their axon terminals, causing the penile veins to constrict, decreasing blood flow from the the penis.
- ☐ d) sympathetic neurons releasing nitric oxide at their axon terminals, causing the penile veins to constrict, decreasing blood flow from the penis.

**30** Estrogens are produced by

- ☐ a) the zona pellucida and the corona radiata.
- ☐ b) primordial follicles and the zona pellucida.
- ☐ c) primary oocytes and cells of the corpus luteum.
- ☐ d) thecal cells and granulosa cells.

**31** About 60 percent of the volume of semen is produced by the

- ☐ a) seminal glands.
- ☐ b) prostate.
- ☐ c) bulbo-urethral glands.
- ☐ d) ductus deferens.

## Short answer

**32** Explain how birth control pills that contain estrogen and progesterone, or only progesterone, prevent ovulation.

**33** What happens to the tertiary ovarian follicle after ovulation?

**34** Identify the sterilization procedures for men and women. How is sexual function affected as a result of each?

---

MasteringA&P™

Access more chapter study tools online in the MasteringA&P Study Area:

- Chapter Quizzes, Chapter Practice Test, and Clinical Case Studies

- Practice Anatomy Lab™ 3.0 — PAL™3.0
- Interactive Physiology™ — iP2™
- A&P Flix™ — A&PFlix™
- PhysioEx™ 9.1 — PhysioEx™ 9.1

---

# Chapter Integration • Applying what you have learned

## Exercise and the absence of menstruation

Exercise-induced amenorrhea, the abnormal absence of menstruation, occurs in 5–25 percent of women athletes. The variability depends on the level of competition and type of sport. For example, female bodybuilders and ballet dancers, both of whom are likely to have low body fat, commonly experience amenorrhea. Among well-nourished female athletes, hard training and exercise may cause the release of stress hormones, which then interfere with the pituitary gland's production of the hormones necessary for maintaining the menstrual cycle.

Although temporary cessation of menstrual cycles itself is not dangerous, there are long-term health consequences for prolonged exercise-induced amenorrhea in premenopausal women. As a result of long-term exercise-induced amenorrhea, women become estrogen deficient, which can lead to other health-related consequences.

Once the diagnosis is confirmed, treatment involves increasing caloric intake and restoring estrogen levels to the normal range. In most cases, exercise-induced amenorrhea is reversible with treatment.

**1** What does amenorrhea in female athletes suggest about the relationship between body fat and menstruation?

**2** How might exercise-induced amenorrhea affect pregnancy?

**3** What are some health-related consequences of estrogen deficiency?

# 27

# Development and Inheritance

## Module 27.1
# Gestation and development are marked by various stages

 **1** **Gestation** (jes-TĀ-shun) is the time that a developing embryo and fetus spends within the uterus, or womb. **Pregnancy** is the state of the female during this time. The process in which gradual anatomical and physiological changes occur from one stage to the next is called **development**. In a mere 9 months, all the tissues, organs, and organ systems we have studied so far take shape and begin to function. What begins as a single cell slightly larger than the period at the end of this sentence becomes an individual with trillions of cells organized into highly specialized structures.

**?** **REVIEW**

**A.** Define *gestation*.

**B.** Distinguish among embryonic, fetal, and prenatal development.

---

**Prenatal Development**

**Embryonic development** comprises the events that occur during the first 2 months after fertilization. The study of these events is called **embryology** (em-brē-OL-ō-jē).

4 weeks

8 weeks

**Fetal development** begins at the start of the 9th week and continues until birth. Embryonic and fetal development are sometimes referred to collectively as **prenatal** (*natus*, birth) **development**, the primary focus of this chapter.

16 weeks

---

**2** For convenience, we consider the gestational period as three integrated *trimesters*, each 3 months long.

| Gestation | | | Postnatal Development |
|---|---|---|---|
| **First Trimester** | **Second Trimester** | **Third Trimester** | **Postnatal development** begins at birth and continues to **maturity**, the state of full development or completed growth. A basic understanding of prenatal and postnatal development gives important insights into anatomical structures. Also, many of the processes of development and growth are similar to those of tissue repair following injury. |
| The **first trimester** is the time of embryonic and early fetal development. During this time, the rudiments (beginning structures) of all the major organ systems appear. | The **second trimester** is dominated by the development of organs and organ systems, a process that nears completion by the end of the 6th month. During this time, body shape and proportions change. By the end of this trimester, the fetus looks distinctively human. | The **third trimester** is characterized by the largest gain in fetal weight. Early in the third trimester, most of the fetus's major organ systems become fully functional. An infant born 1 month or even 2 months prematurely has a reasonable chance of survival if appropriate medical care is available. | |

 **?** **REVIEW**

**C.** What is postnatal development?

 **LO** **LEARNING OUTCOME**

Describe the various stages of gestation and development.

# Module 27.2
# At fertilization, an ovum and a sperm form a zygote that prepares for cell division

**1**    **Fertilization**, or *conception*, is the fusion of 2 haploid gametes (a sperm and a secondary oocyte). This fusion produces a **zygote** (ZĪ-gŏt) with 46 chromosomes—the normal complement in a somatic cell (see Module 3.19). Fertilization typically occurs near the junction between the ampulla and isthmus of the uterine tube, generally within a day after ovulation.

**? REVIEW**
**A.** Define *fertilization*.
**B.** How many chromosomes are contained within a human zygote?

## Fertilization

**Step 1**
**Oocyte at Ovulation**
Ovulation releases a secondary oocyte and the first polar body; both are surrounded by the corona radiata. The oocyte is suspended in metaphase of meiosis II (see Module 26.10).

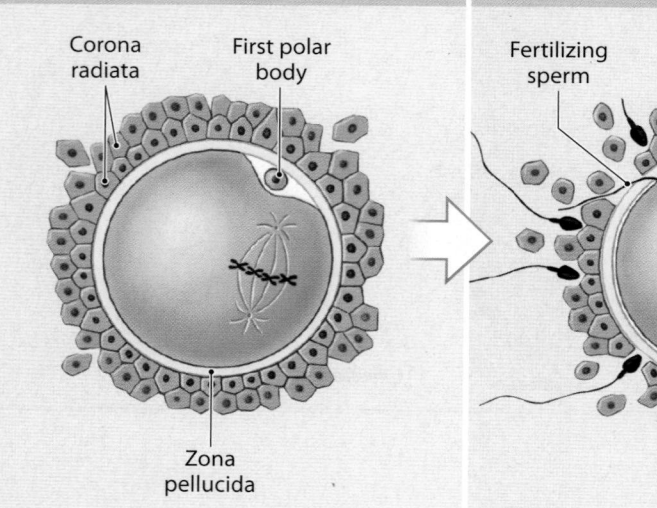

Corona radiata    First polar body

Zona pellucida

**Step 2**
**Fertilization and Oocyte Activation**
Acrosomal enzymes from multiple sperm are required to create gaps between the cells of the corona radiata. A single sperm then makes contact with the oocyte membrane, and membrane fusion occurs, triggering **oocyte activation** and the completion of meiosis. The secondary oocyte is now an **ovum**.

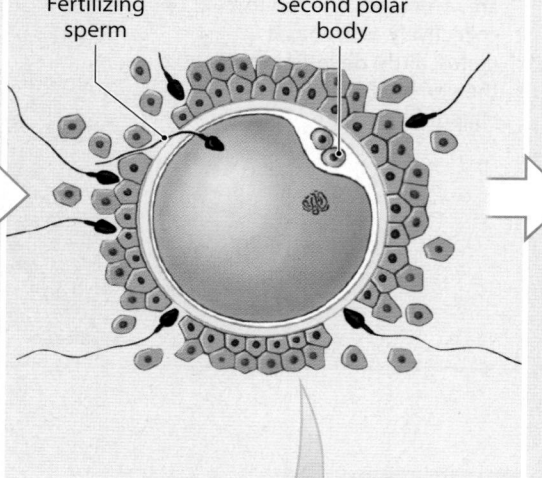

Fertilizing sperm    Second polar body

**Step 3**
**Pronuclei Develop and DNA Synthesis Occurs**
The sperm is absorbed into the cytoplasm, and the female nuclear material within the ovum reorganizes as the **female pronucleus**. As the **male pronucleus** develops, the rest of the sperm breaks down.

Female pronucleus

Nucleus of fertilizing sperm forms the male pronucleus

**2**    This is a photograph of a secondary oocyte surrounded by sperm. The function of a sperm is to deliver the paternal (father's) chromosomes to the secondary oocyte and then facilitate the process of fertilization. In contrast, the secondary oocyte provides all the cellular organelles and inclusions, nourishment, and genetic programming necessary to support embryo development for nearly a week after conception. The volume of the secondary oocyte is now 2000 times greater than that of a sperm.

Oocyte activation is a series of changes in the metabolic activity of the oocyte leading to $Ca^{2+}$ release from the smooth ER. The rise in $Ca^{2+}$ has three effects:
(1) Release of enzymes that harden the zona pellucida and prevent fertilization by more than one sperm
(2) Completion of meiosis II and formation of the second polar body
(3) Activation of enzymes that cause a rapid increase in the ovum's metabolic rate.

**Step 4**
**Spindle Formation Begins**
Male and female pronuclei chromatin condenses into chromosomes, and spindle fibers appear. This is the start of the process of **cleavage** (KLĒV-ij), a series of mitotic cell divisions that produces an increasing number of smaller and smaller daughter cells.

**Step 5**
**Amphimixis Occurs and Cleavage Begins**
The male and female pronuclei then fuse in a process called **amphimixis** (am-fi-MIK-sis; "both mixed together"). The cell is now a zygote that contains the normal complement of 46 chromosomes (each consisting of two duplicate chromatids). Fertilization is complete.

**Step 6**
**First Cleavage Forms Two Blastomeres**
The first cleavage division nears completion about 30 hours after fertilization. This division produces two daughter cells, each one-half the size of the original zygote. These cells are called **blastomeres** (BLAS-tō-mērz).

Male pronucleus    Female pronucleus

Sperm-donated centrioles have duplicated and formed spindle fibers

Metaphase of first cleavage division

This is the "moment of conception." Almost immediately, the maternal and paternal chromosomes line up on a metaphase plate, and the cell prepares to divide (see Module 3.20).

Blastomeres

---

 **Clinical Note**

## Low sperm counts cause male sterility

Of the nearly 200 million sperm introduced into the vagina in a typical ejaculation, only about 10,000 enter a uterine tube, and fewer than 100 reach the isthmus. In general, a male with a sperm count below 20 million per milliliter is functionally sterile because too few sperm survive to reach and fertilize an oocyte. Although it is true that only one sperm fertilizes an oocyte, dozens of sperm are required for successful fertilization. The additional sperm are essential because one sperm does not contain enough acrosomal enzymes to disrupt the corona radiata that surrounds the secondary oocyte.

 **REVIEW**
**C.** Why are numerous sperm required to fertilize a secondary oocyte?

 **LEARNING OUTCOME**
Describe the process of fertilization, and explain the significance of multiple sperm to ensuring its success.

# Module 27.3

# Cleavage continues until the blastocyst implants in the uterine wall

During cleavage, the cytoplasm of the zygote becomes subdivided among an increasing number of progressively smaller blastomeres. A group of blastomeres created by cleavage divisions is called a **pre-embryo**.

**1** Cleavage lasts about 7 days. During that time, the pre-embryo travels the length of the uterine tube. After 3 days, the pre-embryo is a solid ball of cells known as a **morula** (MOR-ū-la; *morus*, mulberry). The morula typically reaches the uterus on day 4. Over the next 2 days, the blastomeres form a **blastocyst** (BLAS-tō-sist), a hollow ball with an inner cavity known as the **blastocoele** (BLAS-tō-sēl). At this stage, the blastomeres are no longer identical in size and shape.

**REVIEW**

**A.** A pre-embryo that consists of a solid ball of cells is known as what type of structure?

| First Cleavage Division | Day 1: Two-Cell Stage | Day 2: Four-Cell Stage | Day 3: Early Morula | Day 4: Advanced Morula |
|---|---|---|---|---|
| Polar bodies | Blastomeres | Blastomeres | | Zona pellucida |

**Day 0**
Fertilization

Ovulation

**Day 5**
Loss of zona pellucida and transport to uterus

**Days 6–9**
Implantation in uterine wall

**Day 10**
Yolk sac formation
(see Module 27.4)

### Day 6: Blastocyst

Loss of the zona pellucida ("hatching") and transport to the uterus leaves the blastocyst freely exposed to the fluid within the uterine cavity. The uterine cavity contains the glycogen-rich secretions of the uterine glands. The rate of growth and cell division now accelerates, and the blastocyst enlarges rapidly.

**FUNCTIONAL LAYER OF ENDOMETRIUM**

Uterine glands

**UTERINE CAVITY**

Blastocyst

The **trophoblast** (TRŌ-fō-blast; *trophos*, food + *blast*, precursor) is the cell layer surrounding the blastocyst. The cells in this layer nourish the embryo and later form part of the placenta.

### Day 7: Implantation

Cleavage ends when the fully formed blastocyst contacts the uterine lining, or endometrium. **Implantation** begins with the attachment of the blastocyst to the endometrium. Implantation proceeds as the blastocyst erodes the endometrial lining and becomes enclosed within the endometrium by day 10.

Blastocoele

The **inner cell mass** lies clustered at one end of the blastocyst. These cells are exposed to the blastocoele but are insulated from contact with the intrauterine environment by the trophoblast. In time, the inner cell mass will form the embryo.

### Day 8: Trophoblast Development

At the point of contact, the trophoblast cells divide rapidly, making the trophoblast several layers thick. The cells closest to the blastocoele remain intact, forming an inner cellular layer of the trophoblast called the **cytotrophoblast**. Near the endometrial wall, the plasma membranes separating the trophoblast cells disappear, creating a layer of cytoplasm containing multiple nuclei. This layer is called the **syncytiotrophoblast**.

Endometrial capillary

Cytotrophoblast

Syncytiotrophoblast

Fingerlike **villi** extend away from the trophoblast into the surrounding endometrium, gradually increasing in size and complexity.

Lacuna

### Day 9: Formation of Amniotic Cavity

As implantation proceeds, the syncytiotrophoblast continues to enlarge and erode the surrounding endometrium. Nutrients released by the eroding uterine glands are absorbed by the syncytiotrophoblast and distributed by diffusion through the underlying cytotrophoblast to the inner cell mass. These nutrients provide the energy needed to support the early stages of embryo formation. Trophoblastic extensions (villi) grow around endometrial capillaries. As the capillary walls are destroyed, maternal blood begins to percolate through trophoblastic channels known as **lacunae**.

At the time of implantation, the inner cell mass has separated from the trophoblast. The separation gradually increases, creating a fluid-filled chamber called the **amniotic** (am-nē-OT-ik) **cavity**.

 **REVIEW**

**B.** Describe the trophoblast.

 **REVIEW**

**C.** Describe the blastocyst and its role in implantation.

**LEARNING OUTCOME**

Explain the processes of cleavage, blastocyst formation, and blastocyst implantation in the uterine wall.

# Module 27.4
# Gastrulation produces three germ layers: ectoderm, endoderm, and mesoderm

**1** Between days 9 and 15 after fertilization, the extra-embryonic membranes begin forming, and the three germ layers appear.

### Day 9: Formation of Amniotic Cavity (continued)

When the amniotic cavity first appears, the cells of the inner cell mass are organized into an oval sheet known as the **blastodisc**. The early blastodisc is two layers thick: a superficial layer that faces the amniotic cavity, and a deeper layer that is exposed to the fluid of the blastocoele. Cells of the superficial layer migrate along the walls of the amniotic cavity and separate the amniotic cavity from the trophoblast. This is the first step in the formation of the **amnion**, one of four **extra-embryonic membranes** we consider further in Module 27.5. At this stage, nutrients released into the amnion and blasto-coele by the advancing trophoblast are absorbed directly by the cells of the blastodisc.

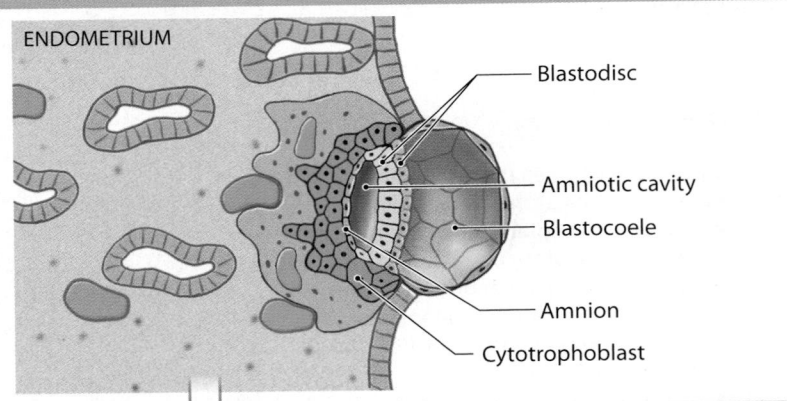

ENDOMETRIUM
- Blastodisc
- Amniotic cavity
- Blastocoele
- Amnion
- Cytotrophoblast

### Day 10: Yolk Sac Formation

While cells from the superficial layer of the inner cell mass migrate around the amniotic cavity, forming the amnion, cells from the deeper layer migrate around the outer edges of the blastocoele. This is the first step in the formation of the **yolk sac**, a second extra-embryonic membrane. For roughly the next 2 weeks, the yolk sac is the primary nutrient source for the inner cell mass. It absorbs and distributes nutrients released into the blastocoele by the cytotrophoblast.

- Syncytiotrophoblast
- Cytotrophoblast
- Yolk sac
- Lacuna

### Day 15: Gastrulation

By day 15, superficial cells of the blastodisc are migrating toward a central line known as the **primitive streak**. Here, the migrating cells leave the surface and move between the two existing layers. This movement forms three distinct embryonic layers: (1) the **ectoderm**, consisting of superficial cells that did not migrate into the interior of the blastodisc; (2) the **endoderm**, consisting of the cells that face the yolk sac; and (3) the **mesoderm**, consisting of the poorly organized layer of migrating cells between the ectoderm and the endoderm. Collectively, these three embryonic layers are called **germ layers**, and the migration process is called **gastrulation** (gas-troo-LĀ-shun). Gastrulation produces an oval, three-layered sheet known as the **embryonic disc**. This disc will form the body of the embryo, whereas all other cells of the blastocyst will be part of the extra-embryonic membranes.

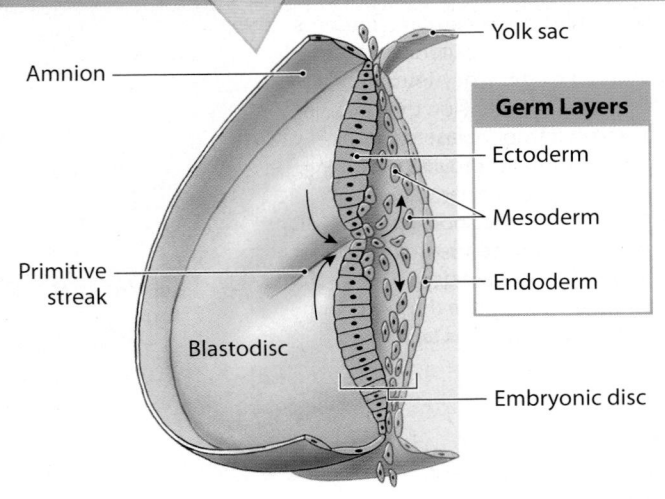

- Yolk sac
- Amnion

**Germ Layers**
- Ectoderm
- Mesoderm
- Endoderm

- Primitive streak
- Blastodisc
- Embryonic disc

**2** The table below summarizes the fates of the germ
layers as the different body systems develop.

## The Fates of the Germ Layers

| Body System | Ectodermal Contributions |
|---|---|
| Integumentary system | Epidermis, hair follicles and hairs, nails, and glands communicating with the skin (sweat glands, mammary glands, and sebaceous glands) |
| Skeletal system | Pharyngeal cartilages and their derivatives in adults (portion of sphenoid, the auditory ossicles, the styloid processes of the temporal bones, the cornu and superior rim of the hyoid bone) |
| Nervous system | All nervous tissue, including brain and spinal cord |
| Endocrine system | Pituitary gland and adrenal medullae |
| Respiratory system | Mucous epithelium of nasal passageways |
| Digestive system | Mucous epithelium of mouth and anus, salivary glands |
| | **Mesodermal Contributions** |
| Integumentary system | Dermis and subcutaneous layer (hypodermis) |
| Skeletal system | All structures except some pharyngeal derivatives |
| Muscular system | All structures |
| Endocrine system | Adrenal cortex, endocrine tissues of heart, kidneys, and gonads |
| Cardiovascular system | All structures |
| Lymphatic system | All structures |
| Urinary system | The kidneys, including the nephrons and the initial portions of the collecting system |
| Reproductive system | The gonads and the adjacent portions of the duct systems |
| Miscellaneous | The lining of the pleural, pericardial, and peritoneal cavities and the connective tissues that support all organ systems |
| | **Endodermal Contributions** |
| Endocrine system | Thymus, thyroid gland, and pancreas |
| Respiratory system | Respiratory epithelium (except nasal passageways) and associated mucous glands |
| Digestive system | Mucous epithelium (except mouth and anus), exocrine glands (except salivary glands), liver, and pancreas |
| Urinary system | Urinary bladder and distal portions of the duct system |
| Reproductive system | Distal portions of the duct system, stem cells that produce gametes |

 **Clinical Note**

## On rare occasions, the trophoblast can become cancerous

The trophoblast undergoes repeated nuclear divisions, shows extensive and rapid growth, has a very high demand for energy, invades and spreads through adjacent tissues, yet fails to activate the maternal immune system—in short, the trophoblast has many of the characteristics of cancer cells. In about 0.1% of pregnancies, something goes wrong with the regulatory mechanisms, and instead of developing normally, the syncytio-trophoblast behaves like a tumor. This condition is called **gestational trophoblastic neoplasia**. Approximately 20% of gestational trophoblastic neoplasias metastasize to other tissues, with potentially fatal results. Consequently, prompt surgical removal of the mass is essential. The surgery is sometimes followed by chemotherapy.

 **INTEGRATION**

**A.** Describe the process that forms the primary germ layers and cite the layer that contributes to nearly all body systems.

 **LEARNING OUTCOME**

Describe gastrulation and the formation of the three germ layers.

# Module 27.5
# The extra-embryonic membranes form the placenta that supports fetal growth and development

Germ layers form four extra-embryonic membranes: the yolk sac (endoderm + mesoderm), the amnion (ectoderm + mesoderm), the allantois (endoderm + mesoderm), and the chorion (mesoderm + cytotrophoblast). Although these membranes support embryonic and fetal development, few traces of their existence remain in adult systems.

**REVIEW**

**A.** Name the four extra-embryonic membranes.

## Formation of the Yolk Sac

## Formation of the Amnion

Endometrium

Syncytiotrophoblast

Cytotrophoblast

The **yolk sac** begins as a layer of cells spread out around the outer edges of the blastocoele to form a complete pouch. This pouch is already visible 10 days after fertilization. The yolk sac is an important site of blood cell formation.

The **amnion** (AM-nē-on) begins as cells from the blastodisc migrate around a cavity that separates the inner cell mass from the cytotrophoblast.

Blastocoele

**Day 10**

As gastrulation proceeds, mesodermal cells migrate around the pouch and complete the formation of the yolk sac. Blood vessels, which appear first in the mesoderm of the yolk sac, collect and distribute nutrients absorbed from the blastocoele. The vascular network of the yolk sac plays a role in early embryonic circulation.

As gastrulation proceeds, mesodermal cells migrate away from the embryonic disc, moving between the ectodermal lining of the amniotic cavity and the surrounding cellular trophoblast. This combination of mesoderm and ectoderm is the completed amnion.

Uterine cavity

The amniotic cavity contains **amniotic fluid**, which surrounds and cushions the developing embryo or fetus.

**Chorion**

Cytotrophoblast

Mesoderm

Blastocoele

Syncytiotrophoblast

**Day 14**

# Formation of the Allantois

# Formation of the Chorion

Endometrium

The **allantois** (a-LAN-tō-is) begins as an outpocket of the endoderm near the base of the yolk sac. The free endodermal tip then grows toward the wall of the blastocyst, surrounded by a mass of mesodermal cells. Its fluid cushions the embryo and fetus.

The mesoderm associated with the allantois spreads around the blastocyst, separating the cytotrophoblast from the blastocoele. This combination of mesoderm and cytotrophoblast is the **chorion** (KŌ-rē-on). It forms part of the fetal portion of the placenta.

Yolk sac

Amniotic cavity

Embryo

Uterine cavity

Blastocoele

Syncytiotrophoblast

**Week 3**

The allantois extends partway into the umbilical stalk. The base of the allantois will form the urinary bladder.

Umbilical stalk

The **placenta** forms the interface between embryonic/fetal and maternal systems. It develops as villi of the chorion invade the endometrium and break down maternal blood vessels. The placenta becomes the primary support structure for the developing embryo (and later, the fetus). The placenta is the site for nutrient, gas, and waste exchange between maternal and embryonic/fetal bloodstreams.

Amniotic cavity

Blastocoele

Embryo

Uterus

Uterine cavity

Endometrium

Yolk sac

**Week 5**

**? REVIEW**

**B.** From which germ layers do the extra-embryonic membranes form, and what are each membrane's functions?

**? REVIEW**

**C.** Describe the placenta.

**Lo LEARNING OUTCOME**

Identify and describe the formation, location, and functions of the extra-embryonic membranes.

# Module 27.6

# The formation of extra-embryonic membranes is associated with major changes in the shape and complexity of the embryo

## Week 2

**1** Migration of mesoderm around the inner surface of the cytotrophoblast forms the chorion. Mesodermal migration around the outside of the amniotic cavity, between the ectodermal cells and the cytotrophoblast, forms the amnion. Mesodermal migration around the endodermal pouch forms the yolk sac.

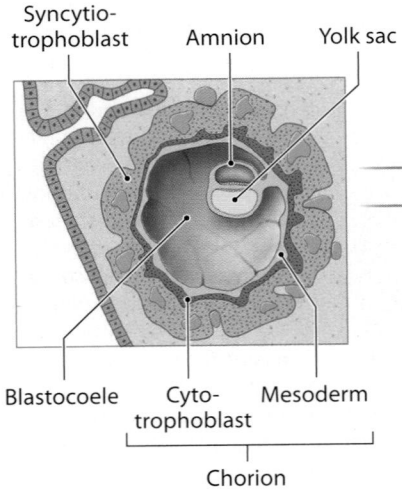

Syncytio-trophoblast    Amnion    Yolk sac

Blastocoele    Cyto-trophoblast    Mesoderm

Chorion

## Week 3

**2** The embryonic disc bulges into the amniotic cavity at the **head fold**. The allantois, an endo-dermal extension surrounded by mesoderm, extends toward the trophoblast.

Head fold of embryo    Amniotic cavity (containing amniotic fluid)    Allantois

Yolk sac

Chorion    Syncytio-trophoblast

Mesoderm extends along the core of each trophoblastic villus, forming **chorionic villi** in contact with maternal tissues. Embryonic blood vessels develop within each villus. Blood flow through those chorionic vessels begins early in the 3rd week of development, when the embryonic heart starts beating.

 **REVIEW**

**A.** Describe the chorionic villi.

## Week 4

**3** The embryo now has a **tail fold** as well as a head fold. The anterior/posterior, left/right, and superior/inferior axes of the developing embryo are now clearly established. The connections between the embryo and the surrounding trophoblast begin to constrict.

The **body stalk**, the connection between embryo and chorion, contains the distal portions of the allantois and blood vessels that carry blood to and from the placenta.

The narrow connection between the endoderm of the embryo and the yolk sac is called the **yolk stalk**.

Tail fold

Embryonic gut    Head fold    Yolk sac

As the chorionic villi enlarge, more maternal blood vessels are eroded. Maternal blood now moves slowly through complex lacunae lined by the syncytial trophoblast. Chorionic blood vessels pass close by, and gases and nutrients diffuse between the embryonic and maternal circulations across the layers of the trophoblast.

 **REVIEW**

**B.** Compare the body stalk with the yolk stalk.

## Week 5

**4** The developing embryo and extra-embryonic membranes bulge into the uterine cavity. The trophoblast pushing out into the uterine cavity remains covered by endometrium but no longer participates in nutrient absorption and embryo support. The embryo moves away from the placenta, and the body stalk and yolk stalk fuse to form an **umbilical stalk**.

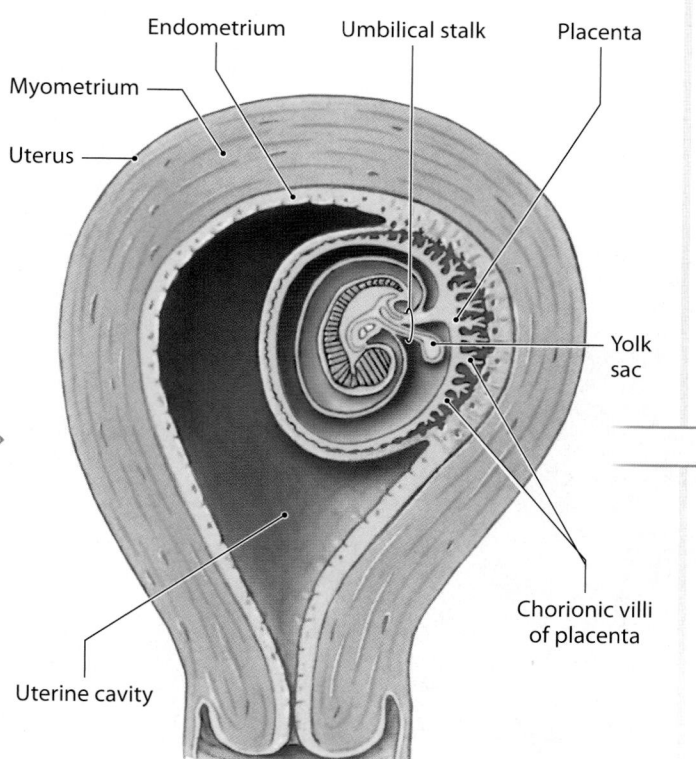

Endometrium  Umbilical stalk  Placenta

Myometrium

Uterus

Yolk sac

Chorionic villi of placenta

Uterine cavity

## Week 10

**5** The amnion has expanded greatly, filling the uterine cavity. The fetus is connected to the placenta by an elongated **umbilical cord** that contains a portion of the allantois, blood vessels, and the remnants of the yolk stalk. A mucus plug forms, preventing bacteria from entering the uterus.

Amniotic cavity  Umbilical cord  Placenta

Amnion

Chorion

Mucus plug

**? REVIEW**

**C.** Identify the structure connecting the fetus to the placenta, and name the extra-embryonic membrane from which it is derived.

**Lo LEARNING OUTCOME**

Describe the roles of the extra-embryonic membranes in embryonic development and placenta formation.

# Module 27.7
# The placenta performs many vital functions during prenatal development

**1** This illustration gives a closer look at the structure of the placenta. The chorionic villi provide the surface area for active and passive exchange of gases, nutrients, and wastes between the fetal and maternal bloodstreams. Blood flowing to the placenta through the paired **umbilical arteries** is deoxygenated (shown in blue) and contains wastes generated by fetal tissues. At the placenta, oxygen supplies are replenished, organic nutrients are added, and carbon dioxide and other organic wastes are removed. Oxygenated blood (shown in red) then returns to the fetus within a single **umbilical vein**. Recall that arteries carry blood away from the heart, and veins carry blood toward the heart.

Area filled with maternal blood

Syncytio-trophoblast

Cytotrophoblast

Fetal blood vessels

Embryonic connective tissue

**Chorionic villus, cross section**　　LM × 280

Endometrium

Umbilical cord (cut)

Yolk sac

Placenta

Amnion

Chorion

Chorionic villi

Area filled with maternal blood

Endometrium

Myometrium

Uterine cavity

Mucus plug in cervical canal

Fetal capillaries

External os

Cervix

Vagina

Umbilical vein

Umbilical arteries

Amnion

Trophoblast (cellular and syncytial layers)

Maternal blood vessels (arrows indicate direction of blood flow)

**2** In addition to supplying nutrients to the fetus, the placenta acts as an endocrine organ. Several hormones are synthesized by the syncytiotrophoblast and released into the maternal bloodstream.

## Placental Hormones

### Human Chorionic Gonadotropin

**Human chorionic gonadotropin (hCG)** appears in the maternal bloodstream soon after implantation. The presence of hCG in blood or urine samples is a reliable indication of pregnancy. Home pregnancy tests detect this hormone in urine. Functionally, hCG resembles luteinizing hormone (LH) because it maintains the integrity of the corpus luteum and promotes the continued secretion of progesterone. As a result, in pregnancy the endometrial lining remains perfectly functional, and menstruation does not occur. In the presence of hCG, the corpus luteum persists for 3–4 months before gradually decreasing in size and secretory function. The decline in luteal function does not trigger the return of uterine cycles, because by the end of the first trimester, the placenta is secreting progesterone.

### Human Placental Lactogen and Placental Prolactin

**Human placental lactogen (hPL)** helps prepare the mammary glands for milk production. At the mammary glands, the conversion from inactive to active status requires placental hormones (hPL, **placental prolactin**, estrogens, and progesterone) and several maternal hormones (growth hormone, prolactin, and thyroid hormones).

### Relaxin

**Relaxin** is a peptide hormone that is secreted by the placenta and the corpus luteum during pregnancy. Relaxin increases the flexibility of the pubic symphysis, permitting the pelvis to expand during delivery; causes the cervix to dilate, making it easier for the fetus to enter the vaginal canal; and suppresses the release of oxytocin by the hypothalamus and delays the onset of labor contractions.

### Progesterone and Estrogens

After the first trimester, the placenta produces sufficient **progesterone** to maintain the endometrial lining and continue the pregnancy. As the end of the third trimester approaches, production of **estrogens** by the placenta accelerates. As we will see shortly, the rising estrogen levels play a role in stimulating labor and delivery (see Module 27.10).

**? REVIEW**

**A.** Name the hormones synthesized by the syncytiotrophoblast.

**B.** The presence of which hormone in the urine provides a reliable indicator of pregnancy, as detected by home pregnancy tests?

**C.** When does the placenta become sufficiently functional to continue the pregnancy?

 **LEARNING OUTCOME**

Discuss the importance of the placenta to the fetus and as an endocrine organ.

# Module 27.8

# Organ systems form in the first trimester and become functional in the second and third trimesters

The first trimester is the most dangerous prenatal stage because the timing of complex processes such as cleavage, implantation, and placenta formation (*placentation*) is vital to forming a viable embryo (*embryogenesis*). These critical events in the first 12 weeks establish the basis for **organogenesis**, the process of organ formation.

**? REVIEW**

**A.** Define *organogenesis*.

**1** After 3 weeks of development, the CNS is forming as a deep groove develops in a thick ectodermal band called the **neural plate** that lies along the posterior midline of the embryo. Somites develop on both sides of the neural plate.

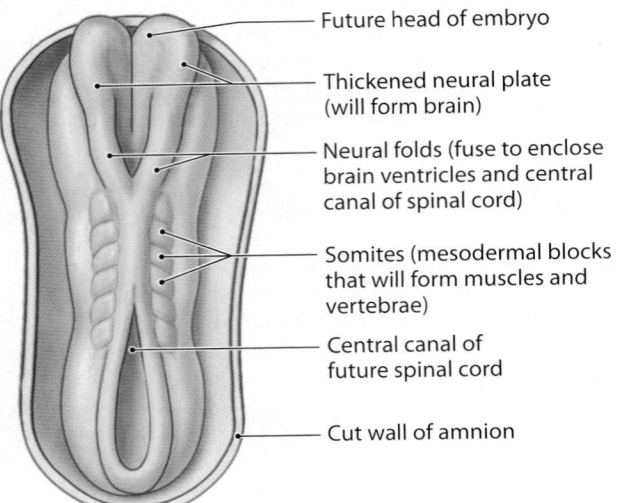

- Future head of embryo
- Thickened neural plate (will form brain)
- Neural folds (fuse to enclose brain ventricles and central canal of spinal cord)
- Somites (mesodermal blocks that will form muscles and vertebrae)
- Central canal of future spinal cord
- Cut wall of amnion

**2** This fiberoptic view of human development at week 4 shows many recognizable features. The heart is beating, pushing blood to and from the placenta, providing the nutrients needed to promote additional growth and development.

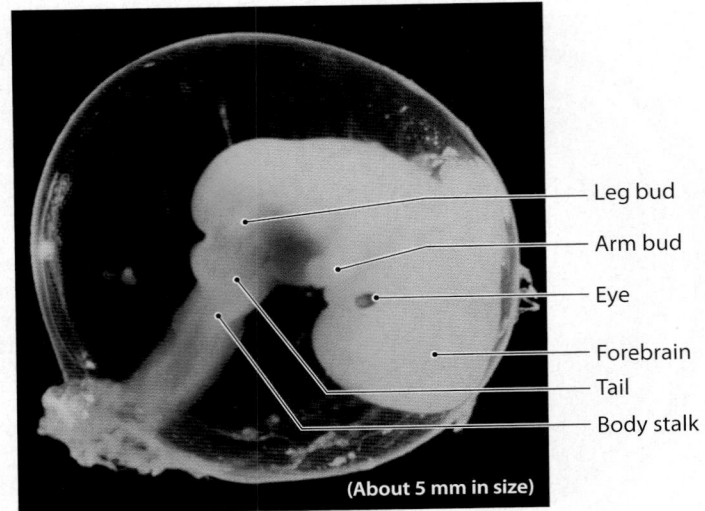

- Leg bud
- Arm bud
- Eye
- Forebrain
- Tail
- Body stalk

(About 5 mm in size)

**3** By week 6, the placenta has formed, and the embryo floats within the amniotic cavity. Body proportions are changing, the limbs are growing longer, and the skull bones are organizing around the already formed brain and eyes.

**4** At the end of the first trimester, the fetus is considerably larger and its human features better defined. The axial and appendicular muscles are forming, and fetal movements will soon begin.

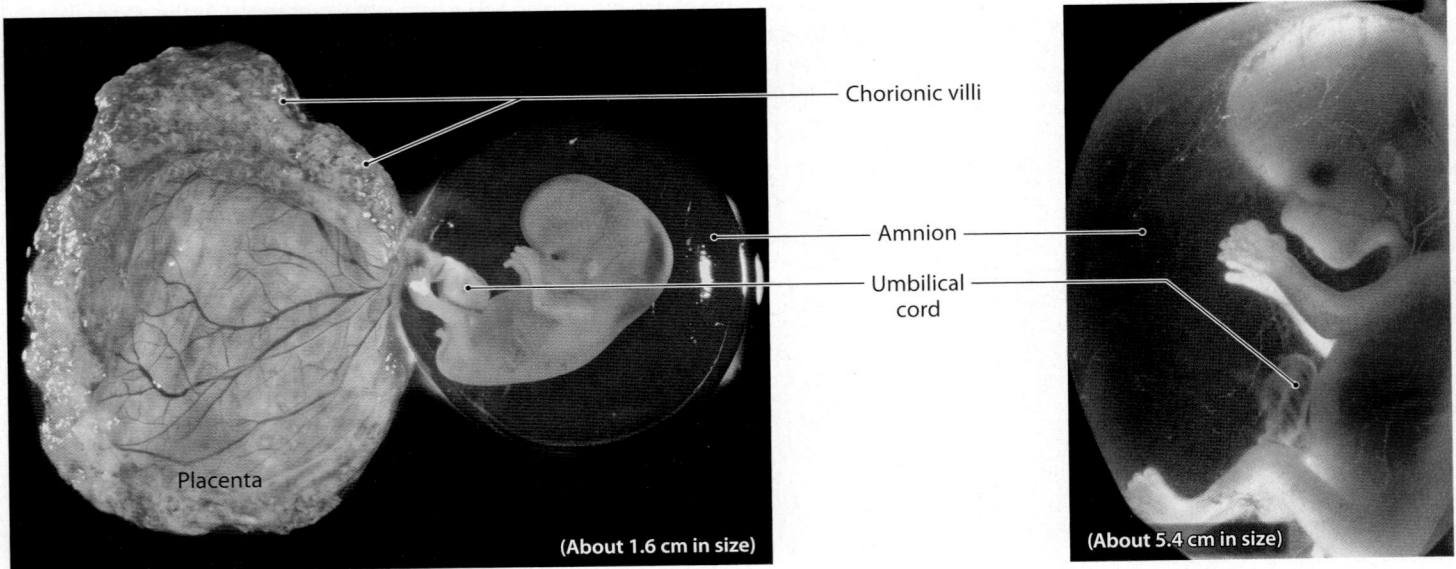

- Chorionic villi
- Amnion
- Umbilical cord
- Placenta

(About 1.6 cm in size)

(About 5.4 cm in size)

**5** This fiberoptic view shows a fetus at 4 months gestation. In the second trimester, the organ systems increase in complexity. The face and palate have their proper form; the cerebral hemispheres are enlarging. Hair follicles are present, and hair growth begins. Nerves have formed, and the fetus moves frequently. The first 8 weeks of fetal growth are the most rapid, with the fetus increasing in weight some 25-fold. By the end of the second trimester, the fetus will have grown to a weight of about 0.64 kg (1.4 lb).

**6** This is an ultrasound of a fetus after 6 months of gestation. During the third trimester, most of the organ systems become ready to function normally without maternal assistance. The rate of growth starts to slow, but in absolute terms the largest weight gain occurs in this trimester. In the final 3 months of gestation, the fetus gains about 2.6 kg (5.7 lb), reaching a full-term weight of approximately 3.2 kg (7 lb).

(About 13.3 cm in size)

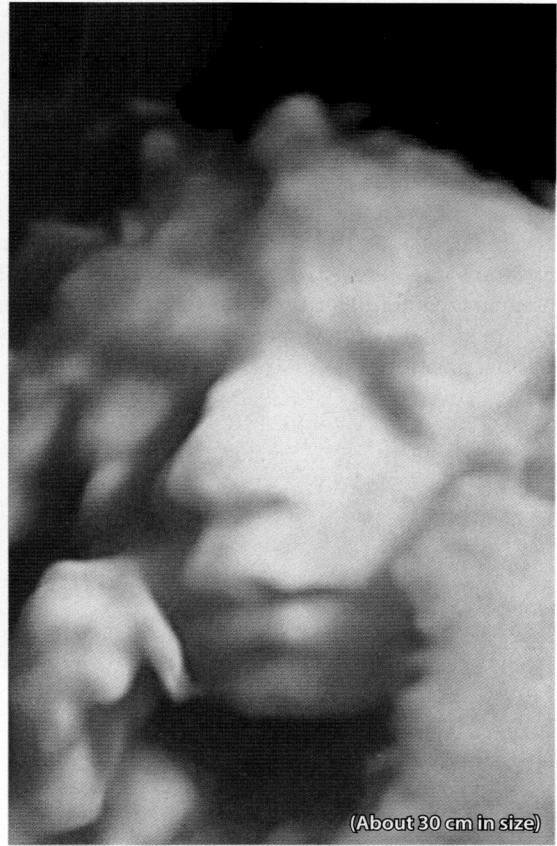

(About 30 cm in size)

**7** Multiple births (twins, triplets, and so forth) can take place for several reasons, including the use of fertility drugs and assisted reproductive technology. According to the U.S. Centers for Disease Control and Prevention (CDC), the rate of twin births in the United States is 33.7 out of every 1000 deliveries. "Fraternal," or **dizygotic** (dī-zī-GOT-ick), twins develop when two separate oocytes are ovulated and fertilized.

"Identical," or **monozygotic**, twins result either from the separation of blastomeres early in cleavage or from the splitting of the inner cell mass before gastrulation. In either event, the genetic makeup of the twins is identical because both formed from the same pair of gametes.

Multiple births can result from multiple ovulations, blastomere splitting, or some combination of the two. Naturally occurring twins occur in 1 of every 89 births.

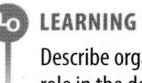

**? REVIEW**

**B.** Identify the general events of fetal development during the second and third trimesters.

**C.** A fetus undergoes its largest absolute weight gain during which trimester?

**LO LEARNING OUTCOME**
Describe organogenesis and its role in the developing fetus.

# Module 27.9

# Pregnancy places anatomical and physiological stresses on maternal systems

Pregnancy places tremendous demands on the mother. The developing fetus is totally dependent on maternal organ systems for nourishment, respiration, and waste removal. Maternal systems perform these functions in addition to their normal operations. For example, the mother must absorb enough oxygen, nutrients, and vitamins for herself and for her fetus, and she must eliminate all the wastes that are generated. Although this is not a burden over the initial weeks of gestation, the demands placed on the mother become significant as the fetus grows.

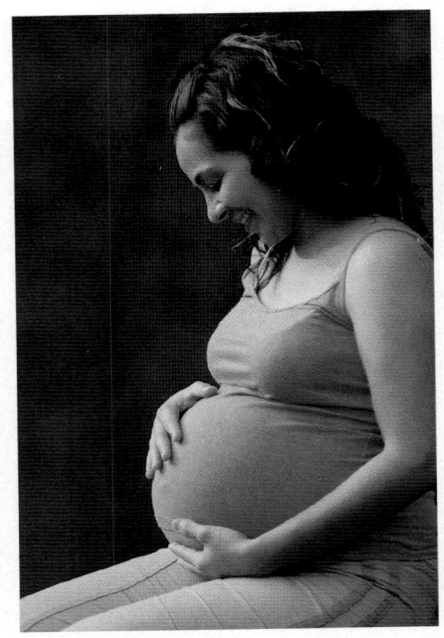

**1** The physical stresses of pregnancy are considerable. It is not unusual for a woman to gain 11–16 kg (25–35 lb) in weight during pregnancy, and the weight is not aligned with the body axis. This means that moving and maintaining balance use additional energy. The sectional views below compare the organ positions in nonpregnant and pregnant females. When the pregnancy is at full term, the uterus and fetus are so large that they push many of the maternal abdominal organs out of their normal positions.

Diaphragm
Liver
Stomach
Pancreas
Transverse colon
Small intestine

Fundus of uterus
Umbilical cord
Placenta
Mucus plug in cervical canal

Uterus
Urinary bladder
Pubic symphysis
Rectum
Urethra
Vagina

External os

**Nonpregnant female**

**Pregnant female (full-term infant)**

**2**   The physiological stresses are even more extreme than the physical ones. This is a summary of the physiological changes that have occurred in maternal systems by the end of the third trimester.

The mother's lungs must deliver the extra oxygen required, and remove the excess carbon dioxide generated, by the fetus. As a result, the maternal respiratory rate increases, and tidal volume increases.

Mammary gland development requires a combination of hormones, including human placental lactogen and prolactin (PRL), estrogens, progesterone, GH, and thyroxine from the maternal endocrine organs. By the end of the 6th month of pregnancy, the mammary glands are fully developed and begin to produce clear secretions that are stored in the duct system of those glands and may be expressed from the nipple.

The kidneys must eliminate the wastes produced by maternal and fetal systems. As a result, the maternal glomerular filtration rate (GFR) increases by nearly 50%.

Because the volume of urine produced increases and the weight of the uterus presses down on the urinary bladder, pregnant women need to urinate frequently.

The volume of blood flowing into the placenta decreases the volume in the systemic circuit of the mother. At the same time, fetal metabolic activity "steals" maternal oxygen and increases maternal $CO_2$ levels. This combination stimulates the production of renin and erythropoietin, leading to an increase in maternal blood volume. By the end of gestation, maternal blood volume has increased by almost 50%.

Pregnant women must nourish both themselves and their fetus, and so they tend to have increased hunger. Maternal requirements for nutrients can increase by up to 30% above normal.

At the end of gestation, a typical uterus has grown from 7.5 cm (3 in.) in length and 30–40 g (1–1.4 oz) in weight to 30 cm (12 in.) in length and 1100 g (2.4 lb) in weight. The uterus may then contain 2 liters of fluid, plus fetus and placenta, for a total weight of roughly 6–7 kg (13–15 lb). This remarkable expansion occurs through the enlargement (hypertrophy) of existing cells, especially smooth muscle fibers, rather than by an increase in the total number of cells.

## Clinical Note

### Medical risks associated with pregnancy increase with age

Although pregnancy is a natural phenomenon, the physical and physiological demands on maternal systems make it potentially dangerous. At any age, the risks associated with pregnancy are significantly greater than those associated with the use of birth control pills. (The notable exception involves women who smoke while on the pill.) For pregnant women over age 35, the chances of dying from pregnancy-related complications are almost twice as great as the chances of being killed in an automobile accident.

 **REVIEW**

**A.** List the major changes that occur in maternal systems during pregnancy.

**B.** Why does a mother's blood volume increase during pregnancy?

 **INTEGRATION**

**C.** Based on the illustrations showing the locations of the internal organs in nonpregnant and pregnant women, explain why some pregnant women experience difficulty breathing.

 **LEARNING OUTCOME**

Describe the interplay between maternal organ systems and the developing fetus.

# Module 27.10

# Multiple factors initiate and accelerate labor and delivery

The tremendous stretching of the uterus is associated with a gradual increase in the rate of spontaneous smooth muscle contractions in the myometrium. In the early stages of pregnancy, progesterone released by the placenta inhibits the uterine smooth muscle, preventing powerful contractions. Late in pregnancy, some women experience occasional spasms in the uterine musculature, but these contractions are not regular or persistent. Such contractions are called **Braxton Hicks contractions** (*false labor*). Placental and fetal factors are involved in the initiation of **true labor**.

**1** This sagittal section shows the position of the fetus at the onset of true labor.

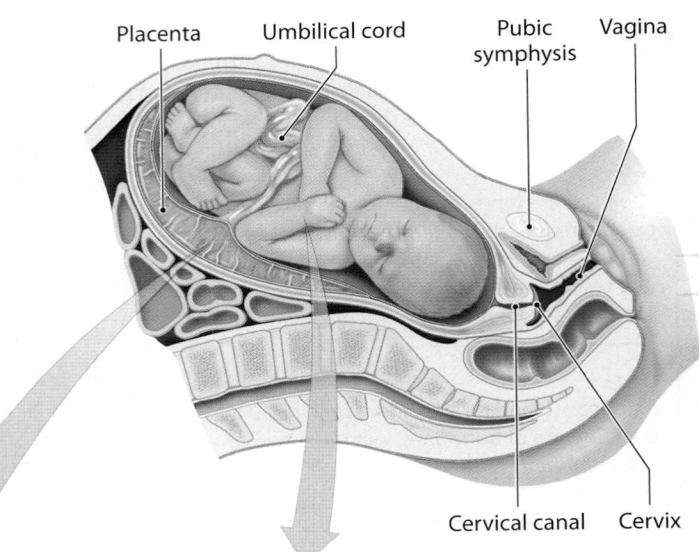

Placenta    Umbilical cord    Pubic symphysis    Vagina

Cervical canal    Cervix

**2** After 9 months of gestation, multiple factors interact to initiate true labor. Once labor contractions have begun in the myometrium, positive feedback ensures that they will continue until delivery has been completed.

## Placental Factors

Placental estrogens increase the sensitivity of the smooth muscle cells of the myometrium and make contractions more likely. As delivery approaches, the production of estrogens accelerates. Estrogens also increase the sensitivity of smooth muscle fibers to oxytocin.

## Fetal Factors

Relaxin produced by the placenta relaxes the pelvic articulations and dilates the cervix.

Growth and the increase in fetal weight stretch the myometrium.

The fetal pituitary gland releases oxytocin in response to estrogens.

## Myometrium Stretches

Stretching of the myometrium increases the sensitivity of the smooth muscle layers, promoting spontaneous contractions that get stronger and more frequent as labor advances.

Labor contractions move the fetus and stretch the myometrium. This stretching stimulates additional oxytocin and prostaglandin release. This **positive feedback** continues until delivery is completed.

## Maternal Oxytocin Release

Maternal oxytocin release is stimulated by high estrogen levels and cervix stretching.

## Prostaglandin Production

Estrogens and oxytocin stimulate the production of prostaglandins in the endometrium. These prostaglandins further stimulate smooth muscle contractions.

## Increased Excitability of the Myometrium

Oxytocin and prostaglandins both stimulate the myometrium. In addition, the sensitivity of the uterus to oxytocin increases dramatically. The smooth muscle in a late-term uterus is 100 times more sensitive to oxytocin than the smooth muscle in a nonpregnant uterus.

## ? REVIEW

**A.** List and describe the factors involved in initiating labor contractions.

**B.** Which chemicals are primarily responsible for initiating contractions of true labor?

**LABOR CONTRACTIONS OCCUR**

**3** The goal of labor is **parturition** (par-tū-RISH-un), or childbirth, the forcible expulsion of the fetus, followed by the placenta. Labor is divided into three stages: the dilation stage, the expulsion stage, and the placental stage.

## Dilation Stage

The **dilation stage** begins with the onset of true labor as the cervix dilates and the fetus shifts toward the cervical canal, moved by gravity and uterine contractions. This stage typically lasts 8 or more hours. At the start of the dilation stage, labor contractions last up to half a minute and occur once every 10–30 minutes; their frequency increases steadily. Late in this stage, the amnion ruptures, an event sometimes referred to as "having one's water break."

## Expulsion Stage

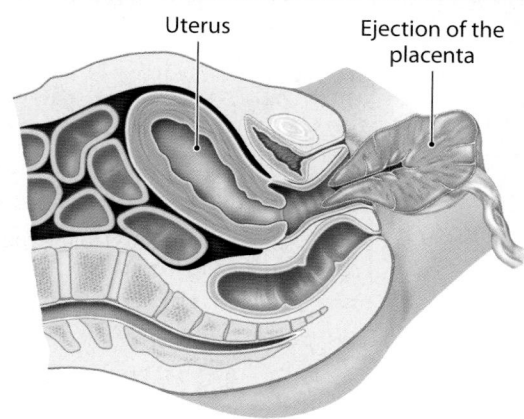

The **expulsion stage** begins as the cervix, pushed open by the approaching fetus, completes dilation (about 10 cm). In this stage, contractions reach maximum intensity, occurring at perhaps 2- or 3-minute intervals and lasting a full minute. Expulsion continues until the fetus has emerged from the vagina; in most cases, the expulsion stage lasts less than 2 hours. The arrival of the newborn infant into the outside world is **delivery**, or birth.

## Placental Stage

Uterus          Ejection of the placenta

During the **placental stage**, muscle tension builds in the walls of the partially empty uterus, which gradually decreases in size. This uterine contraction tears the connections between the endometrium and the placenta, and the placenta, or **afterbirth**, is ejected. The disruption of the placenta is accompanied by a loss of blood, but associated uterine contractions compress the uterine vessels and usually restrict this flow.

 **Clinical Note**

### Premature labor requires medical intervention

**Premature labor** occurs when true labor begins before the fetus has completed normal development. The newborn's chances of surviving are directly related to its body weight at delivery. Even with massive supportive efforts, newborns weighing less than 400 g (14 oz) at birth will not survive, primarily because their respiratory, cardiovascular, and urinary systems are unable to support life without aid from maternal systems. Most fetuses born at 25–27 weeks of gestation (a birth weight under 600 g or 21.1 oz) die despite intensive neonatal care, and survivors have a high risk of developmental abnormalities. **Premature delivery** usually refers to birth at 28–36 weeks (a birth weight over 1 kg or 2.2 lb). With care, these newborns have a good chance of surviving and developing normally.

**? REVIEW**

**C.** Name the three stages of labor, and describe the events that characterize each stage.

 **LEARNING OUTCOME**

List and discuss the events that occur during labor and delivery.

# Module 27.11
# After delivery, development initially requires nourishment by maternal systems

Developmental processes do not end at delivery, because newborns have few of the anatomical, functional, or physiological characteristics of mature adults. During an initial **neonatal period**, the newborn is dependent on the mother for nourishment and protection from environmental hazards, such as extreme changes in temperature. The neonatal period lasts from birth through the first 28 days of life.

**1** Milk is provided to infants through the **milk ejection (milk let-down) reflex**, which is diagrammed here. By the end of the 6th month of pregnancy, the mammary glands are fully developed, and the gland cells begin to produce a secretion known as **colostrum** (kō-LOS-trum). Colostrum contains antibodies that may help the infant ward off infections until its own immune system becomes fully functional. After the first few days of nursing, the mammary glands produce breast milk, which has a much higher fat content than colostrum but still contains antibodies. It also contains **lysozyme**, an enzyme with antibiotic properties.

**?** REVIEW
A. Explain the difference between colostrum and breast milk.

**3 Oxytocin Secretion**

The stimulation of tactile receptors in the nipple leads to the stimulation of secretory neurons in the paraventricular nucleus of the maternal hypothalamus.

Posterior lobe of the pituitary gland

**4 Oxytocin Release**

The hypothalamic neurons release oxytocin by the posterior lobe of the pituitary gland. Oxytocin enters the bloodstream and is distributed throughout the body.

**5 Milk Ejected**

When circulating oxytocin reaches the mammary gland, this hormone causes the contraction of myoepithelial cells in the walls of the lactiferous ducts and sinuses. The result is milk ejection, or milk let-down.

**?** REVIEW
B. Which hormone causes the milk ejection (milk let-down) reflex?

**Start**

**1 Stimulation of Tactile Receptors**

Mammary gland secretion is triggered when the infant sucks on the nipple.

**2 Neural Impulse Transmission**

Impulses are propagated to the spinal cord and then to the brain.

**LO** LEARNING OUTCOME
Describe the neonatal period and the milk ejection reflex.

# Module 27.12
# Postnatal development includes five life stages

**1** Postnatal development includes five **life stages**: (1) the **neonatal period**, (2) **infancy**, (3) **childhood**, (4) **adolescence**, and (5) **maturity**. Once maturity has been reached, the individual is subject to the gradual changes that accompany **senescence** (*senesco*, to grow old), or aging. Growth during infancy and childhood occurs under the direction of circulating hormones, notably growth hormone, adrenal steroids, and thyroid hormones. These hormones affect each tissue and organ in specific ways, depending on cellular sensitivity.

## Postnatal Development

| Neonatal | Infancy | Childhood | Adolescence | Maturity |
|---|---|---|---|---|

Through the neonatal period and infancy, the newborn is dependent on nutrients contained in milk, typically breast milk secreted by the mother's mammary glands. Nutrients may also come from infant formula.

In early childhood, the child is weaned from breast milk or formula. Because growth does not occur uniformly, body proportions gradually change. The head, for example, is relatively large at birth but decreases in proportion with the rest of the body as the child grows to adulthood.

Adolescence begins at **puberty**, the period of sexual maturation, and ends when growth is completed.

5 ft

4 ft

3 ft

2 ft

1 ft

0

1 month          1 year                    Puberty                      18 years
                                          (between
                                       10 and 15 years)

**REVIEW**

**A.** Name the stages of postnatal development, and describe the time frame involved for each stage.

**LEARNING OUTCOME**

Identify the features of and the functions associated with the various life stages.

# Module 27.13

# At puberty, male and female sex hormones have differing effects on most body systems

Many body systems alter their activities in response to changes in circulating levels of sex hormones at **puberty** (see Start arrow at right). The most important sex-related changes are summarized below (males) and at right (females). The effects of testosterone and estrogens are facilitated and enhanced by growth hormone, thyroid hormones, prolactin, and adrenocortical hormones.

**?** **REVIEW**

**B.** Why does a man have a deeper voice and a larger larynx than a woman?

## Responses to Testosterone in Males

### Integumentary System

Testosterone stimulates the development of terminal hairs on the face and chest and stimulates terminal hair growth in the axillae and in the genital area. Adipose tissues respond differently to testosterone than to estrogens, and this difference produces the distinct distributions of subcutaneous body fat in males versus females.

### Skeletal System

Testosterone accelerates bone deposition and skeletal growth. In the process, it promotes closure of the epiphyseal cartilages and thus places a limit on growth in height.

### Muscular System

Testosterone stimulates the growth of skeletal muscle fibers, and the increased muscle mass accounts for significant sex differences in body mass, even for males and females of the same height.

### Nervous System

A surge in testosterone secretion at puberty activates the central nervous system centers concerned with male sexual drive and sexual behaviors.

### Cardiovascular System

Testosterone stimulates erythropoiesis, thereby increasing blood volume and the hematocrit (see Module 17.2).

### Respiratory System

Testosterone stimulates growth of the larynx and a thickening and lengthening of the vocal cords. These changes cause a gradual deepening of the voice in males.

### Reproductive System

Testosterone stimulates the functional development of the accessory reproductive glands, such as the prostate and seminal glands, and helps promote spermatogenesis.

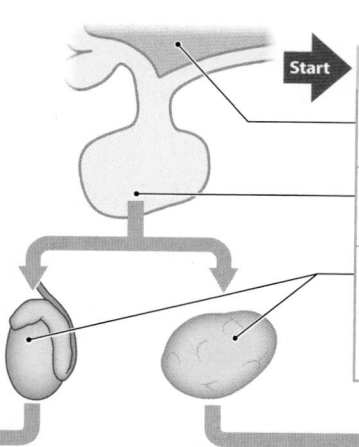

**Start**

## At Puberty

The hypothalamus increases its production of gonadotropin-releasing hormone (GnRH).

Endocrine cells in the anterior lobe of the pituitary gland become more sensitive to GnRH, and circulating levels of FSH and LH rise rapidly.

Testicular or ovarian cells become more sensitive to FSH and LH, initiating (1) gamete production; (2) the secretion of sex hormones, which stimulate the appearance of secondary sex characteristics and behaviors; and (3) a sudden acceleration in the growth rate, ending with closure of the epiphyseal cartilages.

 **REVIEW**

**A.** Name the three major interacting hormonal events associated with the onset of puberty.

## Responses to Estrogens in Females

### Integumentary System

Estrogens stimulate the hair follicles to continue to produce fine vellus hairs and stimulate terminal hair growth in the axillae and in the genital area. The combination of estrogens, prolactin, growth hormone, and thyroid hormones promotes the initial development of the mammary glands.

### Skeletal System

Estrogens cause more rapid epiphyseal closure than does testosterone. In addition, the period of skeletal growth ends at an earlier age in girls than in boys, and so females generally do not grow as tall as males.

### Muscular System

Estrogens stimulate the growth of skeletal muscle fibers, but not to the extent that testosterone does in males.

### Nervous System

A surge in estrogen secretion at puberty activates central nervous system centers involved in female sexual drive and sexual behaviors.

### Cardiovascular System

The iron loss associated with menstruation increases the risk of developing iron-deficiency anemia. Estrogens decrease plasma cholesterol levels and slow plaque formation within arteries. As a result, premenopausal women have a lower risk of atherosclerosis than do adult men.

### Respiratory System

Estrogens do not cause excessive growth of the larynx and vocal cords, so females typically have higher-pitched voices than males.

### Reproductive System

Estrogens target the uterus, promoting a thickening of the myometrium and increasing blood flow to the endometrium. Estrogens also promote the functional development of accessory reproductive structures in females.

**REVIEW**

**C.** Why are premenopausal women at lesser risk of atherosclerosis than men?

**LEARNING OUTCOME**

Explain the roles of hormones in males and females at puberty.

# Section 1 Review
## Overview of Development

**Matching:** Match each lettered term with the most closely related description.

a. hCG

b. conception

c. chorion

d. syncytiotrophoblast

e. colostrum

f. amnion

g. amphimixis

h. morula

i. embryonic disc

j. neonate

k. gestation

l. inner cell mass

m. relaxin

n. blastocyst

| **1** | Fertilization | **1** |_____|
| **2** | Newborn infant | **2** |_____|
| **3** | Pronuclei fuse | **3** |_____|
| **4** | Period of prenatal development | **4** |_____|
| **5** | Pregnancy test | **5** |_____|
| **6** | Softens pubic symphysis | **6** |_____|
| **7** | Mammary gland secretion | **7** |_____|
| **8** | Mesoderm and ectoderm | **8** |_____|
| **9** | Hollow ball of cells | **9** |_____|
| **10** | Mesoderm and cytotrophoblast | **10** |_____|
| **11** | Forms the embryo | **11** |_____|
| **12** | Cytoplasm with many nuclei | **12** |_____|
| **13** | Solid ball of cells | **13** |_____|
| **14** | Gastrulation product | **14** |_____|

**Multiple choice:** Select the correct answer from the list provided.

**15** Fertilization typically occurs in the
- a) lower part of the uterine tube.
- b) upper part of the uterus.
- c) junction between the ampulla and isthmus of the uterine tube.
- d) cervix.

**16** Fetal development begins at the start of the
- a) implantation process.
- b) 2nd month after fertilization.
- c) 9th week after fertilization.
- d) 6th month after fertilization.

**17** Organs and organ systems complete most of their development by the end of the
- a) first trimester.
- b) second trimester.
- c) third trimester.
- d) expulsion stage.

**18** The four critical events that occur during the first trimester include
- a) blastomere, blastocyst, morula, and trophoblast.
- b) cleavage, implantation, placentation, and embryogenesis.
- c) placentation, dilation, expulsion, and organogenesis.
- d) yolk sac, amnion, allantois, and chorion.

**19** The most dangerous period in prenatal or neonatal life is the
- a) first trimester.
- b) second trimester.
- c) third trimester.
- d) expulsion stage.

**20** The systems that were relatively nonfunctional during the fetal period that must become functional at birth are the
- a) cardiovascular, muscular, and skeletal systems.
- b) integumentary, reproductive, and nervous systems.
- c) respiratory, digestive, and urinary systems.
- d) endocrine, nervous, and digestive systems.

**Section integration:** Tina gives birth to a baby with a deformity of the stomach. Tina thinks her baby's condition resulted from a viral infection she had during her third trimester. Explain whether the viral infection likely caused the baby's condition.

**21** _____

_____

## Module 27.14

# A person may be described in terms of genotype and phenotype

Although everyone goes through the same developmental stages, differences in both genetic structure and local environments produce distinctive individual characteristics. Traditionally, **inheritance** or **heredity** refers to the transfer of genetically determined characteristics from generation to generation. The study of the mechanisms responsible for inheritance is called **genetics**. We begin our study of genetics by examining two important concepts: *genotype* and *phenotype*.

**1** One way to understand genotype and phenotype is to compare them to the architecture of a house. The house plan is the genotype, and how the finished house looks is its phenotype.

Every nucleated somatic cell in your body carries copies of the original 46 chromosomes present when you were a zygote. Those chromosomes and their component genes make up your **genotype** (JĒN-ō-tīp; *geno*, gene). In architectural terms, the genotype is a set of plans, like the blueprints for a house.

The collective expression of your genes determines the anatomical and physiological characteristics that make up your **phenotype** (FĒ-nō-tīp; *phaino*, to display). The phenotype results from the interaction between the person's genotype and the environment. Observable features, such as hair or eye color, are called phenotypic *characters*. Character variants, such as brown hair or eyes, are called phenotypic *traits*.

**? REVIEW**

**A.** Distinguish between genotype and phenotype.

**2** This photograph shows the **karyotype** (*karyon*, nucleus + *typos*, mark), or entire set of chromosomes, of a normal male. At amphimixis, one member of each of the 23 chromosome pairs was contributed by the sperm, and the other by the oocyte. The two members of 22 pairs have similar sizes and genes and are known as **homologous** (huh-MOL-ō-gus) **chromosomes**. Those 22 pairs are called **autosomes** (aw-TŌ-sōmz), or non-sex chromosomes. Most of the genes of the autosomes affect somatic characteristics, such as hair color and skin pigmentation. The chromosomes of the 23rd pair are not truly homologous and are called the **sex chromosomes**. One of their functions is to determine whether the individual is genetically male or female. The sex chromosomes of a male consist of an **X chromosome** and a shorter **Y chromosome**, whereas females have two X chromosomes.

**? REVIEW**

**B.** How do autosomes and sex chromosomes differ?

**C.** How can you tell that the karyotype shown here is male?

In this section, we consider basic genetics as it applies to inherited characteristics, such as sex, hair color, and various disorders.

**LO LEARNING OUTCOME**

Describe genetics and inheritance and the relationship between genotype and phenotype.

# Module 27.15

# Genes and chromosomes determine patterns of inheritance

**1** The two chromosomes in a homologous autosomal pair have the same structure and carry genes that affect the same traits. The two chromosomes in a pair may carry the same form or different forms of each gene. The various forms of a given gene are called **alleles** (uh-LĒLZ). The genes are also located at equivalent positions on their respective chromosomes. A gene's position on a chromosome is called a **locus** (LŌ-kus; plural, *loci*).

**2** The phenotype that results from a heterozygous genotype depends on the nature of the interaction between the corresponding alleles. The most common form of interaction among autosomal genes is called **simple inheritance**. In one kind of simple inheritance—**strict dominance**—any dominant allele will be expressed in the phenotype, regardless of any conflicting instructions carried by the other allele. For instance, a person with only one allele for freckles will have freckles, because that allele is dominant over the "nonfreckle" allele. An allele that is **recessive** will be expressed in the phenotype only if that same allele is present on both chromosomes of a homologous pair.

If the two chromosomes of a homologous pair carry the same allele of a particular gene, you are **homozygous** (hō-mō-ZĪ-gus; *homos*, the same) for the trait affected by that gene. That allele will then be expressed in your phenotype. For example, if you receive a gene for curly hair from your father and a gene for curly hair from your mother, you will be homozygous for curly hair—and you will have curly hair.

When you have two different alleles for the same gene, you are **heterozygous** (het-er-ō-ZĪ-gus; *heteros*, other) for the trait determined by that gene.

The human genome contains approximately 22,500 genes. Thus an "average" autosomal pair contains about 1000 pairs of alleles.

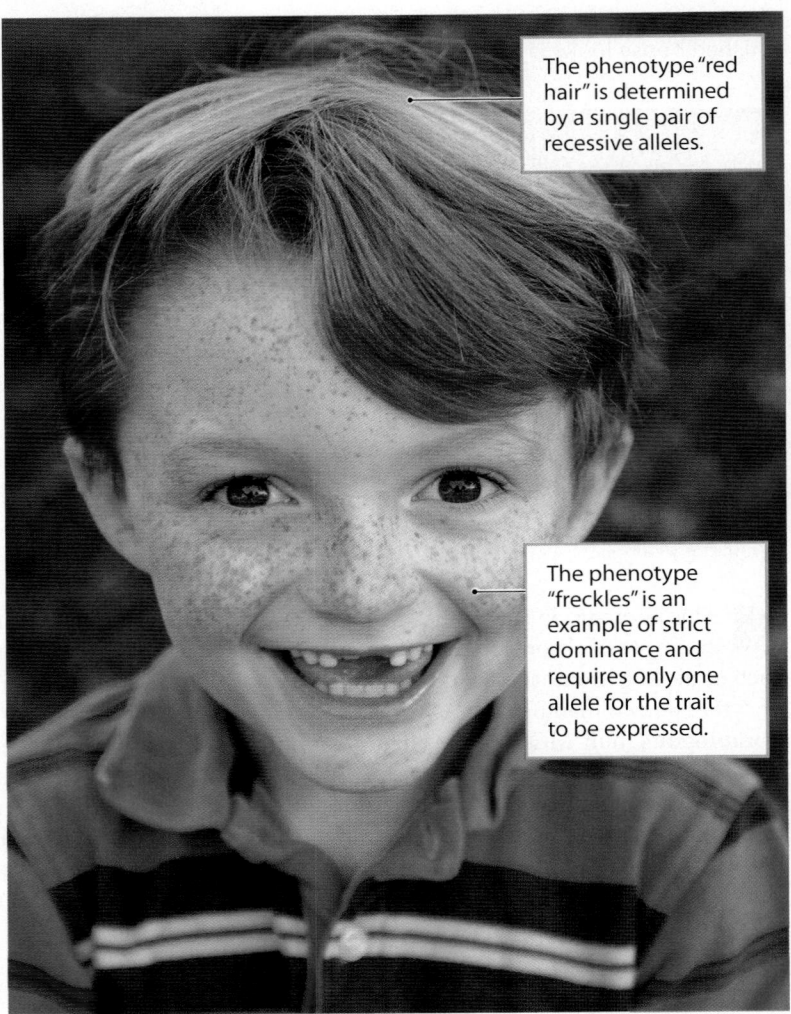

The phenotype "red hair" is determined by a single pair of recessive alleles.

The phenotype "freckles" is an example of strict dominance and requires only one allele for the trait to be expressed.

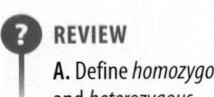 **REVIEW**

**A.** Define *homozygous* and *heterozygous*.

**3** By convention, dominant alleles are indicated by capital letters, and recessive alleles by lowercase letters. Thus for a given trait *T*, the possible genotypes are *TT* (homozygous dominant), *Tt* (heterozygous), and *tt* (homozygous recessive). The **Punnett squares** shown below indicate the possible offspring of a mother with albinism (no skin pigmentation) and a father with normal skin pigmentation. Because albinism is a recessive trait, the maternal genotype is designated *aa*. The father has normal pigmentation, a dominant trait, so his genotype could be either *AA* or *Aa*.

Maternal alleles (contributed by the ovum). Every ovum will carry the recessive gene *a*.

Paternal alleles (contributed by the sperm). Every sperm produced by a homozygous dominant (**AA**) father will carry the **A** allele.

|  | *a* | *a* |
|---|---|---|
| *A* | *Aa* | *Aa* |
| *A* | *Aa* | *Aa* |

All have normal skin pigmentation

If the father is homozygous for normal pigmentation, all of the children will have the genotype *Aa*, and all will have normal skin pigmentation.

Maternal alleles

Half of the sperm produced by a heterozygous (**Aa**) father will carry the dominant allele **A**, and the other half will carry the recessive allele *a*.

|  | *a* | *a* |
|---|---|---|
| *A* | *Aa* | *Aa* |
| *a* | *aa* | *aa* |

50% of the children are heterozygous and have normal pigmentation

50% of the children are homozygous recessive and exhibit albinism

If the father is heterozygous for normal skin pigmentation, the probability that a child will have normal pigmentation is now 50%.

**4** Many phenotypic characteristics are determined by interactions among several genes. Such interactions constitute **polygenic inheritance**. The resulting phenotype depends not only on the nature of the alleles, but also on how those alleles interact. For this reason, you cannot predict the presence or absence of phenotypic characters using a simple Punnett square. An example of polygenic inheritance is brown or black hair color.

**? REVIEW**

**B.** Differentiate between simple inheritance and polygenic inheritance.

**? INTEGRATION**

**C.** The trait "freckles" operates through strict dominance. What would be the phenotype of a person who is heterozygous for this trait?

**Lo LEARNING OUTCOME**

Relate the basic principles of genetics to the inheritance of human traits.

# Module 27.16
# There are several different patterns of inheritance

Autosomes and sex chromosomes show different patterns of inheritance.
This flowchart summarizes the major patterns of inheritance.

## Major Patterns of Inheritance

**1** Autosomes can show two major patterns
of inheritance: *simple* and *polygenic*.
Simple inheritance can be subdivided into *strict
dominance, incomplete dominance,* or *codominance.*

### Inheritance Involving Autosomes

#### Simple Inheritance

The phenotype is determined by a single pair of alleles.
Approximately 80% of your genotype falls within this
category.

#### Polygenic Inheritance

The phenotype is determined by interactions among the
alleles of several genes.

*Examples*
- Hair color (other than blond or red, which are recessive traits)
- Skin color
- Eye color
- Height

#### Strict Dominance

One allele dominates the other allele
and determines the phenotype.

*Examples of dominant traits*
- Normal skin pigmentation
- Freckles
- Nearsightedness
- Farsightedness
- Astigmatism
- Free earlobes
- Tongue-rolling
- Rh factor
- Type A or B blood
- Huntington's disease

*Examples of recessive traits*
- Albino pigmentation
- Absence of freckles
- Normal vision
- Attached earlobes
- Inability to roll tongue
- Rh factor absent
- Type O blood
- Sickle cell disease
- Cystic fibrosis
- Tay–Sachs disease
- Phenylketonuria

#### Incomplete Dominance

In **incomplete dominance**,
heterozygous alleles produce a
phenotype that is intermediate
(not completely dominant) to
the phenotypes of individuals
who are homozygous for
one allele or the other. One
example would be the sickling
gene for hemoglobin. Homozygous people have sickle cell
disease, whereas heterozygous
people have RBCs that sickle
only with low oxygen levels.

*Examples*
- Hemoglobin A (normal Hb)
- Hemoglobin S (abnormal Hb)

#### Codominance

In **codominance**, an individual
who is heterozygous (has
different alleles) for a given
trait exhibits both of the
phenotypes for that trait. Blood
type in humans is determined
by codominance. The alleles
for type A and type B blood
are dominant over the allele
for type O blood, but a person
with one type A allele and one
type B allele has type AB blood.

*Examples*
- Type AB blood
- Structure of albumins

 **REVIEW**

**A.** Compare strict dominance
with codominance.

**2** Sex-linked genes are carried by either sex chromosome (X or Y). X chromosomes are considerably larger and have many more genes than do Y chromosomes. Only a few of these sex-linked genes are homologous. Genes that are found on the X chromosome but not on the Y chromosome are called **X-linked**. In contrast, those genes found only on the Y chromosome are called **Y-linked**. A Y-linked example is the SRY gene for "maleness." It triggers male development in the embryo. A number of clinical disorders are X-linked traits.

### Inheritance Involving Sex Chromosomes

**Sex-linked inheritance** involves genes on the sex chromosomes. Most examples involve the X chromosome. An X-linked allele determines the phenotype in males since there is no corresponding allele on the Y chromosome to mask or affect its expression. Most known cases involve alleles that are recessive in females.

*Examples*
- Red–green color blindness
- Hemophilia (some forms)
- Duchenne muscular dystrophy

X-linked allele (allele not present on Y chromosome)

Y

X

**? REVIEW**
**B.** Why are sex-linked traits expressed more frequently in males than in females?

**3** The inheritance of color blindness illustrates the differences between sex-linked inheritance and autosomal inheritance. In males, the presence of a dominant allele, *C*, on the X chromosome ($X^C$) results in normal color vision; a recessive allele, *c*, on the X chromosome ($X^c$) results in red–green color blindness. This Punnett square reveals that each son of a father with normal vision and a heterozygous (*carrier*) mother has a 50% chance of being red–green color blind, whereas any daughters will have normal color vision.

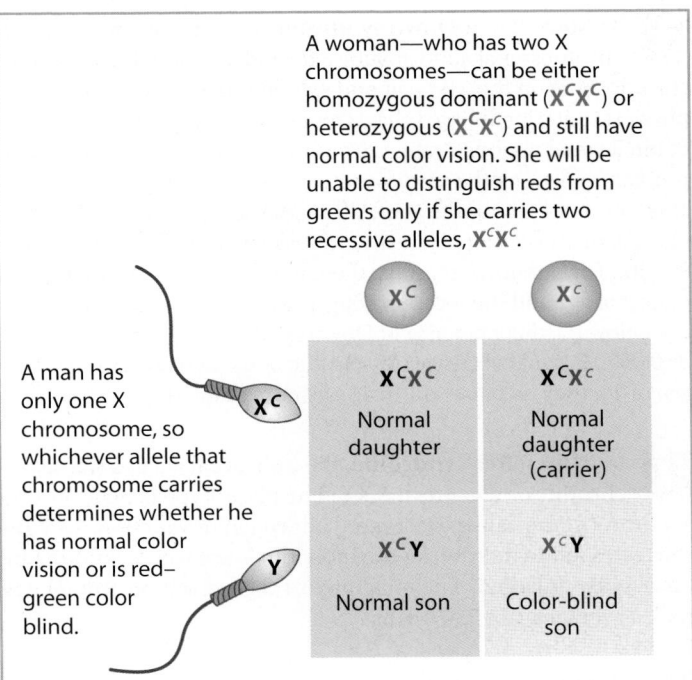

A woman—who has two X chromosomes—can be either homozygous dominant ($X^C X^C$) or heterozygous ($X^C X^c$) and still have normal color vision. She will be unable to distinguish reds from greens only if she carries two recessive alleles, $X^c X^c$.

A man has only one X chromosome, so whichever allele that chromosome carries determines whether he has normal color vision or is red–green color blind.

$X^C$  $X^c$

$X^C$

$X^C X^C$
Normal daughter

$X^C X^c$
Normal daughter (carrier)

Y

$X^C Y$
Normal son

$X^c Y$
Color-blind son

**? INTEGRATION**

**C.** Indicate the type of inheritance involved in each of the following situations: (1) children who exhibit the trait have at least one parent who also exhibits it; (2) children exhibit the trait even though neither parent exhibits it; and (3) the trait is expressed equally in daughters and sons.

**LO LEARNING OUTCOME**

Describe the relationships among the various forms of inheritance, and give examples of representative phenotypic characters, both normal and abnormal.

# Module 27.17
# Many clinical disorders are linked to individual chromosomes or their genes

Chromosomal abnormalities can involve thousands of genes, and as a result they are usually lethal. Most of the time, the embryo or fetus dies before delivery. However, there are a few autosome abnormalities that do not invariably result in prenatal mortality (death). In contrast, variations in the structure of individual genes are relatively common. Although more than 99% of human nucleotide bases are the same in all people, there are about 1.4 million single-base differences, or **single nucleotide polymorphisms (SNPs)**. Some of these SNPs are inconsequential, but others are associated with specific diseases.

**?  REVIEW**
**A.** Define *single nucleotide polymorphism*.

**1** **Trisomy 21**, or **Down syndrome**, is the most common viable chromosomal abnormality. Affected individuals have a third chromosome on the 21st pair and exhibit intellectual disability and certain physical traits, including a characteristic facial appearance. The degree of intellectual disability ranges from moderate to severe, and anatomical problems affecting the cardiovascular system often prove fatal during childhood or early adulthood. Although some people survive to moderate old age, many develop Alzheimer disease while still relatively young (before age 40). For unknown reasons, there is a direct correlation between maternal age and the risk of having a child with trisomy 21. For a maternal age below 25, the incidence of Down syndrome approaches 1 in 2000 births, or 0.05%. For maternal ages 30–34, the odds increase to 1 in 900; and by the age of 44, they increase to 1 in 46, or more than 2%.

**2** In **Klinefelter syndrome**, the individual carries the sex chromosome pattern XXY. The phenotype is male, but the extra X chromosome causes decreased androgen production. As a result, the testes fail to mature, so the individuals are sterile, and the breasts are slightly enlarged. The incidence of this condition among newborn males averages 1 in 750 births.

**3** Individuals with **Turner syndrome** have only a single, female sex chromosome; their sex chromosome complement is designated XO. This kind of chromosomal deletion is known as **monosomy**. The incidence of this condition at delivery has been estimated as 1 in 10,000 live births. The condition may not be recognized at birth, because the phenotype is normal female. But maturational changes do not appear at puberty. The ovaries are nonfunctional, and estrogen production occurs at negligible levels.

**?  REVIEW**
**B.** Name the disorder characterized by each of the following chromosome patterns: (1) XO and (2) XXY.

4 The **human genome**—the full set of genetic material (DNA) in our chromosomes, listed nucleotide by nucleotide—contains an estimated 20,000–25,000 protein-coding genes. Roughly 10,000 different single-gene disorders have been described. Most are very rare, but collectively they may affect 1 in every 200 births. Over 900 of these disorders have been mapped on the genome, and several examples are included in this chromosome map. Many of these conditions were noted in earlier chapters.

Color blindness (multiple forms)

Hemophilia

Neurofibromatosis, type 2 (tumors of the auditory nerves and tissues surrounding the brain)

Down syndrome

Amyotrophic lateral sclerosis*

ADA deficiency (enzyme deficiency that affects the immune system)

Familial hypercholesterolemia (extremely high cholesterol)

Myotonic dystrophy (form of muscular dystrophy)

Amyloidosis (accumulation of an insoluble fibrillar protein in the tissues)

Breast cancer*

Polycystic kidney disease

Tay–Sachs disease (lysosomal storage disease affecting neural tissue)

Marfan syndrome

Alzheimer's disease*

α₁-antitrypsin deficiency (causes a predisposition to develop emphysema)

Retinoblastoma (relatively common tumor of the eye, accounting for 2% of childhood malignancies)

CHROMOSOME PAIRS

22 X Y 1 2 3 4 5 6 7 8 9 10 11 12 13 14 15 16 17 18 19 20 21

Muscular dystrophy

Prostate cancer

Gaucher disease (lysosomal storage disease)

Familial colon cancer*

Retinitis pigmentosa*

Huntington's disease

Familial polyposis of the colon (abnormal tissue growths that commonly lead to colon cancer)

Spinocerebellar ataxia (destroys neurons in the brain and spinal cord, resulting in loss of muscle control)

Cystic fibrosis

Burkitt lymphoma (cancer of lymphocytes)

Retinitis pigmentosa*

Epilepsy, progressive

Malignant melanoma

Ovarian cancer

Multiple endocrine neoplasia, type 2 (tumors in endocrine glands and other tissues)

SCID (severe combined immunodeficiency disease)

Diabetes mellitus, type 1

Sickle cell disease

PKU (phenylketonuria)

\* One form of the disease

**REVIEW**

**C.** Identify the chromosome pair involved in each of the following disorders: (1) ovarian cancer, (2) Tay–Sachs disease, and (3) spinocerebellar ataxia.

**Epigenetics** is the study of inherited traits that do not involve a change in a person's genotype or DNA sequences. Epigenetics acts at a level "above" the DNA by chemically marking DNA to turn specific genes "on" or "off." Understanding the role of epigenetics in human development and disease is an area of active study.

**LEARNING OUTCOME**

Identify several chromosomal disorders, and describe the human genome.

# Section 2 Review
## Genetics and Inheritance

**Matching:** Match each lettered term with the most closely related description.

a. genotype
b. heterozygous
c. locus
d. autosomes
e. simple inheritance
f. homozygous
g. alleles
h. polygenic inheritance
i. homologous
j. genetics
k. karyotype
l. phenotype

**1** Expressed characteristics
**2** Alternate forms of a gene
**3** Refers to the two members of a pair of chromosomes
**4** Array of the entire set of chromosomes in a cell
**5** An individual's chromosomes and genes
**6** Gene's position on a chromosome
**7** Two different alleles for the same gene
**8** Study of the mechanisms of inheritance
**9** Two identical alleles for the same gene
**10** Interactions between alleles on several genes
**11** Phenotype determined by a single pair of alleles
**12** Chromosomes affecting somatic characteristics

**1** _____
**2** _____
**3** _____
**4** _____
**5** _____
**6** _____
**7** _____
**8** _____
**9** _____
**10** _____
**11** _____
**12** _____

**Section integration:** Using the templates below, draw Punnett squares to answer each question about the following genetic conditions.

**13** Tongue-rolling is inherited as a strictly dominant trait. Using *T* for tongue-rolling and *t* for non-tongue-rolling, determine the possible genotypic and phenotypic ratios of offspring born to two heterozygous tongue-rolling parents. What is the probability they will have a child unable to roll his/her tongue?

**14** Cystic fibrosis (CF) is an autosomal homozygous recessive disorder (*cc*) that causes the production of excessively thick mucus. Using *c* for the CF trait and *C* for not having the CF trait, determine the possible genotypic and phenotypic ratios of offspring produced by a mother with CF and a father who is a carrier for CF. What is the probability that they will have a child who will be a carrier for CF?

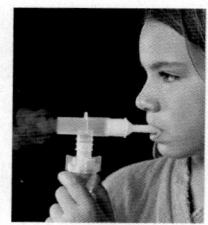

Maternal alleles

Paternal alleles

Maternal alleles

Paternal alleles

**13** _____
_____
_____
_____

**14** _____
_____
_____
_____

# Study Outline

## SECTION 1 • Overview of Development

### Module 27.1

**Gestation and development are marked by various stages p. 1037**

1. **Gestation** is the time that a developing embryo spends within the womb. It occurs during three integrated trimesters.

2. **Embryonic development** occurs during the first 2 months after fertilization.

3. **Fetal development** begins at the start of the 9th week and continues until birth.

4. **Prenatal development** consists of both embryonic and fetal development.

5. **Postnatal development** begins at birth and continues to **maturity**.

### Module 27.2

**At fertilization, an ovum and a sperm form a zygote that prepares for cell division p. 1038**

6. **Fertilization** is the fusion of two haploid gametes, each containing 23 chromosomes, producing a **zygote** with 46 chromosomes. This usually occurs in the uterine tube within a day of ovulation.

7. **Oocyte activation** occurs when a single sperm makes contact with the oocyte membrane, and membrane fusion takes place. Meiosis II is completed, and the secondary oocyte is now an **ovum**.

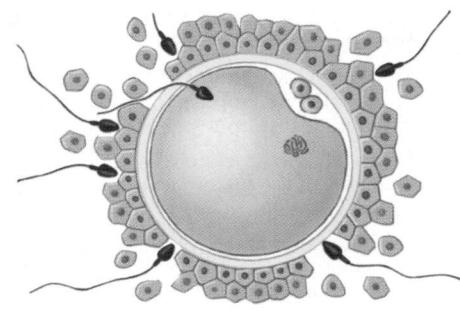

8. The female nuclear material within the ovum reorganizes as the **female pronucleus**. As the **male pronucleus** develops, the rest of the sperm breaks down. DNA synthesis occurs.

9. Male and female pronuclei chromatin condenses into chromosomes, and spindle fibers appear in preparation for cell division. This is the start of **cleavage**, in which mitotic cell divisions produce increasing numbers of smaller and smaller daughter cells.

10. The two pronuclei then fuse in a process called **amphimixis**. The cell is now a zygote that contains the normal complement of 46 chromosomes. Fertilization is complete.

11. Cell division of the zygote produces cells called **blastomeres**.

### Module 27.3

**Cleavage continues until the blastocyst implants in the uterine wall p. 1040**

12. A group of blastomeres created by cleavage divisions is called a **pre-embryo**.

13. After 3 days, the pre-embryo is a solid ball of cells known as a **morula**.

14. By day 6, the blastomeres form a **blastocyst**, a hollow ball with an inner cavity (**blastocoele**), a surrounding cell layer (**trophoblast**), and an inner cluster of cells (**inner cell mass**).

15. **Implantation** (day 7) begins with the attachment of the blastocyst to the uterine endometrium.

16. The cells closest to the blastocoele remain intact, by day 8 forming the **cytotrophoblast**. The **syncytiotrophoblast** forms from the cells near the endometrial wall.

17. **Villi** extend from the syncytiotrophoblast into the endometrium. The **amniotic cavity** forms by day 9 between the inner cell mass and the cytotrophoblast.

### Module 27.4

**Gastrulation produces three germ layers: ectoderm, endoderm, and mesoderm p. 1042**

18. Cells of the inner cell mass are organized into an oval sheet known as the **blastodisc**. Cells of the superficial layer migrate and then separate the amniotic cavity from the trophoblast in the first step of **amnion** formation.

19. The **yolk sac** forms by day 10 from the outer edges of the blastocoele. For the next 2 weeks, the yolk sac is the primary nutrient source for the inner cell mass.

20. By day 15, superficial cells of the blastodisc have begun to migrate toward a central line known as the **primitive streak** and migrate between the existing cell layers. This creates three embryonic layers that are collectively known as **germ layers**: **ectoderm**, **endoderm**, and **mesoderm**.

21. **Gastrulation** produces the three-layered **embryonic disc**. This will form the body of the embryo, whereas all other cells of the blastocyst will be part of the **extra-embryonic membranes**.

### Module 27.5

**The extra-embryonic membranes form the placenta that supports fetal growth and development p. 1044**

22. Germ layers form four extra-embryonic membranes: the yolk sac, the amnion, the allantois, and the chorion.

23. The **yolk sac** begins as a layer of cells spread out around the outer edges of the blastocoele to form a complete pouch.

24. The **amnion** begins as cells from the blastodisc migrate around a cavity that separates the inner cell mass from the trophoblast. The amniotic cavity contains **amniotic fluid** that cushions the developing embryo or fetus.

25. The **allantois** begins as an outpocketing of the endoderm near the base of the yolk sac that develops by week 3.

26. The mesoderm associated with the allantois spreads around the blastocyst, separating the cyto-trophoblast from the blastocoele. By day 14, this combination of mesoderm and cytotrophoblast is the **chorion**.

27. The **placenta** develops by week 5 as villi of the chorion invade the endometrium and break down maternal blood vessels. It becomes the primary support structure for the developing embryo.

## Module 27.6

### The formation of extra-embryonic membranes is associated with major changes in the shape and complexity of the embryo    p. 1046

28. By week 3, the embryonic disc bulges into the amniotic cavity at the **head fold**. By week 4, the embryo also has a **tail fold**.

29. By week 5, the embryo moves away from the placenta, and the **body stalk** and **yolk stalk** fuse to form an **umbilical stalk**.

30. By week 10, the fetus is connected to the placenta by an elongated **umbilical cord** that contains a portion of the allantois, blood vessels, and the remnants of the yolk stalk.

## Module 27.7

### The placenta performs many vital functions during prenatal development    p. 1048

31. Blood flows from the fetus to the placenta through paired **umbilical arteries**.

32. At the placenta, oxygen supplies are replenished, organic nutrients are added, and carbon dioxide and other organic wastes are removed.

33. Blood returns to the fetus within a single **umbilical vein**.

34. The placenta also acts as an endocrine gland, secreting **human chorionic gonadotropin (hCG)**, **human placental lactogen (hPL)**, **placental prolactin**, **relaxin**, **progesterone**, and **estrogens**.

35. The presence of hCG, an indicator of pregnancy, causes the corpus luteum to persist for 3–4 months and continue to secrete progesterone, while hPL and placental prolactin help prepare the mammary glands for milk production.

36. Relaxin increases the flexibility of the pubic symphysis, causes dilation of the cervix, and delays labor contractions until late in pregnancy.

37. The placenta produces sufficient amounts of progesterone to maintain the endometrial lining and continue the pregnancy after the first trimester. Toward the end of the third trimester, estrogen production by the placenta accelerates and assists in stimulating labor and delivery.

## Module 27.8

### Organ systems form in the first trimester and become functional in the second and third trimesters    p. 1050

38. Events in the first 12 weeks establish **organogenesis**, the process of organ formation.

39. The heart is beating by week 4. By week 6, the placenta has formed, and the embryo floats within the amniotic cavity.

40. By the end of the first trimester, the axial and appendicular muscles are forming, and fetal movements will soon begin.

41. By the end of the second trimester, the fetus will have grown to a weight of about 0.64 kg (1.4 lb).

42. During the third trimester, most of the organ systems become ready to function normally without maternal assistance.

43. In the final 3 months of gestation, the fetus gains about 2.6 kg (5.7 lb), reaching a full-term weight of approximately 3.2 kg (7 lb).

44. **Dizygotic** twins develop when two separate oocytes are ovulated and fertilized. Seventy percent of twins are dizygotic. **Monozygotic** twins result either from the separation of blastomeres early in cleavage or from the splitting of the inner cell mass before gastrulation.

## Module 27.9

### Pregnancy places anatomical and physiological stresses on maternal systems    p. 1052

45. The physical strains of pregnancy are considerable. The developing fetus is totally dependent on maternal organ systems that must support the mother as well as the fetus.

46. At full term, the uterus and fetus are so large that they push many of the maternal abdominal organs out of their normal positions.

47. By the end of the third trimester, many physiological stressors related to nourishment, oxygen supply, and waste removal occur in the maternal system.

48. The physical and physiological demands on maternal systems make pregnancy potentially dangerous.

## Module 27.10

### Multiple factors initiate and accelerate labor and delivery  p.    1054

49. Late in pregnancy, some women experience occasional spasms in the uterine musculature called **Braxton Hicks contractions** (*false labor*). **True labor** begins when biochemical and mechanical factors reach a point of no return.

50. Placental factors and fetal factors interact to initiate true labor. **Positive feedback** ensures that contractions of the myometrium will continue until delivery has been completed.

51. Placental factors include placental estrogens that increase the sensitivity of the smooth muscle cells of the myometrium and make contractions more likely, as well as relaxin that dilates the cervix.

52. Fetal factors include fetal stretching of the myometrium that promotes uterine contractions and fetal secretion of oxytocin.

53. **Parturition**, or childbirth, is the forcible expulsion of the fetus and placenta. Labor is divided into three stages: the dilation, expulsion, and placental stages.

54. The **dilation stage** begins with the start of true labor as the cervix dilates. It can last 8 or more hours.

55. The **expulsion stage** begins when the cervix completes dilating, and it continues until the fetus has emerged from the vagina (**delivery**). This usually lasts less than 2 hours.

56. The **placental stage** is the ejection of the placenta, or **afterbirth**.

57. **Premature labor** occurs when true labor begins before the fetus has completed normal development. **Premature delivery** usually refers to birth at 28–36 weeks.

### Module 27.11

**After delivery, development initially requires nourishment by maternal systems** p. 1056

58. The **neonatal period** lasts from birth through the first 28 days of life. At this time, the newborn is dependent on the mother for nourishment.

59. Mammary gland cells produce protein-rich **colostrum** during the neonate's first few days of life; then the gland cells convert to milk production. These secretions are released as a result of the **milk ejection** (milk let-down) **reflex**.

### Module 27.12

**Postnatal development includes five life stages** p. 1057

60. Postnatal development includes five **life stages**: the **neonatal period**, **infancy**, **childhood**, **adolescence**, and **maturity**.

### Module 27.13

**At puberty, male and female sex hormones have differing effects on most body systems** p. 1058

61. At **puberty**, the hypothalamus increases its production of gonadotropin-releasing hormone (GnRH).

62. Endocrine cells in the anterior lobe of the pituitary gland become more sensitive to GnRH, and circulating levels of FSH and LH rise rapidly.

63. Testicular or ovarian cells become more sensitive to FSH and LH. The following responses are initiated: gamete production; the secretion of sex hormones, which stimulate the appearance of secondary sex characteristics; and a sudden acceleration in the growth rate, ending with closure of the epiphyseal cartilages.

## SECTION 2 • Genetics and Inheritance

### Module 27.14

**A person may be described in terms of genotype and phenotype** p. 1061

64. **Genetics** is the study of **inheritance**, or **heredity**, the transfer of genetically determined characteristics from generation to generation.

65. Somatic cells contain copies of the original 46 chromosomes from the zygote. These chromosomes and their genes make up a person's **genotype**. Anatomical and physiological characteristics make up the **phenotype**. *Characters* are observable features, such as hair or eye color, and *traits* are specific variants, such as brown hair or eyes. A **karyotype** is a complete set of a person's chromosomes.

66. The two members of 22 pairs of chromosomes are similar to each other in size and genes and known as **homologous chromosomes**. They are non-sex chromosomes, or **autosomes**. The 23rd pair are called the **sex chromosomes**.

67. The sex chromosomes of a male consist of an **X chromosome** and a shorter **Y chromosome**, whereas females have two X chromosomes.

### Module 27.15

**Genes and chromosomes determine patterns of inheritance** p. 1062

68. The two chromosomes in a homologous autosomal pair have the same structure and carry genes that affect the same traits.

69. The various forms of a given gene are called **alleles**. A gene's position on a chromosome is called a **locus**.

70. If the two chromosomes of a homologous pair carry the same allele of a particular gene, you are **homozygous** for the trait affected by that gene.

*"Red hair" phenotype is determined by a single pair of recessive alleles. "Freckles" phenotype follows strict dominance.*

71. When you have two different alleles for the same gene, you are **heterozygous** for the trait determined by that gene.

72. The most common form of interaction among autosomal genes is **simple inheritance**. In one kind of simple inheritance called **strict dominance**, any dominant allele will be expressed in the phenotype.

73. An allele that is **recessive** will be expressed in the phenotype only if that same allele is present on both chromosomes of a homologous pair.

74. In a **Punnett square**, dominant alleles are indicated by capital letters, and recessive alleles by lowercase letters.

75. Many phenotypic characteristics involve **polygenic inheritance** and are determined by interactions among the alleles of several genes. In these situations, a Punnett square cannot predict the presence or absence of phenotypic characteristics.

### Module 27.16

**There are several different patterns of inheritance** p. 1064

76. Autosomes and sex chromosomes show different patterns of inheritance.

77. Autosomes involve simple inheritance. This includes strict dominance, **incomplete dominance**, and **codominance**. Autosomes also involve polygenic inheritance.

78. **Sex-linked inheritance** involves genes on the sex chromosomes. Most examples involve the X chromosome. An **X-linked** allele determines the phenotype in males since there is no corresponding allele on the Y chromosome to mask its expression. Most known cases involve alleles that are recessive in females.

## Module 27.17

### Many clinical disorders are linked to individual chromosomes or their genes    p. 1066

79. Chromosome abnormalities and **single nucleotide polymorphisms** (**SNPs**) have been linked with various disorders and diseases.

80. **Trisomy 21**, or **Down syndrome**, is the most common viable chromosomal abnormality. Affected individuals exhibit intellectual disability and physical malformations, including a characteristic facial appearance.

81. Individuals with **Klinefelter syndrome** have the sex chromosome pattern XXY. The phenotype is male, but the extra X chromosome causes decreased androgen production, resulting in immature testes, sterility, and enlarged breasts.

82. Individuals with **Turner syndrome** have only a single, female sex chromosome, so their sex chromosome complement is designated XO (**monosomy**). Maturational changes do not appear at puberty; effects include nonfunctional ovaries and negligible estrogen production.

83. The full set of genetic material (DNA) in our chromosomes is called the **human genome** and contains an estimated 20,000–25,000 protein-coding genes. Over 900 single gene disorders have been mapped on the genome.

84. **Epigenetics** is the study of inherited traits that do not involve a change in a person's genotype or DNA sequences.

# Chapter Review Questions

---

**True/False:** Indicate whether each statement is true or false.

**1** Fertilization typically occurs within the uterus.

**2** In the 2 weeks after its development, the yolk sac is the primary nutrient source for the inner cell mass.

**3** Dizygotic twins are also called identical twins.

**4** Relaxin is a hormone released by the placenta as well as the corpus luteum.

**5** There is one umbilical artery and two umbilical veins.

**6** The four extra-embryonic membranes are the yolk sac, amnion, allantois, and chorion.

**7** The structure that implants in the uterine endometrium is an advanced morula.

**8** The health risks associated with oral contraceptives are far greater than the risks associated with pregnancy.

| | |
|---|---|
| **1** | _____ |
| **2** | _____ |
| **3** | _____ |
| **4** | _____ |
| **5** | _____ |
| **6** | _____ |
| **7** | _____ |
| **8** | _____ |

---

**Multiple choice:** Select the correct answer from the list provided.

**9** The hormone detected by pregnancy kits is
- ☐ a) relaxin.
- ☐ b) human chorionic gonadotropin (hCG).
- ☐ c) human placental lactogen (hPL).
- ☐ d) gonadotropin-releasing hormone (GnRH).

**10** X-linked recessive traits
- ☐ a) are more likely to be expressed in females.
- ☐ b) are more likely to be expressed in males.
- ☐ c) are never expressed in females.
- ☐ d) are never expressed in males.

**11** Amphimixis is when

- [ ] a) the blastocyst implants into the uterine endometrium.
- [ ] b) a single sperm contacts the oocyte membrane.
- [ ] c) the male pronucleus develops and spindle fibers appear in preparation for cell division.
- [ ] d) the male and female pronuclei fuse.

**12** For a given trait $T$, how would a homozygous dominant genotype be designated?

- [ ] a) $TT$
- [ ] b) $Tt$
- [ ] c) $tt$
- [ ] d) $tT$

**13** If an allele must be present on both the maternal and paternal chromosomes to affect the phenotype, the allele is said to be

- [ ] a) dominant.
- [ ] b) recessive.
- [ ] c) heterozygous.
- [ ] d) polygenic.

**14** Which of the following sex chromosome patterns represents Klinefelter syndrome?

- [ ] a) YO
- [ ] b) XO
- [ ] c) YYX
- [ ] d) XXY

**15** The hormone responsible for milk ejection is

- [ ] a) prolactin.
- [ ] b) estrogen.
- [ ] c) oxytocin.
- [ ] d) progesterone.

**16** The phases of labor in the correct order are

- [ ] a) dilation, placental, expulsion.
- [ ] b) placental, dilation, expulsion.
- [ ] c) placental, expulsion, dilation.
- [ ] d) dilation, expulsion, placental.

## Short answer

**17** Identify the five life stages of postnatal development and the approximate ages for each stage.

**18** What is trisomy 21, and what are the risk factors for having a baby with this condition?

---

# MasteringA&P™

Access more chapter study tools online in the MasteringA&P Study Area:

- Chapter Quizzes, Chapter Practice Test, and Clinical Case Studies

- Practice Anatomy Lab™ 3.0    PAL 3.0
- Interactive Physiology™    iP2™
- A&P Flix™    A&PFlix™
- PhysioEx™ 9.1    PhysioEx 9.1

---

# Chapter Integration • Applying what you have learned

## The blocks that build a family

Joe and Jane have not been successful in having a child although they have been trying for 2 years. They finally consult a physician and undergo complete physical exams and genetic testing. They discover that Joe's genotype is a normal XY, but he suffers from *oligospermia* (low sperm count). Jane is a carrier (XX$^h$) for hemophilia A, an X-linked recessive trait in which the blood does not clot properly. The doctor informs them that Joe's low sperm count is interfering with their ability to have children and that there is a risk that any child who is born will have hemophilia A.

For most couples, the only sign that the male has a low sperm count is the inability to conceive a child after 1 year of regular intercourse. In most cases, the cause is unknown, but other reasons include damaged sperm ducts, infections, anti-sperm antibodies, or Klinefelter syndrome, which results in a complete lack of sperm in the semen (azoospermia).

Many people with hemophilia maintain active, productive lives.

Signs and symptoms vary and include unexplained bruising, nosebleeds with unknown cause, and painful headaches. Hemophilia A is the most common type of hemophilia and is caused by insufficient amounts of clotting factor VIII.

**1** Since it takes only one sperm to fertilize a secondary oocyte, why is Joe's oligospermia likely to make it difficult for the couple to conceive?

**2** Regarding hemophilia A, what is the probability that the couple will have daughters with hemophilia? What is the probability that the couple will have sons with hemophilia?

## Appendix A Periodic Table of Elements

The **periodic table** presents the known elements in order of their atomic weights. Each horizontal row represents a single electron shell. The number of elements in that row is determined by the maximum number of electrons that can be stored at that energy level. The element at the left end of each row contains a single electron in its outermost electron shell; the element at the right end of the row has a filled outer electron shell. Organizing the elements in this fashion highlights similarities that reflect the composition of the outer electron shell. These similarities are evident when you examine the vertical columns. All the gases of the right-most column—helium, neon, argon, krypton, xenon, and radon—have full electron shells; each is a gas

at normal atmospheric temperature and pressure, and none reacts readily with other elements. These elements, highlighted in blue, are known as the *noble*, or *inert, gases*. In contrast, the elements of the left-most column below hydrogen—lithium, sodium, potassium, rubidium, caesium, and francium—are silvery, soft metals that are so highly reactive that pure forms cannot be found in nature. The fourth and fifth electron levels can hold up to 18 electrons. Table inserts are used for the so-called *lanthanoids* (atomic numbers 57–71) and *actinoids* (atomic numbers 89–103) to save space, as higher levels can store up to 32 electrons. Elements of particular importance to our discussion of human anatomy and physiology are highlighted in pink.

Periodic Table of the Elements:

| 1 H Hydrogen 1.01 | | | | | | | | | | | | 5 B Boron 10.81 | 6 C Carbon 12.01 | 7 N Nitrogen 14.01 | 8 O Oxygen 16.00 | 9 F Fluorine 19.00 | 2 He Helium 4.00 |
|---|---|---|---|---|---|---|---|---|---|---|---|---|---|---|---|---|---|

Key:
Atomic number — 1 H — Chemical symbol
Hydrogen — Element name
Atomic weight — 1.01

\*pending formal approval

# Appendix B Normal Physiological Values

Tables 1 and 2 present normal averages or ranges for the chemical composition of various body fluids. These values are approximations rather than absolute values, because test results vary from laboratory to laboratory due to differences in procedures, equipment, normal solutions, and so forth. Blanks in the tabular data appear where data are not available. The following modules in the text contain additional information about body fluid analysis:

- Module 17.2 (p. 632) discusses the formed elements of whole blood.

- Module 17.4 (p. 637) provides an example of a complete blood count (CBC) and examples of red blood cell (RBC) tests.
- Module 17.9 (p. 646) presents data on the white blood cells (WBCs) in whole blood.
- Module 24.6 (p. 943) compares the normal values for solutes in blood plasma and urine.
- Module 24.12 (p. 955) lists the general characteristics of normal urine.
- Module 24.14 (p. 958) compares the composition of plasma and dialysis fluid.

## Table 1   The Composition of Body Fluids

| Test | Normal Averages or Ranges | | | | | |
|---|---|---|---|---|---|---|
| | **Perilymph** | **Endolymph** | **Synovial Fluid** | **Sweat** | **Saliva** | **Semen** |
| **pH** | | | 7.4 | 4–6.8 | 6.4* | 7.19 |
| **Specific gravity** | | | 1.008–1.015 | 1.001–1.008 | 1.007 | 1.028 |
| **Electrolytes** (mEq/L) | | | | | | |
| Potassium | 5.5–6.3 | 140–160 | 4.0 | 4.3–14.2 | 21 | 31.3 |
| Sodium | 143–150 | 12–16 | 136.1 | 0–104 | 14* | 117 |
| Calcium | 1.3–1.6 | 0.05 | 2.3–4.7 | 0.2–6 | 3 | 12.4 |
| Magnesium | 1.7 | 0.02 | | 0.03–4 | 0.6 | 11.5 |
| Bicarbonate | 17.8–18.6 | 20.4–21.4 | 19.3–30.6 | | 6* | 24 |
| Chloride | 121.5 | 107.1 | 107.1 | 34.3 | 17 | 42.8 |
| **Proteins** (total) (mg/dL) | 200 | 150 | 1.72 g/dL | 7.7 | 386† | 4.5 g/dL |
| **Metabolites** (mg/dL) | | | | | | |
| Amino acids | | | | 47.6 | 40 | 1.26 g/dL |
| Glucose | 104 | | 70–110 | 3.0 | 11 | 224 (fructose) |
| Urea | | | | 26–122 | 20 | 72 |
| Lipids (total) | 12 | | 20.9 | ‡ | 25–500§ | 188 |

*Increases under salivary stimulation.
†Primarily alpha-amylase, with some lysozymes.
‡Not present in eccrine (merocrine) secretions.
§Cholesterol.

## Table 2 The Chemistry of Blood, Cerebrospinal Fluid (CSF), and Urine

| Test | Blood* | CSF | Urine |
|---|---|---|---|
| | | **Normal Averages or Ranges** | |
| **pH** | S: 7.35–7.45 | 7.31–7.34 | 4.5–8.0 |
| **Osmolarity (mOsm/L)** | S: 280–295 | 292–297 | 855–1335 |
| **Electrolytes** | ├————————(mEq/L unless noted)————————┤ | | (urinary loss, mEq per 24-hour period[†]) |
| Bicarbonate | P: 20–28 | 20–24 | 0 |
| Calcium | S: 8.5–10.3 | 2.1–3.0 | 6.5–16.5 |
| Chloride | P: 97–107 | 100–108 | 110–250 |
| Iron | S: 50–150 µg/L | 23–52 µg/L | 40–150 µg |
| Magnesium | S: 1.4–2.1 | 2–2.5 | 6.0–10.0 |
| Phosphorus | S: 1.8–2.9 | 1.2–2.0 | 0.4–1.3 g |
| Potassium | P: 3.5–5.0 | 2.7–3.9 | 25–125 |
| Sodium | P: 135–145 | 137–145 | 40–220 |
| Sulfate | S: 0.2–1.3 | | 1.07–1.3 g |
| **Metabolites** | ├————————(mg/dL unless noted)————————┤ | | (urinary loss, mg per 24-hour period[‡]) |
| Amino acids | P/S: 2.3–5.0 | 10.0–14.7 | 41–133 |
| Ammonia | P: 20–150 µg/dL | 25–80 µg/dL | 340–1200 |
| Bilirubin | S: 0.5–1.0 | < 0.2 | 0 |
| Creatinine | P/S: 0.6–1.5 | 0.5–1.9 | 770–1800 |
| Glucose | P/S: 70–110 | 40–70 | 0 |
| Ketone bodies | S: 0.3–2.0 | 1.3–1.6 | 10–100 |
| Lactic acid | WB: 0.7–2.5 mEq/L[§] | 10–20 | 100–600 |
| Lipids (total) | P: 450–1000 | 0.8–1.7 | 0.002 |
| Cholesterol (total) | S: 150–300 | 0.2–0.8 | 1.2–3.8 |
| Triglycerides | S: 40–150 | 0–0.9 | 0 |
| Urea | P: 8–25 | 12.0 | 1800 |
| Uric acid | P: 2.0–6.0 | 0.2–1.5 | 250–750 |
| **Proteins** | (g/dL) | (mg/dL) | (urinary loss, mg per 24-hour period[‡]) |
| Total | P: 6.0–8.0 | 2.0–4.5 | 0–8 |
| Albumin | S: 3.2–4.5 | 10.6–32.4 | 0–3.5 |
| Globulins (total) | S: 2.3–3.5 | 2.8–15.5 | 7.3 |
| Immunoglobulins | S: 1.0–2.2 | 1.1–1.7 | 3.1 |
| Fibrinogen | P: 0.2–0.4 | 0.65 | 0 |

*S = serum, P = plasma, WB = whole blood.
[†]Because urinary output averages just over 1 liter per day, these electrolyte values are comparable to mEq/L.
[‡]Because urinary metabolite and protein data approximate mg/L or g/L, these data must be divided by 10 for comparison with CSF or blood concentrations.
[§]Venous blood sample.

## Appendix C Codon Chart

A codon is a sequence of three consecutive nucleotides in mRNA that codes for a particular amino acid or signals to stop protein synthesis. Because each mRNA codon consists of three nucleotides, the four nucleotides in mRNA (A, U, G, and C) can produce 64 different combinations. Of these, 61 codons correspond to amino acids and three act as stop signals of protein synthesis. There are only 20 different amino acids, so most amino acids are represented by more than one codon. One codon, AUG, has a dual role. It codes for the amino acid methionine and also as the start signal for protein synthesis. It is always the first codon in a strand of mRNA.

### Genetic Code (mRNA codons)

| First nucleotide | Second nucleotide: U | Second nucleotide: C | Second nucleotide: A | Second nucleotide: G | Third nucleotide |
|---|---|---|---|---|---|
| U | UUU, UUC Phenylalanine; UUA, UUG Leucine | UCU, UCC, UCA, UCG Serine | UAU, UAC Tyrosine; UAA Stop; UAG Stop | UGU, UGC Cysteine; UGA Stop; UGG Tryptophan | U, C, A, G |
| C | CUU, CUC, CUA, CUG Leucine | CCU, CCC, CCA, CCG Proline | CAU, CAC Histidine; CAA, CAG Glutamine | CGU, CGC, CGA, CGG Arginine | U, C, A, G |
| A | AUU, AUC Isoleucine; AUA; AUG Start or Met* | ACU, ACC, ACA, ACG Threonine | AAU, AAC Asparagine; AAA, AAG Lysine | AGU, AGC Serine; AGA, AGG Arginine | U, C, A, G |
| G | GUU, GUC, GUA, GUG Valine | GCU, GCC, GCA, GCG Alanine | GAU, GAC Aspartic acid; GAA, GAG Glutamic acid | GGU, GGC, GGA, GGG Glycine | U, C, A, G |

*Abbreviation for methionine.

# Answers

## CHAPTER 1

### Module Reviews

#### Module 1.1 Review

**A.** *Human A&P* is shorthand for "human antomy and physiology." *Anatomy* refers to structure, and *physiology* refers to function. **B.** Biology is the study of life.

#### Module 1.2 Review

**A.** Eye-tracking studies tell us that the most effective way to learn is to read the text and the art together. **B.** Chances are you discovered how important the text-art integration is to understanding what was described.

#### Module 1.3 Review

**A.** 10,240 square centimeters (cm²)

#### Module 1.4 Review

**A.** The bracelet is found on the woman's right wrist. **B.** The X-marked structure is the left kidney.

#### Module 1.5 Review

**A.** A module is an independent, self-contained unit about a specific topic. Learning outcomes associated with each module appear in the lower right-hand corner of each module. **B.** A learning classification scheme identifies the fundamental levels of learning from lower-order thinking skills to higher-order skills. Common terms associated with levels of learning are *knowing*, *remembering*, *comprehending*, *understanding*, *applying*, *describing*, *analyzing*, *explaining*, *evaluating*, *measuring*, *synthesizing*, and *creating*.

### Section 1 Review

**1.** d **2.** b **3.** e **4.** a **5.** c **6.** the art and text together **7.** key **8.** anatomical position **9.** beginning of each chapter **10.** knowing, remembering **11.** creating, comprehending, analyzing, applying, evaluating **12.** This A&P textbook uses a text-art integration design, thus combining key information in the text and art. Because the information presented in the text is right next to the associated figure, you never have to flip pages to search for the figure.

#### Module 1.6 Review

**A.** "Black Box" refers to missing knowledge. The more you learn about the body, the more you will realize how many other things you don't know. **B.** Several strategies for success in this course include approaching the information in different ways that may include creating your own diagrams, "teaching" fellow students, and asking questions in class; having a set study schedule with a devoted block of time; mastering memorization; avoiding shortcuts; regularly attending and participating in all class sessions; reading assigned material prior to class; avoiding procrastination; and asking for help immediately if you are having any problems.

#### Module 1.7 Review

**A.** The common characteristics shared by all living things are cells, organization, responsiveness, regulation, growth, development, reproduction, and metabolism. **B.** *Growth* refers to an organism's increase in size. *Development* refers to changes in structures and functions as the organism grows. **C.** The basic processes in humans and other animals are respiration, digestion, circulation, and excretion.

#### Module 1.8 Review

**A.** Gross anatomy (also called macroscopic anatomy) is the study of body structures that can be seen with the unaided eye; microscopic anatomy is the study of body structures that cannot be seen without magnification. **B.** Anatomy is the study of internal and external body structures; physiology is the study of how living organisms perform their vital functions.

#### Module 1.9 Review

**A.** The elbow joint moves in a single plane, like the opening and closing of a door on a hinge. **B.** If a structure's anatomy were altered, the structure's function would likely be impaired or perhaps eliminated. **C.** At both the gross anatomical and chemical levels, structure and function are interrelated. The hinge-like movement at the elbow joint depends on the three-dimensional shapes of the joint-forming ends of the humerus and ulna. The opening of a membrane passageway depends on the proper three-dimensional shapes of a chemical messenger and the cell's membrane receptor molecule.

### Section 2 Review

**1.** respiration **2.** reproduction **3.** regulation **4.** circulation **5.** excretion **6.** digestion **7.** metabolism **8.** responsiveness **9.** Anatomy: right atrium, myocardium, left ventricle, endocardium, superior vena cava **10.** Physiology: valve to aorta opens, valve between left atrium and left ventricle closes, pressure in left atrium, electrocardiogram, heartbeat **11.** Both keys and messenger molecules have specific three-dimensional shapes, so both can "fit" into and function with their complementary structures (a lock or a receptor protein) only if the shapes match closely enough. Because messenger molecules need not be flat in two dimensions (as keys tend to be), they can be very complex in shape and thus may be able to bind with a variety of complementary receptor proteins. Moreover, though a lock has moving parts, it does not actually change shape, whereas a receptor protein bound by a messenger can change shape, which in turn affects its function. **12.** Larger organisms simply cannot absorb the amount of needed materials or excrete wastes rapidly enough across their body surfaces as very small organisms can. Their food must be processed, or digested, before being absorbed, and because the processes of absorption, respiration, and excretion occur in different parts of the organism, an efficient internal circulation network for their materials is necessary for survival.

#### Module 1.10 Review

**A.** The cellular level is the simplest level of organization that includes the smallest living units in the body. **B.** An organ is two or more tissues working to perform several functions. **C.** Knowledge about each level of organization contributes to our total understanding of how the body works at the organism level.

#### Module 1.11 Review

**A.** The unit used to measure cell size is the micrometer (μm), which is equal to one-millionth of a meter. **B.** The cell theory holds that (1) cells are the structural building blocks of all plants and animals, (2) new cells are produced through the division of preexisting cells, and (3) cells are the smallest structural units that perform all vital functions. **C.** The oocyte has a diameter about 11 times greater than the red blood cell, or about 88 μm (.088 mm).

#### Module 1.12 Review

**A.** Skeletal muscle moves the bones of the skeleton, cardiac muscle pumps blood through blood vessels, and smooth muscle moves materials along internal passageways. **B.** The body's four primary tissue types that form all body structures are epithelial, connective, muscle, and nervous tissue. Epithelial tissue covers external and internal surfaces and produces secretions; connective tissue fills internal spaces, provides support, and stores energy; muscle tissue is specialized to contract and produce movement; and nervous tissue transmits information.

#### Module 1.13 Review

**A.** The cardiovascular system transports cells and materials, and the lymphatic system returns tissue fluid to the bloodstream. **B.** Falling down a flight of stairs could cause a compound fracture (a broken bone that protrudes through the skin), which could affect (1) the skeletal system (a broken bone), (2) the integumentary system (disruption of skin integrity), (3) the muscular system (broken bone tearing through a muscle), (4) the cardiovascular system (blood loss at the site of injury), (5) the lymphatic system (mobilization of specialized cells to defend against infection), and (6) the nervous system (pain and nerve injury as a result of the trauma).

### Module 1.14 Review
**A.** Your answer may differ but could include the following examples of how a disorder of the nervous system affects the muscular system: A spinal cord injury can cause a person to lose mobility, or a nervous system disorder such as Parkinson's disease can cause a person to experience muscle tremors.

### Module 1.15 Review
**A.** Your answer may differ but could include the following example of how a respiratory system disease affects the cardiovascular system: A patient has an asthma attack and experiences difficulty in breathing; as a result, the blood cannot be fully oxygenated.

### Module 1.16 Review
**A.** Your answer may differ but could include the following example of how a reproductive system disorder could affect the urinary system: A sexually transmitted disease that is acquired through the urethra could affect the urinary system if the bacterium travels up to the urinary bladder and kidneys, causing a more serious infection.

### Section 3 Review
**1.** cells **2.** organs **3.** organ systems **4.** epithelial tissue **5.** external and internal surfaces **6.** glandular secretions **7.** connective tissue **8.** matrix **9.** protein fibers **10.** ground substance **11.** muscle tissue **12.** movement **13.** bones of the skeleton **14.** blood **15.** materials within digestive tract **16.** nervous tissue **17.** neuroglia **18.** In order from simplest to most complex, the levels of organization are: chemical, cellular, tissue, organ, organ system, and organism. **19.** integumentary: protection from environmental hazards, temperature control **20.** skeletal: support, protection of soft tissues, mineral storage, blood formation **21.** muscular: provides skeletal movement, guards entrances and exits to body, produces heat, supports skeleton, protects soft tissues **22.** nervous: directs immediate responses to stimuli, usually by coordinating the activities of other organ systems **23.** endocrine: controls and regulates growth, development, day-night rhythms, fluid balance, calcium, glucose, and water balance, blood cell production, metabolic rate **24.** cardiovascular: propels and distributes blood, maintains blood pressure, transports gases and nutrients, eliminates wastes, assists in temperature regulation and defense against disease **25.** lymphatic: transports lymph and lymphocytes, controls immune response and immune cells **26.** respiratory: filters, warms, and humidifies air; detects smell; conducts air into body; gas exchange between air and blood **27.** digestive: processing of food and absorption of nutrients, minerals, vitamins, and water **28.** urinary: elimination of excess water, salts, and waste products; control of pH **29.** reproductive: produces sperm or oocytes; produces hormones; provides the necessary structure and function for sexual intercourse, fetal growth and development, and nourishment of the newborn **30.** cardiovascular system: blood cells; digestive system: smooth muscle cell; reproductive system: sperm or oocyte; skeletal system: bone cell (osteocyte); nervous system: nerve cell (neuron). (Other cells could be listed for other systems.)

### Module 1.17 Review
**A.** Homeostatic regulation is important to an organism because failure to maintain homeostasis soon leads to illness or death. **B.** The three parts necessary for homeostatic regulation are a receptor or sensor, a control or integration center, and an effector. **C.** Homeostatic control is not precise, and normal ranges are not fixed with absolute values.

### Module 1.18 Review
**A.** Negative feedback systems maintain homeostasis (and provide long-term control over the body's internal conditions and systems) by counteracting any stimulus that moves conditions outside their normal range. An example of homeostatic regulation by negative feedback in the body is temperature regulation. When temperature receptors in the skin and brain detect a rise in body temperature, a negative feedback regulation is initiated. The control centers in the brain receive this input and send commands to effectors, causing skin blood vessels to dilate and sweat glands to secrete. The result is that the body is cooled and homeostasis is achieved. **B.** Positive feedback is useful in processes that, once begun, must move quickly to completion—for example, blood clotting. It is harmful in situations where stable conditions must be maintained because it tends to exaggerate any departure from the desired condition. Thus, positive feedback in the regulation of body temperature would cause a slight fever to spiral out of control, with fatal results.

### Section 4 Review
**1.** positive feedback **2.** homeostatic regulation **3.** homeostasis **4.** receptor or sensor **5.** negative feedback **6.** negative feedback **7.** positive feedback **8.** negative feedback **9.** positive feedback **10.** Blood flow to skin increases, sweating increases, body surface cools, body temperature decreases **11.** Blood flow to skin decreases, shivering occurs, body heat is conserved, body temperature increases **12.** One reason your body temperature may have dropped is that your body may be losing heat faster than it is being produced. (This is more likely to occur on a cool day.) Perhaps hormones have caused a decrease in your metabolic rate, so your body is not producing as much heat as it normally would. Or you may have an infection that has temporarily reset the set point of the body's "thermostat" to a value higher than normal. The last possibility is the most likely explanation given the circumstances.

### Module 1.19 Review
**A.** Greek and Latin are the source of many modern anatomical terms. **B.** An eponym is a term—in our context, a term denoting an anatomical structure or clinical condition—that is named after a person. **C.** Cadaver-based anatomy was established as a discipline studied by medical professionals in Italy at the University of Bologna.

### Module 1.20 Review
**A.** A person in the anatomical position is standing erect, facing the observer, arms at the sides with the palms facing forward, and the feet together. **B.** The anatomical term describing a person lying face down is *prone*. **C.** Clinicians base their descriptions on four abdominopelvic quadrants (determined by the intersection of two imaginary perpendicular lines that cross at the navel, or umbilicus), whereas anatomists use nine abdominopelvic regions. **D.** The purpose of anatomical terms is to provide a standardized frame of reference for describing the human body.

### Module 1.21 Review
**A.** In the anatomical position, an anterior view shows the subject's face, whereas a posterior view shows the subject's back. **B.** (1) sagittal (or midsagittal, if along midline); (2) frontal; (3) transverse, or horizontal **C.** The purpose of directional and sectional terms is to provide a standardized frame of reference for describing the human body.

### Module 1.22 Review
**A.** True body cavities are (1) lined by a serous membrane and (2) share a common embryonic origin. Two essential functions are (1) protecting internal organs and cushioning them from shocks that occur during activity and (2) allowing organs they enclose to change size and shape without disrupting the activities of nearby organs. **B.** The body cavities of the trunk are the thoracic cavity (which contains the pleural and pericardial cavities) and the abdominopelvic cavity (consisting of the peritoneal, abdominal, and

pelvic cavities). **C.** The incision would open the peritoneal portion of the abdominopelvic cavity, which is just inferior to the diaphragm.

### Section 5 Review

**1.** superior **2.** inferior **3.** cranial **4.** caudal **5.** posterior or dorsal **6.** anterior or ventral **7.** lateral **8.** medial **9.** proximal **10.** distal **11.** proximal **12.** distal **13.** thoracic cavity **14.** mediastinum **15.** left lung **16.** trachea, esophagus **17.** heart **18.** diaphragm **19.** abdominopelvic cavity **20.** peritoneal cavity **21.** digestive glands and organs **22.** pelvic cavity **23.** reproductive organs

### Chapter Review Questions

**1.** pericardial cavity encloses the heart **2.** right pleural cavity encloses the right lung **3.** left pleural cavity encloses the left lung **4.** diaphragm is a muscular sheet that separates the thoracic cavity and abdominopelvic cavity **5.** abdominal cavity contains many digestive glands and organs **6.** pelvic cavity contains urinary bladder, reproductive organs, and the last portion of the digestive tract **7.** c **8.** e **9.** k **10.** b **11.** j **12.** h **13.** a **14.** i **15.** n **16.** d **17.** f **18.** m **19.** g **20.** l **21.** b **22.** d **23.** b **24.** b **25.** c **26.** d **27.** c **28.** b **29.** a **30.** c **31.** b **32.** Anatomy is the study of structure. Physiology is the study of function. **33.** The three basic principles of the cell theory are (1) cells are the structural building blocks of all plants and animals; (2) cells are produced by the divisions of preexisting cells; and (3) cells are the smallest structural units that perform all vital functions. **34.** The heart is enclosed by the pericardial cavity; the small intestine is enclosed by the abdominal cavity; the large intestine is enclosed mostly by the abdominal cavity, with the last portion in the pelvic cavity; the lungs are enclosed by the pleural cavities; the kidneys are retroperitoneal. **35.** The four main tissue types are epithelial tissue, connective tissue, muscle tissue, and nervous tissue. Module 1.12 identifies a comprehensive list of where these tissues are found in

the body. **36.** Calcitonin is released when calcium levels are elevated. This hormone should bring about a decrease in blood calcium levels, thus decreasing the stimulus for its release. **37.** A stroke causes damage to the brain, the organ responsible for controlling both voluntary and autonomic activities. Therefore, a patient could lose control of voluntary activities such as walking or speech. Autonomic activities such as bladder control could also be lost.

### Chapter Integration

**1.** The bullet entered the anterior body approximately 2 centimeters inferior to the umbilicus, within the hypogastric (pubic) abdominopelvic region. **2.** Organs found within the abdominal cavity include the liver, stomach, spleen, small intestine, and most of the large intestine. The abdominal aorta and inferior vena cava are two large blood vessels situated posteriorly. If the bullet entered at this location, these structures may have been affected. **3.** It is likely that a massive internal hemorrhage led to this man's death. **4.** Aaron's endocrine system is failing. If he is not treated, long-term high blood glucose could cause damage to the circulatory system, resulting in high blood pressure, which could also damage the urinary system's kidneys and cause damage to the nervous system by depriving the retina of blood flow, causing blindness. **5.** Normally, cellular receptors would detect high levels of blood glucose. Control centers would stimulate the pancreas to secrete insulin. Insulin would have an effect on the tissues to absorb blood glucose. The failing mechanism is likely the response of the pancreas in that it is not producing insulin. **6.** Dietary interventions could include controlling Aaron's diet so that his insulin does not rise rapidly. This could be achieved by eating smaller meals more regularly. Medical interventions could include insulin injections when Aaron's blood glucose is high.

### CHAPTER 2

## Module Reviews

### Module 2.1 Review

**A.** An atom is the smallest stable unit of matter, and matter is anything that takes up space and has mass. **B.** Protons have a positive charge, and neutrons are electrically neutral. Electrons have a negative charge and are much smaller than either protons or neutrons. **C.** Electrons are subatomic particles not in the nucleus. They whirl around the nucleus, creating an electron cloud. **D.** On the moon, the astronaut would have the same mass as on Earth but would weigh 17 pounds ($100 \times 0.17$).

### Module 2.2 Review

**A.** Mass number is larger because it includes the total number of protons and neutrons. **B.** Isotope $^{13}C$ has the same chemical properties but has one more neutron than $^{12}C$. **C.** Isotopes are atoms with the same number of protons but different numbers of neutrons. Hydrogen has three isotopes: hydrogen-1, with a mass number of 1; deuterium, with a mass number of 2; and tritium, with a mass number of 3. The mass number is greater in each isotope because the atoms contain an increasing number of neutrons. The heavier hydrogen sample must contain a higher proportion of one or both of the heavier isotopes. **D.** Trace elements, found in very small amounts in the body, are chemical elements required for normal growth and maintenance. **E.** The six most abundant elements are O, C, H, N, Ca, and P. Together, they account for 99.3% of total body weight.

### Module 2.3 Review

**A.** The maximum number of electrons that can occupy an atom's first three energy levels is 2, 8, and 8, respectively. **B.** Atoms of inert elements are nonreactive because their outermost electron shell (valence shell) contains the maximum number of electrons possible. **C.** A cation is formed

when an atom loses one or more electrons from its outermost electron shell; it has an overall positive charge because it contains more protons than electrons. An anion is formed when an atom gains one or more electrons in its outermost electron shell; it has an overall negative charge because it contains more electrons than protons. **D.** Cations form when an atom loses one or more electrons. When an electron is lost, the repulsion between the remaining electrons decreases, and the protons pull them closer toward the nucleus.

### Module 2.4 Review

**A.** Table salt is made up of more than one element, both sodium and chlorine (as ions) in a fixed one-to-one proportion. **B.** Two pairs, or a total of 4 electrons, are shared in the double covalent bond of an oxygen molecule. **C.** The atoms in a water molecule are held together by polar covalent bonds, in which electrons are shared unequally. **D.** The term *molecule* refers only to chemical structures held together by covalent bonds. Table salt is an ionic compound whose components—sodium ions and chloride ions—are held together by ionic bonds.

### Module 2.5 Review

**A.** Solids have a fixed volume and shape, liquids have a constant volume but no fixed shape, and gases have neither a constant volume nor a fixed shape. **B.** Water molecules are attracted to each other by hydrogen bonds. **C.** The attraction between water molecules at the water's surface creates surface tension, which prevents small objects from penetrating the water. **D.** The low thermal energy of ice slows the vibration of water molecules, resulting in stable hydrogen bonds. Liquid water has more thermal energy than does ice, and hydrogen bonds are less stable as they break and re-form. At the boiling point, hydrogen bonds are broken, and the water molecules are independent of each other.

## Section 1 Review

**1.** f **2.** d **3.** k **4.** a **5.** l **6.** h **7.** i **8.** j
**9.** c **10.** e **11.** g **12.** b **13.** 2 **14.** 4
**15.** 1 **16.** 0 **17.** 6 **18.** 12 **19.** 7
**20.** 7 **21.** 20 **22.** 20 **23.** element
**24.** compound **25.** element
**26.** compound **27.** The outer
energy level of inert elements is
filled with electrons, whereas the
valence shell of reactive elements
is not filled with electrons. Atoms
with unfilled valence shells will
react with other atoms. Atoms
with filled outer energy levels
will not react with other atoms.
**28.** Both polar and nonpolar
molecules are held together by
covalent bonds. However, in a
polar molecule the electrons are
not shared equally, so it carries
small, or partial, positive and
negative charges on its surface;
in a nonpolar molecule, the
electrons are shared equally, so
it is electrically neutral. **29.** Both
covalent bonds and ionic bonds
bind atoms together, but covalent
bonds involve the sharing of
electrons between atoms, whereas
ionic bonds involve the electrical
attraction of oppositely charged
atoms (ions).

### Module 2.6 Review

**A.** Cells are chemical factories
because they use complex
chemical reactions to provide
the energy they need to maintain
homeostasis and to perform
essential functions. **B.** *Work* is
the movement of an object or a
change in the physical structure
of matter. *Energy* is the capacity
to perform work. *Potential
energy* is stored energy that
has the potential (capability)
to do work. *Kinetic energy* is
the energy of motion. **C.** Cells
do *work* when they synthesize
complex molecules and move
materials into and out of cells.
Muscle contraction requires
*energy*. Molecules inside muscle
cells store the *potential energy* of
contraction. The potential energy
of contraction is converted into
*kinetic energy* when a muscle
contracts.

### Module 2.7 Review

**A.** The participants at the start of
a chemical reaction are reactants;

they generate one or more
products. **B.** Chemical reactions
are represented by chemical
equations. An arrow indicates
the direction of the reaction.
**C.** Formula weight is the sum of
the atomic weights of the elements
making up an ionic compound.
**D.** The molecular formula for
glucose, a compound composed
of 6 carbon atoms, 12 hydrogen
atoms, and 6 oxygen atoms, is
$C_6H_{12}O_6$. **E.** The weight of 1 mol
of glucose is 180 grams. (Add the
atomic weights of 6 C = 72 g,
12 H = 12 g, and 6 O = 96 g.)

### Module 2.8 Review

**A.** In hydrolysis reactions,
water is a reactant, whereas in
dehydration synthesis reactions,
water is a product. **B.** Three
types of chemical reactions
important in human physiology
are (1) decomposition reactions,
in which a molecule is broken
down into smaller fragments;
(2) synthesis reactions, in which
small molecules are assembled
into larger ones; and (3) exchange
reactions, in which parts of the
reacting molecules are shuffled
around to produce new products.
**C.** The source of the released
energy is the potential energy
stored in the covalent bonds of
the glucose molecule. Energy
was released when some of the
covalent bonds were broken
during this catabolic reaction.

### Module 2.9 Review

**A.** An enzyme is a protein
that lowers the activation
energy, which is the amount
of energy required to start a
chemical reaction. **B.** Enzymes
promote chemical reactions by
lowering the activation energy
requirements and making it
possible for chemical reactions
to proceed under conditions
compatible with life. **C.** Exergonic
reactions generate heat that helps
maintain body temperature.
**D.** Metabolites are molecules that
can be synthesized or broken
down by chemical reactions inside
our bodies. Nutrients are essential
metabolites normally obtained
from the diet. **E.** Enzymes
are special protein catalysts.

Proteins are one class of organic
compounds.

## Section 2 Review

**1.** h **2.** g **3.** e **4.** f **5.** d **6.** c **7.** a
**8.** b **9.** $H_2$ **10.** 2 H **11.** 6 $H_2O$
**12.** $C_{12}H_{22}O_{11}$ **13.** $C_6H_{12}O_6 + 6 O_2$
$\rightarrow 6 CO_2 + 6 H_2O$ **14.** hydrolysis
reaction **15.** dehydration synthesis
reaction **16.** A decreased amount
of enzyme at the second step
would limit the amount of the
intermediate products in the next
two steps. This would cause a
decrease in the amount of the final
product.

### Module 2.10 Review

**A.** During exercise, water
lubricates the joints for easy
movement. It also functions
in dehydration synthesis and
hydrolysis reactions that occur
during muscle contraction. Water
also cools the body through the
evaporation of perspiration from
the skin. Water dissolves wastes
generated by exercise.

### Module 2.11 Review

**A.** Sodium chloride dissociates
(dissolves) in water as the
positive poles of water molecules
are attracted to the negatively
charged chloride ions, and the
negative poles of water molecules
are attracted to the positively
charged sodium ions. The
ions stay dissolved in solution
because a layer of surrounding
water molecules (a hydration
sphere) separates them from each
other. **B.** Electrolytes are soluble
inorganic molecules whose ions
will conduct an electrical current
in solution. **C.** Hydrophilic
molecules are attracted to water
molecules, whereas hydrophobic
molecules do not interact with
water molecules.

### Module 2.12 Review

**A.** pH is a measure of the
hydrogen ion concentration in a
solution, defined as the negative
logarithm of the hydrogen ion
concentration, expressed in
moles per liter. On the pH scale, 7
represents neutrality, values below
7 indicate acidic solutions, and
values above 7 indicate alkaline
(basic) solutions. **B.** A hydrogen
ion is a hydrogen atom that has

lost its electron, so it is a proton.
**C.** An acid is a compound whose
dissociation in solution releases a
hydrogen ion ($H^+$) and an anion;
a base is a compound whose
dissociation releases a hydroxide
ion ($OH^-$) into the solution or
removes a hydrogen ion ($H^+$) from
the solution; and a salt is an ionic
compound consisting of a cation
other than $H^+$ and an anion other
than $OH^-$. **D.** Buffers stabilize
body pH by removing or replacing
hydrogen ions. The pH of various
body fluids must remain relatively
constant if the body is to maintain
homeostasis and remain healthy.

## Section 3 Review

**1.** e **2.** h **3.** i **4.** d **5.** a **6.** j **7.** c **8.** g
**9.** f **10.** b **11.** effective lubrication
(as between bony surfaces in a
joint) **12.** reactivity (participates
in chemical reactions) **13.** high
heat capacity (readily absorbs
and retains heat) **14.** solubility (is
a solvent for many substances)
**15.** acidic **16.** neutral **17.** alkaline
**18.** The pH 3 solution is 1000
times more acidic than the pH
6 solution—it contains a one-
thousandfold ($10^3$) increase in the
concentration of hydrogen ions
($H^+$). **19.** Three negative effects
of abnormal fluctuations in pH
are cell and tissue damage (due
to broken bonds), changes in the
shapes of proteins, and altered
cellular functions.
**20.** Table salt dissociates (dissolves)
in pure water, but because it does
not release either hydrogen ions
($H^+$) or hydroxide ions ($OH^-$), no
change in pH occurs.

### Module 2.13 Review

**A.** Organic compounds always
contain the elements carbon and
hydrogen, and generally oxygen
as well. **B.** A functional group is
a grouping of atoms that confer
specific chemical properties to
the rest of a molecule to which
it is attached. **C.** Amino groups
are the functional groups of
amino acids. Carboxyl groups are
functional groups of fatty acids
and amino acids. Hydroxyl groups
are the functional groups that
link molecules by dehydration
synthesis and affect a molecule's
solubility. Phosphate groups

are functional groups found in nucleic acids and high-energy compounds. **D.** An amino group acts as a base because it can accept hydrogen ions. A carboxyl group acts as an acid because it releases a hydrogen ion.

### Module 2.14 Review

**A.** Carbohydrates are important as an energy source. **B.** The ring form of glucose is more common. **C.** The three structural classes of carbohydrates are monosaccharides (glucose), disaccharides (sucrose), and polysaccharides (starch). **D.** Muscle cells make (synthesize) glycogen by linking numerous glucose molecules in a series of dehydration synthesis reactions.

### Module 2.15 Review

**A.** Lipids are a diverse group of water-insoluble organic compounds that contain carbon, hydrogen, and oxygen in a ratio that does not approximate 1:2:1 (they contain less oxygen than do carbohydrates). **B.** Examples of lipids found in the body are fatty acids, glycerides, eicosanoids, steroids, phospholipids, and glycolipids. **C.** All fatty acids consist of a hydrocarbon chain and a carboxyl group. In saturated fatty acids, each carbon atom in the hydrocarbon chain has four single covalent bonds that bind the maximum number of hydrogen atoms possible. In unsaturated fatty acids, one or more of the carbon atoms in the hydrocarbon chain has double covalent bonds, so fewer hydrogen atoms are bonded. **D.** In the hydrolysis of a triglyceride, the reactants are a triglyceride and three water molecules. The products are a glycerol molecule and three fatty acids. **E.** Lipids are both a source of energy and a means of energy storage and insulation and physical protection. Some lipids act as chemical messengers between cells. Lipids are essential components of plasma membranes.

### Module 2.16 Review

**A.** Prostaglandins are one type of the eicosanoid lipids derived from arachidonic acid (a fatty acid).

Prostaglandins are released by cells to coordinate local cellular activities. They also produce pain sensations when released by damaged tissues. **B.** Cholesterol is a component of plasma membranes and is important for cell growth and division. **C.** When phospholipids and glycolipids form a micelle, the hydrophobic tails of both molecules are inside the micelle, and their hydrophilic heads form the surface of the micelle. **D.** Cholesterol, phospholipids, and glycolipids are structural lipids that form membranes of cells.

### Module 2.17 Review

**A.** Proteins are organic compounds formed from amino acids. Each amino acid contains a carbon atom, a hydrogen atom, an amino group ($-NH_2$), a carboxyl group ($-COOH$), and a variable group, known as an R group or side chain. **B.** During the dehydration synthesis of two amino acids, a peptide bond links the amino group of one amino acid with the carboxyl group of the other amino acid. **C.** The heat of boiling breaks bonds that maintain the protein's tertiary structure, quaternary structure, or both. The resulting change in shape affects the ability of the protein molecule to perform its normal biological functions. These changes are known as denaturation.

### Module 2.18 Review

**A.** The reactants in an enzymatic reaction are called substrates. **B.** An enzyme's specificity results from the unique shape of its active site, which permits only a substrate with a complementary shape to bind.

### Module 2.19 Review

**A.** ATP (adenosine triphosphate) is a compound consisting of adenosine to which three phosphate groups are attached; high-energy bonds attach the second and third phosphates. **B.** Adenosine monophosphate (AMP) is a nucleotide consisting of adenosine plus a phosphate group ($PO_4^{3-}$), and adenosine diphosphate (ADP) is a compound consisting of adenosine with two

phosphate groups attached. Adding a phosphate group to AMP creates ADP. Breaking one phosphate linkage of ATP to form ADP provides the primary source of energy for physiological processes. **C.** Cells obtain energy from the high-energy bonds of compounds such as ATP. **D.** The products of ATP hydrolysis are ADP, a phosphate group, and energy.

### Module 2.20 Review

**A.** Nucleic acids are large organic molecules composed of carbon, hydrogen, oxygen, nitrogen, and phosphorus that regulate the synthesis of proteins and make up the genetic material in cells. The two classes of nucleic acids are ribonucleic acid (RNA) and deoxyribonucleic acid (DNA). **B.** The complementary strands of DNA are held together through complementary base pairing between adenine and thymine, and between guanine and cytosine. **C.** DNA is composed of a pair of nucleotide chains, and RNA is made up of a single chain. The nitrogenous bases adenine, guanine, and cytosine are common to both DNA and RNA nucleotides. The nitrogenous base thymine is found only in DNA, and uracil is found only in RNA.

### Section 4 Review

**1.** carbohydrates
**2.** polysaccharides **3.** disaccharides
**4.** monosaccharides **5.** lipids
**6.** fatty acids **7.** glycerol **8.** proteins
**9.** amino acids **10.** nucleic acids
**11.** RNA **12.** DNA **13.** nucleotides
**14.** ATP **15.** phosphate groups **16.** polysaccharide, polyunsaturated, polypeptide
**17.** triglyceride **18.** disaccharide, diglyceride, dipeptide
**19.** glycogen, glycolipids **20.** g
**21.** f **22.** c **23.** h **24.** d **25.** e **26.** b
**27.** j **28.** k **29.** i **30.** a

### Chapter Review Questions

**1.** c **2.** c **3.** b **4.** e **5.** c **6.** a **7.** c
**8.** b **9.** d **10.** a **11.** b **12.** d **13.** c
**14.** b **15.** d **16.** a **17.** b **18.** d
**19.** protons, neutrons, and electrons **20.** carbohydrates, lipids, proteins, and nucleic acids **21.** Lipids perform many

functions in the body, including energy reserve, insulation and heat conservation, organ protection, chemical messengers, structural components of plasma (cell) membranes, and digestive secretions. **22.** Proteins are long chains of amino acids bound together by peptide bonds. Progressive folding and pleating of the chain into more complex shapes modifies the primary linear arrangement of amino acids. The increasing levels of complexity are primary, secondary, tertiary, and quaternary structures. **23. (a)** DNA: deoxyribose, phosphate, and nitrogenous bases (A, T, C, G); **(b)** RNA: ribose, phosphate, and nitrogenous bases (A, U, C, G) **24.** Enzymes are specialized protein catalysts that lower the activation energy for chemical reactions. Enzymes speed up chemical reactions but are not used up or changed in the process. **25.** A salt is an ionic compound consisting of any cations other than hydrogen ions and any anions other than hydroxide ions. Acids dissociate and release hydrogen ions, whereas bases remove hydrogen ions from solution (usually by releasing hydroxide ions). **26.** Nonpolar covalent bonds involve an equal sharing of electrons. Polar covalent bonds involve an unequal sharing of electrons. Ionic bonds result from the attraction of oppositely charged ions. (Ions are atoms that have either lost or gained electrons.) **27.** The molecule is a nucleic acid. Carbohydrates and lipids do not contain nitrogen. Although both proteins and nucleic acids contain nitrogen, only nucleic acids normally contain phosphorus. **28.** The insect can walk across the surface of the pond because its small mass is not sufficient to break the surface tension created by the hydrogen bonds between the water molecules. **29.** All of the foods and drink that the student consumed have a pH below 7 and are therefore acidic. These acidic foods, in conjunction with the hydrochloric acid in his stomach necessary for digestion, are likely the cause of his upset stomach. Consuming foods that are slightly alkaline could buffer his stomach contents and relieve his symptoms.

**30. (a)**

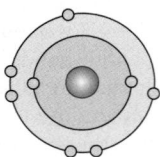

Oxygen atom

**(b)** Two more electrons can fit into the outermost energy level (valence shell) of an oxygen atom.

## Chapter Integration

**1.** Digestion involves breaking down molecules into smaller fragments; therefore, the *digestivo* would be promoting decomposition reactions. **2.** The decomposition reactions occurring in Claire's digestive tract are releasing some of the kinetic energy stored in the covalent bonds of the food she just consumed. This energy can be used to do work, but some of that energy is lost to heat, which could result in her feeling warmer and wanting to remove her jacket. **3.** Nutrients, the essential metabolites obtained from our diet, are digested by enzymes. If the *digestivo* is truly aiding in digestion, it would have to be rich in digestive enzymes. **4.** The polysaccharides would be broken down, or catabolized, into monosaccharides. Lipids such as triglycerides would be broken down into simpler molecules such as diglycerides, monoglycerides, fatty acids, and glycerol. The proteins would be broken down into amino acids.

# CHAPTER 3

## Module Reviews

### Module 3.1 Review

**A.** The cell theory states that cells are the building blocks of all plants and animals, all new cells come from the division of pre-existing cells, and cells are the smallest living units that carry out all vital physiological functions. **B.** All cells are descendants of the fertilized ovum, or zygote. **C.** Cellular differentiation is the development of specific cellular characteristics and functions that are different from the original cell. It results from specific genes in the cell being turned on or off.

### Module 3.2 Review

**A.** Cytoplasm is the material between the plasma membrane and the nuclear membrane; cytosol is the fluid portion of the cytoplasm. **B.** Membranous organelles and their functions are as follows: Endoplasmic reticulum (ER) synthesizes secretory products and participates in intracellular storage and transport; the nucleus controls metabolism, stores and processes genetic information, and controls protein synthesis; rough ER modifies and packages newly synthesized proteins; smooth ER synthesizes lipids and carbohydrates; Golgi apparatus stores, alters, and packages secretory products and lysosomal enzymes; lysosomes remove damaged organelles or intracellular pathogens; mitochondria produce 95 percent of the ATP required by the cell; and peroxisomes neutralize toxic compounds. **C.** The cytoskeleton strengthens and supports the cells and enables movement of cellular structures and materials. **D.** Our cells are surrounded by a watery extracellular fluid. The extracellular fluid in most tissues is called interstitial fluid.

### Module 3.3 Review

**A.** The phospholipid bilayer of the plasma membrane is mostly responsible for isolating a cell from its external environment. **B.** The general functions of the plasma membrane are physical isolation of the cell from its environment, regulation of exchange with the environment, sensitivity to the environment, and structural support. **C.** Channel proteins are integral proteins that allow water and small ions to pass through the plasma membrane. **D.** Phospholipid molecules are amphipathic. Such molecules have a hydrophilic ("water-loving") end and a hydrophobic ("water-fearing") end. As a result, the hydrophilic ends are attracted to the watery extracellular and intracellular environments, whereas the hydrophobic ends are repelled by water and face each other within the plasma membrane.

### Module 3.4 Review

**A.** The three basic components of the cytoskeleton are microfilaments, intermediate filaments, and microtubules. **B.** Microtubules are common to both centrioles and cilia. **C.** Motile cilia propel materials across cell surfaces. **D.** Only male sperm have a flagellum.

### Module 3.5 Review

**A.** Newly synthesized proteins from free ribosomes enter the cytosol; those from fixed ribosomes enter the ER. **B.** Both SER and RER have cisternae. SER lacks ribosomes, and its cisternae are tubular. RER has ribosomes attached to cisternae in the shape of flattened sheets. **C.** The SER functions in the synthesis of lipids such as steroids. Ovaries and testes produce large volumes of steroid hormones, which are lipids, and thus contain large amounts of SER. **D.** The ER is connected to and continuous with the nuclear envelope, which surrounds the nucleus.

### Module 3.6 Review

**A.** The Golgi apparatus (1) renews or modifies the plasma membrane; (2) modifies and packages cellular secretions, such as hormones or enzymes; and (3) packages special enzymes within vesicles (lysosomes) for use within the cell. **B.** Lysosomes contain digestive enzymes. **C.** Lysosomes may (1) remove damaged organelles, breaking them down to enhance recycling and membrane flow; (2) fuse with vesicles containing fluids or solids from the external environment to obtain nutrients or destroy pathogens; and (3) break down and release their digestive enzymes, resulting in the destruction of the cell (autolysis).

### Module 3.7 Review

**A.** A double membrane encloses a mitochondrion; the outer membrane surrounds the organelle, whereas the inner membrane contains folds called cristae and encloses a fluid, enzyme-filled matrix. **B.** Mitochondria require oxygen ($O_2$) to produce ATP, and they release carbon dioxide ($CO_2$) in the process. **C.** Mitochondria produce energy in the form of ATP molecules. A great number of mitochondria in a cell indicates a high demand for energy.

### Section 1 Review

**1.** glycolysis **2.** aerobic (metabolism) **3.** microvilli, microfilaments, microtubules **4.** lysosome, autolysis **5.** microvilli: increase surface area to facilitate absorption of extracellular materials **6.** Golgi apparatus: stores, alters, and packages newly synthesized proteins **7.** lysosome: intracellular removal of pathogens or damaged organelles **8.** mitochondrion: mitochondria produce 95 percent of the ATP required by the cell **9.** peroxisome: neutralizes toxic compounds **10.** nucleus: controls metabolism, stores and processes genetic information, controls protein synthesis **11.** rough endoplasmic reticulum (RER): has ribosomes bound to the membranes; modifies and packages newly synthesized proteins **12.** ribosomes: synthesize proteins **13.** cytoskeleton: provides strength and support; enables movement of cellular structures and materials **14.** Similar to the role of a plasma membrane around a cell, an organelle membrane physically isolates the organelle's contents from the cytosol, regulates exchange with the cytosol, and provides structural support.

### Module 3.8 Review

**A.** The genetic information in the cell is coded in the sequence of nucleotides in DNA. **B.** Most body cells contain a single nucleus, but there are exceptions. Skeletal muscle cells have many nuclei; red blood cells have none. **C.** The nucleus is a cell control center because it contains the genetic information, or directions, for making more than 100,000 proteins. The proteins fulfill structural and functional roles that permit the cell to respond to external changes.

### Module 3.9 Review

**A.** The nucleus is a cellular organelle containing DNA, RNA, and proteins. Surrounding the nucleus is the double-membraned nuclear envelope; the gap within this double membrane is the perinuclear space. Nuclear pores allow for chemical communication between the nucleus and the cytosol. **B.** The DNA in the nucleus stores the cell's instructions for synthesizing proteins. **C.** The nuclei of human somatic cells contain 23 pairs of chromosomes. **D.** The DNA strands are coiled around histone proteins forming nucleosomes, rather than remaining straight double helices. Further packing of DNA occurs as long chains of nucleosomes coil around other proteins.

### Module 3.10 Review

**A.** A gene is a portion of a DNA strand that functions as a hereditary unit and codes for a specific protein. **B.** The genetic code is described as a triplet code because a sequence of three nitrogenous bases—a triplet—specifies the identity of a specific amino acid. **C.** The three types of RNA involved in protein synthesis are messenger RNA (mRNA), transfer RNA (tRNA), and ribosomal RNA (rRNA). **D.** Messenger RNA is the link. It is composed of codons complementary to specific DNA triplets and passes from the nucleus through nuclear pores to ribosomes in the cytoplasm.

### Module 3.11 Review

**A.** Transcription is the synthesis of RNA from a DNA template. The RNA may be rRNA, tRNA, or mRNA. Messenger RNA (mRNA) carries the information needed to synthesize proteins. **B.** The DNA template strand is the strand of DNA that will be used to synthesize RNA. **C.** The substrates are RNA nucleotides, and the product is a strand of mRNA.

**D.** A cell that lacked the enzyme RNA polymerase would not be able to transcribe RNA from DNA.

### Module 3.12 Review

**A.** Translation is the synthesis of a protein using the information provided by the sequence of codons along an mRNA strand. **B.** The second codon is GCA, and its complementary anticodon sequence is CGU. **C.** Deletion of a base during transcription will change the makeup of all subsequent codons in the mRNA base sequence. The altered mRNA sequence will then result in the inclusion of a different series of amino acids into the protein during translation. Almost certainly the protein so produced will not be functional.

### Section 2 Review

**1.** d **2.** h **3.** l **4.** a **5.** k **6.** n **7.** m **8.** b **9.** j **10.** e **11.** i **12.** c **13.** g **14.** f **15.** AUG/UUU/UGU/GCC/ GCC/UUA **16.** UAC/AAA/ACA/ CGG/CGG/AAU **17.** Methionine-Phenylalanine-Cysteine-Alanine-Alanine-Leucine **18.** The nucleus contains the information for synthesizing proteins in the nucleotide sequence of its DNA. Changes in the extracellular fluid can affect cells through the binding of molecules to plasma membrane receptors or by the diffusion of molecules through membrane channels. Such stimuli may result in alterations of genetic activity in the nucleus. These alterations may change biochemical processes and metabolic pathways through the synthesis of additional, fewer, or different enzymes. Altered genetic activity may also change the physical structure of the cell by synthesizing additional, fewer, or different structural proteins.

### Module 3.13 Review

**A.** Permeability is the plasma membrane property that precisely determines which substances can enter or leave the cytoplasm**.** **B.** Three different types of membranes based on permeability are freely permeable membranes, which permit any substance to pass without difficulty; selectively permeable membranes, such

as plasma membranes, which permit some substances to pass but prevent other substances from passing; and impermeable membranes, which permit nothing to pass through. **C.** Passive processes do not require ATP for a substance to pass through the plasma membrane. Active processes require ATP for a substance to pass through the plasma membrane. **D.** Carrier proteins in the plasma membrane are involved in both active and passive processes of membrane passage.

### Module 3.14 Review

**A.** Diffusion is the passive movement of molecules from an area of higher concentration to an area of lower concentration until equilibrium is reached. **B.** Collisions are occurring among and between sugar, dye, and water molecules. **C.** Factors that influence diffusion rates include distance, molecule or ion size, temperature, steepness of the concentration gradient, and electrical forces. **D.** Diffusion is driven by a concentration gradient. The steeper the concentration gradient, the faster the rate of diffusion; the lower the concentration gradient, the slower the rate of diffusion. If the concentration of oxygen in the lungs were to decrease, the concentration gradient between oxygen in the lungs and oxygen in the blood would decrease (as long as the oxygen level of the blood remained constant). Thus, oxygen would diffuse more slowly into the blood.

### Module 3.15 Review

**A.** Osmosis is the passive movement of water across a selectively permeable membrane from a solution with a lower solute concentration to a solution with a higher solute concentration. **B.** Osmotic pressure of a solution refers to the force with which water moves into that solution as a result of its solute concentration. The osmotic pressure of solution B is greater than the osmotic pressure of solution A. **C.** A hypotonic solution would

cause an osmotic flow into the cell, leading to swelling and eventual bursting of the red blood cell (hemolysis). A hypertonic solution would cause an osmotic flow out of the cell, resulting in a shriveled or collapsed red blood cell (crenation). **D.** The 10 percent salt solution is hypertonic with respect to the cells lining the nasal cavity (the solution contains a higher salt concentration than do the cells). The hypertonic solution would draw water out of the cells (which would shrink), adding water to (diluting) the mucus. Thinner mucus would help relieve the congestion.

### Module 3.16 Review

**A.** In carrier-mediated transport, integral membrane proteins bind specific ions or molecules and transport them across the plasma membrane. **B.** The two rate-limiting factors are (1) the presence of a substance's concentration gradient, and (2) the number of carrier proteins. **C.** In both facilitated diffusion and active transport, a carrier molecule transports materials across the plasma membrane. **D.** Transporting hydrogen ions against their concentration gradient—that is, from a site where they are less concentrated (the cells lining the stomach, which produce the hydrogen ions) to a region where they are more concentrated (the digesting stomach contents)—requires energy. Thus, an active transport process must be involved.

### Module 3.17 Review

**A.** Endocytosis is the process in which a membranous vesicle forms at the cell surface, encloses a relatively large volume of extracellular material, and then moves into the cytoplasm. **B.** The three types of endocytosis are receptor-mediated endocytosis, the vesicle-mediated movement of specific target molecules; pinocytosis, the vesicle-mediated movement of extracellular fluid and its contents; and phagocytosis, the vesicle-mediated movement of extracellular solids, especially bacteria and debris. **C.** Exocytosis is the release of cytoplasmic materials into the extracellular environment after a membranous vesicle formed inside the cell fuses with the plasma membrane. **D.** The engulfment of bacteria and their transport into the cell is called phagocytosis.

### Section 3 Review

**1.** diffusion **2.** facilitated diffusion **3.** molecule or ion size **4.** net diffusion of water **5.** active transport **6.** specificity **7.** vesicular transport **8.** exocytosis **9.** pinocytosis **10.** "cell eating" **11.** diffusion **12.** neither **13.** osmosis **14.** diffusion **15.** osmosis

### Module 3.18 Review

**A.** Cell division is important because it is essential to survival. Aging and environmental stressors damage cells. Cellular division replaces old and damaged cells. **B.** Apoptosis is the genetically programmed death of a cell. **C.** Cell division begins with mitosis and ends with cytokinesis.

### Module 3.19 Review

**A.** Interphase is the portion of a cell's life cycle during which the chromosomes are uncoiled and all normal cellular functions except mitosis are under way. Its stages are $G_1$, S, $G_2$, and $G_0$. **B.** This cell is likely in the $G_1$ phase of interphase. **C.** Important enzymes for DNA replication are DNA helicases, DNA polymerase, and DNA ligases. **D.** DNA replication occurs in cells preparing for either mitosis or meiosis.

### Module 3.20 Review

**A.** Mitosis is the division of a cell nucleus into two identical daughter cell nuclei, an essential step in cell division. Its four stages are prophase, metaphase, anaphase, and telophase. **B.** A chromatid is a copy of a duplicated chromosome. Human cells normally contain 23 pairs of chromosomes, so during mitosis a human cell would contain 92 chromatids. **C.** If spindle fibers failed to form during mitosis, the chromosomes would not be able to separate into two sets. If cytokinesis occurred, the result would be one cell with two

sets of chromosomes, and one cell with none.

### Module 3.21 Review

**A.** Cancer is an illness characterized by mutations that disrupt normal cell control mechanisms and produce malignant cells. **B.** A tumor (or neoplasm) is a mass or swelling produced by abnormal growth and division of cells. The cells of a benign tumor typically remain within their tissue of origin and seldom threaten life. **C.** Metastasis is the spread of cancer cells from one site to another, producing secondary tumors. **D.** Angiogenesis refers to the growth of new blood vessels. Malignant tumor cells promote angiogenesis, which provides additional nutrients to speed up the division and growth of malignant cells.

### Section 4 Review

**1.** somatic cells **2.** $G_1$ phase **3.** $G_2$ phase **4.** DNA replication **5.** mitosis **6.** metaphase **7.** telophase **8.** cytokinesis **9.** telophase **10.** prophase **11.** centromere **12.** A cell that undergoes repeated rounds of the cell cycle without cytokinesis could result in a large, multinucleated cell.

### Chapter Review Questions

**1.** b **2.** c **3.** b **4.** b **5.** a **6.** b **7.** a **8.** b **9.** d **10.** b **11.** c **12.** c **13.** anchoring proteins, recognition proteins, receptor proteins, carrier proteins, and channels **14.** Mitosis specifically refers to the division and duplication of the cell's nucleus. Cytokinesis is the division of the cytoplasm into two distinct new cells. **15.** Diffusion is the movement of solutes by concentration differences. Osmosis is the net diffusion of water across a selectively permeable membrane. **16.** Mitochondria contain their own DNA, called mtDNA. It codes for small numbers of RNA and polypeptide molecules. The polypeptides are used in enzymes required for energy production. **17.** The

nonmembranous organelles are the cytoskeleton, microvilli, centrioles, cilia, flagella, and ribosomes. The membranous organelles are the mitochondria, nucleus, endoplasmic reticulum, Golgi apparatus, lysosomes, and peroxisomes. **18.** Prophase: chromosomes coil, forming visible chromosomes; each copy called a chromatid. Metaphase: chromatids move to the metaphase plate. Anaphase: centromere of each chromatid pair splits, and the chromatids separate. Telophase: each new cell prepares to return to interphase, nuclear membranes re-form, nuclei enlarge, and chromosomes uncoil; **19.** If a cell has microvilli on its plasma membrane, it is likely to be actively engaged in absorbing materials from the extracellular fluid. **20.** Water would flow into the cell, and the cell would swell up like a balloon and may eventually burst, releasing its contents. This event is called hemolysis. **21.** e, b, a, g, f, c, d **22.** Phagocytosis begins when cytoplasmic extensions called pseudopodia surround targeted material. The pseudopodia fuse to form a phagosome containing the targeted material. This vesicle fuses with lysosomes. Lysosomal enzymes digest the targeted material. Nutrients are released from the vesicle, and the residue is ejected from the cell through exocytosis. **23.** All somatic cells contain 46 chromosomes. Since mitosis produces two identical nuclei, the resulting cells will each have 46 chromosomes. Meiosis produces sex cells (sperm or oocytes) containing 23 chromosomes, which can unite to form a new somatic cell of 46 chromosomes. **24.** Lipid molecules such as steroid hormones are lipid soluble and therefore easily diffuse across the lipid bilayer of the plasma membrane. **25.** Malignant tumors divide rapidly and release chemicals that stimulate the growth of new blood vessels in a process called angiogenesis. The availability of additional nutrients from these new vessels accelerates tumor growth and metastasis.

**1.** They would want to read in the lab report that the tumor is benign. This would mean the tumor is localized and likely to be easily removed surgically. They would not want to read the word *malignant* because these types of cells divide rapidly and are more likely to threaten life. **2.** The tumor is likely caused by a mutation in a single cell. This mutated cell continued to undergo mitosis, producing more abnormal cells and resulting in a tumor, or mass of abnormal cells. **3.** If this tumor is malignant, it is likely to undergo metastasis and spread to other organs. **4.** If they could, Vito and his doctor would like to switch on apoptosis, or programmed cell death, in the tumor cells. **5.** Serious exercisers have high rates of energy consumption, and over time their mitochondria respond to this high demand for energy by dividing and increasing in number. **6.** Sweating during exercise results in losses of water from the extracellular fluid. As the extracellular fluid osmolarity increases, osmosis moves water of the cells, which begin to dehydrate. Energy drinks contain a solution of water, electrolytes (especially sodium chloride, which is lost in sweat), and glucose. As the digestive tract absorbs the sodium ions and glucose, water follows by osmosis. As the water influx restores normal extracellular fluid volume and osmolarity, water moves back into the cells, rehydrating them.

## CHAPTER 4 ANSWERS

### Module 4.1 Review

**A.** Histology is the study of tissues. **B.** A tissue is a group of similar cells and their cell products. **C.** The four basic tissue types are epithelial tissue, connective tissue, muscle tissue, and nervous tissue. Epithelial tissue covers exposed surfaces, lines passageways, and forms secretory glands. Connective tissue fills spaces, provides support, and stores energy. Muscle tissue contracts to produce movement. Nervous tissue propagates electrical impulses and carries information.

**Module 4.2 Review**

**A.** Early microscopes used only light and could magnify only 10 to 20 times. Today's modern compound light microscopes can magnify up to 1000 times, and modern microscopes using beams of electrons instead of light can magnify over 1 million times. **B.** A light microscope (LM) passes visible light through a thin section of tissue. A transmission electron microscope (TEM) uses magnets to direct a beam of electrons through a finely sectioned object. A scanning electron microscope (SEM) creates a three-dimensional image by first coating a surface with electron-dense material, then bombarding it with electron beams. **C.** Answer: 40×. (Total magnification = 10 × 40 = 400.)

**Module 4.3 Review**

**A.** Epithelial tissue provides physical protection, controls permeability, provides sensation, and produces specialized secretions. **B.** The three characteristic shapes of epithelial cells are squamous (flat), cuboidal (cube-shaped), and columnar (appearing tall and rectangular). A single layer of epithelial cells is a simple epithelium, whereas multiple layers of epithelial cells constitute a stratified epithelium. **C.** Motile cilia on epithelial cell surfaces aid the movement of substances over the epithelial surface.

**Module 4.4 Review**

**A.** Types of epithelial intercellular connections are tight (occluding) junctions, adhesion belts, gap junctions, and desmosomes. Hemidesmosomes connect the epithelium to the noncellular basement membrane. **B.** Gap junctions help to coordinate the functions of adjacent cells by allowing ions and small molecules to pass from cell to cell. In epithelial cells, gap junctions help coordinate functions such as cilia movement. In cardiac and smooth muscle tissues, they are essential for coordinated muscle cell contractions. **C.** Epithelia rely

on blood vessels in underlying tissues to supply needed nutrients. **D.** Epithelial and connective tissues contribute to the formation and maintenance of the basement membrane.

**Module 4.5 Review**

**A.** Both mesothelium and endothelium refer to a simple squamous epithelium, but in different locations. **B.** Because all these sites are subject to physical stresses and abrasion: by food (pharynx and esophagus), by feces (anus), or during intercourse or childbirth (vagina). **C.** Keratinized epithelia are both tough (have physical strength) and water resistant. **D.** No, this is not a sample of the skin surface. A simple squamous epithelium provides too little protection against infection, abrasion, or dehydration to function effectively as the skin surface.

**Module 4.6 Review**

**A.** In a sectional view, simple cuboidal epithelial cells are square and have central nuclei, and the distance between adjacent nuclei is roughly equal to the height of the epithelium. **B.** The epithelium that lines the urinary bladder is a transitional epithelium. It can tolerate repeated cycles of stretching and recoiling back to its original shape. **C.** In an empty or relaxed bladder, the superficial cells of a transitional epithelium appear plump and cuboidal; when stretched, the epithelium appears more like a stratified squamous epithelium. **D.** Simple cuboidal epithelium is associated with secretion and absorption, and transitional epithelium is associated with stretching in structures where large changes in volume occur.

**Module 4.7 Review**

**A.** In a sectional view, simple columnar epithelial cells appear as tall, slender rectangles containing elongated nuclei close to the basement membrane. Cell height is several times the distance between adjacent nuclei. **B.** A pseudostratified columnar epithelium is not truly stratified—even though its nuclei

make it appear so—because all its epithelial cells contact the basement membrane. **C.** Simple columnar epithelia may have microvilli for absorptive functions or motile cilia for the movement of substances. Pseudostratified columnar epithelia typically have motile cilia.

**Module 4.8 Review**

**A.** The two types of glands are called endocrine glands and exocrine glands. Endocrine glands release their secretions into interstitial fluid. Exocrine glands release their secretions into ducts that open onto an epithelial surface. **B.** The three secretory methods used by exocrine glandular epithelial cells are (1) merocrine secretion, (2) apocrine secretion, or (3) holocrine secretion.

**Module 4.9 Review**

**A.** The three characteristics are the structure of the duct, the shape of the secretory area, and the relationship between the duct and secretory areas. **B.** The simplest type of unicellular exocrine gland is a goblet cell. Goblet cells are scattered among the absorptive cells in the epithelium of intestines.

**Section 1 Review**

**1.** simple squamous epithelium **2.** simple cuboidal epithelium **3.** simple columnar epithelium **4.** stratified squamous epithelium **5.** stratified cuboidal epithelium **6.** stratified columnar epithelium **7.** goblet cells **8.** mucin **9.** mucus **10.** ducts **11.** epithelial surfaces **12.** exocrine glands **13.** merocrine secretion **14.** apocrine secretion **15.** holocrine secretion **16.** interstitial fluid **17.** endocrine glands **18.** avascular **19.** alveolar (acinar) gland **20.** transitional epithelium **21.** tight (occluding) junction **22.** basement membrane **23.** simple gland **24.** mesothelium (simple squamous epithelium) **25.** pseudostratified columnar epithelium **26.** urinary bladder, ureters, urine-collecting chambers in the kidney **27.** stratified squamous epithelium **28.** simple columnar epithelium **29.** lining of the peritoneum and pericardium,

exchange surfaces (alveoli) within the lungs **30.** lining of exocrine glands and ducts, kidney tubules **31.** stratified cuboidal epithelium

**Module 4.10 Review**

**A.** The three basic components of connective tissue are specialized cells, extracellular protein fibers, and a fluid known as ground substance. **B.** Connective tissue provides a structural framework for the body, transports fluids and dissolved materials, protects organs, supports and interconnects other tissue types, stores energy, and defends against invading microorganisms. **C.** Connective tissue proper consists of connective tissues with many types of cells and extracellular fibers in a viscous (syrupy) ground substance. Fluid connective tissues have distinct cells in a watery matrix containing dissolved proteins. Supporting connective tissues have less diverse cells and a matrix with more densely packed fibers.

**Module 4.11 Review**

**A.** Cells found in connective tissue proper are melanocytes, fixed macrophages, mast cells, fibroblasts, fibrocytes, adipocytes (fat cells), plasma cells, free macrophages, mesenchymal cells, neutrophils, eosinophils, and lymphocytes. **B.** Fibroblasts secrete hyaluronic acid and the proteins that form collagen, elastic fibers, and reticular fibers in the matrix, and fibroblasts also maintain these connective tissue fibers. **C.** The loose connective tissue that contains primarily lipids is adipose (fat) tissue. **D.** The stroma is a supporting network formed of reticular fibers. **E.** Phagocytic cells found in connective tissue proper include fixed macrophages and free macrophages, neutrophils, and eosinophils.

**Module 4.12 Review**

**A.** The packing or volume of fibers in a connective tissue makes the connective tissue "loose" or "dense." **B.** Continuous recirculation of extracellular fluid (plasma, interstitial fluid, and lymph) helps eliminate local differences in the levels of nutrients, wastes, or toxins; maintains blood volume; and alerts the immune system to infectious agents or infections. **C.** Lack of vitamin C would impair the production of collagen fibers, which add strength to connective tissue. The deficiency would result in tissue that is weak and prone to damage.

### Module 4.13 Review

**A.** Fibers characteristic of cartilage supporting the ear are elastic fibers. **B.** The perichondrium is made up of an outer fibrous layer of dense irregular connective tissue containing fibroblasts and an inner cellular layer. **C.** Appositional growth adds new layers to the outside of a cartilage. Interstitial growth enlarges a cartilage from within.

### Module 4.14 Review

**A.** Bone matrix includes ground substance (a small volume of liquid surrounding insoluble calcium salts) and collagen fibers. **B.** Mature bone cells in lacunae are called osteocytes. **C.** The functional unit of compact bone is an osteon. **D.** Bone contains only collagen fibers, whereas cartilage may contain collagen, elastic, and reticular fibers.

### Module 4.15 Review

**A.** The body cavities lined by serous membranes are the pleural, peritoneal, and pericardial cavities. **B.** The four types of tissue membranes found in the body are mucous membranes, serous membranes, the cutaneous membrane, and synovial membranes. **C.** The three layers of fascia are superficial fascia (areolar and adipose tissue), deep fascia (dense regular connective tissue), and subserous fascia (areolar tissue). **D.** The cutaneous membrane is relatively waterproof and usually dry.

### Section 2 Review

**1.** loose connective tissue **2.** adipose **3.** regular **4.** tendons **5.** ligaments **6.** fluid connective tissue **7.** blood **8.** hyaline **9.** chondrocytes in lacunae **10.** bone **11.** perichondrium,

periosteum, peritoneum, pericardium **12.** osteocytes, osteon **13.** chondrocyte, perichondrium **14.** interstitial growth **15.** lacunae **16.** chondrocyte **17.** osseous tissue **18.** fibrocartilage **19.** adipocytes **20.** synovial membrane **21.** cutaneous membrane **22.** perichondrium

### Module 4.16 Review

**A.** The relative percentage body weight of each of the four tissue types is as follows: muscle tissue 50%, connective tissue 45%, epithelial tissue 3%, and nervous tissue 2%. **B.** The three classifications of muscle tissue are skeletal muscle tissue, cardiac muscle tissue, and smooth muscle tissue. Skeletal muscle tissue functions to pull on bones for movement; cardiac muscle tissue moves blood within the heart and through blood vessels; and smooth muscle tissue moves substances through the digestive tract and regulates the diameter of respiratory passageways and blood vessels, among other functions.

### Module 4.17 Review

**A.** Smooth muscle tissue regulates blood vessel diameter. **B.** Neurons transfer and process information that is in the form of electrical impulses. Neuroglia protect, support, and repair nervous tissue. **C.** Skeletal muscles are organs that contain skeletal muscle tissue, connective tissue, and nervous tissue.

### Module 4.18 Review

**A.** The two processes in the response to tissue injury are inflammation and regeneration. **B.** Inflammation produces swelling, redness, heat (warmth), and pain. **C.** Inflammation can occur in any organ in the body because all organs have connective tissues.

### Section 3 Review

**1.** cardiac **2.** smooth **3.** nonstriated **4.** multinucleate **5.** cell body **6.** dendrites **7.** neuroglia **8.** maintain physical structure of nervous tissue **9.** repair nervous

tissue framework after injury **10.** perform phagocytosis **11.** provide nutrients to neurons **12.** regulate composition of interstitial fluid surrounding neurons **13.** axon **14.** intercalated disc **15.** neuroglia **16.** skeletal muscle tissue **17.** smooth muscle tissue **18.** regeneration **19.** inflammation **20.** Increased blood flow and blood vessel permeability enhance the delivery of oxygen and nutrients and the migration of additional phagocytes into the area and enhance the removal of toxins and waste products from the area.

### Chapter Review Questions

**1.** hyaline cartilage **2.** elastic cartilage **3.** fibrocartilage **4.** cardiac muscle tissue **5.** smooth muscle tissue **6.** skeletal muscle tissue **7.** Bone, or osseous, tissue **8.** adipose tissue **9.** dense regular connective tissue **10.** reticular tissue **11.** areolar tissue **12.** fluid connective tissues (blood or lymph) **13.** c **14.** b **15.** c **16.** a **17.** d **18.** b **19.** b **20.** a **21.** b **22.** d **23.** The skin and the lining of the mouth and throat are areas regularly exposed to severe physical or chemical stresses. The superficial cells, therefore, are continually worn away as a result of these stressors and must be replaced by cells in deeper layers. Stratified squamous epithelium contains a series of layers that allow for deeper cells to replace the lost superficial cells. **24.** Endocrine glands secrete hormones onto the surface of the gland or directly into the surrounding fluid. Exocrine glands secrete directly into ducts. **25.** Nervous tissue contains (1) neurons, which transmit information in the form of electrical impulses; and (2) neuroglia, which protect, support, and repair nervous tissue and maintain the nutrient supply to neurons. **26.** Normally, the extensive interlocking connections between the epithelial skin cells protect underlying tissues from infection by blocking pathogen access. After

a severe skin burn or abrasion, this mechanism can no longer provide protection, and pathogens easily enter the deeper tissues, resulting in an infection. **27.** When viewed with a light microscope, cardiac muscle tissue would demonstrate branched cells, striations, a single nucleus, and intercalated discs. **28.** Skeletal muscle tissue is made up of densely packed, parallel fibers that have striations, but because muscle fibers are composed of cells, they also have many visible nuclei. Jason is probably looking at a tendon (dense regular connective tissue). The small nuclei would be those of fibroblasts. **29.** The skin in the injured area will become red, warm, swollen, and painful. These changes occur as a result of inflammation, the body's first response to injury. Injury to the epithelium and underlying connective tissue will trigger the release of chemicals such as histamine and heparin from mast cells in the area. These chemicals, in turn, initiate the changes identified above.

### Chapter Integration

**1.** The reason for the discrepancy between her microscope and her text is that her microscope is focused at a total magnification of 40×, the typical lowest power on light microscopes found in the common anatomy and physiology lab, but the image in her text is focused at a total magnification of 350×. That is a difference of almost 10 times, so the images will be greatly different. That's like the difference in the view from an airplane window at 1000 feet versus 10,000 feet. **2.** The best way to help Alexis is to advise her to continue to focus her slide with progressively greater magnification to get as close as possible to the image in the book text. Most student microscopes can magnify at 400×, which is a good magnification for her to use because that will allow for the two images to look very similar. The best microscope technique is to first focus an image with the lowest power objective and then focus at progressively greater power. In most cases, you will be able to focus an image that is very close to the magnification identified in the text.

## CHAPTER 5

### Module Reviews

**Module 5.1 Review**
**A.** The two major components of the cutaneous membrane are the epidermis and the dermis. The subcutaneous layer (hypodermis) is not part of the integument. **B.** The major functions of the integumentary system are to protect underlying tissues and organs, excrete wastes by integumentary glands, maintain body temperature, produce melanin and keratin, synthesize vitamin $D_3$, store lipids, and detect touch, pressure, pain, and temperature.

**Module 5.2 Review**
**A.** Thin skin and thick skin refer to the relative thickness of the epidermis, not the cutaneous membrane. Thin skin covers most of the body surface and contains four strata (layers) of cells. Thick skin covers the palms of the hand and soles of the feet and contains five strata of cells. **b.** The five layers of the epidermis from deep to superficial are the stratum basale, stratum spinosum, stratum granulosum, stratum lucidum, and stratum corneum. **C.** Dandruff consists of cells from the stratum corneum. **D.** A splinter that penetrates to the third epidermal layer of the palm is lodged in the stratum granulosum.

**Module 5.3 Review**
**A.** The two pigments in the epidermis are carotene (an orange-yellow pigment) and melanin (a brown, yellow-brown, or black pigment). **B.** Exposure to sunlight or sunlamps darkens skin because the ultraviolet radiation they emit stimulates melanocytes in the epidermis to synthesize the pigment melanin. **C.** When the skin gets warm, arriving oxygenated blood is diverted to the superficial dermis (papillary layer) to eliminate heat. The oxygenated blood imparts a reddish coloration to the light skin. **D.** Cancerous basal cells rarely become malignant and metastasize, or spread, into other tissues throughout the body, as do melanoma cancer cells. Malignant melanoma is extremely dangerous because the cancerous melanocytes spread throughout the lymphatic system. **E.** The child probably has a fondness for vegetables, such as carrots, squash, and sweet potatoes, which have a high amount of carotene, a yellow-orange pigment. If a child eats large amounts of food containing carotene, the pigment is stored in the skin, turning the skin a yellow-orange color.

**Module 5.4 Review**
**A.** The dermis is made up of two layers of connective tissue: a papillary layer and a reticular layer. The superficial papillary layer consists of areolar tissue containing capillaries, lymphatic vessels, and sensory neurons that supply the surface of the skin. The deeper reticular layer consists of dense irregular connective tissue containing collagen and elastic fibers. **b.** Such a cut would cross the tension lines at a right angle, so a greater, rather than lesser, amount of scarring would be predicted.

**Module 5.5 Review**
**A.** A first-degree burn affects only the surface of the skin. This would include most sunburns. A second-degree burn damages the entire epidermis and perhaps some of the dermis. Although blistering occurs, accessory structures such as hair follicles and glands are generally not affected. Third-degree burns, also called full-thickness burns, destroy the epidermis and dermis and extend into the subcutaneous layer (hypodermis). **B.** Third-degree burns often require skin grafting because the tissue damage is so great that the tissue cannot repair itself. **C.** In an autograft, the patient's undamaged skin is used as a graft source. An allograft uses skin from a donor, and a xenograft uses skin from an animal, such as a pig. Autografts are best because they are not rejected by the patient's immune system. **D.** Using the rule of nines, the right leg (a lower limb) is 18% of surface area, the right arm (an upper limb) is 9% of surface area, and the back of the trunk is 18% of surface area (half of 36% trunk total, front and back). Adding these separate values gives a total surface burn area estimate of 45%.

**Section 1 Review**
**1.** accessory structures
**2.** epidermis **3.** granulosum
**4.** papillary layer **5.** nerves
**6.** reticular layer **7.** collagen
**8.** hypodermis **9.** connective
**10.** fat **11.** dermis: the connective tissue layer beneath the epidermis
**12.** epidermis: the protective epithelium covering the surface of the skin **13.** papillary layer: vascularized areolar tissue containing capillaries, lymphatic vessels, and sensory neurons that supply the skin surface
**14.** reticular layer: interwoven meshwork of dense irregular connective tissue containing collagen fibers and elastic fibers **15.** subcutaneous layer (or hypodermis): layer of loose connective tissue below the dermis; **16.** Malignant melanoma is often fatal because melanocytes are located close to the dermal layer, so if they become malignant, they can easily metastasize through the blood vessels and lymphatic vessels in nearby connective tissues.
**17.** The firefighter's comment that the burn is not painful should concern the doctors because it indicates that this is a third-degree burn. This type of burn is full-thickness and can destroy the sensory neurons, which is why the firefighter cannot detect pain despite being seriously burned.

**Module 5.6 Review**
**A.** Epidermal derivatives are accessory structures that originate from the epidermis during embryonic development. They include hair follicles, sebaceous and sweat glands, and nails. **B.** Sweat glands and sebaceous glands are the exocrine glands in the integument.

**Module 5.7 Review**
**A.** A typical hair is a keratinous strand produced by basal cells within a hair follicle. **B.** When arrector pili muscles contract, they pull the hair follicle erect. The overall effect is known as "goose bumps." **C.** Hair follicles transition between (1) an active phase of continuous hair growth; (2) regression; (3) a resting phase, when the hair becomes a club hair; and (4) reactivation of hair growth and shedding of club hair. **D.** Pulling a hair is painful because its root is attached deep within the hair follicle, the base of which is surrounded by a root hair plexus consisting of sensory nerves. Cutting a hair is painless because a hair shaft contains no sensory nerves.

**Module 5.8 Review**
**A.** The exocrine glands in the skin are sebaceous and sweat glands. Together, they assist in thermoregulation, excrete wastes, and lubricate hairs and the epidermis. **B.** Sebum is secreted by holocrine secretion. Sebum lubricates and protects the keratin of the hair shaft, lubricates and conditions the surrounding skin, and inhibits the growth of bacteria. **C.** Deodorants are used to mask the odor of apocrine sweat gland secretions. **D.** Contractile myoepithelial cells aid merocrine secretion discharge. They are found in apocrine and eccrine sweat glands, mammary glands, lacrimal (tear) glands, and salivary glands.

**Module 5.9 Review**
**A.** Nail production occurs at the nail root, an epidermal fold that is not visible from the surface. **B.** The hyponychium is the thickened stratum corneum underlying the free edge of a nail. **C.** The distal free edge of the nail appears white because there are no underlying blood vessels.

**Module 5.10 Review**
**A.** With advancing age, melanocyte activity decreases,

leading to white or gray hair. **B.** As a person ages, the blood supply to the dermis decreases, and eccrine sweat glands become less active. Both changes make it more difficult for older people to cool themselves in hot weather.

### Module 5.11 Review

**A.** Hormones that are necessary for maintaining healthy skin include growth hormone, sex hormones, growth factors (including epidermal growth factor [EGF]), glucocorticoids (which are steroid hormones), and thyroid hormones. **B.** In the presence of UV radiation in sunlight, epidermal cells in the stratum spinosum and stratum basale convert a cholesterol-related steroid into cholecalciferol, also known as vitamin $D_3$. **C.** The hormone cholecalciferol (vitamin $D_3$) is needed to form strong bones and teeth. When the body surface is covered, UV radiation cannot reach the skin to stimulate cholecalciferol (vitamin $D_3$) production, so bones can become fragile.

### Module 5.12 Review

**A.** The first step in tissue repair is inflammation. Inflammation produces swelling, redness, heat, and pain. **B.** Granulation tissue is the combination of blood clots, fibroblasts, and the extensive network of capillaries in healing tissue. **C.** Skin can regenerate effectively even after undergoing considerable damage because stem cells persist in both the epithelial and connective tissue components of skin. In response to injury, cells of the stratum basale replace epithelial cells, and mesenchymal cells replace lost dermal cells.

### Section 2 Review

**1.** free edge **2.** lateral nail fold **3.** nail body **4.** lunula **5.** proximal nail fold **6.** eponychium **7.** eponychium **8.** proximal nail fold **9.** nail root **10.** lunula **11.** nail body **12.** hyponychium **13.** phalanx **14.** dermis **15.** epidermis **16.** hair shaft **17.** sebaceous gland **18.** arrector pili muscle **19.** connective tissue sheath of hair bulb **20.** root hair

plexus **21.** h **22.** c **23.** e **24.** g **25.** i **26.** b **27.** d **28.** a **29.** j **30.** f **31.** The chemicals in hair dyes break the protective covering of the cortex, allowing the dyes to stain the medulla of the shaft. This is not permanent because the cortex remains damaged, allowing shampoo and UV rays from the sun to enter the medulla and affect the color. Also, the viable portion of the hair remains unaffected, so that when the shaft is replaced, the color or any change in curl will be lost.

### Chapter Review Questions

**1.** stratum corneum **2.** stratum lucidum **3.** stratum granulosum **4.** stratum spinosum **5.** stratum basale **6.** false **7.** true **8.** false **9.** false **10.** true **11.** true **12.** false **13.** true **14.** b **15.** b **16.** b **17.** d **18.** a **19.** b **20.** c **21.** a **22.** epidermis, dermis **23.** subcutaneous **24.** Apocrine **25.** basal cell carcinoma **26.** melanin **27.** Epidermal cell division occurs in the stratum basale. **28.** These smooth muscles cause hairs to stand erect when stimulated. **29.** Variations in hair color are due to differences in the pigment produced by the melanocytes in the hair papillae. With age, pigment production decreases, and hair color lightens. White hair is due to the combination of lack of pigment and the presence of air bubbles in the hair shaft medulla. **30.** Lack of vitamin $D_3$ can result in rickets in children. In adults, it can lead to decreased bone density and a greater risk for bone fractures. **31.** Compared to the thin skin of the rest of the body, the thick skin of the palms of the hands has a much thicker stratum corneum and also contains a stratum lucidum. The heat from the campfire, therefore, has less tissue to penetrate on the thin skin of your face before reaching the sensory neurons. **32.** A cut parallel to the tension lines will usually remain closed and heal with little scarring. A cut at right angles to a tension line will be pulled open as severed elastic fibers recoil and

will result in greater scarring. **33.** Fluid and electrolyte balance, thermoregulation, and protection from infection are affected.

### Chapter Integration

**1.** The tattoo ink stays in place because it is deposited into the dermis. The dermal layer is composed mostly of connective tissues with widely scattered cells. The dermal layer does not contain cells that are shed, as are cells in the epidermis. Also, the ink particles are too large to be removed by the body's normal defense mechanisms. **2.** A tattoo can fade over time for a variety of reasons, including the tendency of the ink to disperse and the effect of sunlight on the ink. **3.** Tattoos can be painful because the papillary layer of the dermis contains many sensory neurons. The tattoo needle is damaging to these neurons, causing pain. **4.** The reticular and papillary layers of the dermis contain networks of blood vessels and lymphatics. If the tattoo artist is not hygienic in his or her practice, the risk for infection is substantial.

## CHAPTER 6

### Module Reviews

### Module 6.1 Review

**A.** The axial skeleton is composed of the bones of the skull, thorax, and vertebral column. It contains a total of 80 bones. The appendicular skeleton includes the bones of the limbs and the pectoral and pelvic girdles. It contains a total of 126 bones. **B.** The skeletal system supports the body, stores minerals and lipids, produces blood cells, protects delicate organs and tissues, and functions as levers for movement.

### Module 6.2 Review

**A.** The six broad categories for classifying bones according to shape are flat bones, sutural (Wormian) bones, long bones, irregular bones, sesamoid bones, and short bones. **B.** Bone

markings (surface features) are characteristics of a bone's surface that have certain functions, such as forming a joint; serving as a site for muscle, tendon, or ligament attachment; or allowing the passage of nerves and blood vessels. **C.** Various terms for projections include process, ramus, head, tubercle, tuberosity, trochanter, crest, line, and spine.

### Module 6.3 Review

**A.** The major parts of a long bone are the epiphysis, metaphysis, diaphysis, and medullary cavity. **B.** The medullary cavity—the space within a bone—contains the red bone marrow, which is the site of blood cell production, and the yellow bone marrow, which is adipose tissue that is an important site for energy reserves. **C.** Articular cartilage covers the epiphyseal portions of bone that articulate with other bones. Cartilage is avascular in general, and articular cartilage obtains nourishment by diffusion from synovial fluid within the joint. **D.** Bone injuries are usually painful because of the extensive presence of sensory nerves within the periosteum, diaphysis, medullary cavity, and epiphyses.

### Module 6.4 Review

**A.** Osteogenic cells are stem cells that form osteoblasts. Osteoblasts are immature bone cells that secrete the organic components of bone matrix. **B.** Osteocytes are mature bone cells that maintain the bone matrix. They also take part in repairing damaged bone. **C.** If the ratio of collagen to hydroxyapatite in a bone increased, the bone's compressive strength would decrease, and the bone would also become more flexible. **D.** If the activity of osteoclasts (which demineralize bone) exceeded osteoblast activity (production of new bone), then the bone's mineral content (and thus its mass) would decrease, making the bone weaker.

### Module 6.5 Review

**A.** An osteon is the basic functional unit of mature compact bone; it consists of osteocytes organized around a central canal

and separated by concentric lamellae. **B.** Compact bone, which lies over spongy bone and makes up most of a bone's diaphysis, consists of compactly arranged osteons (Haversian systems); it protects, supports, and resists stress. Spongy bone makes up most of the volume of short, flat, and irregular bones and is also found at the epiphyses of long bones; it stores marrow and provides some support. **C.** The sample is from an epiphysis. The presence of lamellae that are not arranged in osteons is indicative of spongy bone, which is found in epiphyses.

## Module 6.6 Review
**A.** Appositional growth is an increase in bone diameter that occurs through the addition of bony matrix at the bone's surface. **B.** As a bone increases in diameter, the medullary cavity also increases in diameter. **C.** The periosteum is the superficial layer that surrounds a bone. It consists of an outer fibrous layer and an inner cellular layer. The endosteum is an incomplete cellular lining on the inner (medullary) surfaces of bones.

## Module 6.7 Review
**A.** Endochondral ossification is the replacement of a cartilage model with bone. **B.** In endochondral ossification, the original source of osteoblasts is the differentiation of cells in the inner layer of the perichondrium. **C.** X-rays of long bones, such as the femur, can reveal the presence or absence of the epiphyseal cartilage, which separates the epiphysis from the diaphysis so long as the bone is still lengthening. If the epiphyseal cartilage is present, growth is still occurring; if it is not present, the bone has reached its full length, and thus the person has reached his or her full height.

## Module 6.8 Review
**A.** Intramembranous ossification is bone formation within connective tissue without the prior development of a cartilage model. **B.** During intramembranous ossification, embryonic or fibrous connective tissue is replaced by bone. **C.** The primary difference between the two types of ossification is that in intramembranous ossification, bone develops from mesenchymal cells, whereas in endochondral ossification, bone develops from a cartilage model.

## Module 6.9 Review
**A.** Pituitary growth failure is less common today in the United States because children can be treated with synthetic human GH, which provides adequate or nearly adequate amounts of growth hormone for normal growth and development. **B.** Marfan syndrome is a hereditary disorder of connective tissue that results in abnormally long and thin limbs and digits. The condition usually causes life-threatening cardiovascular problems **C.** Gigantism results from overproduction of growth hormone *before* puberty, causing extreme height, whereas acromegaly results from overproduction of growth hormone *after* puberty, causing abnormally thick bones.

## Section 1 Review
**1.** intramembranous ossification
**2.** osteoid **3.** osteocytes
**4.** lacunae **5.** hyaline cartilage
**6.** periosteum **7.** compact bone
**8.** irregular bones **9.** epiphyses
**10.** fossa **11.** medullary cavity
**12.** trabeculae **13.** osteoclasts
**14.** sesamoid bones **15.** ossification or osteogenesis **16.** osteon
**17.** appositional growth
**18.** endochondral ossification
**19.** The fracture might have damaged the epiphyseal cartilage in Rebecca's right leg. Even though the bone healed properly, the damaged leg did not produce as much cartilage as did the undamaged leg. The result would be a shorter bone on the side of the injury.

## Module 6.10 Review
**A.** Organic compounds make up 33% of bone; inorganic components make up 67%.
**B.** The three organ systems that coordinate blood calcium level are the skeletal system, the digestive system, and the urinary system. **C.** If a person's blood calcium were abnormally low, neuron and muscle cell functions would be disrupted. Skeletal muscles might not function properly, so the person's movement would be altered. Cardiac muscle could also be affected, so heart rate is likely to be irregular.

## Module 6.11 Review
**A.** The hormone involved in stimulating the release of calcium ions from bone matrix is parathyroid hormone (PTH). PTH increases the level of blood calcium by causing osteoclasts to release stored calcium from bone. The mechanism of action is indirect because only osteoblasts have PTH receptors. Osteoblasts respond to PTH by releasing RANKL, a hormone that stimulates the differentiation and maturation of bone-eroding osteoclasts, whose activity releases calcium ions from bone matrix. **B.** PTH stimulates an increase in the production of the hormone calcitriol by the kidneys. The additional calcitriol stimulates calcium reabsorption by the kidneys. PTH and calcitriol enhance the absorption of calcium ions by the intestines. **C.** Calcitonin lowers blood calcium level by decreasing osteoclast activity, calcitriol production, and intestinal absorption of calcium ions and by increasing the rate of calcium excretion by the kidneys.

## Module 6.12 Review
**A.** Immediately after a fracture, extensive bleeding occurs at the injury site. After several hours, a large blood clot called a fracture hematoma develops. Next, an internal callus forms as a network of spongy bone unites the inner edges, and an external callus of cartilage and bone stabilizes the outer edges. The cartilaginous external callus is eventually replaced by bone, and the struts of spongy bone then unite the broken ends. With time, the swelling that initially marked the location of the fracture subsides, and the fracture site is remodeled, leaving little evidence that a break occurred. **B.** An open (compound) fracture is a break in the bone in which bone pierces the skin; a closed (simple) fracture is a break in the bone in which no bone penetrates the skin.

## Section 2 Review
**1.** calcitonin **2.** $\downarrow$ $Ca^{2+}$ intestinal absorption **3.** $\downarrow$ blood $Ca^{2+}$ level **4.** parathyroid glands **5.** $\uparrow$ $Ca^{2+}$ intestinal absorption **6.** release of stored $Ca^{2+}$ from bone **7.** homeostasis **8.** spiral fracture (tibia) **9.** transverse fracture (ulna) **10.** greenstick fracture (radius) **11.** comminuted fracture (femur) **12.** compression fracture (vertebra) **13.** Colles fracture (distal radius)

### Chapter Review Questions
**1.** epiphysis **2.** metaphysis
**3.** diaphysis **4.** epiphysis **5.** spongy bone or trabecular bone
**6.** compact bone **7.** medullary cavity **8.** true **9.** false **10.** true **11.** true **12.** false **13.** true **14.** false
**15.** true **16.** b **17.** a **18.** a **19.** b **20.** d **21.** c **22.** a **23.** b **24.** c
**25.** 80, 126 **26.** Sesamoid **27.** sinus
**28.** osteon or Haversian system
**29.** Osteoclasts **30.** (1) support, (2) storage of minerals and lipids, (3) blood cell production, (4) protection, (5) leverage. **31.** The twisting forces of a pirouette are likely to cause a spiral fracture.
**32.** Pituitary growth failure causes reduced epiphyseal cartilage activity and therefore abnormally short bones. This is rare because children can be treated with synthetic growth hormone. Achondroplasia is caused by reduced epiphyseal activity of the long bones, resulting in short, stocky limbs. The trunk, however, is of normal size. **33.** In intramembranous ossification, bone replaces embryonic or fibrous connective tissue. In endochondral ossification, bone replaces a cartilage model. **34.** An epiphyseal fracture is of particular concern because the cartilage can stop growing as it heals. This could result in the fractured limb being shorter than the opposite limb. **35.** Because the osteons of compact bone run parallel to the

long axis of the shaft, impacts perpendicular to the length of the bone are more likely to cause a fracture than stresses parallel to the length of the bone.

**36.** Gigantism is overproduction of growth hormone (GH) before puberty. People with gigantism can be very tall, exceeding 2.7 m (8 ft 11 in.). In acromegaly, GH rises after puberty, so the person does not grow taller, but the bones become thicker. Facial contours change in a characteristic manner.

### Chapter Integration

**1.** The worker forcefully fell on his buttocks in a seated position, which would have resulted in vertical forces on his vertebrae. causing a compression fracture. As confirmed by his coworkers, no bones were piercing the skin; therefore, he has closed fractures. **2.** When the doctor shaves the surface of the worker's hip bone, she will be collecting the osteoblasts located in the cellular layer of the periosteum. As these cells are deposited, they will continue ossification (osteogenesis) and will produce new bone. The grafted material that fills the space between the damaged and healthy vertebrae will continue the ossification process until it forms into a single fused bone. **3.** The region from which the graft is removed will undergo a healing process similar to what occurs in a fracture. A hematoma will form, cells will form a callus, new bone will be deposited, and the graft region will be remodeled with osteoblasts and osteoclasts until it is completely healed.

### CHAPTER 7

### Module Reviews

#### Module 7.1 Review
**A.** The skull is made up of 22 bones (8 cranial bones and 14 facial bones). There are 7 associated bones (6 auditory ossicles and 1 hyoid bone). Therefore, there are 29 bones in the skull. **B.** The axial skeleton provides a framework that supports and protects the brain, spinal cord, and organs in the body cavities of the trunk. It also provides surface area for the attachment of muscles. **C.** The muscles that attach to the axial skeleton position the head, neck, and trunk; perform breathing movements; and stabilize or position the limbs.

#### Module 7.2 Review
**A.** The bones of the cranium are the occipital bone, frontal bone, sphenoid, ethmoid, and the paired parietal and temporal bones. **B.** The facial bones protect and support the entrances to the digestive and respiratory tracts. These bones also provide attachment points for the muscles of facial expression. **C.** A suture is a fibrous connective tissue immovable joint between the flat bones of the skull. **D.** The coronal, squamous, and lambdoid sutures form the boundaries of each parietal bone.

#### Module 7.3 Review
**A.** The facial bones are the paired nasal, lacrimal, palatine, zygomatic, maxillae, and inferior nasal conchae, and the unpaired vomer and mandible. **B.** The vomer and inferior nasal conchae are facial bones; the ethmoid, sphenoid, and temporal bones are cranial bones. **C.** The mandible articulates with the left and right temporal bones. **D.** The right parietal bone is fractured.

#### Module 7.4 Review
**A.** The alveolar process of each maxilla is a projecting ridge that contains the tooth sockets for the upper teeth. **B.** The zygomatic arch consists of the zygomatic process of the temporal bone and the temporal process of the zygomatic bone. **C.** Both the external acoustic meatus and the internal acoustic meatus are found in the temporal bone. **D.** The internal acoustic meatus is a passageway for blood vessels and the facial and vestibulocochlear nerves.

#### Module 7.5 Review
**A.** The temporal bone contains the carotid canal, and the internal carotid artery runs through this passageway. **B.** The foramen ovale provides a passageway for nerves innervating the teeth and jaw. **C.** The foramen magnum is located in the occipital bone. This opening surrounds the connection between the brain and spinal cord.

#### Module 7.6 Review
**A.** The optic canals are found in the sphenoid. The optic nerve and ophthalmic artery pass through this opening. **B.** The sphenoid bone contains the sella turcica. The hypophyseal fossa within the sella turcica encloses the pituitary gland. **C.** The ethmoid contains the cribriform plate. In addition to forming the floor of the cranium and the roof of the nasal cavity, olfactory foramina in this structure serve as passageways for olfactory nerves. **D.** The sphenoid bone unites the cranial and facial bones of the skull.

#### Module 7.7 Review
**A.** The bones of the orbital complex are the frontal, sphenoid, zygomatic, palatine, maxilla, lacrimal, and ethmoid. **B.** The paranasal sinuses lighten the weight of the skull bones, allow the voice to resonate, and provide an extensive area of mucous epithelium. **C.** Frontal: both; maxilla: both; palatine: both; nasal: nasal.

#### Module 7.8 Review
**A.** The foramina of the mandible are the two mental foramina and the two mandibular foramina. **B.** Your lab partner is correct because the hyoid bone does not join directly to any other bone; instead, it supports the larynx and is the attachment site for muscles of the larynx, pharynx, and tongue. **C.** Three auditory ossicles are located within each middle ear cavity, found within the petrous part of the temporal bone. The ossicles play a key role in hearing by conducting vibrations produced by sound waves arriving at the tympanic membrane to the internal ear.

#### Module 7.9 Review
**A.** The major fontanelles are the anterior fontanelle, posterior fontanelle, sphenoidal fontanelle, and mastoid fontanelle. **B.** Because fontanelles are not ossified at birth, they permit flexibility of the skull during childbirth and allow room for the brain to enlarge during infancy and early childhood. **C.** An infant's cranial bones are larger than the facial bones to accommodate the size of the brain. The facial bones develop after birth.

#### Module 7.10 Review
**A.** The secondary curves of the spine allow us to balance our body weight to permit an upright posture with minimal muscular effort. Without the secondary curves, we would not be able to stand upright for extended periods of time. **B.** The major components of a typical vertebra are the articular processes, vertebral arch, and vertebral body. The vertebral arch is composed of laminae, transverse processes, pedicles, and a spinous process. **C.** The intervertebral discs attach to the body of the vertebra.

#### Module 7.11 Review
**A.** The dens is part of the axis, or second cervical vertebra, which is located in the cervical (neck) region of the vertebral column. **B.** The presence of transverse foramina indicates that this vertebra is a cervical vertebra located in the cervical region. **C.** When you run your finger down a person's spine, you can feel the spinous processes of the vertebrae.

#### Module 7.12 Review
**A.** There are five vertebrae in the lumbar region of the vertebral column. **B.** The lumbar vertebrae must support a great deal more weight than do vertebrae that are superior in the spinal column. The large vertebral bodies allow the weight to be distributed over a larger are.

#### Module 7.13 Review
**A.** The sacrum of the axial skeleton joins with the hip bones of the appendicular skeleton. **B.** The sacral and coccygeal regions are made up of fused vertebrae, not individual vertebrae.

#### Module 7.14 Review
**A.** Vertebrosternal ribs attach directly to the sternum by their

own individual costal cartilage. Vertebrochondral ribs do not attach directly to the sternum but rather by means of a shared costal cartilage. **B.** Ribs 11 and 12 lack any connection to the sternum. They are attached only to the vertebrae and body wall muscles. **C.** Along with the ribs and sternum, the 12 thoracic vertebrae make up the thoracic cage.

## Section 1 Review

**1.** frontal **2.** sphenoid **3.** zygomatic **4.** ethmoid **5.** lacrimal **6.** palatine **7.** maxilla **8.** sphenoidal fontanelle **9.** squamous suture **10.** lambdoid suture **11.** mastoid fontanelle **12.** anterior fontanelle **13.** sagittal suture **14.** posterior fontanelle **15.** Region: thoracic **16.** Characteristics: found in chest, heart-shaped body, smaller vertebral foramen, long slender spinous process that points inferiorly, all but two have facets for rib articulations **17.** Region: cervical **18.** Characteristics: found in neck, small oval body, large vertebral foramen, bifid spinous process, transverse foramina **19.** Region: lumbar **20.** Characteristics: found in inferior back, massive vertebral body, smallest vertebral foramen, spinous process blunt and projects posteriorly, no articular facets for ribs or transverse foramina.

## Module 7.15 Review

**A.** There are 126 bones in the appendicular skeleton. **B.** The pectoral girdles attach the upper limbs to the axial skeleton. The pelvic girdle attaches the lower limbs to the axial skeleton.

## Module 7.16 Review

**A.** The bones of the pectoral girdles are two clavicles (collarbones) and two scapulae (shoulder blades). **B.** The clavicle attaches the scapula to the sternum and restricts the scapula's range of motion (mobility). When the clavicle is broken, the scapula has a greater than normal range of motion and is less stable. **C.** The humerus articulates with the scapula at the glenoid cavity. **D.** The pectoral girdles of the appendicular skeleton are attached to the axial skeleton at the sternoclavicular joints.

## Module 7.17 Review

**A.** The two rounded projections on either side of the elbow are the lateral and medial epicondyles of the humerus. **B.** The radius is positioned laterally when the forearm is in the anatomical position. **C.** The four different bone markings associated with the proximal portion of the ulna include the olecranon (elbow point), trochlear notch, coronoid process, and radial notch.

## Module 7.18 Review

**A.** The carpal bones are the scaphoid, lunate, triquetrum, pisiform, trapezium, trapezoid, capitate, and hamate. **B.** Phalanges are bones of the fingers (or toes). **C.** Bill has broken the tip of his thumb, also known as the pollex.

## Module 7.19 Review

**A.** The three bones that fuse to make up a hip bone (coxal bone) are the ilium, ischium, and pubis. **B.** The acetabulum is the concave socket (fossa) on the lateral aspect of the pelvis that articulates with the head of the femur. **C.** When you are seated, the ischial tuberosities bear your body weight. **D.** The auricular surfaces of both the sacrum and the ilium articulate at the sacro-iliac joint.

## Module 7.20 Review

**A.** The bones of the pelvis are the two hip (coxal) bones, the sacrum, and the coccyx. **B.** The two pubic bones are joined anteriorly by the pubic symphysis.

## Module 7.21 Review

**A.** Compared with the male skull, the female skull is lighter and smoother; it is about 10% smaller; the forehead is more vertical; the sinuses are smaller; the teeth are smaller; and the mandible is smaller, narrower, and less robust.

## Module 7.22 Review

**A.** The bones of the lower limb are the femur (thigh), patella (kneecap), tibia and fibula (leg), tarsal bones, metatarsal bones, and phalanges. **B.** The head of the femur articulates with the acetabulum. **C.** The patella is a sesamoid bone. It forms within the tendon of the quadriceps femoris. **D.** The fibula both stabilizes the ankle joint and is an important point of attachment for muscles that move the foot and toes. When the fibula is fractured, those muscles cannot function properly, so walking becomes difficult—and painful.

## Module 7.23 Review

**A.** The tarsal bones are the calcaneus, talus, navicular, cuboid, medial cuneiform, intermediate cuneiform, and lateral cuneiform. **B.** The talus transmits the weight of the body from the tibia toward the toes. **C.** Joey most likely fractured his calcaneus (heel bone).

## Section 2 Review

**1.** clavicle **2.** scapula **3.** humerus **4.** radius **5.** ulna **6.** carpal bones **7.** metacarpal bones **8.** phalanges **9.** hip bone (coxal bone) **10.** femur **11.** patella **12.** tibia **13.** fibula **14.** tarsal bones **15.** metatarsal bones **16.** phalanges **17.** Sex: female **18.** Differences: a. broader, smoother; b. less vertical; c. shallower; d. wide; e. triangular; f. ≥100°; **19.** Sex: male; **20.** Differences: a. narrower, rougher; b. more vertical; c. deeper; d. narrow; e. oval; f. < 90°.

## Chapter Review Questions

**1.** occipital bone **2.** parietal bone **3.** frontal bone **4.** temporal bone **5.** sphenoid **6.** ethmoid **7.** vomer **8.** mandible **9.** lacrimal bone **10.** nasal bone **11.** zygomatic bone **12.** maxilla **13.** true **14.** false **15.** false **16.** true **17.** c **18.** c **19.** b **20.** d **21.** b **22.** b **23.** b **24.** a **25.** Fontanelles permit the skull to distort without damage during delivery, helping to ease the infant through the birth canal. **26.** Movement of the ribs affects the width and depth of the thoracic cage, increasing or decreasing its volume accordingly. This change in volume assists in breathing by increasing and decreasing the pressure inside the thoracic cavity. **27.** The three bones that fuse to form the hip bone (or coxal bone) are the ilium, ischium, and pubis. They meet at the acetabulum.

**28.** Primary curves develop before birth, and the secondary curves develop after birth. The thoracic curve and sacral curves are primary curves. The cervical curve is a secondary curve that forms as an infant begins to balance the weight of the head. The lumbar curve is a secondary curve that develops with the ability to stand. **29.** Arches assist with transferring the weight of the body to the feet and eventually to the ground. The longitudinal arch runs from the calcaneus to the distal portions of the metatarsals. The transverse arch runs perpendicular to the longitudinal arch. It describes the degree of curvature from the medial border to the lateral border of the foot.

## Chapter Integration

**1.** Donny is likely experiencing difficulty breathing because he has two fractured ribs. Breathing involves rib movements that change the volume of the thoracic cavity; broken ribs cause difficulty breathing because the pain hinders full expansion of the rib cage. **2.** Donny probably dislocated his shoulder, a common injury due to the weak nature of the shoulder joint (that is, the articulation between the head of the humerus and the glenoid cavity of the scapula). **3.** Both the appendicular and axial skeleton have been affected in Donny's injuries. The ribs are part of the axial skeleton, and the shoulder joint is part of the appendicular skeleton. **4.** In cleft palate, the bones that do not properly form a suture are the horizontal plates of the palatine bones and maxillae. These bones are paired bones, so a total of four bones are involved. **5.** If you were to look into the mouth of a baby with cleft palate, you would see a visible gap in the hard palate. If there is a gap, or cleft, between these bones, there is no single hard palate in the roof of the mouth. It is therefore possible to see into the nasal cavity because it is immediately superior to the oral cavity. **6.** Babies with cleft palate are likely to have problems with nursing and eating. Later on, these babies could also have speech difficulties. The teeth could be misaligned, requiring orthodontia. If the lip is involved, the condition is called *cleft lip*.

### CHAPTER 8

**Module 8.1 Review**
**A.** Range of motion (ROM) is the amount of movement at a joint. **B.** Synovial joints allow the greatest range of motion.

**Module 8.2 Review**
**A.** The key components of a synovial joint are a fibrous joint (articular) capsule, which surrounds the joint; articular cartilage, which resembles hyaline cartilage and covers the articulating bone surfaces; a synovial membrane, which lines the articular capsule and secretes synovial fluid that lubricates, distributes nutrients, and absorbs shocks; and a joint cavity that is filled with synovial fluid. **B.** Accessory structures include bursae, which are pockets filled with synovial fluid that reduce friction and absorb shocks; fat pads, which protect the articular cartilage; menisci, which are fibrocartilage articular discs that allow for variation in the shapes of the articulating surfaces; ligaments, which are cords of fibrous tissue that support, strengthen, and reinforce the joint; and tendons, which pass across or around a joint, limit the range of motion, and provide mechanical support. **C.** Articular cartilage lacks a blood supply and thus relies on synovial fluid to supply nutrients and remove wastes. If the circulation of synovial fluid were impaired, the cartilage would no longer receive nutrients, and wastes would accumulate. This could cause the cartilage to degenerate, and cells in the tissue may die.

**Module 8.3 Review**
**A.** Possible types of motion at a synovial joint include gliding, angular, circumduction, and rotation (partial). **B.** The types of synovial joints are plane (gliding),

hinge, pivot, condylar, saddle, and ball-and-socket. **C.** A ball-and-socket joint permits the greatest range of motion. **D.** Shoulder: ball-and-socket; elbow: hinge; ankle: plane (gliding); thumb: saddle.

**Module 8.4 Review**
**A.** Flexion and extension are possible movements at hinge joints. **B.** The ankle joint is not at extension in the anatomical position, so dorsiflexion, or flexion at the ankle, is the upward movement of the foot or toes. Plantar extension, or extension at the ankle, is the opposite movement, as when standing on tiptoe. **C.** When a person does jumping jacks, both the upper and lower limbs must perform abduction (when the limbs are spread apart) and adduction (when they are brought back together again).

**Module 8.5 Review**
**A.** The rotation of the head of the radius at the proximal radio-ulnar joint allows pronation and supination of the hand. **B.** Snapping your fingers involves opposition of the thumb and flexion at the third metacarpophalangeal joint. **C.** Protraction, supination, and pronation occur while a person wriggles into tight-fitting gloves.

**Section 1 Review**
**1.** medullary cavity **2.** spongy bone **3.** periosteum **4.** synovial membrane **5.** articular cartilage **6.** joint cavity (containing synovial fluid) **7.** joint capsule **8.** compact bone **9.** flexion **10.** extension **11.** hyperextension **12.** flexion **13.** hyperextension **14.** abduction **15.** adduction **16.** head rotation **17.** pronation **18.** abduction **19.** adduction **20.** opposition **21.** e **22.** d **23.** f **24.** g **25.** b **26.** c **27.** h **28.** a

**Module 8.6 Review**
**A.** Joints cannot be both strong and highly mobile. As a result, the more movable a joint, the weaker it is. **B.** The joints of the appendicular skeleton have a greater range of motion than those of the axial skeleton. **C.** The pectoral girdle and upper limb

are attached to the axial skeleton by the sternoclavicular joint. **D.** The sacrum participates in the sacro-iliac joints (left and right), the lumbrosacral joint, and the sacrococcygeal joint.

**Module 8.7 Review**
**A.** The nucleus pulposus is the gelatinous central region of an intervertebral disc. The anulus fibrosus is the tough layer of fibrocartilage encircling the nucleus pulposus. **B.** The primary vertebral ligaments are the paired ligamenta flava, the posterior longitudinal ligament, interspinous ligament, supraspinous ligament, and anterior longitudinal ligament.

**Module 8.8 Review**
**A.** A bulging disc is an intervertebral disc that is displaced or partly protruding as a result of a compressed nucleus pulposus that distorts the anulus fibrosus. In a herniated disc, the nucleus pulposus breaks through the anulus fibrosus, causing it to protrude into the vertebral canal. **B.** Age-related factors that may lead to vertebral fractures include reduced cushioning by intervertebral discs due to water loss (disc degeneration) and inadequate ossification due to osteopenia. Osteoporosis is a further complicating factor.

**Module 8.9 Review**
**A.** The glenoid labrum, rotator cuff muscles, and associated tendons and ligaments provide most of the stability for the shoulder joint. **B.** The iliofemoral, pubofemoral, and ischiofemoral ligaments are at the hip joint. **C.** The upward force of the head of the humerus could cause partial or complete dislocation of the acromioclavicular joint. This is called a shoulder separation. The bones involved include the clavicle, scapula, and humerus, and the stabilizing ligaments involved are the coracoclavicular, acromioclavicular, coraco-acromial, coracohumeral, and glenohumeral ligaments.

**Module 8.10 Review**
**A.** The ulnar collateral ligament stabilizes the medial surface of

the elbow joint, and the radial collateral ligament stabilizes the lateral surface of the elbow joint. **B.** Damage to the menisci of the knee joint decreases the joint's lateral stability, so the person would have difficulty locking the knee in place while standing and would have to use muscle contractions to stabilize the joint. If the person had to stand for a long time, the muscles would fatigue, and the knee would "give out." It is also likely that the person would feel pain. **C.** A severely hyperextended knee would damage the ACL.

**Module 8.11 Review**
**A.** Rheumatism is a general term describing any painful condition of joints, muscles, or both that is not caused by infection or injury. Osteoarthritis is a form of rheumatism characterized by degeneration of the joint cartilage and the underlying bone. Osteoarthritis results from cumulative wear and tear or genetic factors affecting collagen formation. **B.** A person can slow the progression of arthritis by engaging in regular exercise, doing physical therapy, and taking anti-inflammatory drugs.

**Section 2 Review**
**1.** coracoclavicular ligaments **2.** acromioclavicular ligament **3.** tendon of supraspinatus muscle **4.** acromion **5.** articular capsule **6.** subdeltoid bursa **7.** synovial membrane **8.** humerus **9.** clavicle **10.** coraco-acromial ligament **11.** coracoid process **12.** scapula **13.** articular cartilages **14.** joint cavity **15.** glenoid labrum **16.** patellar surface of femur **17.** fibular collateral ligament or lateral collateral ligament **18.** lateral condyle **19.** lateral meniscus **20.** tibia **21.** fibula **22.** posterior cruciate ligament (PCL) **23.** medial condyle **24.** tibial collateral ligament or medial collateral ligament **25.** medial meniscus **26.** anterior cruciate ligament (ACL) **27.** b **28.** e **29.** h **30.** g **31.** c **32.** a **33.** f **34.** d **35.** The sternoclavicular joints are the only joints between the pectoral girdles and the axial

skeleton. The sacro-iliac joints are the joints between the pelvic girdle and the axial skeleton.

## Chapter Review Questions

**1.** true **2.** false **3.** false **4.** true **5.** true **6.** false **7.** d **8.** c **9.** c **10.** d **11.** d **12.** c **13.** b **14.** b **15.** b **16.** b **17.** d **18.** a **19.** b **20.** c **21.** People tend to get shorter with advancing age in part as a result of the decrease in water content of the nucleus pulposus region of intervertebral discs, and in part as a result of a combination of osteopenia and osteoporosis in the vertebrae. **22.** Articular cartilage lacks a perichondrium, and its matrix contains more water than does the matrix of other cartilages. **23.** (1) plane joint: sternoclavicular joint; (2) hinge joint: elbow joint; (3) pivot joint: proximal radio-ulnar joint; (4) condylar joint: radiocarpal joints; (5) saddle joint: first carpometacarpal joint; (6) ball-and-socket joint: glenohumeral joint; **24.** Menisci may subdivide a synovial cavity, channel the flow of synovial fluid, or allow for variations in the shapes of articular surfaces to assist with joint stability and cushioning. **25.** A loss of bone mass as a result of aging is called osteopenia. When the reduction of bone mass is sufficient to compromise normal function, the condition is called osteoporosis. **26.** Shoulder dislocations occur more frequently than hip dislocations because the shoulder is a more mobile joint. Because the shoulder joint is not bound tightly by ligaments or other structures, it is easier to dislocate when excessive forces are applied. By contrast, the hip joint, although mobile, is less easily dislocated because it is stabilized by four heavy ligaments, and the bones fit together snugly in the joint. Additionally, the synovial capsule of the hip joint is larger than that of the shoulder, and its range of motion is decreased.

## Chapter Integration

**1.** The bones composing the knee joint are the femur, patella, and tibia. The fibula is not part of the joint. **2.** Physical therapy may improve ROM at the joint and increase blood flow to the tissue. **3.** The knee joint is a diarthrotic synovial joint that functions as a hinge; as such, it allows for flexion and extension at the knee. **4.** People who have undergone a total knee arthroplasty may be restricted from running/jogging, high-impact aerobics, and jumping—that is, any activity that places sudden, extreme loads on the artificial joint. **5.** In this scenario, a downward force onto the arm would force the head of the humerus inferiorly out of its articulation with the glenoid cavity. Therefore, Tommy suffered a dislocation of the shoulder joint (glenohumeral joint). A shoulder separation involves the acromioclavicular joint, which is not involved in Tommy's injury. **6.** The affected bones would be the humerus and scapula. The affected ligaments would be the coracohumeral ligament and the glenohumeral ligaments. The affected joint structures would be the articular capsule and the glenoid labrum. The tendons of the muscles surrounding the humeral head are also likely to be involved. **7.** Although a full recovery is possible, it is more likely that he will have long-term joint instability if the ligaments are permanently stretched or ruptured. If the glenoid labrum is damaged, it will no longer help to hold the humeral head in the glenoid cavity. Physical therapy to strengthen the shoulder joint muscles is critical to his recovery. Surgery to tighten the ruptured ligaments may also be required.

## CHAPTER 9

### Module Reviews

#### Module 9.1 Review

**A.** The three types of muscle tissue are skeletal, cardiac, and smooth. Skeletal muscle is directly or indirectly attached to bones and enables movement of the body. Cardiac muscle forms the heart and propels blood. Smooth muscle is found throughout the body and, among other functions, moves substances along the digestive tract and regulates the diameter of small arteries. **B.** Skeletal muscle tissue is voluntary. Cardiac and smooth muscle tissue are involuntary.

#### Module 9.2 Review

**A.** Tendons and aponeuroses attach skeletal muscles to bones. A tendon is a collagenous bundle that connects a skeletal muscle to a bone, whereas an aponeurosis is a broad collagenous sheet that connects skeletal muscle to a broader area that may involve more than one bone. **B.** The connective tissue layers are the epimysium, perimysium, and endomysium. The epimysium is a dense layer of collagen fibers that surrounds the entire muscle; the perimysium divides the skeletal muscle into a series of compartments, each containing a bundle of muscle fibers called a fascicle; and the endomysium surrounds individual skeletal muscle cells (fibers). The collagen fibers of the epimysium, perimysium, and endomysium come together to form either bundles known as tendons or broad sheets called aponeuroses. **C.** The plasma membrane is called the sarcolemma, and the cytoplasm is called the sarcoplasm. **D.** Because tendons attach muscles to bones, severing the tendon would disconnect the muscle from the bone, so the muscle could not move that limb.

#### Module 9.3 Review

**A.** Sarcomeres, the smallest contractile units of a striated muscle cell, are segments of myofibrils. Each sarcomere has a dark A band and light I bands. The A band contains the M line, the H band, and the zone of overlap. Each I band contains thin filaments, but not thick filaments. Z lines mark the boundaries between adjacent sarcomeres. **B.** Transverse tubules, or T tubules, are tubular extensions of the sarcolemma that extend deep into the sarcoplasm, contacting terminal cisternae of the sarcoplasmic reticulum. **C.** Within a resting skeletal muscle fiber, the greatest concentration of $Ca^{2+}$ is in the terminal cisternae of the sarcoplasmic reticulum.

#### Module 9.4 Review

**A.** F-actin (filamentous actin) is a twisted strand made up of two rows of G-actin (globular actin). Each G-actin molecule has an active site to which myosin can bind. **B.** A thick filament is made up of about 300 myosin protein molecules surrounding a strand of elastic protein called titin. **C.** The sliding filament theory describes the process of sarcomere shortening caused by the sliding of thin and thick filaments past one another. **D.** The zone of overlap is important because it is where the free head of the myosin molecule can interact with the thin filaments.

#### Module 9.5 Review

**A.** Sodium–potassium ion pumps maintain the cell's resting membrane potential by exporting $3 Na^+$ from the cell in exchange for $2 K^+$. Open sodium and potassium leak channels in the plasma membrane of cells allow $Na^+$ and $K^+$ to move down their concentration gradients, enabling the pump to work efficiently. **B.** A cell is depolarized when a threshold potential is reached ($-55$ mV for skeletal muscle) and sodium ions ($Na^+$) rush into the cell. The depolarization peaks at $+30$ mV. $Na^+$ inflow then abruptly stops, and repolarization begins as potassium ions ($K^+$) rush out of the cell. This loss of positive ions causes the membrane potential to become negative again. **C.** An excitable membrane permits rapid communication between different parts of a cell. **D.** An action potential travels in only one direction because during the refractory period, the excitable membrane cannot respond to another stimulus. That is, after opening, the sodium channels are inactivated and cannot open again unless the membrane potential changes. This period of sodium channel inactivity—called the refractory period—prevents the action potential from traveling in the other direction, thus restricting its propagation to only one direction.

**Module 9.6 Review**

**A.** The neuromuscular junction (NMJ) is a specialized intercellular connection that enables a motor neuron to communicate with a skeletal muscle fiber. It is made up of an axon terminal, a motor end plate, and a synaptic cleft. **B.** Acetylcholine release is necessary for skeletal muscle contraction because it serves as the first step in the process. A drug that blocks acetylcholine release would prevent binding of ACh with ACh-receptor membrane channels on the motor end plate; therefore, sodium ions would not rush into the muscle fiber's sarcoplasm, and no action potential would be generated in the sarcolemma. As a result, muscle contraction could not occur. **C.** Without AChE, the motor end plate would be continuously stimulated by ACh, locking the muscle in a state of contraction.

**Module 9.7 Review**

**A.** ATP is the molecule that supplies the energy for a muscle fiber contraction. **B.** Once the contraction cycle begins, the steps that occur are as follows: (1) active sites are exposed, (2) cross-bridges form, (3) myosin heads pivot (power stroke), (4) cross-bridges detach, and (5) myosin reactivates (myosin head "recocks").

**Section 1 Review**

1. mitochondrion 2. sarcolemma 3. myofibril 4. thin filament 5. thick filament 6. triad 7. sarcoplasmic reticulum 8. T tubule 9. terminal cisterna 10. sarcoplasm 11. myofibril 12. I band 13. A band 14. H band 15. Z line 16. titin 17. zone of overlap 18. M line 19. thin filament 20. thick filament 21. sarcomere 22. skeletal muscle 23. muscle fascicle 24. muscle fiber 25. myofibril 26. myofilament 27. sarcomere

**Module 9.8 Review**

**A.** The action potential in the sarcolemma travels along the T tubules to the triads, where it triggers the release of $Ca^{2+}$ from the terminal cisternae of the sarcoplasmic reticulum.

**Module 9.9 Review**

**A.** The length of individual sarcomeres is a factor that affects the amount of tension produced when a skeletal muscle fiber contracts. **B.** According to the sarcomere length–tension relationship, (1) the greater the zone of overlap in the sarcomere, the greater the tension the muscle can develop; and (2) there is an optimum range of actin and myosin overlap that will produce the greatest amount of tension. **C.** In an initial latent period (after the stimulus arrives and before tension begins to increase), an action potential generated in the muscle fiber triggers the release of calcium ions from the SR. In the contraction phase, calcium ions bind to troponin (cross-bridges form), and tension begins to increase. In the relaxation phase, tension decreases because cross-bridges have detached and because calcium ion levels have fallen; the active sites are once again covered by the troponin–tropomyosin complex.

**Module 9.10 Review**

**A.** Incomplete tetanus refers to a muscle producing near-peak tension during rapid cycles of contraction and relaxation. Wave summation refers to the addition of one twitch to another. **B.** A motor unit is a motor neuron and all the muscle fibers that it controls. **C.** The finer and more precise the movement produced by a particular muscle, the fewer the number of muscle fibers in the motor unit.

**Module 9.11 Review**

**A.** The greater the load on a muscle, the longer it takes for the muscle to begin to shorten and the less the muscle shortens. **B.** In a concentric contraction, the muscle tension remains constant at a value that just exceeds the load, and the muscle shortens. In an eccentric contraction, the muscle tension is constant, but less than the load, and the muscle elongates. **C.** Yes, a skeletal muscle can contract without shortening; this occurs during an isometric contraction. Whether a contracting muscle shortens (a concentric contraction), elongates (an eccentric contraction), or remains the same length (an isometric contraction) depends on the relationship between the load and the tension produced by actin–myosin interactions.

**Module 9.12 Review**

**A.** Mitochondria absorb oxygen, ADP, phosphate ions, and organic substrates (such as pyruvate molecules generated by glycolysis) from the cytosol to synthesize ATP. **B.** Three sources of energy utilized by muscle fibers are ATP, creatine phosphate (CP), and glycogen. **C.** Muscle fibers produce lactate under conditions of anaerobic metabolism (when there is a lack of oxygen). These conditions occur at peak levels of muscle activity.

**Module 9.13 Review**

**A.** After moderate activity, it may take several hours for a muscle fiber to recover. After sustained activity at higher levels, complete recovery can take a week. **B.** Oxygen debt (or excess postexercise oxygen consumption) is the amount of oxygen intake required after strenuous activity to produce the ATP needed to restore normal, pre-exertion conditions in the body. **C.** Much of the lactate produced during peak exertion diffuses out of the muscle fibers and into the bloodstream. The liver absorbs this lactate and begins converting it into pyruvate.

**Module 9.14 Review**

**A.** Compared with slow fibers, fast fibers have a larger diameter, larger glycogen reserves, less myoglobin, and relatively fewer mitochondria. **B.** A sprinter requires large amounts of ATP for a short burst of activity. To supply this ATP, the sprinter's muscles rely on large glycogen reserves and anaerobic metabolism. Because the muscle fibers have relatively few mitochondria to generate additional ATP, fatigue resistance is also low. Conversely, marathon runners derive most of their ATP from the more efficient process of aerobic metabolism. **C.** Athletes who excel at endurance activities have a higher-than-normal percentage of slow fibers. Slow fibers are physiologically better adapted to this type of activity than are fast fibers, which are less vascular and fatigue faster.

**Module 9.15 Review**

**A.** Muscle hypertrophy is enlargement of a muscle, often in response to activities such as bodybuilding. The number of muscle fibers does not increase; instead, individual muscle fibers enlarge from increases in myofibrils, mitochondria, glycolytic enzymes, and glycogen reserves. Muscle atrophy is the reduction in muscle size, tone, and power from lack of use or from disease in which the muscle is not stimulated by a motor neuron. **B.** A murder victim's time of death can be estimated according to the body's flexibility or rigidity because rigor mortis typically begins a few hours after death, reaches maximum rigidity some 2–7 hours after death, and subsides about 1–6 days later or when decomposition begins. Thus, for example, a victim whose body lacks any signs of rigor mortis likely died within the past few hours. At the molecular level, the membranes of the dead cells are no longer selectively permeable, and the SR is no longer able to retain calcium ions. As calcium ions enter the sarcoplasm, a sustained contraction develops, making the body extremely stiff. Contraction persists because the dead muscle cells can no longer make the ATP required for cross-bridge detachment from the active sites. Rigor mortis lasts until the lysosomal enzymes released by autolysis break down the myofilaments. **C.** While Fred's leg was immobilized, its muscles did not receive sufficient neural stimulation to maintain normal mass, tone, and strength—that is, his leg muscles atrophied—and the muscles were thus unable to support his weight.

## Section 2 Review

**1.** fatty acids **2.** $O_2$ **3.** glucose
**4.** glycogen **5.** CP **6.** creatine **7.** b
**8.** c **9.** a **10.** d **11.** large **12.** small
**13.** intermediate **14.** white **15.** pink **16.** low **17.** high **18.** dense
**19.** intermediate **20.** few **21.** many **22.** intermediate **23.** rapid
**24.** medium **25.** fast **26.** slow **27.** fast **28.** low **29.** high **30.** low **31.** high

### Chapter Review Questions

**1.** true **2.** false **3.** true **4.** false **5.** b
**6.** g **7.** i **8.** c **9.** a **10.** h **11.** j **12.** f
**13.** e **14.** d **15.** d **16.** b **17.** a **18.** b
**19.** d **20.** a **21.** b **22.** c **23.** a **24.** d
**25.** The T tubules carry action potentials into the interior of the cell. **26.** When contracting to lift the coffee cup, the muscles of elbow joint flexion are recruiting few motor units. The greater tension required for curling a dumbbell would be achieved by recruitment of more motor units, including motor units of more powerful muscle fibers. **27.** Throughout the recovery period, oxygen demand remains above normal resting levels as the liver absorbs and recycles the lactate released by muscle fibers during intense exercise. The liver converts the lactate to glucose, which is released into the bloodstream. The transport of lactate to the liver and of glucose back to muscle cells is called the Cori cycle. The additional oxygen consumed by the liver for glucose synthesis is called the oxygen debt, or excess postexercise oxygen consumption. The muscle fibers use the glucose to restore ATP, creatine phosphate, and glycogen concentrations to pre-exertion levels. **28.** Glycolysis in the cytosol and aerobic metabolism in the mitochondria generate ATP from glucose in muscle cells. **29.** When a muscle experiences repeated, exhaustive stimulation, as in exercise, muscle fibers develop more mitochondria, a higher concentration of glycolytic enzymes, larger glycogen reserves, and more myofibrils. Each myofibril contains more thick and thin filaments. The net effect is hypertrophy, or enlargement of the stimulated muscle.

### Chapter Integration

**1.** Sean's muscles are likely rich in slow fibers (type I, red, slow-twitch oxidative fibers). These fibers help him excel because they have a high resistance to fatigue due to their high myoglobin content, dense capillary supply, and high number of mitochondria. This fiber type is excellent for endurance activities such as soccer, but it is not well suited for sprinting from home plate to first base. **2.** Stuart's skeletal muscles are probably rich in fast fibers (type II-B, white, fast-twitch glycolytic fibers). Fast fibers provide Stuart with the speed necessary to get him from home plate to first base because they have a large diameter, and they rapidly reach peak tension, contract fast, and are high in glycolytic enzymes. This fiber type is not well suited for endurance activities and probably explains why he does not perform well in a sport such as soccer. **3.** The percentage of fast versus slow fibers in an individual is genetically determined. Endurance training, however, can cause fast fibers to take on the appearance and characteristics of intermediate fibers (type II-A, fast resistant, fast-twitch oxidative fibers). Sean the soccer player is more likely to have experienced this training benefit.

## CHAPTER 10

### Module 10.1 Review

**A.** The muscular system is divided into axial and appendicular divisions. Axial muscles support and position the axial skeleton. Appendicular muscles support and move the limbs. **B.** There are approximately 700 skeletal muscles in the body. **C.** Factors that influence the performance of a skeletal muscle are muscle fiber organization and how the muscle attaches to the bone.

### Module 10.2 Review

**A.** Contraction of a pennate muscle generates more tension than would contraction of a parallel muscle of the same size because a pennate muscle contains more muscle fibers and thus more myofibrils and sarcomeres. **B.** A lever is a rigid structure—such as a board, a pry bar, or a bone—that moves on a fixed point called the fulcrum. There are three classes of levers: In a first-class lever, the fulcrum lies between the applied force and the load; in a second-class lever, the load lies between the fulcrum and applied force; and in a third-class lever, the applied force is between the fulcrum and the load. Third-class levers are the most common type in the body. **C.** The atlanto-occipital joint between the occipital bone and the first cervical vertebra is part of a first-class lever system. The joint between the two bones acts as the fulcrum because it lies between the skull, which provides the load, and the neck muscles, which provide the applied force.

### Module 10.3 Review

**A.** A synergist is a muscle that helps a larger agonist (or prime mover—a muscle that is responsible for a specific movement) perform its actions more efficiently. **B.** Muscles A and B are antagonists because they perform opposite actions. **C.** The name flexor carpi radialis longus tells you that this muscle is a long (longus) muscle that lies next to the radius (radialis) and flexes (flexor) the wrist (carpi).

### Module 10.4 Review

**A.** Axial muscles arise on the axial skeleton and position the head and spinal column and also move the rib cage. This arrangement makes breathing possible. **B.** The following structures labeled in these figures are not muscles: linea alba, flexor retinaculum, iliotibial tract, patella, tibia, clavicle, sternum, superior extensor retinaculum, inferior extensor retinaculum, lateral malleolus of fibula, medial malleolus of tibia, calcaneal tendon, calcaneus. **C.** biceps brachii = appendicular; external oblique = axial; temporalis = axial; vastus medialis = appendicular.

### Section 1 Review

**1.** unipennate **2.** convergent
**3.** bipennate **4.** circular or sphincter **5.** multipennate
**6.** sternocleidomastoid **7.** deltoid
**8.** biceps brachii **9.** external oblique **10.** pronator teres
**11.** brachioradialis **12.** flexor carpi radialis **13.** rectus femoris
**14.** vastus lateralis **15.** vastus medialis **16.** gastrocnemius
**17.** soleus **18.** pectoralis major
**19.** rectus abdominis **20.** iliopsoas
**21.** tensor fascia latae **22.** gracilis
**23.** sartorius **24.** tibialis anterior
**25.** extensor digitorum longus

### Module 10.5 Review

**A.** The axial muscles stabilize and position the head, neck, and trunk.

### Module 10.6 Review

**A.** The muscles associated with the mouth are the buccinator, depressor labii inferioris, levator labii superioris, levator anguli oris, mentalis, orbicularis oris, risorius, depressor anguli oris, zygomaticus major, and zygomaticus minor. The orbicularis oris is the muscle involved in whistling. **B.** Buccinator = mouth; corrugator supercilii = eye; mentalis = mouth; nasalis = nose; platysma = neck; procerus = nose; risorius = mouth. **C.** A person is able to consciously move the skin on the scalp because the fibers of the epimysium are woven into those of the superficial fascia (subcutaneous layer) and the dermis, whereas muscle fibers (epimysium) of the thigh insert into bone and tendons, not skin tissue.

### Module 10.7 Review

**A.** The extrinsic eye muscles are the superior oblique, superior rectus, lateral rectus, medial rectus, inferior oblique, and inferior rectus. **B.** The medial and lateral pterygoids originate on the lateral pterygoid plates and insert at the medial surface of the ramus of the mandible. **C.** Contracting the masseter elevates the mandible, and relaxing the masseter depresses the mandible. So you would probably be eating or chewing something.

### Module 10.8 Review
**A.** The muscles of the tongue are the genioglossus, hyoglossus, palatoglossus, and styloglossus. **B.** The muscles that elevate the soft palate are the palatal muscles. **C.** Muscles associated with the hyoid that form the floor of the mouth include the mylohyoid, geniohyoid, and digastric.

### Module 10.9 Review
**A.** The spinal flexor muscles are the longus capitis, longus colli, and quadratus lumborum. **B.** The splenius capitis and spinalis cervicus enable you to extend your neck. **C.** The vertebral column does not need a massive series of flexor muscles because (1) many of the large trunk muscles flex the vertebral column when they contract, and (2) most of the body weight lies anterior to the vertebral column, so gravity tends to flex the spine.

### Module 10.10 Review
**A.** The transversus abdominis forms the deepest layer of the abdominal wall muscles. **B.** The rectus abdominis connects the ribs and sternum to the pubic bones. **C.** The external oblique tenses the abdominal wall and compresses the contents of the abdominal cavity, depresses the ribs, and flexes or bends the spine.

### Module 10.11 Review
**A.** The levator ani (pubococcygeus and illiococcygeus), external anal sphincter, and coccygeus make up the pelvic diaphragm. **B.** The bulbospongiosus compresses and stiffens the clitoris and narrows the vaginal opening. **C.** The coccygeus extends from the sacrum and coccyx to the ischial spine.

### Section 2 Review
1. occipitofrontalis (frontal belly)
2. temporalis 3. orbicularis oculi
4. levator labii superioris
5. zygomaticus minor
6. zygomaticus major 7. buccinator
8. orbicularis oris 9. risorius
10. depressor labii inferioris

11. depressor anguli oris
12. masseter 13. mylohyoid
14. digastric 15. geniohyoid
16. omohyoid 17. stylohyoid
18. thyrohyoid 19. sternothyroid
20. sternohyoid
21. sternocleidomastoid

### Module 10.12 Review
**A.** Pectoral girdle muscles originate on the axial skeleton.

### Module 10.13 Review
**A.** The rectus abdominis is the axial muscle that is known as the "six-pack" in physically fit people. **B.** As you move proximally to distally, the muscles become smaller and more numerous, enabling precise movements. **C.** deltoid = appendicular; external oblique = axial; gluteus maximus = appendicular; pectoralis major = appendicular; platysma = axial; rectus femoris = appendicular.

### Module 10.14 Review
**A.** The largest of the superficial muscles that position the pectoral girdle is the trapezius. **B.** The levator scapulae enable you to shrug your shoulders. **C.** The subclavius originates on the first rib and inserts on the inferior border of the clavicle.

### Module 10.15 Review
**A.** The subscapularis originates on the anterior surface of the scapula and inserts on the lesser tubercle of the humerus. **B.** The line of action is the line of force produced when a muscle contracts. **C.** The deltoid abducts the arm.

### Module 10.16 Review
**A.** A retinaculum is a wide band of connective tissue that stabilizes tendons. **B.** The wrist extensors are located on the posterior surface of the forearm. **C.** Supinator and pronator muscles are involved in turning a doorknob.

### Module 10.17 Review
**A.** Muscles that extend the fingers are the extensor digitorum and extensor digiti minimi. **B.** The abductor pollicis longus, extensor pollicis brevis, and extensor pollicis longus are muscles that abduct the wrist. **C.** The names of muscles associated with the

thumb frequently include the term pollicis; pollex is the anatomical term for the thumb.

### Module 10.18 Review
**A.** The intrinsic muscles of the thumb are the adductor pollicis, flexor pollicis brevis, opponens pollicis, and abductor pollicis brevis. **B.** None. No muscles originate on the phalanges. **C.** We are able to move our fingers because tendons from forearm muscles extend across the distal finger joints. When the forearm muscles contract, our fingers move.

### Module 10.19 Review
**A.** The muscles of the gluteal group are the gluteus maximus, gluteus medius, gluteus minimus, and tensor fasciae lata. **B.** Injury to the obturator muscles would impair lateral rotation at the hip.

### Module 10.20 Review
**A.** The biceps femoris, semimembranosus, semitendinosus, sartorius, and popliteus flex the knee. **B.** The muscles of the quadriceps femoris are the rectus femoris, vastus intermedius, vastus lateralis, and vastus medialis. **C.** The muscle whose origin is on the lateral condyle of the femur is the popliteus. **D.** The sartorius crosses the hip and knee joints; it produces flexion at the knee and flexion and lateral rotation at the hip.

### Module 10.21 Review
**A.** The muscles that extend the ankle are the plantaris, gastrocnemius, soleus, tibialis posterior, fibularis longus, and fibularis brevis. **B.** The muscles involved in flexing the toes are the flexor digitorum longus and flexor hallucis longus. **C.** A torn calcaneal tendon would make extension (plantar flexion) at the ankle difficult because this tendon attaches the soleus and gastrocnemius to the calcaneus (heel bone).

### Module 10.22 Review
**A.** The superior and inferior retinacula stabilize the positions of the tendons descending from the leg. **B.** The flexor hallucis

brevis flexes the great toe. **C.** The four lumbrical muscles originate on and insert on tendons.

### Module 10.23 Review
**A.** The lateral and medial intermuscular septa divide the arm into anterior and posterior compartments. **B.** Compartment syndrome can be life threatening because prolonged ischemia leads to tissue death.

### Section 3 Review
1. triceps brachii, long head
2. anconeus 3. extensor carpi ulnaris 4. extensor carpi radialis longus 5. extensor digitorum 6. flexor carpi ulnaris 7. gluteus medius 8. tensor fasciae latae 9. gluteus maximus 10. adductor magnus 11. gracilis 12. biceps femoris 13. semitendinosus 14. semimembranosus 15. sartorius 16. popliteus 17. gastrocnemius 18. tibialis anterior 19. fibularis longus 20. soleus 21. extensor digitorum longus 22. fibularis brevis 23. superior extensor retinaculum 24. inferior extensor retinaculum 25. calcaneal tendon

### Chapter Review Questions

1. second-class lever 2. first-class lever 3. third-class lever 4. true 5. false 6. true 7. true 8. false 9. j 10. h 11. f 12. a 13. g 14. b 15. e 16. c 17. d 18. i 19. b 20. b 21. c 22. a 23. a 24. c 25. d 26. a 27. b 28. b 29. The muscles of the pelvic floor support the organs of the pelvic cavity, flex the coccygeal joints, and control the movement of materials through the anus and urethra. 30. The muscles of the rotator cuff are supraspinatus, infraspinatus, teres minor, and subscapularis. 31. The muscles of the quadriceps femoris are rectus femoris, vastus intermedius, vastus medialis, and vastus lateralis. These muscles produce extension at the knee. Rectus femoris also produces flexion at the hip.

### Chapter Integration

1. The triceps brachii is an antagonist to the biceps brachii. This means that the triceps brachii opposes the movement

of the biceps brachii. To achieve muscle hypertrophy and tone of both muscles, the countering exercises must be performed. **2.** Exercises to tone the abdominal muscle groups include flexing and twisting the trunk. The muscles that the person should target include the rectus abdominis, the external and internal oblique muscles, and the latissimus dorsi. Sit-ups and twisting movements will stimulate these muscles. Placing a weight on the chest while doing sit-ups will produce faster results because the rectus abdominis would be working against a greater load. **3.** When a person is jumping, the hip and knee joints are performing extension, and the ankle joint is plantar flexing. **4.** The primary muscles for hip joint extension are gluteus maximus, biceps femoris, semitendinosus, and semimembranosus. The muscles of the quadriceps femoris (rectus femoris, vastus intermedius, vastus lateralis, and vastus medialis) extend the knee. Plantar flexion is performed by the gastrocnemius, soleus, fibularis longus, and fibularis brevis. **5.** Weight-room exercises that increase jumping strength are best to help Jennifer. Such exercises include squats, knee extensions, knee curls, and heel raises.

## CHAPTER 11

### Module Reviews

### Module 11.1 Review

**A.** The central nervous system (CNS) consists of the brain and spinal cord. The CNS integrates, processes, and coordinates sensory data and motor commands. The peripheral nervous system (PNS) includes all the nervous tissue outside the CNS (except for the ENS, which is within the wall of the gastrointestinal tract). **B.** The sensory division of the PNS brings information to the CNS from peripheral receptors. **C.** The effectors of the ANS are smooth muscle, cardiac muscle, glands, and adipose tissue.

### Module 11.2 Review

**A.** Structural components of a typical neuron include a cell body (including the nucleus and perikaryon, plus neurofilaments and neurotubules), an axon (including the axon hillock, axoplasm, axolemma, telodendria, collateral branches, and axon terminals), and dendrites (including dendritic branches and dendritic spines). **B.** A synapse is the site where a neuron communicates with another cell. It involves a presynaptic cell and a postsynaptic cell whose plasma membranes are separated by a synaptic cleft, where neurotransmitters are released. **C.** A presynaptic cell is usually a neuron. The postsynaptic cell can be a neuron or other type of cell, such as a skeletal muscle fiber or a gland cell. **D.** Most CNS neurons lack centrioles, which organize the microtubules of the spindle apparatus during mitosis, so these cells cannot divide and replace themselves.

### Module 11.3 Review

**A.** Neurons are structurally classified as anaxonic, bipolar, unipolar, or multipolar. **B.** Neurons are functionally classified as sensory neurons, interneurons, or motor neurons. **C.** Because most sensory neurons of the PNS are unipolar, these neurons are more likely have a sensory function. Motor neurons, which are multipolar, have a motor function and are found in the CNS.

### Module 11.4 Review

**A.** Neuroglia of the CNS are ependymal cells, microglia, astrocytes, and oligodendrocytes. **B.** Microglia occur in increased numbers in a person with a brain infection. **C.** Astrocytes protect the CNS from circulating chemicals and hormones by maintaining the blood brain barrier. **D.** White matter of the CNS is dominated by myelinated axons. The gray matter of the CNS includes areas containing neuron cell bodies, dendrites, and unmyelinated axons.

### Module 11.5 Review

**A.** Neuroglia of the PNS are satellite cells and Schwann cells. **B.** Wallerian degeneration occurs in axons of PNS neurons, where Schwann cells help in repairing damaged nerves. **C.** The neurilemma of a myelinating Schwann cell is the outer surface of the myelin sheath covering an internode of one axon. The neurilemma of a nonmyelinating Schwann cell is folded separately around multiple axons.

### Section 1 Review

**1.** dendrite **2.** Nissl bodies
**3.** mitochondrion **4.** nucleus
**5.** nucleolus **6.** cell body
**7.** axon hillock **8.** initial segment
**9.** axolemma **10.** axon
**11.** telodendrion **12.** axon terminal
**13.** multipolar **14.** unipolar
**15.** anaxonic **16.** bipolar
**17.** neuron, neuroglia, neurofilaments, neurofibrils, neurotubules, neurilemma
**18.** dendrite, dendritic spines, telodendria, oligodendrocytes
**19.** efferent fibers **20.** afferent fibers

### Module 11.6 Review

**A.** The membrane potential is the unequal charge distribution between the inner and outer surfaces of the plasma membrane. **B.** Graded potentials are temporary, localized changes in resting membrane potentials that decrease with distance away from the stimulus. In response to sufficiently large graded potentials, action potentials are generated. Action potentials begin in one location and spread along the surface of an axon toward the axon terminals. **C.** Information processing is the integration of stimuli by an individual cell.

### Module 11.7 Review

**A.** The resting membrane potential is the membrane potential of an unstimulated, or "undisturbed," cell. **B.** The sodium-potassium exchange pump maintains the cell's resting membrane potential by ejecting 3 sodium ions from the cell for every 2 potassium ions it recovers from the extracellular fluid. **C.** Decreasing the concentration of extracellular potassium ions would cause more potassium ions to leave the cell through potassium leak channels, which would make the resting membrane potential of the neuron more negative.

### Module 11.8 Review

**A.** Gated ion channels are active channels in the plasma membrane that open or close in response to specific stimuli. **B.** Chemically gated ion channels operate when they bind specific chemicals (such as ACh); voltage-gated ion channels operate in response to changes in the membrane potential; and mechanically gated ion channels operate in response to physical distortion of the membrane surface. **C.** If the voltage-gated sodium ion channels in a neuron's plasma membrane could not open, sodium ions could not rush into the cell, and its membrane potential would not change.

### Module 11.9 Review

**A.** A graded potential (also called a local potential) is a change in the membrane potential that cannot spread far from the site of stimulation. **B.** Depolarization is a shift from the resting membrane potential toward a more positive (less negative) value. Repolarization is the return of the depolarized membrane potential to the resting membrane potential. Hyperpolarization is a shift from the resting membrane potential toward a more negative value. **C.** Movement of sodium ions parallel to the inner and outer surfaces of the plasma membrane accounts for the local currents associated with graded potentials. (As the plasma membrane depolarizes, extracellular sodium ions move toward the open chemically gated sodium channels, replacing ions that have entered the cell.)

### Module 11.10 Review

**A.** An action potential is a propagated change in the membrane potential of excitable cells, initiated by a change in the plasma membrane's permeability to sodium ions. **B.** The all-or-none principle refers to the fact that all stimuli that bring the membrane to threshold generate identical action potentials. A given stimulus triggers either a typical action potential or no action potential at all. **C.** The refractory period is the time between the initiation of an action potential and the restoration of the normal resting membrane potential. The absolute refractory period is the portion of the refractory period during which the membrane cannot respond to further stimulation, no matter its magnitude. The relative refractory period is the time during which the membrane can respond only to a larger-than-normal stimulus. **D.** The events involved in the generation of an action potential are (1) depolarization to threshold, (2) activation of voltage-gated sodium ion channels and rapid depolarization, (3) inactivation of voltage-gated sodium ion channels and activation of voltage-gated potassium ion channels, and (4) closing of voltage-gated potassium ion channels.

**Module 11.11 Review**
**A.** Continuous propagation is the propagation of an action potential along an unmyelinated axon, where the action potential affects every portion of the membrane surface. Saltatory propagation is the relatively rapid propagation of an action potential between successive nodes of a myelinated axon. **B.** The presence of myelin greatly increases the propagation speed of action potentials. However, the propagation speed does vary with axon diameter. The larger the diameter, the faster the action potential travels.

**Module 11.12 Review**
**A.** The components of a chemical synapse—the site where a neuron communicates with another neuron or with a cell of a different type—are a presynaptic cell and a postsynaptic cell, whose plasma membranes are separated by a narrow gap called the synaptic cleft. **B.** The membrane of the postsynaptic cell must reach threshold to generate an action potential. **C.** Synaptic fatigue

occurs in an axon terminal when neutrotransmitters are released into the synaptic cleft at a faster rate than they can be reabsorbed or reycled. When this occurs, the synapse will not function normally. Function returns when the ACh receptor is able to bind ACh again. **D.** In chemical synapses, a neurotransmitter crosses a narrow synaptic cleft. In electrical synapses, the membranes of the presynaptic and postsynaptic cells are joined together by gap junctions.

**Module 11.13 Review**
**A.** An excitatory postsynaptic potential (EPSP) is a graded depolarization of a postsynaptic membrane by a neurotransmitter released by a presynaptic cell. An inhibitory postsynaptic potential (IPSP) is a graded hyperpolarization of a postsynaptic membrane after the arrival of a neurotransmitter. **B.** Temporal summation is the addition of a rapid succession of stimuli occurring at a single synapse. Spatial summation involves the addition of simultaneous stimuli applied at different locations; that is, it involves multiple synapses that are active all at the same time. **C.** No action potential will be generated, because the depolarization did not reach threshold.

**Module 11.14 Review**
**A.** Regulatory neurons facilitate or inhibit the activities of presynaptic neurons by affecting the plasma membrane of the cell body or by altering the sensitivity of axon terminals. **B.** The degree of sustained depolarization at the axon hillock determines the frequency of action potential generation. **C.** The greater the degree of sustained depolarization at the axon hillock, the higher the frequency of generation of action potentials. **D.** Acetylcholine has a direct effect on ion channels through its binding to chemically gated ion channels. The effect of the neurotransmitter serotonin has an indirect effect on ion channels. It acts as a first messenger by binding to

membrane receptors coupled with G proteins. The activated G protein produces or releases a second messenger that then affects ion channels, altering conditions or activities of the postsynaptic cell.

**Section 2 Review**
**1.** action potential **2.** electrical synapse **3.** resting membrane potential **4.** gated ion channels **5.** cholinergic synapses **6.** hyperpolarization **7.** local current **8.** depolarization **9.** acetylcholine (ACh) **10.** calcium ions ($Ca^{2+}$) **11.** synaptic vesicle **12.** acetylcholinesterase (AChE) **13.** ACh gated sodium ion channel **14.** sodium ions ($Na^+$) **15.** An action potential depolarizes the axon terminal. **16.** Calcium ions enter the cytosol of the axon terminal. **17.** ACh is released through exocytosis. **18.** ACh binds to ACh gated sodium ion channel on the postsynaptic membrane, producing a graded depolarization. **19.** The depolarization ends as ACh is broken down into acetate and choline by AChE. **20.** The axon terminal reabsorbs choline from the synaptic cleft and uses it to synthesize new molecules of ACh. **21.** In myelinated fibers, saltatory propagation transmits nerve impulses to the neuromuscular junctions rapidly enough to initiate muscle contractions and promote normal movements. In axons that have become demyelinated, nerve impulses cannot be propagated, so the muscles are not stimulated to contract. Eventually, the muscles atrophy from lack of stimulation (a condition termed *disuse atrophy*).

**Chapter Review Questions**

**1.** ependymal cell **2.** neuron cell body **3.** astrocyte **4.** microglia **5.** myelinated axon **6.** oligodendrocyte **7.** axon **8.** node (node of Ranvier) **9.** false **10.** true **11.** false **12.** true **13.** true **14.** c **15.** e **16.** a **17.** b **18.** f **19.** g **20.** h **21.** d **22.** c **23.** b **24.** b **25.** c **26.** d **27.** a **28.** b **29.** The central nervous system

(CNS) is composed of the brain and spinal cord. The peripheral nervous system (PNS) is made up of all the nervous tissue outside the CNS (except for the ENS) and is divided into the sensory and motor divisions. **30.** The three functional classes of neurons in the nervous system are sensory neurons, interneurons, and motor neurons. Sensory neurons transmit impulses from the PNS to the CNS. Interneurons analyze sensory inputs and coordinate motor outputs. Motor neurons transmit impulses from the CNS to the PNS. **31.** In continuous propagation, which occurs in unmyelinated axons, an action potential appears to move across the membrane surface in a series of tiny steps. In saltatory propagation, which occurs in myelinated axons, only the nodes can respond to a depolarizing stimulus. Saltatory propagation is much faster than continuous propagation. **32.** Graded potentials are localized changes in membrane potential that cannot spread far from the site of stimulation. Action potentials occur in excitable membranes when threshold is reached, causing a depolarization that propagates, or spreads, along the axon, resulting in synaptic activity at the axon terminal. **33.** Temporal summation of EPSPs is the addition of stimuli that arrive at a single synapse in rapid succession. Spatial summation of EPSPs occurs when simultaneous stimuli at multiple synapses have a cumulative effect on membrane potential.

**Chapter Integration**

**1.** Demyelination is the loss or destruction of the myelin sheaths that insulate nerve fibers. **2.** The loss of myelin slows nerve impulse propagation. As a result, information about limb movement and body position moves slowly within the CNS, and motor commands move slowly and erratically. **3.** The glial cells affected by MS are oligodendrocytes, which form the myelin sheaths around axons in the CNS.

### CHAPTER 12

### Module 12.1 Review

**A.** Sensory input travels toward the spinal cord, whereas motor commands travel away from the spinal cord. **B.** A reflex is a rapid, automatic response triggered by specific stimuli. **C.** A spinal reflex is an automatic response triggered by a stimulus that is controlled in the spinal cord and can function without input from the brain.

### Module 12.2 Review

**A.** A typical spinal cord has 31 pairs of spinal nerves, and the spinal cord ends at the level of lumbar vertebra 1 or 2 ($L_1$ or $L_2$). **B.** The pair of spinal nerves that emerges between vertebrae $C_7$ and $T_1$ are named $C_8$. **C.** Gross anatomical features of the cross-sectioned spinal cord include the anterior median fissure (a deep groove along the anterior or ventral surface); the posterior median sulcus (a shallow longitudinal groove); white matter (dominated by myelinated axons); gray matter (composed of cell bodies of neurons, neuroglia, and unmyelinated axons); the central canal (a passageway containing cerebrospinal fluid); a posterior root of each spinal nerve (axons of neurons whose cell bodies are in the posterior root ganglion); the anterior root of each spinal nerve (the axons of motor neurons that extend into the periphery to control somatic and visceral effectors); posterior root ganglia (contain cell bodies of sensory neurons); and spinal nerves (contain the axons of sensory and motor neurons). **D.** The gray matter of the spinal cord is composed of the cell bodies of neurons, neuroglia, and unmyelinated axons. **E.** The cauda equina is made up of extensions of the posterior and anterior roots of spinal segments $L_2$ to $S_5$. It forms as the vertebral column continues to elongate after longitudinal growth of the spinal cord ceases at about age 4.

### Module 12.3 Review

**A.** Cerebrospinal fluid (CSF) is found in the subarachnoid space, which lies deep to the epithelium of the arachnoid mater and superficial to the pia mater. **B.** In a lumbar puncture, the lumbar puncture needle penetrates the epidermis, dermis, subcutaneous layer (hypodermis), and then skeletal muscle before reaching the protective spinal coverings: the dura mater, then the arachnoid matter, and finally the subarachnoid space, which contains cerebrospinal fluid.

### Module 12.4 Review

**A.** Sensory nuclei receive and relay sensory information from peripheral receptors; motor nuclei issue motor commands to peripheral effectors. **B.** The poliovirus-infected neurons would be in the anterior gray horns of the spinal cord, where the cell bodies of somatic motor neurons are located. **C.** A disease that damages myelin sheaths would affect the white matter columns of the spinal cord, which are composed of bundles of myelinated axons.

### Module 12.5 Review

**A.** The three layers of connective tissue of a spinal nerve are the outer epineurium, middle perineurium, and inner endoneurium; the major peripheral branches of a spinal nerve are the posterior ramus, the anterior ramus, and the rami communicantes. **B.** A dermatome is a specific bilateral sensory region monitored by a single pair of spinal nerves. **C.** Shingles is caused by reactivation of the varicella-zoster virus (VZV), the same herpes virus that causes chickenpox. Once reactivated, the virus stimulates painful inflammation of the nerve ganglia and causes skin eruptions in a pattern that corresponds to the affected dermatome.

### Module 12.6 Review

**A.** The anterior ramus carries sensory information from the limbs. **B.** Visceral sensory information reaches the posterior root of a spinal nerve through the posterior and anterior rami and the white ramus communicans. **C.** The posterior ramus provides sensory and motor innervation to the skin and skeletal muscles of the back. **D.** The gray ramus communicans is a bundle of postganglionic sympathetic nerve fibers that are distributed to effectors (glands and smooth muscles) in the body wall and limbs by way of a spinal nerve. The white ramus communicans contains both motor and sensory fibers. It contains the myelinated preganglionic axons of sympathetic motor neurons en route to sympathetic ganglia and abdominopelvic viscera and sensory fibers carrying sensory information from the visceral organs.

### Module 12.7 Review

**A.** A nerve plexus is a complex, interwoven network of nerves. The major plexuses are the cervical, brachial, lumbar, and sacral plexuses. **B.** Spinal nerves $T_2 - T_{11}$ and $Co_1$ are not part of a nerve plexus.

### Module 12.8 Review

**A.** Damage to the phrenic nerves, which innervate the diaphragm, would interfere with the ability to breathe. **B.** An anesthetic that blocks the function of the anterior rami of the cervical spinal nerves would affect the skin and muscles of the back of the neck and of the shoulders.

### Module 12.9 Review

**A.** The brachial plexus is a network of nerves formed by branches of spinal nerve segments $C_5 - T_1$, en route to innervating the upper limb. **B.** A nerve plexus trunk is a large bundle of axons from several spinal nerves; a nerve plexus cord is a smaller branch of nerves that originates at a trunk. **C.** The major nerves associated with the brachial plexus are the dorsal scapular, long thoracic, suprascapular, medial and lateral pectoral, subscapular, thoracodorsal, axillary, medial antebrachial cutaneous, radial, musculocutaneous, median, and ulnar nerves.

### Module 12.10 Review

**A.** The lumbar plexus is a nerve network formed by axons from the anterior rami of spinal nerve segments $T_{12}-L_4$; the sacral plexus is a nerve network formed by the anterior rami of spinal nerve segments $L_4-S_4$. **B.** The major nerves of the sacral plexus are the superior and inferior gluteal, posterior femoral cutaneous, sciatic, and pudendal nerves. **C.** The sciatic nerve divides into the tibial nerve and the common fibular nerve.

### Section 1 Review

**1.** white matter **2.** posterior root ganglion **3.** lateral white column **4.** posterior gray horn **5.** lateral gray horn **6.** anterior gray horn **7.** posterior median sulcus **8.** central canal **9.** sensory nuclei **10.** motor nuclei **11.** anterior gray commissure **12.** anterior root **13.** anterior white commissure **14.** anterior median fissure **15.** anterior view **16.** radial nerve **17.** ulnar nerve **18.** median nerve **19.** posterior view **20.** columns **21.** conus medullaris **22.** nerves **23.** meninges **24.** cauda equina **25.** brachial plexus **26.** dura mater **27.** perineurium **28.** gray ramus communicans

### Module 12.11 Review

**A.** In divergent neural circuits, information spreads from one neuron to several neurons. In convergent neural circuits, several neurons synapse with a single postsynaptic neuron. **B.** In serial processing, information is processed in a stepwise fashion, one neuron to another. **C.** The simplest neural processing occurs in the PNS and spinal cord.

### Module 12.12 Review

**A.** All reflexes are rapid, unconscious patterned responses to a physical stimulus that restore or maintain homeostasis. **B.** The components of a reflex arc include a receptor, a sensory neuron, a motor neuron, and a peripheral effector; interneurons may also be present. **C.** Neural reflexes are classified according to their development (innate reflexes versus acquired reflexes), the nature of the resulting motor response (somatic reflexes versus visceral reflexes), the complexity of the neural circuit involved (monosynaptic reflexes versus polysynaptic reflexes), and the site of information processing (spinal reflexes versus cranial reflexes).

## Module 12.13 Review

**A.** A stretch reflex is a monosynaptic reflex that provides automatic regulation of skeletal muscle length. **B.** In the patellar reflex, the response observed is leg extension, and the effectors involved are the quadriceps femoris. **C.** When stretch receptors are stimulated by gamma motor neurons, the muscle spindles become more sensitive. As a result, little (if any) stretching stimulus is needed to stimulate the contraction of the quadriceps femoris. Thus, the reflex response occurs more quickly.

## Module 12.14 Review

**A.** The flexor reflex is an example of a withdrawal reflex that contracts the flexor muscles of a limb in response to a painful stimulus; so, it has a protective function. **B.** During a withdrawal reflex of the foot, the limb on the opposite side is extended. This response is called a crossed extensor reflex. **C.** All polysynaptic reflexes involve pools of interneurons, involve more than one spinal segment, involve reciprocal inhibition, and have reverberating circuits.

## Module 12.15 Review

**A.** Reinforcement is an enhancement of a spinal reflex through the facilitation of motor neurons involved in reflexes. **B.** Reflex testing provides information about the nervous system's functional status. **C.** A Babinski reflex is abnormal in adults; it indicates possible damage of descending tracts in the spinal cord.

## Section 2 Review

1. divergence 2. convergence 3. serial processing 4. parallel processing 5. reverberation 6. receptor 7. sensory neuron 8. interneuron 9. spinal cord (CNS) 10. motor neuron 11. effector 12. ipsilateral reflex 13. withdrawal reflexes 14. gamma motor neuron

15. flexor reflex 16. visceral reflexes 17. acquired reflexes 18. contralateral reflex 19. reciprocal inhibition 20. reinforcement 21. The withdrawal reflex illustrated in the reflex arc illustration is an innate, somatic, polysynaptic, spinal reflex.

### Chapter Review Questions

1. gray matter 2. white matter 3. anterior root 4. spinal nerve 5. posterior root ganglion 6. posterior root 7. sympathetic ganglion 8. posterior ramus 9. pia mater 10. arachnoid mater 11. dura mater 12. true 13. false 14. false 15. true 16. false 17. c 18. c 19. b 20. d 21. d 22. a 23. b 24. b 25. b 26. a 27. Shingles is caused by the varicella-zoster virus (VZV), the same virus that causes chickenpox. Symptoms include a painful rash and blisters whose distribution corresponds to the affected sensory nerve and its associated dermatome. 28. The student bumped her ulnar nerve as it passed behind the medial epicondyle of the humerus. She is experiencing pain in the fourth and fifth finger, as well as the medial hand. 29. Spinal nerve $C_1$ exits superior to vertebra $C_1$. Spinal nerve $C_8$ exits inferior to vertebra $C_7$. Therefore, there are eight cervical nerves but only seven cervical vertebrae.
30. Cerebrospinal fluid is located in the subarachnoid space. It functions as a shock absorber and a diffusion medium for dissolved gases, nutrients, chemical messengers, and wastes.

### Chapter Integration

1. Snapping of the head to the left would cause a stretching of the spinal nerves on the right side of the neck. The majority of these nerves travel into the right upper limb, which is why the pain would extend only into the right side. The pain was in his neck as well because these nerves originate from the spinal cord in the cervical region. 2. The spinal nerves involved in Dominic's stinger would be those of the brachial plexus, which

are spinal nerves $C_4$–$T_1$. 3. The burning, numbing pain is the result of the stretch placed on the brachial plexus as Dominic's head snapped to the opposite side. These nerves innervate the entire upper limb, which is why the pain extended all the way to his fingertips.
4. The anterior horn in the lumbar region of the spinal cord contains somatic motor neurons that direct the activity of skeletal muscles of the hip, lower limb, and foot. As a result of the injury, Karen would be expected to have poor control of most muscles of the lower limbs, causing difficulty walking (if she could walk at all) and problems maintaining balance (if she could stand). 5. The person would still exhibit defecation (bowel) and urination (urinary bladder) reflexes because these spinal reflexes are processed at the level of the spinal cord. Afferent impulses from the organs would stimulate specific interneurons in the sacral region that synapse with the motor neurons controlling the sphincters, thus bringing about emptying when the organs began to fill. However, a person with a spinal cord transected at $L_1$ would lose voluntary control of the bowel and urinary bladder because these functions rely on impulses carried by motor neurons in the brain that must travel down the spinal cord and synapse with the interneurons and motor neurons involved in the reflex. 6. The effects of a transection of the spinal cord at $L_1$ in an adult are the same as the situation in newborns, in whom the descending tracts required for conscious control of urination have not yet fully developed.

### CHAPTER 13

### Module Reviews

#### Module 13.1 Review

**A.** The three primary brain vesicles are the prosencephalon (forebrain), mesencephalon (midbrain), and rhombencephalon (hindbrain). **B.** The prosencephalon and rhombencephalon form the

scondary brain vesicles. **C.** The telencephalon develops into the cerebrum, which is the largest region of the adult brain.

#### Module 13.2 Review

**A.** The medulla oblongata relays sensory information to other parts of the brainstem and to the thalamus. It also contains centers that regulate autonomic functions, such as heart rate and blood pressure. **B.** The cerebrum contains two lateral ventricles, one in each cerebral hemisphere. **C.** The corpus callosum is a tract of white matter that links the left and right cerebral hemispheres, whereas the septum pellucidum is a partition that separates the two lateral ventricles. **D.** The diencephalon is enclosed by the cerebrum.

#### Module 13.3 Review

**A.** The layers that make up the cranial meninges, from superficial to deep, are the outer dura mater, the middle arachnoid mater, and the inner pia mater. **B.** Tight junctions are characteristic of ependymal cells of each choroid plexus. **C.** Decreased diffusion across the arachnoid granulations would increase the volume of CSF in the ventricles because less CSF would reenter the bloodstream. The increased pressure within the brain due to accumulated CSF could damage the brain.

#### Module 13.4 Review

**A.** The pyramids contain tracts of motor fibers that originate at the cerebral cortex. Some of the motor fibers cross to opposite sides of the medulla oblongata, and that crossing over is called a decussation. **B.** The gracile nucleus and cuneate nucleus are components of the medulla oblongata that relay somatic sensory information to the thalamus. **C.** The ascending and descending tracts of white matter in the medulla oblongata link the brain with the spinal cord.

#### Module 13.5 Review

**A.** The four groups of pontine structures are the (1) tracts that interconnect other portions of the CNS, (2) nuclei involved with the control of respiration, (3) sensory and motor nuclei of cranial nerves, and (4) nuclei and tracts that process and relay information sent to and from the cerebellum.

### Module 13.6 Review
**A.** The components of the cerebellar gray matter are the cerebellar cortex and the cerebellar nuclei. **B.** The arbor vitae is the white matter of the cerebellum, and it connects the cerebellar cortex and cerebellar nuclei with the cerebellar peduncles. **C.** Ataxia is the inability to coordinate muscle movement that can result from damage to the cerebellum from trauma, stroke, or certain drugs, including alcohol.

### Module 13.7 Review
**A.** The two pairs of sensory nuclei contained within the corpora quadrigemina are the superior colliculi and inferior colliculi. **B.** The superior colliculi control the reflex movements of the eyes, head, and neck in response to visual stimuli, such as a bright light. The inferior colliculi control reflex movements of the head, neck, and trunk in response to auditory stimuli, such as a loud noise. **C.** Cranial nerves III, IV, V, VI, VII, VIII, IX, X, XI, and XII arise from the brainstem. **D.** The part of the midbrain that influences the activity of the entire nervous system is the reticular activating system (RAS), a part of the reticular formation. Stimulation of the RAS causes alertness, whereas damage to the RAS produces unconsciousness.

### Module 13.8 Review
**A.** Damage to the lateral geniculate bodies of the thalamus would interfere with the flow of visual information and thus affect the sense of sight. **B.** The preoptic area of the hypothalamus, a component of the diencephalon, is stimulated by changes in body temperature.

### Module 13.9 Review
**A.** The limbic system establishes emotional states; links the conscious, intellectual functions of the cerebral cortex with the unconscious and autonomic functions of the brainstem; facilitates memory storage and retrieval; and affects motivation. **B.** The amygdaloid body plays a role in regulating heart rate, controlling the "fight or flight" response, and in linking emotions with specific memories. **C.** The hippocampus is important in the storage and retrieval of long-term memories.

### Module 13.10 Review
**A.** The basal nuclei are masses of cerebral gray matter that function in the subconscious control of skeletal muscle activity and help coordinate learned movement patterns. **B.** The caudate nucleus is one of the basal nuclei involved with the subconscious adjustment and modification of skeletal muscular activity. **C.** A person who has damage to the basal nuclei would exhibit decreased muscle tone and the loss of coordinated movement.

### Module 13.11 Review
**A.** The lobes of the cerebrum—the frontal lobe, parietal lobe, occipital lobe, and temporal lobe—are named for the overlying bones of the skull. **B.** The insula is an island of cerebral cortex located medial to the lateral sulcus. **C.** Damage to the left postcentral gyrus would interfere with sensory information awareness from the right side of the body.

### Module 13.12 Review
**A.** The primary motor cortex is located in the precentral gyrus of the frontal lobe of the cerebrum. **B.** Damage to the temporal lobes of the cerebrum interferes with the processing of olfactory (smell) and auditory (sound) sensations. **C.** The stroke has damaged Broca's area, located in the frontal lobe.

### Module 13.13 Review
**A.** The longitudinal fasciculi connect the frontal lobe to the other lobes of the same hemisphere. **B.** The axons in the cerebral white matter are called association fibers, commissural fibers, and projection fibers.

### Module 13.14 Review
**A.** The four wave types associated with an EEG are alpha waves (characteristic of normal resting adults), beta waves (characteristic of a person who is concentrating), theta waves (observed in children and frustrated adults), and delta waves (found in a person who is sleeping deeply, in infants, or in people with damaged portions of the brain). **B.** A seizure is a temporary cerebral disorder accompanied by abnormal movements, unusual sensations, inappropriate behavior, or some combination of these signs and symptoms. Epilepsy is a clinical condition characterized by seizures.

### Module 13.15 Review
**A.** The optic, abducens, oculomotor, and trochlear cranial nerves are involved with the eye. **B.** The cranial nerves with motor functions only are the oculomotor (III), trochlear (IV), abducens (VI), accessory (XI), and hypoglossal (XII).

### Section 1 Review
**1.** precentral gyrus **2.** frontal lobe **3.** lateral sulcus **4.** temporal lobe **5.** pons **6.** central sulcus **7.** postcentral gyrus **8.** parietal lobe **9.** occipital lobe **10.** cerebellum **11.** medulla oblongata **12.** olfactory bulb (associated with cranial nerve I, olfactory), S. **13.** oculomotor (III), M. **14.** trigeminal (V), B. **15.** facial (VII), B. **16.** glossopharyngeal (IX), B. **17.** vagus (X), B. **18.** optic (II), S. **19.** trochlear (IV), M. **20.** abducens (VI), M. **21.** vestibulocochlear (VIII), S. **22.** hypoglossal (XII), M. **23.** accessory (XI), M. **24.** thalamus **25.** arcuate fibers **26.** fornix **27.** commissural fibers **28.** basal nuclei **29.** The sensory innervation of the nasal lining, or nasal mucosa, is by way of the maxillary maxillary nerve ($V_2$), a branch of the trigeminal nerve (V). Irritation of the nasal lining increases the frequency of action potentials along the maxillary nerve ($V_2$) through the semilunar ganglion to reach centers in the midbrain, which in turn excite the neurons of the reticular activating system (RAS). Increased activity by the RAS can raise the cerebrum back to consciousness.

### Module 13.16 Review
**A.** The term general senses describes our sensitivity to temperature, pain, touch, pressure, vibration, and proprioception (position). **B.** The size of the receptive field is what enables you to distinguish the exact spot of touch more precisely in certain areas. The smaller the receptive field, the greater your ability to localize an area of stimulus.

### Module 13.17 Review
**A.** The four types of general sensory receptors (and the stimuli that excite them) are nociceptors (pain), thermoreceptors (temperature), mechanoreceptors (physical distortion), and chemoreceptors (chemicals dissolved in body fluids). **B.** The three classes of mechanoreceptors are proprioceptors, which monitor the positions of bones, joints, and muscles; baroreceptors, which detect changes in pressure; and tactile receptors, which respond to the sense of touch. **C.** Adaptation is a decrease in receptor sensitivity in the presence of constant stimulation. Peripheral adaptation reduces the amount of information from receptors that reaches the central nervous system (CNS). In central adaptation, awareness of a stimulus virtually disappears, even though sensory neurons in the CNS remain active.

### Module 13.18 Review
**A.** Tactile receptors found only in the dermis are tactile (Meissner's) corpuscles, lamellar (Pacinian) corpuscles, and bulbous (Ruffini) corpuscles. **B.** A bulbous corpuscle is more sensitive to continuous deep pressure because, unlike a lamellar corpuscle, it undergoes little adaptation.

### Module 13.19 Review
**A.** A somatotopy is a functional map of the somatosensory cortex. **B.** The lateral spinothalamic tracts

carry action potentials generated by nociceptors. **C.** The left cerebral hemisphere (specifically, the primary somatosensory cortex in that hemisphere) receives impulses conducted by the right gracile fasciculus.

### Module 13.20 Review

**A.** The corticobulbar tracts are descending tracts that carry information or commands from the cerebral cortex to motor nuclei in the brainstem. **B.** Corticospinal tracts are descending tracts that carry motor commands from the cerebral cortex to the anterior gray horns of the spinal cord. **C.** Increased stimulation of the motor neurons of the red nucleus would increase stimulation of the skeletal muscles in the upper limbs, thereby increasing their muscle tone.

### Module 13.21 Review

**A.** The thalamus and midbrain control reflexes in response to visual and auditory stimuli experienced while viewing a movie. **B.** The basic reflex motor patterns related to eating and drinking are controlled by the hypothalamus. **C.** As you decide to hit the ball, the premotor cortex receives information from the frontal lobes and then relays that information to the basal nuclei and cerebellum. As the hitting movement begins, the premotor cortex sends additional information to the primary motor cortex.

### Module 13.22 Review

**A.** Referred pain is a sensation felt in a part of the body other than its actual source. **B.** Rabies is contracted by a bite from a rabid animal. The bite injects the rabies virus into peripheral tissues, and the virus infects axons. Retrograde flow then carries the viral particles into the CNS. **C.** Amyotrophic lateral sclerosis (ALS), commonly called Lou Gehrig's disease, is a progressive degeneration of the motor neurons of the CNS, leading to muscle atrophy and eventual paralysis.

### Section 2 Review

**1.** free nerve ending **2.** root hair plexus **3.** tactile discs **4.** tactile

(Meissner's) corpuscle **5.** bulbous (Ruffini) corpuscle **6.** lamellar (Pacinian) corpuscle **7.** anterior **8.** posterior **9.** lateral corticospinal tract of corticospinal pathway (conscious control of skeletal muscles throughout the body) **10.** rubrospinal tract of lateral pathway (subconscious regulation of muscle tone; movement of distal limb muscles) **11.** reticulospinal tract of medial pathway (subconscious regulation of muscle tone; movements of the neck, trunk, and proximal limb muscles) **12.** vestibulospinal tract of medial pathway (subconscious regulation of muscle tone; movements of the neck, trunk, and proximal limb muscles) **13.** tectospinal tract of medial pathway (subconscious regulation of muscle tone; movements of the neck, trunk, and proximal limb muscles in response to bright lights, sudden movements, and loud noises) **14.** anterior corticospinal tract of corticospinal pathway (conscious control of skeletal muscles throughout the body) **15.** posterior column pathway (carries sensations of fine touch, pressure, vibration, and proprioception) **16.** posterior spinocerebellar tract of spinocerebellar pathway (carries proprioceptive information about the position of skeletal muscles, tendons, and joints) **17.** lateral spinothalamic tract of spinothalamic pathway (carries pain and temperature sensations) **18.** anterior spinocerebellar tract of spinocerebellar pathway (carries proprioceptive information about the position of skeletal muscles, tendons, and joints) **19.** anterior spinothalamic tract of spinothalamic pathway (carries crude touch and pressure sensations) **20.** Injuries to the primary motor cortex eliminate the ability to exert fine control over motor units, but gross movements may still be produced by cerebral nuclei using the reticulospinal or rubrospinal tracts.

### Chapter Review Questions

**1.** frontal lobe **2.** corpus callosum **3.** thalamus **4.** hypothalamus

**5.** optic chiasm **6.** mammillary body **7.** midbrain **8.** pons **9.** precentral gyrus **10.** central sulcus **11.** postcentral gyrus **12.** parietal lobe **13.** parieto-occipital sulcus **14.** occipital lobe **15.** pineal gland **16.** cerebral aqueduct **17.** fourth ventricle **18.** medulla oblongata **19.** false **20.** true **21.** false **22.** false **23.** true **24.** b **25.** b **26.** a **27.** d **28.** b **29.** a **30.** b **31.** d **32.** In any inflamed tissue, swelling occurs in the area of the inflammation. The accumulation of fluid in the subarachnoid space can cause damage by pressing against neurons. If the intracranial pressure is excessive, brain damage can occur, and if the pressure involves vital autonomic reflex areas, death can occur. **33.** The nerve that is blocked when a dentist works on a tooth in the bottom jaw is the mandibular nerve ($V_3$), a branch of the trigeminal nerve (V). This is a mixed nerve. Other sensory areas that could be affected by the injection are the inferior gums, lips, and portions of the tongue and palate. Thus the injection may also affect activities carried out by the muscles of mastication.

### Chapter Integration

**1.** The cranial nerve involved in Bell's palsy is the facial nerve (VII) on the left side of his face. **2.** The facial nerve is a mixed nerve, so John is likely to be experiencing both sensory and motor symptoms. His entire face is likely to be experiencing unusual sensations, including the loss of taste on the anterior two-thirds of his tongue on the left side. Besides being unable to move his facial muscles, he probably can't close his eye, make tears (from his lacrimal gland), or produce saliva (from his submandibular and sublingual salivary glands). **3.** Because John is not producing tears and he is unable to close his eye, he should probably take care to keep his eye moist. Using artificial tears (eye drops) and wearing an eye patch while sleeping are good ideas. He may

also want to drink with a straw because he is probably having difficulty in controlling his lips to successfully drink from a glass.

## CHAPTER 14

### Module Reviews

#### Module 14.1 Review

**A.** The SNS provides conscious and subconscious control over skeletal muscles. The ANS controls the visceral effectors: smooth muscle, glands, cardiac muscle, and adipocytes. **B.** Preganglionic neurons are part of the visceral reflex arcs. They are involved mostly in direct reflex responses, rather than responses to commands from the hypothalamus. **C.** Ganglionic neurons of autonomic ganglia innervate visceral effectors.

#### Module 14.2 Review

**A.** The two major divisions of the ANS are the sympathetic division and the parasympathetic division. **B.** In the sympathetic division, axons emerge from thoracic and lumbar segments of the spinal cord and innervate ganglia relatively close to the spinal cord. In the parasympathetic division, axons emerge from the brainstem and sacral segments of the spinal cord and innervate ganglia very close to (or within) target organs. **C.** The enteric nervous system (ENS) is a division of the nervous system. It consists of an extensive network of neurons and nerve networks in the walls of the digestive tract. Although it may be influenced by the sympathetic and parasympathetic divisions, the ENS initiates and coordinates complex visceral reflexes locally, without instructions from the CNS. **D.** The sympathetic division of the ANS is responsible for the physiological changes that occur when you are startled by a loud noise.

#### Module 14.3 Review

**A.** preganglionic neurons ($T_1$–$L_2$) → collateral ganglia → ganglionic neurons (postganglionic fibers) → visceral effectors in abdominopelvic cavity

**B.** An intramural ganglion is a group of neurons embedded in the tissue of a target organ. **C.** General responses to increased sympathetic activity include heightened mental alertness, increased metabolic rate, reduced digestive and urinary functions, activation of energy reserves, increased respiratory rate and dilation of respiratory passageways, elevated heart rate and blood pressure, and activation of sweat glands. General responses to increased parasympathetic activity include decreased metabolic rate, decreased heart rate and blood pressure, increased secretion by salivary and digestive glands, increased motility and blood flow in the digestive tract, and stimulation of urination and defecation.

### Module 14.4 Review

**A.** Splanchnic nerves are three groups of nerves—cardiopulmonary, abdominopelvic (greater, lesser, and lumbar), and pelvic (sacral)—that supply the viscera. **B.** The vagus nerve innervates the cardiac plexus, celiac plexus, inferior mesenteric plexus, and the hypogastric plexus. **C.** The vagus nerve (X) carries most of the outflow of the parasympathetic division of the ANS.

### Module 14.5 Review

**A.** In a tense or anxious person, increased sympathetic stimulation typically causes increased alertness (being "on edge"); a feeling of increased energy; increased heart rate, blood pressure, and breathing rate; increase in muscle tone; and mobilization of energy reserves. **B.** Acetylcholine (ACh) is released by all parasympathetic neurons. **C.** The parasympathetic division is sometimes referred to as the anabolic system because parasympathetic stimulation leads to a general increase in the nutrient content of the blood. Cells throughout the body respond to the increase by absorbing the nutrients and using them to support growth and other anabolic activities.

### Module 14.6 Review

**A.** Both alpha receptors and beta receptors are adrenergic receptors on the membranes of target cells. Alpha receptors are more sensitive to NE than beta receptors, but both receptors are stimulated by E. Alpha-1 receptor stimulation typically excites the target cell, and alpha-2 receptor stimulation generally inhibits the target cell. Stimulation of beta receptors may result in excitation or inhibition of the target cell. **B.** Nicotinic receptors are acetylcholine (ACh) receptors on the surfaces of sympathetic and parasympathetic ganglionic cells; muscarinic receptors are ACh membrane receptors that are located at all parasympathetic neuromuscular and neuroglandular junctions, and at a few sympathetic neuromuscular and neuroglandular junctions. **C.** Blocking the beta receptors on cardiac cells would decrease (or prevent) sympathetic stimulation of tissues containing those cells. As a result, heart rate, force of cardiac muscle contraction, and contraction of the smooth muscle in blood vessel walls would decrease, lowering blood pressure.

### Section 1 Review

**1.** cervical sympathetic ganglia **2.** sympathetic chain ganglia **3.** coccygeal ganglia **4.** cardiopulmonary splanchnic nerves **5.** cardiac and pulmonary plexuses **6.** celiac ganglion **7.** superior mesenteric ganglion **8.** splanchnic nerves **9.** inferior mesenteric ganglion **10.** sympathetic division **11.** thoracolumbar division **12.** thoracic nerves **13.** lumbar nerves **14.** parasympathetic division **15.** craniosacral division **16.** cranial nerves III, VII, IX, X; **17.** sacral nerves **18.** enteric nervous system **19.** f **20.** g **21.** h **22.** b **23.** a **24.** c **25.** e **26.** d

### Module 14.7 Review

**A.** The ANS adjusts the activities of virtually every body system, maintaining homeostasis without instructions or interference from the conscious mind. **B.** The two types of motor pathways are somatic and visceral. **C.** The somatic effector is skeletal muscle. The visceral effectors are smooth muscles, glands, cardiac muscle, and adipocytes.

### Module 14.8 Review

**A.** Dual innervation means that a given body structure receives instructions from both the sympathetic and parasympathetic divisions of the ANS. **B.** Autonomic tone is the background level of activity in sympathetic or parasympathetic motor neurons under resting conditions. It provides a mechanism for fine control of visceral function because it allows a resting neuron to be less active or more active rather than to simply switch from off to on. This is particularly important when only one division innervates a visceral organ or when an organ (e.g., the heart) must be precisely controlled over a broad range of activity levels. **C.** The blood vessels of the skin receive only sympathetic innervation. When you go outside into the cold, sympathetic neurons release NE and cause vasoconstriction of superficial blood vessels through stimulation of alpha receptors. When you get angry, sympathetic activation occurs, and large amounts of epinephrine enter the circulation. This stimulates beta receptors in the superficial blood vessels, dilating those vessels, and stimulates the heart, increasing blood pressure and blood flow. As a result, your skin—and most obviously your face—turns red.

### Module 14.9 Review

**A.** A visceral reflex is an autonomic reflex initiated in the viscera. It is an autonomic motor response that can be modified, facilitated, or inhibited by higher centers, especially those in the hypothalamus. **B.** Short reflexes are autonomic responses that bypass the CNS, whereas long reflexes involve interneurons within the CNS and autonomic delivery of motor commands to the effectors. **C.** The solitary nuclei are large masses of gray matter on each side of the medulla oblongata that serve as processing and sorting centers for visceral sensory information.

### Module 14.10 Review

**A.** Baroreceptors are stretch receptors that monitor changes in pressure. They are located along the digestive tract (stomach, intestines, and colon), within the walls of the urinary bladder, in the carotid and aortic sinuses, and in the lungs. **B.** Chemoreceptors are sensitive to changes in blood pH.

### Module 14.11 Review

**A.** The hypothalamus is the brain structure considered to be the headquarters for the ANS. **B.** The thalamus relays sematosensory information. **C.** A brain tumor pressing on the hypothalamus could interfere with autonomic function because the hypothalamus receives visceral sensory information and controls both sympathetic and parasympathetic functions.

### Section 2 Review

**1.** limbic system and thalamus **2.** hypothalamus **3.** pons **4.** spinal cord $T_1$–$L_2$ **5.** complex visceral reflexes **6.** vasomotor **7.** coughing **8.** respiratory **9.** sympathetic visceral reflexes **10.** parasympathetic visceral reflexes **11.** P **12.** P **13.** S **14.** P **15.** S **16.** S **17.** P **18.** S **19.** S **20.** S **21.** P **22.** S **23.** P **24.** Even though most sympathetic postganglionic fibers are adrenergic, releasing norepinephrine, a few are cholinergic, releasing acetylcholine. This distribution of the cholinergic fibers by the sympathetic division provides a method of regulating sweat gland secretion and selectively controlling blood flow to skeletal muscles while reducing the flow to other tissues in a body wall to maintain homeostasis.

### Chapter Review Questions

**1.** preganglionic neurons **2.** preganglionic fibers **3.** sympathetic ganglion **4.** ganglionic neurons **5.** postganglionic fibers **6.** parasympathetic ganglion **7.** acetylcholine **8.** norepinephrine

9. epinephrine 10. false 11. true 12. false 13. false 14. false 15. true 16. c 17. d 18. a 19. c 20. a 21. c 22. b 23. a 24. b 25. c 26. a 27. a 28. Visceral reflex arcs include a receptor, a sensory neuron, a processing center of one or more interneurons, and two or more visceral motor neurons. Short reflexes bypass the CNS entirely and control very simple motor responses in one small part of a target organ. Long reflexes deliver information to the CNS along the posterior roots of spinal nerves, within the sensory branches of cranial nerves, and within the autonomic nerves that innervate visceral effectors. Long reflexes predominate and are responsible for coordinating responses involving multiple organ systems. 29. There are many differences between the somatic and autonomic nervous systems. Three of the differences are the following: the somatic nervous system exerts control only over skeletal muscle, whereas the autonomic nervous system controls smooth muscle, glands, cardiac muscle, and adipocytes; somatic nervous system effector activity involves an upper motor neuron in the CNS and a lower motor neuron in the PNS, whereas autonomic nervous system effector activity involves an upper motor neuron in the CNS and two lower motor neurons in the PNS; and the somatic nervous system uses only ACh at its target organ, whereas the autonomic nervous system uses ACh, epinephrine, and norepinephrine. 30. The cranial nerves associated with the parasympathetic division are cranial nerves III, VII, IX, and X. The sacral nerves associated with the parasympathetic division are $S_2$, $S_3$, and $S_4$.

### Chapter Integration

1. The parasympathetic division of the autonomic nervous system would be affected if muscarinic receptors were blocked. 2. Blocking muscarinic receptors would block the neurotransmitter acetylcholine (ACh). 3. Muscarinic receptors are cholinergic receptors because they bind with ACh. If scopolamine blocks ACh, it would therefore be described as an anticholinergic drug. 4. ACh is the solitary neurotransmitter of the parasympathetic nervous system. If ACh is blocked while Andrew is wearing this patch, he is likely to experience a decrease in the "rest and digest" activities of the PNS. He is likely to experience dry eyes as a result of decreased lacrimal gland secretion, a dry mouth from decreased salivary gland secretion, dilated pupils due to pupillary reflex inhibition, and blurred near vision because accommodation of the lens for close vision is decreased. 5. Epinephrine would be more effective because it binds to the beta-2 receptors of the smooth muscles surrounding the airways, resulting in their relaxation and airway dilation, thus making it easier for Gregor to breathe. 6. A decrease in blood pressure would stimulate baroreceptors in the carotid sinus and aortic sinus to relay information to the cardiac and vasomotor centers in the medulla oblongata. These centers respond by increasing sympathetic impulses that initiate two complementary sympathetic visceral reflexes. The cardiac center increases the heart rate and force of contraction, and the vasomotor center changes the diameter of peripheral blood vessels to increase blood pressure.

## CHAPTER 15

### Module Reviews

#### Module 15.1 Review
A. The special senses originate at sensory receptor cells, which may be neurons or specialized receptor cells that communicate with sensory neurons. B. A generator potential is the depolarization of specialized sensory neurons capable of generating action potentials that are propagated to the CNS. C. Olfactory receptors are the dendrites of specialized olfactory neurons. The receptors for the other senses—taste, vision, equilibrium, and hearing—are specialized receptor cells that have inexcitable membranes and form synapses with the processes of sensory neurons.

#### Module 15.2 Review
A. Olfaction is the sense of smell and involves olfactory receptors in paired olfactory organs that respond to chemical stimuli. B. Axons from the olfactory epithelium collect into bundles that synapse in the olfactory bulb. Axons leaving the olfactory bulb then travel along the olfactory tract to the olfactory cortex, hypothalamus, and portions of the limbic system. C. The depolarization of an olfactory receptor cell begins with the binding of airborne odorant molecules to their receptors on the membranes of olfactory dendrites. Binding results in the activation of the enzyme adenylate cyclase and formation of cyclic AMP (cAMP). The cAMP opens sodium ion channels in the plasma membrane, leading to depolarization of the olfactory receptor cell's axon.

#### Module 15.3 Review
A. Gustation is the sense of taste, provided by taste receptors responding to chemical stimuli. B. Filiform papillae are slender conical projections on the superior surface of the tongue. They provide friction for the tongue to move objects in the mouth, but they do not contain taste buds. C. Taste receptors respond more readily to unpleasant than to pleasant stimuli. Therefore, we are more sensitive to bitter and sour sensations than to sweet and salty sensations. Such sensitivity has survival value by helping us avoid substances such as acids (which may harm mucous membranes) and bitter-tasting biological toxins.

#### Module 15.4 Review
A. Gustducins are G proteins—protein complexes that use second messengers to produce effects—that are associated with sweet, bitter, and umami sensations. B. The cranial nerves that carry gustatory information are the facial nerve (VII), glossopharyngeal nerve (IX), and vagus nerve (X). C. The gustatory pathway: taste receptors → facial nerve (VII), glossopharyngeal nerve (IX), and vagus nerve (X) → synapse in solitary nucleus of medulla oblongata → medial lemniscus → synapse in thalamus → gustatory cortex of the insula.

### Section 1 Review
1. umami 2. sour 3. bitter 4. salty 5. sweet 6. vallate papillae 7. foliate papillae 8. fungiform papillae 9. filiform papillae 10. e 11. i 12. f 13. m 14. a 15. c 16. k 17. j 18. l 19. h 20. g 21. b 22. d 23. Olfactory receptor cells are specialized sensory neurons whose cilia-shaped dendrites contain receptor proteins. The binding of odorant molecules to the receptor proteins results in a depolarization of the receptor cell and a generator potential that generates action potentials that propagate to the CNS. In contrast, the sensory receptor cells for taste, vision, equilibrium, and hearing have inexcitable membranes that undergo a graded depolarization and release neurotransmitters. These cells all form synapses with the processes of sensory neurons, which then depolarize in response to the neurotransmitters, producing a generator potential that generates action potentials that propagate to the CNS.

#### Module 15.5 Review
A. The first structures that form in eye development are the optic vesicles. B. The retina develops from the inner and outer layers of the optic cup. C. The ependymal cells of the outer layer of the optic cup develop into the photoreceptors.

#### Module 15.6 Review
A. Accessory structures associated with the eye include the eyelids, eyelashes, the superficial epithelium of the eye, and the structures associated with the production, secretion, and removal of tears. B. The conjunctiva is the first layer of

the eye that would be affected by inadequate tear production. **C.** Conjunctivitis, or pinkeye, is inflammation of the conjunctiva, generally caused by a pathogenic infection or by physical, allergic, or chemical irritation of the conjunctival surface. The most obvious sign of conjunctivitis, redness, results from the dilation of blood vessels deep to the conjunctival epithelium.

### Module 15.7 Review
**A.** The layers of the eye are the outer fibrous layer, the intermediate vascular layer (uvea), and the deep inner layer (retina). **B.** The color of eyes is largely determined by the density and distribution of melanocytes in the iris. **C.** Aqueous humor is found in the anterior cavity, between the cornea and the lens, and in the posterior cavity, where it is called vitreous humor.

### Module 15.8 Review
**A.** The cornea does not contain blood vessels. **B.** When light intensity decreases, sympathetic stimulation causes the dilator pupillae to contract, resulting in dilated (enlarged) pupils. **C.** Light passing along the visual axis strikes the fovea centralis, a region of the retina containing the highest concentration of photoreceptors.

### Module 15.9 Review
**A.** The focal point is the point at which the light rays from an object intersect on the retina. **B.** When the ciliary muscles are relaxed, you are viewing something in the distance. **C.** The near point of vision typically increases with age because the elasticity of the lens tends to decrease with age.

### Module 15.10 Review
**A.** The blind spot is the site of the optic disc on the retinal surface. The optic disc is called the blind spot because this portion of the retina has no photoreceptors and thus cannot detect an image. **B.** Rods are responsible for vision in dim light, and objects appear in black and white. Cones provide color vision but require intense light. **C.** When you enter a dimly lit room, you are unlikely to be able to see at all. The low-intensity light in the room would be focused on the fovea centralis, which contains only cones, which cannot be stimulated by low-intensity light.

### Module 15.11 Review
**A.** A photoreceptor is made up of an outer segment and inner segment. The tip of the outer segment is enclosed by pigmented epithelial cells. The outer segment contains flattened membranous discs containing visual pigment molecules sensitive to photons of light. The inner segment contains the major organelles of the cell, and it synapses with a bipolar cell. **B.** A dietary vitamin A deficiency would reduce the quantity of retinal the body could produce, thereby interfering with night vision. **C.** The three types of cones are blue cones, green cones, and red cones.

### Module 15.12 Review
**A.** The two configurations of retinal are the curved or bent 11-*cis* form and the more linear 11-*trans* form. **B.** The visual pigments undergo activation, bleaching, and reassembly during photoreception. **C.** ATP is required for converting 11-*trans* retinal back to its original 11-*cis* form.

### Module 15.13 Review
**A.** Optic radiation refers to the bundles of projection fibers linking the lateral geniculate bodies of the thalamus with the visual cortex in each cerebral hemisphere. **B.** Visual images are perceived in the visual cortex of the occipital lobes of the cerebrum.

### Module 15.14 Review
**A.** Emmetropia is the term for normal vision. **B.** A converging lens (one with at least one convex surface) is used to correct hyperopia.

### Section 2 Review
**1.** posterior cavity **2.** choroid **3.** fovea centralis **4.** optic nerve **5.** optic disc **6.** retina **7.** sclera **8.** fornix **9.** palpebral conjunctiva **10.** bulbar conjunctiva **11.** ciliary body **12.** iris **13.** lens **14.** cornea **15.** ciliary zonule (suspensory ligament) **16.** ora serrata **17.** l **18.** k **19.** g **20.** h **21.** o **22.** i **23.** f **24.** d **25.** n **26.** m **27.** j **28.** b **29.** a **30.** c **31.** e **32.** Light falling on the eye passes through the cornea and strikes the photoreceptors of the retina, bleaching (breaking down) many molecules of the pigment rhodopsin into retinal and opsin. After an intense exposure to light, a photoreceptor cannot respond to further stimulation until its rhodopsin molecules have been regenerated by the conversion of retinal molecules to their original shape and then recombined with opsin molecules. The "ghost" image remains until the rhodopsin molecules are regenerated.

### Module 15.15 Review
**A.** In olfaction and gustation, the sensory receptor cells are exposed to the external environment and communicate with the CNS. In equilibrium and hearing, by contrast, the sensory receptors are isolated within the internal ear and protected from the external environment. **B.** The sensory receptors of the internal ear are called hair cells because their free surfaces are covered with specialized processes similar to cilia and microvilli. Certain hair cells contain a kinocilium (a single large cilium) and many stereocilia. **C.** The internal ear can respond to gravity and acceleration, rotation, or sound, depending on which region is stimulated.

### Module 15.16 Review
**A.** From lateral to medial, the bones in the middle ear are the malleus, incus, and stapes. **B.** The auditory tube (also called the pharyngotympanic tube or eustachian tube) connects the nasopharynx with the middle ear and permits pressure equalization on both sides of the tympanic membrane.

### Module 15.17 Review
**A.** The bony labyrinth is a shell of dense bone composed of the semicircular canals, vestibule, and cochlea. **B.** The semicircular canals are part of the bony labyrinth, and the canals surround the semicircular ducts, which are part of the membranous labyrinth. **C.** Within the membranous labyrinth, receptors in the semicircular ducts respond only to rotation, receptors in the vestibule respond to gravity and linear acceleration, and receptors in the cochlear duct respond only to sound.

### Module 15.18 Review
**A.** Otoliths are densely packed calcium carbonate crystals that sit upon the gelatinous otolithic membrane in the maculae. **B.** Sensory receptors in the macula of utricle and macula of saccule are sensitive to changes in horizontal movement and vertical movement, respectively.

### Module 15.19 Review
**A.** The spiral organ is located in the cochlea of the internal ear. **B.** Perilymph fills the scala vestibuli and the scala tympani, and endolymph fills the cochlear duct. **C.** When the basilar membrane moves in response to pressure changes in the perilymph, the hair cells of the spiral organ press against the tectorial membrane.

### Module 15.20 Review
**A.** A decibel is the unit of measurement for the intensity of sound. **B.** Sound waves enter the external acoustic meatus → tympanic membrane → auditory ossicles → oval window → basilar membrane → round window of scala tympani → hair cells vibrate against tectorial membrane → information relayed to CNS by cochlear nerve of cranial nerve VIII. **C.** If the round window could not move, the perilymph would not be moved by the vibration of the stapes at the oval window; this would reduce or eliminate the perception of sound.

### Module 15.21 Review
**A.** The hair cell receptors for equilibrium are in the vestibule and the semicircular ducts. **B.** Cranial nerves III, IV, VI, and XI are involved with eye, head, and neck movements. **C.** The reflexive response to a loud noise is to turn your head and eyes toward the source of the noise.

### Module 15.22 Review

**A.** Cranial nerves VII (facial), IX (glossopharyngeal), and X (vagus) provide taste sensations from the tongue. **B.** Vertigo is caused by any condition that alters the function of the internal ear receptor complex, the vestibular nerve of the vestibulocochlear nerve, or sensory nuclei and pathways in the central nervous system. Common causes include motion sickness, excessive alcohol consumption, and exposure to certain drugs. **C.** Two common classes of hearing-related disorders are conductive hearing loss and sensorineural hearing loss.

### Section 3 Review

**1.** external ear **2.** middle ear **3.** internal ear **4.** auricle **5.** external acoustic meatus **6.** elastic cartilage **7.** tympanic membrane **8.** auditory ossicles (malleus, incus, and stapes) **9.** tympanic cavity **10.** petrous part of temporal bone **11.** vestibulocochlear nerve (VIII) **12.** cochlea **13.** auditory tube **14.** scala vestibuli **15.** vestibular membrane **16.** cochlear duct (scala media) **17.** spiral organ (organ of Corti) **18.** basilar membrane **19.** scala tympani **20.** spiral ganglion **21.** The rapid descent in the elevator causes the otoliths in the macula of the saccule of each vestibule to slide upward, producing the sensation of downward vertical motion. When the elevator abruptly stops, the otoliths do not. It takes a few seconds for them to come to rest in the normal position. As long as the otoliths are displaced, you will perceive movement.

### Chapter Review Questions

**1.** pigmented layer of the retina **2.** central retinal artery **3.** optic nerve **4.** central retinal vein **5.** neural part of the retina **6.** optic disc (blind spot) **7.** ganglion cell **8.** sclera **9.** choroid **10.** false **11.** false **12.** true **13.** false **14.** true **15.** true **16.** c **17.** b **18.** a **19.** c **20.** b **21.** The four primary taste sensations are salty, sweet, sour, and bitter. The two other taste sensations that have been identified are umami and water. **22.** Senile cataracts are a natural consequence of aging and occur when the lens loses its transparency. **23.** Otitis media is a middle ear infection caused by microorganisms traveling along the auditory tube from the nasopharynx to the middle ear.

### Chapter Integration

**1.** Myopia is the medical term for nearsightedness; people with myopia are able to see nearby objects clearly, but distant objects appear blurry or fuzzy. Hence, Mr. Drummond would have difficulty seeing while driving. **2.** Mr. Drummond will likely have a follow-up visit with an optometrist or ophthalmologist, who will prescribe glasses or contact lenses to alleviate the myopia and restore his distance sight to normal, or near-normal, vision. **3.** Vertigo is another term for dizziness. Vertigo is generally associated with a problem in the internal ear. The movement of Mr. Drummond's arms that occurs while he is standing still provides evidence that there may be a problem with his equilibrium centers (saccule and utricle). Often, inflammation due to infection or a cold affects the internal ear (or cranial nerve VIII) and causes vertigo. **4.** When Mr. Drummond closes his eyes, he does not have any visual cues, so his brain must rely solely on proprioception (the sense of relative position of body parts) and information from the equilibrium centers of the internal ear to maintain normal posture. Because either the internal ear receptors or the sensory nerves are not functioning normally, he is unstable. **5.** The most likely reason for Mr. Drummond's arms to drift to the left is that he is getting inappropriate sensations from equilibrium receptors, either at the maculae (affecting his ability to determine which way is "down") or at one of the horizontal semicircular ducts (making him attempt to compensate for a perceived roll to the right). **6.** Daniel is experiencing sensorineural hearing loss caused by his chronic exposure to loud sounds. This type of hearing loss can be caused by a single exposure to an extremely loud sound (as in a bomb blast), or by long-term exposure to sounds above 90 dB. **7.** Within Daniel's cochlea, the stereocilia in the hair cells of the spiral organ are most likely damaged. Although it is likely that he is conducting sound waves from his tympanic membrane, auditory ossicles, and perilymph, the hair cells are not generating action potentials along the spiral ganglion to the cochlear nerve of cranial nerve VIII. It is also possible that the nerve itself has been damaged as a result of Daniel's exposure to loud sounds. 8. It is not likely that Daniel will ever regain his hearing, because the hair cells of the spiral organ, the spiral ganglion, and cranial nerve VIII do not regenerate.

## CHAPTER 16

### Module Reviews

#### Module 16.1 Review

**A.** The common goal of both the nervous and endocrine systems is to preserve homeostasis by coordinating and regulating the activities of other cells, tissues, organs, and organ systems.

#### Module 16.2 Review

**A.** The structural classes of hormones are (1) amino acid derivatives (thyroid hormones, catecholamines, and tryptophan derivatives); (2) peptide hormones (glycoproteins or short polypeptide chains), which are chains of amino acids that are synthesized as prohormones; and (3) lipid derivatives (eicosanoids and steroid hormones), which contain carbon rings and side chains that are built from fatty acids or cholesterol.

#### Module 16.3 Review

**A.** The endocrine system includes organs whose primary function is the production of hormones or paracrines, which are chemical secretions that are transported by the extracellular fluid or bloodstream to target cells in other sites within the body. **B.** Organs of the endocrine system are the hypothalamus, pituitary gland, thyroid gland, adrenal glands, pancreas (pancreatic islets), pineal gland, and parathyroid glands. Organs of other systems that have secondary endocrine functions are the heart, thymus, digestive tract, kidneys, and gonads.

#### Module 16.4 Review

**A.** A hormone receptor is a protein molecule, located either on the plasma membrane (extracellular receptor) or inside the cell (intracellular receptor), that binds with a specific hormone. **B.** A first messenger is a hormone whose binding to a protein receptor in the plasma membrane gives rise to a second messenger in the cytoplasm. A second messenger changes the rate of various metabolic reactions by acting as an enzyme activator, an enzyme inhibitor, or a cofactor. **C.** Steroid hormones diffuse across the plasma membrane and bind to intracellular receptors in the cytoplasm or nucleus.

#### Module 16.5 Review

**A.** A regulatory hormone is a hormone secreted by the hypothalamus that controls endocrine cells in the anterior lobe of the pituitary gland. **B.** The three mechanisms of hypothalamic integration of neural and endocrine function are (1) secretion of antidiuretic hormone (ADH) and oxytocin (OXT), (2) secretion of regulatory hormones that control activity of the anterior lobe of the pituitary gland, and (3) neural (sympathetic) control over the endocrine cells of the adrenal medullae. **C.** The blood vessels of the hypophyseal portal system link the hypothalamus and anterior lobe of the pituitary gland. Unusually permeable fenestrated capillary beds in each structure are connected by portal vessels. This arrangement ensures that hypothalamic regulatory hormones reach the "downstream" endocrine cells of the anterior lobe directly, before mixing with, and being diluted by, the general circulation.

### Module 16.6 Review
**A.** The two lobes of the pituitary gland are the anterior lobe and the posterior lobe. **B.** The hormones produced and released by the anterior lobe of the pituitary gland are (1) thyroid-stimulating hormone (TSH), which targets the thyroid gland; (2) adrenocorticotropic hormone (ACTH), which targets the adrenal cortex; (3) follicle-stimulating hormone (FSH) and (4) luteinizing hormone (LH), which target the testes in males and the ovaries in females; (5) growth hormone (GH), which targets liver cells (which respond by synthesizing somatomedins); (6) prolactin (PRL), which targets mammary glands in females; and (7) melanocyte-stimulating hormone (MSH), which targets melanocytes in the skin. Hormones released by the posterior lobe of the pituitary gland are (8) oxytocin (OXT), which targets the uterus and mammary glands; and (9) antidiuretic hormone (ADH), which targets the kidneys. **C.** In a person who is dehydrated, the amount of ADH released by the posterior lobe of the pituitary gland increases in response to increased blood osmotic pressure resulting from a rise in solute concentration.

### Module 16.7 Review
**A.** The hypothalamic releasing hormones are corticotropin-releasing hormone (CRH), thyrotropin-releasing hormone (TRH), growth hormone–releasing hormone (GH–RH), prolactin-releasing hormone (PRH), and gonadotropin-releasing hormone (GnRH). **B.** Somatomedins mediate the action of growth hormone (GH). Increased levels of GH typically accompany increased levels of somatomedins. **C.** Increased circulating levels of glucocorticoids inhibit the release of CRH by the hypothalamus. The lack of CRH reduces the secretion of ACTH from the pituitary gland, so ACTH levels would decrease.

### Module 16.8 Review
**A.** Calcitonin aids in calcium regulation. **B.** The hormones of the thyroid gland are thyroxine ($T_4$), or tetraiodothyronine; triiodothyronine ($T_3$); and calcitonin. **C.** Most of the body's reserves of thyroid hormone, thyroxine ($T_4$), are bound to transport proteins in the bloodstream including thyroid-binding globulins, transthyretin, or albumin (a plasma protein). Because these compounds represent such a large reserve of thyroxine, it takes several days after removal of the thyroid gland for blood levels of thyroxine to decline.

### Module 16.9 Review
**A.** The parathyroid glands are embedded in the posterior surfaces of the lateral lobes of the thyroid gland. **B.** Parathyroid hormone (PTH) increases blood calcium levels by increasing calcium release from bones, increasing reabsorption of calcium from the blood by the kidneys, and increasing the production of calcitriol by the kidneys, which increases absorption of calcium from the digestive tract. **C.** Decreased blood calcium levels result in increased secretion of PTH.

### Module 16.10 Review
**A.** The regions of an adrenal gland from superficial to deep are the adrenal capsule, adrenal cortex, and adrenal medulla. The zones are the zona glomerulosa, zona fasciculata, and zona reticularis. **B.** The two hormone-secreting regions of an adrenal gland are the adrenal cortex and adrenal medulla. The cortex secretes mineralocorticoids, primarily aldosterone; glucocorticoids, mainly cortisol (or hydrocortisone) and corticosterone; and androgens. The medulla secretes epinephrine and norepinephrine. **C.** Increased cortisol levels would result in elevated blood glucose levels, because cortisol reduces the use of glucose by cells while increasing both the available glucose (by promoting glycogen breakdown)

and the conversion of amino acids to carbohydrates.

### Module 16.11 Review
**A.** The types of cells in the pancreatic islets (and their hormones) are alpha cells (glucagon), beta cells (insulin), delta cells (GH–IH), and pancreatic polypeptide cells, or PP cells (pancreatic polypeptide hormone). **B.** Insulin secretion lowers blood glucose concentration. **C.** Increased levels of glucagon stimulate the conversion of glycogen to glucose in the liver, which in turn reduces the amount of glycogen stored in the liver.

### Module 16.12 Review
**A.** The hormone-secreting cells of the pineal gland are pinealocytes. **B.** Melatonin secretion is influenced by day/night patterns. Increased amounts of light would inhibit the production (and release) of melatonin from the pineal gland, which receives input concerning the presence of light or darkness from visual pathway collaterals. **C.** In humans, melatonin may maintain circadian rhythms, affect the timing of sexual maturation, and protect against damage by free radicals.

### Module 16.13 Review
**A.** Diabetes mellitus is an endocrine disorder characterized by elevated blood glucose levels high enough to overwhelm the reabsorption capabilities of the kidneys. **B.** The two types of diabetes mellitus are type 1, characterized by inadequate insulin production by the pancreatic beta cells, and type 2, characterized by insulin resistance (failure of the body to use insulin properly). **C.** Some clinical problems associated with diabetes mellitus are diabetic retinopathy, blockages in heart circulation and increased risk for heart attack, diabetic nephropathy, diabetic neuropathy, and decreased blood flow to the distal limbs.

### Section 1 Review
**1.** catecholamines **2.** thyroid hormones **3.** tryptophan derivatives **4.** peptide hormones

**5.** short polypeptides **6.** glycoproteins **7.** small proteins **8.** lipid derivatives **9.** eicosanoids **10.** steroid hormones **11.** transport proteins **12.** c **13.** e **14.** f **15.** g **16.** j **17.** h **18.** b **19.** i **20.** a **21.** d **22.** thymus **23.** pineal gland **24.** pancreatic islet **25.** hypothalamus **26.** kidney **27.** adrenal gland **28.** pituitary gland **29.** gonad

### Module 16.14 Review
**A.** A synergistic effect occurs when two hormones have an additive effect so that the net result is greater than the effect each would produce alone. **B.** The hormones in a negative feedback mechanism have antagonistic effects. **C.** Hormones producing different, but complementary, effects are described as having integrative effects.

### Module 16.15 Review
**A.** The heart secretes natriuretic peptides, which ultimately cause blood pressure and blood volume to decrease. **B.** The kidneys secrete the hormone erythropoietin (EPO) and the enzyme renin. EPO increases red blood cell production. Upon its release into the bloodstream, renin functions as an enzyme that activates the renin-angiotensin-aldosterone system. Together, these responses ultimately cause blood pressure and blood volume to increase.

### Module 16.16 Review
**A.** Hormones necessary for normal growth and development include GH, thyroid hormones, insulin, PTH, calcitriol, and reproductive hormones.

### Module 16.17 Review
**A.** The three phases of the stress response are the alarm phase, the resistance phase, and the exhaustion phase. **B.** The resistance phase is characterized by long-term metabolic adjustments, including mobilization of remaining energy reserves (lipds from adipose tissue and amino acids from skeletal muscle), conservation of glucose, elevation of blood glucose concentration, and conservation of salts and water coupled with the loss of $K^+$ and $H^+$.

**C.** The collapse of vital systems occurs during the exhaustion phase of the general adaptation syndrome.

### Module 16.18 Review

**A.** The prefix *hyper-* refers to excessive hormone production, whereas *hypo-* refers to inadequate hormone production. **B.** Three common causes of hormone hyposecretion are metabolic factors, physical damage, and congenital disorders. **C.** Aldosteronism is characterized by increased body weight due to $Na^+$ and water retention and a low blood $K^+$ concentration.

### Section 2 Review

**1.** cardiac muscle cells **2.** release of natriuretic peptides **3.** increased $Na^+$ and water loss in urine **4.** reduced thirst **5.** decreasing blood pressure and volume **6.** specialized kidney endocrine cells; **7.** erythropoietin (EPO) released **8.** renin released **9.** increased red blood cell production **10.** aldosterone secreted **11.** ADH secreted **12.** increasing blood pressure and volume **13.** c **14.** g **15.** a **16.** e **17.** j **18.** d **19.** i **20.** b **21.** f **22.** h **23.** (1) The two hormones may have opposing or antagonistic effects, such as occurs between insulin (decreases blood glucose levels) and glucagon (increases blood glucose levels). (2) The two hormones may have an additive or synergistic effect, in which the net result is greater than the sum of each acting alone. An example is the enhanced glucose-sparing action of GH in the presence of glucocorticoids. (3) One hormone may have a permissive effect on another, in which the first hormone is needed for the second hormone to produce its effect. For example, epinephrine cannot alter the rate of tissue energy consumption without the presence of thyroid hormones. (4) The hormones may have integrative effects, in which the hormones may produce different but complementary results in specific tissues and organs. An example is the differing effects of calcitriol and parathyroid

hormone (PTH) on tissues involved in calcium metabolism; calcitriol increases calcium ion absorption from digestive system, and PTH inhibits osteoblast activity and enhances calcium ion reabsorption by the kidneys.

### Chapter Review Questions

**1.** true **2.** false **3.** true **4.** false **5.** false **6.** corticotropin-releasing hormone (CRH) **7.** thyrotropin-releasing hormone (TRH) **8.** growth hormone–releasing hormone (GH–RH) **9.** growth hormone–inhibiting hormone (GH–IH) **10.** prolactin releasing hormone (PRH) **11.** prolactin-inhibiting hormone (PIH) **12.** gonadotropin-releasing hormone (GnRH) **13.** adrenocorticotropic hormone (ACTH) **14.** glucocorticoids (steroid hormones) **15.** thyroid-stimulating hormone (TSH) **16.** thyroid hormones **17.** growth hormone (GH) **18.** somatomedins **19.** prolactin (PRL) **20.** follicle-stimulating hormone (FSH) **21.** luteinizing hormone (LH) **22.** inhibin **23.** testosterone **24.** estrogen **25.** progesterone **26.** inhibin **27.** melanocyte-stimulating hormone (MSH) **28.** oxytocin (OXT) **29.** antidiuretic hormone (ADH) **30.** b **31.** d **32.** d **33.** a **34.** b **35.** d **36.** The kidneys release erythropoietin to regulate red blood cell production, calcitriol to increase calcium and phosphate absorption by the intestinal tract, and renin to increase blood volume and blood pressure. **37.** Calcitonin decreases blood calcium levels. Parathyroid hormone increases blood calcium levels.

### Chapter Integration

**1.** Sherry's physician suspects hyperthyroidism because she exhibits the classic signs and symptoms of the condition: restlessness, anxiety, irritability, difficulty sleeping, diarrhea, weight loss, rapid heart rate, and tremors. **2.** Hyperthyroidism is the excess production of $T_3$, $T_4$, or both, by the thyroid gland. Excess thyroid hormone results in

a rapid heartbeat and an increased metabolic rate, among other clinical signs and symptoms. **3.** Sherry's physician could order blood tests to assay the levels of TSH, $T_3$, and $T_4$. Results confirming hyperthyroidism are lower-than-normal TSH levels and elevated $T_3$ and $T_4$ levels. Her physician could also determine whether her condition is primary hyperthyroidism (a problem with the thyroid gland) or secondary hyperthyroidism (a problem with hypothalamic-pituitary control of the thyroid gland). **4.** Sherry's signs and symptoms indicating nervous system involvement include tremors and anxiety.

## CHAPTER 17

### Module Reviews

#### Module 17.1 Review

**A.** The components of the cardiovascular system are the heart, blood vessels (arteries, capillaries, and veins), and blood. **B.** Blood functions to transport dissolved gases, nutrients, hormones, and metabolic wastes; regulate pH and ion composition of interstitial fluids; restrict fluid loss at injury sites; defend against toxins and pathogens; and stabilize body temperature.

#### Module 17.2 Review

**A.** Whole blood is composed of plasma (which contains albumins, globulins, fibrinogen, electrolytes, organic nutrients, and organic wastes) and formed elements (which are platelets, white blood cells, and red blood cells). White blood cells include neutrophils, eosinophils, basophils, lymphocytes, and monocytes. **B.** The hematocrit, also called the packed cell volume, is the percentage of whole blood volume contributed by formed elements. **C.** During an infection, you would expect the level of immunoglobulins (antibodies) in the blood to be elevated.

#### Module 17.3 Review

**A.** Hemocytoblasts form from hematopoietic stem cells.

Hemocytoblasts divide in the red bone marrow, producing two types of cells: lympoid stem cells and myeloid stem cells. **B.** Platelets are cell fragments that develop from megakaryocytes. These large cells shed their cytoplasm in small, membrane-enclosed packets. These packets are the platelets that enter the bloodstream. **C.** Lymphoid stem cells originate in red bone marrow and give rise to lymphocytes; these stem cells also produce lymphocytes in the thymus, spleen, and lymph nodes. Myeloid stem cells are cells in red bone marrow that give rise to all the formed elements except lymphocytes.

### Section 1 Review

**1.** f **2.** o **3.** b **4.** g **5.** l **6.** a **7.** i **8.** k **9.** j **10.** m **11.** e **12.** n **13.** h **14.** c **15.** d

### Module 17.4 Review

**A.** Hematology is the medical study of blood, blood-producing organs, and blood disorders. **B.** A complete blood count (CBC) is a diagnostic blood test used to determine underlying medical conditions. A CBC includes the RBC count, WBC count, erythrocyte indices (such as hemoglobin concentration), hematocrit, platelet count, and WBC differential count. **C.** A patient with a depressed hematocrit level would have anemia.

### Module 17.5 Review

**A.** The functional aspects of red blood cells include: (1) large surface area–to–volume ratio that results in a faster exchange of materials between the RBC and its surroundings; (2) the ability to form stacks of cells; and (3) flexibility, which allows RBCs to bend and pass through capillaries that are narrower than their diameter. **B.** Hemoglobin is a protein—composed of four globular subunits, each bound to a heme molecule—that gives RBCs the ability to transport oxygen in the blood. **C.** Oxyhemoglobin is hemoglobin whose iron has bound oxygen; it is bright red. Deoxyhemoglobin is hemoglobin whose iron has not bound oxygen; it is dark red.

## Module 17.6 Review

**A.** Bilirubin would accumulate in the blood, producing jaundice, because diseases that damage the liver impair the liver's ability to excrete bilirubin in the bile.

## Module 17.7 Review

**A.** Surface antigens on RBCs are glycoproteins in the plasma membrane; they determine blood type. **B.** The most common blood type in the U.S. is $O^+$. **C.** Only type $O^-$ blood can be safely transfused into a person whose blood type is $O^-$. **D.** A person with type A blood also has anti-B antibodies, so if the person received a transfusion of type B blood, the transfused red blood cells would clump, or agglutinate, potentially blocking blood flow to various organs and tissues.

## Module 17.8 Review

**A.** Hemolytic disease of the newborn (HDN) is a condition in which maternal antibodies attack and destroy fetal red blood cells, resulting in fetal anemia. It most commonly occurs in a sensitized $Rh^-$ mother who is carrying an $Rh^+$ fetus. **B.** When RhoGAM (which contains anti-Rh antibodies) is injected into a pregnant $Rh^-$ woman, the anti-Rh antibodies circulate in the mother's bloodstream, where they destroy any fetal RBCs there. This prevents the mother's immune system from making antibodies against the developing fetus's red blood cells. **C.** An $Rh^+$ mother carrying an $Rh^-$ fetus does not require a RhoGAM injection because the fetus is not at risk of Rh incompatibility. The fetus is not at risk because its RBCs lack Rh surface antigens, and the mother's plasma lacks anti-Rh antibodies.

## Module 17.9 Review

**A.** The five types of white blood cells are neutrophils, eosinophils, basophils, monocytes, and lymphocytes. **B.** Basophils enter damaged tissues and release a variety of chemicals, including histamine, which promotes inflammation. **C.** An infected cut would contain a large number of neutrophils, phagocytic white blood cells that are generally the first to arrive at the site of an injury.

## Module 17.10 Review

**A.** Hemostasis is the stopping of bleeding. It involves three phases: the vascular phase, the platelet phase, and the coagulation phase. **B.** During the vascular phase, local blood vessel constriction (vascular spasm) occurs at the injury site. In the platelet phase, platelets are activated, aggregate at the site, and adhere to damaged blood vessel surfaces. In the coagulation phase, factors released by platelets and endothelial cells interact with clotting factors (through either the extrinsic pathway, the intrinsic pathway, or the common pathway) to form a blood clot, a process involving the conversion of soluble fibrinogen to insoluble fibers of fibrin. **C.** Clot retraction and fibrinolysis follow the coagulation phase. The clot consists of fibrin, red blood cells, and platelets. The platelets contract, and the entire clot retracts, pulling the cut vessel edges together. As repairs continue, the clot dissoves through the process of fibrinolysis. Fibrinolysis is activated by plasminogen, which, in turn, activates plasmin, which erodes the foundation of the clot.

## Module 17.11 Review

**A.** Pernicious anemia is insufficient red blood cell production that results from a lack of vitamin $B_{12}$; the blood cells that do develop tend to be macrocytic (abnormally large) and abnormally shaped. Iron deficiency anemia results when the dietary intake or absorption of iron is insufficient, impairing normal hemoglobin synthesis; these blood cells are microcytic (abnormally small). **B.** Two types of leukemia are myeloid leukemia and lymphoid leukemia. **C.** In venipuncture, blood is collected from a superficial vein. Venipuncture is a common clinical procedure because superficial veins are easy to locate, the walls of veins are thinner than comparably sized arteries, and blood pressure in the venous system is relatively low, so the puncture wound seals quickly.

### Section 2 Review

**1.** hematocrit **2.** MCV **3.** Hb concentration **4.** MCH **5.** RBC count **6.** reticulocyte count **7.** leukemias **8.** myeloid leukemia **9.** lymphoid leukemia **10.** nutritional **11.** congenital **12.** infections **13.** pernicious anemia **14.** iron-deficiency anemia **15.** sickle cell disease **16.** hemophilia **17.** sepsis **18.** malaria **19.** e **20.** f **21.** a **22.** k **23.** b **24.** j **25.** l **26.** c **27.** d **28.** g **29.** i **30.** h **31.** During differentiation, the red blood cells of humans (and other mammals) lose most of their organelles, including nuclei and ribosomes. As a result, mature circulating RBCs can neither divide nor synthesize the structural proteins and enzymes required for cellular repairs.

### Chapter Review Questions

**1.** monocyte **2.** neutrophil **3.** erythrocyte, or red blood cell **4.** lymphocyte **5.** basophil **6.** eosinophil **7.** false **8.** false **9.** true **10.** false **11.** true **12.** a **13.** c **14.** a **15.** d **16.** b **17.** b **18.** b **19.** The two types of leukemias are characterized by elevated levels of circulating WBCs. Myeloid leukemia is characterized by the presence of abnormal granulocytes (neutrophils, eosinophils, and basophils) or other cells of the bone marrow. Lymphoid leukemia involves lymphocytes and their stem cells. **20.** Blood stabilizes and maintains body temperature by absorbing and redistributing the heat produced by active skeletal muscles. If body temperature is high, the heat will be lost across the surface of the skin. If body temperature is low, the warm blood is directed to the brain and to other temperature-sensitive organs. **21.** Type A blood has surface antigen A and anti-B plasma antibody; type B blood has surface antigen B and anti-A plasma antibody; type AB blood has both A and B surface antigens and neither anti-A nor anti-B plasma antibodies; and type O blood is lacking A and B surface antigens and has both anti-A and anti-B plasma antibodies.

### Chapter Integration

**1.** The drug that Piero is considering performs like the hormone erythropoietin (EPO). **2.** EPO is carried to red bone marrow, where it stimulates stem cells and developing RBCs. With more RBCs in circulation, Piero will be carrying more oxygen in his blood. This will allow his cells to deliver more oxygen to his working muscles and perform at a higher level. **3.** Serious side effects from taking EPO include blood clots, heart attack, and stroke from the increased blood viscosity due to the increased levels of erythrocytes in blood. **4.** Piero could consider moving from Miami to the mountains so he could train at a higher altitude. Engaging in strenuous exercise at a higher altitude would cause Piero to become more hypoxic. As a result, his body might release more EPO, stimulating erythropoiesis in a natural way and within healthy limits. **5.** Normal red blood cells are shaped like biconcave discs. Hemolysis is the rupture or destruction of red blood cells. In most cases, hemolysis occurs outside the bloodstream in the spleen, liver, and bone marrow. **6.** Among its functions, blood transports heat throughout the body and oxygen to cells. Because Ursula's oxygen-carrying red blood cells were being actively destroyed, her heart had to pump faster to supply tissues with the oxygen they need. The resulting increase in circulation brought more warm blood to her body surface, where greater-than-normal heat loss lowered her body temperature. **7.** Anemia and the accompanying low hemoglobin levels reduce the blood's ability to carry oxygen and nutrients to the body's vital organs and tissues. As a result, Ursula experienced fatigue and weakness.

## CHAPTER 18

### Module Reviews

#### Module 18.1 Review

**A.** The heart is located near the anterior chest wall, directly behind the sternum. **B.** The base of the heart lies at the superior border. The largest veins and arteries of the heart are attached to the base. The apex is located on the inferior aspect of the heart. **C.** The four chambers of the heart are the left atrium and left ventricle and the right atrium and right ventricle. **D.** Each circuit receives the same volume of blood from contraction of the ventricles.

#### Module 18.2 Review

**A.** The mediastinum is the mass of connective tissue between the two pleural cavities that contains the heart, along with the great vessels (large arteries and veins attached to the heart), thymus, esophagus, and trachea. **B.** The heart is surrounded by the pericardium in the anterior mediastinum, deep to the sternum and superior to the diaphragm. **C.** Cardiac tamponade can be a life-threatening condition because the accumulating fluid within the pericardial cavity restricts heart movement.

#### Module 18.3 Review

**A.** From superficial to deep, the layers of the heart wall are the pericardium, myocardium, and endocardium. **B.** The pericardium consists of a fibrous pericardium and two-layered serous pericardium. The parietal layer of the serous pericardium lines the outer wall of the pericardial cavity. A mesothelium lines the pericardial cavity, and an underlying layer of areolar tissue is in contact with the fibrous tissue of the fibrous pericardium. The visceral layer of the serous pericardium has an outer mesothelium and an underlying layer of areolar tissue that attaches directly to the myocardium. **C.** Cardiac tissue is metabolically active and dependent on mitochondrial activity for ATP and local capillaries for obtaining oxygen and nutrients.

#### Module 18.4 Review

**A.** The anterior view of the heart is dominated by the right atrium and right ventricle. **B.** Coronary veins collect blood from the myocardium and carry it to the right atrium. **C.** The anterior interventricular sulcus marks the boundary between the left and right ventricles on the heart's anterior surface; the shallower posterior interventricular sulcus marks the boundary between the left and right ventricles on the posterior surface; and the coronary sulcus is a deep groove that marks the border between the atria and the ventricles.

#### Module 18.5 Review

**A.** The right coronary artery supplies blood to the right atrium, portions of both ventricles, and the conducting system of the heart. The left coronary artery supplies blood to the left ventricle, left atrium, and interventricular septum. **B.** The anterior cardiac veins drain the anterior surface of the right ventricle and empty into the right atrium; the posterior vein of left ventricle drains the area (posterior surface of the left ventricle) supplied by the circumflex artery. **C.** Arteries: left coronary artery, anterior interventricular artery, right coronary artery, marginal arteries, circumflex artery, and posterior interventricular artery. Veins: great cardiac vein, anterior cardiac veins, posterior vein of left ventricle, middle cardiac vein, and small cardiac vein. **D.** During elastic rebound, blood in the aorta is driven forward into the systemic circuit, and some is forced back toward the left ventricle and into the coronary arteries.

#### Module 18.6 Review

**A.** The more muscular left ventricle must generate enough force to propel blood throughout the body in the systemic circuit, whereas the right ventricle must generate only enough force to propel blood the short distance to and from the lungs in the pulmonary circuit. **B.** Damage to the semilunar valve on the right side of the heart would affect blood flow to the pulmonary trunk.

#### Module 18.7 Review

**A.** Cardiac regurgitation is the abnormal backflow of blood into the atria when the ventricles contract, and it is prevented by the chordae tendineae and papillary muscles. **B.** Structurally, the cardiac skeleton stabilizes the positions of the heart valves and ventricular muscle cells; functionally, it electrically isolates the atrial and ventricular myocardia. **C.** Semilunar valves prevent the backflow of blood into the ventricles.

#### Module 18.8 Review

**A.** Arteriosclerosis is any thickening and toughening of arterial walls; atherosclerosis is a type of arteriosclerosis characterized by changes in the endothelial lining and the formation of fatty deposits (plaque) in the tunica media. **B.** Coronary ischemia is a condition in which the blood supply of the coronary arteries is reduced. **C.** Stents are wire-mesh tubes that prop open the natural blood vessel, creating a channel to restore blood flow. Without adequate blood flow to the cardiac muscle, the tissue would die.

#### Section 1 Review

**1.** aortic arch **2.** superior vena cava **3.** right pulmonary arteries **4.** ascending aorta **5.** fossa ovalis **6.** opening of coronary sinus **7.** right atrium **8.** pectinate muscles **9.** tricuspid valve cusp **10.** chordae tendineae **11.** papillary muscle **12.** right ventricle **13.** inferior vena cava **14.** pulmonary trunk **15.** pulmonary valve **16.** left pulmonary arteries **17.** left pulmonary veins **18.** left atrium **19.** aortic valve **20.** mitral valve cusp **21.** left ventricle **22.** interventricular septum **23.** trabeculae carneae **24.** moderator band **25.** g **26.** h **27.** j **28.** a **29.** d **30.** i **31.** b **32.** c **33.** f **34.** e **35.** (a) deoxygenated blood flow: right atrium → tricuspid valve (right atrioventricular valve) → right ventricle → pulmonary semilunar valve; (b) oxygenated blood flow: left atrium → mitral valve (left atrioventricular, or bicuspid, valve) → left ventricle → aortic semilunar valve.

#### Module 18.9 Review

**A.** The cardiac cycle is the period between the start of one heartbeat and the beginning of the next. **B.** The alternate term for heart contraction is *systole*, and the term for heart relaxation is *diastole*. **C.** At a representative heart rate of 75 bpm, atrial systole has a duration of 100 msec, and ventricular systole lasts 270 msec.

#### Module 18.10 Review

**A.** The phases of the cardiac cycle are atrial systole, atrial diastole, ventricular systole, and ventricular diastole. **B.** The two phases of ventricular systole are isovolumetric contraction and ventricular ejection. **C.** No. When pressure in the left ventricle first rises, the heart is contracting, but blood is not leaving the heart. During this initial phase of contraction, called the period of isovolumetric contraction, both the AV valves and the semilunar valves are closed. The increase in pressure is the result of the cardiac muscle contracting. When the pressure in the ventricle exceeds that in the aorta, the aortic semilunar valves are forced open, and blood is rapidly ejected from the ventricle.

#### Module 18.11 Review

**A.** Cardiac muscle has a long refractory period that continues until relaxation is well under way. As a result, another action potential cannot arrive quickly enough for summation to occur, and thus tetany cannot occur. **B.** The three stages of an action potential in a cardiac muscle cell are rapid depolarization, plateau, and repolarization. **C.** Slow calcium channels are voltage-gated calcium channels that open slowly and remain open for a relatively long period—about 175 msec. When they are open, the entry of calcium ions into the cell roughly balances the loss of sodium ions through the active transport of sodium ions. As a result, the membrane potential remains near 0 mV for an extended period.

#### Module 18.12 Review

**A.** Autorhythmicity is the ability of cardiac muscle tissue to contract without neural or hormonal stimulation. **B.** If the cells of the SA node failed to function, the heart would continue to beat, but at a slower rate; the AV node would act as the pacemaker. **C.** If the impulses from the atria were not delayed at the AV node, they would be conducted through the ventricles so quickly by the bundle branches and Purkinje cells that the ventricles would begin contracting before the atria had finished contracting. As a result, the ventricles would not be as full of blood as they could be, and the pumping action of the heart would be less efficient.

### Module 18.13 Review

**A.** An electrocardiogram (ECG or EKG) is a recording of the electrical activities of the heart over time. **B.** The important features of an ECG are the P wave (atrial depolarization), the QRS complex (ventricular depolarization), and the T wave (ventricular repolarization). **C.** Ventricular fibrillation, which causes the condition known as cardiac arrest, is fatal because the ventricles merely quiver and do not pump blood into the systemic circulation.

### Module 18.14 Review

**A.** Bradycardia is a heart rate below 60 beats per minute; tachycardia is a heart rate above 100 beats per minute. **B.** The cardioacceleratory center in the medulla oblongata activates sympathetic neurons to increase heart rate; the cardioinhibitory center (also in the medulla oblongata) controls the parasympathetic neurons that slow heart rate. **C.** Like NE, caffeine acts directly on the conducting system and contractile cells of the heart, increasing the rate at which they depolarize. Drinking large amounts of caffeinated beverages would therefore increase the heart rate.

### Module 18.15 Review

**A.** The end-diastolic volume (EDV) is the amount of blood a ventricle contains at the end of diastole, just before a contraction begins; the end-systolic volume (ESV) is the amount of blood

that remains in the ventricle at the end of ventricular systole. **B.** An increase in venous return would stretch the heart muscle. The more the heart muscle is stretched, the more forcefully it will contract (to a point). The more forceful the contraction, the more blood the heart will eject with each beat (stroke volume). Therefore, increased venous return would increase the stroke volume (if all other factors are constant). **C.** An increase in sympathetic stimulation of the heart would increase heart rate and force of contraction. The end-systolic volume (ESV) is the amount of blood that remains in a ventricle after a contraction (systole). The more forcefully the heart contracts, the more blood it ejects. Therefore, increased sympathetic stimulation results in a lower ESV.

### Module 18.16 Review

**A.** Heart failure is a condition in which the heart can no longer meet the oxygen and nutrient demands of peripheral tissues. **B.** SV = EDV – ESV, so SV = 125 mL – 40 mL = 85 mL.

### Section 2 Review

**1.** d **2.** c **3.** a **4.** g **5.** b **6.** j **7.** i **8.** f **9.** e **10.** h **11.** mitral valve closes **12.** increasing **13.** decreasing **14.** less than **15.** aortic valve is forced open **16.** aorta **17.** ventricular systole

### Chapter Review Questions

**1.** true **2.** false **3.** false **4.** true **5.** true **6.** c **7.** a **8.** b **9.** c **10.** b **11.** b **12.** a **13.** c **14.** a **15.** a **16.** The regurgitation of blood into the atria as the ventricles contract is prevented as the papillary muscles contract and pull on the chordae tendineae. **17.** SA node → internodal pathways → AV node → AV bundle → right and left bundle branches → Purkinje fibers. **18.** The epicardium, or visceral layer of the serous pericardium, is a mesothelial serous membrane that covers the outer surface of the heart. The myocardium is the middle layer of the heart and contains cardiac

muscle tissue, blood vessels, and nerves. The endocardium is made of simple squamous epithelium (endothelium) and areolar tissue and lines the inner surface of the heart. **19.** Sympathetic stimulation increases heart rate and force of contraction (contractility). Parasympathetic stimulation decreases heart rate and force of contraction. **20.** The first heart sound is "lubb" ($S_1$) and marks the start of ventricular contraction. Closing of the AV valves produces this sound. The second sound is "dupp" ($S_2$) and marks the start of ventricular diastole. This sound is produced by the closing of the semilunar valves. The third heart sound ($S_3$) is associated with blood flow into the atria. The fourth heart sound ($S_4$) is associated with atrial contraction.

### Chapter Integration

**1.** The mitral valve is closed during ventricular systole. **2.** The mitral valve is prolapsing into the left atrium. There is no regurgitation because, despite the prolapse, the two cusps of the mitral valve are still in contact and are preventing the backflow of blood. **3.** The echocardiogram could reveal abnormally shaped mitral valve cusps or chordae tendineae that are elongated and unable to restrain the mitral valve cusps during ventricular systole.

# CHAPTER 19

### Module Reviews

### Module 19.1 Review

**A.** The pulmonary circuit transports blood from the right ventricle through the pulmonary arteries, capillaries in the lungs, and pulmonary veins and returns it to the left atrium. **B.** The systemic circuit transports blood through the arteries, capillaries, and veins of the body from the left ventricle to the right atrium. Blood returning to the heart from the systemic circuit must complete the pulmonary circuit before it re-enters the

systemic circuit. **C.** The right atrium receives blood from the systemic circuit. **D.** Efferent vessels (arteries) carry blood away from the heart, afferent vessels (veins) carry blood to the heart, and exchange vessels (capillaries) exchange nutrients, dissolved gases, and wastes between the blood and interstitial fluid.

### Module 19.2 Review

**A.** The five general classes of blood vessels are arteries, arterioles, capillaries, venules, and veins. **B.** A capillary is a small blood vessel, located between an arteriole and a venule, whose thin wall permits exchange between blood and interstitial fluid by diffusion.

### Module 19.3 Review

**A.** The two types of capillaries with a complete endothelium are continuous capillaries and fenestrated capillaries. **B.** Fenestrated capillaries are located where solutes as large as small peptides move freely into and out of the blood. These sites include endocrine glands, the choroid plexus of the brain, absorptive areas of the intestine, and filtration areas of the kidneys.

### Module 19.4 Review

**A.** In the arterial system, pressures are high enough to keep the blood moving away from the heart and through arteries and capillaries. In the venous system, blood pressure is too low to keep the blood moving back toward the heart. Valves in veins prevent blood from flowing backward whenever the venous pressure drops. **B.** Assisted by the presence of valves in the veins, which prevent backflow of the blood, the contraction of the surrounding skeletal muscles squeezes venous blood toward the heart. **C.** The systemic venous system and systemic arterial system contain 64 and 13 percent of the total blood volume, respectively. The remaining volume is contained in the heart (7%), pulmonary circulation (9%), and systemic capillaries (7%). **D.** Varicose veins are sagging, swollen superficial veins in the thighs and legs. They result from the pooling of blood due to gravity and the failure of venous valves.

### Section 1 Review

**1.** artery **2.** vein **3.** smooth muscle **4.** internal elastic membrane **5.** endothelium **6.** tunica intima **7.** tunica media **8.** tunica externa, or tunica adventitia **9.** e **10.** i **11.** j **12.** g **13.** a **14.** k **15.** b **16.** h **17.** l **18.** c **19.** d **20.** f **21.** Contraction or relaxation of the smooth muscle cells in precapillary sphincters controls the blood flow into capillaries. During exercise, certain precapillary sphincters will relax to allow more blood flow into the working muscles to meet their metabolic needs, and into the skin to help with cooling. Other precapillary sphincters will contract to decrease blood flow into nonessential organs such as the digestive viscera. In cold temperatures, precapillary sphincters in the periphery and skin will contract to decrease blood flow to these areas and conserve heat.

### Module 19.5 Review

**A.** Neural and endocrine regulatory mechanisms influence heart rate, stroke volume, peripheral resistance, and venous pressure. **B.** Arterial pressure is much higher than venous pressure because it must push blood a greater distance and through progressively smaller and thinner vessels. **C.** Very low capillary pressure is beneficial because it allows time for diffusion between the blood and the surrounding interstitial fluid.

### Module 19.6 Review

**A.** Total peripheral resistance reflects a combination of vascular resistance, vessel length, vessel luminal diameter, blood viscosity, and turbulence. **B.** The equation $R \propto 1/r^4$ states that resistance (R) is inversely proportional to the fourth power of the vessel radius (r). This means that a small change in vessel diameter results in a large change in resistance. **C.** An increase in vessel diameter would reduce peripheral resistance. (An increase in vessel length would increase peripheral resistance.)

### Module 19.7 Review

**A.** Blood flow is the volume of blood flowing per unit of time through a vessel or group of vessels; it is directly proportional to arterial pressure (increased pressure results in increased blood flow) and inversely proportional to peripheral resistance (increased resistance results in decreased blood flow). **B.** In a healthy person, blood pressure is greater in the aorta than in the inferior vena cava. If the pressure were higher in the inferior vena cava than in the aorta, blood would flow in the reverse direction. **C.** Using the formula MAP = diastolic pressure + (pulse pressure)/3, MAP equals 70 + (125 − 70)/3, which equals 70 + 18.3, or 88.3 mm Hg.

### Module 19.8 Review

**A.** Fluid moves into a capillary (reabsorption) whenever blood colloid osmotic pressure (BCOP) is greater than capillary hydrostatic pressure (CHP). **B.** Edema is an abnormal accumulation of interstitial fluid in peripheral tissues. **C.** Any condition that affects either blood pressure or osmotic pressures in the blood or tissues will shift the balance between hydrostatic and osmotic forces.

### Module 19.9 Review

**A.** Tissue perfusion is blood flow to tissues that is sufficient to deliver adequate oxygen and nutrients. **B.** Cardiovascular autoregulation involves local factors changing the pattern of blood flow within capillary beds in response to chemical changes in interstitial fluids. **C.** Baroreceptor reflexes respond to changes in blood pressure. The baroreceptors—located in the walls of the carotid sinuses and aortic arch—monitor the degree of stretch at those sites. When blood pressure increases, the cardiac centers decrease cardiac output and the vasomotor center is inhibited, resulting in vasodilation; when blood pressure decreases, the cardiac centers increase cardiac output and the vasomotor center is stimulated, resulting in vasoconstriction.

### Module 19.10 Review

**A.** Epinephrine and norepinephrine from the adrenal medullae provide short-term regulation of decreasing blood pressure and blood volume. **B.** Vasoconstriction of the renal artery would decrease both blood flow and blood pressure at the kidney. In response, the kidney would release EPO and renin. EPO increases the rate of red blood cell fomation, which leads to an increase in blood volume. The release of renin would lead to an increase in the level of angiotensin II. The angiotensin II would bring about increased blood pressure and increased blood volume. **C.** Excessive stretching of the right atrium during diastole causes the release of atrial natriuretic peptide (ANP). Excessive stretching of the ventricles during diastole causes the release of brain natriuretic peptide (BNP). The roles of these peptides are to trigger responses whose combined effects act to decrease blood volume and blood pressure. As blood volume and blood pressure decrease, natriuretic peptide production ceases.

### Module 19.11 Review

**A.** Chemoreceptors are located in the carotid bodies, in the aortic bodies, and on the ventrolateral surfaces of the medulla oblongata. **B.** Chemoreceptor reflexes respond to decreasing pH and oxygen levels and to increasing $CO_2$ levels in the blood and cerebrospinal fluid (CSF) by adjusting cardiovascular and repiratory activity. These reflexes stimulate responses by the cardiovascular centers to increase blood pressure through vasoconstriction and increased cardiac output and stimulate responses by the respiratory centers to increase the respiratory rate, which increases pH and oxygen levels and decreases $CO_2$ levels. **C.** An increase in the respiratory rate reduces $CO_2$ levels.

### Module 19.12 Review

**A.** The respiratory pump is a mechanism by which a reduction of pressure in the thoracic cavity during inhalation assists venous return to the heart. **B.** During exercise, cardiac output increases, and blood flow to skeletal muscles increases at the expense of blood flow to less essential organs. Unless compensatory vasoconstriction occurs in "less essential" organs, such as those of the digestive system, vasodilation in skeletal muscles would cause a potentially dangerous decrease in blood pressure and blood flow throughout the body during exercise.

### Module 19.13 Review

**A.** Compensatory mechanisms that respond to blood loss include an increase in cardiac output, a mobilization of venous blood reservoir, peripheral vasoconstriction, and the release of hormones that promote the retention of fluids and the maturation of erythrocytes. **B.** The immediate, short-term problem during hemorrhage is to maintain adequate blood pressure and peripheral blood flow; the long-term problem is to restore normal blood volume. **C.** Circulatory shock occurs when blood loss exceeds about 35 percent of the total blood volume. Circulatory shock involves a series of positive feedback loops that are initiated after homeostasis has been disrupted. Progressive shock is the next stage after circulatory shock. It, too, is a series of positive feedback loops that accelerate tissue damage. Irreversible shock is the fatal stage that occurs if the positive feedback loops initiated during progressive shock are not broken.

### Section 2 Review

**1.** f **2.** i **3.** b **4.** a **5.** c **6.** e **7.** l **8.** g **9.** k **10.** h **11.** d **12.** j **13.** d **14.** j **15.** g **16.** i **17.** h **18.** b **19.** e **20.** a **21.** c **22.** f

### Module 19.14 Review

**A.** Blood islands are aggregations of embryonic cells scattered within the yolk sac that form blood vessels and blood cells. During embryonic development, these islands give rise to hematopoietic stem cells and hemangioblasts. **B.** Hemangioblasts remodel blood islands first into capillary networks and then into larger arterial and venous networks.

### Module 19.15 Review

**A.** The two circulatory circuits of the cardiovascular system are the pulmonary circuit and the systemic circuit. The pulmonary circuit carries deoxygenated blood from the right ventricle to the lungs and returns oxygenated blood to the left atrium. The systemic circuit carries oxygenated blood to the organs and tissues of the body and returns deoxygenated blood to the right atrium. **B.** General patterns of blood vessel organization include the following: (1) The peripheral distributions of arteries and veins on the body's left and right sides are generally identical, except near the heart, where the largest vessels connect to the atria or ventricles; (2) a single vessel may have several names as it crosses specific anatomical boundaries, making accurate anatomical descriptions possible; and (3) tissues and organs are usually serviced by several arteries and veins. **C.** right ventricle → pulmonary trunk → right and left pulmonary arteries → pulmonary arterioles → alveoli → pulmonary venules → pulmonary veins → left atrium.

### Module 19.16 Review

**A.** The largest artery in the body is the aorta. **B.** The two large veins that collect blood from the systemic circuit are the superior vena cava and the inferior vena cava. **C.** A major anatomical difference between the arterial and venous systems is the existence of dual (superficial and deep) venous drainage in the neck and limbs.

### Module 19.17 Review

**A.** The two arteries formed by the division of the brachiocephalic trunk are the right common carotid artery and the right subclavian artery. **B.** A blockage of the left subclavian artery would interfere with blood flow to the left arm. **C.** Thor's bulging vein is his external jugular vein.

### Module 19.18 Review

**A.** The branches of the external carotid artery are the superficial temporal, maxillary, occipital, facial, lingual, and external carotid arteries. **B.** The arterial structure in the neck region that contains baroreceptors is the carotid sinus. **C.** The veins that combine to form the brachiocephalic vein are the external jugular, internal jugular, vertebral, and subclavian veins.

### Module 19.19 Review

**A.** The three branches of the internal carotid artery are the ophthalmic, anterior cerebral, and middle cerebral arteries. **B.** The cerebral arterial circle (also known as the circle of Willis) is a ring-shaped anastomosis that encircles the infundibulum of the pituitary gland. Its anatomical arrangement creates alternate pathways in the cerebral circulation, so that if blood flow is interrupted in one area, other blood vessels can continue to perfuse the entire brain with blood. **C.** The internal jugular veins drain the dural sinuses of the brain.

### Module 19.20 Review

**A.** The inferior vena cava collects most of the venous blood inferior to the diaphragm. **B.** The major branches of the inferior vena cava are the lumbar, gonadal, hepatic, renal, adrenal, and phrenic veins. **C.** Rupturing the celiac trunk would most directly affect the stomach, inferior portion of the esophagus, spleen, liver, gallbladder, and proximal portion of the small intestine.

### Module 19.21 Review

**A.** The unpaired branches of the abdominal aorta that supply blood to the visceral organs are the celiac trunk, superior mesenteric artery, and inferior mesenteric artery. **B.** The three veins that merge to form the hepatic portal vein are the superior mesenteric, inferior mesenteric, and splenic veins. **C.** The left and right gastro-epiploic veins carry blood away from the stomach. **D.** The function of the hepatic portal system is to carry blood with absorbed nutrients from the digestive organs to the liver for processing.

The primary vessel of the hepatic portal system is the hepatic portal vein.

### Module 19.22 Review

**A.** The first two branches of the common iliac artery are the internal iliac artery and the external iliac artery. **B.** The plantar venous arch delivers blood to the anterior tibial, posterior tibial, and fibular (peroneal) veins. **C.** A blockage of the popliteal vein would interfere with blood flow in the tibial and fibular (peroneal) veins (which form the popliteal vein) and the small saphenous vein (which joins the popliteal vein).

### Module 19.23 Review

**A.** left ventricle → ascending aorta → aortic arch → thoracic aorta → abdominal aorta → right common iliac → right external iliac → right femoral → right deep femoral → right hip joint. **B.** right forearm → right brachial → right axillary → right subclavian → right brachiocephalic → superior vena cava → right atrium.

### Module 19.24 Review

**A.** Deoxygenated blood flows from the fetus to the placenta through a pair of umbilical arteries, and oxygenated blood returns from the placenta in a single umbilical vein. The umbilical vein then drains into the ductus venosus within the fetal liver. **B.** The six structures vital to fetal circulation are two umbilical arteries, one umbilical vein, the ductus venosus, the foramen ovale, and the ductus arteriosus. After birth, the foramen ovale closes and persists as the fossa ovalis, a shallow depression; the ductus arteriosus persists as the ligamentum arteriosum, a fibrous cord; and the umbilical vessels and ductus venosus persist throughout life as fibrous cords. **C.** Ventricular septal defects are abnormal openings between the left and right ventricles. Tetralogy of Fallot includes a ventricular septal defect plus three other heart defects: a narrowing of the pulmonary trunk, a displaced aorta, and an enlarged right ventricle with corresponding thickened right and left ventricles.

### Section 3 Review

1. common carotid 2. subclavian 3. brachiocephalic trunk 4. brachial 5. radial 6. popliteal 7. fibular (peroneal) 8. aortic arch 9. celiac trunk 10. renal 11. common iliac 12. external iliac 13. femoral 14. anterior tibial 15. vertebral 16. internal jugular 17. brachiocephalic 18. axillary 19. cephalic 20. median antebrachial 21. ulnar 22. great saphenous 23. fibular 24. superior vena cava 25. inferior vena cava 26. internal iliac 27. deep femoral 28. posterior tibial

### Chapter Review Questions

1. internal carotid artery 2. basilar artery 3. vertebral artery 4. anterior communicating artery 5. anterior cerebral artery 6. posterior communicating artery 7. posterior cerebral artery 8. cerebral arterial circle, or circle of Willis 9. d 10. b 11. d 12. c 13. a 14. d 15. a 16. b 17. A brachiocephalic artery is not needed on the left side of the aortic arch because the majority of the heart lies to the left of midline; this allows the left common carotid artery and left subclavian artery to have direct access to the head and left upper limb, respectively. 18. Blood volume is distributed throughout the body as follows: 64% in the systemic venous system, 9% in the pulmonary circuit, 7% in the heart, 13% in the systemic arterial circuit, and 7% in systemic capillaries. 19. The hepatic portal system directs blood with absorbed nutrients from the digestive system to the liver for processing.

### Chapter Integration

1. The catheter would be inserted into Mr. Samuel's femoral artery. It would be guided then to his external iliac artery, common iliac artery, abdominal aorta, and thoracic aorta to the aortic arch. 2. The dye would be released into the aortic arch at the base of the brachiocephalic trunk and left common carotid artery. This is the

best place because the dye would then travel to the brain by the two common carotid arteries and the two vertebral arteries. **3.** Mr. Samuel was lucky to have had his blockage on one of his vertebral arteries because this vessel joins with the vertebral artery from the opposite side to form the basilar artery. Because the cerebral arterial circle (circle of Willis) that is distal to the basilar artery ensured even distribution of blood, Mr. Samuel did not experience a complete interruption of blood flow to his brain. A blockage in the basilar artery would be more serious because it would block the blood from both vertebral arteries, and a blockage in the middle cerebral artery would be serious because it is distal to the cerebral arterial circle.

# CHAPTER 20

## Module Reviews

### Module 20.1 Review

**A.** Lymphocytes are the primary cells of the lymphatic system. These cells respond to pathogens, abnormal body cells, and foreign proteins such as toxins released by some bacteria. Lymph is the interstitial fluid that has entered a lymphatic vessel. **B.** The components of the lymphatic system are lymphocytes, lymph, lymphatic vessels, primary lymphoid tissues and organs (red bone marrow and the thymus), and secondary lymphoid tissues and organs (lymph nodes, tonsils, MALT, the appendix, and the spleen).

### Module 20.2 Review

**A.** Lymphatic vessels transport lymph from peripheral tissues to the venous system. **B.** Overlapping endothelial cells in lymphatic capillaries act as one-way valves that permit the entry of fluids and solutes but prevent their return to the intercellular spaces. **C.** Lymphatic valves prevent the backflow of lymph in some lymphatic vessels. **D.** Lymphatic capillaries in the small intestine are called lacteals, and they

transport lipids. Lymphatic capillaries elsewhere in the body are associated with blood capillary networks. Lymphatic capillaries are not found in the cornea of the eye, where there is no blood supply.

### Module 20.3 Review

**A.** The right lymphatic duct collects lymph from the right side of the body superior to the diaphragm; the thoracic duct collects lymph from the body inferior to the diaphragm and from the left side of the body superior to the diaphragm. **B.** The right jugular, right subclavian, and right bronchomediastinal trunks empty into the right lymphatic duct, and the left bronchomediastinal, left subclavian, and left jugular trunks empty into the thoracic duct. **C.** Lymphedema is the accumulation of interstitial fluids that results from blocked lymphatic drainage. If the condition persists, connective tissues lose their elasticity, and the swelling becomes permanent.

### Module 20.4 Review

**A.** The three main classes of lymphocytes are T cells, B cells, and natural killer (NK) cells. **B.** The red bone marrow, thymus, and peripheral lymphoid tissues are involved in lymphocytopoiesis. **C.** B cells are responsible for antibody-mediated immunity.

### Module 20.5 Review

**A.** Mucosa-associated lymphoid tissue (MALT) is the collection of lymphoid tissue that protects epithelia lining the digestive, respiratory, urinary, and reproductive tracts. **B.** Tonsils are large lymphoid nodules in the walls of the pharynx. The five tonsils are the left and right palatine tonsils, a single pharyngeal tonsil (adenoid), and a pair of lingual tonsils. **C.** Lymph flow through a lymph node: afferent lymphatics → subcapsular space → cortex → paracortex → medullary sinus → efferent lymphatics.

### Module 20.6 Review

**A.** The thymus is located in the anterior mediastinum, posterior

to the sternum. **B.** The thymus is a pink, grainy organ ranging in weight from 40 g at puberty to less than 12 g by age 50. A capsule covers the thymus and divides it into two lobes, and fibrous partitions called septa divide the lobes into lobules. **C.** Epithelial reticular cells in the cortex maintain the blood thymus barrier.

### Module 20.7 Review

**A.** Functions of the spleen include (1) removing abnormal blood cells and other blood components by phagocytosis, (2) storing iron recycled from the red blood cells, and (3) initiating immune responses by B cells and T cells in response to antigens in the bloodstream. **B.** Red pulp contains large numbers of red blood cells; white pulp resembles lymphoid nodules and contains lymphocytes. **C.** Beginning with the trabecular arteries, blood then flows to the central arteries → capillaries → reticular tissue of red pulp → sinusoids → trabecular veins.

### Section 1 Review

**1.** tonsil **2.** cervical lymph nodes **3.** right lymphatic duct **4.** thymus **5.** cisterna chyli **6.** lumbar lymph nodes **7.** appendix **8.** lymphatics of lower limb **9.** lymphatics of upper limb **10.** axillary lymph nodes **11.** thoracic duct **12.** lymphatics of mammary gland **13.** spleen **14.** mucosa-associated lymphoid tissue (MALT) **15.** pelvic lymph nodes **16.** inguinal lymph nodes **17.** red bone marrow **18.** c **19.** h **20.** l **21.** i **22.** m **23.** j **24.** b **25.** e **26.** g **27.** d **28.** f **29.** a **30.** k

### Module 20.8 Review

**A.** Innate immunity is nonspecific and does not distinguish one type of threat from another. Adaptive immunity is specific and protects against particular threats. **B.** Innate immunity protects us from disease by using physical barriers such as skin, phagocytes that engulf pathogens, immune surveillance of abnormal cells, interferons to defend from viruses, complement to assist antibodies, inflammation to limit the spread

of infection, and fever. **C.** Innate (nonspecific) immunity will be activated immediately in a child with a skinned knee.

### Module 20.9 Review

**A.** The integumentary system provides a physical barrier that is the first line of defense in preventing pathogens and toxins from entering body tissues. Skin secretions flush the surface, hair protects against physical abrasion, and the multiple layers of the skin's epithelium create an interlocking barrier.

### Module 20.10 Review

**A.** The body's phagocytes are neutrophils, eosinophils, and macrophages. Fixed macrophages are scattered among connective tissues and do not move; free macrophages are mobile and reach injury sites by migrating through adjacent tissues or traveling in the bloodstream. **B.** Chemotaxis is phagocyte movement in response to chemical stimuli.

### Module 20.11 Review

**A.** Immune surveillance is the constant monitoring of normal tissues by NK cells sensitive to abnormal antigens on the surfaces of cells. **B.** NK cells recognize unusual proteins, called tumor-specific antigens, on the plasma membranes of cancer cells. When these antigens are detected, the NK cells then destroy the abnormal cells. **C.** Cancer cells can mutate such that either they do not display tumor-specific antigens, or they secrete chemicals that can destroy NK cells. This ability to escape detection is called immunological escape.

### Module 20.12 Review

**A.** Interferons are small proteins that are released by activated lymphocytes, macrophages, and cells infected with viruses; they trigger the production of antiviral proteins that interfere with viral replication within tissue cells. **B.** The complement proteins of the complement system interact with each other in chain reactions that ultimately produce activated forms that target bacterial cell walls and plasma membranes, stimulate inflammation, attract phagocytes, or enhance phagocytosis. **C.** Histamine release by mast cells and basophils in tissues increases local inflammation, thereby accelerating blood flow to the region.

### Module 20.13 Review

**A.** Inflammation, also called the inflammatory response, is a localized response to injury in the body. Its four cardinal signs are redness, swelling, heat, and pain. Lost function is also commonly associated with inflammation. **B.** Pyrogens increase body temperature (produce a fever), which can mobilize defenses, accelerate repairs, and inhibit pathogens. **C.** A rise in the level of interferons suggests a viral infection.

### Section 2 Review

**1.** physical barriers **2.** phagocytes **3.** immune surveillance **4.** interferons **5.** complement system **6.** inflammation **7.** fever **8.** d **9.** d **10.** c **11.** d **12.** a **13.** a **14.** The high body temperatures of a fever may inhibit some viruses and bacteria or increase their reproductive rate so that the disease runs its course more quickly. High body temperatures also accelerate the body's metabolic processes, which may help to mobilize tissue defenses and speed the repair process.

### Module 20.14 Review

**A.** The two cells responsible for coordinating adaptive immunity are T cells and B cells. **B.** A child who receives the polio vaccine develops artificially acquired active immunity.

### Module 20.15 Review

**A.** Antigen presentation occurs when an antigen-glycoprotein, or antigen-MHC protein, combination capable of activating T cells appears in a plasma membrane. T cells sensitive to this antigen are activated if they contact the antigen on the plasma membrane of the antigen-presenting cell. **B.** The major histocompatibility complex (MHC) is a portion of

chromosome 6 containing genes that control the synthesis of membrane glycoproteins. **C.** Class I MHC proteins are in the plasma membranes of all nucleated body cells. Class II MHC proteins are only in the plasma membranes of antigen-presenting cells (APCs) and lymphocytes. **D.** Antigen presentation may stimulate both cell-mediated immunity and antibody-mediated immunity.

### Module 20.16 Review

**A.** CD markers refer to a class of membrane proteins. CD is an abbreviation for "cluster of differentiation." T cell plasma membranes contain CD proteins. T cells with CD8 markers respond to antigens presented by class I MHC proteins. T cells with CD4 markers respond to antigens presented by class II MHC proteins. **B.** The three major types of T cells activated by class I MHC proteins are cytotoxic T cells, memory $T_C$ cells, and regulatory T cells. **C.** Abnormal antigens attached to class I MHC proteins of an infected cell are displayed on the surface of the cell's plasma membrane. The recognition of such abnormal antigens by CD8 T cells initiates an immune response.

### Module 20.17 Review

**A.** Cytokines secreted by helper T cells stimulate both cell-mediated and antibody-mediated immunity. **B.** Sensitization is the process by which a B cell prepares to undergo activation after encountering a specific antigen. During sensitization, the specific antigens bound to membrane antibodies are brought into the cell by endocytosis, where the antigens then become bound to class II MHC proteins. Together, they then appear at the cell surface. **C.** Plasma cells produce and secrete antibodies, so observing an elevated number of plasma cells in the lymph would lead you to expect higher-than-normal antibody levels in the blood.

### Module 20.18 Review

**A.** An antibody molecule consists of two parallel pairs of polypeptide chains: a pair of long, heavy chains

and a pair of short, light chains. Each chain contains both constant segments and variable segments. The constant segments of the heavy chains form the base of the antibody molecule; the free tips of each of the two variable segments form the antigen-binding sites. **B.** An antigenic determinant site is the part of an antigen molecule to which an antibody molecule binds. **C.** A lack of memory cells would affect the secondary response more than the primary response, because memory cells are produced during the initial response to an antigen.

### Module 20.19 Review

**A.** Antigen-antibody complexes help destroy antigens through seven processes: neutralization, prevention of pathogen adhesion, activation of complement, opsonization, attraction of phagocytes, stimulation of inflammation, and precipitation and agglutination. **B.** Opsonization is the process by which the coating of pathogens with antibodies and complement proteins makes the pathogens more susceptible to phagocytosis. **C.** Basophils and mast cells are involved in inflammation.

### Module 20.20 Review

**A.** Hypersensitivities, commonly called allergies, are excessive immune responses to an allergen, which is an antigen that triggers an allergic reaction. **B.** Histamines, leukotrienes, and other chemicals that cause pain and inflammation are released when mast cells and basophils are stimulated in an allergic reaction. **C.** Anaphylaxis is an immune response to a circulating antigen that stimulates mast cells throughout the body to release chemicals that prompt the inflammatory response.

### Module 20.21 Review

**A.** CD8 markers are found on cytotoxic T cells, memory $T_C$ cells, and regulatory T cells; CD4 markers are on all helper T cells. **B.** Plasma cells produce antibodies. **C.** Cytotoxic T cells and NK cells can be activated by direct contact with virus-infected cells.

### Module 20.22 Review

**A.** Autoimmune disorders are conditions that result from the production of antibodies (called autoantibodies) directed against normal antigens in the body (self-antigens). **B.** Immunodeficiency diseases are caused by inadequate immune responses. They may result from problems with the embryonic development of lymphoid organs and tissue, a viral infection that depresses immune function, or treatment with or exposure to immunosuppressive agents, such as radiation or drugs. **C.** The increased incidence of cancer in elderly people may result from a decline in immune surveillance, which results in reduced elimination of tumor cells as they arise.

### Section 3 Review

**1.** c **2.** k **3.** l **4.** i **5.** b **6.** a **7.** g **8.** m **9.** d **10.** f **11.** e **12.** h **13.** j **14.** b **15.** f **16.** g **17.** e **18.** c **19.** d **20.** a **21.** h **22.** i

### Chapter Review Questions

**1.** antigen-binding site **2.** variable segment of light and heavy chains **3.** constant segments of light and heavy chains **4.** heavy chain **5.** light chain **6.** disulfide bond **7.** false **8.** false **9.** true **10.** false **11.** false **12.** true **13.** a **14.** b **15.** c **16.** a **17.** c **18.** d **19.** a **20.** b **21.** In transplant rejection, T cells are activated by contact with MHC proteins on plasma membranes in the donated tissues. The cytotoxic T cells that develop then attack and destroy the foreign cells. **22.** A cytotoxic T cell ($T_C$ cell) destroys its target cell by first encountering its target antigen bound to a class I MHC protein. It then attacks it by any one of the following mechanisms: (1) destroying the target cell's membrane through the release of perforins; (2) activating apoptosis in the target cell; (3) disrupting the target cell's metabolism through the release of lymphotoxin. **23.** Multiple injections of certain vaccines are timed to produce primary and then secondary responses of the immune system. In the primary response to the vaccine injection, B cells produce daughter cells that differentiate into plasma cells and memory B cells. The plasma cells produce antibodies, which represent the primary response, but the primary response does not maintain elevated antibody levels for long periods. Subsequent vaccine injections are necessary to trigger secondary responses, when memory B cells differentiate into plasma cells that produce antibody concentrations that remain high much longer.

### Chapter Integration

**1.** The student cannot yet know whether she will come down with chickenpox. Her elevated blood IgM levels indicate that she is in the early stages of a primary response to the chickenpox virus. If her immune response proves unable to control and then eliminate the virus, she will develop chickenpox. **2.** Because chickenpox is highly contagious, people who have not had chickenpox previously and have not been vaccinated are susceptible to it. **3.** The causative agent for chickenpox and shingles is the varicella-zoster virus (VZV). **4.** IgM antibodies are the first class of immunoglobulins secreted after an antigen is encountered. IgG antibodies are the most common class of immunoglobulins. **5.** Antigenic specificity results from the arrangement of genes controlling the amino acid sequences of the variable segments of antibodies. **6.** Both responses involve the production of antibodies as stimulated by antigens. The primary response develops relatively slowly, and levels of IgG rise more slowly than do levels of IgM. The secondary response develops much more rapidly, is more prolonged, and produces much more IgG than IgM. **7.** Any vaccine, including that for chickenpox, stimulates artificially acquired active immunity.

## CHAPTER 21

### Module Reviews

#### Module 21.1 Review
**A.** Gas exchange between air and the lungs occurs in the alveoli. **B.** The conducting portion of the respiratory tract begins at the nasal cavity and extends through the pharynx, larynx, trachea, bronchi, and larger bronchioles. The respiratory portion includes the respiratory bronchioles and alveoli.

#### Module 21.2 Review
**A.** The respiratory defense system is a series of filtration mechanisms that prevent airway contamination by debris and pathogens. **B.** The respiratory mucosa lines the conducting portion of the respiratory tract. **C.** Cystic fibrosis causes production of dense mucus that restricts respiratory passages and accumulates in the lungs. Harmful bacterial infection of the lungs may also develop, leading to death.

#### Module 21.3 Review
**A.** The structures of the upper respiratory system are the nose, nasal cavity, paranasal sinuses, and pharynx. **B.** Pathway of entering air: nostrils → nasal cavity → nasal vestibule (guarded by hairs that screen out large particles) → superior, middle, and inferior meatuses (air bounces off the conchal surfaces) → choanae (openings between the nasal cavity and nasopharynx) → nasopharynx → oropharynx → laryngopharynx → larynx. **C.** The rich vascularization of the nasal cavity by the expandable veins in the lamina propria radiates body heat, so inhaled air is warmed before it leaves the nasal cavity. The heat also evaporates moisture from the epithelium to humidify the incoming air.

#### Module 21.4 Review
**A.** The paired cartilages are the arytenoid, corniculate, and cuneiform. The unpaired cartilages are the thyroid cartilage, cricoid cartilage, and epiglottis. **B.** The glottis is the "voicebox"

of the larynx. It is made up of the vocal folds (structures that contain the vocal ligaments and are involved with sound production) and the rima glottidis (the opening between the vocal folds and the arytenoid cartilages).**C.** Phonation is the production of sound and is one component of speech; articulation is the modification of sound by the tongue, teeth, and lips for clear speech.

#### Module 21.5 Review
**A.** The right main bronchus is larger in diameter than the left main bronchus, and it descends toward the lung at a steeper angle than the left primary bronchus. **B.** The C-shaped tracheal cartilages allow room for the esophagus to expand when food or liquids are swallowed. **C.** Pathway of airflow along the lower respiratory tract: trachea → main bronchi → lobar bronchi → segmental bronchi → terminal bronchioles → pulmonary lobule.

#### Module 21.6 Review
**A.** A bronchopulmonary segment is a specific region of a lung supplied by a segmental bronchus. **B.** Within the thoracic cavity, the left and right lungs are surrounded by the left and right pleural cavities, respectively. **C.** The curving anterior and inferior borders of each lung follow the coutours of the rib cage. The apex of each lung extends superiorly to the first rib, and the base of each lung rests on the superior surface of the diaphragm. **D.** The left lung is divided into a superior lobe and an inferior lobe by the oblique fissure; in the right lung, the horizontal fissure separates the superior lobe from the middle lobe, while the oblique fissure separates the superior and middle lobes from the inferior lobe.

#### Module 21.7 Review
**A.** Pulmonary lobules are the smallest subdivisions of the lungs; branches of the pulmonary arteries, pulmonary veins, and segmental bronchi supply each lobule. **B.** The blood

air barrier is made up of the fused basement membranes of the alveolar epithelium and capillary endothelium. Because it is very thin and oxygen and carbon dioxide are lipid soluble, diffusion occurs rapidly across the membrane. **C.** Without surfactant, the alveoli would collapse because of the normally high surface tension in the thin layer of water coating the alveolar surfaces.

#### Section 1 Review
**1.** nasal cavity **2.** hard palate **3.** pharynx **4.** glottis **5.** trachea **6.** right lung **7.** nostrils **8.** larynx **9.** main bronchus **10.** lobar bronchus **11.** segmental bronchi **12.** visceral pleura **13.** pulmonary lobule **14.** bronchioles **15.** respiratory bronchiole; **16.** alveolus **17.** h **18.** j **19.** g **20.** d **21.** e **22.** n **23.** a **24.** l **25.** c **26.** k **27.** m **28.** i **29.** b **30.** f

#### Module 21.8 Review
**A.** External respiration is all the processes involved in the exchange of oxygen and carbon dioxide between the blood, lungs, and the external environment. Gas diffusion is the movement of $O_2$ and $CO_2$ across the blood air barrier between alveolar air spaces and alveolar capillaries and across capillary walls between blood and other tissues. Internal respiration is the abosrption of $O_2$ from blood and the release of $CO_2$ by tissue cells. **B.** Hypoxia is low tissue oxygen levels. Anoxia is the complete cutoff of oxygen supply.

#### Module 21.9 Review
**A.** Boyle's law states that at a constant temperature, the pressure of a gas is inversely proportional to its volume. **B.** The movements of the diaphragm and rib cage affect the volume of the lungs. **C.** The intrapulmonary pressure (the pressure inside the respiratory tract) and the atmospheric pressure (the pressure outside the respiratory tract) determine the direction of airflow. Air moves from the area with the higher pressure to the area with the lower pressure.

#### Module 21.10 Review
**A.** The primary inspiratory muscles are the diaphragm and

the external intercostal muscles. **B.** The accessory respiratory muscles become active whenever the primary respiratory muscles are unable to move enough air to meet the oxygen demands of tissues. **C.** The measurable pulmonary volumes are the tidal volume, expiratory reserve volume (ERV), residual volume, and inspiratory reserve volume (IRV).

### Module 21.11 Review

**A.** The respiratory rate is the number of breaths taken each minute. **B.** The respiratory minute volume is the amount of air moved into and out of the respiratory tract each minute, whereas alveolar ventilation is the amount of air reaching the alveoli each minute. Because some of the air never reaches the alveoli but instead remains in the anatomic dead space, alveolar ventilation is lower than respiratory minute volume. **C.** Slow, deep breaths ventilate alveoli more effectively because a smaller amount of the tidal volume of each breath is spent moving air into and out of the anatomic dead space of the lungs.

### Module 21.12 Review

**A.** Dalton's law states that in a mixture of gases, the individual gases exert a pressure proportional to their abundance in the mixture. **B.** Henry's law states that at a given temperature, the amount of a particular gas that dissolves in a liquid is directly proportional to the partial pressure of that gas. Henry's law underlies the diffusion of gases between capillaries and alveoli, and between capillaries and interstitial fluid. **C.** The $P_{O_2}$ decreases from about 100 mm Hg to 95 mm Hg in the pulmonary veins as it mixes with venous blood from capillaries surrounding the conducting passageways. The blood arriving at the peripheral capillaries has a $P_{O_2}$ of 95 mm Hg.

### Module 21.13 Review

**A.** Oxyhemoglobin is hemoglobin to which oxygen molecules have bound. **B.** Both increased temperature and decreased pH (from heat and acidic wastes generated by active skeletal muscles) cause hemoglobin to release more oxygen during exercise than when the muscles are at rest. **C.** BPG (2,3-bisphosphoglycerate) is a compound generated by RBCs during glycolysis that decreases hemoglobin's affinity for oxygen. For any partial pressure of oxygen, if the concentration of BPG increases, the amount of oxygen released by hemoglobin will increase.

### Module 21.14 Review

**A.** Carbon dioxide is transported in the bloodstream dissolved in the plasma, bound to hemoglobin, or as bicarbonate ions in the plasma. **B.** Driven by differences in partial pressure, oxygen enters the blood at the lungs and leaves it in peripheral tissues; similar forces drive carbon dioxide into the blood at the tissues and into the alveoli at the lungs. **C.** Blockage of the trachea would interfere with the body's ability to take in oxygen and eliminate carbon dioxide. Because most carbon dioxide is transported in blood as bicarbonate ions formed from the dissociation of carbonic acid, an inability to eliminate carbon dioxide would result in a buildup of excess hydrogen ions, which would decrease blood pH.

### Module 21.15 Review

**A.** Compliance is the ease with which the lungs expand and recoil. Resistance is an indication of how much force is required to inflate or deflate the lungs. **B.** Three COPDs are asthma, chronic bronchitis, and emphysema. **C.** Chronic bronchitis is long-term inflammation of the mucous membranes in the bronchial tubes; emphysema is a condition in which the alveolar surfaces of the lungs are destroyed and alveoli merge, which reduces respiratory surface area and oxygen absorption, causing breathlessness. People who have chronic bronchitis are sometimes called "blue bloaters"; those with emphysema are called "pink puffers."

### Module 21.16 Review

**A.** The pairs of central nervous system nuclei that adjust the pace of respiration are the apneustic centers and the pneumotaxic centers in the pons. **B.** The respiratory rhythmicity centers in the medulla oblongata generate the respiratory pace. **C.** The pH, $P_{O_2}$, and $P_{CO_2}$ in blood and cerebrospinal fluid stimulate the respiratory centers.

### Module 21.17 Review

**A.** Chemoreceptors are more sensitive to blood carbon dioxide levels than they are to blood oxygen levels. **B.** Hypercapnia is an increase in the $P_{CO_2}$ of arterial blood above the normal range. Hypocapnia is an abnormally low arterial $P_{CO_2}$. **C.** This is unlikely to happen. When Johnny holds his breath, carbon dioxide levels in his blood increase, causing increased stimulation of the inspiratory centers, forcing him to breathe again.

### Module 21.18 Review

**A.** Aging results in deterioration of elastic tissue, arthritic changes that stiffen rib articulations, decreased flexibility at costal cartilages, decreased vital capacity, and some degree of emphysema. **B.** Lung cancer is an aggressive class of malignancies that affect the epithelial cells lining the conducting passageways, mucous glands, and alveoli. More people die from lung cancer than from any other type of cancer. **C.** Dysplasia is the development of abnormal cells; metaplasia is the development of abnormal changes in tissue structure; neoplasia is the conversion of normal cells to tumor (cancerous) cells; and anaplasia is the spread of the malignant cells (metastasis) throughout the body.

### Section 2 Review

**1.** n **2.** i **3.** f **4.** g **5.** h **6.** c **7.** d **8.** e **9.** b **10.** l **11.** m **12.** k **13.** j **14.** a **15.** inspiratory reserve volume (IRV): the amount of air that can be taken in above the tidal volume. **16.** tidal volume ($V_T$): the amount of air inhaled and exhaled during a single respiratory cycle while resting. **17.** expiratory reserve volume (ERV): the amount of air that can be expelled after a completely normal, quiet respiratory cycle. **18.** minimal volume: the amount of air remaining in the lungs if they were to collapse. **19.** inspiratory capacity: the amount of air that can be drawn into the lungs after completing a quiet respiratory cycle. **20.** total lung capacity: the total volume of the lungs. **21.** vital capacity: the maximum amount of air that can be moved into or out of the lungs in a single respiratory cycle. **22.** residual volume: the amount of air remaining in the lungs after a maximal exhalation. **23.** functional residual capacity (FRC): the amount of air that remains in the lungs after completing a quiet respiratory cycle. **24.** External respiration includes all the processes involved in the exchange of oxygen and carbon dioxide between blood, lungs, and the external environment. Pulmonary ventilation, or breathing, is a process of external respiration that involves the physical movement of air into and out of the lungs. Internal respiration is the absorption of oxygen and the release of carbon dioxide by tissue cells.

### Chapter Review Questions

**1.** 40 mm Hg **2.** 45 mm Hg **3.** 100 mm Hg **4.** 40 mm Hg **5.** 100 mm Hg **6.** 40 mm Hg **7.** 95 mm Hg **8.** 40 mm Hg **9.** 40 mm Hg; **10.** 45 mm Hg **11.** 40 mm Hg **12.** 45 mm Hg **13.** e **14.** l **15.** c **16.** i **17.** j **18.** b **19.** a **20.** d **21.** k **22.** f **23.** h **24.** g **25.** d **26.** a **27.** b **28.** b **29.** d **30.** c **31.** The regions of the pharynx are (1) the nasopharynx, located between the soft palate and the choanae; (2) the oropharynx, located between the soft palate and the base of the tongue at the level of the hyoid bone; and (3) the laryngopharynx, which is the portion of the pharynx located between the hyoid bone and entrance to the larynx and esophagus. **32.** The three ways that carbon dioxide is transported in blood are (1) as bicarbonate ions, (2) bound to the protein portion of hemoglobin molecules within the RBC, or (3) dissolved in plasma as $CO_2$.

**33.** Breathing through the nasal cavity is more desirable than breathing through the mouth because the nasal cavity cleanses, moistens, and warms inhaled air, whereas the mouth does not. Breathing through the mouth eliminates much of the conditioning of inhaled air and increases heat and water loss at every exhalation.

### Chapter Integration

**1.** Alveolar ventilation rate (AVR) = respiratory rate × (tidal volume – anatomic dead space). In this case, the dead space is 200 mL (the anatomic dead space plus the volume of the snorkel); therefore, AVR = respiratory rate × (500 – 200). To maintain an AVR of 6.0 L/min, or 6000 mL/minute, the respiratory rate must be 6000/ (500 – 200), or 20 breaths per minute. **2.** Jake's hyperventilation resulted in abnormally low $P_{CO_2}$. This reduced his urge to breathe, so he stayed underwater longer, unaware that his $P_{O_2}$ was so low that he would lose consciousness.

## CHAPTER 22

### Module Reviews

#### Module 22.1 Review

**A.** The respiratory, cardiovascular, and urinary systems all work with the digestive system to support the cells and tissues of the human body. **B.** Major organs of the digestive tract are the oral cavity (mouth), pharynx (throat), esophagus, stomach, small intestine, and large intestine. **C.** The accessory organs of the digestive system are the teeth, tongue, salivary glands, liver, gallbladder, and pancreas.

#### Module 22.2 Review

**A.** The mesenteries—sheets consisting of two layers of serous membrane separated by areolar connective tissue—support and stabilize the organs in the abdominal cavity and provide a route for the passage of associated blood vessels, nerves, and lymphatic vessels. **B.** The four layers of the digestive tract, beginning from the lumen, are the mucosa (adjacent to the lumen), submucosa, muscular layer, and serosa. **C.** The submucosal neural plexus is a nerve network that contains sensory neurons, parasympathetic ganglionic neurons, and sympathetic postganglionic fibers that innervate the mucosa and submucosa. The deeper myenteric plexus is a network of parasympathetic neurons, interneurons, and sympathetic postganglionic fibers that lies between the circular and longitudinal muscle layers of the muscular layer.

#### Module 22.3 Review

**A.** Smooth muscle fibers in either the circular or longitudinal layers of the muscular layer lie parallel to each other. In a longitudinal section of the digestive tract, the fibers of the superficial circular layer appear as little round balls, whereas the fibers of the deeper longitudinal layer are spindle shaped. **B.** Smooth muscle fibers are spindle shaped; they lack T tubules, myofibrils, and sarcomeres; and the sarcoplasmic reticulum forms a loose network throughout the sarcoplasm. Because the tissue lacks sarcomeres, it is nonstriated. Additionally, the thin filaments are anchored to dense bodies. **C.** Smooth muscle can contract over a wider range of resting lengths than can skeletal muscle because the actin and myosin filaments are more loosely organized in smooth muscle.

#### Module 22.4 Review

**A.** The waves of contractions that constitute peristalsis are more efficient in propelling intestinal contents along the digestive tract than segmentation. The reason is that segmentation is basically a churning action that mixes intestinal contents with digestive fluids. **B.** The major mechanisms that regulate and control digestive activities are local factors (stimuli for digestive activities, such as changes in pH or distortion of the intestinal lumen), neural mechanisms (myenteric reflexes), and hormonal mechanisms (involving neuroendocrine cells). **C.** Enteroendocrine cells are endocrine cells scattered among the epithelial cells lining the digestive tract that secrete peptide hormones important to digestion.

#### Section 1 Review

**1.** mesenteric artery and vein **2.** mesentery **3.** circular fold **4.** mucosa **5.** submucosa **6.** muscular layer **7.** serosa **8.** d **9.** e **10.** a **11.** g **12.** b **13.** c **14.** i **15.** m **16.** l **17.** f **18.** n **19.** j **20.** k **21.** h **22.** With a decrease in smooth muscle tone, general motility along the digestive tract decreases, and peristaltic contractions are weaker.

#### Module 22.5 Review

**A.** Ingestion is the entry of foods and liquids into the digestive tract through the mouth. **B.** Chemical digestion is the chemical and enzymatic breakdown of food into small molecules. Absorption is the movement of molecules and other substances across the digestive epithelium and into the interstitial fluid of the digestive tract. **C.** The function of the large intestine is the dehydration and compaction of indigestible materials and organic wastes prior to elimination from the body.

#### Module 22.6 Review

**A.** The oral cavity is lined by stratified squamous epithelium, which provides protection against friction and abrasion. **B.** The hard palate forms the roof of the mouth. **C.** The fauces is the arched opening between the oral cavity and the oropharynx. **D.** A shortened frenulum of tongue could restrict tongue movements, making eating or talking difficult.

#### Module 22.7 Review

**A.** A typical tooth has a crown, a neck, and a root. **B.** The third set of molars is sometimes called the wisdom teeth.

#### Module 22.8 Review

**A.** The pharynx is an anatomical space divided into the nasopharynx, oropharynx, and laryngopharynx that serves as a common passageway for food, liquids, and air. Food passes through the oropharynx and laryngopharynx on its way to the esophagus. **B.** The muscular layer of the esophagus has the usual inner circular and outer longitudinal layers. However, in the superior third of the esophagus, these layers contain skeletal muscle fibers. The middle third contains a mixture of skeletal and smooth muscle tissue. Along the inferior third, only smooth muscle occurs. **C.** During the buccal phase, food is formed into a bolus; during the pharyngeal phase, the bolus contacts the palatal arches and moves into the esophagus; during the esophageal phase, swallowing begins as pharyngeal muscles contract and the bolus is moved toward the stomach by peristaltic waves.

#### Module 22.9 Review

**A.** The falciform ligament is a sheet of mesentery that is a remnant of the ventral mesentery between the liver and the anterior wall of the peritoneal cavity. **B.** The lesser omentum stabilizes the position of the stomach and provides a route for blood vessels (and other structures) entering or leaving the liver. **C.** Peritoneal fluid covers the parietal and visceral layers to prevent friction and subsequent irritation during sliding movements of organs covered by the peritoneum.

#### Module 22.10 Review

**A.** The four major regions of the stomach are the cardia, fundus, body, and pyloric part. **B.** The muscular layer of the stomach has an additional inner, oblique layer of smooth muscle. The longitudinal, circular, and oblique orientations of the muscle fibers in the three layers of the muscular layer allow for the mixing and churning actions necessary for chyme formation. **C.** The lining of the stomach contains rugae, mucosal folds in the lining of the empty stomach that disappear as gastric expansion occurs.

## Module 22.11 Review

**A.** The alkaline mucous layer protects epithelial cells against the acid and enzymes in the gastric lumen. **B.** Parietal cells secrete intrinsic factor and the ions of hydrochloric acid (HCl). **C.** The alkaline tide is a sudden influx of bicarbonate ions into the bloodstream from active parietal cells; it causes a temporary increase in blood pH.

## Module 22.12 Review

**A.** The layers of the small intestine, from superficial to deep, are the mucosa, submucosa, muscular layer, and serosa. **B.** The intestinal mucosa has transverse folds called circular folds (plicae circulares) and small projections called intestinal villi. Pockets called intestinal glands are lined by goblet, epithelial, stem, and paneth cells. **C.** Lacteals are lymphatic capillaries in the intestinal villi that transport materials that cannot enter blood capillaries, such as absorbed fatty acids combined with proteins.

## Module 22.13 Review

**A.** The three segments of the small intestine, from proximal to distal, are the duodenum, jejunum, and ileum. **B.** The proximal portion of the duodenum is found within the epigastric region. **C.** The primary function of the duodenum is to receive chyme from the stomach and neutralize its acids to prevent damage to the absorptive surfaces of the remaining regions of the small intestine, especially the jejunum. **D.** A traumatic injury to the umbilical region would affect all three segments of the small intestine.

## Module 22.14 Review

**A.** The six major hormones that regulate digestive activities are gastrin, secretin, gastric inhibitory peptide (GIP), cholecystokinin (CCK), vasoactive intestinal peptide (VIP), and enterocrinin. **B.** If the small intestine did not produce secretin, the pH of the intestinal contents would be lower (more acidic) than normal. Secretin stimulates the pancreas to release a fluid high in buffers that neutralizes the acidic chyme entering the duodenum from the stomach. **C.** A high-fat meal raises the cholecystokinin (CCK) level in the blood.

## Module 22.15 Review

**A.** The three phases of gastric secretion (named according to the location of the control center involved) are the cephalic phase, which prepares the stomach to receive ingested materials; the gastric phase, which begins with the arrival of food in the stomach; and the intestinal phase, which controls the rate of gastric emptying. **B.** The two central reflexes triggered by stimulation of stretch receptors in the stomach wall are the gastroenteric reflex, which stimulates motility and secretion along the entire small intestine, and the gastroileal reflex, which triggers the opening of the ileocecal valve to allow passage of materials from the small intestine into the large intestine. **C.** Severing the branches of the vagus nerves that supply the stomach would interrupt parasympathetic stimulation of gastric secretions, and the consequent reduction in acid secretions would provide some relief from the pain of gastric ulcers.

## Module 22.16 Review

**A.** The major functions of the large intestine are (1) reabsorbing water and compacting material into feces, (2) absorbing vitamins produced by bacteria, and (3) storing fecal material prior to defecation. **B.** The four regions of the colon are the ascending colon, transverse colon, descending colon, and sigmoid colon. **C.** Mass movements are powerful peristaltic contractions that occur a few times daily in response to distension of the stomach and duodenum.

## Module 22.17 Review

**A.** The mucosa of the large intestine does not produce digestive enzymes. Digestion that does occur in the large intestine results from enzymes secreted into the small intestine or from bacterial action. **B.** Hemorrhoids are distended (swollen) veins in the distal portion of the rectum that may result from pregnancy or straining during defecation. **C.** The two positive feedback loops in the defecation reflex are (1) the short intrinsic myenteric defecation reflex, whereby stretch receptors in the rectal walls promote a series of peristaltic contractions in the sigmoid colon and rectum, moving feces toward the anus; and (2) the long parasympathetic defecation reflex, whereby parasympathetic motor neurons in the sacral spinal cord, also activated by stretch receptors, stimulate mass movements that push feces toward the rectum from the descending colon and sigmoid colon. Defecation occurs with the voluntary relaxation of the external anal sphincter.

## Section 2 Review

**1.** crown **2.** neck **3.** root **4.** enamel **5.** dentin **6.** pulp cavity **7.** gingiva **8.** gingival sulcus **9.** cement **10.** periodontal ligament **11.** root canal **12.** apical foramen **13.** material in jejunum **14.** gastrin **15.** GIP **16.** secretin and CCK **17.** VIP **18.** inhibits **19.** acid production **20.** insulin **21.** bile **22.** intestinal capillaries **23.** gallbladder **24.** nutrient utilization by tissues **25.** Both parietal cells and chief cells are secretory cells found in the gastric glands of the wall of the stomach. However, parietal cells secrete intrinsic factor and, indirectly, hydrochloric acid, whereas chief cells secrete pepsinogen, an inactive proenzyme.

## Module 22.18 Review

**A.** The salivary glands produce secretions that contain mucin and enzymes that are essential to normal digestive function. **B.** The exocrine secretions of the pancreas are buffers and digestive enzymes. The endocrine secretions are insulin, glucagon, pancreatic polypeptide (PP), and GH–IH. **C.** The liver is the accessory organ of the digestive system that has almost 200 known functions.

## Module 22.19 Review

**A.** The three major pairs of salivary glands are the parotid, sublingual, and submandibular glands. **B.** Damage to the parotid glands, which secrete the enzyme salivary amylase, would interfere with the digestion of starches (complex carbohydrates). **C.** The submandibular glands secrete about 70 percent of the saliva. **D.** The submandibular glands secrete substances (antibodies and lysozyme) that reduce oral bacteria populations.

## Module 22.20 Review

**A.** The liver has a tough fibrous capsule, which is covered by a layer of visceral peritoneum. **B.** The falciform ligament marks the division between the left lobe and the right lobe of the liver. **C.** The liver is divided into left, right, caudate, and quadrate lobes. **D.** The gallbladder temporarily stores bile produced by the liver.

## Module 22.21 Review

**A.** A hepatocyte is a liver cell, which is found in a liver lobule. **B.** A portal triad is located at each of the six corners of a liver lobule; each portal triad contains (1) an interlobular vein (a branch of the hepatic portal vein), (2) an interlobular artery (a branch of the hepatic artery proper), and (3) an interlobular bile duct (a small branch of the bile duct). **C.** Stellate macrophages (Kupffer cells) are liver macrophages that engulf pathogens, cell debris, and damaged blood cells.

## Module 22.22 Review

**A.** Emulsification is the breakdown of lipid droplets by bile salts. **B.** The pathway of a drop of bile: hepatic ducts → common hepatic duct → bile duct → duodenal ampulla and papilla → duodenal lumen.

## Module 22.23 Review

**A.** The pancreas produces pancreatic juice that contains buffers and several digestive enzymes necessary for the breakdown of starches, lipids, nucleic acids, and proteins.

## Module 22.24 Review

**A.** Hepatitis is inflammation of the liver. It can be caused by alcohol abuse, drugs, or infection. Cirrhosis is a form of hepatitis that destroys liver cells and replaces them with nonfunctional fibrous connective tissue. In viral hepatitis, hepatitis viruses attack and destroy liver cells. **B.** Cholecystitis is inflammation of the gallbladder, usually resulting from blockage of the cystic duct or the bile duct by gallstones. **C.** The bacterium responsible for most peptic ulcers is *Helicobacter pylori*.

### Section 3 Review

**1.** bile ductule **2.** hepatocytes **3.** central vein **4.** interlobular septum **5.** sinusoid **6.** interlobular artery **7.** interlobular vein **8.** interlobular bile duct **9.** portal triad **10.** g **11.** h **12.** m **13.** j **14.** l **15.** n **16.** b **17.** e **18.** f **19.** a **20.** i **21.** d **22.** k **23.** c **24.** Saliva (1) continuously flushes and cleans oral surfaces, (2) contains buffers that prevent the buildup of acids produced by bacterial action, and (3) contains antibodies (IgA) and lysozyme, which help control the growth of oral bacterial populations. **25.** Such a blockage would interfere with the release of secretions into the duodenum by the pancreas, gallbladder, and liver. Each day, the pancreas normally secretes about 1 liter of pancreatic juice, a mixture of a variety of digestive enzymes and buffer solution. The blockage of pancreatic juice would lead to pancreatitis, an inflammation of the pancreas. Extensive damage to exocrine cells by the blocked digestive enzymes would lead to autolysis that could destroy the pancreas and result in the person's death. If secretion of bile from the bile duct were to become blocked, gallstones could form, leading to damage of the gallbladder wall. Jaundice could also result because bilirubin from the liver would not be excreted in the bile and, instead, would accumulate in body fluids.

### Chapter Review Questions

**1.** gallbladder **2.** fundus **3.** body **4.** neck **5.** cystic duct **6.** bile duct

**7.** pancreatic duct **8.** left and right hepatic ducts **9.** common hepatic duct **10.** hepatic portal vein **11.** hepatic artery proper **12.** true **13.** false **14.** false **15.** true **16.** false **17.** true **18.** a **19.** b **20.** b **21.** a **22.** d **23.** b **24.** The four different types of teeth are incisors, canines (cuspids), premolars (bicuspids), and molars. There are 20 deciduous teeth and 32 permanent teeth. **25.** The three phases of gastric secretion are the cephalic phase, the gastric phase, and the intestinal phase. The cephalic phase of gastric secretion begins when you see, smell, taste, or think of food. The CNS is preparing the stomach to receive food. The gastric phase begins when food arrives in the stomach. It includes distension of the stomach, an increase in pH, and the presence of undigested materials. The intestinal phase begins when chyme first enters the small intestine. **26.** The hepatic portal system is a venous portal system in which the hepatic portal vein receives blood from most of the abdominal viscera and drains into the liver sinusoids.

### Chapter Integration

**1.** Gastric stapling would result in weight loss because the person feels full after eating only a small amount. Consuming a smaller volume of food means consuming fewer calories. **2.** After gastric stapling surgery, the person can eat only a small amount of food before the stretch receptors in the gastric wall become stimulated and a feeling of fullness results. **3.** The smooth muscle in the gastric muscular layer of the functional portion of the stomach gradually becomes increasingly tolerant of distension, and the operation may have to be repeated to achieve the same results. **4.** Gastric bypass and lap-band surgeries, like gastric stapling, reduce gastric capacity, thereby decreasing the amount of food a person can ingest before feeling full.

### CHAPTER 23

## Module Reviews

### Module 23.1 Review

**A.** Catabolism is the breakdown of large molecules into smaller ones, whereas anabolism converts smaller molecules into larger ones. **B.** Metabolic turnover is the process in which cells continuously replace their cellular components.

### Module 23.2 Review

**A.** The reactants required by mitochondria are a continuous supply of 2-carbon substrates and $O_2$; the mitochondrial products are $CO_2$, $H_2O$, and ATP.

### Module 23.3 Review

**A.** The products of glycolysis are 2 molecules each of pyruvate, ATP, and NADH.

### Module 23.4 Review

**A.** The citric acid cycle is the reaction sequence that occurs in the matrix of mitochondria. In the process, organic molecules are broken down, carbon dioxide molecules are released, and hydrogen atoms are transferred to coenzymes that deliver them to the electron transport chain. **B.** The common substrate for the citric acid cycle in mitochondria is the 2-carbon acetate ion, $CH_3COO^-$, which is attached to coenzyme A to form acetyl-CoA. **C.** Coenzymes NAD and FAD transfer hydrogen atoms to the electron transport chain.

### Module 23.5 Review

**A.** Oxidative phosphorylation is the generation of ATP as the result of the transfer of electrons from the coenzymes NADH and $FADH_2$ to oxygen by a sequence of electron carriers within mitochondria. **B.** Oxidation is a gain of oxygen, or loss of hydrogen or electrons from an atom or molecule. Reduction is the loss of oxygen, or gain of hydrogen or electrons to an atom or molecule. **C.** Hydrogen ion channels are the passageways for the diffusion of hydrogen ions from the inner membrane space

of the mitochondria to the matrix. This movement of hydrogen ions powers the production of ATP by ATP synthase.

### Module 23.6 Review

**A.** Most (two-thirds) of the $CO_2$ released in the complete catabolism of glucose occurs during the citric acid cycle. **B.** Glycolysis is important when ATP must be produced anaerobically, for example, during peak levels of physical activity, at all times in red blood cells, or when a tissue is temporarily deprived of oxygen.

### Module 23.7 Review

**A.** Glycogenesis is the formation of glycogen from glucose. Glycogenolysis is the breakdown of glycogen to glucose. **B.** Cells make new compounds to maintain and repair structures, to support growth, and to build up nutrient reserves.

### Section 1 Review

**1.** d **2.** a **3.** e **4.** k **5.** f **6.** j **7.** g **8.** b **9.** c **10.** h **11.** i **12.** c **13.** e **14.** f **15.** m **16.** b **17.** a **18.** j **19.** l **20.** n **21.** g **22.** k **23.** h **24.** d **25.** i **26.** During fasting or starvation, other tissues shift to fatty acid catabolism or amino acid catabolism to conserve glucose for nervous tissue.

### Module 23.8 Review

**A.** Digestion is important because cells throughout the body rely on the organic molecules from the food we eat to produce energy and to replenish the intracellular nutrient pool. **B.** Most nutrient absorption occurs in the small intestine, primarily in the jejunum. **C.** Most nutrients enter into a branch of the hepatic portal vein and are transported to the liver.

### Module 23.9 Review

**A.** Intestinal gas, or flatus, is generated by bacterial activities in the colon when indigestible carbohydrates stimulate bacterial gas production. **B.** Glycogen is synthesized from excess glucose molecules by liver and muscle cells, and it serves as an intracellular glucose reserve. **C.** Carbohydrates are the preferred energy source because proteins and lipids are more important as structural components of cells and tissues.

## Module 23.10 Review

**A.** Micelles are lipid–bile salt complexes (containing fatty acids, glycerol, and monoglycerides) formed in the intestinal lumen. Chylomicrons are lipoproteins formed in intestinal epithelial cells and contain newly synthesized triglycerides, cholesterol, and other lipids surrounded by phospholipids and proteins. **B.** The liver absorbs chylomicrons, removes the triglycerides, combines the cholesterol from the chylomicron with recycled cholesterol, and alters the surface proteins. Newly synthesized complexes are released into the bloodstream as low-density lipoproteins (LDLs) or very low density lipoproteins (VLDLs). **C.** Low-density lipoproteins (LDLs) deliver cholesterol to body tissues, and high-density lipoproteins (HDLs) absorb unused cholesterol from body tissues, returning it to the liver, where it may be packaged into new LDLs or excreted with bile salts in bile.

## Module 23.11 Review

**A.** Beta-oxidation is fatty acid catabolism that produces molecules of acetyl-CoA. **B.** Acetyl-CoA is a reactant molecule in ATP production and in the synthesis of most types of lipids. **C.** Fatty acids may become a source of energy or a component of triglycerides, glycolipids, phospholipids, prostaglandins, cholesterol, and steroids.

## Module 23.12 Review

**A.** The arrival of acidic chyme in the duodenum triggers the release of CCK, which stimulates the production and release of inactive pancreatic proenzymes. Enteropeptidase, released from the duodenum, converts the pancreatic proenzyme trypsinogen into the proteolytic enzyme trypsin. Trypsin then converts other proenzymes to yield chymotrypsin, carboxypeptidase, and elastase.

Each of these enzymes attacks peptide bonds that link specific amino acids while ignoring others. As a result, they break down proteins into a mixture of dipeptides, tripeptides, and amino acids. **B.** The amino group is removed by deamination or transamination. **C.** The ammonium ions combine with carbon dioxide to form urea (in the urea cycle), which is ultimately excreted in the urine.

## Module 23.13 Review

**A.** The absorptive state, lasting about 4 hours, is the period following a meal, when nutrient absorption is under way. During this time, insulin stimulates glucose uptake and glycogenesis, amino acid upate, and protein synthesis. Androgens, estrogens, and growth hormone stimulate protein synthesis. Glycolysis and aerobic metabolism provide the necessary ATP. **B.** Ketone bodies form during the postabsorptive state, when lipids and amino acids are broken down in the liver. The increased concentration of acetyl-CoA that results from their breakdown forms ketone bodies. Ketone bodies are not catabolized by liver cells, and they diffuse into the circulation. **C.** During the absorptive state, insulin prevents a large surge in blood glucose after a meal by stimulating the liver to remove glucose from the circulation. During the postabsorptive state, blood glucose begins to decrease, triggering the release of glucagon, which stimulates the liver to release glucose into the circulation.

## Module 23.14 Review

**A.** Nutrition is the absorption of nutrients from food. **B.** The two classes of vitamins are fat-soluble vitamins and water-soluble vitamins. **C.** Vitamins play an important role in metabolic pathways by serving as coenzymes.

## Module 23.15 Review

**A.** A balanced diet contains all the ingredients needed to maintain homeostasis and prevent malnutrition. **B.** A complete

protein meets the body's amino acid requirements; an incomplete protein is deficient in one or more of the essential amino acids. **C.** The catabolism of lipids releases the greatest amount of energy per gram.

## Module 23.16 Review

**A.** Eating disorders are psychological problems that result in inadequate food consumption (anorexia nervosa) or excessive food consumption followed by purging (bulimia). **B.** Phenylketonuria (PKU) is an inherited metabolic disorder resulting from an inability to convert phenylalanine to tyrosine. **C.** Protein deficiency diseases are nutritional disorders resulting from a lack of one or more essential amino acids. Kwashiorkor is an example of a protein deficiency disease.

## Section 2 Review

**1.** m **2.** n **3.** j **4.** g **5.** a **6.** k **7.** d **8.** l **9.** f **10.** c **11.** i **12.** h **13.** e **14.** b **15.** d **16.** a **17.** d **18.** b **19.** Essential amino acids are necessary in the diet because the body cannot synthesize them. The body can synthesize nonessential amino acids on demand. **20.** (1) Proteins are difficult to break apart because of their complex structure. (2) The energy yield of proteins (4.32 Cal/g) is less than that of lipids (9.46 Cal/g). (3) The by-products of protein or amino acid catabolism are ammonium ions, a toxin that can damage cells. (4) Proteins form the most important structural and functional components of cells. Excessive protein catabolism would threaten homeostasis from the cellular level to the system level of organization. **21.** During the absorptive state, the intestinal mucosa is absorbing nutrients from the digested food. In the postabsorptive state, metabolic activity centers on the mobilization of energy reserves and the maintenance of normal blood glucose levels. **22.** Liver cells can break down or synthesize most carbohydrates, lipids, and amino acids. The liver has an extensive blood supply

and thus can easily monitor and regulate the blood levels of these nutrients. The liver also stores energy in the form of glycogen. **23.** It appears that Claudia is suffering from ketoacidosis as a consequence of her anorexia. Because she is literally starving herself, her body is metabolizing large amounts of fatty acids and amino acids to provide energy and, in the process, is producing large quantities of ketone bodies (normal metabolites from these catabolic processes). One of the ketones formed is acetone, which can be eliminated through the lungs. Acetone has a fruity aroma, so her breath would also smell fruity. The ketones are organic acids, and when they accumulate in the blood, they decrease the blood pH, and when the ketone bodies spill over into the urine, they decrease the pH in urine. (In severe cases of ketoacidosis, the ketone bodies may lower blood pH below 7.05, which may cause coma, cardiac arrhythmias, and death.)

## Module 23.17 Review

**A.** Energetics is the study of the flow of energy and its change from one form to another. **B.** Basal metabolic rate is the minimum resting energy expenditure of an awake, alert person. **C.** Thermoregulation is the homeostatic control of body temperature.

## Module 23.18 Review

**A.** Ghrelin, a hormone secreted by the gastric mucosa when the stomach is not full, inhibits the satiety center and stimulates appetite. **B.** Leptin is a peptide hormone produced by adipose tissue during the synthesis of triglycerides. It stimulates the satiety center and suppresses appetite. **C.** Lack of neuropeptide Y, a hypothalamic neurotransmitter, would probably decrease appetite because it normally stimulates the feeding center.

## Module 23.19 Review

**A.** Insensible perspiration is the evaporation of water from the skin and alveolar surfaces of the lungs. **B.** Radiation accounts for about one-half of a person's heat loss indoors. **C.** Conduction is the direct transfer of heat through physical contact. Convection is heat loss to the cooler air in contact with the skin. The air warmed by the skin rises, and it is repeatedly replaced by cooler air until there is no difference in temperature.

### Module 22.20 Review

**A.** The vasodilation of peripheral vessels would increase blood flow to the skin and thus the amount of heat the body can lose. As a result, body temperature would decrease. **B.** Nonshivering thermogenesis involves the release of hormones that increase the metabolic activity of all tissues, resulting in an increase in body temperature. **C.** Countercurrent exchange is the heat conservation mechanism that results in the conduction of heat from deep arteries to adjacent deep veins in the limbs.

### Section 3 Review

**1.** j **2.** d **3.** i **4.** f **5.** m **6.** l **7.** h **8.** k **9.** b **10.** a **11.** c **12.** n **13.** e **14.** g **15.** d **16.** b **17.** c **18.** c **19.** a **20.** a **21.** The amount of energy used or expended at rest is powered by mitochondrial energy production, which requires oxygen. Using the relationship of 4.825 Calories/L of oxygen, a measurement of the oxygen consumed provides an estimate of the energy expenditure. **22.** The heat-gain center functions in preventing hypothermia, or below-normal body temperature, by conserving body heat and increasing the rate of heat production by the body. **23.** Nonshivering thermogenesis increases the metabolic rate of most tissues through the actions of two hormones, epinephrine and thyroid-stimulating hormone (TSH). In the short term, the heat-gain center stimulates the adrenal medullae to release epinephrine by the sympathetic division of the ANS. Epinephrine quickly increases the breakdown of glycogen (glycogenolysis) in the liver and skeletal muscle, and it increases the metabolic rate of most tissues. The long-

term increase in metabolism occurs primarily in children as the heat-gain center adjusts the rate of thyrotropin-releasing hormone (TRH) release by the hypothalamus. When body temperature is low, additional TRH is released, which stimulates the release of TSH by the anterior lobe of the pituitary gland. The thyroid gland then increases its rate of thyroid hormone release, and these hormones increase the rate of catabolism throughout the body.

### Chapter Review Questions

**1.** false **2.** true **3.** true **4.** false **5.** false **6.** true **7.** d **8.** a **9.** c **10.** b **11.** a **12.** b **13.** d **14.** b **15.** a **16.** a **17.** Some of the energy released during catabolism is captured as ATP, and this ATP can be used as the energy source for anabolism. **18.** A cholesterol level above 200 mg/dL is considered elevated. Elevated cholesterol plus a high LDL:HDL ratio can result in atherosclerotic plaques that can cause heart attacks and strokes. **19.** Excess intake of a vitamin can produce harmful effects in a condition called hypervitaminosis. This condition most commonly involves one of the fat-soluble vitamins. **20.** A calorie is the energy needed to raise the temperature of 1 g of water 1°C. A kilocalorie is the amount of energy needed to raise the temperature of 1 kilogram of water 1°C. A Calorie is equal to a kilocalorie.

### Chapter Integration

**1.** Vitamins and minerals are essential components of the diet because the body cannot synthesize most of the vitamins and minerals it requires. **2.** These terms refer to the lipoproteins in the blood that transport cholesterol. The "good cholesterol" (high-density lipoproteins, or HDLs) transports excess cholesterol to the liver for storage and breakdown, whereas "bad cholesterol" (low-density lipoproteins, or LDLs) transports cholesterol to peripheral tissues,

which includes the arteries. The buildup of cholesterol in arterial walls is linked to cardiovascular disease. **3.** HDLs are considered beneficial because they reduce the amount of cholesterol in the bloodstream by transporting it to the liver for storage or excretion in bile. **4.** Glycogenesis (the synthesis of glycogen from glucose) in the liver increases after a high-carbohydrate meal. **5.** A diet deficient in pyridoxine (vitamin B$_6$) would interfere with the body's ability to metabolize proteins. Subsequent effects of the deficiency include retarded growth, anemia, convulsions, and epithelial changes.

## CHAPTER 24

### Module Reviews

#### Module 24.1 Review

**A.** The kidneys are the major excretory organs of the urinary system. **B.** The functions of the urinary system include removing wastes generated by body cells, adjusting blood volume and pressure, regulating blood plasma concentrations of ions, stabilizing blood pH, conserving nutrients by preventing their loss in urine, and removing drugs and toxins from the bloodstream.

#### Module 24.2 Review

**A.** The renal artery and nerves enter the kidney at the hilum, and the renal vein and ureter exit the kidney at the hilum. **B.** The layers of connective tissue that protect and anchor the kidney are the fibrous capsule, the perinephric fat (perirenal fat capsule), and the renal fascia. The fibrous capsule covers the surface of the kidney; the perinephric fat is a cushioning thick layer of adipose tissue surrounding the fibrous capsule; and the renal fascia is a dense, fibrous outer layer that anchors the kidney to surrounding structures. **C.** The perinephric fat helps hold the kidneys in position against the posterior body wall. If it were to be depleted and the collagen fibers of the fibrous capsule were to become detached,

the kidneys could drop. This may cause pain or distortion of other structures.

#### Module 24.3 Review

**A.** The renal pyramid is a conical mass within the renal medulla that ends at the papilla. **B.** The minor calyx collects urine that is produced by a kidney lobe. A major calyx forms from the fusion of 4–5 minor calyces. **C.** Juxtamedullary nephrons are essential for water conservation and concentrated urine production.

#### Module 24.4 Review

**A.** Filtrate is similar to blood plasma, but filtrate does not contain proteins. **B.** The renal corpuscle consists of the glomerular (Bowman's) capsule and the glomerulus, a capillary network. **C.** The main structures of the collecting system are the collecting duct and papillary duct. The collecting system is a series of tubes that carry tubular fluid away from the nephron. The collecting duct, composed of intercalated cells and principal cells, carries tubular fluid through the osmotic gradient in the renal medulla. The papillary duct collects tubular fluid from multiple collecting ducts and delivers it to a minor calyx.

#### Module 24.5 Review

**A.** Pathway of renal blood flow: renal artery → segmental arteries → interlobar arteries → arcuate arteries → cortical radiate arteries → afferent arterioles → glomerulus → cortical radiate veins → arcuate veins → interlobar veins → renal vein. **B.** Blood enters the glomerulus by the afferent arteriole and leaves by the efferent arteriole. **C.** The vasa recta are long, straight peritubular capillaries that parallel the ascending and descending segments of the nephron loop of a juxtamedullary nephron.

### Section 1 Review

**1.** urinary bladder **2.** nephrons **3.** renal tubules **4.** glomerulus **5.** proximal convoluted tubule **6.** papillary ducts **7.** renal medulla **8.** renal sinus **9.** major calyces **10.** ureter **11.** renal sinus: cavity within kidney that contains calyces, pelvis of the ureter, and segmental arteries and interlobar veins **12.** renal pelvis: funnel-shaped expansion of the superior portion of ureter. **13.** hilum: depression on the medial border of kidney; and site of the apex of the renal pelvis and the passage of segmental arteries, interlobar veins, and renal nerves. **14.** renal papilla: tip of the renal pyramid that projects into a minor calyx. **15.** ureter: tube that conducts urine from the renal pelvis to the urinary bladder. **16.** renal cortex: the outer portion of the kidney, containing renal lobules, renal columns (extensions between the pyramids), renal corpuscles, and the proximal and distal convoluted tubules. **17.** renal medulla: the inner, darker portion of the kidney that contains the renal pyramids.; **18.** renal pyramid: conical mass of the kidney projecting into the medullary region, containing part of the secreting tubules and collecting tubules. **19.** minor calyx: subdivision of major calyces into which urine enters from the renal papillae. **20.** major calyx: primary subdivision of renal pelvis formed from the merging of four or five minor calyces. **21.** kidney lobe: portion of kidney consisting of a renal pyramid and its associated cortical tissue. **22.** renal columns: cortical tissue separating renal pyramids. **23.** fibrous capsule (outer layer): covering of the kidney's outer surface and lining of the renal sinus.

## Module 24.6 Review

**A.** During filtration, fluids and solutes move from the glomerular capillaries and into the capsular space. During reabsorption, fluids and solutes move from the tubular fluid and into the peritubular fluid. During secretion, fluids and solutes move from the peritubular fluid and into the tubular fluid.

## Module 24.7 Review

**A.** The three distinct processes of urine formation in the kidney are filtration, reabsorption, and secretion. **B.** Filtration occurs exclusively across the filtration membrane in the renal corpuscle.

**C.** The nephron loop is the nephron segment solely involved in the reabsorption of water and sodium and chloride ions.

## Module 24.8 Review

**A.** The filtration membrane is formed by the fenestrated endothelium, basement membrane, and foot processes of the podocyte. **B.** Blood pressure is higher in glomerular capillaries than in systemic capillaries because the efferent arteriole has a smaller diameter than the afferent arteriole. **C.** Blood colloid osmotic pressure tends to draw water out of the filtrate and into the plasma because the solute concentration within the blood exceeds that within the filtrate. This action confers an important advantage: it helps conserve body water.

## Module 24.9 Review

**A.** Autoregulation maintains an adequate GFR despite changes in local blood pressure and blood flow. Dilation of afferent arterioles, constriction of efferent arterioles, and contraction of intraglomerular mesangial cells maintain GFR. **B.** In response to decreased filtration pressure, the juxtaglomerular complex increases renin production and release into the bloodstream. **C.** Angiotensin II triggers CNS responses, including enhancing the sensation of thirst, increasing antidiuretic hormone (ADH) production, and increasing sympathetic motor tone.

## Module 24.10 Review

**A.** The distal convoluted tubule (DCT) makes final adjustments to the composition of tubular fluid. **B.** Increased amounts of aldosterone, which promotes $Na^+$ retention and $K^+$ secretion by the kidneys, would increase the $K^+$ concentration of urine. **C.** If the concentration of $Na^+$ in the filtrate decreased, fewer hydrogen ions could be secreted by the countertransport mechanism involving these two ions. As a result, the pH of the tubular fluid would increase.

## Module 24.11 Review

**A.** Countercurrent multiplication in the kidneys is the exchange

of substances between two adjacent nephron loop limbs containing tubular fluid moving in opposite directions; the process enables the kidney tubules to concentrate urine. **B.** The thick ascending limb actively pumps sodium and chloride ions into the peritubular fluid. **C.** An increase in sodium and chloride ions in the peritubular fluid increases the osmotic concentrations around the descending thin limb, resulting in an osmotic outflow of water.

## Module 24.12 Review

**A.** The permeability of the PCT cannot change, and water reabsorption occurs whenever the osmotic concentration of the peritubular fluid exceeds that of the tubular fluid. However, the permeability of the DCT to water increases in response to antidiuretic hormone (ADH). **B.** Obligatory water reabsorption is the water movement that cannot be prevented. This reabsorption usually recovers 85 percent of the volume of filtrate. Facultative water reabsorption is the amount of water that can be precisely controlled. **C.** Increased ADH levels cause the appearance of more water channels, or aquaporins, in the DCT; as a result, more water is reabsorbed into the peritubular fluid, which decreases the volume of water in the urine. **D.** Decreased ADH levels in the DCT reduce the urine osmotic concentration because there is more water in the urine; the result is a larger volume of more dilute urine. **E.** ADH causes the insertion of aquaporins into the apical surfaces of the plasma membranes lining the DCT and collecting duct.

## Module 24.13 Review

**A.** The active removal of ions and organic substrates from the PCT causes an osmotic flow of water out of the tubular fluid.

## Module 24.14 Review

**A.** Chronic renal failure is a gradual loss of renal function, whereas acute renal failure is a sudden loss of renal function. **B.** Dialysis is the process of using an artificial semipermeable

membrane to remove wastes and retain plasma proteins from the blood of a person whose kidneys are not functioning properly. **C.** Patients on dialysis are often given Epogen or Procrit (synthetic forms of erythropoietin) to treat anemia, which occurs because their malfunctioning kidneys produce too little erythropoietin. Erythropoietin is the hormone that stimulates the development of red blood cells in the red bone marrow.

## Section 2 Review

**1.** f **2.** j **3.** h **4.** i **5.** g **6.** a **7.** e **8.** c **9.** d **10.** b **11.** nephron: functional unit of the kidney that filters and excretes waste materials from the blood and forms urine **12.** proximal convoluted tubule: reabsorbs water, ions, and all organic nutrients. **13.** distal convoluted tubule: important site of active secretion. **14.** renal corpuscle: expanded chamber that encloses the glomerulus. **15.** nephron loop: portion of the nephron that produces the concentration gradient in the renal medulla. **16.** collecting system: series of tubes that carry tubular fluid away from the nephron. **17.** collecting duct: portion of collecting system that receives fluid from many nephrons and performs variable reabsorption of water and reabsorption or secretion of sodium, potassium, hydrogen, and bicarbonate ions. **18.** papillary duct: delivers urine to the minor calyx. **19.** The presence of plasma proteins and numerous WBCs in Marissa's urine indicates an increased permeability of the filtration membrane. This condition usually results from inflammation of the filtration membrane within the renal corpuscle. If the condition is temporary, it is probably an acute glomerular nephritis usually associated with a bacterial infection (such as streptococcal sore throat). If the condition is long term, resulting in a nonfunctional kidney, it is referred to as chronic glomerular nephritis. The urine volume would be greater than normal because the plasma proteins increase the osmolarity of the filtrate.

### Module 24.15 Review
**A.** Urine production ends when the fluid enters the renal pelvis. **B.** A pyelogram is an image of the urinary system that is obtained by taking an x-ray after radiopaque dye has been administered intravenously. **C.** The male urethra is longer and transports not only urine, but also semen.

### Module 24.16 Review
**A.** Urine is transported by the ureters, stored within the urinary bladder, and eliminated through the urethra. **B.** The external urethral sphincter must be consciously relaxed to allow urination. **C.** The specialized smooth muscle in the wall of the urinary bladder is the detrusor; its contraction compresses the urinary bladder and expels urine into the urethra.

### Module 24.17 Review
**A.** The two reflexes that control urination are the urine storage reflex and the urine voiding reflex. **B.** The ability to consciously control urination depends on your ability to control the external urethral sphincter. Voluntary relaxation of the external urethral sphincter causes relaxation of the internal urethral sphincter, enabling you to urinate.

### Module 24.18 Review
**A.** Dysuria is the term for painful or difficult urination. **B.** Urinary obstruction at the urethra is more dangerous than urinary obstruction in the ureter because urine would be prevented from exiting the body and would build up in the urinary bladder, leading to rupture of the urinary bladder. Obstruction of a ureter would interfere with the flow of urine between the kidney and the urinary bladder. Obstruction of only one of the two ureters is less hazardous than obstruction of two urethrae because the kidney could still excrete wastes that could be voided.

### Section 3 Review
**1.** rugae **2.** ureter **3.** detrusor **4.** ureteric orifices **5.** trigone **6.** neck

of the urinary bladder **7.** urethra **8.** external urethral sphincter **9.** e **10.** f **11.** b **12.** j **13.** d **14.** k **15.** c **16.** g **17.** a **18.** h **19.** i **20.** Four primary signs and symptoms of urinary disorders are (1) changes in the volume of urine, (2) changes in the appearance of urine, (3) changes in the frequency of urination, and (4) pain. **21.** cystitis/pyelonephritis: Both conditions involve inflammation and infections of the urinary system, but cystitis refers to the urinary bladder, whereas pyelonephritis refers to the kidney. **22.** stress incontinence/overflow incontinence: Both conditions involve an inability to control urination, but stress incontinence involves periodic involuntary leakage, whereas overflow incontinence involves a continual, slow trickle of urine. **23.** polyuria/proteinuria: Both are abnormal urine conditions, but polyuria is the production of excessive amounts of urine, whereas proteinuria is the presence of protein in the urine.

### Chapter Review Questions
**1.** efferent arteriole **2.** juxtamedullary complex **3.** afferent arteriole **4.** glomerular capsule **5.** capsular space **6.** initial segment of renal tubule **7.** parietal layer of glomerular capsule **8.** visceral layer of glomerular capsule **9.** false **10.** true **11.** true **12.** false **13.** false **14.** true **15.** d **16.** a **17.** b **18.** b **19.** c **20.** b **21.** a **22.** a **23.** The processes of urine production are (1) filtration: blood pressure forces water and solutes across the membranes of the glomerular capillaries and into the capsular space; (2) reabsorption: the transport of water and solutes from the tubular fluid, across the tubular epithelium, and into the peritubular fluid; and (3) secretion: the transport of solutes from the peritubular fluid, across the tubular epithelium, and into the tubular fluid. **24.** The primary process involved in glomerular filtration is basically the same as that regulating fluid and solute

movement across capillaries throughout the body: the balance between hydrostatic pressure (fluid pressure) and colloid osmotic pressure (pressure due to materials in solution) on either side of the capillary membrane. **25.** The descending thin limb (DTL) and the thick ascending limb (TAL) of the loop of the nephron are very close together, and the tubular fluid flows in opposite directions. The exchange that occurs between these segments, called *countercurrent multiplication*, results in a higher osmotic concentration in the peritubular fluid compared to the tubular fluid in the ascending limb. This process is responsible for creating the concentration gradient in the renal medulla that enables the kidney to produce highly concentrated urine.

### Chapter Integration

**1.** The formation of crystalline masses can lead to kidney stones, which can cause urinary obstruction. The kidney stones can cause tiny ruptures along the urinary tract, leading to blood in the urine (hematuria). The metabolism of nitrogen-containing melamine releases ammonia, which may cause electrolyte imbalance and renal failure. **2.** Administration of intravenous fluids and dialysis would flush the body of the harmful chemical. Medication to control blood pressure, nausea, and anemia might also be necessary. Frequent blood tests and urinalysis are also performed to monitor kidney function.

## CHAPTER 25

### Module Reviews

### Module 25.1 Review
**A.** The ECF, or extracellular fluid, is made up of the interstitial fluid of peripheral tissues and the plasma of circulating blood. The ICF, or intracellular fluid, consists of the cytosol inside cells. **B.** Fluid compartments are the ECF and

ICF, which behave as separate sections and maintain different ionic compositions. **C.** Proteins make up most of the body mass.

### Module 25.2 Review
**A.** The major routes of fluid loss are urination, evaporation at the skin and at the lungs, and water lost in feces. **B.** A fluid shift is a rapid movement of water between the ECF and ICF in response to an osmotic gradient. **C.** Dehydration is a reduction in the water content of the body that develops when water losses outpace water gains. In dehydration, the osmotic concentration of blood plasma increases.

### Module 25.3 Review
**A.** Mineral balance is the state that exists when the body's ion gains and losses are equal. **B.** The ions absorbed by active transport are sodium, calcium, magnesium, iron, phosphate, and sulfate. **C.** Sodium is a major cation that is essential for normal membrane function. Calcium is a cation that is essential for normal muscle and neuron function and for normal bone structure.

### Module 25.4 Review
**A.** Osmoreceptor inhibition leads to decreased ADH secretion and suppressed thirst. **B.** Aldosterone causes increased urinary sodium retention and thus increases the sodium ion concentration in the ECF.

### Module 25.5 Review
**A.** The kidneys are primarily responsible for regulating the potassium ion concentration in the ECF. **B.** Potassium excretion increases as potassium concentrations increase in the ECF, under aldosterone stimulation, and when the pH of the ECF increases. **C.** Hypokalemia is a condition characterized by blood K$^+$ levels below 3.5 mEq/L. Hyperkalemia is a condition characterized by blood K$^+$ levels above 5.0 mEq/L.

### Section 1 Review
**1.** i **2.** d **3.** h **4.** n **5.** l **6.** m **7.** b **8.** c **9.** a **10.** j **11.** e **12.** k **13.** f **14.** g **15.** b **16.** c **17.** b **18.** a **19.** d **20.** c

**21.** When tissues are burned, cells are destroyed, and the contents of their cytoplasm leak into the interstitial fluid and then move into the blood. Since potassium ions are normally found within cells, damage to a large number of cells releases relatively large amounts of potassium ions into the blood. The elevated potassium level would stimulate cells of the adrenal cortex to produce aldosterone. The elevated levels of aldosterone would promote sodium retention and potassium secretion by the kidneys, thereby accounting for the elevated potassium levels in the patients' urine.

### Module 25.6 Review

**A.** Your body is in acid-base balance when the production of hydrogen ions is precisely offset by their loss and when the pH of body fluids remains within normal limits. **B.** The primary challenge to acid-base homeostasis is that your body generates a variety of acids during normal metabolic operations, and a significant decrease in body fluid pH must be prevented. **C.** Volatile acids can leave solution and enter the atmosphere. Fixed acids remain in body fluids until excreted by the kidneys. Metabolic acids are participants in, or by-products of, cellular metabolism.

### Module 25.7 Review

**A.** Acidemia is the condition in which blood pH decreases below 7.35. Alkalemia is the condition in which blood pH increases above 7.45. **B.** The intermediate compound formed from water and carbon dioxide that directly affects the pH of the ECF is carbonic acid ($H_2CO_3$). **C.** An inverse relationship exists between $P_{CO_2}$ and pH.

### Module 25.8 Review

**A.** The body's three major buffer systems are the phosphate buffer system, the protein buffer systems, and the carbonic acid–bicarbonate buffer system. **B.** The phosphate buffer system plays an important role in buffering the pH of the ICF and the urine. **C.** The carbonic acid–bicarbonate buffer

system prevents pH changes caused by metabolic acids and by fixed acids generated by metabolic activity. It uses the $H^+$ released by these acids to generate carbonic acid, which dissociates into $H_2O$ and $CO_2$, the latter of which is exhaled from the lungs.

### Module 25.9 Review

**A.** Metabolic acidosis results from the production of large numbers of fixed and metabolic acids, from the depletion of the bicarbonate reserve caused by an inability to excrete hydrogen ions at the kidneys, or from bicarbonate loss. **B.** Metabolic alkalosis results from the removal of large numbers of hydrogen ions from body fluids when bicarbonate ion concentrations become elevated. **C.** The kidneys conserve $HCO_3^-$ and eliminate $H^+$ in the urine during metabolic acidosis.

### Module 25.10 Review

**A.** The blood $P_{CO_2}$ of a patient with an airway obstruction would increase, resulting in respiratory acidosis. **B.** A decrease in the pH of body fluids would cause an increase in the respiratory rate.

### Section 2 Review

**1.** increasing $P_{CO_2}$ **2.** blood pH decrease **3.** arterial and CSF chemoreceptors; **4.** increased **5.** $H^+$ **6.** $HCO_3^-$ **7.** decreasing $P_{CO_2}$ **8.** decreased **9.** increased **10.** decreasing $P_{CO_2}$ **11.** blood pH increase **12.** arterial and CSF chemoreceptors **13.** $H^+$ **14.** $HCO_3^-$ **15.** decreased **16.** increased **17.** decreased **18.** increased **19.** The young boy has metabolic and respiratory acidosis. The metabolic acidosis resulted primarily from lactic acidosis (lactate ions and hydrogen ions) generated by the boy's muscles as he struggled in the water. Sustained hypoventilation during drowning contributed to both tissue hypoxia and respiratory acidosis. Respiratory acidosis developed as the $P_{CO_2}$ increased in the ECF, increasing the production of carbonic acid and its dissociation into $H^+$ and $HCO_3^-$. Prompt emergency treatment is essential; the usual

procedure involves some form of artificial or mechanical respiratory assistance (to increase the respiratory rate and decrease the $P_{CO_2}$ in the ECF) coupled with the intravenous infusion of a buffered isotonic solution that would absorb the hydrogen ions in the ECF and increase body fluid pH.

### Chapter Review Questions

**1.** false **2.** true **3.** false **4.** true **5.** false **6.** false **7.** true **8.** false **9.** true **10.** true **11.** d **12.** f **13.** g **14.** a **15.** c **16.** b **17.** h **18.** e **19.** d **20.** b **21.** b **22.** a **23.** d **24.** c **25.** c **26.** b **27.** Fluid balance is the state that exists when water content in the body remains stable over time. Water gains from the digestive tract must balance water losses through urination, evaporation at the skin and lungs, feces, and varying degrees of secretion from sweat glands. Mineral balance is the balance between ion absorption, which occurs across the lining of the small intestine and colon, and ion excretion, which is done primarily by the kidneys. Acid-base balance is the state that exists when the production of hydrogen ions is precisely offset by their loss and when the pH of body fluids remains within normal limits. **28.** Metabolic alkalosis may result from prolonged vomiting because the stomach continues to generate stomach acids to replace those that are lost. As a result the $HCO_3^-$ concentration of the ECF continues to rise. Compensation for metabolic alkalosis involves a reduction in pulmonary ventilation coupled with the increased loss of bicarbonates in the urine.

### Chapter Integration

**1.** Digestive secretions contain high levels of bicarbonate, so people with diarrhea can lose significant amounts of this important ion, leading to acidosis. We would expect the blood pH to be lower than 7.35 and the urine pH to be low (because of increased renal excretion of hydrogen ions).

We would also expect an increase in the rate and depth of breathing as the respiratory system tries to compensate for the lowered pH by eliminating carbon dioxide. **2.** The hypertonic solution will cause fluid to move from the ICF to the ECF, further aggravating Tom's dehydration. The slight increase in pressure and osmolarity of the blood should lead to an increase in ADH, even though ADH levels are probably quite high already. Despite the high ADH levels, urine volume would probably increase, because the kidneys could not reabsorb much of the glucose. The remaining glucose would increase the osmolarity of the tubular fluid, decreasing water reabsorption and increasing urine volume.

## CHAPTER 26

### Module Reviews

#### Module 26.1 Review

**A.** The gonads produce and store reproductive cells called gametes (sperm). **B.** The structures of the male external genitalia are the penis, urethra, and scrotum. **C.** The structures of the male internal genitalia are the ductus deferens (paired), seminal glands, prostate, bulbo-urethral glands, epididymis (paired), spermatic cord, and testes.

#### Module 26.2 Review

**A.** The complex network of channels that is connected to the seminiferous tubules is the rete testis. **B.** On a warm day, the cremaster (as well as the dartos muscle) would be relaxed so that the scrotum could descend away from the warmth of the body, thereby cooling the testes.

#### Module 26.3 Review

**A.** Spermatogenesis is the production of sperm and involves mitosis, meiosis, and spermiogenesis. **B.** Four haploid sperm will be produced from each diploid primary spermatocyte. **C.** A typical sperm has an acrosome, a head, neck, middle piece, and tail (flagellum).

## Module 26.4 Review

**A.** Interstitial endocrine cells produce male androgens, the male sex hormones, including testosterone and androstenedione. **B.** During spermiation, a sperm loses its attachment to the nurse cell and enters the lumen of the seminiferous tubule. **C.** Nurse cells provide a microenvironment that supports spermatogenesis and form the blood testis barrier that isolates sperm from the blood.

## Module 26.5 Review

**A.** The ductus deferens transports sperm from the epididymis to the ejaculatory duct and stores sperm in the ampulla. **B.** The bulbo-urethral glands secrete fluid that lubricates the tip of the penis and neutralizes any urinary acids that may remain in the urethra. **C.** The tail of each epididymis connects with the ductus deferens, which passes through the inguinal canal as part of the spermatic cord. Near the prostate, each ductus deferens enlarges to form an ampulla. The junction of the base of the seminal gland and the ampulla creates the ejaculatory duct, which empties into the urethra.

## Module 26.6 Review

**A.** The three columns of erectile tissue are the corpus spongiosum and the paired corpora cavernosa. **B.** An inability to contract the ischiocavernosus and bulbospongiosus would interfere with the ejaculation phase, including the contractions that are part of the male orgasm.

## Module 26.7 Review

**A.** Important regulatory hormones that establish and maintain male sexual function are follicle-stimulating hormone (FSH), luteinizing hormone (LH), and gonadotropin-releasing hormone (GnRH). Testosterone is the most important androgen, although some tissues, such as the external genitalia and prostate, are more responsive or sensitive to dihydrotestosterone (DHT). **B.** The testes, hypothalamus, and the anterior lobe of the pituitary gland secrete the hormones that control male sexual and reproductive functions. **C.** Low FSH levels would lead to low levels of testosterone in the seminiferous tubules, decreasing both sperm production rate and sperm count.

## Section 1 Review

**1.** prostatic urethra **2.** ductus deferens **3.** penile urethra **4.** penis **5.** epididymis **6.** testis **7.** external urethral orifice **8.** scrotum **9.** seminal gland **10.** prostate **11.** ejaculatory duct **12.** bulbo-urethral gland **13.** j **14.** h **15.** i **16.** l **17.** m **18.** k **19.** n **20.** b **21.** c **22.** a **23.** f **24.** d **25.** e **26.** g **27.** Normal levels of testosterone (1) promote the functional maturation of sperm, (2) maintain the accessory glands and organs of the male reproductive tract, (3) are responsible for establishing and maintaining male secondary sex characteristics, (4) stimulate bone and muscle growth, and (5) stimulate sexual behaviors and sexual drive (libido).

## Module 26.8 Review

**A.** The main organs of the female reproductive system are the external genitalia (vulva and urethra) and the internal genitalia (ovaries, uterine tubes, uterus, and vagina). **B.** The female external genitalia include the mons pubis, clitoris, and labia. **C.** Fertilization normally occurs in the uterine tubes.

## Module 26.9 Review

**A.** The vesico-uterine pouch is the pocket between the uterus and posterior wall of the urinary bladder. The recto-uterine pouch is the pocket between the uterus and the anterior surface of the colon. **B.** The ovaries produce immature female gametes called oocytes; secrete female sex hormones, including estrogens and progesterone; and secrete inhibin, which is involved in the feedback control of FSH production. **C.** The ovaries, uterine tubes, and uterus are enclosed within the broad ligament. The mesovarium supports and stabilizes each ovary.

## Module 26.10 Review

**A.** An oocyte is an immature female gamete whose meiotic divisions will produce a single ovum and two or three polar bodies. **B.** Follicle development may take a year to complete. Important events include (1) formation of primary ovarian follicles from primordial ovarian follicles in an egg nest, (2) formation of secondary ovarian follicles, and (3) formation of tertiary ovarian follicles. During the monthly ovarian cycle one tertiary ovarian follicle matures, releases a secondary oocyte at ovulation, and forms a corpus luteum, which will degenerate into a corpus albicans if fertilazation doesn't occur. **C.** Males produce gametes from puberty until death; females produce gametes only from puberty to menopause. Males produce many gametes at a time; females typically produce one or two per 28-day ovarian cycle. Males release mature gametes that have completed meiosis; females release secondary oocytes suspended in metaphase of meiosis II.

## Module 26.11 Review

**A.** Recently released secondary oocytes reach the uterine tube with the aid of ciliary currents on the inner surfaces of the fimbriae of the infundibulum. **B.** The perimetrium is the outer, incomplete serosal layer; the myometrium is the middle, muscular layer; and the endometrium is the inner, glandular layer. **C.** The regions of the uterus are the fundus (upper, rounded portion), body (ending at the internal os), and cervix (inferior portion).

## Module 26.12 Review

**A.** The endometrium contains the deeper basal layer and the more superficial functional layer. **B.** Menstruation typically lasts from 1 to 7 days. **C.** The uterine cycle begins with the menstrual phase, an interval marked by the destruction of the functional layer and menstruation (menses). After menstruation, the proliferative phase begins, during which the functional layer undergoes repair and thickens. Following the proliferative phase is the secretory phase, during which uterine (endometrial) glands enlarge.

## Module 26.13 Review

**A.** The vagina is a muscular tube extending between the uterus and external genitalia; its lining forms folds called rugae. The proximal portion of the vagina is marked by the cervix, which dips into the vaginal canal, and the shallow recess known as the vaginal fornix. The hymen, a thin epithelial fold, partially blocks the entrance to the vagina until physical distortion ruptures it. **B.** The vagina serves as a passageway for the elimination of menstrual fluids; receives the penis during sexual intercourse and holds sperm prior to their passage into the uterus; and forms the inferior portion of the birth canal, through which the fetus passes during delivery. **C.** The greater vestibular glands in females are similar to the bulbo-urethral glands in males; and both the male penis and the female clitoris have erectile tissue.

## Module 26.14 Review

**A.** Lactation is the secretion of milk by the mammary glands. **B.** Blockage of a single lactiferous sinus would not interfere with the delivery of milk to the nipple, because each mammary gland has several lactiferous sinuses. **C.** Route of milk flow: secretory alveoli of the secretory lobules → ducts within a lobe → lactiferous duct of the lobe → lactiferous sinus → surface of the nipple.

## Module 26.15 Review

**A.** If the LH surge did not occur during an ovarian cycle, meiosis I completion, ovulation, and corpus luteum formation could not take place. **B.** Destruction of the functional layer of the endometrium and menstruation (menses) occur when estrogen and progesteron levels decrease.

## Module 26.16 Review

**A.** Condoms provide some protection against sexually transmitted diseases. **B.** Depo-Provera injections (a progesterone-only form of birth control) result in the cessation of the uterine cycle in 50 percent of women using this product. (Uterine cycles eventually resume after use of the product is discontinued.) **C.** A vasectomy is the surgical removal of a segment of each ductus deferens and the tying or cauterizing of the cut ends, preventing sperm from reaching the distal portions of the male reproductive tract.

## Module 26.17 Review

**A.** Benign prostate hypertrophy (BPH) is an enlargement of the prostate. Prostate cancer is a malignancy, and prostatectomy may be part of treatment. **B.** The human papillomavirus (HPV) causes most cases of cervical cancer. **C.** Sexually transmitted diseases are infections that are transferred from person to person primarily or exclusively by sexual contact or intercourse.

## Section 2 Review

1. infundibulum 2. ovary 3. uterine tube 4. perimetrium 5. myometrium 6. endometrium 7. uterus 8. clitoris 9. labium minus 10. labium majus 11. vaginal fornix 12. cervix 13. external os 14. vagina 15. j 16. d 17. l 18. i 19. b 20. f 21. n 22. m 23. e 24. a 25. h 26. c 27. k 28. g 29. The endometrial cells have receptors for estrogens and progesterone and respond to these hormones as if the cells were still in the body of the uterus. Under the influence of estrogens, the endometrial cells proliferate at the beginning of the uterine (menstrual) cycle and begin to develop glands and blood vessels, which then further develop under the control of progesterone. This dramatic increase in tissue size exerts pressure on neighboring tissues or in some other way interferes with their function. It is the recurring expansion of tissue in an abnormal location that causes periodic pain.

1. ampulla 2. ductus deferens 3. seminal gland 4. prostate prostate 5. corpus spongiosum 6. corpus cavernosum 7. epididymis 8. testis 9. external urethral orifice 10. bulbo-urethral gland; 11. glans penis 12. mons pubis 13. prepuce 14. clitoris 15. labium majus 16. urethral opening 17. labium minus 18. hymen 19. vaginal entrance 20. false 21. true 22. true 23. false 24. true 25. true 26. c 27. c 28. b 29. a 30. d 31. a 32. Birth control pills that contain estrogen and progesterone or only progesterone inhibit GnRH release at the hypothalamus and thus, through negative feedback, inhibit FSH and LH release from the anterior lobe of the pituitary gland. Without FSH, tertiary ovarian follicles do not begin to develop, and levels of estrogen remain low. An LH surge, triggered by the peaking of estrogen, is necessary for ovulation to occur. If the level of estrogen is not allowed to increase above the critical level, the LH surge will not occur, and thus ovulation will not occur, even if a tertiary ovarian follicle managed to develop to a stage at which it could ovulate. Any mature ovarian follicles would ultimately degenerate, and no new ovarian follicles would mature to take their place. Although the ovarian cycle is interrupted, the level of hormones is still adequate to regulate a normal menstrual cycle. 33. After ovulation, the tertiary ovarian follicle initially collapses, and the remaining granulosa cells proliferate to create the corpus luteum that secretes progesterone. 34. The sterilization procedure for men is vasectomy; for women, it is tubal ligation. After vasectomy, male sexual function is not affected. Sperm continue to develop, but they are not present in semen because the ductus deferens is now blocked. A man continues to experience an erection as he did before. After tubal ligation, a woman still experiences the menstrual cycle, but the sperm are unable to contact the secondary oocyte because the uterine tubes are now blocked.

1. The presence of exercise-induced amenorrhea suggests that a certain amount of body fat is necessary for menstrual cycles to occur. If body fat levels fall below some set point, menstruation ceases. 2. Because a woman lacking adequate body fat might not have the energy reserves needed to have a successful pregnancy, compensatory mechanisms in her body prevent pregnancy by shutting down the ovarian cycle and thus the menstrual cycle. When her body subsequently accumulates sufficient energy reserves in body fat, the cycles begin again. 3. Estrogen deficiency can lead to infertility, vaginal and breast atrophy, and osteoporosis. It may also increase the risk of heart attacks later in life.

# CHAPTER 27

## Module Reviews

### Module 27.1 Review

**A.** Gestation is the time that a developing embryo and fetus spends within the uterus, or womb. It is a period of three integrated trimesters, each 3 months long. **B.** Embryonic development refers to the events that occur during the first 2 months after fertilization. Fetal development begins at the start of the 9th week and continues until birth. Prenatal development consists of both embryonic and fetal development. **C.** Postnatal development begins at birth and continues until maturity.

### Module 27.2 Review

**A.** Fertilization is the fusion of a secondary oocyte and a sperm to form a zygote. **B.** A normal zygote contains 46 chromosomes. **C.** Many sperm are needed to achieve fertilization because one sperm does not contain enough acrosomal enzymes to erode the corona radiata surrounding the secondary oocyte.

### Module 27.3 Review

**A.** The morula is the pre-embryo stage that is a solid ball of cells. **B.** The trophoblast is the cell layer surrounding the blastocyst. These cells nourish the embryo and later form part of the placenta. **C.** The blastocyst consists of an outer trophoblast and an inner cell mass. Implantation begins about 6 to 9 days after fertilization, when the blastocyst adheres to the uterine lining, or endometrium. The trophoblast cells in contact with the uterine lining divide and form a syncytiotrophoblast that erodes the endometrial lining. Nutrients from uterine glands are absorbed by the syncytiotrophoblast and distributed by diffusion through the cytotrophoblast to the inner cell mass. The blastocyst becomes enclosed within the endometrium by about 10 days after fertilization.

### Module 27.4 Review

**A.** Gastrulation is the process that forms the primary germ layers—the endoderm, ectoderm, and mesoderm—from the embryonic disc. It is from these germ layers that the body systems differentiate. The mesoderm contributes to nearly all body systems except the nervous and respiratory systems.

### Module 27.5 Review

**A.** The four extra-embryonic membranes are the yolk sac, amnion, allantois, and chorion. **B.** The yolk sac forms from the endoderm and mesoderm; it is an important site of blood cell formation. The amnion forms from the ectoderm and mesoderm; it encloses the fluid that surrounds and cushions the developing embryo and fetus. The allantois forms from the endoderm and mesoderm; its base gives rise to the urinary bladder. The chorion forms from the mesoderm and cytotrophoblast; it surrounds the blastocoele. **C.** The placenta forms the interface between the embryonic/fetal system and the maternal system. It becomes the primary support structure for the developing embryo and fetus. The placenta is the site for nutrient, gas, and waste exchange between maternal and embryonic/fetal bloodstreams.

### Module 27.6 Review

**A.** The chorionic villi are structures that extend outward into the maternal tissues, forming an intricate, branching network through which maternal blood flows. Embryonic blood vessels extend into each chorionic villus. **B.** The body stalk is the connection between the embryo and the chorion; it contains portions of the allantois and blood vessels that carry blood to and from the placenta. The yolk stalk is the connection between the endoderm of the embryo and the yolk sac. **C.** The umbilical cord connects the fetus to the placenta, and it is derived from the allantois.

### Module 27.7 Review

**A.** The hormones synthesized by the syncytiotrophoblast are human chorionic gonadotropin (hCG), human placental lactogen (hPL), placental prolactin, relaxin, progesterone, and estrogens. **B.** The presence of human chorionic gonadotropin (hCG) in the urine is a reliable indicator of pregnancy. **C.** After the first trimester, the placenta is sufficiently functional to maintain the pregnancy.

### Module 27.8 Review

**A.** Organogenesis is the process of organ formation. **B.** In the second trimester, the organ systems increase in complexity. During the third trimester, many of the organ systems become fully functional. **C.** During the third trimester, the fetus undergoes its largest absolute weight gain.

### Module 27.9 Review

**A.** The major changes that occur in maternal systems during pregnancy are increases in respiratory rate and tidal volume, blood volume, nutrient requirements, glomerular filtration rate (GFR), and the size of the uterus and mammary glands. **B.** A mother's blood

volume increases during pregnancy to compensate for the reduction in maternal blood volume resulting from blood flow through the placenta. **C.** Some pregnant women experience difficulty breathing because the enlarged uterus presses up against abdominal organs and the diaphragm, crowding the lungs.

### Module 27.10 Review

**A.** Relaxin, produced by the placenta, softens the pubic symphysis and dilates the cervix, and the weight of the fetus distorts the myometrium of the uterus. Distortion of the cervix and rising estrogen levels promote the release of oxytocin, and the already stretched smooth muscles of the myometrium become even more excitable. **B.** Estrogens and oxytocin stimulate the production of prostaglandins, which are then primarily responsible for initiating true labor. **C.** The dilation stage begins with the onset of true labor as the cervix dilates and the fetus begins to move toward the cervical canal; late in this stage, the amnion ruptures. The expulsion stage begins as the cervix dilates completely and continues until the fetus has completely emerged from the vagina (delivery). In the placental stage, the uterus gradually contracts, tearing the connections between the endometrium and the placenta and ejecting the placenta (also referred to as the afterbirth).

### Module 27.11 Review

**A.** Colostrum contains antibodies and is produced by the mammary glands from the end of the 6th month of pregnancy until a few days after delivery. After that, the glands begin producing breast milk, which contains antibodies and lysozyme but has a higher fat content than colostrum. **B.** Oxytocin causes the milk ejection (milk let-down) reflex.

### Module 27.12 Review

**A.** The postnatal stages of development are the neonatal period, from birth to 1 month; infancy, from 1 month to age 1 year; childhood, from age 1 year

until sexual maturation begins; adolescence, which begins with the onset of sexual maturation (puberty) between ages 10 and 15 years and ends when growth in body size is completed (around 18 years); and maturity, which includes the rest of the person's life. A final stage called senescence, or aging, overlaps with maturity.

### Module 27.13 Review

**A.** The three interacting hormonal events associated with the onset of puberty are (1) increased GnRH production by the hypothalamus; (2) increased sensitivity to GnRH by the anterior lobe of the pituitary gland, and a rapid increase in circulating levels of FSH and LH; and (3) increased sensitivity to FSH and LH by ovarian and testicular cells. **B.** A man has a deeper voice and a larger larynx than a woman because testosterone stimulates laryngeal development in males to a greater extent than estrogen stimulates laryngeal development in females. **C.** The higher estrogen levels in premenopausal women (compared with adult men) decrease plasma cholesterol levels, thereby slowing plaque formation and lowering the risk of atherosclerosis.

### Section 1 Review

**1.** b **2.** j **3.** g **4.** k **5.** a **6.** m **7.** e **8.** f **9.** n **10.** c **11.** l **12.** d **13.** h **14.** i **15.** c **16.** c **17.** b **18.** b **19.** a **20.** c **21.** It is very unlikely that the baby's condition is the result of a viral infection contracted during the third trimester. The development of organ systems occurs during the first trimester, and by the end of the second trimester, most organ systems are fully formed. During the third trimester, the fetus undergoes tremendous growth, but very little new organ formation occurs.

### Module 27.14 Review

**A.** Chromosomes and their component genes make up the genotype. Anatomical and physiological characteristics from the expression of genes make up the phenotype.

**B.** Autosomes (non-sex chromosomes) are the 22 homologous pairs that affect somatic characteristics. The chromosomes of the 23rd pair are the sex chromosomes that determine whether an individual is male (XY) or female (XX). **C.** The karyotype shown is male because it has an XY for the 23rd pair of chromosomes.

### Module 27.15 Review

**A.** Homozygous means that homologous chromosomes carry the same allele of a given gene. Heterozygous means that homologous chromosomes carry different alleles of a given gene. **B.** In simple inheritance, phenotypic characteristics are determined by interactions between a single pair of alleles. Polygenic inheritance involves interactions among alleles of several genes. **C.** The phenotype of a person who is heterozygous for "freckles"—that is, a person with one dominant allele and one recessive allele for that trait—would be "freckles."

### Module 27.16 Review

**A.** In strict dominance, one allele dominates the other, so an individual who is heterozygous for a given trait exhibits the dominant phenotype. In codominance, an individual who is heterozygous for a given trait exhibits both of the phenotypes for that trait. **B.** Sex-linked traits are expressed more frequently in males because the small Y chromosome of males contains few genes that correspond to those on the larger X chromosome. **C.** (1) simple inheritance (dominant); (2) simple inheritance (recessive); (3) autosomal inheritance.

### Module 27.17 Review

**A.** A single nucleotide polymorphism is a variation in a single base pair in a DNA sequence. **B.** (1) XO: Turner syndrome; (2) XXY: Klinefelter syndrome. **C.** (1) ovarian cancer: chromosome pair 9; (2) Tay–Sachs disease: chromosome pair 15; (3) spinocerebellar ataxia: chromosome pair 6.

## Section 2 Review

**1.** l **2.** g **3.** i **4.** k **5.** a **6.** c **7.** b
**8.** j **9.** f **10.** h **11.** e **12.** d

**13.** Tongue-rolling

|  |  | Maternal alleles | Maternal alleles |
|---|---|---|---|
|  |  | $T$ | $t$ |
| Paternal alleles | $T$ | $TT$ | $Tt$ |
| Paternal alleles | $t$ | $Tt$ | $tt$ |

Mom = $Tt$
Dad = $Tt$
The possible genotypic and phenotypic ratios of offspring are the following:

    1 $TT$ = tongue-roller;
      1 in 4 = 25%
    2 $Tt$ = tongue-rollers;
      2 in 4 = 50%
    1 $tt$ = non-tongue-roller;
      1 in 4 = 25%

The probability they will have a child not able to roll his/her tongue is 1 in 4, or 25%.

**14.** Cystic fibrosis

|  |  | Maternal alleles | Maternal alleles |
|---|---|---|---|
|  |  | $c$ | $c$ |
| Paternal alleles | $C$ | $Cc$ | $Cc$ |
| Paternal alleles | $c$ | $cc$ | $cc$ |

Mom = $cc$
Dad = $Cc$
The possible genotypic and phenotypic ratios are the following:

    2 $Cc$ = normal carriers;
      2 in 4 = 50%
    2 $cc$ = cystic fibrosis;
      2 in 4 = 50%

The probability that they will have a child who will be a carrier for CF is 2/4, or 50%. (A child has to inherit two copies of the faulty allele to be born with CF.)

### Chapter Review Questions

**1.** false **2.** true **3.** false **4.** true **5.** false **6.** true **7.** false **8.** false **9.** b **10.** b **11.** d **12.** a **13.** b **14.** d **15.** c **16.** d **17.** The five stages of postnatal development are neonatal period (1 month), infancy (1 year), childhood (up to age 10), adolescence (between 10 and 15 years), and maturity (18 years). **18.** Trisomy 21, or Down syndrome, is a chromosomal abnormality in which the affected person has a third chromosome on the 21st pair. Affected people exhibit intellectual disability and certain physical traits, including a characteristic facial appearance. Anatomical problems affecting the cardiovascular system may prove fatal during childhood or early adulthood. There is a direct correlation between maternal age and the risk of having a child with trisomy 21. For a maternal age below 25, the incidence of Down syndrome approaches 1 in 2000 births, or 0.05 percent. For maternal ages 30–34, the odds increase to 1 in 900; and during ages 35–44, they increase to 1 in 46, or more than 2 percent.

### Chapter Integration

**1.** Although technically it takes only one sperm to fertilize a secondary oocyte, Joe's oligospermia means that too few sperm reach the secondary oocyte in a uterine tube to produce sufficient acrosomal enzymes to enable one sperm to penetrate the corona radiata and fuse with the oocyte membrane. **2.** Hemophilia A is an X-linked recessive trait. The probability that this couple's daughters will have hemophilia is zero, because each daughter will receive a dominant normal allele from her father. There is a 50 percent chance that a son will have hemophilia, because each son has a 50 percent chance of receiving the mother's normal allele and a 50 percent chance of receiving the mother's recessive allele.

# Glossary

**A**

**abdomen:** The region of the trunk between the inferior margin of the rib cage and the superior margin of the pelvis.

**abdominopelvic cavity:** The term used to refer to the general region bounded by the abdominal wall and the pelvis; it contains the peritoneal cavity and visceral organs.

**abducens:** Cranial nerve VI, which innervates the lateral rectus muscle of the eye.

**abduction:** Movement away from the midline of the body, as viewed in the anatomical position.

**abscess:** A localized collection of pus within a damaged tissue.

**absorption:** The active or passive uptake of gases, fluids, or solutes.

**accommodation:** An alteration in the curvature of the lens of the eye to focus an image on the retina.

**acetabulum:** The fossa on the lateral aspect of the pelvis that accommodates the head of the femur.

**acetylcholine (ACh):** A chemical neurotransmitter in the brain and peripheral nervous system; the dominant neurotransmitter in the peripheral nervous system, released at neuromuscular junctions and synapses of the parasympathetic division.

**acetylcholinesterase (AChE):** An enzyme found in the synaptic cleft, bound to the postsynaptic membrane, and in tissue fluids; breaks down and inactivates acetylcholine molecules.

**acetyl-CoA:** An acetyl group bound to coenzyme A, a participant in the anabolic and catabolic pathways for carbohydrates, lipids, and many amino acids.

**acetyl group:** —$CH_3CO$.

**Achilles tendon:** The large tendon that inserts on the calcaneus; tension on this tendon produces extension (plantar flexion) of the foot; also called *calcaneal tendon*.

**acid:** A compound whose dissociation in solution releases a hydrogen ion and an anion; an acidic solution has a pH below 7.0 and contains an excess of hydrogen ions.

**acidosis:** An abnormal physiological state characterized by a blood pH below 7.35.

**acinus/acini:** A histological term referring to a blind pocket, pouch, or sac.

**acoustic:** Pertaining to sound or the sense of hearing.

**acromion:** A continuation of the scapular spine that projects superior to the capsule of the shoulder joint.

**acrosome:** A membranous sac at the tip of a sperm that contains hyaluronidase; also called *acrosomal cap*.

**actin:** The protein component of microfilaments that forms thin filaments in skeletal muscles and produces contractions of all muscles through interaction with thick (myosin) filaments.

**action potential:** A propagated change in the membrane potential of excitable cells, initiated by a change in the membrane permeability to sodium ions.

**active transport:** The ATP-dependent absorption or secretion of solutes across a plasma membrane.

**acute:** Sudden in onset, severe in intensity, and brief in duration.

**adaptation:** A change in pupillary size in response to changes in light intensity; a decrease in receptor sensitivity or perception after chronic stimulation.

**adaptive immunity:** Type of defense that is not present at birth and is related to previous antigen exposure; also called *specific immunity*.

**Addison disease:** A condition resulting from the hyposecretion of glucocorticoids; characterized by lethargy, weakness, hypotension, and increased skin pigmentation.

**adduction:** Movement toward the axis or midline of the body, as viewed in the anatomical position.

**adenine:** A purine; one of the nitrogenous bases in the nucleic acids RNA and DNA.

**adenohypophysis:** The anterior lobe of the pituitary gland.

**adenosine:** A compound consisting of adenine and ribose.

**adenosine diphosphate (ADP):** A compound consisting of adenosine with two phosphate groups attached.

**adenosine monophosphate (AMP):** A nucleotide consisting of adenosine plus a phosphate group ($PO_4^{3-}$); also called *adenosine phosphate*.

**adenosine triphosphate (ATP):** A high-energy compound consisting of adenosine with three phosphate groups attached; the third is attached by a high-energy bond.

**adenylate cyclase:** An enzyme bound to the inner surfaces of plasma membranes that can convert ATP to cyclic AMP; formerly called *adenylyl cyclase*.

**adipocyte:** A fat cell.

**adipose tissue:** Loose connective tissue dominated by adipocytes.

**adrenal cortex:** The superficial portion of the adrenal gland that produces steroid hormones; also called *suprarenal cortex*.

**adrenal gland:** A small endocrine gland that secretes steroids and catecholamines and is located superior to each kidney; also called *suprarenal gland*.

**adrenal medulla:** The core of the adrenal gland; a modified sympathetic ganglion that secretes catecholamines into the blood during sympathetic activation; also called *suprarenal medulla*.

**adrenergic:** An axon terminal that, when stimulated, releases norepinephrine.

**adrenocortical hormone:** Any steroid produced by the adrenal cortex.

**adrenocorticotropic hormone (ACTH):** The hormone that stimulates the production and secretion of glucocorticoids by the zona fasciculata of the adrenal cortex; released by the anterior lobe of the pituitary gland in response to corticotropin-releasing hormone.

**adventitia:** The superficial layer of connective tissue surrounding an internal organ; fibers are continuous with those of surrounding tissues, providing support and stabilization.

**aerobic:** Requiring the presence of oxygen.

**aerobic metabolism:** The complete breakdown of organic substrates into carbon dioxide and water; a process that yields large amounts of ATP but requires mitochondria and oxygen.

**afferent:** Toward a center.

**afferent arteriole:** An arteriole that carries blood to a glomerulus of the kidney.

**afferent fiber:** An axon that carries sensory information to the central nervous system.

**agglutination:** The aggregation of red blood cells due to interactions between surface antigens and plasma antibodies.

**aggregated lymphoid nodules:** Lymphoid nodules beneath the epithelium of the small intestine; also called *Peyer's patches*.

**agonist:** A muscle responsible for a specific movement; also called a *prime mover*.

**agranular:** Without granules; *agranular leukocytes* are monocytes and lymphocytes.

**alba:** White.

**albicans:** White.

**albuginea:** White.

**aldosterone:** A mineralocorticoid produced by the zona glomerulosa of the adrenal cortex; stimulates sodium and water conservation at the kidneys; secreted in response to the presence of angiotensin II.

**alkalosis:** The condition characterized by blood pH greater than 7.45; associated with a relative deficiency of hydrogen ions or an excess of bicarbonate ions.

**allograft:** Tissue transplant between members of the same species.

**alpha (α) receptors:** Membrane receptors sensitive to norepinephrine or epinephrine; stimulation normally results in the excitation of the target cell.

**alveolar sac:** An air-filled chamber that supplies air to several alveoli.

**alveolus/alveoli:** Blind pockets at the end of the respiratory tree, lined by a simple squamous epithelium and surrounded by a capillary network; sites of gas exchange with the blood; a bony socket that holds the root of a tooth.

**Alzheimer's disease:** A disorder resulting from degenerative changes in populations of neurons in the cerebrum, causing dementia characterized by problems with attention, short-term memory, and emotions.

**amination:** The attachment of an amino group to a carbon chain; performed by a variety of cells and important in the synthesis of amino acids.

**amino acid:** Organic compound made up of carbon, hydrogen, oxygen, and nitrogen; building blocks of protein.

**amino group:** —$NH_2$.

**amnion:** One of the four extra-embryonic membranes; surrounds the developing embryo or fetus.

**amniotic fluid:** Fluid that fills the amniotic cavity; cushions and supports the embryo or fetus.

**amphiarthrosis:** A joint that permits a small degree of independent movement; *see* **interosseous membrane** and **pubic symphysis**.

**amphipathic:** Molecule containing both hydrophobic and hydrophilic portions.

**ampulla/ampullae:** A localized dilation in the lumen of a canal or passageway.

**amygdaloid body:** A basal nucleus that is a component of the limbic system and acts as an interface between that system, the cerebrum, and sensory systems; also called *amygdala*.

**amylase:** An enzyme that breaks down polysaccharides; produced by the salivary glands and pancreas.

**anabolism:** The synthesis of complex organic compounds from simpler precursors.

**anaerobic:** Without oxygen.

**anal triangle:** The posterior subdivision of the perineum.

**anaphase:** The mitotic stage in which the paired chromatids separate and move toward opposite ends of the spindle apparatus.

**anaphylaxis:** A hypersensitivity reaction due to the binding of antigens to immunoglobulins (IgE) on the surfaces of mast cells; the release of histamine, serotonin, and prostaglandins by mast cells then causes widespread inflammation; a sudden decrease in blood pressure may occur, producing anaphylactic shock.

**anastomosis:** The joining of two tubes, usually referring to a connection between two peripheral vessels without an intervening capillary bed.

**anatomical position:** An anatomical reference position; the body viewed from the anterior surface with the palms facing forward.

**anatomy:** The study of the structure of the body.

**androgen:** A steroid sex hormone primarily produced by the interstitial cells of the testis and manufactured in small quantities by the adrenal cortex in both sexes.

**anemia:** The condition marked by a decrease in the hematocrit, the hemoglobin content of the blood, or both.

**angiogenesis:** Formation of new blood vessels.

**angiotensin I:** The hormone produced by the activation of angiotensinogen by renin; angiotensin-converting enzyme (ACE) converts angiotensin I into angiotensin II in lung capillaries.

**angiotensin II:** A hormone that causes an increase in systemic blood pressure, stimulates the secretion of aldosterone, promotes thirst, and causes the release of antidiuretic hormone; angiotensin-converting enzyme in lung capillaries converts angiotensin I into angiotensin II.

**angiotensinogen:** The blood protein produced by the liver that is converted to angiotensin I by the enzyme renin.

**anion:** An ion having a negative charge.

**anoxia:** Tissue oxygen deprivation.

**antagonist:** A muscle that opposes the movement of an agonist.

**antebrachium:** The forearm.

**anterior:** On or near the front, or ventral surface, of the body.

**antibiotic:** A chemical agent that selectively kills pathogens, primarily bacteria.

**antibody:** A globular protein produced by plasma cells that will bind to specific antigens and promote their destruction or removal from the body.

**antibody-mediated immunity:** The form of immunity resulting from the presence of circulating antibodies produced by plasma cells; also called *humoral immunity.*

**anticodon:** Three nitrogenous bases on a tRNA molecule that interact with a complementary codon on a strand of mRNA.

**antidiuretic hormone (ADH):** A hormone synthesized in the hypothalamus and secreted by the posterior lobe of the pituitary gland (neurohypophysis); causes water retention by the kidneys and an increase of blood pressure.

**antigen:** A substance capable of inducing an immune response that may include the production of antibodies.

**antigen-antibody complex:** The combination of an antigen and a specific antibody.

**antigenic determinant site:** A portion of an antigen that can interact with an antibody molecule.

**antigen-presenting cell (APC):** A cell that processes antigens and displays them, bound to class II MHC proteins; essential to the initiation of a normal immune response.

**antihistamine:** A chemical agent that blocks the action of histamine on peripheral tissues.

**antrum:** A chamber or pocket.

**anulus:** A cartilage or bone shaped like a ring; also spelled *annulus.*

**anus:** The external opening of the anal canal.

**aorta:** The large, elastic artery that carries blood away from the left ventricle and into the systemic circuit.

**apocrine secretion:** A mode of secretion in which the glandular cell sheds portions of its cytoplasm.

**aponeurosis/aponeuroses:** Broad tendinous sheet(s) that may serve as the origin or insertion of a skeletal muscle.

**appendicular:** Pertaining to the upper or lower limbs.

**appendix:** A blind sac connected to the cecum of the large intestine.

**appositional growth:** The enlargement of a cartilage or bone by the addition of cartilage or bony matrix at its surface.

**aqueous humor:** A fluid similar to perilymph or cerebrospinal fluid that fills the anterior chamber of the eye.

**arachidonic acid:** One of the essential fatty acids.

**arachnoid granulations:** Processes of the arachnoid mater that project into the superior sagittal sinus; sites where cerebrospinal fluid enters the venous circulation.

**arachnoid mater:** The middle meninx that encloses cerebrospinal fluid and protects the central nervous system.

**arbor vitae:** The central, branching mass of white matter inside the cerebellum.

**arcuate:** Curving.

**areolar:** Containing minute spaces, as in areolar tissue.

**areolar tissue:** Loose connective tissue with an open framework.

**arrector pili:** Smooth muscles whose contractions force hairs to stand erect.

**arrhythmias:** Abnormal patterns of cardiac contractions.

**arteriole:** A small arterial branch that delivers blood to a capillary network.

**artery:** A blood vessel that carries blood away from the heart and toward a peripheral capillary.

**articular:** Pertaining to a joint.

**articular capsule:** The dense collagen fiber sleeve that surrounds a joint and provides protection and stabilization.

**articular cartilage:** The cartilage pad that covers the surface of a bone inside a joint cavity.

**articulation:** A joint.

**ascending tract:** A tract carrying information from the spinal cord to the brain.

**association areas:** Cortical areas of the cerebrum that are responsible for the integration of sensory inputs and/or motor commands.

**astrocyte:** One of the four types of neuroglia in the central nervous system; responsible for maintaining the blood brain barrier by the stimulation of endothelial cells.

**atherosclerosis:** The formation of fatty plaques in the walls of arteries, restricting blood flow to deep tissues.

**atom:** The smallest stable unit of matter.

**atomic number:** The number of protons in the nucleus of an atom.

**atomic weight:** The average of the different atomic masses and isotope proportions of a particular element.

**atria:** Thin-walled chambers of the heart that receive venous blood from the pulmonary or systemic circuit.

**atrial natriuretic peptide (ANP):** *See* **natriuretic peptides.**

**atrioventricular (AV) bundle:** Specialized conducting cells in the interventricular septum that carry the contracting stimulus from the AV node to bundle branches and then to Purkinjie fibers; also called *bundle of His.*

**atrioventricular (AV) node:** Specialized cardiac muscle cells that relay the contractile stimulus to the bundle of His, the bundle branches, the Purkinje fibers, and the ventricular myocardium; located at the boundary between the atria and ventricles.

**atrioventricular (AV) valve:** One of the valves that prevents backflow into the atria during ventricular systole (contraction).

**atrophy:** The wasting away of tissues from a lack of use, ischemia, or nutritional abnormalities.

**auditory:** Pertaining to the sense of hearing.

**auditory ossicles:** The bones of the middle ear: malleus, incus, and stapes.

**auditory tube:** A passageway that connects the nasopharynx with the middle ear cavity.

**auricle:** A broad, flattened process that resembles the external ear; in the ear, the expanded, projecting portion that surrounds the external auditory meatus, also called *pinna;* in the heart, the externally visible flap formed by the collapse of the outer wall of a relaxed atrium.

**autoantibodies:** Antibodies that react with antigens on the surfaces of a person's own cells and tissues.

**autograft:** Tissue transplant from the same person.

**autoimmunity:** The immune system's sensitivity to normal cells and tissues, resulting in the production of autoantibodies.

**autolysis:** The destruction of a cell due to the rupture of lysosomal membranes in its cytoplasm.

**automaticity:** The spontaneous depolarization to threshold, characteristic of cardiac pacemaker cells.

**autonomic ganglion:** A collection of visceral motor neurons outside the central nervous system.

**autonomic nerve:** A peripheral nerve consisting of preganglionic or postganglionic autonomic fibers.

**autonomic nervous system (ANS):** Centers, nuclei, tracts, ganglia, and nerves involved in the unconscious regulation of visceral functions; includes components of the central nervous system and the peripheral nervous system.

**autopsy:** The detailed examination of a body after death.

**autoregulation:** Changes in activity that maintain homeostasis in direct response to changes in the local environment; does not require neural or endocrine control.

**autosomal:** Chromosomes other than the X or Y sex chromosome.

**avascular:** Without blood vessels.

**axilla:** The armpit.

**axolemma:** The plasma membrane of an axon, continuous with the plasma membrane of the cell body and dendrites and distinct from any neuroglial coverings.

**axon:** The elongated extension of a neuron that conducts an action potential.

**axon hillock:** In a multipolar neuron, the portion of the cell body adjacent to the initial segment.

**axon terminal:** The club-shaped ending of an axon that contains neuro-transmitters and makes synaptic contact with an effector; also called a *synaptic terminal.*

**axoplasm:** The cytoplasm within an axon.

### B

**bacteria:** Single-celled microorganisms, some pathogenic, that are common in the environment and in and on the body.

**Bainbridge reflex:** The reflexive increase in heart rate after an increase in venous return; due to mechanical and neural factors; also called *atrial reflex.*

**baroreception:** The ability to detect changes in pressure.

**baroreceptor reflex:** A reflexive change in cardiac activity in response to changes in blood pressure.

**baroreceptors:** The receptors responsible for baroreception.

**basal lamina:** A layer of filaments and fibers that attaches an epithelium to the underlying connective tissue.

**basal nuclei:** Nuclei of the cerebrum that are important in the subconscious control of skeletal muscles.

**base:** A compound whose dissociation releases a hydroxide ion ($OH^-$) or removes a hydrogen ion ($H^+$) from the solution.

**basement membrane:** A layer of filaments and fibers that attaches an epithelium to the underlying connective tissue.

**basophils:** Circulating granulocytes (white blood cells) similar in size and function to tissue mast cells.

**B cells:** Lymphocytes capable of differentiating into plasma cells, which produce antibodies.

**benign:** Not malignant.

**beta (β) cells:** Cells of the pancreatic islets that secrete insulin in response to increased blood sugar concentrations.

**beta–oxidation:** Fatty acid catabolism that produces molecules of acetyl-CoA.

**beta (β) receptors:** Membrane receptors sensitive to epinephrine; stimulation may result in the excitation or inhibition of the target cell.

**bicarbonate ion:** $HCO_3^-$; anion component of the carbonic acid-bicarbonate buffer system.

**bicuspid:** Having two cusps or points; refers to a premolar tooth, which has two roots, or to the left AV valve, which has two cusps.

**bicuspid valve:** *See* **mitral valve**.

**bifurcate:** To branch into two parts.

**bile:** The exocrine secretion of the liver; stored in the gallbladder and ejected into the duodenum.

**bile duct:** The duct formed by the union of the cystic duct from the gallbladder and the bile ducts from the liver; terminates at the duodenal ampulla, where it meets the pancreatic duct.

**bile salts:** Steroid derivatives in bile; responsible for the emulsification of ingested lipids.

**bilirubin:** A pigment that is the by-product of hemoglobin catabolism.

**biopsy:** The removal of a small tissue sample for pathological analysis.

**bladder:** A muscular sac that distends as fluid is stored and whose contraction ejects the fluid at an appropriate time; used alone, the term usually refers to the urinary bladder.

**blastocyst:** An early stage in the developing embryo, consisting of an outer trophoblast and an inner cell mass.

**blood brain barrier:** The isolation of the central nervous system from the general circulation; primarily the result of astrocyte regulation of capillary permeabilities.

**blood CSF barrier:** The isolation of the cerebrospinal fluid from the capillaries of the choroid plexus; primarily the result of specialized ependymal cells.

**blood pressure:** A force exerted against vessel walls by the blood in the vessels, due to the push exerted by cardiac contraction and the elasticity of the vessel walls; usually measured along one of the muscular arteries, with systolic pressure measured during ventricular systole (contraction) and diastolic pressure during ventricular diastole (relaxation).

**blood testis barrier:** The isolation of the interior of the seminiferous tubules from the general circulation, due to the activities of the nurse (sustentacular) cells.

**Bohr effect:** The increased oxygen release by hemoglobin due to increased carbon dioxide levels.

**bolus:** A compact mass; usually refers to compacted ingested material (food) on its way to the stomach.

**bone tissue:** A string connective tissue containing specialized cells and a mineralized matrix of crystalline calcium phospahate and calcium carbonate; also called *osseous tissue*.

**bowel:** The intestinal tract.

**brachial:** Pertaining to the arm.

**brachial plexus:** A network formed by branches of spinal nerves $C_5$—$T_1$ en route to innervating the upper limb.

**brachium:** The arm.

**bradycardia:** An abnormally slow heart rate, usually below 50 bpm.

**brain natriuretic peptide (BNP):** *See* **natriuretic peptides**.

**brainstem:** The brain minus the cerebrum, diencephalon, and cerebellum.

**brevis:** Short.

**bronchial tree:** The trachea, bronchi, and bronchioles.

**bronchodilation:** The dilation of the bronchial passages; can be caused by sympathetic stimulation.

**bronchus/bronchi:** Branch/branches of the bronchial tree between the trachea and bronchioles.

**buccal:** Pertaining to the cheeks.

**buffer:** A compound that stabilizes the pH of a solution by removing or releasing hydrogen ions.

**buffer system:** Interacting compounds that prevent increases or decreases in the pH of body fluids; includes the carbonic acid-bicarbonate buffer

system, the phosphate buffer system, and the protein buffer system.

**bulbar:** Pertaining to the brainstem.

**bulbourethral glands:** Mucous glands at the base of the penis that secrete into the penile urethra; the equivalent of the greater vestibular glands of females; also called *Cowper's glands*.

**bundle branches:** Specialized conducting cells in the ventricles that carry the contractile stimulus from the atrioventricular bundle (bundle of His) to the Purkinje fibers.

**bundle of His:** *See* **atrioventricular (AV) bundle**.

**bursa:** A small sac filled with synovial fluid that cushions adjacent structures and reduces friction.

## C

**calcaneal tendon:** The large tendon that inserts on the calcaneus; tension on this tendon produces extension (plantar flexion) of the foot; also called *Achilles tendon*.

**calcaneus:** The heel bone, the largest of the tarsal bones.

**calcification:** The deposition of calcium salts within a tissue.

**calcitonin:** The hormone secreted by C cells of the thyroid when calcium ion concentrations are abnormally high; restores homeostasis by increasing the rate of bone deposition (mostly in childhood) and the rate of calcium loss by the kidneys.

**calculus/calculi:** Solid mass/masses of insoluble materials that form within body fluids, especially the gallbladder, kidneys, or urinary bladder.

**callus:** A localized thickening of the epidermis due to chronic mechanical stresses; a thickened area that forms at the site of a bone break as part of the repair process.

**canaliculi:** Microscopic passageways between cells; bile canaliculi carry bile to bile ducts in the liver; in bone, canaliculi permit the diffusion of nutrients and wastes to and from osteocytes.

**cancer:** An illness caused by mutations leading to the uncontrolled growth and replication of the affected cells.

**cannula:** A tube that can be inserted into the body; commonly placed in blood vessels prior to transfusion or dialysis.

**capacitation:** The activation process that must occur before a sperm can successfully fertilize an oocyte; occurs in the vagina after ejaculation.

**capillary:** A small blood vessel, located between an arteriole and a venule, whose thin wall permits the diffusion of gases, nutrients, and wastes between plasma and interstitial fluids.

**capitulum:** A general term for a small, elevated articular process; refers to the rounded distal surface of the humerus that articulates with the head of the radius.

**caput:** The head.

**carbaminohemoglobin:** Hemoglobin bound to carbon dioxide molecules.

**carbohydrase:** An enzyme that breaks down carbohydrate molecules.

**carbohydrate:** An organic compound containing carbon, hydrogen, and oxygen in a ratio that approximates 1:2:1.

**carbon dioxide:** $CO_2$; a compound produced by the citric acid cycle reactions of aerobic metabolism.

**carbonic anhydrase:** An enzyme that catalyzes the reaction $H_2O + CO_2 \rightarrow H_2CO_3$; important in carbon dioxide transport, gastric acid secretion, and renal pH regulation.

**carcinogenic:** Stimulating cancer formation in affected tissues.

**cardia:** The area of the stomach surrounding its connection with the esophagus.

**cardiac:** Pertaining to the heart.

**cardiac cycle:** One complete heartbeat, including atrial and ventricular systole and diastole.

**cardiac output:** The amount of blood ejected by the left ventricle each minute.

**cardiac reserve:** The potential percentage increase in cardiac output above resting levels.

**cardiac tamponade:** A compression of the heart due to fluid accumulation in the pericardial cavity.

**cardiocyte:** A cardiac muscle cell.

**cardiovascular:** Pertaining to the heart, blood, and blood vessels.

**cardiovascular center:** Poorly localized area in the reticular formation of the medulla oblongata of the brain; includes cardioacceleratory, cardioin-hibitory, and vasomotor centers.

**cardium:** The heart.

**carotene:** A yellow-orange pigment, found in carrots and in green and orange leafy vegetables, that the body can convert to vitamin A.

**carotid artery:** The principal artery of the neck, servicing cervical and cranial structures; one branch, the internal carotid, provides a major blood supply to the brain.

**carotid body:** A group of receptors, adjacent to the carotid sinus, that are sensitive to changes in the carbon dioxide levels, pH, and oxygen concentrations of arterial blood.

**carotid sinus:** A dilated segment at the base of the internal carotid artery whose walls contain baroreceptors sensitive to changes in blood pressure.

**carotid sinus reflex:** Reflexive changes in blood pressure that maintain homeostatic pressures at the carotid sinus, stabilizing blood flow to the brain.

**carpus/carpal:** The wrist.

**cartilage:** A connective tissue with a gelatinous matrix that contains an abundance of fibers.

**catabolism:** The breakdown of complex organic molecules into simpler components, accompanied by the release of energy.

**catalyst:** A substance that accelerates a specific chemical reaction but that is not altered by the reaction.

**catecholamines:** Epinephrine, norepinephrine, dopamine, and related compounds.

**catheter:** A tube surgically inserted into a body cavity or along a blood vessel or excretory passageway for the collection of body fluids, monitoring of blood pressure, or introduction of medications or radiographic dyes.

**cation:** An ion that has a positive charge.

**cauda equina:** Spinal nerve roots distal to the tip of the adult spinal cord; they extend caudally inside the vertebral canal en route to lumbar and sacral segments.

**caudal/caudally:** Closest to or toward the tail (coccyx).

**caudate nucleus:** One of the basal nuclei involved with the subconscious control of skeletal muscles.

**cell:** The smallest living structural unit in the human body.

**cell-mediated immunity:** Resistance to disease by sensitized T cells that destroy antigen-bearing cells by direct contact or through the release of lymphotoxins; also called *cellular immunity*.

**center of ossification:** The site in a connective tissue where bone formation begins.

**central canal:** Longitudinal canal in the center of an osteon that contains blood vessels and nerves, a passageway along the longitudinal axis of the spinal cord that contains cerebrospinal fluid.

**central nervous system (CNS):** The brain and spinal cord.

**centriole:** A cylindrical intracellular organelle composed of nine groups of microtubules, three in each group; functions in mitosis or meiosis by organizing the microtubules of the spindle apparatus.

**centromere:** The localized region where two chromatids remain connected after the chromosomes have replicated; site of spindle fiber attachment.

**centrosome:** A region of cytoplasm that contains a pair of centrioles oriented at right angles to one another.

**cephalic:** Pertaining to the head.

**cerebellum:** The posterior portion of the metencephalon, containing the cerebellar hemispheres; includes the arbor vitae, cerebellar nuclei, and cerebellar cortex.

**cerebral cortex:** An extensive area of neural cortex covering the surfaces of the cerebral hemispheres.

**cerebral hemispheres:** A pair of expanded portions of the cerebrum covered in neural cortex.

**cerebrospinal fluid (CSF):** Fluid bathing the internal and external surfaces of the central nervous system; secreted by the choroid plexus.

**cerebrovascular accident (CVA):** The occlusion of a blood vessel that supplies a portion of the brain, resulting in damage to the dependent neurons; also called *stroke.*

**cerebrum:** The largest portion of the brain, composed of the cerebral hemispheres; includes the cerebral cortex, the basal nuclei, and the internal capsule.

**cerumen:** The waxy secretion of the ceruminous glands along the external acoustic meatus.

**ceruminous glands:** Integumentary glands that secrete cerumen.

**cervix:** The inferior portion of the uterus.

**chemoreception:** The detection of changes in the concentrations of dissolved compounds or gases.

**chemotaxis:** The attraction of phagocytes to the source of abnormal chemicals in tissue fluids.

**chloride shift:** The movement of plasma chloride ions into red blood cells in exchange for bicarbonate ions generated by the intracellular dissociation of carbonic acid.

**cholecystokinin (CCK):** A duodenal hormone that stimulates the contraction of the gallbladder and the secretion of enzymes by the exocrine pancreas.

**cholesterol:** A steroid component of plasma membranes and a substrate for the synthesis of steroid hormones and bile salts.

**choline:** A breakdown product or precursor of acetylcholine.

**cholinergic synapse:** A synapse where the presynaptic membrane releases acetylcholine on stimulation.

**cholinesterase:** The enzyme that breaks down and inactivates acetylcholine.

**chondrocyte:** A cartilage cell.

**chondroitin sulfate:** The predominant proteoglycan in cartilage, responsible for the gelatinous consistency of the matrix.

**chordae tendineae:** Fibrous cords that stabilize the position of the AV valves in the heart, preventing backflow during ventricular systole.

**chorion/chorionic:** An extraembryonic membrane, consisting of the trophoblast and underlying mesoderm, that forms the placenta.

**choroid:** The middle, vascular layer in the wall of the eye.

**choroid plexus:** The vascular complex in the roof of the third and fourth ventricles of the brain, responsible for the production of cerebrospinal fluid.

**chromatid:** One complete copy of a DNA strand and its associated nucleoproteins.

**chromatin:** A histological term referring to the grainy material visible in cell nuclei during interphase; the appearance of the DNA content of the nucleus when the chromosomes are uncoiled.

**chromosomes:** Dense structures, composed of tightly coiled DNA strands and associated histones, that become visible in the nucleus when a cell prepares to undergo mitosis or meiosis; normal human somatic cells each contain 46 chromosomes.

**chronic:** Habitual or long term.

**chylomicrons:** Relatively large droplets that may contain triglycerides, phospholipids, and cholesterol in association with proteins; synthesized and released by intestinal cells and transported to the venous blood by the lymphatic system.

**ciliary body:** A thickened region of the choroid that encircles the lens of the eye; includes the ciliary muscle and the ciliary processes that support the *ciliary zonule* (suspensory ligament) of the lens.

**cilium/cilia:** Slender organelle(s) extending from the free surface of an epithelial cell and generally undergoes cycles of movement; composed of a basal body and microtubules in a 9 + 2 array.

**circulatory system:** The network of blood vessels and lymphatic vessels that facilitate the distribution and circulation of extracellular fluid.

**circumduction:** A movement at a synovial joint in which the distal end of the bone moves in a circular direction, but the shaft does not rotate.

**cisterna:** An expanded or flattened chamber derived from and associated with the endoplasmic reticulum.

**citric acid cycle:** The reaction sequence that occurs in the matrix of mitochondria; in the process, organic molecules are broken down, carbon dioxide molecules are released, and hydrogen atoms are transferred to coenzymes that deliver them to the electron transport system; also called the *Krebs cycle.*

**clot:** A network of fibrin fibers and trapped blood cells; also called a *thrombus* if it occurs within the cardiovascular system.

**clotting factors:** Plasma proteins, synthesized by the liver, that are essential to the clotting response.

**clotting response:** The series of events that result in the formation of a clot.

**coccygeal ligament:** The fibrous extension of the dura mater and filum terminale; provides longitudinal stabilization to the spinal cord.

**coccyx:** The terminal portion of the spinal column, consisting of relatively tiny, fused vertebrae.

**cochlea:** The spiral portion of the bony labyrinth of the internal ear that surrounds the organ of hearing.

**cochlear duct:** The central membranous tube within the cochlea that is filled with endolymph and contains the organ of Corti; also called *scala media.*

**codon:** A sequence of three nitrogenous bases along an mRNA strand that will specify the location of a single amino acid in a peptide chain.

**coelom:** The ventral body cavity, lined by a serous membrane and subdivided during fetal development into the pleural, pericardial, and abdominopelvic (peritoneal) cavities.

**coenzymes:** Complex organic cofactors; most are structurally related to vitamins.

**cofactor:** Ions or molecules that must be attached to the active site before an enzyme can function; examples include mineral ions and several vitamins.

**collagen:** A strong, insoluble protein fiber common in connective tissues.

**collateral ganglion:** A sympathetic ganglion located anterior to the spinal column and separate from the sympathetic chain.

**colliculus/colliculi:** Little mound(s); in the brain, refers to one of the thickenings in the roof of the mesencephalon; the superior colliculi are associated with the visual system, and the inferior colliculi with the auditory system.

**colloid/colloidal suspension:** A solution containing large organic molecules in suspension.

**colon:** The large intestine.

**comminuted:** Broken or crushed into small pieces.

**commissure:** A crossing over from one side to another.

**compact bone:** Dense bone that contains parallel osteons.

**complement:** A group of distinct plasma proteins that interact in a chain reaction after exposure to activated antibodies or the surfaces of certain pathogens; complement proteins promote cell lysis, phagocytosis, and other defense mechanisms.

**compliance:** Distensibility; the ability of certain organs to tolerate changes in volume; indicates the presence of elastic fibers and smooth muscles.

**compound:** A pure chemical substance made up of atoms of two or more different elements in a fixed proportion, regardless of the type of chemical bond joining them.

**concentration:** The amount (in grams) or number of atoms, ions, or molecules (in moles) per unit volume.

**concentration gradient:** Regional differences in the concentration of a particular substance.

**conception:** Fertilization.

**concha/conchae:** Three pairs of thin, scroll-like bones that project into the nasal cavities; the superior and middle conchae are part of the ethmoid, and the inferior conchae articulate with the ethmoid, lacrimal, maxilla, and palatine bones.

**condyle:** A rounded articular projection on the surface of a bone.

**congenital:** Present at birth.

**congestive heart failure (CHF):** The failure to maintain adequate cardiac output due to cardiovascular problems or myocardial damage.

**conjunctiva:** A layer of stratified squamous epithelium that covers the inner surfaces of the eyelids and the anterior surface of the eye to the edges of the cornea.

**connective tissue:** One of the four primary tissue types; provides a structural framework that stabilizes the relative positions of the other tissue types; includes connective tissue proper, cartilage, bone, and blood; contains cell products, cells, and ground substance.

**continuous propagation:** The propagation of an action potential along an unmyelinated axon or a muscle plasma membrane, wherein the action potential affects every portion of the membrane surface.

**contractility:** The ability to contract; possessed by skeletal, smooth, and cardiac muscle cells.

**contralateral reflex:** A reflex that affects the opposite side of the body from the stimulus.

**conus medullaris:** The conical tip of the spinal cord that gives rise to the filum terminale.

**convergence:** In the nervous system, the innervation of a single neuron by axons from several neurons; most common along motor pathways.

**coracoid process:** A hook-shaped process of the scapula that projects above the anterior surface of the capsule of the shoulder joint.

**Cori cycle:** The metabolic exchange of lactate from skeletal muscle for glucose from the liver; performed during the recovery period after muscular exertion.

**cornea:** The transparent portion of the fibrous layer of the anterior surface of the eye.

**corniculate cartilages:** A pair of small laryngeal cartilages.

**cornu:** Horn-shaped.

**coronoid:** Hooked or curved.

**corpora quadrigemina:** The superior and inferior colliculi of the mesencephalic tectum (roof) in the brain.

**corpus/corpora:** Body/bodies.

**corpus callosum:** A large bundle of axons that links centers in the left and right cerebral hemispheres.

**corpus luteum:** The progesterone-secreting mass of follicle cells that develops in the ovary after ovulation.

**cortex:** The outer layer or region of an organ.

**corticobulbar tracts:** Descending tracts that carry information or commands from the cerebral cortex to nuclei and centers in the brainstem.

**corticospinal tracts:** Descending tracts that carry motor commands from the cerebral cortex to the anterior gray horns of the spinal cord.

**corticosteroid:** A steroid hormone produced by the adrenal cortex.

**corticosterone:** A corticosteroid secreted by the zona fasciculata of the adrenal cortex; a glucocorticoid.

**corticotropin-releasing hormone (CRH):** The releasing hormone, secreted by the hypothalamus, that stimulates secretion of adrenocorticotropic hormone (ACTH) by the anterior lobe of the pituitary gland (adenohypophysis).

**cortisol:** A corticosteroid secreted by the zona fasciculata of the adrenal cortex; a glucocorticoid.

**costa/costae:** Rib(s).

**cotransport:** The membrane transport of a nutrient, such as glucose, in company with the movement of an ion, normally sodium; transport requires a carrier protein but does not involve direct ATP expenditure and can occur regardless of the concentration gradient for the nutrient.

**countercurrent exchange:** The transfer of heat, water, or solutes between two fluids that travel in opposite directions.

**countercurrent multiplication:** Active transport between two limbs of a loop that contains a fluid moving in one direction; responsible for the concentration of urine in the kidney tubules.

**covalent bond:** A chemical bond between atoms that involves the sharing of electrons.

**coxal bone:** Hip bone.

**cranial:** Pertaining to the head.

**cranial nerves:** Peripheral nerves originating at the brain.

**craniosacral division:** *See* **parasympathetic division**.

**cranium:** The braincase; the skull bones that surround and protect the brain.

**creatine:** A nitrogenous compound, synthesized in the body, that can form a high-energy bond by connecting to a phosphate group and that serves as an energy reserve.

**creatine phosphate:** A high-energy compound in muscle cells; during muscle activity, the phosphate group is donated to ADP, regenerating ATP; also called *phosphorylcreatine*.

**creatinine:** A breakdown product of creatine metabolism.

**crenation:** Cellular shrinkage due to an osmotic movement of water out of the cytoplasm.

**cribriform plate:** A portion of the ethmoid bone that contains the foramina used by the axons of olfactory receptors en route to the olfactory bulbs of the cerebrum.

**cricoid cartilage:** A ring-shaped cartilage that forms the inferior margin of the larynx.

**crista/cristae:** A ridge-shaped collection of hair cells in the ampulla of a semicircular duct; the crista and cupula form a receptor complex sensitive to movement along the plane of the semicircular canal.

**cross-bridge:** A myosin head that projects from the surface of a thick filament and that can bind to an active site of a thin filament in the presence of calcium ions.

**cuneiform cartilages:** A pair of small cartilages in the larynx.

**cupula:** A gelatinous mass that is located in the ampulla of a semicircular duct in the internal ear and whose movement stimulates the hair cells of the crista.

**Cushing disease:** A condition caused by the oversecretion of adrenal steroids.

**cutaneous membrane:** The epidermis and papillary layer of the dermis.

**cuticle:** The layer of dead, keratinized cells that surrounds the shaft of a hair; for nails, *see* **eponychium**.

**cyanosis:** An abnormal bluish coloration of the skin due to the presence of deoxygenated blood in vessels near the body surface.

**cystic duct:** A duct that carries bile between the gallbladder and the bile duct.

**cytochrome:** A protein-pigment molecule of the electron transport chain.

**cytokinesis:** The cytoplasmic movement that separates two daughter cells at the completion of mitosis.

**cytology:** The study of cells.

**cytoplasm:** The material between the plasma membrane and the nuclear membrane; cell contents.

**cytosine:** A pyrimidine; one of the nitrogenous bases in the nucleic acids RNA and DNA.

**cytoskeleton:** A network of microtubules and microfilaments in the cytoplasm.

**cytosol:** The fluid portion of the cytoplasm.

**cytotoxic:** Poisonous to cells.

**cytotoxic T cells:** Lymphocytes involved in cell-mediated immunity that kill target cells by direct contact or by the secretion of lymphotoxins; also called *killer T cells* and $T_C$ *cells*.

## D

**daughter cells:** Genetically identical cells produced by somatic cell division.

**deamination:** The removal of an amino group from an amino acid.

**decomposition reaction:** A chemical reaction that breaks a molecule into smaller fragments.

**decussate:** To cross over to the opposite side, usually referring to the crossover of the descending tracts of the corticospinal pathway on the ventral surface of the medulla oblongata.

**defecation:** The elimination of fecal wastes.

**degradation:** Breakdown, catabolism.

**dehydration:** A reduction in the water content of the body that threatens homeostasis.

**dehydration synthesis:** The joining of two molecules associated with the removal of a water molecule.

**demyelination:** The loss of the myelin sheath of an axon, normally due to chemical or physical damage to Schwann cells or oligodendrocytes.

**denaturation:** A temporary or permanent change in the three-dimensional structure of a protein.

**dendrite:** A sensory process of a neuron.

**deoxyribonucleic acid (DNA):** A nucleic acid consisting of a double chain of nucleotides that contains the sugar deoxyribose and the nitrogenous bases adenine, guanine, cytosine, and thymine.

**deoxyribose:** A five-carbon sugar resembling ribose but lacking an oxygen atom.

**depolarization:** A change in the membrane potential from a negative value toward 0 mV.

**depression:** Inferior (downward) movement of a body part.

**dermatome:** A sensory region monitored by the dorsal rami of a single spinal segment.

**dermis:** The connective tissue layer beneath the epidermis of the skin.

**detrusor:** Collectively, the three layers of smooth muscle in the wall of the urinary bladder.

**development:** Growth and the acquisition of increasing structural and functional complexity; includes the period from conception to maturity.

**diabetes mellitus:** Polyuria and glycosuria, most commonly due to the inadequate production or diminished sensitivity to insulin with a resulting increase of blood glucose levels.

**diapedesis:** The movement of white blood cells through the walls of blood vessels by migration between adjacent endothelial cells.

**diaphragm:** Any muscular partition; the respiratory muscle that separates the thoracic cavity from the abdominopelvic cavity.

**diaphysis:** The shaft of a long bone.

**diarthrosis:** A synovial joint.

**diastolic pressure:** Pressure measured in the walls of a muscular artery when the left ventricle is in diastole (relaxation).

**diencephalon:** A division of the brain that includes the epithalamus, thalamus, and hypothalamus.

**differential count:** The determination of the relative abundance of each type of white blood cell on the basis of a random sampling of 100 white blood cells.

**differentiation:** The gradual appearance of characteristic cellular specializations during development as the result of gene activation or repression.

**diffusion:** Passive molecular movement from an area of higher concentration to an area of lower concentration.

**digestion:** The chemical breakdown of ingested food into simple molecules that can be absorbed by the cells of the digestive tract.

**digestive system:** The digestive tract and associated glands.

**digestive tract:** An internal passageway that begins at the oral cavity (mouth), ends at the anus, and is lined by a mucous membrane; also called *gastrointestinal* (GI) *tract* or *alimentary canal*.

**dilate:** To increase in diameter; to enlarge or expand.

**disaccharide:** A compound formed by the joining of two simple sugars by dehydration synthesis.

**dissociation:** *See* **ionization**.

**distal:** A direction away from the point of attachment or origin; for a limb, away from its attachment to the trunk.

**distal convoluted tubule (DCT):** The segment of the nephron closest to the connecting tubules and collecting duct; an important site of active secretion.

**diuresis:** Fluid loss by the kidneys; the production of unusually large volumes of urine.

**divergence:** In neural tissue, the spread of information from one neuron to many neurons; an organizational pattern common along sensory pathways of the central nervous system.

**diverticulum:** A sac or pouch in the wall of the colon or other organ.

**DNA molecule:** Two DNA strands wound in a double helix and held together by hydrogen bonds between complementary nitrogenous base pairs.

**dopamine:** An important neurotransmitter in the central nervous system.

**dorsal:** Toward the back, posterior.

**dorsal root ganglion:** A peripheral nervous system ganglion containing the cell bodies of sensory neurons.

**dorsiflexion:** Upward movement of the foot through flexion at the ankle.

**Down syndrome:** A genetic abnormality resulting from the presence of three copies of chromosome 21; people with this condition have characteristic physical and intellectual deficits; also called *trisomy 21*.

**duct:** A passageway that delivers exocrine secretions to an epithelial surface.

**ductus arteriosus:** A vascular connection between the pulmonary trunk and the aorta that functions throughout fetal life; normally closes at birth or shortly thereafter and persists as the ligamentum arteriosum.

**ductus deferens:** A passageway that carries sperm from the epididymis to the ejaculatory duct; also called the *vas deferens*.

**duodenal ampulla:** A chamber that receives bile from the common bile duct and pancreatic secretions from the pancreatic duct.

**duodenal papilla:** A conical projection from the inner surface of the duodenum that contains the opening of the duodenal ampulla.

**duodenum:** The proximal 25 cm (9.8 in.) of the small intestine that contains short villi and submucosal glands.

**dura mater:** The outermost layer of the cranial and spinal meninges.

## E

**eccrine glands:** Sweat glands of the skin that produce a watery secretion.

**ectoderm:** One of the three primary germ layers; covers the surface of the embryo and gives rise to the nervous system, the epidermis and associated glands, and a variety of other structures.

**ectopic:** Outside the normal location.

**effector:** A peripheral gland or muscle cell innervated by a motor neuron.

**efferent:** Conducting a fluid or nerve impulse away from an organ or structure.

**efferent arteriole:** An arteriole carrying blood away from a glomerulus of the kidney.

**efferent fiber:** An axon that carries impulses away from the central nervous system.

**ejaculation:** The ejection of semen from the penis as the result of muscular contractions of the bulbospongiosus and ischiocavernosus muscles.

**ejaculatory ducts:** Short ducts that pass within the walls of the prostate gland and connect the ductus deferens with the prostatic urethra.

**elastin:** Connective tissue fibers that stretch and recoil, providing elasticity to connective tissues.

**electrical coupling:** A connection between adjacent cells that permits the movement of ions and the transfer of graded or propagated changes in the membrane potential from cell to cell.

**electrocardiogram (ECG, EKG):** A graphic record of the electrical activities of the heart, as monitored at specific locations on the body surface.

**electroencephalogram (EEG):** A graphic record of the electrical activities of the brain.

**electrolytes:** Soluble inorganic compounds whose ions will conduct an electrical current in solution.

**electron:** One of the three fundamental subatomic particles; has a negative charge and normally orbits the protons of the nucleus.

**electron transport chain (ETC):** The system made up of four respiratory protein complexes, coenzyme Q, and electron carriers made up of a series of *cytochrome* molecules responsible for aerobic energy production in cells; bound to the inner mitochondrial membrane.

**element:** All the atoms with the same atomic number.

**elevation:** Movement in a superior, or upward, direction.

**embryo:** The developmental stage beginning at fertilization and ending at the start of the third developmental month.

**embryology:** The study of embryonic development, focusing on the first two months after fertilization.

**endocardium:** The simple squamous epithelium that lines the heart and is continuous with the endothelium of the great vessels.

**endochondral ossification:** The replacement of a cartilaginous model by bone; the characteristic mode of formation for skeletal elements other than the bones of the cranium, the clavicles, and sesamoid bones.

**endocrine gland:** A gland that secretes hormones into the blood.

**endocrine system:** The endocrine (ductless) glands and organs of the body.

**endocytosis:** The movement of relatively large volumes of extracellular material into the cytoplasm by the formation of a membranous vesicle at the cell surface; includes pinocytosis and phagocytosis.

**endoderm:** One of the three primary germ layers; the layer on the undersurface of the embryonic disc; gives rise to the epithelia and glands of the digestive system, the respiratory system, and portions of the urinary system.

**endogenous:** Produced within the body.

**endolymph:** The fluid contents of the membranous labyrinth (the saccule, utricle, semicircular ducts, and cochlear duct) of the internal ear.

**endometrium:** The mucous membrane lining the uterus.

**endomysium:** A delicate network of connective tissue fibers that surrounds individual muscle cells.

**endoneurium:** A delicate network of connective tissue fibers that surrounds individual nerve fibers.

**endoplasmic reticulum:** A network of membranous channels in the cytoplasm of a cell that function in intracellular transport, synthesis, storage, packaging, and secretion.

**endosteum:** An incomplete cellular lining on the inner (medullary) surfaces of bones.

**endothelium:** The simple squamous epithelial cells that line blood and lymphatic vessels.

**enteric nervous system (ENS):** One of the three main divisions of the nervous system; it is an extensive nerve network in the walls of the digestive tract that initiates and coordinates digestive motility and secretions.

**enteroendocrine cells:** Endocrine cells scattered among the epithelial cells that line the digestive tract.

**enterogastric reflex:** The reflexive inhibition of gastric secretion; initiated by the arrival of chyme in the small intestine.

**enterokinase:** An enzyme in the lumen of the small intestine that activates the proenzymes secreted by the pancreas.

**enzyme:** A protein that catalyzes a specific biochemical reaction.

**eosinophil:** A type of white blood cell with a lobed nucleus and red-staining granules; participates in the immune response and is especially important during allergic reactions.

**ependyma:** The layer of cells lining the ventricles and central canal of the central nervous system.

**epicardium:** A serous membrane covering the outer surface of the heart; also called *visceral pericardium*.

**epidermis:** The epithelium covering the surface of the skin.

**epididymis:** A coiled duct that connects the rete testis to the ductus deferens; site of functional maturation of sperm.

**epidural space:** The space between the spinal dura mater and the walls of the vertebral foramen; contains blood vessels and adipose tissue; a common site of injection for regional anesthesia.

**epigenetics:** Study of gene expression due to alterations in the reading of DNA sequences.

**epiglottis:** A blade-shaped flap of tissue, reinforced by cartilage, that is attached to the dorsal and superior surface of the thyroid cartilage; folds over the entrance to the larynx during swallowing.

**epimysium:** A dense layer of collagen fibers that surrounds a skeletal muscle and is continuous with the tendons/aponeuroses of the muscle and with the perimysium.

**epineurium:** A dense layer of collagen fibers that surrounds a peripheral nerve.

**epiphyseal cartilage:** The cartilaginous region between the epiphysis and diaphysis of a growing bone.

**epiphysis:** The head of a long bone.

**epithelium:** One of the four primary tissue types; a layer of cells that forms a superficial covering or an internal lining of a body cavity or vessel.

**eponychium:** A narrow zone of stratum corneum that extends across the surface of a nail at its exposed base; also called the *cuticle*.

**equilibrium:** A dynamic state in which two opposing forces or processes are in balance.

**erection:** The stiffening of the penis due to the engorgement of the erectile tissues of the corpora cavernosa and corpus spongiosum.

**erythema:** Redness and inflammation at the surface of the skin.

**erythrocyte:** A red blood cell; has no nucleus and contains large quantities of hemoglobin.

**erythropoietin:** A hormone released by most tissues, and especially by the kidneys, when oxygen levels decrease; stimulates erythropoiesis (red blood cell formation) in red bone marrow.

***Escherichia coli:*** A normal bacterial resident of the large intestine.

**esophagus:** A muscular tube that connects the pharynx to the stomach.

**essential amino acids:** Amino acids that cannot be synthesized in the body in adequate amounts and must be obtained from the diet.

**essential fatty acids:** Fatty acids that cannot be synthesized in the body and must be obtained from the diet.

**estrogens:** A class of steroid sex hormones that includes estradiol.

**evaporation:** A movement of molecules from the liquid state to the gaseous state.

**eversion:** A turning outward.

**excitable membranes:** Membranes that propagate action potentials, a characteristic of muscle cells and nerve cells.

**excitatory postsynaptic potential (EPSP):** The depolarization of a postsynaptic membrane by a chemical neurotransmitter released by the presynaptic cell.

**excretion:** The removal of wastes from the blood, tissues, or organs.

**exocrine gland:** A gland that secretes onto the body surface or into a passageway connected to the exterior.

**exocytosis:** The ejection of cytoplasmic materials by the fusion of a membranous vesicle with the plasma membrane.

**expiration:** Exhalation; breathing out.

**extension:** An increase in the angle between two articulating bones; the opposite of flexion.

**external acoustic meatus:** A passageway in the temporal bone that leads to the tympanic membrane of the internal ear.

**external ear:** The auricle, external acoustic meatus, and tympanic membrane.

**external nares:** The nostrils; the external openings into the nasal cavity.

**external respiration:** The diffusion of gases between the alveolar air and the alveolar capillaries and between the systemic capillaries and peripheral tissues.

**exteroceptors:** General sensory receptors in the skin, mucous membranes, and special sense organs that provide information about the external environment and about our position within it.

**extracellular fluid:** All body fluids other than that contained within cells; includes plasma and interstitial fluid.

**extra-embryonic membranes:** The yolk sac, amnion, chorion, and allantois.

**extrinsic pathway:** A clotting pathway that begins with damage to blood vessels or surrounding tissues and ends with the formation of tissue factor complex.

## F

**facilitated:** Brought closer to threshold, as in the depolarization of a nerve plasma membrane toward threshold; making the cell more sensitive to depolarizing stimuli.

**facilitated diffusion:** The passive movement of a substance across a plasma membrane by means of a protein carrier.

**falciform ligament:** A sheet of mesentery that contains the ligamentum teres, the fibrous remains of the umbilical vein of the fetus.

**falx:** Sickle-shaped.

**falx cerebri:** The curving sheet of dura mater that extends between the two cerebral hemispheres; encloses the superior sagittal sinus.

**fasciae:** Connective tissue fibers, primarily collagen, that form sheets or bands beneath the skin to attach, stabilize, enclose, and separate muscles and other internal organs.

**fasciculus:** A small bundle; usually refers to a collection of nerve axons or muscle fibers.

**fatty acids:** Hydrocarbon chains that end in a carboxyl group.

**fauces:** The passage from the mouth to the pharynx, bounded by the palatal arches, the soft palate, and the uvula.

**feces:** Waste products eliminated by the digestive tract at the anus; contains indigestible residue, bacteria, mucus, and epithelial cells.

**fenestra/fenestrae:** Opening/openings.

**fertilization:** The fusion of a secondary oocyte and a sperm to form a zygote.

**fetus:** The developmental stage lasting from the start of the third developmental month to delivery.

**fibrin:** Insoluble protein fiber that forms the basic framework of a blood clot.

**fibrinogen:** A plasma protein that is the soluble precursor of the insoluble protein fibrin.

**fibroblasts:** Cells of connective tissue proper that produce extracellular fibers and secrete the organic compounds of the extracellular matrix.

**fibrous cartilage:** Cartilage containing an abundance of collagen fibers; located around the edges of joints, in the intervertebral discs, and the menisci of the knee.

**fibrous layer:** The outermost layer of the eye, composed of the sclera and cornea.

**fibula:** The lateral, slender bone of the leg.

**filiform papillae:** Slender conical projections from the dorsal surface of the anterior two-thirds of the tongue.

**filtrate:** The fluid produced by filtration at a glomerulus in the kidney.

**filtration:** The movement of a fluid across a membrane whose pores restrict the passage of solutes on the basis of size.

**filtration pressure:** The hydrostatic pressure responsible for filtration.

**filum terminale:** A fibrous extension of the spinal cord, from the conus medullaris to the coccygeal ligament.

**fimbriae:** Fringes; the fingerlike processes that surround the entrance to the uterine tube.

**first-degree burn:** Burn involving only the epidermis and causing redness and swelling; type of partial-thickness burn.

**fissure:** An elongated groove or opening.

**flagellum/flagella:** An organelle that is structurally similar to a cilium but is used to propel a cell through a fluid; found on sperm.

**flatus:** Intestinal gas.

**flexion:** A movement that reduces the angle between two articulating bones; the opposite of extension.

**flexor:** A muscle that produces flexion.

**flexor reflex:** A reflex contraction of the flexor muscles of a limb in response to a painful stimulus.

**flexure:** A bending.

**folia:** Leaflike folds; the slender folds in the surface of the cerebellar cortex.

**foliate papillae:** Taste buds located on the lateral margins of the posterior region of the tongue.

**follicle:** A small secretory sac or gland.

**follicle-stimulating hormone (FSH):** A hormone secreted by the anterior lobe of the pituitary gland (adenohypophysis); stimulates oogenesis (female) and spermatogenesis (male).

**fontanelle:** A relatively soft, flexible, fibrous region between two flat bones in the developing skull; also spelled *fontanel.*

**foramen/foramina:** Opening(s) or passage(s) through a bone.

**forearm:** The distal portion of the upper limb between the elbow and wrist.

**forebrain:** The cerebrum.

**fornix:** An arch or the space bounded by an arch; in the brain, an arching tract that connects the hippocampus with the mammillary bodies; in the eye, a slender pocket situated where the epithelium of the ocular conjunctiva folds back on itself as the palpebral conjunctiva; in the vagina, the shallow recess surrounding the protrusion of the cervix.

**fossa:** A shallow depression or furrow in the surface of a bone.

**fourth ventricle:** An elongated ventricle of the metencephalon (pons and cerebellum) and the myelencephalon (medulla oblongata) of the brain; the roof contains a region of choroid plexus.

**fovea centralis:** The portion of the retina that provides the sharpest vision because it has the highest concentration of cones; also called *fovea.*

**fracture:** A break or crack in a bone.

**frenulum:** A bridle; usually referring to a band of tissue that restricts movement (e.g., *lingual frenulum*).

**frontal plane:** A sectional plane that divides the body into an anterior portion and a posterior portion; also called *coronal plane.*

**fructose:** A hexose (six-carbon simple sugar) in foods and in semen.

**fundus:** The base of an organ such as the stomach, uterus, or gallbladder.

**G**

**gallbladder:** The pear-shaped reservoir for bile after it is secreted by the liver.

**gametes:** Reproductive cells (sperm or oocytes) that contain half the normal chromosome complement.

**gametogenesis:** The formation of gametes.

**gamma aminobutyric acid (GABA):** A neurotransmitter of the central nervous system whose effects are generally inhibitory.

**gamma motor neurons:** Motor neurons that adjust the sensitivities of muscle spindles (intrafusal fibers).

**ganglion/ganglia:** A collection of neuron cell bodies in the peripheral nervous system.

**gap junctions:** Connections between cells that permit electrical coupling.

**gaster:** The stomach; the body, or belly, of a skeletal muscle.

**gastric:** Pertaining to the stomach.

**gastric glands:** The tubular glands of the stomach whose cells produce acid, enzymes, intrinsic factor, and hormones.

**gastrointestinal (GI) tract:** See **digestive tract.**

**gene:** A portion of a DNA strand that functions as a hereditary unit, is located at a particular site on a specific chromosome, and codes for a specific protein or polypeptide.

**genetics:** The study of mechanisms of heredity.

**geniculate:** Like a little knee; the medial geniculate body and lateral geniculate body are in the thalamus of the brain.

**genitalia:** The reproductive organs.

**genotype:** Genetic makeup of an organism.

**germinal centers:** Pale regions in the interior of lymphoid tissues or lymphoid nodules, where cell divisions occur that produce additional lymphocytes.

**gestation:** The period of intrauterine development; pregnancy.

**gland:** Cells that produce exocrine or endocrine secretions.

**glenoid cavity:** A rounded depression that forms the articular surface of the scapula at the shoulder joint.

**glial cells:** *See* **neuroglia.**

**globular proteins:** Proteins whose tertiary structure makes them rounded and compact.

**glomerular capsule:** The expanded initial portion of the nephron that surrounds the glomerulus.

**glomerular filtration rate:** The rate of filtrate formation at the glomerulus.

**glomerulus:** A ball or knot; in the kidneys, a knot of capillaries that projects into the enlarged, proximal end of a nephron; the site of filtration, the first step in the production of urine.

**glossopharyngeal nerve:** Cranial nerve IX.

**glucagon:** A hormone secreted by the alpha cells of the pancreatic islets; increases blood glucose concentrations.

**glucocorticoids:** Hormones secreted by the zona fasciculata of the adrenal cortex to modify glucose metabolism; cortisol and corticosterone are important examples.

**gluconeogenesis:** The synthesis of glucose from protein or lipid precursors.

**glucose:** A six-carbon sugar, $C_6H_{12}O_6$; the preferred energy source for most cells and normally the only energy source for neurons.

**glycerides:** Lipids composed of glycerol bound to fatty acids.

**glycogen:** A polysaccharide that is an important energy reserve; a polymer consisting of a long chain of glucose molecules.

**glycogenesis:** The synthesis of glycogen from glucose molecules.

**glycogenolysis:** Glycogen breakdown and the liberation of glucose molecules.

**glycolipids:** Compounds created by the combination of carbohydrate and lipid components.

**glycolysis:** The anaerobic cytosolic breakdown of glucose into two 3-carbon molecules of pyruvate, with a net gain of two ATP molecules.

**glycoprotein:** A compound containing a relatively small carbohydrate group attached to a large protein.

**glycosuria:** The presence of glucose in urine; also called *glucosuria.*

**Golgi apparatus:** A cellular organelle consisting of a series of membranous plates that give rise to lysosomes and secretory vesicles.

**gomphosis:** A fibrous synarthrosis that binds a tooth to the bone of the jaw.

**gonadotropin-releasing hormone (GnRH):** A hypothalamic releasing hormone that causes the secretion of both follicle-stimulating hormone and luteinizing hormone by the anterior lobe of the pituitary gland (adenohypophysis).

**gonadotropins:** Follicle-stimulating hormone and luteinizing hormone; hormones that stimulate gamete development and sex hormone secretion.

**gonads:** Reproductive organs that produce gametes and sex hormones.

**granulocytes:** White blood cells containing granules that are visible with the light microscope; includes eosinophils, basophils, and neutrophils; also called *granular leukocytes.*

**gray matter:** Areas in the central nervous system that are dominated by neuron cell bodies, neuroglia, and unmyelinated axons.

**gray ramus communicans:** A bundle of postganglionic sympathetic nerve fibers that are distributed to effectors in the body wall, skin, and limbs by way of a spinal nerve.

**greater omentum:** A large fold of the dorsal mesentery of the stomach; hangs anterior to the intestines.

**groin:** The inguinal region.

**gross anatomy:** The study of the structural features of the body without the aid of a microscope.

**growth hormone (GH):** An anterior lobe of the pituitary gland (adenohypophysis) hormone that stimulates tissue growth and anabolism when nutrients are abundant and restricts tissue glucose dependence when nutrients are in short supply.

**growth hormone-inhibiting hormone (GH-IH):** A hypothalamic regulatory hormone that inhibits growth hormone secretion by the anterior lobe of the pituitary gland.

**guanine:** A purine; one of the nitrogenous bases in the nucleic acids RNA and DNA.

**gustation:** The sense of taste.

**gyrus:** A prominent fold or ridge of neural cortex on the surfaces of the cerebral hemispheres.

## H

**hair:** A keratinous strand produced by epithelial cells of the hair follicle.

**hair cells:** Sensory cells of the internal ear.

**hair follicle:** An accessory structure of the integument; a tube lined by a stratified squamous epithelium that begins at the surface of the skin and ends at the hair papilla.

**hallux:** The big toe; also called the *great toe*.

**haploid:** Possessing half the normal number of chromosomes; a characteristic of gametes.

**hard palate:** The bony roof of the oral cavity, formed by the maxillae and palatine bones.

**helper T cells:** Lymphocytes whose secretions and other activities coordinate cell-mediated and antibody-mediated immunity; also called $T_H$ cells.

**hematocrit:** The percentage of formed elements in a sample of blood; also called *volume of packed red cells (VPRC)* or *packed cell volume (PCV)*.

**hematoma:** An abnormal collection of clotted or partially clotted blood outside a blood vessel.

**hemocytoblasts:** Form from hematopoietic stems cells and whose divisions produce each of the various populations of blood cells.

**hematuria:** The abnormal presence of red blood cells in urine.

**heme:** A porphyrin ring containing a central iron atom that can reversibly bind oxygen molecules; a component of the hemoglobin molecule.

**hemoglobin:** A protein composed of four globular subunits, each bound to a heme molecule; gives red blood cells the ability to transport oxygen in the blood.

**hemolysis:** The breakdown of red blood cells.

**hemopoiesis:** Blood cell formation and differentiation.

**hemorrhage:** Blood loss; to bleed.

**hemostasis:** The stoppage of bleeding.

**heparin:** An anticoagulant released by activated basophils and mast cells.

**hepatic duct:** The duct that carries bile away from the liver lobes and toward the union with the cystic duct.

**hepatic portal vein:** The vessel that carries blood between the intestinal capillaries and the sinusoids of the liver.

**hepatocyte:** A liver cell.

**heterozygous:** Having two different alleles at corresponding sites on a chromosome pair.

**hiatus:** A gap, cleft, or opening.

**high density lipoprotein (HDL):** A lipoprotein with a relatively small lipid content; responsible for the movement of cholesterol from peripheral tissues to the liver.

**hilum:** A localized region where blood vessels, lymphatic vessels, nerves, and/or other anatomical structures are attached to an organ.

**hippocampus:** A region, deep to the floor of a lateral ventricle, involved with emotional states and the conversion of short-term to long-term memories.

**histamine:** The chemical released by stimulated mast cells or basophils to initiate or enhance inflammation.

**histology:** The study of tissues.

**histones:** Proteins associated with the DNA of the nucleus; the DNA strands are wound around them.

**holocrine:** A form of exocrine secretion in which the secretory cell becomes swollen with vesicles and then ruptures.

**homeostasis:** The maintenance of a relatively constant internal environment.

**homozygous:** Having identical alleles at corresponding sites on a chromosome pair.

**hormone:** A compound that is secreted by one cell and travels through the bloodstream to affect the activities of cells in another part of the body.

**human chorionic gonadotropin (hCG):** The placental hormone that maintains the corpus luteum for the first three months of pregnancy.

**human immunodeficiency virus (HIV):** The infectious agent that causes acquired immunodeficiency syndrome (AIDS).

**human leukocyte antigen (HLA):** *See* **MHC protein**.

**human placental lactogen (hPL):** The placental hormone that stimulates the functional development of the mammary glands.

**humoral immunity:** *See* **antibody-mediated immunity**.

**hyaluronic acid:** A carbohydrate component of proteoglycans in the matrix of many connective tissues.

**hyaluronidase:** An enzyme that breaks down the bonds between adjacent follicle cells; produced by some bacteria and found in the acrosome of a sperm.

**hydrogen bond:** A weak interaction between the hydrogen atom on one molecule and a negatively charged portion of another molecule.

**hydrolysis:** The breakage of a chemical bond through the addition of a water molecule; the reverse of dehydration synthesis.

**hydrophilic:** Freely associating with water; readily entering into solution; water-loving.

**hydrophobic:** Incapable of freely associating with water molecules; insoluble; water-fearing.

**hydrostatic pressure:** Fluid pressure.

**hydroxide ion:** $OH^-$.

**hypercapnia:** High blood carbon dioxide concentrations, commonly as a result of hypoventilation or inadequate tissue perfusion.

**hyperpolarization:** The movement of the membrane potential away from the normal resting potential and farther from 0 mV.

**hypersecretion:** The overactivity of glands that produce exocrine or endocrine secretions.

**hypertension:** Abnormally high blood pressure.

**hypertonic:** In comparing two solutions, the solution with the higher osmolarity.

**hypertrophy:** An increase in tissue size without cell division.

**hyperventilation:** A rate of respiration sufficient to decrease blood $P_{CO_2}$ to levels below normal.

**hypocapnia:** An abnormally low blood $P_{CO_2}$ concentration commonly as a result of hyperventilation.

**hypodermic needle:** A needle inserted through the skin to introduce drugs into the subcutaneous layer.

**hypodermis:** *See* **subcutaneous layer**.

**hypophyseal portal system:** The network of vessels that carries blood from capillaries in the hypothalamus to capillaries in the anterior lobe of the pituitary gland.

**hypophysis:** The pituitary gland.

**hyposecretion:** Abnormally low exocrine or endocrine secretion.

**hypothalamus:** The floor of the diencephalon; the region of the brain containing centers involved with the subconscious regulation of visceral functions, emotions, drives, and the coordination of neural and endocrine functions.

**hypotonic:** In comparing two solutions, the solution with the lower osmolarity.

**hypoventilation:** A respiratory rate that is insufficient to keep blood $P_{CO_2}$ within normal levels.

**hypoxia:** A low tissue oxygen concentration.

## I

**ileum:** The distal 2.5 m of the small intestine.

**ilium:** The largest of the three bones whose fusion creates a coxal bone.

**immunity:** Resistance to injuries and diseases caused by foreign substances, toxins, or pathogens.

**immunization:** The production of immunity by the deliberate exposure to antigens under conditions that prevent the development of illness but stimulate the production of memory B cells.

**immunoglobulin (lg):** A circulating antibody.

**implantation:** The attachment of a blastocyst into the endometrium of the uterine wall.

**inclusions:** Aggregations of insoluble pigments, nutrients, or other materials in cytoplasm.

**incus:** The central auditory ossicle, located between the malleus and the stapes in the middle ear cavity.

**inexcitable:** Incapable of propagating an action potential.

**infarct:** An area of dead cells that results from an interruption of blood flow.

**infection:** The invasion and colonization of body tissues by pathogens.

**inferior:** Below, in reference to a particular structure, with the body in the anatomical position.

**inferior vena cava:** The vein that carries blood from the parts of the body inferior to the heart to the right atrium.

**infertility:** The inability to conceive; also called *sterility*.

**inflammation:** A nonspecific defense mechanism that operates at the tissue level; characterized by swelling, redness, heat (warmth), pain, and sometimes loss of function.

**infundibulum:** A tapering, funnel-shaped structure; in the brain, the connection between the pituitary gland and the hypothalamus; in the uterine tube, the entrance bounded by fimbriae that receives the oocyte at ovulation.

**ingestion:** The introduction of food into the digestive tract by way of the mouth; eating.

**inguinal canal:** A passage through the abdominal wall that marks the path of testicular descent and that contains the testicular arteries, veins, and ductus deferens.

**inguinal region:** The area of the abdominal wall near the junction of the trunk and the thighs that contains the external genitalia; the groin.

**inhibin:** A hormone, produced by nurse (sustentacular) cells of the testes and follicular cells of the ovaries, that inhibits the secretion of follicle-stimulating hormone by the anterior lobe of the pituitary gland (adenohypophysis).

**inhibitory postsynaptic potential (IPSP):** A hyperpolarization of the postsynaptic membrane after the arrival of a neurotransmitter.

**initial segment:** The proximal portion of the axon where an action potential first appears.

**injection:** The forcing of fluid into a body part or organ.

**innate immunity:** Type of defense that is genetically determined, present at birth, and is not related to previous exposure to a particular antigen; also called *nonspecific immunity*.

**inner cell mass:** Cells of the blastocyst that will form the body of the embryo.

**innervation:** The distribution of sensory and motor nerves to a specific region or organ.

**insensible perspiration:** Evaporative water loss by diffusion across the epithelium of the skin or evaporation across the alveolar surfaces of the lungs.

**insertion:** A point of attachment of a muscle; the end that is easily movable.

**insoluble:** Incapable of dissolving in solution.

**inspiration:** Inhalation; the movement of air into the respiratory system.

**insulin:** A hormone secreted by beta cells of the pancreatic islets; causes a decrease in blood glucose concentrations.

**integument:** The skin.

**intercalated discs cadiac muscle cells:** Regions where adjacent cardiocytes interlock and where gap junctions permit electrical coupling between the cells.

**intercellular fluid:** *See* **interstitial fluid**.

**interferons:** Peptides released by virus-infected cells, especially lymphocytes, that slow viral replication and make other cells more resistant to viral infection.

**interleukins:** Peptides, released by activated monocytes and lymphocytes, that assist in the coordination of cell-mediated and antibody-mediated immunities.

**internal capsule:** The collection of afferent and efferent fibers of the white matter of the cerebral hemispheres, visible on gross dissection of the brain.

**internal ear:** The membranous labyrinth that contains the organs of hearing and equilibrium.

**internal nares:** The entrance to the nasopharynx from the nasal cavity.

**internal respiration:** The diffusion of gases between interstitial fluid and cytoplasm.

**interneuron:** An association neuron; central nervous system neurons that are between sensory and motor neurons.

**interoceptors:** Sensory receptors monitoring the functions and status of internal organs and systems.

**interosseous membrane:** The fibrous connective tissue membrane between the shafts of the tibia and fibula and between the radius and ulna; an example of a fibrous amphiarthrosis.

**interphase:** The stage in the life cycle of a cell during which the chromosomes are uncoiled and all normal cellular functions except mitosis are under way.

**intersegmental reflex:** A reflex that involves several segments of the spinal cord.

**interstitial fluid:** The fluid in the tissues that fills the spaces between cells.

**interstitial growth:** Type of cartilage growth that occurs during development in which the cartilage expands from within as chondrocytes divide and daughter cells secrete additional matrix.

**interventricular foramen:** The opening that permits fluid movement between the lateral and third ventricles of the brain.

**intervertebral disc:** A fibrocartilage pad between the bodies of successive vertebrae that absorbs shocks.

**intestinal crypt:** A tubular epithelial pocket that is lined by secretory cells and opens into the lumen of the digestive tract; also called *intestinal gland*.

**intestine:** The tubular organ of the digestive tract.

**intracellular fluid:** The cytosol.

**intramembranous ossification:** The formation of bone within a connective tissue without the prior development of a cartilaginous model.

**intrinsic factor:** A glycoprotein, secreted by the parietal cells of the stomach, that facilitates the intestinal absorption of vitamin B$_{12}$.

**intrinsic pathway:** A pathway of the clotting system that begins with the activation of platelets and ends with the formation of factor X activator complex.

**inversion:** A turning inward.

**in vitro:** Outside the body, in an artificial environment.

**in vivo:** In the living body.

**involuntary:** Not under conscious control.

**ion:** An atom or molecule having a positive or negative charge due to the loss or gain, respectively, of one or more electrons.

**ionic bond:** A molecular bond created by the attraction between ions with opposite charges.

**ionization:** Dissociation; the breakdown of a molecule in solution to form ions.

**ipsilateral:** A reflex response that affects the same side as the stimulus.

**iris:** A contractile structure, made up of smooth muscle, that forms the colored portion of the eye.

**ischemia:** An inadequate blood supply to a region of the body.

**ischium:** One of the three bones whose fusion creates a coxal bone.

**islets of Langerhans:** *See* **pancreatic islets**.

**isotonic:** A solution with an osmolarity that does not result in water movement across plasma membranes.

**isotopes:** Forms of an element whose atoms contain the same number of protons but different numbers of neutrons (and thus differ in atomic mass).

**isthmus:** A narrow band of tissue connecting two larger masses.

### J

**jejunum:** The middle part of the small intestine.

**joint:** An area where adjacent bones interact; also called *articulation*.

**juxtaglomerular complex:** Specialized cells in between the walls of the distal convoluted tubule and afferent and efferent arterioles adjacent to the glomerulus; a complex responsible for the release of renin and erythropoietin.

### K

**keratin:** The tough, fibrous protein component of nails, hair, calluses, and the general integumentary surface.

**keto acid:** The carbon chain that remains after the deamination or transamination of an amino acid.

**ketoacidosis:** A condition characterized by a decrease in blood pH due to the presence of large numbers of ketone bodies.

**ketone bodies:** Keto acids produced during the catabolism of lipids and some amino acids; specifically, acetone, acetoacetate, and beta-hydroxybutyrate.

**kidney:** A component of the urinary system; an organ functioning in the regulation of blood composition, including the excretion of wastes and the maintenance of normal fluid and electrolyte balances.

**killer T cells:** *See* **cytotoxic T cells**.

**Krebs cycle:** *See* **citric acid cycle**.

**Kupffer cell:** *See* **stellate macrophage**.

### L

**labium/labia:** Lip(s); the labia majora and labia minora are structures of the female external genitalia.

**labrum:** A lip or rim.

**labyrinth:** A maze of passageways; the structures of the internal ear.

**lacrimal gland:** A tear gland on the dorsolateral surface of the eye.

**lactase:** An enzyme that breaks down the milk sugar lactose.

**lactate:** An anion released by the dissociation of lactic acid or formed from the reduction of pyruvate by NADH.

**lactation:** The production of milk by the mammary glands.

**lacteal:** A terminal lymphatic within an intestinal villus.

**lacuna:** A small pit or cavity.

**lambdoid suture:** The synarthrosis between the parietal and occipital bones of the cranium.

**lamellae:** Concentric layers; the concentric layers of bone within an osteon.

**lamellated corpuscle:** A receptor sensitive to vibration.

**lamina:** A thin sheet or layer.

**lamina propria:** The areolar tissue that underlies a mucous epithelium and forms part of a mucous membrane.

**Langerhans cells:** Cells in the epithelium of the skin and digestive tract that participate in the immune response by presenting antigens to T cells; also called *dendritic cells*.

**large intestine:** The terminal portion of the intestinal tract, consisting of the colon, the rectum, and the anal canal.

**laryngopharynx:** The division of the pharynx that is inferior to the epiglottis and superior to the esophagus.

**larynx:** A complex cartilaginous structure that surrounds and protects the glottis and vocal cords; the superior margin is bound to the hyoid bone, and the inferior margin is bound to the trachea.

**latent period:** The time between the stimulation of a muscle and the start of the contraction phase.

**lateral apertures:** Openings in the roof of the fourth ventricle that permit the circulation of cerebrospinal fluid into the subarachnoid space.

**lateral ventricle:** A fluid-filled chamber within a cerebral hemisphere.

**lateral:** Pertaining to the side.

**lens:** The transparent refractive structure that is between the iris and the vitreous humor.

**lesser omentum:** A double-layered peritoneal fold formed by the mesentery that connects the lesser curvature of the stomach to the liver.

**leukocyte:** A white blood cell.

**ligament:** A dense band of connective tissue fibers that attaches one bone to another.

**ligamentum arteriosum:** The fibrous strand in adults that is the remnant of the ductus arteriosus of the fetal stage.

**ligamentum nuchae:** An elastic ligament between the vertebra prominens and the occipital bone.

**ligamentum teres:** The fibrous strand in the falciform ligament of adults that is the remnant of the umbilical vein; also called the *round ligament of liver*.

**ligate:** To tie off.

**limbic system:** The group of nuclei and centers in the cerebrum and diencephalon that are involved with emotional states, memories, and behavioral drives.

**lingual:** Pertaining to the tongue.

**lipid:** An organic compound containing carbon, hydrogen, and oxygen in a ratio that does not approximate 1:2:1; includes fats, oils, and waxes.

**lipogenesis:** The synthesis of lipids from nonlipid precursors.

**lipolysis:** The catabolism of lipids as a source of energy.

**lipoprotein:** A compound containing a relatively small lipid bound to a protein.

**liver:** An organ of the digestive system that has varied and vital functions, including the production of plasma proteins, the excretion of bile, the storage of energy reserves, the detoxification of poisons, and the interconversion of nutrients.

**lobule:** Histologically, the basic organizational unit of the liver.

**local hormone:** *See* **prostaglandin**.

**loop of Henle:** *See* **nephron loop**.

**loose connective tissue:** A loosely organized, easily distorted connective tissue that contains several fiber types, a varied population of cells, and a viscous ground substance.

**low density lipoprotein (LDL):** Form of lipoprotein that transports cholesterol in the blood to tissues other than the liver.

**lumbar:** Pertaining to the lower back.

**lumen:** The central space within a duct or other internal passageway.

**lungs:** The paired organs of breathing located in the pleural cavities.

**luteinizing hormone (LH):** A hormone produced by the anterior lobe of the pituitary gland (adenohypophysis). In females, it assists FSH in follicle stimulation, triggers ovulation, and promotes the maintenance and secretion of endometrial glands. In males, it stimulates testosterone secretion by the interstitial cells of the testes.

**lymph:** The fluid contents of lymphatic vessels, similar in composition to interstitial fluid.

**lymphatic vessels:** The vessels of the lymphatic system; also called *lymphatics.*

**lymph nodes:** Lymphoid organs that monitor the composition of lymph.

**lymphocyte:** A cell of the lymphatic system that plays a role in the immune response.

**lymphokines:** Chemicals secreted by activated lymphocytes.

**lymphopoiesis:** The production of lymphocytes from lymphoid stem cells.

**lymphotoxin:** A secretion of lymphocytes that kills the target cells.

**lysis:** The destruction of a cell through the rupture of its plasma membrane.

**lysosome:** An intracellular vesicle containing digestive enzymes.

**lysozyme:** An enzyme, present in some exocrine secretions that has antibiotic properties.

**M**

**macrophage:** A phagocytic cell of the monocyte-macrophage system.

**macula:** A receptor complex, located in the saccule or utricle of the internal ear, that responds to linear acceleration or gravity.

**macula** (of retina): An oval area in the retina whose center (the fovea centralis) is the region of sharpest vision; also called *macula lutea.*

**major histocompatibility complex:** *See* **MHC protein**.

**malignant tumor:** A form of cancer characterized by rapid cell growth and the spread of cancer cells throughout the body.

**malleus:** The first auditory ossicle, bound to the tympanic membrane and the incus.

**malnutrition:** An unhealthy state produced by inadequate dietary intake or absorption of nutrients.

**mammary glands:** Milk-producing glands of the female breasts.

**mammillary bodies:** Nuclei in the hypothalamus that affect eating reflexes and behaviors; a component of the limbic system.

**manus:** The hand; manual.

**marrow:** A tissue that fills the internal cavities in bone; dominated by hemopoietic cells (red bone marrow) or by adipose tissue (yellow bone marrow).

**mast cell:** A connective tissue cell that, when stimulated, releases histamine, serotonin, and heparin, initiating the inflammatory response.

**mastication:** Chewing.

**mastoid sinus:** Air-filled spaces in the mastoid process of the temporal bone.

**matrix:** The extracellular fibers and ground substance of a connective tissue.

**maxillary sinus:** One of the paranasal sinuses; an air-filled chamber lined by a respiratory epithelium that is located in a maxilla and opens into the nasal cavity.

**meatus:** An opening or entrance into a passageway.

**mechanoreception:** The detection of mechanical stimuli, such as touch, pressure, or vibration.

**medial:** Toward the midline of the body.

**mediastinum:** The central tissue mass that divides the thoracic cavity into two pleural cavities.

**medulla:** The inner layer or core of an organ.

**medulla oblongata:** The most caudal of the brain regions, also called the *myelencephalon.*

**medullary cavity:** The space within a bone that contains the marrow.

**medullary rhythmicity center:** The center in the medulla oblongata that sets the background pace of respiration; includes inspiratory and expiratory centers.

**megakaryocytes:** Bone marrow cells responsible for the formation of platelets.

**meiosis:** Cell division that produces gametes with half the normal somatic chromosome number.

**melanin:** The yellow-brown, brown, or black pigment produced by the melanocytes of the skin.

**melanocyte:** A specialized cell in the deeper layers of the stratified squamous epithelium of the skin; responsible for the production of melanin.

**melanocyte-stimulating hormone (MSH):** A hormone, produced by the pars intermedia of the anterior lobe of the pituitary gland (adenohypophysis) that stimulates melanin production.

**melatonin:** A hormone secreted by the pineal gland.

**membrane:** Any sheet or partition; a layer consisting of an epithelium and the underlying connective tissue.

**membrane flow:** The movement of sections of membrane surface to and from the cell surface and components of the endoplasmic reticulum, the Golgi apparatus, and vesicles.

**membrane potential:** The potential difference, measured across a plasma membrane and expressed in millivolts, that results from the uneven distribution of positive and negative ions across the plasma membrane.

**membranous labyrinth:** Endolymph-filled tubes that enclose the receptors of the internal ear.

**meninges:** Three membranes that surround the surfaces of the central nervous system; the dura mater, the pia mater, and the arachnoid.

**meniscus:** A fibrocartilage pad between opposing surfaces in a joint.

**menstruation:** The phase of the uterine cycle in which the endometrial functional layer is shed; also called *menses.*

**merocrine:** A method of secretion in which the cell ejects materials from secretory vesicles through exocytosis.

**mesencephalon:** The midbrain; the region between the diencephalon and pons.

**mesenchyme:** Embryonic or fetal connective tissue.

**mesentery:** A double layer of serous membrane that supports and stabilizes the position of an organ in the abdominopelvic cavity and provides a route for the associated blood vessels, nerves, and lymphatic vessels.

**mesoderm:** The middle germ layer, between the ectoderm and endoderm of the embryo.

**mesothelium:** A simple squamous epithelium that lines the body cavities enclosing the lungs, heart, and abdominal organs.

**messenger RNA (mRNA):** RNA formed at transcription to direct protein synthesis in the cytoplasm.

**metabolic turnover:** The continuous breakdown and replacement of organic materials within cells.

**metabolism:** The sum of all biochemical processes under way within the human body at any moment; includes anabolism and catabolism.

**metabolites:** Compounds produced in the body as a result of metabolic reactions.

**metacarpals:** One of the five bones of the palm of the hand.

**metaphase:** The stage of mitosis in which the chromosomes line up along the equatorial plane of the cell.

**metaphysis:** The region of a long bone between the epiphysis and diaphysis, corresponding to the location of the epiphyseal cartilage of the developing bone.

**metastasis:** The spread of cancer cells from one organ to another, leading to the establishment of secondary tumors.

**metatarsal:** One of the five bones of the foot that articulate with the tarsal bones (proximally) and the phalanges (distally).

**metencephalon:** The pons and cerebellum of the brain.

**mHC protein:** A surface antigen that is important to the recognition of foreign antigens and that plays a role in the coordination and activation of the immune response; also called *major histocompatibility complex (MHC) protein* or *human leukocyte antigen (HLA).*

**micelle:** A droplet with hydrophilic portions on the outside; a spherical aggregation of bile salts, monoglycerides, and fatty acids in the lumen of the intestinal tract.

**microfilaments:** Fine protein filaments visible with the electron microscope; components of the cytoskeleton.

**microglia:** Phagocytic neuroglia in the central nervous system.

**microtubules:** Microscopic tubules that are part of the cytoskeleton and are a component in cilia, flagella, the centrioles, and spindle fibers.

**microvilli:** Small, fingerlike extensions of the exposed plasma membrane of an epithelial cell.

**micturition:** Urination.

**midbrain:** The mesencephalon.

**middle ear:** The space between the external and internal ears that contains auditory ossicles.

**midsagittal plane:** A plane passing through the midline of the body that divides it into left and right halves.

**mineralocorticoids:** Corticosteroids produced by the zona glomerulosa of the adrenal cortex; steroids such as aldosterone that affect mineral metabolism.

**mitochondrion:** An intracellular organelle responsible for generating most of the ATP required for cellular operations.

**mitosis:** The division of a single cell nucleus that produces two identical daughter cell nuclei; an essential step in cell division.

**mitral valve:** The left atrioventricular (AV) valve; also called *bicuspid valve.*

**mixed gland:** A gland that contains exocrine and endocrine cells, or an exocrine gland that produces serous and mucous secretions.

**mixed nerve:** A peripheral nerve that contains sensory and motor fibers.

**mole:** A quantity of an element or compound having a mass in grams equal to the element's atomic weight or to the compound's molecular weight.

**molecular weight:** The sum of the atomic weights of all the atoms in a molecule or compound.

**molecule:** A chemical structure containing two or more atoms that are held together by covalent chemical bonds.

**monocytes:** Phagocytic agranulocytes (white blood cells) in the circulating blood.

**monoglyceride:** A lipid consisting of a single fatty acid bound to a molecule of glycerol.

**monosaccharide:** A simple sugar, such as glucose or ribose.

**monosynaptic reflex:** A reflex in which the sensory afferent neuron synapses directly on the motor efferent neuron.

**motor unit:** All of the muscle cells controlled by a single motor neuron.

**mucins:** Proteoglycans responsible for the lubricating properties of mucus.

**mucosa:** A mucous membrane; the epithelium plus the lamina propria.

**mucosa-associated lymphoid tissue (MALT):** The extensive collection of lymphoid tissues linked with the epithelia of the digestive, respiratory, urinary, and reproductive tracts.

**mucous** (adjective): Indicating the presence or production of mucus.

**mucous (goblet) cell:** A goblet-shaped, mucus-producing, unicellular gland in certain epithelia of the digestive and respiratory tracts; they are called mucous cells when they are found in the stomach mucosa, respiratory mucosa, and salivary glands; they are called goblet cells when they are found in the mucosa of the small intestine, large intestine, terminal bronchioles, and conjunctiva.

**mucus** (noun): A lubricating fluid that is composed of water and mucins and is produced by unicellular and multicellular glands along the digestive, respiratory, urinary, and reproductive tracts.

**multipolar neuron:** A neuron with many dendrites and a single axon; the typical form of a motor neuron.

**multi-unit smooth muscle:** A smooth muscle tissue whose muscle cells are innervated in motor units.

**muscarinic receptors:** Membrane receptors sensitive to acetylcholine and to muscarine, a toxin produced by certain mushrooms; located at all parasympathetic neuromuscular and neuroglandular junctions and at a few sympathetic neuromuscular and neuroglandular junctions.

**muscle:** A contractile organ composed of muscle tissue, blood vessels, nerves, connective tissues, and lymphatic vessels.

**muscle tissue:** A tissue characterized by the presence of cells capable of contraction; includes skeletal, cardiac, and smooth muscle tissues.

**muscular layer:** Concentric layers of smooth muscle responsible for peristalsis.

**muscularis mucosae:** The layer of circular and longitudinal smooth muscle beneath the lamina propria; responsible for moving the mucosal surface.

**mutagens:** Chemical agents that induce mutations and may be carcinogenic.

**mutation:** A change in the nucleotide sequence of the DNA in a cell.

**myelencephalon:** *See* **medulla oblongata**.

**myelin:** An insulating sheath around an axon; consists of multiple layers of neuroglial membrane; significantly increases the impulse propagation rate along the axon.

**myelination:** The formation of myelin.

**myenteric plexus:** Parasympathetic motor neurons and sympathetic postganglionic fibers located between the circular and longitudinal layers of the muscularis externa.

**myocardial infarction:** A heart attack; damage to the heart muscle due to an interruption of regional coronary circulation.

**myocardium:** The cardiac muscle tissue of the heart.

**myofibril:** Organized collections of myofilaments in skeletal and cardiac muscle cells.

**myofilaments:** Fine protein filaments composed primarily of the proteins actin (thin filaments) and myosin (thick filaments).

**myoglobin:** An oxygen-binding pigment that is especially common in slow skeletal muscle fibers and cardiac muscle cells.

**myogram:** A recording of the tension produced by muscle fibers when stimulated.

**myometrium:** The thick layer of smooth muscle in the wall of the uterus.

**myosin:** The protein component of thick filaments.

## N

**nail:** A keratinous structure produced by epithelial cells of the nail root.

**nares, external:** The entrance from the exterior to the nasal cavity.

**nares, internal:** The entrance from the nasal cavity to the nasopharynx.

**nasal cavity:** A chamber in the skull that is bounded by the internal and external nares.

**nasolacrimal duct:** The passageway that transports tears from the nasolacrimal sac to the nasal cavity.

**nasolacrimal sac:** A chamber that receives tears from the lacrimal ducts.

**nasopharynx:** A region that is posterior to the internal nares and superior to the soft palate and ends at the oropharynx.

**natriuretic peptides (NP):** Hormones released by specialized cardiac muscle cells when they are stretched by an abnormally large venous return; promote fluid loss and reductions in blood pressure and in venous return. Include atrial natriuretic peptide (ANP) and brain natriuretic peptide (BNP).

**necrosis:** The death of cells or tissues from disease or injury.

**negative feedback:** A corrective mechanism that opposes or negates a variation from normal limits.

**neonate:** A newborn infant, or baby.

**neoplasm:** A tumor, or mass of abnormal tissue.

**nephron:** The basic functional unit of the kidney.

**nephron loop:** The portion of the nephron that creates the concentration gradient in the renal medulla; also called *loop of Henle.*

**nerve impulse:** An action potential in a neuron plasma membrane.

**neural cortex:** An area of gray matter at the surface of the central nervous system.

**neurilemma:** The outer surface of a Schwann cell that encircles an axon in the peripheral nervous system (PNS).

**neurofibrils:** Microfibrils in the cytoplasm of a neuron.

**neurofilaments:** Microfilaments in the cytoplasm of a neuron.

**neuroglandular junction:** A cell junction at which a neuron controls or regulates the activity of a secretory (gland) cell.

**neuroglia:** Cells of the central nervous system and peripheral nervous system that support and protect neurons; also called *glial cells.*

**neurohypophysis:** The posterior lobe of the pituitary gland; contains the axons of hypothalamic neurons, and these axons release OXT and ADH.

**neuromuscular junction:** A synapse between a neuron and a muscle cell.

**neuron:** A cell in nervous tissue that is specialized for intercellular communication through (1) changes in membrane potential and (2) synaptic connections.

**neurotransmitter:** A chemical compound released by one neuron to affect the membrane potential of another.

**neurotubules:** Microtubules in the cytoplasm of a neuron.

**neutron:** A fundamental particle that does not carry a positive or a negative charge.

**neutrophil:** A white blood cell that is very numerous and normally the first of the mobile phagocytic cells to arrive at an area of injury or infection.

**nicotinic receptors:** Acetylcholine receptors on the surfaces of sympathetic and parasympathetic ganglion cells; respond to the compound nicotine.

**nipple:** An elevated epithelial projection on the surface of the breast; contains the openings of the lactiferous sinuses.

**Nissl bodies:** The ribosomes, Golgi apparatus, rough endoplasmic reticulum, and mitochondria of the perikaryon of a typical neuron.

**nitrogenous wastes:** Organic waste products of metabolism that contain nitrogen, such as urea, uric acid, and creatinine.

**nociception:** Pain perception.

**node:** A gap in the myelin sheath of an axon between adjacent Schwann cells or oligodendrocytes; also called *node of Ranvier.*

**nodose ganglion:** A sensory ganglion of cranial nerve X; also called *inferior ganglion.*

**noradrenaline:** *See* **norepinephrine**.

**norepinephrine (NE):** A catecholamine neurotransmitter in the peripheral nervous system and central nervous system, released at most sympathetic neuromuscular and neuroglandular junctions, and a hormone secreted by the adrenal medulla; also called *noradrenaline.*

**nucleic acid:** A polymer of nucleotides that contains a pentose sugar, a phosphate group, and one of four nitrogenous bases that regulate the synthesis of proteins and make up the genetic material in cells.

**nucleolus:** The dense region in the nucleus that is the site of ribosomal RNA synthesis.

**nucleoplasm:** The fluid content of the nucleus.

**nucleoproteins:** Proteins of the nucleus that are generally associated with DNA.

**nucleotide:** A compound consisting of a nitrogenous base, a simple sugar, and a phosphate group.

**nucleus:** A cellular organelle that contains DNA, RNA, and proteins; in the central nervous system, a mass of gray matter.

**nucleus pulposus:** The gelatinous central region of an intervertebral disc.

**nurse cells:** Supporting cells of the seminiferous tubules of the testis; responsible for the differentiation of spermatids, the maintenance of the blood testis barrier, and the secretion of inhibin, and androgen-binding protein; also called *sustentacular cells.*

**nutrient:** An inorganic or organic substance that can be used by the body as a cofactor or to produce energy.

## O

**occlusal surface:** The opposing surfaces of the teeth that come into contact when chewing food.

**ocular:** Pertaining to the eye.

**oculomotor nerve:** Cranial nerve III, which controls the extra-ocular muscles other than the superior oblique and the lateral rectus muscles.

**olecranon:** The proximal end of the ulna that forms the prominent point of the elbow.

**olfaction:** The sense of smell.

**olfactory bulb:** The expanded ends of the olfactory tracts; the sites where the axons of the first cranial nerves (I) synapse on central nervous system interneurons that lie inferior to the frontal lobes of the cerebrum.

**oligodendrocytes:** Central nervous system neuroglia that maintain cellular organization within gray matter and provide a myelin sheath in areas of white matter.

**oocyte:** A cell whose meiotic divisions will produce a single ovum and three polar bodies.

**oogenesis:** Formation and development of an oocyte.

**opsonization:** An effect of coating an object with antibodies; the attraction and enhancement of phagocytosis.

**optic chiasm:** The crossing point of the optic nerves.

**optic nerve:** The second cranial nerve (II), which carries signals from the retina of the eye to the optic chiasm.

**optic tract:** The tract over which nerve impulses from the retina are transmitted between the optic chiasm and the thalamus.

**orbit:** The bony recess of the skull that contains the eyeball.

**organelle:** An intracellular structure with a specific function or group of functions.

**organic compound:** A compound containing carbon, hydrogen, and in most cases oxygen.

**organogenesis:** The formation of organs during embryonic and fetal development.

**organs:** Combinations of tissues that perform complex functions.

**origin:** In a skeletal muscle, the point of attachment that does not change position when the muscle contracts; usually defined in terms of movements from the anatomical position.

**oropharynx:** The middle portion of the pharynx, bounded superiorly by the nasopharynx, anteriorly by the oral cavity, and inferiorly by the laryngopharynx.

**osmolarity:** The total concentration of dissolved materials in a solution, regardless of their specific identities, expressed in moles; also called *osmotic concentration.*

**osmoreceptor:** A receptor sensitive to changes in the osmolarity of plasma.

**osmosis:** The movement of water across a selectively permeable membrane from one solution to another solution that contains a higher solute concentration.

**osmotic pressure:** The force of osmotic water movement; the pressure that must be applied to prevent osmosis across a membrane.

**osseous tissue:** *See* **bone tissue.**

**ossicles:** Small bones.

**ossification:** The formation of bone; osteogenesis.

**osteoblast:** A cell that produces the fibers and matrix of bone.

**osteoclast:** A cell that dissolves the fibers and matrix of bone.

**osteocyte:** A bone cell responsible for the maintenance and turnover of the mineral content of the surrounding bone.

**osteogenic layer:** The inner, cellular layer of the periosteum that aids in bone growth and repair.

**osteolysis:** The breakdown of the mineral matrix of bone.

**osteon:** The basic histological unit of compact bone, consisting of osteocytes organized around a central canal and separated by concentric lamellae.

**otic:** Pertaining to the ear.

**otoliths:** Calcium carbonate crystals embedded in a gelatinous matrix; located on each macula of the vestibule.

**oval window:** An opening in the bony labyrinth where the stapes attaches to the membranous wall of the scala vestibuli (vestibular duct).

**ovarian cycle:** The monthly process of the maturation, ovulation, and degeneration of a tertiary ovarian follicle.

**ovary:** The female reproductive organ that produces oocytes (gametes).

**ovulation:** The release of a secondary oocyte, surrounded by cells of the corona radiata, after the rupture of the wall of a tertiary ovarian follicle; in females, the periodic release of an oocyte from an ovary.

**ovum/ova:** The functional product of meiosis II, produced after the fertilization of a secondary oocyte.

**oxidation:** The gain of oxygen, or the loss of hydrogen or electrons.

**oxytocin (OXT):** A hormone produced by hypothalamic cells and secreted into capillaries at the posterior lobe of the pituitary gland (neurohypophysis); stimulates smooth muscle contractions of the uterus or mammary glands in females and the prostate gland in males.

## P

**pacemaker cells:** Cells of the sinoatrial node that set the pace of cardiac contraction.

**palate:** The horizontal partition separating the oral cavity from the nasal cavity and nasopharynx; divided into an anterior bony (hard) palate and a posterior fleshy (soft) palate.

**palatine:** Pertaining to the palate.

**palpate:** To examine by touch.

**palpebrae:** Eyelids.

**pancreas:** A digestive organ containing exocrine and endocrine tissues; the exocrine portion secretes pancreatic juice, and the endocrine portion secretes hormones, including insulin and glucagon.

**pancreatic duct:** A tubular duct that carries pancreatic juice from the pancreas to the duodenum.

**pancreatic islets:** Aggregations of endocrine cells in the pancreas; also called *islets of Langerhans.*

**pancreatic juice:** A mixture of buffers and digestive enzymes that is discharged into the duodenum under the stimulation of the enzymes secretin and cholecystokinin.

**papilla:** A small, conical projection.

**paralysis:** The loss of voluntary motor control over a portion of the body.

**paranasal sinuses:** Bony chambers, lined by respiratory epithelium, that open into the nasal cavity; the frontal, ethmoidal, sphenoidal, and maxillary sinuses.

**parasagittal:** A section or plane that parallels the midsagittal plane but that does not pass along the midline.

**parasympathetic division:** One of the two divisions of the autonomic nervous system; generally responsible for activities that conserve energy and lower the metabolic rate; also called *craniosacral division.*

**parathyroid glands:** Four small glands embedded in the posterior surface of the thyroid gland; secrete parathyroid hormone.

**parathyroid hormone (PTH):** A hormone secreted by the parathyroid glands when blood calcium levels decrease below the normal range; causes increased osteoclast activity, increased intestinal calcium uptake, and decreased calcium ion loss at the kidneys.

**parenchyma:** The cells of a tissue or organ that are responsible for fulfilling its functional role; distinguished from the stroma of that tissue or organ.

**parietal:** Relating to the parietal bone; referring to the wall of a cavity.

**parietal cells:** Cells of the gastric glands that secrete hydrochloric acid and intrinsic factor.

**Parkinson's disease:** A neurological disorder resulting from dopamine deficiency characterized by rhythmic muscular tremors, rigidity, and a mask-like appearance.

**parotid salivary glands:** Large salivary glands that secrete a saliva containing high concentrations of salivary (alpha) amylase.

**patella:** The sesamoid bone of the knee; also called the *kneecap.*

**pathogen:** A disease-causing organism.

**pathogenic:** Disease-causing.

**pathologist:** A physician specializing in the identification of diseases on the basis of characteristic structural and functional changes in tissues and organs.

**pelvic cavity:** The inferior subdivision of the abdominopelvic cavity; encloses the urinary bladder, the sigmoid colon and rectum, and male or female reproductive organs.

**pelvis:** A bony complex created by the articulations among the coxal bones, the sacrum, and the coccyx.

**penis:** A component of the male external genitalia; a copulatory organ that surrounds the urethra and introduces semen into the female vagina; the developmental equivalent of the female clitoris.

**peptide:** A chain of amino acids linked by peptide bonds.

**peptide bond:** A covalent bond between the amino group of one amino acid and the carboxyl group of another.

**pericardial cavity:** The space between the parietal pericardium and the epicardium (visceral pericardium) that covers the outer surface of the heart.

**pericardium:** The fibrous sac that surrounds the heart; its inner, serous lining is continuous with the epicardium.

**perichondrium:** The layer that surrounds a cartilage, consisting of an outer fibrous region and an inner cellular region.

**perikaryon:** The cytoplasm that surrounds the nucleus in the cell body of a neuron.

**perilymph:** A fluid similar in composition to cerebrospinal fluid; located in the spaces between the bony labyrinth and the membranous labyrinth of the internal ear.

**perimetrium:** The incomplete serosa of the uterus.

**perimysium:** A connective tissue partition that separates adjacent fasciculi in a skeletal muscle.

**perineum:** The pelvic floor and its associated structures.

**perineurium:** A connective tissue partition that separates adjacent bundles of nerve fibers in a peripheral nerve.

**periodontal ligament:** Collagen fibers that bind the cementum of a tooth to the periosteum of the surrounding alveolus.

**periosteum:** The layer that surrounds a bone, consisting of an outer fibrous region and inner cellular region.

**peripheral nervous system (PNS):** All nervous tissue (CNS) outside the central nervous system and the enteric nervous system (ENS).

**peripheral resistance:** The resistance to blood flow; primarily caused by friction with the vessel walls.

**peristalsis:** A wave of smooth muscle contractions that propels materials along the lumen of a tube such as the digestive tract, the ureters, or the ductus deferens.

**peritoneum:** The serous membrane that lines the peritoneal cavity.

**peritubular capillaries:** A network of capillaries that surrounds the proximal and distal convoluted tubules of the kidneys.

**permeability:** The ease with which dissolved materials can cross a membrane; if the membrane is freely permeable, any molecule can cross it; if impermeable, nothing can cross; most biological membranes are selectively permeable.

**peroxisome:** A membranous vesicle containing enzymes that break down hydrogen peroxide ($H_2O_2$).

**pes:** The foot.

**petrosal ganglion:** A sensory ganglion of the glossopharyngeal nerve (IX).

**petrous:** Stony; usually refers to the thickened portion of the temporal bone that encloses the internal ear.

**pH:** The negative exponent (negative logarithm) of the hydrogen ion concentration, expressed in moles per liter.

**phagocyte:** A cell that performs phagocytosis.

**phagocytosis:** The engulfing of extracellular materials or pathogens; the movement of extracellular materials into the cytoplasm by enclosure in a membranous vesicle.

**phalanx/phalanges:** Bone(s) of the finger(s) or toe(s).

**pharmacology:** The study of drugs, their physiological effects, and their clinical uses.

**pharynx:** The throat; a muscular passageway shared by the digestive and respiratory tracts.

**phenotype:** Physical characteristics that are genetically determined.

**phosphate group:** $PO_4^{3-}$; a functional group that can be attached to an organic molecule; required for the formation of high-energy bonds.

**phospholipid:** An important membrane lipid whose structure includes both hydrophilic and hydrophobic regions.

**phosphorylation:** The addition of a high-energy phosphate group to a molecule.

**photoreception:** Sensitivity to light.

**physiology:** The study of function; deals with the ways organisms perform vital activities.

**pia mater:** The innermost layer of the meninges bound to the underlying neural tissue.

**pineal gland:** Neural tissue in the posterior portion of the roof of the diencephalon; secretes melatonin.

**pinna:** *See* auricle.

**pinocytosis:** The introduction of fluids into the cytoplasm by enclosing them in membranous vesicles at the cell surface.

**pituitary gland:** An endocrine organ that is located in the sella turcica of the sphenoid and is connected to the hypothalamus by the infundibulum; includes the posterior lobe (neurohypophysis) and the anterior lobe (adenohypophysis); also called the *hypophysis.*

**placenta:** A temporary structure in the uterine wall that permits diffusion between the fetal and maternal circulatory systems.

**plantar:** Referring to the sole of the foot.

**plantar flexion:** Ankle extension; toe pointing.

**plasma:** The fluid ground substance of whole blood; what remains after the cells have been removed from a sample of whole blood.

**plasma cell:** An activated B cell that secretes antibodies.

**plasma membrane:** A cell membrane; plasmalemma.

**platelets:** Small packets of cytoplasm that contain enzymes important in the clotting response; manufactured in bone marrow by megakaryocytes.

**pleura:** The serous membrane that lines the pleural cavities.

**pleural cavities:** Body cavities of the thoracic region that surround the lungs.

**plexus:** A network or braid.

**polar body:** A nonfunctional packet of cytoplasm that contains chromosomes eliminated from an oocyte during meiosis.

**polar bond:** A covalent bond in which electrons are shared unequally.

**polarized:** Referring to cells that have regional differences in organelle distribution or cytoplasmic composition along a specific axis, such as between the basement membrane and free surface of an epithelial cell.

**pollex:** The thumb.

**polypeptide:** A chain of amino acids strung together by peptide bonds; those containing more than 100 peptides are called *proteins.*

**polysaccharide:** A complex sugar, such as glycogen or a starch.

**polysynaptic reflex:** A reflex in which interneurons are interposed between the sensory fiber and the motor neuron(s).

**polyunsaturated fats:** Fatty acids containing carbon atoms that are linked by double bonds.

**pons:** The portion of the metencephalon that is anterior to the cerebellum.

**popliteal:** Pertaining to the back of the knee.

**positive feedback:** A mechanism that increases a deviation from normal limits after an initial stimulus.

**postcentral gyrus:** The primary sensory cortex, where touch, vibration, pain, temperature, and taste sensations arrive and are consciously perceived.

**posterior:** Toward the back; dorsal.

**postganglionic neuron:** An autonomic neuron in a peripheral ganglion, whose activities control peripheral effectors.

**postsynaptic membrane:** The portion of the plasma membrane of a postsynaptic cell that is part of a synapse.

**potential difference:** The separation of opposite charges; requires a barrier that prevents ion migration.

**precentral gyrus:** The primary motor cortex of a cerebral hemisphere, located anterior to the central sulcus.

**prefrontal cortex:** The anterior portion of each cerebral hemisphere; thought to be involved with higher intellectual functions, predictions, and calculations.

**preganglionic neuron:** A visceral motor neuron in the central nervous system whose output controls one or more ganglionic motor neurons in the peripheral nervous system.

**premotor cortex:** The motor association area between the precentral gyrus and the prefrontal area.

**pre-optic nucleus:** The hypothalamic nucleus that coordinates thermoregulatory activities.

**presynaptic membrane:** The synaptic surface where neurotransmitter release occurs.

**prevertebral ganglion:** *See* collateral ganglion.

**prime mover:** A muscle that performs a specific action.

**proenzyme:** An inactive enzyme secreted by an epithelial cell.

**progesterone:** The most important hormone secreted by the corpus luteum after ovulation.

**prognosis:** A prediction about the possible course or outcome from a specific disease.

**projection fibers:** Axons carrying information from the thalamus to the cerebral cortex.

**prolactin (PRL):** The hormone that stimulates functional development of the mammary glands in females; a secretion of the anterior lobe of the pituitary gland (adenohypophysis).

**pronation:** The rotation of the forearm that makes the palm face posteriorly.

**prone:** Lying face down with the palms facing the floor.

**pronucleus:** An enlarged ovum or sperm nucleus that forms after fertilization but before amphimixis.

**prophase:** The initial phase of mitosis; characterized by the appearance of chromosomes, the breakdown of the nuclear membrane, and the formation of the spindle apparatus.

**proprioception:** The awareness of the positions of bones, joints, and muscles.

**prostaglandin:** A fatty acid secreted by one cell that alters the metabolic activities or sensitivities of adjacent cells; also called *local hormone.*

**prostate gland:** An accessory gland of the male reproductive tract, contributing about one-third of the volume of semen.

**prosthesis:** An artificial substitute for a body part.

**protease:** *See* proteinase.

**protein:** A large polypeptide with a complex structure.

**proteinase:** An enzyme that breaks down proteins into peptides and amino acids.

**proteoglycan:** A substance containing a large polysaccharide complex linked by polypeptide chains; examples include hyaluronic acid and chondroitin sulfate.

**proton:** A fundamental particle having a positive charge.

**protraction:** Movement anteriorly in the horizontal plane.

**proximal:** Toward the attached base of an organ or structure.

**proximal convoluted tubule (PCT):** The segment of the nephron that is located between the glomerular capsule (Bowman's capsule) and the nephron loop; the major site of active reabsorption from filtrate.

**pseudopodia:** Temporary cytoplasmic extensions typical of mobile or phagocytic cells.

**pseudostratified epithelium:** An epithelium that contains several layers of nuclei but whose cells are all in contact with the underlying basement membrane.

**puberty:** A period of rapid growth, sexual maturation, and the appearance of secondary sexual characteristics; normally occurs at ages 10–15 years.

**pubic symphysis:** The fibrocartilaginous amphiarthrosis between the pubic bones of the coxal bones.

**pubis:** The anterior, inferior component of the hip bone.

**pudendum:** The external genitalia.

**pulmonary circuit:** Blood vessels between the pulmonary semilunar valve of the right ventricle and the entrance to the left atrium; the blood flow through the lungs.

**pulmonary ventilation:** The movement of air into and out of the lungs.

**pulvinar nucleus:** The thalamic nucleus involved in the integration of sensory information prior to projection to the cerebral hemispheres.

**pupil:** The opening in the center of the iris through which light enters the eye.

**purine:** A nitrogen compound with a double ring-shaped structure; examples include adenine and guanine, two nitrogenous bases that are common in nucleic acids.

**Purkinje cell:** A large, branching neuron of the cerebellar cortex.

**Purkinje fibers:** Specialized conducting cardiocytes in the ventricles of the heart.

**pyloric part:** The gastric region between the body of the stomach and the duodenum; includes the muscular pylorus and the pyloric sphincter.

**pyloric sphincter:** A ring of smooth muscle that regulates the passage of chyme from the stomach to the duodenum.

**pyrimidine:** A nitrogen compound with a single ring-shaped structure; examples include cytosine, thymine, and uracil, nitrogenous bases that are common in nucleic acids.

**pyruvate:** The anion formed by the dissociation of pyruvic acid, a three-carbon compound produced by glycolysis.

**Q**

**quaternary structure:** The three-dimensional protein structure produced by interactions between protein subunits.

**R**

**rami communicantes:** Paired axon bundles that link the spinal nerves with the ganglia of the sympathetic chain.

**ramus/rami:** A branch/branches.

**raphe:** A seam.

**receptive field:** The area monitored by a single sensory receptor.

**rectum:** The inferior 15 cm (6 in.) of the digestive tract.

**rectus:** Straight.

**red blood cell (RBC):** *See* **erythrocyte**.

**reduction:** The loss of oxygen, or the gain of hydrogen or electrons.

**reductional division:** The first meiotic division, which reduces the chromosome number from 46 to 23.

**reflex:** A rapid, automatic response to a stimulus.

**reflex arc:** The receptor, sensory neuron, motor neuron, and effector involved in a particular reflex; interneurons may be present, depending on the reflex considered.

**refractory period:** The period between the initiation of an action potential and the restoration of the normal resting membrane potential; during this period, the membrane will not respond normally to stimulation.

**regulatory T cells:** Population of T lymphocytes that suppress the immune response.

**relaxation phase:** The period after a contraction when the tension in the muscle fiber returns to resting levels.

**relaxin:** A hormone that loosens the pubic symphysis; secreted by the placenta.

**renal:** Pertaining to the kidneys.

**renal corpuscle:** The initial segment of the nephron, consisting of an expanded chamber that encloses the glomerulus.

**renin:** The enzyme released by cells of the juxtaglomerular complex when renal blood flow decreases; converts angiotensinogen to angiotensin I.

**repolarization:** The movement of the membrane potential away from a positive value and toward the resting potential.

**respiration:** The exchange of gases between cells and the environment; includes pulmonary ventilation, external respiration, internal respiration, and cellular respiration.

**respiratory minute volume** ($V_E$)**:** The amount of air moved into and out of the respiratory system each minute.

**respiratory pump:** A mechanism by which changes in the intrapleural pressures during the respiratory cycle assist the venous return to the heart.

**resting membrane potential:** The membrane potential of an undisturbed cell, or simply the *resting potential.*

**resting potential:** The membrane potential of a normal cell under homeostatic conditions.

**rete:** An interwoven network of blood vessels or passageways.

**reticular activating system (RAS):** The mesencephalic portion of the reticular formation; responsible for arousal and the maintenance of consciousness.

**reticular formation:** A diffuse network of gray matter that extends the entire length of the brainstem (medulla oblongata to the midbrain).

**reticulospinal tracts:** Descending tracts of the medial pathway that carry involuntary motor commands issued by neurons of the reticular formation.

**retina:** The innermost layer of the eye, lining the vitreous chamber; also called *neural layer.*

**retinal:** A visual pigment derived from vitamin A.

**retraction:** Movement posteriorly in the horizontal plane.

**retroperitoneal:** Behind or outside the peritoneal cavity.

**reverberation:** A positive feedback along a chain of neurons such that they remain active once stimulated.

**rheumatism:** A general term used to describe pain in muscles, tendons, bones, or joints.

**Rh factor:** A surface antigen that may be present (Rh-positive) or absent (Rh-negative) from the surfaces of red blood cells.

**rhodopsin:** The visual pigment in the membrane discs of the distal segments of rods.

**rhythmicity center:** A medullary center responsible for the pace of respiration; includes inspiratory and expiratory centers.

**ribonucleic acid:** A nucleic acid consisting of a chain of nucleotides that contain the sugar ribose and the nitrogenous bases adenine, guanine, cytosine, and uracil.

**ribose:** A five-carbon sugar that is a structural component of RNA.

**ribosome:** An organelle that contains rRNA and proteins and is essential to mRNA translation and protein synthesis.

**rod:** A photoreceptor responsible for vision in dim lighting.

**rough endoplasmic reticulum (RER):** A membranous organelle that is a site of protein synthesis and storage.

**round window:** An opening in the bony labyrinth of the internal ear that exposes the membranous wall of the scala tympani (tympanic duct) to the air of the middle ear cavity.

**rubrospinal tracts:** Descending tracts of the lateral pathway that carry involuntary motor commands issued by the red nucleus of the midbrain.

**rugae:** Mucosal folds in the lining of the empty stomach that disappear as gastric distension occurs; folds in the urinary bladder.

**rule of nines:** Method used to calculate the total percentage surface area involved in burns in which the body surface of an adult is divided into multiples of 9.

**S**

**saccule:** A portion of the vestibular apparatus of the internal ear; contains a macula important for providing sensations of gravity and linear acceleration in a vertical dimension.

**sagittal plane:** A sectional plane that divides the body into left and right portions.

**salt:** An inorganic compound consisting of a cation other than H1 and an anion other than OH$^-$.

**saltatory propagation:** The relatively rapid propagation of an action potential between successive nodes of a myelinated axon.

**sarcomere:** The smallest contractile unit of a striated muscle cell.

**sarcoplasm:** The cytoplasm of a muscle cell.

**scala media:** *See* **cochlear duct**.

**scala tympani:** The perilymph-filled chamber of the internal ear, adjacent to the basilar membrane; pressure changes there distort the round window; also called *tympanic duct.*

**scala vestibuli:** A coiled tube filled with perilymph that lies within the bony labyrinth; it is continuous with the scala tympani (tympanic duct) at the tip of the cochlear spiral; also called *vestibular duct.*

**scar tissue:** The thick, collagenous tissue that forms at an injury site.

**Schwann cells:** Neuroglia responsible for the neurilemma that surrounds axons in the peripheral nervous system.

**sciatic nerve:** A nerve innervating the posteromedial portions of the thigh and leg.

**sclera:** The fibrous, outer layer of the eye that forms the white area of the anterior surface; a portion of the fibrous layer of the eye.

**sclerosis:** A hardening and thickening that commonly occurs secondary to tissue inflammation.

**scrotum:** The loose-fitting, fleshy pouch that encloses the testes of the male.

**sebaceous glands:** Glands that secrete sebum; normally associated with hair follicles.

**sebum:** A waxy secretion that coats the surfaces of hairs.

**secondary sex characteristics:** Physical characteristics that appear at puberty in response to sex hormones.

**second-degree burn:** Burn involving the epidermis and dermis that usually forms blisters; type of partial-thickness burn.

**secretin:** A hormone, secreted by the duodenum, that stimulates the production of buffers by the pancreas and inhibits gastric activity.

**semen:** The fluid ejaculate that contains sperm and the secretions of accessory glands of the male reproductive tract.

**semicircular ducts:** The tubular components of the membranous labyrinth of the internal ear; responsible for dynamic equilibrium.

**semilunar valve:** A three-cusped valve guarding the exit from one of the cardiac ventricles; the pulmonary and aortic valves.

**seminal glands:** Glands of the male reproductive tract that produce roughly 60 percent of the volume of semen; also called *seminal vesicles.*

**seminiferous tubules:** Coiled tubules where sperm production occurs in the testes.

**senescence:** Aging.

**sensible perspiration:** Water loss due to secretion by sweat glands.

**septa:** Partitions that subdivide an organ.

**serous cell:** A cell that produces a serous secretion.

**serous membrane:** A squamous epithelium and the underlying loose connective tissue; the lining of the pericardial, pleural, and peritoneal cavities; also called a *serosa.*

**serous secretion:** A watery secretion that contains high concentrations of enzymes.

**serum:** The ground substance of blood plasma from which clotting agents have been removed.

**sesamoid bone:** A bone that forms within a tendon.

**sigmoid colon:** The S-shaped segment of the colon between the descending colon and the rectum.

**sign:** The visible, objective evidence of the presence of a disease.

**simple epithelium:** An epithelium containing a single layer of cells above the basal lamina.

**sinoatrial (SA) node:** The natural pacemaker of the heart; situated in the wall of the right atrium.

**sinus:** A chamber or hollow in a tissue; a large, dilated vein.

**sinusoid:** An exchange vessel that is similar in general structure to a fenestrated capillary. the two differ in size (sinusoids are larger and more irregular in cross section), continuity (sinusoids have gaps between endothelial cells), and support (sinusoids have thin basement membranes, if present at all).

**skeletal muscle:** A contractile organ of the muscular system.

**skeletal muscle tissue:** A contractile tissue dominated by skeletal muscle fibers; characterized as striated, voluntary muscle.

**sliding filament theory:** The concept that a sarcomere shortens as the thick and thin filaments slide past one another.

**small intestine:** The duodenum, jejunum, and ileum; the digestive tract between the stomach and the large intestine.

**smooth endoplasmic reticulum (SER):** A membranous organelle in which lipid and carbohydrate synthesis and storage occur.

**smooth muscle tissue:** Muscle tissue in the walls of many visceral organs; characterized as nonstriated, involuntary muscle.

**soft palate:** The fleshy posterior extension of the hard palate, separating the nasopharynx from the oral cavity.

**solute:** Any materials dissolved in a solution.

**solution:** A fluid containing dissolved materials.

**somatic:** Pertaining to the body.

**somatic nervous system (SNS):** The efferent division of the nervous system that innervates skeletal muscles.

**somatomedins:** Substances stimulating tissue growth; released by the liver after the secretion of growth hormone; also called *insulin-like growth factors.*

**sperm:** Male gamete(s); also called *spermatozoon/ spermatozoa.*

**spermatic cord:** Collectively, the spermatic vessels, nerves, lymphatic vessels, and the ductus deferens, extending between the testes and the proximal end of the inguinal canal.

**spermatocyte:** A cell of the seminiferous tubules that is engaged in meiosis.

**spermatogenesis:** Sperm production.

**spermatozoon/spermatozoa:** *See* **sperm.**

**sphincter:** A muscular ring that contracts to close the entrance or exit of an internal passageway.

**spinal nerve:** One of 31 pairs of nerves that originate on the spinal cord from anterior and posterior roots.

**spindle apparatus:** Microtubule-based structure that distributes duplicated chromosomes to opposite ends of a dividing cell during mitosis.

**spinocerebellar tracts:** Ascending tracts that carry sensory information to the cerebellum.

**spinothalamic tracts:** Ascending tracts that carry poorly localized touch, pressure, pain, vibration, and temperature sensations to the thalamus.

**spinous process:** The prominent posterior projection of a vertebra; formed by the fusion of two laminae.

**spiral organ:** A receptor complex in the scala media of the cochlea that includes the inner and outer hair cells, supporting cells and structures, and the tectorial membrane; provides the sensation of hearing; also called the *organ of Corti.*

**spleen:** A lymphoid organ important for the phagocytosis of red blood cells, the immune response, and lymphocyte production.

**spongy bone:** Bone that consists of an open network of struts and plates that resembles a latticework; also called *trabecular bone.*

**squama:** A broad, flat surface.

**squamous:** Flattened.

**squamous epithelium:** An epithelium whose superficial cells are flattened and platelike.

**stapes:** The auditory ossicle attached to the tympanic membrane.

**stellate macrophage:** Phagocytic cell of the liver sinusoids; also called *Kupffer cell.*

**stenosis:** A constriction or narrowing of a passageway.

**stereocilia:** Elongate microvilli characteristic of the epithelium of the epididymis, portions of the ductus deferens, and the internal ear.

**steroid:** A ring-shaped lipid structurally related to cholesterol.

**stimulus:** An environmental change that produces a change in cellular activities; often used to refer to events that alter the membrane potentials of excitable cells.

**stratified:** Containing several layers.

**stratum:** A layer.

**stretch receptors:** Sensory receptors that respond to stretching of the surrounding tissues.

**stroma:** The connective tissue framework of an organ; distinguished from the functional cells (parenchyma) of that organ.

**subarachnoid space:** A meningeal space containing cerebrospinal fluid; the area between the arachnoid membrane and the pia mater.

**subclavian:** Pertaining to the region immediately posterior and inferior to the clavicle.

**subcutaneous layer:** The layer of loose connective tissue below the dermis; also called *subcutaneous layer* or *superficial fascia.*

**submucosa:** The region between the muscularis mucosae and the muscularis externa.

**subserous fascia:** The loose connective tissue layer beneath the serous membrane that lines the ventral body cavity.

**substrate:** A participant (product or reactant) in an enzyme-catalyzed reaction.

**sulcus:** A groove or furrow.

**summation:** The temporal or spatial addition of contractile force or neural stimuli.

**superior:** Above, in reference to a portion of the body in the anatomical position.

**superior vena cava (SVC):** The vein that carries blood to the right atrium from parts of the body that are superior to the heart.

**supination:** The rotation of the forearm such that the palm faces anteriorly.

**supine:** Lying face up, with palms facing anteriorly.

**surfactant:** A lipid secretion that coats the alveolar surfaces of the lungs and prevents their collapse.

**sustentacular cells:** *See* **nurse cells.**

**sutural bones:** Irregular bones that form in fibrous tissue between the flat bones of the developing cranium; also called *Wormian bones.*

**suture:** A fibrous joint between flat bones of the skull.

**sympathetic division:** The division of the autonomic nervous system that is responsible for "fight or flight" reactions; primarily concerned with the elevation of metabolic rate and increased alertness; also called *thoracolumbar division.*

**symphysis:** A fibrous amphiarthrosis, such as that between adjacent vertebrae or between the pubic bones of the coxal bones.

**symptom:** An abnormality of function as a result of disease; subjective experience of patient.

**synapse:** The site of communication between a nerve cell and some other cell; if the other cell is not a neuron, the term *neuromuscular junction* or *neuroglandular junction* is often used.

**synaptic delay:** The period between the arrival of an impulse at the presynaptic membrane and the initiation of an action potential in the postsynaptic membrane.

**syncytium:** A multinucleate mass of cytoplasm, produced by the fusion of cells or repeated mitoses without cytokinesis.

**syndrome:** A discrete set of signs and symptoms that occur together.

**synergist:** A muscle that assists a prime mover in performing its primary action.

**synovial cavity:** A fluid-filled chamber in a synovial joint.

**synovial fluid:** The substance secreted by synovial membranes that lubricates joints.

**synovial joint:** A freely movable joint where the opposing bone surfaces are separated by synovial fluid; a diarthrosis.

**synovial membrane:** An incomplete layer of fibroblasts facing the synovial cavity, plus the underlying loose connective tissue.

**synthesis:** Manufacture; anabolism.

**system:** An interacting group of organs that performs one or more specific functions.

**systemic circuit:** The vessels between the aortic valve and the entrance to the right atrium; the system other than the vessels of the pulmonary circuit.

**systole:** A period of contraction in a chamber of the heart, as part of the cardiac cycle.

**systolic pressure:** The peak arterial pressure measured during ventricular systole.

**T**

**tarsus:** The ankle.

**tactile:** Pertaining to the sense of touch.

**tarsal bones:** The bones of the ankle (the talus, calcaneus, navicular, and cuneiform bones).

**T cells:** Lymphocytes responsible for cell-mediated immunity and for the coordination and regulation of the immune response; includes cytotoxic T cells, helper T cells, regulatory T cells, and memory T cells.

**tectospinal tracts:** Descending tracts of the medial pathway that carry involuntary motor commands issued by the colliculi.

**telodendria:** Terminal axonal branches that end in axon terminals.

**telophase:** The final stage of mitosis, characterized by the disappearance of the spindle apparatus, the reappearance of the nuclear membrane, the disappearance of the chromosomes, and the completion of cytokinesis.

**temporal:** Pertaining to time (temporal summation) or to the temples (temporal bone).

**tendon:** A collagenous band that connects a skeletal muscle to an element of the skeleton.

**teres:** Round.

**terminal:** Toward the end.

**tertiary structure:** The protein structure that results from interactions among distant portions of the same molecule; complex coiling and folding.

**testes:** The male gonads, sites of gamete production and hormone secretion.

**testosterone:** The main androgen produced by the interstitial endocrine cells of the testes.

**tetraiodothyronine:** $T_4$, or thyroxine, a thyroid hormone.

**thalamus:** The walls of the diencephalon.

**therapy:** The treatment of disease.

**thermoreception:** Sensitivity to temperature changes.

**thermoregulation:** Homeostatic maintenance of body temperature.

**thick filament:** A cytoskeletal filament in a skeletal or cardiac muscle cell; composed of myosin, with a core of titin.

**thin filament:** A cytoskeletal filament in a skeletal or cardiac muscle cell; consists of actin, troponin, and tropomyosin.

**third-degree burn:** Burn involving the entire thickness of the skin and extends into the subcutaneous tissue, muscle, and bone causing scarring; full-thickness burn.

**thoracolumbar division:** The sympathetic division of the autonomic nervous system.

**thorax:** The chest.

**threshold:** The membrane potential at which an action potential begins.

**thrombin:** The enzyme that converts (soluble) fibrinogen to (insoluble) fibrin.

**thymine:** A pyrimidine; one of the nitrogenous bases in the nucleic acid DNA.

**thymosins:** Thymic hormones essential to the development and differentiation of T cells.

**thymus:** A lymphoid organ, the site of T cell development and maturation.

**thyroglobulin:** A circulating transport globulin that binds thyroid hormones.

**thyroid gland:** An endocrine gland whose lobes are lateral to the thyroid cartilage of the larynx.

**thyroid hormones:** Thyroxine ($T_4$) and $T_3$, hormones of the thyroid gland; stimulate tissue metabolism, energy utilization, and growth.

**thyroid-stimulating hormone(TSH):** The hormone, produced by the anterior lobe of the pituitary gland, that triggers the secretion of thyroid hormones by the thyroid gland.

**thyroxine:** A thyroid hormone; also called $T_4$ or *tetraiodothyronine*.

**tidal volume:** The volume of air moved into and out of the lungs during a normal quiet respiratory cycle.

**tissue:** A collection of specialized cells and cell products that performs a specific function.

**titin:** Fibrous protein that connects thick myosin filaments to the z lines in the sarcomere.

**tonsil:** A lymphoid nodule in the wall of the pharynx; the palatine, pharyngeal, and lingual tonsils.

**topical:** Applied to the body surface.

**toxic:** Poisonous.

**trabecula:** A connective tissue partition that subdivides an organ.

**trabecular bone:** *See* **spongy bone**.

**trachea:** The windpipe; an airway extending from the larynx to the primary bronchi.

**tract:** A bundle of axons in the central nervous system.

**transcription:** The encoding of genetic instructions on a strand of mRNA.

**transection:** The severing or cutting of an object in the transverse plane.

**translation:** The process of peptide formation from the instructions carried by an mRNA strand.

**transudate:** A fluid that diffuses across a serous membrane and lubricates opposing surfaces.

**transverse tubules:** The transverse, tubular extensions of the sarcolemma that extend deep into the sarcoplasm, contacting cisternae of the sarcoplasmic reticulum; also called *T tubules*.

**tricuspid valve:** The right atrioventricular valve, which prevents the backflow of blood into the right atrium during ventricular systole.

**trigeminal nerve:** Cranial nerve V, which provides sensory information from the lower portions of the face (including the upper and lower jaws) and delivers motor commands to the muscles of mastication.

**triglyceride:** A lipid that is composed of a molecule of glycerol attached to three fatty acids.

**trochanter:** Large process near the head of the femur.

**trochlea:** A pulley; the spool-shaped medial portion of the condyle of the humerus.

**trochlear nerve:** Cranial nerve IV, controlling the superior oblique muscle of the eye.

**trunk:** The thoracic and abdominopelvic regions; a major arterial branch.

**T tubules:** *See* **transverse tubules**.

**tuberculum:** A small, localized elevation on a bony surface.

**tuberosity:** A large, roughened elevation on a bony surface.

**tumor:** A tissue mass formed by the abnormal growth and replication of cells.

**tunica:** A layer or covering.

**twitch:** A single stimulus-contraction-relaxation cycle in a skeletal muscle fiber.

**tympanic membrane:** The membrane that separates the external acoustic meatus from the middle ear; the membrane whose vibrations are transferred to the auditory ossicles and ultimately to the oval window; also called *eardrum* or *tympanum*.

### U

**umbilical:** Pertaining to the navel.

**umbilical cord:** The connecting stalk between the fetus and the placenta; contains the allantois, the umbilical arteries, and the umbilical vein.

**unipolar neuron:** A sensory neuron whose cell body is in a dorsal root ganglion or a sensory ganglion of a cranial nerve.

**unmyelinated axon:** An axon whose neurolemma does not contain myelin and across which continuous propagation occurs.

**uracil:** A pyrimidine; one of the nitrogenous bases in the nucleic acid RNA.

**ureters:** Muscular tubes, lined by transitional epithelium, that carry urine from the renal pelvis to the urinary bladder.

**urethra:** A muscular tube that carries urine from the urinary bladder to the exterior.

**urinary bladder:** The muscular, distensible sac that stores urine prior to micturition.

**urination:** The voiding of urine; micturition.

**uterus:** The muscular organ of the female reproductive tract in which implantation, placenta formation, and fetal development occur.

**utricle:** The largest chamber of the vestibular apparatus of the internal ear; contains a macula important for providing sensations of gravity and linear acceleration in a horizontal dimension.

### V

**vagina:** A muscular tube extending between the uterus and the vestibule.

**vallate papilla:** One of the large, dome-shaped papillae on the superior surface of the tongue that forms a V, separating the body of the tongue from the root.

**vascular:** Pertaining to blood vessels.

**vasculogenesis:** Formation of the vascular system.

**vasoconstriction:** A decrease in the diameter of arterioles due to the contraction of smooth muscles in the tunica media; increases peripheral resistance; may occur in response to local factors, through the action of hormones, or from the stimulation of the vasomotor center.

**vasodilation:** An increase in the diameter of arterioles due to the relaxation of smooth muscles in the tunica media; decreases peripheral resistance; may occur in response to local factors, through the action of hormones, or after decreased stimulation of the vasomotor center.

**vasomotion:** Rhythmic changes in the pattern of blood flow through a capillary bed due to the alternate contraction and relaxation of precapillary sphincters.

**vasomotor center:** The center in the medulla oblongata whose stimulation produces vasoconstriction and an increase of peripheral resistance.

**vein:** A blood vessel carrying blood from a capillary bed toward the heart.

**vena cava:** One of the major veins delivering systemic blood to the right atrium; superior and inferior venae cavae.

**ventilation:** Air movement into and out of the lungs.

**ventral:** Pertaining to the anterior surface.

**ventricle:** A fluid-filled chamber; in the heart, one of the large chambers discharging blood into the pulmonary or systemic circuits; in the brain, one of four fluid-filled interior chambers.

**venule:** Thin-walled veins that receive blood from capillaries.

**vertebral canal:** The passageway that encloses the spinal cord; a tunnel bounded by the neural arches of adjacent vertebrae.

**vertebral column:** The cervical, thoracic, and lumbar vertebrae, the sacrum, and the coccyx.

**very low density lipoprotein (VLDL):** Form of lipoprotein that transports triglycerides from the intestine and liver to muscle and adipose tissue.

**vesicle:** A membranous sac in the cytoplasm of a cell.

**vestibular nucleus:** The processing center for sensations that arrive from the vestibular apparatus of the internal ear, located near the border between the pons and the medulla oblongata.

**vestibulospinal tracts:** Descending tracts of the medial pathway that carry involuntary motor commands issued by the vestibular nucleus to stabilize the position of the head.

**villus/villi:** Slender, finger-shaped projection(s) of a mucous membrane.

**virus:** A noncellular pathogen.

**viscera:** Internal organs of the thoracic and abdominopelvic cavities.

**visceral:** Pertaining to viscera (internal organs) or their outer coverings.

**visceral smooth muscle:** A smooth muscle tissue that forms sheets or layers in the walls of visceral organs; the cells may not be innervated, and the layers often show automaticity (rhythmic contractions).

**viscosity:** The resistance to flow that a fluid exhibits as a result of molecular interactions within the fluid.

**viscous:** Thick, syrupy.

**vitamin:** An essential organic nutrient that functions as a coenzyme in vital enzymatic reactions.

**vitreous humor:** The fluid portion of the vitreous body, the gelatinous mass in the posterior cavity (vitreous chamber) of the eye.

**voluntary:** Controlled by conscious thought processes.

### W

**white blood cells (WBCs):** The granulocytes and agranulocytes of whole blood.

**white matter:** Regions in the central nervous system that are dominated by myelinated axons.

**white ramus:** A nerve bundle containing the myelinated preganglionic axons of sympathetic motor neurons en route to the sympathetic chain or to a collateral ganglion.

**Wormian bones:** *See* **sutural bones**.

### X

**X chromosome:** One of two sex chromosomes; females have two X chromosomes.

**xenograft:** Tissue transplant from another species, for example, a pig heart valve is transplanted into a human.

**xiphoid process:** The slender, inferior extension of the sternum.

### Y

**Y chromosome:** The sex chromosome whose presence indicates that the individual is a genetic male.

### Z

**zona fasciculata:** The region of the adrenal cortex that secretes glucocorticoids.

**zona glomerulosa:** The region of the adrenal cortex that secretes mineralocorticoids.

**zona reticularis:** The region of the adrenal cortex that secretes androgens.

**zygote:** The fertilized ovum, prior to the start of cleavage.

# Credits

## Illustration Credits

Unless otherwise stated, all illustrations are © Pearson Education. Illustration on p. 916 courtesy of U.S. Department of Agriculture.

## Photo Credits

**Front Matter** p. iv (top to bottom) PK Martini. Amy Heflinger. Cesere Brothers Photography. Tim Mantoani; p. v (top to bottom) William C. Ober and Claire E. Ober. William C. Ober and Claire E. Ober. PK Martini. Ralph T. Hutchings.; VWT-01 Artjazz/Shutterstock; VWT-08 Nik Merkulov/Shutterstock; VWT-09 Designsstock/Shutterstock; VWT-10 Artjazz/Shutterstock; VWT-11 Nik Merkulov/Shutterstock; VWT-13 Designsstock/Shutterstock; VWT-18 Nik Merkulov/Shutterstock; VWT-19 Designsstock/Shutterstock; VWT-20 Artjazz/Shutterstock.

**Chapter 1** Chapter opener David Marchal/E+/Getty Images; p. 3 01-01-01 Isak55/Shutterstock; p. 3 01-01-01 inset Sciencepics/Shutterstock; p. 3 01-01-02 Wavebreakmedia/Shutterstock; p. 4 01-02-01 Graça Victoria/123RF; p. 4 01-02-02 Don Kohlbauer/U-T San Diego/ZUMAPRESS.com/Newscom; p. 4 01-02-03 Wavebreakmedia/Pearson Asset Library/Shutterstock; p. 5 01-02-04 Frederic H. Martini; p. 5 01-02-05 Frederic H. Martini; p. 8 01-04-01 Science Source; p. 11 01-06-01 Darren Baker/Fotolia; p. 11 01-06-02 Science Source; p. 16 01-09-01 Ralph T. Hutchings; p. 34 01-18-01 Pearson Education, Inc.; p. 37 01-19-01 The Granger Collection; p. 38 01-20-01 Ed Fisher/New Yorker Collection/The Cartoon Bank; p. 39 01-20-02 Dr. Andre Kempe/Sciencefoto.De/Getty Images; p. 39 01-20-03 Dr. Andre Kempe/Sciencefoto.De/Getty Images; p. 39 01-20-04 Dr. Andre Kempe/Sciencefoto.De/Getty Images; p. 42 01-22-01 Pearson Education, Inc.; p. 43 01-22-02 Pearson Education, Inc.; p. 43 01-22-03 Pearson Education, Inc.; p. 43 01-22-04 Pearson Education, Inc.; p. 46 01-CR-01 The Granger Collection; p. 47 01-CR-02 Pearson Education, Inc.; p. 49 01-CI-01 Michelle Del Guercio/Science Source; p. 49 01-CI-02 Ben44/Shutterstock.

**Chapter 2** Chapter opener Rrocio/Getty Images; p. 56 02-04-01 GYSSELS/BSIP/Alamy Stock Photo; p. 58 02-05-01 Fotolia; p. 58 02-05-02 Pabijan/Fotolia; p. 58 02-05-03 Khlongwangchao/Shutterstock; p. 58 02-05-04a Meunierd/Shutterstock; p. 58 02-05-04b EpicStockMedia/Shutterstock; p. 58 02-05-04c ArtMarie/iStock/Getty Images; p. 59 02-05-05 Markus Gayda; p. 59 02-05-06 Seelevel; p. 61 02-06-01 Theartofphoto/Fotolia; p. 67 02-09-01a Serg64/Shutterstock; p. 67 02-09-01b Aaron Amat/Shutterstock; p. 69 02-10-01 Photo Works/Shutterstock; p. 69 02-10-02 Trevor Chriss/Alamy Stock Photo; p. 93 02-CI-01 Phil Date/Shutterstock.

**Chapter 3** Chapter opener Felix Mackel/iStock/Getty Images; p. 100 03-04-01 Science Source; p. 104 03-06-01 Birgit Satir; p.p. 106 03-07-01 Bill Longcore/Science Source; p. 110 03-09-01 Don W. Fawcett/Science Source; p.p. 123 03-15-01 Steve Gschmeissner/Science Source; p. 123 03-15-02 David M. Phillips/Science Source; p. 123 03-15-03 Steve Gschmeissner/Science Source; p. 127 03-17-01 Don W. Fawcett/Science Source; p. 132 03-20-01 Ed Reschke/Getty Images; p. 132 03-20-02 Ed Reschke/Getty Images; p. 132 03-20-03 Ed Reschke/Getty Images; p. 133 03-20-04 Ed Reschke/Getty Images; p. 133 03-20-05 Ed Reschke/Getty Images; p. 133 03-20-06 Ed Reschke/Getty Images; p. 133 03-20-07 Ed Reschke/Getty Images; p. 135 03-21-01 Scott Camazine/Newscom; p. 141 03-CI-01 EPSTOCK/Shutterstock; p. 141 03-CI-02 Darren Hubley/Shutterstock.

**Chapter 4** Chapter opener Medicshots/Alamy Stock Photo; p. 144 04-02-01 Alexander Raths/Shutterstock; p. 144 04-02-02 Ariwasabi/Shutterstock; p. 144 04-02-03 O. Bellini/Shutterstock; p. 145 04-02-04 Brian Maudsley/Shutterstock; p. 145 04-02-05 Coffeeyu/E+/Getty Images; p. 145 04-02-06 Avid Creative/Getty Images; p. 145 04-02-07 Pearson Education, Inc.; p. 145 04-02-08 Biophoto Associates/Science Source; p. 145 04-02-09 Prof. P.M. Motta / Univ. "La Sapienza", Rome/Science Source; p. 145 04-02-10 Brian Maudsley/Shutterstock; p. 150 04-05-01 Pearson Education, Inc.; p. 151 04-05-02 Frederic H. Martini; p. 152 04-06-01 Pearson Education, Inc.; p. 152 04-06-02 Courtesy of Dr. Gregory N. Fuller, Chief Neuropathologist, U T MD Anderson Cancer Center.; p. 153 04-06-03 Pearson Education, Inc.; p. 153 04-06-04 Pearson Education, Inc.; p. 154 04-07-01 Pearson Education, Inc.; p. 154 04-07-02 Frederic H. Martini; p. 155 04-07-03 Pearson Education, Inc.; p. 155 04-07-04 Pearson Education, Inc.; p. 155 04-07-05 William C. Ober; p. 155 04-07-06 Courtesy of Dr. Gregory N. Fuller, Chief Neuropathologist, U T MD Anderson Cancer Center.; p. 156 04-08-01 image H. Jastrow from Dr. Jastrow's electron microscopic atlas on the www, http://www.drjastrow.de; p. 161 04-11-01 Pearson Education, Inc.; p. 161 04-11-02 Pearson Education, Inc.; p. 161 04-11-03 Ward's Science; p. 162 04-12-01 Pearson Education, Inc.; p. 162 04-12-02 Pearson Education, Inc.; p. 162 04-12-03 Frederic H. Martini; p. 162 04-12-04 Pearson Education, Inc.; p. 164 04-13-01 Pearson Education, Inc.; p. 164 04-13-02 Pearson Education, Inc.; p. 164 04-13-04 Science Source; p. 164 04-13-05 Pearson Education, Inc.; p. 165 04-13-06 Microscape/Science Source; p. 167 04-14-01 Pearson Education, Inc.; p. 169 04-15-01 Pearson Education, Inc.; p. 171 04-16-01 Ward's Science; p. 172 04-17-01 Pearson Education, Inc.; p. 172 04-17-02 Pearson Education, Inc.; p. 172 04-17-03 Pearson Education, Inc.; p. 172 04-17-04 Pearson Education, Inc.; p. 174 04-18-01 Pixlmaker/Fotolia; p. 174 04-18-02 Leah-Anne Thompson/Shutterstock; p. 179 04-CR-01 Pearson Education, Inc.; p. 179 04-CR-02 Pearson Education, Inc.; p. 179 04-CR-03 Pearson Education, Inc.; p. 179 04-CR-04 Pearson Education, Inc.; p. 179 04-CR-05 Pearson Education, Inc.; p. 179 04-CR-06 Pearson Education, Inc.; p. 179 04-CR-07 Pearson Education, Inc.; p. 179 04-CR-08 Pearson Education, Inc.; p. 04-CR-09 Pearson Education, Inc.; p. 179 04-CR-10 Pearson Education, Inc.; p. 181 04-CI-01 Frederic H. Martini; p. 181 04-CI-02 Pearson Education, Inc.; p. 181 04-CI-03 Alamy.

**Chapter 5** Chapter opener Zdenka Darula/iStock/Getty Images; p. 184 05-02-01 Pearson Education, Inc.; p. 184 05-02-02 Pearson Education, Inc.; p. 185 05-02-03 Pearson Education, Inc.; p. 184 05-02-04 CLOUD HILL IMAGING; p. 186 05-03-01 Pearson Education, Inc.; p. 187 05-03-02 Elizabeth Abel; p. 187 05-03-03 Elizabeth Abel; p. 188 05-04-01 Steve Gschmeissner/Science Source; p. 188 05-04-02 Science Source; p. 191 05-05-01 United States Air Force; p. 196 05-08-01 Frederic H. Martini; p. 197 05-08-02 Frederic H. Martini; p. 197 05-08-03 Frederic H. Martini; p. 198 05-09-01 Thomas, Jim; p. 201 05-11-01 Science Source; p. 202 05-12-01 William C. Ober; p. 203 05-12-02 Kristine Hannon; p. 207 05-CR-01 Pearson Education, Inc.; p. 209 05-CI-01 LEDPIX/Alamy Stock Photo; p. 209 05-CI-02 Superfly Images/E+/Getty Images.

**Chapter 6** Chapter opener Philip Date/Shutterstock; p. 214 06-03-01 Ralph T. Hutchings; p. 217 06-04-01 Seelevel; p. 218 06-05-01 Pearson Education, Inc.; p. 219 06-05-02 Ralph T. Hutchings; p. 223 06-07-01 Alamy; p. 224 06-08-01 Frederic H. Martini; p. 225 06-08-02 Ralph T. Hutchings; p. 225 06-08-03 Ralph T. Hutchings; p. 226 06-09-01 Dennis MacDonald/Alamy Stock Photo; p. 226 06-09-02 Parker, Gary ; p. 226 06-09-03 Joeff Davis; p. 226 06-09-04 Jim Stevenson/Science Source; p. 227 06-09-05 Bettmann/Corbis; p. 227 06-09-06 Kaplan, Frederick; p. 227 06-09-07 Gould, Laura; p. 232 06-12-01 kameel/Fotolia; p. 233 06-12-02 Frederic H. Martini; p. 233 06-12-03 Dr. P. Marazzi/Science Source; p. 233 06-12-04 Scott Camazine/Science Source; p. 233 06-12-05 Frederic H. Martini; p. 233 06-12-06 Frederic H. Martini; p. 233 06-12-07 Mark Aiken; p. 233 06-12-08 Zika, Dan; p. 233 06-12-09 Living Art Enterprises, LLC/Science Source; p. 233 06-12-10 Scott Camazine/Science Source; p. 234 06-SR2-02 Frederic H. Martini; p. 234 06-SR2-03 Frederic H. Martini; p. 234 06-SR2-04 Mark Aiken; p. 234 06-SR2-05 Frederic H. Martini; p. 234 06-SR2-06 Scott Camazine/Science Source; p. 236 06-CR-01 Ralph T. Hutchings; p. 236 06-CR-02 Ralph T. Hutchings; p. 236 06-CR-03 kameel/Fotolia; p. 237 06-CR-04 Ralph T. Hutchings; p. 239 06-CI-01 Jabejon/iStock/Getty Images.

**Chapter 7** Chapter opener Science Picture Co/Getty Images; p. 243 07-02-01 Ralph T. Hutchings; p. 243 07-02-02 Ralph T. Hutchings; p. 251 07-06-01 Ralph T. Hutchings; p. 251 07-06-02 Ralph T. Hutchings; p. 254 07-08-01 Ralph T. Hutchings; p. 254 07-08-02 Ralph T. Hutchings; p. 255 07-08-03 Pearson Education, Inc.; p. 255 07-08-04 Ralph T. Hutchings; p. 255 07-08-05 Michael Wood; p. 256 07-09-01 Timmons Michael; p. 256 07-09-02 Ralph T. Hutchings; p. 258 07-10-01 Pearson Education, Inc.; p. 260 07-11-01 Ralph T. Hutchings; p. 260 07-11-02 Ralph T. Hutchings; p. 261 07-11-03 Robert B. Tallitsch; p. 261 07-11-04

CREDITS

common iliac vein, 725, 733, 737, 739
common pathway, blood clotting, 649, 650
communicating rami, 444
compact bone, 166, 214–219, 223, 232–233
compaction, defined, 868
compartment syndrome, 395
compartments, deep fascia and, 394–395
compartments, leg, 395
compatible blood types, 643
complement, 765
complement system, 770–771, 773, 784
complementary base pairs, 87
complementary strands, 87
complete antigen, 782–784
complete blood count (CBC), 637
complete proteins, 917
complete tetanus, 332
compliance, lungs, 824–825
compound, defined, 56
compound alveolar (acinar) glands, 157
compound fractures, 233
compound glands, 157
compound microscope, 144–145
compound tubular glands, 157
compound tubuloalveolar glands, 157
compression forces, bones, 214
compression fracture, 233, 301
concentration gradient, 120–121
concentric contraction, 334
concentric lamellae, osteon, 218
conception, 1038
condensation reactions, 65
condoms, 1026
conducting portion, respiratory tract, 797
conducting system, heart rate, 680–681
conduction, heat transfer, 923
conductive hearing loss, 585
condylar joints, 293
condylar process, 254
condyle, bone structure, 213
condyle, humerus, 270
cones, eye, 562–565
confluence of sinuses, 731
congenital heart defects, 741
congenital megacolon, 845
congenital talipes equinovarus, 226
conjunctiva, 554
conjunctivitis, 555
connective tissue
    blood vessels, 698–699
    bone abnormalities, 226–227
    cartilage, 164–165
    cuboidal epithelium, 152
    defined, 143
    dermis, 183
    differentiation, 95
    fluid connective tissue, 163
    introduction to, 22, 159
    mediastinum, 43
    percentage of body weight, 171
    sclera, eye, 558
    stratified squamous epithelium, 151
    testes, 1005
connective tissue proper, 159
connexons, 149
conoid tubercle, clavicle, 268
constant segments, antibodies, 782–783
constipation, 883
continuous capillary, 700
continuous propagation, action potentials, 422
contractility, heart, 687
contraction cycle, skeletal muscles, 326–327, 329
contraction phase, muscle tension, 331
contractions, labor and delivery, 1054–1055
contractions, muscle. *See* muscle contractions
contralateral reflex arc, 461
control center, homeostasis, 33, 34
control segment, 114
conus medullaris, 438

convection, heat transfer, 923
convergence, neural circuits, 455
convergent muscles, 352
COPD (chronic obstructive pulmonary disease), 825
copper (Cu), 53, 977
coracoacromial ligament, 302
coracobrachialis muscle, 374–379, 394, 450–451
coracoclavicular ligament, 302
coracohumeral ligament, 302
coracoid process, scapula, 269, 380–381
coracoid process, shoulder joint, 302
cords, brachial plexus nerves, 450–451
Cori cycle, 339
cornea, 554, 556–557, 558
corneal epithelium, 554
corneal limbus, 556–557
corneoscleral junction, 556
corniculate cartilage, 802–803
corona radiata, 1017, 1038
coronal plane, 41
coronal section, 41
coronal suture, skull, 243, 246, 247, 256, 257
coronary arteries, 665, 666
coronary artery disease, 672–673
coronary ischemia, 673
coronary ligament, liver, 877
coronary sinus, 665, 667
coronary sulcus, 664, 665
coronoid fossa, humerus, 270
coronoid process, mandible, 254
coronoid process, radius and ulna, 271
corpora cavernosa, 1008
corpora quadrigemina, 480–481
corpus albicans, 1017
corpus callosum, 473, 484–485, 486, 492
corpus cavernosum, 1008–1009
corpus luteum, 1017
corpus spongiosum, 1008, 1009
corrugator supercilii muscle, 360–361
cortex, hair, 194
cortex, ovary, 1015
cortex, thymus, 761
cortical nephrons, 937, 941
cortical radiate arteries, 940
cortical radiate veins, 940
corticobulbar tracts, 504
corticospinal pathway, 504–505
corticospinal tracts, 504
corticosteroids, 596, 610–611
corticosterone, 611
corticotropin, 602
corticotropin-releasing hormone (CRH), 602, 604–605
cortisol, 611
cortisone, 611
costal, muscle terminology, 355
costal cartilages, 241, 264
costal facets, ribs, 265
costal facets, thoracic vertebrae, 261
costal groove, rib, 265
costal process, 260
costal tuberosity, clavicle, 268
costimulation, 778
cotransport, 124–125
coughing reflex, 533, 827, 829
countercurrent exchange, 925
countercurrent multiplication, 952–953
countertransport, 124–125
covalent bonds
    defined, 57
    energy from, 64
    lipids, 78–79
    organic compounds, 75
Cowper's gland. *See* bulbo-urethral gland
coxal bones, 274–275
cranial, directional term, 40
cranial bones, 242–245
cranial cavity, 242–243
cranial meninges, 474

cranial nerves
    autonomic nervous system, 518, 521
    balance, sense of, 582
    of brain stem, 481
    branches and functions, 494–495
    cervical plexus, overview, 449
    chemoreceptors, 535
    gustatory pathway, 551
    hearing, sense of, 583
    medulla oblongata and, 476
    pons and, 477
    salivary reflex, 874
    somatic motor pathway, 504–505
    sympathetic innervation, 522
    visceral reflexes, 533
cranial reflexes, 456–457
cranial region, 38
craniosacral division, nervous system, 518
cranium. *See also* skull
    axial skeleton, 241
    blood vessels, 728–729
    bones of, 242–243
    gender differences, 277
creatine phosphate (CP), 336–339
creatinine, 943, 945
cremaster muscle, 1001
cremasteric reflex, 463
crenation, 123
crest, bone structure, 213
cribriform plate, 249, 251, 546
cricoid cartilage, 364, 802–803
crista ampullaris, 576
crista galli, 249, 251, 474
cristae, 106–107
cross section, defined, 41
cross-bridges, myosin, 326–327, 329, 330
crossed extensor reflexes, 461
cross-reaction, blood typing, 642
crown, tooth, 850–851
cruciate ligament, knee joint, 291
crude touch and pressure receptors, 498–503
crural region, 38
crus of penis, 1008
crypts of Lieberkühn, 860
crystallins, 558
cuboid bone, 280–281
cuboidal epithelia, 147, 152, 799
cuneate nucleus, 476, 503
cuneiform bones, foot, 280–281
cuneiform cartilage, larynx, 803
cupula, 576
Cushing's disease, 623
cuspids, teeth, 850
cusps, heart valves, 668
cutaneous, muscle terminology, 355
cutaneous membrane, skin, 26, 168, 183, 193
cutaneous plexus, 183, 188
cuticle, hair, 194, 195
cuticle, nail, 198
cyanide, 899
cyanosis, 187
cyclosporin A, 788
cylomicrons, 861
cystic artery, 734
cystic duct, 880
cystic fibrosis (CF), 799
cystic vein, 735
cystitis, 965
cysts, 1029
cytochromes, 898–899
cytokines, 770–771, 780–781
cytokinesis, 129, 130, 132–133, 1039
cytoplasm, 96, 156
cytosine, 86–87, 112–117
cytoskeleton
    epithelial tissues, 148
    introduction to, 96, 97
    structure and function, 98, 100–101

renal corpuscle, 938, 944–947
renal cortex, 936, 937
renal failure, 958–959
renal fascia, 935
renal medulla, 936, 937
renal nerves, 941
renal papilla, 936, 937
renal pelvis, 936
renal pyramid, 936
renal sinus, 936
renal tubules, 938–939
renal vein, 725, 733, 739, 940
renin, 618, 714, 941, 949, 979
renin-angiotensin-aldosterone system (RAAS), 618, 621
rennin, 859
repolarization, action potential, 323, 418–419, 421, 679
reproduction, cell division, 129–133
reproductive system
  cilia, 147
  epithelial tissue, 22
  fetal development, 1043
  hormonal control of, 602
  introduction to, 25
  lymphatic structures, 751
  melatonin, 614
  microvilli, 147
  mucous membranes, 168
  nervous system, 407, 531, 533
  percentage of body weight, 351
  pituitary gland and, 600
  regulation of, 28
  reproductive cells (sex cells), 21
  smooth muscle tissue, 23
reproductive system, female
  anatomic nervous system, 531, 533
  autonomic nervous system effects, 531
  birth control strategies, 1026–1027
  cilia, 147
  epithelial tissue, 22
  gross anatomy, 1013
  hormones, cycle regulation, 1024–1025
  introduction to, 31
  mammary glands, 1023
  microvilli, 147
  mucous membranes, 168
  oogenesis, 1016–1017
  ovarian cycle, 1017, 1024–1025
  ovaries, 1014–1015
  puberty, 1059
  smooth muscle tissue, 23
  structures of, 31
  uterine cycle, 1020–1021, 1024–1025
  uterine tubes, 1018–1019
  vagina, 1022
reproductive system, male
  autonomic nervous system, 531, 533
  autonomic nervous system effects, 531
  birth control strategies, 1026–1027
  cilia, 147
  disorders of, 1028
  ejaculation reflex, 533
  epithelial tissue, 22
  gland structures, 157
  glandular secretions, 1006–1007
  gross anatomy, 999
  introduction to, 31
  microvilli, 147
  penis, 1008–1009
  pseudostratified columnar epithelium, 155
  puberty, 1058
  seminiferous tubules, testes, 1000–1001, 1004–1005
  sexual response, 1009
  smooth muscle tissue, 23
  spermatogenesis, 1002–1005
  sterility, 1039
  structures of, 31
  testosterone, functions of, 1010–1011

RER (rough endoplasmic reticulum), 97, 103
residual volume, lungs, 815
resistance, lungs, 824–825
resistance, peripheral, 705
resistance phase, stress response, 621
resolution, microscopes, 144–145
resonance, sound, 581
respiration
  airflow pattern, 804–805
  alveoli, gas exchange, 808–809
  asthma, 805
  gas diffusion, 818–819
  gas pressures and, 812–813
  gas transport, 820–823
  internal and external, 811
  lungs, lobes of, 806–807
  muscles of, 814
  pulmonary disease, 824–825
  pulmonary volumes and capacities, 815
  regulation of, 816–817, 823, 826–829
  upper respiratory structures, 797–805
respiratory acid-base disorders, 987
respiratory acidosis and alkalosis, 990–991
respiratory bronchiole, 808–809
respiratory centers, pons, 477, 524, 533, 924–925
respiratory defense system, 798
respiratory distress syndrome, 824
respiratory membrane, 809
respiratory minute volume, 816
respiratory mucosa, 788
respiratory pump, 717
respiratory rate, 816
respiratory reflexes, 827
respiratory rhythmicity center, medulla oblongata, 476, 534, 826–829
respiratory system. *See also* respiration
  acid-base balance and, 983, 988–991
  aging and, 830–831
  airflow pattern, 804–805
  alveoli, gas exchange, 808–809
  asthma, 805
  autonomic nervous system effects, 530
  baroreceptors, 534
  body temperature regulation, 924–925
  cells, light micrograph, 145
  chemoreceptors, 535
  cilia, 147
  endocrine system and, 28
  epithelial tissue, 22
  fetal development, 1043
  fluid balance and, 974–975
  gland structures, 157
  heart, gross anatomy, 14
  hyaline cartilage, 164
  introduction to, 25, 29
  larynx, 802–803
  lungs, lobes of, 806–807
  lymphatic structures, 751
  mucous membranes, 168, 798
  nasal cavities, 800–801
  nervous system, 407
  paranasal sinuses, 800–801
  parasympathetic activation, 525
  percentage of body weight, 351
  pharynx, 801
  pregnancy and, 1052–1053
  puberty, changes in, 1058–1059
  regulation of, 476, 477
  sympathetic activation, 524
  thoracic cage, 264–265
  trachea, 804–805
  upper and lower tracts, functions of, 797
respiratory tract, 797
resting heart rate, 684–685
resting membrane potential, nervous tissue, 412, 414–415
resting phase, hair growth cycle, 195
resting membrane potentials, skeletal muscles, 322–323

rete testis, 1001, 1004, 1006
reticular activating system (RAS), 480–481, 498–499
reticular fibers, 160
reticular formation, 476, 477, 504, 583
reticular lamina, 148
reticular layer, dermis, 183, 188
reticular tissue, 161
reticulocyte count, 637
reticulocytes, 635, 640, 641
reticulospinal tracts, 504
retina, 405, 428, 553, 556–557, 568
retinal, eye pigment, 564
retinal artery, 562
retraction, 297
retrograde flow, 404
retroperitoneal, defined, 43, 934
reverberation, neural circuits, 455
revolving nosepiece, microscope, 145
Rh antigens, blood, 643, 644–645
rheumatism, 306–307
rheumatoid arthritis, 306–307, 788
rhodopsin, 564, 566–567
RhoGAM, 645
rhombencephalon, 471
rhomboid, muscle terminology, 355
rhomboid major muscle, 375–377, 450–451
rhomboid minor muscle, 375–377
rhomboid muscle, 357, 375–377
rhythm method, 1027
riboflavin, 915
ribonucleic acid. *See* RNA (ribonucleic acid)
ribose, 85, 905
ribosomal RNA. *See* rRNA (ribosomal RNA)
ribosomes
  fixed ribosomes, 103
  introduction to, 96, 97
  mitochondria, 106–107
  in neurons, 404
  protein synthesis, 113, 116–117
  structure and function, 102–103
ribs
  arm movement, muscles of, 378–379
  axial skeleton, 241
  bone structure, 212
  hyaline cartilage, 164
  muscle terminology, 355
  oblique and rectus muscles, 368–369
  pectoral girdle, muscles of, 376–377
  respiration and, 814
  structures of, 265
  synchondrosis, 289
  thoracic and rib cage, 264–265
  thoracic vertebrae, 261
  vertebral column muscles, 366–367
rickets, 201
right atrioventricular (AV) valve, 668
right atrium, 14, 659, 697, 739
right brachiocephalic vein, 729, 739
right bronchomediastinal lymphatic trunk, 755
right colic artery, 734
right colic flexure, 868
right colic vein, 735
right common carotid artery, 726, 738
right common iliac artery, 736, 738
right common iliac vein, 739
right coronary artery, 666
right external iliac artery, 736, 738
right gastric artery, 734
right gastric vein, 735
right gastroepiploic artery, 734
right gastroepiploic vein, 735
right hepatic duct, 880
right hypochondriac region, 39, 862
right inguinal region, 39, 862
right internal iliac artery, 738
right jugular lymphatic trunk, 755
right lobe, liver, 877
right lower quadrant (RLQ), 39
right lumbar region, 39, 862

water (H₂O) (continued)
  states of matter, 58–59
  urine volume regulation, 954–955
water loss, respiratory system and, 801
water receptors, 548
water-soluble vitamins, 915
wave summation, muscle tension, 332
wavelength, light, 565
wavelength, sound, 580–581
waxes, 78–79
WBC differential count, 637
WBCs. *See* white blood cells (WBCs)
weight, body. *See* body weight
weight, molecular weight and moles, 63
Wernicke's area, 490–491
white, muscle terminology, 355
white blood cells (WBCs)
  bones, synthesis in, 211
  cell types and functions, 20, 633, 646–647
  fluid connective tissue, 163
  leukemia, 651
white matter, brain
  cerebellum, 479
  function of, 492
  medulla oblongata, 476
  midbrain, 481
  pons, 477
white matter, central nervous system, 408, 409
white matter, spinal cord, 438–439, 443
white muscles, 340

white pulp, spleen, 763
white ramus communicans, 446–447
whole blood, 632–633
wide, muscle terminology, 355
wisdom teeth, 851
withdrawal reflex, 460
work, 61
Wormian bones, 243. *See also* bones (osseous tissue)
wound repair, 200, 202–203
wrist joint
  abduction and adduction, 295
  appendicular skeleton, 299
  body regions, 38
  bone structures, 212, 272–273
  carpal tunnel syndrome, 383
  muscle terminology, 355
  muscles of, 373, 380–385
  nerves of, 450–451
  pronation, 296
  radius and ulna bones, 271
  supination, 296

**X**

X chromosomes, 1061–1067
xenograft, 191
xiphoid process, 264, 368–369
X-linked traits, 1065

**Y**

Y chromosomes, 1061–1067
yellow bone marrow, 214–215, 641
Y-linked traits, 1065
yolk sac, 1042–1047
yolk stalk, 1046

**Z**

Z lines, cardiac muscle tissue, 663
Z lines, skeletal muscles, 318, 320–321, 330
zinc (Zn), 53
zona fasciculata, 611
zona glomerulosa, 611
zona pellucida, 1017, 1038
zona reticularis, 611
zone of overlap, skeletal muscles, 318, 321, 330
zwitterions, 82, 987
zygomatic arch, 246, 248, 363
zygomatic bones, 242–243, 244, 246, 248, 252, 253, 360–361
zygomatic process, temporal bone, 246
zygomaticofacial foramen, 252
zygomaticus major, 360–361
zygomaticus minor, 360–361
zygote, 1038–1039
zymogenic cells, 867

# COMMON ABBREVIATIONS USED IN SCIENCE

| Abbreviation | Meaning |
|---|---|
| ACh | acetylcholine |
| AChE | acetylcholinesterase |
| ACTH | adrenocorticotropic hormone |
| ADH | antidiuretic hormone |
| ADP | adenosine diphosphate |
| AIDS | acquired immunodeficiency syndrome |
| ALS | amyotrophic lateral sclerosis |
| AMP | adenosine monophosphate |
| ANP | atrial natriuretic peptide |
| ANS | autonomic nervous system |
| AP | arterial pressure |
| ARDS | adult respiratory distress syndrome |
| atm | atmospheric pressure |
| ATP | adenosine triphosphate |
| ATPase | adenosine triphosphatase |
| AV | atrioventricular |
| AVP | arginine vasopressin |
| BCOP | blood colloid osmotic pressure |
| BMR | basal metabolic rate |
| BPG | bisphosphoglycerate |
| bpm | beats per minute |
| BUN | blood urea nitrogen |
| C | large calorie; Celsius |
| CABG | coronary artery bypass graft |
| CAD | coronary artery disease |
| cAMP | cyclic-AMP |
| CAPD | continuous ambulatory peritoneal dialysis |
| CCK | cholecystokinin |
| CD | cluster of differentiation |
| CF | cystic fibrosis |
| CHF | congestive heart failure |
| CHP | capillary hydrostatic pressure |
| CsHP | capsular hydrostatic pressure |
| CNS | central nervous system |
| CO | cardiac output; carbon monoxide |
| CoA | coenzyme A |
| COMT | catechol-O-methyltransferase |
| COPD | chronic obstructive pulmonary disease |
| CP | creatine phosphate |
| CPK, CK | creatine phosphokinase |
| CPM | continuous passive motion |
| CPR | cardiopulmonary resuscitation |
| CRF | chronic renal failure |
| CRH | corticotropin-releasing hormone |
| CSF | cerebrospinal fluid; colony-stimulating factors |
| CT | computerized tomography; calcitonin |
| CVA | cerebrovascular accident |
| CVS | cardiovascular system |
| DAG | diacylglycerol |
| DC | Doctor of Chiropractic |
| DCT | distal convoluted tubule |
| DDST | Denver Developmental Screening Test |
| DIC | disseminated intravascular coagulation |
| DJD | degenerative joint disease |
| DMD | Duchenne's muscular dystrophy |
| DNA | deoxyribonucleic acid |
| DO | Doctor of Osteopathy |
| DPM | Doctor of Podiatric Medicine |
| DSA | digital subtraction angiography |
| E | epinephrine |
| ECF | extracellular fluid |
| ECG | electrocardiogram |
| EDV | end-diastolic volume |
| EEG | electroencephalogram |
| EKG | electrocardiogram |
| ELISA | enzyme-linked immunosorbent assay |
| EPSP | excitatory postsynaptic potential |
| ERV | expiratory reserve volume |
| ESV | end-systolic volume |
| ETS | electron transport system |
| FAD | flavin adenine dinucleotide |
| FAS | fetal alcohol syndrome |
| FES | functional electrical stimulation |
| FMN | flavin mononucleotide |
| FRC | functional residual capacity |
| FSH | follicle-stimulating hormone |
| GABA | gamma aminobutyric acid |
| GAS | general adaptation syndrome |
| GC | glucocorticoid |
| GFR | glomerular filtration rate |
| GH | growth hormone |
| GH–IH | growth hormone–inhibiting hormone |
| GHP | glomerular hydrostatic pressure |
| GH–RH | growth hormone–releasing hormone |
| GIP | gastric inhibitory peptide |
| GnRH | gonadotropin-releasing hormone |
| GTP | guanosine triphosphate |
| Hb | hemoglobin |
| hCG | human chorionic gonadotropin |
| HCl | hydrochloric acid |
| HDL | high-density lipoprotein |
| HDN | hemolytic disease of the newborn |
| hGH | human growth hormone |
| HIV | human immunodeficiency virus |
| HLA | human leukocyte antigen |
| HMD | hyaline membrane disease |
| HP | hydrostatic pressure |
| hPL | human placental lactogen |
| HR | heart rate |
| Hz | Hertz |
| ICF | intracellular fluid |
| ICOP | interstitial fluid colloid osmotic pressure |
| IGF | insulin-like growth factor |
| IH | inhibiting hormone |
| IM | intramuscular |
| IP₃ | inositol triphosphate |
| IPSP | inhibitory postsynaptic potential |
| IRV | inspiratory reserve volume |
| ISF | interstitial fluid |
| IUD | intrauterine device |
| IVC | inferior vena cava |
| IVF | in vitro fertilization |
| kcal | kilocalorie |
| LDH | lactate dehydrogenase |
| LDL | low density lipoprotein |
| L-DOPA | levodopa |
| LH | luteinizing hormone |
| LLQ | left lower quadrant |
| LM | light micrograph |
| LSD | lysergic acid diethylamide |
| LUQ | left upper quadrant |
| MAO | monoamine oxidase |
| MAP | mean arterial pressure |
| MC | mineralocorticoid |
| MD | Doctor of Medicine |
| mEq | millequivalent; ($10^{-3}$) |
| MHC | major histocompatibility complex |
| MI | myocardial infarction |
| mm Hg | millimeters of mercury |
| mmol | millimole |
| mOsm | milliosmole |
| MRI | magnetic resonance imaging |
| mRNA | messenger RNA |
| MS | multiple sclerosis |
| MSH | melanocyte-stimulating hormone |
| MSH–IH | melanocyte-stimulating hormone–inhibiting hormone |
| NAD | nicotinamide adenine dinucleotide |
| NE | norepinephrine |
| NFP | net filtration pressure |
| NHP | net hydrostatic pressure |
| NO | nitric oxide |
| NRDS | neonatal respiratory distress syndrome |
| OP | osmotic pressure |
| Osm | osmoles |
| OXT | oxytocin |
| PAC | premature atrial contraction |
| PAT | paroxysmal atrial tachycardia |
| PCT | proximal convoluted tubule |
| PCV | packed cell volume |
| PEEP | positive end-expiratory pressure |
| PET | positron emission tomography |
| PFC | perfluorochemical emulsion |
| PG | prostaglandin |
| PID | pelvic inflammatory disease |
| PIH | prolactin-inhibiting hormone |
| PIP | phosphatidylinositol |
| PKC | protein kinase C |
| PKU | phenylketonuria |
| PLC | phospholipase C |
| PMN | polymorphonuclear leukocyte |
| PNS | peripheral nervous system |
| PR | peripheral resistance |
| PRF | prolactin-releasing factor |
| PRL | prolactin |
| psi | pounds per square inch |
| PT | prothrombin time |
| PTA | post-traumatic amnesia; plasma thromboplastin antecedent |
| PTC | phenylthiocarbamide |
| PTH | parathyroid hormone |
| PVC | premature ventricular contraction |
| RAS | reticular activating system |
| RBC | red blood cell |
| RDA | recommended daily allowance |
| RDS | respiratory distress syndrome |
| REM | rapid eye movement |
| RER | rough endoplasmic reticulum |
| RH | releasing hormone |
| RHD | rheumatic heart disease |
| RLQ | right lower quadrant |
| RNA | ribonucleic acid |
| rRNA | ribosomal RNA |
| RUQ | right upper quadrant |
| SA | sinoatrial |
| SCA | sickle cell anemia |
| SCID | severe combined immunodeficiency disease |
| SEM | scanning electron micrograph |
| SER | smooth endoplasmic reticulum |
| SGOT | serum glutamic oxaloacetic transaminase |
| SIADH | syndrome of inappropriate ADH secretion |
| SIDS | sudden infant death syndrome |
| SLE | systemic lupus erythematosus |
| SNS | somatic nervous system |
| STD | sexually transmitted disease |
| SV | stroke volume |
| SVC | superior vena cava |
| T₃ | triiodothyronine |
| T₄ | tetraiodothyronine, or thyroxine |
| TB | tuberculosis |
| TBG | thyroid-binding globulin |
| TEM | transmission electron micrograph |
| TIA | transient ischemic attack |
| Tₘ | transport (tubular) maximum |
| TMJ | temporomandibular joint |
| t-PA | tissue plasminogen activator |
| TRH | thyrotropin-releasing hormone |
| tRNA | transfer RNA |
| TSH | thyroid-stimulating hormone |
| TSS | toxic shock syndrome |
| U.S. | United States |
| UTI | urinary tract infection |
| UTP | uridine triphosphate |
| UV | ultraviolet |
| $\dot{V}_A$ | alveolar ventilation |
| $\dot{V}_D$ | anatomic dead space |
| $\dot{V}_E$ | respiratory minute volume |
| $V_T$ | tidal volume |
| VF | ventricular fibrillation |
| VLDL | very low-density lipoprotein |
| VPRC | volume of packed red cells |
| VT | ventricular tachycardia |
| WBC | white blood cell |

# FOREIGN WORD ROOTS, PREFIXES, SUFFIXES, AND COMBINING FORMS

Each entry starts with the commonly used form or forms of the prefix, suffix, or combining form followed by the word root (shown in italics) and its English translation. One example is also given to illustrate the use of each entry.

**a-,** *a-,* without: avascular
**ab-,** *ab,* from: abduct
**-ac,** *-akos,* pertaining to: cardiac
**acr-,** *akron,* extremity: acromegaly
**ad-,** *ad,* to, toward: adduct
**aden-, adeno-,** *adenos,* gland: adenoid
**adip-,** *adipos,* fat: adipocytes
**aer-,** *aeros,* air: aerobic metabolism
**-al,** *-alis,* pertaining to: brachial
**alb-,** *albicans,* white: albino
**-algia,** *algos,* pain: neuralgia
**allo-,** *allos,* other: allograft
**ana-,** *ana,* up, back: anaphase
**andro-,** *andros,* male: androgen
**angio-,** *angeion,* vessel: angiogram
**ante-,** *ante,* before: antebrachial
**anti-, ant-,** *anti,* against: antibiotic
**apo-,** *apo,* from: apocrine
**arachn-,** *arachne,* spider: arachnoid
**arter-,** *arteria,* artery: arterial
**arthro-,** *arthros,* joint: arthroscopy
**astro-,** *aster,* star: astrocyte
**atel-,** *ateles,* imperfect: atelectasis
**aur-,** *auris,* ear: auricle
**auto-,** *auto,* self: autonomic
**baro-,** *baros,* pressure: baroreceptor
**bi-,** *bi-* two: bifurcate
**bio-,** *bios,* life: biology
**-blast,** *blastos,* precursor: osteoblast
**brachi-,** *brachium,* arm: brachiocephalic
**brachy-,** *brachys,* short: brachydactyly
**brady-,** *bradys,* slow: bradycardia
**bronch-,** *bronchus,* windpipe, airway: bronchial
**carcin-,** *karkinos,* cancer: carcinoma
**cardi-, cardio-,** *kardia,* heart: cardiac
**-cele,** *kele,* tumor, hernia, or swelling: blastocele
**-centesis,** *kentesis,* puncture: thoracocentesis
**cephal-,** *cephalos,* head: brachiocephalic
**cerebr-,** *cerebrum,* brain: cerebral hemispheres
**cerebro-,** *cerebros,* brain: cerebrospinal fluid
**cervic-,** *cervicis,* neck: cervical vertebrae
**chole-,** *chole,* bile: cholecystitis
**-chondrion,** *chondrion,* granule: mitochondrion
**chondro-,** *chondros,* cartilage: chondrocyte
**chrom-, chromo-,** *chroma,* color: chromatin
**circum-,** *circum,* around: circumduction
**-clast,** *klastos,* broken: osteoclast
**coel-, -coel,** *koila,* cavity: coelom
**colo-,** *kolon,* colon: colonoscopy
**contra-,** *contra,* against: contralateral
**corp-,** *corpus,* body: corpuscle
**cortic-,** *cortex,* rind or bark: corticospinal
**cost-,** *costa,* rib: costal
**cranio-,** *cranium,* skull: craniosacral
**cribr-,** *cribrum,* sieve: cribriform
**-crine,** *krinein,* to separate: endocrine
**cut-,** *cutis,* skin: cutaneous
**cyan-,** *kyanos,* blue: cyanosis
**cyst-, -cyst,** *kystis,* sac: blastocyst
**cyt-, cyto-,** *kytos,* a hollow cell: cytology
**de-,** *de,* from, away: deactivation

**dendr-,** *dendron,* tree: dendrite
**dent-,** *dentes,* teeth: dentition
**derm-,** *derma,* skin: dermatome
**desmo-,** *desmos,* band: desmosome
**di-,** *dis,* twice: disaccharide
**dia-,** *dia,* through: diameter
**digit-,** *digit,* a finger or toe: digital
**dipl-,** *diploos,* double: diploid
**dis-,** *dis,* apart, away from: disability
**diure-,** *diourein,* to urinate: diuresis
**dys-,** *dys,* painful: dysmenorrhea
**-ectasis,** *ektasis,* expansion: atelectasis
**ecto-,** *ektos,* outside: ectoderm
**-ectomy,** *ektome,* excision: appendectomy
**ef-,** *ex,* away from: efferent
**emmetro-,** *emmetros,* in proper measure: emmetropia
**encephalo-,** *enkephalos,* brain: encephalitis
**end-, endo-,** *endon,* within: endometrium
**entero-,** *enteron,* intestine: enteric
**epi-,** *epi,* upon: epimysium
**erythema-,** *erythema,* flushed (skin): erythematosis
**erythro-,** *erythros,* red: erythrocyte
**ex-,** *ex,* out of, away from: exocytosis
**extra-,** *exter,* outside of, beyond, in addition: extracellular
**ferr-,** *ferrum,* iron: transferrin
**fil-,** *filum,* thread: filament
**-form,** *-formis,* shape: fusiform
**gastr-,** *gaster,* stomach: gastrointestinal
**-gen, -genic,** *gennan,* to produce: mutagen
**genicula-,** *geniculum,* kneelike structure: geniculates
**genio-,** *geneion,* chin: geniohyoid
**gest-,** *gesto,* to bear: gestation
**glosso-, -glossus,** *glossus,* tongue: hypoglossal
**glyco-,** *glykys,* sugar: glycogen
**-gram,** *gramma,* record: myogram
**gran-,** *granulum,* grain: granulocyte
**-graph, -graphia,** *graphein,* to write, record: electroencephalograph
**gyne-, gyno-,** *gynaikos,* woman: gynecologist
**hem-, hemo-,** *haima,* blood: hemopoiesis
**hemi-,** *hemi-,* one half: hemisphere
**hepato-,** *hepaticus,* liver: hepatocyte
**hetero-,** *heteros,* other: heterozygous
**histo-,** *histos,* tissue: histology
**holo-,** *holos,* entire: holocrine
**homeo-,** *homoios,* similar: homeostasis
**homo-,** *homos,* same: homozygous
**hyal-, hyalo-,** *hyalos,* glass: hyaline
**hydro-,** *hydros,* water: hydrolysis
**hyo-,** *hyoeides,* U-shaped: hyoid bone
**hyper-,** *hyper,* above: hyperpolarization
**hypo-,** *hypo,* under: hypothyroid
**hyster-,** *hystera,* uterus: hysterectomy
**-ia,** *-ia,* state or condition: insomnia
**idi-,** *idios,* one's own: idiopathic
**in-,** *in-,* in, within, or denoting negative effect: inactivate
**infra-,** *infra,* beneath: infra-orbital
**inter-,** *inter,* between: interventricular
**intra-,** *intra,* within: intracapsular
**ipsi-,** *ipse,* itself: ipsilateral
**iso-,** *isos,* equal: isotonic
**-itis,** *-itis,* inflammation: dermatitis